# Complete Solutions Manual

for

Stewart, Redlin, and Watson's

# Algebra and Trigonometry

Second Edition

**Andrew Bulman-Fleming**

Australia • Brazil • Canada • Mexico • Singapore • Spain • United Kingdom • United States

Printed in the United States of America

1 2 3 4 5 6 7 09 08 07 06 05

Printer: Thomson/West

0-495-01360-9

For more information about our products, contact us at:
**Thomson Learning Academic Resource Center**
**1-800-423-0563**

For permission to use material from this text or product, submit a request online at **http://www.thomsonrights.com.**
Any additional questions about permissions can be submitted by email to **thomsonrights@thomson.com.**

**Thomson Higher Education**
**10 Davis Drive**
**Belmont, CA 94002-3098**
**USA**

# Contents

## 2 Coordinates and Graphs 103



## 3 Functions 165



## 4 Polynomial and Rational Functions 243

## 8 Analytic Trigonometry 449



## 9 Polar Coordinates and Vectors 491



## 10 Systems of Equations and Inequalities 531

# 13 Counting and Probability 737

# P Prerequisites

## P.1 Modeling the Real World

**1. (a)** $M = \frac{N}{G} = \frac{230}{5.4} = 42.6$ miles/gallon
**(b)** $25 = \frac{185}{G} \Leftrightarrow G = \frac{185}{25} = 7.4$ gallons

**2. (a)** $T = 70 - 0.003h = 70 - 0.003\,(1500) = 65.5°$ F
**(b)** $64 = 70 - 0.003h \Leftrightarrow 0.003h = 6 \Leftrightarrow h = 2000$ ft

**3. (a)** $P = 0.45d + 14.7 = 0.45\,(180) + 14.7 = 95.7$ lb/in$^2$
**(b)** $P = 0.45d + 14.7 = 0.45\,(330) + 14.7 = 163.2$ lb/in$^2$

**4. (a)** $V = 9.5S = 9.5\left(4.5 \text{ km}^3\right) = 42.75 \text{ km}^3$
**(b)** $8 \text{ km}^3 = 9.5S \Leftrightarrow S = 0.84 \text{ km}^3$

**5. (a)** $P = 0.06s^3 = 0.06\left(12^3\right) = 103.7$ hp
**(b)** $7.5 = 0.06s^3 \Leftrightarrow s^3 = 125$ so $s = 5$ knots

**6.** $C = 500 + 0.35\,(1000) = \$850$.

**7.** $P = 0.8\,(1000) - 500 = \$300$.

**8.** $V = \pi r^2 h = \pi\left(3^2\right)(5) = 45\pi \approx 141.4 \text{ in}^3$

**9.** The number $N$ of days in $w$ weeks is $N = 7w$.

**10.** The number $N$ of cents in $q$ quarters is $N = 25q$.

**11.** The average $A$ of two numbers, $a$ and $b$, is $A = \frac{a+b}{2}$.

**12.** The average $A$ of three numbers, $a$, $b$, and $c$, is $A = \frac{a+b+c}{3}$.

**13.** The sum $S$ of two consecutive integers is $S = n + (n+1) = 2n + 1$, where $n$ is the first integer.

**14.** The sum of three consecutive integers is $S = (n-1) + n + (n+1) = 3n$, where $n$ is the middle integer.

**15.** The sum $S$ of a number $n$ and its square is $S = n + n^2$.

**16.** The sum $S$ of the squares of two numbers $n$ and $m$ is $S = n^2 + m^2$.

**17.** The product of two consecutive integers is $P = n\,(n+1) = n^2 + n$, where $n$ is the smaller integer.

**18.** The product $P$ of an integer $n$ and twice the integer is $P = n\,(2n) = 2n^2$.

**19.** The sum $S$ of an integer $n$ and twice the integer $S = n + 2n = 3n$.

**20.** The distance $d$ in a miles that a car travels in $t$ hour at a speed of $r$ miles per hour is $d = rt$.

**21.** The time it takes an airplane to travel $d$ miles at $r$ miles per hour is $t = \frac{d}{r}$.

**22.** The speed $r$ of a boat that travels $d$ miles in $t$ hours is $r = \frac{d}{t}$.

**23.** The area $A$ of a square of side $x$ is $A = x^2$.

**24.** The length is $4 + x$, so the area $A$ is $A = (4 + x)\,x = 4x + x^2$.

**25.** The length is $4 + x$, so the perimeter $P$ is $P = 2\,(4 + x) + 2x = 8 + 2x + 2x = 8 + 4x$.

**26.** The perimeter $P$ of an equilateral triangle of side $a$ is $P = a + a + a = 3a$.

**27.** The volume $V$ of a cube of side $x$ is $V = x^3$.

**28.** The volume $V$ of a box with a square base of side $x$ and height $2x$ is $V = x \cdot x \cdot 2x = 2x^3$.

**29.** The surface area $A$ of a cube of side $x$ is the sum of the areas of the 6 sides, each with area $x^2$ ; so $A = 6 \cdot x^2 = 6x^2$.

**30.** A box of length $l$, width $w$, and height $h$ has two sides of area $l \cdot w$, two sides of area $l \cdot h$, and two ends with area $w \cdot h$. So the surface area $A$ is $A = 2lw + 2lh + 2wh$.

**31.** The length of one side of the square is $2r$, so its area is $(2r)^2 = 4r^2$. Since the area of the circle is $\pi r^2$, the area of what remains is $A = 4r^2 - \pi r^2 = (4 - \pi)\,r^2$.

**32.** The race track consists of a rectangle of length $x$ and width $2r$ and two semicircles of radius $r$. Each semi circle has length $\pi r$ and two straight runs of length $x$. So the length $L$ is $L = 2x + 2\pi r$.

**33.** The race track consists of a rectangle of length $x$ and width $2r$ and two semicircles of radius $r$. The area of the rectangle is $2rx$ and the area of a circle of radius $r$ is $\pi r^2$. So the area enclosed is $A = 2rx + \pi r^2$.

**34.** The area $A$ of a triangle is $A = \frac{1}{2}$ base· height. Since the base is twice the height $h$ we get $A = \left(\frac{1}{2}\right)(2h)(h) = h^2$.

**35.** This is the volume of the large ball minus the volume of the inside ball. The volume of a ball (sphere) is $\frac{4}{3}\pi\,(\text{radius})^3$. Thus the volume is $V = \frac{4}{3}\pi R^3 - \frac{4}{3}\pi r^3 = \frac{4}{3}\pi\left(R^3 - r^3\right)$.

**36.** The volume $V$ of a cylinder is $V = \pi r^2 h$. Since the radius is $x$ and the height is $2x$ we get $V = \pi\left(x^2\right)(2x) = 2\pi x^3$.

**37. (a)** $\$7.95 + 2\,(\$1.25) = \$7.95 + \$3.75 = \$11.70$

**(b)** The cost $C$, in dollars, of a pizza with $n$ toppings is $C = 7.95 + 1.25n$.

**(c)** Using the model $C = 7.95 + 1.25$ with $C = 14.20$, we get $14.20 = 7.95 + 1.25n \quad\Leftrightarrow\quad 1.25n = 6.25 \quad\Leftrightarrow\quad n = 5$. So the pizza has 5 toppings.

**38. (a)** $3\,(38) + 280\,(0.15) = 114 + 42 = \$156$

**(b)** The cost is $\begin{pmatrix}\text{daily}\\ \text{rental}\end{pmatrix} \times \begin{pmatrix}\text{days}\\ \text{rented}\end{pmatrix} + \begin{pmatrix}\text{cost}\\ \text{per mile}\end{pmatrix} \times \begin{pmatrix}\text{miles}\\ \text{driven}\end{pmatrix}$, so $C = 38n + 0.15m$.

**(c)** We have $C = 149.25$ and $n = 3$. Substituting, we get $149.25 = 38\,(3) + 0.15m \quad\Leftrightarrow\quad 149.25 = 114 + 0.15m \quad\Leftrightarrow\quad 35.25 = m \quad\Leftrightarrow\quad m = 235$. So the rental was driven 235 miles.

**39. (a)** $4\text{ ft/s}\cdot 20\text{ s} = 80\text{ ft}$

**(b)** $d = 4\text{ ft/s}\cdot t\text{ s} = 4t\text{ ft}$

**(c)** Since 1 mile $= 5280$ ft, solve $d = 4t = 5280$ for $t$. We have $4t = 5280 \quad\Leftrightarrow\quad t = \frac{5280}{4} = 1320\text{ s} = 22$ minutes.

**40. (a)** The area that she can mow is $150\text{ ft}^2/\text{min}\times 30\text{ min} = 4{,}500\text{ ft}^2$.

**(b)** The area is $150\text{ ft}^2/\text{min}\times T\text{ min} = 150T\text{ ft}^2$.

**(c)** The area of the lawn is $80\text{ ft}\times 120\text{ ft} = 9{,}600\text{ ft}^2$. If $T$ is the time required, then from part (b), $150T = 9600$, so $T = \frac{9600}{150} = 64$ min.

**(d)** If $R$ is the rate of mowing, then $60R = 9600$, so $R = \frac{9600}{60} = 160\text{ ft}^2/\text{min}$.

**41. (a)** If the width is 20, then the length is 40, so the volume is $20\cdot 20\cdot 40 = 16{,}000\text{ in}^3$.

**(b)** In terms of width, $V = x\cdot x\cdot 2x = 2x^3$.

**(c)** Solve $V = 2x^3 = 6750$ for $x$. We have $2x^3 = 6750 \quad\Leftrightarrow\quad x^3 = 3375 \quad\Leftrightarrow\quad x = 15$. So the width is 15 in. and the length is $2\,(15) = 30$ in. Thus, the dimensions are 15 in. $\times$ 15 in. $\times$ 30 in.

**42. (a)** The 10-minute call cost $0.25 + 0.08\,(10) = 0.25 + 0.80 = \$1.05$.

**(b)** The cost is $\begin{pmatrix}\text{connect}\\ \text{fee}\end{pmatrix} + \begin{pmatrix}\text{cost}\\ \text{per minute}\end{pmatrix} \times (\text{minutes})$, so $C = 0.25 + 0.08t$.

**(c)** When $C = 73$, we have $73 = 25 + 8t \quad\Leftrightarrow\quad 48 = 8t \quad\Leftrightarrow\quad t = 6$. The call lasted 6 minutes.

**(d)** The cost is $\begin{pmatrix}\text{connect}\\ \text{fee}\end{pmatrix} + \begin{pmatrix}\text{cost}\\ \text{per minute}\end{pmatrix} \times (\text{minutes})$, so when the connection fee is $F$ cents and the rate is $r$ cents then the cost (in cents) is given by $C = F + rt$.

**43. (a)** The GPA is $\dfrac{4a + 3b + 2c + 1d + 0f}{a + b + c + d + f} = \dfrac{4a + 3b + 2c + d}{a + b + c + d + f}$.

**(b)** Using $a = 2\cdot 3 = 6$, $b = 4$, $c = 3\cdot 3 = 9$, and $d = f = 0$ in the formula from part (a), we find the GPA to be $\dfrac{4\cdot 6 + 3\cdot 4 + 2\cdot 9}{6 + 4 + 9} = \dfrac{54}{19} \approx 2.84$.

# P.2 Real Numbers

**1. (a)** Natural number: 50
**(b)** Integers: $0, -10, 50$
**(c)** Rational numbers: $0, -10, 50, \frac{22}{7}, 0.538, 1.2\overline{3}, -\frac{1}{3}$
**(d)** Irrational numbers: $\sqrt{7}, \sqrt[3]{2}$

**2. (a)** Natural numbers: $11, \sqrt{16}$
**(b)** Integers: $-11, 11, \sqrt{16}$
**(c)** Rational numbers: $1.001, 0.333\ldots, -11, 11, \frac{13}{15}, \sqrt{16}, 3.14, \frac{15}{3}$
**(d)** Irrational number: $-\pi$

**3.** Commutative Property for addition

**4.** Commutative Property for multiplication

**5.** Associative Property for addition

**6.** Distributive Property

**7.** Distributive Property

**8.** Distributive Property

**9.** Commutative Property for multiplication

**10.** Distributive Property

**11.** $x + 3 = 3 + x$

**12.** $7(3x) = (7 \cdot 3)x$

**13.** $4(A + B) = 4A + 4B$

**14.** $5x + 5y = 5(x + y)$

**15.** $3(x + y) = 3x + 3y$

**16.** $(a - b)8 = 8a - 8b$

**17.** $4(2m) = (4 \cdot 2)m = 8m$

**18.** $\frac{4}{3}(-6y) = \left[\frac{4}{3}(-6)\right]y = -8y$

**19.** $-\frac{5}{2}(2x - 4y) = -\frac{5}{2}(2x) + \frac{5}{2}(4y) = -5x + 10y$

**20.** $(3a)(b + c - 2d) = 3ab + 3ac - 6ad$

**21. (a)** $\frac{3}{10} + \frac{4}{15} = \frac{9}{30} + \frac{8}{30} = \frac{17}{30}$
**(b)** $\frac{1}{4} + \frac{1}{5} = \frac{5}{20} + \frac{4}{20} = \frac{9}{20}$

**22. (a)** $\frac{2}{3} - \frac{3}{5} = \frac{10}{15} - \frac{9}{15} = \frac{1}{15}$
**(b)** $1 + \frac{5}{8} - \frac{1}{6} = \frac{24}{24} + \frac{15}{24} - \frac{4}{24} = \frac{35}{24}$

**23. (a)** $\frac{2}{3}\left(6 - \frac{3}{2}\right) = \frac{2}{3} \cdot 6 - \frac{2}{3} \cdot \frac{3}{2} = 4 - 1 = 3$
**(b)** $0.25\left(\frac{8}{9} + \frac{1}{2}\right) = \frac{1}{4}\left(\frac{16}{18} + \frac{9}{18}\right) = \frac{1}{4} \cdot \frac{25}{18} = \frac{25}{72}$

**24. (a)** $\left(3 + \frac{1}{4}\right)\left(1 - \frac{4}{5}\right) = \left(\frac{12}{4} + \frac{1}{4}\right)\left(\frac{5}{5} - \frac{4}{5}\right) = \frac{13}{4} \cdot \frac{1}{5} = \frac{13}{20}$
**(b)** $\left(\frac{1}{2} - \frac{1}{3}\right)\left(\frac{1}{2} + \frac{1}{3}\right) = \left(\frac{3}{6} - \frac{2}{6}\right)\left(\frac{3}{6} + \frac{2}{6}\right) = \frac{1}{6} \cdot \frac{5}{6} = \frac{5}{36}$

**25. (a)** $\dfrac{2}{\frac{2}{3}} - \dfrac{\frac{2}{3}}{2} = 2 \cdot \frac{3}{2} - \frac{2}{3} \cdot \frac{1}{2} = 3 - \frac{1}{3} = \frac{9}{3} - \frac{1}{3} = \frac{8}{3}$
**(b)** $\dfrac{\frac{1}{12}}{\frac{1}{8} - \frac{1}{9}} = \dfrac{\frac{1}{12}}{\frac{1}{8} - \frac{1}{9}} \cdot \frac{72}{72} = \frac{6}{9-8} = \frac{6}{1} = 6$

**26. (a)** $\dfrac{2 - \frac{3}{4}}{\frac{1}{2} - \frac{1}{3}} = \dfrac{2 - \frac{3}{4}}{\frac{1}{2} - \frac{1}{3}} \cdot \frac{12}{12} = \frac{24-9}{6-4} = \frac{15}{2}$
**(b)** $\dfrac{\frac{2}{5} + \frac{1}{2}}{\frac{1}{10} + \frac{3}{15}} = \dfrac{\frac{2}{5} + \frac{1}{2}}{\frac{1}{10} + \frac{1}{5}} = \dfrac{\frac{2}{5} + \frac{1}{2}}{\frac{1}{10} + \frac{1}{5}} \cdot \frac{10}{10} = \frac{4+5}{1+2} = \frac{9}{3} = 3$

**27. (a)** $2 \cdot 3 = 6$ and $2 \cdot \frac{7}{2} = 7$, so $3 < \frac{7}{2}$
**(b)** $-6 > -7$
**(c)** $3.5 = \frac{7}{2}$

**28. (a)** $3 \cdot \frac{2}{3} = 2$ and $3 \cdot 0.67 = 2.01$, so $\frac{2}{3} < 0.67$
**(b)** $\frac{2}{3} > -0.67$
**(c)** $|0.67| = |-0.67|$

**29. (a)** False
**(b)** True

**30. (a)** True. $\frac{10}{11} = \frac{10}{11} \cdot \frac{13}{13} = \frac{130}{143}$ and $\frac{12}{13} = \frac{12}{13} \cdot \frac{11}{11} = \frac{132}{143}$. Therefore $\frac{10}{11} < \frac{12}{13}$, because $\frac{130}{143} < \frac{132}{143}$.
**(b)** False

**31. (a)** False
**(b)** True

**32. (a)** False
**(b)** True

**33.** **(a)** $x > 0$

**(b)** $t < 4$

**(c)** $a \geq \pi$

**(d)** $-5 < x < \frac{1}{3}$

**(e)** $|p - 3| \leq 5$

**34.** **(a)** $y < 0$

**(b)** $z > 1$

**(c)** $b \leq 8$

**(d)** $0 < w \leq 17$

**(e)** $|y - \pi| \geq 2$

**35.** **(a)** $A \cup B = \{1, 2, 3, 4, 5, 6, 7, 8\}$

**(b)** $A \cap B = \{2, 4, 6\}$

**36.** **(a)** $B \cup C = \{2, 4, 6, 7, 8, 9, 10\}$

**(b)** $B \cap C = \{8\}$

**37.** **(a)** $A \cup C = \{1, 2, 3, 4, 5, 6, 7, 8, 9, 10\}$

**(b)** $A \cap C = \{7\}$

**38.** **(a)** $A \cup B \cup C = \{1, 2, 3, 4, 5, 6, 7, 8, 9, 10\}$

**(b)** $A \cap B \cap C = \varnothing$

**39.** **(a)** $B \cup C = \{x \mid x \leq 5\}$

**(b)** $B \cap C = \{x \mid -1 < x < 4\}$

**40.** **(a)** $A \cap C = \{x \mid -1 < x \leq 5\}$

**(b)** $A \cap B = \{x \mid -2 \leq x < 4\}$

**41.** $(-3, 0) = \{x \mid -3 < x < 0\}$

**42.** $(2, 8] = \{x \mid 2 < x \leq 8\}$

**43.** $[2, 8) = \{x \mid 2 \leq x < 8\}$

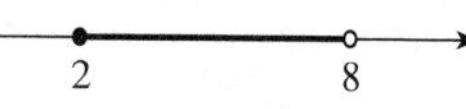


**44.** $\left[-6, -\frac{1}{2}\right] = \left\{x \mid -6 \leq x \leq -\frac{1}{2}\right\}$

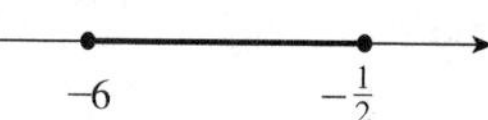


**45.** $[2, \infty) = \{x \mid x \geq 2\}$

**46.** $(-\infty, 1) = \{x \mid x < 1\}$

**47.** $x \leq 1 \quad \Leftrightarrow \quad x \in (-\infty, 1]$

**48.** $1 \leq x \leq 2 \quad \Leftrightarrow \quad x \in [1, 2]$

**49.** $-2 < x \leq 1 \quad \Leftrightarrow \quad x \in (-2, 1]$

−2 1

**50.** $x \geq -5 \quad \Leftrightarrow \quad x \in [-5, \infty)$

**51.** $x > -1 \quad \Leftrightarrow \quad x \in (-1, \infty)$

**52.** $-5 < x < 2 \quad \Leftrightarrow \quad x \in (-5, 2)$

−5 2

**53.** **(a)** $[-3, 5]$

**(b)** $(-3, 5]$

**54.** **(a)** $[0, 2)$

**(b)** $(-2, 0]$

**55.** $(-2, 0) \cup (-1, 1) = (-2, 1)$

**56.** $(-2, 0) \cap (-1, ) = (-1, 0)$

**57.** $[-4, 6] \cap [0, 8) = [0, 6]$

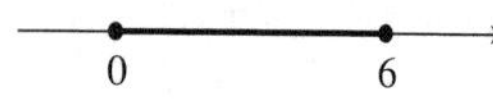


**58.** $[-4, 6] \cup [0, 8) = [-4, 8)$

**59.** $(-\infty, -4) \cup (4, \infty)$

**60.** $(-\infty, 6] \cap (2, 10) = (2, 6]$

**61. (a)** $|100| = 100$

**(b)** $|-73| = 73$

**62. (a)** $\left|\sqrt{5} - 5\right| = -\left(\sqrt{5} - 5\right) = 5 - \sqrt{5}$, since $5 > \sqrt{5}$.

**(b)** $|10 - \pi| = 10 - \pi$, since $10 > \pi$.

**63. (a)** $||-6| - |-4|| = |6 - 4| = |2| = 2$

**(b)** $\frac{-1}{|-1|} = \frac{-1}{1} = -1$

**64. (a)** $|2 - |-12|| = |2 - 12| = |-10| = 10$

**(b)** $-1 - |1 - |-1|| = -1 - |1 - 1| = -1 - |0| = -1$

**65. (a)** $|(-2) \cdot 6| = |-12| = 12$

**(b)** $\left|\left(-\frac{1}{3}\right)(-15)\right| = |5| = 5$

**66. (a)** $\left|\frac{-6}{24}\right| = \left|-\frac{1}{4}\right| = \frac{1}{4}$

**(b)** $\left|\frac{7-12}{12-7}\right| = \left|-\frac{5}{5}\right| = |-1| = 1$

**67.** $|(-2) - 3| = |-5| = 5$

**68.** $|-2.5 - 1.5| = |-4| = 4$

**69. (a)** $|17 - 2| = 15$

**(b)** $|21 - (-3)| = |21 + 3| = |24| = 24$

**(c)** $\left|-\frac{3}{10} - \frac{11}{8}\right| = \left|-\frac{12}{40} - \frac{55}{40}\right| = \left|-\frac{67}{40}\right| = \frac{67}{40}$

**70. (a)** $\left|\frac{7}{15} - \left(-\frac{1}{21}\right)\right| = \left|\frac{49}{105} + \frac{5}{105}\right| = \left|\frac{54}{105}\right| = \left|\frac{18}{35}\right| = \frac{18}{35}$

**(b)** $|-38 - (-57)| = |-38 + 57| = |19| = 19$.

**(c)** $|-2.6 - (-1.8)| = |-2.6 + 1.8| = |-0.8| = 0.8$.

**71. (a)** Let $x = 0.777\ldots$. Then $10x = 7.7777\ldots$ and $x = 0.7777\ldots$. Subtracting, we get $9x = 7$. Thus, $x = \dfrac{7}{9}$.

**(b)** Let $x = 0.2888\ldots$. Then $100x = 28.8888\ldots$ and $10x = 2.8888\ldots$. Subtracting, we get $90x = 26$. Thus, $x = \dfrac{26}{90} = \dfrac{13}{45}$.

**(c)** Let $x = 0.575757\ldots$. Then $100x = 57.5757\ldots$ and $x = 0.5757\ldots$. Subtracting, we get $99x = 57$. Thus, $x = \dfrac{57}{99} = \dfrac{19}{33}$.

**72. (a)** Let $x = 5.2323\ldots$. Then $100x = 523.2323\ldots$ and $x = 5.2323\ldots$. Subtracting, we get $99x = 518$. Thus, $x = \dfrac{518}{99}$.

**(b)** Let $x = 1.3777\ldots$. Then $100x = 137.7777\ldots$ and $10x = 13.7777\ldots$. Subtracting, we get $90x = 124$. Thus, $x - \dfrac{124}{90} = \dfrac{62}{45}$

**(c)** Let $x = 2.13535\ldots$. Then $1000x = 2135.3535\ldots$ and $10x = 21.3535\ldots$. Subtracting, we get $990x = 2114$. Thus, $x = \dfrac{2114}{990} = \dfrac{1057}{495}$.

**73.** Distributive Property

**74.**

| Day | $T_O$ | $T_G$ | $T_O - T_G$ | $\lvert T_O - T_G \rvert$ |
|---|---|---|---|---|
| Sunday | 68 | 77 | −9 | 9 |
| Monday | 72 | 75 | −3 | 3 |
| Tuesday | 74 | 74 | 0 | 0 |
| Wednesday | 80 | 75 | 5 | 5 |
| Thursday | 77 | 69 | 8 | 8 |
| Friday | 71 | 70 | 1 | 1 |
| Saturday | 70 | 71 | −1 | 1 |

$T_O - T_G$ gives more information because it tells us which city had the higher temperature.

**75. (a)** Is $\frac{1}{28}x + \frac{1}{34}y \leq 15$ when $x = 165$ and $y = 230$? We have $\frac{1}{28}(165) + \frac{1}{34}(230) = 5.89 + 6.76 = 12.65 \leq 15$, so indeed the car can travel 165 city miles and 230 highway miles without running out of gas.

**(b)** Here we must solve for $y$ when $x = 280$. So $\frac{1}{28}(280) + \frac{1}{34}y = 15 \quad \Leftrightarrow \quad 10 + \frac{1}{34}y = 15 \quad \Leftrightarrow \quad \frac{1}{34}y = 5 \quad \Leftrightarrow$ $y = 170$. Thus, the car can travel up to 170 highway miles without running out of gas.

**76. (a)** When $L = 60$, $x = 8$, and $y = 6$, we have $L + 2(x + y) = 60 + 2(8 + 6) = 60 + 28 = 88$. Because $88 \leq 108$ the post office will accept this package.
When $L = 48$, $x = 24$, and $y = 24$, we have $L + 2(x + y) = 48 + 2(24 + 24) = 48 + 96 = 144$, and since $144 \not\leq 108$, the post office will *not* accept this package.

**(b)** If $x = y = 9$, then $L + 2(9 + 9) \leq 108 \quad \Leftrightarrow \quad L + 36 \leq 108 \quad \Leftrightarrow \quad L \leq 72$. So the length can be as long as 72 in. = 6 ft.

**77. (a)** Negative since $a > 0 \quad \Leftrightarrow \quad -a < 0$.

**(b)** Positive since $b < 0 \quad \Leftrightarrow \quad -b > 0$.

**(c)** Positive since the product of two negative numbers is positive.

**(d)** Positive since $a - b = a + (-b)$ is the sum of two positive numbers.

**(e)** Negative $c - a = c + (-a)$ is the sum of two negative numbers.

**(f)** Positive since $a$ is positive and $bc$ is positive by part (c).

**(g)** Negative. $ab$ is negative (the product of a positive and a negative) and $ac$ is negative for the same reason. Finally, the sum of two negative numbers is negative.

**(h)** Negative, since $bc$ is positive by Part (c) and $a(bc)$ is the product of two positive number, therefore positive, hence and $-abc$ is negative.

**(i)** Positive since $a$ is known to be positive and $b^2$ is also positive, being a square. The product of two positive numbers is positive.

**78.** Let $x = \dfrac{m_1}{n_1}$ and $y = \dfrac{m_2}{n_2}$ be rational numbers. Then $x + y = \dfrac{m_1}{n_1} + \dfrac{m_2}{n_2} = \dfrac{m_1 n_2 + m_2 n_1}{n_1 n_2}$, $x - y = \dfrac{m_1}{n_1} - \dfrac{m_2}{n_2} = \dfrac{m_1 n_2 - m_2 n_1}{n_1 n_2}$, and $x \cdot y = \dfrac{m_1}{n_1} \cdot \dfrac{m_2}{n_2} = \dfrac{m_1 m_2}{n_1 n_2}$. This shows that the sum, difference, and product of two rational numbers are again rational numbers. However the product of two irrational numbers is not necessarily irrational; for example, $\sqrt{2} \cdot \sqrt{2} = 2$, which is rational. Also, the sum of two irrational numbers is not necessarily irrational; for example, $\sqrt{2} + (-\sqrt{2}) = 0$ which is rational.

**79.** $\frac{1}{2} + \sqrt{2}$ is irrational. Since the sum of two rational numbers is rational (see Exercise 78), if $\frac{1}{2} + \sqrt{2}$ were rational, then sum $\left(\frac{1}{2} + \sqrt{2}\right) + \left(-\frac{1}{2}\right) = \sqrt{2}$ is rational. But this is a contradiction. In general the sum of a rational number and an irrational number is irrational. Zero is rational and the product of zero and any real number is zero. If the rational number is not zero, $\dfrac{a}{b} \neq 0$, then the product of it and any irrational number $w$ is irrational. Again, if we assume the product is rational, say $\dfrac{a}{b} \cdot w = \dfrac{c}{d}$, then since $\dfrac{a}{b} \neq 0$, we have $\dfrac{b}{a} \neq 0$ so $\dfrac{b}{a}\left(\dfrac{a}{b} \cdot w\right) = \dfrac{b}{a} \cdot \dfrac{c}{d} \quad \Leftrightarrow \quad w = \dfrac{bc}{ad}$ which is rational. But $w$ is irrational.

**80.**

| $x$ | 1 | 2 | 10 | 100 | 1000 |
|---|---|---|---|---|---|
| $\frac{1}{x}$ | 1 | $\frac{1}{2}$ | $\frac{1}{10}$ | $\frac{1}{100}$ | $\frac{1}{1000}$ |

As $x$ gets large, the fraction $\dfrac{1}{x}$ gets small. Mathematically, we say that $\dfrac{1}{x}$ goes to zero.

| $x$ | 1 | 0.5 | 0.1 | 0.01 | 0.001 |
|---|---|---|---|---|---|
| $\frac{1}{x}$ | 1 | $\frac{1}{0.5} = 2$ | $\frac{1}{0.1} = 10$ | $\frac{1}{0.01} = 100$ | $\frac{1}{0.001} = 1000$ |

As $x$ gets small, the fraction $\dfrac{1}{x}$ gets large. Mathematically, we say that $\dfrac{1}{x}$ goes to infinity.

**81. (a)** Construct the number $\sqrt{2}$ on the number line by transferring the length of the hypotenuse of a right triangle with legs of length 1 and 1.

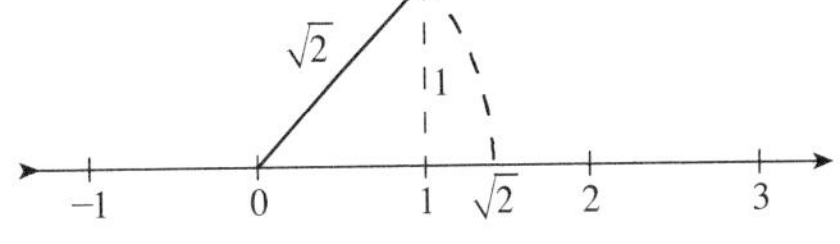


**(b)** Construct a right triangle with legs of length 1 and 2. By the Pythagorean Theorem, the length of the hypotenuse is $\sqrt{1^2 + 2^2} = \sqrt{5}$. Then transfer the length of the hypotenuse to the number line.

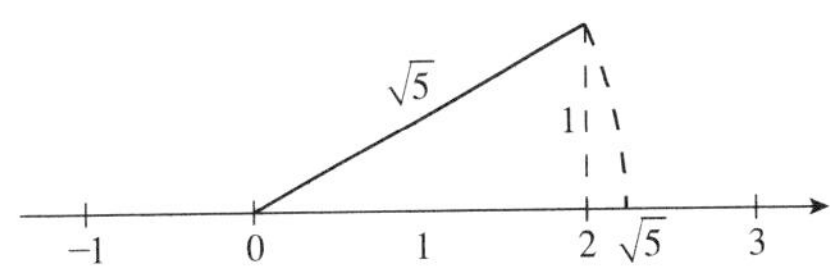


**(c)** Construct a right triangle with legs of length $\sqrt{2}$ and 2 [construct $\sqrt{2}$ as in part (a)]. By the Pythagorean Theorem, the length of the hypotenuse is $\sqrt{\left(\sqrt{2}\right)^2 + 2^2} = \sqrt{6}$. Then transfer the length of the hypotenuse to the number line.

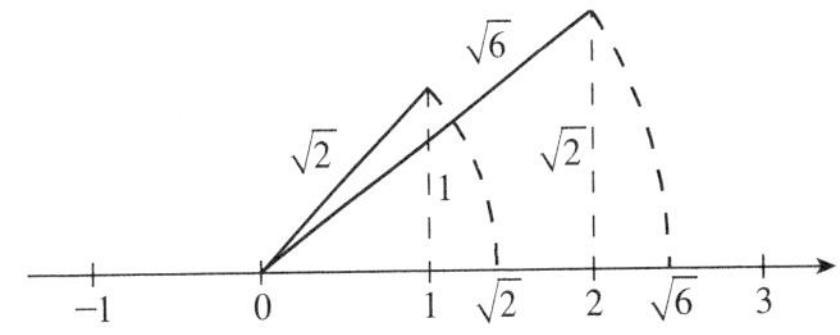


**82. (a)** Subtraction is not commutative. For example, $5 - 1 \neq 1 - 5$.

**(b)** Division is not commutative. For example, $5 \div 1 \neq 1 \div 5$.

**(c)** Putting on socks and putting on shoes are not commutative. (Wearing your socks over your shoes is not the same as wearing your shoes over your socks!)

**(d)** These actions commute, since you get the same result when you put on your coat and then your hat as when you put on your hat and then your coat.

**(e)** No, you must wash the laundry before drying it.

**(f)** Answers will vary.

**(g)** Answers will vary.

## P.3 Integer Exponents

**1.** $5^2 \cdot 5 = 5^3 = 125$

**2.** $2^3 \cdot 2^2 = 2^{3+2} = 2^5 = 32$

**3.** $\left(2^3\right)^2 = 2^6 = 64$

**4.** $\left(2^3\right)^0 = 1$

**5.** $(-6)^0 = 1$

**6.** $-6^0 = -\left(6^0\right) = -(1) = -1$

**7.** $-3^2 = -9$

**8.** $(-3)^2 = 9$

**9.** $\left(\frac{1}{3}\right)^4 \cdot 3^6 = \dfrac{1}{3^4} \cdot 3^6 = 3^{-4} \cdot 3^6 = 3^2 = 9$

**10.** $5^2 \cdot \left(\frac{1}{5}\right)^3 = 5^2 \cdot 5^{-3} = 5^{2-3} = 5^{-1} = \frac{1}{5}$

**11.** $\dfrac{10^7}{10^4} = 10^{7-4} = 10^3 = 1{,}000$

**12.** $\dfrac{3}{3^{-2}} = 3^{1-(-2)} = 3^3 = 27$

**13.** $\dfrac{4^{-3}}{2^{-8}} = \dfrac{\left(2^2\right)^{-3}}{2^{-8}} = \dfrac{2^{-6}}{2^{-8}} = 2^{-6-(-8)} = 2^{-6+8} = 2^2 = 4$

**14.** $\dfrac{3^{-2}}{9} = \dfrac{3^{-2}}{3^2} = \dfrac{1}{3^{2-(-2)}} = \dfrac{1}{3^4} = \frac{1}{81}$

**15.** $\left(\frac{1}{4}\right)^{-2} = (4)^2 = 16$

**16.** $\left(\dfrac{2}{3}\right)^{-3} = \left(\dfrac{3}{2}\right)^3 = \dfrac{3^3}{2^3} = \dfrac{27}{8}$

**17.** $\left(\dfrac{3}{2}\right)^{-2} \cdot \dfrac{9}{16} = \left(\dfrac{2}{3}\right)^2 \cdot \dfrac{9}{16} = \dfrac{2^2}{3^2} \cdot \dfrac{9}{16} = \dfrac{4}{9} \cdot \dfrac{9}{16} = \dfrac{1}{4}$

**18.** $\left(\dfrac{1}{2}\right)^4 \cdot \left(\dfrac{5}{2}\right)^{-2} = \left(\dfrac{1}{2}\right)^4 \cdot \left(\dfrac{2}{5}\right)^2 = \dfrac{1}{2^4} \cdot \dfrac{2^2}{5^2} = \dfrac{1}{2^{4-2}} \cdot \dfrac{1}{5^2} = \dfrac{1}{2^2 \cdot 5^2} = \dfrac{1}{100}$

**19.** $\left(\dfrac{1}{13}\right)^0 \left(\dfrac{2}{3}\right)^6 \left(\dfrac{4}{9}\right)^{-3} = 1 \cdot \left(\dfrac{2^6}{3^6}\right)\left(\dfrac{9}{4}\right)^3 = \left(\dfrac{2^6}{3^6}\right)\left(\dfrac{3^2}{2^2}\right)^3 = \left(\dfrac{2^6}{3^6}\right)\left(\dfrac{3^6}{2^6}\right) = 1$

**20.** $\dfrac{3^2 \cdot 4^{-2} \cdot 5}{2^{-4} \cdot 3^3 \cdot 25} = \dfrac{3^2 \cdot \left(2^2\right)^{-2} \cdot 5}{2^{-4} \cdot 3^3 \cdot 5^2} = \dfrac{3^2 \cdot 2^{-4} \cdot 5}{2^{-4} \cdot 3^3 \cdot 5^2} = 2^{-4-(-4)} \cdot 3^{2-3} \cdot 5^{1-2} = 2^0 \cdot 3^{-1} \cdot 5^{-1} = \dfrac{1}{3 \cdot 5} = \dfrac{1}{15}$

**21.** $2^{-2} + 2^{-3} = \dfrac{1}{2^2} + \dfrac{1}{2^3} = \dfrac{1}{4} + \dfrac{1}{8} = \dfrac{2}{8} + \dfrac{1}{8} = \dfrac{3}{8}$

**22.** $3^{-1} - 3^{-3} = \dfrac{1}{3} - \dfrac{1}{3^3} = \dfrac{1}{3} - \dfrac{1}{27} = \dfrac{9}{27} - \dfrac{1}{27} = \dfrac{8}{27}$

**23.** $(2x)^4 x^3 = 2^4 x^4 x^3 = 16x^{4+3} = 16x^7$

**24.** $\frac{1}{6}x\left(3x^2\right)^3 = \frac{1}{6}x\left(3^3 x^{2\cdot 3}\right) = \dfrac{3^3}{2 \cdot 3} \cdot x \cdot x^6 = \dfrac{3^2}{2} \cdot x^{1+6} = \frac{9}{2}x^7$

**25.** $(-3y)^4 = (-3)^4 y^4 = 3^4 y^4 = 81y^4$

**26.** $\left(5x^2\right)^3 \left(\frac{1}{25}x^4\right)^2 = \left(5^3 x^6\right)\left[\left(\dfrac{1}{5^2}\right)^2 x^8\right] = \left(5^3 x^6\right)\left(\dfrac{1}{5^4}x^8\right) = 5^{3-4} \cdot x^{6+8} = 5^{-1}x^{14} = \frac{1}{5}x^{14}$

**27.** $(3z)^2 \left(6z^2\right)^{-3} = \left(3^2 z^2\right)\left(6^{-3} z^{2(-3)}\right) = \left(3^2 z^2\right)\left(6^{-3} z^{-6}\right) = 3^2 (2 \cdot 3)^{-3} z^{2-6} = 3^2 \left(2^{-3} \cdot 3^{-3}\right) z^{2-6}$
$= 3^{2-3} 2^{-3} z^{-4} = 3^{-1} 2^{-3} z^{-4} = \dfrac{1}{3 \cdot 2^3 z^4} = \dfrac{1}{24z^4}$

**28.** $\left(2z^2\right)^{-5} z^{10} = 2^{-5} z^{-10} z^{10} = 2^{-5} z^{-10+10} = \dfrac{1}{2^5} z^0 = \dfrac{1}{32}$

**29.** $\dfrac{5x^2}{25x^5} = \frac{5}{25} \cdot \dfrac{x^2}{x^5} = \frac{1}{5} \cdot \dfrac{1}{x^{5-2}} = \frac{1}{5} \cdot \dfrac{1}{x^3} = \dfrac{1}{5x^3}$

**30.** $\left(\dfrac{3}{x}\right)^4 \left(\dfrac{4}{x}\right)^{-2} = \left(\dfrac{3}{x}\right)^4 \left(\dfrac{x}{4}\right)^2 = \dfrac{3^4}{x^4} \cdot \dfrac{x^2}{4^2} = \dfrac{3^4}{4^2} \cdot x^{2-4} = \frac{81}{16}x^{-2} = \dfrac{81}{16x^2}$

**31.** $\dfrac{10(x+y)^4}{5(x+y)^3} = \dfrac{10}{5} \cdot \dfrac{(x+y)^4}{(x+y)^3} = 2 \cdot (x+y)^{4-3} = 2(x+y) = 2x + 2y$

**32.** $\dfrac{[2(r-s)]^2}{(r-s)^3} = \dfrac{2^2 (r-s)^2}{(r-s)^3} = \dfrac{2^2}{(r-s)^{3-2}} = \dfrac{4}{r-s}$

**33.** $a^9 a^{-5} = a^{9-5} = a^4$

**34.** $\left(3y^2\right)\left(4y^5\right) = 3 \cdot 4y^{2+5} = 12y^7$

**35.** $\left(12x^2y^4\right)\left(\frac{1}{2}x^5y\right) = \left(12 \cdot \frac{1}{2}\right) x^{2+5} y^{4+1} = 6x^7y^5$

**36.** $(6y)^3 = 6^3 y^3 = 216y^3$

**37.** $\dfrac{x^9 (2x)^4}{x^3} = 2^4 \cdot x^{9+4-3} = 16x^{10}$

**38.** $\dfrac{a^{-3}b^4}{a^{-5}b^5} = a^{-3-(-5)} b^{4-5} = a^2 b^{-1} = \dfrac{a^2}{b}$

**39.** $b^4\left(\frac{1}{3}b^2\right)\left(12b^{-8}\right) = \dfrac{12}{3} b^{4+2-8} = 4b^{-2} = \dfrac{4}{b^2}$

**40.** $\left(2s^3t^{-1}\right)\left(\frac{1}{4}s^6\right)\left(16t^4\right) = \dfrac{2 \cdot 16}{4} s^{3+6} t^{-1+4} = 8s^9t^3$

**41.** $(rs)^3 (2s)^{-2} (4r)^4 = r^3 s^3 2^{-2} s^{-2} 4^4 r^4 = r^3 s^3 2^{-2} s^{-2} 2^{2\cdot 4} r^4 = 2^{-2+8} r^{3+4} s^{3-2} = 2^6 r^7 s = 64r^7 s$

**42.** $\left(2u^2v^3\right)^3 \left(3u^3v\right)^{-2} = 2^3 u^6 v^9 \cdot 3^{-2} u^{-6} v^{-2} = \dfrac{2^3}{3^2} u^{6-6} v^{9-2} = \frac{8}{9}v^7$

**43.** $\dfrac{(6y^3)^4}{2y^5} = \dfrac{6^4 y^{3\cdot 4}}{2y^5} = \dfrac{6^4}{2}\, y^{12-5} = 648y^7$

**44.** $\dfrac{(2x^3)^2(3x^4)}{(x^3)^4} = \dfrac{2^2x^6 \cdot 3x^4}{x^{12}} = 12x^{6+4-12} = 12x^{-2} = \dfrac{12}{x^2}$

**45.** $\dfrac{(x^2y^3)^4(xy^4)^{-3}}{x^2y} = \dfrac{x^8y^{12}x^{-3}y^{-12}}{x^2y} = x^{8-3-2}y^{12-12-1} = x^3y^{-1} = \dfrac{x^3}{y}$

**46.** $\left(\dfrac{c^4d^3}{cd^2}\right)\left(\dfrac{d^2}{c^3}\right)^3 = \dfrac{c^4d^3}{cd^2} \cdot \dfrac{d^6}{c^9} = c^{4-1-9}d^{3+6-2} = c^{-6}d^7 = \dfrac{d^7}{c^6}$

**47.** $\dfrac{(xy^2z^3)^4}{(x^3y^2z)^3} = \dfrac{x^4y^8z^{12}}{x^9y^6z^3} = x^{4-9}y^{8-6}z^{12-3} = x^{-5}y^2z^9 = \dfrac{y^2z^9}{x^5}$

**48.** $\left(\dfrac{xy^{-2}z^{-3}}{x^2y^3z^{-4}}\right)^{-3} = \dfrac{x^{-3}y^6z^9}{x^{-6}y^{-9}z^{12}} = x^{-3-(-6)}y^{6-(-9)}z^{9-12} = x^3y^{15}z^{-3} = \dfrac{x^3y^{15}}{z^3}$

**49.** $\left(\dfrac{q^{-1}rs^{-2}}{r^{-5}sq^{-8}}\right)^{-1} = \dfrac{qr^{-1}s^2}{r^5s^{-1}q^8} = q^{1-8}r^{-1-5}s^{2-(-1)} = q^{-7}r^{-6}s^3 = \dfrac{s^3}{q^7r^6}$

**50.** $(3ab^2c)\left(\dfrac{2a^2b}{c^3}\right)^{-2} = 3ab^2c \cdot \dfrac{2^{-2}a^{-4}b^{-2}}{c^{-6}} = \frac{3}{4}a^{1-4}b^{2-2}c^{1-(-6)} = \frac{3}{4}a^{-3}b^0c^7 = \dfrac{3c^7}{4a^3}$

**51.** $69{,}300{,}000 = 6.93 \times 10^7$

**52.** $7{,}200{,}000{,}000{,}000 = 7.2 \times 10^{12}$

**53.** $0.000028536 = 2.8536 \times 10^{-5}$

**54.** $0.0001213 = 1.213 \times 10^{-4}$

**55.** $129{,}540{,}000 = 1.2954 \times 10^8$

**56.** $7{,}259{,}000{,}000 = 7.259 \times 10^9$

**57.** $0.0000000014 = 1.4 \times 10^{-9}$

**58.** $0.0007029 = 7.029 \times 10^{-4}$

**59.** $3.19 \times 10^5 = 319{,}000$

**60.** $2.721 \times 10^8 = 272{,}100{,}000$

**61.** $2.670 \times 10^{-8} = 0.00000002670$

**62.** $9.999 \times 10^{-9} = 0.000000009999$

**63.** $7.1 \times 10^{14} = 710{,}000{,}000{,}000{,}000$

**64.** $6 \times 10^{12} = 6{,}000{,}000{,}000{,}000$

**65.** $8.55 \times 10^{-3} = 0.00855$

**66.** $6.257 \times 10^{-10} = 0.0000000006257$

**67.** **(a)** $5{,}900{,}000{,}000{,}000 \text{ mi} = 5.9 \times 10^{12}$ mi

**(b)** $0.0000000000004 \text{ cm} = 4 \times 10^{-13}$ cm

**(c)** 33 billion billion molecules $= 33 \times 10^9 \times 10^9 = 3.3 \times 10^{19}$ molecules

**68.** **(a)** $93{,}000{,}000 \text{ mi} = 9.3 \times 10^7$ mi

**(b)** $0.000000000000000000000053 \text{ g} = 5.3 \times 10^{-23}$ g

**(c)** $5{,}970{,}000{,}000{,}000{,}000{,}000{,}000{,}000 \text{ kg} = 5.97 \times 10^{24}$ kg

**69.** $(7.2 \times 10^{-9})(1.806 \times 10^{-12}) = 7.2 \times 1.806 \times 10^{-9} \times 10^{-12} \approx 13.0 \times 10^{-21} = 1.3 \times 10^{-20}$

**70.** $(1.062 \times 10^{24})(8.61 \times 10^{19}) = 1.062 \times 8.61 \times 10^{24} \times 10^{19} \approx 9.14 \times 10^{43}$

**71.** $\dfrac{1.295643 \times 10^9}{(3.610 \times 10^{-17})(2.511 \times 10^6)} = \dfrac{1.295643}{3.610 \times 2.511} \times 10^{9+17-6} \approx 0.1429 \times 10^{19} = 1.429 \times 10^{19}$

**72.** $\dfrac{(73.1)(1.6341 \times 10^{28})}{0.0000000019} = \dfrac{(7.31 \times 10)(1.6341 \times 10^{28})}{1.9 \times 10^{-9}} = \dfrac{7.31 \times 1.6341}{1.9} \times 10^{1+28-(-9)} \approx 6.3 \times 10^{38}$

**73.** $\dfrac{(0.0000162)(0.01582)}{(594621000)(0.0058)} = \dfrac{(1.62 \times 10^{-5})(1.582 \times 10^{-2})}{(5.94621 \times 10^8)(5.8 \times 10^{-3})} = \dfrac{1.62 \times 1.582}{5.94621 \times 5.8} \times 10^{-5-2-8+3} = 0.074 \times 10^{-12}$

$= 7.4 \times 10^{-14}$

**74.** $\dfrac{(3.542 \times 10^{-6})^9}{(5.05 \times 10^4)^{12}} = \dfrac{(3.542)^9 \times 10^{-54}}{(5.05)^{12} \times 10^{48}} = \dfrac{87747.96}{275103767.10} \times 10^{-54-48} \approx 3.19 \times 10^{-4} \times 10^{-102} \approx 3.19 \times 10^{-106}$

**75. (a)** $b^5$ is negative since a negative number raised to an odd power is negative.

**(b)** $b^{10}$ is positive since a negative number raised to an even power is positive.

**(c)** $ab^2c^3$ we have (positive) $(\text{negative})^2$ $(\text{negative})^3$ = (positive) (positive) (negative) which is negative.

**(d)** Since $b - a$ is negative, $(b-a)^3 = (\text{negative})^3$ which is negative.

**(e)** Since $b - a$ is negative, $(b-a)^4 = (\text{negative})^4$ which is positive.

**(f)** $\dfrac{a^3c^3}{b^6c^6} = \dfrac{(\text{positive})^3\,(\text{negative})^3}{(\text{negative})^6\,(\text{negative})^6} = \dfrac{(\text{positive})\,(\text{negative})}{(\text{positive})\,(\text{positive})} = \dfrac{\text{negative}}{\text{positive}}$ which is negative.

**76. (a)** $\dfrac{a^m}{a^n} = \dfrac{a \cdot a \cdot a \cdot \cdots \cdot a\ (m \text{ factors})}{a \cdot a \cdot a \cdot \cdots \cdot a\ (n \text{ factors})} = a \cdot a \cdot a \cdot \cdots \cdot a\ (m-n \text{ factors}) = a^{m-n}, a \neq 0$

**(b)** $\left(\dfrac{a}{b}\right)^n = \dfrac{a}{b} \cdot \dfrac{a}{b} \cdot \cdots \cdot \dfrac{a}{b}\ (n \text{ factors}) = \dfrac{a \cdot a \cdot a \cdot \cdots \cdot a\ (n \text{ factors})}{b \cdot b \cdot b \cdot \cdots \cdot b\ (n \text{ factors})} = \dfrac{a^n}{b^n}, b \neq 0$

**(c)** $\left(\dfrac{a}{b}\right)^{-n} = \left[\left(\dfrac{a}{b}\right)^{-1}\right]^n = \left[\dfrac{1}{\left(\frac{a}{b}\right)}\right]^n = \left(\dfrac{b}{a}\right)^n = \dfrac{b^n}{a^n}$, $a \neq 0$, $b \neq 0$. The first equality follows from Law 3. The second equality is given by the Negative Exponents Law. The third equality is simplification, and the fourth equality follows from Law 5.

**77.** Since one light year is $5.9 \times 10^{12}$ miles, Centauri is about $4.3 \times 5.9 \times 10^{12} \approx 2.54 \times 10^{13}$ miles away or 25,400,000,000,000 miles away.

**78.** $9.3 \times 10^7 \text{ mi} = 186{,}000 \dfrac{\text{mi}}{\text{s}} \times t \text{ s} \quad \Leftrightarrow \quad t = \dfrac{9.3 \times 10^7}{186{,}000} \text{ s} = 500 \text{ s} = 8\tfrac{1}{3} \text{ min.}$

**79.** Volume = (average depth) (area) $= (3.7 \times 10^3 \text{ m})\,(3.6 \times 10^{14} \text{ m}^2)\left(\dfrac{10^3 \text{ liters}}{\text{m}^3}\right) \approx 1.33 \times 10^{21}$ liters

**80.** Each person's share is equal to $\dfrac{\text{national debt}}{\text{population}} = \dfrac{5.736 \times 10^{12}}{2.83 \times 10^8} \approx 2.03 \times 10^4 = \$20{,}300.$

**81.** The number of molecules is equal to

$$(\text{volume}) \cdot \left(\frac{\text{liters}}{\text{m}^3}\right) \cdot \left(\frac{\text{molecules}}{22.4 \text{ liters}}\right) = (5 \cdot 10 \cdot 3) \cdot (10^3) \cdot \left(\frac{6.02 \times 10^{23}}{22.4}\right) \approx 4.03 \times 10^{27}$$

**82. (a)**

| Person | Weight | Height | BMI $= 703\dfrac{W}{H^2}$ | Result |
|---|---|---|---|---|
| Brian | 295 lb | 5 ft 10 in. = 70 in. | 42.32 | obese |
| Linda | 105 lb | 5 ft 6 in. = 66 in. | 16.95 | underweight |
| Larry | 220 lb | 6 ft 4 in. = 76 in. | 26.78 | overweight |
| Helen | 110 lb | 5 ft 2 in. = 62 in. | 20.12 | normal |

**(b)** Answers will vary.

**83.**

| Year | Total interest |
|---|---|
| 1 | \$152.08 |
| 2 | 308.79 |
| 3 | 470.26 |
| 4 | 636.64 |
| 5 | 808.08 |

**84.** Since $10^6 = 10^3 \cdot 10^3$ it would take 1000 days $\approx 2.74$ years to spend the million dollars.
Since $10^9 = 10^3 \cdot 10^6$ it would take $10^6 = 1{,}000{,}000$ days $\approx 2739.72$ years to spend the billion dollars.

**85. (a)** $\dfrac{18^5}{9^5} = \left(\dfrac{18}{9}\right)^5 = 2^5 = 32$

**(b)** $20^6 \cdot (0.5)^6 = (20 \cdot 0.5)^6 = 10^6 = 1{,}000{,}000$

**86.** $\left|10^{50} - 10^{10}\right| < 10^{50}$, whereas $\left|10^{101} - 10^{100}\right| = 10^{100}\left|10 - 1\right| = 9 \times 10^{100} > 10^{50}$. So $10^{10}$ is closer to $10^{50}$ than $10^{100}$ is to $10^{101}$.

# P.4 Rational Exponents and Radicals

**1.** $\dfrac{1}{\sqrt{5}} = 5^{-1/2}$

**2.** $\sqrt[3]{7^2} = 7^{2/3}$

**3.** $4^{2/3} = \sqrt[3]{4^2} = \sqrt[3]{16}$

**4.** $11^{-3/2} = \left(11^{3/2}\right)^{-1} = \left(\sqrt{11^3}\right)^{-1} = \dfrac{1}{\sqrt{11^3}}$

**5.** $\sqrt[5]{5^3} = 5^{3/5}$

**6.** $2^{-1.5} = 2^{-3/2} = \dfrac{1}{\sqrt{2^3}} = \dfrac{1}{\sqrt{8}}$

**7.** $a^{2/5} = \sqrt[5]{a^2}$

**8.** $\dfrac{1}{\sqrt{x^5}} = \dfrac{1}{x^{5/2}} = x^{-5/2}$

**9. (a)** $\sqrt{16} = \sqrt{4^2} = 4$

**(b)** $\sqrt[4]{16} = \sqrt[4]{2^4} = 2$

**(c)** $\sqrt[4]{\dfrac{1}{16}} = \sqrt[4]{\left(\dfrac{1}{2}\right)^4} = \dfrac{1}{2}$

**10. (a)** $\sqrt{64} = \sqrt{8^2} = 8$

**(b)** $\sqrt[3]{-64} = \sqrt[3]{(-4)^3} = -4$

**(c)** $\sqrt[5]{-32} = \sqrt[5]{(-2)^5} = -2$

**11. (a)** $\sqrt{\dfrac{4}{9}} = \sqrt{\left(\dfrac{2}{3}\right)^2} = \dfrac{2}{3}$

**(b)** $\sqrt[4]{256} = \sqrt[4]{4^4} = 4$

**(c)** $\sqrt[6]{\dfrac{1}{64}} = \sqrt[6]{\dfrac{1}{2^6}} = \dfrac{\sqrt[6]{1}}{\sqrt[6]{2^6}} = \dfrac{1}{2}$

**12. (a)** $\sqrt{7}\sqrt{28} = \sqrt{7}\sqrt{4 \cdot 7} = \sqrt{7} \cdot 2 \cdot \sqrt{7} = 2 \cdot 7 = 14$

**(b)** $\dfrac{\sqrt{48}}{\sqrt{3}} = \sqrt{\dfrac{48}{3}} = \sqrt{16} = 4$

**(c)** $\sqrt[4]{24}\sqrt[4]{54} = \sqrt[4]{8 \cdot 3}\sqrt[4]{2 \cdot 27} = \sqrt[4]{2^3 \cdot 3 \cdot 2 \cdot 3^3}$
$= \sqrt[4]{2^4 \cdot 3^4} = 2 \cdot 3 = 6$

**13. (a)** $\left(\dfrac{4}{9}\right)^{-1/2} = \left(\dfrac{2^2}{3^2}\right)^{-1/2} = \dfrac{2^{-1}}{3^{-1}} = \dfrac{3}{2}$

**(b)** $(-32)^{2/5} = \left[(-2)^5\right]^{2/5} = (-2)^2 = 4$

**(c)** $(-125)^{-1/3} = \left[(-5)^3\right]^{-1/3} = (-5)^{-1} = \dfrac{1}{-5} = -\dfrac{1}{5}$

**14. (a)** $1024^{-0.1} = \left(2^{10}\right)^{-0.1} = 2^{-1} = \frac{1}{2}$

**(b)** $\left(-\frac{27}{8}\right)^{2/3} = \left(\sqrt[3]{-\frac{27}{8}}\right)^2 = \left(-\frac{3}{2}\right)^2 = \frac{9}{4}$

**(c)** $\left(\frac{25}{64}\right)^{3/2} = \left(\sqrt{\frac{25}{64}}\right)^3 = \left(\frac{5}{8}\right)^3 = \frac{125}{512}$

**15.** **(a)** $\left(\frac{1}{32}\right)^{2/5} = \left(\frac{1}{2^5}\right)^{2/5} = \left(\frac{1}{2}\right)^{5\cdot(2/5)} = \left(\frac{1}{2}\right)^2 = \frac{1}{4}$

**(b)** $(27)^{-4/3} = (3^3)^{-4/3} = (3)^{3\cdot(-4/3)} = 3^{-4} = \frac{1}{3^4} = \frac{1}{81}$

**(c)** $\left(\frac{1}{8}\right)^{-2/3} = \left(\frac{1}{2^3}\right)^{-2/3} = (2^{-3})^{-2/3} = 2^{(-3)\cdot(-2/3)} = 2^2 = 4$

**16.** **(a)** $(-1000)^{-2/3} = \left[(-10)^3\right]^{-2/3} = (-10)^{-2} = \frac{1}{100}$

**(b)** $(10{,}000)^{-3/2} = (10^4)^{-3/2} = 10^{4\cdot(-3/2)} = 10^{-6} = \frac{1}{1{,}000{,}000}$

**(c)** $(-8000)^{4/3} = \left[(-20)^3\right]^{4/3} = (-20)^{3\cdot(4/3)} = (-20)^4 = 160{,}000$

**17.** **(a)** $100^{-1.5} = (10^2)^{-1.5} = 10^{2\cdot(-1.5)} = 10^{-3} = \frac{1}{10^3} = \frac{1}{1000}$

**(b)** $4^{2/3} \cdot 6^{2/3} \cdot 9^{2/3} = (2^2)^{2/3} \cdot (2 \cdot 3)^{2/3} \cdot (3^2)^{2/3} = 2^{4/3} \cdot 2^{2/3} \cdot 3^{2/3} \cdot 3^{4/3} = 2^2 \cdot 3^2 = 4 \cdot 9 = 36$

**(c)** $0.001^{-2/3} = (10^{-3})^{-2/3} = 10^2 = 100$

**18.** **(a)** $\left(\frac{1}{16}\right)^{-0.75} = \left(\frac{1}{2^4}\right)^{-3/4} = (2^{-4})^{-3/4} = 2^{(-4)\cdot(-3/4)} = 2^3 = 8$

**(b)** $0.25^{-0.5} = \left(\frac{1}{4}\right)^{-1/2} = \left(\frac{1}{2^2}\right)^{-1/2} = (2^{-2})^{-1/2} = 2^{(-2)\cdot(-1/2)} = 2^1 = 2$

**(c)** $9^{1/3} \cdot 15^{1/3} \cdot 25^{1/3} = (3^2)^{1/3} \cdot (3 \cdot 5)^{1/3} \cdot (5^2)^{1/3} = 3^{2/3} \cdot \left(3^{1/3} \cdot 5^{1/3}\right) \cdot 5^{2/3} = 3^{2/3+1/3} \cdot 5^{2/3+1/3} = 3^1 \cdot 5^1 = 15$

**19.** When $x = 3$, $y = 4$, $z = -1$ we have $\sqrt{x^2 + y^2} = \sqrt{3^2 + 4^2} = \sqrt{9 + 16} = \sqrt{25} = 5$.

**20.** When $x = 3$, $y = 4$, $z = -1$ we have $\sqrt[4]{x^3 + 14y + 2z} = \sqrt[4]{3^3 + 14(4) + 2(-1)} = \sqrt[4]{27 + 56 - 2} = \sqrt[4]{81} = \sqrt[4]{3^4} = 3$.

**21.** When $x = 3$, $y = 4$, $z = -1$ we have

$$\begin{aligned}(9x)^{2/3} + (2y)^{2/3} + z^{2/3} &= (9 \cdot 3)^{2/3} + (2 \cdot 4)^{2/3} + (-1)^{2/3} = (3^3)^{2/3} + (2^3)^{2/3} + (1)^{1/3} \\ &= 3^2 + 2^2 + 1 = 9 + 4 + 1 = 14.\end{aligned}$$

**22.** When $x = 3$, $y = 4$, $z = -1$ we have $(xy)^{2z} = (3 \cdot 4)^{2\cdot(-1)} = 12^{-2} = \frac{1}{144}$.

**23.** $\sqrt{32} + \sqrt{18} = \sqrt{16 \cdot 2} + \sqrt{9 \cdot 2} = \sqrt{4^2 \cdot 2} + \sqrt{3^2 \cdot 2} = 4\sqrt{2} + 3\sqrt{2} = 7\sqrt{2}$

**24.** $\sqrt{75} + \sqrt{48} = \sqrt{25 \cdot 3} + \sqrt{16 \cdot 3} = \sqrt{5^2 \cdot 3} + \sqrt{4^2 \cdot 3} = 5\sqrt{3} + 4\sqrt{3} = 9\sqrt{3}$

**25.** $\sqrt{125} - \sqrt{45} = \sqrt{25 \cdot 5} - \sqrt{9 \cdot 5} = \sqrt{5^2 \cdot 5} - \sqrt{3^2 \cdot 5} = 5\sqrt{5} - 3\sqrt{5} = 2\sqrt{5}$

**26.** $\sqrt[3]{54} - \sqrt[3]{16} = \sqrt[3]{2 \cdot 3^3} - \sqrt[3]{2^3 \cdot 2} = 3\sqrt[3]{2} - 2\sqrt[3]{2} = \sqrt[3]{2}$

**27.** $\sqrt[3]{108} - \sqrt[3]{32} = 3\sqrt[3]{4} - 2\sqrt[3]{4} = \sqrt[3]{4}$

**28.** $\sqrt{8} + \sqrt{50} = \sqrt{2^2 \cdot 2} + \sqrt{5^2 \cdot 2} = 2\sqrt{2} + 5\sqrt{2} = 7\sqrt{2}$

**29.** $\sqrt{245} - \sqrt{125} = 7\sqrt{5} - 5\sqrt{5} = 2\sqrt{5}$

**30.** $\sqrt[3]{24} - \sqrt[3]{81} = \sqrt[3]{2^3 \cdot 3} - \sqrt[3]{3^4} = 2\sqrt[3]{3} - 3\sqrt[3]{3} = -\sqrt[3]{3}$

**31.** $\sqrt[5]{96} + \sqrt[5]{3} = \sqrt[5]{32 \cdot 3} + \sqrt[5]{3} = \sqrt[5]{2^5 \cdot 3} + \sqrt[5]{3} = 2\sqrt[5]{3} + \sqrt[5]{3} = 3\sqrt[5]{3}$

**32.** $\sqrt[4]{48} - \sqrt[4]{3} = \sqrt[4]{16 \cdot 3} - \sqrt[4]{2^3 \cdot 2} = \sqrt[4]{2^4 \cdot 3} - \sqrt[4]{3} = 2\sqrt[4]{3} - \sqrt[4]{3} = \sqrt[4]{3}$

**33.** $\sqrt[4]{x^4} = |x|$

**34.** $\sqrt[5]{x^{10}} = (x^{10})^{1/5} = x^2$

**35.** $\sqrt[4]{16x} = \sqrt[4]{2^4 x^8} = 2x^2$

**36.** $\sqrt[3]{x^3 y^6} = (x^3 y^6)^{1/3} = xy^2$

**37.** $\sqrt[3]{x^3 y} = (x^3)^{1/3} y^{1/3} = x\sqrt[3]{y}$

**38.** $\sqrt{x^4 y^4} = (x^4 y^4)^{1/2} = x^2 y^2$

**39.** $\sqrt[5]{a^6 b^7} = a^{6/5} b^{7/5} = a \cdot a^{1/5} b \cdot b^{2/5} = ab\sqrt[5]{ab^2}$

**40.** $\sqrt[3]{a^2 b}\sqrt[3]{a^4 b} = \sqrt[3]{a^6 b^2} = (a^6 b^2)^{1/3} = a^2 b^{2/3} = a^2\sqrt[3]{b^2}$

**41.** $\sqrt[3]{\sqrt{64x^6}} = (8|x^3|)^{1/3} = 2|x|$

**42.** $\sqrt[4]{x^4 y^2 z^2} = \sqrt[4]{x^4}\sqrt[4]{y^2 z^2} = |x|\sqrt[4]{y^2 z^2}$

**43.** $x^{2/3}x^{1/5} = x^{(10/15+3/15)} = x^{13/15}$

**44.** $\left(2x^{3/2}\right)(4x)^{-1/2} = \left(2x^{3/2}\right)\left(2^2x\right)^{-1/2} = \left(2x^{3/2}\right)\left(2^{-1}x^{-1/2}\right) = 2^{1-1}\cdot x^{3/2-1/2} = 2^0\cdot x = x$

**45.** $\left(-3a^{1/4}\right)(9a)^{-3/2} = \left(-1\cdot 3a^{1/4}\right)\left(3^2a\right)^{-3/2} = \left(-1\cdot 3a^{1/4}\right)\left(3^{-3}a^{-3/2}\right) = -1\cdot 3^{1-3}\cdot a^{1/4-3/2}$
$= -1\cdot 3^{-3}a^{-5/4} = -\dfrac{1}{3^3\cdot a^{5/4}} = -\dfrac{1}{9a^{5/4}}$

**46.** $\left(-2a^{3/4}\right)\left(5a^{3/2}\right) = -10a^{3/4}a^{3/2} = -10a^{(3/4+6/4)} = -10a^{9/4}$

**47.** $(4b)^{1/2}\left(8b^{2/5}\right) = \sqrt{4}\cdot 8b^{1/2}b^{2/5} = 16b^{(5/10+4/10)} = 16b^{9/10}$

**48.** $\left(8x^6\right)^{-2/3} = \left(2^3\right)^{-2/3}\left(x^6\right)^{-2/3} = 2^{-6/3}x^{-12/3} = 2^{-2}x^{-4} = \dfrac{1}{4x^4}$

**49.** $\left(c^2d^3\right)^{-1/3} = c^{-2/3}d^{-1} = \dfrac{1}{c^{2/3}d}$

**50.** $\left(4x^6y^8\right)^{3/2} = \left(2^2\right)^{3/2}\left(x^6\right)^{3/2}\left(y^8\right)^{3/2} = 2^{6/2}x^{18/2}y^{24/2} = 2^3x^9y^{12} = 8x^9y^{12}$

**51.** $\left(y^{3/4}\right)^{2/3} = y^{(3/4)\cdot(2/3)} = y^{1/2}$

**52.** $\left(a^{2/5}\right)^{-3/4} = a^{(2/5)\cdot(-3/4)} = a^{-6/20} = a^{-3/10} = \dfrac{1}{a^{3/10}}$

**53.** $\left(2x^4y^{-4/5}\right)^3\left(8y^2\right)^{2/3} = 2^3x^{12}y^{-12/5}8^{2/3}y^{4/3} = 2^{3+2}x^{12}y^{(-12/5+4/3)} = \dfrac{32x^{12}}{y^{16/15}}$. Note that $8^{2/3} = \left(8^{1/3}\right)^2 = 2^2$.

**54.** $\left(x^{-5}y^3z^{10}\right)^{-3/5} = x^{15/5}y^{-9/5}z^{-30/5} = x^3y^{-9/5}z^{-6} = \dfrac{x^3}{y^{9/5}z^6}$

**55.** $\left(\dfrac{x^6y}{y^4}\right)^{5/2} = \dfrac{x^{15}y^{5/2}}{y^{10}} = x^{15}y^{5/2-10} = x^{15}y^{-15/2} = \dfrac{x^{15}}{y^{15/2}}$

**56.** $\left(\dfrac{-2x^{1/3}}{y^{1/2}z^{1/6}}\right)^4 = \dfrac{(-2)^4x^{4/3}}{y^{4/2}z^{4/6}} = \dfrac{16x^{4/3}}{y^2z^{2/3}}$

**57.** $\left(\dfrac{3a^{-2}}{4b^{-1/3}}\right)^{-1} = \dfrac{3^{-1}a^2}{4^{-1}b^{1/3}} = \dfrac{4a^2}{3b^{1/3}}$

**58.** $\dfrac{\left(y^{10}z^{-5}\right)^{1/5}}{\left(y^{-2}z^3\right)^{1/3}} = \dfrac{y^2z^{-1}}{y^{-2/3}z} = y^{2-(-2/3)}z^{-1-1} = y^{8/3}z^{-2} = \dfrac{y^{8/3}}{z^2}$

**59.** $\dfrac{(9st)^{3/2}}{\left(27s^3t^{-4}\right)^{2/3}} = \dfrac{27s^{3/2}t^{3/2}}{9s^2t^{-8/3}} = 3s^{3/2-2}t^{3/2+8/3} = 3s^{-1/2}t^{25/6} = \dfrac{3t^{25/6}}{s^{1/2}}$

**60.** $\left(\dfrac{a^2b^{-3}}{x^{-1}y^2}\right)^3\left(\dfrac{x^{-2}b^{-1}}{a^{3/2}y^{1/3}}\right) = \dfrac{a^6b^{-9}}{x^{-3}y^6}\cdot\dfrac{x^{-2}b^{-1}}{a^{3/2}y^{1/3}} = a^{6-(3/2)}b^{-9-1}x^{-2-(-3)}y^{-6-1/3} = a^{9/2}b^{-10}x^1y^{-19/3} = \dfrac{a^{9/2}x}{b^{10}y^{19/3}}$

**61.** **(a)** $\dfrac{1}{\sqrt{6}} = \dfrac{1}{\sqrt{6}}\cdot\dfrac{\sqrt{6}}{\sqrt{6}} = \dfrac{\sqrt{6}}{6}$

**(b)** $\dfrac{3}{\sqrt{2}} = \dfrac{3}{\sqrt{2}}\cdot\dfrac{\sqrt{2}}{\sqrt{2}} = \dfrac{3\sqrt{2}}{2}$

**(c)** $\dfrac{9}{\sqrt{3}} = \dfrac{9}{\sqrt{3}}\cdot\dfrac{\sqrt{3}}{\sqrt{3}} = \dfrac{9\sqrt{3}}{3} = 3\sqrt{3}$

**62.** **(a)** $\dfrac{12}{\sqrt{3}} = \dfrac{12}{\sqrt{3}}\cdot\dfrac{\sqrt{3}}{\sqrt{3}} = \dfrac{12\sqrt{3}}{3} = 4\sqrt{3}$

**(b)** $\dfrac{5}{\sqrt{2}} = \dfrac{5}{\sqrt{2}}\cdot\dfrac{\sqrt{2}}{\sqrt{2}} = \dfrac{5\sqrt{2}}{2}$

**(c)** $\dfrac{2}{\sqrt{6}} = \dfrac{2}{\sqrt{6}}\cdot\dfrac{\sqrt{6}}{\sqrt{6}} = \dfrac{2\sqrt{6}}{6} = \dfrac{\sqrt{6}}{3}$

**63. (a)** $\dfrac{1}{\sqrt[3]{4}} = \dfrac{1}{\sqrt[3]{2^2}} \cdot \dfrac{\sqrt[3]{2}}{\sqrt[3]{2}} = \dfrac{\sqrt[3]{2}}{2}$

**(b)** $\dfrac{1}{\sqrt[4]{3}} = \dfrac{1}{\sqrt[4]{3}} \cdot \dfrac{\sqrt[4]{3^3}}{\sqrt[4]{3^3}} = \dfrac{\sqrt[4]{3^3}}{3} = \dfrac{\sqrt[4]{27}}{3}$

**(c)** $\dfrac{8}{\sqrt[5]{2}} = \dfrac{8}{\sqrt[5]{2}} \cdot \dfrac{\sqrt[5]{2^4}}{\sqrt[5]{2^4}} = \dfrac{8\sqrt[5]{2^4}}{2} = 4\sqrt[5]{2^4} = 4\sqrt[5]{16}$

**64. (a)** $\dfrac{1}{\sqrt[5]{2^3}} = \dfrac{1}{\sqrt[5]{2^3}} \cdot \dfrac{\sqrt[5]{2^2}}{\sqrt[5]{2^2}} = \dfrac{\sqrt[5]{2^2}}{2} = \dfrac{\sqrt[5]{4}}{2}$

**(b)** $\dfrac{2}{\sqrt[4]{3}} = \dfrac{2}{\sqrt[4]{3}} \cdot \dfrac{\sqrt[4]{3^3}}{\sqrt[4]{3^3}} = \dfrac{2\sqrt[4]{3^3}}{3} = \dfrac{2\sqrt[4]{27}}{3}$

**(c)** $\dfrac{3}{\sqrt[4]{2^3}} = \dfrac{3}{\sqrt[4]{2^3}} \cdot \dfrac{\sqrt[4]{2}}{\sqrt[4]{2}} = \dfrac{3\sqrt[4]{2}}{2}$

**65. (a)** $\dfrac{1}{\sqrt[3]{x}} = \dfrac{1}{\sqrt[3]{x}} \cdot \dfrac{\sqrt[3]{x^2}}{\sqrt[3]{x^2}} = \dfrac{\sqrt[3]{x^2}}{x}$

**(b)** $\dfrac{1}{\sqrt[5]{x^2}} = \dfrac{1}{\sqrt[5]{x^2}} \cdot \dfrac{\sqrt[5]{x^3}}{\sqrt[5]{x^3}} = \dfrac{\sqrt[5]{x^3}}{x}$

**(c)** $\dfrac{1}{\sqrt[7]{x^3}} = \dfrac{1}{\sqrt[7]{x^3}} \cdot \dfrac{\sqrt[7]{x^4}}{\sqrt[7]{x^4}} = \dfrac{\sqrt[7]{x^4}}{x}$

**66. (a)** $\dfrac{1}{\sqrt[3]{x^2}} = \dfrac{1}{\sqrt[3]{x^2}} \cdot \dfrac{\sqrt[3]{x}}{\sqrt[3]{x}} = \dfrac{\sqrt[3]{x}}{x}$

**(b)** $\dfrac{1}{\sqrt[4]{x^3}} = \dfrac{1}{\sqrt[4]{x^3}} \cdot \dfrac{\sqrt[4]{x}}{\sqrt[4]{x}} = \dfrac{\sqrt[4]{x}}{x}$

**(c)** $\dfrac{1}{\sqrt[3]{x^4}} = \dfrac{1}{\sqrt[3]{x^3 \cdot x}} = \dfrac{1}{x\sqrt[3]{x}} = \dfrac{1}{x\sqrt[3]{x}} \cdot \dfrac{\sqrt[3]{x^2}}{\sqrt[3]{x^2}} = \dfrac{\sqrt[3]{x^2}}{x\sqrt[3]{x^3}} = \dfrac{\sqrt[3]{x^2}}{x^2}$

**67.** First convert 1135 feet to miles. This gives $1135 \text{ ft} = 1135 \cdot \dfrac{1 \text{ mile}}{5280 \text{ feet}} = 0.215$ mi. Thus the distance you can see is given by $D = \sqrt{2rh + h^2} = \sqrt{2(3960)(0.215) + (0.215)^2} \approx \sqrt{1702.8} \approx 41.3$ miles.

**68. (a)** Using $f = 0.4$ and substituting $d = 65$, we obtain $s = \sqrt{30fd} = \sqrt{30 \times 0.4 \times 65} \approx 28$ mi/h.

**(b)** Using $f = 0.5$ and substituting $s = 50$, we find $d$. This gives $s = \sqrt{30fd} \Leftrightarrow 50 = \sqrt{30 \cdot (0.5)\,d} \Leftrightarrow 50 = \sqrt{15d} \Leftrightarrow 2500 = 15d \Leftrightarrow d = \frac{500}{3} \approx 167$ feet.

**69. (a)** Substituting, we get $0.30(60) + 0.38(3400)^{1/2} - 3(650)^{1/3} \approx 18 + 0.38(58.31) - 3(8.66) \approx 18 + 22.16 - 25.98 \approx 14.18$. Since this value is less than 16, the sailboat qualifies for the race.

**(b)** Solve for $A$ when $L = 65$and $V = 600$. Substituting, we get $0.30(65) + 0.38A^{1/2} - 3(600)^{1/3} \le 16 \Leftrightarrow 19.5 + 0.38A^{1/2} - 25.30 \le 16 \Leftrightarrow 0.38A^{1/2} - 5.80 \le 16 \Leftrightarrow 0.38A^{1/2} \le 21.80 \Leftrightarrow A^{1/2} \le 57.38 \Leftrightarrow A \le 3292.0$. Thus, the largest possible sail is 3292 ft$^2$.

**70.**

| Planet | $d$ | $\sqrt{d}$ | $T$ | $\sqrt[3]{T}$ |
|---|---|---|---|---|
| Mercury | 0.387 | 0.622 | 0.241 | 0.622 |
| Venus | 0.723 | 0.850 | 0.615 | 0.850 |
| Earth | 1.000 | 1.000 | 1.000 | 1.000 |
| Mars | 1.523 | 1.234 | 1.881 | 1.234 |
| Jupiter | 5.203 | 2.281 | 11.861 | 2.281 |
| Saturn | 9.541 | 3.089 | 29.457 | 3.088 |

$\sqrt{d} \approx \sqrt[3]{T}$ or $d^3 \approx T^2$

**71.** Since 1 day = 86,400 s, 365.25 days = 31,557,600 s. Substituting, we obtain $d = \left(\dfrac{6.67 \times 10^{-11} \times 1.99 \times 10^{30}}{4\pi^2}\right)^{1/3} \cdot (3.15576 \times 10^7)^{2/3} \approx 1.5 \times 10^{11} \text{ m} = 1.5 \times 10^8$ km.

**72. (a)** Substituting the given values we get $V = 1.486\dfrac{75^{2/3} \cdot 0.050^{1/2}}{24.1^{2/3} \cdot 0.040} \approx 17.707$ ft/s.

**(b)** Since the volume of the flow is $V \cdot A$, the canal discharge is $17.707 \cdot 75 \approx 1328.0$ ft$^3$/s.

**73. (a)**

| $n$ | 1 | 2 | 5 | 10 | 100 |
|---|---|---|---|---|---|
| $2^{1/n}$ | $2^{1/1}=2$ | $2^{1/2}=1.414$ | $2^{1/5}=1.149$ | $2^{1/10}=1.072$ | $2^{1/100}=1.007$ |

So when $n$ gets large, $2^{1/n}$ decreases to 1.

**(b)**

| $n$ | 1 | 2 | 5 | 10 | 100 |
|---|---|---|---|---|---|
| $\left(\frac{1}{2}\right)^{1/n}$ | $\left(\frac{1}{2}\right)^{1/1}=0.5$ | $\left(\frac{1}{2}\right)^{1/2}=0.707$ | $\left(\frac{1}{2}\right)^{1/5}=0.871$ | $\left(\frac{1}{2}\right)^{1/10}=0.933$ | $\left(\frac{1}{2}\right)^{1/100}=0.993$ |

So when $n$ gets large, $\left(\frac{1}{2}\right)^{1/n}$ increases to 1.

**74. (a)** Since $\frac{1}{2}>\frac{1}{3}$, $2^{1/2}>2^{1/3}$.

**(b)** $\left(\frac{1}{2}\right)^{1/2}=2^{-1/2}$ and $\left(\frac{1}{2}\right)^{1/3}=2^{-1/3}$. Since $-\frac{1}{2}<-\frac{1}{3}$, we have $\left(\frac{1}{2}\right)^{1/2}<\left(\frac{1}{2}\right)^{1/3}$.

**(c)** We find a common root: $7^{1/4}=7^{3/12}=\left(7^3\right)^{1/12}=343^{1/12}$; $4^{1/3}=4^{4/12}=\left(4^4\right)^{1/12}=256^{1/12}$. So $7^{1/4}>4^{1/3}$.

**(d)** We find a common root: $\sqrt[3]{5}=5^{1/3}=5^{2/6}=\left(5^2\right)^{1/6}=25^{1/6}$; $\sqrt{3}=3^{1/2}=3^{3/6}=\left(3^3\right)^{1/6}=27^{1/6}$. So $\sqrt[3]{5}<\sqrt{3}$.

**75.** Using $v=c/10$ in the formula $m=\dfrac{m_0}{\sqrt{1-v^2/c^2}}$, we get $m=\dfrac{m_0}{\sqrt{1-(c/10)^2/c^2}}=\dfrac{m_0}{\sqrt{1-\frac{1}{100}}}$. Thus $m=\dfrac{1}{\sqrt{\frac{99}{100}}}m_0=\dfrac{10\sqrt{11}}{33}m_0$. So the rest mass of the spaceship is multiplied by $\dfrac{10\sqrt{11}}{33}\approx 1.005$.

Using $v=c/2$ in the formula, we get $m=\dfrac{m_0}{\sqrt{1-(c/2)^2/c^2}}=\dfrac{m_0}{\sqrt{1-\frac{1}{4}}}$. Thus $m=\dfrac{1}{\sqrt{\frac{3}{4}}}m_0=\dfrac{2\sqrt{3}}{3}m_0$. So the rest mass of the spaceship is multiplied by $\frac{2\sqrt{3}}{3}\approx 1.15$.

Using $v=0.9c$ in the formula, we get $m=\dfrac{m_0}{\sqrt{1-(0.9c)^2/c^2}}=\dfrac{m_0}{\sqrt{1-0.81}}$. Thus $m=\dfrac{1}{\sqrt{0.19}}m_0\approx 2.29m_0$. So the rest mass of the spaceship is multiplied by $\frac{1}{\sqrt{0.19}}\approx 2.29$.

As the spaceship travels closer to the speed of light, $v\to c$, so the term $\dfrac{v^2}{c^2}$ approaches 1. So $\sqrt{1-\dfrac{v^2}{c^2}}$ approaches 0, and we obtain $m=\dfrac{m_0}{\text{very small number}}$, which means $m=(\text{very large number})\,m_0$. Hence the mass of the space ship becomes arbitrarily large as its speed approaches the speed of light. The actual value of the speed of light does not affect the calculations.

## P.5 Algebraic Expressions

**1.** Trinomial, terms $x^2$, $-3x$, and 7, degree 2

**2.** Binomial, terms $2x^5$ and $4x^2$, degree 5

**3.** Monomial, term $-8$, degree 0

**4.** Monomial, term $\frac{1}{2}x^7$, degree 7

**5.** Polynomial, terms $x$, $-x^2$, $x^3$, and $-x^4$, degree 4

**6.** Binomial, terms $\sqrt{2}x$ and $-\sqrt{3}$, degree 1

**7.** Not a polynomial

**8.** Not a polynomial

**9.** Polynomial, degree 3

**10.** Polynomial, degree 5

**11.** Not a polynomial

**12.** Polynomial, degree 1

**13.** $(12x - 7) - (5x - 12) = 12x - 7 - 5x + 12 = 7x + 5$

**14.** $(5 - 3x) + (2x - 8) = -x - 3$

**15.** $\left(3x^2 + x + 1\right) + \left(2x^2 - 3x - 5\right) = 5x^2 - 2x - 4$

**16.** $\left(3x^2 + x + 1\right) - \left(2x^2 - 3x - 5\right) = 3x^2 + x + 1 - 2x^2 + 3x + 5 = x^2 + 4x + 6$

**17.** $\left(x^3 + 6x^2 - 4x + 7\right) - \left(3x^2 + 2x - 4\right) = x^3 + 6x^2 - 4x + 7 - 3x^2 - 2x + 4 = x^3 + 3x^2 - 6x + 11$

**18.** $3(x - 1) + 4(x + 2) = 3x - 3 + 4x + 8 = 7x + 5$

**19.** $8(2x + 5) - 7(x - 9) = 16x + 40 - 7x + 63 = 9x + 103$

**20.** $4\left(x^2 - 3x + 5\right) - 3\left(x^2 - 2x + 1\right) = 4x^2 - 12x + 20 - 3x^2 + 6x - 3 = x^2 - 6x + 17$

**21.** $2(2 - 5t) + t^2(t - 1) - \left(t^4 - 1\right) = 4 - 10t + t^3 - t^2 - t^4 + 1 = -t^4 + t^3 - t^2 - 10t + 5$

**22.** $5(3t - 4) - \left(t^2 + 2\right) - 2t(t - 3) = 15t - 20 - t^2 - 2 - 2t^2 + 6t = -3t^2 + 21t - 22$

**23.** $x^2\left(2x^2 - x + 1\right) = x^2\left(2x^2\right) - x^2(x) + x^2(1) = 2x^4 - x^3 + x^2$

**24.** $3x^3\left(x^4 - 4x^2 + 5\right) = 3x^3x^4 - 3x^3\left(4x^2\right) + 3x^3(5) = 3x^7 - 12x^5 + 15x^3$

**25.** $\sqrt{x}\,(x - \sqrt{x}) = x^{1/2}\left(x - x^{1/2}\right) = x^{1/2}x - x^{1/2}x^{1/2} = x^{3/2} - x$

**26.** $x^{3/2}\left(\sqrt{x} - 1/\sqrt{x}\right) = x^{3/2}\left(x^{1/2} - x^{-1/2}\right) = x^{3/2}x^{1/2} - x^{3/2}x^{-1/2} = x^2 - x$

**27.** $y^{1/3}\left(y^2 - 1\right) = y^{1/3}y^2 - y^{1/3} = y^{7/3} - y^{1/3}$

**28.** $y^{1/4}\left(y^{1/2} + 2y^{3/4}\right) = \left(y^{1/4}\right)\left(y^{2/4}\right) + y^{1/4}\left(2y^{3/4}\right) = y^{3/4} + 2y$

**29.** $(3t - 2)(7t - 5) = 21t^2 - 15t - 14t + 10 = 21t^2 - 29t + 10$

**30.** $(4x - 1)(3x + 7) = 12x^2 + 28x - 3x - 7 = 12x^2 + 25x - 7$

**31.** $(x + 2y)(3x - y) = 3x^2 - xy + 6xy - 2y^2 = 3x^2 + 5xy - 2y^2$

**32.** $(4x - 3y)(2x + 5y) = 8x^2 + 20xy - 6xy - 15y^2 = 8x^2 + 14xy - 15y^2$

**33.** $(1 - 2y)^2 = 1 - 4y + 4y^2$

**34.** $(3x + 4)^2 = (3x)^2 + 2(3x)(4) + (4)^2 = 9x^2 + 24x + 16$

**35.** $\left(2x^2 + 3y^2\right)^2 = \left(2x^2\right)^2 + 2\left(2x^2\right)\left(3y^2\right) + \left(3y^2\right)^2 = 4x^4 + 12x^2y^2 + 9y^4$

**36.** $\left(c + \frac{1}{c}\right)^2 = (c)^2 + 2(c)\left(\frac{1}{c}\right) + \left(\frac{1}{c}\right)^2 = c^2 + 2 + \frac{1}{c^2}$

**37.** $(2x - 5)\left(x^2 - x + 1\right) = 2x^3 - 2x^2 + 2x - 5x^2 + 5x - 5 = 2x^3 - 7x^2 + 7x - 5$

**38.** $(1 + 2x)\left(x^2 - 3x + 1\right) = x^2 - 3x + 1 + 2x^3 - 6x^2 + 2x = 2x^3 - 5x^2 - x + 1$

**39.** $\left(x^2 - a^2\right)\left(x^2 + a^2\right) = \left(x^2\right)^2 - \left(a^2\right)^2 = x^4 - a^4$ (difference of squares)

**40.** $\left(x^{1/2} + y^{1/2}\right)\left(x^{1/2} - y^{1/2}\right) = \left(x^{1/2}\right)^2 - \left(y^{1/2}\right)^2 = x - y$

**41.** $\left(\sqrt{a} - \frac{1}{b}\right)\left(\sqrt{a} + \frac{1}{b}\right) = (\sqrt{a})^2 - \left(\frac{1}{b}\right)^2 = a - \frac{1}{b^2}$ (difference of squares)

**42.** $\left(\sqrt{h^2 + 1} + 1\right)\left(\sqrt{h^2 + 1} - 1\right) = \left(\sqrt{h^2 + 1}\right)^2 - (1)^2 = h^2 + 1 - 1 = h^2$

**43.** $\left(1 + a^3\right)^3 = 1 + 3\left(a^3\right) + 3\left(a^3\right)^2 + \left(a^3\right)^3 = 1 + 3a^3 + 3a^6 + a^9$ (perfect cube)

**44.** $(1 - 2y)^3 = (1)^3 - 3(1)^2(2y) + 3(1)(2y)^2 - (2y)^3 = 1 - 6y + 12y^2 - 8y^3$

**45.** $\left(x^2 + x - 1\right)\left(2x^2 - x + 2\right) = x^2\left(2x^2 - x + 2\right) + x\left(2x^2 - x + 2\right) - \left(2x^2 - x + 2\right)$

$= 2x^4 - x^3 + 2x^2 + 2x^3 - x^2 + 2x - 2x^2 + x - 2 = 2x^4 + x^3 - x^2 + 3x - 2$

**46.** $(3x^3 + x^2 - 2)(x^2 + 2x - 1) = 3x^3(x^2 + 2x - 1) + x^2(x^2 + 2x - 1) - 2(x^2 + 2x - 1)$
$= 3x^5 + 6x^4 - 3x^3 + x^4 + 2x^3 - x^2 - 2x^2 - 4x + 2 = 3x^5 + 7x^4 - x^3 - 3x^2 - 4x + 2$

**47.** $(x^2 + x - 2)(x^3 - x + 1) = x^5 - x^3 + x^2 + x^4 - x^2 + x - 2x^3 + 2x - 2 = x^5 + x^4 - 3x^3 + 3x - 2$

**48.** $(1 + x + x^2)(1 - x + x^2) = [(x^2 + 1) + x][(x^2 + 1) - x] = (x^2 + 1)^2 - (x)^2 = x^4 + 2x^2 + 1 - x^2 = x^4 + x^2 + 1$

**49.** $\left(1 + x^{4/3}\right)\left(1 - x^{2/3}\right) = 1 - x^{2/3} + x^{4/3} - x^{6/3} = 1 - x^{2/3} + x^{4/3} - x^2$

**50.** $(1 - b)^2(1 + b)^2 = [(1 - b)(1 + b)]^2 = (1 - b^2)^2 = 1 - 2b^2 + b^4$

**51.** $(3x^2y + 7xy^2)(x^2y^3 - 2y^2) = 3x^4y^4 - 6x^2y^3 + 7x^3y^5 - 14xy^4 = 3x^4y^4 + 7x^3y^5 - 6x^2y^3 - 14xy^4$ (arranging in decreasing powers of $x$)

**52.** $(x^4y - y^5)(x^2 + xy + y^2) = x^6y + x^5y^2 + x^4y^3 - x^2y^5 - xy^6 - y^7$

**53.** $(2x + y - 3)(2x + y + 3) = [(2x + y) - 3][(2x + y) + 3] = (2x + y)^2 - 3^2 = 4x^2 + 4xy + y^2 - 9$

**54.** $(x^2 + y - 2)(x^2 + y + 2) = [(x^2 + y) - 2][(x^2 + y) + 2] = (x^2 + y)^2 - 2^2 = x^4 + 2x^2y + y^2 - 4$

**55.** $(x + y + z)(x - y - z) = [x + (y + z)][x - (y + z)] = x^2 - (y + z)^2 = x^2 - (y^2 + 2yz + z^2) = x^2 - y^2 - 2yz - z^2$

**56.** $(x^2 - y + z)(x^2 + y - z) = [x^2 - (y - z)][x^2 + (y - z)] = (x^2)^2 - (y - z)^2 = x^4 - (y^2 - 2yz + z^2)$
$= x^4 - y^2 + 2yz - z^2$

**57. (a)** The height of the box is $x$, its width is $6 - 2x$, and its length is $10 - 2x$. Since Volume = height × width × length, we have $V = x(6 - 2x)(10 - 2x)$.

**(b)** $V = x(60 - 32x + 4x^2) = 60x - 32x^2 + 4x^3$, degree 3.

**(c)** When $x = 1$, the volume is $V = 60(1) - 32(1^2) + 4(1^3) = 32$, and when $x = 2$, the volume is $V = 60(2) - 32(2^2) + 4(2^3) = 24$.

**58. (a)** The width is the width of the lot minus the setbacks of 10 feet each. Thus width $= x - 20$ and length $= y - 20$. Since Area = width × length, we get $A = (x - 20)(y - 20)$.

**(b)** $A = (x - 20)(y - 20) = xy - 20x - 20y + 400$

**(c)** For the $100 \times 400$ lot, the building envelope has $A = (100 - 20)(400 - 20) = 80(380) = 30{,}400$. For the $200 \times 200$, lot the building envelope has $A = (200 - 20)(200 - 20) = 180(180) = 32{,}400$. The $200 \times 200$ lot has a larger building envelope.

**59. (a)** $A = 2000(1 + r)^3 = 2000(1 + 3r + 3r^2 + r^3) = 2000 + 6000r + 6000r^2 + 2000r^3$, degree 3.

**(b)** Remember that % means divide by 100, so $2\% = 0.02$.

| Interest rate $r$ | 2% | 3% | 4.5% | 6% | 10% |
|---|---|---|---|---|---|
| Amount $A$ | \$2122.42 | \$2185.45 | \$2282.33 | \$2382.03 | \$2662.00 |

**60. (a)** $P = R - C = (50x - 0.05x^2) - (50 + 30x - 0.1x^2) = 50x - 0.05x^2 - 50 - 30x + 0.1x^2 = 0.05x^2 + 20x - 50$.

**(b)** The profit on 10 calculators is $P = 0.05(10^2) + 20(10) - 50 = \$155$. The profit on 20 calculators is $P = 0.05(20^2) + 20(20) - 50 = \$370$ .

**61. (a)** When $x = 1$, $(x + 5)^2 = (1 + 5)^2 = 36$ and $x^2 + 25 = 1^2 + 25 = 26$.

**(b)** $(x + 5)^2 = x^2 + 10x + 25$

**62. (a)** The degree of the product is the sum of the degrees.

**(b)** The degree of a sum is at most the largest of the degrees — it could be lower than either.

**(c)** Product: $(2x^3 + x - 3)(-2x^3 - x + 7) = -4x^6 - 2x^4 + 14x^3 - 2x^4 - x^2 + 7x + 6x^3 + 3x - 21$
$= -4x^6 - 4x^4 + 20x^3 - x^2 + 10x - 21$

Sum: $(2x^3 + x - 3) + (-2x^3 - x + 7) = 4$.

# P.6 Factoring

**1.** $5a - 20 = 5(a - 4)$

**2.** $-3b + 12 = -3(b - 4)$

**3.** $-2x^3 + 16x = -2x(x^2 - 8)$

**4.** $2x^4 + 4x^3 - 14x^2 = 2x^2(x^2 + 2x - 7)$

**5.** $y(y - 6) + 9(y - 6) = (y - 6)(y + 9)$

**6.** $(z + 2)^2 - 5(z + 2) = (z + 2)[(z + 2) - 5]$
$= (z + 2)(z - 3)$

**7.** $2x^2y - 6xy^2 + 3xy = xy(2x - 6y + 3)$

**8.** $-7x^4y^2 + 14xy^3 + 21xy^4 = 7xy^2(-x^3 + 2y + 3y^2)$

**9.** $x^2 + 2x - 3 = (x - 1)(x + 3)$

**10.** $x^2 - 6x + 5 = (x - 5)(x - 1)$

**11.** $6 + y - y^2 = (3 - y)(2 + y)$

**12.** $24 - 5t - t^2 = (8 + t)(3 - t)$

**13.** $8x^2 - 14x - 15 = (2x - 5)(4x + 3)$

**14.** $6y^2 + 11y - 21 = (y + 3)(6y - 7)$

**15.** $(3x + 2)^2 + 8(3x + 2) + 12 = [(3x + 2) + 2][(3x + 2) + 6] = (3x + 4)(3x + 8)$

**16.** $2(a + b)^2 + 5(a + b) - 3 = [(a + b) + 3][2(a + b) - 1] = (a + b + 3)(2a + 2b - 1)$

**17.** $9a^2 - 16 = (3a)^2 - 4^2 = (3a - 4)(3a + 4)$

**18.** $(x + 3)^2 - 4 = (x + 3)^2 - 2^2 = [(x + 3) - 2][(x + 3) + 2] = (x + 1)(x + 5)$

**19.** $27x^3 + y^3 = (3x)^3 + y^3 = (3x + y)[(3x)^2 + 3xy + y^2] = (3x + y)(9x^2 - 3xy + y^2)$

**20.** $a^3 - b^6 = a^3 - (b^2)^3 = (a - b^2)\left[a^2 + ab^2 + (b^2)^2\right] = (a - b^2)(a^2 + ab^2 + b^4)$

**21.** $8s^3 - 125t^3 = (2s)^3 - (5t)^3 = (2s - 5t)[(2s)^2 + (2s)(5t) + (5t)^2] = (2s - 5t)(4s^2 + 10st + 25t^2)$

**22.** $1 + 1000y^3 = 1 + (10y)^3 = (1 + 10y)[1 - 10y + (10y)^2] = (1 + 10y)(1 - 10y + 100y^2)$

**23.** $x^2 + 12x + 36 = x^2 + 2(6x) + 6^2 = (x + 6)^2$

**24.** $16z^2 - 24z + 9 = (4z)^2 - 2(4z)(3) + 3^2 = (4z - 3)^2$

**25.** $x^3 + 4x^2 + x + 4 = x^2(x + 4) + 1(x + 4) = (x + 4)(x^2 + 1)$

**26.** $3x^3 - x^2 + 6x - 2 = x^2(3x - 1) + 2(3x - 1) = (3x - 1)(x^2 + 2)$

**27.** $2x^3 + x^2 - 6x - 3 = x^2(2x + 1) - 3(2x + 1) = (2x + 1)(x^2 - 3)$. If irrational coefficients are permitted, then this can be further factored as $(2x + 1)(x - \sqrt{3})(x - \sqrt{3})$.

**28.** $-9x^3 - 3x^2 + 3x + 1 = -3x^2(3x + 1) + 1(3x + 1) = (3x + 1)(-3x^2 + 1) = (3x + 1)(1 - 3x^2)$. If irrational coefficients are permitted, then this can be further factored as $(3x + 1)(1 - \sqrt{3}x)(1 + \sqrt{3}x)$

**29.** $x^3 + x^2 + x + 1 = x^2(x + 1) + 1(x + 1) = (x + 1)(x^2 + 1)$

**30.** $x^5 + x^4 + x + 1 = x^4(x + 1) + 1(x + 1) = (x + 1)(x^4 + 1)$

**31.** $12x^3 + 18x = 6x(2x^2 + 3)$

**32.** $30x^3 + 15x^4 = 15x^3(2 + x)$

**33.** $6y^4 - 15y^3 = 3y^3(2y - 5)$

**34.** $5ab - 8abc = ab(5 - 8c)$

**35.** $x^2 - 2x - 8 = (x - 4)(x + 2)$

**36.** $x^2 - 14x + 48 = (x - 8)(x - 6)$

**37.** $y^2 - 8y + 15 = (y - 3)(y - 5)$

**38.** $z^2 + 6z - 16 = (z - 2)(z + 8)$

**39.** $2x^2+5x+3=(2x+3)(x+1)$

**40.** $2x^2+7x-4=(2x-1)(x+4)$

**41.** $9x^2-36x-45=9\left(x^2-4x-5\right)=9(x-5)(x+1)$

**42.** $8x^2+10x+3=(4x+3)(2x+1)$

**43.** $6x^2-5x-6=(3x+2)(2x-3)$

**44.** $6+5t-6t^2=(3-2t)(2+3t)$

**45.** $4t^2-12t+9=(2t-3)^2$

**46.** $4x^2+4xy+y^2=(2x+y)^2$

**47.** $r^2-6rs+9s^2=(r-3s)^2$

**48.** $25s^2-10st+t^2=(5s-t)^2$

**49.** $x^2-36=(x-6)(x+6)$

**50.** $4x^2-25=(2x-5)(2x+5)$

**51.** $49-4y^2=(7-2y)(7+2y)$

**52.** $4t^2-9s^2=(2t-3s)(2t+3s)$

**53.** $(a+b)^2-(a-b)^2=[(a+b)-(a-b)][(a+b)+(a-b)]=(2b)(2a)=4ab$

**54.** $$\left(1+\frac{1}{x}\right)^2-\left(1-\frac{1}{x}\right)^2=\left[\left(1+\frac{1}{x}\right)-\left(1-\frac{1}{x}\right)\right]\left[\left(1+\frac{1}{x}\right)+\left(1-\frac{1}{x}\right)\right]$$
$$=\left(1+\frac{1}{x}-1+\frac{1}{x}\right)\left(1+\frac{1}{x}+1-\frac{1}{x}\right)=\left(\frac{2}{x}\right)(2)=\frac{4}{x}$$

**55.** $x^2\left(x^2-1\right)-9\left(x^2-1\right)=\left(x^2-1\right)\left(x^2-9\right)=(x-1)(x+1)(x-3)(x+3)$

**56.** $\left(a^2-1\right)b^2-4\left(a^2-1\right)=\left(a^2-1\right)\left(b^2-4\right)=(a-1)(a+1)(b-2)(b+2)$

**57.** $t^3+1=(t+1)\left(t^2-t+1\right)$

**58.** $x^3-27=x^3-3^3=(x-3)\left(x^2+3x+9\right)$

**59.** $8x^3-125=(2x)^3-5^3=(2x-5)\left[(2x)^2+(2x)(5)+5^2\right]=(2x-5)\left(4x^2+10x+25\right)$

**60.** $x^6+64=x^6+2^6=\left(x^2\right)^3+(4)^3=\left(x^2+4\right)\left[\left(x^2\right)^2-4\left(x^2\right)+(4)^2\right]=\left(x^2+4\right)\left(x^4-4x^2+16\right)$

**61.** $x^6-8y^3=\left(x^2\right)^3-(2y)^3=\left(x^2-2y\right)\left[\left(x^2\right)^2+\left(x^2\right)(2y)+(2y)^2\right]=\left(x^2-2y\right)\left(x^4+2x^2y+4y^2\right)$

**62.** $27a^3+b^6=(3a)^3+\left(b^2\right)^3=\left(3a+b^2\right)\left[(3a)^2-(3a)\left(b^2\right)+\left(b^2\right)^2\right]=\left(3a+b^2\right)\left(9a^2-3ab^2+b^4\right)$

**63.** $x^3+2x^2+x=x\left(x^2+2x+1\right)=x(x+1)^2$

**64.** $3x^3-27x=3x\left(x^2-9\right)=3x(x-3)(x+3)$

**65.** $x^4+2x^3-3x^2=x^2\left(x^2+2x-3\right)=x^2(x-1)(x+3)$

**66.** $x^3+3x^2-x-3=\left(x^3+3x^2\right)+(-x-3)=x^2(x+3)-(x+3)=(x+3)\left(x^2-1\right)=(x+3)(x-1)(x+1)$

**67.** $y^3-3y^2-4y+12=\left(y^3-3y^2\right)+(-4y+12)=y^2(y-3)+(-4)(y-3)=(y-3)\left(y^2-4\right)$
$=(y-3)(y-2)(y+2)$ (factor by grouping)

**68.** $y^3-y^2+y-1=y^2(y-1)+1(y-1)=\left(y^2+1\right)(y-1)$

**69.** $2x^3+4x^2+x+2=\left(2x^3+4x^2\right)+(x+2)=2x^2(x+2)+(1)(x+2)=(x+2)\left(2x^2+1\right)$ (factor by grouping)

**70.** $3x^3+5x^2-6x-10=x^2(3x+5)-2(3x+5)=\left(x^2-2\right)(3x+5)$

**71.** $(x-1)(x+2)^2-(x-1)^2(x+2)=(x-1)(x+2)[(x+2)-(x-1)]=3(x-1)(x+2)$

**72.** $(x+1)^3x-2(x+1)^2x^2+x^3(x+1)=x(x+1)\left[(x+1)^2-2(x+1)x+x^2\right]=x(x+1)[(x+1)-x]^2$
$=x(x+1)(1)^2=x(x+1)$

**73.** $y^4(y+2)^3+y^5(y+2)^4=y^4(y+2)^3[(1)+y(y+2)]=y^4(y+2)^3\left(y^2+2y+1\right)=y^4(y+2)^3(y+1)^2$

**74.** $n(x-y)+(n-1)(y-x)=n(x-y)-(n-1)(x-y)=(x-y)[n-(n-1)]=x-y$

**75.** Start by factoring $y^2-7y+10$, and then substitute $a^2+1$ for $y$. This gives

$$(a^2+1)^2-7(a^2+1)+10=[(a^2+1)-2][(a^2+1)-5]=(a^2-1)(a^2-4)=(a-1)(a+1)(a-2)(a+2)$$

**76.** $$(a^2+2a)^2-2(a^2+2a)-3=[(a^2+2a)-3][(a^2+2a)+1]=(a^2+2a-3)(a^2+2a+1)$$
$$=(a-1)(a+3)(a+1)^2$$

**77.** $x^{5/2}-x^{1/2}=x^{1/2}(x^2-1)=\sqrt{x}(x-1)(x+1)$

**78.** $3x^{-1/2}+4x^{1/2}+x^{3/2}=x^{-1/2}(3+4x+x^2)=\left(\frac{1}{\sqrt{x}}\right)(3+x)(1+x)$

**79.** Start by factoring out the power of $x$ with the smallest exponent, that is, $x^{-3/2}$. So

$$x^{-3/2}+2x^{-1/2}+x^{1/2}=x^{-3/2}(1+2x+x^2)=\frac{(1+x)^2}{x^{3/2}}.$$

**80.** $$(x-1)^{7/2}-(x-1)^{3/2}=(x-1)^{3/2}\left[(x-1)^2-1\right]=(x-1)^{3/2}[(x-1)-1][(x-1)+1]$$
$$=(x-1)^{3/2}(x-2)(x)$$

**81.** Start by factoring out the power of $(x^2+1)$ with the smallest exponent, that is, $(x^2+1)^{-1/2}$. So

$$(x^2+1)^{1/2}+2(x^2+1)^{-1/2}=(x^2+1)^{-1/2}\left[(x^2+1)+2\right]=\frac{x^2+3}{\sqrt{x^2+1}}.$$

**82.** $x^{-1/2}(x+1)^{1/2}+x^{1/2}(x+1)^{-1/2}=x^{-1/2}(x+1)^{-1/2}[(x+1)+x]=\frac{2x+1}{\sqrt{x}\sqrt{x+1}}$

**83.** $$2x^{1/3}(x-2)^{2/3}-5x^{4/3}(x-2)^{-1/3}=x^{1/3}(x-2)^{-1/3}[2(x-2)-5x]=x^{1/3}(x-2)^{-1/3}(2x-4-5x)$$
$$=x^{1/3}(x-2)^{-1/3}(-3x-4)=\frac{(-3x-4)\sqrt[3]{x}}{\sqrt[3]{x-2}}$$

**84.** $$3x^{-1/2}(x^2+1)^{5/4}-x^{3/2}(x^2+1)^{1/4}=x^{-1/2}(x^2+1)^{1/4}\left[3(x^2+1)-x^2(1)\right]$$
$$=x^{-1/2}(x^2+1)^{1/4}(3x^2+3-x^2)=x^{-1/2}(x^2+1)^{1/4}(2x^2+3)=\frac{\sqrt[4]{x^2+1}(2x^2+3)}{\sqrt{x}}$$

**85.** $$3x^2(4x-12)^2+x^3(2)(4x-12)(4)=x^2(4x-12)[3(4x-12)+x(2)(4)]=4x^2(x-3)(12x-36+8x)$$
$$=4x^2(x-3)(20x-36)=16x^2(x-3)(5x-9)$$

**86.** $$5(x^2+4)^4(2x)(x-2)^4+(x^2+4)^5(4)(x-2)^3=2(x^2+4)^4(x-2)^3\left[(5)(x)(x-2)+(x^2+4)(2)\right]$$
$$=2(x^2+4)^4(x-2)^3(5x^2-10x+2x^2+8)=2(x^2+4)^4(x-2)^3(7x^2-10x+8)$$

**87.** $$3(2x-1)^2(2)(x+3)^{1/2}+(2x-1)^3\left(\tfrac{1}{2}\right)(x+3)^{-1/2}=(2x-1)^2(x+3)^{-1/2}\left[6(x+3)+(2x-1)\left(\tfrac{1}{2}\right)\right]$$
$$=(2x-1)^2(x+3)^{-1/2}\left(6x+18+x-\tfrac{1}{2}\right)=(2x-1)^2(x+3)^{-1/2}\left(7x+\tfrac{35}{2}\right)$$

**88.** $$\tfrac{1}{3}(x+6)^{-2/3}(2x-3)^2+(x+6)^{1/3}(2)(2x-3)(2)=\tfrac{1}{3}(x+6)^{-2/3}(2x-3)[(2x-3)+(3)(x+6)(4)]$$
$$=\tfrac{1}{3}(x+6)^{-2/3}(2x-3)[2x-3+12x+72]=\tfrac{1}{3}(x+6)^{-2/3}(2x-3)(14x+69)$$

**89.** $(x^2+3)^{-1/3}-\tfrac{2}{3}x^2(x^2+3)^{-4/3}=(x^2+3)^{-4/3}\left[(x^2+3)-\tfrac{2}{3}x^2\right]=(x^2+3)^{-4/3}\left(\tfrac{1}{3}x^2+3\right)=\frac{\frac{1}{3}x^2+3}{(x^2+3)^{4/3}}$

**90.** $$\tfrac{1}{2}x^{-1/2}(3x+4)^{1/2}-\tfrac{3}{2}x^{1/2}(3x+4)^{-1/2}=\tfrac{1}{2}x^{-1/2}(3x+4)^{-1/2}[(3x+4)-3x]=\tfrac{1}{2}x^{-1/2}(3x+4)^{-1/2}(4)$$
$$=2x^{-1/2}(3x+4)^{-1/2}$$

**91. (a)** $\frac{1}{2}\left[(a+b)^2-\left(a^2+b^2\right)\right]=\frac{1}{2}\left[a^2+2ab+b^2-a^2-b^2\right]=\frac{1}{2}(2ab)=ab.$

**(b)** $$\begin{aligned}\left(a^2+b^2\right)^2-\left(a^2-b^2\right)^2&=\left[\left(a^2+b^2\right)-\left(a^2-b^2\right)\right]\left[\left(a^2+b^2\right)+\left(a^2-b^2\right)\right]\\&=\left(a^2+b^2-a^2+b^2\right)\left(a^2+b^2+a^2-b^2\right)=\left(2b^2\right)\left(2a^2\right)=4a^2b^2\end{aligned}$$

**(c)** $\text{LHS}=\left(a^2+b^2\right)\left(c^2+d^2\right)=a^2c^2+a^2d^2+b^2c^2+b^2d^2.$

$\text{RHS}=(ac+bd)^2+(ad-bc)^2=a^2c^2+2abcd+b^2d^2+a^2d^2-2abcd+b^2c^2=a^2c^2+a^2d^2+b^2c^2+b^2d^2.$

So LHS = RHS, that is, $\left(a^2+b^2\right)\left(c^2+d^2\right)=(ac+bd)^2+(ad-bc)^2.$

**(d)** $$\begin{aligned}4a^2c^2-\left(c^2-b^2+a^2\right)^2&=(2ac)^2-\left(c^2-b^2+a^2\right)\\&=\left[(2ac)-\left(c^2-b^2+a^2\right)\right]\left[(2ac)+\left(c^2-b^2+a^2\right)\right]\text{ (difference of squares)}\\&=\left(2ac-c^2+b^2-a^2\right)\left(2ac+c^2-b^2+a^2\right)\\&=\left[b^2-\left(c^2-2ac+a^2\right)\right]\left[\left(c^2+2ac+a^2\right)-b^2\right]\text{ (regrouping)}\\&=\left[b^2-(c-a)^2\right]\left[(c+a)^2-b^2\right]\text{ (perfect squares)}\\&=[b-(c-a)][b+(c-a)][(c+a)-b][(c+a)+b]\text{ (each factor is a difference of squares)}\\&=(b-c+a)(b+c-a)(c+a-b)(c+a+b)\\&=(a+b-c)(-a+b+c)(a-b+c)(a+b+c)\end{aligned}$$

**(e)** $$\begin{aligned}x^4+3x^2+4&=\left(x^4+4x^2+4\right)-x^2=\left(x^2+2\right)^2-x^2=\left[\left(x^2+2\right)-x\right]\left[\left(x^2+2\right)+x\right]\\&=\left(x^2-x+2\right)\left(x^2+x+2\right)\end{aligned}$$

**92.** Difference of Cubes: $(A-B)\left(A^2+AB+B^2\right)=A^3+A^2B+AB^2-A^2B-AB^2-B^3=A^3-B^3$

Sum of Cubes: $(A+B)\left(A^2-AB+B^2\right)=A^3-A^2B+AB^2+A^2B-AB^2+B^3=A^3+B^3$

**93.** The volume of the shell is the difference between the volumes of the outside cylinder (with radius $R$) and the inside cylinder (with radius $r$). Thus $V=\pi R^2h-\pi r^2h=\pi\left(R^2-r^2\right)h=\pi(R-r)(R+r)h=2\pi\cdot\dfrac{R+r}{2}\cdot h\cdot(R-r)$. The average radius is $\dfrac{R+r}{2}$ and $2\pi\cdot\dfrac{R+r}{2}$ is the average circumference (length of the rectangular box), $h$ is the height, and $R-r$ is the thickness of the rectangular box. Thus $V=\pi R^2h-\pi r^2h=2\pi\cdot\dfrac{R+r}{2}\cdot h\cdot(R-r)=2\pi\cdot(\text{average radius})\cdot(\text{height})\cdot(\text{thickness})$

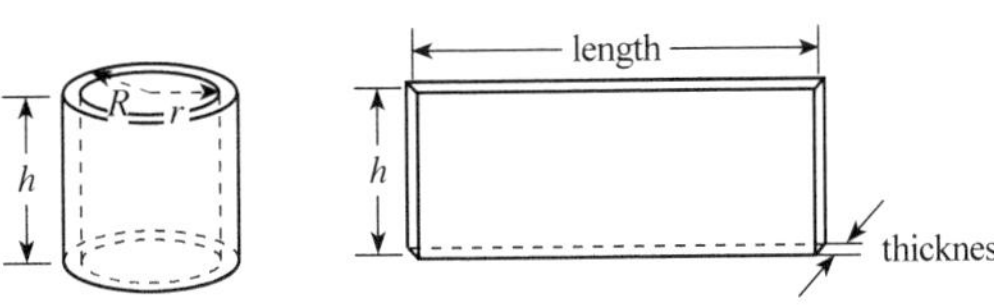


**94. (a)** Moved portion = field − habitat

**(b)** Using the difference of squares, we get $b^2-(b-2x)^2=[b-(b-2x)][b+(b-x)]=2x(2b-2x)=4x(b-x).$

**95. (a)** $528^2-527^2=(528-527)(528+527)=1(1055)=1055$

**(b)** $122^2-120^2=(122-120)(122+120)=2(242)=484$

**(c)** $1020^2-1010^2=(1020-1010)(1020+1010)=10(2030)=20{,}300$

**(d)** $49\cdot51=(50-1)(50+1)=50^2-1=2500-1=2499$

**(e)** $998\cdot1002=(1000-2)(1000+2)=1000^2-2^2=1{,}000{,}000-4=999{,}996$

**96. (a)** $A^4 - B^4 = (A^2 - B^2)(A^2 + B^2) = (A - B)(A + B)(A^2 + B^2)$

$A^6 - B^6 = (A^3 - B^3)(A^3 + B^3)$ (difference of squares)

$= (A - B)(A^2 + AB + B^2)(A + B)(A^2 - AB + B^2)$ (difference and sum of cubes)

**(b)** $12^4 - 7^4 = 20{,}736 - 2{,}401 = 18{,}335$; $12^6 - 7^6 = 2{,}985{,}984 - 117{,}649 = 2{,}868{,}335$

**(c)** $18{,}335 = 12^4 - 7^4 = (12 - 7)(12 + 7)(12^2 + 7^2) = 5(19)(144 + 49) = 5(19)(193)$

$2{,}868{,}335 = 12^6 - 7^6 = (12 - 7)(12 + 7)\left[12^2 + 12(7) + 7^2\right]\left[12^2 - 12(7) + 7^2\right]$

$= 5(19)(144 + 84 + 49)(144 - 84 + 49) = 5(19)(277)(109)$

**97.**

$$\begin{array}{rrrr} & & A & +\ 1 \\ \times & & A & -\ 1 \\ \hline & & -A & -\ 1 \\ A^2 & + & A & \\ \hline A^2 & & & -\ 1 \end{array} \qquad \begin{array}{rrrrr} & & A^2 & +\ A & +\ 1 \\ \times & & & A & -\ 1 \\ \hline & & -A^2 & -\ A & -\ 1 \\ A^3 & + & A^2 & +\ A & \\ \hline A^3 & & & & -\ 1 \end{array} \qquad \begin{array}{rrrrrr} & & A^3 & +\ A^2 & +\ A & +\ 1 \\ \times & & & & A & -\ 1 \\ \hline & & -A^3 & -\ A^2 & -\ A & -\ 1 \\ A^4 & + & A^3 & +\ A^2 & +\ A & \\ \hline A^4 & & & & & -\ 1 \end{array}$$

Based on the pattern, we suspect that $A^5 - 1 = (A - 1)(A^4 + A^3 + A^2 + A + 1)$. Check:

$$\begin{array}{rrrrrrr} & & A^4 & +\ A^3 & +\ A^2 & +\ A & +\ 1 \\ \times & & & & & A & -\ 1 \\ \hline & & -A^4 & -\ A^3 & -\ A^2 & -\ A & -\ 1 \\ A^5 & + & A^4 & +\ A^3 & +\ A^2 & +\ A & \\ \hline A^5 & & & & & & -\ 1 \end{array}$$

The general pattern is $A^n - 1 = (A - 1)(A^{n-1} + A^{n-2} + \cdots + A^2 + A + 1)$, where $n$ is a positive integer.

**98. (a)** $x^4 + x^2 - 2 = (x^2 - 1)(x^2 + 2) = (x - 1)(x + 1)(x^2 + 2)$

**(b)** $x^4 + 2x^2 + 9 = (x^4 + 6x^2 + 9) - 4x^2 = (x^2 + 3)^2 - (2x)^2 = \left[(x^2 + 3) - 2x\right]\left[(x^2 + 3) + 2x\right]$

$= (x^2 - 2x + 3)(x^2 + 2x + 3)$

**(c)** $x^4 + 4x^2 + 16 = (x^4 + 8x^2 + 16) - 4x^2 = (x^2 + 4)^2 - (2x)^2$

$= \left[(x^2 + 4) - 2x\right]\left[(x^2 + 4) + 2x\right] = (x^2 - 2x + 4)(x^2 + 2x + 4)$

**(d)** $x^4 + 2x^2 + 1 = (x^2 + 1)^2$

## P.7 Rational Expressions

**1. (a)** When $x = 5$, we get $4(5^2) - 10(5) + 3 = 53$.

**(b)** The domain is all real numbers.

**2. (a)** When $x = -1$, we get $-(-1)^4 + (-1)^3 + 9(-1) = -1 - 1 - 9 = -11$.

**(b)** The domain is all real numbers.

**3. (a)** When $x = 7$, we get $\dfrac{2(7) + 1}{7 - 4} = \dfrac{15}{3} = 5$.

**(b)** Since $x - 4 \neq 0$, we have $x \neq 4$, so the domain is $\{x \mid x \neq 4\}$.

**4. (a)** When $x = 1$, we get $\dfrac{2(1^2) - 5}{3(1) + 6} = \dfrac{-3}{9} = -\dfrac{1}{3}$.

**(b)** Since $3x + 6 \neq 0$, we have $x \neq -2$, so the domain is $\{x \mid x \neq -2\}$.

**5. (a)** When $x = 6$, we get $\sqrt{6+3} = \sqrt{9} = 3$.

**(b)** $x + 3 \geq 0 \quad \Leftrightarrow \quad x \geq -3$, so the domain is $\{x \mid x \geq -3\}$.

**6. (a)** When $x = 5$, we get $\dfrac{1}{\sqrt{5-1}} = \dfrac{1}{\sqrt{4}} = \dfrac{1}{2}$.

**(b)** $x - 1 > 0 \quad \Leftrightarrow \quad x > 1$, so the domain is $\{x \mid x > 1\}$.

**7.** $\dfrac{12x}{6x^2} = \dfrac{6x \cdot 2}{6x \cdot x} = \dfrac{2}{x}$

**8.** $\dfrac{81x^3}{18x} = \dfrac{9x \cdot 9x^2}{9x \cdot 2} = \dfrac{9x^2}{2} = \frac{9}{2}x^2$

**9.** $\dfrac{5y^2}{10y + y^2} = \dfrac{y \cdot 5y}{y \cdot (10+y)} = \dfrac{5y}{10+y}$

**10.** $\dfrac{14t^2 - t}{7t} = \dfrac{t \cdot (14t - 1)}{t \cdot 7} = \dfrac{14t - 1}{7} = 2t - \dfrac{1}{7}$

**11.** $\dfrac{3(x+2)(x-1)}{6(x-1)^2} = \dfrac{3(x-1)\cdot(x+2)}{3(x-1)\cdot 2(x-1)} = \dfrac{x+2}{2(x-1)}$

**12.** $\dfrac{4(x^2-1)}{12(x+2)(x-1)} = \dfrac{4(x+1)(x-1)}{12(x+2)(x-1)} = \dfrac{x+1}{3(x+2)}$

**13.** $\dfrac{x-2}{x^2-4} = \dfrac{x-2}{(x-2)(x+2)} = \dfrac{1}{x+2}$

**14.** $\dfrac{x^2-x-2}{x^2-1} = \dfrac{(x-2)(x+1)}{(x-1)(x+1)} = \dfrac{x-2}{x-1}$

**15.** $\dfrac{x^2+6x+8}{x^2+5x+4} = \dfrac{(x+2)(x+4)}{(x+1)(x+4)} = \dfrac{x+2}{x+1}$

**16.** $\dfrac{x^2-x-12}{x^2+5x+6} = \dfrac{(x-4)(x+3)}{(x+2)(x+3)} = \dfrac{x-4}{x+2}$

**17.** $\dfrac{y^2+y}{y^2-1} = \dfrac{y(y+1)}{(y-1)(y+1)} = \dfrac{y}{y-1}$

**18.** $\dfrac{y^2-3y-18}{2y^2+5y+3} = \dfrac{(y-6)(y+3)}{(2y+3)(y+1)}$

**19.** $\dfrac{2x^3-x^2-6x}{2x^2-7x+6} = \dfrac{x(2x^2-x-6)}{(2x-3)(x-2)} = \dfrac{x(2x+3)(x-2)}{(2x-3)(x-2)} = \dfrac{x(2x+3)}{2x-3}$

**20.** $\dfrac{1-x^2}{x^3-1} = \dfrac{(1-x)(1+x)}{(x-1)(x^2+x+1)} = \dfrac{-(x-1)(1+x)}{(x-1)(x^2+x+1)} = \dfrac{-(x+1)}{x^2+x+1}$

**21.** $\dfrac{4x}{x^2-4} \cdot \dfrac{x+2}{16x} = \dfrac{4x}{(x-2)(x+2)} \cdot \dfrac{x+2}{16x} = \dfrac{1}{4(x-2)}$

**22.** $\dfrac{x^2-25}{x^2-16} \cdot \dfrac{x+4}{x+5} = \dfrac{(x-5)(x+5)}{(x-4)(x+4)} \cdot \dfrac{x+4}{x+5} = \dfrac{x-5}{x-4}$

**23.** $\dfrac{x^2-x-12}{x^2-9} \cdot \dfrac{3+x}{4-x} = \dfrac{(x-4)(x+3)}{(x-3)(x+3)} \cdot \dfrac{x+3}{-(x-4)} = \dfrac{x+3}{-(x-3)} = \dfrac{3+x}{3-x}$

**24.** $\dfrac{x^2+2x-3}{x^2-2x-3} \cdot \dfrac{3-x}{3+x} = \dfrac{(x+3)(x-1)}{(x-3)(x+1)} \cdot \dfrac{-(x-3)}{x+3} = \dfrac{-(x-1)}{x+1} = \dfrac{-x+1}{x+1} = \dfrac{1-x}{1+x}$

**25.** $\dfrac{t-3}{t^2+9} \cdot \dfrac{t+3}{t^2-9} = \dfrac{(t-3)(t+3)}{(t^2+9)(t-3)(t+3)} = \dfrac{1}{t^2+9}$

**26.** $\dfrac{x^2-x-6}{x^2+2x} \cdot \dfrac{x^3+x^2}{x^2-2x-3} = \dfrac{(x-3)(x+2)}{x(x+2)} \cdot \dfrac{x^2(x+1)}{(x-3)(x+1)} = x$

**27.** $\dfrac{x^2+7x+12}{x^2+3x+2} \cdot \dfrac{x^2+5x+6}{x^2+6x+9} = \dfrac{(x+3)(x+4)}{(x+1)(x+2)} \cdot \dfrac{(x+2)(x+3)}{(x+3)(x+3)} = \dfrac{x+4}{x+1}$

**28.** $\dfrac{x^2+2xy+y^2}{x^2-y^2} \cdot \dfrac{2x^2-xy-y^2}{x^2-xy-2y^2} = \dfrac{(x+y)(x+y)}{(x-y)(x+y)} \cdot \dfrac{(x-y)(2x+y)}{(x-2y)(x+y)} = \dfrac{2x+y}{x-2y}$

**29.** $\dfrac{2x^2+3x+1}{x^2+2x-15} \div \dfrac{x^2+6x+5}{2x^2-7x+3} = \dfrac{2x^2+3x+1}{x^2+2x-15} \cdot \dfrac{2x^2-7x+3}{x^2+6x+5} = \dfrac{(2x+1)(x+1)}{(x-3)(x+5)} \cdot \dfrac{(2x-1)(x-3)}{(x+1)(x+5)}$

$= \dfrac{(2x+1)(2x-1)}{(x+5)(x+5)} = \dfrac{(2x+1)(2x-1)}{(x+5)^2}$

**30.** $\dfrac{4y^2-9}{2y^2+9y-18} \div \dfrac{2y^2+y-3}{y^2+5y-6} = \dfrac{4y^2-9}{2y^2+9y-18} \cdot \dfrac{y^2+5y-6}{2y^2+y-3} = \dfrac{(2y-3)(2y+3)}{(2y-3)(y+6)} \cdot \dfrac{(y-1)(y+6)}{(y-1)(2y+3)} = 1$

**31.** $\dfrac{\dfrac{x^3}{x+1}}{\dfrac{x}{x^2+2x+1}} = \dfrac{x^3}{x+1} \cdot \dfrac{x^2+2x+1}{x} = \dfrac{x^3(x+1)(x+1)}{(x+1)x} = x^2(x+1)$

**32.** $\dfrac{\dfrac{2x^2-3x-2}{x^2-1}}{\dfrac{2x^2+5x+2}{x^2+x-2}} = \dfrac{2x^2-3x-2}{x^2-1} \cdot \dfrac{x^2+x-2}{2x^2+5x+2} = \dfrac{(x-2)(2x+1)}{(x-1)(x+1)} \cdot \dfrac{(x-1)(x+2)}{(x+2)(2x+1)} = \dfrac{x-2}{x+1}$

**33.** $\dfrac{x/y}{z} = \dfrac{x}{y} \cdot \dfrac{1}{z} = \dfrac{x}{yz}$

**34.** $\dfrac{x}{y/z} = x \div \dfrac{y}{z} = \dfrac{x}{1} \cdot \dfrac{z}{y} = \dfrac{xz}{y}$

**35.** $2 + \dfrac{x}{x+3} = \dfrac{2(x+3)}{x+3} + \dfrac{x}{x+3} = \dfrac{2x+6+x}{x+3} = \dfrac{3x+6}{x+3} = \dfrac{3(x+2)}{(x+3)}$

**36.** $\dfrac{2x-1}{x+4} - 1 = \dfrac{2x-1}{x+4} + \dfrac{-(x+4)}{x+4} = \dfrac{2x-1-x-4}{x+4} = \dfrac{x-5}{x+4}$

**37.** $\dfrac{1}{x+5} + \dfrac{2}{x-3} = \dfrac{x-3}{(x+5)(x-3)} + \dfrac{2(x+5)}{(x+5)(x-3)} = \dfrac{x-3+2x+10}{(x+5)(x-3)} = \dfrac{3x+7}{(x+5)(x-3)}$

**38.** $\dfrac{1}{x+1} + \dfrac{1}{x-1} = \dfrac{x-1}{(x+1)(x-1)} + \dfrac{x+1}{(x+1)(x-1)} = \dfrac{x-1+x+1}{(x+1)(x-1)} = \dfrac{2x}{(x+1)(x-1)}$

**39.** $\dfrac{1}{x+1} - \dfrac{1}{x+2} = \dfrac{x+2}{(x+1)(x+2)} + \dfrac{-(x+1)}{(x+1)(x+2)} = \dfrac{x+2-x-1}{(x+1)(x+2)} = \dfrac{1}{(x+1)(x+2)}$

**40.** $\dfrac{x}{x-4} - \dfrac{3}{x+6} = \dfrac{x(x+6)}{(x-4)(x+6)} + \dfrac{-3(x-4)}{(x-4)(x+6)} = \dfrac{x^2+6x-3x+12}{(x-4)(x+6)} = \dfrac{x^2+3x+12}{(x-4)(x+6)}$

**41.** $\dfrac{x}{(x+1)^2} + \dfrac{2}{x+1} = \dfrac{x}{(x+1)^2} + \dfrac{2(x+1)}{(x+1)(x+1)} = \dfrac{x+2x+2}{(x+1)^2} = \dfrac{3x+2}{(x+1)^2}$

**42.** $\dfrac{5}{2x-3} - \dfrac{3}{(2x-3)^2} = \dfrac{5(2x-3)}{(2x-3)^2} - \dfrac{3}{(2x-3)^2} = \dfrac{10x-15-3}{(2x-3)^2} = \dfrac{10x-18}{(2x-3)^2} = \dfrac{2(5x-9)}{(2x-3)^2}$

**43.** $u+1+\dfrac{u}{u+1} = \dfrac{(u+1)(u+1)}{u+1} + \dfrac{u}{u+1} = \dfrac{u^2+2u+1+u}{u+1} = \dfrac{u^2+3u+1}{u+1}$

**44.** $\dfrac{2}{a^2} - \dfrac{3}{ab} + \dfrac{4}{b^2} = \dfrac{2b^2}{a^2b^2} - \dfrac{3ab}{a^2b^2} + \dfrac{4a^2}{a^2b^2} = \dfrac{2b^2-3ab+4a^2}{a^2b^2}$

**45.** $\dfrac{1}{x^2} + \dfrac{1}{x^2+x} = \dfrac{1}{x^2} + \dfrac{1}{x(x+1)} = \dfrac{x+1}{x^2(x+1)} + \dfrac{x}{x^2(x+1)} = \dfrac{2x+1}{x^2(x+1)}$

**46.** $\dfrac{1}{x} + \dfrac{1}{x^2} + \dfrac{1}{x^3} = \dfrac{x^2}{x^3} + \dfrac{x}{x^3} + \dfrac{1}{x^3} = \dfrac{x^2+x+1}{x^3}$

**47.** $\dfrac{2}{x+3} - \dfrac{1}{x^2+7x+12} = \dfrac{2}{x+3} - \dfrac{1}{(x+3)(x+4)} = \dfrac{2(x+4)}{(x+3)(x+4)} + \dfrac{-1}{(x+3)(x+4)}$

$= \dfrac{2x+8-1}{(x+3)(x+4)} = \dfrac{2x+7}{(x+3)(x+4)}$

**48.** $\dfrac{x}{x^2-4} + \dfrac{1}{x-2} = \dfrac{x}{(x-2)(x+2)} + \dfrac{1}{x-2} = \dfrac{x}{(x-2)(x+2)} + \dfrac{x+2}{(x-2)(x+2)}$

$= \dfrac{2x+2}{(x-2)(x+2)} = \dfrac{2(x+1)}{(x-2)(x+2)}$

**49.** $\dfrac{1}{x+3} + \dfrac{1}{x^2-9} = \dfrac{1}{x+3} + \dfrac{1}{(x-3)(x+3)} = \dfrac{x-3}{(x-3)(x+3)} + \dfrac{1}{(x-3)(x+3)} = \dfrac{x-2}{(x-3)(x+3)}$

**50.** $\dfrac{x}{x^2+x-2} - \dfrac{2}{x^2-5x+4} = \dfrac{x}{(x-1)(x+2)} + \dfrac{-2}{(x-1)(x-4)}$

$= \dfrac{x(x-4)}{(x-1)(x+2)(x-4)} + \dfrac{-2(x+2)}{(x-1)(x+2)(x-4)} = \dfrac{x^2-4x-2x-4}{(x-1)(x+2)(x-4)} = \dfrac{x^2-6x-4}{(x-1)(x+2)(x-4)}$

**51.** $\dfrac{2}{x} + \dfrac{3}{x-1} - \dfrac{4}{x^2-x} = \dfrac{2}{x} + \dfrac{3}{x-1} - \dfrac{4}{x(x-1)} = \dfrac{2(x-1)}{x(x-1)} + \dfrac{3x}{x(x-1)} + \dfrac{-4}{x(x-1)} = \dfrac{2x-2+3x-4}{x(x-1)} = \dfrac{5x-6}{x(x-1)}$

**52.** $\dfrac{x}{x^2-x-6}-\dfrac{1}{x+2}-\dfrac{2}{x-3}=\dfrac{x}{(x-3)(x+2)}+\dfrac{-1}{x+2}+\dfrac{-2}{x-3}$

$$=\frac{x}{(x-3)(x+2)}+\frac{-1(x-3)}{(x-3)(x+2)}+\frac{-2(x+2)}{(x-3)(x+2)}=\frac{x-x+3-2x-4}{(x-3)(x+2)}=\frac{-2x-1}{(x-3)(x+2)}$$

**53.** $\dfrac{1}{x^2+3x+2}-\dfrac{1}{x^2-2x-3}=\dfrac{1}{(x+2)(x+1)}-\dfrac{1}{(x-3)(x+1)}$

$$=\frac{x-3}{(x-3)(x+2)(x+1)}+\frac{-(x+2)}{(x-3)(x+2)(x+1)}=\frac{x-3-x-2}{(x-3)(x+2)(x+1)}=\frac{-5}{(x-3)(x+2)(x+1)}$$

**54.** $\dfrac{1}{x+1}-\dfrac{2}{(x+1)^2}+\dfrac{3}{x^2-1}=\dfrac{1}{x+1}+\dfrac{-2}{(x+1)^2}+\dfrac{3}{(x-1)(x+1)}$

$$=\frac{(x+1)(x-1)}{(x-1)(x+1)^2}+\frac{-2(x-1)}{(x-1)(x+1)^2}+\frac{3(x+1)}{(x-1)(x+1)^2}$$

$$=\frac{x^2-1}{(x-1)(x+1)^2}+\frac{-2x+2}{(x-1)(x+1)^2}+\frac{3x+3}{(x-1)(x+1)^2}=\frac{x^2-1-2x+2+3x+3}{(x-1)(x+1)^2}=\frac{x^2+x+4}{(x-1)(x+1)^2}$$

**55.** $\dfrac{\dfrac{x}{y}-\dfrac{y}{x}}{\dfrac{1}{x^2}-\dfrac{1}{y^2}}=\dfrac{\dfrac{x^2-y^2}{xy}}{\dfrac{y^2-x^2}{x^2y^2}}=\dfrac{x^2-y^2}{xy}\cdot\dfrac{x^2y^2}{y^2-x^2}=\dfrac{xy}{-1}=-xy$. An alternative method is to multiply the numerator and denominator by the common denominator of both the numerator and denominator, in this case $x^2y^2$:

$$\frac{\dfrac{x}{y}-\dfrac{y}{x}}{\dfrac{1}{x^2}-\dfrac{1}{y^2}}=\frac{\left(\dfrac{x}{y}-\dfrac{y}{x}\right)}{\left(\dfrac{1}{x^2}-\dfrac{1}{y^2}\right)}\cdot\frac{x^2y^2}{x^2y^2}=\frac{x^3y-xy^3}{y^2-x^2}=\frac{xy\left(x^2-y^2\right)}{y^2-x^2}=-xy.$$

**56.** $x-\dfrac{y}{\dfrac{x}{y}+\dfrac{y}{x}}=x-\dfrac{y}{\dfrac{x}{y}+\dfrac{y}{x}}\cdot\dfrac{xy}{xy}=x-\dfrac{xy^2}{x^2+y^2}=\dfrac{x\left(x^2+y^2\right)}{x^2+y^2}-\dfrac{xy^2}{x^2+y^2}=\dfrac{x^3+xy^2-xy^2}{x^2+y^2}=\dfrac{x^3}{x^2+y^2}$

**57.** $\dfrac{1+\dfrac{1}{c-1}}{1-\dfrac{1}{c-1}}=\dfrac{\dfrac{c-1}{c-1}+\dfrac{1}{c-1}}{\dfrac{c-1}{c-1}+\dfrac{-1}{c-1}}=\dfrac{\dfrac{c}{c-1}}{\dfrac{c-2}{c-1}}=\dfrac{c}{c-1}\cdot\dfrac{c-1}{c-2}=\dfrac{c}{c-2}$. Using the alternative method, we obtain

$$\frac{1+\dfrac{1}{c-1}}{1-\dfrac{1}{c-1}}=\frac{\left(1+\dfrac{1}{c-1}\right)}{\left(1-\dfrac{1}{c-1}\right)}\cdot\frac{c-1}{c-1}=\frac{c-1+1}{c-1-1}=\frac{c}{c-2}.$$

**58.** $1+\dfrac{1}{1+\dfrac{1}{1+x}}=1+\dfrac{1}{1+\dfrac{1}{1+x}}\cdot\dfrac{1+x}{1+x}=1+\dfrac{1+x}{1+x+1}=1+\dfrac{1+x}{2+x}=\dfrac{2+x}{2+x}+\dfrac{1+x}{2+x}=\dfrac{3+2x}{2+x}$

**59.** $\dfrac{\dfrac{5}{x-1}-\dfrac{2}{x+1}}{\dfrac{x}{x-1}+\dfrac{1}{x+1}}=\dfrac{\dfrac{5(x+1)}{(x-1)(x+1)}+\dfrac{-2(x-1)}{(x-1)(x+1)}}{\dfrac{x(x+1)}{(x-1)(x+1)}+\dfrac{x-1}{(x-1)(x+1)}}=\dfrac{\dfrac{5x+5-2x+2}{(x-1)(x+1)}}{\dfrac{x^2+x+x-1}{(x-1)(x+1)}}$

$$=\frac{3x+7}{(x-1)(x+1)}\cdot\frac{(x-1)(x+1)}{x^2+2x-1}=\frac{3x+7}{x^2+2x-1}$$

Alternatively,

$$\frac{\dfrac{5}{x-1}-\dfrac{2}{x+1}}{\dfrac{x}{x-1}+\dfrac{1}{x+1}}=\frac{\left(\dfrac{5}{x-1}-\dfrac{2}{x+1}\right)}{\left(\dfrac{x}{x-1}+\dfrac{1}{x+1}\right)}\cdot\frac{(x-1)(x+1)}{(x-1)(x+1)}=\frac{5(x+1)-2(x-1)}{x(x+1)+(x-1)}$$

$$=\frac{5x+5-2x+2}{x^2+x+x-1}=\frac{3x+7}{x^2+2x-1}$$

**60.** $$\frac{\dfrac{a-b}{a}-\dfrac{a+b}{b}}{\dfrac{a-b}{b}+\dfrac{a+b}{a}}=\frac{\dfrac{a-b}{a}-\dfrac{a+b}{b}}{\dfrac{a-b}{b}+\dfrac{a+b}{a}}\cdot\frac{ab}{ab}=\frac{(a-b)\,b-(a+b)\,a}{(a-b)\,a+(a+b)\,b}=\frac{ab-b^2-a^2-ab}{a^2-ab+ab+b^2}$$
$$=\frac{-a^2-b^2}{a^2+b^2}=\frac{-\left(a^2+b^2\right)}{a^2+b^2}=-1$$

**61.** $$\frac{x^{-2}-y^{-2}}{x^{-1}+y^{-1}}=\frac{\dfrac{1}{x^2}-\dfrac{1}{y^2}}{\dfrac{1}{x}+\dfrac{1}{y}}=\frac{\dfrac{y^2}{x^2y^2}-\dfrac{x^2}{x^2y^2}}{\dfrac{y}{xy}+\dfrac{x}{xy}}=\frac{y^2-x^2}{x^2y^2}\cdot\frac{xy}{y+x}=\frac{(y-x)\,(y+x)\,xy}{x^2y^2\,(y+x)}=\frac{y-x}{xy}$$

Alternatively, $$\frac{x^{-2}-y^{-2}}{x^{-1}+y^{-1}}=\frac{\left(\dfrac{1}{x^2}-\dfrac{1}{y^2}\right)}{\left(\dfrac{1}{x}+\dfrac{1}{y}\right)}\cdot\frac{x^2y^2}{x^2y^2}=\frac{y^2-x^2}{xy^2+x^2y}=\frac{(y-x)\,(y+x)}{xy\,(y+x)}=\frac{y-x}{xy}.$$

**62.** $$\frac{x^{-1}+y^{-1}}{(x+y)^{-1}}=\frac{\dfrac{1}{x}+\dfrac{1}{y}}{\dfrac{1}{x+y}}=\frac{\dfrac{1}{x}+\dfrac{1}{y}}{\dfrac{1}{x+y}}\cdot\frac{xy\,(x+y)}{xy\,(x+y)}=\frac{y\,(x+y)+x\,(x+y)}{xy}$$
$$=\frac{xy+y^2+x^2+xy}{xy}=\frac{x^2+2xy+y^2}{xy}=\frac{(x+y)^2}{xy}$$

**63.** $$\frac{1}{1+a^n}+\frac{1}{1+a^{-n}}=\frac{1}{1+a^n}+\frac{1}{1+a^{-n}}\cdot\frac{a^n}{a^n}=\frac{1}{1+a^n}+\frac{a^n}{a^n+1}=\frac{1+a^n}{1+a^n}=1$$

**64.** $$\frac{\left(a+\dfrac{1}{b}\right)^m\left(a-\dfrac{1}{b}\right)^n}{\left(b+\dfrac{1}{a}\right)^m\left(b-\dfrac{1}{a}\right)^n}=\left(\frac{a+\dfrac{1}{b}}{b+\dfrac{1}{a}}\right)^m\left(\frac{a-\dfrac{1}{b}}{b-\dfrac{1}{a}}\right)^n=\left(\frac{\dfrac{ab+1}{b}}{\dfrac{ab+1}{a}}\right)^m\left(\frac{\dfrac{ab-1}{b}}{\dfrac{ab-1}{a}}\right)^n$$
$$=\left(\frac{a}{b}\right)^m\left(\frac{a}{b}\right)^n=\left(\frac{a}{b}\right)^{m+n}=\frac{a^{m+n}}{b^{m+n}}$$

**65.** $$\frac{\dfrac{1}{a+h}-\dfrac{1}{a}}{h}=\frac{\dfrac{a}{a\,(a+h)}-\dfrac{a+h}{a\,(a+h)}}{h}=\frac{\dfrac{a-a-h}{a\,(a+h)}}{h}=\frac{-h}{a\,(a+h)}\cdot\frac{1}{h}=\frac{-1}{a\,(a+h)}$$

**66.** $$\frac{(x+h)^{-3}-x^{-3}}{h}=\frac{\dfrac{1}{(x+h)^3}-\dfrac{1}{x^3}}{h}=\left(\frac{1}{(x+h)^3}-\frac{1}{x^3}\right)\cdot\frac{1}{h}=\left(\frac{x^3}{(x+h)^3\,x^3}+\frac{-(x+h)^3}{(x+h)^3\,x^3}\right)\cdot\frac{1}{h}$$
$$=\frac{x^3-\left(x^3+3x^2h+3xh^2+h^3\right)}{(x+h)^3\,x^3}\cdot\frac{1}{h}=\frac{x^3-x^3-3x^2h-3xh^2-h^3}{h\,(x+h)^3\,x^3}$$
$$=\frac{-3x^2h-3xh^2-h^3}{h\,(x+h)^3\,x^3}=\frac{-h\left(3x^2+3xh+h^2\right)}{h\,(x+h)^3\,x^3}=\frac{-\left(3x^2+3xh+h^2\right)}{(x+h)^3\,x^3}$$

**67.** $$\frac{\dfrac{1-(x+h)}{2+(x+h)}-\dfrac{1-x}{2+x}}{h}=\frac{\dfrac{(2+x)\,(1-x-h)}{(2+x)\,(2+x+h)}-\dfrac{(1-x)\,(2+x+h)}{(2+x)\,(2+x+h)}}{h}$$
$$=\frac{\dfrac{2-x-x^2-2h-xh}{(2+x)\,(2+x+h)}-\dfrac{2-x-x^2+h-xh}{(2+x)\,(2+x+h)}}{h}=\frac{-3h}{(2+x)\,(2+x+h)}\cdot\frac{1}{h}=\frac{-3}{(2+x)\,(2+x+h)}$$

**68.** $$\frac{(x+h)^3-7\,(x+h)-\left(x^3-7x\right)}{h}=\frac{x^3+3x^2h+3xh^2+h^3-7x-7h-x^3+7x}{h}$$
$$=\frac{3x^2h+3xh^2+h^3-7h}{h}=\frac{h\left(3x^2+3xh+h^2-7\right)}{h}=3x^2+3xh+h^2-7$$

**69.** $\sqrt{1+\left(\frac{x}{\sqrt{1-x^2}}\right)^2}=\sqrt{1+\frac{x^2}{1-x^2}}=\sqrt{\frac{1-x^2}{1-x^2}+\frac{x^2}{1-x^2}}=\sqrt{\frac{1}{1-x^2}}=\frac{1}{\sqrt{1-x^2}}$

**70.** $\sqrt{1+\left(x^3-\frac{1}{4x^3}\right)^2}=\sqrt{1+x^6-\frac{2x^3}{4x^3}+\frac{1}{16x^6}}=\sqrt{1+x^6-\frac{1}{2}+\frac{1}{16x^6}}=\sqrt{x^6+\frac{1}{2}+\frac{1}{16x^6}}$

$=\sqrt{\left(x^3+\frac{1}{4x^3}\right)^2}=\left|x^3+\frac{1}{4x^3}\right|$

**71.** $\frac{3(x+2)^2(x-3)^2-(x+2)^3(2)(x-3)}{(x-3)^4}=\frac{(x+2)^2(x-3)[3(x-3)-(x+2)(2)]}{(x-3)^4}$

$=\frac{(x+2)^2(3x-9-2x-4)}{(x-3)^3}=\frac{(x+2)^2(x-13)}{(x-3)^3}$

**72.** $\frac{2x(x+6)^4-x^2(4)(x+6)^3}{(x+6)^8}=\frac{(x+6)^3\left[2x(x+6)-4x^2\right]}{(x+6)^8}=\frac{2x^2+12x-4x^2}{(x+6)^5}=\frac{12x-2x^2}{(x+6)^5}=\frac{2x(6-x)}{(x+6)^5}$

**73.** $\frac{2(1+x)^{1/2}-x(1+x)^{-1/2}}{1+x}=\frac{(1+x)^{-1/2}[2(1+x)-x]}{1+x}=\frac{x+2}{(1+x)^{3/2}}$

**74.** $\frac{\left(1-x^2\right)^{1/2}+x^2\left(1-x^2\right)^{-1/2}}{1-x^2}=\frac{\left(1-x^2\right)^{-1/2}\left(1-x^2+x^2\right)}{1-x^2}=\frac{1}{\left(1-x^2\right)^{3/2}}$

**75.** $\frac{3(1+x)^{1/3}-x(1+x)^{-2/3}}{(1+x)^{2/3}}=\frac{(1+x)^{-2/3}[3(1+x)-x]}{(1+x)^{2/3}}=\frac{2x+3}{(1+x)^{4/3}}$

**76.** $\frac{(7-3x)^{1/2}+\frac{3}{2}x(7-3x)^{-1/2}}{7-3x}=\frac{(7-3x)^{-1/2}\left(7-3x+\frac{3}{2}x\right)}{7-3x}=\frac{7-\frac{3}{2}x}{(7-3x)^{3/2}}$

**77.** $\frac{1}{2-\sqrt{3}}=\frac{1}{2-\sqrt{3}}\cdot\frac{2+\sqrt{3}}{2+\sqrt{3}}=\frac{2+\sqrt{3}}{4-3}=\frac{2+\sqrt{3}}{1}=2+\sqrt{3}$

**78.** $\frac{2}{3-\sqrt{5}}=\frac{2}{3-\sqrt{5}}\cdot\frac{3+\sqrt{5}}{3+\sqrt{5}}=\frac{2(3+\sqrt{5})}{9-5}=\frac{2(3+\sqrt{5})}{4}=\frac{3+\sqrt{5}}{2}$

**79.** $\frac{2}{\sqrt{2}+\sqrt{7}}=\frac{2}{\sqrt{2}+\sqrt{7}}\cdot\frac{\sqrt{2}-\sqrt{7}}{\sqrt{2}-\sqrt{7}}=\frac{2(\sqrt{2}-\sqrt{7})}{2-7}=\frac{2(\sqrt{2}-\sqrt{7})}{-5}=\frac{2(\sqrt{7}-\sqrt{2})}{5}$

**80.** $\frac{1}{\sqrt{x}+1}=\frac{1}{\sqrt{x}+1}\cdot\frac{\sqrt{x}-1}{\sqrt{x}-1}=\frac{\sqrt{x}-1}{x-1}$

**81.** $\frac{y}{\sqrt{3}+\sqrt{y}}=\frac{y}{\sqrt{3}+\sqrt{y}}\cdot\frac{\sqrt{3}-\sqrt{y}}{\sqrt{3}-\sqrt{y}}=\frac{y(\sqrt{3}-\sqrt{y})}{3-y}=\frac{y\sqrt{3}-y\sqrt{y}}{3-y}$

**82.** $\frac{2(x-y)}{\sqrt{x}-\sqrt{y}}=\frac{2(x-y)}{\sqrt{x}-\sqrt{y}}\cdot\frac{\sqrt{x}+\sqrt{y}}{\sqrt{x}+\sqrt{y}}=\frac{2(x-y)(\sqrt{x}+\sqrt{y})}{x-y}=2(\sqrt{x}+\sqrt{y})=2\sqrt{x}+2\sqrt{y}$

**83.** $\frac{1-\sqrt{5}}{3}=\frac{1-\sqrt{5}}{3}\cdot\frac{1+\sqrt{5}}{1+\sqrt{5}}=\frac{1-5}{3(1+\sqrt{5})}=\frac{-4}{3(1+\sqrt{5})}$

**84.** $\frac{\sqrt{3}+\sqrt{5}}{2}=\frac{\sqrt{3}+\sqrt{5}}{2}\cdot\frac{\sqrt{3}-\sqrt{5}}{\sqrt{3}-\sqrt{5}}=\frac{3-5}{2(\sqrt{3}-\sqrt{5})}=\frac{-2}{2(\sqrt{3}-\sqrt{5})}=\frac{-1}{\sqrt{3}-\sqrt{5}}$

**85.** $\frac{\sqrt{r}+\sqrt{2}}{5}=\frac{\sqrt{r}+\sqrt{2}}{5}\cdot\frac{\sqrt{r}-\sqrt{2}}{\sqrt{r}-\sqrt{2}}=\frac{r-2}{5(\sqrt{r}-\sqrt{2})}$

**86.** $\frac{\sqrt{x}-\sqrt{x+h}}{h\sqrt{x}\sqrt{x+h}}=\frac{\sqrt{x}-\sqrt{x+h}}{h\sqrt{x}\sqrt{x+h}}\cdot\frac{\sqrt{x}+\sqrt{x+h}}{\sqrt{x}+\sqrt{x+h}}=\frac{x-(x+h)}{h\sqrt{x}\sqrt{x+h}(\sqrt{x}+\sqrt{x+h})}$

$=\frac{-h}{h\sqrt{x}\sqrt{x+h}(\sqrt{x}+\sqrt{x+h})}=-\frac{1}{\sqrt{x}\sqrt{x+h}(\sqrt{x}+\sqrt{x+h})}$

**87.** $\sqrt{x^2+1}-x=\dfrac{\sqrt{x^2+1}-x}{1}\cdot\dfrac{\sqrt{x^2+1}+x}{\sqrt{x^2+1}+x}=\dfrac{x^2+1-x^2}{\sqrt{x^2+1}+x}=\dfrac{1}{\sqrt{x^2+1}+x}$

**88.** $\sqrt{x+1}-\sqrt{x}=\dfrac{\sqrt{x+1}-\sqrt{x}}{1}\cdot\dfrac{\sqrt{x+1}+\sqrt{x}}{\sqrt{x+1}+\sqrt{x}}=\dfrac{x+1-x}{\sqrt{x+1}+\sqrt{x}}=\dfrac{1}{\sqrt{x+1}+\sqrt{x}}$

**89.** $\dfrac{16+a}{16}=\frac{16}{16}+\dfrac{a}{16}=1+\dfrac{a}{16}$, so the statement is true.

**90.** This statement is false. For example, take $b=2$ and $c=1$. Then LHS $=\dfrac{b}{b-c}=\dfrac{2}{2-1}=2$, while RHS $=1-\dfrac{b}{c}=1-\frac{2}{1}=-1$, and $2\neq-1$.

**91.** This statement is false. For example, take $x=2$. Then LHS $=\dfrac{2}{4+x}=\dfrac{2}{4+2}=\dfrac{2}{6}=\dfrac{1}{3}$, while RHS $=\dfrac{1}{2}+\dfrac{2}{x}=\dfrac{1}{2}+\dfrac{2}{2}=\dfrac{3}{2}$, and $\dfrac{1}{3}\neq\dfrac{3}{2}$.

**92.** This statement is false. For example, take $x=5$ and $y=2$. Substituting, we obtain LHS $=\dfrac{x+1}{y+1}=\dfrac{5+1}{2+1}=\dfrac{6}{3}=2$, while RHS $=\dfrac{x}{y}=\dfrac{5}{2}$, and $2\neq\dfrac{5}{2}$.

**93.** This statement is false. For example, take $x=0$ and $y=1$. Then LHS $=\dfrac{x}{x+y}=\dfrac{0}{0+1}=0$, while RHS $=\dfrac{1}{1+y}=\dfrac{1}{1+1}=\dfrac{1}{2}$, and $0\neq\dfrac{1}{2}$.

**94.** This statement is false. For example, take $x=1$ and $y=1$. Then LHS $=2\left(\frac{a}{b}\right)=2\left(\frac{1}{1}\right)=2$, while RHS $=\dfrac{2a}{2b}=\dfrac{2}{2}=1$, and $2\neq1$.

**95.** This statement is true: $\dfrac{-a}{b}=(-a)\left(\dfrac{1}{b}\right)=(-1)(a)\left(\dfrac{1}{b}\right)=(-1)\left(\frac{a}{b}\right)=-\dfrac{a}{b}$.

**96.** This statement is true: $\dfrac{1+x+x^2}{x}=\dfrac{1}{x}+\dfrac{x}{x}+\dfrac{x^2}{x}=\dfrac{1}{x}+1+x$.

**97. (a)** $R=\dfrac{1}{\dfrac{1}{R_1}+\dfrac{1}{R_2}}=\dfrac{1}{\dfrac{1}{R_1}+\dfrac{1}{R_2}}\cdot\dfrac{R_1R_2}{R_1R_2}=\dfrac{R_1R_2}{R_2+R_1}$

**(b)** Substituting $R_1=10$ ohms and $R_2=20$ ohms gives $R=\dfrac{(10)(20)}{(20)+(10)}=\dfrac{200}{30}\approx6.7$ ohms.

**98. (a)** The average cost $A=\dfrac{\text{Cost}}{\text{number of shirts}}=\dfrac{500+6x+0.01x^2}{x}$.

**(b)**

| $x$ | 10 | 20 | 50 | 100 | 200 | 500 | 1000 |
|---|---|---|---|---|---|---|---|
| Average cost | \$56.10 | \$31.20 | \$16.50 | \$12.00 | \$10.50 | \$12.00 | \$16.50 |

**99.**

| $x$ | 2.80 | 2.90 | 2.95 | 2.99 | 2.999 | 3 | 3.001 | 3.01 | 3.05 | 3.10 | 3.20 |
|---|---|---|---|---|---|---|---|---|---|---|---|
| $\dfrac{x^2-9}{x-3}$ | 5.80 | 5.90 | 5.95 | 5.99 | 5.999 | ? | 6.001 | 6.01 | 6.05 | 6.10 | 6.20 |

From the table, we see that the expression $\dfrac{x^2-9}{x-3}$ approaches 6 as $x$ approaches 3. We simplify the expression: $\dfrac{x^2-9}{x-3}=\dfrac{(x-3)(x+3)}{x-3}=x+3$, $x\neq3$. Clearly as $x$ approaches 3, $x+3$ approaches 6. This explains the result in the table.

**100.** No, squaring $\dfrac{2}{\sqrt{x}}$ changes its value by a factor of $\dfrac{2}{\sqrt{x}}$.

**101.** Answers will vary.

| Algebraic Error | Counterexample |
|---|---|
| $\frac{1}{a}+\frac{1}{b}\neq\frac{1}{a+b}$ | $\frac{1}{2}+\frac{1}{2}\neq\frac{1}{2+2}$ |
| $(a+b)^2\neq a^2+b^2$ | $(1+3)^2\neq 1^2+3^2$ |
| $\sqrt{a^2+b^2}\neq a+b$ | $\sqrt{5^2+12^2}\neq 5+12$ |
| $\frac{a+b}{a}\neq b$ | $\frac{2+6}{2}\neq 6$ |
| $\left(a^3+b^3\right)^{1/3}\neq a+b$ | $\left(2^3+2^3\right)^{1/3}\neq 2+2$ |
| $\frac{a^m}{a^n}\neq a^{m/n}$ | $\frac{3^5}{3^2}\neq 3^{5/2}$ |
| $a^{-1/n}\neq\frac{1}{a^n}$ | $64^{-1/3}\neq\frac{1}{64^3}$ |

**102. (a)** $A+B$ **(b)** $AB$ **(c)** $A^{1/3}$ **(d)** $A/B$

# Chapter P Review

**1.** Commutative Property for addition.

**2.** Commutative Law for multiplication.

**3.** Distributive Property.

**4.** Distributive Property.

**5.** $x\in[-2,6) \quad\Leftrightarrow\quad -2\leq x<6$

**6.** $x\in(0,10] \quad\Leftrightarrow\quad 0<x\leq 10$

**7.** $x\in(-\infty,4] \quad\Leftrightarrow\quad x\leq 4$

**8.** $x\in[-2,\infty) \quad\Leftrightarrow\quad -2\leq x$

**9.** $x\geq 5 \quad\Leftrightarrow\quad x\in[5,\infty)$

**10.** $x<-3 \quad\Leftrightarrow\quad x\in(-\infty,-3)$

**11.** $-1<x\leq 5 \quad\Leftrightarrow\quad x\in(-1,5]$

**12.** $0\leq x\leq\frac{1}{2} \quad\Leftrightarrow\quad x\in\left[0,\frac{1}{2}\right]$

**13.** $|3-|-9||=|3-9|=|-6|=6$

**14.** $1-|1-|-1||=1-|1-1|=1-|0|=1$

**15.** $2^{-3}-3^{-2}=\frac{1}{8}-\frac{1}{9}=\frac{9}{72}-\frac{8}{72}=\frac{1}{72}$

**16.** $\sqrt[3]{-125}=\sqrt[3]{(-5)^3}=-5$

**17.** $216^{-1/3}=\dfrac{1}{216^{1/3}}=\dfrac{1}{\sqrt[3]{216}}=\dfrac{1}{6}$

**18.** $64^{2/3}=\left(4^3\right)^{2/3}=4^2=16$

**19.** $\dfrac{\sqrt{242}}{\sqrt{2}} = \sqrt{\dfrac{242}{2}} = \sqrt{121} = 11$

**20.** $\sqrt[4]{4}\sqrt[4]{324} = \sqrt[4]{4}\sqrt[4]{4 \cdot 81} = \sqrt[4]{2^2}\sqrt[4]{2^2 \cdot 3^4} = 2 \cdot 3 = 6$

**21.** $2^{1/2}8^{1/2} = \sqrt{2} \cdot \sqrt{8} = \sqrt{16} = 4$

**22.** $\sqrt{2}\sqrt{50} = \sqrt{100} = 10$

**23.** $\dfrac{1}{x^2} = x^{-2}$

**24.** $x\sqrt{x} = x^1 \cdot x^{1/2} = x^{3/2}$

**25.** $x^2x^m\left(x^3\right)^m = x^{2+m+3m} = x^{4m+2}$

**26.** $\left(\left(x^m\right)^2\right)^n = \left(x^{2m}\right)^n = x^{2mn}$

**27.** $x^ax^bx^c = x^{a+b+c}$

**28.** $\left(\left(x^a\right)^b\right)^c = \left(x^{ab}\right)^c = x^{abc}$

**29.** $x^{c+1}\left(x^{2c-1}\right)^2 = x^{(c+1)+2(2c-1)} = x^{c+1+4c-2} = x^{5c-1}$

**30.** $\dfrac{\left(x^2\right)^n x^5}{x^n} = \dfrac{x^{2n} \cdot x^5}{x^n} = x^{2n+5-n} = x^{n+5}$

**31.** $\left(2x^3y\right)^2\left(3x^{-1}y^2\right) = 4x^6y^2 \cdot 3x^{-1}y^2 = 4 \cdot 3x^{6-1}y^{2+2}$
$= 12x^5y^4$

**32.** $\left(a^2\right)^{-3}\left(a^3b\right)^2\left(b^3\right)^4 = a^{-6} \cdot a^6b^2 \cdot b^{12}$
$= a^{-6+6}b^{2+12} = b^{14}$

**33.** $\dfrac{x^4\left(3x\right)^2}{x^3} = \dfrac{x^4 \cdot 9x^2}{x^3} = 9x^{4+2-3} = 9x^3$

**34.** $\left(\dfrac{r^2s^{4/3}}{r^{1/3}s}\right)^6 = \dfrac{r^{12}s^8}{r^2s^6} = r^{12-2}s^{8-6} = r^{10}s^2$

**35.** $\sqrt[3]{\left(x^3y\right)^2y^4} = \sqrt[3]{x^6y^4y^2} = \sqrt[3]{x^6y^6} = x^2y^2$

**36.** $\sqrt{x^2y^4} = \sqrt{x^2} \cdot \sqrt{\left(y^2\right)^2} = |x|\,y^2$

**37.** $\dfrac{x}{2+\sqrt{x}} = \dfrac{x}{2+\sqrt{x}} \cdot \dfrac{2-\sqrt{x}}{2-\sqrt{x}} = \dfrac{x\left(2-\sqrt{x}\right)}{4-x}$. Here simplify means to rationalize the denominator.

**38.** $\dfrac{\sqrt{x}+1}{\sqrt{x}-1} = \dfrac{\sqrt{x}+1}{\sqrt{x}-1} \cdot \dfrac{\sqrt{x}+1}{\sqrt{x}+1} = \dfrac{\left(\sqrt{x}\right)^2 + 2\sqrt{x} + 1}{x-1} = \dfrac{x + 2\sqrt{x} + 1}{x-1}$. Here simplify means to rationalize the denominator.

**39.** $\dfrac{8r^{1/2}s^{-3}}{2r^{-2}s^4} = 4r^{(1/2)-(-2)}s^{-3-4} = 4r^{5/2}s^{-7} = \dfrac{4r^{5/2}}{s^7}$

**40.** $\left(\dfrac{ab^2c^{-3}}{2a^3b^{-4}}\right)^{-2} = \dfrac{a^{-2}b^{-4}c^6}{2^{-2}a^{-6}b^8} = 2^2a^{-2-(-6)}b^{-4-8}c^6 = 4a^4b^{-12}c^6 = \dfrac{4a^4c^6}{b^{12}}$

**41.** $78{,}250{,}000{,}000 = 7.825 \times 10^{10}$

**42.** $2.08 \times 10^{-8} = 0.0000000208$

**43.** $\dfrac{ab}{c} \approx \dfrac{(0.00000293)\left(1.582 \times 10^{-14}\right)}{2.8064 \times 10^{12}} = \dfrac{\left(2.93 \times 10^{-6}\right)\left(1.582 \times 10^{-14}\right)}{2.8064 \times 10^{12}} = \dfrac{2.93 \cdot 1.582}{2.8064} \times 10^{-6-14-12}$
$\approx 1.65 \times 10^{-32}$

**44.** $80\,\dfrac{\text{times}}{\text{minute}} \cdot \dfrac{60\text{ minutes}}{\text{hour}} \cdot \dfrac{24\text{ hours}}{\text{day}} \cdot \dfrac{365\text{ days}}{\text{year}} \cdot 90\text{ years} \approx 3.8 \times 10^9$ times

**45.** $2x^2y - 6xy^2 = 2xy\left(x - 3y\right)$

**46.** $12x^2y^4 - 3xy^5 + 9x^3y^2 = 3xy^2\left(4xy^2 - y^3 + 3x^2\right)$

**47.** $x^2 - 9x + 18 = \left(x-6\right)\left(x-3\right)$

**48.** $x^2 + 3x - 10 = \left(x+5\right)\left(x-2\right)$

**49.** $3x^2 - 2x - 1 = \left(3x+1\right)\left(x-1\right)$

**50.** $6x^2 + x - 12 = \left(3x-4\right)\left(2x+3\right)$

**51.** $4t^2 - 13t - 12 = \left(4t+3\right)\left(t-4\right)$

**52.** $x^4 - 2x^2 + 1 = \left(x^2-1\right)^2 = \left[\left(x-1\right)\left(x+1\right)\right]^2 = \left(x-1\right)^2\left(x+1\right)^2$

**53.** $25 - 16t^2 = \left(5-4t\right)\left(5+4t\right)$

**54.** $2y^6 - 32y^2 = 2y^2\left(y^4-16\right) = 2y^2\left(y^2+4\right)\left(y^2-4\right) = 2y^2\left(y^2+4\right)\left(y+2\right)\left(y-2\right)$

**55.** $x^6 - 1 = \left(x^3 - 1\right)\left(x^3 + 1\right) = (x - 1)\left(x^2 + x + 1\right)(x + 1)\left(x^2 - x + 1\right)$

**56.** $y^3 - 2y^2 - y + 2 = y^2(y - 2) - 1(y - 2) = (y - 2)\left(y^2 - 1\right) = (y - 2)(y - 1)(y + 1)$

**57.** $x^{-1/2} - 2x^{1/2} + x^{3/2} = x^{-1/2}\left(1 - 2x + x^2\right) = x^{-1/2}(1 - x)^2$

**58.** $a^4b^2 + ab^5 = ab^2\left(a^3 + b^3\right) = ab^2(a + b)\left(a^2 - ab + b^2\right)$

**59.** $4x^3 - 8x^2 + 3x - 6 = 4x^2(x - 2) + 3(x - 2) = \left(4x^2 + 3\right)(x - 2)$

**60.** $8x^3 + y^6 = (2x)^3 + \left(y^2\right)^3 = \left(2x + y^2\right)\left(4x^2 - 2xy^2 + y^4\right)$

**61.** $\left(x^2 + 2\right)^{5/2} + 2x\left(x^2 + 2\right)^{3/2} + x^2\sqrt{x^2 + 2} = \left(x^2 + 2\right)^{1/2}\left(\left(x^2 + 2\right)^2 + 2x\left(x^2 + 2\right) + x^2\right)$

$= \sqrt{x^2 + 2}\left(x^4 + 4x^2 + 4 + 2x^3 + 4x + x^2\right) = \sqrt{x^2 + 2}\left(x^4 + 2x^3 + 5x^2 + 4x + 4\right) = \sqrt{x^2 + 2}\left(x^2 + x + 2\right)^2$

**62.** $3x^3 - 2x^2 + 18x - 12 = x^2(3x - 2) + 6(3x - 2) = (3x - 2)\left(x^2 + 6\right)$

**63.** $a^2y - b^2y = y\left(a^2 - b^2\right) = y(a - b)(a + b)$

**64.** $ax^2 + bx^2 - a - b = x^2(a + b) - 1(a + b) = (a + b)\left(x^2 - 1\right) = (a + b)(x - 1)(x + 1)$

**65.** $(x + 1)^2 - 2(x + 1) + 1 = [(x + 1) - 1]^2 = x^2$. You can also obtain this result by expanding each term and then simplifying.

**66.** $(a + b)^2 + 2(a + b) - 15 = [(a + b) - 3][(a + b) + 5] = (a + b - 3)(a + b + 5)$

**67.** $(2x + 1)(3x - 2) - 5(4x - 1) = 6x^2 - 4x + 3x - 2 - 20x + 5 = 6x^2 - 21x + 3$

**68.** $(2y - 7)(2y + 7) = 4y^2 - 49$

**69.** $\left(2a^2 - b\right)^2 = \left(2a^2\right)^2 - 2\left(2a^2\right)(b) + (b)^2 = 4a^4 - 4a^2b + b^2$

**70.** $(1 + x)(2 - x) - (3 - x)(3 + x) = 2 + x - x^2 - \left(9 - x^2\right) = 2 + x - x^2 - 9 + x^2 = -7 + x$

**71.** $(x - 1)(x - 2)(x - 3) = (x - 1)\left(x^2 - 5x + 6\right) = x^3 - 5x^2 + 6x - x^2 + 5x - 6 = x^3 - 6x^2 + 11x - 6$

**72.** $(2x + 1)^3 = (2x)^3 + 3(2x)^2(1) + 3(2x)(1)^2 + (1)^3 = 8x^3 + 12x^2 + 6x + 1$

**73.** $\sqrt{x}\left(\sqrt{x} + 1\right)\left(2\sqrt{x} - 1\right) = \left(x + \sqrt{x}\right)\left(2\sqrt{x} - 1\right) = 2x\sqrt{x} - x + 2x - \sqrt{x} = 2x^{3/2} + x - x^{1/2}$

**74.** $x^3(x - 6)^2 + x^4(x - 6) = x^3(x - 6)[(x - 6) + x] = x^3(x - 6)(2x - 6) = x^3\left(2x^2 - 18x + 36\right)$

$= 2x^5 - 18x^4 + 36x^3$

**75.** $x^2(x - 2) + x(x - 2)^2 = x^3 - 2x^2 + x\left(x^2 - 4x + 4\right) = x^3 - 2x^2 + x^3 - 4x^2 + 4x = 2x^3 - 6x^2 + 4x$

**76.** $\dfrac{x^3 + 2x^2 + 3x}{x} = \dfrac{x\left(x^2 + 2x + 3\right)}{x} = x^2 + 2x + 3$

**77.** $\dfrac{x^2 - 2x - 3}{2x^2 + 5x + 3} = \dfrac{(x - 3)(x + 1)}{(2x + 3)(x + 1)} = \dfrac{x - 3}{2x + 3}$

**78.** $\dfrac{t^3 - 1}{t^2 - 1} = \dfrac{(t - 1)\left(t^2 + t + 1\right)}{(t - 1)(t + 1)} = \dfrac{t^2 + t + 1}{t + 1}$

**79.** $\dfrac{x^2 + 2x - 3}{x^2 + 8x + 16} \cdot \dfrac{3x + 12}{x - 1} = \dfrac{(x + 3)(x - 1)}{(x + 4)(x + 4)} \cdot \dfrac{3(x + 4)}{(x - 1)} = \dfrac{3(x + 3)}{x + 4}$

**80.** $\dfrac{x^3/(x - 1)}{x^2/\left(x^3 - 1\right)} = \dfrac{x^3}{x - 1} \cdot \dfrac{x^3 - 1}{x^2} = \dfrac{x^3}{x - 1} \cdot \dfrac{(x - 1)\left(x^2 + x + 1\right)}{x^2} = x\left(x^2 + x + 1\right) = x^3 + x^2 + x$

**81.** $\dfrac{x^2 - 2x - 15}{x^2 - 6x + 5} \div \dfrac{x^2 - x - 12}{x^2 - 1} = \dfrac{(x - 5)(x + 3)}{(x - 5)(x - 1)} \cdot \dfrac{(x - 1)(x + 1)}{(x - 4)(x + 3)} = \dfrac{x + 1}{x - 4}$

**82.** $x - \dfrac{1}{x + 1} = \dfrac{x(x + 1)}{x + 1} - \dfrac{1}{x + 1} = \dfrac{x^2 + x - 1}{x + 1}$

**83.** $\dfrac{1}{x - 1} - \dfrac{x}{x^2 + 1} = \dfrac{x^2 + 1}{(x - 1)\left(x^2 + 1\right)} - \dfrac{x(x - 1)}{(x - 1)\left(x^2 + 1\right)} = \dfrac{x^2 + 1 - x^2 + x}{(x - 1)\left(x^2 + 1\right)} = \dfrac{x + 1}{(x - 1)\left(x^2 + 1\right)}$

**84.** $\dfrac{2}{x}+\dfrac{1}{x-2}+\dfrac{3}{(x-2)^2}=\dfrac{2(x-2)^2}{x(x-2)^2}+\dfrac{x(x-2)}{x(x-2)^2}+\dfrac{3x}{x(x-2)^2}$

$=\dfrac{2(x^2-4x+4)+x^2-2x+3x}{x(x-2)^2}=\dfrac{2x^2-8x+8+x^2-2x+3x}{x(x-2)^2}=\dfrac{3x^2-7x+8}{x(x-2)^2}$

**85.** $\dfrac{1}{x-1}-\dfrac{2}{x^2-1}=\dfrac{1}{x-1}-\dfrac{2}{(x-1)(x+1)}=\dfrac{x+1}{(x-1)(x+1)}-\dfrac{2}{(x-1)(x+1)}$

$=\dfrac{x+1-2}{(x-1)(x+1)}=\dfrac{x-1}{(x-1)(x+1)}=\dfrac{1}{x+1}$

**86.** $\dfrac{1}{x+2}+\dfrac{1}{x^2-4}-\dfrac{2}{x^2-x-2}=\dfrac{1}{x+2}+\dfrac{1}{(x-2)(x+2)}-\dfrac{2}{(x-2)(x+1)}$

$=\dfrac{(x-2)(x+1)}{(x-2)(x+1)(x+2)}+\dfrac{x+1}{(x-2)(x+1)(x+2)}-\dfrac{2(x+2)}{(x-2)(x+1)(x+2)}$

$=\dfrac{x^2-x-2+x+1-2x-4}{(x-2)(x+1)(x+2)}=\dfrac{x^2-2x-5}{(x-2)(x+1)(x+2)}$

**87.** $\dfrac{\frac{1}{x}-\frac{1}{2}}{x-2}=\dfrac{\frac{2}{2x}-\frac{x}{2x}}{x-2}=\dfrac{2-x}{2x}\cdot\dfrac{1}{x-2}=\dfrac{-1(x-2)}{2x}\cdot\dfrac{1}{x-2}=\dfrac{-1}{2x}$

**88.** $\dfrac{\frac{1}{x}-\frac{1}{x+1}}{\frac{1}{x}+\frac{1}{x+1}}=\dfrac{\frac{1}{x}-\frac{1}{x+1}}{\frac{1}{x}+\frac{1}{x+1}}\cdot\dfrac{x(x+1)}{x(x+1)}=\dfrac{(x+1)-x}{(x+1)+x}=\dfrac{1}{2x+1}$

**89.** $\dfrac{3(x+h)^2-5(x+h)-(3x^2-5x)}{h}=\dfrac{3x^2+6xh+3h^2-5x-5h-3x^2+5x}{h}=\dfrac{6xh+3h^2-5h}{h}$

$=\dfrac{h(6x+3h-5)}{h}=6x+3h-5$

**90.** $\dfrac{\sqrt{x+h}-\sqrt{x}}{h}=\dfrac{\sqrt{x+h}-\sqrt{x}}{h}\cdot\dfrac{\sqrt{x+h}+\sqrt{x}}{\sqrt{x+h}+\sqrt{x}}=\dfrac{(x+h)-x}{h\left(\sqrt{x+h}+\sqrt{x}\right)}=\dfrac{h}{h\left(\sqrt{x+h}+\sqrt{x}\right)}=\dfrac{1}{\sqrt{x+h}+\sqrt{x}}$

**91.** This statement is false. For example, take $x=1$ and $y=1$. Then LHS $=(x+y)^3=(1+1)^3=2^3=8$, while RHS $=x^3+y^3=1^3+1^3=1+1=2$, and $8\neq 2$.

**92.** This statement is true for $a\neq 1$: $\dfrac{1+\sqrt{a}}{1-a}=\dfrac{1+\sqrt{a}}{1-a}\cdot\dfrac{1-\sqrt{a}}{1-\sqrt{a}}=\dfrac{1-a}{(1-a)(1-\sqrt{a})}=\dfrac{1}{1-\sqrt{a}}$.

**93.** This statement is true: $\dfrac{12+y}{y}=\dfrac{12}{y}+\dfrac{y}{y}=\dfrac{12}{y}+1$.

**94.** This statement is false. For example, take $a=1$ and $b=1$. Then LHS $=\sqrt[3]{a+b}=\sqrt[3]{1+1}=\sqrt[3]{2}$, while RHS $=\sqrt[3]{a}+\sqrt[3]{b}=\sqrt[3]{1}+\sqrt[3]{1}=1+1=2$, and $\sqrt[3]{2}\neq 2$.

**95.** This statement is false. For example, take $a=-1$. Then LHS $=\sqrt{a^2}=\sqrt{(-1)^2}=\sqrt{1}=1$, which does not equal $a=-1$. The true statement is $\sqrt{a^2}=|a|$.

**96.** This statement is false. For example, take $x=1$. Then LHS $=\dfrac{1}{x+4}=\dfrac{1}{1+4}=\dfrac{1}{5}$, while RHS $=\dfrac{1}{x}+\dfrac{1}{4}=\dfrac{1}{1}+\dfrac{1}{4}=\dfrac{5}{4}$, and $\dfrac{1}{5}\neq\dfrac{5}{4}$.

**97.** This statement is false. For example, take $x=1$ and $y=1$. Then LHS $=x^3+y^3=1^3+1^3=1+1=2$, while RHS $=(x+y)(x^2+xy+y^2)=(1+1)\left[1^2+(1)(1)+1^2\right]=2(3)=6$.

**98.** This statement is false. For example, take $x=1$. Then LHS $=\dfrac{x^2+1}{x^2+2x+1}=\dfrac{1+1}{1+2+1}=\dfrac{2}{5}$, while RHS $=\dfrac{1}{2x+1}=\dfrac{1}{2+1}=\dfrac{1}{3}$, and $\dfrac{2}{5}\neq\dfrac{1}{3}$.

**99.** Substituting for $a$ and $b$, we obtain

$$a^2+b^2=(2mn)^2+\left(m^2-n^2\right)^2=4m^2n^2+m^4-2m^2n^2+n^4=m^4+2m^2n^2+n^4=\left(m^2+n^2\right)^2$$

Since this last expression is $c^2$, we have $a^2+b^2=c^2$ for these values.

**100.** Substituting, we obtain

$$\begin{aligned}\sqrt{1+t^2}&=\sqrt{1+\left[\frac{1}{2}\left(x^3-\frac{1}{x^3}\right)\right]^2}=\sqrt{1+\frac{1}{4}\left(x^6-2\frac{x^3}{x^3}+\frac{1}{x^6}\right)}=\sqrt{1+\frac{x^6}{4}-\frac{1}{2}+\frac{1}{4x^6}}\\&=\sqrt{\frac{x^6}{4}+\frac{1}{2}+\frac{1}{4x^6}}=\sqrt{\left(\frac{x^3}{2}+\frac{1}{2x^3}\right)^2}\end{aligned}$$

Since $x>0$, $\sqrt{\left(\frac{x^3}{2}+\frac{1}{2x^3}\right)^2}=\frac{x^3}{2}+\frac{1}{2x^3}=\frac{1}{2}\left(x^3+\frac{1}{x^3}\right)$.

# Chapter P Test

**1. (a)**

$[-2,5)$ $\qquad$ $(7,\infty)$

**(b)** $x\le 3 \Leftrightarrow x\in(-\infty,3]$; $0<x\le 8 \Leftrightarrow x\in(0,8]$

**(c)** The distance is $|-12-39|=|-51|=51$.

**2. (a)** $(-3)^4=81$

**(b)** $-3^4=-81$

**(c)** $3^{-4}=\frac{1}{3^4}=\frac{1}{81}$

**(d)** $\frac{5^{23}}{5^{21}}=5^{23-21}=5^2=25$

**(e)** $\left(\frac{2}{3}\right)^{-2}=\frac{3^2}{2^2}=\frac{9}{4}$

**(f)** $\frac{\sqrt[6]{64}}{\sqrt{32}}=\frac{\sqrt[6]{2^6}}{\sqrt{16\cdot 2}}=\frac{2}{4\sqrt{2}}=\frac{1}{2\sqrt{2}}$

**(g)** $16^{-3/4}=\left(2^4\right)^{-3/4}=2^{-3}=\frac{1}{8}$

**3. (a)** $186{,}000{,}000{,}000=1.86\times10^{11}$

**(b)** $0.0000003965=3.965\times10^{-7}$

**4. (a)** $\sqrt{200}-\sqrt{32}=10\sqrt{2}-4\sqrt{2}=6\sqrt{2}$

**(b)** $\left(3a^3b^3\right)\left(4ab^2\right)^2=3a^3b^3\cdot4^2a^2b^4=48a^5b^7$

**(c)** $\sqrt[3]{\frac{125}{x^{-9}}}=\sqrt[3]{5^3\cdot x^9}=5x^3$

**(d)** $\left(\frac{3x^{3/2}y^3}{x^2y^{-1/2}}\right)^{-2}=\frac{3^{-2}x^{-3}y^{-6}}{x^{-4}y^1}=\frac{1}{9}x^{-3-(-4)}y^{-6-1}=\frac{1}{9}xy^{-7}=\frac{x}{9y^7}$

**5. (a)** $3(x+6)+4(2x-5)=3x+18+8x-20=11x-2$

**(b)** $(x+3)(4x-5)=4x^2-5x+12x-15=4x^2+7x-15$

**(c)** $\left(\sqrt{a}+\sqrt{b}\right)\left(\sqrt{a}-\sqrt{b}\right)=\left(\sqrt{a}\right)^2-\left(\sqrt{b}\right)^2=a-b$

**(d)** $(2x+3)^2=(2x)^2+2(2x)(3)+(3)^2=4x^2+12x+9$

**(e)** $(x+2)^3=(x)^3+3(x)^2(2)+3(x)(2)^2+(2)^3=x^3+6x^2+12x+8$

**6. (a)** $4x^2-25=(2x-5)(2x+5)$

**(b)** $2x^2+5x-12=(2x-3)(x+4)$

**(c)** $x^3 - 3x^2 - 4x + 12 = x^2(x-3) - 4(x-3) = (x-3)(x^2-4) = (x-3)(x-2)(x+2)$

**(d)** $x^4 + 27x = x(x^3+27) = x(x+3)(x^2-3x+9)$

**(e)** $3x^{3/2} - 9x^{1/2} + 6x^{-1/2} = 3x^{-1/2}(x^2-3x+2) = 3x^{-1/2}(x-2)(x-1)$

**(f)** $x^3y - 4xy = xy(x^2-4) = xy(x-2)(x+2)$

**7. (a)** $\dfrac{x^2+3x+2}{x^2-x-2} = \dfrac{(x+1)(x+2)}{(x+1)(x-2)} = \dfrac{x+2}{x-2}$

**(b)** $\dfrac{2x^2-x-1}{x^2-9} \cdot \dfrac{x+3}{2x+1} = \dfrac{(2x+1)(x-1)}{(x-3)(x+3)} \cdot \dfrac{x+3}{2x+1} = \dfrac{x-1}{x-3}$

**(c)** $\dfrac{x^2}{x^2-4} - \dfrac{x+1}{x+2} = \dfrac{x^2}{(x-2)(x+2)} - \dfrac{x+1}{x+2} = \dfrac{x^2}{(x-2)(x+2)} + \dfrac{-(x+1)(x-2)}{(x-2)(x+2)}$

$= \dfrac{x^2-(x^2-x-2)}{(x-2)(x+2)} = \dfrac{x+2}{(x-2)(x+2)} = \dfrac{1}{x-2}$

**(d)** $\dfrac{\frac{y}{x}-\frac{x}{y}}{\frac{1}{y}-\frac{1}{x}} = \dfrac{\frac{y}{x}-\frac{x}{y}}{\frac{1}{y}-\frac{1}{x}} \cdot \dfrac{xy}{xy} = \dfrac{y^2-x^2}{x-y} = \dfrac{(y-x)(y+x)}{x-y} = \dfrac{-(x-y)(y+x)}{x-y} = -(y+x)$

**8. (a)** $\dfrac{6}{\sqrt[3]{4}} = \dfrac{6}{\sqrt[3]{2^2}} = \dfrac{6}{\sqrt[3]{2^2}} \cdot \dfrac{\sqrt[3]{2}}{\sqrt[3]{2}} = \dfrac{6\sqrt[3]{2}}{2} = 3\sqrt[3]{2}$

**(b)** $\dfrac{\sqrt{6}}{2+\sqrt{3}} = \dfrac{\sqrt{6}}{2+\sqrt{3}} \cdot \dfrac{2-\sqrt{3}}{2-\sqrt{3}} = \dfrac{2\sqrt{6}-\sqrt{18}}{4-3} = \dfrac{2\sqrt{6}-\sqrt{9\cdot 2}}{1} = 2\sqrt{6}-3\sqrt{2}$

## Focus on Problem Solving: General Principles

**1.** Let $d$ be the distance traveled to and from work. Let $t_1$ and $t_2$ be the times for the trip from home to work and the trip from work to home, respectively. Using time$=\dfrac{\text{distance}}{\text{rate}}$, we get $t_1 = \dfrac{d}{50}$ and $t_2 = \dfrac{d}{30}$. Since average speed $= \dfrac{\text{distance traveled}}{\text{total time}}$, we have average speed $= \dfrac{2d}{t_1+t_2} = \dfrac{2d}{\dfrac{d}{50}+\dfrac{d}{30}}$ $\Leftrightarrow$

$$\text{average speed} = \frac{150\,(2d)}{150\left(\dfrac{d}{50}+\dfrac{d}{30}\right)} = \frac{300d}{3d+5d} = \tfrac{300}{8} = 37.5 \text{ mi/h}.$$

**2.** Let $r$ be the rate of the descent. We use the formula time $= \dfrac{\text{distance}}{\text{rate}}$; the ascent takes $\frac{1}{15}$ h, the descent takes $\dfrac{1}{r}$ h, and the total trip should take $\dfrac{2}{30} = \dfrac{1}{15}$ h. Thus we have $\dfrac{1}{15} + \dfrac{1}{r} = \dfrac{1}{15} \quad \Leftrightarrow \quad \dfrac{1}{r} = 0$, which is impossible. So the car cannot go fast enough to average 30 mi/h for the 2-mile trip.

**3.** We use the formula $d = rt$. Since the car and the van each travel at a speed of 40 mi/h, they approach each other at a combined speed of 80 mi/h. (The distance between them decreases at a rate of 80 mi/h.) So the time spent driving till they meet is $t = \dfrac{d}{r} = \frac{120}{80} = 1.5$ hours. Thus, the fly flies at a speed of 100 mi/h for 1.5 hours, and therefore travels a distance of $d = rt = (100)\,(1.5) = 150$ miles.

**4.** Let us start with a given price $P$. After a discount of 40%, the price decreases to $0.6P$. After a discount of 20%, the price decreases to $0.8P$, and after another 20% discount, it becomes $0.8\,(0.8P) = 0.64P$. Since $0.6P < 0.64P$, a 40% discount is better.

**5.** We continue the pattern. Three parallel cuts produce 10 pieces. Thus, each new cut produces an additional 3 pieces. Since the first cut produces 4 pieces, we get the formula $f\,(n) = 4 + 3\,(n-1)$, $n \geq 1$. Since $f\,(142) = 4 + 3\,(141) = 427$, we see that 142 parallel cuts produce 427 pieces.

**6.** By placing two amoebas into the vessel, we skip the first simple division which took 3 minutes. Thus when we place two amoebas into the vessel, it will take $60 - 3 = 57$ minutes for the vessel to be full of amoebas.

**7.** George's speed is $\frac{1}{50}$ lap/s, and Sue's is $\frac{1}{30}$ lap/s. So after $t$ seconds, George has run $\dfrac{t}{50}$ laps, and Sue has run $\dfrac{t}{30}$ laps. They will next be side by side when Sue has run exactly one more lap than George, that is, when $\dfrac{t}{30} = \dfrac{t}{50} + 1$. We solve for $t$ by first multiplying by 150, so $50t = 30t + 150 \quad \Leftrightarrow \quad 20t = 150 \quad \Leftrightarrow \quad t = 75$. Therefore, they will be even after 75 s.

**8.** The statement is false. Here is one particular counterexample:

| | Player A | Player B |
|---|---|---|
| First half | 1 hit in 99 at-bats: average $= \frac{1}{99}$ | 0 hit in 1 at-bat: average $= \frac{0}{1}$ |
| Second half | 1 hit in 1 at-bat: average $= \frac{1}{1}$ | 98 hits in 99 at-bats: average $= \frac{98}{99}$ |
| Entire season | 2 hits in 100 at-bats: average $= \frac{2}{100}$ | 99 hits in 100 at-bats: average $= \frac{99}{100}$ |

**9.** *Method 1:* After the exchanges, the volume of liquid in the pitcher and in the cup is the same as it was to begin with. Thus, any coffee in the pitcher of cream must be replacing an equal amount of cream that has ended up in the coffee cup.

*Method 2:* Alternatively, look at the drawing of the spoonful of coffee and cream mixture being returned to the pitcher of cream. Suppose it is possible to separate the cream and the coffee, as shown. Then you can see that the coffee going into the cream occupies the same volume as the cream that was left in the coffee.

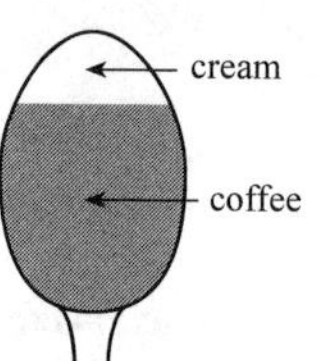


*Method 3 (an algebraic approach):* Suppose the cup of coffee has $y$ spoonfuls of coffee. When one spoonful of cream is added to the coffee cup, the resulting mixture has the following ratios: $\dfrac{\text{cream}}{\text{mixture}} = \dfrac{1}{y+1}$ and $\dfrac{\text{coffee}}{\text{mixture}} = \dfrac{y}{y+1}$.

So, when we remove a spoonful of the mixture and put it into the pitcher of cream, we are really removing $\dfrac{1}{y+1}$ of a spoonful of cream and $\dfrac{y}{y+1}$ spoonful of coffee. Thus the amount of cream left in the mixture (cream in the coffee) is $1 - \dfrac{1}{y+1} = \dfrac{y}{y+1}$ of a spoonful. This is the same as the amount of coffee we added to the cream.

**10.** It remains the same. The weight of the ice cube is the same as the weight of the volume of water it displaces. Thus as the ice cube melts, it becomes water that fits exactly into the space displaced by the floating ice cube.

**11.** Let $r$ be the radius of the earth in feet. Then the circumference (length of the ribbon) is $2\pi r$. When we increase the radius by 1 foot, the new radius is $r + 1$, so the new circumference is $2\pi(r+1)$. Thus you need $2\pi(r+1) - 2\pi r = 2\pi$ extra feet of ribbon.

**12.** We know that $\sqrt{2}$ is irrational. Now $\sqrt{2}^{\sqrt{2}}$ is either rational or irrational. If $\sqrt{2}^{\sqrt{2}}$ is rational, we would be done because we would have found an irrational number raised to an irrational number which is rational. If $\sqrt{2}^{\sqrt{2}}$ is not rational, then consider $\left(\sqrt{2}^{\sqrt{2}}\right)^{\sqrt{2}} = \sqrt{2}^{\sqrt{2}\cdot\sqrt{2}} = \sqrt{2}^{2} = 2$. So, again we would have found an irrational number (namely $\sqrt{2}^{\sqrt{2}}$) raised to an irrational number (namely $\sqrt{2}$) which is rational (namely 2).

**13.** Let $r_1 = 8$. $r_2 = \frac{1}{2}\left(8 + \frac{72}{8}\right) = \frac{1}{2}(8+9) = 8.5$, $r_3 = \frac{1}{2}\left(8.5 + \dfrac{72}{8.5}\right) = \frac{1}{2}(8.5 + 8.471) = 8.485$, and $r_4 = \frac{1}{2}\left(8.485 + \dfrac{72}{8.485}\right) = \frac{1}{2}(8.485 + 8.486) = 8.485$. Thus $\sqrt{72} \approx 8.49$.

**14.** Let $n$, $n+1$, and $n+2$ be three consecutive integers. Then the product of three consecutive integers plus the middle is $n(n+1)(n+2) + (n+1) = n^3 + 3n^2 + 2n + n + 1 = n^3 + 3n^2 + 3n + 1 = (n+1)^3$.

**15.** The first few powers of 3 are$3^1 = 3$, $3^2 = 9$, $3^3 = 27$, $3^4 = 81$, $3^5 = 243$, $3^6 = 729$. It appears that the final digit cycles in a pattern, namely $3 \to 9 \to 7 \to 1 \to 3 \to 9 \to 7 \to 1$, of length 4. Since$459 = 4 \times 114 + 3$, the final digit is the third in the cycle, namely 7.

**16.** Let us see what happens when we square similar numbers with fewer 9 's: $39^2 = 1521$, $399^2 = 159201$, $3999^2 = 15992001$, $39999^2 = 1599920001$.

The pattern is that the square always seems to start with 15 and end with 1, and if $39\cdots9$ has $n$ 9 's, then the 2 in the middle of its square is preceded by $(n-1)$ 9 's and followed by $(n-1)$ 0 's. From this pattern, we make the guess that $3{,}999{,}999{,}999{,}999^2 = 15{,}999{,}999{,}999{,}992{,}000{,}000{,}000{,}001$

This can be verified by writing the number as follows:

$$\begin{aligned} 3{,}999{,}999{,}999{,}999^2 &= (4{,}000{,}000{,}000{,}000 - 1)^2 = 4{,}000{,}000{,}000{,}000^2 - 2\cdot 4{,}000{,}000{,}000{,}000 + 1 \\ &= 16{,}000{,}000{,}000{,}000{,}000{,}000{,}000{,}000 - 8{,}000{,}000{,}000{,}000 + 1 \\ &= 15{,}999{,}999{,}999{,}992{,}000{,}000{,}000{,}001 \end{aligned}$$

**17.** Let $p$ be an odd prime and label the other integer sides of the triangle as shown. Then $p^2 + b^2 = c^2 \quad \Leftrightarrow \quad p^2 = c^2 - b^2 \quad \Leftrightarrow \quad p^2 = (c-b)(c+b)$. Now $c - b$ and $c + b$ are integer factors of $p^2$ where $p$ is a prime.

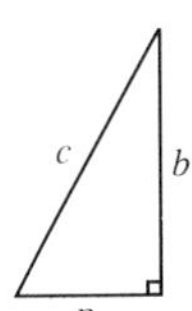


*Case 1:* $c - b = p$ and $c + b = p$ Subtracting gives $b = 0$ which contradicts that $b$ is the length of a side of a triangle.

*Case 2:* $c - b = 1$ and $c + b = p^2$ Adding gives $2c = 1 + p^2 \quad \Leftrightarrow \quad c = \dfrac{1+p^2}{2}$ and $b = \dfrac{p^2-1}{2}$. Since $p$ is odd, $p^2$ is odd. Thus $\dfrac{1+p^2}{2}$ and $\dfrac{p^2-1}{2}$ both define integers and are the only solutions. (For example, if we choose $p = 3$, then $c = 5$ and $b = 4$.)

**18.** $x^2 + y^2 = 4z + 3$. We consider all possible cases for $x$ and $y$ even or odd.

*Case 1: $x$ and $y$ are both odd* Since the square of an odd number is odd, and the sum of two odd numbers is even, the left hand side is even but the right hand side is odd. Thus the left hand side cannot equal the right hand side.

*Case 2: $x$ and $y$ are both even* Since the square of an even number is even, and the sum of two even numbers is even, the left hand side is even but the right hand side is odd. Thus the left hand side cannot equal the right hand side.

*Case 3: $x$ is odd and $y$ is even* Let $x = 2n + 1$ and let $y = 2m$ for some integers $n$ and $m$. Then $(2n+1)^2 + (2m)^2 = 4z + 3 \quad \Leftrightarrow \quad 4n^2 + 4n + 1 + 4m^2 = 4z + 3 \quad \Leftrightarrow \quad 4\left(n^2 + n + m^2\right) + 1 = 4z + 3$. However, when the left hand side is divided by 4 the remainder is 1, but when the right hand side is divided by 4 the remainder is 3. Thus the left hand side cannot equal the right hand side.

*Case 4: $x$ is even and $y$ is odd* This is handled in the same way as Case 3.

Thus, there is no integer solution to $x^2 + y^2 = 4z + 3$.

**19.** The north pole is such a point. And there are others: Consider a point $a_1$ near the south pole such that the parallel passing through $a_1$ forms a circle $C_1$ with circumference exactly one mile. Any point $P_1$ exactly one mile north of the circle $C_1$ along a meridian is a point satisfying the conditions in the problem: starting at $P_1$ she walks one mile south to the point $a_1$ on the circle $C_1$, then one mile east along $C_1$ returning to the point $a_1$, then north for one mile to $P_1$. That's not all. If a point $a_2$ (or $a_3, a_4, a_5, \ldots$) is chosen near the south pole so that the parallel passing through it forms a circle $C_2$ ($C_3$, $C_4$, $C_5, \ldots$) with a circumference of exactly $\frac{1}{2}$ mile ( $\frac{1}{3}$ mi, $\frac{1}{4}$ mi, $\frac{1}{5}$ mi, $\ldots$), then the point $P_2$ ($P_3$, $P_4$, $P_5, \ldots$) one mile north of $a_2$ ($a_3, a_4, a_5, \ldots$) along a meridian satisfies the conditions of the problem: she walks one mile south from $P_2$ ($P_3$, $P_4$, $P_5, \ldots$) arriving at $a_2$ ( $a_3, a_4, a_5, \ldots$) along the circle $C_2$ ($C_3, C_4, C_5, \ldots$), walks east along the circle for one mile thus traversing the circle twice (three times, four times, five times, …) returning to $a_2$ ($a_3, a_4, a_5, \ldots$), and then walks north one mile to $P_2$ ( $P_3, P_4, P_5, \ldots$).

**20.** Let $h_1$ be the height of the pyramid whose square base length is $a$. Then $h_1 - h$ is the height of the pyramid whose square base length is $b$. Thus the volume of the truncated pyramid is$V = \frac{1}{3}h_1 a^2 - \frac{1}{3}(h_1 - h)\, b^2 = \frac{1}{3}h_1 a^2 - \frac{1}{3}h_1 b^2 + \frac{1}{3}hb^2 = \frac{1}{3}h_1\left(a^2 - b^2\right) + \frac{1}{3}hb^2$. Next we must find a relationship between $h_1$ and the other variables. Using geometry, the ratio of the height to the base of the two pyramids must be the same. Thus $\dfrac{h_1}{a} = \dfrac{h_1 - h}{b} \quad \Leftrightarrow \quad bh_1 = ah_1 - ah \quad \Leftrightarrow \quad ah = ah_1 - bh_1 = (a-b)\, h_1 \quad \Leftrightarrow \quad \dfrac{ah}{a-b} = h_1$. Substituting for $h_1$ we have $V = \frac{1}{3}\left(\dfrac{ah}{a-b}\right)\left(a^2 - b^2\right) + \frac{1}{3}hb^2 = \frac{1}{3}(ah)(a+b) + \frac{1}{3}hb^2 = \frac{1}{3}ha^2 + \frac{1}{3}hab + \frac{1}{3}hb^2 = \frac{1}{3}h\left(a^2 + ab + b^2\right)$.

**21.** Let $r$ and $R$ be the radius of the smaller circle and larger circle, respectively. Since the line is tangent to the smaller circle, using the Pythagorean Theorem, we get the following relationship between the radii: $R^2 = 1^2 + r^2$.Thus the area of the shaded region is $\pi R^2 - \pi r^2 = \pi\left(1 + r^2\right) - \pi r^2 = \pi$.

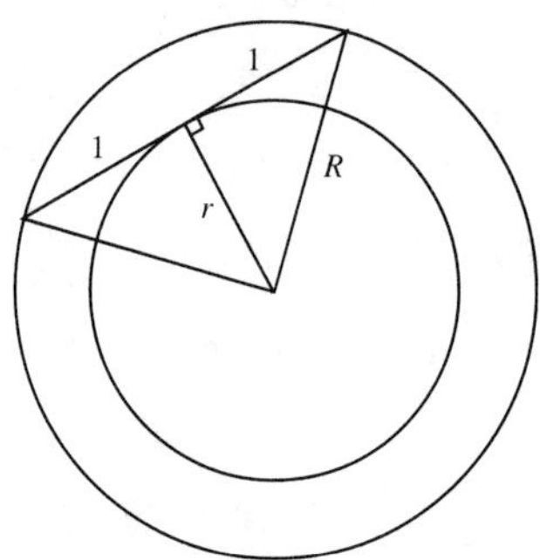


**22.** Label the figures as shown. The unshaded area is $c^2$ in the first figure and $a^2 + b^2$ in the second figure. These unshaded areas are equal, since in each case they equal the area of the large square minus the areas of the four shaded triangles. Thus $a^2 + b^2 = c^2$. Finally, note that $a$, $b$, and $c$ are the legs and hypotenuse of the shaded triangle.

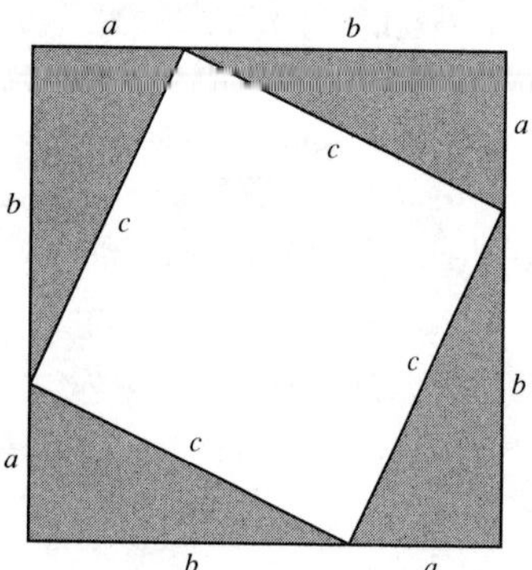


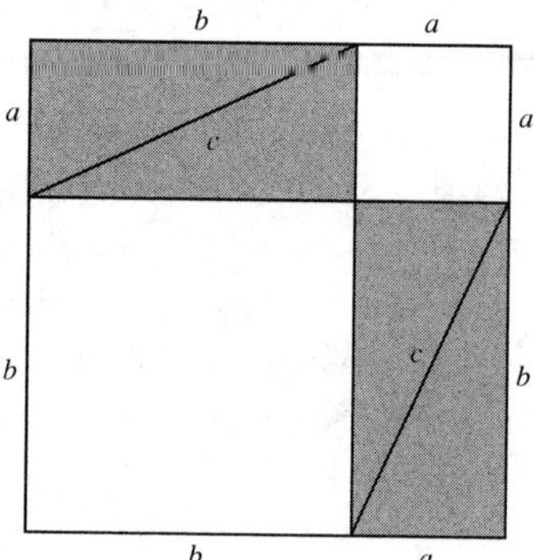


**23.** Since $\sqrt[3]{1729} \approx 12.0023$, we start with $n = 12$ and find the other perfect cube.

| $n$ | $1729 - n^3$ |
|---|---|
| 12 | $1729 - (12)^3 = 1$ |
| 11 | $1729 - (11)^3 = 398$ |
| 10 | $1729 - (10)^3 = 729$ |

Since 1 and 729 are perfect cubes, the two representations we seek are $1^3 + 12^3 = 1729$ and $9^3 + 10^3 = 1729$.

**24. (a)** $\left(\sqrt{3 + 2\sqrt{2}} - \sqrt{3 - 2\sqrt{2}}\right)^2 = 3 + 2\sqrt{2} - 2 \cdot \sqrt{(3 + 2\sqrt{2})\,(3 - 2\sqrt{2})} + 3 - 2\sqrt{2} = 6 - 2 \cdot \sqrt{9 - 8} = 6 - 2 = 4$.

Therefore, $\sqrt{3 + 2\sqrt{2}} - \sqrt{3 - 2\sqrt{2}} = \sqrt{4} = 2$.

**(b)** $\dfrac{\sqrt{2} + \sqrt{6}}{\sqrt{2 + \sqrt{3}}} = \sqrt{\dfrac{(\sqrt{2} + \sqrt{6})^2}{2 + \sqrt{3}}} = \sqrt{\dfrac{2 + 2\sqrt{12} + 6}{2 + \sqrt{3}}} = \sqrt{\dfrac{8 + 4\sqrt{3}}{2 + \sqrt{3}}} = \sqrt{\dfrac{4\,(2 + \sqrt{3})}{2 + \sqrt{3}}} = 2$

**25. (a)** Imagine that the rooms are colored alternately black and white (like a checkerboard) with the entrance being a white room. So there are 18 black rooms and 18 white rooms. If a tourist is in a white room, he must next go to a black room, and if he is in a black room, he must next go to a white room.

**(b)** Since the entrance is white, any path that the tourist takes must be of the form W, B, W, B, W, B, . . .. Since the museum contains an even number of rooms and the tourist starts in a white room, he must end his tour in a black room. But the exit is in a white room, so the proposed tour is impossible.

# 1 Equations and Inequalities

## 1.1 Basic Equations

**1. (a)** When $x = -2$, LHS $= 4(-2) + 7 = -8 + 7 = -1$ and RHS $= 9(-2) - 3 = -18 - 3 = -21$. Since LHS $\neq$ RHS, $x = -2$ is not a solution.

**(b)** When $x = 2$, LHS $= 4(-2) + 7 = 8 + 7 = 15$ and RHS $= 9(2) - 3 = 18 - 3 = 15$. Since LHS $=$ RHS, $x = 2$ is a solution.

**2. (a)** When $x = -1$, LHS $= 2 - 5(-1) = 2 + 5 = 7$ and RHS $= 8 + (-1) = 7$. Since LHS $=$ RHS, $x = -1$ is a solution.

**(b)** When $x = 1$, LHS $= 2 - 5(1) = 2 - 5 = -3$ and RHS $= 8 + (1) = 9$. Since LHS $\neq$ RHS, $x = 1$ is not a solution.

**3. (a)** When $x = 2$, LHS $= 1 - [2 - (3 - (2))] = 1 - [2 - 1] = 1 - 1 = 0$ and RHS $= 4(2) - (6 + (2)) = 8 - 8 = 0$. Since LHS $=$ RHS, $x = 2$ is a solution.

**(b)** When $x = 4$ LHS $= 1 - [2 - (3 - (4))] = 1 - [2 - (-1)] = 1 - 3 = -2$ and RHS $= 4(4) - (6 + (4)) = 16 - 10 = 6$. Since LHS $\neq$ RHS, $x = 4$ is not a solution.

**4. (a)** When $x = 2$, LHS $= \frac{1}{2} - \dfrac{1}{2 - 4} = \frac{1}{2} - \dfrac{1}{-2} = \frac{1}{2} + \frac{1}{2} = 1$ and RHS $= 1$. Since LHS $=$ RHS, $x = 2$ is a solution.

**(b)** When $x = 4$ the expression $\dfrac{1}{4 - 4}$ is not defined, so $x = 4$ is not a solution.

**5. (a)** When $x = -1$, LHS $= 2(-1)^{1/3} - 3 = 2(-1) - 3 = -2 - 3 = -5$. Since LHS $\neq 1$, $x = -1$ is not a solution.

**(b)** When $x = 8$ LHS $= 2(8)^{1/3} - 3 = 2(2) - 3 = 4 - 3 = 1 =$ RHS. So $x = 8$ is a solution.

**6. (a)** When $x = 4$, LHS $= \dfrac{4^{3/2}}{4 - 6} = \dfrac{2^3}{-2} = \dfrac{8}{-2} = -4$ and RHS $= (4) - 8 = -4$. Since LHS $=$ RHS, $x = 4$ is a solution.

**(b)** When $x = 8$, LHS $= \dfrac{8^{3/2}}{8 - 6} = \dfrac{\left(2^3\right)^{3/2}}{2} = \dfrac{2^{9/2}}{2} = 2^{7/2}$ and RHS $= (8) - 8 = 0$. Since LHS $\neq$ RHS, $x = 8$ is not a solution.

**7. (a)** When $x = 0$, LHS $= \dfrac{0 - a}{0 - b} = \dfrac{-a}{-b} = \dfrac{a}{b} =$ RHS. So $x = 0$ is a solution.

**(b)** When $x = b$, LHS $= \dfrac{b - a}{b - b} = \dfrac{b - a}{0}$ is not defined, so $x = b$ is not a solution.

**8. (a)** When $x = b + \sqrt{b^2 - 2}$,

$$\begin{aligned} \text{LHS} &= \tfrac{1}{2}\left(b + \sqrt{b^2 - 2}\right)^2 - b\left(b + \sqrt{b^2 - 2}\right) + 1 = \tfrac{1}{2}\left[b^2 + 2b\sqrt{b^2 - 2} + \left(b^2 - 2\right)\right] - b^2 - b\sqrt{b^2 - 2} + 1 \\ &= \tfrac{1}{2}b^2 + b\sqrt{b^2 - 2} + \tfrac{1}{2}b^2 - 1 - b^2 - b\sqrt{b^2 - 2} + 1 = 0 = \text{RHS} \end{aligned}$$

Since LHS $=$ RHS, $x = b + \sqrt{b^2 - 2}$ is a solution.

**(b)** When $x = b - \sqrt{b^2 - 2}$,

$$\begin{aligned} \text{LHS} &= \tfrac{1}{2}\left(b - \sqrt{b^2 - 2}\right)^2 - b\left(b - \sqrt{b^2 - 2}\right) + 1 = \tfrac{1}{2}\left[b^2 - 2b\sqrt{b^2 - 2} + \left(b^2 - 2\right)\right] - b^2 + b\sqrt{b^2 - 2} + 1 \\ &= \tfrac{1}{2}b^2 - b\sqrt{b^2 - 2} + \tfrac{1}{2}b^2 - 1 - b^2 + b\sqrt{b^2 - 2} + 1 = 0 = \text{RHS} \end{aligned}$$

Since LHS $=$ RHS, $x = b - \sqrt{b^2 - 2}$ is a solution.

**9.** $2x + 7 = 31 \quad\Leftrightarrow\quad 2x = 24 \quad\Leftrightarrow\quad x = 12$

**10.** $5x - 3 = 4 \quad\Leftrightarrow\quad 5x = 7 \quad\Leftrightarrow\quad x = \frac{7}{5}$

**11.** $\frac{1}{2}x - 8 = 1 \quad\Leftrightarrow\quad \frac{1}{2}x = 9 \quad\Leftrightarrow\quad x = 18$

**12.** $3 + \frac{1}{3}x = 5 \quad\Leftrightarrow\quad \frac{1}{3}x = 2 \quad\Leftrightarrow\quad x = 6$

**13.** $x - 3 = 2x + 6 \quad \Leftrightarrow \quad -9 = x$

**14.** $4x + 7 = 9x - 13 \quad \Leftrightarrow \quad 20 = 5x \quad \Leftrightarrow \quad x = 4$

**15.** $-7w = 15 - 2w \quad \Leftrightarrow \quad -5w = 15 \quad \Leftrightarrow \quad w = -3$

**16.** $5t - 13 = 12 - 5t \quad \Leftrightarrow \quad 10t = 25 \quad \Leftrightarrow \quad t = \frac{5}{2}$

**17.** $\frac{1}{2}y - 2 = \frac{1}{3}y \quad \Leftrightarrow \quad 3y - 12 = 2y$ (multiply both sides by the LCD, 6) $\quad \Leftrightarrow \quad y = 12$

**18.** $\dfrac{z}{5} = \frac{3}{10}z + 7 \quad \Leftrightarrow \quad 2z = 3z + 70 \quad \Leftrightarrow \quad z = -70$

**19.** $2(1 - x) = 3(1 + 2x) + 5 \quad \Leftrightarrow \quad 2 - 2x = 3 + 6x + 5 \quad \Leftrightarrow \quad 2 - 2x = 8 + 6x \quad \Leftrightarrow \quad -6 = 8x \quad \Leftrightarrow \quad x = -\frac{3}{4}$

**20.** $5(x + 3) + 9 = -2(x - 2) - 1 \quad \Leftrightarrow \quad 5x + 15 + 9 = -2x + 4 - 1 \quad \Leftrightarrow \quad 5x + 24 = -2x + 3 \quad \Leftrightarrow \quad 7x = -21 \quad \Leftrightarrow$ $x = -3$

**21.** $4\left(y - \frac{1}{2}\right) - y = 6(5 - y) \quad \Leftrightarrow \quad 4y - 2 - y = 30 - 6y \quad \Leftrightarrow \quad 3y - 2 = 30 - 6y \quad \Leftrightarrow \quad 9y = 32 \quad \Leftrightarrow \quad y = \frac{32}{9}$

**22.** $\frac{2}{3}y + \frac{1}{2}(y - 3) = \dfrac{y + 1}{4} \quad \Leftrightarrow \quad 8y + 6(y - 3) = 3(y + 1) \quad \Leftrightarrow \quad 8y + 6y - 18 = 3y + 3 \quad \Leftrightarrow \quad 14y - 18 = 3y + 3$ $\Leftrightarrow \quad 11y = 21 \quad \Leftrightarrow \quad y = \frac{21}{11}$

**23.** $\dfrac{1}{x} = \dfrac{4}{3x} + 1 \quad \Rightarrow \quad 3 = 4 + 3x$ (multiply both sides by the LCD, $3x$) $\quad \Leftrightarrow \quad -1 = 3x \quad \Leftrightarrow \quad x = -\frac{1}{3}$

**24.** $\dfrac{2x - 1}{x + 2} = \dfrac{4}{5} \quad \Rightarrow \quad 5(2x - 1) = 4(x + 2) \quad \Leftrightarrow \quad 10x - 5 = 4x + 8 \quad \Leftrightarrow \quad 6x = 13 \quad \Leftrightarrow \quad x = \frac{13}{6}$

**25.** $\dfrac{2}{t + 6} = \dfrac{3}{t - 1} \quad \Rightarrow \quad 2(t - 1) = 3(t + 6)$ [multiply both sides by the LCD, $(t - 1)(t + 6)$] $\quad \Leftrightarrow \quad 2t - 2 = 3t + 18$ $\Leftrightarrow \quad -20 = t$

**26.** $\dfrac{1}{t - 1} + \dfrac{t}{3t - 2} = \frac{1}{3} \quad \Rightarrow \quad 3(3t - 2) + 3t(t - 1) = (t - 1)(3t - 2) \quad \Leftrightarrow \quad 9t - 6 + 3t^2 - 3t = 3t^2 - 5t + 2 \quad \Leftrightarrow$ $11t = 8 \quad \Leftrightarrow \quad t = \frac{8}{11}$

**27.** $r - 2[1 - 3(2r + 4)] = 61 \quad \Leftrightarrow \quad r - 2(1 - 6r - 12) = 61 \quad \Leftrightarrow \quad r - 2(-6r - 11) = 61 \Leftrightarrow \quad r + 12r + 22 = 61$ $\Leftrightarrow \quad 13r = 39 \quad \Leftrightarrow \quad r = 3$

**28.** $(t - 4)^2 = (t + 4)^2 + 32 \quad \Leftrightarrow \quad t^2 - 8t + 16 = t^2 + 8t + 16 + 32 \quad \Leftrightarrow \quad -16t = 32 \quad \Leftrightarrow \quad t = -2$

**29.** $\sqrt{3}x + \sqrt{12} = \dfrac{x + 5}{\sqrt{3}} \quad \Leftrightarrow \quad 3x + 6 = x + 5$ (multiply both sides by $\sqrt{3}$) $\quad \Leftrightarrow \quad 2x = -1 \quad \Leftrightarrow \quad x = -\frac{1}{2}$

**30.** $\frac{2}{3}x - \frac{1}{4} = \frac{1}{6}x - \frac{1}{9} \quad \Leftrightarrow \quad 24x - 9 = 6x - 4 \quad \Leftrightarrow \quad 18x = 5 \quad \Leftrightarrow \quad x = \frac{5}{18}$

**31.** $\dfrac{2}{x} - 5 = \dfrac{6}{x} + 4 \quad \Rightarrow \quad 2 - 5x = 6 + 4x \quad \Leftrightarrow \quad -4 = 9x \quad \Leftrightarrow \quad -\frac{4}{9} = x$

**32.** $\dfrac{6}{x - 3} = \dfrac{5}{x + 4} \quad \Rightarrow \quad 6(x + 4) = 5(x - 3) \quad \Leftrightarrow \quad 6x + 24 = 5x - 15 \quad \Leftrightarrow \quad x = -39$

**33.** $\dfrac{3}{x + 1} - \frac{1}{2} = \dfrac{1}{3x + 3} \quad \Rightarrow \quad 3(6) - (3x + 3) = 2$ [multiply both sides by $6(x + 1)$] $\quad \Leftrightarrow \quad 18 - 3x - 3 = 2 \quad \Leftrightarrow$ $-3x + 15 = 2 \quad \Leftrightarrow \quad -3x = -13 \quad \Leftrightarrow \quad x = \frac{13}{3}$

**34.** $\dfrac{4}{x - 1} + \dfrac{2}{x + 1} = \dfrac{35}{x^2 - 1} \quad \Rightarrow \quad 4(x + 1) + 2(x - 1) = 35 \quad \Leftrightarrow \quad 4x + 4 + 2x - 2 = 35 \quad \Leftrightarrow \quad 6x + 2 = 35 \quad \Leftrightarrow$ $6x = 33 \quad \Leftrightarrow \quad x = \frac{11}{2}$

**35.** $\dfrac{2x - 7}{2x + 4} = \frac{2}{3} \quad \Rightarrow \quad (2x - 7)3 = 2(2x + 4)$ (cross multiply) $\quad \Leftrightarrow \quad 6x - 21 = 4x + 8 \quad \Leftrightarrow \quad 2x = 29 \Leftrightarrow \quad x = \frac{29}{2}$

**36.** $\dfrac{12x - 5}{6x + 3} = 2 - \dfrac{5}{x} \quad \Rightarrow \quad (12x - 5)x = 2x(6x + 3) - 5(6x + 3) \quad \Leftrightarrow \quad 12x^2 - 5x = 12x^2 + 6x - 30x - 15 \quad \Leftrightarrow$ $12x^2 - 5x = 12x^2 - 24x - 15 \quad \Leftrightarrow \quad 19x = -15 \quad \Leftrightarrow \quad x = -\frac{15}{19}$

**37.** $x - \frac{1}{3}x - \frac{1}{2}x - 5 = 0 \quad \Leftrightarrow \quad 6x - 2x - 3x - 30 = 0$ (multiply both sides by 6) $\quad \Leftrightarrow \quad x = 30$

**38.** $2x - \dfrac{x}{2} + \dfrac{x + 1}{4} = 6x \quad \Leftrightarrow \quad 8x - 2x + x + 1 = 24x \quad \Leftrightarrow \quad 7x + 1 = 24x \quad \Leftrightarrow \quad 1 = 17x \quad \Leftrightarrow \quad x = \frac{1}{17}$

**39.** $\frac{1}{z} - \frac{1}{2z} - \frac{1}{5z} = \frac{10}{z+1} \Rightarrow 10(z+1) - 5(z+1) - 2(z+1) = 10(10z)$ [multiply both sides by $10z(z+1)$] $\Leftrightarrow$ $3(z+1) = 100z \Leftrightarrow 3z+3 = 100z \Leftrightarrow 3 = 97z \Leftrightarrow \frac{3}{97} = z$

**40.** Multiply the LHS by $\frac{2+w}{2+w}$: $\frac{1}{1-\frac{3}{2+w}} = 60 \Leftrightarrow \frac{2+w}{2+w-3} = 60$. Thus, $\frac{2+w}{w-1} = 60 \Rightarrow 2+w = 60(w-1)$ $\Leftrightarrow 2+w = 60w - 60 \Leftrightarrow 62 = 59w \Leftrightarrow w = \frac{62}{59}$.

**41.** $\frac{u}{u - \frac{u+1}{2}} = 4 \Rightarrow u = 4\left(u - \frac{u+1}{2}\right)$ (cross multiply) $\Leftrightarrow u = 4u - 2u - 2 \Leftrightarrow u = 2u - 2 \Leftrightarrow 2 = u$

**42.** $\frac{1}{3-t} + \frac{4}{3+t} + \frac{16}{9-t^2} = 0 \Rightarrow (3+t) + 4(3-t) + 16 = 0 \Leftrightarrow 3+t+12-4t+16 = 0 \Leftrightarrow -3t+31 = 0$ $\Leftrightarrow -3t = -31 \Leftrightarrow t = \frac{31}{3}$

**43.** $\frac{x}{2x-4} - 2 = \frac{1}{x-2} \Rightarrow x - 2(2x-4) = 2$ [multiply both sides by $2(x-2)$] $\Leftrightarrow x - 4x + 8 = 2 \Leftrightarrow$ $-3x = -6 \Leftrightarrow x = 2$. But substituting $x = 2$ into the original equation does not work, since we cannot divide by 0. Thus there is no solution.

**44.** $\frac{1}{x+3} + \frac{5}{x^2-9} = \frac{2}{x-3} \Rightarrow (x-3) + 5 = 2(x+3) \Leftrightarrow x+2 = 2x+6 \Leftrightarrow x = -4$

**45.** $\frac{3}{x+4} = \frac{1}{x} + \frac{6x+12}{x^2+4x} \Rightarrow 3(x) = (x+4) + 6x + 12$ (multiply both sides by $x(x+4)$] $\Leftrightarrow 3x = 7x + 16 \Leftrightarrow$ $-4x = 16 \Leftrightarrow x = -4$. But substituting $x = -4$ into the original equation does not work, since we cannot divide by 0. Thus, there is no solution.

**46.** $\frac{1}{x} - \frac{2}{2x+1} = \frac{1}{2x^2+x} \Rightarrow (2x+1) - 2(x) = 1 \Leftrightarrow 1 = 1$. This is an identity for $x \neq 0$ and $x \neq -\frac{1}{2}$, so the solutions are all real numbers except 0 and $-\frac{1}{2}$.

**47.** $x^2 = 49 \Rightarrow x = \pm 7$

**48.** $x^2 = 18 \Rightarrow x = \pm\sqrt{18} = \pm 3\sqrt{2}$

**49.** $x^2 - 24 = 0 \Leftrightarrow x^2 = 24 \Rightarrow x = \pm\sqrt{24} = \pm 2\sqrt{6}$

**50.** $x^2 - 7 = 0 \Leftrightarrow x^2 = 7 \Rightarrow x = \pm\sqrt{7}$

**51.** $8x^2 - 64 = 0 \Leftrightarrow x^2 - 8 = 0 \Leftrightarrow x^2 = 8 \Rightarrow x = \pm\sqrt{8} - \pm 2\sqrt{2}$

**52.** $5x^2 - 125 = 0 \Leftrightarrow 5(x^2 - 25) = 0 \Leftrightarrow x^2 = 25 \Rightarrow x = \pm 5$

**53.** $x^2 + 16 = 0 \Leftrightarrow x^2 = -16$ which has no real solution.

**54.** $6x^2 + 100 = 0 \Leftrightarrow 6x^2 = -100 \Leftrightarrow x^2 = -\frac{50}{3}$, which has no real solution.

**55.** $(x+2)^2 = 4 \Leftrightarrow (x+2)^2 = 4 \Rightarrow x+2 = \pm 2$. If $x+2 = 2$, then $x = 0$. If $x+2 = -2$, then $x = -4$. The solutions are $-4$ and 0.

**56.** $3(x-5)^2 = 15 \Leftrightarrow (x-5)^2 = 5 \Rightarrow x - 5 = \pm\sqrt{5} \Leftrightarrow x = 5 \pm \sqrt{5}$

**57.** $x^3 = 27 \Leftrightarrow x = 27^{1/3} = 3$

**58.** $x^5 + 32 = 0 \Leftrightarrow x^5 = -32 \Leftrightarrow x = -32^{1/5} = -2$

**59.** $0 = x^4 - 16 = (x^2+4)(x^2-4) = (x^2+4)(x-2)(x+2) .x^2 + 4 = 0$ has no real solution. If $x - 2 = 0$, then $x = 2$. If $x + 2 = 0$, then $x = -2$. The solutions are $\pm 2$.

**60.** $64x^6 = 27 \Leftrightarrow x^6 - \frac{27}{64} \Rightarrow x = +\left(\frac{27}{64}\right)^{1/6} = \pm\frac{27^{1/6}}{64^{1/6}} = \pm\frac{\sqrt{3}}{2}$

**61.** $x^4 + 64 = 0 \Leftrightarrow x^4 = -64$ which has no real solution.

**62.** $(x-1)^3 + 8 = 0 \Leftrightarrow (x-1)^3 = -8 \Leftrightarrow x - 1 = (-8)^{1/3} = -2 \Leftrightarrow x = -1$.

**63.** $(x+2)^4 - 81 = 0 \Leftrightarrow (x+2)^4 = 81 \Leftrightarrow [(x+2)^4]^{1/4} = \pm 81^{1/4} \Leftrightarrow x+2 = \pm 3$. So $x+2 = 3$, then $x = 1$. If $x+2 = -3$, then $x = -5$. The solutions are $-5$ and 1.

**64.** $(x+1)^4+16=0 \Leftrightarrow (x+1)^4=-16$, which has no real solution.

**65.** $3(x-3)^3=375 \Leftrightarrow (x-3)^3=125 \Leftrightarrow (x-3)=125^{1/3}=5 \Leftrightarrow x=3+5=8$

**66.** $4(x+2)^5=1 \Leftrightarrow (x+2)^5=\frac{1}{4} \Rightarrow x+2=\sqrt[5]{\frac{1}{4}} \Leftrightarrow x=-2+\sqrt[5]{\frac{1}{4}}$

**67.** $\sqrt[3]{x}=5 \Leftrightarrow x=5^3=125$

**68.** $x^{4/3}-16=0 \Leftrightarrow x^{4/3}=16=2^4 \Leftrightarrow \left(x^{4/3}\right)^3=\left(2^4\right)^3=2^{12} \Leftrightarrow x^4=2^{12} \Leftrightarrow$
$x=\pm\left(2^{12}\right)^{1/4}=\pm 2^3=\pm 8$

**69.** $2x^{5/3}+64=0 \Leftrightarrow 2x^{5/3}=-64 \Leftrightarrow x^{5/3}=-32 \Leftrightarrow x=(-32)^{3/5}=\left(-2^5\right)^{1/5}=(-2)^3=-8$

**70.** $6x^{2/3}-216=0 \Leftrightarrow 6x^{2/3}=216 \Leftrightarrow x^{2/3}=36=(\pm 6)^2 \Leftrightarrow \left(x^{2/3}\right)^{3/2}=\left[(\pm 6)^2\right]^{3/2} \Leftrightarrow$
$x=(\pm 6)^3=\pm 216$

**71.** $3.02x+1.48=10.92 \Leftrightarrow 3.02x=9.44 \Leftrightarrow x=\dfrac{9.44}{3.02}\approx 3.13$

**72.** $8.36-0.95x=9.97 \Leftrightarrow -0.95x=1.61 \Leftrightarrow x=\dfrac{1.61}{-0.95}\approx -1.69$

**73.** $2.15x-4.63=x+1.19 \Leftrightarrow 1.15x=5.82 \Leftrightarrow x=\dfrac{5.82}{1.15}\approx 5.06$

**74.** $3.95-x=2.32x+2.00 \Leftrightarrow 1.95=3.32x \Leftrightarrow x=\dfrac{1.95}{3.32}\approx 0.59$

**75.** $3.16(x+4.63)=4.19(x-7.24) \Leftrightarrow 3.16x+14.63=4.19x-30.34 \Leftrightarrow$
$44.97=1.03x \Leftrightarrow x=\dfrac{44.97}{1.03}\approx 43.66$

**76.** $2.14(x-4.06)=2.27-0.11x \Leftrightarrow 2.14x-8.6684=2.27-0.11x \Leftrightarrow 2.25x=10.9584 \Leftrightarrow$
$x=4.8704\approx 4.87$

**77.** $\dfrac{0.26x-1.94}{3.03-2.44x}=1.76 \Rightarrow 0.26x-1.94=1.76(3.03-2.44x) \Leftrightarrow 0.26x-1.94=5.33-4.29x \Leftrightarrow$
$4.55x=7.27 \Leftrightarrow x=\dfrac{7.27}{4.55}\approx 1.60$

**78.** $\dfrac{1.73x}{2.12+x}=1.51 \Leftrightarrow 1.73x=1.51(2.12+x) \Leftrightarrow 1.73x=3.20+1.51x \Leftrightarrow 0.22x=3.20 \Leftrightarrow$
$x=\dfrac{3.20}{0.22}\approx 14.55$

**79.** $PV=nRT \Leftrightarrow R=\dfrac{PV}{nT}$

**80.** $F=G\dfrac{mM}{r^2} \Leftrightarrow m=\dfrac{Fr^2}{GM}$

**81.** $\dfrac{1}{R}=\dfrac{1}{R_1}+\dfrac{1}{R_2} \Leftrightarrow R_1R_2=RR_2+RR_1$(multiply both sides by the LCD, $RR_1R_2$). Thus $R_1R_2-RR_1=RR_2$
$\Leftrightarrow R_1(R_2-R)=RR_2 \Leftrightarrow R_1=\dfrac{RR_2}{R_2-R}$.

**82.** $P=2l+2w \Leftrightarrow 2w=P-2l \Leftrightarrow w=\dfrac{P-2l}{2}$

**83.** $\dfrac{ax+b}{cx+d}=2 \Leftrightarrow ax+b=2(cx+d) \Leftrightarrow ax+b=2cx+2d \Leftrightarrow ax-2cx=2d-b \Leftrightarrow (a-2c)x=2d-b$
$\Leftrightarrow x=\dfrac{2d-b}{a-2c}$

**84.** $a-2[b-3(c-x)]=6 \Leftrightarrow a-2(b-3c+3x)=6 \Leftrightarrow a-2b+6c-6x=6 \Leftrightarrow -6x=6-a+2b-6c$
$\Leftrightarrow x=\dfrac{6-a+2b-6c}{-6}=-\dfrac{6-a+2b-6c}{6}$

**85.** $a^2x + (a-1) = (a+1)x \Leftrightarrow a^2x - (a+1)x = -(a-1) \Leftrightarrow (a^2 - (a+1))x = -a+1 \Leftrightarrow$
$(a^2 - a - 1)x = -a + 1 \Leftrightarrow x = \dfrac{-a+1}{a^2-a-1}$

**86.** $\dfrac{a+1}{b} = \dfrac{a-1}{b} + \dfrac{b+1}{a} \Leftrightarrow a(a+1) = a(a-1) + b(b+1) \Leftrightarrow a^2 + a = a^2 - a + b^2 + b \Leftrightarrow 2a = b^2 + b$
$\Leftrightarrow a = \frac{1}{2}(b^2+b)$

**87.** $V = \frac{1}{3}\pi r^2 h \Leftrightarrow r^2 = \dfrac{3V}{\pi h} \Rightarrow r = \pm\sqrt{\dfrac{3V}{\pi h}}$

**88.** $F = G\dfrac{mM}{r^2} \Leftrightarrow r^2 = G\dfrac{mM}{F} \Rightarrow r = \pm\sqrt{G\dfrac{mM}{F}}$

**89.** $a^2 + b^2 = c^2 \Leftrightarrow b^2 = c^2 - a^2 \Rightarrow b = \pm\sqrt{c^2 - a^2}$

**90.** $A = P\left(1 + \dfrac{i}{100}\right)^2 \Leftrightarrow \dfrac{A}{P} = \left(1 + \dfrac{i}{100}\right)^2 \Rightarrow 1 + \dfrac{i}{100} = \pm\sqrt{\dfrac{A}{P}} \Leftrightarrow \dfrac{i}{100} = -1 \pm \sqrt{\dfrac{A}{P}} \Leftrightarrow$
$i = -100 \pm 100\sqrt{\dfrac{A}{P}}$

**91.** $V = \frac{4}{3}\pi r^3 \Leftrightarrow r^3 = \dfrac{3V}{4\pi} \Leftrightarrow r = \sqrt[3]{\dfrac{3V}{4\pi}}$

**92.** $x^4 + y^4 + z^4 = 100 \Leftrightarrow x^4 = 100 - y^4 - z^4 \Rightarrow x = \pm\sqrt[4]{100 - y^4 - z^4}$

**93. (a)** The shrinkage factor when $w = 250$ is $S = \dfrac{0.032(250) - 2.5}{10{,}000} = \dfrac{8 - 2.5}{10{,}000} = 0.00055$. So the beam shrinks $0.00055 \times 12.025 \approx 0.007$ m, so when it dries it will be $12.025 - 0.007 = 12.018$ m long.

**(b)** Substituting $S = 0.00050$ we get $0.00050 = \dfrac{0.032w - 2.5}{10{,}000} \Leftrightarrow 5 = 0.032w - 2.5 \Leftrightarrow 7.5 = 0.032w \Leftrightarrow$
$w = \dfrac{7.5}{0.032} \approx 234.375$. So the water content should be $234.375$ kg/m$^3$.

**94.** Substituting $C = 3600$ we get $3600 = 450 + 3.75x \Leftrightarrow 3150 = 3.75x \Leftrightarrow x = \dfrac{3150}{3.75} = 840$. So the toy manufacturer can manufacture 840 toy trucks.

**95. (a)** Solving for $v$ when $P = 10{,}000$ we get $10{,}000 = 15.6v^3 \Leftrightarrow v^3 \approx 641.02 \Leftrightarrow v \approx 8.6$ km/h.

**(b)** Solving for $v$ when $P = 50{,}000$ we get $50{,}000 = 15.6v^3 \Leftrightarrow v^3 \approx 3205.13 \Leftrightarrow v \approx 14.7$ km/h.

**96.** Substituting $F = 300$ we get $300 = 0.3x^{3/4} \Leftrightarrow 1000 = 10^3 = x^{3/4} \Leftrightarrow x^{1/4} = 10 \Leftrightarrow x = 10^4 - 10{,}000$ lb.

**97. (a)** $3(0) + k - 5 = k(0) - k + 1 \Leftrightarrow k - 5 = -k + 1 \Leftrightarrow 2k = 6 \Leftrightarrow k = 3$

**(b)** $3(1) + k - 5 = k(1) - k + 1 \Leftrightarrow 3 + k - 5 = k - k + 1 \Leftrightarrow k - 2 = 1 \Leftrightarrow k = 3$

**(c)** $3(2) + k - 5 = k(2) - k + 1 \Leftrightarrow 6 + k - 5 = 2k - k + 1 \Leftrightarrow k + 1 = k + 1$. $x = 2$ is a solution for every value of $k$. That is, $x = 2$ is a solution to every member of this family of equations.

**98.** When we multiplied by $x$, we introduced $x = 0$ as a solution. When we divided by $x - 1$, we are really dividing by 0, since $x = 1 \Leftrightarrow x - 1 = 0$.

**99. (a)** Using the formulas for volume we have: for the sphere, $V = \frac{4}{3}\pi r^3$; for the cylinder, $V = \pi r^2 h_1$; and for the cone, $V = \frac{1}{3}\pi r^2 h_2$. Since these have the same volume we must have $\frac{4}{3}\pi r^3 = \pi r^2 h_1$ and $\frac{4}{3}\pi r^3 = \frac{1}{3}\pi r^2 h_2$.

**(b)** Setting the volume of the sphere equal to the volume of the cylinder, we have $\frac{4}{3}\pi r^3 = \pi r^2 h_1 \Leftrightarrow \frac{4}{3}r = h_1$. So the height of the cylinder is $h_1 = \frac{4}{3}r$.

Likewise, setting the volume of the sphere equal to the volume of the cone, we have $\frac{4}{3}\pi r^3 = \frac{1}{3}\pi r^2 h_2 \Leftrightarrow 4r = h_2$. So the height of the cone is $h_2 = 4r$.

# 1.2 Modeling with Equations

**1.** If $n$ is the first integer, then $n+1$ is the middle integer, and $n+2$ is the third integer. So the sum of the three consecutive integers is $n+(n+1)+(n+2)=3n+3$.

**2.** If $n$ is the middle integer, then $n-1$ is the first integer, and $n+1$ is the third integer. So the sum of the three consecutive integers is $(n-1)+n+(n+1)=3n$.

**3.** If $s$ is the third test score, then since the other test scores are 78 and 82, the average of the three test scores is $\dfrac{78+82+s}{3}=\dfrac{160+s}{3}$.

**4.** If $q$ is the fourth quiz score, then since the other quiz scores are 8, 8, and 8, the average of the four quiz scores is $\dfrac{8+8+8+q}{4}=\dfrac{24+q}{4}$.

**5.** If $x$ dollars are invested at $2\frac{1}{2}\%$ simple interest, then the first year you will receive $0.025x$ dollars in interest.

**6.** If $n$ is the number of months the apartment is rented, and each month the rent is \$795, then the total rent paid is $795n$.

**7.** Since $w$ is the width of the rectangle, the length is three times the width, or $3w$. Then area $=$ length $\times$ width $=3w\times w=3w^2$ ft$^2$.

**8.** Since $w$ is the width of the rectangle, the length is $w+5$. The perimeter is

$$2\times\text{length}+2\times\text{width}=2(w+5)+2(w)=4w+10$$

**9.** Since distance $=$ rate$\times$ time we have distance $=s\times(45\text{ min})\dfrac{1\text{ h}}{60\text{ min}}=\frac{3}{4}s$ mi.

**10.** If $d$ is the given distance, in miles, and distance $=$ rate $\times$ time, we have time $=\dfrac{\text{distance}}{\text{rate}}=\dfrac{d}{55}$.

**11.** If $x$ is the quantity of pure water added, the mixture will contain 25 oz of salt and $3+x$ gallons of water. Thus the concentration is $\dfrac{25}{3+x}$.

**12.** If $p$ is the number of pennies in the purse, then the number of nickels is $2p$, the number of dimes is $4+2p$, and the number of quarters is $(2p)+(4+2p)=4p+4$. Thus the value (in cents) of the change in the purse is $1\cdot p+5\cdot 2p+10\cdot(4+2p)+25\cdot(4p+4)=p+10p+40+20p+100p+100=131p+140$.

**13.** Let $x$ be the first integer. Then $x+1$ and $x+2$ are the next consecutive integers. So $x+(x+1)+(x+2)=156$ $\Leftrightarrow$ $3x+3=156$ $\Leftrightarrow$ $3x=153$ $\Leftrightarrow$ $x=51$. Thus the consecutive integers are 51, 52, and 53.

**14.** Let $x$ be the first consecutive odd integer. Then $x+2$, $x+4$, and $x+6$ are the next consecutive odd integers. So $x+(x+2)+(x+4)+(x+6)=416$ $\Leftrightarrow$ $4x+12=416$ $\Leftrightarrow$ $4x=404$ $\Leftrightarrow$ $x=101$. Thus the consecutive odd integers are 101, 103, 105, and 107.

**15.** Let $m$ be the amount invested at $4\frac{1}{2}\%$. Then $12{,}000-m$ is the amount invested at 4%.
Since the total interest is equal to the interest earned at $4\frac{1}{2}\%$ plus the interest earned at 4%, we have $525=0.045m+0.04(12{,}000-m)$ $\Leftrightarrow$ $525=0.045m+480-0.04m$ $\Leftrightarrow$ $45=0.005m$ $\Leftrightarrow$ $m=\dfrac{45}{0.005}=9000$. Thus \$9000 is invested at $4\frac{1}{2}\%$, and $\$12{,}000-9000=\$3000$ is invested at 4%.

**16.** Let $m$ be the amount invested at $5\frac{1}{2}\%$. Then $4000 + m$ is the total amount invested. Thus

$$4\tfrac{1}{2}\% \text{ of the total investment} = \text{interest earned at } 4\% + \text{interest earned at } 5\tfrac{1}{2}\%$$

So $0.045\,(4000 + m) = 0.04\,(4000) + 0.055m \quad\Leftrightarrow\quad 180 + 0.045m = 160 + 0.055m \quad\Leftrightarrow\quad 20 = 0.01m \quad\Leftrightarrow$ $m = \dfrac{20}{0.01} = 2000$. Thus \$2,000 needs to be invested at $5\frac{1}{2}\%$.

**17.** Let $x$ be her monthly salary. Since her annual salary $=$ 12 $\times$ (monthly salary) $+$ (Christmas bonus) we have $97{,}300 = 12x + 8{,}500 \quad\Leftrightarrow\quad 88{,}800 = 12x \quad\Leftrightarrow\quad x \approx 7{,}400$. Her monthly salary is \$7,400.

**18.** Let $s$ be the husband's annual salary. Then her annual salary is $1.15s$. Since husband's annual salary+wife's annual salary $=$ total annual income, we have $s + 1.15s = 69{,}875 \quad\Leftrightarrow\quad 2.15s = 69{,}875 \quad\Leftrightarrow$ $s = 32{,}500$. Thus the husband's annual salary is \$32,500.

**19.** Let $h$ be the amount that Craig inherits. So $(x + 22{,}000)$ is the amount that he invests and doubles. Thus $2\,(x + 22{,}000) = 134{,}000 \quad\Leftrightarrow\quad 2x + 44{,}000 = 134{,}000 \quad\Leftrightarrow\quad 2x = 90{,}000 \quad\Leftrightarrow\quad x = 45{,}000$. So Craig inherits \$45,000.

**20.** Let $x$ be the overtime hours Helen works. Since gross pay $=$ regular salary $+$ overtime pay, we obtain the equation $352.50 = 7.50 \times 35 + 7.50 \times 1.5 \times x \quad\Leftrightarrow\quad 352.50 = 262.50 + 11.25x \quad\Leftrightarrow\quad 90 = 11.25x \quad\Leftrightarrow\quad x = \dfrac{90}{11.25} = 8$. Thus Helen worked 8 hours of overtime.

**21.** Let $x$ be the hours the assistant worked. Then $2x$ is the hours the plumber worked. Since the labor charge is equal to the plumber's labor plus the assistant's labor, we have $4025 = 45\,(2x) + 25x \quad\Leftrightarrow\quad 4025 = 90x + 25x \quad\Leftrightarrow$ $4025 = 115x \quad\Leftrightarrow\quad x = \frac{4025}{115} = 35$. Thus the assistant works for 35 hours, and the plumber works for $2 \times 35 = 70$ hours.

**22.** Let $a$ be the daughter's age now. Then her father's age now is $4a$. In 6 years, her age will $a + 6$, and her father's age will be $3\,(a + 6)$. But her father's age in 6 years is also $4a + 6$. So $3\,(a + 6) = 4a + 6 \quad\Leftrightarrow\quad 3a + 18 = 4a + 6 \quad\Leftrightarrow\quad 12 = a$. Thus the daughter is 12 years old.

**23.** All ages are in terms of the daughter's age 7 years ago. Let $y$ be age of the daughter 7 years ago. Then $11y$ is the age of the movie star 7 years ago. Today, the daughter is $y + 7$, and the movie star is $11y + 7$. But the movie star is also 4 times his daughter's age today. So $4\,(y + 7) = 11y + 7 \quad\Leftrightarrow\quad 4y + 28 = 11y + 7 \quad\Leftrightarrow\quad 21 = 7y \quad\Leftrightarrow\quad y = 3$. Thus the movie star's age today is $11\,(3) + 7 = 40$ years.

**24.** Let $h$ be number of home runs Babe Ruth hit. Then $h + 41$ is the number of home runs that Hank Aaron hit. So $1469 = h + h + 41 \quad\Leftrightarrow\quad 1428 = 2h \quad\Leftrightarrow\quad h = 714$. Thus Babe Ruth hit 714 home runs.

**25.** Let $p$ be the number of pennies. Then $p$ is the number of nickels and $p$ is the number of dimes. So the value of the coins in the purse is the value of the pennies plus the value of the nickels plus the value of the dimes. Thus $1.44 = 0.01p + 0.05p + 0.10p \quad\Leftrightarrow\quad 1.44 = 0.16p \quad\Leftrightarrow\quad p = \frac{1.44}{0.16} = 9$. So the purse contains 9 pennies, 9 nickels, and 9 dimes.

**26.** Let $q$ be the number of quarters. Then $2q$ is the number of dimes, and $2q + 5$ is the number of nickels. Thus $3.00 =$ value of the nickels+ value of the dimes+ value of the quarters. So $3.00 = 0.05\,(2q + 5) + 0.10\,(2q) + 0.25q \quad\Leftrightarrow\quad 3.00 = 0.10q + 0.25 + 0.20q + 0.25q \quad\Leftrightarrow\quad 2.75 = 0.55q \quad\Leftrightarrow$ $q = \frac{2.75}{0.55} = 5$. Thus Mary has 5 quarters, $2\,(5) = 10$ dimes, and $2\,(5) + 5 = 15$ nickels.

**27.** Let $l$ be the length of the garden. Since area $=$ width $\cdot$ length, we obtain the equation $1125 = 25l \quad\Leftrightarrow\quad l = \frac{1125}{25} = 45$ ft. So the garden is 45 feet long.

**28.** Let $w$ be the width of the pasture. Then the length of the pasture is $2w$. Since area$=$ length$\times$ width we have $115{,}200 = w\,(2w) = 2w^2 \quad\Leftrightarrow\quad w^2 = 57{,}600 \quad\Rightarrow\quad w = \pm 240$. Thus the width of the pasture is 240 feet.

**29.** Let $x$ be the length of a side of the square plot. As shown in the figure, area of the plot = area of the building + area of the parking lot. Thus, $x^2 = 60\,(40) + 12{,}000 = 2{,}400 + 12{,}000 = 14{,}400 \quad \Rightarrow \quad x = \pm 120$. So the plot of land measures 120 feet by 120 feet.

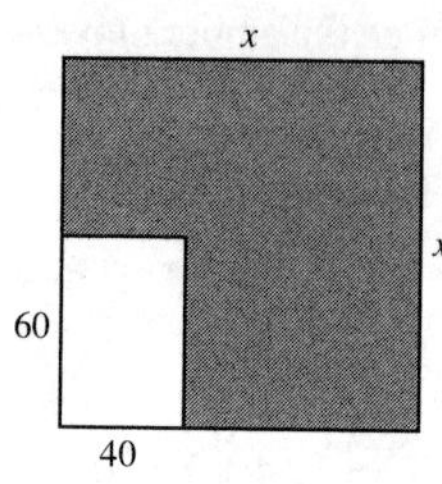


**30.** Let $w$ be the width of the building lot. Then the length of the building lot is $5w$. Since a half-acre is $\frac{1}{2} \cdot 43{,}560 = 21{,}780$ and area is length times width, we have $21{,}780 = w\,(5w) = 5w^2 \quad \Leftrightarrow \quad w^2 = 4{,}356 \quad \Rightarrow \quad w = \pm 66$. Thus the width of the building lot is 66 feet and the length of the building lot is $5\,(66) = 330$ feet.

**31.** The figure is a trapezoid, so its area is $\dfrac{\text{base}_1 + \text{base}_2}{2}$ (height). Putting in the known quantities, we have $120 = \dfrac{y + 2y}{2}\,(y) = \frac{3}{2}y^2 \quad \Leftrightarrow \quad y^2 = 80 \quad \Rightarrow \quad y = \pm\sqrt{80} = \pm 4\sqrt{5}$. Since length is positive, $y = 4\sqrt{5} \approx 8.94$.

**32.** First we write a formula for the area of the figure in terms of $x$. Region $A$ has dimensions 10 cm and $x$ cm and region $B$ has dimensions 6 cm and $x$ cm. So the shaded region has area $(10 \cdot x) + (6 \cdot x) = 16x$ cm$^2$. We are given that this is equal to 144 cm$^2$, so $144 = 16x \quad \Leftrightarrow \quad x = \frac{144}{16} = 9$ cm.

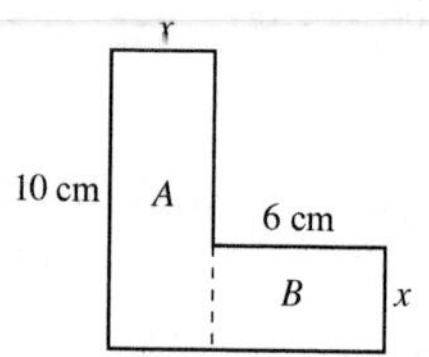


**33.** Let $x$ be the width of the strip. Then the length of the mat is $20 + 2x$, and the width of the mat is $15 + 2x$. Now the perimeter is twice the length plus twice the width, so $102 = 2\,(20 + 2x) + 2\,(15 + 2x) \quad \Leftrightarrow \quad 102 = 40 + 4x + 30 + 4x \quad \Leftrightarrow \quad 102 = 70 + 8x \quad \Leftrightarrow \quad 32 = 8x \quad \Leftrightarrow \quad x = 4$. Thus the strip of mat is 4 inches wide.

**34.** Let $x$ be the amount (in mL) of 60% acid solution to be used. Then $(300 - x)$ mL of 30% solution would have to be used to yield a total of 300 mL of solution.

| | 60% acid | 30% acid | Mixture |
|---|---|---|---|
| mL | $x$ | $300 - x$ | 300 |
| Rate (% acid) | 0.60 | 0.30 | 0.50 |
| Value | $0.60x$ | $0.30\,(300 - x)$ | $0.50\,(300)$ |

Thus the total amount of pure acid used is $0.60x + 0.30\,(300 - x) = 0.50\,(300) \quad \Leftrightarrow \quad 0.3x + 90 = 150 \quad \Leftrightarrow \quad x = \dfrac{60}{0.3} = 200$. So 200 mL of 60% acid solution must be mixed with 100 mL of 30% solution to get 300 mL of 50% acid solution.

**35.** Let $x$ be the grams of silver added. The weight of the rings is $5 \times 18$ g $= 90$ g.

| | 5 rings | Pure silver | Mixture |
|---|---|---|---|
| Grams | 90 | $x$ | $90 + x$ |
| Rate (% gold) | 0.90 | 0 | 0.75 |
| Value | $0.90\,(90)$ | $0x$ | $0.75\,(90 + x)$ |

So $0.90\,(90) + 0x = 0.75\,(90 + x) \quad \Leftrightarrow \quad 81 = 67.5 + 0.75x \quad \Leftrightarrow \quad 0.75x = 13.5 \quad \Leftrightarrow \quad x = \frac{13.5}{0.75} = 18$. Thus 18 grams of silver must be added to get the required mixture.

**36.** Let $x$ be the number of liters of water to be boiled off. The result will contain $6 - x$ liters.

| | Original | Water | Final |
|---|---|---|---|
| Liters | 6 | $-x$ | $6 - x$ |
| Concentration | 120 | 0 | 200 |
| Amount | $120(6)$ | 0 | $200(6 - x)$ |

So $120(6) + 0 = 200(6 - x) \Leftrightarrow 720 = 1200 - 200x \Leftrightarrow 200x = 480 \Leftrightarrow x = 2.4$. Thus 2.4 liters need to be boiled off.

**37.** Let $x$ be the liters of coolant removed and replaced by water.

| | 60% antifreeze | 60% antifreeze (removed) | Water | Mixture |
|---|---|---|---|---|
| Liters | 3.6 | $x$ | $x$ | 3.6 |
| Rate (% antifreeze) | 0.60 | 0.60 | 0 | 0.50 |
| Value | $0.60(3.6)$ | $-0.60x$ | $0x$ | $0.50(3.6)$ |

so $0.60(3.6) - 0.60x + 0x = 0.50(3.6) \Leftrightarrow 2.16 - 0.6x = 1.8 \Leftrightarrow -0.6x = -0.36 \Leftrightarrow x = \dfrac{-0.36}{-0.6} = 0.6$. Thus 0.6 liters must be removed and replaced by water.

**38.** Let $x$ be the number of gallons of 2% bleach removed from the tank. This is also the number of gallons of pure bleach added to make the 5% mixture.

| | Original 2% | Pure bleach | 5% mixture |
|---|---|---|---|
| Gallons | $100 - x$ | $x$ | 100 |
| Concentration | 0.02 | 1 | 0.05 |
| Bleach | $0.02(100 - x)$ | $1x$ | $0.05 \cdot 100$ |

So $0.02(100 - x) + x = 0.05 \cdot 100 \Leftrightarrow 2 - 0.02x + x = 5 \Leftrightarrow 0.98x = 3 \Leftrightarrow x = 3.06$. Thus 3.06 gallons need to removed and replaced with pure bleach.

**39.** Let $c$ be the concentration of fruit juice in the cheaper brand. The new mixture that Jill makes will consist of 650 mL of the original fruit punch and 100 ml of the cheaper fruit punch.

| | Original Fruit Punch | Cheaper Fruit Punch | Mixture |
|---|---|---|---|
| mL | 650 | 100 | 750 |
| Concentration | 0.50 | $c$ | 0.48 |
| Juice | $0.50 \cdot 650$ | $100c$ | $0.48 \cdot 750$ |

So $0.50 \cdot 650 + 100c = 0.48 \cdot 750 \Leftrightarrow 325 + 100c = 360 \Leftrightarrow 100c = 35 \Leftrightarrow c = 0.35$. Thus the cheaper brand is only 35% fruit juice.

**40.** Let $x$ be the number of pounds of \$3.00/lb tea Then $80 - x$ is the number of pounds of \$2.75/lb tea.

| | \$3.00 tea | \$2.75 tea | Mixture |
|---|---|---|---|
| Pounds | $x$ | $80 - x$ | 80 |
| Rate (cost per pound) | 3.00 | 2.75 | 2.90 |
| Value | $3.00x$ | $2.75(80 - x)$ | $2.90(80)$ |

So $3.00x + 2.75(80 - x) = 2.90(80) \Leftrightarrow 3.00x + 220 - 2.75x = 232 \Leftrightarrow 0.25x = 12 \Leftrightarrow x = 48$. The mixture uses 48 pounds of \$3.00/lb tea and $80 - 48 = 32$ pounds of \$2.75/lb tea.

**41.** Let $x$ be the length of the man's shadow, in meters. Using similar triangles, $\frac{10+x}{6}=\frac{x}{2}$ $\Leftrightarrow$ $20+2x=6x$ $\Leftrightarrow$ $4x=20$ $\Leftrightarrow$ $x=5$. Thus the man's shadow is 5 meters long.

**42.** Let $x$ be the height of the tall tree. Here we use the property that corresponding sides in similar triangles are proportional. The base of the similar triangles starts at eye level of the woodcutter, 5 feet. Thus we obtain the proportion $\frac{x-5}{15}=\frac{150}{25}$

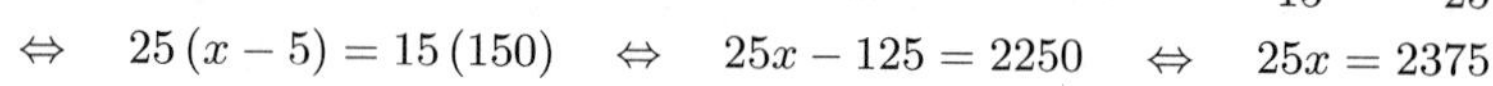

$\Leftrightarrow$ $25(x-5)=15(150)$ $\Leftrightarrow$ $25x-125=2250$ $\Leftrightarrow$ $25x=2375$

$\Leftrightarrow$ $x=95$. Thus the tree is 95 feet tall.

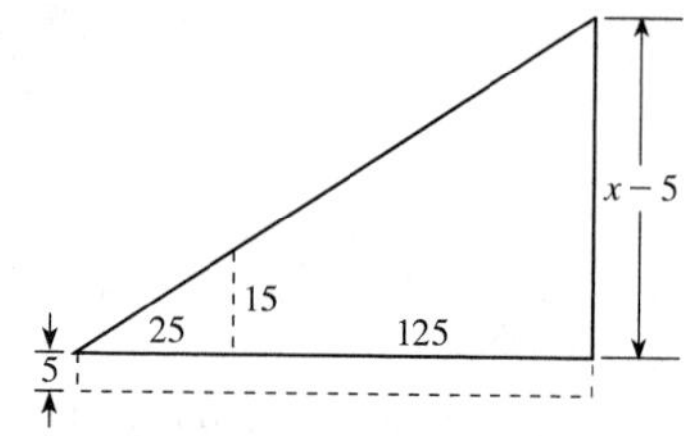


**43.** Let $x$ be the distance from the fulcrum to where the mother sits. Then substituting the known values into the formula given, we have $100(8)=125x$ $\Leftrightarrow$ $800=125x$ $\Leftrightarrow$ $x=6.4$. So the mother should sit 6.4 feet from the fulcrum.

**44.** Let $w$ be the largest weight that can be hung. In this exercise, the edge of the building acts as the fulcrum, so the 240 lb man is sitting 25 feet from the fulcrum. Then substituting the known values into the formula given in Exercise 43, we have $240(25)=5w$ $\Leftrightarrow$ $6000=5w$ $\Leftrightarrow$ $w=1200$. Therefore, 1200 pounds is the largest weight that can be hung.

**45.** Let $t$ be the time in minutes it would take Candy and Tim if they work together. Candy delivers the papers at a rate of $\frac{1}{70}$ of the job per minute, while Tim delivers the paper at a rate of $\frac{1}{80}$ of the job per minute. The sum of the fractions of the job that each can do individually in one minute equals the fraction of the job they can do working together. So we have $\frac{1}{t}=\frac{1}{70}+\frac{1}{80}$ $\Leftrightarrow$ $560=8t+7t$ $\Leftrightarrow$ $560=15t$ $\Leftrightarrow$ $t=37\frac{1}{3}$ minutes. Since $\frac{1}{3}$ of a minute is 20 seconds, it would take them 37 minutes 20 seconds if they worked together.

**46.** Let $t$ be the time, in minutes, it takes Hilda to mow the lawn. Since Hilda is twice as fast as Stan, it takes Stan $2t$ minutes to mow the lawn by himself. Thus $40\cdot\frac{1}{t}+40\cdot\frac{1}{2t}=1$ $\Leftrightarrow$ $40+20=t$ $\Leftrightarrow$ $t=60$. So it would take Stan $2(60)=120$ minutes to mow the lawn.

**47.** Let $t$ be the time, in hours, it takes Karen to paint a house alone. Then working together, Karen and Betty can paint a house in $\frac{2}{3}t$ hours. The sum of their individual rates equals their rate working together, so $\frac{1}{t}+\frac{1}{6}=\frac{1}{\frac{2}{3}t}$ $\Leftrightarrow$ $\frac{1}{t}+\frac{1}{6}=\frac{3}{2t}$ $\Leftrightarrow$ $6+t=9$ $\Leftrightarrow$ $t=3$. Thus it would take Karen 3 hours to paint a house alone.

**48.** Let $h$ be the time, in hours, to fill the swimming pool using Jim's hose alone. Since Bob's hose takes 20% less time, it uses only 80% of the time, or $0.8h$. Thus $18\cdot\frac{1}{h}+18\cdot\frac{1}{0.8h}=1$ $\Leftrightarrow$ $18\cdot 0.8+18=0.8h$ $\Leftrightarrow$ $14.4+18=0.8h$ $\Leftrightarrow$ $32.4=0.8h$ $\Leftrightarrow$ $h=40.5$. Jim's hose takes 40.5 hours, and Bob's hose takes 32.4 hours to fill the pool alone.

**49.** Let $t$ be the time in hours that Wendy spent on the train. Then $\frac{11}{2}-t$ is the time in hours that Wendy spent on the bus. We construct a table:

| | Rate | Time | Distance |
|---|---|---|---|
| By train | 40 | $t$ | $40t$ |
| By bus | 60 | $\frac{11}{2}-t$ | $60\left(\frac{11}{2}-t\right)$ |

The total distance traveled is the sum of the distances traveled by bus and by train, so $300=40t+60\left(\frac{11}{2}-t\right)$ $\Leftrightarrow$ $300=40t+330-60t$ $\Leftrightarrow$ $-30=-20t$ $\Leftrightarrow$ $t=\frac{-30}{-20}=1.5$ hours. So the time spent on the train is $5.5-1.5=4$ hours.

**50.** Let $r$ be the speed of the slower cyclist, in mi/h. Then the speed of the faster cyclist is $2r$.

| | Rate | Time | Distance |
|---|---|---|---|
| Slower cyclist | $r$ | 2 | $2r$ |
| Faster cyclist | $2r$ | 2 | $4r$ |

When they meet, they will have traveled a total of 90 miles, so $2r + 4r = 90 \quad \Leftrightarrow \quad 6r = 90 \quad \Leftrightarrow \quad r = 15$. The speed of the slower cyclist is 15 mi/h, while the speed of the faster cyclist is $2(15) = 30$ mi/h.

**51.** Let $r$ be the speed of the plane from Montreal to Los Angeles. Then $r + 0.20r = 1.20r$ is the speed of the plane from Los Angeles to Montreal.

| | Rate | Time | Distance |
|---|---|---|---|
| Montreal to L.A. | $r$ | $\dfrac{2500}{r}$ | 2500 |
| L.A. to Montreal | $1.2r$ | $\dfrac{2500}{1.2r}$ | 2500 |

The total time is the sum of the times each way, so $9\frac{1}{6} = \dfrac{2500}{r} + \dfrac{2500}{1.2r} \quad \Leftrightarrow \quad \dfrac{55}{6} = \dfrac{2500}{r} + \dfrac{2500}{1.2r} \quad \Leftrightarrow$ $55 \cdot 1.2r = 2500 \cdot 6 \cdot 1.2 + 2500 \cdot 6 \quad \Leftrightarrow \quad 66r = 18{,}000 + 15{,}000 \quad \Leftrightarrow \quad 66r = 33{,}000 \quad \Leftrightarrow \quad r = \frac{33{,}000}{66} = 500.$ Thus the plane flew at a speed of 500 mi/h on the trip from Montreal to Los Angeles.

**52.** Let $x$ be the speed of the car in mi/h. Since a mile contains 5280 ft and an hour contains 3600 s, 1 mi/h $= \frac{5280 \text{ ft}}{3600 \text{ s}} = \frac{22}{15}$ ft/s. The truck is traveling at $50 \cdot \frac{22}{15} = \frac{220}{3}$ ft/s. So in 6 seconds, the truck travels $6 \cdot \frac{220}{3} = 440$ feet. Thus the back end of the car must travel the length of the car, the length of the truck, and the 440 feet in 6 seconds, so its speed must be $\frac{14+30+440}{6} = \frac{242}{3}$ ft/s. Converting to mi/h, we have that the speed of the car is $\frac{242}{3} \cdot \frac{15}{22} = 55$ mi/h.

**53.** Let $l$ be the length of the lot in feet. Then the length of the diagonal is $l + 10$. We apply the Pythagorean Theorem with the hypotenuse as the diagonal. So $l^2 + 50^2 = (l + 10)^2 \quad \Leftrightarrow \quad l^2 + 2500 = l^2 + 20l + 100 \quad \Leftrightarrow \quad 20l = 2400$ $\Leftrightarrow \quad l = 120$. Thus the length of the lot is 120 feet.

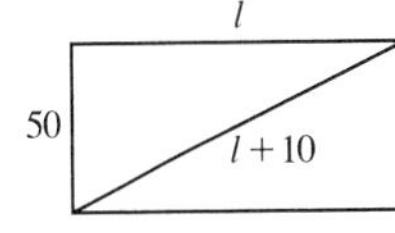


**54.** Let $r$ be the radius of the running track. The running track consists of two semicircles and two straight sections 110 yards long, so we get the equation $2\pi r + 220 = 440 \quad \Leftrightarrow \quad 2\pi r = 220 \quad \Leftrightarrow \quad r = \frac{110}{\pi} = 35.03$. Thus the radius of the semicircle is about 35 yards.

**55.** Let $h$ be the height in feet of the structure. The structure is composed of a right cylinder with radius 10 and height $\frac{2}{3}h$ and a cone with base radius 10 and height $\frac{1}{3}h$. Using the formulas for the volume of a cylinder and that of a cone, we obtain the equation $1400\pi = \pi(10)^2\left(\frac{2}{3}h\right) + \frac{1}{3}\pi(10)^2\left(\frac{1}{3}h\right) \quad \Leftrightarrow \quad 1400\pi = \frac{200\pi}{3}h + \frac{100\pi}{9}h \quad \Leftrightarrow \quad 126 = 6h + h$ (multiply both sides by $\dfrac{9}{100\pi}$) $\quad \Leftrightarrow \quad 126 = 7h \quad \Leftrightarrow \quad h = 18$. Thus the height of the structure is 18 feet.

**56.** Let $h$ be the height of the break, in feet. Then the portion of the bamboo above the break is $10 - h$. Applying the Pythagorean Theorem, we obtain $h^2 + 3^2 = (10 - h)^2 \quad \Leftrightarrow \quad h^2 + 9 = 100 - 20h + h^2 \quad \Leftrightarrow \quad -91 = -20h$ $\Leftrightarrow \quad h = \frac{91}{20} = 4.55$. Thus the break is 4.55 ft above the ground.

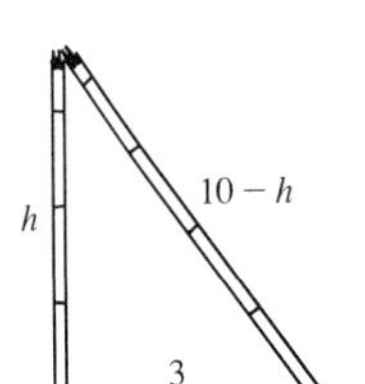


**57.** Pythagoras was born about 569 BC in Samos, Ionia and died about 475 BC.
Euclid was born about 325 BC and died about 265 BC in Alexandria, Egypt.
Archimedes was born in 287 BC in Syracuse, Sicily and died in 212 BC in Syracuse.

## 1.3 Quadratic Equations

**1.** $x^2 + x - 12 = 0 \Leftrightarrow (x-3)(x+4) = 0 \Leftrightarrow x - 3 = 0$ or $x + 4 = 0$. Thus, $x = 3$ or $x = -4$.

**2.** $x^2 + 3x - 4 = 0 \Leftrightarrow (x-1)(x+4) = 0 \Leftrightarrow x - 1 = 0$ or $x + 4 = 0$. Thus, $x = 1$ or $x = -4$.

**3.** $x^2 - 7x + 12 = 0 \Leftrightarrow (x-4)(x-3) = 0 \Leftrightarrow x - 4 = 0$ or $x - 3 = 0$. Thus, $x = 4$ or $x = 3$.

**4.** $x^2 + 8x + 12 = 0 \Leftrightarrow (x+2)(x+6) = 0 \Leftrightarrow x + 2 = 0$ or $x + 6 = 0$. Thus, $x = -2$ or $x = -6$.

**5.** $3x^2 - 5x - 2 = 0 \Leftrightarrow (3x+1)(x-2) = 0 \Leftrightarrow 3x + 1 = 0$ or $x - 2 = 0$. Thus, $x = -\dfrac{1}{3}$ or $x = 2$.

**6.** $4x^2 - 4x - 15 = 0 \Leftrightarrow (2x+3)(2x-5) = 0 \Leftrightarrow 2x + 3 = 0$ or $2x - 5 = 0$. Thus, $x = -\frac{3}{2}$ or $x = \frac{5}{2}$.

**7.** $2y^2 + 7y + 3 = 0 \Leftrightarrow (2y+1)(y+3) = 0 \Leftrightarrow 2y + 1 = 0$ or $y + 3 = 0$. Thus, $y = -\dfrac{1}{2}$ or $y = -3$.

**8.** $4w^2 = 4w + 3 \Leftrightarrow 4w^2 - 4w - 3 = 0 \Leftrightarrow (2w+1)(2w-3) = 0 \Leftrightarrow 2w + 1 = 0$ or $2w - 3 = 0$. If $2w + 1 = 0$, then $w = -\frac{1}{2}$; if $2w - 3 = 0$, then $w = \frac{3}{2}$.

**9.** $6x^2 + 5x = 4 \Leftrightarrow 6x^2 + 5x - 4 = 0 \Leftrightarrow (2x-1)(3x+4) = 0 \Leftrightarrow 2x - 1 = 0$ or $3x + 4 = 0$. If $2x - 1 = 0$, then $x = \frac{1}{2}$; if $3x + 4 = 0$, then $x = -\frac{4}{3}$.

**10.** $3x^2 + 1 = 4x \Leftrightarrow 3x^2 - 4x + 1 = 0 \Leftrightarrow (3x-1)(x-1) = 0 \Leftrightarrow 3x - 1 = 0$ or $x - 1 = 0$. If $3x - 1 = 0$, then $x = \frac{1}{3}$; if $x - 1 = 0$, then $x = 1$.

**11.** $x^2 = 5(x+100) \Leftrightarrow x^2 = 5x + 500 \Leftrightarrow x^2 - 5x - 500 = 0 \Leftrightarrow (x-25)(x+20) = 0 \Leftrightarrow x - 25 = 0$ or $x + 20 = 0$. Thus, $x = 25$ or $x = -20$.

**12.** $6x(x-1) = 21 - x \Leftrightarrow 6x^2 - 6x = 21 - x \Leftrightarrow 6x^2 - 5x - 21 = 0 \Leftrightarrow (2x+3)(3x-7) = 0 \Leftrightarrow$ $2x + 3 = 0$ or $3x - 7 = 0$. If $2x + 3 = 0$, then $x = -\frac{3}{2}$; if $3x - 7 = 0$, then $x = \frac{7}{3}$.

**13.** $x^2 + 2x - 5 = 0 \Leftrightarrow x^2 + 2x = 5 \Leftrightarrow x^2 + 2x + 1 = 5 + 1 \Leftrightarrow (x+1)^2 = 6 \Rightarrow$ $x + 1 = \pm\sqrt{6} \Leftrightarrow x = -1 \pm \sqrt{6}$.

**14.** $x^2 - 4x + 2 = 0 \Leftrightarrow x^2 - 4x = -2 \Leftrightarrow x^2 - 4x + 4 = -2 + 4 \Leftrightarrow (x-2)^2 = 2 \Rightarrow x - 2 = \pm\sqrt{2} \Leftrightarrow$ $x = 2 \pm \sqrt{2}$.

**15.** $x^2 - 6x - 11 = 0 \Leftrightarrow x^2 - 6x = 11 \Leftrightarrow x^2 - 6x + 9 = 11 + 9 \Leftrightarrow (x-3)^2 = 20 \Rightarrow x - 3 = \pm 2\sqrt{5} \Leftrightarrow$ $x = 3 \pm 2\sqrt{5}$.

**16.** $x^2 + 3x - \frac{7}{4} = 0 \Leftrightarrow x^2 + 3x = \dfrac{7}{4} \Leftrightarrow x^2 + 3x + \frac{9}{4} = \frac{7}{4} + \frac{9}{4} \Leftrightarrow \left(x + \frac{3}{2}\right)^2 = \frac{16}{4} = 4 \Rightarrow x + \frac{3}{2} = \pm 2 \Leftrightarrow$ $x = -\frac{3}{2} \pm 2 \Leftrightarrow x = \frac{1}{2}$ or $x = -\frac{7}{2}$.

**17.** $x^2 + x - \frac{3}{4} = 0 \Leftrightarrow x^2 + x = \dfrac{3}{4} \Leftrightarrow x^2 + x + \frac{1}{4} = \frac{3}{4} + \frac{1}{4} \Leftrightarrow \left(x + \frac{1}{2}\right)^2 = 1 \Rightarrow x + \frac{1}{2} = \pm 1 \Leftrightarrow x = -\frac{1}{2} \pm 1$. So $x = -\frac{1}{2} - 1 = -\frac{3}{2}$ and $x = -\frac{1}{2} + 1 = \frac{1}{2}$.

**18.** $x^2 - 5x + 1 = 0 \Leftrightarrow x^2 - 5x = -1 \Leftrightarrow x^2 - 5x + \frac{25}{4} = -1 + \frac{25}{4} \Leftrightarrow \left(x - \frac{5}{2}\right)^2 = \frac{21}{4} \Rightarrow$ $x - \frac{5}{2} = \pm\sqrt{\frac{21}{4}} = \pm\frac{\sqrt{21}}{2} \Leftrightarrow x = \frac{5}{2} \pm \frac{\sqrt{21}}{2}$.

**19.** $x^2 + 22x + 21 = 0 \Leftrightarrow x^2 + 22x = -21 \Leftrightarrow x^2 + 22x + 11^2 = -21 + 11^2 = -21 + 121 \Leftrightarrow (x+11)^2 = 100$ $\Rightarrow x + 11 = \pm 10 \Leftrightarrow x = -11 \pm 10$. Thus, $x = -1$ or $x = -21$.

**20.** $x^2 - 18x = 19 \Leftrightarrow x^2 - 18x + (-9)^2 = 19 + (-9)^2 = 19 + 81 \Leftrightarrow (x-9)^2 = 100 \Rightarrow x - 9 = \pm 10 \Leftrightarrow$ $x = 9 \pm 10$, so $x = -1$ or $x = 19$.

**21.** $2x^2 + 8x + 1 = 0 \Leftrightarrow x^2 + 4x + \frac{1}{2} = 0 \Leftrightarrow x^2 + 4x = -\frac{1}{2} \Leftrightarrow x^2 + 4x + 4 = -\frac{1}{2} + 4 \Leftrightarrow (x+2)^2 = \frac{7}{2}$ $\Rightarrow x + 2 = \pm\sqrt{\frac{7}{2}} \Leftrightarrow x = -2 \pm \frac{\sqrt{14}}{2}$.

**22.** $3x^2 - 6x - 1 = 0 \Leftrightarrow x^2 - 2x - \frac{1}{3} = 0 \Leftrightarrow x^2 - 2x = \frac{1}{3} \Leftrightarrow x^2 - 2x + 1 = \frac{1}{3} + 1 \Leftrightarrow (x-1)^2 = \frac{4}{3} \Rightarrow$ $x - 1 = \pm\sqrt{\frac{4}{3}} \Leftrightarrow x = 1 \pm \frac{2\sqrt{3}}{3}$.

**23.** $4x^2 - x = 0 \Leftrightarrow x^2 - \frac{1}{4}x = 0 \Leftrightarrow x^2 - \frac{1}{4}x + \frac{1}{64} = \frac{1}{64} \Leftrightarrow \left(x - \frac{1}{8}\right)^2 = \frac{1}{64} \Rightarrow x - \frac{1}{8} = \pm\frac{1}{8} \Leftrightarrow$ $x = \frac{1}{8} \pm \frac{1}{8}$, so $x = \frac{1}{8} - \frac{1}{8} = 0$ or $x = \dfrac{1}{8} + \frac{1}{8} = \frac{1}{4}$.

**24.** $x^2 = \frac{3}{4}x - \frac{1}{8} \Leftrightarrow x^2 - \dfrac{3}{4}x = -\frac{1}{8} \Leftrightarrow x^2 - \frac{3}{4}x + \frac{9}{64} = -\frac{1}{8} + \frac{9}{64} \Leftrightarrow \left(x - \frac{3}{8}\right)^2 = \frac{1}{64} \Rightarrow x - \frac{3}{8} = \pm\frac{1}{8} \Leftrightarrow$ $x = \frac{3}{8} \pm \frac{1}{8}$, so $x = \frac{1}{2}$ or $x = \frac{1}{4}$.

**25.** $x^2 - 2x - 15 = 0 \Leftrightarrow (x+3)(x-5) = 0 \Leftrightarrow x + 3 = 0$ or $x - 5 = 0$.Thus, $x = -3$ or $x = 5$.

**26.** $x^2 + 5x - 6 = 0 \Leftrightarrow (x-1)(x+6) = 0 \Leftrightarrow x - 1 = 0$ or $x + 6 = 0$. Thus, $x = 1$ or $x = -6$.

**27.** $x^2 - 7x + 10 = 0 \Leftrightarrow (x-5)(x-2) = 0 \Leftrightarrow x - 5 = 0$ or $x - 2 = 0$.Thus, $x = 5$ or $x = 2$.

**28.** $x^2 + 30x + 200 = 0 \Leftrightarrow (x+10)(x+20) = 0 \Leftrightarrow x + 10 = 0$ or $x + 20 = 0$. Thus, $x = -10$ or $x = -20$.

**29.** $2x^2 + x - 3 = 0 \Leftrightarrow (x-1)(2x+3) = 0 \Leftrightarrow x - 1 = 0$ or $2x + 3 = 0$. If $x - 1 = 0$, then $x = 1$; if $2x + 3 = 0$, then $x = -\frac{3}{2}$.

**30.** $3x^2 + 7x + 4 = 0 \Leftrightarrow (3x+4)(x+1) = 0 \Leftrightarrow 3x + 4 = 0$ or $x + 1 = 0$. Thus, $x = -\frac{4}{3}$ or $x = -1$.

**31.** $x^2 + 3x + 1 = 0 \Leftrightarrow x^2 + 3x = -1 \Leftrightarrow x^2 + 3x + \frac{9}{4} = -1 + \frac{9}{4} = \frac{5}{4} \Leftrightarrow \left(x + \frac{3}{2}\right)^2 = \frac{5}{4} \Rightarrow$ $x + \frac{3}{2} = \pm\sqrt{\frac{5}{4}} = \pm\dfrac{\sqrt{5}}{2} \Leftrightarrow x = -\frac{3}{2} \pm \dfrac{\sqrt{5}}{2}$.

**32.** $x^2 - 4x + 2 = 0 \Leftrightarrow x^2 - 4x = -2 \Leftrightarrow x^2 - 4x + 4 = -2 + 4 \Leftrightarrow (x-2)^2 = 2 \Leftrightarrow x - 2 = \pm\sqrt{2} \Leftrightarrow$ $x = 2 \pm \sqrt{2}$.

**33.** $x^2 + 12x - 27 = 0 \Leftrightarrow x^2 + 12x = 27 \Leftrightarrow x^2 + 12x + 36 = 27 + 36 \Leftrightarrow (x+6)^2 = 63 \Rightarrow$ $x + 6 = \pm 3\sqrt{7} \Leftrightarrow x = -6 \pm 3\sqrt{7}$.

**34.** $8x^2 - 6x - 9 = 0 \Leftrightarrow (2x-3)(4x+3) = 0 \Leftrightarrow 2x - 3 = 0$ or $4x + 3 = 0$. If $2x - 3 = 0$, then $x = \frac{3}{2}$; if $4x + 3 = 0$, then $x = -\frac{3}{4}$.

**35.** $3x^2 + 6x - 5 = 0 \Leftrightarrow x^2 + 2x - \frac{5}{3} = 0 \Leftrightarrow x^2 + 2x = \frac{5}{3} \Leftrightarrow x^2 + 2x + 1 = \frac{5}{3} + 1 \Leftrightarrow (x+1)^2 = \frac{8}{3} \Rightarrow$ $x + 1 = \pm\sqrt{\frac{8}{3}} \Leftrightarrow x = -1 \pm \frac{2\sqrt{6}}{3}$.

**36.** $x^2 - 6x + 1 = 0 \Rightarrow$

$$x = \frac{-b \pm \sqrt{b^2 - 4ac}}{2a} = \frac{-(-6) \pm \sqrt{(-6)^2 - 4(1)(1)}}{2(1)} = \frac{6 \pm \sqrt{36-4}}{2} = \frac{6 \pm \sqrt{32}}{2} = \frac{6 \pm 4\sqrt{2}}{2} = 3 \pm 2\sqrt{2}.$$

**37.** $2y^2 - y - \frac{1}{2} = 0 \Rightarrow y = \dfrac{-b \pm \sqrt{b^2 - 4ac}}{2a} = \dfrac{-(-1) \pm \sqrt{(-1)^2 - 4(2)\left(-\frac{1}{2}\right)}}{2(2)} = \dfrac{1 \pm \sqrt{1+4}}{4} = \dfrac{1 \pm \sqrt{5}}{4}$.

**38.** $\theta^2 - \frac{3}{2}\theta + \frac{9}{16} = 0 \Leftrightarrow \left(\theta - \frac{3}{4}\right)^2 = 0 \Leftrightarrow \theta - \dfrac{3}{4} = 0 \Leftrightarrow \theta = \frac{3}{4}$

**39.** $4x^2 + 16x - 9 = 0 \Leftrightarrow (2x-1)(2x+9) = 0 \Leftrightarrow 2x - 1 = 0$ or $2x + 9 = 0$. If $2x - 1 = 0$, then $x = \frac{1}{2}$; if $2x + 9 = 0$, then $x = -\frac{9}{2}$.

**40.** $0 = x^2 - 4x + 1 = 0 \Rightarrow$

$$x = \frac{-b \pm \sqrt{b^2 - 4ac}}{2a} = \frac{-(-4) \pm \sqrt{(-4)^2 - 4(1)(1)}}{2(1)} = \frac{4 \pm \sqrt{16-4}}{2} = \frac{4 \pm \sqrt{12}}{2} = \frac{4 \pm 2\sqrt{3}}{2} = 2 \pm \sqrt{3}.$$

**41.** $3 + 5z + z^2 = 0 \Rightarrow z = \dfrac{-b \pm \sqrt{b^2 - 4ac}}{2a} = \dfrac{-(5) \pm \sqrt{(5)^2 - 4(1)(3)}}{2(1)} = \dfrac{-5 \pm \sqrt{25-12}}{2} = \dfrac{-5 \pm \sqrt{13}}{2}$.

**42.** $w^2 = 3(w-1) \Leftrightarrow w^2 - 3w + 3 = 0 \Rightarrow w = \dfrac{-(-3) \pm \sqrt{(-3)^2 - 4(1)(3)}}{2(1)} = \dfrac{3 \pm \sqrt{9-12}}{2} = \dfrac{3 \pm \sqrt{-3}}{2}$.

Since the discriminant is less than 0, the equation has no real solution.

**43.** $x^2 - \sqrt{5}x + 1 = 0 \quad \Rightarrow \quad x = \dfrac{-b \pm \sqrt{b^2 - 4ac}}{2a} = \dfrac{(-\sqrt{5}) \pm \sqrt{(-\sqrt{5})^2 - 4(1)(1)}}{2(1)} = \dfrac{\sqrt{5} \pm \sqrt{5-4}}{2} = \dfrac{\sqrt{5} \pm 1}{2}$.

**44.** *Method 1:* Use the quadratic formula immediately. $\sqrt{6}x^2 + 2x - \sqrt{\frac{3}{2}} = 0 \quad \Rightarrow$

$$x = \frac{-b \pm \sqrt{b^2 - 4ac}}{2a} = \frac{-(2) \pm \sqrt{(2)^2 - 4(\sqrt{6})\left(-\sqrt{\frac{3}{2}}\right)}}{2(\sqrt{6})} = \frac{-2 \pm \sqrt{4+12}}{2\sqrt{6}} = \frac{-2 \pm \sqrt{16}}{2\sqrt{6}} = \frac{-2 \pm 4}{2\sqrt{6}}.$$

So $x = -\frac{\sqrt{6}}{2}$ or $x = \frac{\sqrt{6}}{6}$.

*Method 2:* Multiply by $\sqrt{6}$ and use the result in the quadratic formula. $\sqrt{6}x^2 + 2x - \sqrt{\frac{3}{2}} = 0 \Leftrightarrow 6x^2 + 2\sqrt{6}x - 3 = 0 \Rightarrow$

$$x = \frac{-b \pm \sqrt{b^2 - 4ac}}{2a} = \frac{-(2\sqrt{6}) \pm \sqrt{(2\sqrt{6})^2 - 4(6)(-3)}}{2(6)} = \frac{-2\sqrt{6} \pm \sqrt{24+72}}{12} = \frac{-2\sqrt{6} \pm 4\sqrt{6}}{12}.$$

So $x = -\frac{\sqrt{6}}{2}$ or $x = \frac{\sqrt{6}}{6}$.

**45.** $10y^2 - 16y + 5 = 0 \quad \Rightarrow$

$$x = \frac{-b \pm \sqrt{b^2 - 4ac}}{2a} = \frac{-(-16) \pm \sqrt{(-16)^2 - 4(10)(5)}}{2(10)} = \frac{16 \pm \sqrt{256 - 200}}{20} = \frac{16 \pm \sqrt{56}}{20} = \frac{8 \pm \sqrt{14}}{10}.$$

**46.** $25x^2 + 70x + 49 = 0 \quad \Leftrightarrow \quad (5x+7)^2 = 0 \quad \Leftrightarrow \quad 5x + 7 = 0 \quad \Leftrightarrow \quad 5x = -7 \quad \Leftrightarrow \quad x = -\frac{7}{5}$.

**47.** $3x^2 + 2x + 2 = 0 \quad \Rightarrow \quad x = \dfrac{-b \pm \sqrt{b^2 - 4ac}}{2a} = \dfrac{-(2) \pm \sqrt{(2)^2 - 4(3)(2)}}{2(3)} = \dfrac{-2 \pm \sqrt{4-24}}{6} = \dfrac{-2 \pm \sqrt{-20}}{6}$.

Since the discriminant is less than 0, the equation has no real solution.

**48.** $5x^2 - 7x + 5 \quad \Rightarrow \quad x = \dfrac{-b \pm \sqrt{b^2 - 4ac}}{2a} = \dfrac{-(-7) \pm \sqrt{(-7)^2 - 4(5)(5)}}{2(5)} = \dfrac{7 \pm \sqrt{49 - 100}}{10} = \dfrac{7 \pm \sqrt{-51}}{10}$.

Since the discriminant is less than 0, the equation has no real solution.

**49.** $x^2 - 0.011x - 0.064 = 0 \quad \Rightarrow$

$$x = \frac{-(-0.011) \pm \sqrt{(-0.011)^2 - 4(1)(-0.064)}}{2(1)} = \frac{0.011 \pm \sqrt{0.000121 + 0.256}}{2} \approx \frac{0.011 \pm 0.506}{2}.$$

Thus, $x \approx \dfrac{0.011 + 0.506}{2} = 0.259$ or $x \approx \dfrac{0.011 - 0.506}{2} = -0.248$.

**50.** $x^2 - 2.450x + 1.500 = 0 \quad \Rightarrow$

$$x = \frac{-(-2.450) \pm \sqrt{(-2.450)^2 - 4(1)(1.500)}}{2(1)} = \frac{2.450 \pm \sqrt{6.0025 - 6}}{2} = \frac{2.450 \pm \sqrt{0.0025}}{2} = \frac{2.450 \pm 0.050}{2}.$$

Thus, $x = \dfrac{2.450 + 0.050}{2} = 1.250$ or $x = \dfrac{2.450 - 0.050}{2} = 1.200$.

**51.** $x^2 - 2.450x + 1.501 = 0 \quad \Rightarrow$

$$x = \frac{-(-2.450) \pm \sqrt{(-2.450)^2 - 4(1)(1.501)}}{2(1)} = \frac{2.450 \pm \sqrt{6.0025 - 6.004}}{2} = \frac{2.450 \pm \sqrt{-0.0015}}{2}.$$

Thus, there is no real solution.

**52.** $x^2 - 1.800x + 0.810 = 0 \quad \Rightarrow$

$x = \dfrac{-(-1.800) \pm \sqrt{(-1.800)^2 - 4(1)(0.810)}}{2(1)} = \dfrac{1.800 \pm \sqrt{3.24 - 3.24}}{2} = \dfrac{1.800 \pm \sqrt{0}}{2} = 0.900$. Thus the only solution is $x = 0.900$.

**53.** $2.232x^2 - 4.112x = 6.219 \quad\Leftrightarrow\quad 2.232x^2 - 4.112x - 6.219 = 0 \quad\Rightarrow$

$$x = \frac{-(-4.112) \pm \sqrt{(-4.112)^2 - 4(2.232)(-6.219)}}{2(2.232)} = \frac{4.112 \pm \sqrt{16.9085 + 55.5232}}{4.464} \approx \frac{4.112 \pm 8.511}{4.464}.$$ Thus,

$$x = \frac{4.112 - 8.511}{4.464} = -0.985 \text{ or } x = \frac{4.112 + 8.511}{4.464} = 2.828.$$

**54.** $12.714x^2 + 7.103x = 0.987 \quad\Leftrightarrow\quad 12.714x^2 + 7.103x - 0.987 = 0 \quad\Rightarrow$

$$x = \frac{-(7.103) \pm \sqrt{(7.103)^2 - 4(12.714)(-0.987)}}{2(12.714)} = \frac{-7.103 \pm \sqrt{50.4526 + 50.1949}}{25.428} \approx \frac{-7.103 \pm 10.032}{25.428}.$$ Thus,

$$x \approx \frac{-7.103 - 10.032}{25.428} = -0.674 \text{ or } x \approx \frac{-7.103 + 10.032}{25.428} = 0.115.$$

**55.** $h = \frac{1}{2}gt^2 + v_0 t \quad\Leftrightarrow\quad \frac{1}{2}gt^2 + v_0 t - h = 0$. Using the quadratic formula,

$$t = \frac{-(v_0) \pm \sqrt{(v_0)^2 - 4\left(\frac{1}{2}g\right)(-h)}}{2\left(\frac{1}{2}g\right)} = \frac{-v_0 \pm \sqrt{v_0^2 + 2gh}}{g}.$$

**56.** $S = \dfrac{n(n+1)}{2} \quad\Leftrightarrow\quad 2S = n^2 + n \quad\Leftrightarrow\quad n^2 + n - 2S = 0$. Using the quadratic formula,

$$n = \frac{-1 \pm \sqrt{(1)^2 - 4(1)(-2S)}}{2(1)} = \frac{-1 \pm \sqrt{1 + 8S}}{2}.$$

**57.** $A = 2x^2 + 4xh \quad\Leftrightarrow\quad 2x^2 + 4xh - A = 0$. Using the quadratic formula,

$$x = \frac{-(4h) \pm \sqrt{(4h)^2 - 4(2)(-A)}}{2(2)} = \frac{-4h \pm \sqrt{16h^2 + 8A}}{4} = \frac{-4h \pm \sqrt{4(4h^2 + 2A)}}{4} = \frac{-4h \pm 2\sqrt{4h^2 + 2A}}{4}$$

$$= \frac{2\left(-2h \pm \sqrt{4h^2 + 2A}\right)}{4} = \frac{-2h \pm \sqrt{4h^2 + 2A}}{2}$$

**58.** $A = 2\pi r^2 + 2\pi rh \quad\Leftrightarrow\quad 2\pi r^2 + 2\pi rh - A = 0$. Using the quadratic formula,

$$r = \frac{-(2\pi h) \pm \sqrt{(2\pi h)^2 - 4(2\pi)(-A)}}{2(2\pi)} = \frac{-2\pi h \pm \sqrt{4\pi^2 h^2 + 8\pi A}}{4\pi} = \frac{-\pi h \pm \sqrt{\pi^2 h^2 + 2\pi A}}{2\pi}.$$

**59.** $\dfrac{1}{s+a} + \dfrac{1}{s+b} = \dfrac{1}{c} \quad\Leftrightarrow\quad c(s+b) + c(s+a) = (s+a)(s+b) \quad\Leftrightarrow\quad cs + bc + cs + ac = s^2 + as + bs + ab \quad\Leftrightarrow$ $s^2 + (a + b - 2c)s + (ab - ac - bc) = 0$. Using the quadratic formula,

$$s = \frac{-(a + b - 2c) \pm \sqrt{(a + b - 2c)^2 - 4(1)(ab - ac - bc)}}{2(1)}$$

$$= \frac{-(a + b - 2c) \pm \sqrt{a^2 + b^2 + 4c^2 + 2ab - 4ac - 4bc - 4ab + 4ac + 4bc}}{2}$$

$$= \frac{-(a + b - 2c) \pm \sqrt{a^2 + b^2 + 4c^2 - 2ab}}{2}$$

**60.** $\dfrac{1}{r} + \dfrac{2}{1-r} = \dfrac{4}{r^2} \quad\Leftrightarrow\quad r^2(1-r)\left(\dfrac{1}{r} + \dfrac{2}{1-r}\right) = r^2(1-r)\left(\dfrac{4}{r^2}\right) \quad\Leftrightarrow\quad r(1-r) + 2r^2 = 4(1-r)$ $\Leftrightarrow\quad r - r^2 + 2r^2 = 4 - 4r \quad\Leftrightarrow\quad r^2 + 5r - 4 = 0$. Using the quadratic formula,

$$r = \frac{-(5) \pm \sqrt{(5)^2 - 4(1)(-4)}}{2(1)} = \frac{-5 \pm \sqrt{25 + 16}}{2} = \frac{-5 \pm \sqrt{41}}{2}.$$

**61.** $D = b^2 - 4ac = (-6)^2 - 4(1)(1) = 32$. Since $D$ is positive, this equation has two real solutions.

**62.** $x^2 = 6x - 9 \quad \Leftrightarrow \quad x^2 - 6x + 9$, so $D = b^2 - 4ac = (-6)^2 - 4(1)(9) = 36 - 36 = 0$. Since $D = 0$, this equation has one real solution.

**63.** $D = b^2 - 4ac = (2.20)^2 - 4(1)(1.21) = 4.84 - 4.84 = 0$. Since $D = 0$, this equation has one real solution.

**64.** $D = b^2 - 4ac = (2.21)^2 - 4(1)(1.21) = 4.8841 - 4.84 = 0.0441$. Since $D \neq 0$, this equation has two real solutions.

**65.** $D = b^2 - 4ac = (5)^2 - 4(4)\left(\frac{13}{8}\right) = 25 - 26 = -1$. Since $D$ is negative, this equation has no real solution.

**66.** $D = b^2 - 4ac = (-4)^2 - 4(9)\left(\frac{4}{9}\right) = 16 - 16 = 0$. Since $D = 0$, this equation has one real solution.

**67.** $D = b^2 - 4ac = (r)^2 - 4(1)(-s) = r^2 + 4s$. Since $D$ is positive, this equation has two real solutions.

**68.** $D = b^2 - 4ac = (-r)^2 - 4(1)(s) = r^2 - 4s > 0$ (since $r > 2\sqrt{s}$). Since $D$ is positive, this equation has two real solutions.

**69.** $a^2x^2 + 2ax + 1 = 0 \quad \Leftrightarrow \quad (ax+1)^2 = 0 \quad \Leftrightarrow \quad ax + 1 = 0$. So $ax + 1 = 0$ then $ax = -1 \quad \Leftrightarrow \quad x = -\dfrac{1}{a}$.

**70.** $b^2x^2 - 5bx + 4 = 0 \quad \Leftrightarrow \quad (bx-4)(bx-1) = 0 \quad \Leftrightarrow \quad bx - 4 = 0$ or $bx - 1 = 0$. If $bx - 4 = 0$, then $bx = 4 \quad \Leftrightarrow \quad x = \dfrac{4}{b}$. If $bx - 1 = 0$, then $bx = 1 \quad \Leftrightarrow \quad x = \dfrac{1}{b}$.

**71.** $ax^2 - (2a+1)x + (a+1) = 0 \quad \Leftrightarrow \quad [ax - (a+1)](x-1) = 0 \quad \Leftrightarrow \quad ax - (a+1) = 0$ or $x - 1 = 0$. If $ax - (a+1) = 0$, then $x = \dfrac{a+1}{a}$; if $x - 1 = 0$, then $x = 1$.

**72.** $bx^2 + 2x + \dfrac{1}{b} = 0 \quad \Leftrightarrow \quad b^2x^2 + 2bx + 1 = 0 \quad \Leftrightarrow \quad (bx+1)^2 = 0 \quad \Leftrightarrow \quad bx + 1 = 0 \quad \Leftrightarrow \quad \Leftrightarrow \quad x = -\dfrac{1}{b}$.

**73.** We want to find the values of $k$ that make the discriminant 0. Thus $k^2 - 4(4)(25) = 0 \quad \Leftrightarrow \quad k^2 = 400 \quad \Leftrightarrow \quad k = \pm 20$.

**74.** We want to find the values of $k$ that make the discriminant 0. Thus $D = 36^2 - 4(k)(k) = 0 \quad \Leftrightarrow \quad 4k^2 = 36^2 \quad \Rightarrow \quad 2k = \pm 36 \quad \Leftrightarrow \quad k = \pm 18$.

**75.** Let $n$ be one number. Then the other number must be $55 - n$,since $n + (55 - n) = 55$. Because the product is 684, we have $(n)(55-n) = 684 \quad \Leftrightarrow \quad 55n - n^2 = 684 \quad \Leftrightarrow \quad n^2 - 55n + 684 = 0 \quad \Rightarrow \quad n = \frac{-(-55) \pm \sqrt{(-55)^2 - 4(1)(684)}}{2(1)} = \frac{55 \pm \sqrt{3025 - 2736}}{2} = \frac{55 \pm \sqrt{289}}{2} = \frac{55 \pm 17}{2}$. So $n = \frac{55+17}{2} = \frac{72}{2} = 36$ or $n = \frac{55-17}{2} = \frac{38}{2} = 19$. In either case, the two numbers are 19 and 36.

**76.** Let $n$ be one even number. Then the next even number is $n + 2$. Thus we get the equation $n^2 + (n+2)^2 = 1252 \quad \Leftrightarrow \quad n^2 + n^2 + 4n + 4 = 1252 \quad \Leftrightarrow \quad 0 = 2n^2 + 4n - 1248 = 2\left(n^2 + 2n - 624\right) = 2(n-24)(n+26)$. So $n = 24$ or $n = -26$. Thus the consecutive even integers are 24 and 26 or $-26$ and $-24$.

**77.** Let $w$ be the width of the garden in feet. Then the length is $w + 10$. Thus $875 = w(w+10) \quad \Leftrightarrow \quad w^2 + 10w - 875 = 0 \quad \Leftrightarrow \quad (w+35)(w-25) = 0$. So $w + 35 = 0$ in which case $w = -35$, which is not possible, or $w - 25 = 0$ and so $w = 25$. Thus the width is 25 feet and the length is 35 feet.

**78.** Let $w$ be the width of the bedroom. Then its length is $w + 7$. Since area is length times width, we have $228 = (w+7)w = w^2 + 7w \quad \Leftrightarrow \quad w^2 + 7w - 228 = 0 \quad \Leftrightarrow \quad (w+19)(w-12) = 0 \quad \Leftrightarrow \quad w + 19 = 0$ or $w - 12 = 0$. Thus $w = -19$ or $w = 12$. Since the width must be positive, the width is 12 feet.

**79.** Let $w$ be the width of the garden in feet. We use the perimeter to express the length $l$ of the garden in terms of width. Since the perimeter is twice the width plus twice the length, we have $200 = 2w + 2l \quad \Leftrightarrow \quad 2l = 200 - 2w \quad \Leftrightarrow \quad l = 100 - w$. Using the formula for area, we have $2400 = w(100 - w) = 100w - w^2 \quad \Leftrightarrow \quad w^2 - 100w + 2400 = 0 \quad \Leftrightarrow \quad (w-40)(w-60) = 0$. So $w - 40 = 0 \quad \Leftrightarrow \quad w = 40$, or $w - 60 = 0 \quad \Leftrightarrow \quad w = 60$. If $w = 40$, then $l = 100 - 40 = 60$. And if $w = 60$, then $l = 100 - 60 = 40$. So the length is 60 feet and the width is 40 feet.

**80.** First we write a formula for the area of the figure in terms of $x$. Region $A$ has dimensions 14 in. and $x$ in. and region $B$ has dimensions $(13 + x)$ in. and $x$ in. So the area of the figure is $(14 \cdot x) + [(13 + x)\, x] = 14x + 13x + x^2 = x^2 + 27x$. We are given that this is equal to 160 in$^2$, so $160 = x^2 + 27x \quad \Leftrightarrow \quad x^2 + 27x - 160 = 0 \quad \Leftrightarrow \quad (x + 32)(x - 5) \quad \Leftrightarrow \quad x = -32$ or $x = 5$. $x$ must be positive, so $x = 5$ in.

x
A
14 in.
13 in.
B
x

**81.** The shaded area is the sum of the area of a rectangle and the area of a triangle. So $A = y\,(1) + \frac{1}{2}\,(y)\,(y) = \frac{1}{2}y^2 + y$. We are given that the area is 1200 cm$^2$, so $1200 = \frac{1}{2}y^2 + y \quad \Leftrightarrow \quad y^2 + 2y - 2400 = 0 \quad \Leftrightarrow \quad (y + 50)(y - 48) = 0$. $y$ is positive, so $y = 48$ cm.

**82.** Setting $P = 1250$ and solving for $x$, we have $1250 = \frac{1}{10}x\,(300 - x) = 30x - \frac{1}{10}x^2 \quad \Leftrightarrow \quad \frac{1}{10}x^2 - 30x + 1250 = 0$.

Using the quadratic formula, $x = \dfrac{-(-30) \pm \sqrt{(-30)^2 - 4\left(\frac{1}{10}\right)(1250)}}{2\left(\frac{1}{10}\right)} = \dfrac{30 \pm \sqrt{900 - 500}}{0.2} = \dfrac{30 \pm 20}{0.2}$. Thus $x = \dfrac{30 - 20}{0.2} = 50$ or $x = \dfrac{30 + 20}{0.2} = 250$. Since he must have $0 \le x \le 200$, he should make 50 ovens per week.

**83.** Let $x$ be the length of one side of the cardboard, so we start with a piece of cardboard $x$ by $x$. When 4 inches are removed from each side, the base of the box is $x - 8$ by $x - 8$. Since the volume is 100 in$^3$, we get $4\,(x - 8)^2 = 100 \quad \Leftrightarrow \quad x^2 - 16x + 64 = 25 \quad \Leftrightarrow \quad x^2 - 16x + 39 = 0 \quad \Leftrightarrow \quad (x - 3)(x - 13) = 0$.So $x = 3$ or $x = 13$. But $x = 3$ is not possible, since then the length of the base would be $3 - 8 = -5$, and all lengths must be positive. Thus $x = 13$, and the piece of cardboard is 13 inches by 13 inches.

**84.** Let $r$ be the radius of the can. Now using the formula $V = \pi r^2 h$ with $V = 40\pi$ cm$^3$ and $h = 10$, we solve for $r$. Thus $40\pi = \pi r^2\,(10) \quad \Leftrightarrow \quad 4 = r^2 \quad \Rightarrow \quad r = \pm 2$. Since $r$ represents radius, $r > 0$. Thus $r = 2$, and the diameter is 4 cm.

**85.** Let $w$ be the width of the lot in feet. Then the length is $w + 6$. Using the Pythagorean Theorem, we have $w^2 + (w + 6)^2 = (174)^2 \quad \Leftrightarrow \quad w^2 + w^2 + 12w + 36 = 30{,}276 \quad \Leftrightarrow \quad 2w^2 + 12w - 30240 = 0 \quad \Leftrightarrow \quad w^2 + 6w - 15120 = 0 \quad \Leftrightarrow \quad (w + 126)(w - 120) = 0$. So either $w + 126 = 0$ in which case $w = -126$, which is not possible, or $w - 120 = 0$ in which case $w = 120$. Thus the width is 120 feet and the length is 126 feet.

**86.** Let $h$ be the height of the flagpole, in feet. Then the length of each guy wire is $h + 5$. Since the distance between the points where the wires are fixed to the ground is equal to one guy wire, the triangle is equilateral, and the flagpole is the perpendicular bisector of the base. Thus from the Pythagorean Theorem, we get $\left[\frac{1}{2}(h + 5)\right]^2 + h^2 = (h + 5)^2 \quad \Leftrightarrow \quad h^2 + 10h + 25 + 4h^2 = 4h^2 + 40h + 100 \quad \Leftrightarrow \quad h^2 - 30h - 75 = 0 \quad \Rightarrow$ $h = \frac{-(-30) \pm \sqrt{(-30)^2 - 4(1)(-75)}}{2(1)} = \frac{30 \pm \sqrt{900 + 300}}{2} = \frac{30 \pm \sqrt{1200}}{2} = \frac{30 \pm 20\sqrt{3}}{2}$. Since $h = \frac{30 - 20\sqrt{3}}{2} < 0$, we reject it. Thus the height is $h = \frac{30 + 20\sqrt{3}}{2} = 15 + 10\sqrt{3} \approx 32.32$ ft $\approx 32$ ft 4 in.

**87.** Using $h_0 = 288$, we solve $0 = -16t^2 + 288$, for $t \ge 0$. So $0 = -16t^2 + 288 \quad \Leftrightarrow \quad 16t^2 = 288 \quad \Leftrightarrow \quad t^2 = 18 \quad \Rightarrow \quad t = \pm\sqrt{18} = \pm 3\sqrt{2}$. Thus it takes $3\sqrt{2} \approx 4.24$ seconds for the ball the hit the ground.

**88.** **(a)** Using $h_0 = 96$, half the distance is 48, so we solve the equation $48 = -16t^2 + 96 \quad \Leftrightarrow \quad -48 = -16t^2 \quad \Leftrightarrow \quad 3 = t^2 \quad \Rightarrow \quad t = \pm\sqrt{3}$. Since $t \ge 0$, it takes $\sqrt{3} \approx 1.732$ s.

**(b)** The ball hits the ground when $h = 0$, so we solve the equation $0 = -16t^2 + 96 \quad \Leftrightarrow \quad 16t^2 = 96 \quad \Leftrightarrow \quad t^2 = 6 \quad \Rightarrow \quad t = \pm\sqrt{6}$. Since $t \ge 0$, it takes $\sqrt{6} \approx 2.449$ s.

**89.** We are given $v_o = 40$ ft/s.

**(a)** Setting $h = 24$, we have $24 = -16t^2 + 40t \Leftrightarrow 16t^2 - 40t + 24 = 0 \Leftrightarrow 8(2t^2 - 5t + 3) = 0 \Leftrightarrow 8(2t-3)(t-1) = 0 \Leftrightarrow t = 1$ or $t = 1\frac{1}{2}$.Therefore, the ball reaches 24 feet in 1 second (ascending) and again after $1\frac{1}{2}$ seconds (descending).

**(b)** Setting $h = 48$, we have $48 = -16t^2 + 40t \Leftrightarrow 16t^2 - 40t + 48 = 0 \Leftrightarrow 2t^2 - 5t + 6 = 0 \Leftrightarrow t = \dfrac{5 \pm \sqrt{25 - 48}}{4} = \dfrac{5 \pm \sqrt{-23}}{4}$. However, since the discriminant $D < 0$, there is no real solution, and hence the ball never reaches a height of 48 feet.

**(c)** The greatest height $h$ is reached only once. So $h = -16t^2 + 40t \Leftrightarrow 16t^2 - 40t + h = 0$ has only one solution. Thus $D = (-40)^2 - 4(16)(h) = 0 \Leftrightarrow 1600 - 64h = 0 \Leftrightarrow h = 25$. So the greatest height reached by the ball is 25 feet.

**(d)** Setting $h = 25$, we have $25 = -16t^2 + 40t \Leftrightarrow 16t^2 - 40t + 25 = 0 \Leftrightarrow (4t - 5)^2 = 0 \Leftrightarrow t = 1\frac{1}{4}$. Thus the ball reaches the highest point of its path after $1\frac{1}{4}$ seconds.

**(e)** Setting $h = 0$ (ground level), we have $0 = -16t^2 + 40t \Leftrightarrow 2t^2 - 5t = 0 \Leftrightarrow t(2t - 5) = 0 \Leftrightarrow t = 0$ (start) or $t = 2\frac{1}{2}$. So the ball hits the ground in $2\frac{1}{2}$ s.

**90.** If the maximum height is 100 feet, then the discriminant of the equation, $16t^2 - v_o t + 100 = 0$, must equal zero. So $0 = b^2 - 4ac = (-v_o)^2 - 4(16)(100) \Leftrightarrow v_o^2 = 6400 \Rightarrow v_o = \pm 80$. Since $v_o = -80$ does not make sense, we must have $v_o = 80$ ft/s.

**91. (a)** The fish population on January 1, 2002 corresponds to $t = 0$, so $F = 1000\left(30 + 17(0) - (0)^2\right) = 30,000$. To find when the population will again reach this value, we set $F = 30,000$, giving $30000 = 1000\left(30 + 17t - t^2\right) = 30000 + 17000t - 1000t^2 \Leftrightarrow 0 = 17000t - 1000t^2 = 1000t(17 - t) \Leftrightarrow t = 0$ or $t = 17$. Thus the fish population will again be the same 17 years later, that is, on January 1, 2019.

**(b)** Setting $F = 0$, we have $0 = 1000\left(30 + 17t - t^2\right) \Leftrightarrow t^2 - 17t - 30 = 0 \Leftrightarrow t = \dfrac{17 \pm \sqrt{289 + 120}}{-2} = \dfrac{17 \pm \sqrt{409}}{-2} = \dfrac{17 \pm 20.22}{2}$. Thus $t \approx -1.612$ or $t \approx 18.612$.Since $t < 0$ is inadmissible, it follows that the fish in the lake will have died out 18.612 years after January 1, 2002, that is on August 12, 2020.

**92.** Let $y$ be the circumference of the circle, so $360 - y$ is the perimeter of the square. Use the circumference to find the radius, $r$, in terms of $y$: $y = 2\pi r \Rightarrow r = y/(2\pi)$. Thus the area of the circle is $\pi\left[y/(2\pi)\right]^2 = y^2/(4\pi)$. Now if the perimeter of the square is $360 - y$, the length of each side is $\frac{1}{4}(360 - y)$, and the area of the square is $\left[\frac{1}{4}(360 - y)\right]^2$. Setting these areas equal, we obtain $y^2/(4\pi) = \left[\frac{1}{4}(360 - y)\right]^2 \Leftrightarrow y/(2\sqrt{\pi}) = \frac{1}{4}(360 - y) \Leftrightarrow 2y = 360\sqrt{\pi} - \sqrt{\pi}y \Leftrightarrow (2 + \sqrt{\pi})y = 360\sqrt{\pi}$. Therefore, $y = 360\sqrt{\pi}/(2 + \sqrt{\pi}) \approx 169.1$. Thus one wire is 169.1 in. long and the other is 190.9 in. long.

**93.** Let $x$ be the rate, in mi/h, at which the salesman drove between Ajax and Barrington.

| Cities | Distance | Rate | Time |
|---|---|---|---|
| Ajax $\to$ Barrington | 120 | $x$ | $\frac{120}{x}$ |
| Barrington $\to$ Collins | 150 | $x+10$ | $\frac{150}{x+10}$ |

We have used the equation time $= \dfrac{\text{distance}}{\text{rate}}$ to fill in the "Time" column of the table. Since the second part of the trip took 6 minutes (or $\frac{1}{10}$ hour) more than the first, we can use the time column to get the equation $\dfrac{120}{x} + \dfrac{1}{10} = \dfrac{150}{x+10}$ $\Rightarrow$ $120(10)(x+10)+x(x+10) = 150(10x)$ $\Leftrightarrow$ $1200x+12{,}000+x^2+10x = 1500x$ $\Leftrightarrow$ $x^2-290x+12{,}000 = 0$ $\Leftrightarrow$ $x = \frac{-(-290)\pm\sqrt{(-290)^2-4(1)(12{,}000)}}{2} = \frac{290\pm\sqrt{84{,}100-48{,}000}}{2} = \frac{290\pm\sqrt{36{,}100}}{2} = \frac{290\pm190}{2} = 145 \pm 95$. Hence, the salesman drove either 50 mi/h or 240 mi/h between Ajax and Barrington. (The first choice seems more likely!)

**94.** Let $x$ be the rate, in mi/h, at which Kiran drove from Tortula to Cactus.

| Cities | Distance | Rate | Time |
|---|---|---|---|
| Tortula $\to$ Cactus | 250 | $x$ | $\frac{250}{x}$ |
| Cactus $\to$ Dry Junction | 360 | $x+10$ | $\frac{360}{x+10}$ |

We have used time $= \dfrac{\text{distance}}{\text{rate}}$ to fill in the time column of the table. We are given that the sum of the times is 11 hours. Thus we get the equation $\dfrac{250}{x} + \dfrac{360}{x+10} = 11$ $\Leftrightarrow$ $250(x+10) + 360x = 11x(x+10)$ $\Leftrightarrow$ $250x + 2500 + 360x = 11x^2 + 110x$ $\Leftrightarrow$ $11x^2 - 500x - 2500 = 0$ $\Rightarrow$

$$x = \frac{-(-500) \pm \sqrt{(-500)^2 - 4(11)(-2500)}}{2(11)} = \frac{500 \pm \sqrt{250{,}000 + 110{,}000}}{22} = \frac{500 \pm \sqrt{360{,}000}}{22} = \frac{500 \pm 600}{22}.$$

Hence, Kiran drove either $-4.54$ mi/h (impossible) or 50 mi/h between Tortula and Cactus.

**95.** Let $r$ be the rowing rate in km/h of the crew in still water. Then their rate upstream was $r - 3$ km/h, and their rate downstream was $r + 3$ km/h.

| | Distance | Rate | Time |
|---|---|---|---|
| Upstream | 6 | $r-3$ | $\frac{6}{r-3}$ |
| Downstream | 6 | $r+3$ | $\frac{6}{r+3}$ |

Since the time to row upstream plus the time to row downstream was 2 hours 40 minutes $= \frac{8}{3}$ hour, we get the equation $\dfrac{6}{r-3} + \dfrac{6}{r+3} = \dfrac{8}{3}$ $\Leftrightarrow$ $6(3)(r+3)+6(3)(r-3) = 8(r-3)(r+3)$ $\Leftrightarrow$ $18r+54+18r-54 = 8r^2-72$ $\Leftrightarrow$ $0 = 8r^2 - 36r - 72 = 4\left(2r^2 - 9r - 18\right) = 4(2r+3)(r-6)$. Since $2r+3=0$ $\Leftrightarrow$ $r = -\frac{3}{2}$ is impossible, the solution is $r - 6 = 0$ $\Leftrightarrow$ $r = 6$. So the rate of the rowing crew in still water is 6 km/h.

**96.** Let $r$ be the speed of the southbound boat. Then $r + 3$ is the speed of the eastbound boat. In two hours the southbound boat has traveled $2r$ miles and the eastbound boat has traveled $2(r+3) = 2r+6$ miles. Since they are traveling is directions with are $90^\circ$ apart, we can use the Pythagorean Theorem to get $(2r)^2 + (2r+6)^2 = 30^2$ $\Leftrightarrow$ $4r^2 + 4r^2 + 24r + 36 = 900$ $\Leftrightarrow$ $8r^2 + 24r - 864 = 0$ $\Leftrightarrow$ $8\left(r^2 + 3r - 108\right) = 0$ $\Leftrightarrow$ $8(r+12)(r-9) = 0$. So $r = -12$ or $r = 9$. Since speed is positive, the speed of the southbound boat is 9 mi/h.

**97.** Let $w$ be the uniform width of the lawn. With $w$ cut off each end, the area of the factory is $(240 - 2w)(180 - 2w)$. Since the lawn and the factory are equal in size this area, is $\frac{1}{2} \cdot 240 \cdot 180$. So $21{,}600 = 43{,}200 - 480w - 360w + 4w^2 \Leftrightarrow 0 = 4w^2 - 840w + 21{,}600 = 4(w^2 - 210w + 5400) = 4(w - 30)(w - 180) \Rightarrow w = 30$ or $w = 180$. Since 180 ft is too wide, the width of the lawn is 30 ft, and the factory is 120 ft by 180 ft.

**98.** Let $h$ be the height the ladder reaches (in feet). Using the Pythagorean Theorem we have $\left(7\frac{1}{2}\right)^2 + h^2 = \left(19\frac{1}{2}\right)^2 \Leftrightarrow \left(\frac{15}{2}\right)^2 + h^2 = \left(\frac{39}{4}\right)^2 \Leftrightarrow h^2 = \left(\frac{39}{4}\right)^2 - \left(\frac{15}{2}\right)^2 = \frac{1521}{4} - \frac{225}{4} = \frac{1296}{4} = 324$. So $h = \sqrt{324} = 18$.

**99.** Let $t$ be the time, in hours it takes Irene to wash all the windows. Then it takes Henry $t + \frac{3}{2}$ hours to wash all the windows, and the sum of the fraction of the job per hour they can do individually equals the fraction of the job they can do together. Since 1 hour 48 minutes $= 1 + \frac{48}{60} = 1 + \frac{4}{5} = \frac{9}{5}$, we have $\dfrac{1}{t} + \dfrac{1}{t + \frac{3}{2}} = \dfrac{1}{\frac{9}{5}} \Leftrightarrow \dfrac{1}{t} + \dfrac{2}{2t+3} = \frac{5}{9}$

$\Rightarrow 9(2t+3) + 2(9t) = 5t(2t+3) \Leftrightarrow 18t + 27 + 18t = 10t^2 + 15t \Leftrightarrow 10t^2 - 21t - 27 = 0$

$\Leftrightarrow t = \dfrac{-(-21) \pm \sqrt{(-21)^2 - 4(10)(-27)}}{2(10)} = \dfrac{21 \pm \sqrt{441 + 1080}}{20} = \dfrac{21 \pm 39}{20}$. So $t = \dfrac{21 - 39}{20} = -\dfrac{9}{10}$ or $t = \dfrac{21 + 39}{20} = 3$. Since $t < 0$ is impossible, all the windows are washed by Irene alone in 3 hours and by Henry alone in $3 + \frac{3}{2} = 4\frac{1}{2}$ hours.

**100.** Let $t$ be the time, in hours, it takes Kay to deliver all the flyers alone. Then it takes Lynn $t + 1$ hours to deliver all the flyers alone, and it takes the group $0.4t$ hours to do it together. Thus $\frac{1}{4} + \dfrac{1}{t} + \dfrac{1}{t+1} = \dfrac{1}{0.4t} \Leftrightarrow$ $\frac{1}{4}(0.4t) + \dfrac{1}{t}(0.4t) + \dfrac{1}{t+1}(0.4t) = 1 \Leftrightarrow t + 4 + \dfrac{4t}{t+1} = 10 \Leftrightarrow t(t+1) + 4(t+1) + 4t = 10(t+1) \Leftrightarrow$ $t^2 + t + 4t + 4 + 4t = 10t + 10 \Leftrightarrow t^2 - t - 6 = 0 \Leftrightarrow (t-3)(t+2) = 0$. So $t = 3$ or $t = -2$. Since $t = -2$ is impossible, it takes Kay 3 hours to deliver all the flyers alone.

**101.** Let $x$ be the distance from the center of the earth to the dead spot (in thousands of miles). Now setting $F = 0$, we have $0 = -\dfrac{K}{x^2} + \dfrac{0.012K}{(239 - x)^2} \Leftrightarrow \dfrac{K}{x^2} = \dfrac{0.012K}{(239 - x)^2} \Leftrightarrow K(239 - x)^2 = 0.012Kx^2 \Leftrightarrow$ $57121 - 478x + x^2 = 0.012x^2 \Leftrightarrow 0.988x^2 - 478x + 57121 = 0$. Using the quadratic formula, we obtain $x = \frac{-(-478) \pm \sqrt{(-478)^2 - 4(0.988)(57121)}}{2(0.988)} = \frac{478 \pm \sqrt{228484 - 225742.192}}{1.976} = \frac{478 \pm \sqrt{2741.808}}{1.976} \approx \frac{478 \pm 52.362}{1.976} \approx 241.903 \pm 26.499$. So either $x \approx 241.903 + 26.499 \approx 268$ or $x \approx 241.903 - 26.499 \approx 215$. Since 268 is greater than the distance from the earth to the moon, we reject it; thus $x \approx 215{,}000$ miles.

**102.** $x^2 - 9x + 20 = (x-4)(x-5) = 0$, so $x = 4$ or $x = 5$. The roots are 4 and 5. The product is $4 \cdot 5 = 20$, and the sum is $4 + 5 = 9$. $x^2 - 2x - 8 = (x-4)(x+2) = 0$, so $x = 4$ or $x = -2$. The roots are 4 and $-2$. The product is $4 \cdot (-2) = -8$, and the sum is $4 + (-2) = 2$. $x^2 + 4x + 2 = 0$, so using the quadratic formula, $x = \dfrac{-4 \pm \sqrt{4^2 - 4(1)(2)}}{2(1)} = \dfrac{-4 \pm \sqrt{8}}{2} = \dfrac{-4 \pm 2\sqrt{2}}{2} = -2 \pm \sqrt{2}$. The roots are $-2 - \sqrt{2}$ and $-2 + \sqrt{2}$. The product is $(-2 - \sqrt{2}) \cdot (-2 + \sqrt{2}) = 4 - 2 = 2$, and the sum is $(-2 - \sqrt{2}) + (-2 + \sqrt{2}) = -4$. In general, if $x = r_1$ and $x = r_2$ are roots, then $x^2 + bx + c = (x - r_1)(x - r_2) = x^2 - r_1x - r_2x + r_1r_2 = x^2 - (r_1 + r_2)x + r_1r_2$. Equating the coefficients, we get $c = r_1r_2$ and $b = -(r_1 + r_2)$.

**103.** Let $x$ equal the original length of the reed in cubits. Then $x - 1$ is the piece that fits 60 times along the length of the field, that is, the length is $60(x-1)$. The width is $30x$. Then converting cubits to ninda, we have $375 = 60(x-1) \cdot 30x \cdot \dfrac{1}{12^2} = \frac{25}{2}x(x-1) \Leftrightarrow 30 = x^2 - x \Leftrightarrow x^2 - x - 30 = 0 \Leftrightarrow (x-6)(x+5) = 0$. So $x = 6$ or $x = -5$. Since $x$ must be positive, the original length of the reed is 6 cubits.

# 1.4 Complex Numbers

**1.** $5-7i$: real part 5, imaginary part $-7$.

**2.** $-6+4i$: real part $-6$, imaginary part 4.

**3.** $\dfrac{-2-5i}{3}=-\frac{2}{3}-\frac{5}{3}i$: real part $-\frac{2}{3}$, imaginary part $-\frac{5}{3}$.

**4.** $\dfrac{4+7i}{2}=2+\frac{7}{2}i$: real part 2, imaginary part $\frac{7}{2}$.

**5.** 3: real part 3, imaginary part 0.

**6.** $-\frac{1}{2}$: real part $-\frac{1}{2}$, imaginary part 0.

**7.** $-\frac{2}{3}i$: real part 0, imaginary part $-\frac{2}{3}$.

**8.** $i\sqrt{3}$: real part 0, imaginary part $\sqrt{3}$.

**9.** $\sqrt{3}+\sqrt{-4}=\sqrt{3}+2i$: real part $\sqrt{3}$, imaginary part 2.

**10.** $2-\sqrt{-5}=2-i\sqrt{5}$: real part 2, imaginary part $-\sqrt{5}$.

**11.** $(2-5i)+(3+4i)=(2+3)+(-5+4)\,i=5-i$

**12.** $(2+5i)+(4-6i)=(2+4)+(5-6)\,i=6-i$

**13.** $(-6+6i)+(9-i)=(-6+9)+(6-1)\,i=3+5i$

**14.** $(3-2i)+\left(-5-\frac{1}{3}i\right)=(3-5)+\left(-2-\frac{1}{3}\right)i=-2-\frac{7}{3}i$

**15.** $\left(7-\frac{1}{2}i\right)-\left(5+\frac{3}{2}i\right)=(7-5)+\left(-\frac{1}{2}-\frac{3}{2}\right)i=2-2i$

**16.** $(-4+i)-(2-5i)=-4+i-2+5i=(-4-2)+(1+5)\,i=-6+6i$

**17.** $(-12+8i)-(7+4i)=-12+8i-7-4i=(-12-7)+(8-4)\,i=-19+4i$

**18.** $6i-(4-i)=6i-4+i=(-4)+(6+1)\,i=-4+7i$

**19.** $4\,(-1+2i)=-4+8i$

**20.** $2i\left(\frac{1}{2}-i\right)=i-2i^2=2+i$

**21.** $(7-i)\,(4+2i)=28+14i-4i-2i^2=(28+2)+(14-4)\,i=30+10i$

**22.** $(5-3i)\,(1+i)=5+5i-3i-3i^2=(5+3)+(5-3)\,i=8+2i$

**23.** $(3-4i)\,(5-12i)=15-36i-20i+48i^2=(15-48)+(-36-20)\,i=-33-56i$

**24.** $\left(\frac{2}{3}+12i\right)\left(\frac{1}{6}+24i\right)=\frac{1}{9}+16i+2i+288i^2=\left(\frac{1}{9}-288\right)+(16+2)\,i=-\frac{2591}{9}+18i$

**25.** $(6+5i)\,(2-3i)=12-18i+10i-15i^2=(12+15)+(-18+10)\,i=27-8i$

**26.** $(-2+i)\,(3-7i)=-6+14i+3i-7i^2=(-6+7)+(14+3)\,i=1+17i$

**27.** $\dfrac{1}{i}=\dfrac{1}{i}\cdot\dfrac{i}{i}=\dfrac{i}{i^2}=\dfrac{i}{-1}=-i$

**28.** $\dfrac{1}{1+i}=\dfrac{1}{1+i}\cdot\dfrac{1-i}{1-i}=\dfrac{1-i}{1-i^2}=\dfrac{1-i}{1+1}=\dfrac{1-i}{2}=\frac{1}{2}-\frac{1}{2}i$

**29.** $\dfrac{2-3i}{1-2i}=\dfrac{2-3i}{1-2i}\cdot\dfrac{1+2i}{1+2i}=\dfrac{2+4i-3i-6i^2}{1-4i^2}=\dfrac{(2+6)+(4-3)\,i}{1+4}=\dfrac{8+i}{5}$ or $\frac{8}{5}+\frac{1}{5}i$

**30.** $\dfrac{5-i}{3+4i}=\dfrac{5-i}{3+4i}\cdot\dfrac{3-4i}{3-4i}=\dfrac{15-20i-3i+4i^2}{9-16i^2}=\dfrac{(15-4)+(-20-3)\,i}{9+16}=\dfrac{11-23i}{25}=\frac{11}{25}-\frac{23}{25}i$

**31.** $\dfrac{26+39i}{2-3i}=\dfrac{26+39i}{2-3i}\cdot\dfrac{2+3i}{2+3i}=\dfrac{52+78i+78i+117i^2}{4-9i^2}=\dfrac{(52-117)+(78+78)\,i}{4+9}=\dfrac{-65+156i}{13}$
$=\dfrac{13\,(-5+12i)}{13}=-5+12i$

**32.** $\dfrac{25}{4-3i}=\dfrac{25}{4-3i}\cdot\dfrac{4+3i}{4+3i}=\dfrac{100+75i}{16-9i^2}=\dfrac{100+75i}{16+9}=\dfrac{100+75i}{25}=\dfrac{25\,(4+3i)}{25}=4+3i$

**33.** $\dfrac{10i}{1-2i}=\dfrac{10i}{1-2i}\cdot\dfrac{1+2i}{1+2i}=\dfrac{10i+20i^2}{1-4i^2}=\dfrac{-20+10i}{1+4}=\dfrac{5\,(-4+2i)}{5}=-4+2i$

**34.** $(2-3i)^{-1}=\dfrac{1}{2-3i}=\dfrac{1}{2-3i}\cdot\dfrac{2+3i}{2+3i}=\dfrac{2+3i}{4-9i^2}=\dfrac{2+3i}{4+9}=\dfrac{2+3i}{13}=\frac{2}{13}+\frac{3}{13}i$

**35.** $\dfrac{4+6i}{3i}=\dfrac{4+6i}{3i}\cdot\dfrac{i}{i}=\dfrac{4i+6i^2}{3i^2}=\dfrac{-6+4i}{-3}=\dfrac{-6}{-3}+\dfrac{4}{-3}i=2-\frac{4}{3}i$

**36.** $\dfrac{-3+5i}{15i}=\dfrac{-3+5i}{15i}\cdot\dfrac{i}{i}=\dfrac{-3i+5i^2}{15i^2}=\dfrac{-5-3i}{-15}=\dfrac{-5}{-15}+\dfrac{-3}{-15}i=\frac{1}{3}+\frac{1}{5}i$

**37.** $\dfrac{1}{1+i}-\dfrac{1}{1-i}=\dfrac{1}{1+i}\cdot\dfrac{1-i}{1-i}-\dfrac{1}{1-i}\cdot\dfrac{1+i}{1+i}=\dfrac{1-i}{1-i^2}-\dfrac{1+i}{1-i^2}=\dfrac{1-i}{2}+\dfrac{-1-i}{2}=-i$

**38.** $\dfrac{(1+2i)(3-i)}{2+i}=\dfrac{3-i+6i-2i^2}{2+i}=\dfrac{5+5i}{2+i}\cdot\dfrac{2-i}{2-i}=\dfrac{10-5i+10i-5i^2}{4-i^2}=\dfrac{(10+5)+(-5+10)i}{5}$
$=\dfrac{15+5i}{5}=\frac{15}{5}+\frac{5}{5}i=3+i$

**39.** $i^3=i^2\cdot i=-1\cdot i=-i$

**40.** $(2i)^4=2^4\cdot i^4=16\cdot(1)=16$

**41.** $i^{100}=\left(i^4\right)^{25}=(1)^{25}=1$

**42.** $i^{1002}=\left(i^4\right)^{250}\cdot i^2=(1)^{250}\cdot(-1)=-1$

**43.** $\sqrt{-25}=5i$

**44.** $\sqrt{-\frac{9}{4}}=\frac{3}{2}i$

**45.** $\sqrt{-3}\sqrt{-12}=i\sqrt{3}\cdot 2i\sqrt{3}=6i^2=-6$

**46.** $\sqrt{\frac{1}{3}}\sqrt{-27}=\sqrt{\frac{1}{3}}\cdot 3\sqrt{3}i=3i$

**47.** $\left(3-\sqrt{-5}\right)\left(1+\sqrt{-1}\right)=\left(3-i\sqrt{5}\right)(1+i)=3+3i-i\sqrt{5}-i^2\sqrt{5}=\left(3+\sqrt{5}\right)+\left(3-\sqrt{5}\right)i$

**48.** $\dfrac{1-\sqrt{-1}}{1+\sqrt{-1}}=\dfrac{1-i}{1+i}=\dfrac{1-i}{1+i}\cdot\dfrac{1-i}{1-i}=\dfrac{1-i-i+i^2}{1-i^2}=\dfrac{(1-1)+(-1-1)i}{1+1}=\dfrac{-2i}{2}=-i$

**49.** $\dfrac{2+\sqrt{-8}}{1+\sqrt{-2}}=\dfrac{2+2i\sqrt{2}}{1+i\sqrt{2}}=\dfrac{2+2i\sqrt{2}}{1+i\sqrt{2}}\cdot\dfrac{1-i\sqrt{2}}{1-i\sqrt{2}}=\dfrac{2-2i\sqrt{2}+2i\sqrt{2}-4i^2}{1-2i^2}=\dfrac{(2+4)+\left(-2\sqrt{2}+2\sqrt{2}\right)i}{1+2}=\frac{6}{3}=2$

**50.** $\left(\sqrt{3}-\sqrt{-4}\right)\left(\sqrt{6}-\sqrt{-8}\right)=\left(\sqrt{3}-2i\right)\left(\sqrt{6}-2\sqrt{2}i\right)=\sqrt{18}-2\sqrt{6}i-2\sqrt{6}i+4\sqrt{2}i^2$
$=\left(3\sqrt{2}-4\sqrt{2}\right)+\left(-2\sqrt{6}-2\sqrt{6}\right)i=-\sqrt{2}-4\sqrt{6}i$

**51.** $\dfrac{\sqrt{-36}}{\sqrt{-2}\sqrt{-9}}=\dfrac{6i}{i\sqrt{2}\cdot 3i}=\dfrac{2}{i\sqrt{2}}\cdot\dfrac{i\sqrt{2}}{i\sqrt{2}}=\dfrac{2i\sqrt{2}}{2i^2}=\dfrac{i\sqrt{2}}{-1}=-i\sqrt{2}$

**52.** $\dfrac{\sqrt{-7}\sqrt{-49}}{\sqrt{28}}=\dfrac{\left(\sqrt{7}i\right)(7i)}{2\sqrt{7}}=\dfrac{7i^2}{2}=-\frac{7}{2}$

**53.** $x^2+9=0\quad\Leftrightarrow\quad x^2=-9\quad\Rightarrow\quad x=\pm 3i$

**54.** $9x^2+4=0\quad\Leftrightarrow\quad 9x^2=-4\quad\Leftrightarrow\quad x^2=-\frac{4}{9}\quad\Rightarrow\quad x=\pm\frac{2}{3}i$

**55.** $x^2-4x+5=0\quad\Rightarrow\quad x=\dfrac{-(-4)\pm\sqrt{(-4)^2-4(1)(5)}}{2(1)}=\dfrac{4\pm\sqrt{16-20}}{2}=\dfrac{4\pm\sqrt{-4}}{2}=\dfrac{4\pm 2i}{2}=2\pm i$

**56.** $x^2+2x+2=0\quad\Rightarrow\quad x=\dfrac{-(2)\pm\sqrt{(2)^2-4(1)(2)}}{2(1)}=\dfrac{-2\pm\sqrt{4-8}}{2}=\dfrac{-2\pm\sqrt{-4}}{2}=\dfrac{-2\pm 2i}{2}=-1\pm i$

**57.** $x^2+x+1=0\quad\Rightarrow\quad x=\dfrac{-(1)\pm\sqrt{(1)^2-4(1)(1)}}{2(1)}=\dfrac{-1\pm\sqrt{1-4}}{2}=\dfrac{-1\pm\sqrt{-3}}{2}=\dfrac{-1\pm i\sqrt{3}}{2}=-\frac{1}{2}\pm\dfrac{i\sqrt{3}}{2}$

**58.** $x^2-3x+3=0\quad\Rightarrow\quad x=\dfrac{-(-3)\pm\sqrt{(-3)^2-4(1)(3)}}{2(1)}=\dfrac{3\pm\sqrt{9-12}}{2}=\dfrac{3\pm\sqrt{-3}}{2}=\dfrac{3\pm\sqrt{3}i}{2}=\frac{3}{2}\pm\dfrac{\sqrt{3}}{2}i$

**59.** $2x^2-2x+1=0\quad\Rightarrow\quad x=\dfrac{-(-2)\pm\sqrt{(-2)^2-4(2)(1)}}{2(2)}=\dfrac{2\pm\sqrt{4-8}}{4}=\dfrac{2\pm\sqrt{-4}}{4}=\dfrac{2\pm 2i}{4}=\frac{1}{2}\pm\frac{1}{2}i$

**60.** $2x^2+3=2x \quad \Leftrightarrow \quad 2x^2-2x+3=0 \quad \Rightarrow$

$x=\frac{-(-2)\pm\sqrt{(-2)^2-4(2)(3)}}{2(2)}=\frac{2\pm\sqrt{4-24}}{4}=\frac{2\pm\sqrt{-20}}{4}=\frac{2\pm2\sqrt{5}i}{4}=\frac{1}{2}\pm\frac{\sqrt{5}}{2}i$

**61.** $t+3+\frac{3}{t}=0 \quad \Leftrightarrow \quad t^2+3t+3=0 \quad \Rightarrow \quad t=\frac{-(3)\pm\sqrt{(3)^2-4(1)(3)}}{2(1)}=\frac{-3\pm\sqrt{9-12}}{2}=\frac{-3\pm\sqrt{-3}}{2}=\frac{-3\pm i\sqrt{3}}{2}=-\frac{3}{2}\pm i\frac{\sqrt{3}}{2}$

**62.** $z+4+\frac{12}{z}=0 \quad \Leftrightarrow \quad z^2+4z+12=0 \quad \Rightarrow$

$z=\frac{-(4)\pm\sqrt{(4)^2-4(1)(12)}}{2(1)}=\frac{-4\pm\sqrt{16-48}}{2}=\frac{-4\pm\sqrt{-32}}{2}=\frac{-4\pm4\sqrt{2}i}{2}=-2\pm2\sqrt{2}i$

**63.** $6x^2+12x+7=0 \quad \Rightarrow$

$x=\frac{-(12)\pm\sqrt{(12)^2-4(6)(7)}}{2(6)}=\frac{-12\pm\sqrt{144-168}}{12}=\frac{-12\pm\sqrt{-24}}{12}=\frac{-12\pm2i\sqrt{6}}{12}=\frac{-12}{12}\pm\frac{2i\sqrt{6}}{12}=-1\pm\frac{\sqrt{6}}{6}i$

**64.** $4x^2-16x+19=0 \quad \Rightarrow$

$x=\frac{-(-16)\pm\sqrt{(-16)^2-4(4)(19)}}{2(4)}=\frac{16\pm\sqrt{256-304}}{8}=\frac{16\pm\sqrt{-48}}{8}=\frac{16\pm4\sqrt{3}i}{8}=\frac{16}{8}\pm\frac{4\sqrt{3}}{8}i=2\pm i\frac{\sqrt{3}}{2}$

**65.** $\frac{1}{2}x^2-x+5=0 \quad \Rightarrow \quad x=\frac{-(-1)\pm\sqrt{(-1)^2-4\left(\frac{1}{2}\right)(5)}}{2\left(\frac{1}{2}\right)}=\frac{1\pm\sqrt{1-10}}{1}=1\pm\sqrt{-9}=1\pm3i$

**66.** $x^2+\frac{1}{2}x+1=0 \quad \Rightarrow$

$$x=\frac{-\left(\frac{1}{2}\right)\pm\sqrt{\left(\frac{1}{2}\right)^2-4(1)(1)}}{2(1)}=\frac{-\frac{1}{2}\pm\sqrt{\frac{1}{4}-4}}{2}=\frac{-\frac{1}{2}\pm\sqrt{-\frac{15}{4}}}{2}=\frac{-\frac{1}{2}\pm\frac{1}{2}\sqrt{15}i}{2}=-\frac{1}{4}\pm\frac{\sqrt{15}}{4}i$$

**67.** $\text{LHS}=\overline{z}+\overline{w}=\overline{(a+bi)}+\overline{(c+di)}=a-bi+c-di=(a+c)+(-b-d)i=(a+c)-(b+d)i.$

$\text{RHS}=\overline{z+w}=\overline{(a+bi)+(c+di)}=\overline{(a+c)+(b+d)i}=(a+c)-(b+d)i.$

Since LHS = RHS, this proves the statement.

**68.** $\text{LHS}=\overline{zw}=\overline{(a+bi)(c+di)}=\overline{ac+adi+bci+bdi^2}=\overline{(ac-bd)+(ad+bc)i}=(ac-bd)-(ad+bc)i.$

$\text{RHS}=\overline{z}\cdot\overline{w}=\overline{a+bi}\cdot\overline{c+di}=(a-bi)(c-di)=ac-adi-bci+bdi^2=(ac-bd)-(ad+bc)i.$

Since LHS = RHS, this proves the statement.

**69.** $\text{LHS}=(\overline{z})^2=\left(\overline{(a+bi)}\right)^2=(a-bi)^2=a^2-2abi+b^2i^2=\left(a^2-b^2\right)-2abi.$

$\text{RHS}=\overline{z^2}=\overline{(a+bi)^2}=\overline{a^2+2abi+b^2i^2}=\overline{(a^2-b^2)+2abi}=\left(a^2-b^2\right)-2abi.$

Since LHS = RHS, this proves the statement.

**70.** $\overline{\overline{z}}=\overline{\overline{a+bi}}=\overline{a-bi}=a+bi=z.$

**71.** $z+\overline{z}=(a+bi)+\overline{(a+bi)}=a+bi+a-bi=2a$, which is a real number.

**72.** $z-\overline{z}=(a+bi)-\overline{(a+bi)}=a+bi-(a-bi)=a+bi-a+bi=2bi$, which is a pure imaginary number.

**73.** $z\cdot\overline{z}=(a+bi)\cdot\overline{(a+bi)}=(a+bi)\cdot(a-bi)=a^2-b^2i^2=a^2+b^2$, which is a real number.

**74.** Suppose $z=\overline{z}$. Then we have $(a+bi)=\overline{(a+bi)} \quad \Rightarrow \quad a+bi=a-bi \quad \Rightarrow \quad 0=-2bi \quad \Rightarrow \quad b=0$, so $z$ is real. Now if $z$ is real, then $z=a+0i$(where $a$ is real). Since $\overline{z}=a-0i$, we have $z=\overline{z}$.

**75.** Using the quadratic formula, the solutions to the equation are $x=\frac{-b\pm\sqrt{b^2-4ac}}{2a}$. Since both solutions are imaginary, we have $b^2-4ac<0 \quad \Leftrightarrow \quad 4ac-b^2>0$, so the solutions are $x=\frac{-b}{2a}\pm\frac{\sqrt{4ac-b^2}}{2a}i$, where $\sqrt{4ac-b^2}$ is a real number. Thus the solutions are complex conjugates of each other.

**76.** $i=i,\ i^5=i^4\cdot i=i,\ i^9=i^8\cdot i=i;\quad i^2=-1,\ i^6=i^4\cdot i^2=-1,\ i^{10}=i^8\cdot i^2=-1;$ $i^3=-i,\ i^7=i^4\cdot i^3=-i,\ i^{11}=i^8\cdot i^3=-i;\quad i^4=1,\ i^8=i^4\cdot i^4=1,\ i^{12}=i^8\cdot i^4=1$. Because $i^4=1$, we have $i^n=i^r$, where $r$ is the remainder when $n$ is divided by 4, that is, $n=4\cdot k+r$, where $k$ is an integer and $0\leq r<4$. Since $4446=4\cdot1111+2$, we must have $i^{4446}=i^2=-1$.

# 1.5 Other Types of Equations

**1.** $x^3 = 16x \Leftrightarrow 0 = x^3 - 16x = x(x^2 - 16) = x(x-4)(x+4)$. So $x = 0$, $x - 4 = 0 \Leftrightarrow x = 4$, or $x + 4 = 0 \Leftrightarrow x = -4$. The solutions are 0 and $\pm 4$.

**2.** $x^5 = 27x^2 \Leftrightarrow 0 = x^5 - 27x^2 = x^2(x^3 - 27) \Leftrightarrow x^2 = 0$ or $x^3 - 27 = 0$. If $x^2 = 0$, then $x = 0$. If $x^3 - 27 = 0$, then $x^3 = 27 \Leftrightarrow x = 3$. The solutions are 0 and 3.

**3.** $0 = x^6 - 81x^2 = x^2(x^4 - 81) = x^2(x^2 - 9)(x^2 + 9) = x^2(x-3)(x+3)(x^2+9)$. So $x^2 = 0 \Leftrightarrow x = 0$, or $x - 3 = 0 \Leftrightarrow x = 3$, or $x + 3 = 0 \Leftrightarrow x = -3.x^2 + 9 = 0 \Leftrightarrow x^2 = -9$ which has no real solution. The solutions are 0 and $\pm 3$.

**4.** $0 = x^5 - 16x = x(x^4 - 16) = x(x^2+4)(x^2-4) = x(x^2+4)(x-2)(x+2)$. Since $x^2 + 4 = 0$ has no real solution, thus either $x\ x = 0$, or $x = 2$, or $x = -2$. The solutions are 0, 2, and $-2$.

**5.** $0 = x^5 + 8x^2 = x^2(x^3 + 8) = x^2(x+2)(x^2 - 2x + 4) \Leftrightarrow x^2 = 0, x + 2 = 0$, or $x^2 - 2x + 4 = 0$. If $x^2 = 0$, then $x = 0$; if $x + 2 = 0$, then $x = -2$, and $x^2 - 2x + 4 = 0$ has no real solution. Thus the solutions are $x = 0$ and $x = -2$.

**6.** $0 = x^4 + 64x = x(x^3 + 64) \Leftrightarrow x = 0$ or $x^3 + 64 = 0$. If $x^3 + 64 = 0$, then $x^3 = -64 \Leftrightarrow x = -4$. The solutions are 0 and $-4$.

**7.** $0 = x^3 - 5x^2 + 6x = x(x^2 - 5x + 6) = x(x-2)(x-3) \Leftrightarrow x = 0, x - 2 = 0$, or $x - 3 = 0$. Thus $x = 0$, or $x = 2$, or $x = 3$. The solutions are $x = 0$, $x = 2$, and $x = 3$.

**8.** $0 = x^4 - x^3 - 6x^2 = x^2(x^2 - x - 6) = x^2(x-3)(x+2)$. Thus either $x^2 = 0$, so $x = 0$,or $x = 3$, or $x = -2$. The solutions are 0,3, and $-2$.

**9.** $0 = x^4 + 4x^3 + 2x^2 = x^2(x^2 + 4x + 2)$. So either $x^2 = 0 \Leftrightarrow x = 0$, or using the quadratic formula on $x^2 + 4x + 2 = 0$, we have $x = \frac{-4\pm\sqrt{4^2-4(1)(2)}}{2(1)} = \frac{-4\pm\sqrt{16-8}}{2} = \frac{-4\pm\sqrt{8}}{2} = \frac{-4\pm2\sqrt{2}}{2} = -2\pm\sqrt{2}$. The solutions are 0, $-2-\sqrt{2}$, and $-2+\sqrt{2}$.

**10.** $0 = (x-2)^5 - 9(x-2)^3 = (x-2)^3\left[(x-2)^2 - 9\right] = (x-2)^3\left[(x-2)-3\right]\left[(x-2)+3\right] = (x-2)^3(x-5)(x+1)$. Thus either $x = 2$,or $x = 5$, or $x = -1$. The solutions are 2, 5, and $-1$.

**11.** $0 = x^3 - 5x^2 - 2x + 10 = x^2(x-5) - 2(x-5) = (x-5)(x^2-2)$. If $x - 5 = 0$, then $x = 5$. If $x^2 - 2 = 0$, then $x^2 = 2 \Leftrightarrow x = \pm\sqrt{2}$. The solutions are 5 and $\pm\sqrt{2}$.

**12.** $0 = 2x^3 + x^2 - 18x - 9 = x^2(2x+1) - 9(2x+1) = (2x+1)(x^2-9) = (2x+1)(x-3)(x+3)$. The solutions are $-\frac{1}{2}$, 3, and $-3$.

**13.** $x^3 - x^2 + x - 1 = x^2 + 1 \Leftrightarrow 0 = x^3 - 2x^2 + x - 2 = x^2(x-2) + (x-2) = (x-2)(x^2+1)$. Since $x^2 + 1 = 0$ has no real solution, the only solution comes from $x - 2 = 0 \Leftrightarrow x = 2$.

**14.** $7x^3 - x + 1 = x^3 + 3x^2 + x \Leftrightarrow 0 = 6x^3 - 3x^2 - 2x + 1 = 3x^2(2x-1) - (2x-1) = (2x-1)(3x^2-1) \Leftrightarrow 2x - 1 = 0$ or $3x^2 - 1 = 0$. If $2x - 1 = 0$, then $x = \frac{1}{2}$. If $3x^2 - 1 = 0$, then $3x^2 = 1 \Leftrightarrow x^2 = \frac{1}{3} \Rightarrow x = \pm\sqrt{\frac{1}{3}}$. The solutions are $\frac{1}{2}$ and $\pm\sqrt{\frac{1}{3}}$.

**15.** $\dfrac{1}{x-1} + \dfrac{1}{x+2} = \frac{5}{4} \Leftrightarrow 4(x-1)(x+2)\left(\dfrac{1}{x-1} + \dfrac{1}{x+2}\right) = 4(x-1)(x+2)\left(\frac{5}{4}\right) \Leftrightarrow 4(x+2) + 4(x-1) = 5(x-1)(x+2) \Leftrightarrow 4x + 8 + 4x - 4 = 5x^2 + 5x - 10 \Leftrightarrow 5x^2 - 3x - 14 = 0 \Leftrightarrow (5x+7)(x-2) = 0$. If $5x + 7 = 0$, then $x = -\frac{7}{5}$; if $x - 2 = 0$, then $x = 2$. The solutions are $-\dfrac{7}{5}$ and 2.

**16.** $\dfrac{10}{x} - \dfrac{12}{x-3} + 4 = 0 \Leftrightarrow x(x-3)\left(\dfrac{10}{x} - \dfrac{12}{x-3} + 4\right) = 0 \Leftrightarrow (x-3)10 - 12x + 4x(x-3) = 0 \Leftrightarrow 10x - 30 - 12x + 4x^2 - 12x = 0 \Leftrightarrow 4x^2 - 14x - 30 = 0$. Using the quadratic formula, we have $x = \frac{-(-14)\pm\sqrt{(-14)^2-4(4)(-30)}}{2(4)} = \frac{14\pm\sqrt{196+480}}{8} = \frac{14\pm\sqrt{676}}{8} = \frac{14\pm26}{8}$. So the solutions are 5 and $-\frac{3}{2}$.

**17.** $\dfrac{x^2}{x+100}=50 \;\Rightarrow\; x^2=50(x+100)=50x+5000 \;\Leftrightarrow\; x^2-50x-5000=0 \;\Leftrightarrow\; (x-100)(x+50)=0$ $\Leftrightarrow\; x-100=0$ or $x+50=0$. Thus $x=100$ or $x=-50$. The solutions are 100 and $-50$.

**18.** $1+\dfrac{2x}{(x+3)(x+4)}=\dfrac{2}{x+3}+\dfrac{4}{x+4} \;\Rightarrow\; (x+3)(x+4)+2x=2(x+4)+4(x+3) \;\Leftrightarrow$ $x^2+7x+12+2x=2x+8+4x+12 \;\Leftrightarrow\; x^2+3x-8=0$. Using the quadratic formula, we have $x=\frac{-(3)\pm\sqrt{(3)^2-4(1)(-8)}}{2(1)}=\frac{-3\pm\sqrt{9+32}}{2}=\frac{-3\pm\sqrt{41}}{2}$. The solutions are $\frac{-3\pm\sqrt{41}}{2}$.

**19.** $\dfrac{x+5}{x-2}=\dfrac{5}{x+2}+\dfrac{28}{x^2-4} \;\Rightarrow\; (x+2)(x+5)=5(x-2)+28 \;\Leftrightarrow\; x^2+7x+10=5x-10+28 \;\Leftrightarrow$ $x^2+2x-8=0 \;\Leftrightarrow\; (x-2)(x+4)=0 \;\Leftrightarrow\; x-2=0$ or $x+4=0 \;\Leftrightarrow\; x=2$ or $x=-4$. However, $x=2$ is inadmissible since we can't divide by 0 in the original equation, so the only solution is $-4$.

**20.** $\dfrac{x}{2x+7}-\dfrac{x+1}{x+3}=1 \;\Leftrightarrow\; x(x+3)-(x+1)(2x+7)=(2x+7)(x+3) \;\Leftrightarrow\; x^2+3x-2x^2-9x-7=2x^2+$ $13x+21 \;\Leftrightarrow\; 3x^2+19x+28=0 \;\Leftrightarrow\; (3x+7)(x+4)=0$. Thus either $3x+7=0$, so $x=-\frac{7}{3}$, or $x=-4$. The solutions are $-\frac{7}{3}$ and $-4$.

**21.** $\dfrac{1}{x-1}-\dfrac{2}{x^2}=0 \;\Leftrightarrow\; x^2-2(x-1)=0 \;\Leftrightarrow\; x^2-2x+2=0 \;\Rightarrow$

$$x=\frac{-(-2)\pm\sqrt{(-2)^2-4(1)(2)}}{2(1)}=\frac{2\pm\sqrt{4-8}}{2}=\frac{2\pm\sqrt{-4}}{2}.$$ Since the radicand is negative, there is no real solution.

**22.** $\dfrac{x+\frac{2}{x}}{3+\frac{4}{x}}=5x \;\Rightarrow\; \left(\dfrac{x+\frac{2}{x}}{3+\frac{4}{x}}\right)\cdot\dfrac{x}{x}=\dfrac{x^2+2}{3x+4}=5x \;\Rightarrow\; x^2+$ $2=5x(3x+4) \;\Leftrightarrow\; x^2+2=15x^2+20x \;\Leftrightarrow\; 0=14x^2+20x-2 \;\Rightarrow$

$$x=\frac{-(20)\pm\sqrt{(20)^2-4(14)(-2)}}{2(14)}=\frac{-20\pm\sqrt{400+112}}{28}=\frac{-20\pm\sqrt{512}}{28}=\frac{-20\pm16\sqrt{2}}{28}=\frac{-5\pm4\sqrt{2}}{7}.$$ The solutions are $\dfrac{-5\pm4\sqrt{2}}{7}$.

**23.** $0=(x+5)^2-3(x+5)-10=[(x+5)-5][(x+5)+2]=x(x+7) \;\Leftrightarrow\; x=0$ or $x=-7$. The solutions are 0 and $-7$.

**24.** Let $w=\dfrac{x+1}{x}$. Then $0=\left(\dfrac{x+1}{x}\right)^2+4\left(\dfrac{x+1}{x}\right)+3$ becomes $0=w^2+4w+3=(w+1)(w+3)$. Now if $w+1=0$, then $\dfrac{x+1}{x}+1=0 \;\Leftrightarrow\; \dfrac{x+1}{x}=-1 \;\Leftrightarrow\; x+1=-x \;\Leftrightarrow\; x=-\frac{1}{2}$, and if $w+3=0$, then $\dfrac{x+1}{x}+3=0 \;\Leftrightarrow\; \dfrac{x+1}{x}=-3 \;\Leftrightarrow\; x+1=-3x \;\Leftrightarrow\; x=-\frac{1}{4}$. The solutions are $-\frac{1}{2}$ and $-\frac{1}{4}$.

**25.** Let $w=\dfrac{1}{x+1}$. Then $\left(\dfrac{1}{x+1}\right)^2-2\left(\dfrac{1}{x+1}\right)-8=0$ becomes $w^2-2w-8=0 \;\Leftrightarrow\; (w-4)(w+2)=0$. So $w-4=0 \;\Leftrightarrow\; w=4$, and $w+2=0 \;\Leftrightarrow\; w=-2$. When $w=4$, we have $\dfrac{1}{x+1}=4 \;\Leftrightarrow\; 1=4x+4 \;\Leftrightarrow$ $-3=4x \;\Leftrightarrow\; x=-\frac{3}{4}$. When $w=-2$, we have $\dfrac{1}{x+1}=-2 \;\Leftrightarrow\; 1=-2x-2 \;\Leftrightarrow\; 3=-2x \;\Leftrightarrow\; x=-\frac{3}{2}$. Solutions are $-\frac{3}{4}$ and $-\frac{3}{2}$.

**26.** Let $w=\dfrac{x}{x+2}$. Then $\left(\dfrac{x}{x+2}\right)^2=\dfrac{4x}{x+2}-4$ becomes $w^2=4w-4 \;\Leftrightarrow\; 0=w^2-4w+4=(w-2)^2$. Now if $w-2=0$, then $\dfrac{x}{x+2}-2=0 \;\Leftrightarrow\; \dfrac{x}{x+2}=2 \;\Leftrightarrow\; x=2x+4 \;\Leftrightarrow\; x=-4$. The solution is $-4$.

**27.** Let $w = x^2$. Then $x^4 - 13x^2 + 40 = (x^2)^2 - 13x^2 + 40 = 0$ becomes $w^2 - 13w + 40 = 0 \Leftrightarrow (w-5)(w-8) = 0$. So $w - 5 = 0 \Leftrightarrow w = 5$, and $w - 8 = 0 \Leftrightarrow w = 8$. When $w = 5$, we have $x^2 = 5 \Rightarrow x = \pm\sqrt{5}$. When $w = 8$, we have $x^2 = 8 \Rightarrow x = \pm\sqrt{8} = \pm 2\sqrt{2}$. The solutions are $\pm\sqrt{5}$ and $\pm 2\sqrt{2}$.

**28.** $0 = x^4 - 5x^2 + 4 = (x^2 - 4)(x^2 - 1) = (x-2)(x+2)(x-1)(x+1)$ .So $x = 2$, $x = -2$, or $x = 1$, or $x = -1$. The solutions are $-2$, 2, $-1$, and 1.

**29.** $2x^4 + 4x^2 + 1 = 0$. The LHS is the sum of two nonnegative numbers and a positive number, so $2x^4 + 4x^2 + 1 \geq 1 \neq 0$. This equation has no real solution.

**30.** $0 = x^6 - 2x^3 - 3 = (x^3 - 3)(x^3 + 1)$. If $x^3 - 3 = 0$, then $x^3 = 3 \Leftrightarrow x = \sqrt[3]{3}$, or if $x^3 = -1 \Leftrightarrow x = -1$. Thus $x = \sqrt[3]{3}$ or $x = -1$. The solutions are $\sqrt[3]{3}$ and $-1$.

**31.** $0 = x^6 - 26x^3 - 27 = (x^3 - 27)(x^3 + 1)$. If $x^3 - 27 = 0 \Leftrightarrow x^3 = 27$, so $x = 3$. If $x^3 + 1 = 0 \Leftrightarrow x^3 = -1$, so $x = -1$. The solutions are 3 and $-1$.

**32.** $x^8 + 15x^4 = 16 \Leftrightarrow 0 = x^8 + 15x^4 - 16 = (x^4 + 16)(x^4 - 1)$. If $x^4 + 16 = 0$, then $x^4 = -16$ which is impossible(for real numbers). If $x^4 - 1 = 0 \Leftrightarrow x^4 = 1$, so $x = \pm 1$. The solutions are 1 and $-1$.

**33.** Let $u = x^{2/3}$. Then $0 = x^{4/3} - 5x^{2/3} + 6$ becomes $u^2 - 5u + 6 = 0 \Leftrightarrow (u-3)(u-2) = 0 \Leftrightarrow u - 3 = 0$ or $u - 2 = 0$. If $u - 3 = 0$, then $x^{2/3} - 3 = 0 \Leftrightarrow x^{2/3} = 3 \Rightarrow x = \pm 3^{3/2} = \pm 3\sqrt{3}$. If $u - 2 = 0$, then $x^{2/3} - 2 = 0$ $\Leftrightarrow x^{2/3} = 2 \Rightarrow x = \pm 2^{3/2} = 2\sqrt{2}$. The solutions are $\pm 3\sqrt{3}$ and $\pm 2\sqrt{2}$.

**34.** Let $u = \sqrt[4]{x}$; then$0 = \sqrt{x} - 3\sqrt[4]{x} - 4 = u^2 - 3u - 4 = (u-4)(u+1)$. So $u - 4 = \sqrt[4]{x} - 4 = 0 \Leftrightarrow \sqrt[4]{x} = 4 \Rightarrow$ $x = 4^4 = 256$, or $u + 1 = \sqrt[4]{x} + 1 = 0 \Leftrightarrow \sqrt[4]{x} = -1$. However, $\sqrt[4]{x}$ is the positive fourth root, so this cannot equal $-1$. The only solution is 256.

**35.** $4(x+1)^{1/2} - 5(x+1)^{3/2} + (x+1)^{5/2} = 0 \Leftrightarrow \sqrt{x+1}\left[4 - 5(x+1) + (x+1)^2\right] = 0 \Leftrightarrow$ $\sqrt{x+1}\,(4 - 5x - 5 + x^2 + 2x + 1) = 0 \Leftrightarrow \sqrt{x+1}\,(x^2 - 3x) = 0 \Leftrightarrow \sqrt{x+1} \cdot x(x-3) = 0 \Leftrightarrow x = -1$ or $x = 0$ or $x = 3$. The solutions are $-1$, 0, and 3.

**36.** Let $u = x - 4$; then $0 = 2(x-4)^{7/3} - (x-4)^{4/3} - (x-4)^{1/3} = 2u^{7/3} - u^{4/3} - u^{1/3} =$ $u^{1/3}(2u^2 - u - 1) = u^{1/3}(2u+1)(u-1)$. So $u^{1/3} = x - 4 = 0 \Leftrightarrow x = 4$, or $2u + 1 = 2(x-4) + 1 = 2x - 7 = 0$ $\Leftrightarrow 2x = 7 \Leftrightarrow x = \frac{7}{2}$, or $u - 1 = (x-4) - 1 = x - 5 = 0 \Leftrightarrow x = 5$. The solutions are 4, $\frac{7}{2}$,and 5.

**37.** $0 = x^{3/2} + 8x^{1/2} + 16x^{-1/2} = x^{-1/2}(x^2 + 8x + 16) = x^{-1/2}(x+4)^2$. Now $x^{-1/2} = \dfrac{1}{\sqrt{x}}$ so $x \neq 0$. So $x + 4 = 0$ $\Leftrightarrow x = -4$. But $\sqrt{-4}$ is not a real number, so this equation has no real solution. Alternatively, we see that this is the sum of three positive real numbers (remember $x \neq 0$), so it never equals zero.

**38.** $x^{1/2} + 3x^{-1/2} = 10x^{-3/2} \Leftrightarrow 0 = x^{1/2} + 3x^{-1/2} - 10x^{-3/2} = x^{-3/2}(x^2 + 3x - 10) = x^{-3/2}(x-2)(x+5)$. Now $x^{-3/2}$ never equals 0, and no solution can be negative, because we cannot take the $\frac{1}{2}$ power of a negative number. Thus 2 is the only solution.

**39.** Let $u = x^{1/6}$. (We choose the exponent $\frac{1}{6}$ because the LCD of 2, 3, and 6 is 6.) Then $x^{1/2} - 3x^{1/3} = 3x^{1/6} - 9 \Leftrightarrow x^{3/6} - 3x^{2/6} = 3x^{1/6} - 9 \Leftrightarrow u^3 - 3u^2 = 3u - 9 \Leftrightarrow$ $0 = u^3 - 3u^2 - 3u + 9 = u^2(u-3) - 3(u-3) = (u-3)(u^2-3)$. So $u - 3 = 0$ or $u^2 - 3 = 0$. If $u - 3 = 0$, then $x^{1/6} - 3 = 0 \Leftrightarrow x^{1/6} = 3 \Leftrightarrow x = 3^6 = 729$. If $u^2 - 3 = 0$, then $x^{1/3} - 3 = 0 \Leftrightarrow x^{1/3} = 3 \Leftrightarrow$ $x = 3^3 = 27$.The solutions are 729 and 27.

**40.** Let $u = \sqrt{x}$. Then $0 = x - 5\sqrt{x} + 6$ becomes $u^2 - 5u + 6 = (u-3)(u-2) = 0$. If $u - 3 = 0$, then $\sqrt{x} - 3 = 0 \Leftrightarrow$ $\sqrt{x} = 3 \Rightarrow x = 9$. If $u - 2 = 0$, then $\sqrt{x} - 2 = 0 \Leftrightarrow \sqrt{x} = 2 \Rightarrow x = 4$. The solutions are 9 and 4.

**41.** $\dfrac{1}{x^3} + \dfrac{4}{x^2} + \dfrac{4}{x} = 0 \Rightarrow 1 + 4x + 4x^2 = 0 \Leftrightarrow (1+2x)^2 = 0 \Leftrightarrow 1 + 2x = 0 \Leftrightarrow 2x = -1 \Leftrightarrow x = -\frac{1}{2}$. The solution is $-\frac{1}{2}$.

**42.** $0 = 4x^{-4} - 16x^{-2} + 4$. Multiplying by $\dfrac{x^4}{4}$ we get, $0 = 1 - 4x^2 + x^4$. Substituting $u = x^2$, we get $0 = 1 - 4u + u^2$, and using the quadratic formula, we get $u = \frac{-(-4)\pm\sqrt{(-4)^2-4(1)(1)}}{2(1)} = \frac{4\pm\sqrt{16-4}}{2} = \frac{4\pm\sqrt{12}}{2} = \frac{4\pm2\sqrt{3}}{2} = 2\pm\sqrt{3}$. Substituting back, we have $x^2 = 2\pm\sqrt{3}$, and since $2+\sqrt{3}$ and $2-\sqrt{3}$ are both positive we have $x = \pm\sqrt{2+\sqrt{3}}$ or $x = \pm\sqrt{2-\sqrt{3}}$. Thus the solutions are $-\sqrt{2-\sqrt{3}}$, $\sqrt{2-\sqrt{3}}$, $-\sqrt{2+\sqrt{3}}$, and $\sqrt{2+\sqrt{3}}$.

**43.** $\sqrt{2x+1}+1 = x \quad\Leftrightarrow\quad \sqrt{2x+1} = x-1 \quad\Rightarrow\quad 2x+1 = (x-1)^2 \quad\Leftrightarrow\quad 2x+1 = x^2-2x+1 \quad\Leftrightarrow\quad 0 = x^2-4x = x(x-4)$. Potential solutions are $x=0$ and $x-4 \quad\Leftrightarrow\quad x=4$. These are only potential solutions since squaring is not a reversible operation. We must check each potential solution in the original equation.

Checking $x=0$: $\sqrt{2(0)+1}+1 = (0) \quad\Leftrightarrow\quad \sqrt{1}+1 = 0$ is false.

Checking $x=4$: $\sqrt{2(4)+1}+1 = (4) \quad\Leftrightarrow\quad \sqrt{9}+1 = 4 \quad\Leftrightarrow\quad 3+1=4$ is true. The only solution is $x=4$.

**44.** $x-\sqrt{9-3x} = 0 \quad\Leftrightarrow\quad x = \sqrt{9-3x} \quad\Rightarrow\quad x^2 = 9-3x \quad\Leftrightarrow\quad 0 = x^2+3x-9$. Using the quadratic formula to find the potential solutions, we have $x = \frac{-3\pm\sqrt{3^2-4(1)(-9)}}{2(1)} = \frac{-3\pm\sqrt{45}}{2} = \frac{-3\pm3\sqrt{5}}{2}$. Substituting each of these solutions into the original equation, we see that $x = \frac{-3+3\sqrt{5}}{2}$ is a solution, but $x = \frac{-3-3\sqrt{5}}{2}$ is not. Thus $x = \frac{-3+3\sqrt{5}}{2}$ is the only solution.

**45.** $\sqrt{5-x}+1 = x-2 \quad\Leftrightarrow\quad \sqrt{5-x} = x-3 \quad\Rightarrow\quad 5-x = (x-3)^2 \quad\Leftrightarrow\quad 5-x = x^2-6x+9 \quad\Leftrightarrow\quad 0 = x^2-5x+4 = (x-4)(x-1)$. Potential solutions are $x=4$ and $x=1$. We must check each potential solution in the original equation.

Checking $x=4$: $\sqrt{5-(4)}+1 = (4)-2 \quad\Leftrightarrow\quad \sqrt{1}+1 = 4-2 \quad\Leftrightarrow\quad 1+1=2$ is true.

Checking $x=1$: $\sqrt{5-(1)}+1 = (1)-2 \quad\Leftrightarrow\quad \sqrt{4}+1 = -1 \quad\Leftrightarrow\quad 2+1=-1$ is false. The only solution is $x=4$.

**46.** $2x+\sqrt{x+1} = 8 \quad\Leftrightarrow\quad \sqrt{x+1} = 8-2x \quad\Rightarrow\quad x+1 = (8-2x)^2 \quad\Leftrightarrow\quad x+1 = 64-32x+4x^2 \quad\Leftrightarrow\quad 0 = 4x^2-33x+63 = (4x-21)(x-3)$. Potential solutions are $x=\frac{21}{4}$ and $x=3$. Substituting each of these solutions into the original equation, we see that $x=3$ is a solution, but $x=\frac{21}{4}$ is not. Thus 3 is the only solution.

**47.** $x-\sqrt{x+3} = \dfrac{x}{2} \quad\Leftrightarrow\quad \dfrac{x}{2} = \sqrt{x+3} \quad\Rightarrow\quad \left(\frac{1}{2}x\right)^2 = x+3 \quad\Leftrightarrow\quad \frac{1}{4}x^2 = x+3 \quad\Leftrightarrow\quad x^2 = 4x+12 \quad\Leftrightarrow\quad 0 = x^2-4x-12 = (x-6)(x+2)$. Potential solutions are $x=6$ and $x=-2$. We must check each potential solution in the original equation. Checking $x=6$: $6-\sqrt{6+3} = \frac{6}{2} \quad\Leftrightarrow\quad 6-\sqrt{6+3} = 6-\sqrt{9} = 6-3$, which is true. Checking $x=-2$: $-2-\sqrt{-2+3} = -\frac{2}{2} \quad\Leftrightarrow\quad -2-\sqrt{-2+3} = -2-\sqrt{1} = -2-1 = -3 = -\frac{2}{2} \quad\Leftrightarrow\quad -3=-1$, hence this not a solution. The only solution is $x=6$.

**48.** $x+2\sqrt{x-7} = 10 \quad\Leftrightarrow\quad 2\sqrt{x-7} = 10-x \quad\Rightarrow\quad 4(x-7) = (10-x)^2 \quad\Leftrightarrow\quad 4x-28 = 100-20x+x^2 \quad\Leftrightarrow\quad 0 = x^2-24x+128 = (x-8)(x+16)$. Potential solutions are $x=8$ and $x=16$. Substituting each of these solutions into the original equation, we see that $x=8$ is a solution, but $x=16$ is not. Thus 8 is the only solution.

**49.** $\sqrt{\sqrt{x-5}+x} = 5$. Squaring both sides, we get $\sqrt{x-5}+x = 25 \quad\Leftrightarrow\quad \sqrt{x-5} = 25-x$. Squaring both sides again, we get $x-5 = (25-x)^2 \quad\Leftrightarrow\quad x-5 = 625-50x+x^2 \quad\Leftrightarrow\quad 0 = x^2-51x+630 = (x-30)(x-21)$. Potential solutions are $x=30$ and $x=21$. We must check each potential solution in the original equation.

Checking $x=30$: $\sqrt{\sqrt{(30)-5}+(30)} = 5 \quad\Leftrightarrow\quad \sqrt{\sqrt{(30)-5}+(30)} = \sqrt{\sqrt{25}+30} = \sqrt{35} = 5$, hence $x=30$ is not a solution.

Checking $x=21$: $\sqrt{\sqrt{(21)-5}+21} = 5 \quad\Leftrightarrow\quad \sqrt{\sqrt{(21)-5}+21} = \sqrt{\sqrt{16}+21} = \sqrt{25} = 5$, hence $x=21$ is the only solution.

**50.** $\sqrt[3]{4x^2-4x} = x \quad\Leftrightarrow\quad 4x^2-4x = x^3 \quad\Leftrightarrow\quad 0 = x^3-4x^2+4x = x\left(x^2-4x+4\right) = x(x-2)^2$. So $x=0$ or $x=2$. The solutions are 0 and 2.

**51.** $x^2\sqrt{x+3} = (x+3)^{3/2} \Leftrightarrow 0 = x^2\sqrt{x+3} - (x+3)^{3/2} \Leftrightarrow 0 = \sqrt{x+3}\left[\left(x^2\right) - (x+3)\right] \Leftrightarrow$ $0 = \sqrt{x+3}\left(x^2 - x - 3\right)$. If $(x+3)^{1/2} = 0$, then $x+3=0 \Leftrightarrow x=-3$. If $x^2 - x - 3 = 0$, then using the quadratic formula $x = \frac{1\pm\sqrt{13}}{2}$. The solutions are $-3$ and $\frac{1\pm\sqrt{13}}{2}$.

**52.** Let $u = \sqrt{11-x^2}$. By definition of $u$ we require it to be nonnegative. Now $\sqrt{11-x^2} - \dfrac{2}{\sqrt{11-x^2}} = 1 \Leftrightarrow$ $u - \dfrac{2}{u} = 1$. Multiplying both sides by $u$ we obtain $u^2 - 2 = u \Leftrightarrow 0 = u^2 - u - 2 = (u-2)(u+1)$. So $u=2$ or $u=-1$. But since $u$ must be nonnegative, we only have $u = 2 \Leftrightarrow \sqrt{11-x^2} = 2 \Rightarrow 11 - x^2 = 4 \Leftrightarrow x^2 = 7$ $\Leftrightarrow x = \pm\sqrt{7}$. The solutions are $\pm\sqrt{7}$.

**53.** $\sqrt{x+\sqrt{x+2}} = 2$. Squaring both sides, we get $x + \sqrt{x+2} = 4 \Leftrightarrow \sqrt{x+2} = 4 - x$. Squaring both sides again, we get $x + 2 = (4-x)^2 = 16 - 8x + x^2 \Leftrightarrow 0 = x^2 - 9x + 14 \Leftrightarrow 0 = (x-7)(x-2)$. If $x - 7 = 0$, then $x = 7$. If $x - 2 = 0$, then $x = 2$. So $x = 2$ is a solution but $x = 7$ is not, since it does not satisfy the original equation.

**54.** $\sqrt{1+\sqrt{x+\sqrt{2x+1}}} = \sqrt{5+\sqrt{x}}$. We square both sides to get $1 + \sqrt{x+\sqrt{2x+1}} = 5 + \sqrt{x} \Rightarrow x + \sqrt{2x+1} = (4+\sqrt{x})^2 = 16 + 8\sqrt{x} + x \Leftrightarrow \sqrt{2x+1} = 16 + 8\sqrt{x}$. Again, squaring both sides, we obtain $2x + 1 = (16+8\sqrt{x})^2 = 256 + 256\sqrt{x} + 64x \Leftrightarrow -62x - 255 = 256\sqrt{x}$. We could continue squaring both sides until we found possible solutions; however, consider the last equation. Since we are working with real numbers, for $\sqrt{x}$ to be defined, we must have $x \geq 0$. Then $-62x - 255 < 0$ while $256\sqrt{x} \geq 0$, so there is no solution.

**55.** $x^3 = 1 \Leftrightarrow x^3 - 1 = 0 \Leftrightarrow (x-1)\left(x^2+x+1\right) = 0 \Leftrightarrow x - 1 = 0$ or $x^2 + x + 1 = 0$. If $x - 1 = 0$, then $x = 1$. If $x^2 + x + 1 = 0$, then using the quadratic formula $x = \frac{-1\pm i\sqrt{3}}{2}$. The solutions are 1 and $\frac{-1\pm i\sqrt{3}}{2}$.

**56.** $0 = x^4 - 16 = \left(x^2-4\right)\left(x^2+4\right) = (x-2)(x+2)(x-2i)(x+2i)$. Setting each factor in turn equal to zero, we see that the solutions are $\pm2$ and $\pm2i$.

**57.** $x^3 + x^2 + x = 0 \Leftrightarrow x\left(x^2+x+1\right) = 0 \Leftrightarrow x = 0$ or $x = \frac{-1\pm i\sqrt{3}}{2}$. The solutions are 0 and $\frac{-1\pm i\sqrt{3}}{2}$.

**58.** $x^4 + x^3 + x^2 + x = 0 \Leftrightarrow$
$0 = x\left(x^3+x^2+x+1\right) = x\left[x^2(x+1) + (x+1)\right] = x(x+1)\left(x^2+1\right) = x(x+1)(x-i)(x+i)$. Setting each factor in turn equal to zero, we see that the solutions are 0, $-1$, and $\pm i$.

**59.** $x^4 - 6x^2 + 8 = 0 \Leftrightarrow \left(x^2-4\right)\left(x^2-2\right) = 0 \Leftrightarrow x = \pm2$ or $x = \pm\sqrt{2}$. The solutions are $\pm2$ and $\pm\sqrt{2}$.

**60.** $0 = x^3 + 3x^2 + 9x + 27 = x^2(x+3) + 9(x+3) = (x+3)\left(x^2+9\right) = (x+3)(x-3i)(x+3i)$. Setting each factor in turn equal to zero, we see that the solutions are $-3$ and $\pm3\,i$.

**61.** $x^6 - 9x^3 + 8 = 0 \Leftrightarrow \left(x^3-8\right)\left(x^3-1\right) = 0 \Leftrightarrow (x-2)\left(x^2+2x+4\right)(x-1)\left(x^2+x+1\right) = 0 \Leftrightarrow x = 2$ or $x = \dfrac{-2\pm2i\sqrt{3}}{2} = -1 \pm i\sqrt{3}$ or $x = 1$ or $x = \dfrac{-1\pm i\sqrt{3}}{2}$. The solutions are 2, $-1\pm i\sqrt{3}$, 1, and $\dfrac{-1\pm i\sqrt{3}}{2}$.

**62.** $0 = x^6 + 9x^4 - 4x^2 - 36 = x^4\left(x^2+9\right) - 4\left(x^2+9\right) = \left(x^2+9\right)\left(x^4-4\right)$
$= (x-3i)(x+3i)\left(x^2-2\right)\left(x^2+2\right) = (x-3i)(x+3i)\left(x-\sqrt{2}\right)\left(x+\sqrt{2}\right)\left(x-\sqrt{2}i\right)\left(x-\sqrt{2}i\right)$
The six solutions are $\pm3\,i$, $\pm\sqrt{2}$, and $\pm\sqrt{2}i$.

**63.** $\sqrt{x^2+1}+\dfrac{8}{\sqrt{x^2+1}}=\sqrt{x^2+9}$. Squaring both sides, we have $\left(x^2+1\right)+16+\dfrac{64}{x^2+1}=x^2+9 \Leftrightarrow \dfrac{64}{x^2+1}=-8$ $\Leftrightarrow \dfrac{8}{x^2+1}=-1 \Leftrightarrow x^2+1=-8 \Leftrightarrow x^2=-9 \Leftrightarrow x=\pm 3i$. We must check each potential solution in the original equation. Checking $x=\pm 3i$: $\sqrt{(\pm 3i)^2+1}+\dfrac{8}{\sqrt{(\pm 3i)^2+1}} \stackrel{?}{=} \sqrt{(\pm 3i)^2+9}$,

$\text{LHS}=\sqrt{-8}+\dfrac{8}{\sqrt{-8}}=2\sqrt{2}i+\dfrac{8}{2\sqrt{2}i}=2\sqrt{2}i-2\sqrt{2}i=0$. $\text{RHS}=\sqrt{(\pm 3i)^2+9}=\sqrt{-9+9}=0$. Since LHS = RHS, $-3i$ and $3i$ are solutions.

**64.** $1-\sqrt{x^2+7}=6-x^2 \Leftrightarrow x^2-5=\sqrt{x^2+7}$. Squaring both sides, we get $x^4-10x^2+25=x^2+7 \Leftrightarrow$ $0=x^4-11x^2+18=\left(x^2-9\right)\left(x^2-2\right)=(x-3)(x+3)\left(x-\sqrt{2}\right)\left(x+\sqrt{2}\right)$. The possible solutions are $x=\pm 3$ and $x=\pm\sqrt{2}$.

Checking $x=\pm 3$, we have LHS $=1-\sqrt{(\pm 3)^2+7}=1-\sqrt{16}=-3$; RHS $=6-(\pm 3)^2=-3$. Since LHS = RHS these are solutions.

Checking $\pm\sqrt{2}$,we have LHS $=1-\sqrt{\left(\pm\sqrt{2}\right)^2+7}=1-\sqrt{9}=-2$; RHS $=6-\left(\pm\sqrt{2}\right)^2=4$. Since LHS $\neq$ RHS these are not solutions. Thus the only solutions are $\pm 3$.

**65.** $0=x^4+5ax^2+4a^2=\left(x^2+a\right)\left(x^2+4a\right)$. Since $a$ is positive, $x^2+a=0 \Leftrightarrow x^2=-a \Leftrightarrow x=\pm i\sqrt{a}$. Again, since $a$ is positive, $x^2+4a=0 \Leftrightarrow x^2=-4a \Leftrightarrow x=\pm 2i\sqrt{a}$. Thus the four solutions are: $\pm i\sqrt{a}$, $\pm 2i\sqrt{a}$.

**66.** $0=a^3x^3+b^3=(ax+b)\left(a^2x^2-abx+b^2\right)$. So $ax+b=0 \Leftrightarrow ax=-b \Leftrightarrow x=-\dfrac{b}{a}$ or

$$x=\frac{-(-ab)\pm\sqrt{(-ab)^2-4\left(a^2\right)\left(b^2\right)}}{2\left(a^2\right)}=\frac{ab\pm\sqrt{-3a^2b^2}}{2a^2}=\frac{ab\pm\sqrt{3}abi}{2a^2}=\frac{b\pm\sqrt{3}bi}{2a}.$$

Thus, the three solutions are $-\dfrac{b}{a}$ and $\dfrac{b\pm\sqrt{3}bi}{2a}$.

**67.** $\sqrt{x+a}+\sqrt{x-a}=\sqrt{2}\sqrt{x+6}$. Squaring both sides, we have $x+a+2\left(\sqrt{x+a}\right)\left(\sqrt{x-a}\right)+x-a=2(x+6) \Leftrightarrow 2x+2\left(\sqrt{x+a}\right)\left(\sqrt{x-a}\right)=2x+12 \Leftrightarrow$ $2\left(\sqrt{x+a}\right)\left(\sqrt{x-a}\right)=12 \Leftrightarrow \left(\sqrt{x+a}\right)\left(\sqrt{x-a}\right)=6$. Squaring both sides again we have $(x+a)(x-a)=36$ $\Leftrightarrow x^2-a^2=36 \Leftrightarrow x^2=a^2+36 \Leftrightarrow x=\pm\sqrt{a^2+36}$. Checking these answers, we see that $x=-\sqrt{a^2+36}$ is not a solution (for example, try substituting $a=8$), but $x=\sqrt{a^2+36}$ is a solution.

**68.** Let $w=x^{1/6}$. Then $x^{1/3}=w^2$ and $x^{1/2}=w^3$, and so $0=w^3+aw^2+bw+ab=w^2(w+a)+b(w+a)=(w+a)\left(w^2+b\right) \Leftrightarrow \left(\sqrt[6]{x}+a\right)\left(\sqrt[3]{x}+b\right)$. So $\sqrt[6]{x}+a=0$ $\Leftrightarrow \sqrt[6]{x}=-a$; however, since $\sqrt[6]{x}$ is positive by definition is positive and $-a$ is negative, this is impossible. Setting the other factor equal to zero, we have $\sqrt[3]{x}+b=0 \Leftrightarrow \sqrt[3]{x}=-b \Rightarrow x=-b^3$. Checking $x=-b^3$, we have $\sqrt{-b^3}+a\sqrt[3]{-b^3}+b\sqrt[6]{-b^3}+ab=b\sqrt{b}i+a(-b)+b\left(\sqrt{-b}\right)+ab=2b\sqrt{b}i\neq 0$, so this is not a solution either. Therefore, there are no solutions to this equation.

**69.** Let $x$ be the number of people originally intended to take the trip. Then originally, the cost of the trip is $\dfrac{900}{x}$. After 5 people cancel, there are now $x-5$ people, each paying $\dfrac{900}{x}+2$. Thus $900=(x-5)\left(\dfrac{900}{x}+2\right) \Leftrightarrow$ $900=900+2x-\dfrac{4500}{x}-10 \Leftrightarrow 0=2x-10-\dfrac{4500}{x} \Leftrightarrow 0=2x^2-10x-4500=(2x-100)(x+45)$. Thus either $2x-100=0$, so $x=50$, or $x+45=0$, $x=-45$. Since the number of people on the trip must be positive, originally 50 people intended to take the trip.

**70.** Let $n$ be the number of people in the group, so each person now pays $\dfrac{120{,}000}{n}$. If one person joins the group, then there would be $n+1$ members in the group, and each person would pay $\dfrac{120{,}000}{n} - 6000$. So $(n+1)\left(\dfrac{120{,}000}{n} - 6000\right) = 120{,}000 \quad\Leftrightarrow\quad \left[\left(\dfrac{n}{6000}\right)\left(\dfrac{120{,}000}{n} - 6000\right)\right](n+1) = \left(\dfrac{n}{6000}\right)120{,}000 \quad\Leftrightarrow$ $(20-n)(n+1) = 20n \quad\Leftrightarrow\quad -n^2+19n+20 = 20n \quad\Leftrightarrow\quad 0 = n^2+n-20 = (n-4)(n+5)$. Thus $n=4$ or $n=-5$. Since $n$ must be positive, there are now 4 friends in the group.

**71.** We want to solve for $t$ when $P = 500$. Letting $u = \sqrt{t}$ and substituting, we have $500 = 3t + 10\sqrt{t} + 140 \quad\Leftrightarrow$ $500 = 3u^2 + 10u + 140 \quad\Leftrightarrow\quad 0 = 3u^2 + 10u - 360 \quad\Rightarrow\quad u = \dfrac{-5 \pm \sqrt{1105}}{3}$. Since $u = \sqrt{t}$, we must have $u \geq 0$. So $\sqrt{t} = u = \dfrac{-5+\sqrt{1105}}{3} \approx 9.414 \quad\Rightarrow\quad t =\approx 88.62$. So it will take 89 days for the fish population to reach 500.

**72.** Let $d$ be the distance from the lens to the object. Then the distance from the lens to the image is $d-4$. So substituting $F = 4.8$, $x = d$, and $y = d-4$, and then solving for $x$, we have $\dfrac{1}{4.8} = \dfrac{1}{d} + \dfrac{1}{d-4}$. Now we multiply by the LCD, $4.8d(d-4)$, to get $d(d-4) = 4.8(d-4) + 4.8d \quad\Leftrightarrow\quad d^2 - 4d = 9.6d - 19.2 \quad\Leftrightarrow\quad 0 = d^2 - 13.6d + 19.2 \quad\Rightarrow$ $d = \dfrac{13.6 \pm 10.4}{2}$. So $d = 1.6$ or $d = 12$. Since $d-4$ must also be positive, the object is 12 cm from the lens.

**73.** We have that the volume is 180 ft$^3$, so $x(x-4)(x+9) = 180 \quad\Leftrightarrow\quad x^3 + 5x^2 - 36x = 180 \quad\Leftrightarrow$ $x^3 + 5x^2 - 36x - 180 = 0 \quad\Leftrightarrow\quad x^2(x+5) - 36(x+5) = 0 \quad\Leftrightarrow\quad (x+5)(x^2-36) = 0 \quad\Leftrightarrow$ $(x+5)(x+6)(x-6) = 0 \quad\Rightarrow\quad x = 6$ is the only positive solution. So the box is 2 feet by 6 feet by 15 feet.

**74.** Let $r$ be the radius of the larger sphere, in mm. Equating the volumes, we have $\frac{4}{3}\pi r^3 = \frac{4}{3}\pi\left(2^3+3^3+4^3\right) \quad\Leftrightarrow$ $r^3 = 2^3+3^3+4^4 \quad\Leftrightarrow\quad r^3 = 99 \quad\Leftrightarrow\quad r = \sqrt[3]{99} \approx 4.63$. Therefore, the radius of the larger sphere is about 4.63 mm.

**75.** Let $x$ be the length, in miles, of the abandoned road to be used. Then the length of the abandoned road not used is $40-x$, and the length of the new road is $\sqrt{10^2+(40-x)^2}$ miles, by the Pythagorean Theorem. Since the cost of the road is cost per mile $\times$ number of miles, we have $100{,}000x + 200{,}000\sqrt{x^2-80x+1700} = 6{,}800{,}000$ $\Leftrightarrow\quad 2\sqrt{x^2-80x+1700} = 68 - x$.Squaring both sides, we get $4x^2 - 320x + 6800 = 4624 - 136x + x^2 \quad\Leftrightarrow$ $3x^2 - 184x + 2176 = 0 \quad\Leftrightarrow\quad x = \frac{184\pm\sqrt{33856-26112}}{6} = \frac{184\pm 88}{6} \quad\Leftrightarrow\quad x = \frac{136}{3}$ or $x = 16$. Since $45\frac{1}{3}$ is longer than the existing road, 16 miles of the abandoned road should be used. A completely new road would have length $\sqrt{10^2+40^2}$ (let $x=0$) and would cost $\sqrt{1700} \times 200{,}000 \approx 8.3$ million dollars. So no, it would not be cheaper.

**76.** Let $x$ be the distance, in feet, that he goes on the boardwalk before veering off onto the sand. The distance along the boardwalk from where he started to the point on the boardwalk closest to the umbrella is $\sqrt{750^2 - 210^2} = 720$ ft. Thus the distance that he walks on the sand is $\sqrt{(720 - x)^2 + 210^2} = \sqrt{518{,}400 - 1440x + x^2 + 44{,}100} = \sqrt{x^2 - 1440x + 562{,}500}$.

| | Distance | Rate | Time |
|---|---|---|---|
| Along boardwalk | $x$ | 4 | $\frac{x}{4}$ |
| Across sand | $\sqrt{x^2 - 1440x + 562{,}500}$ | 2 | $\frac{\sqrt{x^2 - 1440x + 562{,}500}}{2}$ |

Since 4 minutes 45 seconds = 285 seconds, we equate the time it takes to walk along the boardwalk and across the sand to the total time to get $285 = \frac{x}{4} + \frac{\sqrt{x^2 - 1440x + 562{,}500}}{2} \Leftrightarrow 1140 - x = 2\sqrt{x^2 - 1440x + 562{,}500}$. Squaring both sides, we get $(1140 - x)^2 = 4\left(x^2 - 1440x + 562{,}500\right) \Leftrightarrow 1{,}299{,}600 - 2280x + x^2 = 4x^2 - 5760x + 2{,}250{,}000$ $\Leftrightarrow 0 = 3x^2 - 3480x + 950{,}400 = 3\left(x^2 - 1160x + 316{,}800\right) = 3(x - 720)(x - 440)$. So $x - 720 = 0$ $\Leftrightarrow x = 720$, and $x - 440 = 0 \Leftrightarrow x = 440$. Checking $x = 720$, the distance across the sand is 210 feet. So $\frac{720}{4} + \frac{210}{2} = 180 + 105 = 285$ seconds. Checking $x = 440$, the distance across the sand is $\sqrt{(720 - 440)^2 + 210^2} = 350$ feet. So $\frac{440}{4} + \frac{350}{2} = 110 + 175 = 285$ seconds. Since both solutions are less than or equal to 720 feet, we have two solutions: he walks 440 feet down the boardwalk and then heads towards his umbrella, or he walks 720 feet down the boardwalk and then heads toward his umbrella.

**77.** Let $x$ be the height of the pile in feet. Then the diameter is $3x$ and the radius is $\frac{3}{2}x$ feet. Since the volume of the cone is 1000 ft$^3$, we have $\frac{\pi}{3}\left(\frac{3x}{2}\right)^2 x = 1000 \Leftrightarrow \frac{3\pi x^3}{4} = 1000 \Leftrightarrow x^3 = \frac{4000}{3\pi} \Leftrightarrow x = \sqrt[3]{\frac{4000}{3\pi}} \approx 7.52$ feet.

**78.** Let $r$ be the radius of the tank, in feet. The volume of the spherical tank is $\frac{4}{3}\pi r^3$ and is also $750 \times 0.1337 = 100.275$. So $\frac{4}{3}\pi r^3 = 100.275 \Leftrightarrow r^3 = 23.938 \Leftrightarrow r = 2.88$ feet.

**79.** Let $x$ be the length of the hypotenuse of the triangle, in feet. Then one of the other sides has length $x - 7$ feet, and since the perimeter is 392 feet, the remaining side must have length $392 - x - (x - 7) = 399 - 2x$. From the Pythagorean Theorem, we get $(x - 7)^2 + (399 - 2x)^2 = x^2 \Leftrightarrow$ $4x^2 - 1610x + 159250 = 0$. Using the quadratic formula, we get

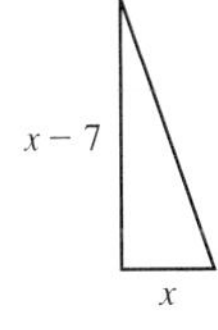


$x = \frac{1610 \pm \sqrt{1610^2 - 4(4)(159250)}}{2(4)} = \frac{1610 \pm \sqrt{44100}}{8} = \frac{1610 \pm 210}{8}$, and so $x = 227.5$ or $x = 175$. But if $x = 227.5$, then the side of length $x - 7$ combined with the hypotenuse already exceeds the perimeter of 392 feet, and so we must have $x = 175$. Thus the other sides have length $175 - 7 = 168$ and $399 - 2(175) = 49$. The lot has sides of length 49 feet, 168 feet, and 175 feet.

**80.** Let $h$ be the height of the screens in inches. Hence, the width of the smaller screen is $h + 5$ inches, and the width of the bigger screen is $1.8h$ inches. The diagonal measure of the smaller screen is $\sqrt{h^2 + (h + 5)^2}$, and the diagonal measure of the larger screen is $\sqrt{h^2 + (1.8h)^2} = \sqrt{4.24h^2} \approx 2.06h$. Thus $\sqrt{h^2 + (h + 5)^2} + 14 = 2.06h \Leftrightarrow \sqrt{h^2 + (h + 5)^2} = 2.06h - 14$. Squaring both sides yields $h^2 + h^2 + 10h + 25 = 4.24h^2 - 57.68h + 196 \Leftrightarrow 0 = 2.24h^2 - 67.68h + 171$. Applying the quadratic formula, we obtain $h = \frac{67.68 \pm \sqrt{(-67.68)^2 - 4(2.24)(171)}}{2(2.24)} = \frac{67.68 \pm \sqrt{3048.4224}}{4.48} = \frac{67.68 \pm 55.21}{4.48}$. So $h = 27.4$ or $h = 2.8$. Since high definition television monitors don't come in heights of 2.8 inches, the height of the screens is 27.4 inches.

**81.** Since the total time is 3 s, we have $3 = \dfrac{\sqrt{d}}{4} + \dfrac{d}{1090}$. Letting $w = \sqrt{d}$, we have $3 = \frac{1}{4}w + \frac{1}{1090}w^2 \Leftrightarrow \frac{1}{1090}w^2 + \frac{1}{4}w - 3 = 0 \Leftrightarrow 2w^2 + 545w - 6540 = 0 \Rightarrow w = \dfrac{-545 \pm 591.054}{4}$. Since $w \geq 0$, we have $\sqrt{d} = w \approx 11.51$, so $d = 132.56$. The well is 132.6 ft deep.

**82. (a)** *Method 1:* Let $u = \sqrt{x}$, so $u^2 = x$. Thus $x - \sqrt{x} - 2 = 0$ becomes $u^2 - u - 2 = 0 \Leftrightarrow (u-2)(u+1) = 0$. So $u = 2$ or $u = -1$. If $u = 2$, then $\sqrt{x} = 2 \Rightarrow x = 4$. If $u = -1$, then $\sqrt{x} = -1 \Rightarrow x = 1$. So the possible solutions are 4 and 1. Checking $x = 4$ we have $4 - \sqrt{4} - 2 = 4 - 2 - 2 = 0$. Checking $x = 1$ we have $1 - \sqrt{1} - 2 = 1 - 1 - 2 \neq 0$. The only solution is 4.

*Method 2:* $x - \sqrt{x} - 2 = 0 \Leftrightarrow x - 2 = \sqrt{x} \Rightarrow x^2 - 4x + 4 = x \Leftrightarrow x^2 - 5x + 4 = 0 \Leftrightarrow (x-4)(x-1) = 0$. So the possible solutions are 4 and 1. Checking will result in the same solution.

**(b)** *Method 1:* Let $u = \dfrac{1}{x-3}$, so $u^2 = \dfrac{1}{(x-3)^2}$. Thus $\dfrac{12}{(x-3)^2} + \dfrac{10}{x-3} + 1 = 0$ becomes $12u^2 + 10u + 1 = 0$. Using the quadratic formula, we have $u = \frac{-10 \pm \sqrt{10^2 - 4(12)(1)}}{2(12)} = \frac{-10 \pm \sqrt{52}}{24} = \frac{-10 \pm 2\sqrt{13}}{24} = \frac{-5 \pm \sqrt{13}}{12}$. If $u = \frac{-5-\sqrt{13}}{12}$, then $\dfrac{1}{x-3} = \dfrac{-5-\sqrt{13}}{12} \Leftrightarrow x - 3 = \frac{12}{-5-\sqrt{13}} \cdot \frac{-5+\sqrt{13}}{-5+\sqrt{13}} = \frac{12\left(-5+\sqrt{13}\right)}{12} = -5 + \sqrt{13}$. So $x = -2 + \sqrt{13}$.

If $u = \frac{-5+\sqrt{13}}{12}$, then $\dfrac{1}{x-3} = \dfrac{-5+\sqrt{13}}{12} \Leftrightarrow x - 3 = \frac{12}{-5+\sqrt{13}} \cdot \frac{-5-\sqrt{13}}{-5-\sqrt{13}} = \frac{12\left(-5-\sqrt{13}\right)}{12} = -5 - \sqrt{13}$. So $x = -2 - \sqrt{13}$.

The solutions are $-2 \pm \sqrt{13}$.

*Method 2:* Multiplying by the LCD, $(x-3)^2$, we get $(x-3)^2\left(\dfrac{12}{(x-3)^2} + \dfrac{10}{x-3} + 1\right) = 0 \cdot (x-3)^2 \Leftrightarrow 12 + 10(x-3) + (x-3)^2 = 0 \Leftrightarrow 12 + 10x - 30 + x^2 - 6x + 9 = 0 \Leftrightarrow x^2 + 4x - 9 = 0$. Using the quadratic formula, we have $u = \frac{-4 \pm \sqrt{4^2 - 4(1)(-9)}}{2} = \frac{-4 \pm \sqrt{52}}{2} = \frac{-4 \pm 2\sqrt{13}}{22} = -2 \pm \sqrt{13}$. The solutions are $-2 \pm \sqrt{13}$.

## 1.6 Inequalities

**1.** $x = -2$: $-2 - 3 \overset{?}{>} 0$. No, $-5 \not> 0$. $x = -1$: $-1 - 3 \overset{?}{>} 0$. No, $-4 \not> 0$. $x = 0$: $0 - 3 \overset{?}{>} 0$. No, $-3 \not> 0$. $x = \frac{1}{2}$: $\frac{1}{2} - 3 \overset{?}{>} 0$. No, $-\frac{5}{2} \not> 0$. $x = 1$: $1 - 3 \overset{?}{>} 0$. No, $-2 \not> 0$. $x = \sqrt{2}$: $\sqrt{2} - 3 \overset{?}{>} 0$. No, $\sqrt{2} - 3 \not> 0$. $x = 2$: $2 - 3 \overset{?}{>} 0$. No, $-1 \not> 0$. $x = 4$: $4 - 3 \overset{?}{>} 0$. Yes, $1 > 0$. Only 4 satisfies the inequality.

**2.** $x = -2$: $-2 + 1 \overset{?}{<} 2$. Yes, $-1 < 2$. $x = -1$: $-1 + 1 \overset{?}{<} 2$. Yes, $0 < 2$. $x = 0$: $0 + 1 \overset{?}{<} 2$. Yes, $1 < 2$. $x = \frac{1}{2}$: $\frac{1}{2} + 1 \overset{?}{<} 2$. Yes, $\frac{3}{2} < 2$. $x = 1$: $1 + 1 \overset{?}{<} 2$. No, $2 \not< 2$. $x = \sqrt{2}$: $\sqrt{2} + 1 \overset{?}{<} 2$. No, since $\sqrt{2} > 1$. $x = 2$: $2 + 1 \overset{?}{<} 2$. No, $3 \not< 2$. $x = 4$: $4 + 1 \overset{?}{<} 2$. No, $5 \not< 2$. The elements $-2$, $-1$, 0, and $\frac{1}{2}$ all satisfy the inequality.

**3.** $x = -2$: $3 - 2(-2) \overset{?}{\leq} \frac{1}{2}$. No, $7 \not\leq \frac{1}{2}$. $x = -1$: $3 - 2(-1) \overset{?}{\leq} \frac{1}{2}$. No, $6 \not\leq \frac{1}{2}$. $x = 0$: $3 - 2(0) \overset{?}{\leq} \frac{1}{2}$. No, $3 \not\leq \frac{1}{2}$. $x = \frac{1}{2}$: $3 - 2\left(\frac{1}{2}\right) \overset{?}{\leq} \frac{1}{2}$. No, $2 \not\leq \frac{1}{2}$. $x = 1$: $3 - 2(1) \overset{?}{\leq} \frac{1}{2}$. No, $1 \not\leq \frac{1}{2}$. $x = \sqrt{2}$: $3 - 2\left(\sqrt{2}\right) \overset{?}{\leq} \frac{1}{2}$. Yes, $3 - 2\sqrt{2} \leq \frac{1}{2}$. $x = 2$: $3 - 2(2) \overset{?}{\leq} \frac{1}{2}$. Yes, $-1 \leq \frac{1}{2}$. $x = 4$: $3 - 2(4) \overset{?}{\leq} \frac{1}{2}$. Yes, $5 \leq \frac{1}{2}$. The elements $\sqrt{2}$, 2, and 4 all satisfy the inequality.

**4.** $x=-2$: $2(-2)-1\overset{?}{\geq}-2$. No, $-5\not\geq-2$. $x=-1$: $2(-1)-1\overset{?}{\geq}\frac{1}{2}$. No, $6-1\not\geq\frac{1}{2}$.

$x=0$: $2(0)-1\overset{?}{\geq}0$. No, $-1\not\geq 0$. $x=\frac{1}{2}$: $2\left(\frac{1}{2}\right)-1\overset{?}{\geq}\frac{1}{2}$. No, $0\not\geq\frac{1}{2}$.

$x=1$: $2(1)-1\overset{?}{\geq}1$. Yes since $1\geq 1$. $x=\sqrt{2}$: $2\left(\sqrt{2}\right)-1\overset{?}{\geq}\sqrt{2}$. Yes, since $2\sqrt{2}-1\approx 1.8$ and $\sqrt{2}\approx 1.4$.

$x=2$: $2(2)-1\overset{?}{\geq}2$. Yes, $3\geq 2$. $x=4$: $2(4)-1\overset{?}{\geq}4$. Yes, $7\geq 4$.

The elements 1, $\sqrt{2}$, 2, and 4 all satisfy the inequality.

**5.** $x=-2$: $1\overset{?}{<}2(-2)-4\overset{?}{\leq}7$. No, $2(-2)-4=-8$ and $1\not<-8$.

$x=-1$: $1\overset{?}{<}2(-1)-4\overset{?}{\leq}7$. No, $2(-1)-4=-6$ and $1\not<-6$.

$x=0$: $1\overset{?}{<}2(0)-4\overset{?}{\leq}7$. No, $2(0)-4=-4$ and $1\not<-4$.

$x=\frac{1}{2}$: $1\overset{?}{<}2\left(\frac{1}{2}\right)-4\overset{?}{\leq}7$. No, $2\left(\frac{1}{2}\right)-4=-3$ and $1\not<-3$.

$x=1$: $1\overset{?}{<}2(1)-4\overset{?}{\leq}7$. No, $2(1)-4=-2$ and $1\not<-2$.

$x=\sqrt{2}$: $1\overset{?}{<}2\left(\sqrt{2}\right)-4\overset{?}{\leq}7$. No, $2\sqrt{2}-4<0$ and $1\not<0$.

$x=2$: $1\overset{?}{<}2(2)-4\overset{?}{\leq}7$. No, $2(1)-4=-2$ and $1\not<-2$.

$x=4$: $1\overset{?}{<}2(4)-4\overset{?}{\leq}7$. Yes, $2(4)-4=4$ and $1<4\leq 7$. Only 4 satisfies the inequality.

**6.** $x=-2$: $2\overset{?}{\leq}3-(-2)\overset{?}{<}2$. No, $3-(-2)=5\not<2$. $x=-1$: $-2\overset{?}{\leq}3-(-1)\overset{?}{<}2$. No, $3-(-1)=4\not<2$.

$x=0$: $-2\overset{?}{\leq}3-0\overset{?}{<}2$. No, $3-(0)=3\not<2$. $x=\frac{1}{2}$: $-2\overset{?}{\leq}3-\frac{1}{2}\overset{?}{<}2$. No, $3-\left(\frac{1}{2}\right)=\frac{5}{2}\not<2$.

$x=1$: $-2\overset{?}{\leq}3-1\overset{?}{<}2$ No, $3-(1)=2\not<2$. $x=\sqrt{2}$: $-2\overset{?}{\leq}3-\sqrt{2}\overset{?}{<}2$. Yes, $3-\sqrt{2}\approx 1.5$ and $-2\leq 1.5<2$.

$x=2$: $-2\overset{?}{\leq}3-2\overset{?}{<}2$. Yes, $3-2=1$ and $-2\leq 1<2$.

$x=4$: $-2\overset{?}{\leq}3-4\overset{?}{<}2$. Yes, $3-4=-1$ and $-2\leq-1<2$. The elements $\sqrt{2}$, 2, and 4 all satisfy the inequality.

**7.** $x=-2$: $\frac{1}{(-2)}\overset{?}{\leq}\frac{1}{2}$. Yes, $-\frac{1}{2}\leq\frac{1}{2}$. $x=-1$: $\frac{1}{(-1)}\overset{?}{\leq}\frac{1}{2}$. Yes, $-1\leq\frac{1}{2}$. $x=0$: $\frac{1}{0}\overset{?}{\leq}\frac{1}{2}$. No, $\frac{1}{0}$ is not defined.

$x=\frac{1}{2}$: $\frac{1}{1/2}\overset{?}{\leq}\frac{1}{2}$. No, $2\not\leq\frac{1}{2}$. $x=1$: $\frac{1}{1}\overset{?}{\leq}\frac{1}{2}$. No, $1\not\leq\frac{1}{2}$. $x=\sqrt{2}$: $\frac{1}{\sqrt{2}}\overset{?}{\leq}\frac{1}{2}$. No, $2\not\leq\sqrt{2}$.

$x=2$: $\frac{1}{2}\overset{?}{\leq}\frac{1}{2}$. Yes, $\frac{1}{2}\leq\frac{1}{2}$. $x=4$: $\frac{1}{4}\overset{?}{\leq}\frac{1}{2}$. Yes. The elements $-2$, $-1$, 2, and 4 all satisfy the inequality.

**8.** $x=-2$: $(-2)^2+2\overset{?}{<}4$. No, $(-2)^2+2=6$ and $6\not<4$. $x=-1$: $(-1)^2+2\overset{?}{<}4$. Yes, $(-1)^2+2=3$ and $3<4$.

$x=0$: $0^2+2\overset{?}{<}4$. Yes, $0^2+2=2$ and $2<4$. $x=\frac{1}{2}$: $\left(\frac{1}{2}\right)^2+2\overset{?}{<}4$. Yes, $\left(\frac{1}{2}\right)^2+2=\frac{9}{4}$ and $\frac{9}{4}<4$.

$x=1$: $1^2+2\overset{?}{<}4$. Yes, $1^2+2=3$ and $3<4$. $x=\sqrt{2}$: $\left(\sqrt{2}\right)^2+2\overset{?}{<}4$. No, $\left(\sqrt{2}\right)^2+2=4$ and $4\not<4$.

$x=2$: $2^2+2\overset{?}{<}4$. No, $2^2+2=6$ and $6\not<4$. $x=4$: $4^2+2\overset{?}{<}4$. No, $4^2+2=18$ and $18\not<4$.

The elements $-1$, 0, $\frac{1}{2}$, and 1 all satisfy the inequality.

**9.** $2x\leq 7$ $\Leftrightarrow$ $x\leq\frac{7}{2}$. Interval: $\left(-\infty,\frac{7}{2}\right]$

Graph: $\frac{7}{2}$

**10.** $-4x\geq 10$ $\Leftrightarrow$ $x\leq-\frac{5}{2}$. Interval: $\left(-\infty,-\frac{5}{2}\right]$

Graph: $-\frac{5}{2}$

**11.** $2x-5>3$ $\Leftrightarrow$ $2x>8$ $\Leftrightarrow$ $x>4$

Interval: $(4,\infty)$

Graph: 4

**12.** $3x+11<5$ $\Leftrightarrow$ $3x<-6$ $\Leftrightarrow$ $x<-2$

Interval: $(-\infty,-2)$

Graph:

**13.** $7 - x \geq 5 \quad\Leftrightarrow\quad -x \geq -2 \quad\Leftrightarrow\quad x \leq 2$
Interval: $(-\infty, 2]$

Graph: [number line, closed dot at 2, shaded left]

**14.** $5 - 3x \leq -16 \quad\Leftrightarrow\quad -3x \leq -21 \quad\Leftrightarrow\quad x \geq 7$
Interval: $[7, \infty)$

Graph: [number line, closed dot at 7, shaded right]

**15.** $2x + 1 < 0 \quad\Leftrightarrow\quad 2x < -1 \quad\Leftrightarrow\quad x < -\frac{1}{2}$
Interval: $\left(-\infty, -\frac{1}{2}\right)$

Graph: [number line, open dot at $-\frac{1}{2}$, shaded left]

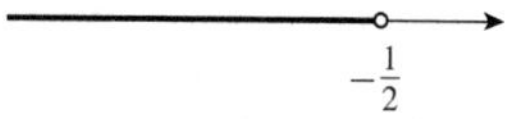

**16.** $0 < 5 - 2x \quad\Leftrightarrow\quad 2x < 5 \quad\Leftrightarrow\quad x < \frac{5}{2}$
Interval: $\left(-\infty, \frac{5}{2}\right)$

Graph: [number line, open dot at $\frac{5}{2}$, shaded left]

**17.** $3x + 11 \leq 6x + 8 \quad\Leftrightarrow\quad 3 \leq 3x \quad\Leftrightarrow\quad 1 \leq x$
Interval: $[1, \infty)$

Graph: [number line, closed dot at 1, shaded right]

**18.** $6 - x \geq 2x + 9 \quad\Leftrightarrow\quad -3 \geq 3x \quad\Leftrightarrow\quad -1 \geq x$
Interval: $(-\infty, -1]$

Graph: [number line, closed dot at −1, shaded left]

**19.** $\frac{1}{2}x - \frac{2}{3} > 2 \quad\Leftrightarrow\quad \frac{1}{2}x > \frac{8}{3} \quad\Leftrightarrow\quad x > \frac{16}{3}$
Interval: $\left(\frac{16}{3}, \infty\right)$

Graph: [number line, open dot at $\frac{16}{3}$, shaded right]

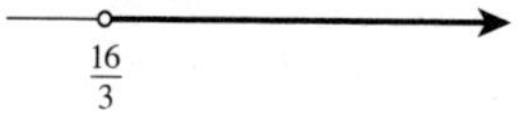

**20.** $\frac{2}{5}x + 1 < \frac{1}{5} - 2x \quad\Leftrightarrow\quad \frac{12}{5}x < -\frac{4}{5} \quad\Leftrightarrow\quad x < -\frac{1}{3}$
Interval: $\left(-\infty, -\frac{1}{3}\right)$

Graph: [number line, open dot at $-\frac{1}{3}$, shaded left]

**21.** $\frac{1}{3}x + 2 < \frac{1}{6}x - 1 \quad\Leftrightarrow\quad \frac{1}{6}x < -3 \quad\Leftrightarrow\quad x < -18$
Interval: $(-\infty, -18)$

Graph: [number line, open dot at −18, shaded left]

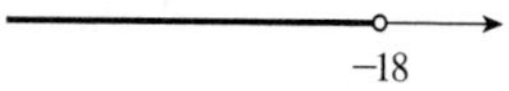

**22.** $\frac{2}{3} - \frac{1}{2}x \geq \frac{1}{6} + x$ (multiply both sides by 6) $\Leftrightarrow$
$4 - 3x \geq 1 + 6x \quad\Leftrightarrow\quad 3 \geq 9x \quad\Leftrightarrow\quad \frac{1}{3} \geq x$
Interval: $\left(-\infty, \frac{1}{3}\right]$

Graph: [number line, closed dot at $\frac{1}{3}$, shaded left]

**23.** $4 - 3x \leq -(1 + 8x) \quad\Leftrightarrow\quad 4 - 3x \leq -1 - 8x \quad\Leftrightarrow$
$5x \leq -5 \quad\Leftrightarrow\quad x \leq -1$
Interval: $(-\infty, -1]$

Graph: [number line, closed dot at −1, shaded left]

**24.** $2(7x - 3) \leq 12x + 16 \quad\Leftrightarrow\quad 14x - 6 \leq 12x + 16$
$\Leftrightarrow\quad 2x \leq 22 \quad\Leftrightarrow\quad x \leq 11$
Interval: $(-\infty, 11]$

Graph: [number line, closed dot at 11, shaded left]

**25.** $2 \leq x + 5 < 4 \quad\Leftrightarrow\quad -3 \leq x < -1$
Interval: $[-3, -1)$

Graph: [number line, closed dot at −3, open dot at −1, shaded between]

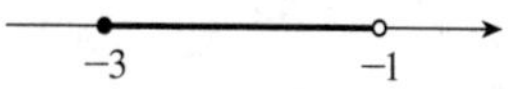

**26.** $5 \leq 3x - 4 \leq 14 \quad\Leftrightarrow\quad 9 \leq 3x \leq 18 \quad\Leftrightarrow$
$3 \leq x \leq 6$
Interval: $[3, 6]$

Graph: [number line, closed dots at 3 and 6, shaded between]

**27.** $-1 < 2x - 5 < 7 \quad\Leftrightarrow\quad 4 < 2x < 12 \quad\Leftrightarrow$
$2 < x < 6$
Interval: $(2, 6)$

Graph: [number line, open dots at 2 and 6, shaded between]

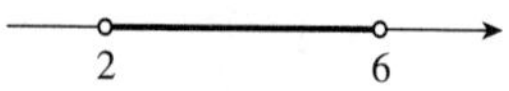

**28.** $1 < 3x + 4 \leq 16 \quad\Leftrightarrow\quad -3 < 3x \leq 12 \quad\Leftrightarrow$
$-1 < x \leq 4$
Interval: $(-1, 4]$

Graph: [number line, open dot at −1, closed dot at 4, shaded between]

**29.** $-2 < 8 - 2x \le -1 \quad \Leftrightarrow \quad -10 < -2x \le -9 \quad \Leftrightarrow$ $5 > x \ge \frac{9}{2} \quad \Leftrightarrow \quad \frac{9}{2} \le x < 5$

Interval: $\left[\frac{9}{2}, 5\right)$

Graph: 

**30.** $-3 \le 3x + 7 \le \frac{1}{2} \quad \Leftrightarrow \quad -10 \le 3x \le -\frac{13}{2} \quad \Leftrightarrow$ $-\frac{10}{3} \le x \le -\frac{13}{6}$

Interval: $\left[-\frac{10}{3}, -\frac{13}{6}\right]$

Graph: 

**31.** $\frac{2}{3} \ge \frac{2x - 3}{12} > \frac{1}{6} \quad \Leftrightarrow \quad 8 \ge 2x - 3 > 2$ (multiply each expression by 12) $\Leftrightarrow \quad 11 \ge 2x > 5 \quad \Leftrightarrow \quad \frac{11}{2} \ge x > \frac{5}{2}$.

Interval: $\left(\frac{5}{2}, \frac{11}{2}\right]$. Graph: 

**32.** $-\frac{1}{2} < \frac{4 - 3x}{5} \le \frac{1}{4} \quad \Leftrightarrow$ (multiply each expression by 20) $-10 < 4(4 - 3x) \le 5 \quad \Leftrightarrow \quad -10 < 16 - 12x \le 5 \quad \Leftrightarrow$

$-26 < -12x \le -11 \quad \Leftrightarrow \quad \frac{13}{6} > x \ge \frac{11}{12} \quad \Leftrightarrow \quad \frac{11}{12} \le x < \frac{13}{6}$. Interval: $\left[\frac{11}{12}, \frac{13}{6}\right)$. Graph: 

**33.** $(x + 2)(x - 3) < 0$. The expression on the left of the inequality changes sign where $x = -2$ and where $x = 3$. Thus we must check the intervals in the following table.

| Interval | $(-\infty, -2)$ | $(-2, 3)$ | $(3, \infty)$ |
|---|---|---|---|
| Sign of $x + 2$ | $-$ | $+$ | $+$ |
| Sign of $x - 3$ | $-$ | $-$ | $+$ |
| Sign of $(x + 2)(x - 3)$ | $+$ | $-$ | $+$ |

From the table, the solution set is $\{x \mid -2 < x < 3\}$. Interval: $(-2, 3)$.

Graph: 

**34.** $(x - 5)(x + 4) \ge 0$. The expression on the left of the inequality changes sign when $x = 5$ and $x = -4$. Thus we must check the intervals in the following table.

| Interval | $(-\infty, -4)$ | $(-4, 5)$ | $(5, \infty)$ |
|---|---|---|---|
| Sign of $x - 5$ | $-$ | $-$ | $+$ |
| Sign of $x + 4$ | $-$ | $+$ | $+$ |
| Sign of $(x - 5)(x + 4)$ | $+$ | $-$ | $+$ |

From the table, the solution set is $\left\{x \mid x \le -\frac{1}{3} \text{ or } 1 \le x\right\}$.

Interval: $(-\infty, -4] \cup [5, \infty)$.

Graph: 

**35.** $x(2x + 7) \ge 0$. The expression on the left of the inequality changes sign where $x = 0$ and where $x = -\frac{7}{2}$. Thus we must check the intervals in the following table.

| Interval | $\left(-\infty, -\frac{7}{2}\right)$ | $\left(-\frac{7}{2}, 0\right)$ | $(0, \infty)$ |
|---|---|---|---|
| Sign of $x$ | $-$ | $-$ | $+$ |
| Sign of $2x + 7$ | $-$ | $+$ | $+$ |
| Sign of $x(2x + 7)$ | $+$ | $-$ | $+$ |

From the table, the solution set is $\left\{x \mid x \le -\frac{7}{2} \text{ or } 0 \le x\right\}$.

Interval: $\left(-\infty, -\frac{7}{2}\right] \cup [0, \infty)$.

Graph: 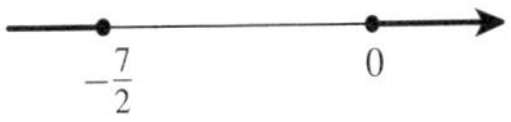

**36.** $x(2-3x) \le 0$. The expression on the left of the inequality changes sign when $x = 0$ and $x = \frac{2}{3}$. Thus we must check the intervals in the following table.

| Interval | $(-\infty, 0)$ | $\left(0, \frac{2}{3}\right)$ | $\left(\frac{2}{3}, \infty\right)$ |
|---|---|---|---|
| Sign of $x$ | $-$ | $+$ | $+$ |
| Sign of $2-3x$ | $+$ | $+$ | $-$ |
| Sign of $x(2-3x)$ | $-$ | $+$ | $-$ |

From the table, the solution set is $\left\{x \mid x \le 0 \text{ or } \frac{2}{3} \le x\right\}$. Interval: $(-\infty, 0] \cup \left[\frac{2}{3}, \infty\right)$.

Graph: 0, $\frac{2}{3}$

**37.** $x^2 - 3x - 18 \le 0 \quad \Leftrightarrow \quad (x+3)(x-6) \le 0$. The expression on the left of the inequality changes sign where $x = 6$ and where $x = -3$. Thus we must check the intervals in the following table.

| Interval | $(-\infty, -3)$ | $(-3, 6)$ | $(6, \infty)$ |
|---|---|---|---|
| Sign of $x+3$ | $-$ | $+$ | $+$ |
| Sign of $x-6$ | $-$ | $-$ | $+$ |
| Sign of $(x+3)(x-6)$ | $+$ | $-$ | $+$ |

From the table, the solution set is $\{x \mid -3 \le x \le 6\}$. Interval: $[-3, 6]$.

Graph: $-3$, 6

**38.** $x^2 + 5x + 6 > 0 \quad \Leftrightarrow \quad (x+3)(x+2) > 0$. The expression on the left of the inequality changes sign when $x = -3$ and $x = -2$. Thus we must check the intervals in the following table.

| Interval | $(-\infty, -3)$ | $(-3, -2)$ | $(-2, \infty)$ |
|---|---|---|---|
| Sign of $x+3$ | $-$ | $+$ | $+$ |
| Sign of $x+2$ | $-$ | $-$ | $+$ |
| Sign of $(x+3)(x+2)$ | $+$ | $-$ | $+$ |

From the table, the solution set is $\{x \mid x < -3 \text{ or } -2 < x\}$. Interval: $(-\infty, -3) \cup (-2, \infty)$.

Graph: $-3$, $-2$

**39.** $2x^2 + x \ge 1 \quad \Leftrightarrow \quad 2x^2 + x - 1 \ge 0 \quad \Leftrightarrow \quad (x+1)(2x-1) \ge 0$. The expression on the left of the inequality changes sign where $x = -1$ and where $x = \frac{1}{2}$. Thus we must check the intervals in the following table.

| Interval | $(-\infty, -1)$ | $\left(-1, \frac{1}{2}\right)$ | $\left(\frac{1}{2}, \infty\right)$ |
|---|---|---|---|
| Sign of $x+1$ | $-$ | $+$ | $+$ |
| Sign of $2x-1$ | $-$ | $-$ | $+$ |
| Sign of $(x+1)(2x-1)$ | $+$ | $-$ | $+$ |

From the table, the solution set is $\left\{x \mid x \le -1 \text{ or } \frac{1}{2} \le x\right\}$. Interval: $(-\infty, -1] \cup \left[\frac{1}{2}, \infty\right)$.

Graph: $-1$, $\frac{1}{2}$

**40.** $x^2 < x + 2 \quad \Leftrightarrow \quad x^2 - x - 2 < 0 \quad \Leftrightarrow \quad (x+1)(x-2) < 0$. The expression on the left of the inequality changes sign when $x = -1$ and $x = 2$. Thus we must check the intervals in the following table.

| Interval | $(-\infty, -1)$ | $(-1, 2)$ | $(2, \infty)$ |
|---|---|---|---|
| Sign of $x+1$ | $-$ | $+$ | $+$ |
| Sign of $x-2$ | $-$ | $-$ | $+$ |
| Sign of $(x+1)(x-2)$ | $+$ | $-$ | $+$ |

From the table, the solution set is $\{x \mid -1 < x < 2\}$. Interval: $(-1, 2)$.

Graph: $-1$, 2

**41.** $3x^2 - 3x < 2x^2 + 4 \quad \Leftrightarrow \quad x^2 - 3x - 4 < 0 \quad \Leftrightarrow \quad (x+1)(x-4) < 0$. The expression on the left of the inequality changes sign where $x = -1$ and where $x = 4$. Thus we must check the intervals in the following table.

| Interval | $(-\infty, -1)$ | $(-1, 4)$ | $(4, \infty)$ |
|---|---|---|---|
| Sign of $x + 1$ | $-$ | $+$ | $+$ |
| Sign of $x - 4$ | $-$ | $-$ | $+$ |
| Sign of $(x+1)(x-4)$ | $+$ | $-$ | $+$ |

From the table, the solution set is $\{x \mid -1 < x < 4\}$. Interval: $(-1, 4)$.

Graph: −1 4

**42.** $5x^2 + 3x \geq 3x^2 + 2 \quad \Leftrightarrow \quad 2x^2 + 3x - 2 \geq 0 \quad \Leftrightarrow \quad (2x-1)(x+2) \geq 0$. The expression on the left of the inequality changes sign when $x = \frac{1}{2}$ and $x = -2$. Thus we must check the intervals in the following table.

| Interval | $(-\infty, -2)$ | $\left(-2, \frac{1}{2}\right)$ | $\left(\frac{1}{2}, \infty\right)$ |
|---|---|---|---|
| Sign of $2x - 1$ | $-$ | $-$ | $+$ |
| Sign of $x + 2$ | $-$ | $+$ | $+$ |
| Sign of $(2x-1)(x+2)$ | $+$ | $-$ | $+$ |

From the table, the solution set is $\{x \mid x \leq -2 \text{ or } \frac{1}{2} \leq x\}$. Interval: $(-\infty, -2] \cup \left[\frac{1}{2}, \infty\right)$.

Graph: −2 $\frac{1}{2}$

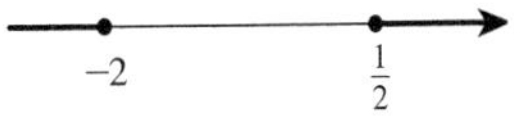

**43.** $x^2 > 3(x+6) \quad \Leftrightarrow \quad x^2 - 3x - 18 > 0 \quad \Leftrightarrow \quad (x+3)(x-6) > 0$. The expression on the left of the inequality changes sign where $x = 6$ and where $x = -3$. Thus we must check the intervals in the following table.

| Interval | $(-\infty, -3)$ | $(-3, 6)$ | $(6, \infty)$ |
|---|---|---|---|
| Sign of $x + 3$ | $-$ | $+$ | $+$ |
| Sign of $x - 6$ | $-$ | $-$ | $+$ |
| Sign of $(x+3)(x-6)$ | $+$ | $-$ | $+$ |

From the table, the solution set is $\{x \mid x < -3 \text{ or } 6 < x\}$. Interval: $(-\infty, -3) \cup (6, \infty)$.

Graph: −3 6

**44.** $x^2 + 2x > 3 \quad \Leftrightarrow \quad x^2 + 2x - 3 > 0 \quad \Leftrightarrow \quad (x+3)(x-1) > 0$. The expression on the left of the inequality changes sign when $x = -3$ and $x = 1$. Thus we must check the intervals in the following table.

| Interval | $(-\infty, -3)$ | $(-3, 1)$ | $(1, \infty)$ |
|---|---|---|---|
| Sign of $x + 3$ | $-$ | $+$ | $+$ |
| Sign of $x - 1$ | $-$ | $-$ | $+$ |
| Sign of $(x+3)(x-1)$ | $+$ | $-$ | $+$ |

From the table, the solution set is $\{x \mid x < -3 \text{ or } 1 < x\}$. Interval: $(-\infty, -3) \cup (1, \infty)$.

Graph: −3 1

**45.** $x^2 < 4 \quad \Leftrightarrow \quad x^2 - 4 < 0 \quad \Leftrightarrow \quad (x+2)(x-2) < 0$. The expression on the left of the inequality changes sign where $x = -2$ and where $x = 2$. Thus we must check the intervals in the following table.

| Interval | $(-\infty, -2)$ | $(-2, 2)$ | $(2, \infty)$ |
|---|---|---|---|
| Sign of $x + 2$ | $-$ | $+$ | $+$ |
| Sign of $x - 2$ | $-$ | $-$ | $+$ |
| Sign of $(x+2)(x-2)$ | $+$ | $-$ | $+$ |

From the table, the solution set is $\{x \mid -2 < x < 2\}$. Interval: $(-2, 2)$.

Graph: −2 2

**46.** $x^2 \geq 9 \quad \Leftrightarrow \quad x^2 - 9 \geq 0 \quad \Leftrightarrow \quad (x+3)(x-3) \geq 0$. The expression on the left of the inequality changes sign when $x = -3$ and $x = 3$. Thus we must check the intervals in the following table.

| Interval | $(-\infty, -3)$ | $(-3, 3)$ | $(3, \infty)$ |
|---|---|---|---|
| Sign of $x+3$ | $-$ | $+$ | $+$ |
| Sign of $x-3$ | $-$ | $-$ | $+$ |
| Sign of $(x+3)(x-3)$ | $+$ | $-$ | $+$ |

From the table, the solution set is $\{x \mid x \leq -3 \text{ or } 3 \leq x\}$.
Interval: $(-\infty, -3] \cup [3, \infty)$.

Graph: −3 3

**47.** $-2x^2 \leq 4 \quad \Leftrightarrow \quad -2x^2 - 4 \leq 0 \quad \Leftrightarrow \quad -2(x^2+1) \leq 0$. Since $x^2 + 1 > 0$, $-2(x^2+1) \leq 0$ for all $x$.

Interval: $(-\infty, \infty)$. Graph:

**48.** $(x+2)(x-1)(x-3) \leq 0$. The expression on the left of the inequality changes sign when $x = -2$, $x = 1$, and $x = 3$. Thus we must check the intervals in the following table.

| Interval | $(-\infty, -2)$ | $(-2, 1)$ | $(1, 3)$ | $(3, \infty)$ |
|---|---|---|---|---|
| Sign of $x+2$ | $-$ | $+$ | $+$ | $+$ |
| Sign of $x-1$ | $-$ | $-$ | $+$ | $+$ |
| Sign of $x-3$ | $-$ | $-$ | $-$ | $+$ |
| Sign of $(x+2)(x-1)(x-3)$ | $-$ | $+$ | $-$ | $+$ |

From the table, the solution set is $\{x \mid x \leq -2 \text{ or } 1 \leq x \leq 3\}$. Interval: $(-\infty, -2] \cup [1, 3]$.

Graph: −2 1 3

**49.** $x^3 - 4x > 0 \quad \Leftrightarrow \quad x(x^2 - 4) > 0 \quad \Leftrightarrow \quad x(x+2)(x-2) > 0$. The expression on the left of the inequality changes sign where $x = 0$, $x = -2$ and where $x = 4$. Thus we must check the intervals in the following table.

| Interval | $(-\infty, -2)$ | $(-2, 0)$ | $(0, 2)$ | $(2, \infty)$ |
|---|---|---|---|---|
| Sign of $x$ | $-$ | $-$ | $+$ | $+$ |
| Sign of $x+2$ | $-$ | $+$ | $+$ | $+$ |
| Sign of $x-2$ | $-$ | $-$ | $-$ | $+$ |
| Sign of $x(x+2)(x-2)$ | $-$ | $+$ | $-$ | $+$ |

From the table, the solution set is $\{x \mid -2 < x < 0 \text{ or } x > 2\}$. Interval: $(-2, 0) \cup (2, \infty)$.

Graph: −2 0 2

**50.** $16x \leq x^3 \quad \Leftrightarrow \quad 0 \leq x^3 - 16x = x(x^2 - 16) = x(x-4)(x+4)$. The expression on the left of the inequality changes sign when $x = -4$, $x = 0$, and $x = 4$. Thus we must check the intervals in the following table.

| Interval | $(-\infty, -4)$ | $(-4, 0)$ | $(0, 4)$ | $(4, \infty)$ |
|---|---|---|---|---|
| Sign of $x+4$ | $-$ | $+$ | $+$ | $+$ |
| Sign of $x$ | $-$ | $-$ | $+$ | $+$ |
| Sign of $x-4$ | $-$ | $-$ | $-$ | $+$ |
| Sign of $x(x+4)(x-4)$ | $-$ | $+$ | $-$ | $+$ |

From the table, the solution set is $\{x \mid -4 \leq x \leq 0 \text{ or } 4 \leq x\}$. Interval: $[-4, 0] \cup [4, \infty)$. Graph: −4 0 4

**51.** $\dfrac{x-3}{x+1} \geq 0$. The expression on the left of the inequality changes sign where $x = -1$ and where $x = 3$. Thus we must check the intervals in the following table.

| Interval | $(-\infty, -1)$ | $(-1, 3)$ | $(3, \infty)$ |
|---|---|---|---|
| Sign of $x+1$ | $-$ | $+$ | $+$ |
| Sign of $x-3$ | $-$ | $-$ | $+$ |
| Sign of $\dfrac{x-3}{x+1}$ | $+$ | $-$ | $+$ |

From the table, the solution set is $\{x \mid x < -1 \text{ or } x \leq 3\}$. Since the denominator cannot equal 0 we must have $x \neq -1$. Interval: $(-\infty, -1) \cup [3, \infty)$.

Graph: 

**52.** $\dfrac{2x+6}{x-2} < 0$. The expression on the left of the inequality changes sign when $x = -3$ and $x = 2$. Thus we must check the intervals in the following table.

| Interval | $(-\infty, -3)$ | $(-3, 2)$ | $(2, \infty)$ |
|---|---|---|---|
| Sign of $2x+6$ | $-$ | $+$ | $+$ |
| Sign of $x-2$ | $-$ | $-$ | $+$ |
| Sign of $\dfrac{2x+6}{x-2}$ | $+$ | $-$ | $+$ |

From the table, the solution set is $\{x \mid -3 < x < 2\}$. Interval: $(-3, 2)$.

Graph: 

**53.** $\dfrac{4x}{2x+3} > 2 \quad\Leftrightarrow\quad \dfrac{4x}{2x+3} - 2 > 0 \quad\Leftrightarrow\quad \dfrac{4x}{2x+3} - \dfrac{2(2x+3)}{2x+3} > 0 \quad\Leftrightarrow\quad \dfrac{-6}{2x+3} > 0$. The expression on the left of the inequality changes sign where $x = -\frac{3}{2}$. Thus we must check the intervals in the following table.

| Interval | $\left(-\infty, -\frac{3}{2}\right)$ | $\left(-\frac{3}{2}, \infty\right)$ |
|---|---|---|
| Sign of $-6$ | $-$ | $-$ |
| Sign of $2x+3$ | $-$ | $+$ |
| Sign of $\dfrac{-6}{2x+3}$ | $+$ | $-$ |

From the table, the solution set is $\left\{x \mid x < -\frac{3}{2}\right\}$. Interval: $\left(-\infty, -\frac{3}{2}\right)$.

Graph: 

**54.** $-2 < \dfrac{x+1}{x-3} \quad\Leftrightarrow\quad 0 < \dfrac{x+1}{x-3} + 2 \quad\Leftrightarrow\quad 0 < \dfrac{x+1}{x-3} + \dfrac{2(x-3)}{x-3} \quad\Leftrightarrow\quad 0 < \dfrac{3x-5}{x-3}$. The expression on the left of the inequality changes sign when $x = \frac{5}{3}$ and $x = 3$. Thus we must check the intervals in the following table.

| Interval | $\left(-\infty, \frac{5}{3}\right)$ | $\left(\frac{5}{3}, 3\right)$ | $(3, \infty)$ |
|---|---|---|---|
| Sign of $3x-5$ | $-$ | $+$ | $+$ |
| Sign of $x-3$ | $-$ | $-$ | $+$ |
| Sign of $\dfrac{3x-5}{x-3}$ | $+$ | $-$ | $+$ |

From the table, the solution set is $\left\{x \mid x < \frac{5}{3} \text{ or } 3 < x < \infty\right\}$. Interval: $\left(-\infty, \frac{5}{3}\right) \cup (3, \infty)$.

Graph: 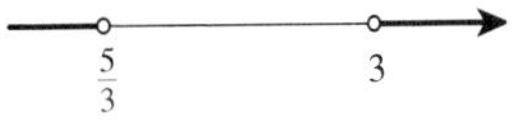


**55.** $\dfrac{2x+1}{x-5} \leq 3 \quad\Leftrightarrow\quad \dfrac{2x+1}{x-5} - 3 \leq 0 \quad\Leftrightarrow\quad \dfrac{2x+1}{x-5} - \dfrac{3(x-5)}{x-5} \leq 0 \quad\Leftrightarrow\quad \dfrac{-x+16}{x-5} \leq 0$. The expression on the left of the inequality changes sign where $x = 16$ and where $x = 5$. Thus we must check the intervals in the following table.

| Interval | $(-\infty, 5)$ | $(5, 16)$ | $(16, \infty)$ |
|---|---|---|---|
| Sign of $-x+16$ | $+$ | $+$ | $-$ |
| Sign of $x-5$ | $-$ | $+$ | $+$ |
| Sign of $\dfrac{-x+16}{x-5}$ | $-$ | $+$ | $-$ |

From the table, the solution set is $\{x \mid x < 5 \text{ or } x \geq 16\}$. Since the denominator cannot equal 0, we must have $x \neq 5$. Interval: $(-\infty, 5) \cup [16, \infty)$.

Graph: 

**56.** $\dfrac{3+x}{3-x} \geq 1 \quad \Leftrightarrow \quad \dfrac{3+x}{3-x} - 1 \geq 0 \quad \Leftrightarrow \quad \dfrac{3+x}{3-x} - \dfrac{3-x}{3-x} \geq 0 \quad \Leftrightarrow \quad \dfrac{2x}{3-x} \geq 0$. The expression on the left of the inequality changes sign when $x = 0$ and $x = 3$. Thus we must check the intervals in the following table.

| Interval | $(-\infty, 0)$ | $(0, 3)$ | $(3, \infty)$ |
|---|---|---|---|
| Sign of $3 - x$ | $+$ | $+$ | $-$ |
| Sign of $2x$ | $-$ | $+$ | $+$ |
| Sign of $\dfrac{2x}{3-x}$ | $-$ | $+$ | $-$ |

Since the denominator cannot equal 0, we must have $x \neq 3$. The solution set is $\{x \mid 0 \leq x < 3\}$. Interval: $[0, 3)$.

Graph: 0 3

**57.** $\dfrac{4}{x} < x \quad \Leftrightarrow \quad \dfrac{4}{x} - x < 0 \quad \Leftrightarrow \quad \dfrac{4}{x} - \dfrac{x \cdot x}{x} < 0 \quad \Leftrightarrow \quad \dfrac{4 - x^2}{x} < 0 \quad \Leftrightarrow \quad \dfrac{(2-x)(2+x)}{x} < 0$. The expression on the left of the inequality changes sign where $x = 0$, where $x = -2$, and where $x = 2$. Thus we must check the intervals in the following table.

| Interval | $(-\infty, -2)$ | $(-2, 0)$ | $(0, 2)$ | $(2, \infty)$ |
|---|---|---|---|---|
| Sign of $2 + x$ | $-$ | $+$ | $+$ | $+$ |
| Sign of $x$ | $-$ | $-$ | $+$ | $+$ |
| Sign of $2 - x$ | $+$ | $+$ | $+$ | $-$ |
| Sign of $\dfrac{(2-x)(2+x)}{x}$ | $+$ | $-$ | $+$ | $-$ |

From the table, the solution set is $\{x \mid -2 < x < 0 \text{ or } 2 < x\}$. Interval: $(-2, 0) \cup (2, \infty)$.

Graph: −2 0 2

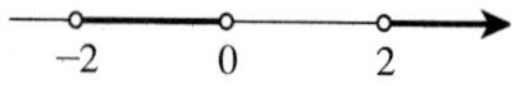

**58.** $\dfrac{x}{x+1} > 3x \quad \Leftrightarrow \quad \dfrac{x}{x+1} - 3x > 0 \quad \Leftrightarrow \quad \dfrac{x}{x+1} - \dfrac{3x(x+1)}{x+1} > 0 \quad \Leftrightarrow \quad \dfrac{-2x - 3x^2}{x+1} > 0 \quad \Leftrightarrow$ $\dfrac{-x(2+3x)}{x+1} > 0$. The expression on the left of the inequality changes sign when$x = 0$, $x = -\frac{2}{3}$, and $x = -1$. Thus we must check the intervals in the following table.

| Interval | $(-\infty, -1)$ | $(-1, -\frac{2}{3})$ | $(-\frac{2}{3}, 0)$ | $(0, \infty)$ |
|---|---|---|---|---|
| Sign of $-x$ | $+$ | $+$ | $+$ | $-$ |
| Sign of $2 + 3x$ | $-$ | $-$ | $+$ | $+$ |
| Sign of $x + 1$ | $-$ | $+$ | $+$ | $+$ |
| Sign of $\dfrac{(2-x)(2+x)}{x}$ | $+$ | $-$ | $+$ | $-$ |

From the table, the solution set is $\{x \mid x < -1 \text{ or } -\frac{2}{3} < x < 0\}$. Interval: $(-\infty, -1) \cup (-\frac{2}{3}, 0)$.

Graph: −1 $-\frac{2}{3}$ 0

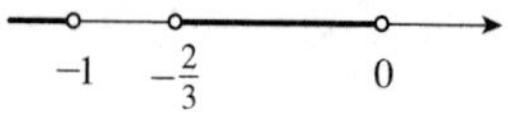

**59.** $1+\dfrac{2}{x+1}\le\dfrac{2}{x}$ $\Leftrightarrow$ $1+\dfrac{2}{x+1}-\dfrac{2}{x}\le 0$ $\Leftrightarrow$ $\dfrac{x(x+1)}{x(x+1)}+\dfrac{2x}{x(x+1)}-\dfrac{2(x+1)}{x(x+1)}\le 0$ $\Leftrightarrow$

$\dfrac{x^2+x+2x-2x-2}{x(x+1)}\le 0$ $\Leftrightarrow$ $\dfrac{x^2+x-2}{x(x+1)}\le 0$ $\Leftrightarrow$ $\dfrac{(x+2)(x-1)}{x(x+1)}\le 0$. The expression on the left of the inequality changes sign where $x=-2$, where $x=-1$, where $x=0$, and where $x=1$. Thus we must check the intervals in the following table.

| Interval | $(-\infty,-2)$ | $(-2,-1)$ | $(-1,0)$ | $(0,1)$ | $(1,\infty)$ |
|---|---|---|---|---|---|
| Sign of $x+2$ | $-$ | $+$ | $+$ | $+$ | $+$ |
| Sign of $x-1$ | $-$ | $-$ | $-$ | $-$ | $+$ |
| Sign of $x$ | $-$ | $-$ | $-$ | $+$ | $+$ |
| Sign of $x+1$ | $-$ | $-$ | $+$ | $+$ | $+$ |
| Sign of $\dfrac{(x+2)(x-1)}{x(x+1)}$ | $+$ | $-$ | $+$ | $-$ | $+$ |

Since $x=-1$ and $x=0$ yield undefined expressions, we cannot include them in the solution. From the table, the solution set is $\{x \mid -2\le x<-1 \text{ or } 0<x\le 1\}$. Interval: $[-2,-1)\cup(0,1]$. Graph:

−2 −1 0 1

**60.** $\dfrac{3}{x-1}-\dfrac{4}{x}\ge 1$ $\Leftrightarrow$ $\dfrac{3}{x-1}-\dfrac{4}{x}-1\ge 0$ $\Leftrightarrow$ $\dfrac{3x}{x(x-1)}-\dfrac{4(x-1)}{x(x-1)}-\dfrac{x(x-1)}{x(x-1)}\ge 0$ $\Leftrightarrow$

$\dfrac{3x-4x+4-x^2+x}{x(x-1)}\ge 0$ $\Leftrightarrow$ $\dfrac{4-x^2}{x(x-1)}\ge 0$ $\Leftrightarrow$ $\dfrac{(2-x)(2+x)}{x(x-1)}\ge 0$. The expression on the left of the inequality changes sign when $x=2$, $x=-2$, $x=0$, and $x=1$. Thus we must check the intervals in the following table.

| Interval | $(-\infty,-2)$ | $(-2,0)$ | $(0,1)$ | $(1,2)$ | $(2,\infty)$ |
|---|---|---|---|---|---|
| Sign of $2-x$ | $+$ | $+$ | $+$ | $+$ | $-$ |
| Sign of $2+x$ | $-$ | $+$ | $+$ | $+$ | $+$ |
| Sign of $x$ | $-$ | $-$ | $+$ | $+$ | $+$ |
| Sign of $x-1$ | $-$ | $-$ | $-$ | $+$ | $+$ |
| Sign of $\dfrac{(2-x)(2+x)}{x(x-1)}$ | $-$ | $+$ | $-$ | $+$ | $-$ |

Since $x=0$ and $x=1$ give undefined expressions, we cannot include them in the solution. From the table, the solution set is $\{x \mid -2\le x<0 \text{ or } 1<x\le 2\}$. Interval: $[-2,0)\cup(1,2]$. Graph:

−2 0 1 2

**61.** $\dfrac{6}{x-1}-\dfrac{6}{x}\geq 1 \quad\Leftrightarrow\quad \dfrac{6}{x-1}-\dfrac{6}{x}-1\geq 0 \quad\Leftrightarrow\quad \dfrac{6x}{x(x-1)}-\dfrac{6(x-1)}{x(x-1)}-\dfrac{x(x-1)}{x(x-1)}\geq 0 \quad\Leftrightarrow$

$\dfrac{6x-6x+6-x^2+x}{x(x-1)}\geq 0 \quad\Leftrightarrow\quad \dfrac{-x^2+x+6}{x(x-1)}\geq 0 \quad\Leftrightarrow\quad \dfrac{(-x+3)(x+2)}{x(x-1)}\geq 0$. The expression on the left of the inequality changes sign where $x=3$, where $x=-2$, where $x=0$, and where $x=1$. Thus we must check the intervals in the following table.

| Interval | $(-\infty,-2)$ | $(-2,0)$ | $(0,1)$ | $(1,3)$ | $(3,\infty)$ |
|---|---|---|---|---|---|
| Sign of $-x+3$ | $+$ | $+$ | $+$ | $+$ | $-$ |
| Sign of $x+2$ | $-$ | $+$ | $+$ | $+$ | $+$ |
| Sign of $x$ | $-$ | $-$ | $+$ | $+$ | $+$ |
| Sign of $x-1$ | $-$ | $-$ | $-$ | $+$ | $+$ |
| Sign of $\dfrac{(-x+3)(x+2)}{x(x-1)}$ | $-$ | $+$ | $-$ | $+$ | $-$ |

From the table, the solution set is $\{x \mid -2\leq x<0 \text{ or } 1<x\leq 3\}$. The points $x=0$ and $x=1$ are excluded from the solution set because they make the denominator zero. Interval: $[-2,0)\cup(1,3]$. Graph:

–2 0 1 3

**62.** $\dfrac{x}{2}\geq\dfrac{5}{x+1}+4 \quad\Leftrightarrow\quad \dfrac{x}{2}-\dfrac{5}{x+1}-4\geq 0 \quad\Leftrightarrow\quad \dfrac{x(x+1)}{2(x+1)}-\dfrac{2\cdot 5}{2(x+1)}-\dfrac{4(2)(x+1)}{2(x+1)}\geq 0 \quad\Leftrightarrow$

$\dfrac{x^2+x-10-8x-8}{2(x+1)}\geq 0 \quad\Leftrightarrow\quad \dfrac{x^2-7x-18}{2(x+1)}\geq 0 \quad\Leftrightarrow\quad \dfrac{(x-9)(x+2)}{2(x+1)}\geq 0$. The expression on the left of the inequality changes sign when $x=9$, $x=-2$, and $x=-1$. Thus we must check the intervals in the following table.

| Interval | $(-\infty,-2)$ | $(-2,-1)$ | $(-1,9)$ | $(9,\infty)$ |
|---|---|---|---|---|
| Sign of $x-9$ | $-$ | $-$ | $-$ | $+$ |
| Sign of $x+2$ | $-$ | $+$ | $+$ | $+$ |
| Sign of $x+1$ | $-$ | $-$ | $+$ | $+$ |
| Sign of $\dfrac{(x-9)(x+2)}{2(x+1)}$ | $-$ | $+$ | $-$ | $+$ |

From the table, the solution set is $\{x \mid -2\leq x<-1 \text{ or } 9\leq x\}$. The point $x=-1$ is excluded from the solution set because it makes the expression undefined. Interval: $[-2,-1)\cup[9,\infty)$. Graph:

–1 –2 9

**63.** $\dfrac{x+2}{x+3} < \dfrac{x-1}{x-2} \quad\Leftrightarrow\quad \dfrac{x+2}{x+3} - \dfrac{x-1}{x-2} < 0 \quad\Leftrightarrow\quad \dfrac{(x+2)(x-2)}{(x+3)(x-2)} - \dfrac{(x-1)(x+3)}{(x-2)(x+3)} < 0 \quad\Leftrightarrow$

$\dfrac{x^2-4-x^2-2x+3}{(x+3)(x-2)} < 0 \quad\Leftrightarrow\quad \dfrac{-2x-1}{(x+3)(x-2)} < 0$. The expression on the left of the inequality changes sign where $x = -\frac{1}{2}$, where $x = -3$, and where $x = 2$. Thus we must check the intervals in the following table.

| Interval | $(-\infty,-3)$ | $\left(-3,-\frac{1}{2}\right)$ | $\left(-\frac{1}{2},2\right)$ | $(2,\infty)$ |
|---|---|---|---|---|
| Sign of $-2x-1$ | $+$ | $+$ | $-$ | $-$ |
| Sign of $x+3$ | $-$ | $+$ | $+$ | $+$ |
| Sign of $x-2$ | $-$ | $-$ | $-$ | $+$ |
| Sign of $\dfrac{-2x-1}{(x+3)(x-2)}$ | $+$ | $-$ | $+$ | $-$ |

From the table, the solution set is $\left\{x \mid -3 < x < -\frac{1}{2} \text{ or } 2 < x\right\}$. Interval: $\left(-3,-\frac{1}{2}\right) \cup (2,\infty)$.

Graph: $-3$ $-\frac{1}{2}$ $2$

**64.** $\dfrac{1}{x+1} + \dfrac{1}{x+2} \le 0 \quad\Leftrightarrow\quad \dfrac{x+2}{(x+1)(x+2)} + \dfrac{x+1}{(x+1)(x+2)} \le 0 \quad\Leftrightarrow\quad \dfrac{x+2+x+1}{(x+1)(x+2)} \le 0 \quad\Leftrightarrow$

$\dfrac{2x+3}{(x+1)(x+2)} \le 0$. The expression on the left of the inequality changes sign when $x = -\frac{3}{2}$, $x = -1$, and $x = -2$. Thus we must check the intervals in the following table.

| Interval | $(-\infty,-2)$ | $\left(-2,-\frac{3}{2}\right)$ | $\left(-\frac{3}{2},-1\right)$ | $(-1,\infty)$ |
|---|---|---|---|---|
| Sign of $2x+3$ | $-$ | $-$ | $+$ | $+$ |
| Sign of $x+1$ | $-$ | $-$ | $-$ | $+$ |
| Sign of $x+2$ | $-$ | $+$ | $+$ | $+$ |
| Sign of $\dfrac{2x+3}{(x+1)(x+2)}$ | $-$ | $+$ | $-$ | $+$ |

From the table, the solution set is $\left\{x \mid x < -2 \text{ or } -\frac{3}{2} \le x < -1\right\}$.The points $x = -2$ and $x = -1$ are excluded from the solution because the expression is undefined at those values. Interval: $(-\infty,-2) \cup \left[-\frac{3}{2},-1\right)$.

Graph: $-2$ $-\frac{3}{2}$ $-1$

**65.** $x^4 > x^2 \quad \Leftrightarrow \quad x^4 - x^2 > 0 \quad \Leftrightarrow \quad x^2\left(x^2 - 1\right) > 0 \quad \Leftrightarrow \quad x^2\left(x-1\right)\left(x+1\right) > 0$. The expression on the left of the inequality changes sign where $x = 0$, where $x = 1$, and where $x = -1$. Thus we must check the intervals in the following table.

| Interval | $(-\infty, -1)$ | $(-1, 0)$ | $(0, 1)$ | $(1, \infty)$ |
|---|---|---|---|---|
| Sign of $x^2$ | $+$ | $+$ | $+$ | $+$ |
| Sign of $x - 1$ | $-$ | $-$ | $-$ | $+$ |
| Sign of $x + 1$ | $-$ | $+$ | $+$ | $+$ |
| Sign of $x^2\left(x-1\right)\left(x+1\right)$ | $+$ | $-$ | $-$ | $+$ |

From the table, the solution set is $\{x \mid x < -1 \text{ or } 1 < x\}$. Interval: $(-\infty, -1) \cup (1, \infty)$. Graph:

−1 1

**66.** $x^5 > x^2 \quad \Leftrightarrow \quad x^5 - x^2 > 0 \quad \Leftrightarrow \quad x^2\left(x^3 - 1\right) > 0 \quad \Leftrightarrow \quad x^2\left(x-1\right)\left(x^2+x+1\right) > 0$. The expression on the left of the inequality changes sign when $x = 0$ and $x = 1$. But the solution of $x^2 + x + 1 = 0$ are

$$x = \frac{-1 \pm \sqrt{(1)^2 - 4(1)(1)}}{2(1)} = \frac{-1 \pm \sqrt{-3}}{2}.$$

Since these are not real solutions. The expression $x^2 + x + 1$ does not changes signs, so we must check the intervals in the following table.

| Interval | $(-\infty, 0)$ | $(0, 1)$ | $(1, \infty)$ |
|---|---|---|---|
| Sign of $x^2$ | $+$ | $+$ | $+$ |
| Sign of $x - 1$ | $-$ | $-$ | $+$ |
| Sign of $x^2 + x + 1$ | $+$ | $+$ | $+$ |
| Sign of $x^2\left(x-1\right)\left(x^2+x+1\right)$ | $-$ | $-$ | $+$ |

From the table, the solution set is $\{x \mid 1 < x\}$. Interval: $(1, \infty)$. Graph:

1

**67.** For $\sqrt{16 - 9x^2}$ to be defined as a real number we must have $16 - 9x^2 \geq 0 \quad \Leftrightarrow \quad (4 - 3x)(4 + 3x) \geq 0$. The expression in the inequality changes sign at $x = \frac{4}{3}$ and $x = -\frac{4}{3}$.

| Interval | $\left(-\infty, -\frac{4}{3}\right)$ | $\left(-\frac{4}{3}, \frac{4}{3}\right)$ | $\left(\frac{4}{3}, \infty\right)$ |
|---|---|---|---|
| Sign of $4 - 3x$ | $+$ | $+$ | $-$ |
| Sign of $4 + 3x$ | $-$ | $+$ | $+$ |
| Sign of $(4 - 3x)(4 + 3x)$ | $-$ | $+$ | $-$ |

Thus $-\frac{4}{3} \leq x \leq \frac{4}{3}$.

**68.** For $\sqrt{3x^2 - 5x + 2}$ to be defined as a real number, we must have $3x^2 - 5x + 2 \geq 0 \quad \Leftrightarrow \quad (3x - 2)(x - 1) \geq 0$. The expression on the left of the inequality changes sign when$x = \frac{2}{3}$ and $x = 1$. Thus we must check the intervals in the following table.

| Interval | $\left(-\infty, \frac{2}{3}\right)$ | $\left(\frac{2}{3}, 1\right)$ | $(1, \infty)$ |
|---|---|---|---|
| Sign of $3x - 2$ | $-$ | $+$ | $+$ |
| Sign of $x - 1$ | $-$ | $-$ | $+$ |
| Sign of $(3x - 2)(x - 1)$ | $+$ | $-$ | $+$ |

Thus $x \leq \frac{2}{3}$ or $1 \leq x$.

**69.** For $\left(\dfrac{1}{x^2 - 5x - 14}\right)^{1/2}$ to be defined as a real number we must have $x^2 - 5x - 14 > 0 \quad \Leftrightarrow \quad (x - 7)(x + 2) > 0$.

The expression in the inequality changes sign at $x = 7$ and $x = -2$.

| Interval | $(-\infty, -2)$ | $(-2, 7)$ | $(7, \infty)$ |
|---|---|---|---|
| Sign of $x - 7$ | $-$ | $-$ | $+$ |
| Sign of $x + 2$ | $-$ | $+$ | $+$ |
| Sign of $(x - 7)(x + 2)$ | $+$ | $-$ | $+$ |

Thus $x < -2$ or $7 < x$, and the solution set is $(-\infty, -2) \cup (7, \infty)$.

**70.** For $\sqrt[4]{\dfrac{1 - x}{2 + x}}$ to be defined as a real number we must have $\dfrac{1 - x}{2 + x} \geq 0$.The expression on the left of the inequality changes sign when $x = 1$ and $x = -2$. Thus we must check the intervals in the following table.

| Interval | $(-\infty, -2)$ | $(-2, 1)$ | $(1, \infty)$ |
|---|---|---|---|
| Sign of $1 - x$ | $+$ | $+$ | $-$ |
| Sign of $2 + x$ | $-$ | $+$ | $+$ |
| Sign of $\dfrac{1 - x}{2 + x}$ | $-$ | $+$ | $-$ |

Thus $-2 < x \leq 1$. Note that $x = -2$ has been excluded from the solution set because the expression is undefined at that value.

**71. (a)** $a(bx - c) \geq bc$ (where $a, b, c > 0$) $\quad \Leftrightarrow \quad bx - c \geq \dfrac{bc}{a} \quad \Leftrightarrow \quad bx \geq \dfrac{bc}{a} + c \quad \Leftrightarrow \quad x \geq \dfrac{1}{b}\left(\dfrac{bc}{a} + c\right) = \dfrac{c}{a} + \dfrac{c}{b}$

$\Leftrightarrow \quad x \geq \dfrac{c}{a} + \dfrac{c}{b}$.

**(b)** We have $a \leq bx + c < 2a$, where $a, b, c > 0 \quad \Leftrightarrow \quad a - c \leq bx < 2a - c \quad \Leftrightarrow \quad \dfrac{a - c}{b} \leq x < \dfrac{2a - c}{b}$.

**72.** We are given $\frac{a}{b} < \frac{c}{d}$, where $a$, $b$, $c$, and $d$ are all positive. So multiplying by $d$, we get $\frac{ad}{b} < c$ and so by Rule 1 for Inequalities, $a + \frac{ad}{b} < a + c \quad \Leftrightarrow \quad \frac{ab + ad}{b} < a + c \quad \Leftrightarrow \quad \frac{a}{b}(b + d) < a + c \quad \Leftrightarrow \quad \frac{a}{b} < \frac{a + c}{b + d}$, since $b + d$ is positive. Similarly, $\frac{a}{b} < \frac{c}{d} \quad \Leftrightarrow \quad a < \frac{bc}{d} \quad \Leftrightarrow \quad a + c < \frac{bc}{d} + c \quad \Leftrightarrow \quad a + c < \frac{bc}{d} + \frac{dc}{d} \quad \Leftrightarrow \quad a + c < \frac{c}{d}(b + d)$ $\Leftrightarrow \quad \frac{a + c}{b + d} < \frac{c}{d}$, since $b + d > 0$. Combining the two inequalities, $\frac{a}{b} < \frac{a + c}{b + d} < \frac{c}{d}$.

**73.** Inserting the relationship $C = \frac{5}{9}(F - 32)$, we have $20 \le C \le 30 \quad \Leftrightarrow \quad 20 \le \frac{5}{9}(F - 32) \le 30 \quad \Leftrightarrow$ $36 \le F - 32 \le 54 \quad \Leftrightarrow \quad 68 \le F \le 86$.

**74.** Inserting the relationship $F = \frac{9}{5}C + 32$, we have $50 \le F \le 95 \quad \Leftrightarrow \quad 50 \le \frac{9}{5}C + 32 \le 95 \quad \Leftrightarrow \quad 18 \le \frac{9}{5}C \le 63$ $\Leftrightarrow \quad 10 \le C \le 35$.

**75.** Let $x$ be the average number of miles driven per day. Each day the cost of Plan A is $30 + 0.10x$, and the cost of Plan B is 50. Plan B saves money when $50 < 30 + 0.10x \quad \Leftrightarrow \quad 20 < 0.1x \quad \Leftrightarrow \quad 200 < x$. So Plan B saves money when you average more than 200 miles a day.

**76.** Let $m$ be the number of minutes of long-distance calls placed per month. Then under Plan A, the cost will be $25 + 0.05m$, and under Plan B, the cost will be $5 + 0.12m$. To determine when Plan B is advantageous, we must solve $25 + 0.05m > 5 + 0.12m \quad \Leftrightarrow \quad 20 > 0.07m \quad \Leftrightarrow \quad 285.7 > m$. So Plan B is advantageous if a person places fewer than 286 minutes of long-distance calls during a month.

**77.** We need to solve $6400 \le 0.35m + 2200 \le 7100$ for $m$. So $6400 \le 0.35m + 2200 \le 7100 \quad \Leftrightarrow \quad 4200 \le 0.35m \le 4900$ $\Leftrightarrow \quad 12{,}000 \le m \le 14{,}000$. She plans on driving between 12,000 and 14,000 miles.

**78. (a)** $T = 20 - \frac{h}{100}$, where $T$ is the temperature in °C, and $h$ is the height in meters.

**(b)** Solving the expression in part (a) for $h$, we get $h = 100(20 - T)$. So $0 \le h \le 5000 \quad \Leftrightarrow$ $0 \le 100(20 - T) \le 5000 \quad \Leftrightarrow \quad 0 \le 20 - T \le 50 \quad \Leftrightarrow \quad -20 \le -T \le 30 \quad \Leftrightarrow \quad 20 \ge T \ge -30$. Thus the range of temperature is from 20° C down to $-30$° C.

**79. (a)** Let $x$ be the number of \$3 increases. Then the number of seats sold is $120 - x$. So $P = 200 + 3x \quad \Leftrightarrow$ $3x = P - 200 \quad \Leftrightarrow \quad x = \frac{1}{3}(P - 200)$. Substituting for $x$ we have that the number of seats sold is $120 - x = 120 - \frac{1}{3}(P - 200) = -\frac{1}{3}P + \frac{560}{3}$.

**(b)** $90 \le -\frac{1}{3}P + \frac{560}{3} \le 115 \quad \Leftrightarrow \quad 270 \le 360 - P + 200 \le 345 \quad \Leftrightarrow \quad 270 \le -P + 560 \le 345 \quad \Leftrightarrow$ $-290 \le -P \le -215 \quad \Leftrightarrow \quad 290 \ge P \ge 215$. Putting this into standard order, we have $215 \le P \le 290$. So the ticket prices are between \$215 and \$290.

**80.** If the customer buys $x$ pounds of coffee at \$6.50 per pound, then his cost $c$ will be $6.50x$. Thus $x = \frac{c}{6.5}$. Since the scale's accuracy is $\pm 0.03$ lb, and the scale shows 3 lb, we have $3 - 0.03 \le x \le 3 + 0.03 \quad \Leftrightarrow \quad 2.97 \le \frac{c}{6.5} \le 3.03 \quad \Leftrightarrow$ $(6.50)\,2.97 \le c \le (6.50)\,3.03 \quad \Leftrightarrow \quad 19.305 \le c \le 19.695$. Since the customer paid \$19.50, he could have been over- or undercharged by as much as 19.5 cents.

**81.** $0.0004 \leq \dfrac{4{,}000{,}000}{d^2} \leq 0.01$. Since $d^2 \geq 0$ and $d \neq 0$, we can multiply each expression by $d^2$ to obtain $0.0004d^2 \leq 4{,}000{,}000 \leq 0.01d^2$. Solving each pair, we have $0.0004d^2 \leq 4{,}000{,}000 \quad \Leftrightarrow \quad d^2 \leq 10{,}000{,}000{,}000$ $\Rightarrow \quad d \leq 100{,}000$ (recall that $d$ represents distance, so it is always nonnegative). Solving $4{,}000{,}000 \leq 0.01d^2 \quad \Leftrightarrow$ $400{,}000{,}000 \leq d^2 \Rightarrow 20{,}000 \leq d$. Putting these together, we have $20{,}000 \leq d \leq 100{,}000$.

**82.** $\dfrac{600{,}000}{x^2+300} < 500 \quad \Leftrightarrow \quad 600{,}000 < 500\left(x^2+300\right)$ (Note that $x^2 + 300 \geq 300 > 0$, so we can multiply both sides by the denominator and not worry that we might be multiplying both sides by a negative number or by zero.) $1200 < x^2 + 300$ $\Leftrightarrow \quad 0 < x^2 - 900 \quad \Leftrightarrow \quad 0 < (x-30)(x+30)$. The expression in the inequality changes sign at $x = 30$ and $x = -30$. However, since $x$ represents distance, we must have $x > 0$.

| Interval | $(0, 30)$ | $(30, \infty)$ |
|---|---|---|
| Sign of $x - 30$ | $-$ | $+$ |
| Sign of $x + 30$ | $+$ | $+$ |
| Sign of $(x-30)(x+30)$ | $-$ | $+$ |

So $x > 30$ and you must stand at least 30 meters from the center of the fire.

**83.** $128 + 16t - 16t^2 \geq 32 \quad \Leftrightarrow \quad -16t^2 + 16t + 96 \geq 0 \quad \Leftrightarrow \quad -16\left(t^2 - t - 6\right) \geq 0 \quad \Leftrightarrow \quad -16(t-3)(t+2) \geq 0$. The expression on the left of the inequality changes sign at $x = -2$, at $t = 3$, and at $t = -2$. However, $t \geq 0$, so the only endpoint is $t = 3$.

| Interval | $(0, 3)$ | $(3, \infty)$ |
|---|---|---|
| Sign of $-16$ | $-$ | $-$ |
| Sign of $t - 3$ | $-$ | $+$ |
| Sign of $t + 2$ | $+$ | $+$ |
| Sign of $-16(t-3)(t+2)$ | $+$ | $-$ |

So $0 \leq t \leq 3$.

**84.** Solve $30 \leq 10 + 0.9v - 0.01v^2$ for $10 \leq v \leq 75$. We have $30 \leq 10 + 0.9v - 0.01v^2 \quad \Leftrightarrow \quad 0.01v^2 - 0.9v + 20 \leq 0$ $\Leftrightarrow \quad (0.1v - 4)(0.1v - 5) \leq 0$. The possible endpoints are $0.1v - 4 = 0 \quad \Leftrightarrow \quad 0.1v = 4 \quad \Leftrightarrow \quad v = 40$ and $0.1v - 5 = 0 \quad \Leftrightarrow \quad 0.1v = 5 \quad \Leftrightarrow \quad v = 50$.

| Interval | $(10, 40)$ | $(40, 50)$ | $(50, 75)$ |
|---|---|---|---|
| Sign of $0.1v - 4$ | $-$ | $+$ | $+$ |
| Sign of $0.1v - 5$ | $-$ | $-$ | $+$ |
| Sign of $(0.1v-4)(0.1v-5)$ | $+$ | $-$ | $+$ |

Thus he must drive between 40 and 50 mi/h.

**85.** $240 > v + \dfrac{v^2}{20} \quad\Leftrightarrow\quad \frac{1}{20}v^2 + v - 240 < 0 \quad\Leftrightarrow\quad \left(\frac{1}{20}v - 3\right)(v + 80) < 0$. The expression in the inequality changes sign at $v = 60$ and $v = -80$. However, since $v$ represents the speed, we must have $v \geq 0$.

| Interval | $(0, 60)$ | $(60, \infty)$ |
|---|---|---|
| Sign of $\frac{1}{20}v - 3$ | $-$ | $+$ |
| Sign of $v + 80$ | $+$ | $+$ |
| Sign of $\left(\frac{1}{20}v - 3\right)(v + 80)$ | $-$ | $+$ |

So Kerry must drive less than 60 mi/h.

**86.** Solve $2400 \leq 20x - \left(2000 + 8x + 0.0025x^2\right) \quad\Leftrightarrow\quad 2400 \leq 20x - 2000 - 8x - 0.0025x^2 \quad\Leftrightarrow$ $0.0025x^2 - 12x + 4400 \leq 0 \quad\Leftrightarrow\quad (0.0025x - 1)(x - 4400) \leq 0$. The expression on the left of the inequality changes sign when $x = 400$ and $x = 4400$. Since the manufacturer can only sell positive units, we check the intervals in the following table.

| Interval | $(0, 400)$ | $(400, 4400)$ | $(4400, \infty)$ |
|---|---|---|---|
| Sign of $0.0025x - 1$ | $-$ | $+$ | $+$ |
| Sign of $x - 4400$ | $-$ | $-$ | $+$ |
| Sign of $(0.0025x - 1)(x - 4400)$ | $+$ | $-$ | $+$ |

So the manufacturer must sell between 400 and 4400 units to enjoy a profit of at least \$2400.

**87.** Let $n$ be the number of people in the group. Then the bus fare is $\dfrac{360}{n}$, and the cost of the theater tickets is $30 - 0.25n$. We want the total cost to be less than \$39 per person, that is, $\dfrac{360}{n} + (30 - 0.25n) < 39$. If we multiply this inequality by $n$, we will not change the direction of the inequality; $n$ is positive since it represents the number of people. So we get $360 + n(30 - 0.25n) < 39n \quad\Leftrightarrow\quad -0.25n^2 + 30n + 360 < 39n \quad\Leftrightarrow\quad -0.25n^2 - 9n + 360 < 0 \quad\Leftrightarrow$ $(0.25n + 15)(-n + 24)$. The expression on the left of the inequality changes sign when $n = -60$ and $n = 24$. Since $n > 0$, we check the intervals in the following table.

| Interval | $(0, 24)$ | $(24, \infty)$ |
|---|---|---|
| Sign of $0.25n + 15$ | $+$ | $+$ |
| Sign of $-n + 24$ | $+$ | $-$ |
| Sign of $(0.25n + 15)(-n + 24)$ | $+$ | $-$ |

So the group must have more than 24 people in order that the cost of the theater tour is less than \$39.

**88.** Let $x$ be the length of the garden and $w$ its width. Using the fact that the perimeter is 120 ft, we must have $2x + 2w = 120$ $\Leftrightarrow$ $w = 60 - x$. Now since the area must be at least 800 ft$^2$, we have $800 < x(60 - x)$ $\Leftrightarrow$ $800 < 60x - x^2$ $\Leftrightarrow$ $x^2 - 60x + 800 < 0$ $\Leftrightarrow$ $(x - 20)(x - 40) < 0$. The expression in the inequality changes sign at $x = 20$ and $x = 40$. However, since $x$ represents length, we must have $x > 0$.

| Interval | $(0, 20)$ | $(20, 40)$ | $(40, \infty)$ |
|---|---|---|---|
| Sign of $x - 20$ | $-$ | $+$ | $+$ |
| Sign of $x - 40$ | $-$ | $-$ | $+$ |
| Sign of $(x - 20)(x - 40)$ | $+$ | $-$ | $+$ |

The length of the garden should be between 20 and 40 feet.

**89.** *Case 1:* $a < b < 0$ We have $a \cdot a > a \cdot b$, since $a < 0$, and $b \cdot a > b \cdot b$, since $b < 0$. So $a^2 > a \cdot b > b^2$, that is $a < b < 0$ $\Rightarrow$ $a^2 > b^2$. Continuing, we have $a \cdot a^2 < a \cdot b^2$, since $a < 0$ and $b^2 \cdot a < b^2 \cdot b$, since $b^2 > 0$. So $a^3 < ab^2 < b^3$. Thus $a < b < 0$ $\Rightarrow$ $a^3 > b^3$. So $a < b < 0$ $\Rightarrow$ $a^n > b^n$, if $n$ is even, and $a^n < b$, if $n$ is odd.

*Case 2:* $0 < a < b$ We have $a \cdot a < a \cdot b$, since $a > 0$, and $b \cdot a < b \cdot b$, since $b < 0$. So $a^2 < a \cdot b < b^2$. Thus $0 < a < b$ $\Rightarrow$ $a^2 < b^2$. Likewise, $a^2 \cdot a < a^2 \cdot b$ and $b \cdot a^2 < b \cdot b^2$, thus $a^3 < b^3$. So $0 < a < b$ $\Rightarrow$ $a^n < b^n$, for all positive integers $n$.

*Case 3:* $a < 0 < b$ If $n$ is odd, then $a^n < b^n$, because $a^n$ is negative and $b^n$ is positive. If $n$ is even, then we could have either $a^n < b^n$ or $a^n > b^n$. For example, $-1 < 2$ and $(-1)^2 < 2^2$, but $-3 < 2$ and $(-3)^2 > 2^2$.

**90.** The rule we want to apply here is "$a < b \Rightarrow ac < bc$ if $c > 0$ and $a < b \Rightarrow ac > bc$ if $c < 0$". Thus we cannot simply multiply by $x$, since we don't yet know if $x$ is positive or negative, so in solving $1 < \frac{3}{x}$, we must consider two cases.

*Case 1:* $x > 0$ Multiplying both sides by $x$, we have $x < 3$. Together with our initial condition, we have $0 < x < 3$.

*Case 2:* $x < 0$ Multiplying both sides by $x$, we have $x > 3$. But $x < 0$ and $x > 3$ have no elements in common, so this gives no additional solution. Hence, the only solutions are $0 < x < 3$.

## 1.7 Absolute Value Equations and Inequalities

**1.** $|4x| = 24$ $\Leftrightarrow$ $4x = \pm 24$ $\Leftrightarrow$ $x = \pm 6$.

**2.** $|6x| = 15$ $\Leftrightarrow$ $6x = \pm 15$ $\Leftrightarrow$ $x = \pm\frac{5}{2}$.

**3.** $5|x| + 3 = 28$ $\Leftrightarrow$ $5|x| = 25$ $\Leftrightarrow$ $|x| = 5$ $\Leftrightarrow$ $x = \pm 5$.

**4.** $\frac{1}{2}|x| - 7 = 2$ $\Leftrightarrow$ $\frac{1}{2}|x| = 9$ $\Leftrightarrow$ $|x| = 18$ $\Leftrightarrow$ $x = \pm 18$.

**5.** $|x - 3| = 2$ is equivalent to $x - 3 = \pm 2$ $\Leftrightarrow$ $x = 3 \pm 2$ $\Leftrightarrow$ $x = 1$ or $x = 5$.

**6.** $|2x - 3| = 7$ is equivalent to either $2x - 3 = 7$ $\Leftrightarrow$ $2x = 10$ $\Leftrightarrow$ $x = 5$; or $2x - 3 = -7$ $\Leftrightarrow$ $2x = -4$ $\Leftrightarrow$ $x = -2$. The two solutions are $x = 5$ and $x = -2$.

**7.** $|x + 4| = 0.5$ is equivalent to $x + 4 = \pm 0.5$ $\Leftrightarrow$ $x = -4 \pm 0.5$ $\Leftrightarrow$ $x = -4.5$ or $x = -3.5$.

**8.** $|x + 4| = -3$. Since the absolute value is always nonnegative, there is no solution.

**9.** $|4x+7|=9$ is equivalent to either $4x+7=9 \iff 4x=2 \iff x=\dfrac{1}{2}$; or $4x+7=-9 \iff 4x=-16 \iff x=-4$. The two solutions are $x=\frac{1}{2}$ and $x=-4$.

**10.** $\left|\frac{1}{2}x-2\right|=1$ is equivalent to either $\frac{1}{2}x-2=1 \iff \frac{1}{2}x=3 \iff x=6$; or $\frac{1}{2}x-2=-1 \iff \frac{1}{2}x=1 \iff x=2$. The two solutions are $x=6$ and $x=2$.

**11.** $4-|3x+6|=1 \iff -|3x+6|=-3 \iff |3x+6|=3$, which is equivalent to either $3x+6=3 \iff 3x=-3 \iff x=-1$; or $3x+6=-3 \iff 3x=-9 \iff x=-3$. The two solutions are $x=-1$ and $x=-3$.

**12.** $|5-2x|+6=14 \iff |5-2x|=8$ which is equivalent to either $5-2x=8 \iff -2x=3 \iff x=-\frac{3}{2}$; or $5-2x=-8 \iff -2x=-13 \iff x=\frac{13}{2}$. The two solutions are $x=-\frac{3}{2}$ and $x=\frac{13}{2}$.

**13.** $3|x+5|+6=15 \iff 3|x+5|=9 \iff |x+5|=3$, which is equivalent to either $x+5=3 \iff x=-2$; or $x+5=-3 \iff x=-8$. The two solutions are $x=-2$ and $x=-8$.

**14.** $20+|2x-4|=15 \iff |2x-4|=-5$. Since the absolute value is always nonnegative, there is no solution.

**15.** $8+5\left|\frac{1}{3}x-\frac{5}{6}\right|=33 \iff 5\left|\frac{1}{3}x-\frac{5}{6}\right|=25 \iff \left|\frac{1}{3}x-\frac{5}{6}\right|=5$, which is equivalent to either $\frac{1}{3}x-\frac{5}{6}=5 \iff \frac{1}{3}x=\frac{35}{6} \iff x=\frac{35}{2}$; or $\frac{1}{3}x-\frac{5}{6}=-5 \iff \frac{1}{3}x=-\frac{25}{6} \iff x=-\frac{25}{2}$. The two solutions are $x=-\frac{25}{2}$ and $x=\frac{35}{2}$.

**16.** $\left|\frac{3}{5}x+2\right|-\frac{1}{2}=4 \iff \left|\frac{3}{5}x+2\right|=\frac{9}{2}$ which is equivalent to either $\frac{3}{5}x+2=\frac{9}{2} \iff \frac{3}{5}x=\frac{5}{2} \iff x=\frac{25}{6}$; or $\frac{3}{5}x+2=-\frac{9}{2} \iff \frac{3}{5}x=-\frac{13}{2} \iff x=-\frac{65}{6}$. The two solutions are $x=\frac{25}{6}$ and $x=-\frac{65}{6}$.

**17.** $|x-1|=|3x+2|$, which is equivalent to either $x-1=3x+2 \iff -2x=3 \iff x=-\frac{3}{2}$; or $x-1=-(3x+2) \iff x-1=-3x-2 \iff 4x=-1 \iff x=-\frac{1}{4}$. The two solutions are $x=-\frac{3}{2}$ and $x=-\frac{1}{4}$.

**18.** $|x+3|=|2x+1|$ is equivalent to either $x+3=2x+1 \iff -x=-2 \iff x=2$ or to $x+3=-(2x+1) \iff x+3=-2x-1 \iff 3x=-4 \iff x=-\frac{4}{3}$. The two solutions are $x=2$ and $x=-\frac{4}{3}$.

**19.** $|x|\le 4 \iff -4\le x\le 4$. Interval: $[-4,4]$.

**20.** $|3x|<15 \iff -15<3x<15 \iff -5<x<5$. Interval: $(-5,5)$.

**21.** $|2x|>7$ is equivalent to $2x>7 \iff x>\frac{7}{2}$; or $2x<7 \iff x<-\frac{7}{2}$. Interval: $\left(-\infty,-\frac{7}{2}\right)\cup\left(\frac{7}{2},\infty\right)$.

**22.** $\frac{1}{2}|x|\ge 1 \iff |x|\ge 2$ is equivalent to $x\ge 2$ or $x\le -2$. Interval: $(-\infty,-2]\cup[2,\infty)$.

**23.** $|x-5|\le 3 \iff -3\le x-5\le 3 \iff 2\le x\le 8$. Interval: $[2,8]$.

**24.** $|x-9|>9$ is equivalent to $x-9>9 \iff x>18$, or $x-9<-9 \iff x<0$. Interval: $(-\infty,0)\cup(18,\infty)$.

**25.** $|x+1|\ge 1$ is equivalent to $x+1\ge 1 \iff x\ge 0$; or $x+1\le -1 \iff x\le -2$. Interval: $(-\infty,-2]\cup[0,\infty)$.

**26.** $|x+4|\le 0$ is equivalent to $|x+4|=0 \iff x+4=0 \iff x=-4$. The solution is just the point $x=-4$.

**27.** $|x+5|\ge 2$ is equivalent to $x+5\ge 2 \iff x\ge -3$; or $x+5\le -2 \iff x\le -7$. Interval: $(-\infty,-7]\cup[-3,\infty)$.

**28.** $|x+1|\ge 3$ is equivalent either to $x+1\ge 3 \iff x\ge 2$, or to $x+1\le -3 \iff x\le -4$. Interval: $(-\infty,-4]\cup[2,\infty)$.

**29.** $|2x-3|\le 0.4 \iff -0.4\le 2x-3\le 0.4 \iff 2.6\le 2x\le 3.4 \iff 1.3\le x\le 1.7$. Interval: $[1.3,1.7]$.

**30.** $|5x-2|<6 \iff -6<5x-2<6 \iff -4<5x<8 \iff -\frac{4}{5}<x<\frac{8}{5}$. Interval: $\left(-\frac{4}{5},\frac{8}{5}\right)$.

**31.** $\left|\dfrac{x-2}{3}\right|<2 \iff -2<\dfrac{x-2}{3}<2 \iff -6<x-2<6 \iff -4<x<8$. Interval: $(-4,8)$.

**32.** $\left|\dfrac{x+1}{2}\right|\ge 4 \iff \left|\frac{1}{2}(x+1)\right|\ge 4 \iff \frac{1}{2}|x+1|\ge 4 \iff |x+1|\ge 8$ which is equivalent either to $x+1\ge 8 \iff x\ge 7$, or to $x+1\le -8 \iff x\le -9$. Interval: $(-\infty,-9]\cup[7,\infty)$.

**33.** $|x+6|<0.001 \iff -0.001<x+6<0.001 \iff -6.001<x<-5.999$. Interval: $(-6.001,-5.999)$.

**34.** $|x-a|<d \iff -d<x-a<d \iff a-d<x<a+d$. Interval: $(a-d,a+d)$.

**35.** $4|x+2|-3<13 \Leftrightarrow 4|x+2|<16 \Leftrightarrow |x+2|<4 \Leftrightarrow -4<x+2<4 \Leftrightarrow -6<x<2$. Interval: $(-6,2)$.

**36.** $3-|2x+4|\le 1 \Leftrightarrow -|2x+4|\le -2 \Leftrightarrow |2x+4|\ge 2$ which is equivalent to either $2x+4\ge 2 \Leftrightarrow 2x\ge -2 \Leftrightarrow x\ge -1$; or $2x+4\le -2 \Leftrightarrow 2x\le -6 \Leftrightarrow x\le -3$. Interval: $(-\infty,-3]\cup[-1,\infty)$.

**37.** $8-|2x-1|\ge 6 \Leftrightarrow -|2x-1|\ge -2 \Leftrightarrow |2x-1|\le 2 \Leftrightarrow -2\le 2x-1\le 2 \Leftrightarrow -1\le 2x\le 3 \Leftrightarrow -\frac{1}{2}\le x\le \frac{3}{2}$. Interval: $\left[-\frac{1}{2},\frac{3}{2}\right]$.

**38.** $7|x+2|+5>4 \Leftrightarrow 7|x+2|>-1 \Leftrightarrow |x+2|>-\frac{1}{7}$. Since the absolute value is always nonnegative, the inequality is true for all real numbers. In interval notation, we have $(-\infty,\infty)$.

**39.** $\frac{1}{2}\left|4x+\frac{1}{3}\right|>\frac{5}{6} \Leftrightarrow \left|4x+\frac{1}{3}\right|>\frac{5}{3}$, which is equivalent to $4x+\frac{1}{3}>\frac{5}{3} \Leftrightarrow 4x>\frac{4}{3} \Leftrightarrow x>\frac{1}{3}$, or $4x+\dfrac{1}{3}<-\frac{5}{3} \Leftrightarrow 4x<-2 \Leftrightarrow x<-\frac{1}{2}$. Interval: $\left(-\infty,-\frac{1}{2}\right)\cup\left(\frac{1}{3},\infty\right)$.

**40.** $2\left|\frac{1}{2}x+3\right|+3\le 51 \Leftrightarrow 2\left|\frac{1}{2}x+3\right|\le 48 \Leftrightarrow \left|\frac{1}{2}x+3\right|\le 24 \Leftrightarrow -24<\frac{1}{2}x+3<24 \Leftrightarrow -27<\frac{1}{2}x<21 \Leftrightarrow -54<x<42$. Interval: $(-54,42)$.

**41.** $1\le|x|\le 4$. If $x\ge 0$, then this is equivalent to $1\le x\le 4$. If $x<0$, then this is equivalent to $1\le -x\le 4 \Leftrightarrow -1\ge x\ge -4 \Leftrightarrow -4\le x\le -1$. Interval: $[-4,-1]\cup[1,4]$.

**42.** $0<|x-5|\le\frac{1}{2}$. For $x\ne 5$, this is equivalent to $-\frac{1}{2}\le x-5\le\frac{1}{2} \Leftrightarrow \frac{9}{2}\le x\le\frac{11}{2}$. Since $x=5$ is excluded, the solution is $\left[\frac{9}{2},5\right)\cup\left(5,\frac{11}{2}\right]$.

**43.** $\dfrac{1}{|x+7|}>2 \Leftrightarrow 1>2|x+7|\ (x\ne -7) \Leftrightarrow |x+7|<\frac{1}{2} \Leftrightarrow -\frac{1}{2}<x+7<\frac{1}{2} \Leftrightarrow -\frac{15}{2}<x<-\frac{13}{2}$ and $x\ne -7$. Interval: $\left(-\frac{15}{2},-7\right)\cup\left(-7,-\frac{13}{2}\right)$.

**44.** $\dfrac{1}{|2x-3|}\le 5 \Leftrightarrow \frac{1}{5}\le|2x-3|$, since $|2x-3|>0$, provided $2x-3\ne 0 \Leftrightarrow x\ne\frac{3}{2}$. Now for $x\ne\frac{3}{2}$, we have $\frac{1}{5}\le|2x-3|$ is equivalent to $\frac{1}{5}\le 2x-3 \Leftrightarrow \frac{16}{5}\le 2x \Leftrightarrow \frac{8}{5}\le x$; or $2x-3\le-\frac{1}{5} \Leftrightarrow 2x\le\frac{14}{5} \Leftrightarrow x\le\frac{7}{5}$. Interval: $\left(-\infty,\frac{7}{5}\right]\cup\left[\frac{8}{5},\infty\right)$.

**45.** $|x|<3$

**46.** $|x|>2$

**47.** $|x-7|\ge 5$

**48.** $|x-2|\le 4$

**49.** $|x|\le 2$

**50.** $|x|\ge 1$

**51.** $|x|>3$

**52.** $|x|<4$

**53.** **(a)** Let $x$ be the thickness of the laminate. Then $|x-0.020|\le 0.003$.

**(b)** $|x-0.020|\le 0.003 \Leftrightarrow -0.003\le x-0.020\le 0.003 \Leftrightarrow 0.017\le x\le 0.023$.

**54.** $\left|\dfrac{h-68.2}{2.9}\right|\le 2 \Leftrightarrow -2\le\dfrac{h-68.2}{2.9}\le 2 \Leftrightarrow -5.8\le h-68.2\le 5.8 \Leftrightarrow 62.4\le h\le 74.0$. Thus 95% of the adult males are between 62.4 in and 74.0 in.

**55.** $|x-1|$ is the distance between $x$ and 1; $|x-3|$ is the distance between $x$ and 3. So $|x-1|<|x-3|$ represents those points closer to 1 than to 3, and the solution is $x<2$, since 2 is the point halfway between 1 and 3. If $a<b$, then the solution to $|x-a|<|x-b|$ is $x<\dfrac{a+b}{2}$.

# Chapter 1 Review

**1.** $5x+11=36 \Leftrightarrow 5x=25 \Leftrightarrow x=5$

**2.** $3-x=5+3x \Leftrightarrow -2=4x \Leftrightarrow x=-\frac{1}{2}$

**3.** $3x+12=24 \Leftrightarrow 3x=12 \Leftrightarrow x=4$

**4.** $5x-7=42 \Leftrightarrow 5x=49 \Leftrightarrow x=\frac{49}{5}$

**5.** $7x-6=4x+9 \Leftrightarrow 3x=15 \Leftrightarrow x=5$

**6.** $8-2x=14+x \Leftrightarrow -3x=6 \Leftrightarrow x=-2$

**7.** $\frac{1}{3}x - \frac{1}{2} = 2 \quad\Leftrightarrow\quad 2x - 3 = 12 \quad\Leftrightarrow\quad 2x = 15 \quad\Leftrightarrow\quad x = \frac{15}{2}$

**8.** $\frac{2}{3}x + \frac{3}{5} = \frac{1}{5} - 2x \quad\Leftrightarrow\quad 10x + 9 = 3 - 30x \quad\Leftrightarrow\quad 40x = -6 \quad\Leftrightarrow\quad x = -\frac{6}{40} = -\frac{3}{20}$

**9.** $2(x+3) - 4(x-5) = 8 - 5x \quad\Leftrightarrow\quad 2x + 6 - 4x + 20 = 8 - 5x \quad\Leftrightarrow\quad -2x + 26 = 8 - 5x \quad\Leftrightarrow\quad 3x = -18 \quad\Leftrightarrow$
$x = -6$

**10.** $\dfrac{x-5}{2} - \dfrac{2x+5}{3} = \dfrac{5}{6} \quad\Leftrightarrow\quad 3(x-5) - 2(2x+5) = 5 \quad\Leftrightarrow\quad 3x - 15 - 4x - 10 = 5 \quad\Leftrightarrow\quad -x = 30 \quad\Leftrightarrow\quad x = -30$

**11.** $\dfrac{x+1}{x-1} = \dfrac{2x-1}{2x+1} \quad\Leftrightarrow\quad (x+1)(2x+1) = (2x-1)(x-1) \quad\Leftrightarrow\quad 2x^2 + 3x + 1 = 2x^2 - 3x + 1 \quad\Leftrightarrow\quad 6x = 0 \quad\Leftrightarrow$
$x = 0$

**12.** $\dfrac{x}{x+2} - 3 = \dfrac{1}{x+2} \quad\Leftrightarrow\quad x - 3(x+2) = 1 \quad\Leftrightarrow\quad x - 3x - 6 = 1 \quad\Leftrightarrow\quad -2x = 7 \quad\Leftrightarrow\quad x = -\dfrac{7}{2}$

**13.** $x^2 = 144 \quad\Rightarrow\quad x = \pm 12$

**14.** $4x^2 = 49 \quad\Leftrightarrow\quad x^2 = \frac{49}{4} \quad\Rightarrow\quad x = \pm\frac{7}{2}$

**15.** $5x^4 - 16 = 0 \quad\Leftrightarrow\quad 5x^4 = 16 \quad\Leftrightarrow\quad x^4 = \frac{16}{5} \quad\Rightarrow\quad x^2 = \pm\frac{4}{\sqrt{5}}$. Now $x^2 = -\frac{4}{\sqrt{5}}$ has no real solution, so we solve
$x^2 = \frac{4}{\sqrt{5}} \quad\Rightarrow\quad x = \pm\frac{2}{\sqrt[4]{5}} = \pm\frac{2\sqrt[4]{5^3}}{5}$

**16.** $x^3 - 27 = 0 \quad\Leftrightarrow\quad x^3 = 27 \quad\Rightarrow\quad x = 3.$

**17.** $5x^6 - 15 = 0 \quad\Leftrightarrow\quad 5x^6 = 15 \quad\Leftrightarrow\quad x^6 = 3 \quad\Leftrightarrow\quad x = \sqrt[6]{3}.$

**18.** $6x^4 + 15 = 0 \quad\Leftrightarrow\quad 6x^4 = -15 \quad\Leftrightarrow\quad x^4 = -\frac{5}{2}$. Since $x^4$ must be nonnegative, there is no real solution.

**19.** $(x+1)^3 = -64 \quad\Leftrightarrow\quad x + 1 = -4 \quad\Leftrightarrow\quad x = -1 - 4 = -5.$

**20.** $(x+2)^2 - 2 = 0 \quad\Leftrightarrow\quad (x+2)^2 = 2 \quad\Leftrightarrow\quad x + 2 = \pm\sqrt{2} \quad\Leftrightarrow\quad x = -2 \pm \sqrt{2}.$

**21.** $\sqrt[3]{x} = -3 \quad\Leftrightarrow\quad x = (-3)^3 = -27.$

**22.** $x^{2/3} - 4 = 0 \quad\Leftrightarrow\quad \left(x^{1/3}\right)^2 = 4 \quad\Rightarrow\quad x^{1/3} = \pm 2 \quad\Leftrightarrow\quad x = \pm 8.$

**23.** $4x^{3/4} - 500 = 0 \quad\Leftrightarrow\quad 4x^{3/4} = 500 \quad\Leftrightarrow\quad x^{3/4} = 125 \quad\Leftrightarrow\quad x = 125^{4/3} = 5^4 = 625.$

**24.** $(x-2)^{1/5} = 2 \quad\Leftrightarrow\quad x - 2 = 2^5 = 32 \quad\Leftrightarrow\quad x = 2 + 32 = 34.$

**25.** $\dfrac{x+1}{x-1} = \dfrac{3x}{3x-6} = \dfrac{3x}{3(x-2)} = \dfrac{x}{x-2} \quad\Leftrightarrow\quad (x+1)(x-2) = x(x-1) \quad\Leftrightarrow\quad x^2 - x - 2 = x^2 - x \quad\Leftrightarrow\quad -2 = 0.$
Since this last equation is never true, there is no real solution to the original equation.

**26.** $(x+2)^2 = (x-4)^2 \quad\Leftrightarrow\quad (x+2)^2 - (x-4)^2 = 0 \quad\Leftrightarrow\quad [(x+2) - (x-4)][(x+2) + (x-4)] = 0 \quad\Leftrightarrow$
$[x + 2 - x + 4][x + 2 + x - 4] = 6(2x - 2) = 0 \quad\Leftrightarrow\quad 2x - 2 = 0 \quad\Leftrightarrow\quad x = 1.$

**27.** $x^2 - 9x + 14 = 0 \quad\Leftrightarrow\quad (x-7)(x-2) = 0 \quad\Leftrightarrow\quad x = 7$or $x = 2.$

**28.** $x^2 + 24x + 144 = 0 \quad\Leftrightarrow\quad (x+12)^2 = 0 \quad\Leftrightarrow\quad x + 12 = 0 \quad\Leftrightarrow\quad x = -12.$

**29.** $2x^2 + x = 1 \quad\Leftrightarrow\quad 2x^2 + x - 1 = 0 \quad\Leftrightarrow\quad (2x-1)(x+1) = 0$. So either $2x - 1 = 0 \quad\Leftrightarrow\quad 2x = 1 \quad\Leftrightarrow\quad x = \frac{1}{2}$; or
$x + 1 = 0 \quad\Leftrightarrow\quad x = -1.$

**30.** $3x^2 + 5x - 2 = 0 \quad\Leftrightarrow\quad (3x-1)(x+2) = 0 \quad\Leftrightarrow\quad x = \frac{1}{3}$ or $x = -2.$

**31.** $0 = 4x^3 - 25x = x\left(4x^2 - 25\right) = x(2x-5)(2x+5) = 0$. So either $x = 0$; or $2x - 5 = 0 \quad\Leftrightarrow\quad 2x = 5 \quad\Leftrightarrow\quad x = \frac{5}{2}$;
or $2x + 5 = 0 \quad\Leftrightarrow\quad 2x = -5 \quad\Leftrightarrow\quad x = -\frac{5}{2}.$

**32.** $x^3 - 2x^2 - 5x + 10 = 0 \quad\Leftrightarrow\quad x^2(x-2) - 5(x-2) = 0 \quad\Leftrightarrow\quad (x-2)\left(x^2 - 5\right) = 0 \quad\Leftrightarrow\quad x = 2$ or $x = \pm\sqrt{5}.$

**33.** $3x^2 + 4x - 1 = 0 \Rightarrow$
$x = \frac{-b\pm\sqrt{b^2-4ac}}{2a} = \frac{-(4)\pm\sqrt{(4)^2-4(3)(-1)}}{2(3)} = \frac{-4\pm\sqrt{16+12}}{6} = \frac{-4\pm\sqrt{28}}{6} = \frac{-4\pm2\sqrt{7}}{6} = \frac{2\left(-2\pm\sqrt{7}\right)}{6} = \frac{-2\pm\sqrt{7}}{3}.$

**34.** $x^2 - 3x + 9 = 0 \quad\Rightarrow\quad x = \frac{-b\pm\sqrt{b^2-4ac}}{2a} = \frac{-(-3)\pm\sqrt{(-3)^2-4(1)(9)}}{2(1)} = \frac{3\pm\sqrt{9-36}}{2} = \frac{3\pm\sqrt{-27}}{2}$, which are not real numbers.
There is no real solution.

**35.** $\dfrac{1}{x}+\dfrac{2}{x-1}=3 \;\Leftrightarrow\; (x-1)+2(x)=3(x)(x-1) \;\Leftrightarrow\; x-1+2x=3x^2-3x \;\Leftrightarrow\; 0=3x^2-6x+1 \;\Rightarrow$
$x=\frac{-b\pm\sqrt{b^2-4ac}}{2a}=\frac{-(-6)\pm\sqrt{(-6)^2-4(3)(1)}}{2(3)}=\frac{6\pm\sqrt{36-12}}{6}=\frac{6\pm\sqrt{24}}{6}=\frac{6\pm2\sqrt{6}}{6}=\frac{2(3\pm\sqrt{6})}{6}=\frac{3\pm\sqrt{6}}{3}$.

**36.** $\dfrac{x}{x-2}+\dfrac{1}{x+2}=\dfrac{8}{x^2-4} \;\Leftrightarrow\; x(x+2)+(x-2)=8 \;\Leftrightarrow\; x^2+2x+x-2=8 \;\Leftrightarrow\; x^2+3x-10=0 \;\Leftrightarrow$
$(x-2)(x+5)=0 \;\Leftrightarrow\; x=2$ or $x=-5$. However, since $x=2$ makes the expression undefined, we reject this solution. Hence the only solution is $x=-5$.

**37.** $x^4-8x^2-9=0 \;\Leftrightarrow\; (x^2-9)(x^2+1)=0 \;\Leftrightarrow\; (x-3)(x+3)(x^2+1)=0 \;\Rightarrow\; x-3=0 \;\Leftrightarrow\; x=3$, or $x+3=0 \;\Leftrightarrow\; x=-3$, however $x^2+1=0$ has no real solution. The solutions are $x=\pm3$.

**38.** $x-4\sqrt{x}=32$. Let $u=\sqrt{x}$. Then $u^2-4u=32 \;\Leftrightarrow\; u^2-4u-32=0 \;\Leftrightarrow\; (u-8)(u+4)=0$. So $u-8=0$ $\Rightarrow\; \sqrt{x}-8=0 \;\Leftrightarrow\; \sqrt{x}=8 \;\Rightarrow\; x=64$, or $u+4=0 \;\Rightarrow\; \sqrt{x}+4=0 \;\Leftrightarrow\; \sqrt{x}=-4$, which has no real solution. The only solution is $x=64$.

**39.** $x^{-1/2}-2x^{1/2}+x^{3/2}=0 \;\Leftrightarrow\; x^{-1/2}(1-2x+x^2)=0 \;\Leftrightarrow\; x^{1/2}(1-x)^2=0$. Since $x^{-1/2}\neq0$, the only solution comes from $(1-x)^2=0 \;\Leftrightarrow\; 1-x=0 \;\Leftrightarrow\; x=1$.

**40.** $(1+\sqrt{x})^2-2(1+\sqrt{x})-15=0$. Let $u=1+\sqrt{x}$, then the equation becomes $u^2-2u-15=0 \;\Leftrightarrow$ $(u-5)(u+3)=0$. So $u-5=1+\sqrt{x}-5=0 \;\Leftrightarrow\; \sqrt{x}=4 \;\Rightarrow\; x=16$; or $u+3=1+\sqrt{x}+3=0 \;\Leftrightarrow$ $\sqrt{x}=-4$, which has no real solution. The only solution is $x=16$.

**41.** $|x-7|=4 \;\Leftrightarrow\; x-7=\pm4 \;\Leftrightarrow\; x=7\pm4$, so $x=11$ or $x=3$.

**42.** $|3x|=18$ is equivalent to $3x=\pm18 \;\Leftrightarrow\; x=\pm6$.

**43.** $|2x-5|=9$ is equivalent to $2x-5=\pm9 \;\Leftrightarrow\; 2x=5\pm9 \;\Leftrightarrow\; x=\frac{5\pm9}{2}$. So $x=-2$ or $x=7$.

**44.** $4|3-x|+3=15 \;\Leftrightarrow\; 4|3-x|=12 \;\Leftrightarrow\; |3-x|=3 \;\Leftrightarrow\; 3-x=\pm3$. So $3-x=3 \;\Leftrightarrow\; -x=0 \;\Leftrightarrow$ $x=0$; or $3-x=-3 \;\Leftrightarrow\; -x=-6 \;\Leftrightarrow\; x=6$. The solutions are $x=0$ and $x=6$.

**45.** Let $x$ be the number of pounds of raisins. Then the number of pounds of nuts is $50-x$.

| | Raisins | Nuts | Mixture |
|---|---|---|---|
| Pounds | $x$ | $50-x$ | 50 |
| Rate (cost per pound) | 3.20 | 2.40 | 2.72 |

So $3.20x+2.40(50-x)=2.72(50) \;\Leftrightarrow\; 3.20x+120-2.40x=136 \;\Leftrightarrow\; 0.8x=16 \;\Leftrightarrow\; x=20$. Thus the mixture uses 20 pounds of raisins and $50-20=30$ pounds of nuts.

**46.** Let $t$ be the number of hours that Anthony drives. Then Helen drives for $t-\frac{1}{4}$ hours.

| | Rate | Time | Distance |
|---|---|---|---|
| Anthony | 45 | $t$ | $45t$ |
| Helen | 40 | $t-\frac{1}{4}$ | $40\left(t-\frac{1}{4}\right)$ |

When they pass each other, they will have traveled a total of 160 miles. So $45t+40\left(t-\frac{1}{4}\right)=160 \;\Leftrightarrow$ $45t+40t-10=160 \;\Leftrightarrow\; 85t=170 \;\Leftrightarrow\; t=2$. Since Anthony leaves at 2:00 P.M. and travels for 2 hours, they pass each other at 4:00 P.M.

**47.** Let $r$ be the rate the woman runs in mi/h. Then she cycles at $r+8$ mi/h.

| | Rate | Time | Distance |
|---|---|---|---|
| Cycle | $r+8$ | $\dfrac{4}{r+8}$ | 4 |
| Run | $r$ | $\dfrac{2.5}{r}$ | 2.5 |

Since the total time of the workout is 1 hour, we have $\dfrac{4}{r+8}+\dfrac{2.5}{r}=1$. Multiplying by $2r\,(r+8)$, we get $4\,(2r)+2.5\,(2)\,(r+8)=2r\,(r+8) \quad\Leftrightarrow\quad 8r+5r+40=2r^2+16r \quad\Leftrightarrow\quad 0=2r^2+3r-40 \quad\Rightarrow$ $r=\frac{-3\pm\sqrt{(3)^2-4(2)(-40)}}{2(2)}=\frac{-3\pm\sqrt{9+320}}{4}=\frac{-3\pm\sqrt{329}}{4}$. Since $r\geq 0$, we reject the negative value. She runs at $r=\frac{-3+\sqrt{329}}{4}\approx 3.78$ mi/h.

**48.** Substituting 75 for $d$, we have $75=x+\dfrac{x^2}{20} \quad\Leftrightarrow\quad 1500=20x+x^2 \quad\Leftrightarrow\quad x^2+20x-1500=0 \quad\Leftrightarrow$ $(x-30)\,(x+50)=0$.So $x=30$ or $x=-50$. The speed of the car was 30 mi/h.

**49.** Let $x$ be the length of one side in cm. Then $28-x$ is the length of the other side. Using the Pythagorean Theorem, we have $x^2+(28-x)^2=20^2 \quad\Leftrightarrow\quad x^2+784-56x+x^2=400 \quad\Leftrightarrow\quad 2x^2-56x+384=0 \quad\Leftrightarrow$ $2\left(x^2-28x+192\right)=0 \quad\Leftrightarrow\quad 2\,(x-12)\,(x-16)=0$. So $x=12$ or $x=16$. If $x=12$, then the other side is $28-12=16$. Similarly, if $x=16$, then the other side is 12. The sides are 12 cm and 16 cm.

**50.** Let $t$ be the time it would take Abbie to paint a living room if she works alone. It would take Beth $2t$ hours to paint the living room alone, and it would take $3t$ hours for Cathie to paint the living room. Thus Abbie does $\dfrac{1}{t}$ of the job per hour, Beth does $\dfrac{1}{2t}$ of the job per hour, and Cathie does $\dfrac{1}{3t}$ of the job per hour. So $\dfrac{1}{t}+\dfrac{1}{2t}+\dfrac{1}{3t}=1 \quad\Leftrightarrow\quad 6+3+2=6t$ $\quad\Leftrightarrow\quad 6t=11 \quad\Leftrightarrow\quad t=\frac{11}{6}$. So it would Abbie 1 hour 50 minutes to paint the living room alone.

**51.** Let $w$ be width of the pool. Then the length of the pool is $2w$, and its volume is $8\,(w)\,(2w)=8464 \quad\Leftrightarrow\quad 16w^2=8464$ $\quad\Leftrightarrow\quad w^2=529 \quad\Rightarrow\quad w=\pm 23$. Since $w>0$, we reject the negative value. The pool is 23 feet wide, $2\,(23)=46$ feet long, and 8 feet deep.

**52.** Let $l$ be length of each garden plot. The width of each plot is then $\dfrac{80}{l}$ and the total amount of fencing material is $4\,(l)+6\left(\dfrac{80}{l}\right)=88$. Thus $4l+\dfrac{480}{l}=88 \quad\Leftrightarrow\quad 4l^2+480=88l \quad\Leftrightarrow\quad 4l^2-88l+480=0 \quad\Leftrightarrow$ $4\left(l^2-22l+120\right)=0 \quad\Leftrightarrow\quad 4\,(l-10)\,(l-12)=0$. So $l=10$ or $l=12$. If $l=10$ ft, then the width of each plot is $\frac{80}{10}=8$ ft. If $l=12$ ft, then the width of each plot is $\frac{80}{12}=6.67$ ft. Both solutions are possible.

**53.** $(3-5i)-(6+4i)=3-5i-6-4i=-3-9i$

**54.** $(-2+3i)+\left(\frac{1}{2}-i\right)=-2+3i+\frac{1}{2}-i=-\frac{3}{2}+2i$

**55.** $(2+7i)\,(6-i)=12-2i+42i-7i^2=12+40i+7=19+40i$

**56.** $3\,(5-2i)\,\frac{i}{5}=(15-6i)\,\frac{i}{5}=3i-\frac{6}{5}i^2=\frac{6}{5}+3i$

**57.** $\dfrac{2-3i}{2+3i}=\dfrac{2-3i}{2+3i}\cdot\dfrac{2-3i}{2-3i}=\dfrac{4-12i+9i^2}{4-9i^2}=\dfrac{4-12i-9}{4+9}=\dfrac{-5-12i}{13}=-\frac{5}{13}-\frac{12}{13}i$

**58.** $\dfrac{2+i}{4-3i}=\dfrac{2+i}{4-3i}\cdot\dfrac{4+3i}{4+3i}=\dfrac{8+6i+4i+3i^2}{16-9i^2}=\dfrac{8+10i-3}{16+9}=\dfrac{5+10i}{25}=\frac{1}{5}+\frac{2}{5}i$

**59.** $i^{45}=i^{44}i=\left(i^4\right)^{11}i=(1)^{11}\,i=i$

**60.** $(3-i)^3=(3)^3-3\,(3)^2\,(i)+3\,(3)\,(i)^2-(i)^3=27-27i+9i^2-i^3=27-27i-9+i=18-26i$

**61.** $\left(1-\sqrt{-3}\right)\left(2+\sqrt{-4}\right)=\left(1-\sqrt{3}i\right)(2+2i)=2+2i-2\sqrt{3}i-2\sqrt{3}i^2=2+\left(2-2\sqrt{3}\right)i+2\sqrt{3}$
$=\left(2+2\sqrt{3}\right)+\left(2-2\sqrt{3}\right)i$

**62.** $\sqrt{-5}\cdot\sqrt{-20}=\sqrt{5}i\cdot 2\sqrt{5}i=10i^2=-10$

**63.** $x^2+16=0 \quad\Leftrightarrow\quad x^2=-16 \quad\Rightarrow\quad x=\pm\sqrt{-16}=\pm 4i$

**64.** $x^2=-12 \quad\Rightarrow\quad x=\pm\sqrt{-12}=\pm 2\sqrt{3}i$

**65.** $x^2+6x+10=0 \quad\Rightarrow\quad x=\frac{-6\pm\sqrt{6^2-4(1)(10)}}{2(1)}=\frac{-6\pm\sqrt{36-40}}{2}=\frac{-6\pm\sqrt{-4}}{2}=\frac{-6\pm 2i}{2}=-3\pm i$

**66.** $2x^2-3x+2=0 \quad\Rightarrow\quad x=\frac{-(-3)\pm\sqrt{(-3)^2-4(2)(2)}}{2(2)}=\frac{3\pm\sqrt{9-16}}{4}=\frac{3\pm\sqrt{-7}}{4}=\frac{3}{4}\pm\frac{\sqrt{7}}{4}i$

**67.** $x^4-256=0 \quad\Leftrightarrow\quad \left(x^2+16\right)\left(x^2-16\right)=0$. Thus either $x^2+16=0 \quad\Leftrightarrow\quad x^2=-16 \quad\Rightarrow\quad x=\pm\sqrt{-16} \quad\Leftrightarrow$ $x=\pm 4i$; or $x^2-16=0 \quad\Leftrightarrow\quad (x-4)(x+4)=0 \quad\Rightarrow\quad x-4=0 \quad\Leftrightarrow\quad x=4$; or $x+4=0 \quad\Leftrightarrow\quad x=-4$. The solutions are $\pm 4i$ and $\pm 4$.

**68.** $x^3-2x^2+4x-8=0 \quad\Leftrightarrow\quad x^2(x-2)+4(x-2)=0 \quad\Leftrightarrow\quad (x-2)\left(x^2+4\right)=0 \quad\Leftrightarrow\quad x=2$, or $x^2+4=0$ $\Leftrightarrow\quad x^2=-4 \quad\Rightarrow\quad x=\pm\sqrt{-4}=\pm 2i$. The solutions are 2 and $\pm 2i$.

**69.** $x^2+4x=(2x+1)^2 \quad\Leftrightarrow\quad x^2+4x=4x^2+4x+1 \quad\Leftrightarrow\quad 0=3x^2+1 \quad\Leftrightarrow\quad 3x^2=-1 \quad\Leftrightarrow$ $x^2=-\frac{1}{3} \Rightarrow \quad x=\pm\sqrt{-\frac{1}{3}}=\pm\frac{\sqrt{3}}{3}i$.

**70.** $x^3=125 \quad\Leftrightarrow\quad x^3-125=0 \quad\Leftrightarrow\quad (x-5)\left(x^2+5x+25\right)=0$. So $x=5$, or $x=\frac{-5\pm\sqrt{5^2-4(1)(25)}}{2(1)}=\frac{-5\pm\sqrt{25-100}}{2}=\frac{-5\pm\sqrt{-75}}{2}=\frac{-5\pm 5\sqrt{3}i}{2}$. The solutions are 5 and $-\frac{5}{2}\pm\frac{5\sqrt{3}}{2}i$.

**71.** $3x-2>-11 \quad\Leftrightarrow\quad 3x>-9 \quad\Leftrightarrow\quad x>-3$.
Interval: $(-3,\infty)$

Graph: −3

**72.** $12-x\geq 7x \quad\Leftrightarrow\quad 12\geq 8x \quad\Leftrightarrow\quad \frac{3}{2}\geq x$.
Interval: $\left(-\infty,\frac{3}{2}\right]$

Graph: $\frac{3}{2}$

**73.** $-1<2x+5\leq 3 \quad\Leftrightarrow\quad -6<2x\leq -2 \quad\Leftrightarrow$ $-3<x\leq -1$
Interval: $(-3,-1]$

Graph: −3 −1

**74.** $3-x\leq 2x-7 \quad\Leftrightarrow\quad 10\leq 3x \quad\Leftrightarrow\quad \frac{10}{3}\leq x$
Interval: $\left[\frac{10}{3},\infty\right)$

Graph: $\frac{10}{3}$

**75.** $x^2+4x-12>0 \quad\Leftrightarrow\quad (x-2)(x+6)>0$. The expression on the left of the inequality changes sign where $x=2$ and where $x=-6$. Thus we must check the intervals in the following table.

| Interval | $(-\infty,-6)$ | $(-6,2)$ | $(2,\infty)$ |
|---|---|---|---|
| Sign of $x-2$ | − | − | + |
| Sign of $x+6$ | − | + | + |
| Sign of $(x-2)(x+6)$ | + | − | + |

Interval: $(-\infty,-6)\cup(2,\infty)$.

Graph: 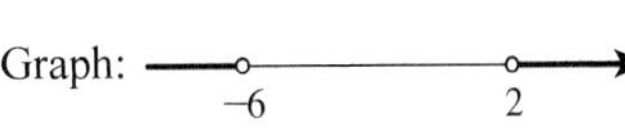


**76.** $x^2\leq 1 \quad\Leftrightarrow\quad x^2-1\leq 0 \quad\Leftrightarrow\quad (x-1)(x+1)\leq 0$. The expression on the left of the inequality changes sign when $x=-1$ and $x=1$. Thus we must check the intervals in the following table.

| Interval | $(-\infty,-1)$ | $(-1,1)$ | $(1,\infty)$ |
|---|---|---|---|
| Sign of $x-1$ | − | − | + |
| Sign of $x+1$ | − | + | + |
| Sign of $(x-1)(x+1)$ | + | − | + |

Interval: $[-1,1]$

Graph: 

**77.** $\dfrac{2x+5}{x+1} \le 1 \Leftrightarrow \dfrac{2x+5}{x+1} - 1 \le 0 \Leftrightarrow \dfrac{2x+5}{x+1} - \dfrac{x+1}{x+1} \le 0 \Leftrightarrow \dfrac{x+4}{x+1} \le 0$. The expression on the left of the inequality changes sign where $x = -1$ and where $x = -4$. Thus we must check the intervals in the following table.

| Interval | $(-\infty, -4)$ | $(-4, -1)$ | $(-1, \infty)$ |
|---|---|---|---|
| Sign of $x+4$ | $-$ | $+$ | $+$ |
| Sign of $x+1$ | $-$ | $-$ | $+$ |
| Sign of $\dfrac{x+4}{x+1}$ | $+$ | $-$ | $+$ |

We exclude $x = -1$, since the expression is not defined at this value. Thus the solution is $[-4, -1)$.

Graph: –4 1

**78.** $2x^2 \ge x+3 \Leftrightarrow 2x^2 - x - 3 \ge 0 \Leftrightarrow (2x-3)(x+1) \ge 0$. The expression on the left of the inequality changes sign when $-1$ and $\frac{3}{2}$. Thus we must check the intervals in the following table.

| Interval | $(-\infty, -1)$ | $\left(-1, \frac{3}{2}\right)$ | $\left(\frac{3}{2}, \infty\right)$ |
|---|---|---|---|
| Sign of $2x-3$ | $-$ | $-$ | $+$ |
| Sign of $x+1$ | $-$ | $+$ | $+$ |
| Sign of $(2x-3)(x+1)$ | $+$ | $-$ | $+$ |

Interval: $(-\infty, -1] \cup \left[\frac{3}{2}, \infty\right)$

Graph: –1 $\frac{3}{2}$

**79.** $\dfrac{x-4}{x^2-4} \le 0 \Leftrightarrow \dfrac{x-4}{(x-2)(x+2)} \le 0$. The expression on the left of the inequality changes sign where $x = -2$, where $x = 2$, and where $x = 4$. Thus we must check the intervals in the following table.

| Interval | $(-\infty, -2)$ | $(-2, 2)$ | $(2, 4)$ | $(4, \infty)$ |
|---|---|---|---|---|
| Sign of $x-4$ | $-$ | $-$ | $-$ | $+$ |
| Sign of $x-2$ | $-$ | $-$ | $+$ | $+$ |
| Sign of $x+2$ | $-$ | $+$ | $+$ | $+$ |
| Sign of $\dfrac{x-4}{(x-2)(x+2)}$ | $-$ | $+$ | $-$ | $+$ |

Since the expression is not defined when $x = \pm 2$,we exclude these values and the solution is $(-\infty, -2) \cup (2, 4]$.

Graph: –2 2 4

**80.** $\dfrac{5}{x^3 - x^2 - 4x + 4} < 0 \Leftrightarrow \dfrac{5}{x^2(x-1) - 4(x-1)} < 0 \Leftrightarrow \dfrac{5}{(x-1)(x^2-4)} < 0 \Leftrightarrow$
$\dfrac{5}{(x-1)(x-2)(x+2)} < 0$. The expression on the left of the inequality changes sign when $-2$, $1$,and $2$. Thus we must check the intervals in the following table.

| Interval | $(-\infty, -2)$ | $(-2, 1)$ | $(1, 2)$ | $(2, \infty)$ |
|---|---|---|---|---|
| Sign of $x-1$ | $-$ | $-$ | $+$ | $+$ |
| Sign of $x-2$ | $-$ | $-$ | $-$ | $+$ |
| Sign of $x+2$ | $-$ | $+$ | $+$ | $+$ |
| Sign of $\dfrac{5}{(x-1)(x-2)(x+2)}$ | $-$ | $+$ | $-$ | $+$ |

Interval: $(-\infty, -2) \cup (1, 2)$

Graph: –2 1 2

**81.** $|x-5| \le 3 \quad \Leftrightarrow \quad -3 \le x-5 \le 3 \quad \Leftrightarrow \quad 2 \le x \le 8$.

Interval: $[2, 8]$

Graph:

2 8

**82.** $|x-4| < 0.02 \quad \Leftrightarrow \quad -0.02 < x-4 < 0.02 \quad \Leftrightarrow$

$3.98 < x < 4.02$

Interval: $(3.98, 4.02)$

Graph:

3.98 4.02

**83.** $|2x+1| \ge 1$ is equivalent to $2x+1 \ge 1$ or $2x+1 \le -1$.

*Case 1:* $2x+1 \ge 1 \quad \Leftrightarrow \quad 2x \ge 0 \quad \Leftrightarrow \quad x \ge 0$

*Case 2:* $2x+1 \le -1 \quad \Leftrightarrow \quad 2x \le -2 \quad \Leftrightarrow \quad x \le -1$

Interval: $(-\infty, -1] \cup [0, \infty)$.

Graph:

−1 0

**84.** $|x-1|$ is the distance between $x$ and 1 on the number line, and $|x-3|$ is the distance between $x$ and 3. We want those points that are closer to 1 than to 3. Since 2 is midway between 1 and 3, we get $x \in (-\infty, 2)$ as the solution.

Graph:

2

**85. (a)** For $\sqrt{24-x-3x^2}$ to define a real number, we must have $24-x-3x^2 \ge 0 \quad \Leftrightarrow \quad (8-3x)(3+x) \ge 0$. The expression on the left of the inequality changes sign where $8-3x=0 \quad \Leftrightarrow \quad -3x=-8 \quad \Leftrightarrow \quad x=\frac{8}{3}$; or where $x=-3$. Thus we must check the intervals in the following table.

| Interval | $(-\infty, -3)$ | $\left(-3, \frac{8}{3}\right)$ | $\left(\frac{8}{3}, \infty\right)$ |
|---|---|---|---|
| Sign of $8-3x$ | + | + | − |
| Sign of $3+x$ | − | + | + |
| Sign of $(8-3x)(3+x)$ | − | + | − |

Interval: $\left[-3, \frac{8}{3}\right]$.

Graph:

−3 $\frac{8}{3}$

**(b)** For $\dfrac{1}{\sqrt[4]{x-x^4}}$ to define a real number we must have $x-x^4 > 0 \quad \Leftrightarrow \quad x\left(1-x^3\right) > 0 \quad \Leftrightarrow$ $x(1-x)\left(1+x+x^2\right) > 0$. The expression on the left of the inequality changes sign where $x=0$; or where $x=1$; or where $1+x+x^2=0 \quad \Rightarrow \quad x=\frac{-1\pm\sqrt{1^2-4(1)(1)}}{2(1)}=\frac{1\pm\sqrt{1-4}}{2}$ which is imaginary. We check the intervals in the following table.

| Interval | $(-\infty, 0)$ | $(0, 1)$ | $(1, \infty)$ |
|---|---|---|---|
| Sign of $x$ | − | + | + |
| Sign of $1-x$ | + | + | − |
| Sign of $1+x+x^2$ | + | + | + |
| Sign of $x(1-x)\left(1+x+x^2\right)$ | − | + | − |

Interval: $(0, 1)$.

Graph:

0 1

**86.** We have $8 \le \frac{4}{3}\pi r^3 \le 12 \quad \Leftrightarrow \quad \dfrac{6}{\pi} \le r^3 \le \dfrac{9}{\pi} \quad \Leftrightarrow \quad \sqrt[3]{\dfrac{6}{\pi}} \le r \le \sqrt[3]{\dfrac{9}{\pi}}$. Thus $r \in \left[\sqrt[3]{\dfrac{6}{\pi}}, \sqrt[3]{\dfrac{9}{\pi}}\right]$.

# Chapter 1 Test

**1. (a)** $3x - 7 = \frac{1}{2}x \Leftrightarrow -7 = -\frac{5}{2}x \Leftrightarrow x = \frac{14}{5}$

**(b)** $x^{2/3} - 9 = 0 \Leftrightarrow x^{2/3} = 9 \Leftrightarrow x^{1/3} = \pm 3 \Leftrightarrow x = \pm 27$

**(c)** $\dfrac{2x}{2x-5} = \dfrac{x+2}{x-1} \Leftrightarrow 2x(x-1) = (2x-5)(x+2) \Leftrightarrow 2x^2 - 2x = 2x^2 - x - 10 \Leftrightarrow -x = -10 \Leftrightarrow x = 10$

**2.** Let $d$ be the distance between Ajax and Bixby. Then we have the following table.

| | Rate | Time | Distance |
|---|---|---|---|
| Ajax to Bixby | 50 | $\dfrac{d}{50}$ | $d$ |
| Bixby to Ajax | 60 | $\dfrac{d}{60}$ | $d$ |

We use the fact that the total time is $4\frac{2}{5}$ hours to get the equation $\dfrac{d}{50} + \dfrac{d}{60} = \dfrac{22}{5} \Leftrightarrow 6d + 5d = 1320 \Leftrightarrow 11d = 1320 \Leftrightarrow d = 120$. Thus Ajax and Bixby are 120 miles apart.

**3. (a)** $(6 - 2i) - (7 - \frac{1}{2}i) = 6 - 2i - 7 + \frac{1}{2}i = -1 - \frac{3}{2}i$

**(b)** $(1 + i)(3 - 2i) = 3 - 2i + 3i - 2i^2 = 3 + i + 2 = 5 + i$

**(c)** $\dfrac{5+10i}{3-4i} = \dfrac{5+10i}{3-4i} \cdot \dfrac{3+4i}{3+4i} = \dfrac{15 + 20i + 30i + 40i^2}{9 - 16i^2} = \dfrac{15 + 50i - 40}{9 + 16} = \dfrac{-25 + 50i}{25} = -1 + 2i$

**(d)** $i^{50} = i^{48} \cdot i^2 = \left(i^4\right)^{12} \cdot i^2 = (1)^{12} \cdot (-1) = -1$

**(e)** $\left(2 - \sqrt{-2}\right)\left(\sqrt{8} + \sqrt{-4}\right) = \left(2 - \sqrt{2}i\right)\left(2\sqrt{2} + 2i\right) = 4\sqrt{2} + 4i - 4i - 2\sqrt{2}i^2 = 4\sqrt{2} + 2\sqrt{2} = 6\sqrt{2}$

**4. (a)** $x^2 - x - 12 = 0 \Leftrightarrow (x-4)(x+3) = 0$. So $x = 4$ or $x = -3$.

**(b)** $2x^2 + 4x + 3 = 0 \Rightarrow x = \frac{-4 \pm \sqrt{4^2 - 4(2)(3)}}{2(2)} = \frac{-4 \pm \sqrt{16-24}}{4} = \frac{-4 \pm \sqrt{-8}}{4} = \frac{-4 \pm 2\sqrt{2}i}{4} = -1 \pm \frac{\sqrt{2}}{2}i$

**(c)** $\sqrt{3 - \sqrt{x+5}} = 2 \Rightarrow 3 - \sqrt{x+5} = 4 \Leftrightarrow -1 = \sqrt{x+5}$. (Note that this is impossible, so there can be no solution.) Squaring both sides again, we get $1 = x + 5 \Leftrightarrow x = -4$. But this does not satisfy the original equation, so there is no solution. (You must always check your final answers if you have squared both sides when solving an equation, since extraneous answers may be introduced, as here.)

**(d)** $x^{1/2} - 3x^{1/4} + 2 = 0$. Let $u = x^{1/4}$, then we have $u^2 - 3u + 2 = 0 \Leftrightarrow (u-2)(u-1) = 0$. So $u - 2 = x^{1/4} - 2 = 0 \Leftrightarrow x^{1/4} = 2 \Rightarrow x = 2^4 = 16$; or $u - 1 = x^{1/4} - 1 = 0 \Leftrightarrow x^{1/4} = 1 \Rightarrow x = 1^4 = 1$. Thus the solutions are $x = 16$ and $x = 1$.

**(e)** $x^4 - 16x^2 = 0 \Leftrightarrow x^2(x^2 - 16) = 0 \Leftrightarrow x^2(x-4)(x+4) = 0$. So $x = 0, -4$, or $4$.

**(f)** $3|x-4| - 10 = 0 \Leftrightarrow 3|x-4| = 10 \Leftrightarrow |x-4| = \frac{10}{3} \Leftrightarrow x - 4 = \pm\frac{10}{3} \Leftrightarrow x = 4 \pm \frac{10}{3}$. So $x = 4 - \frac{10}{3} = \frac{2}{3}$ or $x = 4 + \frac{10}{3} = \frac{22}{3}$. Thus the solutions are $x = \frac{2}{3}$ and $x = \frac{22}{3}$.

**5.** Let $w$ be the width of the parcel of land. Then $w + 70$ is the length of the parcel of land. Then $w^2 + (w+70)^2 = 130^2 \Leftrightarrow w^2 + w^2 + 140w + 4900 = 16{,}900 \Leftrightarrow 2w^2 + 140w - 12{,}000 = 0 \Leftrightarrow w^2 + 70w - 6000 = 0 \Leftrightarrow (w-50)(w+120) = 0$. So $w = 50$ or $w = -120$. Since $w \geq 0$, the width is $w = 50$ ft and the length is $w + 70 = 120$ ft.

**6. (a)** $-1 \leq 5 - 2x < 10 \quad \Leftrightarrow \quad -6 \leq -2x < 5 \quad \Leftrightarrow \quad 3 \geq x > -\frac{5}{2}$ or $-\frac{5}{2} < x \leq 3$.

Interval: $\left(-\frac{5}{2}, 3\right]$. Graph: 

**(b)** $x(x-1)(x-2) > 0$.The expression on the left of the inequality changes sign when $x = 0$, $x = 1$, and $x = 2$. Thus we must check the intervals in the following table.

| Interval | $(-\infty, 0)$ | $(0, 1)$ | $(1, 2)$ | $(2, \infty)$ |
|---|---|---|---|---|
| Sign of $x$ | $-$ | $+$ | $+$ | $+$ |
| Sign of $x - 1$ | $-$ | $-$ | $+$ | $+$ |
| Sign of $x - 2$ | $-$ | $-$ | $-$ | $+$ |
| Sign of $x(x-1)(x-2)$ | $-$ | $+$ | $-$ | $+$ |

From the table, the solution set is $\{x \mid 0 < x < 1 \text{ or } 2 < x\}$.

Interval: $(0, 1) \cup (2, \infty)$. Graph: 

**(c)** $|x - 3| < 2$ is equivalent to $-2 < x - 3 < 2 \quad \Leftrightarrow \quad 1 < x < 5$.

Interval: $(1, 5)$. Graph: 

**(d)** $\dfrac{2x+5}{x+1} \leq 1 \quad \Leftrightarrow \quad \dfrac{2x+5}{x+1} - 1 \leq 0 \quad \Leftrightarrow \quad \dfrac{2x+5}{x+1} - \dfrac{x+1}{x+1} \leq 0 \quad \Leftrightarrow \quad \dfrac{x+4}{x+1} \leq 0$. The expression on the left of the inequality changes sign where $x = -4$ and where $x = -1$. Thus we must check the intervals in the following table.

| Interval | $(-\infty, -4)$ | $(-4, -1)$ | $(-1, \infty)$ |
|---|---|---|---|
| Sign of $x + 4$ | $-$ | $+$ | $+$ |
| Sign of $x + 1$ | $-$ | $-$ | $+$ |
| Sign of $\dfrac{x+4}{x+1}$ | $+$ | $-$ | $+$ |

Since $x = -1$ makes the expression in the inequality undefined, we exclude this value.

Interval: $[-4, -1)$. Graph: 

**7.** $5 \leq \frac{5}{9}(F - 32) \leq 10 \quad \Leftrightarrow \quad 9 \leq F - 32 \leq 18 \quad \Leftrightarrow \quad 41 \leq F \leq 50$. Thus the medicine is to be stored at a temperature between $41°$ F and $50°$ F.

**8.** For $\sqrt{4x - x^2}$ to be defined as a real number $4x - x^2 \geq 0 \quad \Leftrightarrow \quad x(4 - x) \geq 0$. The expression on the left of the inequality changes sign when $x = 0$ and $x = 4$. Thus we must check the intervals in the following table.

| Interval | $(-\infty, 0)$ | $(0, 4)$ | $(4, \infty)$ |
|---|---|---|---|
| Sign of $x$ | $-$ | $+$ | $+$ |
| Sign of $4 - x$ | $+$ | $+$ | $-$ |
| Sign of $x(4 - x)$ | $-$ | $+$ | $-$ |

From the table, we see that $\sqrt{4x - x^2}$ is defined when $0 \leq x \leq 4$.

# Focus on Modeling: Making the Best Decisions

**1. (a)** The total cost is $\begin{pmatrix}\text{cost of}\\\text{copier}\end{pmatrix}+\begin{pmatrix}\text{maintenance}\\\text{cost}\end{pmatrix}\begin{pmatrix}\text{number}\\\text{of months}\end{pmatrix}+\begin{pmatrix}\text{copy}\\\text{cost}\end{pmatrix}\begin{pmatrix}\text{number}\\\text{of months}\end{pmatrix}$. Each month the copy cost is $8000 \cdot 0.03 = 240$. Thus we get $C_1(n) = 5800 + 25n + 240n = 5800 + 265n$.

**(b)** In this case the cost is $\begin{pmatrix}\text{rental}\\\text{cost}\end{pmatrix}\begin{pmatrix}\text{number}\\\text{of months}\end{pmatrix}+\begin{pmatrix}\text{copy}\\\text{cost}\end{pmatrix}\begin{pmatrix}\text{number}\\\text{of months}\end{pmatrix}$. Each month the copy cost is $8000 \cdot 0.06 = 480$. Thus we get $C_2(n) = 95n + 480n = 575n$.

**(c)**

| Years | $n$ | Purchase | Rental |
|---|---|---|---|
| 1 | 12 | 8,980 | 6,900 |
| 2 | 24 | 12,160 | 13,800 |
| 3 | 36 | 15,340 | 20,700 |
| 4 | 48 | 18,520 | 27,600 |
| 5 | 60 | 21,700 | 34,500 |
| 6 | 72 | 24,880 | 41,400 |

**(d)** The cost is the same when $C_1(n) = C_2(n)$. So $5800 + 265n = 575n \quad\Leftrightarrow\quad 5800 = 310n \quad\Leftrightarrow\quad n \approx 18.71$ months.

**2. (a)** The cost of Plan 1 is $3 \cdot \begin{pmatrix}\text{daily}\\\text{cost}\end{pmatrix}+\begin{pmatrix}\text{cost per}\\\text{mile}\end{pmatrix}\begin{pmatrix}\text{number}\\\text{of miles}\end{pmatrix} = 3 \cdot 65 + 0.15x = 195 + 0.15x$.

The cost of Plan 2 is $3 \cdot \begin{pmatrix}\text{daily}\\\text{cost}\end{pmatrix} = 3 \cdot 90 = 270$.

**(b)** When $x = 400$, Plan 1 costs $195 + 0.15(400) = \$255$ and Plan 2 costs \$270, so Plan 1 is cheaper. When $x = 800$, Plan 1 costs $195 + 0.15(800) = \$315$ and Plan 2 costs \$270, so Plan 2 is cheaper.

**(c)** The cost is the same when $195 + 0.15x = 270 \quad\Leftrightarrow\quad 0.15 = 75x \quad\Leftrightarrow\quad x = 500$. So both plans cost \$270 when the businessman drives 500 miles.

**3. (a)** The total cost is $\begin{pmatrix}\text{setup}\\\text{cost}\end{pmatrix}+\begin{pmatrix}\text{cost per}\\\text{tire}\end{pmatrix}\begin{pmatrix}\text{number}\\\text{of tires}\end{pmatrix}$. So $C(x) = 8000 + 22x$.

**(b)** The revenue is $\begin{pmatrix}\text{price per}\\\text{tire}\end{pmatrix}\begin{pmatrix}\text{number}\\\text{of tires}\end{pmatrix}$. So $R(x) = 49x$.

**(c)** Profit = Revenue – Cost. So $P(x) = R(x) - C(x) = 49x - (8000 + 22x) = 27x - 8000$.

**(d)** Break even is when profit is zero. Thus $27x - 8000 = 0 \quad\Leftrightarrow\quad 27x = 8000 \quad\Leftrightarrow\quad x \approx 296.3$. So they need to sell at least 297 tires to break even.

**4. (a)** *Option 1:* In this option the width is constant at 100. Let $x$ be the increase in length. Then the additional area is width $\times \begin{pmatrix} \text{increase} \\ \text{in length} \end{pmatrix} = 100x$. The cost is the sum of the costs of moving the old fence, and of installing the new one. The cost of moving is $\$6 \cdot 100 = \$600$ and the cost of installation is $2 \cdot 10 \cdot x = 20x$, so the total cost is $C = 20x + 600$. Solving for $x$, we get $C = 20x + 600 \quad\Leftrightarrow\quad 20x = C - 600 \quad\Leftrightarrow\quad x = \dfrac{C - 600}{20}$. Substituting in the area we have $A_1 = 100\left(\dfrac{C - 600}{20}\right) = 5\,(C - 600) = 5C - 3{,}000$.

*Option 2:* In this option the length is constant at 180. Let $y$ be the increase in the width. Then the additional area is length $\times \begin{pmatrix} \text{increase} \\ \text{in width} \end{pmatrix} = 180y$. The cost of moving the old fence is $6 \cdot 180 = \$1080$ and the cost of installing the new one is $2 \cdot 10 \cdot y = 20x$, so the total cost is $C = 20y + 1080$. Solving for $y$, we get $C = 20y + 1080 \quad\Leftrightarrow\quad 20y = C - 1080 \quad\Leftrightarrow\quad y = \dfrac{C - 1080}{20}$. Substituting in the area we have $A_2 = 180\left(\dfrac{C - 1080}{20}\right) = 9\,(C - 1080) = 9C - 9{,}720$.

**(b)**

| Cost, $C$ | Area gain $A_1$ from Option 1 | Area gain $A_2$ from Option 2 |
|---|---|---|
| \$1100 | 2,500 ft$^2$ | 180 ft$^2$ |
| \$1200 | 3,000 ft$^2$ | 1,080 ft$^2$ |
| \$1500 | 4,500 ft$^2$ | 3,780 ft$^2$ |
| \$2000 | 7,000 ft$^2$ | 8,280 ft$^2$ |
| \$2500 | 9,500 ft$^2$ | 12,780 ft$^2$ |
| \$3000 | 12,000 ft$^2$ | 17,280 ft$^2$ |

**(c)** If the farmer has only \$1200, Option 1 gives him the greatest gain. If the farmer has only \$2000, Option 2 gives him the greatest gain.

**5. (a)** Design 1 is a square and the perimeter of a square is four times the length of a side. $24 = 4x$, so each side is $x = 6$ feet long. Thus the area is $6^2 = 36$ ft$^2$.

Design 2 is a circle with perimeter $2\pi r$ and area $\pi r^2$. Thus we must solve $2\pi r = 24 \quad\Leftrightarrow\quad r = \dfrac{12}{\pi}$. Thus, the area is $\pi\left(\dfrac{12}{\pi}\right)^2 = \dfrac{144}{\pi} \approx 45.8$ ft$^2$. Design 2 gives the largest area.

**(b)** In Design 1, the cost is \$3 times the perimeter $p$, so $120 = 3p$ and the perimeter is 40 feet. By part (a), each side is then $\frac{40}{4} = 10$ feet long. So the area is $10^2 = 100$ ft$^2$.

In Design 2, the cost is \$4 times the perimeter $p$. Because the perimeter is $2\pi r$, we get $120 = 4\,(2\pi r)$ so $r = \dfrac{120}{8\pi} = \dfrac{15}{\pi}$. The area is $\pi r^2 = \pi\left(\dfrac{15}{\pi}\right)^2 = \dfrac{225}{\pi} \approx 71.6$ ft$^2$. Design 1 gives the largest area.

**6. (a)** Plan 1: Tomatoes every year. Profit = acres × (Revenue − cost) = $100\,(1600 - 300) = 130{,}000$. Then for $n$ years the profit is $P_1(n) = 130{,}000n$.

**(b)** Plan 2: Soybeans followed by tomatoes. The profit for two years is Profit = acres × $\left[\begin{pmatrix}\text{soybean}\\\text{revenue}\end{pmatrix} + \begin{pmatrix}\text{tomato}\\\text{revenue}\end{pmatrix}\right] = 100\,(1200 + 1600) = 280{,}000$. Remember that no fertilizer is needed in this plan. Then for $2k$ years, the profit is $P_2(2k) = 280{,}000k$.

**(c)** $P_1(10) = 130{,}000\,(10) = 1{,}300{,}000$. Since $2k = 10$ when $k = 5$. Thus $P_2(10) = 280{,}000\,(5) = 1{,}400{,}000$. So Plan B is more profitable.

**7. (a)**

| Minutes | Plan A | Plan B | Plan C |
|---|---|---|---|
| 500 | $39 + 100\,(0.20) = \$59$ | $49 | $69 |
| 600 | $39 + 200\,(0.20) = \$79$ | $49 | $69 |
| 700 | $39 + 300\,(0.20) = \$99$ | $49 + 100\,(0.15) = \$64$ | $69 |
| 800 | $39 + 400\,(0.20) = \$119$ | $49 + 200\,(0.15) = \$79$ | $69 |
| 900 | $39 + 500\,(0.20) = \$139$ | $49 + 300\,(0.15) = \$94$ | $69 |
| 1000 | $39 + 600\,(0.20) = \$159$ | $49 + 400\,(0.15) = \$109$ | $69 |
| 1100 | $39 + 700\,(0.20) = \$179$ | $49 + 500\,(0.15) = \$124$ | $69 + 100\,(0.15) = \$84$ |

**(b)** If Jane uses either 500 or 700 minutes, Plan B is the cheapest, but if she uses 1100 minutes, Plan C is the cheapest.

**8. (a)** In this plan, Company A gets \$3.2 million and Company B gets \$3.2 million. Company A's investment is \$1.4 million, so they make a profit of $3.2 - 1.4 = \$1.8$ million. Company B's investment is \$2.6 million, so they make a profit of $3.2 - 2.6 = \$0.6$ million. So Company A makes three times the profit that Company B does.

**(b)** The original investment is $1.4 + 2.6 = \$4$ million. So after giving the original investment back, they then share the profit of \$2.4 million. So each gets an additional \$1.2 million. So Company A gets a total of $1.4 + 1.2 = \$2.6$ million and Company B gets $2.6 + 1.2 = \$3.8$ million. So even though Company B invests more, they make the same profit as Company A.

**(c)** The original investment is \$4 million, so Company A gets $\frac{1.4}{4} \cdot 6.4 = \$2.24$ million and Company B gets $\frac{2.6}{4} \cdot 6.4 = \$4.16$ million. This seems the fairest.

# 2 Coordinates and Graphs

## 2.1 The Coordinate Plane

**1.**

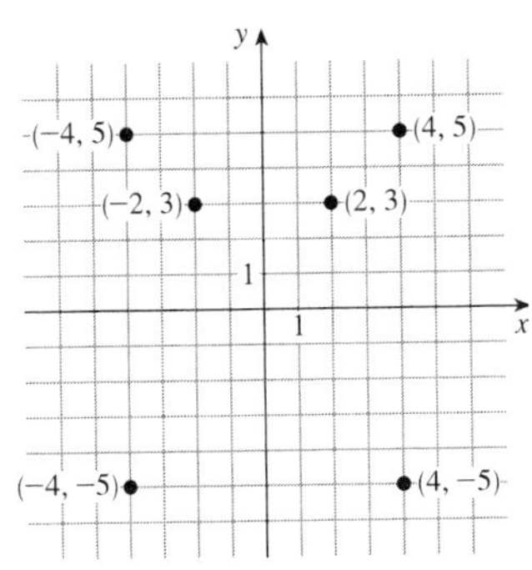


**2.** $A(5,1)$, $B(1,2)$, $C(-2,6)$, $D(-6,2)$, $E(-4,-1)$, $F(-2,0)$, $G(-1,-3)$, $H(2,-2)$

**3.** The two points are $(0,2)$ and $(3,0)$.

**(a)** $d=\sqrt{(3-0)^2+(0-(-2))^2}=\sqrt{3^2+2^2}=\sqrt{9+4}=\sqrt{13}$

**(b)** midpoint: $\left(\dfrac{3+0}{2},\dfrac{0+2}{2}\right)=\left(\frac{3}{2},1\right)$

**4.** The two points are $(-2,-1)$ and $(2,2)$.

**(a)** $d=\sqrt{(-2-2)^2+(-1-2)^2}=\sqrt{(-4)^2+(-3)^2}=\sqrt{16+9}=\sqrt{25}=5$

**(b)** midpoint: $\left(\dfrac{-2+2}{2},\dfrac{-1+2}{2}\right)=\left(0,\frac{1}{2}\right)$

**5.** The two points are $(-3,3)$ and $(5,-3)$.

**(a)** $d=\sqrt{(-3-5)^2+(3-(-3))^2}=\sqrt{(-8)^2+6^2}=\sqrt{64+36}=\sqrt{100}=10$

**(b)** midpoint: $\left(\dfrac{-3+5}{2},\dfrac{3+(-3)}{2}\right)=(1,0)$

**6.** The two points are $(-2,-3)$ and $(4,-1)$.

**(a)** $d=\sqrt{(-2-4)^2+(-3-(-1))^2}=\sqrt{(-6)^2+(-2)^2}=\sqrt{36+4}=\sqrt{50}=2\sqrt{10}$

**(b)** midpoint: $\left(\dfrac{-2+4}{2},\dfrac{-3+(-1)}{2}\right)=(1,-2)$

**7. (a)**

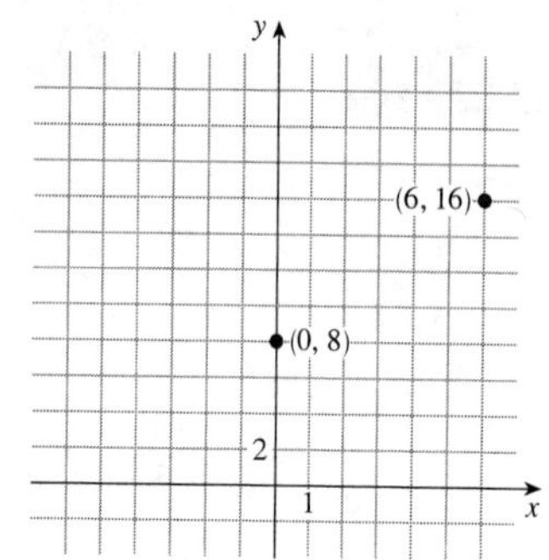


**(b)** $d = \sqrt{(0-6)^2 + (8-16)^2}$

$= \sqrt{(-6)^2 + (-8)^2} = \sqrt{100} = 10$

**(c)** Midpoint: $\left(\frac{0+6}{2}, \frac{8+16}{2}\right) = (3, 12)$

**8. (a)**

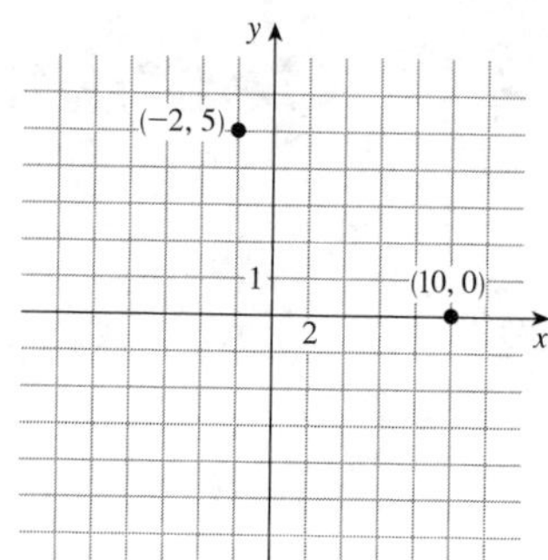


**(b)** $d = \sqrt{(-2-10)^2 + (5-0)^2}$

$= \sqrt{(-12)^2 + (5)^2} = \sqrt{169} = 13$

**(c)** Midpoint: $\left(\frac{-2+10}{2}, \frac{5+0}{2}\right) = \left(4, \frac{5}{2}\right)$

**9. (a)**

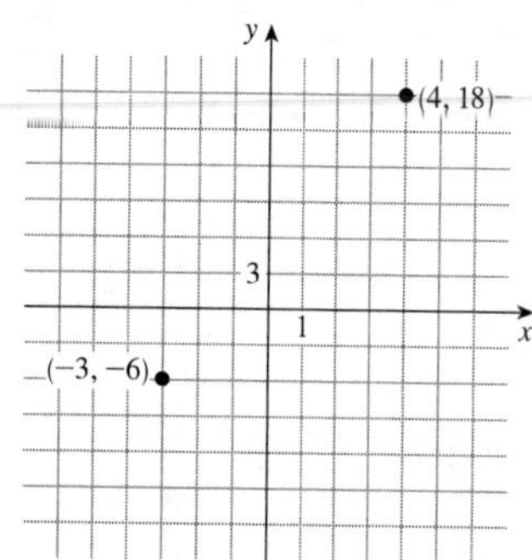


**(b)** $d = \sqrt{(-3-4)^2 + (-6-18)^2}$

$= \sqrt{(-7)^2 + (-24)^2} = \sqrt{49+576} = \sqrt{625} = 25$

**(c)** Midpoint: $\left(\frac{-3+4}{2}, \frac{-6+18}{2}\right) = \left(\frac{1}{2}, 6\right)$

**10. (a)**

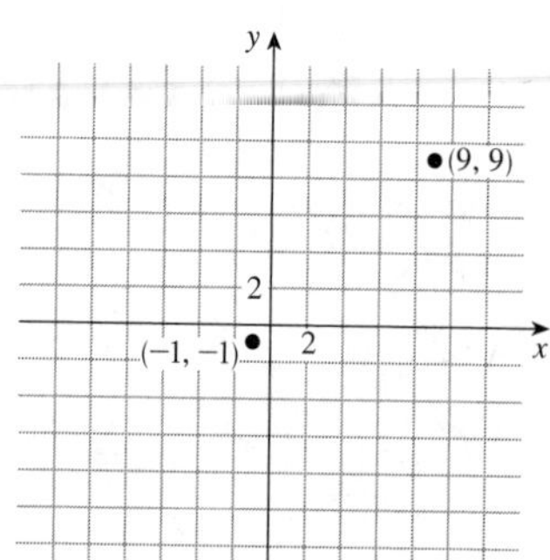


**(b)** $d = \sqrt{(-1-9)^2 + (-1-9)^2}$

$= \sqrt{(-10)^2 + (-10)^2} = \sqrt{200} = 10\sqrt{2}$

**(c)** Midpoint: $\left(\frac{-1+9}{2}, \frac{-1+9}{2}\right) = (4, 4)$

**11. (a)**

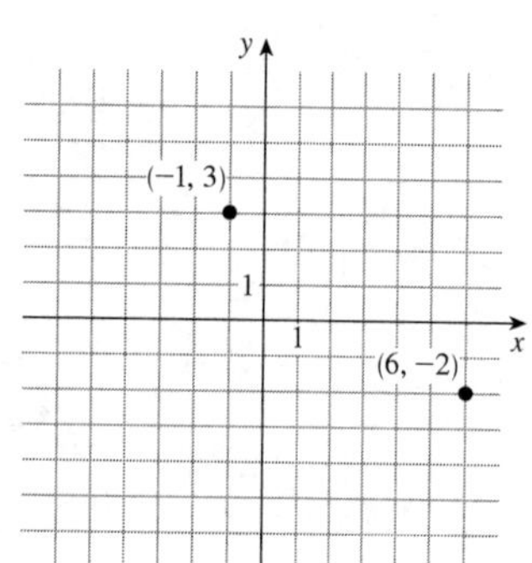


**(b)** $d = \sqrt{(6-(-1))^2 + (-2-3)^2}$

$= \sqrt{7^2 + (-5)^2} = \sqrt{49+25} = \sqrt{74}$

**(c)** Midpoint: $\left(\frac{6-1}{2}, \frac{-2+3}{2}\right) = \left(\frac{5}{2}, \frac{1}{2}\right)$

**12. (a)**

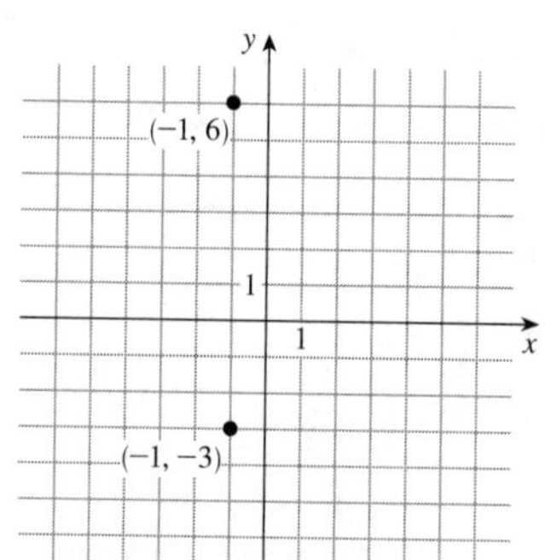


**(b)** $d = \sqrt{(-1-(-1))^2 + (6-(-3))^2}$

$= \sqrt{0^2 + 9^2} = 9$

**(c)** Midpoint: $\left(\frac{-1+(-1)}{2}, \frac{6+(-3)}{2}\right) = \left(-1, \frac{3}{2}\right)$

**13. (a)**

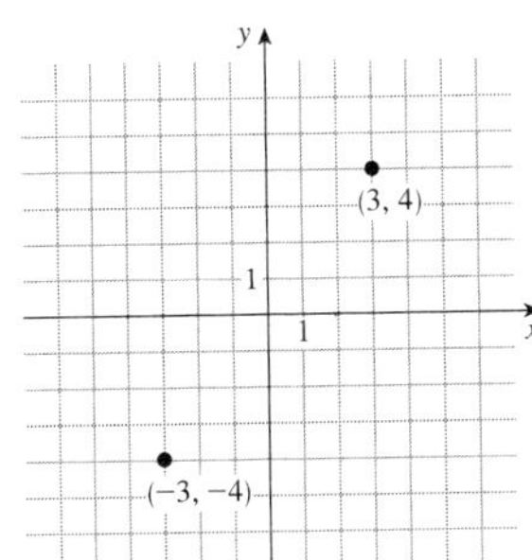


**(b)** $d = \sqrt{(3-(-3))^2 + (4-(-4))^2}$
$= \sqrt{6^2 + 8^2} = \sqrt{36+64} = \sqrt{100} = 10$

**(c)** Midpoint: $\left(\frac{3+(-3)}{2}, \frac{4+(-4)}{2}\right) = (0, 0)$

**14. (a)**

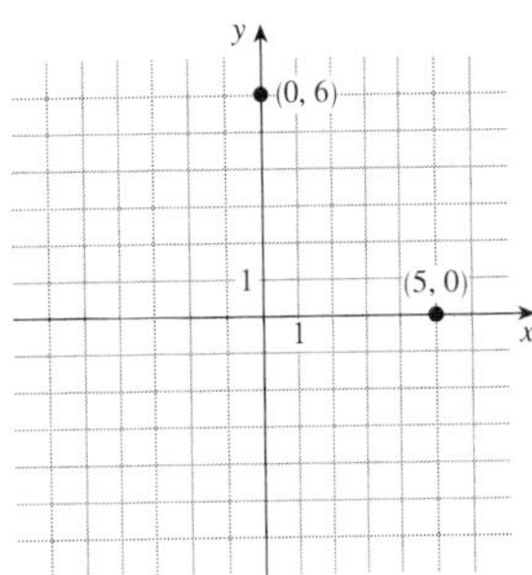


**(b)** $d = \sqrt{(5-0)^2 + (0-6)^2}$
$= \sqrt{5^2 + (-6)^2} = \sqrt{25+36} = \sqrt{61}$

**(c)** Midpoint: $\left(\frac{5+0}{2}, \frac{0+6}{2}\right) = \left(\frac{5}{2}, 3\right)$

**15.** $d(A, B) = \sqrt{(1-5)^2 + (3-3)^2} = \sqrt{(-4)^2} = 4.$
$d(A, C) = \sqrt{(1-1)^2 + (3-(-3))^2} = \sqrt{(6)^2} = 6.$ So the area is $4 \cdot 6 = 24$.

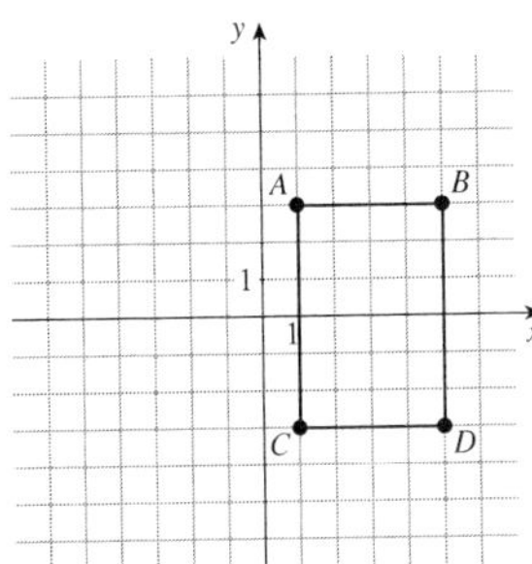


**16.** The area of a parallelogram is its base times its height. Since two sides are parallel to the $x$-axis, we use the length of one of these as the base. Thus, the base is $d(A, B) = \sqrt{(1-5)^2 + (2-2)^2} = \sqrt{(-4)^2} = 4$. The height is the change in the $y$ coordinates, thus, the height is $6 - 2 = 4$. So the area of the parallelogram is base $\cdot$ height $= 4 \cdot 4 = 16$.

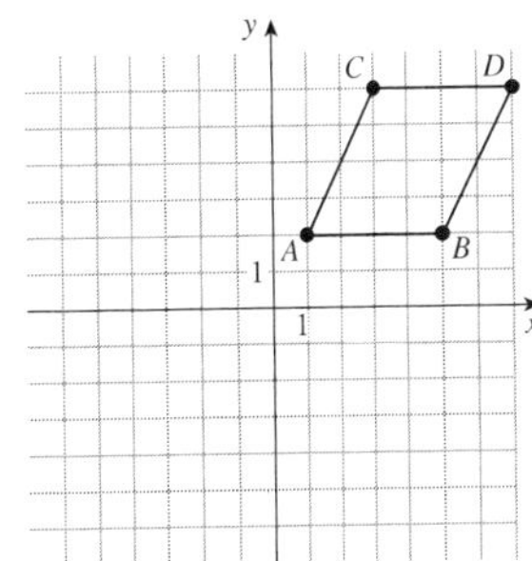


**17.** From the graph, the quadrilateral $ABCD$ has a pair of parallel sides, so $ABCD$ is a trapezoid. The area is $\frac{1}{2}(b_1 + b_2)h$. From the graph we see that
$b_1 = d(A, B) = \sqrt{(1-5)^2 + (0-0)^2} = \sqrt{4^2} = 4;$
$b_2 = d(C, D) = \sqrt{(4-2)^2 + (3-3)^2} = \sqrt{2^2} = 2;$
and $h$ is the difference in $y$-coordinates is $|3-0| = 3$.
Thus the area of the trapezoid is $\frac{1}{2}(4+2)3 = 9$.

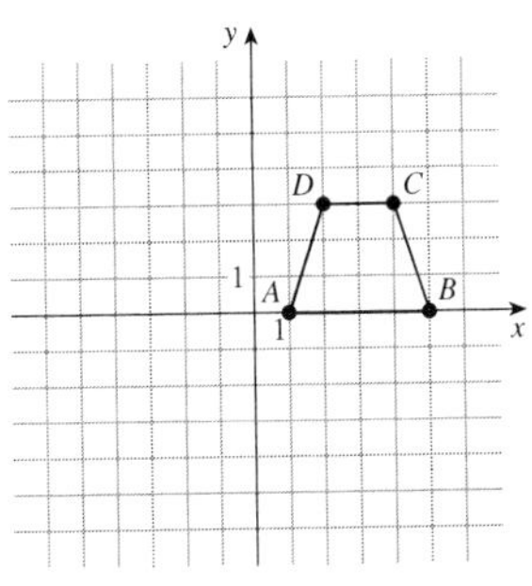


**18.** The point $S$ must be located at $(0, -4)$. To find the area, we find the length of one side and square it. This gives

$$d(Q, R) = \sqrt{(-5-0)^2 + (1-6)^2}$$
$$= \sqrt{(-5)^2 + (-5)^2} = \sqrt{25+25} = \sqrt{50}$$

So the area is $\left(\sqrt{50}\right)^2 = 50$.

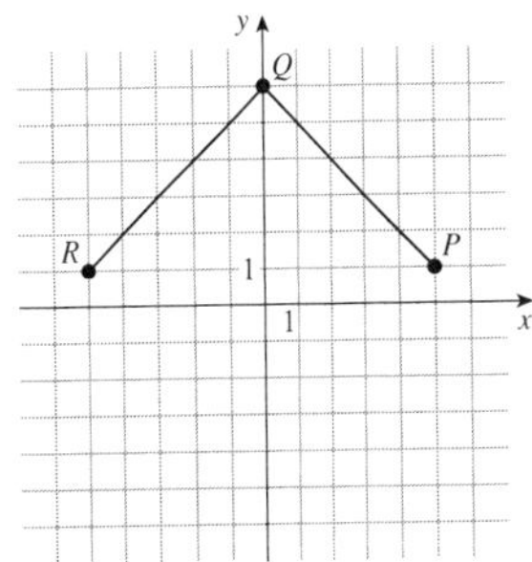

**19.** $\{(x, y) \mid x \leq 0\}$

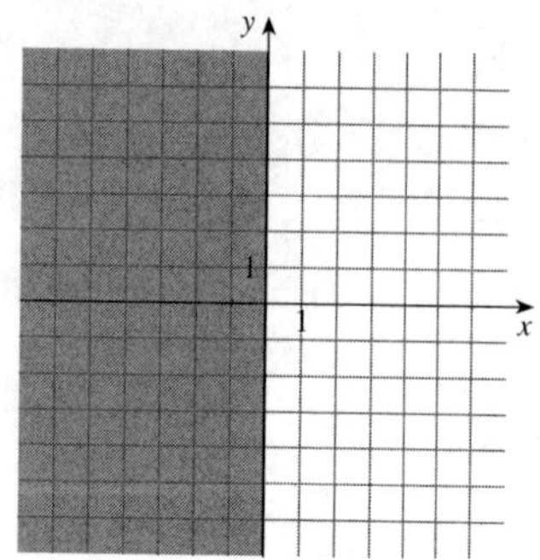

**20.** $\{(x, y) \mid y \geq 0\}$

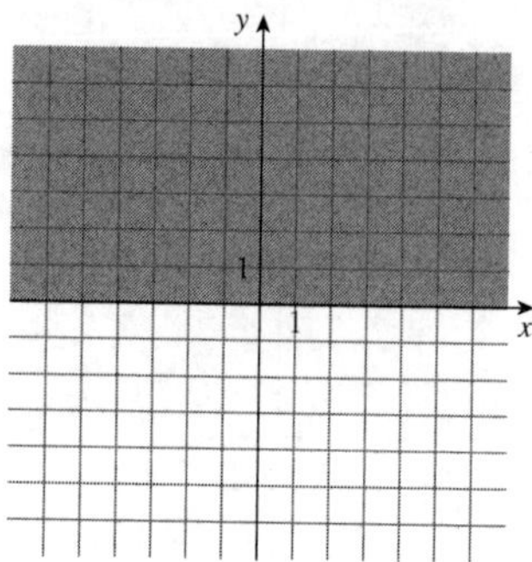

**21.** $\{(x, y) \mid x = 3\}$

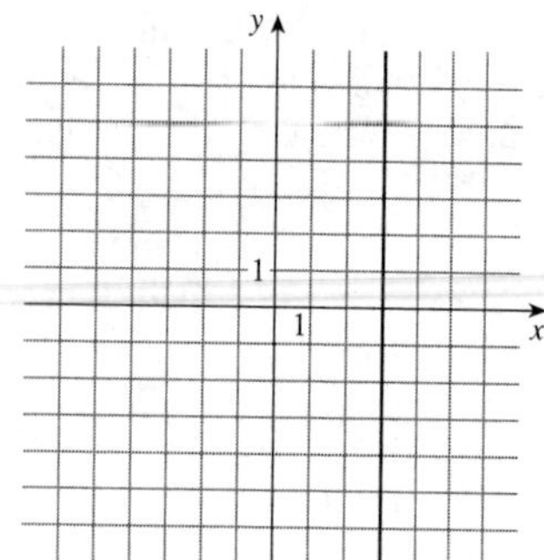

**22.** $\{(x, y) \mid y = -2\}$

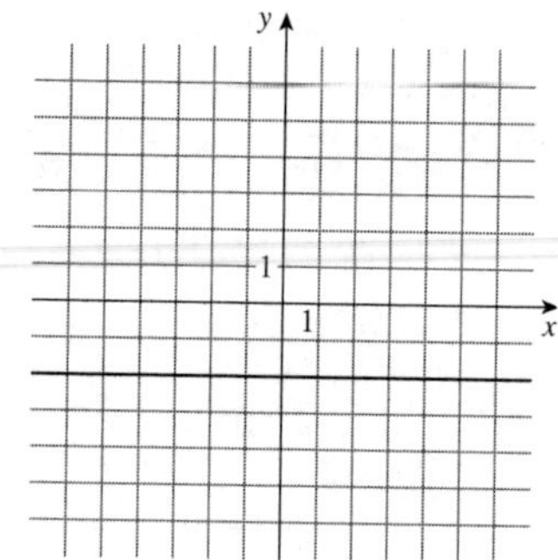

**23.** $\{(x, y) \mid 1 < x < 2\}$

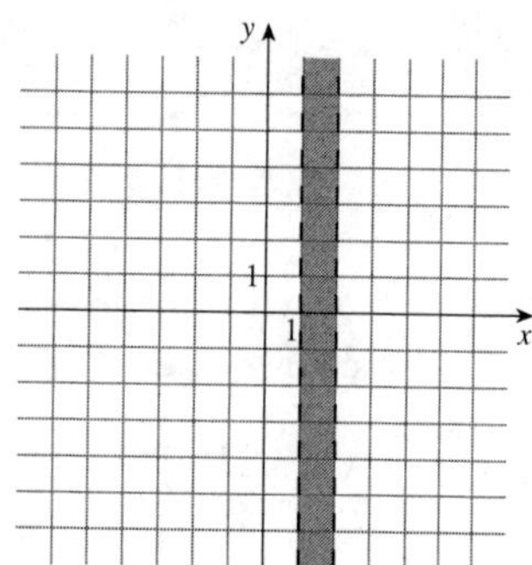

**24.** $\{(x, y) \mid 0 \leq y \leq 4\}$

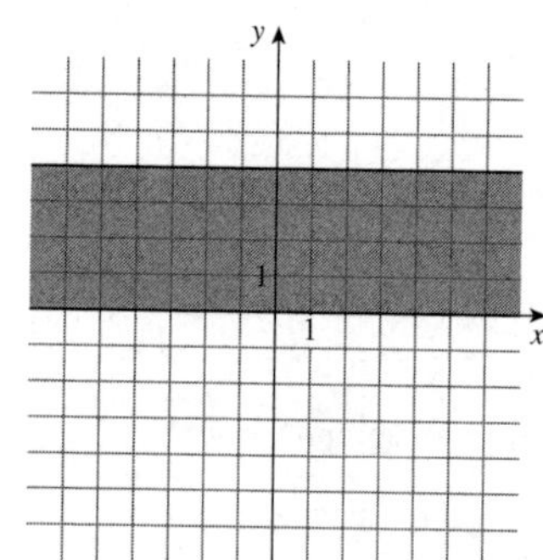

**25.** $\{(x, y) \mid xy < 0\}$
$= \{(x, y) \mid x < 0 \text{ and } y > 0 \text{ or } x > 0 \text{ and } y < 0\}$

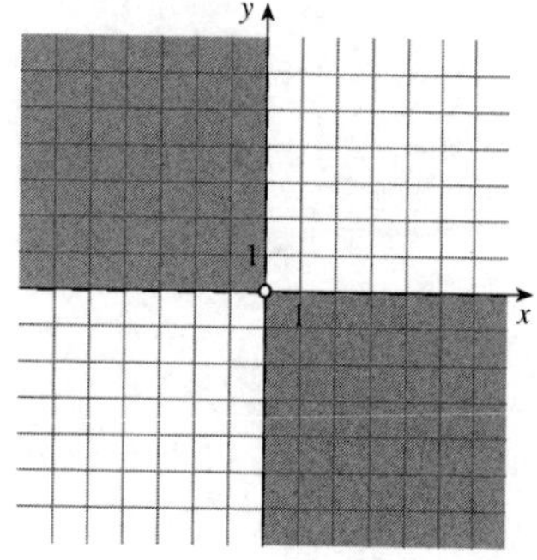

**26.** $\{(x, y) \mid xy > 0\}$
$= \{(x, y) \mid x < 0 \text{ and } y < 0 \text{ or } x > 0 \text{ and } y > 0\}$

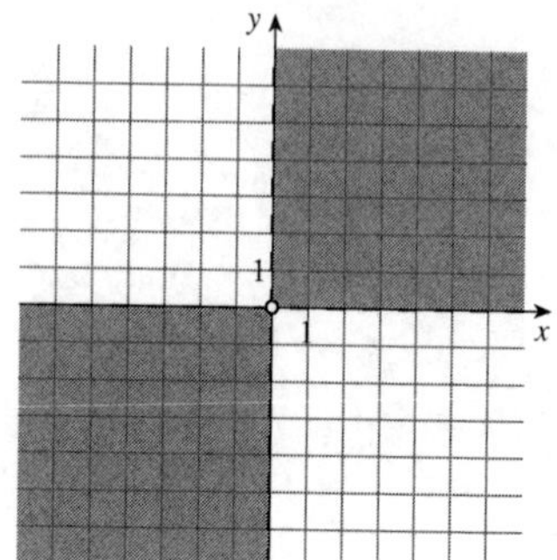

**27.** $\{(x, y) \mid x \geq 1 \text{ and } y < 3\}$

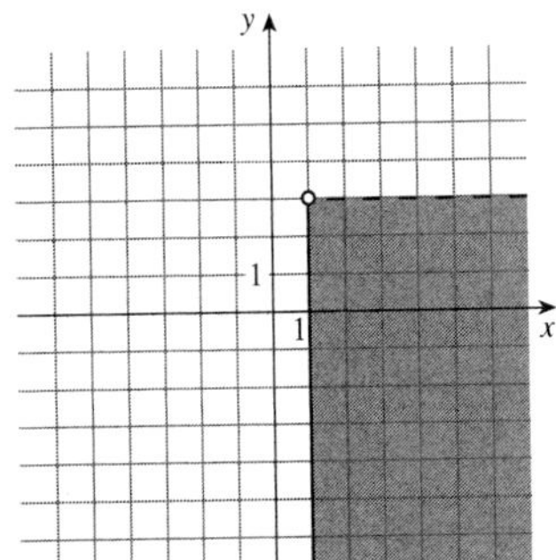

**28.** $\{(x, y) \mid -2 < x < 2 \text{ and } y \geq 3\}$

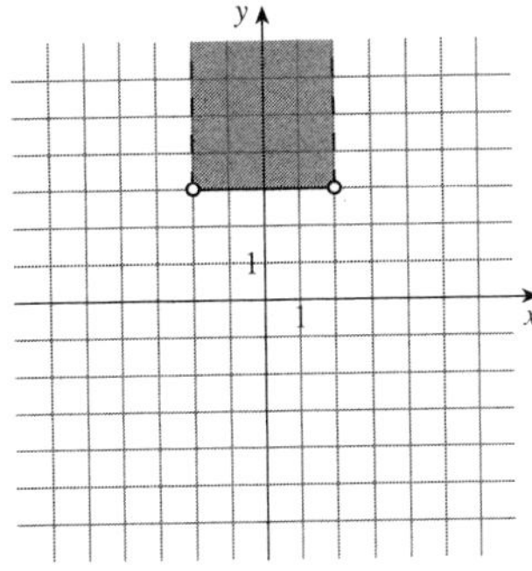

**29.** $\{(x, y) \mid |x| > 4\}$

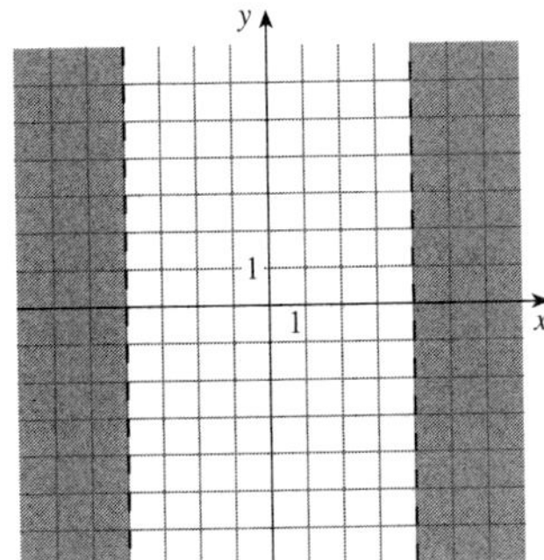

**30.** $\{(x, y) \mid |y| \leq 2\}$

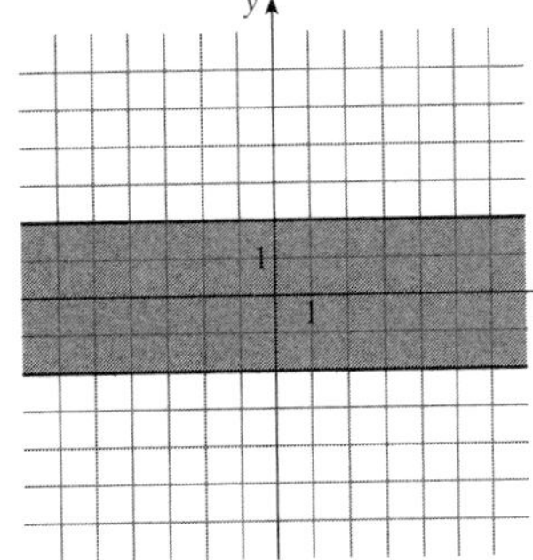

**31.** $\{(x, y) \mid |x| \leq 2 \text{ and } |y| \leq 3\}$

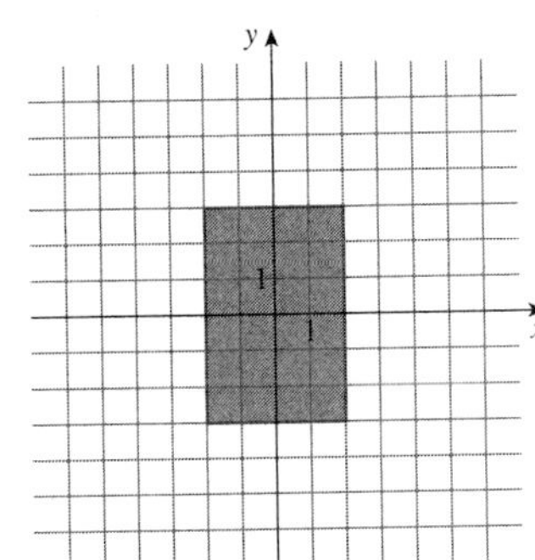

**32.** $\{(x, y) \mid |x| > 2 \text{ and } |y| > 3\}$

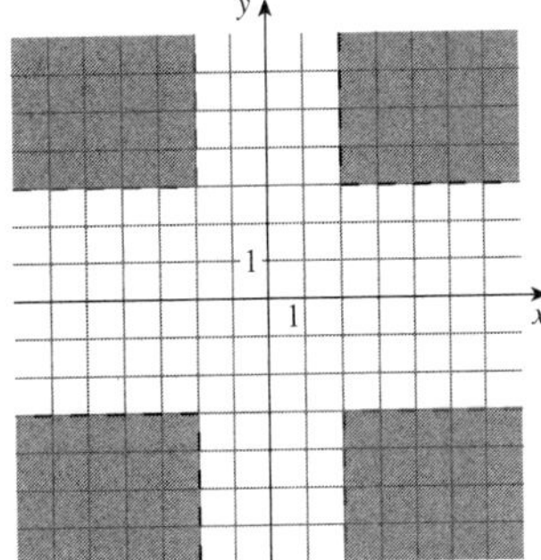

**33.** $d(0, A) = \sqrt{(6-0)^2 + (7-0)^2} = \sqrt{6^2 + 7^2} = \sqrt{36 + 49} = \sqrt{85}$.

$d(0, B) = \sqrt{(-5-0)^2 + (8-0)^2} = \sqrt{(-5)^2 + 8^2} = \sqrt{25 + 64} = \sqrt{89}$.

Thus point $A(6, 7)$ is closer to the origin.

**34.** $d(E, C) = \sqrt{(-6-(-2))^2 + (3-1)^2} = \sqrt{(-4)^2 + 2^2} = \sqrt{16 + 4} = \sqrt{20}$.

$d(E, D) = \sqrt{(3-(-2))^2 + (0-1)^2} = \sqrt{5^2 + (-1)^2} = \sqrt{25 + 1} = \sqrt{26}$.

Thus point $C$ is closer to point $E$.

**35.** $d(P, R) = \sqrt{(-1-3)^2 + (-1-1)^2} = \sqrt{(-4)^2 + (-2)^2} = \sqrt{16 + 4} = \sqrt{20} = 2\sqrt{5}$.

$d(Q, R) = \sqrt{(-1-(-1))^2 + (-1-3)^2} = \sqrt{0 + (-4)^2} = \sqrt{16} = 4$. Thus point $Q(-1, 3)$ is closer to point $R$.

**36. (a)** The distance from $(7,3)$ to the origin is $\sqrt{(7-0)^2+(3-0)^2}=\sqrt{7^2+3^2}=\sqrt{49+9}=\sqrt{58}$. The distance from $(3,7)$ to the origin is $\sqrt{(3-0)^2+(7-0)^2}=\sqrt{3^2+7^2}=\sqrt{9+49}=\sqrt{58}$. So the points are the same distance from the origin.

**(b)** The distance from $(a,b)$ to the origin is $\sqrt{(a-0)^2+(b-0)^2}=\sqrt{a^2+b^2}$. The distance from $(b,a)$ to the origin is $\sqrt{(b-0)^2+(a-0)^2}=\sqrt{b^2+a^2}=\sqrt{a^2+b^2}$. So the points are the same distance from the origin.

**37.** Since we do not know which pair are isosceles, we find the length of all three sides.

$d(A,B)=\sqrt{(-3-0)^2+(-1-2)^2}=\sqrt{(-3)^2+(-3)^2}=\sqrt{9+9}=\sqrt{18}=3\sqrt{2}$.

$d(C,B)=\sqrt{(-3-(-4))^2+(-1-3)^2}=\sqrt{1^2+(-4)^2}=\sqrt{1+16}=\sqrt{17}$.

$d(A,C)=\sqrt{(0-(-4))^2+(2-3)^2}=\sqrt{4^2+(-1)^2}=\sqrt{16+1}=\sqrt{17}$. So sides $AC$ and $CB$ have the same length.

**38.** Since the side $AB$ is parallel to the $x$-axis, we use this as the base in the formula area $=\frac{1}{2}$ (base $\cdot$ height). The height is the change in the $y$-coordinates. Thus, the base is $|-2-4|=6$ and the height is $|4-1|=3$. So the area is $\frac{1}{2}(6\cdot 3)=9$.

**39. (a)** Here we have $A=(2,2)$, $B=(3,-1)$, and $C=(-3,-3)$. So

$d(A,B)=\sqrt{(3-2)^2+(-1-2)^2}=\sqrt{1^2+(-3)^2}=\sqrt{1+9}=\sqrt{10}$;

$d(C,B)=\sqrt{(3-(-3))^2+(-1-(-3))^2}=\sqrt{6^2+2^2}=\sqrt{36+4}=\sqrt{40}=2\sqrt{10}$;

$d(A,C)=\sqrt{(-3-2)^2+(-3-2)^2}=\sqrt{(-5)^2+(-5)^2}=\sqrt{25+25}=\sqrt{50}=5\sqrt{2}$.

Since $[d(A,B)]^2+[d(C,B)]^2=[d(A,C)]^2$, we conclude that the triangle is a right triangle.

**(b)** The area of the triangle is $\frac{1}{2}\cdot d(C,B)\cdot d(A,B)=\frac{1}{2}\cdot\sqrt{10}\cdot 2\sqrt{10}=10$.

**40.** $d(A,B)=\sqrt{(11-6)^2+(-3-(-7))^2}=\sqrt{5^2+4^2}=\sqrt{25+16}=\sqrt{41}$;

$d(A,C)=\sqrt{(2-6)^2+(-2-(-7))^2}=\sqrt{(-4)^2+5^2}=\sqrt{16+25}=\sqrt{41}$;

$d(B,C)=\sqrt{(2-11)^2+(-2-(-3))^2}=\sqrt{(-9)^2+1^2}=\sqrt{81+1}=\sqrt{82}$.

Since $[d(A,B)]^2+[d(A,C)]^2=[d(B,C)]^2$, we conclude that the triangle is a right triangle. The area is $\frac{1}{2}\left(\sqrt{41}\cdot\sqrt{41}\right)=\frac{41}{2}$.

**41.** We show that all sides are the same length (its a rhombus) and then show that the diagonals are equal. Here we have $A=(-2,9)$, $B=(4,6)$, $C=(1,0)$, and $D=(-5,3)$. So

$d(A,B)=\sqrt{(4-(-2))^2+(6-9)^2}=\sqrt{6^2+(-3)^2}=\sqrt{36+9}=\sqrt{45}$;

$d(B,C)=\sqrt{(1-4)^2+(0-6)^2}=\sqrt{(-3)^2+(-6)^2}=\sqrt{9+36}=\sqrt{45}$;

$d(C,D)=\sqrt{(-5-1)^2+(3-0)^2}=\sqrt{(-6)^2+(-3)^2}=\sqrt{36+9}=\sqrt{45}$;

$d(D,A)=\sqrt{(-2-(-5))^2+(9-3)^2}=\sqrt{3^2+6^2}=\sqrt{9+36}=\sqrt{45}$. So the points form a rhombus. Also $d(A,C)=\sqrt{(1-(-2))^2+(0-9)^2}=\sqrt{3^2+(-9)^2}=\sqrt{9+81}=\sqrt{90}=3\sqrt{10}$, and $d(B,D)=\sqrt{(-5-4)^2+(3-6)^2}=\sqrt{(-9)^2+(-3)^2}=\sqrt{81+9}=\sqrt{90}=3\sqrt{10}$. Since the diagonals are equal, the rhombus is a square.

**42.** $d(A,B) = \sqrt{(3-(-1))^2+(11-3)^2} = \sqrt{4^2+8^2} = \sqrt{16+64} = \sqrt{80} = 4\sqrt{5}$.

$d(B,C) = \sqrt{(5-3)^2+(15-11)^2} = \sqrt{2^2+4^2} = \sqrt{4+16} = \sqrt{20} = 2\sqrt{5}$.

$d(A,C) = \sqrt{(5-(-1))^2+(15-3)^2} = \sqrt{6^2+12^2} = \sqrt{36+144} = \sqrt{180} = 6\sqrt{5}$. So $d(A,B)+d(B,C)=d(A,C)$, and the points are collinear.

**43.** Let $P=(0,y)$ be such a point. Setting the distances equal we get

$\sqrt{(0-5)^2+(y-(-5))^2} = \sqrt{(0-1)^2+(y-1)^2} \quad\Leftrightarrow$

$\sqrt{25+y^2+10y+25} = \sqrt{1+y^2-2y+1} \quad\Rightarrow\quad y^2+10y+50 = y^2-2y+2 \quad\Leftrightarrow\quad 12y=-48 \quad\Leftrightarrow$

$y=-4$. Thus, the point is $P=(0,-4)$. Check:

$\sqrt{(0-5)^2+(-4-(-5))^2} = \sqrt{(-5)^2+1^2} = \sqrt{25+1} = \sqrt{26}$;

$\sqrt{(0-1)^2+(-4-1)^2} = \sqrt{(-1)^2+(-5)^2} = \sqrt{25+1} = \sqrt{26}$.

**44.** The midpoint of $AB$ is $C' = \left(\frac{1+3}{2},\frac{0+6}{2}\right) = (2,3)$. So the length of the median $CC'$ is $d(C,C') = \sqrt{(2-8)^2+(3-2)^2} = \sqrt{37}$. The midpoint of $AC$ is $B' = \left(\frac{1+8}{2},\frac{0+2}{2}\right) = \left(\frac{9}{2},1\right)$. So the length of the median $BB'$ is $d(B,B') = \sqrt{\left(\frac{9}{2}-3\right)^2+(1-6)^2} = \frac{\sqrt{109}}{2}$. The midpoint of $BC$ is $A' = \left(\frac{3+8}{2},\frac{6+2}{2}\right) = \left(\frac{11}{2},4\right)$. So the length of the median $AA'$ is $d(A,A') = \sqrt{\left(\frac{11}{2}-1\right)^2+(4-0)^2} = \frac{\sqrt{145}}{2}$.

**45.** We find the midpoint $M$ of $PQ$, and then the midpoint of $PM$. Now $M = \left(\frac{-1+7}{2},\frac{3+5}{2}\right) = (3,4)$, and the midpoint of $PM$ is thus $\left(\frac{-1+3}{2},\frac{3+4}{2}\right) = \left(1,\frac{7}{2}\right)$.

**46.** Points on a perpendicular bisector of $PQ$ are the same distance from the points $P$ and $Q$. For point $A$,

$d(P,A) = \sqrt{(5-(-2))^2+(-7-1)^2} = \sqrt{7^2+(-8)^2} = \sqrt{113}$ and

$d(Q,A) = \sqrt{(5-12)^2+(-7-(-1))^2} = \sqrt{(-7)^2+(-6)^2} = \sqrt{85}$. Since

$d(P,A) \neq d(Q,A)$, point $A$ does not lie on the perpendicular bisector of $PQ$.

For point $B$, $d(P,B) = \sqrt{(6-(-2))^2+(7-1)^2} = \sqrt{8^2+6^2} = 10$, and

$d(Q,B) = \sqrt{(6-12)^2+(7-(-1))^2} = \sqrt{(-6)^2+8^2} = 10$. Since

$d(P,B) = d(Q,B)$, point $B(6,7)$ lies on the perpendicular bisector of $PQ$.

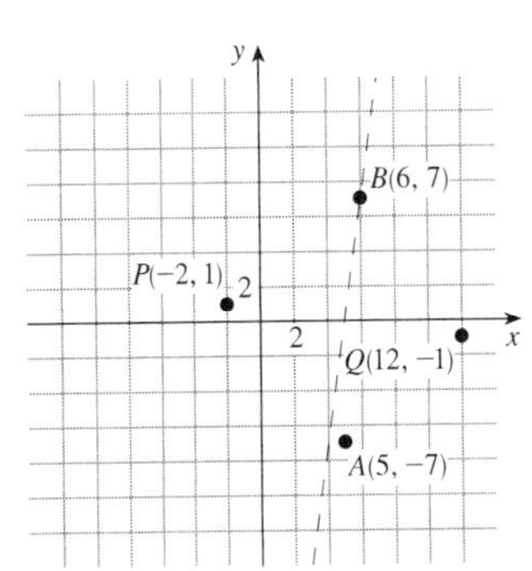

**47.** As indicated by Example 5, we must find a point $S(x_1, y_1)$ such that the midpoints of $PR$ and $QS$ are the same. Thus $\left(\frac{4+(-1)}{2}, \frac{2+(-4)}{2}\right) = \left(\frac{x_1+1}{2}, \frac{y_1+1}{2}\right)$. Setting the $x$-coordinates equal, we get $\frac{4+(-1)}{2} = \frac{x_1+1}{2} \Leftrightarrow 4-1 = x_1+1 \Leftrightarrow x_1 = 2$. Setting the $y$-coordinates equal, we get $\frac{2+(-4)}{2} = \frac{y_1+1}{2} \Leftrightarrow 2-4 = y_1+1 \Leftrightarrow$ $y_1 = -3$. Thus $S = (2, -3)$.

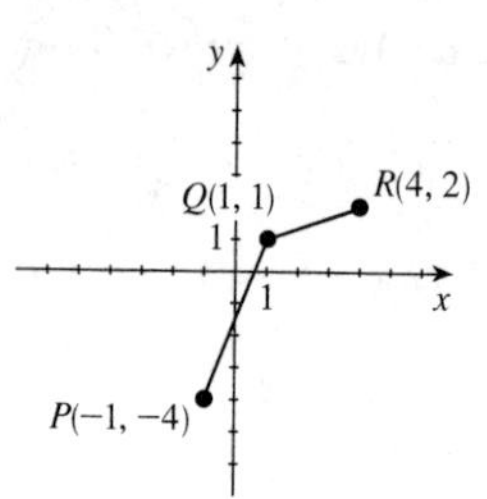


**48.** We solve the equation $6 = \frac{2+x}{2}$ to find the $x$ coordinate of $B$. This gives $6 = \frac{2+x}{2} \Leftrightarrow 12 = 2+x \Leftrightarrow x = 10$. Likewise, $8 = \frac{3+y}{2} \Leftrightarrow 16 = 3+y \Leftrightarrow y = 13$. Thus, $B = (10, 13)$.

**49. (a)**

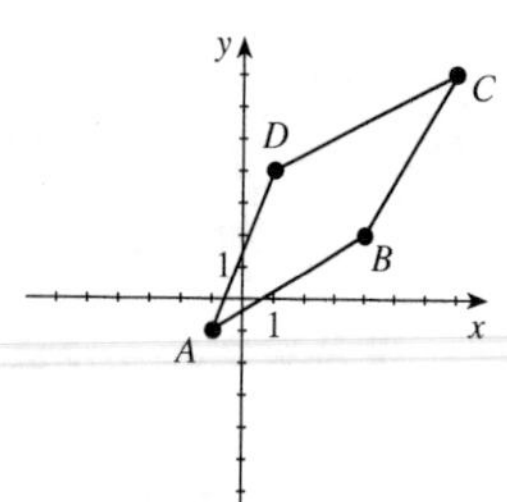


**(b)** The midpoint of $AC$ is $\left(\frac{-2+7}{2}, \frac{-1+7}{2}\right) = \left(\frac{5}{2}, 3\right)$, the midpoint of $BD$ is $\left(\frac{4+1}{2}, \frac{2+4}{2}\right) = \left(\frac{5}{2}, 3\right)$.

**(c)** Since the they have the same midpoint, we conclude that the diagonals bisect each other.

**50.** We have $M = \left(\frac{a+0}{2}, \frac{b+0}{2}\right) = \left(\frac{a}{2}, \frac{b}{2}\right)$. Thus,

$$d(C, M) = \sqrt{\left(\frac{a}{2}-0\right)^2 + \left(\frac{b}{2}-0\right)^2} = \sqrt{\frac{a^2}{4} + \frac{b^2}{4}} = \frac{\sqrt{a^2+b^2}}{2};$$

$$d(A, M) = \sqrt{\left(\frac{a}{2}-a\right)^2 + \left(\frac{b}{2}-0\right)^2} = \sqrt{\left(-\frac{a}{2}\right)^2 + \left(\frac{b}{2}\right)^2} = \sqrt{\frac{a^2}{4} + \frac{b^2}{4}} = \frac{\sqrt{a^2+b^2}}{2};$$

$$d(B, M) = \sqrt{\left(\frac{a}{2}-0\right)^2 + \left(\frac{b}{2}-b\right)^2} = \sqrt{\left(\frac{a}{2}\right)^2 + \left(-\frac{b}{2}\right)^2} = \sqrt{\frac{a^2}{4} + \frac{b^2}{4}} = \frac{\sqrt{a^2+b^2}}{2}.$$

**51. (a)** $d(A, B) = \sqrt{3^2+4^2} = \sqrt{25} = 5$.

**(b)** We want the distances from $C = (4, 2)$ to $D = (11, 26)$. The walking distance is $|4-11| + |2-26| = 7+24 = 31$ blocks. Straight-line distance is $\sqrt{(4-11)^2 + (2-26)^2} = \sqrt{7^2+24^2} = \sqrt{625} = 25$ blocks.

**(c)** The two points are on the same avenue or the same street.

**52. (a)** The midpoint is at $\left(\frac{3+27}{2}, \frac{7+17}{2}\right) = (15, 12)$, which is at the intersection of 15th Street and 12th Avenue.

**(b)** They each must walk $|15-3| + |12-7| = 12+5 = 17$ blocks.

**53.** The midpoint of the line segment is $(66, 45)$. The pressure experienced by an ocean diver at a depth of 66 feet is 45 lb/in$^2$.

**54. (a)** The point $(5, 3)$ is shifted to $(5+3, 3+2) = (8, 5)$.

**(b)** The point $(a, b)$ is shifted to $(a+3, b+2)$.

**(c)** Let $(x, y)$ be the point that is shifted to $(3, 4)$. Then $(x+3, y+2) = (3, 4)$. Setting the $x$-coordinates equal, we get $x+3 = 3 \Leftrightarrow x = 0$. Setting the $y$-coordinates equal, we get $y+2 = 4 \Leftrightarrow y = 2$.So the point is $(0, 2)$.

**(d)** $A = (-5, -1)$, so $A' = (-5+3, -1+2) = (-2, 1)$; $B = (-3, 2)$, so $B' = (-3+3, 2+2) = (0, 4)$; and $C = (2, 1)$, so $C' = (2+3, 1+2) = (5, 3)$.

**55. (a)** The point $(3, 7)$ is reflected to the point $(-3, 7)$.

**(b)** The point $(a, b)$ is reflected to the point $(-a, b)$.

**(c)** Since the point $(-a, b)$ is the reflection of $(a, b)$, the point $(-4, -1)$ is the reflection of $(4, -1)$.

**(d)** $A = (3, 3)$, so $A' = (-3, 3)$; $B = (6, 1)$, so $B' = (-6, 1)$; and $C = (1, -4)$, so $C' = (-1, -4)$.

**56.** We solve the equation $6 = \dfrac{2+x}{2}$ to find the $x$ coordinate of $B$: $6 = \dfrac{2+x}{2} \quad \Leftrightarrow \quad 12 = 2 + x \quad \Leftrightarrow \quad x = 10$.

Likewise, for the $y$ coordinate of $B$, we have $8 = \dfrac{3+y}{2} \quad \Leftrightarrow \quad 16 = 3 + y \quad \Leftrightarrow \quad y = 13$. Thus $B = (10, 13)$.

**57.** We need to find a point $S(x_1, y_1)$ such that $PQRS$ is a parallelogram. As indicated by Example 3, this will be the case if the diagonals $PR$ and $QS$ bisect each other. So the midpoints of $PR$ and $QS$ are the same. Thus $\left(\dfrac{0+5}{2}, \dfrac{-3+3}{2}\right) = \left(\dfrac{x_1+2}{2}, \dfrac{y_1+2}{2}\right)$. Setting the $x$-coordinates equal, we get $\dfrac{0+5}{2} = \dfrac{x_1+2}{2} \quad \Leftrightarrow \quad 0 + 5 = x_1 + 2 \quad \Leftrightarrow \quad x_1 = 3$.

Setting the $y$-coordinates equal, we get $\dfrac{-3+3}{2} = \dfrac{y_1+2}{2} \quad \Leftrightarrow$

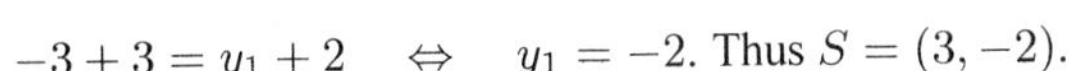
$-3 + 3 = y_1 + 2 \quad \Leftrightarrow \quad y_1 = -2$. Thus $S = (3, -2)$.

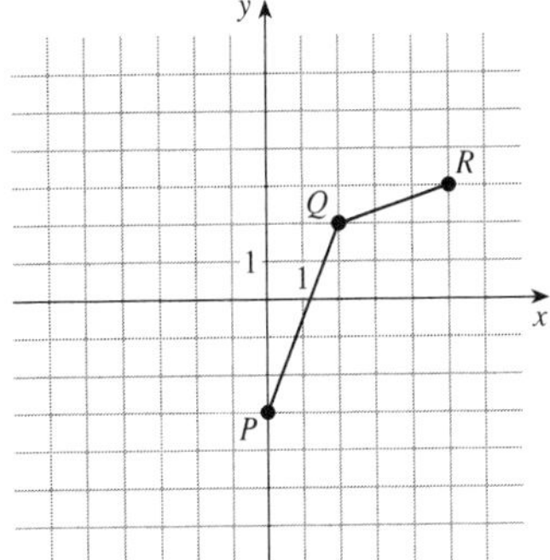


## 2.2 Graphs of Equations in Two Variables

**1.** $(0, 2)$: $2 \stackrel{?}{=} 3(0) - 2 \quad \Leftrightarrow \quad 2 \stackrel{?}{=} -2$. No. $\left(\frac{1}{3}, 1\right)$: $1 \stackrel{?}{=} 3\left(\frac{1}{3}\right) - 2 \quad \Leftrightarrow \quad 1 \stackrel{?}{=} 1 - 2$. No.

$(1, 1)$: $1 \stackrel{?}{=} 3(1) - 2 \quad \Leftrightarrow \quad 1 \stackrel{?}{=} 3 - 2$. Yes. So $(1, 1)$ is on the graph of this equation.

**2.** $(1, 0)$: $0 \stackrel{?}{=} \sqrt{(1)+1} \quad \Leftrightarrow \quad 0 \stackrel{?}{=} \sqrt{2}$. No. $(0, 1)$: $1 \stackrel{?}{=} \sqrt{(0)+1} \quad \Leftrightarrow \quad 1 \stackrel{?}{=} \sqrt{1}$. Yes.

$(3, 2)$: $2 \stackrel{?}{=} \sqrt{(3)+1} \quad \Leftrightarrow \quad 2 \stackrel{?}{=} \sqrt{4}$. Yes. So $(0, 1)$ and $(3, 2)$ are points on the graph of this equation.

**3.** $(0, 0)$: $0 - 2(0) - 1 \stackrel{?}{=} 0 \quad \Leftrightarrow \quad -1 \stackrel{?}{=} 0$. No. $(1, 0)$: $1 - 2(0) - 1 \stackrel{?}{=} 0 \quad \Leftrightarrow \quad -1 + 1 \stackrel{?}{=} 0$. Yes.

$(-1, 1)$: $(-1) - 2(-1) - 1 \stackrel{?}{=} 0 \quad \Leftrightarrow \quad -1 + 2 - 1 \stackrel{?}{=} 0$. Yes.

So $(1, 0)$ and $(-1, -1)$ are points on the graph of this equation.

**4.** $(1, 1)$: $(1)\left[(1)^2 + 1\right] \stackrel{?}{=} 1 \quad \Leftrightarrow \quad 1(2) \stackrel{?}{=} 1$. No. $\left(1, \frac{1}{2}\right)$: $\left(\frac{1}{2}\right)\left[(1)^2 + 1\right] \stackrel{?}{=} 1 \quad \Leftrightarrow \quad \frac{1}{2}(2) \stackrel{?}{=} 1$. Yes.

$\left(-1, \frac{1}{2}\right)$: $\left(\frac{1}{2}\right)\left[(-1)^2 + 1\right] \stackrel{?}{=} 1 \quad \Leftrightarrow \quad \frac{1}{2}(2) \stackrel{?}{=} 1$. Yes.

So both $\left(1, \frac{1}{2}\right)$ and $\left(-1, \frac{1}{2}\right)$ are points on the graph of this equation.

**5.** $(0, -2)$: $(0)^2 + (0)(-2) + (-2)^2 \stackrel{?}{=} 4 \quad \Leftrightarrow \quad 0 + 0 + 4 \stackrel{?}{=} 4$. Yes.

$(1, -2)$: $(1)^2 + (1)(-2) + (-2)^2 \stackrel{?}{=} 4 \quad \Leftrightarrow \quad 1 - 2 + 4 \stackrel{?}{=} 4$. No.

$(2, -2)$: $(2)^2 + (2)(-2) + (-2)^2 \stackrel{?}{=} 4 \quad \Leftrightarrow \quad 4 - 4 + 4 \stackrel{?}{=} 4$. Yes.

So $(0, -2)$ and $(2, -2)$ are points on the graph of this equation.

**6.** $(0, 1)$: $(0)^2 + (1)^2 - 1 \stackrel{?}{=} 0 \quad \Leftrightarrow \quad 0 + 1 - 1 \stackrel{?}{=} 0$. Yes.

$\left(\frac{1}{\sqrt{2}}, \frac{1}{\sqrt{2}}\right)$: $\left(\frac{1}{\sqrt{2}}\right)^2 + \left(\frac{1}{\sqrt{2}}\right)^2 - 1 \stackrel{?}{=} 0 \quad \Leftrightarrow \quad \frac{1}{2} + \frac{1}{2} - 1 \stackrel{?}{=} 0$. Yes.

$\left(\frac{3}{\sqrt{2}}, \frac{1}{2}\right)$: $\left(\frac{3}{\sqrt{2}}\right)^2 + \left(\frac{1}{2}\right)^2 - 1 \stackrel{?}{=} 0 \quad \Leftrightarrow \quad \frac{3}{4} + \frac{1}{4} - 1 \stackrel{?}{=} 0$. Yes.

So $(0, 1)$, $\left(\frac{1}{\sqrt{2}}, \frac{1}{\sqrt{2}}\right)$ and $\left(\frac{3}{\sqrt{2}}, \frac{1}{2}\right)$, are all points on the graph of this equation.

**7.** To find $x$-intercepts, set $y = 0$. This gives $0 = 4x - x^2 \quad \Leftrightarrow \quad 0 = x(4 - x) \quad \Leftrightarrow \quad 0 = x$ or $x = 4$, so the $x$-intercept are 0 and 4. To find $y$-intercepts, set $x = 0$. This gives $y = 4(0) - 0^2 \quad \Leftrightarrow \quad y = 0$, so the $y$-intercept is 0.

**8.** To find $x$-intercepts, set $y = 0$. This gives $\dfrac{x^2}{9} + \dfrac{0^2}{4} = 1 \quad \Leftrightarrow \quad \dfrac{x^2}{9} = 1 \quad \Leftrightarrow \quad x^2 = 9 \quad \Leftrightarrow \quad x = \pm 3$, so the $x$-intercept are $-3$ and 3.

To find $y$-intercepts, set $x = 0$. This gives $\dfrac{0^2}{9} + \dfrac{y^2}{4} = 1 \quad \Leftrightarrow \quad \dfrac{y^2}{4} = 1 \quad \Leftrightarrow \quad y^2 = 4 \quad \Leftrightarrow \quad x = \pm 2$, so the $y$-intercept are $-2$ and 2.

**9.** To find $x$-intercepts, set $y = 0$. This gives $x^4 + 0^2 - x(0) = 16 \quad \Leftrightarrow \quad x^4 = 16 \quad \Leftrightarrow \quad x = \pm 2$. So the $x$-intercept are $-2$ and 2.

To find $y$-intercepts, set $x = 0$. This gives $0^4 + y^2 - (0)y = 16 \quad \Leftrightarrow \quad y^2 = 16 \quad \Leftrightarrow \quad y = \pm 4$. So the $y$-intercept are $-4$ and 4.

**10.** To find $x$-intercepts, set $y = 0$. This gives $x^2 + 0^3 - x^2(0)^2 = 64 \quad \Leftrightarrow \quad x^2 = 64 \quad \Leftrightarrow \quad x = \pm 8$. So the $x$-intercept are $-8$ and 8.

To find $y$-intercepts, set $x = 0$. This gives $0^2 + y^3 - (0)^2 y^2 - 64 \quad \Leftrightarrow \quad y^3 = 64 \quad \Leftrightarrow \quad y = 4$. So the $y$-intercept is 4.

**11.** To find $x$-intercepts, set $y = 0$. This gives $0 = x - 3 \quad \Leftrightarrow \quad x = 3$, so the $x$-intercept is 3.

To find $y$-intercepts, set $x = 0$. This gives $y = 0 - 3 \quad \Leftrightarrow \quad y = -3$, so the $y$-intercept is $-3$.

**12.** To find $x$-intercepts, set $y = 0$. This gives $0 = x^2 - 5x + 6 \quad \Leftrightarrow 0 = (x - 2)(x - 3)$. So $x - 2 = 0$ and $x = 2$ or $x - 3 = 0$ and $x = 3$, and the $x$-intercepts are at 2 and 3.

To find $y$-intercepts, set $x = 0$. This gives $y = 0^2 - 0 + 6 \quad \Leftrightarrow \quad y = 6$, so the $y$-intercept is 6.

**13.** To find $x$-intercepts, set $y = 0$. This gives $0 = x^2 - 9 \quad \Leftrightarrow \quad x^2 = 9 \quad \Rightarrow \quad x = \pm 3$, so the $x$-intercept are $\pm 3$.

To find $y$-intercepts, set $x = 0$. This gives $y = (0)^2 - 9 \quad \Leftrightarrow \quad y = -9$, so the $y$-intercept is $-9$.

**14.** To find $x$-intercepts, set $y = 0$. This gives $0 - 2x(0) + 2x = 1 \quad \Leftrightarrow \quad 2x = 1 \quad \Leftrightarrow \quad x = \dfrac{1}{2}$, so the $x$-intercept is $\frac{1}{2}$.

To find $y$-intercepts, set $x = 0$. This gives $y - 2(0)y + 2(0) = 1 \quad \Leftrightarrow \quad y = 1$, so the $y$-intercept is 1.

**15.** To find $x$-intercepts, set $y = 0$. This gives $x^2 + (0)^2 = 4 \quad \Leftrightarrow \quad x^2 = 4 \quad \Rightarrow \quad x = \pm 2$, so the $x$-intercept are $\pm 2$.

To find $y$-intercepts, set $x = 0$. This gives $(0)^2 + y^2 = 4 \quad \Leftrightarrow \quad y^2 = 4 \quad \Rightarrow \quad y = \pm 2$, so the $y$-intercept are $\pm 2$.

**16.** To find $x$-intercepts, set $y = 0$. This gives $0 = \sqrt{x + 1} \quad \Leftrightarrow \quad 0 = x + 1 \quad \Leftrightarrow \quad x = -1$, so the $x$-intercept is $-1$.

To find $y$-intercepts, set $x = 0$. This gives $y = \sqrt{0 + 1} \quad \Leftrightarrow \quad y = 1$, so the $y$-intercept is 1.

**17.** To find $x$-intercepts, set $y = 0$. This gives $x(0) = 5 \quad \Leftrightarrow \quad 0 = 5$, which is impossible, so there is no $x$-intercept.

To find $y$-intercepts, set $x = 0$. This gives $(0)y = 5 \quad \Leftrightarrow \quad 0 = 5$, which is again impossible, so there is no $y$-intercept.

**18.** To find $x$-intercepts, set $y = 0$. This gives $x^2 - x(0) + (0) = 1 \quad \Leftrightarrow \quad x^2 = 1 \quad \Rightarrow \quad x = \pm 1$, so the $x$-intercepts are $-1$ and 1.

To find $y$-intercepts, set $x = 0$. This gives $y = (0)^2 - (0)y + y = 1 \quad \Leftrightarrow \quad y = 1$, so the $y$-intercept is 1.

**19.** $y = -x$

| $x$ | $y$ |
|---|---|
| $-4$ | 4 |
| $-2$ | 2 |
| 0 | 0 |
| 1 | $-1$ |
| 2 | $-2$ |
| 3 | $-3$ |
| 4 | $-4$ |

When $y = 0$, $x = 0$. So the $x$-intercept is 0, and the $y$-intercept is also 0.

$x$-axis symmetry: $(-y) = -x \quad \Leftrightarrow \quad y = x$, which is not the same as $y = -x$, so the graph is not symmetric with respect to the $x$-axis.

$y$-axis symmetry: $y = -(-x) \quad \Leftrightarrow \quad y = x$, which is not the same as $y = -x$, so the graph is not symmetric with respect to the $y$-axis.

Origin symmetry: $(-y) = -(-x)$ is the same as $y = -x$, so the graph is symmetric with respect to the origin.

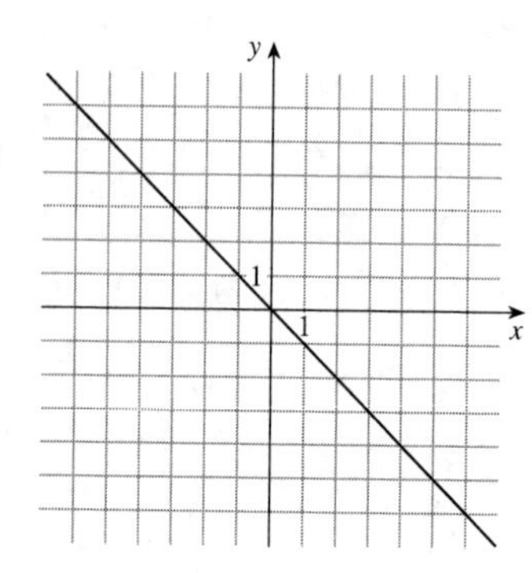

**20.** $y = 2x$

| $x$ | $y$ |
|---|---|
| $-4$ | $4$ |
| $-2$ | $2$ |
| $0$ | $0$ |
| $1$ | $-1$ |
| $2$ | $-2$ |
| $3$ | $-3$ |
| $4$ | $-4$ |

$y = 0 \quad \Leftrightarrow \quad 2x = 0$, so the $x$- and $y$-intercepts are $0$.
$x$-axis symmetry: $(-y) = 2x \quad \Leftrightarrow \quad -y = 2x$, which is not the same, so the graph is not symmetric with respect to the $x$-axis.
$y$-axis symmetry: $y = 2(-x) \quad \Leftrightarrow \quad y = -2x$, which is not the same, so the graph is not symmetric with respect to the $y$-axis.
Origin symmetry: $(-y) = 2(-x)$ is the same as $y = 2x$, so the graph is symmetric with respect to the origin.

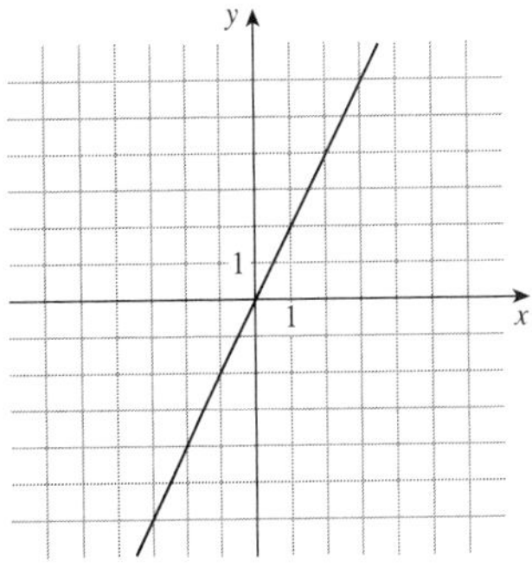


**21.** $y = -x + 4$

| $x$ | $y$ |
|---|---|
| $-4$ | $8$ |
| $-2$ | $6$ |
| $0$ | $4$ |
| $1$ | $3$ |
| $2$ | $2$ |
| $3$ | $1$ |
| $4$ | $0$ |

When $y = 0$ we get $x = 4$. So the $x$-intercept is $4$, and $x = 0 \quad \Rightarrow \quad y = 4$, so the $y$-intercept is $4$.
$x$-axis symmetry: $(-y) = -x + 4 \quad \Leftrightarrow \quad y = x - 4$, which is not the same as $y = -x + 4$, so the graph is not symmetric with respect to the $x$-axis.
$y$-axis symmetry: $y = -(-x) + 4 \quad \Leftrightarrow \quad y = x + 4$, which is not the same as $y = -x + 4$, so the graph is not symmetric with respect to the $y$-axis.
Origin symmetry: $(-y) = -(-x) + 4 \quad \Leftrightarrow \quad y = -x - 4$, which is not the same as $y = -x + 4$, so the graph is not symmetric with respect to the origin.

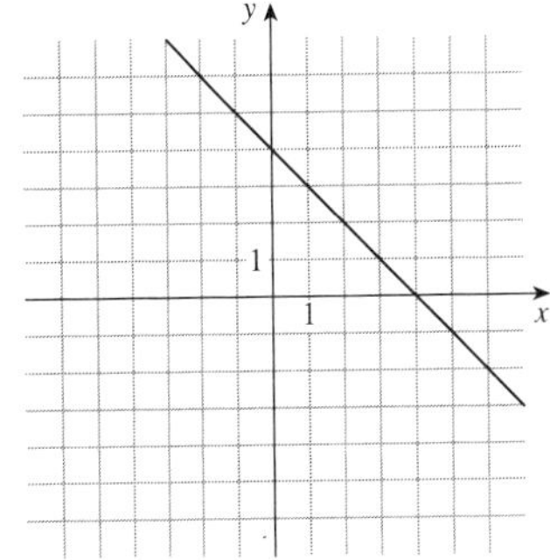


**22.** $y = 3x + 3$

| $x$ | $y$ |
|---|---|
| $-3$ | $-6$ |
| $-2$ | $-3$ |
| $-1$ | $0$ |
| $0$ | $3$ |
| $1$ | $6$ |
| $2$ | $9$ |
| $3$ | $12$ |

$y = 0 \quad \Rightarrow \quad 0 = 3x + 3 \quad \Leftrightarrow \quad 3x = -3 \quad \Leftrightarrow \quad x = -1$, so the $x$-intercept is $-1$, and $x = 0 \quad \Rightarrow \quad y = 3(0) + 3 = 3$, so the $y$-intercept is $3$.
$x$-axis symmetry: $(-y) = 3x + 3 \quad \Leftrightarrow \quad y = -3x - 3$, which is not the same as $y = 3x + 3$, so the graph is not symmetric with respect to the $x$-axis.
$y$-axis symmetry: $y = 3(-x) + 3 \quad \Leftrightarrow \quad y = -3x + 3$, which is not the same as $y = 3x + 3$, so the graph is not symmetric with respect to the $y$-axis.
Origin symmetry: $(-y) = 3(-x) + 3 \quad \Leftrightarrow \quad y = 3x - 3$, which is not the same as $y = 3x + 3$, so the graph is not symmetric with respect to the origin.

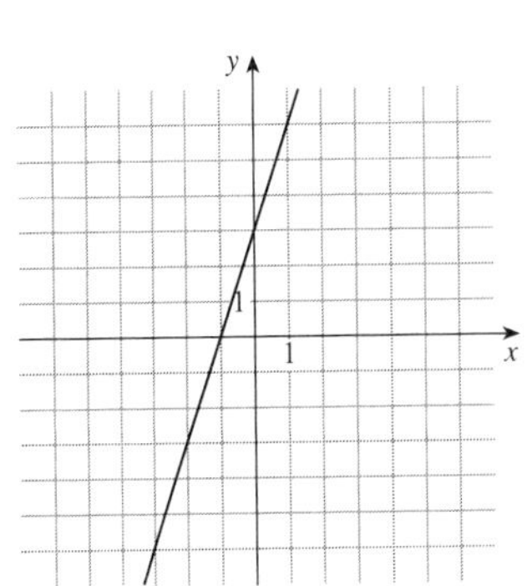

**23.** $2x - y = 6$

| $x$ | $y$ |
|---|---|
| $-1$ | $-8$ |
| 0 | $-6$ |
| 1 | $-4$ |
| 2 | $-2$ |
| 3 | 0 |
| 4 | 2 |
| 5 | 4 |

When $y = 0$ we get $2x = 6$ So the $x$-intercept is 3. When $x = 0$ we get $-y = 6$ so the $y$-intercept is $-6$.

$x$-axis symmetry: $2x - (-y) = 6 \quad \Leftrightarrow \quad 2x + y = 6$, which is not the same, so the graph is not symmetric with respect to the $x$-axis.

$y$-axis symmetry: $2(-x) - y = 6 \quad \Leftrightarrow \quad 2x + y = -6$, so the graph is not symmetric with respect to the $y$-axis.

Origin symmetry: $2(-x) - (-y) = 6 \quad \Leftrightarrow \quad -2x + y = 6$, which not he same, so the graph is not symmetric with respect to the origin.

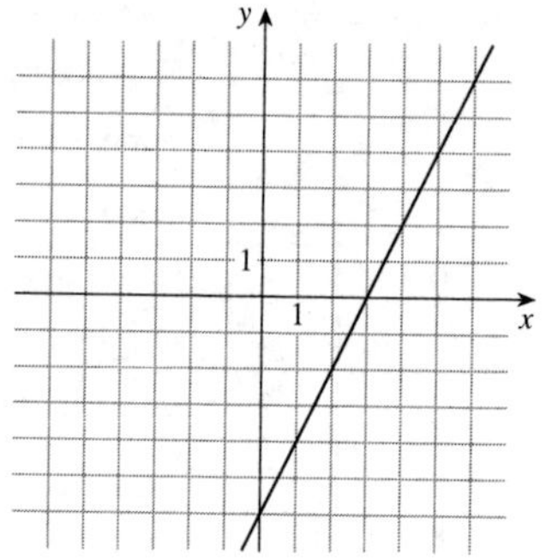


**24.** Solve for $y$: $x + y = 3 \quad \Leftrightarrow \quad y = -x + 3$.

| $x$ | $y$ |
|---|---|
| $-2$ | 5 |
| $-1$ | 4 |
| 0 | 3 |
| 1 | 2 |
| 2 | 1 |
| 3 | 0 |
| 4 | $-1$ |

$y = 0 \quad \Rightarrow \quad 0 = -x + 3 \quad \Leftrightarrow \quad x = 3$, so the $x$-intercept is 3, and $x = 0 \quad \Rightarrow \quad y = -(0) + 3 = 3$, so the $y$-intercept is 3.

$x$-axis symmetry: $x + (-y) = 3 \quad \Leftrightarrow \quad x - y = 3$, which is not the same as $x + y = 3$, so the graph is not symmetric with respect to the $x$-axis.

$y$-axis symmetry: $(-x) + y = 5 \quad \Leftrightarrow \quad -x + y = 3$, which is not the same as $x + y = 3$, so the graph is not symmetric with respect to the $y$-axis.

Origin symmetry: $(-x) + (-y) = 3 \quad \Leftrightarrow \quad -x - y = 3$, which is not the same as $x + y = 3$, so the graph is not symmetric with respect to the origin.

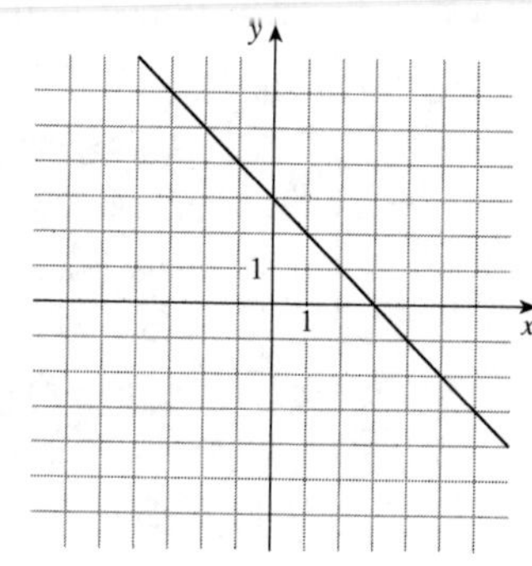


**25.** $y = 1 - x^2$

| $x$ | $y$ |
|---|---|
| $-3$ | $-8$ |
| $-2$ | $-3$ |
| $-1$ | 0 |
| 0 | 1 |
| 1 | 0 |
| 2 | $-3$ |
| 3 | $-8$ |

$y = 0 \quad \Rightarrow \quad 0 = 1 - x^2 \quad \Leftrightarrow \quad x^2 = 1 \quad \Rightarrow \quad x = \pm 1$, so the $x$-intercepts are 1 and $-1$, and $x = 0 \quad \Rightarrow \quad y = 1 - (0)^2 = 1$, so the $y$-intercept is 1.

$x$-axis symmetry: $(-y) = 1 - x^2 \quad \Leftrightarrow \quad -y = 1 - x^2$,which is not the same as $y = 1 - x^2$, so the graph is not symmetric with respect to the $x$-axis.

$y$-axis symmetry: $y = 1 - (-x)^2 \quad \Leftrightarrow \quad y = 1 - x^2$, so the graph is symmetric with respect to the $y$-axis.

Origin symmetry: $(-y) = 1 - (-x)^2 \quad \Leftrightarrow \quad -y = 1 - x^2$ which is not the same as $y = 1 - x^2$. The graph is not symmetric with respect to the origin.

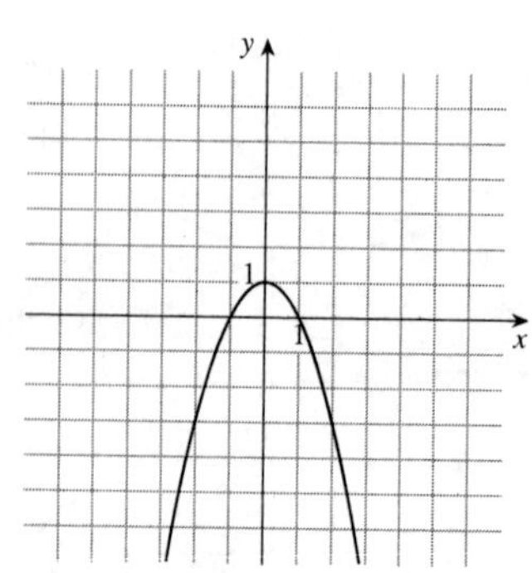

**26.** $y = x^2 + 2$

| $x$ | $y$ |
|---|---|
| $-3$ | 11 |
| $-2$ | 6 |
| $-1$ | 3 |
| 0 | 2 |
| 1 | 3 |
| 2 | 6 |
| 3 | 11 |

$y = 0 \quad \Rightarrow \quad 0 = x^2 + 2 \quad \Leftrightarrow \quad -2 = x^2$, since $x^2 \geq 0$, there is no $x$-intercept, and $x = 0 \quad \Rightarrow \quad y = (0)^2 + 2 = 2$, so the $y$-intercept is 2.

$x$-axis symmetry: $(-y) = x^2 + 2 \quad \Leftrightarrow \quad -y = x^2 + 2$, which is not the same, so the graph is not symmetric with respect to the $x$-axis.

$y$-axis symmetry: $y = (-x)^2 + 2 = x^2 + 2$, so the graph is symmetric with respect to the $y$-axis.

Origin symmetry: $(-y) = (-x)^2 + 2 \quad \Leftrightarrow \quad -y = x^2 + 2$, which is not the same, so the graph is not symmetric with respect to the origin.

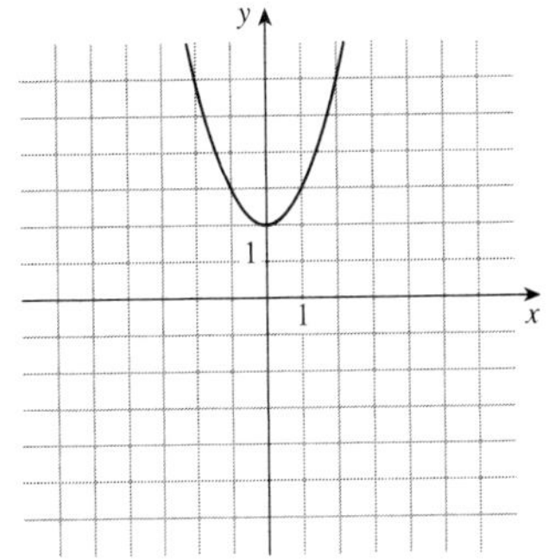


**27.** $4y = x^2 \quad \Leftrightarrow \quad y = \frac{1}{4}x^2$

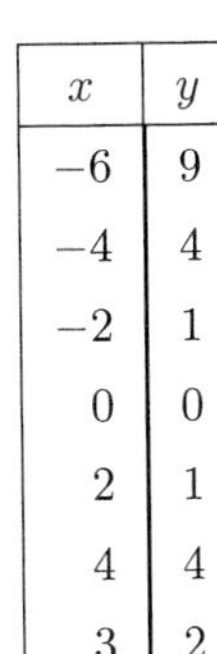

| $x$ | $y$ |
|---|---|
| $-6$ | 9 |
| $-4$ | 4 |
| $-2$ | 1 |
| 0 | 0 |
| 2 | 1 |
| 4 | 4 |
| 3 | 2 |

$y = 0 \quad \Rightarrow \quad 0 = \frac{1}{4}x^2 \quad \Leftrightarrow \quad x^2 = 0 \quad \Rightarrow \quad x = 0$, so the $x$-intercept is 0, and $x = 0 \quad \Rightarrow \quad y = \frac{1}{4}(0)^2 = 0$, so the $y$-intercept is 0.

$x$-axis symmetry: $(-y) = \frac{1}{4}x^2$, which is not the same as $y = \frac{1}{4}x^2$, so the graph is not symmetric with respect to the $x$-axis.

$y$-axis symmetry: $y = \frac{1}{4}(-x)^2 \quad \Leftrightarrow \quad y = \frac{1}{4}x^2$, so the graph is symmetric with respect to the $y$-axis.

Origin symmetry: $(-y) = \frac{1}{4}(-x)^2 \quad \Leftrightarrow \quad -y = \frac{1}{4}x^2$, which is not the same as $y = \frac{1}{4}x^2$, so the graph is not symmetric with respect to the origin.

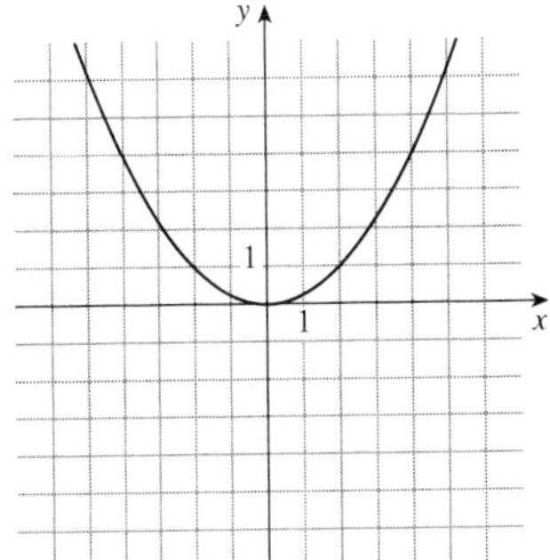


**28.** $8y = x^3 \quad \Leftrightarrow \quad y = \frac{1}{8}x^3$

| $x$ | $y$ |
|---|---|
| $-6$ | $-27$ |
| $-4$ | $-8$ |
| $-2$ | $-1$ |
| 0 | 0 |
| 2 | 1 |
| 4 | 8 |
| 3 | 27 |

$y = 0 \quad \Rightarrow \quad 0 = \frac{1}{8}x^3 \quad \Leftrightarrow \quad x^3 = 0 \quad \Rightarrow \quad x = 0$, so the $x$-intercept is 0, and $x = 0 \quad \Rightarrow \quad y = \frac{1}{8}(0)^3 = 0$, so the $y$-intercept is 0.

$x$-axis symmetry: $(-y) = \frac{1}{8}x^3$, which is not the same as $y = \frac{1}{8}x^3$, so the graph is not symmetric with respect to the $x$-axis.

$y$-axis symmetry: $y = \frac{1}{8}(-x)^3 = -\frac{1}{8}x^3$, which is not the same as $y = \frac{1}{8}x^3$, so the graph is not symmetric with respect to the $y$-axis.

Origin symmetry: $(-y) = \frac{1}{8}(-x)^3 \quad \Leftrightarrow \quad -y = -\frac{1}{8}x^3$, which is the same as $y = \frac{1}{8}x^3$, so the graph is symmetric with respect to the origin.

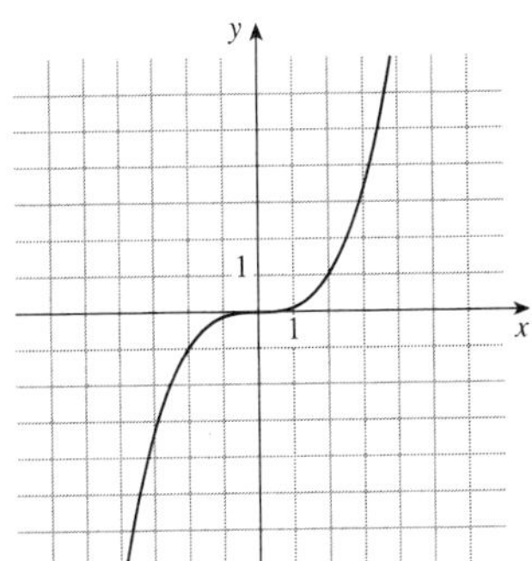

**29.** $y = x^2 - 9$

| $x$ | $y$ |
|---|---|
| $-4$ | $7$ |
| $-3$ | $0$ |
| $-2$ | $-5$ |
| $-1$ | $-8$ |
| $0$ | $-9$ |
| $1$ | $-8$ |
| $2$ | $-5$ |
| $3$ | $0$ |
| $4$ | $7$ |

$y = 0 \Rightarrow 0 = x^2 - 9 \Leftrightarrow x^2 = 9 \Rightarrow x = \pm 3$, so the $x$-intercepts are 3 and $-3$, and $x = 0 \Rightarrow y = (0)^2 - 9 = -9$, so the $y$-intercept is $-9$.

$x$-axis symmetry: $(-y) = x^2 - 9$, which is not the same as $y = x^2 - 9$, so the graph is not symmetric with respect to the $x$-axis.

$y$-axis symmetry: $y = (-x)^2 - 9 \Leftrightarrow y = x^2 - 9$, so the graph is symmetric with respect to the $y$-axis.

Origin symmetry: $(-y) = (-x)^2 - 9 \Leftrightarrow -y = x^2 - 9$, which is not the same as $y = x^2 - 9$, so the graph is not symmetric with respect to the origin.

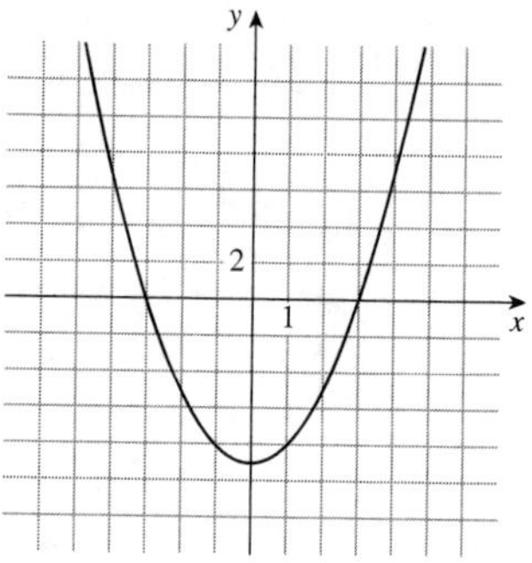


**30.** $y = 9 - x^2$

| $x$ | $y$ |
|---|---|
| $-4$ | $-7$ |
| $-3$ | $0$ |
| $-2$ | $5$ |
| $-1$ | $8$ |
| $0$ | $9$ |
| $1$ | $8$ |
| $2$ | $5$ |
| $3$ | $0$ |
| $4$ | $-7$ |

$y = 0 \Rightarrow 0 = 9 - x^2 \Leftrightarrow x^2 = 9 \Rightarrow x = \pm 3$, so the $x$-intercepts are $-3$ and 3, and $x = 0 \Rightarrow y = 9 - (0)^2 = 9$, so the $y$-intercept is 9.

$x$-axis symmetry: $(-y) = 9 - x^2$, which is not the same as $y = 9 - x^2$, so the graph is not symmetric with respect to the $x$-axis.

$y$-axis symmetry: $y = 9 - (-x)^2 = 9 - x^2$, so the graph is symmetric with respect to the $y$-axis.

Origin symmetry: $(-y) = 9 - (-x)^2 \Leftrightarrow -y = 9 - x^2$, which is not the same as $y = 9 - x^2$, so the graph is not symmetric with respect to the origin.

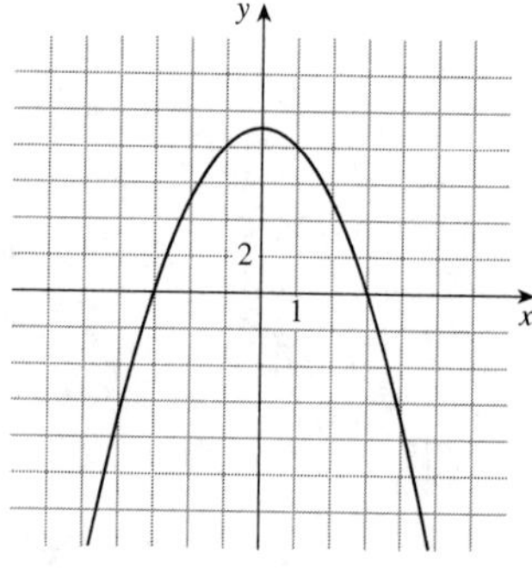


**31.** $xy = 2 \Leftrightarrow y = \dfrac{2}{x}$

| $x$ | $y$ |
|---|---|
| $-4$ | $-\frac{1}{2}$ |
| $-2$ | $-1$ |
| $-1$ | $-2$ |
| $-\frac{1}{2}$ | $-4$ |
| $-\frac{1}{4}$ | $-8$ |
| $\frac{1}{4}$ | $8$ |
| $\frac{1}{2}$ | $4$ |
| $1$ | $2$ |
| $2$ | $1$ |
| $4$ | $\frac{1}{2}$ |

$y = 0$ or $x = 0 \Rightarrow 0 = 2$, which is impossible, so this equation has no $x$-intercept and no $y$-intercept.

$x$-axis symmetry: $x(-y) = 2 \Leftrightarrow -xy = 2$, which is not the same as $xy = 2$, so the graph is not symmetric with respect to the $x$-axis.

$y$-axis symmetry: $(-x)y = 2 \Leftrightarrow -xy = 2$, which is not the same as $xy = 2$, so the graph is not symmetric with respect to the $y$-axis.

Origin symmetry: $(-x)(-y) = 2 \Leftrightarrow xy = 2$, so the graph is symmetric with respect to the origin.

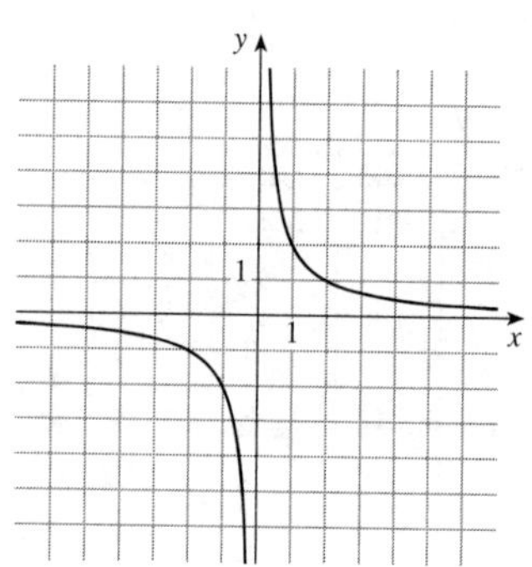

**32.** Solve for $x$ in terms of $y$: $x + y^2 = 4 \quad \Leftrightarrow \quad x = 4 - y^2$

| $x$ | $y$ |
|---|---|
| $-12$ | $-4$ |
| $-5$ | $-3$ |
| $0$ | $-2$ |
| $3$ | $-1$ |
| $4$ | $0$ |
| $3$ | $1$ |
| $0$ | $2$ |
| $-5$ | $3$ |
| $-12$ | $4$ |

$y = 0 \quad \Rightarrow \quad x + 0^2 = 4 \quad \Leftrightarrow \quad x = 4$, so the $x$-intercept is 4, and $x = 0 \quad \Rightarrow \quad 0 + y^2 = 4 \quad \Rightarrow \quad y = \pm 2$, so the $y$-intercepts are $-2$ and 2.

$x$-axis symmetry: $x + (-y)^2 = 4 \quad \Leftrightarrow \quad x + y^2 = 4$, so the graph is symmetric with respect to the $x$-axis.

$y$-axis symmetry: $(-x) + y^2 = 4 \quad \Leftrightarrow \quad -x + y^2 = 4$, which is not the same, so the graph is not symmetric with respect to the $y$-axis.

Origin symmetry: $(-x) + (-y)^2 = 4 \quad \Leftrightarrow \quad -x + y^2 = 4$, which is not the same as $x + y^2 = 4$, so the graph is not symmetric with respect to the origin.

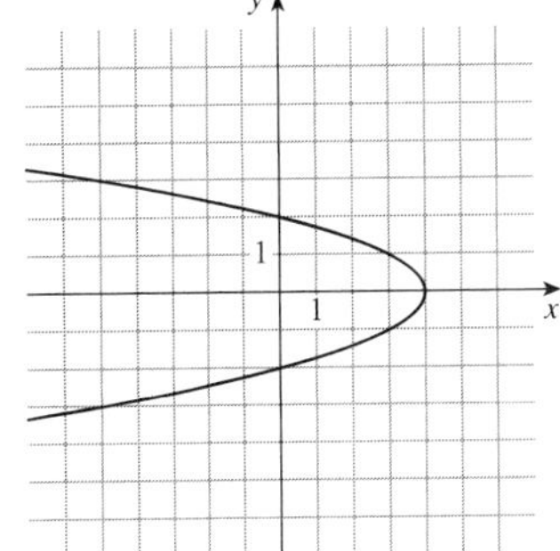


**33.** $y = \sqrt{x}$

| $x$ | $y$ |
|---|---|
| $0$ | $0$ |
| $\frac{1}{4}$ | $\frac{1}{2}$ |
| $1$ | $1$ |
| $2$ | $\sqrt{2}$ |
| $4$ | $2$ |
| $9$ | $3$ |
| $16$ | $4$ |

$y = 0 \quad \Rightarrow \quad 0 = \sqrt{x} \quad \Leftrightarrow \quad x = 0$. so the $x$-intercept is 0, and $x = 0 \quad \Rightarrow \quad y = \sqrt{0} = 0$, so the $y$-intercept is 0. Since we are graphing real numbers and $\sqrt{x}$ is defined to be a nonnegative number, the equation is not symmetric with respect to the $x$-axis nor with respect to the $y$-axis. Also, the equation is not symmetric with respect to the origin.

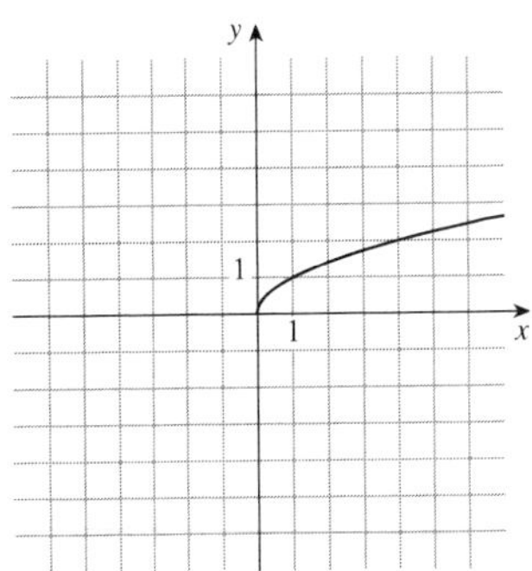


**34.** $x^2 + y^2 = 9$

| $x$ | $y$ |
|---|---|
| $-3$ | $0$ |
| $-2$ | $\pm\sqrt{5}$ |
| $-1$ | $\pm 2\sqrt{2}$ |
| $0$ | $\pm 3$ |
| $1$ | $\pm 2\sqrt{2}$ |
| $2$ | $\pm\sqrt{5}$ |
| $3$ | $0$ |

$y = 0 \quad \Rightarrow \quad x^2 + 0^2 = 9 \quad \Leftrightarrow \quad x = \pm 3$, so the $x$-intercepts are $-3$ and 3, and $x = 0 \quad \Rightarrow \quad 0^2 + y^2 = 9 \quad \Leftrightarrow \quad y = \pm 3$, so the $y$-intercepts are $-3$ and 3.

$x$-axis symmetry: $x^2 + (-y)^2 = 9 \quad \Leftrightarrow \quad x^2 + y^2 = 9$, so it is symmetric with respect to the $x$-axis.

$y$-axis symmetry: $(-x)^2 + y^2 = 9 \quad \Leftrightarrow \quad x^2 + y^2 = 9$, so it is symmetric with respect to the $y$-axis.

Origin symmetry: $(-x)^2 + (-y)^2 = 9 \quad \Leftrightarrow \quad x^2 + y^2 = 9$, so it is symmetric with respect to the origin.

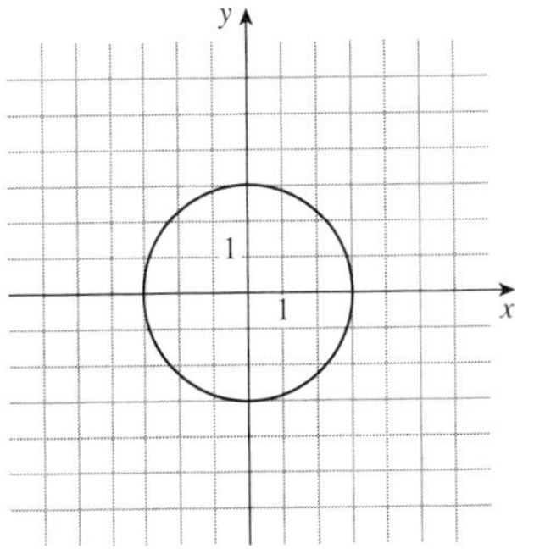

**35.** $y = \sqrt{4 - x^2}$. Since the radicand (the inside of the square root) cannot be negative, we must have $4 - x^2 \geq 0 \quad \Leftrightarrow \quad x^2 \leq 4 \quad \Leftrightarrow \quad |x| \leq 2$.

| $x$ | $y$ |
|---|---|
| $-2$ | $0$ |
| $-1$ | $\sqrt{3}$ |
| $0$ | $4$ |
| $1$ | $\sqrt{3}$ |
| $2$ | $0$ |

$y = 0 \quad \Rightarrow \quad 0 = \sqrt{4 - x^2} \quad \Leftrightarrow \quad 4 - x^2 = 0 \quad \Leftrightarrow \quad x^2 = 4 \quad \Rightarrow \quad x = \pm 2$, so the $x$-intercept are $-2$ and $2$, and $x = 0 \quad \Rightarrow \quad y = \sqrt{4 - (0)^2} = \sqrt{4} = 2$, so the $y$-intercept is 2. Since $y \geq 0$, the graph is not symmetric with respect to the $x$-axis.

$y$-axis symmetry: $y = \sqrt{4 - (-x)^2} = \sqrt{4 - x^2}$, so the graph is symmetric with respect to the $y$-axis. Also, since $y \geq 0$ the graph is not symmetric with respect to the origin.

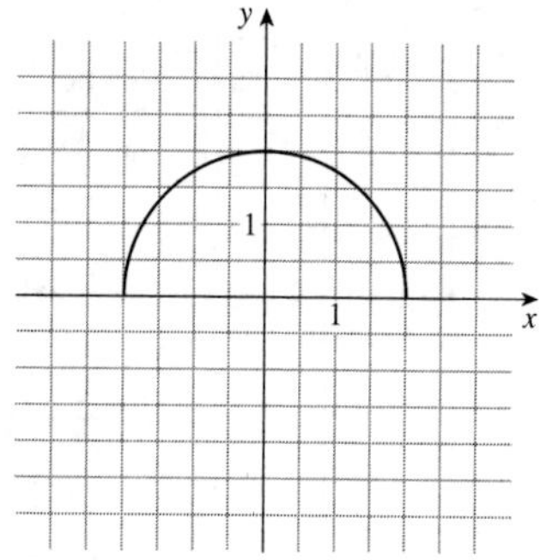


**36.** $y = -\sqrt{4 - x^2}$Since the radicand (the inside of the square root) cannot be negative, we must have $4 - x^2 \geq 0 \quad \Leftrightarrow \quad x^2 \leq 4 \quad \Leftrightarrow \quad |x| \leq 2$. So $-2 \leq x \leq 2$ is the only portion of the $x$-axis where this equation is defined.

| $x$ | $y$ |
|---|---|
| $-2$ | $0$ |
| $-1$ | $-\sqrt{3}$ |
| $0$ | $-4$ |
| $1$ | $-\sqrt{3}$ |
| $2$ | $0$ |

$y = 0 \quad \Rightarrow \quad 0 = -\sqrt{4 - x^2} \quad \Leftrightarrow \quad 4 - x^2 = 0 \quad \Leftrightarrow \quad x^2 = 4 \quad \Rightarrow \quad x = \pm 2$, so the $x$-intercepts are $-2$ and $2$, and $x = 0 \quad \Rightarrow \quad y = -\sqrt{4 - (0)^2} = -\sqrt{4} = -2$, so the $y$-intercept is $-2$.

$x$-axis symmetry: Since $y \leq 0$, this graph is not symmetric with respect to the $x$-axis.

$y$-axis symmetry: $y = -\sqrt{4 - (-x)^2} = -\sqrt{4 - x^2}$, so the graph is symmetric with respect to the $y$-axis.

Origin symmetry: Since $y \leq 0$, the graph is not symmetric with respect to the origin.

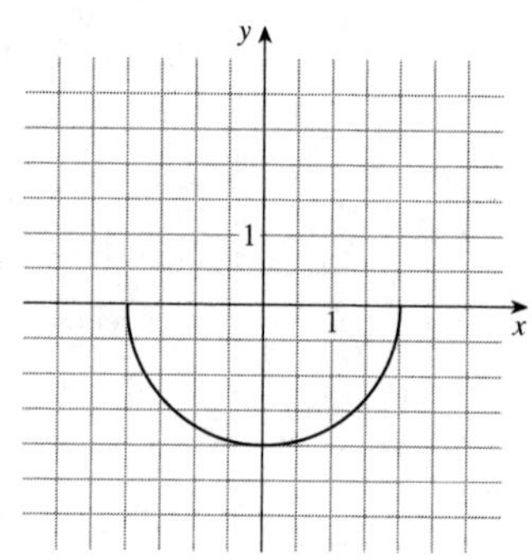


**37.** $y = |x|$

| $x$ | $y$ |
|---|---|
| $-3$ | $3$ |
| $-2$ | $2$ |
| $-1$ | $1$ |
| $0$ | $0$ |
| $1$ | $1$ |
| $2$ | $2$ |
| $3$ | $3$ |

$y = 0 \quad \Rightarrow \quad 0 = |x| \quad \Leftrightarrow \quad x = 0$, so the $x$-intercept is 0, and $x = 0 \quad \Rightarrow \quad y = |0| = 0$, so the $y$-intercept is 0.

Since $y \geq 0$, the graph is not symmetric with respect to the $x$-axis.

$y$-axis symmetry: $y = |-x| = |x|$, so the graph is symmetric with respect to the $y$-axis.

Since $y \geq 0$, the graph is not symmetric with respect to the origin.

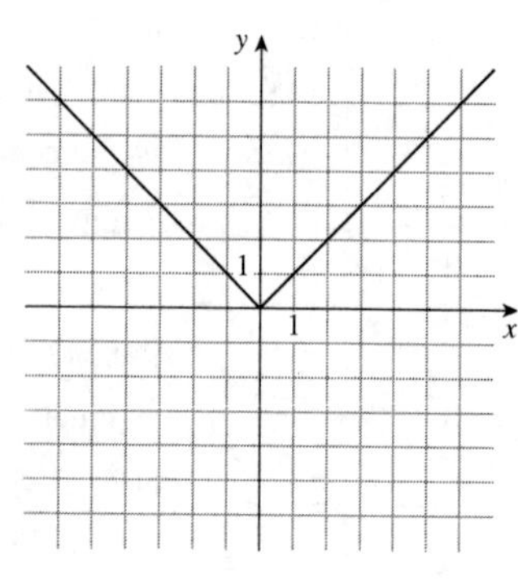

**38.** $x = |y|$. Here we insert values of $y$ and find the corresponding value of $x$.

| $x$ | $y$ |
|---|---|
| 3 | $-3$ |
| 2 | $-2$ |
| 1 | $-1$ |
| 0 | 0 |
| 1 | 1 |
| 2 | 2 |
| 3 | 3 |

$y = 0 \quad \Rightarrow \quad x = |0| = 0$, so the $x$-intercept is 0, and $x = 0 \Rightarrow 0 = |y| \quad \Leftrightarrow \quad y = 0$, so the $y$-intercept is 0.

$x$-axis symmetry: $x = |-y| = |y|$, so the graph is symmetric with respect to the $x$-axis. $y$-axis symmetry: Since $x \geq 0$, the graph is not symmetric with respect to the $y$-axis. Origin symmetry: Since $x \geq 0$, the graph is not symmetric with respect to the origin.

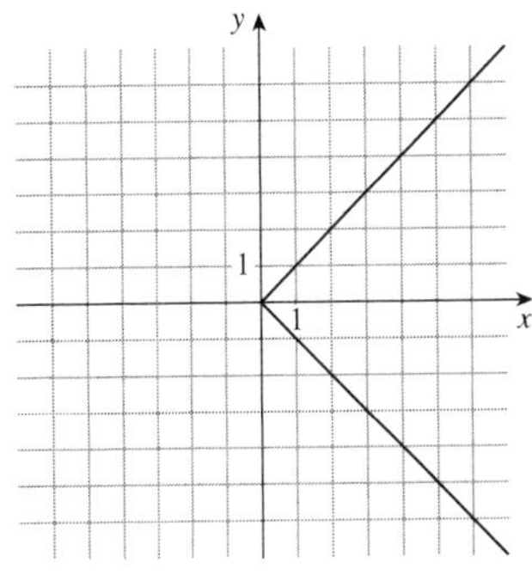


**39.** $y = 4 - |x|$

| $x$ | $y$ |
|---|---|
| $-6$ | $-2$ |
| $-4$ | 0 |
| $-2$ | 2 |
| 0 | 4 |
| 2 | 2 |
| 4 | 0 |
| 6 | $-2$ |

$y = 0 \quad \Rightarrow \quad 0 = 4 - |x| \quad \Leftrightarrow \quad |x| = 4 \quad \Rightarrow \quad x = \pm 4$, so the $x$-intercepts are $-4$ and 4, and $x = 0 \quad \Rightarrow \quad y = 4 - |0| = 4$, so the $y$-intercept is 4.

$x$-axis symmetry: $(-y) = 4 - |x| \quad \Leftrightarrow \quad y = -4 + |x|$, which is not the same as $y = 4 - |x|$, so the graph is not symmetric with respect to the $x$-axis.

$y$-axis symmetry: $y = 4 - |-x| = 4 - |x|$, so the graph is symmetric with respect to the $y$-axis.

Origin symmetry: $(-y) = 4 - |-x| \quad \Leftrightarrow \quad y = -4 + |x|$,which is not the same as $y = 4 - |x|$, so the graph is not symmetric with respect to the origin.

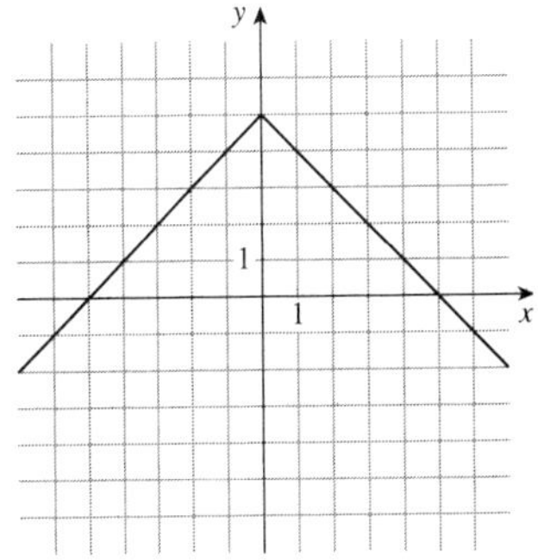


**40.** $y = |4 - x|$

| $x$ | $y$ |
|---|---|
| $-6$ | 10 |
| $-4$ | 8 |
| $-2$ | 6 |
| 0 | 4 |
| 2 | 2 |
| 4 | 0 |
| 6 | 2 |
| 8 | 4 |
| 10 | 6 |

$y = 0 \quad \Rightarrow \quad 0 = |4 - x| \quad \Leftrightarrow \quad 4 - x = 0 \quad \Rightarrow \quad x = 4$, so the $x$-intercept is 4, and $x = 0 \quad \Rightarrow \quad y = |4 - 0| = |4| = 4$, so the $y$-intercept is 4.

$x$-axis symmetry: $y = |4 - (-x)| = |4 + x|$, which is not the same as $y = |4 - x|$, so the graph is not symmetric with respect to the $x$-axis.

$y$-axis symmetry: $(-y) = |4 - x| \quad \Leftrightarrow \quad -y = |4 - x|$, which is not the same as $y = |4 - x|$, so the graph is not symmetric with respect to the $y$-axis.

Origin symmetry: $(-y) = |4 - (-x)| \quad \Leftrightarrow \quad -y = |4 + x|$, which is not the same as $y = |4 - x|$, so the graph is not symmetric with respect to the origin.

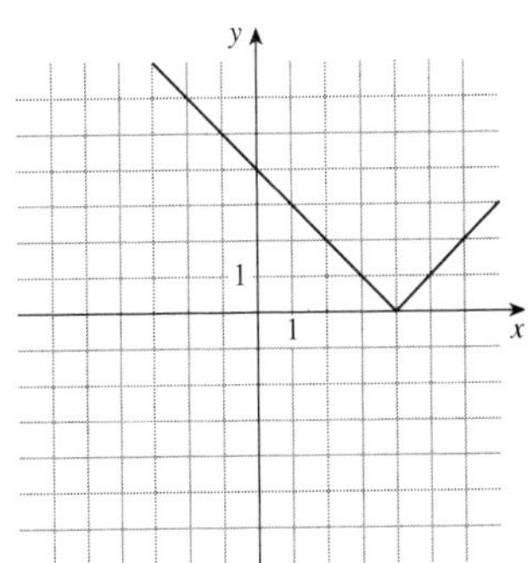

**41.** Since $x = y^3$ is solved for $x$ in terms of $y$, we insert values for $y$ and find the corresponding values of $x$.

| $x$ | $y$ |
|---|---|
| $-27$ | $-3$ |
| $-8$ | $-2$ |
| $-1$ | $-1$ |
| $0$ | $0$ |
| $1$ | $1$ |
| $8$ | $2$ |
| $27$ | $3$ |

$y = 0 \quad \Rightarrow \quad x = (0)^3 = 0$, so the $x$-intercept is 0, and $x = 0 \quad \Rightarrow$ $0 = y^3 \quad \Rightarrow \quad y = 0$, so the $y$-intercept is 0.

$x$-axis symmetry: $x = (-y)^3 = -y^3$, which is not the same as $x = y^3$, so the graph is not symmetric with respect to the $x$-axis.

$y$-axis symmetry: $(-x) = y^3 \quad \Leftrightarrow \quad x = -y^3$, which is not the same as $x = y^3$, so the graph is not symmetric with respect to the $y$-axis.

Origin symmetry: $(-x) = (-y)^3 \quad \Leftrightarrow \quad -x = -y^3 \quad \Leftrightarrow$ $x = y^3$, so the graph is symmetric with respect to the origin.

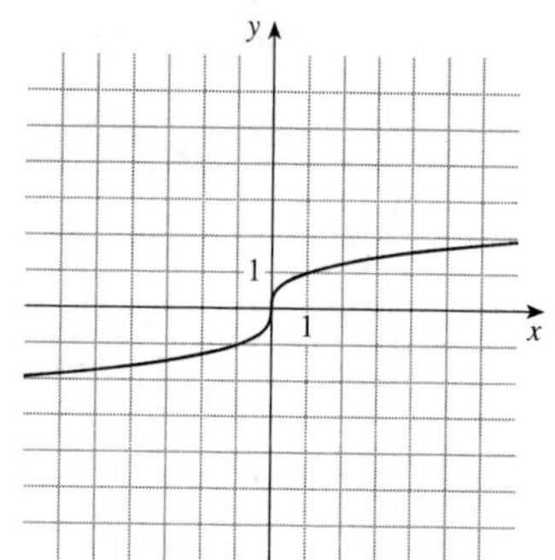


**42.** $y = x^3 - 1$

| $x$ | $y$ |
|---|---|
| $-3$ | $-28$ |
| $-2$ | $-9$ |
| $-1$ | $-2$ |
| $0$ | $-1$ |
| $1$ | $1$ |
| $2$ | $7$ |
| $3$ | $26$ |

$y = 0 \quad \Rightarrow \quad 0 = x^3 - 1 = (x-1)\left(x^2 + x + 1\right)$. So $x - 1 = 0$ $\Leftrightarrow \quad x = 1$, while $x^2 + x + 1 = 0$ has no real solution, so the $x$-intercept is 1, and $x = 0 \quad \Rightarrow y = (0)^3 - 1 = -1$, so the $y$-intercept is $-1$.

$x$-axis symmetry: $-y = x^3 - 1$, which is not the same as $y = x^3 - 1$, so the graph is not symmetric with respect to the $x$-axis.

$y$-axis symmetry: $y = (-x)^3 - 1 = -x^3 - 1$, which is not the same as $y = x^3 - 1$, so the graph is not symmetric with respect to the $y$-axis.

Origin symmetry: $(-y) = (-x)^3 - 1 \quad \Leftrightarrow \quad -y = -x^3 - 1 \quad \Leftrightarrow$ $y = x^3 + 1$ which is not the same. Not symmetric with respect to the origin.

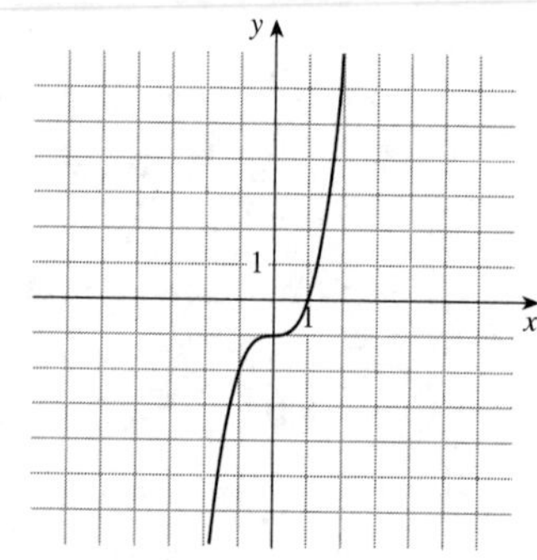


**43.** $y = x^4$

| $x$ | $y$ |
|---|---|
| $-3$ | $81$ |
| $-2$ | $16$ |
| $-1$ | $1$ |
| $0$ | $0$ |
| $1$ | $1$ |
| $2$ | $16$ |
| $3$ | $81$ |

$y = 0 \quad \Rightarrow \quad 0 = x^4 \quad \Rightarrow \quad x = 0$, so the $x$-intercept is 0, and $x = 0$ $\Rightarrow \quad y = 0^4 = 0$, so the $y$-intercept is 0.

$x$-axis symmetry: $(-y) = x^4 \quad \Leftrightarrow \quad y = -x^4$, which is not the same as $y = x^4$, so the graph is not symmetric with respect to the $x$-axis.

$y$-axis symmetry: $y = (-x)^4 = x^4$, so the graph is symmetric with respect to the $y$-axis.

Origin symmetry: $(-y) = (-x)^4 \quad \Leftrightarrow \quad -y = x^4$, which is not the same as $y = x^4$, so the graph is not symmetric with respect to the origin.

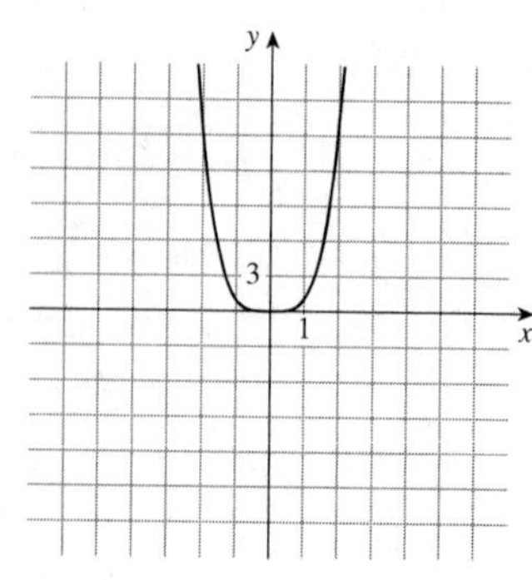

**44.** $y = 16 - x^4$

| $x$ | $y$ |
|---|---|
| $-3$ | $-65$ |
| $-2$ | $0$ |
| $-1$ | $15$ |
| $0$ | $16$ |
| $1$ | $15$ |
| $2$ | $0$ |
| $3$ | $-65$ |

$y = 0 \quad \Rightarrow \quad 0 = 16 - x^4 \quad \Rightarrow \quad x^4 = 16 \quad \Rightarrow \quad x^2 = 4 \Rightarrow$ $x = \pm 2$, so the $x$-intercepts are $\pm 2$, and so $x = 0 \quad \Rightarrow$ $y = 16 - 0^4 = 16$, so the $y$-intercept is 16.

$x$-axis symmetry: $(-y) = 16 - x^4 \quad \Leftrightarrow \quad y = -16 + x^4$, which is not the same as $y = 16 - x^4$, so the graph is not symmetric with respect to the $x$-axis. $y$-axis symmetry: $y = 16 - (-x)^4 = 16 - x^4$, so the graph is symmetric with respect to the $y$-axis.

Origin symmetry: $(-y) = 16 - (-x)^4 \quad \Leftrightarrow \quad -y = 16 - x^4$, which is not the same as $y = 16 - x^4$, so the graph is not symmetric with respect to the origin.

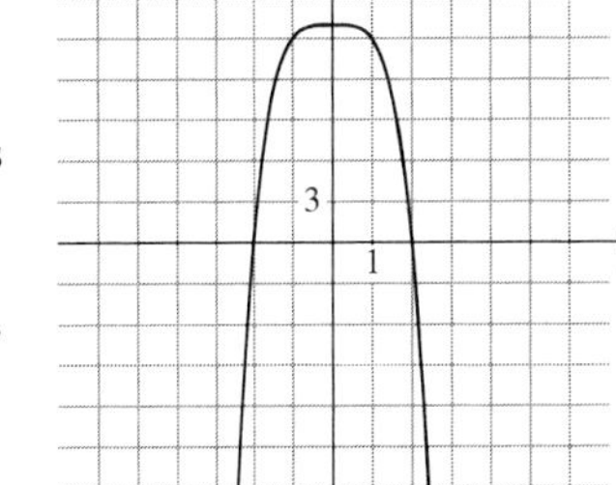


**45.** $x$-axis symmetry: $(-y) = x^4 + x^2 \quad \Leftrightarrow \quad y = -x^4 - x^2$, which is not the same as $y = x^4 + x^2$, so the graph is not symmetric with respect to the $x$-axis.

$y$-axis symmetry: $y = (-x)^4 + (-x)^2 = x^4 + x^2$, so the graph is symmetric with respect to the $y$-axis.

Origin symmetry: $(-y) = (-x)^4 + (-x)^2 \quad \Leftrightarrow \quad -y = x^4 + x^2$, which is not the same as $y = x^4 + x^2$, so the graph is not symmetric with respect to the origin.

**46.** $x$-axis symmetry: $x = (-y)^4 - (-y)^2 = y^4 - y^2$, so the graph is symmetric with respect to the $x$-axis.

$y$-axis symmetry: $(-x) = y^4 - y^2$, which is not the same as $x = y^4 - y^2$, so the graph is not symmetric with respect to the $y$-axis.

Origin symmetry: $(-x) = (-y)^4 - (-y)^2 \quad \Leftrightarrow \quad -x = y^4 - y^2$, which is not the same as $x = y^4 - y^2$, so the graph is not symmetric with respect to the origin.

**47.** $x$-axis symmetry: $x^2(-y)^2 + x(-y) = 1 \quad \Leftrightarrow \quad x^2y^2 - xy = 1$, which is not the same as $x^2y^2 + xy = 1$, so the graph is not symmetric with respect to the $x$-axis.

$y$-axis symmetry: $(-x)^2 y^2 + (-x) y = 1 \quad \Leftrightarrow \quad x^2y^2 - xy = 1$, which is not the same as $x^2y^2 + xy = 1$, so the graph is not symmetric with respect to the $y$-axis.

Origin symmetry: $(-x)^2(-y)^2 + (-x)(-y) = 1 \quad \Leftrightarrow \quad x^2y^2 + xy = 1$, so the graph is symmetric with respect to the origin.

**48.** $x$-axis symmetry: $x^4(-y)^4 + x^2(-y)^2 = 1 \quad \Leftrightarrow \quad x^4y^4 + x^2y^2 = 1$, so the graph is symmetric with respect to the $x$-axis.

$y$-axis symmetry: $(-x)^4 y^4 + (-x)^2 y^2 = 1 \quad \Leftrightarrow \quad x^4y^4 + x^2y^2 = 1$, so the graph is symmetric with respect to the $y$-axis.

Origin symmetry: $(-x)^4(-y)^4 + (-x)^2(-y)^2 = 1 \quad \Leftrightarrow \quad x^4y^4 + x^2y^2 = 1$, so the graph is symmetric with respect to the origin.

**49.** $x$-axis symmetry: $(-y) = x^3 + 10x \quad \Leftrightarrow \quad y = -x^3 - 10x$, which is not the same as $y = x^3 + 10x$, so the graph is not symmetric with respect to the $x$-axis.

$y$-axis symmetry: $y = (-x)^3 + 10(-x) \quad \Leftrightarrow \quad y = -x^3 - 10x$, which is not the same as $y = x^3 + 10x$, so the graph is not symmetric with respect to the $y$-axis.

Origin symmetry: $(-y) = (-x)^3 + 10(-x) \quad \Leftrightarrow \quad -y = -x^3 - 10x \quad \Leftrightarrow \quad y = x^3 + 10x$, so the graph is symmetric with respect to the origin.

**50.** $x$-axis symmetry: $(-y) = x^2 + |x| \quad \Leftrightarrow \quad y = -x^2 - |x|$, which is not the same as $y = x^2 + |x|$, so the graph is not symmetric with respect to the $x$-axis.

$y$-axis symmetry: $y = (-x)^2 + |-x| \quad \Leftrightarrow \quad y = x^2 + |x|$, so the graph is symmetric with respect to the $y$-axis. Note that $|-x| = |x|$.

Origin symmetry: $(-y) = (-x)^2 + |-x| \quad \Leftrightarrow \quad -y = x^2 + |x| \quad \Leftrightarrow \quad y = -x^2 - |x|$, which is not the same as $y = x^2 + |x|$, so the graph is not symmetric with respect to the origin.

**51.** Symmetric with respect to the $y$-axis.

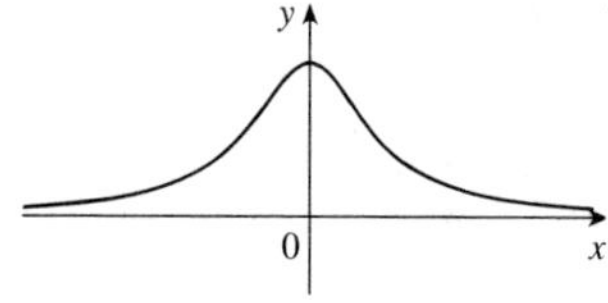


**52.** Symmetric with respect to the $x$-axis.

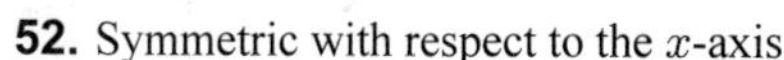

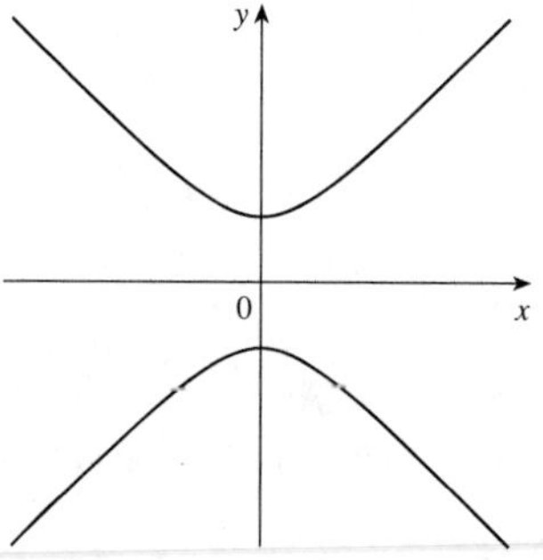


**53.** Symmetric with respect to the origin.

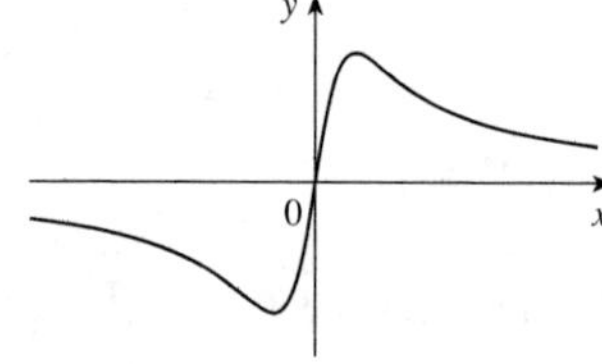


**54.** Symmetric with respect to the origin.

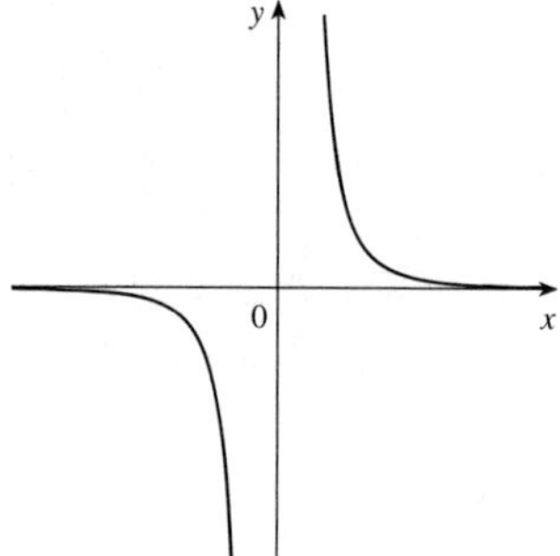


**55.** Using $h = 2$, $k = -1$, and $r = 3$, we get $(x-2)^2 + (y-(-1))^2 = 3^2 \quad \Leftrightarrow \quad (x-2)^2 + (y+1)^2 = 9$.

**56.** Using $h = -1$, $k = -4$, and $r = 8$ we get $(x-(-1))^2 + (y-(-4))^2 = 8^2 \quad \Leftrightarrow \quad (x+1)^2 + (y+4)^2 = 64$.

**57.** The equation of a circle centered at the origin is $x^2 + y^2 = r^2$. Using the point $(4, 7)$ we solve for $r^2$. This gives $(4)^2 + (7)^2 = r^2 \quad \Leftrightarrow \quad 16 + 49 = 65 = r^2$. Thus, the equation of the circle is $x^2 + y^2 = 65$.

**58.** Using $h = -1$ and $k = 5$, we get $(x-(-1))^2 + (y-5)^2 = r^2 \quad \Leftrightarrow \quad (x+1)^2 + (y-5)^2 = r^2$. Next, using the point $(-4, -6)$, we solve for $r^2$. This gives $(-4+1)^2 + (-6-5)^2 = r^2 \quad \Leftrightarrow \quad 130 = r^2$. Thus, an equation of the circle is $(x+1)^2 + (y-5)^2 = 130$.

**59.** The center is at the midpoint of the line segment, which is $\left(\dfrac{-1+5}{2}, \dfrac{1+5}{2}\right) = (2, 3)$. The radius is one half the diameter, so $r = \frac{1}{2}\sqrt{(-1-5)^2 + (1-5)^2} = \frac{1}{2}\sqrt{36+16} = \frac{1}{2}\sqrt{52} = \sqrt{13}$. Thus, an equation of the circle is $(x-2)^2 + (y-3)^2 = \left(\sqrt{13}\right)^2$ or $(x-2)^2 + (y-3)^2 = 13$.

**60.** The center is at the midpoint of the line segment, which is $\left(\dfrac{-1+7}{2}, \dfrac{3+-5}{2}\right) = (3, -1)$. The radius is one half the diameter, so $r = \frac{1}{2}\sqrt{(-1-7)^2 + (3-(-5))^2} = 4\sqrt{2}$. Thus, an equation of the circle is $(x-3)^2 + (y+1)^2 = 32$ .

**61.** Since the circle is tangent to the $x$-axis, it must contain the point $(7, 0)$, so the radius is the change in the $y$-coordinates. That is, $r = |-3-0| = 3$. So the equation of the circle is $(x-7)^2 + (y-(-3))^2 = 3^2$, which is $(x-7)^2 + (y+3)^2 = 9$.

**62.** Since the circle with $r = 5$ lies in the first quadrant and is tangent to both the $x$-axis and the $y$-axis, the center of the circle is at $(5, 5)$. Therefore, the equation of the circle is $(x - 5)^2 + (y - 5)^2 = 25$.

**63.** From the figure, the center of the circle is at $(-2, 2)$. The radius is the change in the $y$-coordinates, so $r = |2 - 0| = 2$. Thus the equation of the circle is $(x - (-2))^2 + (y - 2)^2 = 2^2$, which is $(x + 2)^2 + (y - 2)^2 = 4$.

**64.** From the figure, the center of the circle is at $(-1, 1)$. The radius is the distance from the center to the point $(2, 0)$. Thus $r = \sqrt{(-1 - 2)^2 + (1 - 0)^2} = \sqrt{9 + 1} = \sqrt{10}$, and the equation of the circle is $(x + 1)^2 + (y - 1)^2 = 10$.

**65.** Completing the square gives $x^2 + y^2 - 2x + 4y + 1 = 0 \Leftrightarrow x^2 - 2x + \left(\frac{-2}{2}\right)^2 + y^2 + 4y + \left(\frac{4}{2}\right)^2 = -1 + \left(\frac{-2}{2}\right)^2 + \left(\frac{4}{2}\right)^2$ $\Leftrightarrow x^2 - 2x + 1 + y^2 + 4y + 4 = -1 + 1 + 4 \Leftrightarrow (x - 1)^2 + (y + 2)^2 = 4$. Thus, the center is $(1, -2)$, and the radius is 2.

**66.** Completing the square gives $x^2 + y^2 - 2x - 2y = 2 \quad \Leftrightarrow \quad x^2 - 2x + \left(\frac{-2}{2}\right)^2 + y^2 - 2y + \left(\frac{-2}{2}\right)^2 = 2 + \left(\frac{-2}{2}\right)^2 + \left(\frac{-2}{2}\right)^2$ $\Leftrightarrow \quad x^2 - 2x + 1 + y^2 - 2y + 1 = 2 + 1 + 1 \quad \Leftrightarrow \quad (x - 1)^2 + (y - 1)^2 = 4$. Thus, the center is $(1, 1)$, and the radius is 2.

**67.** Completing the square gives $x^2 + y^2 - 4x + 10y + 13 = 0 \quad \Leftrightarrow \quad x^2 - 4x + \left(\frac{-4}{2}\right)^2 + y^2 + 10y + \left(\frac{10}{2}\right)^2 = -13 + \left(\frac{4}{2}\right)^2 + \left(\frac{10}{2}\right)^2$ $\Leftrightarrow \quad x^2 - 4x + 4 + y^2 + 10y + 25 = -13 + 4 + 25 \quad \Leftrightarrow \quad (x - 2)^2 + (y + 5)^2 = 16$.
Thus, the center is $(2, -5)$, and the radius is 4.

**68.** Completing the square gives $x^2 + y^2 + 6y + 2 = 0 \quad \Leftrightarrow \quad x^2 + y^2 + 6y + \left(\frac{6}{2}\right)^2 = -2 + \left(\frac{6}{2}\right)^2 \quad \Leftrightarrow$ $x^2 + y^2 + 6y + 9 = -2 + 9 \quad \Leftrightarrow \quad x^2 + (y + 3)^2 = 7$. Thus, the circle has center $(0, -3)$ and radius $\sqrt{7}$.

**69.** Completing the square gives $x^2 + y^2 + x = 0 \quad \Leftrightarrow \quad x^2 + x + \left(\frac{1}{2}\right)^2 + y^2 = \left(\frac{1}{2}\right)^2 \quad \Leftrightarrow \quad x^2 + x + \frac{1}{4} + y^2 = \frac{1}{4} \quad \Leftrightarrow$ $\left(x + \frac{1}{2}\right)^2 + y^2 = \frac{1}{4}$. Thus, the circle has center $\left(-\frac{1}{2}, 0\right)$ and radius $\frac{1}{2}$.

**70.** Completing the square gives $x^2 + y^2 + 2x + y + 1 = 0 \quad \Leftrightarrow \quad x^2 + 2x + \left(\frac{2}{2}\right)^2 + y^2 + y + \left(\frac{1}{2}\right)^2 = -1 + 1 + \left(\frac{1}{2}\right)^2 \quad \Leftrightarrow$ $x^2 + 2x + 1 + y^2 + y + \frac{1}{4} = \frac{1}{4} \quad \Leftrightarrow \quad (x + 1)^2 + \left(y + \frac{1}{2}\right)^2 = \frac{1}{4}$. Thus, the circle has center $\left(-1, -\frac{1}{2}\right)$ and radius $\frac{1}{2}$.

**71.** Completing the square gives $x^2 + y^2 - \frac{1}{2}x + \frac{1}{2}y = \frac{1}{8} \quad \Leftrightarrow \quad x^2 - \frac{1}{2}x + \left(\frac{-1/2}{2}\right)^2 + y^2 + \frac{1}{2}y + \left(\frac{1/2}{2}\right)^2 = \frac{1}{8} +$ $\left(\frac{-1/2}{2}\right)^2 + \left(\frac{1/2}{2}\right)^2 \quad \Leftrightarrow \quad x^2 - \frac{1}{2}x + \frac{1}{16} + y^2 + \frac{1}{2}y + \frac{1}{16} = \frac{1}{8} + \frac{1}{16} + \frac{1}{16} = \frac{2}{8} = \frac{1}{4} \quad \Leftrightarrow \quad \left(x - \frac{1}{4}\right)^2 + \left(y + \frac{1}{4}\right)^2 = \frac{1}{4}$.
Thus, the circle has center $\left(\frac{1}{4}, -\frac{1}{4}\right)$ and radius $\frac{1}{2}$.

**72.** Completing the square gives $x^2 + y^2 + \frac{1}{2}x + 2y + \frac{1}{16} = 0 \quad \Leftrightarrow \quad x^2 + \frac{1}{2}x + \left(\frac{1/2}{2}\right)^2 + y^2 + 2y + \left(\frac{2}{2}\right)^2 = -\frac{1}{16} + \left(\frac{1/2}{2}\right)^2 + \left(\frac{2}{2}\right)^2$ $\Leftrightarrow \quad \left(x + \frac{1}{4}\right)^2 + (y + 1)^2 = 1$. Thus, the circle has center $\left(-\frac{1}{4}, -1\right)$ and radius 1.

**73.** Completing the square gives $x^2 + y^2 + 4x - 10y = 21$ $\Leftrightarrow \quad x^2 + 4x + \left(\frac{4}{2}\right)^2 + y^2 - 10y + \left(\frac{-10}{2}\right)^2 = 21 +$ $\left(\frac{4}{2}\right)^2 + \left(\frac{-10}{2}\right)^2 \quad \Leftrightarrow$ $(x + 2)^2 + (y - 5)^2 = 21 + 4 + 25 = 50$. Thus, the circle has center $(-2, 5)$ and radius $\sqrt{50} = 5\sqrt{2}$.

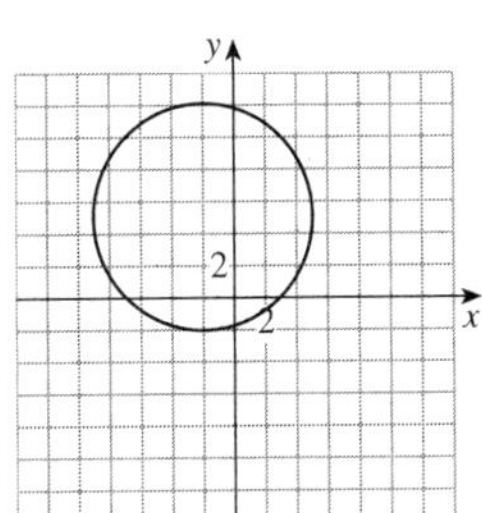


**74.** First divide by 4, then complete the square. This gives $4x^2 + 4y^2 + 2x = 0 \quad \Leftrightarrow$ $x^2 + y^2 + \frac{1}{2}x = 0 \quad \Leftrightarrow x^2 + \frac{1}{2}x + \_\_ + y^2 = 0 \quad \Leftrightarrow$ $x^2 + \frac{1}{2}x + \left(\frac{1/2}{2}\right)^2 + y^2 = \left(\frac{1/2}{2}\right)^2 \quad \Leftrightarrow \left(x + \frac{1}{4}\right)^2 +$ $y^2 = \frac{1}{16}$. Thus, the circle has center $\left(-\frac{1}{4}, 0\right)$ and radius $\frac{1}{4}$.

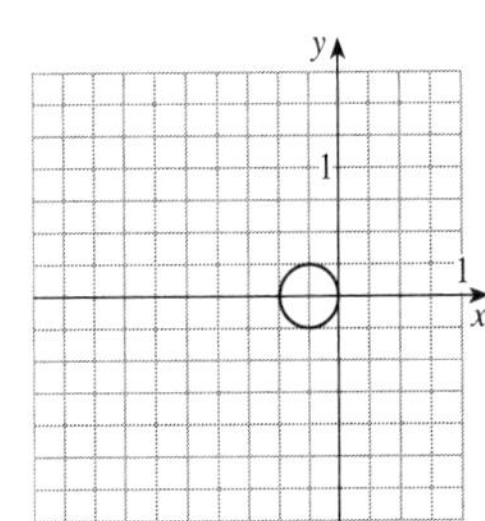

**75.** Completing the square gives
$x^2 + y^2 + 6x - 12y + 45 = 0 \Leftrightarrow$
$(x+3)^2 + (y-6)^2 = -45 + 9 + 36 = 0$. Thus, the center is $(-3, 6)$, and the radius is 0. This is a degenerate circle whose graph consists only of the point $(-3, 6)$.

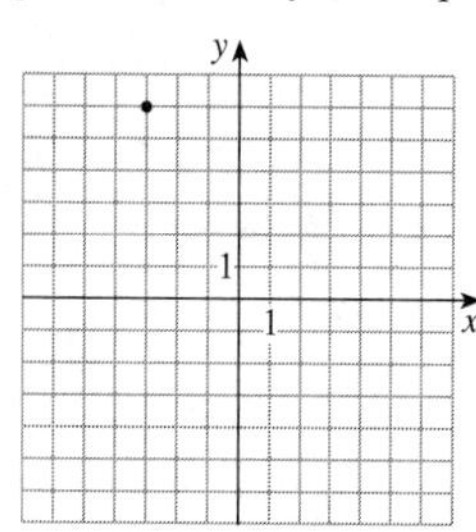


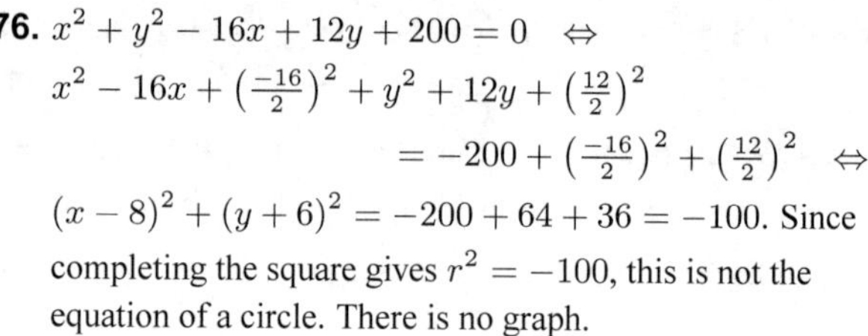

**76.** $x^2 + y^2 - 16x + 12y + 200 = 0 \Leftrightarrow$
$x^2 - 16x + \left(\frac{-16}{2}\right)^2 + y^2 + 12y + \left(\frac{12}{2}\right)^2$
$= -200 + \left(\frac{-16}{2}\right)^2 + \left(\frac{12}{2}\right)^2 \Leftrightarrow$
$(x-8)^2 + (y+6)^2 = -200 + 64 + 36 = -100$. Since completing the square gives $r^2 = -100$, this is not the equation of a circle. There is no graph.

**77.** $\{(x,y) \mid x^2 + y^2 \le 1\}$. This is the set of points inside (and on) the circle $x^2 + y^2 = 1$.

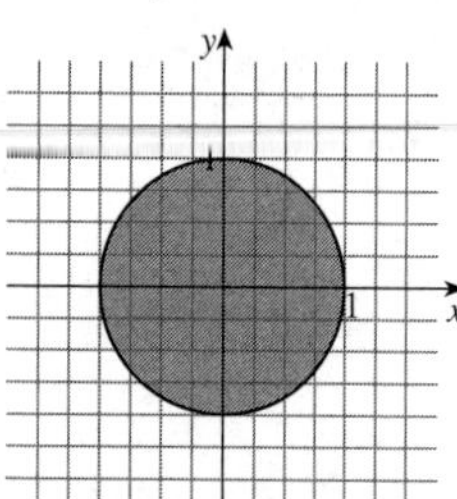


**78.** $\{(x,y) \mid x^2 + y^2 > 4\}$. This is the set of points outside the circle $x^2 + y^2 = 4$.

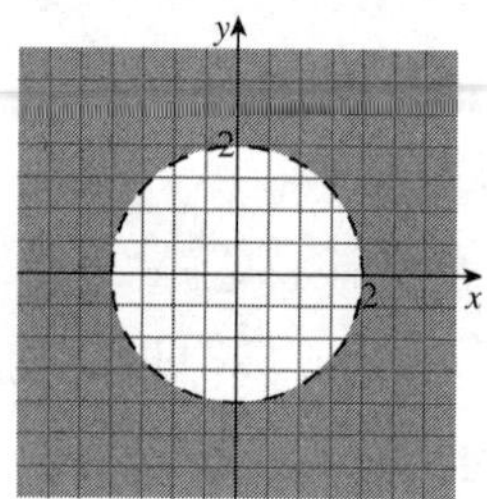


**79.** $\{(x,y) \mid 1 \le x^2 + y^2 < 9\}$. This is the set of points outside the circle $x^2 + y^2 = 1$ and inside (but not on) the circle $x^2 + y^2 = 9$.

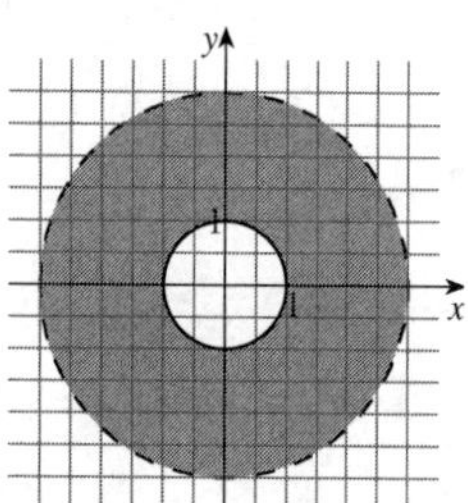


**80.** $\{(x,y) \mid 2x < x^2 + y^2 \le 4\}$. Completing the square gives $2x < x^2 + y^2 \Leftrightarrow 1 < (x+1)^2 + y^2$. Thus, this is the set of points outside the circle $(x+1)^2 + y^2 = 1$ and inside (or on) the circle $x^2 + y^2 = 4$.

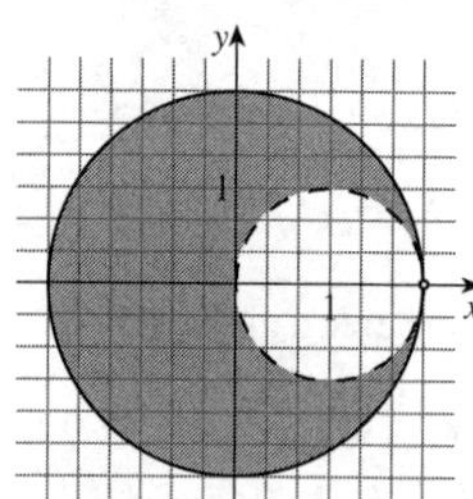


**81.** Completing the square gives $x^2 + y^2 - 4y - 12 = 0 \Leftrightarrow$
$x^2 + y^2 - 4y + \left(\frac{-4}{2}\right)^2 = 12 + \left(\frac{-4}{2}\right)^2 \Leftrightarrow$
$x^2 + (y-2)^2 = 16$. Thus, the center is $(0, 2)$, and the radius is 4. So the circle $x^2 + y^2 = 4$, with center $(0,0)$ and radius 2, sits completely inside the larger circle. Thus, the area is $\pi 4^2 - \pi 2^2 = 16\pi - 4\pi = 12\pi$.

**82.** This is the top quarter of the circle of radius 3. Thus, the area is $\frac{1}{4}(9\pi) = \frac{9\pi}{4}$.

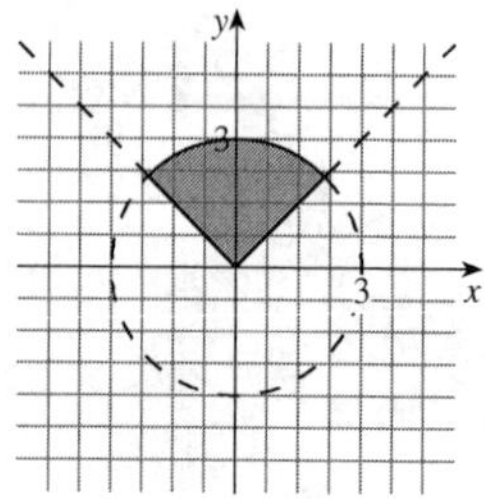

**83. (a)** In 1980 inflation was 14%; in 1990, it was 6%; in 1999, it was 2%.

**(b)** Inflation exceeded 6% from 1975 to 1976 and from 1978 to 1982.

**(c)** Between 1980 and 1985 the inflation rate generally decreased. Between 1987 and 1992 the inflation rate generally increased.

**(d)** The highest rate was about 14% in 1980. The lowest was about 1% in 2002.

**84. (a)** Closest: 2 Mm. Farthest: 8 Mm.

**(b)** When $y = 2$ we have $\dfrac{(x-3)^2}{25} + \dfrac{2^2}{16} = 1 \quad\Leftrightarrow\quad \dfrac{(x-3)^2}{25} + \frac{1}{4} = 1 \quad\Leftrightarrow\quad \dfrac{(x-3)^2}{25} = \frac{3}{4} \quad\Leftrightarrow$ $(x-3)^2 = \frac{75}{4}$. Taking the square root of both sides we get $x - 3 = \pm\sqrt{\frac{75}{4}} = \pm\frac{5\sqrt{3}}{2} \quad\Leftrightarrow\quad x = 3 \pm \frac{5\sqrt{3}}{2}$. So $x = 3 - \frac{5\sqrt{3}}{2} \approx -1.33$ or $x = 3 + \frac{5\sqrt{3}}{2} \approx 7.33$. The distance from $(-1.33, 2)$ to the center $(0,0)$ is $d = \sqrt{(-1.33-0)^2 + (2-0)^2} = \sqrt{5.7689} \approx 2.40$. The distance from $(7.33, 2)$ to the center $(0,0)$ is $d = \sqrt{(7.33-0)^2 + (2-0)^2} = \sqrt{57.7307} \approx 7.60$.

**85.** Completing the square gives $x^2 + y^2 + ax + by + c = 0 \quad\Leftrightarrow\quad x^2 + ax + \left(\frac{a}{2}\right)^2 + y^2 + by + \left(\frac{b}{2}\right)^2 = -c + \left(\frac{a}{2}\right)^2 + \left(\frac{b}{2}\right)^2$ $\Leftrightarrow \quad \left(x + \frac{a}{2}\right)^2 + \left(y + \frac{b}{2}\right)^2 = -c + \dfrac{a^2+b^2}{4}$. This equation represents a circle only when $-c + \dfrac{a^2+b^2}{4} > 0$. This equation represents a point when $-c + \dfrac{a^2+b^2}{4} = 0$, and this equation represents the empty set when $-c + \dfrac{a^2+b^2}{4} < 0$. When the equation represents a circle, the center is $\left(-\frac{a}{2}, -\frac{b}{2}\right)$, and the radius is $\sqrt{-c + \dfrac{a^2+b^2}{4}} = \frac{1}{2}\sqrt{a^2+b^2-4ac}$.

**86. (a) (i)** $(x-2)^2 + (y-1)^2 = 9$, the center is at $(2,1)$, and the radius is 3. $(x-6)^2 + (y-4)^2 = 16$, the center is at $(6,4)$, and the radius is 4. The distance between centers is $\sqrt{(2-6)^2 + (1-4)^2} = \sqrt{(-4)^2 + (-3)^2} = \sqrt{16+9} = \sqrt{25} = 5$. Since $5 < 3 + 4$, these circles intersect.

**(ii)** $x^2 + (y-2)^2 = 4$, the center is at $(0,2)$, and the radius is 2. $(x-5)^2 + (y-14)^2 = 9$, the center is at $(5,14)$, and the radius is 3. The distance between centers is $\sqrt{(0-5)^2 + (2-14)^2} = \sqrt{(-5)^2 + (-12)^2} = \sqrt{25+144} = \sqrt{169} = 13$. Since $13 > 2 + 3$, these circles do not intersect.

**(iii)** $(x-3)^2 + (y+1)^2 = 1$, the center is at $(3,-1)$, and the radius is 1. $(x-2)^2 + (y-2)^2 = 25$, the center is at $(2,2)$, and the radius is 5. The distance between centers is $\sqrt{(3-2)^2 + (-1-2)^2} = \sqrt{1^2 + (-3)^2} = \sqrt{1+9} = \sqrt{10}$. Since $\sqrt{10} < 1 + 5$, these circles intersect.

**(b)** As shown in the diagram, if two circles intersect, then the centers of the circles and one point of intersection form a triangle. So because in any triangle each side has length less than the sum of the other two, the two circles will intersect only if the distance between their centers, $d$, is less than or equal to the sum of the radii, $r_1$ and $r_2$. That is, the circles will intersect if $d \le r_1 + r_2$.

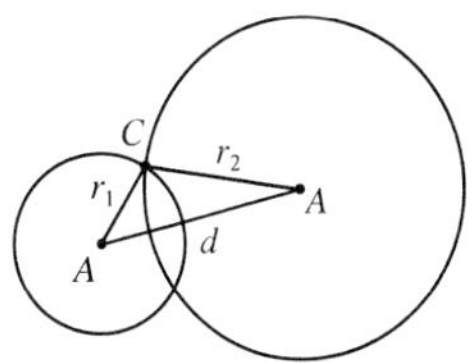

**87. (a)** Symmetric about the $x$-axis.

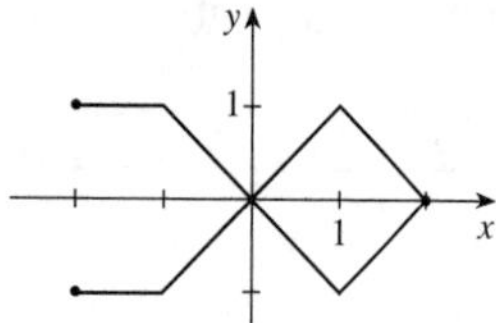

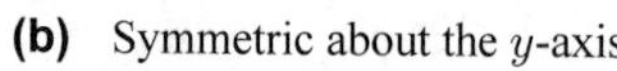
**(b)** Symmetric about the $y$-axis.

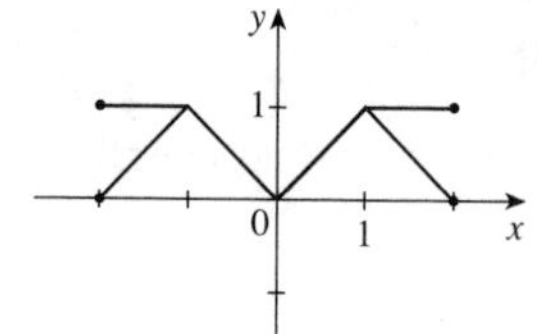

**(c)** Symmetric about the origin.

**88. (a)** In geometry: The line runs through the center of the circle and intersects it at the points $A$ and $B$.

In coordinate geometry: The line $y = \frac{4}{3}x$ runs through the center of the circle $x^2 + y^2 = 25$ and intersects it at the points $A(-3,-4)$ and $B(3,4)$.

**(b)**

It is difficult to attach geometric significance to points on this curve, unlike the circle where the points on it are a fixed distance from the center.

**(c)** In coordinate geometry we can use equations to describe curves. In geometry without coordinates, we are restricted to geometric descriptions (involving, for example, distances between points) to describe curves.

## 2.3 Graphing Calculators; Solving Equations and Inequalities Graphically

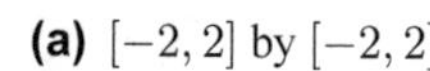
**1.** $y = x^4 + 2$

**(a)** $[-2,2]$ by $[-2,2]$

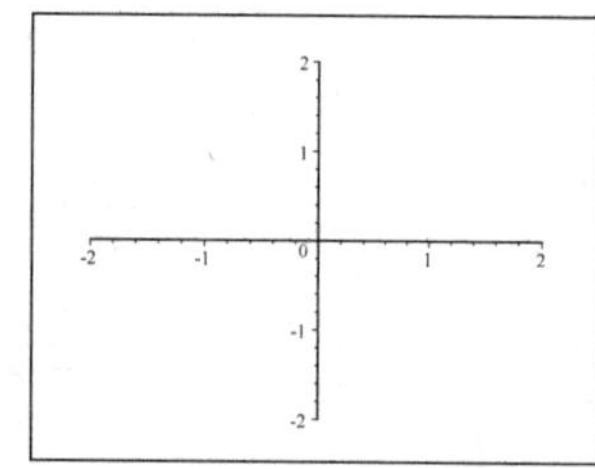

**(b)** $[0,4]$ by $[0,4]$

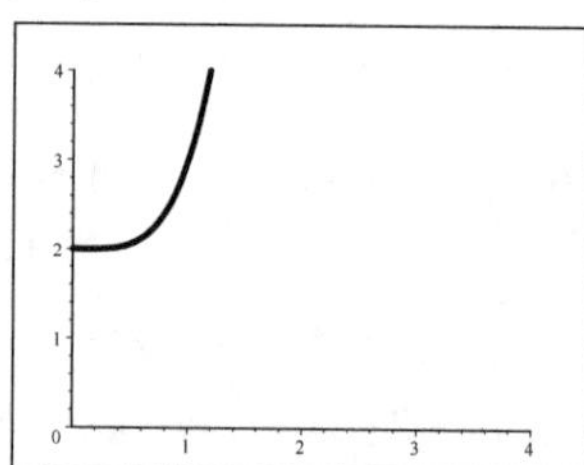

**(c)** $[-8,8]$ by $[-4,40]$

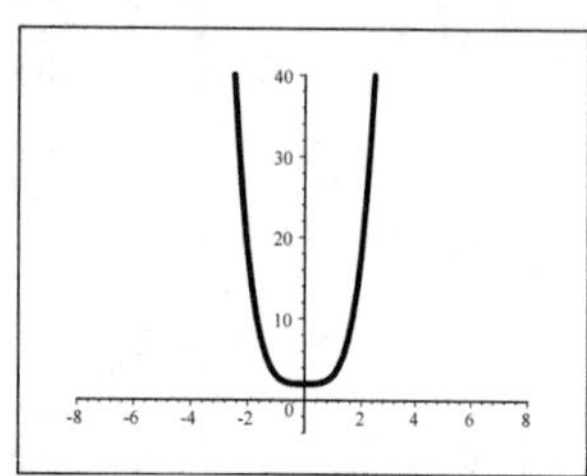

**(d)** $[-40,40]$ by $[-80,800]$

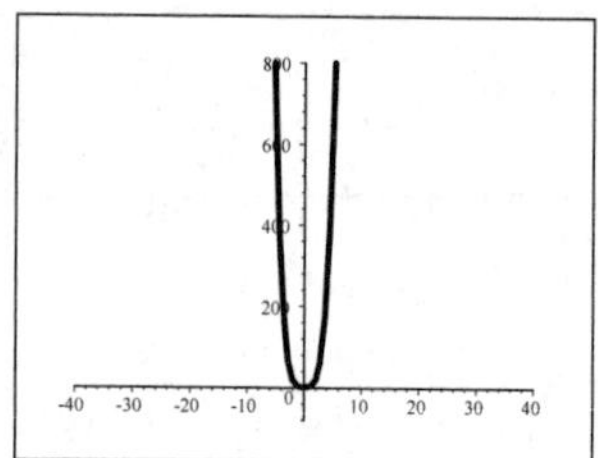

The viewing rectangle in part (c) produces the most appropriate graph of the equation.

**2.** $y = x^2 + 7x + 6$

**(a)** $[-5, 5]$ by $[-5, 5]$

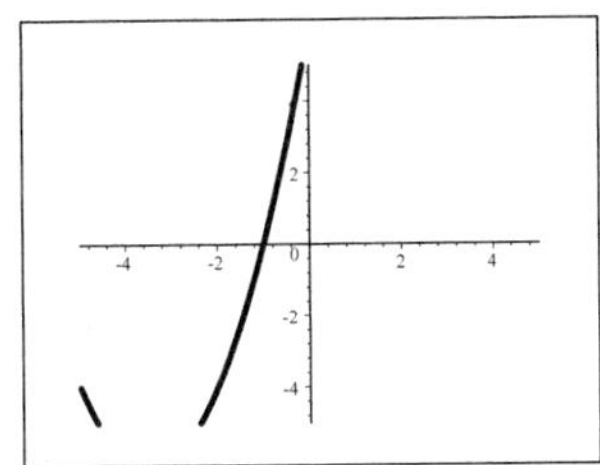

**(b)** $[0, 10]$ by $[-20, 100]$

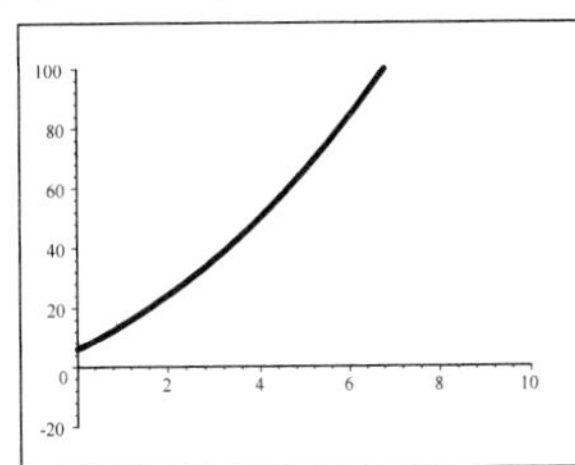

**(c)** $[-15, 8]$ by $[-20, 100]$

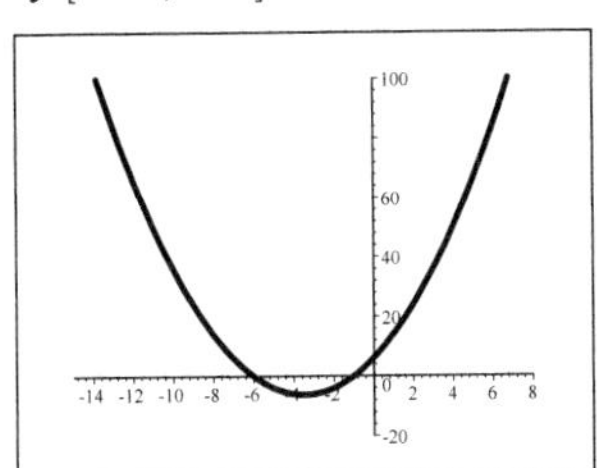

**(d)** $[-10, 3]$ by $[-100, 20]$

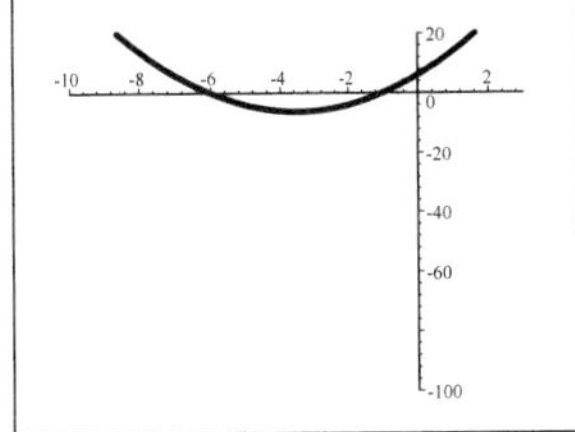

The viewing rectangle in part (c) produces the most appropriate graph of the equation.

**3.** $y = 100 - x^2$

**(a)** $[-4, 4]$ by $[-4, 4]$

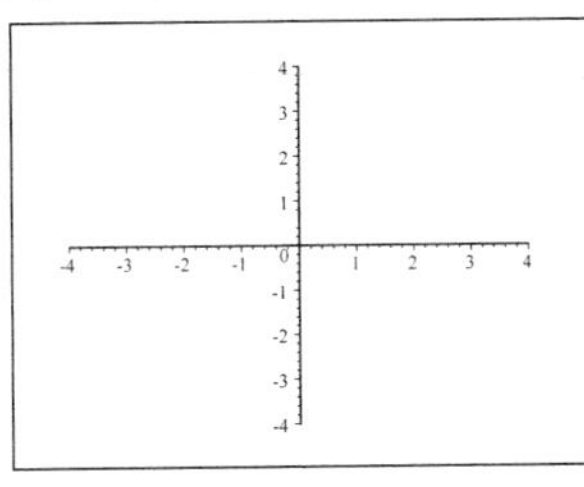

**(b)** $[-10, 10]$ by $[-10, 10]$

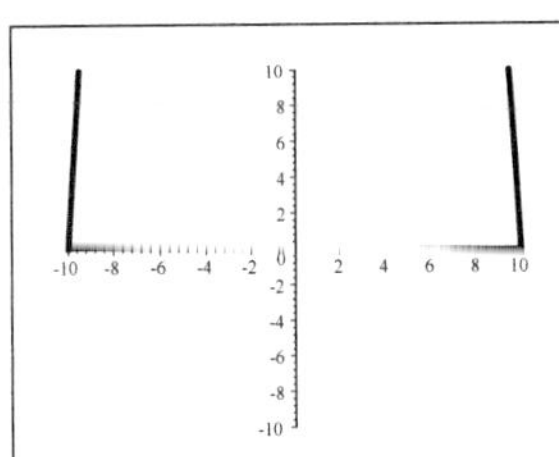

**(c)** $[-15, 15]$ by $[-30, 110]$

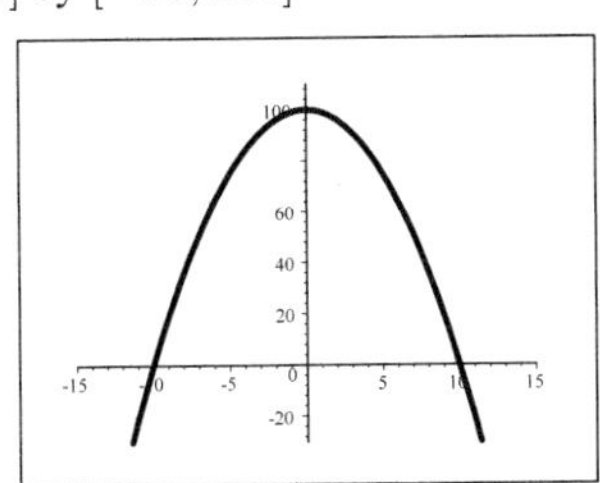

**(d)** $[-4, 4]$ by $[-30, 110]$

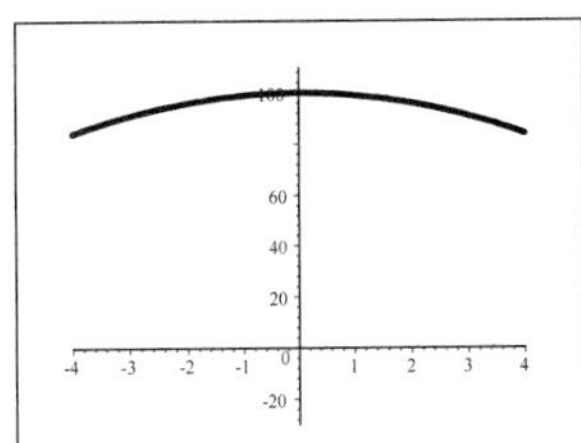

The viewing rectangle in part (c) produces the most appropriate graph of the equation.

**4.** $y = 2x^2 - 1000$

**(a)** $[-10, 10]$ by $[-10, 10]$

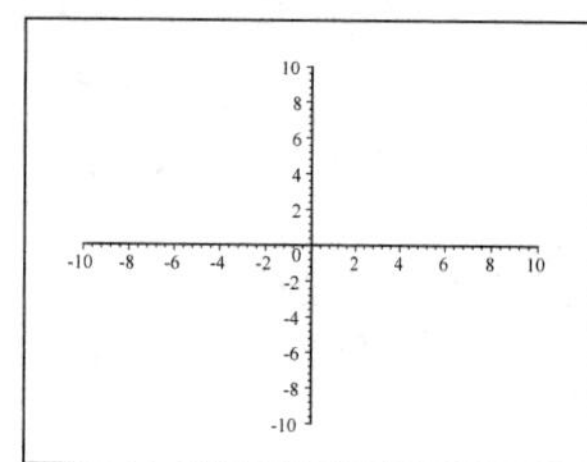

**(b)** $[-10, 10]$ by $[-100, 100]$

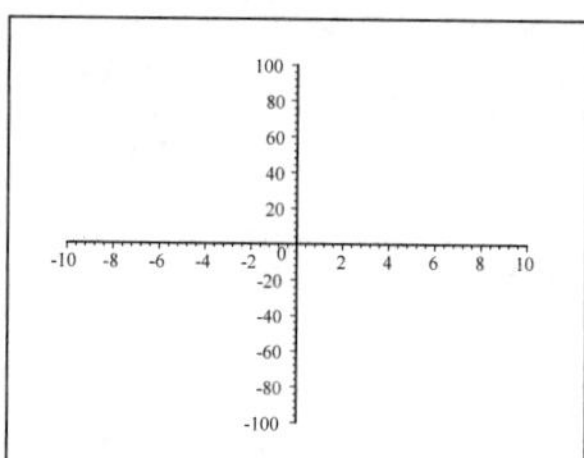

**(c)** $[-10, 10]$ by $[-1000, 1000]$

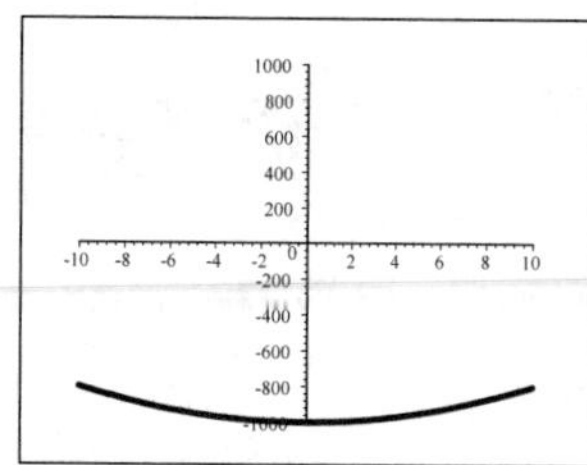

**(d)** $[-25, 25]$ by $[-1200, 200]$

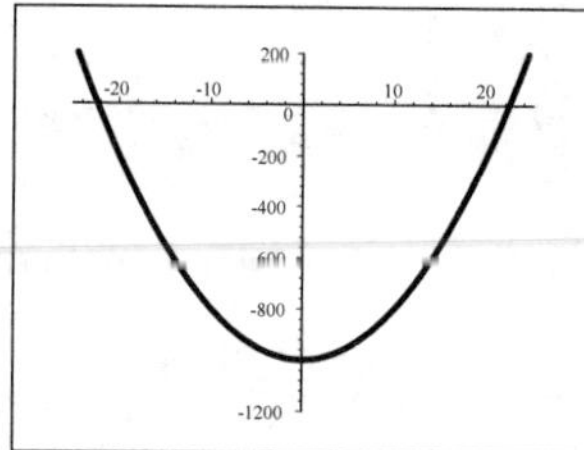

The viewing rectangle in part (d) produces the most appropriate graph of the equation.

**5.** $y = 10 + 25x - x^3$

**(a)** $[-4, 4]$ by $[-4, 4]$

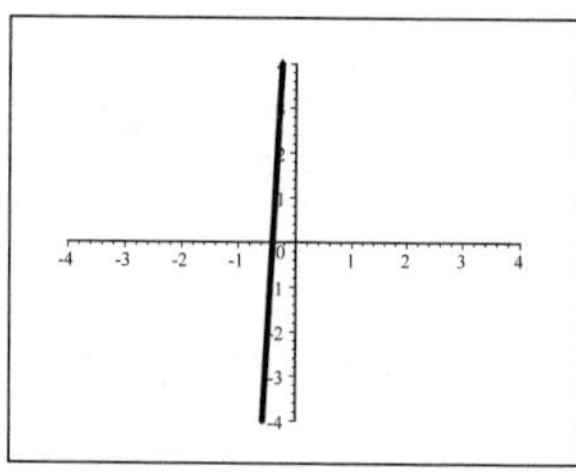

**(b)** $[-10, 10]$ by $[-10, 10]$

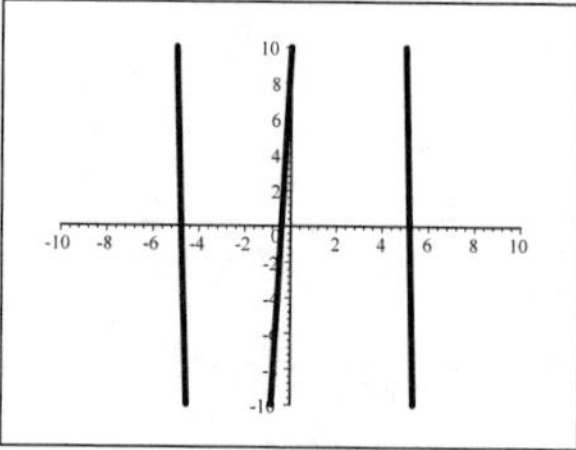

**(c)** $[-20, 20]$ by $[-100, 100]$

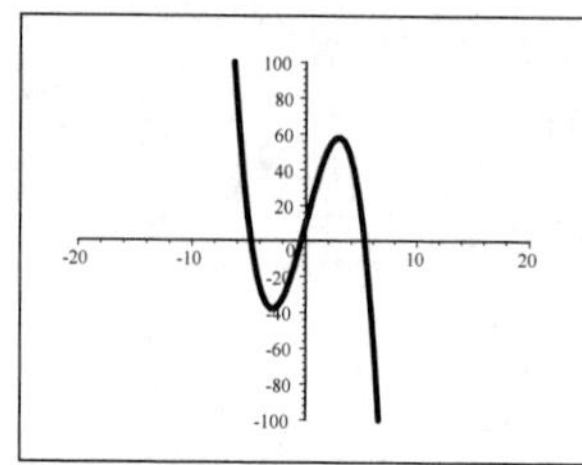

**(d)** $[-100, 100]$ by $[-200, 200]$

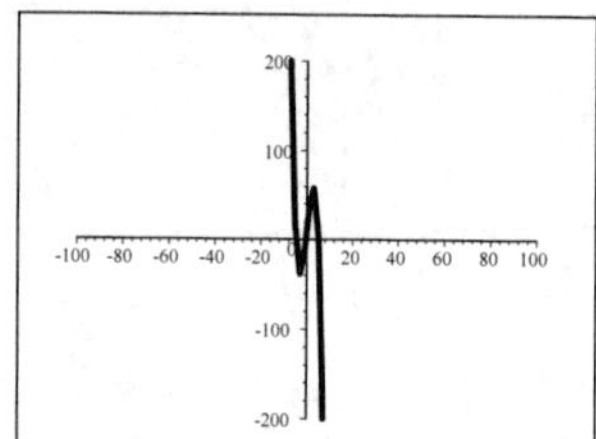

The viewing rectangle in part (c) produces the most appropriate graph of the equation.

**6.** $y = \sqrt{8x - x^2}$

**(a)** $[-4, 4]$ by $[-4, 4]$

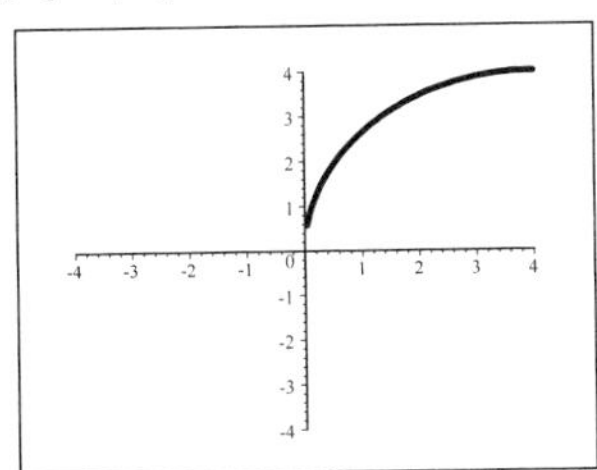

**(b)** $[-5, 5]$ by $[0, 100]$

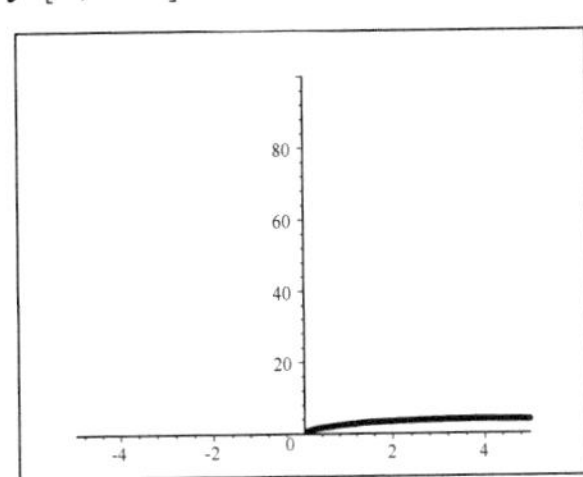

**(c)** $[-10, 10]$ by $[-10, 40]$

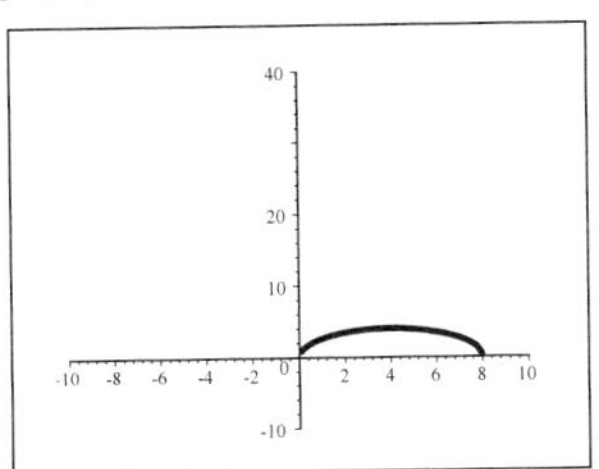

**(d)** $[-2, 10]$ by $[-2, 6]$

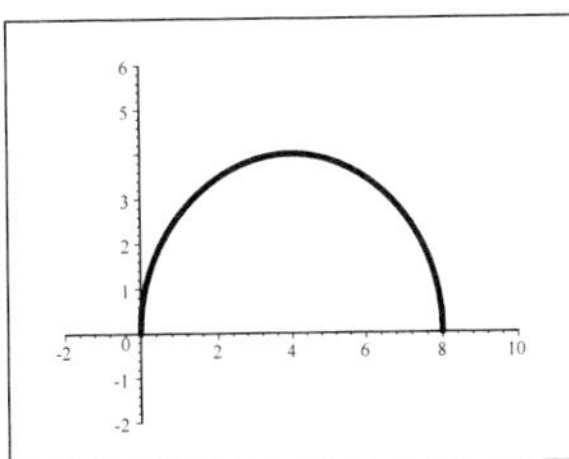

From the graphs we see that the viewing rectangle in (d) produces the most appropriate graph of the equation.
*Note:* Squaring both sides yields the equation $y^2 = 8x - x^2 \quad \Leftrightarrow \quad (x-4)^2 + y^2 = 16$. Since this gives a circle, the original equation represents the top half of a circle.

**7.** $y = 100x^2$, $[-2, 2]$ by $[-10, 400]$

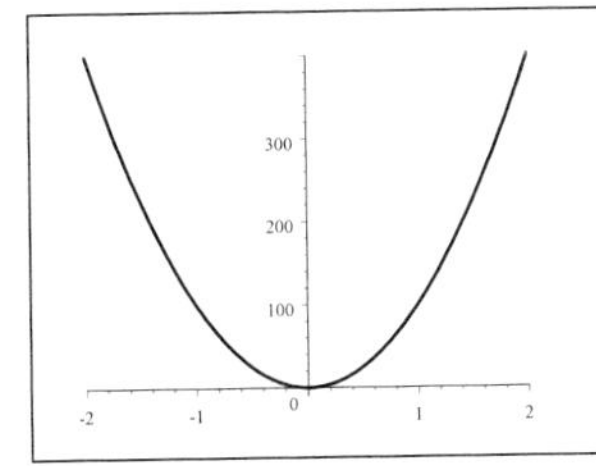

**8.** $y = -100x^2$, $[-2, 2]$ by $[-400, 10]$

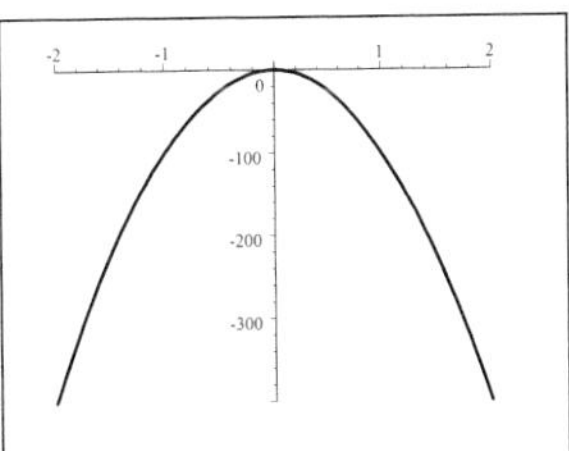

**9.** $y = 4 + 6x - x^2$, $[-4, 10]$ by $[-10, 20]$

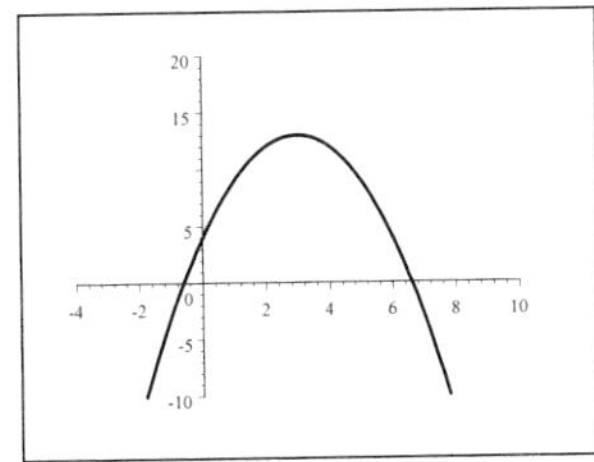

**10.** $y = 0.3x^2 + 1.7x - 3$, $[-15, 10]$ by $[-10, 20]$

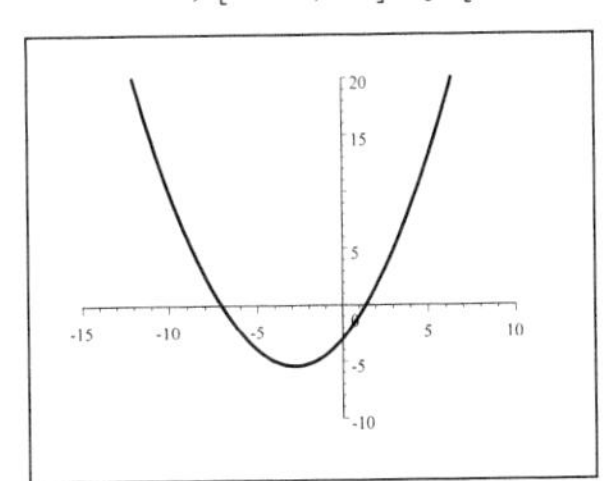

**11.** $y = \sqrt[4]{256 - x^2}$. We require that $256 - x^2 \geq 0 \quad \Rightarrow$ $-16 \leq x \leq 16$, so we graph $y = \sqrt[4]{256 - x^2}$ in the viewing rectangle $[-20, 20]$ by $[-1, 5]$.

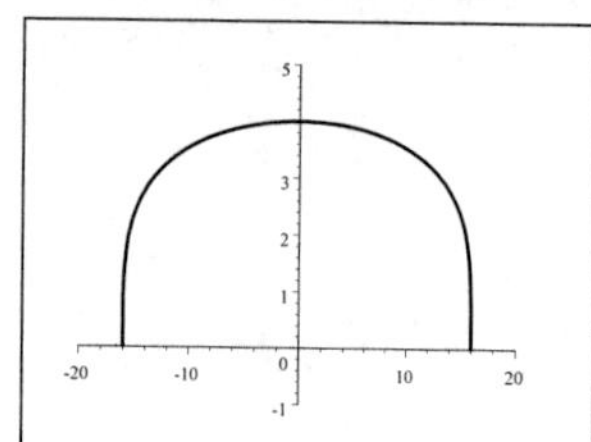

**12.** $y = \sqrt{12x - 17}$, $[0, 10]$ by $[0, 20]$

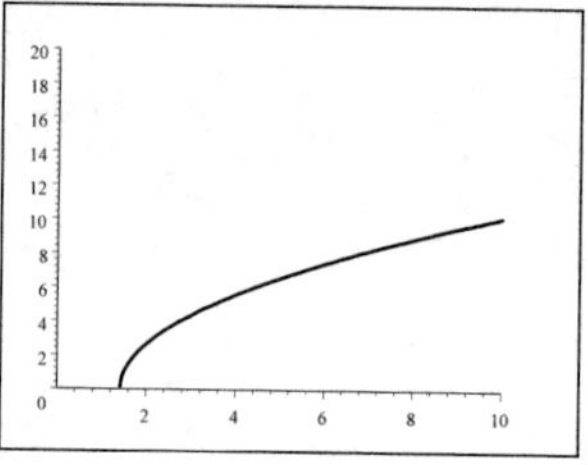

**13.** $y = 0.01x^3 - x^2 + 5$, $[-50, 150]$ by $[-2000, 2000]$

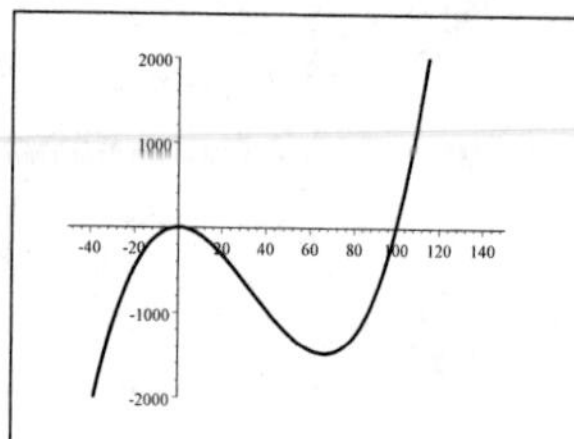

**14.** $y = x(x + 6)(x - 9)$, $[-10, 10]$ by $[-250, 150]$

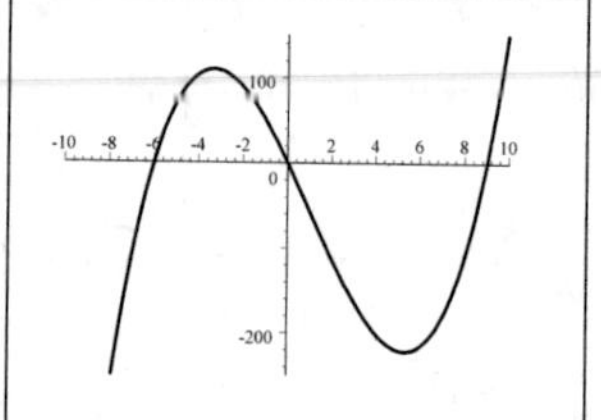

**15.** $y = x^4 - 4x^3$, $[-4, 6]$ by $[-50, 100]$

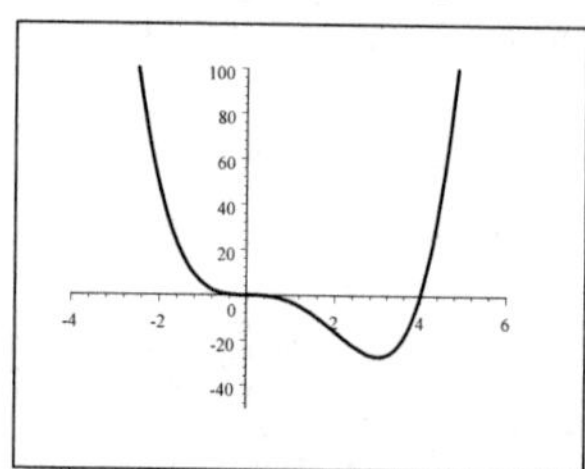

**16.** $y = \dfrac{x}{x^2 + 25}$, $[-10, 10]$ by $[-0.2, 0.2]$

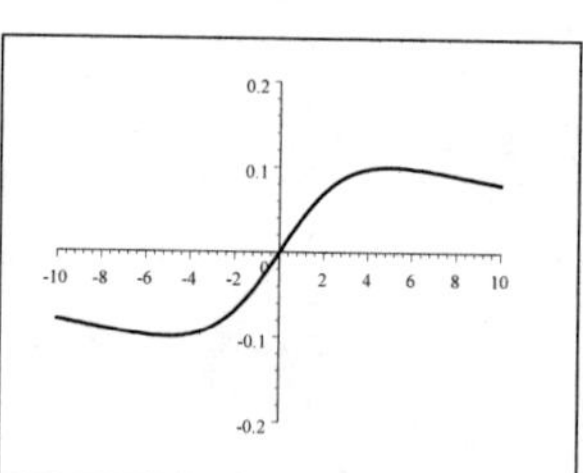

**17.** $y = 1 + |x - 1|$, $[-3, 5]$ by $[-1, 5]$

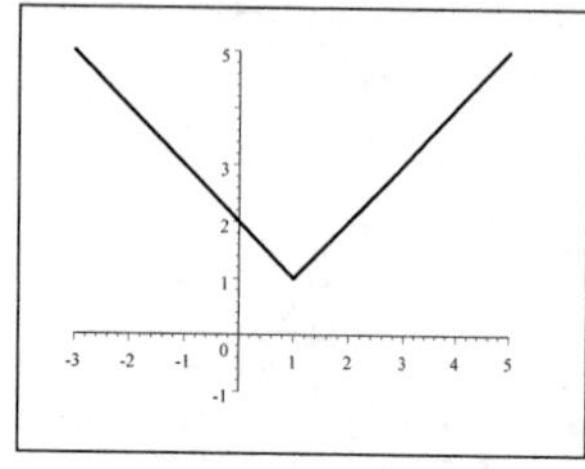

**18.** $y = 2x - |x^2 - 5|$, $[-10, 10]$ by $[-10, 10]$

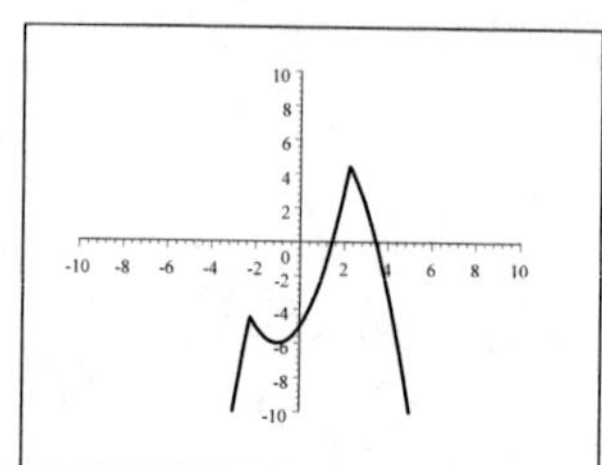

**19.** $x^2 + y^2 = 9 \quad \Leftrightarrow \quad y^2 = 9 - x^2 \quad \Rightarrow$ $y = \pm\sqrt{9 - x^2}$. So we graph the functions $y_1 = \sqrt{9 - x^2}$ and $y_2 = -\sqrt{9 - x^2}$ in the viewing rectangle $[-8, 8]$ by $[-6, 6]$.

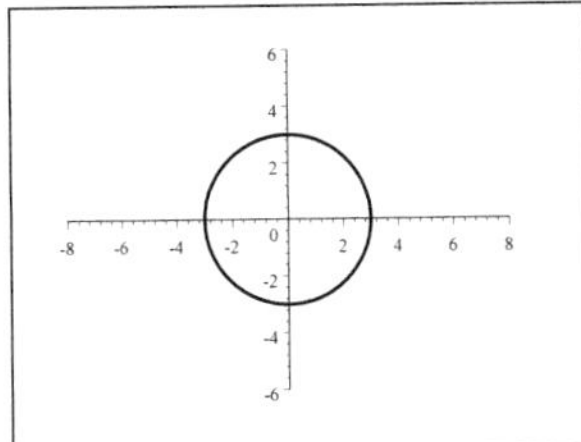

**20.** $(y-1)^2 + x^2 = 1 \quad \Leftrightarrow \quad (y-1)^2 = 1 - x^2 \quad \Rightarrow$ $y - 1 = \pm\sqrt{1 - x^2} \quad \Leftrightarrow \quad y = 1 \pm \sqrt{1 - x^2}$. So we graph the functions $y_1 = 1 + \sqrt{1 - x^2}$ and $y_2 = 1 - \sqrt{1 - x^2}$ in the viewing rectangle $[-2, 2]$ by $[0, 3]$.

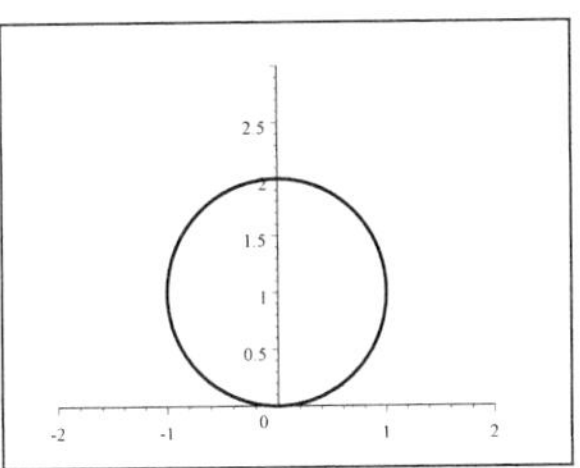

**21.** $4x^2 + 2y^2 = 1 \quad \Leftrightarrow \quad 2y^2 = 1 - 4x^2 \quad \Leftrightarrow$ $y^2 = \dfrac{1 - 4x^2}{2} \quad \Rightarrow \quad y = \pm\sqrt{\dfrac{1 - 4x^2}{2}}$. So we graph the functions $y_1 = \sqrt{\dfrac{1 - 4x^2}{2}}$ and $y_2 = -\sqrt{\dfrac{1 - 4x^2}{2}}$ in the viewing rectangle $[-1.2, 1.2]$ by $[-0.8, 0.8]$.

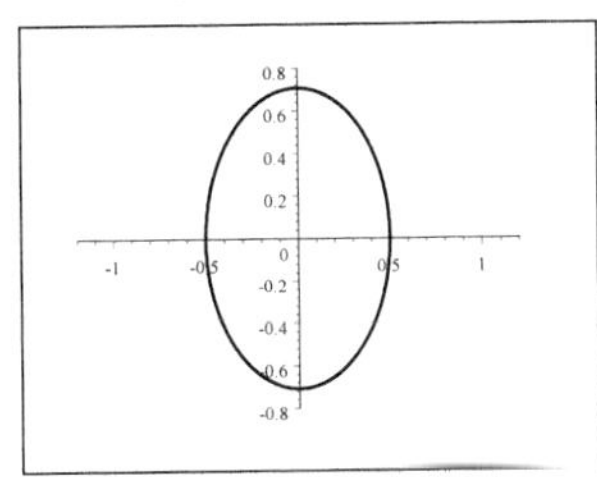

**22.** $y^2 - 9x^2 = 1 \quad \Leftrightarrow \quad y^2 = 1 + 9x^2 \quad \Rightarrow$ $y = \pm\sqrt{1 + 9x^2}$. So we graph the functions $y_1 = \sqrt{1 + 9x^2}$ and $y_2 = -\sqrt{1 + 9x^2}$ in the viewing rectangle $[-5, 5]$ by $[-5, 5]$.

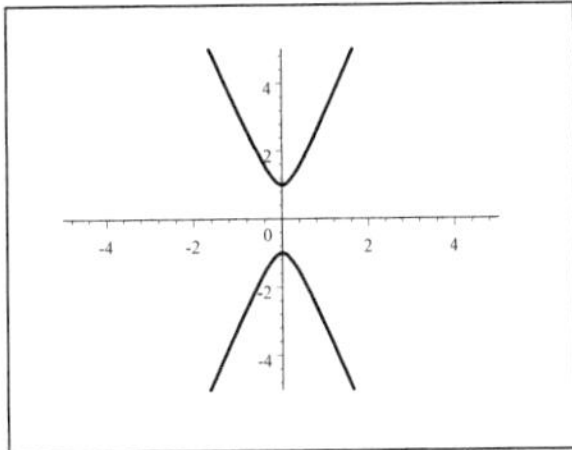

**23.** Although the graphs of $y = -3x^2 + 6x - \frac{1}{2}$ and $y = \sqrt{7 - \frac{7}{12}x^2}$ appear to intersect in the viewing rectangle $[-4, 4]$ by $[-1, 3]$, there is no point of intersection. You can verify this by zooming in.

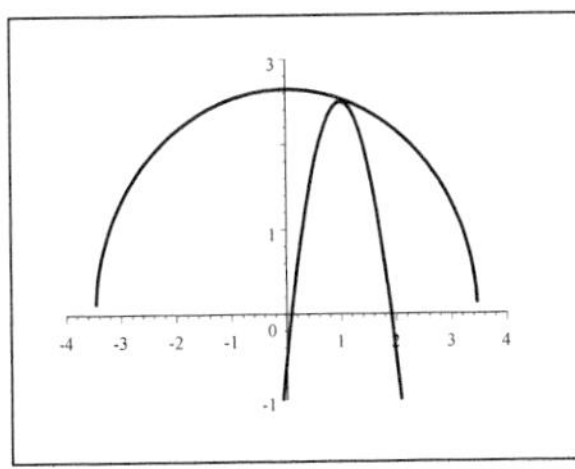

**24.** Although the graphs of $y = \sqrt{49 - x^2}$ and $y = \frac{1}{5}(41 - 3x)$ appear to intersect in the viewing rectangle $[-8, 8]$ by $[-1, 8]$, there is no point of intersection. You can verify this by zooming in.

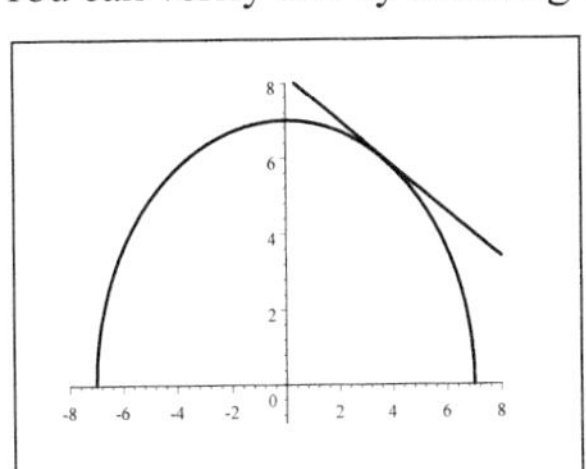

**25.** The graphs of $y = 6 - 4x - x^2$ and $y = 3x + 18$ appear to have two points of intersection in the viewing rectangle $[-6, 2]$ by $[-5, 20]$. You can verify that $x = -4$ and $x = -3$ are exact solutions.

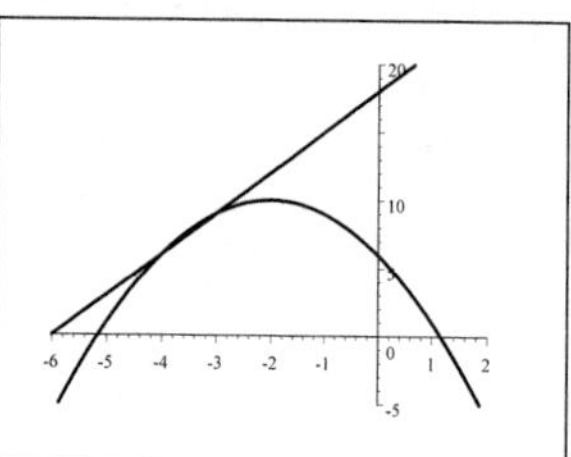

**26.** The graphs of $y = x^3 - 4x$ and $y = x + 5$ appear to have one point of intersection in the viewing rectangle $[-4, 4]$ by $[-15, 15]$. The solution is $x \approx 2.627$.

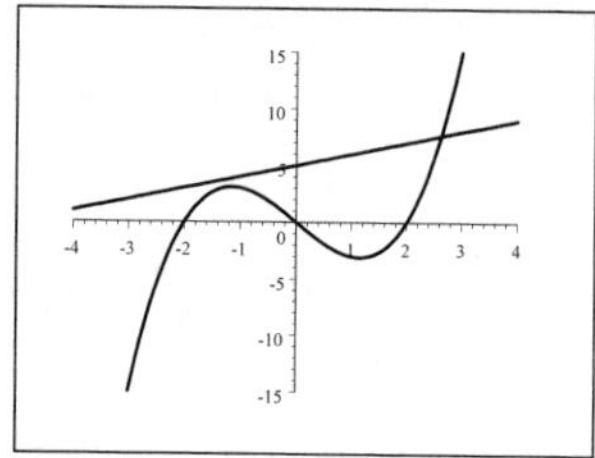

**27.** Algebraically: $x - 4 = 5x + 12 \quad \Leftrightarrow \quad -16 = 4x \quad \Leftrightarrow \quad x = -4$.

Graphically: We graph the two equations $y_1 = x - 4$ and $y_2 = 5x + 12$ in the viewing rectangle $[-6, 4]$ by $[-10, 2]$. Zooming in, we see that the solution is $x = -4$.

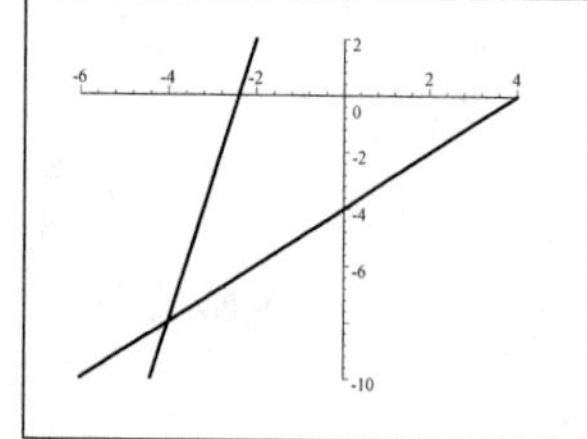

**28.** Algebraically: $\frac{1}{2}x - 3 = 6 + 2x \quad \Leftrightarrow \quad -9 = \frac{3}{2}x \quad \Leftrightarrow \quad x = -6$.

Graphically: We graph the two equations $y_1 = \frac{1}{2}x - 3$ and $y_2 = 6 + 2x$ in the viewing rectangle $[-10, 5]$ by $[-10, 5]$. Zooming in, we see that the solution is $x = -6$.

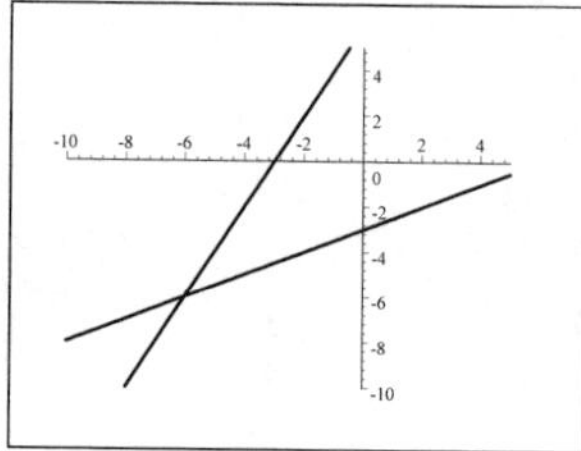

**29.** Algebraically: $\dfrac{2}{x} + \dfrac{1}{2x} = 7 \quad \Leftrightarrow$

$2x\left(\dfrac{2}{x} + \dfrac{1}{2x}\right) = 2x\,(7) \quad \Leftrightarrow \quad 4 + 1 = 14x \quad \Leftrightarrow$

$x = \frac{5}{14}$.

Graphically: We graph the two equations $y_1 = \dfrac{2}{x} + \dfrac{1}{2x}$ and $y_2 = 7$ in the viewing rectangle $[-2, 2]$ by $[-2, 8]$. Zooming in, we see that the solution is $x \approx 0.36$.

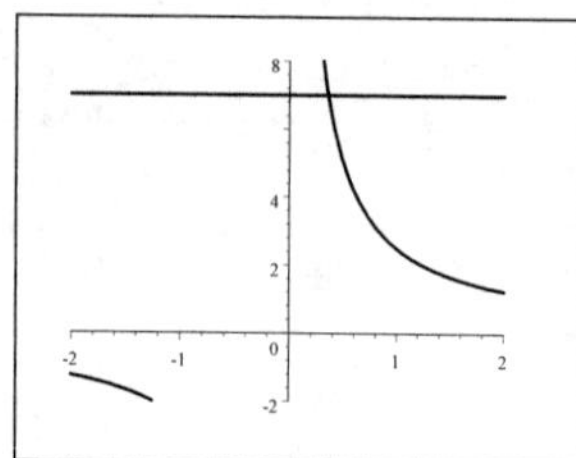

**30.** Algebraically: $\dfrac{4}{x+2} - \dfrac{6}{2x} = \dfrac{5}{2x+4} \quad \Leftrightarrow$

$2x\,(x+2)\left(\dfrac{4}{x+2} - \dfrac{6}{2x}\right) = 2x\,(x+2)\left(\dfrac{5}{2x+4}\right)$

$\Leftrightarrow \quad 2x\,(4) - (x+2)\,(6) = x\,(5) \quad \Leftrightarrow$

$8x - 6x - 12 = 5x \quad \Leftrightarrow \quad -12 = 3x \quad \Leftrightarrow \quad -4 = x$.

Graphically: We graph the two equations $y_1 = \dfrac{4}{x+2} - \dfrac{6}{2x}$ and $y_2 = \dfrac{5}{2x+4}$ in the viewing rectangle $[-5, 5]$ by $[-10, 10]$. Zooming in, we see that there is only one solution at $x = -4$.

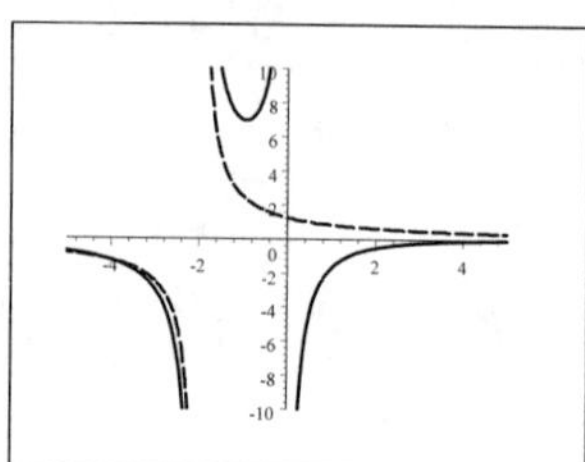

**31.** Algebraically: $x^2 - 32 = 0 \quad \Leftrightarrow \quad x^2 = 32 \quad \Rightarrow$ $x = \pm\sqrt{32} = \pm 4\sqrt{2}$.

Graphically: We graph the equation $y_1 = x^2 - 32$ and determine where this curve intersects the $x$-axis. We use the viewing rectangle $[-10, 10]$ by $[-5, 5]$. Zooming in, we see that solutions are $x \approx 5.66$ and $x \approx -5.66$.

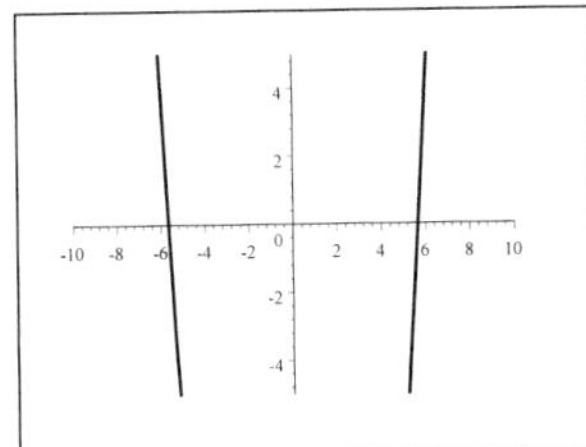

**32.** Algebraically: $x^3 + 16 = 0 \quad \Leftrightarrow \quad x^3 = -16 \quad \Leftrightarrow$ $x = -2\sqrt[3]{2}$.

Graphically: We graph the equation $y = x^3 + 16$ and determine where this curve intersects the $x$-axis. We use the viewing rectangle $[-5, 5]$ by $[-5, 5]$. Zooming in, we see that the solution is $x \approx -2.52$.

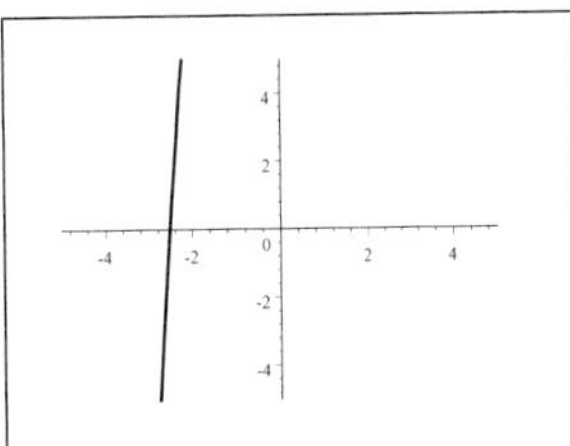

**33.** Algebraically: $16x^4 = 625 \quad \Leftrightarrow \quad x^4 = \frac{625}{16} \quad \Rightarrow$ $x = \pm\frac{5}{2} = \pm 2.5$.

Graphically: We graph the two equations $y_1 = 16x^4$ and $y_2 = 625$ in the viewing rectangle $[-5, 5]$ by $[610, 640]$. Zooming in, we see that solutions are $x = \pm 2.5$.

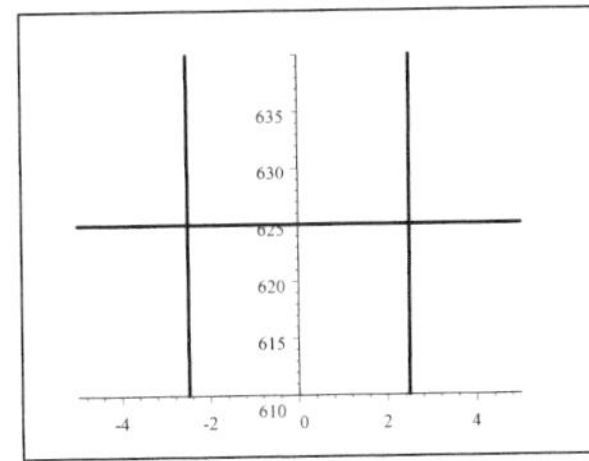

**34.** Algebraically: $2x^5 - 243 = 0 \quad \Leftrightarrow \quad 2x^5 = 243 \quad \Leftrightarrow$ $x^5 = \frac{243}{2} \quad \Leftrightarrow \quad x = \sqrt[5]{\frac{243}{2}} = \frac{3}{2}\sqrt[5]{16}$.

Graphically: We graph the equation $y = 2x^5 - 243$ and determine where this curve intersects the $x$-axis. We use the viewing rectangle $[-5, 10]$ by $[-5, 5]$. Zooming in, we see that the solution is $x \approx 2.61$.

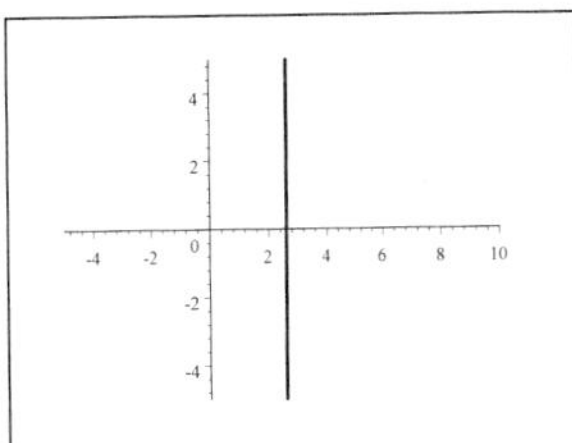

**35.** Algebraically: $(x - 5)^4 - 80 = 0 \quad \Leftrightarrow \quad (x - 5)^4 = 80$ $\Rightarrow \quad x - 5 = \pm\sqrt[4]{80} = \pm 2\sqrt[4]{5} \quad \Leftrightarrow \quad x = 5 \pm 2\sqrt[4]{5}$.

Graphically: We graph the equation $y_1 = (x - 5)^4 - 80$ and determine where this curve intersects the $x$-axis. We use the viewing rectangle $[-1, 9]$ by $[-5, 5]$. Zooming in, we see that solutions are $x \approx 2.01$ and $x \approx 7.99$.

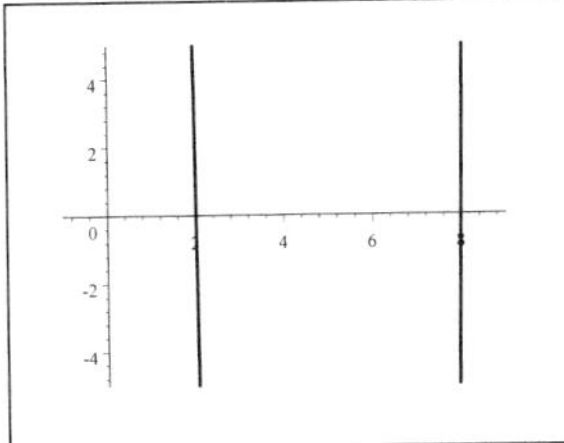

**36.** Algebraically: $6(x + 2)^5 = 64 \quad \Leftrightarrow$ $(x + 2)^5 = \frac{64}{6} = \frac{32}{3} \quad \Leftrightarrow \quad x + 2 = \sqrt[5]{\frac{32}{3}} = \frac{2}{3}\sqrt[5]{81}$ $\Leftrightarrow \quad x = -2 + \frac{2}{3}\sqrt[5]{81}$.

Graphically: We graph the two equations $y_1 = 6(x + 2)^5$ and $y_2 = 64$ in the viewing rectangle $[-5, 5]$ by $[50, 70]$. Zooming in, we see that the solution is $x \approx -0.39$.

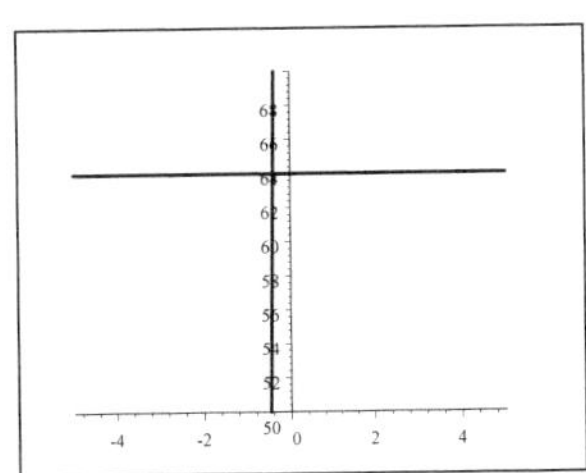

**37.** We graph $y = x^2 - 7x + 12$ in the viewing rectangle $[0, 6]$ by $[-0.1, 0.1]$. The solutions appear to be exactly $x = 3$ and $x = 4$. [In fact $x^2 - 7x + 12 = (x - 3)(x - 4)$.]

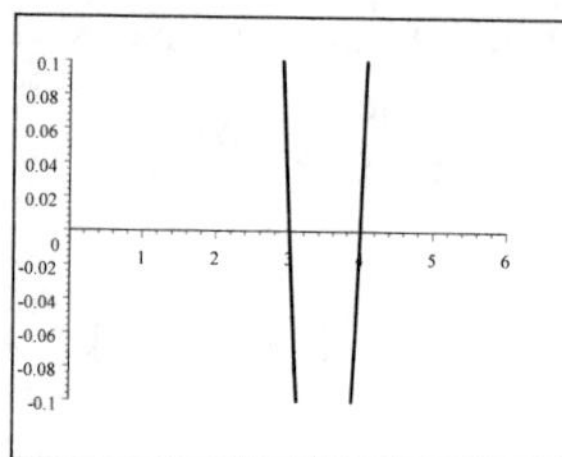

**38.** We graph $y = x^2 - 0.75x + 0.125$ in the viewing rectangle $[-2, 2]$ by $[-0.1, 0.1]$. The solutions are $x = 0.25$ and $x = 0.50$.

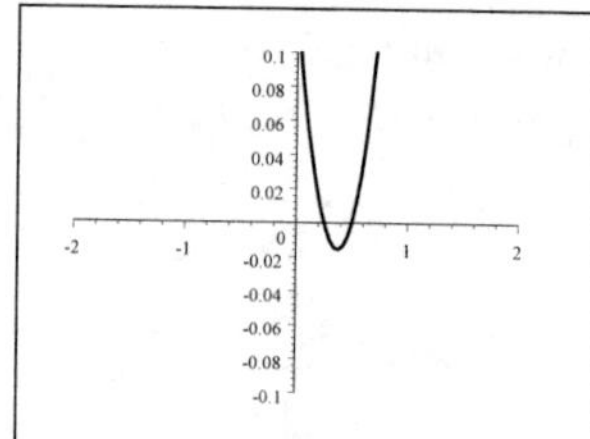

**39.** We graph $y = x^3 - 6x^2 + 11x - 6$ in the viewing rectangle $[-1, 4]$ by $[-0.1, 0.1]$. The solutions are $x = 1.00$, $x = 2.00$,and $x = 3.00$.

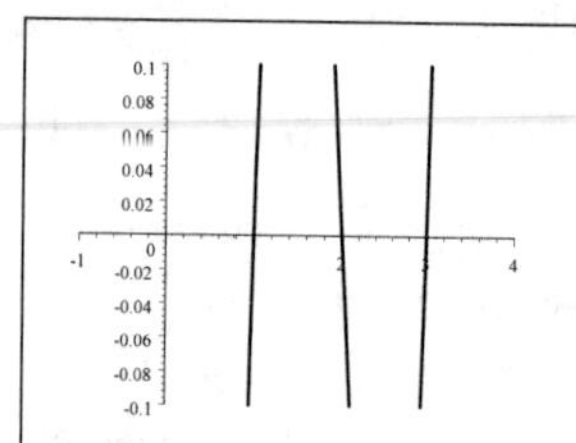

**40.** Since $16x^3 + 16x^2 = x + 1 \quad \Leftrightarrow$ $16x^3 + 16x^2 - x - 1 = 0$, we graph $y = 16x^3 + 16x^2 - x - 1$ in the viewing rectangle $[-2, 2]$ by $[-0.1, 0.1]$. The solutions are: $x = -1.00$, $x = -0.25$, and $x = 0.25$.

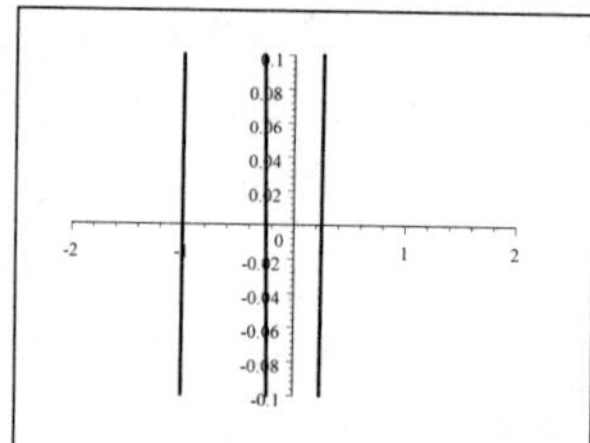

**41.** We first graph $y = x - \sqrt{x + 1}$ in the viewing rectangle $[-1, 5]$ by $[-0.1, 0.1]$ and find that the solution is near 1.6. Zooming in, we see that solutions is $x \approx 1.62$.

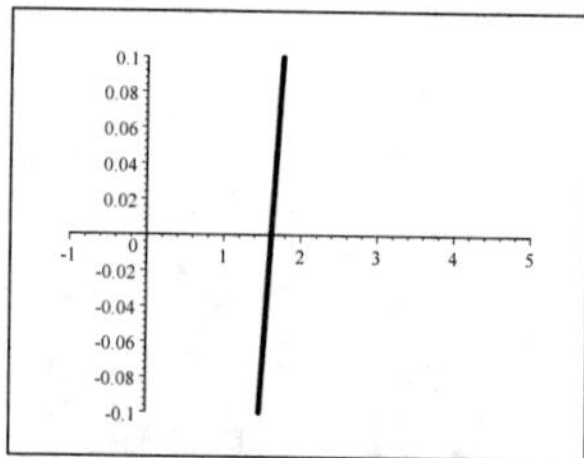

**42.** $1 + \sqrt{x} = \sqrt{1 + x^2} \quad \Leftrightarrow$ $1 + \sqrt{x} - \sqrt{1 + x^2} = 0$.Since $\sqrt{x}$ is only defined for $x \geq 0$, we start with the viewing rectangle $[-1, 5]$ by $[-1, 1]$. In this rectangle, there appears to be an exact solution at $x = 0$ and another solution between $x = 2$ and $x = 2.5$. We then use the viewing rectangle $[2.3, 2.35]$ by $[-0.01, 0.01]$, and isolate the second solution as $x \approx 2.314$. Thus the solutions are $x = 0$ and $x \approx 2.31$.

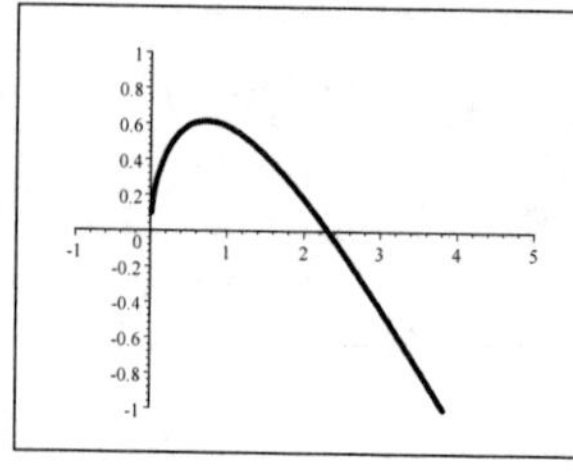

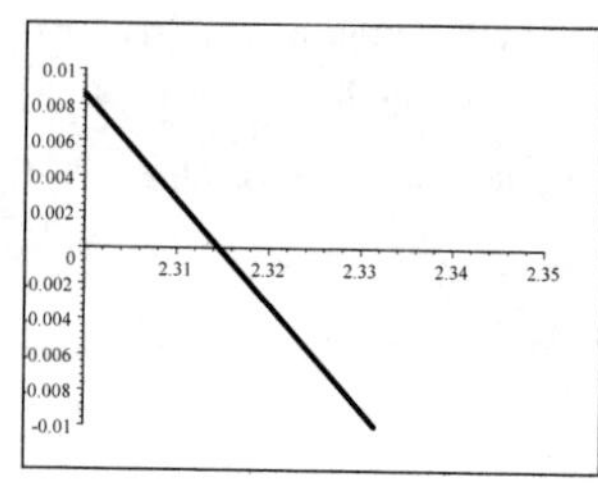

**43.** We graph $y = x^{1/3} - x$ in the viewing rectangle $[-3, 3]$ by $[-1, 1]$. The solutions are $x = -1$, $x = 0$, and $x = 1$, as can be verified by substitution.

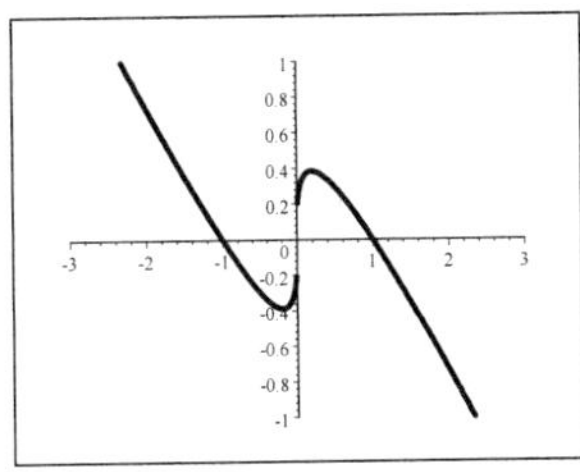

**44.** Since $x^{1/2}$ is only defined for $x \geq 0$, we start by graphing $y = x^{1/2} + x^{1/3} - x$ in the viewing rectangle $[-1, 5]$ by $[-1, 1]$ .We see a solution at $x = 0$ and another one between $x = 3$ and $x = 3.5$. We then use the viewing rectangle $[3.3, 3.4]$ by $[-0.01, 0.01]$, and isolate the second solution as $x \approx 3.31$. Thus, the solutions are $x = 0$ and $x \approx 3.31$.

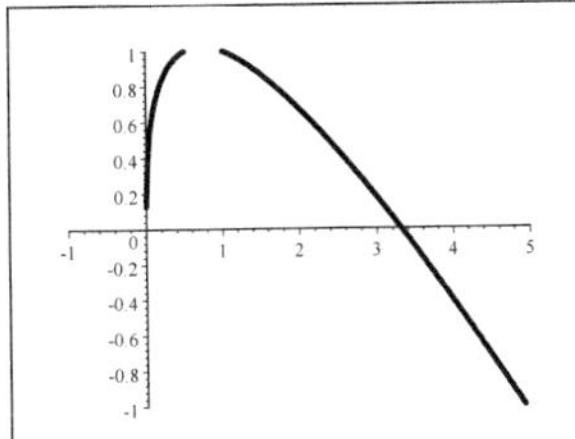

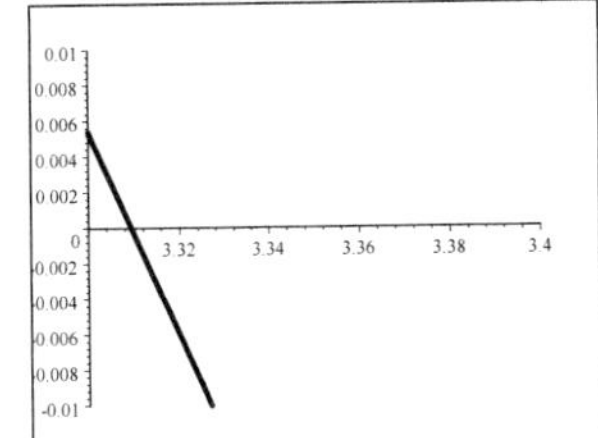

**45.** $x^3 - 2x^2 - x - 1 = 0$, so we start by graphing the function $y = x^3 - 2x^2 - x - 1$ in the viewing rectangle $[-10, 10]$ by $[-100, 100]$. There appear to be two solutions, one near $x = 0$ and another one between $x = 2$ and $x = 3$. We then use the viewing rectangle $[-1, 5]$ by $[-1, 1]$ and zoom in on the only solution, $x \approx 2.55$.

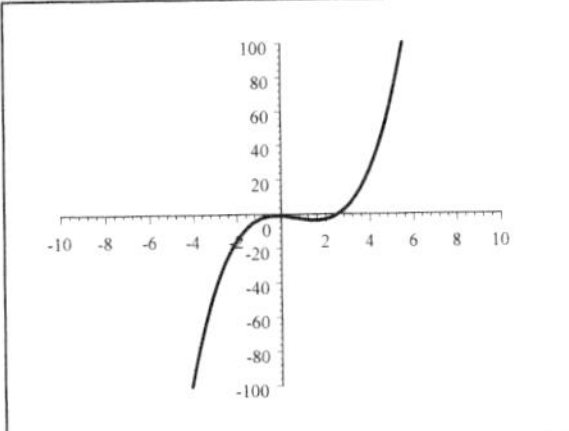

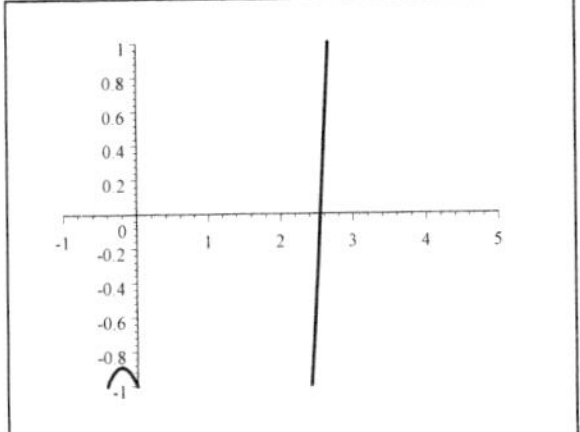

**46.** $x^4 - 8x^2 + 2 = 0$. We start by graphing the function $y = x^4 - 8x^2 + 2$ in the viewing rectangle $[-10, 10]$ by $[-10, 10]$. There appear to be four solutions between $x = -3$ and $x = 3$. We then use the viewing rectangle $[-5, 5]$ by $[-1, 1]$, and zoom to find the four solutions $x \approx -2.78$, $x \approx -0.51$, $x \approx 0.51$, and $x \approx 2.78$.

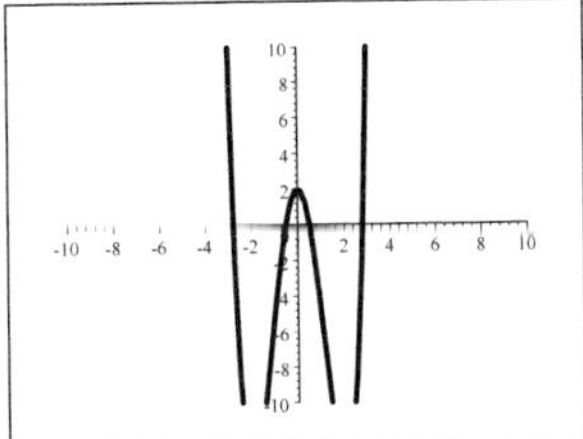

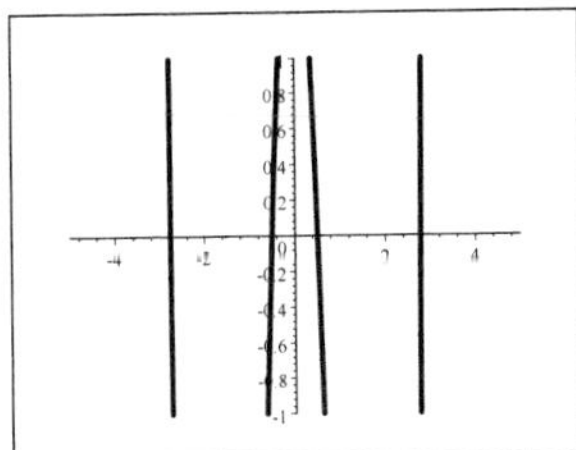

**47.** $x(x-1)(x+2) = \frac{1}{6}x \quad \Leftrightarrow$ $x(x-1)(x+2) - \frac{1}{6}x = 0$. We start by graphing the function $y = x(x-1)(x+2) - \frac{1}{6}x$ in the viewing rectangle $[-5, 5]$ by $[-10, 10]$. There appear to be three solutions. We then use the viewing rectangle $[-2.5, 2.5]$ by $[-1, 1]$ and zoom into the solutions at $x \approx -2.05$, $x = 0.00$, and $x \approx 1.05$.

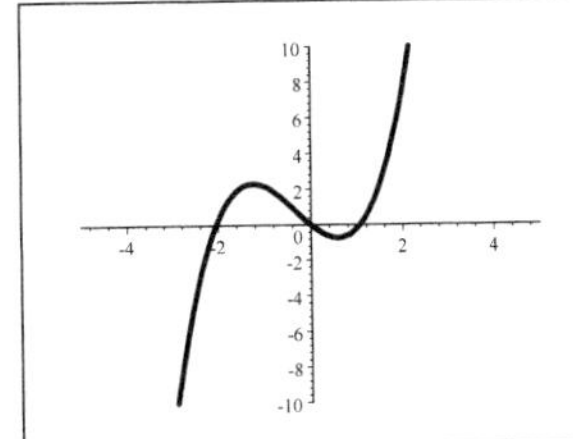

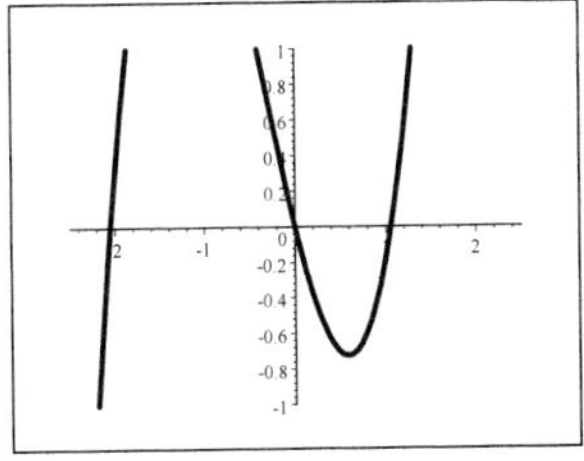

**48.** $x^4 = 16 - x^3$. We start by graphing the functions $y_1 = x^4$ and $y_2 = 16 - x^3$ in the viewing rectangle $[-10, 10]$ by $[-5, 40]$. There appears to be two solutions, one near $x = -2$ and another one near $x = 2$. We then use the viewing rectangle $[-2.4, -2.2]$ by $[27, 29]$, and zoom in to find the solution at $x \approx -2.31$. We then use the viewing rectangle $[1.7, 1.8]$ by $[9.5, 10.5]$, and zoom in to find the solution at $x \approx 1.79$.

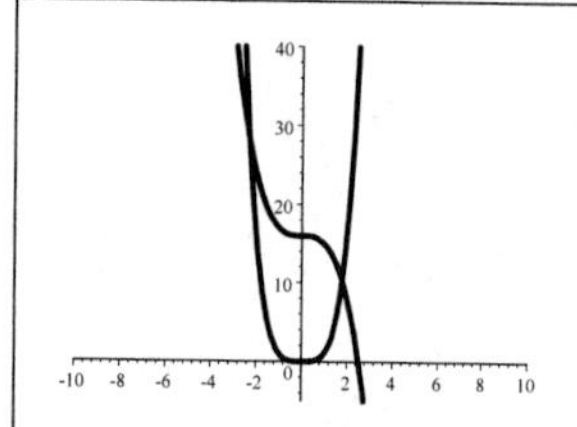


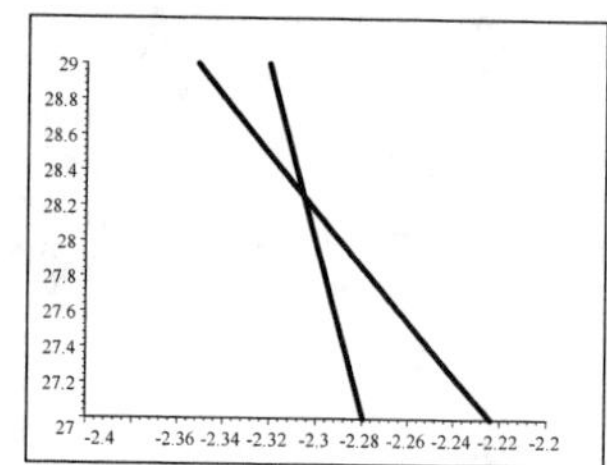


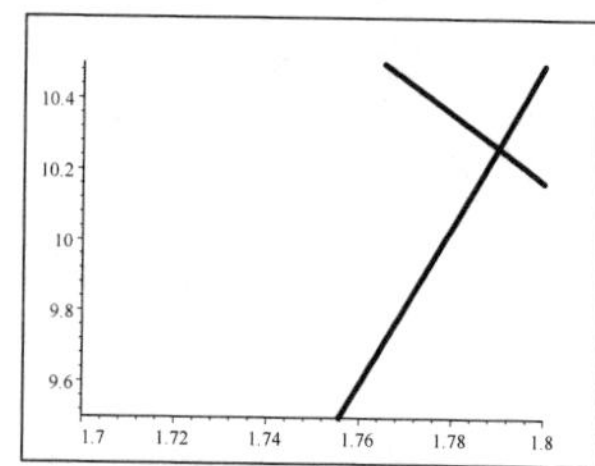


**49.** We graph $y = x^2 - 3x - 10$ in the viewing rectangle $[-5, 6]$ by $[-14, 2]$. The the solution to the inequality is $[-2, 5]$.

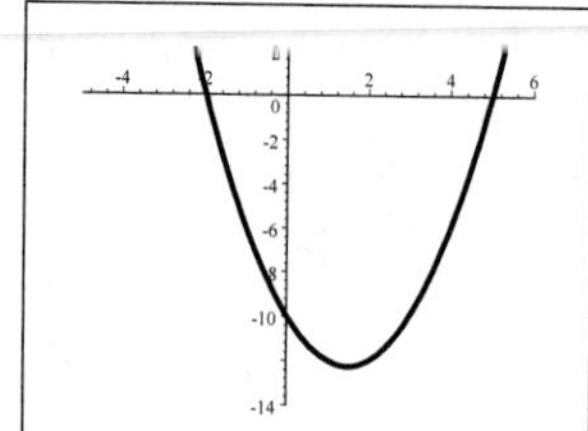


**50.** Since $0.5x^2 + 0.875x \leq 0.25 \quad \Leftrightarrow$ $0.5x^2 + 0.875x - 0.25 \leq 0$, we graph $y = 0.5x^2 + 0.875x - 0.25$ in the viewing rectangle $[-3, 1]$ by $[-5, 5]$. Thus the solution to the inequality is $-2 \leq x \leq 0.25$.

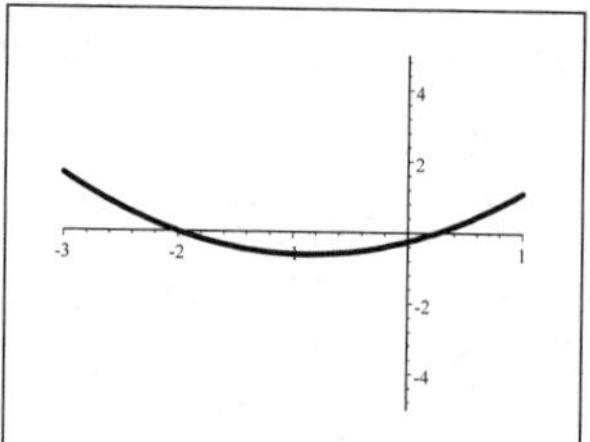


**51.** Since $x^3 + 11x \leq 6x^2 + 6 \quad \Leftrightarrow \quad x^3 - 6x^2 + 11x - 6 \leq 0$, we graph $y = x^3 - 6x^2 + 11x - 6$ in the viewing rectangle $[0, 5]$ by $[-5, 5]$. The solution set is $(-\infty, 1.0] \cup [2.0, 3.0]$.

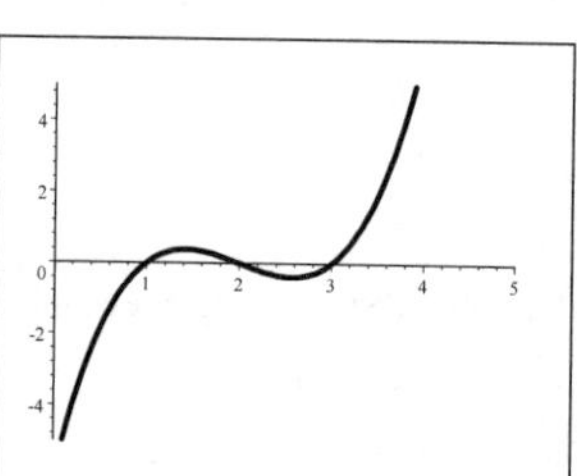


**52.** Since $16x^3 + 24x^2 > -9x - 1 \quad \Leftrightarrow$ $16x^3 + 24x^2 + 9x + 1 > 0$, we graph $y = 16x^3 + 24x^2 + 9x + 1$ in the viewing rectangle $[-3, 1]$ by $[-5, 5]$. From this rectangle, we see that $x = -1$ is an $x$-intercept, but it is unclear what is occurring between $x = -0.5$ and

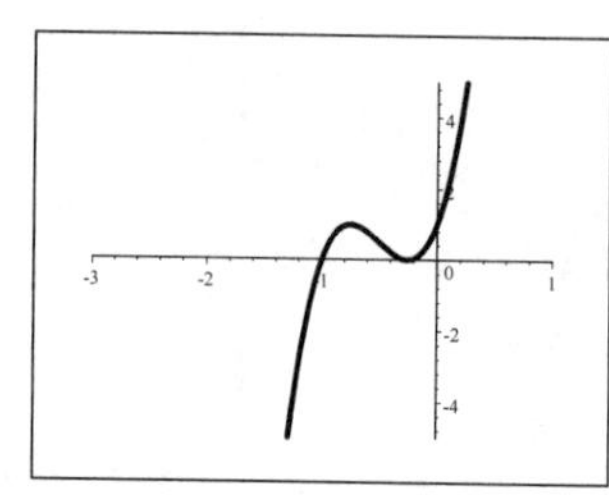


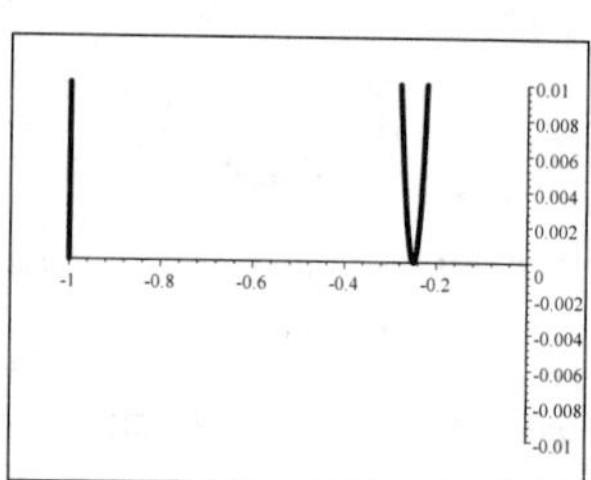


$x = 0$. We then use the viewing rectangle $[-1, 0]$ by $[-0.01, 0.01]$. It shows $y = 0$ at $x = -0.25$. Thus in interval notation, the solution is $(-1, -0.25) \cup (-0.25, \infty)$.

**53.** Since $x^{1/3} \le x \quad \Leftrightarrow \quad x^{1/3} - x < 0$, we graph $y = x^{1/3} - x$ in the viewing rectangle $[-3, 3]$ by $[-1, 1]$. From this, we find that the solution set is $(-1, 0) \cup (1, \infty)$.

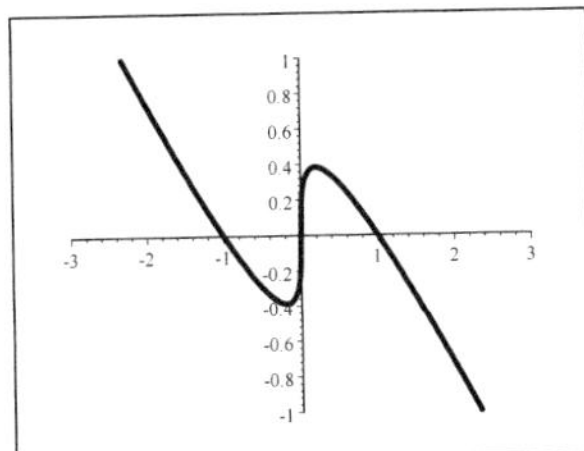


**54.** Since $\sqrt{0.5x^2 + 1} \le 2|x| \Leftrightarrow \sqrt{0.5x^2 + 1} - 2|x| \le 0$, we graph $y = \sqrt{0.5x^2 + 1} - 2|x|$ in the viewing rectangle $[-1, 1]$ by $[-1, 1]$. We locate the $x$-intercepts at $x \approx \pm 0.535$. Thus in interval notation, the solution is approximately $(-\infty, -0.535] \cup [0.535, \infty)$.

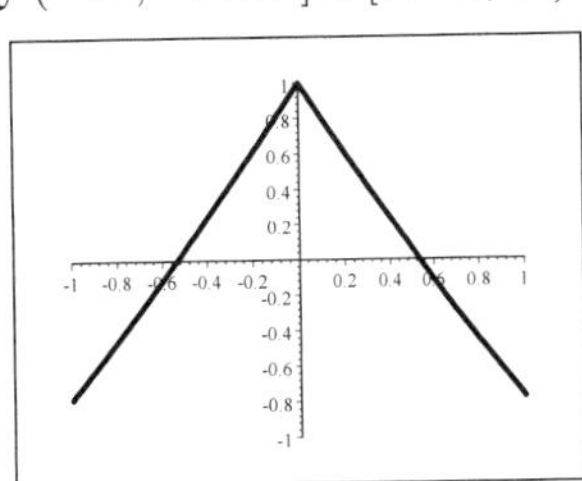


**55.** Since $(x+1)^2 < (x-1)^2 \quad \Leftrightarrow$ $(x+1)^2 - (x-1)^2 < 0$, we graph $y = (x+1)^2 - (x-1)^2$ in the viewing rectangle $[-2, 2]$ by $[-5, 5]$. The solution set is $(-\infty, 0)$.

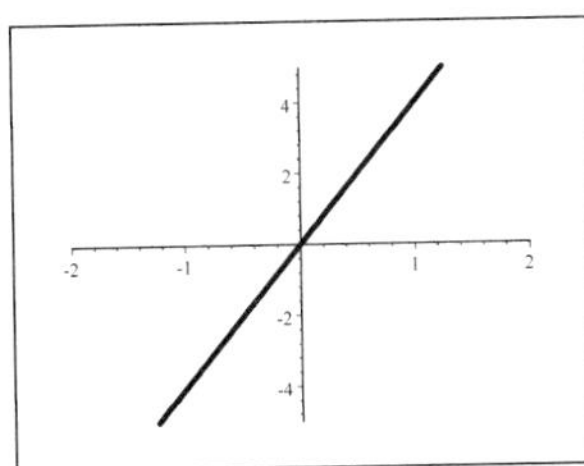


**56.** Since $(x+1)^2 \le x^3 \quad \Leftrightarrow \quad (x+1)^2 - x^3 \le 0$, we graph $y = (x+1)^2 - x^3$ in the viewing rectangle $[-4, 4]$ by $[-1, 1]$. The $x$-intercept is close to $x = 2$. Using a trace function, we obtain $x \approx 2.148$. Thus the solution is $[2.148, \infty)$.

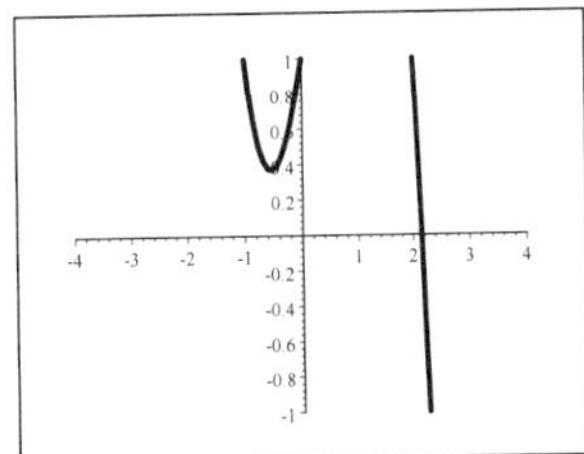


**57.** As in Example 6, we graph the equation $y = x^3 - 6x^2 + 9x - \sqrt{x}$ in the viewing rectangle $[0, 10]$ by $[-2, 15]$. We see the two solutions found in Example 6 and what appears to be an additional solution at $x = 0$. In the viewing rectangle $[0, 0.05]$ by $[-0.25, 0.25]$, we find yet another solution at $x \approx 0.01$. We can verify that $x = 0$ is an exact solution by substitution.

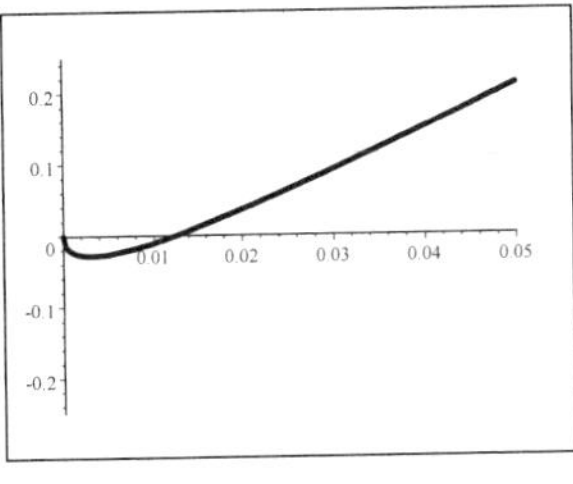


**58. (a)**

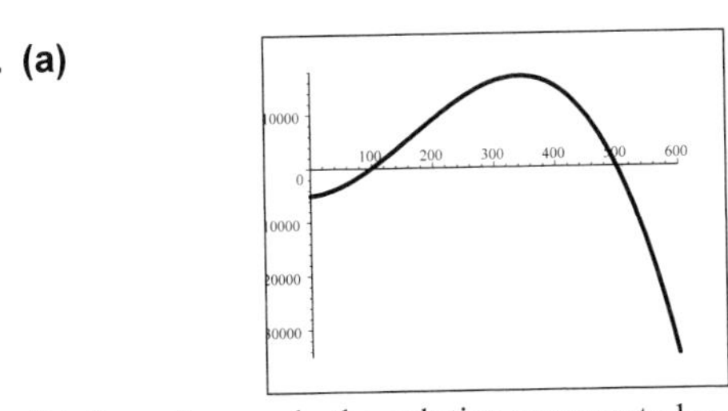


**(b)** From the graph, the solution appears to be 100 cooktops. Verifying it by substituting $x = 100$ in $y = 10x + 0.5x^2 - 0.001x^3 - 5000$ we get $y = 0$.

**(c)** Using a zoom or trace function on a calculator, we find that the company's profits are greater than \$15,000 for $280 < x < 400$.

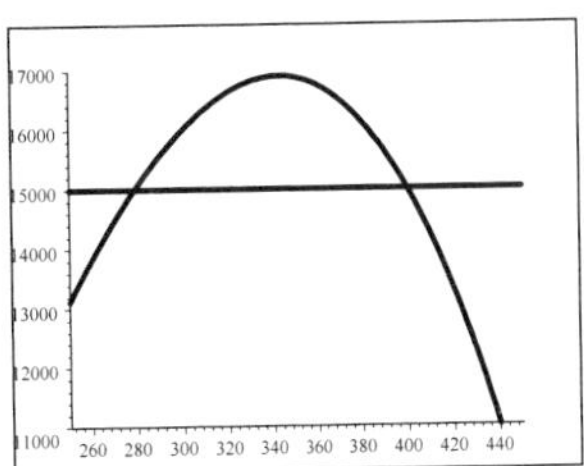

**59.** Using a zoom or trace function, we find that $y \geq 10$ for $x \geq 66.7$. We could estimate this since if $x < 100$, then $\left(\frac{x}{5280}\right)^2 \leq 0.00036$. So for $x < 100$ we have $\sqrt{1.5x + \left(\frac{x}{5280}\right)^2} \approx \sqrt{1.5x}$. Solving $\sqrt{1.5x} > 10$ we get $1.5 > 100$ or $x > \frac{100}{1.5} = 66.7$ mi.

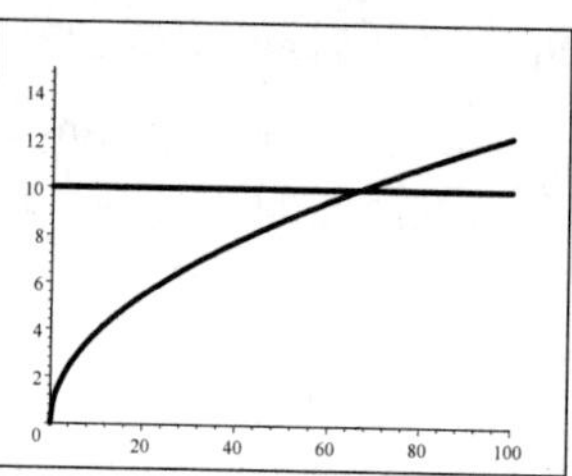

**60.** Answers may vary. Possible answers are given.

**(a)** Absolute value must be chosen from a menu or written as "abs() ".

**(b)** Some calculators or computer graphing programs will not take the odd root of a negative number, even though it is defined. Since $\sqrt[5]{-x} = -\sqrt[5]{x}$, you may have to enter two functions: `Y_1=x^(1/5)` and `Y_2=-(-x)^(1/5)`.

**(c)** Here you must remember to use parentheses and enter $x/(x-1)$.

**(d)** As in part (b), some graphing devices may not give the entire graph. When this is the case, you will need to determine where the radicand (inside of the radical) is negative. Here you might need to define two functions: `Y_1=x^3+(x+2)^(1/3)` and `Y_2=x^3-(-x-2)^(1/3)`.

**61.** Calculators perform operations in the following order: exponents are applied before division and division is applied before addition. Therefore, `Y_1=x^1/3` is interpreted as $y = \frac{x^1}{3} = \frac{x}{3}$, which is the equation of a line. Likewise, `Y_2=x/x+4` is interpreted as $y = \frac{x}{x} + 4 = 1 + 4 = 5$. Instead, enter the following: `Y_1=x^(1/3), Y_2=x/(x+4)`.

**62.** Answers will vary.

**63. (a)** We graph $y_1 = x^3 - 3x$ and $y_2 = k$ for $k = -4, -2, 0, 2,$ and $4$ in the viewing rectangle $[-5, 5]$ by $[-10, 10]$. The number of solutions and the solutions are shown in the table below.

| $k$ | Number of solutions | Solutions |
|---|---|---|
| $-4$ | 1 | $x \approx -2.20$ |
| $-2$ | 2 | $x = -2, x = 1$ |
| $0$ | 3 | $x \approx \pm 1.73, x = 0$ |
| $2$ | 2 | $x = -1, x = 2$ |
| $4$ | 1 | $x \approx 2.20$ |

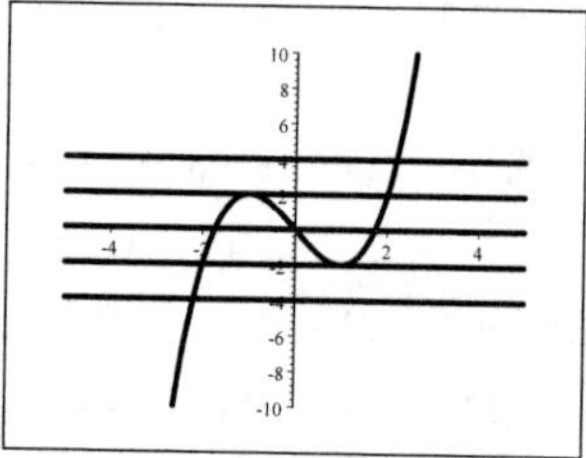

**(b)** The equation $x^3 - 3x = k$ will have one solution for all $k < -2$ or $k > 2$, it will have exactly two solutions when $k = \pm 2$, and it will have three solutions for $-2 < k < 2$.

## 2.4 Lines

**1.** $m = \frac{y_2 - y_1}{x_2 - x_1} = \frac{2-0}{4-0} = \frac{2}{4} = \frac{1}{2}$

**2.** $m = \frac{y_2 - y_1}{x_2 - x_1} = \frac{0-(-6)}{0-(2)} = \frac{6}{-2} = -3$

**3.** $m = \frac{y_2 - y_1}{x_2 - x_1} = \frac{0-2}{-10-2} = \frac{-2}{-12} = \frac{1}{6}$

**4.** $m = \frac{y_2 - y_1}{x_2 - x_1} = \frac{2-(3)}{1-(3)} = \frac{-1}{-2} = \frac{1}{2}$

**5.** $m = \frac{y_2 - y_1}{x_2 - x_1} = \frac{4-3}{2-4} = \frac{1}{-2} = -\frac{1}{2}$

**6.** $m = \frac{y_2 - y_1}{x_2 - x_1} = \frac{3-(-5)}{(-4)-(2)} = \frac{8}{-6} = -\frac{4}{3}$

**7.** $m = \frac{y_2 - y_1}{x_2 - x_1} = \frac{6-(-3)}{-1-1} = \frac{9}{-2} = -\frac{9}{2}$

**8.** $m = \frac{y_2 - y_1}{x_2 - x_1} = \frac{0-(-4)}{6-(-1)} = \frac{4}{7}$

**9.** For $\ell_1$, we find two points, $(-1, 2)$ and $(0, 0)$ that lie on the line. Thus the slope of $\ell_1$ is $m = \dfrac{y_2 - y_1}{x_2 - x_1} = \dfrac{2-0}{-1-0} = -2$.

For $\ell_2$, we find two points $(0, 2)$ and $(2, 3)$. Thus, the slope of $\ell_2$ is $m = \dfrac{y_2 - y_1}{x_2 - x_1} = \dfrac{3-2}{2-0} = \frac{1}{2}$. For $\ell_3$ we find the points $(2, -2)$ and $(3, 1)$. Thus, the slope of $\ell_3$ is $m = \dfrac{y_2 - y_1}{x_2 - x_1} = \dfrac{1-(-2)}{3-2} = 3$. For $\ell_4$, we find the points $(-2, -1)$ and $(2, -2)$. Thus, the slope of $\ell_4$ is $m = \dfrac{y_2 - y_1}{x_2 - x_1} = \dfrac{-2-(-1)}{2-(-2)} = \dfrac{-1}{4} = -\frac{1}{4}$.

**10. (a)**

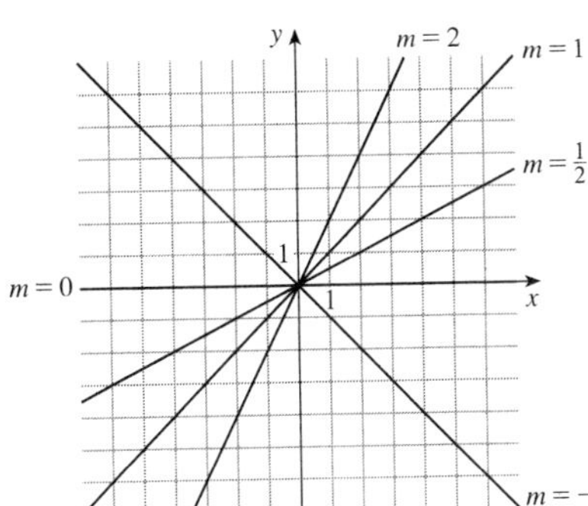


**(b)**

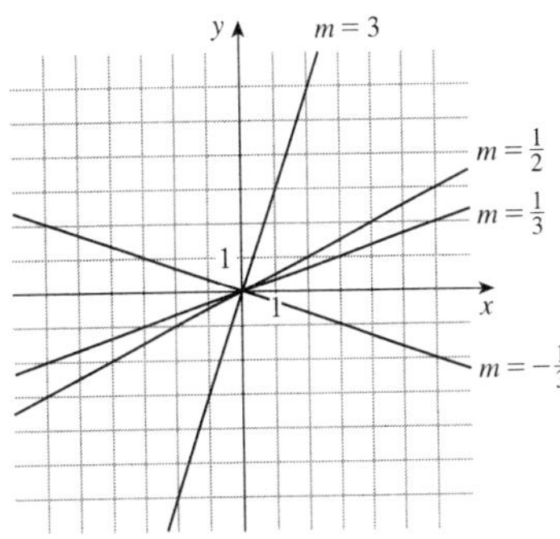


**11.** First we find two points $(0, 4)$ and $(4, 0)$ that lie on the line. So the slope is $m = \dfrac{0-4}{4-0} = -1$. Since the $y$-intercept is 4, the equation of the line is $y = mx + b = -1x + 4$. So $y = -x + 4$, or $x + y - 4 = 0$.

**12.** We find two points on the graph, $(0, 4)$ and $(-2, 0)$. So the slope is $m = \dfrac{0-4}{-2-0} = 2$. Since the $y$-intercept is 4, the equation of the line is $y = mx + b = 2x + 4$, so $y = 2x + 4 \quad \Leftrightarrow \quad 2x - y + 4 = 0$.

**13.** We choose the two intercepts as points, $(0, -3)$ and $(2, 0)$. So the slope is $m = \dfrac{0-(-3)}{2-0} = \frac{3}{2}$. Since the $y$-intercept is $-3$, the equation of the line is $y = mx + b = \frac{3}{2}x - 3$, or $3x - 2y - 6 = 0$.

**14.** We choose the two intercepts, $(0, -4)$ and $(-3, 0)$. So the slope is $m = \dfrac{0-(-4)}{-3-0} = -\frac{4}{3}$. Since the $y$-intercept is $-4$, the equation of the line is $y = mx + b = -\frac{4}{3}x - 4 \quad \Leftrightarrow \quad 4x + 3y + 12 = 0$.

**15.** Using the equation $y - y_1 = m(x - x_1)$, we get $y - 3 = 1(x - 2) \quad \Leftrightarrow \quad -x + y = 1 \quad \Leftrightarrow \quad x - y + 1 = 0$.

**16.** Using the equation $y - y_1 = m(x - x_1)$, we get $y - 4 = -1(x - (-2)) \quad \Leftrightarrow \quad y - 4 = -x - 2 \quad \Leftrightarrow \quad x + y - 2 = 0$.

**17.** Using the equation $y - y_1 = m(x - x_1)$, we get $y - 7 = \frac{2}{3}(x - 1) \quad \Leftrightarrow \quad 3y - 21 = 2x - 2 \quad \Leftrightarrow \quad -2x + 3y = 19$ $\Leftrightarrow \quad 2x - 3y + 19 = 0$.

**18.** Using the equation $y - y_1 = m(x - x_1)$, we get $y - (-5) = -\frac{7}{2}(x - (-3)) \quad \Leftrightarrow \quad 2y + 10 = -7x - 21 \quad \Leftrightarrow$ $7x + 2y + 31 = 0$.

**19.** First we find the slope, which is $m = \dfrac{y_2 - y_1}{x_2 - x_1} = \dfrac{6-1}{1-2} = \dfrac{5}{-1} = -5$. Substituting into $y - y_1 = m(x - x_1)$, we get $y - 6 = -5(x - 1) \quad \Leftrightarrow \quad y - 6 = -5x + 5 \quad \Leftrightarrow \quad 5x + y - 11 = 0$.

**20.** First we find the slope, which is $m = \dfrac{y_2 - y_1}{x_2 - x_1} = \dfrac{3-(-2)}{4-(-1)} = \frac{5}{5} = 1$. Substituting into $y - y_1 = m(x - x_1)$, we get $y - 3 = 1(x - 4) \quad \Leftrightarrow \quad y - 3 = x - 4 \quad \Leftrightarrow \quad x - y - 1 = 0$.

**21.** Using $y = mx + b$, we have $y = 3x + (-2)$ or $3x - y - 2 = 0$.

**22.** Using $y = mx + b$, we have $y = \frac{2}{5}x + 4 \quad \Leftrightarrow \quad 2x - 5y + 20 = 0$.

**23.** We are given two points, $(1, 0)$ and $(0, -3)$. Thus, the slope is $m = \dfrac{y_2 - y_1}{x_2 - x_1} = \dfrac{-3-0}{0-1} = \dfrac{-3}{-1} = 3$. Using the $y$-intercept, we have $y = 3x + (-3)$ or $y = 3x - 3$ or $3x - y - 3 = 0$.

**24.** We are given two points, $(-8, 0)$ and $(0, 6)$. Thus, the slope is $m = \frac{y_2 - y_1}{x_2 - x_1} = \frac{6 - 0}{0 - (-8)} = \frac{6}{8} = \frac{3}{4}$. Using the $y$-intercept we have $y = \frac{3}{4}x + 6 \quad \Leftrightarrow \quad 3x - 4y + 24 = 0$.

**25.** Since the equation of a horizontal line passing through $(a, b)$ is $y = b$,the equation of the horizontal line passing through $(4, 5)$ is $y = 5$.

**26.** Any line parallel to the $y$-axis will have undefined slope and be of the form $x = a$. Since the graph of the line passes through the point $(4, 5)$, the equation of the line is $x = 4$.

**27.** Since $x + 2y = 6 \quad \Leftrightarrow \quad 2y = -x + 6 \quad \Leftrightarrow \quad y = -\frac{1}{2}x + 3$, the slope of this line is $-\frac{1}{2}$. Thus, the line we seek is given by $y - (-6) = -\frac{1}{2}(x - 1) \quad \Leftrightarrow \quad 2y + 12 = -x + 1 \quad \Leftrightarrow \quad x + 2y + 11 = 0$.

**28.** Since $2x + 3y + 4 = 0 \quad \Leftrightarrow \quad 3y = -2x - 4 \quad \Leftrightarrow \quad y = -\frac{2}{3}x - \frac{4}{3}$, the slope of this line is $m = -\frac{2}{3}$. Substituting $m = -\frac{2}{3}$ and $b = 6$ into the slope intercept formula, the line we seek is given by $y = -\frac{2}{3}x + 6 \quad \Leftrightarrow \quad 2x + 3y - 18 = 0$.

**29.** Any line parallel to $x = 5$ will have undefined slope and be of the form $x = a$. Thus the equation of the line is $x = -1$.

**30.** Any line perpendicular to $y = 1$ has undefined slope and is of the form $x = a$. Since the graph of the line passes through the point $(2, 6)$, the equation of the line is $x = 2$.

**31.** First find the slope of $2x + 5y + 8 = 0$. This gives $2x + 5y + 8 = 0 \quad \Leftrightarrow \quad 5y = -2x - 8 \quad \Leftrightarrow \quad y = -\frac{2}{5}x - \frac{8}{5}$. So the slope of the line that is perpendicular to $2x + 5y + 8 = 0$ is $m = -\frac{1}{-2/5} = \frac{5}{2}$. The equation of the line we seek is $y - (-2) = \frac{5}{2}(x - (-1)) \quad \Leftrightarrow \quad 2y + 4 = 5x + 5 \quad \Leftrightarrow \quad 5x - 2y + 1 = 0$.

**32.** First find the slope of the line $4x - 8y = 1$. This gives $4x - 8y = 1 \quad \Leftrightarrow \quad -8y = -4x + 1 \quad \Leftrightarrow \quad y = \frac{1}{2}x - \frac{1}{8}$. So the slope of the line that is perpendicular to $4x - 8y = 1$ is $m = -\frac{1}{1/2} = -2$. The equation of the line we seek is $y - \left(-\frac{2}{3}\right) = -2\left(x - \frac{1}{2}\right) \quad \Leftrightarrow \quad y + \frac{2}{3} = -2x + 1 \quad \Leftrightarrow \quad 6x + 3y - 1 = 0$.

**33.** First find the slope of the line passing through $(2, 5)$ and $(-2, 1)$. This gives $m = \frac{1 - 5}{-2 - 2} = \frac{-4}{-4} = 1$, and so the equation of the line we seek is $y - 7 = 1(x - 1) \quad \Leftrightarrow \quad x - y + 6 = 0$.

**34.** First find the slope of the line passing through $(1, 1)$ and $(5, -1)$. This gives $m = \frac{-1 - 1}{5 - 1} = \frac{-2}{4} = -\frac{1}{2}$, and so the slope of the line that is perpendicular is $m = -\frac{1}{-1/2} = 2$. Thus the equation of the line we seek is $y + 11 = 2(x + 2) \quad \Leftrightarrow$ $2x - y - 7 = 0$.

**35. (a)**

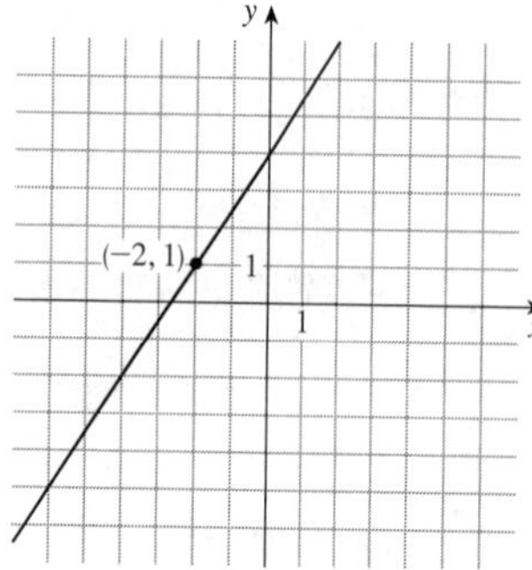


**(b)** $y - 1 = \frac{3}{2}(x - (-2)) \quad \Leftrightarrow \quad 2y - 2 = 3(x + 2)$
$\Leftrightarrow \quad 2y - 2 = 3x + 6 \quad \Leftrightarrow \quad 3x - 2y + 8 = 0$.

**36. (a)**

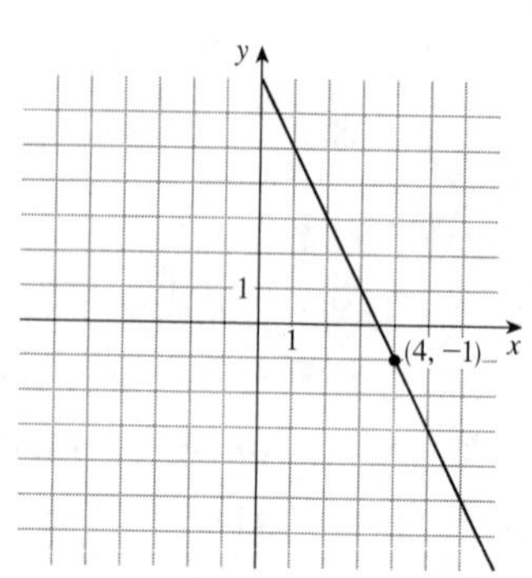


**(b)** $y - (-1) = -2(x - 4) \quad \Leftrightarrow \quad y + 1 = -2x + 8$
$\Leftrightarrow \quad 2x + y - 7 = 0$.

**37.**

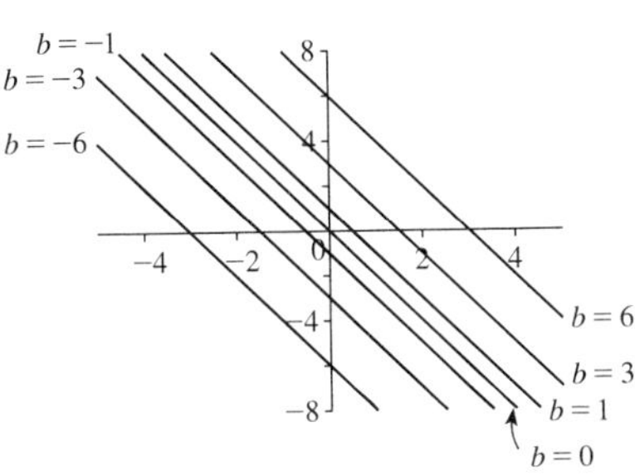


$y = -2x + b$, $b = 0, \pm 1, \pm 3, \pm 6$. They have the same slope, so they are parallel.

**38.**

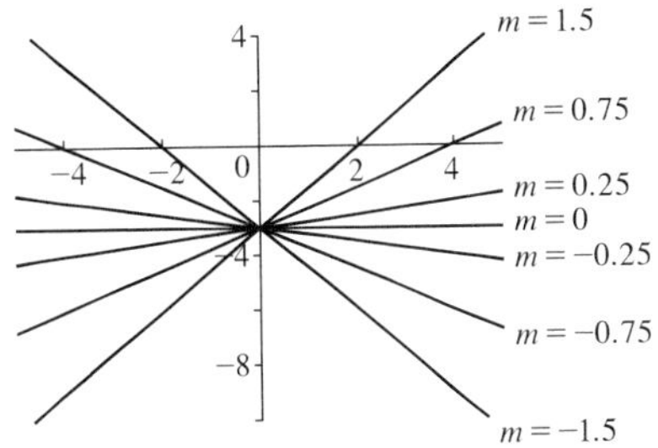


$y = mx - 3$, $m = 0, \pm 0.25, \pm 0.75, \pm 1.5$. Each of the lines contains the point $(0, -3)$ because the point $(0, -3)$ satisfies each equation $y = mx - 3$. Since $(0, -3)$ is on the $y$-axis, they all have the same $y$-intercept.

**39.**

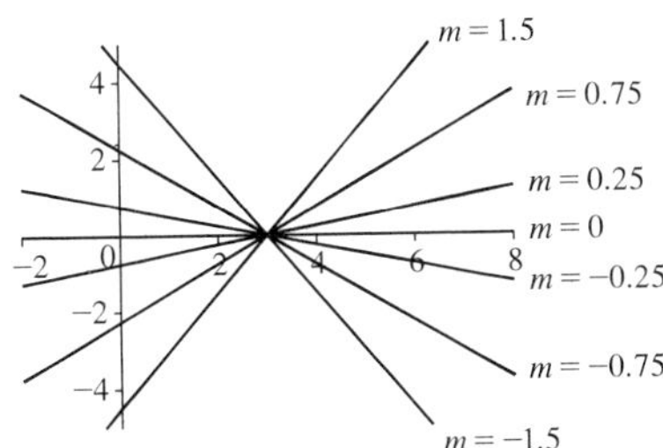


$y = m(x - 3)$, $m = 0, \pm 0.25, \pm 0.75, \pm 1.5$. Each of the lines contains the point $(3, 0)$ because the point $(3, 0)$ satisfies each equation $y = m(x - 3)$. Since $(3, 0)$ is on the $x$-axis, we could also say that they all have the same $x$-intercept.

**40.**

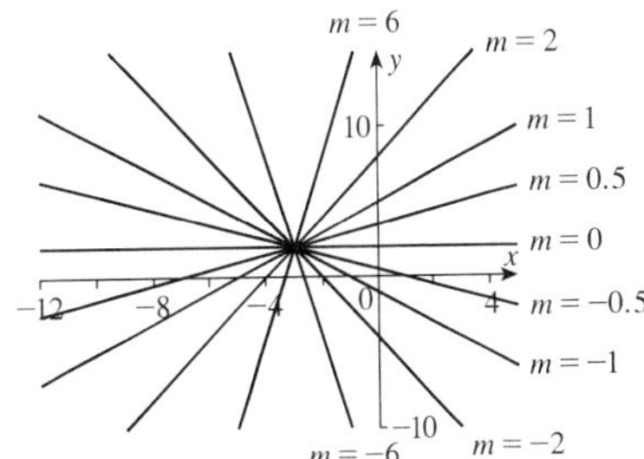


$y = 2 + m(x + 3)$, $m = 0, \pm 0.5, \pm 1, \pm 2, \pm 6$. Each of the lines contains the point $(-3, 2)$ because the point $(-3, 2)$ satisfies each equation $y = 2 + m(x + 3)$.

**41.** $x + y = 3 \quad \Leftrightarrow \quad y = -x + 3$. So the slope is $-1$, and the $y$-intercept is 3.

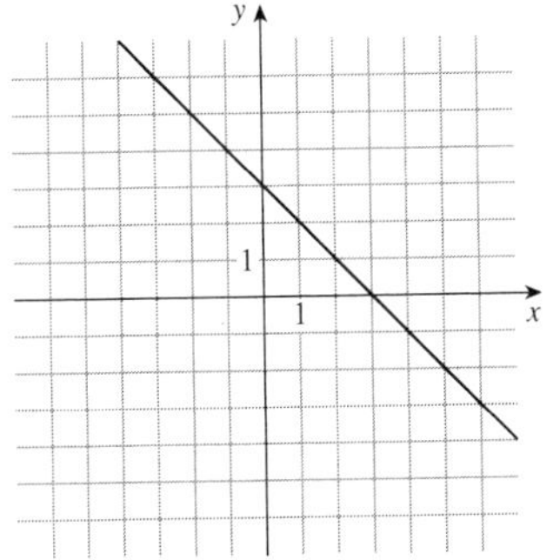

**42.** $3x - 2y = 12 \quad \Leftrightarrow \quad -2y = -3x + 12 \quad \Leftrightarrow$ $y = \frac{3}{2}x - 6$. So the slope is $\frac{3}{2}$, and the $y$-intercept is $-6$.

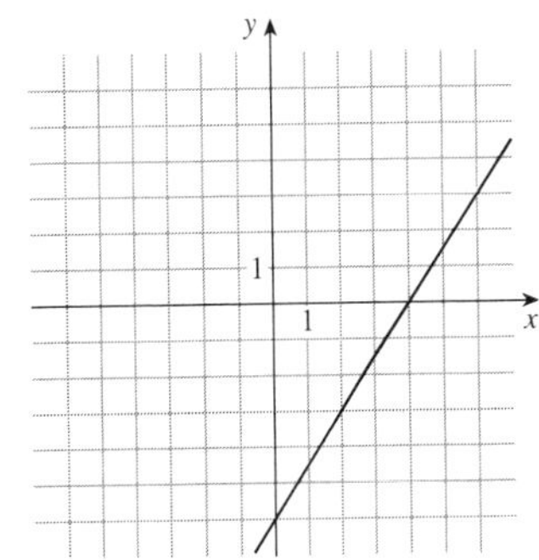

**43.** $x + 3y = 0 \quad \Leftrightarrow \quad 3y = -x \quad \Leftrightarrow \quad y = -\frac{1}{3}x$. So the slope is $-\frac{1}{3}$, and the $y$-intercept is 0.

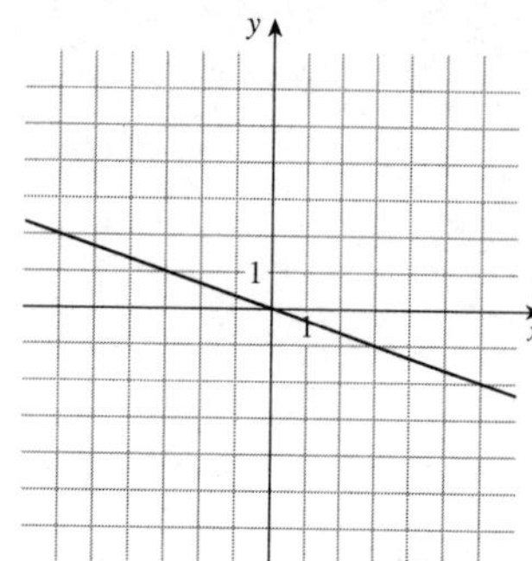


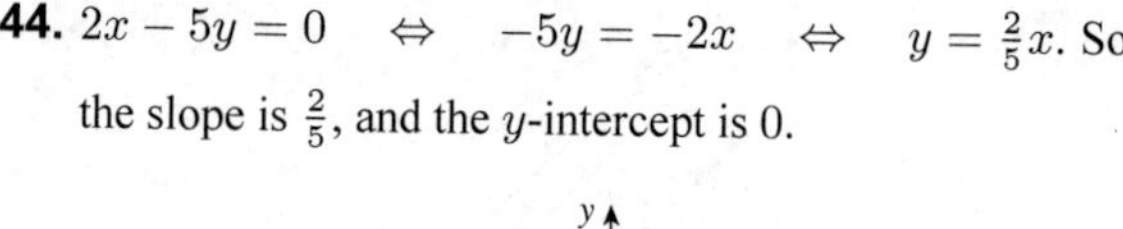

**44.** $2x - 5y = 0 \quad \Leftrightarrow \quad -5y = -2x \quad \Leftrightarrow \quad y = \frac{2}{5}x$. So the slope is $\frac{2}{5}$, and the $y$-intercept is 0.

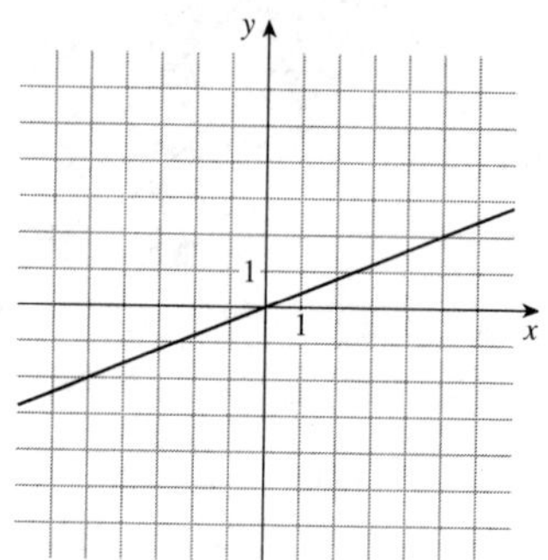


**45.** $\frac{1}{2}x - \frac{1}{3}y + 1 = 0 \quad \Leftrightarrow \quad -\frac{1}{3}y = -\frac{1}{2}x - 1 \quad \Leftrightarrow$ $y = \frac{3}{2}x + 3$. So the slope is $\frac{3}{2}$, and the $y$-intercept is 3.

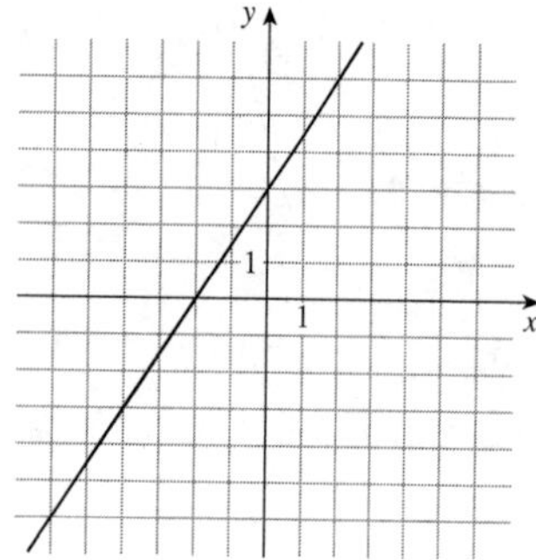


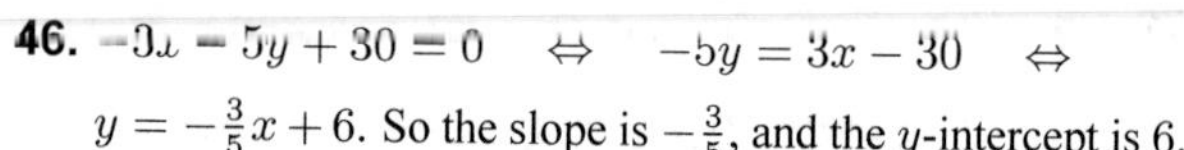

**46.** $-3x - 5y + 30 = 0 \quad \Leftrightarrow \quad -5y = 3x - 30 \quad \Leftrightarrow$ $y = -\frac{3}{5}x + 6$. So the slope is $-\frac{3}{5}$, and the $y$-intercept is 6.

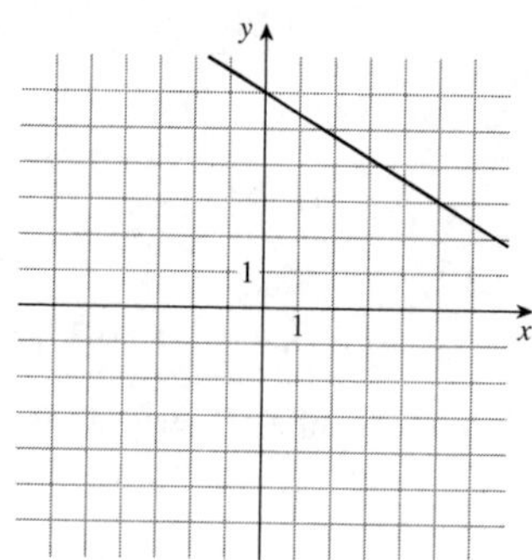


**47.** $y = 4$ can also be expressed as $y = 0x + 4$. So the slope is 0, and the $y$-intercept is 4.

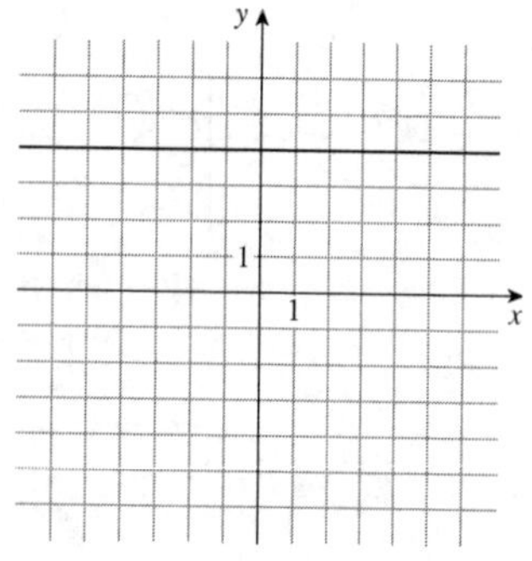


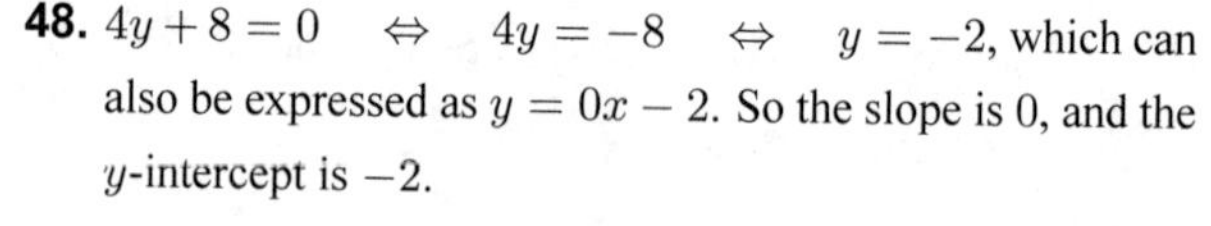

**48.** $4y + 8 = 0 \quad \Leftrightarrow \quad 4y = -8 \quad \Leftrightarrow \quad y = -2$, which can also be expressed as $y = 0x - 2$. So the slope is 0, and the $y$-intercept is $-2$.

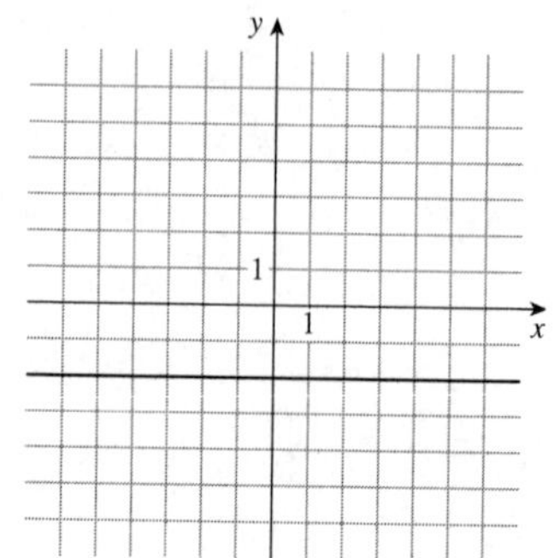

**49.** $3x - 4y = 12 \quad \Leftrightarrow \quad -4y = -3x + 12 \quad \Leftrightarrow$ $y = \frac{3}{4}x - 3$. So the slope is $\frac{3}{4}$, and the $y$-intercept is $-3$.

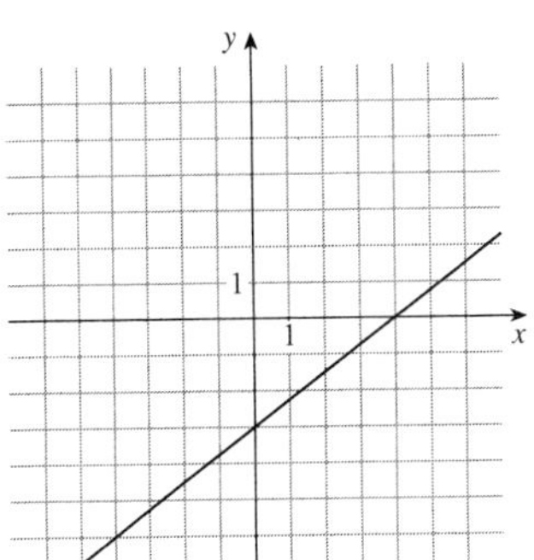

**50.** $x = -5$ cannot be expressed in the form $y = mx + b$. So the slope is undefined, and there is no $y$-intercept. This is a vertical line.

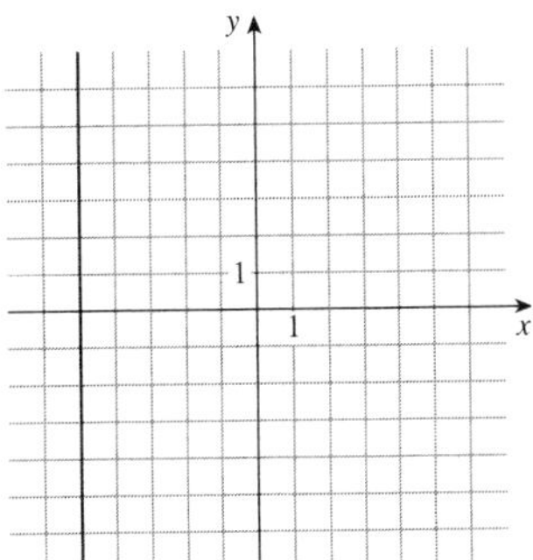

**51.** $3x + 4y - 1 = 0 \quad \Leftrightarrow \quad 4y = -3x + 1 \quad \Leftrightarrow$ $y = -\frac{3}{4}x + \frac{1}{4}$. So the slope is $-\frac{3}{4}$, and the $y$-intercept is $\frac{1}{4}$.

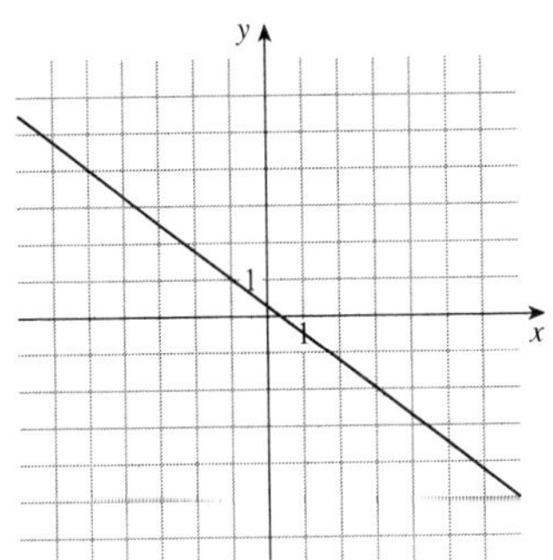

**52.** $4x + 5y = 10 \quad \Leftrightarrow \quad 5y = -4x + 10 \quad \Leftrightarrow$ $y = -\frac{4}{5}x + 2$. So the slope is $-\frac{4}{5}$, and the $y$-intercept is 2.

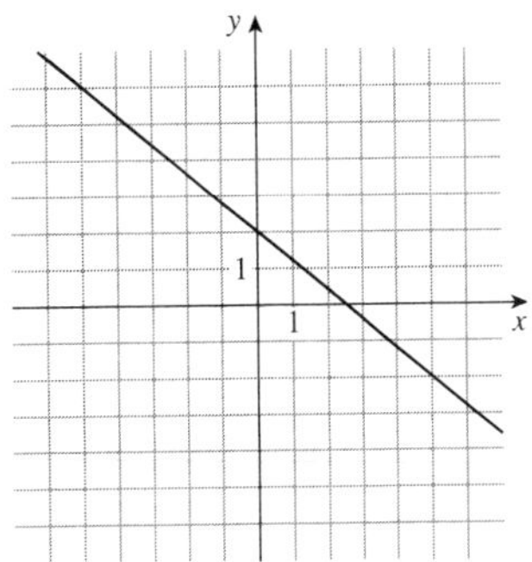

**53.** We first plot the points to find the pairs of points that determine each side. Next we find the slopes of opposite sides. The slope of $AB$ is $\dfrac{4-1}{7-1} = \dfrac{3}{6} = \dfrac{1}{2}$, and the slope of $DC$ is $\dfrac{10-7}{5-(-1)} = \dfrac{3}{6} = \dfrac{1}{2}$. Since these slope are equal, these two sides are parallel. The slope of $AD$ is $\dfrac{7-1}{-1-1} = \dfrac{6}{-2} = -3$, and the slope of $BC$ is $\dfrac{10-4}{5-7} = \dfrac{6}{-2} = -3$. Since these slope are equal, these two sides are parallel. Hence $ABCD$ is a parallelogram.

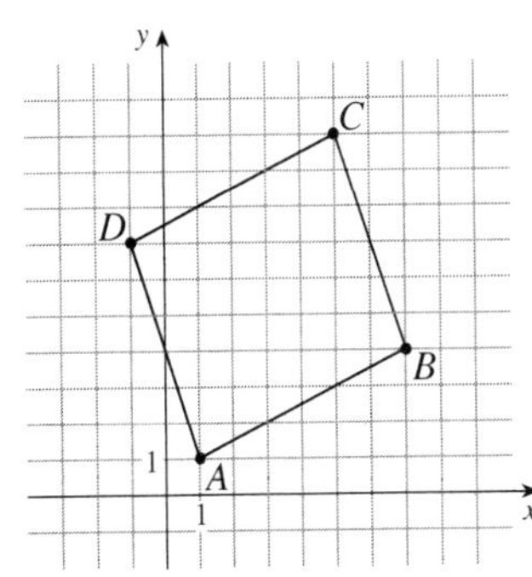

**54.** We first plot the points to determine the perpendicular sides. Next find the slopes of the sides. The slope of $AB$ is $\dfrac{3-(-1)}{3-(-3)} = \dfrac{4}{6} = \dfrac{2}{3}$, and the slope of $AC$ is $\dfrac{8-(-1)}{-9-(-3)} = \dfrac{9}{-6} = -\dfrac{3}{2}$. Since (slope of $AB$) $\times$ (slope of $AC$) $= \left(\frac{2}{3}\right)\left(-\frac{3}{2}\right) = -1$, the sides are perpendicular, and $ABC$ is a right triangle.

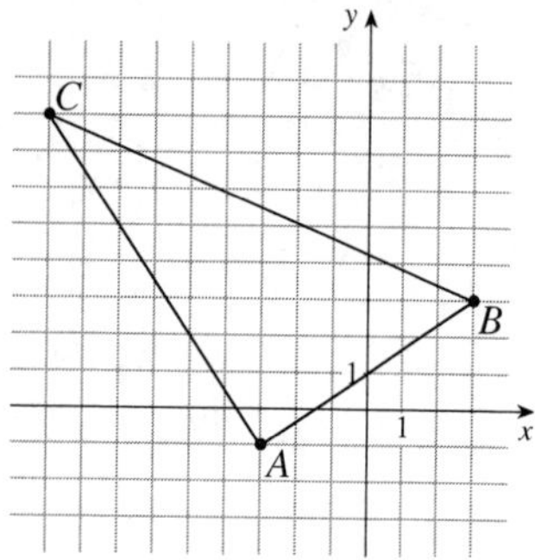


**55.** We first plot the points to find the pairs of points that determine each side. Next we find the slopes of opposite sides. The slope of $AB$ is $\dfrac{3-1}{11-1} = \dfrac{2}{10} = \dfrac{1}{5}$ and the slope of $DC$ is $\dfrac{6-8}{0-10} = \dfrac{-2}{-10} = \dfrac{1}{5}$. Since these slope are equal, these two sides are parallel. Slope of $AD$ is $\dfrac{6-1}{0-1} = \dfrac{5}{-1} = -5$, and the slope of $BC$ is $\dfrac{3-8}{11-10} = \dfrac{-5}{1} = -5$. Since these slope are equal, these two sides are parallel. Since (slope of $AB$) $\times$ (slope of $AD$) $= \frac{1}{5} \times (-5) = -1$, the first two sides are each perpendicular to the second two sides. So the sides form a rectangle.

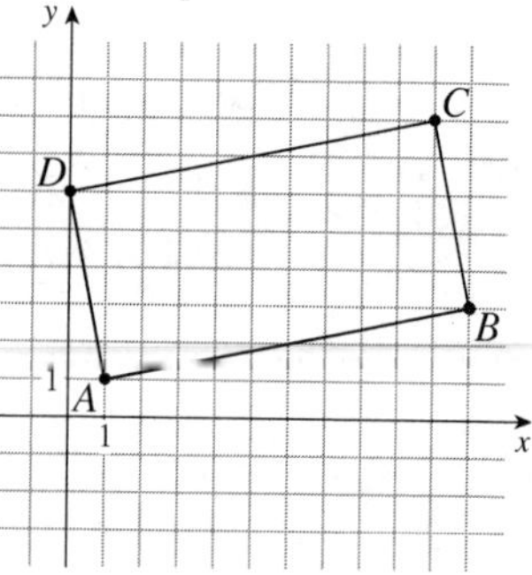


**56. (a)** The slope of the line passing through $(1,1)$ and $(3,9)$ is $\dfrac{9-1}{3-1} = \dfrac{8}{2} = 4$. The slope of the line passing through $(1,1)$ and $(6,21)$ is $\dfrac{21-1}{6-1} = \dfrac{20}{5} = 4$. Since the slopes are equal, the points are collinear.

**(b)** The slope of the line passing through $(-1,3)$ and $(1,7)$ is $\dfrac{7-3}{1-(-1)} = \dfrac{4}{2} = 2$. The slope of the line passing through $(-1,3)$ and $(4,15)$ is $\dfrac{15-3}{4-(-1)} = \dfrac{12}{5}$. Since the slopes are not equal, the points are not collinear.

**57.** We need the slope and the midpoint of the line $AB$. The midpoint of $AB$ is $\left(\dfrac{1+7}{2}, \dfrac{4-2}{2}\right) = (4,1)$, and the slope of $AB$ is $m = \dfrac{-2-4}{7-1} = \dfrac{-6}{6} = -1$. The slope of the perpendicular bisector will have slope $\dfrac{-1}{m} = \dfrac{-1}{-1} = 1$. Using the point-slope form, the equation of the perpendicular bisector is $y - 1 = 1(x-4)$ or $x - y - 3 = 0$.

**58.** We find the intercepts (the length of the sides). When $x = 0$, we have $2y + 3(0) - 6 = 0 \quad \Leftrightarrow \quad 2y = 6 \quad \Leftrightarrow \quad y = 3$, and when $y = 0$, we have $2(0) + 3x - 6 = 0 \quad \Leftrightarrow \quad 3x = 6 \quad \Leftrightarrow \quad x = 2$. Thus, the area of the triangle is $\frac{1}{2}(3)(2) = 3$.

**59. (a)** We start with the two points $(a,0)$ and $(0,b)$. The slope of the line that contains them is $\dfrac{b-0}{0-a} = -\dfrac{b}{a}$. So the equation of the line containing them is $y = -\dfrac{b}{a}x + b$ (using the slope-intercept form). Dividing by $b$ (since $b \neq 0$) gives $\dfrac{y}{b} = -\dfrac{x}{a} + 1 \quad \Leftrightarrow \quad \dfrac{x}{a} + \dfrac{y}{b} = 1$.

**(b)** Setting $a = 6$ and $b = -8$, we get $\dfrac{x}{6} + \dfrac{y}{-8} = 1 \quad \Leftrightarrow \quad 4x - 3y = 24 \quad \Leftrightarrow \quad 4x - 3y - 24 = 0$.

**60. (a)** The line tangent at $(3,-4)$ will be perpendicular to the line passing through the points $(0,0)$ and $(3,-4)$. The slope of this line is $\dfrac{-4-0}{3-0}=-\dfrac{4}{3}$. Thus, the slope of the tangent line will be $-\dfrac{1}{(-4/3)}=\dfrac{3}{4}$. Then the equation of the tangent line is $y-(-4)=\frac{3}{4}(x-3) \quad\Leftrightarrow\quad 4(y+4)=3(x-3) \quad\Leftrightarrow\quad 3x-4y-25=0$.

**(b)** Since diametrically opposite points on the circle have parallel tangent lines, the other point is $(-3,4)$.

**61.** Let $h$ be the change in your horizontal distance, in feet. Then $-\frac{6}{100}=\dfrac{-1000}{h} \quad\Leftrightarrow\quad h=\dfrac{100,000}{6}\approx 16{,}667$. So the change in your horizontal distance is about 16,667 feet.

**62. (a)** The slope represents the increase in the average surface temperature in degrees in $^\circ$ C per year. The $T$ -intercept is the average surface temperature in 1900, or $8.5^\circ$ C.

**(b)** In 2100, $T=2100-1900=200$, so $T=0.02\,(200)+8.50=12.5^\circ$ C.

**63. (a)** The slope is $0.0417D=0.0417\,(200)=8.34$. It represents the increase in dosage for each one-year increase in the child's age.

**(b)** When $a=0$, $c=8.34\,(0+1)=8.34$ mg.

**64. (a)**

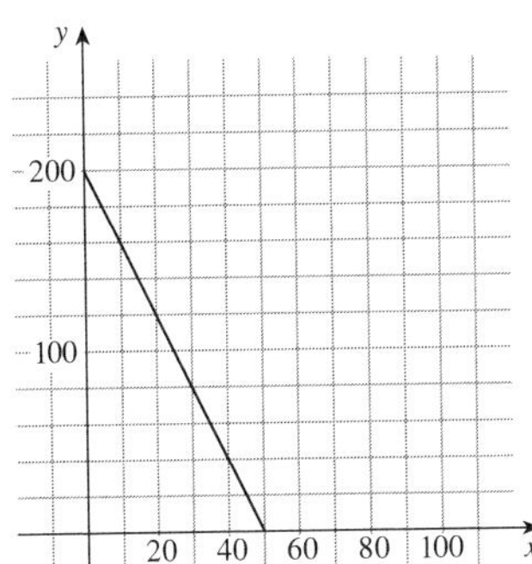


**(b)** The slope, $-4$, represents the decline in number of spaces sold for each \$1 increase in rent. The $y$-intercept is the number of spaces at the flea market, 200, and the $x$-intercept is the cost per space when the manager rents no spaces, \$50.

**65. (a)**

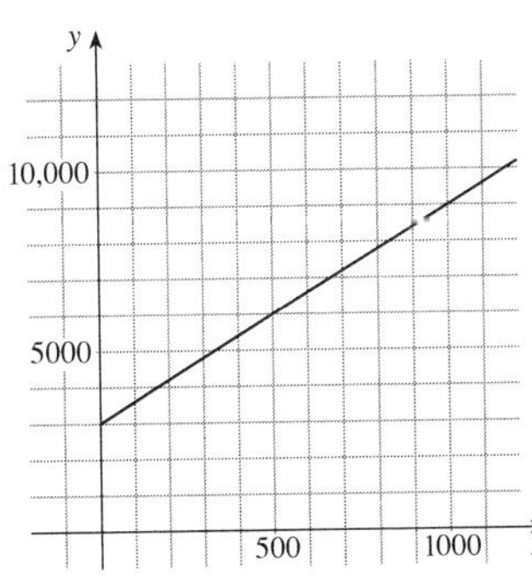


**(b)** The slope is the cost per toaster oven, \$6. The $y$-intercept, \$3000, is the monthly fixed cost — the cost that is incurred no matter how many toaster ovens are produced.

**66. (a)**

| C | $-30^\circ$ | $-20^\circ$ | $-10^\circ$ | $0^\circ$ | $10^\circ$ | $20^\circ$ | $30^\circ$ |
|---|---|---|---|---|---|---|---|
| F | $-22^\circ$ | $-4^\circ$ | $14^\circ$ | $32^\circ$ | $50^\circ$ | $68^\circ$ | $86^\circ$ |

**(b)** Substituting $a$ for both F and C, we have $a=\frac{9}{5}a+32 \quad\Leftrightarrow\quad -\frac{4}{5}a=32 \quad\Leftrightarrow\quad a=-40^\circ$. Thus both scales agree at $-40^\circ$.

**67. (a)** Using $n$ in place of $x$ and $t$ in place of $y$, we find that the slope is $\dfrac{t_2-t_1}{n_2-n_1}=\dfrac{80-70}{168-120}=\dfrac{10}{48}=\dfrac{5}{24}$. So the linear equation is $t-80=\frac{5}{24}(n-168) \quad\Leftrightarrow\quad t-80=\frac{5}{24}n-35 \quad\Leftrightarrow\quad t=\frac{5}{24}n+45$.

**(b)** When $n=150$, the temperature is approximately given by $t=\frac{5}{24}(150)+45=76.25^\circ\text{ F}\approx 76^\circ$ F.

**68. (a)** Using $t$ in place of $x$ and $V$ in place of $y$, we find the slope of the line using the points $(0, 4000)$ and $(4, 200)$. Thus, the slope is $m = \dfrac{200 - 4000}{4 - 0} = \dfrac{-3800}{4} = -950$. Using the $V$-intercept, the linear equation is $V = -950t + 4000$.

**(b)**

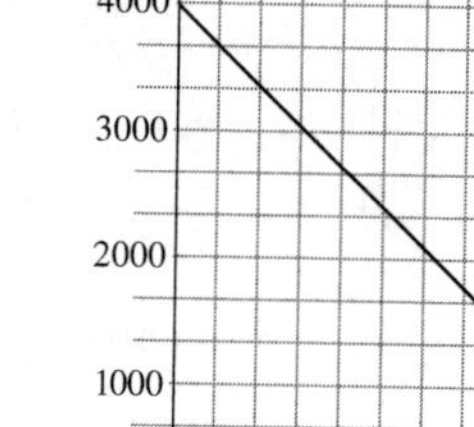


**(c)** The slope represents the rate of depreciation of the computer. The $V$ -intercept represents the cost of the computer.

**(d)** When $t = 3$, the value of the computer is given by $V = -950\,(3) + 4000 = 1150$.

**69. (a)** We are given $\dfrac{\text{change in pressure}}{\text{10 feet change in depth}} = \dfrac{4.34}{10} = 0.434$. Using $P$ for pressure and $d$ for depth, and using the point $P = 15$ when $d = 0$, we have $P - 15 = 0.434\,(d - 0) \quad \Leftrightarrow \quad P = 0.434d + 15$.

**(b)**

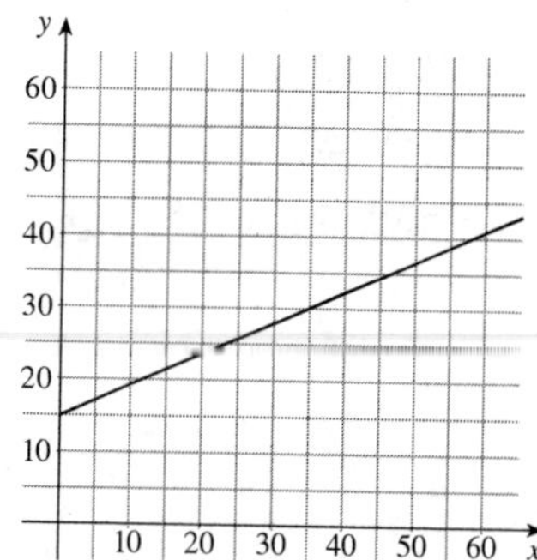


**(c)** The slope represents the increase in pressure per foot of descent. The $y$-intercept represents the pressure at the surface.

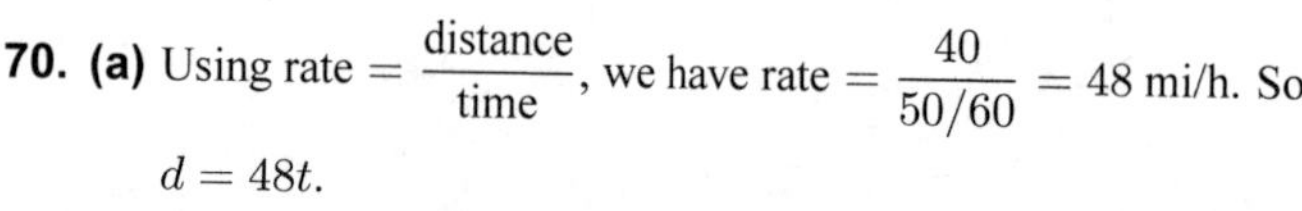

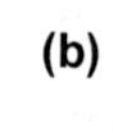

**(d)** When $P = 100$, then $100 = 0.434d + 15 \quad \Leftrightarrow \quad 0.434d = 85 \quad \Leftrightarrow \quad d = 195.9$ ft. Thus the pressure is 100 lb/in$^3$ at a depth of approximately 196 ft.

**70. (a)** Using rate $= \dfrac{\text{distance}}{\text{time}}$, we have rate $= \dfrac{40}{50/60} = 48$ mi/h. So $d = 48t$.

**(b)**

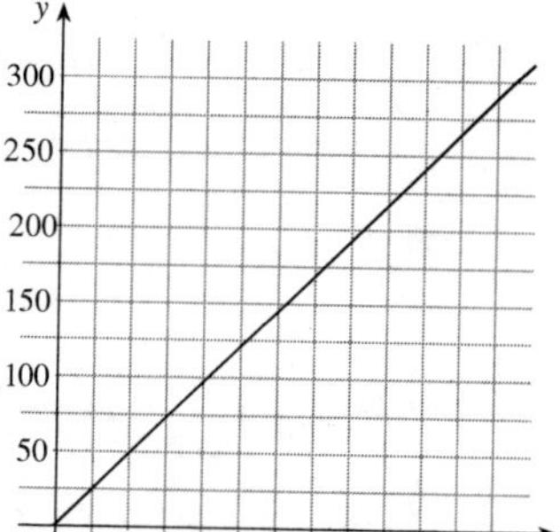


**(c)** The slope of the line is 48 and it represents the speed in mi/h.

**71. (a)** Using $d$ in place of $x$ and $C$ in place of $y$, we find the slope to be $\dfrac{C_2 - C_1}{d_2 - d_1} = \dfrac{460 - 380}{800 - 480} = \dfrac{80}{320} = \dfrac{1}{4}$. So the linear equation is $C - 460 = \frac{1}{4}\,(d - 800) \quad \Leftrightarrow \quad C - 460 = \frac{1}{4}d - 200 \quad \Leftrightarrow \quad C = \frac{1}{4}d + 260$.

**(b)** Substituting $d = 1500$ we get $C = \frac{1}{4}\,(1500) + 260 = 635$. Thus, the cost of driving 1500 miles is \$635.

**(c)**

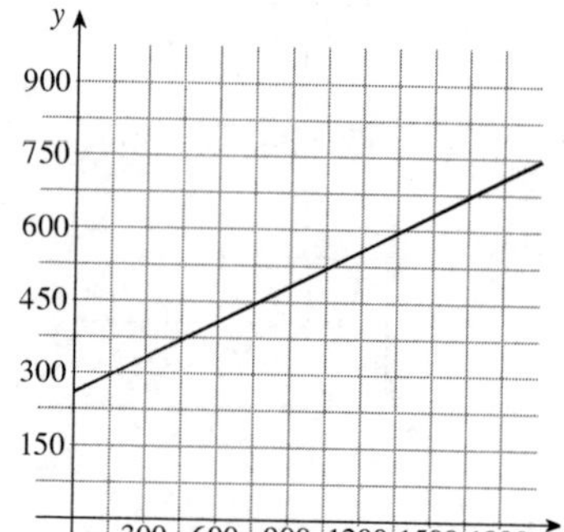


The slope of the line represents the cost per mile, \$0.25.

**(d)** The $y$-intercept represents the fixed cost, \$260.

**(e)** It is a suitable model because you have fixed monthly costs such as insurance and car payments, as well as costs that occur as you drive, such as gasoline, oil, tires, etc., and the cost of these for each additional mile driven is a constant.

**72.** **(a)** Using the points $(100, 2200)$ and $(300, 4800)$, we find that the slope is $\dfrac{4800-2200}{300-100} = \frac{2600}{200} = 13$. So $y - 2200 = 13(x - 100) \quad \Leftrightarrow$ $y = 13x + 900$.

**(b)** The slope of the line in part (a) is 13, and it represents the cost of producing each additional chair.

**(c)** The $y$-intercept is 900, and it represents the fixed daily costs of operating the factory.

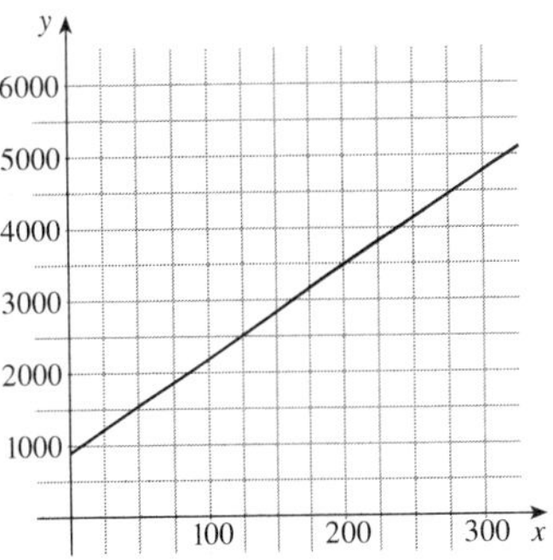


**73.** Slope is the rate of change of one variable per unit change in another variable. So if the slope is positive, then the temperature is rising. Likewise, if the slope is negative then the temperature is decreasing. If the slope is 0, then the temperature is not changing.

**74.** We label the three points $A$, $B$, and $C$. If the slope of the line segment $\overline{AB}$ is equal to the slope of the line segment $\overline{BC}$, then the points $A$, $B$, and $C$ are collinear. Using the distance formula, we find the distance between $A$ and $B$, between $B$ and $C$, and between $A$ and $C$. If the sum of the two smaller distances equals the largest distance, the points $A$, $B$, and $C$ are collinear.

*Another method:* Find an equation for the line through $A$ and $B$. Then check if $C$ satisfies the equation. If so, the points are collinear.

## 2.5 Modeling Variation

**1.** $T = kx$, where $k$ is constant.

**2.** $P = kw$, where $k$ is constant.

**3.** $v = \dfrac{k}{z}$, where $k$ is constant.

**4.** $w = kmn$, where $k$ is constant.

**5.** $y = \dfrac{ks}{t}$, where $k$ is constant.

**6.** $P = \dfrac{k}{T}$, where $k$ is constant.

**7.** $z = k\sqrt{y}$, where $k$ is constant.

**8.** $A = k\dfrac{t^2}{x^3}$, where $k$ is constant.

**9.** $V = klwh$, where $k$ is constant.

**10.** $S = kr^2\theta^2$, where $k$ is constant.

**11.** $R = \dfrac{ki}{Pt}$, where $k$ is constant.

**12.** $A = k\sqrt{xy}$, where $k$ is constant.

**13.** Since $y$ is directly proportional to $x$, $y = kx$. Since $y = 42$ when $x = 6$, we have $42 = k(6) \quad \Leftrightarrow \quad k = 7$. So $y = 7x$.

**14.** $z = \dfrac{k}{t}$. Since $z = 5$ when $t = 3$, we have $5 = \dfrac{k}{3} \quad \Leftrightarrow \quad k = 15$, so $z = \dfrac{15}{t}$.

**15.** Since $M$ varies directly as $x$ and inversely as $y$, $M = \dfrac{kx}{y}$. Since $M = 5$ when $x = 2$ and $y = 6$, we have $5 = \dfrac{k(2)}{6} \quad \Leftrightarrow$ $k = 15$. Therefore $M = \dfrac{15x}{y}$.

**16.** $S = kpq$. Since $S = 180$ when $p = 4$ and $q = 5$, we have $180 = k(4)(5) \quad \Leftrightarrow \quad 180 = 20k \quad \Leftrightarrow \quad k = 9$. So $S = 9pq$.

**17.** Since $W$ is inversely proportional to the square of $r$, $W = \dfrac{k}{r^2}$. Since $W = 10$ when $r = 6$, we have $10 = \dfrac{k}{(6)^2} \quad \Leftrightarrow$ $k = 360$. So $W = \dfrac{360}{r^2}$.

**18.** $t = k\dfrac{xy}{r}$. Since $t = 25$ when $x = 2$, $y = 3$, and $r = 12$, we have $25 = k\dfrac{(2)(3)}{12} \Leftrightarrow k = 50$. So $t = 50\dfrac{xy}{r}$.

**19.** Since $C$ is jointly proportional to $l$, $w$, and $h$, we have $C = klwh$. Since $C = 128$ when $l = w = h = 2$, we have $128 = k(2)(2)(2) \Leftrightarrow 128 = 8k \Leftrightarrow k = 16$. Therefore, $C = 16lwh$.

**20.** $H = kl^2w^2$. Since $H = 36$ when $l = 2$ and $w = \frac{1}{3}$, we have $36 = k(2)^2\left(\frac{1}{3}\right)^2 \Leftrightarrow 36 = \frac{4}{9}k \Leftrightarrow k = 81$. So $H = 81l^2w^2$.

**21.** Since $s$ is inversely proportional to the square root of $t$, we have $s = \dfrac{k}{\sqrt{t}}$. Since $s = 100$ when $t = 25$, we have $100 = \dfrac{k}{\sqrt{25}} \Leftrightarrow 100 = \dfrac{k}{5} \Leftrightarrow k = 500$. So $s = \dfrac{500}{\sqrt{t}}$.

**22.** $M = k\dfrac{abc}{d}$. Since $M = 128$ when $a = d$ and $b = c = 2$, we have $128 = k\dfrac{a(2)(2)}{a} = 4k \Leftrightarrow k = 32$. So $M = 32\dfrac{abc}{d}$.

**23.** **(a)** The force $F$ needed is $F = kx$.

**(b)** Since $F = 40$ when $x = 5$, we have $40 = k(5) \Leftrightarrow k = 8$.

**(c)** From part (b), we have $F = 8x$. Substituting $x = 4$ into $F = 8x$ gives $F = 8(4) = 32$ N.

**24.** **(a)** Let $T$ and $l$ be the period and the length of the pendulum, respectively. Then $T = k\sqrt{l}$.

**(b)** $T = k\sqrt{l} \Rightarrow T^2 = k^2 l \Leftrightarrow l = \dfrac{T^2}{k^2}$. If the period is doubled, the new length is $\dfrac{(2T)^2}{k^2} = 4\dfrac{T^2}{k^2} = 4l$. So we would quadruple the length $l$ to double the period $T$.

**25.** **(a)** $C = kpm$

**(b)** Since $C = 60{,}000$ when $p = 120$ and $m = 4000$, we get $60{,}000 = k(120)(4000) \Leftrightarrow k = \frac{1}{8}$. So $C = \frac{1}{8}pm$.

**(c)** Substituting $p = 92$ and $m = 5{,}000$, we get $C = \frac{1}{8}(92)(5{,}000) = \$57{,}500$ and $k = \frac{1}{8}$.

**26.** **(a)** $P = \dfrac{kT}{V}$.

**(b)** Substituting $P = 33.2$, $T = 400$, and $V = 100$, we get $33.2 = \dfrac{k(400)}{100} \Leftrightarrow k = 8.3$. Thus $k = 8.3$ and the equation is $P = \dfrac{8.3T}{V}$.

**(c)** Substituting $T = 500$ and $V = 80$, we have $P = \dfrac{8.3(500)}{80} = 51.875$ kPa. Hence the pressure of the sample of gas is about 51.9 kPa.

**27.** **(a)** $P = ks^3$.

**(b)** Since $P = 96$ when $s = 20$, we get $96 = k \cdot 20^3 \Leftrightarrow k = 0.012$. So $P = 0.012s^3$.

**(c)** Substituting $x = 30$, we get $P = 0.012 \cdot 30^3 = 324$ watts.

**28.** $P = ks^3$. Since $P = 80$ when $s = 10$ we have $80 = k(10)^3$ so $k = 0.08$. Thus $P = 0.08s^3$. When $s = 15$ we have $P = 0.08(15)^3 = 270$. So 270 hp is needed to power the boat at 15 knots.

**29.** $L = \dfrac{k}{d^2}$. Since $L = 70$ when $d = 10$, we have $70 = \dfrac{k}{10^2}$ so $k = 7{,}000$. Thus $L = \dfrac{7{,}000}{d^2}$. When $d = 100$ we get $L = \dfrac{7{,}000}{100^2} = 0.7$ dB.

**30.** $D = ks^2$. Since $D = 240$ when $s = 50$ we have $240 = k(50)^2$ so $k = 0.096$. Thus $D = 0.096s^2$. When $D = 160$ then $160 = 0.096s^2 \Leftrightarrow s^2 = 1666.7$ so $s \approx 40$ mi/h (for safety reasons we round down).

**31.** $P = kAv^3$. If $A = \frac{1}{2}A_0$ and $v = 2v_0$, then $P = k\left(\frac{1}{2}A_0\right)(2v_0)^3 = \frac{1}{2}kA_0\left(8v_0^3\right) = 4kA_0v_0^3$. The power is increased by a factor of 4.

**32.** $L = ks^2A$. Since $L = 1700$ when $s = 50$ and $A = 500$, we have $1700 = k\left(50^2\right)(500) \quad\Leftrightarrow\quad k = 0.00136$. Thus $L = 0.00136s^2A$. When $A = 600$ and $s = 40$ we get the lift is $L = 0.00136\left(40^2\right)(600) = 1305.6$ lb.

**33.** $F = kAs^2$. Since $F = 220$ when $A = 40$ and $s = 5$. Solving for $k$ we have $220 = k(40)(5)^2 \quad\Leftrightarrow\quad 220 = 1000k \quad\Leftrightarrow\quad k = 0.22$. Now when $A = 28$ and $F = 175$ we get $175 = 0.220(28)s^2 \quad\Leftrightarrow\quad 28.4090 = s^2$ so $s = \sqrt{28.4090} = 5.33$ mi/h.

**34. (a)** $F = k\dfrac{ws^2}{r}$

**(b)** For the first car we have $w_1 = 1600$ and $s_1 = 60$ and for the second car we have $w_2 = 2500$. Since the forces are equal we have $k\dfrac{1600 \cdot 60^2}{r} = k\dfrac{2500 \cdot s_2^2}{r} \quad\Leftrightarrow\quad \dfrac{16 \cdot 60^2}{25} = s_2^2$, so $s_2 = 48$ mi/h.

**35. (a)** $R = \dfrac{kL}{d^2}$

**(b)** Since $R = 140$ when $L = 1.2$ and $d = 0.005$, we get $140 = \dfrac{k(1.2)}{(0.005)^2} \quad\Leftrightarrow\quad k = \frac{7}{2400} = 0.00291\overline{6}$.

**(c)** Substituting $L = 3$ and $d = 0.008$, we have $R = \dfrac{7}{2400} \cdot \dfrac{3}{(0.008)^2} = \dfrac{4375}{32} \approx 137\ \Omega$.

**36. (a)** $T^2 = kd^3$

**(b)** Substituting $T = 365$ and $d = 93 \times 10^6$, we get $365^2 = k \cdot \left(93 \times 10^6\right)^3 \quad\Leftrightarrow\quad k = 1.66 \times 10^{-19}$.

**(c)** $T^2 = 1.66 \times 10^{-19}\left(2.79 \times 10^9\right)^3 = 3.60 \times 10^9 \quad\Rightarrow\quad T = 6.00 \times 10^4$. Hence the period of Neptune is $6.00 \times 10^4$ days$\approx$ 164 years.

**37. (a)** For the sun, $E_S = k6000^4$ and for earth $E_E = k300^4$. Thus $\dfrac{E_S}{E_E} = \dfrac{k6000^4}{k300^4} = \left(\frac{6000}{300}\right)^4 = 20^4 = 160{,}000$. So the sun produces 160,000 times the radiation energy per unit area than the Earth.

**(b)** The surface area of the sun is $4\pi(435{,}000)^2$ and the surface area of the Earth is $4\pi(3{,}960)^2$. So the sun has $\dfrac{4\pi(435{,}000)^2}{4\pi(3{,}960)^2} = \left(\dfrac{435{,}000}{3{,}960}\right)^2$ times the surface area of the Earth. Thus the total radiation emitted by the sun is $160{,}000 \times \left(\dfrac{435{,}000}{3{,}960}\right)^2 = 1{,}930{,}670{,}340$ times the total radiation emitted by the Earth.

**38.** Let $V$ be the value of a building lot on Galiano Island, $A$ the area of the lot, and $q$ the quantity of the water produced. Since $V$ is jointly proportional to the area and water quantity, we have $V = kAq$. When $A = 200 \cdot 300 = 60{,}000$ and $q = 10$, we have $V = \$48{,}000$, so $48{,}000 = k(60{,}000)(10) \quad\Leftrightarrow\quad k = 0.08$. Thus $V = 0.08Aq$. Now when $A = 400 \cdot 400 = 160{,}000$ and $q = 4$, the value is $V = 0.08(160{,}000)(4) = \$51{,}200$.

**39.** Let $S$ be the final size of the cabbage, in pounds, let $N$ be the amount of nutrients it receives, in ounces, and let $c$ be the number of other cabbages around it. Then $S = k\dfrac{N}{c}$. When $N = 20$ and $c = 12$, we have $S = 30$, so substituting, we have $30 = k\frac{20}{12} \quad\Leftrightarrow\quad k = 18$. Thus $S = 18\dfrac{N}{c}$. When $N = 10$ and $c = 5$, the final size is $S = 18\left(\frac{10}{5}\right) = 36$ lb.

**40.** Let $H$ be the heat experienced by a hiker at a campfire, let $A$ be the amount of wood, and let $d$ be the distance from campfire. So $H = k\dfrac{A}{d^3}$. When the hiker is 20 feet from the fire, the heat experienced is $H = k\dfrac{A}{20^3}$, and when the amount of wood is doubled, the heat experienced is $H = k\dfrac{2A}{d^3}$. So $k\dfrac{A}{8{,}000} = k\dfrac{2A}{d^3} \quad\Leftrightarrow\quad d^3 = 16{,}000 \quad\Leftrightarrow\quad d = 20\sqrt[3]{2} \approx 25.2$ feet.

**41. (a)** Since $f$ is inversely proportional to $L$, we have $f = \frac{k}{L}$, where $k$ is a positive constant.

**(b)** If we replace $L$ by $2L$ we have $\frac{k}{2L} = \frac{1}{2} \cdot \frac{k}{L} = \frac{1}{2}f$. So the frequency of the vibration is cut in half.

**42. (a)** Since $r$ is jointly proportional to $x$ and $P - x$, we have $r = kx\,(P - x)$, where $k$ is a positive constant.

**(b)** When 10 people are infected the rate is $r = k10\,(5000 - 10) = 49{,}900k$. When 1000 people are infected the rate is $r = k \cdot 1000 \cdot (5000 - 1000) = 4{,}000{,}000k$. So the rate is much higher when 1000 people are infected. Comparing these rates, we find that $\frac{1000 \text{ people infected}}{10 \text{ people infected}} = \frac{4{,}000{,}000k}{49{,}900k} \approx 80$. So the infection rate when 1000 people are infected is about 80 times as large as when 10 people are infected.

**(c)** When the entire population is infected the rate is $r = k\,(5000)\,(5000 - 5000) = 0$. This makes sense since there are no more people who can be infected.

**43.** Examples include radioactive decay and exponential growth in biology.

# Chapter 2 Review

**1. (a)**

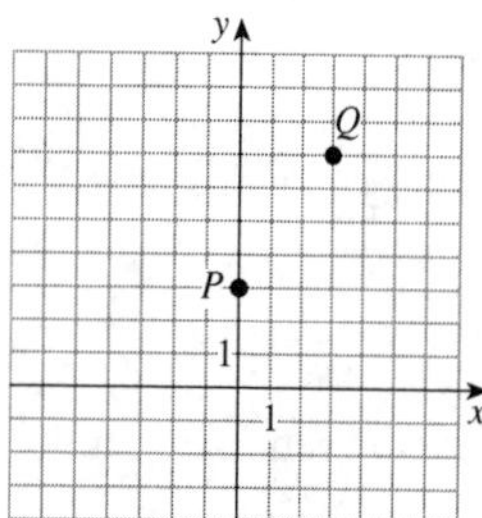


**(b)** The distance from $P$ to $Q$ is

$$d\,(P, Q) = \sqrt{(0 - 3)^2 + (3 - 7)^2} = \sqrt{9 + 16} = \sqrt{25} = 5$$

**(c)** The midpoint is $\left(\frac{0+3}{2}, \frac{3+7}{2}\right) = \left(\frac{3}{2}, 5\right)$.

**(d)** The line has slope $m = \frac{3 - 7}{0 - 3} = \frac{4}{3}$ and equation $y - 3 = \frac{4}{3}\,(x - 0) \Leftrightarrow y - 3 = \frac{4}{3}x \Leftrightarrow y = \frac{4}{3}x + 3$.

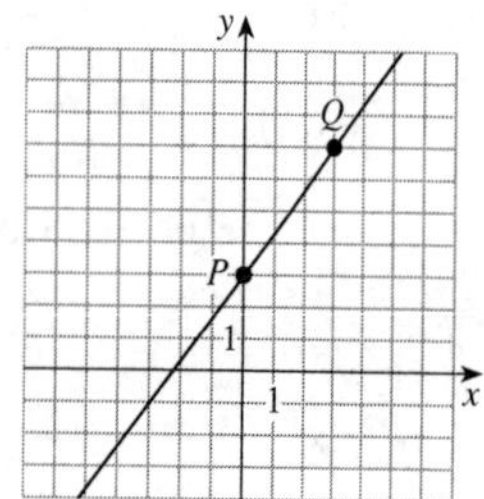


**(e)** The radius of this circle was found in part (b). It is $r = d\,(P, Q) = 5$. So an equation is $(x - 0)^2 + (y - 3)^2 = (5)^2 \Leftrightarrow x^2 + (y - 3)^2 = 25$.

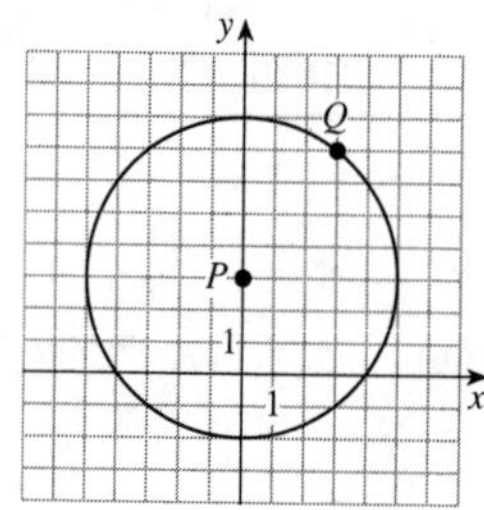

**2. (a)**

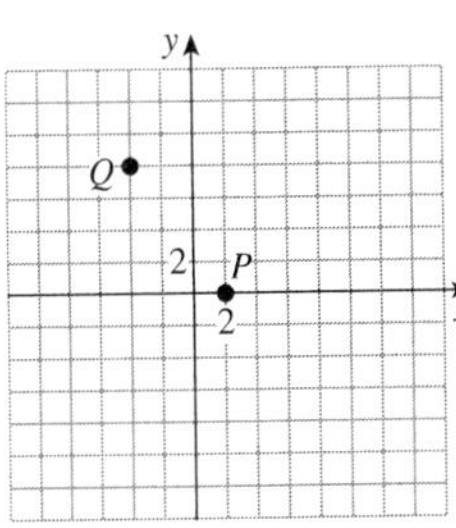


**(b)** The distance from $P$ to $Q$ is

$$d(P,Q) = \sqrt{[2-(-4)]^2 + (0-8)^2} = \sqrt{36+64} = \sqrt{100} = 10$$

**(c)** The midpoint is $\left(\dfrac{2+(-4)}{2}, \dfrac{0+8}{2}\right) = (-1, 4)$.

**(d)** The line has slope $m = \dfrac{0-8}{2-(-4)} = \dfrac{-8}{6} = -\frac{4}{3}$, and equation $y - 0 = -\frac{4}{3}(x-2) \Leftrightarrow y = -\frac{4}{3}x + \frac{8}{3}$.

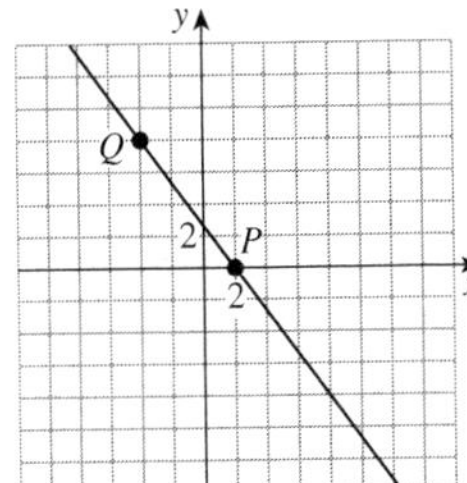


**(e)** The radius of this circle was found in part (b). It is $r = d(P,Q) = 10$. So an equation is $(x-2)^2 + (y-0)^2 = (10)^2 \Leftrightarrow (x-2)^2 + y^2 = 100$.

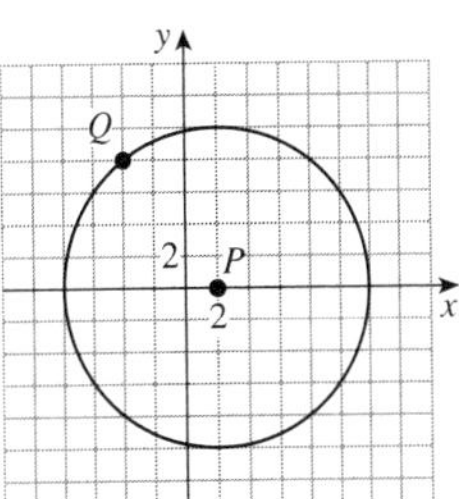


**3. (a)**

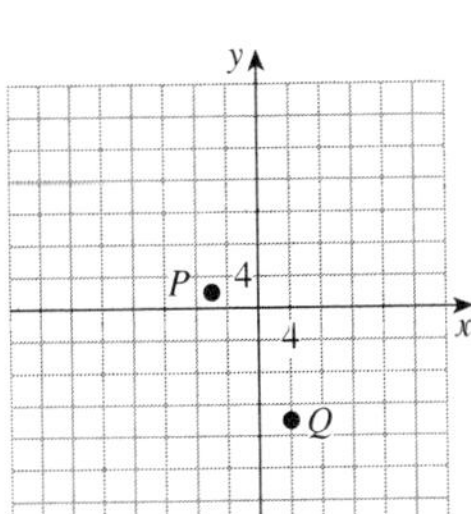


**(b)** The distance from $P$ to $Q$ is

$$d(P,Q) = \sqrt{(-6-4)^2 + [2-(-14)]^2} = \sqrt{100+256} = \sqrt{356} = 2\sqrt{89}$$

**(c)** The midpoint is $\left(\dfrac{-6+4}{2}, \dfrac{2+(-14)}{2}\right) = (-1, -6)$.

**(d)** The line has slope $m = \dfrac{2-(-14)}{-6-4} = \dfrac{16}{-10} = -\dfrac{8}{5}$ and equation $y - 2 = -\frac{8}{5}(x+6) \Leftrightarrow y - 2 = -\frac{8}{5}x - \frac{48}{5} \Leftrightarrow y = -\frac{8}{5}x - \frac{38}{5}$.

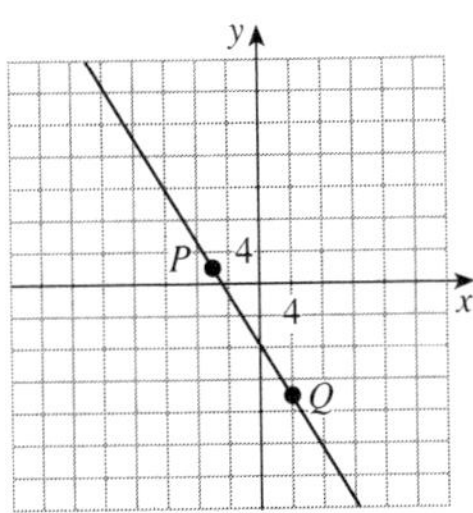


**(e)** The radius of this circle was found in part (b). It is $r = d(P,Q) = 2\sqrt{89}$. So an equation is $[x-(-6)]^2 + (y-2)^2 = \left(2\sqrt{89}\right)^2 \Leftrightarrow (x+6)^2 + (y-2)^2 = 356$.

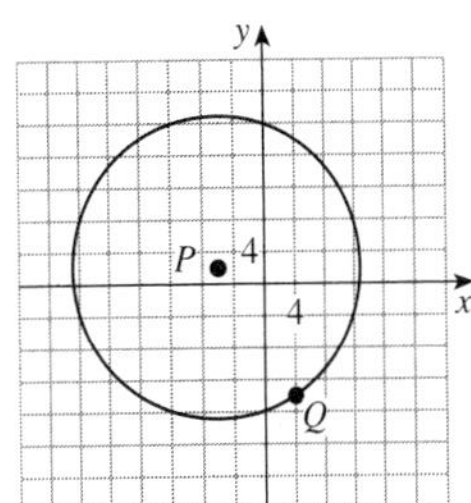

**4. (a)**

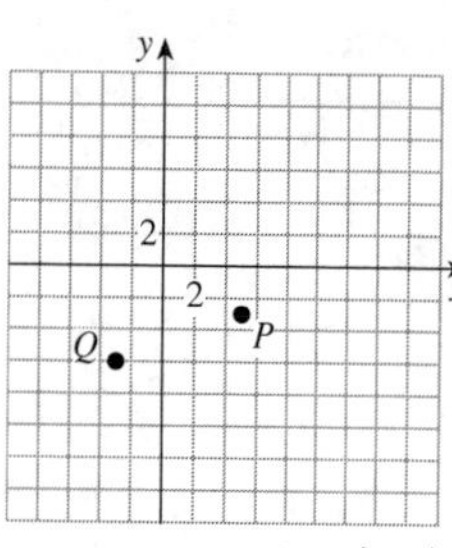


**(b)** The distance from $P$ to $Q$ is

$$d(P,Q) = \sqrt{[5-(-3)]^2 + [-2-(-6)]^2} = \sqrt{64+16} = \sqrt{80} = 4\sqrt{5}.$$

**(c)** The midpoint is $\left(\dfrac{5+(-3)}{2}, \dfrac{-2+(-6)}{2}\right) = (1,-4)$.

**(d)** The line has slope $m = \dfrac{-2-(-6)}{5-(-3)} = \frac{4}{8} = \frac{1}{2}$, and has equation $y-(-2) = \frac{1}{2}(x-5) \Leftrightarrow y+2 = \frac{1}{2}x - \frac{5}{2} \Leftrightarrow y = \frac{1}{2}x - \frac{9}{2}$.

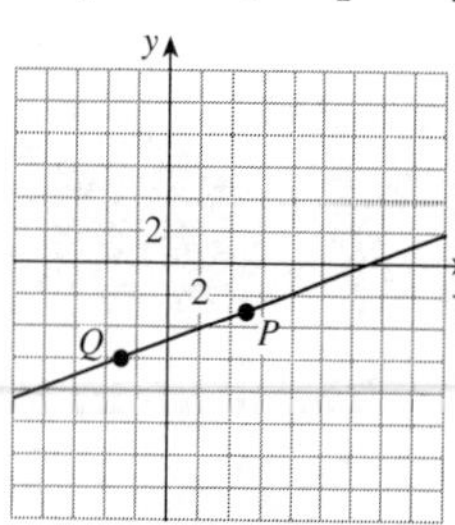


**(e)** The radius of this circle was found in part (b). It is $r = d(P,Q) = 4\sqrt{5}$. So an equation is $(x-5)^2 + [y-(-2)]^2 = \left(4\sqrt{5}\right)^2 \Leftrightarrow (x-5)^2 + (y+2)^2 = 80$.

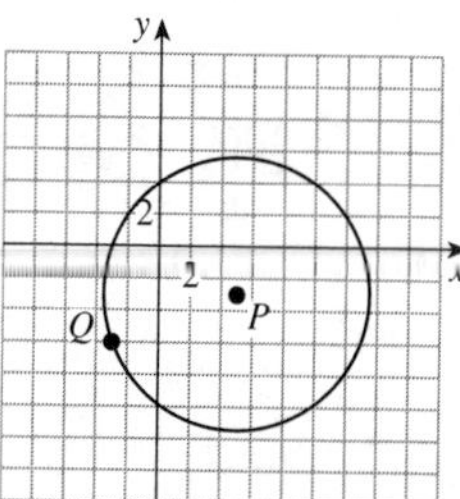


**5.**

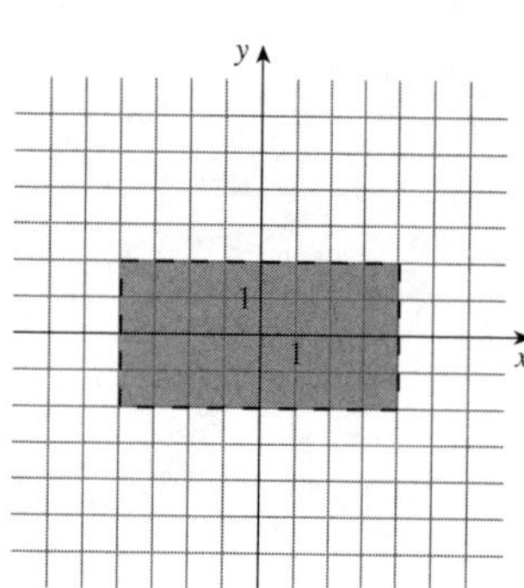


**6.** $\{(x,y) \mid x \geq 4 \text{ or } y \geq 2\}$

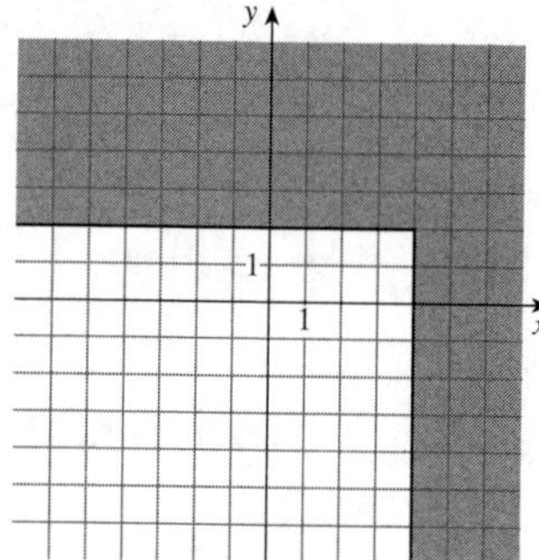


**7.** $d(A,C) = \sqrt{(4-(-1))^2 + (4-(-3))^2} = \sqrt{(4+1)^2 + (4+3)^2} = \sqrt{74}$ and $d(B,C) = \sqrt{(5-(-1))^2 + (3-(-3))^2} = \sqrt{(5+1)^2 + (3+3)^2} = \sqrt{72}$. Therefore, $B$ is closer to $C$.

**8.** The circle with center at $(2,-5)$ and radius $\sqrt{2}$ has equation $(x-2)^2 + (y+5)^2 = \left(\sqrt{2}\right)^2 \Leftrightarrow (x-2)^2 + (y+5)^2 = 2$.

**9.** The center is $C = (-5,-1)$, and the point $P = (0,0)$ is on the circle. The radius of the circle is $r = d(P,C) = \sqrt{(0-(-5))^2 + (0-(-1))^2} = \sqrt{(0+5)^2 + (0+1)^2} = \sqrt{26}$. Thus, an equation of the circle is $(x+5)^2 + (y+1)^2 = 26$.

**10.** The midpoint of segment $PQ$ is $\left(\dfrac{2-1}{2}, \dfrac{3+8}{2}\right) = \left(\dfrac{1}{2}, \dfrac{11}{2}\right)$, and the radius is $\frac{1}{2}$ of the distance from $P$ to $Q$, or $r = \frac{1}{2} \cdot d(P,Q) = \frac{1}{2}\sqrt{(2-(-1))^2 + (3-8)^2} = \frac{1}{2}\sqrt{(2+1)^2 + (3-8)^2} \Leftrightarrow r = \frac{1}{2}\sqrt{34}$. Thus the equation is $\left(x - \frac{1}{2}\right)^2 + \left(y - \frac{11}{2}\right)^2 = \frac{17}{2}$.

**11.** $x^2+y^2+2x-6y+9=0 \quad\Leftrightarrow\quad (x^2+2x)+(y^2-6y)=-9 \quad\Leftrightarrow\quad (x^2+2x+1)+(y^2-6y+9)=-9+1+9$ $\Leftrightarrow\quad (x+1)^2+(y-3)^2=1$. This equation represents a circle with center at $(-1,3)$ and radius 1.

**12.** $2x^2+2y^2-2x+8y=\frac{1}{2} \quad\Leftrightarrow\quad x^2-x+y^2+4y=\frac{1}{4} \quad\Leftrightarrow\quad \left(x^2-x+\frac{1}{4}\right)+\left(y^2+4y+4\right)=\frac{1}{4}+\frac{1}{4}+4 \quad\Leftrightarrow$ $\left(x-\frac{1}{2}\right)^2+(y+2)^2=\frac{9}{2}$. This is the equation of a circle whose center is $\left(\frac{1}{2},-2\right)$and radius is $\dfrac{3}{\sqrt{2}}$.

**13.** $x^2+y^2+72=12x \quad\Leftrightarrow\quad (x^2-12x)+y^2=-72 \quad\Leftrightarrow\quad (x^2-12x+36)+y^2=-72+36 \quad\Leftrightarrow$ $(x-6)^2+y^2=-36$. Since the left side of this equation must be greater than or equal to zero, this equation has no graph.

**14.** $x^2+y^2-6x-10y+34=0 \quad\Leftrightarrow\quad x^2-6x+y^2-10y=-34 \quad\Leftrightarrow\quad (x^2-6x+9)+(y^2-10y+25)=-34+9+25$ $\Leftrightarrow\quad (x-3)^2+(y-5)^2=0$. This is the equation of the point $(3,5)$.

**15.** $y=2-3x$

| $x$ | $y$ |
|---|---|
| $-2$ | 8 |
| 0 | 2 |
| $\frac{2}{3}$ | 0 |

$x$-axis symmetry: $(-y)=2-3x \quad\Leftrightarrow\quad y=-2+3x$, which is not the same as the original equation, so the graph is not symmetric with respect to the $x$-axis.

$y$-axis symmetry: $y=2-3(-x) \quad\Leftrightarrow\quad y=2+3x$, which is not the same as the original equation, so the graph is not symmetric with respect to the $y$-axis.

Origin symmetry: $(-y)=2-3(-x) \quad\Leftrightarrow\quad -y=2+3x \quad\Leftrightarrow$ $y=-2-3x$, which is not the same as the original equation, so the graph is not symmetric with respect to the origin.

Hence the graph has no symmetry.

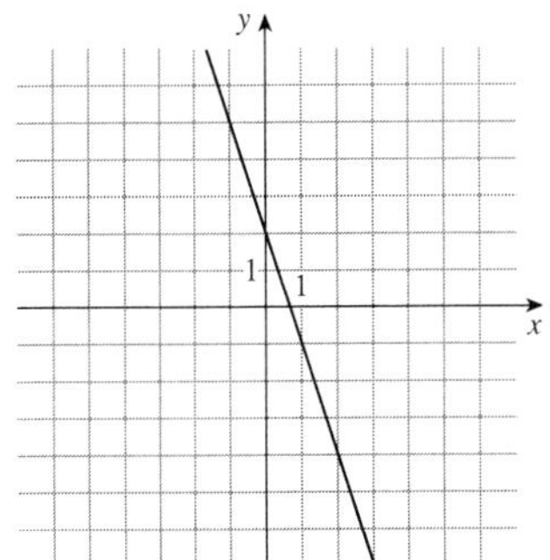


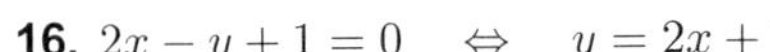

**16.** $2x-y+1=0 \quad\Leftrightarrow\quad y=2x+1$

| $x$ | $y$ |
|---|---|
| $-2$ | $-3$ |
| 0 | 1 |
| $-\frac{1}{2}$ | 0 |

$x$-axis symmetry: $(-y)=2x+1 \quad\Leftrightarrow\quad y=-2x-1$, which is not the same as the original equation, so the graph is not symmetric with respect to the $x$-axis.

$y$-axis symmetry: $y=2(-x)+1 \quad\Leftrightarrow\quad y=-2x+1$, which is not the same as the original equation, so the graph is not symmetric with respect to the $y$-axis.

Origin symmetry: $(-y)=2(-x)+1 \quad\Leftrightarrow\quad -y=-2x+1$ $\Leftrightarrow\quad y=2x-1$, which is not the same as the original equation, so the graph is not symmetric with respect to the origin.

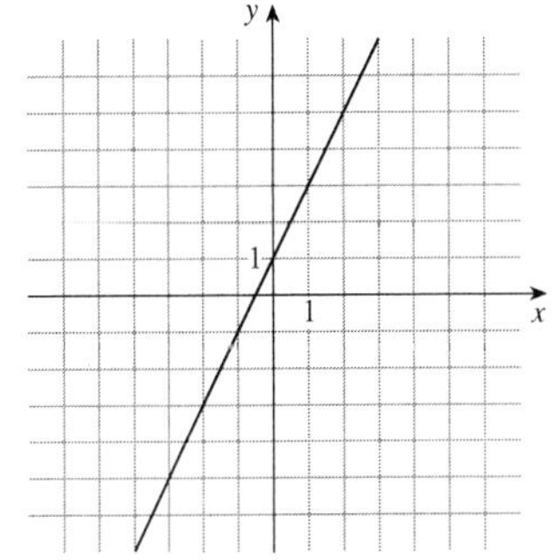

**17.** $x+3y=21 \quad\Leftrightarrow\quad y=-\frac{1}{3}x+7$

| $x$ | $y$ |
|---|---|
| $-3$ | 8 |
| 0 | 7 |
| 21 | 0 |

$x$-axis symmetry: $x+3(-y)=21 \quad\Leftrightarrow\quad x-3y=21$, which is not the same as the original equation, so the graph is not symmetric with respect to the $x$-axis.

$y$-axis symmetry: $(-x)+3y=21 \quad\Leftrightarrow\quad x-3y=-21$, which is not the same as the original equation, so the graph is not symmetric with respect to the $y$-axis.

Origin symmetry: $(-x)+3(-y)=21 \quad\Leftrightarrow\quad x+3y=-21$, which is not the same as the original equation, so the graph is not symmetric with respect to the origin. Hence the graph has no symmetry.

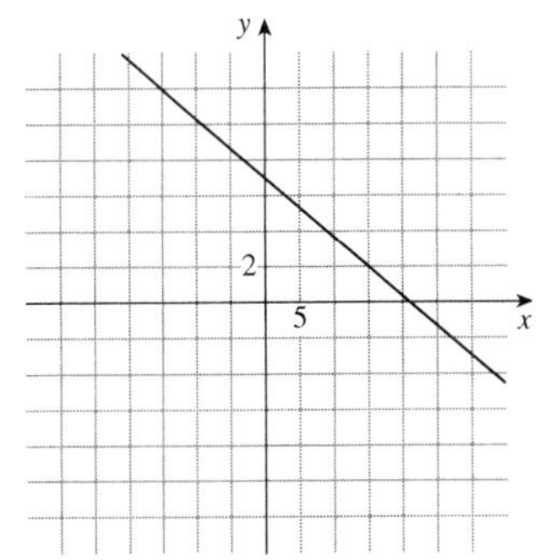

**18.** $x = 2y + 12 \quad \Leftrightarrow \quad 2y = x - 12 \quad \Leftrightarrow \quad y = \frac{1}{2}x - 6$

| $x$ | $y$ |
|---|---|
| $-4$ | $-8$ |
| $0$ | $-6$ |
| $12$ | $0$ |

$x$-axis symmetry: $(-y) = \frac{1}{2}x - 6 \quad \Leftrightarrow \quad y = -\frac{1}{2}x + 6$, which is not the same as the original equation, so the graph is not symmetric with respect to the $x$-axis.

$y$-axis symmetry: $y = \frac{1}{2}(-x) - 6 \quad \Leftrightarrow \quad y = -\frac{1}{2}x - 6$, which is not the same as the original equation, so the graph is not symmetric with respect to the $y$-axis.

Origin symmetry: $(-y) = \frac{1}{2}(-x) - 6 \quad \Leftrightarrow \quad y = \frac{1}{2}x + 6$, which is not the same as the original equation, so the graph is not symmetric with respect to the origin.

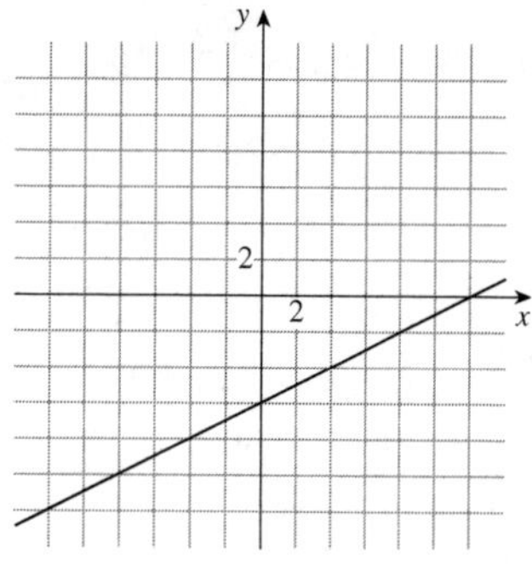


**19.** $\frac{x}{2} - \frac{y}{7} = 1 \quad \Leftrightarrow \quad y = \frac{7}{2}x - 7$

| $x$ | $y$ |
|---|---|
| $-2$ | $-14$ |
| $0$ | $-7$ |
| $2$ | $0$ |

$x$-axis symmetry: $\frac{x}{2} - \frac{(-y)}{7} = 1 \quad \Leftrightarrow \quad \frac{x}{2} + \frac{y}{7} = 1$, which is not the same as the original equation, so the graph is not symmetric with respect to the $x$-axis.

$y$-axis symmetry: $\frac{(-x)}{2} - \frac{y}{7} = 1 \quad \Leftrightarrow \quad \frac{x}{2} + \frac{y}{7} = -1$, which is not the same as the original equation, so the graph is not symmetric with respect to the $y$-axis.

Origin symmetry: $\frac{(-x)}{2} - \frac{(-y)}{7} = 1 \quad \Leftrightarrow \quad \frac{x}{2} - \frac{y}{7} = -1$, which is not the same as the original equation, so the graph is not symmetric with respect to the origin. Hence, the graph has no symmetry.

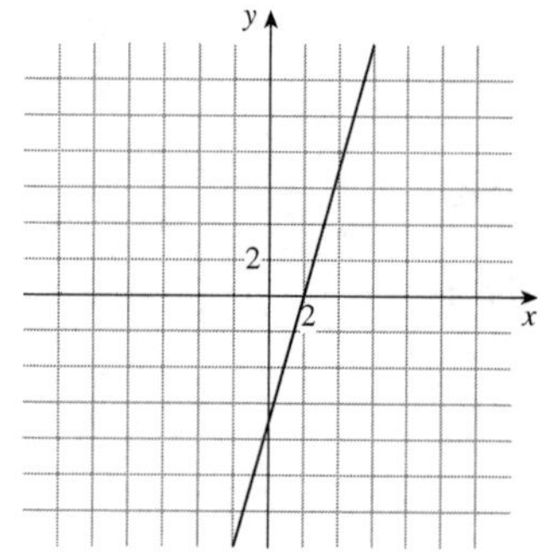


**20.** $\frac{x}{4} + \frac{y}{5} = 0 \quad \Leftrightarrow \quad 5x + 4y = 0$

| $x$ | $y$ |
|---|---|
| $-4$ | $5$ |
| $0$ | $0$ |
| $4$ | $-5$ |

$x$-axis symmetry: $5x + 4(-y) = 0 \quad \Leftrightarrow \quad 5x - 4y = 0$, which is not the same as the original equation, so the graph is not symmetric with respect to the $x$-axis.

$y$-axis symmetry: $5(-x) + 4y = 0 \quad \Leftrightarrow \quad -5x + 4y = 0$, which is not the same as the original equation, so the graph is not symmetric with respect to the $y$-axis.

Origin symmetry: $5(-x) + 4(-y) = 0 \quad \Leftrightarrow \quad -5x - 4y = 0 \quad \Leftrightarrow$ $5x + 4y = 0$, which is the original equation, so the graph is symmetric with respect to the origin.

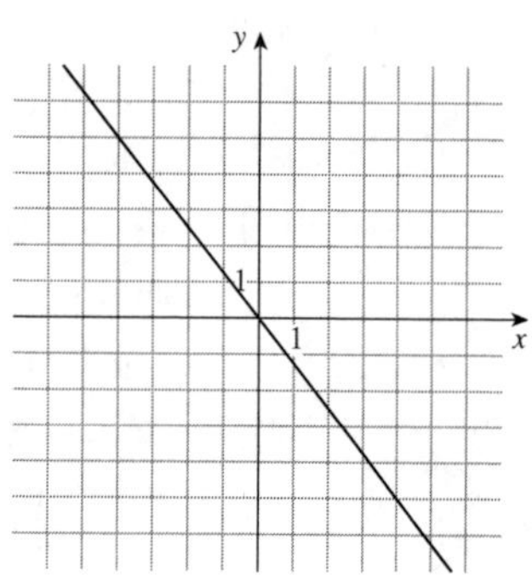

**21.** $y = 16 - x^2$

| $x$ | $y$ |
|---|---|
| $-3$ | 7 |
| $-1$ | 15 |
| 0 | 16 |
| 1 | 15 |
| 3 | 7 |

$x$-axis symmetry: $(-y) = 16 - x^2 \quad \Leftrightarrow \quad y = -16 + x^2$, which is not the same as the original equation, so the graph is not symmetric with respect to the $x$-axis.

$y$-axis symmetry: $y = 16 - (-x)^2 \quad \Leftrightarrow \quad y = 16 - x^2$, which is the same as the original equation, so its is symmetric with respect to the $y$-axis.

Origin symmetry: $(-y) = 16 - (-x)^2 \quad \Leftrightarrow \quad y = -16 + x^2$, which is not the same as the original equation, so the graph is not symmetric with respect to the origin. Hence, the graph is symmetric with respect to the $y$-axis.

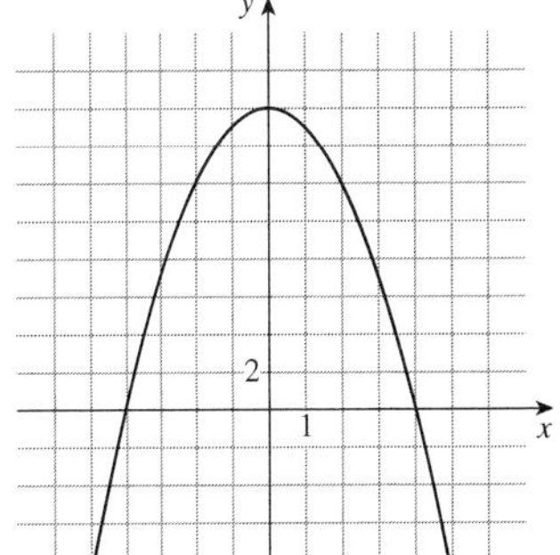


**22.** $8x + y^2 = 0 \quad \Leftrightarrow \quad y^2 = -8x$

| $x$ | $y$ |
|---|---|
| $-8$ | $\pm 8$ |
| $-2$ | $\pm 4$ |
| 0 | 0 |

$x$-axis symmetry: $(-y)^2 = -8x \quad \Leftrightarrow \quad y^2 = -8x$, which is the same as the original equation, so the graph is symmetric with respect to the $x$-axis.

$y$-axis symmetry: $y^2 = -8(-x) \quad \Leftrightarrow \quad y^2 = 8x$, which is not the same as the original equation, so the graph is not symmetric with respect to the $y$-axis.

Origin symmetry: $(-y)^2 = -8(-x) \quad \Leftrightarrow \quad y^2 = 8x$, which is not the same as the original equation, so the graph is not symmetric with respect to the origin.

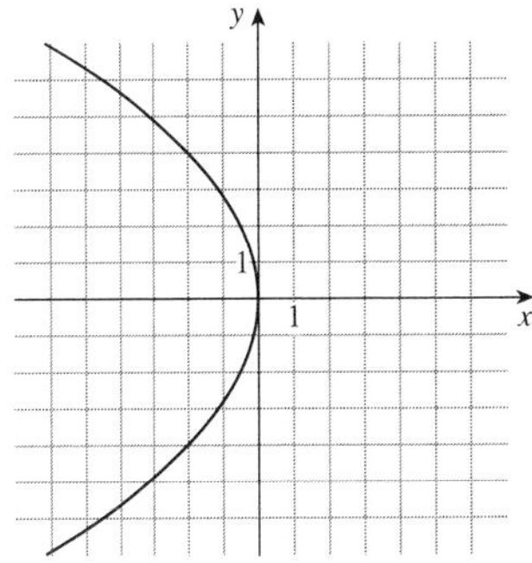


**23.** $x = \sqrt{y}$

| $x$ | $y$ |
|---|---|
| 0 | 0 |
| 1 | 1 |
| 2 | 4 |
| 3 | 9 |

$x$-axis symmetry: $x = \sqrt{-y}$, which is not the same as the original equation, so the graph is not symmetric with respect to the $x$-axis.

$y$-axis symmetry: $(-x) = \sqrt{y} \quad \Leftrightarrow \quad x = -\sqrt{y}$, which is not the same as the original equation, so the graph is not symmetric with respect to the $y$-axis.

Origin symmetry: $(-x) = \sqrt{-y}$, which is not the same as the original equation, so the graph is not symmetric with respect to the origin. Hence, the graph has no symmetry.

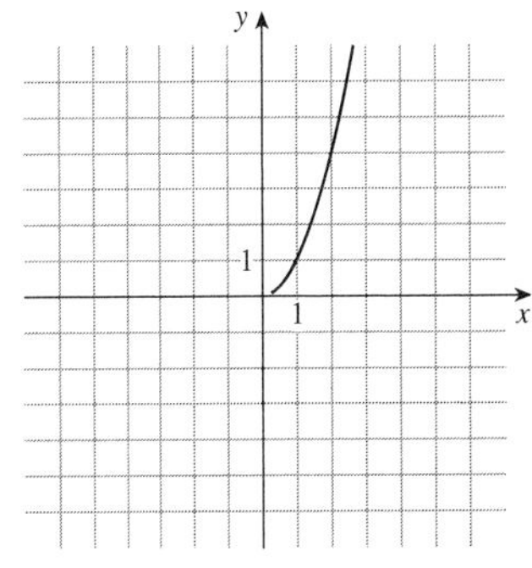

**24.** $y = -\sqrt{1 - x^2}$

| $x$ | $y$ |
|---|---|
| $-1$ | $0$ |
| $\frac{1}{2}$ | $-\frac{\sqrt{3}}{2}$ |
| $0$ | $-1$ |
| $1$ | $0$ |

$x$-axis symmetry: $(-y) = -\sqrt{1 - x^2} \quad \Leftrightarrow \quad y = \sqrt{1 - x^2}$, which is not the same as the original equation, so the graph is not symmetric with respect to the $x$-axis.

$y$-axis symmetry: $y = -\sqrt{1 - (-x)^2} \quad \Leftrightarrow \quad y = -\sqrt{1 - x^2}$, which is the same as the original equation, so the graph is symmetric with respect to the $y$-axis.

Origin symmetry: $(-y) = -\sqrt{1 - (-x)^2} \quad \Leftrightarrow \quad y = \sqrt{1 - x^2}$, which is not the same as the original equation, so the graph is not symmetric with respect to the origin.

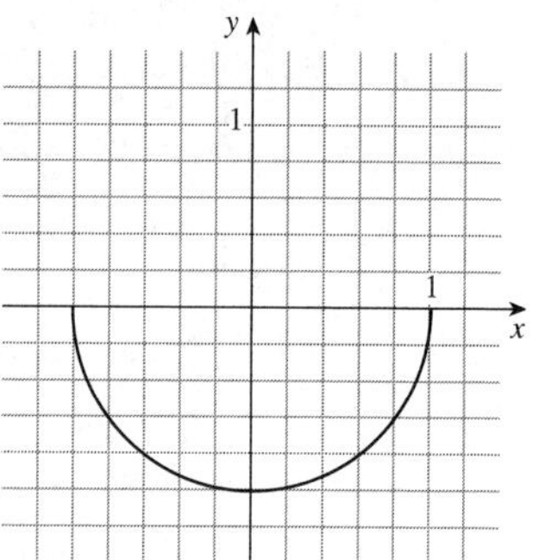

**25.** $y = x^2 - 6x$. Viewing rectangle $[-10, 10]$ by $[-10, 10]$.

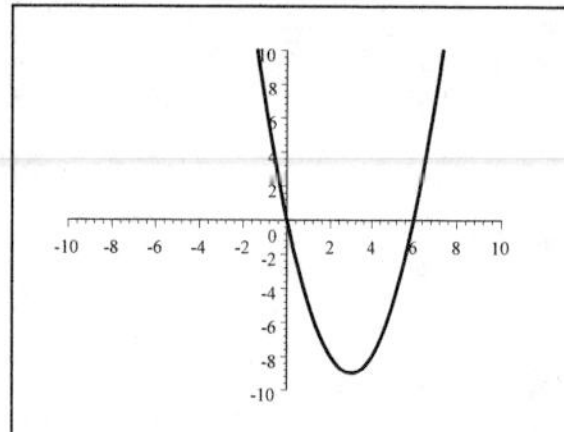

**26.** $y = \sqrt{5 - x}$; Viewing rectangle $[-10, 6]$ by $[-1, 5]$.

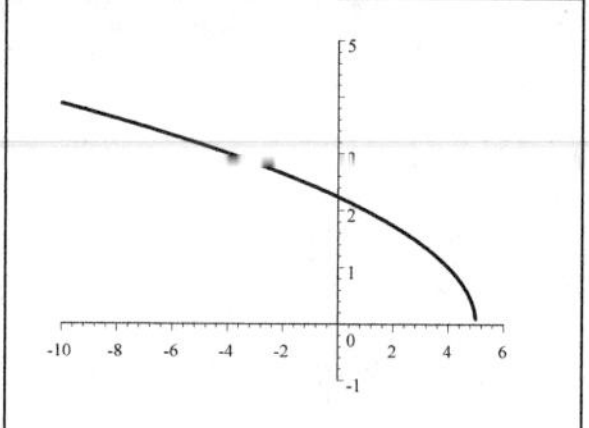

**27.** $y = x^3 - 4x^2 - 5$. Viewing rectangle $[-4, 10]$ by $[-30, 20]$.

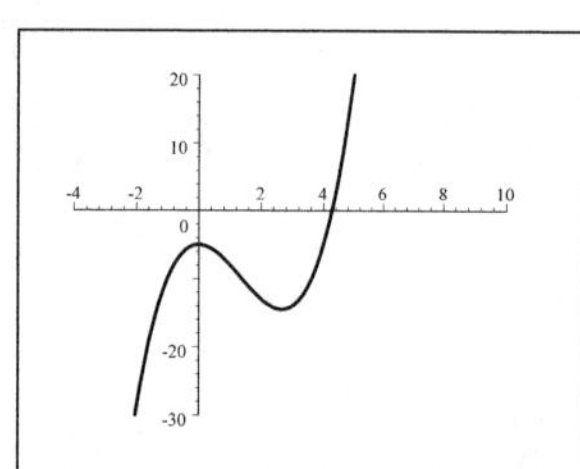

**28.** $\dfrac{x^2}{4} + y^2 = 1 \quad \Leftrightarrow \quad y^2 = 1 - \dfrac{x^2}{4} \quad \Rightarrow$

$y = \pm\sqrt{1 - \dfrac{x^2}{4}}$. Viewing rectangle $[-3, 3]$ by $[-2, 2]$.

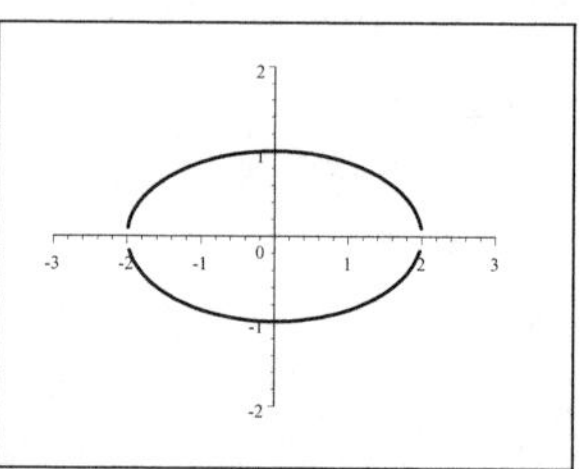

**29.** $x^2 - 4x = 2x + 7$. We graph the equations $y_1 = x^2 - 4x$ and $y_2 = 2x + 7$ in the viewing rectangle rectangle $[-10, 10]$ by $[-5, 25]$. Using a zoom or trace function, we get the solutions $x = -1$ and $x = 7$.

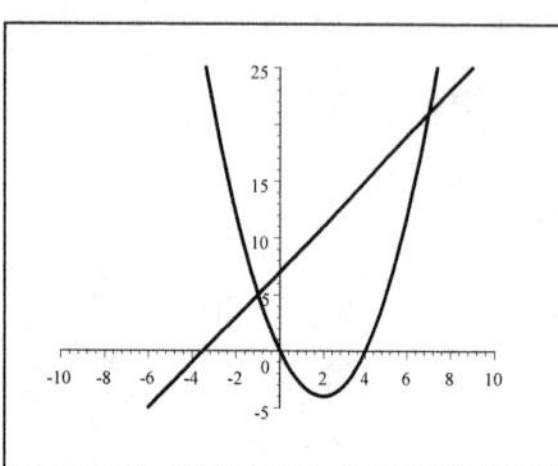

**30.** $\sqrt{x + 4} = x^2 - 5$. We graph the equations $y_1 = \sqrt{x + 4}$ and $y_2 = x^2 - 5$ in the viewing rectangle $[-4, 5]$ by $[0, 10]$. Using a zoom or trace function, we get the solutions $x \approx -2.50$ and $x \approx 2.76$.

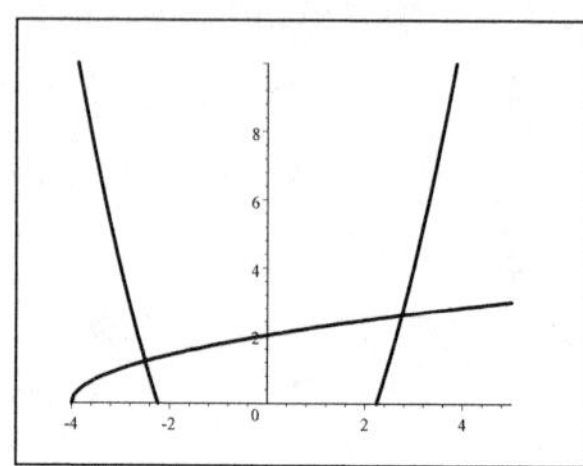

**31.** $x^4 - 9x^2 = x - 9$. We graph the equations $y_1 = x^4 - 9x^2$ and $y_2 = x - 9$ in the viewing rectangle $[-5, 5]$ by $[-25, 10]$. Using a zoom or trace function, we get the solutions $x \approx -2.72$, $x \approx -1.15$, $x = 1.00$, and $x \approx 2.87$.

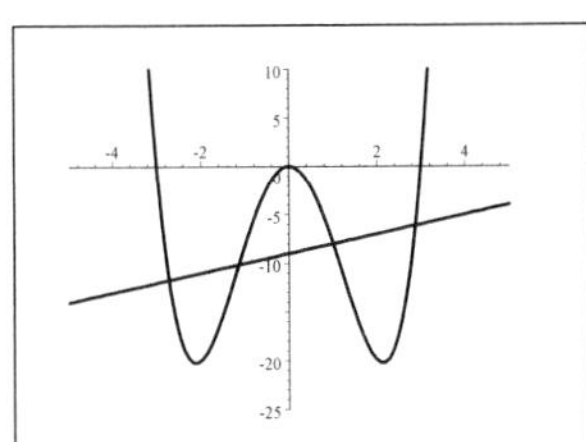

**32.** $||x + 3| - 5| = 2$. We graph the equations $y_1 = ||x + 3| - 5|$ and $y_2 = 2$ in the viewing rectangle $[-20, 20]$ by $[0, 10]$. Using Zoom and/or Trace, we get the solutions $x = -10$, $x = -6$, $x = 0$, and $x = 4$.

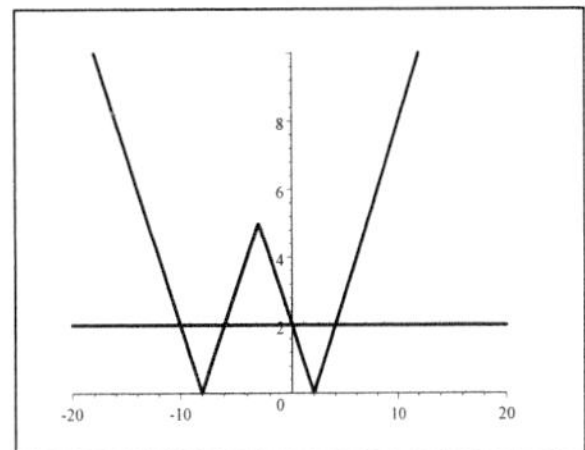

**33.** $4x - 3 \geq x^2$. We graph the equations $y_1 = 4x - 3$ and $y_2 = x^2$ in the viewing rectangle $[-5, 5]$ by $[0, 15]$. Using a zoom or trace function, we find the points of intersection are at $x = 1$ and $x = 3$. Since we want $4x - 3 \geq x^2$, the solution is the interval $[1, 3]$.

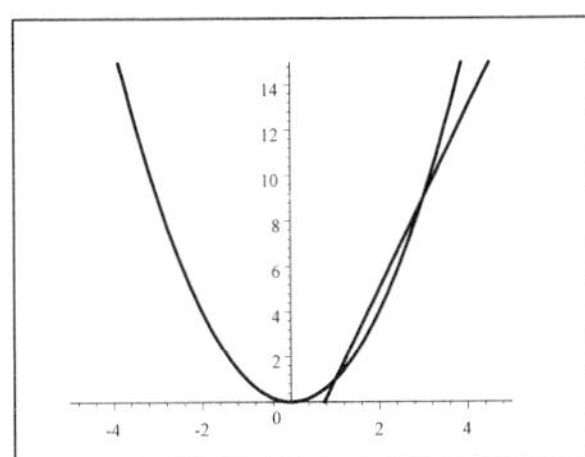

**34.** $x^3 - 4x^2 - 5x > 2$. We graph the equations $y_1 = x^3 - 4x^2 - 5x$ and $y_2 = 2$ in the viewing rectangle $[-10, 10]$ by $[-5, 5]$. We find that the point of intersection is at $x \approx 5.07$. Since we want $x^3 - 4x^2 - 5x > 2$, the solution is the interval $(5.07, \infty)$.

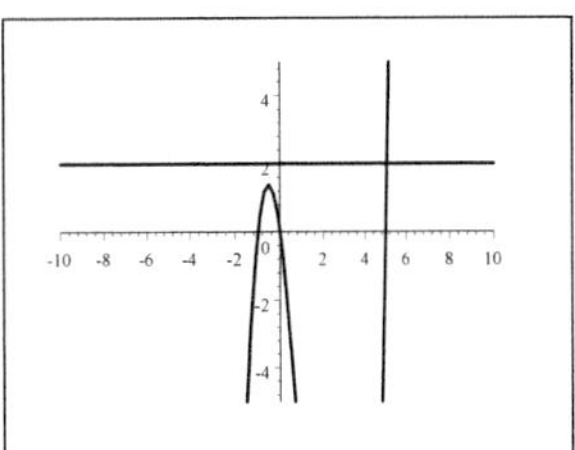

**35.** $x^4 - 4x^2 < \frac{1}{2}x - 1$. We graph the equations $y_1 = x^4 - 4x^2$ and $y_2 = \frac{1}{2}x - 1$ in the viewing rectangle $[-5, 5]$ by $[-5, 5]$. We find the points of intersection are at $x \approx -1.85$, $x \approx -0.60$, $x \approx 0.45$, and $x = 2.00$. Since we want $x^4 - 4x^2 < \frac{1}{2}x - 1$, the solution is $(-1.85, -0.60) \cup (0.45, 2.00)$.

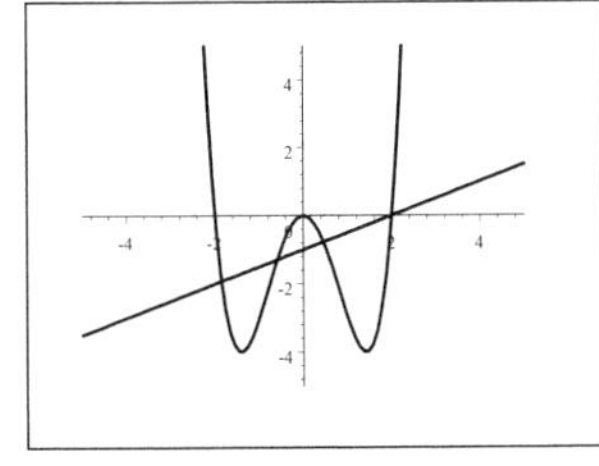

**36.** $\left|x^2 - 16\right| - 10 \geq 0$. We graph the equation $y = \left|x^2 - 16\right| - 10$ in the viewing rectangle $[-10, 10]$ by $[-10, 10]$. Using a zoom or trace function, we find that the $x$-intercepts are $x \approx \pm 5.10$ and $x \approx \pm 2.45$. Since we want $\left|x^2 - 16\right| - 10 \geq 0$, the solution is approximately $(-\infty, -5.10] \cup [-2.45, 2.45] \cup [5.10, \infty)$.

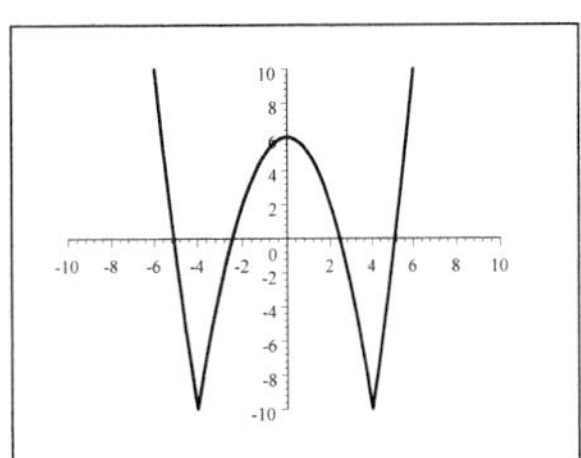

**37.** The line has slope $m = \dfrac{-4+6}{2+1} = \dfrac{2}{3}$, and so, by the point-slope formula, the equation is $y + 4 = \frac{2}{3}(x - 2) \quad\Leftrightarrow\quad y = \frac{2}{3}x - \frac{16}{3} \quad\Leftrightarrow\quad 2x - 3y - 16 = 0$.

**38.** Using the point-slope formula, the equation is $y + 3 = -\frac{1}{2}(x - 6) \quad\Leftrightarrow\quad y + 3 = -\frac{1}{2}x + 3 \quad\Leftrightarrow\quad y = -\frac{1}{2}x$ or $x + 2y = 0$.

**39.** The $x$-intercept is 4, and the $y$-intercept is 12, so the slope is $m = \dfrac{12-0}{0-4} = -3$. Therefore, by the slope-intercept formula, the equation of the line is $y = -3x + 12 \quad \Leftrightarrow \quad 3x + y - 12 = 0$.

**40.** $x - 3y + 16 = 0 \quad \Leftrightarrow \quad -3y = -x - 16 \quad \Leftrightarrow \quad y = \frac{1}{3}x + \frac{16}{3}$. So the slope of the perpendicular line we seek is $m = -\dfrac{1}{(1/3)} = -3$. Then the equation of the perpendicular line passing through the point $(1,7)$ is $y - 7 = -3(x-1)$ $\Leftrightarrow \quad y = -3x + 10$ or $3x + y - 10 = 0$.

**41.** We first find the slope of the line $3x + 15y = 22$. This gives $3x + 15y = 22 \quad \Leftrightarrow \quad 15y = -3x + 22 \quad \Leftrightarrow$ $y = -\frac{1}{5}x + \frac{22}{15}$. So this line has slope $m = -\frac{1}{5}$, as does any line parallel to it. Then the parallel line passing through the origin has equation $y - 0 = -\frac{1}{5}(x-0) \quad \Leftrightarrow \quad x + 5y = 0$.

**42.** The line passing through the points $(-1,-3)$ and $(3,2)$ has slope $m = \dfrac{2+3}{3+1} = \dfrac{5}{4}$. Therefore the equation of the parallel line passing through the point $(5,2)$ is $y - 2 = \frac{5}{4}(x-5) \quad \Leftrightarrow \quad 4y - 8 = 5x - 25 \quad \Leftrightarrow \quad 5x - 4y - 17 = 0$.

**43.** **(a)** The slope, $0.3$, represents the increase in length of the spring for each unit increase in weight $w$. The $S$ -intercept is the resting or natural length of the spring.

**(b)** When $w = 5$, $S = 0.3(5) + 2.5 = 1.5 + 2.5 = 4.0$ inches.

**44.** **(a)** We use the information to find two points, $(0, 60000)$ and $(3, 70500)$. Then the slope is $m = \dfrac{70{,}500 - 60{,}000}{3-0} = \dfrac{10{,}500}{3} = 3{,}500$. So $S = 3{,}500t + 60{,}000$.

**(b)** The slope represents her annual increase, \$3500, and the $S$-intercept represents her initial salary, \$60,000.

**(c)** When $t = 12$, her salary will be $S = 3500(12) + 60{,}000 = 42{,}000 + 60{,}000 = \$102{,}000$.

**45.** Since $M$ varies directly as $z$ we have $M = kz$. Substituting $M = 120$ when $z = 15$, we find $120 = k(15) \quad \Leftrightarrow \quad k = 8$. Therefore, $M = 8z$.

**46.** Since $z$ is inversely proportional to $y$, we have $z = \dfrac{k}{y}$. Substituting $z = 12$ when $y = 16$, we find $12 = \dfrac{k}{16} \quad \Leftrightarrow$ $k = 192$. Therefore $z = \dfrac{192}{y}$.

**47.** **(a)** The intensity $I$ varies inversely as the square of the distance $d$, so $I = \dfrac{k}{d^2}$.

**(b)** Substituting $I = 1000$ when $d = 8$, we get $1000 = \dfrac{k}{(8)^2} \quad \Leftrightarrow \quad k = 64{,}000$.

**(c)** From parts (a) and (b), we have $I = \dfrac{64{,}000}{d^2}$. Substituting $d = 20$, we get $I = \dfrac{64{,}000}{(20)^2} = 160$ candles.

**48.** Let $f$ be the frequency of the string and $l$ be the length of the string. Since the frequency is inversely proportional to the length, we have $f = \dfrac{k}{l}$. Substituting $l = 12$ when $k = 440$, we find $440 = \dfrac{k}{12} \quad \Leftrightarrow \quad k = 5280$. Therefore $f = \dfrac{5280}{l}$. For $f = 660$, we must have $660 = \dfrac{5280}{l} \quad \Leftrightarrow \quad l = \frac{5280}{660} = 8$. So the string needs to be shortened to 8 inches.

**49.** Let $v$ be the terminal velocity of the parachutist in mi/h and $w$ be his weight in pounds. Since the terminal velocity is directly proportional to the square root of the weight, we have $v = k\sqrt{w}$. Substituting $v = 9$ when $w = 160$, we solve for $k$. This gives $9 = k\sqrt{160} \quad \Leftrightarrow \quad k = \dfrac{9}{\sqrt{160}} \approx 0.712$. Thus $v = 0.712\sqrt{w}$. When $w = 240$, the terminal velocity is $v = 0.712\sqrt{240} \approx 11$ mi/h.

**50.** Let $r$ be the maximum range of the baseball and $v$ be the velocity of the baseball. Since the maximum range is directly proportional to the square of the velocity, we have $r = lv^2$. Substituting $v = 60$ and $r = 242$, we find $242 = k(60)^2 \quad \Leftrightarrow$ $k \approx 0.0672$. If $v = 70$, then we have a maximum range of $r = 0.0672(70)^2 = 329.4$ feet.

**51.** Here the center is at $(0,0)$, and the circle passes through the point $(-5,12)$, so the radius is $r=\sqrt{(-5-0)^2+(12-0)^2}=\sqrt{25+144}=\sqrt{169}=13$. The equation of the circle is $x^2+y^2=13^2 \Leftrightarrow x^2+y^2=169$. The line shown is the tangent that passes through the point $(-5,12)$, so it is perpendicular to the line through the points $(0,0)$ and $(-5,12)$. This line has slope $m_1=\dfrac{12-0}{-5-0}=-\dfrac{12}{5}$. The slope of the line we seek is $m_2=-\dfrac{1}{m_1}=-\dfrac{1}{-12/5}=\dfrac{5}{12}$. Thus, an equation of the tangent line is $y-12=\frac{5}{12}(x+5) \Leftrightarrow y-12=\frac{5}{12}x+\frac{25}{12} \Leftrightarrow y=\frac{5}{12}x+\frac{169}{12} \Leftrightarrow 5x-12y+169=0$.

**52.** Since the circle is tangent to the $x$-axis at the point $(5,0)$ and tangent to the $y$-axis at the point $(0,5)$, the center is at $(5,5)$ and the radius is 5. Thus an equation is $(x-5)^2+(y-5)^2=5^2 \Leftrightarrow (x-5)^2+(y-5)^2=25$. The slope of the line passing through the points $(8,1)$ and $(5,5)$ is $m=\dfrac{5-1}{5-8}=\dfrac{4}{-3}=-\dfrac{4}{3}$, so an equation of the line we seek is $y-1=-\frac{4}{3}(x-8) \Leftrightarrow 4x+3y-35=0$.

# Chapter 2 Test

**1. (a)**

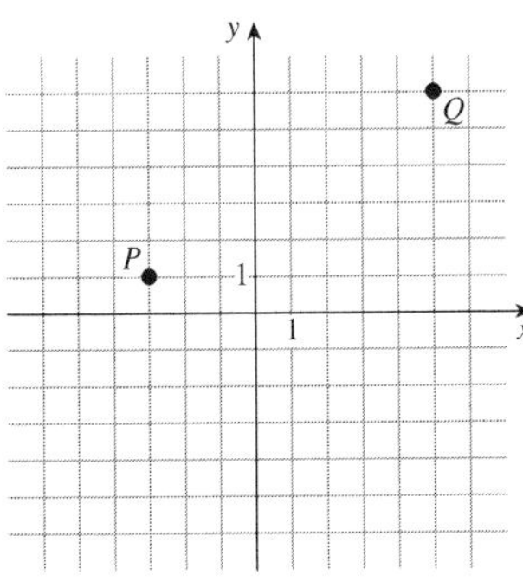


**(b)** The distance between $P$ and $Q$ is

$$d(P,Q)=\sqrt{(-3-5)^2+(1-6)^2}=\sqrt{64+25}=\sqrt{89}.$$

**(c)** The midpoint is $\left(\dfrac{-3+5}{2},\dfrac{1+6}{2}\right)=\left(1,\frac{7}{2}\right)$.

**(d)** The slope of the line is $\dfrac{1-6}{-3-5}=\dfrac{-5}{-8}=\dfrac{5}{8}$.

**(e)** The perpendicular bisector of $PQ$ contains the midpoint, $\left(1,\frac{7}{2}\right)$, and it slope is the negative reciprocal of $\frac{5}{8}$. Thus the slope is $-\frac{1}{5/8}=-\frac{8}{5}$. Hence the equation is $y-\frac{7}{2}=-\frac{8}{5}(x-1) \Leftrightarrow$ $y=-\frac{8}{5}x+\frac{8}{5}+\frac{7}{2}=-\frac{8}{5}x+\frac{51}{10}$. That is, $y=-\frac{8}{5}x+\frac{51}{10}$.

**(f)** The center of the circle is the midpoint, $\left(1,\frac{7}{2}\right)$, and the length of the radius is $\frac{1}{2}\sqrt{89}$. Thus the equation of the circle whose diameter is $PQ$ is $(x-1)^2+\left(y-\frac{7}{2}\right)^2=\left(\frac{1}{2}\sqrt{89}\right)^2 \Leftrightarrow (x-1)^2+\left(y-\frac{7}{2}\right)^2=\frac{89}{4}$.

**2. (a)** $x^2+y^2=25=5^2$ has center $(0,0)$ and radius 5.

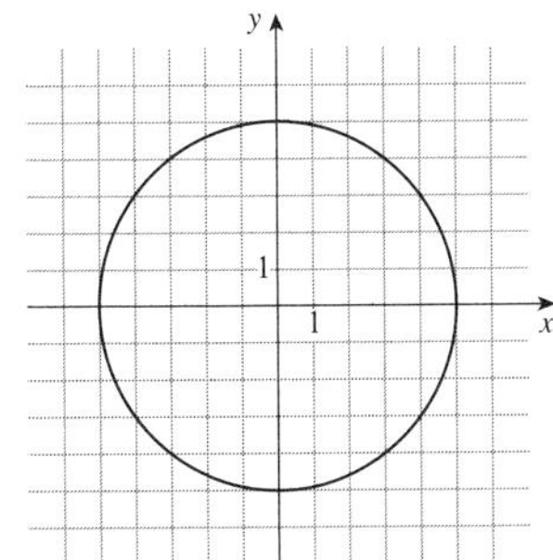


**(b)** $(x-2)^2+(y+1)^2=9=3^2$ has center $(2,-1)$ and radius 3.

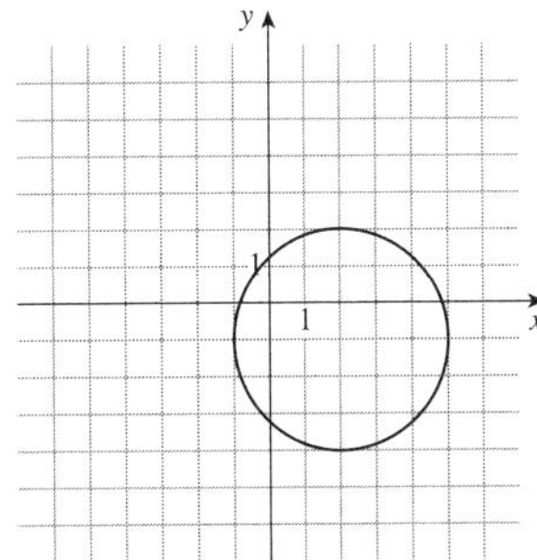


**(c)** $x^2+6x+y^2-2y+6=0 \Leftrightarrow x^2+6x+9+y^2-2y+1=4 \Leftrightarrow (x+3)^2+(y-1)^2=4=2^2$ has center $(-3,1)$ and radius 2.

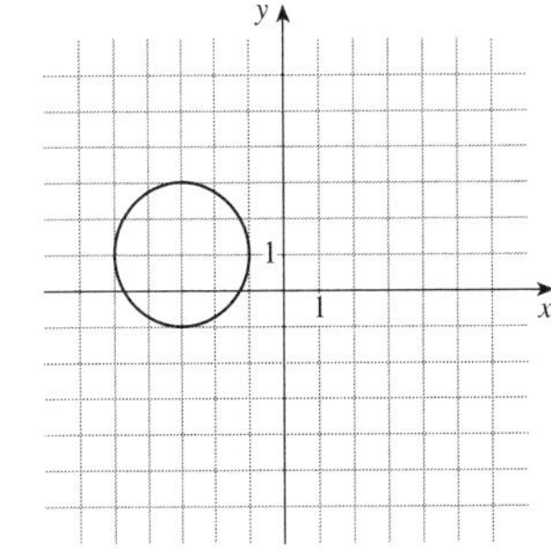

**3.** $2x - 3y = 15 \quad \Leftrightarrow$

$-3y = -2x + 15 \quad \Leftrightarrow \quad y = \frac{2}{3}x - 5.$

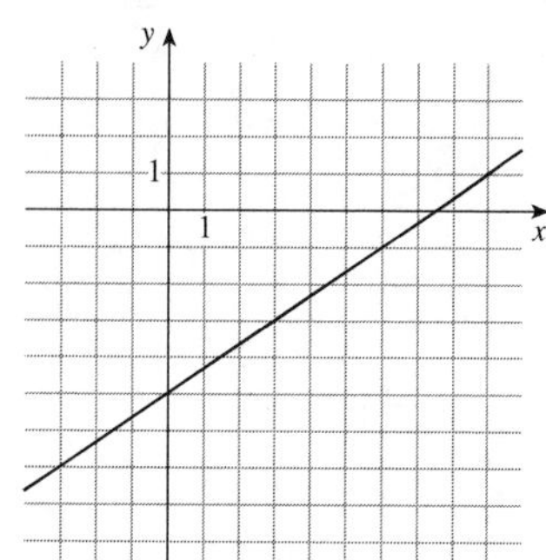


**4. (a)** $3x + y - 10 = 0 \quad \Leftrightarrow \quad y = -3x + 10$, so the slope of the line we seek is $-3$. Using the point-slope, $y - (-6) = -3(x - 3) \quad \Leftrightarrow$ $y + 6 = -3x + 9 \quad \Leftrightarrow \quad 3x + y - 3 = 0.$

**(b)** Using the intercept form we get $\dfrac{x}{6} + \dfrac{y}{4} = 1 \quad \Leftrightarrow \quad 2x + 3y = 12$ $\Leftrightarrow \quad 2x + 3y - 12 = 0.$

**5. (a)** When $x = 100$ we have $T = 0.08(100) - 4 = 8 - 4 = 4$, so the temperature at one meter is 4° C.

**(b)**

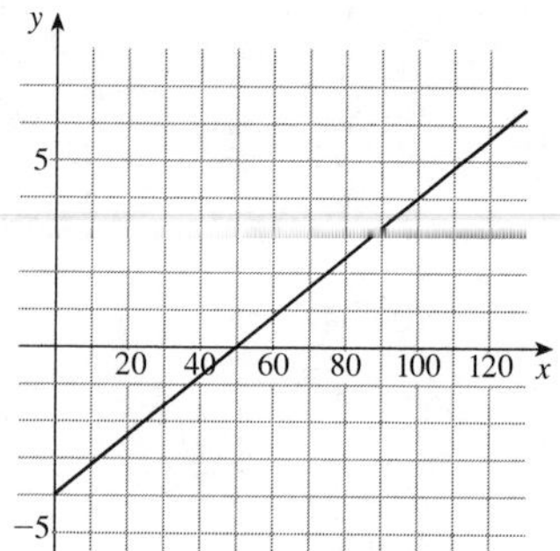


**(c)** The slope represents the raise in temperature as the depth increase. The $T$-intercept is the surface temperature of the soil and the $x$-intercept represents the depth of the "frost line", where the soil below is not frozen.

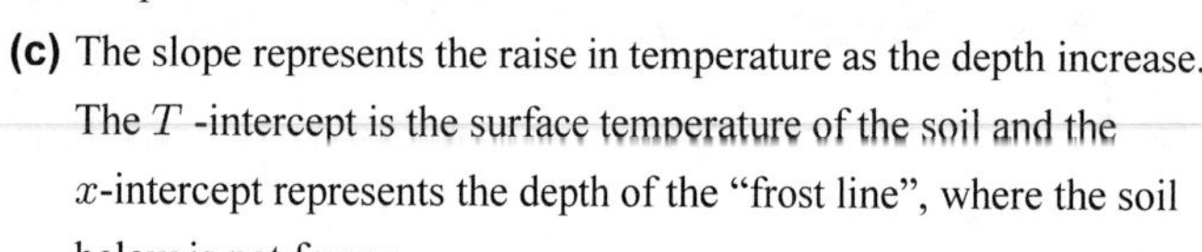

**6. (a)** $x^3 - 9x - 1 = 0$. We graph the equation $y = x^3 - 9x - 1$ in the viewing rectangle $[-5, 5]$ by $[-10, 10]$. We find that the roots of the equation are $x \approx -2.94, -0.11, 3.05$.

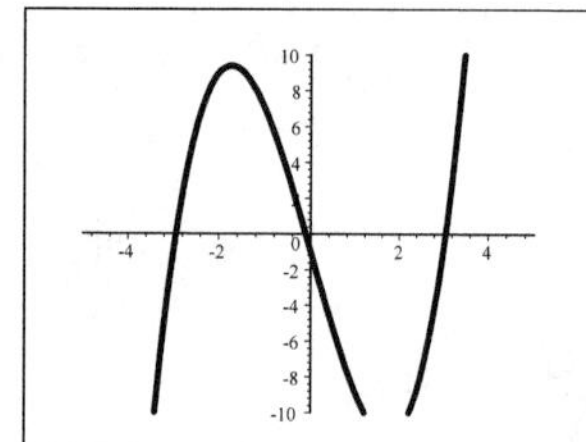

**(b)** $\frac{1}{2}x + 2 \geq \sqrt{x^2 + 1}$. We graph the equations $y_1 = \frac{1}{2}x + 2$ and $y_2 = \sqrt{x^2 + 1}$ in the viewing rectangle $[-5, 5]$ by $[0, 5]$. We find that the points of intersection occur at $x \approx -0.78$ and $x \approx 1.28$. Since we want $\frac{1}{2}x + 2 \geq \sqrt{x^2 + 1}$, the solution is approximately the interval $[-0.78, 1.28]$.

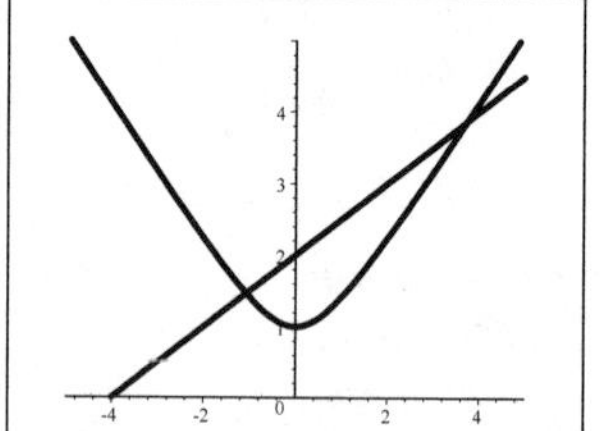

**7. (a)** $M = k\dfrac{wh^2}{L}$

**(b)** Substituting $w = 4$, $h = 6$, $L = 12$, and $M = 4800$, we have $4800 = k\dfrac{(4)(6^2)}{12} \quad \Leftrightarrow \quad k = 400$. Thus $M = 400\dfrac{wh^2}{L}$.

**(c)** Now if $L = 10$, $w = 3$, and $h = 10$, then $M = 400\dfrac{(3)(10^2)}{10} = 12{,}000$. So the beam can support 12,000 pounds.

# Focus on Modeling: Fitting Lines to Data

**1. (a)**

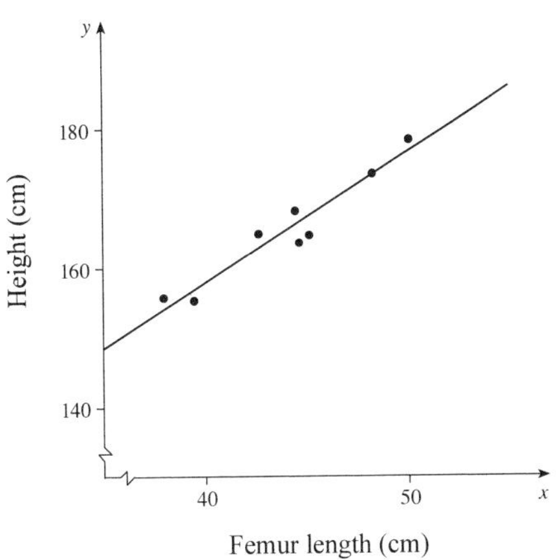


**(b)** Using a graphing calculator, we obtain the regression line $y = 1.8807x + 82.65$.

**(c)** Using $x = 58$ in the equation $y = 1.8807x + 82.65$, we get $y = 1.8807\,(58) + 82.65 \approx 191.7$ cm.

**2. (a)**

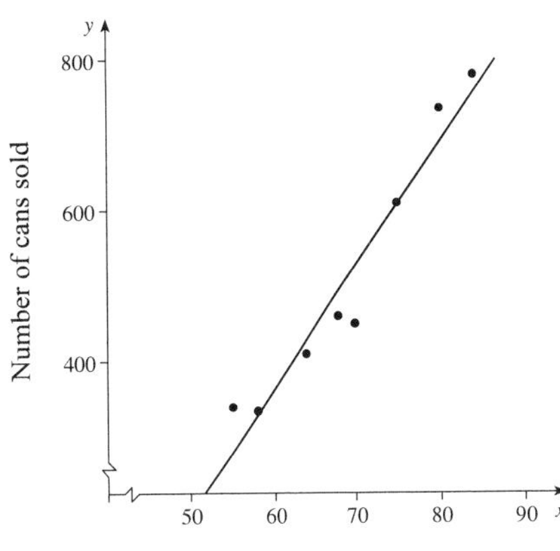


**(b)** Using a graphing calculator, we obtain the regression line $y = 16.4163x - 621.83$.

**(c)** Using $x = 95$ in the equation $y = 16.4163x - 621.83$, we get $y = 16.4163\,(95) - 621.83 \approx 938$ cans.

**3. (a)**

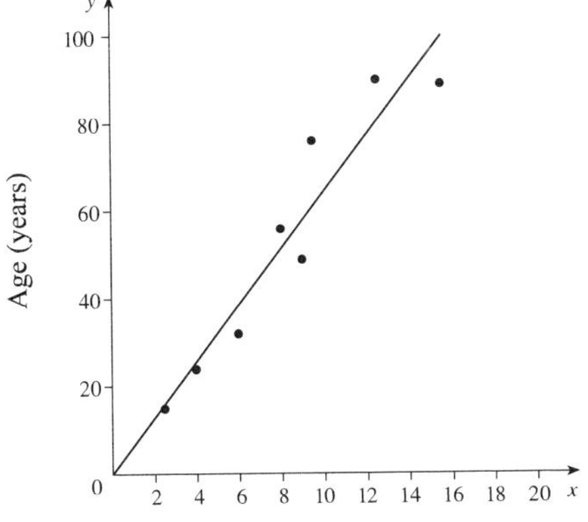


**(b)** Using a graphing calculator, we obtain the regression line $y = 6.451x - 0.1523$.

**(c)** Using $x = 18$ in the equation $y = 6.451x - 0.1523$, we get $y = 6.451\,(18) - 0.1523 \approx 116$ years.

**4. (a)**

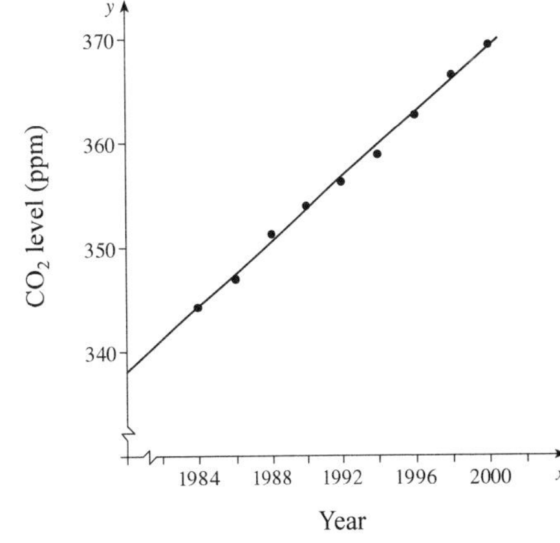


**(b)** Using a graphing calculator, we obtain the regression line $y = 1.555x - 2740.8$.

**(c)** Using $x = 2001$ in the equation $y = 1.555x - 2740.8$, we get $y = 1.555\,(2001) - 2740.8 \approx 370.8$ ppm $CO_2$.

**5. (a)**

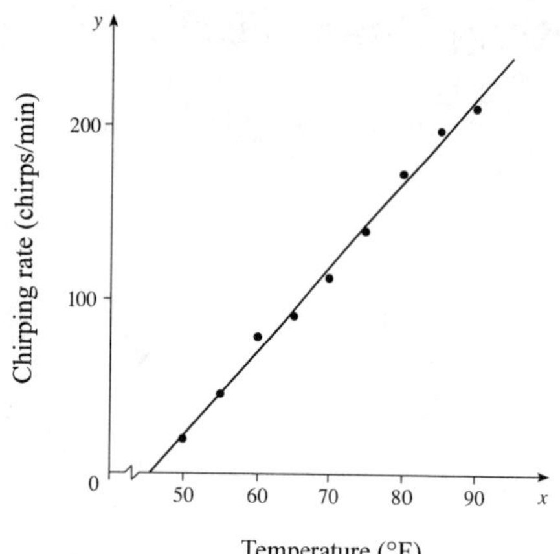


**(b)** Using a graphing calculator, we obtain the regression line $y = 4.857x - 220.97$.

**(c)** Using $x = 100°$ F in the equation $y = 4.857x - 220.97$, we get $y \approx 265$ chirps per minute.

**6. (a)**

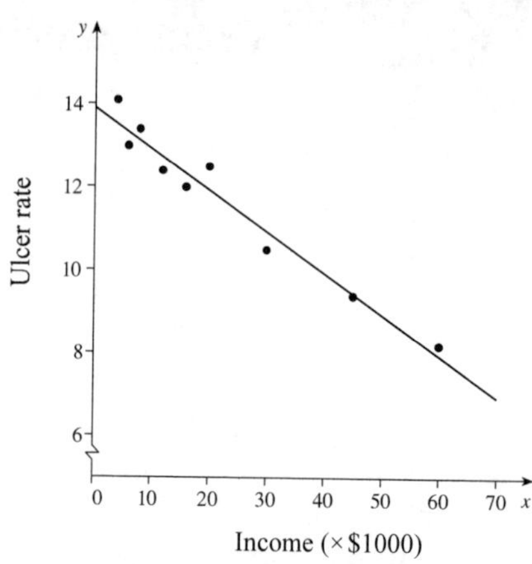


**(b)** Using a graphing calculator, we obtain the regression line $y = -0.0995x + 13.9$, where $x$ is measured in thousands of dollars.

**(c)** Using the regression line equation, $y = -0.0995x + 13.9$, we get an estimated $y = 11.4$ ulcers per 100 population when $x = \$25,000$.

**(d)** Again using $y = -0.0995x + 13.9$, we get an estimated $y = 5.8$ ulcers per 100 population when $x = \$80,000$.

**7. (a)**

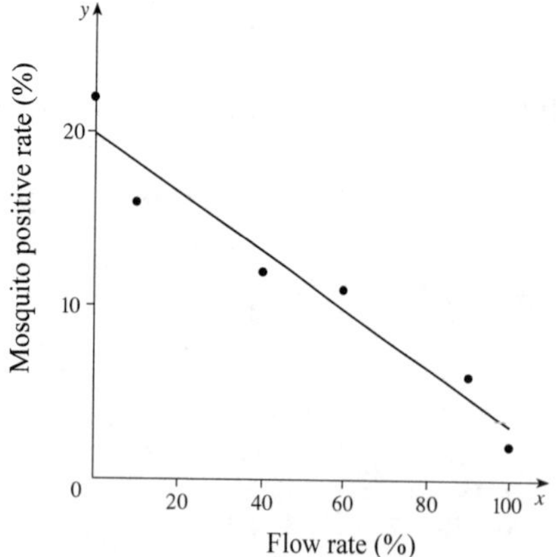


**(b)** Using a graphing calculator, we obtain the regression line $y = -0.168x + 19.89$.

**(c)** Using the regression line equation $y = -0.168x + 19.89$, we get $y \approx 8.13\%$ when $x = 70\%$.

**8. (a)**

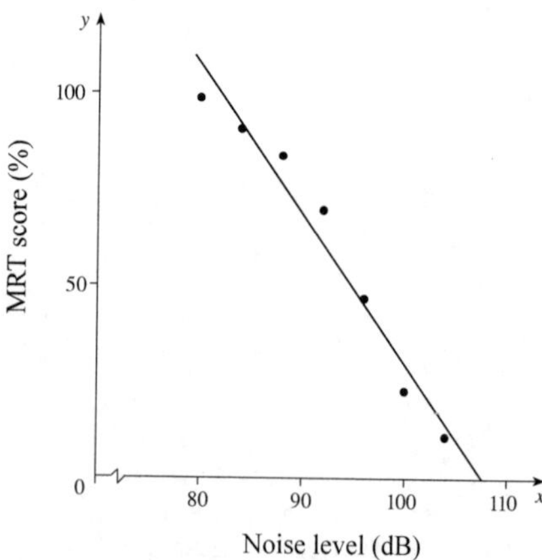


**(b)** Using a graphing calculator, we obtain $y = -3.9018x + 419.7$.

**(c)** The correlation coefficient is $r = -0.98$, so linear model is appropriate for $x$ between 80 dB and 104 dB.

**(d)** Substituting $x = 94$ into the regression equation, we get $y = -3.9018\,(94) + 419.7 \approx 53$. So the intelligibility is about 53%.

**9. (a)**

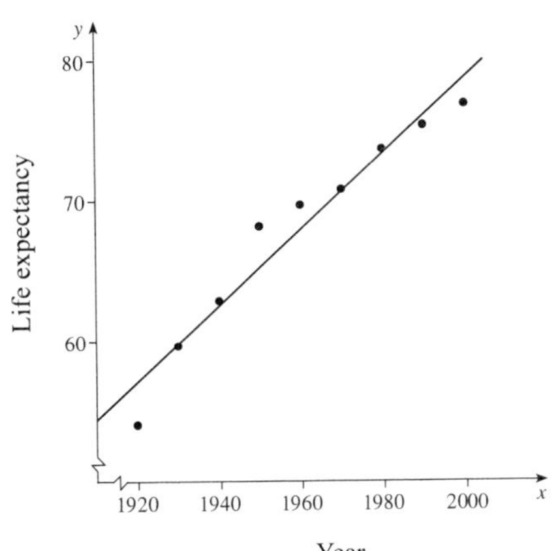


**(b)** Using a graphing calculator, we obtain $y = 0.27083x - 462.9$.

**(c)** We substitute $x = 2004$ in the model $y = 0.27083x - 462.9$ to get $y = 79.8$, that is, a life expectancy of 79.8 years.

**(d)** As of this writing, data for 2004 are not yet available. The life expectancy of a child born in the US in 2003 is 77.6 years.

**10. (a)**

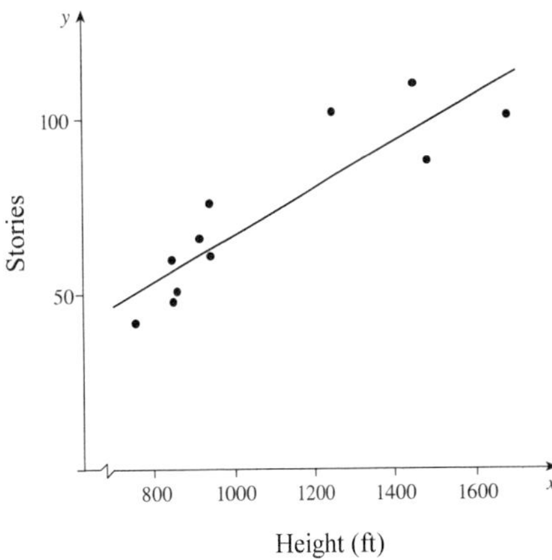


**(b)** Using a graphing calculator, we obtain the regression line $y = 0.0669x + 0.23$.

**(c)** The slope is 0.0669. It indicates the fraction of a storey represented by one foot of height in the building. Since $\dfrac{1}{0.0669} \approx 14.94$, each storey is about 15 feet.

**11. (a)** If we take $x = 0$ in 1900 for both men and women, then the regression equation for the men's data is $y = -0.173x + 64.72$ and the regression equation for the women's data is $y = -0.269x + 78.67$.

**(b)**

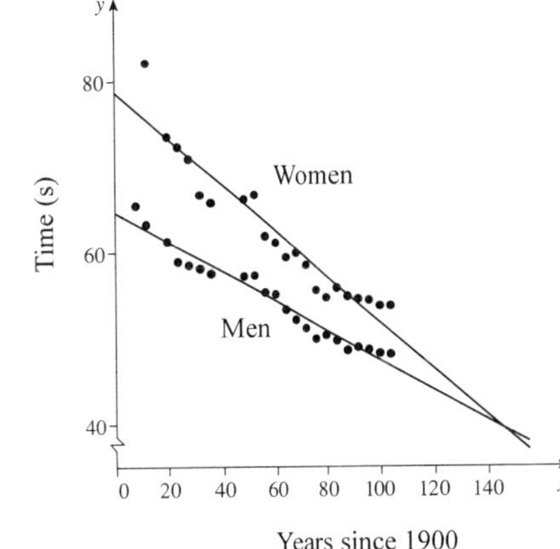


These lines predict that the women will overtake the men in this event when $-0.173x + 64.72 = -0.269x + 78.67 \quad \Leftrightarrow \quad 0.096x = 13.95 \quad \Leftrightarrow \quad x = 145.31$, or in 2045. This seems unlikely, but who knows?

**12. (a)** Using a graphing calculator, we obtain the regression line $y = 0.7804x + 15.5$.

**(b)** Answers will vary.

**13.** The students should find a fairly strong correlation between shoe size and height.

**14.** Results will depend on student surveys in each class.

# 3 Functions

## 3.1 What is a Function?

**1.** $f(x) = 2(x+3)$

**2.** $f(x) = \frac{x}{7} - 4$

**3.** $f(x) = (x-5)^2$

**4.** $f(x) = \frac{1}{3}(\sqrt{x}+8)$

**5.** Subtract 4, then divide by 3.

**6.** Divide by 3, then subtract 4.

**7.** Square, then add 2.

**8.** Add 2, then take the square root.

**9.** Machine diagram for $f(x) = \sqrt{x-1}$.

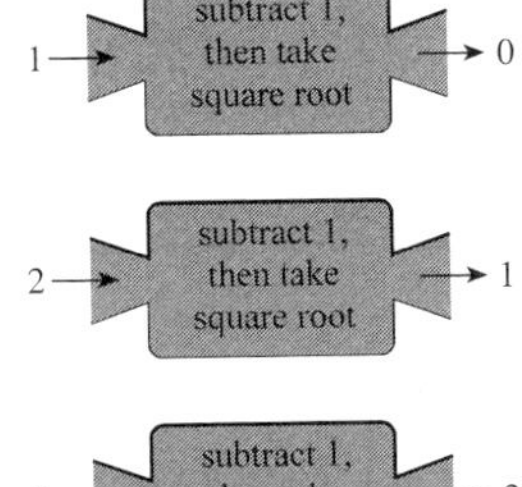


**10.** Machine diagram for $f(x) = \dfrac{3}{x-2}$.

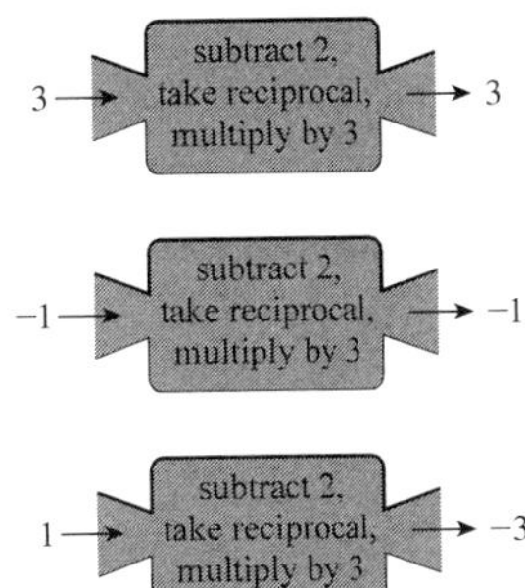


**11.** $f(x) = 2(x-1)^2$

| $x$ | $f(x)$ |
|---|---|
| $-1$ | $2(-1-1)^2 = 8$ |
| $0$ | $2(-1)^2 = 2$ |
| $1$ | $2(1-1)^2 = 0$ |
| $2$ | $2(2-1)^2 = 2$ |
| $3$ | $2(3-1)^2 = 8$ |

**12.** $g(x) = |2x+3|$

| $x$ | $g(x)$ |
|---|---|
| $-3$ | $\lvert 2(-3)+3\rvert = 3$ |
| $-2$ | $\lvert 2(-2)+3\rvert = 1$ |
| $0$ | $\lvert 2(0)+3\rvert = 3$ |
| $1$ | $\lvert 2(1)+3\rvert = 5$ |
| $3$ | $\lvert 2(3)+3\rvert = 9$ |

**13.** $f(1) = 2(1)+1 = 3$; $f(-2) = 2(-2)+1 = -3$; $f\left(\frac{1}{2}\right) = 2\left(\frac{1}{2}\right)+1 = 2$; $f(a) = 2(a)+1 = 2a+1$; $f(-a) = 2(-a)+1 = -2a+1$; $f(a+b) = 2(a+b)+1 = 2a+2b+1$.

**14.** $f(0) = 0^2+2(0) = 0$; $f(3) = 3^2+2(3) = 9+6 = 15$; $f(-3) = (-3)^2+2(-3) = 9-6 = 3$; $f(a) = a^2+2(a) = a^2+2a$; $f(-x) = (-x)^2+2(-x) = x^2-2x$; $f\left(\frac{1}{a}\right) = \left(\frac{1}{a}\right)^2+2\left(\frac{1}{a}\right) = \frac{1}{a^2}+\frac{2}{a}$.

**15.** $g(2) = \dfrac{1-(2)}{1+(2)} = \dfrac{-1}{3} = -\dfrac{1}{3}$; $g(-2) = \dfrac{1-(-2)}{1+(-2)} = \dfrac{3}{-1} = -3$; $g\left(\frac{1}{2}\right) = \dfrac{1-\left(\frac{1}{2}\right)}{1+\left(\frac{1}{2}\right)} = \dfrac{\frac{1}{2}}{\frac{3}{2}} = \dfrac{1}{3}$; $g(a) = \dfrac{1-(a)}{1+(a)} = \dfrac{1-a}{1+a}$; $g(a-1) = \dfrac{1-(a-1)}{1+(a-1)} = \dfrac{1-a+1}{1+a-1} = \dfrac{2-a}{a}$; $g(-1) = \dfrac{1-(-1)}{1+(-1)} = \dfrac{2}{0}$, so $g(-1)$ is not defined.

**16.** $h(1) = (1) + \frac{1}{(1)} = 2$; $h(-1) = (-1) + \frac{1}{-1} = -1 - 1 = -2$; $h(2) = 2 + \frac{1}{2} = \frac{5}{2}$; $h\left(\frac{1}{2}\right) = \dfrac{1}{2} + \dfrac{1}{\frac{1}{2}} = \dfrac{1}{2} + 2 = \dfrac{5}{2}$;
$h(x) = x + \dfrac{1}{x}$; $h\left(\dfrac{1}{x}\right) = \dfrac{1}{x} + \dfrac{1}{\frac{1}{x}} = \dfrac{1}{x} + x$.

**17.** $f(0) = 2(0)^2 + 3(0) - 4 = -4$; $f(2) = 2(2)^2 + 3(2) - 4 = 8 + 6 - 4 = 10$;
$f(-2) = 2(-2)^2 + 3(-2) - 4 = 8 - 6 - 4 = -2$; $f(\sqrt{2}) = 2(\sqrt{2})^2 + 3(\sqrt{2}) - 4 = 4 + 3\sqrt{2} - 4 = 3\sqrt{2}$;
$f(x+1) = 2(x+1)^2 + 3(x+1) - 4 = 2x^2 + 4x + 2 + 3x + 3 - 4 = 2x^2 + 7x + 1$;
$f(-x) = 2(-x)^2 + 3(-x) - 4 = 2x^2 - 3x - 4$.

**18.** $f(0) = 0^3 - 4(0)^2 = 0 + 0 = 0$; $f(1) = 1^3 - 4(1)^2 = 1 - 4 = -3$; $f(-1) = (-1)^3 - 4(-1)^2 = -1 - 4 = -5$;
$f\left(\frac{3}{2}\right) = \left(\frac{3}{2}\right)^3 - 4\left(\dfrac{3}{2}\right)^2 = \frac{27}{8} - 9 = -\frac{45}{8}$; $f\left(\dfrac{x}{2}\right) = \left(\dfrac{x}{2}\right)^3 - 4\left(\dfrac{x}{2}\right)^2 = \dfrac{x^3}{8} - x^2$; $f(x^2) = (x^2)^3 - 4(x^2)^2 = x^6 - 4x^4$.

**19.** $f(-2) = 2|-2-1| = 2(3) = 6$; $f(0) = 2|0-1| = 2(1) = 2$;
$f\left(\frac{1}{2}\right) = 2\left|\frac{1}{2} - 1\right| = 2\left(\frac{1}{2}\right) = 1$; $f(2) = 2|2-1| = 2(1) = 2$; $f(x+1) = 2|(x+1) - 1| = 2|x|$;
$f(x^2+2) = 2|(x^2+2) - 1| = 2|x^2+1| = 2x^2 + 2$ (since $x^2 + 1 > 0$).

**20.** $f(-2) = \dfrac{|-2|}{-2} = \dfrac{2}{-2} = -1$; $f(-1) = \dfrac{|-1|}{-1} = \dfrac{1}{-1} = -1$; $f(x)$ is not defined at $x = 0$; $f(5) = \dfrac{|5|}{5} = \dfrac{5}{5} = 1$;
$f(x^2) = \dfrac{|x^2|}{x^2} = \dfrac{x^2}{x^2} = 1$ since $x^2 > 0$, $x \neq 0$; $f\left(\dfrac{1}{x}\right) = \dfrac{|1/x|}{1/x} = \dfrac{x}{|x|}$.

**21.** Since $-2 < 0$, we have $f(-2) = (-2)^2 = 4$. Since $-1 < 0$, we have $f(-1) = (-1)^2 = 1$. Since $0 \geq 0$, we have $f(0) = 0 + 1 = 1$. Since $1 \geq 0$, we have $f(1) = 1 + 1 = 2$. Since $2 \geq 0$, we have $f(2) = 2 + 1 = 3$.

**22.** Since $-3 \leq 2$, we have $f(-3) = 5$. Since $0 \leq 2$, we have $f(0) = 5$. Since $2 \leq 2$, we have $f(2) = 5$. Since $3 > 2$, we have $f(3) = 2(3) - 3 = 3$. Since $5 > 2$, we have $f(5) = 2(5) - 3 = 7$.

**23.** Since $-4 \leq -1$, we have $f(-4) = (-4)^2 + 2(-4) = 16 - 8 = 8$. Since $-\frac{3}{2} \leq -1$, we have $f\left(-\frac{3}{2}\right) = \left(-\frac{3}{2}\right)^2 + 2\left(-\frac{3}{2}\right) = \frac{9}{4} - 3 = -\frac{3}{4}$. Since $-1 \leq -1$, we have $f(-1) = (-1)^2 + 2(-1) = 1 - 2 = -1$. Since $-1 < 0 \leq 1$, we have $f(0) = 0$. Since $25 > 1$, we have $f(25) = -1$.

**24.** Since $-5 < 0$, we have $f(-5) = 3(-5) = -15$. Since $0 \leq 0 \leq 2$, we have $f(0) = 0 + 1 = 1$. Since $0 \leq 1 \leq 2$, we have $f(1) = 1 + 1 = 2$. Since $0 \leq 2 \leq 2$, we have $f(2) = 2 + 1 = 3$. Since $5 > 2$, we have $f(5) = (5-2)^2 = 9$.

**25.** $f(x+2) = (x+2)^2 + 1 = x^2 + 4x + 4 + 1 = x^2 + 4x + 5$; $f(x) + f(2) = x^2 + 1 + (2)^2 + 1 = x^2 + 1 + 4 + 1 = x^2 + 6$.

**26.** $f(2x) = 3(2x) - 1 = 6x - 1$; $2f(x) = 2(3x - 1) = 6x - 2$.

**27.** $f(x^2) = x^2 + 4$; $[f(x)]^2 = [x+4]^2 = x^2 + 8x + 16$.

**28.** $f\left(\dfrac{x}{3}\right) - 6\left(\dfrac{x}{3}\right) - 18 = 2x - 18$; $\dfrac{f(x)}{3} = \dfrac{6x - 18}{3} = \dfrac{3(2x-6)}{3} = 2x - 6$.

**29.** $f(a) = 3(a) + 2 = 3a + 2$; $f(a+h) = 3(a+h) + 2 = 3a + 3h + 2$;
$$\frac{f(a+h) - f(a)}{h} = \frac{(3a + 3h + 2) - (3a + 2)}{h} = \frac{3a + 3h + 2 - 3a - 2}{h} = \frac{3h}{h} = 3.$$

**30.** $f(a) = (a)^2 + 1 = a^2 + 1$; $f(a+h) = (a+h)^2 + 1 = a^2 + 2ah + h^2 + 1$;
$$\begin{aligned}\frac{f(a+h) - f(a)}{h} &= \frac{(a^2 + 2ah + h^2 + 1) - (a^2 + 1)}{h} = \frac{a^2 + 2ah + h^2 + 1 - a^2 - 1}{h} \\ &= \frac{2ah + h^2}{h} = \frac{h(2a+h)}{h} = 2a + h.\end{aligned}$$

**31.** $f(a) = 5$; $f(a+h) = 5$; $\dfrac{f(a+h) - f(a)}{h} = \dfrac{5 - 5}{h} = 0$.

**32.** $f(a)=\dfrac{1}{a+1}$; $f(a+h)=\dfrac{1}{a+h+1}$;

$$\frac{f(a+h)-f(a)}{h}=\frac{\dfrac{1}{a+h+1}-\dfrac{1}{a+1}}{h}=\frac{\dfrac{a+1}{(a+1)(a+h+1)}-\dfrac{a+h+1}{(a+1)(a+h+1)}}{h}$$

$$=\frac{\dfrac{-h}{(a+1)(a+h+1)}}{h}=\frac{-1}{(a+1)(a+h+1)}.$$

**33.** $f(a)=\dfrac{a}{a+1}$; $f(a+h)=\dfrac{a+h}{a+h+1}$;

$$\frac{f(a+h)-f(a)}{h}=\frac{\dfrac{a+h}{a+h+1}-\dfrac{a}{a+1}}{h}=\frac{\dfrac{(a+h)(a+1)}{(a+h+1)(a+1)}-\dfrac{a(a+h+1)}{(a+h+1)(a+1)}}{h}$$

$$=\frac{\dfrac{(a+h)(a+1)-a(a+h+1)}{(a+h+1)(a+1)}}{h}=\frac{a^2+a+ah+h-\left(a^2+ah+a\right)}{h(a+h+1)(a+1)}$$

$$=\frac{1}{(a+h+1)(a+1)}$$

**34.** $f(a)=\dfrac{2a}{a-1}$; $f(a+h)=\dfrac{2(a+h)}{a+h-1}$;

$$\frac{f(a+h)-f(a)}{h}=\frac{\dfrac{2a+h}{a+h-1}-\dfrac{2a}{a-1}}{h}=\frac{\dfrac{(2a+h)(a-1)}{(a+h-1)(a-1)}-\dfrac{2a(a+h-1)}{(a+h-1)(a-1)}}{h}$$

$$=\frac{\dfrac{(2a+h)(a-1)-2a(a+h-1)}{(a+h-1)(a-1)}}{h}=\frac{2a^2+ah-2a-h-\left(2a^2+2ah-2a\right)}{h(a+h-1)(a-1)}$$

$$=\frac{-ah-h}{h(a+h-1)(a-1)}=-\frac{a+1}{(a+h-1)(a-1)}$$

**35.** $f(a)=3-5a+4a^2$;

$$f(a+h)=3-5(a+h)+4(a+h)^2=3-5a-5h+4\left(a^2+2ah+h^2\right)$$
$$=3-5a-5h+4a^2+8ah+4h^2;$$

$$\frac{f(a+h)-f(a)}{h}=\frac{\left(3-5a-5h+4a^2+8ah+4h^2\right)-\left(3-5a+4a^2\right)}{h}$$

$$=\frac{3-5a-5h+4a^2+8ah+4h^2-3+5a-4a^2}{h}=\frac{-5h+8ah+4h^2}{h}$$

$$=\frac{h(-5+8a+4h)}{h}=-5+8a+4h.$$

**36.** $f(a)=a^3$; $f(a+h)=(a+h)^3=a^3+3a^2h+3ah^2+h^3$;

$$\frac{f(a+h)-f(a)}{h}=\frac{\left(a^3+3a^2h+3ah^2+h^3\right)-\left(a^3\right)}{h}=\frac{3a^2h+3ah^2+h^3}{h}$$

$$=\frac{h\left(3a^2+3ah+h^2\right)}{h}=3a^2+3ah+h^2.$$

**37.** $f(x)=2x$. Since there is no restrictions, the domain is the set of real numbers, $(-\infty,\infty)$.

**38.** $f(x)=x^2+1$. Since there is no restrictions, the domain is all real numbers, $(-\infty,\infty)$.

**39.** $f(x)=2x$. The domain is restricted by the exercise to $[-1,5]$.

**40.** $f(x)=x^2+1$. The domain is restricted by the exercise to $[0,5]$.

**41.** $f(x)=\dfrac{1}{x-3}$. Since the denominator cannot equal 0 we have $x-3\neq 0 \quad\Leftrightarrow\quad x\neq 3$. Thus the domain is $\{x \mid x\neq 3\}$. In interval notation, the domain is $(-\infty,3)\cup(3,\infty)$.

**42.** $f(x) = \frac{1}{3x-6}$. Since the denominator cannot equal 0, we have $3x - 6 \neq 0 \quad \Leftrightarrow \quad 3x \neq 6 \quad \Leftrightarrow \quad x \neq 2$. In interval notation, the domain is $(-\infty, 2) \cup (2, \infty)$.

**43.** $f(x) = \frac{x+2}{x^2-1}$. Since the denominator cannot equal 0 we have $x^2 - 1 \neq 0 \quad \Leftrightarrow \quad x^2 \neq 1 \quad \Rightarrow \quad x \neq \pm 1$. Thus the domain is $\{x \mid x \neq \pm 1\}$. In interval notation, the domain is $(-\infty, -1) \cup (-1, 1) \cup (1, \infty)$.

**44.** $f(x) = \frac{x^4}{x^2+x-6}$. Since the denominator cannot equal 0, $x^2 + x - 6 \neq 0 \quad \Leftrightarrow \quad (x+3)(x-2) \neq 0 \quad \Rightarrow$ $x \neq -3$ or $x \neq 2$. In interval notation, the domain is $(-\infty, -3) \cup (-3, 2) \cup (2, \infty)$.

**45.** $f(x) = \sqrt{x-5}$. We require $x - 5 \geq 0 \quad \Leftrightarrow \quad x \geq 5$. Thus the domain is $\{x \mid x \geq 5\}$. The domain can also be expressed in interval notation as $[5, \infty)$.

**46.** $f(x) = \sqrt[4]{x+9}$. Since even roots are only defined for nonnegative numbers, we must have $x + 9 \geq 0 \quad \Leftrightarrow \quad x \geq -9$, so the domain is $[-9, \infty)$.

**47.** $f(t) = \sqrt[3]{t-1}$. Since the odd root is defined for all real numbers, the domain is the set of real numbers, $(-\infty, \infty)$.

**48.** $g(x) = \sqrt{7-3x}$. For the square root to be defined, we must have $7 - 3x \geq 0 \quad \Leftrightarrow \quad 7 \geq 3x \quad \Leftrightarrow \quad \frac{7}{3} \geq x$. Thus the domain is $\left(-\infty, \frac{7}{3}\right]$.

**49.** $h(x) = \sqrt{2x-5}$. Since the square root is defined as a real number only for nonnegative numbers, we require that $2x - 5 \geq 0 \quad \Leftrightarrow \quad 2x \geq 5 \quad \Leftrightarrow \quad x \geq \frac{5}{2}$. So the domain is $\{x \mid x \geq \frac{5}{2}\}$. In interval notation, the domain is $\left[\frac{5}{2}, \infty\right)$.

**50.** $G(x) = \sqrt{x^2-9}$. We must have $x^2 - 9 \geq 0 \quad \Leftrightarrow \quad x^2 \geq 9 \quad \Leftrightarrow \quad |x| \geq 3 \quad \Rightarrow \quad x \geq 3$ or $x \leq -3$. Thus the domain is $(-\infty, -3] \cup [3, \infty)$.

**51.** $g(x) = \frac{\sqrt{2+x}}{3-x}$. We require $2 + x \geq 0$, and the denominator cannot equal 0. Now $2 + x \geq 0 \quad \Leftrightarrow \quad x \geq -2$, and $3 - x \neq 0 \quad \Leftrightarrow \quad x \neq 3$. Thus the domain is $\{x \mid x \geq -2 \text{ and } x \neq 3\}$, which can be expressed in interval notation as $[-2, 3) \cup (3, \infty)$.

**52.** $g(x) = \frac{\sqrt{x}}{2x^2+x-1}$. We must have $x \geq 0$ for the numerator and $2x^2 + x - 1 \neq 0$ for the denominator. So $2x^2 + x - 1 \neq 0 \quad \Leftrightarrow \quad (2x-1)(x+1) \neq 0 \quad \Rightarrow \quad 2x - 1 \neq 0$ or $x + 1 \neq 0 \quad \Leftrightarrow \quad x \neq \frac{1}{2}$ or $x \neq -1$. Thus the domain is $\left[0, \frac{1}{2}\right) \cup \left(\frac{1}{2}, \infty\right)$.

**53.** $g(x) = \sqrt[4]{x^2-6x}$. Since the input to an even root must be nonnegative, we have $x^2 - 6x \geq 0 \quad \Leftrightarrow \quad x(x-6) \geq 0$. We make a table:

| | $(-\infty, 0)$ | $(0, 6)$ | $(6, \infty)$ |
|---|---|---|---|
| Sign of $x$ | $-$ | $+$ | $+$ |
| Sign of $x - 6$ | $-$ | $-$ | $+$ |
| Sign of $x(x-6)$ | $+$ | $-$ | $+$ |

Thus the domain is $(-\infty, 0] \cup [6, \infty)$.

**54.** $g(x) = \sqrt{x^2-2x-8}$. We must have $x^2 - 2x - 8 \geq 0 \quad \Leftrightarrow \quad (x-4)(x+2) \geq 0$. We make a table:

| | $(-\infty, -2)$ | $(-2, 4)$ | $(4, \infty)$ |
|---|---|---|---|
| Sign of $x - 4$ | $-$ | $-$ | $+$ |
| Sign of $x + 2$ | $-$ | $+$ | $+$ |
| Sign of $(x-4)(x+2)$ | $+$ | $-$ | $+$ |

Thus the domain is $(-\infty, -2] \cup [4, \infty)$.

**55.** $f(x) = \dfrac{3}{\sqrt{x-4}}$. Since the input to an even root must be nonnegative and the denominator cannot equal 0, we have $x - 4 > 0 \quad \Leftrightarrow \quad x > 4$. Thus the domain is $(4, \infty)$.

**56.** $f(x) = \dfrac{x^2}{\sqrt{6-x}}$. Since the input to an even root must be nonnegative and the denominator cannot equal 0, we have $6 - x > 0 \quad \Leftrightarrow \quad 6 > x$. Thus the domain is $(-\infty, 6)$.

**57.** $f(x) = \dfrac{(x+1)^2}{\sqrt{2x-1}}$. Since the input to an even root must be nonnegative and the denominator cannot equal 0, we have $2x - 1 > 0 \quad \Leftrightarrow \quad x > \frac{1}{2}$. Thus the domain is $\left(\frac{1}{2}, \infty\right)$.

**58.** $f(x) = \dfrac{x}{\sqrt[4]{9-x^2}}$. Since the input to an even root must be nonnegative and the denominator cannot equal 0, we have $9 - x^2 > 0 \quad \Leftrightarrow \quad (3-x)(3+x) > 0$. We make a table:

| Interval | $(-\infty, -3)$ | $(-3, 3)$ | $(3, \infty)$ |
|---|---|---|---|
| Sign of $3 - x$ | $+$ | $+$ | $-$ |
| Sign of $3 + x$ | $-$ | $+$ | $+$ |
| Sign of $(x-4)(x+2)$ | $-$ | $+$ | $-$ |

Thus the domain is $(-3, 3)$.

**59. (a)** $C(10) = 1500 + 3(10) + 0.02(10)^2 + 0.0001(10)^3 = 1500 + 30 + 2 + 0.1 = 1532.1$
$C(100) = 1500 + 3(100) + 0.02(100)^2 + 0.0001(100)^3 = 1500 + 300 + 200 + 100 = 2100$

**(b)** $C(10)$ represents the cost of producing 10 yards of fabric and $C(100)$ represents the cost of producing 100 yards of fabric.

**(c)** $C(0) = 1500 + 3(0) + 0.02(0)^2 + 0.0001(0)^3 = 1500$

**60. (a)** $S(2) = 4\pi(2)^2 = 16\pi \approx 50.27$, $S(3) = 4\pi(3)^2 = 36\pi \approx 113.10$.

**(b)** $S(2)$ represents the surface area of a sphere of radius 2, and $S(3)$ represents the surface area of a sphere of radius 3.

**61. (a)** $D(0.1) = \sqrt{2(3960)(0.1) + (0.1)^2} = \sqrt{792.01} \approx 28.1$ miles
$D(0.2) = \sqrt{2(3960)(0.2) + (0.2)^2} = \sqrt{1584.04} \approx 39.8$ miles

**(b)** 1135 feet $= \frac{1135}{5280}$ miles $\approx 0.215$ miles. $D(0.215) = \sqrt{2(3960)(0.215) + (0.215)^2} = \sqrt{1702.846} \approx 41.3$ miles

**(c)** $D(7) = \sqrt{2(3960)(7) + (7)^2} = \sqrt{55489} \approx 235.6$ miles

**62. (a)** $V(0) = 50\left(1 - \frac{0}{20}\right)^2 = 50$ and $V(20) = 50\left(1 - \frac{20}{20}\right)^2 = 0$.

**(b)** $V(0) = 50$ represents the volume of the full tank at time $t = 0$, and $V(20) = 0$ represents the volume of the empty tank twenty minutes later.

**(c)**

| $x$ | $V(x)$ |
|---|---|
| 0 | 50 |
| 5 | 28.125 |
| 10 | 12.5 |
| 15 | 3.125 |
| 20 | 0 |

**63. (a)** $v(0.1) = 18500(0.25 - 0.1^2) = 4440$,

$v(0.4) = 18500(0.25 - 0.4^2) = 1665$.

**(b)** They tell us that the blood flows much faster (about 2.75 times faster) 0.1 cm from the center than 0.1 cm from the edge.

**(c)**

| $r$ | $v(r)$ |
|---|---|
| 0 | 4625 |
| 0.1 | 4440 |
| 0.2 | 3885 |
| 0.3 | 2960 |
| 0.4 | 1665 |
| 0.5 | 0 |

**64. (a)** $R(1) = \sqrt{\dfrac{13 + 7(1)^{0.4}}{1 + 4(1)^{0.4}}} = \sqrt{\dfrac{20}{5}} = 2$,

$R(10) = \sqrt{\dfrac{13 + 7(10)^{0.4}}{1 + 4(10)^{0.4}}} \approx 1.66$, and

$R(100) = \sqrt{\dfrac{13 + 7(100)^{0.4}}{1 + 4(100)^{0.4}}} \approx 1.48$.

**(b)**

| $x$ | $R(x)$ |
|---|---|
| 1 | 2 |
| 10 | 1.66 |
| 100 | 1.48 |
| 200 | 1.44 |
| 500 | 1.41 |
| 1000 | 1.39 |

**65. (a)** $L(0.5c) = 10\sqrt{1 - \dfrac{(0.5c)^2}{c^2}} \approx 8.66$ m, $L(0.75c) = 10\sqrt{1 - \dfrac{(0.75c)^2}{c^2}} \approx 6.61$ m, and $L(0.9c) = 10\sqrt{1 - \dfrac{(0.9c)^2}{c^2}} \approx 4.36$ m.

**(b)** It will appear to get shorter.

**66. (a)** Since $0 \leq 5{,}000 \leq 10{,}000$ we have $T(5{,}000) = 0$. Since $10{,}000 < 12{,}000 \leq 20{,}000$ we have $T(12{,}000) = 0.08(12{,}000) = 960$. Since $20{,}000 < 25{,}000$ we have $T(25{,}000) = 1600 + 0.15(25{,}000) = 5350$.

**(b)** There is no tax on \$5000, a tax of \$960 on \$12,000 income, and a tax of \$5350 on \$25,000.

**67. (a)** $C(75) = 75 + 15 = \$90$; $C(90) = 90 + 15 = \$105$; $C(100) = \$100$; and $C(105) = \$105$.

**(b)** The total price of the books purchased, including shipping.

**68. (a)** $T(x) = \begin{cases} 75x & \text{if } 0 \leq x \leq 2 \\ 50x + 50 & \text{if } x > 2 \end{cases}$

**(b)** $T(2) = 75(2) = 150$; $T(3) = 50(3) + 50 = 200$; and $T(5) = 5(50) + 50 = 300$.

**(c)** The total cost of the lodgings.

**69. (a)** $F(x) = \begin{cases} 15(40 - x) & \text{if } 0 < x < 40 \\ 0 & \text{if } 40 \leq x \leq 65 \\ 15(x - 65) & \text{if } x > 65 \end{cases}$.

**(b)** $F(30) = 15(40 - 10) = 15 \cdot 10 = \$150$; $F(50) = \$0$; and $F(75) = 15(75 - 65)\,15 \cdot 10 = \$150$.

**(c)** The fines for violating the speed limits on the freeway.

**70.** We assume the grass grows linearly.

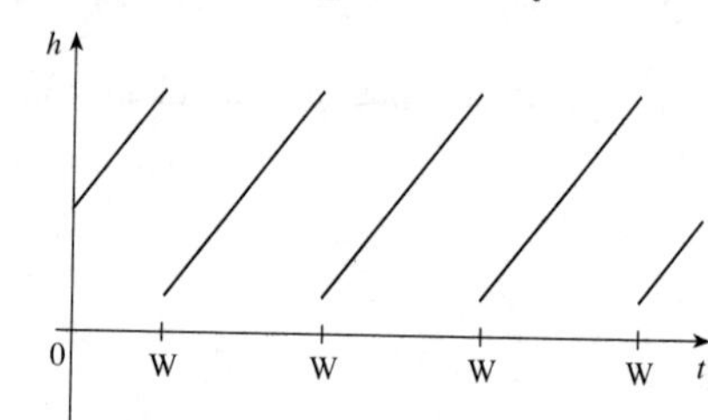

**71.**

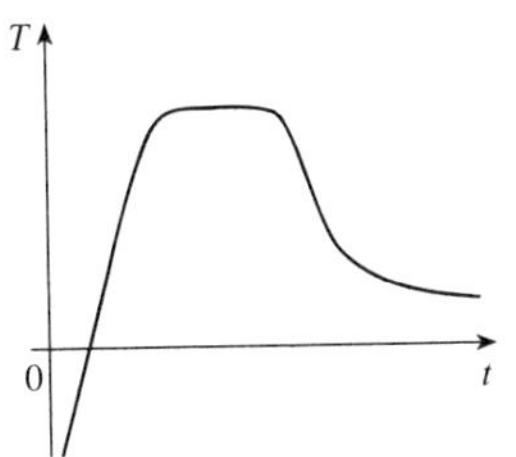


**72.**

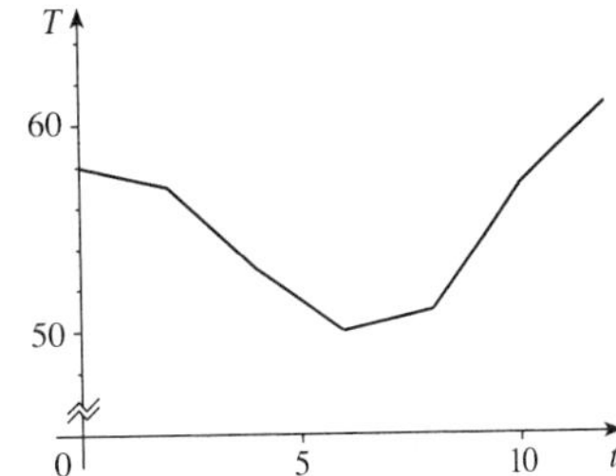


**73.**

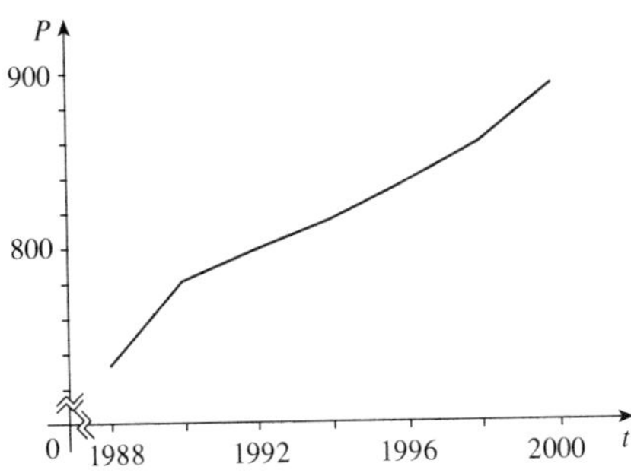


**74.** Answers will vary.

**75.** Answers will vary.

# 3.2 Graphs of Functions

**1.**

| $x$ | $f(x) = 2$ |
|---|---|
| $-9$ | $2$ |
| $-6$ | $2$ |
| $-3$ | $2$ |
| $0$ | $2$ |
| $3$ | $2$ |
| $6$ | $2$ |

**2.**

| $x$ | $f(x) = -3$ |
|---|---|
| $-4$ | $-3$ |
| $-2$ | $-3$ |
| $0$ | $-3$ |
| $2$ | $-3$ |
| $4$ | $-3$ |
| $6$ | $-3$ |

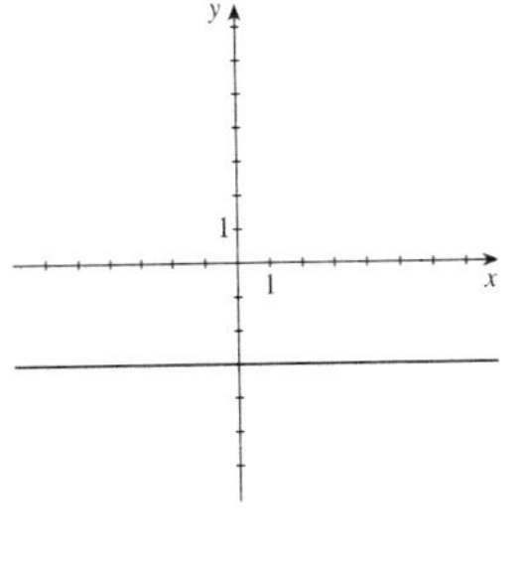


**3.**

| $x$ | $f(x) = 2x - 4$ |
|---|---|
| $-1$ | $-6$ |
| $0$ | $-4$ |
| $1$ | $-2$ |
| $2$ | $0$ |
| $3$ | $2$ |
| $4$ | $4$ |
| $5$ | $6$ |

**4.**

| $x$ | $f(x) = 6 - 3x$ |
|---|---|
| $-2$ | $12$ |
| $-1$ | $9$ |
| $0$ | $6$ |
| $1$ | $3$ |
| $2$ | $0$ |
| $3$ | $-3$ |

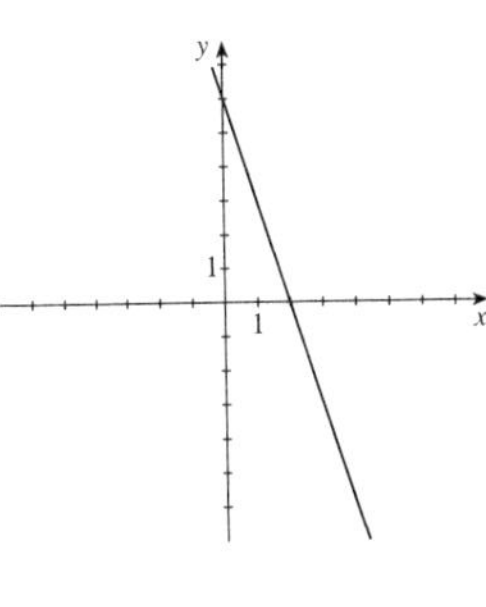

**5.**

| $x$ | $f(x) = -x + 3,$ $-3 \le x \le 3$ |
|---|---|
| $-3$ | 6 |
| $-2$ | 5 |
| 0 | 3 |
| 1 | 2 |
| 2 | 1 |
| 3 | 0 |

**6.**

| $x$ | $f(x) = \dfrac{x-3}{2},$ $0 \le x \le 5$ |
|---|---|
| 0 | $-1.5$ |
| 1 | $-1$ |
| 2 | $-0.5$ |
| 3 | 0 |
| 4 | 0.5 |
| 5 | 1 |

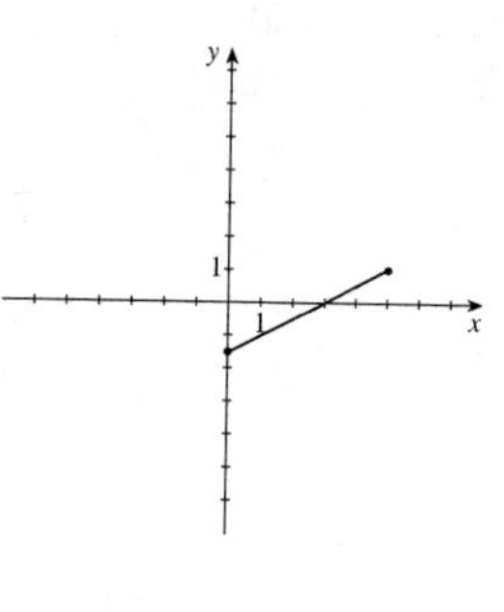

**7.**

| $x$ | $f(x) = -x^2$ |
|---|---|
| $\pm 4$ | $-16$ |
| $\pm 3$ | $-9$ |
| $\pm 2$ | $-4$ |
| $\pm 1$ | $-1$ |
| 0 | 0 |

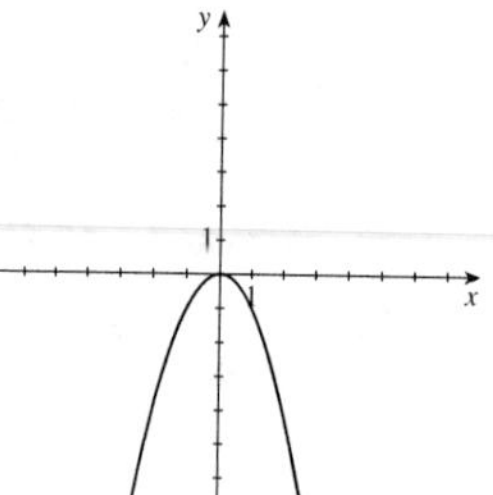

**8.**

| $x$ | $f(x) = x^2 - 4$ |
|---|---|
| $\pm 5$ | 21 |
| $\pm 4$ | 12 |
| $\pm 3$ | 5 |
| $\pm 2$ | 0 |
| $\pm 1$ | $-3$ |
| 0 | $-4$ |

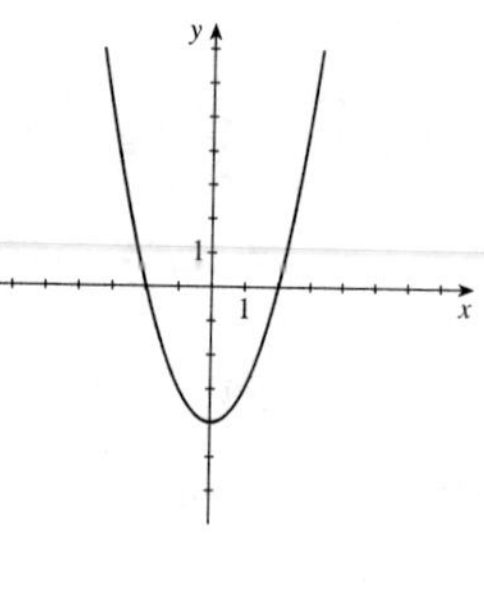

**9.**

| $x$ | $g(x) = x^3 - 8$ |
|---|---|
| $-2$ | $-16$ |
| $-1$ | $-9$ |
| 0 | $-8$ |
| 1 | $-7$ |
| 2 | 0 |
| 3 | 19 |

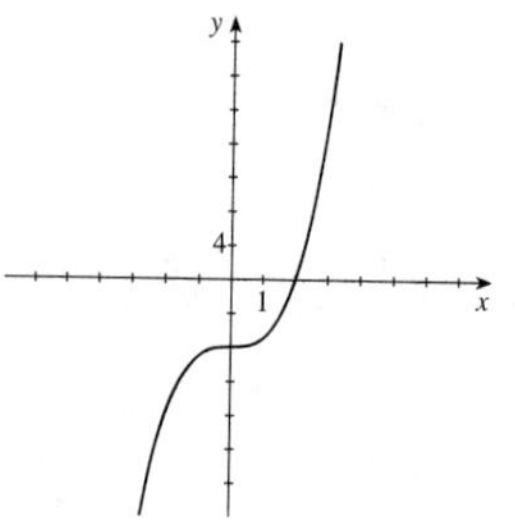

**10.**

| $x$ | $g(x) = 4x^2 - x^4$ |
|---|---|
| $\pm 4$ | $-64$ |
| $\pm 3$ | $-45$ |
| $\pm 2$ | 0 |
| $\pm 1$ | 3 |
| 0 | 0 |

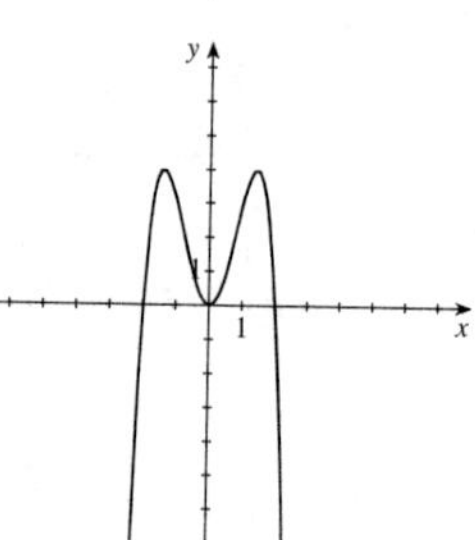

**11.**

| $x$ | $g(x) = \sqrt{x+4}$ |
|---|---|
| $-4$ | 0 |
| $-3$ | 1 |
| $-2$ | 1.414 |
| $-1$ | 1.732 |
| 0 | 2 |
| 1 | 2.236 |

**12.**

| $x$ | $g(x) = \sqrt{-x}$ |
|---|---|
| $-9$ | 3 |
| $-5$ | 2.236 |
| $-4$ | 2 |
| $-2$ | 1.414 |
| $-1$ | 1 |
| 0 | 0 |

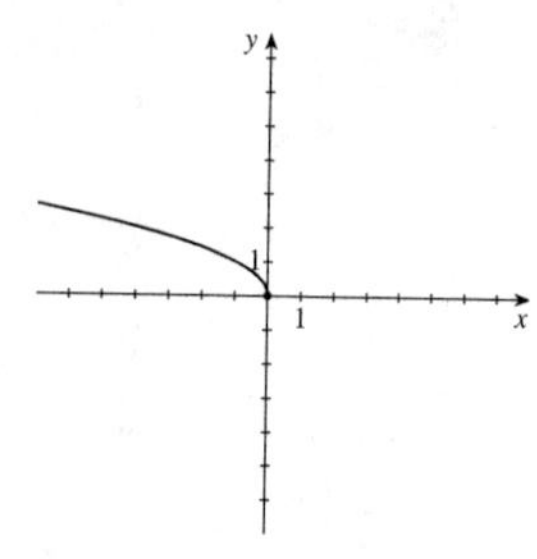

**13.**

| $x$ | $F(x) = \frac{1}{x}$ |
|---|---|
| $-5$ | $-0.2$ |
| $-1$ | $-1$ |
| $-0.1$ | $-10$ |
| $0.1$ | $10$ |
| $1$ | $1$ |
| $5$ | $0.2$ |

**14.**

| $x$ | $F(x) = \frac{1}{x+4}$ |
|---|---|
| $-6$ | $-0.5$ |
| $-5$ | $-1$ |
| $-4.5$ | $-2$ |
| $-4.1$ | $-10$ |
| $-3.9$ | $10$ |
| $-3.5$ | $2$ |
| $-3$ | $1$ |
| $-2$ | $0.5$ |

**15.**

| $x$ | $H(x) = \lvert 2x \rvert$ |
|---|---|
| $\pm 5$ | $10$ |
| $\pm 4$ | $8$ |
| $\pm 3$ | $6$ |
| $\pm 2$ | $4$ |
| $\pm 1$ | $2$ |
| $0$ | $0$ |

**16.**

| $x$ | $H(x) = \lvert x+1 \rvert$ |
|---|---|
| $-5$ | $4$ |
| $-4$ | $3$ |
| $-3$ | $2$ |
| $-2$ | $1$ |
| $-1$ | $0$ |
| $0$ | $1$ |
| $1$ | $2$ |

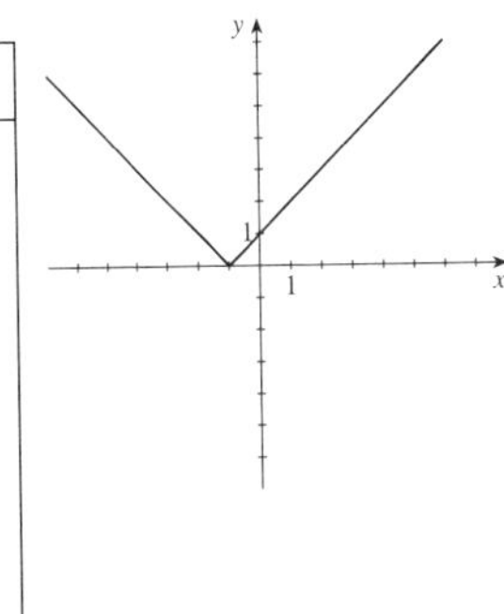

**17.**

| $x$ | $G(x) = \lvert x \rvert + x$ |
|---|---|
| $-5$ | $0$ |
| $-2$ | $0$ |
| $0$ | $0$ |
| $1$ | $2$ |
| $2$ | $4$ |
| $5$ | $10$ |

**18.**

| $x$ | $G(x) = \lvert x \rvert - x$ |
|---|---|
| $-5$ | $10$ |
| $-2$ | $4$ |
| $-1$ | $2$ |
| $0$ | $0$ |
| $1$ | $0$ |
| $3$ | $0$ |

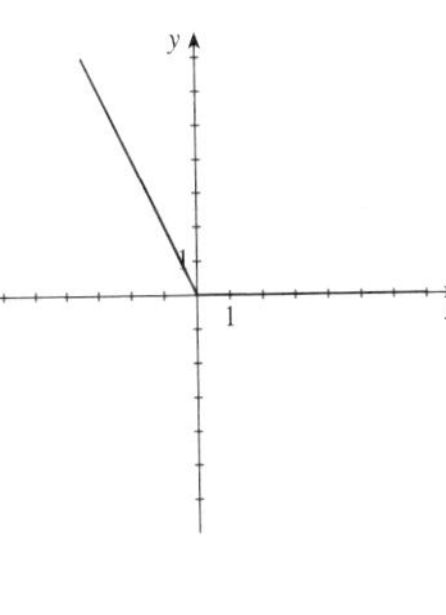

**19.**

| $x$ | $f(x) = \lvert 2x-2 \rvert$ |
|---|---|
| $-5$ | $12$ |
| $-2$ | $8$ |
| $0$ | $2$ |
| $1$ | $0$ |
| $2$ | $2$ |
| $5$ | $8$ |

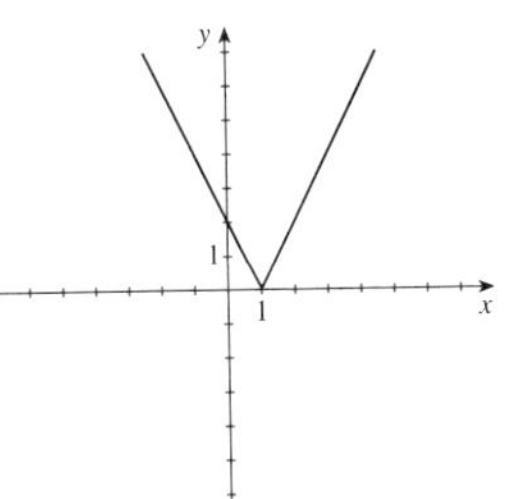

**20.**

| $x$ | $f(x) = \frac{x}{\lvert x \rvert}$ |
|---|---|
| $-3$ | $-1$ |
| $-2$ | $-1$ |
| $-1$ | $-1$ |
| $0$ | undefined |
| $1$ | $0$ |
| $2$ | $0$ |
| $3$ | $0$ |

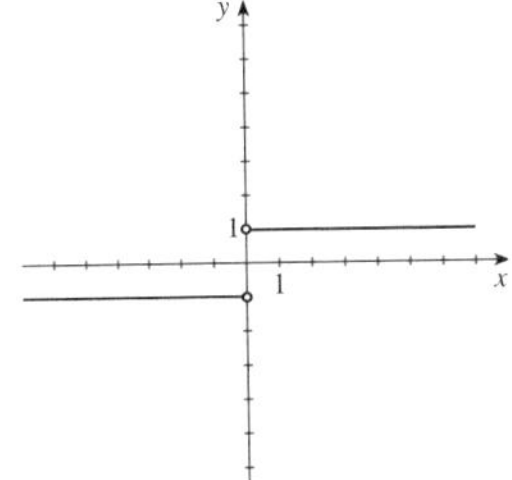

**21.**

| $x$ | $g(x) = \frac{2}{x^2}$ |
|---|---|
| $\pm 5$ | 0.08 |
| $\pm 4$ | 0.125 |
| $\pm 3$ | 0.2 |
| $\pm 2$ | 0.5 |
| $\pm 1$ | 2 |
| $\pm 0.5$ | 8 |

**22.**

| $x$ | $g(x) = \frac{\lvert x \rvert}{x^2}$ |
|---|---|
| $\pm 4$ | 0.25 |
| $\pm 2$ | 0.5 |
| $\pm 1$ | 1 |
| $\pm 0.5$ | 2 |
| $\pm 0.25$ | 4 |
| $\pm 0.10$ | 10 |

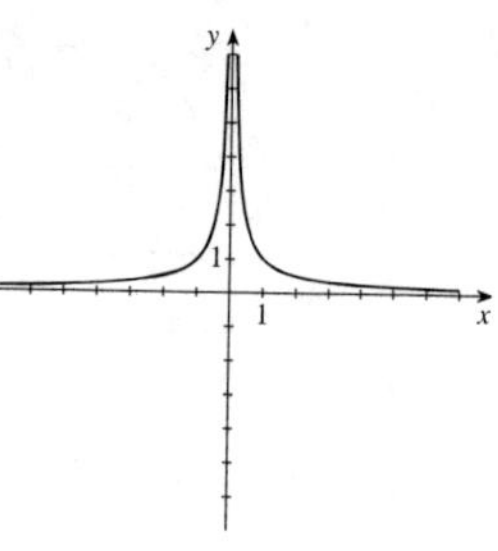

**23. (a)** $h(-2) = 1;\ h(0) = -1;\ h(2) = 3;\ h(3) = 4.$

**(b)** Domain: $[-3, 4]$. Range: $[-1, 4]$.

**24. (a)** $g(-4) = 3;\ g(-2) = 2;\ g(0) = -2;\ g(2) = 1;\ g(4) = 0.$

**(b)** Domain: $[-4, 4]$. Range: $[-2, 3]$.

**25. (a)** $f(0) = 3 > \frac{1}{2} = g(0)$. So $f(0)$ is larger.

**(b)** $f(-3) \approx -\frac{3}{2} < 2 = g(-3)$. So $g(-3)$ is larger.

**(c)** For $x = -2$ and $x = 2$.

**26. (a)** $f(0.5) \approx 1.2$

**(b)** $f(3) \approx 2.1$

**(c)** $x = 0.4$ and $x = 3.6$ are the only solutions to the equation $f(x) = 1$.

**27. (a)**

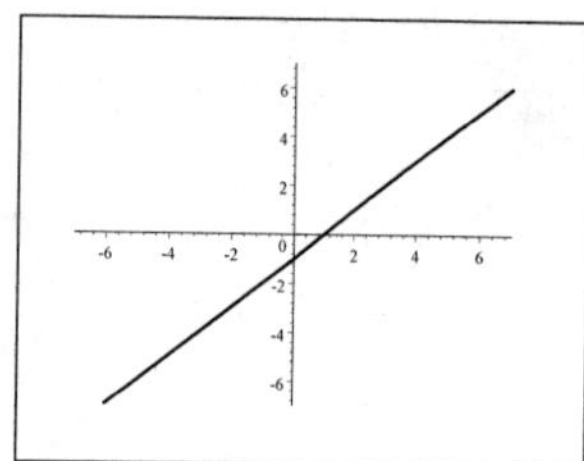

**(b)** Domain: $(-\infty, \infty)$; Range: $(-\infty, \infty)$

**28. (a)**

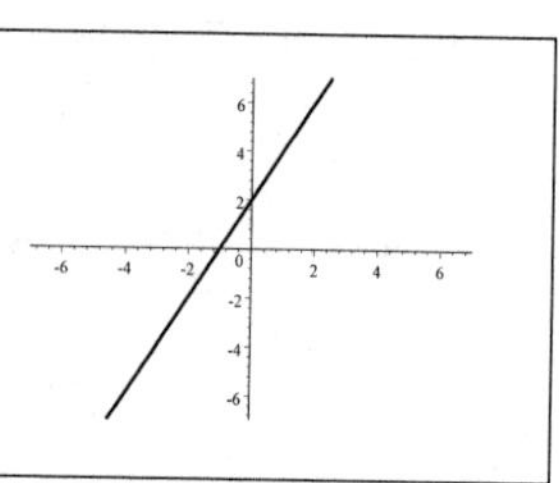

**(b)** Domain: $(-\infty, \infty)$; Range: $(-\infty, \infty)$

**29. (a)**

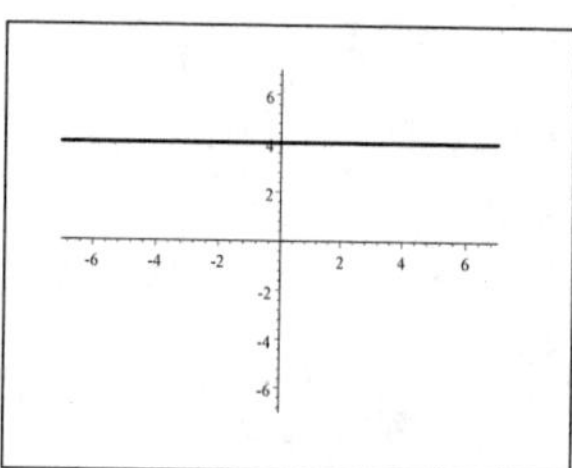

**(b)** Domain: $(-\infty, \infty)$; Range: $\{4\}$

**30. (a)**

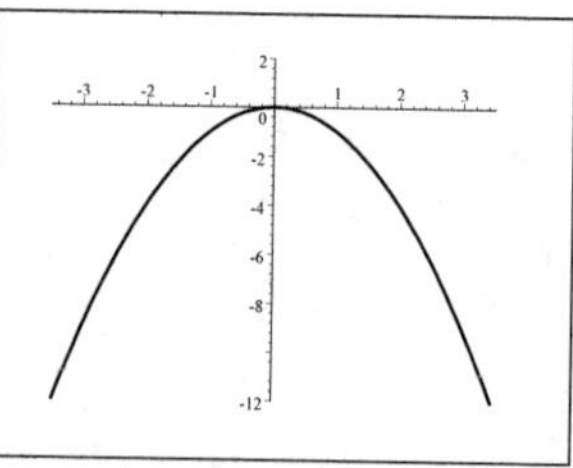

**(b)** Domain: $(-\infty, \infty)$; Range: $(-\infty, 0]$

**31. (a)**

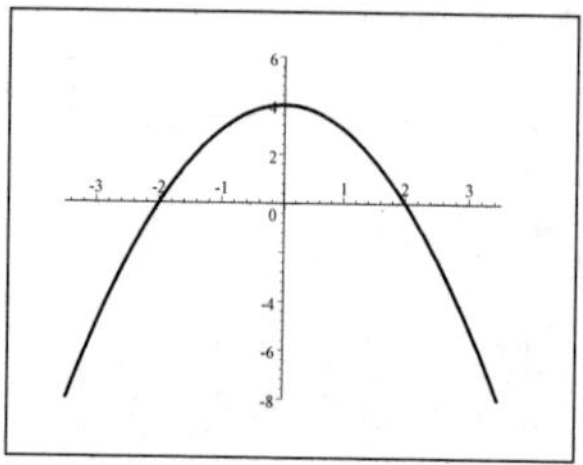

**(b)** Domain: $(-\infty, \infty)$; Range: $(-\infty, 4]$

**32. (a)**

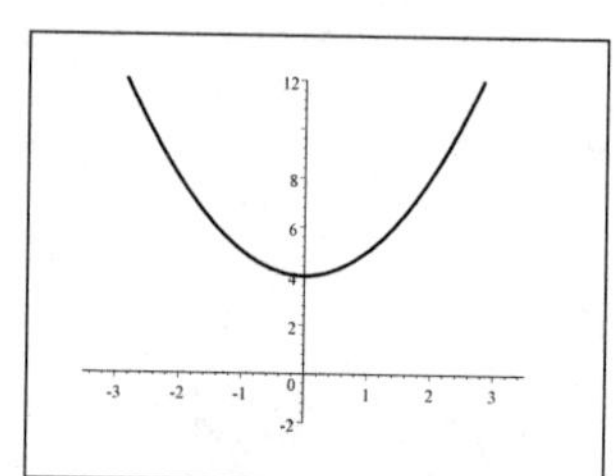

**(b)** Domain: $(-\infty, \infty)$; Range: $[4, \infty)$

**33. (a)**

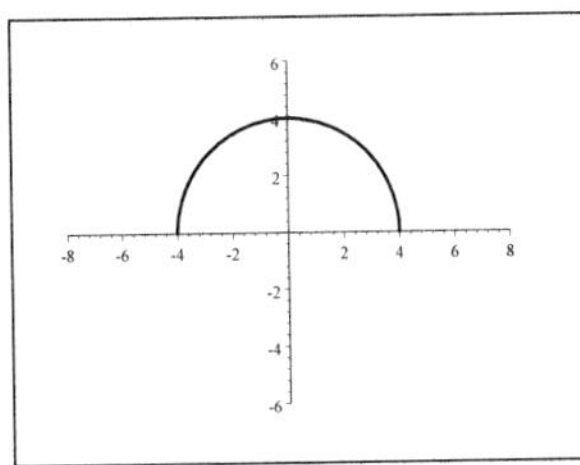

**(b)** Domain: $[-4,4]$; Range: $[0,4]$

**34. (a)**

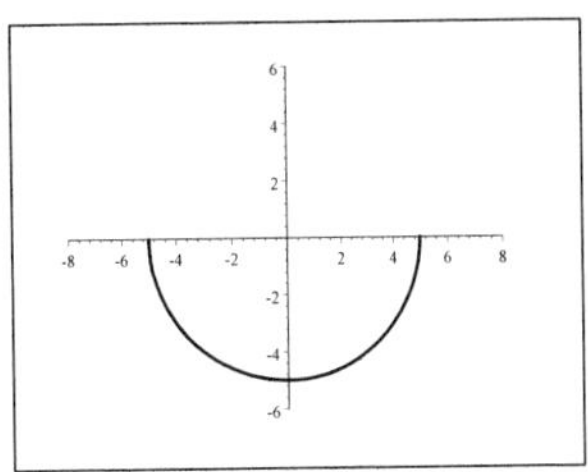

**(b)** Domain: $[-5,5]$; Range: $[-5,0]$

**35. (a)**

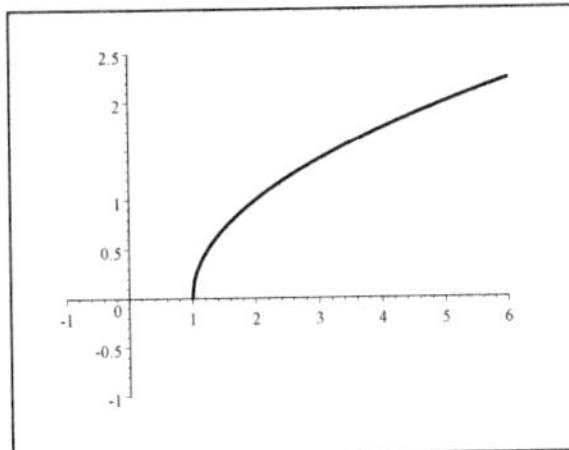

**(b)** Domain: $[1,\infty)$; Range: $[0,\infty)$

**36. (a)**

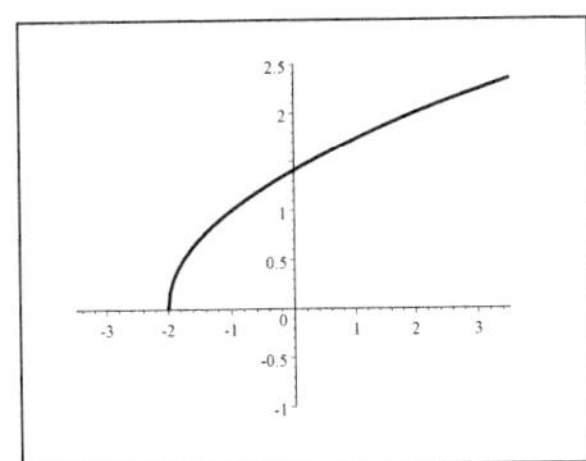

**(b)** Domain: $[-2,\infty)$; Range: $[0,\infty)$

**37.** $f(x)=\begin{cases} 0 & \text{if } x<2 \\ 1 & \text{if } x\ge 2 \end{cases}$

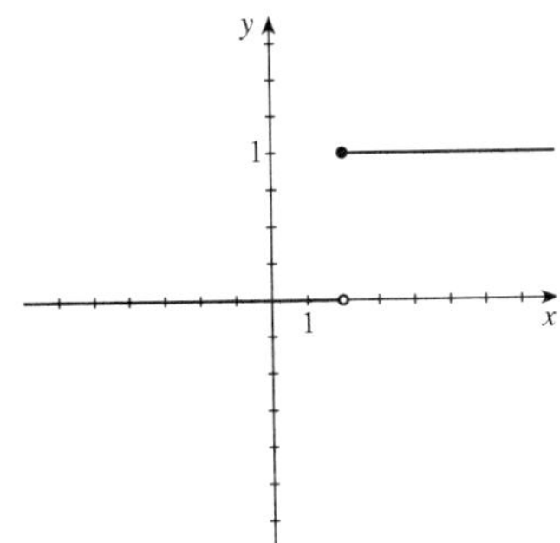


**38.** $f(x)=\begin{cases} 1 & \text{if } x\le 1 \\ x+1 & \text{if } x>1 \end{cases}$

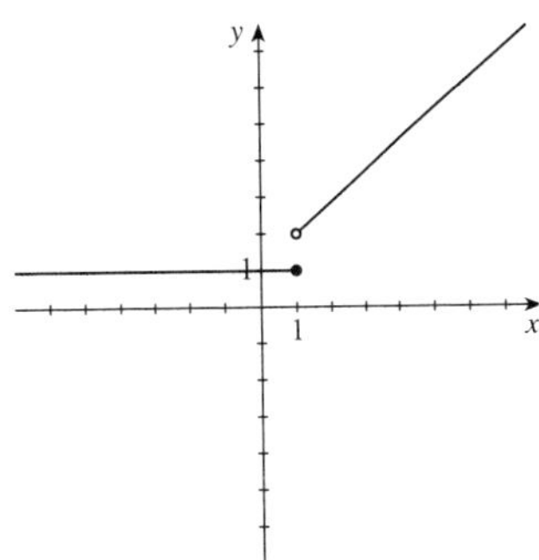


**39.** $f(x)=\begin{cases} 3 & \text{if } x<2 \\ x-1 & \text{if } x\ge 2 \end{cases}$

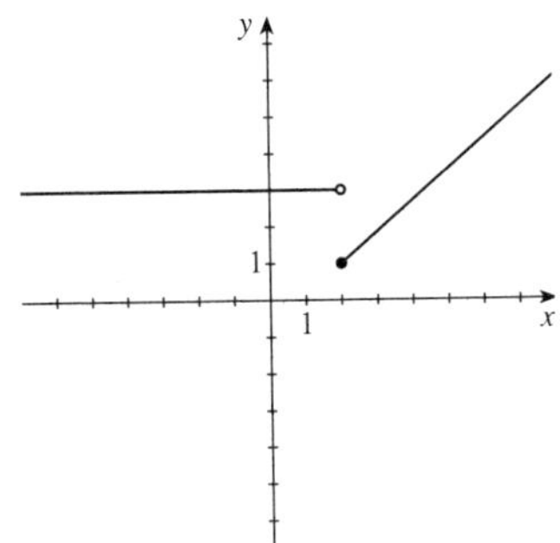


**40.** $f(x)=\begin{cases} 1-x & \text{if } x<-2 \\ 5 & \text{if } x\ge -2 \end{cases}$

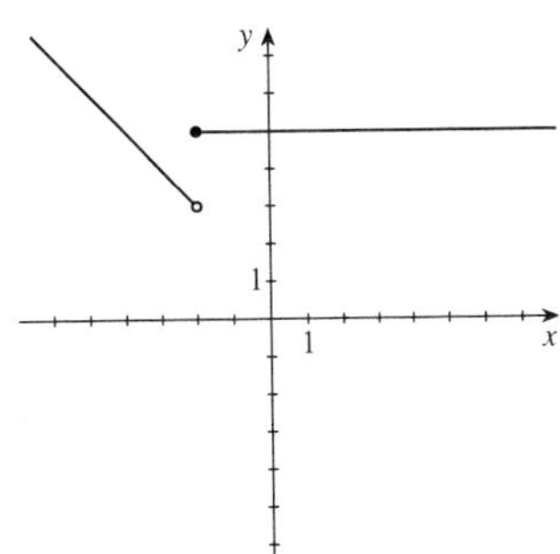

**41.** $f(x) = \begin{cases} x & \text{if } x \le 0 \\ x+1 & \text{if } x > 0 \end{cases}$

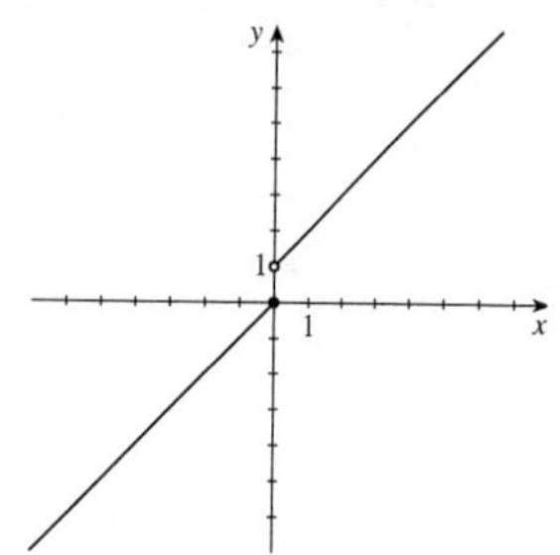

**42.** $f(x) = \begin{cases} 2x+3 & \text{if } x < -1 \\ 3-x & \text{if } x \ge -1 \end{cases}$

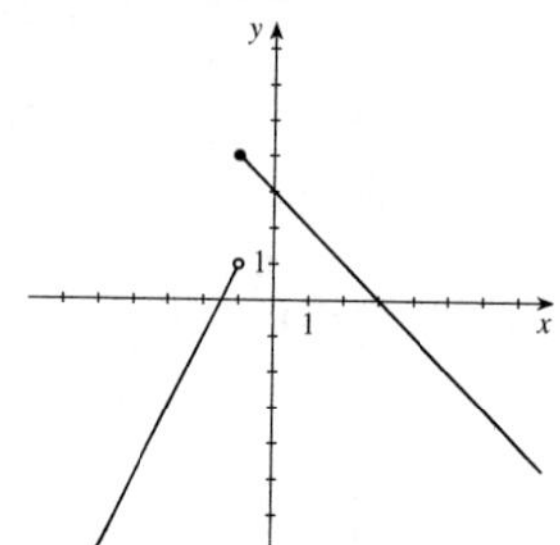

**43.** $f(x) = \begin{cases} -1 & \text{if } x < -1 \\ 1 & \text{if } -1 \le x \le 1 \\ -1 & \text{if } x > 1 \end{cases}$

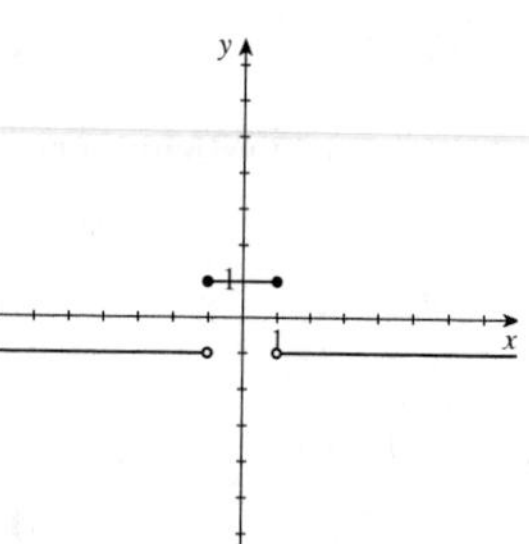

**44.** $f(x) = \begin{cases} -1 & \text{if } x < -1 \\ x & \text{if } -1 \le x \le 1 \\ 1 & \text{if } x > 1 \end{cases}$

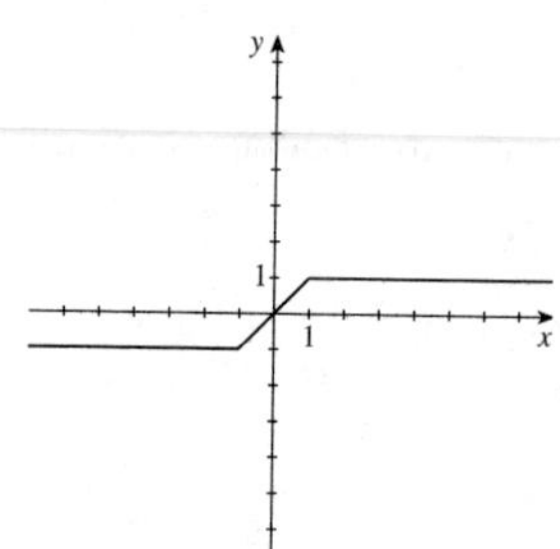

**45.** $f(x) = \begin{cases} 2 & \text{if } x \le -1 \\ x^2 & \text{if } x > -1 \end{cases}$

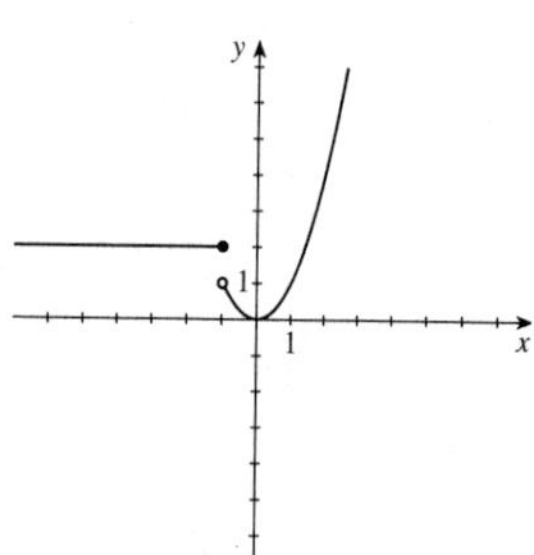

**46.** $f(x) = \begin{cases} 1-x^2 & \text{if } x \le 2 \\ x & \text{if } x > 2 \end{cases}$

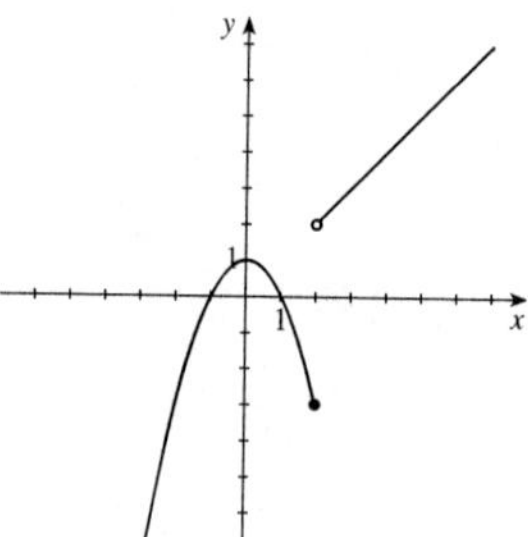

**47.** $f(x) = \begin{cases} 0 & \text{if } |x| \le 2 \\ 3 & \text{if } |x| > 2 \end{cases}$

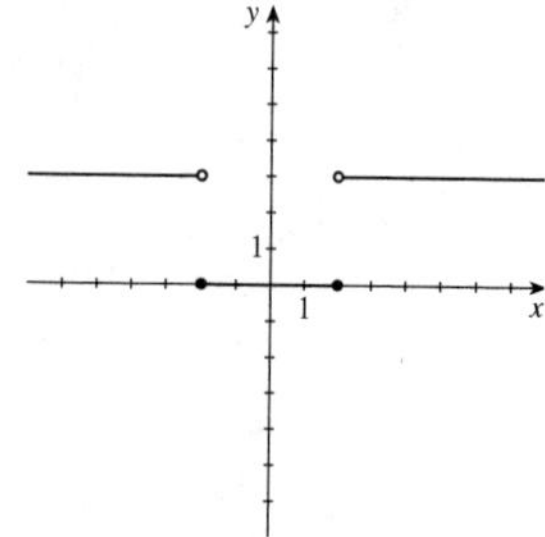

**48.** $f(x) = \begin{cases} x^2 & \text{if } |x| \le 1 \\ 1 & \text{if } |x| > 1 \end{cases}$

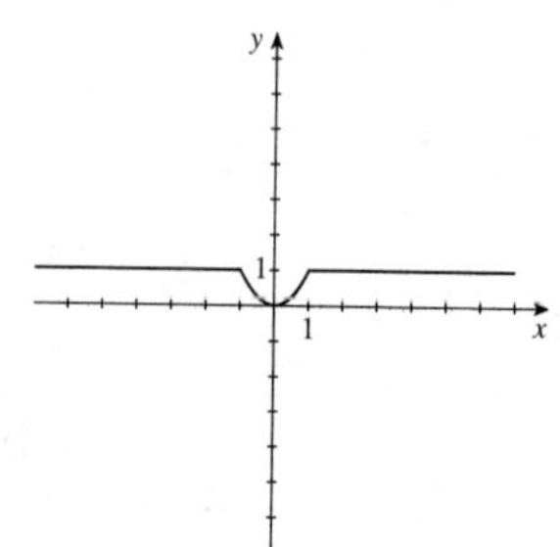

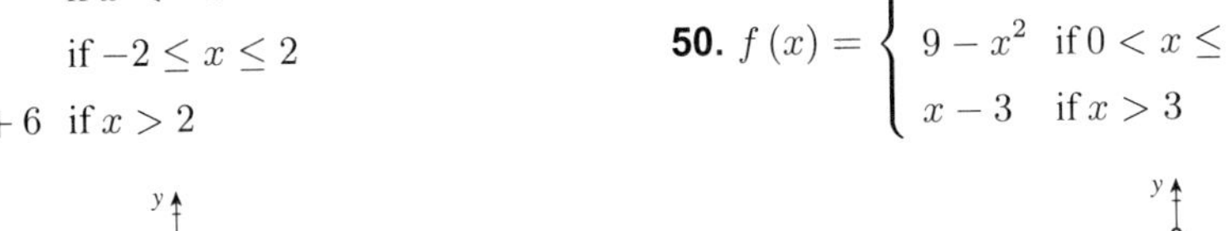

**49.** $f(x) = \begin{cases} 4 & \text{if } x < -2 \\ x^2 & \text{if } -2 \le x \le 2 \\ -x + 6 & \text{if } x > 2 \end{cases}$

**50.** $f(x) = \begin{cases} -x & \text{if } x \le 0 \\ 9 - x^2 & \text{if } 0 < x \le 3 \\ x - 3 & \text{if } x > 3 \end{cases}$

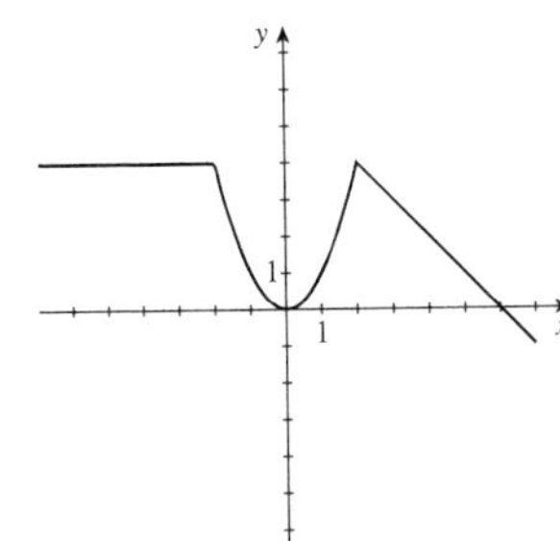


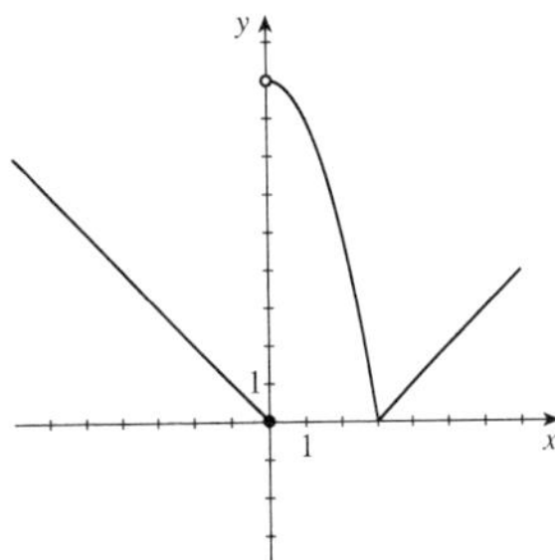


**51.** $f(x) = \begin{cases} x + 2 & \text{if } x \le -1 \\ x^2 & \text{if } x > -1 \end{cases}$

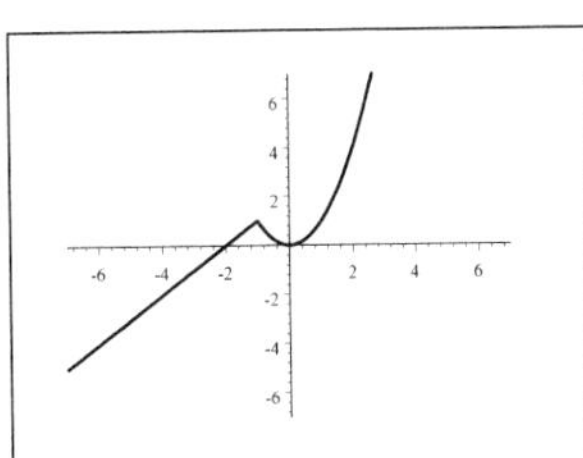


**52.** $f(x) = \begin{cases} 2x - x^2 & \text{if } x > 1 \\ (x - 1)^3 & \text{if } x \le 1 \end{cases}$ The first graph shows the output of a typical graphing device. However, the actual graph of this function is also shown, and its difference from the graphing device's version should be noted.

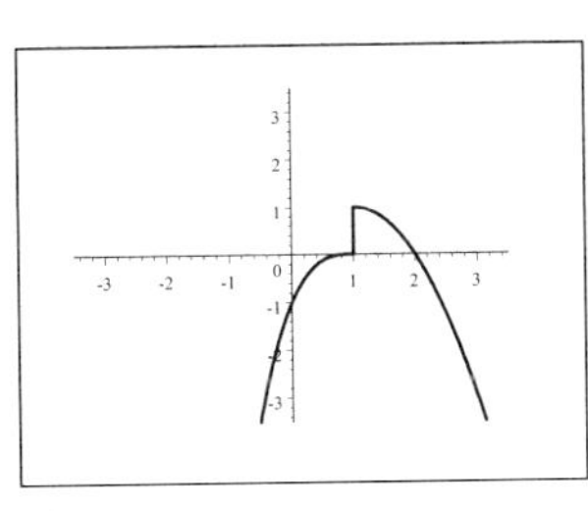


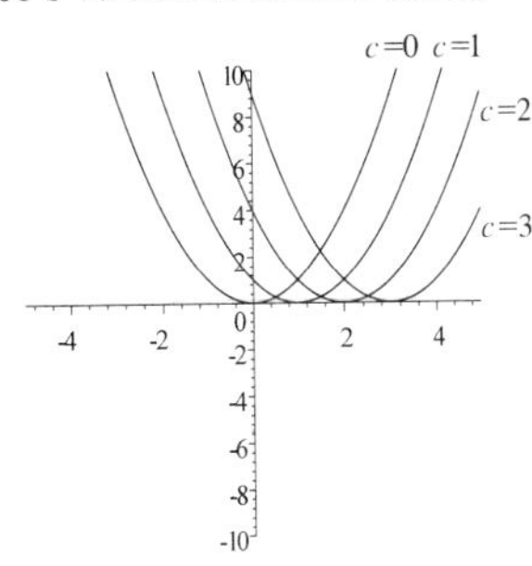


**53.** $f(x) = \begin{cases} -2 & \text{if } x < -2 \\ x & \text{if } -2 \le x \le 2 \\ 2 & \text{if } x > 2 \end{cases}$

**54.** $f(x) = \begin{cases} 1 & \text{if } x \le -1 \\ 1 - x & \text{if } -1 < x \le 2 \\ -2 & \text{if } x > 2 \end{cases}$

**55.** The curves in parts (a) and (c) are graphs of a function of $x$, by the Vertical Line Test.

**56.** The curves in parts (b) and (c) are graphs of functions of $x$.

**57.** The given curve is the graph of a function of $x$. Domain: $[-3, 2]$. Range: $[-2, 2]$.

**58.** No, the given curve is not the graph of a function of $x$.

**59.** No, the given curve is not the graph of a function of $x$, by the Vertical Line Test.

**60.** The given curve is the graph of a function of $x$. Domain: $[-3, 2]$. Range: $\{-2\} \cup (0, 3]$.

**61.** Solving for $y$ in terms of $x$ gives $x^2 + 2y = 4 \quad \Leftrightarrow \quad 2y = 4 - x^2 \quad \Leftrightarrow \quad y = 2 - \frac{1}{2}x^2$. This defines $y$ as a function of $x$.

**62.** Solving for $y$ in terms of $x$ gives $3x + 7y = 21 \quad \Leftrightarrow \quad 7y = -3x + 21 \quad \Leftrightarrow \quad y = -\frac{3}{7}x + 3$. This defines $y$ as a function of $x$.

**63.** Solving for $y$ in terms of $x$ gives $x = y^2 \Leftrightarrow y = \pm\sqrt{x}$. The last equation gives two values of $y$ for a given value of $x$. Thus, this equation does not define $y$ as a function of $x$.

**64.** Solving for $y$ in terms of $x$ gives $x^2 + (y-1)^2 = 4 \Leftrightarrow (y-1)^2 = 4 - x^2 \Leftrightarrow y - 1 = \pm\sqrt{4-x^2} \Leftrightarrow y = 1 \pm \sqrt{4-x^2}$. The last equation gives two values of $y$ for a given value of $x$. Thus, this equation does not define $y$ as a function of $x$.

**65.** Solving for $y$ in terms of $x$ gives $x + y^2 = 9 \Leftrightarrow y^2 = 9 - x \Leftrightarrow y = \pm\sqrt{9-x}$. The last equation gives two values of $y$ for a given value of $x$. Thus, this equation does not define $y$ as a function of $x$.

**66.** Solving for $y$ in terms of $x$ gives $x^2 + y = 9 \Leftrightarrow y = 9 - x^2$. This defines $y$ as a function of $x$.

**67.** Solving for $y$ in terms of $x$ gives $x^2y + y = 1 \Leftrightarrow y\left(x^2+1\right) = 1 \Leftrightarrow y = \dfrac{1}{x^2+1}$. This defines $y$ as a function of $x$.

**68.** Solving for $y$ in terms of $x$ gives $\sqrt{x} + y = 12 \Leftrightarrow y = 12 - \sqrt{x}$. This defines $y$ as a function of $x$.

**69.** Solving for $y$ in terms of $x$ gives $2|x| + y = 0 \Leftrightarrow y = -2|x|$. This defines $y$ as a function of $x$.

**70.** Solving for $y$ in terms of $x$ gives $2x + |y| = 0 \Leftrightarrow |y| = -2x$. Since $|a| = |-a|$, the last equation gives two values of $y$ for a given value of $x$. Thus, this equation does not define $y$ as a function of $x$.

**71.** Solving for $y$ in terms of $x$ gives $x = y^3 \Leftrightarrow y = \sqrt[3]{x}$. This defines $y$ as a function of $x$.

**72.** Solving for $y$ in terms of $x$ gives $x = y^4 \Leftrightarrow y = \pm\sqrt[4]{x}$. The last equation gives two values of $y$ for a given value of $x$. Thus, this equation does not define $y$ as a function of $x$.

**73. (a)** $f(x) = x^2 + c$, for $c = 0$, 2, 4, and 6.

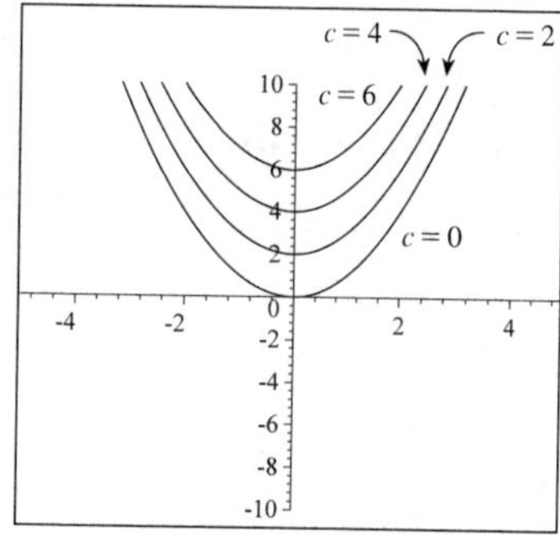


**(b)** $f(x) = x^2 + c$, for $c = 0, -2, -4$, and $-6$.

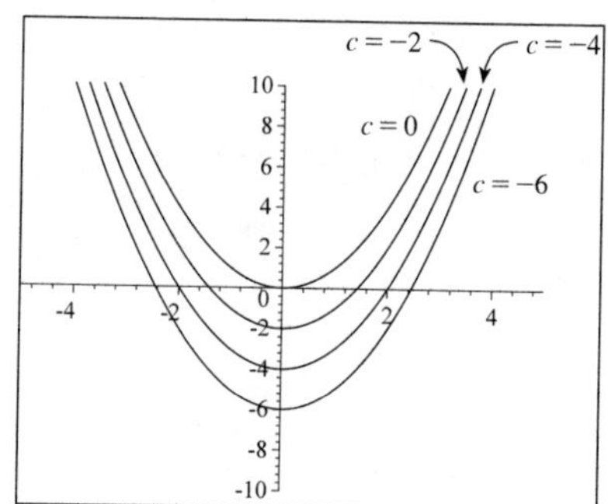


**(c)** The graphs in part (a) are obtained by shifting the graph of $f(x) = x^2$ upward $c$ units, $c > 0$. The graphs in part (b) are obtained by shifting the graph of $f(x) = x^2$ downward $c$ units.

**74. (a)** $f(x) = (x-c)^2$, for $c = 0$, 1, 2, and 3.

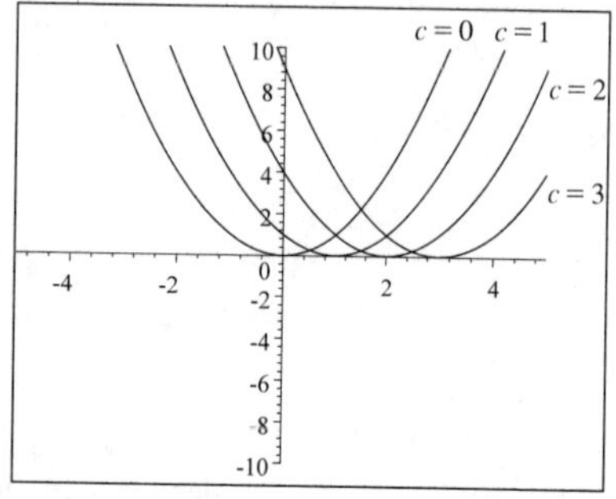


**(b)** $f(x) = (x-c)^2$, for $c = 0, -1, -2$, and $-3$.

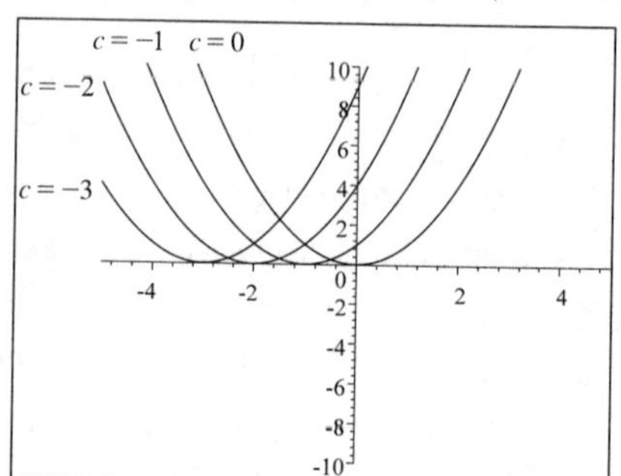


**(c)** The graphs in part (a) are obtained by shifting the graph of $y = x^2$ to the right 1, 2, and 3 units, while the graphs in part (b) are obtained by shifting the graph of $y = x^2$ to the left 1, 2, and 3 units.

**75. (a)** $f(x) = (x - c)^3$, for $c = 0, 2, 4$, and $6$.

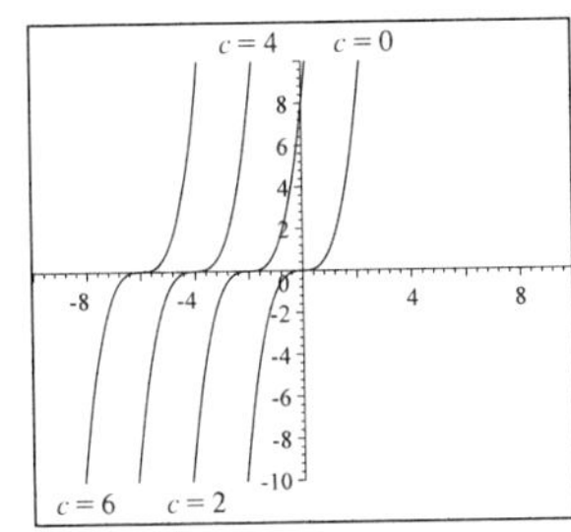


**(b)** $f(x) = (x - c)^3$, for $c = 0, -2, -4$, and $-6$.

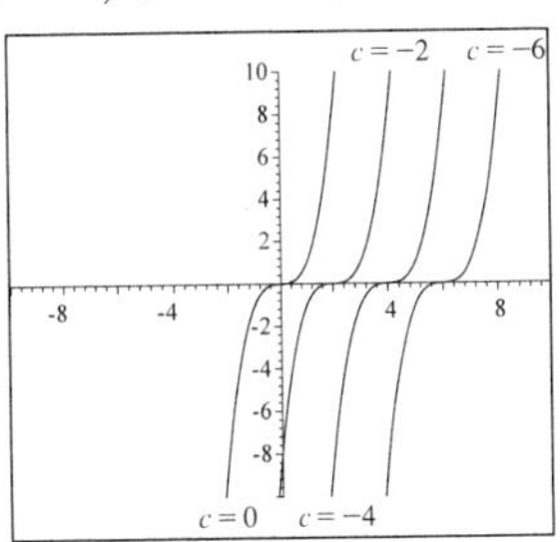


**(c)** The graphs in part (a) are obtained by shifting the graph of $f(x) = x^3$ to the right $c$ units, $c > 0$. The graphs in part (b) are obtained by shifting the graph of $f(x) = x^3$ to the left $|c|$ units, $c < 0$.

**76. (a)** $f(x) = cx^2$, for $c = 1, \frac{1}{2}, 2$, and $4$.

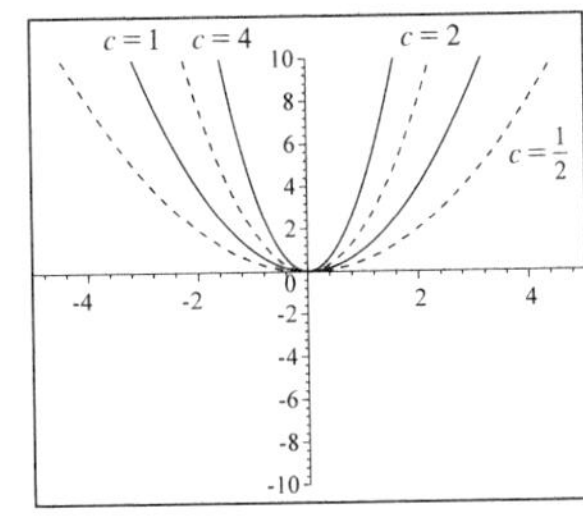


**(b)** $f(x) = cx^2$, for $c = 1, -1, -\frac{1}{2}$, and $-2$.

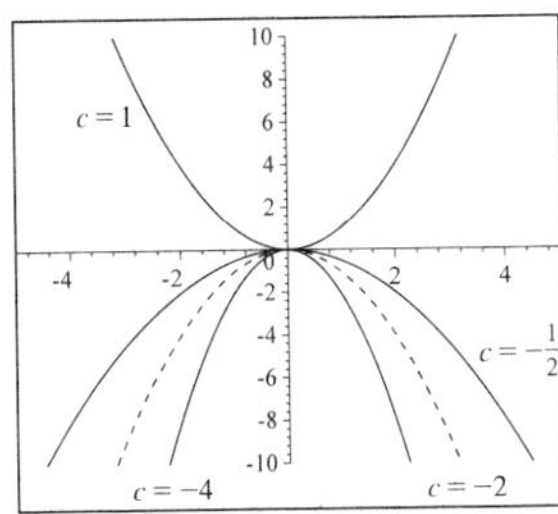


**(c)** As $|c|$ increases, the graph of $f(x) = cx^2$ is stretched vertically. As $|c|$ decreases, the graph of $f$ is flattened. When $c < 0$, the graph is reflected about the $x$-axis.

**77. (a)** $f(x) = x^c$, for $c = \frac{1}{2}, \frac{1}{4}$, and $\frac{1}{6}$.

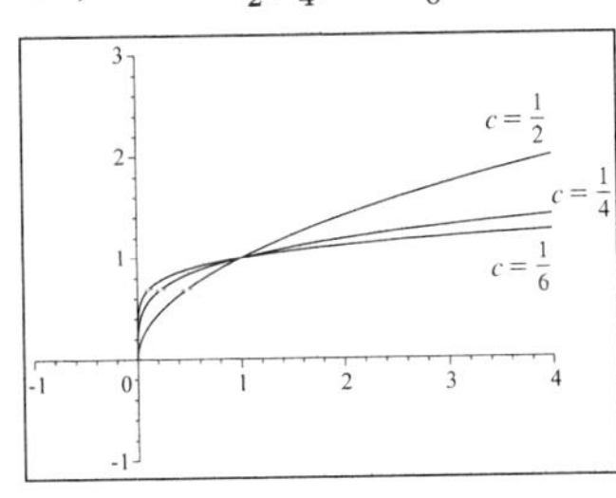


**(b)** $f(x) = x^c$, for $c = 1, \frac{1}{3}$, and $\frac{1}{5}$.

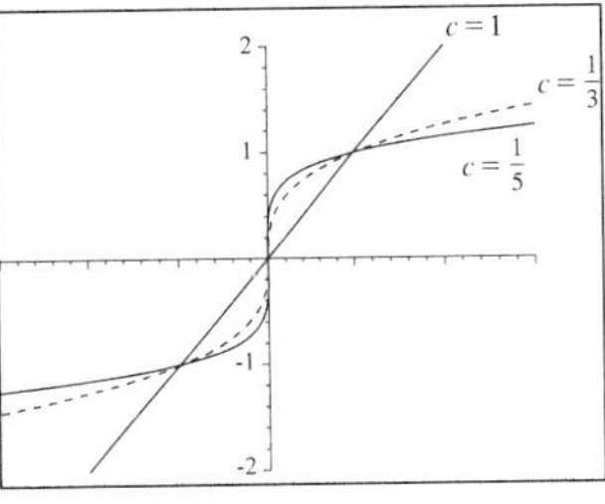


**(c)** Graphs of even roots are similar to $y = \sqrt{x}$, graphs of odd roots are similar to $y = \sqrt[3]{x}$. As $c$ increases, the graph of $y = \sqrt[c]{x}$ becomes steeper near $x = 0$ and flatter when $x > 1$.

**78. (a)** $f(x) = \dfrac{1}{x^n}$, for $n = 1$ and $3$.

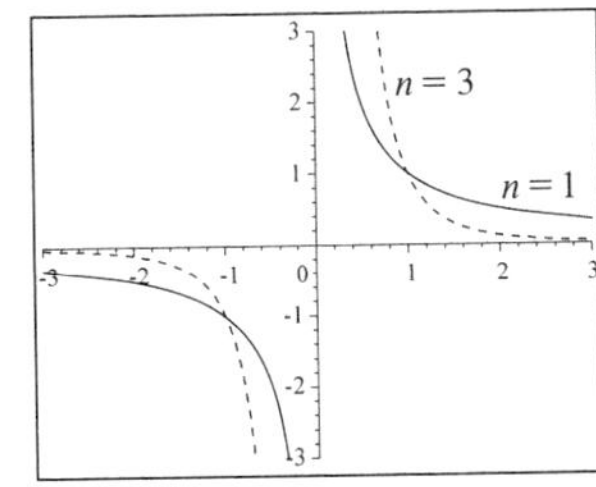


**(b)** $f(x) = \dfrac{1}{x^n}$, for $n = 2$ and $4$.

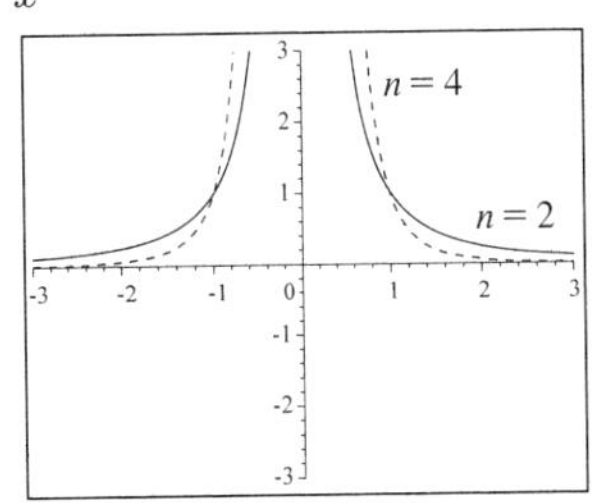


**(c)** As $n$ increases, the graphs of $y = 1/x^n$ go to zero faster for $x$ large. Also, as $n$ increases and $x$ goes to 0, the graphs of $y = 1/x^n$ go to infinity faster. The graphs of $y = 1/x^n$ for $n$ odd are similar to each other. Likewise, the graphs for $n$ even are similar to each other.

**79.** The slope of the line segment joining the points $(-2, 1)$ and $(4, -6)$ is $m = \dfrac{-6-1}{4-(-2)} = -\frac{7}{6}$. Using the point-slope form, we have $y - 1 = -\frac{7}{6}(x+2) \Leftrightarrow y = -\frac{7}{6}x - \frac{7}{3} + 1 \Leftrightarrow y = -\frac{7}{6}x - \frac{4}{3}$. Thus the function is $f(x) = -\frac{7}{6}x - \frac{4}{3}$ for $-2 \le x \le 4$.

**80.** The slope of the line containing the points $(-3, -2)$ and $(6, 3)$ is $m = \dfrac{-2-3}{-3-6} = \dfrac{-5}{-9} = \frac{5}{9}$. Using the point-slope equation of the line, we have $y - 3 = \frac{5}{9}(x-6) \Leftrightarrow y = \frac{5}{9}x - \frac{10}{3} + 3 = \dfrac{5}{9}x - \frac{1}{3}$. Thus the function is $f(x) = \frac{5}{9}x - \frac{1}{3}$, for $-3 \le x \le 6$.

**81.** First solve the circle for $y$: $x^2 + y^2 = 9 \Leftrightarrow y^2 = 9 - x^2 \Rightarrow y = \pm\sqrt{9-x^2}$. Since we seek the top half of the circle, we choose $y = \sqrt{9-x^2}$. So the function is $f(x) = \sqrt{9-x^2}$, $-3 \le x \le 3$.

**82.** First solve the circle for $y$: $x^2 + y^2 = 9 \Leftrightarrow y^2 = 9 - x^2 \Rightarrow y = \pm\sqrt{9-x^2}$. Since we seek the bottom half of the circle, we choose $y = -\sqrt{9-x^2}$. So the function is $f(x) = -\sqrt{9-x^2}$, $-3 \le x \le 3$.

**83.** This person appears to gain weight steadily until the age of 21 when this person's weight gain slows down. At age 30, this person experiences a sudden weight loss, but recovers, and by about age 40, this person's weight seems to level off at around 200 pounds. It then appears that the person dies at about age 68. The sudden weight loss could be due to a number of reasons, among them major illness, a weight loss program, etc.

**84.** The salesman travels away from home and stops to make a sales call between 9 A.M. and 10 A.M., and then travels further from home for a sales call between 12 noon and 1 P.M.. Next he travels along a route that takes him closer to home before taking him further away from home. He then makes a final sales call between 5 P.M. and 6 P.M. and then returns home.

**85.** Runner A won the race. All runners finished the race. Runner B fell, but got up and finished the race.

**86.** **(a)** At 6 A.M. the graph shows that the power consumption is about 500 megawatts. Since $t = 18$ represents 6 P.M., the graph shows that the power consumption at 6 P.M. is about 725 megawatts.

**(b)** The power consumption is lowest between 3 A.M. and 4 A.M..

**(c)** The power consumption is highest just before 12 noon.

**87.** **(a)** The first noticeable movements occurred at time $t = 5$ seconds.

**(b)** It seemed to end at time $t = 30$ seconds.

**(c)** Maximum intensity was reached at $t = 17$ seconds.

**88.** **(a)** $E(x) = \begin{cases} 6.00 + 0.10x & \text{if } 0 \le x \le 300 \\ 36.00 + 0.06(x - 300) & \text{if } 300 < x \end{cases}$

**(b)**

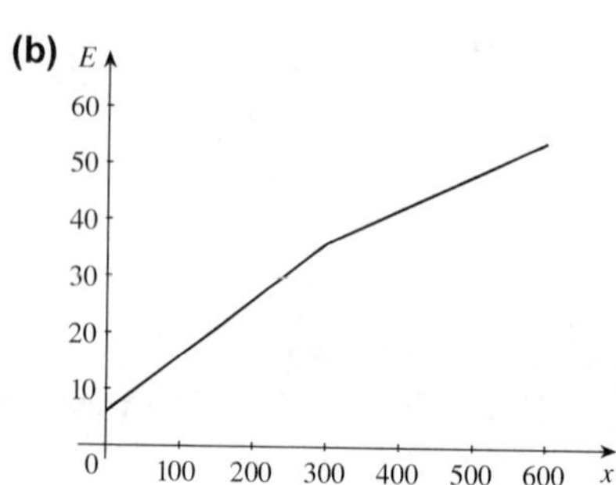

**89.** $C(x) = \begin{cases} 2.00 & \text{if } 0 < x \le 1 \\ 2.20 & \text{if } 1 < x \le 1.1 \\ 2.40 & \text{if } 1.1 < x \le 1.2 \\ \vdots & \\ 4.00 & \text{if } 1.9 < x < 2 \end{cases}$

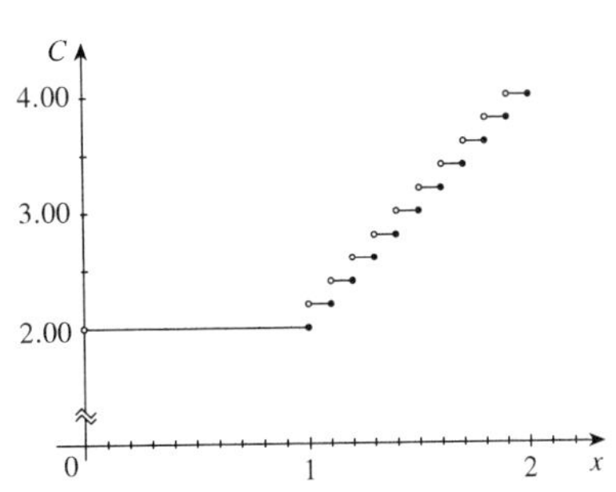

**90.** $P(x) = \begin{cases} 0.37 & \text{if } 0 < x \le 1 \\ 0.60 & \text{if } 1 < x \le 2 \\ 0.83 & \text{if } 2 < x \le 3 \\ \vdots & \\ 2.90 & \text{if } 11 < x \le 12 \end{cases}$

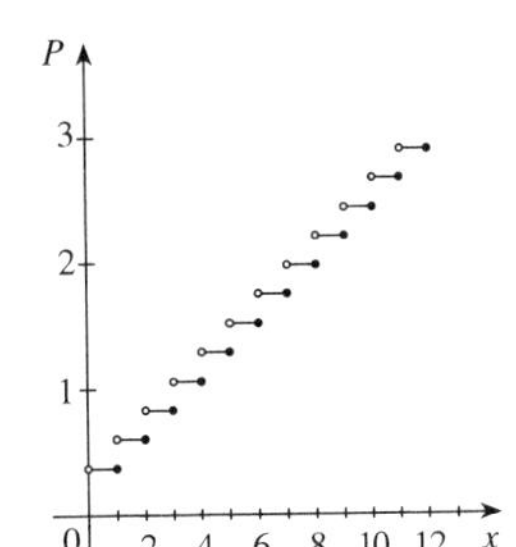

**91.** The graph of $x = y^2$ is not the graph of a function because both $(1, 1)$ and $(-1, 1)$ satisfy the equation $x = y^2$. The graph of $x = y^3$ is the graph of a function because $x = y^3 \quad \Leftrightarrow \quad x^{1/3} = y$. If $n$ is even, then both $(1, 1)$ and $(-1, 1)$ satisfies the equation $x = y^n$, so the graph of $x = y^n$ is not the graph of a function. When $n$ is odd, $y = x^{1/n}$ is defined for all real numbers, and since $y = x^{1/n} \quad \Leftrightarrow \quad x = y^n$, the graph of $x = y^n$ is the graph of a function.

**92.** Answers vary. Some examples are almost anything we purchase based on weight, volume, length, or time, for example gasoline. Although the amount delivered by the pump is continuous, the amount we pay is rounded to the penny. An example involving time would be the cost of a telephone call.

**93.**

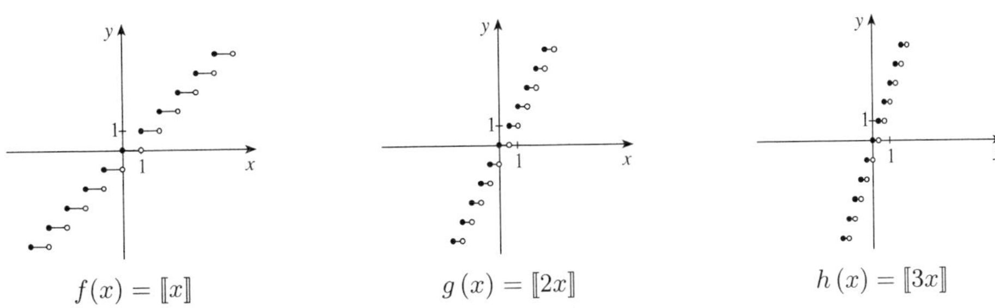

$f(x) = [\![x]\!]$ $g(x) = [\![2x]\!]$ $h(x) = [\![3x]\!]$

The graph of $k(x) = [\![nx]\!]$ is a step function whose steps are each $\frac{1}{n}$ wide.

**94. (a)** The graphs of $f(x) = x^2 + x - 6$ and $g(x) = \left|x^2 + x - 6\right|$ are shown in the viewing rectangle $[-10, 10]$ by $[-10, 10]$.

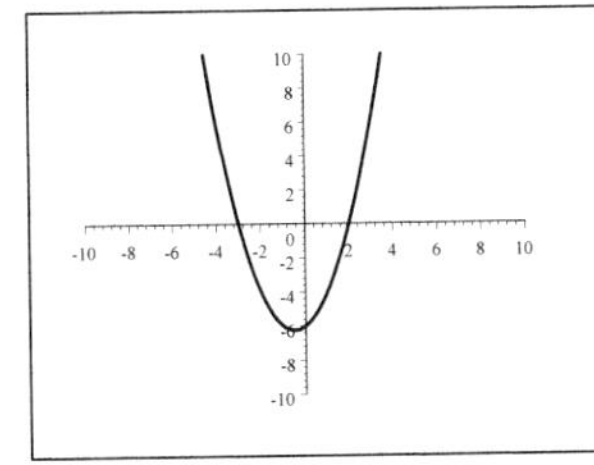

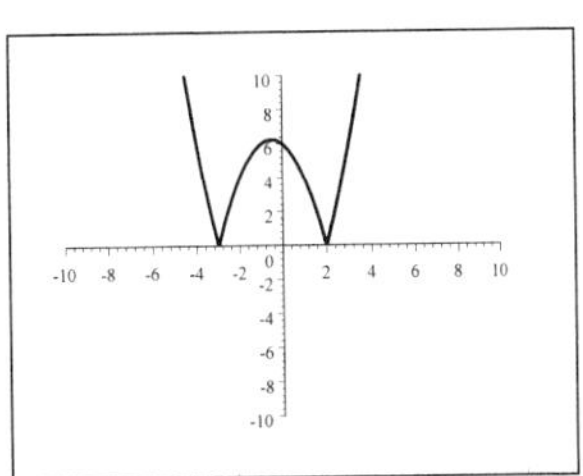

For those values of $x$ where $f(x) \ge 0$, the graphs of $f$ and $g$ coincide, and for those values of $x$ where $f(x) < 0$, the graph of $g$ is obtained from that of $f$ by reflecting the part below the $x$-axis about the $x$-axis.

**(b)** The graphs of $f(x) = x^4 - 6x^2$ and $g(x) = |x^4 - 6x^2|$ are shown in the viewing rectangle $[-5, 5]$ by $[-10, 15]$.

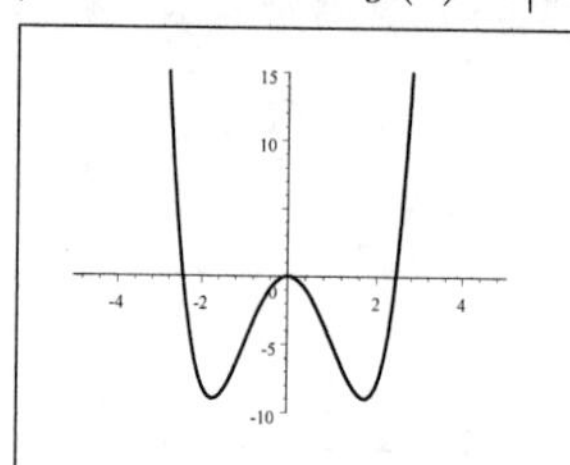

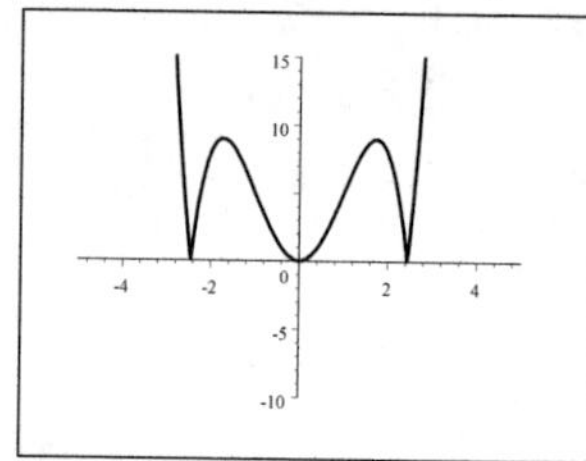

For those values of $x$ where $f(x) \geq 0$, the graphs of $f$ and $g$ coincide, and for those values of $x$ where $f(x) < 0$, the graph of $g$ is obtained from that of $f$ by reflecting the part below the $x$-axis above the $x$-axis.

**(c)** In general, if $g(x) = |f(x)|$, then for those values of $x$ where $f(x) \geq 0$, the graphs of $f$ and $g$ coincide, and for those values of $x$ where $f(x) < 0$, the graph of $g$ is obtained from that of $f$ by reflecting the part below the $x$-axis above the $x$-axis.

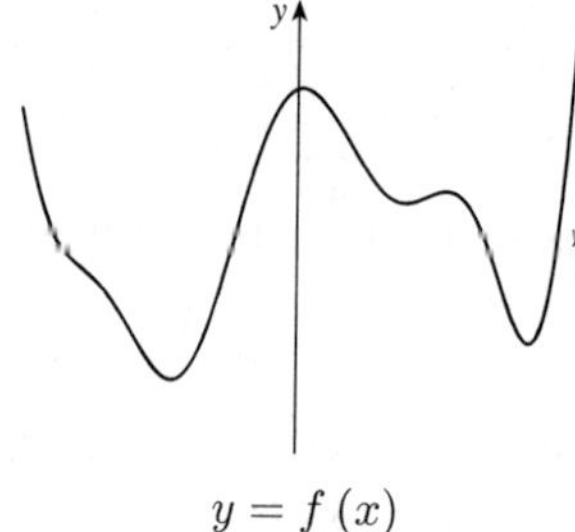

$y = f(x)$

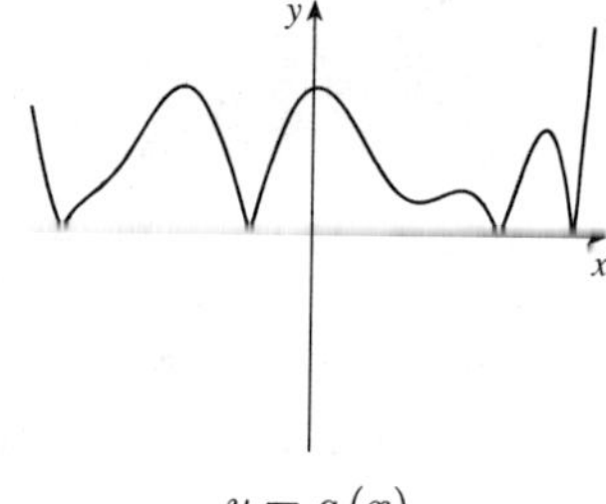

$y = g(x)$

# 3.3 Increasing and Decreasing Functions; Average Rate of Change

1. **(a)** The function is increasing on $[-1, 1]$ and $[2, 4]$. **(b)** The function is decreasing on $[1, 2]$.

2. **(a)** The function is increasing on $[0, 1]$. **(b)** The function is decreasing on $[-2, 0]$ and $[1, 3]$.

3. **(a)** The function is increasing on $[-2, -1]$ and $[1, 2]$. **(b)** The function is decreasing on $[-3, -2]$, $[-1, 1]$, and $[2, 3]$.

4. **(a)** The function is increasing on $[-1, 1]$. **(b)** The function is decreasing on $[-2, -1]$ and $[1, 2]$.

5. **(a)** $f(x) = x^{2/5}$ is graphed in the viewing rectangle $[-10, 10]$ by $[-5, 5]$.

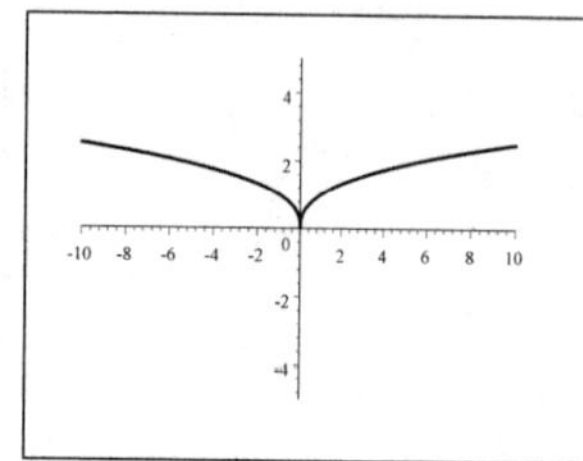

**(b)** The function is increasing on $[0, \infty)$. It is decreasing on $(-\infty, 0]$.

6. **(a)** $f(x) = 4 - x^{2/3}$ is graphed in the viewing rectangle $[-10, 10]$ by $[-10, 10]$.

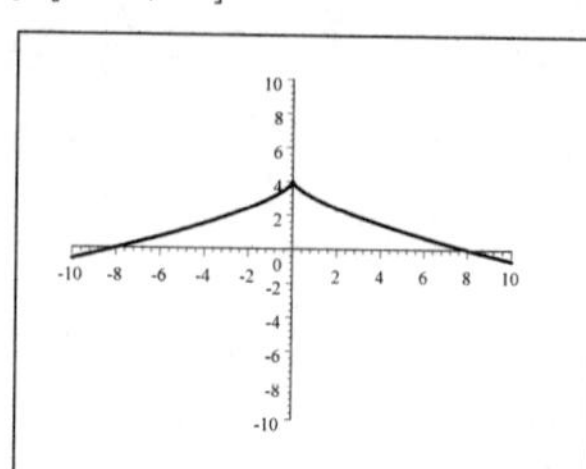

**(b)** The function is increasing on $(-\infty, 0]$. It is decreasing on $[0, \infty)$.

**7. (a)** $f(x) = x^2 - 5x$ is graphed in the viewing rectangle $[-2, 7]$ by $[-10, 10]$.

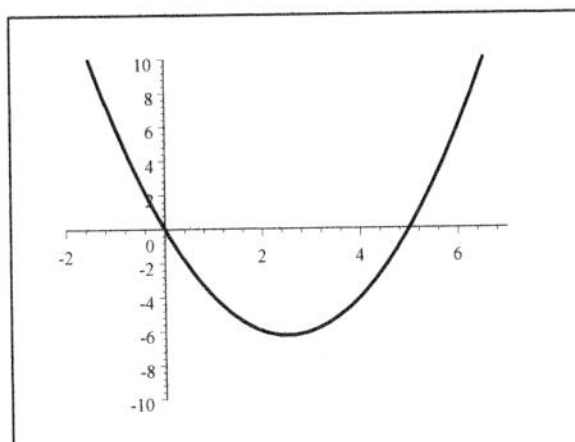

**(b)** The function is increasing on $[2.5, \infty)$. It is decreasing on $(-\infty, 2.5]$.

**8. (a)** $f(x) = x^3 - 4x$ is graphed in the viewing rectangle $[-10, 10]$ by $[-10, 10]$.

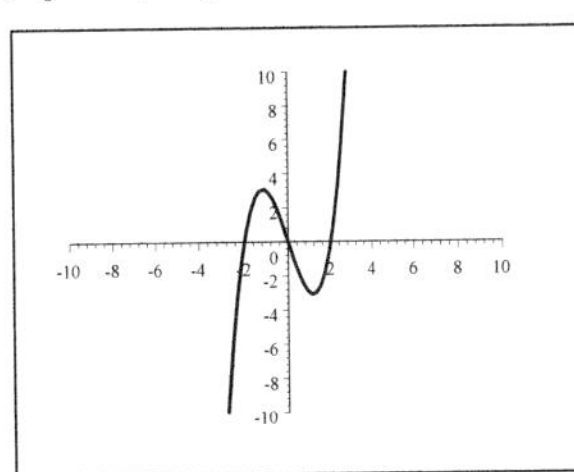

**(b)** The function is increasing on $(-\infty, -1.15]$ and $[1.15, \infty)$. It is decreasing on $[-1.15, 1.15]$.

**9. (a)** $f(x) = 2x^3 - 3x^2 - 12x$ is graphed in the viewing rectangle $[-3, 5]$ by $[-25, 20]$.

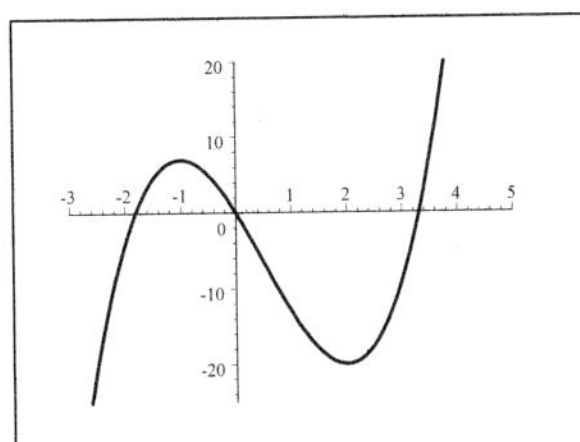

**(b)** The function is increasing on $(-\infty, -1]$ and $[2, \infty)$. It is decreasing on $[-1, 2]$.

**10. (a)** $f(x) = x^4 - 16x^2$ is graphed in the viewing rectangle $[-10, 10]$ by $[-70, 10]$.

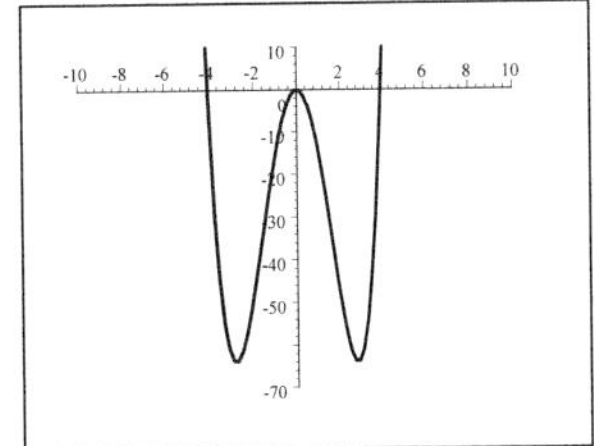

**(b)** The function is increasing on $[-2.83, 0]$ and $[2.83, \infty)$. It is decreasing on $(-\infty, -2.83]$ and $[0, 2.83]$.

**11. (a)** $f(x) = x^3 + 2x^2 - x - 2$ is graphed in the viewing rectangle $[-5, 5]$ by $[-3, 3]$.

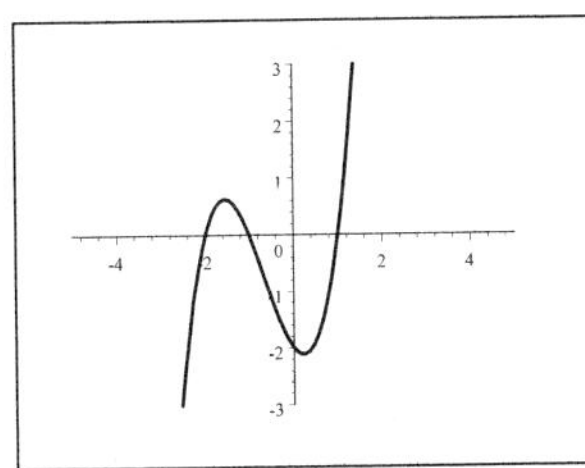

**(b)** The function is increasing on $(-\infty, -1.55]$ and $[0.22, \infty)$. It is decreasing on $[-1.55, 0.22]$.

**12. (a)** $f(x) = x^4 - 4x^3 + 2x^2 + 4x - 3$ is graphed in the viewing rectangle $[-3, 5]$ by $[-5, 5]$.

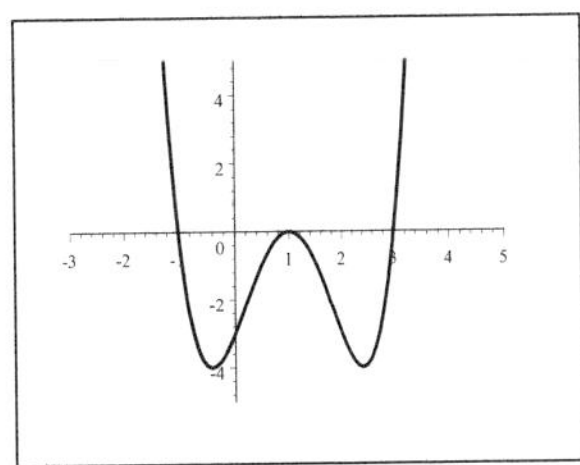

**(b)** The function is increasing on $(-0.4, 1]$ and $[2.4, \infty)$. It is decreasing on $[-\infty, -0.4]$ and $[1, 2.4]$.

**13.** We use the points $(1, 3)$ and $(4, 5)$, so the average rate of change is $\dfrac{5-3}{4-1} = \frac{2}{3}$.

**14.** We use the points $(1, 4)$ and $(5, 2)$, so the average rate of change is $\dfrac{2-4}{5-1} = \dfrac{-2}{4} = -\frac{1}{2}$.

**15.** We use the points $(0, 6)$ and $(5, 2)$, so the average rate of change is $\dfrac{2-6}{5-0} = \dfrac{-4}{5}$.

**16.** We use the points $(-1, 0)$ and $(5, 4)$, so the average rate of change is $\dfrac{4-0}{5-(-1)} = \frac{4}{6} = \frac{2}{3}$.

**17.** The average rate of change is $\dfrac{f(3) - f(2)}{3-2} = \dfrac{[3(3) - 2] - [3(2) - 2]}{1} = 7 - 4 = 3$.

**18.** The average rate of change is $\dfrac{g(5)-g(1)}{5-1}=\dfrac{\left[5+\frac{1}{2}(5)\right]-\left[5+\frac{1}{2}(1)\right]}{4}=\dfrac{\frac{15}{2}-\frac{11}{2}}{4}=\frac{2}{4}=\frac{1}{2}$.

**19.** The average rate of change is $\dfrac{h(4)-h(-1)}{4-(-1)}=\dfrac{\left[4^2+2(4)\right]-\left[(-1)^2+2(-1)\right]}{5}=\dfrac{24-(-1)}{5}=5$.

**20.** The average rate of change is $\dfrac{f(0)-f(-2)}{0-(-2)}=\dfrac{\left[1-3(0)^2\right]-\left[1-3(-2)^2\right]}{2}=\dfrac{1-(-11)}{2}=\dfrac{12}{2}=6$.

**21.** The average rate of change is $\dfrac{f(10)-f(0)}{10-0}=\dfrac{\left[10^3-4\left(10^2\right)\right]-\left[0^3-4\left(0^2\right)\right]}{10-0}=\dfrac{600-0}{10}=60$.

**22.** The average rate of change is $\dfrac{f(3)-f(-1)}{3-(-1)}=\dfrac{\left[3+3^4\right]-\left[(-1)+(-1)^4\right]}{4}=\dfrac{3+81}{4}=\frac{84}{4}=21$.

**23.** The average rate of change is

$$\frac{f(2+h)-f(2)}{(2+h)-2}=\frac{\left[3(2+h)^2\right]-\left[3\left(2^2\right)\right]}{h}=\frac{12+12h+3h^2-12}{h}=\frac{12h+3h^2}{h}=\frac{h(12+3h)}{h}=12+3h.$$

**24.** The average rate of change is

$$\frac{f(1+h)-f(1)}{(1+h)-1}=\frac{\left[4-(1+h)^2\right]-\left[4-1^2\right]}{h}=\frac{3-2h-h^2-3}{h}=\frac{-2h-h^2}{h}=\frac{h(-2-h)}{h}=-2-h.$$

**25.** The average rate of change is $\dfrac{g(1)-g(a)}{1-a}=\dfrac{\frac{1}{1}-\frac{1}{a}}{1-a}\cdot\dfrac{a}{a}=\dfrac{a-1}{a(1-a)}=\dfrac{-1(1-a)}{a(1-a)}=\dfrac{-1}{a}$.

**26.** The average rate of change is $\dfrac{g(h)-g(0)}{h-0}=\dfrac{\frac{2}{h+1}-\frac{2}{0+1}}{h}\cdot\dfrac{h+1}{h+1}=\dfrac{2-2(h+1)}{h(h+1)}=\dfrac{-2h}{h(h+1)}=\dfrac{-2}{h+1}$.

**27.** The average rate of change is

$$\frac{f(a+h)-f(a)}{(a+h)-a}=\frac{\frac{2}{a+h}-\frac{2}{a}}{h}\cdot\frac{a(a+h)}{a(a+h)}=\frac{2a-2(a+h)}{ah(a+h)}=\frac{-2h}{ah(a+h)}=\frac{-2}{a(a+h)}.$$

**28.** The average rate of change is

$$\frac{f(a+h)-f(a)}{(a+h)-a}=\frac{\sqrt{a+h}-\sqrt{a}}{h}\cdot\frac{\sqrt{a+h}+\sqrt{a}}{\sqrt{a+h}+\sqrt{a}}=\frac{(a+h)-a}{h\left(\sqrt{a+h}+\sqrt{a}\right)}=\frac{h}{h\left(\sqrt{a+h}+\sqrt{a}\right)}=\frac{1}{\sqrt{a+h}+\sqrt{a}}.$$

**29. (a)** The average rate of change is

$$\frac{f(a+h)-f(a)}{(a+h)-a}=\frac{\left[\frac{1}{2}(a+h)+3\right]-\left[\frac{1}{2}a+3\right]}{h}=\frac{\frac{1}{2}a+\frac{1}{2}h+3-\frac{1}{2}a-3}{h}=\frac{\frac{1}{2}h}{h}=\frac{1}{2}.$$

**(b)** The slope of the line $f(x)=\frac{1}{2}x+3$ is $\frac{1}{2}$, which is also the average rate of change.

**30. (a)** The average rate of change is

$$\frac{g(a+h)-g(a)}{(a+h)-a}=\frac{[-4(a+h)+2]-[-4a+2]}{h}=\frac{-4a-4h+2+4a-2}{h}=\frac{-4h}{h}=-4.$$

**(b)** The slope of the line $g(x)=-4x+2$ is $-4$, which is also the average rate of change.

**31. (a)** The function $P$ is increasing on $[0,150]$ and $[300,365]$ and decreasing on $[150,300]$.

**(b)** The average rate of change is $\dfrac{W(200)-W(100)}{200-100}=\dfrac{50-75}{200-100}=\dfrac{-25}{100}=-\dfrac{1}{4}$ ft/day.

**32. (a)** The function $P$ is increasing on $[0,25]$ and decreasing on $[25,50]$.

**(b)** The average rate of change is $\dfrac{P(40)-P(20)}{40-20}=\dfrac{40-40}{40-20}=\dfrac{0}{20}=0$.

**(c)** The population increased and decreased the same amount during the 20 years.

**33. (a)** The average rate of change of population is $\frac{1{,}591 - 856}{2001 - 1998} = \frac{735}{3} = 245$ persons/yr.

**(b)** The average rate of change of population is $\frac{826 - 1{,}483}{2004 - 2002} = \frac{-657}{2} = -328.5$ persons/yr.

**(c)** The population was increasing from 1997 to 2001.

**(d)** The population was decreasing from 2001 to 2006.

**34. (a)** The average speed is $\frac{800 - 400}{152 - 68} = \frac{400}{84} = \frac{100}{21} \approx 4.76$ m/s.

**(b)** The average speed is $\frac{1{,}600 - 1{,}200}{412 - 263} = \frac{400}{149} \approx 2.68$ m/s.

**(c)**

| Lap | Length of time to run lap | Average speed of lap. |
|---|---|---|
| 1 | 32 | 6.25 m/s |
| 2 | 36 | 5.56 m/s |
| 3 | 40 | 5.00 m/s |
| 4 | 44 | 4.55 m/s |
| 5 | 51 | 3.92 m/s |
| 6 | 60 | 3.33 m/s |
| 7 | 72 | 2.78 m/s |
| 8 | 77 | 2.60 m/s |

The man is slowing down throughout the run.

**35. (a)** The average rate of change of sales is $\frac{584 - 512}{2003 - 1993} = \frac{72}{10} = 7.2$ units/yr.

**(b)** The average rate of change of sales is $\frac{520 - 512}{1994 - 1993} = \frac{8}{1} = 8$ units/yr.

**(c)** The average rate of change of sales is $\frac{410 - 520}{1996 - 1994} = \frac{-110}{2} = -55$ units/yr.

**(d)**

| Year | CD players sold | Change in sales from previous year |
|---|---|---|
| 1993 | 512 | – |
| 1994 | 520 | 8 |
| 1995 | 413 | −107 |
| 1996 | 410 | −3 |
| 1997 | 468 | 58 |
| 1998 | 510 | 42 |
| 1999 | 590 | 80 |
| 2000 | 607 | 17 |
| 2001 | 732 | 125 |
| 2002 | 612 | −120 |
| 2003 | 584 | −28 |

Sales increased most quickly between 2000 and 2001. Sales decreased most quickly between 2001 and 2002.

**36.**

| Year | Number of books |
|---|---|
| 1980 | 420 |
| 1981 | 460 |
| 1982 | 500 |
| 1985 | 620 |
| 1990 | 820 |
| 1992 | 900 |
| 1995 | 1020 |
| 1997 | 1100 |
| 1998 | 1140 |
| 1999 | 1180 |
| 2000 | 1220 |

**37. (a)** For all three runners, the average rate of change is $\frac{d(10) - d(0)}{10 - 0} = \frac{100}{10} = 10$.

**(b)** Runner A gets a great jump out of the blocks but tires at the end of the race. Runner B runs a steady race. Runner C is slow at the beginning but accelerates down the track.

**38. (a)**

| Time (s) | Average speed (ft/s) |
|---|---|
| 0–2 | 17 |
| 2–4 | 18 |
| 4–6 | 63 |
| 6–8 | 147 |
| 8–10 | 237 |

The average speed is increasing, so the car is accelerating. From the shape of the graph, we see that the slope continues to get steeper.

**(b)**

| Time (s) | Average speed (ft/s) |
|---|---|
| 30–32 | 263 |
| 32–34 | 144 |
| 34–36 | 91 |
| 36–38 | 74 |
| 38–40 | 48 |

The average speed is decreasing, so the car is decelerating. From the shape of the graph, we see that the slope continues to get flatter.

**39. (a)** $f(x)$ is always increasing, and $f(x) > 0$ for all $x$.

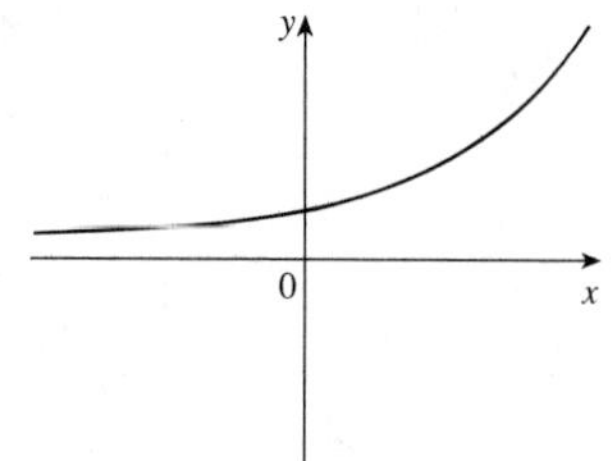


**(b)** $f(x)$ is always decreasing, and $f(x) > 0$ for all $x$.

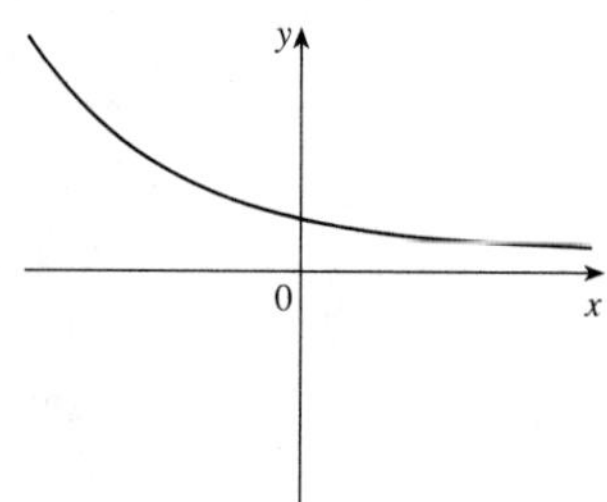

**(c)** $f(x)$ is always increasing, and $f(x) < 0$ for all $x$.

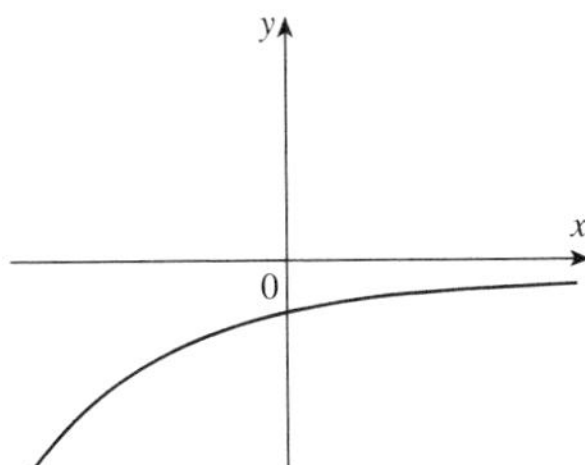


**(d)** $f(x)$ is always decreasing, and $f(x) < 0$ for all $x$.

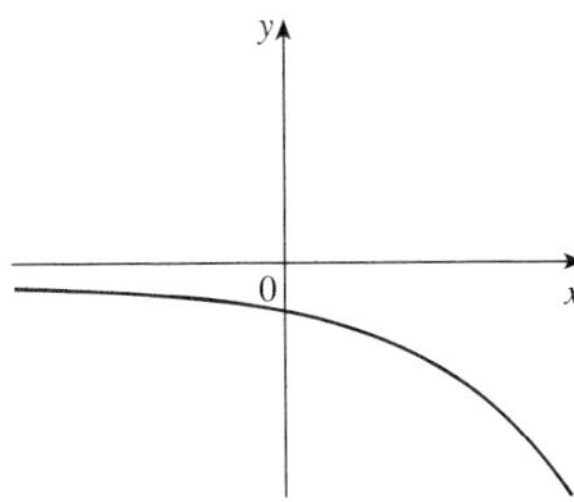


## 3.4 Transformations of Functions

**1. (a)** Shift the graph of $y = f(x)$ downward 5 units.
**(b)** Shift the graph of $y = f(x)$ to the right 5 units.

**2. (a)** Shift the graph of $y = f(x)$ to the left 7 units.
**(b)** Shift the graph of $y = f(x)$ upward 7 units.

**3. (a)** Shift the graph of $f$ to the left $\frac{1}{2}$ unit.
**(b)** Shift the graph of $f$ upward $\frac{1}{2}$ unit.

**4. (a)** Reflect the graph of $y = f(x)$ about the $x$-axis.
**(b)** Reflect the graph of $y = f(x)$ about the $y$-axis.

**5. (a)** Reflect the graph of $y = f(x)$ about the $x$-axis, and stretch vertically by a factor of 2.
**(b)** Reflect the graph of $y = f(x)$ about the $x$-axis, and shrink vertically by a factor of $\frac{1}{2}$.

**6. (a)** Reflect the graph of $y = f(x)$ about the $x$-axis, and then shift upward 5 units.
**(b)** Stretch the graph of $y = f(x)$ vertically by a factor of 3, and then shift downward 5 units.

**7. (a)** Shift the graph of $y = f(x)$ to the right 4 units, and then upward $\frac{3}{4}$ of a unit.
**(b)** Shift the graph of $y = f(x)$ to the left 4 units, and then downward $\frac{3}{4}$ of a unit.

**8. (a)** Shift the graph of $y = f(x)$ to the left 2 units, stretch vertically by a factor of 2, and then shift downward 2 units.
**(b)** Shift the graph of $y = f(x)$ to the right 2 units, stretch vertically by a factor of 2,and then shift upward 2 units.

**9. (a)** Shrink the graph of $y = f(x)$ horizontally by a factor of $\frac{1}{4}$.
**(b)** Stretch the graph of $y = f(x)$ horizontally by a factor of 4.

**10. (a)** Shrink the graph of $y = f(x)$ horizontally by a factor of $\frac{1}{2}$, and then reflect the resulting graph about the $x$-axis.
**(b)** Shrink the graph of $y = f(x)$ horizontally by a factor of $\frac{1}{2}$ and then shift downward 1 unit.

**11.** $g(x) = f(x-2) = (x-2)^2 = x^2 - 4x + 4$

**12.** $g(x) = f(x) + 3 = x^3 + 3$

**13.** $g(x) = f(x+1) + 2 = |x+1| + 2$

**14.** $g(x) = 2f(x) = 2|x|$

**15.** $g(x) = -f(x+2) = -\sqrt{x+2}$

**16.** $g(x) = -f(x-2)+1 = -(x-2)^2+1 = -x^2+4x-3$

**17. (a)** $y = f(x-4)$ is graph #3.
**(b)** $y = f(x) + 3$ is graph #1.
**(c)** $y = 2f(x+6)$ is graph #2.
**(d)** $y = -f(2x)$ is graph #4.

**18. (a)** $y = \frac{1}{3}f(x)$ is graph #2.
**(b)** $y = -f(x+4)$ is graph #3.
**(c)** $y = f(x-5) + 3$ is graph #1.
**(d)** $y = f(-x)$ is graph #4.

**19. (a)** $y = f(x-2)$

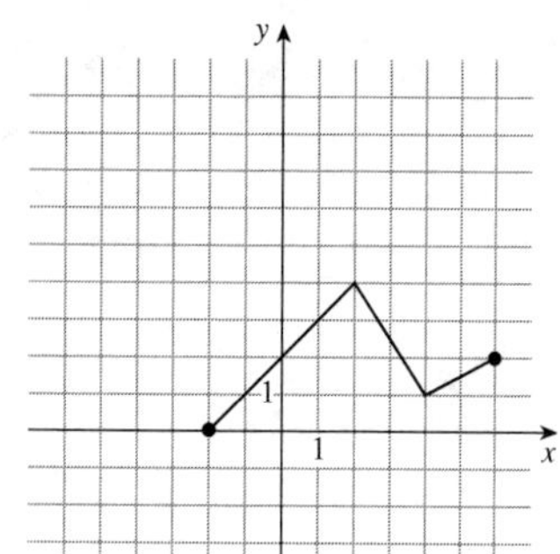


**(b)** $y = f(x) - 2$

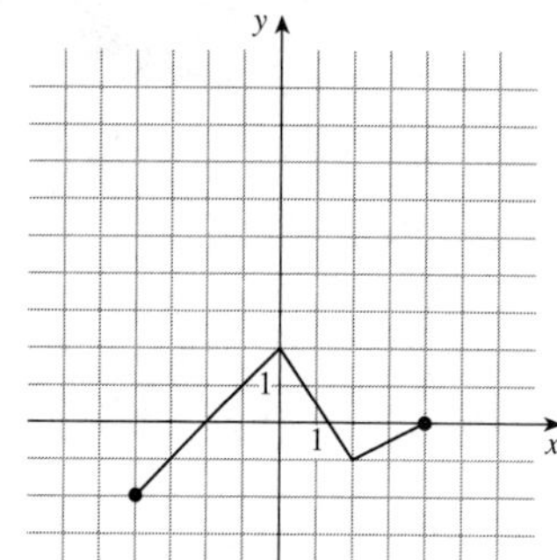


**(c)** $y = 2f(x)$

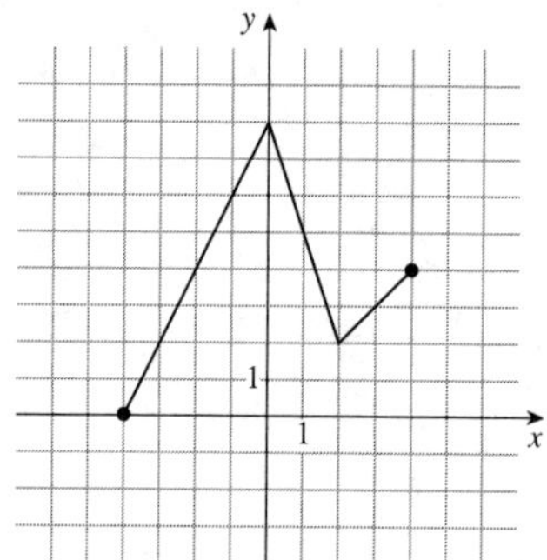


**(d)** $y = -f(x) + 3$

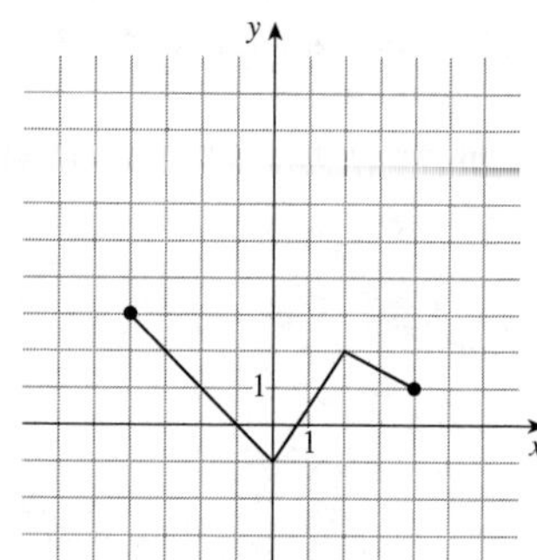


**(e)** $y = f(-x)$

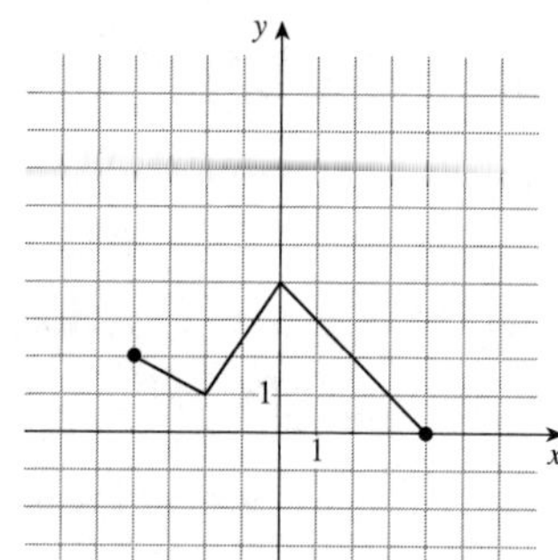


**(f)** $y = \frac{1}{2}f(x-1)$

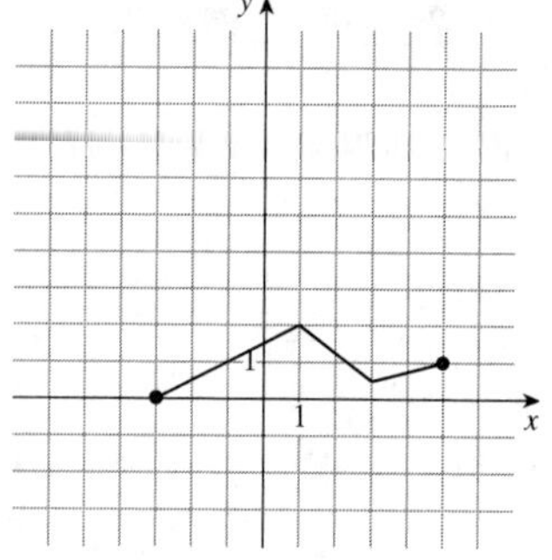


**20. (a)** $y = g(x+1)$

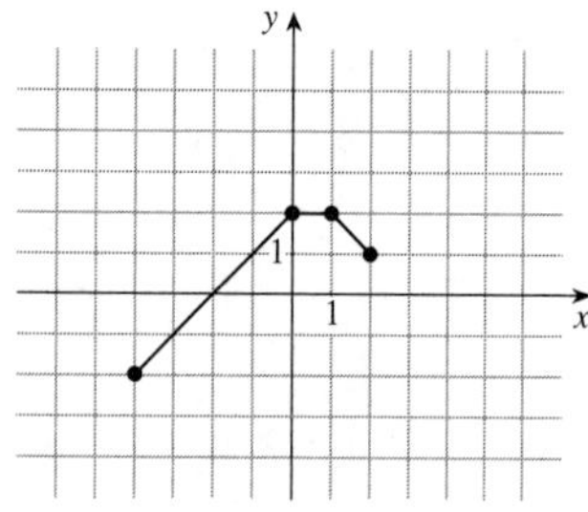


**(b)** $y = -g(x+1)$

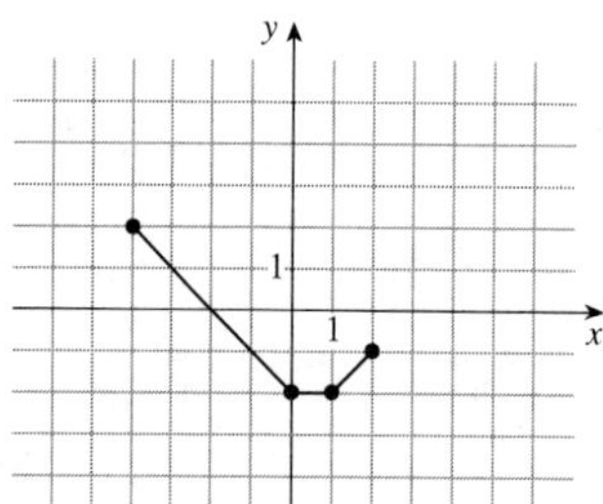


**(c)** $y = g(x-2)$

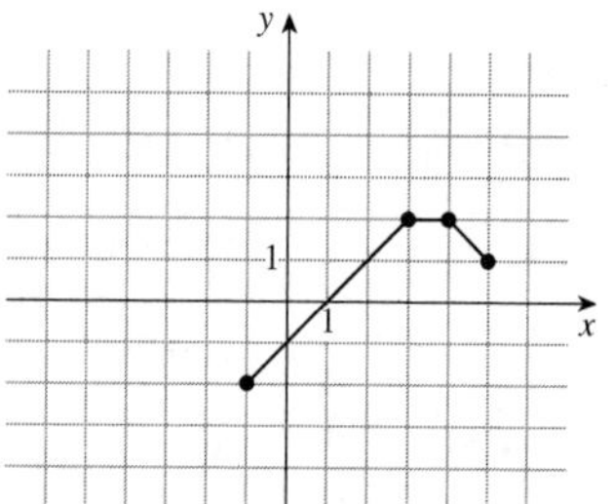


**(d)** $y = g(x) - 2$

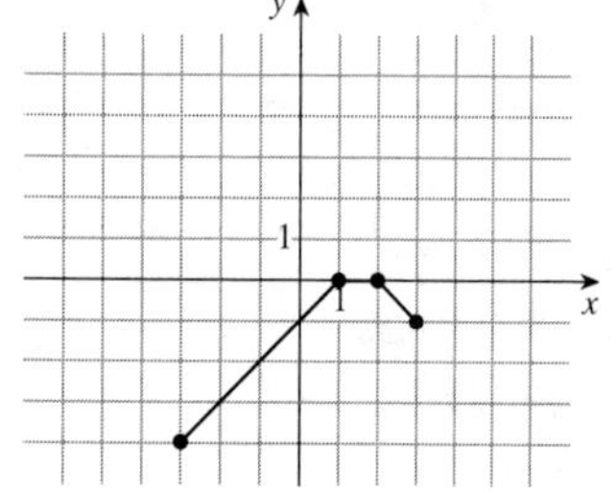


**(e)** $y = -g(x) + 2$

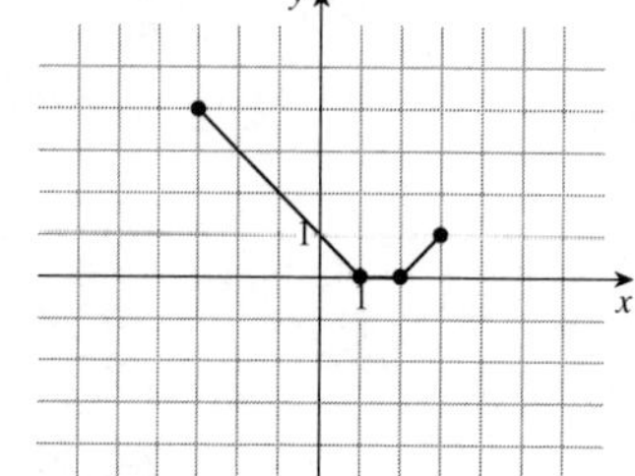


**(f)** $y = 2g(x)$

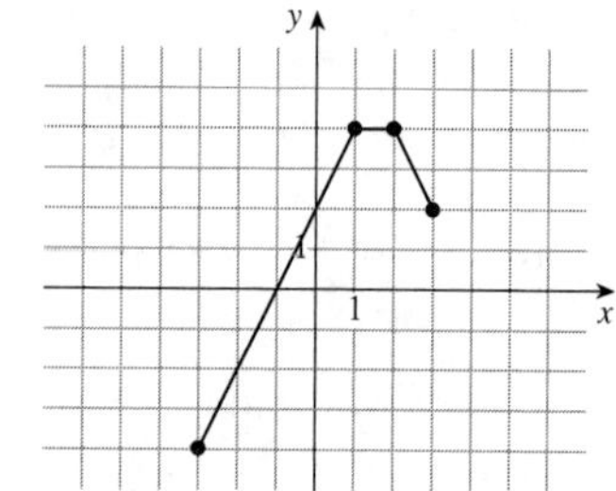

**21. (a)** $f(x) = \frac{1}{x}$

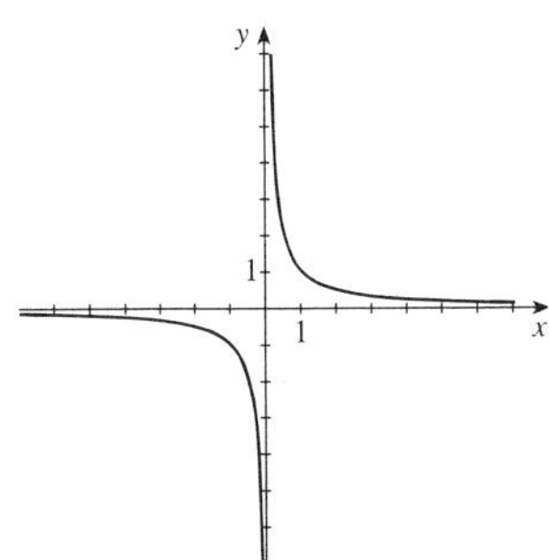

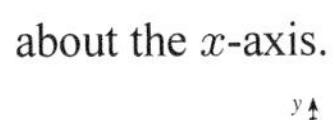

**(b) (i)** $y = -\frac{1}{x}$. Reflect the graph of $f$ about the $x$-axis.

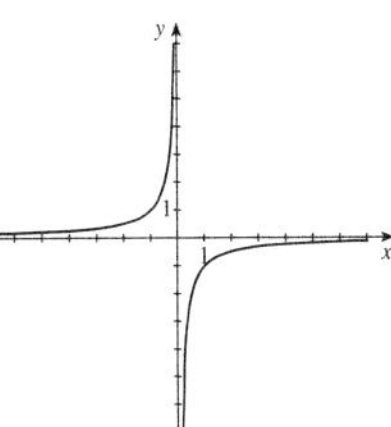

**(ii)** $y = \frac{1}{x-1}$. Shift the graph of $f$ to the right 1 unit.

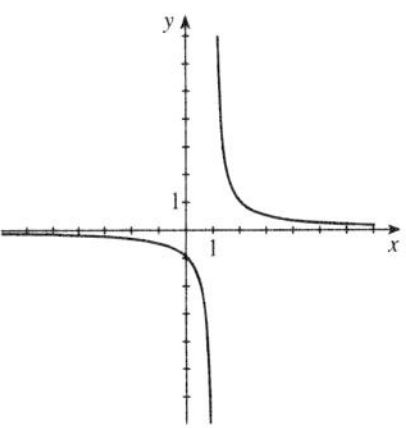

**(iii)** $y = \frac{2}{x+2}$. Shift the graph of $f$ to the left 2 units and stretch vertically by a factor of 2.

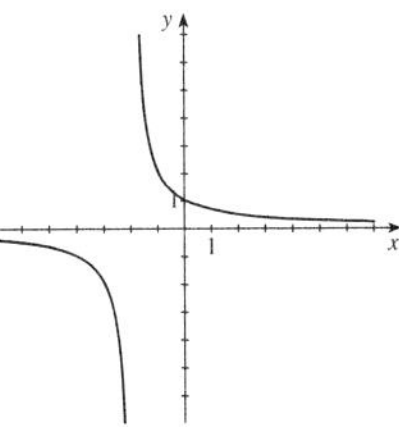

**(iv)** $y = 1 + \frac{1}{x-3}$. Shift the graph of $f$ to the right 3 units and upward 1 unit.

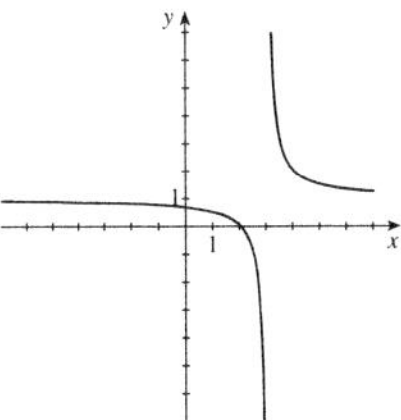

**22. (a)** $f(x) = \sqrt[3]{x}$

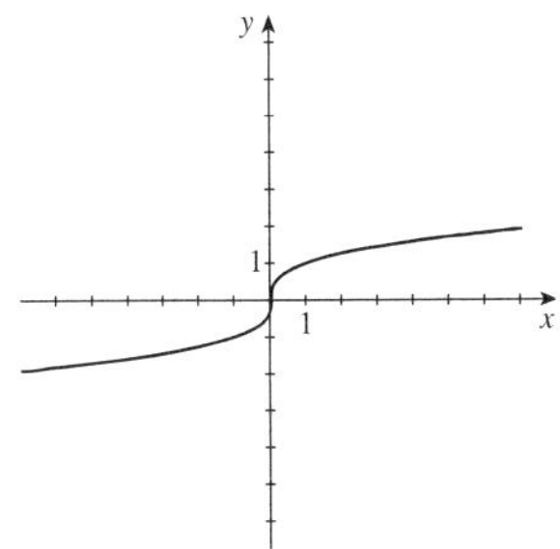

**(b) (i)** $\sqrt[3]{x-2}$. Shift the graph of $f(x) = \sqrt[3]{x}$ to the right 2 units.

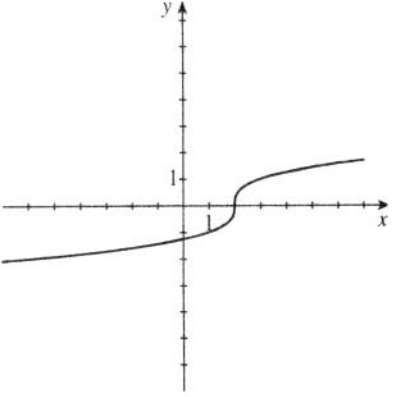

**(ii)** $y = \sqrt[3]{x+2} + 2$. Shift the graph of $f(x) = \sqrt[3]{x}$ to the left 2 units and upward 2 units.

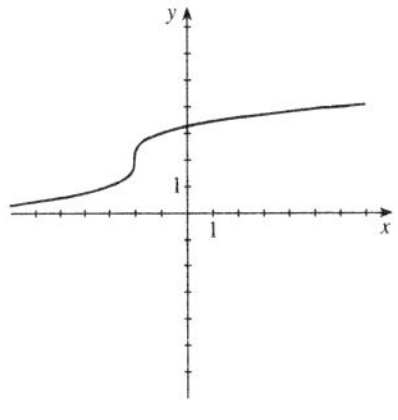

**(iii)** $y = 1 - \sqrt[3]{x}$. Reflect the graph of $f(x) = \sqrt[3]{x}$ about the $x$-axis, and then shift it upward 1 unit.

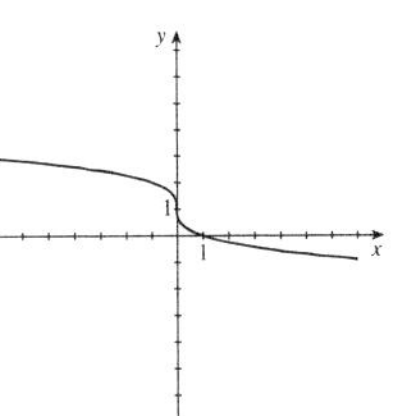

**(iv)** $y = 2\sqrt[3]{x}$. Stretch the graph of $f(x) = \sqrt[3]{x}$ vertically by a factor of 2.

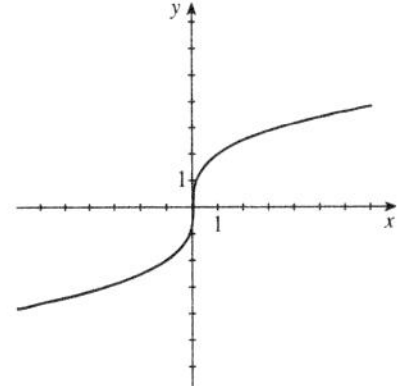

**23. (a)** The graph of $g(x) = (x+2)^2$ is obtained by shifting the graph of $f(x)$ to the left 2 units.

**(b)** The graph of $g(x) = x^2 + 2$ is obtained by shifting the graph of $f(x)$ upward 2 units.

**24. (a)** The graph of $g(x) = (x-4)^3$ is obtained by shifting the graph of $f(x)$ to the right 4 units.

**(b)** The graph of $g(x) = x^3 - 4$ is obtained by shifting the graph of $f(x)$ downward 4 units.

**25.** **(a)** The graph of $g(x) = 2\sqrt{x}$ is obtained by stretching the graph of $f(x)$ vertically by a factor of 2.

**(b)** The graph of $g(x) = \frac{1}{2}\sqrt{x-2}$ is obtained by shifting the graph of $f(x)$ to the right 2 units, and then shrinking the graph vertically by a factor of $\frac{1}{2}$.

**26.** **(a)** The graph of $g(x) = 3|x| + 1$ is obtained by stretching the graph of $f(x)$ vertically by a factor of 3, then shifting the graph upward 1 unit.

**(b)** The graph of $g(x) = -|x+1|$ is obtained by shifting the graph of $f(x)$ left 1 unit, then reflecting the graph about the $x$-axis.

**27.** $y = f(x-2) + 3$. When $f(x) = x^2$, we get $y = (x-2)^2 + 3 = x^2 - 4x + 4 + 3 = x^2 - 4x + 7$.

**28.** $y = f(x+4) - 1$. When $f(x) = x^3$, we get $y = (x+4)^3 - 1 = x^3 + 12x^2 + 48x + 64 - 1 = x^3 + 12x^2 + 48x + 63$.

**29.** $y = -5f(x+3)$. When $f(x) = \sqrt{x}$, we get $y = -5\sqrt{x+3}$

**30.** $y = \frac{1}{2}f(-x) + \frac{3}{5}$. When $f(x) = \sqrt[3]{x}$, we get $y = \frac{1}{2}\sqrt[3]{-x} + \frac{3}{5}$.

**31.** $y = 0.1f\left(x - \frac{1}{2}\right) - 2$. When $f(x) = |x|$, we get $y = 0.1\left|x - \frac{1}{2}\right| - 2$

**32.** $y = 3f(x+1) + 10$. When $f(x) = |x|$, we get $y = 3|x+1| + 10$

**33.** $f(x) = (x-2)^2$. Shift the graph of $y = x^2$ to the right 2 units.

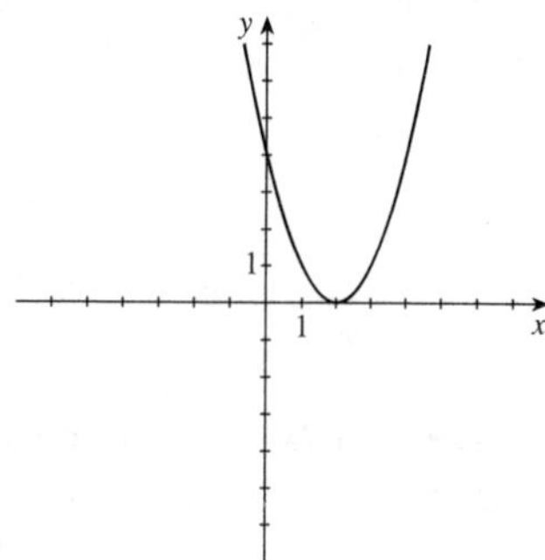


**34.** $f(x) = (x+7)^2$. Shift the graph of $y = x^2$ to the left 7 units.

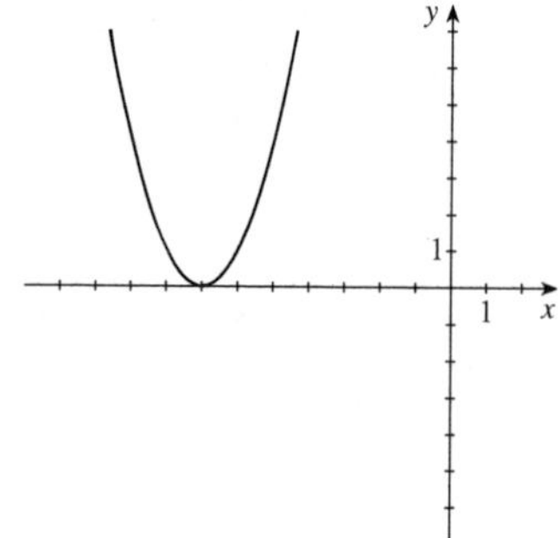


**35.** $f(x) = -(x+1)^2$. Shift the graph of $y = x^2$ to the left 1 unit, then reflect about the $x$-axis.

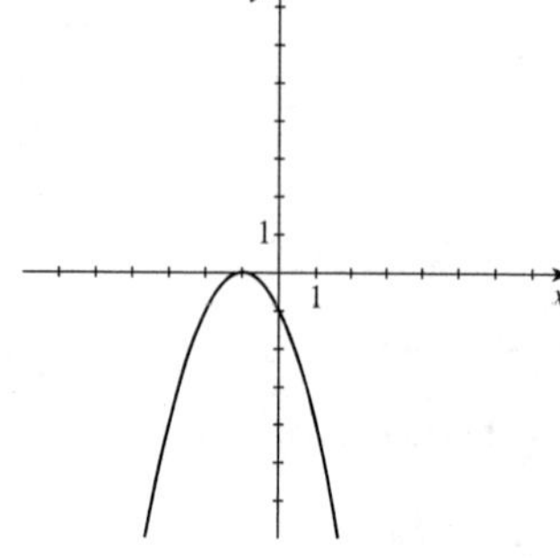


**36.** $f(x) = 1 - x^2$. Reflect the graph of $y = x^2$ about the $x$-axis, then shift it upward 1 unit.

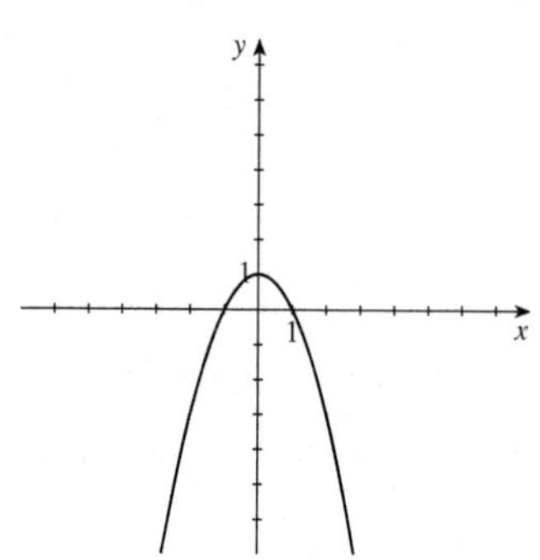

**37.** $f(x) = x^3 + 2$. Shift the graph of $y = x^3$ upward 2 units.

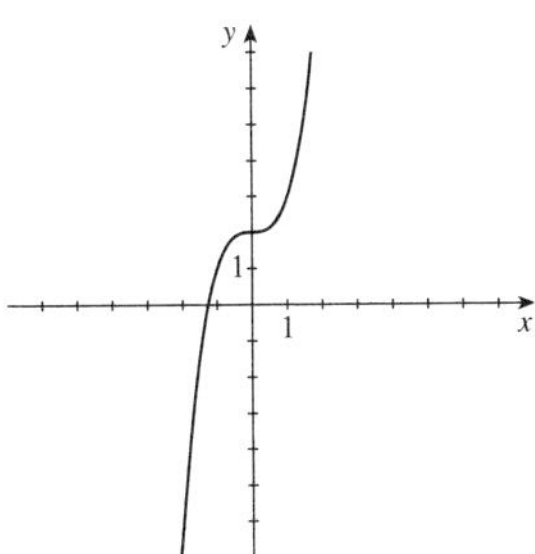


**38.** $f(x) = -x^3$. Reflect the graph of $y = x^3$ about the $x$-axis.

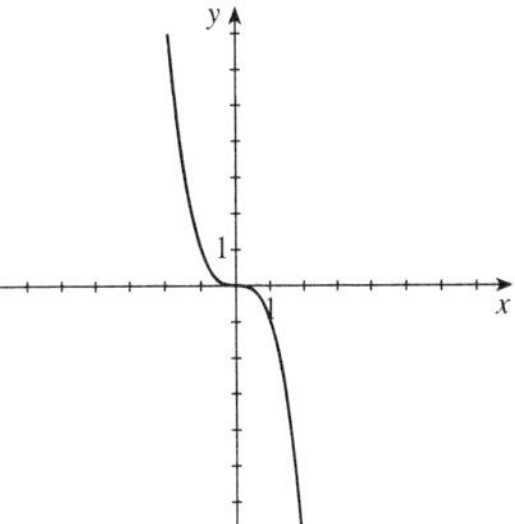


**39.** $y = 1 + \sqrt{x}$. Shift the graph of $y = \sqrt{x}$ upward 1 unit.

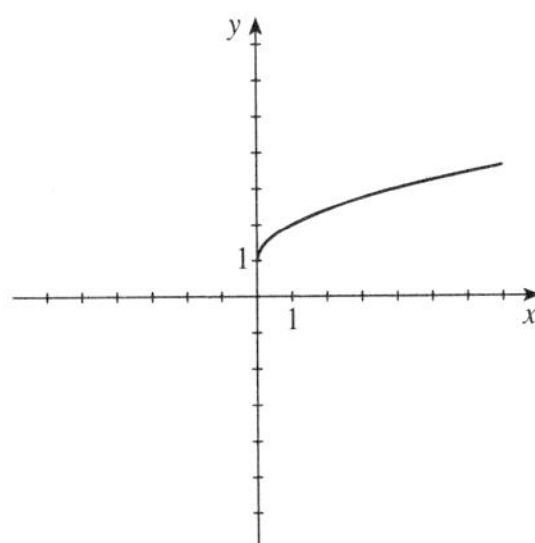


**40.** $y = 2 - \sqrt{x+1}$. Shift the graph of $y = \sqrt{x}$ to the left 1 unit, reflect it about the $x$-axis, and finally shift it upward 2 units.

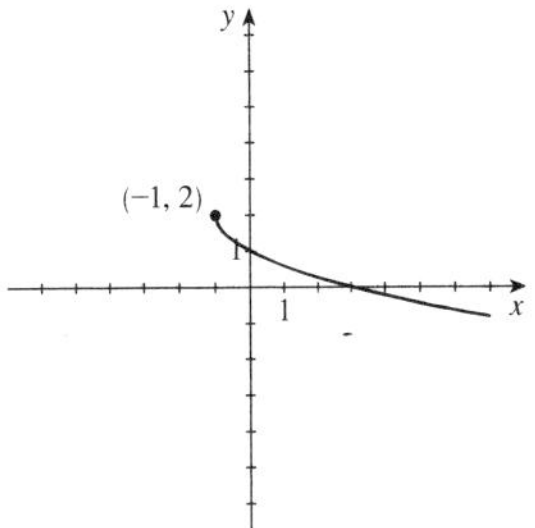


**41.** $y = \frac{1}{2}\sqrt{x+4} - 3$. Shift the graph of $y = \sqrt{x}$ to the left 4 units, shrink vertically by a factor of $\frac{1}{2}$, and then shift it downward 3 units.

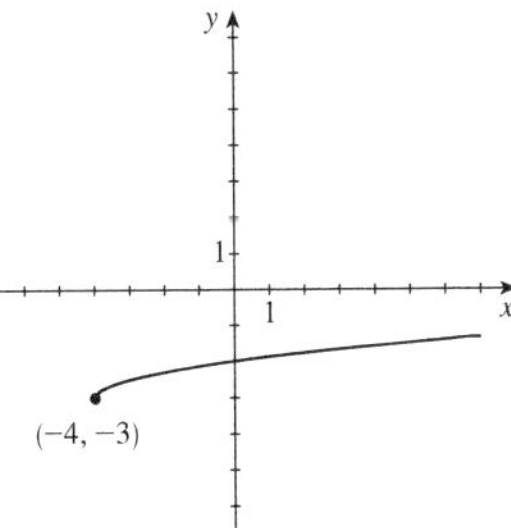


**42.** $y = 3 - 2(x-1)^2$. Shift the graph of $y = x^2$ to the right 1 unit, reflect it about the $x$-axis, stretch it vertically by a factor of 2, and then shift it upward 3 units.

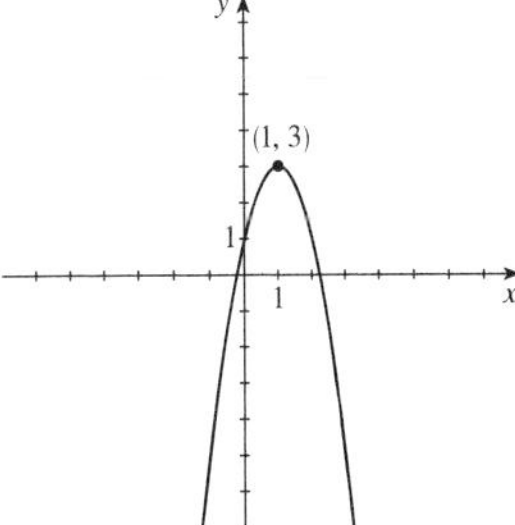


**43.** $y = 5 + (x+3)^2$. Shift the graph of $y = x^2$ to the left 3 units, then upward 5 units.

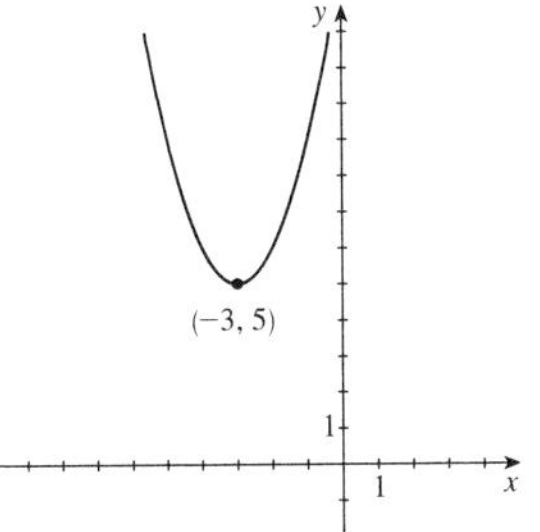


**44.** $y = \frac{1}{3}x^3 - 1$. Shrink the graph of $y = x^3$ vertically by a factor of $\frac{1}{3}$, then shift it downward 1 unit.

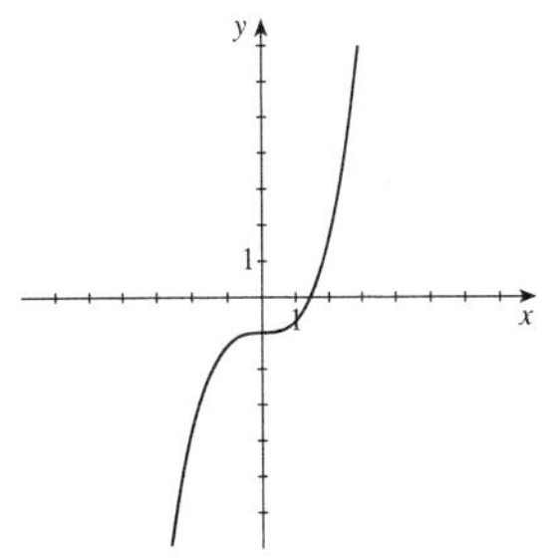

**45.** $y = |x| - 1$. Shift the graph of $y = |x|$ downward 1 unit.

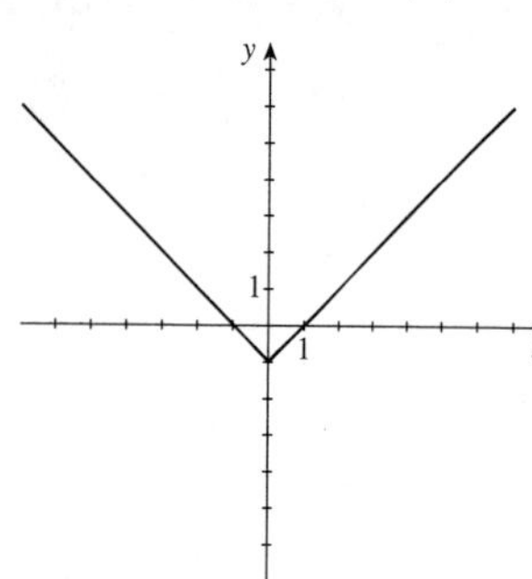


**46.** $y = |x - 1|$. Shift the graph of $y = |x|$ to the right 1 unit.

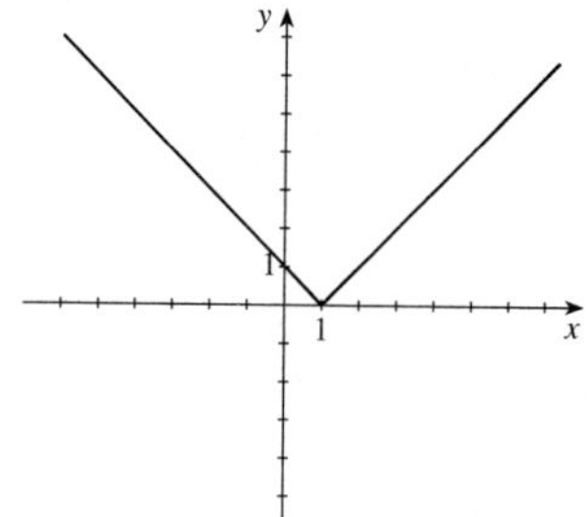
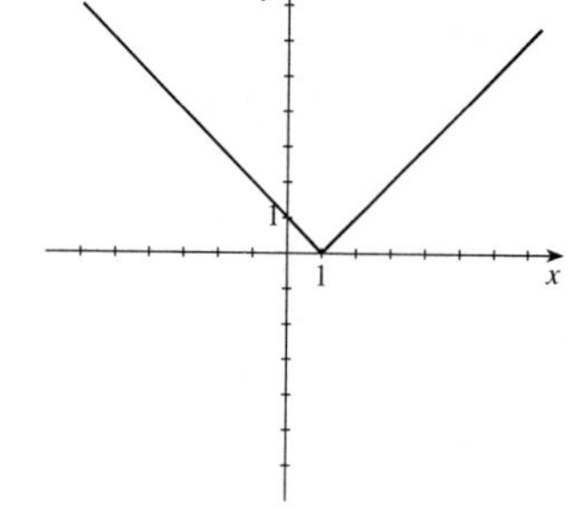


**47.** $y = |x + 2| + 2$. Shift the graph of $y = |x|$ to the left 2 units and upward 2 units.

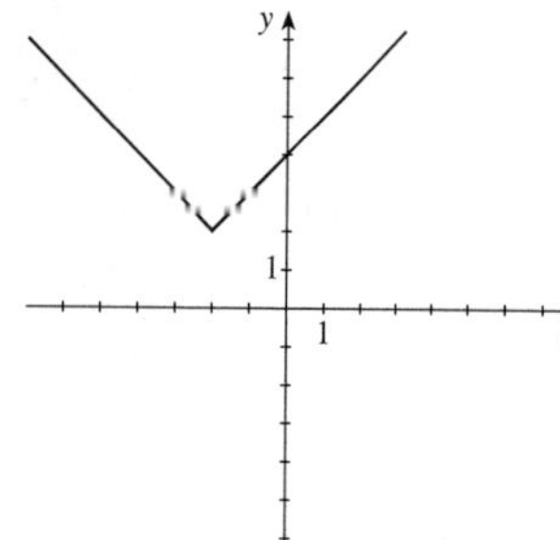


**48.** $y = 2 - |x|$. Reflect the graph of $y = |x|$ about the $x$-axis, and then shift it upward 2 units.

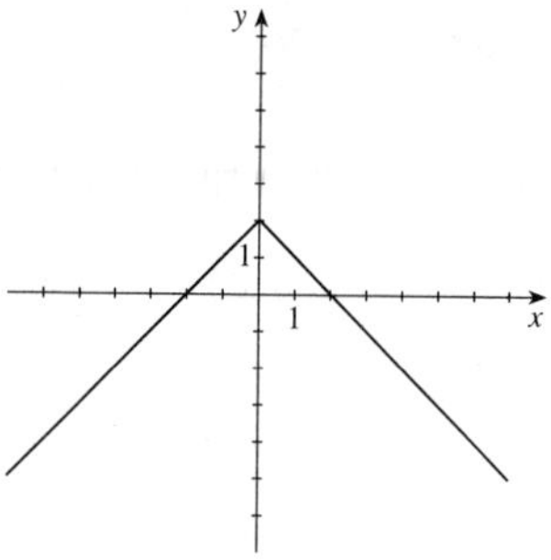


**49.**

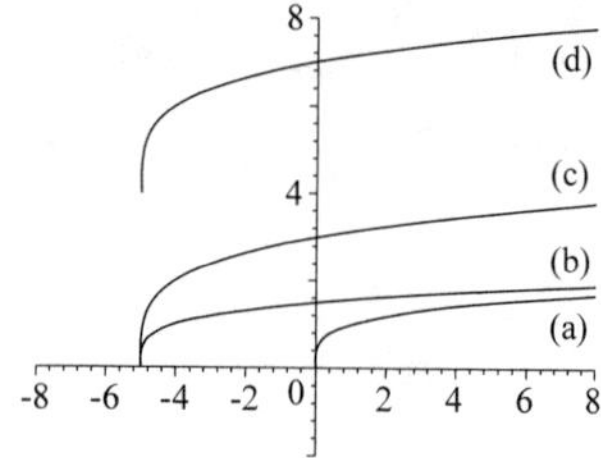


For part (b), shift the graph in (a) to the left 5 units; for part (c), shift the graph in (a) to the left 5 units, and stretch it vertically by a factor of 2; for part (d), shift the graph in (a) to the left 5 units, stretch it vertically by a factor of 2, and then shift it upward 4 units.

**50.**

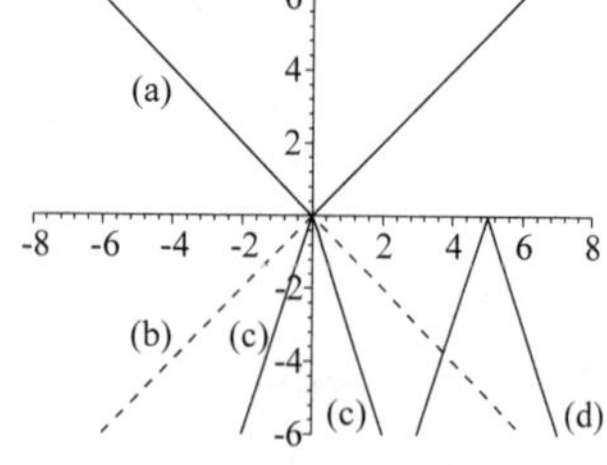


For (b), reflect the graph in (a) about the $x$-axis; for (c), stretch the graph in (a) vertically by a factor of 3 and reflect about the $x$-axis; for (d), shift the graph in (a) to the right 5 units, stretch it vertically by a factor of 3, and reflect it about the $x$-axis. The order in which each operation is applied to the graph in (a) is not important to obtain the graphs in part (c) and (d).

**51.**

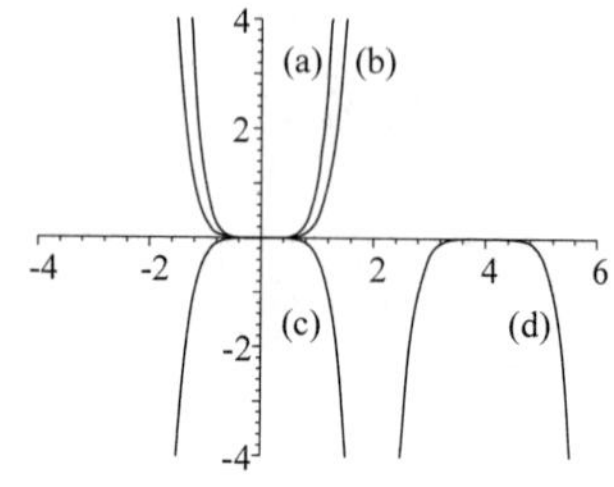


For part (b), shrink the graph in (a) vertically by a factor of $\frac{1}{3}$; for part (c), shrink the graph in (a) vertically by a factor of $\frac{1}{3}$, and reflect it about the $x$-axis; for part (d), shift the graph in (a) to the right 4 units, shrink vertically by a factor of $\frac{1}{3}$, and then reflect it about the $x$-axis.

**52.** 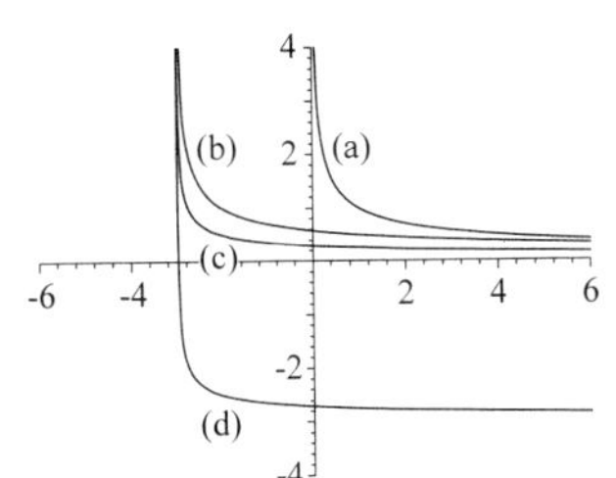


For (b), shift the graph in (a) to the left 3 units; for (c), shift the graph in (a) to the left 3 units and shrink it vertically by a factor of $\frac{1}{2}$; for (d), shift the graph in (a) to the left 3 units, shrink it vertically by a factor of $\frac{1}{2}$, and then shift it downward 3 units. The order in which each operation is applied to the graph in (a) is not important to sketch (c), while it is important in (d).

**53. (a)** $y = g(2x)$

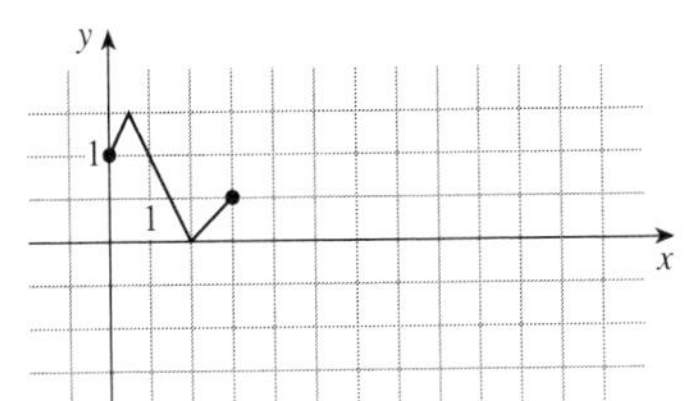

**(b)** $y = g\left(\frac{1}{2}x\right)$

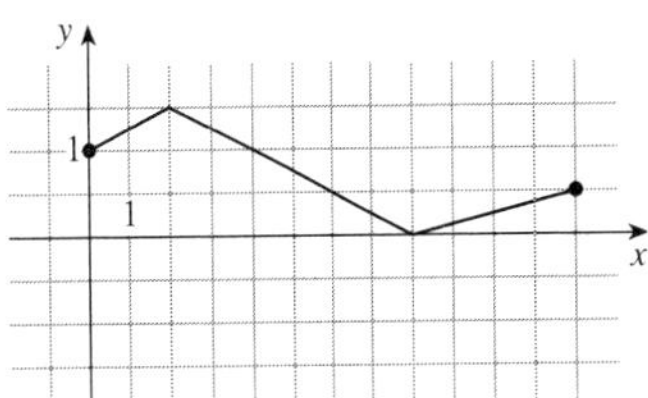

**54. (a)** $y = h(3x)$

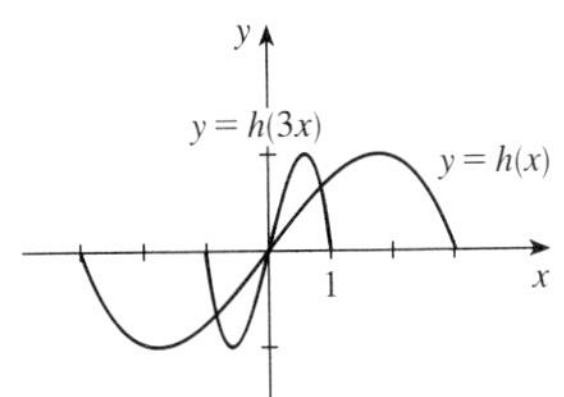


**(b)** $y = h\left(\frac{1}{3}x\right)$

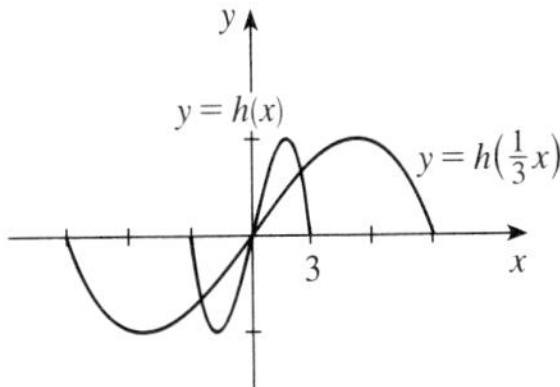


**55. (a)** Even

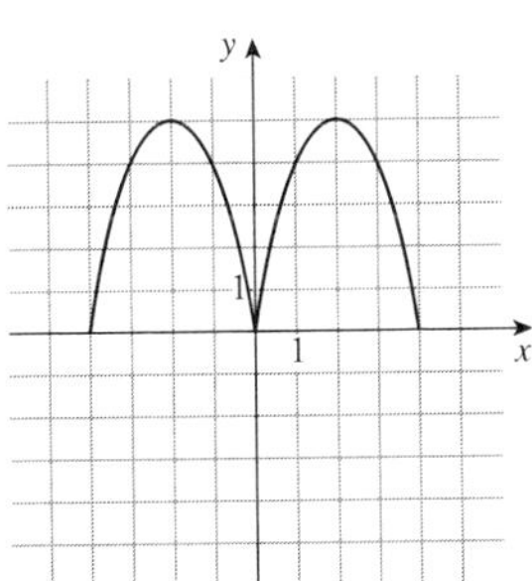

**(b)** Odd

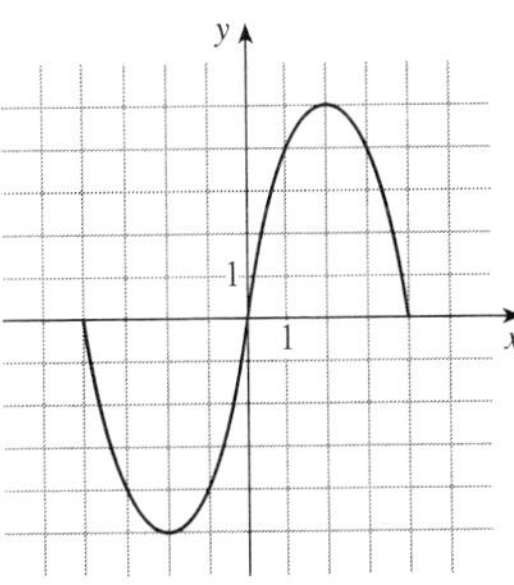

**56. (a)** Even

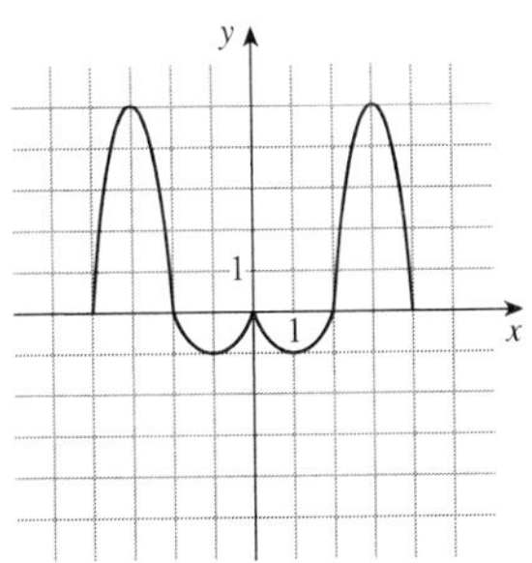

**(b)** Odd

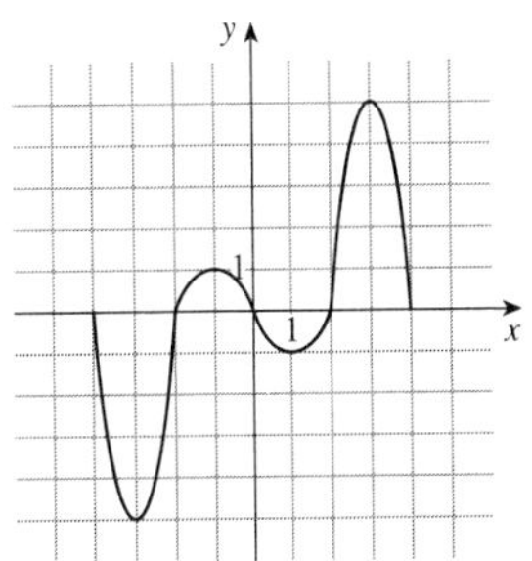

**57.** $y = [\![2x]\!]$

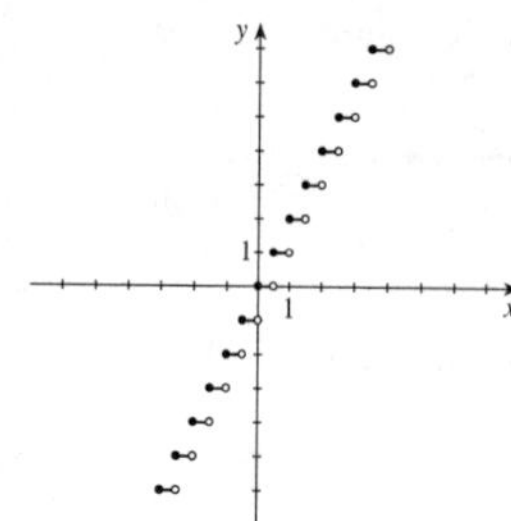

**58.** $y = [\![\frac{1}{4}x]\!]$

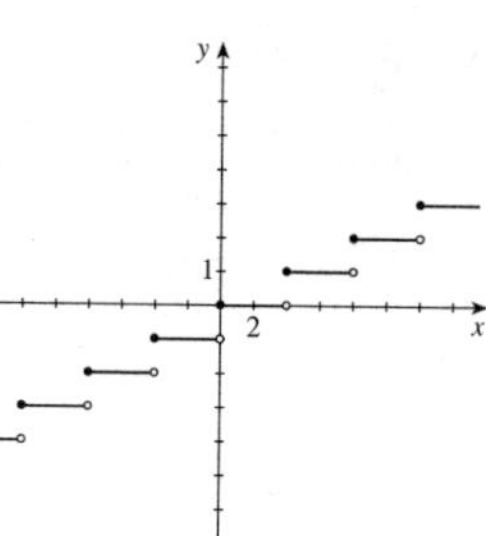

**59. (a)** $y = f(x) = \sqrt{2x - x^2}$

**(b)** $y = f(2x) = \sqrt{2(2x) - (2x)^2}$
$= \sqrt{4x - 4x^2}$

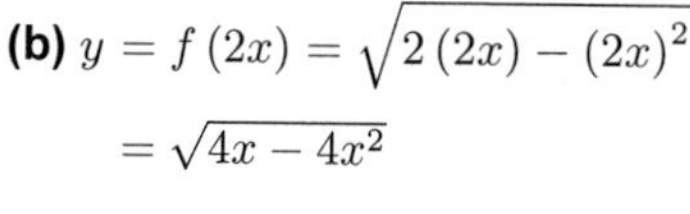

**(c)** $y = f\left(\frac{1}{2}x\right) = \sqrt{2\left(\frac{1}{2}x\right) - \left(\frac{1}{2}x\right)^2}$
$= \sqrt{x - \frac{1}{4}x^2}$

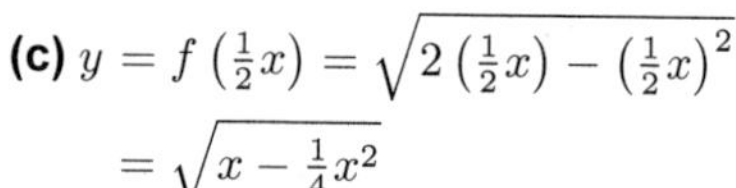

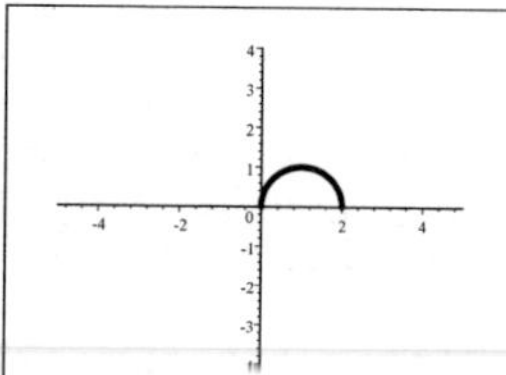

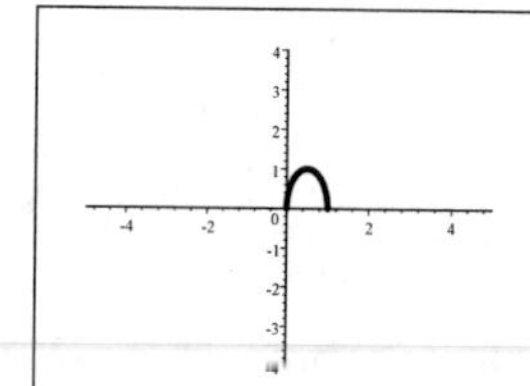

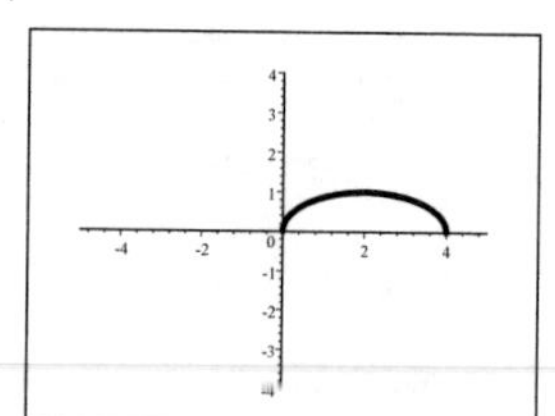

The graph in part (b) is obtained by horizontally shrinking the graph in part (a) by a factor of 2 (so the graph is half as wide). The graph in part (c) is obtained by horizontally stretching the graph in part (a) by a factor of 2 (so the graph is twice as wide).

**60. (a)** $y = f(x) = \sqrt{2x - x^2}$

**(b)** $y = f(-x) = \sqrt{2(-x) - (-x)^2}$
$= \sqrt{-2x - x^2}$

**(c)** $y = -f(-x) = -\sqrt{2(-x) - (-x)^2}$
$= -\sqrt{-2x = x^2}$

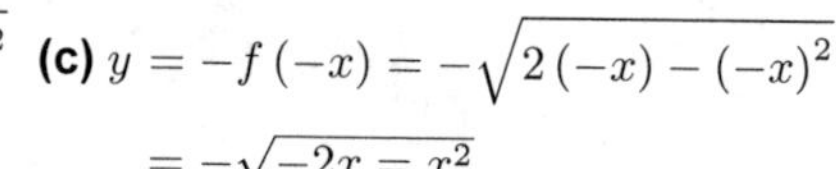

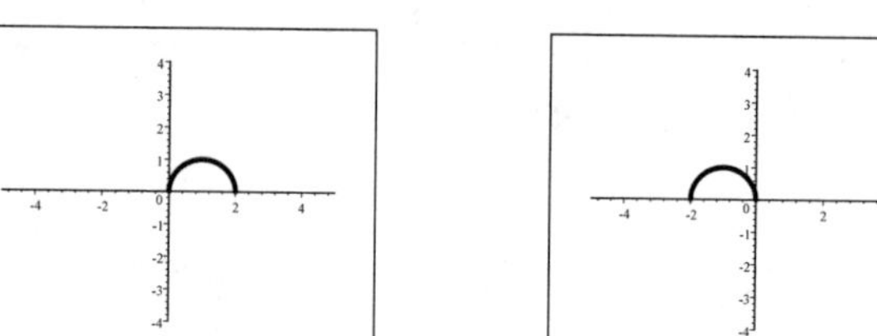

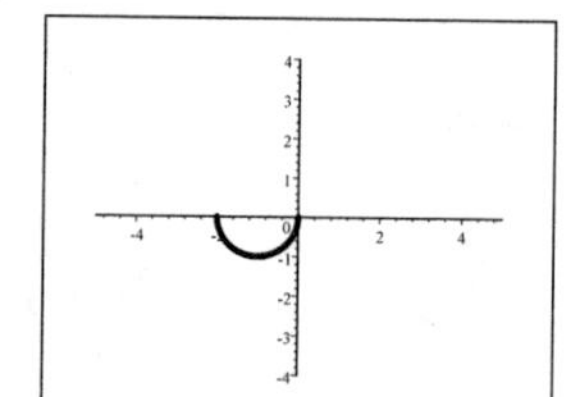

**(d)** $y = f(-2x) = \sqrt{2(-2x) - (-2x)^2}$
$= -\sqrt{-2x - x^2} = \sqrt{-4x - 4x^2}$

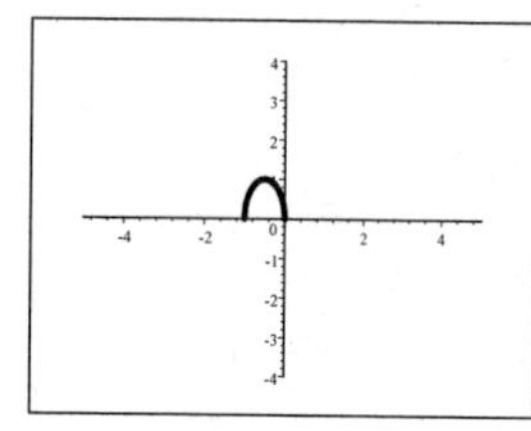

**(e)** $y = f\left(-\frac{1}{2}x\right) = \sqrt{2\left(-\frac{1}{2}x\right) - \left(-\frac{1}{2}x\right)^2}$
$= \sqrt{-x - \frac{1}{4}x^2}$

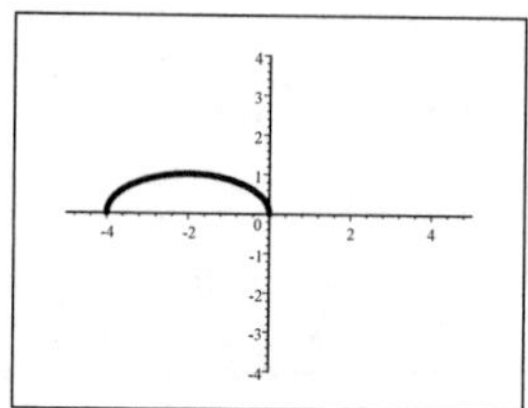

The graph in part (b) is obtained by reflecting the graph in part (a) about the $y$-axis. The graph in part (c) is obtained by reflecting the graph in part (a) about the origin. The graph in part (d) is obtained by reflecting the graph in part (a) about the $y$-axis and then horizontally shrinking the graph by a factor of 2 (so the graph is $\frac{1}{2}$ as wide). The graph in part (e) is obtained by reflecting the graph is part (a) about the $y$-axis and then horizontally stretching the graph by a factor of 2 (so the graph is twice as wide).

**61.** $f(x) = x^{-2}$. $f(-x) = (-x)^{-2} = x^{-2} = f(x)$. Thus $f(x)$ is even.

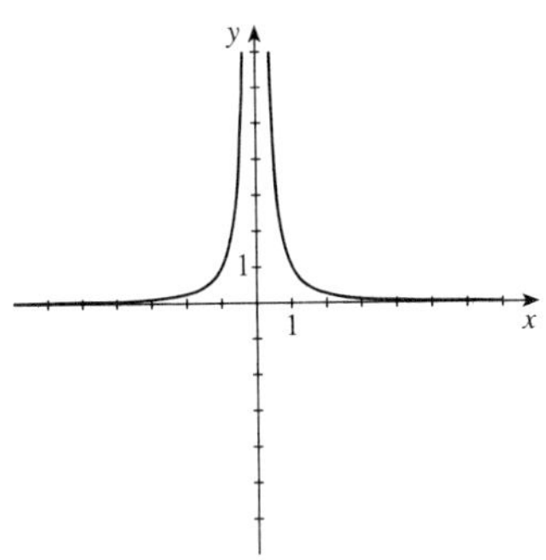


**62.** $f(x) = x^{-3}$. $f(-x) = (-x)^{-3} = -x^{-3} = -f(x)$. Thus $f(x)$ is odd.

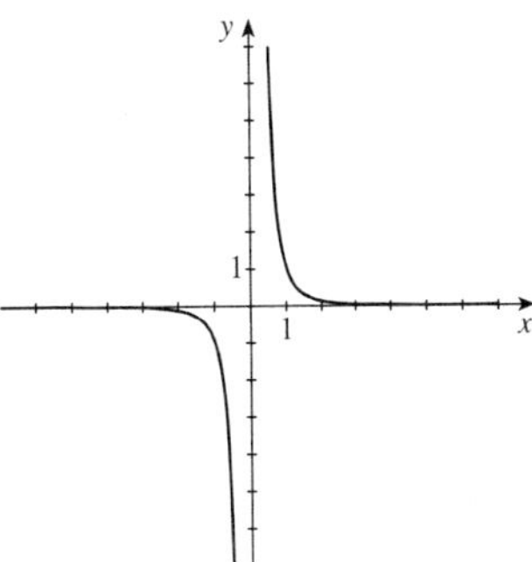


**63.** $f(x) = x^2 + x$. $f(-x) = (-x)^2 + (-x) = x^2 - x$. Thus $f(-x) \neq f(x)$. Also, $f(-x) \neq -f(x)$, so $f(x)$ is neither odd nor even.

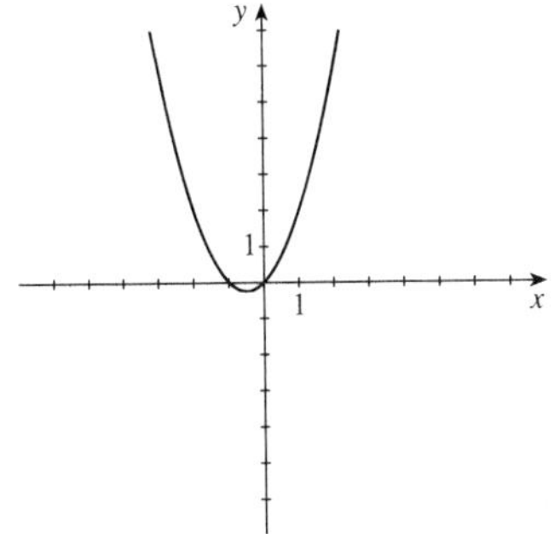


**64.** $f(x) = x^4 - 4x^2$.

$f(-x) = (-x)^4 - 4(-x)^2 = x^4 - 4x^2 = f(x)$. Thus $f(x)$ is even.

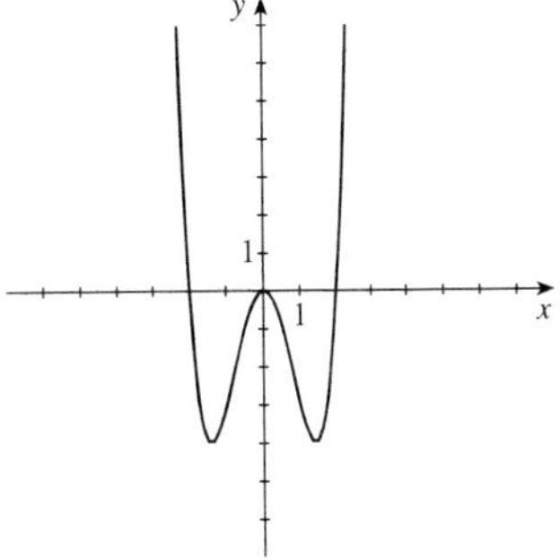


**65.** $f(x) = x^3 - x$.

$$\begin{aligned} f(-x) &= (-x)^3 - (-x) = -x^3 + x \\ &= -\left(x^3 - x\right) = -f(x). \end{aligned}$$

Thus $f(x)$ is odd.

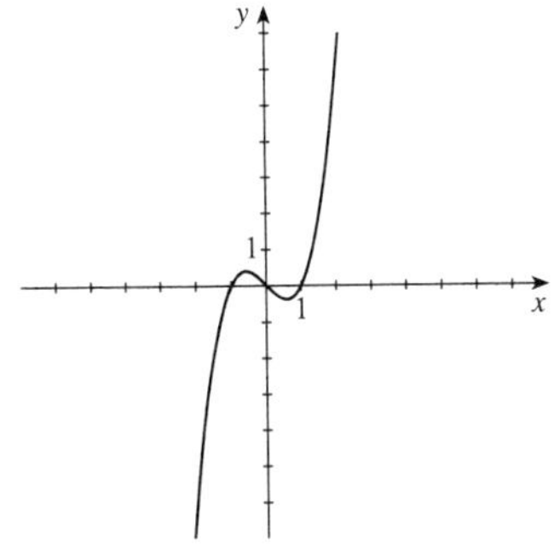


**66.** $f(x) = 3x^3 + 2x^2 + 1$.

$f(-x) = 3(-x)^3 + 2(-x)^2 + 1 = -3x^3 + 2x^2 + 1$. Thus $f(-x) \neq f(x)$. Also $f(-x) \neq -f(x)$, so $f(x)$ is neither odd nor even.

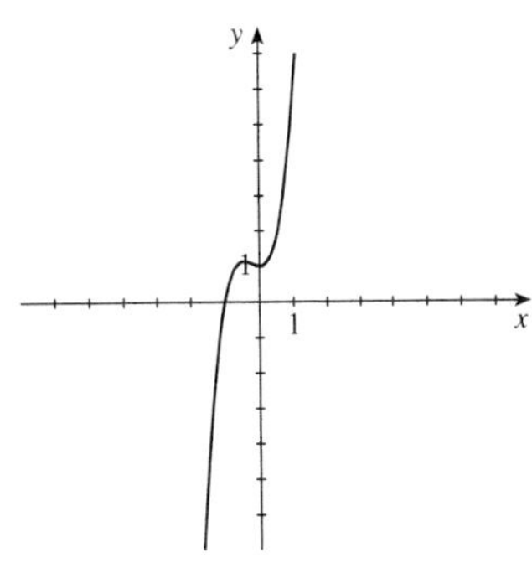

**67.** $f(x) = 1 - \sqrt[3]{x}$. $f(-x) = 1 - \sqrt[3]{(-x)} = 1 + \sqrt[3]{x}$. Thus $f(-x) \neq f(x)$. Also $f(-x) \neq -f(x)$, so $f(x)$ is neither odd nor even.

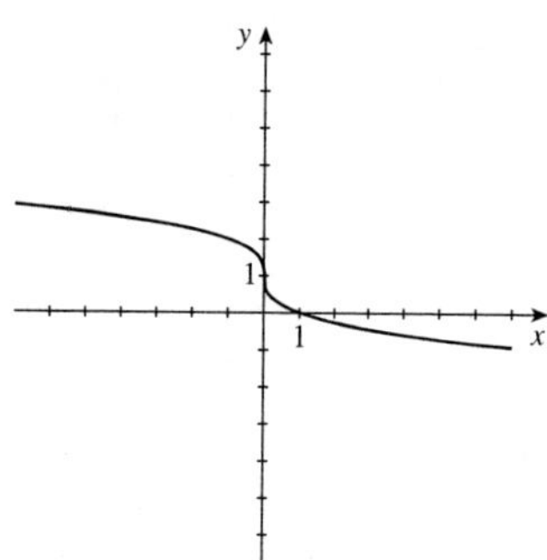


**68.** $f(x) = x + 1/x$.

$$f(-x) = (-x) + 1/(-x) = -x - 1/x$$
$$= -(x + 1/x) = -f(x).$$

Thus $f(x)$ is odd.

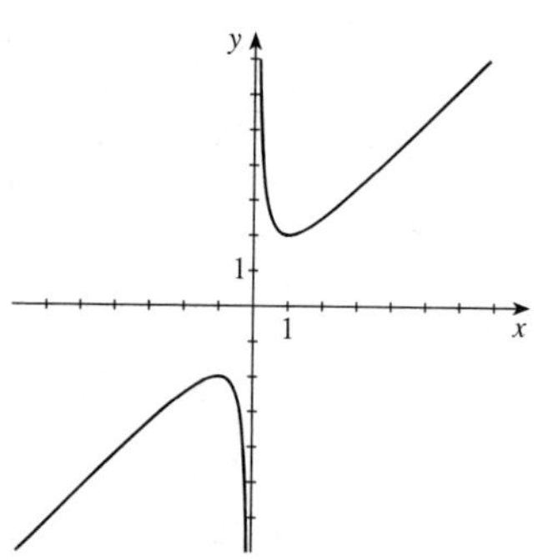


**69.** Since $f(x) = x^2 - 4 < 0$, for $-2 < x < 2$, the graph of $y = g(x)$ is found by sketching the graph of $y = f(x)$ for $x \leq -2$ and $x \geq 2$, then reflecting about the $x$-axis the part of the graph of $y = f(x)$ for $-2 < x < 2$.

**70.** $g(x) = |x^4 - 4x^2|$

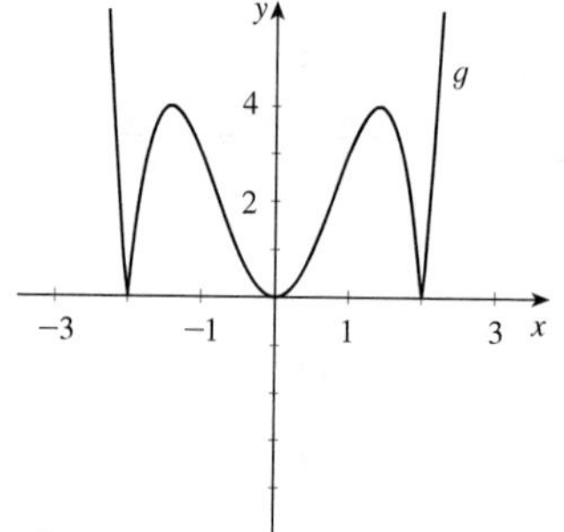


**71. (a)** $f(x) = 4x - x^2$

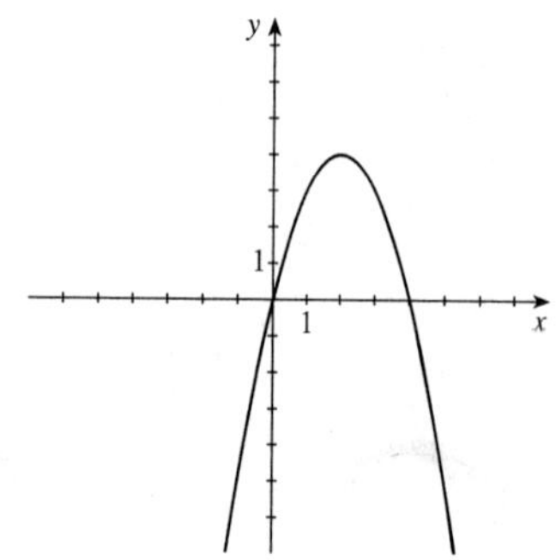


**(b)** $f(x) = |4x - x^2|$

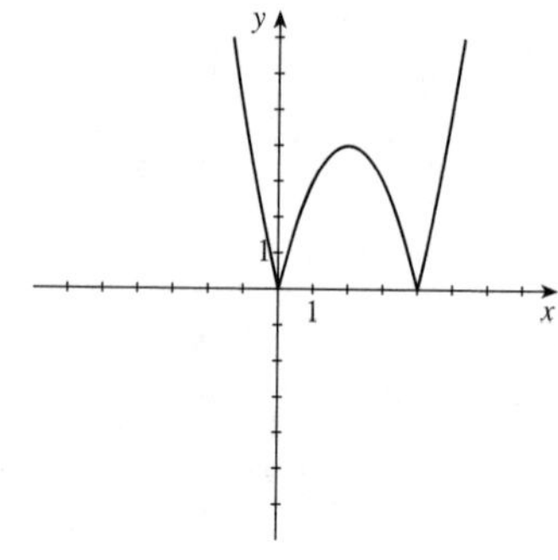

**72. (a)** $f(x) = x^3$

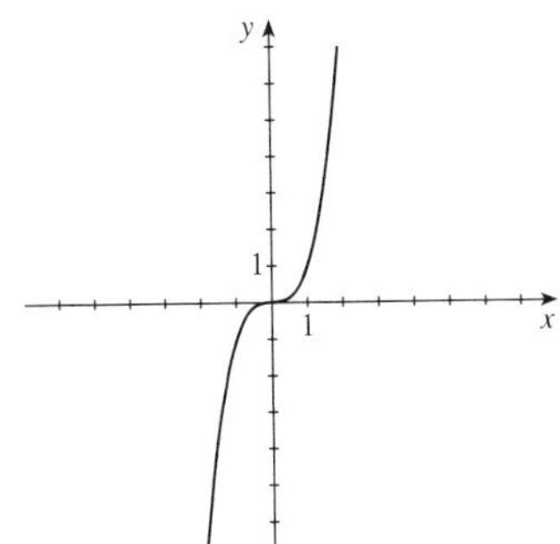


**(b)** $g(x) = |x^3|$

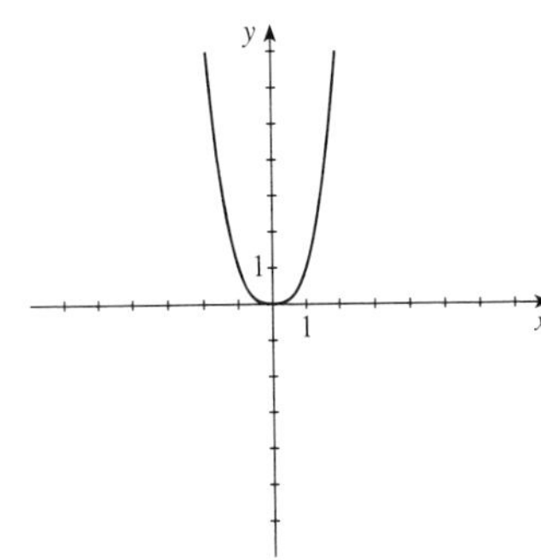


**73. (a)** The graph of $y = t^2$ must be shrunk vertically by a factor of $0.01$ and shifted vertically 4 units up to obtain the graph of $y = f(t)$.

**(b)** The graph of $y = f(t)$ must be shifted horizontally 10 units to the right to obtain the graph of $y = g(t)$. So $g(t) = f(t - 10) = 4 + 0.01(t - 10)^2 = 5 - 0.02t + 0.01t^2$.

**74. (a)** The graph of $y = t^2$ must be shrunk vertically by a factor of $\frac{1}{2}$ and shifted up 2 units to obtain the graph of $y = C(t)$.

**(b)** The graph of $y = C(t)$ must be stretched vertically by a factor of $\frac{9}{5}$ and shifted up 32 units to obtain the graph of $y = F(t)$. So $F(t) = \frac{9}{5}C(t) + 32 = \frac{9}{5}\left(\frac{1}{2}t^2 + 2\right) + 32 = \frac{9}{10}t^2 + \frac{178}{5}$.

**75.** $f$ even implies $f(-x) = f(x)$; $g$ even implies $g(-x) = g(x)$; $f$ odd implies $f(-x) = -f(x)$; and $g$ odd implies $g(-x) = -g(x)$

If $f$ and $g$ are both even, then $(f + g)(-x) = f(-x) + g(-x) = f(x) + g(x) = (f + g)(x)$ and $f + g$ is even.

If $f$ and $g$ are both odd, then $(f + g)(-x) = f(-x) + g(-x) = -f(x) - g(x) = -(f + g)(x)$ and $f + g$ is odd.

If $f$ odd and $g$ even, then $(f + g)(-x) = f(-x) + g(-x) = -f(x) + g(x)$, which is neither odd nor even.

**76.** $f$ even implies $f(-x) = f(x)$; $g$ even implies $g(-x) = g(x)$; $f$ odd implies $f(-x) = -f(x)$; and $g$ odd implies $g(-x) = -g(x)$.

If $f$ and $g$ are both even, then $(fg)(-x) = f(-x) \cdot g(-x) = f(x) \cdot g(x) = (fg)(x)$. Thus $fg$ is even.

If $f$ and $g$ are both odd, then $(fg)(-x) = f(-x) \cdot g(-x) = -f(x) \cdot (-g(x)) = f(x) \cdot g(x) = (fg)(x)$. Thus $fg$ is even

If $f$ if odd and $g$ is even, then $(fg)(-x) = f(-x) \cdot g(-x) = f(x) \cdot (-g(x)) = -f(x) \cdot g(x) = -(fg)(x)$. Thus $fg$ is odd.

**77.** $f(x) = x^n$ is even when $n$ is an even integer and $f(x) = x^n$ is odd when $n$ is an odd integer.

These names were chosen because polynomials with only terms with odd powers are odd functions, and polynomials with only terms with even powers are even functions.

## 3.5 Quadratic Functions; Maxima and Minima

**1. (a)** Vertex: $(3, 4)$

**(b)** Maximum value of $f$: 4.

**2. (a)** Vertex: $(-2, 8)$

**(b)** Maximum value of $f$: 8.

**3. (a)** Vertex: $(1, -3)$

**(b)** Minimum value of $f$: $-3$.

**4. (a)** Vertex: $(-1, -4)$

**(b)** Minimum value of $f$: $-4$.

**5.** **(a)** $f(x) = x^2 - 6x = (x-3)^2 - 9$

**(b)** Vertex: $y = x^2 - 6x = x^2 - 6x + 9 - 9 = (x-3)^2 - 9$. So the vertex is at $(3, -9)$.

$x$-intercepts: $y = 0 \quad \Rightarrow \quad 0 = x^2 - 6x = x(x-6)$. So $x = 0$ or $x = 6$. The $x$-intercepts are $x = 0$ and $x = 6$.

$y$-intercept: $x = 0 \quad \Rightarrow \quad y = 0$. The $y$-intercept is $y = 0$.

**(c)**

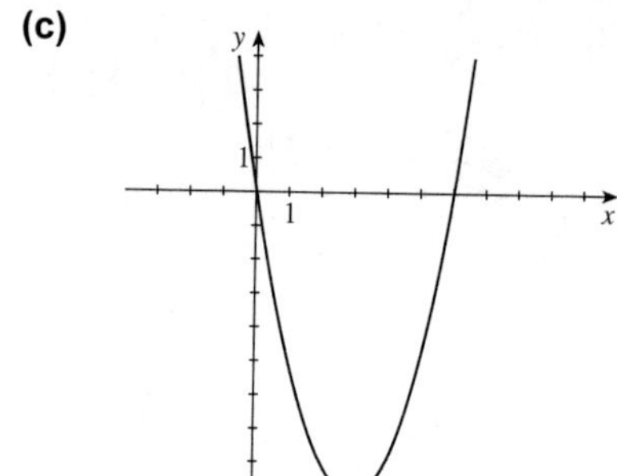


**6.** **(a)** $f(x) = x^2 + 8x = (x+4)^2 - 16$

**(b)** The vertex is $(-4, -16)$. $x$-intercepts: $y = 0 \quad \Rightarrow$ $0 = x^2 + 8x = x(x+8)$. So $x = 0$ or $x = -8$. The $x$-intercepts are $x = 0$ and $x = -8$. $y$-intercept: $x = 0 \quad \Rightarrow \quad y = 0$. The $y$-intercept is $y = 0$.

**(c)**

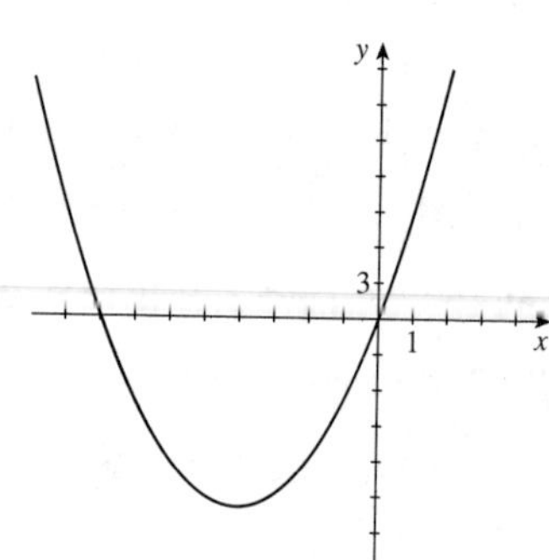


**7.** **(a)** $f(x) = 2x^2 + 6x = 2\left(x + \frac{3}{2}\right)^2 - \frac{9}{2}$

**(b)** The vertex is $\left(-\frac{3}{2}, -\frac{9}{2}\right)$.

$x$-intercepts: $y = 0 \quad \Rightarrow \quad 0 = 2x^2 + 6x = 2x(x+3) \quad \Rightarrow$ $x = 0$ or $x = -3$. The $x$-intercepts are $x = 0$ and $x = -3$.

$y$-intercept: $x = 0 \quad \Rightarrow \quad y = 0$. The $y$-intercept is $y = 0$.

**(c)**

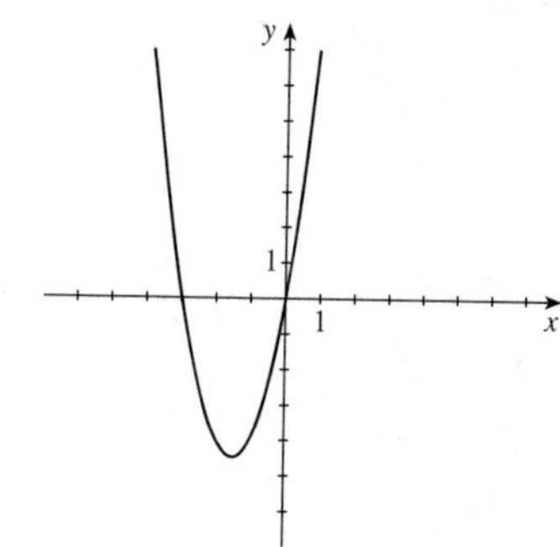


**8.** **(a)** $f(x) = -x^2 + 10x = -(x-5)^2 + 25$

**(b)** The vertex is $(5, 25)$.

$x$-intercepts: $y = 0 \quad \Rightarrow \quad 0 = -x^2 + 10x = -x(x-10) = 0$ $\Rightarrow x = 0$ or $x = 10$. The $x$-intercepts are $x = 0$ and $x = 10$.

$y$-intercept: $x = 0 \quad \Rightarrow \quad y = 0$. The $y$-intercept is $y = 0$.

**(c)**

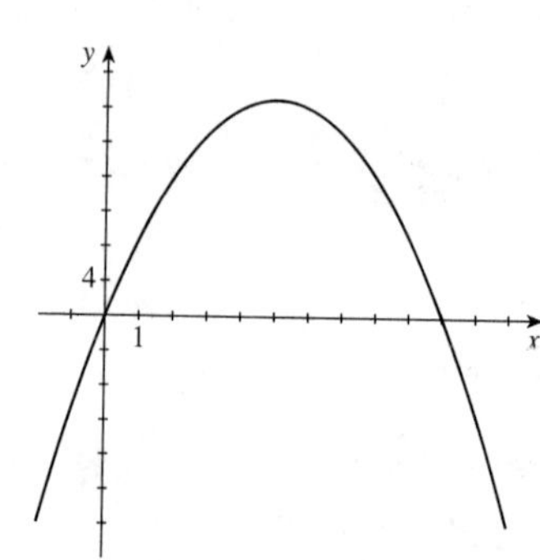

**9. (a)** $f(x) = x^2 + 4x + 3 = (x+2)^2 - 1$

**(b)** The vertex is $(-2, -1)$.

$x$-intercepts: $y = 0 \quad \Rightarrow \quad 0 = x^2 + 4x + 3 = (x+1)(x+3)$. So $x = -1$ or $x = -3$. The $x$-intercepts are $x = -1$ and $x = -3$.

$y$-intercept: $x = 0 \quad \Rightarrow \quad y = 3$. The $y$-intercept is $y = 3$.

**(c)**

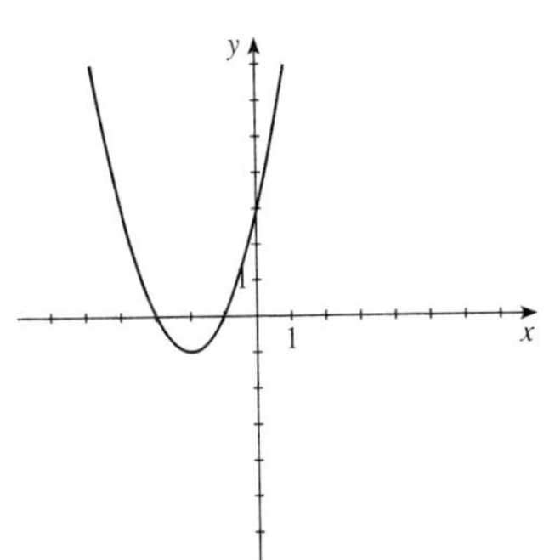


**10. (a)** $f(x) = x^2 - 2x + 2 = (x-1)^2 + 1$

**(b)** The vertex is $(1, 1)$.

$x$-intercepts: $y = 0 \quad \Rightarrow \quad (x-1)^2 + 1 = 0 \quad \Leftrightarrow$ $(x-1)^2 = -1$. Since this last equation has no real solution, there is no $x$-intercept.

$y$-intercept: $x = 0 \quad \Rightarrow \quad y = 2$. The $y$-intercept is $y = 2$.

**(c)**

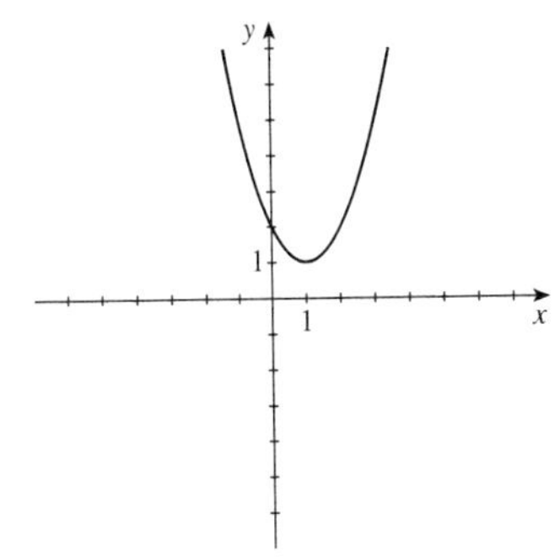


**11. (a)** $f(x) = -x^2 + 6x + 4 = -(x-3)^2 + 13$

**(b)** The vertex is $(3, 13)$.

$x$-intercepts: $y = 0 \quad \Rightarrow \quad 0 = -(x-3)^2 + 13 \quad \Leftrightarrow$ $(x-3)^2 = 13 \Rightarrow x - 3 = \pm\sqrt{13} \quad \Leftrightarrow \quad x = 3 \pm \sqrt{13}$. The $x$-intercepts are $x = 3 - \sqrt{13}$ and $x = 3 + \sqrt{13}$.

$y$-intercept: $x = 0 \quad \Rightarrow \quad y = 4$. The $y$-intercept is $y = 4$.

**(c)**

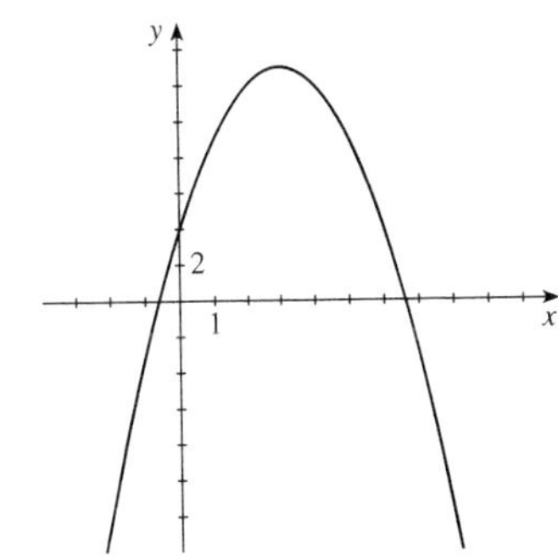


**12. (a)** $f(x) = -x^2 - 4x + 4 = -(x+2)^2 + 8$

**(b)** The vertex is $(-2, 8)$.

$x$-intercepts: $y = 0 \quad \Rightarrow \quad 0 = -x^2 - 4x + 4 \Leftrightarrow 0 = x^2 + 4x - 4$.

Using the quadratic formula,

$$x = \frac{-4 \pm \sqrt{(4)^2 - 4(1)(-4)}}{2(1)} = \frac{-4 \pm \sqrt{32}}{2} = \frac{2\left(-2 \pm 2\sqrt{2}\right)}{2} = -2 \pm 2\sqrt{2}.$$

The $x$-intercepts are $x = -2 + 2\sqrt{2}$ and $x = -2 - 2\sqrt{2}$.

$y$-intercept: $x = 0 \quad \Rightarrow \quad y = 4$. The $y$-intercept is $y = 4$.

**(c)**

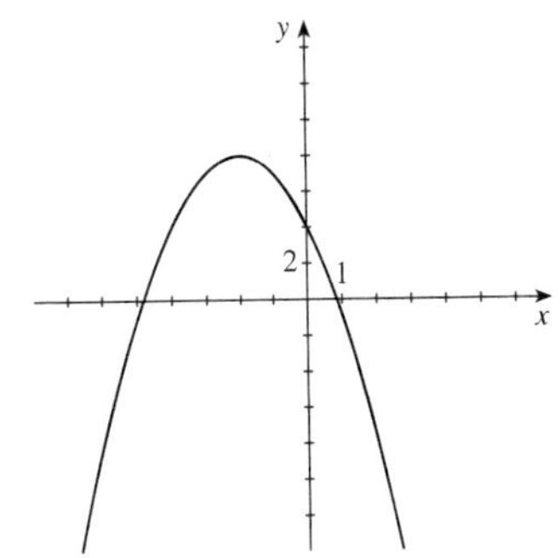

**13. (a)** $f(x) = 2x^2 + 4x + 3 = 2(x+1)^2 + 1$

**(b)** The vertex is $(-1, 1)$.

$x$-intercepts: $y = 0 \quad \Rightarrow \quad 0 = 2x^2 + 4x + 3 = 2(x+1)^2 + 1$ $\Leftrightarrow \quad 2(x+1)^2 = -1$. Since this last equation has no real solution, there is no $x$-intercept.

$y$-intercept: $x = 0 \quad \Rightarrow \quad y = 3$. The $y$-intercept is $y = 3$.

**(c)**

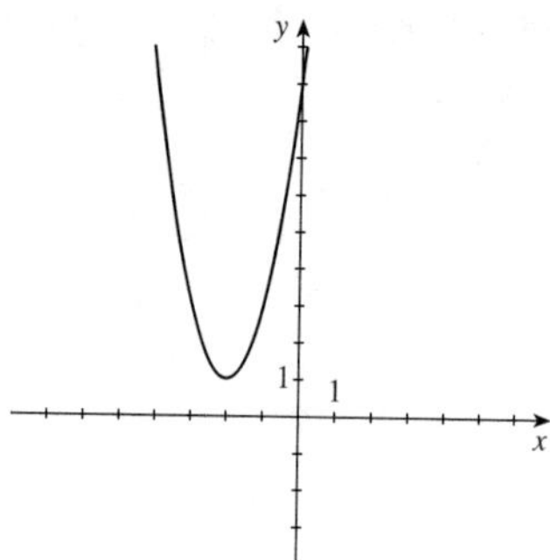


**14. (a)** $f(x) = -3x^2 + 6x - 2 = -3(x-1)^2 + 1$

**(b)** The vertex is $(1, 1)$.

$x$-intercepts: $y = 0 \quad \Rightarrow \quad 0 = -3(x-1)^2 + 1 = 0$ $\Leftrightarrow (x-1)^2 = \frac{1}{3} \quad \Rightarrow \quad x - 1 = \pm\sqrt{\frac{1}{3}} \quad \Leftrightarrow \quad x = 1 \pm \sqrt{\frac{1}{3}}$. The $x$-intercepts are $x = 1 + \sqrt{\frac{1}{3}}$ and $x = 1 - \sqrt{\frac{1}{3}}$.

$y$-intercept: $x = 0 \quad \Rightarrow \quad y = -2$. The $y$-intercept is $y = -2$.

**(c)**

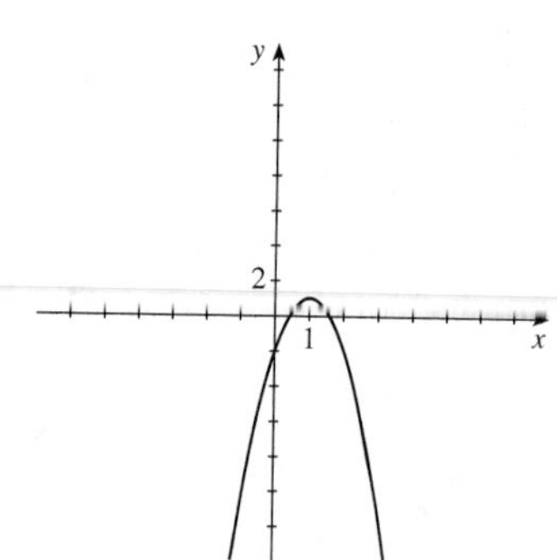


**15. (a)** $f(x) = 2x^2 - 20x + 57 = 2(x-5)^2 + 7$

**(b)** The vertex is $(5, 7)$.

$x$-intercepts: $y = 0 \quad \Rightarrow$ $0 = 2x^2 - 20x + 57 = 2(x-5)^2 + 7 \Leftrightarrow 2(x-5)^2 = -7$. Since this last equation has no real solution, there is no $x$-intercept.

$y$-intercept: $x = 0 \quad \Rightarrow \quad y = 57$. The $y$-intercept is $y = 57$.

**(c)**

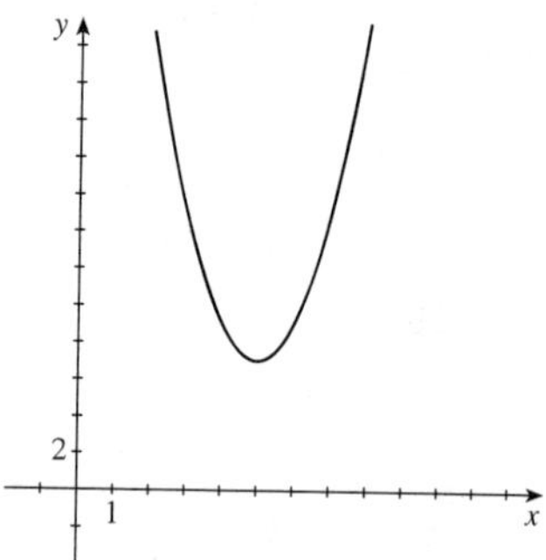


**16. (a)** $f(x) = 2x^2 + x - 6 = 2\left(x + \frac{1}{4}\right)^2 - \frac{49}{8}$

**(b)** The vertex is $\left(-\frac{1}{4}, -\frac{49}{8}\right)$.

$x$-intercepts: $y = 0 \quad \Rightarrow \quad 0 = 2x^2 + x - 6 = (2x-3)(x+2)$ $\Rightarrow x = \frac{3}{2}$ or $x = -2$. The $x$-intercepts are $x = \frac{3}{2}$ and $x = -2$.

$y$-intercept: $x = 0 \quad \Rightarrow \quad y = -6$. The $y$-intercept is $y = -6$.

**(c)**

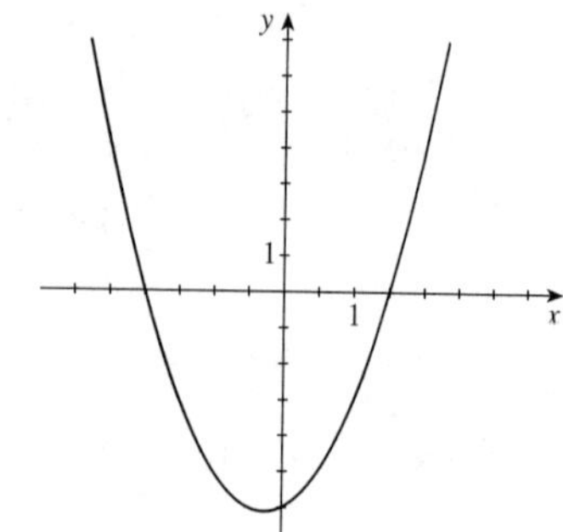

**17. (a)** $f(x) = -4x^2 - 16x + 3 = -4(x+2)^2 + 19$

**(b)** The vertex is $(-2, 19)$.

$x$-intercepts: $y = 0 \quad \Rightarrow$

$0 = -4x^2 - 16x + 3 = -4(x+2)^2 + 19 \Leftrightarrow 4(x+2)^2 = 19$

$\Leftrightarrow \quad (x+2)^2 = \frac{19}{4} \Rightarrow x + 2 = \pm\sqrt{\frac{19}{4}} = \pm\frac{\sqrt{19}}{2} \quad \Leftrightarrow$

$x = -2 \pm \frac{\sqrt{19}}{2}$. The $x$-intercepts are $x = -2 - \frac{\sqrt{19}}{2}$ and $x = -2 + \frac{\sqrt{19}}{2}$.

$y$-intercept: $x = 0 \quad \Rightarrow \quad y = 3$. The $y$-intercept is $y = 3$.

**(c)**

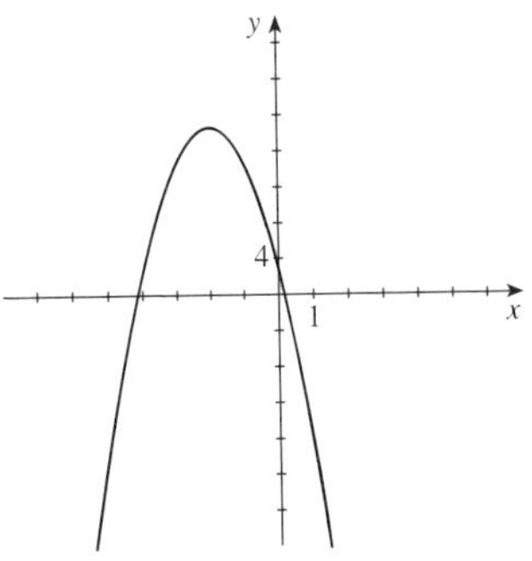


**18. (a)** $f(x) = 6x^2 + 12x - 5 = 6(x+1)^2 - 11$

**(b)** The vertex is $(-1, -11)$.

$x$-intercepts: $y = 0 \quad \Rightarrow \quad 0 = 6x^2 + 12x - 5$. Using the quadratic formula, $x = \frac{-12 \pm \sqrt{(12)^2 - 4(6)(-5)}}{2(6)} = \frac{-12 \pm \sqrt{264}}{12} = \frac{-6 \pm \sqrt{66}}{6}$. The $x$-intercepts are $x = \frac{-6 \pm \sqrt{66}}{6}$.

$y$-intercept: $x = 0 \quad \Rightarrow \quad y = -5$. The $y$-intercept is $y = -5$.

**(c)**

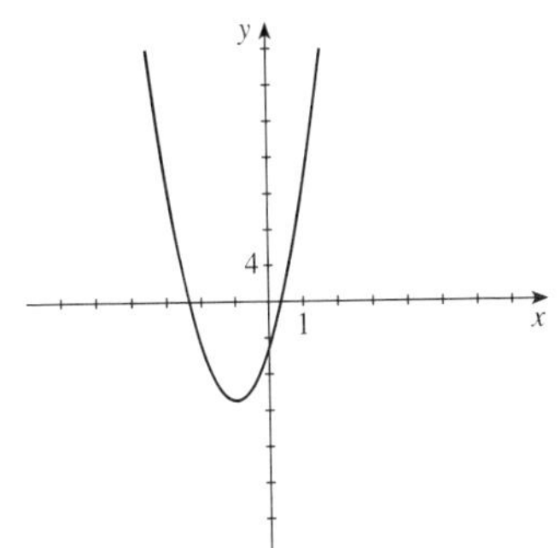


**19. (a)** $f(x) = 2x - x^2 = -(x^2 - 2x)$
$= -(x^2 - 2x + 1) + 1 = -(x-1)^2 + 1$

**(b)**

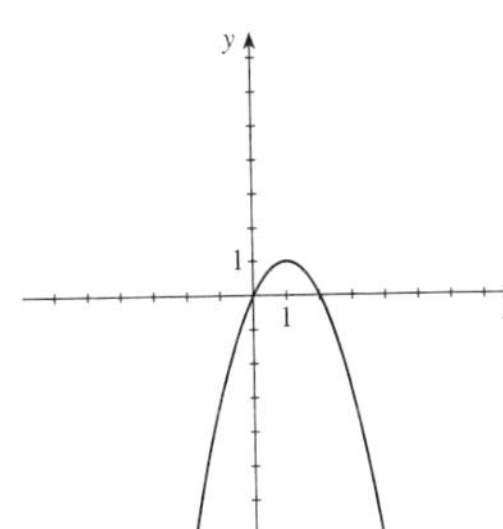


**(c)** The maximum value is $f(1) = 1$.

**20. (a)** $f(x) = x + x^2 = \left(x^2 + x + \frac{1}{4}\right) - \frac{1}{4}$
$= \left(x + \frac{1}{2}\right)^2 - \frac{1}{4}$

**(b)**

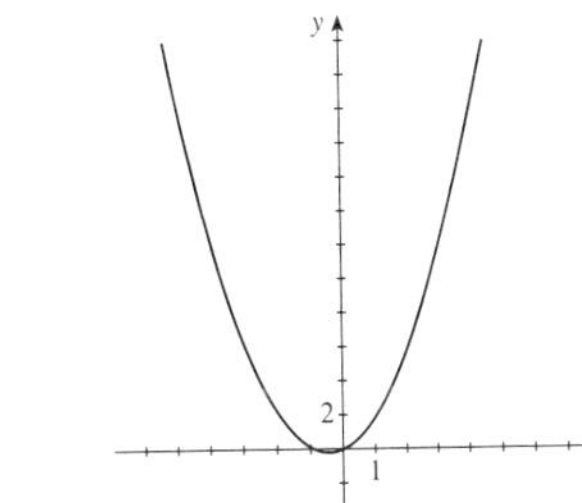


**(c)** The minimum value is $f\left(-\frac{1}{2}\right) = -\frac{1}{4}$.

**21. (a)** $f(x) = x^2 + 2x - 1 = (x^2 + 2x) - 1$
$= (x^2 + 2x + 1) - 1 - 1 = (x+1)^2 - 2$

**(b)**

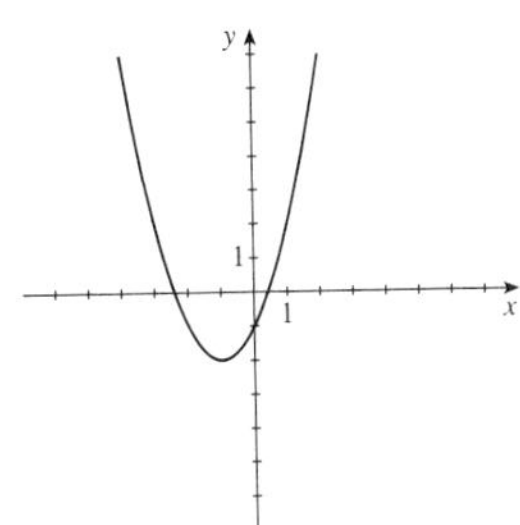


**(c)** The minimum value is $f(-1) = -2$.

**22. (a)** $f(x) = x^2 - 8x + 8 = (x^2 - 8x + 16) + 8 - 16$
$= (x-4)^2 - 8$

**(b)**

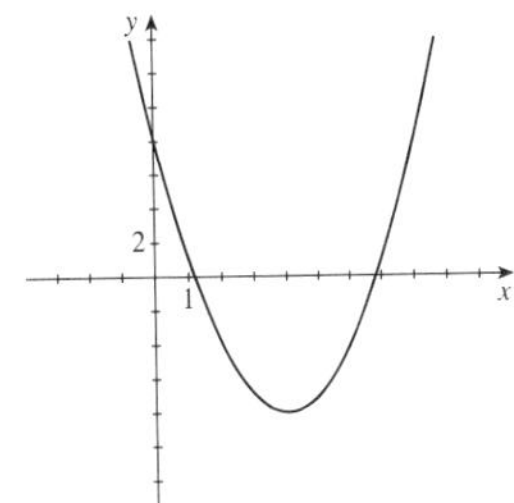


**(c)** The minimum value is $f(4) = -8$.

**23. (a)** $f(x) = -x^2 - 3x + 3 = -(x^2 + 3x) + 3$
$= -(x^2 + 3x + \frac{9}{4}) + 3 + \frac{9}{4}$
$= -(x + \frac{3}{2})^2 + \frac{21}{4}$

**(b)**

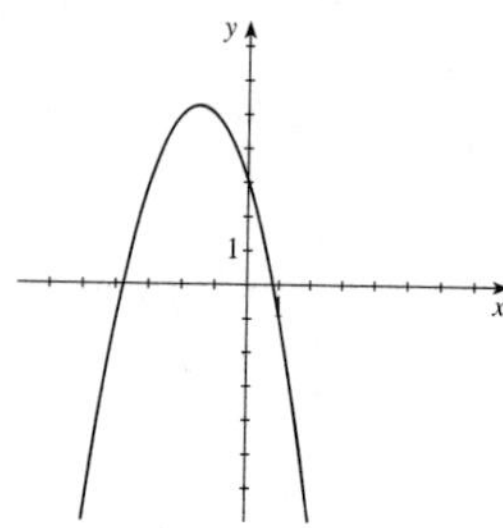

**(c)** The maximum value is $f(-\frac{3}{2}) = \frac{21}{4}$.

**24. (a)** $f(x) = 1 - 6x - x^2 = -(x^2 + 6x) + 1$
$= -(x^2 + 6x + 9) + 1 + 9$
$= -(x + 3)^2 + 10$

**(b)**

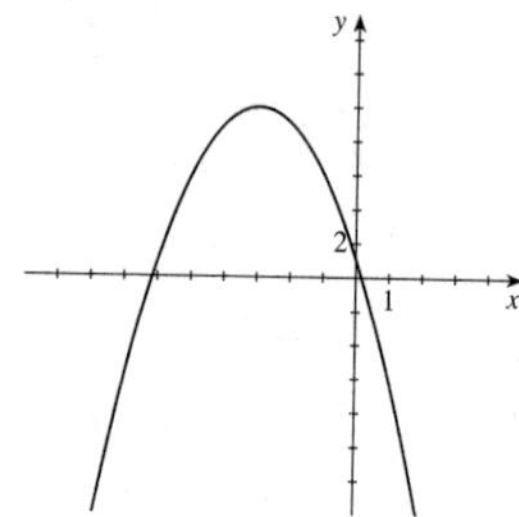

**(c)** The maximum value is $f(-3) = 10$.

**25. (a)** $g(x) = 3x^2 - 12x + 13 = 3(x^2 - 4x) + 13$
$= 3(x^2 - 4x + 4) + 13 - 12$
$= 3(x - 2)^2 + 1$

**(b)**

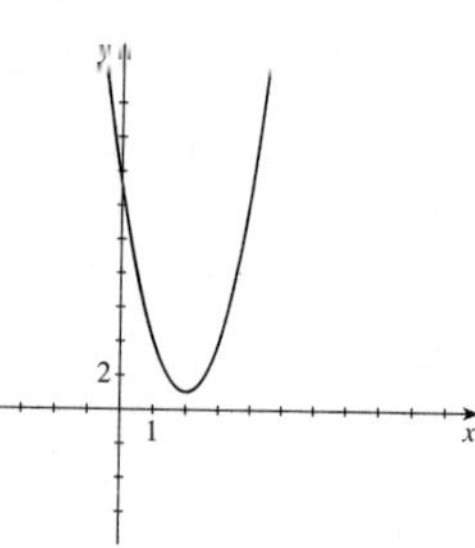

**(c)** The minimum value is $g(2) = 1$.

**26. (a)** $g(x) = 2x^2 + 8x + 11 = 2(x^2 + 4x) + 11$
$= 2(x^2 + 4x + 4) + 11 - 8$
$= 2(x + 2)^2 + 3$

**(b)**

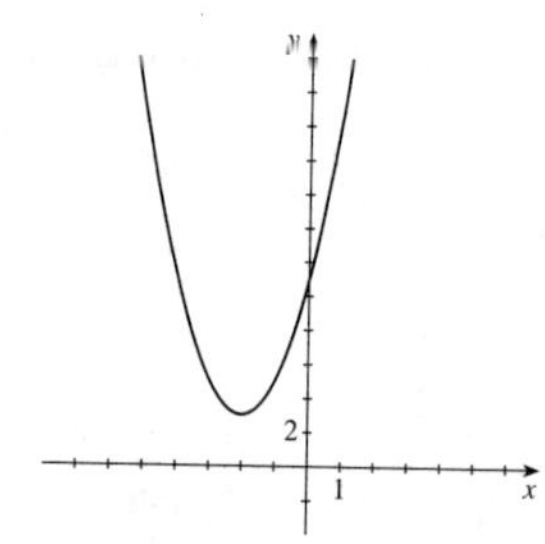

**(c)** The minimum value is $g(-2) = 3$.

**27. (a)** $h(x) = 1 - x - x^2 = -(x^2 + x) + 1$
$= -(x^2 + x + \frac{1}{4}) + 1 + \frac{1}{4}$
$= -(x + \frac{1}{2})^2 + \frac{5}{4}$

**(b)**

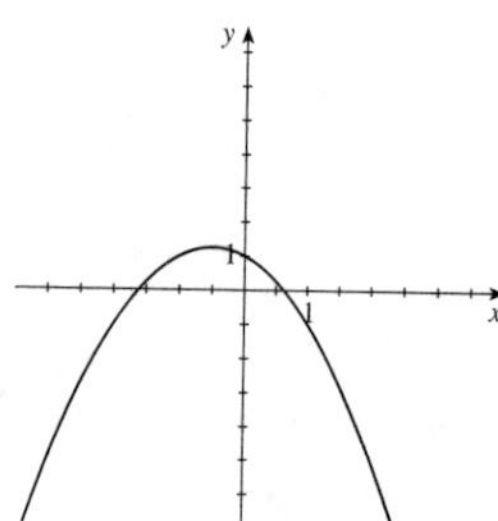

**(c)** The maximum value is $h(-\frac{1}{2}) = \frac{5}{4}$.

**28. (a)** $h(x) = 3 - 4x - 4x^2 = -4(x^2 + x) + 3$
$= -4(x^2 + x + \frac{1}{4}) + 3 + 1$
$= -4(x + \frac{1}{2})^2 + 4$

**(b)**

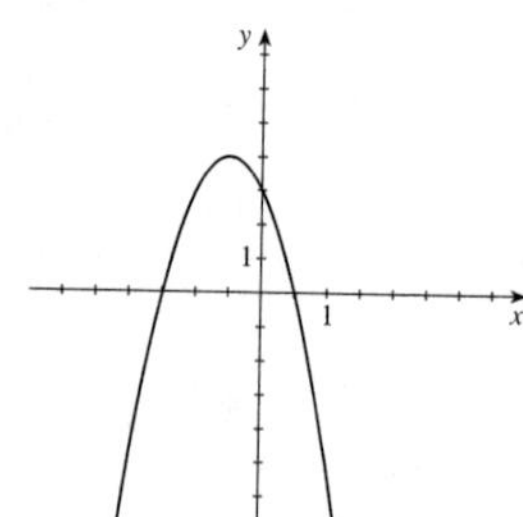

**(c)** The maximum value is $h(-\frac{1}{2}) = 4$.

**29.** $f(x) = x^2 + x + 1 = (x^2 + x) + 1 = (x^2 + x + \frac{1}{4}) + 1 + \frac{1}{4} = (x + \frac{1}{2})^2 + \frac{3}{4}$.
Therefore, the minimum value is $f(-\frac{1}{2}) = \frac{3}{4}$.

**30.** $f(x) = 1 + 3x - x^2 = -(x^2 - 3x) + 1 = -(x^2 - 3x + \frac{9}{4}) + 1 + \frac{9}{4} = -(x - \frac{3}{2})^2 + \frac{13}{4}$. Therefore, the maximum value is $f(\frac{3}{2}) = \frac{13}{4}$.

**31.** $f(t) = 100 - 49t - 7t^2 = -7(t^2 + 7t) + 100 = -7(t^2 + 7t + \frac{49}{4}) + 100 + \frac{343}{4} = -7(t + \frac{7}{2})^2 + \frac{743}{4}$.
Therefore, the maximum value is $f(-\frac{7}{2}) = \frac{743}{4} = 185.75$.

**32.** $f(t) = 10t^2 + 40t + 113 = 10(t^2 + 4t) + 113 = 10(t^2 + 4t + 4) + 113 - 40 = 10(t+2)^2 + 73$. Therefore, the minimum value is $f(-2) = 73$.

**33.** $f(s) = s^2 - 1.2s + 16 = (s^2 - 1.2s) + 16 = (s^2 - 1.2s + 0.36) + 16 - 0.36 = (s - 0.6)^2 + 15.64$. Therefore, the minimum value is $f(0.6) = 15.64$.

**34.** $g(x) = 100x^2 - 1500x = 100(x^2 - 15x) = 100\left(x^2 - 15x + \frac{225}{4}\right) - 5625 = 10\left(x - \frac{15}{2}\right)^2 - 5625$. Therefore, the minimum value is $g\left(\frac{15}{2}\right) = -5625$.

**35.** $h(x) = \frac{1}{2}x^2 + 2x - 6 = \frac{1}{2}(x^2 + 4x) - 6 = \frac{1}{2}(x^2 + 4x + 4) - 6 - 2 = \frac{1}{2}(x+2)^2 - 8$. Therefore, the minimum value is $h(-2) = -8$.

**36.** $f(x) = -\dfrac{x^2}{3} + 2x + 7 = -\frac{1}{3}(x^2 + 6x) + 7 = -\frac{1}{3}(x^2 + 6x + 9) + 7 + 3 = -\frac{1}{3}(x+3)^2 + 10$. Therefore, the maximum value is $f(-3) = 10$.

**37.** $f(x) = 3 - x - \frac{1}{2}x^2 = -\frac{1}{2}(x^2 + 2x) + 3 = -\frac{1}{2}(x^2 + 2x + 1) + 3 + \frac{1}{2} = -\frac{1}{2}(x+1) + \frac{7}{2}$. Therefore, the maximum value is $f(-1) = \frac{7}{2}$.

**38.** $g(x) = 2x(x-4) + 7 = 2x^2 - 8x + 7 = 2(x^2 - 4x) + 7 = 2(x^2 - 4x + 4) + 7 - 8 = 2(x-2)^2 - 1$. Therefore, the minimum value is $g(2) = -1$.

**39.** Since the vertex is at $(1, -2)$, the function is of the form $f(x) = a(x-1)^2 - 2$. Substituting the point $(4, 16)$, we get $16 = a(4-1)^2 - 2 \quad\Leftrightarrow\quad 16 = 9a - 2 \quad\Leftrightarrow\quad 9a = 18 \quad\Leftrightarrow\quad a = 2$. So the function is $f(x) = 2(x-1)^2 - 2 = 2x^2 - 4x$.

**40.** Since the vertex is $(3, 4)$, the function is of the form $y = a(x-3)^2 + 4$. Since the parabola passes through the point $(1, -8)$, it must satisfy $-8 = a(1-3)^2 + 4 \quad\Leftrightarrow\quad -8 = 4a + 4 \quad\Leftrightarrow\quad 4a = -12 \quad\Leftrightarrow\quad a = -3$. So the function is $y = -3(x-3)^2 + 4 = -3x^2 + 18x - 23$.

**41.** $f(x) = -x^2 + 4x - 3 = -(x^2 - 4x) - 3 = -(x^2 - 4x + 4) - 3 + 4 = -(x-2)^2 + 1$. So the domain of $f(x)$ is $(-\infty, \infty)$. Since $f(x)$ has a maximum value of 1, the range is $(-\infty, 1]$.

**42.** $f(x) = x^2 - 2x - 3 = (x^2 - 2x + 1) - 3 - 1 = (x-1)^2 - 4$. Then the domain of the function is all real numbers, and since the minimum value of the function is $f(1) = -4$, the range of the function is $[-4, \infty)$.

**43.** $f(x) = 2x^2 + 6x - 7 = 2\left(x + \frac{3}{2}\right)^2 - 7 - \frac{9}{2} = 2\left(x + \frac{3}{2}\right)^2 - \frac{23}{2}$. The domain of the function is all real numbers, and since the minimum value of the function is $f\left(-\frac{3}{2}\right) = -\frac{23}{2}$, the range of the function is $\left[-\frac{23}{2}, \infty\right)$.

**44.** $f(x) = -3x^2 + 6x + 4 = -3(x-1)^2 + 4 + 3 = -3(x-1)^2 + 7$. The domain of the function is all real numbers, and since the maximum value of the function is $f(1) = 7$, the range of the function is $(-\infty, 7]$.

**45. (a)** The graph of $f(x) = x^2 + 1.79x - 3.21$ is shown. The minimum value is $f(x) \approx -4.01$.

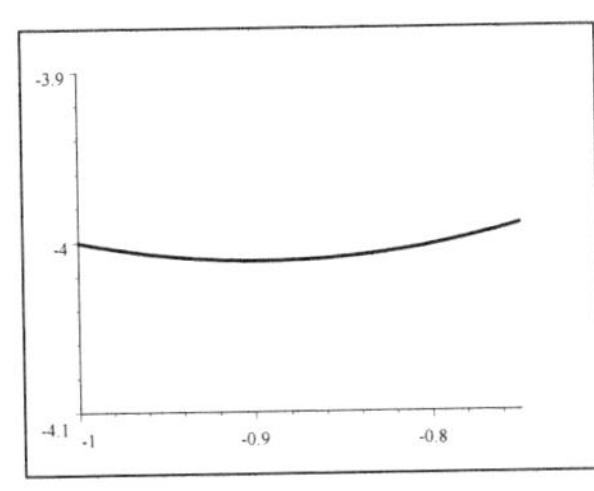


**(b)**
$$\begin{aligned} f(x) &= x^2 + 1.79x - 3.21 \\ &= \left[x^2 + 1.79x + \left(\tfrac{1.79}{2}\right)^2\right] - 3.21 - \left(\tfrac{1.79}{2}\right)^2 \\ &= (x + 0.895)^2 - 4.011025 \end{aligned}$$

Therefore, the exact minimum of $f(x)$ is $-4.011025$.

**46. (a)** The graph of $f(x) = 1 + x - \sqrt{2}x^2$ is shown. The maximum value is $f(x) \approx 1.18$.

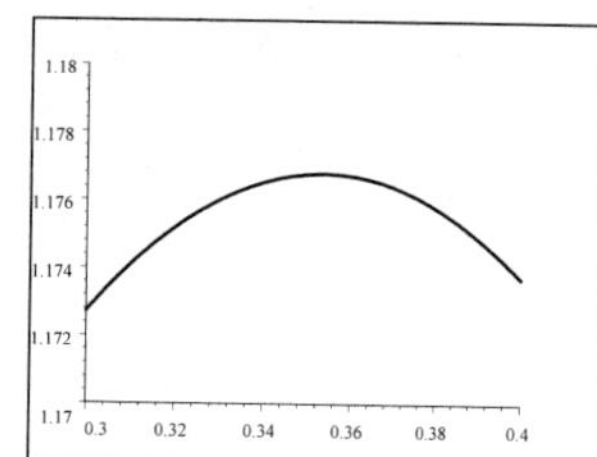

**(b)** $f(x) = 1 + x - \sqrt{2}\,x^2 = -\sqrt{2}\left(x^2 - \frac{\sqrt{2}}{2}x\right) + 1$
$= -\sqrt{2}\left[x^2 - \frac{\sqrt{2}}{2}x + \left(\frac{\sqrt{2}}{4}\right)^2\right] + 1 + \frac{\sqrt{2}}{8}$
$= -\sqrt{2}\left(x - \frac{\sqrt{2}}{4}\right)^2 + \frac{8+\sqrt{2}}{8}$

Therefore, the exact maximum of $f(x)$ is $\frac{8+\sqrt{2}}{8}$.

**47.** Local maximum: 2 at $x = 0$. Local minimum: $-1$ at $x = -2$ and 0 at $x = 2$.

**48.** Local maximum: 2 at $x = -2$ and $x = 1$ at $x = 2$. Local minimum: $-1$ at $x = 0$.

**49.** Local maximum: 0 at $x = 0$ and 1 at $x = 3$. Local minimum: $-2$ at $x = -2$ and $-1$ at $x = 1$.

**50.** Local maximum: 3 at $x = -2$ and 2 at $x = 1$. Local minimum: 0 at $x = -1$ and $-1$ at $x = 2$.

**51.** In the first graph, we see that $f(x) = x^3 - x$ has a local minimum and a local maximum. Smaller $x$- and $y$-ranges show that $f(x)$ has a local maximum of about 0.38 when $x \approx -0.58$ and a local minimum of about $-0.38$ when $x \approx 0.58$.

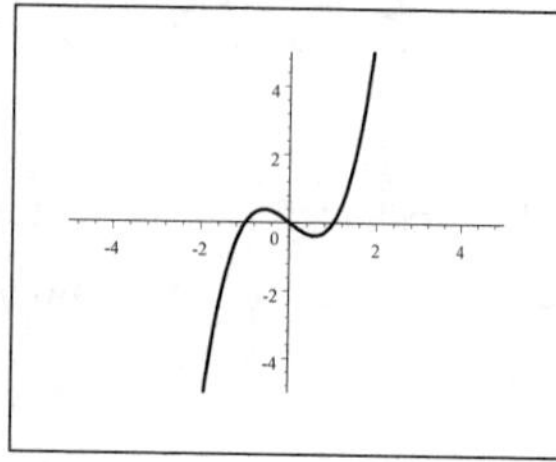
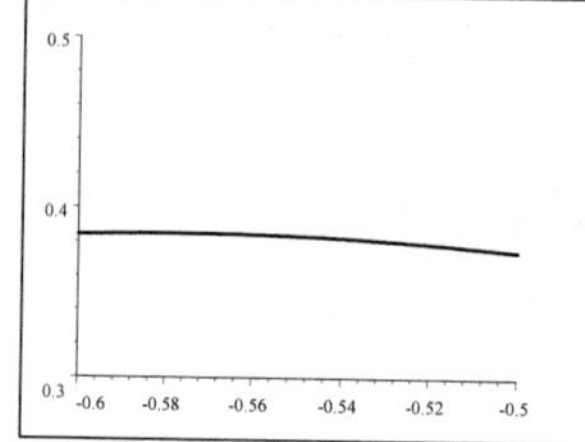
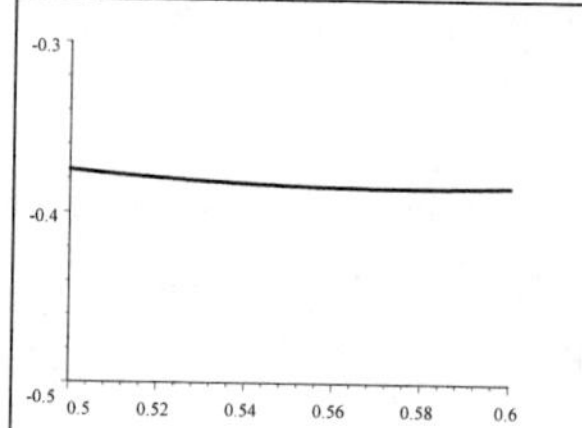

**52.** In the first graph, we see that $f(x) = 3 + x + x^2 - x^3$ has a local minimum and a local maximum. Smaller $x$- and $y$-ranges show that $f(x)$ has a local maximum of about 4.00 when $x \approx 1.00$ and a local minimum of about 2.81 when $x \approx -0.33$.

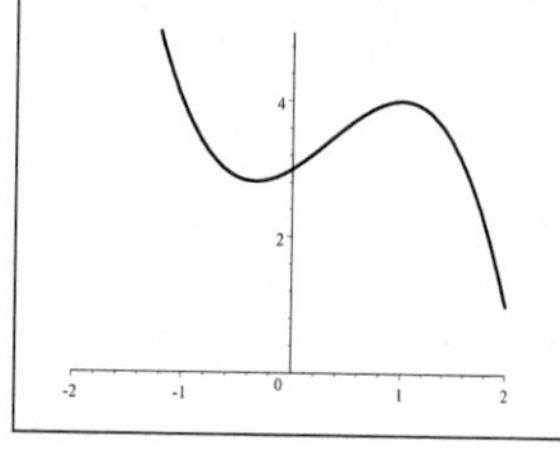
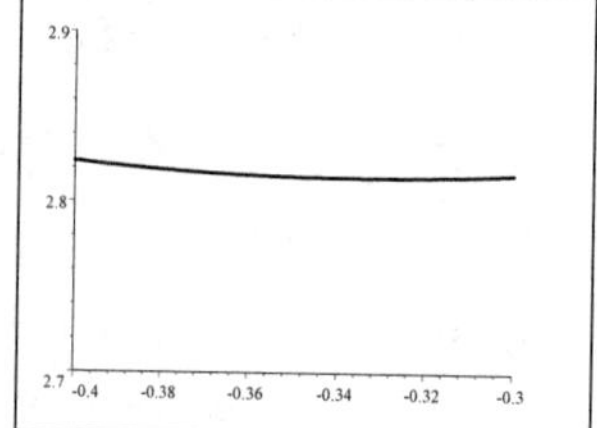
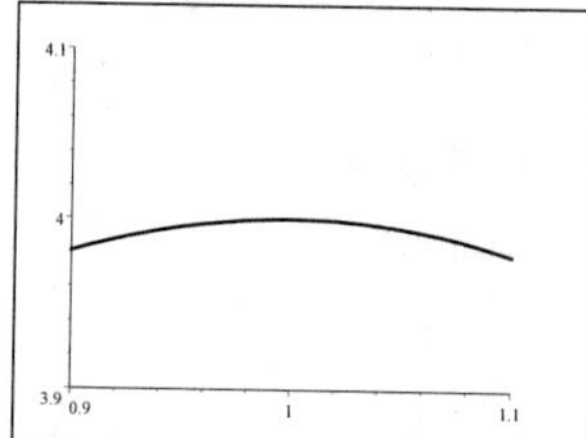

**53.** In the first graph, we see that $g(x) = x^4 - 2x^3 - 11x^2$ has two local minimums and a local maximum. The local maximum is $g(x) = 0$ when $x = 0$. Smaller $x$- and $y$-ranges show that local minima are $g(x) \approx -13.61$ when $x \approx -1.71$ and $g(x) \approx -73.32$ when $x \approx 3.21$.

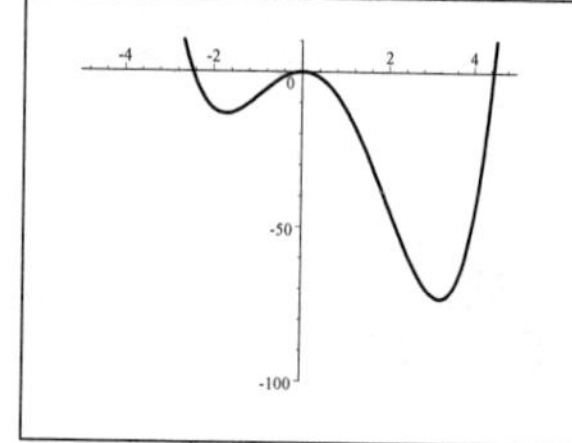
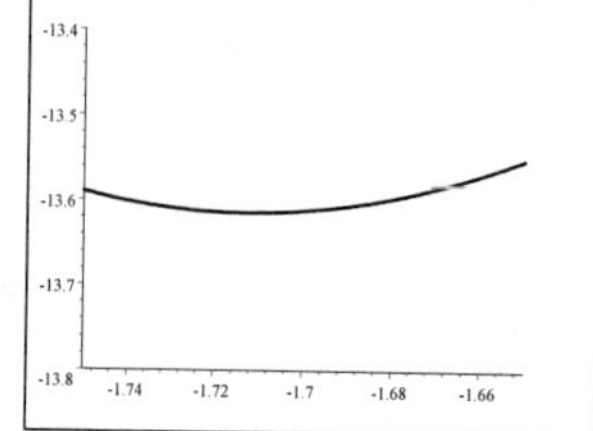
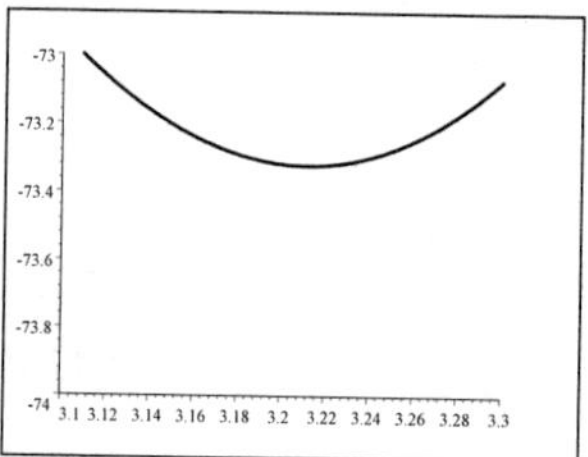

**54.** In the first graph, we see that $g(x) = x^5 - 8x^3 + 20x$ has two local minimums and two local maximums. The local maximums are $g(x) \approx -7.87$ when $x \approx -1.93$ and $g(x) \approx 13.02$ when $x = 1.04$. Smaller $x$- and $y$-ranges show that local minimums are $g(x) \approx -13.02$ when $x = -1.04$ and $g(x) \approx 7.87$ when $x \approx 1.93$. Notice that since $g(x)$ is odd, the local maxima and minima are related.

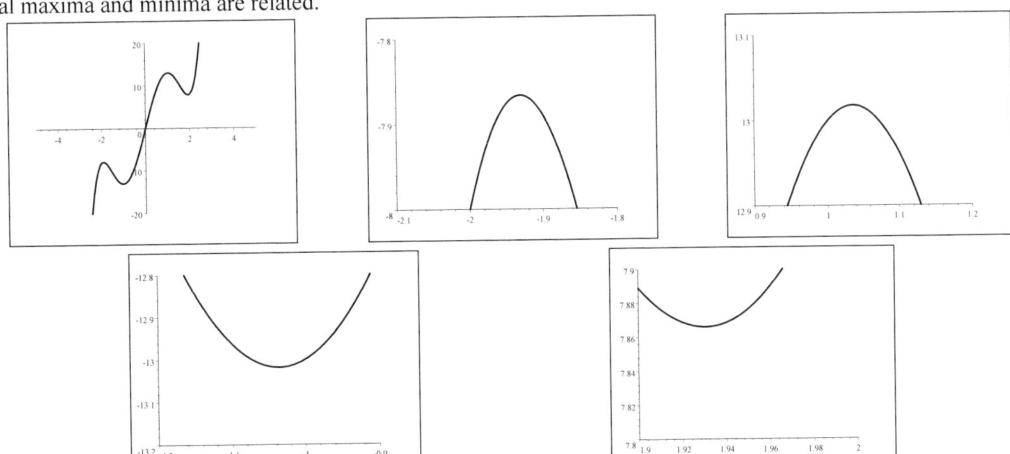

**55.** In the first graph, we see that $U(x) = x\sqrt{6-x}$ only has a local maximum. Smaller $x$- and $y$-ranges show that $U(x)$ has a local maximum of about 5.66 when $x \approx 4.00$.

**56.** In the first viewing rectangle below, we see that $U(x) = x\sqrt{x - x^2}$ has only a local maximum. Smaller $x$- and $y$-ranges show that $U(x)$ has a local maximum of about 0.32 when $x \approx 0.75$.

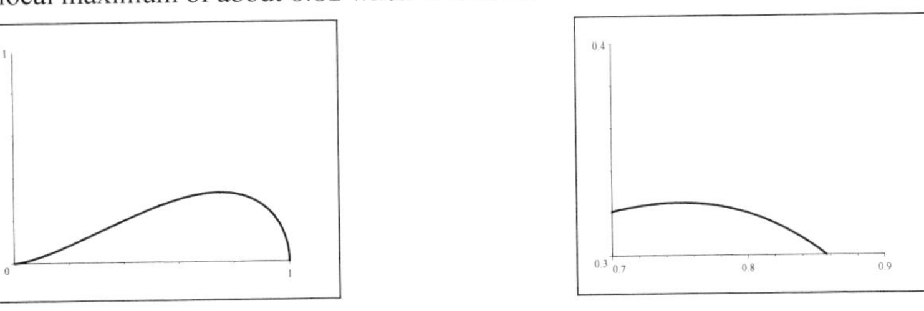

**57.** In the first graph, we see that $V(x) = \dfrac{1-x^2}{x^3}$ has a local minimum and a local maximum. Smaller $x$- and $y$-ranges show that $V(x)$ has a local maximum of about 0.38 when $x \approx -1.73$ and a local minimum of about $-0.38$ when $x \approx 1.73$.

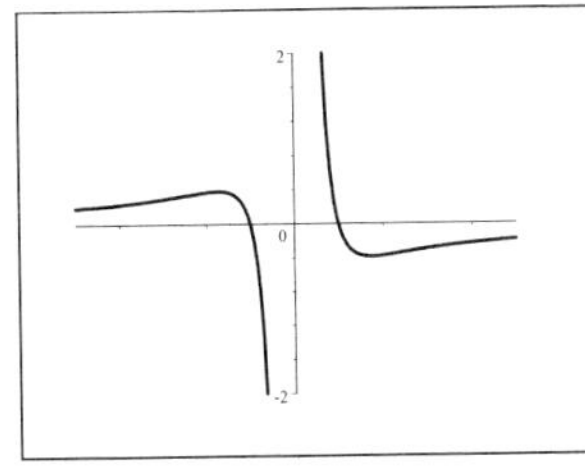
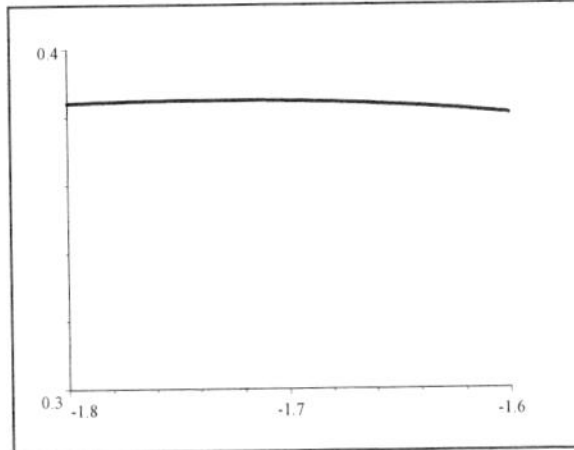
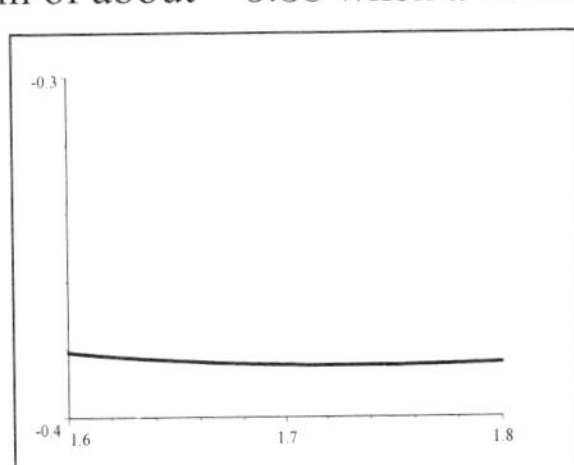

**58.** In the first viewing rectangle below, we see that $V(x) = \dfrac{1}{x^2 + x + 1}$ only has a local maximum. Smaller $x$- and $y$-ranges show that $V(x)$ has a local maximum of about $1.33$ when $x \approx -0.50$.

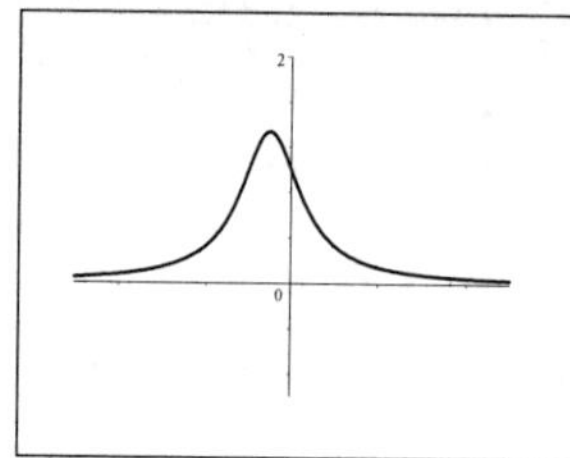


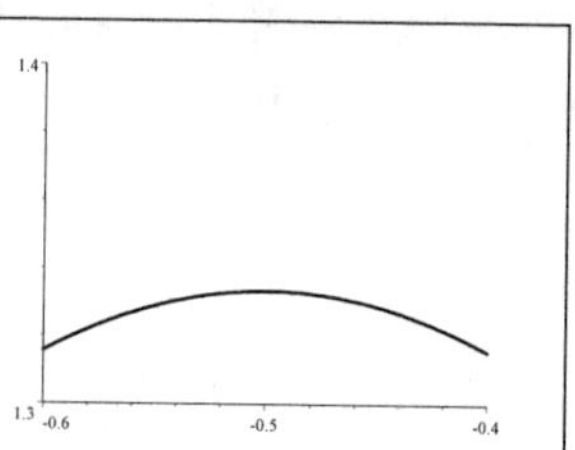


**59.** $y = f(t) = 40t - 16t^2 = -16\left(t^2 - \frac{5}{2}\right) = -16\left[t^2 - \frac{5}{2}t + \left(\frac{5}{4}\right)^2\right] + 16\left(\frac{5}{4}\right)^2 = -16\left(t - \frac{5}{4}\right)^2 + 25$. Thus the maximum height attained by the ball is $f\left(\frac{5}{4}\right) = 25$ feet.

**60. (a)** $y = -0.005x^2 + x + 5 = -0.005\left(x^2 - 200x\right) + 5 = -0.005\left(x^2 - 200x + 10{,}000\right) + 5 + 50 = -0.005(x - 100)^2 + 55$. Thus the maximum height attained by the football is 55 ft.

**(b)** We solve for $y = 0$: $0 = -0.005x^2 + x + 5$. Using the quadratic formula, we have $x = \frac{-1 \pm \sqrt{1^2 - (4)(-0.005)(5)}}{2(-0.005)} = \frac{-1 \pm \sqrt{1.1}}{-0.01} = 100 \pm 100\sqrt{1.1}$. Since he throws the football downfield, we take the positive root, so $x = 100 + 100\sqrt{1.1} \approx 204.9$ feet.

**61.** $R(x) = 80x - 0.4x^2 = -0.4\left(x^2 - 200x\right) = -0.4\left(x^2 - 200x + 10{,}000\right) + 4{,}000 = -0.4(x - 100)^2 + 4{,}000$. So revenue is maximized at \$4,000 when 100 units are sold.

**62.** $P(x) = -0.001x^2 + 3x - 1800 = -0.001\left(x^2 - 3000x\right) - 1800 = -0.001\left(x^2 - 3000x + 2{,}250{,}000\right) - 1800 + 2250 = -0.001(x - 1500)^2 + 450$. The vendor's maximum profit occurs when he sells 1500 cans and the profit is \$450.

**63.** $E(n) = \frac{2}{3}n - \frac{1}{90}n^2 = -\frac{1}{90}\left(n^2 - 60n\right) = -\frac{1}{90}\left(n^2 - 60n + 900\right) + 10 = -\dfrac{1}{90}(n - 30)^2 + 10$. Since the maximum of the function occurs when $n = 30$, the viewer should watch the commercial 30 times for maximum effectiveness.

**64.** $C(t) = 0.06t - 0.0002t^2 = -0.0002\left(t^2 - 300t\right) = -0.0002\left(t^2 - 300t + 22{,}500\right) + 4.5 = -0.0002(t - 150)^2 + 4.5$. The maximum concentration of 4.5 mg/L occurs after 150 minutes.

**65.** Graphing $A(n) = n(900 - 9n)$ in the viewing rectangle $[0, 100]$ by $[0, 25000]$, we see that maximum yield of apples occurs when there are 50 trees per acre.

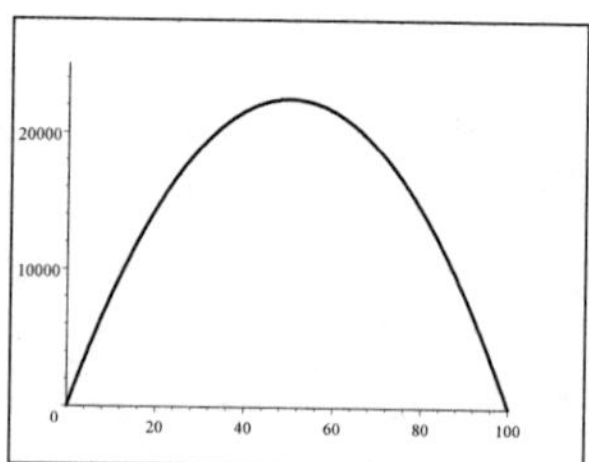


**66.** In the first graph, we see the general location of the minimum of $E(v) = 2.73v^3 \dfrac{10}{v - 5}$. In the second graph, we isolate the minimum, and from this graph, we see that energy is minimized when $v \approx 7.5$ mi/h.

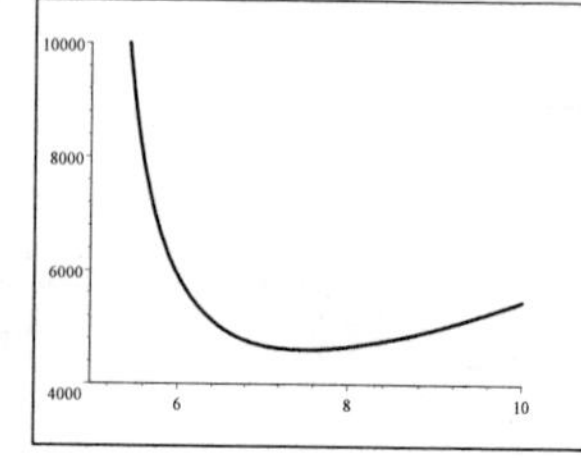


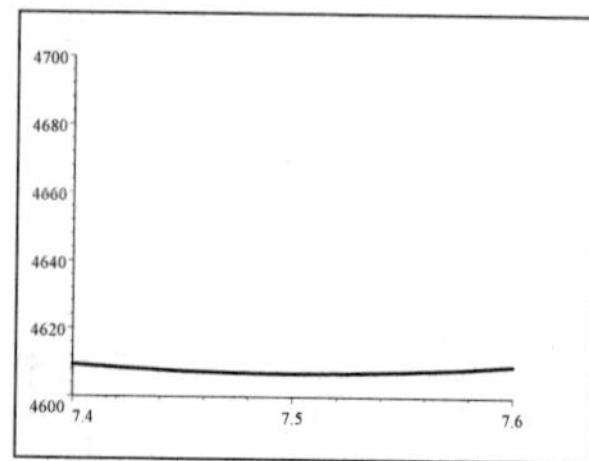

**67.** In the first graph, we see the general location of the maximum of $N(s) = \dfrac{88s}{17 + 17\left(\frac{s}{20}\right)^2}$. In the second graph we isolate the maximum, and from this graph we see that at the speed of 20 mi/h the largest number of cars that can use the highway safely is 52.

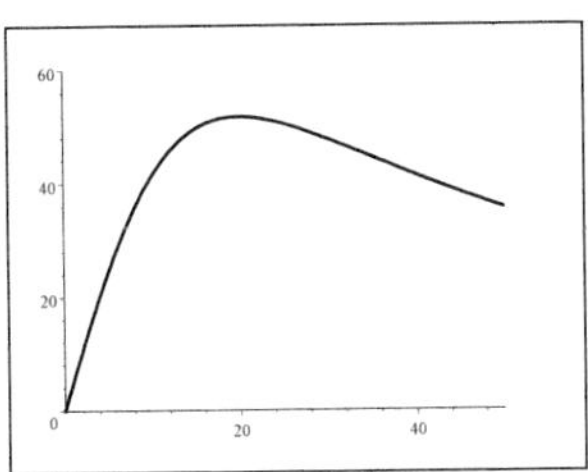

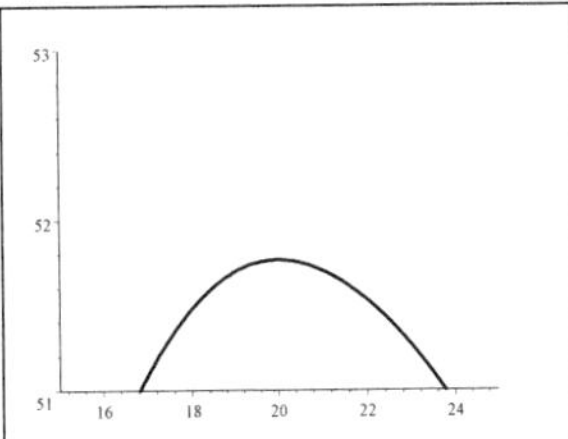

**68.** In the first graph, we see the general location of the minimum of $V = 999.87 - 0.06426T + 0.0085043T^2 - 0.0000679T^3$ is around $T = 4$. In the second graph, we isolate the minimum, and from this graph, we see that the minimum volume of 1 kg of water occurs at $T \approx 3.96°$ C.

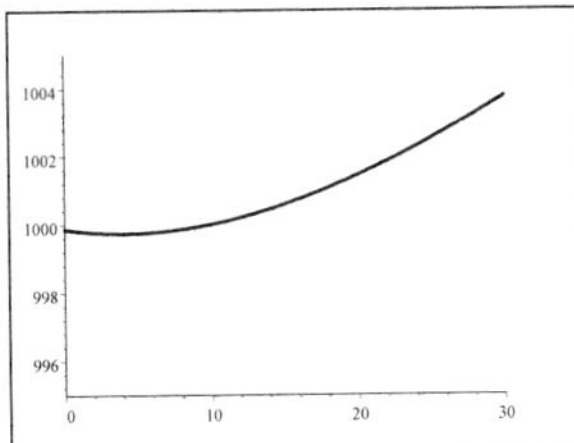

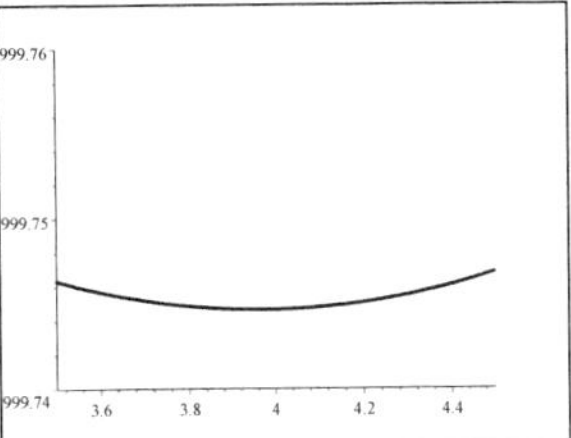

**69.** In the first graph, we see the general location of the maximum of $v(r) = 3.2(1 - r)r^2$ is around $r = 0.7$ cm. In the second graph, we isolate the maximum, and from this graph we see that at the maximum velocity is $0.47$ when $r \approx 0.67$ cm.

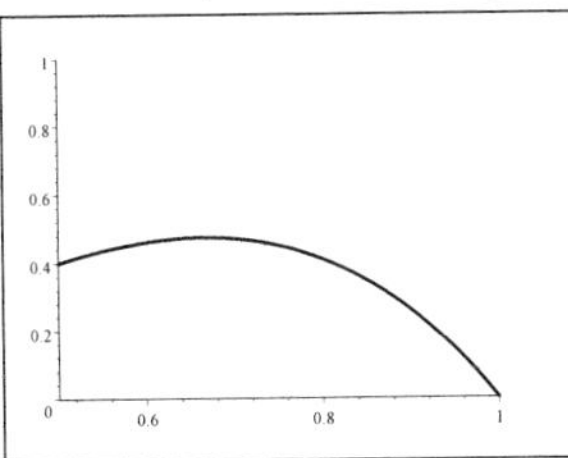

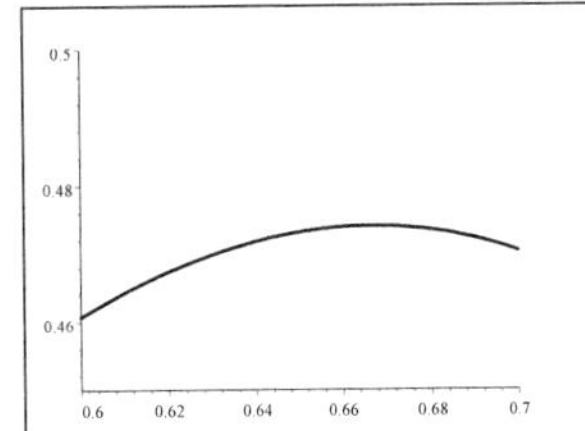

**70.** Numerous answers are possible.

**71.** **(a)** If $x = a$ is a local maximum of $f(x)$ then $f(a) \geq f(x) \geq 0$ for all $x$ around $x = a$. So $[g(a)]^2 \geq [g(x)]^2$ and thus $g(a) \geq g(x)$. Similarly, if $x = b$ is a local minimum of $f(x)$, then $f(x) \geq f(b) \geq 0$ for all $x$ around $x = b$. So $[g(x)]^2 \geq [g(b)]^2$ and thus $g(x) \geq g(b)$.

**(b)** Using the distance formula,

$$g(x) = \sqrt{(x-3)^2 + (x^2 - 0)^2} = \sqrt{x^4 + x^2 - 6x + 9}$$

**(c)** Let $f(x) = x^4 + x^2 - 6x + 9$. From the graph, we see that $f(x)$ has a minimum at $x = 1$. Thus $g(x)$ also has a minimum at $x = 1$ and this minimum value is $g(1) = \sqrt{1^4 + 1^2 - 6(1) + 9} = \sqrt{5}$.

**72.** $f(x) = 3 + 4x^2 - x^4$. Substituting $t = x^2$, we have $f(\sqrt{t}) = 3 + 4t - t^2 = -(t^2 - 4t) + 3 = -(t^2 - 4t + 4) + 3 + 4 = -(t-2)^2 + 7$. Therefore, the function $f(\sqrt{t}) = f(x)$ has a maximum value of 7.

# 3.6 Combining Functions

**1.** $f(x) = x - 3$ has domain $(-\infty, \infty)$. $g(x) = x^2$ has domain $(-\infty, \infty)$. The intersection of the domains of $f$ and $g$ is $(-\infty, \infty)$.
$(f + g)(x) = (x - 3) + (x^2) = x^2 + x - 3$, and the domain is $(-\infty, \infty)$.
$(f - g)(x) = (x - 3) - (x^2) = -x^2 + x - 3$, and the domain is $(-\infty, \infty)$.
$(fg)(x) = (x - 3)(x^2) = x^3 - 3x^2$, and the domain is $(-\infty, \infty)$.
$\left(\frac{f}{g}\right)(x) = \frac{x-3}{x^2}$, and the domain is $\{x \mid x \neq 0\}$.

**2.** $f(x) = x^2 + 2x$ has domain $(-\infty, \infty)$. $g(x) = 3x^2 - 1$ has domain $(-\infty, \infty)$. The intersection of the domains of $f$ and $g$ is $(-\infty, \infty)$.
$(f + g)(x) = x^2 + 2x + (3x^2 - 1) = 4x^2 + 2x - 1$, and the domain is $(-\infty, \infty)$.
$(f - g)(x) = x^2 + 2x - (3x^2 - 1) = -2x^2 + 2x + 1$, and the domain is $(-\infty, \infty)$.
$(fg)(x) = (x^2 + 2x)(3x^2 - 1) = 3x^4 + 6x^3 - x^2 - 2x$, and the domain is $(-\infty, \infty)$.
$\left(\frac{f}{g}\right)(x) = \frac{x^2 + 2x}{3x^2 - 1}$, $3x^2 - 1 \neq 0 \Rightarrow x \neq \pm\frac{\sqrt{3}}{3}$, and the domain is $\left\{x \mid x \neq \pm\frac{\sqrt{3}}{3}\right\}$.

**3.** $f(x) = \sqrt{4 - x^2}$, has domain $[-2, 2]$. $g(x) = \sqrt{1 + x}$, has domain $[-1, \infty)$. The intersection of the domains of $f$ and $g$ is $[-1, 2]$.
$(f + g)(x) = \sqrt{4 - x^2} + \sqrt{1 + x}$, and the domain is $[-1, 2]$.
$(f - g)(x) = \sqrt{4 - x^2} - \sqrt{1 + x}$, and the domain is $[-1, 2]$.
$(fg)(x) = \sqrt{4 - x^2}\sqrt{1 + x} = \sqrt{-x^3 - x^2 + 4x + 4}$, and the domain is $[-1, 2]$.
$\left(\frac{f}{g}\right)(x) = \frac{\sqrt{4 - x^2}}{\sqrt{1 + x}} = \sqrt{\frac{4 - x^2}{1 + x}}$, and the domain is $(-1, 2]$.

**4.** $f(x) = \sqrt{9 - x^2}$ has domain $[-3, 3]$. $g(x) = \sqrt{x^2 - 4}$ has domain $(-\infty, -2] \cup [2, \infty)$. The intersection of the domains of $f$ and $g$ is $[-3, -2] \cup [2, 3]$.
$(f + g)(x) = \sqrt{9 - x^2} + \sqrt{x^2 - 4}$, and the domain is $[-3, -2] \cup [2, 3]$.
$(f - g)(x) = \sqrt{9 - x^2} - \sqrt{x^2 - 4}$, and the domain is $[-3, -2] \cup [2, 3]$.
$(fg)(x) = \sqrt{9 - x^2} \cdot \sqrt{x^2 - 4} = \sqrt{-x^4 + 13x^2 - 36}$, and the domain is $[-3, -2] \cup [2, 3]$.
$\left(\frac{f}{g}\right)(x) = \frac{\sqrt{9 - x^2}}{\sqrt{x^2 - 4}} = \sqrt{\frac{9 - x^2}{x^2 - 4}}$, and the domain is $[-3, -2) \cup (2, 3]$.

**5.** $f(x) = \frac{2}{x}$ has domain $x \neq 0$. $g(x) = \frac{4}{x + 4}$, has domain $x \neq -4$. The intersection of the domains of $f$ and $g$ is $\{x \mid x \neq 0, -4\}$; in interval notation, this is $(-\infty, -4) \cup (-4, 0) \cup (0, \infty)$.
$(f + g)(x) = \frac{2}{x} + \frac{4}{x + 4} = \frac{2}{x} + \frac{4}{x + 4} = \frac{2(3x + 4)}{x(x + 4)}$, and the domain is $(-\infty, -4) \cup (-4, 0) \cup (0, \infty)$.
$(f - g)(x) = \frac{2}{x} - \frac{4}{x + 4} = -\frac{2(x - 4)}{x(x + 4)}$, and the domain is $(-\infty, -4) \cup (-4, 0) \cup (0, \infty)$.
$(fg)(x) = \frac{2}{x} \cdot \frac{4}{x + 4} = \frac{8}{x(x + 4)}$, and the domain is $(-\infty, -4) \cup (-4, 0) \cup (0, \infty)$.
$\left(\frac{f}{g}\right)(x) = \frac{\frac{2}{x}}{\frac{4}{x + 4}} = \frac{x + 4}{2x}$, and the domain is $(-\infty, -4) \cup (-4, 0) \cup (0, \infty)$.

**6.** $f(x) = \dfrac{2}{x+1}$ has domain $x \neq -1$. $g(x) = \dfrac{x}{x+1}$ has domain $x \neq -1$. The intersection of the domains of $f$ and $g$ is $\{x \mid x \neq -1\}$; in interval notation, this is $(-\infty, -1) \cup (-1, \infty)$.

$(f+g)(x) = \dfrac{2}{x+1} + \dfrac{x}{x+1} = \dfrac{x+2}{x+1}$, and the domain is $(-\infty, -1) \cup (-1, \infty)$.

$(f-g)(x) = \dfrac{2}{x+1} - \dfrac{x}{x+1} = \dfrac{2-x}{x+1}$, and the domain is $(-\infty, -1) \cup (-1, \infty)$.

$(fg)(x) = \dfrac{2}{x+1} \cdot \dfrac{x}{x+1} = \dfrac{2x}{(x+1)^2}$, and the domain is $(-\infty, -1) \cup (-1, \infty)$.

$\left(\dfrac{f}{g}\right)(x) = \dfrac{\frac{2}{x+1}}{\frac{x}{x+1}} = \dfrac{2}{x}$, so $x \neq 0$ as well. Thus the domain is $(-\infty, -1) \cup (-1, 0) \cup (0, \infty)$.

**7.** $f(x) = \sqrt{x} + \sqrt{1-x}$. The domain of $\sqrt{x}$ is $[0, \infty)$, and the domain of $\sqrt{1-x}$ is $(-\infty, 1]$. Thus the domain is $(-\infty, 1] \cap [0, \infty) = [0, 1]$.

**8.** $g(x) = \sqrt{x+1} + \dfrac{1}{x}$. The domain of $\sqrt{x+1}$ is $[-1, \infty)$, and the domain of $\dfrac{1}{x}$ is $x \neq 0$. Since $x \neq 0$ is $(-\infty, 0) \cup (0, \infty)$, the domain is $[-1, \infty) \cap \{(-\infty, 0) \cup (0, \infty)\} = [-1, 0) \cup (0, \infty)$.

**9.** $h(x) = (x-3)^{-1/4} = \dfrac{1}{(x-3)^{1/4}}$. Since $1/4$ is an even root and the denominator can not equal $0$, $x - 3 > 0 \quad \Leftrightarrow$ $x > 3$. So the domain is $(3, \infty)$.

**10.** $k(x) = \dfrac{\sqrt{x+3}}{x-1}$. The domain of $\sqrt{x+3}$ is $[-3, \infty)$, and the domain of $\dfrac{1}{x-1}$ is $x \neq 1$. Since $x \neq 1$ is $(-\infty, 1) \cup (1, \infty)$, the domain is $[-3, \infty) \cap \{(-\infty, 1) \cup (1, \infty)\} = [-3, 1) \cup (1, \infty)$.

**11.**

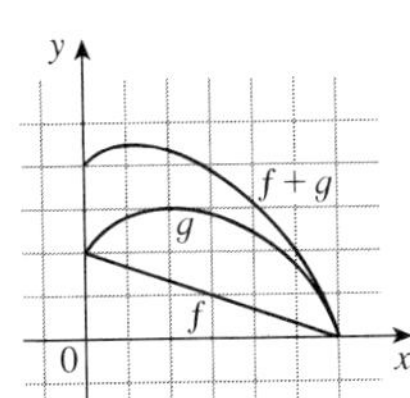


**12.**

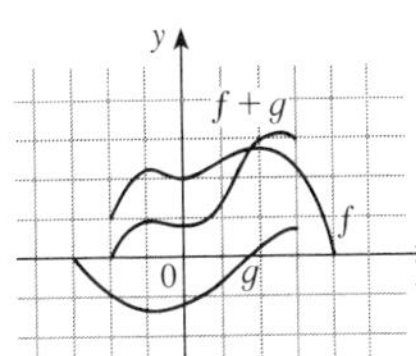


**13.**

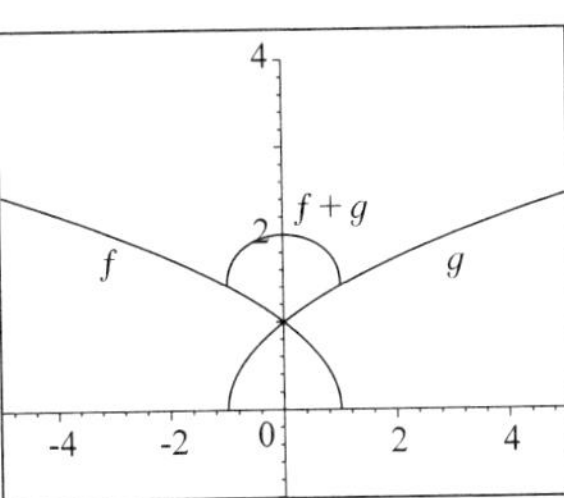


**14.**

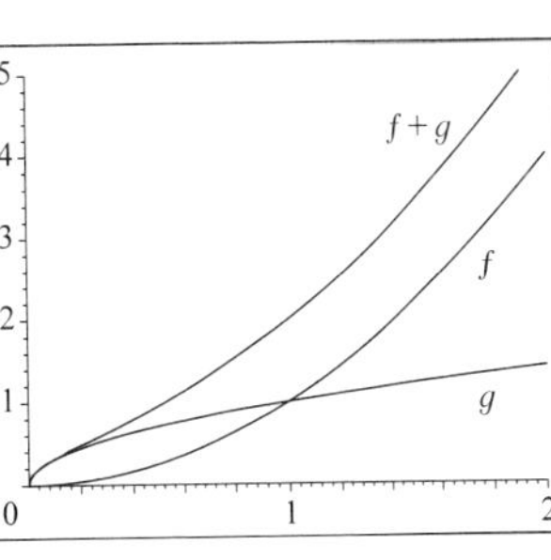


**15.**

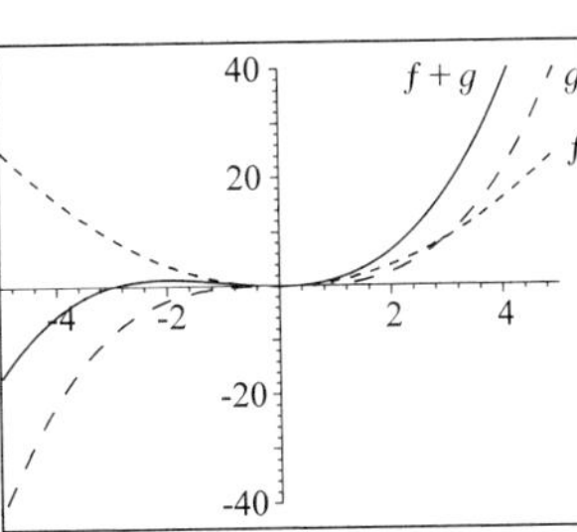


**16.**

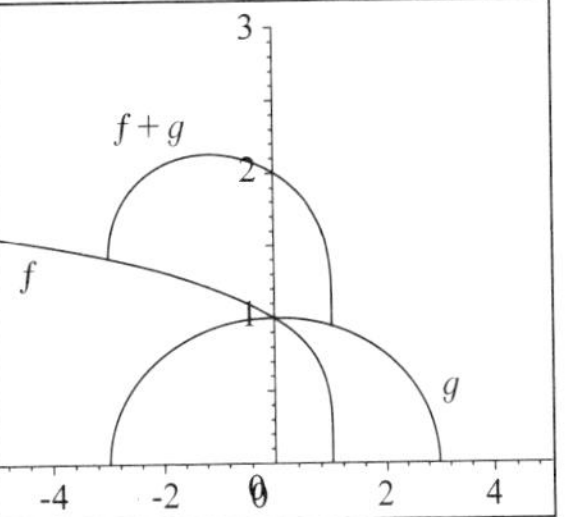


**17. (a)** $f(g(0)) = f(2 - (0)^2) = f(2) = 3(2) - 5 = 1$

**(b)** $g(f(0)) = g(3(0) - 5) = g(-5) = 2 - (-5)^2 = -23$

**18. (a)** $f(f(4)) = f(3(4) - 5) = f(7) = 3(7) - 5 = 16$

**(b)** $g(g(3)) = g\left(2 - (3)^2\right) = g(-7) = 2 - (-7)^2 = -47$

**19. (a)** $(f \circ g)(-2) = f(g(-2)) = f\left(2 - (-2)^2\right) = f(-2) = 3(-2) - 5 = -11$

**(b)** $(g \circ f)(-2) = g(f(-2)) = g(3(-2) - 5) = g(-11) = 2 - (-11)^2 = -119$

**20. (a)** $(f \circ f)(-1) = f(f(-1)) = f(3(-1) - 5) = f(-8) = 3(-8) - 5 = -29$

**(b)** $(g \circ g)(2) = g(g(2)) = g\left(2 - (2)^2\right) = g(-2) = 2 - (-2)^2 = -2$

**21. (a)** $(f \circ g)(x) = f(g(x)) = f\left(2 - x^2\right) = 3\left(2 - x^2\right) - 5 = 6 - 3x^2 - 5 = 1 - 3x^2$

**(b)** $(g \circ f)(x) = g(f(x)) = g(3x - 5) = 2 - (3x - 5)^2 = 2 - \left(9x^2 - 30x + 25\right) = -9x^2 + 30x - 23$

**22. (a)** $(f \circ f)(x) = f(f(x)) = f(3x - 5) = 3(3x - 5) - 5 = 9x - 15 - 5 = 9x - 20$

**(b)** $(g \circ g)(x) = g(g(x)) = g\left(2 - x^2\right) = 2 - \left(2 - x^2\right)^2 = 2 - \left(4 - 4x^2 + x^4\right) = -x^4 + 4x^2 - 2$

**23.** $f(g(2)) = f(5) = 4$

**24.** $f(0) = 0$, so $g(f(0)) = g(0) = 3$.

**25.** $(g \circ f)(4) = g(f(4)) = g(2) = 5$

**26.** $g(0) = 3$, so $(f \circ g)(0) = f(3) = 0$

**27.** $(g \circ g)(-2) = g(g(-2)) = g(1) = 4$

**28.** $f(4) = 2$, so $(f \circ f)(4) = f(2) = -2$.

**29.** $f(x) = 2x + 3$, has domain $(-\infty, \infty)$; $g(x) = 4x - 1$, has domain $(-\infty, \infty)$.
$(f \circ g)(x) = f(4x - 1) = 2(4x - 1) + 3 = 8x + 1$, and the domain is $(-\infty, \infty)$.
$(g \circ f)(x) = g(2x + 3) = 4(2x + 3) - 1 = 8x + 11$, and the domain is $(-\infty, \infty)$.
$(f \circ f)(x) = f(2x + 3) = 2(2x + 3) + 3 = 4x + 9$, and the domain is $(-\infty, \infty)$.
$(g \circ g)(x) = g(4x - 1) = 4(4x - 1) - 1 = 16x - 5$, and the domain is $(-\infty, \infty)$.

**30.** $f(x) = 6x - 5$ has domain $(-\infty, \infty)$. $g(x) = \dfrac{x}{2}$ has domain $(-\infty, \infty)$.

$(f \circ g)(x) = f\left(\dfrac{x}{2}\right) = 6\left(\dfrac{x}{2}\right) - 5 = 3x - 5$, and the domain is $(-\infty, \infty)$.

$(g \circ f)(x) = g(6x - 5) = \dfrac{6x - 5}{2} = 3x - \frac{5}{2}$, and the domain is $(-\infty, \infty)$.

$(f \circ f)(x) = f(6x - 5) = 6(6x - 5) - 5 = 36x - 35$, and the domain is $(-\infty, \infty)$.

$(g \circ g)(x) = g\left(\dfrac{x}{2}\right) = \dfrac{\frac{x}{2}}{2} = \dfrac{x}{4}$, and the domain is $(-\infty, \infty)$.

**31.** $f(x) = x^2$, has domain $(-\infty, \infty)$; $g(x) = x + 1$, has domain $(-\infty, \infty)$.
$(f \circ g)(x) = f(x + 1) = (x + 1)^2 = x^2 + 2x + 1$, and the domain is $(-\infty, \infty)$.
$(g \circ f)(x) = g\left(x^2\right) = \left(x^2\right) + 1 = x^2 + 1$, and the domain is $(-\infty, \infty)$.
$(f \circ f)(x) = f\left(x^2\right) = \left(x^2\right)^2 = x^4$, and the domain is $(-\infty, \infty)$.
$(g \circ g)(x) = g(x + 1) = (x + 1) + 1 = x + 2$, and the domain is $(-\infty, \infty)$.

**32.** $f(x) = x^3 + 2$ has domain $(-\infty, \infty)$. $g(x) = \sqrt[3]{x}$ has domain $(-\infty, \infty)$.
$(f \circ g)(x) = f\left(\sqrt[3]{x}\right) = \left(\sqrt[3]{x}\right)^3 + 2 = x + 2$, and the domain is $(-\infty, \infty)$.
$(g \circ f)(x) = g\left(x^3 + 2\right) = \sqrt[3]{x^3 + 2}$ and the domain is $(-\infty, \infty)$.
$(f \circ f)(x) = f\left(x^3 + 2\right) = \left(x^3 + 2\right)^3 + 2 = x^9 + 6x^6 + 12x^3 + 8 + 2 = x^9 + 6x^6 + 12x^3 + 10$, and the domain is $(-\infty, \infty)$.

$(g \circ g)(x) = g\left(\sqrt[3]{x}\right) = \sqrt[3]{\sqrt[3]{x}} = \left(x^{1/3}\right)^{1/3} = x^{1/9}$, and the domain is $(-\infty, \infty)$.

**33.** $f(x) = \frac{1}{x}$, has domain $\{x \mid x \neq 0\}$; $g(x) = 2x + 4$, has domain $(-\infty, \infty)$.

$(f \circ g)(x) = f(2x+4) = \frac{1}{2x+4}$. $(f \circ g)(x)$ is defined for $2x + 4 \neq 0 \quad \Leftrightarrow \quad x \neq -2$. So the domain is $\{x \mid x \neq -2\} = (-\infty, -2) \cup (-2, \infty)$.

$(g \circ f)(x) = g\left(\frac{1}{x}\right) = 2\left(\frac{1}{x}\right) + 4 = \frac{2}{x} + 4$, the domain is $\{x \mid x \neq 0\} = (-\infty, 0) \cup (0, \infty)$.

$(f \circ f)(x) = f\left(\frac{1}{x}\right) = \frac{1}{\left(\frac{1}{x}\right)} = x$. $(f \circ f)(x)$ is defined whenever both $f(x)$ and $f(f(x))$ are defined; that is, whenever $\{x \mid x \neq 0\} = (-\infty, 0) \cup (0, \infty)$.

$(g \circ g)(x) = g(2x+4) = 2(2x+4) + 4 = 4x + 8 + 4 = 4x + 12$, and the domain is $(-\infty, \infty)$.

**34.** $f(x) = x^2$ has domain $(-\infty, \infty)$. $g(x) = \sqrt{x-3}$ has domain $[3, \infty)$.

$(f \circ g)(x) = f\left(\sqrt{x-3}\right) = \left(\sqrt{x-3}\right)^2 = x - 3$, and the domain is $[3, \infty)$.

$(g \circ f)(x) = g\left(x^2\right) = \sqrt{x^2 - 3}$. For the domain we must have $x^2 \geq 3 \quad \Rightarrow \quad x \leq -\sqrt{3}$ or $x \geq \sqrt{3}$. Thus the domain is $\left(-\infty, -\sqrt{3}\right] \cup \left[\sqrt{3}, \infty\right)$.

$(f \circ f)(x) = f\left(x^2\right) = \left(x^2\right)^2 = x^4$, and the domain is $(-\infty, \infty)$.

$(g \circ g)(x) = g\left(\sqrt{x-3}\right) = \sqrt{\sqrt{x-3} - 3}$. For the domain we must have $\sqrt{x-3} \geq 3 \quad \Rightarrow \quad x - 3 \geq 9 \quad \Rightarrow$ $x \geq 12$, so the domain is $[12, \infty)$.

**35.** $f(x) = |x|$, has domain $(-\infty, \infty)$; $g(x) = 2x + 3$, has domain $(-\infty, \infty)$

$(f \circ g)(x) = f(2x+4) = |2x+3|$, and the domain is $(-\infty, \infty)$.

$(g \circ f)(x) = g(|x|) = 2|x| + 3$, and the domain is $(-\infty, \infty)$.

$(f \circ f)(x) = f(|x|) = ||x|| = |x|$, and the domain is $(-\infty, \infty)$.

$(g \circ g)(x) = g(2x+3) = 2(2x+3) + 3 = 4x + 6 + 3 = 4x + 9$. Domain is $(-\infty, \infty)$.

**36.** $f(x) = x - 4$ has domain $(-\infty, \infty)$. $g(x) = |x+4|$ has domain $(-\infty, \infty)$.

$(f \circ g)(x) = f(|x+4|) = |x+4| - 4$, and the domain is $(-\infty, \infty)$.

$(g \circ f)(x) = g(x-4) = |(x-4)+4| = |x|$, and the domain is $(-\infty, \infty)$.

$(f \circ f)(x) = f(x-4) = (x-4) - 4 = x - 8$, and the domain is $(-\infty, \infty)$.

$(g \circ g)(x) = g(|x+4|) = ||x+4| + 4| = |x+4| + 4$ ($|x+4| + 4$ is always positive). The domain is $(-\infty, \infty)$.

**37.** $f(x) = \frac{x}{x+1}$, has domain $\{x \mid x \neq -1\}$; $g(x) = 2x - 1$, has domain $(-\infty, \infty)$

$(f \circ g)(x) = f(2x-1) = \frac{2x-1}{(2x-1)+1} = \frac{2x-1}{2x}$, and the domain is $\{x \mid x \neq 0\} = (-\infty, 0) \cup (0, \infty)$.

$(g \circ f)(x) = g\left(\frac{x}{x+1}\right) = 2\left(\frac{x}{x+1}\right) - 1 = \frac{2x}{x+1} - 1$, and the domain is $\{x \mid x \neq -1\} = (-\infty, -1) \cup (-1, \infty)$

$(f \circ f)(x) = f\left(\frac{x}{x+1}\right) = \frac{\frac{x}{x+1}}{\frac{x}{x+1} + 1} \cdot \frac{x+1}{x+1} = \frac{x}{x + x + 1} = \frac{x}{2x+1}$. $(f \circ f)(x)$ is defined whenever both $f(x)$ and $f(f(x))$ are defined; that is, whenever $x \neq -1$ and $2x + 1 \neq 0 \quad \Rightarrow \quad x \neq -\frac{1}{2}$, which is $(-\infty, -1) \cup \left(-1, -\frac{1}{2}\right) \cup \left(-\frac{1}{2}, \infty\right)$.

$(g \circ g)(x) = g(2x-1) = 2(2x-1) - 1 = 4x - 2 - 1 = 4x - 3$, and the domain is $(-\infty, \infty)$.

**38.** $f(x)=\dfrac{1}{\sqrt{x}}$ has domain $\{x\mid x>0\}$; $g(x)=x^2-4x$ has domain $(-\infty,\infty)$.

$(f\circ g)(x)=f(x^2-4x)=\dfrac{1}{\sqrt{x^2-4x}}$. $(f\circ g)(x)$ is defined whenever $0<x^2-4x=x(x-4)$. The product of two numbers is positive either when both numbers are negative or when both numbers are positive. So the domain of $f\circ g$ is $\{x\mid x<0 \text{ and } x<4\}\cup\{x\mid x>0 \text{ and } x>4\}$ which is $(-\infty,0)\cup(4,\infty)$.

$(g\circ f)(x)=g\left(\dfrac{1}{\sqrt{x}}\right)=\left(\dfrac{1}{\sqrt{x}}\right)^2-4\left(\dfrac{1}{\sqrt{x}}\right)=\dfrac{1}{x}-\dfrac{4}{\sqrt{x}}$. $(g\circ f)(x)$ is defined whenever both $f(x)$ and $g(f(x))$ are defined, that is, whenever $x>0$. So the domain of $g\circ f$ is $(0,\infty)$.

$(f\circ f)(x)=f\left(\dfrac{1}{\sqrt{x}}\right)=\dfrac{1}{\sqrt{\dfrac{1}{\sqrt{x}}}}=x^{1/4}$. $(f\circ f)(x)$ is defined whenever both $f(x)$ and $f(f(x))$ are defined, that is, whenever $x>0$. So the domain of $f\circ f$ is $(0,\infty)$.

$(g\circ g)(x)=g(x^2-4x)=(x^2-4x)^2-4(x^2-4x)=x^4-8x^3+16x^2-4x^2+16x=x^4-8x^3+12x^2+16x$, and the domain is $(-\infty,\infty)$.

**39.** $f(x)=\sqrt[3]{x}$, has domain $(-\infty,\infty)$; $g(x)=\sqrt[4]{x}$, has domain $[0,\infty)$.

$(f\circ g)(x)=f(\sqrt[4]{x})=\sqrt[3]{\sqrt[4]{x}}=\sqrt[12]{x}$. $(f\circ g)(x)$ is defined whenever both $g(x)$ and $f(g(x))$ are defined. Since $f(x)$ has no restriction, the domain is $[0,\infty)$.

$(g\circ f)(x)=g(\sqrt[3]{x})=\sqrt[4]{\sqrt[3]{x}}=\sqrt[12]{x}$. $(g\circ f)(x)$ is defined whenever both $f(x)$ and $g(f(x))$ are defined; that is, whenever $x\geq 0$. So the domain is $[0,\infty)$.

$(f\circ f)(x)=f(\sqrt[3]{x})=\sqrt[3]{\sqrt[3]{x}}=\sqrt[9]{x}$. $(f\circ f)(x)$ is defined whenever both $f(x)$ and $f(f(x))$ are defined. Since $f(x)$ is defined everywhere, the domain is $(-\infty,\infty)$.

$(g\circ g)(x)=g(\sqrt[4]{x})=\sqrt[4]{\sqrt[4]{x}}=\sqrt[16]{x}$. $(g\circ g)(x)$ is defined whenever both $g(x)$ and $g(g(x))$ are defined; that is, whenever $x\geq 0$. So the domain is $[0,\infty)$.

**40.** $f(x)=\dfrac{2}{x}$ has domain $\{x\mid x\neq 0\}$ $g(x)=\dfrac{x}{x+2}$ has domain $\{x\mid x\neq -2\}$.

$(f\circ g)(x)=f\left(\dfrac{x}{x+2}\right)=\dfrac{2}{\dfrac{x}{x+2}}=\dfrac{2x+4}{x}$. $(f\circ g)(x)$ is defined whenever both $g(x)$ and $f(g(x))$ are defined; that is, whenever $x\neq 0$ and $x\neq -2$. So the domain is $\{x\mid x\neq 0,-2\}$.

$(g\circ f)(x)=g\left(\dfrac{2}{x}\right)=\dfrac{\dfrac{2}{x}}{\dfrac{2}{x}+2}=\dfrac{2}{2+2x}=\dfrac{1}{1+x}$. $(g\circ f)(x)$ is defined whenever both $f(x)$ and $g(f(x))$ are defined; that is, whenever $x\neq 0$ and $x\neq -1$. So the domain is $\{x\mid x\neq 0,-1\}$.

$(f\circ f)(x)=f\left(\dfrac{2}{x}\right)=\dfrac{2}{\dfrac{2}{x}}=x$. $(f\circ f)(x)$ is defined whenever both $f(x)$ and $f(f(x))$ are defined; that is, whenever $x\neq 0$. So the domain is $\{x\mid x\neq 0\}$.

$(g\circ g)(x)=g\left(\dfrac{x}{x+2}\right)=\dfrac{\dfrac{x}{x+2}}{\dfrac{x}{x+2}+2}=\dfrac{x}{x+2(x+2)}=\dfrac{x}{3x+4}$. $(g\circ g)(x)$ is defined whenever both $g(x)$ and $g(g(x))$ are defined; that is whenever $x\neq -2$ and $x\neq -\frac{4}{3}$. So the domain is $\{x\mid x\neq -2,-\frac{4}{3}\}$.

**41.** $(f\circ g\circ h)(x)=f(g(h(x)))=f(g(x-1))=f(\sqrt{x-1})=\sqrt{x-1}-1$

**42.** $(g\circ h)(x)=g(x^2+2)=(x^2+2)^3=x^6+6x^4+12x^2+8$.

$(f\circ g\circ h)(x)=f(x^6+6x^4+12x^2+8)=\dfrac{1}{x^6+6x^4+12x^2+8}$.

**43.** $(f \circ g \circ h)(x) = f(g(h(x))) = f(g(\sqrt{x})) = f(\sqrt{x} - 5) = (\sqrt{x} - 5)^4 + 1$

**44.** $(g \circ h)(x) = g(\sqrt[3]{x}) = \dfrac{\sqrt[3]{x}}{\sqrt[3]{x} - 1}$. $(f \circ g \circ h)(x) = f\left(\dfrac{\sqrt[3]{x}}{\sqrt[3]{x} - 1}\right) = \sqrt{\dfrac{\sqrt[3]{x}}{\sqrt[3]{x} - 1}}$.

**For Exercises 45–54, many answers are possible.**

**45.** $F(x) = (x - 9)^5$. Let $f(x) = x^5$ and $g(x) = x - 9$, then $F(x) = (f \circ g)(x)$.

**46.** $F(x) = \sqrt{x} + 1$. If $f(x) = x + 1$ and $g(x) = \sqrt{x}$, then $F(x) = (f \circ g)(x)$.

**47.** $G(x) = \dfrac{x^2}{x^2 + 4}$. Let $f(x) = \dfrac{x}{x + 4}$ and $g(x) = x^2$, then $G(x) = (f \circ g)(x)$.

**48.** $G(x) = \dfrac{1}{x + 3}$. If $f(x) = \dfrac{1}{x}$ and $g(x) = x + 3$, then $G(x) = (f \circ g)(x)$.

**49.** $H(x) = \left|1 - x^3\right|$. Let $f(x) = |x|$ and $g(x) = 1 - x^3$, then $H(x) = (f \circ g)(x)$.

**50.** $H(x) = \sqrt{1 + \sqrt{x}}$. If $f(x) = \sqrt{1 + x}$ and $g(x) = \sqrt{x}$, then $H(x) = (f \circ g)(x)$.

**51.** $F(x) = \dfrac{1}{x^2 + 1}$. Let $f(x) = \dfrac{1}{x}$, $g(x) = x + 1$, and $h(x) = x^2$, then $F(x) = (f \circ g \circ h)(x)$.

**52.** $F(x) = \sqrt[3]{\sqrt{x} - 1}$. If $g(x) = x - 1$ and $h(x) = \sqrt{x}$, then $(g \circ h)(x) = \sqrt{x} - 1$, and if $f(x) = \sqrt[3]{x}$, then $F(x) = (f \circ g \circ h)(x)$.

**53.** $G(x) = (4 + \sqrt[3]{x})^9$. Let $f(x) = x^9$, $g(x) = 4 + x$, and $h(x) = \sqrt[3]{x}$, then $G(x) = (f \circ g \circ h)(x)$.

**54.** $G(x) = \dfrac{2}{(3 + \sqrt{x})^2}$. If $g(x) = 3 + x$ and $h(x) = \sqrt{x}$, then $(g \circ h)(x) = 3 + \sqrt{x}$, and if $f(x) = \dfrac{2}{x^2}$, then $G(x) = (f \circ g \circ h)(x)$.

**55.** The price per sticker is $0.15 - 0.000002x$ and the number sold is $x$, so the revenue is $R(x) = (0.15 - 0.000002x)x = 0.15x - 0.000002x^2$.

**56.** As found in Exercise 55, the revenue is $R(x) = 0.15x - 0.000002x^2$, and the cost is $0.095x - 0.0000005x^2$, so the profit is $P(x) = 0.15x - 0.000002x^2 - (0.095x - 0.0000005x^2) = 0.055x - 0.0000015x^2$.

**57. (a)** Since the ripple travels at a speed of 60 cm/s, the distance traveled in $t$ seconds is the radius, so $g(t) = 60t$.

**(b)** The area of a circle is $\pi r^2$, so $f(r) = \pi r^2$.

**(c)** $f \circ g = \pi(g(t))^2 = \pi(60t)^2 = 3600\pi t^2$ cm$^2$. This function represents the area of the ripple as a function of time.

**58. (a)** Let $f(t)$ be the radius of the spherical balloon in centimeters. Since the radius is increasing at a rate of 1 cm/s, the radius is $f(t) = t$ after $t$ seconds.

**(b)** The volume of the balloon can be written as $g(r) = \frac{4}{3}\pi r^3$.

**(c)** $g \circ f = \frac{4}{3}\pi(t)^3 = \frac{4}{3}\pi t^3$. $g \circ f$ represents the volume as a function of time.

**59.** Let $r$ be the radius of the spherical balloon in centimeters. Since the radius is increasing at a rate of 2 cm/s, the radius is $r = 2t$ after $t$ seconds. Therefore, the surface area of the balloon can be written as $S = 4\pi r^2 = 4\pi(2t)^2 = 4\pi(4t^2) = 16\pi t^2$.

**60. (a)** $f(x) = 0.80x$

**(b)** $g(x) = x - 50$

**(c)** $f \circ g = f(x - 50) = 0.80(x - 50) = 0.80x - 40$. $f \circ g$ represents applying the \$50 coupon, then the 20% discount. $g \circ f = g(0.80x) = 0.80x - 50$. $g \circ f$ represents applying the 20% discount, then the \$50 coupon. So applying the 20% discount, then the \$50 coupon gives the lower price.

**61. (a)** $f(x) = 0.90x$

**(b)** $g(x) = x - 100$

**(c)** $f \circ g = f(x - 100) = 0.90(x - 100) = 0.90x - 90$. $f \circ g$ represents applying the \$100 coupon, then the 10% discount. $g \circ f = g(0.90x) = 0.90x - 100$. $g \circ f$ represents applying the 10% discount, then the \$100 coupon. So applying the 10% discount, then the \$100 coupon gives the lower price.

**62.** Let $t$ be the time since the plane flew over the radar station.

**(a)** Let $s$ be the distance in miles between the plane and the radar station, and let $d$ be the horizontal distance that the plane has flown. Using the Pythagorean theorem, $s = f(d) = \sqrt{1+d^2}$.

**(b)** Since distance = rate × time, we have $d = g(t) = 350t$.

**(c)** $s(t) = (f \circ g)(t) = f(350t) = \sqrt{1+(350t)^2} = \sqrt{1+122{,}500t^2}$.

**63.** $A(x) = 1.05x$. $(A \circ A)(x) = A(A(x)) = A(1.05x) = 1.05(1.05x) = (1.05)^2 x$. $(A \circ A \circ A)(x) = A(A \circ A(x)) = A((1.05)^2 x) = 1.05\left[(1.05)^2 x\right] = (1.05)^3 x$. $(A \circ A \circ A \circ A)(x) = A(A \circ A \circ A(x)) = A((1.05)^3 x) = 1.05\left[(1.05)^3 x\right] = (1.05)^4 x$. $A$ represents the amount in the account after 1 year; $A \circ A$ represents the amount in the account after 2 years; $A \circ A \circ A$ represents the amount in the account after 3 years; and $A \circ A \circ A \circ A$ represents the amount in the account after 4 years. We can see that if we compose $n$ copies of $A$, we get $(1.05)^n x$.

**64.** Yes. If $f(x) = m_1x + b_1$ and $g(x) = m_2x + b_2$, then $(f \circ g)(x) = f(m_2x + b_2) = m_1(m_2x + b_2) + b_1 = m_1m_2x + m_1b_2 + b_1$, which is a linear function, because it is of the form $y = mx + b$. The slope is $m_1m_2$.

**65.** $g(x) = 2x + 1$ and $h(x) = 4x^2 + 4x + 7$.

*Method 1:* Notice that $(2x+1)^2 = 4x^2 + 4x + 1$. We see that adding 6 to this quantity gives $(2x+1)^2 + 6 = 4x^2 + 4x + 1 + 6 = 4x^2 + 4x + 7$, which is $h(x)$. So let $f(x) = x^2 + 6$, and we have $(f \circ g)(x) = (2x+1)^2 + 6 = h(x)$.

*Method 2:* Since $g(x)$ is linear and $h(x)$ is a second degree polynomial, $f(x)$ must be a second degree polynomial, that is, $f(x) = ax^2 + bx + c$ for some $a$, $b$, and $c$. Thus $f(g(x)) = f(2x+1) = a(2x+1)^2 + b(2x+1) + c \Leftrightarrow 4ax^2 + 4ax + a + 2bx + b + c = 4ax^2 + (4a+2b)x + (a+b+c) = 4x^2 + 4x + 7$. Comparing this with $f(g(x))$, we have $4a = 4$ (the $x^2$ coefficients), $4a + 2b = 4$ (the $x$ coefficients), and $a + b + c = 7$ (the constant terms) $\Leftrightarrow a = 1$ and $2a + b = 2$ and $a + b + c = 7 \Leftrightarrow a = 1, b = 0, c = 6$. Thus $f(x) = x^2 + 6$.

$f(x) = 3x + 5$ and $h(x) = 3x^2 + 3x + 2$.

Note since $f(x)$ is linear and $h(x)$ is quadratic, $g(x)$ must also be quadratic. We can then use trial and error to find $g(x)$. Another method is the following: We wish to find $g$ so that $(f \circ g)(x) = h(x)$. Thus $f(g(x)) = 3x^2 + 3x + 2 \Leftrightarrow 3(g(x)) + 5 = 3x^2 + 3x + 2 \Leftrightarrow 3(g(x)) = 3x^2 + 3x - 3 \Leftrightarrow g(x) = x^2 + x - 1$.

**66.** If $g(x)$ is even, then $h(-x) = f(g(-x)) = f(g(x)) = h(x)$. So yes, $h$ is always an even function.

If $g(x)$ is odd, then $h$ is not necessarily an odd function. For example, if we let $f(x) = x - 1$ and $g(x) = x^3$, $g$ is an odd function, but $h(x) = (f \circ g)(x) = f(x^3) = x^3 - 1$ is not an odd function.

If $g(x)$ is odd and $f$ is also odd, then $h(-x) = (f \circ g)(-x) = f(g(-x)) = f(-g(x)) = -f(g(x)) = -(f \circ g)(x) = -h(x)$. So in this case, $h$ is also an odd function.

If $g(x)$ is odd and $f$ is even, then $h(-x) = (f \circ g)(-x) = f(g(-x)) = f(-g(x)) = f(g(x)) = (f \circ g)(x) = h(x)$, so in this case, $h$ is an even function.

## 3.7 One-to-One Functions and Their Inverses

**1.** By the Horizontal Line Test, $f$ is not one-to-one.

**2.** By the Horizontal Line Test, $f$ is one-to-one.

**3.** By the Horizontal Line Test, $f$ is one-to-one.

**4.** By the Horizontal Line Test, $f$ is not one-to-one.

**5.** By the Horizontal Line Test, $f$ is not one-to-one.

**6.** By the Horizontal Line Test, $f$ is one-to-one.

**7.** $f(x) = -2x + 4$. If $x_1 \neq x_2$, then $-2x_1 \neq -2x_2$ and $-2x_1 + 4 \neq -2x_2 + 4$. So $f$ is a one-to-one function.

**8.** $f(x) = 3x - 2$. If $x_1 \neq x_2$, then $3x_1 \neq 3x_2$ and $3x_1 - 2 \neq 3x_2 - 2$. So $f$ is a one-to-one function.

**9.** $g(x) = \sqrt{x}$. If $x_1 \neq x_2$, then $\sqrt{x_1} \neq \sqrt{x_2}$ because two different numbers cannot have the same square root. Therefore, $g$ is a one-to-one function.

**10.** $g(x) = |x|$. Since every number and its negative have the same absolute value, that is, $|-1| = 1 = |1|$, $g$ is not a one-to-one function.

**11.** $h(x) = x^2 - 2x$. Since $h(0) = 0$ and $h(2) = (2) - 2(2) = 0$ we have $h(0) = h(2)$. So $f$ is not a one-to-one function.

**12.** $h(x) = x^3 + 8$. If $x_1 \neq x_2$, then $x_1^3 \neq x_2^3$ and $x_1^3 + 8 \neq x_2^3 + 8$. So $f$ is a one-to-one function.

**13.** $f(x) = x^4 + 5$. Every nonzero number and its negative have the same fourth power. For example, $(-1)^4 = 1 = (1)^4$, so $f(-1) = f(1)$. Thus $f$ is not a one-to-one function.

**14.** $f(x) = x^4 + 5$, $0 \leq x \leq 2$. If $x_1 \neq x_2$, then $x_1^4 \neq x_2^4$ because two different positive numbers cannot have the same fourth power. Thus, $x_1^4 + 5 \neq x_2^4 + 5$. So $f$ is a one-to-one function.

**15.** $f(x) = \dfrac{1}{x^2}$. Every nonzero number and its negative have the same square. For example, $\dfrac{1}{(-1)^2} = 1 = \dfrac{1}{(1)^2}$, so $f(-1) = f(1)$. Thus $f$ is not a one-to-one function.

**16.** $f(x) = \dfrac{1}{x}$. If $x_1 \neq x_2$, then $\dfrac{1}{x_1} \neq \dfrac{1}{x_2}$. So $f$ is a one-to-one function.

**17. (a)** $f(2) = 7$. Since $f$ is one-to-one, $f^{-1}(7) = 2$.

**(b)** $f^{-1}(3) = -1$. Since $f$ is one-to-one, $f(-1) = 3$.

**18. (a)** $f(5) = 18$. Since $f$ is one-to-one, $f^{-1}(18) = 5$.

**(b)** $f^{-1}(4) = 2$. Since $f$ is one-to-one, $f(2) = 4$.

**19.** $f(x) = 5 - 2x$. Since $f$ is one-to-one and $f(1) = 5 - 2(1) = 3$, then $f^{-1}(3) = 1$. (Find 1 by solving the equation $5 - 2x = 3$.)

**20.** To find $g^{-1}(5)$, we find the $x$ value such that $g(x) = 5$; that is, we solve the equation $g(x) = x^2 + 4x = 5$. Now $x^2 + 4x = 5 \quad \Leftrightarrow \quad x^2 + 4x - 5 = 0 \quad \Leftrightarrow \quad (x-1)(x+5) = 0 \quad \Leftrightarrow \quad x = 1$ or $x = -5$. Since the domain of $g$ is $[-2, \infty)$, $x = 1$ is the only value where $g(x) = 5$. Therefore, $g^{-1}(5) = 1$.

**21.** $f(g(x)) = f(x+6) = (x+6) - 6 = x$ for all $x$.
$g(f(x)) = g(x-6) = (x-6) + 6 = x$ for all $x$. Thus $f$ and $g$ are inverses of each other.

**22.** $f(g(x)) = f\left(\dfrac{x}{3}\right) = 3\left(\dfrac{x}{3}\right) = x$ for all $x$.
$g(f(x)) = g(3x) = \dfrac{3x}{3} = x$ for all $x$. Thus $f$ and $g$ are inverses of each other.

**23.** $f(g(x)) = f\left(\dfrac{x+5}{2}\right) = 2\left(\dfrac{x+5}{2}\right) - 5 = x + 5 - 5 = x$ for all $x$.
$g(f(x)) = g(2x-5) = \dfrac{(2x-5)+5}{2} = x$ for all $x$. Thus $f$ and $g$ are inverses of each other.

**24.** $f(g(x)) = f(3-4x) = \dfrac{3-(3-4x)}{4} = \dfrac{3-3+4x}{4} = x$ for all $x$.
$g(f(x)) = g\left(\dfrac{3-x}{4}\right) = 3 - 4\left(\dfrac{3-x}{4}\right) = 3 - 3 + x = x$ for all $x$. Thus $f$ and $g$ are inverses of each other.

**25.** $f(g(x)) = f\left(\dfrac{1}{x}\right) = \dfrac{1}{1/x} = x$ for all $x$. Since $f(x) = g(x)$, we also have $g(f(x)) = x$. Thus $f$ and $g$ are inverses of each other.

**26.** $f(g(x)) = f(\sqrt[5]{x}) = (\sqrt[5]{x})^5 = x$ for all $x$.
$g(f(x)) = g(x^5) = \sqrt[5]{x^5} = x$ for all $x$. Thus $f$ and $g$ are inverses of each other.

**27.** $f(g(x)) = f(\sqrt{x+4}) = (\sqrt{x+4})^2 - 4 = x + 4 - 4 = x$ for all $x \geq -4$.
$g(f(x)) = g(x^2 - 4) = \sqrt{(x^2-4)+4} = \sqrt{x^2} = x$ for all $x \geq 0$. Thus $f$ and $g$ are inverses of each other.

**28.** $f(g(x)) = f\left((x-1)^{1/3}\right) = \left((x-1)^{1/3}\right)^3 + 1 = x - 1 + 1 = x$ for all $x$.

$g(f(x)) = g(x^3+1) = \left[(x-1)^{1/3}\right]^3 + 1 = x - 1 + 1 = x$ for all $x$. Thus $f$ and $g$ are inverses of each other.

**29.** $f(g(x)) = f\left(\frac{1}{x}+1\right) = \dfrac{1}{\left(\frac{1}{x}+1\right)-1} = x$ for all $x \neq 0$.

$g(f(x)) = g\left(\frac{1}{x-1}\right) = \dfrac{1}{\left(\frac{1}{x-1}\right)} + 1 = (x-1)+1 = x$ for all $x \neq 1$. Thus $f$ and $g$ are inverses of each other.

**30.** $f(g(x)) = f\left(\sqrt{4-x^2}\right) = \sqrt{4-\left(\sqrt{4-x^2}\right)^2} = \sqrt{4-4+x^2} = \sqrt{x^2} = x$, for all $0 \le x \le 2$. (Note that the last equality is possible since $x \ge 0$.)

$g(f(x)) = g\left(\sqrt{4-x^2}\right) = \sqrt{4-\left(\sqrt{4-x^2}\right)^2} = \sqrt{4-4+x^2} = \sqrt{x^2} = x$, for all $0 \le x \le 2$. (Again, the last equality is possible since $x \ge 0$.) Thus $f$ and $g$ are inverses of each other.

**31.** $f(x) = 2x+1$. $y = 2x+1 \quad\Leftrightarrow\quad 2x = y-1 \quad\Leftrightarrow\quad x = \frac{1}{2}(y-1)$. So $f^{-1}(x) = \frac{1}{2}(x-1)$.

**32.** $f(x) = 6-x$. $y = 6-x \quad\Leftrightarrow\quad x = 6-y$. So $f^{-1}(x) = 6-x$.

**33.** $f(x) = 4x+7$. $y = 4x+7 \quad\Leftrightarrow\quad 4x = y-7 \quad\Leftrightarrow\quad x = \frac{1}{4}(y-7)$. So $f^{-1}(x) = \frac{1}{4}(x-7)$.

**34.** $f(x) = 3-5x$. $y = 3-5x \quad\Leftrightarrow\quad -5x = y-3 \quad\Leftrightarrow\quad x = -\frac{1}{5}(y-3) = \dfrac{1}{5}(3-y)$. So $f^{-1}(x) = \frac{1}{5}(3-x)$.

**35.** $f(x) = \dfrac{x}{2}$. $y = \dfrac{x}{2} \quad\Leftrightarrow\quad x = 2y$. So $f^{-1}(x) = 2x$.

**36.** $f(x) = \dfrac{1}{x^2}$, $x > 0$. Since the function $f$ is restricted to positive values, $f$ is one-to-one. So $y = \dfrac{1}{x^2} \quad\Leftrightarrow\quad \sqrt{y} = \dfrac{1}{x}$ $\quad\Leftrightarrow\quad x = \dfrac{1}{\sqrt{y}}$. Thus $f^{-1}(x) = \dfrac{1}{\sqrt{x}}$, $x > 0$.

**37.** $f(x) = \dfrac{1}{x+2}$. $y = \dfrac{1}{x+2} \quad\Leftrightarrow\quad x+2 = \dfrac{1}{y} \quad\Leftrightarrow\quad x = \dfrac{1}{y} - 2$. So $f^{-1}(x) = \dfrac{1}{x} - 2$.

**38.** $f(x) = \dfrac{x-2}{x+2}$. $y = \dfrac{x-2}{x+2} \quad\Leftrightarrow\quad y(x+2) = x-2 \quad\Leftrightarrow\quad xy+2y = x-2 \quad\Leftrightarrow\quad xy - x = -2-2y \quad\Leftrightarrow$ $x(y-1) = -2(y+1) \quad\Leftrightarrow\quad x = \dfrac{-2(y+1)}{y-1}$. So $f^{-1}(x) = \dfrac{-2(x+1)}{x-1}$.

**39.** $f(x) = \dfrac{1+3x}{5-2x}$. $y = \dfrac{1+3x}{5-2x} \quad\Leftrightarrow\quad y(5-2x) = 1+3x \quad\Leftrightarrow\quad 5y - 2xy = 1+3x \quad\Leftrightarrow\quad 3x + 2xy = 5y-1$ $\quad\Leftrightarrow\quad x(3+2y) = 5y-1 \quad\Leftrightarrow\quad x = \dfrac{5y-1}{2y+3}$. So $f^{-1}(x) = \dfrac{5x-1}{2x+3}$.

**40.** $f(x) = 5-4x^3$. $y = 5-4x^3 \quad\Leftrightarrow\quad 4x^3 = 5-y \quad\Leftrightarrow\quad x^3 = \frac{1}{4}(5-y) \quad\Leftrightarrow\quad x = \sqrt[3]{\frac{1}{4}(5-y)}$. So $f^{-1}(x) = \sqrt[3]{\frac{1}{4}(5-x)}$.

**41.** $f(x) = \sqrt{2+5x}, x \ge -\frac{2}{5}$. $y = \sqrt{2+5x}, y \ge 0 \quad\Leftrightarrow\quad y^2 = 2+5x \quad\Leftrightarrow\quad 5x = y^2-2 \quad\Leftrightarrow\quad x = \frac{1}{5}(y^2-2)$ and $y \ge 0$. So $f^{-1}(x) = \frac{1}{5}(x^2-2)$, $x \ge 0$.

**42.** $f(x) = x^2+x = \left(x^2+x+\frac{1}{4}\right) - \frac{1}{4} = \left(x+\frac{1}{2}\right)^2 - \frac{1}{4}$, $x \ge -\dfrac{1}{2}$. $y = \left(x+\frac{1}{2}\right)^2 - \frac{1}{4} \quad\Leftrightarrow\quad \left(x+\frac{1}{2}\right)^2 = y + \frac{1}{4} \quad\Leftrightarrow$ $x + \frac{1}{2} = \sqrt{y+\frac{1}{4}} \quad\Leftrightarrow\quad x = \sqrt{y+\frac{1}{4}} - \frac{1}{2}$, $y \ge -\frac{1}{4}$. So $f^{-1}(x) = \sqrt{x+\frac{1}{4}} - \frac{1}{2}$, $x \ge -\frac{1}{4}$. (Note that $x \ge -\frac{1}{2}$, so that $x + \frac{1}{2} \ge 0$, and hence $\left(x+\frac{1}{2}\right)^2 = y + \frac{1}{4} \quad\Leftrightarrow\quad x + \dfrac{1}{2} = \sqrt{y+\frac{1}{4}}$. Also, since $x \ge -\frac{1}{2}$, $y = \left(x+\frac{1}{2}\right)^2 - \frac{1}{4} \ge -\frac{1}{4}$ so that $y + \frac{1}{4} \ge 0$, and hence $\sqrt{y+\frac{1}{4}}$ is defined.)

**43.** $f(x) = 4 - x^2$, $x \geq 0$. $y = 4 - x^2 \quad \Leftrightarrow \quad x^2 = 4 - y \quad \Leftrightarrow \quad x = \sqrt{4 - y}$. So $f^{-1}(x) = \sqrt{4 - x}$. Note: $x \geq 0 \quad \Rightarrow$ $f(x) \leq 4$.

**44.** $f(x) = \sqrt{2x - 1}$. $y = \sqrt{2x - 1} \quad \Leftrightarrow \quad 2x - 1 = y^2 \quad \Leftrightarrow \quad x = \frac{1}{2}\left(y^2 + 1\right)$. Since the range of $f$ is $f(x) \geq 0$ so $f^{-1}(x) = \frac{1}{2}\left(x^2 + 1\right)$ for $x \geq 0$.

**45.** $f(x) = 4 + \sqrt[3]{x}$. $y = 4 + \sqrt[3]{x} \quad \Leftrightarrow \quad \sqrt[3]{x} = y - 4 \quad \Leftrightarrow \quad x = (y - 4)^3$. So $f^{-1}(x) = (x - 4)^3$.

**46.** $f(x) = \left(2 - x^3\right)^5$. $y = \left(2 - x^3\right)^5 \quad \Leftrightarrow \quad 2 - x^3 = \sqrt[5]{y} \quad \Leftrightarrow \quad x^3 = 2 - \sqrt[5]{y} \quad \Leftrightarrow \quad x = \sqrt[3]{2 - \sqrt[5]{y}}$. So $f^{-1}(x) = \sqrt[3]{2 - \sqrt[5]{x}}$.

**47.** $f(x) = 1 + \sqrt{1 + x}$. $y = 1 + \sqrt{1 + x}$, $y \geq 1 \quad \Leftrightarrow \quad \sqrt{1 + x} = y - 1 \quad \Leftrightarrow \quad 1 + x = (y - 1)^2$ $\Leftrightarrow x = (y - 1)^2 - 1 = y^2 - 2y$. So $f^{-1}(x) = x^2 - 2x$, $x \geq 1$.

**48.** $f(x) = \sqrt{9 - x^2}$, $0 \leq x \leq 3$. $y = \sqrt{9 - x^2} \quad \Leftrightarrow \quad y^2 = 9 - x^2 \quad \Leftrightarrow \quad x^2 = 9 - y^2 \quad \Rightarrow \quad x = \sqrt{9 - y^2}$ (since we must have $x \geq 0$ ). So $f^{-1}(x) = \sqrt{9 - x^2}$, $0 \leq x \leq 3$.

**49.** $f(x) = x^4$, $x \geq 0$. $y = x^4$, $y \geq 0 \quad \Leftrightarrow \quad x = \sqrt[4]{y}$. So $f^{-1}(x) = \sqrt[4]{x}$, $x \geq 0$.

**50.** $f(x) = 1 - x^3$. $y = 1 - x^3 \quad \Leftrightarrow \quad x^3 = 1 - y \quad \Leftrightarrow \quad x = \sqrt[3]{1 - y}$. So $f^{-1}(x) = \sqrt[3]{1 - x}$.

**51. (a), (b)** $f(x) = 3x - 6$

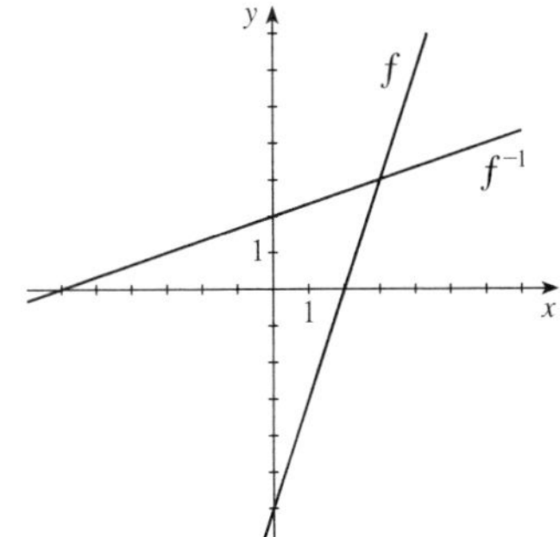


**(c)** $f(x) = 3x - 6$. $y = 3x - 6 \quad \Leftrightarrow$ $3x = y + 6 \quad \Leftrightarrow \quad x = \frac{1}{3}(y + 6)$. So $f^{-1}(x) = \frac{1}{3}(x + 6)$.

**52. (a), (b)** $f(x) = 16 - x^2$, $x \geq 0$

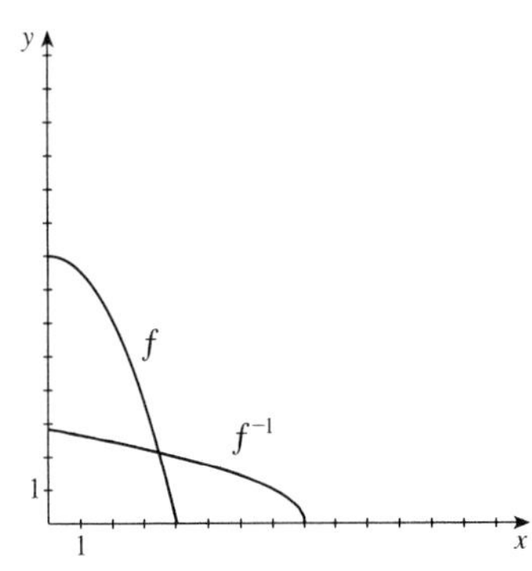


**(c)** $f(x) = 16 - x^2$, $x \geq 0$. $y = 16 - x^2 \quad \Leftrightarrow$ $x^2 = 16 - y \quad \Leftrightarrow \quad x = \sqrt{16 - y}$. So $f^{-1}(x) = \sqrt{16 - x}$, $x \leq 16$. (*Note:* $x \geq 0$ $\Rightarrow \quad f(x) = 16 - x^2 \leq 16$.)

**53. (a), (b)** $f(x) = \sqrt{x+1}$

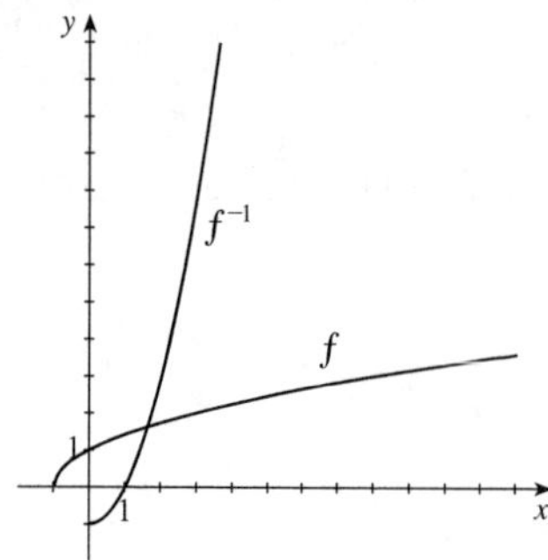


**(c)** $f(x) = \sqrt{x+1}, x \geq -1$.
$y = \sqrt{x+1}, y \geq 0 \quad \Leftrightarrow \quad y^2 = x+1 \quad \Leftrightarrow$
$x = y^2 - 1$ and $y \geq 0$. So $f^{-1}(x) = x^2 - 1$,
$x \geq 0$.

**54. (a), (b)** $f(x) = x^3 - 1$

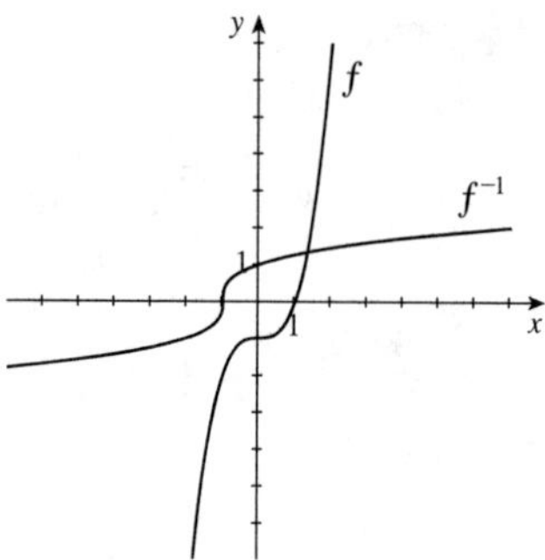


**(c)** $f(x) = x^3 - 1 \quad \Leftrightarrow \quad y = x^3 - 1 \quad \Leftrightarrow$
$x^3 = y + 1 \quad \Leftrightarrow \quad x = \sqrt[3]{y+1}$. So
$f^{-1}(x) = \sqrt[3]{x+1}$.

**55.** $f(x) = x^3 - x$. Using a graphing device and the Horizontal Line Test, we see that $f$ is not a one-to-one function. For example, $f(0) = 0 = f(-1)$.

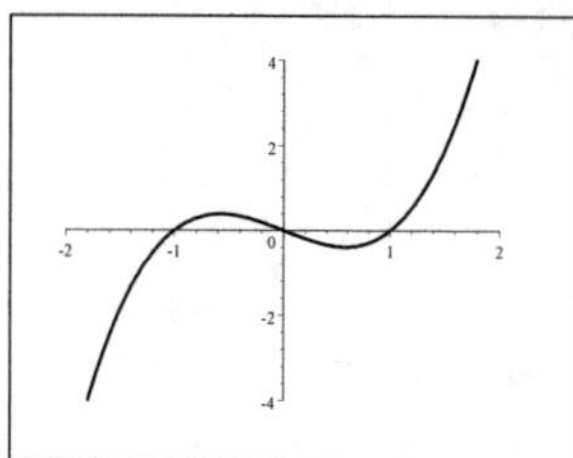

**56.** $f(x) = x^3 + x$. Using a graphing device and the Horizontal Line Test, we see that $f$ is a one-to-one function.

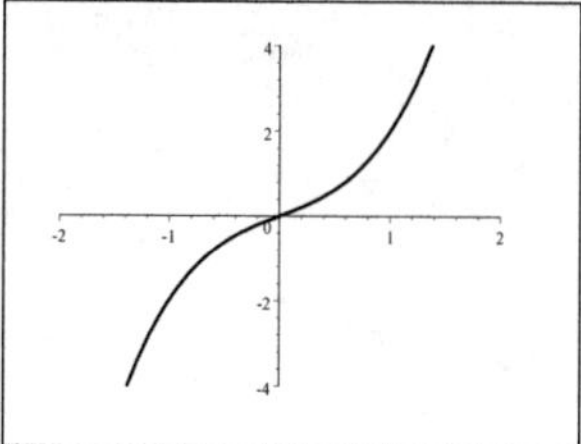

**57.** $f(x) = \dfrac{x+12}{x-6}$. Using a graphing device and the Horizontal Line Test, we see that $f$ is a one-to-one function.

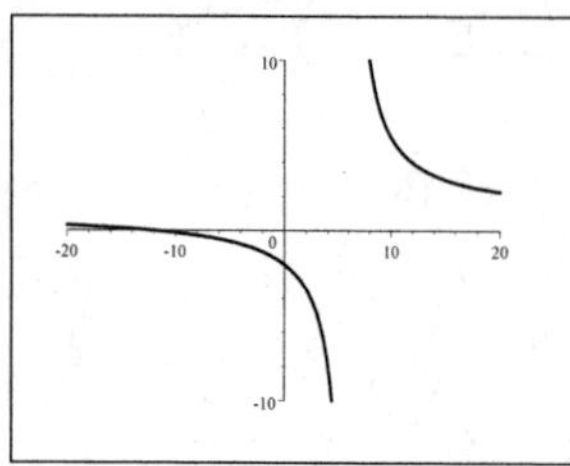

**58.** $f(x) = \sqrt{x^3 - 4x + 1}$. Using a graphing device and the Horizontal Line Test, we see that $f$ is not a one-to-one function. For example, $f(0) = 1 = f(2)$.

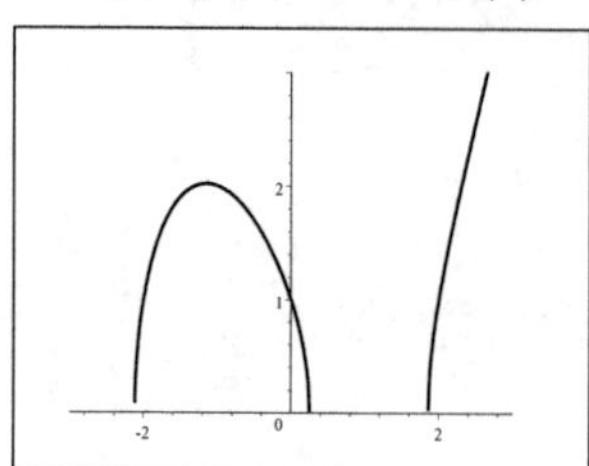

**59.** $f(x) = |x| - |x-6|$. Using a graphing device and the Horizontal Line Test, we see that $f$ is not a one-to-one function. For example $f(0) = -6 = f(-2)$.

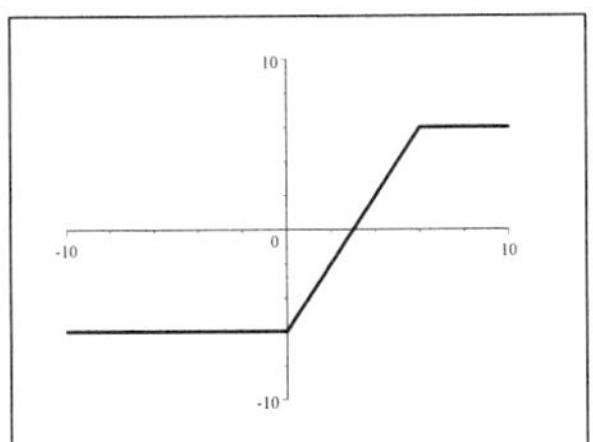

**60.** $f(x) = x \cdot |x|$. Using a graphing device and the Horizontal Line Test, we see that $f$ is a one-to-one function.

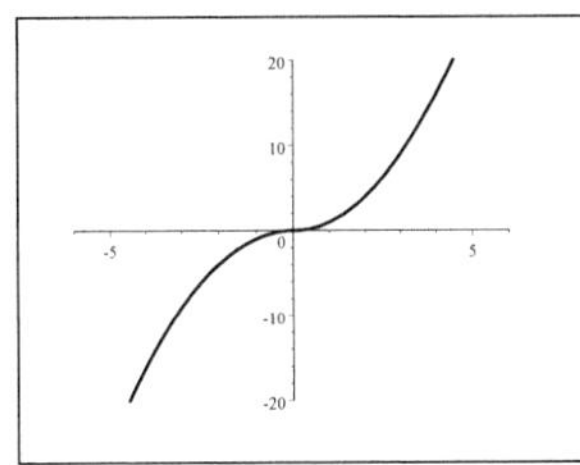

**61. (a)** $y = f(x) = 2 + x \quad \Leftrightarrow \quad x = y - 2$. So $f^{-1}(x) = x - 2$.

**(b)**

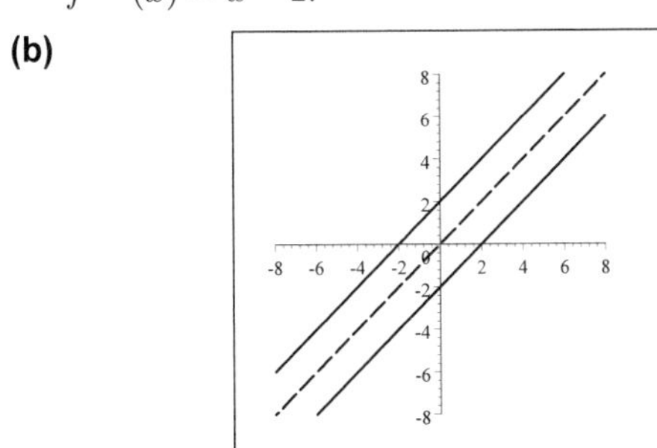

**62. (a)** $y = f(x) = 2 - \frac{1}{2}x \quad \Leftrightarrow \quad \frac{1}{2}x = 2 - y \quad \Leftrightarrow \quad x = 4 - 2y$. So $f^{-1}(x) = 4 - 2x$.

**(b)**

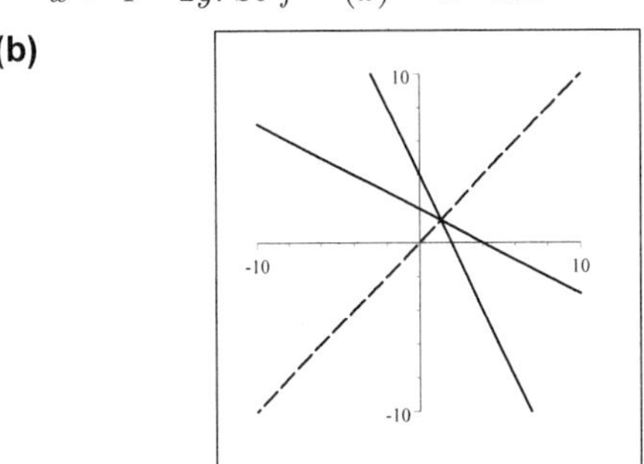

**63. (a)** $y = g(x) = \sqrt{x+3}, y \geq 0 \quad \Leftrightarrow \quad x + 3 = y^2$, $y \geq 0 \quad \Leftrightarrow \quad x = y^2 - 3, y \geq 0$. So $g^{-1}(x) = x^2 - 3, x \geq 0$.

**(b)**

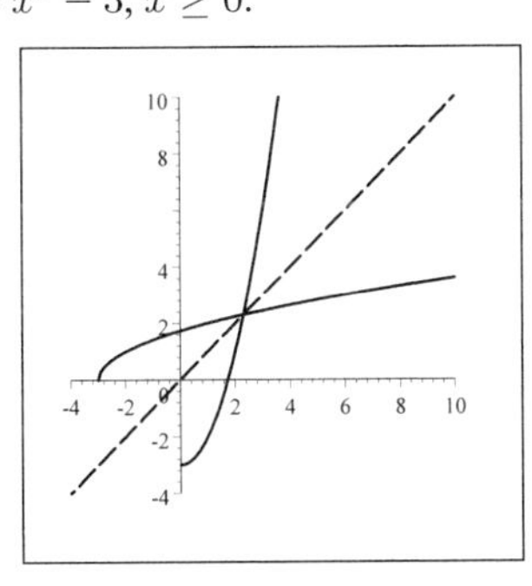

**64. (a)** $y = g(x) = x^2 + 1, x \geq 0 \quad \Leftrightarrow \quad x^2 = y - 1$, $x \geq 0 \quad \Leftrightarrow \quad x = \sqrt{y-1}$. So $g^{-1}(x) = \sqrt{x-1}$.

**(b)**

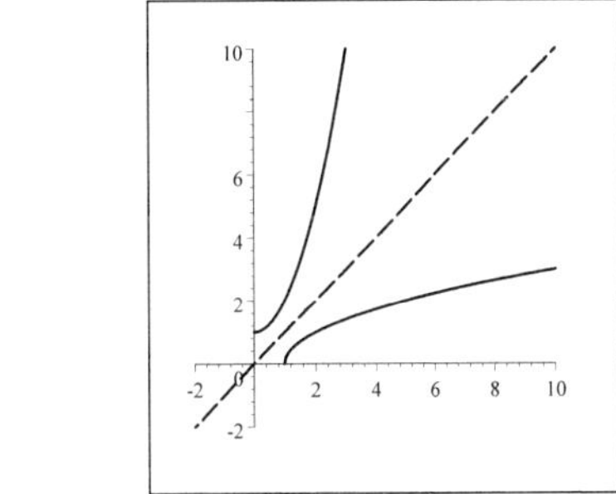

**65.** If we restrict the domain of $f(x)$ to $[0, \infty)$, then $y = 4 - x^2 \quad \Leftrightarrow \quad x^2 = 4 - y \quad \Rightarrow \quad x = \sqrt{4-y}$ (since $x \geq 0$, we take the positive square root). So $f^{-1}(x) = \sqrt{4-x}$.
If we restrict the domain of $f(x)$ to $(-\infty, 0]$, then $y = 4 - x^2 \quad \Leftrightarrow \quad x^2 = 4 - y \quad \Rightarrow \quad x = -\sqrt{4-y}$ (since $x \leq 0$, we take the negative square root). So $f^{-1}(x) = -\sqrt{4-x}$.

**66.** If we restrict the domain of $g(x)$ to $[1, \infty)$, then $y = (x-1)^2 \quad \Rightarrow \quad x - 1 = \sqrt{y}$ (since $x \geq 1$ we take the positive square root) $\Leftrightarrow x = 1 + \sqrt{y}$. So $g^{-1}(x) = 1 + \sqrt{x}$.
If we restrict the domain of $g(x)$ to $(-\infty, 1]$, then $y = (x-1)^2 \quad \Rightarrow \quad x - 1 = -\sqrt{y}$ (since $x \leq 1$ we take the negative square root) $\Leftrightarrow x = 1 - \sqrt{y}$. So $g^{-1}(x) = 1 - \sqrt{x}$.

**67.** If we restrict the domain of $h(x)$ to $[-2,\infty)$, then $y=(x+2)^2 \quad \Rightarrow \quad x+2=\sqrt{y}$ (since $x\geq -2$, we take the positive square root) $\Leftrightarrow x=-2+\sqrt{y}$. So $h^{-1}(x)=-2+\sqrt{x}$.

If we restrict the domain of $h(x)$ to $(-\infty,-2]$, then $y=(x+2)^2 \quad \Rightarrow \quad x+2=-\sqrt{y}$ (since $x\leq -2$, we take the negative square root) $\Leftrightarrow x=-2-\sqrt{y}$. So $h^{-1}(x)=-2-\sqrt{x}$.

**68.** 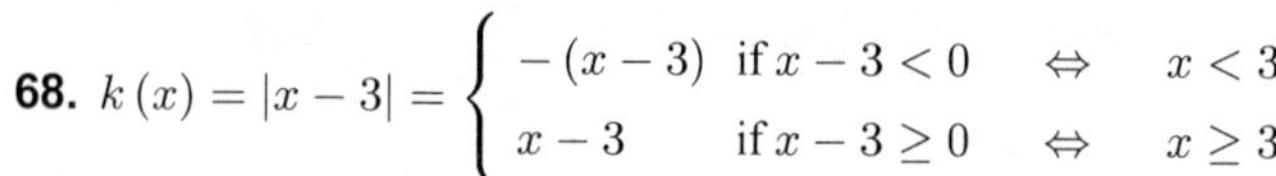

$$k(x)=|x-3|=\begin{cases}-(x-3) & \text{if } x-3<0 \quad \Leftrightarrow \quad x<3\\ x-3 & \text{if } x-3\geq 0 \quad \Leftrightarrow \quad x\geq 3\end{cases}$$

If we restrict the domain of $k(x)$ to $[3,\infty)$, then $y=x-3 \quad \Leftrightarrow \quad x=3+y$. So $k^{-1}(x)=3+x$.

If we restrict the domain of $k(x)$ to $(-\infty,3]$, then $y=-(x-3) \quad \Leftrightarrow \quad y=-x+3 \quad \Leftrightarrow \quad x=3-y$. So $k^{-1}(x)=3-x$.

**69.**

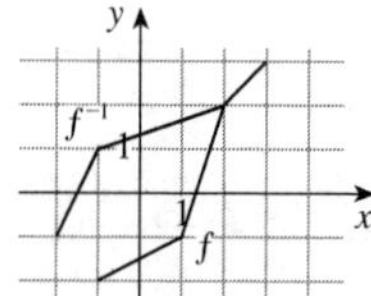


**70.**

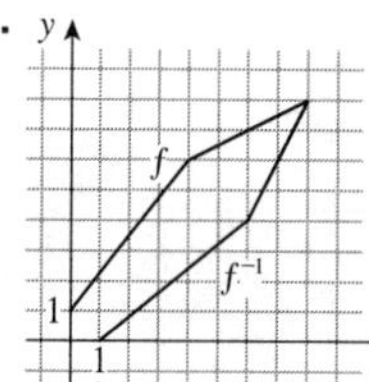


**71. (a)** $f(x)=500+80x$.

**(b)** $f(x)=500+80x$. $y=500+80x \quad \Leftrightarrow \quad 80x=y-500 \quad \Leftrightarrow \quad x=\dfrac{y-500}{80}$. So $f^{-1}(x)=\dfrac{x-500}{80}$. $f^{-1}$ represents the number of hours of investigation the investigate spends on a case for $x$ dollars.

**(c)** $f^{-1}(1220)=\dfrac{1220-500}{80}=\dfrac{720}{80}=9$. The investigator spent 9 hours investigating this case.

**72. (a)** $V(t)=100\left(1-\dfrac{t}{40}\right)^2$, $0\leq t\leq 40$. $y=100\left(1-\dfrac{t}{40}\right)^2 \quad \Leftrightarrow \quad \dfrac{y}{100}=\left(1-\dfrac{t}{40}\right)^2 \quad \Rightarrow$

$1-\dfrac{t}{40}=\pm\sqrt{\dfrac{y}{100}} \quad \Leftrightarrow \quad \dfrac{t}{40}=1\pm\dfrac{\sqrt{y}}{10} \quad \Leftrightarrow \quad t=40\pm 4\sqrt{y}$. Since $t\leq 40$ we must have $V^{-1}(t)=40-4\sqrt{t}$.

$V^{-1}$ represents time that has elapsed since the tank started to leak.

**(b)** $V^{-1}(15)=40-4\sqrt{15}\approx 24.5$ minutes. In 24.5 minutes the tank has drained to just 15 gallons of water.

**73. (a)** $v(r)=18{,}500\left(0.25-r^2\right)$. $t=18{,}500\left(0.25-r^2\right) \quad \Leftrightarrow \quad t=4625-18{,}500r^2 \quad \Leftrightarrow \quad 18500r^2=4625-t$

$\Leftrightarrow \quad r^2=\dfrac{4625-t}{18{,}500} \quad \Rightarrow \quad r=\pm\sqrt{\dfrac{4625-t}{18{,}500}}$. Since $r$ represents a distance, $r\geq 0$, so $v^{-1}(t)=\sqrt{\dfrac{4625-t}{18{,}500}}$.

$v^{-1}$ represents the radius in the vein that has the velocity $v$.

**(b)** $v^{-1}(30)=\sqrt{\dfrac{4625-30}{18{,}500}}\approx 0.498$ cm. The velocity is 30 at 0.498 cm from the center of the artery or vein.

**74. (a)** $D(p)=-3p+150$. $y=-3p+150 \quad \Leftrightarrow \quad 3p=150-y \Leftrightarrow p=50-\frac{1}{3}y$. So $D^{-1}(p)=50-\frac{1}{3}p$. $D^{-1}$ represents the price that is associated with demand $D$.

**(b)** $D^{-1}(30)=50-\frac{1}{3}(30)=40$. So when the demand is 30 units the price per unit is \$40.

**75. (a)** $F(x)=\frac{9}{5}x+32$. $y=\frac{9}{5}x+32 \quad \Leftrightarrow \quad \frac{9}{5}x=y-32 \quad \Leftrightarrow \quad x=\frac{5}{9}(y-32)$. So $F^{-1}(x)=\frac{5}{9}(x-32)$. $F^{-1}$ represents the Celsius temperature that corresponds to the Fahrenheit temperature of $F$.

**(b)** $F^{-1}(86)=\frac{5}{9}(86-32)=\frac{5}{9}(54)=30$. So $86^\circ$ Fahrenheit is the same as $30^\circ$ Celsius.

**76. (a)** $f(x)=0.8159x$.

**(b)** $f(x)=0.8159x$. $y=0.8159x \quad \Leftrightarrow \quad x=1.2256y$. So $f^{-1}(x)=1.2256x$. $f^{-1}$ represents the exchange rate from US dollars to Canadian dollars.

**(c)** $f^{-1}(12{,}250)=1.2256(12{,}250)=15{,}013.60$. So \$12,250 in US currency is worth \$15,013.60 in Canadian currency.

**77. (a)** $f(x) = \begin{cases} 0.1x, & \text{if } 0 \le x \le 20{,}000 \\ 2000 + 0.2(x - 20{,}000) & \text{if } x > 20{,}000 \end{cases}$

**(b)** We will find the inverse of each piece of the function $f$.

$f_1(x) = 0.1x$. $y = 0.1x \quad\Leftrightarrow\quad x = 10y$. So $f_1^{-1}(x) = 10x$.

$f_2(x) = 2000 + 0.2(x - 20{,}000) = 0.2x - 2000$. $y = 0.2x - 2000 \quad\Leftrightarrow\quad 0.2x = y + 2000 \quad\Leftrightarrow$ $x = 5y + 10{,}000$. So $f_2^{-1}(x) = 5x + 10{,}000$.

Since $f(0) = 0$ and $f(20{,}000) = 2000$ we have $f^{-1}(x) = \begin{cases} 10x, & \text{if } 0 \le x \le 2000 \\ 5x + 10{,}000 & \text{if } x > 2000 \end{cases}$ It represents the taxpayer's income.

**(b)** $f^{-1}(10{,}000) = 5(10{,}000) + 10{,}000 = 60{,}000$. The required income is \$60,000.

**78. (a)** $f(x) = 0.85x$.

**(b)** $g(x) = x - 1000$.

**(c)** $H = f \circ g = f(x - 1000) = 0.85(x - 1000) = 0.85x - 850$.

**(d)** $H(x) = 0.85x - 850$. $y = 0.85x - 850 \quad\Leftrightarrow\quad 0.85x = y + 850 \quad\Leftrightarrow\quad x = 1.176y + 1000$. So $H^{-1}(x) = 1.176x + 1000$. The function $H^{-1}$ represents the original sticker price for a given discounted price.

**(e)** $H^{-1}(13{,}000) = 1.176(13{,}000) + 1000 = 16{,}266$. So the original price of the car is \$16,288 when the discounted price (\$1000 rebate, then 15% off) is \$13,000.

**79.** $f(x) = 7 + 2x$. $y = 7 + 2x \quad\Leftrightarrow\quad 2x = y - 7 \quad\Leftrightarrow\quad x = \dfrac{y-7}{2}$. So $f^{-1}(x) = \dfrac{x-7}{2}$. $f^{-1}$ is the number of toppings on a pizza that costs $x$ dollars.

**80.** $f(x) = mx + b$. Notice that $f(x_1) = f(x_2) \quad\Leftrightarrow\quad mx_1 + b = mx_2 + b \quad\Leftrightarrow\quad mx_1 = mx_2$. We can conclude that $x_1 = x_2$ if and only if $m \neq 0$. Therefore $f$ is one-to-one if and only if $m \neq 0$. If $m \neq 0$, $f(x) = mx + b \quad\Leftrightarrow$ $y = mx + b \quad\Leftrightarrow\quad mx = y - b \quad\Leftrightarrow\quad x = \dfrac{y-b}{m}$. So, $f^{-1}(x) = \dfrac{x-b}{m}$.

**81. (a)** $f(x) = \dfrac{2x+1}{5}$ is "multiply by 2, add 1, and then divide by 5 ". So the reverse is "multiply by 5, subtract 1, and then divide by 2 " or $f^{-1}(x) = \dfrac{5x-1}{2}$. Check: $f \circ f^{-1}(x) = f\left(\dfrac{5x-1}{2}\right) = \dfrac{2\left(\dfrac{5x-1}{2}\right)+1}{5} = \dfrac{5x-1+1}{5} = \dfrac{5x}{5} = x$

and $f^{-1} \circ f(x) = f^{-1}\left(\dfrac{2x+1}{5}\right) = \dfrac{5\left(\dfrac{2x+1}{5}\right)-1}{2} = \dfrac{2x+1-1}{2} = \dfrac{2x}{2} = x$.

**(b)** $f(x) = 3 - \dfrac{1}{x} = \dfrac{-1}{x} + 3$ is "take the negative reciprocal and add 3 ". Since the reverse of "take the negative reciprocal" is "take the negative reciprocal ", $f^{-1}(x)$ is "subtract 3 and take the negative reciprocal ", that is, $f^{-1}(x) = \dfrac{-1}{x-3}$. Check: $f \circ f^{-1}(x) = f\left(\dfrac{-1}{x-3}\right) = 3 - \dfrac{1}{\dfrac{-1}{x-3}} = 3 - \left(1 \cdot \dfrac{x-3}{-1}\right) = 3 + x - 3 = x$ and

$f^{-1} \circ f(x) = f^{-1}\left(3 - \dfrac{1}{x}\right) = \dfrac{-1}{\left(3 - \dfrac{1}{x}\right) - 3} = \dfrac{-1}{-\dfrac{1}{x}} = -1 \cdot \dfrac{x}{-1} = x$.

**(c)** $f(x) = \sqrt{x^3+2}$ is "cube, add 2, and then take the square root". So the reverse is "square, subtract 2, then take the cube root " or $f^{-1}(x) = \sqrt[3]{x^2-2}$. Domain for $f(x)$ is $[-\sqrt[3]{2}, \infty)$; domain for $f^{-1}(x)$ is $[0, \infty)$. Check: $f \circ f^{-1}(x) = f\left(\sqrt[3]{x^2-2}\right) = \sqrt{\left(\sqrt[3]{x^2-2}\right)^3+2} = \sqrt{x^2-2+2} = \sqrt{x^2} = x$ (on the appropriate domain) and $f^{-1} \circ f(x) = f^{-1}\left(\sqrt{x^3+2}\right) = \sqrt[3]{\left(\sqrt{x^3+2}\right)^2-2} = \sqrt[3]{x^3+2-2} = \sqrt[3]{x^3} = x$ (on the appropriate domain).

**(d)** $f(x) = (2x-5)^3$ is "double, subtract 5, and then cube". So the reverse is "take the cube root, add 5, and divide by 2" or $f^{-1}(x) = \dfrac{\sqrt[3]{x}+5}{2}$ Domain for both $f(x)$ and $f^{-1}(x)$ is $(-\infty, \infty)$. Check: $f \circ f^{-1}(x) = f\left(\dfrac{\sqrt[3]{x}+5}{2}\right) = \left[2\left(\dfrac{\sqrt[3]{x}+5}{2}\right)-5\right]^3 = \left(\sqrt[3]{x}+5-5\right)^3 = \left(\sqrt[3]{x}\right)^3 = \sqrt[3]{x^3} = x$ and

$f^{-1} \circ f(x) = f^{-1}\left((2x-5)^3\right) = \dfrac{\sqrt[x]{(2x-5)^3}+5}{2} = \dfrac{(2x-5)+5}{2} = \dfrac{2x}{2} = x.$

In a function like $f(x) = 3x-2$, the variable occurs only once and it easy to see how to reverse the operations step by step. But in $f(x) = x^3+2x+6$, you apply two different operations to the variable $x$ (cubing and multiplying by 2) and then add 6, so it is not possible to reverse the operations step by step.

**82.** $f(I(x)) = f(x)$; therefore $f \circ I = f$. $I(f(x)) = f(x)$; therefore $I \circ f = f$.

By definition, $f \circ f^{-1}(x) = x = I(x)$; therefore $f \circ f^{-1} = I$. Similarly, $f^{-1} \circ f(x) = x = I(x)$; therefore $f^{-1} \circ f = I$.

**83. (a)** We find $g^{-1}(x)$: $y = 2x+1 \quad \Leftrightarrow \quad 2x = y-1 \quad \Leftrightarrow \quad x = \frac{1}{2}(y-1)$. So $g^{-1}(x) = \frac{1}{2}(x-1)$. Thus $f(x) = h \circ g^{-1}(x) = h\left(\frac{1}{2}(x-1)\right) = 4\left[\frac{1}{2}(x-1)\right]^2 + 4\left[\frac{1}{2}(x-1)\right] + 7 = x^2-2x+1+2x-2+7 = x^2+6$.

**(b)** $f \circ g = h \quad \Leftrightarrow \quad f^{-1} \circ f \circ g = f^{-1} \circ h \quad \Leftrightarrow \quad I \circ g = f^{-1} \circ h \quad \Leftrightarrow \quad g = f^{-1} \circ h$.
Note that we compose with $f^{-1}$ on the left on each side of the equation. We find $f^{-1}$:
$y = 3x+5 \quad \Leftrightarrow \quad 3x = y-5 \quad \Leftrightarrow \quad x = \dfrac{1}{3}(y-5)$. So $f^{-1}(x) = \frac{1}{3}(x-5)$. Thus $g(x) = f^{-1} \circ h(x) = f^{-1}(3x^2+3x+2) = \frac{1}{3}\left[(3x^2+3x+2)-5\right] = \frac{1}{3}\left[3x^2+3x-3\right] = x^2+x-1$.

# Chapter 3 Review

**1.** $f(x) = x^2-4x+6$; $f(0) = (0)^2-4(0)+6 = 6$; $f(2) = (2)^2-4(2)+6 = 2$; $f(-2) = (-2)^2-4(-2)+6 = 18$; $f(a) = (a)^2-4(a)+6 = a^2-4a+6$; $f(-a) = (-a)^2-4(-a)+6 = a^2+4a+6$; $f(x+1) = (x+1)^2-4(x+1)+6 = x^2+2x+1-4x-4+6 = x^2-2x+3$; $f(2x) = (2x)^2-4(2x)+6 = 4x^2-8x+6$; $2f(x)-2 = 2(x^2-4x+6)-2 = 2x^2-8x+12-2 = 2x^2-8x+10$.

**2.** $f(x) = 4-\sqrt{3x-6}$; $f(5) = 4-\sqrt{15-6} = 3$; $f(9) = 4-\sqrt{27-6} = 4-\sqrt{21}$; $f(a+2) = 4-\sqrt{3a+6-6} = 4-\sqrt{3a}$; $f(-x) = 4-\sqrt{3(-x)-6} = 4-\sqrt{-3x-6}$; $f(x^2) = 4-\sqrt{3x^2-6}$; $[f(x)]^2 = \left(4-\sqrt{3x-6}\right)^2 = 16-8\sqrt{3x-6}+3x-6 = 10-8\sqrt{3x-6}+3x$.

**3. (a)** $f(-2) = -1$. $f(2) = 2$.

**(b)** The domain of $f$ is $[-4, 5]$.

**(b)** The range of $f$ is $[-4, 4]$.

**(d)** $f$ is increasing on $[-4, -2]$ and $[-1, 4]$; $f$ is decreasing on $[-2, -1]$ and $[4, 5]$.

**(e)** $f$ is not a one-to-one, for example, $f(-2) = -1 = f(0)$. There are many more examples.

**4.** By the Vertical Line Test, figures (b) and (c) are graphs of functions. By the Horizontal Line Test, figure (c) is the graph of a one-to-one function.

**5.** Domain: We must have $x + 3 \geq 0 \quad \Leftrightarrow \quad x \geq -3$. In interval notation, the domain is $[-3, \infty)$.
Range: For $x$ in the domain of $f$, we have $x \geq -3 \quad \Leftrightarrow \quad x + 3 \geq 0 \quad \Leftrightarrow \quad \sqrt{x+3} \geq 0 \quad \Leftrightarrow \quad f(x) \geq 0$. So the range is $[0, \infty)$.

**6.** $F(t) = t^2 + 2t + 5 = \left(t^2 + 2t + 1\right) + 5 - 1 = (t+1)^2 + 4$. Therefore $F(t) \geq 4$ for all $t$. Since there are no restrictions on $t$, the domain of $F$ is $(-\infty, \infty)$, and the range is $[4, \infty)$.

**7.** $f(x) = 7x + 15$. The domain is all real numbers, $(-\infty, \infty)$.

**8.** $f(x) = \dfrac{2x+1}{2x-1}$. Then $2x - 1 \neq 0 \quad \Leftrightarrow \quad x \neq \frac{1}{2}$. So the domain of $f$ is $\left\{x \mid x \neq \frac{1}{2}\right\}$.

**9.** $f(x) = \sqrt{x+4}$. We require $x + 4 \geq 0 \quad \Leftrightarrow \quad x \geq -4$. Thus the domain is $[-4, \infty)$.

**10.** $f(x) = 3x - \dfrac{2}{\sqrt{x+1}}$. The domain of $f$ is the set of $x$ where $x + 1 > 0 \quad \Leftrightarrow \quad x > -1$. So the domain is $(-1, \infty)$.

**11.** $f(x) = \dfrac{1}{x} + \dfrac{1}{x+1} + \dfrac{1}{x+2}$. The denominators cannot equal 0, therefore the domain is $\{x \mid x \neq 0, -1, -2\}$.

**12.** $g(x) = \dfrac{2x^2 + 5x + 3}{2x^2 - 5x - 3} = \dfrac{2x^2 + 5x + 3}{(2x+1)(x-3)}$. The domain of $g$ is the set of all $x$ where the denominator is not 0. So the domain is $\{x \mid 2x + 1 \neq 0 \text{ and } x - 3 \neq 0\} = \left\{x \mid x \neq -\frac{1}{2} \text{ and } x \neq 3\right\}$.

**13.** $h(x) = \sqrt{4-x} + \sqrt{x^2-1}$. We require the expression inside the radicals be nonnegative. So $4 - x \geq 0 \quad \Leftrightarrow \quad 4 \geq x$; also $x^2 - 1 \geq 0 \quad \Leftrightarrow \quad (x-1)(x+1) \geq 0$. We make a table:

| Interval | $(-\infty, -1)$ | $(-1, 1)$ | $(1, \infty)$ |
|---|---|---|---|
| Sign of $x - 1$ | $-$ | $-$ | $+$ |
| Sign of $x + 1$ | $-$ | $+$ | $+$ |
| Sign of $(x-1)(x+1)$ | $+$ | $-$ | $+$ |

Thus the domain is $(-\infty, 4] \cap \{(-\infty, -1] \cup [1, \infty)\} = (-\infty, -1] \cup [1, 4]$.

**14.** $f(x) = \dfrac{\sqrt[3]{2x+1}}{\sqrt[3]{2x}+2}$. Since we have an odd root, the domain is the set of all $x$ where the denominator is not 0. Now $\sqrt[3]{2x} + 2 \neq 0 \quad \Leftrightarrow \quad \sqrt[3]{2x} \neq -2 \quad \Leftrightarrow \quad 2x \neq -8 \quad \Leftrightarrow \quad x \neq -4$. Thus the domain of $f$ is $\{x \mid x \neq -4\}$.

**15.** $f(x) = 1 - 2x$

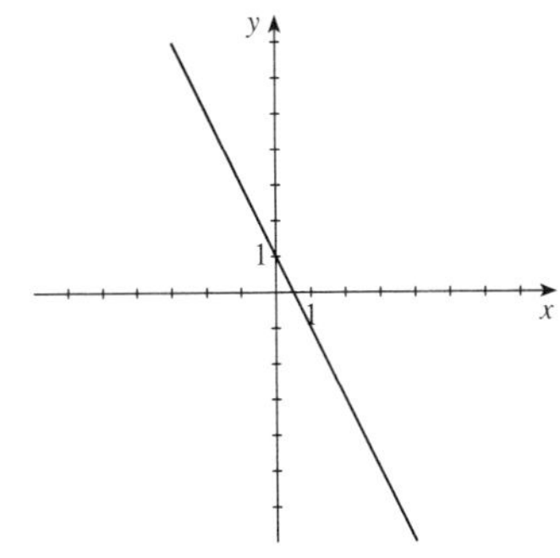


**16.** $f(x) = \frac{1}{3}(x-5)$, $2 \leq x \leq 8$

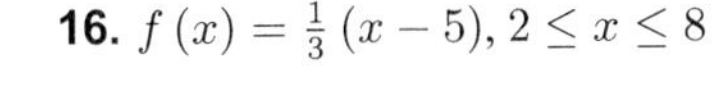

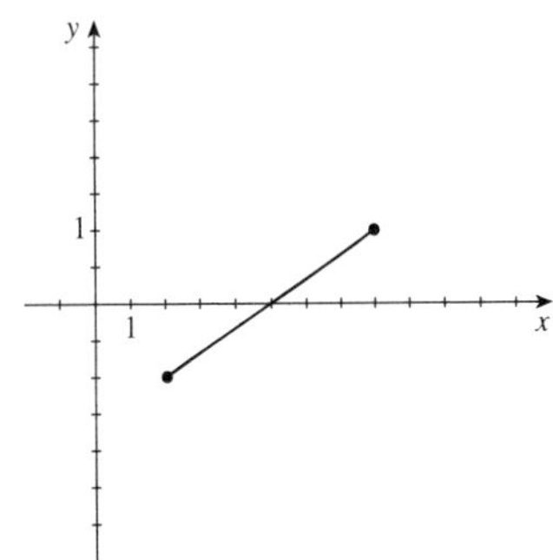

**17.** $f(t) = 1 - \frac{1}{2}t^2$

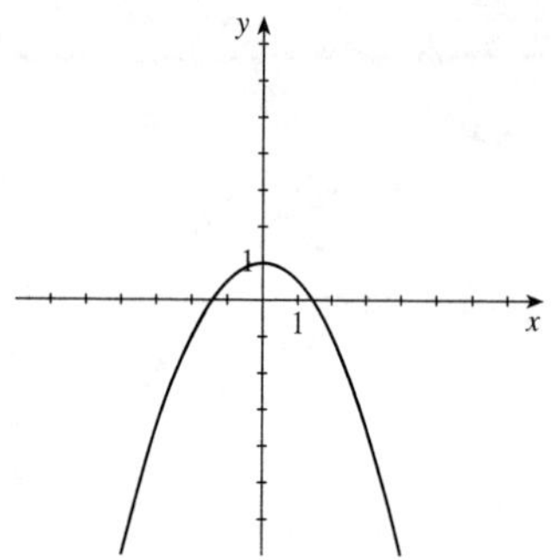

**18.** $g(t) = t^2 - 2t = (t-1)^2 - 1$

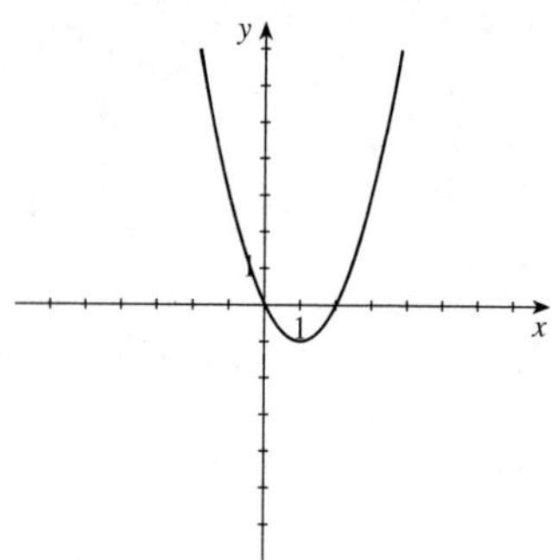

**19.** $f(x) = x^2 - 6x + 6$

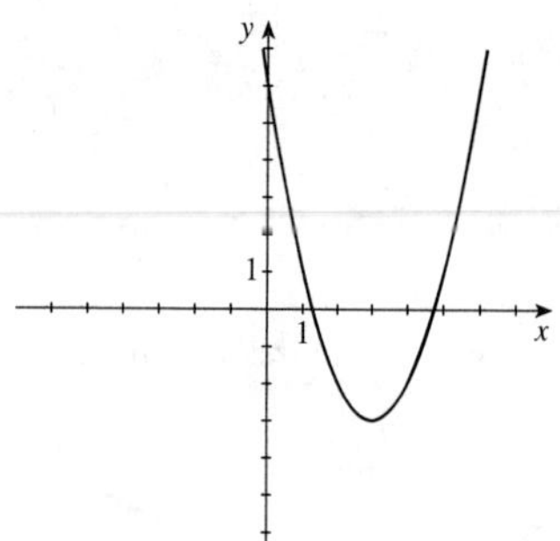

**20.** $f(x) = 3 - 8x - 2x^2 = -2(x+2)^2 + 11$

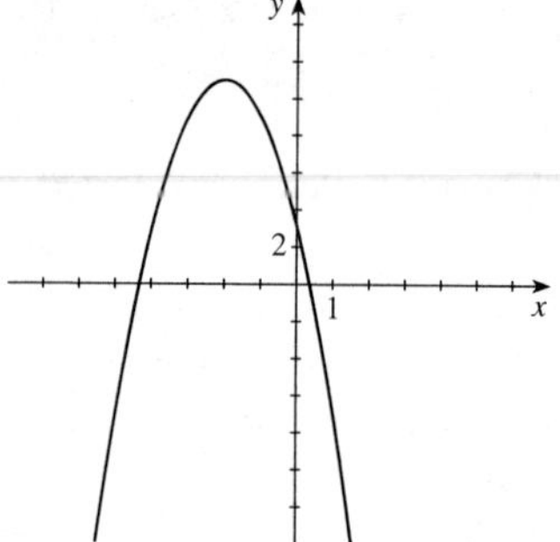

**21.** $g(x) = 1 - \sqrt{x}$

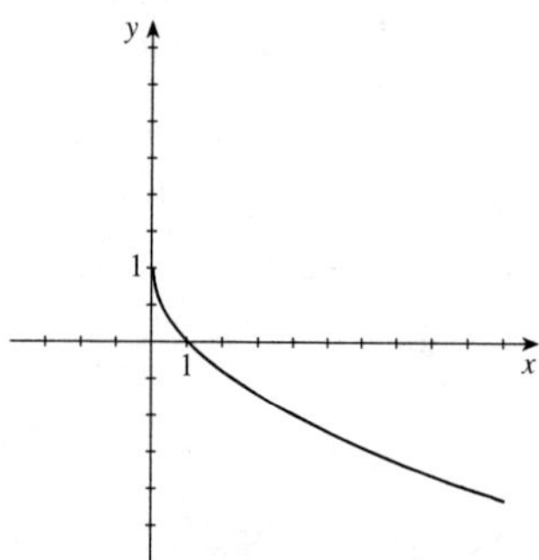

**22.** $g(x) = -|x|$

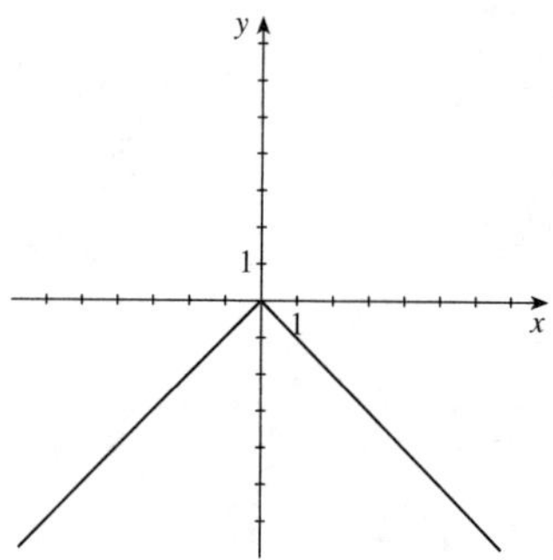

**23.** $h(x) = \frac{1}{2}x^3$

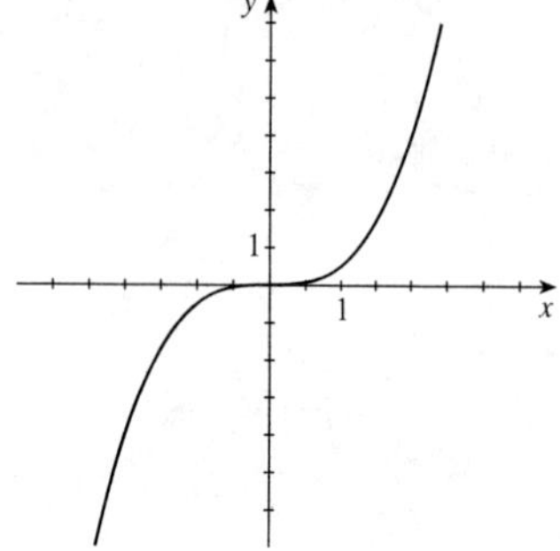

**24.** $h(x) = \sqrt{x+3}$

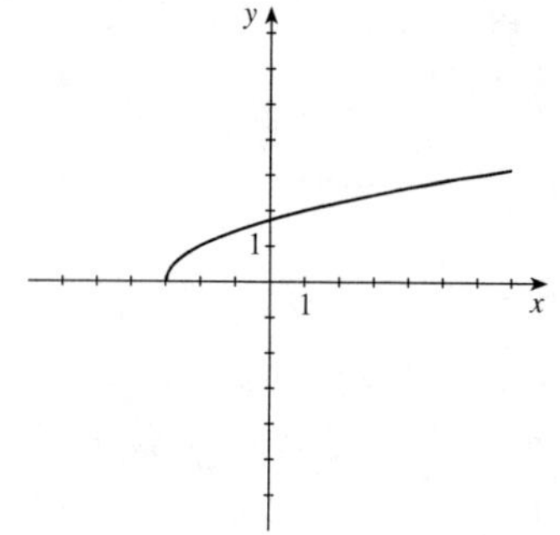

**25.** $h(x) = \sqrt[3]{x}$

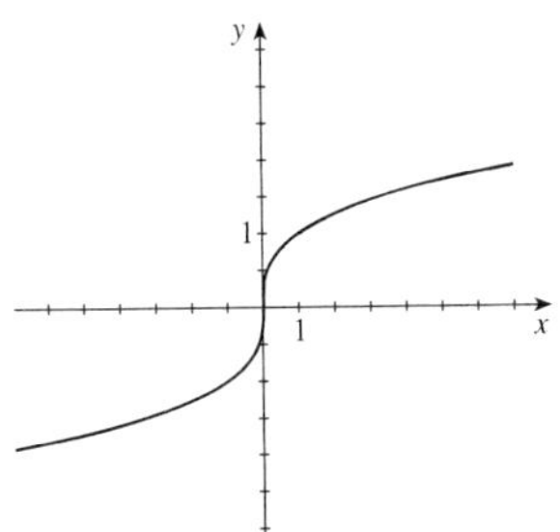

**26.** $H(x) = x^3 - 3x^2 = x^2(x-3)$

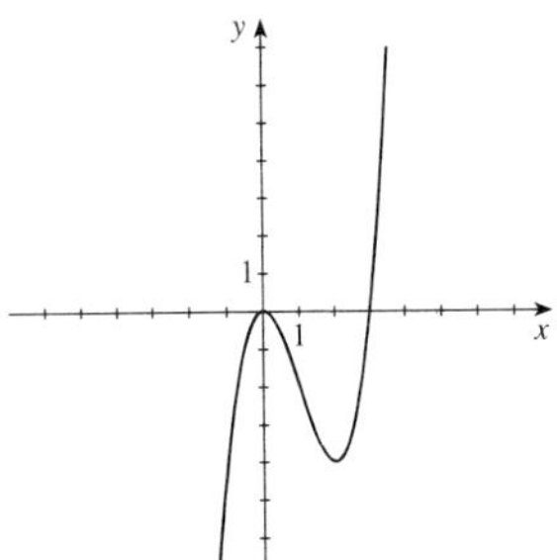

**27.** $g(x) = \dfrac{1}{x^2}$

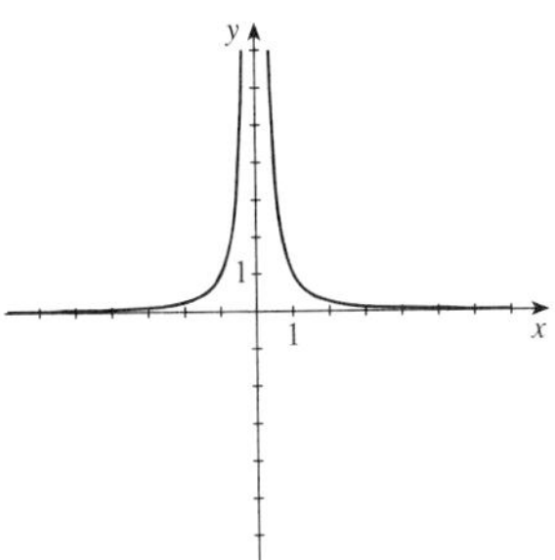

**28.** $G(x) = \dfrac{1}{(x-3)^2}$

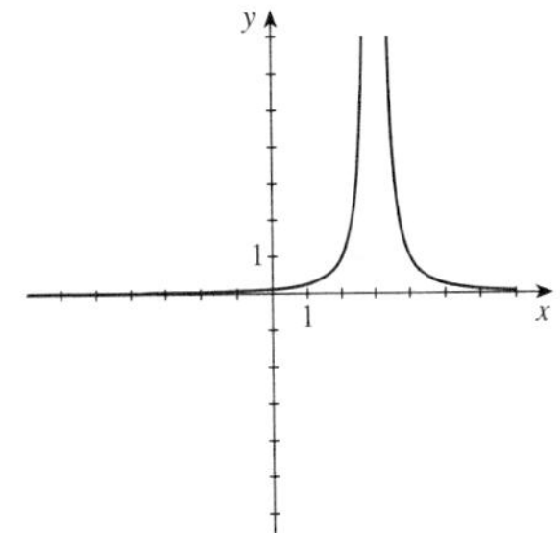

**29.** $f(x) = \begin{cases} 1 - x & \text{if } x < 0 \\ 1 & \text{if } x \geq 0 \end{cases}$

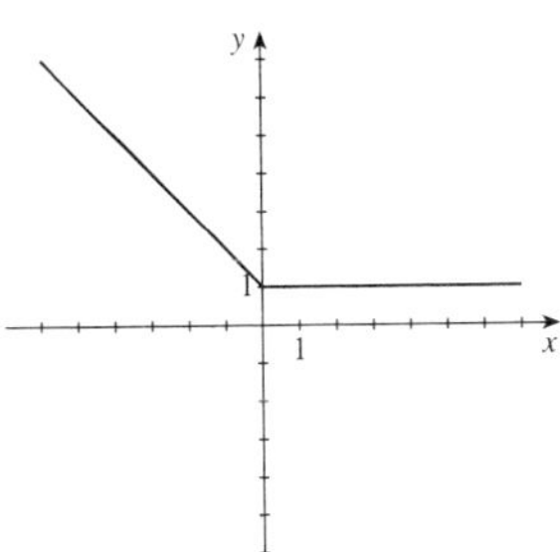

**30.** $f(x) = \begin{cases} 1 - 2x & \text{if } x \leq 0 \\ 2x - 1 & \text{if } x > 0 \end{cases}$

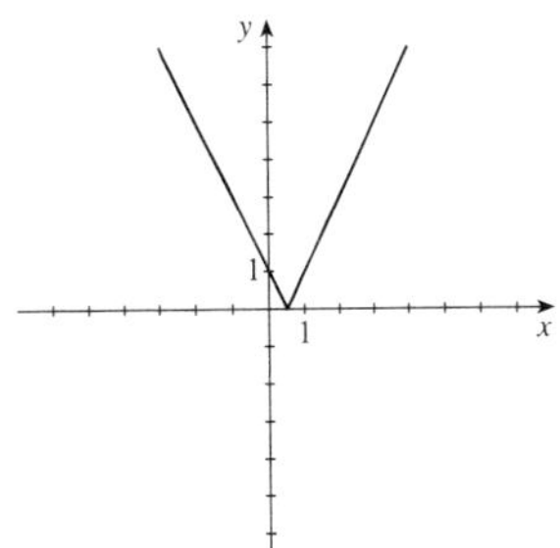

**31.** $f(x) = \begin{cases} x + 6 & \text{if } x < -2 \\ x^2 & \text{if } x \geq -2 \end{cases}$

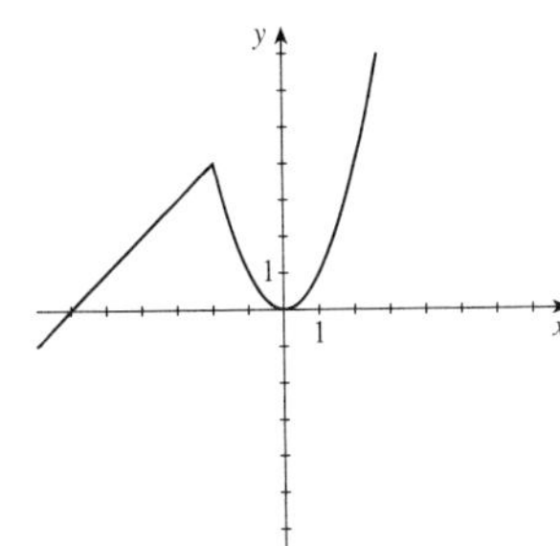

**32.** $f(x) = \begin{cases} -x & \text{if } x < 0 \\ x^2 & \text{if } 0 \leq x < 2 \\ 1 & \text{if } x \geq 2 \end{cases}$

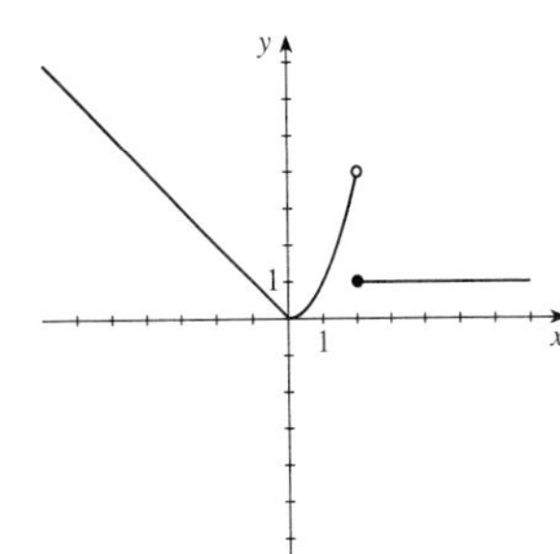

**33.** $f(x) = 6x^3 - 15x^2 + 4x - 1$

**(i)** $[-2, 2]$ by $[-2, 2]$

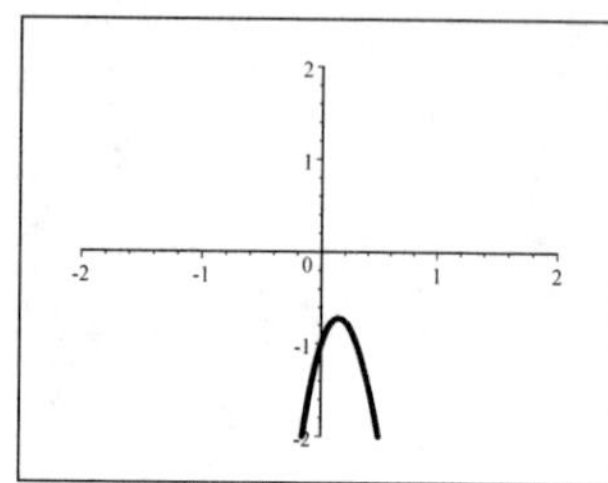

**(ii)** $[-8, 8]$ by $[-8, 8]$

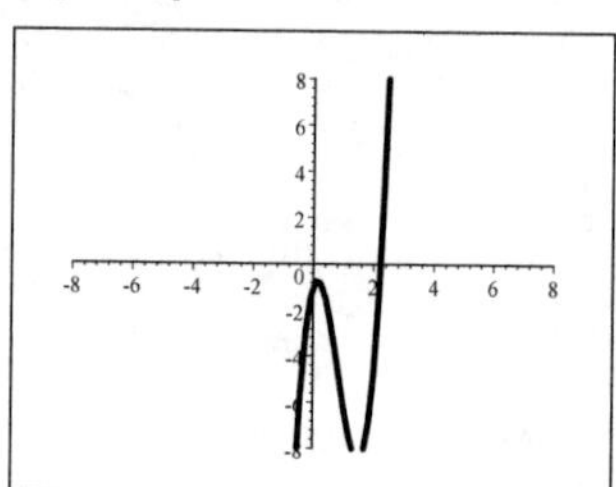

**(iii)** $[-4, 4]$ by $[-12, 12]$

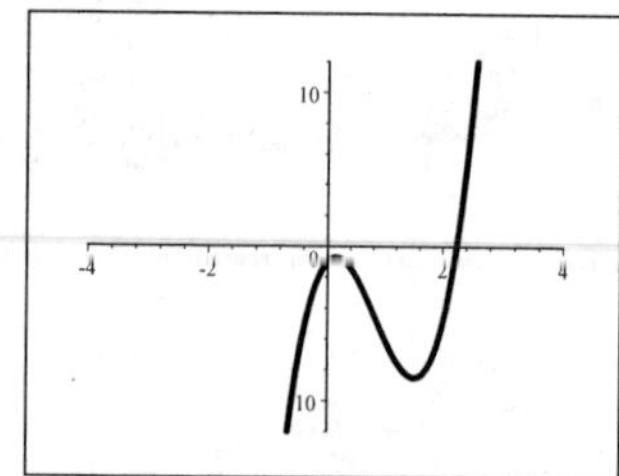

**(iv)** $[-100, 100]$ by $[-100, 100]$

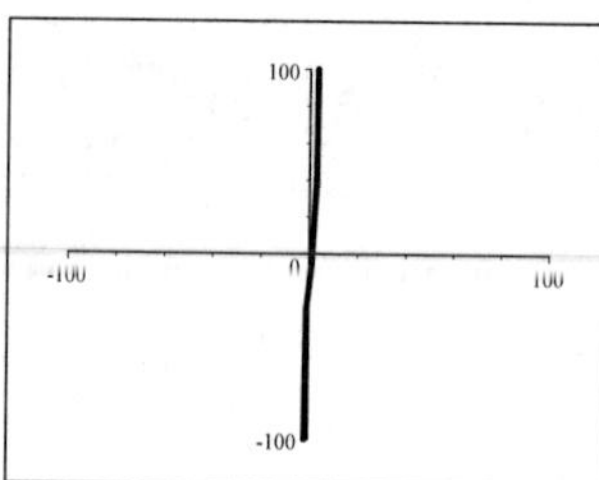

From the graphs, we see that the viewing rectangle in (iii) produces the most appropriate graph.

**34.** $f(x) = \sqrt{100 - x^3}$

**(i)** $[-4, 4]$ by $[-4, 4]$

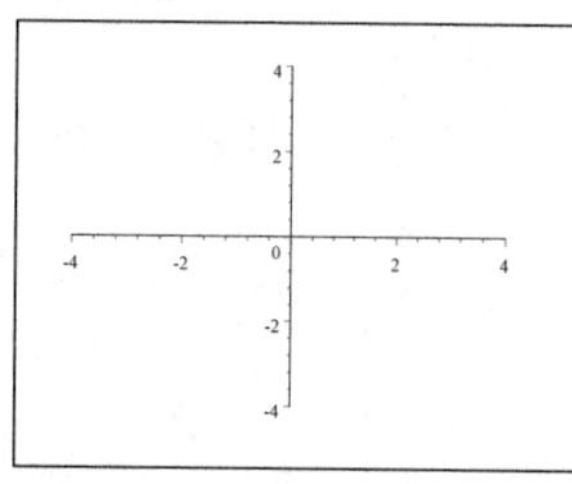

**(ii)** $[-10, 10]$ by $[-10, 10]$

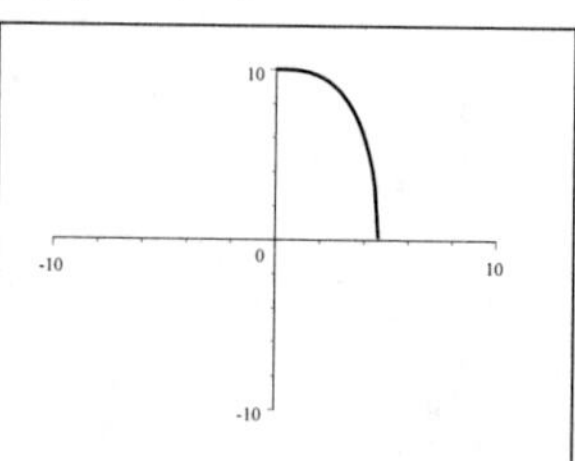

**(iii)** $[-10, 10]$ by $[-10, 40]$

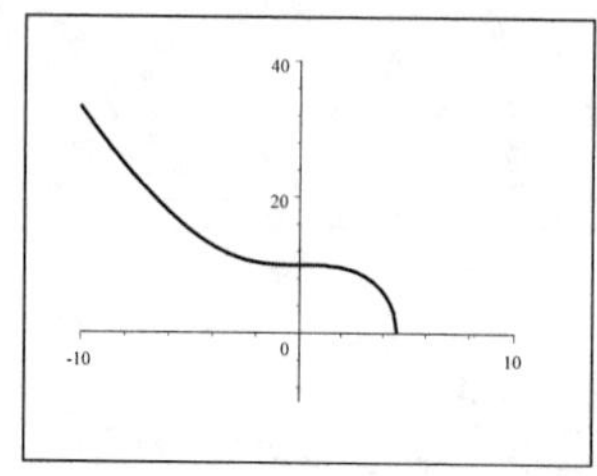

**(iv)** $[-100, 100]$ by $[-100, 100]$

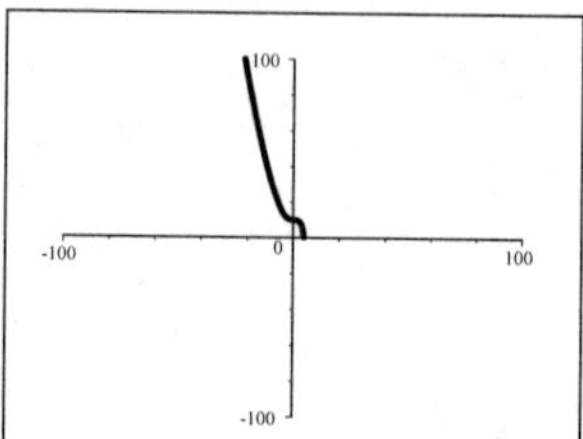

From the graphs, we see that the viewing rectangle in (iii) produces the most appropriate graph of $f$.

**35.** $f(x) = x^2 + 25x + 173$
$= \left(x^2 + 25x + \frac{625}{4}\right) + 173 - \frac{625}{4}$
$= \left(x + \frac{25}{2}\right)^2 + \frac{67}{4}$

We use the viewing rectangle $[-30, 5]$ by $[-20, 250]$.

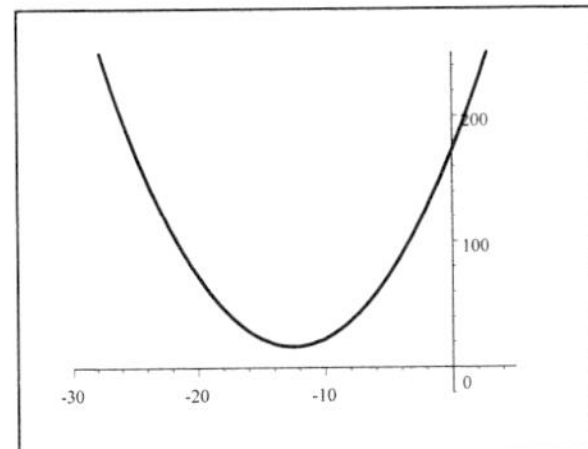


**36.** $f(x) = 1.1x^3 - 9.6x^2 - 1.4x + 3.2$. Here we experiment to find an appropriate viewing rectangle. The viewing rectangle shown is $[-10, 10]$ by $[-125, 10]$.

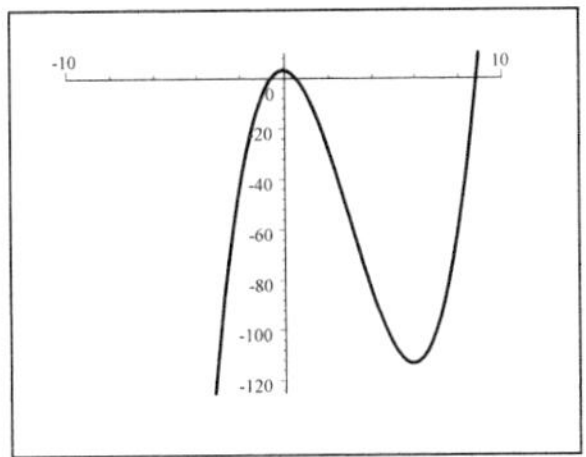


**37.** $f(x) = \dfrac{x}{\sqrt{x^2 + 16}}$. Since $\sqrt{x^2 + 16} \geq \sqrt{x^2} = |x|$, it follows that $y$ should behave like $\dfrac{x}{|x|}$. Thus we use the viewing rectangle $[-20, 20]$ by $[-2, 2]$.

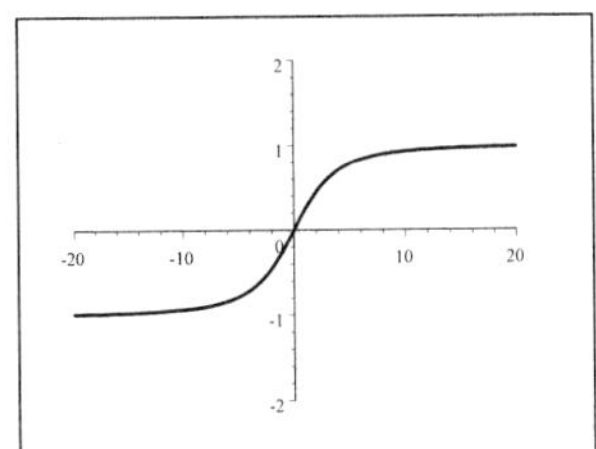


**38.** $f(x) = |x(x + 2)(x + 4)|$. Here, the interesting parts of the function occur near $y = 0$. Thus we choose a viewing rectangle that includes $x = -4, -2$, and $0$, where $f(x)$ is close to $0$. So we choose the viewing rectangle $[-5, 5]$ by $[-1, 15]$.

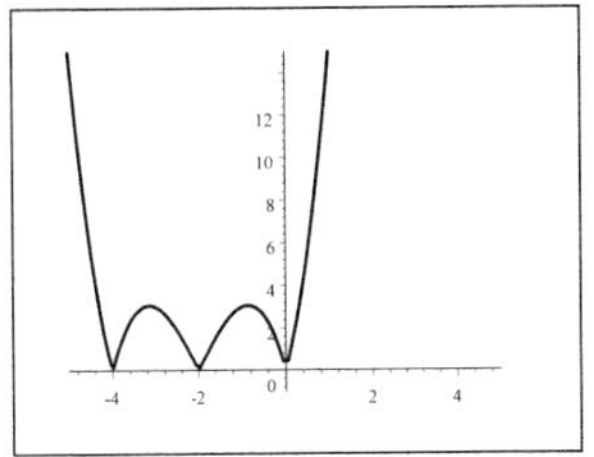


**39.** $f(x) = \sqrt{x^3 - 4x + 1}$. The domain consists of all $x$ where $x^3 - 4x + 1 \geq 0$. Using a graphing device, we see that the domain is approximately $[-2.1, 0.2] \cup [1.9, \infty)$.

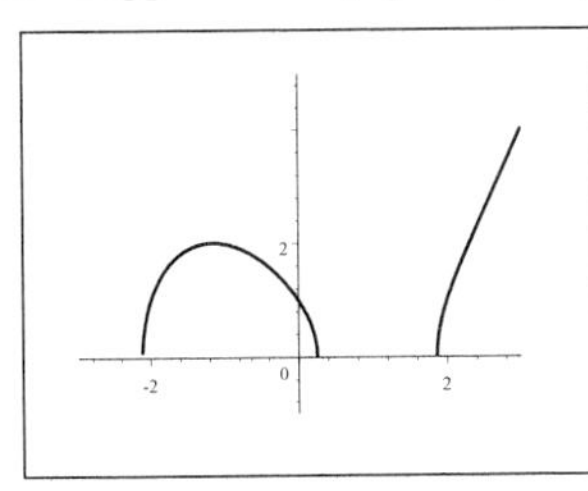


**40.** We find that the range of $f(x) = x^4 - x^3 + x^2 + 3x - 6$ is about $[-7.10, \infty)$.

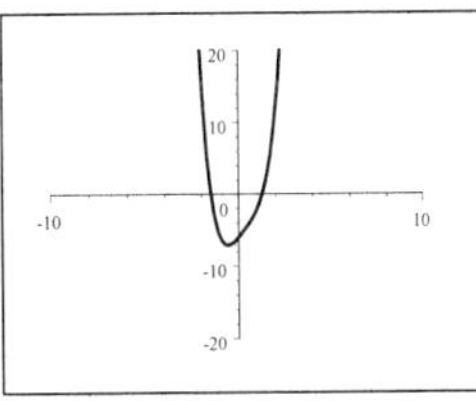


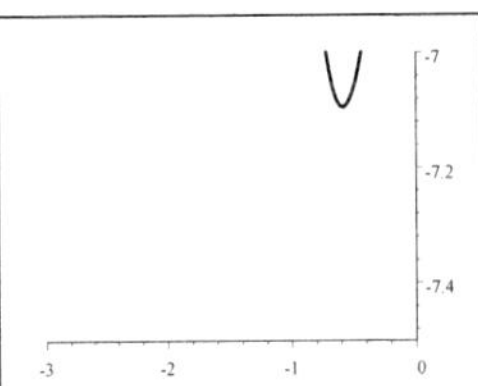


**41.** The average rate of change is $\dfrac{f(2) - f(0)}{2 - 0} = \dfrac{\left[(2)^2 + 3(2)\right] - \left[0^2 + 3(0)\right]}{2} = \dfrac{4 + 6 - 0}{2} = 5.$

**42.** The average rate of change is $\dfrac{f(8) - f(4)}{8 - 4} = \dfrac{\frac{1}{8-2} - \frac{1}{4-2}}{4} = \dfrac{\frac{1}{6} - \frac{1}{2}}{4} \cdot \dfrac{6}{6} = \dfrac{1 - 3}{24} = -\dfrac{1}{12}.$

**43.** The average rate of change is $\dfrac{f(3 + h) - f(3)}{(3 + h) - 3} = \dfrac{\dfrac{1}{3 + h} - \dfrac{1}{3}}{h} \cdot \dfrac{3(3 + h)}{3(3 + h)} = \dfrac{3 - (3 + h)}{3h(3 + h)} = \dfrac{-h}{3h(3 + h)} = -\dfrac{1}{3(3 + h)}.$

**44.** The average rate of change is

$$\frac{f(a+h)-f(a)}{(a+h)-a} = \frac{(a+h+1)^2-(a+1)^2}{h} = \frac{a^2+2ah+h^2+2a+2h+1-a^2-2a-1}{h}$$
$$= \frac{2ah+h^2+2h}{h} = \frac{h(2a+h+2)}{h} = 2a+h+2.$$

**45.** $f(x) = x^3 - 4x^2$ is graphed in the viewing rectangle $[-5, 5]$ by $[-20, 10]$. $f(x)$ is increasing on $(-\infty, 0]$ and $[2.67, \infty)$. It is decreasing on $[0, 2.67]$.

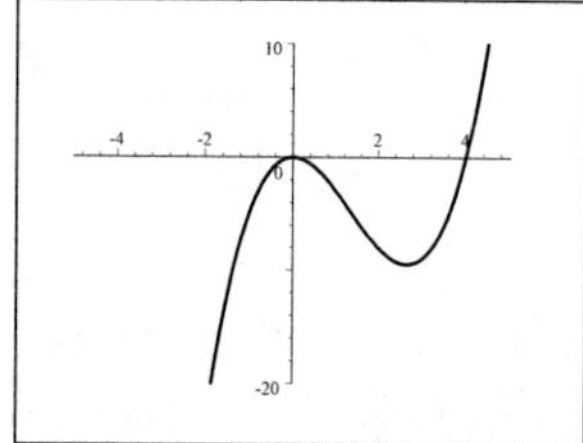


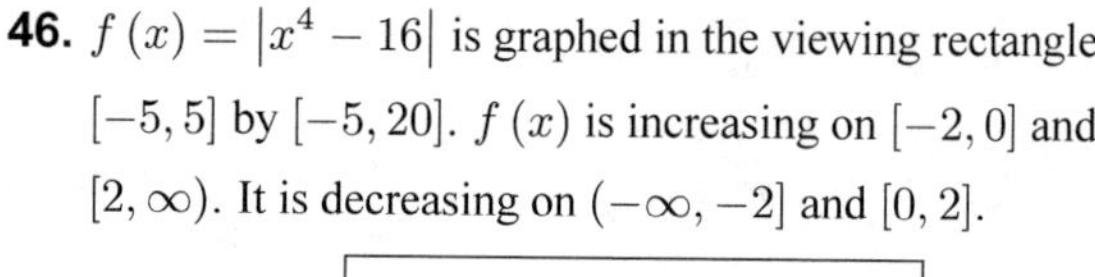

**46.** $f(x) = |x^4 - 16|$ is graphed in the viewing rectangle $[-5, 5]$ by $[-5, 20]$. $f(x)$ is increasing on $[-2, 0]$ and $[2, \infty)$. It is decreasing on $(-\infty, -2]$ and $[0, 2]$.

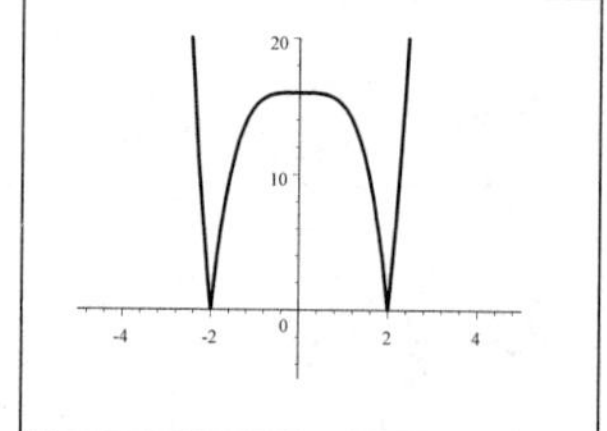


**47.** **(a)** $y = f(x) + 8$. Shift the graph of $f(x)$ upward 8 units.

**(b)** $y = f(x+8)$. Shift the graph of $f(x)$ to the left 8 units.

**(c)** $y = 1 + 2f(x)$. Stretch the graph of $f(x)$ vertically by a factor of 2, then shift it upward 1 unit.

**(d)** $y = f(x-2) - 2$. Shift the graph of $f(x)$ to the right 2 units, then downward 2 units.

**(e)** $y = f(-x)$. Reflect the graph of $f(x)$ about the $y$-axis.

**(f)** $y = -f(-x)$. Reflect the graph of $f(x)$ first about the $y$-axis, then reflect about the $x$-axis.

**(g)** $y = -f(x)$. Reflect the graph of $f(x)$ about the $x$-axis.

**(h)** $y = f^{-1}(x)$. Reflect the graph of $f(x)$ about the line $y = x$.

**48.** **(a)** $y = f(x-2)$

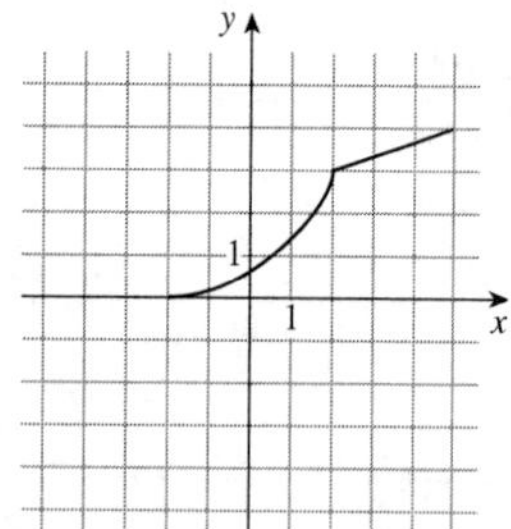


**(b)** $y = -f(x)$

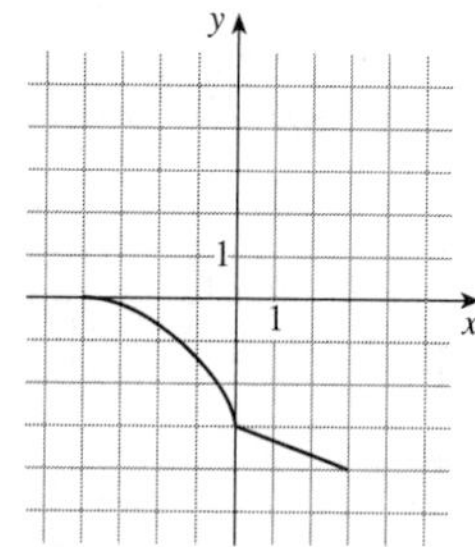


**(c)** $y = 3 - f(x)$

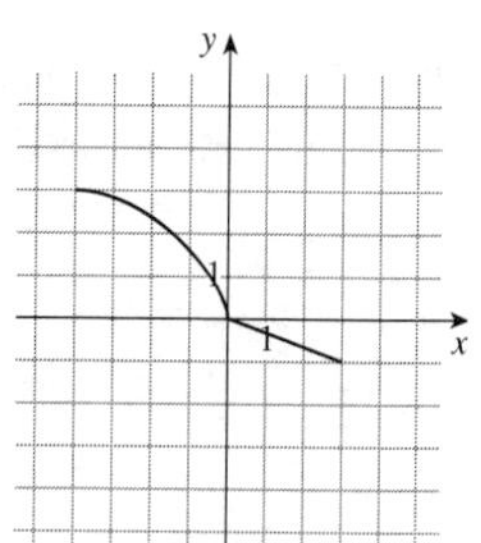


**(d)** $y = \frac{1}{2}f(x) - 1$

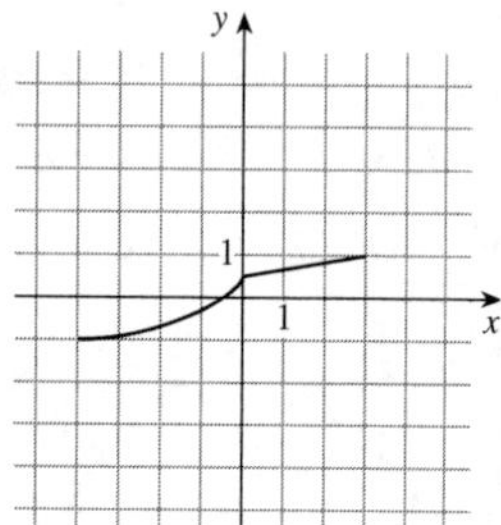


**(e)** $y = f^{-1}(x)$

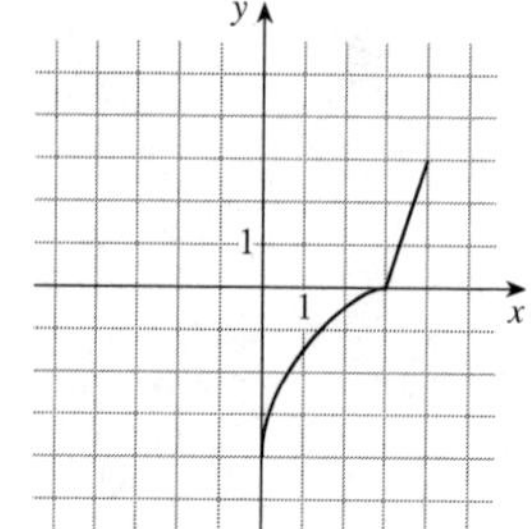


**(f)** $y = f(-x)$

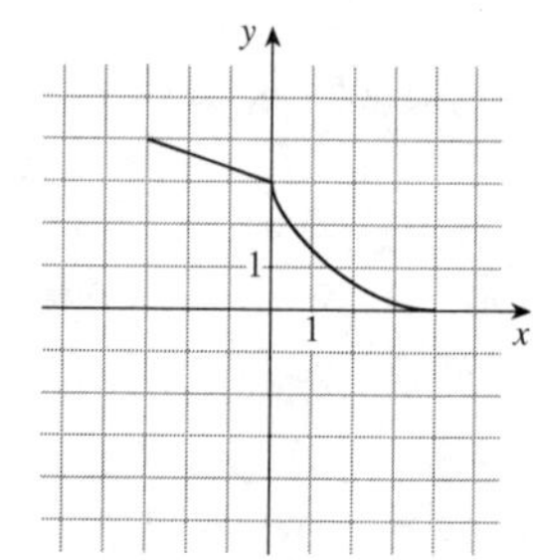

**49. (a)** $f(x) = 2x^5 - 3x^2 + 2$. $f(-x) = 2(-x)^5 - 3(-x)^2 + 2 = -2x^5 - 3x^2 + 2$. Since $f(x) \neq f(-x)$, $f$ is not even. $-f(x) = -2x^5 + 3x^2 - 2$. Since $-f(x) \neq f(-x)$, $f$ is not odd.

**(b)** $f(x) = x^3 - x^7$. $f(-x) = (-x)^3 - (-x)^7 = -(x^3 - x^7) = -f(x)$, hence $f$ is odd.

**(c)** $f(x) = \dfrac{1-x^2}{1+x^2}$. $f(-x) = \dfrac{1-(-x)^2}{1+(-x)^2} = \dfrac{1-x^2}{1+x^2} = f(x)$. Since $f(x) = f(-x)$, $f$ is even.

**(d)** $f(x) = \dfrac{1}{x+2}$. $f(-x) = \dfrac{1}{(-x)+2} = \dfrac{1}{2-x}$. $-f(x) = -\dfrac{1}{x+2}$. Since $f(x) \neq f(-x)$, $f$ is not even, and since $f(-x) \neq -f(x)$, $f$ is not odd.

**50. (a)** This function is odd.

**(b)** This function is neither even nor odd.

**(c)** This function is even.

**(d)** This function is neither even nor odd.

**51.** $f(x) = x^2 + 4x + 1 = (x^2 + 4x + 4) + 1 - 4 = (x+2)^2 - 3$.

**52.** $f(x) = -2x^2 + 12x + 12 = -2(x^2 - 6x) + 12 = -2(x^2 - 6x + 9) + 12 + 18 = -2(x-3)^2 + 30$

**53.** $g(x) = 2x^2 + 4x - 5 = 2(x^2 + 2x) - 5 = 2(x^2 + 2x + 1) - 5 - 2 = 2(x+1)^2 - 7$. So the minimum value is $g(-1) = -7$.

**54.** $f(x) = 1 - x - x^2 = -(x^2 + x) + 1 = -(x^2 + x + \frac{1}{4}) + 1 + \frac{1}{4} = -(x + \frac{1}{2})^2 + \frac{5}{4}$. So the maximum value of $f$ is $\frac{5}{4}$.

**55.** $h(t) = -16t^2 + 48t + 32 = -16(t^2 - 3t) + 32 = -16(t^2 - 3t + \frac{9}{4}) + 32 + 36$
$= -16(t^2 - 3t + \frac{9}{4}) + 68 = -16(t - \frac{3}{2})^2 + 68$

The stone reaches a maximum height of 68 feet.

**56.** $P(x) = -1500 + 12x - 0.0004x^2 = -0.0004(x^2 - 30{,}000x) - 1500 = -0.0004(x^2 - 30{,}000x + 225{,}000{,}000) - 1500 + 90{,}000 = -0.0004(x - 15{,}000)^2 + 88{,}500$

The maximum profit occurs when 15,000 units are sold, and the maximum profit is \$88,500.

**57.** $f(x) = 3.3 + 1.6x - 2.5x^3$. In the first viewing rectangle, $[-2, 2]$ by $[-4, 8]$, we see that $f(x)$ has a local maximum and a local minimum. In the next viewing rectangle, $[0.4, 0.5]$ by $[3.78, 3.80]$, we isolate the local maximum value as approximately 3.79 when $x \approx 0.46$. In the last viewing rectangle, $[-0.5, -0.4]$ by $[2.80, 2.82]$, we isolate the local minimum value as 2.81 when $x \approx -0.46$.

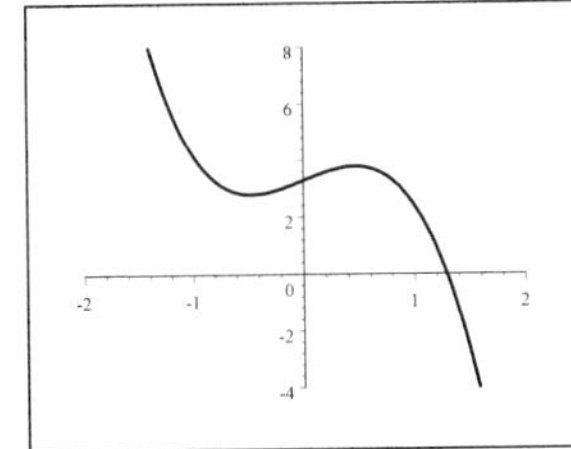


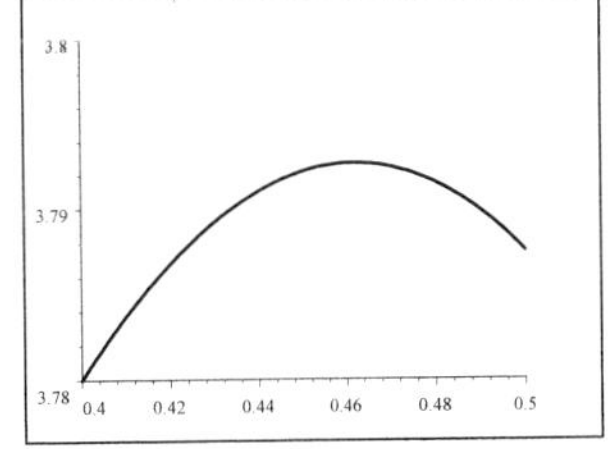


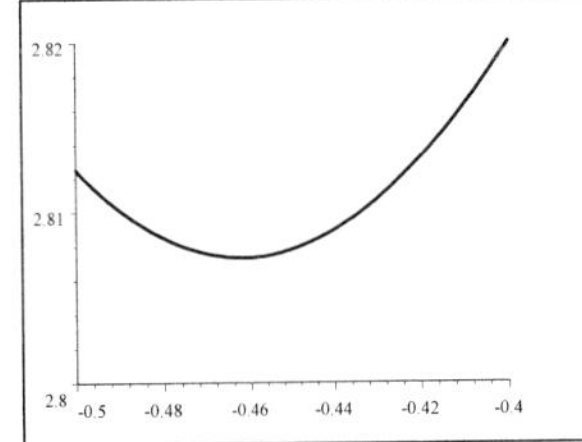


**58.** $f(x) = x^{2/3}(6-x)^{1/3}$. In the first viewing rectangle, $[-10, 10]$ by $[-10, 10]$, we see that $f(x)$ has a local maximum and a local minimum. The local minimum is 0 at $x = 0$ (and is easily verified). In the next viewing rectangle, $[3.95, 4.05]$ by $[3.16, 3.18]$, we isolate the local maximum value as approximately 3.175 when $x \approx 4.00$.

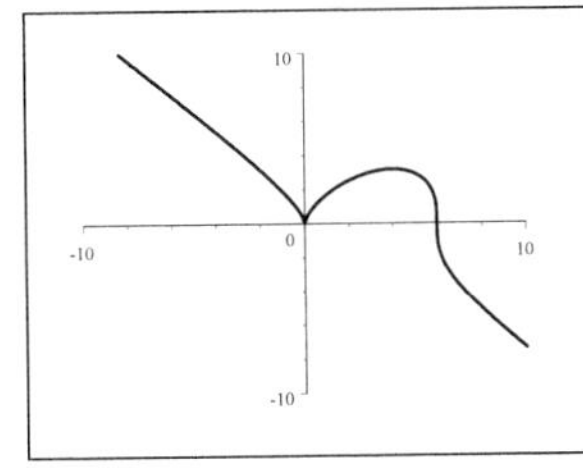


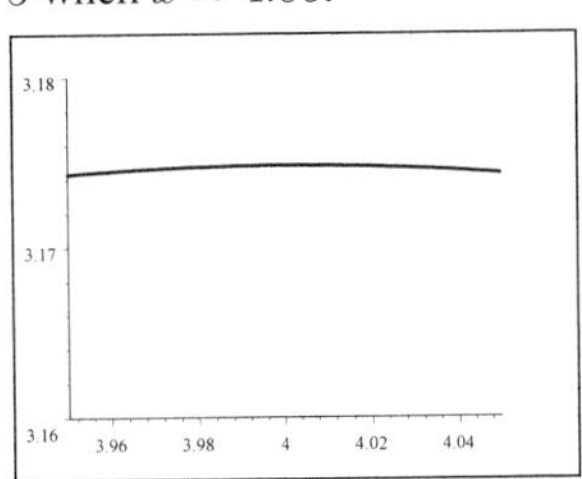

**59.** $f(x) = x^2 - 3x + 2$ and $g(x) = 4 - 3x$.

**(a)** $(f + g)(x) = (x^2 - 3x + 2) + (4 - 3x) = x^2 - 6x + 6$

**(b)** $(f - g)(x) = (x^2 - 3x + 2) - (4 - 3x) = x^2 - 2$

**(c)** $(fg)(x) = (x^2 - 3x + 2)(4 - 3x) = 4x^2 - 12x + 8 - 3x^3 + 9x^2 - 6x = -3x^3 + 13x^2 - 18x + 8$

**(d)** $\left(\frac{f}{g}\right)(x) = \frac{x^2 - 3x + 2}{4 - 3x}, x \neq \frac{4}{3}$

**(e)** $(f \circ g)(x) = f(4 - 3x) = (4 - 3x)^2 - 3(4 - 3x) + 2 = 16 - 24x + 9x^2 - 12 + 9x + 2 = 9x^2 - 15x + 6$

**(f)** $(g \circ f)(x) = g(x^2 - 3x + 2) = 4 - 3(x^2 - 3x + 2) = -3x^2 + 9x - 2$

**60.** $f(x) = 1 + x^2$ and $g(x) = \sqrt{x - 1}$. (Remember that the proper domains must apply.)

**(a)** $(f \circ g)(x) = f(\sqrt{x - 1}) = 1 + (\sqrt{x - 1})^2 = 1 + x - 1 = x$

**(b)** $(g \circ f)(x) = g(1 + x^2) = \sqrt{(1 + x^2) - 1} = \sqrt{x^2} = |x|$

**(c)** $(f \circ g)(2) = f(g(2)) = f\left(\sqrt{(2) - 1}\right) = f(1) = 1 + (1)^2 = 2.$

**(d)** $(f \circ f)(2) = f(f(2)) = f(1 + (2)^2) = f(5) = 1 + (5)^2 = 26.$

**(e)** $(f \circ g \circ f)(x) = f((g \circ f)(x)) = f(|x|) = 1 + (|x|)^2 = 1 + x^2$. Note that $(g \circ f)(x) = |x|$ by part (b).

**(f)** $(g \circ f \circ g)(x) = g((f \circ g)(x)) = g(x) = \sqrt{x - 1}$. Note that $(f \circ g)(x) = x$ by part (a).

**61.** $f(x) = 3x - 1$ and $g(x) = 2x - x^2$.

$(f \circ g)(x) = f(2x - x^2) = 3(2x - x^2) - 1 = -3x^2 + 6x - 1$, and the domain is $(-\infty, \infty)$.

$(g \circ f)(x) = g(3x - 1) = 2(3x - 1) - (3x - 1)^2 = 6x - 2 - 9x^2 + 6x - 1 = -9x^2 + 12x - 3$ , and the domain is $(-\infty, \infty)$

$(f \circ f)(x) = f(3x - 1) = 3(3x - 1) - 1 = 9x - 4$, and the domain is $(-\infty, \infty)$.

$(g \circ g)(x) = g(2x - x^2) = 2(2x - x^2) - (2x - x^2)^2 = 4x - 2x^2 - 4x^2 + 4x^3 - x^4 = -x^4 + 4x^3 - 6x^2 + 4x$, and domain is $(-\infty, \infty)$.

**62.** $f(x) = \sqrt{x}$, has domain $\{x \mid x \geq 0\}$. $g(x) = \frac{2}{x - 4}$, has domain $\{x \mid x \neq 4\}$.

$(f \circ g)(x) = f\left(\frac{2}{x - 4}\right) = \sqrt{\frac{2}{x - 4}}$. $(f \circ g)(x)$ is defined whenever both $g(x)$ and $f(g(x))$ are defined; that is, whenever $x \neq 4$ and $\frac{2}{x - 4} \geq 0$. Now $\frac{2}{x - 4} \geq 0 \quad \Leftrightarrow \quad x - 4 > 0 \quad \Leftrightarrow \quad x > 4$. So the domain of $f \circ g$ is $(4, \infty)$.

$(g \circ f)(x) = g(\sqrt{x}) = \frac{2}{\sqrt{x} - 4}$. $(g \circ f)(x)$ is defined whenever both $f(x)$ and $g(f(x))$ are defined; that is, whenever $x \geq 0$ and $\sqrt{x} - 4 \neq 0$. Now $\sqrt{x} - 4 \neq 0 \quad \Leftrightarrow \quad x \neq 16$. So the domain of $g \circ f$ is $[0, 16) \cup (16, \infty)$.

$(f \circ f)(x) = f(\sqrt{x}) = \sqrt{\sqrt{x}} = x^{1/4}$. $(f \circ f)(x)$ is defined whenever both $f(x)$ and $f(f(x))$ are defined; that is, whenever $x \geq 0$. So the domain of $f \circ f$ is $[0, \infty)$.

$(g \circ g)(x) = g\left(\frac{2}{x - 4}\right) = \frac{2}{\frac{2}{x - 4} - 4} = \frac{2(x - 4)}{2 - 4(x - 4)} = \frac{x - 4}{9 - 2x}$. $(g \circ g)(x)$ is defined whenever both $g(x)$ and $g(g(x))$ are defined; that is, whenever $x \neq 4$ and $9 - 2x \neq 0$. Now $9 - 2x \neq 0 \quad \Leftrightarrow \quad 2x \neq 9 \quad \Leftrightarrow \quad x \neq \frac{9}{2}$. So the domain of $g \circ g$ is $\{x \mid x \neq \frac{9}{2}, 4\}$.

**63.** $f(x) = \sqrt{1 - x}$, $g(x) = 1 - x^2$ and $h(x) = 1 + \sqrt{x}$.

$$(f \circ g \circ h)(x) = f(g(h(x))) = f(g(1 + \sqrt{x})) = f\left(1 - (1 + \sqrt{x})^2\right) = f(1 - (1 + 2\sqrt{x} + x))$$
$$= f(-x - 2\sqrt{x}) = \sqrt{1 - (-x - 2\sqrt{x})} = \sqrt{1 + 2\sqrt{x} + x} = \sqrt{(1 + \sqrt{x})^2} = 1 + \sqrt{x}$$

**64.** If $h(x) = \sqrt{x}$ and $g(x) = 1 + x$ , then $(g \circ h)(x) = g(\sqrt{x}) = 1 + \sqrt{x}$. If $f(x) = \dfrac{1}{\sqrt{x}}$, then

$$(f \circ g \circ h)(x) = f(1 + \sqrt{x}) = \frac{1}{\sqrt{1 + \sqrt{x}}} = T(x).$$

**65.** $f(x) = 3 + x^3$. If $x_1 \neq x_2$, then $x_1^3 \neq x_2^3$ (unequal numbers have unequal cubes), and therefore $3 + x_1^3 \neq 3 + x_2^3$. Thus $f$ is a one-to-one function.

**66.** $g(x) = 2 - 2x + x^2 = (x^2 - 2x + 1) + 1 = (x-1)^2 + 1$. Since $g(0) = 2 = g(2)$ , as is true for all pairs of numbers equidistant from 1, $g$ is not a one-to-one function.

**67.** $h(x) = \dfrac{1}{x^4}$. Since the fourth powers of a number and its negative are equal, $h$ is not one-to-one. For example, $h(-1) = \dfrac{1}{(-1)^4} = 1$ and $h(1) = \dfrac{1}{(1)^4} = 1$, so $h(-1) = h(1)$.

**68.** $r(x) = 2 + \sqrt{x+3}$. If $x_1 \neq x_2$, then $x_1 + 3 \neq x_2 + 3$, so $\sqrt{x_1 + 3} \neq \sqrt{x_2 + 3}$ and $2 + \sqrt{x_1 + 3} \neq 2 + \sqrt{x_2 + 3}$. Thus $r$ is one-to-one.

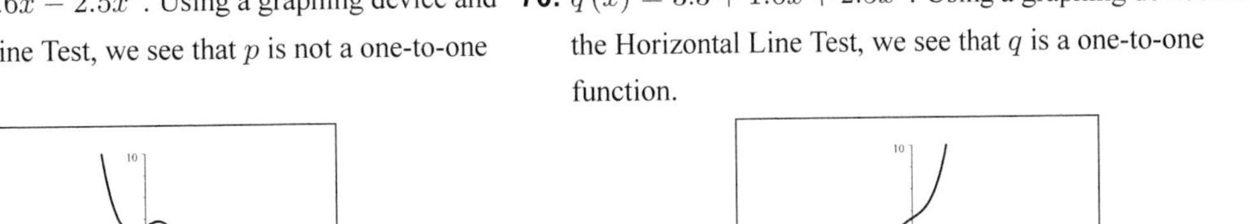

**69.** $p(x) = 3.3 + 1.6x - 2.5x^3$. Using a graphing device and the Horizontal Line Test, we see that $p$ is not a one-to-one function.

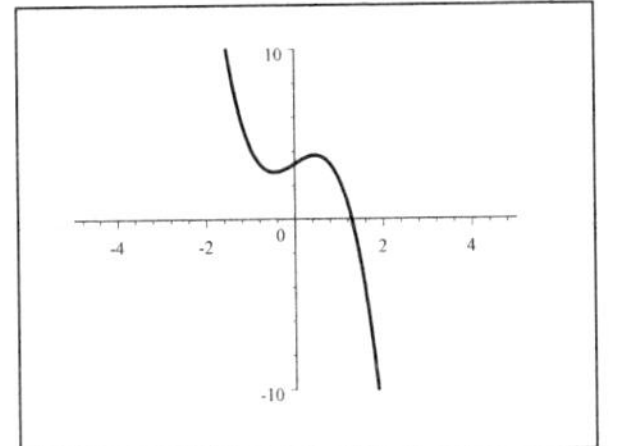


**70.** $q(x) = 3.3 + 1.6x + 2.5x^3$. Using a graphing device and the Horizontal Line Test, we see that $q$ is a one-to-one function.

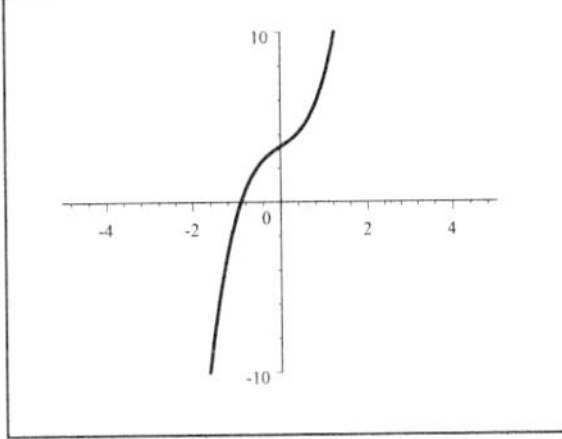


**71.** $f(x) = 3x - 2 \quad \Leftrightarrow \quad y = 3x - 2 \quad \Leftrightarrow \quad 3x = y + 2 \quad \Leftrightarrow \quad x = \frac{1}{3}(y+2)$. So $f^{-1}(x) = \frac{1}{3}(x+2)$.

**72.** $f(x) = \dfrac{2x+1}{3}$. $y = \dfrac{2x+1}{3} \Leftrightarrow 2x + 1 = 3y \quad \Leftrightarrow \quad 2x = 3y - 1 \quad \Leftrightarrow \quad x = \frac{1}{2}(3y-1)$. So $f^{-1}(x) = \frac{1}{2}(3x-1)$.

**73.** $f(x) = (x+1)^3 \quad \Leftrightarrow \quad y = (x+1)^3 \quad \Leftrightarrow \quad x + 1 = \sqrt[3]{y} \quad \Leftrightarrow \quad x = \sqrt[3]{y} - 1$. So $f^{-1}(x) = \sqrt[3]{x} - 1$.

**74.** $f(x) = 1 + \sqrt[5]{x-2}$. $y = 1 + \sqrt[5]{x-2} \quad \Leftrightarrow \quad y - 1 = \sqrt[5]{x-2} \quad \Leftrightarrow \quad x - 2 = (y-1)^5 \quad \Leftrightarrow \quad x = 2 + (y-1)^5$. So $f^{-1}(x) = 2 + (x-1)^5$.

**75. (a), (b)** $f(x) = x^2 - 4, x \geq 0$

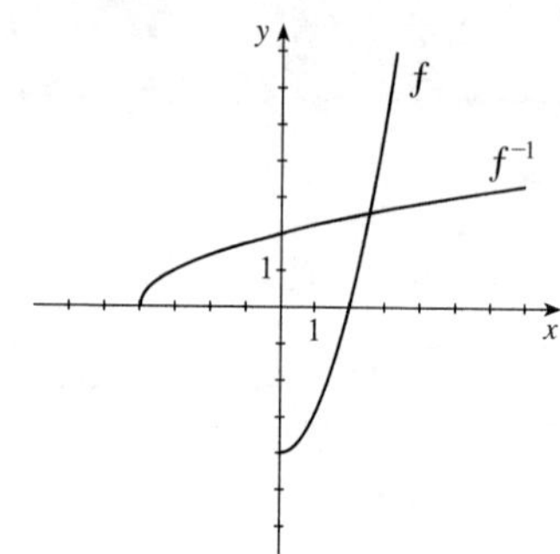


**(c)** $f(x) = x^2 - 4, x \geq 0 \quad \Leftrightarrow \quad y = x^2 - 4,$ $y \geq -4 \quad \Leftrightarrow \quad x^2 = y + 4 \quad \Leftrightarrow$ $x = \sqrt{y+4}$. So $f^{-1}(x) = \sqrt{x+4}$, $x \geq -4$.

**76.** $f(x) = 1 + \sqrt[4]{x}$

**(a)** If $x_1 \neq x_2$, then $\sqrt[4]{x_1} \neq \sqrt[4]{x_2}$, and so $1 + \sqrt[4]{x_1} \neq 1 + \sqrt[4]{x_2}$. Therefore, $f$ is a one-to-one function.

**(b), (c)**

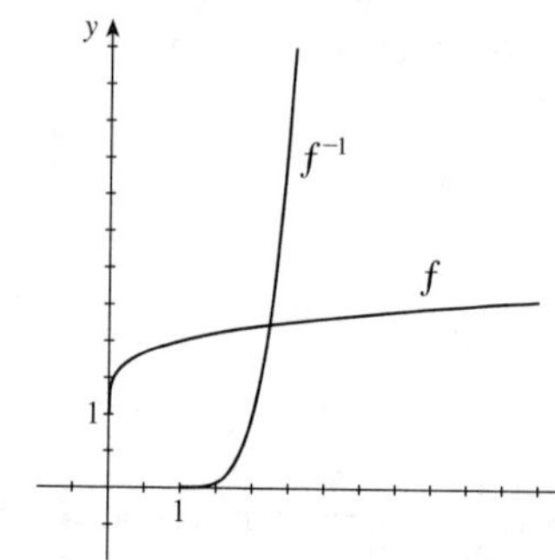


**(d)** $f(x) = 1 + \sqrt[4]{x}$. $y = 1 + \sqrt[4]{x} \quad \Leftrightarrow$ $\sqrt[4]{x} = y - 1 \quad \Leftrightarrow \quad x = (y-1)^4$. So $f^{-1}(x) = (x-1)^4$, $x \geq 1$. Note that the domain of $f$ is $[0, \infty)$, so $y = 1 + \sqrt[4]{x} \geq 1$. Hence, the domain of $f^{-1}$ is $[1, \infty)$.

# Chapter 3 Test

**1.** By the Vertical Line Test, figures (a) and (b) are graphs of functions. By the Horizontal Line Test, only figure (a) is the graph of a one-to-one function.

**2. (a)** $f(3) = \dfrac{\sqrt{3+1}}{3} = \dfrac{\sqrt{4}}{3} = \frac{2}{3}$; $f(5) = \dfrac{\sqrt{5+1}}{5} = \dfrac{\sqrt{6}}{5}$; $f(a - +1) = \dfrac{\sqrt{(a-1)+1}}{a-1} = \dfrac{\sqrt{a}}{a-1}$.

**(b)** $f(x) = \dfrac{\sqrt{x+1}}{x}$. Our restrictions are that the input to the radical is nonnegative, and the denominator must not be equal to zero. Thus $x + 1 \geq 0 \quad \Leftrightarrow \quad x \geq -1$ and $x \neq 0$. In interval notation, the domain is $[-1, 0) \cup (0, \infty)$.

**3.** The average rate of change is $\dfrac{f(2) - f(5)}{2 - 5} = \dfrac{[2^2 - 2(2)] - [5^2 - 2(5)]}{-3} = \dfrac{4 - 4 - (25 - 10)}{-3} = \dfrac{-15}{-3} = 5$.

**4. (a)** $f(x) = x^3$

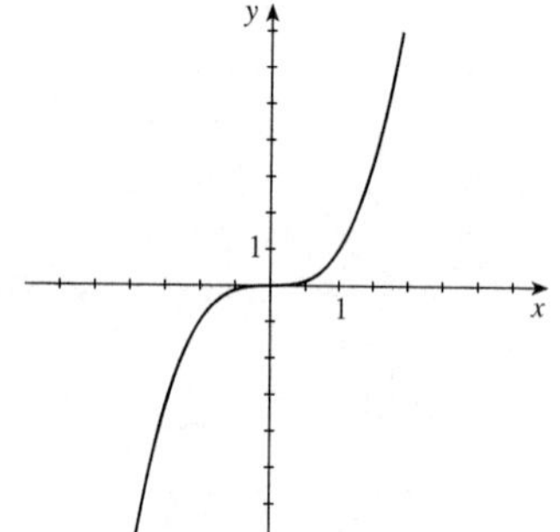


**(b)** $g(x) = (x-1)^3 - 2$. To obtain the graph of $g$, shift the graph of $f$ to the right 1 unit and downward 2 units.

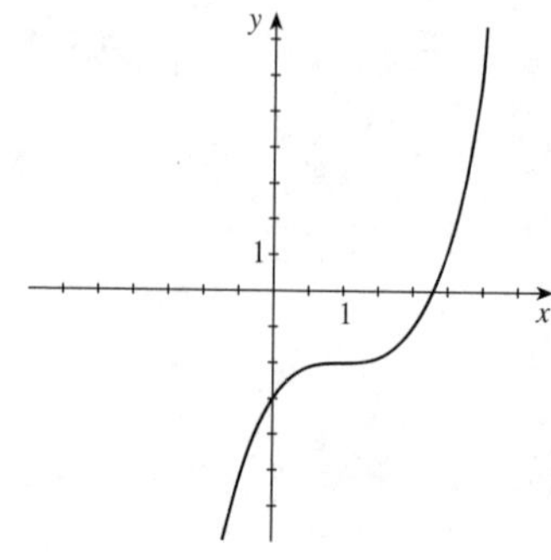

**5. (a)** $y = f(x-3) + 2$. Shift the graph of $f(x)$ to the right 3 units, then shift the graph upward 2 units.

**(b)** $y = f(-x)$. Reflect the graph of $f(x)$ about the $y$-axis.

**6. (a)**
$$\begin{aligned} f(x) &= 2x^2 - 8x + 13 \\ &= 2\left(x^2 - 4x\right) + 13 \\ &= 2\left(x^2 - 4x + 4\right) + 13 - 8 \\ &= 2(x-2)^2 + 5 \end{aligned}$$

**(b)**

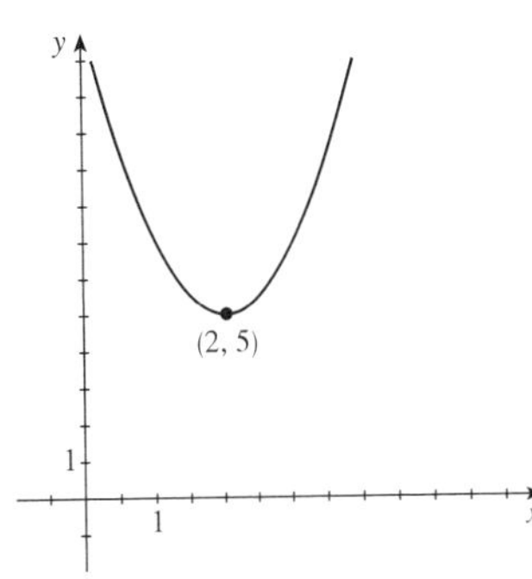


**(c)** Since $f(x) = 2(x-2)^2 + 5$ is in standard form, the minimum value of $f$ is $f(2) = 5$.

**7. (a)** $f(-2) = 1 - (-2)^2 = 1 - 4 = -3$ (since $-2 \leq 0$).
$f(1) = 2(1) + 1 = 2 + 1 = 3$ (since $1 > 0$).

**(b)**

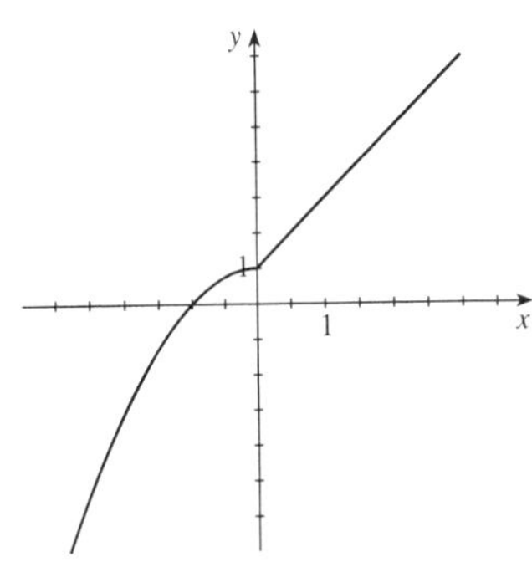


**8.** $f(x) = x^2 + 1$; $g(x) = x - 3$.

**(a)** $(f \circ g)(x) = f(g(x)) = f(x-3) = (x-3)^2 + 1 = x^2 - 6x + 9 + 1 = x^2 - 6x + 10$

**(b)** $(g \circ f)(x) = g(f(x)) = g\left(x^2+1\right) = \left(x^2+1\right) - 3 = x^2 - 2$

**(c)** $f(g(2)) = f(-1) = (-1)^2 + 1 = 2$. (We have used the fact that $g(2) = (2) - 3 = -1$.)

**(d)** $g(f(2)) = g(5) = 5 - 3 = 2$. (We have used the fact that $f(2) = 2^2 + 1 = 5$.)

**(e)** $(g \circ g \circ g)(x) = g(g(g(x))) = g(g(x-3)) = g(x-6) = (x-6) - 3 = x - 9$. (We have used the fact that $g(x-3) = (x-3) - 3 = x - 6$.)

**9. (a)** $f(x) = \sqrt{3-x}$, $x \leq 3$ $\Leftrightarrow$ $y = \sqrt{3-x}$ $\Leftrightarrow$ $y^2 = 3 - x$ $\Leftrightarrow$ $x = 3 - y^2$. Thus $f^{-1}(x) = 3 - x^2$, $x \geq 0$.

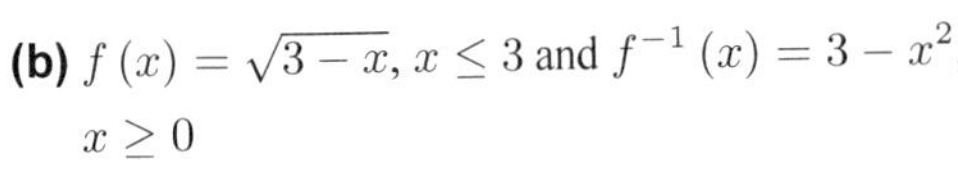
**(b)** $f(x) = \sqrt{3-x}$, $x \leq 3$ and $f^{-1}(x) = 3 - x^2$, $x \geq 0$

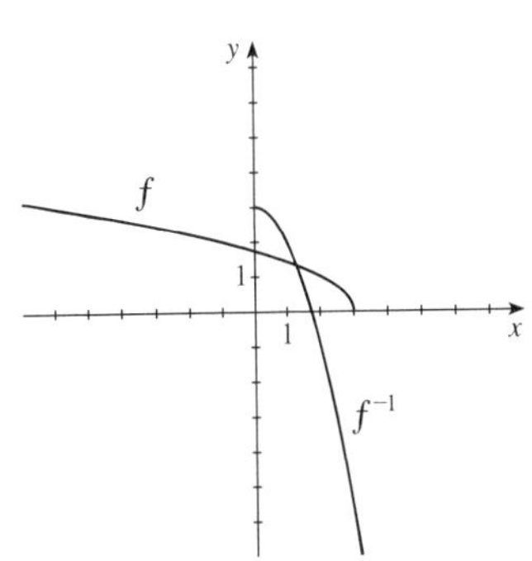

**10. (a)** The domain of $f$ is $[0, 6]$, and the range of $f$ is $[1, 7]$.

**(b)**

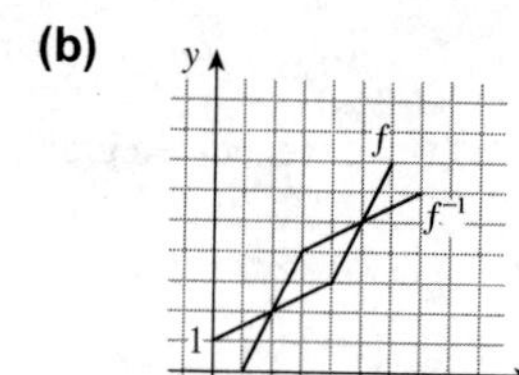


**(c)** The average rate of change is $\dfrac{f(6) - f(2)}{6 - 2} = \dfrac{7 - 2}{4} = \dfrac{5}{4}$.

**11. (a)** $f(x) = 3x^4 - 14x^2 + 5x - 3$. The graph is shown in the viewing rectangle $[-10, 10]$ by $[-30, 10]$.

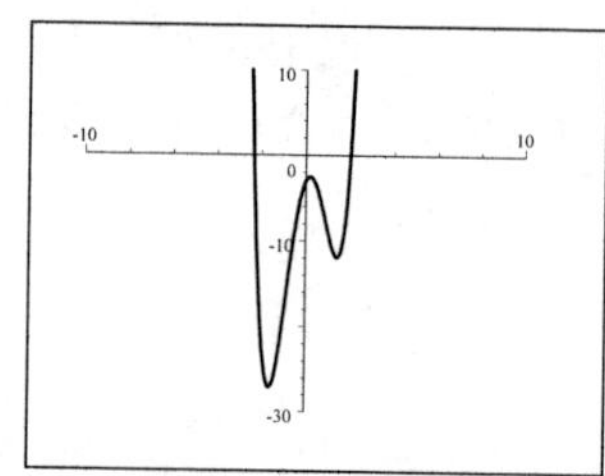


**(b)** No, by the Horizontal Line Test.

**(c)** The local maximum is approximately $-2.55$ when $x \approx 0.18$, as shown in the first viewing rectangle $[0.15, 0.25]$ by $[-2.6, -2.5]$. One local minimum is approximately $-27.18$ when $x \approx -1.61$, as shown in the second viewing rectangle $[-1.65, -1.55]$ by $[-27.5, -27]$. The other local minimum is approximately $-11.93$ when $x \approx 1.43$, as shown is the viewing rectangle $[1.4, 1.5]$ by $[-12, -11.9]$.

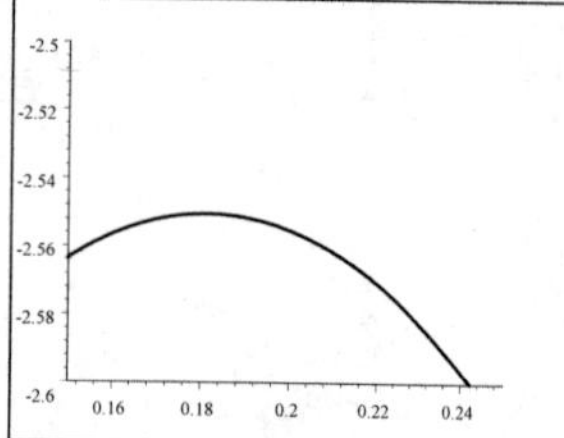


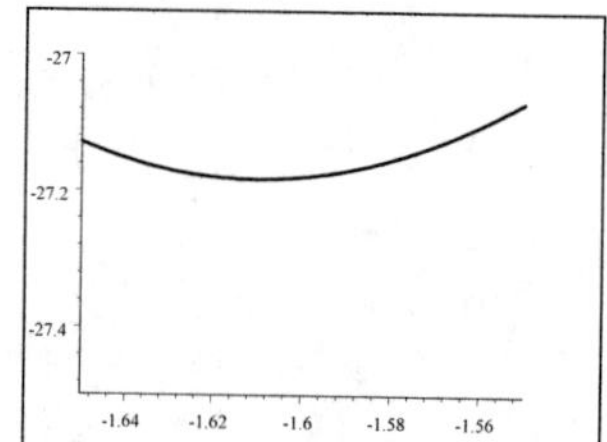


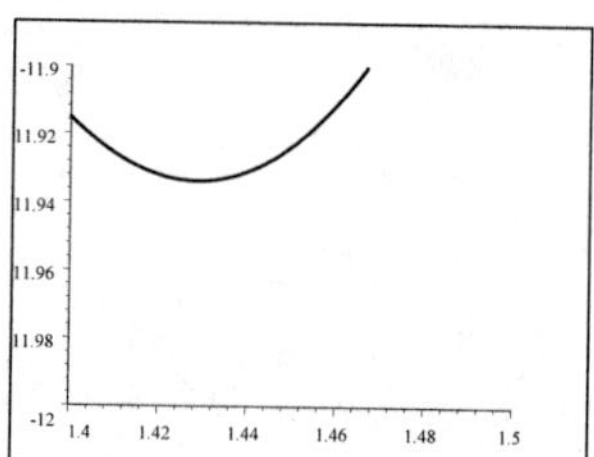


**(d)** Using the graph in part (a) and the local minimum, $-27.18$, found in part (c), we see that the range is $[-27.18, \infty)$.

**(e)** Using the information from part (c) and the graph in part (a), $f(x)$ is increasing on the intervals $[-1.61, 0.18]$ and $[1.43, \infty)$ and decreasing on the intervals $(-\infty, -1.61]$ and $[0.18, 1.43]$.

## Focus on Modeling: Modeling with Functions

**1.** Let $w$ be the width of the building lot. Then the length of the lot is $3w$. So the area of the building lot is $A(w) = 3w^2$, $w > 0$.

**2.** Let $w$ be the width of the poster. Then the length of the poster is $w + 10$. So the area of the poster is $A(w) = w(w + 10) = w^2 + 10w$.

**3.** Let $w$ be the width of the base of the rectangle. Then the height of the rectangle is $\frac{1}{2}w$. Thus the volume of the box is given by the function $V(w) = \frac{1}{2}w^3$, $w > 0$.

**4.** Let $r$ be the radius of the cylinder. Then the height of the cylinder is $4r$. Since for a cylinder $V = \pi r^2 h$, the volume of the cylinder is given by the function $V(r) = \pi r^2 (4r) = 4\pi r^3$.

**5.** Let $P$ be the perimeter of the rectangle and $y$ be the length of the other side. Since $P = 2x + 2y$ and the perimeter is 20, we have $2x + 2y = 20 \quad \Leftrightarrow \quad x + y = 10 \quad \Leftrightarrow \quad y = 10 - x$. Since area is $A = xy$, substituting gives $A(x) = x(10 - x) = 10x - x^2$, and since $A$ must be positive, the domain is $0 < x < 10$.

**6.** Let $A$ be the area and $y$ be the length of the other side. Then $A = xy = 16 \quad \Leftrightarrow \quad y = \dfrac{16}{x}$. Substituting into $P = 2x + 2y$ gives $P = 2x + 2 \cdot \dfrac{16}{x} = 2x + \dfrac{32}{x}$, where $x > 0$.

**7.**

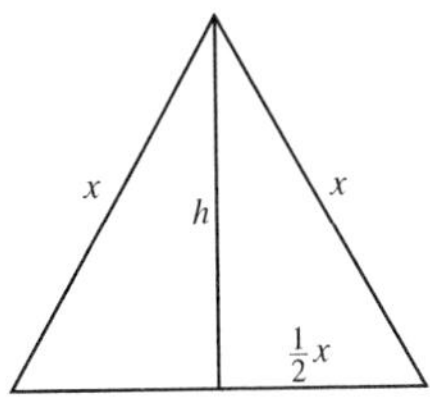


Let $h$ be the height of an altitude of the equilateral triangle whose side has length $x$, as shown in the diagram. Thus the area is given by $A = \frac{1}{2}xh$. By the Pythagorean Theorem, $h^2 + \left(\frac{1}{2}x\right)^2 = x^2 \quad \Leftrightarrow \quad h^2 + \frac{1}{4}x^2 = x^2 \quad \Leftrightarrow$ $h^2 = \frac{3}{4}x^2 \Leftrightarrow h = \frac{\sqrt{3}}{2}x$. Substituting into the area of a triangle, we get $A(x) = \frac{1}{2}xh = \frac{1}{2}x\left(\frac{\sqrt{3}}{2}x\right) = \frac{\sqrt{3}}{4}x^2$, $x > 0$.

**8.** Let $d$ represent the length of any side of a cube. Then the surface area is $S = 6d^2$, and the volume is $V = d^3 \quad \Leftrightarrow$ $d = \sqrt[3]{V}$. Substituting for $d$ gives $S(V) = 6\left(\sqrt[3]{V}\right)^2 = 6V^{2/3}$, $V > 0$.

**9.** We solve for $r$ in the formula for the area of a circle. This gives $A = \pi r^2 \quad \Leftrightarrow \quad r^2 = \dfrac{A}{\pi} \quad \Rightarrow \quad r = \sqrt{\dfrac{A}{\pi}}$, so the model is $r(A) = \sqrt{\dfrac{A}{\pi}}$, $A > 0$.

**10.** Let $r$ be the radius of a circle. Then the area is $A = \pi r^2$, and the circumference is $C = 2\pi r \quad \Leftrightarrow \quad r = \dfrac{C}{2\pi}$. Substituting for $r$ gives $A(C) = \pi\left(\dfrac{C}{2\pi}\right)^2 = \dfrac{C^2}{4\pi}$, $C > 0$.

**11.** Let $h$ be the height of the box in feet. The volume of the box is $V = 60$. Then $x^2 h = 60 \quad \Leftrightarrow \quad h = \dfrac{60}{x^2}$. The surface area, $S$, of the box is the sum of the area of the 4 sides and the area of the base and top. Thus $S = 4xh + 2x^2 = 4x\left(\dfrac{60}{x^2}\right) + 2x^2 = \dfrac{240}{x} + 2x^2$, so the model is $S(x) = \dfrac{240}{x} + 2x^2$, $x > 0$.

**12.** By similar triangles, $\dfrac{5}{L} = \dfrac{12}{L + d} \quad \Leftrightarrow \quad 5(L + d) = 12L \quad \Leftrightarrow \quad 5d = 7L \quad \Leftrightarrow \quad L = \dfrac{5d}{7}$. The model is $L(d) = \frac{5}{7}d$.

**13.** 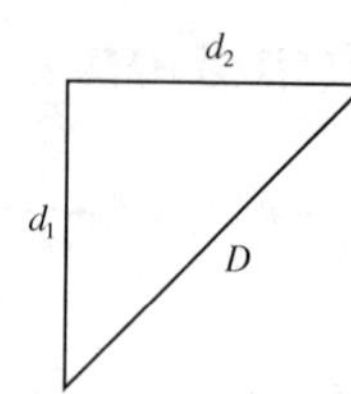


Let $d_1$ be the distance traveled south by the first ship and $d_2$ be the distance traveled east by the second ship. The first ship travels south for $t$ hours at 5 mi/h, so $d_1 = 15t$ and, similarly, $d_2 = 20t$. Since the ships are traveling at right angles to each other, we can apply the Pythagorean Theorem to get $D^2 = d_1^2 + d_2^2 = (15t)^2 + (20t)^2 = 225t^2 + 400t^2 = 625t^2$.

**14.** Let $n$ be one of the numbers. Then the other number is $60 - n$, so the product is given by the function $P(n) = n(60 - n) = 60n - n^2$.

**15.** 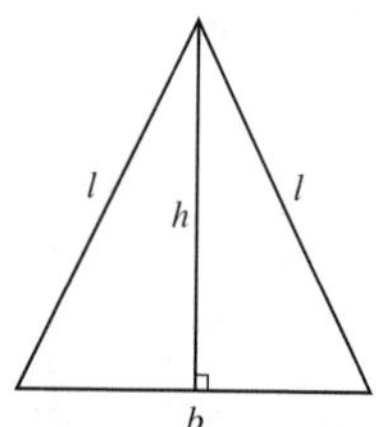


Let $b$ be the length of the base, $l$ be the length of the equal sides, and $h$ be the height in centimeters. Since the perimeter is 8, $2l + b = 8 \quad \Leftrightarrow \quad 2l = 8 - b$ $\Leftrightarrow \quad l = \frac{1}{2}(8 - b)$. By the Pythagorean Theorem, $h^2 + \left(\frac{1}{2}b\right)^2 = l^2 \quad \Leftrightarrow$ $h = \sqrt{l^2 - \frac{1}{4}b^2}$. Therefore the area of the triangle is

$$A = \tfrac{1}{2} \cdot b \cdot h = \tfrac{1}{2} \cdot b\sqrt{l^2 - \tfrac{1}{4}b^2} = \frac{b}{2}\sqrt{\tfrac{1}{4}(8-b)^2 - \tfrac{1}{4}b^2}$$
$$= \frac{b}{4}\sqrt{64 - 16b + b^2 - b^2} = \frac{b}{4}\sqrt{64 - 16b} = \frac{b}{4} \cdot 4\sqrt{4 - b} = b\sqrt{4 - b}$$

so the model is $A(b) = b\sqrt{4 - b}$, $0 < b < 4$.

**16.** Let $x$ be the length of the shorter leg of the right triangle. Then the length of the other triangle is $2x$. Since it is a right triangle, the length of the hypotenuse is $\sqrt{x^2 + (2x)^2} = \sqrt{5x^2} = \sqrt{5}\,x$ (since $x \geq 0$ ). Thus the perimeter of the triangle is $P(x) = x + 2x + \sqrt{5}\,x = \left(3 + \sqrt{5}\right)x$.

**17.** Let $w$ be the length of the rectangle. By the Pythagorean Theorem, $\left(\frac{1}{2}w\right)^2 + h^2 = 10^2 \quad \Leftrightarrow \quad \frac{w^2}{4} + h^2 = 10^2$ $\Leftrightarrow \quad w^2 = 4\left(100 - h^2\right) \quad \Leftrightarrow \quad w = 2\sqrt{100 - h^2}$ (since $w > 0$ ). Therefore, the area of the rectangle is $A = wh = 2h\sqrt{100 - h^2}$, so the model is $A(h) = 2h\sqrt{100 - h^2}$, $0 < h < 10$.

**18.** Using the formula for the volume of a cone, $V = \frac{1}{3}\pi r^2 h$, we substitute $V = 100$ and solve for $h$. Thus $100 = \frac{1}{3}\pi r^2 h$ $\Leftrightarrow \quad h(r) = \dfrac{300}{\pi r^2}$.

**19. (a)** We complete the table.

| First number | Second number | Product |
|---|---|---|
| 1 | 18 | 18 |
| 2 | 17 | 34 |
| 3 | 16 | 48 |
| 4 | 15 | 60 |
| 5 | 14 | 70 |
| 6 | 13 | 78 |
| 7 | 12 | 84 |
| 8 | 11 | 88 |
| 9 | 10 | 90 |
| 10 | 9 | 90 |
| 11 | 8 | 88 |

From the table we conclude that the numbers is still increasing, the numbers whose product is a maximum should both be 9.5.

**(b)** Let $x$ be one number: then $19 - x$ is the other number, and so the product, $p$, is

$p(x) = x(19 - x) = 19x - x^2$.

**(c)** $p(x) = 19x - x^2 = -\left(x^2 - 19x\right)$

$$= -\left[x^2 - 19x + \left(\tfrac{19}{2}\right)^2\right] + \left(\tfrac{19}{2}\right)^2$$
$$= -(x - 9.5)^2 + 90.25$$

So the product is maximized when the numbers are both 9.5.

**20.** Let the positive numbers be $x$ and $y$. Since their sum is 100, we have $x+y=100 \Leftrightarrow y=100-x$. We wish to minimize the sum of squares, which is $S=x^2+y^2=x^2+(100-x)^2$. So $S(x)=x^2+(100-x)^2=x^2+10{,}000-200x+x^2=2x^2-200x+10{,}000=2(x^2-100x)+10{,}000=2(x^2-100x+2500)+10{,}000-5000=2(x-50)^2+5000$. Thus the minimum sum of squares occurs when $x=50$. Then $y=100-50=50$. Therefore both numbers are 50.

**21.** Let $x$ and $y$ be the two numbers. Since their sum is $-24$, we have $x+y=-24 \Leftrightarrow y=-x-24$. The product of the two numbers is $P=xy=x(-x-24)=-x^2-24x$, which we wish to maximize. So $P=-x^2-24x=-(x^2+24x)=-(x^2+24x+144)+144=-(x+12)^2+144$. Thus the maximum product is 144, and it occurs when $x=-12$ and $y=-(-12)-24=-12$. Thus the two numbers are $-12$ and $-12$.

**22.** Let $w$ and $l$ be the width and the length of the rectangle in feet. We want all rectangles with perimeter equal to 20, so we have $2w+2l=20 \Leftrightarrow l=10-w$. The area of a rectangle is given by $A(w)=l\cdot w=(10-w)w=10w-w^2=-(w^2-10w)=-(w^2-10w+25)+25=-(w-5)^2+25$. So the area is maximized when $w=5$, and hence the largest rectangle is a square where the dimension of each side is 5 feet.

**23. (a)** Let $x$ be the width of the field (in feet) and $l$ be the length of the field (in feet). Since the farmer has 2400 ft of fencing we must have $2x+l=2400$.

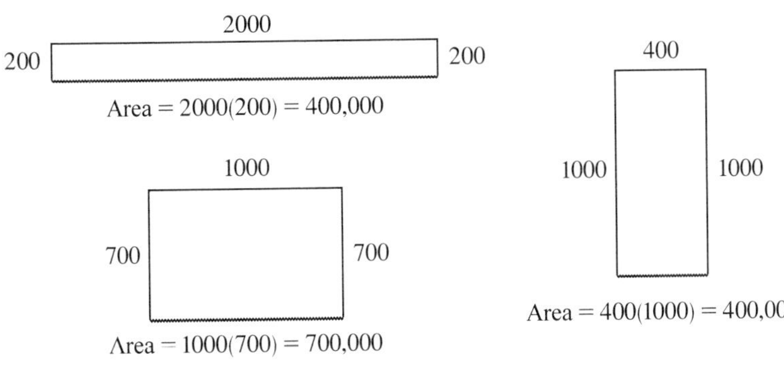


| Width | Length | Area |
|---|---|---|
| 200 | 2000 | 400,000 |
| 300 | 1800 | 540,000 |
| 400 | 1600 | 640,000 |
| 500 | 1400 | 700,000 |
| 600 | 1200 | 720,000 |
| 700 | 1000 | 700,000 |
| 800 | 800 | 640,000 |

It appears that the field of largest area is about 600 ft $\times$ 1200 ft.

**(b)** Let $x$ be the width of the field (in feet) and $l$ be the length of the field (in feet). Since the farmer has 2400 ft of fencing we must have $2x+l=2400 \Leftrightarrow l=2400-2x$. The area of the fenced-in field is given by $A(x)=l\cdot x=(2400-2x)x=-2x^2+2400x=-2(x^2-1200x)$.

**(c)** The area is $A(x)=-2(x^2-1200x+600^2)+2(600^2)=-2(x-600)^2+720000$. So the maximum area occurs when $x=600$ feet and $l=2400-2(600)=1200$ feet.

**24. (a)** Let $w$ be the width of the rectangular area (in feet) and $l$ be the length of the field (in feet). Since the farmer has 750 feet of fencing, we must have $5w+2l=750 \Leftrightarrow 2l=750-5w \Leftrightarrow l=\frac{5}{2}(150-w)$. Thus the total area of the four pens is $A(w)=l\cdot w=\frac{5}{2}w(150-w)=-\frac{5}{2}(w^2-150w)$.

**(b)** We complete the square to get $A(w)=-\frac{5}{2}(w^2-150w)=-\frac{5}{2}(w^2-150w+75^2)+(\frac{5}{2})\cdot 75^2=-\frac{5}{2}(w-75)^2+14062.5$. Therefore, the largest possible total area of the four pens is 14,062.5 square feet.

**25. (a)** Let $x$ be the length of the fence along the road. If the area is 1200, we have $1200 = x\cdot$ width, so the width of the garden is $\dfrac{1200}{x}$. Then the cost of the fence is given by the function $C(x) = 5(x) + 3\left[x + 2\cdot\dfrac{1200}{x}\right] = 8x + \dfrac{7200}{x}$.

**(b)** We graph the function $y = C(x)$ in the viewing rectangle $[0, 75] \times [0, 800]$. From this we get the cost is minimized when $x = 30$ ft. Then the width is $\frac{1200}{30} = 40$ ft. So the length is 30 ft and the width is 40 ft.

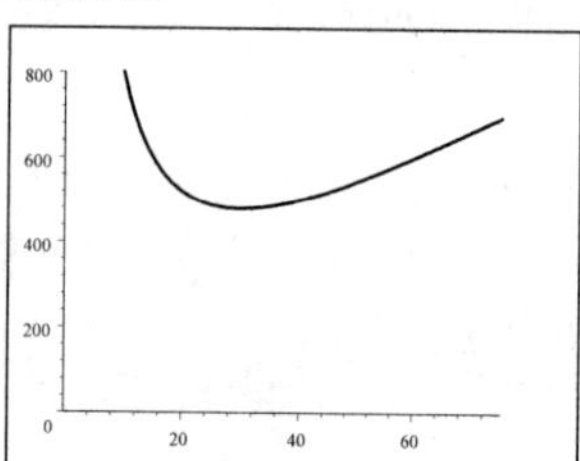


**(c)** We graph the function $y = C(x)$ and $y = 600$ in the viewing rectangle $[10, 65] \times [450, 650]$. From this we get that the cost is at most \$600 when $15 \le x \le 60$. So the range of lengths he can fence along the road is 15 feet to 60 feet.

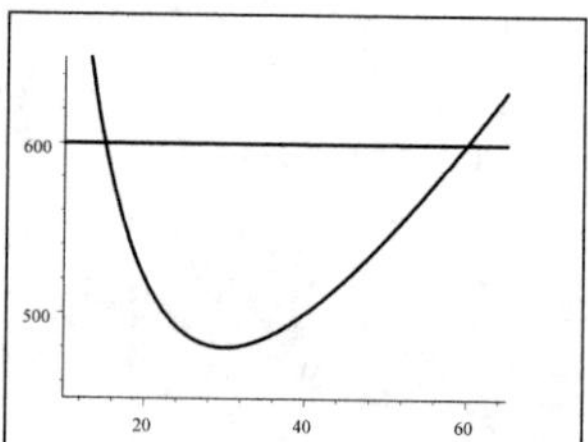


**26. (a)** Let $x$ be the length of wire in cm that is bent into a square. So $10 - x$ is the length of wire in cm that is bent into the second square. The width of each square is $\dfrac{x}{4}$ and $\dfrac{10-x}{4}$, and the area of each square is $\left(\dfrac{x}{4}\right)^2 = \dfrac{x^2}{16}$ and $\left(\dfrac{10-x}{4}\right)^2 = \dfrac{100 - 20x + x^2}{16}$. Thus the sum of the areas is $A(x) = \dfrac{x^2}{16} + \dfrac{100 - 20x + x^2}{16} = \dfrac{100 - 20x + 2x^2}{16} = \frac{1}{8}x^2 - \frac{5}{4}x + \frac{25}{4}$.

**(b)** We complete the square. $A(x) = \frac{1}{8}x^2 - \frac{5}{4}x + \frac{25}{4} = \frac{1}{8}(x^2 - 10x) + \frac{25}{4} = \frac{1}{8}(x^2 - 10x + 25) + \dfrac{25}{4} - \frac{25}{8} = \frac{1}{8}(x-5)^2 + \frac{25}{8}$ So the minimum area is $\frac{25}{8}$ cm$^2$ when each piece is 5 cm long.

**27. (a)** Let $p$ be the price of the ticket. So $10 - p$ is the difference in ticket price and therefore the number of tickets sold is $27{,}000 + 3000(10 - p) = 57{,}000 - 3000p$. Thus the revenue is $R(p) = p(57{,}000 - 3000p) = 57{,}000p - 3000p^2$.

**(b)** $R(p) = 0 = p(57{,}000 - 3000p)$. So $p = 0$ *or* $= \dfrac{57{,}000}{3000} = 19$. So at \$19 no one will come.

**(c)** We complete the square:

$$\begin{aligned} R(p) &= 57{,}000p - 3000p^2 = -3000(p^2 - 19p) = -3000\left(p^2 - 19p + \tfrac{19^2}{4}\right) + 270{,}750 \\ &= -3000\left(p - \tfrac{19}{2}\right)^2 + 270{,}750 \end{aligned}$$

The revenue is maximized when $p = \frac{19}{2}$, and so the price should be set at \$9.50.

**28. (a)** Let $x$ be the number of one dollar increases in the price of a bird feeder. So the selling price will be $10+x$ dollars, and the number of bird feeders sold will be $20 - 2x$. The revenue from the sales will be $(10 + x)(20 - 2x)$, and the cost will be $6(20 - 2x)$. The profits will be $P(x) = (10 + x)(20 - 2x) - 6(20 - 2x) = (4 + x)(20 - 2x) = 80 + 12x - 2x^2$.

**(b)** Completing the square we get $P(x) = 80 + 12x - 2x^2 = -2(x^2 - 6x) + 80 = -2(x^2 - 6x + 9) + 80 + 18 = -2(x-3)^2 + 98$. Thus the profit would be maximized at \$98 when $x = 3$. So the bird society should set the selling price at $10 + x = \$13$.

**29. (a)** Let $h$ be the height in feet of the straight portion of the window. The circumference of the semicircle is $C = \frac{1}{2}\pi x$. Since the perimeter of the window is 30 feet, we have $x + 2h + \frac{1}{2}\pi x = 30$. Solving for $h$, we get $2h = 30 - x - \frac{1}{2}\pi x \quad \Leftrightarrow \quad h = 15 - \frac{1}{2}x - \frac{1}{4}\pi x$. The area of the window is $A(x) = xh + \frac{1}{2}\pi\left(\frac{1}{2}x\right)^2 = x\left(15 - \frac{1}{2}x - \frac{1}{4}\pi x\right) + \frac{1}{8}\pi x^2 = 15x - \frac{1}{2}x^2 - \frac{1}{8}\pi x^2$.

**(b)** $A(x) = 15x - \frac{1}{2}x^2 - \frac{1}{8}\pi x^2 = 15x - \frac{1}{8}(\pi+4)x^2 = -\frac{1}{8}(\pi+4)\left[x^2 - \frac{120}{\pi+4}x\right]$

$$= -\frac{1}{8}(\pi+4)\left[x^2 - \frac{120}{\pi+4}x + \left(\frac{60}{\pi+4}\right)^2\right] + \frac{450}{\pi+4} = -\frac{1}{8}(\pi+4)\left(x - \frac{60}{\pi+4}\right)^2 + \frac{450}{\pi+4}$$

The area is maximized when $x = \dfrac{60}{\pi+4} \approx 8.40$, and hence $h \approx 15 - \frac{1}{2}(8.40) - \dfrac{1}{4}\pi(8.40) \approx 4.20$.

**30. (a)** The height of the box is $x$, the width of the box is $12 - 2x$, and the length of the box is $20 - 2x$. Therefore, the volume of the box is

$$\begin{aligned} V(x) &= x(12-2x)(20-2x) \\ &= 4x^3 - 64x^2 + 240x, 0 < x < 6 \end{aligned}$$

**(b)** We graph the function $y = V(x)$ in the viewing rectangle $[0, 6] \times [200, 270]$.

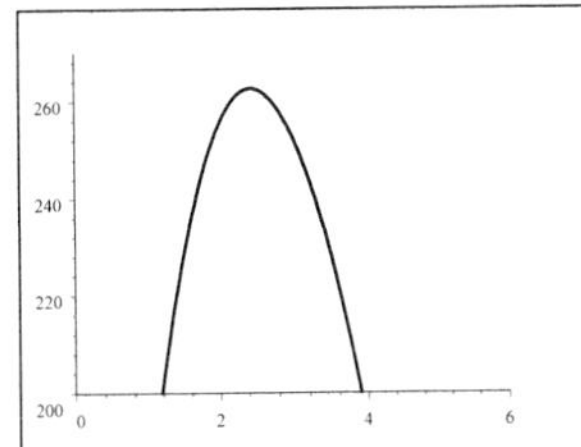


From the calculator we get that the volume of the box is greater than 200 in$^3$ for $1.174 \le x \le 3.898$ (accurate to 3 decimal places).

**(c)** From the graph, the volume of the box with the largest volume is 262.682 in$^3$ when $x \approx 2.427$.

**31. (a)** Let $x$ be the length of one side of the base and let $h$ be the height of the box in feet. Since the volume of the box is $V = x^2h = 12$, we have $x^2h = 12 \quad \Leftrightarrow \quad h = \dfrac{12}{x^2}$. The surface area, $A$, of the box is sum of the area of the four sides and the area of the base. Thus the surface area of the box is given by the formula $A(x) = 4xh + x^2 = 4x\left(\dfrac{12}{x^2}\right) + x^2 = \dfrac{48}{x} + x^2, x > 0$.

**(b)** The function $y = A(x)$ is shown in the first viewing rectangle below. In the second viewing rectangle, we isolate the minimum, and we see that the amount of material is minimized when $x$ (the length and width) is 2.88 ft. Then the height is $h = \dfrac{12}{x^2} \approx 1.44$ ft.

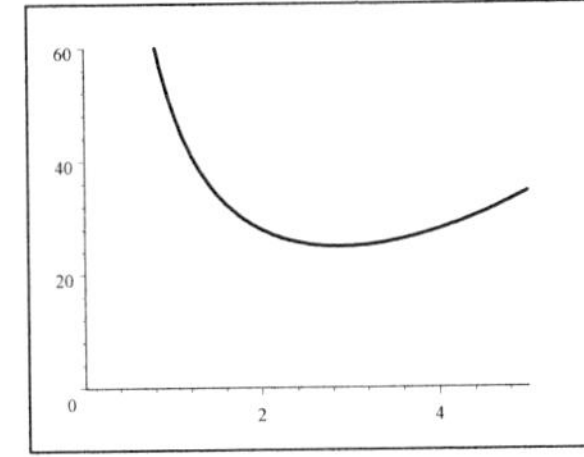


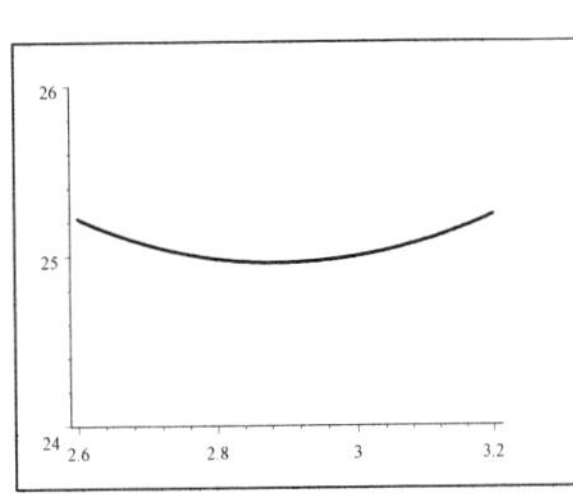

**32.** Let $A, B, C$, and $D$ be the vertices of a rectangle with base $AB$ on the $x$-axis and its other two vertices $C$ and $D$ above the $x$-axis and lying on the parabola $y = 8 - x^2$. Let $C$ have the coordinates $(x, y)$, $x > 0$. By symmetry, the coordinates of $D$ must be $(-x, y)$. So the width of the rectangle is $2x$, and the length is $y = 8 - x^2$. Thus the area of the rectangle is $A(x) = \text{length} \cdot \text{width} = 2x(8 - x^2) = 16x - 2x^3$. The graphs of $A(x)$ below show that the area is maximized when $x \approx 1.63$. Hence the maximum area occurs when the width is 3.26 and the length is 5.33.

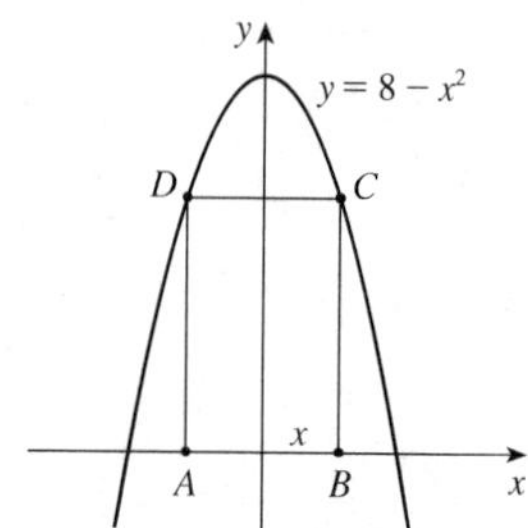


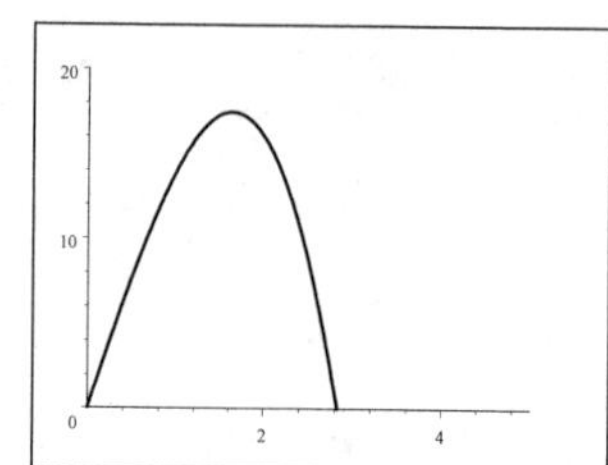


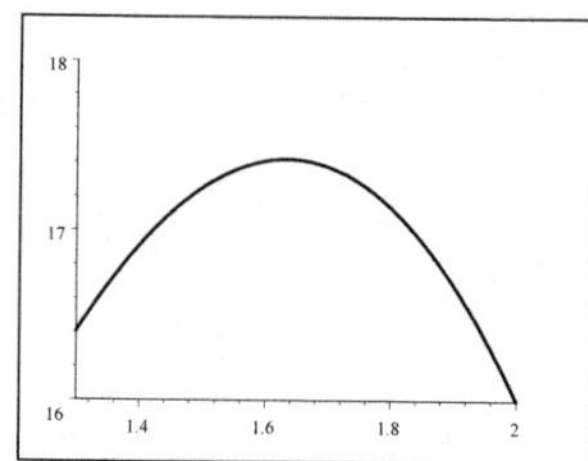


**33. (a)** Let $w$ be the width of the pen and $l$ be the length in meters. We use the area to establish a relationship between $w$ and $l$. Since the area is 100 m$^2$, we have $l \cdot w = 100 \quad \Leftrightarrow \quad l = \dfrac{100}{w}$. So the amount of fencing used is

$$F = 2l + 2w = 2\left(\frac{100}{w}\right) + 2w = \frac{200 + 2w^2}{w}.$$

**(b)** Using a graphing device, we first graph $F$ in the viewing rectangle $[0, 40]$ by $[0, 100]$, and locate the approximate location of the minimum value. In the second viewing rectangle, $[8, 12]$ by $[39, 41]$, we see that the minimum value of $F$ occurs when $w = 10$. Therefore the pen should be a square with side 10 m.

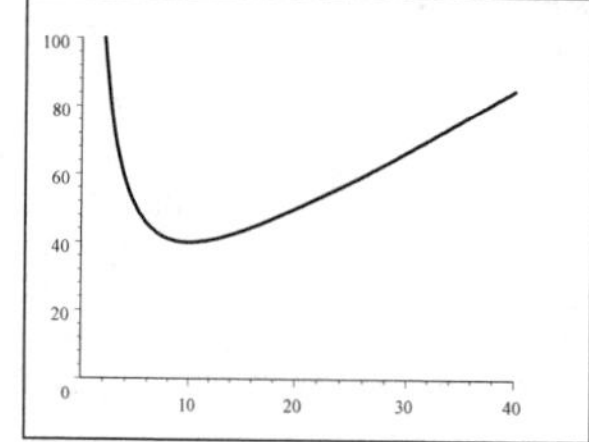


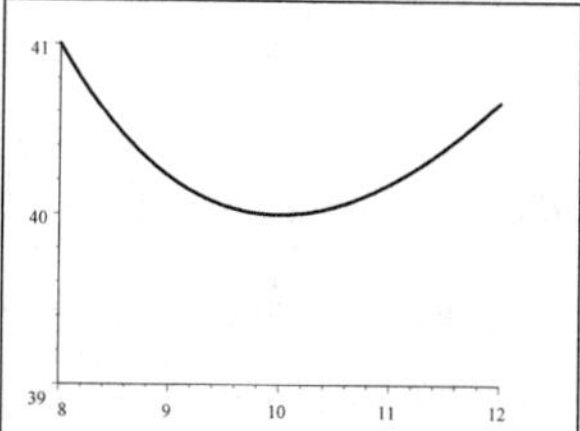


**34. (a)** Let $t_1$ represent the time, in hours, spent walking, and let $t_2$ represent the time spent rowing. Since the distance walked is $x$ and the walking speed is 5 mi/h, the time spent walking is $t_1 = \frac{1}{5}x$. By the Pythagorean Theorem, the distance rowed is

$d = \sqrt{2^2 + (7 - x)^2} = \sqrt{x^2 - 14x + 53}$, and so the time spent rowing is $t_2 = \frac{1}{2} \cdot \sqrt{x^2 - 14x + 53}$. Thus the total time is $T(x) = \frac{1}{2}\sqrt{x^2 - 14x + 53} + \frac{1}{5}x$.

**(b)** We graph $y = T(x)$. Using the zoom function, we see that $T$ is minimized when $x \approx 6.13$. He should land at a point 6.13 miles from point $B$.

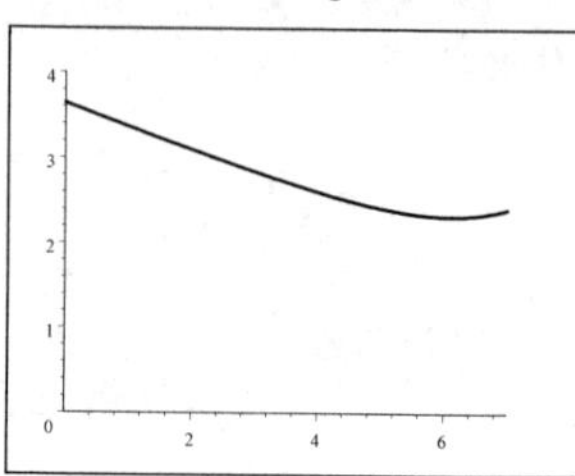

**35. (a)** Let $x$ be the distance from point $B$ to $C$, in miles. Then the distance from $A$ to $C$ is $\sqrt{x^2+25}$, and the energy used in flying from $A$ to $C$ then $C$ to $D$ is $f(x) = 14\sqrt{x^2+25} + 10(12-x)$.

**(b)** By using a graphing device, the energy expenditure is minimized when the distance from $B$ to $C$ is about 5.1 miles.

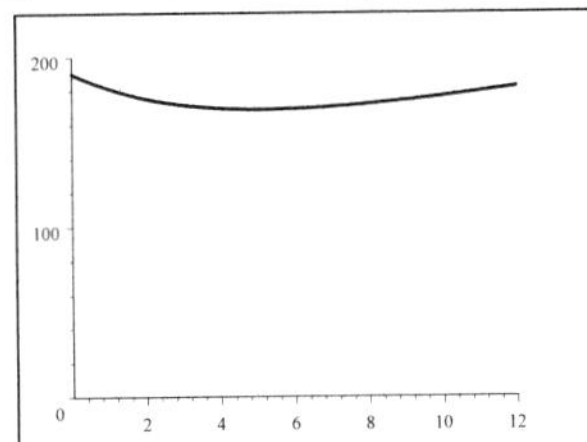


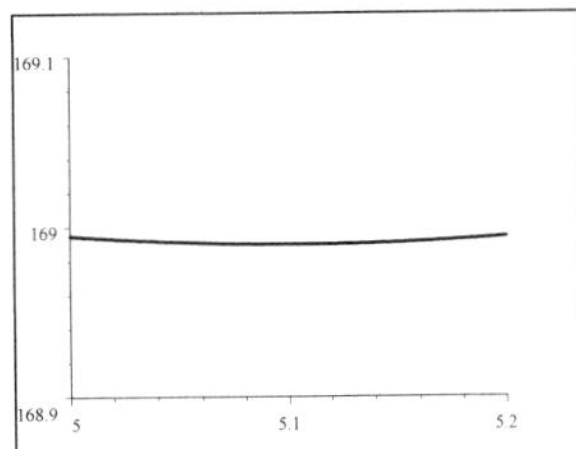


**36. (a)** Using the Pythagorean Theorem, we have that the height of the upper triangles is $\sqrt{25-x^2}$ and the height of the lower triangles is $\sqrt{144-x^2}$. So the area of the each of the upper triangles is $\frac{1}{2}x\sqrt{25-x^2}$, and the area of the each of the lower triangles is $\frac{1}{2}x\sqrt{144-x^2}$. Since there are two upper triangles and two lower triangles, we get that the total area is $A(x) = 2 \cdot \left[\frac{1}{2}x\sqrt{25-x^2}\right] + 2 \cdot \left[\frac{1}{2}x\sqrt{144-x^2}\right] = x\left(\sqrt{25-x^2} + \sqrt{144-x^2}\right)$.

**(b)** The function $y = A(x) = x\left(\sqrt{25-x^2} + \sqrt{144-x^2}\right)$ is shown in the first viewing rectangle below. In the second viewing rectangle, we isolate the maximum, and we see that the area of the kite is maximized when $x \approx 4.615$. So the length of the horizontal crosspiece must be $2 \cdot 4.615 = 9.23$. The length of the vertical crosspiece is $\sqrt{5^2-(4.615)^2} + \sqrt{12^2-(4.615)^2} \approx 13.00$.

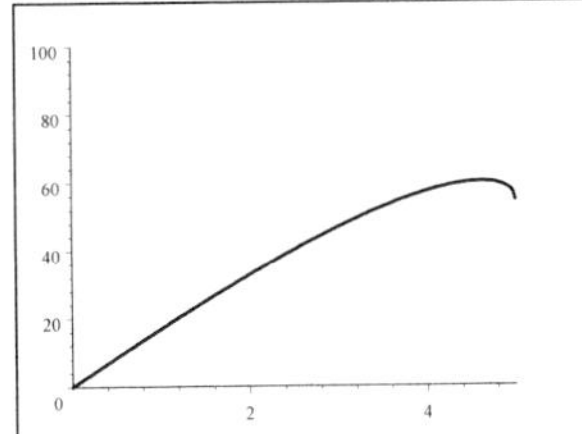


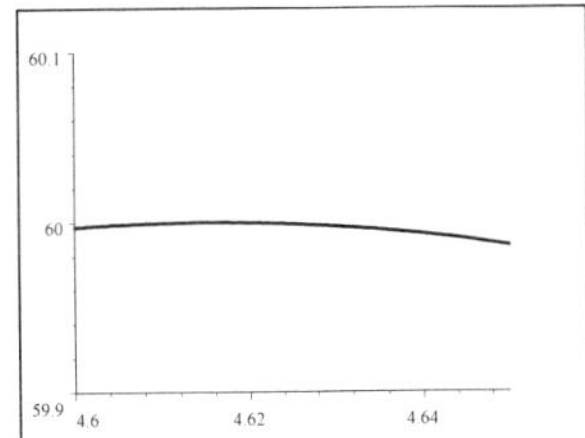

# 4 Polynomial and Rational Functions

## 4.1 Polynomial Functions and Their Graphs

**1. (a)** $P(x) = x^2 - 4$

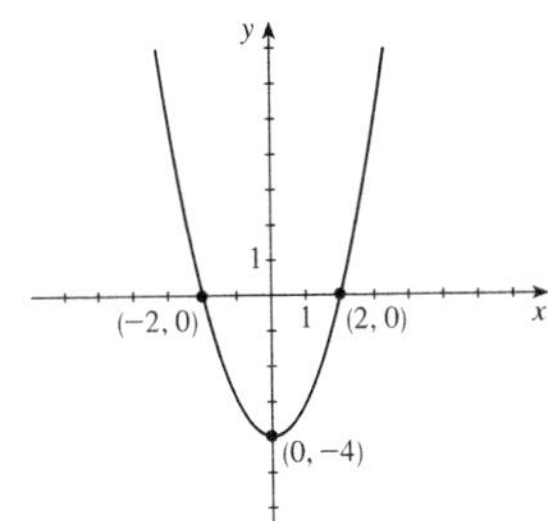


**(b)** $Q(x) = (x - 4)^2$

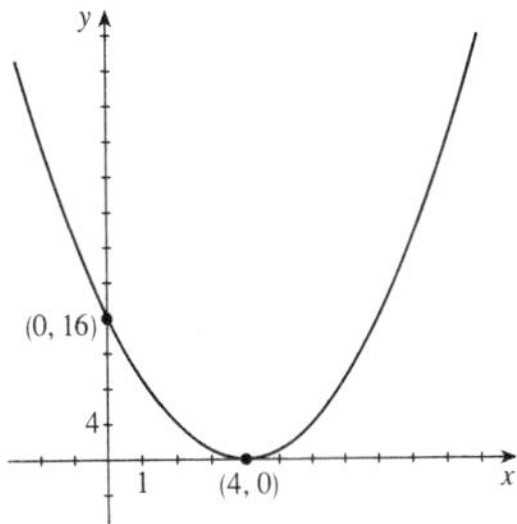


**(c)** $R(x) = 2x^2 - 2$

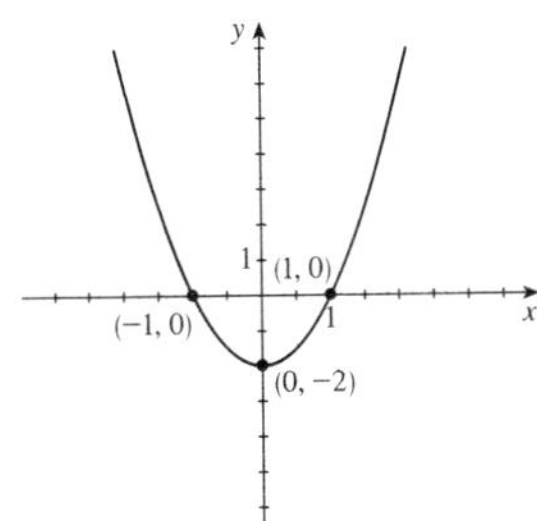


**(d)** $S(x) = 2(x - 2)^2$

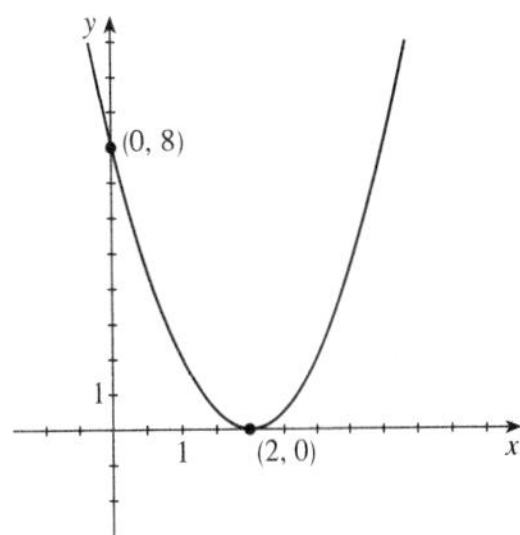


**2. (a)** $P(x) = x^4 - 16$

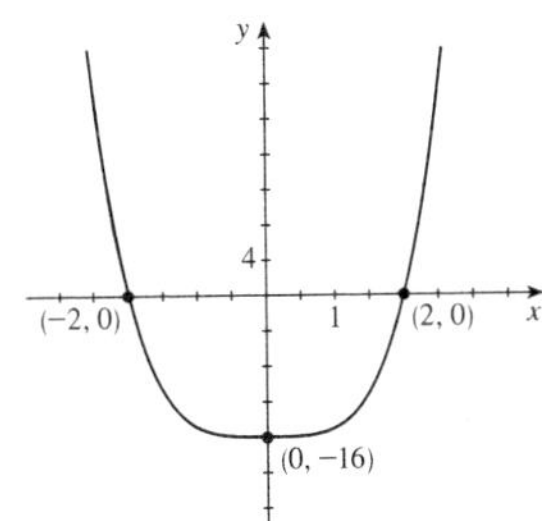


**(b)** $Q(x) = (x + 2)^4$

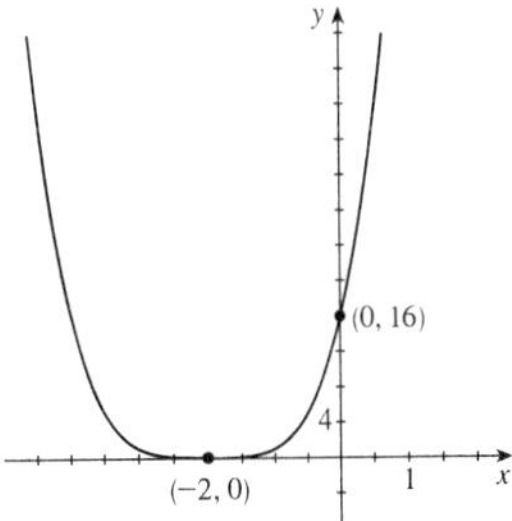


**(c)** $R(x) = (x + 2)^4 - 16$

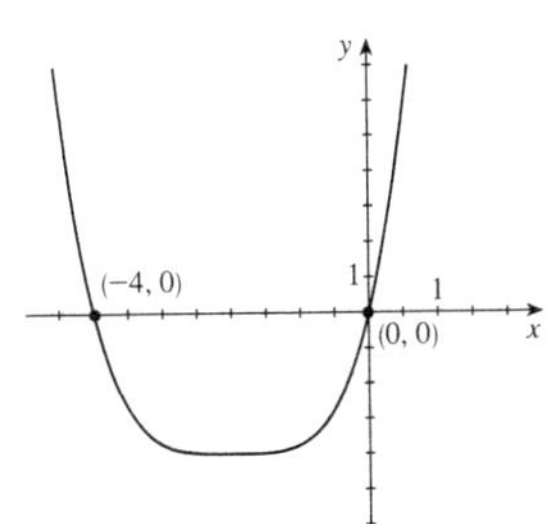


**(d)** $S(x) = -2(x + 2)^4$

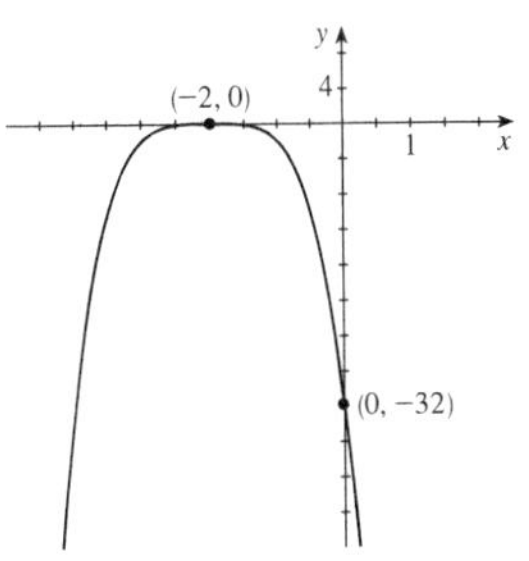

**3. (a)** $P(x) = x^3 - 8$

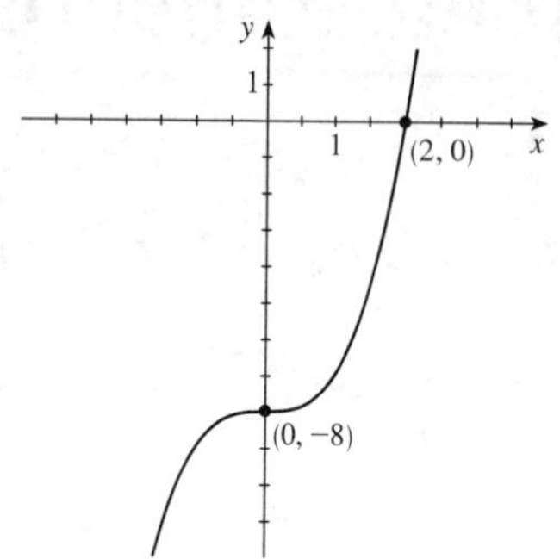


**(b)** $Q(x) = -x^3 + 27$

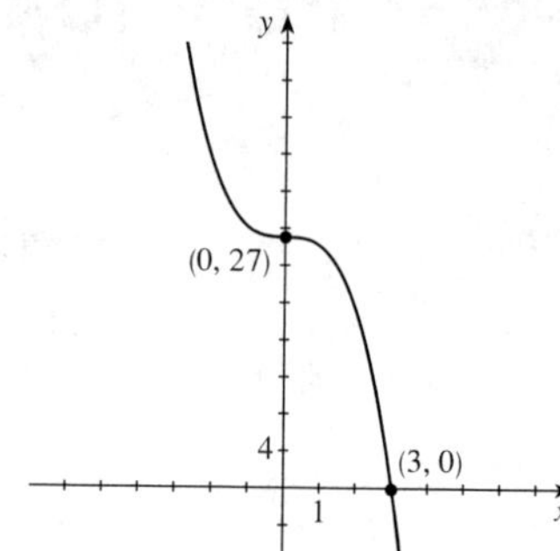


**(c)** $R(x) = -(x+2)^3$

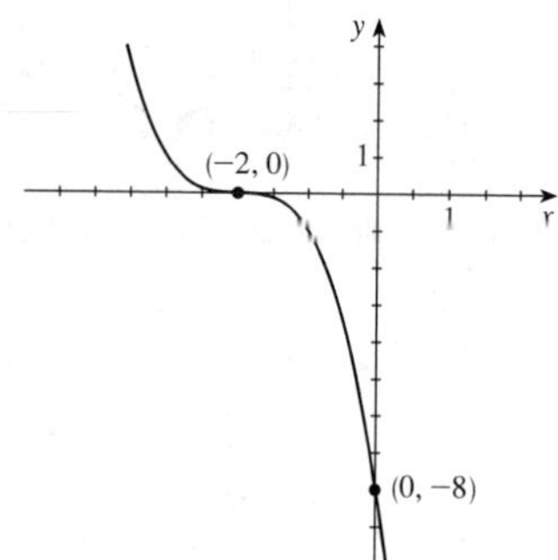


**(d)** $S(x) = \frac{1}{2}(x-1)^3 + 4$

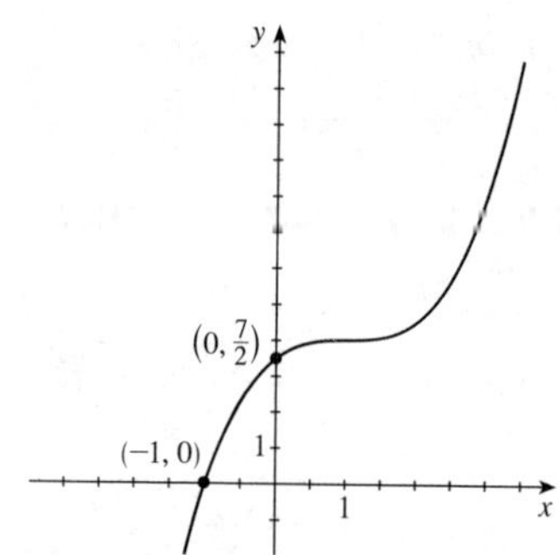


**4. (a)** $P(x) = (x+3)^5$

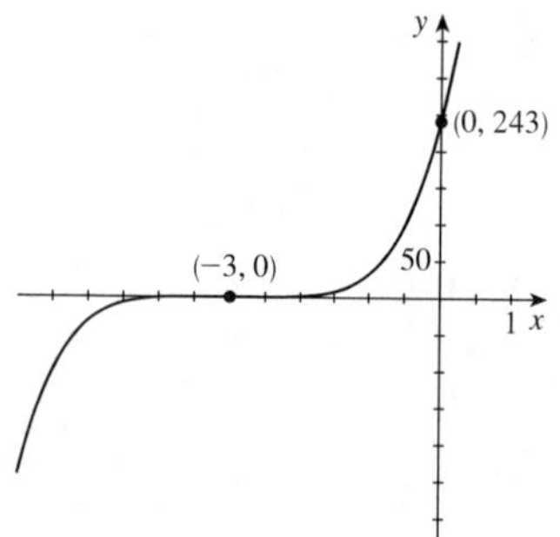


**(b)** $Q(x) = 2(x+3)^5 - 64$

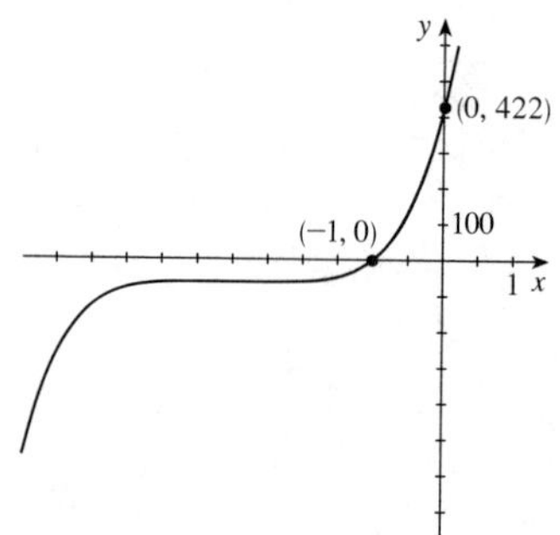


**(c)** $R(x) = -\frac{1}{2}(x-2)^5$

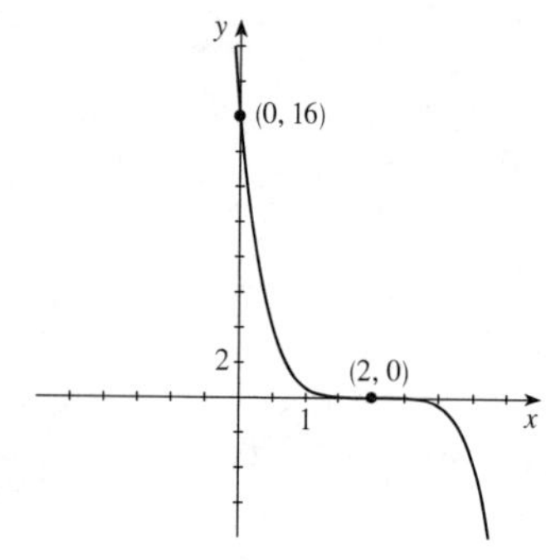


**(d)** $S(x) = -\frac{1}{2}(x-2)^5 + 16$

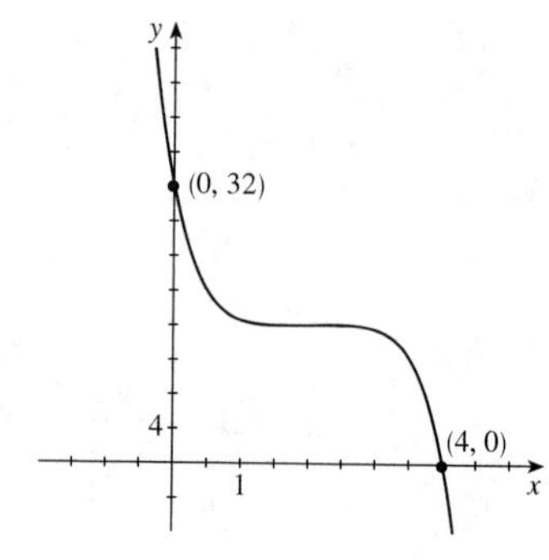

**5.** III **6.** I **7.** V **8.** II **9.** VI **10.** IV

**11.** $P(x) = (x-1)(x+2)$

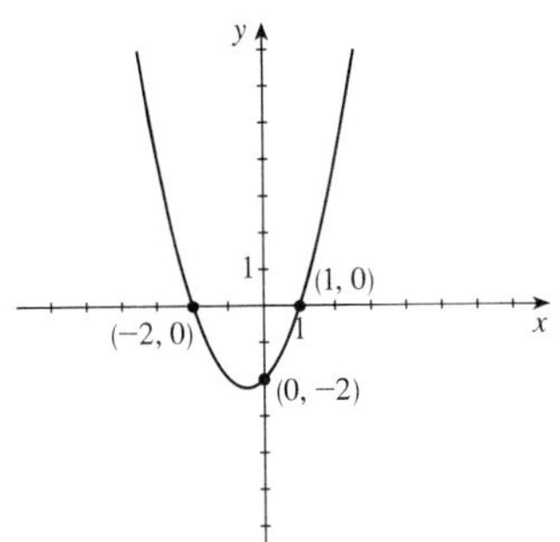


**12.** $P(x) = (x-1)(x+1)(x-2)$

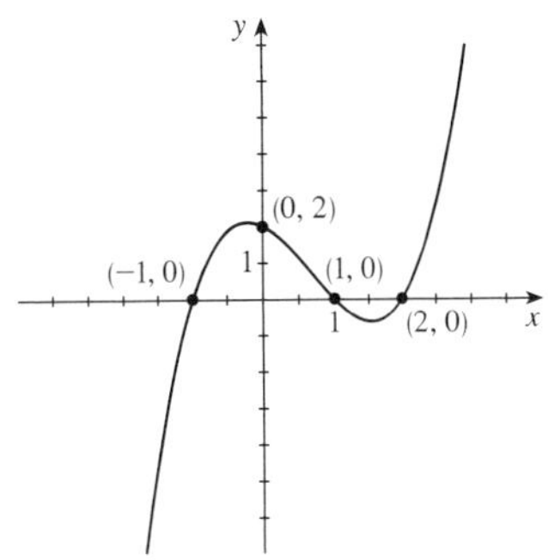


**13.** $P(x) = x(x-3)(x+2)$

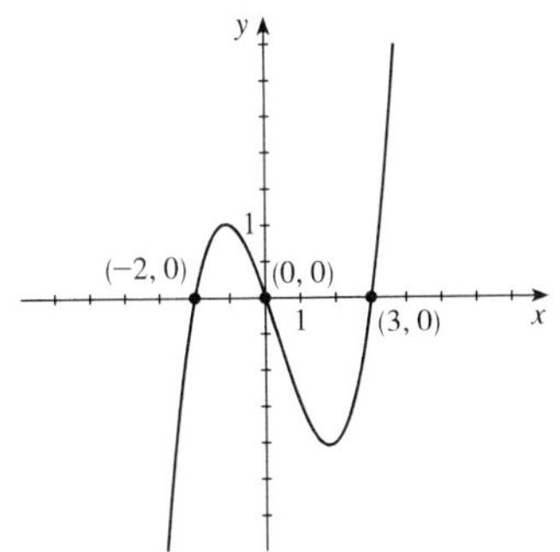


**14.** $P(x) = (2x-1)(x+1)(x+3)$

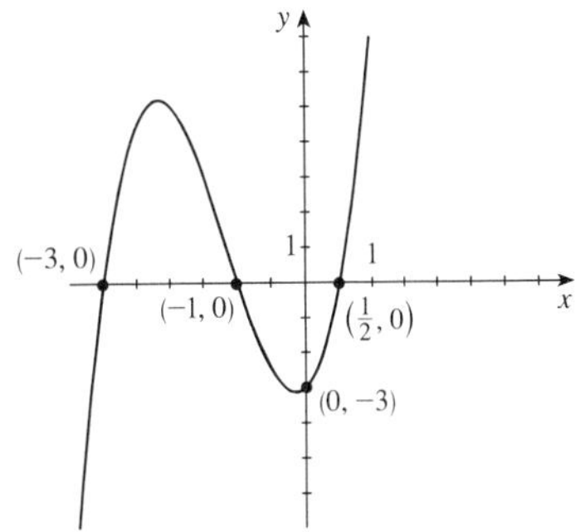


**15.** $P(x) = (x-3)(x+2)(3x-2)$

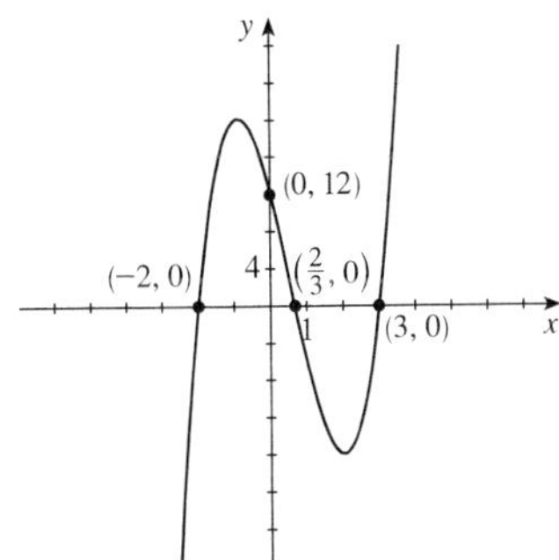


**16.** $P(x) = \frac{1}{5}x(x-5)^2$

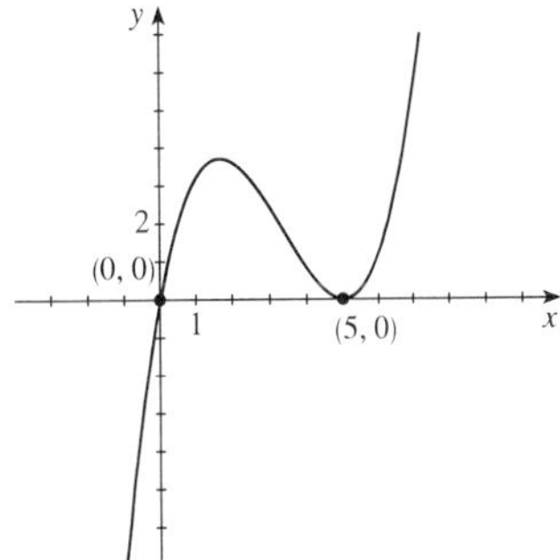


**17.** $P(x) = (x-1)^2(x-3)$

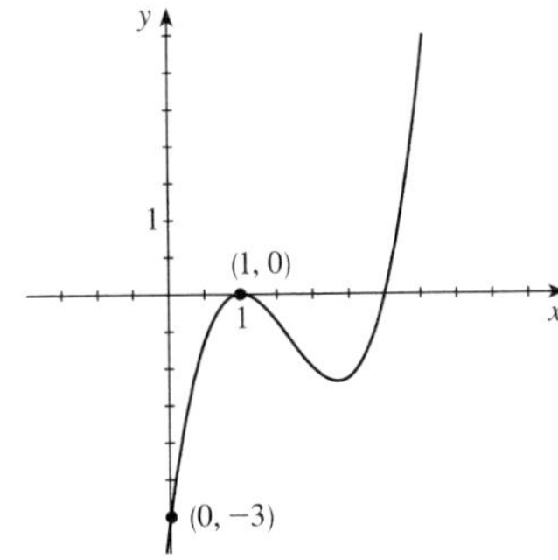


**18.** $P(x) = \frac{1}{4}(x+1)^3(x-3)$

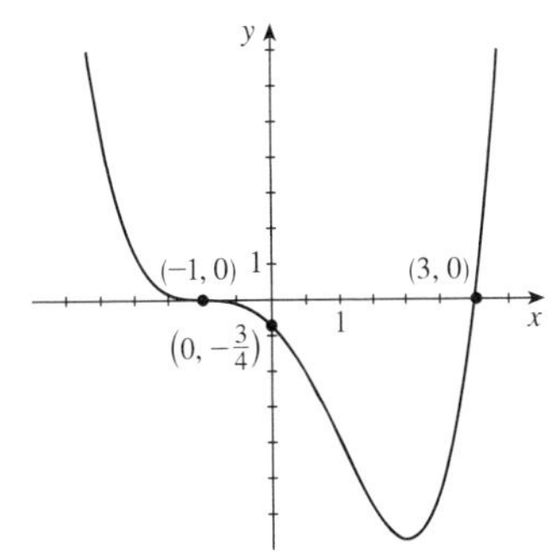

**19.** $P(x) = \frac{1}{12}(x+2)^2(x-3)^2$

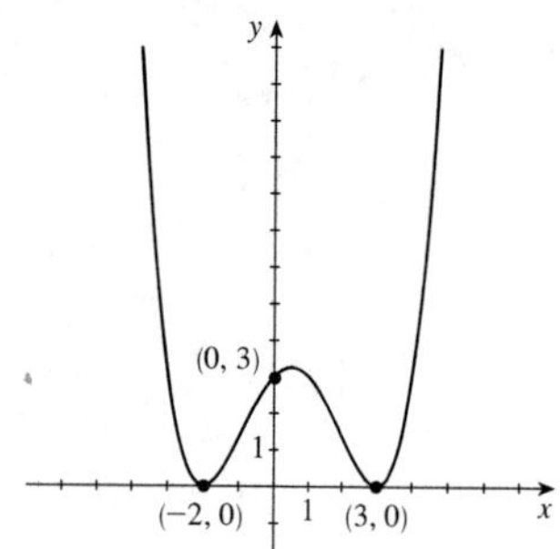


**20.** $P(x) = (x-1)^2(x+2)^3$

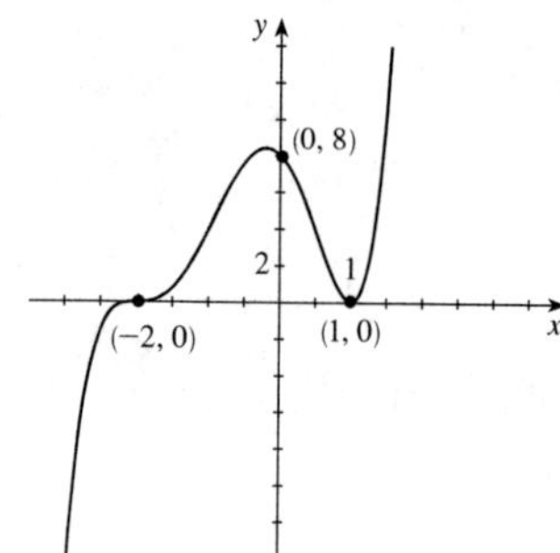


**21.** $P(x) = x^3(x+2)(x-3)^2$

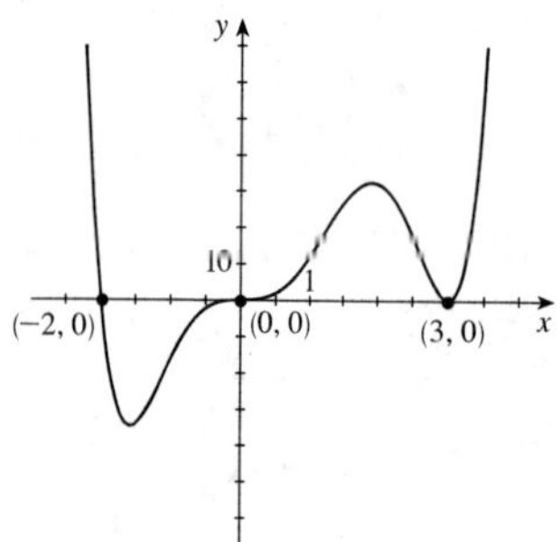


**22.** $P(x) = (x-3)^2(x+1)^2$

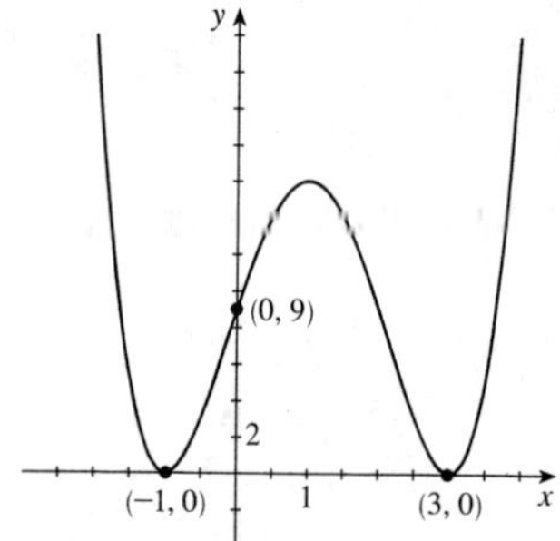


**23.** $P(x) = x^3 - x^2 - 6x = x(x+2)(x-3)$

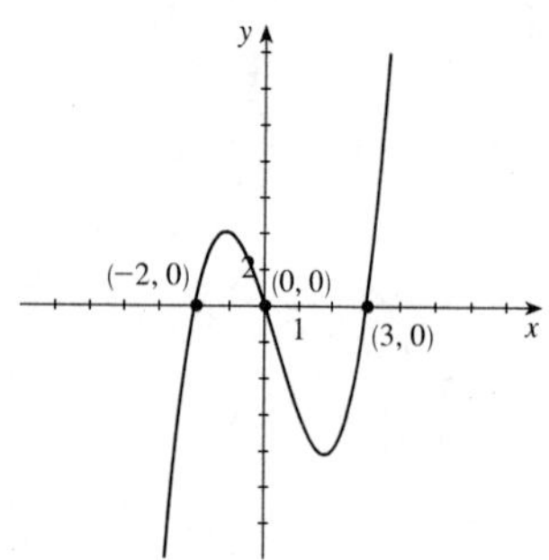


**24.** $P(x) = x^3 + 2x^2 - 8x = x(x-2)(x+4)$

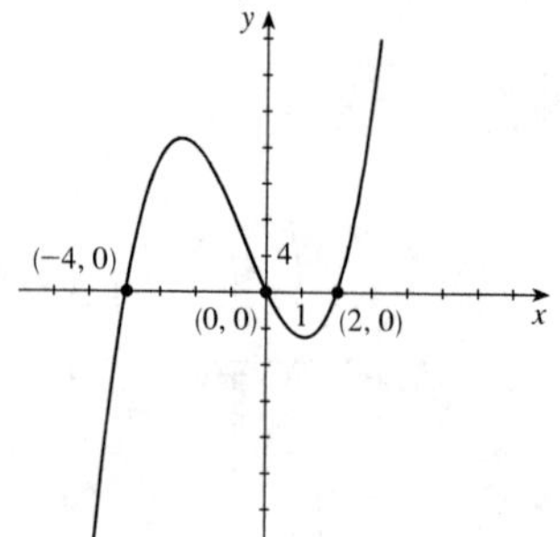


**25.** $P(x) = -x^3 + x^2 + 12x = -x(x+3)(x-4)$

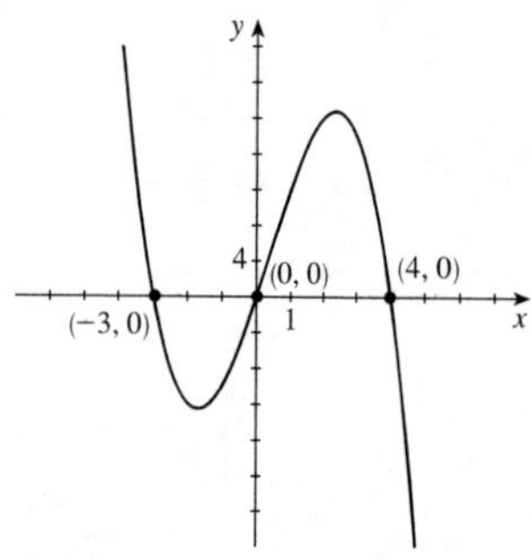


**26.** $P(x) = -2x^3 - x^2 + x = -x\left(2x^2 + x - 1\right)$
$= -x(2x-1)(x+1)$

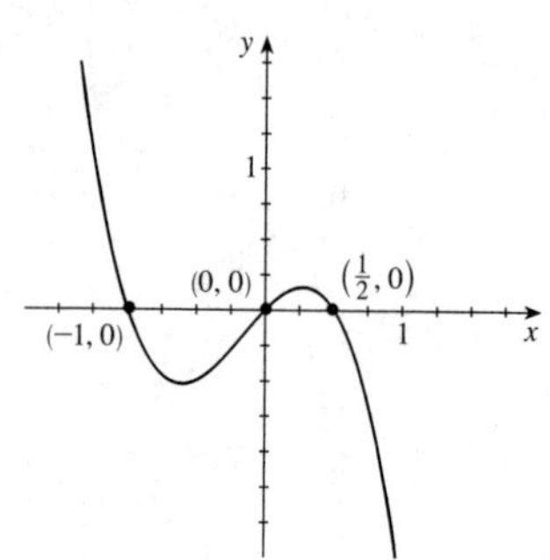

**27.** $P(x) = x^4 - 3x^3 + 2x^2 = x^2(x-1)(x-2)$

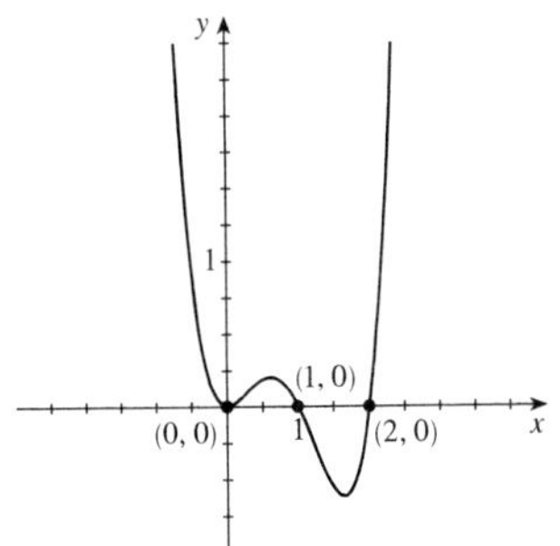


**28.** $P(x) = x^5 - 9x^3 = x^3(x+3)(x-3)$

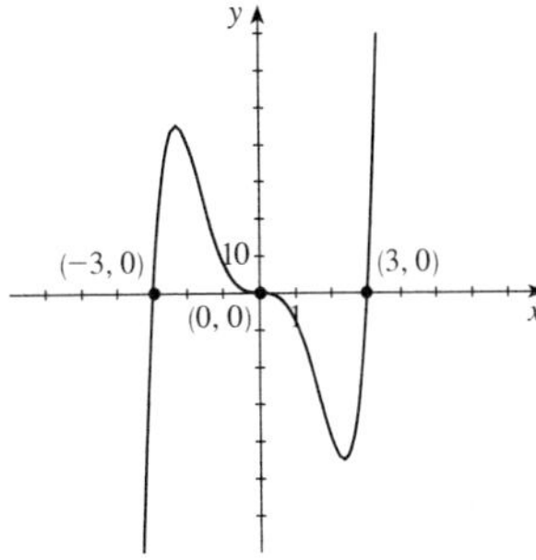


**29.** $P(x) = x^3 + x^2 - x - 1 = (x-1)(x+1)^2$

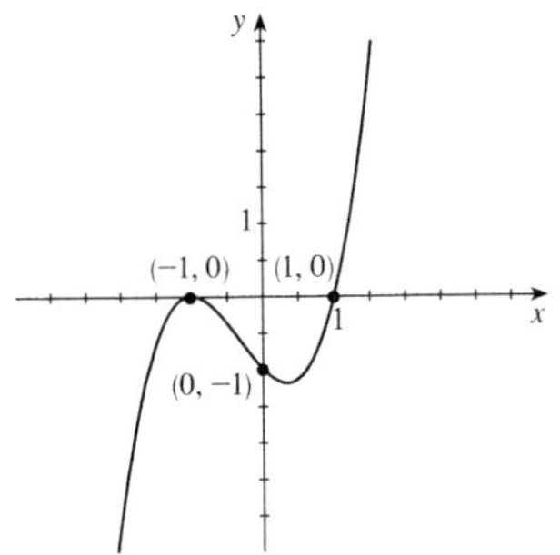


**30.** $P(x) = x^3 + 3x^2 - 4x - 12 = (x+3)(x-2)(x+2)$

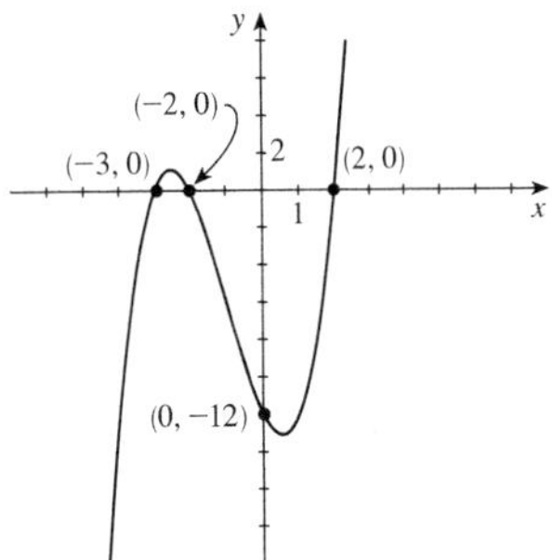


**31.** $P(x) = 2x^3 - x^2 - 18x + 9$
$= (x-3)(2x-1)(x+3)$

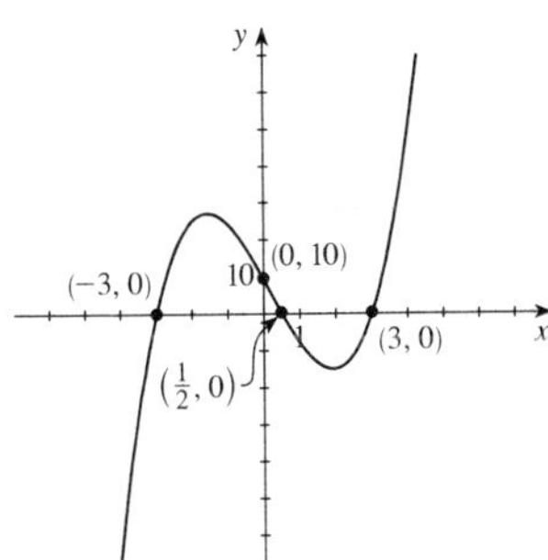


**32.** $P(x) = \frac{1}{8}\left(2x^4 + 3x^3 - 16x - 24\right)^2$
$= \frac{1}{8}(x-2)^2(2x+3)^2\left(x^2+2x+4\right)^2$

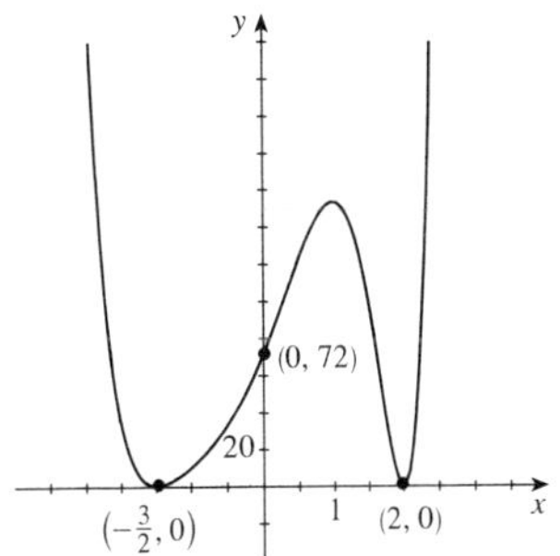


**33.** $P(x) = x^4 - 2x^3 - 8x + 16$
$= (x-2)^2\left(x^2+2x+4\right)$

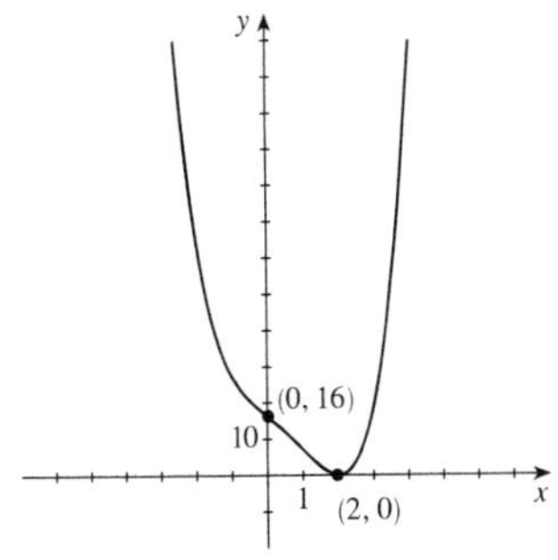


**34.** $P(x) = x^4 - 2x^3 + 8x - 16$
$= (x+2)(x-2)\left(x^2-2x+4\right)$

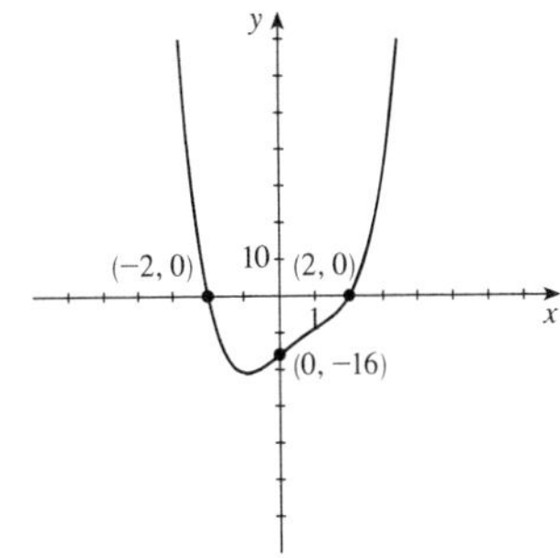

**35.** $P(x) = x^4 - 3x^2 - 4 = (x-2)(x+2)(x^2+1)$

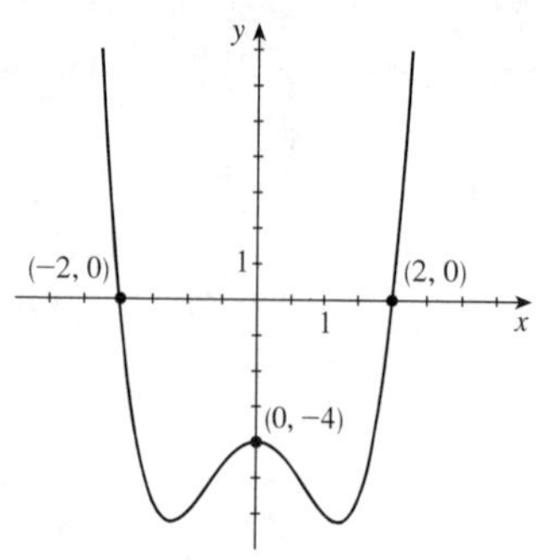


**36.** $P(x) = x^6 - 2x^3 + 1 = (x^3-1)^2 = (x-1)^2(x^2+x+1)^2$

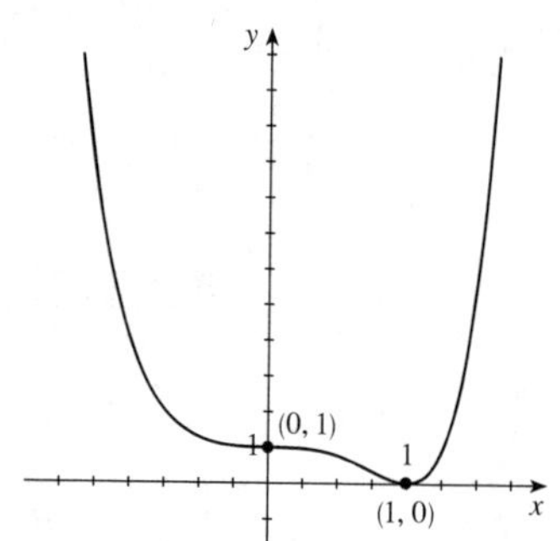


**37.** $P(x) = 3x^3 - x^2 + 5x + 1$; $Q(x) = 3x^3$. Since $P$ has odd degree and positive leading coefficient, it has the following end behavior: $y \to \infty$ as $x \to \infty$ and $y \to -\infty$ as $x \to -\infty$.

On a large viewing rectangle, the graphs of $P$ and $Q$ look almost the same. On a small viewing rectangle, we see that the graphs of $P$ and $Q$ have different intercepts.

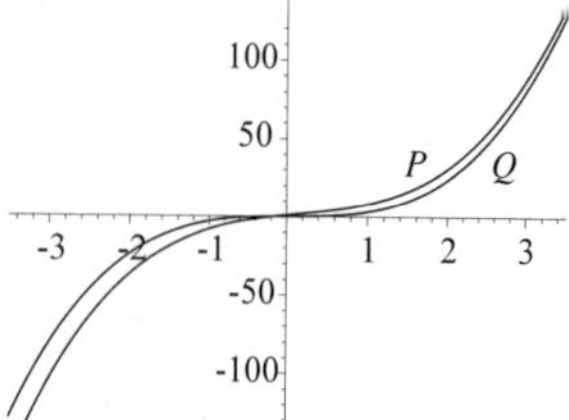


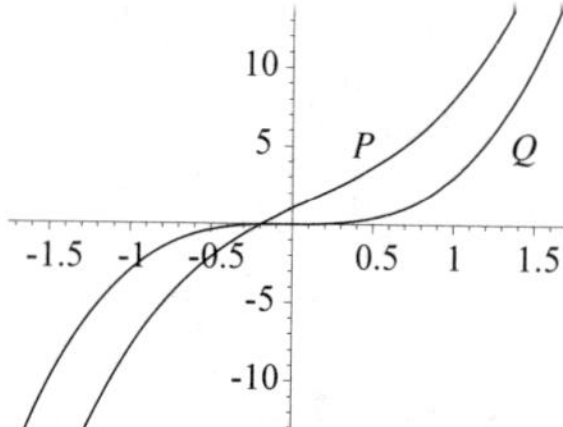


**38.** $P(x) = -\frac{1}{8}x^3 + \frac{1}{4}x^2 + 12x$; $Q(x) = -\frac{1}{8}x^3$. Since $P$ has odd degree and negative leading coefficient, it has the following end behavior: $y \to -\infty$ as $x \to \infty$ and $y \to \infty$ as $x \to -\infty$.

On a large viewing rectangle, the graphs of $P$ and $Q$ look almost the same. On a small viewing rectangle, the graphs of $P$ and $Q$ look very different and seem (wrongly) to have different end behavior.

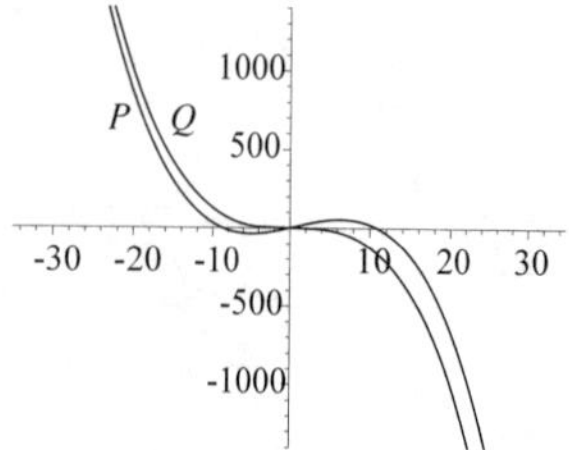


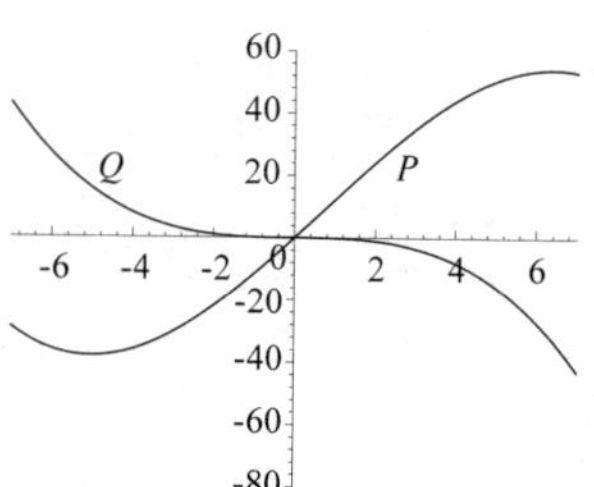


**39.** $P(x) = x^4 - 7x^2 + 5x + 5$; $Q(x) = x^4$. Since $P$ has even degree and positive leading coefficient, it has the following end behavior: $y \to \infty$ as $x \to \infty$ and $y \to \infty$ as $x \to -\infty$.

On a large viewing rectangle, the graphs of $P$ and $Q$ look almost the same. On a small viewing rectangle, the graphs of $P$ and $Q$ look very different and we see that they have different intercepts.

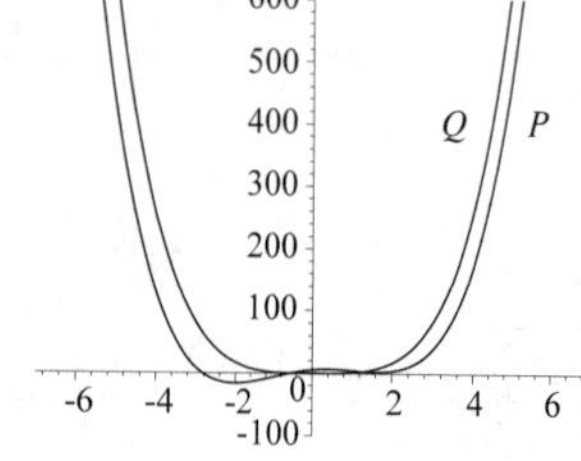


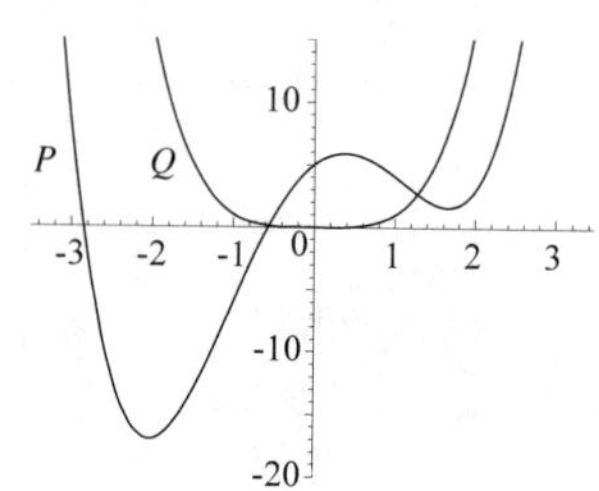

**40.** $P(x) = -x^5 + 2x^2 + x$; $Q(x) = -x^5$. Since $P$ has odd degree and negative leading coefficient, it has the following end behavior: $y \to -\infty$ as $x \to \infty$ and $y \to \infty$ as $x \to -\infty$.

On a large viewing rectangle, the graphs of $P$ and $Q$ look almost the same. On a small viewing rectangle, the graphs of $P$ and $Q$ look very different.

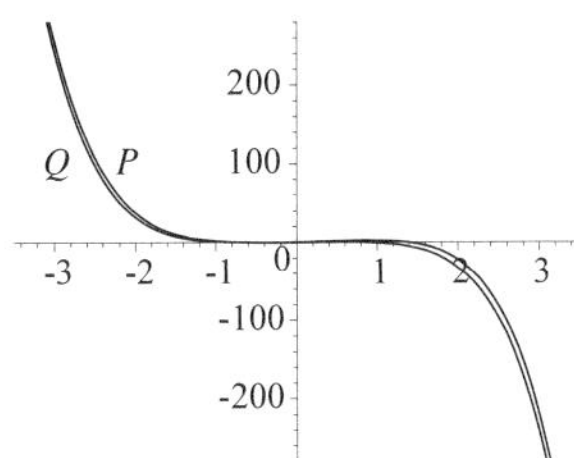


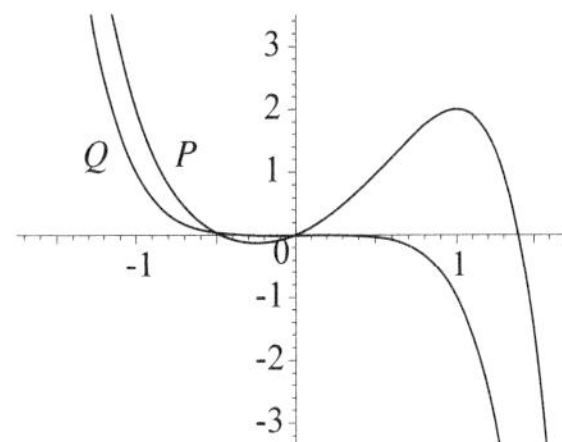


**41.** $P(x) = x^{11} - 9x^9$; $Q(x) = x^{11}$. Since $P$ has odd degree and positive leading coefficient, it has the following end behavior: $y \to \infty$ as $x \to \infty$ and $y \to -\infty$ as $x \to -\infty$.

On a large viewing rectangle, the graphs of $P$ and $Q$ look like they have the same end behavior. On a small viewing rectangle, the graphs of $P$ and $Q$ look very different and seem (wrongly) to have different end behavior.

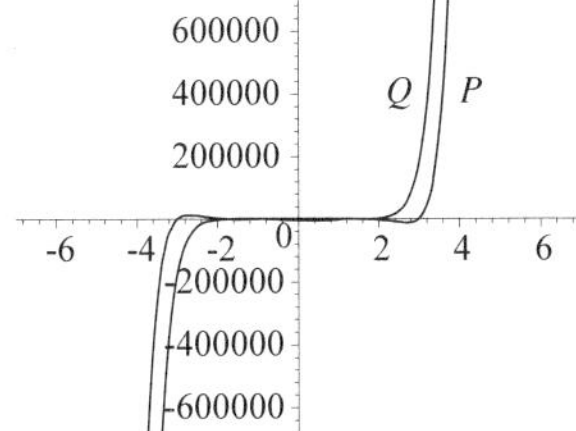


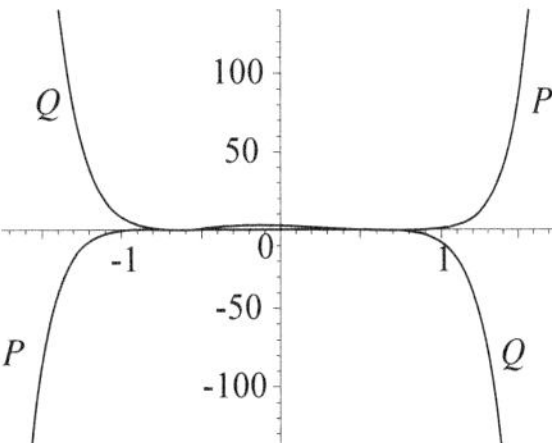


**42.** $P(x) = 2x^2 - x^{12}$; $Q(x) = -x^{12}$. Since $P$ has even degree and negative leading coefficient, it has the following end behavior: $y \to -\infty$ as $x \to \infty$ and $y \to -\infty$ as $x \to -\infty$.

On a large viewing rectangle, the graphs of $P$ and $Q$ look almost the same. On a small viewing rectangle, the graphs of $P$ and $Q$ look very different.

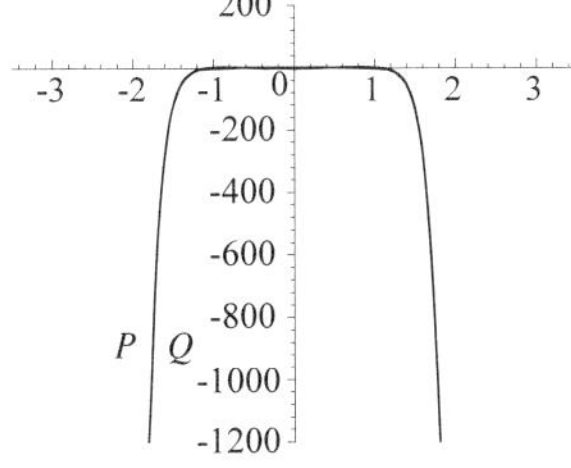


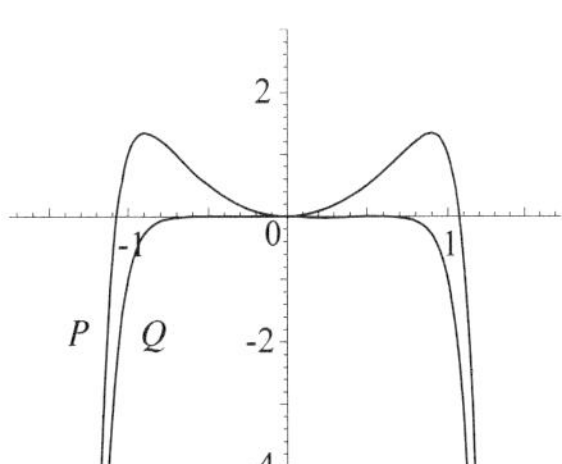


**43.** **(a)** $x$-intercepts at 0 and 4, $y$-intercept at 0.

**(b)** Local maximum at $(2, 4)$, no local minimum.

**44.** **(a)** $x$-intercepts at 0 and 4.5, $y$-intercept at 0.

**(b)** Local maximum at $(0, 0)$, local minimum at $(3, -3)$.

**45.** **(a)** $x$-intercepts at $-2$ and 1, $y$-intercept at $-1$.

**(b)** Local maximum at $(1, 0)$, local minimum at $(-1, -2)$.

**46.** **(a)** $x$-intercepts at 0 and 4, $y$-intercept at 0.

**(b)** No local maximum, local minimum at $(3, -3)$.

**47.** $y = -x^2 + 8x$, $[-4, 12]$ by $[-50, 30]$

No local minimum. Local maximum at $(4, 16)$.

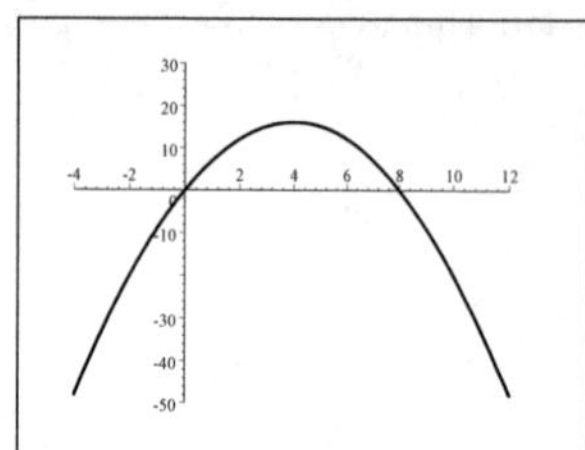

**48.** $y = x^3 - 3x^2$, $[-2, 5]$ by $[-10, 10]$

Local minimum at $(2, -4)$. Local maximum at $(0, 0)$.

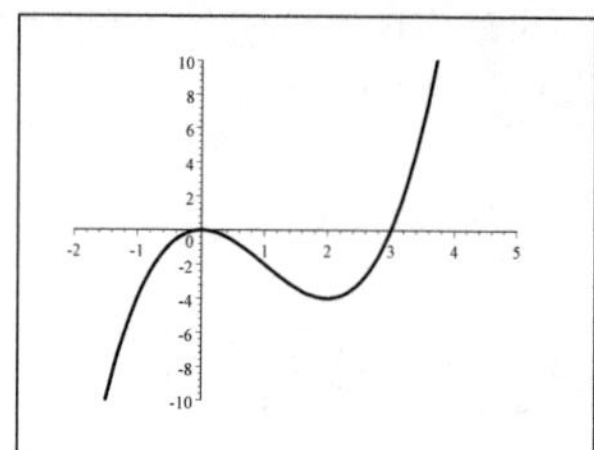

**49.** $y = x^3 - 12x + 9$, $[-5, 5]$ by $[-30, 30]$

Local minimum at $(-2, 25)$, Local maximum at $(2, -7)$.

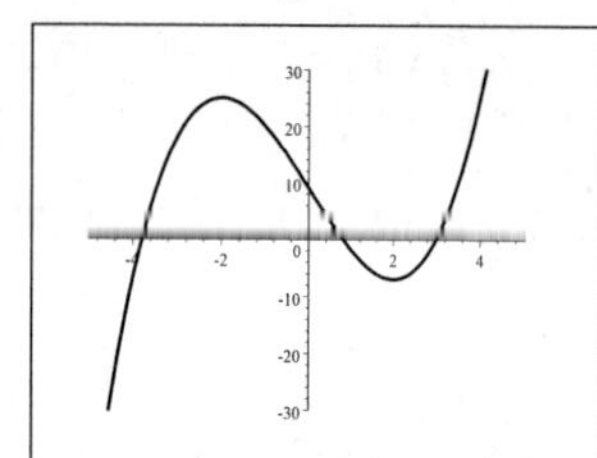

**50.** $y = 2x^3 - 3x^2 - 12x - 32$, $[-5, 5]$ by $[-60, 30]$

Local minimum at $(2, -52)$. Local maximum at $(-1, -25)$.

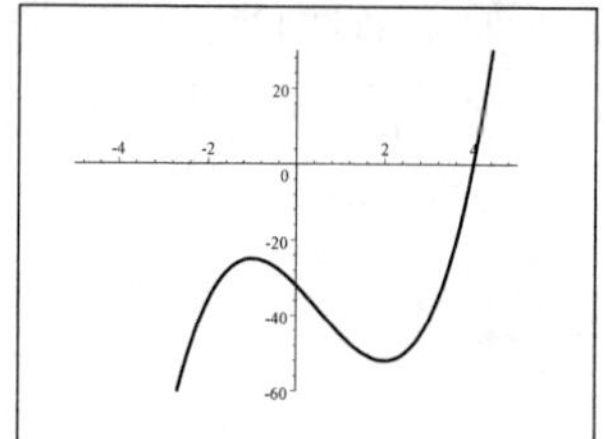

**51.** $y = x^4 + 4x^3$, $[-5, 5]$ by $[-30, 30]$

Local minimum at $(-3, -27)$. No local maximum.

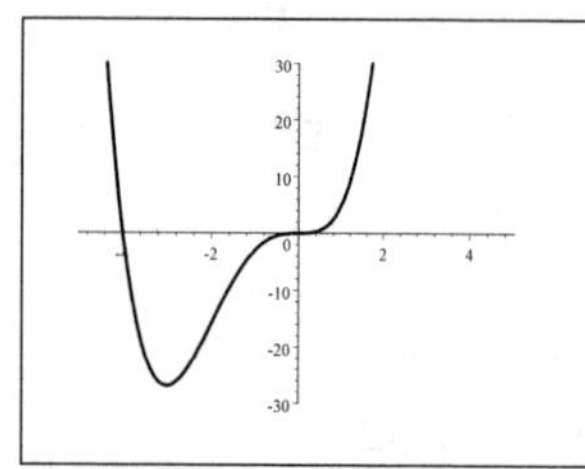

**52.** $y = x^4 - 18x^2 + 32$, $[-5, 5]$ by $[-100, 100]$

Local minima at $(-3, -49)$ and $(3, -49)$. Local maximum at $(0, 32)$.

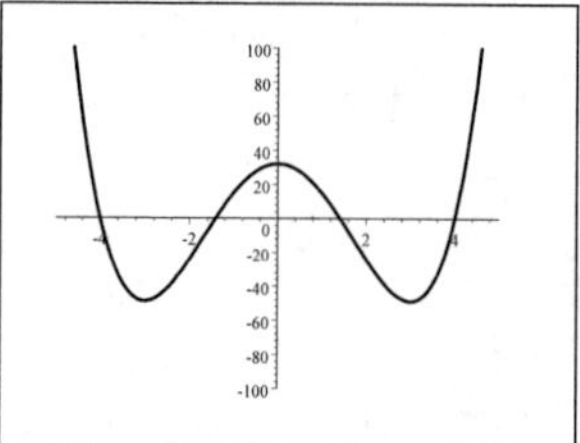

**53.** $y = 3x^5 - 5x^3 + 3$, $[-3, 3]$ by $[-5, 10]$

Local maximum at $(-1, 5)$. Local minimum at $(1, 1)$.

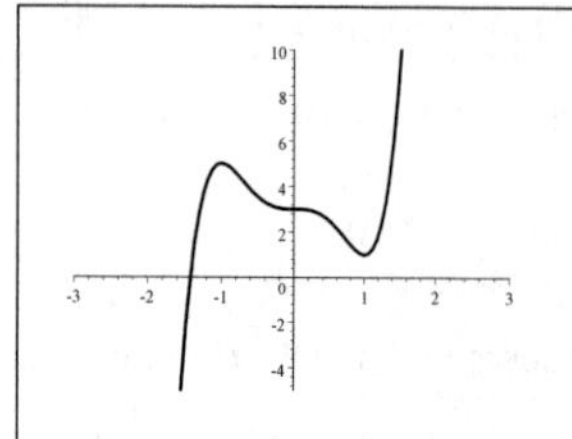

**54.** $y = x^5 - 5x^2 + 6$, $[-3, 3]$ by $[-5, 10]$

Local minimum at $(1.26, 1.24)$. Local maximum at $(0, 6)$.

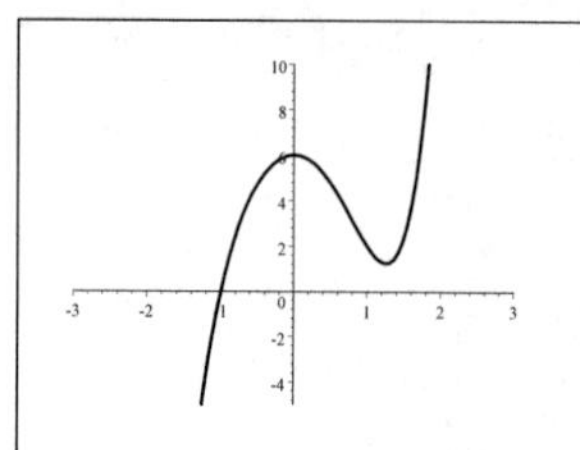

**55.** $y = -2x^2 + 3x + 5$ has one local maximum at $(0.75, 6.13)$.

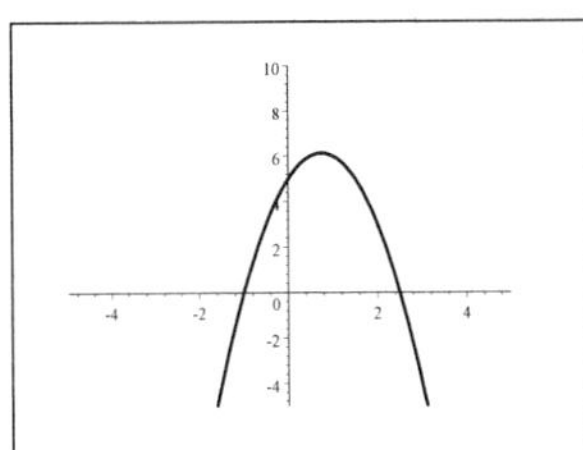

**56.** $y = x^3 + 12x$ has no local maximum or minimum.

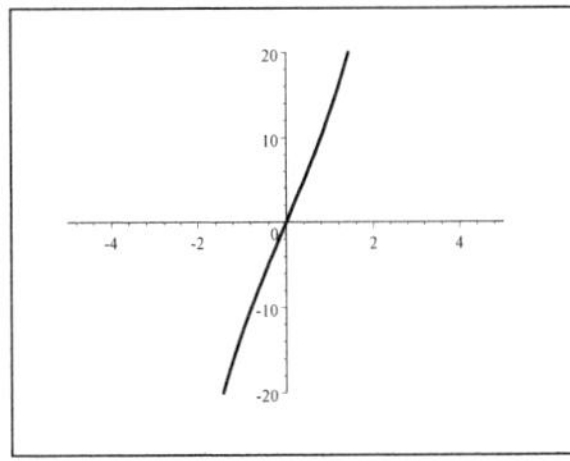

**57.** $y = x^3 - x^2 - x$ has one local maximum at $(-0.33, 0.19)$ and one local minimum at $(1.00, -1.00)$.

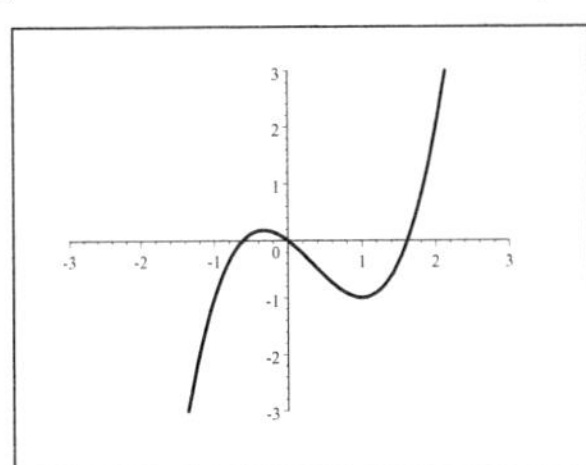

**58.** $y = 6x^3 + 3x + 1$ has no local maximum or minimum.

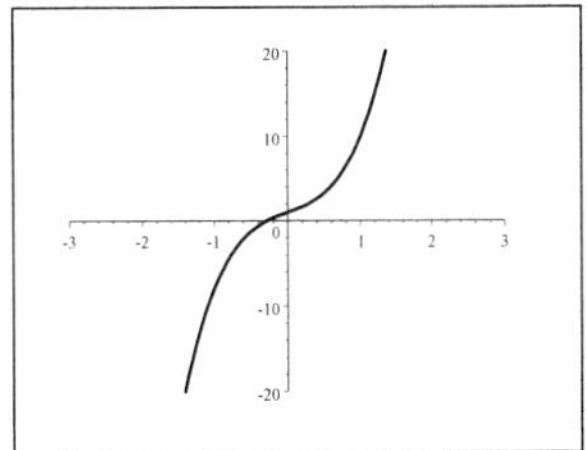

**59.** $y = x^4 - 5x^2 + 4$ has one local maximum at $(0, 4)$ and two local minima at $(-1.58, -2.25)$ and $(1.58, -2.25)$.

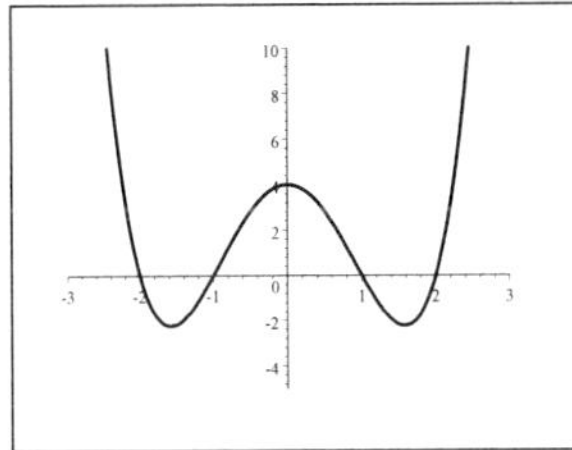

**60.** $y = 1.2x^5 + 3.75x^4 - 7x^3 - 15x^2 + 18x$ has two local maxima at $(0.50, 4.65)$ and $(-2.97, 12.10)$ and two local minima at $(-1.40, -27.44)$ and $(1.40, -2.54)$.

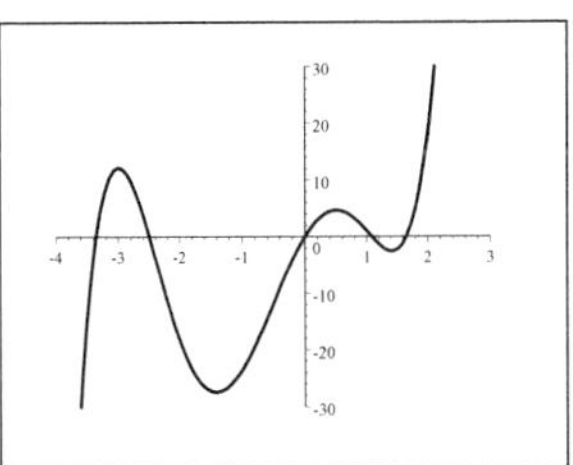

**61.** $y = (x - 2)^5 + 32$ has no maximum or minimum.

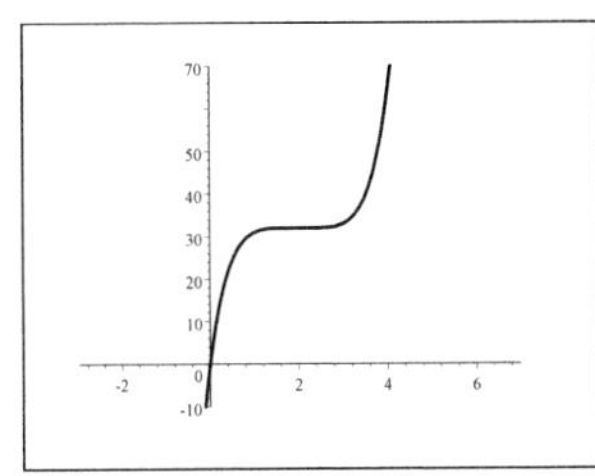

**62.** $y = \left(x^2 - 2\right)^3$ has one local minimum at $(0, -8)$.

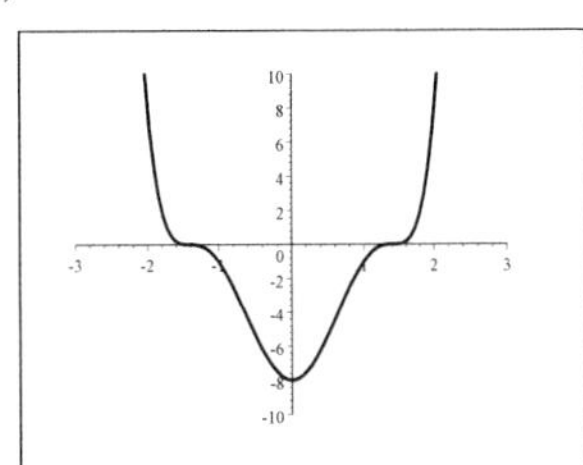

**63.** $y = x^8 - 3x^4 + x$ has one local maximum at $(0.44, 0.33)$ and two local minima at $(1.09, -1.15)$ and $(-1.12, -3.36)$.

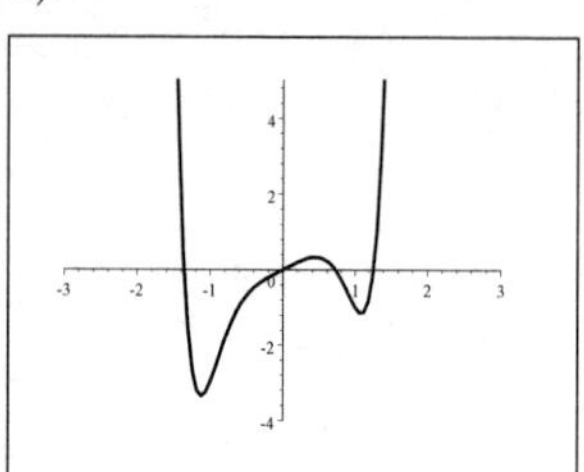


**64.** $y = \frac{1}{3}x^7 - 17x^2 + 7$ has one local maximum at $(0, 7)$ and one local minimum at $(1.71, -28.46)$.

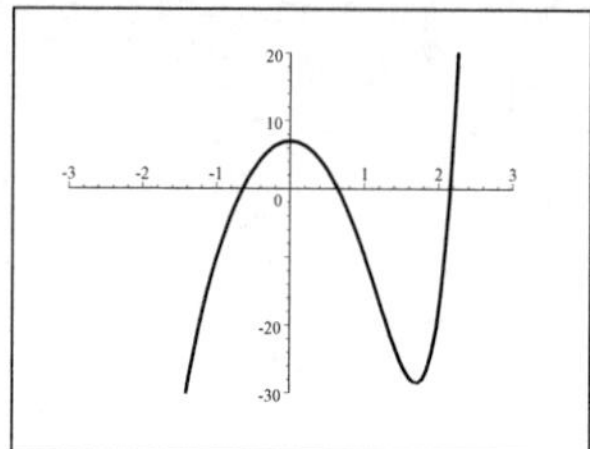


**65.** $y = cx^3$; $c = 1, 2, 5, \frac{1}{2}$. Increasing the value of $c$ stretches the graph vertically.

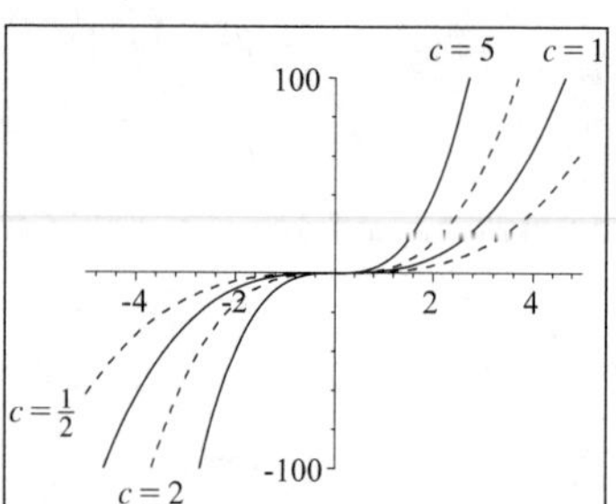


**66.** $P(x) = (x - c)^4$; $c = -1, 0, 1, 2$. Increasing the value of $c$ shifts the graph to the right.

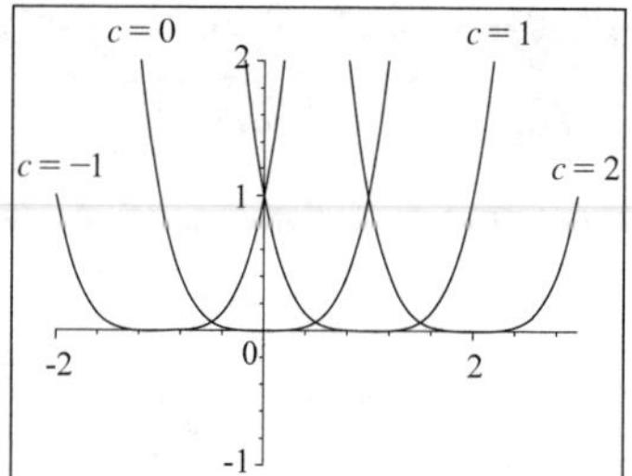


**67.** $P(x) = x^4 + c$; $c = -1, 0, 1$, and $2$. Increasing the value of $c$ moves the graph up.

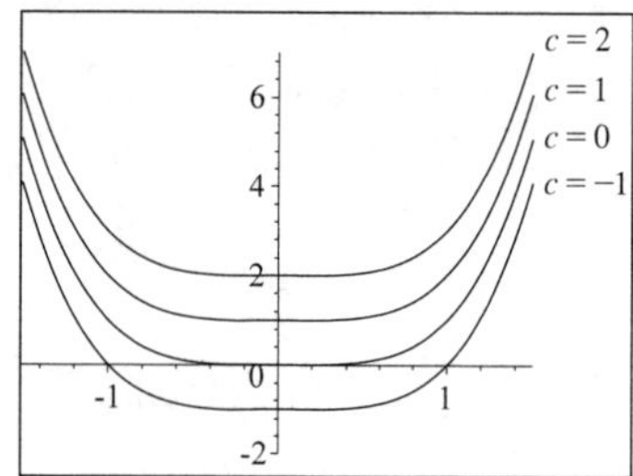


**68.** $P(x) = x^3 + cx$; $c = 2, 0, -2, -4$. Increasing the value of $c$ makes the "bumps" in the graph flatter.

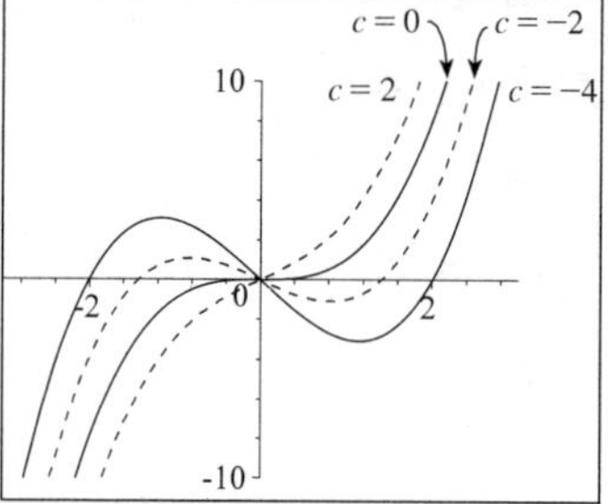


**69.** $P(x) = x^4 - cx$; $c = 0, 1, 8$, and $27$. Increasing the value of $c$ causes a deeper dip in the graph, in the fourth quadrant, and moves the positive $x$-intercept to the right.

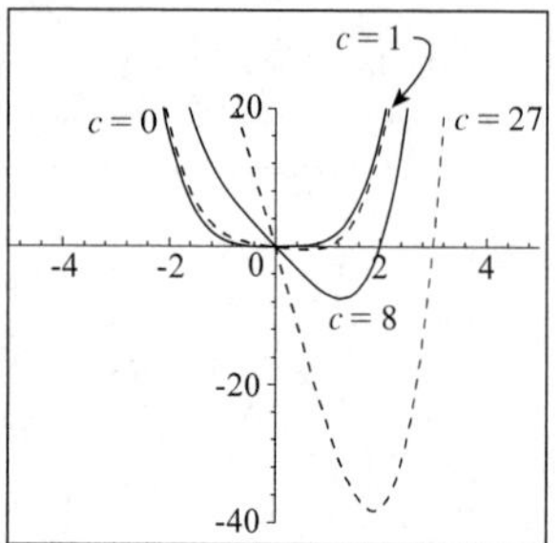


**70.** $P(x) = x^c$; $c = 1, 3, 5, 7$. The larger $c$ gets, the flatter the graph is near the origin, and the steeper it is away from the origin.

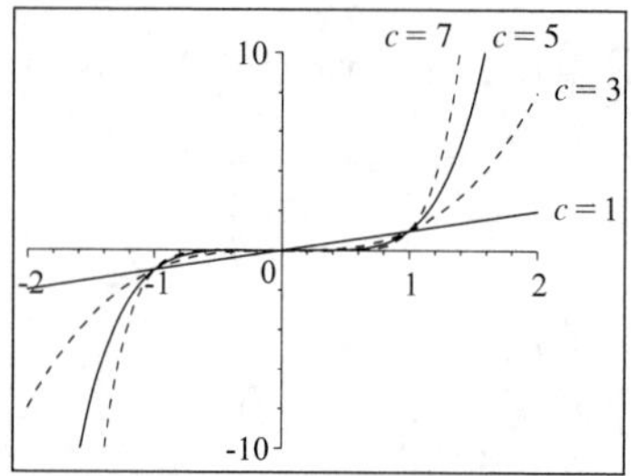

**71. (a)**

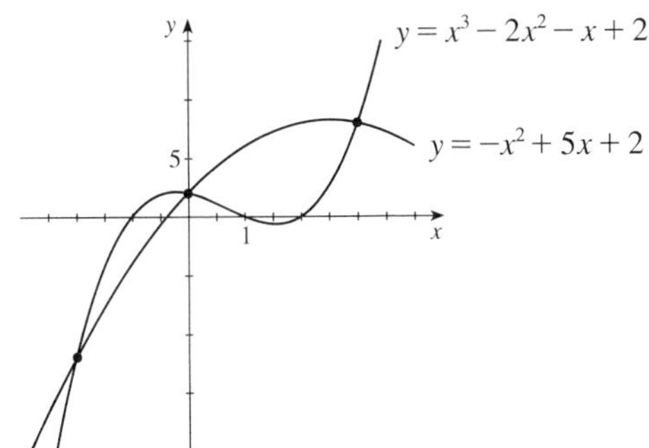


**(b)** The two graphs appear to intersect at 3 points.

**(c)** $x^3 - 2x^2 - x + 2 = -x^2 + 5x + 2 \quad \Leftrightarrow \quad x^3 - x^2 - 6x = 0 \quad \Leftrightarrow \quad x\left(x^2 - x - 6\right) = 0 \Leftrightarrow$ $x(x-3)(x+2) = 0$. Then either $x = 0$, $x = 3$, or $x = -2$. If $x = 0$, then $y = 2$; if $x = 3$ then $y = 8$; if $x = -2$, then $y = -12$. Hence the points where the two graphs intersect are $(0, 2)$, $(3, 8)$, and $(-2, -12)$.

**72.** Graph 1 belongs to $y = x^4$. Graph 2 belongs to $y = x^2$. Graph 3 belongs to $y = x^6$. Graph 4 belongs to $y = x^3$. Graph 5 belongs to $y = x^5$.

**73. (a)** Let $P(x)$ be a polynomial containing only odd powers of $x$. Then each term of $P(x)$ can be written as $Cx^{2n+1}$, for some constant $C$ and integer $n$. Since $C(-x)^{2n+1} = -Cx^{2n+1}$, each term of $P(x)$ is an odd function. Thus by part (a), $P(x)$ is an odd function.

**(b)** Let $P(x)$ be a polynomial containing only even powers of $x$. Then each term of $P(x)$ can be written as $Cx^{2n}$, for some constant $C$ and integer $n$. Since $C(-x)^{2n} = Cx^{2n}$, each term of $P(x)$ is an even function. Thus by part (b), $P(x)$ is an even function.

**(c)** Since $P(x)$ contains both even and odd powers of $x$, we can write it in the form $P(x) = R(x) + Q(x)$, where $R(x)$ contains all the even-powered terms in $P(x)$ and $Q(x)$ contains all the odd-powered terms. By part (d), $Q(x)$ is an odd function, and by part (e), $R(x)$ is an even function. Thus, since neither $Q(x)$ nor $R(x)$ are constantly 0 (by assumption), by part (c), $P(x) = R(x) + Q(x)$ is neither even nor odd.

**(d)** $P(x) = x^5 + 6x^3 - x^2 - 2x + 5 = \left(x^5 + 6x^3 - 2x\right) + \left(-x^2 + 5\right) = P_O(x) + P_E(x)$ where $P_O(x) = x^5 + 6x^3 - 2x$ and $P_E(x) = -x^2 + 5$. Since $P_O(x)$ contains only odd powers of $x$, it is an odd function, and since $P_E(x)$ contains only even powers of $x$, it is an even function.

**74. (a)** $P(x) = (x-1)(x-3)(x-4)$. Local maximum at $(1.8, 2.1)$. Local minimum at $(3.6, -0.6)$.

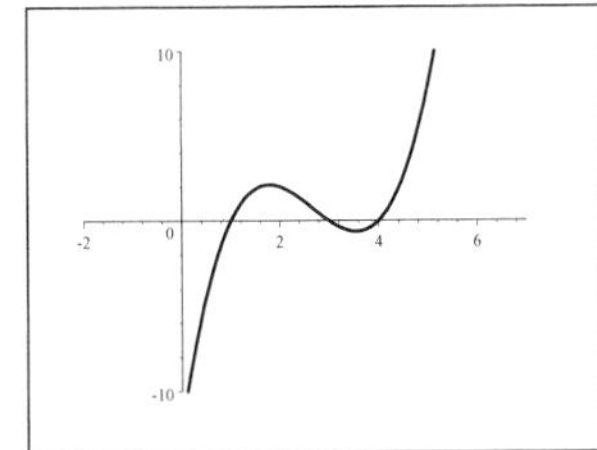


**(b)** Since $Q(x) = P(x) + 5$, each point on the graph of $Q$ has $y$-coordinate 5 units more than the corresponding point on the graph of $P$. Thus $Q$ has a local maximum at $(1.8, 7.1)$ and a local minimum at $(3.5, 4.4)$.

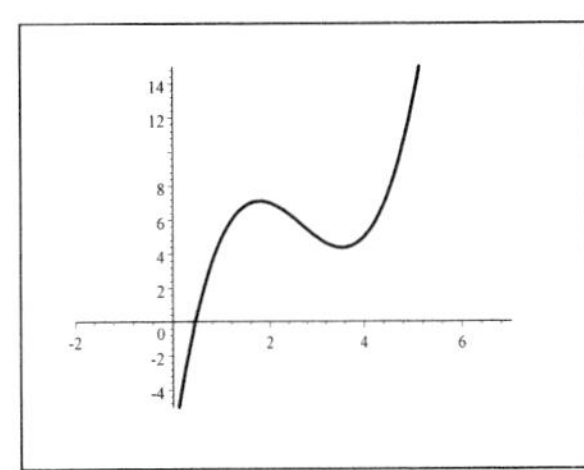

**75. (a)** $P(x) = (x-2)(x-4)(x-5)$ has one local maximum and one local minimum.

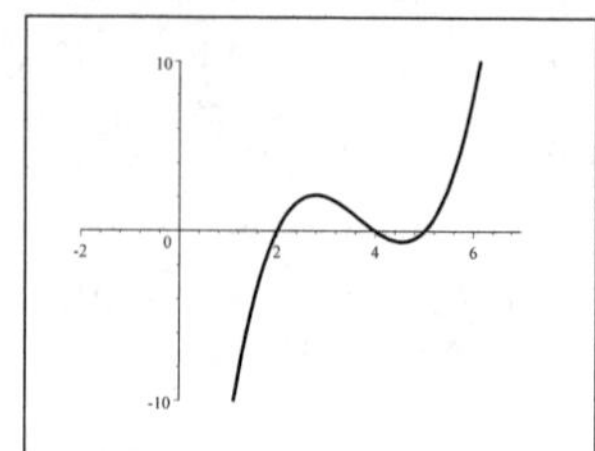

**(b)** Since $P(a) = P(b) = 0$, and $P(x) > 0$ for $a < x < b$ (see the table below), the graph of $P$ must first rise and then fall on the interval $(a, b)$, and so $P$ must have at least one local maximum between $a$ and $b$. Using similar reasoning, the fact that $P(b) = P(c) = 0$ and $P(x) < 0$ for $b < x < c$ shows that $P$ must have at least one local minimum between $b$ and $c$. Thus $P$ has at least two local extrema.

| Interval | $(-\infty, a)$ | $(a, b)$ | $(b, c)$ | $(c, \infty)$ |
|---|---|---|---|---|
| Sign of $x-a$ | $-$ | $+$ | $+$ | $+$ |
| Sign of $x-b$ | $-$ | $-$ | $+$ | $+$ |
| Sign of $x-c$ | $-$ | $-$ | $-$ | $+$ |
| Sign of $(x-a)(x-b)(x-c)$ | $-$ | $+$ | $-$ | $+$ |

**76. (a)** From the graph, $P(x) = x^3 - 4x = x(x-2)(x+2)$ has three $x$-intercepts, one local maximum, and one local minimum.

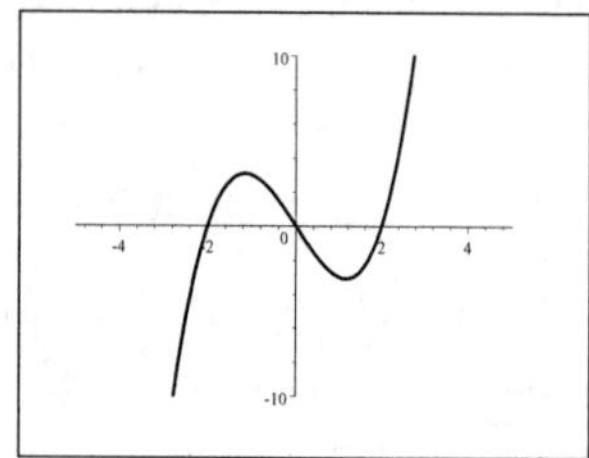

**(b)** From the graph, $Q(x) = x^3 + 4x = x(x^2+4)$ has one $x$ intercept and no local maximum or minimum.

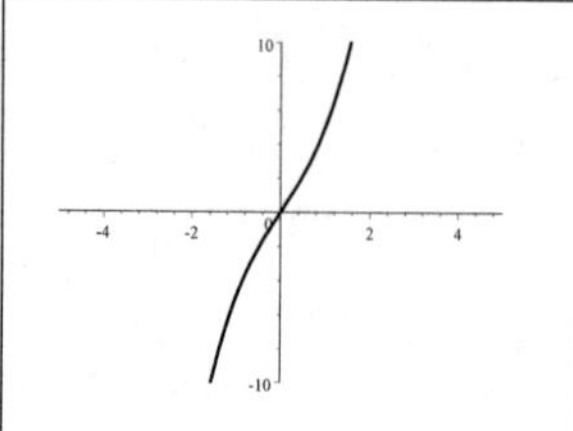

**(c)** For the $x$-intercepts of $P(x) = x^3 - ax$, we solve $x^3 - ax = 0$. Then we have $x(x^2 - a) = 0 \quad \Leftrightarrow \quad x = 0$ or $x^2 = a$. If $x^2 = a$, then $x = \pm\sqrt{a}$. So $P$ has 3 $x$-intercepts. Since $P(x) = x(x^2 - a) = x(x+\sqrt{a})(x-\sqrt{a})$, by part (c) of problem 67, $P$ has 2 local extrema. For the $x$-intercepts of $Q(x) = x^3 + ax$, we solve $x^3 + ax = 0$. Then we have $x(x^2 + a) = 0 \Leftrightarrow x = 0$ or $x^2 = -a$. The equation $x^2 = -a$ has no real solutions because $a > 0$. So $Q$ has 1 $x$-intercept. We now show that $Q$ is always increasing and hence has no extrema. If$x_1 < x_2$, then $ax_1 < ax_2$ (because $a > 0$) and $x_1^3 < x_2^3$. So we have $x_1^3 + ax_1 < x_2^3 + ax_2$, and hence $Q(x_1) < Q(x_2)$. Thus $Q$ is increasing, that is, its graph always rises, and so it has no local extrema.

**77.** $P(x) = 8x + 0.3x^2 - 0.0013x^3 - 372$

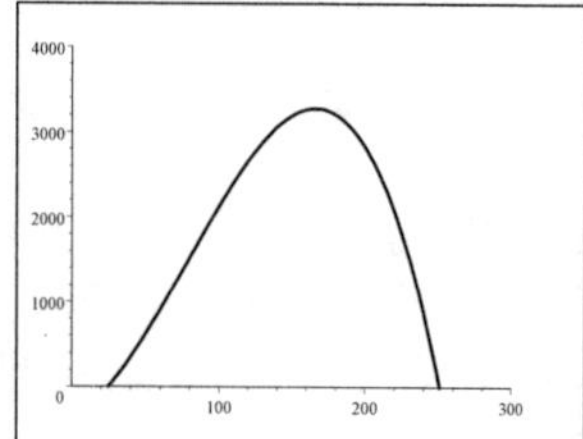

**(a)** For the firm to break even, $P(x) = 0$. From the graph, we see that $P(x) = 0$ when $x \approx 25.2$. Of course, the firm cannot produce fractions of a blender, so the manufacturer must produce at least 26 blenders a year.

**(b)** No, the profit does not increase indefinitely. The largest profit is approximately \$ 3276.22, which occurs when the firm produces 166 blenders per year.

**78.** $P(t) = 120t - 0.4t^4 + 1000$

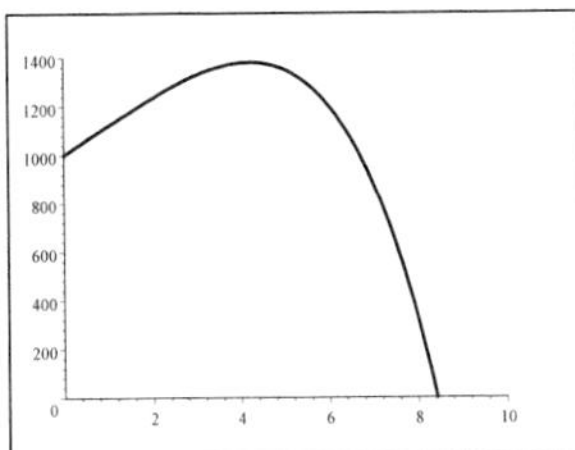

**(a)** A maximum population of approximately 1380 is attained after 4.22 months.

**(b)** The rabbit population disappears after approximately 8.42 months.

**79.** **(a)** The length of the bottom is $40 - 2x$, the width of the bottom is $20 - 2x$, and the height is $x$, so the volume of the box is $V = x(20 - 2x)(40 - 2x) = 4x^3 - 120x^2 + 800x$.

**(b)** Since the height and width must be positive, we must have $x > 0$ and $20 - 2x > 0$, and so the domain of $V$ is $0 < x < 10$.

**(c)** Using the domain from part (b), we graph $V$ in the viewing rectangle $[0, 10]$ by $[0, 1600]$. The maximum volume is $V \approx 1539.6$ when $x = 4.23$.

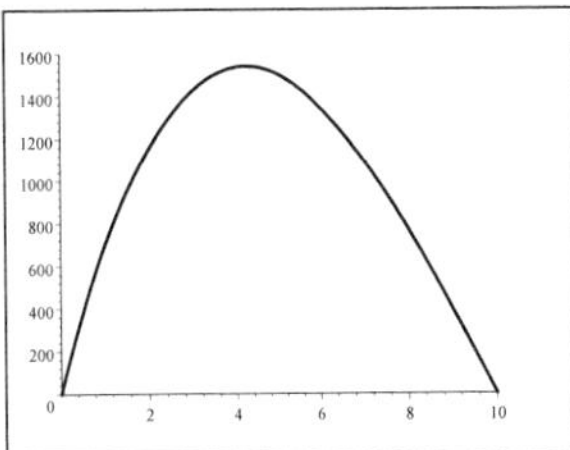

**80.** **(a)** Let $h$ be the height of the box. Then the total length of all 12 edges is $8x + 4h = 144$ in. Thus, $8x + 4h = 144 \quad \Leftrightarrow \quad 2x + h = 36 \quad \Leftrightarrow \quad h = 36 - 2x$. The volume of the box is equal to (area of base) $\times$ (height) $= \left(x^2\right) \times (36 - 2x) = -2x^3 + 36x^2$. Therefore, the volume of the box is $V(x) = -2x^3 + 36x^2 = 2x^2(18 - x)$.

**(b)** Since the length of the base is $x$, we must have $x > 0$. Likewise, the height must be positive so $36 - 2x > 0 \quad \Leftrightarrow \quad x < 18$. Putting these together, we get that the domain of $V$ is $0 < x < 18$.

**(c)** Using the domain from part (b), we graph $V$ in the viewing rectangle $[0, 18]$ by $[0, 2000]$. The maximum volume is $V = 1728$ in$^3$ when $x = 12$ in.

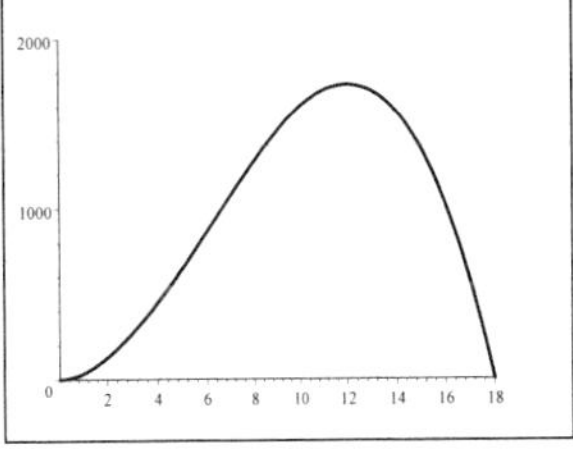

**81.**

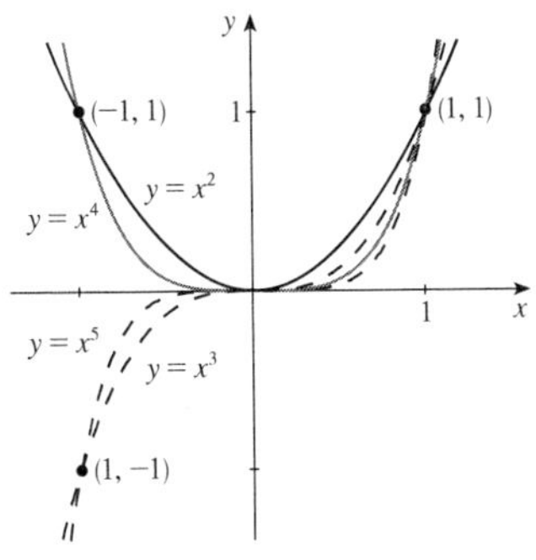


The graph of $y = x^{100}$ is close to the $x$-axis for $|x| < 1$, but passes through the points $(1, 1)$ and $(-1, 1)$. The graph of $y = x^{101}$ behaves similarly except that the $y$-values are negative for negative values of $x$, and it passes through $(-1, -1)$ instead of $(-1, 1)$.

**82.** Since the polynomial shown has five zeros, it has at least five factors, and so the degree of the polynomial is greater than or equal to 5.

**83.** No, it is impossible. The end behavior of a third degree polynomial is the same as that of $y = kx^3$, and for this function, the values of $y$ go off in opposite directions as $x \to \infty$ and $x \to -\infty$. But for a function with just one extremum, the values of $y$ would head off in the same direction (either both up or both down) on either side of the extremum. An $n$th-degree polynomial can have $n-1$ extrema or $n-3$ extrema or $n-5$ extrema, and so on (decreasing by 2). A polynomial that has six local extrema must be of degree 7 or higher. For example, $P(x) = (x-1)(x-2)(x-3)(x-4)(x-5)(x-6)(x-7)$ has six local extrema.

**84.** No, it is not possible. Clearly a polynomial must have a local minimum between any two local maxima.

## 4.2 Dividing Polynomials

**1.**

$$\begin{array}{r|l} & 3x - 4 \\ \hline x+3 & 3x^2+5x-4 \\ & 3x^2+9x \\ \hline & \quad -4x-4 \\ & \quad -4x-12 \\ \hline & \qquad 8 \end{array}$$

Thus the quotient is $3x - 4$ and the remainder is 8, so $P(x) = 3x^2 + 5x - 4 = (x+3) \cdot (3x-4) + 8$.

**2.** *Long division:*

$$\begin{array}{r|l} & x^2 + 5x - 1 \\ \hline x-1 & x^3 + 4x^2 - 6x + 1 \\ & x^3 - x^2 \\ \hline & \quad 5x^2 - 6x \\ & \quad 5x^2 - 5x \\ \hline & \qquad -x + 1 \\ & \qquad -x + 1 \\ \hline & \qquad\quad 0 \end{array}$$

Thus $P(x) = x^3 + 4x^2 - 6x + 1 = (x-1) \cdot (x^2 + 5x - 1)$.

*Synthetic division:*

$$\begin{array}{r|rrrr} 1 & 1 & 4 & -6 & 1 \\ & & 1 & 5 & -1 \\ \hline & 1 & 5 & -1 & 0 \end{array}$$

Thus the quotient is $x^2 + 5x - 1$ and the remainder is 0, as above.

**3.**

$$\begin{array}{r|l} & x^2 \qquad\quad - 1 \\ \hline 2x-3 & 2x^3 - 3x^2 - 2x \\ & 2x^3 - 3x^2 \\ \hline & \qquad\quad -2x \\ & \qquad\quad -2x + 3 \\ \hline & \qquad\qquad -3 \end{array}$$

Thus the quotient is $x^2 - 1$ and the remainder is $-3$, and $P(x) = 2x^3 - 3x^2 - 2x = (x^2 - 1)(2x - 3) - 3$.

**4.**

$$\begin{array}{r|l} & 2x^2 - x + 4 \\ \hline 2x+1 & 4x^3 \qquad + 7x + 9 \\ & 4x^3 + 2x^2 \\ \hline & \quad -2x^2 + 7x \\ & \quad -2x^2 - x \\ \hline & \qquad\quad 8x + 9 \\ & \qquad\quad 8x + 4 \\ \hline & \qquad\qquad 5 \end{array}$$

Thus $P(x) = 4x^3 + 7x + 9 = (2x^2 - x + 4)(2x + 1) + 5$.

**5.**

$$\begin{array}{r|l}
 & x^2 - x - 3 \\ \hline
x^2+3 & x^4 - x^3 + 0x^2 + 4x + 2 \\
 & x^4 \qquad + 3x^2 \\ \hline
 & \quad - x^3 - 3x^2 + 4x \\
 & \quad - x^3 \qquad - 3x \\ \hline
 & \qquad -3x^2 + 7x + 2 \\
 & \qquad -3x^2 \qquad - 9 \\ \hline
 & \qquad\qquad 7x + 11
\end{array}$$

Thus the quotient is $x^2 - x - 3$ and the remainder is $7x + 11$, and

$$\begin{aligned} P(x) &= x^3 + 4x^2 - 6x + 1 \\ &= \left(x^2+3\right)\cdot\left(x^2 - x - 3\right) + (7x + 11) \end{aligned}$$

**6.**

$$\begin{array}{r|l}
 & 2x^3 + 4x^2 \qquad + 8 \\ \hline
x^2-2 & x^5 + 4x^4 - 4x^3 + 0x^2 - x - 3 \\
 & x^5 \qquad - 4x^3 \\ \hline
 & \quad 4x^4 \qquad + 0x^2 \\
 & \quad 4x^4 \qquad - 8x^2 \\ \hline
 & \qquad\qquad 8x^2 - x - 3 \\
 & \qquad\qquad 8x^2 \qquad - 16 \\ \hline
 & \qquad\qquad\quad -x + 13
\end{array}$$

Thus

$$\begin{aligned} P(x) &= 2x^5 + 4x^4 - 4x^3 - x - 3 \\ &= \left(x^2 - 2\right)\cdot\left(2x^3 + 4x^2 + 8\right) + (-x + 13) \end{aligned}$$

**7.**

$$\begin{array}{r|rrr}
-3 & 1 & 4 & -8 \\
 & & -3 & -3 \\ \hline
 & 1 & 1 & -11
\end{array}$$

Thus the quotient is $x + 1$ and the remainder is $-11$, and

$$\frac{P(x)}{D(x)} = \frac{x^2 + 4x - 8}{x + 3} = (x + 1) + \frac{-11}{x + 3}.$$

**8.**

$$\begin{array}{r|rrrr}
4 & 1 & 0 & 6 & 5 \\
 & & 4 & 16 & 88 \\ \hline
 & 1 & 4 & 22 & 93
\end{array}$$

Thus the quotient is $x^2 + 4x + 22$ and the remainder is 93, and $\dfrac{P(x)}{D(x)} = \dfrac{x^3 + 6x + 5}{x - 4} = \left(x^2 + 4x + 22\right) + \dfrac{93}{x - 4}$.

**9.**

$$\begin{array}{r|l}
 & 2x - \frac{1}{2} \\ \hline
2x-1 & 4x^2 - 3x - 7 \\
 & 4x^2 - 2x \\ \hline
 & \qquad -x - 7 \\
 & \qquad -x + \frac{1}{2} \\ \hline
 & \qquad\quad -\frac{15}{2}
\end{array}$$

Thus the quotient is $2x - \frac{1}{2}$ and the remainder is $-\frac{15}{2}$, and

$$\frac{P(x)}{D(x)} = \frac{4x^2 - 3x - 7}{2x - 1} = \left(2x - \tfrac{1}{2}\right) + \frac{-\frac{15}{2}}{2x - 1}.$$

**10.**

$$\begin{array}{r|l}
 & 2x^2 + 3x \\ \hline
3x-4 & 6x^3 + x^2 - 12x + 5 \\
 & 6x^3 - 8x^2 \\ \hline
 & \qquad 9x^2 - 12x \\
 & \qquad 9x^2 - 12x \\ \hline
 & \qquad\qquad\quad 5
\end{array}$$

Thus the quotient is $2x^2 + 3x$ and the remainder is 5, and

$$\frac{P(x)}{D(x)} = \frac{6x^3 + x^2 - 12x + 5}{3x - 4} = \left(2x^2 + 3x\right) + \frac{5}{3x - 4}.$$

**11.**

$$\begin{array}{r|l}
 & 2x^2 - x + 1 \\ \hline
x^2+4 & 2x^4 - x^3 + 9x^2 \\
 & 2x^4 \qquad + 8x^2 \\ \hline
 & \quad -x^3 + x^2 \\
 & \quad -x^3 \qquad - 4x \\ \hline
 & \qquad\quad x^2 + 4x \\
 & \qquad\quad x^2 \qquad + 4 \\ \hline
 & \qquad\qquad 4x - 4
\end{array}$$

Thus the quotient is $2x^2 - x + 1$ and the remainder is $4x - 4$, and

$$\frac{P(x)}{D(x)} = \frac{2x^4 - x^3 + 9x^2}{x^2 + 4} = \left(2x^2 - x + 1\right) + \frac{4x - 4}{x^2 + 4}.$$

**12.**

$$\begin{array}{r|l}
 & x^3 \qquad - x + 1 \\ \hline
x^2+x-1 & x^5 + x^4 - 2x^3 + 0x^2 + x + 1 \\
 & x^5 + x^4 - x^3 \\ \hline
 & \qquad\quad -x^3 \qquad + x \\
 & \qquad\quad -x^3 - x^2 + x \\ \hline
 & \qquad\qquad\quad x^2 \qquad + 1 \\
 & \qquad\qquad\quad x^2 + x - 1 \\ \hline
 & \qquad\qquad\qquad -x + 2
\end{array}$$

Thus the quotient is $x^3 - x + 1$ and the remainder is $-x + 2$, and

$$\begin{aligned} \frac{P(x)}{D(x)} &= \frac{x^5 + x^4 - 2x^3 + x + 1}{x^2 + x - 1} \\ &= \left(x^3 - x + 1\right) + \frac{-x + 2}{x^2 + x - 1} \end{aligned}$$

**13.**

$$\begin{array}{r|l} & x - 2 \\ \hline x-4 & x^2 - 6x - 8 \\ & x^2 - 4x \\ \hline & \quad -2x - 8 \\ & \quad -2x + 8 \\ \hline & \qquad -16 \end{array}$$

Thus the quotient is $x - 2$ and the remainder is $-16$.

**14.**

$$\begin{array}{r|l} & x^2 + x \\ \hline x-2 & x^3 - x^2 - 2x + 6 \\ & x^3 - 2x^2 \\ \hline & \quad x^2 - 2x \\ & \quad x^2 - 2x \\ \hline & \qquad 6 \end{array}$$

Thus the quotient is $x^2 + x$, and the remainder is 6.

**15.**

$$\begin{array}{r|l} & 2x^2 - 1 \\ \hline 2x+1 & 4x^3 + 2x^2 - 2x - 3 \\ & 4x^3 + 2x^2 \\ \hline & \qquad -2x - 3 \\ & \qquad -2x - 1 \\ \hline & \qquad\quad -2 \end{array}$$

Thus the quotient is $2x^2 - 1$ and the remainder is $-2$.

**16.**

$$\begin{array}{r|l} & \frac{1}{3}x^2 + \frac{1}{3}x + \frac{2}{3} \\ \hline 3x+6 & x^3 + 3x^2 + 4x + 3 \\ & x^3 + 2x^2 \\ \hline & \quad x^2 + 4x \\ & \quad x^2 + 2x \\ \hline & \qquad 2x + 3 \\ & \qquad 2x + 4 \\ \hline & \qquad\quad -1 \end{array}$$

Thus the quotient is $\frac{1}{3}x^2 + \frac{1}{3}x + \frac{2}{3}$, and the remainder is $-1$.

**17.**

$$\begin{array}{r|l} & x + 2 \\ \hline x^2-2x+2 & x^3 + 0x^2 + 6x + 3 \\ & x^3 - 2x^2 + 2x \\ \hline & \quad 2x^2 + 4x + 3 \\ & \quad 2x^2 - 4x + 4 \\ \hline & \qquad 8x - 1 \end{array}$$

Thus the quotient is $x + 2$, and the remainder is $8x - 1$.

**18.**

$$\begin{array}{r|l} & 3x^2 - 8x - 1 \\ \hline x^2+x+3 & 3x^4 - 5x^3 + 0x^2 - 20x - 5 \\ & 3x^4 + 3x^3 + 9x^2 \\ \hline & \quad -8x^3 - 9x^2 - 20x \\ & \quad -8x^3 - 8x^2 - 24x \\ \hline & \qquad -x^2 + 4x - 5 \\ & \qquad -x^2 - x - 3 \\ \hline & \qquad\quad 5x - 2 \end{array}$$

Thus the quotient is $3x^2 - 8x - 1$, and the remainder is $5x - 2$.

**19.**

$$\begin{array}{r|l} & 3x + 1 \\ \hline 2x^2+0x+5 & 6x^3 + 2x^2 + 22x + 0 \\ & 6x^3 + 15x \\ \hline & \quad 2x^2 + 7x + 0 \\ & \quad 2x^2 + 5 \\ \hline & \qquad 7x - 5 \end{array}$$

Thus the quotient is $3x + 1$, and the remainder is $7x - 5$.

**20.**

$$\begin{array}{r|l} & 3 \\ \hline 3x^2-7x & 9x^2 - x + 5 \\ & 9x^2 - 21x \\ \hline & \quad 20x + 5 \end{array}$$

Thus the quotient is 3, and the remainder is $20x + 5$.

**21.**

$$\begin{array}{r|l}
 & x^4 \qquad\qquad\qquad\qquad\qquad + 1 \\ \hline
x^2+1 & x^6 + 0x^5 + x^4 + 0x^3 + x^2 + 0x + 1 \\
 & x^6 \qquad + x^4 \\ \hline
 & \qquad\qquad 0 \qquad + x^2 \qquad + 1 \\
 & \qquad\qquad\qquad\quad x^2 \qquad + 1 \\ \hline
 & \qquad\qquad\qquad\qquad\qquad\quad 0
\end{array}$$

Thus the quotient is $x^4 + 1$, and the remainder is 0.

**22.**

$$\begin{array}{r|l}
 & \tfrac{1}{2}x^3 - x^2 - \tfrac{5}{2}x - \tfrac{7}{4} \\ \hline
4x^2-6x+8 & 2x^5 - 7x^4 + 0x^3 + 0x^2 + 0x - 13 \\
 & 2x^5 - 3x^4 + 4x^3 \\ \hline
 & \quad -4x^4 - 4x^3 + 0x^2 \\
 & \quad -4x^4 + 6x^3 - 8x^2 \\ \hline
 & \qquad -10x^3 + 8x^2 + 0x \\
 & \qquad -10x^3 + 15x^2 - 20x \\ \hline
 & \qquad\qquad -7x^2 + 20x - 13 \\
 & \qquad\qquad -7x^2 + \tfrac{21}{2}x - 14 \\ \hline
 & \qquad\qquad\qquad \tfrac{19}{2}x + 1
\end{array}$$

Thus the quotient is $\frac{1}{2}x^3 - x^2 - \frac{5}{2}x - \frac{7}{4}$, and the remainder is $\frac{19}{2}x + 1$.

**23.** The synthetic division table for this problem takes the following form.

$$\begin{array}{r|rrr}
3 & 1 & -5 & 4 \\
 & & 3 & -6 \\ \hline
 & 1 & -2 & -2
\end{array}$$

Thus the quotient is $x - 2$, and the remainder is $-2$.

**24.** The synthetic division table for this problem takes the following form.

$$\begin{array}{r|rrr}
1 & 1 & -5 & 4 \\
 & & 1 & -4 \\ \hline
 & 1 & -4 & 0
\end{array}$$

Thus the quotient is $x - 4$, and the remainder is 0.

**25.** The synthetic division table for this problem takes the following form.

$$\begin{array}{r|rrr}
6 & 3 & 5 & 0 \\
 & & 18 & 138 \\ \hline
 & 3 & 23 & 138
\end{array}$$

Thus the quotient is $3x + 23$, and the remainder is 138.

**26.** The synthetic division table for this problem takes the following form.

$$\begin{array}{r|rrr}
-5 & 4 & 0 & -3 \\
 & & -20 & 100 \\ \hline
 & 4 & -20 & 97
\end{array}$$

Thus the quotient is $4x - 20$,and the remainder is 97.

**27.** Since $x + 2 = x - (-2)$, the synthetic division table for this problem takes the following form.

$$\begin{array}{r|rrrr}
-2 & 1 & 2 & 2 & 1 \\
 & & -2 & 0 & -4 \\ \hline
 & 1 & 0 & 2 & -3
\end{array}$$

Thus the quotient is $x^2 + 2$, and the remainder is $-3$.

**28.** The synthetic division table for this problem takes the following form.

$$\begin{array}{r|rrrr}
5 & 3 & -12 & -9 & 1 \\
 & & 15 & 15 & 30 \\ \hline
 & 3 & 3 & 6 & 31
\end{array}$$

Thus the quotient is $3x^2 + 3x + 6$, and the remainder is 31.

**29.** Since $x + 3 = x - (-3)$ and $x^3 - 8x + 2 = x^3 + 0x^2 - 8x + 2$, the synthetic division table for this problem takes the following form.

$$\begin{array}{r|rrrr}
-3 & 1 & 0 & -8 & 2 \\
 & & -3 & 9 & -3 \\ \hline
 & 1 & -3 & 1 & -1
\end{array}$$

Thus the quotient is $x^2 - 3x + 1$, and the remainder is $-1$.

**30.** The synthetic division table for this problem takes the following form.

$$\begin{array}{r|rrrrr}
2 & 1 & -1 & 1 & -1 & 2 \\
 & & 2 & 2 & 6 & 10 \\ \hline
 & 1 & 1 & 3 & 5 & 12
\end{array}$$

Thus the quotient is $x^3 + x^2 + 3x + 5$, and the remainder is 12.

**31.** Since $x^5 + 3x^3 - 6 = x^5 + 0x^4 + 3x^3 + 0x^2 + 0x - 6$, the synthetic division table for this problem takes the following form.

$$\begin{array}{r|rrrrrr} 1 & 1 & 0 & 3 & 0 & 0 & -6 \\ & & 1 & 1 & 4 & 4 & 4 \\ \hline & 1 & 1 & 4 & 4 & 4 & -2 \end{array}$$

Thus the quotient is $x^4 + x^3 + 4x^2 + 4x + 4$, and the remainder is $-2$.

**32.** The synthetic division table for this problem takes the following form.

$$\begin{array}{r|rrrr} 3 & 1 & -9 & 27 & -27 \\ & & 3 & -18 & 27 \\ \hline & 1 & -6 & 9 & 0 \end{array}$$

Thus the quotient is $x^2 - 6x + 9$, and the remainder is 0.

**33.** The synthetic division table for this problem takes the following form.

$$\begin{array}{r|rrrr} \frac{1}{2} & 2 & 3 & -2 & 1 \\ & & 1 & 2 & 0 \\ \hline & 2 & 4 & 0 & 1 \end{array}$$

Thus the quotient is $2x^2 + 4x$, and the remainder is 1.

**34.** The synthetic division table for this problem takes the following form.

$$\begin{array}{r|rrrrr} -\frac{2}{3} & 6 & 10 & 5 & 1 & 1 \\ & & -4 & -4 & -\frac{2}{3} & -\frac{2}{9} \\ \hline & 6 & 6 & 1 & \frac{1}{3} & \frac{7}{9} \end{array}$$

Thus the quotient is $6x^3 + 6x^2 + x + \frac{1}{3}$, and the remainder is $\frac{7}{9}$.

**35.** Since $x^3 - 27 = x^3 + 0x^2 + 0x - 27$, the synthetic division table for this problem takes the following form.

$$\begin{array}{r|rrrr} 3 & 1 & 0 & 0 & -27 \\ & & 3 & 9 & 27 \\ \hline & 1 & 3 & 9 & 0 \end{array}$$

Thus the quotient is $x^2 + 3x + 9$, and the remainder is 0.

**36.** The synthetic division table for this problem takes the following form.

$$\begin{array}{r|rrrrr} -2 & 1 & 0 & 0 & 0 & -16 \\ & & -2 & 4 & -8 & 16 \\ \hline & 1 & -2 & 4 & -8 & 0 \end{array}$$

Thus the quotient is $x^3 - 2x^2 + 4x - 8$, and the remainder is 0.

**37.** $P(x) = 4x^2 + 12x + 5$, $c = -1$

$$\begin{array}{r|rrr} -1 & 4 & 12 & 5 \\ & & -4 & -8 \\ \hline & 4 & 8 & -3 \end{array}$$

Therefore, by the Remainder Theorem, $P(-1) = -3$.

**38.** $P(x) = 2x^2 + 9x + 1$, $c = \frac{1}{2}$

$$\begin{array}{r|rrr} \frac{1}{2} & 2 & 9 & 1 \\ & & 1 & 5 \\ \hline & 2 & 10 & 6 \end{array}$$

Therefore, by the Remainder Theorem, $P\left(\frac{1}{2}\right) = 6$.

**39.** $P(x) = x^3 + 3x^2 - 7x + 6$, $c = 2$

$$\begin{array}{r|rrrr} 2 & 1 & 3 & -7 & 6 \\ & & 2 & 10 & 6 \\ \hline & 1 & 5 & 3 & 12 \end{array}$$

Therefore, by the Remainder Theorem, $P(2) = 12$.

**40.** $P(x) = x^3 - x^2 + x + 5$, $c = -1$

$$\begin{array}{r|rrrr} -1 & 1 & -1 & 1 & 5 \\ & & -1 & 2 & -3 \\ \hline & 1 & -2 & 3 & 2 \end{array}$$

Therefore, by the Remainder Theorem, $P(-1) = 2$.

**41.** $P(x) = x^3 + 2x^2 - 7$, $c = -2$

$$\begin{array}{r|rrrr} -2 & 1 & 2 & 0 & -7 \\ & & -2 & 0 & 0 \\ \hline & 1 & 0 & 0 & -7 \end{array}$$

Therefore, by the Remainder Theorem, $P(-2) = -7$.

**42.** $P(x) = 2x^3 - 21x^2 + 9x - 200$, $c = 11$

$$\begin{array}{r|rrrr} 11 & 2 & -21 & 9 & -200 \\ & & 22 & 11 & 220 \\ \hline & 2 & 1 & 20 & 20 \end{array}$$

Therefore, by the Remainder Theorem, $P(11) = 20$.

**43.** $P(x) = 5x^4 + 30x^3 - 40x^2 + 36x + 14$, $c = -7$

$$\begin{array}{r|rrrrr} -7 & 5 & 30 & -40 & 36 & 14 \\ & & -35 & 35 & 35 & -497 \\ \hline & 5 & -5 & -5 & 71 & -483 \end{array}$$

Therefore, by the Remainder Theorem, $P(-7) = -483$.

**44.** $P(x) = 6x^5 + 10x^3 + x + 1$, $c = -2$

$$\begin{array}{r|rrrrrr} -2 & 6 & 0 & 10 & 0 & 1 & 1 \\ & & -12 & 24 & -68 & 136 & -274 \\ \hline & 6 & -12 & 34 & -68 & 137 & -273 \end{array}$$

Therefore, by the Remainder Theorem, $P(-2) = -273$.

**45.** $P(x) = x^7 - 3x^2 - 1$

$= x^7 + 0x^6 + 0x^5 + 0x^4 + 0x^3 - 3x^2 + 0x - 1$

$c = 3$

$$\begin{array}{r|rrrrrrrr} 3 & 1 & 0 & 0 & 0 & 0 & -3 & 0 & -1 \\ & & 3 & 9 & 27 & 81 & 243 & 720 & 2160 \\ \hline & 1 & 3 & 9 & 27 & 81 & 240 & 720 & 2159 \end{array}$$

Therefore by the Remainder Theorem, $P(3) = 2159$.

**46.** $P(x) = -2x^6 + 7x^5 + 40x^4 - 7x^2 + 10x + 112$, $c = -3$

$$\begin{array}{r|rrrrrrr} -3 & -2 & 7 & 40 & 0 & -7 & 10 & 112 \\ & & 6 & -39 & -3 & 9 & -6 & -12 \\ \hline & -2 & 13 & 1 & -3 & 2 & 4 & 100 \end{array}$$

Therefore, by the Remainder Theorem, $P(-3) = 100$.

**47.** $P(x) = 3x^3 + 4x^2 - 2x + 1$, $c = \frac{2}{3}$

$$\begin{array}{r|rrrr} \frac{2}{3} & 3 & 4 & -2 & 1 \\ & & 2 & 4 & \frac{4}{3} \\ \hline & 3 & 6 & 2 & \frac{7}{3} \end{array}$$

Therefore, by the Remainder Theorem, $P\left(\frac{2}{3}\right) = \frac{7}{3}$.

**48.** $P(x) = x^3 - x + 1$, $c = \frac{1}{4}$

$$\begin{array}{r|rrrr} \frac{1}{4} & 1 & 0 & -1 & 1 \\ & & \frac{1}{4} & \frac{1}{16} & -\frac{15}{64} \\ \hline & 1 & \frac{1}{4} & -\frac{15}{16} & \frac{49}{64} \end{array}$$

Therefore, by the Remainder Theorem, $P\left(\frac{1}{4}\right) = \frac{49}{64}$.

**49.** $P(x) = x^3 + 2x^2 - 3x - 8$, $c = 0.1$

$$\begin{array}{r|rrrr} 0.1 & 1 & 2 & -3 & -8 \\ & & 0.1 & 0.21 & -0.279 \\ \hline & 1 & 2.1 & -2.79 & -8.279 \end{array}$$

Therefore, by the Remainder Theorem, $P(0.1) = -8.279$.

**50. (a)** $P(x) = 6x^7 - 40x^6 + 16x^5 - 200x^4 - 60x^3 - 69x^2 + 13x - 139$, $c = 7$

$$\begin{array}{r|rrrrrrrr} 7 & 6 & -40 & 16 & -200 & -60 & -69 & 13 & -139 \\ & & 42 & 14 & 210 & 70 & 70 & 7 & 140 \\ \hline & 6 & 2 & 30 & 10 & 10 & 1 & 20 & 1 \end{array}$$

Therefore, by the Remainder Theorem, $P(7) = 1$.

**(b)** $P(7) = 6(7)^7 - 40(7)^6 + 16(7)^5 - 200(7)^4 - 60(7)^3 - 69(7)^2 + 13(7) - 139$

$= 6(823{,}543) - 40(117{,}649) + 16(16{,}807) - 200(2401) - 60(343) - 69(49) + 13(7) - 139 = 1$

which agrees with the value obtained by synthetic division, but requires more work.

**51.** $P(x) = x^3 - 3x^2 + 3x - 1$, $c = 1$

$$\begin{array}{r|rrrr} 1 & 1 & -3 & 3 & -1 \\ & & 1 & -2 & 1 \\ \hline & 1 & -2 & 1 & 0 \end{array}$$

Since the remainder is 0, $x - 1$ is a factor.

**52.** $P(x) = x^3 + 2x^2 - 3x - 10$, $c = 2$

$$\begin{array}{r|rrrr} 2 & 1 & 2 & -3 & -10 \\ & & 2 & 8 & 10 \\ \hline & 1 & 4 & 5 & 0 \end{array}$$

Since the remainder is 0, $x - 2$ is a factor.

**53.** $P(x) = 2x^3 + 7x^2 + 6x - 5$, $c = \frac{1}{2}$

$$\begin{array}{r|rrrr} \frac{1}{2} & 2 & 7 & 6 & -5 \\ & & 1 & 4 & 5 \\ \hline & 2 & 8 & 10 & 0 \end{array}$$

Since the remainder is 0, $x - \frac{1}{2}$ is a factor.

**54.** $P(x) = x^4 + 3x^3 - 16x^2 - 27x + 63$, $c = 3, -3$

$$\begin{array}{r|rrrrr} 3 & 1 & 3 & -16 & -27 & 63 \\ & & 3 & 18 & 6 & -63 \\ \hline & 1 & 6 & 2 & -21 & 0 \end{array}$$

Since the remainder is 0, $x - 3$ is a factor. We next show that $x + 3$ is also a factor by using synthetic division on the quotient of the above synthetic division, $x^3 + 6x^2 - 2x - 21$.

$$\begin{array}{r|rrrr} -3 & 1 & 6 & 2 & -21 \\ & & -3 & -9 & 21 \\ \hline & 1 & 3 & -7 & 0 \end{array}$$

Since the remainder is 0, $x + 3$ is a factor.

**55.** $P(x) = x^3 - x^2 - 11x + 15$, $c = 3$

$$\begin{array}{r|rrrr} 3 & 1 & -1 & -11 & 15 \\ & & 3 & 6 & -15 \\ \hline & 1 & 2 & -5 & 0 \end{array}$$

Since the remainder is 0, we know that 3 is a zero $x^3 - x^2 - 11x + 15 = (x - 3)(x^2 + 2x - 5)$. Now $x^2 + 2x - 5 = 0$ when $x = \frac{-2 \pm \sqrt{2^2 + 4(1)(5)}}{2} = -1 \pm \sqrt{6}$. Hence, the zeros are $-1 - \sqrt{6}$, $-1 + \sqrt{6}$, and 3.

**56.** $P(x) = 3x^4 - x^3 - 21x^2 - 11x + 6$, $c = \frac{1}{3}, -2$.

$$\begin{array}{r|rrrrr} \frac{1}{3} & 3 & -1 & -21 & -11 & 6 \\ & & 1 & 0 & -7 & -6 \\ \hline & 3 & 0 & -21 & -18 & 0 \end{array}$$

Since the remainder is 0, it follows that $\frac{1}{3}$ is a zero. Thus $P(x) = \left(x - \frac{1}{3}\right)\left(3x^3 - 21x - 18\right)$. Next, we use synthetic division on the quotient.

$$\begin{array}{r|rrrr} -2 & 3 & 0 & -21 & -18 \\ & & -6 & 12 & 18 \\ \hline & 3 & -6 & -9 & 0 \end{array}$$

Since the remainder is 0, it follows that 2 is a zero. So

$$\begin{aligned} P(x) &= \left(x - \tfrac{1}{3}\right)(x + 2)\left(3x^2 - 6x - 9\right) \\ &= 3\left(x - \tfrac{1}{3}\right)(x + 2)\left(x^2 - 2x - 3\right) \\ &= 3\left(x - \tfrac{1}{3}\right)(x + 2)(x - 3)(x + 1) \end{aligned}$$

Hence, the zeros are $\frac{1}{3}$, $-2$, $-1$, and 3.

**57.** Since the zeros are $x = -1$, $x = 1$, and $x = 3$, the factors are $x + 1$, $x - 1$, and $x - 3$. Thus $P(x) = (x + 1)(x - 1)(x - 3) = x^3 - 3x^2 - x + 3$.

**58.** Since the zeros are $x = -2$, $x = 0$, $x = 2$, and $x = 4$, the factors are $x + 2$, $x$, $x - 2$, and $x - 4$. Thus $P(x) = c(x + 2)x(x - 2)(x - 4)$. If we let $c = 1$, then $P(x) = x^4 - 4x^3 - 4x^2 + 16x$.

**59.** Since the zeros are $x = -1$, $x = 1$, $x = 3$, and $x = 5$, the factors are $x + 1$, $x - 1$, $x - 3$, and $x - 5$. Thus $P(x) = (x + 1)(x - 1)(x - 3)(x - 5) = x^4 - 8x^3 + 14x^2 + 8x - 15$.

**60.** Since the zeros are $x=-2$, $x=-1$, $x=0$, $x=1$, and $x=2$, the factors are $x+2$, $x+1$, $x$, $x-1$, and $x-2$. Thus $P(x)=c(x+2)(x+1)x(x-1)(x-2)$. If we let $c=1$, then $P(x)=x^5-5x^3+4x$.

**61.** Since the zeros of the polynomial are 1, $-2$, and 3, it follows that $P(x)=C(x-1)(x+2)(x-3)=C(x^3-2x^2-5x+6)=Cx^3-2Cx^2-5Cx+6C$. Since the coefficient of $x^2$ is to be 3, $-2C=3$ so $C=-\frac{3}{2}$. Therefore, $P(x)=-\frac{3}{2}(x^3-2x^2-5x+6)=-\frac{3}{2}x^3+3x^2+\frac{15}{2}x-9$ is the polynomial.

**62.** Since the zeros of the polynomial are 1, $-1$, 2, and $\frac{1}{2}$, it follows that $P(x)=C(x-1)(x+1)(x-2)(x-\frac{1}{2})=C(x^2-1)(x^2-\frac{5}{2}x+1)$. To ensure integer coefficients, we choose $C$ to be any nonzero multiple of 2. When $C=2$, we have $P(x)=2(x^2-1)(x^2-\frac{5}{2}x+1)=(x^2-1)(2x^2-5x+2)=2x^4-5x^3+5x-2$. (Note: This is just one of many polynomials with the desired zeros and integer coefficients. We can choose $C$ to be any even integer.)

**63.** The $y$-intercept is 2 and the zeros of the polynomial are $-1$, 1, and 2. It follows that $P(x)=C(x+1)(x-1)(x-2)=C(x^3-2x^2-x+2)$. Since $P(0)=2$ we have $2=C[(0)^3-2(0)^2-(0)+2]$ $\Leftrightarrow$ $2=2C\Leftrightarrow C=1$ and $P(x)=(x+1)(x-1)(x-2)=x^3-2x^2-x+2$.

**64.** The $y$-intercept is 4 and the zeros of the polynomial are $-1$ and 2 with 2 being degree two. It follows that $P(x)=C(x+1)(x-2)^2=C(x^3-3x^2+4)$. Since $P(0)=4$ we have $4=C[(0)^3-3(0)^2+4]$ $\Leftrightarrow$ $4=4C$ $\Leftrightarrow$ $C=1$ and $P(x)=x^3-3x^2+4$.

**65.** The $y$-intercept is 4 and the zeros of the polynomial are $-2$ and 1 both being degree two. It follows that $P(x)=C(x+2)^2(x-1)^2=C(x^4+2x^3-3x^2-4x+4)$. Since $P(0)=4$ we have $4=C[(0)^4+2(0)^3-3(0)^2-4(0)+4]$ $\Leftrightarrow$ $4=4C\Leftrightarrow C=1$. Thus $P(x)=(x+2)^2(x-1)^2=x^4+2x^3-3x^2-4x+4$.

**66.** The $y$-intercept is 2 and the zeros of the polynomial are $-2$, $-1$, and 1 with 1 being degree two. It follows that $P(x)=C(x+2)(x+1)(x-1)^2=C(x^4+x^3-3x^2-x+2)$. Since $P(0)=2$ we have $4=C[(0)^4+(0)^3-3(0)^2-(0)+2]$ $\Leftrightarrow$ $2=2C$ so $C=1$ and $P(x)=x^4+x^3-3x^2-x+2$.

**67.** **A.** By the Remainder Theorem, the remainder when $P(x)=6x^{1000}-17x^{562}+12x+26$ is divided by $x+1$ is $P(-1)=6(-1)^{1000}-17(-1)^{562}+12(-1)+26=6-17-12+26=3$.

**B.** If $x-1$ is a factor of $Q(x)=x^{567}-3x^{400}+x^9+2$, then $Q(1)$ must equal 0. $Q(1)=(1)^{567}-3(1)^{400}+(1)^9+2=1-3+1+2=1\neq 0$, so $x-1$ is not a factor.

**68.** $R(x)=x^5-2x^4+3x^3-2x^2+3x+4=(x^4-2x^3+3x^2-2x+3)x+4=((x^3-2x^2+3x-2)x+3)x+4=(((x^2-2x+3)x-2)x+3)x+4=((((x-2)x+3)x-2)x+3)x+4$

So to calculate $R(3)$, we start with 3, then subtract 2, multiply by 3, add 3, multiply by 3, subtract 2, multiply by 3, add 3, multiply by 3, and add 4, to get 157.

## 4.3 Real Zeros of Polynomials

**1.** $P(x)=x^3-4x^2+3$ has possible rational zeros $\pm1$ and $\pm3$.

**2.** $Q(x)=x^4-3x^3-6x+8$ has possible rational zeros $\pm1$, $\pm2$, $\pm4$, $\pm8$.

**3.** $R(x)=2x^5+3x^3+4x^2-8$ has possible rational zeros $\pm1$, $\pm2$, $\pm4$, $\pm8$, $\pm\frac{1}{2}$.

**4.** $S(x)=6x^4-x^2+2x+12$ has possible rational zeros $\pm1$, $\pm2$, $\pm3$, $\pm4$, $\pm6$, $\pm12$, $\pm\frac{1}{2}$, $\pm\frac{3}{2}$, $\pm\frac{1}{3}$, $\pm\frac{2}{3}$, $\pm\frac{4}{3}$, $\pm\frac{1}{6}$.

**5.** $T(x)=4x^4-2x^2-7$ has possible rational zeros $\pm1$, $\pm7$, $\pm\frac{1}{2}$, $\pm\frac{7}{2}$, $\pm\frac{1}{4}$, $\pm\frac{7}{4}$.

**6.** $U(x)=12x^5+6x^3-2x-8$ has possible rational zeros $\pm1$, $\pm2$, $\pm4$, $\pm8$, $\pm\frac{1}{2}$, $\pm\frac{1}{3}$, $\pm\frac{2}{3}$, $\pm\frac{4}{3}$, $\pm\frac{8}{3}$, $\pm\frac{1}{4}$, $\pm\frac{1}{6}$, $\pm\frac{1}{12}$.

**7.** **(a)** $P(x) = 5x^3 - x^2 - 5x + 1$ has possible rational zeros $\pm 1, \pm \frac{1}{5}$.

**(b)** From the graph, the actual zeroes are $-1$, $\frac{1}{5}$, and 1.

**8.** **(a)** $P(x) = 3x^3 + 4x^2 - x - 2$ has possible rational zeros $\pm 1, \pm 2, \pm \frac{1}{3}, \pm \frac{2}{3}$.

**(b)** From the graph, the actual zeroes are $-1$ and $\frac{2}{3}$.

**9.** **(a)** $P(x) = 2x^4 - 9x^3 + 9x^2 + x - 3$ has possible rational zeros $\pm 1, \pm 3, \pm \frac{1}{2}, \pm \frac{3}{2}$.

**(b)** From the graph, the actual zeroes are $-\frac{1}{2}$, 1, and 3.

**10.** **(a)** $P(x) = 4x^4 - x^3 - 4x + 1$ has possible rational zeros $\pm 1, \pm \frac{1}{2}, \pm \frac{1}{4}$.

**(b)** From the graph, the actual zeroes are $\frac{1}{4}$ and 1.

**11.** $P(x) = x^3 + 3x^2 - 4$. The possible rational zeros are $\pm 1, \pm 2, \pm 4$. $P(x)$ has 1 variation in sign and hence 1 positive real zero. $P(-x) = -x^3 + 3x^2 - 4$ has 2 variations in sign and hence 0 or 2 negative real zeros.

$$\begin{array}{r|rrrr} 1 & 1 & 3 & 0 & -4 \\ & & 1 & 4 & 4 \\ \hline & 1 & 4 & 4 & 0 \end{array} \quad \Rightarrow x = 1 \text{ is a zero.}$$

$$\begin{aligned} P(x) &= x^3 + 3x^2 - 4 = (x - 1)\left(x^2 + 4x + 4\right) \\ &= (x - 1)(x + 2)^2 \end{aligned}$$

Therefore, the zeros are $x = -2, 1$.

**12.** $P(x) = x^3 - 7x^2 + 14x - 8$. The possible rational zeros are $\pm 1, \pm 2, \pm 4, \pm 8$. $P(x)$ has 3 variations in sign and hence 1 or 3 positive real zeros.

$P(-x) = -x^3 - 7x^2 - 14x - 8$ has 0 variations in sign and hence no negative real zeros.

$$\begin{array}{r|rrrr} 1 & 1 & -7 & 14 & -8 \\ & & 1 & -6 & 8 \\ \hline & 1 & -6 & 8 & 0 \end{array} \quad \Rightarrow x = 1 \text{ is a zero.}$$

So

$$\begin{aligned} P(x) &= x^3 - 7x^2 + 14x - 8 = (x - 1)\left(x^2 - 6x + 8\right) \\ &= (x - 1)(x - 2)(x - 4) \end{aligned}$$

Therefore, the zeros are $x = 1, 2, 4$.

**13.** $P(x) = x^3 - 3x - 2$. The possible rational zeros are $\pm 1, \pm 2$. $P(x)$ has 1 variation in sign and hence 1 positive real zero. $P(-x) = -x^3 + 3x - 2$ has 2 variations in sign and hence 0 or 2 negative real zeros.

$$\begin{array}{r|rrrr} 1 & 1 & 0 & -3 & -2 \\ & & 1 & 1 & -2 \\ \hline & 1 & 1 & -2 & -4 \end{array} \quad \Rightarrow x = 1 \text{ is not a zero.} \qquad \begin{array}{r|rrrr} 2 & 1 & 0 & -3 & -2 \\ & & 2 & 4 & 2 \\ \hline & 1 & 2 & 1 & 0 \end{array} \quad \Rightarrow x = 2 \text{ is a zero.}$$

$P(x) = x^3 - 3x - 2 = (x - 2)\left(x^2 + 2x + 1\right) = (x - 2)(x + 1)^2$. Therefore, the zeros are $x = 2, -1$.

**14.** $P(x) = x^3 + 4x^2 - 3x - 18$. The possible rational zeros are $\pm 1, \pm 2, \pm 3, \pm 6, \pm 9, \pm 18$. $P(x)$ has 1 variation in sign and hence 1 positive real zero. $P(-x) = -x^3 + 4x^2 + 3x - 18$ has 2 variations in sign and hence 0 or 2 negative real zeros.

$$\begin{array}{r|rrrr} 1 & 1 & 4 & -3 & -18 \\ & & 1 & 5 & 2 \\ \hline & 1 & 5 & 2 & -16 \end{array} \qquad \begin{array}{r|rrrr} 2 & 1 & 4 & -3 & -18 \\ & & 2 & 12 & 18 \\ \hline & 1 & 6 & 9 & 0 \end{array} \quad \Rightarrow x = 2 \text{ is a zero.}$$

$P(x) = x^3 + 4x^2 - 3x - 18 = (x - 2)\left(x^2 + 6x + 9\right) = (x - 2)(x + 3)^2$. Therefore, the zeros are $x = -3, 2$.

**15.** $P(x) = x^3 - 6x^2 + 12x - 8$. The possible rational zeros are $\pm 1, \pm 2, \pm 4, \pm 8$. $P(x)$ has 3 variations in sign and hence 1 or 3 positive real zeros. $P(-x) = -x^3 - 6x^2 - 12x - 8$ has no variations in sign and hence 0 negative real zeros.

$$\begin{array}{r|rrrr} 1 & 1 & -6 & 12 & -8 \\ & & 1 & -5 & 7 \\ \hline & 1 & -5 & 7 & -1 \end{array} \quad \Rightarrow x = 1 \text{ is not a zero.} \qquad \begin{array}{r|rrrr} 2 & 1 & -6 & 12 & -8 \\ & & 2 & -8 & 8 \\ \hline & 1 & -4 & 4 & 0 \end{array} \quad \Rightarrow x = 2 \text{ is a zero.}$$

$P(x) = x^3 - 6x^2 + 12x - 8 = (x - 2)\left(x^2 - 4x + 4\right) = (x - 2)^3$. Therefore, the zero is $x = 2$.

**16.** $P(x) = x^3 - x^2 - 8x + 12$. The possible rational zeros are $\pm1, \pm2, \pm3, \pm4, \pm6, \pm12$. $P(x)$ has 2 variations in sign and hence 0 or 2 positive real zeros. $P(-x) = -x^3 - x^2 + 8x + 12$ has 1 variation in sign and hence 1 negative real zero.

$$\begin{array}{r|rrrr} 1 & 1 & -1 & -8 & 12 \\ & & 1 & 0 & -8 \\ \hline & 1 & 0 & -8 & 4 \end{array} \qquad \begin{array}{r|rrrr} 2 & 1 & -1 & -8 & 12 \\ & & 2 & 2 & -12 \\ \hline & 1 & 1 & -6 & 0 \end{array} \Rightarrow x = 2 \text{ is a zero.}$$

$P(x) = x^3 - x^2 - 8x + 12 = (x-2)(x^2 + x - 6) = (x-2)(x+3)(x-2)$. Therefore, the zeros are $x = -3, 2$.

**17.** $P(x) = x^3 - 4x^2 + x + 6$. The possible rational zeros are $\pm1, \pm2, \pm3, \pm6$. $P(x)$ has 2 variations in sign and hence 0 or 2 positive real zeros. $P(-x) = -x^3 - 4x^2 - x + 6$ has 1 variation in sign and hence 1 negative real zeros.

$$\begin{array}{r|rrrr} -1 & 1 & -4 & 1 & 6 \\ & & -1 & 5 & -6 \\ \hline & 1 & -5 & 6 & 0 \end{array} \Rightarrow x + 1 \text{ is a factor.}$$

So

$$\begin{aligned} P(x) = x^3 - 4x^2 + x + 6 &= (x+1)(x^2 - 5x + 6) \\ &= (x+1)(x-3)(x-2) \end{aligned}$$

Therefore, the zeros are $x = -1, 2, 3$.

**18.** $P(x) = x^3 - 4x^2 - 7x + 10$. The possible rational zeros are $\pm1, \pm2, \pm5, \pm10$. $P(x)$ has 2 variations in sign and hence 0 or 2 positive real zeros.

$P(-x) = -x^3 - 4x^2 + 7x + 10$ has 1 variation in sign and hence 1 negative real zero.

$$\begin{array}{r|rrrr} 1 & 1 & -4 & -7 & 10 \\ & & 1 & -3 & -10 \\ \hline & 1 & -3 & -10 & 0 \end{array} \Rightarrow x = 1 \text{ is a zero.}$$

So

$$\begin{aligned} P(x) = x^3 - 4x^2 - 7x + 10 &= (x-1)(x^2 - 3x - 10) \\ &= (x-1)(x-5)(x+2) \end{aligned}$$

Therefore, the zeros are $x = -2, 1, 5$.

**19.** $P(x) = x^3 + 3x^2 + 6x + 4$. The possible rational zeros are $\pm1, \pm2, \pm4$. $P(x)$ has no variation in sign and hence no positive real zeros. $P(-x) = -x^3 + 3x^2 - 6x + 4$ has 3 variations in sign and hence 1 or 3 negative real zeros.

$$\begin{array}{r|rrrr} -1 & 1 & 3 & 6 & 4 \\ & & -1 & -2 & -4 \\ \hline & 1 & 2 & 4 & 0 \end{array} \Rightarrow x + 1 \text{ is a factor.}$$

So $P(x) = x^3 + 3x^2 + 6x + 4 = (x+1)(x^2 + 2x + 4)$. Now, $Q(x) = x^2 + 2x + 4$ has no real zeros, since the discriminant of this quadratic is $b^2 - 4ac = (2)^2 - 4(1)(4) = -12 < 0$. Thus, the only real zero is $x = -1$.

**20.** $P(x) = x^3 - 2x^2 - 2x - 3$. The possible rational zeros are $\pm1, \pm3$. $P(x)$ has 1 variation in sign and hence 1 positive real zero. $P(-x) = -x^3 - 2x^2 + 2x - 3$ has 2 variations in sign and hence 0 or 2 negative real zeros.

$$\begin{array}{r|rrrr} 1 & 1 & -2 & -2 & -3 \\ & & 1 & -1 & -3 \\ \hline & 1 & -1 & -3 & -6 \end{array} \qquad \begin{array}{r|rrrr} 3 & 1 & -2 & -2 & -3 \\ & & 3 & 3 & 3 \\ \hline & 1 & 1 & 1 & 0 \end{array} \Rightarrow x = 3 \text{ is a zero.}$$

So $P(x) = x^3 - 2x^2 - 2x - 3 = (x-3)(x^2 + x + 1)$. Now, $Q(x) = x^2 + x + 1$ has no real zeros, since the discriminant is $b^2 - 4ac = (1)^2 - 4(1)(1) = -3 < 0$. Thus, the only real zero is $x = 3$.

**21.** *Method 1:* $P(x) = x^4 - 5x^2 + 4$. The possible rational zeros are $\pm1$, $\pm2$, $\pm4$. $P(x)$ has 1 variation in sign and hence 1 positive real zero. $P(-x) = x^4 - 5x^2 + 4$ has 2 variations in sign and hence 0 or 2 negative real zeros.

$$\begin{array}{r|rrrrr} 1 & 1 & 0 & -5 & 0 & 4 \\ & & 1 & 1 & -4 & -4 \\ \hline & 1 & 1 & -4 & -4 & 0 \end{array} \Rightarrow x = 1 \text{ is a zero.}$$

Thus $P(x) = x^4 - 5x^2 + 4 = (x-1)(x^3 + x^2 - 4x - 4)$. Continuing with the quotient we have:

$$\begin{array}{r|rrrr} -1 & 1 & 1 & -4 & -4 \\ & & -1 & 0 & 4 \\ \hline & 1 & 0 & -4 & 0 \end{array} \Rightarrow x = -1 \text{ is a zero.}$$

$P(x) = x^4 - 5x^2 + 4 = (x-1)(x+1)(x^2-4) = (x-1)(x+1)(x-2)(x+2)$. Therefore, the zeros are $x = \pm1$, $\pm2$.

*Method 2:* Substituting $u = x^2$, the polynomial becomes $P(u) = u^2 - 5u + 4$, which factors: $u^2 - 5u + 4 = (u-1)(u-4) = (x^2-1)(x^2-4)$, so either $x^2 = 1$ or $x^2 = 4$. If $x^2 = 1$, then $x = \pm1$; if $x^2 = 4$, then $x = \pm2$. Therefore, the zeros are $x = \pm1, \pm2$.

**22.** $P(x) = x^4 - 2x^3 - 3x^2 + 8x - 4$. Using synthetic division, we see that $(x-1)$ is a factor of $P(x)$:

$$\begin{array}{r|rrrrr} 1 & 1 & -2 & -3 & 8 & -4 \\ & & 1 & -1 & -4 & 4 \\ \hline & 1 & -1 & -4 & 4 & 0 \end{array} \Rightarrow x = 1 \text{ is a zero.}$$

We continue by factoring the quotient, and we see that $(x-1)$ is again a factor:

$$\begin{array}{r|rrrr} 1 & 1 & -1 & -4 & 4 \\ & & 1 & 0 & -4 \\ \hline & 1 & 0 & -4 & 0 \end{array} \Rightarrow x = 1 \text{ is a zero.}$$

$P(x) = x^4 - 2x^3 - 3x^2 + 8x - 4 = (x-1)(x-1)(x^2-4) = (x-1)^2(x-2)(x+2)$

Therefore, the zeros are $x = 1, \pm2$.

**23.** $P(x) = x^4 + 6x^3 + 7x^2 - 6x - 8$. The possible rational zeros are $\pm1$, $\pm2$, $\pm4$, $\pm8$. $P(x)$ has 1 variation in sign and hence 1 positive real zero. $P(-x) = x^4 - 6x^3 + 7x^2 + 6x - 8$ has 3 variations in sign and hence 1 or 3 negative real zeros.

$$\begin{array}{r|rrrrr} 1 & 1 & 6 & 7 & -6 & -8 \\ & & 1 & 7 & 14 & 8 \\ \hline & 1 & 7 & 14 & 8 & 0 \end{array} \Rightarrow x = 1 \text{ is a zero}$$

and there are no other positive zeros. Thus $P(x) = x^4 + 6x^3 + 7x^2 - 6x - 8 = (x-1)(x^3 + 7x^2 + 14x + 8)$. Continuing by factoring the quotient, we have:

$$\begin{array}{r|rrrr} -1 & 1 & 7 & 14 & 8 \\ & & -1 & -6 & -8 \\ \hline & 1 & 6 & 8 & 0 \end{array} \Rightarrow x = -1 \text{ is a zero.}$$

So $P(x) = x^4 + 6x^3 + 7x^2 - 6x - 8 = (x-1)(x+1)(x^2+6x+8) = (x-1)(x+1)(x+2)(x+4)$. Therefore, the zeros are $x = -4, -2, \pm1$.

**24.** $P(x) = x^4 - x^3 - 23x^2 - 3x + 90$. The possible rational zeros are $\pm1, \pm2, \pm3, \pm5, \pm6, \pm9, \pm10, \pm15, \pm18, \pm30, \pm45, \pm90$. Since $P(x)$ has 2 variations in sign, $P$ has 0 or 2 positive real zeros. Since $P(-x) = x^4 + x^3 - 23x^2 + 3x + 90$ has 2 variations in sign, $P$ has 0 or 2 negative real zeros.

$$\begin{array}{r|rrrrr} 1 & 1 & -1 & -23 & -3 & 90 \\ & & 1 & 0 & -23 & -26 \\ \hline & 1 & 0 & -23 & -26 & 64 \end{array} \qquad \begin{array}{r|rrrrr} 2 & 1 & -1 & -23 & -3 & 90 \\ & & 2 & 2 & -42 & -90 \\ \hline & 1 & 1 & -21 & -45 & 0 \end{array} \Rightarrow x = 2 \text{ is a zero.}$$

$P(x) = (x-2)(x^3 + x^2 - 21x - 45)$. Continuing with the quotient we have:

$$\begin{array}{r|rrrr} 3 & 1 & 1 & -21 & -45 \\ & & 3 & 12 & -27 \\ \hline & 1 & 4 & -9 & -72 \end{array} \qquad \begin{array}{r|rrrr} 5 & 1 & 1 & -21 & -45 \\ & & 5 & 30 & 45 \\ \hline & 1 & 6 & 9 & 0 \end{array} \Rightarrow x = 5 \text{ is a zero.}$$

$P(x) = (x-2)(x-5)(x^2 + 6x + 9) = (x-2)(x-5)(x+3)^2$. Therefore, the zeros are $x = -3, 2, 5$.

**25.** $P(x) = 4x^4 - 25x^2 + 36$ has possible rational zeros $\pm1, \pm2, \pm3, \pm4, \pm6, \pm9, \pm12, \pm18, \pm36, \pm\frac{1}{2}, \pm\frac{1}{4}, \pm\frac{3}{2}, \pm\frac{3}{4}, \pm\frac{9}{2}, \pm\frac{9}{4}$. Since $P(x)$ has 2 variations in sign, there are 0 or 2 positive real zeros. Since $P(-x) = 4x^4 - 25x^2 + 36$ has 2 variations in sign, there are 0 or 2 negative real zeros.

$$\begin{array}{r|rrrrr} 1 & 4 & 0 & -25 & 0 & 36 \\ & & 4 & 4 & -21 & -21 \\ \hline & 4 & 4 & -21 & -21 & 15 \end{array} \qquad \begin{array}{r|rrrrr} 2 & 4 & 0 & -25 & 0 & 36 \\ & & 8 & 16 & -18 & -36 \\ \hline & 4 & 8 & -9 & -18 & 0 \end{array} \Rightarrow x = 2 \text{ is a zero.}$$

$P(x) = (x-2)(4x^3 + 8x^2 - 9x - 18)$

$$\begin{array}{r|rrrr} 2 & 4 & 8 & -9 & -18 \\ & & 8 & 32 & 46 \\ \hline & 4 & 16 & 23 & 28 \end{array} \Rightarrow \text{all positive, } x = 2 \text{ is an upper bound.} \qquad \begin{array}{r|rrrr} \frac{1}{2} & 4 & 8 & -9 & -18 \\ & & 2 & 5 & -2 \\ \hline & 4 & 10 & -4 & -20 \end{array}$$

$$\begin{array}{r|rrrr} \frac{1}{4} & 4 & 8 & -9 & -18 \\ & & 1 & \frac{9}{4} & -\frac{27}{16} \\ \hline & 4 & 9 & -\frac{27}{4} & -\frac{315}{16} \end{array} \qquad \begin{array}{r|rrrr} \frac{3}{2} & 4 & 8 & -9 & -18 \\ & & 6 & 21 & 18 \\ \hline & 4 & 14 & 12 & 0 \end{array} \Rightarrow x = \frac{3}{2} \text{ is a zero.}$$

$P(x) = (x-2)(2x-3)(2x^2 + 7x + 6) = (x-2)(2x-3)(2x+3)(x+2)$. Therefore, the zeros are $x = \pm2, \pm\frac{3}{2}$.
*Note:* Since $P(x)$ has only even terms, factoring by substitution also works. Let $x^2 = u$; then $P(u) = 4u^2 - 25u + 36 = (u-4)(4u-9) = (x^2 - 4)(4x^2 - 9)$, which gives the same results.

**26.** $P(x) = x^4 - x^3 - 5x^2 + 3x + 6$. The possible rational zeros are $\pm1, \pm2, \pm3, \pm6$. Since $P(x)$ has 2 variations in sign, $P$ has 0 or 2 positive real zeros. Since $P(-x) = x^4 + x^3 - 5x^2 - 3x + 6$ has 2 variations in sign, $P$ has 0 or 2 negative real zeros.

$$\begin{array}{r|rrrrr} 1 & 1 & -1 & -5 & 3 & 6 \\ & & 1 & 0 & -5 & -2 \\ \hline & 1 & 0 & -5 & -2 & 4 \end{array} \qquad \begin{array}{r|rrrrr} 2 & 1 & -1 & -5 & 3 & 6 \\ & & 2 & 2 & -6 & -6 \\ \hline & 1 & 1 & -3 & -3 & 0 \end{array} \Rightarrow x = 2 \text{ is a zero.}$$

$P(x) = (x-2)(x^3 + x^2 - 3x - 3)$. Continuing with the quotient we have:

$$\begin{array}{r|rrrr} 3 & 1 & 1 & -3 & -3 \\ & & 3 & 12 & 27 \\ \hline & 1 & 4 & 9 & 24 \end{array} \Rightarrow x = 3 \text{ is an upper bound.} \qquad \begin{array}{r|rrrr} -1 & 1 & 1 & -3 & -3 \\ & & -1 & 0 & 3 \\ \hline & 1 & 0 & -3 & 0 \end{array} \Rightarrow x = -1 \text{ is a zero.}$$

$P(x) = (x-2)(x+1)(x^2 - 3)$. Therefore, the rational zeros are $x = -1, 2$.

**27.** $P(x) = x^4 + 8x^3 + 24x^2 + 32x + 16$. The possible rational zeros are $\pm 1, \pm 2, \pm 4, \pm 8, \pm 16$. $P(x)$ has no variations in sign and hence no positive real zero. $P(-x) = x^4 - 8x^3 + 24x^2 - 32x + 16$ has 4 variations in sign and hence 0 or 2 or 4 negative real zeros.

$$\begin{array}{r|rrrrr} -1 & 1 & 8 & 24 & 32 & 16 \\ & & -1 & -7 & -17 & -15 \\ \hline & 1 & 7 & 17 & 15 & 1 \end{array} \Rightarrow x = -1 \text{ is not a zero.}$$

$$\begin{array}{r|rrrrr} -2 & 1 & 8 & 24 & 32 & 16 \\ & & -2 & -12 & -24 & -16 \\ \hline & 1 & 6 & 12 & 8 & 0 \end{array} \Rightarrow x = -2 \text{ is a zero.}$$

Thus $P(x) = x^4 + 8x^3 + 24x^2 + 32x + 16 = (x+2)(x^3 + 6x^2 + 12x + 8)$. Continuing by factoring the quotient, we have

$$\begin{array}{r|rrrr} -2 & 1 & 6 & 12 & 8 \\ & & -2 & -8 & -8 \\ \hline & 1 & 4 & 4 & 0 \end{array} \Rightarrow x = -2 \text{ is a zero.}$$

Thus $P(x) = (x+2)^2 (x^2 + 4x + 4) = (x+2)^4$. Therefore, the zero is $x = -2$

**28.** $P(x) = 2x^3 + 7x^2 + 4x - 4$. The possible rational zeros are $\pm 1, \pm 2, \pm 4, \pm\frac{1}{2}$. Since $P(x)$ has 1 variation in sign, $P$ has 1 positive real zero. Since $P(-x) = -2x^3 + 7x^2 - 4x - 4$ has 2 variations in sign, $P$ has 0 or 2 negative real zeros.

$$\begin{array}{r|rrrr} 1 & 2 & 7 & 4 & -4 \\ & & 2 & 9 & 13 \\ \hline & 2 & 9 & 13 & 9 \end{array} \Rightarrow x = 1 \text{ is an upper bound.}$$

$$\begin{array}{r|rrrr} \frac{1}{2} & 2 & 7 & 4 & -4 \\ & & 1 & 4 & 4 \\ \hline & 2 & 8 & 8 & 0 \end{array} \Rightarrow x = \tfrac{1}{2} \text{ is a zero.}$$

$P(x) = (x - \frac{1}{2})(2x^2 + 8x + 8) = 2(x - \frac{1}{2})(x^2 + 4x + 4) = 2(x - \frac{1}{2})(x+2)^2$. Therefore, the zeros are $x = -2, \frac{1}{2}$.

**29.** Factoring by grouping can be applied to this exercise. $4x^3 + 4x^2 - x - 1 = 4x^2(x+1) - (x+1) = (x+1)(4x^2 - 1) = (x+1)(2x+1)(2x-1)$. Therefore, the zeros are $x = -1, \pm\frac{1}{2}$.

**30.** We use factoring by grouping: $P(x) = 2x^3 - 3x^2 - 2x + 3 = 2x(x^2 - 1) - 3(x^2 - 1) = (x^2 - 1)(2x - 3) = (x-1)(x+1)(2x-3)$. Therefore, the zeros are $x = \frac{3}{2}, \pm 1$.

**31.** $P(x) = 4x^3 - 7x + 3$. The possible rational zeros are $\pm 1, \pm 3, \pm\frac{1}{2}, \pm\frac{3}{2}, \pm\frac{1}{4}, \pm\frac{3}{4}$. Since $P(x)$ has 2 variations in sign, there are 0 or 2 positive zeros. Since $P(-x) = -4x^3 + 7x + 3$ has 1 variation in sign, there is 1 negative zero.

$$\begin{array}{r|rrrr} \frac{1}{2} & 4 & 0 & -7 & 3 \\ & & 2 & 1 & -3 \\ \hline & 4 & 2 & -6 & 0 \end{array} \Rightarrow x = \tfrac{1}{2} \text{ is a zero.}$$

$P(x) = (x - \frac{1}{2})(4x^2 + 2x - 6) = (2x - 1)(2x^2 + x - 3) = (2x-1)(x-1)(2x+3) = 0$. Thus, the zeros are $x = -\frac{3}{2}, \frac{1}{2}, 1$.

**32.** $P(x) = 8x^3 + 10x^2 - x - 3$. The possible rational zeros are $\pm 1, \pm 3, \pm\frac{1}{2}, \pm\frac{3}{2}, \pm\frac{1}{4}, \pm\frac{3}{4}, \pm\frac{1}{8}, \pm\frac{3}{8}$. Since $P(x)$ has 1 variation in sign, $P$ has 1 positive real zero. Since $P(-x) = -8x^3 + 10x^2 + x - 3$ has 2 variations in sign, $P$ has 0 or 2 negative real zeros.

$$\begin{array}{r|rrrr} 1 & 8 & 10 & -1 & -3 \\ & & 8 & 18 & 17 \\ \hline & 8 & 18 & 17 & 14 \end{array} \Rightarrow x = 1 \text{ is an upper bound.}$$

$$\begin{array}{r|rrrr} \frac{1}{2} & 8 & 10 & -1 & -3 \\ & & 4 & 7 & 3 \\ \hline & 8 & 14 & 6 & 0 \end{array} \Rightarrow x = \tfrac{1}{2} \text{ is a zero.}$$

$$\begin{aligned} P(x) &= 8x^3 + 10x^2 - x - 3 = (x - \tfrac{1}{2})(8x^2 + 14x + 6) \\ &= 2(x - \tfrac{1}{2})(4x^2 + 7x + 3) = (2x-1)(x+1)(4x+3) = 0 \end{aligned}$$

Therefore, the zeros are $x = -1, -\frac{3}{4}, \frac{1}{2}$.

**33.** $P(x) = 4x^3 + 8x^2 - 11x - 15$. The possible rational zeros are $\pm 1, \pm 3, \pm 5, \pm\frac{1}{2}, \pm\frac{1}{4}, \pm\frac{3}{2}, \pm\frac{3}{4}, \pm\frac{5}{2}, \pm\frac{5}{4}$. $P(x)$ has 1 variation in sign and hence 1 positive real zero. $P(-x) = -4x^3 + 8x^2 + 11x - 15$ has 2 variations in sign, so $P$ has 0 or 2 negative real zeros.

$$\begin{array}{r|rrrr} 1 & 4 & 8 & -11 & -15 \\ & & 4 & 12 & 1 \\ \hline & 4 & 12 & 1 & -14 \end{array} \Rightarrow x = 1 \text{ is not a zero.}$$

$$\begin{array}{r|rrrr} 3 & 4 & 8 & -11 & -15 \\ & & 12 & 60 & 147 \\ \hline & 4 & 20 & 49 & 132 \end{array} \Rightarrow x = 3 \text{ is not a zero.}$$

$$\begin{array}{r|rrrr} 5 & 4 & 8 & -11 & -15 \\ & & 20 & 140 & 645 \\ \hline & 4 & 28 & 129 & 630 \end{array} \Rightarrow x = 5 \text{ is not a zero.}$$

$$\begin{array}{r|rrrr} 3 & 4 & 8 & -11 & -15 \\ & & 12 & 60 & 147 \\ \hline & 4 & 20 & 49 & 132 \end{array} \Rightarrow x = 3 \text{ is not a zero.}$$

$$\begin{array}{r|rrrr} \frac{1}{2} & 4 & 8 & -11 & -15 \\ & & 2 & 5 & -3 \\ \hline & 4 & 10 & -6 & -18 \end{array} \Rightarrow x = \tfrac{1}{2} \text{ is not a zero.}$$

$$\begin{array}{r|rrrr} \frac{1}{4} & 4 & 8 & -11 & -15 \\ & & 1 & \frac{9}{4} & -\frac{35}{16} \\ \hline & 4 & 9 & -\frac{35}{4} & -\frac{275}{16} \end{array} \Rightarrow x = \tfrac{1}{4} \text{ is not a zero.}$$

$$\begin{array}{r|rrrr} \frac{3}{2} & 4 & 8 & -11 & -15 \\ & & 6 & 21 & 15 \\ \hline & 4 & 14 & 10 & 0 \end{array} \Rightarrow x = \tfrac{3}{2} \text{ is a zero.}$$

Thus $P(x) = 4x^3 + 8x^2 - 11x - 15 = \left(x - \frac{3}{2}\right)\left(4x^2 + 14x + 10\right)$. Continuing by factoring the quotient, whose possible rational zeros are $-1, -5, -\frac{1}{2}, -\frac{1}{4}, -\frac{5}{2}$, and $-\frac{5}{4}$, we have

$$\begin{array}{r|rrr} -1 & 4 & 14 & 10 \\ & & -4 & -10 \\ \hline & 4 & 10 & 0 \end{array} \Rightarrow x = -1 \text{ is a zero.}$$

Thus $P(x) = \left(x - \frac{3}{2}\right)(x+1)(4x+10)$ has zeros $\frac{3}{2}$, $-1$, and $-\frac{5}{2}$.

**34.** $P(x) = 6x^3 + 11x^2 - 3x - 2$. The possible rational zeros are $\pm 1, \pm 2, \pm\frac{1}{2}, \pm\frac{1}{3}, \pm\frac{1}{6}, \pm\frac{2}{3}$. $P(x)$ has 1 variation in sign and hence 1 positive real zero. $P(-x) = -6x^3 + 11x^2 + 3x - 2$ has 2 variations in sign and hence 0 or 2 negative real zeros.

$$\begin{array}{r|rrrr} 1 & 6 & 11 & -3 & -2 \\ & & 6 & 17 & 14 \\ \hline & 6 & 17 & 14 & 12 \end{array} \Rightarrow x = 1 \text{ is not a zero.}$$

$$\begin{array}{r|rrrr} 2 & 6 & 11 & -3 & -2 \\ & & 12 & 46 & 86 \\ \hline & 6 & 23 & 43 & 84 \end{array} \Rightarrow x = 2 \text{ is not a zero.}$$

$$\begin{array}{r|rrrr} \frac{1}{2} & 6 & 11 & -3 & -2 \\ & & 3 & 7 & 2 \\ \hline & 6 & 14 & 4 & 0 \end{array} \Rightarrow x = \tfrac{1}{2} \text{ is a zero.}$$

Thus, $P(x) = 6x^3 + 11x^2 - 3x - 2 = \left(x - \frac{1}{2}\right)\left(6x^2 + 14x + 4\right)$. Continuing:

$$\begin{array}{r|rrr} -1 & 6 & 14 & 4 \\ & & -6 & -8 \\ \hline & 6 & 8 & -4 \end{array} \Rightarrow x = -1 \text{ is not a zero.}$$

$$\begin{array}{r|rrr} -2 & 6 & 14 & 4 \\ & & -12 & -4 \\ \hline & 6 & 2 & 0 \end{array} \Rightarrow x = -2 \text{ is a zero.}$$

Thus, $P(x) = \left(x - \frac{1}{2}\right)(x+2)(6x+2)$ has zeros $\frac{1}{2}$, $-2$, and $-\frac{1}{3}$.

**35.** $P(x) = 2x^4 - 7x^3 + 3x^2 + 8x - 4$. The possible rational zeros are $\pm 1, \pm 2, \pm 4, \pm\frac{1}{2}$. $P(x)$ has 3 variations in sign and hence 1 or 3 positive real zeros. $P(-x) = 2x^4 + 7x^3 + 3x^2 - 8x - 4$ has 1 variation in sign and hence 1 negative real zero.

$$\begin{array}{r|rrrrr} 1 & 2 & -7 & 3 & 8 & -4 \\ & & 2 & -5 & -2 & 6 \\ \hline & 2 & -5 & -2 & 6 & 2 \end{array} \Rightarrow x = 1 \text{ is not a zero.} \qquad \begin{array}{r|rrrrr} \frac{1}{2} & 2 & -7 & 3 & 8 & -4 \\ & & 1 & -3 & 0 & 4 \\ \hline & 2 & -6 & 0 & 8 & 0 \end{array} \Rightarrow x = \tfrac{1}{2} \text{ is a zero.}$$

Thus $P(x) = 2x^4 - 7x^3 + 3x^2 + 8x - 4 = \left(x - \frac{1}{2}\right)\left(2x^3 - 6x^2 + 8\right)$. Continuing by factoring the quotient, we have:

$$\begin{array}{r|rrrr} 2 & 2 & -6 & 0 & 8 \\ & & 4 & -4 & -8 \\ \hline & 2 & -2 & -4 & 0 \end{array} \Rightarrow x = 2 \text{ is a zero.}$$

$P(x) = \left(x - \frac{1}{2}\right)(x - 2)\left(2x^2 - 2x - 4\right) = 2\left(x - \frac{1}{2}\right)(x - 2)\left(x^2 - x - 2\right) = 2\left(x - \frac{1}{2}\right)(x - 2)^2(x + 1)$. Thus, the zeros are $x = \frac{1}{2}, 2, -1$.

**36.** $P(x) = 6x^4 - 7x^3 - 12x^2 + 3x + 2$. The possible rational zeros are $\pm 1, \pm 2, \pm\frac{1}{2}, \pm\frac{1}{3}, \pm\frac{2}{3}, \pm\frac{1}{6}$. Since $P(x)$ has 2 variations in sign, $P$ has 0 or 2 positive real zeros. Since $P(-x) = 6x^4 + 7x^3 - 12x^2 - 3x + 2$ has 2 variations in sign, $P$ has 0 or 2 negative real zeros.

$$\begin{array}{r|rrrrr} 1 & 6 & -7 & -12 & 3 & 2 \\ & & 6 & -1 & -13 & -10 \\ \hline & 6 & -1 & -13 & -10 & -8 \end{array} \qquad \begin{array}{r|rrrrr} 2 & 6 & -7 & -12 & 3 & 2 \\ & & 12 & 10 & -4 & -2 \\ \hline & 6 & 5 & -2 & -1 & 0 \end{array} \Rightarrow x = 2 \text{ is a zero.}$$

$P(x) = 6x^4 - 7x^3 - 12x^2 + 3x + 2 = (x - 2)\left(6x^3 + 5x^2 - 2x - 1\right)$. Continuing by factoring the quotient, we first note that the possible rational zeros are $-1, \pm\frac{1}{2}, \pm\frac{1}{3}, \pm\frac{1}{6}$. We have:

$$\begin{array}{r|rrrr} \frac{1}{2} & 6 & 5 & -2 & -1 \\ & & 3 & 4 & 1 \\ \hline & 6 & 8 & 2 & 0 \end{array} \Rightarrow x = \tfrac{1}{2} \text{ is a zero.}$$

$P(x) = (x - 2)\left(x - \frac{1}{2}\right)\left(6x^2 + 8x + 2\right) = 2(x - 2)\left(x - \frac{1}{2}\right)\left(3x^2 + 4x + 1\right) = (x - 2)\left(x - \frac{1}{2}\right)(x + 1)(3x + 1)$. Therefore, the zeros are $x = -1, -\frac{1}{3}, \frac{1}{2}, 2$.

**37.** $P(x) = x^5 + 3x^4 - 9x^3 - 31x^2 + 36$. The possible rational zeros are $\pm 1, \pm 2, \pm 3, \pm 4, \pm 6, \pm 8, \pm 9, \pm 12, \pm 18$. $P(x)$ has 2 variations in sign and hence 0 or 2 positive real zeros. $P(-x) = -x^5 + 3x^4 + 9x^3 - 31x^2 + 36$ has 3 variations in sign and hence 1 or 3 negative real zeros.

$$\begin{array}{r|rrrrrr} 1 & 1 & 3 & -9 & -31 & 0 & 36 \\ & & 1 & 4 & -5 & -36 & -36 \\ \hline & 1 & 4 & -5 & -36 & -36 & 0 \end{array} \Rightarrow x = 1 \text{ is a zero.}$$

So $P(x) = x^5 + 3x^4 - 9x^3 - 31x^2 + 36 = (x - 1)\left(x^4 + 4x^3 - 5x^2 - 36x - 36\right)$. Continuing by factoring the quotient, we have:

$$\begin{array}{r|rrrrr} 1 & 1 & 4 & -5 & -36 & -36 \\ & & 1 & 5 & 0 & -36 \\ \hline & 1 & 1 & 0 & -36 & -72 \end{array} \qquad \begin{array}{r|rrrrr} 2 & 1 & 4 & -5 & -36 & -36 \\ & & 2 & 12 & 14 & -44 \\ \hline & 1 & 6 & 7 & -22 & -80 \end{array}$$

$$\begin{array}{r|rrrrr} 3 & 1 & 4 & -5 & -36 & -36 \\ & & 3 & 21 & 48 & 36 \\ \hline & 1 & 7 & 16 & 12 & 0 \end{array} \Rightarrow x = 3 \text{ is a zero.}$$

So $P(x) = (x-1)(x-3)(x^3+7x^2+16x+12)$. Since we have 2 positive zeros, there are no more positive zeros, so we continue by factoring the quotient with possible negative zeros.

$$\begin{array}{r|rrrr} -1 & 1 & 7 & 16 & 12 \\ & & -1 & -6 & -10 \\ \hline & 1 & 6 & 10 & 2 \end{array} \qquad \begin{array}{r|rrrr} -2 & 1 & 7 & 16 & 12 \\ & & -2 & -10 & -12 \\ \hline & 1 & 5 & 6 & 0 \end{array} \Rightarrow x=-2 \text{ is a zero.}$$

Then $P(x) = (x-1)(x-3)(x+2)(x^2+5x+6) = (x-1)(x-3)(x+2)^2(x+3)$. Thus, the zeros are $x = 1, 3, -2, -3$.

**38.** $P(x) = x^5 - 4x^4 - 3x^3 + 22x^2 - 4x - 24$ has possible rational zeros $\pm1, \pm2, \pm3, \pm4, \pm6, \pm8, \pm12, \pm24$. Since $P(x)$ has 3 variations in sign, there are 1 or 3 positive real zeros. Since $P(-x) = -x^5 - 4x^4 + 3x^3 + 22x^2 + 4x - 24$ has 2 variations in sign, there are 0 or 2 negative real zeros.

$$\begin{array}{r|rrrrrr} 1 & 1 & -4 & -3 & 22 & -4 & -24 \\ & & 1 & -3 & -6 & 16 & 12 \\ \hline & 1 & -3 & -6 & 16 & 12 & -12 \end{array} \qquad \begin{array}{r|rrrrrr} 2 & 1 & -4 & -3 & 22 & -4 & -24 \\ & & 2 & -4 & -14 & 16 & 24 \\ \hline & 1 & -2 & -7 & 8 & 12 & 0 \end{array} \Rightarrow x=2 \text{ is a zero.}$$

$P(x) = (x-2)(x^4 - 2x^3 - 7x^2 + 8x + 12)$

$$\begin{array}{r|rrrrr} 2 & 1 & -2 & -7 & 8 & 12 \\ & & 2 & 0 & -14 & -12 \\ \hline & 1 & 0 & -7 & -6 & 0 \end{array} \Rightarrow x=2 \text{ is a zero again.}$$

$P(x) = (x-2)^2(x^3 - 7x - 6)$

$$\begin{array}{r|rrrr} 2 & 1 & 0 & -7 & -6 \\ & & 2 & 4 & -6 \\ \hline & 1 & 2 & -3 & -12 \end{array} \qquad \begin{array}{r|rrrr} 3 & 1 & 0 & -7 & -6 \\ & & 3 & 9 & 6 \\ \hline & 1 & 3 & 2 & 0 \end{array} \Rightarrow x=3 \text{ is a zero.}$$

$P(x) = (x-2)^2(x-3)(x^2+3x+2) = (x-2)^2(x-3)(x+1)(x+2) = 0$. Therefore, the zeros are $x = -1, \pm2, 3$.

**39.** $P(x) = 3x^5 - 14x^4 - 14x^3 + 36x^2 + 43x + 10$ has possible rational zeros $\pm1, \pm2, \pm5, \pm10, \pm\frac{1}{3}, \pm\frac{2}{3}, \pm\frac{5}{3}, \pm\frac{10}{3}$. Since $P(x)$ has 2 variations in sign, there are 0 or 2 positive real zeros. Since $P(-x) = -3x^5 - 14x^4 + 14x^3 + 36x^2 - 43x + 10$ has 3 variations in sign, there are 1 or 3 negative real zeros.

$$\begin{array}{r|rrrrrr} 1 & 3 & -14 & -14 & 36 & 43 & 10 \\ & & 3 & -11 & -25 & 11 & 54 \\ \hline & 3 & -11 & -25 & 11 & 54 & 64 \end{array} \qquad \begin{array}{r|rrrrrr} 2 & 3 & -14 & -14 & 36 & 43 & 10 \\ & & 6 & -16 & -60 & -48 & -10 \\ \hline & 3 & -8 & -30 & -24 & -5 & 0 \end{array} \Rightarrow x=2 \text{ is a zero.}$$

$P(x) = (x-2)(3x^4 - 8x^3 - 30x^2 - 24x - 5)$

$$\begin{array}{r|rrrrr} 2 & 3 & -8 & -30 & -24 & -5 \\ & & 6 & -4 & -68 & -184 \\ \hline & 3 & -2 & -34 & -92 & -189 \end{array} \qquad \begin{array}{r|rrrrr} 5 & 3 & -8 & -30 & -24 & -5 \\ & & 15 & 35 & 25 & 5 \\ \hline & 3 & 7 & 5 & 1 & 0 \end{array} \Rightarrow x=5 \text{ is a zero.}$$

$P(x) = (x-2)(x-5)(3x^3+7x^2+5x+1)$. Since $3x^3+7x^2+5x+1$ has no variation in sign, there are no more positive zeros.

$$\begin{array}{r|rrrr} -1 & 3 & 7 & 5 & 1 \\ & & -3 & -4 & -1 \\ \hline & 3 & 4 & 1 & 0 \end{array} \Rightarrow x=-1 \text{ is a zero.}$$

$P(x) = (x-2)(x-5)(x+1)(3x^2+4x+1) = (x-2)(x-5)(x+1)(x+1)(3x+1)$. Therefore, the zeros are $x = -1, -\frac{1}{3}, 2, 5$.

**40.** $P(x) = 2x^6 - 3x^5 - 13x^4 + 29x^3 - 27x^2 + 32x - 12$ has possible rational zeros $\pm1$, $\pm2$, $\pm3$, $\pm4$, $\pm6$, $\pm12$, $\pm\frac{1}{2}$, $\pm\frac{3}{2}$. Since $P(x)$ has 5 variations in sign, there are 1 or 3 or 5 positive real zeros. Since $P(-x) = 2x^6 + 3x^5 - 13x^4 - 29x^3 + 27x^2 - 32x - 12$ has 3 variations in sign, there are 1 or 3 negative real zeros.

$$\begin{array}{r|rrrrrrr} 1 & 2 & -3 & -13 & 29 & -27 & 32 & -12 \\ & & 2 & -1 & -14 & 15 & -12 & 20 \\ \hline & 2 & -1 & -14 & 15 & -12 & 20 & 8 \end{array}$$

$$\begin{array}{r|rrrrrrr} 2 & 2 & -3 & -13 & 29 & -27 & 32 & -12 \\ & & 4 & 2 & -22 & 14 & -26 & 12 \\ \hline & 2 & 1 & -11 & 7 & -13 & 6 & 0 \end{array} \Rightarrow x = 2 \text{ is a zero.}$$

$P(x) = (x-2)(2x^5 + x^4 - 11x^3 + 7x^2 - 13x + 6)$. We continue with the quotient:

$$\begin{array}{r|rrrrrr} 2 & 2 & 1 & -11 & 7 & -13 & 6 \\ & & 4 & 10 & -2 & 10 & -6 \\ \hline & 2 & 5 & -1 & 5 & -3 & 0 \end{array} \Rightarrow x = 2 \text{ is a zero again.}$$

$P(x) = (x-2)^2 (2x^4 + 5x^3 - x^2 + 5x - 3)$. We continue with the quotient, first noting 2 is no longer a possible rational solution.

$$\begin{array}{r|rrrrr} 3 & 2 & 5 & -1 & 5 & -3 \\ & & 6 & 22 & 42 & 94 \\ \hline & 2 & 11 & 21 & 47 & 91 \end{array} \Rightarrow x = 3 \text{ is an upper bound.}$$

We know that there is at least 1 more positive zero.

$$\begin{array}{r|rrrrr} \frac{1}{2} & 2 & 5 & -1 & 5 & -3 \\ & & 1 & 2 & 1 & 3 \\ \hline & 2 & 6 & 2 & 6 & 0 \end{array} \Rightarrow x = \tfrac{1}{2} \text{ is a zero.}$$

$P(x) = (x-2)^2 \left(x - \frac{1}{2}\right)(2x^3 + 6x^2 + 2x + 6)$. We can factor $2x^3 + 6x^2 + 2x + 6$ by grouping; $2x^3 + 6x^2 + 2x + 6 = (2x^3 + 6x^2) + (2x + 6) = (2x + 6)(x^2 + 1)$. So $P(x) = 2(x-2)^2 \left(x - \frac{1}{2}\right)(x+3)(x^2+1)$. Since $x^2 + 1$ has no real zeros, the zeros of $P$ are $x = -3, 2, \frac{1}{2}$.

**41.** $P(x) = x^3 + 4x^2 + 3x - 2$. The possible rational zeros are $\pm1$, $\pm2$. $P(x)$ has 1 variation in sign and hence 1 positive real zero. $P(-x) = -x^3 + 4x^2 - 3x - 2$ has 2 variations in sign and hence 0 or 2 negative real zeros.

$$\begin{array}{r|rrrr} 1 & 1 & 4 & 3 & -2 \\ & & 1 & 5 & 8 \\ \hline & 1 & 5 & 8 & 6 \end{array} \Rightarrow x = 1 \text{ is an upper bound.}$$

$$\begin{array}{r|rrrr} -1 & 1 & 4 & 3 & -2 \\ & & -1 & -3 & 0 \\ \hline & 1 & 3 & 0 & -2 \end{array}$$

$$\begin{array}{r|rrrr} -2 & 1 & 4 & 3 & -2 \\ & & -2 & -4 & 2 \\ \hline & 1 & 2 & -1 & 0 \end{array} \Rightarrow x = -2 \text{ is a zero.}$$

So $P(x) = (x+2)(x^2 + 2x - 1)$. Using the quadratic formula on the second factor, we have: $x = \frac{-2 \pm \sqrt{2^2 - 4(1)(-1)}}{2(1)} = \frac{-2 \pm \sqrt{8}}{2} = \frac{-2 \pm 2\sqrt{2}}{2} = -1 \pm \sqrt{2}$. Therefore, the zeros are $x = -2, -1 + \sqrt{2}, -1 - \sqrt{2}$.

**42.** $P(x) = x^3 - 5x^2 + 2x + 12$. The possible rational zeros are $\pm1, \pm2, \pm3, \pm4, \pm6, \pm12$. $P(x)$ has 2 variations in sign and hence 0 or 2 positive real zeros. $P(-x) = -x^3 - 5x^2 - 2x + 12$ has 1 variation in sign and hence 1 negative real zero.

$$\begin{array}{r|rrrr} 1 & 1 & -5 & 2 & 12 \\ & & 1 & -4 & -2 \\ \hline & 1 & -4 & -2 & 10 \end{array} \qquad \begin{array}{r|rrrr} 2 & 1 & -5 & 2 & 12 \\ & & 2 & -6 & -8 \\ \hline & 1 & -3 & -4 & 4 \end{array}$$

$$\begin{array}{r|rrrr} 3 & 1 & -5 & 2 & 12 \\ & & 3 & -6 & -12 \\ \hline & 1 & -2 & -4 & 0 \end{array} \Rightarrow x = 3 \text{ is a zero.}$$

So $P(x) = (x-3)(x^2 - 2x - 4)$. Using the quadratic formula on the second factor, we have: $x = \frac{-(-2)\pm\sqrt{(-2)^2-4(1)(-4)}}{2(1)} = \frac{2\pm\sqrt{20}}{2} = \frac{2\pm2\sqrt{5}}{2} = 1 \pm \sqrt{5}$. Therefore, the zeros are $x = 3, 1 \pm \sqrt{5}$.

**43.** $P(x) = x^4 - 6x^3 + 4x^2 + 15x + 4$. The possible rational zeros are $\pm1, \pm2, \pm4$. $P(x)$ has 2 variations in sign and hence 0 or 2 positive real zeros. $P(-x) = x^4 + 6x^3 + 4x^2 - 15x + 4$ has 2 variations in sign and hence 0 or 2 negative real zeros.

$$\begin{array}{r|rrrrr} 1 & 1 & -6 & 4 & 15 & 4 \\ & & 1 & -5 & -1 & 14 \\ \hline & 1 & -5 & -1 & 14 & 18 \end{array} \qquad \begin{array}{r|rrrrr} 2 & 1 & -6 & 4 & 15 & 4 \\ & & 2 & -8 & -8 & 14 \\ \hline & 1 & -4 & -4 & 7 & 18 \end{array}$$

$$\begin{array}{r|rrrrr} 4 & 1 & -6 & 4 & 15 & 4 \\ & & 4 & -8 & -16 & -4 \\ \hline & 1 & -2 & -4 & -1 & 0 \end{array} \Rightarrow x = 4 \text{ is a zero.}$$

So $P(x) = (x-4)(x^3 - 2x^2 - 4x - 1)$. Continuing by factoring the quotient, we have:

$$\begin{array}{r|rrrr} 4 & 1 & -2 & -4 & -1 \\ & & 4 & 8 & 16 \\ \hline & 1 & 2 & 4 & 15 \end{array} \Rightarrow x = 4 \text{ is an upper bound.} \qquad \begin{array}{r|rrrr} -1 & 1 & -2 & -4 & -1 \\ & & -1 & 3 & 1 \\ \hline & 1 & -3 & -1 & 0 \end{array} \Rightarrow x = -1 \text{ is a zero.}$$

So $P(x) = (x-4)(x+1)(x^2 - 3x - 1)$. Using the quadratic formula on the third factor, we have: $x = \frac{-(-3)\pm\sqrt{(-3)^2-4(1)(-1)}}{2(1)} = \frac{3\pm\sqrt{13}}{2}$. Therefore, the zeros are $x = 4, -1, \frac{3\pm\sqrt{13}}{2}$.

**44.** $P(x) = x^4 + 2x^3 - 2x^2 - 3x + 2$. The possible rational zeros are $\pm1, \pm2$. $P(x)$ has 2 variations in sign and hence 0 or 2 positive real zeros. $P(-x) = x^4 - 2x^3 - 2x^2 + 3x + 2$ has 2 variations in sign and hence 0 or 2 negative real zeros.

$$\begin{array}{r|rrrrr} 1 & 1 & 2 & -2 & -3 & 2 \\ & & 1 & 3 & 1 & -2 \\ \hline & 1 & 3 & 1 & -2 & 0 \end{array} \Rightarrow x = 1 \text{ is a zero.}$$

$P(x) = (x-1)(x^3 + 3x^2 + x - 2)$. Continuing with the quotient:

$$\begin{array}{r|rrrr} 1 & 1 & 3 & 1 & -2 \\ & & 1 & 4 & 5 \\ \hline & 1 & 4 & 5 & 3 \end{array} \Rightarrow x = 1 \text{ is an upper bound.} \qquad \begin{array}{r|rrrr} -1 & 1 & 3 & 1 & -2 \\ & & -1 & -2 & 1 \\ \hline & 1 & 2 & -1 & -1 \end{array} \qquad \begin{array}{r|rrrr} -2 & 1 & 3 & 1 & -2 \\ & & -2 & -2 & 2 \\ \hline & 1 & 1 & -1 & 0 \end{array} \Rightarrow x = -2 \text{ is a zero.}$$

So $P(x) = (x-1)(x+2)(x^2 + x - 1)$. Using the quadratic formula on the third factor, we have: $x = \dfrac{-1 \pm \sqrt{1^2 - 4(1)(-1)}}{2(1)} = \dfrac{1 \pm \sqrt{5}}{2}$. Therefore, the zeros are $x = 1, -2, \frac{1\pm\sqrt{5}}{2}$.

**45.** $P(x) = x^4 - 7x^3 + 14x^2 - 3x - 9$. The possible rational zeros are $\pm1, \pm3, \pm9$. $P(x)$ has 3 variations in sign and hence 1 or 3 positive real zeros. $P(-x) = x^4 + 7x^3 + 14x^2 + 3x - 4$ has 1 variation in sign and hence 1 negative real zero.

$$\begin{array}{r|rrrrr} 1 & 1 & -7 & 14 & -3 & -9 \\ & & 1 & -6 & 8 & 5 \\ \hline & 1 & -6 & 8 & 5 & 4 \end{array} \qquad \begin{array}{r|rrrrr} 3 & 1 & -7 & 14 & -3 & -9 \\ & & 3 & -12 & 6 & 9 \\ \hline & 1 & -4 & 2 & 3 & 0 \end{array} \Rightarrow x = 3 \text{ is a zero.}$$

So $P(x) = (x-3)(x^3 - 4x^2 + 2x + 3)$. Since the constant term of the second term is 3, $\pm9$ are no longer possible zeros. Continuing by factoring the quotient, we have:

$$\begin{array}{r|rrrr} 3 & 1 & -4 & 2 & 3 \\ & & 3 & -3 & -3 \\ \hline & 1 & -1 & -1 & 0 \end{array} \Rightarrow x = 3 \text{ is a zero again.}$$

So $P(x) = (x-3)^2(x^2 - x - 1)$. Using the quadratic formula on the second factor, we have: $x = \frac{-(-1)\pm\sqrt{(-1)^2-4(1)(-1)}}{2(1)} = \frac{1\pm\sqrt{5}}{2}$. Therefore, the zeros are $x = 3, \frac{1\pm\sqrt{5}}{2}$.

**46.** $P(x) = x^5 - 4x^4 - x^3 + 10x^2 + 2x - 4$. The possible rational zeros are $\pm1, \pm2, \pm4$. $P(x)$ has 3 variations in sign and hence 1 or 3 positive real zeros. $P(-x) = -x^5 - 4x^4 + x^3 + 10x^2 - 2x - 4$ has 2 variations in sign and hence 0 or 2 negative real zeros.

$$\begin{array}{r|rrrrrr} 1 & 1 & -4 & -1 & 10 & 2 & -4 \\ & & 1 & -3 & -4 & 6 & 8 \\ \hline & 1 & -3 & -4 & 6 & 8 & 4 \end{array} \qquad \begin{array}{r|rrrrrr} 2 & 1 & -4 & -1 & 10 & 2 & -4 \\ & & 2 & -4 & -10 & 0 & 4 \\ \hline & 1 & -2 & -5 & 0 & 2 & 0 \end{array} \Rightarrow x = 2 \text{ is a zero.}$$

So $P(x) = (x-2)(x^4 - 2x^3 - 5x^2 + 2)$. Since the constant term of the second factor is 2, $\pm4$ are no longer possible zeros. Continuing by factoring the quotient, we have:

$$\begin{array}{r|rrrrr} 2 & 1 & -2 & -5 & 0 & 2 \\ & & 2 & 0 & -10 & -20 \\ \hline & 1 & 0 & -5 & -10 & 18 \end{array} \qquad \begin{array}{r|rrrrr} -1 & 1 & -2 & -5 & 0 & 2 \\ & & -1 & 3 & 2 & -2 \\ \hline & 1 & -3 & -2 & 2 & 0 \end{array} \Rightarrow x = -1 \text{ is a zero.}$$

So $P(x) = (x-2)(x+1)(x^3 - 3x^2 - 2x - 2)$. Continuing by factoring the quotient, we have:

$$\begin{array}{r|rrrr} -1 & 1 & -3 & -2 & 2 \\ & & -1 & 4 & -2 \\ \hline & 1 & -4 & 2 & 0 \end{array} \Rightarrow x = -1 \text{ is a zero again.}$$

So $P(x) = (x-2)(x+1)^2(x^2 - 4x + 2)$. Using the quadratic formula on the second factor, we have: $x = \frac{-(-4)\pm\sqrt{(-4)^2-4(1)(2)}}{2(1)} = \frac{4\pm\sqrt{8}}{2} = \frac{4\pm2\sqrt{2}}{2} = 2\pm\sqrt{2}$. Therefore, the zeros are $x = -1, 2, 2\pm\sqrt{2}$.

**47.** $P(x) = 4x^3 - 6x^2 + 1$. The possible rational zeros are $\pm1, \pm\frac{1}{2}, \pm\frac{1}{4}$. $P(x)$ has 2 variations in sign and hence 0 or 2 positive real zeros. $P(-x) = -4x^3 - 6x^2 + 1$ has 1 variation in sign and hence 1 negative real zero.

$$\begin{array}{r|rrrr} 1 & 4 & -6 & 0 & 1 \\ & & 4 & -2 & -2 \\ \hline & 4 & -2 & -2 & -1 \end{array} \qquad \begin{array}{r|rrrr} \frac{1}{2} & 4 & -6 & 0 & 1 \\ & & 2 & -2 & -1 \\ \hline & 4 & -4 & -2 & 0 \end{array} \Rightarrow x = \tfrac{1}{2} \text{ is a zero.}$$

So $P(x) = (x - \frac{1}{2})(4x^2 - 4x - 2)$. Using the quadratic formula on the second factor, we have: $x = \frac{-(-4)\pm\sqrt{(-4)^2-4(4)(-2)}}{2(4)} = \frac{4\pm\sqrt{48}}{8} = \frac{4\pm4\sqrt{3}}{8} = \frac{1\pm\sqrt{3}}{2}$. Therefore, the zeros are $x = \frac{1}{2}, \frac{1\pm\sqrt{3}}{2}$.

**48.** $P(x) = 3x^3 - 5x^2 - 8x - 2$. The possible rational zeros are $\pm 1, \pm 2, \pm\frac{1}{3}, \pm\frac{2}{3}$. $P(x)$ has 1 variation in sign and hence 1 positive real zero. $P(-x) = -3x^3 - 5x^2 + 8x - 2$ has 2 variations in sign and hence 0 or 2 negative real zeros.

$$\begin{array}{r|rrrr} 1 & 3 & -5 & -8 & -2 \\ & & 3 & -2 & -10 \\ \hline & 3 & -2 & -10 & -12 \end{array} \qquad \begin{array}{r|rrrr} 2 & 3 & -5 & -8 & -2 \\ & & 6 & 2 & -12 \\ \hline & 3 & 1 & -6 & -14 \end{array}$$

$$\begin{array}{r|rrrr} \frac{1}{3} & 3 & -5 & -8 & -2 \\ & & 1 & -\frac{4}{3} & -\frac{28}{9} \\ \hline & 3 & -4 & -\frac{28}{3} & -\frac{46}{9} \end{array} \qquad \begin{array}{r|rrrr} \frac{2}{3} & 3 & -5 & -8 & -2 \\ & & 2 & -2 & -\frac{20}{3} \\ \hline & 3 & -3 & -10 & -\frac{26}{3} \end{array}$$

Thus we have tried all the positive rational zeros, so we try the negative zeros.

$$\begin{array}{r|rrrr} -1 & 3 & -5 & -8 & -2 \\ & & -3 & 8 & 0 \\ \hline & 3 & -8 & 0 & -2 \end{array} \qquad \begin{array}{r|rrrr} -2 & 3 & -5 & -8 & -2 \\ & & -6 & 22 & -28 \\ \hline & 3 & -11 & 14 & -30 \end{array}$$

$$\begin{array}{r|rrrr} -\frac{1}{3} & 3 & -5 & -8 & -2 \\ & & -1 & 2 & 2 \\ \hline & 3 & -6 & -6 & 0 \end{array} \quad \Rightarrow x = -\tfrac{1}{3} \text{ is a zero.}$$

So $P(x) = \left(x + \frac{1}{3}\right)\left(3x^2 - 6x - 6\right) = 3\left(x + \frac{1}{3}\right)\left(x^2 - 2x - 2\right)$. Using the quadratic formula on the second factor, we have: $x = \frac{-(-2) \pm \sqrt{(-2)^2 - 4(1)(-2)}}{2(1)} = \frac{2 \pm \sqrt{12}}{2} = \frac{2 \pm 2\sqrt{3}}{2} = 1 \pm \sqrt{3}$. Therefore, the zeros are $x = -\frac{1}{3}, 1 \pm \sqrt{3}$.

**49.** $P(x) = 2x^4 + 15x^3 + 17x^2 + 3x - 1$. The possible rational zeros are $\pm 1, \pm\frac{1}{2}$. $P(x)$ has 1 variation in sign and hence 1 positive real zero. $P(-x) = 2x^4 - 15x^3 + 17x^2 - 3x - 1$ has 3 variations in sign and hence 1 or 3 negative real zeros.

$$\begin{array}{r|rrrrr} \frac{1}{2} & 2 & 15 & 17 & 3 & -1 \\ & & 1 & 8 & \frac{25}{2} & \frac{31}{4} \\ \hline & 2 & 16 & 25 & \frac{31}{2} & \frac{27}{4} \end{array} \quad \Rightarrow x = \tfrac{1}{2} \text{ is an upper bound.}$$

$$\begin{array}{r|rrrrr} -\frac{1}{2} & 2 & 15 & 17 & 3 & -1 \\ & & -1 & -7 & -5 & 1 \\ \hline & 2 & 14 & 10 & -2 & 0 \end{array} \quad \Rightarrow x = -\tfrac{1}{2} \text{ is a zero.}$$

So $P(x) = \left(x + \frac{1}{2}\right)\left(2x^3 + 14x^2 + 10x - 2\right) = 2\left(x + \frac{1}{2}\right)\left(x^3 + 7x^2 + 5x - 1\right)$.

$$\begin{array}{r|rrrr} -1 & 1 & 7 & 5 & -1 \\ & & -1 & -6 & 1 \\ \hline & 1 & 6 & -1 & 0 \end{array} \quad \Rightarrow x = -1 \text{ is a zero.}$$

So $P(x) = \left(x + \frac{1}{2}\right)\left(2x^3 + 14x^2 + 10x - 2\right) = 2\left(x + \frac{1}{2}\right)(x + 1)\left(x^2 + 6x - 1\right)$ Using the quadratic formula on the third factor, we have $x = \frac{-(6) \pm \sqrt{(6)^2 - 4(1)(-1)}}{2(1)} = \frac{-6 \pm \sqrt{40}}{2} = \frac{-6 \pm 2\sqrt{10}}{2} = -3 \pm \sqrt{10}$. Therefore, the zeros are $x = -1$, $-\frac{1}{2}, -3 \pm \sqrt{10}$.

**50.** $P(x) = 4x^5 - 18x^4 - 6x^3 + 91x^2 - 60x + 9$. The possible rational zeros are $\pm1, \pm3, \pm9, \pm\frac{1}{2}, \pm\frac{3}{2}, \pm\frac{9}{2}, \pm\frac{1}{4}, \pm\frac{3}{4}, \pm\frac{9}{4}$. $P(x)$ has 4 variations in sign and hence 0 or 2 or 4 positive real zeros. $P(-x) = -4x^5 - 18x^4 + 6x^3 + 91x^2 + 60x + 9$ has 1 variation in sign and hence 1 negative real zero.

$$\begin{array}{r|rrrrrr} 1 & 4 & -18 & -6 & 91 & -60 & 9 \\ & & 4 & -14 & -20 & 71 & 1 \\ \hline & 4 & -14 & -20 & 71 & 11 & 10 \end{array} \qquad \begin{array}{r|rrrrrr} 3 & 4 & -18 & -6 & 91 & -60 & 9 \\ & & 12 & -18 & -72 & 57 & -9 \\ \hline & 4 & -6 & -24 & 19 & -3 & 0 \end{array} \Rightarrow x = 3 \text{ is a zero.}$$

So $P(x) = (x-3)\left(4x^4 - 6x^3 - 24x^2 + 19x - 3\right)$. Continuing by factoring the quotient, we have:

$$\begin{array}{r|rrrrr} 3 & 4 & -6 & -24 & 19 & -3 \\ & & 12 & 18 & -18 & 3 \\ \hline & 4 & 6 & -6 & 1 & 0 \end{array} \Rightarrow x = 3 \text{ is a zero again.}$$

So $P(x) = (x-3)^2\left(4x^3 + 6x^2 - 6x + 1\right)$. Continuing by factoring the quotient, we have:

$$\begin{array}{r|rrrr} 3 & 4 & 6 & -6 & 1 \\ & & 12 & 54 & 144 \\ \hline & 4 & 18 & 48 & 1445 \end{array} \Rightarrow x = 3 \text{ is an upper bound.} \qquad \begin{array}{r|rrrr} \frac{1}{2} & 4 & 6 & -6 & 1 \\ & & 2 & 4 & -1 \\ \hline & 4 & 8 & -2 & 0 \end{array} \Rightarrow x = \tfrac{1}{2} \text{ is a zero.}$$

So $P(x) = (x-3)^2\left(x - \frac{1}{2}\right)\left(4x^2 + 8x - 2\right) = 2(x-3)^2\left(x - \frac{1}{2}\right)\left(2x^2 + 4x - 1\right)$. Using the quadratic formula on the second factor, we have $x = \frac{-4 \pm \sqrt{4^2 - 4(2)(-1)}}{2(2)} = \frac{-4 \pm \sqrt{24}}{4} = -1 \pm \frac{\sqrt{6}}{2}$. Therefore, the zeros are $x = \frac{1}{2}, 3, -1 \pm \frac{\sqrt{6}}{2}$.

**51. (a)** $P(x) = x^3 - 3x^2 - 4x + 12$ has possible rational zeros $\pm1, \pm2, \pm3, \pm4, \pm6, \pm12$.

$$\begin{array}{r|rrrr} 1 & 1 & -3 & -4 & 12 \\ & & 1 & -2 & -6 \\ \hline & 1 & -2 & -6 & 6 \end{array}$$

$$\begin{array}{r|rrrr} 2 & 1 & -3 & -4 & 12 \\ & & 2 & -2 & -12 \\ \hline & 1 & -1 & -6 & 0 \end{array} \Rightarrow x = 2 \text{ is a zero.}$$

**(b)**

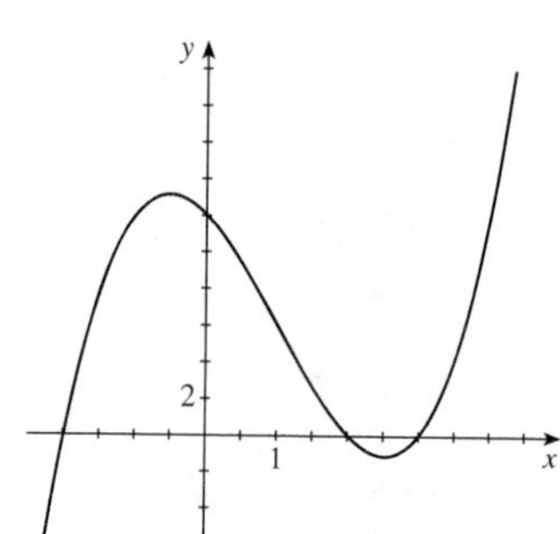


So $P(x) = (x-2)\left(x^2 - x - 6\right) = (x-2)(x+2)(x-3)$. The real zeros of $P$ are $-2, 2, 3$.

**52. (a)** $P(x) = -x^3 - 2x^2 + 5x + 6$ has possible rational zeros $\pm1, \pm2, \pm3, \pm6$.

$$\begin{array}{r|rrrr} 1 & -1 & -2 & 5 & 6 \\ & & -1 & -3 & 2 \\ \hline & -1 & -3 & 2 & 8 \end{array}$$

$$\begin{array}{r|rrrr} 2 & -1 & -2 & 5 & 6 \\ & & -2 & -8 & -6 \\ \hline & -1 & -4 & -3 & 0 \end{array} \Rightarrow x = 2 \text{ is a zero.}$$

**(b)**

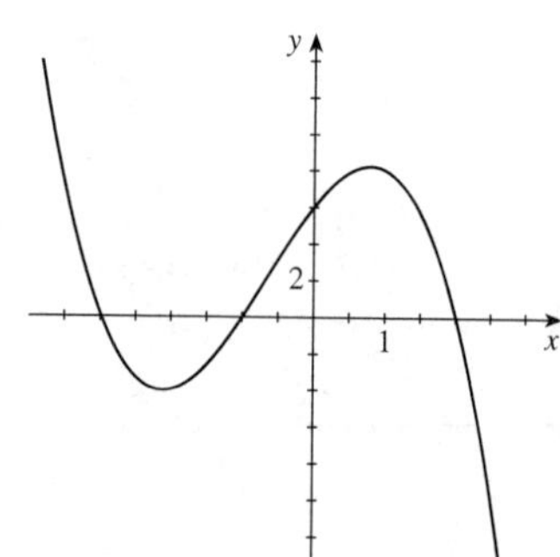


So $P(x) = (x-2)\left(-x^2 - 4x - 3\right) = -(x-2)\left(x^2 + 4x + 3\right) = -(x-2)(x+1)(x+3)$. The real zeros of $P$ are $2, -1, -3$.

**53. (a)** $P(x) = 2x^3 - 7x^2 + 4x + 4$ has possible rational zeros $\pm 1$, $\pm 2$, $\pm 4$, $\pm \frac{1}{2}$.

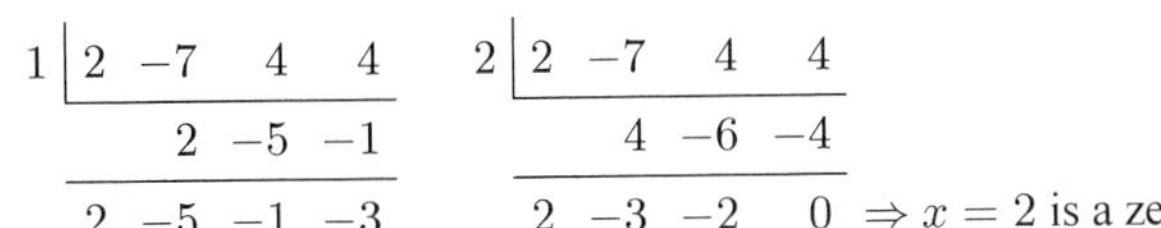

$$\begin{array}{r|rrrr} 1 & 2 & -7 & 4 & 4 \\ & & 2 & -5 & -1 \\ \hline & 2 & -5 & -1 & -3 \end{array} \qquad \begin{array}{r|rrrr} 2 & 2 & -7 & 4 & 4 \\ & & 4 & -6 & -4 \\ \hline & 2 & -3 & -2 & 0 \end{array} \Rightarrow x = 2 \text{ is a zero.}$$

So $P(x) = (x-2)\left(2x^2 - 3x - 2\right)$. Continuing:

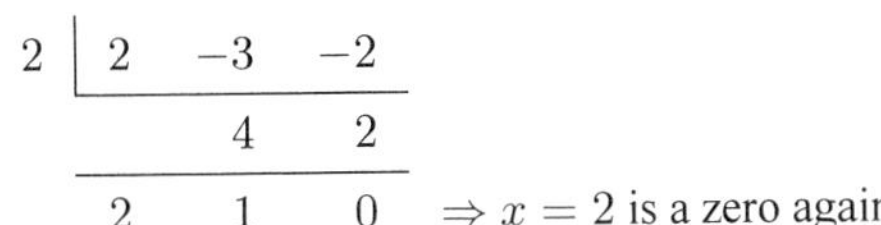

$$\begin{array}{r|rrr} 2 & 2 & -3 & -2 \\ & & 4 & 2 \\ \hline & 2 & 1 & 0 \end{array} \Rightarrow x = 2 \text{ is a zero again.}$$

Thus $P(x) = (x-2)^2(2x+1)$. The real zeros of $P$ are 2 and $-\frac{1}{2}$.

**(b)**

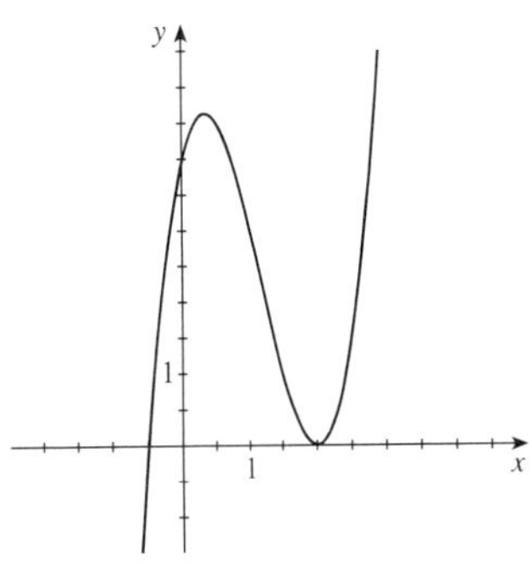


**54. (a)** $P(x) = 3x^3 + 17x^2 + 21x - 9$ has possible rational zeros $\pm 1$, $\pm 3$, $\pm 9$, $\pm \frac{1}{3}$, $\pm \frac{2}{3}$.

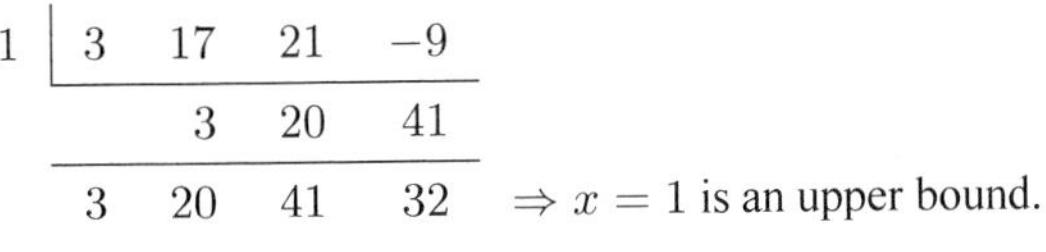

$$\begin{array}{r|rrrr} 1 & 3 & 17 & 21 & -9 \\ & & 3 & 20 & 41 \\ \hline & 3 & 20 & 41 & 32 \end{array} \Rightarrow x = 1 \text{ is an upper bound.}$$

$$\begin{array}{r|rrrr} \frac{1}{3} & 3 & 17 & 21 & -9 \\ & & 1 & 6 & 9 \\ \hline & 3 & 18 & 27 & 0 \end{array} \Rightarrow x = \tfrac{1}{3} \text{ is a zero.}$$

**(b)**

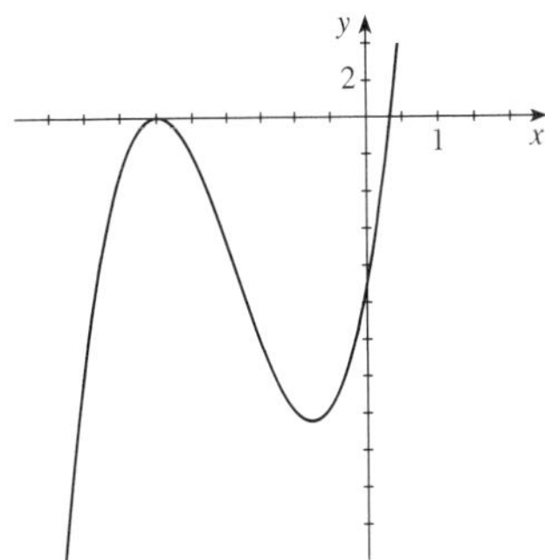


So $P(x) = \left(x - \frac{1}{3}\right)\left(3x^2 + 18x + 27\right) = 3\left(x - \frac{1}{3}\right)\left(x^2 + 6x + 9\right) = 3\left(x - \frac{1}{3}\right)(x+3)^2$. The real zeros of $P$ are $-3$, $\frac{1}{3}$.

**55. (a)** $P(x) = x^4 - 5x^3 + 6x^2 + 4x - 8$ has possible rational zeros $\pm 1$, $\pm 2$, $\pm 4$, $\pm 8$.

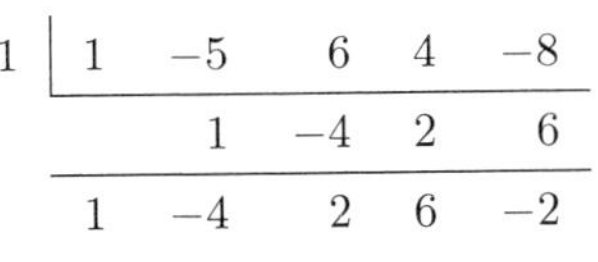

$$\begin{array}{r|rrrrr} 1 & 1 & -5 & 6 & 4 & -8 \\ & & 1 & -4 & 2 & 6 \\ \hline & 1 & -4 & 2 & 6 & -2 \end{array}$$

$$\begin{array}{r|rrrrr} 2 & 1 & -5 & 6 & 4 & -8 \\ & & 2 & -6 & 0 & 8 \\ \hline & 1 & -3 & 0 & 4 & 0 \end{array} \Rightarrow x = 2 \text{ is a zero.}$$

**(b)**

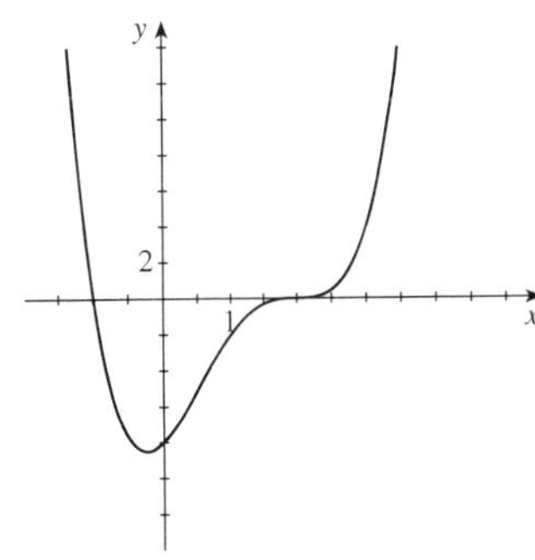


So $P(x) = (x-2)\left(x^3 - 3x^2 + 4\right)$ and the possible rational zeros are restricted to $-1$, $\pm 2$, $\pm 4$.

$$\begin{array}{r|rrrr} 2 & 1 & -3 & 0 & 4 \\ & & 2 & -2 & -4 \\ \hline & 1 & -1 & -2 & 0 \end{array} \Rightarrow x = 2 \text{ is a zero again.}$$

$P(x) = (x-2)^2\left(x^2 - x - 2\right) = (x-2)^2(x-2)(x+1) = (x-2)^3(x+1)$. So the real zeros of $P$ are $-1$ and 2.

**56. (a)** $P(x) = -x^4 + 10x^2 + 8x - 8$ has possible rational zeros $\pm 1, \pm 2, \pm 4, \pm 8$.

| 1 | −1 | 0 | 10 | 8 | −8 |
|---|---|---|---|---|---|
| | | −1 | −1 | 9 | 17 |
| | −1 | −1 | 9 | 17 | 9 |

| 2 | −1 | 0 | 10 | 8 | −8 |
|---|---|---|---|---|---|
| | | −2 | −4 | 12 | 40 |
| | −1 | −2 | 6 | 20 | 32 |

| 4 | −1 | 0 | 10 | 8 | −8 |
|---|---|---|---|---|---|
| | | −4 | −16 | −24 | −64 |
| | −1 | −4 | −6 | −16 | −72 |

$\Rightarrow x = 4$ is an upper bound.

| −1 | −1 | 0 | 10 | 8 | −8 |
|---|---|---|---|---|---|
| | | 1 | −1 | −9 | 1 |
| | −1 | 1 | 9 | −1 | −7 |

| −2 | −1 | 0 | 10 | 8 | −8 |
|---|---|---|---|---|---|
| | | 2 | −4 | −12 | 8 |
| | −1 | 2 | 6 | −4 | 0 |

$\Rightarrow x = -2$ is a zero.

**(b)**

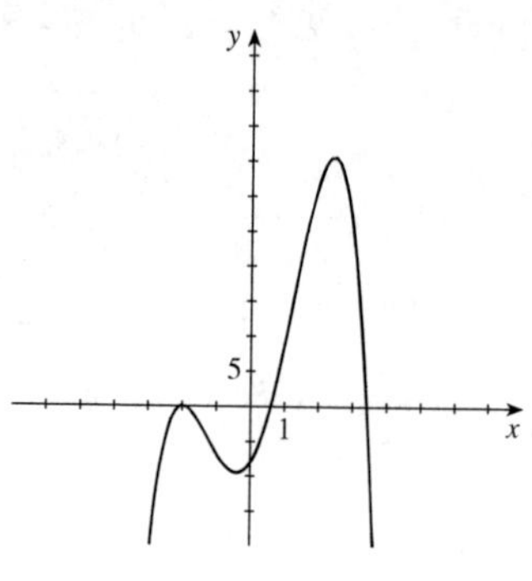


So $P(x) = (x+2)\left(-x^3 + 2x^2 + 6x - 4\right)$. Continuing, we have:

| −2 | −1 | 2 | 6 | −4 |
|---|---|---|---|---|
| | | 2 | −8 | 4 |
| | −1 | 4 | −2 | 0 |

$\Rightarrow x = -2$ is a zero again.

$P(x) = (x+2)^2\left(-x^2 + 4x - 2\right)$. Using the quadratic formula on the second factor, we have $x = \frac{-4 \pm \sqrt{4^2 - 4(-1)(-2)}}{2(-1)} = \frac{-4 \pm \sqrt{8}}{-2} = \frac{-4 \pm 2\sqrt{2}}{-2} = 2 \pm \sqrt{2}$. So the real zeros of $P$ are $-2, 2 \pm \sqrt{2}$.

**57. (a)** $P(x) = x^5 - x^4 - 5x^3 + x^2 + 8x + 4$ has possible rational zeros $\pm 1, \pm 2, \pm 4$.

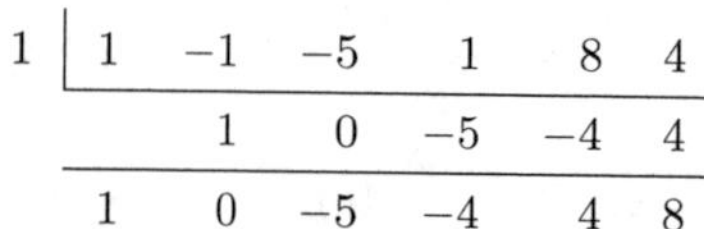

| 1 | 1 | −1 | −5 | 1 | 8 | 4 |
|---|---|---|---|---|---|---|
| | | 1 | 0 | −5 | −4 | 4 |
| | 1 | 0 | −5 | −4 | 4 | 8 |

| 2 | 1 | −1 | −5 | 1 | 8 | 4 |
|---|---|---|---|---|---|---|
| | | 2 | 2 | −6 | −10 | −4 |
| | 1 | 1 | −3 | −5 | −2 | 0 |

$\Rightarrow x = 2$ is a zero.

**(b)**

So $P(x) = (x-2)\left(x^4 + x^3 - 3x^2 - 5x - 2\right)$, and the possible rational zeros are restricted to $-1, \pm 2$.

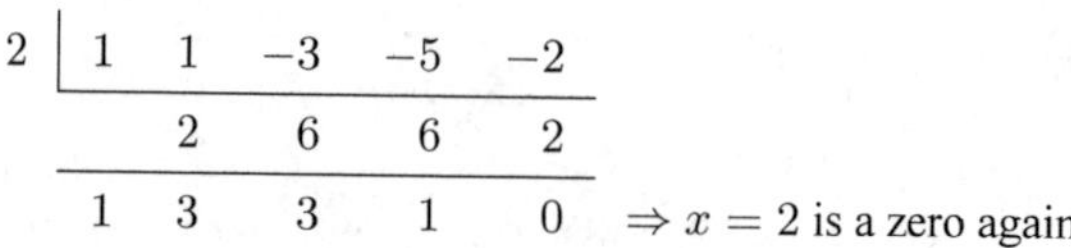

| 2 | 1 | 1 | −3 | −5 | −2 |
|---|---|---|---|---|---|
| | | 2 | 6 | 6 | 2 |
| | 1 | 3 | 3 | 1 | 0 |

$\Rightarrow x = 2$ is a zero again.

So $P(x) = (x-2)^2\left(x^3 + 3x^2 + 3x + 1\right)$, and the possible rational zeros are restricted to $-1$.

| −1 | 1 | 3 | 3 | 1 |
|---|---|---|---|---|
| | | −1 | −2 | −1 |
| | 1 | 2 | 1 | 0 |

$\Rightarrow x = -1$ is a zero.

So $P(x) = (x-2)^2(x+1)\left(x^2 + 2x + 1\right) = (x-2)^2(x+1)^3$., and the real zeros of $P$ are $-1$ and $2$.

**58. (a)** $P(x) = x^5 - x^4 - 6x^3 + 14x^2 - 11x + 3$ has possible rational zeros $\pm 1, \pm 3$.

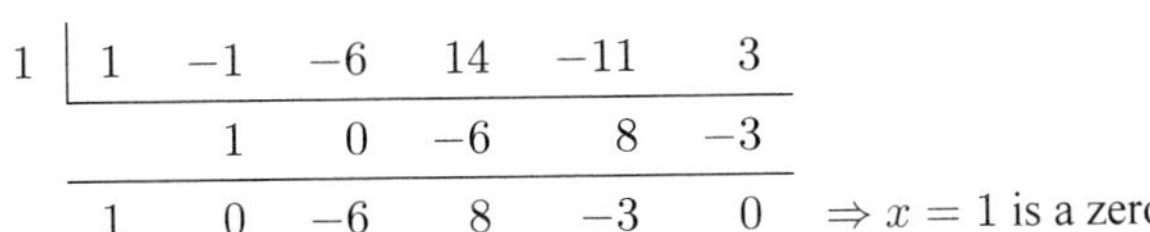

| 1 | 1 | −1 | −6 | 14 | −11 | 3 |
|---|---|---|---|---|---|---|
| | | 1 | 0 | −6 | 8 | −3 |
| | 1 | 0 | −6 | 8 | −3 | 0 |

$\Rightarrow x = 1$ is a zero.

So $P(x) = (x-1)(x^4 - 6x^2 + 8x - 3)$:

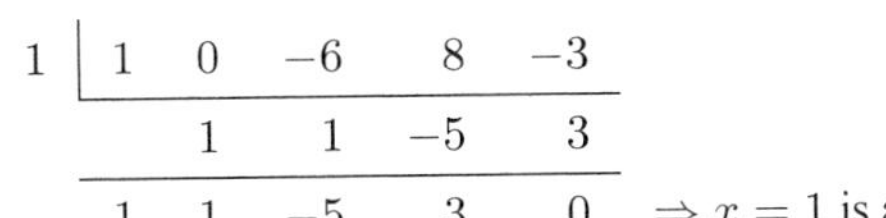

| 1 | 1 | 0 | −6 | 8 | −3 |
|---|---|---|---|---|---|
| | | 1 | 1 | −5 | 3 |
| | 1 | 1 | −5 | 3 | 0 |

$\Rightarrow x = 1$ is a zero again.

So $P(x) = (x-1)^2(x^3 + x^2 - 5x + 3)$:

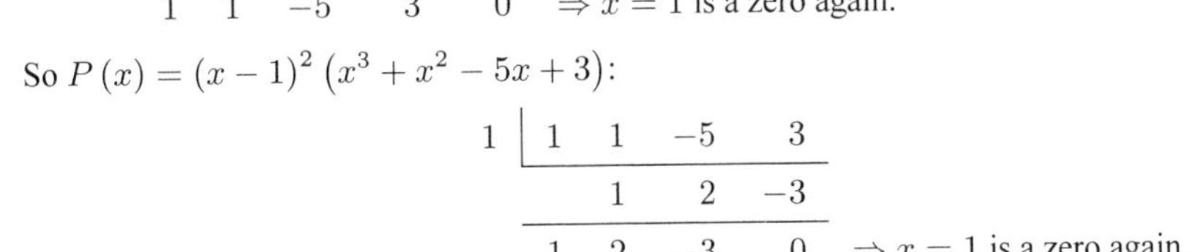

| 1 | 1 | 1 | −5 | 3 |
|---|---|---|---|---|
| | | 1 | 2 | −3 |
| | 1 | 2 | −3 | 0 |

$\Rightarrow x = 1$ is a zero again.

So $P(x) = (x-1)^3(x^2 + 2x - 3) = (x-1)^4(x+3)$, and the real zeros of $P$ are 1 and $-3$.

**(b)**

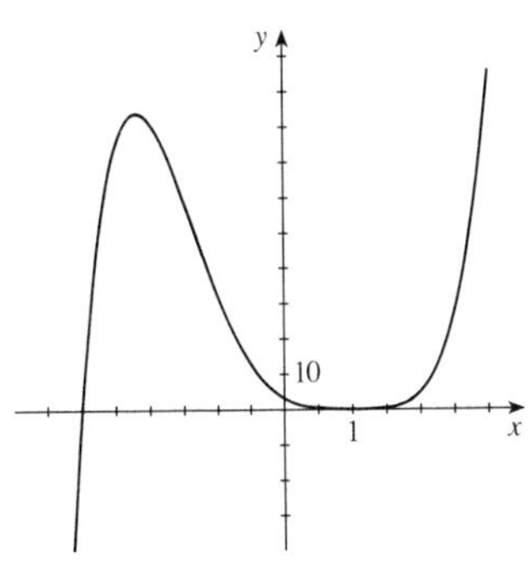


**59.** $P(x) = x^3 - x^2 - x - 3$. Since $P(x)$ has 1 variation in sign, $P$ has 1 positive real zero. Since $P(-x) = -x^3 - x^2 + x - 3$ has 2 variations in sign, $P$ has 2 or 0 negative real zeros. Thus, $P$ has 1 or 3 real zeros.

**60.** $P(x) = 2x^3 - x^2 + 4x - 7$. Since $P(x)$ has 3 variations in signs, $P$ has 3 or 1 positive real zeros. Since $P(-x) = -2x^3 - x^2 - 4x - 7$ has no variation in sign, there is no negative real zero. Thus, $P$ has 1 or 3 real zeros.

**61.** $P(x) = 2x^6 + 5x^4 - x^3 - 5x - 1$. Since $P(x)$ has 1 variation in sign, $P$ has 1 positive real zero. Since $P(-x) = 2x^6 + 5x^4 + x^3 + 5x - 1$ has 1 variation in sign, $P$ has 1 negative real zero. Therefore, $P$ has 2 real zeros.

**62.** $P(x) = x^4 + x^3 + x^2 + x + 12$. Since $P(x)$ has no variations in sign, $P$ has no positive real zeros. Since $P(-x) = x^4 - x^3 + x^2 - x + 12$ has 4 variations in sign, $P$ has 4, 2, or 0 negative real zeros. Therefore, $P(x)$ has 0, 2, or 4 real zeros.

**63.** $P(x) = x^5 + 4x^3 - x^2 + 6x$. Since $P(x)$ has 2 variations in sign, $P$ has 2 or 0 positive real zeros. Since $P(-x) = -x^5 - 4x^3 - x^2 - 6x$ has no variation in sign, $P$ has no negative real zero. Therefore, $P$ has a total of 1 or 3 real zeros (since $x = 0$ is a zero, but is neither positive nor negative).

**64.** $P(x) = x^8 - x^5 + x^4 - x^3 + x^2 - x + 1$. Since $P(x)$ has 6 variations in sign, the polynomial has 6, 4, 2, or 0 positive real zeros. Since $P(-x)$ has no variation in sign, the polynomial has no negative real zeros. Therefore, $P$ has 6, 4, 2, or 0 real zeros.

**65.** $P(x) = 2x^3 + 5x^2 + x - 2$; $a = -3$, $b = 1$

| −3 | 2 | 5 | 1 | −2 |
|---|---|---|---|---|
| | | −6 | 3 | −12 |
| | 2 | −1 | 4 | −14 |

alternating signs $\Rightarrow$ lower bound.

| 1 | 2 | 5 | 1 | −2 |
|---|---|---|---|---|
| | | 2 | 7 | 8 |
| | 2 | 7 | 8 | 6 |

all nonnegative $\Rightarrow$ upper bound.

Therefore $a = -3$ and $b = 1$ are lower and upper bounds.

**66.** $P(x) = x^4 - 2x^3 - 9x^2 + 2x + 8$; $a = -3$, $b = 5$

$$\begin{array}{r|rrrrr} -3 & 1 & -2 & -9 & 2 & 8 \\ & & -3 & 15 & -18 & 48 \\ \hline & 1 & -5 & 6 & -16 & 56 \end{array} \quad \text{Alternating signs} \Rightarrow \text{lower bound.}$$

$$\begin{array}{r|rrrrr} 5 & 1 & -2 & -9 & 2 & 8 \\ & & 5 & 15 & 30 & 160 \\ \hline & 1 & 3 & 6 & 32 & 168 \end{array} \quad \text{All nonnegative} \Rightarrow \text{upper bound.}$$

Therefore $a = -3$ and $b = 5$ are lower and upper bounds.

**67.** $P(x) = 8x^3 + 10x^2 - 39x + 9$; $a = -3$, $b = 2$

$$\begin{array}{r|rrrr} -3 & 8 & 10 & -39 & 9 \\ & & -24 & 42 & -9 \\ \hline & 8 & -14 & 3 & 0 \end{array} \quad \text{alternating signs} \Rightarrow \text{lower bound.}$$

$$\begin{array}{r|rrrr} 2 & 8 & 10 & -39 & 9 \\ & & 16 & 52 & 26 \\ \hline & 8 & 26 & 13 & 35 \end{array} \quad \text{all nonnegative} \Rightarrow \text{upper bound.}$$

Therefore $a = -3$ and $b = 2$ are lower and upper bounds. Note that $x = -3$ is also a zero.

**68.** $P(x) = 3x^4 - 17x^3 + 24x^2 - 9x + 1$; $a = 0$, $b = 6$

$$\begin{array}{r|rrrrr} 0 & 3 & -17 & 24 & -9 & 1 \\ & & 0 & 0 & 0 & 0 \\ \hline & 3 & -17 & 24 & -9 & 1 \end{array} \quad \text{Alternating signs} \Rightarrow \text{lower bound.}$$

$$\begin{array}{r|rrrrr} 6 & 3 & -17 & 24 & -9 & 1 \\ & & 18 & 6 & 180 & 1026 \\ \hline & 3 & 1 & 30 & 171 & 1027 \end{array} \quad \text{All nonnegative} \Rightarrow \text{upper bound.}$$

Therefore $a = 0$, $b = 6$ are lower and upper bounds, respectively. Note, since $P(x)$ alternates in sign, by Descartes' Rule of Signs, 0 is automatically a lower bound.

**69.** $P(x) = x^3 - 3x^2 + 4$ and use the Upper and Lower Bounds Theorem:

$$\begin{array}{r|rrrr} -1 & 1 & -3 & 0 & 4 \\ & & -1 & 4 & -4 \\ \hline & 1 & -4 & 4 & 0 \end{array} \quad \text{alternating signs} \Rightarrow \text{lower bound.}$$

$$\begin{array}{r|rrrr} 3 & 1 & -3 & 0 & 4 \\ & & 3 & 0 & 0 \\ \hline & 1 & 0 & 0 & 4 \end{array} \quad \text{all nonnegative} \Rightarrow \text{upper bound.}$$

Therefore $-1$ is a lower bound (and a zero) and 3 is an upper bound. (There are many possible solutions.)

**70.** $P(x) = 2x^3 - 3x^2 - 8x + 12$ and using the Upper and Lower Bounds Theorem:

$$\begin{array}{r|rrrr} -2 & 2 & -3 & -8 & 12 \\ & & -4 & 14 & -12 \\ \hline & 2 & -7 & 6 & 0 \end{array} \quad \text{Alternating signs} \quad \Rightarrow \quad x = -2 \text{ is a lower bound (and a zero).}$$

$$\begin{array}{r|rrrr} 3 & 2 & -3 & -8 & 12 \\ & & 6 & 9 & 3 \\ \hline & 2 & 3 & 1 & 15 \end{array} \quad \text{All nonnegative} \quad \Rightarrow \quad x = 3 \text{ is an upper bound.}$$

(There are many possible solutions.)

**71.** $P(x) = x^4 - 2x^3 + x^2 - 9x + 2$.

$$\begin{array}{r|rrrrr} 1 & 1 & -2 & 1 & -9 & 2 \\ & & 1 & -1 & 0 & -9 \\ \hline & 1 & -1 & 0 & -9 & -7 \end{array} \qquad \begin{array}{r|rrrrr} 2 & 1 & -2 & 1 & -9 & 2 \\ & & 2 & 0 & 2 & -14 \\ \hline & 1 & 0 & 1 & -7 & -12 \end{array}$$

$$\begin{array}{r|rrrrr} 3 & 1 & -2 & 1 & -9 & 2 \\ & & 3 & 3 & 12 & 9 \\ \hline & 1 & 1 & 4 & 3 & 11 \end{array} \quad \text{all positive} \quad \Rightarrow \quad \text{upper bound.}$$

$$\begin{array}{r|rrrrr} -1 & 1 & -2 & 1 & -9 & 2 \\ & & -1 & 3 & -4 & 13 \\ \hline & 1 & -3 & 4 & -13 & 15 \end{array} \quad \text{alternating signs} \quad \Rightarrow \quad \text{lower bound.}$$

Therefore $-1$ is a lower bound and 3 is an upper bound. (There are many possible solutions.)

**72.** Set $P(x) = x^5 - x^4 + 1$.

$$\begin{array}{r|rrrrrr} 1 & 1 & -1 & 0 & 0 & 0 & 1 \\ & & 1 & 0 & 0 & 0 & 0 \\ \hline & 1 & 0 & 0 & 0 & 0 & 1 \end{array} \quad \text{All nonnegative} \quad \Rightarrow \quad x = 1 \text{ is an upper bound.}$$

$$\begin{array}{r|rrrrrr} -1 & 1 & -1 & 0 & 0 & 0 & 1 \\ & & -1 & 2 & -2 & 2 & -2 \\ \hline & 1 & -2 & 2 & -2 & 2 & -1 \end{array} \quad \text{Alternating signs} \quad \Rightarrow \quad x = -1 \text{ is a lower bound.}$$

(There are many possible solutions.)

**73.** $P(x) = 2x^4 + 3x^3 - 4x^2 - 3x + 2$.

$$\begin{array}{r|rrrrr} 1 & 2 & 3 & -4 & -3 & 2 \\ & & 2 & 5 & 1 & -2 \\ \hline & 2 & 5 & 1 & -2 & 0 \end{array} \quad \Rightarrow x = 1 \text{ is a zero.}$$

$P(x) = (x - 1)(2x^3 + 5x^2 + x - 2)$

$$\begin{array}{r|rrrr} -1 & 2 & 5 & 1 & -2 \\ & & -2 & -3 & 2 \\ \hline & 2 & 3 & -2 & 0 \end{array} \quad \Rightarrow x = -1 \text{ is a zero.}$$

$P(x) = (x - 1)(x + 1)(2x^2 + 3x - 2) = (x - 1)(x + 1)(2x - 1)(x + 2)$. Therefore, the zeros are $x = -2, \frac{1}{2}, \pm 1$.

**74.** $P(x) = 2x^4 + 15x^3 + 31x^2 + 20x + 4$. The possible rational zeros are $\pm1, \pm2, \pm4, \pm\frac{1}{2}$. Since all of the coefficients are positive, there are no positive zeros. Since $P(-x) = 2x^4 - 15x^3 + 31x^2 - 20x + 4$ has 4 variations in sign, there are 0, 2, or 4 negative real zeros.

$$\begin{array}{r|rrrrr} -1 & 2 & 15 & 31 & 20 & 4 \\ & & -2 & -13 & -18 & -2 \\ \hline & 2 & 13 & 18 & 2 & 2 \end{array} \qquad \begin{array}{r|rrrrr} -2 & 2 & 15 & 31 & 20 & 4 \\ & & -4 & -22 & -18 & -4 \\ \hline & 2 & 11 & 9 & 2 & 0 \end{array} \Rightarrow x = -2 \text{ is a zero.}$$

$P(x) = (x+2)(2x^3 + 11x^2 + 9x + 2)$:

$$\begin{array}{r|rrrr} -2 & 2 & 11 & 9 & 2 \\ & & -4 & -14 & 10 \\ \hline & 2 & 7 & -5 & 12 \end{array} \qquad \begin{array}{r|rrrr} -4 & 2 & 11 & 9 & 2 \\ & & -8 & -12 & 12 \\ \hline & 2 & 3 & -3 & 14 \end{array} \qquad \begin{array}{r|rrrr} -\frac{1}{2} & 2 & 11 & 9 & 2 \\ & & -1 & -5 & -2 \\ \hline & 2 & 10 & 4 & 0 \end{array}\ x = -\tfrac{1}{2} \text{ is a zero.}$$

$P(x) = (x+2)(2x+1)(x^2 + 5x + 2)$. Now if $x^2 + 5x + 2 = 0$, then $x = \frac{-5\pm\sqrt{25-4(1)(2)}}{2} = \frac{-5\pm\sqrt{17}}{2}$. Thus, the zeros are $-2, -\frac{1}{2}$, and $\frac{-5\pm\sqrt{17}}{2}$.

**75.** *Method 1:* $P(x) = 4x^4 - 21x^2 + 5$ has 2 variations in sign, so by Descartes' rule of signs there are either 2 or 0 positive zeros. If we replace $x$ with $(-x)$, the function does not change, so there are either 2 or 0 negative zeros. Possible rational zeros are $\pm1, \pm\frac{1}{2}, \pm\frac{1}{4}, \pm5, \pm\frac{5}{2}, \pm\frac{5}{4}$.By inspection, $\pm1$ and $\pm5$ are not zeros, so we must look for non-integer solutions:

$$\begin{array}{r|rrrrr} \frac{1}{2} & 4 & 0 & -21 & 0 & 5 \\ & & 2 & 1 & -10 & -5 \\ \hline & 4 & 2 & -20 & -10 & 0 \end{array} \Rightarrow x = \tfrac{1}{2} \text{ is a zero.}$$

$P(x) = \left(x - \frac{1}{2}\right)(4x^3 + 2x^2 - 20x - 10)$, continuing with the quotient, we have:

$$\begin{array}{r|rrrr} -\frac{1}{2} & 4 & 2 & -20 & -10 \\ & & -2 & 0 & 10 \\ \hline & 4 & 0 & -20 & 0 \end{array} \Rightarrow x = -\tfrac{1}{2} \text{ is a zero.}$$

$P(x) = \left(x - \frac{1}{2}\right)\left(x + \frac{1}{2}\right)(4x^2 - 20) = 0$. If $4x^2 - 20 = 0$, then $x = \pm\sqrt{5}$. Thus the zeros are $x = \pm\frac{1}{2}, \pm\sqrt{5}$.

*Method 2:* Substituting $u = x^2$, the equation becomes $4u^2 - 21u + 5 = 0$, which factors: $4u^2 - 21u + 5 = (4u - 1)(u - 5) = (4x^2 - 1)(x^2 - 5)$. Then either we have $x^2 = 5$, so that $x = \pm\sqrt{5}$, or we have $x^2 = \frac{1}{4}$, so that $x = \pm\sqrt{\frac{1}{4}} = \pm\frac{1}{2}$. Thus the zeros are $x = \pm\frac{1}{2}, \pm\sqrt{5}$.

**76.** $P(x) = 6x^4 - 7x^3 - 8x^2 + 5x = x(6x^3 - 7x^2 - 8x + 5)$. So $x = 0$ is a zero. Continuing with the quotient, $Q(x) = 6x^3 - 7x^2 - 8x + 5$. The possible rational zeros are $\pm1, \pm5, \pm\frac{1}{2}, \pm\frac{5}{2}, \pm\frac{1}{3}, \pm\frac{5}{3}, \pm\frac{1}{6}, \pm\frac{5}{6}$. Since $Q(x)$ has 2 variations in sign, there are 0 or 2 positive real zeros. Since $Q(-x) = 6x^4 + 7x^3 - 8x^2 - 5x$ has 1 variation in sign, there is 1 negative real zero.

$$\begin{array}{r|rrrr} 1 & 6 & -7 & -8 & 5 \\ & & 6 & -1 & -9 \\ \hline & 6 & -1 & -9 & -4 \end{array} \qquad \begin{array}{r|rrrr} 5 & 6 & -7 & -8 & 5 \\ & & 30 & 115 & 535 \\ \hline & 6 & 23 & 107 & 540 \end{array} \quad \text{All positive} \Rightarrow \text{upper bound.}$$

$$\begin{array}{r|rrrr} \frac{1}{2} & 6 & -7 & -8 & 5 \\ & & 3 & -2 & -5 \\ \hline & 6 & -4 & -10 & 0 \end{array} \Rightarrow x = \tfrac{1}{2} \text{ is a zero.}$$

$P(x) = x(2x-1)(3x^2 - 2x - 5) = x(2x-1)(3x-5)(x+1)$. Therefore, the zeros are $0, -1, \frac{1}{2}$ and $\frac{5}{3}$.

**77.** $P(x) = x^5 - 7x^4 + 9x^3 + 23x^2 - 50x + 24$. The possible rational zeros are $\pm1, \pm2, \pm3, \pm4, \pm6, \pm8, \pm12, \pm24$. $P(x)$ has 4 variations in sign and hence 0, 2, or 4 positive real zeros. $P(-x) = -x^5 - 7x^4 - 9x^3 + 23x^2 + 50x + 24$ has 1 variation in sign, and hence 1 negative real zero.

$$\begin{array}{r|rrrrrr} 1 & 1 & -7 & 9 & 23 & -50 & 24 \\ & & 1 & -6 & 3 & 26 & -24 \\ \hline & 1 & -6 & 3 & 26 & -24 & 0 \end{array} \Rightarrow x = 1 \text{ is a zero.}$$

$P(x) = (x-1)(x^4 - 6x^3 + 3x^2 + 26x - 24)$; continuing with the quotient, we try 1 again.

$$\begin{array}{r|rrrrr} 1 & 1 & -6 & 3 & 26 & -24 \\ & & 1 & -5 & -2 & 24 \\ \hline & 1 & -5 & -2 & 24 & 0 \end{array} \Rightarrow x = 1 \text{ is a zero again.}$$

$P(x) = (x-1)^2(x^3 - 5x^2 - 2x + 24)$; continuing with the quotient, we start by trying 1 again.

$$\begin{array}{r|rrrr} 1 & 1 & -5 & -2 & 24 \\ & & 1 & -4 & -6 \\ \hline & 1 & -4 & -6 & 18 \end{array} \quad \begin{array}{r|rrrr} 2 & 1 & -5 & -2 & 24 \\ & & 2 & -6 & -16 \\ \hline & 1 & -3 & -8 & 8 \end{array} \quad \begin{array}{r|rrrr} 3 & 1 & -5 & -2 & 24 \\ & & 3 & -6 & -24 \\ \hline & 1 & -2 & -8 & 0 \end{array} \Rightarrow x = 3 \text{ is a zero.}$$

$P(x) = (x-1)^2(x-3)(x^2 - 2x - 8) = (x-1)^2(x-3)(x-4)(x+2)$. Therefore, the zeros are $x = -2, 1, 3, 4$.

**78.** $P(x) = 8x^5 - 14x^4 - 22x^3 + 57x^2 - 35x + 6$. The possible rational zeros are $\pm1, \pm2, \pm3, \pm6, \pm\frac{1}{2}$, $\pm\frac{3}{2}, \pm\frac{1}{4}, \pm\frac{3}{4}, \pm\frac{1}{8}, \pm\frac{3}{8}$. Since $P(x)$ has 4 variations in sign, there are 0, 2, or 4 positive real zeros. Since $P(-x) = -8x^5 - 14x^4 + 22x^3 + 57x^2 + 35x + 6$ has 1 variation in sign, there is 1 negative real zero.

$$\begin{array}{r|rrrrrr} -1 & 8 & -14 & -22 & 57 & -35 & 6 \\ & & -8 & 22 & 0 & -57 & 92 \\ \hline & 8 & -22 & 0 & 57 & -92 & 98 \end{array} \quad \begin{array}{r|rrrrrr} -2 & 8 & -14 & -22 & 57 & -35 & 6 \\ & & -16 & 60 & -76 & 38 & -6 \\ \hline & 8 & -30 & 38 & -19 & 3 & 0 \end{array} \Rightarrow x = -2 \text{ is a zero.}$$

$P(x) = (x+2)(8x^4 - 30x^3 + 38x^2 - 19x + 3)$. All the other real zeros are positive.

$$\begin{array}{r|rrrrr} 1 & 8 & -30 & 38 & -19 & 3 \\ & & 8 & -22 & 16 & -3 \\ \hline & 8 & -22 & 16 & -3 & 0 \end{array} \Rightarrow x = 1 \text{ is a zero.}$$

$P(x) = (x+2)(x-1)(8x^3 - 22x^2 + 16x - 3)$.

$$\begin{array}{r|rrrr} 1 & 8 & -22 & 16 & -3 \\ & & 8 & -14 & 2 \\ \hline & 8 & -14 & 2 & -1 \end{array} \qquad \begin{array}{r|rrrr} \frac{1}{2} & 8 & -22 & 16 & -3 \\ & & 4 & -9 & \frac{7}{2} \\ \hline & 8 & -18 & 7 & \frac{1}{2} \end{array}$$

Since $f\left(\frac{1}{2}\right) > 0 > f(1)$, there must be a zero between $\frac{1}{2}$ and 1. We try $\frac{3}{4}$:

$$\begin{array}{r|rrrr} \frac{3}{4} & 8 & -22 & 16 & -3 \\ & & 6 & -12 & 3 \\ \hline & 8 & -16 & 4 & 0 \end{array} \Rightarrow x = \tfrac{3}{4} \text{ is a zero.}$$

$P(x) = (x+2)(x-1)(4x-3)(2x^2 - 4x + 1)$. Now, $2x^2 - 4x + 1 = 0$ when $x = \frac{4\pm\sqrt{16-4(2)(1)}}{2(2)} = \frac{2\pm\sqrt{2}}{2}$. Thus, the zeros are $1, \frac{3}{4}, -2$, and $\frac{2\pm\sqrt{2}}{2}$.

**79.** $P(x) = x^3 - x - 2$. The only possible rational zeros of $P(x)$ are $\pm 1$ and $\pm 2$.

$$\begin{array}{r|rrrr} 1 & 1 & 0 & -1 & -2 \\ & & 1 & 1 & 0 \\ \hline & 1 & 1 & 0 & -2 \end{array} \qquad \begin{array}{r|rrrr} 2 & 1 & 0 & -1 & -2 \\ & & 2 & 4 & 6 \\ \hline & 1 & 2 & 3 & 4 \end{array} \qquad \begin{array}{r|rrrr} -1 & 1 & 0 & -1 & -2 \\ & & -1 & 1 & 0 \\ \hline & 1 & -1 & 0 & -2 \end{array}$$

Since the row that contains $-1$ alternates between nonnegative and nonpositive, $-1$ is a lower bound and there is no need to try $-2$. Therefore, $P(x)$ does not have any rational zeros.

**80.** $P(x) = 2x^4 - x^3 + x + 2$. The only possible rational zeros of $P(x)$ are $\pm 1, \pm 2, \pm\frac{1}{2}$.

$$\begin{array}{r|rrrrr} \frac{1}{2} & 2 & -1 & 0 & 1 & 2 \\ & & 1 & 0 & 0 & \frac{1}{2} \\ \hline & 2 & 0 & 0 & 1 & \frac{5}{2} \end{array} \quad \text{All nonnegative} \quad \Rightarrow \quad x = \tfrac{1}{2} \text{ is an upper bound.}$$

$$\begin{array}{r|rrrrr} -1 & 2 & -1 & 0 & 1 & 2 \\ & & -2 & 3 & -3 & 2 \\ \hline & 2 & -3 & 3 & -2 & 4 \end{array} \quad \text{Alternating signs} \quad \Rightarrow \quad x = -1 \text{ is a lower bound.}$$

$$\begin{array}{r|rrrrr} -\frac{1}{2} & 2 & -1 & 0 & 1 & 2 \\ & & -1 & 1 & -\frac{1}{2} & -\frac{1}{4} \\ \hline & 2 & -2 & 1 & \frac{1}{2} & \frac{7}{4} \end{array}$$

Therefore, there is no rational zero.

**81.** $P(x) = 3x^3 - x^2 - 6x + 12$ has possible rational zeros $\pm 1, \pm 2, \pm 3, \pm 4, \pm 6, \pm 12, \pm\frac{1}{3}, \pm\frac{2}{3}, \pm\frac{4}{3}$.

$$\begin{array}{r|rrrr} & 3 & -1 & -6 & 12 \\ \hline 1 & 3 & 2 & -4 & 8 \\ 2 & 3 & 5 & 4 & 20 \\ -1 & 3 & -4 & -2 & 14 \\ -2 & 3 & -7 & 8 & -4 \end{array}$$

Row 2: all positive $\Rightarrow$ $x = 2$ is an upper bound

Row $-2$: alternating signs $\Rightarrow$ $x = -2$ is a lower bound

$$\begin{array}{r|rrrr} & 3 & -1 & -6 & 12 \\ \hline \frac{1}{3} & 3 & 0 & -6 & 10 \\ \frac{2}{3} & 3 & 1 & -\frac{16}{3} & \frac{76}{9} \\ \frac{4}{3} & 3 & 3 & -2 & \frac{28}{3} \\ -\frac{1}{3} & 3 & -2 & -\frac{16}{3} & \frac{124}{9} \\ -\frac{2}{3} & 3 & -3 & -4 & \frac{44}{3} \\ -\frac{4}{3} & 3 & -5 & \frac{2}{3} & \frac{100}{9} \end{array}$$

Therefore, there is no rational zero.

**82.** $P(x) = x^{50} - 5x^{25} + x^2 - 1$. The only possible rational zeros of $P(x)$ are $\pm 1$. Since $P(1) = (1)^{50} - 5(1)^{25} + (1)^2 - 1 = -4$ and $P(-1) = (-1)^{50} - 5(-1)^{25} + (-1)^2 - 1 = 6$, $P(x)$ does not have a rational zero.

**83.** $P(x) = x^3 - 3x^2 - 4x + 12$, $[-4, 4]$ by $[-15, 15]$. The possible rational zeros are $\pm 1, \pm 2, \pm 3, \pm 4, \pm 6, \pm 12$. By observing the graph of $P$, the rational zeros are $x = -2$, 2, 3.

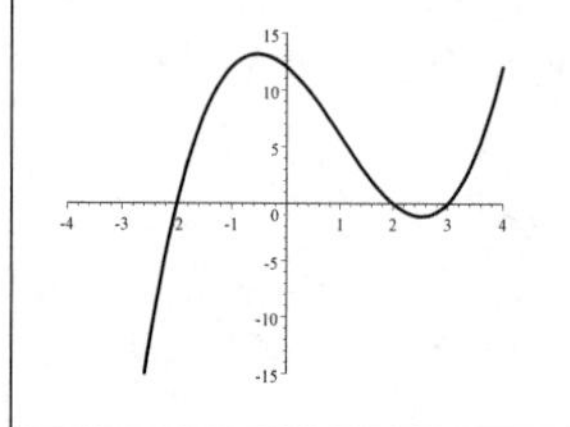

**84.** $P(x) = x^4 - 5x^2 + 4$, $[-4, 4]$ by $[-30, 30]$. The possible rational solutions are $\pm 1, \pm 2, \pm 4$. By observing the graph of the equation, the solutions of the given equation are $x = \pm 1, \pm 2$.

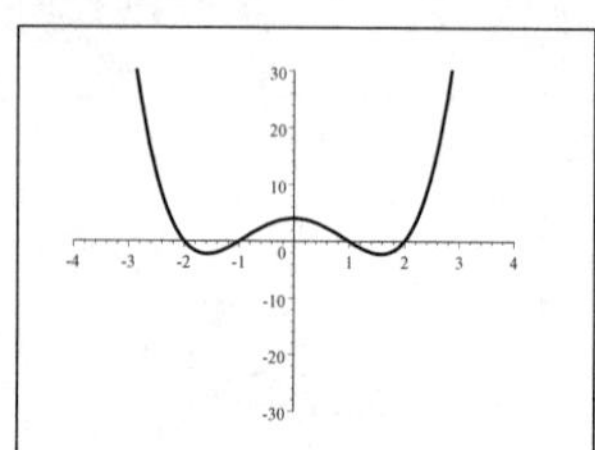

**85.** $P(x) = 2x^4 - 5x^3 - 14x^2 + 5x + 12$, $[-2, 5]$ by $[-40, 40]$. The possible rational zeros are $\pm 1, \pm 2, \pm 3, \pm 4, \pm 6, \pm 12, \pm \frac{1}{2}, \pm \frac{3}{2}$. By observing the graph of $P$, the zeros are $x = -\frac{3}{2}, -1, 1, 4$.

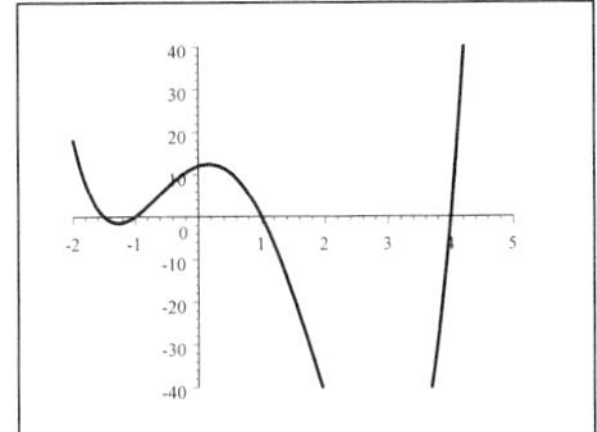


**86.** $P(x) = 3x^3 + 8x^2 + 5x + 2$, $[-3, 3]$ by $[-10, 10]$ The possible rational solutions are $\pm 1$, $\pm 2$, $\pm \frac{1}{3}$, $\pm \frac{2}{3}$. By observing the graph of the equation, the only real solution of the given equation is $x = -2$.

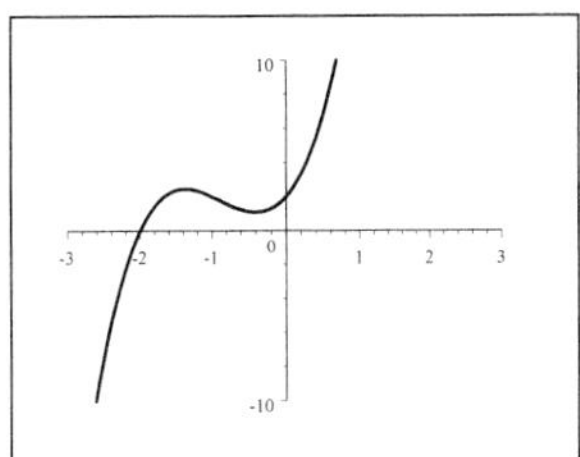


**87.** $x^4 - x - 4 = 0$. Possible rational solutions are $\pm 1$, $\pm 2$, $\pm 4$.

$$\begin{array}{r|rrrrr} 1 & 1 & 0 & 0 & -1 & -4 \\ & & 1 & 1 & 1 & 0 \\ \hline & 1 & 1 & 1 & 0 & -4 \end{array} \qquad \begin{array}{r|rrrrr} 2 & 1 & 0 & 0 & -1 & -4 \\ & & 2 & 4 & 8 & 14 \\ \hline & 1 & 2 & 4 & 7 & 10 \end{array} \Rightarrow x = 2 \text{ is an upper bound.}$$

$$\begin{array}{r|rrrrr} -1 & 1 & 0 & 0 & -1 & -4 \\ & & -1 & 1 & -1 & 2 \\ \hline & 1 & -1 & 1 & -2 & -2 \end{array} \qquad \begin{array}{r|rrrrr} -2 & 1 & 0 & 0 & -1 & -4 \\ & & -2 & 4 & -8 & 18 \\ \hline & 1 & -2 & 4 & -9 & 14 \end{array} \Rightarrow x = -2 \text{ is a lower bound.}$$

Therefore, we graph the function $P(x) = x^4 - x - 4$ in the viewing rectangle $[-2, 2]$ by $[-5, 20]$ and see there are two solutions. In the viewing rectangle $[-1.3, -1.25]$ by $[-0.1, 0.1]$, we find the solution $x \approx -1.28$. In the viewing rectangle $[1.5.1.6]$ by $[-0.1, 0.1]$, we find the solution $x \approx 1.53$. Thus the solutions are $x \approx -1.28, 1.53$.

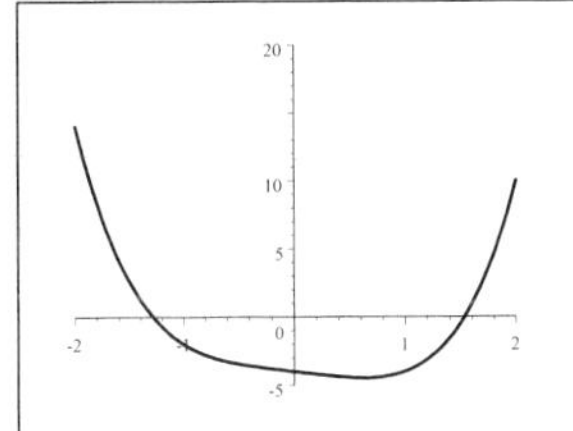


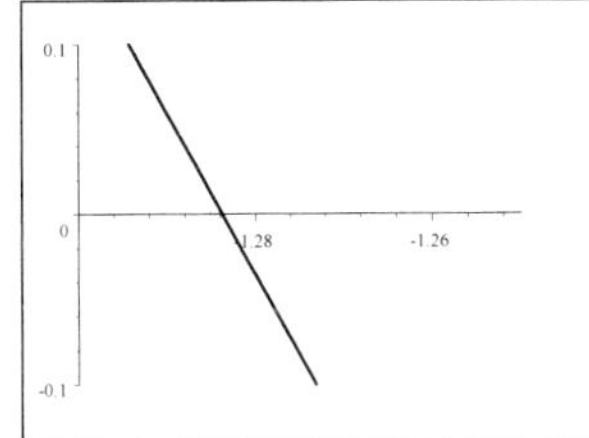


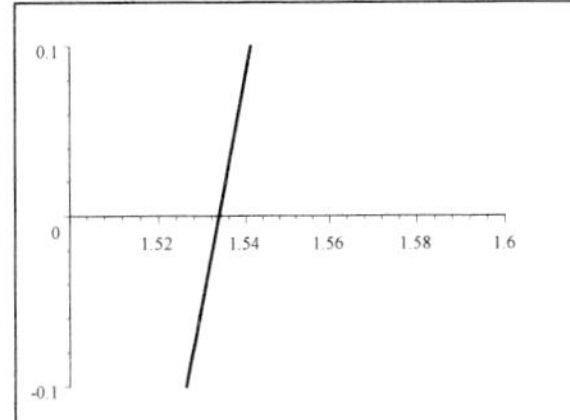


**88.** $2x^3 - 8x^2 + 9x - 9 = 0$. Possible rational solutions are $\pm 1$, $\pm 3$, $\pm 9$, $\pm \frac{1}{2}$, $\pm \frac{3}{2}$, $\pm \frac{9}{2}$.

$$\begin{array}{r|rrrr} 1 & 2 & -8 & 9 & -9 \\ & & 2 & -6 & 3 \\ \hline & 2 & -6 & 3 & -6 \end{array} \qquad \begin{array}{r|rrrr} 3 & 2 & -8 & 9 & -9 \\ & & 6 & -6 & 9 \\ \hline & 2 & -2 & 3 & 0 \end{array} \Rightarrow x = 3 \text{ is a zero.}$$

We graph $P(x) = 2x^3 - 8x^2 + 9x - 9$ in the viewing rectangle $[-4, 6]$ by $[-40, 40]$. It appears that the equation has no other real solution. We can factor $2x^3 - 8x^2 + 9x - 9 = (x - 3)(2x^2 - 2x + 3)$. Since the quotient is a quadratic expression, we can use the quadratic formula to locate the other possible solutions: $x = \frac{2 \pm \sqrt{2^2 - 4(2)(3)}}{2(2)}$, which are not real. So the only solution is $x = 3$.

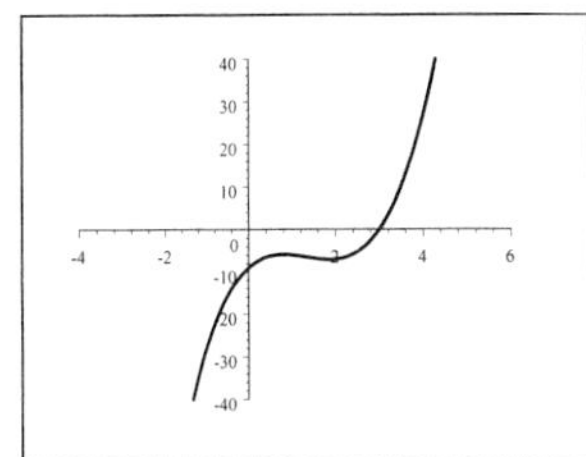

**89.** $4.00x^4 + 4.00x^3 - 10.96x^2 - 5.88x + 9.09 = 0$.

$$\begin{array}{r|rrrrr} 1 & 4 & 4 & -10.96 & -5.88 & 9.09 \\ & & 4 & 8 & -2.96 & -8.84 \\ \hline & 4 & 8 & -2.96 & -8.84 & 0.25 \end{array} \qquad \begin{array}{r|rrrrr} 2 & 4 & 4 & -10.96 & -5.88 & 9.09 \\ & & 8 & 24 & 26.08 & 40.40 \\ \hline & 4 & 12 & 13.04 & 20.2 & 49.49 \end{array} \Rightarrow x = 2 \text{ is an upper bound.}$$

$$\begin{array}{r|rrrrr} -2 & 4 & 4 & -10.96 & -5.88 & 9.09 \\ & & -8 & 8 & 5.92 & -0.08 \\ \hline & 4 & -4 & -2.96 & 0.04 & 9.01 \end{array} \qquad \begin{array}{r|rrrrr} -3 & 4 & 4 & -10.96 & -5.88 & 9.09 \\ & & -12 & 24 & -39.12 & 135 \\ \hline & 4 & -8 & 13.04 & -45 & 144.09 \end{array} \Rightarrow x = -3 \text{ is a lower bound.}$$

Therefore, we graph the function $P(x) = 4.00x^4 + 4.00x^3 - 10.96x^2 - 5.88x + 9.09$ in the viewing rectangle $[-3, 2]$ by $[-10, 40]$. There appear to be two solutions. In the viewing rectangle $[-1.6, -1.4]$ by $[-0.1, 0.1]$, we find the solution $x \approx -1.50$. In the viewing rectangle $[0.8, 1.2]$ by $[0, 1]$, we see that the graph comes close but does not go through the $x$-axis. Thus there is no solution here. Therefore, the only solution is $x \approx -1.50$.

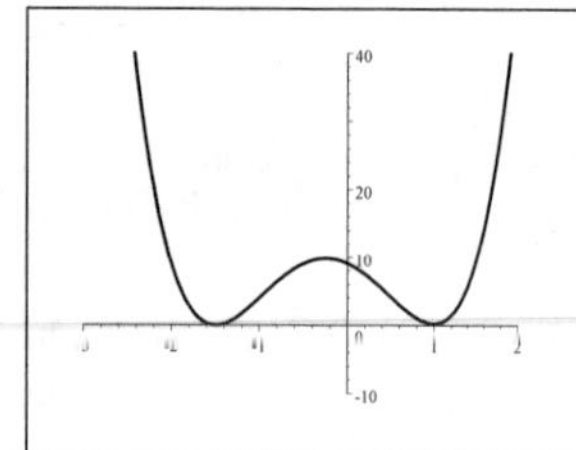


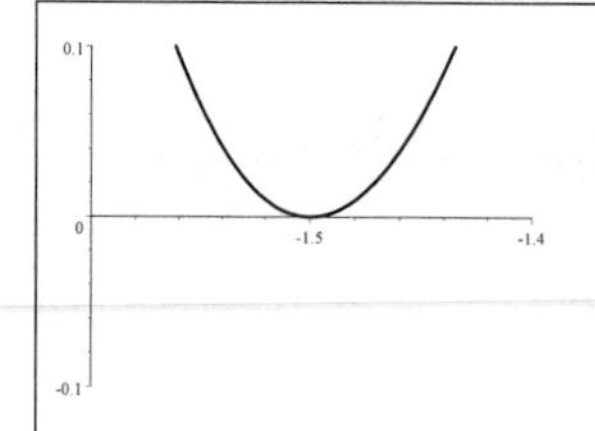


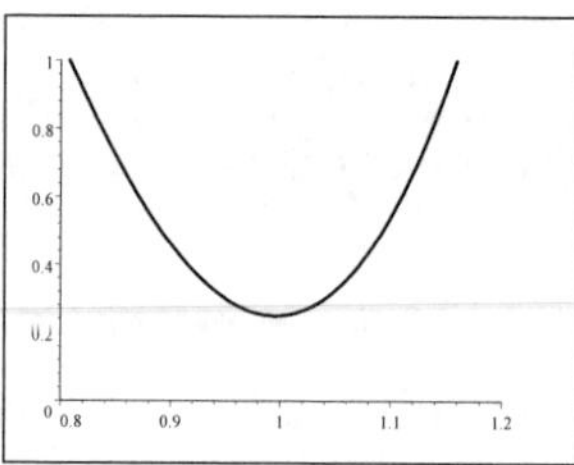


**90.** $x^5 + 2x^4 + 0.96x^3 + 5x^2 + 10x + 4.8 = 0$. Since all the coefficients are positive, there is no positive solution. So $x = 0$ is an upper bound.

$$\begin{array}{r|rrrrrr} -2 & 1 & 2 & 0.96 & 5 & 10 & 4.8 \\ & & -2 & 0 & -1.92 & -6.16 & -7.68 \\ \hline & 1 & 0 & 0.96 & 3.08 & 3.84 & -2.88 \end{array} \qquad \begin{array}{r|rrrrrr} -3 & 1 & 2 & 0.96 & 5 & 10 & 4.8 \\ & & -3 & 3 & -11.88 & 20.64 & -91.92 \\ \hline & 1 & -1 & 3.96 & -6.88 & 30.64 & -87.12 \end{array} \Rightarrow x = -3 \text{ is a lower bound.}$$

Therefore, we graph $P(x) = x^5 + 2x^4 + 0.96x^3 + 5x^2 + 10x + 4.8$ in the viewing rectangle $[-3, 0]$ by $[-10, 5]$ and see that there are three possible solutions. In the viewing rectangle $[-1.75, -1.7]$ by $[-0.1, 0.1]$, we find the solution $x \approx -1.71$. In the viewing rectangle $[-1.25, -1.15]$ by $[-0.1, 0.1]$, we find the solution $x \approx -1.20$. In the viewing rectangle $[-0.85, -0.75]$ by $[-0.1, 0.1]$, we find the solution $x \approx -0.80$. So the solutions are $x \approx -1.71, -1.20, -0.80$.

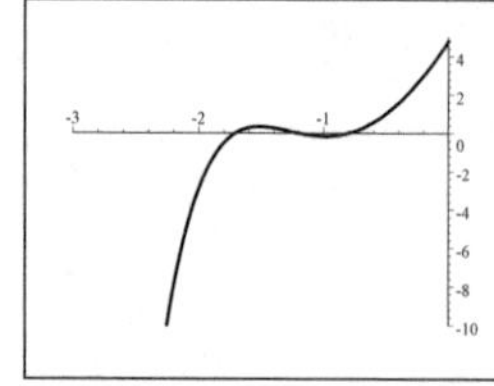


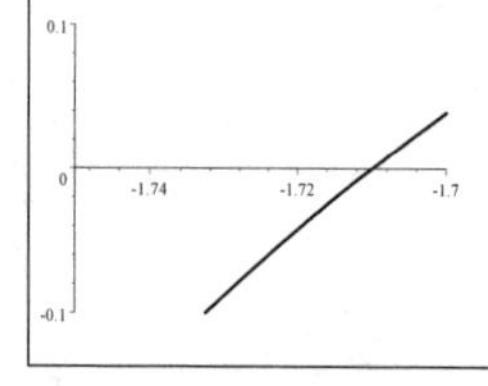


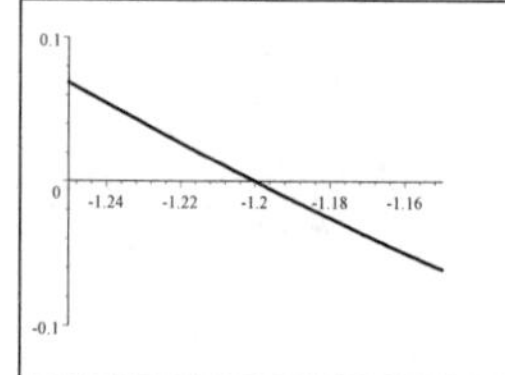


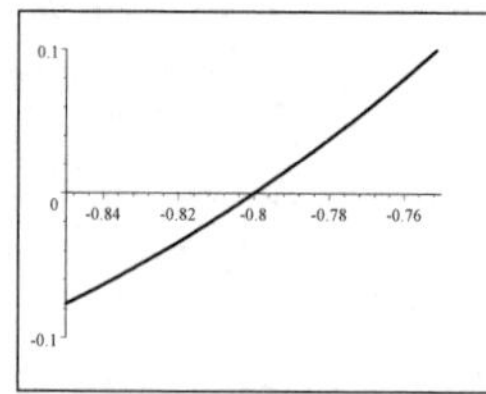


**91. (a)** Since $z > b$, we have $z - b > 0$. Since all the coefficients of $Q(x)$ are nonnegative, and since $z > 0$, we have $Q(z) > 0$ (being a sum of positive terms). Thus, $P(z) = (z - b) \cdot Q(z) + r > 0$, since the sum of a positive number and a nonnegative number.

**(b)** In part (a), we showed that if $b$ satisfies the conditions of the first part of the Upper and Lower Bounds Theorem and $z > b$, then $P(z) > 0$. This means that no real zero of $P$ can be larger than $b$, so $b$ is an upper bound for the real zeros.

**(c)** Suppose $-b$ is a negative lower bound for the real zeros of $P(x)$. Then clearly $b$ is an upper bound for $P_1(x) = P(-x)$. Thus, as in Part (a), we can write $P_1(x) = (x - b) \cdot Q(x) + r$, where $r > 0$ and the coefficients of $Q$ are all nonnegative, and $P(x) = P_1(-x) = (-x - b) \cdot Q(-x) + r = (x + b) \cdot [-Q(-x)] + r$. Since the coefficients of $Q(x)$ are all nonnegative, the coefficients of $-Q(-x)$ will be alternately nonpositive and nonnegative, which proves the second part of the Upper and Lower Bounds Theorem.

**92.** $P(x) = x^5 - x^4 - x^3 - 5x^2 - 12x - 6$ has possible rational zeros $\pm1, \pm2, \pm3, \pm6$. Since $P(x)$ has 1 variation in sign, there is 1 positive real zero. Since $P(-x) = -x^5 - x^4 + x^3 - 5x^2 + 12x - 6$ has 4 variations in sign, there are 0, 2, or 4 negative real zeros.

$$\begin{array}{c|cccccc} 1 & 1 & -1 & -1 & -5 & -12 & -6 \\ & & 1 & 0 & -1 & -6 & -18 \\ \hline & 1 & 0 & -1 & -6 & -18 & -24 \end{array}$$

$$\begin{array}{c|cccccc} 2 & 1 & -1 & -1 & -5 & -12 & -6 \\ & & 2 & 2 & 2 & -6 & -36 \\ \hline & 1 & 1 & 1 & -3 & -18 & -42 \end{array}$$

$$\begin{array}{c|cccccc} 3 & 1 & -1 & -1 & -5 & -12 & -6 \\ & & 3 & 6 & 15 & 30 & 54 \\ \hline & 1 & 2 & 5 & 10 & 18 & 48 \end{array} \Rightarrow 3 \text{ is an upper bound.}$$

$$\begin{array}{c|cccccc} -1 & 1 & -1 & -1 & -5 & -12 & -6 \\ & & -1 & 2 & -1 & 6 & 6 \\ \hline & 1 & -2 & 1 & -6 & -6 & 0 \end{array} \Rightarrow x = -1 \text{ is a zero.}$$

$P(x) = (x+1)(x^4 - 2x^3 + x^2 - 6x - 6)$, continuing with the quotient we have

$$\begin{array}{c|ccccc} -1 & 1 & -2 & 1 & -6 & -6 \\ & & -1 & 3 & -4 & 10 \\ \hline & 1 & -3 & 4 & -10 & 4 \end{array} \Rightarrow -1 \text{ is a lower bound.}$$

Therefore, there is 1 rational zero, namely $-1$. Since there are 1, 3 or 5 real zeros, and we found 1 rational zero, there must be 0, 2 or 4 irrational zeros. However, since 1 zero must be positive, there cannot be 0 irrational zeros. Therefore, there is exactly 1 rational zero and 2 or 4 irrational zeros.

**93.** Let $r$ be the radius of the silo. The volume of the hemispherical roof is $\frac{1}{2}\left(\frac{4}{3}\pi r^3\right) = \frac{2}{3}\pi r^3$. The volume of the cylindrical section is $\pi\left(r^2\right)(30) = 30\pi r^2$. Because the total volume of the silo is 15,000 ft$^3$, we get the following equation: $\frac{2}{3}\pi r^3 + 30\pi r^2 = 15000 \quad \Leftrightarrow \quad \frac{2}{3}\pi r^3 + 30\pi r^2 - 15000 = 0 \Leftrightarrow \pi r^3 + 45\pi r^2 - 22500 = 0$. Using a graphing device, we first graph the polynomial in the viewing rectangle $[0, 15]$ by $[-10000, 10000]$. The solution, $r \approx 11.28$ ft., is shown in the viewing rectangle $[11.2, 11.4]$ by $[-1, 1]$.

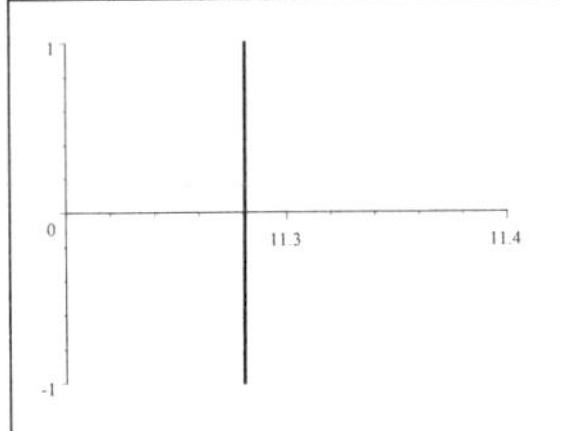


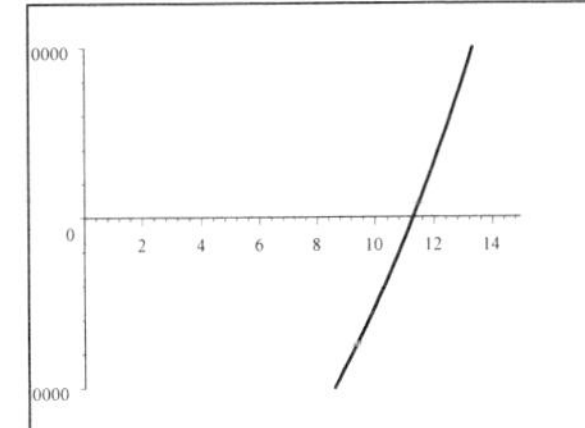


**94.** Given that $x$ is the length of a side of the rectangle, we have that the length of the diagonal is $x+10$, and the length of the other side of the rectangle is $\sqrt{(x+10)^2 - x^2}$. Hence $x\sqrt{(x+10)^2 - x^2} = 5000 \quad \Rightarrow \quad x^2(20x + 100) = 25{,}000{,}000$ $\Leftrightarrow \quad 2x^3 + 10x^2 - 2{,}500{,}000 = 0 \quad \Leftrightarrow \quad x^3 + 5x^2 - 1{,}250{,}000 = 0$. The first viewing rectangle, $[0, 120]$ by $[-100, 500]$, shows there is one solution. The second viewing rectangle, $[106, 106.1]$ by $[-0.1, 0.1]$, shows the solution is $x = 106.08$. Therefore, the dimensions of the rectangle are 47 ft by 106 ft.

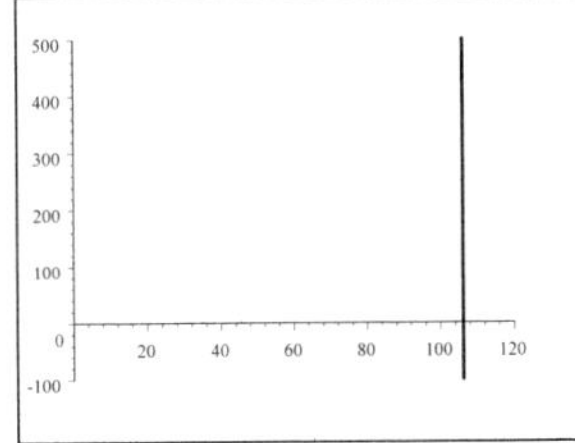


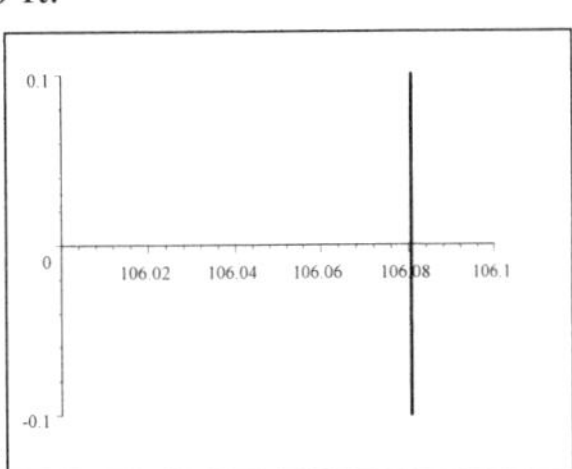

**95.** $h(t) = 11.60t - 12.41t^2 + 6.20t^3 - 1.58t^4 + 0.20t^5 - 0.01t^6$ is shown in the viewing rectangle $[0, 10]$ by $[0, 6]$.

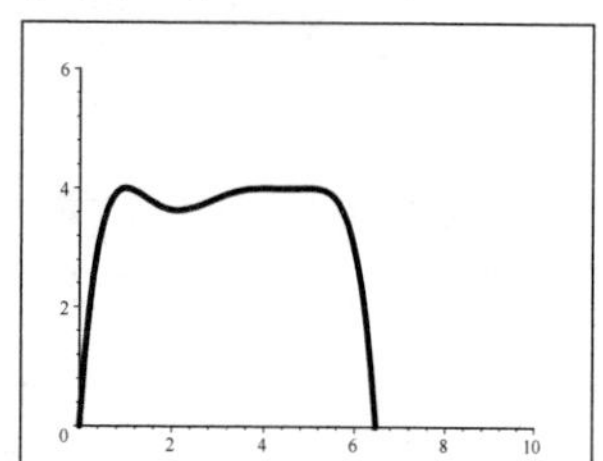


**(a)** It started to snow again.

**(b)** No, $h(t) \leq 4$.

**(c)** The function $h(t)$ is shown in the viewing rectangle $[6, 6.5]$ by $[0, 0.5]$. The $x$-intercept of the function is a little less than 6.5, which means that the snow melted just before midnight on Saturday night.

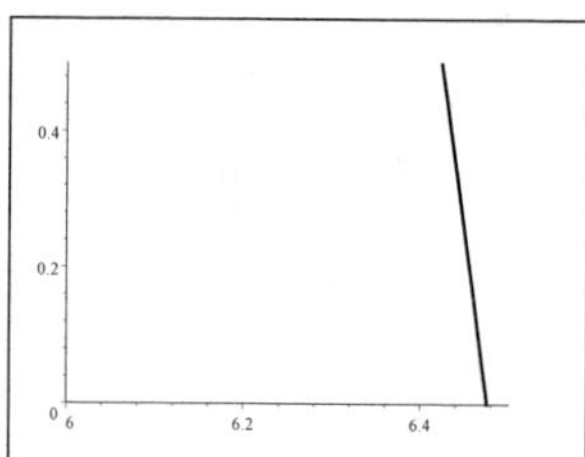


**96.** The volume of the box is $V = 1500 = x(20 - 2x)(40 - 2x) = 4x^3 - 120x^2 + 800x \Leftrightarrow 4x^3 - 120x^2 + 800x - 1500 = 4(x^3 - 30x^2 + 200x - 375) = 0$. Clearly, we must have $20 - 2x > 0$, and so $0 < x < 10$.

| 5 | 1 | $-30$ | 200 | $-375$ | |
|---|---|---|---|---|---|
| | | 5 | $-125$ | 375 | |
| | 1 | $-25$ | 75 | 0 | $\Rightarrow x = 5$ is a zero. |

$x^3 - 30x^2 + 200x - 375 = (x - 5)(x^2 - 25x + 75) = 0$. Using the quadratic formula, we find the other zeros: $x = \frac{25 \pm \sqrt{625 - 4(1)(75)}}{2} = \frac{25 \pm \sqrt{325}}{2} = \frac{25 \pm 5\sqrt{13}}{2}$. Since $\frac{25 + 5\sqrt{13}}{2} > 10$, the two answers are: $x =$ height $= 5$ cm, width $= 20 - 2(5) = 10$ cm, and length $= 40 - 2(5) = 30$ cm; and $x =$ height $= \frac{25 - 5\sqrt{13}}{2} \approx 3.49$ cm, width $= 20 - (25 - 5\sqrt{13}) = 5\sqrt{13} - 5 \approx 13.03$ cm, and length $= 40 - (25 - 5\sqrt{13}) = 15 + 5\sqrt{13} \approx 33.03$ cm.

**97.** Let $r$ be the radius of the cone and cylinder and let $h$ be the height of the cone. Since the height and diameter are equal, we get $h = 2r$. So the volume of the cylinder is $V_1 = \pi r^2 \cdot (\text{cylinder height}) = 20\pi r^2$, and the volume of the cone is $V_2 = \frac{1}{3}\pi r^2 h = \frac{1}{3}\pi r^2 (2r) = \frac{2}{3}\pi r^3$. Since the total volume is $\dfrac{500\pi}{3}$, it follows that $\frac{2}{3}\pi r^3 + 20\pi r^2 = \dfrac{500\pi}{3} \Leftrightarrow r^3 + 30r^2 - 250 = 0$. By Descartes' Rule of Signs, there is 1 positive zero. Since $r$ is between 2.76 and 2.765 (see the table), the radius should be 2.76 m (correct to two decimals).

| $r$ | $r^3 + 30r^2 - 250$ |
|---|---|
| 1 | $-219$ |
| 2 | $-122$ |
| 3 | 47 |
| 2.7 | $-11.62$ |
| 2.76 | $-2.33$ |
| 2.77 | 1.44 |
| 2.765 | 1.44 |
| 2.8 | 7.15 |

**98. (a)** Let $x$ be the length, in ft, of each side of the base and let $h$ be the height. The volume of the box is $V = 2\sqrt{2} = hx^2$, and so $hx^2 = 2\sqrt{2}$. The length of the diagonal on the base is $\sqrt{x^2 + x^2} = \sqrt{2x^2}$, and hence the length of the diagonal between opposite corners is $\sqrt{2x^2 + h^2} = x + 1$. Squaring both sides of the equation, we have $2x^2 + h^2 = x^2 + 2x + 1 \Leftrightarrow h^2 = -x^2 + 2x + 1 \Leftrightarrow h = \sqrt{-x^2 + 2x + 1}$. Therefore, $2\sqrt{2} = hx^2 = \left(\sqrt{-x^2 + 2x + 1}\right)x^2 \Leftrightarrow (-x^2 + 2x + 1)x^4 = 8 \Leftrightarrow x^6 - 2x^5 - x^4 + 8 = 0$.

**(b)** We graph $y = x^6 - 2x^5 - x^4 + 8$ in the viewing rectangle $[0, 5]$ by $[-10, 10]$, and we see that there are two solutions. In the second viewing rectangle, $[1.4, 1.5]$ by $[-1, 1]$, we see the solution $x \approx 1.45$. The third viewing rectangle, $[2.25, 2.35]$ by $[-1, 1]$, shows the solution $x \approx 2.31$.

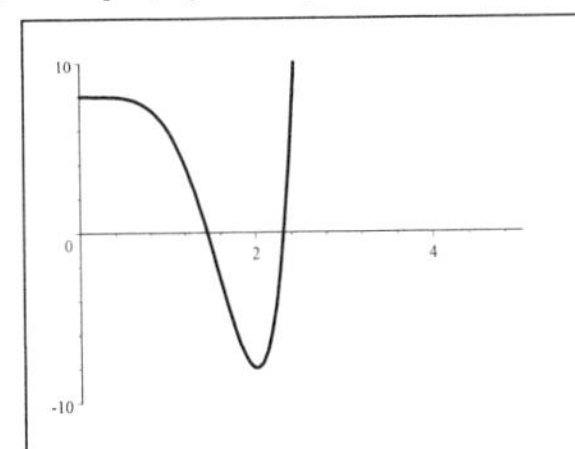

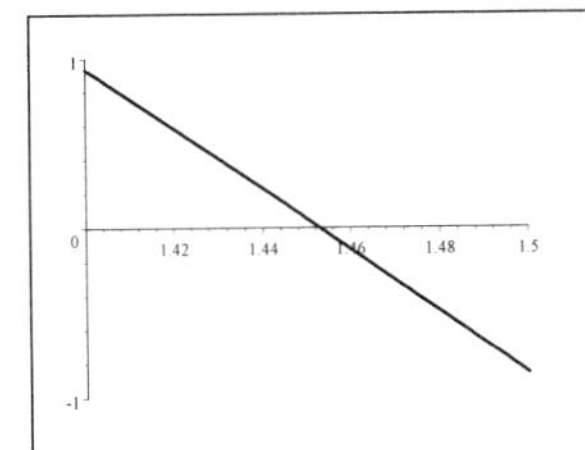

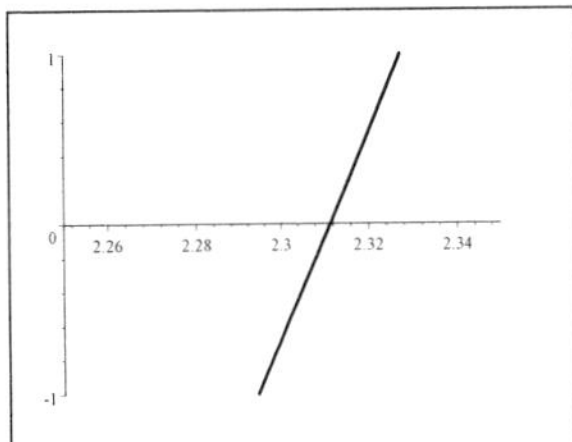


If $x = \text{width} = \text{length} = 1.45$ ft, then height $= \sqrt{-x^2 + 2x + 1} = 1.34$ ft, and if $x = \text{width} = \text{length} = 2.31$ ft, then height $= \sqrt{-x^2 + 2x + 1} = 0.53$ ft.

**99.** Let $b$ be the width of the base, and let $l$ be the length of the box. Then the length plus girth is $l + 4b = 108$, and the volume is $V = lb^2 = 2200$. Solving the first equation for $l$ and substituting this value into the second equation yields $l = 108 - 4b \Rightarrow V = (108 - 4b)\, b^2 = 2200 \quad \Leftrightarrow \quad 4b^3 - 108b^2 + 2200 = 0 \Leftrightarrow 4\left(b^3 - 27b^2 + 550\right) = 0$. Now $P(b) = b^3 - 27b^2 + 550$ has two variations in sign, so there are 0 or 2 positive real zeros. We also observe that since $l > 0, b < 27$, so $b = 27$ is an upper bound. Thus the possible positive rational real zeros are $1, 2, 3, 10, 11, 22, 25$.

| 1 | 1 | −27 | 0 | 550 |
|---|---|---|---|---|
| | | 1 | −26 | −26 |
| | 1 | −26 | −26 | 524 |

| 2 | 1 | −27 | 0 | 550 |
|---|---|---|---|---|
| | | 2 | −50 | −100 |
| | 1 | −25 | −50 | 450 |

| 5 | 1 | −27 | 0 | 550 |
|---|---|---|---|---|
| | | 5 | −110 | −550 |
| | 1 | −22 | −110 | 0 |

$\Rightarrow b = 5$ is a zero.

$P(b) = (b - 5)\left(b^2 - 22b - 110\right)$. The other zeros are $b = \frac{22 \pm \sqrt{484 - 4(1)(-110)}}{2} = \frac{22 \pm \sqrt{924}}{2} = \frac{22 \pm 30.397}{2}$. The positive answer from this factor is $b \approx 26.20$.Thus we have two possible solutions, $b = 5$ or $b \approx 26.20$. If $b = 5$, then $l = 108 - 4(5) = 88$; if $b \approx 26.20$, then $l = 108 - 4(26.20) = 3.20$. Thus the length of the box is either 88 in. or 3.20 in.

**100. (a)** An odd-degree polynomial must have a real zero. The end behavior of such a polynomial requires that the graph of the polynomial heads off in opposite directions as $x \to \infty$ and $x \to -\infty$. Thus the graph must cross the $x$-axis.

**(b)** There are many possibilities one of which is $P(x) = x^4 + 1$.

**(c)** $P(x) = x\left(x - \sqrt{2}\right)\left(x + \sqrt{2}\right) = x^3 - 2x$.

**(d)** $P(x) = \left(x - \sqrt{2}\right)\left(x + \sqrt{2}\right)\left(x - \sqrt{3}\right)\left(x + \sqrt{3}\right) = x^4 - 5x^2 + 6$. If a polynomial with integer coefficients has no real zeroes, then the polynomial must have even degree.

**101. (a)** Substituting $X - \frac{a}{3}$ for $x$ we have

$$\begin{aligned} x^3 + ax^2 + bx + c &= \left(X - \frac{a}{3}\right)^3 + a\left(X - \frac{a}{3}\right)^2 + b\left(X - \frac{a}{3}\right) + c \\ &= X^3 - aX^2 + \frac{a^2}{3}X + \frac{a^3}{27} + a\left(X^2 - \frac{2a}{3}X + \frac{a^2}{9}\right) + bX - \frac{ab}{3} + c \\ &= X^3 - aX^2 + \frac{a^2}{3}X + \frac{a^3}{27} + aX^2 - \frac{2a^2}{3}X + \frac{a^3}{9} + bX - \frac{ab}{3} + c \\ &= X^3 + (-a + a)X^2 + \left(-\frac{a^2}{3} - \frac{2a^2}{3} + b\right)X + \left(\frac{a^3}{27} + \frac{a^3}{9} - \frac{ab}{3} + c\right) \\ &= X^3 + \left(b - a^2\right)X + \left(\frac{4a^3}{27} - \frac{ab}{3} + c\right) \end{aligned}$$

**(b)** $x^3 + 6x^2 + 9x + 4 = 0$. Setting $a = 6$, $b = 9$, and $c = 4$, we have: $X^3 + \left(9 - 6^2\right)X + (32 - 18 + 4) = X^3 - 27X + 18$.

**102. (a)** Using the cubic formula, $x = \sqrt[3]{\frac{-2}{2} + \sqrt{\frac{2^2}{4} + \frac{(-3)^3}{27}}} + \sqrt[3]{\frac{-2}{2} - \sqrt{\frac{2^2}{4} + \frac{(-3)^3}{27}}} = \sqrt[3]{-1} + \sqrt[3]{-1} = -1 - 1 = -2$.

$$\begin{array}{r|rrrr} -2 & 1 & 0 & -3 & 2 \\ & & -2 & 4 & -2 \\ \hline & 1 & -2 & 1 & 0 \end{array}$$

So $(x + 2)\left(x^2 - 2x + 1\right) = (x + 2)(x - 1)^2 = 0 \quad \Rightarrow \quad x = -2, 1$. Using the methods from this section, we have

$$\begin{array}{r|rrrr} 1 & 1 & 0 & -3 & 2 \\ & & 1 & 1 & -2 \\ \hline & 1 & 1 & -2 & 0 \end{array}$$

So $x^3 - 3x + 2 = (x - 1)\left(x^2 + x - 2\right) = (x - 1)^2 (x + 2) = 0 \quad \Leftrightarrow \quad x = -2, 1$.
Since this factors easily, the factoring method was easier.

**(b)** Using the cubic formula,

$$\begin{aligned} x &= \sqrt[3]{\frac{-(-54)}{2} + \sqrt{\frac{(-54)^2}{4} + \frac{(-27)^3}{27}}} + \sqrt[3]{\frac{-(-54)}{2} - \sqrt{\frac{(-54)^2}{4} + \frac{(-27)^3}{27}}} \\ &= \sqrt[3]{\frac{54}{2} + \sqrt{27^2 - 27^2}} + \sqrt[3]{\frac{54}{2} - \sqrt{27^2 - 27^2}} = \sqrt[3]{27} + \sqrt[3]{27} = 3 + 3 = 6 \end{aligned}$$

$$\begin{array}{r|rrrr} 6 & 1 & 0 & -27 & -54 \\ & & 6 & 36 & 54 \\ \hline & 1 & 6 & 9 & 0 \end{array}$$

$x^3 - 27x - 54 = (x - 6)\left(x^2 + 6x + 9\right) = (x - 6)(x + 3)^2 = 0 \quad \Rightarrow \quad x = -3, 6$.
Using methods from this section,

$$\begin{array}{r|rrrr} -1 & 1 & 0 & -27 & -54 \\ & & -1 & 1 & 26 \\ \hline & 1 & -1 & -26 & -28 \end{array} \qquad \begin{array}{r|rrrr} -2 & 1 & 0 & -27 & -54 \\ & & -2 & 4 & 46 \\ \hline & 1 & -2 & -23 & -8 \end{array} \qquad \begin{array}{r|rrrr} -3 & 1 & 0 & -27 & -54 \\ & & -3 & 9 & 54 \\ \hline & 1 & -3 & -18 & 0 \end{array}$$

So $x^3 - 27x - 54 = (x + 3)\left(x^2 - 3x - 18\right) = (x - 6)(x + 3)^2 = 0 \quad \Leftrightarrow \quad x = -3, 6$.
Since this factors easily, the factoring method was easier.

**(c)** Using the cubic formula,

$$x = \sqrt[3]{\frac{-4}{2}+\sqrt{\frac{4^2}{4}+\frac{3^3}{27}}}+\sqrt[3]{\frac{-4}{2}-\sqrt{\frac{4^2}{4}+\frac{3^3}{27}}}$$
$$= \sqrt[3]{-2+\sqrt{4+1}}+\sqrt[3]{-2-\sqrt{4+1}} = \sqrt[3]{-2+\sqrt{5}}+\sqrt[3]{-2-\sqrt{5}}$$

From the graphing calculator, we see that $P(x) = x^3 + 3x + 4$ has one zero.
Using methods from this section, $P(x)$ has possible rational zeros $\pm 1, \pm 2, \pm 4$.

$$\begin{array}{c|cccc} 1 & 1 & 0 & 3 & 4 \\ & & 1 & 1 & 4 \\ \hline & 1 & 1 & 4 & 4 \end{array} \Rightarrow 1 \text{ is an upper bound.}$$

$$\begin{array}{c|cccc} -1 & 1 & 0 & 3 & 4 \\ & & -1 & 1 & -4 \\ \hline & 1 & -1 & 4 & 0 \end{array} \Rightarrow x = -1 \text{ is a zero.}$$

$P(x) = x^3 + 3x + 4 = (x+1)(x^2 - x + 4)$. Using the quadratic formula we have: $x = \frac{-(-1)\pm\sqrt{(-1)^2-4(1)(4)}}{2} = \frac{1\pm\sqrt{-15}}{2}$ which is not a real number. Since it is not easy to see that $\sqrt[3]{-2+\sqrt{5}}+\sqrt[3]{-2-\sqrt{5}} = -1$, we see that the factoring method was much easier.

# 4.4 Complex Zeros and the Fundamental Theorem of Algebra

**1. (a)** $x^4 + 4x^2 = 0 \quad\Leftrightarrow\quad x^2(x^2+4) = 0$. So $x = 0$ or $x^2 + 4 = 0$. If $x^2 + 4 = 0$ then $x^2 = -4 \Leftrightarrow x = \pm 2i$. Therefore, the solutions are $x = 0$ and $\pm 2i$.

**(b)** To get the complete factorization, we factor the remaining quadratic factor $P(x) = x^2(x+4) = x^2(x-2i)(x+2i)$.

**2. (a)** $x^5 + 9x^3 = 0 \quad\Leftrightarrow\quad x^3(x^2+9) = 0$. So $x = 0$ or $x^2 + 9 = 0$. If $x^2 + 9 = 0$ then $x = \pm 3i$. Therefore, the zeros of $P$ are $x = 0, \pm 3i$.

**(b)** Since $-3i$ and $3i$ are the zeros from $x^2 + 9 = 0$, $x + 3i$ and $x - 3i$ are the factors of $x^2 + 9$. Thus the complete factorization is $P(x) = x^3(x^2+9) = x^3(x+3i)(x-3i)$.

**3. (a)** $x^3 - 2x^2 + 2x = 0 \quad\Leftrightarrow\quad x(x^2 - 2x + 2) = 0$. So $x = 0$ or $x^2 - 2x + 2 = 0$. If $x^2 - 2x + 2 = 0$ then $x = \frac{-(-2)\pm\sqrt{(-2)^2-4(1)(2)}}{2} = \frac{2\pm\sqrt{-4}}{2} = \frac{2\pm 2i}{2} = 1 \pm i$. Therefore, the solutions are $x = 0, 1 \pm i$.

**(b)** Since $1 - i$ and $1 + i$ are zeros, $x - (1-i) = x - 1 + i$ and $x - (1+i) = x - 1 - i$ are the factors of $x^2 - 2x + 2$. Thus the complete factorization is $P(x) = x(x^2 - 2x + 2) = x(x-1+i)(x-1-i)$.

**4. (a)** $x^3 + x^2 + x = 0 \; x(x^2+x+1) = 0$ . So $x = 0$ or $x^2 + x + 1 = 0$. If $x^2 + x + 1 = 0$ then $x = \frac{-(1)\pm\sqrt{(1)^2-4(1)(1)}}{2(1)} = \frac{-1\pm\sqrt{-3}}{2} = -\frac{1}{2} \pm i\frac{\sqrt{3}}{2}$. Therefore, the zeros of $P$ are $x = 0, -\frac{1}{2} \pm i\frac{\sqrt{3}}{2}$.

**(b)** The zeros of $x^2 + x + 1 = 0$ are $-\frac{1}{2} - i\frac{\sqrt{3}}{2}$ and $-\frac{1}{2} + i\frac{\sqrt{3}}{2}$, so factoring we get $x^2 + x + 1 = \left[x - \left(-\frac{1}{2} - i\frac{\sqrt{3}}{2}\right)\right]\left[x - \left(-\frac{1}{2} + i\frac{\sqrt{3}}{2}\right)\right] = \left(x + \frac{1}{2} + i\frac{\sqrt{3}}{2}\right)\left(x + \frac{1}{2} - i\frac{\sqrt{3}}{2}\right)$. Thus the complete factorization is $P(x) = x(x^2 + x + 1) = x\left(x + \frac{1}{2} + i\frac{\sqrt{3}}{2}\right)\left(x + \frac{1}{2} - i\frac{\sqrt{3}}{2}\right)$.

**5. (a)** $x^4+2x^2+1=0 \quad \Leftrightarrow \quad (x^2+1)^2=0 \quad \Leftrightarrow \quad x^2+1=0 \quad \Leftrightarrow \quad x^2=-1 \Leftrightarrow x=\pm i$. Therefore the zeros of $P$ are $x=\pm i$.

**(b)** Since $-i$ and $i$ are zeros, $x+i$ and $x-i$ are the factors of $x^2+1$. Thus the complete factorization is $P(x)=(x^2+1)^2=[(x+i)(x-i)]^2=(x+i)^2(x-i)^2$.

**6. (a)** $x^4-x^2-2=0 \quad \Leftrightarrow \quad (x^2-2)(x^2+1)=0$. So $x^2-2=0$ or $x^2+1=0$. If $x^2-2=0$ then $x^2=2 \quad \Leftrightarrow \quad x=\pm\sqrt{2}$. And if $x^2+1=0$ then $x^2=-1 \quad \Leftrightarrow \quad x=\pm i$. Therefore, the zeros of $P$ are $x=\pm\sqrt{2}, \pm i$.

**(b)** To get the complete factorization, we factor the quadratic factors to get $P(x)=(x^2-2)(x^2+1)=(x-\sqrt{2})(x+\sqrt{2})(x-i)(x+i)$.

**7. (a)** $x^4-16=0 \quad \Leftrightarrow \quad 0=(x^2-4)(x^2+4)=(x-2)(x+2)(x^2+4)$. So $x=\pm 2$ or $x^2+4=0$. If $x^2+4=0$ then $x^2=-4 \quad \Rightarrow \quad x=\pm 2i$. Therefore the zeros of $P$ are $x=\pm 2, \pm 2i$.

**(b)** Since $-i$ and $i$ are zeros, $x+i$ and $x-i$ are the factors of $x^2+1$. Thus the complete factorization is $P(x)=(x-2)(x+2)(x^2+4)=(x-2)(x+2)(x-2i)(x+2i)$.

**8. (a)** $x^4+6x^2+9=0 \quad \Leftrightarrow \quad (x^2+3)^2=0 \quad \Leftrightarrow \quad x^2=-3$. So $x=\pm i\sqrt{3}$ are the only zeros of $P$ (each of multiplicity 2).

**(b)** To get the complete factorization, we factor the quadratic factor to get $P(x)=(x^2+3)^2=\left[(x-i\sqrt{3})(x+i\sqrt{3})\right]^2=(x-i\sqrt{3})^2(x+i\sqrt{3})^2$.

**9. (a)** $x^3+8=0 \quad \Leftrightarrow \quad (x+2)(x^2-2x+4)=0$. So $x=-2$ or $x^2-2x+4=0$. If $x^2-2x+4=0$ then $x=\frac{-(-2)\pm\sqrt{(-2)^2-4(1)(4)}}{2}=\frac{2\pm\sqrt{-12}}{2}=\frac{2\pm 2i\sqrt{3}}{2}=1\pm i\sqrt{3}$. Therefore, the zeros of $P$ are $x=-2, 1\pm i\sqrt{3}$.

**(b)** Since $1-i\sqrt{3}$ and $1+i\sqrt{3}$ are the zeros from the $x^2-2x+4=0$, $x-(1-i\sqrt{3})$ and $x-(1+i\sqrt{3})$ are the factors of $x^2-2x+4$. Thus the complete factorization is

$$\begin{aligned}P(x)&=(x+2)(x^2-2x+4)=(x+2)\left[x-(1-i\sqrt{3})\right]\left[x-(1+i\sqrt{3})\right]\\&=(x+2)(x-1+i\sqrt{3})(x-1-i\sqrt{3})\end{aligned}$$

**10. (a)** $x^3-8=0 \quad \Leftrightarrow \quad (x-2)(x^2+2x+4)=0$. So $x=2$ or $x^2+2x+4=0$. If $x^2+2x+4=0$ then $x=\frac{-(2)\pm\sqrt{(2)^2-4(1)(4)}}{2}=\frac{-2\pm\sqrt{-12}}{2}=\frac{-2\pm 2i\sqrt{3}}{2}=-1\pm i\sqrt{3}$. Therefore, the zeros of $P$ are $x=2, -1\pm i\sqrt{3}$.

**(b)** Since $-1-i\sqrt{3}$ and $-1+i\sqrt{3}$ are the zeros from $x^2+2x+4=0$, $x-(-1-i\sqrt{3})$ and $x-(-1+i\sqrt{3})$ are the factors of $x^2-2x+4$. Thus the complete factorization is

$$\begin{aligned}P(x)&=(x-2)(x^2+2x+4)=(x-2)\left[x-(-1-i\sqrt{3})\right]\left[x-(-1+i\sqrt{3})\right]\\&=(x-2)(x+1+i\sqrt{3})(x+1-i\sqrt{3})\end{aligned}$$

**11. (a)** $x^6 - 1 = 0 \quad \Leftrightarrow \quad 0 = (x^3 - 1)(x^3 + 1) = (x - 1)(x^2 + x + 1)(x + 1)(x^2 - x + 1)$. Clearly, $x = \pm 1$ are solutions. If $x^2 + x + 1 = 0$, then $x = \frac{-1 \pm \sqrt{1 - 4(1)(1)}}{2} = \frac{-1 \pm \sqrt{-3}}{2} = -\frac{1}{2} \pm \frac{\sqrt{-3}}{2}$ so $x = -\frac{1}{2} \pm i\frac{\sqrt{3}}{2}$. And if $x^2 - x + 1 = 0$, then $x = \frac{1 \pm \sqrt{1 - 4(1)(1)}}{2} = \frac{1 \pm \sqrt{-3}}{2} = \frac{1}{2} \pm \frac{\sqrt{-3}}{2} = \frac{1}{2} \pm i\frac{\sqrt{3}}{2}$. Therefore, the zeros of $P$ are $x = \pm 1$, $-\frac{1}{2} \pm i\frac{\sqrt{3}}{2}$, $\frac{1}{2} \pm i\frac{\sqrt{3}}{2}$.

**(b)** The zeros of $x^2 + x + 1 = 0$ are $-\frac{1}{2} - i\frac{\sqrt{3}}{2}$ and $-\frac{1}{2} + i\frac{\sqrt{3}}{2}$, so $x^2 + x + 1$ factors as $\left[x - \left(-\frac{1}{2} - i\frac{\sqrt{3}}{2}\right)\right]\left[x - \left(-\frac{1}{2} + i\frac{\sqrt{3}}{2}\right)\right] = \left(x + \frac{1}{2} + i\frac{\sqrt{3}}{2}\right)\left(x + \frac{1}{2} - i\frac{\sqrt{3}}{2}\right)$. Similarly,since the zeros of $x^2 - x + 1 = 0$ are $\frac{1}{2} - i\frac{\sqrt{3}}{2}$ and $\frac{1}{2} + i\frac{\sqrt{3}}{2}$, so $x^2 - x + 1$ factors as $\left[x - \left(\frac{1}{2} - i\frac{\sqrt{3}}{2}\right)\right]\left[x - \left(\frac{1}{2} + i\frac{\sqrt{3}}{2}\right)\right] = \left(x - \frac{1}{2} + i\frac{\sqrt{3}}{2}\right)\left(x - \frac{1}{2} - i\frac{\sqrt{3}}{2}\right)$. Thus the complete factorization is

$$\begin{aligned} P(x) &= (x - 1)(x^2 + x + 1)(x + 1)(x^2 - x + 1) \\ &= (x - 1)(x + 1)\left(x + \tfrac{1}{2} + i\tfrac{\sqrt{3}}{2}\right)\left(x + \tfrac{1}{2} - i\tfrac{\sqrt{3}}{2}\right)\left(x - \tfrac{1}{2} + i\tfrac{\sqrt{3}}{2}\right)\left(x - \tfrac{1}{2} - i\tfrac{\sqrt{3}}{2}\right) \end{aligned}$$

**12. (a)** $x^6 - 7x^3 - 8 = 0 \quad \Leftrightarrow \quad 0 = (x^3 - 8)(x^3 + 1) = (x - 2)(x^2 + 2x + 4)(x + 1)(x^2 - x + 1)$. Clearly, $x = -1$ and $x = 2$ are solutions. If $x^2 + 2x + 4 = 0$, then $x = \frac{-2 \pm \sqrt{4 - 4(1)(4)}}{2} = \frac{-2 \pm \sqrt{-12}}{2} = -\frac{2}{2} \pm \frac{2\sqrt{-3}}{2}$ so $x = -1 \pm \sqrt{3}i$. If $x^2 - x + 1 = 0$, then $x = \frac{1 \pm \sqrt{1 - 4(1)(1)}}{2} = \frac{1 \pm \sqrt{-3}}{2} = \frac{1}{2} \pm \frac{\sqrt{-3}}{2} = \frac{1}{2} \pm i\frac{\sqrt{3}}{2}$. Therefore, the zeros of $P$ are $x = -1, 2, -1 \pm \sqrt{3}i, \frac{1}{2} \pm i\frac{\sqrt{3}}{2}$.

**(b)** From Exercise 10, $x^2 + 2x + 4 = (x + 1 + i\sqrt{3})(x + 1 - i\sqrt{3})$ and from Exercise 11, $x^2 - x + 1 = \left(x - \frac{1}{2} + i\frac{\sqrt{3}}{2}\right)\left(x - \frac{1}{2} - i\frac{\sqrt{3}}{2}\right)$. Thus the complete factorization is

$$\begin{aligned} P(x) &= (x - 2)(x^2 + 2x + 4)(x + 1)(x^2 - x + 1) \\ &= (x - 2)(x + 1)\left(x + 1 + i\sqrt{3}\right)\left(x + 1 - i\sqrt{3}\right)\left(x - \tfrac{1}{2} + i\tfrac{\sqrt{3}}{2}\right)\left(x - \tfrac{1}{2} - i\tfrac{\sqrt{3}}{2}\right) \end{aligned}$$

**13.** $P(x) = x^2 + 25 = (x - 5i)(x + 5i)$. The zeros of $P$ are $5i$ and $-5i$, both multiplicity 1.

**14.** $P(x) = 4x^2 + 9 = (2x - 3i)(2x + 3i)$. The zeros of $P$ are $\frac{3}{2}i$ and $-\frac{3}{2}i$, both multiplicity 1.

**15.** $Q(x) = x^2 + 2x + 2$. Using the quadratic formula $x = \frac{-(2) \pm \sqrt{(2)^2 - 4(1)(2)}}{2(1)} = \frac{-2 \pm \sqrt{-4}}{2} = \frac{-2 \pm 2i}{2} = -1 \pm i$. So $Q(x) = (x + 1 - i)(x + 1 + i)$. The zeros of $Q$ are $-1 - i$ (multiplicity 1) and $-1 + i$ (multiplicity 1).

**16.** $Q(x) = x^2 - 8x + 17 = (x^2 - 8x + 16) + 1 = (x - 4)^2 + 1 = [(x - 4) - i][(x - 4) + i] = (x - 4 - i)(x - 4 + i)$. The zeros of $Q$ are $4 + i$ and $4 - i$, both multiplicity 1.

**17.** $P(x) = x^3 + 4x = x(x^2 + 4) = x(x - 2i)(x + 2i)$. The zeros of $P$ are 0, $2i$, and $-2i$ (all multiplicity 1).

**18.** $P(x) = x^3 + x^2 + x = x(x^2 + x + 1)$. Using the quadratic formula, we have $x = \frac{-(1) \pm \sqrt{(1)^2 - 4(1)(1)}}{2(1)} = \frac{1 \pm \sqrt{-3}}{2} = \frac{1}{2} \pm i\frac{\sqrt{3}}{2}$. The zeros of $P$ are 0, $\frac{1}{2} + i\frac{\sqrt{3}}{2}$, and $\frac{1}{2} - i\frac{\sqrt{3}}{2}$, all of multiplicity 1. And $P(x) = x\left(x - \frac{1}{2} - i\frac{\sqrt{3}}{2}\right)\left(x - \frac{1}{2} + i\frac{\sqrt{3}}{2}\right)$.

**19.** $Q(x) = x^4 - 1 = (x^2 - 1)(x^2 + 1) = (x - 1)(x + 1)(x^2 + 1) = (x - 1)(x + 1)(x - i)(x + i)$. The zeros of $Q$ are $1, -1, i$, and $-i$ (all of multiplicity 1).

**20.** $Q(x) = x^4 - 625 = (x^2 - 25)(x^2 + 25) = (x - 5)(x + 5)(x^2 + 25) = (x - 5)(x + 5)(x - 5i)(x + 5i)$. The zeros of $Q$ are $5, -5, 5i$, and $5i$, all multiplicity 1.

**21.** $P(x) = 16x^4 - 81 = (4x^2 - 9)(4x^2 + 9) = (2x - 3)(2x + 3)(2x - 3i)(2x + 3i)$. The zeros of $P$ are $\frac{3}{2}$, $-\frac{3}{2}$, $\frac{3}{2}i$, and $-\frac{3}{2}i$ (all of multiplicity 1).

**22.** $P(x) = x^3 - 64 = (x - 4)(x^2 + 4x + 16)$. Using the quadratic formula, we have $x = \frac{-4 \pm \sqrt{16 - 4(1)(16)}}{2} = \frac{-4 \pm \sqrt{-48}}{2} = \frac{-4 \pm 4\sqrt{3}i}{2} = -2 \pm 2\sqrt{3}i$. The zeros of $P$ are 4, $-2 + 2\sqrt{3}i$, and $-2 - 2\sqrt{3}i$, all of multiplicity 1. And $P(x) = (x - 4)(x + 2 - 2\sqrt{3}i)(x + 2 + 2\sqrt{3}i)$.

**23.** $P(x) = x^3 + x^2 + 9x + 9 = x^2(x + 1) + 9(x + 1) = (x + 1)(x^2 + 9) = (x + 1)(x - 3i)(x + 3i)$. The zeros of $P$ are $-1$, $3i$, and $-3i$ (all of multiplicity 1).

**24.** $P(x) = x^6 - 729 = (x^3 - 27)(x^3 + 27) = (x - 3)(x^2 + 3x + 9)(x + 3)(x^2 - 3x + 9)$. Using the quadratic formula on $x^2 + 3x + 9$ we have $x = \frac{-3 \pm \sqrt{9 - 4(1)(9)}}{2} = \frac{-3 \pm \sqrt{-27}}{2} = -\frac{3}{2} \pm \frac{\sqrt{-27}}{2}$. Using the quadratic formula on $x^2 - 3x + 9$ we have $x = \frac{3 \pm \sqrt{9 - 4(1)(9)}}{2} = \frac{3 \pm \sqrt{-27}}{2} = \frac{3}{2} \pm \frac{\sqrt{-27}}{2} = \frac{3}{2} \pm \frac{3\sqrt{3}}{2}i$. The zeros of $P$ are 3, $-3$, $-\frac{3}{2} + \frac{3\sqrt{3}}{2}i$, $-\frac{3}{2} - \frac{3\sqrt{3}}{2}i$, $\frac{3}{2} + \frac{3\sqrt{3}}{2}i$, and $\frac{3}{2} - \frac{3\sqrt{3}}{2}i$, all multiplicity 1. And $P(x) = (x - 3)(x + 3)\left(x + \frac{3}{2} - \frac{3\sqrt{3}}{2}i\right)\left(x + \frac{3}{2} + \frac{3\sqrt{3}}{2}i\right)\left(x - \frac{3}{2} - \frac{3\sqrt{3}}{2}i\right)\left(x - \frac{3}{2} + \frac{3\sqrt{3}}{2}i\right)$.

**25.** $Q(x) = x^4 + 2x^2 + 1 = (x^2 + 1)^2 = (x - i)^2(x + i)^2$. The zeros of $Q$ are $i$ and $-i$ (both of multiplicity 2).

**26.** $Q(x) = x^4 + 10x^2 + 25 = (x^2 + 5)^2 = [(x - i\sqrt{5})(x + i\sqrt{5})]^2 = (x - i\sqrt{5})^2(x + i\sqrt{5})^2$. The zeros of $Q$ are $i\sqrt{5}$ and $-i\sqrt{5}$, both of multiplicity 2.

**27.** $P(x) = x^4 + 3x^2 - 4 = (x^2 - 1)(x^2 + 4) = (x - 1)(x + 1)(x - 2i)(x + 2i)$. The zeros of $P$ are 1, $-1$, $2i$, and $-2i$ (all of multiplicity 1).

**28.** $P(x) = x^5 + 7x^3 = x^3(x^2 + 7) = x^3(x - i\sqrt{7})(x + i\sqrt{7})$. The zeros of $P$ are 0 (multiplicity 3), $i\sqrt{7}$ and $-i\sqrt{7}$, both of multiplicity 1.

**29.** $P(x) = x^5 + 6x^3 + 9x = x(x^4 + 6x^2 + 9) = x(x^2 + 3)^2 = x(x - \sqrt{3}i)^2(x + \sqrt{3}i)^2$. The zeros of $P$ are 0 (multiplicity 1), $\sqrt{3}i$ (multiplicity 2), and $-\sqrt{3}i$ (multiplicity 2).

**30.** $P(x) = x^6 + 16x^3 + 64 = (x^3 + 8)^2 = (x + 2)^2(x^2 - 2x + 4)^2$. Using the quadratic formula, on $x^2 - 2x + 4$ we have $x = \frac{-(-2) \pm \sqrt{(-2)^2 - 4(1)(4)}}{2} = \frac{2 \pm \sqrt{-12}}{2} = \frac{2 \pm 2i\sqrt{3}}{2} = 1 \pm i\sqrt{3}$. The zeros of $P$ are $-2$, $1 + i\sqrt{3}$, and $1 - i\sqrt{3}$, all multiplicity 2.

**31.** Since $1 + i$ and $1 - i$ are conjugates, the factorization of the polynomial must be $P(x) = a(x - [1 + i])(x - [1 - i]) = a(x^2 - 2x + 2)$. If we let $a = 1$, we get $P(x) = x^2 - 2x + 2$.

**32.** Since $1 + i\sqrt{2}$ and $1 - i\sqrt{2}$ are conjugates, the factorization of the polynomial must be $P(x) = c\left(x - [1 + i\sqrt{2}]\right)\left(x - [1 - i\sqrt{2}]\right) = c(x^2 - 2x + 3)$. If we let $c = 1$, we get $P(x) = x^2 - 2x + 3$.

**33.** Since $2i$ and $-2i$ are conjugates, the factorization of the polynomial must be $Q(x) = b(x - 3)(x - 2i)(x + 2i] = b(x - 3)(x^2 + 4) = b(x^3 - 3x^2 + 4x - 12)$. If we let $b = 1$, we get $Q(x) = x^3 - 3x^2 + 4x - 12$.

**34.** Since $i$ is a zero, by the Conjugate Roots Theorem, $-i$ is also a zero. So the factorization of the polynomial must be $Q(x) = b(x + 0)(x - i)(x + i) = bx(x^2 + 1) = b(x^3 + x)$. If we let $b = 1$, we get $Q(x) = x^3 + x$.

**35.** Since $i$ is a zero, by the Conjugate Roots Theorem, $-i$ is also a zero. So the factorization of the polynomial must be $P(x) = a(x - 2)(x - i)(x + i) = a(x^3 - 2x^2 + x - 2)$. If we let $a = 1$, we get $P(x) = x^3 - 2x^2 + x - 2$.

**36.** Since $1+i$ is a zero, by the Conjugate Roots Theorem, $1-i$ is also a zero. So the factorization of the polynomial must be $Q(x)=a(x+3)(x-[1+i])(x-[1-i])=a(x+3)(x^2-2x+2)=a(x^3+x^2-4x+6)$. If we let $a=1$, we get $Q(x)=x^3+x^2-4x+6$.

**37.** Since the zeros are $1-2i$ and $1$ (with multiplicity 2), by the Conjugate Roots Theorem, the other zero is $1+2i$. So a factorization is

$$\begin{aligned} R(x) &= c(x-[1-2i])(x-[1+2i])(x-1)^2 = c([x-1]+2i)([x-1]-2i)(x-1)^2 \\ &= c([x-1]^2-[2i]^2)(x^2-2x+1) = c(x^2-2x+1+4)(x^2-2x+1) = c(x^2-2x+5)(x^2-2x+1) \\ &= c(x^4-2x^3+x^2-2x^3+4x^2-2x+5x^2-10x+5) = c(x^4-4x^3+10x^2-12x+5) \end{aligned}$$

If we let $c=1$ we get $R(x)=x^4-4x^3+10x^2-12x+5$.

**38.** Since $S(x)$ has zeros $2i$ and $3i$, by the Conjugate Roots Theorem, the other zeros of $S(x)$ are $-2i$ and $-3i$. So a factorization of $S(x)$ is

$$S(x)=C(x-2i)(x+2i)(x-3i)(x+3i)=C(x^2-4i^2)(x^2-9i^2)=C(x^2+4)(x^2+9)=C(x^4+13x^2+36)$$

If we let $C=1$, we get $S(x)=x^4+13x^2+36$.

**39.** Since the zeros are $i$ and $1+i$, by the Conjugate Roots Theorem, the other zeros are $-i$ and $1-i$. So a factorization is

$$\begin{aligned} T(x) &= C(x-i)(x+i)(x-[1+i])(x-[1-i]) \\ &= C(x^2-i^2)([x-1]-i)([x-1]+i) = C(x^2+1)(x^2-2x+1-i^2) = C(x^2+1)(x^2-2x+2) \\ &= C(x^4-2x^3+2x^2+x^2-2x+2) = C(x^4-2x^3+3x^2-2x+2) = Cx^4-2Cx^3+3Cx^2-2Cx+2C \end{aligned}$$

Since the constant coefficient is 12, it follows that $2C=12 \Leftrightarrow C=6$, and so $T(x)=6(x^4-2x^3+3x^2-2x+2)=6x^4-12x^3+18x^2-12x+12$.

**40.** Since $U(x)$ has zeros $\frac{1}{2}$, $-1$ (with multiplicity two), and $-i$, by the Conjugate Roots Theorem, the other zero is $i$. So a factorization of $U(x)$ is

$$U(x)=c\left(x-\tfrac{1}{2}\right)(x+1)^2(x+i)(x-i)=\tfrac{1}{2}c(2x-1)(x^2+2x+1)(x^2+1)=\tfrac{1}{2}c(2x^5+3x^4+2x^3+2x^2-1)$$

Since the leading coefficient is 4, we have $4=\frac{1}{2}c(2)=c$. Thus we have $U(x)=\frac{1}{2}(4)(2x^5+3x^4+2x^3+2x^2-1)=4x^5+6x^4+4x^3+4x^2-2$.

**41.** $P(x)=x^3+2x^2+4x+8=x^2(x+2)+4(x+2)=(x+2)(x^2+4)=(x+2)(x-2i)(x+2i)$. Thus the zeros are $-2$ and $\pm 2i$.

**42.** $P(x)=x^3-7x^2+17x-15$. We start by trying the possible rational factors of the polynomial:

$$\begin{array}{r|rrrr} 1 & 1 & -7 & 17 & -15 \\ & & 1 & -6 & 11 \\ \hline & 1 & -6 & 11 & -4 \end{array} \qquad \begin{array}{r|rrrr} 3 & 1 & -7 & 17 & -15 \\ & & 3 & -12 & 15 \\ \hline & 1 & -4 & 5 & 0 \end{array} \Rightarrow x=3 \text{ is a zero.}$$

So $P(x)=(x-3)(x^2-4x+5)$. Using the quadratic formula on the second factor, we have $x=\frac{4\pm\sqrt{16-4(1)(5)}}{2}=\frac{4\pm\sqrt{-4}}{2}=\frac{4\pm 2i}{2}=2\pm i$. Thus the zeros are $3$, $2\pm i$.

**43.** $P(x) = x^3 - 2x^2 + 2x - 1$. By inspection, $P(1) = 1 - 2 + 2 - 1 = 0$, and hence $x = 1$ is a zero.

$$\begin{array}{r|rrrr} 1 & 1 & -2 & 2 & -1 \\ & & 1 & -1 & 1 \\ \hline & 1 & -1 & 1 & 0 \end{array}$$

Thus $P(x) = (x-1)(x^2 - x + 1)$. So $x = 1$ or $x^2 - x + 1 = 0$.

Using the quadratic formula, we have $x = \frac{1 \pm \sqrt{1-4(1)(1)}}{2} = \frac{1 \pm i\sqrt{3}}{2}$. Hence, the zeros are 1 and $\frac{1 \pm i\sqrt{3}}{2}$.

**44.** $P(x) = x^3 + 7x^2 + 18x + 18$ has possible rational zeros $\pm1, \pm2, \pm3, \pm6, \pm9, \pm18$. Since all of the coefficients are positive, there are no positive real zeros.

$$\begin{array}{r|rrrr} -1 & 1 & 7 & 18 & 18 \\ & & -1 & -6 & -12 \\ \hline & 1 & 6 & 12 & 6 \end{array} \qquad \begin{array}{r|rrrr} -2 & 1 & 7 & 18 & 18 \\ & & -2 & -10 & -16 \\ \hline & 1 & 5 & 8 & 2 \end{array} \qquad \begin{array}{r|rrrr} -3 & 1 & 7 & 18 & 18 \\ & & -3 & -12 & -18 \\ \hline & 1 & 4 & 6 & 0 \end{array} \Rightarrow x = -3 \text{ is a zero.}$$

So $P(x) = (x-3)(x^2 + 4x + 6)$. Using the quadratic formula on the second factor, we have $x = \frac{-4 \pm \sqrt{16-4(1)(6)}}{2} = \frac{-4 \pm \sqrt{-8}}{2} = \frac{-4 \pm 2\sqrt{2}i}{2} = -2 \pm \sqrt{2}i$. Thus the zeros are $-3, -2 \pm \sqrt{2}i$.

**45.** $P(x) = x^3 - 3x^2 + 3x - 2$.

$$\begin{array}{r|rrrr} 2 & 1 & -3 & 3 & -2 \\ & & 2 & -2 & 2 \\ \hline & 1 & -1 & 1 & 0 \end{array}$$

Thus $P(x) = (x-2)(x^2 - x + 1)$. So $x = 2$ or $x^2 - x + 1 = 0$

Using the quadratic formula we have $x = \frac{1 \pm \sqrt{1-4(1)(1)}}{2} = \frac{1 \pm i\sqrt{3}}{2}$. Hence, the zeros are 2, and $\frac{1 \pm i\sqrt{3}}{2}$.

**46.** $P(x) = x^3 - x - 6$ has possible zeros $\pm1, \pm2, \pm3$.

$$\begin{array}{r|rrrr} 1 & 1 & 0 & -1 & -6 \\ & & 1 & 1 & 0 \\ \hline & 1 & 1 & 0 & -6 \end{array} \qquad \begin{array}{r|rrrr} 2 & 1 & 0 & -1 & -6 \\ & & 2 & 4 & 6 \\ \hline & 1 & 2 & 3 & 0 \end{array} \Rightarrow x = 2 \text{ is a zero.}$$

$P(x) = (x-2)(x^2 + 2x + 3)$. Now $x^2 + 2x + 3$ has zeros $x = \frac{-2 \pm \sqrt{4-4(1)(3)}}{2} = \frac{-2 \pm 2i\sqrt{2}}{2} = -1 \pm i\sqrt{2}$. Thus the zeros are $2, -1 \pm i\sqrt{2}$.

**47.** $P(x) = 2x^3 + 7x^2 + 12x + 9$ has possible rational zeros $\pm1, \pm3, \pm9, \pm\frac{1}{2}, \pm\frac{3}{2}, \pm\frac{9}{2}$. Since all coefficients are positive, there are no positive real zeros.

$$\begin{array}{r|rrrr} -1 & 2 & 7 & 12 & 9 \\ & & -2 & -5 & -7 \\ \hline & 2 & 5 & 7 & 2 \end{array} \qquad \begin{array}{r|rrrr} -2 & 2 & 7 & 12 & 9 \\ & & -4 & -6 & -12 \\ \hline & 2 & 3 & 6 & -3 \end{array}$$

There is a zero between $-1$ and $-2$.

$$\begin{array}{r|rrrr} -\frac{3}{2} & 2 & 7 & 12 & 9 \\ & & -3 & -6 & -9 \\ \hline & 2 & 4 & 6 & 0 \end{array} \Rightarrow x = -\tfrac{3}{2} \text{ is a zero.}$$

$P(x) = \left(x + \frac{3}{2}\right)(2x^2 + 4x + 6) = 2\left(x + \frac{3}{2}\right)(x^2 + 2x + 3)$. Now $x^2 + 2x + 3$ has zeros $x = \frac{-2 \pm \sqrt{4-4(3)(1)}}{2} = \frac{-2 \pm 2\sqrt{-2}}{2} = -1 \pm i\sqrt{2}$. Hence, the zeros are $-\frac{3}{2}$ and $-1 \pm i\sqrt{2}$.

**48.** Using synthetic division, we see that $(x-3)$ is a factor of the polynomial:

$$\begin{array}{r|rrrr} 1 & 2 & -8 & 9 & -9 \\ & & 2 & -6 & 3 \\ \hline & 2 & -6 & 3 & -6 \end{array} \qquad \begin{array}{r|rrrr} 3 & 2 & -8 & 9 & -9 \\ & & 6 & -6 & 9 \\ \hline & 2 & -2 & 3 & 0 \end{array} \Rightarrow x = 3 \text{ is a zero.}$$

So $P(x) = 2x^3 - 8x^2 + 9x - 9 = (x-3)\left(2x^2 - 2x + 3\right)$. Using the quadratic formula, we find the other two solutions: $x = \dfrac{2 \pm \sqrt{4 - 4(3)(2)}}{2(2)} = \dfrac{2 \pm \sqrt{-20}}{4} = \frac{1}{2} \pm \dfrac{\sqrt{5}}{2}i$. Thus the zeros are 3, $\frac{1}{2} \pm \dfrac{\sqrt{5}}{2}i$.

**49.** $P(x) = x^4 + x^3 + 7x^2 + 9x - 18$. Since $P(x)$ has one change in sign, we are guaranteed a positive zero, and since $P(-x) = x^4 - x^3 + 7x^2 - 9x - 18$, there are 1 or 3 negative zeros.

$$\begin{array}{r|rrrrr} 1 & 1 & 1 & 7 & 9 & -18 \\ & & 1 & 2 & 9 & 18 \\ \hline & 1 & 2 & 9 & 18 & 0 \end{array}$$

Therefore, $P(x) = (x-1)\left(x^3 + 2x^2 + 9x + 18\right)$. Continuing with the quotient, we try negative zeros.

$$\begin{array}{r|rrrr} -1 & 1 & 2 & 9 & 18 \\ & & -1 & -1 & -8 \\ \hline & 1 & 1 & 8 & 10 \end{array} \qquad \begin{array}{r|rrrr} -2 & 1 & 2 & 9 & 18 \\ & & -2 & 0 & -18 \\ \hline & 1 & 0 & 9 & 0 \end{array}$$

$P(x) = (x-1)(x+2)\left(x^2+9\right) = (x-1)(x+2)(x-3i)(x+3i)$. Therefore,the zeros are 1, $-2$, and $\pm 3i$.

**50.** $P(x) = x^4 - 2x^3 - 2x^2 - 2x - 3$ has possible zeros $\pm 1$, $\pm 3$.

$$\begin{array}{r|rrrrr} 1 & 1 & -2 & -2 & -2 & -3 \\ & & 1 & -1 & -3 & -5 \\ \hline & 1 & -1 & -3 & -5 & -8 \end{array} \qquad \begin{array}{r|rrrrr} 3 & 1 & -2 & -2 & -2 & -3 \\ & & 3 & 3 & 3 & 3 \\ \hline & 1 & 1 & 1 & 1 & 0 \end{array} \Rightarrow x = 3 \text{ is a zero.}$$

$P(x) = (x-3)\left(x^3 + x^2 + x + 1\right)$. If we factor the second factor by grouping, we get $x^3 + x^2 + x + 1 = x^2(x+1) + 1(x+1) = (x+1)\left(x^2+1\right)$. So we have $P(x) = (x-3)(x+1)\left(x^2+1\right) = (x-3)(x+1)(x-i)(x+i)$. Thus the zeros are 3, $-1$, $i$, and $-i$.

**51.** We see a pattern and use it to factor by grouping. This gives

$$\begin{aligned} P(x) &= x^5 - x^4 + 7x^3 - 7x^2 + 12x - 12 = x^4(x-1) + 7x^2(x-1) + 12(x-1) = (x-1)\left(x^4 + 7x^2 + 12\right) \\ &= (x-1)\left(x^2+3\right)\left(x^2+4\right) = (x-1)\left(x - i\sqrt{3}\right)\left(x + i\sqrt{3}\right)(x-2i)(x+2i) \end{aligned}$$

Therefore,the zeros are 1, $\pm i\sqrt{3}$, and $\pm 2i$.

**52.** $P(x) = x^5 + x^3 + 8x^2 + 8 = x^3\left(x^2+1\right) + 8\left(x^2+1\right) = \left(x^2+1\right)\left(x^3+8\right) = \left(x^2+1\right)(x+2)\left(x^2 - 2x + 4\right)$ (factoring a sum of cubes). So $x = -2$, or $x^2 + 1 = 0$. If $x^2 + 1 = 0$, then $x^2 = -1 \Rightarrow x = \pm i$. If $x^2 - 2x + 4 = 0$, then $x = \frac{2 \pm \sqrt{4 - 4(1)(4)}}{2} = 1 \pm \frac{\sqrt{-12}}{2} = 1 \pm \sqrt{3}i$. Thus, the zeros are $-2$, $\pm i$, $1 \pm \sqrt{3}i$.

**53.** $P(x) = x^4 - 6x^3 + 13x^2 - 24x + 36$ has possible rational zeros $\pm 1, \pm 2, \pm 3, \pm 4, \pm 6, \pm 9, \pm 12, \pm 18$. $P(x)$ has 4 variations in sign and $P(-x)$ has no variation in sign.

$$\begin{array}{r|rrrrr} 1 & 1 & -6 & 13 & -24 & 36 \\ & & 1 & -5 & 8 & -16 \\ \hline & 1 & -5 & 8 & -16 & 20 \end{array} \qquad \begin{array}{r|rrrrr} 2 & 1 & -6 & 13 & -24 & 36 \\ & & 2 & -8 & 10 & -28 \\ \hline & 1 & -4 & 5 & -14 & 8 \end{array} \qquad \begin{array}{r|rrrrr} 3 & 1 & -6 & 13 & -24 & 36 \\ & & 3 & -9 & 12 & -36 \\ \hline & 1 & -3 & 4 & -12 & 0 \end{array} \Rightarrow x = 3 \text{ is a zero.}$$

Continuing:

$$\begin{array}{r|rrrr} 3 & 1 & -3 & 4 & -12 \\ & & 3 & 0 & 12 \\ \hline & 1 & 0 & 4 & 0 \end{array} \Rightarrow x = 3 \text{ is a zero.}$$

$P(x) = (x-3)^2 (x^2+4) = (x-3)^2 (x-2i)(x+2i)$. Therefore,the zeros are 3 (multiplicity 2) and $\pm 2i$.

**54.** $P(x) = x^4 - x^2 + 2x + 2$ has possible rational zeros $\pm 1, \pm 2$.

$$\begin{array}{r|rrrrr} 1 & 1 & 0 & -1 & 2 & 2 \\ & & 1 & 1 & 0 & 2 \\ \hline & 1 & 1 & 0 & 2 & 4 \end{array} \; 1 \text{ is an upper bound.} \qquad \begin{array}{r|rrrrr} -1 & 1 & 0 & -1 & 2 & 2 \\ & & -1 & 1 & 0 & -2 \\ \hline & 1 & -1 & 0 & 2 & 0 \end{array} \qquad \begin{array}{r|rrrr} -1 & 1 & -1 & 0 & 2 \\ & & -1 & 2 & -2 \\ \hline & 1 & -2 & 2 & 0 \end{array}$$

$P(x) = (x+1)^2 (x^2 - 2x + 2)$. Using the quadratic formula on $x^2 - 2x + 2$, we have $x = \frac{2 \pm \sqrt{4-8}}{2} = \frac{2 \pm 2i}{2} = 1 \pm i$. Thus, the zeros of $P(x)$ are $-1, 1 \pm i$.

**55.** $P(x) = 4x^4 + 4x^3 + 5x^2 + 4x + 1$ has possible rational zeros $\pm 1, \pm \frac{1}{2}, \pm \frac{1}{4}$. Since there is no variation in sign, all real zeros (if there are any) are negative.

$$\begin{array}{r|rrrrr} -1 & 4 & 4 & 5 & 4 & 1 \\ & & -4 & 0 & -5 & 1 \\ \hline & 4 & 0 & 5 & -1 & 2 \end{array} \qquad \begin{array}{r|rrrrr} -\frac{1}{2} & 4 & 4 & 5 & 4 & 1 \\ & & -2 & -1 & -2 & -1 \\ \hline & 4 & 2 & 4 & 2 & 0 \end{array} \Rightarrow x = -\tfrac{1}{2} \text{ is a zero.}$$

$P(x) = \left(x + \frac{1}{2}\right)(4x^3 + 2x^2 + 4x + 2)$. Continuing:

$$\begin{array}{r|rrrr} -\frac{1}{2} & 4 & 2 & 4 & 2 \\ & & -2 & 0 & -2 \\ \hline & 4 & 0 & 4 & 0 \end{array} \Rightarrow x = -\tfrac{1}{2} \text{ is a zero again.}$$

$P(x) = \left(x + \frac{1}{2}\right)^2 (4x^2 + 4)$. Thus, the zeros of $P(x)$ are $-\frac{1}{2}$ and $\pm i$.

**56.** $P(x) = 4x^4 + 2x^3 - 2x^2 - 3x - 1$ has possible rational zeros $\pm 1, \pm \frac{1}{2}, \pm \frac{1}{4}$. $P$ has one variation in sign, so $P$ has one positive real zero.

$$\begin{array}{r|rrrrr} 1 & 4 & 2 & -2 & -3 & -1 \\ & & 4 & 6 & 4 & 1 \\ \hline & 4 & 6 & 4 & 1 & 0 \end{array} \Rightarrow 1 \text{ is a zero.}$$

$P(x) = (x-1)(4x^3 + 6x^2 + 4x + 1)$. Continuing:

$$\begin{array}{r|rrrr} -1 & 4 & 6 & 4 & 1 \\ & & -4 & -2 & 2 \\ \hline & 4 & 2 & -2 & 3 \end{array} \qquad \begin{array}{r|rrrr} -\frac{1}{2} & 4 & 6 & 4 & 1 \\ & & -2 & -2 & -1 \\ \hline & 4 & 4 & 2 & 0 \end{array} \Rightarrow -\tfrac{1}{2} \text{ is a zero.}$$

$P(x) = (x-1)\left(x + \frac{1}{2}\right)(4x^2 + 4x + 2)$. Using the quadratic formula on $4x^2 + 4x + 2$, we find $x = \frac{-4 \pm \sqrt{16-32}}{8} = -\frac{1}{2} \pm \frac{1}{2}i$. Thus, $P$ has zeros $1, -\frac{1}{2}, -\frac{1}{2} \pm \frac{1}{2}i$.

**57.** $P(x) = x^5 - 3x^4 + 12x^3 - 28x^2 + 27x - 9$ has possible rational zeros $\pm 1, \pm 3, \pm 9$. $P(x)$ has 4 variations in sign and $P(-x)$ has 1 variation in sign.

$$\begin{array}{r|rrrrrr} 1 & 1 & -3 & 12 & -28 & 27 & -9 \\ & & 1 & -2 & 10 & -18 & 9 \\ \hline & 1 & -2 & 10 & -18 & 9 & 0 \end{array} \Rightarrow x = 1 \text{ is a zero.}$$

$$\begin{array}{r|rrrrr} 1 & 1 & -2 & 10 & -18 & 9 \\ & & 1 & -1 & 9 & -9 \\ \hline & 1 & -1 & 9 & -9 & 0 \end{array} \Rightarrow x = 1 \text{ is a zero.}$$

$$\begin{array}{r|rrrr} 1 & 1 & -1 & 9 & -9 \\ & & 1 & 0 & 9 \\ \hline & 1 & 0 & 9 & 0 \end{array} \Rightarrow x = 1 \text{ is a zero.}$$

$P(x) = (x-1)^3 (x^2+9) = (x-1)^3 (x-3i)(x+3i)$. Therefore,the zeros are 1 (multiplicity 3) and $\pm 3i$.

**58.** $P(x) = x^5 - 2x^4 + 2x^3 - 4x^2 + x - 2$ has possible rational zeros $\pm 1, \pm 2$.

$$\begin{array}{r|rrrrrr} 1 & 1 & -2 & 2 & -4 & 1 & -2 \\ & & 1 & -1 & 1 & -3 & -2 \\ \hline & 1 & -1 & 1 & -3 & -2 & -4 \end{array}$$

$$\begin{array}{r|rrrrrr} 2 & 1 & -2 & 2 & -4 & 1 & -2 \\ & & 2 & 0 & 4 & 0 & 2 \\ \hline & 1 & 0 & 2 & 0 & 1 & 0 \end{array} \Rightarrow x = 2 \text{ is a zero.}$$

$P(x) = (x-2)(x^4+2x^2+1) = (x-2)(x^2+1)^2 = (x-2)(x-i)^2(x+i)^2$. Thus, the zeros of $P(x)$ are 2, $\pm i$.

**59. (a)** $P(x) = x^3 - 5x^2 + 4x - 20 = x^2(x-5) + 4(x-5) = (x-5)(x^2+4)$

**(b)** $P(x) = (x-5)(x-2i)(x+2i)$

**60. (a)** $P(x) = x^3 - 2x - 4$

$$\begin{array}{r|rrrr} 1 & 1 & 0 & -2 & -4 \\ & & 1 & 1 & -1 \\ \hline & 1 & 1 & -1 & -5 \end{array} \qquad \begin{array}{r|rrrr} 2 & 1 & 0 & -2 & -4 \\ & & 2 & 4 & 4 \\ \hline & 1 & 2 & 2 & 0 \end{array}$$

$P(x) = x^3 - 2x - 4 = (x-2)(x^2+2x+2)$

**(b)** $P(x) = (x-2)(x+1-i)(x+1+i)$

**61. (a)** $P(x) = x^4 + 8x^2 - 9 = (x^2-1)(x^2+9) = (x-1)(x+1)(x^2+9)$

**(b)** $P(x) = (x-1)(x+1)(x-3i)(x+3i)$

**62. (a)** $P(x) = x^4 + 8x^2 + 16 = (x^2+4)^2$

**(b)** $P(x) = (x-2i)^2(x+2i)^2$

**63. (a)** $P(x) = x^6 - 64 = (x^3-8)(x^3+8) = (x-2)(x^2+2x+4)(x+2)(x^2-2x+4)$

**(b)** $P(x) = (x-2)(x+2)(x+1-i\sqrt{3})(x+1+i\sqrt{3})(x-1-i\sqrt{3})(x-1+i\sqrt{3})$

**64. (a)** $P(x) = x^5 - 16x = x(x^4-16) = x(x^2-4)(x^2+4) = x(x-2)(x+2)(x^2+4)$

**(b)** $P(x) = x(x-2)(x+2)(x-2i)(x+2i)$

**65. (a)** $x^4 - 2x^3 - 11x^2 + 12x = x\left(x^3 - 2x^2 - 11x + 12\right) = 0$. We first find the bounds for our viewing rectangle.

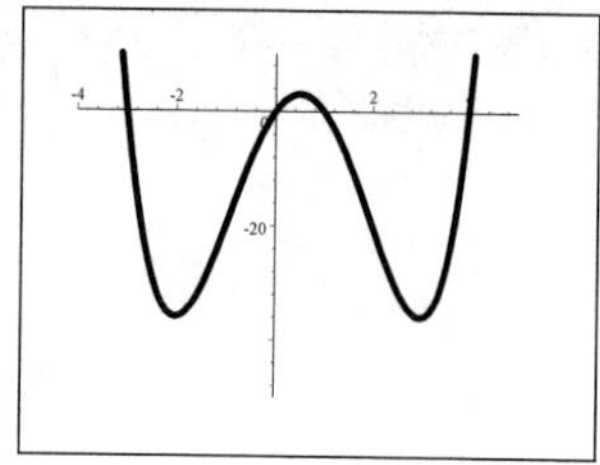

| | 1 | −2 | −11 | 12 | |
|---|---|---|---|---|---|
| 5 | 1 | 3 | 4 | 32 | $\Rightarrow x = 5$ is an upper bound. |
| −4 | 1 | −6 | 13 | −50 | $\Rightarrow x = -4$ is a lower bound. |

We graph $P(x) = x^4 - 2x^3 - 11x^2 + 12x$ in the viewing rectangle $[-4, 5]$ by $[-50, 10]$ and see that it has 4 real solutions. Since this matches the degree of $P(x)$, $P(x)$ has no imaginary solution.

**(b)** $x^4 - 2x^3 - 11x^2 + 12x - 5 = 0$. We use the same bounds for our viewing rectangle, $[-4, 5]$ by $[-50, 10]$, and see that $R(x) = x^4 - 2x^3 - 11x^2 + 12x - 5$ has 2 real solutions. Since the degree of $R(x)$ is 4, $R(x)$ must have 2 imaginary solutions.

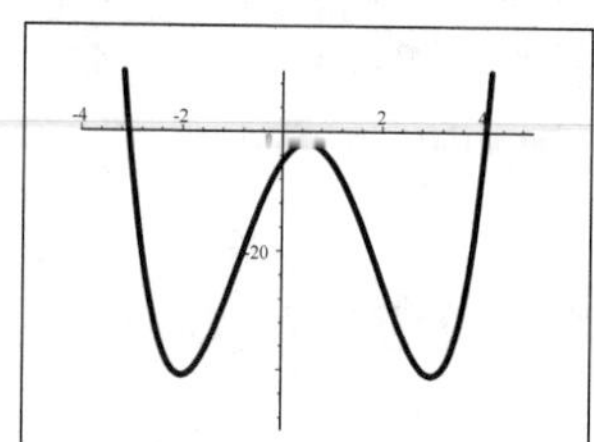

**(c)** $x^4 - 2x^3 - 11x^2 + 12x + 40 = 0$. We graph $T(x) = x^4 - 2x^3 - 11x^2 + 12x + 40$ in the viewing rectangle $[-4, 5]$ by $[-10, 50]$, and see that $T$ has no real solution. Since the degree of $T$ is 4, $T$ must have 4 imaginary solutions.

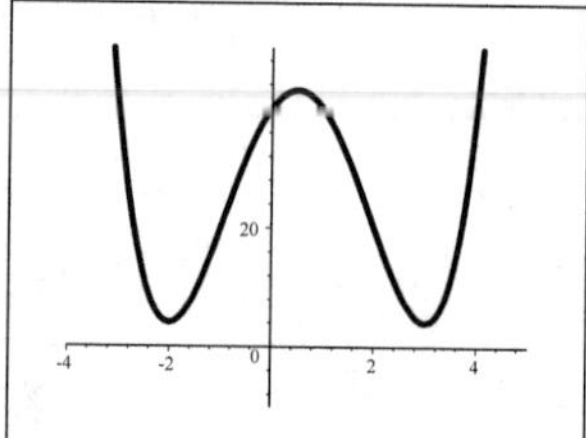

**66. (a)** $2x + 4i = 1 \quad\Leftrightarrow\quad 2x = 1 - 4i \quad\Leftrightarrow\quad x = \frac{1}{2} - 2i$.

**(b)** $x^2 - ix = 0 \quad\Leftrightarrow\quad x(x - i) = 0 \Leftrightarrow x = 0, i$.

**(c)** $x^2 + 2ix - 1 = 0 \quad\Leftrightarrow\quad (x + i)^2 = 0 \Leftrightarrow x = -i$.

**(d)** $ix^2 - 2x + i = 0$. Using the quadratic formula, we get

$x = \frac{2 \pm \sqrt{(-2)^2 - 4(i)(i)}}{2i} = \frac{2 \pm \sqrt{8}}{2i} = \frac{2 \pm 2\sqrt{2}}{2i} = \frac{1 \pm \sqrt{2}}{i} = \left(1 \pm \sqrt{2}\right)(-i) = \left(-1 \pm \sqrt{2}\right)i$.

**67. (a)** $P(x) = x^2 - (1 + i)x + (2 + 2i)$. So $P(2i) = (2i)^2 - (1 + i)(2i) + 2 + 2i = -4 - 2i + 2 + 2 + 2i = 0$, and $P(1 - i) = (1 - i)^2 - (1 + i)(1 - i) + (2 + 2i) = 1 - 2i - 1 - 1 - 1 + 2 + 2i = 0$. Therefore, $2i$ and $1 - i$ are solutions of the equation $x^2 - (1 + i)x + (2 + 2i) = 0$. However, $P(-2i) = (-2i)^2 - (1 + i)(-2i) + 2 + 2i = -4 + 2i - 2 + 2 + 2i = -4 + 4i$, and $P(1 + i) = (1 + i)^2 - (1 + i)(1 + i) + 2 + 2i = 2 + 2i$. Since, $P(-2i) \neq 0$ and $P(1 + i) \neq 0$, $-2i$ and $1 + i$ are not solutions.

**(b)** This does not violate the Conjugate Roots Theorem because the coefficients of the polynomial $P(x)$ are not all real.

**68. (a)** Since $i$ and $1 + i$ are zeros, $-i$ and $1 - i$ are also zeros. So

$$\begin{aligned} P(x) &= C(x - i)(x + i)(x - [1 + i])(x - [1 - i]) = C\left(x^2 + 1\right)\left(x^2 - 2x + 2\right) \\ &= C\left(x^4 - 2x^3 + 2x^2 + x^2 - 2x + 2\right) = C\left(x^4 - 2x^3 + 3x^2 - 2x + 2\right) \end{aligned}$$

Since $C = 1$, the polynomial is $P(x) = x^4 - 2x^3 + 3x^2 - 2x + 2$.

**(b)** Since $i$ and $1 + i$ are zeros,

$$P(x) = C(x - i)(x - [i + 1]) = C\left(x^2 - xi - x - xi - 1 + i\right) = C\left[x^2 - (1 + 2i)x - 1 + i\right]$$

Since $C = 1$, the polynomial is $P(x) = x^2 - (1 + 2i)x - 1 + i$.

**69.** Since $P$ has real coefficients, the imaginary zeros come in pairs: $a \pm bi$ (by the Conjugate Roots Theorem), where $b \neq 0$. Thus there must be an even number of imaginary zeros. Since $P$ is of odd degree, it has an odd number of zeros (counting multiplicity). It follows that $P$ has at least one real zero.

**70.** $x^4 - 1 = 0 \quad \Leftrightarrow \quad (x^2 - 1)(x^2 + 1) = 0 \quad \Leftrightarrow \quad (x - 1)(x + 1)(x + i)(x - i) = 0 \quad \Leftrightarrow \quad x = \pm 1, \pm i$. So there are four fourth roots of 1, two that are real and two that are complex. Consider $P(x) = x^n - 1$, where $n$ is even. $P$ has one change in sign so $P$ has exactly one real positive zero, namely $x = 1$. Since $P(-x) = P(x)$, $P$ also has exactly one real negative zero, namely $x = -1$. Thus $P$ must have $n - 2$ complex roots. As a result, $x^n = 1$ has two real $n$ th zeros and $n - 2$ complex roots.

$x^3 - 1 = 0 \quad \Leftrightarrow \quad (x - 1)(x^2 + x + 1) = 0 \Leftrightarrow x = 1, \frac{-1 \pm \sqrt{3}i}{2}$. So there is one real cube zero of unity and two complex roots. Now consider $Q(x) = x^k - 1$, where $k$ is odd. Since $Q$ has one change in sign, $Q$ has exactly one real positive zero, namely $x = 1$. But $Q(-x) = -x^k - 1$ has no changes in sign, so there are no negative real zeros. As a result, $x^k = 1$ has one real $k$th zero and $k - 1$ complex roots.

# 4.5 Rational Functions

**1.** $r(x) = \dfrac{x}{x - 2}$

**(a)**

| $x$ | $r(x)$ |
|---|---|
| 1.5 | −3 |
| 1.9 | −19 |
| 1.99 | −199 |
| 1.999 | −1999 |

| $x$ | $r(x)$ |
|---|---|
| 2.5 | 5 |
| 2.1 | 21 |
| 2.01 | 201 |
| 2.001 | 2001 |

| $x$ | $r(x)$ |
|---|---|
| 10 | 1.25 |
| 50 | 1.042 |
| 100 | 1.020 |
| 1000 | 1.002 |

| $x$ | $r(x)$ |
|---|---|
| −10 | 0.833 |
| −50 | 0.962 |
| −100 | 0.980 |
| −1000 | 0.998 |

**(b)** $r(x) \to -\infty$ as $x \to 2^-$ and $r(x) \to \infty$ as $x \to 2^+$.

**(c)** $r$ has horizontal asymptote $y = 1$.

**2.** $r(x) = \dfrac{4x + 1}{x - 2}$

**(a)**

| $x$ | $r(x)$ |
|---|---|
| 1.5 | −14 |
| 1.9 | −86 |
| 1.99 | −896 |
| 1.999 | −8996 |

| $x$ | $r(x)$ |
|---|---|
| 2.5 | 22 |
| 2.1 | 94 |
| 2.01 | 904 |
| 2.001 | 9004 |

| $x$ | $r(x)$ |
|---|---|
| 10 | 5.125 |
| 50 | 4.188 |
| 100 | 4.092 |
| 1000 | 4.009 |

| $x$ | $r(x)$ |
|---|---|
| −10 | 3.25 |
| −50 | 3.827 |
| −100 | 3.912 |
| −1000 | 3.991 |

**(b)** $r(x) \to -\infty$ as $x \to 2^-$ and $r(x) \to \infty$ as $x \to 2^+$.

**(c)** $r$ has horizontal asymptote $y = 4$.

**3.** $r(x) = \dfrac{3x - 10}{(x - 2)^2}$

**(a)**

| $x$ | $r(x)$ |
|---|---|
| 1.5 | −22 |
| 1.9 | −430 |
| 1.99 | −40,300 |
| 1.999 | −4,003,000 |

| $x$ | $r(x)$ |
|---|---|
| 2.5 | −10 |
| 2.1 | −370 |
| 2.01 | −39,700 |
| 2.001 | −3,997,000 |

| $x$ | $r(x)$ |
|---|---|
| 10 | 0.3125 |
| 50 | 0.0608 |
| 100 | 0.0302 |
| 1000 | 0.0030 |

| $x$ | $r(x)$ |
|---|---|
| −10 | −0.2778 |
| −50 | −0.0592 |
| −100 | −0.0298 |
| −1000 | −0.0030 |

**(b)** $r(x) \to -\infty$ as $x \to 2$.

**(c)** $r$ has horizontal asymptote $y = 0$.

**4.** $r(x) = \dfrac{3x^2+1}{(x-2)^2}$

**(a)**

| $x$ | $r(x)$ |
|---|---|
| 1.5 | 31 |
| 1.9 | 1183 |
| 1.99 | 128,803 |
| 1.999 | 12,988,003 |

| $x$ | $r(x)$ |
|---|---|
| 2.5 | 79 |
| 2.1 | 1423 |
| 2.01 | 131,203 |
| 2.001 | 13,012,003 |

| $x$ | $r(x)$ |
|---|---|
| 10 | 4.703 |
| 50 | 3.256 |
| 100 | 3.124 |
| 1000 | 3.012 |

| $x$ | $r(x)$ |
|---|---|
| $-10$ | 2.09 |
| $-50$ | 2.774 |
| $-100$ | 2.884 |
| $-1000$ | 2.988 |

**(b)** $r(x) \to \infty$ as $x \to 2$.

**(c)** $r$ has horizontal asymptote $y = 3$.

**5.** $r(x) = \dfrac{x-1}{x+4}$. When $x = 0$, we have $r(0) = -\frac{1}{4}$, so the $y$-intercept is $-\frac{1}{4}$. The numerator is 0 when $x = 1$, so the $x$-intercept is 1.

**6.** $s(x) = \dfrac{3x}{x-5}$. When $x = 0$, we have $s(0) = 0$, so the $y$-intercept is 0. The numerator is zero when $3x = 0$ or $x = 0$, so the $x$-intercept is 0.

**7.** $t(x) = \dfrac{x^2-x-2}{x-6}$. When $x = 0$, we have $t(0) = \dfrac{-2}{-6} = \frac{1}{3}$, so the $y$-intercept is $\frac{1}{3}$. The numerator is 0 when $x^2 - x - 2 = (x-2)(x+1) = 0$ or when $x = 2$ or $x = -1$, so the $x$-intercepts are 2 and $-1$.

**8.** $r(x) = \dfrac{2}{x^2+3x-4}$. When $x = 0$, we have $r(0) = \dfrac{2}{-4} = -\frac{1}{2}$, so the $y$-intercept is $-\frac{1}{2}$. The numerator is never zero, so there is no $x$-intercept.

**9.** $r(x) = \dfrac{x^2-9}{x^2}$. Since 0 is not in the domain of $r(x)$, there is no $y$-intercept. The numerator is 0 when $x^2 - 9 = (x-3)(x+3) = 0$ or when $x = \pm 3$, so the $x$-intercepts are $\pm 3$.

**10.** $r(x) = \dfrac{x^3+8}{x^2+4}$. When $x = 0$, we have $r(0) = \frac{8}{4} = 2$, so the $y$-intercept is 2. The $x$-intercept occurs when $x^3 + 8 = 0$ $\Leftrightarrow$ $(x+2)(x^2-2x+4) = 0$ $\Leftrightarrow$ $x = -2$ or $x = 1 \pm i\sqrt{3}$, which has only one real solution, so the $x$-intercept is $-2$.

**11.** From the graph, the $x$-intercept is 3, the $y$-intercept is 3, the vertical asymptote is $x = 2$, and the horizontal asymptote is $y = 2$.

**12.** From the graph, the $x$-intercept is 0, the $y$-intercept is 0, the horizontal asymptote is $y = 0$, and the vertical asymptotes are $x = -1$ and $x = 2$.

**13.** From the graph, the $x$-intercepts are $-1$ and 1, the $y$-intercept is about $\frac{1}{4}$, the vertical asymptotes are $x = -2$ and $x = 2$, and the horizontal asymptote is $y = 1$.

**14.** From the graph, the $x$-intercepts are $\pm 2$, the $y$-intercept is $-6$, the horizontal asymptote is $y = 2$, and there are no vertical asymptotes

**15.** $r(x) = \dfrac{3}{x+2}$. There is a vertical asymptote where $x+2 = 0$ $\Leftrightarrow$ $x = -2$. We have $r(x) = \dfrac{3}{x+2} = \dfrac{3/x}{1+(2/x)} \to 0$ as $x \to \pm\infty$, so the horizontal asymptote is $y = 0$.

**16.** $s(x) = \dfrac{2x+3}{x-1} = \dfrac{2+(3/x)}{1-(1/x)} \to 2$ as $x \to \pm\infty$. The horizontal asymptote is $y = 2$. There is a vertical asymptote when $x - 1 = 0 \Leftrightarrow x = 1$, so the vertical asymptote is $x = 1$.

**17.** $t(x) = \dfrac{x^2}{x^2 - x - 6} = \dfrac{x^2}{(x-3)(x+2)} = \dfrac{1}{1 - \dfrac{1}{x} - \dfrac{6}{x^2}} \to 1$ as $x \to \pm\infty$. Hence, the horizontal asymptote is $y = 1$.

The vertical asymptotes occur when $(x-3)(x+2) = 0 \quad \Leftrightarrow \quad x = 3$ or $x = -2$, and so the vertical asymptotes are $x = 3$ and $x = -2$.

**18.** $r(x) = \dfrac{2x - 4}{x^2 + 2x + 1} = \dfrac{\dfrac{2}{x} - \dfrac{4}{x^2}}{1 + \dfrac{2}{x} + \dfrac{1}{x^2}} \to 0$ as $x \to \pm\infty$. Thus, the horizontal asymptote is $y = 0$. Also, $y = \dfrac{2(x-2)}{(x+1)^2}$ so there is a vertical asymptote when $x + 1 = 0 \quad \Leftrightarrow \quad x = -1$, so the vertical asymptote is $x = -1$.

**19.** $s(x) = \dfrac{6}{x^2 + 2}$. There is no vertical asymptote since $x^2 + 2$ is never 0. Since $s(x) = \dfrac{6}{x^2 + 2} = \dfrac{\dfrac{6}{x^2}}{1 + \dfrac{2}{x^2}} \to 0$ as $x \to \pm\infty$, the horizontal asymptote is $y = 0$.

**20.** $t(x) = \dfrac{(x-1)(x-2)}{(x-3)(x-4)} = \dfrac{x^2 - 3x + 2}{x^2 - 7x + 12} = \dfrac{1 - \dfrac{3}{x} + \dfrac{2}{x^2}}{1 - \dfrac{7}{x} + \dfrac{12}{x^2}} \to 1$ as $x \to \pm\infty$, so the horizontal asymptote is $y = 1$.

Also, vertical asymptotes occur when $(x-3)(x-4) = 0 \quad \Rightarrow \quad x = 3, 4$, so the two vertical asymptotes are $x = 3$ and $x = 4$.

**21.** $r(x) = \dfrac{6x - 2}{x^2 + 5x - 6}$. A vertical asymptote occurs when $x^2 + 5x - 6 = (x+6)(x-1) = 0 \Leftrightarrow x = 1$ or $x = -6$. Because the degree of the denominator is greater than the degree of the numerator, the horizontal asymptote is $y = 0$.

**22.** $s(x) = \dfrac{3x^2}{x^2 + 2x + 5} = \dfrac{3}{1 + \dfrac{2}{x} + \dfrac{5}{x^2}} \to 3$ as $x \to \pm\infty$, so the horizontal asymptote is $y = 3$. Also, vertical asymptotes occur when $x^2 + 2x + 5 = 0 \quad \Rightarrow \quad x = \frac{-2 \pm \sqrt{4 - 20}}{2} = -1 \pm 2i$. Since there is no real zero, there are no vertical asymptotes.

**23.** $y = \dfrac{x^2 + 2}{x - 1}$. A vertical asymptote occurs when $x - 1 = 0 \Leftrightarrow x = 1$. There are no horizontal asymptotes because the degree of the numerator is greater than the degree of the denominator.

**24.** $r(x) = \dfrac{x^3 + 3x^2}{x^2 - 4} = \dfrac{x^2(x+3)}{(x-2)(x+2)}$. Because the degree of the numerator is greater than the degree of the denominator, the function has no horizontal asymptotes. Two vertical asymptotes occur at $x = 2$ and $x = -2$. By using long division, we see that $r(x) = x + 3 + \dfrac{4x + 12}{x^2 - 4}$ so $y = x + 3$ is a slant asymptote.

**In Exercises 25–32, let** $f(x) = \dfrac{1}{x}$.

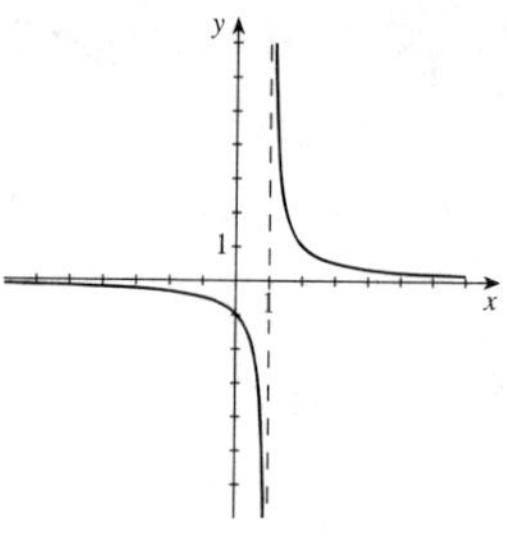


**25.** $r(x) = \dfrac{1}{x-1} = f(x-1)$. From this form we see that the graph of $r$ is obtained from the graph of $f$ by shifting 1 unit to the right. Thus $r$ has vertical asymptote $x = 1$ and horizontal asymptote $y = 0$.

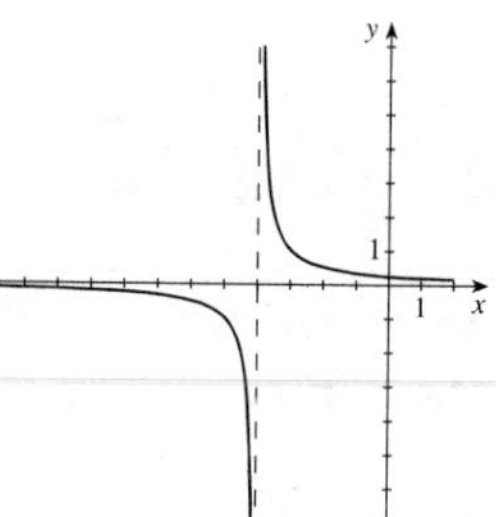


**26.** $r(x) = \dfrac{1}{x+4} = f(x+4)$. From this form we see that the graph of $r$ is obtained from the graph of $f$ by shifting 4 units to the left. Thus $r$ has vertical asymptote $x = -4$ and horizontal asymptote $y = 0$.

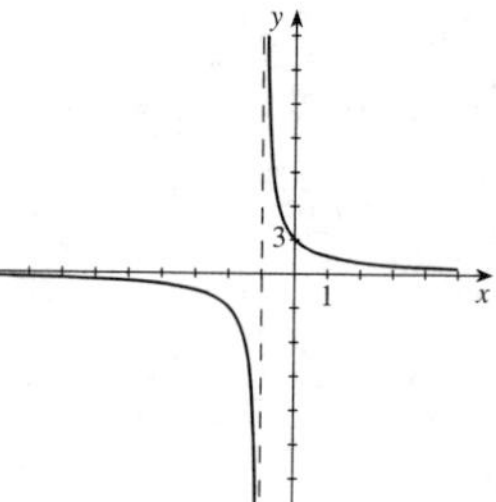


**27.** $s(x) = \dfrac{3}{x+1} = 3\left(\dfrac{1}{x+1}\right) = 3f(x+1)$. From this form we see that the graph of $s$ is obtained from the graph of $f$ by shifting 1 unit to the left and stretching vertically by a factor of 3. Thus $s$ has vertical asymptote $x = -1$ and horizontal asymptote $y = 0$.

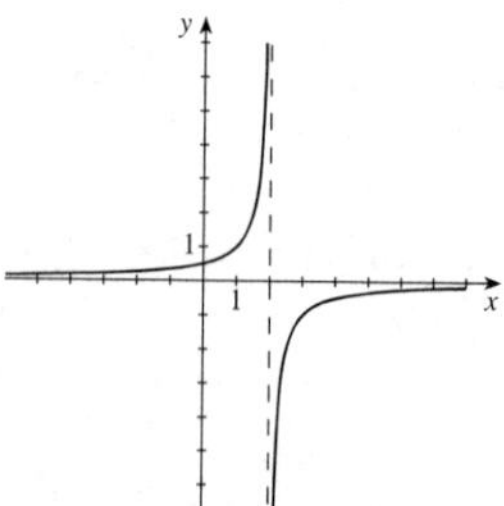


**28.** $s(x) = \dfrac{-2}{x-2} = -2\left(\dfrac{1}{x-2}\right) = -2f(x-2)$. From this form we see that the graph of $s$ is obtained from the graph of $f$ by shifting 2 units to the right, stretching vertically by a factor of 2, and then reflecting about the $x$-axis. Thus $s$ has vertical asymptote $x = 2$ and horizontal asymptote $y = 0$.

**29.** $t(x) = \dfrac{2x-3}{x-2} = 2 + \dfrac{1}{x-2} = f(x-2) + 2$ (see long division below). From this form we see that the graph of $t$ is obtained from the graph of $f$ by shifting 2 units to the right and 2 units vertically. Thus $t$ has vertical asymptote $x = 2$ and horizontal asymptote $y = 2$.

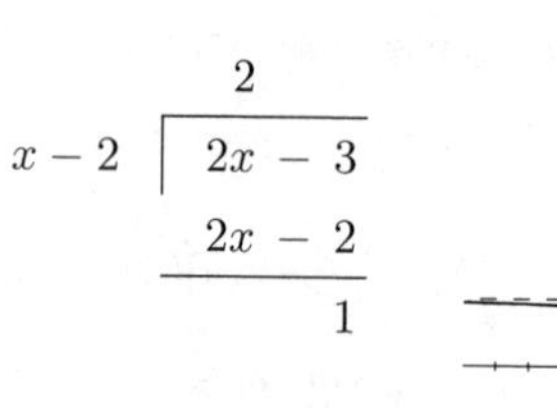


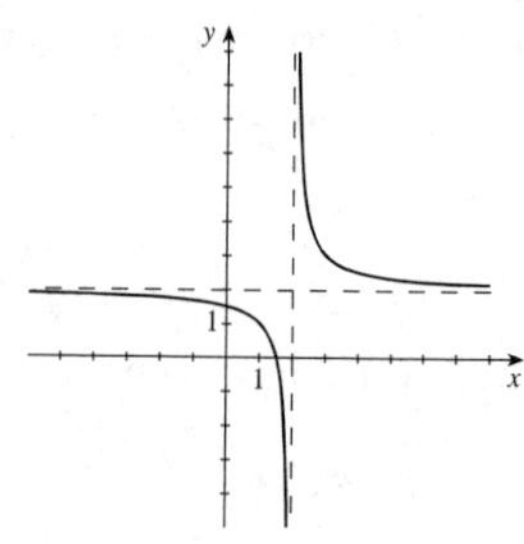

**30.** $t(x) = \dfrac{3x-3}{x+2} = 2 - \dfrac{9}{x+2} = 3 - 9\left(\dfrac{1}{x+2}\right)$
$= -9f(x+2) + 3$

From this form we see that the graph of $t$ is obtained from the graph of $f$ by shifting 2 units to the left, stretching vertically by a factor of 9, reflecting about the $x$-axis, and then shifting 3 units vertically. Thus $t$ has vertical asymptote $x = -2$ and horizontal asymptote $y = 3$.

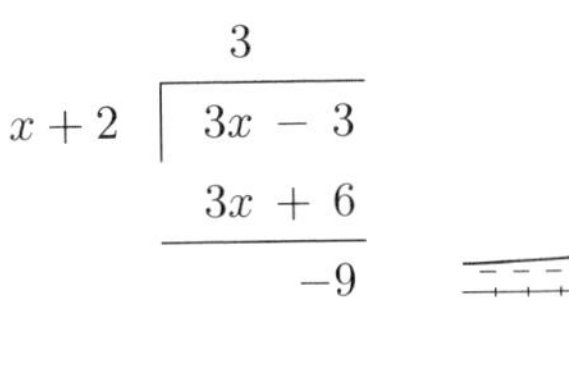

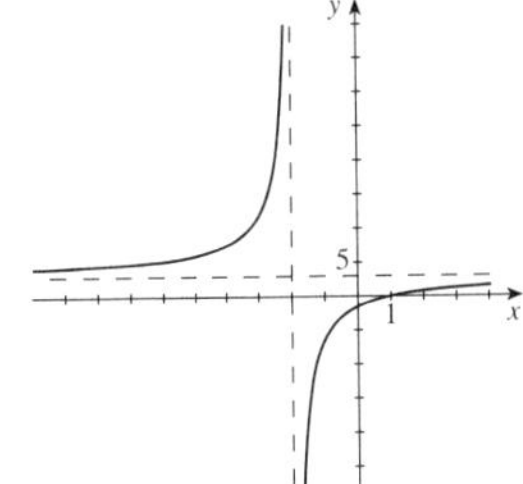


**31.** $r(x) = \dfrac{x+2}{x+3} = 1 - \dfrac{1}{x+3} = -f(x+3) + 1$ (see long division below). From this form we see that the graph of $r$ is obtained from the graph of $f$ by shifting 3 units to the left, reflect about the $x$-axis, and then shifting vertically 1 unit. Thus $r$ has vertical asymptote $x = -3$ and horizontal asymptote $y = 1$.

$$\begin{array}{r|l} & \quad 1 \\ x+3 & x+2 \\ & x+3 \\ \hline & \quad -1 \end{array}$$

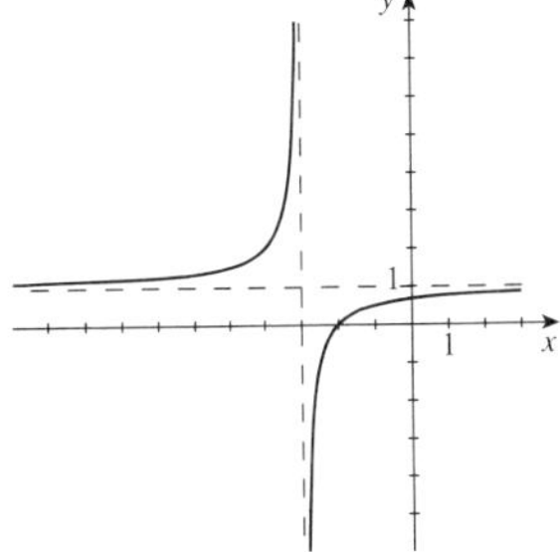


**32.** $t(x) = \dfrac{2x-9}{x-4} = 2 - \dfrac{1}{x-4} = 2 - \left(\dfrac{1}{x-4}\right)$
$= -f(x-4) + 2$

From this form we see that the graph of $t$ is obtained from the graph of $f$ by shifting 4 units to the right, reflecting about the $x$-axis, and then shifting 2 units vertically. Thus $t$ has vertical asymptote $x = 4$ and horizontal asymptote $y = 2$.

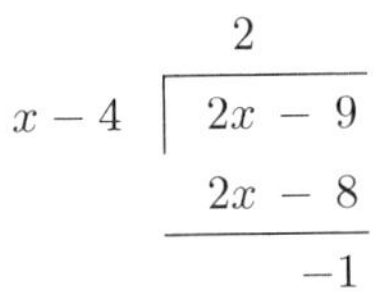

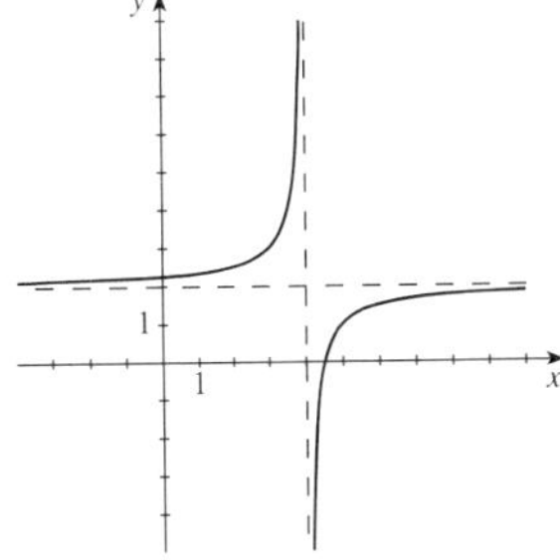


**33.** $y = \dfrac{4x-4}{x+2}$. When $x = 0$, $y = -2$, so the $y$-intercept is $-2$. When $y = 0$, $4x - 4 = 0 \quad \Leftrightarrow \quad x = 1$, so the $x$-intercept is 1. Since the degree of the numerator and denominator are the same the horizontal asymptote is $y = \frac{4}{1} = 4$. A vertical asymptote occurs when $x = -2$. As $x \to -2^+$, $y = \dfrac{4x-4}{x+2} \to -\infty$, and as $x \to -2^-$, $y = \dfrac{4x-4}{x+2} \to \infty$.

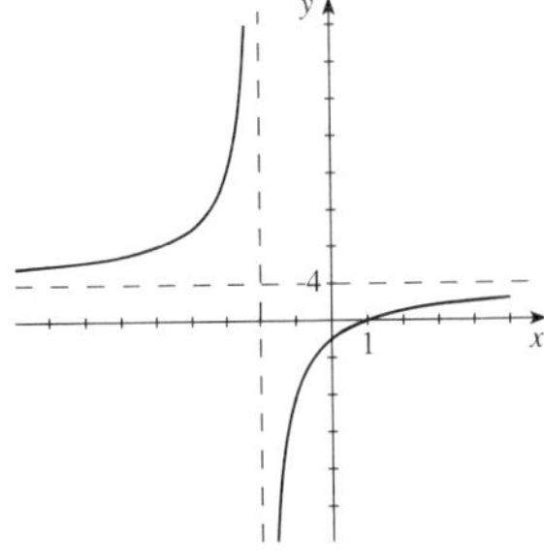


**34.** $r(x) = \dfrac{2x+6}{-6x+3} = \dfrac{2(x+3)}{-3(2x-1)}$. When $x = 0$, we have $y = 2$, so the $y$-intercept is 2. When $y = 0$, we have $x + 3 = 0 \quad \Leftrightarrow \quad x = -3$, so the $x$-intercept is $-3$. A vertical asymptote occurs when $2x - 1 = 0 \Leftrightarrow x = \frac{1}{2}$. Because the degree of the denominator and the numerator are the same, the horizontal asymptote is $y = \frac{2}{-6} = -\frac{1}{3}$.

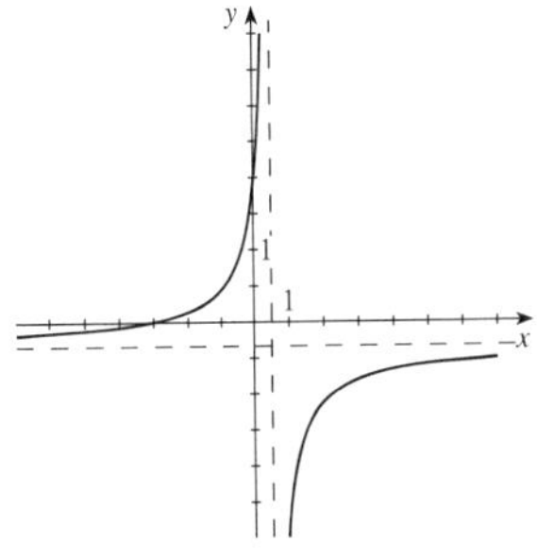

**35.** $s(x) = \dfrac{4 - 3x}{x + 7}$. When $x = 0$, $y = \frac{4}{7}$, so the $y$-intercept is $\frac{4}{7}$. The $x$-intercepts occur when $y = 0 \quad \Leftrightarrow \quad 4 - 3x = 0 \quad \Leftrightarrow \quad x = \frac{4}{3}$. A vertical asymptote occurs when $x = -7$. Since the degree of the numerator and denominator are the same the horizontal asymptote is $y = \dfrac{-3}{1} = -3$.

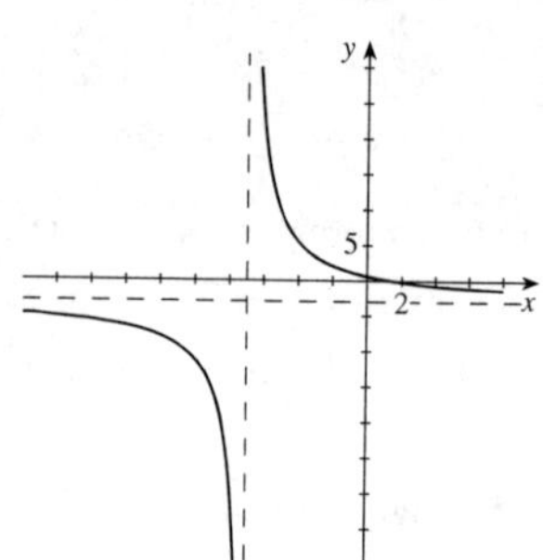


**36.** $s(x) = \dfrac{1 - 2x}{2x + 3}$. When $x = 0$, $y = \frac{1}{3}$, so the $y$-intercept is $\frac{1}{3}$. When $y = 0$, we have $1 - 2x = 0 \Leftrightarrow x = \frac{1}{2}$, so the $x$-intercept is $\frac{1}{2}$. A vertical asymptote occurs when $2x + 3 = 0 \quad \Leftrightarrow \quad x = -\frac{3}{2}$, and because the degree of the denominator and the numerator are the same, the horizontal asymptote is $y = -1$.

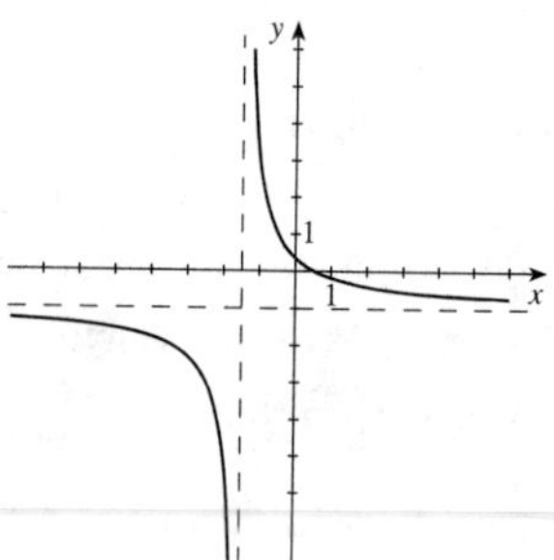


**37.** $r(x) = \dfrac{18}{(x - 3)^2}$. When $x = 0$, $y = \frac{18}{9} = 2$, and so the $y$-intercept is 2. Since the numerator can never be zero, there is no $x$-intercept. There is a vertical asymptote when $x - 3 = 0 \quad \Leftrightarrow \quad x = 3$, and because the degree of the asymptote is $y = 0$.

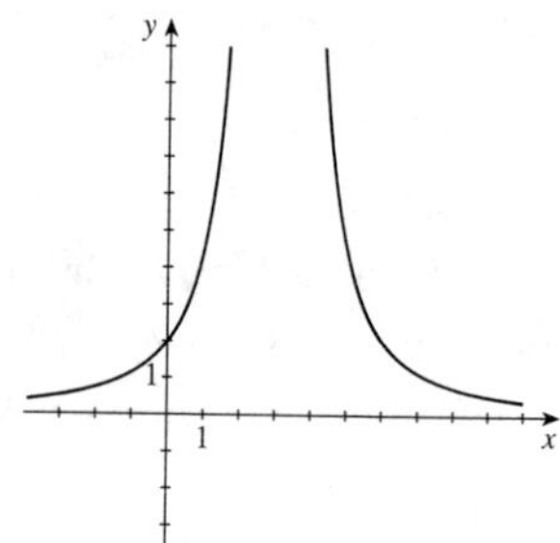


**38.** $r(x) = \dfrac{x - 2}{(x + 1)^2}$. When $x = 0$, we have $y = -2$, so the $y$-intercept is $-2$. When $y = 0$, we have $x - 2 = 0 \quad \Leftrightarrow \quad x = 2$, so the $x$-intercept is 2. A vertical asymptote occurs when $x = -1$, and because the degree of the denominator is greater than the degree of the numerator, the horizontal asymptote is $y = 0$.

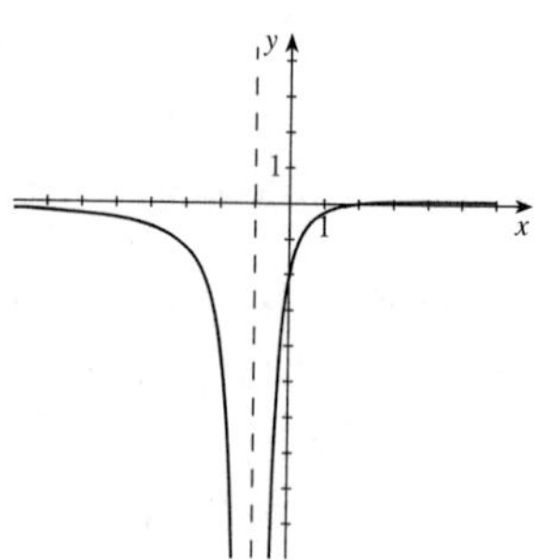


**39.** $s(x) = \dfrac{4x - 8}{(x - 4)(x + 1)}$. When $x = 0$, $y = \dfrac{-8}{(-4)(1)} = 2$, so the $y$-intercept is 2. When $y = 0$, $4x - 8 = 0 \quad \Leftrightarrow \quad x = 2$, so the $x$-intercept is 2. The vertical asymptotes are $x = -1$ and $x = 4$, and because the degree of the numerator is less than the degree of the denominator, the horizontal asymptote is $y = 0$.

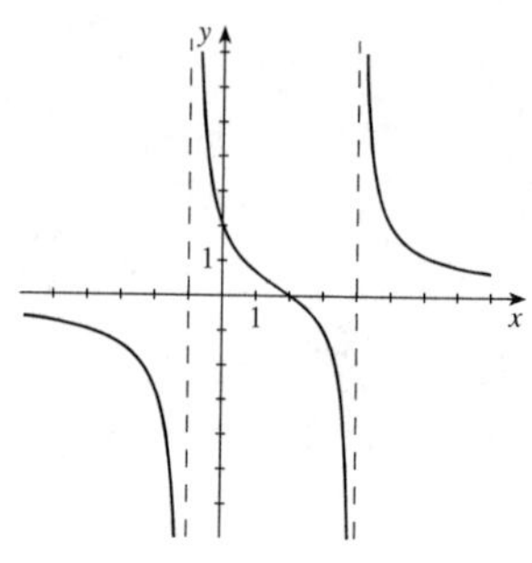

**40.** $s(x) = \dfrac{x+2}{(x+3)(x-1)}$. When $x = 0$, $y = \dfrac{2}{-3}$, so the $y$-intercept is $-\frac{2}{3}$ When $y = 0$, we have $x + 2 = 0 \quad \Leftrightarrow \quad x = -2$, so the $x$-intercept is $-2$. A vertical asymptote occurs when $(x+3)(x-1) = 0 \Leftrightarrow x = -3$ and $x = 1$. Because the degree of the denominator is greater than the degree of the numerator, the horizontal asymptote is $y = 0$.

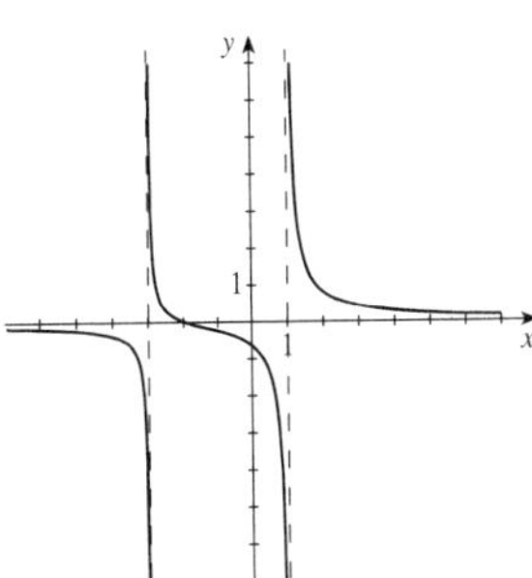


**41.** $s(x) = \dfrac{6}{x^2 - 5x - 6}$. When $x = 0$, $y = \dfrac{6}{-6} = -1$, so the $y$-intercept is $-1$. Since the numerator is never zero, there is no $x$-intercept. The vertical asymptotes occur when $x^2 - 5x - 6 = (x+1)(x-6) \Leftrightarrow x = -1$ and $x = 6$, and because the degree of the numerator is less less than the degree of the denominator, the horizontal asymptote is $y = 0$.

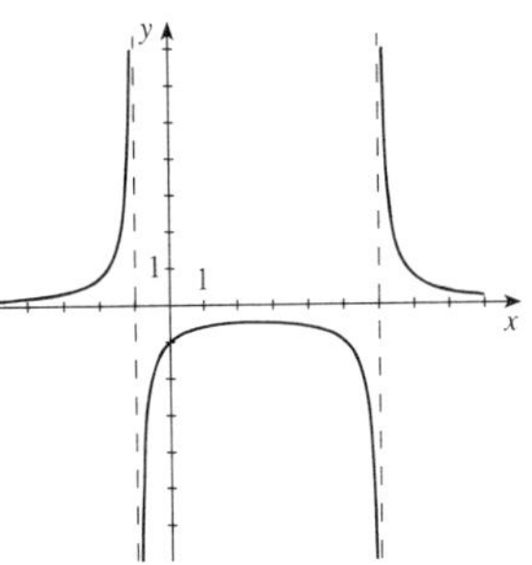


**42.** $s(x) = \dfrac{2x-4}{x^2+x-2} = \dfrac{2(x-2)}{(x-1)(x+2)}$. When $x = 0$, $y = 2$, so the $y$-intercept is 2. When $y = 0$, we have $2x - 4 = 0 \quad \Leftrightarrow \quad x = 2$, so the $x$-intercept is 2. A vertical asymptote occurs when $(x-1)(x+2) = 0 \Leftrightarrow x = 1$ and $x = -2$. Because the degree of the denominator is greater than the degree of the numerator, the horizontal asymptote is $y = 0$.

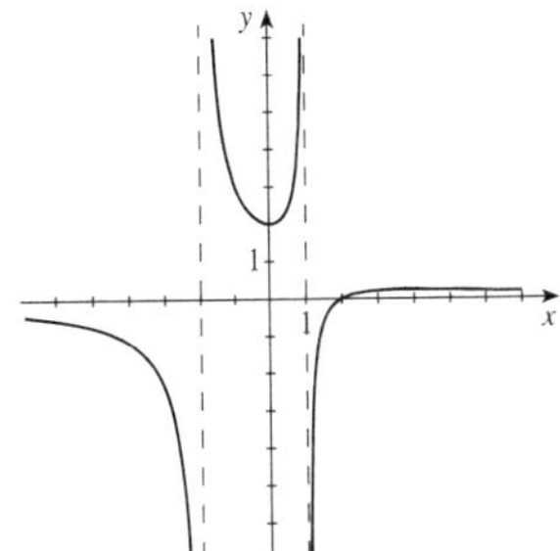


**43.** $t(x) = \dfrac{3x+6}{x^2+2x-8}$. When $x = 0$, $y = \dfrac{6}{-8} = -\dfrac{3}{4}$, so the $y$-intercept is $-\frac{3}{4}$. When $y = 0$, $3x + 6 = 0 \quad \Leftrightarrow \quad x = -2$, so the $x$-intercept is $-2$. The vertical asymptotes occur when $x^2 + 2x - 8 = (x-2)(x+4) = 0 \quad \Leftrightarrow \quad x = 2$ and $x = -4$. Since the degree of the numerator is less than the degree of the denominator, the horizontal asymptote is $y = 0$.

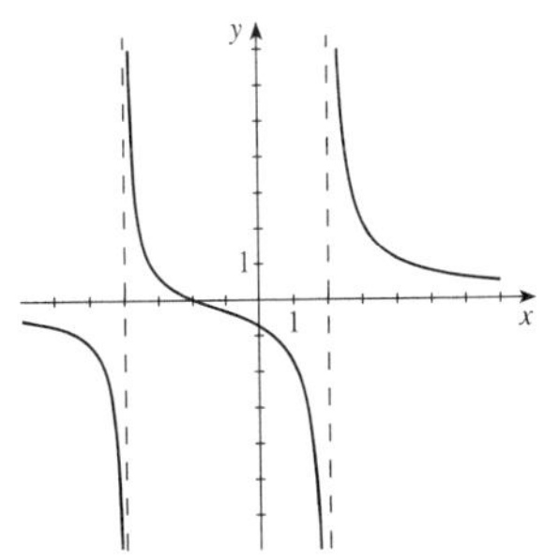


**44.** $t(x) = \dfrac{x-2}{x^2-4x} = \dfrac{x-2}{x(x-4)}$. Since $x = 0$ is not in the domain of $t(x)$, there is no $y$-intercept. When $y = 0$, we have $x - 2 = 0 \Leftrightarrow x = 2$, so the $x$-intercept is 2. A vertical asymptote occurs when $x(x-4) = 0 \quad \Leftrightarrow \quad x = 0$ and $x = 4$. Because the degree of the denominator is greater than the degree of the numerator, the horizontal asymptote is $y = 0$.

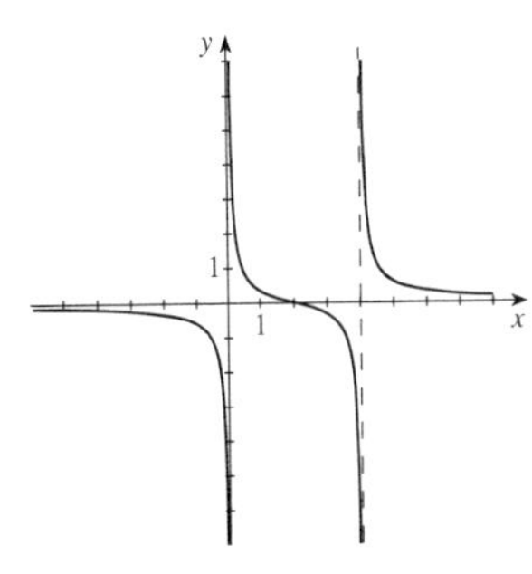

**45.** $r(x) = \dfrac{(x-1)(x+2)}{(x+1)(x-3)}$. When $x = 0$, $y = \frac{2}{3}$, so the $y$-intercept is $\frac{2}{3}$. When $y = 0$, $(x-1)(x+2) = 0 \quad \Rightarrow \quad x = -2, 1$, so, the $x$-intercepts are $-2$ and 1. The vertical asymptotes are $x = -1$ and $x = 3$, and because the degree of the numerator and denominator are the same the horizontal asymptote is $y = \frac{1}{1} = 1$.

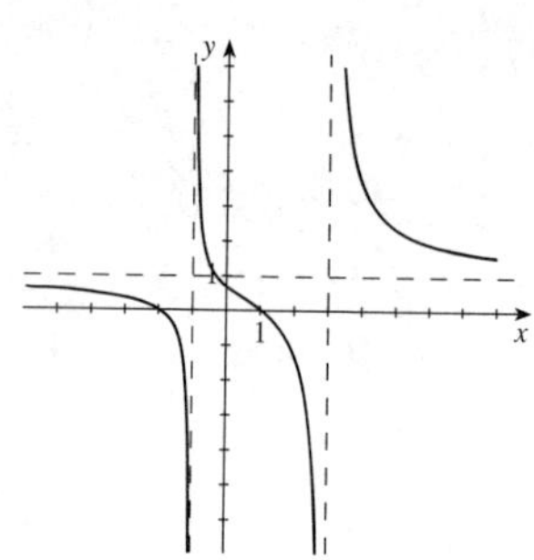


**46.** $r(x) = \dfrac{2x(x+2)}{(x-1)(x-4)}$. When $x = 0$, we have $y = 0$, so the graph passes through the origin. Also, when $y = 0$, we have $2x(x+2) = 0 \Leftrightarrow x = 0, -2$, so the $x$-intercepts are 0 and $-2$. There are two vertical asymptotes, $x = 1$ and $x = 4$. Because the degree of the denominator and numerator are the same, the horizontal asymptote is $y = \frac{2}{1} = 2$.

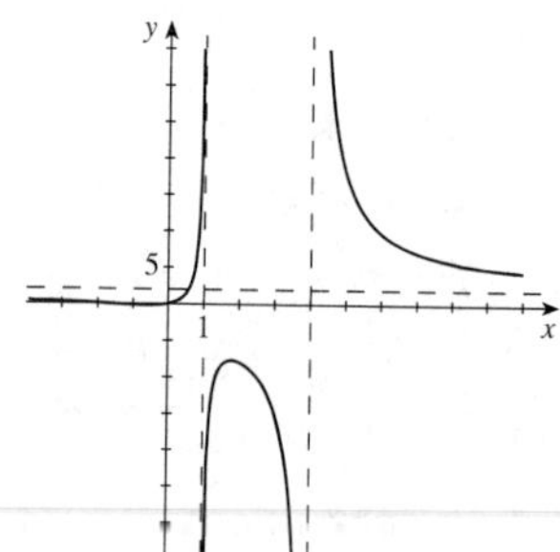


**47.** $r(x) = \dfrac{x^2 - 2x + 1}{x^2 + 2x + 1} = \dfrac{(x-1)^2}{(x+1)^2} = \left(\dfrac{x-1}{x+1}\right)^2$. When $x = 0$, $y = 1$, so the $y$-intercept is 1. When $y = 0$, $x = 1$, so the $x$-intercept is 1 . A vertical asymptote occurs at $x + 1 = 0 \Leftrightarrow x = -1$. Because the degree of the numerator and denominator are the same the horizontal asymptote is $y = \frac{1}{1} = 1$.

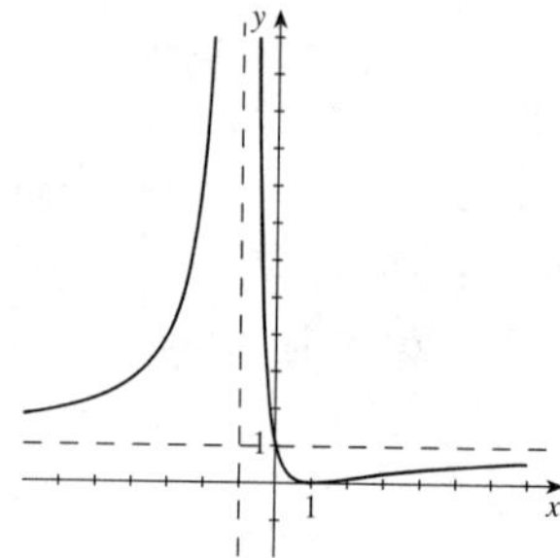


**48.** $r(x) = \dfrac{4x^2}{x^2 - 2x - 3} = \dfrac{4x^2}{(x-3)(x+1)}$. When $x = 0$, we have $y = 0$, so the graph passes through the origin. Vertical asymptotes occur at $x = -1$ and $x = 3$. Because the degree of the denominator and numerator are the same, the horizontal asymptote is $y = \frac{4}{1} = 4$.

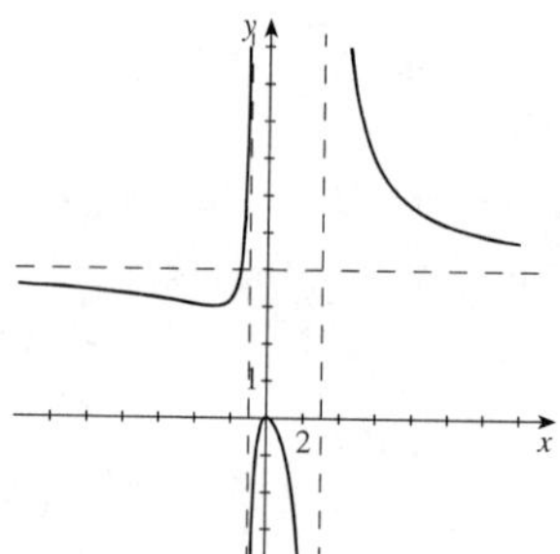


**49.** $r(x) = \dfrac{2x^2 + 10x - 12}{x^2 + x - 6} = \dfrac{2(x-1)(x+6)}{(x-2)(x+3)}$. When $x = 0$, $y = \dfrac{2(-1)(6)}{(-2)(3)} = 2$, so the $y$-intercept is 2. When $y = 0$, $2(x-1)(x+6) = 0$ $\Rightarrow \quad x = -6, 1$, so the $x$-intercepts are $-6$ and 1. Vertical asymptotes occur when $(x-2)(x+3) = 0 \Leftrightarrow x = -3$ or $x = 2$. Because the degree of the numerator and denominator are the same the horizontal asymptote is $y = \frac{2}{1} = 2$.

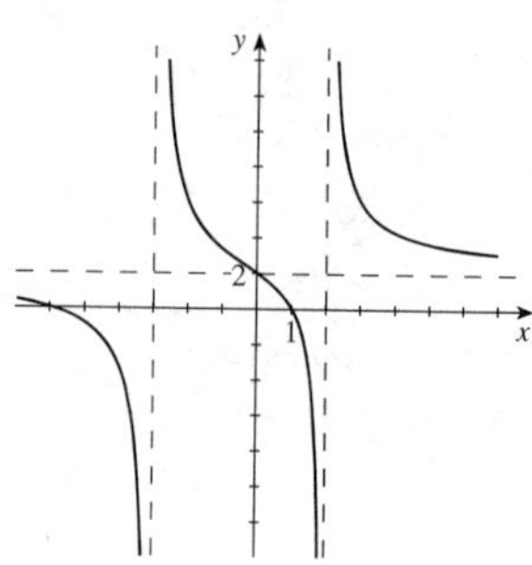

**50.** $r(x) = \dfrac{2x^2 + 2x - 4}{x^2 + x} = \dfrac{2(x+2)(x-1)}{x(x+1)}$. Vertical asymptotes occur at $x = 0$ and $x = -1$. Since $x$ cannot equal zero, there is no $y$-intercept. When $y = 0$, we have $x = -2$ or $1$, so the $x$-intercepts are $-2$ and $1$. Because the degree of the denominator and numerator are the same, the horizontal asymptote is $y = \frac{2}{1} = 2$.

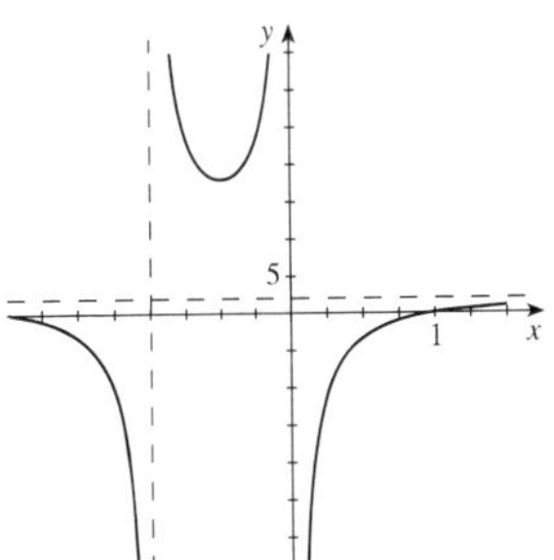


**51.** $y = \dfrac{x^2 - x - 6}{x^2 + 3x} = \dfrac{(x-3)(x+2)}{x(x+3)}$. The $x$-intercept occurs when $y = 0 \Leftrightarrow$ $(x-3)(x+2) = 0 \quad \Rightarrow \quad x = -2, 3$, so the $x$-intercepts are $-2$ and $3$. There is no $y$-intercept because $y$ is undefined when $x = 0$. The vertical asymptotes are $x = 0$ and $x = -3$. Because the degree of the numerator and denominator are the same, the horizontal asymptotes is $y = \frac{1}{1} = 1$.

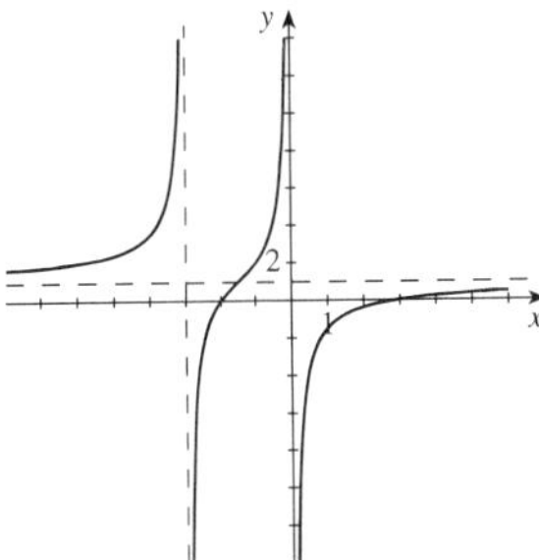


**52.** $r(x) = \dfrac{x^2 + 3x}{x^2 - x - 6} = \dfrac{x(x+3)}{(x-3)(x+2)}$. When $x = 0$, we have $y = 0$, so the graph passes through the origin. When $y = 0$, we have $x = 0$ or $-3$, so the $x$-intercepts are $0$ and $-3$. Vertical asymptotes occur at $x = -2$ and $x = 3$. Because the degree of the denominator and numerator are the same, the horizontal asymptote is $y = \frac{1}{1} = 1$.

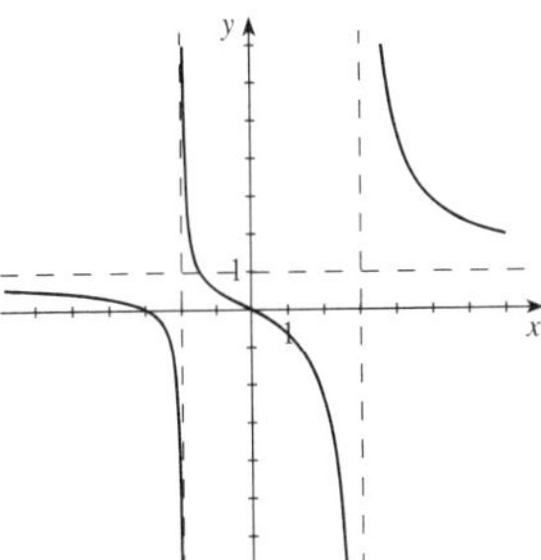


**53.** $r(x) = \dfrac{3x^2 + 6}{x^2 - 2x - 3} = \dfrac{3(x^2 + 2)}{(x-3)(x+1)}$. When $x = 0$, $y = -2$, so the $y$-intercept is $-2$. Since the numerator can never equal zero, there is no $x$-intercept. Vertical asymptotes occur when $x = -1, 3$. Because the degree of the numerator and denominator are the same, the horizontal asymptote is.$y = \frac{3}{1} = 3$.

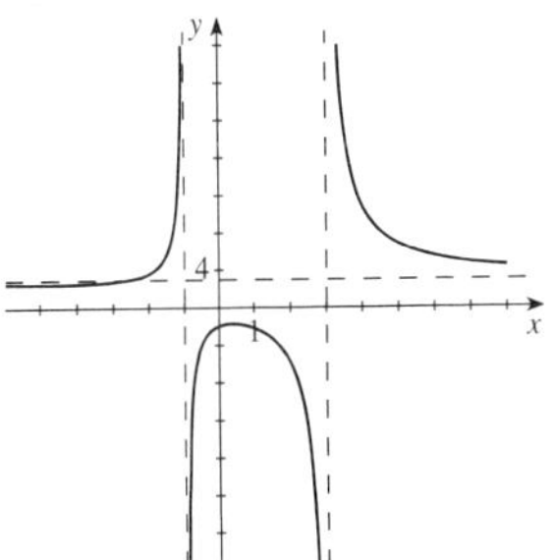


**54.** $r(x) = \dfrac{5x^2 + 5}{x^2 + 4x + 4} = \dfrac{5(x^2 + 1)}{(x+2)^2}$. When $x = 0$, we have $y = \dfrac{5}{4}$, so the $y$-intercept is $\frac{5}{4}$. Since $x^2 + 1 > 0$ for all real $x$, $y$ never equals zero, and there is no $x$-intercept. The vertical asymptote is $x = -2$. Because the degree of the denominator and numerator are the same, the horizontal asymptote occurs at $y = \frac{5}{1} = 5$.

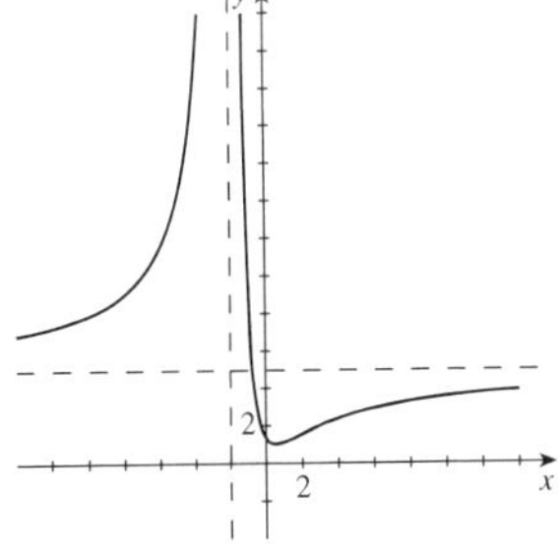

**55.** $s(x) = \dfrac{x^2 - 2x + 1}{x^3 - 3x^2} = \dfrac{(x-1)^2}{x^2(x-3)}$. Since $x = 0$ is not in the domain of $s(x)$, there is no $y$-intercept. The $x$-intercept occurs when $y = 0 \Leftrightarrow$ $x^2 - 2x + 1 = (x-1)^2 = 0 \quad \Rightarrow \quad x = 1$, so the $x$-intercept is 1. Vertical asymptotes occur when $x = 0, 3$. Since the degree of the numerator is less than the degree of the denominator, the horizontal asymptote is $y = 0$.

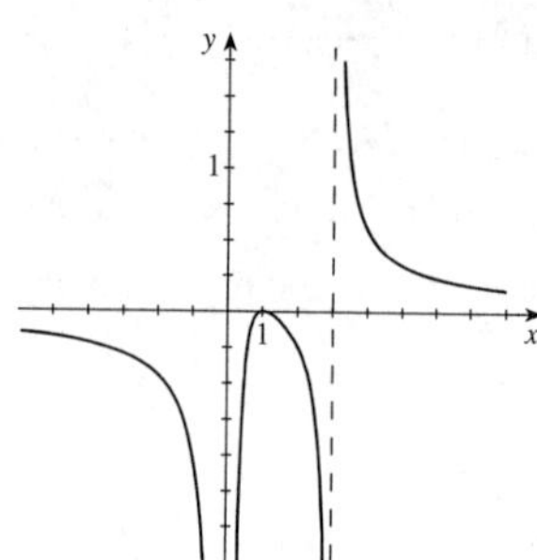


**56.** $r(x) = \dfrac{x^3 - x^2}{x^3 - 3x - 2} = \dfrac{x^2(x-1)}{x^3 - 3x - 2}$. When $x = 0$, we have $y = 0$, so the $y$-intercept is 0. When $y = 0$, we have $x^2(x-1) = 0$, so the $x$-intercepts are 0 and 1. Vertical asymptotes occur when $x^3 - 3x - 2 = 0$. Since $x^3 - 3x - 2 = 0$ when $x = 2$, we can factor $(x-2)(x+1)^2 = 0$, so the vertical asymptotes occur at $x = 2$ and $x = -1$. Because the degree of the denominator and numerator are the same, the horizontal asymptote is $y = \frac{1}{1} = 1$.

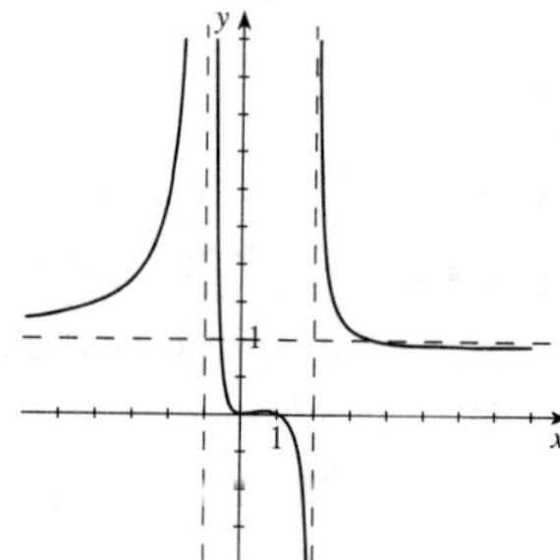


**57.** $r(x) = \dfrac{x^2}{x-2}$. When $x = 0$, $y = 0$, so the graph passes through the origin. There is a vertical asymptote when $x - 2 = 0 \Leftrightarrow x = 2$, with $y \to \infty$ as $x \to 2^+$, and $y \to -\infty$ as $x \to 2^-$. Because the degree of the numerator is greater than the degree of the denominator, there is no horizontal asymptotes. By using long division, we see that $y = x + 2 + \dfrac{4}{x-2}$, so $y = x + 2$ is a slant asymptote.

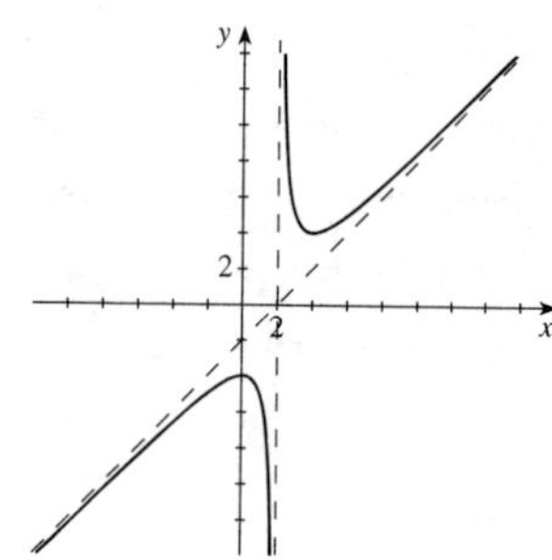


**58.** $r(x) = \dfrac{x^2 + 2x}{x-1} = \dfrac{x(x+2)}{x-1}$. When $x = 0$, we have $y = 0$, so the graph passes through the origin. Also, when $y = 0$, we have $x = 0$ or $-2$, so the $x$-intercepts are $-2$ and 0. The vertical asymptote is $x = 1$. There is no horizontal asymptote, and the line $y = x + 3$ is a slant asymptote because by long division, we have $y = x + 3 + \dfrac{2}{x-1}$.

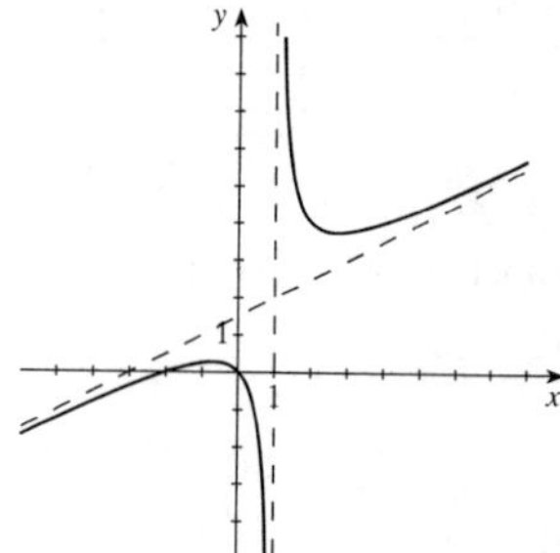


**59.** $r(x) = \dfrac{x^2 - 2x - 8}{x} = \dfrac{(x-4)(x+2)}{x}$. The vertical asymptote is $x = 0$, thus, there is no $y$-intercept. If $y = 0$, then $(x-4)(x+2) = 0 \Rightarrow x = -2, 4$, so the $x$-intercepts are $-2$ and 4. Because the degree of the numerator is greater than the degree of the denominator, there are no horizontal asymptotes. By using long division, we see that $y = x - 2 - \dfrac{8}{x}$, so $y = x - 2$ is a slant asymptote.

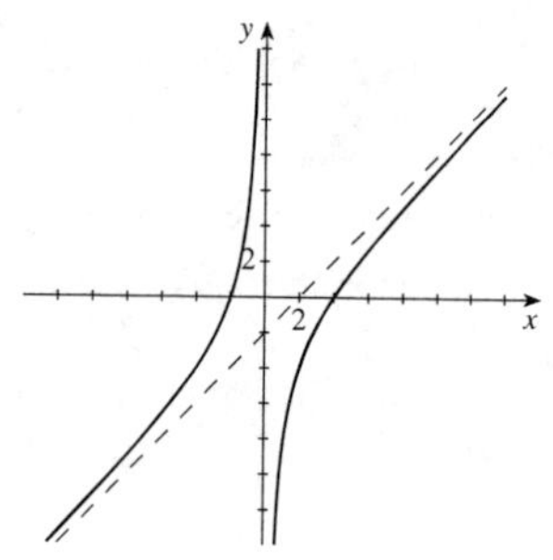

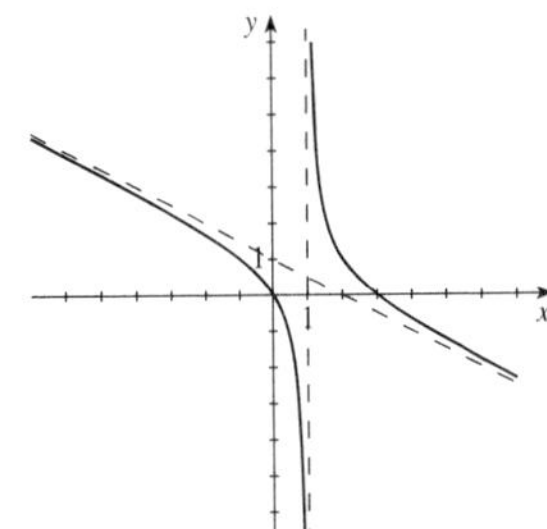


**60.** $r(x) = \dfrac{3x - x^2}{2x - 2} = \dfrac{x(3 - x)}{2(x - 1)}$. When $x = 0$, we have $y = 0$, so the graph passes through the origin. Also, when $y = 0$, we have $x = 0$ or $x = 3$, so the $x$-intercepts are 0 and 3. The vertical asymptote is $x = 1$. There is no horizontal asymptote, and the line $y = -\frac{1}{2}x + 1$ is a slant asymptote because by long division we have $y = -\frac{1}{2}x + 1 + \dfrac{1}{x - 1}$.

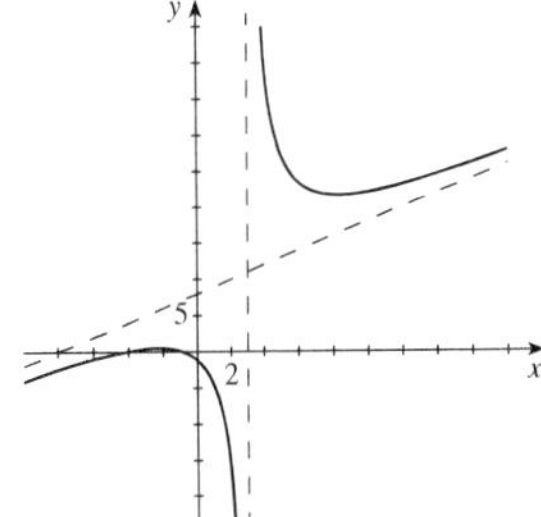


**61.** $r(x) = \dfrac{x^2 + 5x + 4}{x - 3} = \dfrac{(x + 4)(x + 1)}{x - 3}$. When $x = 0$, $y = -\frac{4}{3}$, so the $y$-intercept is $-\frac{4}{3}$. When $y = 0$, $(x + 4)(x + 1) = 0 \quad \Leftrightarrow \quad x = -4, -1$, so the two $x$-intercepts are $-4$ and $-1$. A vertical asymptote occurs when $x = 3$, with $y \to \infty$ as $x \to 3^+$, and $y \to -\infty$ as $x \to 3^-$. Using long division, we see that $y = x + 8 + \dfrac{28}{x - 3}$, so $y = x + 8$ is a slant asymptote.

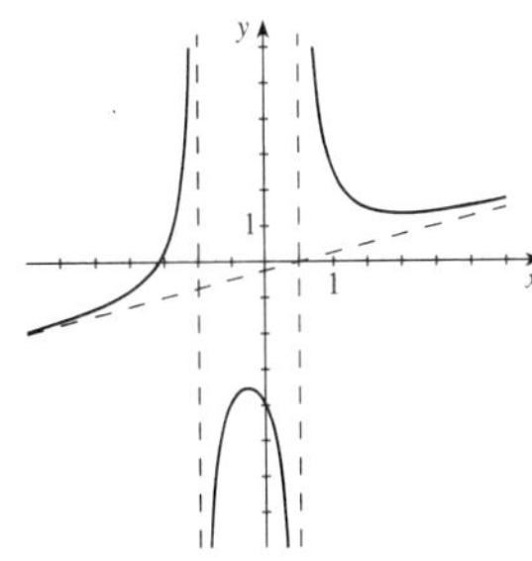


**62.** $r(x) = \dfrac{x^3 + 4}{2x^2 + x - 1} = \dfrac{x^3 + 4}{(2x - 1)(x + 1)}$. When $x = 0$, we have $y = \dfrac{0 + 4}{0 + 0 - 1} = -4$, so the $y$-intercept is $-4$. Since $x^3 + 4 = 0 \quad \Rightarrow$ $x = -\sqrt[3]{4}$, the $x$-intercept is $x = -\sqrt[3]{4}$. There are vertical asymptotes where $(2x - 1)(x + 1) = 0 \Rightarrow x = \frac{1}{2}$ or $x = -1$. Since the degree of the numerator is greater than the degree of the denominator, there is no horizontal asymptote. By long division, we have $y = \frac{1}{2}x - \frac{1}{4} + \dfrac{\frac{3}{4}x + \frac{15}{4}}{2x^2 - x - 1}$, so the line $y = \frac{1}{2}x - \frac{1}{4}$ is a slant asymptote.

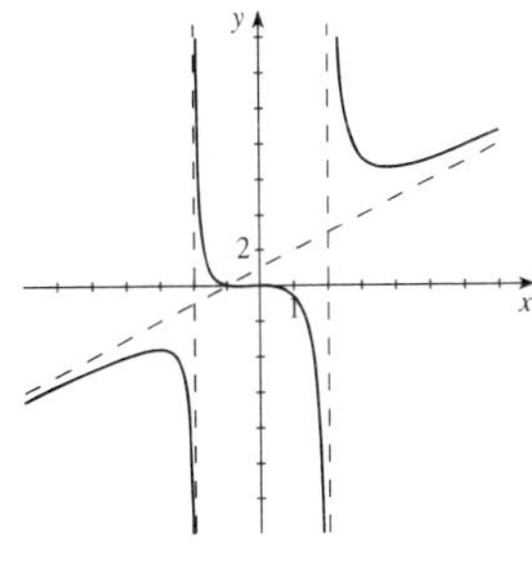


**63.** $r(x) = \dfrac{x^3 + x^2}{x^2 - 4} = \dfrac{x^2(x + 1)}{(x - 2)(x + 2)}$. When $x = 0$, $y = 0$, so the graph passes through the origin. Moreover, when $y = 0$, we have $x^2(x + 1) = 0 \quad \Rightarrow$ $x = 0, -1$, so the $x$-intercepts are 0 and $-1$. Vertical asymptotes occur when $x = \pm 2$; as $x \to \pm 2^-$, $y = -\infty$ and as $x \to \pm 2^+$, $y \to \infty$. Because the degree of the numerator is greater than the degree of the denominator, there is no horizontal asymptote. Using long division, we see that $y = x + 1 + \dfrac{4x + 4}{x^2 - 4}$, so $y = x + 1$ is a slant asymptote.

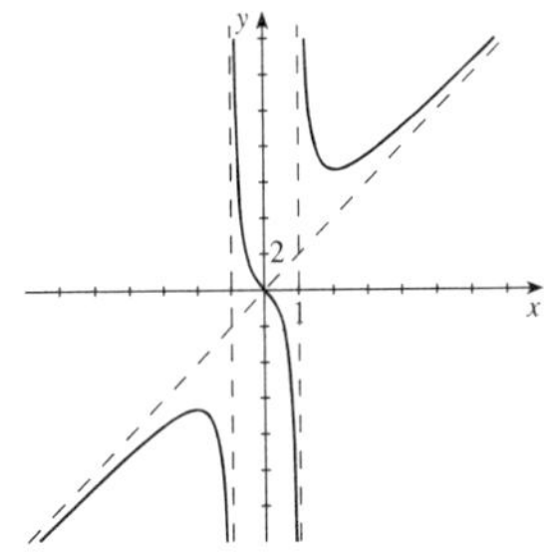


**64.** $r(x) = \dfrac{2x^3 + 2x}{x^2 - 1} = \dfrac{2x(x^2 + 1)}{(x - 1)(x + 1)}$. When $x = 0$, we have $y = 0$, so the graph passes through the origin. Also, note that $x^2 + 1 > 0$, for all real $x$, so the only $x$-intercept is 0. There are two vertical asymptotes at $x = -1$ and $x = 1$. There is no horizontal asymptote, and the line $y = 2x$ is a slant asymptote because by long division, we have $y = 2x + \dfrac{4x}{x^2 - 1}$.

**65.** $f(x) = \dfrac{2x^2 + 6x + 6}{x + 3}$, $g(x) = 2x$. $f$ has vertical asymptote $x = -3$.

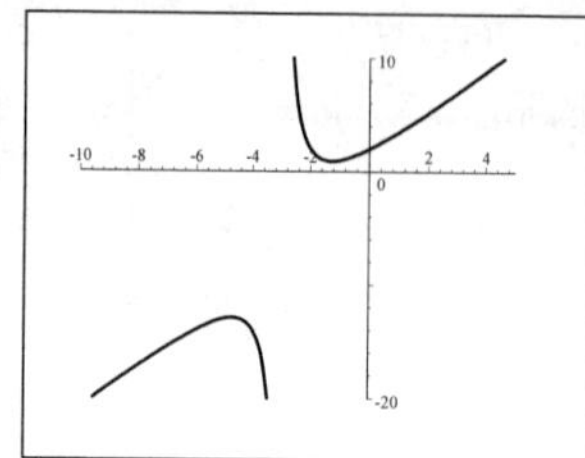

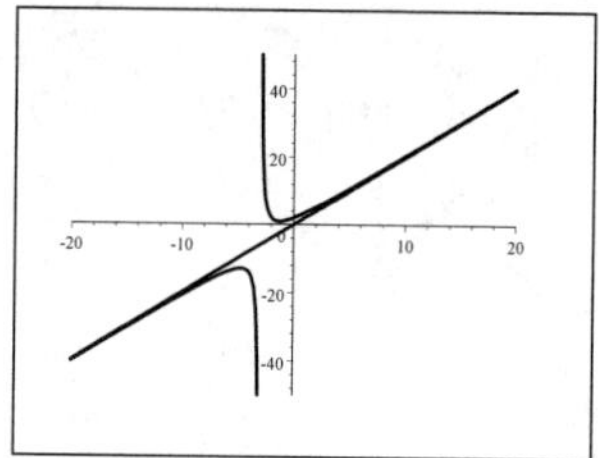

**66.** $f(x) = \dfrac{-x^3 + 6x^2 - 5}{x^2 - 2x}$, $g(x) = -x + 4$. $f$ has vertical asymptotes $x = 0$ and $x = 2$.

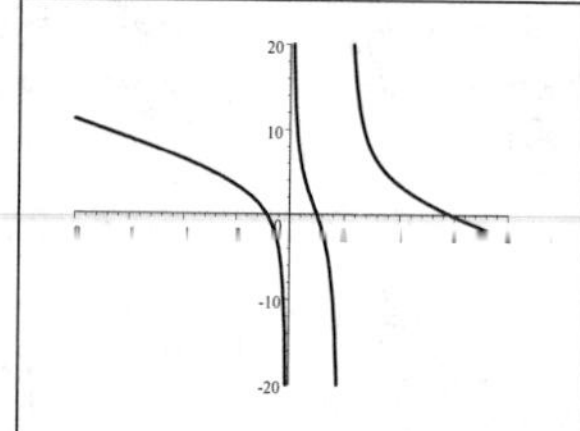

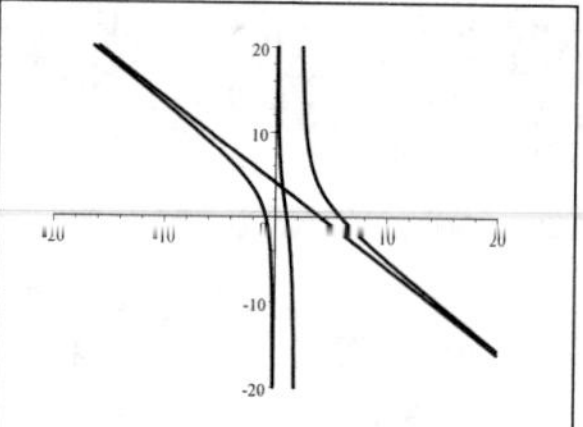

**67.** $f(x) = \dfrac{x^3 - 2x^2 + 16}{x - 2}$, $g(x) = x^2$. $f$ has vertical asymptote $x = 2$.

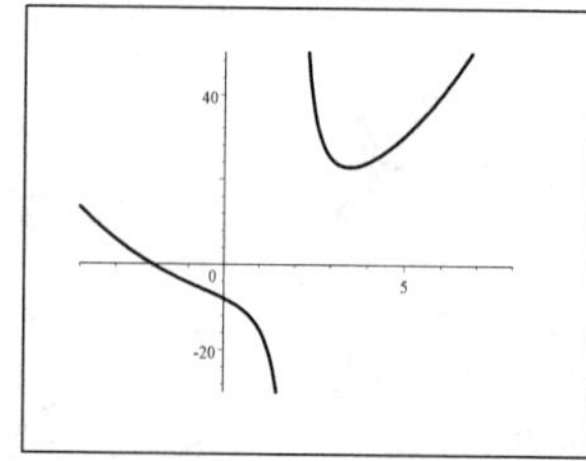

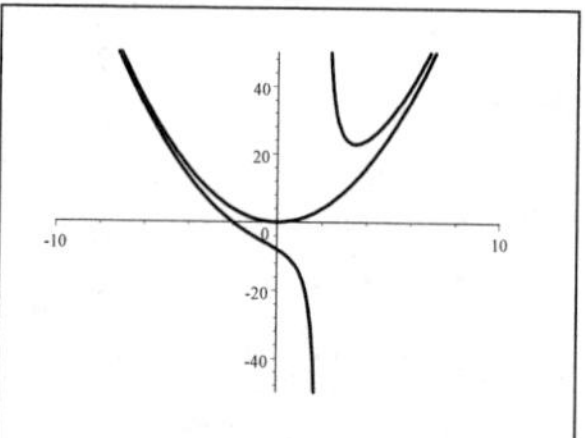

**68.** $f(x) = \dfrac{-x^4 + 2x^3 - 2x}{(x - 1)^2}$, $g(x) = 1 - x^2$. $f$ has vertical asymptote $x = 1$.

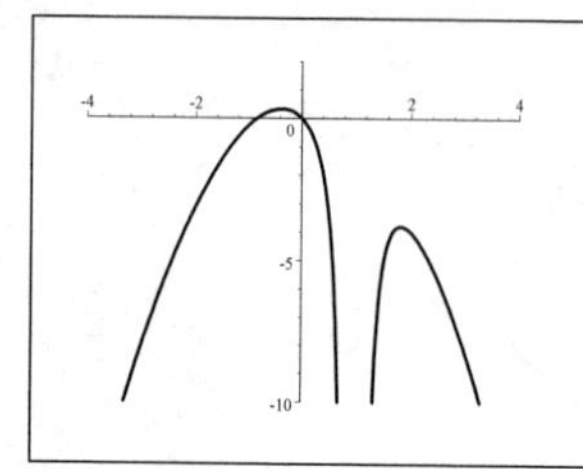

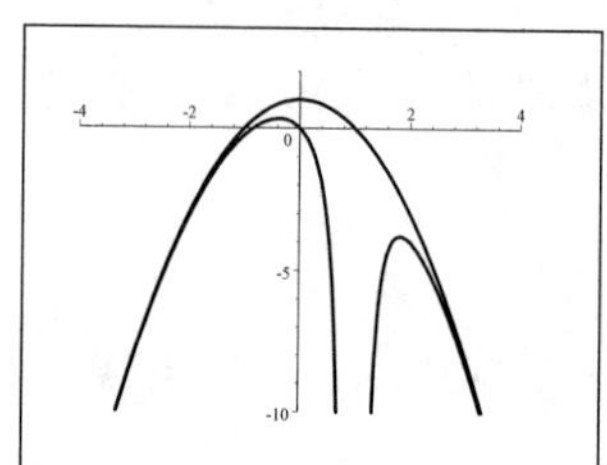

**69.** $f(x) = \dfrac{2x^2 - 5x}{2x+3}$ has vertical asymptote $x = -1.5$, $x$-intercepts 0 and 2.5, $y$-intercept 0, local maximum $(-3.9, -10.4)$, and local minimum $(0.9, -0.6)$. Using long division, we get $f(x) = x - 4 + \dfrac{12}{2x+3}$. From the graph, we see that the end behavior of $f(x)$ is like the end behavior of $g(x) = x - 4$.

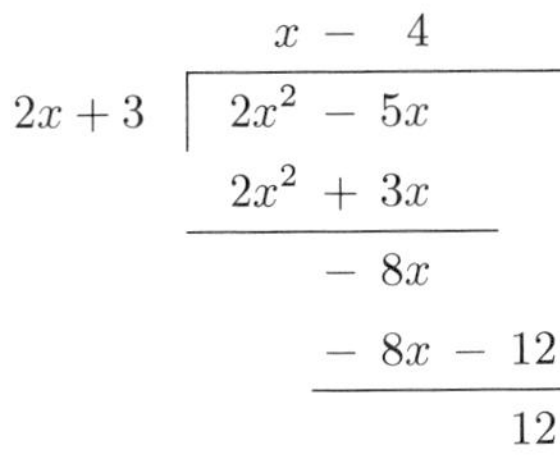

$$\begin{array}{r|l} & x - 4 \\ \hline 2x+3 & 2x^2 - 5x \\ & 2x^2 + 3x \\ \hline & \quad -8x \\ & \quad -8x - 12 \\ \hline & \qquad\quad 12 \end{array}$$

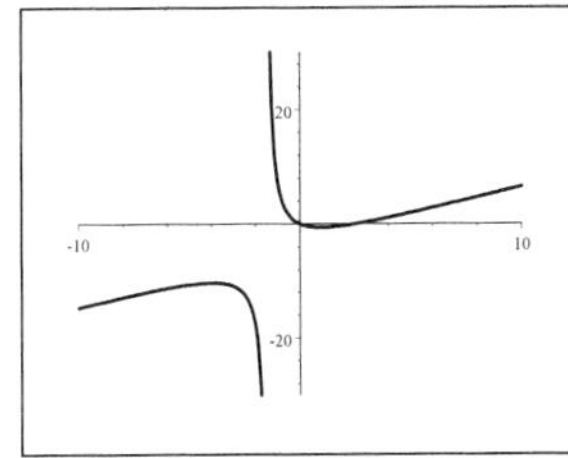


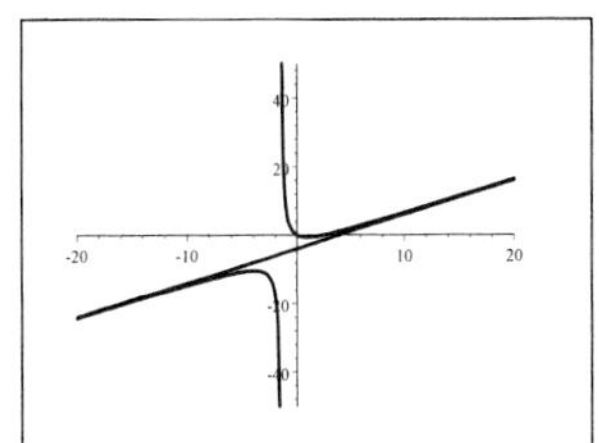


**70.** $f(x) = \dfrac{x^4 - 3x^3 + x^2 - 3x + 3}{x^2 - 3x}$ has vertical asymptotes are $x = 0$, $x = 3$, $x$-intercept 0.82, and no $y$-intercept. The local minima are $(-0.80, 2.63)$ and $(3.38, 14.76)$. The local maximum is $(2.56, 4.88)$. By using long division, we see that $f(x) = x^2 + 1 + \dfrac{3}{x^2 - 3x}$. From the second graph, we see that the end behavior of $f(x)$ is the same as the end behavior of $g(x) = x^2 + 1$.

$$\begin{array}{r|l} & x^2 \qquad\quad + 1 \\ \hline x^2 - 3x & x^4 - 3x^3 + x^2 - 3x + 3 \\ & x^4 - 3x^3 \\ \hline & \qquad 0x^3 + x^2 - 3x \\ & \qquad\qquad x^2 - 3x \\ \hline & \qquad\qquad\qquad\quad 3 \end{array}$$

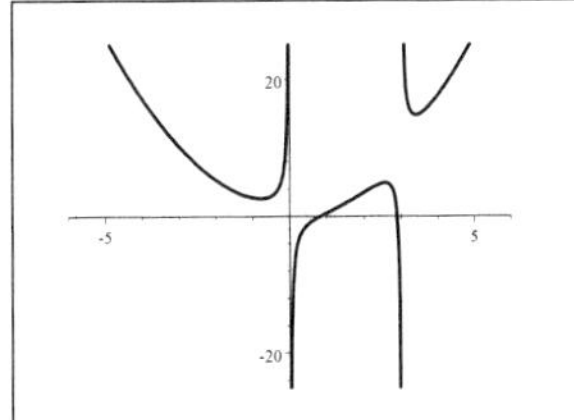


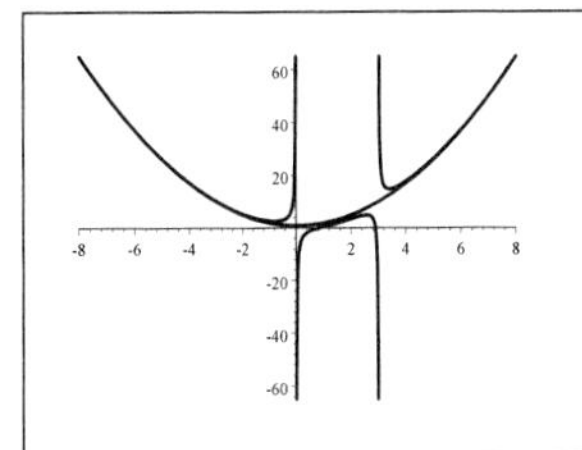


**71.** $f(x) = \dfrac{x^5}{x^3 - 1}$ has vertical asymptote $x = 1$, $x$-intercept 0, $y$-intercept 0, and local minimum $(1.4, 3.1)$. Thus $y = x^2 + \dfrac{x^2}{x^3 - 1}$. From the graph we see that the end behavior of $f(x)$ is like the end behavior of $g(x) = x^2$.

$$\begin{array}{r|l} & x^2 \\ \hline x^3 - 1 & x^5 \\ & x^5 - x^2 \\ \hline & \qquad x^2 \end{array}$$

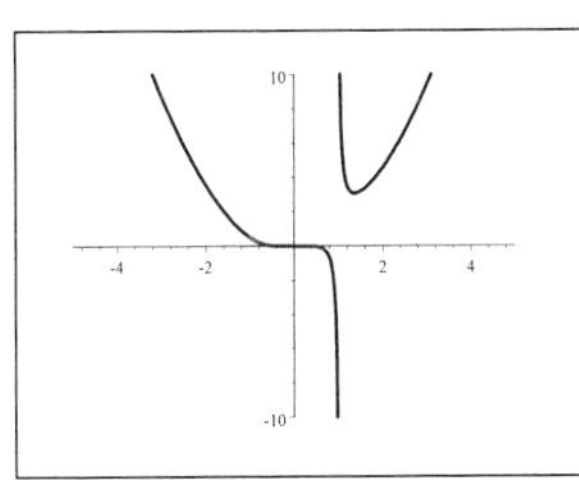


Graph of $f$

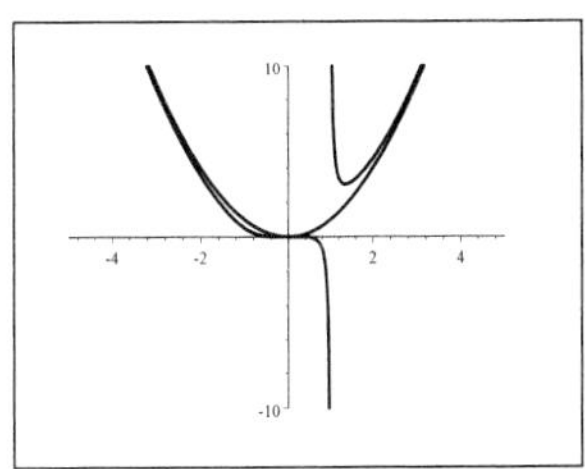


Graph of $f$ and $g$

**72.** $f(x) = \dfrac{x^4}{x^2 - 2}$ has vertical asymptotes $x = \pm 1.41$, $x$-intercept 0, and $y$-intercept 0. The local maximum is $(0, 0)$. The local minima are $(-2, 8)$ and $(2, 8)$. By using long division, we see that $f(x) = x^2 + 2 + \dfrac{4}{x^2 - 2}$. From the second graph, we see that the end behavior of $f(x)$ is the same as the end behavior of $g(x) = x^2 + 2$.

$$\begin{array}{r|l} & x^2 \qquad\quad + 2 \\ \hline x^2 - 2 & x^4 + 0x^3 + 0x^2 + 0x + 0 \\ & x^4 \qquad\quad - 2x^2 \\ \hline & \qquad\qquad 2x^2 \\ & \qquad\qquad 2x^2 \qquad - 4 \\ \hline & \qquad\qquad\qquad\qquad 4 \end{array}$$

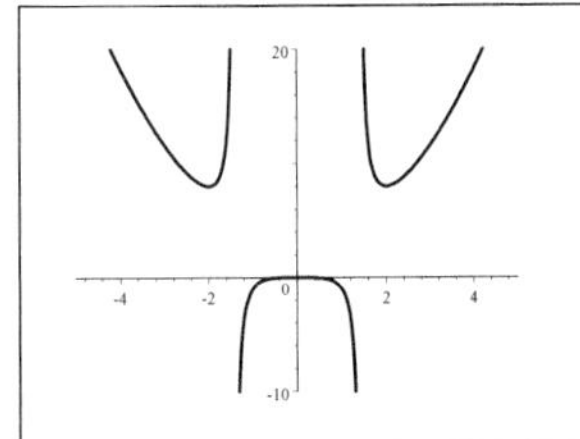


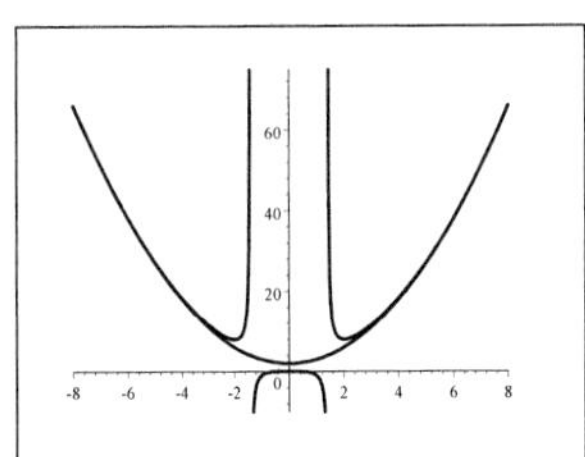

**73.** $f(x) = \dfrac{x^4 - 3x^3 + 6}{x - 3}$ has vertical asymptote $x = 3$, $x$-intercepts 1.6 and 2.7, $y$-intercept $-2$, local maxima $(-0.4, -1.8)$ and $(2.4, 3.8)$, and local minima $(0.6, -2.3)$ and $(3.4, 54.3)$. Thus $y = x^3 + \dfrac{6}{x-3}$. From the graphs, we see that the end behavior of $f(x)$ is like the end behavior of $g(x) = x^3$.

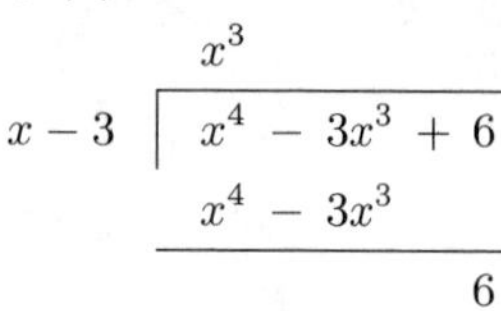


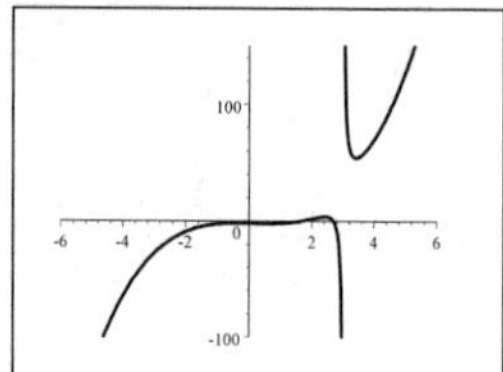

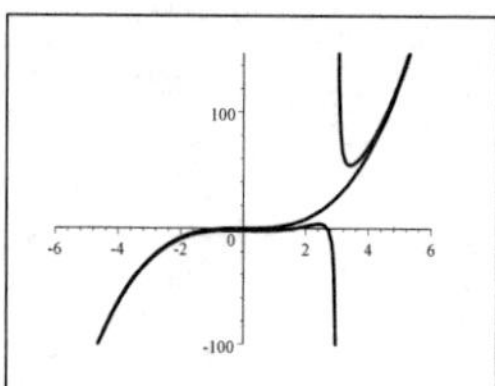

**74.** $r(x) = \dfrac{4 + x^2 - x^4}{x^2 - 1} = \dfrac{-x^4 + x^2 + 4}{(x-1)(x+1)}$ has vertical asymptotes $x \approx \pm 1.41$, $x$-intercepts $\pm 1.6$, and $y$-intercept $-1$. The local maximum is $(0, -0.4)$ and there is no local minimum.

Thus $y = -x^2 + \dfrac{6}{x^2 - 1}$. From the graphs, we see that the end behavior of $f(x)$ is like the end behavior of $g(x) = -x^2$.

$$\begin{array}{r|l} & -x^2 \\ \hline x^2 - 1 & -x^4 + 0x^3 + x^2 + 0x + 4 \\ & -x^4 \quad\quad + x^2 \\ \hline & \quad\quad 0 \quad\quad + 4 \end{array}$$

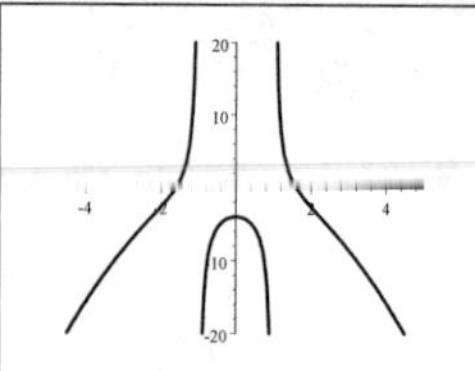

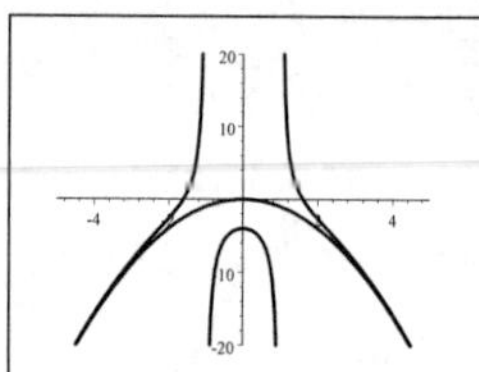

**75. (a)**

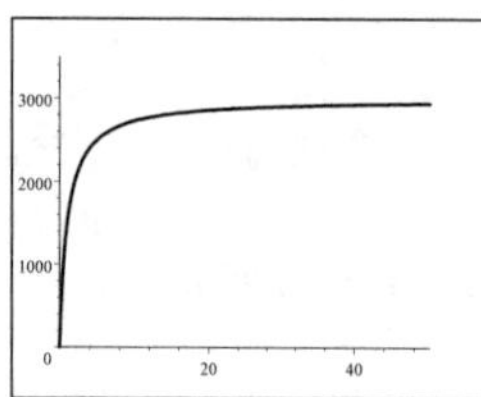

**(b)** $p(t) = \dfrac{3000t}{t+1} = 3000 - \dfrac{3000}{t+1}$. So as $t \to \infty$, we have $p(t) \to 3000$.

**76. (a)**

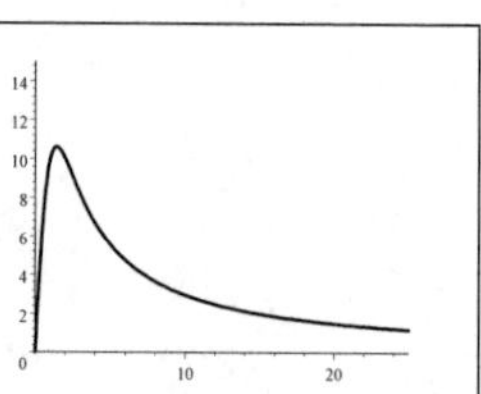

**(b)** $c(t) = \dfrac{30t}{t^2 + 2}$. Since the degree of the denominator is larger than the degree of the numerator, $c(t) \to 0$ as $t \to \infty$.

**77.** $c(t) = \dfrac{5t}{t^2 + 1}$

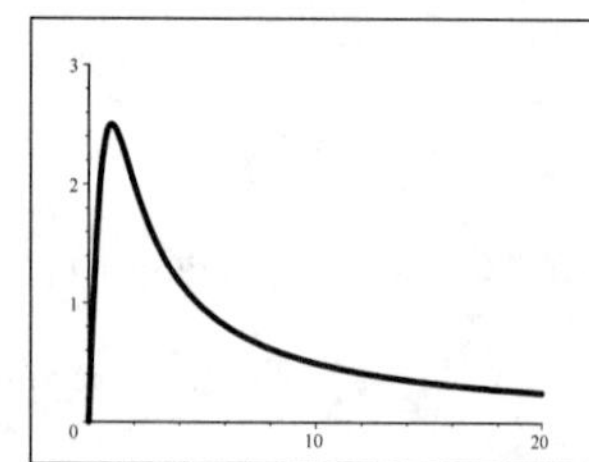

**(a)** The highest concentration of drug is 2.50 mg/L, and it is reached 1 hour after the drug is administered.

**(b)** The concentration of the drug in the bloodstream goes to 0.

**(c)** From the first viewing rectangle, we see that an approximate solution is near $t = 15$. Thus we graph $y = \dfrac{5t}{t^2 + 1}$ and $y = 0.3$ in the viewing rectangle $[14, 18]$ by $[0, 0.5]$. So it takes about 16.61 hours for the concentration to drop below 0.3 mg/L.

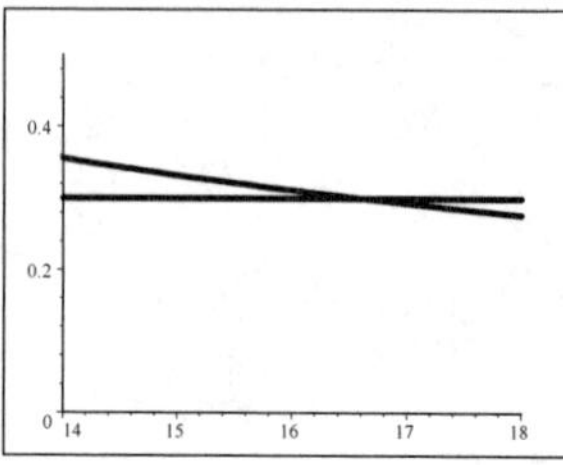

**78.** Substituting for $R$ and $g$, we have $h(v) = \dfrac{(6.4 \times 10^6) v^2}{2(9.8)(6.4 \times 10^6) - v^2}$. The vertical asymptote is $v \approx 11{,}000$, and it represents the escape velocity from the earth's gravitational pull: 11,000 m/s $\approx$ 1900 mi/h.

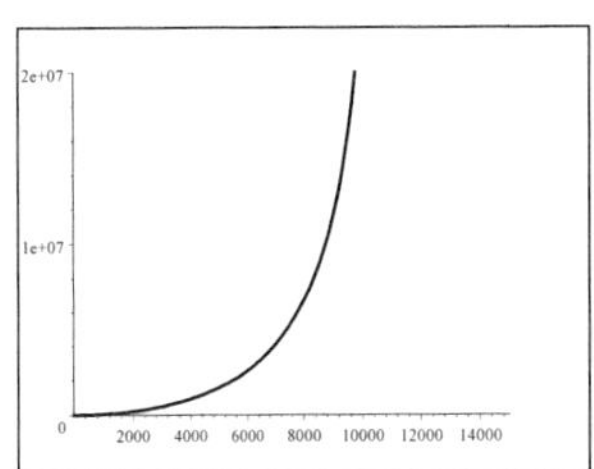


**79.** $P(v) = P_0\left(\dfrac{s_0}{s_0 - v}\right) \quad \Rightarrow \quad P(v) = 440\left(\dfrac{332}{332 - v}\right)$

If the speed of the train approaches the speed of sound, the pitch of the whistle becomes very loud. This would be experienced as a "sonic boom"— an effect seldom heard with trains.

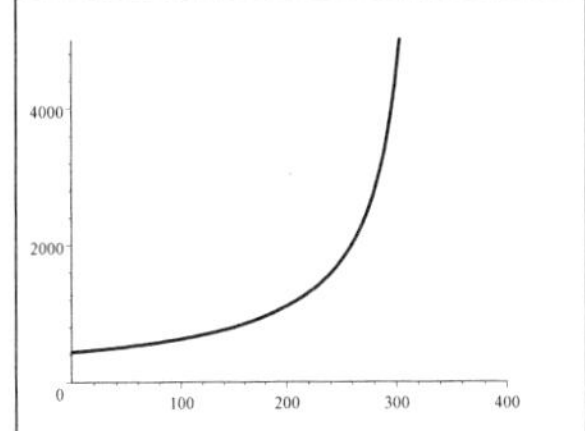


**80. (a)** $\dfrac{1}{x} + \dfrac{1}{y} = \dfrac{1}{F} \quad \Leftrightarrow \quad \dfrac{1}{y} = \dfrac{1}{F} - \dfrac{1}{x} \quad \Leftrightarrow \quad \dfrac{1}{y} = \dfrac{x - F}{xF} \quad \Leftrightarrow$

$y = \dfrac{xF}{x - F}$. Using $F = 55$, we get $y = \dfrac{55x}{x - 55}$. Since $y \geq 0$, we use the viewing rectangle $[0, 1000]$ by $[0, 250]$.

**(b)** $y$ approaches 55 millimeters.

**(c)** $y$ approaches $\infty$.

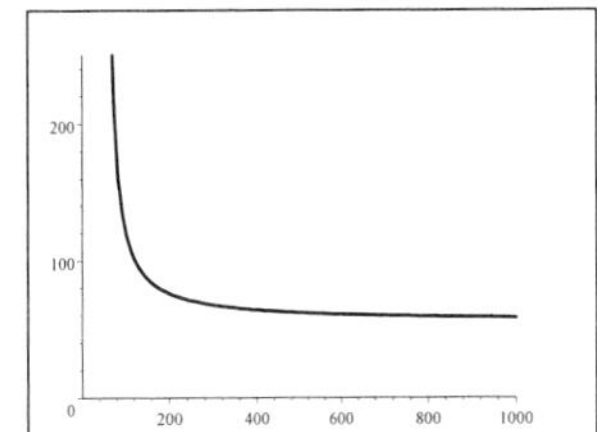


**81.** Vertical asymptote $x = 3$: $p(x) = \dfrac{1}{x - 3}$. Vertical asymptote $x = 3$ and horizontal asymptote $y = 2$: $r(x) = \dfrac{2x}{x - 3}$. Vertical asymptotes $x = 1$ and $x = -1$, horizontal asymptote 0, and $x$-intercept 4: $q(x) = \dfrac{x - 4}{(x - 1)(x + 1)}$. Of course, other answers are possible.

**82.** $r(x) = \dfrac{x^6 + 10}{x^4 + 8x^2 + 15}$ has no $x$-intercept since the numerator has no real roots. Likewise, $r(x)$ has no vertical asymptotes, since the denominator has no real roots. Since the degree of the numerator is two greater than the degree of the denominator, $r(x)$ has no horizontal or slant asymptotes.

**83. (a)** $r(x) = \dfrac{3x^2 - 3x - 6}{x - 2} = \dfrac{3(x - 2)(x + 1)}{x - 2} = 3(x + 1)$, for $x \neq 2$. Therefore, $r(x) = 3x + 3$, $x \neq 2$. Since $3(2) + 3 = 9$, the graph is the line $y = 3x + 3$ with the point $(2, 9)$ removed.

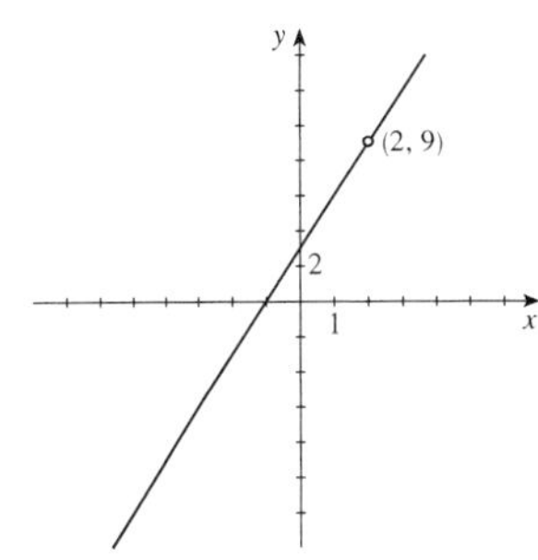

**(b)** $s(x) = \frac{x^2+x-20}{x+5} = \frac{(x-4)(x+5)}{x+5} = x-4$, for $x \neq -5$. Therefore, $s(x) = x-4$, $x \neq -5$. Since $(-5)-4 = -9$, the graph is the line $y = x-4$ with the point $(-5,-9)$ removed.

$t(x) = \frac{2x^2-x-1}{x-1} = \frac{(2x+1)(x-1)}{x-1} = 2x+1$, for $x \neq 1$. Therefore, $t(x) = 2x+1$, $x \neq 1$. Since $2(1)+1 = 3$, the graph is the line $y = 2x+1$ with the point $(1,3)$ removed.

$u(x) = \frac{x-2}{x^2-2x} = \frac{x-2}{x(x-2)} = \frac{1}{x}$, for $x \neq 2$. Therefore, $u(x) = \frac{1}{x}$, $x \neq 2$. When $x = 2$, $\frac{1}{x} = \frac{1}{2}$, so the graph is the curve $y = \frac{1}{x}$ with the point $\left(2, \frac{1}{2}\right)$ removed.

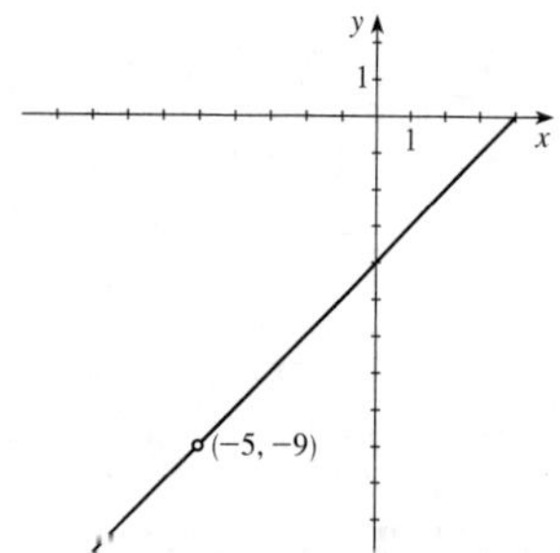


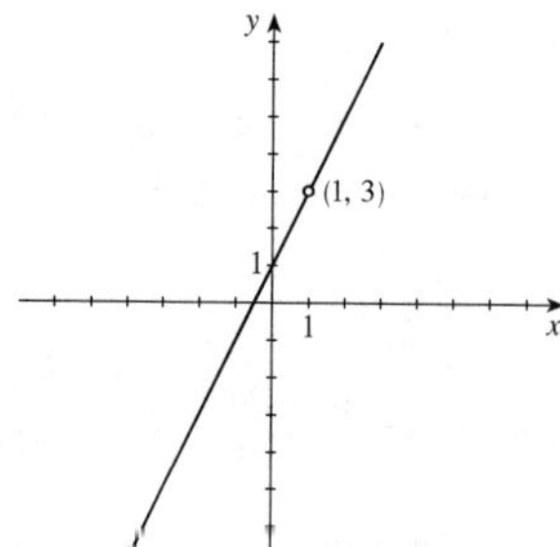


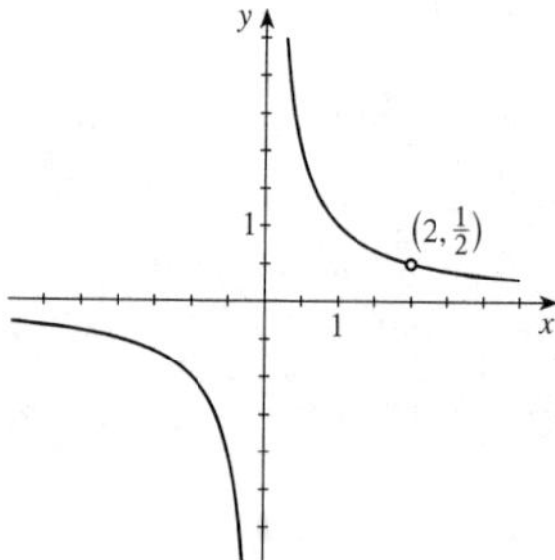


**84. (a)** Let $f(x) = \frac{1}{x^2}$. Then $r(x) = \frac{1}{(x-2)^2} = f(x-2)$. From this form we see that the graph of $r$ is obtained from the graph of $f$ by shifting 2 units to the right. Thus $r$ has vertical asymptote $x = 2$ and horizontal asymptote $y = 0$.

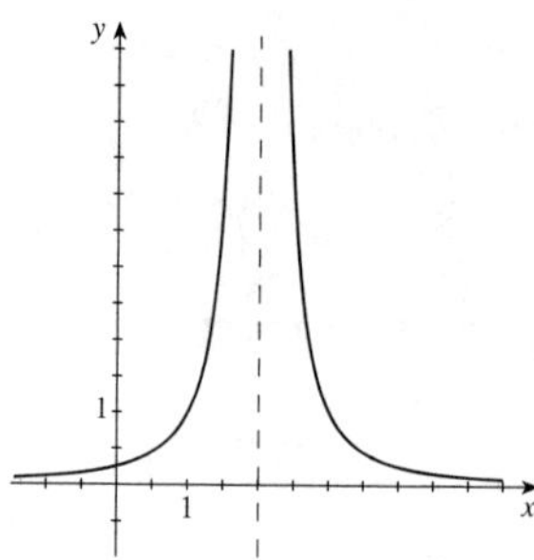


**(b)** $s(x) = \frac{2x^2+4x+5}{x^2+2x+1} = 2 + \frac{3}{(x+1)^2} = 3f(x+1)+2$. From this form we see that the graph of $s$ is obtained from the graph of $f$ by shifting 1 unit to the left, stretching vertically by a factor of 3, and shifting 2 units vertically. Thus $r$ has vertical asymptote $x = -1$ and horizontal asymptote $y = 2$.

$$\begin{array}{r|l} & \quad\quad\;\; 2 \\ x^2+2x+1 & 2x^2+4x+5 \\ & \underline{2x^2+4x+2} \\ & \quad\quad\;\; 3 \end{array}$$

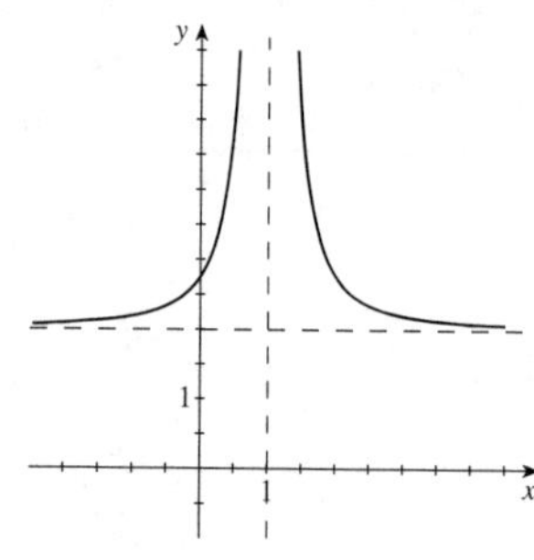

**(c)** Using long division, we see that $p(x) = \dfrac{2-3x^2}{x^2-4x+4} = -3 + \dfrac{-12x+14}{x^2-4x+4}$ which cannot be graphed by transforming $f(x) = \dfrac{1}{x^2}$. Using long division on $q$ we have:

$$\begin{array}{r|l} & -3 \\ x^2-4x+4 & -3x^2+0x+2 \\ & -3x^2+12x-12 \\ \hline & -12x+14 \end{array} \qquad \begin{array}{r|l} & -3 \\ x^2-4x+4 & -3x^2+12x+0 \\ & -3x^2+12x-12 \\ \hline & 12 \end{array}$$

So $q(x) = \dfrac{12x-3x^2}{x^2-4x+4} = -3 + \dfrac{12}{(x-2)^2} = 12f(x-2) - 3$. From this form we see that the graph of $q$ is obtained from the graph of $f$ by shifting 2 units to the right, stretching vertically by a factor of 12, and then shifting 3 units vertically down. Thus the vertical asymptote is $x = 2$ and the horizontal asymptote is $y = -3$. We show $y = p(x)$ just to verify that we cannot obtain $p(x)$ from $y = \dfrac{1}{x^2}$.

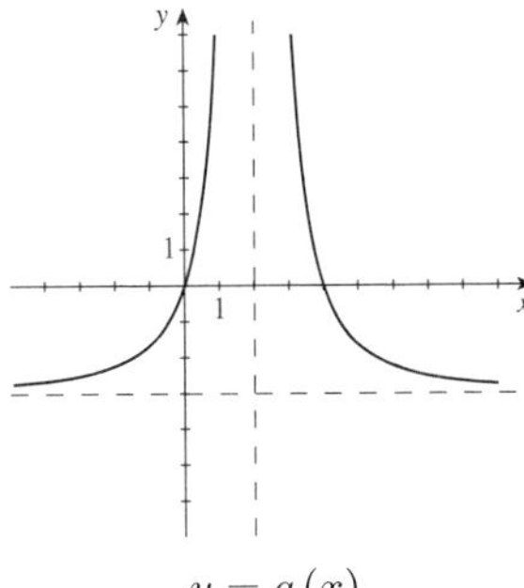

$y = q(x)$

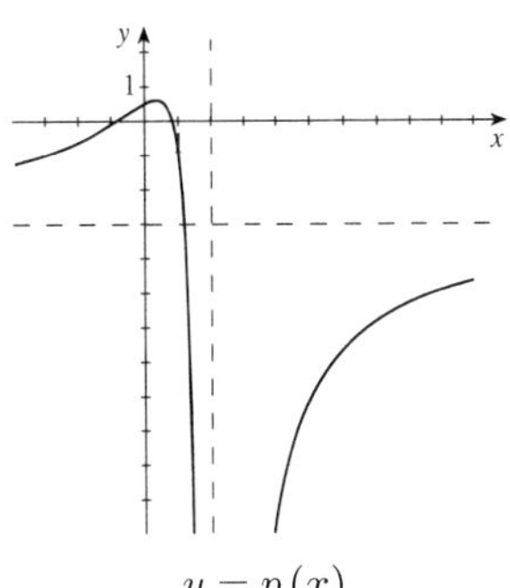

$y = p(x)$

# Chapter 4 Review

**1.** $P(x) = -x^3 + 64$

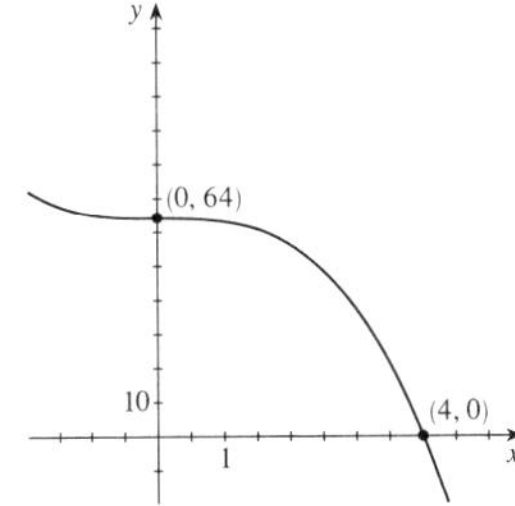


**2.** $P(x) = 2x^3 - 16 = 2(x-2)(x^2+2x+4)$

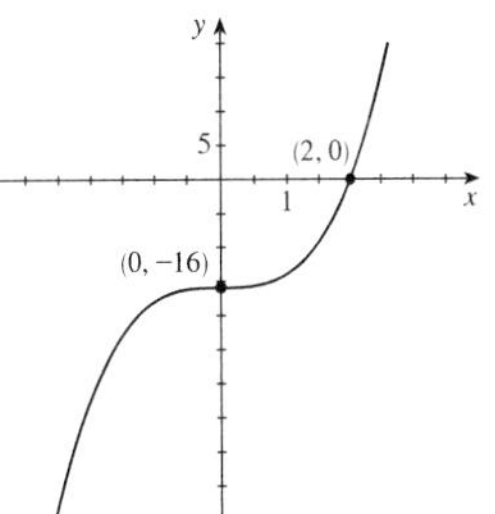


**3.** $P(x) = 2(x-1)^4 - 32$

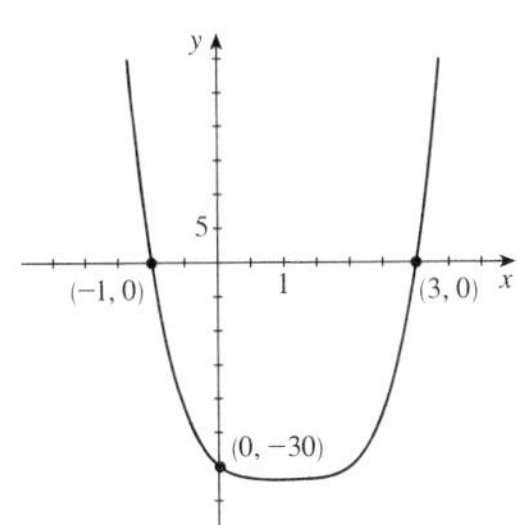


**4.** $P(x) = 64 - (x-3)^4$

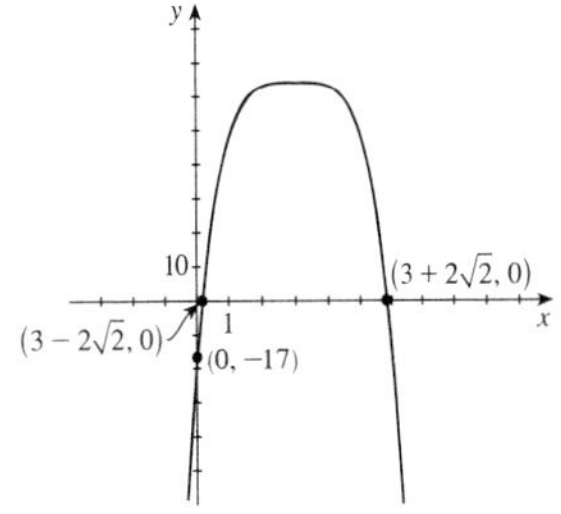

**5.** $P(x) = 32 + (x-1)^5$

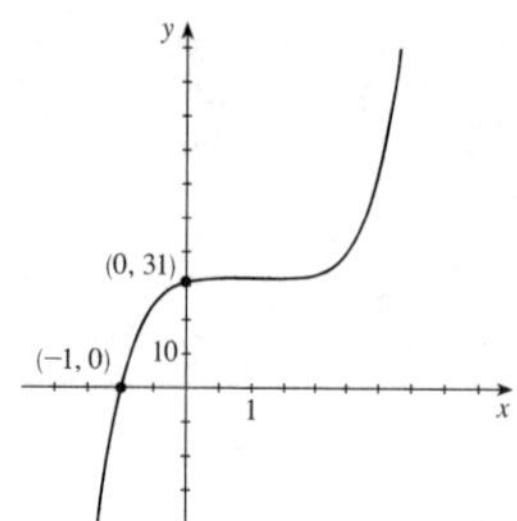


**6.** $P(x) = -3(x+2)^5 + 96$

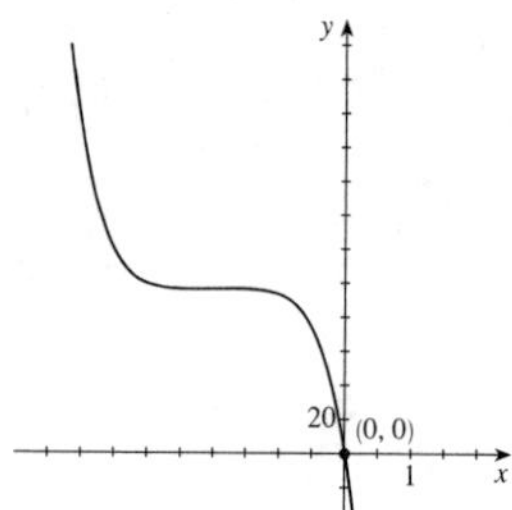


**7.** $P(x) = x^3 - 4x + 1$. $x$-intercepts: $-2.1$, $0.3$, and $1.9$. $y$-intercept: 1. Local maximum is $(-1.2, 4.1)$. Local minimum is $(1.2, -2.1)$. $y \to \infty$ as $x \to \infty$; $y \to -\infty$ as $x \to -\infty$.

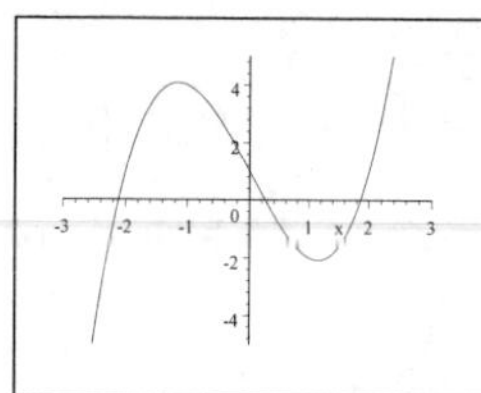

**8.** $P(x) = -2x^3 + 6x^2 - 2$. $x$-intercepts: $-0.5$, $0.7$, and $2.9$. $y$-intercept: $-2$. Local maximum is $(2, 6)$. Local minimum is $(0, -2)$. $y \to \infty$ as $x \to -\infty$ and $y \to -\infty$ as $x \to \infty$.

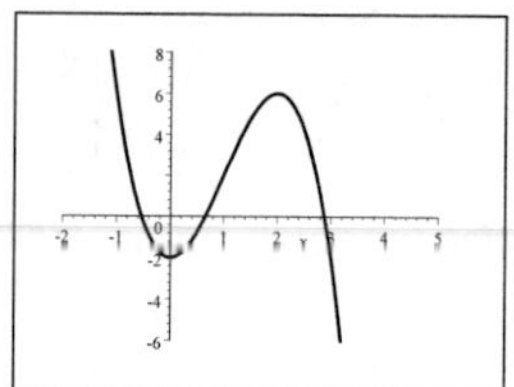

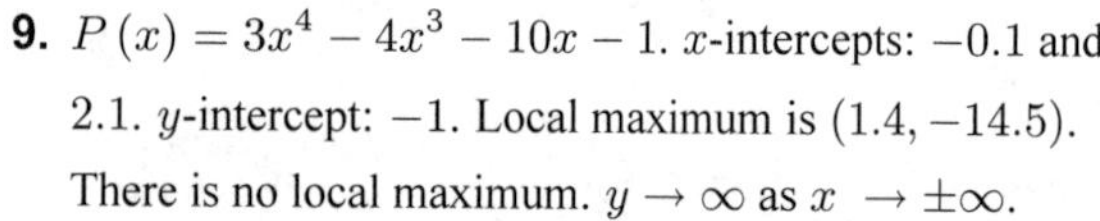

**9.** $P(x) = 3x^4 - 4x^3 - 10x - 1$. $x$-intercepts: $-0.1$ and $2.1$. $y$-intercept: $-1$. Local maximum is $(1.4, -14.5)$. There is no local maximum. $y \to \infty$ as $x \to \pm\infty$.

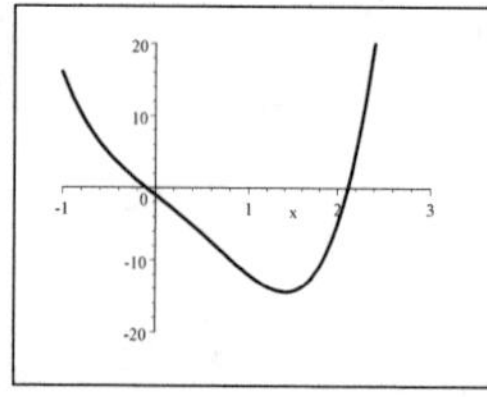

**10.** $P(x) = x^5 + x^4 - 7x^3 - x^2 + 6x + 3$. $x$-intercepts: $-3.0$, $1.3$, and $1.9$. $y$-intercept: $-3$. Local maxima are $(-2.4, 33.2)$ and $(0.5, 5.0)$. Local minima are $(-0.6, 0.6)$ and $(1.6, -1.6)$. $y \to -\infty$ as $x \to -\infty$ and $y \to \infty$ as $x \to \infty$.

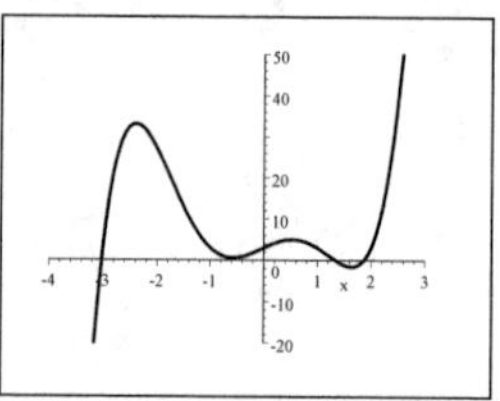

**11. (a)** Use the Pythagorean Theorem and solving for $y^2$ we have, $x^2 + y^2 = 10^2 \quad \Leftrightarrow \quad y^2 = 100 - x^2$. Substituting we get

$S = 13.8x\left(100 - x^2\right) = 1380x - 13.8x^3$.

**(b)** Domain is $[0, 10]$.

**(c)**

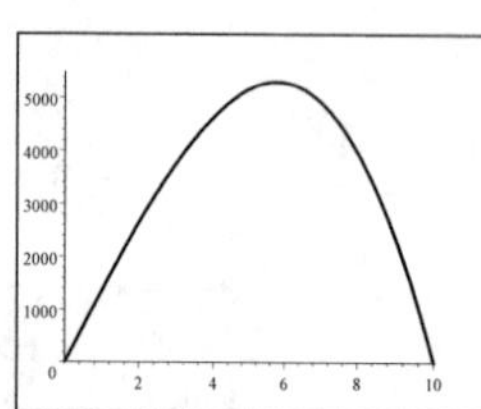

**(d)** The strongest beam has width 5.8 inches.

**12. (a)** The area of the four sides is $2x^2 + 2xy = 1200 \Leftrightarrow$

$$2xy = 1200 - 2x^2 \quad \Leftrightarrow \quad y = \frac{600 - x^2}{x}.$$

Substituting we get

$$V = x^2 y = x^2\left(\frac{600 - x^2}{x}\right) = 600x - x^3.$$

**(b)**

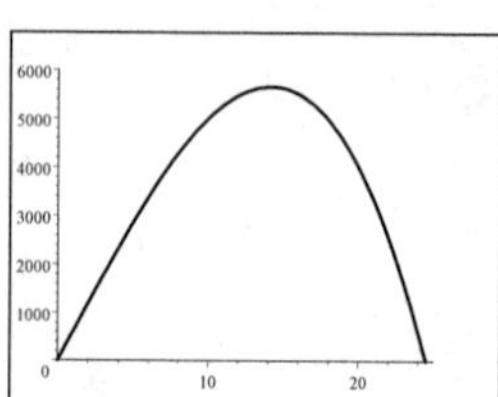

**(c)** $V$ is maximized when $x \approx 14.14$, $y \approx 28.28$

**13.** $\dfrac{x^2 - 3x + 5}{x - 2}$

$$\begin{array}{r|rrr} 2 & 1 & -3 & 5 \\ & & 2 & -2 \\ \hline & 1 & -1 & 3 \end{array}$$

Using synthetic division, we see that $Q(x) = x - 1$ and $R(x) = 3$.

**14.** $\dfrac{x^2 + x - 12}{x - 3} = \dfrac{(x+4)(x-3)}{x-3} = x + 4$. Factoring we have $Q(x) = x + 4$ and $R(x) = 0$. This can be confirmed with synthetic division:

$$\begin{array}{r|rrr} 3 & 1 & 1 & -12 \\ & & 3 & 12 \\ \hline & 1 & 4 & 0 \end{array}$$

**15.** $\dfrac{x^3 - x^2 + 11x + 2}{x - 4}$

$$\begin{array}{r|rrrr} 4 & 1 & -1 & 11 & 2 \\ & & 4 & 12 & 92 \\ \hline & 1 & 3 & 23 & 94 \end{array}$$

Using synthetic division, we see that $Q(x) = x^2 + 3x + 23$ and $R(x) = 94$.

**16.** $\dfrac{x^3 + 2x^2 - 10}{x + 3}$

$$\begin{array}{r|rrrr} -3 & 1 & 2 & 0 & -10 \\ & & -3 & 3 & -9 \\ \hline & 1 & -1 & 3 & -19 \end{array}$$

Using synthetic division, we see that $Q(x) = x^2 - x + 3$ and $R(x) = -19$.

**17.** $\dfrac{x^4 - 8x^2 + 2x + 7}{x + 5}$

$$\begin{array}{r|rrrrr} -5 & 1 & 0 & -8 & 2 & 7 \\ & & -5 & 25 & -85 & 415 \\ \hline & 1 & -5 & 17 & -83 & 422 \end{array}$$

Using synthetic division, we see that $Q(x) = x^3 - 5x^2 + 17x - 83$ and $R(x) = 422$.

**18.** $\dfrac{2x^4 + 3x^3 - 12}{x + 4}$

$$\begin{array}{r|rrrrr} -4 & 2 & 3 & 0 & 0 & -12 \\ & & -8 & 20 & -80 & 320 \\ \hline & 2 & -5 & 20 & -80 & 308 \end{array}$$

Using synthetic division, we see that $Q(x) = 2x^3 - 5x^2 + 20x - 80$ and $R(x) = 308$.

**19.** $\dfrac{2x^3 + x^2 - 8x + 15}{x^2 + 2x - 1}$

$$\begin{array}{r|rrrr} & 2x & -3 & & \\ \hline x^2 + 2x - 1 & 2x^3 & + x^2 & - 8x & + 15 \\ & 2x^3 & + 4x^2 & - 2x & \\ \hline & & -3x^2 & - 6x & + 15 \\ & & -3x^2 & - 6x & + 3 \\ \hline & & & & 12 \end{array}$$

Therefore, $Q(x) = 2x - 3$, and $R(x) = 12$.

**20.** $\dfrac{x^4 - 2x^2 + 7x}{x^2 - x + 3}$

$$\begin{array}{r|rrrrr} & x^2 & + x & - 4 & & \\ \hline x^2 - x + 3 & x^4 & + 0x^3 & - 2x^2 & + 7x & + 0 \\ & x^4 & - x^3 & + 3x^3 & & \\ \hline & & x^3 & - 5x^2 & + 7x & \\ & & x^3 & - x^2 & + 3x & \\ \hline & & & -4x^2 & + 4x & + 0 \\ & & & -4x^2 & + 4x & - 12 \\ \hline & & & & & 12 \end{array}$$

Therefore, $Q(x) = x^2 + x - 4$, and $R(x) = 12$.

**21.** $P(x) = 2x^3 - 9x^2 - 7x + 13$; find $P(5)$.

$$\begin{array}{r|rrrr} 5 & 2 & -9 & -7 & 13 \\ & & 10 & 5 & -10 \\ \hline & 2 & 1 & -2 & 3 \end{array}$$

Therefore, $P(5) = 3$.

**22.** $Q(x) = x^4 + 4x^3 + 7x^2 + 10x + 15$; find $Q(-3)$

$$\begin{array}{r|rrrrr} -3 & 1 & 4 & 7 & 10 & 15 \\ & & -3 & -3 & -12 & 6 \\ \hline & 1 & 1 & 4 & -2 & 21 \end{array}$$

By the Remainder Theorem, we have $Q(-3) = 21$.

**23.** $\frac{1}{2}$ is a zero of $P(x) = 2x^4 + x^3 - 5x^2 + 10x - 4$ if $P\left(\frac{1}{2}\right) = 0$.

$$\begin{array}{r|rrrrr} \frac{1}{2} & 2 & 1 & -5 & 10 & -4 \\ & & 1 & 1 & -2 & 4 \\ \hline & 2 & 2 & -4 & 8 & 0 \end{array}$$

Since $P\left(\frac{1}{2}\right) = 0$, $\frac{1}{2}$ is a zero of the polynomial.

**24.** $x + 4$ is a factor of $P(x) = x^5 + 4x^4 - 7x^3 - 23x^2 + 23x + 12$ if $P(-4) = 0$.

$$\begin{array}{r|rrrrrr} -4 & 1 & 4 & -7 & -23 & 23 & 12 \\ & & -4 & 0 & 28 & -20 & -12 \\ \hline & 1 & 0 & -7 & 5 & 3 & 0 \end{array}$$

Since $P(-4) = 0$, $x + 4$ is a factor of the polynomial.

**25.** $P(x) = x^{500} + 6x^{201} - x^2 - 2x + 4$. The remainder from dividing $P(x)$ by $x - 1$ is $P(1) = (1)^{500} + 6(1)^{201} - (1)^2 - 2(1) + 4 = 8$.

**26.** Let $P(x) = x^{101} - x^4 + 2$. The remainder from dividing $P(x)$ by $x + 1$ is $P(-1) = (-1)^{101} - (-1)^4 + 2 = 0$.

**27. (a)** $P(x) = x^5 - 6x^3 - x^2 + 2x + 18$ has possible rational zeros $\pm 1, \pm 2, \pm 3, \pm 6, \pm 9, \pm 18$.

**(b)** Since $P(x)$ has 2 variations in sign, there are either 0 or 2 positive real zeros. Since $P(-x) = -x^5 + 6x^3 - x^2 - 2x + 18$ has 3 variations in sign, there are 1 or 3 negative real zeros.

**28. (a)** $P(x) = 6x^4 + 3x^3 + x^2 + 3x + 4$ has possible rational zeros $\pm 1, \pm 2, \pm 4, \pm\frac{1}{2}, \pm\frac{1}{3}, \pm\frac{2}{3}, \pm\frac{4}{3}, \pm\frac{1}{6}$.

**(b)** Since $P(x)$ has no variations in sign, there are no positive real zeros. Since $P(-x) = 6x^4 - 3x^3 + x^2 - 3x + 4$ has 4 variations in sign, there are 0, 2, or 4 negative real zeros.

**29. (a)** $P(x) = x^3 - 16x = x(x^2 - 16)$
$= x(x - 4)(x + 4)$

has zeros $-4, 0, 4$ (all of multiplicity 1).

**(b)**

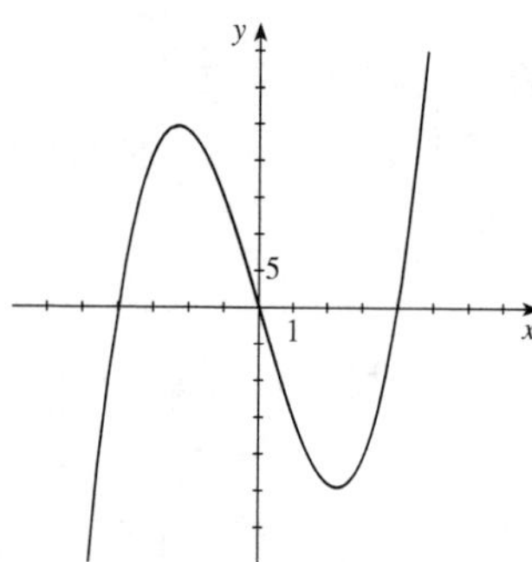


**30. (a)** $P(x) = x^3 - 3x^2 - 4x$
$= x(x + 1)(x - 4)$

has zeros $0, -1$, and $4$ (all of multiplicity 1).

**(b)**

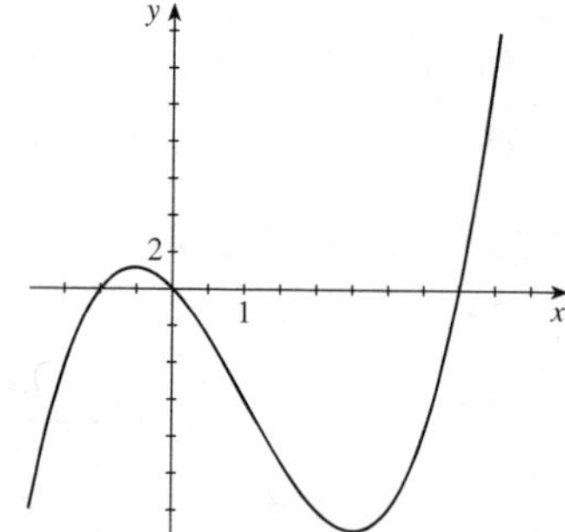


**31. (a)** $P(x) = x^4 + x^3 - 2x^2 = x^2(x^2 + x - 2)$
$= x^2(x + 2)(x - 1)$

The zeros are 0 (multiplicity 2), $-2$ (multiplicity 1), and 1 (multiplicity 1)

**(b)**

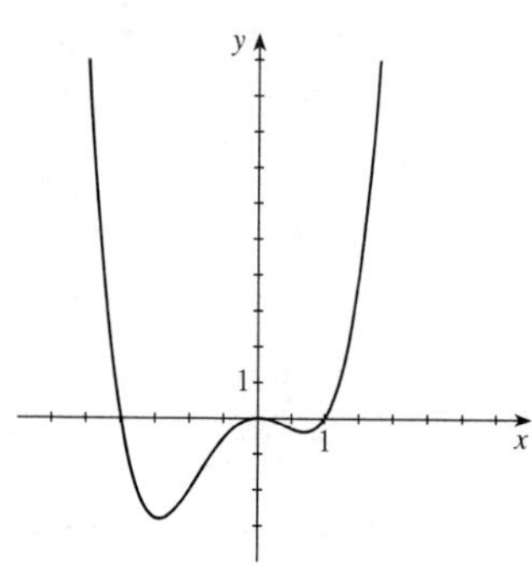


**32. (a)** $P(x) = x^4 - 5x^2 + 4 = (x^2 - 4)(x^2 - 1)$
$= (x - 2)(x + 2)(x - 1)(x + 1)$

Thus, the zeros are $-1, 1, -2, 2$ (all of multiplicity 1).

**(b)**

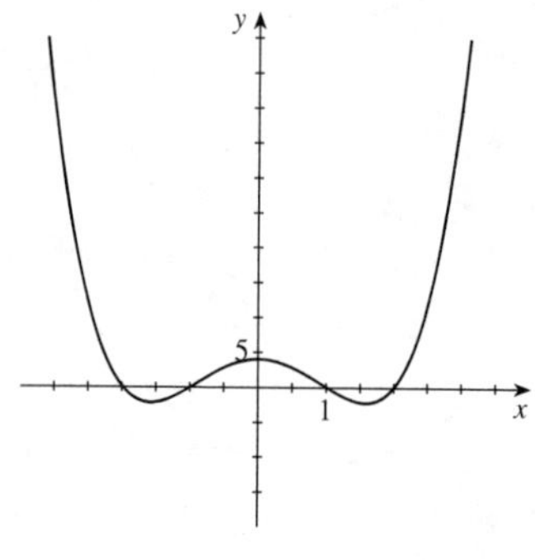

**33. (a)** $P(x) = x^4 - 2x^3 - 7x^2 + 8x + 12$. The possible rational zeros are $\pm 1, \pm 2, \pm 3, \pm 4, \pm 6, \pm 12$. $P$ has 2 variations in sign, so it has either 2 or 0 positive real zeros.

| 1 | 1 | −2 | −7 | 8 | 12 |
|---|---|---|---|---|---|
| | | 1 | −1 | −8 | 0 |
| | 1 | −1 | −8 | 0 | 12 |

| 2 | 1 | −2 | −7 | 8 | 12 |
|---|---|---|---|---|---|
| | | 2 | 0 | −14 | −12 |
| | 2 | 0 | −7 | −6 | 0 |

$\Rightarrow x = 2$ is a root.

$P(x) = x^4 - 2x^3 - 7x^2 + 8x + 12 = (x - 2)(x^3 - 7x - 6)$. Continuing:

| 2 | 1 | 0 | −6 | −6 |
|---|---|---|---|---|
| | | 2 | 4 | −4 |
| | 1 | 2 | −2 | −10 |

| 3 | 1 | 0 | −7 | −6 |
|---|---|---|---|---|
| | | 3 | 9 | 6 |
| | 1 | 3 | 2 | 0 |

so $x = 3$ is a root and

$$\begin{aligned} P(x) &= (x-2)(x-3)(x^2+3x+2) \\ &= (x-2)(x-3)(x+1)(x+2) \end{aligned}$$

Therefore the real roots are $-2, -1, 2$, and $3$ (all of multiplicity 1).

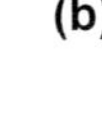
**(b)**

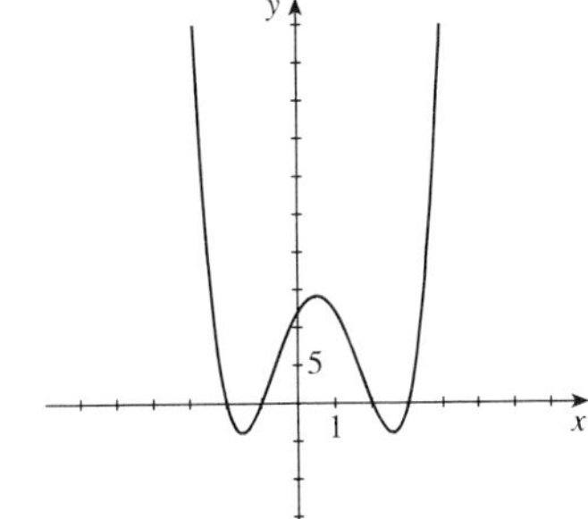


**34. (a)**

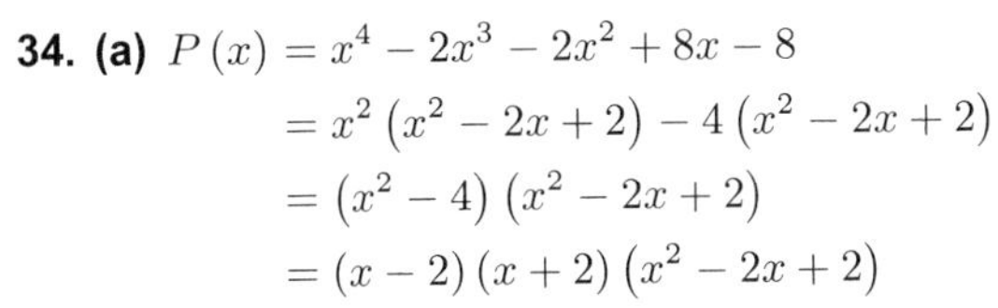

$$\begin{aligned} P(x) &= x^4 - 2x^3 - 2x^2 + 8x - 8 \\ &= x^2(x^2 - 2x + 2) - 4(x^2 - 2x + 2) \\ &= (x^2 - 4)(x^2 - 2x + 2) \\ &= (x-2)(x+2)(x^2 - 2x + 2) \end{aligned}$$

The quadratic is irreducible, so the real roots are $\pm 2$ (each of multiplicity 1).

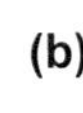
**(b)**

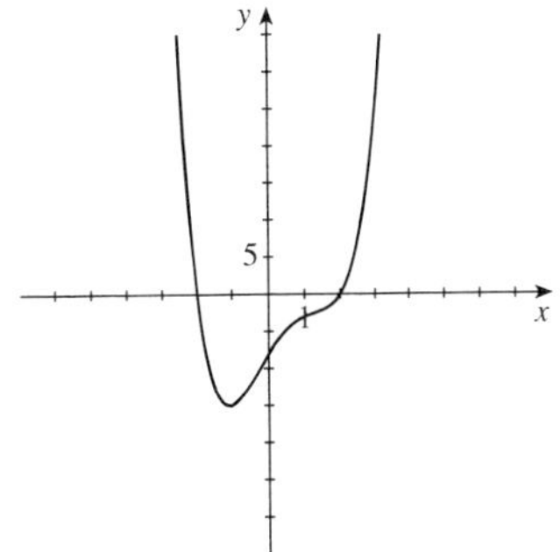

**35. (a)** $P(x) = 2x^4 + x^3 + 2x^2 - 3x - 2$. The possible rational roots are $\pm 1, \pm 2, \pm \frac{1}{2}$. $P$ has one variation in sign, and hence 1 positive real root. $P(-x)$ has 3 variations in sign and hence either 3 or 1 negative real roots.

$$\begin{array}{r|rrrrr} 1 & 2 & 1 & 2 & -3 & -2 \\ & & 2 & 3 & 5 & 2 \\ \hline & 2 & 3 & 5 & 2 & 0 \end{array} \Rightarrow x = 1 \text{ is a zero.}$$

$P(x) = 2x^4 + x^3 + 2x^2 - 3x - 2 = (x-1)(2x^3 + 3x^2 + 5x + 2)$. Continuing:

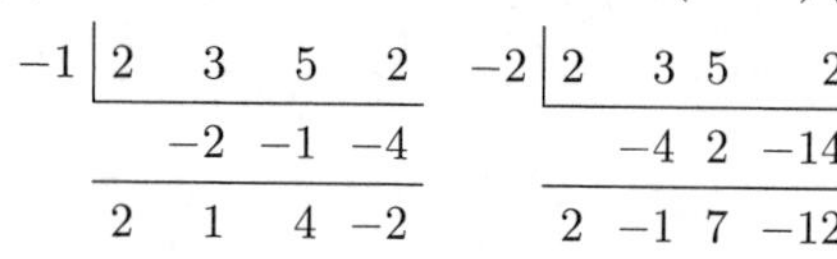

$$\begin{array}{r|rrrr} -1 & 2 & 3 & 5 & 2 \\ & & -2 & -1 & -4 \\ \hline & 2 & 1 & 4 & -2 \end{array} \qquad \begin{array}{r|rrrr} -2 & 2 & 3 & 5 & 2 \\ & & -4 & 2 & -14 \\ \hline & 2 & -1 & 7 & -12 \end{array}$$

$$\begin{array}{r|rrrr} -\frac{1}{2} & 2 & 3 & 5 & 2 \\ & & -1 & -1 & -2 \\ \hline & 2 & 2 & 4 & 0 \end{array} \Rightarrow x = -\tfrac{1}{2} \text{ is a zero.}$$

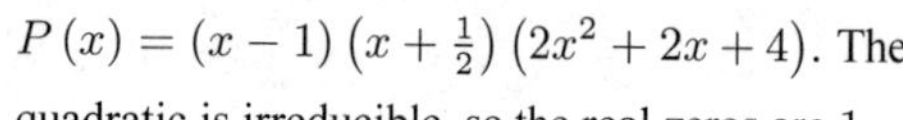

$P(x) = (x-1)\left(x+\frac{1}{2}\right)(2x^2 + 2x + 4)$. The quadratic is irreducible, so the real zeros are 1 and $-\frac{1}{2}$ (each of multiplicity 1).

**(b)**

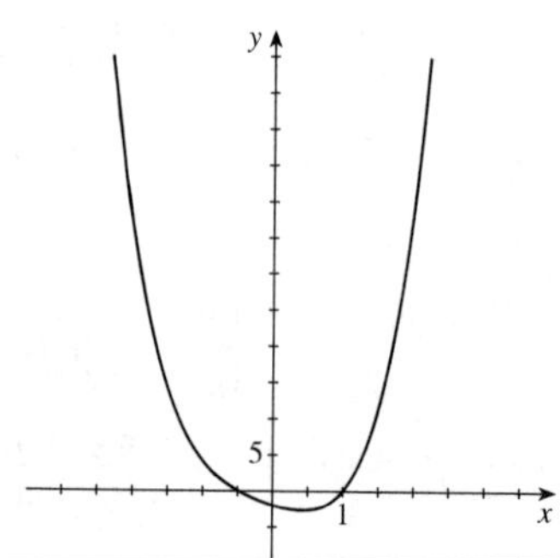


**36. (a)** $P(x) = 9x^5 - 21x^4 + 10x^3 + 6x^2 - 3x - 1$. The possible rational zeros are $\pm 1, \pm \frac{1}{3}, \pm \frac{1}{9}$. $P$ has 3 variations in sign, hence 3 or 1 positive real roots. $P(-x)$ has 2 variations in sign, hence 2 or 0 real negative roots.

$$\begin{array}{r|rrrrrr} 1 & 9 & -21 & 10 & 6 & -3 & -1 \\ & & 9 & -12 & -2 & 4 & 1 \\ \hline & 9 & -12 & -2 & 4 & 1 & 0 \end{array} \Rightarrow x = 1 \text{ is a zero.}$$

$P(x) = (x-1)(9x^4 - 12x^3 - 2x^2 + 4x + 1)$. Continuing:

$$\begin{array}{r|rrrrr} 1 & 9 & -12 & -2 & 4 & 1 \\ & & 9 & -3 & -5 & -1 \\ \hline & 9 & -3 & -5 & -1 & 0 \end{array} \Rightarrow x = 1 \text{ is a zero again.}$$

$P(x) = (x-1)^2(9x^3 - 3x^2 - 5x - 1)$.

Continuing:

$$\begin{array}{r|rrrr} 1 & 9 & -3 & -5 & -1 \\ & & 9 & 6 & 1 \\ \hline & 9 & 6 & 1 & 0 \end{array} \Rightarrow x = 1 \text{ is a zero yet again.}$$

$$\begin{aligned} P(x) &= (x-1)^3(9x^2 + 6x + 1) \\ &= (x-1)^3(3x+1)^2 \end{aligned}$$

So the real zeros of $P$ are 1 (multiplicity 3) and $-\frac{1}{3}$ (multiplicity 2).

**(b)**

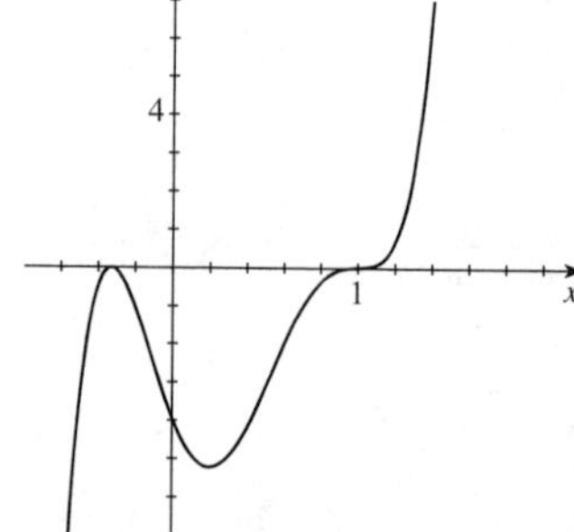


**37.** Since the zeros are $-\frac{1}{2}$, 2, and 3, a factorization is

$$\begin{aligned} P(x) &= C\left(x+\tfrac{1}{2}\right)(x-2)(x-3) = \tfrac{1}{2}C(2x+1)(x^2 - 5x + 6) \\ &= \tfrac{1}{2}C(2x^3 - 10x^2 + 12x + x^2 - 5x + 6) = \tfrac{1}{2}C(2x^3 - 9x^2 + 7x + 6) \end{aligned}$$

Since the constant coefficient is 12, $\frac{1}{2}C(6) = 12 \quad \Leftrightarrow \quad C = 4$, and so the polynomial is $P(x) = 4x^3 - 18x^2 + 14x + 12$.

**38.** Since the zeros are $\pm 3i$ and 4 (which is a double zero), a factorization is $P(x) = C(x-4)^2(x-3i)(x+3i) = C(x^2-8x+16)(x^2+9) = C(x^4-8x^3+25x^2-72x+144)$. Since all of the coefficients are integers, we can choose $C = 1$, so $P(x) = x^4 - 8x^3 + 25x^2 - 72x + 144$.

**39.** No, there is no polynomial of degree 4 with integer coefficients that has zeros $i$, $2i$, $3i$ and $4i$. Since the imaginary zeros of polynomial equations with real coefficients come in complex conjugate pairs, there would have to be 8 zeros, which is impossible for a polynomial of degree 4.

**40.** $P(x) = 3x^4 + 5x^2 + 2 = (3x^2+2)(x^2+1)$. Since $3x^2 + 2 = 0$ and $x^2 + 1 = 0$ have no real zeros, it follows that $3x^4 + 5x^2 + 2$ has no real zeros.

**41.** $P(x) = x^3 - 3x^2 - 13x + 15$ has possible rational zeros $\pm 1, \pm 3, \pm 5, \pm 15$.

$$\begin{array}{r|rrrr} 1 & 1 & -3 & -13 & 15 \\ & & 1 & -2 & -15 \\ \hline & 1 & -2 & -15 & 0 \end{array} \Rightarrow x = 1 \text{ is a zero.}$$

So $P(x) = x^3 - 3x^2 - 13x + 15 = (x-1)(x^2-2x-15) = (x-1)(x-5)(x+3)$. Therefore, the zeros are $-3$, 1, and 5.

**42.** $P(x) = 2x^3 + 5x^2 - 6x - 9$ has possible rational zeros $\pm 1, \pm 3, \pm 9, \pm\frac{1}{2}, \pm\frac{3}{2}, \pm\frac{9}{2}$. Since there is one variation in sign, there is a positive real zero.

$$\begin{array}{r|rrrr} 1 & 2 & 5 & -6 & -9 \\ & & 2 & 7 & 1 \\ \hline & 2 & 7 & 1 & -8 \end{array} \qquad \begin{array}{r|rrrr} 3 & 2 & 5 & -6 & -9 \\ & & 6 & 33 & 81 \\ \hline & 2 & 11 & 27 & 72 \end{array} \Rightarrow x = 3 \text{ is an upper bound} \qquad \begin{array}{r|rrrr} \frac{3}{2} & 2 & 5 & -6 & -9 \\ & & 3 & 12 & 9 \\ \hline & 2 & 8 & 6 & 0 \end{array} \Rightarrow x = \tfrac{3}{2} \text{ is a zero.}$$

So $P(x) = 2x^3 + 5x^2 - 6x - 9 = (2x-3)(x^2+4x+3) = (2x-3)(x+3)(x+1)$. Therefore, the zeros are $-3$, $-1$ and $\frac{3}{2}$.

**43.** $P(x) = x^4 + 6x^3 + 17x^2 + 28x + 20$ has possible rational zeros $\pm1, \pm2, \pm4, \pm5, \pm10, \pm20$. Since all of the coefficients are positive, there are no positive real zeros.

$$\begin{array}{r|rrrrr} -1 & 1 & 6 & 17 & 28 & 20 \\ & & -1 & -5 & -12 & -16 \\ \hline & 1 & 5 & 12 & 16 & 4 \end{array} \qquad \begin{array}{r|rrrrr} -2 & 1 & 6 & 17 & 28 & 20 \\ & & -2 & -8 & -18 & -20 \\ \hline & 1 & 4 & 9 & 10 & 0 \end{array} \Rightarrow x = -2 \text{ is a zero.}$$

$P(x) = x^4 + 6x^3 + 17x^2 + 28x + 20 = (x+2)(x^3 + 4x^2 + 9x + 10)$. Continuing with the quotient, we have

$$\begin{array}{r|rrrr} -2 & 1 & 4 & 9 & 10 \\ & & -2 & -4 & -10 \\ \hline & 1 & 2 & 5 & 0 \end{array} \Rightarrow x = -2 \text{ is a zero.}$$

Thus $P(x) = x^4 + 6x^3 + 17x^2 + 28x + 20 = (x+2)^2(x^2 + 2x + 5)$. Now $x^2 + 2x + 5 = 0$ when $x = \frac{-2\pm\sqrt{4-4(5)(1)}}{2} = \frac{-2\pm4i}{2} = -1 \pm 2i$. Thus, the zeros are $-2$ (multiplicity 2) and $-1 \pm 2i$.

**44.** $P(x) = x^4 + 7x^3 + 9x^2 - 17x - 20$ has possible rational zeros $\pm1, \pm2, \pm4, \pm5, \pm10, \pm20$.

$$\begin{array}{r|rrrrr} 1 & 1 & 7 & 9 & -17 & -20 \\ & & 1 & 8 & 17 & 0 \\ \hline & 1 & 8 & 17 & 0 & -20 \end{array} \qquad \begin{array}{r|rrrrr} 2 & 1 & 7 & 9 & -17 & -20 \\ & & 2 & 18 & 54 & 74 \\ \hline & 1 & 9 & 27 & 37 & 54 \end{array} \Rightarrow x = 2 \text{ is an upper bound.} \qquad \begin{array}{r|rrrrr} -1 & 1 & 7 & 9 & -17 & -20 \\ & & -1 & -6 & -3 & 20 \\ \hline & 1 & 6 & 3 & -20 & 0 \end{array} \Rightarrow x = -1 \text{ is a zero.}$$

So $P(x) = x^4 + 7x^3 + 9x^2 - 17x - 20 = (x+1)(x^3 + 6x^2 + 3x - 20)$. Continuing with the quotient, we have

$$\begin{array}{r|rrrr} -1 & 1 & 6 & 3 & -20 \\ & & -1 & -5 & 2 \\ \hline & 1 & 5 & -2 & -18 \end{array} \qquad \begin{array}{r|rrrr} -2 & 1 & 6 & 3 & -20 \\ & & -2 & -8 & 10 \\ \hline & 1 & 4 & -5 & -10 \end{array} \qquad \begin{array}{r|rrrr} -4 & 1 & 6 & 3 & -20 \\ & & -4 & -8 & 20 \\ \hline & 1 & 2 & -5 & 0 \end{array} \Rightarrow x = -4 \text{ is a zero.}$$

So $P(x) = x^4 + 7x^3 + 9x^2 - 17x - 20 = (x+1)(x+4)(x^2 + 2x - 5)$. Now using the quadratic formula on $x^2 + 2x - 5$ we have: $x = \dfrac{-2 \pm \sqrt{4 - 4(1)(-5)}}{2} = \dfrac{-2 \pm 2\sqrt{6}}{2} = -1 \pm \sqrt{6}$. Thus, the zeros are $-4$, $-1$, and $-1 \pm \sqrt{6}$.

**45.** $P(x) = x^5 - 3x^4 - x^3 + 11x^2 - 12x + 4$ has possible rational zeros $\pm1, \pm2, \pm4$.

$$\begin{array}{r|rrrrrr} 1 & 1 & -3 & -1 & 11 & -12 & 4 \\ & & 1 & -2 & -3 & 8 & -4 \\ \hline & 1 & -2 & -3 & 8 & -4 & 0 \end{array} \Rightarrow x = 1 \text{ is a zero.}$$

$P(x) = x^5 - 3x^4 - x^3 + 11x^2 - 12x + 4 = (x-1)(x^4 - 2x^3 - 3x^2 + 8x - 4)$. Continuing with the quotient, we have

$$\begin{array}{r|rrrrr} 1 & 1 & -2 & -3 & 8 & -4 \\ & & 1 & -1 & -4 & 4 \\ \hline & 1 & -1 & -4 & 4 & 0 \end{array} \Rightarrow x = 1 \text{ is a zero.}$$

$$\begin{aligned} x^5 - 3x^4 - x^3 + 11x^2 - 12x + 4 &= (x-1)^2(x^3 - x^2 - 4x + 4) = (x-1)^3(x^2 - 4) \\ &= (x-1)^3(x-2)(x+2) \end{aligned}$$

Therefore, the zeros are 1 (multiplicity 3), $-2$, and 2.

**46.** $P(x) = x^4 - 81 = (x^2 - 9)(x^2 + 9) = (x-3)(x+3)(x^2+9) = (x-3)(x+3)(x-3i)(x+3i)$. Thus, the zeros are $\pm3, \pm3i$.

**47.** $P(x) = x^6 - 64 = (x^3 - 8)(x^3 + 8) = (x - 2)(x^2 + 2x + 4)(x + 2)(x^2 - 2x + 4)$. Now using the quadratic formula to find the zeros of $x^2 + 2x + 4$, we have $x = \frac{-2 \pm \sqrt{4 - 4(4)(1)}}{2} = \frac{-2 \pm 2\sqrt{3}i}{2} = -1 \pm \sqrt{3}i$, and using the quadratic formula to find the zeros of $x^2 - 2x + 4$, we have $x = \frac{2 \pm \sqrt{4 - 4(4)(1)}}{2} = \frac{2 \pm 2\sqrt{3}i}{2} = 1 \pm \sqrt{3}i$. Therefore, the zeros are 2, $-2$, $1 \pm \sqrt{3}i$, and $-1 \pm \sqrt{3}i$.

**48.** $P(x) = 18x^3 + 3x^2 - 4x - 1$ has possible rational zeros $\pm 1, \pm\frac{1}{2}, \pm\frac{1}{3}, \pm\frac{1}{6}, \pm\frac{1}{9}, \pm\frac{1}{18}$.

$$\begin{array}{r|rrrr} 1 & 18 & 3 & -4 & -1 \\ & & 18 & 21 & 17 \\ \hline & 18 & 21 & 17 & 16 \end{array} \Rightarrow x = 1 \text{ is an upper bound.} \qquad \begin{array}{r|rrrr} \frac{1}{2} & 18 & 3 & -4 & -1 \\ & & 9 & 6 & 1 \\ \hline & 18 & 12 & 2 & 0 \end{array} \Rightarrow x = \tfrac{1}{2} \text{ is a zero.}$$

So $P(x) = 18x^3 + 3x^2 - 4x - 1 = (2x - 1)(9x^2 + 6x + 1) = (2x - 1)(3x + 1)^2$. Thus the zeros are $\frac{1}{2}$ and $-\frac{1}{3}$ (multiplicity 2).

**49.** $P(x) = 6x^4 - 18x^3 + 6x^2 - 30x + 36 - 6(x^4 - 3x^3 + x^2 - 5x + 6)$ has possible rational zeros $\pm 1, \pm 2, \pm 3, \pm 6$.

$$\begin{array}{r|rrrrr} 1 & 6 & -18 & 6 & -30 & 36 \\ & & 6 & -12 & -6 & -36 \\ \hline & 6 & -12 & -6 & -36 & 0 \end{array} \Rightarrow x = 1 \text{ is a zero.}$$

So $P(x) = 6x^4 - 18x^3 + 6x^2 - 30x + 36 = (x - 1)(6x^3 - 12x^2 - 6x - 36) = 6(x - 1)(x^3 - 2x^2 - x - 6)$. Continuing with the quotient we have

$$\begin{array}{r|rrrr} 1 & 1 & -2 & -1 & -6 \\ & & 1 & -1 & -2 \\ \hline & 1 & -1 & -2 & -8 \end{array} \qquad \begin{array}{r|rrrr} 2 & 1 & -2 & -1 & -6 \\ & & 2 & 0 & -2 \\ \hline & 1 & 0 & -1 & -8 \end{array} \qquad \begin{array}{r|rrrr} 3 & 1 & -2 & -1 & -6 \\ & & 3 & 3 & 6 \\ \hline & 1 & 1 & 2 & 0 \end{array} \Rightarrow x = 3 \text{ is a zero.}$$

So $P(x) = 6x^4 - 18x^3 + 6x^2 - 30x + 36 = 6(x - 1)(x - 3)(x^2 + x + 2)$. Now $x^2 + x + 2 = 0$ when $x = \frac{-1 \pm \sqrt{1 - 4(1)(2)}}{2} = \frac{-1 \pm \sqrt{7}i}{2}$, and so the zeros are 1, 3, and $\frac{-1 \pm \sqrt{7}i}{2}$.

**50.** $P(x) = x^4 + 15x^2 + 54 = (x^2 + 9)(x^2 + 6)$. If $x^2 = -9$, then $x = \pm 3i$. If $x^2 = -6$, then $x = \pm\sqrt{6}i$. Therefore, the zeros are $\pm 3i$ and $\pm\sqrt{6}i$.

**51.** $2x^2 = 5x + 3 \quad \Leftrightarrow \quad 2x^2 - 5x - 3 = 0$. The solutions are $x = -0.5, 3$.

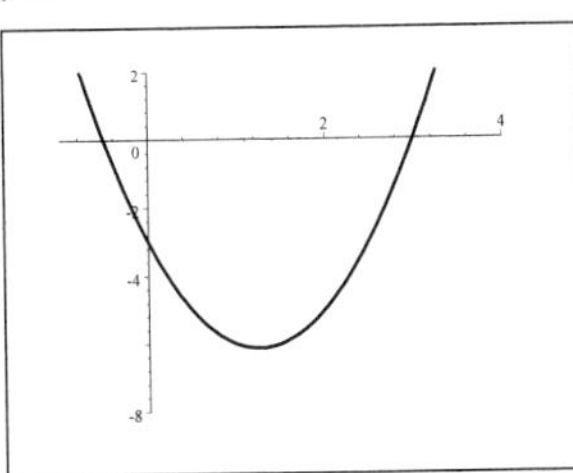

**52.** Let $P(x) = x^3 + x^2 - 14x - 24$. The solutions to $P(x) = 0$ are $x = -3, -2$, and 4.

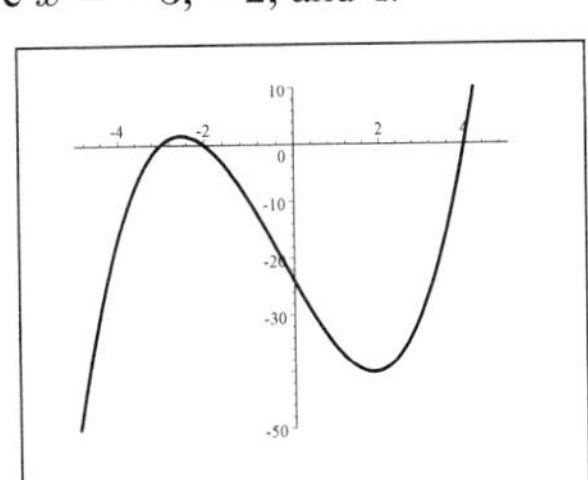

**53.** $x^4 - 3x^3 - 3x^2 - 9x - 2 = 0$ has solutions $x \approx -0.24$, 4.24.

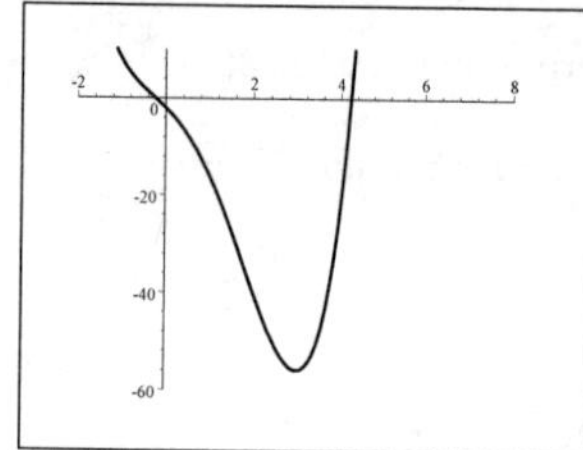

**54.** $x^5 = x + 3 \quad \Leftrightarrow \quad x^5 - x - 3 = 0$. We graph $P(x) = x^5 - x - 3$. The only real solution is 1.34.

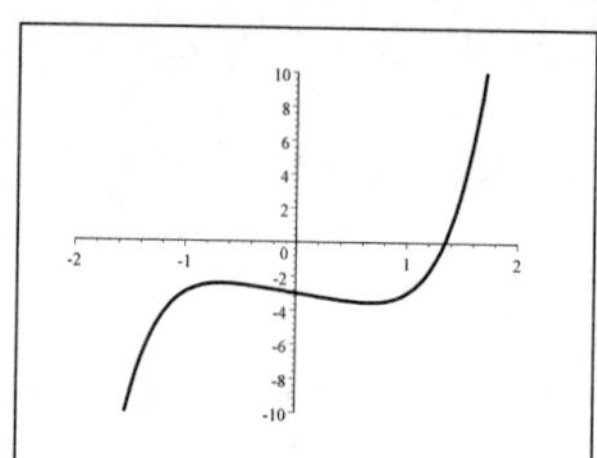

**55.** $r(x) = \dfrac{3x - 12}{x + 1}$. When $x = 0$, we have $r(0) = \dfrac{-12}{1} = -12$, so the $y$-intercept is $-12$. Since $y = 0$, when $3x - 12 = 0 \quad \Leftrightarrow \quad x = 4$, the $x$-intercept is 4. The vertical asymptote is $x = -1$. Because the degree of the denominator and numerator are the same, the horizontal asymptote is $y = \frac{3}{1} = 3$.

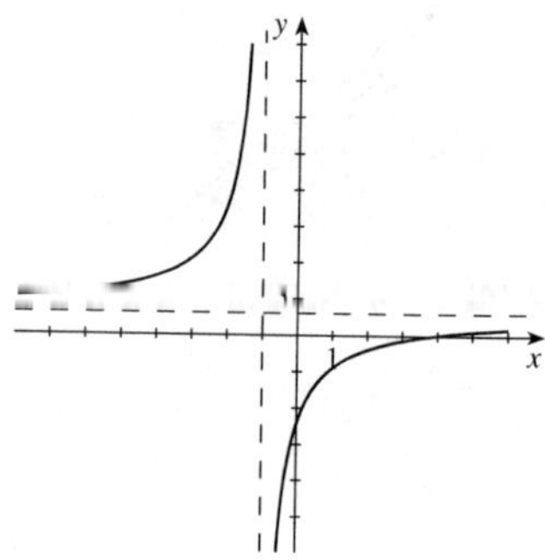

**56.** $r(x) = \dfrac{1}{(x + 2)^2}$. When $x = 0$, we have $r(0) = \dfrac{1}{2^2} = \frac{1}{4}$, so the $y$-intercept is $\frac{1}{4}$. Since the numerator is 1, $y$ never equals zero and there is no $x$-intercept. There is a vertical asymptote at $x = -2$. The horizontal asymptote is $y = 0$ because the degree of the denominator is greater than the degree of the numerator.

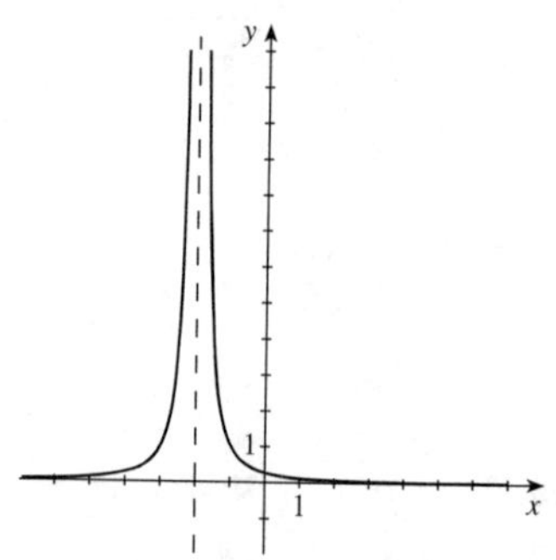

**57.** $r(x) = \dfrac{x - 2}{x^2 - 2x - 8} = \dfrac{x - 2}{(x + 2)(x - 4)}$. When $x = 0$, we have $r(0) = \frac{-2}{-8} = \frac{1}{4}$, so the $y$-intercept is $\frac{1}{4}$. When $y = 0$, we have $x - 2 = 0 \quad \Leftrightarrow$ $x = 2$, so the $x$-intercept is 2. There are vertical asymptotes at $x = -2$ and $x = 4$.

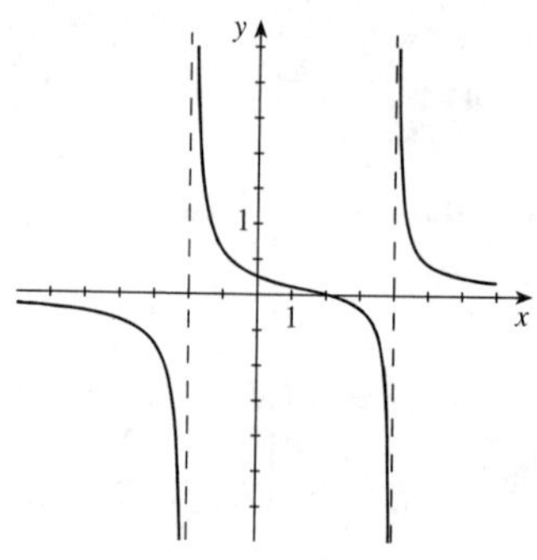

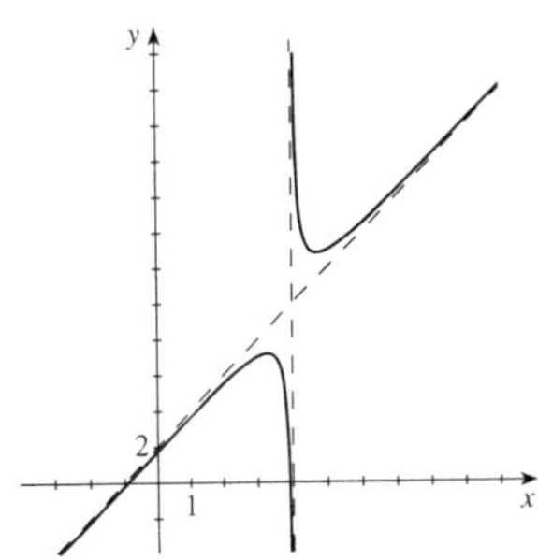


**58.** $r(x) = \dfrac{2x^2 - 6x - 7}{x - 4}$. When $x = 0$, we have $r(0) = \frac{-7}{-4} = \frac{7}{4}$, so the $y$-intercept is $y = \frac{7}{4}$. We use the quadratic formula to find the $x$-intercepts:

$x = \frac{-(-6) \pm \sqrt{(-6)^2 - 4(2)(-7)}}{2(2)} = \frac{6 \pm \sqrt{92}}{4} = \frac{3 \pm \sqrt{23}}{2}$. Thus the $x$-intercepts are $x \approx 3.9$ and $x \approx -0.9$. The vertical asymptote is $x = 4$. Because the degree of the numerator is greater than the degree of the denominator, there is no horizontal asymptote. By long division, we have $r(x) = 2x + 2 + \dfrac{1}{x - 4}$, so the slant asymptote is $r(x) = 2x + 2$.

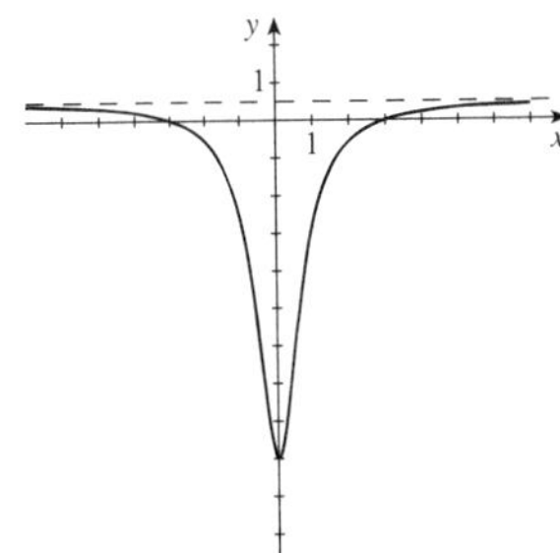


**59.** $r(x) = \dfrac{x^2 - 9}{2x^2 + 1} = \dfrac{(x + 3)(x - 3)}{2x^2 + 1}$. When $x = 0$, we have $r(0) = \frac{-9}{1}$, so the $y$-intercept is $-9$. When $y = 0$, we have $x^2 - 9 = 0 \quad \Leftrightarrow \quad x = \pm 3$ so the $x$-intercepts are $-3$ and $3$. Since $2x^2 + 1 > 0$, the denominator is never zero so there are no vertical asymptotes. The horizontal asymptote is at $y = \frac{1}{2}$ because the degree of the denominator and numerator are the same.

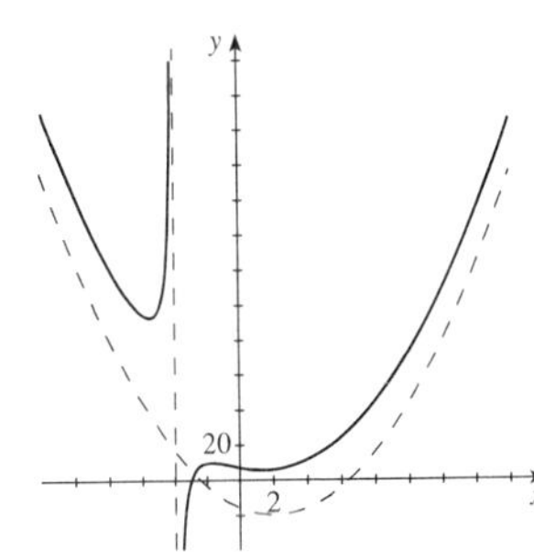


**60.** $r(x) = \dfrac{x^3 + 27}{x + 4}$. When $x = 0$, we have $r(0) = \frac{27}{4}$, so the $y$-intercept is $y = \frac{27}{4}$. When $y = 0$, we have $x^3 + 27 = 0 \quad \Leftrightarrow \quad x^3 = -27 \Rightarrow x = -3$. Thus the $x$-intercept is $x = -3$. The vertical asymptote is $x = -4$. Because the degree of the numerator is greater than the degree of the denominator, there is no horizontal asymptote. By long division, wc have $r(x) = x^2 - 4x + 16 - \dfrac{37}{x + 4}$. So the end behavior of $y$ is like the end behavior of $g(x) = x^2 - 4x - 16$.

**61.** $r(x) = \dfrac{x - 3}{2x + 6}$. From the graph we see that the $x$-intercept is 3, the $y$-intercept is $-0.5$, there is a vertical asymptote at $x = -3$ and a horizontal asymptote at $y = 0.5$, and there is no local extremum.

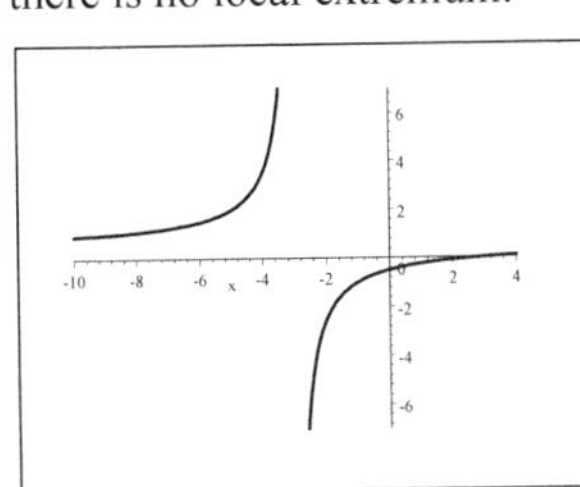


**62.** $r(x) = \dfrac{2x - 7}{x^2 + 9}$. From the graph we see that the $x$-intercept is 3.5, the $y$-intercept is $-0.78$, there is a horizontal asymptote at $y = 0$ and no vertical asymptote, the local minimum is $(-1.11, -0.90)$, and the local maximum is $(8.11, 0.12)$.

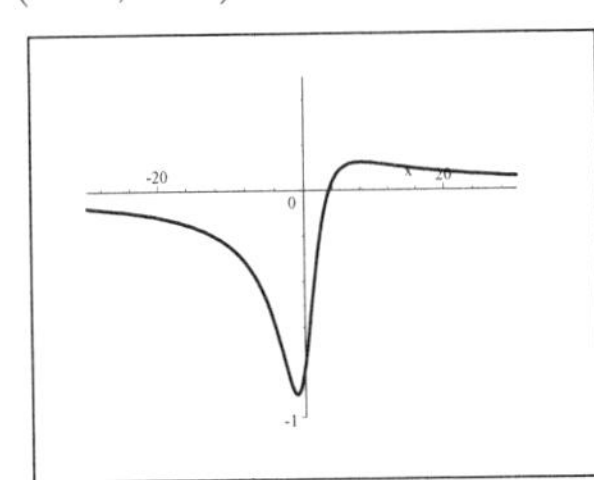

**63.** $r(x) = \dfrac{x^3+8}{x^2-x-2}$. From the graph we see that the $x$-intercept is $-2$, the $y$-intercept is $-4$, there are vertical asymptotes at $x=-1$ and $x=2$ and a horizontal asymptote at $y=0.5$. $r$ has a local maximum of $(0.425, -3.599)$ and a local minimum of $(4.216, 7.175)$. By using long division, we see that $f(x) = x+1+\dfrac{10-x}{x^2-x-2}$, so $f$ has a slant asymptote of $y=x+1$.

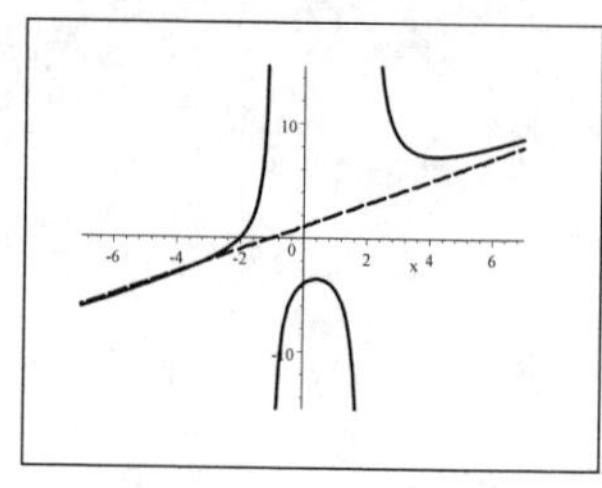

**64.** $r(x) = \dfrac{2x^3-x^2}{x+1}$. From the graph we see that the $x$-intercepts are 0 and $\frac{1}{2}$, the $y$-intercept is 0, there is a vertical asymptote at $x=-1$, the local maximum is $(0,0)$, and the local minima are $(-1.57, 17.90)$ and $(0.32, -0.03)$.

Using long division, we see that $r(x) = 2x^2-3x+3-\dfrac{3}{x+1}$. So the end behavior of $r$ is the same as the end behavior of $g(x) = 2x^2-3x+3$.

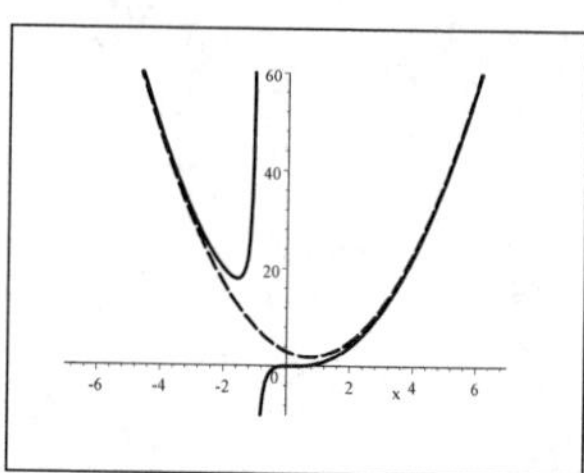

**65.** The graphs of $y = x^4+x^2+24x$ and $y = 6x^3+20$ intersect when $x^4+x^2+24x = 6x^3+20 \Leftrightarrow x^4-6x^3+x^2+24x-20=0$. The possible rational zeros are $\pm1, \pm2, \pm4, \pm5, \pm10, \pm20$.

$$\begin{array}{r|rrrrr} 1 & 1 & -6 & 1 & 24 & -20 \\ & & 1 & -5 & -4 & 20 \\ \hline & 1 & -5 & -4 & 20 & 0 \end{array} \quad \Rightarrow x=1 \text{ is a zero.}$$

So $x^4-6x^3+x^2+24x-20 = (x-1)(x^3-5x^2-4x+20) = 0$. Continuing with the quotient:

$$\begin{array}{r|rrrr} 1 & 1 & -5 & -4 & 20 \\ & & 1 & -4 & -8 \\ \hline & 1 & -4 & -8 & 12 \end{array} \qquad \begin{array}{r|rrrr} 2 & 1 & -5 & -4 & 20 \\ & & 2 & -6 & -20 \\ \hline & 1 & -3 & -10 & 0 \end{array} \quad \Rightarrow x=2 \text{ is a zero.}$$

So $x^4-6x^3+x^2+24x-20 = (x-1)(x-2)(x^2-3x-10) = (x-1)(x-2)(x-5)(x+2) = 0$. Hence, the points of intersection are $(1,26)$, $(2,68)$, $(5,770)$, and $(-2,-28)$.

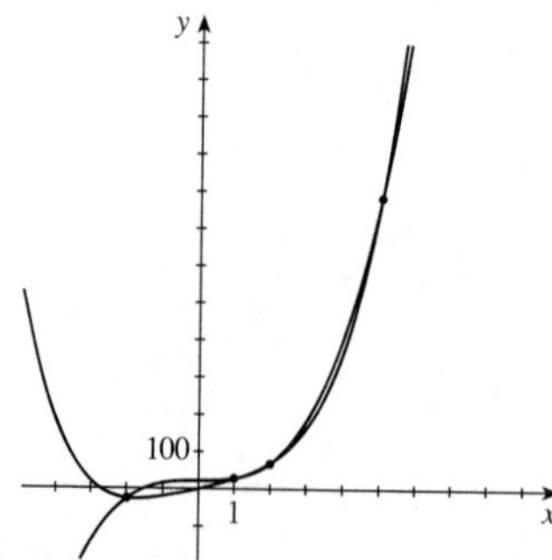

# Chapter 4 Test

**1.** $f(x) = -(x+2)^3 + 27$ has $y$-intercept $y = -2^3 + 27 = 19$ and $x$-intercept where $-(x+2)^3 = -27 \quad \Leftrightarrow \quad x = 1$.

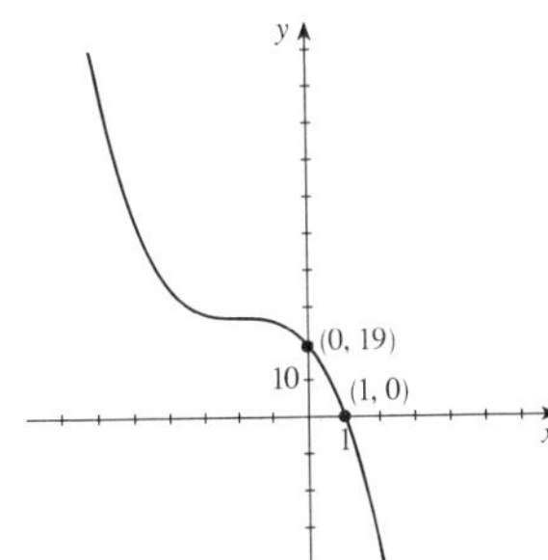


**2. (a)**

$$\begin{array}{r|rrrrr} 2 & 1 & 0 & -4 & 2 & 5 \\ & & 2 & 4 & 0 & 4 \\ \hline & 1 & 2 & 0 & 2 & 9 \end{array}$$

Therefore, the quotient is $Q(x) = x^3 + 2x^2 + 2$, and the remainder is $R(x) = 9$.

**(b)**

$$\begin{array}{r|l} & x^3 + 2x^2 \qquad + \tfrac{1}{2} \\ \hline 2x^2 - 1 & 2x^5 + 4x^4 - x^3 - x^2 + 0x + 7 \\ & 2x^5 \qquad - x^3 \\ \hline & \qquad 4x^4 \qquad - x^2 \\ & \qquad 4x^4 \qquad - 2x^2 \\ \hline & \qquad\qquad x^2 \qquad + 7 \\ & \qquad\qquad x^2 \qquad - \tfrac{1}{2} \\ \hline & \qquad\qquad\qquad \tfrac{15}{2} \end{array}$$

Therefore, the quotient is $Q(x) = x^3 + 2x^2 + \frac{1}{2}$ and the remainder is $R(x) = \frac{15}{2}$.

**3. (a)** Possible rational zeros are: $\pm 1, \pm 3, \pm \frac{1}{2}, \pm \frac{3}{2}$.

**(b)**

$$\begin{array}{r|rrrr} -1 & 2 & -5 & -4 & 3 \\ & & -2 & 7 & -3 \\ \hline & 2 & -7 & 3 & 0 \end{array} \Rightarrow x = -1 \text{ is a zero.}$$

$$\begin{aligned} P(x) &= (x+1)\left(2x^2 - 7x + 3\right) = (x+1)(2x-1)(x-3) \\ &= 2(x+1)\left(x - \tfrac{1}{2}\right)(x-3) \end{aligned}$$

**(c)** The zeros of $P$ are $x = -1, 3, \frac{1}{2}$.

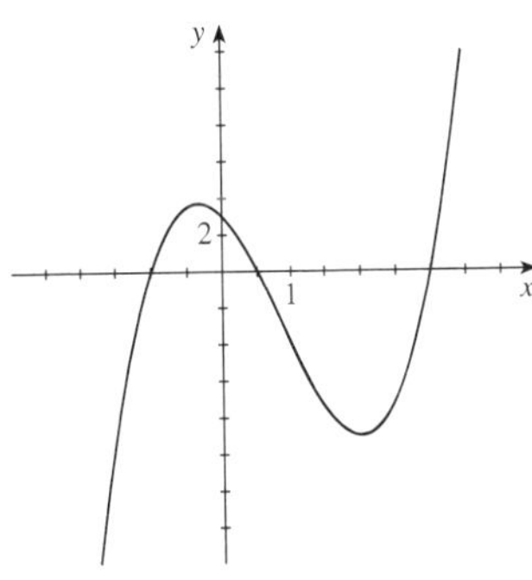


**4.** $P(x) = x^3 - x^2 - 4x - 6$. Possible rational zeros are: $\pm 1, \pm 2, \pm 3, \pm 6$.

$$\begin{array}{r|rrrr} 1 & 1 & -1 & -4 & -6 \\ & & 1 & 0 & -4 \\ \hline & 1 & 0 & -4 & -10 \end{array} \qquad \begin{array}{r|rrrr} 2 & 1 & -1 & -4 & -6 \\ & & 2 & 2 & -4 \\ \hline & 1 & 1 & -2 & -10 \end{array} \qquad \begin{array}{r|rrrr} 3 & 1 & -1 & -4 & -6 \\ & & 3 & 6 & 6 \\ \hline & 1 & 2 & 2 & 0 \end{array} \Rightarrow x = 3 \text{ is a zero.}$$

So $P(x) = (x-3)\left(x^2 + 2x + 2\right)$. Using the quadratic formula on the second factor, we have $x = \dfrac{-2 \pm \sqrt{2^2 - 4(1)(2)}}{2(1)} = \dfrac{-2 \pm \sqrt{-4}}{2} = \dfrac{-2 \pm 2\sqrt{-1}}{2} = -1 \pm i$. So zeros of $P(x)$ are $3$, $-1 - i$, and $-1 + i$.

**5.** $P(x) = x^4 - 2x^3 + 5x^2 - 8x + 4$. The possible rational zeros of $P$ are: $\pm 1$, $\pm 2$, and $\pm 4$. Since there are four changes in sign, $P$ has 4, 2, or 0 positive real zeros.

$$\begin{array}{r|rrrrr} 1 & 1 & -2 & 5 & -8 & 4 \\ & & 1 & -1 & 4 & -4 \\ \hline & 1 & -1 & 4 & -4 & 0 \end{array}$$

So $P(x) = (x-1)(x^3 - x^2 + 4x - 4)$. Factoring the second factor by grouping, we have $P(x) = (x-1)\left[x^2(x-1) + 4(x-1)\right] = (x-1)(x^2+4)(x-1) = (x-1)^2(x-2i)(x+2i)$.

**6.** Since $3i$ is a zero of $P(x)$, $-3i$ is also a zero of $P(x)$. And since $-1$ is a zero of multiplicity 2, $P(x) = (x+1)^2(x-3i)(x+3i) = (x^2+2x+1)(x^2+9) = x^4 + 2x^3 + 10x^2 + 18x + 9$.

**7.** $P(x) = 2x^4 - 7x^3 + x^2 - 18x + 3$.

**(a)** Since $P(x)$ has 4 variations in sign, $P(x)$ can have 4, 2, or 0 positive real zeros. Since $P(-x) = 2x^4 + 7x^3 + x^2 + 18x + 3$ has no variations in sign, there are no negative real zeros.

**(b)**

$$\begin{array}{r|rrrrr} 4 & 2 & -7 & 1 & -18 & 3 \\ & & 8 & 4 & 20 & 8 \\ \hline & 2 & 1 & 5 & 2 & 11 \end{array}$$

Since the last row contains no negative entry, 4 is an upper bound for the real zeros of $P(x)$.

$$\begin{array}{r|rrrrr} -1 & 2 & -7 & 1 & -18 & 3 \\ & & -1 & 9 & -10 & 28 \\ \hline & 2 & -9 & 10 & -28 & 31 \end{array}$$

Since the last row alternates in sign, $-1$ is a lower bound for the real zeros of $P(x)$.

**(c)** Using the upper and lower limit from part (b), we graph $P(x)$ in the viewing rectangle $[-1, 4]$ by $[-1, 1]$. The two real zeros are 0.17 and 3.93.

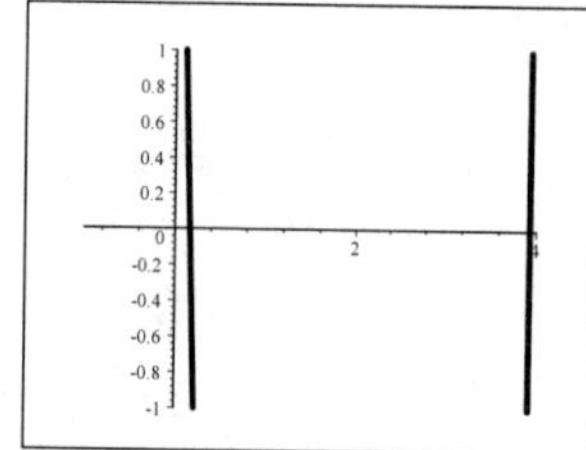


**(d)** Local minimum $(2.8, -70.3)$.

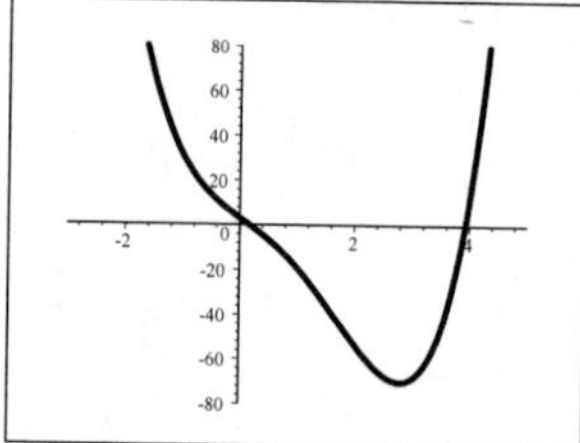


**8.** $r(x) = \dfrac{2x-1}{x^2 - x - 2}$, $s(x) = \dfrac{x^3 + 27}{x^2 + 4}$, $t(x) = \dfrac{x^3 - 9x}{x+2}$ and $u(x) = \dfrac{x^2 + x - 6}{x^2 - 25}$.

**(a)** $r(x)$ has the horizontal asymptote $y = 0$ because the degree of the denominator is greater than the degree of the numerator. $u(x)$ has the horizontal asymptote $y = \frac{1}{1} = 1$ because the degree of the numerator and the denominator are the same.

**(b)** The degree of the numerator of $s(x)$ is one more than the degree of the denominator, so $s$ has a slant asymptote.

**(c)** The denominator of $s(x)$ is never 0, so $s$ has no vertical asymptote.

**(d)** $u(x) = \dfrac{x^2+x-6}{x^2-25} = \dfrac{(x+3)(x-2)}{(x-5)(x+5)}$. When $x=0$, we have $u(x) = \frac{-6}{-25} = \frac{6}{25}$, so the $y$-intercept is $y = \frac{6}{25}$. When $y=0$, we have $x=-3$ or $x=2$, so the $x$-intercepts are $-3$ and $2$. The vertical asymptotes are $x=-5$ and $x=5$. The horizontal asymptote occurs at $y = \frac{1}{1} = 1$ because the degree of the denominator and numerator are the same.

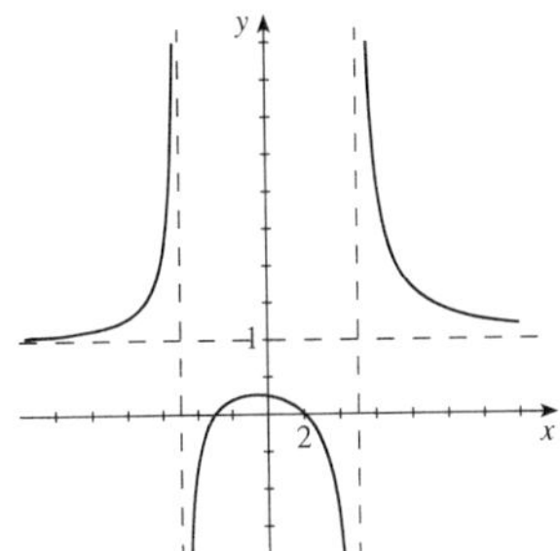


**(e)**

$$
\begin{array}{r|l}
 & x^2 - 2x - 5 \\
\hline
x+2 & x^3 + 0x^2 - 9x + 0 \\
 & x^3 + 2x^2 \\
\hline
 & \quad -2x^2 - 9x \\
 & \quad -2x^2 - 4x \\
\hline
 & \qquad\quad -5x + 0 \\
 & \qquad\quad -5x - 10 \\
\hline
 & \qquad\qquad\quad -10
\end{array}
$$

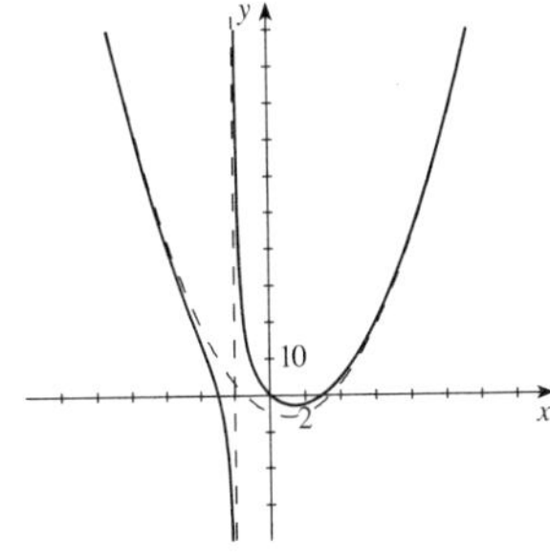


Thus $P(x) = x^2 - 2x - 5$ and $t(x) = \dfrac{x^3 - 9x}{x+2}$ have the same end behavior.

## Focus on Modeling: Fitting Polynomial Curves to Data

**1. (a)** Using a graphing calculator, we obtain the quadratic polynomial $y = -0.275428x^2 + 19.7485x - 273.5523$ (where miles are measured in thousands).

**(b)**

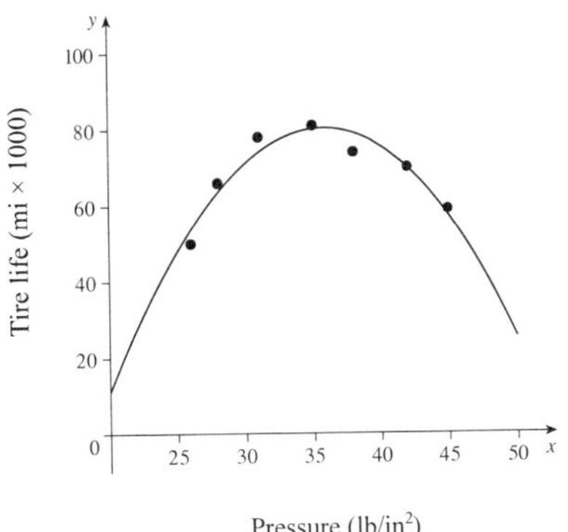


**(c)** Moving the cursor along the path of the polynomial, we find that 35.85 lb/in$^2$ gives the longest tire life.

**2. (a)** Using a graphing calculator, we obtain the quadratic polynomial $y = -0.2783333x^2 + 18.4655x - 166.732$ (where plants/acre are measured in thousands).

**(b)**

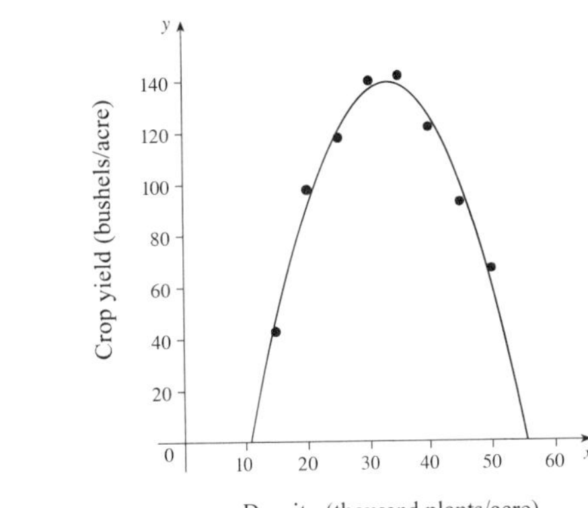


**(c)** Moving the cursor along the path of the polynomial, we find that yield when 37,000 plants are planted per acre is about 135 bushels/acre.

**3. (a)** Using a graphing calculator, we obtain the cubic polynomial

$$y = 0.00203709x^3 - 0.104522x^2 + 1.966206x + 1.45576.$$

**(b)**

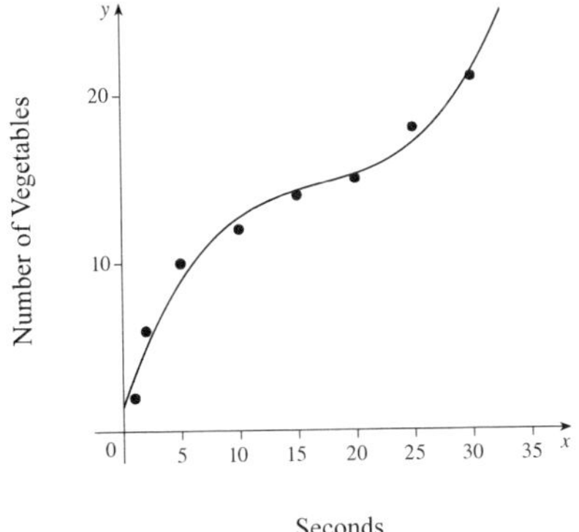


**(c)** Moving the cursor along the path of the polynomial, we find that the subjects could name about 43 vegetables in 40 seconds.

**(d)** Moving the cursor along the path of the polynomial, we find that the subjects could name 5 vegetables in about 2.0 seconds.

**4. (a)** Using a graphing calculator, we obtain the quartic polynomial

$$y = -55.907634x^4 + 1414.8779x^3 - 12,199.094x^2 + 42,577.206x - 25,714.646.$$

**(b)**

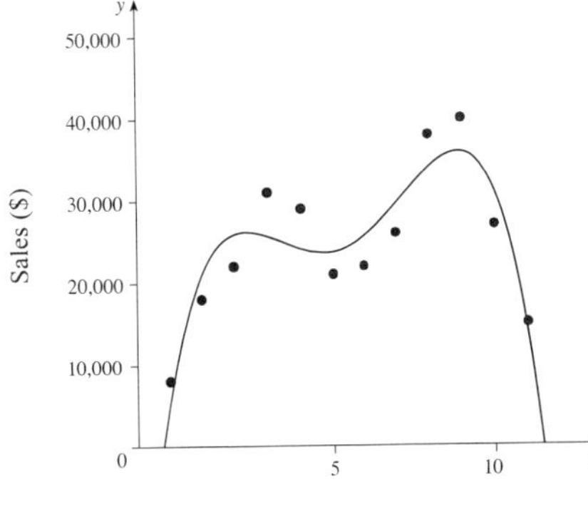


**(c)** Yes since there appear to be two local maxima and one local minimum.

**5. (a)**

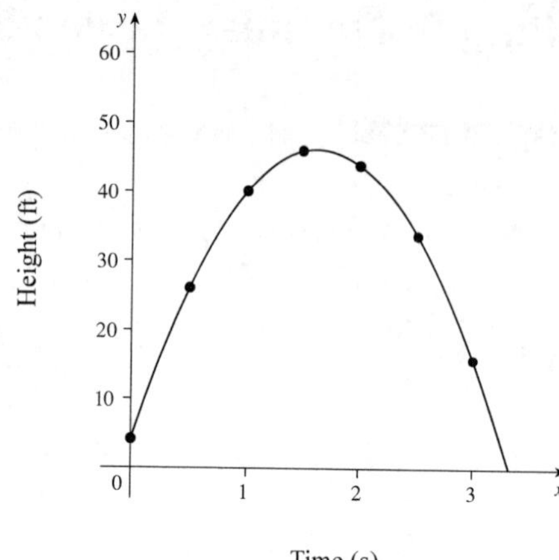


A quadratic model seems appropriate.

**(b)** Using a graphing calculator, we obtain the quadratic polynomial $y = -16.0x^2 + 51.8429x + 4.20714$.

**(c)** Moving the cursor along the path of the polynomial, we find that the ball is 20 ft. above the ground 0.3 seconds and 2.9 seconds after it is thrown upward.

**(d)** Again, moving the cursor along the path of the polynomial, we find that the maximum height is 46.2 ft.

**6. (a)** Using a graphing calculator, we obtain the quadratic polynomial

$y = 0.0120536x^2 - 0.490357x + 4.96571.$

**(b)**

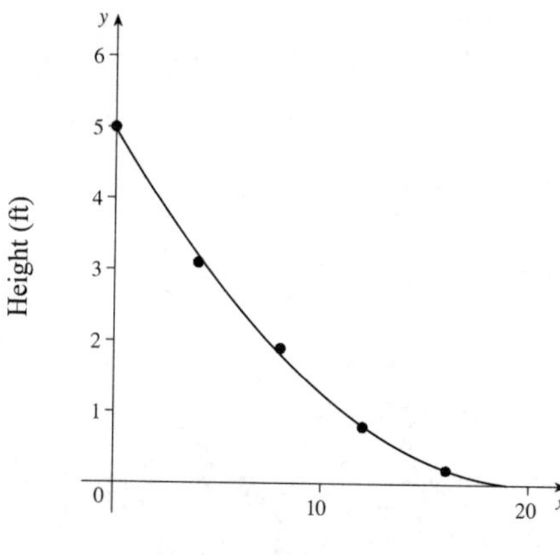


**(c)** Moving the cursor along the path of the polynomial, we find that the tank should drain is 19.0 minutes.

# 5 Exponential and Logarithmic Functions

## 5.1 Exponential Functions

**1.** $f(x) = 4^x$; $f(0.5) = 2$, $f(\sqrt{2}) = 7.103$, $f(\pi) = 77.880$, $f\left(\frac{1}{3}\right) = 1.587$

**2.** $f(x) = 3^{x+1}$; $f(-1.5) = 0.577$, $f(\sqrt{3}) = 20.115$, $f(e) = 59.439$, $f\left(-\frac{5}{4}\right) = 0.760$

**3.** $g(x) = \left(\frac{2}{3}\right)^{x-1}$; $g(1.3) = 0.885$, $g(\sqrt{5}) = 0.606$, $g(2\pi) = 0.117$, $g\left(-\frac{1}{2}\right) = 1.837$

**4.** $g(x) = \left(\frac{3}{4}\right)^{2x}$; $g(0.7) = 0.668$, $g\left(\frac{\sqrt{7}}{2}\right) = 0.467$, $g\left(\frac{1}{\pi}\right) = 0.833$, $g\left(\frac{2}{3}\right) = 0.681$

**5.** $f(x) = 2^x$

| $x$ | $y$ |
|---|---|
| $-4$ | $\frac{1}{16}$ |
| $-2$ | $\frac{1}{4}$ |
| 0 | 1 |
| 2 | 4 |
| 4 | 16 |

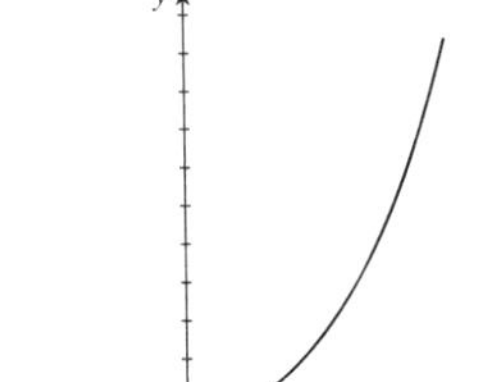


**6.** $g(x) = 8^x$

| $x$ | $y$ |
|---|---|
| $-2$ | $\frac{1}{64}$ |
| $-1$ | $\frac{1}{8}$ |
| 0 | 1 |
| 1 | 8 |
| 2 | 64 |

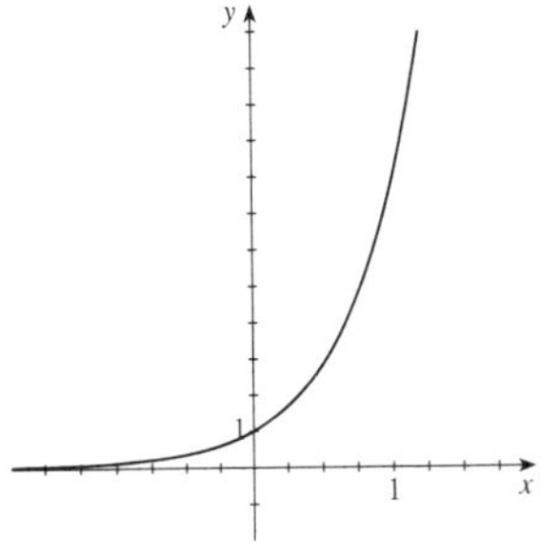


**7.** $f(x) = \left(\frac{1}{3}\right)^x$

| $x$ | $y$ |
|---|---|
| $-2$ | 9 |
| $-1$ | 3 |
| 0 | 1 |
| 1 | $\frac{1}{3}$ |
| 2 | $\frac{1}{9}$ |

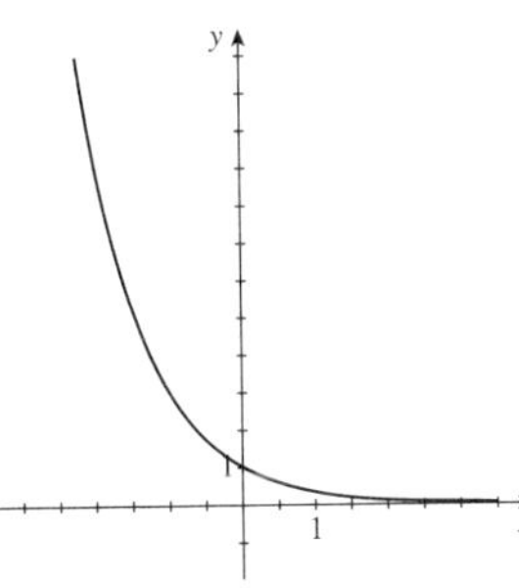


**8.** $h(x) = (1.1)^x$

| $x$ | $y$ |
|---|---|
| $-5$ | 0.620921323 |
| $-1$ | $0.\overline{90}$ |
| 0 | 1 |
| 5 | 1.61051 |
| 10 | 2.59374246 |

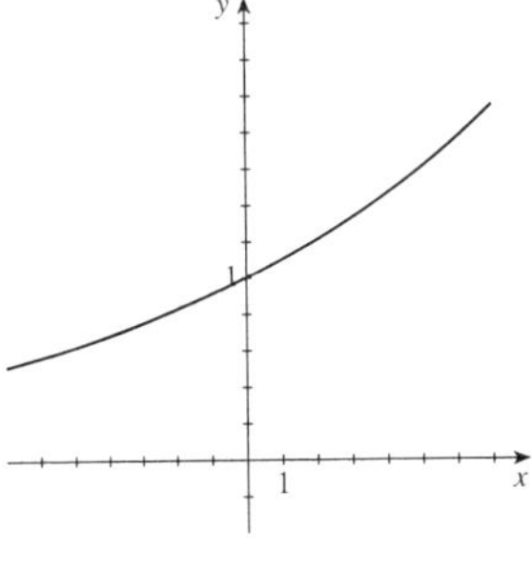


**9.** $f(x) = 3e^x$

| $x$ | $y$ |
|---|---|
| $-2$ | 0.406 |
| $-1$ | 1.104 |
| 0 | 3 |
| 0.5 | 4.946 |
| 1 | 8.155 |
| 1.5 | 13.445 |
| 2 | 22.167 |

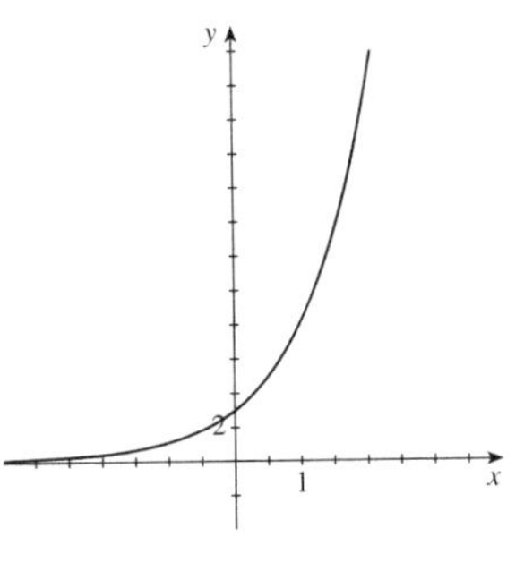


**10.** $g(x) = 2e^{-0.5x}$

| $x$ | $y$ |
|---|---|
| $-4$ | 14.7761 |
| $-3$ | 8.96338 |
| $-2$ | 5.43656 |
| $-1$ | 3.29744 |
| 0 | 2 |
| 1 | 1.21306 |
| 2 | 0.73576 |
| 3 | 0.44626 |
| 4 | 0.27067 |

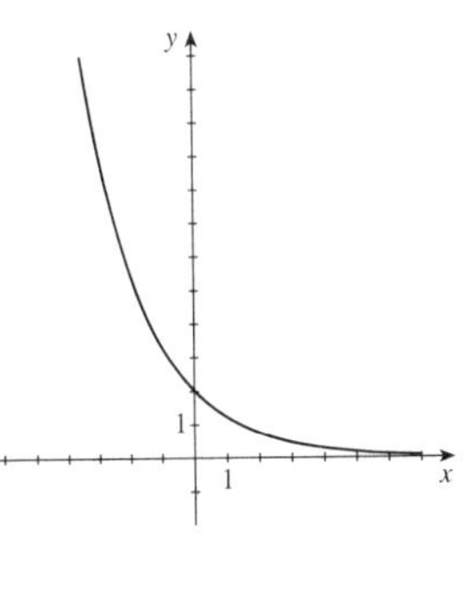

**11.** $f(x) = 2^x$ and $g(x) = 2^{-x}$

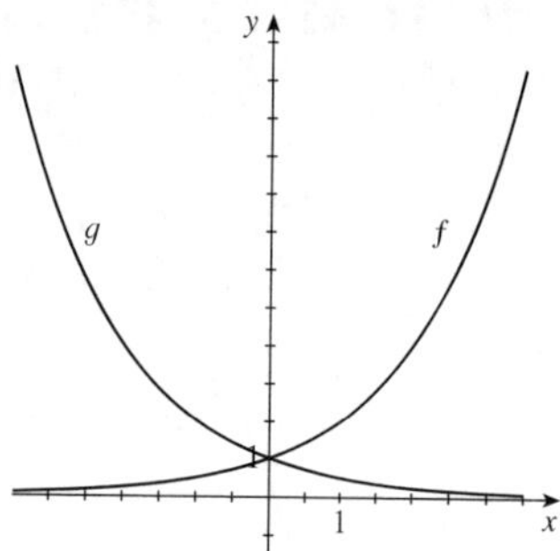


**12.** $f(x) = 3^{-x}$ and $g(x) = \left(\frac{1}{3}\right)^x$

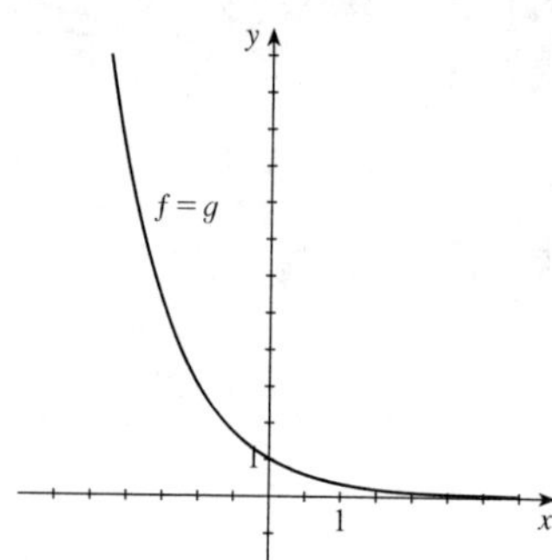


**13.** $f(x) = 4^x$ and $g(x) = 7^x$.

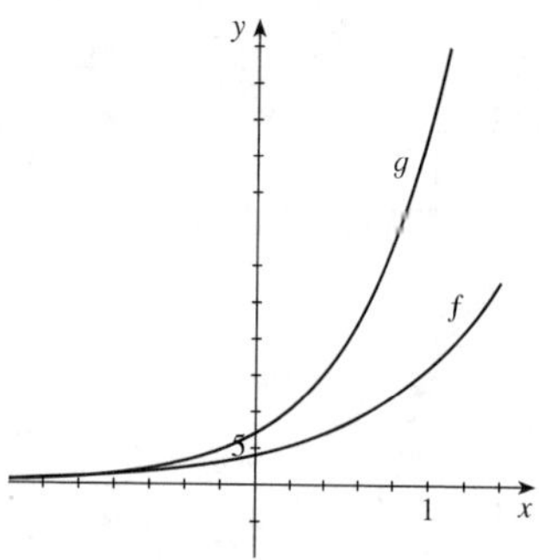


**14.** $f(x) = \left(\frac{2}{3}\right)^x$ and $g(x) = \left(\frac{4}{3}\right)^x$.

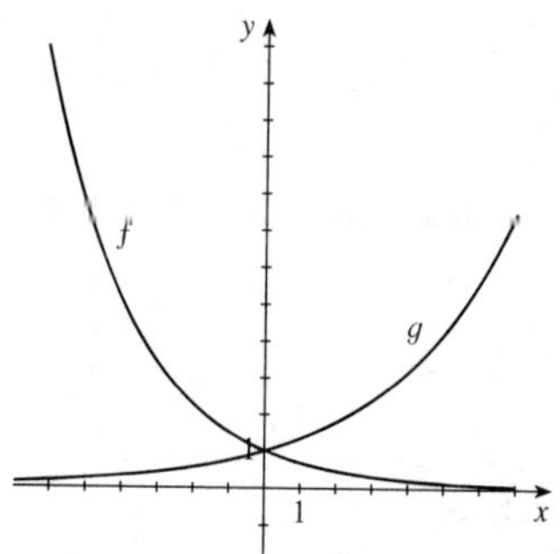


**15.** From the graph, $f(2) = a^2 = 9$, so $a = 3$. Thus $f(x) = 3^x$.

**16.** From the graph, $f(-1) = a^{-1} = \frac{1}{5}$, so $a = 5$. Thus $f(x) = 5^x$.

**17.** From the graph, $f(2) = a^2 = \frac{1}{16}$, so $a = \frac{1}{4}$. Thus $f(x) = \left(\frac{1}{4}\right)^x$.

**18.** From the graph, $f(-3) = a^{-3} = 8$, so $a = \frac{1}{2}$. Thus $f(x) = \left(\frac{1}{2}\right)^x$.

**19.** III **20.** V **21.** I **22.** VI **23.** II **24.** IV

**25.** The graph of $f(x) = -3^x$ is obtained by reflecting the graph of $y = 3^x$ about the $x$-axis. Domain: $(-\infty, \infty)$. Range: $(-\infty, 0)$. Asymptote: $y = 0$.

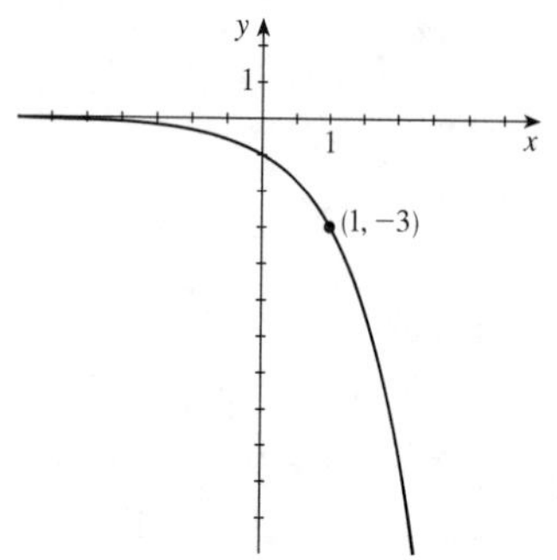


**26.** The graph of $f(x) = 10^{-x}$ is obtained by reflecting the graph of $y = 10^x$ about the $y$-axis. Domain: $(-\infty, \infty)$. Range: $(0, \infty)$. Asymptote: $y = 0$.

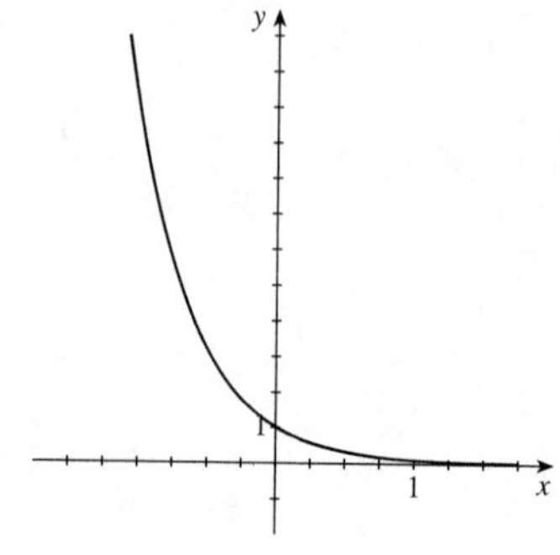

**27.** $g(x) = 2^x - 3$. The graph of $g$ is obtained by shifting the graph of $y = 2^x$ downward 3 units. Domain: $(-\infty, \infty)$. Range: $(-3, \infty)$. Asymptote: $y = -3$.

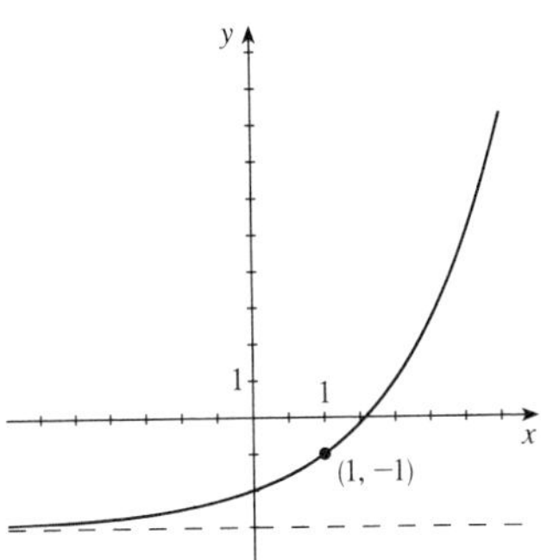


**28.** $g(x) = 2^{x-3}$. The graph of $g$ is obtained by shifting the graph of $y = 2^x$ to the right 3 units. Domain: $(-\infty, \infty)$. Range: $(0, \infty)$. Asymptote: $y = 0$.

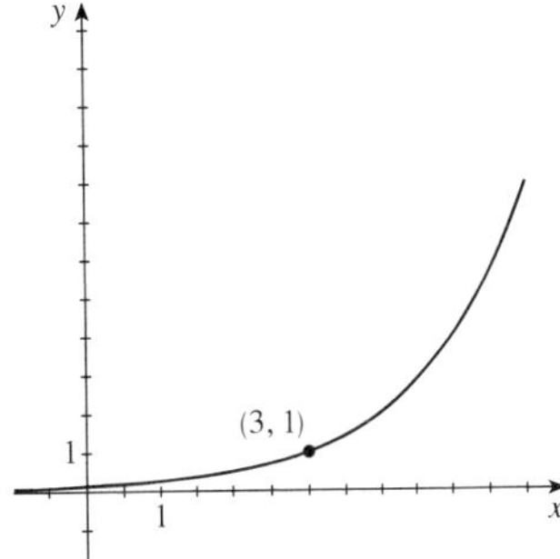


**29.** $h(x) = 4 + \left(\frac{1}{2}\right)^x$. The graph of $h$ is obtained by shifting the graph of $y = \left(\frac{1}{2}\right)^x$ upward 4 units. Domain: $(-\infty, \infty)$. Range: $(4, \infty)$. Asymptote: $y = 4$.

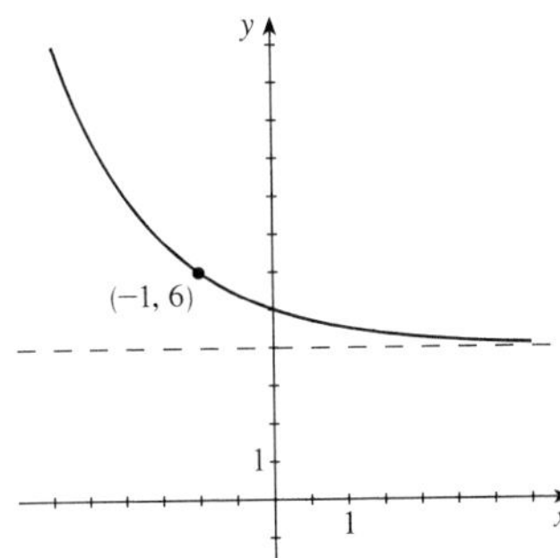


**30.** $h(x) = 6 - 3^x$. The graph of $h$ is obtained by reflecting the graph of $y = 3^x$ about the $x$-axis and shifting upward 6 units. Domain: $(-\infty, \infty)$. Range: $(-\infty, 6)$. Asymptote: $y = 6$.

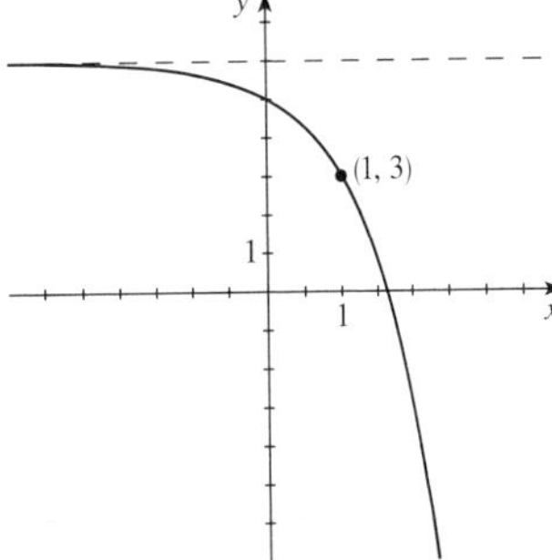


**31.** $f(x) = 10^{x+3}$. The graph of $f$ is obtained by shifting the graph of $y = 10^x$ to the left 3 units. Domain: $(-\infty, \infty)$. Range: $(0, \infty)$. Asymptote: $y = 0$.

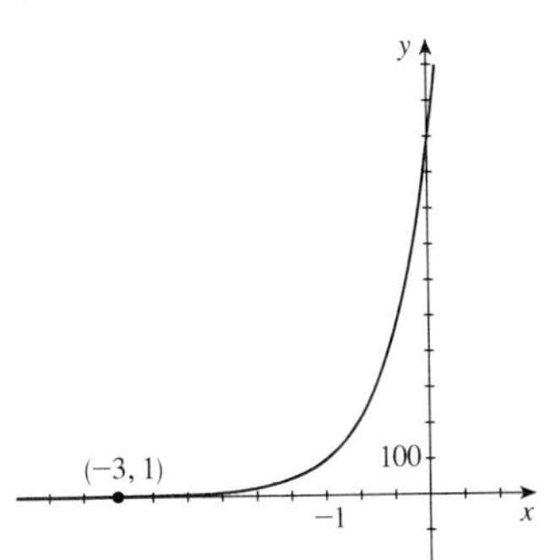


**32.** $f(x) = -\left(\frac{1}{5}\right)^x$. Note that $f(x) = -\left(\frac{1}{5}\right)^x = -5^{-x}$. So the graph of $f$ is obtained by reflecting the graph of $y = 5^x$ about the $y$-axis and about the $x$-axis. Domain: $(-\infty, \infty)$. Range: $(-\infty, 0)$. Asymptote: $y = 0$.

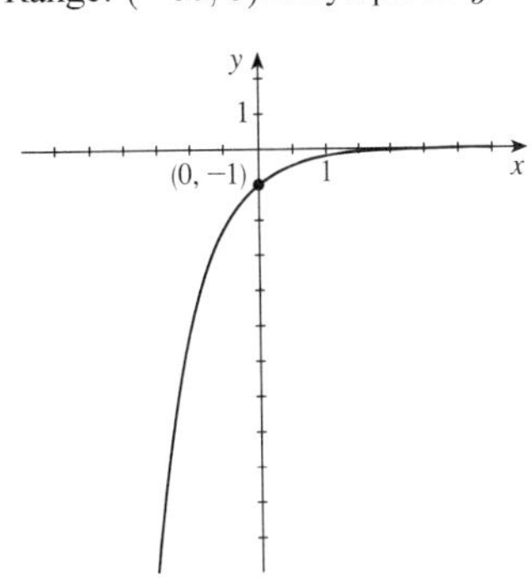

**33.** $y = -e^x$. The graph of $y = -e^x$ is obtained from the graph of $y = e^x$ by reflecting it about the $x$-axis. Domain: $(-\infty, \infty)$. Range: $(-\infty, 0)$. Asymptote: $y = 0$.

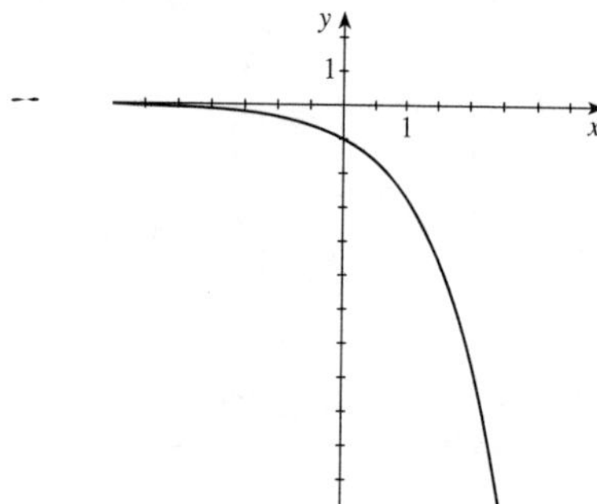

**34.** $f(x) = 1 - e^x$. The graph of $f(x) = 1 - e^x$ is obtained by reflecting the graph of $y = e^x$ about the $x$-axis and then shifting upward 1 unit. Domain: $(-\infty, \infty)$. Range: $(-\infty, 1)$. Asymptote: $y = 1$.

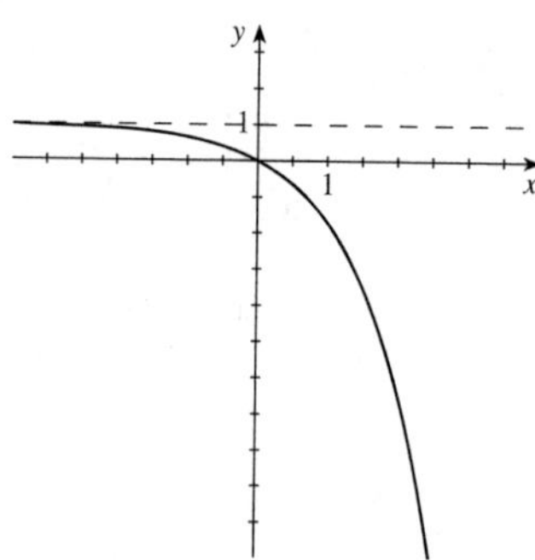

**35.** $y = e^{-x} - 1$. The graph of $y = e^{-x} - 1$ is obtained from the graph of $y = e^x$ by reflecting it about the $y$-axis then shifting downward 1 unit. Domain: $(-\infty, \infty)$. Range: $(-1, \infty)$. Asymptote: $y = -1$.

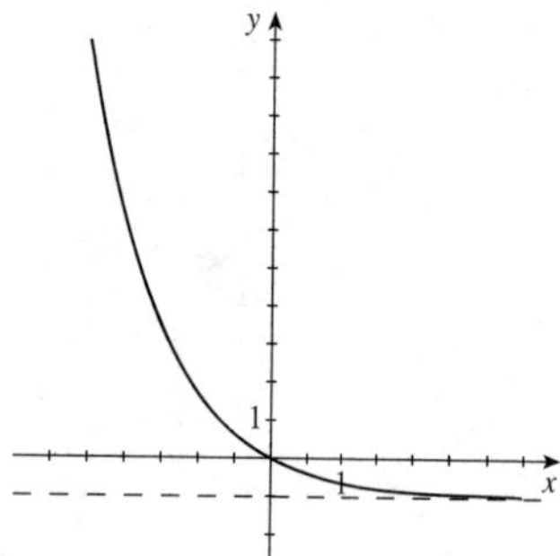

**36.** $f(x) = -e^{-x}$. The graph of $f(x) = -e^{-x}$ is obtained by reflecting the graph of $y = e^x$ about the $y$-axis and then about the $x$-axis. Domain: $(-\infty, \infty)$. Range: $(-\infty, 0)$. Asymptote: $y = 0$.

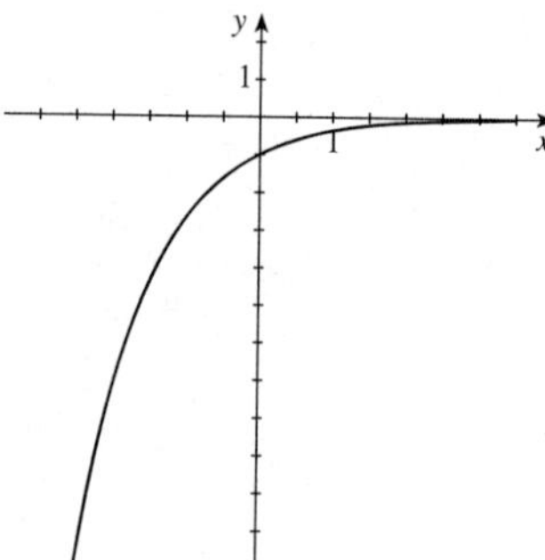

**37.** $y = e^{x-2}$. The graph of $y = e^{x-2}$ is obtained from the graph of $y = e^x$ by shifting it to the right 2 units. Domain: $(-\infty, \infty)$. Range: $(0, \infty)$. Asymptote: $y = 0$.

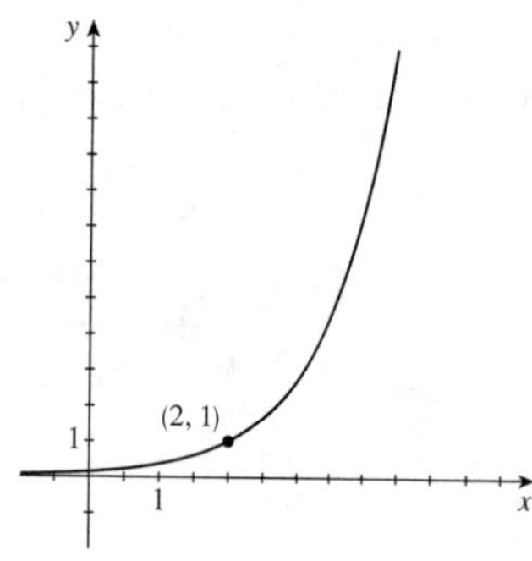


**38.** $f(x) = e^{x-3} + 4$. The graph of $f(x) = e^{x-3} + 4$ is obtained by shifting the graph of $y = e^x$ to the right 3 units, and then upward 4 units. Domain: $(-\infty, \infty)$. Range: $(4, \infty)$. Asymptote: $y = 4$.

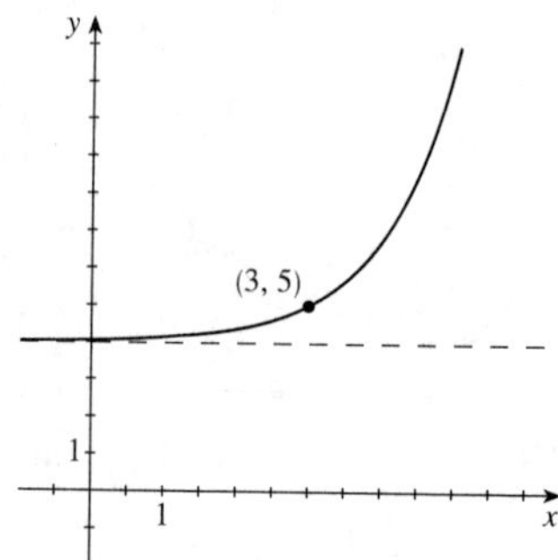


**39.** Using the points $(0, 3)$ and $(2, 12)$, we have $f(0) = Ca^0 = 3 \Leftrightarrow C = 3$. We also have $f(2) = 3a^2 = 12 \Leftrightarrow a^2 = 4 \Leftrightarrow a = 2$ (recall that for an exponential function $f(x) = a^x$ we require $a > 0$). Thus $f(x) = 3 \cdot 2^x$.

**40.** Using the points $(-1,15)$ and $(0,5)$ we have $f(0)=Ca^0=5 \quad\Leftrightarrow\quad C=5$. Then $f(-1)=5a^{-1}=15 \quad\Leftrightarrow\quad a^{-1}=3 \quad\Leftrightarrow\quad a=\frac{1}{3}$. Thus $f(x)=5\left(\frac{1}{3}\right)^x$.

**41. (a)**

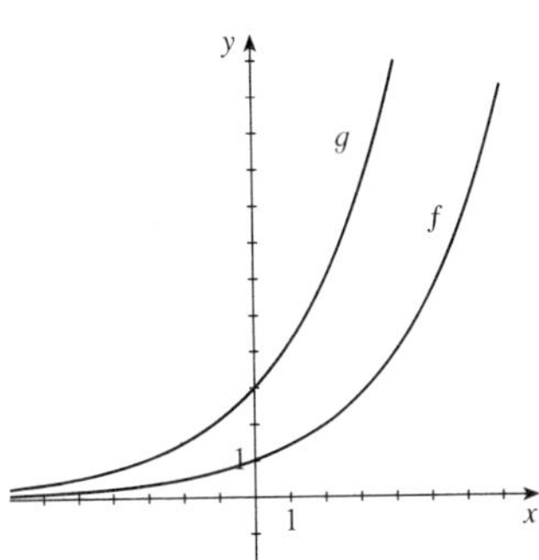


**(b)** Since $g(x)=3(2^x)=3f(x)$ and $f(x)>0$, the height of the graph of $g(x)$ is always three times the height of the graph of $f(x)=2^x$, so the graph of $g$ is steeper than the graph of $f$.

**42. (a)**

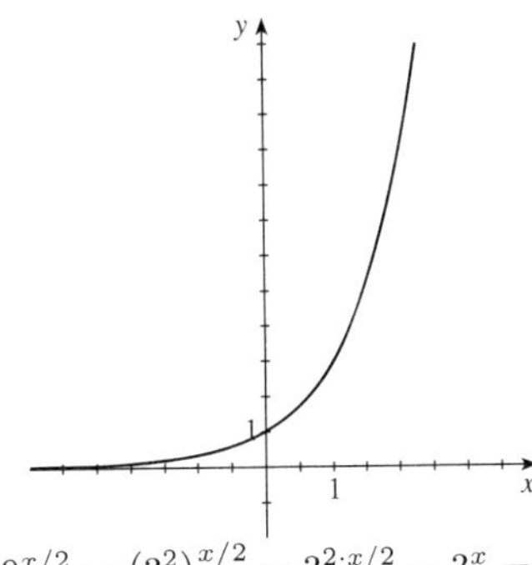


**(b)** $f(x)=9^{x/2}=\left(3^2\right)^{x/2}=3^{2\cdot x/2}=3^x=g(x)$. So $f(x)=g(x)$, and the graphs are the same.

**43.** $f(x)=10^x$, so $\dfrac{f(x+h)-f(x)}{h}=\dfrac{10^{x+h}-10^x}{h}=\dfrac{10^x\cdot 10^h-10^x}{h}=10^x\,\dfrac{10^h-1}{h}$.

**44.**

| $x$ | $f(x)=x^3$ | $g(x)=3^x$ |
|---|---|---|
| 0 | 0 | 1 |
| 1 | 1 | 3 |
| 2 | 8 | 9 |
| 3 | 27 | 27 |
| 4 | 64 | 81 |
| 5 | 125 | 243 |
| 6 | 216 | 729 |
| 7 | 343 | 2187 |
| 8 | 512 | 6561 |
| 9 | 729 | 19,683 |
| 10 | 1000 | 59,049 |
| 15 | 3375 | 14,348,907 |
| 20 | 8000 | 3,486,784,401 |

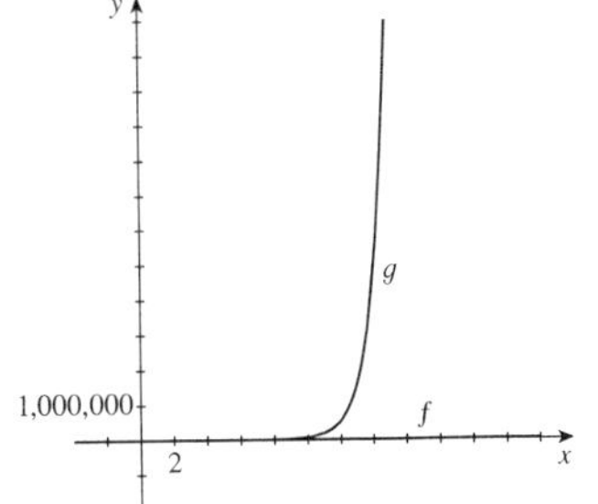


**45.**

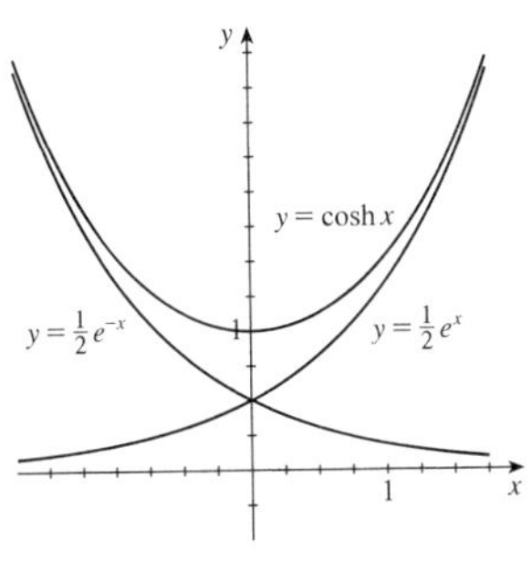


**46.**

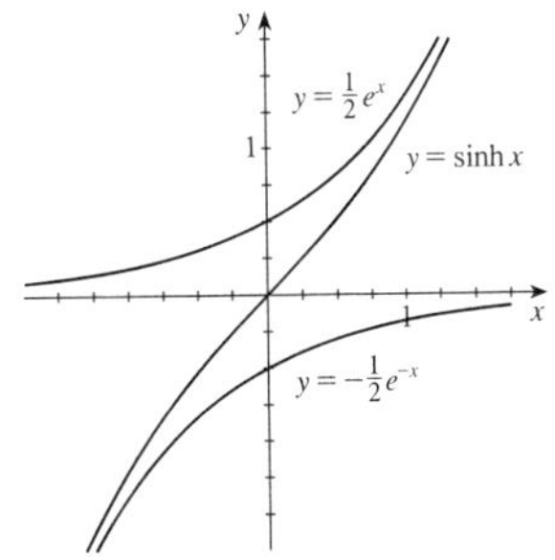

**47.** $\cosh(-x) = \dfrac{e^{-x} + e^{-(-x)}}{2} = \dfrac{e^{-x} + e^{x}}{2} = \dfrac{e^{x} + e^{-x}}{2} = \cosh x$

**48.** $\sinh(-x) = \dfrac{e^{-x} - e^{-(-x)}}{2} = \dfrac{e^{-x} - e^{x}}{2} = -\dfrac{-e^{-x} + e^{x}}{2} = -\dfrac{e^{x} - e^{-x}}{2} = -\sinh x$

**49.** $(\cosh x)^2 - (\sinh x)^2 = \left(\dfrac{e^x + e^{-x}}{2}\right)^2 - \left(\dfrac{e^x - e^{-x}}{2}\right)^2 = \frac{1}{4}\left(e^{2x} + 2 + e^{-2x}\right) - \frac{1}{4}\left(e^{2x} - 2 + e^{-2x}\right) = \frac{2}{4} + \frac{2}{4} = 1$

**50.** $\sinh x \cosh y + \cosh x \sinh y = \dfrac{e^x - e^{-x}}{2} \cdot \dfrac{e^y + e^{-y}}{2} + \dfrac{e^x + e^{-x}}{2} \cdot \dfrac{e^y - e^{-y}}{2}$

$$= \frac{e^{x+y} + e^{x-y} - e^{y-x} - e^{-(x+y)}}{4} + \frac{e^{x+y} - e^{x-y} + e^{y-x} - e^{-(x+y)}}{4} = \frac{e^{x+y} - e^{-(x+y)}}{2} = \sinh(x+y)$$

**51. (a)** From the graphs below, we see that the graph of $f$ ultimately increases much more quickly than the graph of $g$.

**(i)** $[0, 5]$ by $[0, 20]$ **(ii)** $[0, 25]$ by $[0, 10^7]$ **(iii)** $[0, 50]$ by $[0, 10^8]$

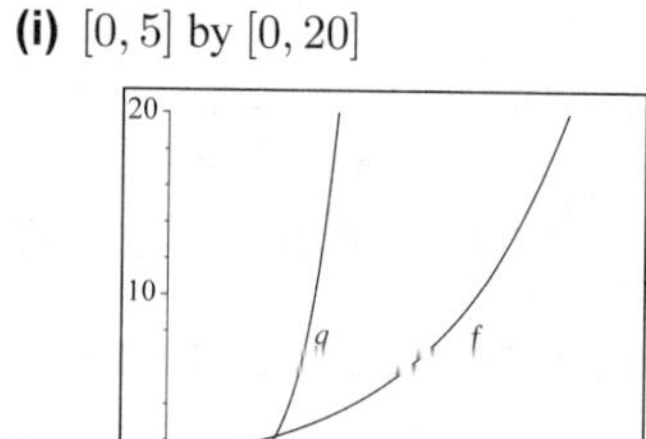


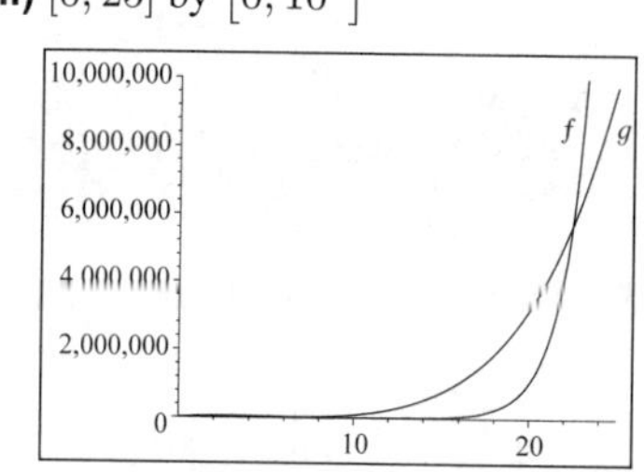


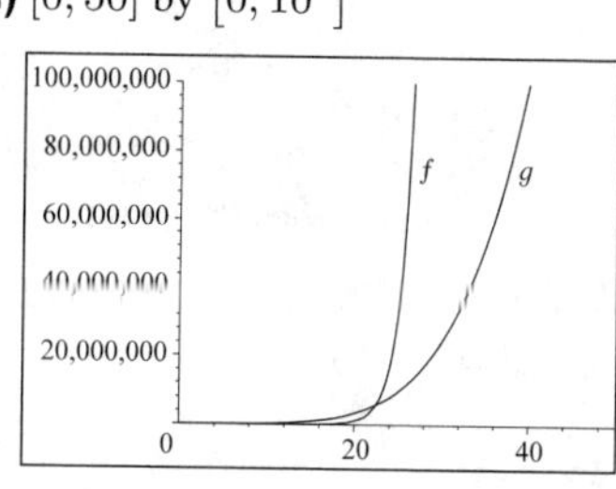


**(b)** From the graphs in parts (a)(i) and (a)(ii), we see that the approximate solutions are $x \approx 1.2$ and $x \approx 22.4$.

**52. (a)** **(i)** $[-4, 4]$ by $[0, 20]$ **(ii)** $[0, 10]$ by $[0, 5000]$ **(iii)** $[0, 20]$ by $[0, 10^5]$

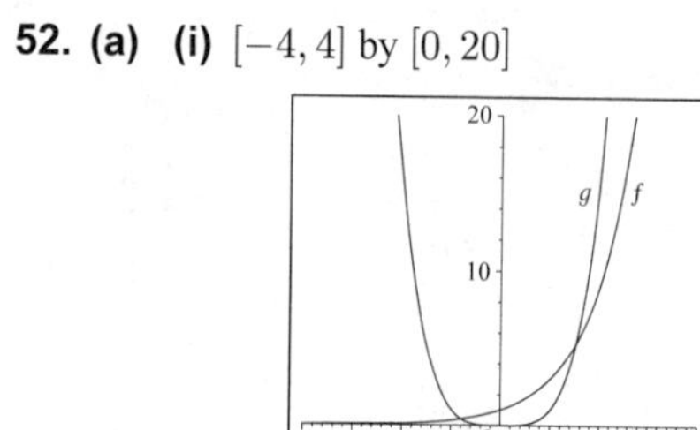


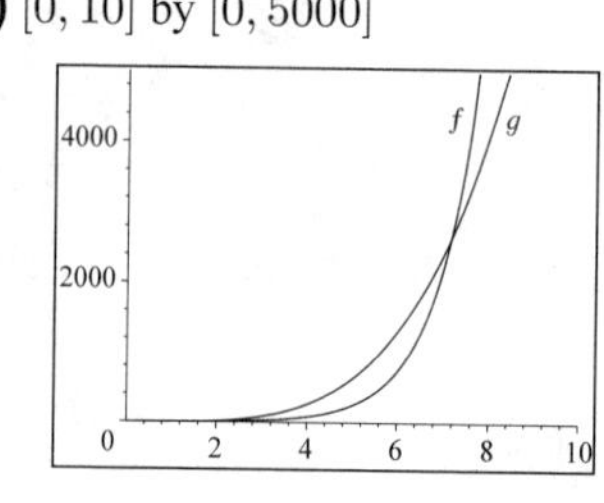


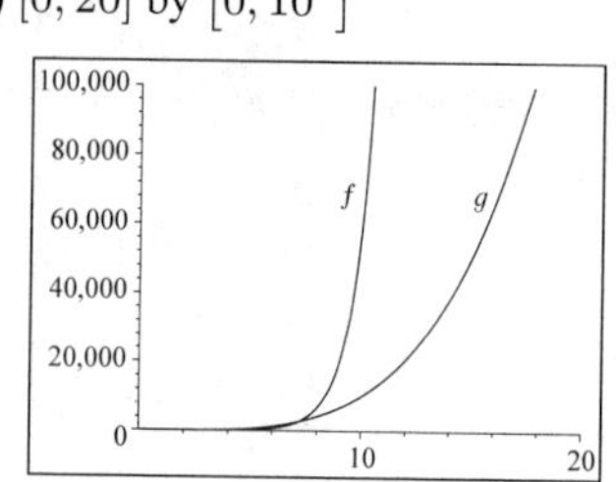


**(b)** From the graphs in parts (i) and (ii), we see that the solutions of $3^x = x^4$ are $x \approx -0.80$, $x \approx 1.52$ and $x \approx 7.17$.

**53.**

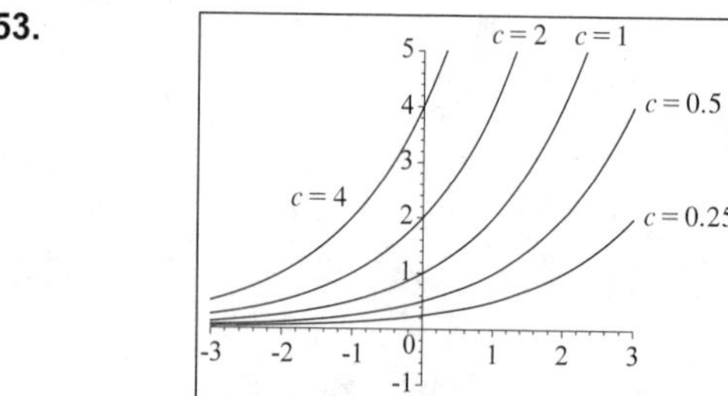

The larger the value of $c$, the more rapidly the graph of $f(x) = c2^x$ increases. Also notice that the graphs are just shifted horizontally 1 unit. This is because of our choice of $c$; each $c$ in this exercise is of the form $2^k$. So $f(x) = 2^k \cdot 2^x = 2^{x+k}$.

**54.**

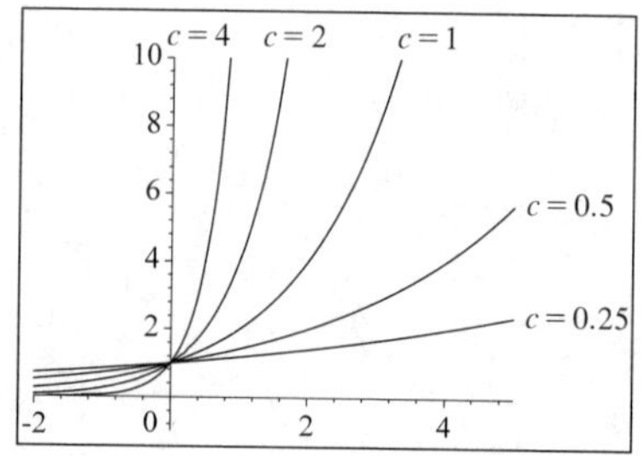


The larger the value of $c$, the more rapidly the graph of $f(x) = 2^{cx}$ increases. In general, $f(x) = 2^{cx} = (2^c)^x$; so, for example, $f(x) = 2^{2x} = (2^2)^x = 4^x$.

**55.**

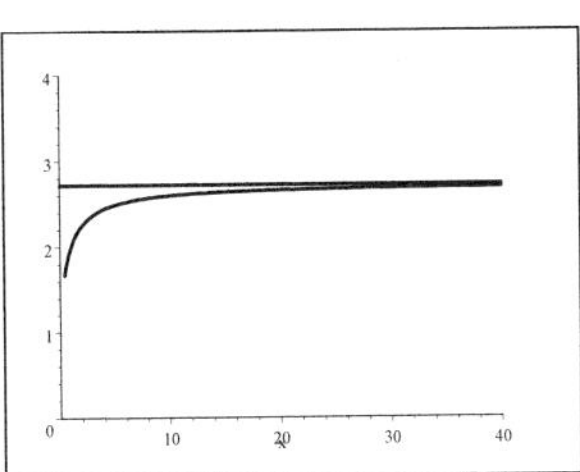

Note from the graph that $y = [1 + (1/x)]^x$ approaches $e$ as $x$ get large.

**56.**

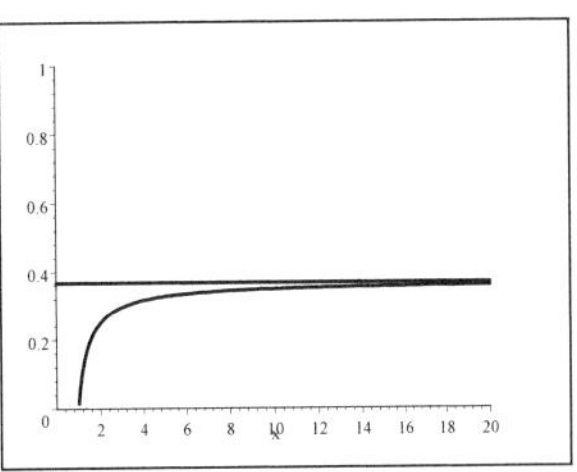

From the graph, we see that $f(x) = [1 - (1/x)]^x$ approaches $1/e \approx 0.368$ as $x$ get large.

**57. (a)**

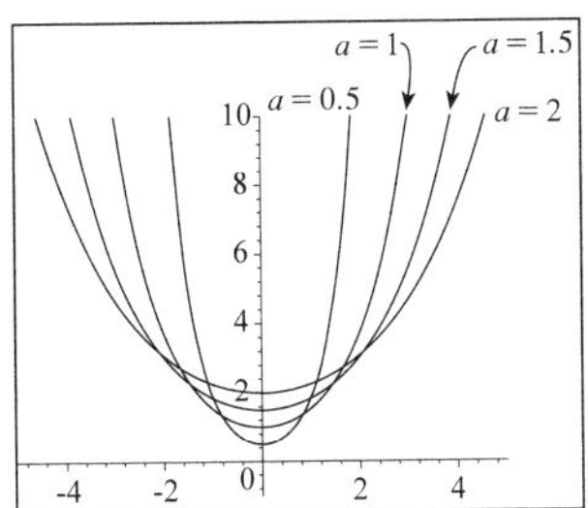


**(b)** As $a$ increases the curve $y = \dfrac{a}{2}\left(e^{x/a} + e^{-x/a}\right)$ flattens out and the $y$ intercept increases.

**58.** $y = 2^{1/x}$ has vertical asymptote $x = 0$ and horizontal asymptote $y = 1$.

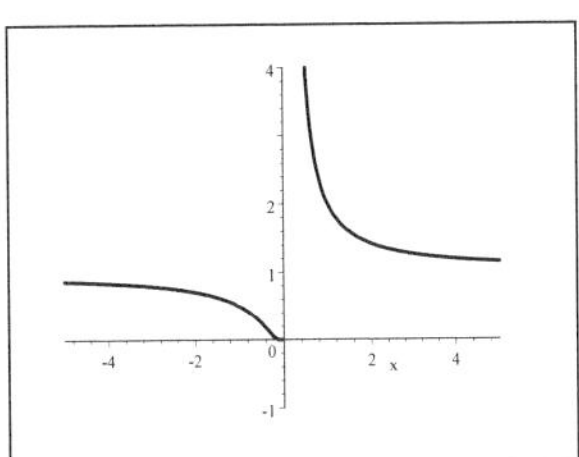

**59.** $y = \dfrac{e^x}{x}$ has vertical asymptote $x = 0$ and horizontal asymptote $y = 0$. As $x \to -\infty$, $y \to 0$, and as $x \to \infty$, $y \to \infty$.

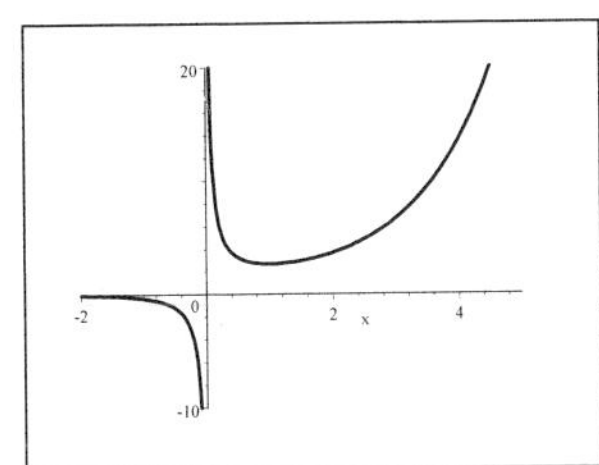

**60.** $g(x) = x^x$. Notice that $g(x)$ is only defined for $x \geq 0$. The graph of $g(x)$ is shown in the viewing rectangle $[0, 1.5]$ by $[0, 1.5]$. From the graph, we see that there is a local minimum of about $0.69$ when $x \approx 0.37$.

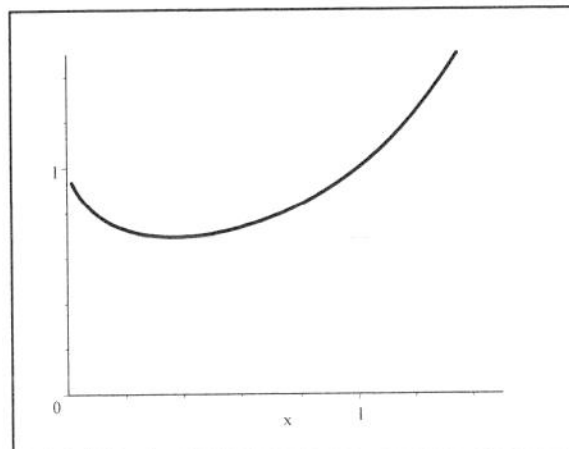

**61.** $g(x) = e^x + e^{-3x}$. The graph of $g(x)$ is shown in the viewing rectangle $[-4, 4]$ by $[0, 20]$. From the graph, we see that there is a local minimum of approximately $1.75$ when $x \approx 0.27$.

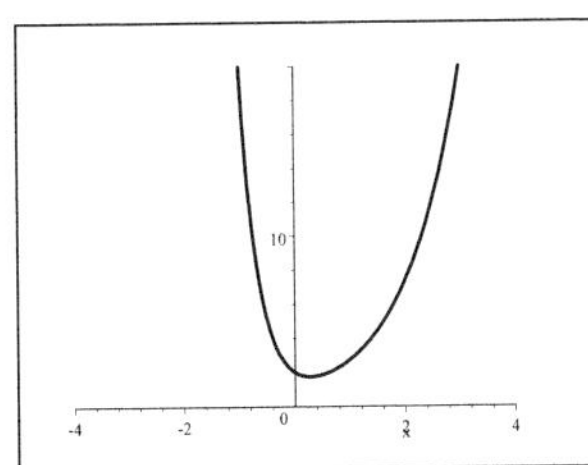

**62.** $y = 10^{x - x^2}$

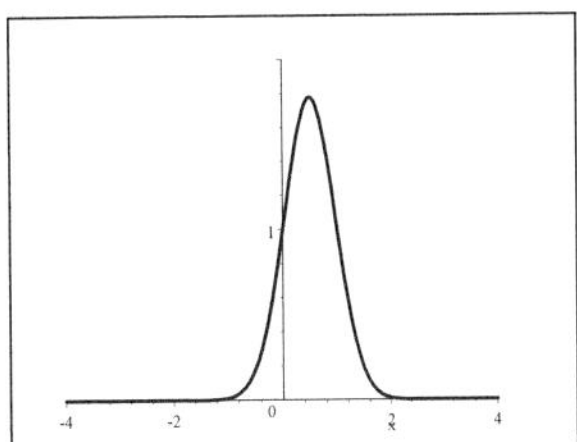

**(a)** From the graph, we see that the function is increasing on $(-\infty, 0.50]$ and decreasing on $[0.50, \infty)$.

**(b)** From the graph, we see that the range is approximately $(0, 1.78]$.

**63.** $y = xe^{-x}$

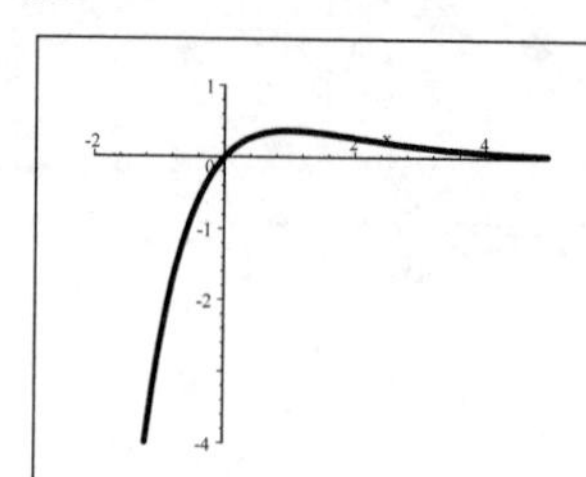

**(a)** From the graph, we see that the function $f(x) = xe^{x}$ is increasing on $(-\infty, 1]$ and decreasing on $[1, \infty)$.

**(b)** From the graph, we see that the range is approximately $(-\infty, 0.37)$.

**64.** $D(t) = 50e^{-0.2t}$. So when $t = 3$ we have $D(3) = 50e^{-0.2(3)} \approx 27.4$ milligrams.

**65.** $m(t) = 13e^{-0.015t}$

**(a)** $m(0) = 13$ kg.

**(b)** $m(45) = 13e^{-0.015(45)} = 13e^{-0.675} = 6.619$ kg. Thus the mass of the radioactive substance after 45 days is about 6.6 kg.

**66. (a)** $m(0) = 6e^{-0.087(0)} \approx 6$ grams

**(b)** $m(20) = 6e^{-0.087(20)} \approx 6(0.1755) = 1.053$. Thus approximately 1 gram of radioactive iodine remains after 20 days.

**67.** $v(t) = 80\left(1 - e^{-0.2t}\right)$

**(a)** $v(0) = 80\left(1 - e^{0}\right) = 80(1 - 1) = 0$.

**(b)** $v(5) = 80\left(1 - e^{-0.2(5)}\right) \approx 80(0.632) = 50.57$ ft/s. So the velocity after 5 s is about 50.6 ft/s.

$v(10) = 80\left(1 - e^{-0.2(10)}\right) \approx 80(0.865) = 69.2$ ft/s. So the velocity after 10 s is about 69.2 ft/s.

**(c)**

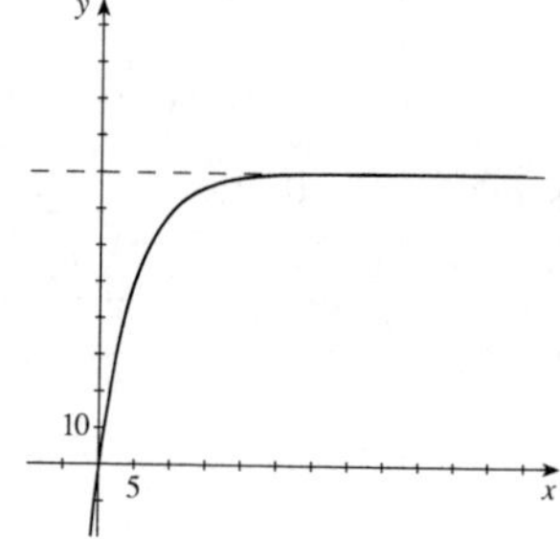


**(d)** The terminal velocity is 80 ft/s.

**68. (a)** $Q(5) = 15\left(1 - e^{-0.04(5)}\right) \approx 15(0.1813) = 2.7345$. Thus approximately 2.7 lb of salt are in the barrel after 5 minutes.

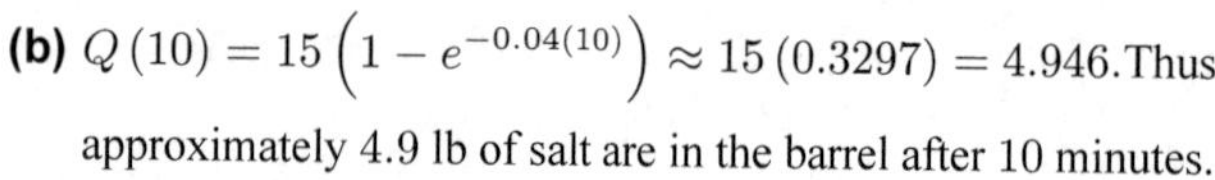

**(b)** $Q(10) = 15\left(1 - e^{-0.04(10)}\right) \approx 15(0.3297) = 4.946$. Thus approximately 4.9 lb of salt are in the barrel after 10 minutes.

**(c)**

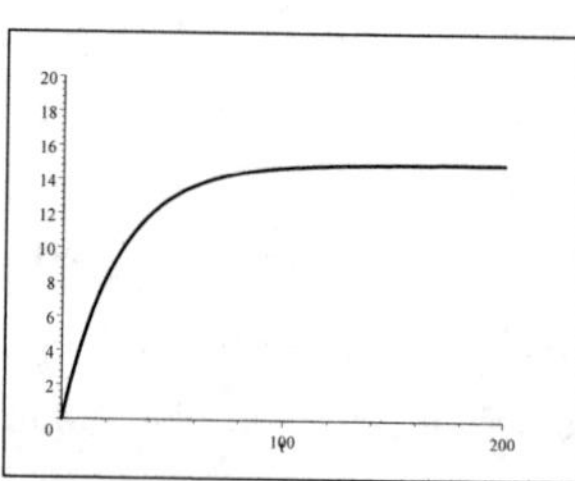


**(d)** The amount of salt approaches 15 lb. This is to be expected, since 50 gal $\times$ 0.3 lb/gal $= 15$ lb.

**69.** $P(t) = \dfrac{1200}{1 + 11e^{-0.2t}}$

**(a)** $P(0) = \dfrac{1200}{1 + 11e^{-0.2(0)}} = \dfrac{1200}{1 + 11} = 100$.

**(b)** $P(10) = \dfrac{1200}{1 + 11e^{-0.2(10)}} \approx 482$. $P(20) = \dfrac{1200}{1 + 11e^{-0.2(20)}} \approx 999$. $P(30) = \dfrac{1200}{1 + 11e^{-0.2(30)}} \approx 1168$.

**(c)** As $t \to \infty$ we have $e^{-0.2t} \to 0$, so $P(t) \to \dfrac{1200}{1 + 0} = 1200$. The graph shown confirms this.

**70.** $n(t) = \dfrac{5600}{0.5 + 27.5e^{-0.044t}}$

**(a)** $n(0) = \frac{5600}{28} = 200$

**(b)**

**(c)** From the graph, we see that $n(t)$ approaches about 11,200 as $t$ gets large.

**71.** $D(t) = \dfrac{5.4}{1 + 2.9e^{-0.01t}}$. So $D(20) = \dfrac{5.4}{1 + 2.9e^{-0.01(20)}} \approx 1.600$ ft.

**72. (a)** Substituting $n_0 = 50$ and $t = 12$, we have

$$n(12) = \frac{300}{0.05 + \left(\frac{300}{50} - 0.05\right)e^{-0.55(12)}} \approx 5164.$$

**(b)**

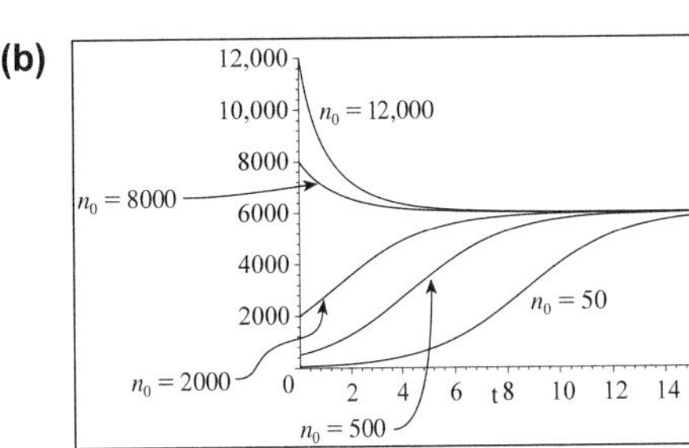


**(c)** The population approaches 6000 rabbits.

**73.** Using the formula $A(t) = P(1+i)^k$ with $P = 5000$, $i = 4\%$ per year $= \dfrac{0.04}{12}$ per month, and $k = 12 \cdot$ number of years, we fill in the table:

| Time (years) | Amount |
|---|---|
| 1 | \$5203.71 |
| 2 | \$5415.71 |
| 3 | \$5636.36 |
| 4 | \$5865.99 |
| 5 | \$6104.98 |
| 6 | \$6353.71 |

**74.** Using the formula $A(t) = P(1+i)^k$ with $P = 5000$, $i = \dfrac{\text{rate per year}}{12}$ per month, and $k = 12 \cdot 5 = 60$ months, we fill in the table:

| Rate per year | Amount |
|---|---|
| 1% | \$5256.25 |
| 2% | \$5525.39 |
| 3% | \$5808.08 |
| 4% | \$6104.98 |
| 5% | \$6416.79 |
| 6% | \$6744.25 |

**75.** $P = 10{,}000$, $r = 0.10$, and $n = 2$. So $A(t) = 10{,}000\left(1 + \frac{0.10}{2}\right)^{2t} = 10{,}000 \cdot 1.05^{2t}$.

**(a)** $A(5) = 10{,}000 \cdot 1.05^{10} \approx 16{,}288.95$, and so the value of the investment is \$16,288.95.

**(b)** $A(10) = 10{,}000 \cdot 1.05^{20} \approx 26{,}532.98$, and so the value of the investment is \$26,532.98.

**(c)** $A(15) = 10{,}000 \cdot 1.05^{30} \approx 43{,}219.42$, and so the value of the investment is \$43,219.42.

**76.** $P = 4{,}000$, $r = 0.16$, and $n = 4$. So $A(t) = 4000\left(1 + \frac{0.16}{4}\right)^{4t} = 4000 \cdot 1.04^{4t}$.

**(a)** $A(4) = 4000 \cdot (1.04)^{16} \approx 7491.92$, and so the amount due is \$7,491.92.

**(b)** $A(6) = 4000 \cdot (1.04)^{24} \approx 10{,}253.22$, and so the amount due is \$10,253.22.

**(c)** $A(8) = 4000 \cdot (1.04)^{32} \approx 14{,}032.23$, and so the amount due is \$14,032.23.

**77.** $P = 3000$ and $r = 0.09$. Then we have $A(t) = 3000\left(1 + \dfrac{0.09}{n}\right)^{nt}$, and so $A(5) = 3000\left(1 + \dfrac{0.09}{n}\right)^{5n}$.

**(a)** If $n = 1$, $A(5) = 3000\left(1 + \frac{0.09}{1}\right)^5 = 3000 \cdot 1.09^5 \approx \$4{,}615.87$.

**(b)** If $n = 2$, $A(5) = 3000\left(1 + \frac{0.09}{2}\right)^{10} = 3000 \cdot 1.045^{10} \approx \$4{,}658.91$.

**(c)** If $n = 12$, $A(5) = 3000\left(1 + \frac{0.09}{12}\right)^{60} = 3000 \cdot 1.0075^{60} \approx \$4{,}697.04$.

**(d)** If $n = 52$, $A(5) = 3000\left(1 + \frac{0.09}{52}\right)^{260} \approx \$4{,}703.11$.

**(e)** If $n = 365$, $A(5) = 3000\left(1 + \frac{0.09}{365}\right)^{1825} \approx \$4{,}704.68$.

**(f)** If $n = 24 \cdot 365 = 8760$, $A(5) = 3000\left(1 + \frac{0.09}{8760}\right)^{43800} \approx \$4{,}704.93$ .

**(g)** If interest is compounded continuously, $A(5) = 3000 \cdot e^{0.45} \approx \$4{,}704.94$.

**78.** $P = 4000$, $n = 4$, and $t = 5$. Then $A(5) = 4000\left(1 + \dfrac{r}{4}\right)^{4(5)}$, and so $A(5) = 4000\left(1 + \dfrac{r}{4}\right)^{20}$.

**(a)** If $r = 0.06$, $A(5) = 4000\left(1 + \frac{0.06}{4}\right)^{20} = 4000 \cdot (1.015)^{20} \approx \$5{,}387.42$.

**(b)** If $r = 0.065$, $A(5) = 4000\left(1 + \frac{0.065}{4}\right)^{20} = 4000 \cdot (1.01625)^{20} \approx \$5{,}521.68$.

**(c)** If $r = 0.07$, $A(5) = 4000\left(1 + \frac{0.07}{4}\right)^{20} = 4000 \cdot (1.0175)^{20} \approx \$5{,}659.11$.

**(d)** If $r = 0.08$, $A(5) = 4000\left(1 + \frac{0.08}{4}\right)^{20} = 4000 \cdot (1.02)^{20} \approx \$5{,}943.79$.

**79.** We find the effective rate with $P = 1$ and $t = 1$. So $A = \left(1 + \dfrac{r}{n}\right)^{n}$

**(i)** $n = 2$, $r = 0.085$; $A(2) = \left(1 + \frac{0.085}{2}\right)^2 = (1.0425)^2 \approx 1.0868$.

**(ii)** $n = 4$, $r = 0.0825$; $A(4) = \left(1 + \frac{0.0825}{4}\right)^4 = (1.020625)^4 \approx 1.0851$.

**(iii)** Continuous compounding: $r = 0.08$; $A(1) = e^{0.08} \approx 1.0833$.

Since (i) is larger than the others, the best investment is the one at 8.5% compounded semiannually.

**80.** We find the effective rate for $P = 1$ and $t = 1$.

**(i)** If $r = 0.0925$ and $n = 2$, then $A(2) = \left(1 + \frac{0.0925}{2}\right)^2 = (1.04625)^2 \approx 1.0946$.

**(ii)** If $r = 0.09$ and interest is compounded continuously, then $A(1) = e^{0.09} \approx 1.0942$.

Since the effective rate in (i) is greater than the effective rate in (ii), we can see that the account paying $9\frac{1}{4}\%$ per year compounded semiannually is the better investment.

**81. (a)** We must solve for $P$ in the equation $10000 = P\left(1 + \frac{0.09}{2}\right)^{2(3)} = P(1.045)^6 \quad\Leftrightarrow\quad 10000 = 1.3023P \quad\Leftrightarrow$ $P = 7678.96$. Thus the present value is \$7,678.96.

**(b)** We must solve for $P$ in the equation $100000 = P\left(1 + \frac{0.08}{12}\right)^{12(5)} = P(1.00667)^{60} \quad\Leftrightarrow\quad 100000 = 1.4898P$ $\Leftrightarrow\quad P = \$67{,}121.04$.

**82. (a)** $A(t) = 5000\left(1 + \dfrac{0.09}{2}\right)^{2t} = 5000(1.045)^{2t} = 5000(1.092025)^t$. **(b)**

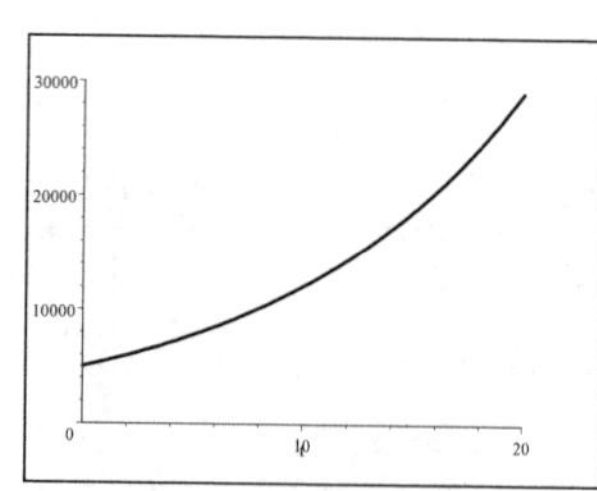


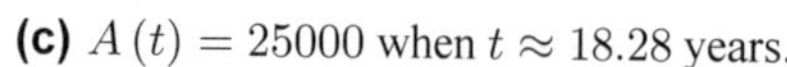

**(c)** $A(t) = 25000$ when $t \approx 18.28$ years.

**83. (a)** In this case the payment is \$1 million.

**(b)** In this case the total pay is $2 + 2^2 + 2^3 + \cdots + 2^{30} > 2^{30}$ cents $= \$10{,}737{,}418.24$. Since this is much more than method (a), method (b) is more profitable.

**84.** Since $f(40) = 2^{40} = 1{,}099{,}511{,}627{,}776$, it would take a sheet of paper 4 inches by 1,099,511,627,776 inches. Since there are 12 inches in a foot and 5,280 feet in a mile, 1,099,511,627,776 inches $\approx$ 1.74 million miles. So the dimensions of the sheet of paper required are 4 inches by about 1.74 million miles.

# 5.2 Logarithmic Functions

**1.**

| Logarithmic form | Exponential form |
|---|---|
| $\log_8 8 = 1$ | $8^1 = 8$ |
| $\log_8 64 = 2$ | $8^2 = 64$ |
| $\log_8 4 = \frac{2}{3}$ | $8^{2/3} = 4$ |
| $\log_8 512 = 3$ | $8^3 = 512$ |
| $\log_8 \frac{1}{8} = -1$ | $8^{-1} = \frac{1}{8}$ |
| $\log_8 \frac{1}{64} = -2$ | $8^{-2} = \frac{1}{64}$ |

**2.**

| Logarithmic form | Exponential form |
|---|---|
| $\log_4 64 = 3$ | $4^3 = 64$ |
| $\log_4 2 = \frac{1}{2}$ | $4^{1/2} = 2$ |
| $\log_4 8 = \frac{3}{2}$ | $4^{3/2} = 8$ |
| $\log_4 \frac{1}{16} = -2$ | $4^{-2} = \frac{1}{16}$ |
| $\log_4 \frac{1}{2} = -\frac{1}{2}$ | $4^{-1/2} = \frac{1}{2}$ |
| $\log_4 \frac{1}{32} = -\frac{5}{2}$ | $4^{-5/2} = \frac{1}{32}$ |

**3.** **(a)** $5^2 = 25$ **(b)** $5^0 = 1$

**4.** **(a)** $10^{-1} = 0.1$ **(b)** $8^3 = 512$

**5.** **(a)** $8^{1/3} = 2$ **(b)** $2^{-3} = \frac{1}{8}$

**6.** **(a)** $3^4 = 81$ **(b)** $8^{2/3} = 4$

**7.** **(a)** $e^x = 5$ **(b)** $e^5 = y$

**8.** **(a)** $e^2 = x + 1$ **(b)** $e^4 = x - 1$

**9.** **(a)** $\log_5 125 = 3$ **(b)** $\log_{10} 0.0001 = -4$

**10.** **(a)** $\log_{10} 1000 = 3$ **(b)** $\log_{81} 9 = \frac{1}{2}$

**11.** **(a)** $\log_8 \frac{1}{8} = -1$ **(b)** $\log_2 \left(\frac{1}{8}\right) = -3$

**12.** **(a)** $\log_4 0.125 = -\frac{3}{2}$ **(b)** $\log_7 343 = 3$

**13.** **(a)** $\ln 2 = x$ **(b)** $\ln y = 3$

**14.** **(a)** $\ln 0.5 = x + 1$ **(b)** $\ln t = 0.5x$

**15.** **(a)** $\log_3 3 = 1$ **(b)** $\log_3 1 = \log_3 3^0 = 0$ **(c)** $\log_3 3^2 = 2$

**16.** **(a)** $\log_5 5^4 = 4$ **(b)** $\log_4 64 = \log_4 4^3 = 3$ **(c)** $\log_9 9 = 1$

**17.** **(a)** $\log_6 36 = \log_6 6^2 = 2$ **(b)** $\log_9 81 = \log_9 9^2 = 2$ **(c)** $\log_7 7^{10} = 10$

**18.** **(a)** $\log_2 32 = \log_2 2^5 = 5$ **(b)** $\log_8 8^{17} = 17$ **(c)** $\log_6 1 = \log_6 6^0 = 0$

**19.** **(a)** $\log_3 \left(\frac{1}{27}\right) = \log_3 3^{-3} = -3$ **(b)** $\log_{10} \sqrt{10} = \log_{10} 10^{1/2} = \frac{1}{2}$ **(c)** $\log_5 0.2 = \log_5 \left(\frac{1}{5}\right) = \log_5 5^{-1} = -1$

**20.** **(a)** $\log_5 125 = \log_5 5^3 = 3$ **(b)** $\log_{49} 7 = \log_{49} 49^{1/2} = \frac{1}{2}$ **(c)** $\log_9 \sqrt{3} = \log_9 3^{1/2} = \log_9 \left(9^{1/2}\right)^{1/2} = \log_9 9^{1/4} = \frac{1}{4}$

**21.** **(a)** $2^{\log_2 37} = 37$ **(b)** $3^{\log_3 8} = 8$ **(c)** $e^{\ln \sqrt{5}} = \sqrt{5}$

**22.** **(a)** $e^{\ln \pi} = \pi$ **(b)** $10^{\log 5} = 5$ **(c)** $10^{\log 87} = 87$

**23.** **(a)** $\log_8 0.25 = \log_8 8^{-2/3} = -\frac{2}{3}$ **(b)** $\ln e^4 = 4$ **(c)** $\ln \left(\dfrac{1}{e}\right) = \ln e^{-1} = -1$

**24.** **(a)** $\log_4 \sqrt{2} = \log_4 2^{1/2} = \log_4 \left(4^{1/2}\right)^{1/2} = \log_4 4^{1/4} = \frac{1}{4}$

**(b)** $\log_4 \left(\frac{1}{2}\right) = \log_4 2^{-1} = \log_4 \left(4^{1/2}\right)^{-1} = \log_4 4^{-1/2} = -\frac{1}{2}$

**(c)** $\log_4 8 = \log_4 2^3 = \log_4 \left(4^{1/2}\right)^3 = \log_4 4^{3/2} = \frac{3}{2}$

**25.** **(a)** $\log_2 x = 5 \quad \Leftrightarrow \quad x = 2^5 = 32$ **(b)** $x = \log_2 16 = \log_2 2^4 = 4$

**26.** **(a)** $\log_5 x = 4 \quad \Leftrightarrow \quad x = 5^4 = 625$ **(b)** $x = \log_{10} (0.1) = \log_{10} 10^{-1} = -1$

**27.** **(a)** $x = \log_3 243 = \log_3 3^5 = 5$
**(b)** $\log_3 x = 3 \quad \Leftrightarrow \quad x = 3^3 = 27$

**28.** **(a)** $x = \log_4 2 = \log_4 4^{1/2} = \frac{1}{2}$
**(b)** $\log_4 x = 2 \quad \Leftrightarrow \quad x = 4^2 = 16$

**29.** **(a)** $\log_{10} x = 2 \quad \Leftrightarrow \quad x = 10^2 = 100$
**(b)** $\log_5 x = 2 \quad \Leftrightarrow \quad x = 5^2 = 25$

**30.** **(a)** $\log_x 1000 = 3 \quad \Leftrightarrow \quad x^3 = 1000 \quad \Leftrightarrow \quad x = 10$
**(b)** $\log_x 25 = 2 \quad \Leftrightarrow \quad x^2 = 25 \quad \Leftrightarrow \quad x = 5$

**31.** **(a)** $\log_x 16 = 4 \quad \Leftrightarrow \quad x^4 = 16 \quad \Leftrightarrow \quad x = 2$
**(b)** $\log_x 8 = \dfrac{3}{2} \quad \Leftrightarrow \quad x^{3/2} = 8 \quad \Leftrightarrow \quad x = 8^{2/3} = 4$

**32.** **(a)** $\log_x 6 = \frac{1}{2} \quad \Leftrightarrow \quad x^{1/2} = 6 \quad \Leftrightarrow \quad x = 36$
**(b)** $\log_x 3 = \frac{1}{3} \quad \Leftrightarrow \quad x^{1/3} = 3 \quad \Leftrightarrow \quad x = 27$

**33.** **(a)** $\log 2 \approx 0.3010$
**(b)** $\log 35.2 \approx 1.5465$
**(c)** $\log\left(\frac{2}{3}\right) \approx -0.1761$

**34.** **(a)** $\log 50 \approx 1.6990$
**(b)** $\log \sqrt{2} \approx 0.1505$
**(c)** $\log\left(3\sqrt{2}\right) \approx 0.6276$

**35.** **(a)** $\ln 5 \approx 1.6094$
**(b)** $\ln 25.3 \approx 3.2308$
**(c)** $\ln\left(1 + \sqrt{3}\right) \approx 1.0051$

**36.** **(a)** $\ln 27 \approx 3.2958$
**(b)** $\ln 7.39 \approx 2.0001$
**(c)** $\ln 54.6 \approx 4.0000$

**37.** Since the point $(5, 1)$ is on the graph, we have $1 = \log_a 5 \quad \Leftrightarrow \quad a^1 = 5$. Thus the function is $y = \log_5 x$.

**38.** Since the point $\left(\frac{1}{2}, -1\right)$is on the graph, we have $-1 = \log_a\left(\frac{1}{2}\right) \quad \Leftrightarrow \quad a^{-1} = \frac{1}{2} \quad \Leftrightarrow \quad a = 2$. Thus the function is $y = \log_2 x$.

**39.** Since the point $\left(3, \frac{1}{2}\right)$ is on the graph, we have $\frac{1}{2} = \log_a 3 \quad \Leftrightarrow \quad a^{1/2} = 3 \quad \Leftrightarrow \quad a = 9$. Thus the function is $y = \log_9 x$.

**40.** Since the point $(9, 2)$ is on the graph, we have $2 = \log_a 9 \quad \Leftrightarrow \quad a^2 = 9 \quad \Leftrightarrow \quad a = 3$. Thus the function is $y = \log_3 x$.

**41.** II **42.** V **43.** III **44.** IV **45.** VI **46.** I

**47.** The graph of $y = \log_4 x$ is obtained from the graph of $y = 4^x$ by reflecting it about the line $y = x$.

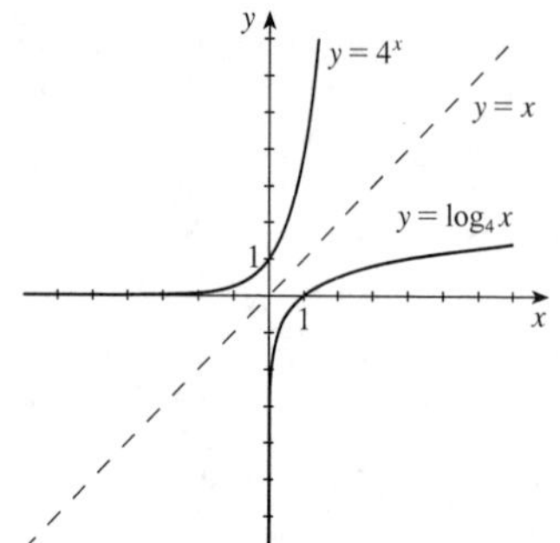


**48.** The graph of $y = \log_3 x$ is obtained from the graph of $y = 3^x$ by reflecting it about the line $y = x$.

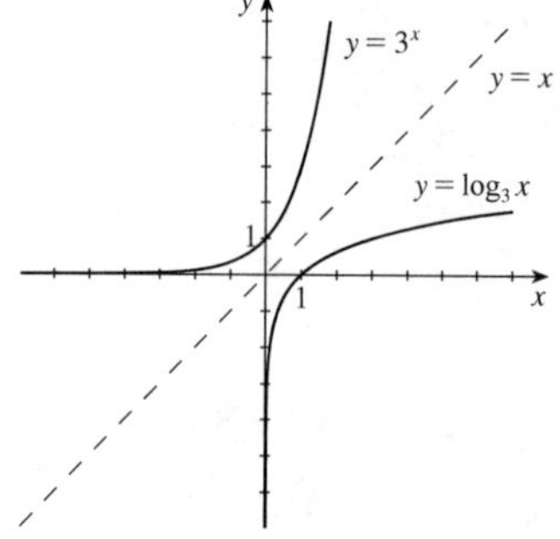

**49.** $f(x) = \log_2(x-4)$. The graph of $f$ is obtained from the graph of $y = \log_2 x$ by shifting it to the right 4 units. Domain: $(4, \infty)$. Range: $(-\infty, \infty)$. Vertical asymptote: $x = 4$.

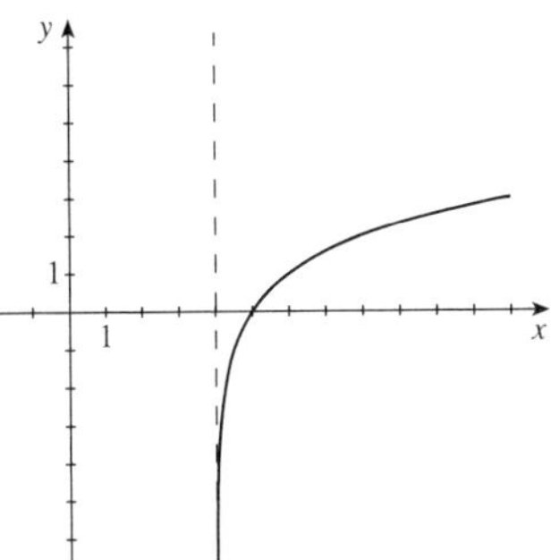


**50.** $f(x) = -\log_{10} x$. The graph of $f$ is obtained from the graph of $y = \log_{10} x$ by reflecting it about the $x$-axis. Domain: $(0, \infty)$. Range: $(-\infty, \infty)$. Vertical asymptote: $x = 0$.

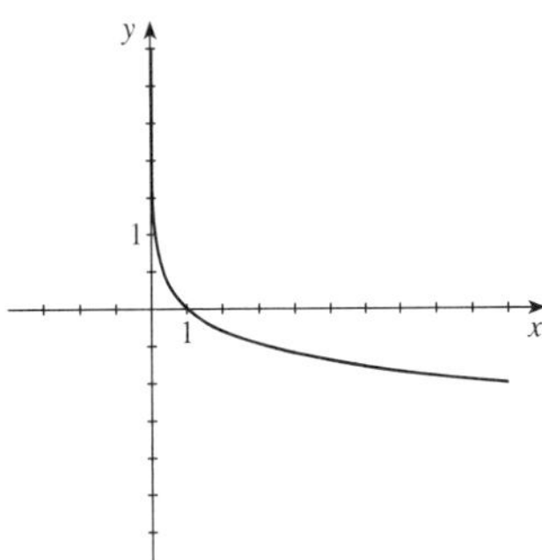


**51.** $g(x) = \log_5(-x)$. The graph of $g$ is obtained from the graph of $y = \log_5 x$ by reflecting it about the $y$-axis. Domain: $(-\infty, 0)$. Range: $(-\infty, \infty)$. Vertical asymptote: $x = 0$.

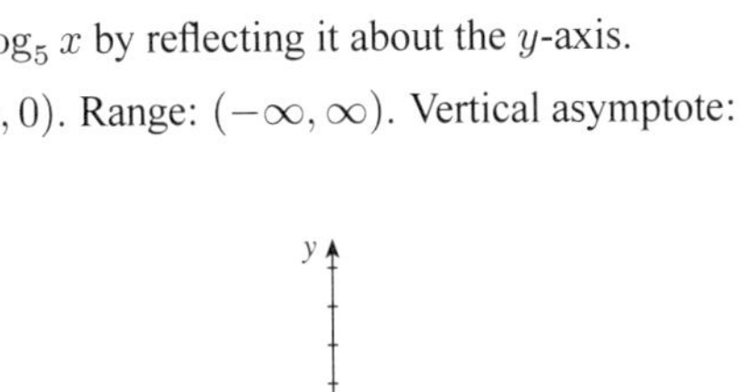


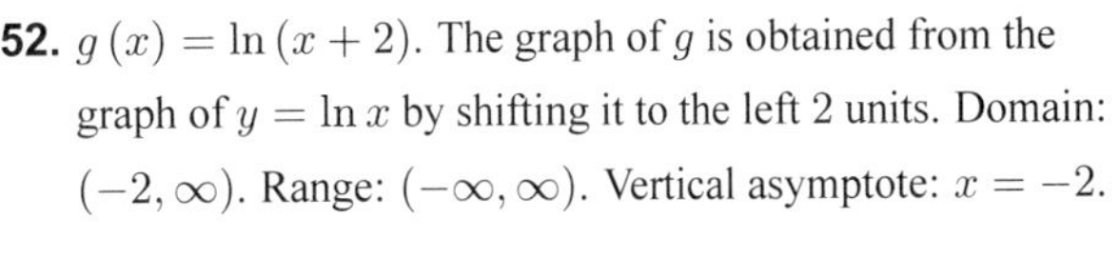

**52.** $g(x) = \ln(x+2)$. The graph of $g$ is obtained from the graph of $y = \ln x$ by shifting it to the left 2 units. Domain: $(-2, \infty)$. Range: $(-\infty, \infty)$. Vertical asymptote: $x = -2$.

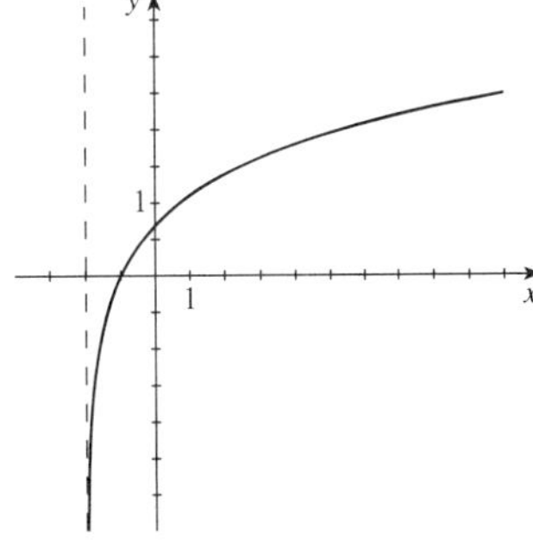


**53.** $y = 2 + \log_3 x$. The graph of $y = 2 + \log_3 x$ is obtained from the graph of $y = \log_3 x$ by shifting it upward 2 units. Domain: $(0, \infty)$. Range: $(-\infty, \infty)$. Vertical asymptote: $x = 0$.

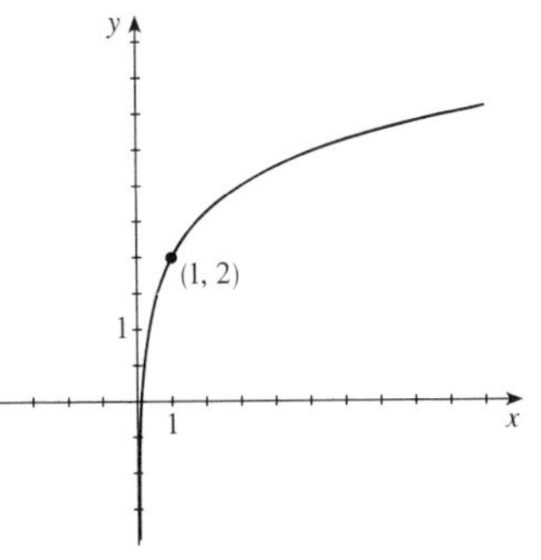


**54.** $y = \log_3(x-1) - 2$. The graph of $y = \log_3(x-1) - 2$ is obtained from the graph of $y = \log_3 x$ by shifting it to the right 1 unit and then downward 2 units. Domain: $(1, \infty)$. Range: $(-\infty, \infty)$. Vertical asymptote: $x = 1$.

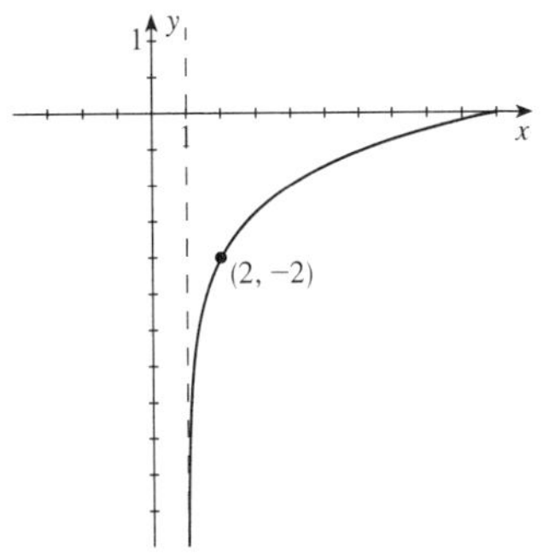

**55.** $y = 1 - \log_{10} x$. The graph of $y = 1 - \log_{10} x$ is obtained from the graph of $y = \log_{10} x$ by reflecting it about the $x$-axis, and then shifting it upward 1 unit. Domain: $(0, \infty)$. Range: $(-\infty, \infty)$. Vertical asymptote: $x = 0$.

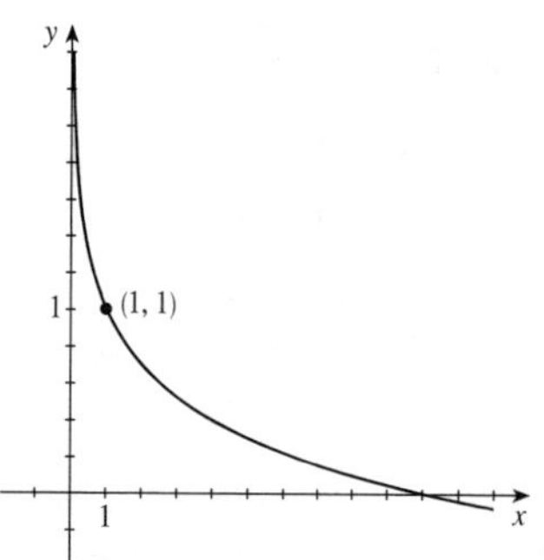


**56.** $y = 1 + \ln(-x)$. The graph of $y = 1 + \ln(-x)$ is obtained from the graph of $y = \ln x$ by reflecting it about the $y$-axis and then shifting it upward 1 unit. Domain: $(-\infty, 0)$. Range: $(-\infty, \infty)$. Vertical asymptote: $x = 0$.

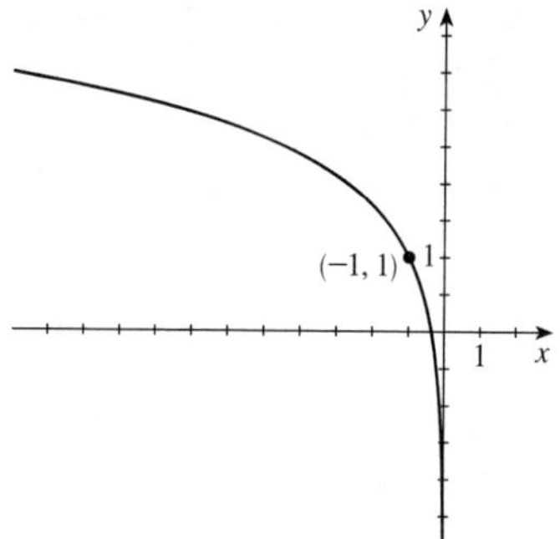


**57.** $y = |\ln x|$. The graph of $y = |\ln x|$ is obtained from the graph of $y = \ln x$ by reflecting the part of the graph for $0 < x < 1$ about the $x$-axis. Domain: $(0, \infty)$. Range: $[0, \infty)$. Vertical asymptote: $x = 0$.

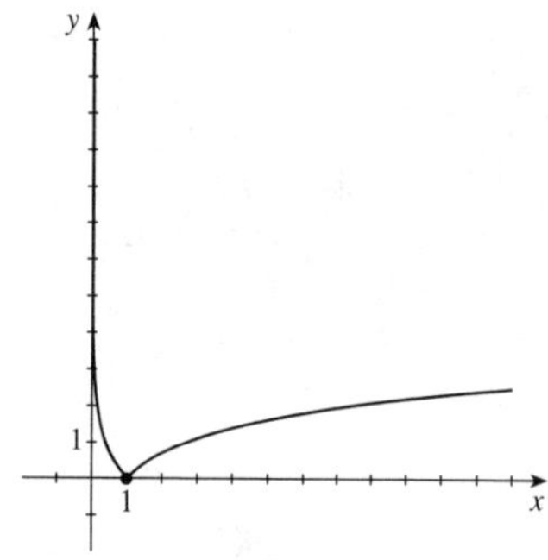


**58.** $y = \ln|x|$. Note that $y = \begin{cases} \ln x & \text{if } x > 0 \\ \ln(-x) & \text{if } x < 0 \end{cases}$ The graph of $y = \ln|x|$ is obtained by reflecting the graph of $y = \ln x$ about the $x$-axis. Domain: $(-\infty, 0) \cup (0, \infty)$. Range: $(-\infty, \infty)$. Vertical asymptote: $x = 0$.

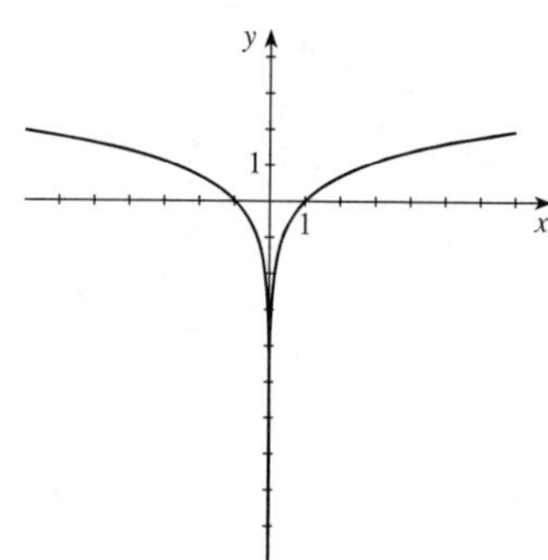


**59.** $f(x) = \log_{10}(x + 3)$. We require that $x + 3 > 0 \Leftrightarrow x > -3$, so the domain is $(-3, \infty)$.

**60.** $f(x) = \log_5(8 - 2x)$. Then we must have $8 - 2x > 0 \Leftrightarrow 8 > 2x \Leftrightarrow 4 > x$, and so the domain is $(-\infty, 4)$.

**61.** $g(x) = \log_3(x^2 - 1)$. We require that $x^2 - 1 > 0 \Leftrightarrow x^2 > 1 \Rightarrow x < -1$ or $x > 1$, so the domain is $(-\infty, -1) \cup (1, \infty)$.

**62.** $g(x) = \ln(x - x^2)$. Then we must have $x - x^2 > 0 \Leftrightarrow x(1 - x) > 0$. Using the methods from Chapter 1 with the endpoints 0 and 1, we get the table at right. Thus the domain is $(0, 1)$.

| Interval | $(-\infty, 0)$ | $(0, 1)$ | $(1, \infty)$ |
|---|---|---|---|
| Sign of $x$ | $-$ | $+$ | $+$ |
| Sign of $1 - x$ | $+$ | $+$ | $-$ |
| Sign of $x(1 - x)$ | $-$ | $+$ | $-$ |

**63.** $h(x) = \ln x + \ln(2 - x)$. We require that $x > 0$ and $2 - x > 0 \Leftrightarrow x > 0$ and $x < 2 \Leftrightarrow 0 < x < 2$, so the domain is $(0, 2)$.

**64.** $h(x) = \sqrt{x - 2} - \log_5(10 - x)$. Then we must have $x - 2 \geq 0$ and $10 - x > 0 \Leftrightarrow x \geq 2$ and $10 > x \Leftrightarrow 2 \leq x < 10$. So the domain is $[2, 10)$.

**65.** $y = \log_{10}\left(1 - x^2\right)$ has domain $(-1, 1)$, vertical asymptotes $x = -1$ and $x = 1$, and local maximum $y = 0$ at $x = 0$.

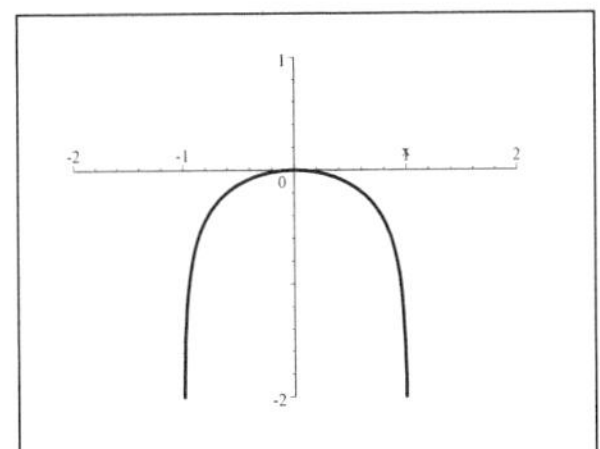

**66.** $y = \ln\left(x^2 - x\right) = \ln\left(x\left(x - 1\right)\right)$ has domain $(-\infty, 0) \cup (1, \infty)$, vertical asymptotes $x = 0$ and $x = 1$, and no local maximum or minimum.

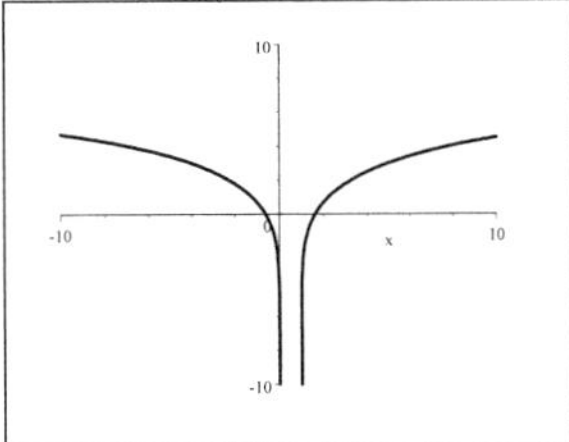


**67.** $y = x + \ln x$ has domain $(0, \infty)$, vertical asymptote $x = 0$, and no local maximum or minimum.

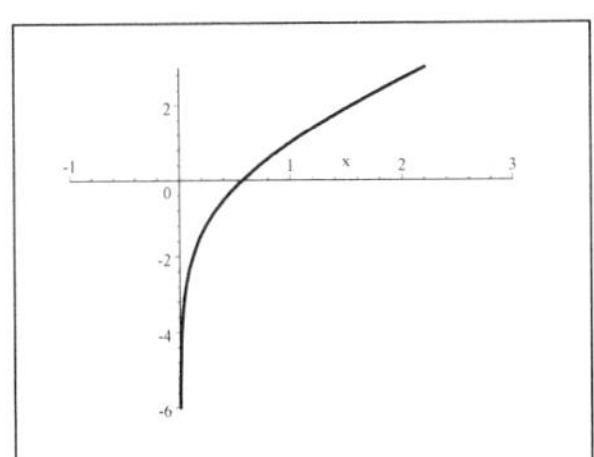

**68.** $y = x\left(\ln x\right)^2$ has domain $(0, \infty)$, no vertical asymptote, local minimum $y = 0$ at $x = 1$, and local maximum $y \approx 0.54$ at $x \approx 0.14$.

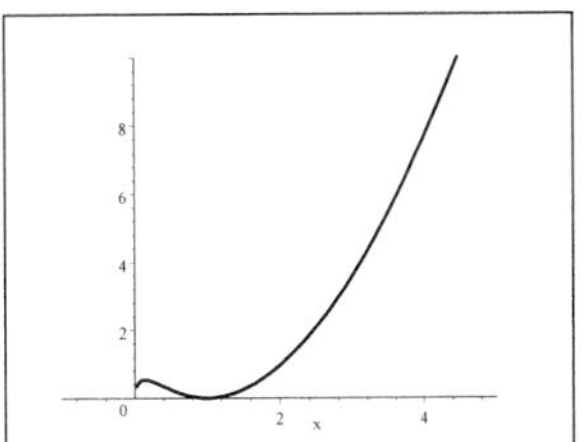

**69.** $y = \dfrac{\ln x}{x}$ has domain $(0, \infty)$, vertical asymptote $x = 0$, horizontal asymptote $y = 0$, and local maximum $y \approx 0.37$ at $x \approx 2.72$.

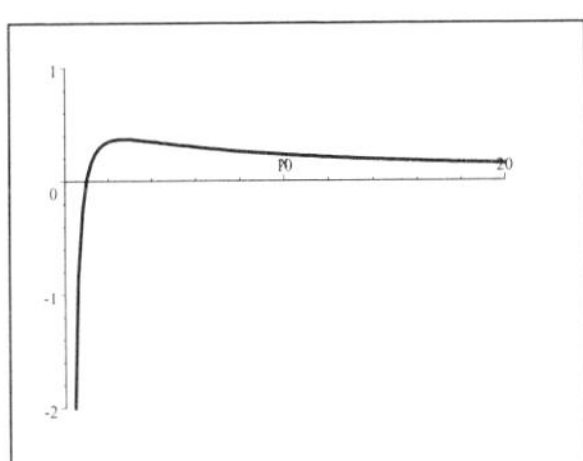

**70.** $y = x\log_{10}\left(x + 10\right)$ has domain $(-10, \infty)$, vertical asymptote $x = -10$, and local minimum $y \approx -3.62$ at $x \approx -5.87$.

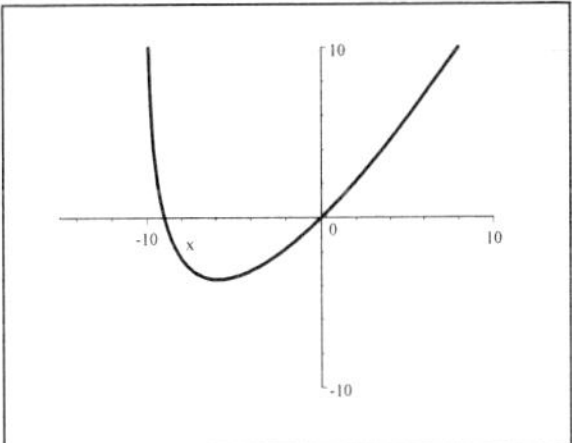


**71.** The graph of $g\left(x\right) = \sqrt{x}$ grows faster than the graph of $f\left(x\right) = \ln x$.

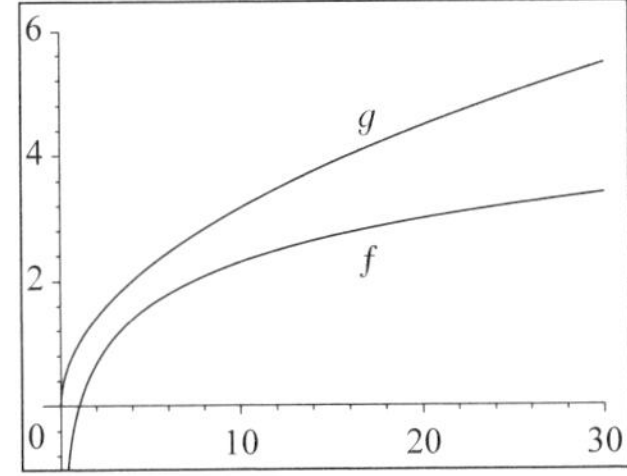


**72. (a)**

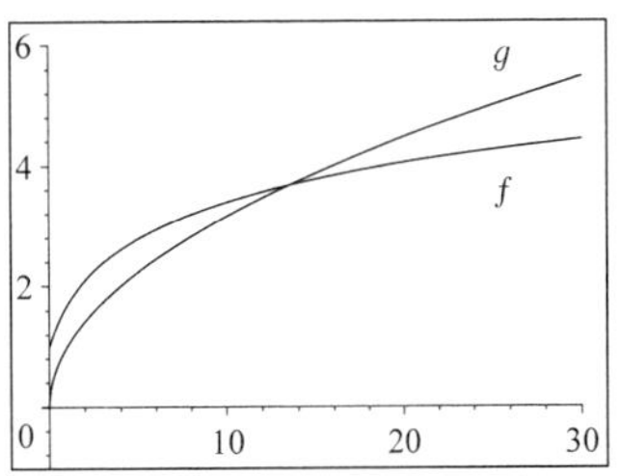


**(b)** From the graph, we see that the solution to the equation $\sqrt{x} = 1 + \ln\left(1 + x\right)$ is $x \approx 13.50$.

**73. (a)**

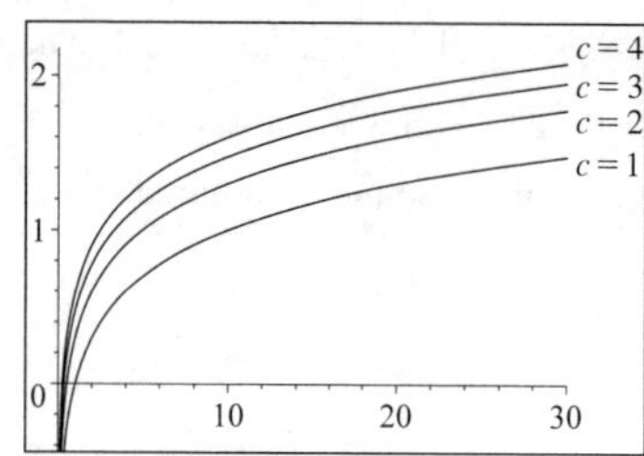


**(b)** Notice that $f(x) = \log(cx) = \log c + \log x$, so as $c$ increases, the graph of $f(x) = \log(cx)$ is shifted upward $\log c$ units.

**74. (a)**

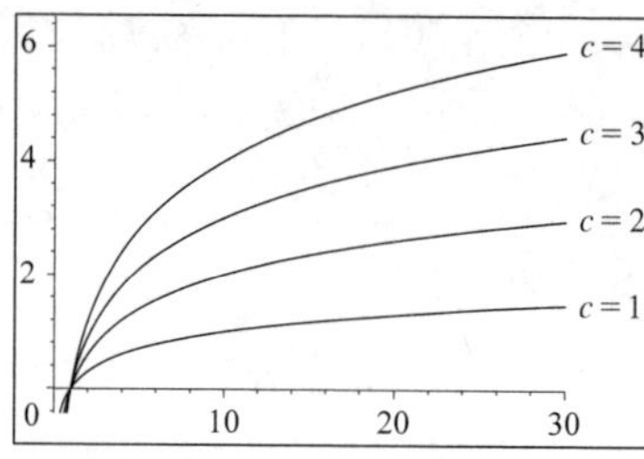


**(b)** As $c$ increases, the graph of $f(x) = c\log(x)$ stretches vertically by a factor of $c$.

**75. (a)** $f(x) = \log_2(\log_{10} x)$. Since the domain of $\log_2 x$ is the positive real numbers, we have: $\log_{10} x > 0 \Leftrightarrow x > 10^0 = 1$. Thus the domain of $f(x)$ is $(1, \infty)$.

**(b)** $y = \log_2(\log_{10} x) \Leftrightarrow 2^y = \log_{10} x \Leftrightarrow 10^{2^y} = x$. Thus $f^{-1}(x) = 10^{2^x}$.

**76. (a)** $f(x) = \ln(\ln(\ln x))$. We must have $\ln(\ln x) > 0 \Leftrightarrow \ln x > 1 \Leftrightarrow x > e$. So the domain of $f$ is $(e, \infty)$.

**(b)** $y = \ln(\ln(\ln x)) \Leftrightarrow e^y = \ln(\ln x) \Leftrightarrow e^{e^y} = \ln x \Leftrightarrow e^{e^{e^y}} = x$. Thus the inverse function is $f^{-1}(x) = e^{e^{e^x}}$.

**77. (a)** $f(x) = \dfrac{2^x}{1+2^x}$. $y = \dfrac{2^x}{1+2^x} \Leftrightarrow$

$y + y2^x = 2^x \Leftrightarrow$

$y = 2^x - y2^x = 2^x(1-y) \Leftrightarrow$

$2^x = \dfrac{y}{1-y} \Leftrightarrow x = \log_2\left(\dfrac{y}{1-y}\right)$. Thus

$f^{-1}(x) = \log_2\left(\dfrac{x}{1-x}\right)$.

**(b)** $\dfrac{x}{1-x} > 0$. Solving this using the methods from Chapter 1, we start with the endpoints, 0 and 1.

| Interval | $(-\infty, 0)$ | $(0, 1)$ | $(1, \infty)$ |
|---|---|---|---|
| Sign of $x$ | $-$ | $+$ | $+$ |
| Sign of $1-x$ | $+$ | $+$ | $-$ |
| Sign of $\dfrac{x}{1-x}$ | $-$ | $+$ | $-$ |

Thus the domain of $f^{-1}(x)$ is $(0, 1)$.

**78.** Using $I = 0.7I_0$ we have $C = -2500\ln\left(\dfrac{I}{I_0}\right) = -2500\ln 0.7 = 891.69$ moles/liter.

**79.** Using $D = 0.73D_0$ we have $A = -8267\ln\left(\dfrac{D}{D_0}\right) = -8267\ln 0.73 \approx 2601$ years.

**80.** Substituting $N = 1{,}000{,}000$ we get $t = 3\dfrac{\log(N/50)}{\log 2} = 3\dfrac{\log 20{,}000}{\log 2} \approx 42.86$ hours.

**81.** When $r = 6\%$ we have $t = \dfrac{\ln 2}{0.06} \approx 11.6$ years. When $r = 7\%$ we have $t = \dfrac{\ln 2}{0.07} \approx 9.9$ years. And when $r = 8\%$ we have $t = \dfrac{\ln 2}{0.08} \approx 8.7$ years.

**82.** Using $k = 0.25$ and substituting $C = 0.9C_0$ we have $t = -0.25\ln\left(1 - \dfrac{C}{C_0}\right) = -0.25\ln(1-0.9) = -0.25\ln 0.1 \approx 0.58$ hours.

**83.** Using $A = 100$ and $W = 5$ we find the ID to be $\dfrac{\log(2A/W)}{\log 2} = \dfrac{\log(2 \cdot 100/5)}{\log 2} = \dfrac{\log 40}{\log 2} \approx 5.32$. Using $A = 100$ and $W = 10$ we find the ID to be $\dfrac{\log(2A/W)}{\log 2} = \dfrac{\log(2 \cdot 100/10)}{\log 2} = \dfrac{\log 20}{\log 2} \approx 4.32$. So the smaller icon is $\dfrac{5.23}{4.32} \approx 1.23$ times harder.

**84. (a)** Since 2 feet= 24 inches, the height of the graph is $2^{24} = 1677216$ inches. Now, since there are 12 inches per foot and 5280 feet per mile, there are $12\,(5280) = 63{,}360$ inches per mile. So the height of the graph is $\frac{1{,}677{,}216}{63360} \approx 264.8$, or about 265 miles.

**(b)** Since $\log_2\left(2^{24}\right) = 24$, we must be about $2^{24}$ inches$\approx$ 265 miles to the right of the origin before the height of the graph of $y = \log_2 x$ reaches 24 inches or 2 feet.

**85.** $\log\left(\log 10^{100}\right) = \log 100 = 2$

$\log\left(\log\left(\log 10^{\text{googol}}\right)\right) = \log\left(\log\left(\text{googol}\right)\right) = \log\left(\log 10^{100}\right) = \log\left(100\right) = 2$

**86.** Notice that $\log_a x$ is increasing for $a > 1$. So we have $\log_4 17 > \log_4 16 = \log_4 4^2 = 2$. Also, we have $\log_5 24 < \log_5 25 = \log_5 5^2 = 2$. Thus, $\log_5 24 < 2 < \log_4 17$.

**87.** The numbers between 1000 and 9999 (inclusive) each have 4 digits, while $\log 1000 = 3$ and $\log 10{,}000 = 4$. Since $[\![\log x]\!] = 3$ for all integers $x$ where $1000 \le x < 10{,}000$, the number of digits is $[\![\log x]\!] + 1$. Likewise, if $x$ is an integer where $10^{n-1} \le x < 10^n$, then $x$ has $n$ digits and $[\![\log x]\!] = n - 1$. Since $[\![\log x]\!] = n - 1 \quad\Leftrightarrow\quad n = [\![\log x]\!] + 1$, the number of digits in $x$ is $[\![\log x]\!] + 1$.

## 5.3 Laws of Logarithms

**1.** $\log_3 \sqrt{27} = \log_3 3^{3/2} = \frac{3}{2}$

**2.** $\log_2 160 - \log_2 5 = \log_2 \frac{160}{5} = \log_2 32 = \log_2 2^5 = 5$

**3.** $\log 4 + \log 25 = \log\left(4 \cdot 25\right) = \log 100 = 2$

**4.** $\log \frac{1}{\sqrt{1000}} = \log 10^{-3/2} = -\frac{3}{2}$

**5.** $\log_4 192 - \log_4 3 = \log_4 \frac{192}{3} = \log_4 64 = \log_4 4^3 = 3$

**6.** $\log_{12} 9 + \log_{12} 16 = \log_{12}\left(9 \cdot 16\right) = \log_{12} 144 = \log_{12} 12^2 = 2$

**7.** $\log_2 6 - \log_2 15 + \log_2 20 = \log_2 \frac{6}{15} + \log_2 20 = \log_2\left(\frac{2}{5} \cdot 20\right) = \log_2 8 = \log_2 2^3 = 3$

**8.** $\log_3 100 - \log_3 18 - \log_3 50 = \log_3\left(\dfrac{100}{18 \cdot 50}\right) = \log_3\left(\frac{1}{9}\right) = \log_3 3^{-2} = -2$

**9.** $\log_4 16^{100} = \log_4\left(4^2\right)^{100} = \log_4 4^{200} = 200$

**10.** $\log_2 8^{33} = \log_2\left(2^3\right)^{33} = \log_2 2^{99} = 99$

**11.** $\log\left(\log 10^{10{,}000}\right) = \log\left(10{,}000 \log 10\right) = \log\left(10{,}000 \cdot 1\right) = \log\left(10{,}000\right) = \log 10^4 = 4 \log 10 = 4$

**12.** $\ln\left(\ln e^{e^{200}}\right) = \ln\left(e^{200} \ln e\right) = \ln e^{200} = 200 \ln e = 200$

**13.** $\log_2 2x = \log_2 2 + \log_2 x = 1 + \log_2 x$

**14.** $\log_3\left(5y\right) = \log_3 5 + \log_3 y$

**15.** $\log_2\left[x\left(x - 1\right)\right] = \log_2 x + \log_2\left(x - 1\right)$

**16.** $\log_5\left(\dfrac{x}{2}\right) = \log_5 x - \log_5 2$

**17.** $\log 6^{10} = 10 \log 6$

**18.** $\ln\left(\sqrt{z}\right) = \ln\left(z^{1/2}\right) = \frac{1}{2} \ln z$

**19.** $\log_2\left(AB^2\right) = \log_2 A + \log_2 B^2 = \log_2 A + 2 \log_2 B$

**20.** $\log_6 \sqrt[4]{17} = \frac{1}{4} \log_6 17$

**21.** $\log_3\left(x\sqrt{y}\right) = \log_3 x + \log_3 \sqrt{y} = \log_3 x + \frac{1}{2} \log_3 y$

**22.** $\log_2\left(xy\right)^{10} = 10 \log_2\left(xy\right) = 10\left(\log_2 x + \log_2 y\right)$

**23.** $\log_5 \sqrt[3]{x^2 + 1} = \frac{1}{3} \log_5\left(x^2 + 1\right)$

**24.** $\log_a\left(\dfrac{x^2}{yz^3}\right) = \log_a x^2 - \log_a\left(yz^3\right) = 2 \log_a x - \left(\log_a y + 3 \log_a z\right)$

**25.** $\ln \sqrt{ab} = \frac{1}{2} \ln ab = \frac{1}{2}\left(\ln a + \ln b\right)$

**26.** $\ln \sqrt[3]{3r^2 s} = \frac{1}{3} \ln\left(3r^2 s\right) = \dfrac{1}{3}\left[\ln 3 + \ln r^2 + \ln s\right] = \frac{1}{3}\left(\ln 3 + 2 \ln r + \ln s\right)$

**27.** $\log\left(\dfrac{x^3 y^4}{z^6}\right) = \log\left(x^3 y^4\right) - \log z^6 = 3 \log x + 4 \log y - 6 \log z$

**28.** $\log \dfrac{a^2}{b^4\sqrt{c}} = \log a^2 - \log\left(b^4\sqrt{c}\right) = 2\log a - \left(4\log b + \frac{1}{2}\log c\right)$

**29.** $\log_2\left(\dfrac{x\left(x^2+1\right)}{\sqrt{x^2-1}}\right) = \log_2 x + \log_2\left(x^2+1\right) - \frac{1}{2}\log_2\left(x^2-1\right)$

**30.** $\log_5\sqrt{\dfrac{x-1}{x+1}} = \frac{1}{2}\log_5\left(\dfrac{x-1}{x+1}\right) = \frac{1}{2}\left[\log_5(x-1) - \log_5(x+1)\right]$

**31.** $\ln\left(x\sqrt{\dfrac{y}{z}}\right) = \ln x + \frac{1}{2}\ln\left(\dfrac{y}{z}\right) = \ln x + \frac{1}{2}(\ln y - \ln z)$

**32.** $\ln\dfrac{3x^2}{(x+1)^{10}} = \ln\left(3x^2\right) - \ln(x+1)^{10} = \ln 3 + 2\ln x - 10\ln(x+1)$

**33.** $\log\sqrt[4]{x^2+y^2} = \frac{1}{4}\log\left(x^2+y^2\right)$

**34.** $\log\dfrac{x}{\sqrt[3]{1-x}} = \log x - \log\sqrt[3]{1-x} = \log x - \frac{1}{3}\log(1-x)$

**35.** $\log\sqrt{\dfrac{x^2+4}{\left(x^2+1\right)\left(x^3-7\right)^2}} = \frac{1}{2}\log\dfrac{x^2+4}{\left(x^2+1\right)\left(x^3-7\right)^2} = \frac{1}{2}\left[\log\left(x^2+4\right) - \log\left(x^2+1\right)\left(x^3-7\right)^2\right]$

$= \frac{1}{2}\left[\log\left(x^2+4\right) - \log\left(x^2+1\right) - 2\log\left(x^3-7\right)\right]$

**36.** $\log\sqrt{x\sqrt{y\sqrt{z}}} = \frac{1}{2}\log\left(x\sqrt{y\sqrt{z}}\right) = \frac{1}{2}\left(\log x + \log\sqrt{y\sqrt{z}}\right) = \frac{1}{2}\left[\log x + \frac{1}{2}\log\left(y\sqrt{z}\right)\right]$

$= \frac{1}{2}\left[\log x + \frac{1}{2}\left(\log y + \frac{1}{2}\log z\right)\right] = \frac{1}{2}\log x + \frac{1}{4}\log y + \frac{1}{8}\log z$

**37.** $\ln\dfrac{x^3\sqrt{x-1}}{3x+4} = \ln\left(x^3\sqrt{x-1}\right) - \ln(3x+4) = 3\ln x + \frac{1}{2}\ln(x-1) - \ln(3x+4)$

**38.** $\log\dfrac{10^x}{x\left(x^2+1\right)\left(x^4+2\right)} = \log 10^x - \log\left[x\left(x^2+1\right)\left(x^4+2\right)\right] = x - \left[\log x + \log\left(x^2+1\right) + \log\left(x^4+2\right)\right]$

**39.** $\log_3 5 + 5\log_3 2 = \log_3 5 + \log_3 2^5 = \log_3\left(5\cdot 2^5\right) = \log_3 160$

**40.** $\log 12 + \frac{1}{2}\log 7 - \log 2 = \log\left(12\sqrt{7}\right) - \log 2 = \log\dfrac{12\sqrt{7}}{2} = \log\left(6\sqrt{7}\right)$

**41.** $\log_2 A + \log_2 B - 2\log_2 C = \log_2(AB) - \log_2\left(C^2\right) = \log_2\left(\dfrac{AB}{C^2}\right)$

**42.** $\log_5\left(x^2-1\right) - \log_5(x-1) = \log_5\dfrac{x^2-1}{x-1} = \log_5\dfrac{(x-1)(x+1)}{x-1} = \log_5(x+1)$

**43.** $4\log x - \frac{1}{3}\log\left(x^2+1\right) + 2\log(x-1) = \log x^4 - \log\sqrt[3]{x^2+1} + \log(x-1)^2$

$= \log\left(\dfrac{x^4}{\sqrt[3]{x^2+1}}\right) + \log(x-1)^2 = \log\left(\dfrac{x^4(x-1)^2}{\sqrt[3]{x^2+1}}\right)$

**44.** $\ln(a+b) + \ln(a-b) - 2\ln c = \ln[(a+b)(a-b)] - \ln\left(c^2\right) = \ln\dfrac{a^2-b^2}{c^2}$

**45.** $\ln 5 + 2\ln x + 3\ln\left(x^2+5\right) = \ln\left(5x^2\right) + \ln\left(x^2+5\right)^3 = \ln\left[5x^2\left(x^2+5\right)^3\right]$

**46.** $2\left[\log_5 x + 2\log_5 y - 3\log_5 z\right] = 2\log_5\dfrac{xy^2}{z^3} = \log_5\left(\dfrac{xy^2}{z^3}\right)^2 = \log_5\dfrac{x^2y^4}{z^6}$

**47.** $\frac{1}{3}\log(2x+1) + \frac{1}{2}\left[\log(x-4) - \log\left(x^4-x^2-1\right)\right] = \log\sqrt[3]{2x+1} + \frac{1}{2}\log\dfrac{x-4}{x^4-x^2-1}$

$= \log\left(\sqrt[3]{2x+1}\cdot\sqrt{\dfrac{x-4}{x^4-x^2-1}}\right)$

**48.** $\log_a b + c\log_a d - r\log_a s = \log_a\left(bd^c\right) - \log_a s^r = \log_a\dfrac{bd^c}{s^r}$

**49.** $\log_2 5 = \dfrac{\log 5}{\log 2} \approx 2.321928$

**50.** $\log_5 2 = \dfrac{\log 2}{\log 5} \approx 0.430677$

**51.** $\log_3 16 = \dfrac{\log 16}{\log 3} \approx 2.523719$

**52.** $\log_6 92 = \dfrac{\log 92}{\log 6} \approx 2.523658$

**53.** $\log_7 2.61 = \dfrac{\log 2.61}{\log 7} \approx 0.493008$

**54.** $\log_6 532 = \dfrac{\log 532}{\log 6} \approx 3.503061$

**55.** $\log_4 125 = \dfrac{\log 125}{\log 4} \approx 3.482892$

**56.** $\log_{12} 2.5 = \dfrac{\log 2.5}{\log 12} \approx 0.368743$

**57.** $\log_3 x = \dfrac{\log_e x}{\log_e 3} = \dfrac{\ln x}{\ln 3} = \dfrac{1}{\ln 3}\ln x$. The graph of $y = \dfrac{1}{\ln 3}\ln x$ is shown in the viewing rectangle $[-1, 4]$ by $[-3, 2]$.

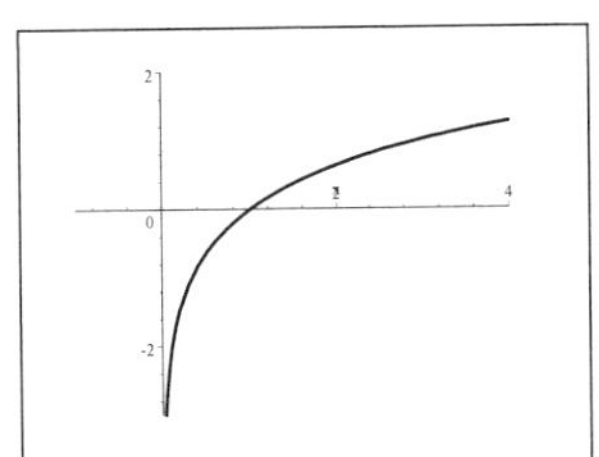

**58.** Note that $\log_c x = \left(\dfrac{1}{\ln c}\right)\ln x$ (by the change of base formula). So the graph of $y = \log_c x$ is obtained from the graph of $y = \ln x$ by either shrinking or stretching vertically by a factor of $\dfrac{1}{\ln c}$ depending on whether $\ln c > 1$ or $\ln c < 1$. All of the graphs pass through $(1, 0)$ because $\log_c 1 = 0$ for all $c$.

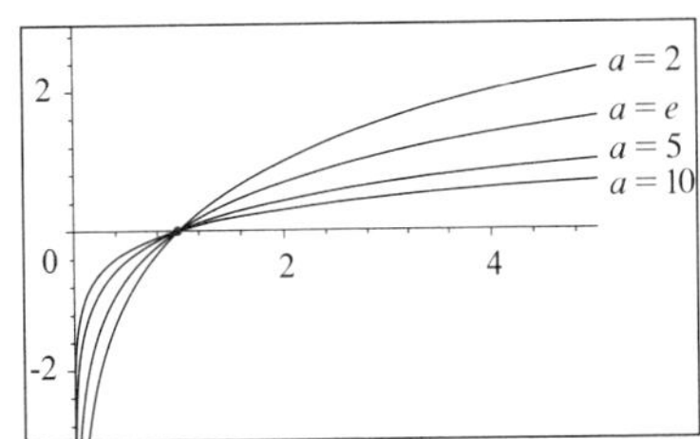


**59.** $\log e = \dfrac{\ln e}{\ln 10} = \dfrac{1}{\ln 10}$

**60.** $(\log_2 5)(\log_5 7) = \dfrac{\log 5}{\log 2} \cdot \dfrac{\log 7}{\log 5} = \dfrac{\log 7}{\log 2} = \log_2 7$

**61.** $$-\ln\left(x - \sqrt{x^2 - 1}\right) = \ln\left(\frac{1}{x - \sqrt{x^2 - 1}}\right) = \ln\left(\frac{1}{x - \sqrt{x^2 - 1}} \cdot \frac{x + \sqrt{x^2 - 1}}{x + \sqrt{x^2 - 1}}\right) = \ln\left(\frac{x + \sqrt{x^2 - 1}}{x^2 - (x^2 - 1)}\right)$$
$$= \ln\left(x + \sqrt{x^2 - 1}\right)$$

**62.** From Example 5(a), $P = \dfrac{P_0}{(t+1)^c}$. Substituting $P_0 = 80$, $t = 24$, and $c = 0.3$ we have $P = \dfrac{80}{(24+1)^{0.3}} \approx 30.5$. So the student should get a score of 30.

**63. (a)** $\log P = \log c - k \log W \quad \Leftrightarrow \quad \log P = \log c - \log W^k \quad \Leftrightarrow \quad \log P = \log\left(\dfrac{c}{W^k}\right) \quad \Leftrightarrow \quad P = \dfrac{c}{W^k}$.

**(b)** Using $k = 2.1$ and $c = 8000$, when $W = 2$ we have $P = \dfrac{8000}{2^{2.1}} \approx 1866$ and when $W = 10$ we have $P = \dfrac{8000}{10^{2.1}} \approx 64$.

**64. (a)** $\log S = \log c + k \log A \quad \Leftrightarrow \quad \log S = \log c + \log A^k \quad \Leftrightarrow \quad \log S = \log\left(cA^k\right) \quad \Leftrightarrow \quad S = cA^k$.

**(b)** If $A = 2A_0$ when $k = 3$ we get $S = c\,(2A_0)^3 = c \cdot 2^3 \cdot A_0^3 = 8 \cdot cA_0^3$. Thus doubling the area increases the species eightfold.

**65. (a)** $M = -2.5\log(B/B_0) = -2.5\log B + 2.5\log B_0$.

**(b)** Suppose $B_1$ and $B_2$ are the brightness of two stars such that $B_1 < B_2$ and let $M_1$ and $M_2$ be their respective magnitudes. Since log is an increasing function, we have $\log B_1 < \log B_2$. Then $\log B_1 < \log B_2 \quad \Leftrightarrow \quad \log B_1 - \log B_0 < \log B_2 - \log B_0 \quad \Leftrightarrow \quad \log(B_1/B_0) < \log(B_2/B_0) \quad \Leftrightarrow$ $-2.5\log(B_1/B_0) > -2.5\log(B_2/B_0) \quad \Leftrightarrow \quad M_1 > M_2$. Thus the brighter star has less magnitudes.

**(c)** Let $B_1$ be the brightness of the star Albiero. Then $100B_1$ is the brightness of Betelgeuse, and its magnitude is
$M = -2.5\log(100B_1/B_0) = -2.5\left[\log 100 + \log(B_1/B_0)\right] = -2.5\left[2 + \log(B_1/B_0)\right] = -5 - 2.5\log(B_1/B_0)$
$= -5 +$ magnitude of Albiero

**66. (a)** False; $\log(x/y) = \log x - \log y \neq (\log x)/(\log y)$.

**(b)** False; $\log_2 x - \log_2 y = \log_2(x/y) \neq \log_2(x - y)$.

**(c)** True; the equation is an identity: $\log_5(a/b^2) = \log_5 a - \log_5 b^2 = \log_5 a - 2\log_5 b$.

**(d)** True; the equation is an identity: $\log 2^z = z \log 2$.

**(e)** False; $\log P + \log Q = \log(PQ) \neq (\log P)(\log Q)$.

**(f)** False; $\log a - \log b = \log(a/b) \neq (\log a)/(\log b)$.

**(g)** False; $x \log_2 7 = \log_2 7^x \neq (\log_2 7)^x$.

**(h)** True; the equation is an identity. $\log_a a^a = a \log_a a = a \cdot 1 = a$.

**(i)** False; $\log(x - y) \neq (\log x)/(\log y)$. For example, $0 = \log(3 - 2) \neq (\log 3)/(\log 2)$.

**(j)** True; the equation is an identity: $-\ln(1/A) = -\ln A^{-1} = -1(-\ln A) = \ln A$.

**67.** The error is on the first line: $\log 0.1 < 0$, so $2\log 0.1 < \log 0.1$.

**68.** Let $f(x) = x^2$. Then $f(2x) = (2x)^2 = 4x^2 = 4f(x)$. Now the graph of $f(2x)$ is the same as the graph of $f$ shrunk horizontally by a factor of $\frac{1}{2}$, whereas the graph of $4f(x)$ is the same as the graph of $f(x)$ stretched vertically by a factor of 4.

Let $g(x) = e^x$. Then $g(x + 2) = e^{x+2} = e^2 e^x = e^2 g(x)$. This shows that a horizontal shift of 2 units to the right is the same as a vertical stretch by a factor of $e^2$.

Let $h(x) = \ln x$. Then $h(2x) = \ln 2x = \ln 2 + \ln x = \ln 2 + h(x)$. This shows that a horizontal shrinking by a factor of $\frac{1}{2}$ is the same as a vertical shift upward by $\ln 2$.

## 5.4 Exponential and Logarithmic Equations

**1.** $10^x = 25 \Leftrightarrow \log 10^x = \log 25 \Leftrightarrow x \log 10 = \log 25 \Leftrightarrow x \approx 1.398$

**2.** $10^{-x} = 4 \Leftrightarrow \log 10^{-x} = \log 4 \Leftrightarrow -x = \log 4 \Leftrightarrow x = -\log 4 \approx -0.6021$

**3.** $e^{-2x} = 7 \Leftrightarrow \ln e^{-2x} = \ln 7 \Leftrightarrow -2x \ln e = \ln 7 \Leftrightarrow -2x = \ln 7 \Leftrightarrow x = -\frac{1}{2}\ln 7 \approx -0.9730$

**4.** $e^{3x} = 12 \Leftrightarrow \ln e^{3x} = \ln 12 \Leftrightarrow 3x = \ln 12 \Leftrightarrow x = \dfrac{\ln 12}{3} \approx 0.8283$

**5.** $2^{1-x} = 3 \Leftrightarrow \log 2^{1-x} = \log 3 \Leftrightarrow (1 - x)\log 2 = \log 3 \Leftrightarrow 1 - x = \dfrac{\log 3}{\log 2} \Leftrightarrow$
$x = 1 - \dfrac{\log 3}{\log 2} \approx -0.5850$

**6.** $3^{2x-1} = 5 \Leftrightarrow \log 3^{2x-1} = \log 5 \Leftrightarrow (2x - 1)\log 3 = \log 5 \Leftrightarrow 2x - 1 = \dfrac{\log 5}{\log 3} \Leftrightarrow 2x = 1 + \dfrac{\log 5}{\log 3}$
$\Leftrightarrow x = \frac{1}{2}\left(1 + \dfrac{\log 5}{\log 3}\right) \approx 1.2325$

**7.** $3e^x = 10 \Leftrightarrow e^x = \frac{10}{3} \Leftrightarrow x = \ln\left(\frac{10}{3}\right) \approx 1.2040$

**8.** $2e^{12x} = 17 \Leftrightarrow e^{12x} = \frac{17}{2} \Leftrightarrow 12x = \ln\left(\frac{17}{2}\right) \Leftrightarrow x = \frac{1}{12}\left[\ln\left(\frac{17}{2}\right)\right] \approx 0.1783$

**9.** $e^{1-4x} = 2 \Leftrightarrow 1 - 4x = \ln 2 \Leftrightarrow -4x = -1 + \ln 2 \Leftrightarrow x = \dfrac{1 - \ln 2}{4} = 0.0767$

**10.** $4(1 + 10^{5x}) = 9 \Leftrightarrow 1 + 10^{5x} = \frac{9}{4} \Leftrightarrow 10^{5x} = \frac{5}{4} \Leftrightarrow 5x = \log\left(\frac{5}{4}\right) \Leftrightarrow$
$x = \frac{1}{5}[\log 5 - \log 4] \approx 0.0194$

**11.** $4 + 3^{5x} = 8 \Leftrightarrow 3^{5x} = 4 \Leftrightarrow \log 3^{5x} = \log 4 \Leftrightarrow 5x \log 3 = \log 4 \Leftrightarrow 5x = \dfrac{\log 4}{\log 3} \Leftrightarrow$
$x = \dfrac{\log 4}{5 \log 3} \approx 0.2524$

**12.** $2^{3x} = 34 \quad \Leftrightarrow \quad \log 2^{3x} = \log 34 \quad \Leftrightarrow \quad 3x \log 2 = \log 34 \quad \Leftrightarrow \quad x = \dfrac{\log 34}{3 \log 2} \approx 1.6958$

**13.** $8^{0.4x} = 5 \quad \Leftrightarrow \quad \log 8^{0.4x} = \log 5 \quad \Leftrightarrow \quad 0.4x \log 8 = \log 5 \quad \Leftrightarrow \quad 0.4x = \dfrac{\log 5}{\log 8} \quad \Leftrightarrow \quad x = \dfrac{\log 5}{0.4 \log 8} \approx 1.9349$

**14.** $3^{x/14} = 0.1 \quad \Leftrightarrow \quad \log 3^{x/14} = \log 0.1 \quad \Leftrightarrow \quad \left(\dfrac{x}{14}\right) \log 3 = \log 0.1 \quad \Leftrightarrow \quad x = \dfrac{14 \log 0.1}{\log 3} \approx -29.3426$

**15.** $5^{-x/100} = 2 \quad \Leftrightarrow \quad \log 5^{-x/100} = \log 2 \quad \Leftrightarrow \quad -\dfrac{x}{100} \log 5 = \log 2 \quad \Leftrightarrow \quad x = -\dfrac{100 \log 2}{\log 5} \approx -43.0677$

**16.** $e^{3-5x} = 16 \quad \Leftrightarrow \quad 3 - 5x = \ln 16 \quad \Leftrightarrow \quad -5x = \ln 16 - 3 \quad \Leftrightarrow \quad x = -\dfrac{1}{5}(\ln 16 - 3) \approx 0.0455$

**17.** $e^{2x+1} = 200 \quad \Leftrightarrow \quad 2x + 1 = \ln 200 \quad \Leftrightarrow \quad 2x = -1 + \ln 200 \quad \Leftrightarrow \quad x = \dfrac{-1 + \ln 200}{2} \approx 2.1492$

**18.** $\left(\frac{1}{4}\right)^x = 75 \quad \Leftrightarrow \quad 4^{-x} = 75 \quad \Leftrightarrow \quad \log 4^{-x} = \log 75 \quad \Leftrightarrow \quad (-x)(\log 4) = \log 75 \quad \Leftrightarrow \quad -x = \dfrac{\log 75}{\log 4} \quad \Leftrightarrow$
$x = -\dfrac{\log 75}{\log 4} \approx -3.1144$

**19.** $5^x = 4^{x+1} \quad \Leftrightarrow \quad \log 5^x = \log 4^{x+1} \quad \Leftrightarrow \quad x \log 5 = (x+1) \log 4 = x \log 4 + \log 4 \quad \Leftrightarrow \quad x \log 5 - x \log 4 = \log 4$
$\Leftrightarrow \quad x(\log 5 - \log 4) = \log 4 \quad \Leftrightarrow \quad x = \dfrac{\log 4}{\log 5 - \log 4} \approx 6.2126$

**20.** $10^{1-x} = 6^x \quad \Leftrightarrow \quad \log 10^{1-x} = \log 6^x \quad \Leftrightarrow \quad 1 - x = x(\log 6) \quad \Leftrightarrow \quad 1 = x(\log 6) + x \quad \Leftrightarrow \quad 1 = x(\log 6 + 1)$
$\Leftrightarrow \quad x = \dfrac{1}{\log 6 + 1} \approx 0.5624$

**21.** $2^{3x+1} = 3^{x-2} \quad \Leftrightarrow \quad \log 2^{3x+1} = \log 3^{x-2} \quad \Leftrightarrow \quad (3x+1) \log 2 = (x-2) \log 3 \quad \Leftrightarrow \quad 3x \log 2 +$
$\log 2 = x \log 3 - 2 \log 3 \quad \Leftrightarrow \quad 3x \log 2 - x \log 3 = -\log 2 - 2 \log 3 \quad \Leftrightarrow \quad x(3 \log 2 - \log 3) = -(\log 2 + 2 \log 3)$
$\Leftrightarrow \quad s = -\dfrac{\log 2 + 2 \log 3}{3 \log 2 - \log 3} \approx -2.9469$

**22.** $7^{x/2} = 5^{1-x} \quad \Leftrightarrow \quad \log 7^{x/2} = \log 5^{1-x} \quad \Leftrightarrow \quad \left(\dfrac{x}{2}\right) \log 7 = (1-x) \log 5 \quad \Leftrightarrow \quad \left(\dfrac{x}{2}\right) \log 7 = \log 5 - x \log 5$
$\Leftrightarrow \quad \left(\dfrac{x}{2}\right) \log 7 + x \log 5 = \log 5 \quad \Leftrightarrow \quad x\left(\frac{1}{2} \log 7 + \log 5\right) = \log 5 \quad \Leftrightarrow \quad x = \dfrac{\log 5}{\frac{1}{2} \log 7 + \log 5} \approx 0.6232$

**23.** $\dfrac{50}{1 + e^{-x}} = 4 \quad \Leftrightarrow \quad 50 = 4 + 4e^{-x} \quad \Leftrightarrow \quad 46 = 4e^{-x} \quad \Leftrightarrow \quad 11.5 = e^{-x} \quad \Leftrightarrow \quad \ln 11.5 = -x \quad \Leftrightarrow$
$x = -\ln 11.5 \approx -2.4423$

**24.** $\dfrac{10}{1 + e^{-x}} = 2 \quad \Leftrightarrow \quad 10 = 2 + 2e^{-x} \quad \Leftrightarrow \quad 8 = 2e^{-x} \quad \Leftrightarrow \quad 4 = e^{-x} \quad \Leftrightarrow \quad \ln 4 = -x \quad \Leftrightarrow$
$x = -\ln 4 \approx -1.3863$

**25.** $100(1.04)^{2t} = 300 \quad \Leftrightarrow \quad 1.04^{2t} = 3 \quad \Leftrightarrow \quad \log 1.04^{2t} = \log 3 \quad \Leftrightarrow \quad 2t \log 1.04 = \log 3 \quad \Leftrightarrow$
$t = \dfrac{\log 3}{2 \log 1.04} \approx 14.0055$

**26.** $(1.00625)^{12t} = 2 \quad \Leftrightarrow \quad \log 1.00625^{12t} = \log 2 \quad \Leftrightarrow \quad 12t \log 1.00625 = \log 2 \quad \Leftrightarrow$
$t = \dfrac{\log 2}{12 \log 1.00625} \approx 9.2708$

**27.** $x^2 2^x - 2^x = 0 \quad \Leftrightarrow \quad 2^x(x^2 - 1) = 0 \Rightarrow 2^x = 0$ (never) or $x^2 - 1 = 0$. If $x^2 - 1 = 0$, then $x^2 = 1 \Rightarrow x = \pm 1$. So the only solutions are $x = \pm 1$.

**28.** $x^2 10^x - x 10^x = 2(10^x) \quad \Leftrightarrow \quad x^2 10^x - x 10^x - 2(10^x) = 0 \quad \Leftrightarrow \quad 10^x(x^2 - x - 2) = 0 \Rightarrow 10^x = 0$(never) or $x^2 - x - 2 = 0$. If $x^2 - x - 2 = 0$, then $(x-2)(x+1) = 0 \Rightarrow x = 2, -1$. So the only solutions are $x = 2, -1$.

**29.** $4x^3 e^{-3x} - 3x^4 e^{-3x} = 0 \quad \Leftrightarrow \quad x^3 e^{-3x}(4 - 3x) = 0 \Rightarrow x = 0$ or $e^{-3x} = 0$ (never) or $4 - 3x = 0$. If $4 - 3x = 0$, then $3x = 4 \quad \Leftrightarrow \quad x = \frac{4}{3}$. So the solutions are $x = 0$ and $x = \frac{4}{3}$.

**30.** $x^2e^x + xe^x - e^x = 0 \quad \Leftrightarrow \quad e^x(x^2 + x - 1) = 0 \Rightarrow e^x = 0$ (impossible) or $x^2 + x - 1 = 0$. If $x^2 + x - 1 = 0$, then $x = \dfrac{-1 \pm \sqrt{5}}{2}$. So the solutions are $x = \dfrac{-1 \pm \sqrt{5}}{2}$.

**31.** $e^{2x} - 3e^x + 2 = 0 \quad \Leftrightarrow \quad (e^x - 1)(e^x - 2) = 0 \Rightarrow e^x - 1 = 0$ or $e^x - 2 = 0$. If $e^x - 1 = 0$, then $e^x = 1 \quad \Leftrightarrow$ $x = \ln 1 = 0$. If $e^x - 2 = 0$, then $e^x = 2 \quad \Leftrightarrow \quad x = \ln 2 \approx 0.6931$. So the solutions are $x = 0$ and$x \approx 0.6931$.

**32.** $e^{2x} - e^x - 6 = 0 \quad \Leftrightarrow \quad (e^x - 3)(e^x + 2) = 0 \Rightarrow e^x + 2 = 0$ (impossible) or $e^x - 3 = 0$. If $e^x - 3 = 0$, then $e^x = 3$ $\Leftrightarrow \quad x = \ln 3 \approx 1.0986$. So the only solution is $x \approx 1.0986$.

**33.** $e^{4x} + 4e^{2x} - 21 = 0 \quad \Leftrightarrow \quad (e^{2x} + 7)(e^{2x} - 3) = 0 \Rightarrow e^{2x} = -7$ or $e^{2x} = 3$. Now $e^{2x} = -7$ has no solution, since $e^{2x} > 0$ for all $x$. But we can solve $e^{2x} = 3 \quad \Leftrightarrow \quad 2x = \ln 3 \quad \Leftrightarrow \quad x = \frac{1}{2}\ln 3 \approx 0.5493$. So the only solution is $x \approx 0.5493$.

**34.** $e^x - 12e^{-x} - 1 = 0 \quad \Leftrightarrow \quad e^x - 1 - 12e^{-x} = 0 \quad \Leftrightarrow \quad e^x(e^x - 1 - 12e^{-x}) = 0 \cdot e^x \quad \Leftrightarrow \quad e^{2x} - e^x - 12 = 0$ $\Leftrightarrow \quad (e^x - 4)(e^x + 3) = 0 \Rightarrow e^x + 3 = 0$ (impossible) or $e^x - 4 = 0$. If $e^x - 4 = 0$, then $e^x = 4 \quad \Leftrightarrow$ $x = \ln 4 \approx 1.3863$. So the only solution is $x \approx 1.3863$.

**35.** $\ln x = 10 \quad \Leftrightarrow \quad x = e^{10} \approx 22026$

**36.** $\ln(2 + x) = 1 \quad \Leftrightarrow \quad 2 + x = e^1 \quad \Leftrightarrow \quad x = e - 2 \approx 0.7183$

**37.** $\log x = -2 \quad \Leftrightarrow \quad x = 10^{-2} = 0.01$

**38.** $\log(x - 4) = 3 \quad \Leftrightarrow \quad x - 4 = 10^3 = 1000 \quad \Leftrightarrow \quad x = 1004$

**39.** $\log(3x + 5) = 2 \quad \Leftrightarrow \quad 3x + 5 = 10^2 = 100 \quad \Leftrightarrow \quad 3x = 95 \quad \Leftrightarrow \quad x = \frac{95}{3} \approx 31.6667$

**40.** $\log_3(2 - x) = 3 \quad \Leftrightarrow \quad 2 - x = 3^3 = 27 \quad \Leftrightarrow \quad -x = 25 \quad \Leftrightarrow \quad x = -25$

**41.** $2 - \ln(3 - x) = 0 \quad \Leftrightarrow \quad 2 = \ln(3 - x) \quad \Leftrightarrow \quad e^2 = 3 - x \quad \Leftrightarrow \quad x = 3 - e^2 \approx -4.3891$

**42.** $\log_2(x^2 - x - 2) = 2 \quad \Leftrightarrow \quad x^2 - x - 2 = 2^2 = 4 \quad \Leftrightarrow \quad x^2 - x - 6 = 0 \quad \Leftrightarrow \quad (x - 3)(x + 2) = 0 \quad \Leftrightarrow$ $x = 3$ or $x = -2$. Thus the solutions are $x = 3$ and $x = -2$.

**43.** $\log_2 3 + \log_2 x = \log_2 5 + \log_2(x - 2) \quad \Leftrightarrow \quad \log_2(3x) = \log_2(5x - 10) \quad \Leftrightarrow \quad 3x = 5x - 10 \quad \Leftrightarrow \quad 2x = 10$ $\Leftrightarrow \quad x = 5$

**44.** $2\log x = \log 2 + \log(3x - 4) \quad \Leftrightarrow \quad \log(x^2) = \log(6x - 8) \quad \Leftrightarrow \quad x^2 = 6x - 8 \quad \Leftrightarrow \quad x^2 - 6x + 8 = 0 \quad \Leftrightarrow$ $(x - 4)(x - 2) = 0 \quad \Leftrightarrow \quad x = 4$ or $x = 2$. Thus the solutions are $x = 4$ and $x = 2$.

**45.** $\log x + \log(x - 1) = \log(4x) \quad \Leftrightarrow \quad \log[x(x - 1)] = \log(4x) \quad \Leftrightarrow \quad x^2 - x = 4x \quad \Leftrightarrow \quad x^2 - 5x = 0 \quad \Leftrightarrow$ $x(x - 5) = 0 \Rightarrow x = 0$ or $x = 5$. So the possible solutions are $x = 0$ and $x = 5$. However, when $x = 0$, $\log x$ is undefined. Thus the only solution is $x = 5$.

**46.** $\log_5 x + \log_5(x + 1) = \log_5 20 \quad \Leftrightarrow \quad \log_5(x^2 + x) = \log_5 20 \quad \Leftrightarrow \quad x^2 + x = 20 \quad \Leftrightarrow \quad x^2 + x - 20 = 0 \quad \Leftrightarrow$ $(x + 5)(x - 4) = 0 \quad \Leftrightarrow \quad x = -5$ or $x = 4$. Since $\log_5(-5)$ is undefined, the only solution is $x = 4$.

**47.** $\log_5(x + 1) - \log_5(x - 1) = 2 \quad \Leftrightarrow \quad \log_5\left(\dfrac{x + 1}{x - 1}\right) = 2 \quad \Leftrightarrow \quad \dfrac{x + 1}{x - 1} = 5^2 \quad \Leftrightarrow \quad x + 1 = 25x - 25 \quad \Leftrightarrow$ $24x = 26 \quad \Leftrightarrow \quad x = \frac{13}{12}$

**48.** $\log x + \log(x - 3) = 1 \quad \Leftrightarrow \quad \log[x(x - 3)] = 1 \quad \Leftrightarrow \quad x^2 - 3x = 10 \quad \Leftrightarrow \quad x^2 - 3x - 10 = 0 \quad \Leftrightarrow$ $(x + 2)(x - 5) = 0 \quad \Leftrightarrow \quad x = -2$ or $x = 5$. Since $\log(-2)$ is undefined, the only solution is $x = 5$.

**49.** $\log_9(x - 5) + \log_9(x + 3) = 1 \quad \Leftrightarrow \quad \log_9[(x - 5)(x + 3)] = 1 \quad \Leftrightarrow \quad (x - 5)(x + 3) = 9^1 \quad \Leftrightarrow$ $x^2 - 2x - 24 = 0 \quad \Leftrightarrow \quad (x - 6)(x + 4) = 0 \Rightarrow x = 6$ or$-4$. However, $x = -4$ is inadmissible, so $x = 6$ is the only solution.

**50.** $\ln(x - 1) + \ln(x + 2) = 1 \quad \Leftrightarrow \quad \ln[(x - 1)(x + 2)] = 1 \quad \Leftrightarrow \quad x^2 + x - 2 = e \quad \Leftrightarrow \quad x^2 + x - (2 + e) = 0 \Rightarrow$ $x = \frac{-1 \pm \sqrt{1 + 4(2 + e)}}{2} = \frac{-1 \pm \sqrt{9 + 4e}}{2}$. Since $x - 1 < 0$ when $x = \frac{-1 - \sqrt{9 + 4e}}{2}$, the only solution is $x = \frac{-1 + \sqrt{9 + 4e}}{2} \approx 1.7290$

**51.** $\log(x + 3) = \log x + \log 3 \quad \Leftrightarrow \quad \log(x + 3) = \log(3x) \quad \Leftrightarrow \quad x + 3 = 3x \quad \Leftrightarrow \quad 2x = 3 \quad \Leftrightarrow \quad x = \frac{3}{2}$

**52.** $(\log x)^3 = 3\log x \quad\Leftrightarrow\quad (\log x)^3 - 3\log x = 0 \quad\Leftrightarrow\quad (\log x)\left((\log x)^2 - 3\right) \quad\Leftrightarrow\quad (\log x) = 0$ or $(\log x)^2 - 3 = 0$. Now $\log x = 0 \quad\Leftrightarrow\quad x = 1$. Also $(\log x)^2 - 3 = 0 \quad\Leftrightarrow\quad (\log x)^2 = 3 \quad\Leftrightarrow\quad \log x = \pm\sqrt{3} \quad\Leftrightarrow\quad x = 10^{\pm\sqrt{3}}$, so $x = 10^{\sqrt{3}} \approx 53.9574$ or $x = 10^{-\sqrt{3}} \approx 0.0185$. Thus the solutions to the equation are $x = 1$, $x = 10^{\sqrt{3}} \approx 53.9574$ and $x = 10^{-\sqrt{3}} \approx 0.0185$.

**53.** $2^{2/\log_5 x} = \frac{1}{16} \quad\Leftrightarrow\quad \log_2 2^{2/\log_5 x} = \log_2\left(\frac{1}{16}\right) \quad\Leftrightarrow\quad \dfrac{2}{\log_5 x} = -4 \quad\Leftrightarrow\quad \log_5 x = -\frac{1}{2} \quad\Leftrightarrow$ $x = 5^{-1/2} = \frac{1}{\sqrt{5}} \approx 0.4472$

**54.** $\log_2(\log_3 x) = 4 \quad\Leftrightarrow\quad \log_3 x = 2^4 = 16 \quad\Leftrightarrow\quad x = 3^{16} = 43{,}046{,}721$

**55.** $\ln x = 3 - x \quad\Leftrightarrow\quad \ln x + x - 3 = 0$. Let $f(x) = \ln x + x - 3$. We need to solve the equation $f(x) = 0$. From the graph of $f$, we get $x \approx 2.21$.

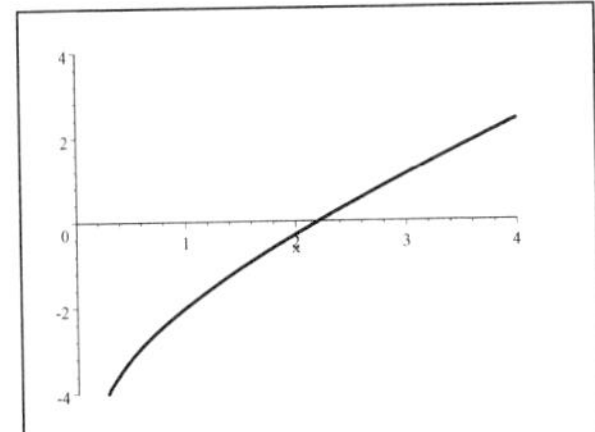

**56.** $\log x = x^2 - 2 \quad\Leftrightarrow\quad \log x - x^2 + 2 = 0$. Let $f(x) = \log x - x^2 + 2$. We need to solve the equation $f(x) = 0$. From the graph of $f$, we get $x \approx 0.01$ or $x \approx 1.47$.

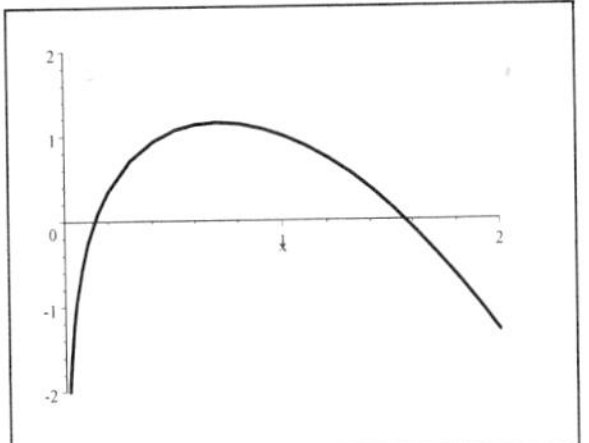

**57.** $x^3 - x = \log_{10}(x+1) \quad\Leftrightarrow$ $x^3 - x - \log_{10}(x+1) = 0$. Let $f(x) = x^3 - x - \log_{10}(x+1)$. We need to solve the equation $f(x) = 0$. From the graph of $f$, we get $x = 0$ or $x \approx 1.14$.

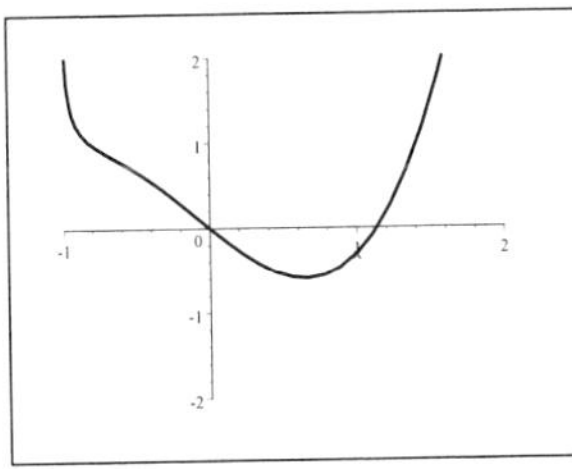

**58.** $x = \ln\left(4 - x^2\right) \quad\Leftrightarrow\quad x - \ln\left(4 - x^2\right) = 0$. Let $f(x) = x - \ln\left(4 - x^2\right)$. We need to solve the equation $f(x) = 0$. From the graph of $f$, we get $xx \approx -1.96$ or $x \approx 1.06$.

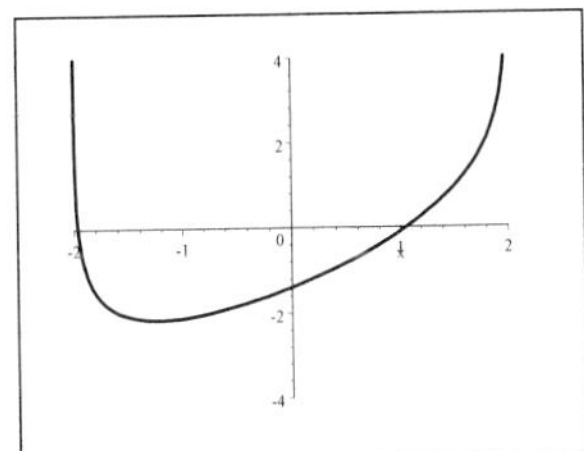

**59.** $e^x = -x \quad\Leftrightarrow\quad e^x + x = 0$. Let $f(x) = e^x + x$. We need to solve the equation $f(x) = 0$. From the graph of $f$, we get $x \approx -0.57$.

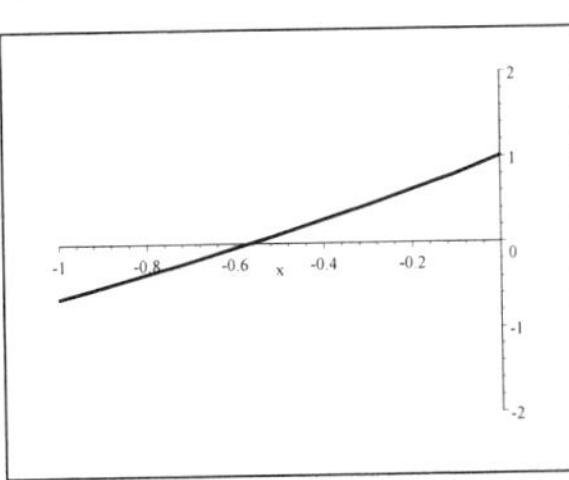

**60.** $2^{-x} = x - 1 \quad\Leftrightarrow\quad 2^{-x} - x + 1 = 0$. Let $f(x) = 2^{-x} - x + 1$. We need to solve the equation $f(x) = 0$. From the graph of $f$, we get $x \approx 1.38$.

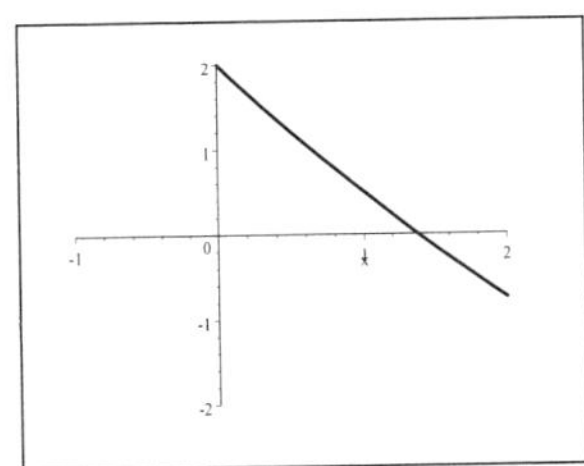

**61.** $4^{-x} = \sqrt{x} \quad \Leftrightarrow \quad 4^{-x} - \sqrt{x} = 0$. Let $f(x) = 4^{-x} - \sqrt{x}$. We need to solve the equation $f(x) = 0$. From the graph of $f$, we get $x \approx 0.36$.

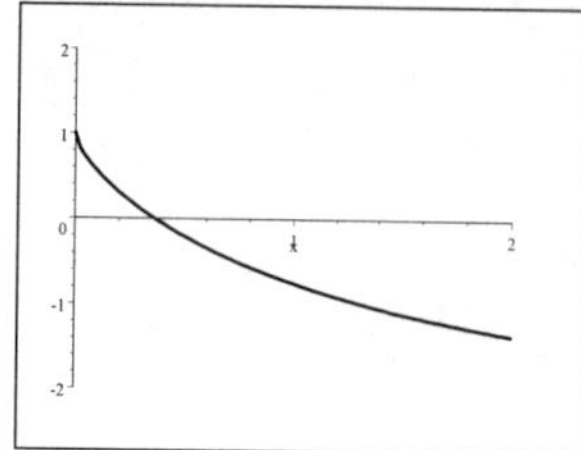

**62.** $e^{x^2} - 2 = x^3 - x \quad \Leftrightarrow \quad e^{x^2} - 2 - x^3 + x = 0$. Let $f(x) = e^{x^2} - 2 - x^3 + x$. We need to solve the equation $f(x) = 0$. From the graph of $f$, we get $x \approx -0.89$ or $x \approx 0.71$.

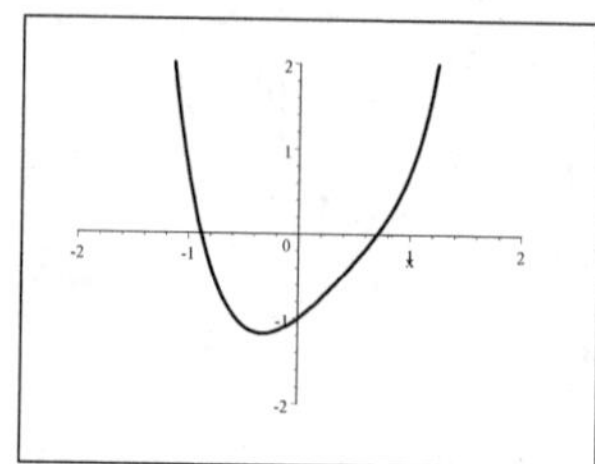

**63.** $\log(x-2) + \log(9-x) < 1 \quad \Leftrightarrow \quad \log[(x-2)(9-x)] < 1 \quad \Leftrightarrow \quad \log(-x^2 + 11x - 18) < 1 \Rightarrow$ $-x^2 + 11x - 18 < 10^1 \quad \Leftrightarrow \quad 0 < x^2 - 11x + 28 \quad \Leftrightarrow \quad 0 < (x-7)(x-4)$. Also, since the domain of a logarithm is positive we must have $0 < -x^2 + 11x - 18 \quad \Leftrightarrow \quad 0 < (x-2)(9-x)$. Using the methods from Chapter 1 with the endpoints 2, 4, 7, 9 for the intervals, we make the following table:

| Interval | $(-\infty, 2)$ | $(2, 4)$ | $(4, 7)$ | $(7, 9)$ | $(9, \infty)$ |
|---|---|---|---|---|---|
| Sign of $x - 7$ | $-$ | $-$ | $-$ | $+$ | $+$ |
| Sign of $x - 4$ | $-$ | $-$ | $+$ | $+$ | $+$ |
| Sign of $x - 2$ | $-$ | $+$ | $+$ | $+$ | $+$ |
| Sign of $9 - x$ | $+$ | $+$ | $+$ | $+$ | $-$ |
| Sign of $(x-7)(x-4)$ | $+$ | $+$ | $-$ | $+$ | $+$ |
| Sign of $(x-2)(9-x)$ | $-$ | $+$ | $+$ | $+$ | $-$ |

Thus the solution is $(2, 4) \cup (7, 9)$.

**64.** $3 \le \log_2 x \le 4 \quad \Leftrightarrow \quad 2^3 \le x \le 2^4 \quad \Leftrightarrow \quad 8 \le x \le 16$.

**65.** $2 < 10^x < 5 \quad \Leftrightarrow \quad \log 2 < x < \log 5 \quad \Leftrightarrow \quad 0.3010 < x < 0.6990$. Hence the solution to the inequality is approximately the interval $(0.3010, 0.6990)$.

**66.** $x^2 e^x - 2e^x < 0 \quad \Leftrightarrow \quad e^x(x^2 - 2) < 0 \quad \Leftrightarrow \quad e^x(x - \sqrt{2})(x + \sqrt{2}) < 0$. We use the methods of Chapter 1 with the endpoints $-\sqrt{2}$ and $\sqrt{2}$, noting that $e^x > 0$ for all $x$. We make a table:

| Interval | $(-\infty, -\sqrt{2})$ | $(-\sqrt{2}, \sqrt{2})$ | $(\sqrt{2}, \infty)$ |
|---|---|---|---|
| Sign of $e^x$ | $+$ | $+$ | $+$ |
| Sign of $(x - \sqrt{2})$ | $-$ | $-$ | $+$ |
| Sign of $(x + \sqrt{2})$ | $-$ | $+$ | $+$ |
| Sign of $e^x(x - \sqrt{2})(x + \sqrt{2})$ | $+$ | $-$ | $+$ |

Thus $-\sqrt{2} < x < \sqrt{2}$.

**67. (a)** $A(3) = 5000\left(1 + \dfrac{0.085}{4}\right)^{4(3)} = 5000(1.02125^{12}) = 6435.09$. Thus the amount after 3 years is \$6,435.09.

**(b)** $10000 = 5000\left(1 + \dfrac{0.085}{4}\right)^{4t} = 5000(1.02125^{4t}) \quad \Leftrightarrow \quad 2 = 1.02125^{4t} \quad \Leftrightarrow \quad \log 2 = 4t \log 1.02125 \quad \Leftrightarrow$ $t = \dfrac{\log 2}{4 \log 1.02125} \approx 8.24$ years. Thus the investment will double in about 8.24 years.

**68. (a)** $A(2) = 6500e^{0.06(2)} \approx \$7328.73$

**(b)** $8000 = 6500e^{0.06t} \quad \Leftrightarrow \quad \frac{16}{13} = e^{0.06t} \quad \Leftrightarrow \quad \ln\left(\frac{16}{13}\right) = 0.06t \quad \Leftrightarrow \quad t = \dfrac{1}{0.06}\ln\left(\frac{16}{13}\right) \approx 3.46$. So the investment doubles in about $3\frac{1}{2}$ years.

**69.** $8000 = 5000\left(1 + \dfrac{0.075}{4}\right)^{4t} = 5000\left(1.01875^{4t}\right) \quad \Leftrightarrow \quad 1.6 = 1.01875^{4t} \quad \Leftrightarrow \quad \log 1.6 = 4t\log 1.01875 \quad \Leftrightarrow$ $t = \dfrac{\log 1.6}{4\log 1.01875} \approx 6.33$ years. The investment will increase to \$8000 in approximately 6 years and 4 months.

**70.** $5000 = 4000\left(1 + \dfrac{0.0975}{2}\right)^{2t} \quad \Leftrightarrow \quad 1.25 = (1.04875)^{2t} \quad \Leftrightarrow \quad \log 1.25 = 2t\log 1.04875 \quad \Leftrightarrow$ $t = \dfrac{\log 1.25}{2\log 1.04875} \approx 2.344$. So it takes about $2\dfrac{1}{3}$ years to save \$5000.

**71.** $2 = e^{0.085t} \quad \Leftrightarrow \quad \ln 2 = 0.085t \quad \Leftrightarrow \quad t = \dfrac{\ln 2}{0.085} \approx 8.15$ years. Thus the investment will double in about 8.15 years.

**72.** $1435.77 = 1000\left(1 + \frac{r}{2}\right)^{2(4)} \quad \Leftrightarrow \quad 1.43577 = \left(1 + \frac{r}{2}\right)^{8} \quad \Leftrightarrow \quad 1 + \dfrac{r}{2} = \sqrt[8]{1.43577} \quad \Leftrightarrow \quad \dfrac{r}{2} = \sqrt[8]{1.43577} - 1$ $\Leftrightarrow \quad r = 2\left(\sqrt[8]{1.43577} - 1\right) \approx 0.0925$. Thus the rate was about 9.25%.

**73.** $r_{\text{eff}} = \left(1 + \frac{r}{n}\right)^{n} - 1$. Here $r = 0.08$ and $n = 12$, so $r_{\text{eff}} = \left(1 + \dfrac{0.08}{12}\right)^{12} - 1 = (1.0066667)^{12} - 1 = 8.30\%$.

**74.** $r_{\text{APY}} = e^{r} - 1$. Here $r = 0.05.5$ so $r_{\text{APY}} = e^{0.055} - 1 \approx 1.0565 - 1 = 0.565$. So the annual percentage yield is about 5.65%.

**75.** $15e^{-0.087t} = 5 \quad \Leftrightarrow \quad e^{-0.087t} = \frac{1}{3} \quad \Leftrightarrow \quad -0.087t = \ln\left(\frac{1}{3}\right) = -\ln 3 \quad \Leftrightarrow \quad t = \dfrac{\ln 3}{0.087} \approx 12.6277$. So only 5 grams remain after approximately 13 days.

**76.** We want to solve for $t$ in the equation $80\left(e^{-0.2t} - 1\right) = -70$ (when motion is downwards, the velocity is negative). Then $80\left(e^{-0.2t} - 1\right) = -70 \quad \Leftrightarrow \quad e^{-0.2t} - 1 = -\frac{7}{8} \quad \Leftrightarrow \quad e^{-0.2t} = \frac{1}{8} \quad \Leftrightarrow \quad -0.2t = \ln\left(\frac{1}{8}\right) \quad \Leftrightarrow$ $t = \dfrac{\ln\left(\frac{1}{8}\right)}{-0.2} \approx 10.4$ seconds. Thus the velocity is 70 ft/sec after about 10 seconds.

**77. (a)** $P(3) = \dfrac{10}{1 + 4e^{-0.8(3)}} = 7.337$, so there are approximately 7337 fish after 3 years.

**(b)** We solve for $t$. $\dfrac{10}{1 + 4e^{-0.8t}} = 5 \quad \Leftrightarrow \quad 1 + 4e^{-0.8t} = \frac{10}{5} = 2 \quad \Leftrightarrow \quad 4e^{-0.8t} = 1 \quad \Leftrightarrow \quad e^{-0.8t} = 0.25 \quad \Leftrightarrow$ $-0.8t = \ln 0.25 \quad \Leftrightarrow \quad t = \dfrac{\ln 0.25}{-0.8} = 1.73$. So the population will reach 5000 fish in about 1 year and 9 months.

**78. (a)** $I = 10e^{-0.008(30)} = 10e^{-0.24} = 7.87$. So at 30 ft the intensity is 7.87 lumens.

**(b)** $5 = 10e^{-0.008x} \quad \Leftrightarrow \quad e^{-0.008x} = \frac{1}{2} \quad \Leftrightarrow \quad -0.008x = \ln\left(\frac{1}{2}\right) \quad \Leftrightarrow \quad x = \dfrac{\ln(1/2)}{-0.008} \approx 86.6$. So the intensity drops to 25 lumens at 86.6 ft.

**79. (a)** $\ln\left(\dfrac{P}{P_0}\right) = -\dfrac{h}{k} \quad \Leftrightarrow \quad \dfrac{P}{P_0} = e^{-h/k} \quad \Leftrightarrow \quad P = P_0e^{-h/k}$. Substituting $k = 7$ and $P_0 = 100$ we get $P = 100e^{-h/7}$.

**(b)** When $h = 4$ we have $P = 100e^{-4/7} \approx 56.47$ kPa.

**80. (a)** $\ln\left(\dfrac{T - 20}{200}\right) = -0.11t \quad \Leftrightarrow \quad \dfrac{T - 20}{200} = e^{-0.11t} \quad \Leftrightarrow \quad T - 20 = 200e^{-0.11t} \quad \Leftrightarrow \quad T = 20 + 200e^{-0.11t.}$

**(b)** When $t = 20$ we have $T = 20 + 200e^{-0.11(20)} = 20 + 200e^{-2.2} \approx 42.2^\circ$ F.

**81.** **(a)** $I = \frac{60}{13}\left(1 - e^{-13t/5}\right) \quad \Leftrightarrow \quad \frac{13}{60}I = 1 - e^{-13t/5} \quad \Leftrightarrow \quad e^{-13t/5} = 1 - \frac{13}{60}I \quad \Leftrightarrow \quad -\frac{13}{5}t = \ln\left(1 - \frac{13}{60}I\right) \quad \Leftrightarrow$ $t = -\frac{5}{13}\ln\left(1 - \frac{13}{60}I\right)$.

**(b)** Substituting $I = 2$, we have $t = -\frac{5}{13}\ln\left[1 - \frac{13}{60}(2)\right] \approx 0.218$ seconds.

**82.** **(a)** $P = M - Ce^{-kt} \quad \Leftrightarrow \quad Ce^{-kt} = M - P \quad \Leftrightarrow$

$e^{-kt} = \dfrac{M-P}{C} \quad \Leftrightarrow \quad -kt = \ln\left(\dfrac{M-P}{C}\right) \quad \Leftrightarrow$

$t = -\dfrac{1}{k}\ln\left(\dfrac{M-P}{C}\right)$

**(b)** $P(t) = 20 - 14e^{-0.024t}$. Substituting $M = 20$, $C = 14$, $k = 0.024$, and $P = 12$ into $t = -\dfrac{1}{k}\ln\left(\dfrac{M-P}{C}\right)$, we have

$t = -\dfrac{1}{0.024}\ln\left(\dfrac{20-12}{14}\right) \approx 23.32$. So it takes about 23 months.

**(c)**

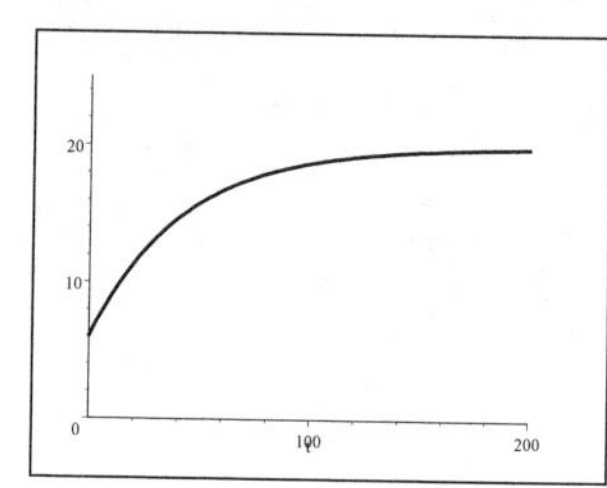

**83.** Since $9^1 = 9$, $9^2 = 81$, and $9^3 = 729$, the solution of $9^x = 20$ must be between 1 and 2 (because 20 is between 9 and 81), whereas the solution to $9^x = 100$ must be between 2 and 3 (because 100 is between 81 and 729).

**84.** Notice that $\log\left(x^{1/\log x}\right) = \dfrac{1}{\log x}\log x = 1$, so $x^{1/\log x} = 10^1$ for all $x > 0$. So $x^{1/\log x} = 5$ has no solution, and $x^{1/\log x} = k$ has a solution only when $k = 10$. This is verified by the graph of $f(x) = x^{1/\log x}$.

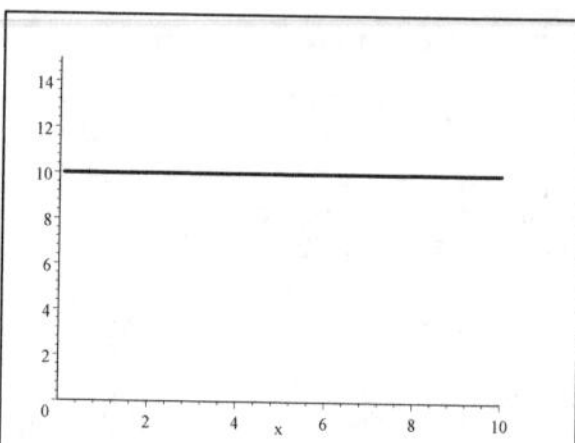

**85.** **(a)** $(x-1)^{\log(x-1)} = 100(x-1) \quad \Leftrightarrow \quad \log\left((x-1)^{\log(x-1)}\right) = \log(100(x-1)) \quad \Leftrightarrow$ $[\log(x-1)]\log(x-1) = \log 100 + \log(x-1) \quad \Leftrightarrow \quad [\log(x-1)]^2 - \log(x-1) - 2 = 0 \quad \Leftrightarrow$ $[\log(x-1) - 2][\log(x-1) + 1] = 0$. Thus either $\log(x-1) = 2 \quad \Leftrightarrow \quad x = 101$ or $\log(x-1) = -1 \quad \Leftrightarrow$ $x = \frac{11}{10}$.

**(b)** $\log_2 x + \log_4 x + \log_8 x = 11 \quad \Leftrightarrow \quad \log_2 x + \log_2\sqrt{x} + \log_2\sqrt[3]{x} = 11 \quad \Leftrightarrow \quad \log_2\left(x\sqrt{x}\sqrt[3]{x}\right) = 11 \quad \Leftrightarrow$ $\log_2\left(x^{11/6}\right) = 11 \quad \Leftrightarrow \quad \frac{11}{6}\log_2 x = 11 \quad \Leftrightarrow \quad \log_2 x = 6 \quad \Leftrightarrow \quad x = 2^6 = 64$

**(c)** $4^x - 2^{x+1} = 3 \quad \Leftrightarrow \quad (2^x)^2 - 2(2^x) - 3 = 0 \quad \Leftrightarrow \quad (2^x - 3)(2^x + 1) = 0 \quad \Leftrightarrow$ either $2^x = 3 \quad \Leftrightarrow$ $x = \dfrac{\ln 3}{\ln 2}$ or $2^x = -1$, which has no real solution. So $x = \dfrac{\ln 3}{\ln 2}$ is the only real solution.

# 5.5 Modeling with Exponential and Logarithmic Functions

**1.** **(a)** $n(0) = 500$.

**(b)** The relative growth rate is $0.45 = 45\%$.

**(c)** $n(3) = 500e^{0.45(3)} \approx 1929$.

**(d)** $10{,}000 = 500e^{0.45t} \quad \Leftrightarrow \quad 20 = e^{0.45t} \quad \Leftrightarrow \quad 0.45t = \ln 20 \quad \Leftrightarrow \quad t = \dfrac{\ln 20}{0.45} \approx 6.66$ hours, or 6 hours 40 minutes.

**2. (a)** The relative growth rate is $0.012 = 1.2\%$.

**(b)** $n(5) = 12e^{0.012(5)} = 12e^{0.06} \approx 12.74$ million fish.

**(c)** $30 = 12e^{0.012t} \Leftrightarrow 2.5 = e^{0.012t} \Leftrightarrow 0.012t = \ln 2.5 \Leftrightarrow$ $t = \dfrac{\ln 2.5}{0.012} \approx 76.36$. Thus the fish population reaches 30 million after about 76 years.

**(d)**

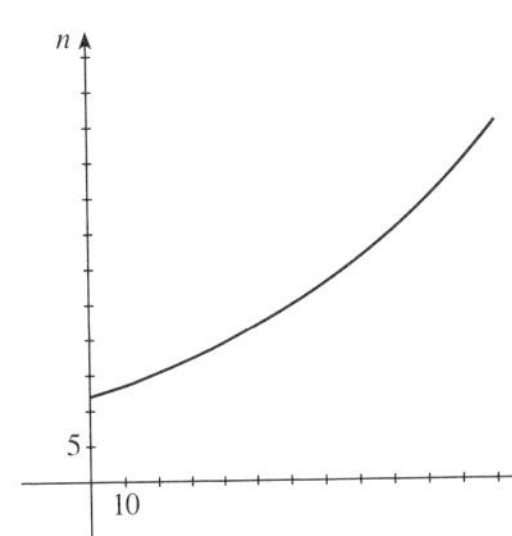


**3. (a)** $r = 0.08$ and $n(0) = 18000$. Thus the population is given by the formula $n(t) = 18{,}000e^{0.08t}$.

**(b)** $t = 2008 - 2000 = 8$. Then we have $n(8) = 18000e^{0.08(8)} = 18000e^{0.64} \approx 34{,}137$. Thus there should be 34,137 foxes in the region by the year 2008.

**(c)**

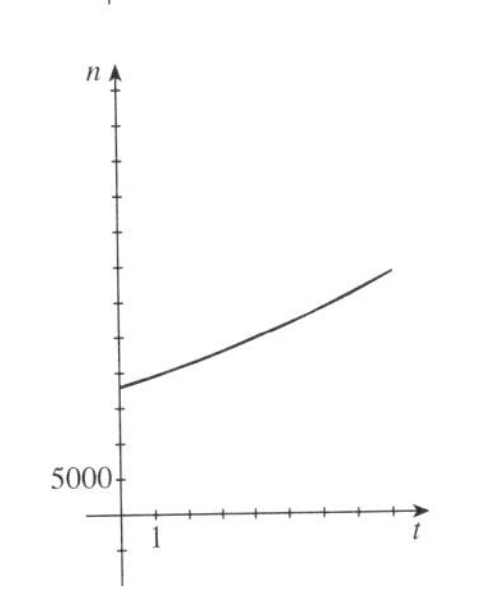


**4.** $n(t) = n_0e^{rt}$; $n_0 = 110$ million, $t = 2020 - 1995 = 25$

**(a)** $r = 0.03$; $n(25) = 110{,}000{,}000e^{0.03(25)} = 110{,}000{,}000e^{0.75} \approx 232{,}870{,}000$. Thus at a 3% growth rate, the projected population will be approximately 233 million people by the year 2020.

**(b)** $r = 0.02$; $n(25) = 110{,}000{,}000e^{0.02(25)} = 110{,}000{,}000e^{0.50} \approx 181{,}359{,}340$. Thus at a 2% growth rate, the projected population will be approximately 181 million people by the year 2020.

**5. (a)** $n(t) = 112{,}000e^{0.04t}$.

**(b)** $t = 2000 - 1994 = 6$ and $n(6) = 112{,}000e^{0.04(6)} \approx 142380$. The projected population is about142,000.

**(c)** $200{,}000 = 112{,}000e^{0.04t} \Leftrightarrow \frac{25}{14} = e^{0.04t} \Leftrightarrow 0.04t = \ln\left(\frac{25}{14}\right) \Leftrightarrow t = 25\ln\left(\frac{25}{14}\right) \approx 14.5$. Since $1994 + 14.5 = 2008.5$, the population will reach 200,000 during the year 2008.

**6. (a)** $n(t) = n_0e^{rt}$ with $n_0 = 85$ and $r = 0.18$. Thus $n(t) = 85e^{0.18t}$.

**(b)** $n(3) = 85e^{0.18(3)} \approx 146$ frogs.

**(c)** $600 = 85e^{0.18t} \Leftrightarrow \frac{120}{17} = e^{0.18t} \Leftrightarrow 0.18t = \ln\left(\frac{120}{17}\right) \Leftrightarrow t = \frac{1}{0.18}\ln\left(\frac{120}{17}\right) \approx 10.86$. So the population will reach 600 frogs in about 11 years.

**7. (a)** The deer population in 1996 was 20,000.

**(b)** Using the model $n(t) = 20{,}000e^{rt}$ and the point $(4, 31000)$, we have $31{,}000 = 20{,}000e^{4r} \Leftrightarrow 1.55 = e^{4r} \Leftrightarrow$ $4r = \ln 1.55 \Leftrightarrow r = \frac{1}{4}\ln 1.55 \approx 0.1096$. Thus $n(t) = 20{,}000e^{0.1096t}$

**(c)** $n(8) = 20{,}000e^{0.1096(8)} \approx 48{,}218$, so the projected deer population in 2004 is about 48,000.

**(d)** $100{,}000 = 20{,}000e^{0.1096t} \Leftrightarrow 5 = e^{0.1096t} \Leftrightarrow 0.1096t = \ln 5 \Leftrightarrow t = \dfrac{\ln 5}{0.1096} \approx 14.63$. Since $1996 + 14.63 = 2010.63$, the deer population will reach 100,000 during the year 2010.

**8. (a)** Since the population grows exponentially, the population is represented by $n(t) = n_0e^{rt}$, with $n_0 = 1500$ and $n(30) = 3000$. Solving for $r$, we have $3000 = 1500e^{30r} \Leftrightarrow 2 = e^{30r} \Leftrightarrow 30r = \ln 2 \Leftrightarrow$ $r = (\ln 2)/30 \approx 0.023$. Thus $n(t) = 1500e^{0.023t}$.

**(b)** Since 2 hours is 120 minutes, the number of bacteria in 2 hours is $n(120) = 1500e^{0.023(120)} \approx 24{,}000$.

**(c)** We need to solve $4000 = 1500e^{0.023t}$ for $t$. So $4000 = 1500e^{0.023t} \Leftrightarrow \frac{8}{3} = e^{0.023t} \Leftrightarrow 0.023t = \ln\frac{8}{3} \Leftrightarrow$ $t = \dfrac{\ln(8/3)}{0.023} \approx 42.6$. Thus the bacteria population will reach 4000 in about 43 minutes.

**9. (a)** Using the formula $n(t) = n_0 e^{rt}$ with $n_0 = 8600$ and $n(1) = 10000$, we solve for $r$, giving $10000 = n(1) = 8600e^r$ $\Leftrightarrow$ $\frac{50}{43} = e^r$ $\Leftrightarrow$ $r = \ln\left(\frac{50}{43}\right) \approx 0.1508$. Thus $n(t) = 8600e^{0.1508t}$.

**(b)** $n(2) = 8600e^{0.1508(2)} \approx 11627$. Thus the number of bacteria after two hours is about 11,600.

**(c)** $17200 = 8600e^{0.1508t}$ $\Leftrightarrow$ $2 = e^{0.1508t}$ $\Leftrightarrow$ $0.1508t = \ln 2$ $\Leftrightarrow$ $t = \dfrac{\ln 2}{0.1508} \approx 4.596$. Thus the number of bacteria will double in about 4.6 hours.

**10. (a)** Using $n(t) = n_0 e^{rt}$ with $n(2) = 400$ and $n(6) = 25{,}600$, we have $n_0 e^{2r} = 400$ and $n_0 e^{6r} = 25{,}600$. Dividing the second equation by the first gives $\dfrac{n_0 e^{6r}}{n_0 e^{2r}} = \dfrac{25{,}600}{400} = 64$ $\Leftrightarrow$ $e^{4r} = 64$ $\Leftrightarrow$ $4r = \ln 64$ $\Leftrightarrow$ $r = \frac{1}{4}\ln 64 \approx 1.04$. Thus the relative rate of growth is about 104%.

**(b)** Since $r = \frac{1}{4}\ln 64 = \frac{1}{2}\ln 8$, we have from part (a) $n(t) = n_0 e^{\left(\frac{1}{2}\ln 8\right)t}$. Since $n(2) = 400$, we have $400 = n_0 e^{\ln 8}$ $\Leftrightarrow$ $n_0 = \dfrac{400}{e^{\ln 8}} = \frac{400}{8} = 50$. So the initial size of the culture was 50.

**(c)** Substituting $n_0 = 50$ and $r = 1.04$, we have $n(t) = n_0 e^{rt} = 50e^{1.04t}$.

**(d)** $n(4.5) = 50e^{1.04(4.5)} = 50e^{4.68} \approx 5388.5$, so the size after 4.5 hours is approximately 5400.

**(e)** $n(t) = 50{,}000 = 50e^{1.04t}$ $\Leftrightarrow$ $e^{1.04t} = 1000$ $\Leftrightarrow$ $1.04t = \ln 1000$ $\Leftrightarrow$ $t = \dfrac{\ln 1000}{1.04} \approx 6.64$. Hence the population will reach 50,000 after roughly $6\frac{2}{3}$ hours.

**11. (a)** $2n_0 = n_0 e^{0.02t}$ $\Leftrightarrow$ $2 = e^{0.02t}$ $\Leftrightarrow$ $0.02t = \ln 2$ $\Leftrightarrow$ $t = 50\ln 2 \approx 34.65$. So we have $t = 1995 + 34.65 = 2029.65$, and hence at the current growth rate the population will double by the year 2029.

**(b)** $3n_0 = n_0 e^{0.02t}$ $\Leftrightarrow$ $3 = e^{0.02t}$ $\Leftrightarrow$ $0.02t = \ln 3$ $\Leftrightarrow$ $t = 50\ln 3 \approx 54.93$. So we have $t = 1995 + 54.93 = 2049.93$, and hence at the current growth rate the population will triple by the year 2050.

**12. (a)** Calculating dates relative to 1950 gives $n_0 = 10{,}586{,}223$ and $n(30) = 23{,}668{,}562$. Then $n(30) = 10{,}586{,}223e^{30r} = 23{,}668{,}562$ $\Leftrightarrow$ $e^{30r} = \frac{23{,}668{,}562}{10{,}586{,}223} \approx 2.2358$ $\Leftrightarrow$ $30r = \ln 2.2358$ $\Leftrightarrow$ $r = \frac{1}{30}\ln 2.2358 \approx 0.0268$. Thus $n(t) = 10{,}586{,}223e^{0.0268t}$.

**(b)** $2(10{,}586{,}223) = 10{,}586{,}223e^{0.0268t}$ $\Leftrightarrow$ $2 = e^{0.0268t}$ $\Leftrightarrow$ $\ln 2 = 0.0268t$ $\Leftrightarrow$ $t = \dfrac{\ln 2}{0.0268} \approx 25.86$. So the population doubles in about 26 years.

**(c)** $t = 2000 - 1950 = 50$; $n(50) \approx 10586223e^{0.0268(50)} \approx 40{,}429{,}246$ and so the population in the year 2000 will be approximately 40,429,000.

**13.** $n(t) = n_0 e^{2t}$. When $n_0 = 1$, the critical level is $n(24) = e^{2(24)} = e^{48}$.We solve the equation $e^{48} = n_0 e^{2t}$, where $n_0 = 10$. This gives$e^{48} = 10e^{2t}$ $\Leftrightarrow$ $48 = \ln 10 + 2t$ $\Leftrightarrow$ $2t = 48 - \ln 10$ $\Leftrightarrow$ $t = \frac{1}{2}(48 - \ln 10) \approx 22.85$ hours.

**14.** From the formula for radioactive decay, we have $m(t) = m_0 e^{-rt}$, where $r = \dfrac{\ln 2}{h}$.

**(a)** We have $m_0 = 22$ and $h = 1600$, so $r = \dfrac{\ln 2}{1600} \approx 0.000433$ and the amount after $t$ years is given by $m(t) = 22e^{-0.000433t}$.

**(b)** $m(4000) = 22e^{-0.000433(4000)} \approx 3.89$, so the amount after 4000 years is about 4 mg.

**(c)** We have to solve for $t$ in the equation $18 = 22\,e^{-0.000433t}$. This gives $18 = 22e^{-0.000433t}$ $\Leftrightarrow$ $\frac{9}{11} = e^{-0.000433t}$ $\Leftrightarrow$ $-0.000433t = \ln\left(\frac{9}{11}\right)$ $\Leftrightarrow$ $t = \dfrac{\ln\left(\frac{9}{11}\right)}{-0.000433} \approx 463.4$, so it takes about 463 years.

**15. (a)** Using $m(t) = m_0 e^{-rt}$ with $m_0 = 10$ and $h = 30$, we have $r = \frac{\ln 2}{h} = \frac{\ln 2}{30} \approx 0.0231$. Thus $m(t) = 10e^{-0.0231t}$.

**(b)** $m(80) = 10e^{-0.0231(80)} \approx 1.6$ grams.

**(c)** $2 = 10e^{-0.0231t} \quad \Leftrightarrow \quad \frac{1}{5} = e^{-0.0231t} \quad \Leftrightarrow \quad \ln\left(\frac{1}{5}\right) = -0.0231t \quad \Leftrightarrow \quad t = \frac{-\ln 5}{-0.0231} \approx 70$ years.

**16. (a)** $m(60) = 40e^{-0.0277(60)} \approx 7.59$, so the mass remaining after 60 days is about 8 g.

**(b)** $10 = 40e^{-0.0277t} \quad \Leftrightarrow \quad 0.25 = e^{-0.0277t} \quad \Leftrightarrow \quad \ln 0.25 = -0.0277t \quad \Leftrightarrow \quad t = -\frac{\ln 0.25}{0.0277} \approx 50.05$, so it takes about 50 days.

**(c)** We need to solve for $t$ in the equation $20 = 40e^{-0.0277t}$. We have $20 = 40e^{-0.0277t} \quad \Leftrightarrow \quad e^{-0.277t} = \frac{1}{2} \quad \Leftrightarrow$ $-0.0277t = \ln \frac{1}{2} \quad \Leftrightarrow \quad t = \frac{\ln \frac{1}{2}}{-0.0277} \approx 25.02$. Thus the half-life of thorium-234 is about 25 days.

**17.** By the formula in the text, $m(t) = m_0 e^{-rt}$ where $r = \frac{\ln 2}{h}$, so $m(t) = 50e^{-[(\ln 2)/28]t}$. We need to solve for $t$ in the equation $32 = 50e^{-[(\ln 2)/28]t}$. This gives $e^{-[(\ln 2)/28]t} = \frac{32}{50} \quad \Leftrightarrow \quad -\frac{\ln 2}{28}t = \ln\left(\frac{32}{50}\right) \quad \Leftrightarrow$ $t = -\frac{28}{\ln 2} \cdot \ln\left(\frac{32}{50}\right) \approx 18.03$, so it takes about 18 years.

**18.** From the formula for radioactive decay, we have $m(t) = m_0 e^{-rt}$, where $r = \frac{\ln 2}{h}$. Since $h = 30$, we have $r = \frac{\ln 2}{30} \approx 0.0231$ and $m(t) = m_0 e^{-0.0231t}$. In this exercise we have to solve for $t$ in the equation $0.05m_0 = m_0 e^{-0.0231t}$ $\Leftrightarrow \quad e^{-0.0231t} = 0.05 \quad \Leftrightarrow \quad -0.0231t = \ln 0.05 \quad \Leftrightarrow \quad t = \frac{\ln 0.05}{-0.0231} \approx 129.7$. So it will take about 130 s.

**19.** By the formula for radioactive decay, we have $m(t) = m_0 e^{-rt}$, where $r = \frac{\ln 2}{h}$, in other words $m(t) = m_0 e^{-[(\ln 2)/h]t}$. In this exercise we have to solve for $h$ in the equation $200 = 250e^{-[(\ln 2)/h] \cdot 48} \quad \Leftrightarrow \quad 0.8 = e^{-[(\ln 2)/h] \cdot 48} \quad \Leftrightarrow$ $\ln(0.8) = -\frac{\ln 2}{h} \cdot 48 \quad \Leftrightarrow \quad h = -\frac{\ln 2}{\ln 0.8} \cdot 48 \approx 149.1$ hours. So the half-life is approximately 149 hours.

**20.** From the formula for radioactive decay, we have $m(t) = m_0 e^{-rt}$, where $r = \frac{\ln 2}{h}$. In other words, $m(t) = m_0 e^{-[(\ln 2)/h]t}$.

**(a)** Using $m(3) = 0.58m_0$, we have to solve for $h$ in the equation $0.58m_0 = m(3) = m_0 e^{-[(\ln 2)/h]3}$. Then $0.58m_0 = m_0 e^{-[(3\ln 2)/h]} \quad \Leftrightarrow \quad e^{-[(3\ln 2)/h]} = 0.58 \quad \Leftrightarrow \quad -\frac{3\ln 2}{h} = \ln 0.58 \quad \Leftrightarrow$ $h = -\frac{3\ln 2}{\ln 0.58} \approx 3.82$ days. Thus the half-life of Radon-222 is about 3.82 days.

**(b)** Here we have to solve for $t$ in the equation $0.2m_0 = m_0 e^{-[(\ln 2)/3.82]t}$. So we have $0.2m_0 = m_0 e^{-[(\ln 2)/3.82]t} \quad \Leftrightarrow$ $0.2 = e^{-[(\ln 2)/3.82]t} \quad \Leftrightarrow \quad -\frac{\ln 2}{3.82}t = \ln 0.2 \quad \Leftrightarrow \quad t = -\frac{3.82\ln 0.2}{\ln 2} \approx 8.87$. So it takes roughly 9 days for a sample of Radon-222 to decay to 20% of its original mass.

**21.** By the formula in the text, $m(t) = m_0 e^{-[(\ln 2)/h] \cdot t}$, so we have $0.65 = 1 \cdot e^{-[(\ln 2)/5730] \cdot t} \quad \Leftrightarrow \quad \ln(0.65) = -\frac{\ln 2}{5730}t$ $\Leftrightarrow \quad t = -\frac{5730\ln 0.65}{\ln 2} \approx 3561$. Thus the artifact is about 3560 years old.

**22.** From the formula for radioactive decay, we have $m(t) = m_0 e^{-rt}$ where $r = \frac{\ln 2}{h}$. Since $h = 5730$, $r = \frac{\ln 2}{5730} \approx 0.000121$ and $m(t) = m_0 e^{-0.000121t}$. We need to solve for $t$ in the equation $0.59m_0 = m_0 e^{-0.000121t} \quad \Leftrightarrow \quad e^{-0.000121t} = 0.59$ $\Leftrightarrow \quad -0.000121t = \ln 0.59 \quad \Leftrightarrow \quad t = \frac{\ln 0.59}{-0.000121} \approx 4360.6$. So the mummy was buried about 4360 years ago.

**23. (a)** $T(0) = 65 + 145e^{-0.05(0)} = 65 + 145 = 210°$ F.

**(b)** $T(10) = 65 + 145e^{-0.05(10)} \approx 152.9$. Thus the temperature after 10 minutes is about 153° F.

**(c)** $100 = 65 + 145e^{-0.05t} \Leftrightarrow 35 = 145e^{-0.05t} \Leftrightarrow 0.2414 = e^{-0.05t} \Leftrightarrow \ln 0.2414 = -0.05t \Leftrightarrow$ $t = -\dfrac{\ln 0.2414}{0.05} \approx 28.4$. Thus the temperature will be 100° F in about 28 minutes.

**24. (a)** We use Newton's Law of Cooling: $T(t) = T_s + D_0 e^{-kt}$ with $k = 0.1947$, $T_s = 60$, and $D_0 = 98.6 - 60 = 38.6$. So $T(t) = 60 + 38.6e^{-0.1947t}$.

**(b)** Solve $T(t) = 72$. So $72 = 60 + 38.6e^{-0.1947t} \Leftrightarrow 38.6e^{-0.1947t} = 12 \Leftrightarrow e^{-0.1947t} = \dfrac{12}{38.6} \Leftrightarrow$ $-0.1947t = \ln\left(\dfrac{12}{38.6}\right) \Leftrightarrow t = -\dfrac{1}{0.1947}\ln\left(\dfrac{12}{38.6}\right) \approx 6.00$, and the time of death was about 6 hours ago.

**25.** Using Newton's Law of Cooling, $T(t) = T_s + D_0 e^{-kt}$ with $T_s = 75$ and $D_0 = 185 - 75 = 110$. So $T(t) = 75 + 110e^{-kt}$.

**(a)** Since $T(30) = 150$, we have $T(30) = 75 + 110e^{-30k} = 150 \Leftrightarrow 110e^{-30k} = 75 \Leftrightarrow e^{-30k} = \frac{15}{22} \Leftrightarrow$ $-30k = \ln\left(\frac{15}{22}\right) \Leftrightarrow k = -\frac{1}{30}\ln\left(\frac{15}{22}\right)$. Thus we have $T(45) = 75 + 110e^{(45/30)\ln(15/22)} \approx 136.9$, and so the temperature of the turkey after 45 minutes is about 137° F.

**(b)** The temperature will be 100°F when $75 + 110e^{(t/30)\ln(15/22)} = 100 \Leftrightarrow e^{(t/30)\ln(15/22)} = \dfrac{25}{110} = \frac{5}{22} \Leftrightarrow$ $\left(\dfrac{t}{30}\right)\ln\left(\frac{15}{22}\right) = \ln\left(\frac{5}{22}\right) \Leftrightarrow t = 30\dfrac{\ln\left(\frac{5}{22}\right)}{\ln\left(\frac{15}{22}\right)} \approx 116.1$. So the temperature will be 100° F after 116 minutes.

**26.** We use Newton's Law of Cooling: $T(t) = T_s + D_0 e^{-kt}$, with $T_s = 20$ and $D_0 = 100 - 20 = 80$. So $T(t) = 20 + 80e^{-kt}$. Since $T(15) = 75$, we have $20 + 80e^{-15k} = 75 \Leftrightarrow 80e^{-15k} = 55 \Leftrightarrow e^{-15k} = \frac{11}{16} \Leftrightarrow$ $-15k = \ln\left(\frac{11}{16}\right) \Leftrightarrow k = -\frac{1}{15}\ln\left(\frac{11}{16}\right)$. Thus $T(25) = 20 + 80e^{(25/15)\cdot\ln(11/16)} \approx 62.8$, and so the temperature after another 10 min is 63° C. The function $T(t) = 20 + 80e^{(1/15)\cdot\ln(11/16)t}$ is shown in the viewing rectangle $[0, 30]$ by $[50, 100]$.

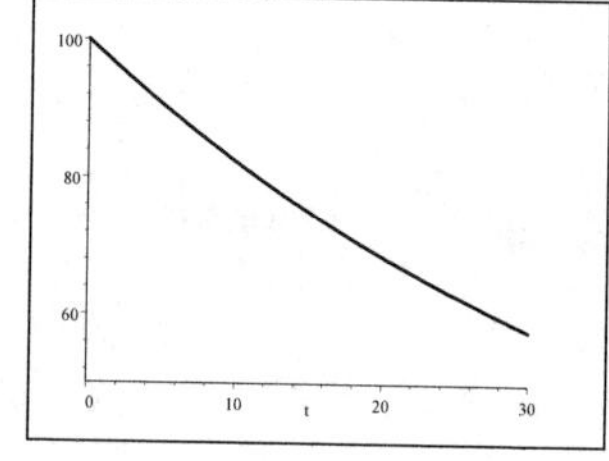


**27. (a)** $\text{pH} = -\log[\text{H}^+] = -\log(5.0 \times 10^{-3}) \approx 2.3$

**(b)** $\text{pH} = -\log[\text{H}^+] = -\log(3.2 \times 10^{-4}) \approx 3.5$

**(c)** $\text{pH} = -\log[\text{H}^+] = -\log(5.0 \times 10^{-9}) \approx 8.3$

**28.** $\text{pH} = -\log[\text{H}^+] = -\log(3.1 \times 10^{-8}) \approx 7.5$ and the substance is basic.

**29. (a)** $\text{pH} = -\log[\text{H}^+] = 3.0 \Leftrightarrow [\text{H}^+] = 10^{-3}$ M

**(b)** $\text{pH} = -\log[\text{H}^+] = 6.5 \Leftrightarrow [\text{H}^+] = 10^{-6.5} \approx 3.2 \times 10^{-7}$ M

**30. (a)** $\text{pH} = -\log[\text{H}^+] = 4.6 \Leftrightarrow [\text{H}^+] = 10^{-4.6}$ M $\approx 2.5 \times 10^{-5}$ M

**(b)** $\text{pH} = -\log[\text{H}^+] = 7.3 \Leftrightarrow [\text{H}^+] = 10^{-7.3}$ M $\approx 5.0 \times 10^{-8}$ M

**31.** $4.0 \times 10^{-7} \le [\text{H}^+] \le 1.6 \times 10^{-5} \Leftrightarrow \log(4.0 \times 10^{-7}) \le \log[\text{H}^+] \le \log(1.6 \times 10^{-5}) \Leftrightarrow$ $-\log(4.0 \times 10^{-7}) \ge \text{pH} \ge -\log(1.6 \times 10^{-5}) \Leftrightarrow 6.4 \ge \text{pH} \ge 4.8$. Therefore the range of pH readings for cheese is approximately 4.8 to 6.4.

**32.** $2.8 \le \text{pH} \le 3.8 \Leftrightarrow -2.8 \ge -\text{pH} \ge -3.8 \Leftrightarrow 10^{-2.8} \ge 10^{-\text{pH}} \ge 10^{-3.8} \Leftrightarrow$ $1.58 \times 10^{-3} \ge [\text{H}^+] \ge 1.58 \times 10^{-4}$. The range of $[\text{H}^+]$ is $1.58 \times 10^{-4}$ to $1.58 \times 10^{-3}$.

**33.** Let $I_0$ be the intensity of the smaller earthquake and $I_1$ the intensity of the larger earthquake. Then $I_1 = 20I_0$. Notice that $M_0 = \log\left(\frac{I_0}{S}\right) = \log I_0 - \log S$ and $M_1 = \log\left(\frac{I_1}{S}\right) = \log\left(\frac{20I_0}{S}\right) = \log 20 + \log I_0 - \log S$. Then $M_1 - M_0 = \log 20 + \log I_0 - \log S - \log I_0 + \log S = \log 20 \approx 1.3$. Therefore the magnitude is 1.3 times larger.

**34.** Let the subscript $S$ represent the San Francisco earthquake and $J$ the Japan earthquake. Then we have $M_S = \log\left(\frac{I_S}{S}\right) = 8.3 \quad\Leftrightarrow\quad I_S = S \cdot 10^{8.3}$ and $M_J = \log\left(\frac{I_J}{S}\right) = 4.9 \quad\Leftrightarrow\quad I_J = S \cdot 10^{4.9}$. So $\frac{I_S}{I_J} = \frac{10^{8.3}}{10^{4.9}} = 10^{3.4} \approx 2511.9$, and so the San Francisco earthquake was 2500 times more intense than the Japan earthquake.

**35.** Let the subscript $A$ represent the Alaska earthquake and $S$ represent the San Francisco earthquake. Then $M_A = \log\left(\frac{I_A}{S}\right) = 8.6 \quad\Leftrightarrow\quad I_A = S \cdot 10^{8.6}$ ; also, $M_S = \log\left(\frac{I_S}{S}\right) = 8.3 \quad\Leftrightarrow\quad I_S = S \cdot 10^{8.6}$.So $\frac{I_A}{I_S} = \frac{S \cdot 10^{8.6}}{S \cdot 10^{8.3}} = 10^{0.3} \approx 1.995$, and hence the Alaskan earthquake was roughly twice as intense as the San Francisco earthquake.

**36.** Let the subscript $N$ represent the Northridge, California earthquake and $K$ the Kobe, Japan earthquake. Then $M_N = \log\left(\frac{I_N}{S}\right) = 6.8 \quad\Leftrightarrow\quad I_N = S \cdot 10^{6.8}$ and $M_K = \log\left(\frac{I_K}{S}\right) = 7.2 \quad\Leftrightarrow\quad I_K = S \cdot 10^{7.2}$. So $\frac{I_K}{I_N} = \frac{10^{7.2}}{10^{6.8}} = 10^{0.4} \approx 2.51$, and so the Kobe, Japan earthquake was 2.5 times more intense than the Northridge, California earthquake.

**37.** Let the subscript $M$ represent the Mexico City earthquake, and $T$ represent the Tangshan earthquake. We have $\frac{I_T}{I_M} = 1.26 \quad\Leftrightarrow\quad \log 1.26 = \log\frac{I_T}{I_M} = \log\frac{I_T/S}{I_M/S} = \log\frac{I_T}{S} - \log\frac{I_M}{S} = M_T - M_M$. Therefore $M_T = M_M + \log 1.26 \approx 8.1 + 0.1 = 8.2$. Thus the magnitude of the Tangshan earthquake was roughly 8.2.

**38.** $\beta = 10\log\left(\frac{I}{I_0}\right) = 10\log\left(\frac{2.0 \times 10^{-5}}{1.0 \times 10^{-12}}\right) = 10\log\left(2 \times 10^7\right) = 10\left(\log 2 + \log 10^7\right) = 10\left(\log 2 + 7\right) \approx 73$. Therefore the intensity level was 73 dB.

**39.** $98 = 10\log\left(\frac{I}{10^{-12}}\right) \quad\Leftrightarrow\quad \log\left(I \cdot 10^{12}\right) = 9.8 \quad\Leftrightarrow\quad \log I = 9.8 - \log 10^{12} = -2.2 \quad\Leftrightarrow$ $I = 10^{-2.2} \approx 6.3 \times 10^{-3}$. So the intensity was $6.3 \times 10^{-3}$ watts/m$^2$.

**40.** Let the subscript $M$ represent the power mower and $C$ the rock concert. Then $106 = 10\log\left(\frac{I_M}{10^{-12}}\right) \quad\Leftrightarrow$ $\log\left(I_M \cdot 10^{12}\right) = 10.6 \quad\Leftrightarrow\quad I_M \cdot 10^{12} = 10^{10.6}$. Also $120 = 10\log\left(\frac{I_C}{10^{-12}}\right) \quad\Leftrightarrow\quad \log\left(I_C \cdot 10^{12}\right) = 12.0 \quad\Leftrightarrow$ $I_C \cdot 10^{12} = 10^{12.0}$. So $\frac{I_C}{I_M} = \frac{10^{12}}{10^{10.6}} = 10^{1.4} \approx 25.12$, and so the ratio of intensity is roughly 25.

**41. (a)** $\beta_1 = 10\log\left(\frac{I_1}{I_0}\right)$ and $I_1 = \frac{k}{d_1^2} \Leftrightarrow \beta_1 = 10\log\left(\frac{k}{d_1^2 I_0}\right) = 10\left[\log\left(\frac{k}{I_0}\right) - 2\log d_1\right] = 10\log\left(\frac{k}{I_0}\right) - 20\log d_1$.

Similarly, $\beta_2 = 10\log\left(\frac{k}{I_0}\right) - 20\log d_2$. Substituting the expression for $\beta_1$ gives

$$\beta_2 = 10\log\left(\frac{k}{I_0}\right) - 20\log d_1 + 20\log d_1 - 20\log d_2 = \beta_1 + 20\log d_1 - 20\log d_2 = \beta_1 + 20\log\left(\frac{d_1}{d_2}\right).$$

**(b)** $\beta_1 = 120$, $d_1 = 2$, and $d_2 = 10$. Then $\beta_2 = \beta_1 + 20\log\left(\frac{d_1}{d_2}\right) = 120 + 20\log\left(\frac{2}{10}\right) = 120 + 20\log 0.2 \approx 106$, and so the intensity level at 10 m is approximately 106 dB.

# Chapter 5 Review

**1.** $f(x) = 2^{-x+1}$. Domain $(-\infty, \infty)$, range $(0, \infty)$, asymptote $y = 0$.

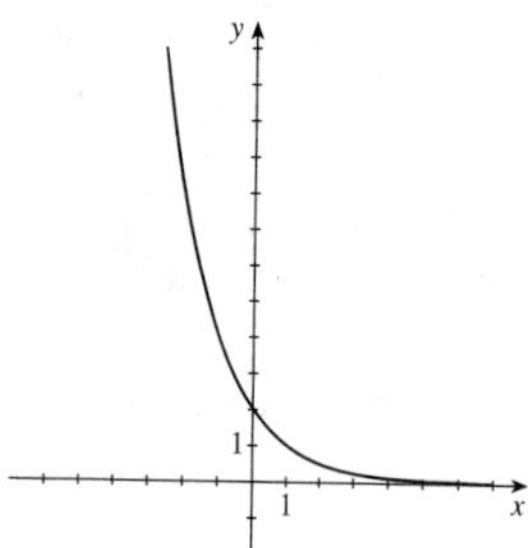


**2.** $f(x) = 3^{x-2}$. domain $(-\infty, \infty)$, range $(0, \infty)$, asymptote $y = 0$.

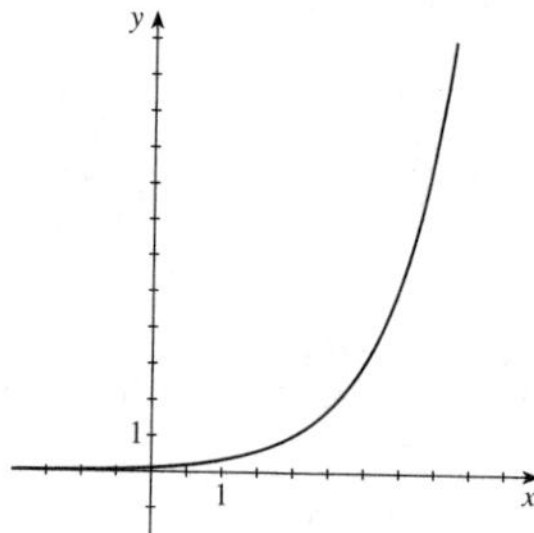


**3.** $g(x) = 3 + 2^x$. Domain $(-\infty, \infty)$, range $(3, \infty)$, asymptote $y = 3$.

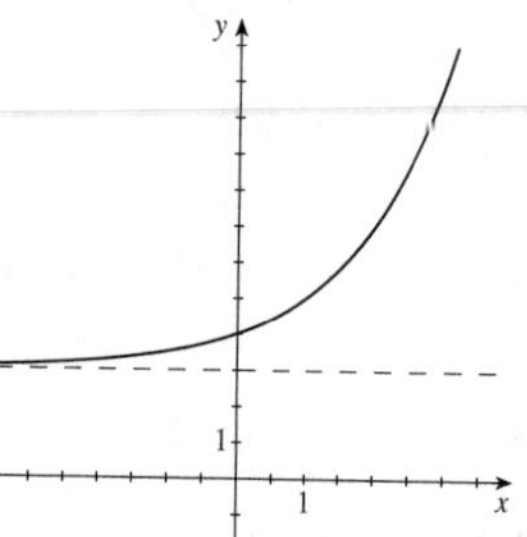


**4.** $g(x) = 5^{-x} - 5$. Domain $(-\infty, \infty)$, range $(-5, \infty)$, asymptote $y = -5$.

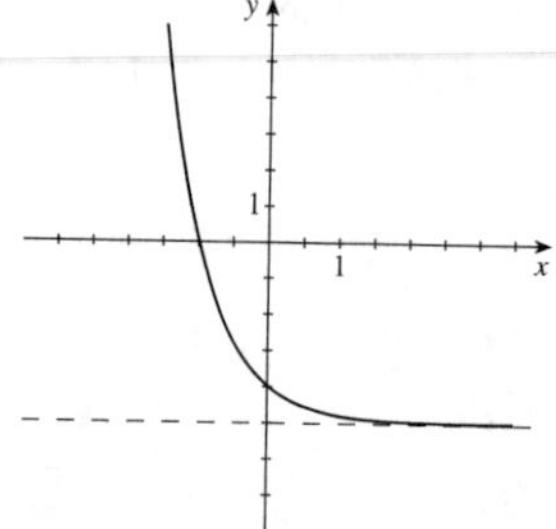


**5.** $f(x) = \log_3(x - 1)$. Domain $(1, \infty)$, range $(-\infty, \infty)$, asymptote $x = 1$.

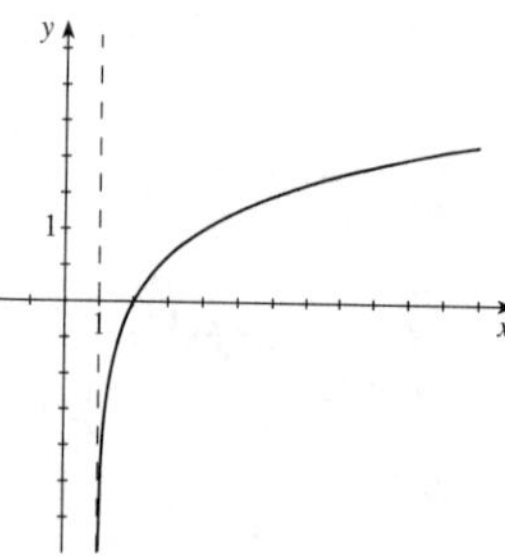


**6.** $g(x) = \log(-x)$. Domain $(-\infty, 0)$, range $(-\infty, \infty)$, asymptote $x = 0$.

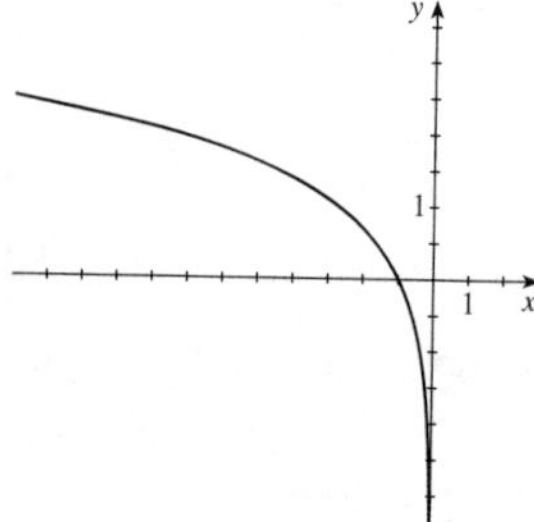


**7.** $f(x) = 2 - \log_2 x$. Domain $(0, \infty)$, range $(-\infty, \infty)$, asymptote $x = 0$.

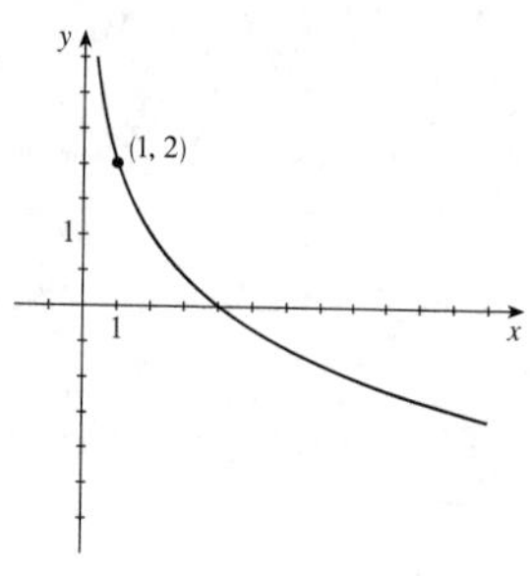


**8.** $f(x) = 3 + \log_5(x + 4)$. Domain $(-4, \infty)$, range $(-\infty, \infty)$, asymptote $x = -4$.

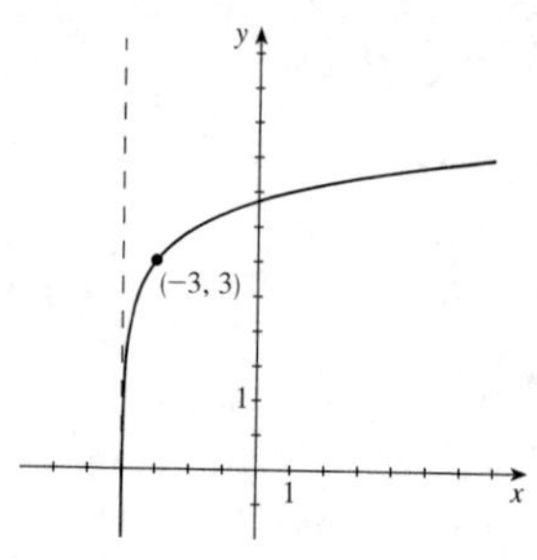

**9.** $F(x) = e^x - 1$. Domain $(-\infty, \infty)$, range $(-1, \infty)$, asymptote $y = -1$.

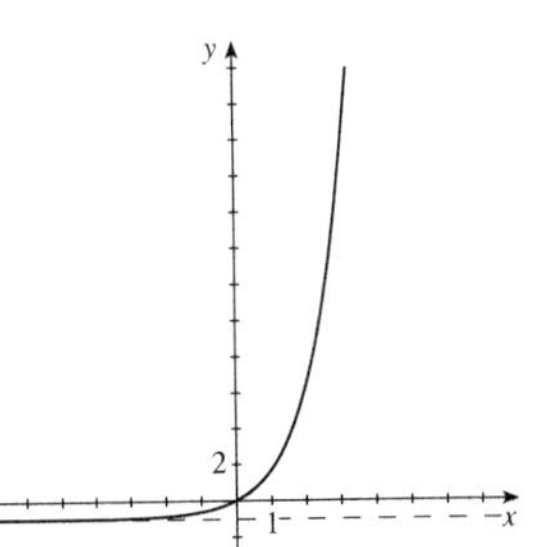


**10.** $G(x) = \frac{1}{2}e^{x-1}$. Domain $(-\infty, \infty)$, range $(0, \infty)$, asymptote $y = 0$.

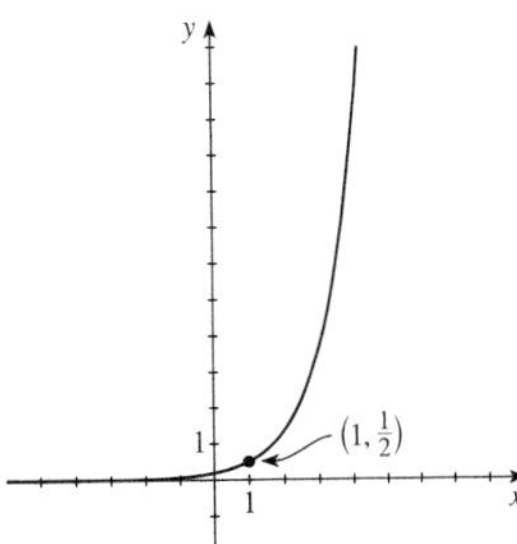


**11.** $g(x) = 2\ln x$. Domain $(0, \infty)$, range $(-\infty, \infty)$, asymptote $x = 0$.

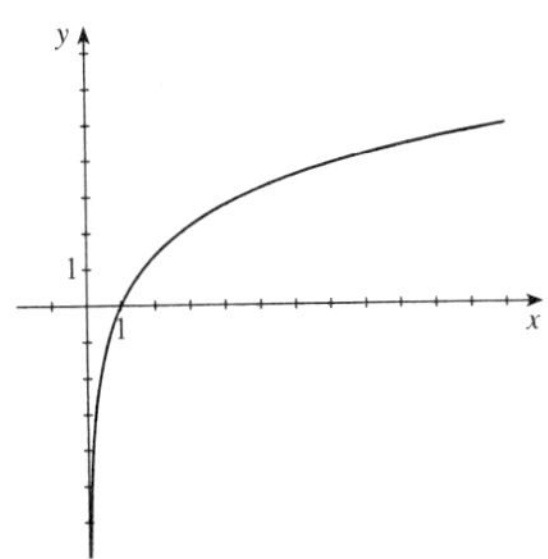


**12.** $g(x) = \ln(x^2)$. Domain $\{x \mid x \neq 0\} = (-\infty, 0) \cup (0, \infty)$, range $(-\infty, \infty)$, asymptote $x = 0$.

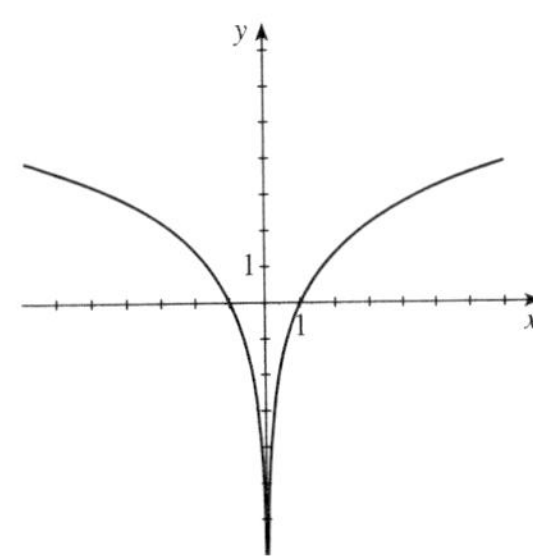


**13.** $f(x) = 10^{x^2} + \log(1 - 2x)$. Since $\log u$ is defined only for $u > 0$, we require $1 - 2x > 0 \Leftrightarrow -2x > -1 \Leftrightarrow x < \frac{1}{2}$, and so the domain is $\left(-\infty, \frac{1}{2}\right)$.

**14.** $g(x) = \ln(2 + x - x^2)$. We must have $2 + x - x^2 > 0$ (since $\ln y$ is defined only for $y > 0$) $\Leftrightarrow x^2 - x - 2 < 0 \Leftrightarrow (x - 2)(x + 1) < 0$. The endpoints of the intervals are 2 and $-1$.

| Interval | $(-\infty, -1)$ | $(-1, 2)$ | $(2, \infty)$ |
|---|---|---|---|
| Sign of $x - 2$ | $-$ | $-$ | $+$ |
| Sign of $x + 1$ | $-$ | $+$ | $+$ |
| Sign of $(x - 2)(x + 1)$ | $+$ | $-$ | $+$ |

Thus the domain is $(-1, 2)$.

**15.** $h(x) = \ln(x^2 - 4)$. We must have $x^2 - 4 > 0$ ( since $\ln y$ is defined only for $y > 0$) $\Leftrightarrow x^2 - 4 > 0 \Leftrightarrow (x - 2)(x + 2) > 0$. The endpoints of the intervals are $-2$ and 2.

| Interval | $(-\infty, -2)$ | $(-2, 2)$ | $(2, \infty)$ |
|---|---|---|---|
| Sign of $x - 2$ | $-$ | $-$ | $+$ |
| Sign of $x + 2$ | $-$ | $+$ | $+$ |
| Sign of $(x - 2)(x + 2)$ | $+$ | $-$ | $+$ |

Thus the domain is $(-\infty, -2) \cup (2, \infty)$.

**16.** $k(x) = \ln|x|$. We must have $|x| > 0$. So $|x| > 0 \Leftrightarrow -x > 0$ or $x > 0$. Since $-x > 0 \Leftrightarrow x < 0$, the domain is $x < 0$ or $x > 0$ which is equivalent to $x \neq 0$. In interval notation, $(-\infty, 0) \cup (0, \infty)$.

**17.** $\log_2 1024 = 10 \quad \Leftrightarrow \quad 2^{10} = 1024$

**18.** $\log_6 37 = x \quad \Leftrightarrow \quad 6^x = 37$

**19.** $\log x = y \quad \Leftrightarrow \quad 10^y = x$

**20.** $\ln c = 17 \quad \Leftrightarrow \quad e^{17} = c$

**21.** $2^6 = 64 \quad \Leftrightarrow \quad \log_2 64 = 6$

**22.** $49^{-1/2} = \frac{1}{7} \quad \Leftrightarrow \quad \log_{49} \frac{1}{7} = -\frac{1}{2}$

**23.** $10^x = 74 \quad \Leftrightarrow \quad \log_{10} 74 = x \quad \Leftrightarrow \quad \log 74 = x$

**24.** $e^k = m \quad \Leftrightarrow \quad \ln m = k$

**25.** $\log_2 128 = \log_2 \left(2^7\right) = 7$

**26.** $\log_8 1 = \log_8 \left(8^0\right) = 0$

**27.** $10^{\log 45} = 45$

**28.** $\log 0.000001 = \log 10^{-6} = -6$

**29.** $\ln \left(e^6\right) = 6$

**30.** $\log_4 8 = \log_4 \left(4^{3/2}\right) = \frac{3}{2}$

**31.** $\log_3 \frac{1}{27} = \log_3 3^{-3} = -3$

**32.** $2^{\log_2 13} = 13$

**33.** $\log_5 \sqrt{5} = \log_5 5^{1/2} = \frac{1}{2}$

**34.** $e^{2 \ln 7} = \left(e^{\ln 7}\right)^2 = 7^2 = 49$

**35.** $\log 25 + \log 4 = \log (25 \cdot 4) = \log 10^2 = 2$

**36.** $\log_3 \sqrt{243} = \log_3 \left(3^{5/2}\right) = \frac{5}{2}$

**37.** $\log_2 \left(16^{23}\right) = \log_2 \left(2^4\right)^{23} = \log_2 2^{92} = 92$

**38.** $\log_5 250 - \log_5 2 = \log_5 \frac{250}{2} = \log_5 125 = \log_5 5^3 = 3$

**39.** $\log_8 6 - \log_8 3 + \log_8 2 = \log_8 \left(\frac{6}{3} \cdot 2\right) = \log_8 4 = \log_8 8^{2/3} = \frac{2}{3}$

**40.** $\log_{10} \left(\log_{10} 10^{100}\right) = \log_{10} 100 = \log_{10} 10^2 = 2$

**41.** $\log \left(AB^2C^3\right) = \log A + 2 \log B + 3 \log C$

**42.** $\log_2 \left(x\sqrt{x^2+1}\right) = \log_2 x + \log_2 \sqrt{x^2+1} = \log_2 x + \frac{1}{2} \log_2 \left(x^2+1\right)$

**43.** $\ln \sqrt{\dfrac{x^2-1}{x^2+1}} = \frac{1}{2} \ln \left(\dfrac{x^2-1}{x^2+1}\right) = \frac{1}{2} \left[\ln \left(x^2-1\right) - \ln \left(x^2+1\right)\right]$

**44.** $\log \left(\dfrac{4x^3}{y^2 (x-1)^5}\right) = \log \left(4x^3\right) - \log \left[y^2 (x-1)^5\right] = \log 4 + 3 \log x - \left[2 \log y + 5 \log (x-1)\right]$

**45.** $\log_5 \left(\dfrac{x^2 (1-5x)^{3/2}}{\sqrt{x^3-x}}\right) = \log_5 x^2 (1-5x)^{3/2} - \log_5 \sqrt{x \left(x^2-1\right)} = 2 \log_5 x + \dfrac{3}{2} \log_5 (1-5x) - \frac{1}{2} \log_5 \left(x^3-x\right)$

**46.** $\ln \left(\dfrac{\sqrt[3]{x^4+12}}{(x+16)\sqrt{x-3}}\right) = \frac{1}{3} \ln \left(x^4+12\right) - \left[\ln (x+16) + \frac{1}{2} \ln (x-3)\right]$

**47.** $\log 6 + 4 \log 2 = \log 6 + \log 2^4 = \log \left(6 \cdot 2^4\right) = \log 96$

**48.** $\log x + \log \left(x^2 y\right) + 3 \log y = \log \left(x \cdot x^2 y \cdot y^3\right) = \log \left(x^3 y^4\right)$

**49.** $\frac{3}{2} \log_2 (x-y) - 2 \log_2 \left(x^2+y^2\right) = \log_2 (x-y)^{3/2} - \log_2 \left(x^2+y^2\right)^2 = \log_2 \left(\dfrac{(x-y)^{3/2}}{\left(x^2+y^2\right)^2}\right)$

**50.** $\log_5 2 + \log_5 (x+1) - \frac{1}{3} \log_5 (3x+7) = \log_5 \left[2 (x+1)\right] - \log_5 (3x+7)^{1/3} = \log_5 \left(\dfrac{2(x+1)}{\sqrt[3]{3x+7}}\right)$

**51.** $\log (x-2) + \log (x+2) - \frac{1}{2} \log \left(x^2+4\right) = \log \left[(x-2)(x+2)\right] - \log \sqrt{x^2+4} = \log \left(\dfrac{x^2-4}{\sqrt{x^2+4}}\right)$

**52.** $\frac{1}{2} \left[\ln (x-4) + 5 \ln \left(x^2+4x\right)\right] = \frac{1}{2} \ln \left[(x-4)\left(x^2+4x\right)^5\right] = \ln \sqrt{(x-4)\left(x^2+4x\right)^5}$

**53.** $\log_2 (1-x) = 4 \quad \Leftrightarrow \quad 1-x = 2^4 \quad \Leftrightarrow \quad x = 1 - 2^4 = -15$

**54.** $2^{3x-5} = 7 \quad \Leftrightarrow \quad \log 2^{3x-5} = \log 7 \quad \Leftrightarrow \quad 3x - 5 = \dfrac{\log 7}{\log 2} \quad \Leftrightarrow \quad x = \dfrac{1}{3}\left(5 + \dfrac{\log 7}{\log 2}\right) \approx 2.60$

**55.** $5^{5-3x} = 26 \quad \Leftrightarrow \quad \log_5 26 = 5 - 3x \quad \Leftrightarrow \quad 3x = 5 - \log_5 26 \quad \Leftrightarrow \quad x = \frac{1}{3} \left(5 - \log_5 26\right) \approx 0.99$

**56.** $\ln(2x-3)=14 \quad\Leftrightarrow\quad e^{\ln(2x-3)}=e^{14} \quad\Leftrightarrow\quad 2x-3=e^{14} \quad\Leftrightarrow\quad x=\frac{1}{2}\left(3+e^{14}\right)\approx 601303.64$

**57.** $e^{3x/4}=10 \quad\Leftrightarrow\quad \ln e^{3x/4}=\ln 10 \quad\Leftrightarrow\quad \dfrac{3x}{4}=\ln 10 \quad\Leftrightarrow\quad x=\frac{4}{3}\ln 10\approx 3.07$

**58.** $2^{1-x}=3^{2x+5} \quad\Leftrightarrow\quad \log 2^{1-x}=\log 3^{2x+5} \quad\Leftrightarrow\quad (1-x)\log 2=(2x+5)\log 3 \quad\Leftrightarrow$

$x(2\log 3+\log 2)=\log 2-5\log 3 \quad\Leftrightarrow\quad x=\dfrac{\log 2-5\log 3}{\log 2+2\log 3}=\dfrac{\log\frac{2}{3^5}}{\log(2\cdot 9)}\approx -1.66$

**59.** $\log x+\log(x+1)=\log 12 \quad\Leftrightarrow\quad \log[x(x+1)]=\log 12 \quad\Leftrightarrow\quad x(x+1)=12 \quad\Leftrightarrow\quad x^2+x-12=0 \quad\Leftrightarrow$ $(x+4)(x-3)=0\Rightarrow x=3$ or $-4$. Because $\log x$ and $\log(x+1)$ are undefined at $x=-4$, it follows that $x=3$ is the only solution.

**60.** $\log_8(x+5)-\log_8(x-2)=1 \quad\Leftrightarrow\quad \log_8\left(\dfrac{x+5}{x-2}\right)=1 \quad\Leftrightarrow\quad \dfrac{x+5}{x-2}=8^1=8 \quad\Leftrightarrow\quad x+5=8x-16 \quad\Leftrightarrow$

$7x=21 \quad\Leftrightarrow\quad x=3$

**61.** $x^2e^{2x}+2xe^{2x}=8e^{2x} \quad\Leftrightarrow\quad e^{2x}\left(x^2+2x-8\right)=0 \quad\Leftrightarrow\quad x^2+2x-8=0$ (since $e^{2x}>0$ for all $x$) $\Leftrightarrow$ $(x+4)(x-2)=0 \quad\Leftrightarrow\quad x=2,-4$

**62.** $2^{3^x}=5 \quad\Leftrightarrow\quad \log 2^{3^x}=\log 5 \quad\Leftrightarrow\quad 3^x\log 2=\log 5 \quad\Leftrightarrow\quad 3^x=\dfrac{\log 5}{\log 2} \quad\Leftrightarrow\quad \log 3^x=\log\left(\dfrac{\log 5}{\log 2}\right) \quad\Leftrightarrow$

$x\log 3=\log\left(\dfrac{\log 5}{\log 2}\right) \quad\Leftrightarrow\quad x=\dfrac{1}{\log 3}\cdot\log\left(\dfrac{\log 5}{\log 2}\right)\approx 0.77$

**63.** $5^{-2x/3}=0.63 \quad\Leftrightarrow\quad \dfrac{-2x}{3}\log 5=\log 0.63 \quad\Leftrightarrow\quad x=-\dfrac{3\log 0.63}{2\log 5}\approx 0.430618$

**64.** $2^{3x-5}=7 \quad\Leftrightarrow\quad (3x-5)\log 2=\log 7 \quad\Leftrightarrow\quad x=\frac{1}{3}\left(5+\dfrac{\log 7}{\log 2}\right)\approx 2.602452$

**65.** $5^{2x+1}=3^{4x-1} \quad\Leftrightarrow\quad (2x+1)\log 5=(4x-1)\log 3 \quad\Leftrightarrow\quad 2x\log 5+\log 5=4x\log 3-\log 3 \quad\Leftrightarrow$

$x(2\log 5-4\log 3)=-\log 3-\log 5 \quad\Leftrightarrow\quad x=\dfrac{\log 3+\log 5}{4\log 3-2\log 5}\approx 2.303600$

**66.** $e^{-15k}=10000 \quad\Leftrightarrow\quad -15k=\ln 10000 \quad\Leftrightarrow\quad k=-\frac{1}{15}\ln 10000\approx -0.614023$

**67.** $y=e^{x/(x+2)}$. Vertical asymptote $x=-2$, horizontal asymptote $y=2.72$, no maximum or minimum.

**68.** $y=2x^2-\ln x$. Vertical asymptote $x=0$, no horizontal asymptote, local minimum of about $1.19$ at $x\approx 0.5$.

**69.** $y=\log\left(x^3-x\right)$. Vertical asymptotes $x=-1$, $x=0$, $x=1$, no horizontal asymptote, local maximum of about $-0.41$ when $x\approx -0.58$.

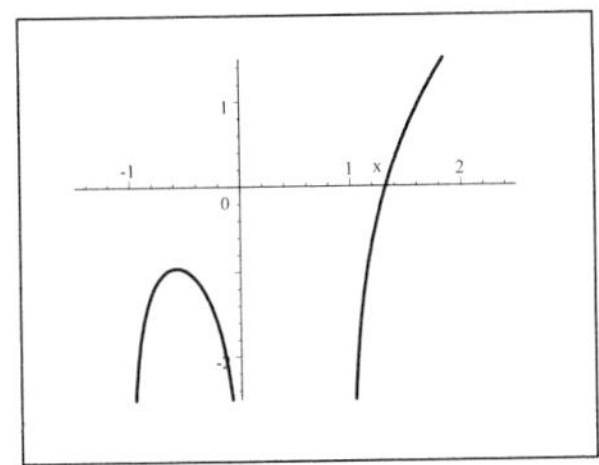

**70.** $y=10^x-5^x$. No vertical asymptote, horizontal asymptote $y=0$, local minimum of about $-0.13$ at $x\approx -0.5$.

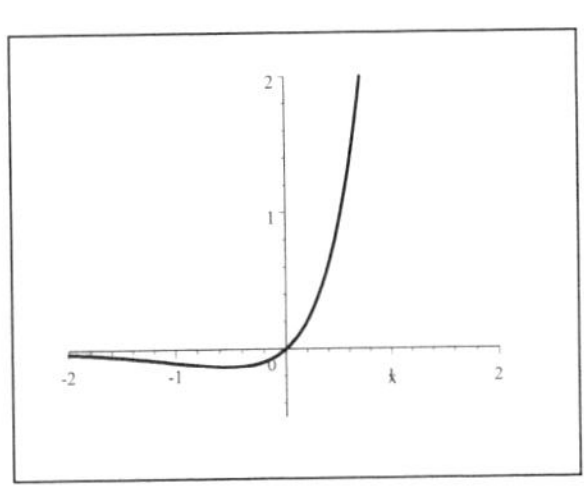

**71.** $3\log x = 6 - 2x$. We graph $y = 3\log x$ and $y = 6 - 2x$ in the same viewing rectangle. The solution occurs where the two graphs intersect. From the graphs, we see that the solution is $x \approx 2.42$.

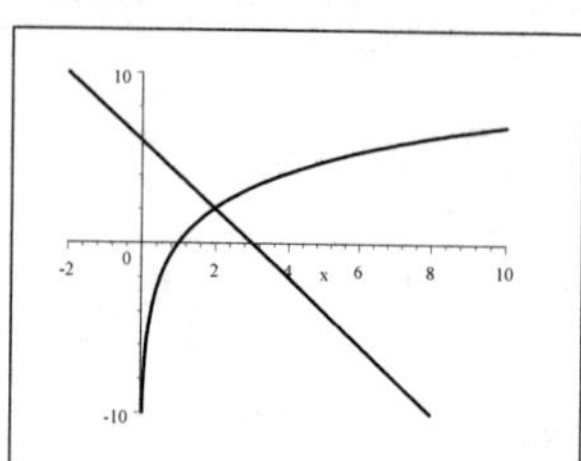

**72.** $4 - x^2 = e^{-2x}$. From the graphs, we see that the solutions are $x \approx -0.64$ and $x \approx 2$.

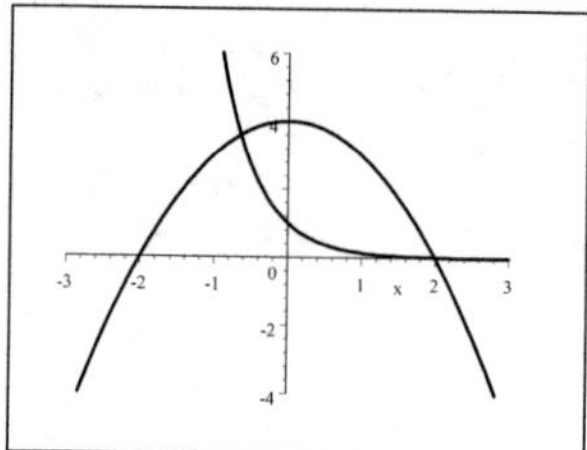

**73.** $\ln x > x - 2$. We graph the function $f(x) = \ln x - x + 2$, and we see that the graph lies above the $x$-axis for $0.16 < x < 3.15$. So the approximate solution of the given inequality is $0.16 < x < 3.15$.

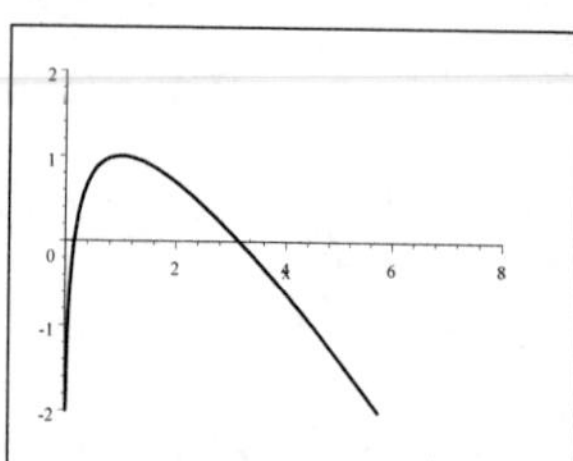

**74.** $e^x < 4x^2 \quad \Leftrightarrow \quad e^x - 4x^2 < 0$. We graph the function $f(x) = e^x - 4x^2$, and we see that the graph lies below the $x$-axis for $(-\infty, -0.41) \cup (0.71, 4.31)$.

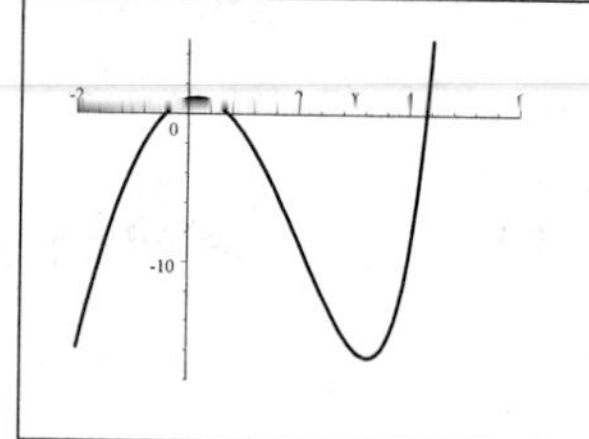

**75.** $f(x) = e^x - 3e^{-x} - 4x$. We graph the function $f(x)$, and we see that the function is increasing on $(-\infty, 0]$ and $[1.10, \infty)$ and that it is decreasing on $[0, 1.10]$.

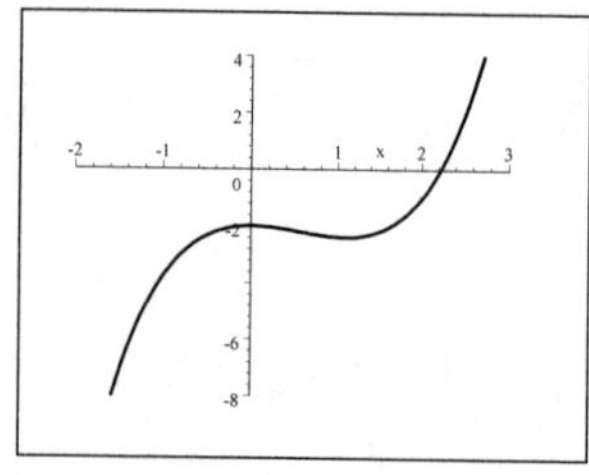

**76.** The line has $x$-intercept at $x = e^0 = 1$. When $x = e^a$, $y = \ln e^a = a$. Therefore, using the point-slope equation, we have $y - 0 = \dfrac{a - 0}{e^a - 1}(x - 1) \quad \Leftrightarrow$

$y = \dfrac{a}{e^a - 1}(x - 1)$.

**77.** $\log_4 15 = \dfrac{\log 15}{\log 4} = 1.953445$

**78.** $0.2 \le \log x < 2 \quad \Leftrightarrow \quad 10^{0.2} \le x < 10^2 \quad \Leftrightarrow \quad \sqrt[5]{10} \le x < 100$.

**79.** Notice that $\log_4 258 > \log_4 256 = \log_4 4^4 = 4$ and so $\log_4 258 > 4$. Also $\log_5 620 < \log_5 625 = \log_5 5^4 = 4$ and so $\log_5 620 < 4$. Then $\log_4 258 > 4 > \log_5 620$ and so $\log_4 258$ is larger.

**80.** $f(x) = 2^{3^x}$. Then $y = 2^{3^x} \quad \Leftrightarrow \quad \log_2 y = 3^x \quad \Leftrightarrow \quad \log_3(\log_2 y) = x$, and so the inverse function is $f^{-1}(x) = \log_3(\log_2 x)$. Since $\log_3 y$ is defined only when $y > 0$, we have $\log_2 x > 0 \quad \Leftrightarrow \quad x > 1$. Therefore the domain is $(1, \infty)$, and the range is $(-\infty, \infty)$.

**81.** $P = 12{,}000$, $r = 0.10$, and $t = 3$. Then $A = P\left(1 + \frac{r}{n}\right)^{nt}$.

**(a)** For $n = 2$, $A = 12{,}000\left(1 + \frac{0.10}{2}\right)^{2(3)} = 12{,}000\left(1.05^6\right) \approx \$16{,}081.15$.

**(b)** For $n = 12$, $A = 12{,}000\left(1 + \frac{0.10}{12}\right)^{12(3)} \approx \$16{,}178.18$.

**(c)** For $n = 365$, $A = 12{,}000\left(1 + \frac{0.10}{365}\right)^{365(3)} \approx \$16{,}197.64$.

**(d)** For $n = \infty$, $A = Pe^{rt} = 12{,}000e^{0.10(3)} \approx \$16{,}198.31$.

**82.** $P = 5000$, $r = 0.085$, and $n = 2$.

**(a)** For $t = 1.5$, $A = 5000\left(1 + \frac{0.085}{2}\right)^{2(1.5)} = 5000 \cdot 1.0425^3 \approx \$5664.98$.

**(b)** We want to find $t$ such that $A = 7000$. Then $A = 5000 \cdot 1.0425^{2t} = 7000 \Leftrightarrow 1.0425^{2t} = \frac{7000}{5000} = \frac{7}{5} \Leftrightarrow 2t = \dfrac{\log\left(\frac{7}{5}\right)}{\log 1.0425} \Leftrightarrow t = \dfrac{\log\left(\frac{7}{5}\right)}{2\log 1.0425} \approx 4.04$, and so the investment will amount to \$7000 after approximately 4 years.

**83. (a)** Using the model $n(t) = n_0e^{rt}$, with $n_0 = 30$ and $r = 0.15$, we have the formula $n(t) = 30e^{0.15t}$.

**(b)** $n(4) = 30e^{0.15(4)} \approx 55$.

**(c)** $500 = 30e^{0.15t} \Leftrightarrow \frac{50}{3} = e^{0.15t} \Leftrightarrow 0.15t = \ln\left(\frac{50}{3}\right) \Leftrightarrow t = \dfrac{1}{0.15}\ln\left(\frac{50}{3}\right) \approx 18.76$. So the stray cat population will reach 500 in about 19 years.

**84.** Using the model $n(t) = n_0e^{rt}$, with $n_0 = 10000$ and $n(1) = 25000$, we have $25000 = n(1) = 10000e^{r\cdot 1} \Leftrightarrow e^r = \frac{5}{2} \Leftrightarrow r = \ln\frac{5}{2} \approx 0.916$. So $n(t) = 10000e^{0.916t}$.

**(a)** Here we must solve the equation $n(t) = 20000$ for $t$. So $n(t) = 10000e^{0.916t} = 20000 \Leftrightarrow e^{0.916t} = 2 \Leftrightarrow 0.916t = \ln 2 \Leftrightarrow t = \dfrac{\ln 2}{0.916} \approx 0.756$. Thus the doubling period is about 45 minutes.

**(b)** $n(3) = 10000e^{0.916\cdot 3} \approx 156250$, so the population after 3 hours is about 156,250.

**85. (a)** From the formula for radioactive decay, we have $m(t) = 10e^{-rt}$, where $r = -\dfrac{\ln 2}{2.7 \times 10^5}$. So after 1000 years the amount remaining is $m(1000) = 10 \cdot e^{\left[-\ln 2/\left(2.7\times 10^5\right)\right]\cdot 1000} = 10e^{-(\ln 2)/\left(2.7\times 10^2\right)} = 10e^{-(\ln 2)/270} \approx 9.97$. Therefore the amount remaining is about 9.97 mg.

**(b)** We solve for $t$ in the equation $7 = 10e^{-\left[\ln 2/\left(2.7\times 10^5\right)\right]\cdot t}$. We have $7 = 10e^{-\left[\ln 2/\left(2.7\times 10^5\right)\right]\cdot t} \Leftrightarrow 0.7 = e^{-\left[\ln 2/\left(2.7\times 10^5\right)\right]\cdot t} \Leftrightarrow \ln 0.7 = -\dfrac{\ln 2}{2.7 \times 10^5}\cdot t \Leftrightarrow t = -\dfrac{\ln 0.7}{\ln 2}\cdot 2.7 \times 10^5 \approx 138{,}934.75$. Thus it takes about 139,000 years.

**86.** From the formula for radioactive decay, we have $m(t) = m_0e^{-rt}$, where $r = \dfrac{\ln 2}{h}$. So $m(t) = m_0e^{-[(\ln 2)/h]t}$.

**(a)** Using $m(8) = 0.33m_0$, we solve for $h$. We have $0.33m_0 = m(8) = m_0e^{-(\ln 2)/h} \Leftrightarrow 0.33 = e^{-(8\ln 2)/h} \Leftrightarrow -\dfrac{8\ln 2}{h} = \ln 0.33 \Leftrightarrow h = -\dfrac{8\ln 2}{\ln 0.33} \approx 5.002$. So the half-life of this element is roughly 5 days.

**(b)** $m(12) = m_0e^{-[(\ln 2)/5]\cdot 12} \approx 0.19m_0$, so about 19% of the original mass remains.

**87. (a)** From the formula for radioactive decay, $r = \dfrac{\ln 2}{1590} \approx 0.0004359$ and $n(t) = 150 \cdot e^{-0.0004359t}$.

**(b)** $n(1000) = 150 \cdot e^{-0.0004359\cdot 1000} \approx 97.00$, and so the amount remaining is about 97.00 mg.

**(c)** Find $t$ so that $50 = 150 \cdot e^{-0.0004359t}$. We have $50 = 150 \cdot e^{-0.0004359t} \Leftrightarrow \frac{1}{3} = e^{-0.0004359t} \Leftrightarrow t = -\dfrac{1}{0.0004359}\ln\left(\frac{1}{3}\right) \approx 2520$. Thus only 50 mg remain after about 2520 years.

**88.** From the formula for radioactive decay, we have $m(t) = m_0 e^{-rt}$, where $r = \dfrac{\ln 2}{h}$. Since $h = 4$, we have $r = \dfrac{\ln 2}{4} \approx 0.173$ and $m(t) = m_0 e^{-0.173t}$.

**(a)** Using $m(20) = 0.375$, we solve for $m_0$. We have $0.375 = m(20) = m_0 e^{-0.173 \cdot 20} \quad \Leftrightarrow \quad 0.03125 m_0 = 0.375$ $\Leftrightarrow \quad m_0 = \dfrac{0.375}{0.03125} \approx 12$. So the initial mass of the sample was about 12 g.

**(b)** $m(t) = 12e^{-0.173t}$.

**(c)** $m(3) = 12e^{-0.173 \cdot 3} \approx 7.135$. So there are about 7.1 g remaining after 3 days.

**(d)** Here we solve $m(t) = 0.15$ for $t$: $0.15 = 12e^{-0.173t} \quad \Leftrightarrow \quad 0.0125 = e^{-0.173t} \quad \Leftrightarrow \quad -0.173t = \ln 0.0125 \quad \Leftrightarrow$ $t = \dfrac{\ln 0.0125}{-0.173} \approx 25.3$. So it will take about 25 days until only 15% of the substance remains.

**89. (a)** Using $n_0 = 1500$ and $n(5) = 3200$ in the formula $n(t) = n_0 e^{rt}$, we have $3200 = n(5) = 1500e^{5r} \quad \Leftrightarrow$ $e^{5r} = \frac{32}{15} \quad \Leftrightarrow \quad 5r = \ln\left(\frac{32}{15}\right) \quad \Leftrightarrow \quad r = \frac{1}{5}\ln\left(\frac{32}{15}\right) \approx 0.1515$. Thus $n(t) = 1500 \cdot e^{0.1515t}$.

**(b)** We have $t = 1999 - 1988 = 11$ so $n(11) = 1500e^{0.1515 \cdot 11} \approx 7940$. Thus in 1999 the bird population should be about 7940.

**90.** We use Newton's Law of Cooling: $T(t) = T_s + D_0 e^{-kt}$ with $k = 0.0341$, $T_s = 60$ and $D_0 = 190 - 60 = 130$. So $90 = T(t) = 60 + 130e^{-0.0341t} \quad \Leftrightarrow \quad 90 = 60 + 130e^{-0.0341t} \quad \Leftrightarrow \quad 130e^{-0.0341t} = 30 \quad \Leftrightarrow \quad e^{-0.0341t} = \frac{3}{13}$ $\Leftrightarrow \quad -0.0341t = \ln\left(\frac{3}{13}\right) \quad \Leftrightarrow \quad t = \dfrac{-\ln(3/13)}{0.0341} \approx 43.0$, so the engine cools to 90° F in about 43 minutes.

**91.** $[H^+] = 1.3 \times 10^{-8}$ M. Then $\text{pH} = -\log[H^+] = -\log(1.3 \times 10^{-8}) \approx 7.9$, and so fresh egg whites are basic.

**92.** $\text{pH} = 1.9 = -\log[H^+]$. Then $[H^+] = 10^{-1.9} \approx 1.26 \times 10^{-2}$ M.

**93.** Let $I_0$ be the intensity of the smaller earthquake and $I_1$ be the intensity of the larger earthquake. Then $I_1 = 35I_0$. Since $M = \log\left(\dfrac{I}{S}\right)$, we have $M_0 = \log\left(\dfrac{I_0}{S}\right) = 6.5$ and $M_1 = \log\left(\dfrac{I_1}{S}\right) = \log\left(\dfrac{35I_0}{S}\right) = \log 35 + \log\left(\dfrac{I_0}{S}\right) = \log 35 +$ $M_0 = \log 35 + 6.5 \approx 8.04$. So the magnitude on the Richter scale of the larger earthquake is approximately 8.0.

**94.** Let the subscript $J$ represent the jackhammer and $W$ the whispering: $\beta_J = 132 = 10\log\left(\dfrac{I_J}{I_0}\right) \quad \Leftrightarrow$ $\log\left(\dfrac{I_J}{I_0}\right) = 13.2 \quad \Leftrightarrow \quad \dfrac{I_J}{I_0} = 10^{13.2}$. Similarly $\dfrac{I_W}{I_0} = 10^{2.8}$. So $\dfrac{I_J}{I_W} = \dfrac{10^{13.2}}{10^{2.8}} = 10^{10.4} \approx 2.51 \times 10^{10}$, and so the ratio of intensities is $2.51 \times 10^{10}$.

# Chapter 5 Test

**1.** $y = 2^x$ and $y = \log_2 x$.

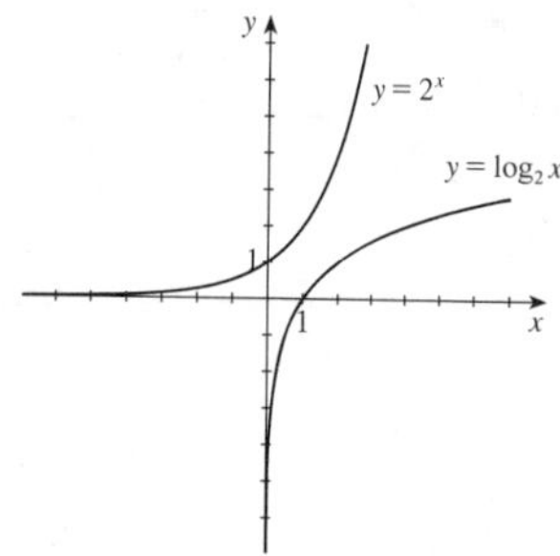


**2.** $f(x) = \log(x + 1)$ has domain $(-1, \infty)$, range $(-\infty, \infty)$, and vertical asymptote $x = -1$.

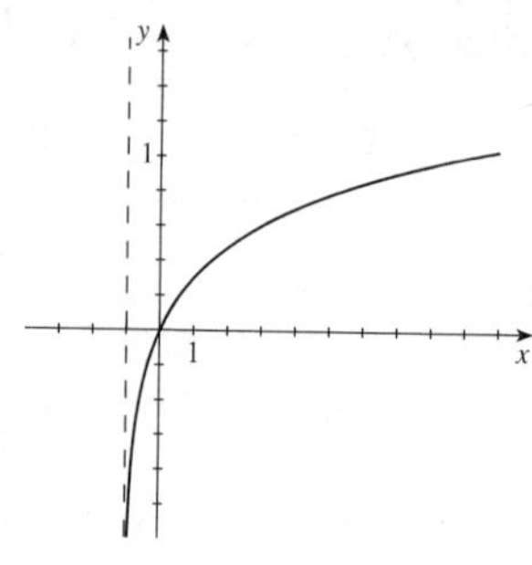

**3. (a)** $\log_3 \sqrt{27} = \log_3 (3^3)^{1/2} = \log_3 3^{3/2} = \frac{3}{2}$

**(b)** $\log_2 80 - \log_2 10 = \log_2 \left(\frac{80}{10}\right) = \log_2 8 = \log_2 2^3 = 3$

**(c)** $\log_8 4 = \log_8 8^{2/3} = \frac{2}{3}$

**(d)** $\log_6 4 + \log_6 9 = \log_6 (4 \cdot 9) = \log_6 6^2 = 2$

**4.** $\log \sqrt[3]{\dfrac{x+2}{x^4 (x^2+4)}} = \dfrac{1}{3} \log \left(\dfrac{x+2}{x^4 (x^2+4)}\right) = \frac{1}{3} \left[\log (x+2) - \left(4 \log x + \log \left(x^2+4\right)\right)\right]$
$= \frac{1}{3} \log (x+2) - \frac{4}{3} \log x - \frac{1}{3} \log \left(x^2+4\right)$

**5.** $\ln x - 2 \ln \left(x^2+1\right) + \frac{1}{2} \ln \left(3-x^4\right) = \ln \left(x\sqrt{3-x^4}\right) - \ln \left(x^2+1\right)^2 = \ln \left(\dfrac{x\sqrt{3-x^4}}{\left(x^2+1\right)^2}\right)$

**6. (a)** $2^{x-1} = 10 \quad \Leftrightarrow \quad \log 2^{x-1} = \log 10 = 1 \quad \Leftrightarrow \quad (x-1) \log 2 = 1 \quad \Leftrightarrow \quad x - 1 = \dfrac{1}{\log 2} \quad \Leftrightarrow$
$x = 1 + \dfrac{1}{\log 2} \approx 4.32$

**(b)** $5 \ln (3-x) = 4 \quad \Leftrightarrow \quad \ln (3-x) = \frac{4}{5} \quad \Leftrightarrow \quad e^{\ln(3-x)} = e^{4/5} \quad \Leftrightarrow \quad 3 - x = e^{4/5} \quad \Leftrightarrow$
$x = 3 - e^{4/5} \approx 0.77$

**(c)** $10^{x+3} = 6^{2x} \quad \Leftrightarrow \quad \log 10^{x+3} = \log 6^{2x} \quad \Leftrightarrow \quad x + 3 = 2x \log 6 \quad \Leftrightarrow \quad 2x \log 6 - x = 3 \quad \Leftrightarrow$
$x (2 \log 6 - 1) = 3 \quad \Leftrightarrow \quad x = \dfrac{3}{2 \log 6 - 1} \approx 5.39$

**(d)** $\log_2 (x+2) + \log_2 (x-1) = 2 \quad \Leftrightarrow \quad \log_2 ((x+2)(x-1)) = 2 \quad \Leftrightarrow \quad x^2 + x - 2 = 2^2 \quad \Leftrightarrow$
$x^2 + x - 6 = 0 \quad \Leftrightarrow \quad (x+3)(x-2) = 0 \Rightarrow x = -3$ or $x = 2$. However, both logarithms are undefined at $x = -3$, so the only solution is $x = 2$.

**7. (a)** From the formula for population growth, we have $8000 = 1000e^{r \cdot 1}$ $\Leftrightarrow \quad 8 = e^r \quad \Leftrightarrow \quad r = \ln 8 \approx 2.07944$. Thus $n(t) = 1000e^{2.07944t}$.

**(b)** $n(1.5) = 1000e^{2.07944(1.5)} \approx 22{,}627$

**(c)** $15000 = 1000e^{2.07944t} \quad \Leftrightarrow \quad 15 = e^{2.07944t} \quad \Leftrightarrow$
$\ln 15 = 2.07944t \quad \Leftrightarrow \quad t = \dfrac{\ln 15}{2.07944} \approx 1.3$. Thus the population will reach 15,000 after approximately 1.3 hours.

**(d)**

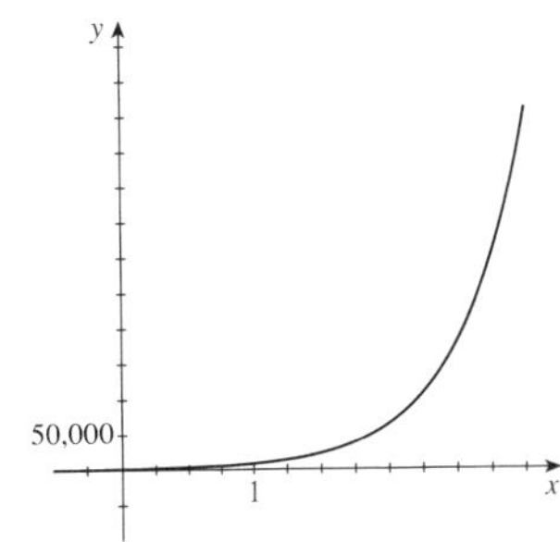


**8. (a)** $A(t) = 12{,}000 \left(1 + \frac{0.056}{12}\right)^{12t}$, where $t$ is in years.

**(b)** $A(t) = 12{,}000 \left(1 + \frac{0.056}{365}\right)^{365t}$. So $A(3) = 12{,}000 \left(1 + \dfrac{0.056}{365}\right)^{365(3)} = \$14{,}195.06$.

**(c)** $A(t) = 12{,}000 \left(1 + \frac{0.056}{2}\right)^{2t} = 12{,}000 (1.028)^{2t}$. So $20{,}000 = 12{,}000 (1.028)^{2t} \quad \Leftrightarrow \quad 1.6667 = 1.028^{2t} \quad \Leftrightarrow$
$\ln 1.6667 = \ln 1.028^{2t} \quad \Leftrightarrow \quad \ln 1.6667 = 2t \ln 1.028 \quad \Leftrightarrow \quad t = \frac{\ln 1.6667}{2 \ln 1.028} \approx 9.25$ years.

**9.** $f(x) = \dfrac{e^x}{x^3}$

**(a)**

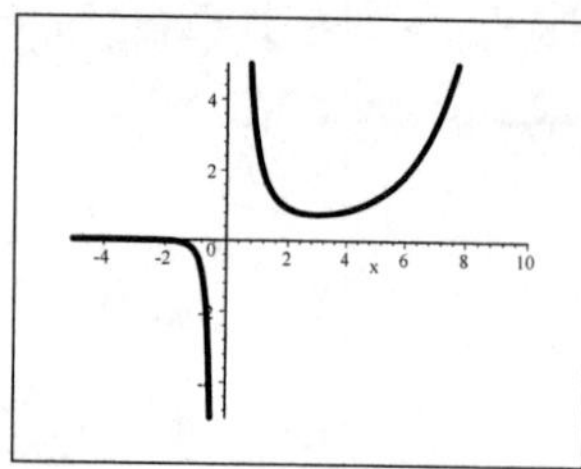

**(b)** $f$ has vertical asymptote $x = 0$ and horizontal asymptote $y = 0$.

**(c)** $f$ has a local minimum of about $0.74$ when $x \approx 3.00$.

**(d)** The range of $f$ is approximately $(-\infty, 0) \cup [0.74, \infty)$.

**(e)** $\dfrac{e^x}{x^3} = 2x + 1$.

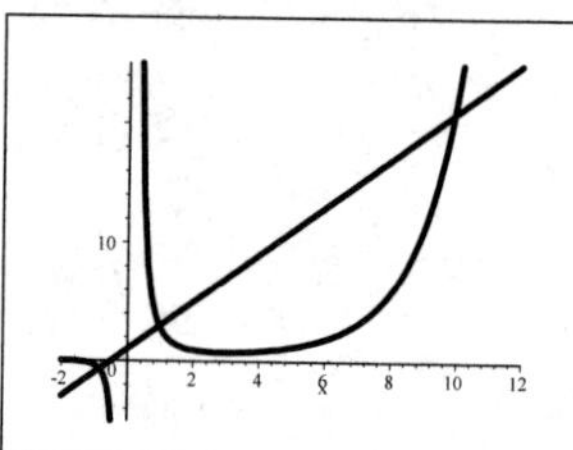

We see that the graphs intersect at $x \approx -0.85$, $0.96$, and $9.92$.

## Focus on Modeling: Fitting Exponential and Power Curves to Data

**1. (a)**

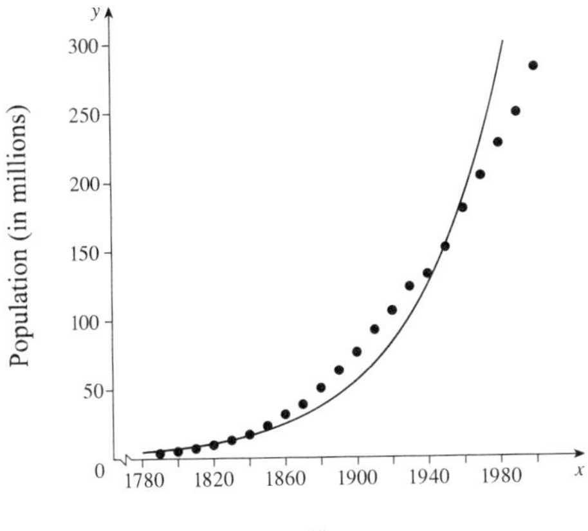


**(b)** Using a graphing calculator, we obtain the model $y = ab^t$, where $a = 1.1806094 \times 10^{-15}$ and $b = 1.0204139$, and $y$ is the population (in millions) in the year $t$.

**(c)** Substituting $t = 2010$ into the model of part (b), we get $y = ab^{2010} \approx 515.9$ million.

**(d)** According to the model, the population in 1965 should have been about $y = ab^{1965} \approx 207.8$ million.

**(e)** The values given by the model are clearly much too large. This means that an exponential model is *not* appropriate for these data.

**2. (a)**

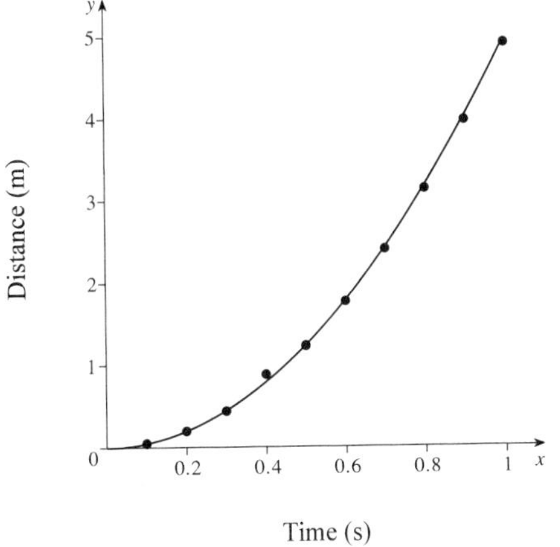


**(b)** We let $t$ represent the time (in seconds) and $y$ the distance fallen (in meters). Using a calculator, we obtain the power model: $y = 4.9622t^{2.0027}$.

**(c)** When $t = 3$ the model predicts that $y = 44.792$ m.

**3. (a)** Yes.

**(b)**

| Year $t$ | Health Expenditures $E$ (\$bn) | $\ln E$ |
|---|---|---|
| 1970 | 74.3 | 4.30811 |
| 1980 | 251.1 | 5.52585 |
| 1985 | 434.5 | 6.07420 |
| 1987 | 506.2 | 6.22693 |
| 1990 | 696.6 | 6.54621 |
| 1992 | 820.3 | 6.70967 |
| 1994 | 937.2 | 6.84290 |
| 1996 | 1039.4 | 6.94640 |
| 1998 | 1150.0 | 7.04752 |
| 2000 | 1310.0 | 7.17778 |
| 2001 | 1424.5 | 7.26158 |

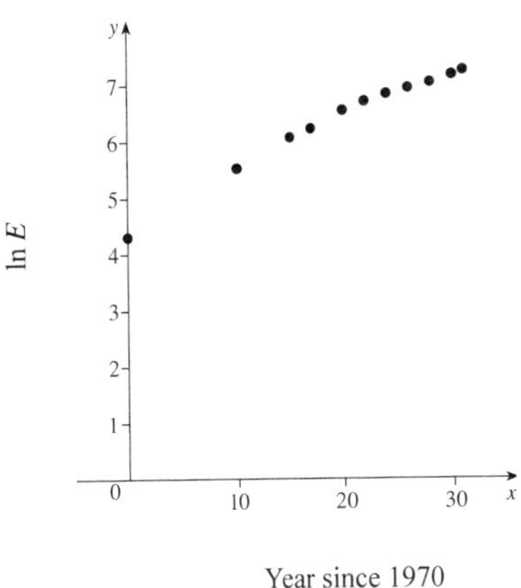


Yes, the scatter plot appears to be roughly linear.

**(c)** Let $t$ be the number of years elapsed since 1970 . Then $\ln E = 4.551437 + 0.09238268t$, where $E$ is expenditure in billions of dollars.

**(d)** $E = e^{4.551437+0.09238268t} = 94.76849e^{0.09238268t}$

**(e)** In 2009 we have $t = 2009 - 1970 = 39$, so the estimated 2009 health-care expenditures are $94.76849e^{0.09238268(39)} \approx 3478.5$ billion dollars.

**4. (a)**

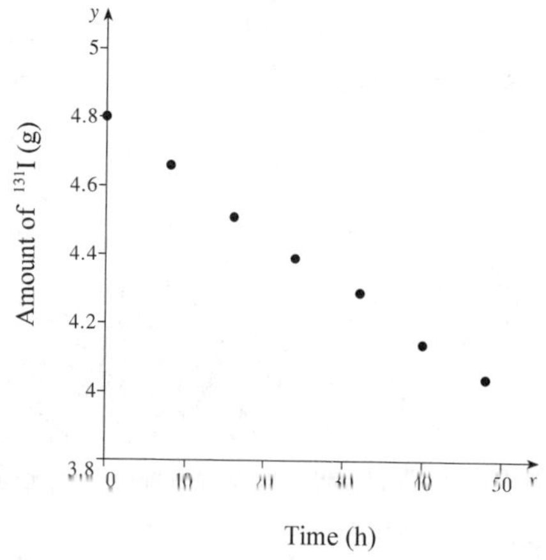


**(b)** Let $t$ be the time (in hours) and $y$ be amount of iodine-131 (in grams). Using a calculator, we obtain the exponential model $y = ab^t$, where $a = 4.79246$ and $b = 0.99642$.

**(c)** To find the half-life of iodine-131, we must find the time when the sample has decayed to half its original mass. Setting $y = 2.40$ g, we get

$2.40 = 4.79246 \cdot (0.99642)^t \quad \Leftrightarrow$

$\ln 2.40 = \ln 4.79246 + t \ln 0.99642 \quad \Leftrightarrow$

$t = \dfrac{\ln 2.40 - \ln 4.79246}{\ln 0.99642} \approx 192.8$ h.

**5. (a)** Using a graphing calculator, we find that $I_0 = 22.7586444$ and $k = 0.1062398$.

**(b)**

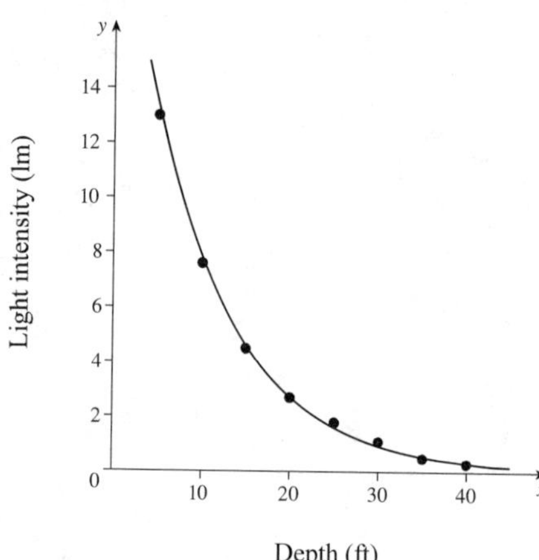


**(c)** We solve $0.15 = 22.7586444e^{-0.1062398x}$ for $x$:

$0.15 = 22.7586444e^{-0.1062398x} \quad \Leftrightarrow$

$0.006590902 = e^{-0.1062398x} \quad \Leftrightarrow$

$-5.022065 = -0.1062398x \quad \Leftrightarrow \quad x \approx 47.27$. So light intensity drops below 0.15 lumens below around 47.27 feet.

**6. (a)** Using a graphing calculator, we find the power function model $y = 49.70030t^{-0.15437}$ and the exponential model $y = 44.82418 \cdot (0.99317^t)$.

**(b)**

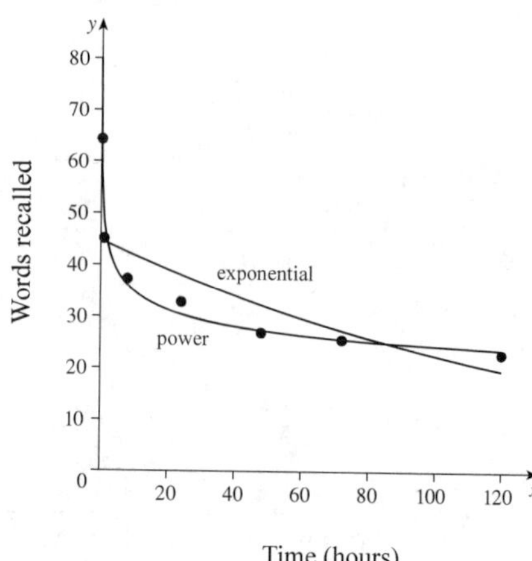


**(c)** The power function, $y = 49.70030t^{-0.15437}$, seems to provide a better model.

**7. (a)** Let $t$ be the number of years elapsed since 1970, and let $y$ be the number of millions of tons of lead emissions in year $t$. Using a graphing calculator, we obtain the exponential model $y = ab^t$, where $a = 301.813054$ and $b = 0.819745$.

**(b)** Using a graphing calculator, we obtain the fourth-degree polynomial model $y = at^4 + bt^3 + ct^2 + dt + e$, where $a = -0.002430$, $b = 0.135159$, $c = -2.014322$, $d = -4.055294$, and $e = 199.092227$.

**(c)** The exponential and polynomial models are shown. From the graph, the polynomial model appears to fit the data better.

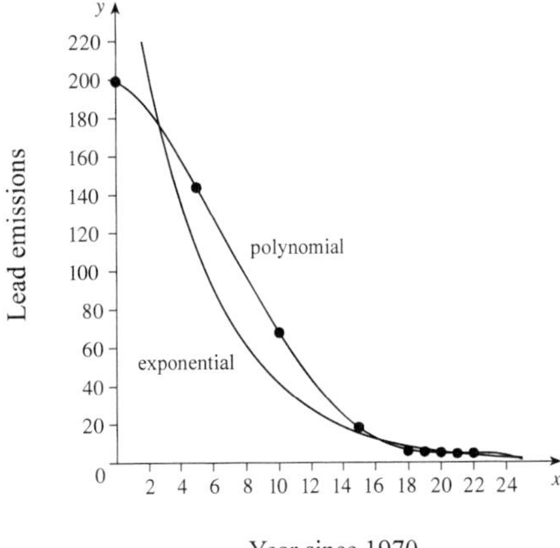


**(d)** Using the exponential model, we estimate emissions in 1972 to be $y(2) \approx 202.8$ million metric tons and emissions in 1982 to be $y(12) \approx 27.8$ million metric tons.
Using the polynomial model, we estimate emissions in 1972 to be $y(2) \approx 184.0$ million metric tons and emissions in 1982 to be $y(12) \approx 43.5$ million metric tons.

**8.** Let $x$ be the reduction in emissions (in percent), and $y$ be the cost (in dollars). First we make a scatter plot of the data. A linear model does not appear appropriate, so we try an exponential model. Using a calculator, we get the model $y = ab^x$, where $a = 2.414$ and $b = 1.05452$. This model is graphed on the scatter plot.

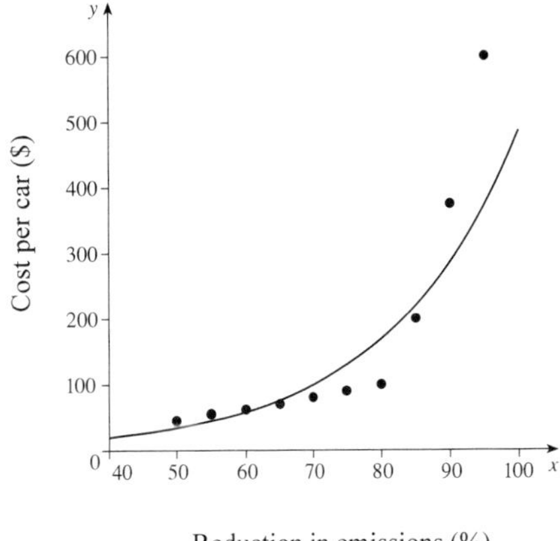


**9. (a)**

**(b)**

| $x$ | $y$ | $\ln x$ | $\ln y$ |
|---|---|---|---|
| 2 | 0.08 | 0.69315 | $-2.52573$ |
| 4 | 0.12 | 1.38629 | $-2.12026$ |
| 6 | 0.18 | 1.79176 | $-1.71480$ |
| 8 | 0.25 | 2.07944 | $-1.38629$ |
| 10 | 0.36 | 2.30259 | $-1.02165$ |
| 12 | 0.52 | 2.48491 | $-0.65393$ |
| 14 | 0.73 | 2.63906 | $-0.31471$ |
| 16 | 1.06 | 2.77259 | 0.05827 |

**(b)**

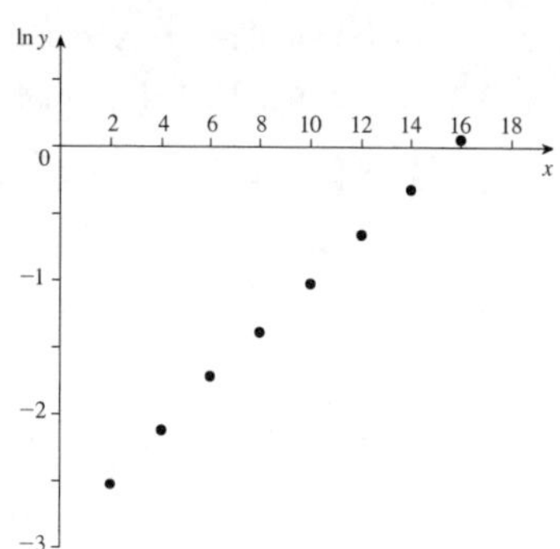


**(c)** The exponential function.

**(d)** $y = a \cdot b^x$ where $a = 0.057697$ and $b = 1.200236$.

**10. (a)**

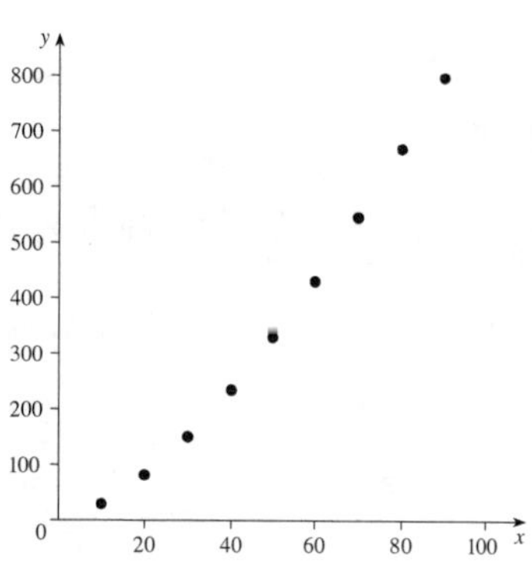


**(b)**

| $x$ | $y$ | $\ln x$ | $\ln y$ |
|---|---|---|---|
| 10 | 29 | 2.30259 | 3.36730 |
| 20 | 82 | 2.99573 | 4.40672 |
| 30 | 151 | 3.40120 | 5.01728 |
| 40 | 235 | 3.68888 | 5.45959 |
| 50 | 330 | 3.91202 | 5.79909 |
| 60 | 430 | 4.09434 | 6.06379 |
| 70 | 546 | 4.24850 | 6.30262 |
| 80 | 669 | 4.38203 | 6.50578 |
| 90 | 797 | 4.49981 | 6.68085 |

**(b)**

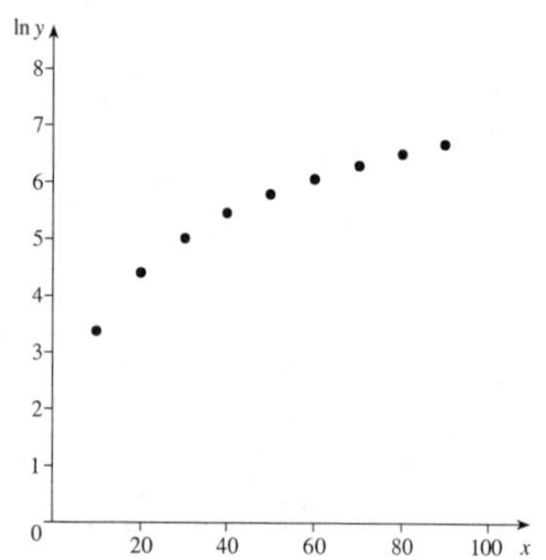


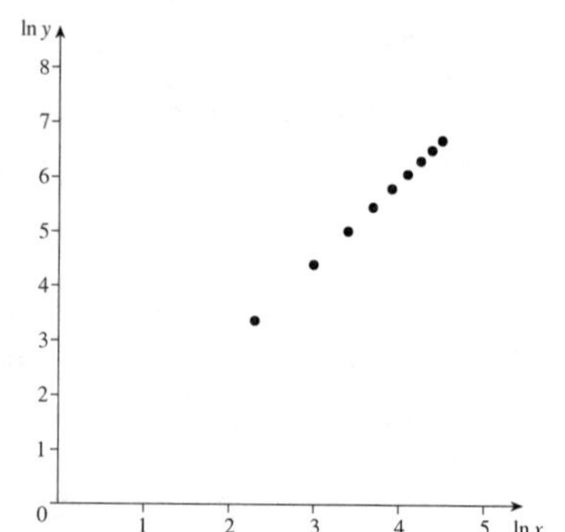


**(c)** The power function is more appropriate.

**(d)** $y = ax^n$ where $a = 0.893421326$ and $n = 1.50983$.

**11. (a)** Using the `Logistic` command on a TI-83 we get $y = \dfrac{c}{1 + ae^{-bx}}$ where $a = 49.10976596$, $b = 0.4981144989$, and $c = 500.855793$.

**(b)** Using the model $N = \dfrac{c}{1 + ae^{-bt}}$ we solve for $t$. So $N = \dfrac{c}{1 + ae^{-bt}} \Leftrightarrow 1 + ae^{-bt} = \dfrac{c}{N}$

$$\Leftrightarrow \quad ae^{-bt} = \left(\frac{c}{N}\right) - 1 = \frac{c - N}{N} \quad \Leftrightarrow \quad e^{-bt} = \frac{c - N}{aN} \quad \Leftrightarrow \quad -bt = \ln(c - N) - \ln aN$$

$\Leftrightarrow \quad t = \dfrac{1}{b}[\ln aN - \ln(c - N)]$. Substituting the values for $a$, $b$, and $c$, with $N = 400$ we have $t = \frac{1}{0.4981144989}(\ln 19643.90638 - \ln 100.855793) \approx 10.58$ days.

**12. (a)** Using the `LnReg` command on a TI-83, we get $y = a + b \ln x$ where $a = -7154.888128$ and $b = 1061.006551$.

**(b)** Using the model $C = -7154.888128 + (1061.006551) \ln 2005 \approx 912$ metric tons.

# 6 Trigonometric Functions of Angles

## 6.1 Angle Measure

**1.** $72^\circ = 72^\circ \cdot \frac{\pi}{180^\circ}$ rad $= \frac{2\pi}{5}$ rad $\approx 1.257$ rad

**2.** $54^\circ = 54^\circ \cdot \frac{\pi}{180^\circ}$ rad $= \frac{3\pi}{10}$ rad $\approx 0.942$ rad

**3.** $-45^\circ = -45^\circ \cdot \frac{\pi}{180^\circ}$ rad $= -\frac{\pi}{4}$ rad $\approx -0.785$ rad

**4.** $-60^\circ = -60^\circ \cdot \frac{\pi}{180^\circ}$ rad $= -\frac{\pi}{3}$ rad $\approx -1.047$ rad

**5.** $-75^\circ = -75^\circ \cdot \frac{\pi}{180^\circ}$ rad $= -\frac{5\pi}{12}$ rad $\approx -1.309$ rad

**6.** $-300^\circ = -300^\circ \cdot \frac{\pi}{180^\circ}$ rad $= -\frac{5\pi}{3}$ rad $\approx -5.236$ rad

**7.** $1080^\circ = 1080^\circ \cdot \frac{\pi}{180^\circ}$ rad $= 6\pi$ rad $\approx 18.850$ rad

**8.** $3960^\circ = 3960^\circ \cdot \frac{\pi}{180^\circ}$ rad $= 22\pi$ rad $\approx 69.115$ rad

**9.** $96^\circ = 96^\circ \cdot \frac{\pi}{180^\circ}$ rad $= \frac{8\pi}{15}$ rad $\approx 1.676$ rad

**10.** $15^\circ = 15^\circ \cdot \frac{\pi}{180^\circ}$ rad $= \frac{\pi}{12}$ rad $\approx 0.262$ rad

**11.** $7.5^\circ = 7.5^\circ \cdot \frac{\pi}{180^\circ}$ rad $= \frac{\pi}{24}$ rad $\approx 0.131$ rad

**12.** $202.5^\circ = 202.5^\circ \cdot \frac{\pi}{180^\circ}$ rad $= \frac{9\pi}{8}$ rad $\approx 3.534$ rad

**13.** $\frac{7\pi}{6} = \frac{7\pi}{6} \cdot \frac{180^\circ}{\pi} = 210^\circ$

**14.** $\frac{11\pi}{3} = \frac{11\pi}{3} \cdot \frac{180^\circ}{\pi} = 660^\circ$

**15.** $-\frac{5\pi}{4} = -\frac{5\pi}{4} \cdot \frac{180^\circ}{\pi} = -225^\circ$

**16.** $-\frac{3\pi}{2} = -\frac{3\pi}{2} \cdot \frac{180^\circ}{\pi} = -270^\circ$

**17.** $3 = 3 \cdot \frac{180^\circ}{\pi} = \frac{540^\circ}{\pi} \approx 171.9^\circ$

**18.** $-2 = -2 \cdot \frac{180^\circ}{\pi} = -\frac{360^\circ}{\pi} \approx -114.6^\circ$

**19.** $-1.2 = -1.2 \cdot \frac{180^\circ}{\pi} = -\frac{216^\circ}{\pi} = -68.8^\circ$

**20.** $3.4 = 3.4 \cdot \frac{180^\circ}{\pi} = \frac{612^\circ}{\pi} \approx 194.8^\circ$

**21.** $\frac{\pi}{10} = \frac{\pi}{10} \cdot \frac{180^\circ}{\pi} = 18^\circ$

**22.** $\frac{5\pi}{18} = \frac{5\pi}{18} \cdot \frac{180^\circ}{\pi} = 50^\circ$

**23.** $-\frac{2\pi}{15} = -\frac{2\pi}{15} \cdot \frac{180^\circ}{\pi} = -24^\circ$

**24.** $-\frac{13\pi}{12} = -\frac{13\pi}{12} \cdot \frac{180^\circ}{\pi} = -195^\circ$

**25.** $50^\circ$ is coterminal with $50^\circ + 360^\circ = 410^\circ$, $50^\circ + 720^\circ = 770^\circ$, $50^\circ - 360^\circ = -310^\circ$, and $50^\circ - 720^\circ = -670^\circ$. (Other answers are possible.)

**26.** $135^\circ$ is coterminal with $135^\circ + 360^\circ = 495^\circ$, $135^\circ + 720^\circ = 855^\circ$, $135^\circ - 360^\circ = -225^\circ$, and $135^\circ - 720^\circ = -585^\circ$. (Other answers are possible.)

**27.** $\frac{3\pi}{4}$ is coterminal with $\frac{3\pi}{4} + 2\pi = \frac{11\pi}{4}$, $\frac{3\pi}{4} + 4\pi = \frac{19\pi}{4}$, $\frac{3\pi}{4} - 2\pi = -\frac{5\pi}{4}$, and $\frac{3\pi}{4} - 4\pi = -\frac{13\pi}{4}$. (Other answers are possible.)

**28.** $\frac{11\pi}{6}$ is coterminal with $\frac{11\pi}{6} + 2\pi = \frac{23\pi}{6}$, $\frac{11\pi}{6} + 4\pi = \frac{35\pi}{6}$, $\frac{11\pi}{6} - 2\pi = -\frac{\pi}{6}$, and $\frac{11\pi}{6} - 4\pi = -\frac{13\pi}{6}$. (Other answers are possible.)

**29.** $-\frac{\pi}{4}$ is coterminal with $-\frac{\pi}{4} + 2\pi = \frac{7\pi}{4}$, $-\frac{\pi}{4} + 4\pi = \frac{15\pi}{4}$, $-\frac{\pi}{4} - 2\pi = -\frac{9\pi}{4}$, and $-\frac{\pi}{4} - 4\pi = -\frac{17\pi}{4}$. (Other answers are possible.)

**30.** $-45^\circ$ is coterminal with $-45^\circ + 360^\circ = 315^\circ$, $-45^\circ + 720^\circ = 675^\circ$, $-45^\circ - 360^\circ = -405^\circ$, and $-45^\circ - 720^\circ = -765^\circ$. (Other answers are possible.)

**31.** Since $430^\circ - 70^\circ = 360^\circ$, the angles are coterminal.

**32.** Since $330^\circ - (-30^\circ) = 360^\circ$, the angles are coterminal.

**33.** Since $\frac{17\pi}{6} - \frac{5\pi}{6} = \frac{12\pi}{6} = 2\pi$; the angles are coterminal.

**34.** Since $\frac{32\pi}{3} - \frac{11\pi}{3} = \frac{21\pi}{3} = 7\pi$ is not a multiple of $2\pi$, the angles are not coterminal.

**35.** Since $875^\circ - 155^\circ = 720^\circ = 2 \cdot 360^\circ$, the angles are coterminal.

**36.** Since $340^\circ - 50^\circ = 290^\circ$ is not a multiple of $360^\circ$, the angles are not coterminal.

**37.** Since $733^\circ - 2 \cdot 360^\circ = 13^\circ$, the angles $733^\circ$ and $13^\circ$ are coterminal.

**38.** Since $361^\circ - 1^\circ = 360^\circ$, the angles $361^\circ$ and $1^\circ$ are coterminal.

**39.** Since $1110^\circ - 3 \cdot 360^\circ = 30^\circ$, the angles $1110^\circ$ and $30^\circ$ are coterminal.

**40.** Since $-100^\circ - 260^\circ = -360^\circ$ is a multiple of $360^\circ$, the angles $-100^\circ$ and $260^\circ$ are coterminal.

**41.** Since $-800^\circ + 3 \cdot 360^\circ = 280^\circ$, the angles $-800^\circ$ and $280^\circ$ are coterminal.

**42.** Since $1270^\circ - 190^\circ = 1080^\circ = 3 \cdot 360^\circ$ is a multiple of $360^\circ$, the angles $1270^\circ$ and $190^\circ$ are coterminal.

**43.** Since $\frac{17\pi}{6} - 2\pi = \frac{5\pi}{6}$, the angles $\frac{17\pi}{6}$ and $\frac{5\pi}{6}$ are coterminal.

**44.** Since $-\frac{7\pi}{3} - \frac{5\pi}{3} = 4\pi$ is a multiple of $2\pi$, the angles $-\frac{7\pi}{3}$ and $\frac{5\pi}{3}$ are coterminal.

**45.** Since $87\pi - 43 \cdot 2\pi = \pi$, the angles $87\pi$ and $\pi$ are coterminal.

**46.** Since $10 - 2\pi \approx 3.717$, the angles 10 and $10 - 2\pi$ are coterminal.

**47.** Since $\frac{17\pi}{4} - 2 \cdot 2\pi = \frac{\pi}{4}$, the angles $\frac{17\pi}{4}$ and $\frac{\pi}{4}$ are coterminal.

**48.** Since $\frac{51\pi}{2} - \frac{3\pi}{2} = 24\pi = 12 \cdot 2\pi$, the angles $\frac{51\pi}{2}$ and $\frac{3\pi}{2}$ are coterminal.

**49.** Using the formula $s = \theta r$, the length of the arc is $s = \left(220^\circ \cdot \frac{\pi}{180^\circ}\right) \cdot 5 = \frac{55\pi}{9} \approx 19.2$.

**50.** $\theta = \dfrac{s}{r} = \frac{10}{5} = 2 \text{ rad} = 2 \cdot \frac{180^\circ}{\pi} \approx 114.6^\circ$

**51.** Solving for $r$ we have $r = \frac{s}{\theta}$, so the radius of the circle is $r = \frac{8}{2} = 4$.

**52.** Using the formula $s = \theta r$, the length of the arc is $s = 45^\circ \cdot \frac{\pi}{180^\circ} \cdot 10 = \frac{5\pi}{2} \approx 7.85$ m

**53.** Using the formula $s = \theta r$, the length of the arc is $s = 2 \cdot 2 = 4$ mi.

**54.** Solving for $\theta$, we have $\theta = \frac{s}{r}$, the measure of the central angle is $\theta = \frac{6}{5} = 1.2 \text{ rad} = 1.2 \cdot \frac{180^\circ}{\pi} \approx 68.8^\circ$.

**55.** Solving for $\theta$, we have $\theta = \frac{s}{r}$, so the measure of the central angle is $\theta = \frac{100}{50} = 2$ rad. Converting to degrees we have $\theta = 2 \cdot \frac{180^\circ}{\pi} \approx 114.6^\circ$

**56.** Solving for $r$, we have $r = \frac{s}{\theta}$, so the radius of the circle is $r = \dfrac{3}{25^\circ \cdot \frac{\pi}{180^\circ}} = 3 \cdot \frac{36}{5\pi} \approx 6.88$ ft.

**57.** Solving for $r$, we have $r = \frac{s}{\theta}$, so the radius of the circle is $r = \frac{6}{\pi/6} = \frac{36}{\pi} \approx 11.46$ m.

**58.** Solving for $r$, we have $r = \frac{s}{\theta}$, so the radius of the circle is $r = \dfrac{4}{135^\circ \cdot \frac{\pi}{180^\circ}} = \frac{16}{3\pi} \approx 1.70$ ft.

**59.** **(a)** $A = \frac{1}{2}r^2\theta = \frac{1}{2} \cdot 8^2 \cdot 80^\circ \cdot \frac{\pi}{180^\circ} = 32 \cdot \frac{4\pi}{9} = \frac{128\pi}{9} \approx 44.68$

**(b)** $A = \frac{1}{2}r^2\theta = \frac{1}{2} \cdot 10^2 \cdot 0.5 = 25$

**60.** **(a)** $A = \frac{1}{2}r^2\theta \Rightarrow r = \sqrt{\dfrac{2A}{\theta}}$, so the radius is $\sqrt{\dfrac{2 \cdot 12}{0.7}} \approx 5.86$.

**(b)** $r = \sqrt{\dfrac{2A}{\theta}} = \sqrt{\dfrac{2 \cdot 12}{150^\circ \cdot \frac{\pi}{180^\circ}}} \approx 3.03$

**61.** $A = \frac{1}{2}r^2\theta = \frac{1}{2} \cdot 10^2 \cdot 1 = 50 \text{ m}^2$

**62.** $\theta = 60^\circ = 60^\circ \cdot \frac{\pi}{180^\circ} = \frac{\pi}{3}$ rad. Then $A = \frac{1}{2}r^2\theta = \frac{1}{2} \cdot 3^2 \cdot \frac{\pi}{3} = \frac{3\pi}{2} \approx 4.7 \text{ mi}^2$

**63.** $\theta = 2$ rad, $A = 16 \text{ m}^2$. Since $A = \frac{1}{2}r^2\theta$, we have $r = \sqrt{2A/\theta} = \sqrt{2 \cdot 16/2} = \sqrt{16} = 4$ m.

**64.** $r = 24$ mi, $A = 288 \text{ mi}^2$. Then $\theta = \dfrac{2A}{r^2} = \dfrac{2 \cdot 288}{24^2} = 1 \text{ rad} \approx 57.3^\circ$.

**65.** Since the area of the circle is $72 \text{ cm}^2$, the radius of the circle is $r = \sqrt{A/\pi} = \sqrt{72/\pi}$. Then the area of the sector is $A = \frac{1}{2}r^2\theta = \frac{1}{2} \cdot \frac{72}{\pi} \cdot \frac{\pi}{6} = 6 \text{ cm}^2$.

**66.** Referring to the figure, we have $AC = 3 + 1 = 4$, $BC = 1 + 2 = 3$, and $AB = 2 + 3 = 5$. Since $AB^2 = AC^2 + BC^2$, then by the Pythagorean Theorem, the triangle is a right triangle. Therefore, $\theta = \frac{\pi}{2}$ and $A = \frac{1}{2}r^2\theta = \frac{1}{2} \cdot 1^2 \cdot \frac{\pi}{2} = \frac{\pi}{4}$ ft$^2$.

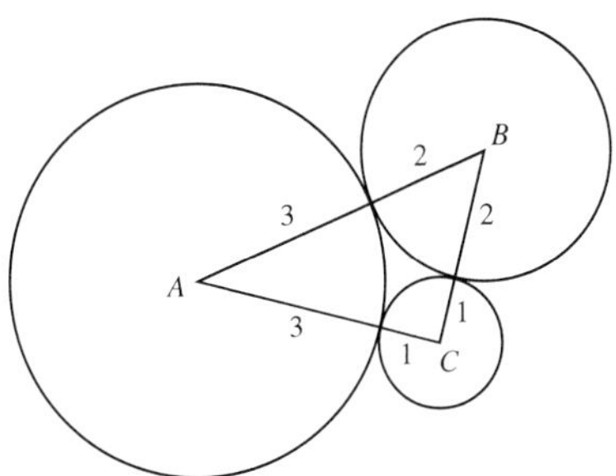


**67.** The circumference of each wheel is $\pi d = 28\pi$ in. If the wheels revolve 10,000 times, the distance traveled is $10{,}000 \cdot 28\pi \text{ in.} \cdot \frac{1 \text{ ft}}{12 \text{ in.}} \cdot \frac{1 \text{ mi}}{5280 \text{ ft}} \approx 13.88$ mi.

**68.** Since the diameter is 30 in., we have $r = 15$ in. In one revolution, the arc length (distance traveled) is $s = \theta r = 2\pi \cdot 15 = 30\pi$ in. The total distance traveled is 1 mi $\cdot$ 5280 ft/mi $\cdot$ 12 in/ft $=$ 63,360 in. $=$ 63,360 in. $\cdot \frac{1 \text{ rev}}{30\pi \text{ in}} \approx 672.27$ rev. Therefore the car wheel will make approximately 672 revolutions.

**69.** We find the measure of the angle in degrees and then convert to radians. $\theta = 40.5° - 25.5° = 15°$ and $15 \cdot \frac{\pi}{180°}$ rad $= \frac{\pi}{12}$ rad. Then using the formula $s = \theta r$, we have $s = \frac{\pi}{12} \cdot 3960 = 330\pi \approx 1036.725$ and so the distance between the two cities is roughly 1037 mi.

**70.** $\theta = 35° - 30° = 5° = 5° \cdot \frac{\pi}{180°}$ rad $= \frac{\pi}{3}$ rad. Then using the formula $s = \theta r$, the length of the arc is $s = \frac{\pi}{3} \cdot 3960 = 110\pi \approx 345.575$. So the distance between the two cities is roughly 346 mi.

**71.** In one day, the earth travels $\frac{1}{365}$ of its orbit which is $\frac{2\pi}{365}$ rad. Then $s = \theta r = \frac{2\pi}{365} \cdot 93{,}000{,}000 \approx 1{,}600{,}911.3$, so the distance traveled is approximately 1.6 million miles.

**72.** Since the sun is so far away, we can assume that the rays of the sun are parallel when striking the earth. Thus, the angle formed at the center of the earth is also $\theta = 7.2°$. So $r = \frac{s}{\theta} = \frac{500}{7.2° \cdot \frac{\pi}{180°}} = \frac{180 \cdot 500}{7.2\pi} \approx 3980$ mi, and the circumference is $c = 2\pi r = \frac{2\pi \cdot 180 \cdot 500}{7.2\pi} = 25{,}000$ mi.

**73.** The central angle is 1 minute $= \left(\frac{1}{60}\right)° = \frac{1}{60} \cdot \frac{\pi}{180°}$ rad $= \frac{\pi}{10{,}800}$ rad. Then $s = \theta r = \frac{\pi}{10{,}800} \cdot 3960 \approx 1.152$, and so a nautical mile is approximately 1.152 mi.

**74.** The area is $A = \frac{1}{2}r^2\theta = \frac{1}{2} \cdot 300^2 \cdot \left(280° \cdot \frac{\pi}{180°}\right) \approx 219{,}900$ ft$^2$.

**75.** The area is equal to the area of the large sector (with radius 34 in.) minus the area of the small sector (with radius 14 in.) Thus, $A = \frac{1}{2}r_1^2\theta - \frac{1}{2}r_2^2\theta = \frac{1}{2}\left(34^2 - 14^2\right)\left(135° \cdot \frac{\pi}{180°}\right) \approx 1131$ in.$^2$.

**76.** The area available to the cow is shown in the diagram. Its area is the sum of four quarter-circles:

$$\begin{aligned} A &= \tfrac{1}{4}\pi\left(100^2 + 50^2 + 40^2 + 30^2\right) \\ &= 3750\pi \\ &\approx 11{,}781 \text{ ft}^2 \end{aligned}$$

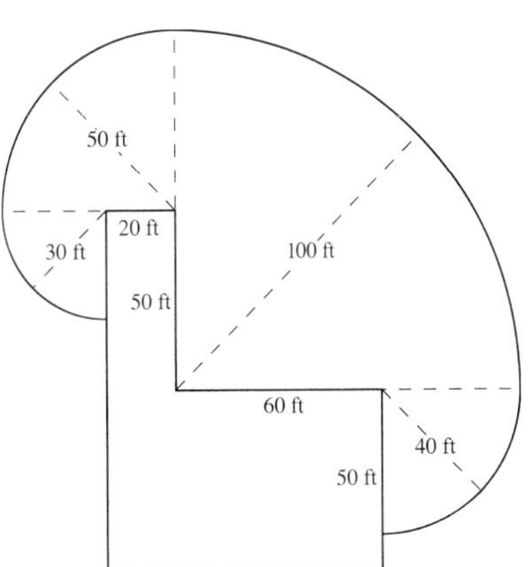


**77.** $v = \frac{8 \cdot 2\pi \cdot 2}{15} = \frac{32\pi}{15} \approx 6.702$ ft/s.

**78.** **(a)** The angular speed is $\omega = \dfrac{45 \cdot 2\pi \text{ rad}}{1 \text{ min}} = 90\pi$ rad/min.

**(b)** The linear speed is $v = \dfrac{45 \cdot 2\pi \cdot 16}{1} = 1440\pi$ in./min $\approx 4523.9$ in./min.

**79.** **(a)** The angular speed is $\omega = \dfrac{1000 \cdot 2\pi \text{ rad}}{1 \text{ min}} = 2000\pi$ rad/min.

**(b)** The linear speed is $v = \dfrac{1000 \cdot 2\pi \cdot \frac{6}{12}}{60} = \dfrac{50\pi}{3}$ ft/s $\approx 52.4$ ft/s.

**80.** 23 h 56 min 4 s $= 23.9344$ hr. So the linear speed is $\dfrac{1 \cdot 2\pi \cdot (3960)}{1 \text{ day}} \cdot \dfrac{1 \text{ day}}{23.9344 \text{ hr}} \approx 1039.57$ mi/h.

**81.** $v = \dfrac{600 \cdot 2\pi \cdot \frac{11}{12} \text{ ft}}{1 \text{ min}} \cdot \dfrac{1 \text{ mi}}{5280 \text{ ft}} \cdot \dfrac{60 \text{ min}}{1 \text{ hr}} = 12.5\pi$ mi/h $\approx 39.3$ mi/h.

**82.** **(a)** The angular speed is $\omega = \dfrac{\text{linear speed}}{\text{radius}} = \dfrac{50 \text{ mi/h}}{4 \text{ ft}} \cdot \dfrac{1 \text{ h}}{60 \text{ min}} \cdot \dfrac{5280 \text{ ft}}{1 \text{ mi}} = 1100$ rad/min.

**(b)** The rate of revolution is $\dfrac{\text{angular speed}}{2\pi} = \dfrac{1100 \text{ rad/min}}{2\pi} \approx 175$ revolutions/min.

**83.** $v = \dfrac{100 \cdot 2\pi \cdot 0.20 \text{ m}}{60 \text{ s}} = \dfrac{2\pi}{3} \approx 2.09$ m/s.

**84.** **(a)** The angular speed is $\omega = \dfrac{\text{linear speed of pedal}}{\text{radius of wheel sprocket}} = \dfrac{40 \cdot 2\pi \cdot 4}{2} = 160$ rad/min.

**(b)** The linear speed of the bicycle is $v =$ angular speed $\cdot$ radius $= (160 \text{ rad/min}) \cdot (13) = 2080\pi$ ft/min $\approx 74.26$ mi/h.

**85.** **(a)** The circumference of the opening is the length of the arc subtended by the angle $\theta$ on the flat piece of paper, that is, $C = s = r\theta = 6 \cdot \frac{5\pi}{3} = 10\pi \approx 31.4$ cm.

**(b)** Solving for $r$, we find $r = \dfrac{C}{2\pi} = \dfrac{10\pi}{2\pi} = 5$ cm.

**(c)** By the Pythagorean Theorem, $h^2 = 6^2 - 5^2 = 11$, so $h = \sqrt{11} \approx 3.3$ cm.

**(d)** The volume of a cone is $V = \frac{1}{3}\pi r^2 h$. In this case $V = \frac{1}{3}\pi \cdot 5^2 \cdot \sqrt{11} \approx 86.8$ cm$^3$.

**86.** **(a)** With an arbitrary angle $\theta$, the circumference of the opening is

$$C = 6\theta,\ r = \frac{C}{2\pi} = \frac{3\theta}{\pi},\ h = \sqrt{6^2 - r^2} = \sqrt{36 - \frac{9\theta^2}{\pi^2}}, \text{ and}$$

$$V = \tfrac{1}{3}\pi r^2 h = \frac{\pi}{3} \cdot \frac{9\theta^2}{\pi^2}\sqrt{36 - \frac{9\theta^2}{\pi^2}} = \frac{9}{\pi^2}\theta^2\sqrt{4\pi^2 - \theta^2}.$$

**(b)**



**(c)** The volume seems to be maximized for $\theta \approx 5.13$ rad or about 293°.

**87.** Answers will vary, although of course everyone prefers radians.

**88.** There are 5 hours 45 minutes between 1:00 P.M. and 6:45 P.M.. Since 45 minutes is 0.75 hour, the minute hand moves through $5.75 \cdot 2\pi = 11.5\pi \approx 36.128$ rad. The hour hand moves through $5.75 \cdot \frac{1}{12} \cdot 2\pi = \frac{23}{24}\pi \approx 3.011$ rad.

## 6.2 Trigonometry of Right Triangles

**1.** $\sin\theta = \frac{4}{5}$, $\cos\theta = \frac{3}{5}$, $\tan\theta = \frac{4}{3}$, $\csc\theta = \frac{5}{4}$, $\sec\theta = \frac{5}{3}$, $\cot\theta = \frac{3}{4}$

**2.** $\sin\theta = \frac{7}{25}$, $\cos\theta = \frac{24}{25}$, $\tan\theta = \frac{7}{24}$, $\csc\theta = \frac{25}{7}$, $\sec\theta = \frac{25}{24}$, $\cot\theta = \frac{24}{7}$

**3.** The remaining side is obtained by the Pythagorean Theorem: $\sqrt{41^2 - 40^2} = \sqrt{81} = 9$. Then $\sin\theta = \frac{40}{41}$, $\cos\theta = \frac{9}{41}$, $\tan\theta = \frac{40}{9}$, $\csc\theta = \frac{41}{40}$, $\sec\theta = \frac{41}{9}$, $\cot\theta = \frac{9}{40}$

**4.** The hypotenuse is obtained by the Pythagorean Theorem: $\sqrt{8^2 + 15^2} = \sqrt{289} = 17$. Then $\sin\theta = \frac{15}{17}$, $\cos\theta = \frac{8}{17}$, $\tan\theta = \frac{15}{8}$, $\csc\theta = \frac{17}{15}$, $\sec\theta = \frac{17}{8}$, $\cot\theta = \frac{8}{15}$

**5.** The remaining side is obtained by the Pythagorean Theorem: $\sqrt{3^2+2^2}=\sqrt{13}$. Then $\sin\theta=\frac{2}{\sqrt{13}}=\frac{2\sqrt{13}}{13}$, $\cos\theta=\frac{3}{\sqrt{13}}=\frac{3\sqrt{13}}{13}$, $\tan\theta=\frac{2}{3}$, $\csc\theta=\frac{\sqrt{13}}{2}$, $\sec\theta=\frac{\sqrt{13}}{3}$, $\cot\theta=\frac{3}{2}$

**6.** The remaining side is obtained by the Pythagorean Theorem: $\sqrt{8^2-7^2}=\sqrt{15}$. Then $\sin\theta=\frac{7}{8}$, $\cos\theta=\frac{\sqrt{15}}{8}$, $\tan\theta=\frac{7}{\sqrt{15}}=\frac{7\sqrt{15}}{15}$, $\csc\theta=\frac{8}{7}$, $\sec\theta=\frac{8}{\sqrt{15}}=\frac{8\sqrt{15}}{15}$, $\cot\theta=\frac{\sqrt{15}}{7}$

**7.** $c=\sqrt{5^2+3^2}=\sqrt{34}$

**(a)** $\sin\alpha=\cos\beta=\frac{3}{\sqrt{34}}=\frac{3\sqrt{34}}{34}$

**(b)** $\tan\alpha=\cot\beta=\frac{3}{5}$

**(c)** $\sec\alpha=\csc\beta=\frac{\sqrt{34}}{5}$

**8.** $b=\sqrt{7^2-4^2}=\sqrt{33}$

**(a)** $\sin\alpha=\cos\beta=\frac{4}{7}$

**(b)** $\tan\alpha=\cot\beta=\frac{4}{\sqrt{33}}$

**(c)** $\sec\alpha=\csc\beta=\frac{7}{\sqrt{33}}$

**9.** Since $\sin 30^\circ=\dfrac{x}{25}$, we have $x=25\sin 30^\circ=25\cdot\frac{1}{2}=\frac{25}{2}$.

**10.** Since $\sin 45^\circ=\dfrac{12}{x}$, we have $x=\dfrac{12}{\sin 45^\circ}=\dfrac{12}{\frac{1}{\sqrt{2}}}=12\sqrt{2}$.

**11.** Since $\sin 60^\circ=\dfrac{x}{13}$, we have $x=13\sin 60^\circ=13\cdot\frac{\sqrt{3}}{2}=\frac{13\sqrt{3}}{2}$.

**12.** Since $\tan 30^\circ=\dfrac{4}{x}$, we have $x=\dfrac{4}{\tan 30^\circ}=\dfrac{4}{\frac{1}{\sqrt{3}}}=4\sqrt{3}$.

**13.** Since $\tan 36^\circ=\dfrac{12}{x}$, we have $x=\dfrac{12}{\tan 36^\circ}\approx 16.51658$.

**14.** Since $\sin 53^\circ=\dfrac{25}{x}$, we have $x=\dfrac{25}{\sin 53^\circ}\approx 31.30339$.

**15.** $\dfrac{x}{28}=\cos\theta \quad\Leftrightarrow\quad x=28\cos\theta$, and $\dfrac{y}{28}=\sin\theta \quad\Leftrightarrow\quad y=28\sin\theta$.

**16.** $\dfrac{x}{4}=\tan\theta \quad\Leftrightarrow\quad x=4\tan\theta$, and $\dfrac{4}{y}=\cos\theta \quad\Leftrightarrow\quad y=\dfrac{4}{\cos\theta}=4\sec\theta$.

**17.** $\sin\theta=\frac{3}{5}$. Then the third side is $x=\sqrt{5^2-3^2}=4$. The other five ratios are $\cos\theta=\frac{4}{5}$, $\tan\theta=\frac{3}{4}$, $\csc\theta=\frac{5}{3}$, $\sec\theta=\frac{5}{4}$, and $\cot\theta=\frac{4}{3}$.

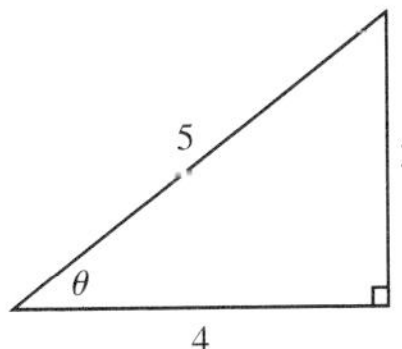


**18.** $\cos\theta=\frac{9}{40}$. The third side is $y=\sqrt{40^2-9^2}=\sqrt{1519}=7\sqrt{31}$. The other five ratios are $\sin\theta=\frac{7\sqrt{31}}{40}$, $\tan\theta=\frac{7\sqrt{31}}{9}$, $\csc\theta=\frac{40}{7\sqrt{31}}=\frac{40\sqrt{31}}{217}$, $\sec\theta=\frac{40}{9}$, and $\cot\theta=\frac{9}{7\sqrt{31}}=\frac{9\sqrt{31}}{217}$.

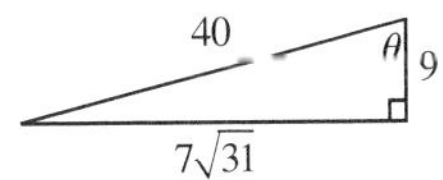


**19.** $\cot\theta=1$. Then the third side is $r=\sqrt{1^2+1^2}=\sqrt{2}$. The other five ratios are $\sin\theta=\frac{1}{\sqrt{2}}=\frac{\sqrt{2}}{2}$, $\cos\theta=\frac{1}{\sqrt{2}}=\frac{\sqrt{2}}{2}$, $\tan\theta=1$, $\csc\theta=\sqrt{2}$, and $\sec\theta=\sqrt{2}$.

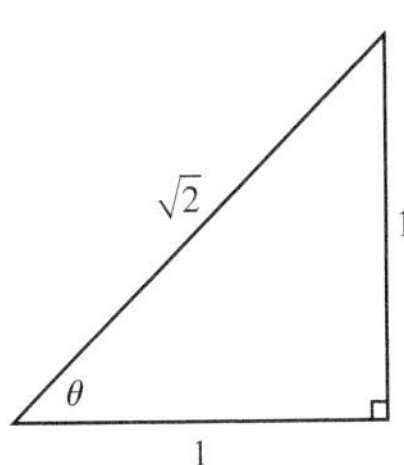


**20.** $\tan\theta=\sqrt{3}$. The third side is $r=\sqrt{1^2+3}=2$. The other five ratios are $\sin\theta=\frac{\sqrt{3}}{2}$, $\cos\theta=\frac{1}{2}$, $\csc\theta=\frac{2}{\sqrt{3}}$, $\sec\theta=2$, and $\cot\theta=\frac{1}{\sqrt{3}}$.

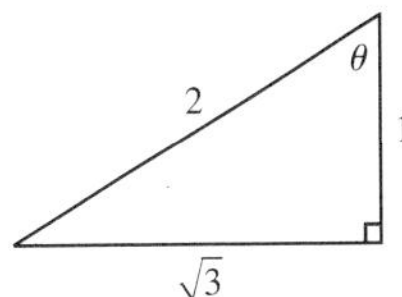

**21.** $\sec\theta = 7$. The third side is $y = \sqrt{7^2 - 2^2} = \sqrt{45} = 3\sqrt{5}$ The other five ratios are $\sin\theta = \frac{3\sqrt{5}}{7}$, $\cos\theta = \frac{2}{7}$, $\tan\theta = \frac{3\sqrt{5}}{2}$, $\csc\theta = \frac{7}{3\sqrt{5}} = \frac{7\sqrt{5}}{15}$, and $\cot\theta = \frac{2}{3\sqrt{5}} = \frac{2\sqrt{5}}{15}$.

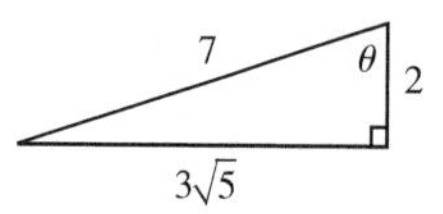


**22.** $\csc\theta = \frac{13}{12}$. The third side is $x = \sqrt{13^2 - 12^2} = 5$. The other five ratios are $\sin\theta = \frac{12}{13}$, $\cos\theta = \frac{5}{13}$, $\tan\theta = \frac{12}{5}$, $\sec\theta = \frac{13}{5}$, and $\cot\theta = \frac{5}{12}$.

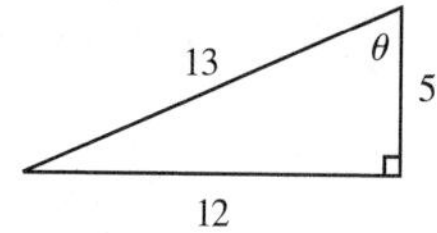


**23.** $\sin\frac{\pi}{6} + \cos\frac{\pi}{6} = \frac{1}{2} + \frac{\sqrt{3}}{2} = \frac{1+\sqrt{3}}{2}$

**24.** $\sin 30^\circ \csc 30^\circ = \sin 30^\circ \cdot \dfrac{1}{\sin 30^\circ} = 1$

**25.** $\sin 30^\circ \cos 60^\circ + \sin 60^\circ \cos 30^\circ = \frac{1}{2}\cdot\frac{1}{2} + \frac{\sqrt{3}}{2}\cdot\frac{\sqrt{3}}{2} = \frac{1}{4} + \frac{3}{4} = 1$

**26.** $(\sin 60^\circ)^2 + (\cos 60^\circ)^2 = \left(\frac{\sqrt{3}}{2}\right)^2 + \left(\frac{1}{2}\right)^2 = \frac{3}{4} + \frac{1}{4} = 1$

**27.** $(\cos 30^\circ)^2 - (\sin 30^\circ)^2 = \left(\frac{\sqrt{3}}{2}\right)^2 - \left(\frac{1}{2}\right)^2 = \frac{3}{4} - \frac{1}{4} = \frac{1}{2}$

**28.** $\left(\sin\frac{\pi}{3}\cos\frac{\pi}{4} - \sin\frac{\pi}{4}\cos\frac{\pi}{3}\right)^2 = \left(\frac{\sqrt{3}}{2}\cdot\frac{1}{\sqrt{2}} - \frac{1}{\sqrt{2}}\cdot\frac{1}{2}\right)^2 = \left[\frac{1}{2\sqrt{2}}\left(\sqrt{3} - 1\right)\right]^2 = \frac{1}{8}\left(\sqrt{3}-1\right)^2 = \frac{1}{8}\left(3 - 2\sqrt{3} + 1\right)$
$= \frac{1}{8}\left(4 - 2\sqrt{3}\right) = \frac{1}{4}\left(2 - \sqrt{3}\right)$

**29.** This is an isosceles right triangle, so the other leg has length $16\tan 45^\circ = 16$, the hypotenuse has length $\dfrac{16}{\sin 45^\circ} = 16\sqrt{2} \approx 22.63$, and the other angle is $90^\circ - 45^\circ = 45^\circ$.

**30.** The other leg has length $100\tan 75^\circ \approx 26.79$, the hypotenuse has length $\dfrac{100}{\sin 75^\circ} \approx 103.52$, and the other angle is $90^\circ - 75^\circ = 15^\circ$.

**31.** The other leg has length $35\tan 52^\circ \approx 44.79$, the hypotenuse has length $\dfrac{35}{\cos 52^\circ} \approx 56.85$, and the other angle is $90^\circ - 52^\circ = 38^\circ$.

**32.** The adjacent leg has length $1000\cos 68^\circ \approx 374.61$, the opposite leg has length $1000\sin 68^\circ \approx 927.18$, and the other angle is $90^\circ - 68^\circ = 22^\circ$.

**33.** The adjacent leg has length $33.5\cos\frac{\pi}{8} \approx 30.95$, the opposite leg has length $33.5\sin\frac{\pi}{8} \approx 12.82$, and the other angle is $\frac{\pi}{2} - \frac{\pi}{8} = \frac{3\pi}{8}$.

**34.** The opposite leg has length $72.3\tan\frac{\pi}{6} \approx 41.74$, the hypotenuse has length $\dfrac{72.3}{\cos\frac{\pi}{6}} \approx 83.48$, and the other angle is $\frac{\pi}{2} - \frac{\pi}{6} = \frac{\pi}{3}$.

**35.** The adjacent leg has length $\dfrac{106}{\tan\frac{\pi}{5}} \approx 145.90$, the hypotenuse has length $\dfrac{106}{\sin\frac{\pi}{5}} \approx 180.34$, and the other angle is $\frac{\pi}{2} - \frac{\pi}{5} = \frac{3\pi}{10}$.

**36.** The adjacent leg has length $425\cos\frac{3\pi}{8} \approx 162.64$, the opposite leg has length $425\sin\frac{3\pi}{8} \approx 392.65$, and the other angle is $\frac{\pi}{2} - \frac{3\pi}{8} = \frac{\pi}{8}$.

**37.** $\sin\theta \approx \frac{1}{2.24} \approx 0.45$. $\cos\theta \approx \frac{2}{2.24} \approx 0.89$, $\tan\theta = \frac{1}{2}$, $\csc\theta \approx 2.24$, $\sec\theta \approx \frac{2.24}{2} \approx 1.12$, $\cot\theta \approx 2.00$.

**38.** 40°

$\sin 40^\circ \approx 0.64$, $\cos 40^\circ \approx 0.77$, $\tan 40^\circ \approx \dfrac{0.64}{0.77} \approx 0.83$, $\csc 40^\circ \approx 1.56$, $\sec 40^\circ \approx 1.39$, $\cot 40^\circ \approx 1.20$.

**39.** $x = \dfrac{100}{\tan 60^\circ} + \dfrac{100}{\tan 30^\circ} \approx 230.9$

**40.** Let $d$ be the length of the base of the $60^\circ$ triangle. Then $\tan 60^\circ = \dfrac{85}{d} \Leftrightarrow d = \dfrac{85}{\tan 60^\circ} \approx 49.075$, and so $\tan 30^\circ = \dfrac{85}{d+x} \Leftrightarrow d + x = \dfrac{85}{\tan 30^\circ} \Leftrightarrow x = \dfrac{85}{\tan 30^\circ} - d \approx 98.1$.

**41.** Let $h$ be the length of the shared side. Then $\sin 60^\circ = \dfrac{50}{h} \Leftrightarrow h = \dfrac{50}{\sin 60^\circ} \approx 57.735 \Leftrightarrow \sin 65^\circ = \dfrac{h}{x} \Leftrightarrow$ $x = \dfrac{h}{\sin 65^\circ} \approx 63.7$

**42.** Let $h$ be the hypotenuse of the top triangle. Then $\sin 30^\circ = \dfrac{5}{h} \Leftrightarrow h = \dfrac{5}{\sin 30^\circ} = 10$, and so $\tan 60^\circ = \dfrac{h}{x} \Leftrightarrow$ $x = \dfrac{h}{\tan 60^\circ} = \dfrac{10}{\tan 60^\circ} \approx 5.8$.

**43.** 



From the diagram, $\sin\theta = \dfrac{x}{y}$ and $\tan\theta = \dfrac{y}{10}$, so $x = y\sin\theta = 10\sin\theta\tan\theta$.

**44.** $\sin\theta = \dfrac{a}{1} \Leftrightarrow a = \sin\theta$, $\tan\theta = \dfrac{b}{1} \Leftrightarrow b = \tan\theta$, $\cos\theta = \dfrac{1}{c} \Leftrightarrow c = \sec\theta$, $\cos\theta = \dfrac{d}{1} \Leftrightarrow d = \cos\theta$

**45.** Let $h$ be the height, in feet, of the Empire State Building. Then $\tan 11^\circ = \dfrac{h}{5280} \Leftrightarrow h = 5280 \cdot \tan 11^\circ \approx 1026$ ft.

**46. (a)** Let $r$ be the distance, in feet, between the plane and the Gateway Arch. Therefore, $\sin 22^\circ = \dfrac{35{,}000}{r} \Leftrightarrow$ $r = \dfrac{35{,}000}{\sin 22^\circ} \approx 93{,}431$ ft.

**(b)** Let $x$ be the distance, also in feet, between a point on the ground directly below the plane and the Gateway Arch. Then $\tan 22^\circ = \dfrac{35{,}000}{x} \Leftrightarrow x = \dfrac{35{,}000}{\tan 22^\circ} \approx 86{,}628$ ft.

**47. (a)** Let $h$ be the distance, in miles, that the beam has diverged. Then $\tan 0.5^\circ = \dfrac{h}{240{,}000} \Leftrightarrow$ $h = 240{,}000 \cdot \tan 0.5^\circ \approx 2100$ mi.

**(b)** Since the deflection is about 2100 mi whereas the radius of the moon is about 1000 mi, the beam will not strike the moon.

**48.** Let $x$ be the distance, in feet, of the ship from the base of the lighthouse. Then $\tan 23^\circ = \dfrac{200}{x} \Leftrightarrow$ $x = \dfrac{200}{\tan 23^\circ} \approx 471$ ft.

**49.** Let $h$ represent the height, in feet, that the ladder reaches on the building. Then $\sin 72^\circ = \dfrac{h}{20} \Leftrightarrow$ $h = 20\sin 72^\circ \approx 19$ ft.

**50.** Let $\theta$ represent the angle of elevation of the ladder. Let $h$ represent the height, in feet, that the ladder reaches on the building. Then $\cos\theta = \frac{6}{20} = 0.3 \Leftrightarrow \theta = \cos^{-1} 0.3 \approx 1.266 \text{ rad} \approx 72.5^\circ$. By the Pythagorean Theorem $h^2 + 6^2 = 20^2 \Leftrightarrow$ $h = \sqrt{400 - 36} = \sqrt{364} \approx 19$ ft.

**51.** Let $\theta$ be the angle of elevation of the sun. Then $\tan\theta = \frac{96}{120} = 0.8 \Leftrightarrow \theta = \tan^{-1} 0.8 \approx 0.675 \approx 38.7^\circ$.

**52.** Let $h$ be the height, in feet, of the communication tower. Then $\sin 65^\circ = \dfrac{h}{600} \Leftrightarrow h = 600\sin 65^\circ \approx 544$ ft.

**53.** Let $h$ be the height, in feet, of the kite above the ground. Then $\sin 50^\circ = \dfrac{h}{450} \Leftrightarrow h = 450\sin 50^\circ \approx 345$ ft.

**54.** 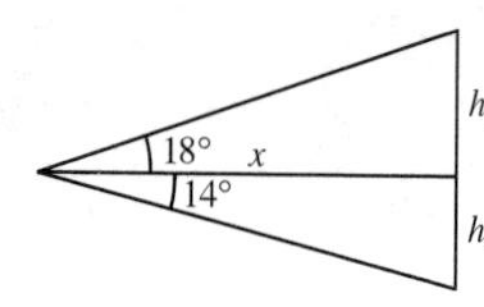


Let $h_1$ be the height of the flagpole above elevation and let $h_2$ be the height below, as shown in the figure. So $\tan 18° = \frac{h_1}{x} \Leftrightarrow h_1 = x \tan 18°$. Similarly, $h_2 = x \tan 14°$. Since the flagpole is 60 feet tall, we have $h_1 + h_2 = 60$, so $x(\tan 18° + \tan 14°) = 60 \Leftrightarrow x = \frac{60}{\tan 18° + \tan 14°} \approx 104.5$ ft.

**55.** Let $h_1$ be the height of the window in feet and $h_2$ be the height from the window to the top of the tower. Then $\tan 25° = \frac{h_1}{325} \Leftrightarrow h_1 = 325 \cdot \tan 25° \approx 152$ ft. Also, $\tan 39° = \frac{h_2}{325} \Leftrightarrow h_2 = 325 \cdot \tan 39° \approx 263$ ft. Therefore, the height of the window is approximately 152 ft and the height of the tower is approximately $152 + 263 = 415$ ft.

**56.** 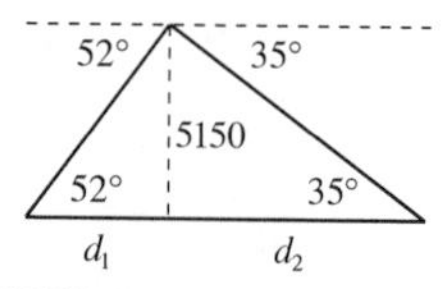


Let $d_1$ be the distance, in feet, between a point directly below the plane and one car, and $d_2$ be the distance, in feet, between the same point and the other car. Then $\tan 52° = \frac{5150}{d_1} \Leftrightarrow d_1 = \frac{5150}{\tan 52°} \approx 4023.62$ ft, and $\tan 35° = \frac{5150}{d_2} \Leftrightarrow$ $d_2 = \frac{5150}{\tan 35°} \approx 7354.96$ ft. So the distance between the two cars is about $d_1 + d_2 \approx 4023.62 + 7354.96 \approx 11{,}379$ ft.

**57.** Let $d_1$ be the distance, in feet, between a point directly below the plane and one car, and $d_2$ be the distance, in feet, between the same point and the other car. Then $\tan 52° = \frac{d_1}{5150} \Leftrightarrow d_1 = 5150 \cdot \tan 52° \approx 6591.7$ ft. Also, $\tan 38° = \frac{d_2}{5150} \Leftrightarrow$ $d_2 = 5150 \cdot \tan 38° \approx 4023.6$ ft. So in this case, the distance between the two cars is about 2570 ft.

**58.** Let $x$ be the distance, in feet, between a point directly below the balloon and the first mile post. Let $h$ be the height, in feet, of the balloon. Then $\tan 22° = \frac{h}{x}$ and $\tan 20° = \frac{h}{x + 5280}$. So $h = x \tan 22° = (x + 5280) \tan 20° \Leftrightarrow$ $x = \frac{5280 \cdot \tan 20°}{\tan 22° - \tan 20°} \approx 47{,}977$ ft. Therefore $h \approx 47{,}976.9 \cdot \tan 22° \approx 19{,}384$ ft $\approx 3.7$ mi.

**59.** Let $x$ be the horizontal distance, in feet, between a point on the ground directly below the top of the mountain and the point on the plain closest to the mountain. Let $h$ be the height, in feet, of the mountain. Then $\tan 35° = \frac{h}{x}$ and $\tan 32° = \frac{h}{x + 1000}$. So $h = x \tan 35° = (x + 1000) \tan 32° \Leftrightarrow x = \frac{1000 \cdot \tan 32°}{\tan 35° - \tan 32°} \approx 8294.2$. Thus $h \approx 8294.2 \cdot \tan 35° \approx 5808$ ft.

**60.** 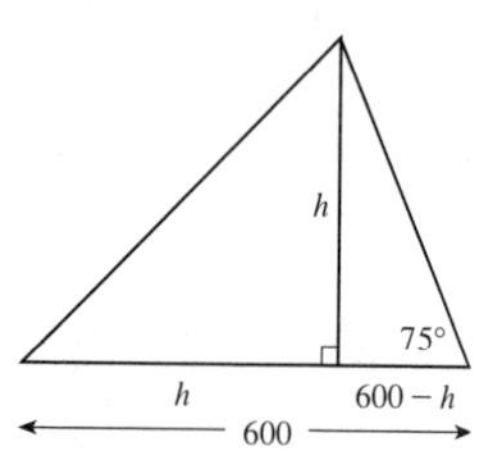


Since the angle of elevation from the observer is 45°, the distance from the observer is $h$, as shown in the figure. Thus, the length of the leg in the smaller right triangle is $600 - h$. Then $\tan 75° = \frac{h}{600 - h} \Leftrightarrow (600 - h) \tan 75° = h \Leftrightarrow$ $600 \tan 75° = h(1 + \tan 75°) \Leftrightarrow h = \frac{600 \tan 75°}{1 + \tan 75°} \approx 473$ m.

**61.** Let $d$ be the distance, in miles, from the earth to the sun. Then $\tan 89.95° = \frac{d}{240{,}000} \Leftrightarrow$ $d = 240{,}000 \cdot \tan 89.95° \approx 91.7$ million miles.

**62. (a)** $s = r\theta \Leftrightarrow \theta = \frac{s}{r} = \frac{6155}{3960} \approx 1.5543$ rad $\approx 89.05°$

**(b)** Let $d$ represent the distance, in miles, from the center of the earth to the moon. Since $\cos \theta = \frac{3960}{d}$, we have $d = \frac{3960}{\cos \theta} \approx \frac{3960}{\cos 89.05°} \approx 239{,}961.5$. So the distance $AC$ is $239{,}961.5 - 3960 \approx 236{,}000$ mi.

**63.** Let $r$ represent the radius, in miles, of the earth. Then $\sin 60.276^\circ = \frac{r}{r+600} \Leftrightarrow (r+600)\sin 60.276^\circ = r \Leftrightarrow 600\sin 60.276^\circ = r(1-\sin 60.276^\circ) \Leftrightarrow r = \frac{600\sin 60.276^\circ}{1-\sin 60.276^\circ} \approx 3960.099$. So the earth's radius is about 3960 mi.

**64.** Let $d$ represent the distance, in miles, from the earth to Alpha Centauri. Since $\sin 0.000211^\circ = \frac{93{,}000{,}000}{d}$, we have $d = \frac{93{,}000{,}000}{\sin 0.000211^\circ} \approx 25{,}253{,}590{,}022{,}410$. So the distance from the earth to Alpha Centauri is about $2.53 \times 10^{13}$ mi.

**65.** Let $d$ be the distance, in AU, between Venus and the sun. Then $\sin 46.3^\circ = \dfrac{d}{1} = d$, so $d = \sin 46.3^\circ \approx 0.723$ AU.

**66.** If two triangles are similar, then their corresponding angles are equal and their corresponding sides are proportional. That is, if triangle $ABC$ is similar to triangle $A'B'C'$ then $|AB| = r\,|A'B'|$, $|AC| = r\,|A'C'|$, and $|BC| = r\,|B'C'|$. Thus when we express any trigonometric ratio of these lengths as a fraction, the factor $r$ cancels out.

# 6.3 Trigonometric Functions of Angles

**1. (a)** The reference angle for $150^\circ$ is $180^\circ - 150^\circ = 30^\circ$.
**(b)** The reference angle for $330^\circ$ is $360^\circ - 330^\circ = 30^\circ$.
**(c)** The reference angle for $-30^\circ$ is $-(-30^\circ) = 30^\circ$.

**2. (a)** The reference angle for $120^\circ$ is $180^\circ - 120^\circ = 60^\circ$.
**(b)** The reference angle for $-210^\circ$ is $210^\circ - 180^\circ = 30^\circ$.
**(c)** The reference angle for $780^\circ$ is $780^\circ - 720^\circ = 60^\circ$.

**3. (a)** The reference angle for $225^\circ$ is $225^\circ - 180^\circ = 45^\circ$.
**(b)** The reference angle for $810^\circ$ is $810^\circ - 720^\circ = 90^\circ$.
**(c)** The reference angle for $-105^\circ$ is $180^\circ - 105^\circ = 75^\circ$.

**4. (a)** The reference angle for $99^\circ$ is $180^\circ - 99^\circ = 81^\circ$.
**(b)** The reference angle for $-199^\circ$ is $199^\circ - 180^\circ = 19^\circ$.
**(c)** The reference angle for $359^\circ$ is $360^\circ - 359^\circ = 1^\circ$.

**5. (a)** The reference angle for $\frac{11\pi}{4}$ is $3\pi - \frac{11\pi}{4} = \frac{\pi}{4}$.
**(b)** The reference angle for $-\frac{11\pi}{6}$ is $2\pi - \frac{11\pi}{6} = \frac{\pi}{6}$.
**(c)** The reference angle for $\frac{11\pi}{3}$ is $4\pi - \frac{11\pi}{3} = \frac{\pi}{3}$.

**6. (a)** The reference angle for $\frac{4\pi}{3}$ is $\frac{4\pi}{3} - \pi = \frac{\pi}{3}$.
**(b)** The reference angle for $\frac{33\pi}{4}$ is $\frac{33\pi}{4} - 8\pi = \frac{\pi}{4}$.
**(c)** The reference angle for $-\frac{23\pi}{6}$ is $4\pi - \frac{23\pi}{6} = \frac{\pi}{6}$.

**7. (a)** The reference angle for $\frac{5\pi}{7}$ is $\pi - \frac{5\pi}{7} = \frac{2\pi}{7}$.
**(b)** The reference angle for $-1.4\pi$ is $1.4\pi - \pi = 0.4\pi$.
**(c)** The reference angle for 1.4 is 1.4 because $1.4 < \frac{\pi}{2}$.

**8. (a)** The reference angle for $2.3\pi$ is $2.3\pi - 2\pi = 0.3\pi$.
**(b)** The reference angle for 2.3 is $\pi - 2.3 \approx 0.84$.
**(c)** The reference angle for $-10\pi$ is $10\pi - 10\pi = 0$.

**9.** $\sin 150^\circ = \sin 30^\circ = \frac{1}{2}$

**10.** $\sin 225^\circ = -\sin 45^\circ = -\frac{1}{\sqrt{2}} = -\frac{\sqrt{2}}{2}$

**11.** $\cos 135^\circ = -\cos 45^\circ = -\frac{1}{\sqrt{2}} = -\frac{\sqrt{2}}{2}$

**12.** $\cos(-60^\circ) = \cos 60^\circ = \frac{1}{2}$

**13.** $\tan(-60^\circ) = -\tan 60^\circ = -\sqrt{3}$

**14.** $\sec 300^\circ = \sec 60^\circ = \frac{1}{\cos 60^\circ} = 2$

**15.** $\csc(-630^\circ) = \csc 90^\circ = \frac{1}{\sin 90^\circ} = 1$

**16.** $\cot 210^\circ = \cot 30^\circ = \frac{1}{\tan 30^\circ} = \sqrt{3}$

**17.** $\cos 570^\circ = -\cos 30^\circ = -\frac{\sqrt{3}}{2}$

**18.** $\sec 120^\circ = -\sec 60^\circ = -\frac{1}{\cos 60^\circ} = -2$

**19.** $\tan 750^\circ = \tan 30^\circ = \frac{1}{\sqrt{3}} = \frac{\sqrt{3}}{3}$

**20.** $\cos 660^\circ = \cos 60^\circ = \frac{1}{2}$

**21.** $\sin\frac{2\pi}{3} = \sin\frac{\pi}{3} = \frac{\sqrt{3}}{2}$

**22.** $\sin\frac{5\pi}{3} = -\sin\frac{\pi}{3} = -\frac{\sqrt{3}}{2}$

**23.** $\sin\frac{3\pi}{2} = -\sin\frac{\pi}{2} = -1$

**24.** $\cos\frac{7\pi}{3} = \cos\frac{\pi}{3} = \frac{1}{2}$

**25.** $\cos\left(-\frac{7\pi}{3}\right) = \cos\frac{\pi}{3} = \frac{1}{2}$

**26.** $\tan\frac{5\pi}{6} = -\tan\frac{\pi}{6} = -\frac{1}{\sqrt{3}} = -\frac{\sqrt{3}}{3}$

**27.** $\sec\frac{17\pi}{3} = \sec\frac{\pi}{3} = \dfrac{1}{\cos\frac{\pi}{3}} = 2$

**28.** $\csc\frac{5\pi}{4} = -\csc\frac{\pi}{4} = -\dfrac{1}{\sin\frac{\pi}{4}} = -\sqrt{2}$

**29.** $\cot\left(-\frac{\pi}{4}\right) = -\cot\frac{\pi}{4} = \dfrac{-1}{\tan\frac{\pi}{4}} = -1$

**30.** $\cos\frac{7\pi}{4} = \cos\frac{\pi}{4} = \frac{1}{\sqrt{2}} = \frac{\sqrt{2}}{2}$

**31.** $\tan\frac{5\pi}{2} = \tan\frac{\pi}{2}$ which is undefined.

**32.** $\sin\frac{11\pi}{6} = -\sin\frac{\pi}{6} = -\frac{1}{2}$

**33.** Since $\sin\theta < 0$ and $\cos\theta < 0$, $\theta$ is in quadrant III.

**34.** Since both $\tan\theta$ and $\sin\theta$ are negative, $\theta$ is in quadrant IV.

**35.** $\sec\theta > 0 \;\Rightarrow\; \cos\theta > 0$. Also $\tan\theta < 0 \;\Rightarrow\; \dfrac{\sin\theta}{\cos\theta} < 0 \quad\Leftrightarrow\quad \sin\theta < 0$ (since $\cos\theta > 0$). Since $\sin\theta < 0$ and $\cos\theta > 0$, $\theta$ is in quadrant IV.

**36.** Since $\csc\theta > 0 \;\Rightarrow\; \sin\theta > 0$ and $\cos\theta < 0$, $\theta$ is in quadrant II.

**37.** $\sec^2\theta = 1 + \tan^2\theta \quad\Leftrightarrow\quad \tan^2\theta = \dfrac{1}{\cos^2\theta} - 1 \quad\Leftrightarrow$

$\tan\theta = \sqrt{\dfrac{1}{\cos^2\theta} - 1} = \sqrt{\dfrac{1-\cos^2\theta}{\cos^2\theta}} = \dfrac{\sqrt{1-\cos^2\theta}}{|\cos\theta|} = \dfrac{\sqrt{1-\cos^2\theta}}{-\cos\theta}$ (since $\cos\theta < 0$ in quadrant III, $|\cos\theta| = -\cos\theta$). Thus $\tan\theta = -\dfrac{\sqrt{1-\cos^2\theta}}{\cos\theta}$.

**38.** $\cot\theta = \dfrac{\cos\theta}{\sin\theta} = \dfrac{-\sqrt{1-\sin^2\theta}}{\sin\theta}$ because $\cos\theta < 0$ in quadrant II.

**39.** $\cos^2\theta + \sin^2\theta = 1 \quad\Leftrightarrow\quad \cos\theta = \sqrt{1-\sin^2\theta}$ because $\cos\theta > 0$ in quadrant IV.

**40.** $\sec\theta = \dfrac{1}{\cos\theta} = \dfrac{1}{\sqrt{1-\sin^2\theta}}$ because all trigonometric functions are positive in quadrant I.

**41.** $\sec^2\theta = 1 + \tan^2\theta \quad\Leftrightarrow\quad \sec\theta = -\sqrt{1+\tan^2\theta}$ because $\sec\theta < 0$ in quadrant II.

**42.** $\csc^2\theta = 1 + \cot^2\theta \quad\Leftrightarrow\quad \csc\theta = -\sqrt{1+\cot^2\theta}$ because $\csc\theta < 0$ in quadrant III.

**43.** $\sin\theta = \frac{3}{5}$. Then $x = -\sqrt{5^2-3^2} = -\sqrt{16} = -4$, since $\theta$ is in quadrant II. Thus, $\cos\theta = -\frac{4}{5}$, $\tan\theta = -\frac{3}{4}$, $\csc\theta = \frac{5}{3}$, $\sec\theta = -\frac{5}{4}$, and $\cot\theta = -\frac{4}{3}$.

**44.** $\cos\theta = -\frac{7}{12}$. Since $\theta$ is in quadrant III, $y = -\sqrt{12^2-7^2} = -\sqrt{95}$, and so $\sin\theta = -\frac{\sqrt{95}}{12}$, $\tan\theta = \frac{\sqrt{95}}{7}$, $\csc\theta = -\frac{12}{\sqrt{95}}$, $\sec\theta = -\frac{12}{7}$, and $\cot\theta = \frac{7}{\sqrt{95}}$.

**45.** $\tan\theta = -\frac{3}{4}$. Then $r = \sqrt{3^2+4^2} = 5$, and so $\sin\theta = -\frac{3}{5}$, $\cos\theta = \frac{4}{5}$, $\csc\theta = -\frac{5}{3}$, $\sec\theta = \frac{5}{4}$, and $\cot\theta = -\frac{4}{3}$.

**46.** $\sec\theta = 5$. Then $\cos\theta = \frac{1}{5}$ and $y = -\sqrt{5^2-1^2} = -2\sqrt{6}$. So $\sin\theta = -\frac{2\sqrt{6}}{5}$, $\cos\theta = \frac{1}{5}$, $\tan\theta = -2\sqrt{6}$, $\csc\theta = -\frac{5}{2\sqrt{6}}$, and $\cot\theta = -\frac{1}{2\sqrt{6}}$.

**47.** $\csc\theta = 2$. Then $\sin\theta = \frac{1}{2}$ and $x = \sqrt{2^2-1^2} = \sqrt{3}$. So $\sin\theta = \frac{1}{2}$, $\cos\theta = \frac{\sqrt{3}}{2}$, $\tan\theta = \frac{1}{\sqrt{3}} = \frac{\sqrt{3}}{3}$, $\sec\theta = \frac{2}{\sqrt{3}} = \frac{2\sqrt{3}}{3}$, and $\cot\theta = \sqrt{3}$.

**48.** $\cot\theta = \frac{1}{4}$. Then $\tan\theta = 4$ and $r = \sqrt{4^2+1^2} = \sqrt{17}$. So $\sin\theta = -\frac{4}{\sqrt{17}}$, $\cos\theta = -\frac{1}{\sqrt{17}}$, $\tan\theta = 4$, $\csc\theta = -\frac{\sqrt{17}}{4}$, and $\sec\theta = -\sqrt{17}$.

**49.** $\cos\theta = -\frac{2}{7}$. Then $y = \sqrt{7^2-2^2} = \sqrt{45} = 3\sqrt{5}$, and so $\sin\theta = \frac{3\sqrt{5}}{7}$, $\tan\theta = -\frac{3\sqrt{5}}{2}$, $\csc\theta = \frac{7}{3\sqrt{5}} = \frac{7\sqrt{5}}{15}$, $\sec\theta = -\frac{7}{2}$, and $\cot\theta = -\frac{2}{3\sqrt{5}} = -\frac{2\sqrt{5}}{15}$.

**50.** $\tan\theta = -4$. Then $r = \sqrt{4^2+1^2} = \sqrt{17}$, and so $\sin\theta = \frac{4}{\sqrt{17}}$, $\cos\theta = -\frac{1}{\sqrt{17}}$, $\csc\theta = \frac{\sqrt{17}}{4}$, $\sec\theta = -\sqrt{17}$, and $\cot\theta = -\frac{1}{4}$.

**51.** **(a)** $\sin 2\theta = \sin\left(2\cdot\frac{\pi}{3}\right) = \sin\frac{2\pi}{3} = \sin\frac{\pi}{3} = \frac{\sqrt{3}}{2}$, while $2\sin\theta = 2\sin\frac{\pi}{3} = 2\cdot\frac{\sqrt{3}}{2} = \sqrt{3}$.

**(b)** $\sin\frac{1}{2}\theta = \sin\left(\frac{1}{2}\cdot\frac{\pi}{3}\right) = \sin\frac{\pi}{6} = \frac{1}{2}$, while $\frac{1}{2}\sin\theta = \frac{1}{2}\sin\frac{\pi}{3} = \frac{1}{2}\cdot\frac{\sqrt{3}}{2} = \frac{\sqrt{3}}{4}$.

**(c)** $\sin^2\theta = \left(\sin\frac{\pi}{3}\right)^2 = \left(\frac{\sqrt{3}}{2}\right)^2 = \frac{3}{4}$, while $\sin\left(\theta^2\right) = \sin\left(\frac{\pi}{3}\right)^2 = \sin\frac{\pi^2}{9} \approx 0.88967$.

**52.** $a = 7$, $b = 9$, and $\theta = 72^\circ$. Thus, $A = \frac{1}{2}(7)(9)\sin 72^\circ = \frac{63}{2}\sin 72^\circ \approx 30.0$.

**53.** $a = 10$, $b = 22$, and $\theta = 10^\circ$. Thus, the area of the triangle is $A = \frac{1}{2}(10)(22)\sin 10^\circ = 110\sin 10^\circ \approx 19.1$.

**54.** Since the triangle is equilateral, $\theta = 60^\circ$ and $a = b = 10$. Thus, $A = \frac{1}{2}(10)^2\sin 60^\circ = 25\sqrt{3} \approx 43.3$.

**55.** $A = 16$, $a = 5$, and $b = 7$. So $\sin\theta = \dfrac{2A}{ab} = \dfrac{2 \cdot 16}{5 \cdot 7} = \dfrac{32}{35} \quad\Leftrightarrow\quad \theta = \sin^{-1}\frac{32}{35} \approx 66.1^\circ$.

**56.** Let the length of each of the equal sides be $x$. Then $A = \frac{1}{2}x^2 \sin\theta \Rightarrow x = \sqrt{\dfrac{2A}{\sin\theta}}$. Substituting, we find

$$x = \sqrt{\frac{48}{\sin\frac{5\pi}{6}}} = 4\sqrt{6} \approx 9.8 \text{ cm}.$$

**57.** For the sector defined by the two sides, $A_1 = \frac{1}{2}r^2\theta = \frac{1}{2} \cdot 2^2 \cdot 120^\circ \cdot \frac{\pi}{180^\circ} = \frac{4\pi}{3}$. For the triangle defined by the two sides, $A_2 = \frac{1}{2}ab\sin\theta = \frac{1}{2} \cdot 2 \cdot 2 \cdot \sin 120^\circ = 2\sin 60^\circ = \sqrt{3}$. Thus the area of the region is $A_1 - A_2 = \frac{4\pi}{3} - \sqrt{3} \approx 2.46$.

**58.** The area of the entire circle is $\pi r^2 = \pi \cdot 12^2 = 144\pi$, the area of the sector is $\frac{1}{2}r^2\theta = \frac{1}{2} \cdot 12^2 \cdot \frac{\pi}{3} = 24\pi$, and the area of the triangle is $\frac{1}{2}ab\sin\theta = \frac{1}{2} \cdot 12^2 \cdot \sin\frac{\pi}{3} = 36\sqrt{3}$, so the area of the shaded region is $A_1 - A_2 + A_3 = 144\pi - 24\pi + 36\sqrt{3} = 120\pi + 36\sqrt{3} \approx 439.3$.

**59.** $\sin^2\theta + \cos^2\theta = 1 \quad\Leftrightarrow\quad (\sin^2\theta + \cos^2\theta) \cdot \dfrac{1}{\cos^2\theta} = 1 \cdot \dfrac{1}{\cos^2\theta} \quad\Leftrightarrow\quad \tan^2\theta + 1 = \sec^2\theta$

**60.** $\sin^2\theta + \cos^2\theta = 1 \quad\Leftrightarrow\quad (\sin^2\theta + \cos^2\theta) \cdot \dfrac{1}{\sin^2\theta} = 1 \cdot \dfrac{1}{\sin^2\theta} \quad\Leftrightarrow\quad 1 + \cot^2\theta = \csc^2\theta$

**61. (a)** $\tan\theta = \dfrac{h}{1 \text{ mile}}$, so $h = \tan\theta \cdot 1 \text{ mile} \cdot \dfrac{5280 \text{ ft}}{1 \text{ mile}} = 5280\tan\theta$ ft.

**(b)**

| $\theta$ | $20^\circ$ | $60^\circ$ | $80^\circ$ | $85^\circ$ |
|---|---|---|---|---|
| $h$ | 1922 | 9145 | 29,944 | 60,351 |

**62. (a)** Let the depth of the water be $h$. Then the cross-sectional area of the gutter is a trapezoid whose height is $h = 10\sin\theta$. The bases are 10 and $10 + 2(10\cos\theta) = 10 + 20\cos\theta$. Thus, the area is $A(\theta) = \dfrac{b_1 + b_2}{2} \cdot h = \dfrac{10 + 10 + 20\cos\theta}{2} \cdot 10\sin\theta = 100\sin\theta + 100\sin\theta\cos\theta$.

**(b)**

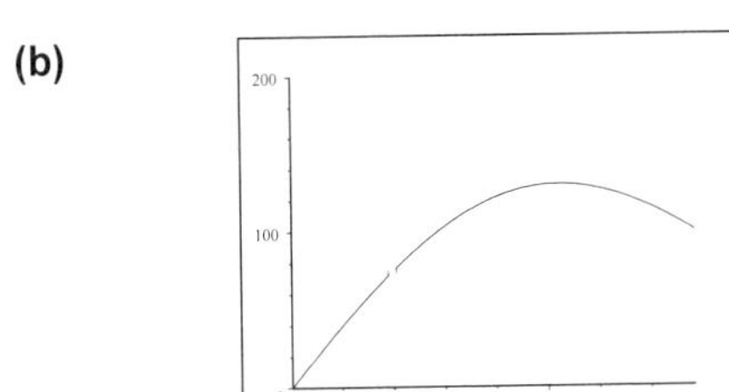


**(c)**

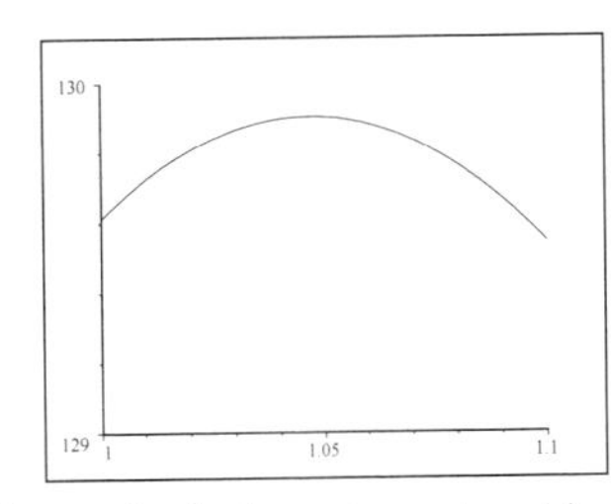


From the graph, the largest area is achieved when $\theta \approx 1.047$ rad $\approx 60^\circ$.

**63. (a)** From the figure in the text, we express depth and width in terms of $\theta$. Since $\sin\theta = \dfrac{\text{depth}}{20}$ and $\cos\theta = \dfrac{\text{width}}{20}$, we have depth $= 20\sin\theta$ and width $= 20\cos\theta$. Thus, the cross-section area of the beam is $A(\theta) = (\text{depth})(\text{width}) = (20\cos\theta)(20\sin\theta) = 400\cos\theta\sin\theta$.

**(b)**

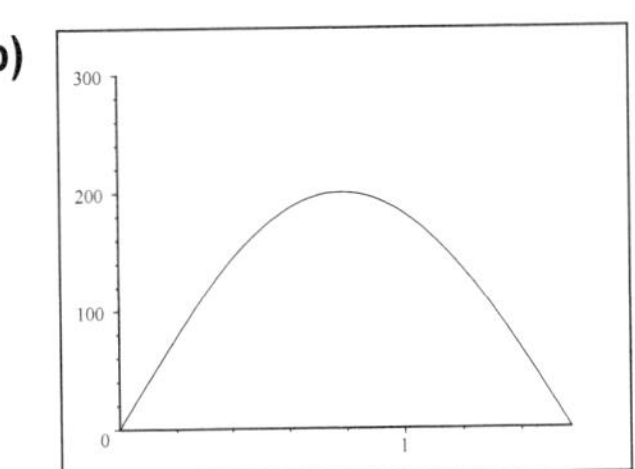


**(c)** The beam with the largest cross-sectional area is the square beam, $10\sqrt{2}$ by $10\sqrt{2}$ (about 14.14 by 14.14).

**64.** Using depth $= 20\sin\theta$ and width $= 20\cos\theta$, from Exercise 63, we have strength $= k(\text{width})(\text{depth})^2$. Thus $S(\theta) = k(20\cos\theta)(20\sin\theta)^2 = 8000k\cos\theta\sin^2\theta$.

**65. (a)** On Earth, the range is $R = \dfrac{v_0^2 \sin(2\theta)}{g} = \dfrac{12^2 \sin\frac{\pi}{3}}{32} = \dfrac{9\sqrt{3}}{4} \approx 3.897$ ft and the height is $H = \dfrac{v_0^2 \sin^2\theta}{2g} = \dfrac{12^2 \sin^2\frac{\pi}{6}}{2 \cdot 32} = \dfrac{9}{16} = 0.5625$ ft.

**(b)** On the moon, $R = \dfrac{12^2 \sin\frac{\pi}{3}}{5.2} \approx 23.982$ ft and $H = \dfrac{12^2 \sin^2\frac{\pi}{6}}{2 \cdot 5.2} \approx 3.462$ ft

**66.** Substituting, $t = \sqrt{\dfrac{2000}{16 \sin 30^\circ}} = 5\sqrt{10} \approx 15.8$ s.

**67. (a)** $W = 3.02 - 0.38 \cot\theta + 0.65 \csc\theta$

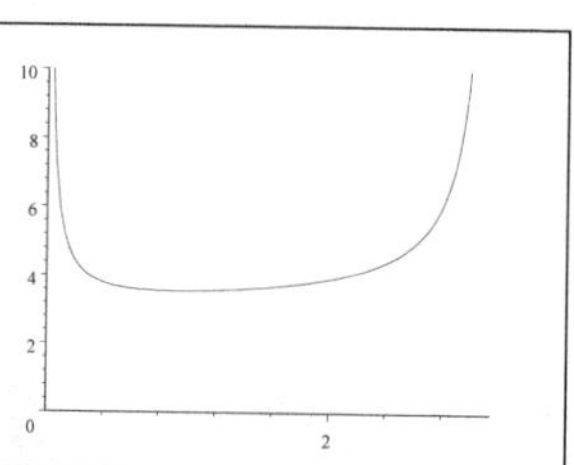

**(b)** From the graph, it appears that $W$ has its minimum value at about $\theta = 0.946 \approx 54.2^\circ$.

**68. (a)** We label the lengths $L_1$ and $L_2$ as shown in the figure. Since $\sin\theta = \dfrac{9}{L_1}$ and $\cos\theta = \dfrac{6}{L_2}$, we have $L_1 = 9\csc\theta$ and $L_2 = 6\sec\theta$. Thus $L(\theta) = L_1 + L_2 = 9\csc\theta + 6\sec\theta$.

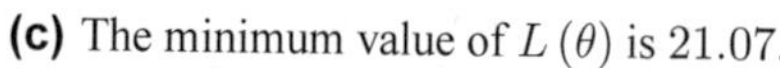

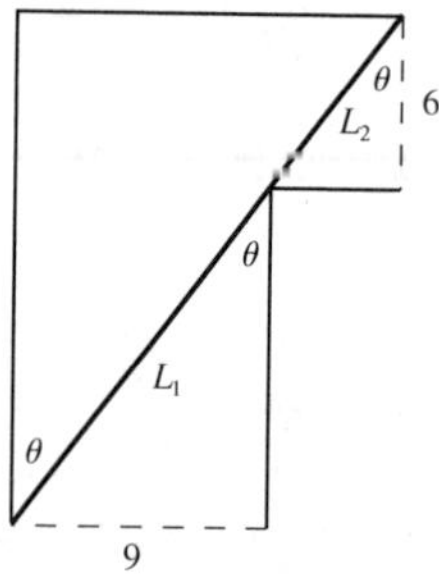


**(b)**

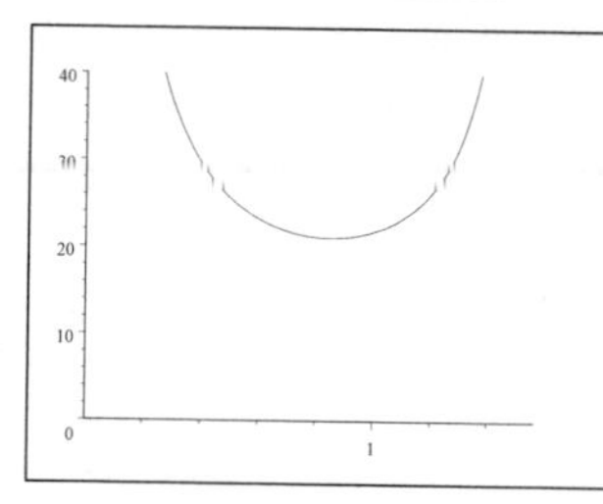

**(c)** The minimum value of $L(\theta)$ is 21.07.

**(d)** The minimum value of $L(\theta)$ is the shortest distance the pipe must pass through.

**69.** We have $\sin\alpha = k \sin\beta$, where $\alpha = 59.4^\circ$ and $k = 1.33$. Substituting, $\sin 59.4^\circ = 1.33 \sin\beta$ $\Rightarrow \sin\beta = \dfrac{\sin 59.4^\circ}{1.33} \approx 0.6472$. Using a calculator, we find that $\beta \approx \sin^{-1} 0.6472 \approx 40.3^\circ$, so $\theta = 4\beta - 2\alpha \approx 4(40.3^\circ) - 2(59.4^\circ) = 42.4^\circ$.

**70.** $\sin 4^\circ = 0.0697564737$ but $\sin 4 = -0.7568024953$. Your partner has found $\sin 4^\circ$ instead of $\sin 4$ radians.

**71.** $\cos\theta = \dfrac{\text{adj}}{\text{hyp}} = \dfrac{|OP|}{|OR|} = \dfrac{|OP|}{1} = |OP|$. Since $QS$ is tangent to the circle at $R$, $\triangle ORQ$ is a right triangle. Then $\tan\theta = \dfrac{\text{opp}}{\text{adj}} = \dfrac{|RQ|}{|OR|} = |RQ|$ and $\sec\theta = \dfrac{\text{hyp}}{\text{adj}} = \dfrac{|OQ|}{|OR|} = |OQ|$. Since $\angle SOQ$ is a right angle $\triangle SOQ$ is a right triangle and $\angle OSR = \theta$. Then $\csc\theta = \dfrac{\text{hyp}}{\text{opp}} = \dfrac{|OS|}{|OR|} = |OS|$ and $\cot\theta = \dfrac{\text{adj}}{\text{opp}} = \dfrac{|SR|}{|OR|} = |SR|$. Summarizing, we have $\sin\theta = |PR|$, $\cos\theta = |OP|$, $\tan\theta = |RQ|$, $\sec\theta = |OQ|$, $\csc\theta = |OS|$, and $\cot\theta = |SR|$.

## 6.4 The Law of Sines

**1.** $\angle C = 180^\circ - 98.4^\circ - 24.6^\circ = 57^\circ$. $x = \dfrac{376 \sin 57^\circ}{\sin 98.4^\circ} \approx 318.75$.

**2.** $\angle C = 180^\circ - 37.5^\circ - 28.1^\circ = 114.4^\circ$. $x = \dfrac{17 \sin 114.4^\circ}{\sin 37.5^\circ} \approx 25.4$.

**3.** $\angle C = 180^\circ - 52^\circ - 70^\circ = 58^\circ$. $x = \dfrac{26.7 \sin 52^\circ}{\sin 58^\circ} \approx 24.8$.

**4.** $\sin\theta = \dfrac{56.3 \sin 67^\circ}{80.2} \approx 0.646$. Then $\theta \approx \sin^{-1} 0.646 \approx 40.3^\circ$.

**5.** $\sin C = \dfrac{36\sin 120^\circ}{45} \approx 0.693 \Leftrightarrow \angle C \approx \sin^{-1} 0.693 \approx 44^\circ$.

**6.** $\angle C = 180^\circ - 102^\circ - 28^\circ = 50^\circ$. $x = \dfrac{185\sin 50^\circ}{\sin 102^\circ} \approx 144.9$.

**7.** $\angle C = 180^\circ - 46^\circ - 20^\circ = 114^\circ$. Then $a = \dfrac{65\sin 46^\circ}{\sin 114^\circ} \approx 51$ and $b = \dfrac{65\sin 20^\circ}{\sin 114^\circ} \approx 24$.

**8.** $\angle B = 180^\circ - 30^\circ - 100^\circ = 50^\circ$. Then $c = \dfrac{2\sin 100^\circ}{\sin 50^\circ} \approx 2.57$ and $a = \dfrac{2\sin 30^\circ}{\sin 50^\circ} \approx 1.31$.

**9.** $\angle B = 68^\circ$, so $\angle A = 180^\circ - 68^\circ - 68^\circ = 44^\circ$ and $a = \dfrac{12\sin 44^\circ}{\sin 68^\circ} \approx 8.99$.

**10.** $\sin B = \dfrac{3.4\sin 80^\circ}{6.5} \approx 0.515$, so $\angle B = \sin^{-1} 0.515 \approx 31^\circ$. Then $\angle C = 180^\circ - 80^\circ - 31^\circ = 69^\circ$ and $c = \dfrac{6.5\sin 69^\circ}{\sin 80^\circ} \approx 6.2$.

**11.** $\angle C = 180^\circ - 50^\circ - 68^\circ = 62^\circ$. Then $a = \dfrac{230\sin 50^\circ}{\sin 62^\circ} \approx 200$ and $b = \dfrac{230\sin 68^\circ}{\sin 62^\circ} \approx 242$.

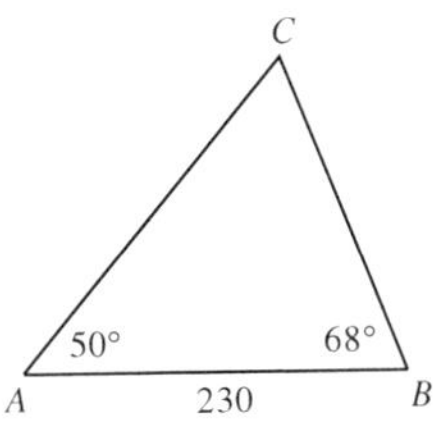


**12.** $\angle C = 180^\circ - 110^\circ - 23^\circ = 47^\circ$. Then $a = \dfrac{50\sin 23^\circ}{\sin 47^\circ} \approx 26.7$ and $b = \dfrac{50\sin 110^\circ}{\sin 47^\circ} \approx 64.2$.

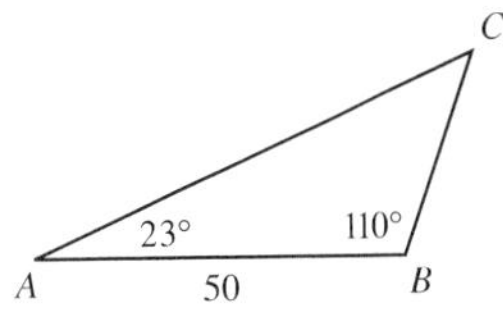


**13.** $\angle B = 180^\circ - 30^\circ - 65^\circ = 85^\circ$. Then $a = \dfrac{10\sin 30^\circ}{\sin 85^\circ} \approx 5.0$ and $c = \dfrac{10\sin 65^\circ}{\sin 85^\circ} \approx 9$.

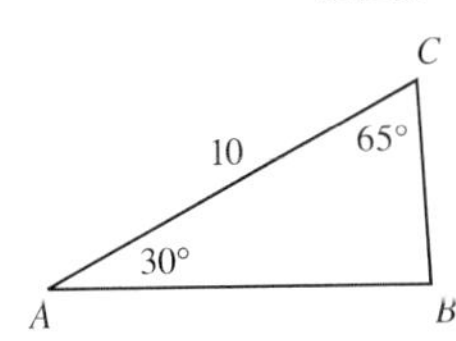


**14.** $\angle C = 180^\circ - 95^\circ - 22^\circ = 63^\circ$. Then $b = \dfrac{420\sin 95^\circ}{\sin 22^\circ} \approx 1116.9$ and $c = \dfrac{420\sin 63^\circ}{\sin 22^\circ} \approx 999.0$.

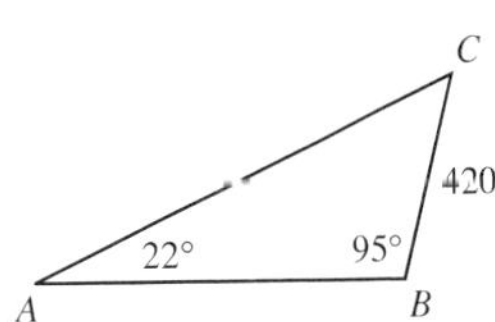


**15.** $\angle A = 180^\circ - 51^\circ - 29^\circ = 100^\circ$. Then $a = \dfrac{44\sin 100^\circ}{\sin 29^\circ} \approx 89$ and $c = \dfrac{44\sin 51^\circ}{\sin 29^\circ} \approx 71$.

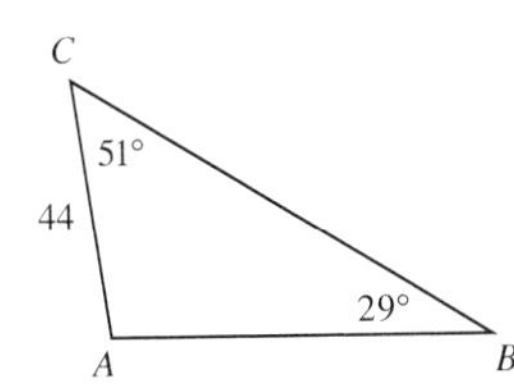


**16.** $\angle A = 180^\circ - 100^\circ - 10^\circ = 70^\circ$. Then $a = \dfrac{115\sin 70^\circ}{\sin 100^\circ} \approx 109.7$ and $b = \dfrac{115\sin 10^\circ}{\sin 100^\circ} \approx 20.3$.

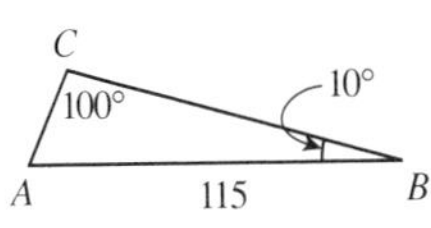


**17.** Since $\angle A > 90^\circ$ there is only one triangle. $\sin B = \dfrac{15\sin 110^\circ}{28} \approx 0.503 \quad \Leftrightarrow \quad \angle B \approx \sin^{-1} 0.503 \approx 30^\circ$. Then $\angle C \approx 180^\circ - 110^\circ - 30^\circ = 40^\circ$, and so $c = \dfrac{28\sin 40^\circ}{\sin 110^\circ} \approx 19$. Thus $\angle B \approx 30^\circ$, $\angle C \approx 40^\circ$, and $c \approx 19$.

**18.** $\sin C = \dfrac{40 \sin 37^\circ}{30} \approx 0.802 \Leftrightarrow \angle C_1 \approx \sin^{-1} 0.822 \approx 53.4^\circ$ or $\angle C_2 \approx 180^\circ - 53.4^\circ \approx 126.6^\circ$.

If $\angle C_1 \approx 53.4^\circ$, then $\angle B_1 \approx 180^\circ - 37^\circ - 53.4^\circ = 89.6^\circ$ and $b_1 = \dfrac{30 \sin 89.6^\circ}{\sin 37^\circ} \approx 49.8$.

If $\angle C_2 \approx 126.6^\circ$, then $\angle B_2 \approx 180^\circ - 37^\circ - 126.6^\circ = 16.4^\circ$ and $b_2 = \dfrac{30 \sin 16.4^\circ}{\sin 37^\circ} \approx 14.1$.

Thus, one triangle has $\angle B_1 \approx 89.6^\circ$, $\angle C_1 \approx 53.4^\circ$, and $b_1 \approx 49.8$; the other has $\angle B_2 \approx 16.4^\circ$, $\angle C_2 \approx 126.6^\circ$, and $b_2 \approx 14.1$.

**19.** $\angle A = 125^\circ$ is the largest angle, but since side $a$ is not the longest side, there can be no such triangle.

**20.** $\sin B = \dfrac{45 \sin 38^\circ}{42} \approx 0.660 \Leftrightarrow \angle B_1 \approx \sin^{-1} 0.660 \approx 41.3^\circ$ or $\angle B_2 \approx 180^\circ - 41.3^\circ \approx 138.7^\circ$.

If $\angle B_1 \approx 41.3^\circ$, then $\angle A_1 \approx 180^\circ - 38^\circ - 41.3^\circ = 100.7^\circ$ and $a_1 = \dfrac{42 \sin 100.7^\circ}{\sin 38^\circ} \approx 67$.

If $\angle B_2 \approx 138.7^\circ$, then $\angle A_2 \approx 180^\circ - 38^\circ - 138.7^\circ = 3.3^\circ$ and $a_2 = \dfrac{42 \sin 3.3^\circ}{\sin 38^\circ} \approx 3.9$.

Thus, one triangle has $\angle A_1 \approx 100.7^\circ$, $\angle B_1 \approx 41.3^\circ$, and $a_1 \approx 67$; the other has $\angle A_2 \approx 3.3^\circ$, $\angle B_2 \approx 138.7^\circ$, and $a_2 \approx 3.9$.

**21.** $\sin C = \dfrac{30 \sin 25^\circ}{25} \approx 0.507 \Leftrightarrow \angle C_1 \approx \sin^{-1} 0.507 \approx 30.47^\circ$ or $\angle C_2 \approx 180^\circ - 30.47^\circ = 149.53^\circ$.

If $\angle C_1 = 30.47^\circ$, then $\angle A_1 \approx 180^\circ - 25^\circ - 30.47^\circ = 124.53^\circ$ and $a_1 = \dfrac{25 \sin 124.53^\circ}{\sin 25^\circ} \approx 48.73$.

If $\angle C_2 = 149.53^\circ$, then $\angle A_2 \approx 180^\circ - 25^\circ - 149.53^\circ = 5.47^\circ$ and $a_2 = \dfrac{25 \sin 5.47^\circ}{\sin 25^\circ} \approx 5.64$.

Thus, one triangle has $\angle A_1 \approx 125^\circ$, $\angle C_1 \approx 30^\circ$, and $a_1 \approx 49$; the other has $\angle A_2 \approx 5^\circ$, $\angle C_2 \approx 150^\circ$, and $a_2 \approx 5.6$.

**22.** $\sin B = \dfrac{100 \sin 30^\circ}{75} = \frac{2}{3} \quad \Leftrightarrow \quad \angle B_1 \approx \sin^{-1} \frac{2}{3} \approx 41.8^\circ$ or $\angle B_2 \approx 180^\circ - 41.8^\circ = 138.2^\circ$.

If $\angle B_1 \approx 41.8^\circ$, then $\angle C_1 \approx 180^\circ - 30^\circ - 41.8^\circ = 108.2^\circ$ and $c_1 \approx \dfrac{75 \sin 108.2^\circ}{\sin 30^\circ} \approx 142.5$.

If $\angle B_2 \approx 138.2^\circ$, then $\angle C_2 \approx 180^\circ - 30^\circ - 138.2^\circ = 11.8^\circ$ and $c_2 \approx \dfrac{75 \sin 11.8^\circ}{\sin 30^\circ} \approx 30.7$.

Thus, one triangle has $\angle B_1 \approx 41.8^\circ$, $\angle C_1 \approx 108.2^\circ$, and $c_1 \approx 142.5$; the other has $\angle B_2 \approx 138.2^\circ$, $\angle C_2 \approx 11.8^\circ$, and $c_2 \approx 30.7$.

**23.** $\sin B = \dfrac{100 \sin 50^\circ}{50} \approx 1.532$. Since $|\sin \theta| \leq 1$ for all $\theta$, there can be no such angle $B$, and thus no such triangle.

**24.** $\sin B = \dfrac{80 \sin 135^\circ}{100} \approx 0.566 \Leftrightarrow \angle B_1 \approx \sin^{-1} 0.566 \approx 34.4^\circ$ or $\angle B_2 \approx 180^\circ - 34.4^\circ = 145.6^\circ$.

If $\angle B_1 \approx 34.4^\circ$, then $\angle C \approx 180^\circ - 135^\circ - 34.4^\circ = 10.6^\circ$ and $c \approx \dfrac{100 \sin 10.6^\circ}{\sin 135^\circ} \approx 25.9$.

If $\angle B_2 \approx 180^\circ - 34.4^\circ = 145.6^\circ$, then $\angle A + \angle B_2 = 135^\circ + 145.6^\circ > 180^\circ$, so there is no such triangle.

Thus, the only possible triangle is $\angle B \approx 34.4^\circ$, $\angle C \approx 10.6^\circ$, and $c \approx 25.9$.

**25.** $\sin A = \dfrac{26 \sin 29^\circ}{15} \approx 0.840 \Leftrightarrow \angle A_1 \approx \sin^{-1} 0.840 \approx 57.2^\circ$ or $\angle A_2 \approx 180^\circ - 57.2^\circ = 122.8^\circ$.

If $\angle A_1 \approx 57.2^\circ$, then $\angle B_1 = 180^\circ - 29^\circ - 57.2^\circ = 93.8^\circ$ and $b_1 \approx \dfrac{15 \sin 93.8^\circ}{\sin 29^\circ} \approx 30.9$.

If $\angle A_2 \approx 122.8^\circ$, then $\angle B_2 = 180 - 29^\circ - 122.8^\circ = 28.2^\circ$ and $b_2 \approx \dfrac{15 \sin 28.1^\circ}{\sin 29^\circ} \approx 14.6$.

Thus, one triangle has $\angle A_1 \approx 57.2^\circ$, $\angle B_1 \approx 93.8^\circ$, and $b_1 \approx 30.9$; the other has $\angle A_2 \approx 122.8^\circ$, $\angle B_2 \approx 28.2^\circ$, and $b_2 \approx 14.6$.

**26.** $\sin C = \dfrac{82 \sin 58^\circ}{73} \approx 0.953$, so $\angle C_1 \approx \sin^{-1} 0.953 \approx 72.4^\circ$ or $\angle C_2 \approx 180 - 72.4^\circ = 107.6^\circ$.

If $\angle C_1 \approx 72.4^\circ$, then $\angle A_1 \approx 180^\circ - 58^\circ - 72.4^\circ = 49.6^\circ$ and $a_1 \approx \dfrac{73 \sin 49.6^\circ}{\sin 58^\circ} \approx 65.6$.

If $\angle C_2 \approx 107.6^\circ$, then $\angle A_2 \approx 180^\circ - 58^\circ - 107.6^\circ = 14.4^\circ$ and $a_2 \approx \dfrac{73 \sin 14.4^\circ}{\sin 58^\circ} \approx 21.4$.

Thus, one triangle has $\angle A_1 \approx 49.6^\circ$, $\angle C_1 \approx 72.4^\circ$, and $a_1 \approx 65.6$; the other has $\angle A_2 \approx 14.4^\circ$, $\angle C_2 \approx 107.6^\circ$, and $a_2 \approx 21.4$.

**27. (a)** From $\triangle ABC$ and the Law of Sines we get $\dfrac{\sin 30^\circ}{20} = \dfrac{\sin B}{28} \quad \Leftrightarrow \quad \sin B = \dfrac{28 \sin 30^\circ}{20} = 0.7$, so $\angle B \approx \sin^{-1} 0.7 \approx 44.427^\circ$. Since $\triangle BCD$ is isosceles, $\angle B = \angle BDC \approx 44.427^\circ$. Thus, $\angle BCD = 180^\circ - 2\angle B \approx 91.146^\circ \approx 91.1^\circ$.

**(b)** From $\triangle ABC$ we get $\angle BCA = 180^\circ - \angle A - \angle B \approx 180^\circ - 30^\circ - 44.427^\circ = 105.573^\circ$. Hence $\angle DCA = \angle BCA - \angle BCD \approx 105.573^\circ - 91.146^\circ = 14.4^\circ$.

**28.** By symmetry, $\angle DCB = 25^\circ$, so $\angle A = 180^\circ - 25^\circ - 50^\circ = 105^\circ$. Then by the Law of Sines, $|AD| = \dfrac{12 \sin 25^\circ}{\sin 105^\circ} \approx 5.25$.

**29. (a)** $\sin B = \dfrac{20 \sin 40^\circ}{15} \approx 0.857 \Leftrightarrow \angle B_1 \approx \sin^{-1} 0.857 \approx 58.99^\circ$ or $\angle B_2 \approx 180^\circ - 58.99^\circ \approx 121.01^\circ$.

If $\angle B_1 = 30.47^\circ$, then $\angle C_1 \approx 180^\circ - 15^\circ - 58.99^\circ = 106.01^\circ$ and $c_1 = \dfrac{15 \sin 106.01^\circ}{\sin 40^\circ} \approx 22.43$.

If $\angle B_2 = 121.01^\circ$, then $\angle C_2 \approx 180^\circ - 15^\circ - 121.01^\circ = 43.99^\circ$ and $c_2 = \dfrac{15 \sin 43.99}{\sin 40^\circ} \approx 16.21$. Thus there are two triangles.

**(b)** By the area formula given in Section 6.3, $\dfrac{\text{Area of } \triangle\ ABC}{\text{Area of } \triangle\ A'B'C'} = \dfrac{\frac{1}{2}ab \sin C}{\frac{1}{2}ab \sin C'} = \dfrac{\sin C}{\sin C'}$, since $a$ and $b$ are the same in both triangles.

**30.** By the formula of Section 6.3, the area of $\triangle ABC$ is $A = \frac{1}{2}ab \sin C$. Since we are given $a$ and the three angles, we need to find $b$ in terms of these. By the Law of Sines, $\dfrac{\sin B}{b} = \dfrac{\sin A}{a} \quad \Leftrightarrow \quad b = \dfrac{a \sin B}{\sin A}$. Thus,
$A = \frac{1}{2}ab \sin C = \frac{1}{2}a \left(\dfrac{a \sin B}{\sin A}\right) \sin C = \dfrac{a^2 \sin B \sin C}{2 \sin A}$.

**31. (a)** Let $a$ be the distance from satellite to the tracking station $A$ in miles. Then the subtended angle at the satellite is $\angle C = 180^\circ - 93^\circ - 84.2^\circ = 2.8^\circ$, and so $a = \dfrac{50 \sin 84.2^\circ}{\sin 2.8^\circ} \approx 1018$ mi.

**(b)** Let $d$ be the distance above the ground in miles. Then $d = 1018.3 \sin 87^\circ \approx 1017$ mi.

**32. (a)** Let $x$ be the distance from the plane to point $A$. Then $\dfrac{x}{AB} = \dfrac{\sin 48^\circ}{\sin(180^\circ - 32^\circ - 48^\circ)} = \dfrac{\sin 48^\circ}{\sin 100^\circ} \Leftrightarrow$

$x = 5 \cdot \dfrac{\sin 48^\circ}{\sin 100^\circ} \approx 3.77$ mi.

**(b)** Let $h$ be the height of the plane. Then $\sin 32^\circ = \dfrac{h}{x} \Rightarrow h = 3.77 \sin 32^\circ \approx 2.00$ mi.

**33.** $\angle C = 180^\circ - 82^\circ - 52^\circ = 46^\circ$, so by the Law of Sines, $\dfrac{|AC|}{\sin 52^\circ} = \dfrac{|AB|}{\sin 46^\circ} \quad \Leftrightarrow \quad |AC| = \dfrac{|AB| \sin 52^\circ}{\sin 46^\circ}$, so substituting we have $|AC| = \dfrac{200 \sin 52^\circ}{\sin 46^\circ} \approx 219$ ft.

**34.** $\sin \angle ABC = \dfrac{312 \sin 48.6^\circ}{527} \approx 0.444 \quad \Leftrightarrow \quad \angle ABC \approx \sin^{-1} 0.444 \approx 26.4^\circ$, and so $\angle BCA \approx 180^\circ - 48.6^\circ - 26.4^\circ = 105^\circ$. Then the distance between $A$ and $B$ is $|AB| = \dfrac{527 \sin 105^\circ}{\sin 48.6^\circ} \approx 678.5$ ft.

**35.** 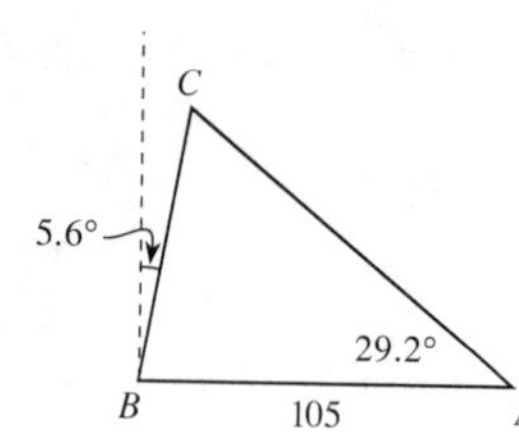


We draw a diagram. $A$ is the position of the tourist and $C$ is the top of the tower. $\angle B = 90° - 5.6° = 84.4°$ and so $\angle C = 180° - 29.2° - 84.4° = 66.4°$. Thus, by the Law of Sines, the length of the tower is $|BC| = \dfrac{105 \sin 29.2°}{\sin 66.4°} \approx 55.9$ m.

**36.** 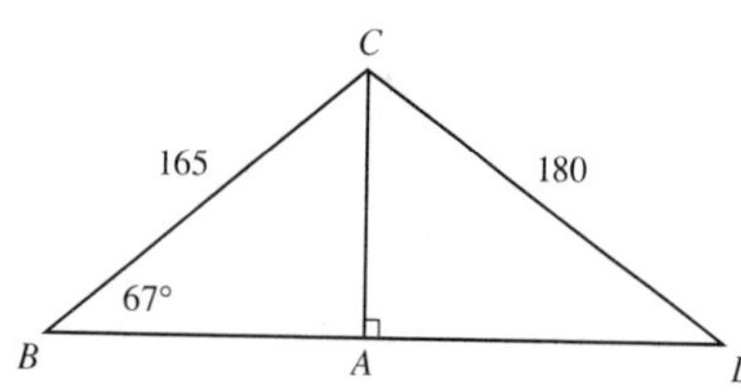


The situation is illustrated in the diagram. $|AC| = 165 \sin 67° \approx 151.9$ ft, so using the Pythagorean Theorem we can calculate $|BA| = \sqrt{165^2 - 151.9^2} \approx 64.4$ ft and $|AD| = \sqrt{180^2 - 151.9^2} \approx 96.6$ ft. Thus the anchor points are $|BA| + |AD| \approx 64.4 + 96.6 = 161$ ft apart.

**37.** The angle subtended by the top of the tree and the sun's rays is $\angle A = 180° - 90° - 52° = 38°$. Thus the height of the tree is $h = \dfrac{215 \sin 30°}{\sin 38°} \approx 175$ ft.

**38.** 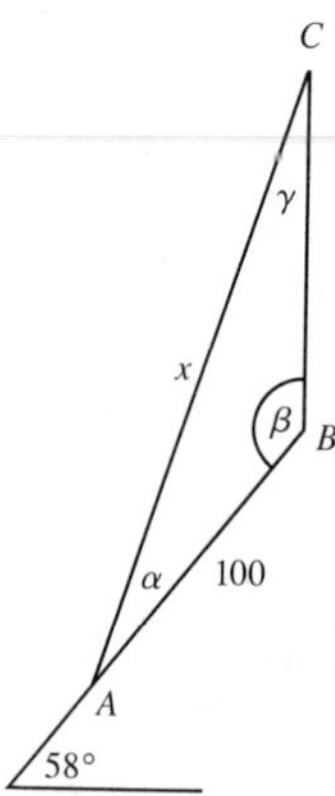


Let $x$ be the length of the wire, as shown in the figure. Since $\alpha = 12°$, other angles in $\Delta ABC$ are $\beta = 90° + 58° = 148°$, and $\gamma = 180° - (12° + 148°) = 20°$, Thus, $\dfrac{x}{\sin 148°} = \dfrac{100}{\sin 20°} \quad \Leftrightarrow \quad x = 100 \cdot \dfrac{\sin 148°}{\sin 20°} \approx 155$ ft.

**39.** Call the balloon's position $R$. Then in $\triangle PQR$, we see that $\angle P = 62° - 32° = 30°$, and $\angle Q = 180° - 71° + 32° = 141°$. Therefore, $\angle R = 180° - 30° - 141° = 9°$. So by the Law of Sines, $\dfrac{|QR|}{\sin 30°} = \dfrac{|PQ|}{\sin 9°} \quad \Leftrightarrow \quad |QR| = 60 \cdot \dfrac{\sin 30°}{\sin 9°} \approx 192$ m.

**40.** 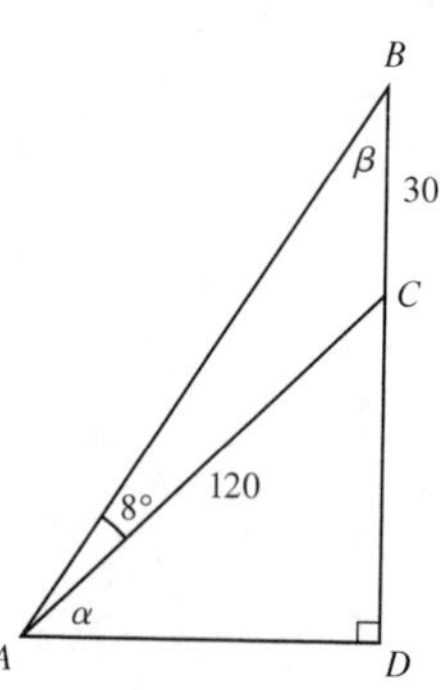


Label the diagram as shown, and let the hill's angle of elevation be $\alpha$. Then applying the Law of Sines to $\Delta ABC$, $\dfrac{\sin \beta}{120} = \dfrac{\sin 8°}{30} \quad \Leftrightarrow$ $\sin \beta = 4 \sin 8° \approx 0.55669 \Rightarrow \beta \approx \sin^{-1} 0.55669 \approx 33.8°$. But from $\Delta ABD$, $\angle BAD + \angle B = (\alpha + 8°) + \beta = 90°$, so $\alpha \approx 90° - 8° - 33.8° = 48.2°$.

**41.** Let $d$ be the distance from the earth to Venus, and let $\beta$ be the angle formed by sun, Venus, and earth. By the Law of Sines, $\dfrac{\sin \beta}{1} = \dfrac{\sin 39.4°}{0.723} \approx 0.878$, so either $\beta \approx \sin^{-1} 0.878 \approx 61.4°$ or $\beta \approx 180° - \sin^{-1} 0.878 \approx 118.6°$. In the first case, $\dfrac{d}{\sin(180° - 39.4° - 61.4°)} = \dfrac{0.723}{\sin 39.4°} \quad \Leftrightarrow \quad d \approx 1.119$ AU; in the second case, $\dfrac{d}{\sin(180° - 39.4° - 118.6°)} = \dfrac{0.723}{\sin 39.4°} \quad \Leftrightarrow \quad d \approx 0.427$ AU.

**42. (a)** Applying the Law of Sines to $\triangle ABC$, we get $\frac{\sin B}{b} = \frac{\sin 60^\circ}{c}$ or $\sin B = \frac{b\sin 60^\circ}{c}$. Similarly, applying the Law of Sines to $\triangle BCD$ gives $\frac{\sin B}{r} = \frac{\sin 120^\circ}{c+d}$ or $\sin B = \frac{r\sin 120^\circ}{c+d}$. Since $\sin 120^\circ = \sin 60^\circ$, we have $\frac{b}{c} = \frac{r}{c+d}$ $\Leftrightarrow$ $\frac{b}{r} = \frac{c}{c+d}$ $(*)$. Similarly, from $\triangle ADC$ and the Law of Sines we have $\frac{\sin D}{b} = \frac{\sin 60^\circ}{d}$ or $\sin D = \frac{b\sin 60^\circ}{d}$, and from $\triangle BDC$ we have $\sin D = \frac{a\sin 120^\circ}{c+d}$. Thus, $\frac{b\sin 60^\circ}{d} = \frac{a\sin 120^\circ}{c+d}$ $\Leftrightarrow$ $\frac{b}{a} = \frac{d}{c+d}$. Combining this with $(*)$, we get $\frac{b}{r} + \frac{b}{a} = \frac{c}{c+d} + \frac{d}{c+d} = \frac{c+d}{c+d} = 1$. Solving for $r$, we find $\frac{b}{r} = 1 - \frac{b}{a} = \frac{a-b}{a}$ $\Leftrightarrow$ $\frac{r}{b} = \frac{a}{a-b}$ $\Leftrightarrow$ $r = \frac{ab}{a-b}$.

**(b)** $r = \frac{4\cdot 3}{4-3} = 12$ cm.

**(c)** If $a = b$, then $r$ is infinite, and so the face is a flat disk.

**43.**

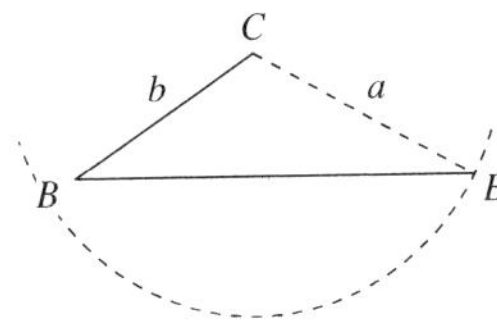


$a \geq b$: One solution

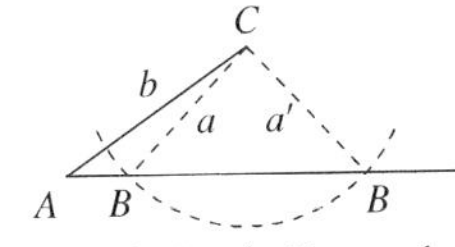


$b > a > b\sin A$: Two solutions

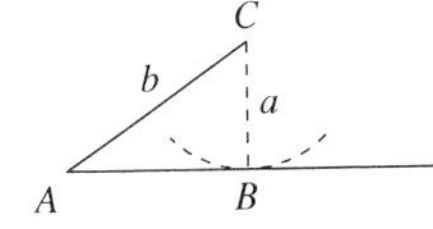


$a = b\sin A$: One solution

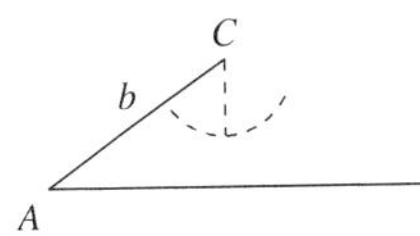


$a < b\sin A$: No solution

$\angle A = 30^\circ$, $b = 100$, $\sin A = \frac{1}{2}$. If $a \geq b = 100$ then there is one triangle. If $100 > a > 100\sin 30^\circ = 50$, then there are two possible triangles. If $a = 50$, then there is one (right) triangle. And if $a < 50$, then no triangle is possible.

## 6.5 The Law of Cosines

**1.** $x^2 = 21^2 + 42^2 - 2\cdot 21\cdot 42\cdot\cos 39^\circ = 441 + 1764 - 1764\cos 39^\circ \approx 834.115$ and so $x \approx \sqrt{834.115} \approx 28.9$.

**2.** $x^2 = 15^2 + 18^2 - 2\cdot 15\cdot 18\cdot\cos 108^\circ = 225 + 324 - 540\cos 108^\circ \approx 715.869$ and so $x \approx \sqrt{715.869} \approx 26.8$.

**3.** $x^2 = 25^2 + 25^2 - 2\cdot 25\cdot 25\cdot\cos 140^\circ = 625 + 625 - 1250\cos 140^\circ \approx 2207.556$ and so $x \approx \sqrt{2207.556} \approx 47$.

**4.** $x^2 = 2^2 + 8^2 - 2\cdot 2\cdot 8\cdot\cos 88^\circ = 4 + 64 - 32\cos 88^\circ \approx 66.883$ and so $x \approx \sqrt{66.883} \approx 8.2$.

**5.** $37.83^2 = 68.01^2 + 42.15^2 - 2\cdot 68.01\cdot 42.15\cdot\cos\theta$. Then $\cos\theta = \frac{37.83^2 - 68.01^2 - 42.15^2}{-2\cdot 68.01\cdot 42.15} \approx 0.867$ $\Leftrightarrow$ $\theta \approx \cos^{-1} 0.867 \approx 29.89^\circ$.

**6.** $154.6^2 = 60.1^2 + 122.5^2 - 2\cdot 60.1\cdot 122.5\cdot\cos\theta$. Then $\cos\theta = \frac{154.6^2 - 60.1^2 - 122.5^2}{-2\cdot 60.1\cdot 122.5} \approx -0.359$ $\Leftrightarrow$ $\theta \approx \cos^{-1}(-0.359) \approx 111^\circ$.

**7.** $x^2 = 24^2 + 30^2 - 2\cdot 24\cdot 30\cdot\cos 30^\circ = 576 + 900 - 1440\cos 30^\circ \approx 228.923$ and so $x \approx \sqrt{228.923} \approx 15$.

**8.** $20^2 = 10^2 + 12^2 - 2\cdot 10\cdot 12\cdot\cos\theta$. Then $\cos\theta = \frac{20^2 - 10^2 - 12^2}{-2\cdot 10\cdot 12} = \frac{156}{-240} = -0.65$ $\Leftrightarrow$ $\theta \approx \cos^{-1}(-0.65) \approx 130.54^\circ$.

**9.** $c^2 = 10^2 + 18^2 - 2\cdot 10\cdot 18\cdot\cos 120^\circ = 100 + 324 - 360\cos 120^\circ = 604$ and so $c \approx \sqrt{604} \approx 24.576$. Then $\sin A \approx \frac{18\sin 120^\circ}{24.576} \approx 0.634295$ $\Leftrightarrow$ $\angle A \approx \sin^{-1} 0.634295 \approx 39.4^\circ$, and $\angle B \approx 180^\circ - 120^\circ - 39.4^\circ = 20.6^\circ$.

**10.** $12^2 = 40^2 + 44^2 - 2\cdot 40\cdot 44\cdot\cos B \Leftrightarrow \cos B = \frac{12^2 - 40^2 - 44^2}{-2\cdot 40\cdot 44} \approx 0.964$ $\Leftrightarrow$ $\angle B \approx \cos^{-1} 0.964 \approx 15^\circ$. Then $\sin A = \frac{40\sin 15.5^\circ}{12} \approx 0.891 \Leftrightarrow \angle A \approx \sin^{-1} 0.891 \approx 63^\circ$, and so $\angle C \approx 180^\circ - 15^\circ - 63^\circ = 102^\circ$.

**11.** $c^2 = 3^2 + 4^2 - 2 \cdot 3 \cdot 4 \cdot \cos 53° = 9 + 16 - 24 \cos 53° \approx 10.556 \quad \Leftrightarrow \quad c \approx \sqrt{10.556} \approx 3.2$. Then $\sin B = \dfrac{4 \sin 53°}{3.25} \approx 0.983 \quad \Leftrightarrow \quad \angle B \approx \sin^{-1} 0.983 \approx 79°$ and $\angle A \approx 180° - 53° - 79° = 48°$.

**12.** $a^2 = 60^2 + 30^2 - 2 \cdot 60 \cdot 30 \cdot \cos 70° = 3600 + 900 - 3600 \cos 70° \approx 3268.73$ $\Leftrightarrow \quad a = \sqrt{3268.73} \approx 57.2$. Then $\sin C \approx \dfrac{30 \sin 70°}{57.2} \approx 0.493 \quad \Leftrightarrow$ $\angle C \approx \sin^{-1} 0.493 \approx 29.5°$, and $\angle B \approx 180° - 70° - 29.5° = 80.5°$.

**13.** $20^2 = 25^2 + 22^2 - 2 \cdot 25 \cdot 22 \cdot \cos A \quad \Leftrightarrow \quad \cos A = \frac{20^2 - 25^2 - 22^2}{-2 \cdot 25 \cdot 22} \approx 0.644 \quad \Leftrightarrow \quad \angle A \approx \cos^{-1} 0.644 \approx 50°$. Then $\sin B \approx \frac{25 \sin 49.9°}{20} \approx 0.956 \quad \Leftrightarrow \quad \angle B \approx \sin^{-1} 0.956 \approx 73°$, and so $\angle C \approx 180° - 50° - 73° = 57°$.

**14.** $10^2 = 12^2 + 16^2 - 2 \cdot 12 \cdot 16 \cdot \cos A \Leftrightarrow \cos A = \frac{10^2 - 12^2 - 16^2}{-2 \cdot 12 \cdot 16} = 0.78125 \quad \Leftrightarrow$ $\angle A \approx \cos^{-1} 0.78125 \approx 38.6°$. Then $\sin B \approx \frac{12 \sin 38.6°}{10} \approx 0.749 \quad \Leftrightarrow$ $\angle B \approx \sin^{-1} 0.749 \approx 48.5°$, and so $\angle C \approx 180° - 38.6° - 48.5° = 92.9°$.

**15.** $\sin C = \frac{162 \sin 40°}{125} \approx 0.833 \Leftrightarrow \angle C_1 \approx \sin^{-1} 0.833 \approx 56.4°$ or $\angle C_2 \approx 180° - 56.4° \approx 123.6°$.
If $\angle C_1 \approx 56.4°$, then $\angle A_1 \approx 180° - 40° - 56.4° = 83.6°$ and $a_1 = \frac{125 \sin 83.6°}{\sin 40°} \approx 193$.
If $\angle C_2 \approx 123.6°$, then $\angle A_2 \approx 180° - 40° - 123.6° = 16.4°$ and $a_2 = \frac{125 \sin 16.4°}{\sin 40°} \approx 54.9$.
Thus, one triangle has $\angle A \approx 83.6°$, $\angle C \approx 56.4°$, and $a \approx 193$; the other has $\angle A \approx 16.4°$, $\angle C \approx 123.6°$, and $a \approx 54.9$.

**16.** $\sin A = \frac{65 \sin 52°}{50} \approx 1.024$. Since $|\sin \theta| \leq 1$ for all $\theta$, there is no such $\angle A$, and hence there is no such triangle

**17.** $\sin B = \frac{65 \sin 55°}{50} \approx 1.065$. Since $|\sin \theta| \leq 1$ for all $\theta$, there is no such $\angle B$, and hence there is no such triangle.

**18.** $\angle A = 180° - 61° - 83° = 36°$. Then $b = \frac{73.5 \sin 61°}{\sin 36°} \approx 109.4$ and $c = \frac{73.5 \sin 83°}{\sin 36°} \approx 124.1$.

**19.** $\angle B = 180° - 35° - 85° = 60°$. Then $x = \frac{3 \sin 35°}{\sin 60°} \approx 2$.

**20.** $x^2 = 10° + 18° - 2 \cdot 10 \cdot 18 \cdot \cos 40° = 100 + 324 - 360 \cos 40° \approx 148.224$ and so $x \approx \sqrt{148.224} \approx 12.2$.

**21.** $x = \frac{50 \sin 30°}{\sin 100°} \approx 25.4$

**22.** $4^2 = 10^2 + 11^2 - 2 \cdot 10 \cdot 11 \cdot \cos \theta$. Then $\cos \theta = \frac{4^2 - 10^2 - 11^2}{-2 \cdot 10 \cdot 11} = \frac{-205}{-220} \approx 0.932 \quad \Leftrightarrow \quad \theta \approx \cos^{-1} 0.932 \approx 21.3°$.

**23.** $b^2 = 110^2 + 138^2 - 2(110)(138) \cdot \cos 38° = 12{,}100 + 19{,}044 - 30{,}360 \cos 38° \approx 7220.0$ and so $b \approx 85.0$. Therefore, using the Law of Cosines again, we have $\cos \theta = \frac{110^2 + 85^2 - 128^2}{2(110)(138)} \quad \Leftrightarrow \quad \theta \approx 89.15°$.

**24.** $\sin \theta = \frac{10 \sin 40°}{8} \approx 0.803 \quad \Leftrightarrow \quad \theta \approx \sin^{-1} 0.803 \approx 53.5°$ or $\theta \approx 180° - 53.5° \approx 126.5°$, but $53.5°$ doesn't fit the picture, so $\theta \approx 126.5°$.

**25.** $x^2 = 38^2 + 48^2 - 2 \cdot 38 \cdot 48 \cdot \cos 30° = 1444 + 2304 - 3648 \cos 30° \approx 588.739$ and so $x \approx 24.3$.

**26.** $\angle A = 180° - 98° - 25° = 57°$. Then $x = \frac{1000 \sin 98°}{\sin 57°} \approx 1180.8$.

**27.** The semiperimeter is $s = \frac{9+12+15}{2} = 18$, so by Heron's Formula the area is $A = \sqrt{18(18-9)(18-12)(18-15)} = \sqrt{2916} = 54$.

**28.** The semiperimeter is $s = \frac{1+2+2}{2} = \frac{5}{2}$, so by Heron's Formula the area is $A = \sqrt{\frac{5}{2}\left(\frac{5}{2} - 1\right)\left(\frac{5}{2} - 2\right)\left(\frac{5}{2} - 2\right)} = \sqrt{\frac{15}{16}} = \frac{\sqrt{15}}{4} \approx 0.968$.

**29.** The semiperimeter is $s = \frac{7+8+9}{2} = 12$, so by Heron's Formula the area is $A = \sqrt{12(12-7)(12-8)(12-9)} = \sqrt{720} = 12\sqrt{5} \approx 26.8$.

**30.** The semiperimeter is $s = \frac{11+100+101}{2} = 106$, so by Heron's Formula the area is $A = \sqrt{106(106-11)(106-100)(106-101)} = \sqrt{302{,}100} = 10\sqrt{3021} \approx 550$.

**31.** The semiperimeter is $s = \frac{3+4+6}{2} = \frac{13}{2}$, so by Heron's Formula the area is $A = \sqrt{\frac{13}{2}\left(\frac{13}{2} - 3\right)\left(\frac{13}{2} - 4\right)\left(\frac{13}{2} - 6\right)} = \sqrt{\frac{455}{16}} = \frac{\sqrt{455}}{4} \approx 5.33$.

**32.** Both of the smaller triangles have the same area. The semiperimeter of each is $s = \frac{2+5+5}{2} = 6$, so by Heron's Formula the area of each is $A = \sqrt{6(6-2)(6-5)(6-5)} = \sqrt{24} = 2\sqrt{6}$, so the shaded area is $4\sqrt{6} \approx 9.80$.

**33.** We draw a diagonal connecting the vertices adjacent to the $100^\circ$ angle. This forms two triangles. Consider the triangle with sides of length 5 and 6 containing the $100^\circ$ angle. The area of this triangle is $A_1 = \frac{1}{2}(5)(6)\sin 100^\circ \approx 14.77$. To use Heron's Formula to find the area of the second triangle, we need to find the length of the diagonal using the Law of Cosines: $c^2 = a^2 + b^2 - 2ab\cos C = 5^2 + 6^2 - 2\cdot 5\cdot 6\cos 100^\circ \approx 71.419 \Rightarrow c \approx 8.45$. Thus the second triangle has semiperimeter $s = \dfrac{8+7+8.45}{2} \approx 11.7255$ and area $A_2 = \sqrt{11.7255\,(11.7255-8)\,(11.7255-7)\,(11.7255-8.45)} \approx 26.00$. The area of the quadrilateral is the sum of the areas of the two triangles: $A = A_1 + A_2 \approx 14.77 + 26.00 = 40.77$.

**34.** We draw a line segment with length $x$ bisecting the $60^\circ$ angle to create two triangles. By the Law of Cosines, $3^2 = 4^2 + x^2 - 2\cdot 4x\cos 30^\circ \quad\Leftrightarrow\quad x^2 - 4\sqrt{3}x + 7 = 0$. Using the quadratic formula, we find $x = 2\sqrt{3} \pm \sqrt{5}$. The minus sign provides the correct length of about 1.23 (the other solution is about 5.7, which corresponds to a convex quadrilateral with the same side lengths), so the semiperimeter of each triangle is $s = \dfrac{3+4+2\sqrt{3}-\sqrt{5}}{2}$ and the total area of the figure is $A = 2\sqrt{s\,(s-3)\,(s-4)\left(s-2\sqrt{3}+\sqrt{5}\right)} \approx 2.46$.

**35.** Label the centers of the circles $A$, $B$, and $C$, as in the figure. By the Law of Cosines, $\cos A = \dfrac{AB^2 + AC^2 - BC^2}{2\,(AB)\,(AC)} = \dfrac{9^2 + 10^2 - 11^2}{2\,(9)\,(10)} = \frac{1}{3} \Rightarrow$ $\angle A \approx 70.53^\circ$. Now, by the Law of Sines, $\dfrac{\sin 70.53^\circ}{11} = \dfrac{\sin B}{AC} = \dfrac{\sin C}{AB}$. So $\sin B = \frac{10}{11}\sin 70.53^\circ \approx 0.85710 \Rightarrow B \approx \sin^{-1} 0.85710 \approx 58.99^\circ$ and $\sin C = \frac{9}{11}\sin 70.53^\circ \approx 0.77139 \Rightarrow C \approx \sin^{-1} 0.77139 \approx 50.48^\circ$. The area of $\Delta ABC$ is $\frac{1}{2}(AB)(AC)\sin A = \frac{1}{2}(9)(10)(\sin 70.53^\circ) \approx 42.426$.

The area of sector $A$ is given by $S_A = \pi R^2 \cdot \dfrac{\theta}{360^\circ} = \pi (4)^2 \cdot \dfrac{70.53^\circ}{360^\circ} \approx 9.848$. Similarly, the areas of sectors $B$ and $C$ are $S_B \approx 12.870$ and $S_C \approx 15.859$. Thus, the area enclosed between the circles is $A = \Delta ABC - S_A - S_B - S_C \Rightarrow A \approx 42.426 - 9.848 - 12.870 - 15.859 \approx 3.85\text{ cm}^2$.

**36.** In any $\Delta ABC$, the Law of Cosines gives $a^2 = b^2 + c^2 - 2bc\cdot\cos A$, $b^2 = a^2 + c^2 - 2ac\cdot\cos B$, and $c^2 = a^2 + b^2 - 2ab\cdot\cos C$. Adding the second and third equations gives

$$\begin{aligned} b^2 &= a^2 + c^2 - 2ac\cdot\cos B \\ c^2 &= a^2 + b^2 - 2ab\cdot\cos C \\ b^2 + c^2 &= 2a^2 + b^2 + c^2 - 2a\,(c\cos B + b\cos C) \end{aligned}$$

Thus

$$\begin{aligned} 2a^2 - 2a\,(c\cos B + b\cos C) &= 0 \\ 2a\,(a - c\cos B + b\cos C) &= 0 \end{aligned}$$

Since $a \neq 0$ we must have $a - (c\cos B + b\cos C) = 0 \Leftrightarrow a = b\cos C + c\cos B$. The other laws follow from the symmetry of $a$, $b$, and $c$.

**37.** Let $c$ be the distance across the lake, in miles. Then $c^2 = 2.82^2 + 3.56^2 - 2\,(2.82)\,(3.56)\cdot\cos 40.3^\circ \approx 5.313 \quad\Leftrightarrow$ $c \approx 2.30$ mi.

**38.** Suppose $ABCD$ is a parallelogram with $AB = DC = 5$, $AD = BC = 3$, and $\angle A = 50°$ (see the figure). Since opposite angles are equal in a parallelogram, it follows that $\angle C = 50°$, and $\angle B + \angle D = 360° - 100° = 260°$. Thus, $\angle B = \angle D = \dfrac{260°}{2} = 130°$.

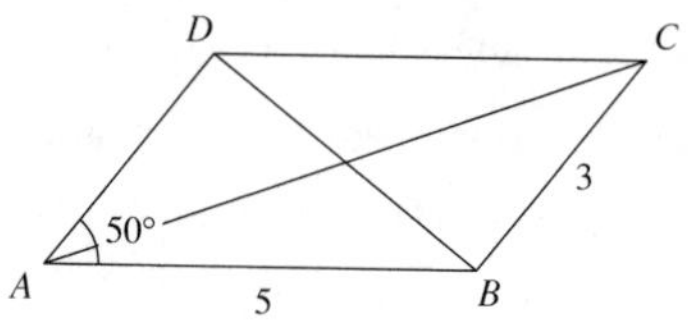


By the Law of Cosines, $AC^2 = 3^2 + 5^2 - 2\,(3)\,(5) \cdot \cos 130° \Rightarrow AC = \sqrt{9 + 25 - 30 \cdot \cos 130°} \approx 7.3$. Similarly, $BD = \sqrt{3^2 + 5^2 - 2\,(3)\,(5) \cdot \cos 50°} \approx 3.8$.

**39.** In half an hour, the faster car travels 25 miles while the slower car travels 15 miles. The distance between them is given by the Law of Cosines: $d^2 = 25^2 + 15^2 - 2\,(25)\,(15) \cdot \cos 65° \Rightarrow$ $d = \sqrt{25^2 + 15^2 - 2\,(25)\,(15) \cdot \cos 65°} = 5\sqrt{25 + 9 - 30 \cdot \cos 65°} \approx 23.1$ mi.

**40.** Let $x$ be the car's distance from its original position. Since the car travels at a constant speed of 40 miles per hour, it must have traveled 40 miles east, and then 20 miles northeast (which is 45° east of "due north"). From the diagram, we see that $\angle\beta = 135°$, so

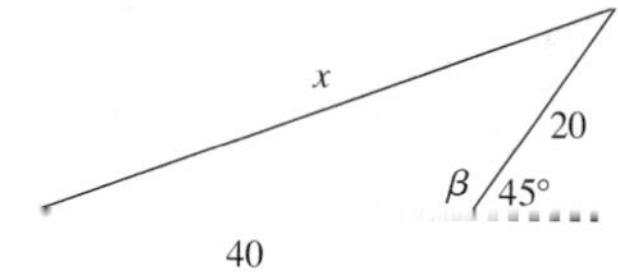


$x = \sqrt{20^2 + 40^2 - 2\,(20)\,(40) \cdot \cos 135°} = 10\sqrt{4 + 16 - 16 \cdot \cos 135°} \approx 56.0$ mi.

**41.** The pilot travels a distance of $625 \cdot 1.5 = 937.5$ miles in her original direction and $625 \cdot 2 = 1250$ miles in the new direction. Since she makes a course correction of 10° to the right, the included angle is $180° - 10° = 170°$. From the figure, we use the Law of Cosines to get

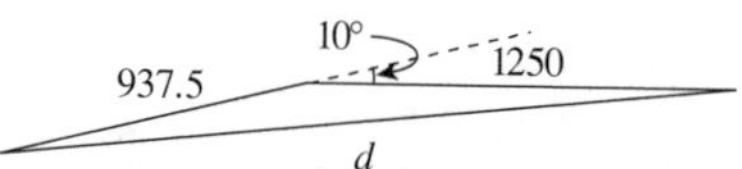


the expression $d^2 = 937.5^2 + 1250^2 - 2\,(937.5)\,(1250) \cdot \cos 170° \approx 4{,}749{,}549.42$, so $d \approx 2179$ miles. Thus, the pilot's distance from her original position is approximately 2179 miles.

**42.** Let $d$ be the distance between the two boats in miles. After one hour, the boats have traveled distances of 30 miles and 26 miles. Also, the angle subtended by their directions is $180° - 50° - 70° = 60°$. Then $d^2 = 30^2 + 26^2 - 2 \cdot 30 \cdot 26 \cdot \cos 60° = 796 \quad \Leftrightarrow \quad d \approx \sqrt{796} \approx 28.2$. Thus the distance between the two boats is about 28 miles.

**43. (a)** The angle subtended at Egg Island is 100°. Thus using the Law of Cosines, the distance from Forrest Island to the fisherman's home port is

$$\begin{aligned} x^2 &= 30^2 + 50^2 - 2 \cdot 30 \cdot 50 \cdot \cos 100° \\ &= 900 + 2500 - 3000 \cos 100° \approx 3920.945 \end{aligned}$$

and so $x \approx \sqrt{3920.945} \approx 62.62$ miles.

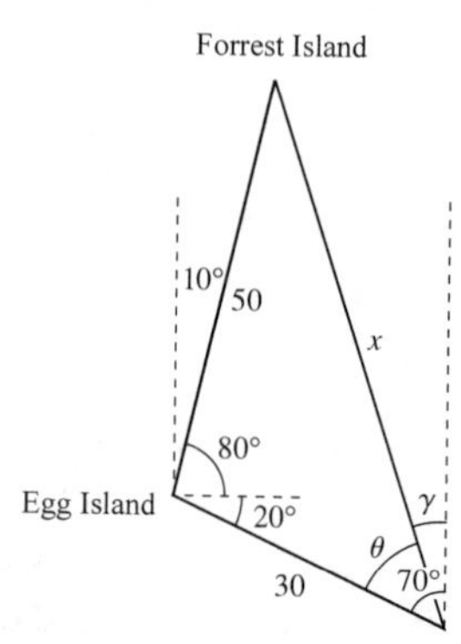


**(b)** Let $\theta$ be the angle shown in the figure. Using the Law of Sines, $\sin\theta = \dfrac{50 \sin 100°}{62.62} \approx 0.7863 \Leftrightarrow \theta \approx \sin^{-1} 0.7863 \approx 51.8°$. Then $\gamma = 90° - 20° - 51.8° = 18.2°$. Thus the bearing to his home port is S 18.2° E.

**44.** **(a)** In the 30 minutes the pilot flies 100 miles due east. So using the Law of Cosines, $x^2 = 100^2 + 300^2 - 2 \cdot 100 \cdot 300 \cdot \cos 40^\circ = 100^2 (1 + 9 - 6\cos 40^\circ) \approx 100^2 (5.404)$ and so $x \approx \sqrt{100^2 (5.404)} \approx 232.5$. Thus the pilot is 232.5 miles from his destination.

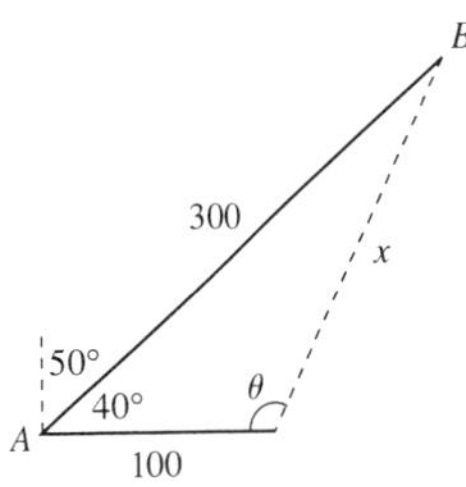


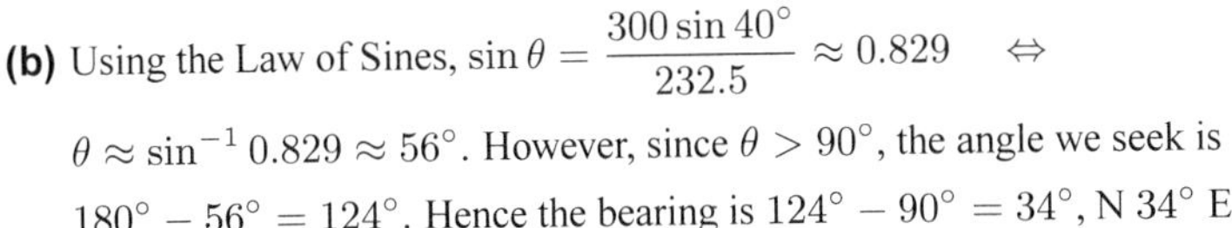

**(b)** Using the Law of Sines, $\sin\theta = \dfrac{300\sin 40^\circ}{232.5} \approx 0.829 \quad \Leftrightarrow$ $\theta \approx \sin^{-1} 0.829 \approx 56^\circ$. However, since $\theta > 90^\circ$, the angle we seek is $180^\circ - 56^\circ = 124^\circ$. Hence the bearing is $124^\circ - 90^\circ = 34^\circ$, N $34^\circ$ E.

**45.** The largest angle is the one opposite the longest side; call this angle $\theta$. Then by the Law of Cosines, $44^2 = 36^2 + 22^2 - 2(36)(22) \cdot \cos\theta \quad \Leftrightarrow \quad \cos\theta = \dfrac{36^2 + 22^2 - 44^2}{2(36)(22)} = -0.09848 \Rightarrow \theta \approx \cos^{-1}(-0.09848) \approx 96^\circ$.

**46.** Let $\theta$ be the angle formed by the cables. The two tugboats and the barge form a triangle: the side opposite $\theta$ has a length of 120 ft and the other two sides have lengths of 212 and 230 ft. Therefore, $120^2 = 212^2 + 230^2 - 2(212)(230) \cdot \cos\theta$ $\Leftrightarrow \quad \cos\theta = \dfrac{212^2 + 230^2 - 120^2}{2(212)(230)} \Rightarrow \cos\theta = 0.8557 \Rightarrow \theta \approx \cos^{-1} 0.8557 \approx 31^\circ$.

**47.** Let $d$ be the distance between the kites; then $d^2 \approx 380^2 + 420^2 - 2(380)(420) \cdot \cos 30^\circ \Rightarrow$ $d \approx \sqrt{380^2 + 420^2 - 2(380)(420) \cdot \cos 30^\circ} \approx 211$ ft.

**48.** Let $x$ be the length of the wire and let $\theta$ be the angle opposite $x$, as shown in the figure. Since the mountain is inclined $32^\circ$, we must have $\theta = 180^\circ - (90^\circ - 32^\circ) = 122^\circ$. Thus, $x = \sqrt{55^2 + 125^2 - 2(55)(125) \cdot \cos 122^\circ} \approx 161$ ft.

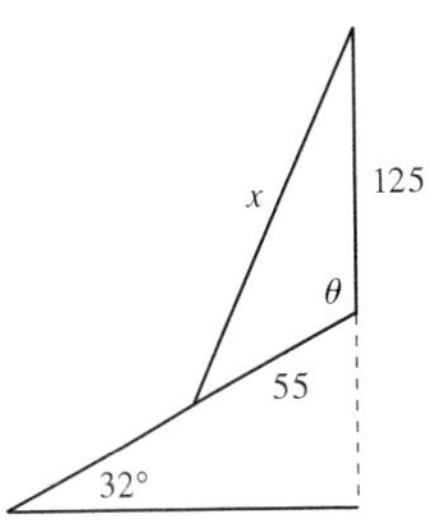


**49.** *Solution 1:* From the figure, we see that $\gamma = 106^\circ$ and $\sin 74^\circ = \dfrac{3400}{b} \quad \Leftrightarrow$ $b = \dfrac{3400}{\sin 74^\circ} \approx 3537$. Thus, $x^2 = 800^2 + 3537^2 - 2(800)(3537)\cos 106^\circ \quad \Rightarrow$ $x = \sqrt{800^2 + 3537^2 - 2(800)(3537)\cos 106^\circ} \quad \Rightarrow \quad x \approx 3835$ ft.

*Solution 2:* Notice that $\tan 74^\circ = \dfrac{3400}{a} \quad \Leftrightarrow \quad a = \dfrac{3400}{\tan 74^\circ} \approx 974.9$. By the Pythagorean theorem, $x^2 = (a + 800)^2 + 3400^2$. So $x = \sqrt{(974.9 + 800)^2 + 3400^2} \approx 3835$ ft.

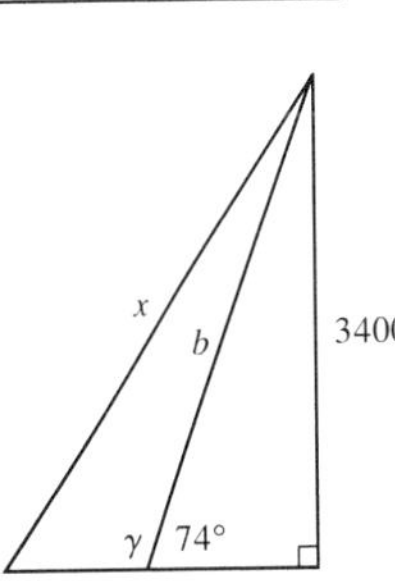


**50.** Let the woman be at point $A$, the first landmark (at $62^\circ$) be at point $B$, and the other landmark be at point $C$. We want to find the length $BC$. Now, $\cos 62^\circ = \dfrac{1150}{AB} \Leftrightarrow AB = \dfrac{1150}{\cos 62^\circ} \approx 2450$. Similarly, $\cos 54^\circ = \dfrac{1150}{AC} \quad \Leftrightarrow$ $AC = \dfrac{1150}{\cos 54^\circ} \approx 1956$. Therefore, by the Law of Cosines, $BC^2 = AB^2 + AC^2 - 2(AB)(AC) \cdot \cos 43^\circ \quad \Rightarrow$ $BC = \sqrt{2450^2 + 1956^2 - 2(2450)(1956) \cdot \cos 43^\circ} \Rightarrow BC \approx 1679$. Thus, the two landmarks are roughly 1679 feet apart.

**51.** By Heron's formula, $A = \sqrt{s(s-a)(s-b)(s-c)}$, where $s = \dfrac{a+b+c}{2} = \dfrac{112+148+190}{2} = 225$. Thus, $A = \sqrt{225(225-112)(225-148)(225-190)} \approx 8277.7 \text{ ft}^2$. Since the land value is \$20 per square foot, the value of the lot is approximately $8277.7 \cdot 20 = \$165{,}554$.

**52.** Having found $a \approx 13.2$ using the Law of Cosines, we use the Law of Sines to find $\angle B$: $\dfrac{\sin B}{10.5} = \dfrac{\sin 46.5^\circ}{13.2} \Leftrightarrow$ $\sin B = \dfrac{10.5 \sin 46.5^\circ}{13.2} \approx 0.577$. Now there are two angles $\angle B$ between $0^\circ$ and $180^\circ$ which have $\sin B = 0.577$, namely $\angle B_1 \approx 35.2^\circ$ and $\angle B_2 \approx 144.8^\circ$. But we must choose $\angle B_1$, since otherwise $\angle A + \angle B > 180^\circ$.

Using the Law of Sines again, $\dfrac{\sin C}{18.0} = \dfrac{\sin 46.5^\circ}{13.2} \Leftrightarrow \sin C = \dfrac{18.0 \sin 46.5^\circ}{13.2} \approx 0.989$, so either $\angle C \approx 81.5^\circ$ or $\angle C \approx 98.5^\circ$. In this case we must choose $\angle C \approx 98.5^\circ$ so that the sum of the angles in the triangle is $\angle A + \angle B + \angle C \approx 46.5^\circ + 35.2^\circ + 98.5^\circ \approx 180^\circ$. (The fact that the angles do not sum to exactly $180^\circ$, and the discrepancies between these results and those of Example 3, are due to roundoff error.)

The method in this exercise is slightly easier computationally, but the method in Example 3 is more foolproof.

# Chapter 6 Review

**1. (a)** $60^\circ = 60 \cdot \frac{\pi}{180} = \frac{\pi}{3} \approx 1.05$ rad

**(b)** $330^\circ = 330 \cdot \frac{\pi}{180} = \frac{11\pi}{6} \approx 5.76$ rad

**(c)** $-135^\circ = -135 \cdot \frac{\pi}{180} = -\frac{3\pi}{4} \approx -3.36$ rad

**(d)** $-90^\circ = -90 \cdot \frac{\pi}{180} = -\frac{\pi}{2} \approx -1.57$ rad

**2. (a)** $24^\circ = 24 \cdot \frac{\pi}{180} = \frac{2\pi}{15} \approx 0.42$ rad

**(b)** $-330^\circ = -330 \cdot \frac{\pi}{180} = -\frac{11\pi}{6} \approx -5.76$ rad

**(c)** $750^\circ = 750 \cdot \frac{\pi}{180} = \frac{25\pi}{6} \approx 13.09$ rad

**(d)** $5^\circ = 5 \cdot \frac{\pi}{180} = \frac{\pi}{36} \approx 0.87$ rad

**3. (a)** $\frac{5\pi}{2}$ rad $= \frac{5\pi}{2} \cdot \frac{180}{\pi} = 450^\circ$

**(b)** $-\frac{\pi}{6}$ rad $= -\dfrac{\pi}{6} \cdot \frac{180}{\pi} = -30^\circ$

**(c)** $\frac{9\pi}{4}$ rad $= \frac{9\pi}{4} \cdot \frac{180}{\pi} = 405^\circ$

**(d)** $3.1$ rad$= 3.1 \cdot \frac{180}{\pi} = \frac{558}{\pi} \approx 177.6^\circ$

**2. (a)** $8$ rad $= 8 \cdot \frac{180}{\pi} = \frac{1440}{\pi} \approx 458.37^\circ$

**(b)** $-\frac{5}{2}$ rad $= -\frac{5}{2} \cdot \frac{180}{\pi} = \frac{450}{\pi} \approx 143.24^\circ$

**(c)** $\frac{11\pi}{6}$ rad $= \frac{11\pi}{6} \cdot \frac{180}{\pi} = 330^\circ$

**(d)** $\frac{3\pi}{5}$ rad $= \frac{3\pi}{5} \cdot \frac{180}{\pi} = 108^\circ$

**5.** $r = 8$ m, $\theta = 1$ rad. Then $s = r\theta = 8 \cdot 1 = 8$ m.

**6.** $s = 7$ ft, $r = 5$ ft. Then $\theta = \dfrac{s}{r} = \frac{7}{5} = 1.4 \text{ rad} \approx 80.2^\circ$

**7.** $s = 100$ ft, $\theta = 70^\circ = 70 \cdot \frac{\pi}{180} = \frac{7\pi}{18}$ rad. Then $r = \frac{s}{\theta} = 100 \cdot \frac{18}{7\pi} = \frac{1800}{7\pi} \approx 82$ ft.

**8.** Since the diameter is 28 in, $r = 14$ in. In one revolution, the arc length (distance traveled) is $s = \theta r = 2\pi \cdot 14 = 28\pi$ in. The total distance traveled is 60 mi/h $\cdot$ 0.5 h $=$ 30 mi $=$ 30 mi $\cdot$ 5280 ft/mi $\cdot$ 12 in./ft $=$ 1,900,800 in. The number of revolution is 1,900,800 in $\cdot \dfrac{1 \text{ rev}}{28\pi \text{ in.}} \approx 21608.7$ rev. Therefore the car wheel will make approximately 21,609 revolutions.

**9.** $r = 3960$ miles, $s = 2450$ miles. Then $\theta = \dfrac{s}{r} = \dfrac{2450}{3960} \approx 0.619 \text{ rad} = 0.619 \cdot \frac{\pi}{180} \approx 35.448^\circ$ and so the angle is approximately $35.4^\circ$.

**10.** $r = 5$ m, $\theta = 2$ rad. Then $A = \frac{1}{2}r^2\theta = \frac{1}{2} \cdot 5^2 \cdot 2 = 25 \text{ m}^2$.

**11.** $A = \frac{1}{2}r^2\theta = \frac{1}{2}(200)^2\left(52^\circ \cdot \frac{\pi}{180^\circ}\right) \approx 18{,}151 \text{ ft}^2$

**12.** $A = 125 \text{ ft}^2$, $r = 25$ ft. Then $\theta = \dfrac{2A}{r^2} = \dfrac{2 \cdot 125}{25^2} = \frac{250}{625} = 0.4 \text{ rad} \approx 22.9^\circ$

**13.** The angular speed is $\omega = \dfrac{150 \cdot 2\pi \text{ rad}}{1 \text{ min}} = 300\pi \text{ rad/min} \approx 942.5 \text{ rad/min}$. The linear speed is $v = \dfrac{150 \cdot 2\pi \cdot 8}{1} = 2400\pi \text{ in./min} \approx 7539.8 \text{ in./min}$.

**14. (a)** The angular speed of the engine is $\omega_e = \frac{3500 \cdot 2\pi \text{ rad}}{1 \text{ min}} = 7000\pi$ rad/min.

**(b)** To find the angular speed $\omega_w$ of the wheels, we calculate $g = \frac{\omega_e}{\omega_w} = \frac{7000\pi \text{ rad/min}}{\omega_w} = 0.9 \Leftrightarrow \omega_w \approx 7777.8\pi$ rad/min.

**(c)** The speed of the car is the angular speed of the wheels times their radius:
$\frac{7777.8\pi \text{ rad}}{\text{min}} \cdot (11 \text{ in}) \cdot \frac{60 \text{ min}}{1 \text{ hr}} \cdot \frac{1 \text{ mile}}{63{,}360 \text{ in.}} \approx 254.5$ mi/h.

**15.** $r = \sqrt{5^2 + 7^2} = \sqrt{74}$. Then $\sin\theta = \frac{5}{\sqrt{74}}$, $\cos\theta = \frac{7}{\sqrt{74}}$, $\tan\theta = \frac{5}{7}$, $\csc\theta = \frac{\sqrt{74}}{5}$, $\sec\theta = \frac{\sqrt{74}}{7}$, and $\cot\theta = \frac{7}{5}$.

**16.** $x = \sqrt{10^2 - 3^2} = \sqrt{91}$. Then $\sin\theta = \frac{3}{10}$, $\cos\theta = \frac{\sqrt{91}}{10}$, $\tan\theta = \frac{3}{\sqrt{91}}$, $\csc\theta = \frac{10}{3}$, $\sec\theta = \frac{10}{\sqrt{91}}$, and $\cot\theta = \frac{\sqrt{91}}{3}$.

**17.** $\frac{x}{5} = \cos 40^\circ \Leftrightarrow x = 5\cos 40^\circ \approx 3.83$, and $\frac{y}{5} = \sin 40^\circ \Leftrightarrow y = 5\sin 40^\circ \approx 3.21$.

**18.** $\cos 35^\circ = \frac{2}{x} \Leftrightarrow x = \frac{2}{\cos 35^\circ} \approx 2.44$, and $\tan 35^\circ = \frac{y}{2} \Leftrightarrow y = 2\tan 35^\circ \approx 1.40$.

**19.** $\frac{1}{x} = \sin 20^\circ \Leftrightarrow x = \frac{1}{\sin 20^\circ} \approx 2.92$, and $\frac{x}{y} = \cos 20^\circ \Leftrightarrow y = \frac{x}{\cos 20^\circ} \approx \frac{2.924}{0.9397} \approx 3.11$.

**20.** $\cos 30^\circ = \frac{x}{4} \Leftrightarrow x = 4\cos 30^\circ \approx 3.46$, and $\sin 30^\circ = \frac{y}{x} \Leftrightarrow y = x\sin 30^\circ = 3.46 \cdot 0.5 \approx 1.73$.

**21.** $w = 3\sin 20^\circ \approx 1.026$,
$v = 3\cos 20^\circ \approx 2.819$.

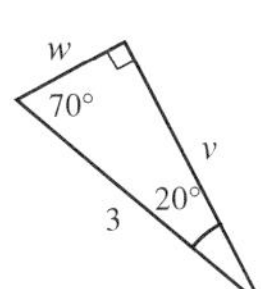


**22.** The other angle is $90^\circ - 60^\circ = 30^\circ$,
$\tan 60^\circ = l/20 \Leftrightarrow l = 20\tan 60^\circ \approx 34.64$,
and $\cos 60^\circ = 20/h \Leftrightarrow$
$h = 20/(\cos 60^\circ) = 40$.

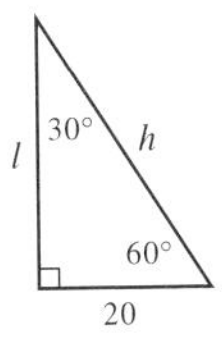


**23.** $\tan\theta = \frac{1}{a} \Leftrightarrow a = \frac{1}{\tan\theta} = \cot\theta$, $\sin\theta = \frac{1}{b} \Leftrightarrow b = \frac{1}{\sin\theta} = \csc\theta$

**24.** Let $h$ be the height of the tower in meters. Then $\tan 28.81^\circ = \frac{h}{1000} \Leftrightarrow h = 1000\tan 28.81^\circ \approx 550$ m.

**25.**

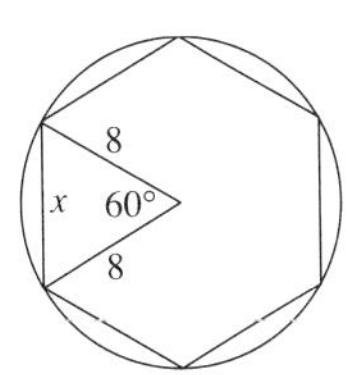


One side of the hexagon together with radial line segments through its endpoints forms a triangle with two sides of length 8 m and subtended angle $60^\circ$. Let $x$ be the length of one such side (in meters). By the Law of Cosines, $x^2 = 8^2 + 8^2 - 2 \cdot 8 \cdot 8 \cdot \cos 60^\circ = 64 \Leftrightarrow x = 8$. Thus the perimeter of the hexagon is $6x = 6 \cdot 8 = 48$ m.

**26.** As the crankshaft moves in its circular pattern, point $Q$ is determined by the angle $\theta$, namely it has coordinates $Q(2\cos\theta, 2\sin\theta)$. We split the triangle into two right triangles $\triangle OQR$ and $\triangle PQR$, as shown in the figure. Let $h$ be the height of the piston. We consider two cases, $0^\circ \le \theta < 180^\circ$ and $180^\circ \le \theta < 360^\circ$.

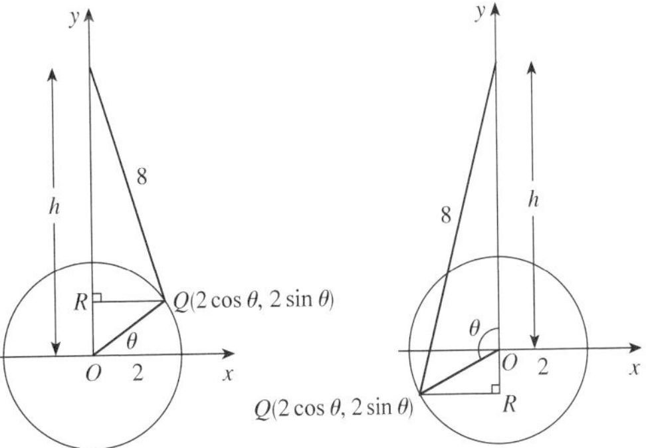


If $0^\circ \le \theta < 180^\circ$, then $h$ is the sum of $OR$ and $RP$. Using the Pythagorean Theorem, we find $RP = \sqrt{8^2 - (2\cos\theta)^2}$, while $OR$ is the $y$-coordinate of the point $Q$, $2\sin\theta$. Thus $h = \sqrt{64 - 4\cos^2\theta} + 2\sin\theta$.

If $180^\circ \le \theta < 360^\circ$, then $h$ is the difference between $RP$ and $RO$. Again, $RP = \sqrt{64 - 4\cos^2\theta}$ and $OR$ is the $y$-coordinate of the point $Q$, $2\sin\theta$. Thus $h = \sqrt{64 - 4\cos^2\theta} - 2|\sin\theta|$. Since $\sin\theta < 0$ for $180^\circ < \theta < 360^\circ$, this also reduces to $h = \sqrt{64 - 4\cos^2\theta} + 2\sin\theta$.

Since we get the same result in both cases, the height of the piston in terms of $\theta$ is $h = \sqrt{64 - 4\cos^2\theta} + 2\sin\theta$.

**27.** Let $r$ represent the radius, in miles, of the moon. Then $\tan\dfrac{\theta}{2} = \dfrac{r}{r+|AB|}$, $\theta = 0.518°$ $\Leftrightarrow$ $r = (r+236{,}900)\cdot\tan 0.259°$ $\Leftrightarrow$ $r(1-\tan 0.259°) = 236{,}900\cdot\tan 0.259°$ $\Leftrightarrow$ $r = \dfrac{236{,}900\cdot\tan 0.259°}{1-\tan 0.259°} \approx 1076$ and so the radius of the moon is roughly 1076 miles.

**28.** Let $d_1$ represent the horizontal distance from a point directly below the plane to the closer ship in feet, and $d_2$ represent the horizontal distance to the other ship in feet. Then $\tan 52° = \dfrac{35{,}000}{d_1}$ $\Leftrightarrow$ $d_1 = \dfrac{35{,}000}{\tan 52°}$, and similarly $\tan 40° = \dfrac{35{,}000}{d_2}$ $\Leftrightarrow$ $d_2 = \dfrac{35{,}000}{\tan 40°}$. So the distance between the two ships is $d_2 - d_1 = \dfrac{35{,}000}{\tan 40°} - \dfrac{35{,}000}{\tan 52°} \approx 14{,}400$ ft.

**29.** $\sin 315° = -\sin 45° = -\frac{1}{\sqrt{2}} = -\frac{\sqrt{2}}{2}$

**30.** $\csc\frac{9\pi}{4} = \csc\frac{\pi}{4} = \sqrt{2}$

**31.** $\tan(-135°) = \tan 45° = 1$

**32.** $\cos\frac{5\pi}{6} = -\cos\frac{\pi}{6} = -\frac{\sqrt{3}}{2}$

**33.** $\cot\left(-\frac{22\pi}{3}\right) = \cot\frac{2\pi}{3} = \cot\frac{\pi}{3} = -\frac{1}{\sqrt{3}} = -\frac{\sqrt{3}}{3}$

**34.** $\sin 405° = \sin 45° = \frac{1}{\sqrt{2}} = \frac{\sqrt{2}}{2}$

**35.** $\cos 585° = \cos 225° = -\cos 45° = -\frac{1}{\sqrt{2}} = -\frac{\sqrt{2}}{2}$

**36.** $\sec\frac{22\pi}{3} = \sec\frac{4\pi}{3} = -\sec\frac{\pi}{3} = -2$

**37.** $\csc\frac{8\pi}{3} = \csc\frac{2\pi}{3} = \csc\frac{\pi}{3} = \frac{2}{\sqrt{3}} = \frac{2\sqrt{3}}{3}$

**38.** $\sec\frac{13\pi}{6} = \sec\frac{\pi}{6} = \frac{2\sqrt{3}}{3}$

**39.** $\cot(-390°) = \cot(-30°) = -\cot 30° = -\sqrt{3}$

**40.** $\tan\frac{23\pi}{4} = \tan\frac{3\pi}{4} = -\tan\frac{\pi}{4} = -1$

**41.** $r = \sqrt{(-5)^2 + 12^2} = \sqrt{169} = 13$. Then $\sin\theta = \frac{12}{13}$, $\cos\theta = -\frac{5}{13}$, $\tan\theta = -\frac{12}{5}$, $\csc\theta = \frac{13}{12}$, $\sec\theta = -\frac{13}{5}$, and $\cot\theta = -\frac{5}{12}$.

**42.** If $\theta$ is in standard position, then the terminal point of $\theta$ on the unit circle is simply $(\cos\theta, \sin\theta)$. Since the terminal point is given as $\left(-\frac{\sqrt{3}}{2}, \frac{1}{2}\right)$, $\sin\theta = \frac{1}{2}$.

**43.** $y - \sqrt{3}x + 1 = 0 \Leftrightarrow y = \sqrt{3}x - 1$, so the slope of the line is $m = \sqrt{3}$. Then $\tan\theta = m = \sqrt{3} \Leftrightarrow \theta = 60°$.

**44.** $4y - 2x - 1 = 0 \Leftrightarrow y = \frac{1}{2}x + \frac{1}{4}$. The slope of the line is $m = \frac{1}{2}$. Then $\tan\theta = m = \frac{1}{2}$and $r = \sqrt{1^2 + 2^2} = \sqrt{5}$. So $\sin\theta = -\frac{1}{\sqrt{5}}$, $\cos\theta = -\frac{2}{\sqrt{5}}$, $\tan\theta = \frac{1}{2}$, $\csc\theta = -\sqrt{5}$, $\sec\theta = -\frac{\sqrt{5}}{2}$, and $\cot\theta = 2$.

**45.** $\sec^2\theta = 1 + \tan^2\theta \Leftrightarrow \tan^2\theta = \dfrac{1}{\cos^2\theta} - 1 \Leftrightarrow$ $\tan\theta = \sqrt{\dfrac{1}{\cos^2\theta} - 1} = \sqrt{\dfrac{1-\cos^2\theta}{\cos^2\theta}} = \dfrac{\sqrt{1-\cos^2\theta}}{|\cos\theta|} = \dfrac{\sqrt{1-\cos^2\theta}}{-\cos\theta}$ (since $\cos\theta < 0$ in quadrant III, $|\cos\theta| = -\cos\theta$). Thus $\tan\theta = -\dfrac{\sqrt{1-\cos^2\theta}}{\cos\theta}$.

**46.** $\sec\theta = \dfrac{1}{\cos\theta} = \dfrac{1}{-\sqrt{1-\sin^2\theta}}$(since $\cos\theta < 0$ in quadrant III).

**47.** $\tan^2\theta = \dfrac{\sin^2\theta}{\cos^2\theta} = \dfrac{\sin^2\theta}{1-\sin^2\theta}$

**48.** $\csc^2\theta\cos^2\theta = \dfrac{1}{\sin^2\theta}\cdot\cos^2\theta = \dfrac{1-\sin^2\theta}{\sin^2\theta} = \dfrac{1}{\sin^2\theta} - 1$

**49.** $\tan\theta = \frac{\sqrt{7}}{3}$, $\sec\theta = \frac{4}{3}$. Then $\cos\theta = \frac{3}{4}$ and $\sin\theta = \tan\theta\cdot\cos\theta = \frac{\sqrt{7}}{4}$, $\csc\theta = \frac{4}{\sqrt{7}} = \frac{4\sqrt{7}}{7}$, and $\cot\theta = \frac{3}{\sqrt{7}} = \frac{3\sqrt{7}}{7}$.

**50.** $\sec\theta = \frac{41}{40}$, $\csc\theta = -\frac{41}{9}$. Then $\sin\theta = -\frac{9}{41}$, $\cos\theta = \frac{40}{41}$, $\tan\theta = \dfrac{\sin\theta}{\cos\theta} = \dfrac{-\frac{9}{41}}{\frac{40}{41}} = -\frac{9}{40}$, and $\cot\theta = -\frac{40}{9}$.

**51.** $\sin\theta = \frac{3}{5}$. Since $\cos\theta < 0$, $\theta$ is in quadrant II. Thus, $x = -\sqrt{5^2 - 3^2} = -\sqrt{16} = -4$ and so $\cos\theta = -\frac{4}{5}$, $\tan\theta = -\frac{3}{4}$, $\csc\theta = \frac{5}{3}$, $\sec\theta = -\frac{5}{4}$, $\cot\theta = -\frac{4}{3}$.

**52.** $\sec\theta = -\frac{13}{5}$and $\tan\theta > 0$. Then $\cos\theta = -\frac{5}{13}$, and $\theta$ must be in quadrant III $\Rightarrow \sin\theta < 0$. Therefore, $\sin\theta = -\sqrt{1-\cos^2\theta} = -\sqrt{1-\frac{25}{169}} = -\frac{12}{13}$, $\csc\theta = -\frac{13}{12}$, $\tan\theta = \dfrac{\sin\theta}{\cos\theta} = \frac{12}{5}$, and $\cot\theta = \frac{5}{12}$.

**53.** $\tan\theta = -\frac{1}{2}$. $\sec^2\theta = 1+\tan^2\theta = 1+\frac{1}{4} = \frac{5}{4} \quad\Leftrightarrow\quad \cos^2\theta = \dfrac{4}{5} \Rightarrow \cos\theta = -\sqrt{\frac{4}{5}} = -\frac{2}{\sqrt{5}}$ since $\cos\theta < 0$ in quadrant II. But $\tan\theta = \dfrac{\sin\theta}{\cos\theta} = -\frac{1}{2} \quad\Leftrightarrow\quad \sin\theta = -\frac{1}{2}\cos\theta = -\frac{1}{2}\left(-\frac{2}{\sqrt{5}}\right) = \frac{1}{\sqrt{5}}$. Therefore, $\sin\theta+\cos\theta = \frac{1}{\sqrt{5}} + \left(-\frac{2}{\sqrt{5}}\right) = -\frac{1}{\sqrt{5}} = -\frac{\sqrt{5}}{5}$.

**54.** $\sin\theta = \frac{1}{2}$for $\theta$ in quadrant I. Then $\tan\theta+\sec\theta = \dfrac{\sin\theta}{\cos\theta}+\dfrac{1}{\cos\theta} = \dfrac{\sin\theta+1}{\cos\theta} = \dfrac{\sin\theta+1}{\sqrt{1-\sin^2\theta}} = \dfrac{\frac{1}{2}+1}{\sqrt{1-\frac{1}{2}}} = \dfrac{2\sqrt{3}}{2} = \sqrt{3}$.

**55.** By the Pythagorean Theorem, $\sin^2\theta+\cos^2\theta = 1$ for any angle $\theta$.

**56.** $\cos\theta = -\frac{\sqrt{3}}{2}$ and $\frac{\pi}{2} < \theta < \pi$. Then $\theta = \frac{5\pi}{6} \Rightarrow 2\theta = \frac{10\pi}{6} = \frac{5\pi}{3}$. So $\sin 2\theta = \sin\frac{5\pi}{3} = -\sin\frac{\pi}{3} = -\frac{\sqrt{3}}{2}$.

**57.** $\angle B = 180^\circ - 30^\circ - 80^\circ = 70^\circ$, and so by the Law of Sines, $x = \dfrac{10\sin 30^\circ}{\sin 70^\circ} \approx 5.32$.

**58.** $x = \dfrac{2\sin 45^\circ}{\sin 105^\circ} \approx 1.46$

**59.** $x^2 = 100^2+210^2 - 2\cdot 100\cdot 210\cdot\cos 40^\circ \approx 21{,}926.133 \quad\Leftrightarrow\quad x \approx 148.07$

**60.** $x^2 = 2^2+8^2-2(2)(8)\cos 120^\circ = 84 \quad\Leftrightarrow\quad x\approx\sqrt{84}\approx 9.17$

**61.** $\sin B = \dfrac{20\sin 60^\circ}{70} \approx 0.247 \Leftrightarrow \angle B \approx \sin^{-1} 0.247 \approx 14.33^\circ$. Then $\angle C \approx 180^\circ - 60^\circ - 14.33^\circ = 105.67^\circ$, and so $x \approx \dfrac{70\sin 105.67^\circ}{\sin 60^\circ} \approx 77.82$.

**62.** $\sin B = \dfrac{4\sin 110^\circ}{6} \approx 0.626 \Leftrightarrow \angle B \approx 38.79^\circ$. Then $\angle C \approx 180^\circ - 110^\circ - 38.79^\circ = 31.21^\circ$, and so $x \approx \dfrac{6\sin 31.21^\circ}{\sin 110^\circ} \approx 3.3$.

**63.** After 2 hours the ships have traveled distances $d_1 = 40$ mi and $d_2 = 56$ mi. The subtended angle is $180^\circ - 32^\circ - 42^\circ = 106^\circ$. Let $d$ be the distance between the two ships in miles. Then by the Law of Cosines, $d^2 = 40^2+56^2-2(40)(56)\cos 106^\circ \approx 5970.855 \quad\Leftrightarrow\quad d\approx 77.3$ miles.

**64.** Let $h$ represent the height of the building in feet, and $x$ the horizontal distance from the building to point $B$. Then $\tan 24.1^\circ = \dfrac{h}{x+600}$ and $\tan 30.2^\circ = \dfrac{h}{x} \quad\Leftrightarrow\quad x = h\cot 30.2^\circ$. Substituting for $x$ gives $\tan 24.1^\circ = \dfrac{h}{h\cot 30.2^\circ+600} \quad\Leftrightarrow\quad h = \tan 24.1^\circ\,(h\cot 30.2^\circ+600) \quad\Leftrightarrow$ $h = \dfrac{600\cdot\tan 24.1^\circ}{1-\tan 24.1^\circ\cot 30.2^\circ} \approx 1160$ ft.

**65.** Let $d$ be the distance, in miles, between the points $A$ and $B$ . Then by the Law of Cosines, $d^2 = 3.2^2+5.6^2-2(3.2)(5.6)\cos 42^\circ \approx 14.966 \quad\Leftrightarrow\quad d\approx 3.9$ mi.

**66.** $\angle C = 180^\circ - 42.3^\circ - 68.9^\circ = 68.8^\circ$. Then $b = \dfrac{120\sin 68.9^\circ}{\sin 68.8^\circ} \approx 120.08$ miles. Let $d$ be the shortest distance, in miles, to the shore. Then $d = b\sin A \approx 120.08\sin 42.3^\circ \approx 80.8$ miles.

**67.** $A = \frac{1}{2}ab\sin\theta = \frac{1}{2}(8)(14)\sin 35^\circ \approx 32.12$

**68.** By Heron's Formula, $A = \sqrt{s(s-a)(s-b)(s-c)}$, where $s = \dfrac{a+b+c}{2} = \dfrac{5+6+8}{2} = 9.5$. Thus, $A = \sqrt{9.5(9.5-5)(9.5-6)(9.5-8)} \approx 14.98$.

# Chapter 6 Test

**1.** $330° = 330 \cdot \frac{\pi}{180} = \frac{11\pi}{6}$ rad. $-135° = -135 \cdot \frac{\pi}{180} = -\frac{3\pi}{4}$ rad.

**2.** $\frac{4\pi}{3}$ rad $= \frac{4\pi}{3} \cdot \frac{180}{\pi} = 240°$. $-1.3$ rad $= -1.3 \cdot \frac{180}{\pi} = -\frac{234}{\pi} \approx -74.5°$

**3. (a)** The angular speed is $\omega = \frac{120 \cdot 2\pi \text{ rad}}{1 \text{ min}} = 240\pi$ rad/min.

**(b)** The linear speed is

$$v = \frac{120 \cdot 2\pi \cdot 16}{1} = 3840\pi \text{ ft/min}$$
$$\approx 12{,}063.7 \text{ ft/min}$$
$$\approx 137 \text{ mi/h.}$$

**4. (a)** $\sin 405° = \sin 45° = \frac{1}{\sqrt{2}} = \frac{\sqrt{2}}{2}$

**(b)** $\tan(-150°) = \tan 30° = \frac{1}{\sqrt{3}} = \frac{\sqrt{3}}{3}$

**(c)** $\sec \frac{5\pi}{3} = \sec \frac{\pi}{3} = 2$

**(d)** $\csc \frac{5\pi}{2} = \csc \frac{\pi}{2} = 1$

**5.** $r = \sqrt{3^2 + 2^2} = \sqrt{13}$. Then $\tan\theta + \sin\theta = \frac{2}{3} + \frac{2}{\sqrt{13}} = \frac{2(\sqrt{13}+3)}{3\sqrt{13}} = \frac{26+6\sqrt{13}}{39}$.

**6.** $\sin\theta = \dfrac{a}{24} \quad\Leftrightarrow\quad a = 24\sin\theta$. Also, $\cos\theta = \dfrac{b}{24} \quad\Leftrightarrow\quad b = 24\cos\theta$.

**7.** $\cos\theta = -\frac{1}{3}$ and $\theta$ is in quadrant III, so $r = 3$, $x = -1$, and $y = -\sqrt{3^2 - 1^2} = -2\sqrt{2}$. Then $\tan\theta\cot\theta + \csc\theta = \tan\theta \cdot \dfrac{1}{\tan\theta} + \csc\theta = 1 - \frac{3}{2\sqrt{2}} = \frac{2\sqrt{2}-3}{2\sqrt{2}} = \frac{4-3\sqrt{2}}{4}$.

**8.** $\sin\theta = \frac{5}{13}$, $\tan\theta = -\frac{5}{12}$. Then $\sec\theta = \dfrac{1}{\cos\theta} = \dfrac{1}{\cos\theta} \cdot \dfrac{\sin\theta}{\sin\theta} = \tan\theta \cdot \dfrac{1}{\sin\theta} = -\frac{5}{12} \cdot \frac{13}{5} = -\frac{13}{12}$.

**9.** $\sec^2\theta = 1 + \tan^2\theta \quad\Leftrightarrow\quad \tan\theta = \pm\sqrt{\sec^2\theta - 1}$. Thus, $\tan\theta = -\sqrt{\sec^2\theta - 1}$ since $\tan\theta < 0$ in quadrant II.

**10.** $\tan 73° = \dfrac{h}{6} \Rightarrow h = 6\tan 73° \approx 19.6$ ft.

**11.** By the Law of Cosines, $x^2 = 10^2 + 12^2 - 2(10)(12) \cdot \cos 48° \approx 8.409 \quad\Leftrightarrow\quad x \approx 9.1$.

**12.** $\angle C = 180° - 52° - 69° = 59°$. Then by the Law of Sines, $x = \frac{230 \sin 69°}{\sin 59°} \approx 250.5$.

**13.** Let $h$ be the height of the shorter altitude. Then $\tan 20° = \dfrac{h}{50} \quad\Leftrightarrow\quad h = 50\tan 20°$ and $\tan 28° = \dfrac{x+h}{50} \quad\Leftrightarrow\quad x + h = 50\tan 28° \quad\Leftrightarrow\quad x = 50\tan 28° - h = 50\tan 28° - 50\tan 20° \approx 8.4$.

**14.** Let $\angle A$ and $\angle X$ be the other angles in the triangle. Then $\sin A = \frac{15\sin 108°}{28} \approx 0.509 \Leftrightarrow \angle A \approx 30.63°$. Then $\angle X \approx 180° - 108° - 30.63° \approx 41.37°$, and so $x \approx \frac{28\sin 41.37°}{\sin 108°} \approx 19.5$.

**15. (a)** $A\,(\text{sector}) = \frac{1}{2}r^2\theta = \frac{1}{2} \cdot 10^2 \cdot 72 \cdot \frac{\pi}{180} = 50 \cdot \frac{72\pi}{180}$. $A\,(\text{triangle}) = \frac{1}{2}r \cdot r\sin\theta = \frac{1}{2} \cdot 10^2 \sin 72°$. Thus, the area of the shaded region is $A\,(\text{shaded}) = A\,(\text{sector}) - A\,(\text{triangle}) = 50\left(\frac{72\pi}{180} - \sin 72°\right) \approx 15.3 \text{ m}^2$.

**(b)** The shaded region is bounded by two pieces: one piece is part of the triangle, the other is part of the circle. The first part has length $l = \sqrt{10^2 + 10^2 - 2(10)(10) \cdot \cos 72°} = 10\sqrt{2 - 2 \cdot \cos 72°}$. The second has length $s = 10 \cdot 72 \cdot \frac{\pi}{180} = 4\pi$. Thus, the perimeter of the shaded region is $p = l + s = 10\sqrt{2 - 2\cos 72°} + 4\pi \approx 24.3$ m.

**16. (a)** If $\theta$ is the angle opposite the longest side, then by the Law of Cosines $\cos\theta = \frac{9^2+13^2-20^2}{2(9)(20)} = -0.6410$. Therefore, $\theta = \cos^{-1}(-0.6410) \approx 129.9°$.

**(b)** From part (a), $\theta \approx 129.9°$, so the area of the triangle is $A = \frac{1}{2}(9)(13)\sin 129.9° \approx 44.9 \text{ units}^2$. Another way to find the area is to use Heron's Formula: $A = \sqrt{s(s-a)(s-b)(s-c)}$, where $s = \frac{a+b+c}{2} = \frac{9+13+20}{2} = 21$. Thus, $A = \sqrt{21(21-20)(21-13)(21-9)} = \sqrt{2016} \approx 44.9 \text{ units}^2$.

**17.** Label the figure as shown. Now $\angle\beta = 85° - 75° = 10°$, so by the Law of Sines,

$$\frac{x}{\sin 75°} = \frac{100}{\sin 10°} \quad\Leftrightarrow\quad x = 100 \cdot \frac{\sin 75°}{\sin 10°}. \text{ Now } \sin 85° = \frac{h}{x} \quad\Leftrightarrow$$

$$h = x\sin 85° = 100 \cdot \frac{\sin 75°}{\sin 10°}\sin 85° \approx 554.$$

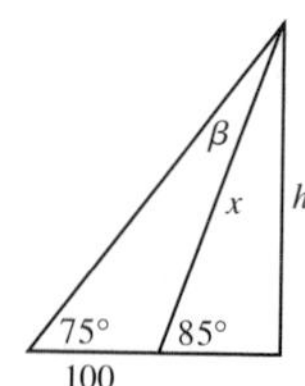

# Focus on Modeling: Surveying

**1.** Let $x$ be the distance between the church and City Hall. To apply the Law of Sines to the triangle with vertices at City Hall, the church, and the first bridge, we first need the measure of the angle at the first bridge, which is $180^\circ - 25^\circ - 30^\circ = 125^\circ$. Then $\dfrac{x}{\sin 125^\circ} = \dfrac{0.86}{\sin 30^\circ} \Leftrightarrow x = \dfrac{0.86 \sin 125^\circ}{\sin 30^\circ} \approx 1.4089$. So the distance between the church and City Hall is about 1.41 miles.

**2.** To find the distance $z$ between the fire hall and the school, we use the distance found in the text between the bank and the cliff. To find $z$ we first need to find the length of the edges labeled $x$ and $y$. In the bank-second bridge-cliff triangle, the third angle is $180^\circ - 50^\circ - 60^\circ = 70^\circ$, so $x = \dfrac{1.55 \sin 50^\circ}{\sin 70^\circ} \approx 1.26$. In the second bridge-school-cliff triangle, the third angle is $180^\circ - 80^\circ - 55^\circ = 45^\circ$, so $y = \dfrac{1.26 \sin 45^\circ}{\sin 55^\circ} \approx 1.09$. Finally, in the school-fire hall-cliff triangle, the third angle is $180^\circ - 45^\circ - 80^\circ = 55^\circ$, so $z = \dfrac{1.09 \sin 80^\circ}{\sin 55^\circ} \approx 1.31$. Thus, the fire hall and the school are about 1.31 miles apart.

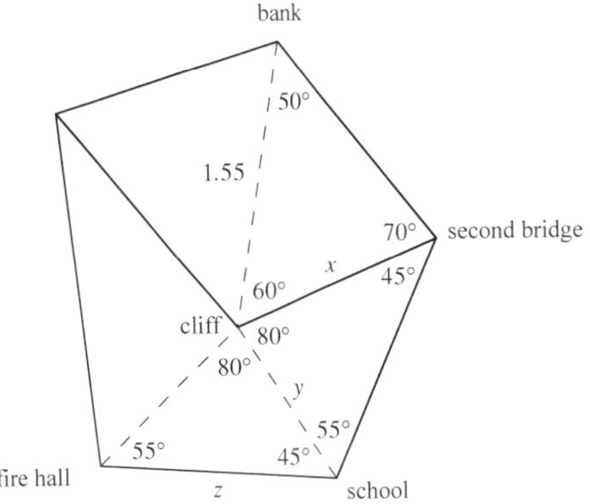


**3.** First notice that $\angle DBC = 180^\circ - 20^\circ - 95^\circ - 65^\circ$ and $\angle DAC = 180^\circ - 60^\circ - 45^\circ = 75^\circ$. From $\triangle ACD$ we get $\dfrac{|AC|}{\sin 45^\circ} = \dfrac{20}{\sin 75^\circ} \Leftrightarrow |AC| = \dfrac{20 \sin 45^\circ}{\sin 75^\circ} \approx 14.6^\circ$. From $\triangle BCD$ we get $\dfrac{|BC|}{\sin 95^\circ} = \dfrac{20}{\sin 65^\circ} \Leftrightarrow |BC| = \dfrac{20 \sin 95^\circ}{\sin 65^\circ} \approx 22.0$. By applying the Law of Cosines to $\triangle ABC$ we get $|AB|^2 = |AC|^2 + |BC|^2 - 2\,|AC|\,|BC| \cos 40^\circ \approx 14.6^2 + 22.0^2 - 2 \cdot 14.6 \cdot 22.0 \cdot \cos 40^\circ \approx 205$, so $|AB| \approx \sqrt{205} \approx 14.3$ m. Therefore, the distance between $A$ and $B$ is approximately 14.3 m.

**4.** Let $h$ represent the height in meters of the cliff, and $d$ the horizontal distance to the cliff. The third horizontal angle is $180^\circ - 69.4^\circ - 51.6^\circ = 59^\circ$ and so $d = \dfrac{200 \sin 51.6^\circ}{\sin 59^\circ} \approx 182.857$. Then $\tan 33.1^\circ = \dfrac{h}{d} \Leftrightarrow$ $h = d \tan 33.1^\circ \approx 182.857 \tan 33.1^\circ \approx 119.2$ m.

**5. (a)** In $\triangle ABC$, $\angle B = 180^\circ - \beta$, so $\angle C = 180^\circ - \alpha - (180^\circ - \beta) = \beta - \alpha$. By the Law of Sines, $\dfrac{|BC|}{\sin \alpha} = \dfrac{|AB|}{\sin (\beta - \alpha)}$

$$\Rightarrow |BC| = |AB| \frac{\sin \alpha}{\sin (\beta - \alpha)} = \frac{d \sin \alpha}{\sin (\beta - \alpha)}.$$

**(b)** From part (a) we know that $|BC| = \dfrac{d \sin \alpha}{\sin (\beta - \alpha)}$. But $\sin \beta = \dfrac{h}{|BC|} \Leftrightarrow |BC| = \dfrac{h}{\sin \beta}$. Therefore,

$$|BC| = \frac{d \sin \alpha}{\sin (\beta - \alpha)} = \frac{h}{\sin \beta} \Rightarrow h = \frac{d \sin \alpha \sin \beta}{\sin (\beta - \alpha)}.$$

**(c)** $h = \dfrac{d \sin \alpha \sin \beta}{\sin (\beta - \alpha)} = \dfrac{800 \sin 25^\circ \sin 29^\circ}{\sin 4^\circ} \approx 2350$ ft

**6.** Let the surveyor be at point $A$, the first landmark (with angle of depression $42^\circ$) be at point $B$, and the other landmark be at point $C$. We want to find $|BC|$. Now $\sin 42^\circ = \dfrac{2430}{|AB|} \Leftrightarrow |AB| = \dfrac{2430}{\sin 42^\circ} \approx 3631.6$. Similarly, $\sin 39^\circ = \dfrac{2430}{|AC|}$ $\Leftrightarrow |AC| = \dfrac{2430}{\sin 39^\circ} \approx 3861.3$. Therefore, by the Law of Cosines, $|BC|^2 = |AB|^2 + |AC|^2 - 2\,|AB|\,|AC| \cos 68^\circ \Rightarrow$ $BC = \sqrt{3631.6^2 + 3861.3^2 - 2\,(3631.6)\,(3861.3) \cos 68^\circ} \approx 4194$. Thus, the two landmarks are approximately 4194 ft apart.

**7.** We start by labeling the edges and calculating the remaining angles, as shown in the first figure. Using the Law of Sines, we find the following: $\dfrac{a}{\sin 29^\circ} = \dfrac{150}{\sin 60^\circ} \Leftrightarrow a = \dfrac{150 \sin 29^\circ}{\sin 60^\circ} \approx 83.97$, $\dfrac{b}{\sin 91^\circ} = \dfrac{150}{\sin 60^\circ} \Leftrightarrow b = \dfrac{150 \sin 91^\circ}{\sin 60^\circ} \approx 173.18$, $\dfrac{c}{\sin 32^\circ} = \dfrac{173.18}{\sin 87^\circ} \Leftrightarrow c = \dfrac{173.18 \sin 32^\circ}{\sin 87^\circ} \approx 91.90$, $\dfrac{d}{\sin 61^\circ} = \dfrac{173.18}{\sin 87^\circ} \Leftrightarrow e = \dfrac{173.18 \sin 61^\circ}{\sin 87^\circ} \approx 151.67$, $\dfrac{e}{\sin 41^\circ} = \dfrac{151.67}{\sin 51^\circ} \Leftrightarrow e = \dfrac{151.67 \sin 41^\circ}{\sin 51^\circ} \approx 128.04$, $\dfrac{f}{\sin 88^\circ} = \dfrac{151.67}{\sin 51^\circ} \Leftrightarrow f = \dfrac{151.67 \sin 88^\circ}{\sin 51^\circ} \approx 195.04$, $\dfrac{g}{\sin 50^\circ} = \dfrac{195.04}{\sin 92^\circ} \Leftrightarrow g = \dfrac{195.04 \sin 50^\circ}{\sin 92^\circ} \approx 149.50$, and $\dfrac{h}{\sin 38^\circ} = \dfrac{195.04}{\sin 92^\circ} \Leftrightarrow h = \dfrac{195.04 \sin 38^\circ}{\sin 92^\circ} \approx 120.15$. Note that we used two decimal places throughout our calculations. Our results are shown (to one decimal place) in the second figure.

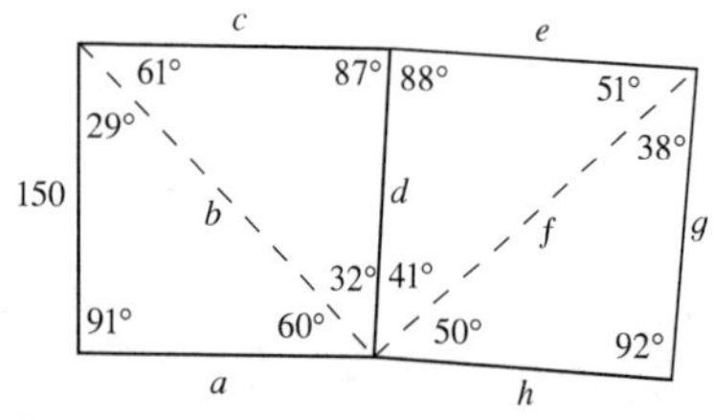


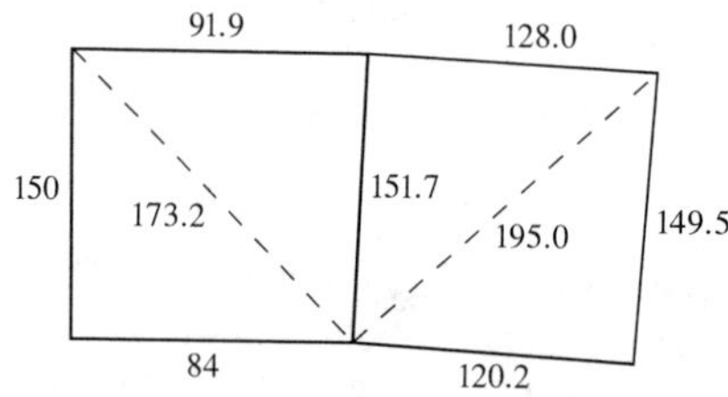


**8.** Answers will vary. Mount Everest was measured to be exactly 29,000 ft high, but the surveyors published the height as 29,002 ft to make it clear that they considered their measurement accurate to within a foot. The accepted figure today is 29,028 ft.

# 7 Trigonometric Functions of Real Numbers

## 7.1 The Unit Circle

**1.** Since $\left(\frac{4}{5}\right)^2+\left(-\frac{3}{5}\right)^2=\frac{16}{25}+\frac{9}{25}=1$, $P\left(\frac{4}{5},-\frac{3}{5}\right)$ lies on the unit circle.

**2.** Since $\left(-\frac{5}{13}\right)^2+\left(\frac{12}{13}\right)^2=\frac{144}{169}+\frac{25}{169}=1$, $P\left(-\frac{5}{13},\frac{12}{13}\right)$ lies on the unit circle.

**3.** Since $\left(\frac{7}{25}\right)^2+\left(\frac{24}{25}\right)^2=\frac{49}{625}+\frac{576}{625}=1$, $P\left(\frac{7}{25},\frac{24}{25}\right)$ lies on the unit circle.

**4.** Since $\left(-\frac{5}{7}\right)^2+\left(-\frac{2\sqrt{6}}{7}\right)^2=\frac{25}{49}+\frac{24}{49}=1$, $P\left(-\frac{5}{7},-\frac{2\sqrt{6}}{7}\right)$ lies on the unit circle.

**5.** Since $\left(-\frac{\sqrt{5}}{3}\right)^2+\left(\frac{2}{3}\right)^2=\frac{5}{9}+\frac{4}{9}=1$, $P\left(-\frac{\sqrt{5}}{3},\frac{2}{3}\right)$ lies on the unit circle.

**6.** Since $\left(\frac{\sqrt{11}}{6}\right)^2+\left(\frac{5}{6}\right)^2=\frac{11}{36}+\frac{25}{36}=1$, $P\left(\frac{\sqrt{11}}{6},\frac{5}{6}\right)$ lies on the unit circle.

**7.** $\left(-\frac{3}{5}\right)^2+y^2=1\Leftrightarrow y^2=1-\frac{9}{25}\Leftrightarrow y^2=\frac{16}{25}\Leftrightarrow y=\pm\frac{4}{5}$. Since $P(x,y)$ is in quadrant III, $y$ is negative, so the point is $P\left(-\frac{3}{5},-\frac{4}{5}\right)$.

**8.** $x^2+\left(-\frac{7}{25}\right)^2=1\Leftrightarrow x^2=1-\frac{49}{625}\Leftrightarrow x^2=\frac{576}{625}\Leftrightarrow x=\pm\frac{24}{25}$. Since $P$ is in quadrant IV, $x$ is positive, so the point is $P\left(\frac{24}{25},-\frac{7}{25}\right)$.

**9.** $x^2+\left(\frac{1}{3}\right)^2=1\Leftrightarrow x^2=1-\frac{1}{9}\Leftrightarrow x^2=\frac{8}{9}\Leftrightarrow x=\pm\frac{2\sqrt{2}}{3}$. Since $P$ is in quadrant II, $x$ is negative, so the point is $P\left(-\frac{2\sqrt{2}}{3},\frac{1}{3}\right)$.

**10.** $\left(\frac{2}{5}\right)^2+y^2=1\Leftrightarrow y^2=1-\frac{4}{25}\Leftrightarrow y^2=\frac{21}{25}\Leftrightarrow y=\pm\frac{\sqrt{21}}{5}$. Since $P$ is in quadrant I, $y$ is positive, so the point is $P\left(\frac{2}{5},\frac{\sqrt{21}}{5}\right)$.

**11.** $x^2+\left(-\frac{2}{7}\right)^2=1\Leftrightarrow x^2=1-\frac{4}{49}\Leftrightarrow x^2=\frac{45}{49}\Leftrightarrow x=\pm\frac{3\sqrt{5}}{7}$. Since $P(x,y)$ is in quadrant IV, $x$ is positive, so the point is $P\left(\frac{3\sqrt{5}}{7},-\frac{2}{7}\right)$.

**12.** $\left(-\frac{2}{3}\right)^2+y^2=1\Leftrightarrow y^2=1-\frac{4}{9}\Leftrightarrow y^2=\frac{5}{9}\Leftrightarrow y=\pm\frac{\sqrt{5}}{3}$. Since $P$ is in quadrant II, $y$ is positive, so the point is $P\left(-\frac{2}{3},\frac{\sqrt{5}}{3}\right)$.

**13.** $\left(\frac{4}{5}\right)^2+y^2=1\Leftrightarrow y^2=1-\frac{16}{25}\Leftrightarrow y^2=\frac{9}{25}\Leftrightarrow y=\pm\frac{3}{5}$. Since its $y$-coordinate is positive, the point is $P\left(\frac{4}{5},\frac{3}{5}\right)$.

**14.** $x^2+\left(-\frac{1}{3}\right)^2=1\Leftrightarrow x^2=1-\frac{1}{9}\Leftrightarrow x^2=\frac{8}{9}\Leftrightarrow x=\pm\frac{2\sqrt{2}}{3}$. Since its $x$-coordinate is positive, the point is $P\left(\frac{2\sqrt{2}}{3},-\frac{1}{3}\right)$.

**15.** $x^2+\left(\frac{2}{3}\right)^2=1\Leftrightarrow x^2=1-\frac{4}{9}\Leftrightarrow x^2=\frac{5}{9}\Leftrightarrow x=\pm\frac{\sqrt{5}}{3}$. Since its $x$-coordinate is negative, the point is $P\left(-\frac{\sqrt{5}}{3},\frac{2}{3}\right)$.

**16.** $x^2+\left(-\frac{\sqrt{5}}{5}\right)^2=1\Leftrightarrow x^2=1-\frac{5}{25}\Leftrightarrow x^2=\frac{20}{25}\Leftrightarrow x=\pm\frac{2\sqrt{5}}{5}$. Since its $x$-coordinate is positive, the point is $P\left(\frac{2\sqrt{5}}{5},-\frac{\sqrt{5}}{5}\right)$.

**17.** $\left(-\frac{\sqrt{2}}{3}\right)^2+y^2=1\Leftrightarrow y^2=1-\frac{2}{9}\Leftrightarrow y^2=\frac{7}{9}\Leftrightarrow y=\pm\frac{\sqrt{7}}{3}$. Since $P$ lies below the $x$-axis, its $y$-coordinate is negative, so the point is $P\left(-\frac{\sqrt{2}}{3},-\frac{\sqrt{7}}{3}\right)$.

**18.** $\left(-\frac{2}{5}\right)^2+y^2=1\Leftrightarrow y^2=1-\frac{4}{25}\Leftrightarrow y^2=\frac{21}{25}\Leftrightarrow y=\pm\frac{\sqrt{21}}{5}$. Since $P$ lies above the $x$-axis, its $y$-coordinate is positive, so the point is $P\left(-\frac{2}{5},\frac{\sqrt{21}}{5}\right)$.

**19.**

| $t$ | Terminal Point |
|---|---|
| $0$ | $(1,0)$ |
| $\frac{\pi}{4}$ | $\left(\frac{\sqrt{2}}{2},\frac{\sqrt{2}}{2}\right)$ |
| $\frac{\pi}{2}$ | $(0,1)$ |
| $\frac{3\pi}{4}$ | $\left(-\frac{\sqrt{2}}{2},\frac{\sqrt{2}}{2}\right)$ |
| $\pi$ | $(-1,0)$ |

| $t$ | Terminal Point |
|---|---|
| $\pi$ | $(-1,0)$ |
| $\frac{5\pi}{4}$ | $\left(-\frac{\sqrt{2}}{2},-\frac{\sqrt{2}}{2}\right)$ |
| $\frac{3\pi}{2}$ | $(0,-1)$ |
| $\frac{7\pi}{4}$ | $\left(\frac{\sqrt{2}}{2},-\frac{\sqrt{2}}{2}\right)$ |
| $2\pi$ | $(1,0)$ |

**20.**

| $t$ | Terminal Point |
|---|---|
| $0$ | $(1,0)$ |
| $\frac{\pi}{6}$ | $\left(\frac{\sqrt{3}}{2},\frac{1}{2}\right)$ |
| $\frac{\pi}{3}$ | $\left(\frac{1}{2},\frac{\sqrt{3}}{2}\right)$ |
| $\frac{\pi}{2}$ | $(0,1)$ |
| $\frac{2\pi}{3}$ | $\left(-\frac{1}{2},\frac{\sqrt{3}}{2}\right)$ |
| $\frac{5\pi}{6}$ | $\left(-\frac{\sqrt{3}}{2},\frac{1}{2}\right)$ |
| $\pi$ | $(-1,0)$ |

| $t$ | Terminal Point |
|---|---|
| $\pi$ | $(-1,0)$ |
| $\frac{7\pi}{6}$ | $\left(-\frac{\sqrt{3}}{2},-\frac{1}{2}\right)$ |
| $\frac{4\pi}{3}$ | $\left(-\frac{1}{2},-\frac{\sqrt{3}}{2}\right)$ |
| $\frac{3\pi}{2}$ | $(0,-1)$ |
| $\frac{5\pi}{3}$ | $\left(\frac{1}{2},-\frac{\sqrt{3}}{2}\right)$ |
| $\frac{11\pi}{6}$ | $\left(\frac{\sqrt{3}}{2},-\frac{1}{2}\right)$ |
| $2\pi$ | $(1,0)$ |

**21.** $P(x,y)=(0,1)$

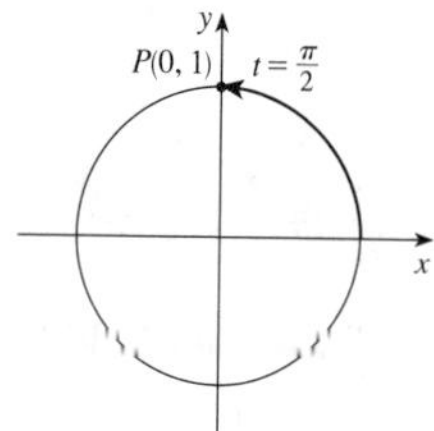


**22.** $P(x,y)=(0,-1)$

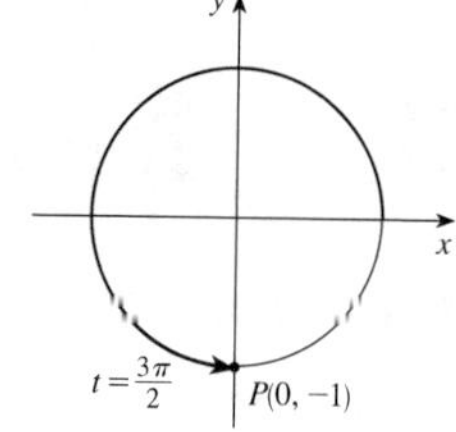


**23.** $P(x,y)=\left(-\frac{\sqrt{3}}{2},\frac{1}{2}\right)$

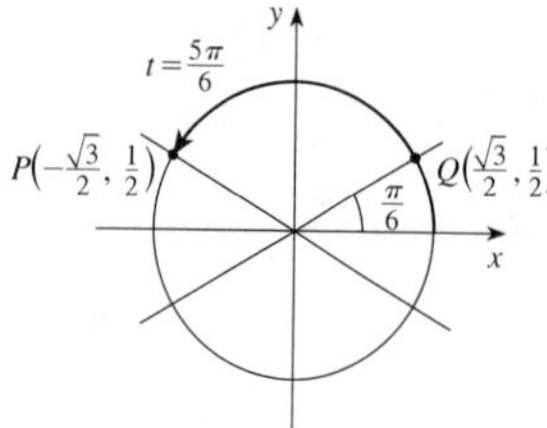


**24.** $P(x,y)=\left(-\frac{\sqrt{3}}{2},-\frac{1}{2}\right)$

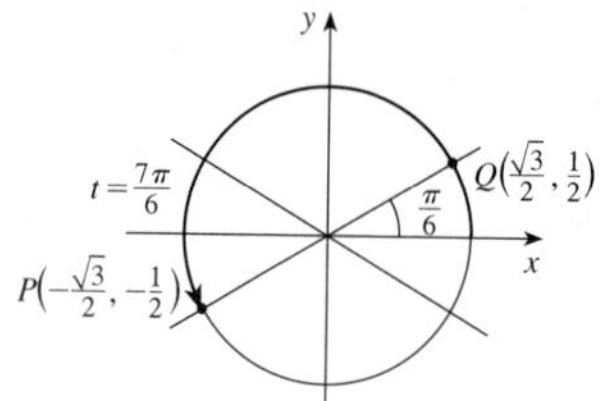


**25.** $P(x,y)=\left(\frac{1}{2},-\frac{\sqrt{3}}{2}\right)$

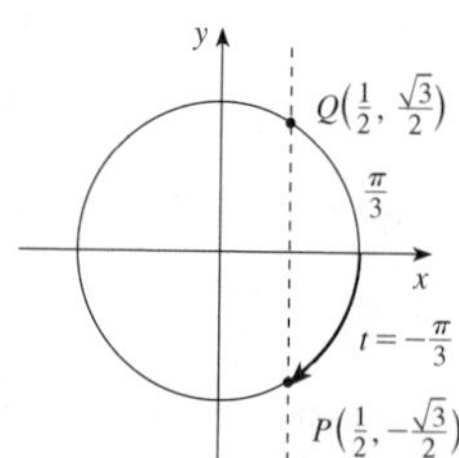


**26.** $P(x,y)=\left(\frac{1}{2},-\frac{\sqrt{3}}{2}\right)$

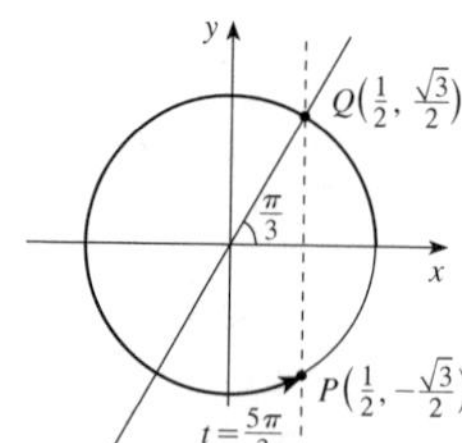


**27.** $P(x,y)=\left(-\frac{1}{2},\frac{\sqrt{3}}{2}\right)$

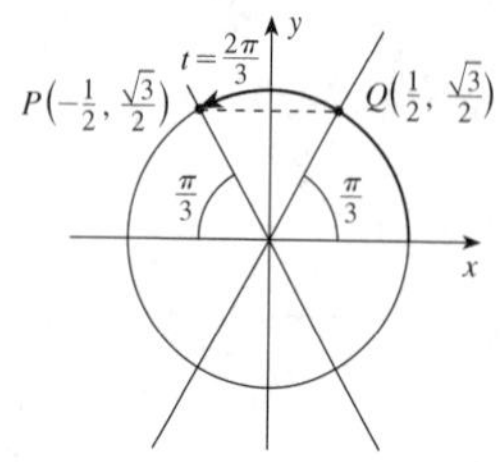


**28.** $P(x,y)=(0,-1)$

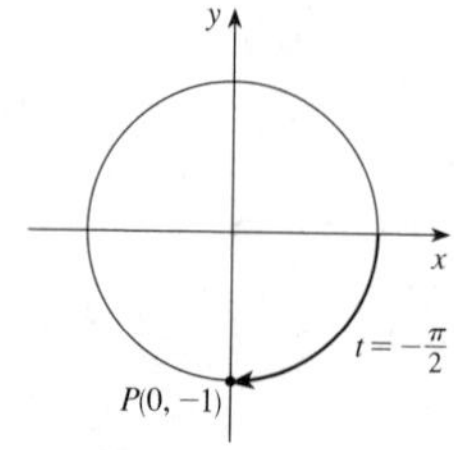

**29.** $P(x,y)=\left(-\frac{\sqrt{2}}{2},-\frac{\sqrt{2}}{2}\right)$

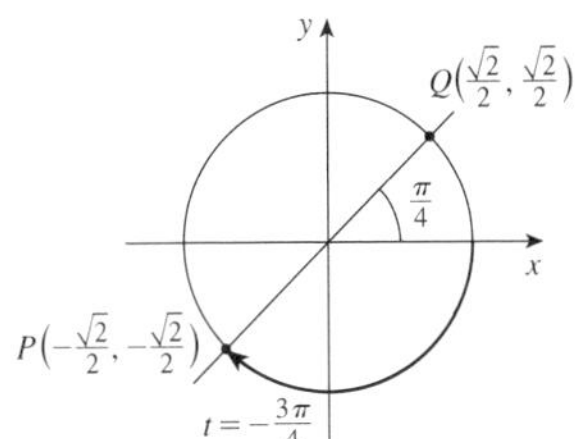

**30.** $P(x,y)=\left(\frac{\sqrt{3}}{2},-\frac{1}{2}\right)$

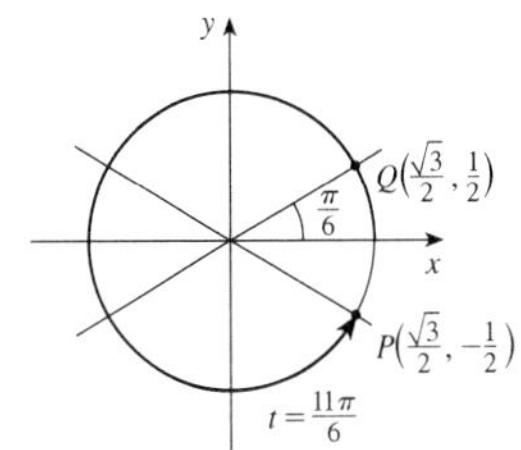

**31.** Let $Q(x,y)=\left(\frac{3}{5},\frac{4}{5}\right)$ be the terminal point determined by $t$.

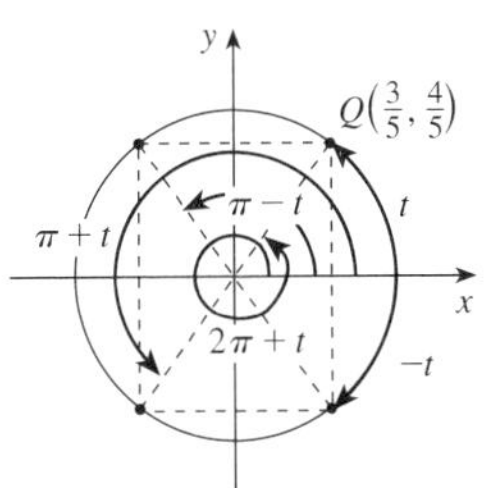

**(a)** $\pi-t$ determines the point $P(-x,y)=\left(-\frac{3}{5},\frac{4}{5}\right)$.
**(b)** $-t$ determines the point $P(x,-y)=\left(\frac{3}{5},-\frac{4}{5}\right)$.
**(c)** $\pi+t$ determines the point $P(-x,-y)=\left(-\frac{3}{5},-\frac{4}{5}\right)$.
**(d)** $2\pi+t$ determines the point $P(x,y)=\left(\frac{3}{5},\frac{4}{5}\right)$.

**32.** Let $Q(x,y)=\left(\frac{3}{4},\frac{\sqrt{7}}{4}\right)$ be the terminal point determined by $t$.

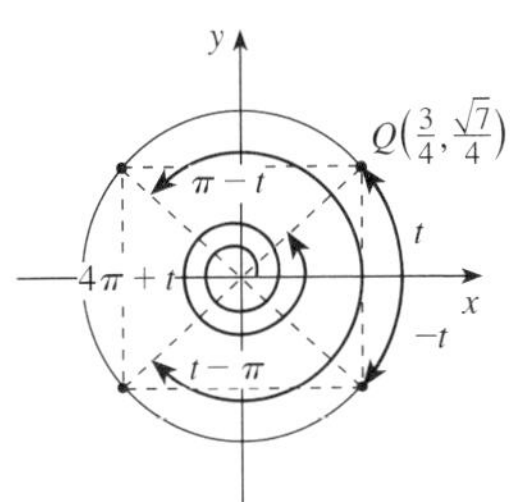

**(a)** $-t$ determines the point $P(x,-y)=\left(\frac{3}{4},-\frac{\sqrt{7}}{4}\right)$.
**(b)** $4\pi+t$ determines the point $P(x,y)=\left(\frac{3}{4},\frac{\sqrt{7}}{4}\right)$.
**(c)** $\pi-t$ determines the point $P(-x,y)=\left(-\frac{3}{4},\frac{\sqrt{7}}{4}\right)$.
**(d)** $t-\pi$ determines the point $P(-x,-y)=\left(-\frac{3}{4},-\frac{\sqrt{7}}{4}\right)$.

**33.** **(a)** $\bar{t}=\frac{5\pi}{4}-\pi=\frac{\pi}{4}$
**(b)** $\bar{t}=\frac{7\pi}{3}-2\pi=\frac{\pi}{3}$
**(c)** $\bar{t}=\frac{4\pi}{3}-\pi=\frac{\pi}{3}$
**(d)** $\bar{t}=\frac{\pi}{6}$

**34.** **(a)** $\bar{t}=\pi-\frac{5\pi}{6}=\frac{\pi}{6}$
**(b)** $\bar{t}=\frac{7\pi}{6}-\pi=\frac{\pi}{6}$
**(c)** $\bar{t}=4\pi-\frac{11\pi}{3}=\frac{\pi}{3}$
**(d)** $\bar{t}=2\pi-\frac{7\pi}{4}=\frac{\pi}{4}$

**35.** **(a)** $\bar{t}=\pi-\frac{5\pi}{7}=\frac{2\pi}{7}$
**(b)** $\bar{t}=\pi-\frac{7\pi}{9}=\frac{2\pi}{9}$
**(c)** $\bar{t}=\pi-3\approx 0.142$
**(d)** $\bar{t}=2\pi-5\approx 1.283$

**36.** **(a)** $\bar{t}=\frac{11\pi}{5}-2\pi=\frac{\pi}{5}$
**(b)** $\bar{t}=\frac{9\pi}{7}-\pi=\frac{2\pi}{7}$
**(c)** $\bar{t}=2\pi-6\approx 0.283$
**(d)** $\bar{t}=7-2\pi\approx 0.717$

**37.** **(a)** $\bar{t}=\pi-\frac{2\pi}{3}=\frac{\pi}{3}$
**(b)** $P\left(-\frac{1}{2},\frac{\sqrt{3}}{2}\right)$

**38.** **(a)** $\bar{t}=\frac{4\pi}{3}-\pi=\frac{\pi}{3}$
**(b)** $P\left(-\frac{1}{2},-\frac{\sqrt{3}}{2}\right)$

**39.** **(a)** $\bar{t}=\pi-\frac{3\pi}{4}=\frac{\pi}{4}$
**(b)** $P\left(-\frac{\sqrt{2}}{2},\frac{\sqrt{2}}{2}\right)$

**40.** **(a)** $\bar{t}=\frac{7\pi}{3}-2\pi=\frac{\pi}{3}$
**(b)** $P\left(\frac{1}{2},\frac{\sqrt{3}}{2}\right)$

**41.** **(a)** $\bar{t} = \pi - \frac{2\pi}{3} = \frac{\pi}{3}$
**(b)** $P\left(-\frac{1}{2}, -\frac{\sqrt{3}}{2}\right)$

**42.** **(a)** $\bar{t} = \frac{7\pi}{6} - \pi = \frac{\pi}{6}$
**(b)** $P\left(-\frac{\sqrt{3}}{2}, \frac{1}{2}\right)$

**43.** **(a)** $\bar{t} = \frac{13\pi}{4} - 3\pi = \frac{\pi}{4}$
**(b)** $P\left(-\frac{\sqrt{2}}{2}, -\frac{\sqrt{2}}{2}\right)$

**44.** **(a)** $\bar{t} = \frac{13\pi}{6} - 2\pi = \frac{\pi}{6}$
**(b)** $P\left(\frac{\sqrt{3}}{2}, \frac{1}{2}\right)$

**45.** **(a)** $\bar{t} = \frac{7\pi}{6} - \pi = \frac{\pi}{6}$
**(b)** $P\left(-\frac{\sqrt{3}}{2}, -\frac{1}{2}\right)$

**46.** **(a)** $\bar{t} = \frac{17\pi}{4} - 4\pi = \frac{\pi}{4}$
**(b)** $P\left(\frac{\sqrt{2}}{2}, \frac{\sqrt{2}}{2}\right)$

**47.** **(a)** $\bar{t} = 4\pi - \frac{11\pi}{3} = \frac{\pi}{3}$
**(b)** $P\left(\frac{1}{2}, \frac{\sqrt{3}}{2}\right)$

**48.** **(a)** $\bar{t} = \frac{31\pi}{6} - 5\pi = \frac{\pi}{6}$
**(b)** $P\left(-\frac{\sqrt{3}}{2}, -\frac{1}{2}\right)$

**49.** **(a)** $\bar{t} = \frac{16\pi}{3} - 5\pi = \frac{\pi}{3}$
**(b)** $P\left(-\frac{1}{2}, -\frac{\sqrt{3}}{2}\right)$

**50.** **(a)** $\bar{t} = -10\pi - \left(-\frac{41\pi}{4}\right) = \frac{\pi}{4}$
**(b)** $P\left(\frac{\sqrt{2}}{2}, -\frac{\sqrt{2}}{2}\right)$

**51.** $t = 1 \Rightarrow (0.5, 0.8)$

**52.** $t = 2.5 \Rightarrow P(-0.8, 0.6)$

**53.** $t = -1.1 \Rightarrow (0.5, -0.9)$

**54.** $t = 4.2 \Rightarrow P(-0.6, -0.9)$

**55.** The distances $PQ$ and $PR$ are equal because they both subtend arcs of length $\frac{\pi}{3}$. Since $P(x, y)$ is a point on the unit circle, $x^2 + y^2 = 1$. Now $d(P, Q) = \sqrt{(x-x)^2 + (y-(-y))^2} = 2y$ and $d(R, S) = \sqrt{(x-0)^2 + (y-1)^2} = \sqrt{x^2 + y^2 - 2y + 1} = \sqrt{2 - 2y}$ (using the fact that $x^2 + y^2 = 1$). Setting these equal gives $2y = \sqrt{2 - 2y} \Rightarrow 4y^2 = 2 - 2y \Leftrightarrow 4y^2 + 2y - 2 = 0 \Leftrightarrow 2(2y-1)(y+1) = 0$. So $y = -1$ or $y = \frac{1}{2}$. Since $P$ is in quadrant I, $y = \frac{1}{2}$ is the only viable solution. Again using $x^2 + y^2 = 1$ we have $x^2 + \left(\frac{1}{2}\right)^2 = 1 \Leftrightarrow x^2 = \frac{3}{4}$ $\Rightarrow x = \pm\frac{\sqrt{3}}{2}$. Again, since $P$ is in quadrant I the coordinates must be $\left(\frac{\sqrt{3}}{2}, \frac{1}{2}\right)$.

**56.** $P$ is the reflection of $Q$ about the line $y = x$. Since $Q$ is the point $Q\left(\frac{\sqrt{3}}{2}, \frac{1}{2}\right)$ it follows that $P$ is the point $P\left(\frac{1}{2}, \frac{\sqrt{3}}{2}\right)$.

## 7.2 Trigonometric Functions of Real Numbers

**1.**

| $t$ | $\sin t$ | $\cos t$ |
|---|---|---|
| $0$ | $0$ | $1$ |
| $\frac{\pi}{4}$ | $\frac{\sqrt{2}}{2}$ | $\frac{\sqrt{2}}{2}$ |
| $\frac{\pi}{2}$ | $1$ | $0$ |
| $\frac{3\pi}{4}$ | $\frac{\sqrt{2}}{2}$ | $-\frac{\sqrt{2}}{2}$ |
| $\pi$ | $0$ | $-1$ |
| $\frac{5\pi}{4}$ | $-\frac{\sqrt{2}}{2}$ | $-\frac{\sqrt{2}}{2}$ |
| $\frac{3\pi}{2}$ | $-1$ | $0$ |
| $\frac{7\pi}{4}$ | $-\frac{\sqrt{2}}{2}$ | $\frac{\sqrt{2}}{2}$ |
| $2\pi$ | $0$ | $1$ |

**2.**

| $t$ | $\sin t$ | $\cos t$ |
|---|---|---|
| $0$ | $0$ | $1$ |
| $\frac{\pi}{6}$ | $\frac{1}{2}$ | $\frac{\sqrt{3}}{2}$ |
| $\frac{\pi}{3}$ | $\frac{\sqrt{3}}{2}$ | $\frac{1}{2}$ |
| $\frac{\pi}{2}$ | $1$ | $0$ |
| $\frac{2\pi}{3}$ | $\frac{\sqrt{3}}{2}$ | $-\frac{1}{2}$ |
| $\frac{5\pi}{6}$ | $\frac{1}{2}$ | $-\frac{\sqrt{3}}{2}$ |
| $\pi$ | $0$ | $-1$ |

| $t$ | $\sin t$ | $\cos t$ |
|---|---|---|
| $\pi$ | $0$ | $-1$ |
| $\frac{7\pi}{6}$ | $-\frac{1}{2}$ | $-\frac{\sqrt{3}}{2}$ |
| $\frac{4\pi}{3}$ | $-\frac{\sqrt{3}}{2}$ | $-\frac{1}{2}$ |
| $\frac{3\pi}{2}$ | $-1$ | $0$ |
| $\frac{5\pi}{3}$ | $-\frac{\sqrt{3}}{2}$ | $\frac{1}{2}$ |
| $\frac{11\pi}{6}$ | $-\frac{1}{2}$ | $\frac{\sqrt{3}}{2}$ |
| $2\pi$ | $0$ | $1$ |

**3. (a)** $\sin\frac{2\pi}{3}=\frac{\sqrt{3}}{2}$
**(b)** $\cos\frac{2\pi}{3}=-\frac{1}{2}$
**(c)** $\tan\frac{2\pi}{3}=-\sqrt{3}$

**4. (a)** $\sin\frac{5\pi}{6}=\frac{1}{2}$
**(b)** $\cos\frac{5\pi}{6}=-\frac{\sqrt{3}}{2}$
**(c)** $\tan\frac{5\pi}{6}=-\frac{\sqrt{3}}{3}$

**5. (a)** $\sin\frac{7\pi}{6}=-\frac{1}{2}$
**(b)** $\sin\left(-\frac{\pi}{6}\right)=-\frac{1}{2}$
**(c)** $\sin\frac{11\pi}{6}=-\frac{1}{2}$

**6. (a)** $\cos\frac{5\pi}{3}=\frac{1}{2}$
**(b)** $\cos\left(-\frac{5\pi}{3}\right)=\frac{1}{2}$
**(c)** $\cos\frac{7\pi}{3}=\frac{1}{2}$

**7. (a)** $\cos\frac{3\pi}{4}=-\frac{\sqrt{2}}{2}$
**(b)** $\cos\frac{5\pi}{4}=-\frac{\sqrt{2}}{2}$
**(c)** $\cos\frac{7\pi}{4}=\frac{\sqrt{2}}{2}$

**8. (a)** $\sin\frac{3\pi}{4}=\frac{\sqrt{2}}{2}$
**(b)** $\sin\frac{5\pi}{4}=-\frac{\sqrt{2}}{2}$
**(c)** $\sin\frac{7\pi}{4}=-\frac{\sqrt{2}}{2}$

**9. (a)** $\sin\frac{7\pi}{3}=\frac{\sqrt{3}}{2}$
**(b)** $\csc\frac{7\pi}{3}=\frac{2\sqrt{3}}{3}$
**(c)** $\cot\frac{7\pi}{3}=\frac{\sqrt{3}}{3}$

**10. (a)** $\cos\left(-\frac{\pi}{3}\right)=\frac{1}{2}$
**(b)** $\sec\left(-\frac{\pi}{3}\right)=2$
**(c)** $\tan\left(-\frac{\pi}{3}\right)=-\sqrt{3}$

**11. (a)** $\sin\left(-\frac{\pi}{2}\right)=-1$
**(b)** $\cos\left(-\frac{\pi}{2}\right)=0$
**(c)** $\cot\left(-\frac{\pi}{2}\right)=0$

**12. (a)** $\sin\left(-\frac{3\pi}{2}\right)=1$
**(b)** $\cos\left(-\frac{3\pi}{2}\right)=0$
**(c)** $\cot\left(-\frac{3\pi}{2}\right)=0$

**13. (a)** $\sec\frac{11\pi}{3}=2$
**(b)** $\csc\frac{11\pi}{3}=-\frac{2\sqrt{3}}{3}$
**(c)** $\sec\left(-\frac{\pi}{3}\right)=2$

**14. (a)** $\cos\frac{7\pi}{6}=-\frac{\sqrt{3}}{2}$
**(b)** $\sec\frac{7\pi}{6}=-\frac{2\sqrt{3}}{3}$
**(c)** $\csc\frac{7\pi}{6}=-2$

**15. (a)** $\tan\frac{5\pi}{6}=-\frac{\sqrt{3}}{3}$
**(b)** $\tan\frac{7\pi}{6}=\frac{\sqrt{3}}{3}$
**(c)** $\tan\frac{11\pi}{6}=-\frac{\sqrt{3}}{3}$

**16. (a)** $\cot\left(-\frac{\pi}{3}\right)=-\frac{\sqrt{3}}{3}$
**(b)** $\cot\frac{2\pi}{3}=-\frac{\sqrt{3}}{3}$
**(c)** $\cot\frac{5\pi}{3}=-\frac{\sqrt{3}}{3}$

**17. (a)** $\cos\left(-\frac{\pi}{4}\right)=\frac{\sqrt{2}}{2}$
**(b)** $\csc\left(-\frac{\pi}{4}\right)=-\sqrt{2}$
**(c)** $\cot\left(-\frac{\pi}{4}\right)=-1$

**18. (a)** $\sin\frac{5\pi}{4}=-\frac{\sqrt{2}}{2}$
**(b)** $\sec\frac{5\pi}{4}=-\sqrt{2}$
**(c)** $\tan\frac{5\pi}{4}=1$

**19. (a)** $\csc\left(-\frac{\pi}{2}\right)=-1$
**(b)** $\csc\frac{\pi}{2}=1$
**(c)** $\csc\frac{3\pi}{2}=-1$

**20. (a)** $\sec(-\pi)=-1$
**(b)** $\sec\pi=-1$
**(c)** $\sec 4\pi=1$

**21. (a)** $\sin 13\pi=0$
**(b)** $\cos 14\pi=1$
**(c)** $\tan 15\pi=0$

**22. (a)** $\sin\frac{25\pi}{2}=\sin\left(\frac{\pi}{2}+24\pi\right)=1$
**(b)** $\cos\frac{25\pi}{2}=\cos\left(\frac{\pi}{2}+24\pi\right)=0$
**(c)** $\cot\frac{25\pi}{2}=\cot\left(\frac{\pi}{2}+24\pi\right)=0$

**23.** $t=0 \Rightarrow \sin t=0$, $\cos t=1$, $\tan t=0$, $\sec t=1$, $\csc t$ and $\cot t$ are undefined.

**24.** $t=\frac{\pi}{2} \Rightarrow \sin t=1$, $\cos t=0$, $\csc t=1$, $\cot t=0$, $\tan t$ and $\sec t$ are undefined.

**25.** $t=\pi \Rightarrow \sin t=0$, $\cos t=-1$, $\tan t=0$, $\sec t=-1$, $\csc t$ and $\cot t$ are undefined.

**26.** $t=\frac{3\pi}{2} \Rightarrow \sin t=-1$, $\cos t=0$, $\csc t=-1$, $\cot t=0$, $\tan t$ and $\sec t$ are undefined.

**27.** $\left(\frac{3}{5}\right)^2+\left(\frac{4}{5}\right)^2=\frac{9}{25}+\frac{1}{2}=1$. So $\sin t=\frac{4}{5}$, $\cos t=\frac{3}{5}$, and $\tan t=\frac{\frac{4}{5}}{\frac{3}{5}}=\frac{4}{3}$.

**28.** $\left(-\frac{3}{5}\right)^2+\left(\frac{4}{5}\right)^2=\frac{9}{25}+\frac{1}{2}=1$. So $\sin t=\frac{4}{5}$, $\cos t=-\frac{3}{5}$, and $\tan t=\frac{\frac{4}{5}}{-\frac{3}{5}}=-\frac{4}{3}$.

**29.** $\left(\frac{\sqrt{5}}{4}\right)^2 + \left(-\frac{\sqrt{11}}{4}\right)^2 = \frac{5}{16} + \frac{11}{16} = 1$. So $\sin t = -\frac{\sqrt{11}}{4}$, $\cos t = \frac{\sqrt{5}}{4}$, and $\tan t = \dfrac{-\frac{\sqrt{11}}{4}}{\frac{\sqrt{5}}{4}} = -\frac{\sqrt{11}}{\sqrt{5}} = -\frac{\sqrt{55}}{5}$.

**30.** $\left(-\frac{1}{3}\right)^2 + \left(-\frac{2\sqrt{2}}{3}\right)^2 = \frac{1}{9} + \frac{8}{9} = 1$. So $\sin t = -\frac{2\sqrt{2}}{3}$, $\cos t = -\frac{1}{3}$, and $\tan t = \dfrac{-\frac{2\sqrt{2}}{3}}{-\frac{1}{3}} = 2\sqrt{2}$.

**31.** $\left(-\frac{6}{7}\right)^2 + \left(\frac{\sqrt{13}}{7}\right)^2 = \frac{36}{49} + \frac{13}{49} = 1$. So $\sin t = \frac{\sqrt{13}}{7}$, $\cos t = -\frac{6}{7}$, and $\tan t = \dfrac{\frac{\sqrt{13}}{7}}{-\frac{6}{7}} = -\frac{\sqrt{13}}{6}$.

**32.** $\left(\frac{40}{41}\right)^2 + \left(\frac{9}{41}\right)^2 = \frac{1600}{1681} + \frac{81}{1681} = 1$. So $\sin t = \frac{9}{41}$, $\cos t = \frac{40}{41}$, and $\tan t = \dfrac{\frac{9}{41}}{\frac{40}{41}} = \frac{9}{40}$.

**33.** $\left(-\frac{5}{13}\right)^2 + \left(-\frac{12}{13}\right)^2 = \frac{25}{169} + \frac{144}{169} = 1$. So $\sin t = -\frac{12}{13}$, $\cos t = -\frac{5}{13}$, and $\tan t \dfrac{-\frac{12}{13}}{-\frac{5}{13}} = \frac{12}{5}$.

**34.** $\left(\frac{\sqrt{5}}{5}\right)^2 + \left(\frac{2\sqrt{5}}{5}\right)^2 = \frac{5}{25} + \frac{20}{25} = 1$. So $\sin t = \frac{2\sqrt{5}}{5}$, $\cos t = \frac{\sqrt{5}}{5}$, and $\tan t = \dfrac{\frac{2\sqrt{5}}{5}}{\frac{\sqrt{5}}{5}} = 2$.

**35.** $\left(-\frac{20}{29}\right)^2 + \left(\frac{21}{29}\right)^2 = \frac{400}{841} + \frac{441}{841} = 1$. So $\sin t = \frac{21}{29}$, $\cos t = -\frac{20}{29}$, and $\tan t = \dfrac{\frac{21}{29}}{-\frac{20}{29}} = -\frac{21}{20}$.

**36.** $\left(\frac{24}{25}\right)^2 + \left(-\frac{7}{25}\right)^2 = \frac{576}{625} + \frac{49}{625} = 1$. So $\sin t = -\frac{7}{25}$, $\cos t = \frac{24}{25}$, and $\tan t = \dfrac{-\frac{7}{25}}{\frac{24}{25}} = -\frac{7}{24}$.

**37.** **(a)** 0.8
**(b)** 0.84147

**38.** **(a)** 0.7
**(b)** 0.69671

**39.** **(a)** 0.9
**(b)** 0.93204

**40.** **(a)** 0.3
**(b)** 0.28366

**41.** **(a)** 1.0
**(b)** 1.02964

**42.** **(a)** $-3.6$
**(b)** $-3.60210$

**43.** **(a)** $-0.6$
**(b)** $-0.57482$

**44.** **(a)** 0.9
**(b)** 0.88345

**45.** $\sin t \cdot \cos t$. Since $\sin t$ is positive in quadrant II and $\cos t$ is negative in quadrant II, their product is negative.

**46.** $\tan t \cdot \sec t$ is negative in quadrant IV because $\tan t$ is negative and $\cos t$ is positive in quadrant IV.

**47.** $\dfrac{\tan t \cdot \sin t}{\cot t} = \tan t \cdot \dfrac{1}{\cot t} \cdot \sin t = \tan t \cdot \tan t \cdot \sin t = \tan^2 t \cdot \sin t$. Since $\tan^2 t$ is always positive and $\sin t$ is negative in quadrant III, the expression is negative in quadrant III.

**48.** $\cos t \cdot \sec t$ is positive in any quadrant, since $\cos t \cdot \sec t = \cos t \cdot \dfrac{1}{\cos t} = 1$, provided $\cos t \neq 0$.

**49.** Quadrant II

**50.** Quadrant III

**51.** Quadrant II

**52.** Quadrant II

**53.** $\sin t = \sqrt{1 - \cos^2 t}$

**54.** $\cos t = \sqrt{1 - \sin^2 t}$

**55.** $\tan t = \dfrac{\sin t}{\cos t} = \dfrac{\sin t}{\sqrt{1 - \sin^2 t}}$

**56.** $\tan t = -\dfrac{\sqrt{1 - \cos^2 t}}{\cos t}$

**57.** $\sec t = -\sqrt{1 + \tan^2 t}$

**58.** $\csc t = -\sqrt{1 + \cot^2 t}$

**59.** $\tan t = \sqrt{\sec^2 t - 1}$

**60.** $\sin t = -\sqrt{1 - \cos^2 t} = -\sqrt{1 - \dfrac{1}{\sec^2 t}}$

**61.** $\tan^2 t = \dfrac{\sin^2 t}{\cos^2 t} = \dfrac{\sin^2 t}{1 - \sin^2 t}$

**62.** $\sec^2 t \cdot \sin^2 t = \dfrac{1}{\cos^2 t} \cdot \left(1 - \cos^2 t\right) = \dfrac{1}{\cos^2 t} - 1$

**63.** $\sin t = \frac{3}{5}$ and $t$ is in quadrant II, so the terminal point determined by $t$ is $P\left(x, \frac{3}{5}\right)$. Since $P$ is on the unit circle $x^2 + \left(\frac{3}{5}\right)^2 = 1$. Solving for $x$ gives $x = \pm\sqrt{1 - \frac{9}{25}} = \pm\sqrt{\frac{1}{2}} = \pm\dfrac{4}{5}$. Since $t$ is in quadrant III, $x = -\frac{4}{5}$. Thus the terminal point is $P\left(-\frac{4}{5}, \frac{3}{5}\right)$. Thus, $\cos t = -\dfrac{4}{5}$, $\tan t = -\frac{3}{4}$, $\csc t = \frac{5}{3}$, $\sec t = -\frac{5}{4}$, $\cot t = -\frac{4}{3}$.

**64.** $\cos t = -\frac{4}{5}$ and $t$ lies in quadrant III, so the terminal point determined by $t$ is $P\left(-\frac{4}{5}, y\right)$. Since $P$ is on the unit circle $\left(-\frac{4}{5}\right)^2 + y^2 = 1$. Solving for $y$ gives $x = \pm\sqrt{1-\frac{1}{2}} = \pm\sqrt{\frac{9}{25}} = \pm\frac{3}{5}$. Since $t$ is in quadrant III, $y = -\frac{3}{5}$. Thus the terminal point is $P\left(-\frac{4}{5}, -\frac{3}{5}\right)$. Thus, $\sin t = -\frac{3}{5}$, $\tan t = \frac{3}{4}$, $\csc t = -\frac{5}{3}$, $\sec t = -\frac{5}{4}$, $\cot t = \frac{4}{3}$.

**65.** $\sec t = 3$ and $t$ lies in quadrant IV. Thus, $\cos t = \frac{1}{3}$ and the terminal point determined by $t$ is $P\left(\frac{1}{3}, y\right)$. Since $P$ is on the unit circle $\left(\frac{1}{3}\right)^2 + y^2 = 1$. Solving for $y$ gives $y = \pm\sqrt{1-\frac{1}{9}} = \pm\sqrt{\frac{8}{9}} = \pm\frac{2\sqrt{2}}{3}$. Since $t$ is in quadrant IV, $y = -\frac{2\sqrt{2}}{3}$. Thus the terminal point is $P\left(\frac{1}{3}, -\frac{2\sqrt{2}}{3}\right)$. Therefore, $\sin t = -\frac{2\sqrt{2}}{3}$, $\cos t = \frac{1}{3}$, $\tan t = -2\sqrt{2}$, $\csc t = -\dfrac{3}{2\sqrt{2}} = -\dfrac{3\sqrt{2}}{4}$, $\cot t = -\dfrac{1}{2\sqrt{2}} = -\dfrac{\sqrt{2}}{4}$.

**66.** $\tan t = \frac{1}{4}$ and $t$ lies in quadrant III. Since $\sec^2 t = \tan^2 t + 1$ we have $\sec^2 t = \left(\frac{1}{4}\right)^2 + 1 = \frac{1}{16} + 1 = \frac{17}{16}$. Thus $\sec t = \pm\sqrt{\frac{17}{16}} = \pm\frac{\sqrt{17}}{4}$. Since $\sec t < 0$ in quadrant III we have $\sec t = -\frac{\sqrt{17}}{4}$, so $\cos t = \dfrac{1}{\sec t} = \dfrac{1}{-\frac{\sqrt{17}}{4}} = -\dfrac{4}{\sqrt{17}} = -\dfrac{4\sqrt{17}}{17}$. Since $\tan t \cdot \cos t = \sin t$ we have $\sin t = \left(\frac{1}{4}\right)\left(-\frac{4}{\sqrt{17}}\right) = -\frac{1}{\sqrt{17}} = -\frac{\sqrt{17}}{17}$. Thus, the terminal point determined by $t$ is $P\left(-\frac{4\sqrt{17}}{17}, -\frac{\sqrt{17}}{17}\right)$. Therefore, $\sin t = -\frac{\sqrt{17}}{17}$, $\cos t = -\frac{4\sqrt{17}}{17}$, $\csc t = -\sqrt{17}$, $\sec t = -\frac{\sqrt{17}}{4}$, $\cot t = 4$.

**67.** $\tan t = -\frac{3}{4}$ and $\cos t > 0$, so $t$ is in quadrant IV. Since $\sec^2 t = \tan^2 t + 1$ we have $\sec^2 t = \left(-\frac{3}{4}\right)^2 + 1 = \frac{9}{16} + 1 = \frac{25}{16}$. Thus $\sec t = \pm\sqrt{\frac{25}{16}} = \pm\frac{5}{4}$. Since $\cos t > 0$, we have $\cos t = \dfrac{1}{\sec t} = \dfrac{1}{\frac{5}{4}} = \frac{4}{5}$. Let $P\left(\frac{4}{5}, y\right)$. Since $\tan t \cdot \cos t = \sin t$ we have $\sin t = \left(-\frac{3}{4}\right)\left(\frac{4}{5}\right) = -\frac{3}{5}$. Thus, the terminal point determined by $t$ is $P\left(\frac{4}{5}, -\frac{3}{5}\right)$, and so $\sin t = -\frac{3}{5}$, $\cos t = \frac{4}{5}$, $\csc t = -\frac{5}{3}$, $\sec t = \frac{5}{4}$, $\cot t = -\frac{4}{3}$.

**68.** $\sec t = 2$ and $\sin t < 0$, so $t$ is in quadrant IV. Thus, $\cos t = \frac{1}{2}$ and the terminal point determined by $t$ is $P\left(\frac{1}{2}, y\right)$. Since $P$ is on the unit circle $\left(\frac{1}{2}\right)^2 + y^2 = 1$. Solving for $y$ gives $y = \pm\sqrt{1-\frac{1}{4}} = \pm\sqrt{\frac{3}{4}} = \pm\frac{\sqrt{3}}{2}$. Since $t$ is in quadrant IV, $y = -\frac{\sqrt{3}}{2}$. Thus the terminal point is $P\left(\frac{1}{2}, -\frac{\sqrt{3}}{2}\right)$, and so $\sin t = -\frac{\sqrt{3}}{2}$, $\cos t = \frac{1}{2}$, $\tan t = -\sqrt{3}$, $\csc t = -\frac{2}{\sqrt{3}} = -\frac{2\sqrt{3}}{3}$, $\cot t = -\frac{1}{\sqrt{3}} = -\frac{\sqrt{3}}{3}$.

**69.** $\sin t = -\frac{1}{4}$, $\sec t < 0$, so $t$ is in quadrant III. So the terminal point determined by $t$ is $P\left(x, -\frac{1}{4}\right)$. Since $P$ is on the unit circle $x^2 + \left(-\frac{1}{4}\right)^2 = 1$. Solving for $x$ gives $x = \pm\sqrt{1-\frac{1}{16}} = \pm\sqrt{\frac{15}{16}} = \pm\frac{\sqrt{15}}{4}$. Since $t$ is in quadrant III, $x = -\dfrac{\sqrt{15}}{4}$. Thus, the terminal point determined by $t$ is $P\left(-\frac{\sqrt{15}}{4}, -\frac{1}{4}\right)$, and so $\cos t = -\frac{\sqrt{15}}{4}$, $\tan t = \frac{1}{\sqrt{15}} = \frac{\sqrt{15}}{15}$, $\csc t = -4$, $\sec t = -\frac{4}{\sqrt{15}} = -\frac{4\sqrt{15}}{15}$, $\cot t = \sqrt{15}$.

**70.** $\tan t = -4$ and $t$ lies in quadrant II. Since $\sec^2 t = \tan^2 t + 1$ we have $\sec^2 t = (-4)^2 + 1 = 16 + 1 = 17$. Thus $\sec t = \pm\sqrt{17}$. Since $\sec t < 0$, we have $\sec t = -\sqrt{17}$and $\cos t = \dfrac{1}{\sec t} = \frac{1}{-\sqrt{17}} = -\dfrac{\sqrt{17}}{17}$. Since $\tan t \cdot \cos t = \sin t$ we have $\sin t = (-4)\left(-\frac{\sqrt{17}}{17}\right) = \frac{4\sqrt{17}}{17}$. Thus, the terminal point determined by $t$ is $P\left(-\frac{\sqrt{17}}{17}, \frac{4\sqrt{17}}{17}\right)$. Thus, $\sin t = \frac{4\sqrt{17}}{17}$, $\cos t = -\frac{\sqrt{17}}{17}$, $\csc t = \frac{\sqrt{17}}{4}$, $\sec t = -\sqrt{17}$, $\cot t = -\frac{1}{4}$.

**71.** $f(-x) = (-x)^2 \sin(-x) = -x^2 \sin x = -f(x)$, so $f$ is odd.

**72.** $f(-x) = (-x)^2 \cos 2(-x) = x^2 \cos 2x = f(x)$, so $f$ is even.

**73.** $f(-x) = \sin(-x)\cos(-x) = -\sin x \cos x = -f(x)$, so $f$ is odd.

**74.** $f(-x) = \sin(-x) + \cos(-x) = -\sin x + \cos x$ which is neither $f(x)$ nor $-f(x)$, so $f$ is neither even nor odd.

**75.** $f(-x) = |-x|\cos(-x) = |x|\cos x = f(x)$, so $f$ is even.

**76.** $f(-x) = -x\sin^3(-x) = -x[\sin(-x)]^3 = -x(-\sin x)^3 = x\sin^3 x = f(x)$, so $f$ is even.

**77.** $f(-x) = (-x)^3 + \cos(-x) = -x^3 + \cos x$ which is neither $f(x)$ nor $-f(x)$, so $f$ is neither even nor odd.

**78.** $f(-x) = \cos(\sin(-x)) = \cos(-\sin x) = \cos(\sin x) = f(x)$, so $f$ is even.

**79.**

| $t$ | $y(t)$ |
|---|---|
| 0 | 4 |
| 0.25 | −2.83 |
| 0.50 | 0 |
| 0.75 | 2.83 |
| 1.00 | −4 |
| 1.25 | 2.83 |

**80.** **(a)** $B(6) = 80 + 7\sin\frac{\pi}{2} = 87$

**(b)** $B(10.5) = 80 + 7\sin\frac{10.5\pi}{12} \approx 82.7$

**(c)** $B(12) = 80 + 7\sin\pi = 80$

**(d)** $B(20) = 80 + 7\sin\frac{20\pi}{12} = 80 - \frac{7\sqrt{3}}{2} \approx 73.9$

**81.** **(a)** $I(0.1) = 0.8e^{-0.3}\sin 1 \approx 0.499\text{ A}$

**(b)** $I(0.5) = 0.8e^{-1.5}\sin 5 \approx -0.171\text{ A}$

**82.**

| $t$ | $H(t)$ |
|---|---|
| 0 | 175 |
| 1 | 150.4 |
| 2 | 100 |
| 4 | 38.6 |
| 6 | 100 |
| 8 | 150.3 |
| 12 | 58.8 |

**83.** Notice that if $P(t) = (x, y)$, then $P(t+\pi) = (-x, -y)$. Thus,

**(a)** $\sin(t+\pi) = -y$ and $\sin t = y$. Therefore, $\sin(t+\pi) = -\sin t$.

**(b)** $\cos(t+\pi) = -x$ and $\cos t = x$. Therefore, $\cos(t+\pi) = -\cos t$.

**(c)** $\tan(t+\pi) = \dfrac{\sin(t+\pi)}{\cos(t+\pi)} = \dfrac{-y}{-x} = \dfrac{y}{x} = \dfrac{\sin t}{\cos t} = \tan t$.

**84.** To prove that $\Delta AOB = \Delta CDO$, first note that $OB = OD = 1$ and $\angle OAB = \angle OCD = \frac{\pi}{2}$. Now $\angle COD + \angle AOB + \frac{\pi}{2} = \pi \quad\Leftrightarrow\quad \angle COD = \frac{\pi}{2} - \angle AOB = \angle ABO$. Since we know two angles and one side to be equal, the triangles are (SAA) congruent. Thus $AB = OC$ and $OA = CD$, so if $B$ has coordinates $(x, y)$, then $D$ has coordinates $(-y, x)$. Therefore,

**(a)** $\sin\left(t + \frac{\pi}{2}\right) = x = \cos t$

**(b)** $\cos\left(t + \frac{\pi}{2}\right) = -y$, and $\sin t = y$. Therefore, $\cos\left(t + \frac{\pi}{2}\right) = -\sin t$.

**(c)** $\tan\left(t + \frac{\pi}{2}\right) = \dfrac{x}{-y} = -\dfrac{x}{y} = -\dfrac{\cos t}{\sin t} = -\cot t$

# 7.3 Trigonometric Graphs

**1.** $f(x) = 1 + \cos x$

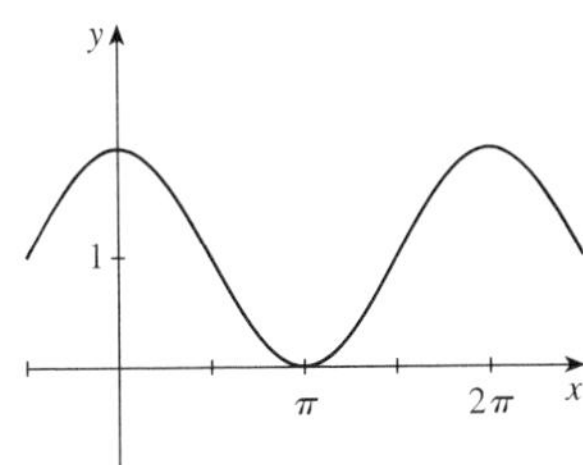


**2.** $f(x) = 3 + \sin x$

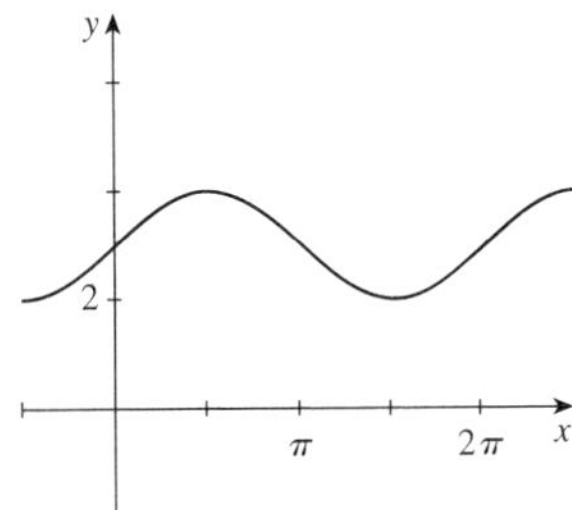


**3.** $f(x) = -\sin x$

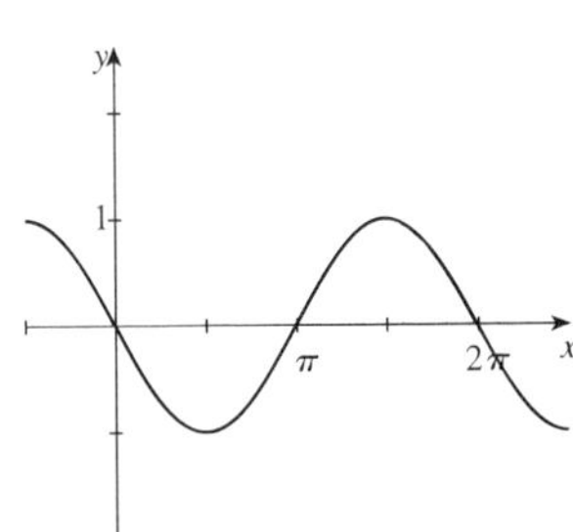


**4.** $f(x) = 2 - \cos x$

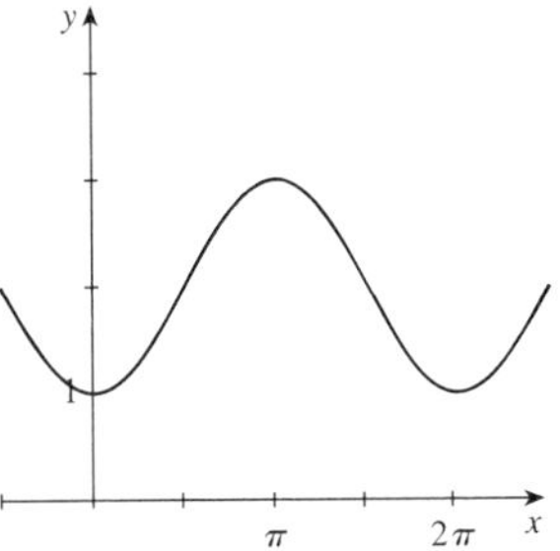


**5.** $f(x) = -2 + \sin x$

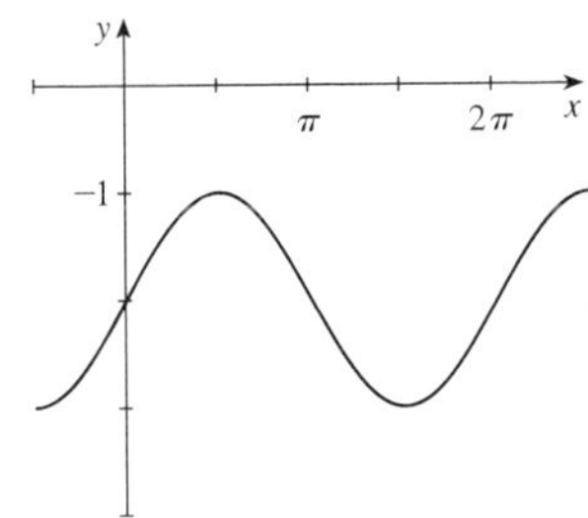


**6.** $f(x) = -1 + \cos x$

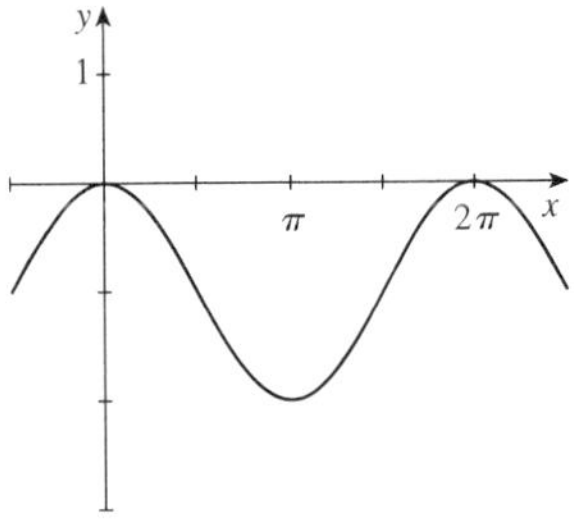


**7.** $g(x) = 3 \cos x$

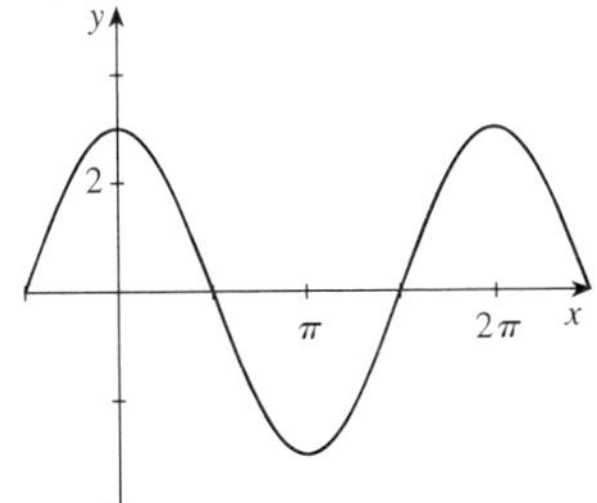


**8.** $g(x) = 2 \sin x$

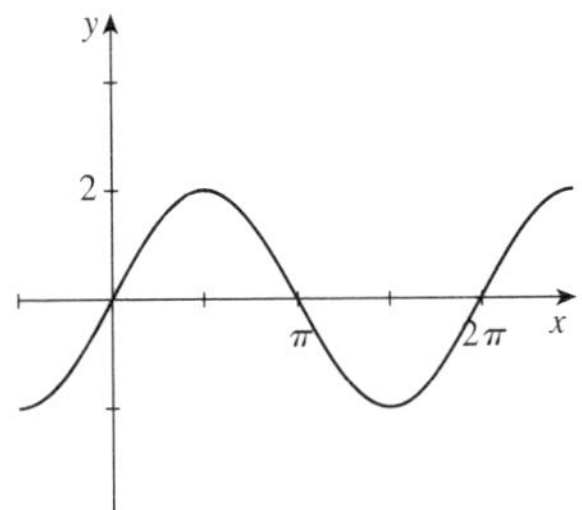

**9.** $g(x) = -\frac{1}{2}\sin x$

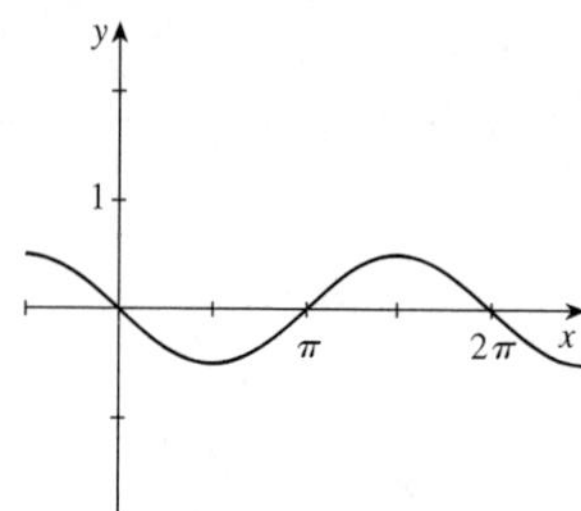


**10.** $g(x) = -\frac{2}{3}\cos x$

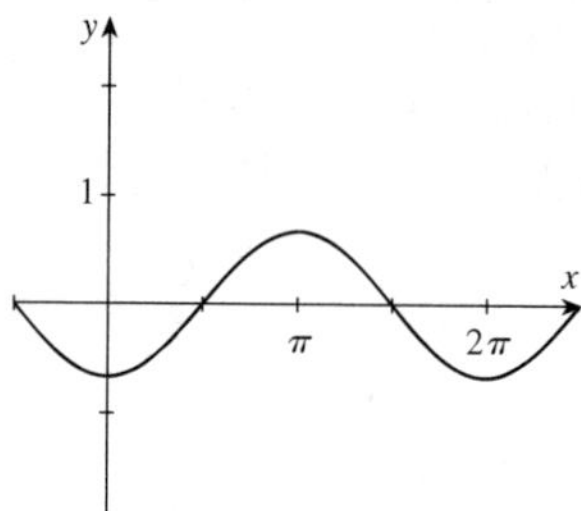


**11.** $g(x) = 3 + 3\cos x$

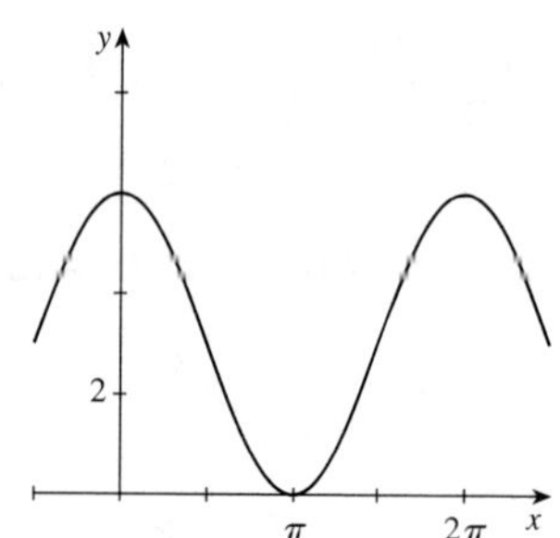


**12.** $g(x) = 4 - 2\sin x$

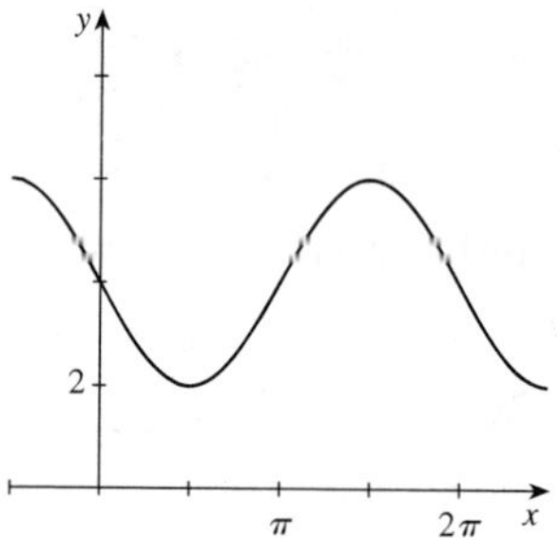


**13.** $h(x) = |\cos x|$

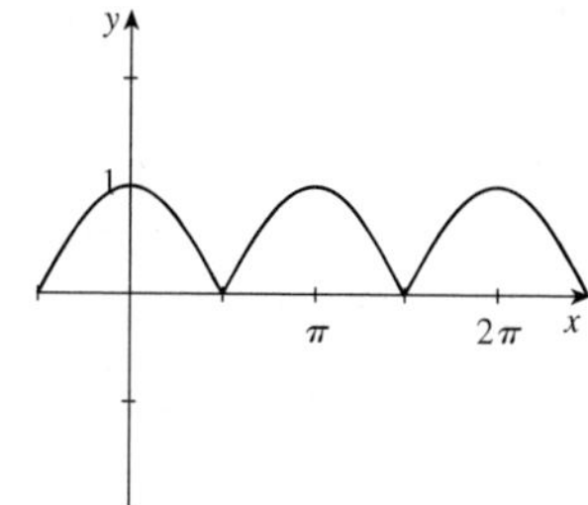


**14.** $h(x) = |\sin x|$

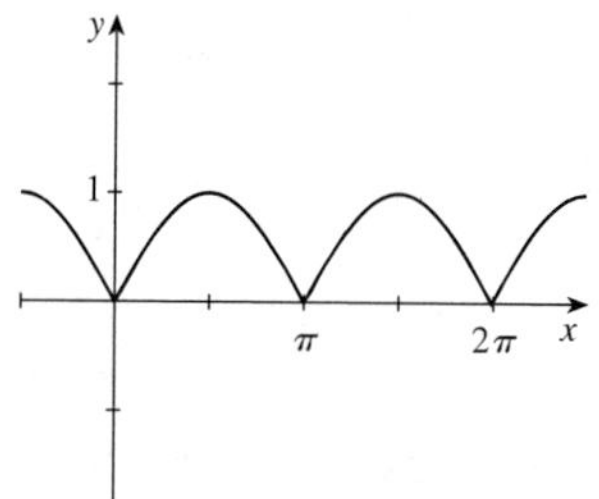


**15.** $y = \cos 2x$ has amplitude 1 and period $\pi$.

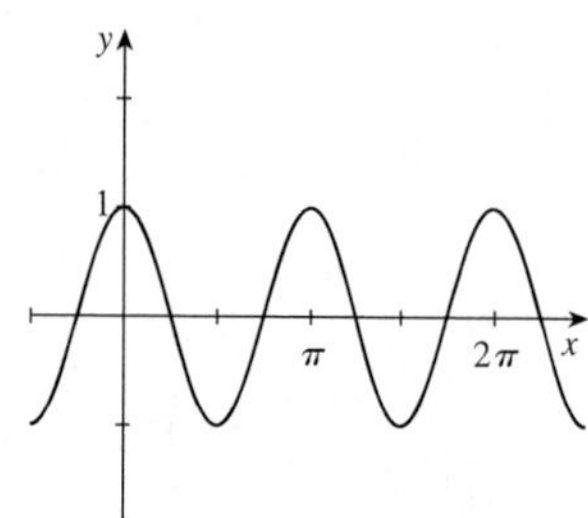


**16.** $y = -\sin 2x$ has amplitude 1 and period $\pi$.

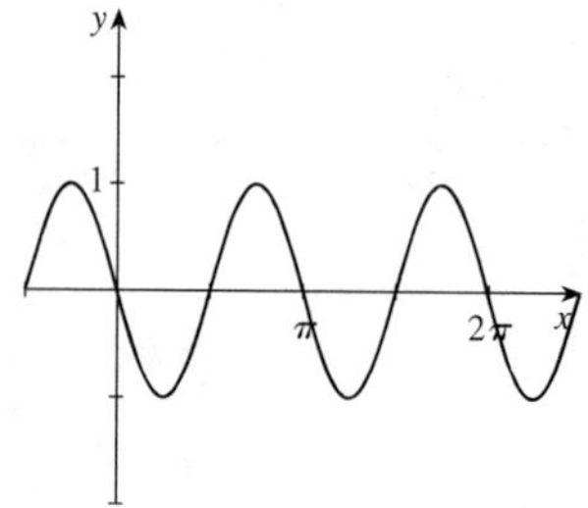

**17.** $y = -3\sin 3x$ has amplitude 3 and period $\frac{2\pi}{3}$.

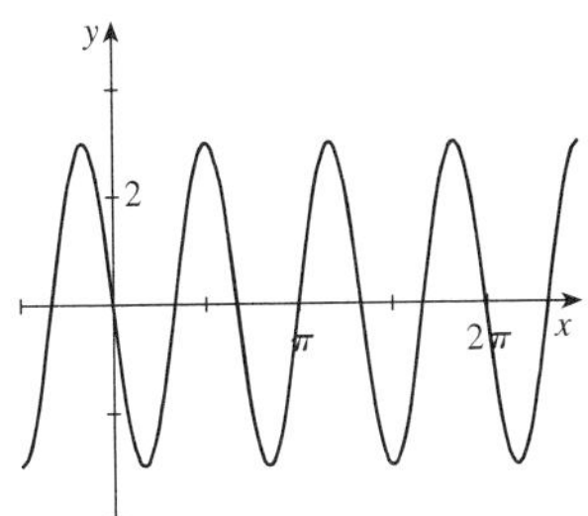


**18.** $y = \frac{1}{2}\cos 4x$ has amplitude $\frac{1}{2}$ and period $\frac{\pi}{2}$.

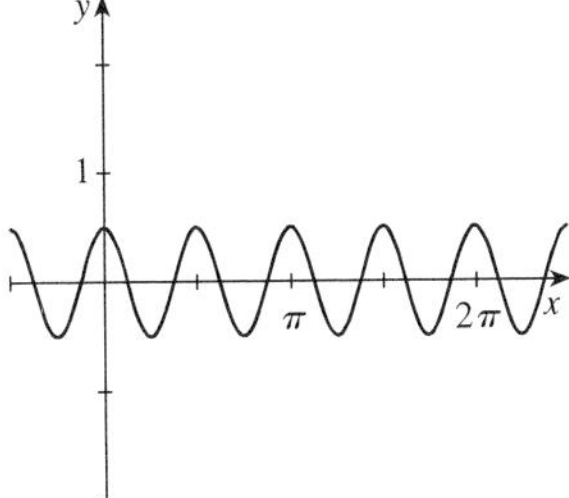


**19.** $y = 10\sin \frac{1}{2}x$ has amplitude 10 and period $4\pi$.

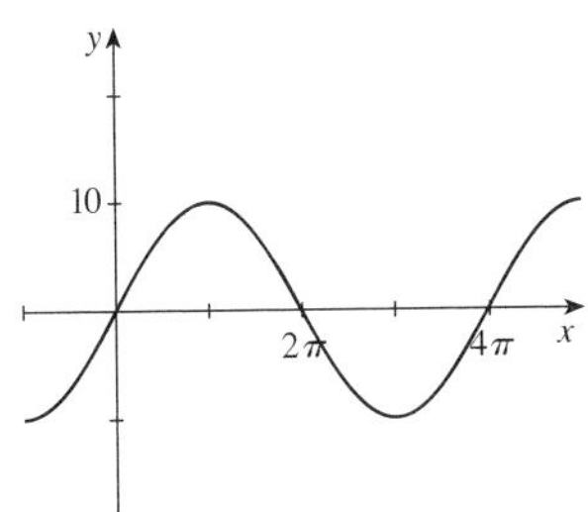


**20.** $y = 5\cos \frac{1}{4}x$ has amplitude 5 and period $8\pi$.

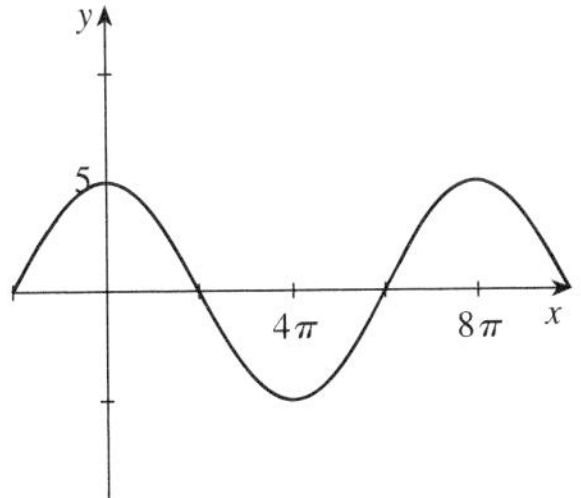


**21.** $y = -\frac{1}{3}\cos \frac{1}{3}x$ has amplitude $\frac{1}{3}$ and period $6\pi$.

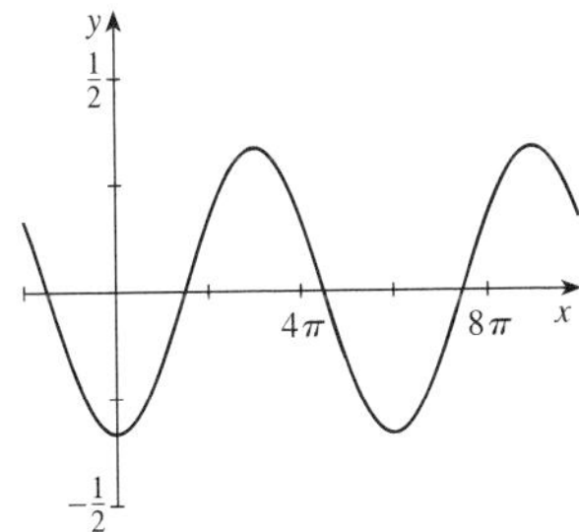


**22.** $y = 4\sin(-2x)$ has amplitude 4 and period $\pi$.

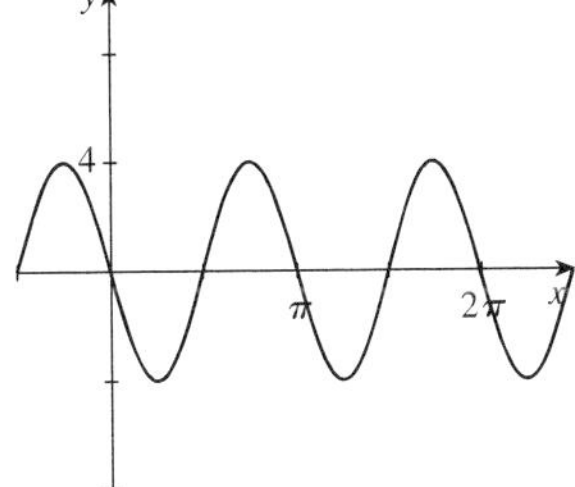


**23.** $y = -2\sin 2\pi x$ has amplitude 2 and period 1.

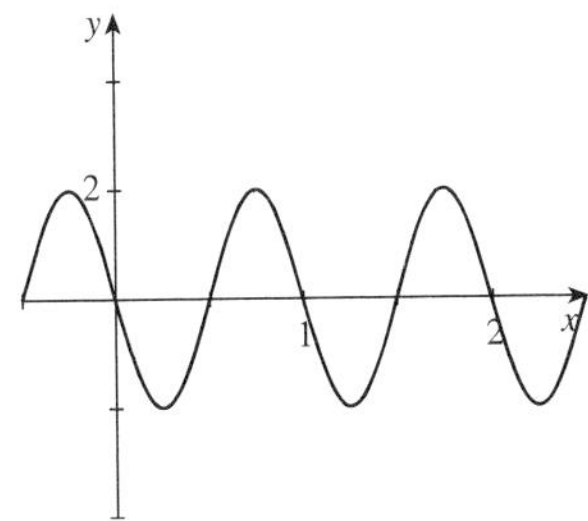


**24.** $y = -3\sin \pi x$ has amplitude 3 and period 2.

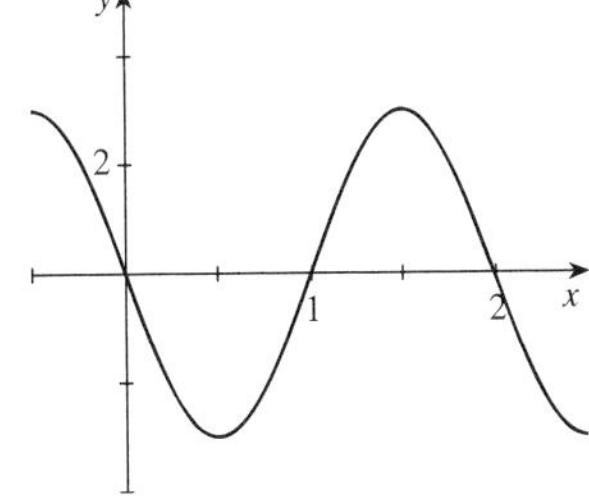

**25.** $y = 1 + \frac{1}{2}\cos \pi x$ has amplitude $\frac{1}{2}$ and period 2.

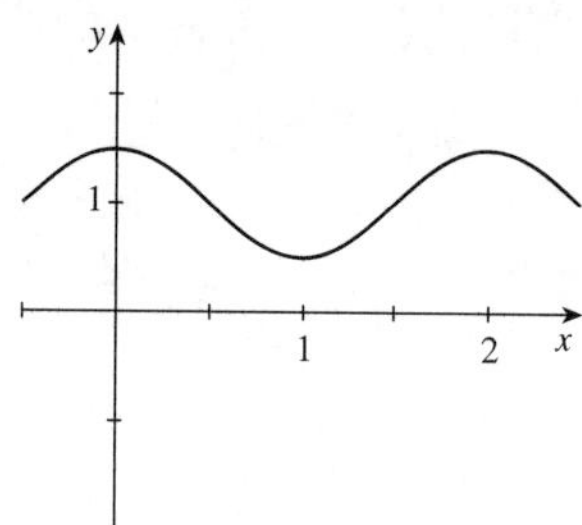


**26.** $y = -2 + \cos 4\pi x$ has amplitude 1 and period $\frac{1}{2}$.

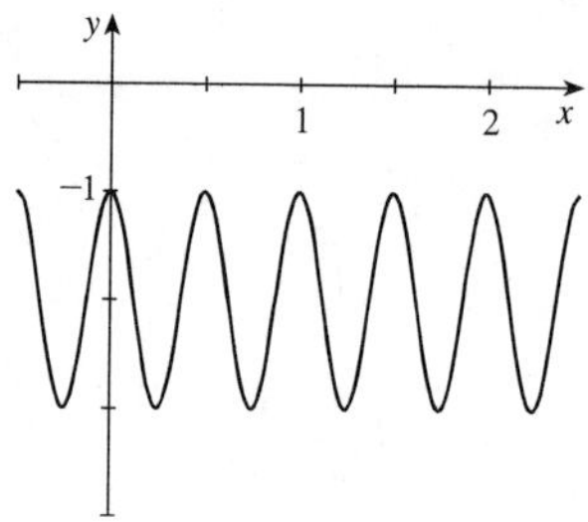


**27.** $y = \cos\left(x - \frac{\pi}{2}\right)$ has amplitude 1, period $2\pi$, and phase shift $\frac{\pi}{2}$.

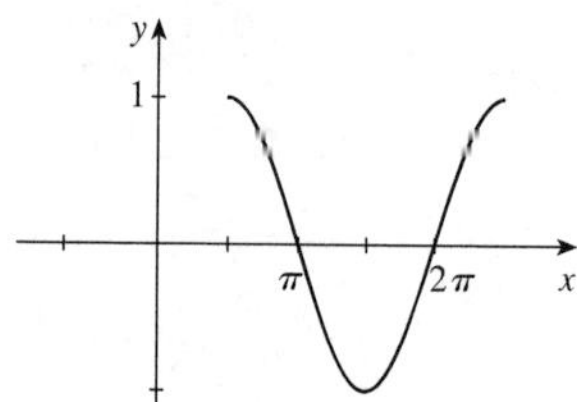


**28.** $y = 2\sin\left(x - \frac{\pi}{3}\right)$ has amplitude 2, period $2\pi$, and phase shift $\frac{\pi}{3}$.

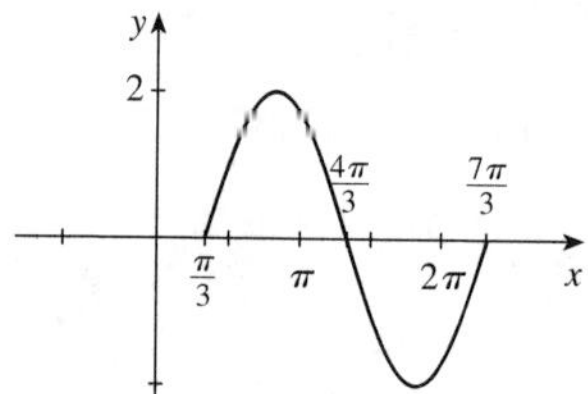


**29.** $y = -2\sin\left(x - \frac{\pi}{6}\right)$ has amplitude 2, period $2\pi$, and phase shift $\frac{\pi}{6}$.

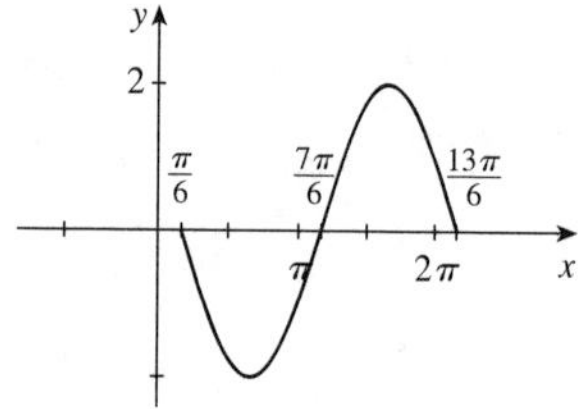


**30.** $y = 3\cos\left(x + \frac{\pi}{4}\right)$ has amplitude 3, period $2\pi$, and phase shift $-\frac{\pi}{4}$.

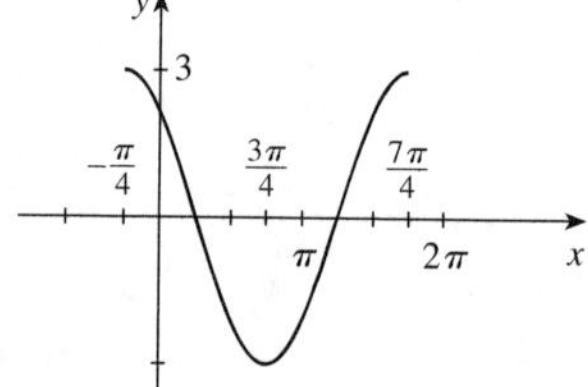


**31.** $y = -4\sin 2\left(x + \frac{\pi}{2}\right)$ has amplitude 4, period $\pi$, and phase shift $-\frac{\pi}{2}$.

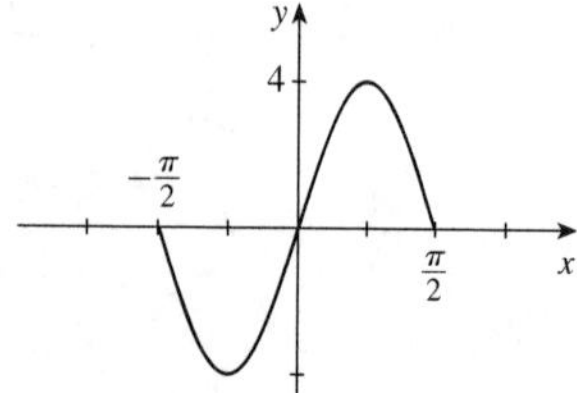


**32.** $y = \sin \frac{1}{2}\left(x + \frac{\pi}{4}\right)$ has amplitude 1, period $4\pi$, and phase shift $-\frac{\pi}{4}$.

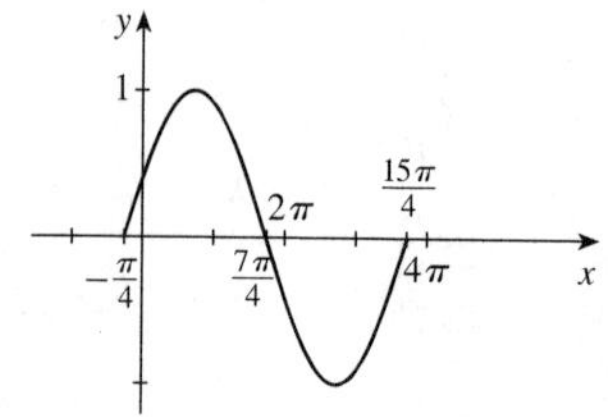

**33.** $y = 5\cos\left(3x - \frac{\pi}{4}\right) = 5\cos 3\left(x - \frac{\pi}{12}\right)$ has amplitude 5, period $\frac{2\pi}{3}$, and phase shift $\frac{\pi}{12}$.

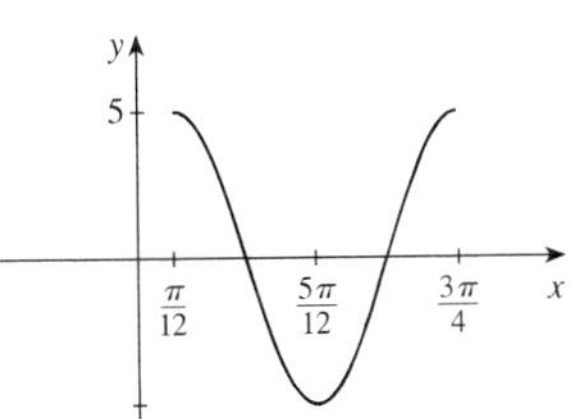

**34.** $y = 2\sin\left(\frac{2}{3}x - \frac{\pi}{6}\right) = 2\sin\frac{2}{3}\left(x - \frac{\pi}{4}\right)$ has amplitude 2, period $3\pi$, and phase shift $\frac{\pi}{4}$.

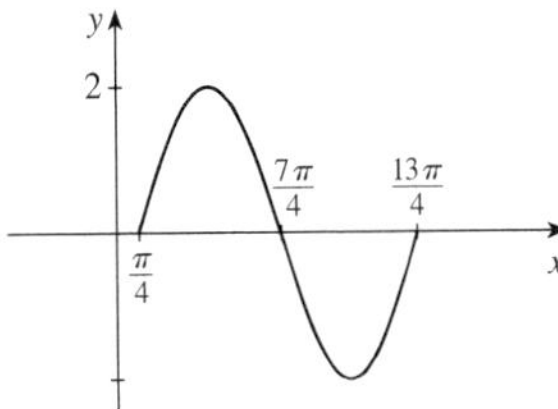

**35.** $y = \frac{1}{2} - \frac{1}{2}\cos\left(2x - \frac{\pi}{3}\right) = \frac{1}{2} - \frac{1}{2}\cos 2\left(x - \frac{\pi}{6}\right)$ has amplitude $\frac{1}{2}$, period $\pi$, and phase shift $\frac{\pi}{6}$.

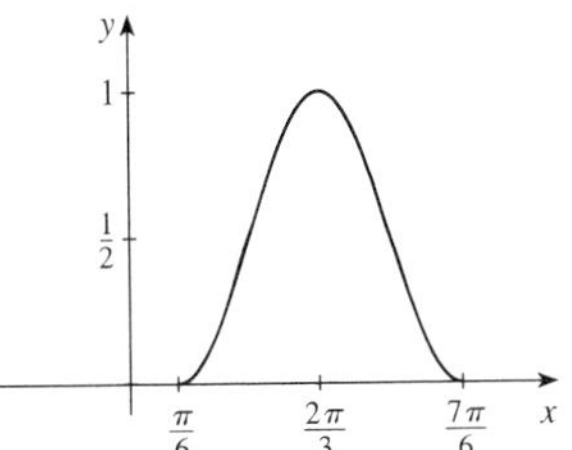

**36.** $y = 1 + \cos\left(3x + \frac{\pi}{2}\right) = 1 + \cos 3\left(x + \frac{\pi}{6}\right)$ has amplitude 1, period $\frac{2\pi}{3}$, and phase shift $-\frac{\pi}{6}$.

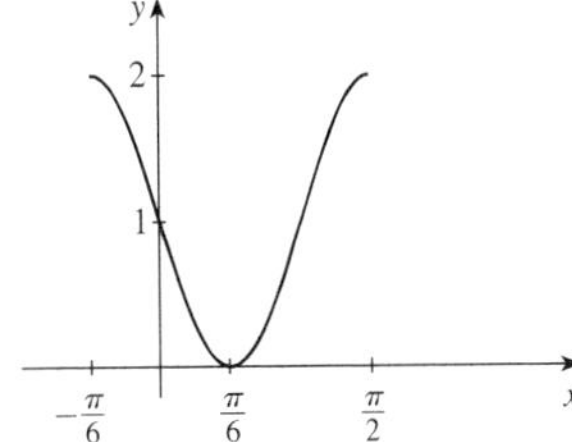

**37.** $y = 3\cos\pi\left(x + \frac{1}{2}\right)$ has amplitude 3, period 2, and phase shift $-\frac{1}{2}$.

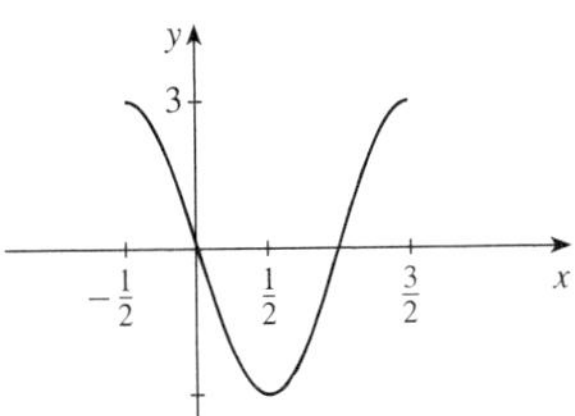

**38.** $y = 3 + 2\sin 3(x + 1)$ has amplitude 2, period $\frac{2\pi}{3}$, and phase shift $-1$.

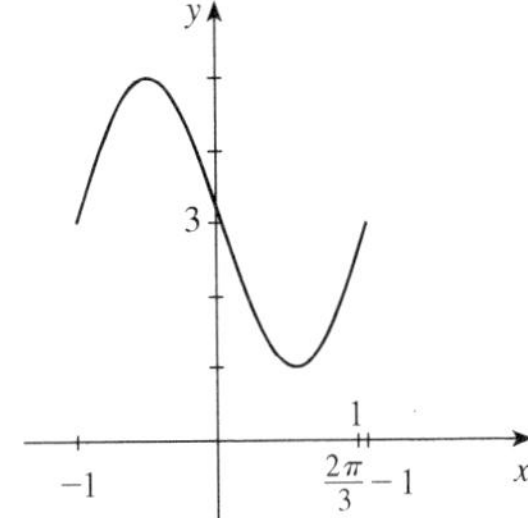

**39.** $y = \sin(3x + \pi) = \sin 3\left(x + \frac{\pi}{3}\right)$ has amplitude 1, period $\frac{2\pi}{3}$, and phase shift $-\frac{\pi}{3}$

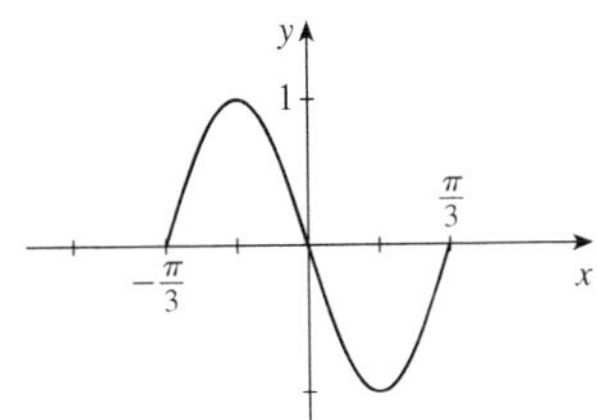

**40.** $y = \cos\left(\frac{\pi}{2} - x\right) = \cos\left(x - \frac{\pi}{2}\right)$ has amplitude 1, period $2\pi$, and phase shift $\frac{\pi}{2}$.

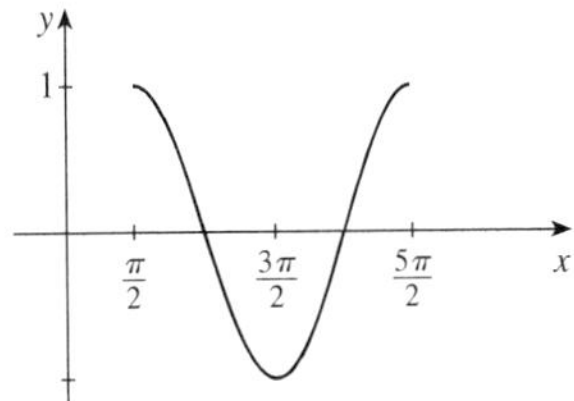

**41. (a)** This function has amplitude $a = 4$, period $\frac{2\pi}{k} = 2\pi$, and phase shift $b = 0$ as a sine curve.

**(b)** $y = a\sin k(x - b) = 4\sin x$

**42. (a)** This curve has amplitude $a = 2$, period $\frac{2\pi}{k} = \pi$, and phase shift $b = 0$ as a cosine curve.

**(b)** $y = a\cos k(x - b) = 2\cos 2x$

**43. (a)** This curve has amplitude $a = \frac{3}{2}$, period $\frac{2\pi}{k} = \frac{2\pi}{3}$, and phase shift $b = 0$ as a cosine curve.

**(b)** $y = a\cos k(x - b) = \frac{3}{2}\cos 3x$

**44. (a)** This curve has amplitude $a = 3$, period $\frac{2\pi}{k} = 4\pi$, and phase shift $b = 0$ as a sine curve.

**(b)** $y = 3\sin\frac{1}{2}x$

**45. (a)** This curve has amplitude $a = \frac{1}{2}$, period $\frac{2\pi}{k} = \pi$, and phase shift $b = -\frac{\pi}{3}$ as a cosine curve.

**(b)** $y = -\frac{1}{2}\cos 2\left(x + \frac{\pi}{3}\right)$

**46. (a)** This curve has amplitude $a = \frac{1}{10}$, period $\frac{2\pi}{k} = \pi$, and phase shift $b = -\frac{\pi}{2}$ as a sine curve.

**(b)** $y = \frac{1}{10}\sin 2\left(x + \frac{\pi}{2}\right)$

**47. (a)** This curve has amplitude $a = 4$, period $\frac{2\pi}{k} = \frac{3}{2}$, and phase shift $b = -\frac{1}{2}$ as a sine curve.

**(b)** $y = 4\sin\frac{4\pi}{3}\left(x + \frac{1}{2}\right)$

**48. (a)** This curve has amplitude $a = 5$, period $\frac{2\pi}{k} = 1$, and phase shift $b = 0$ as a cosine curve.

**(b)** $y = 5\cos 2\pi x$

**49.** $f(x) = \cos 100x$, $[-0.1, 0.1]$ by $[-1.5, 1.5]$

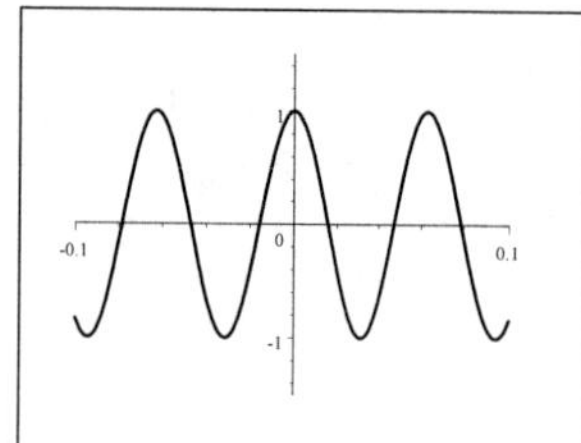


**50.** $f(x) = 3\sin 120x$, $[-0.1, 0.1]$ by $[-4, 4]$

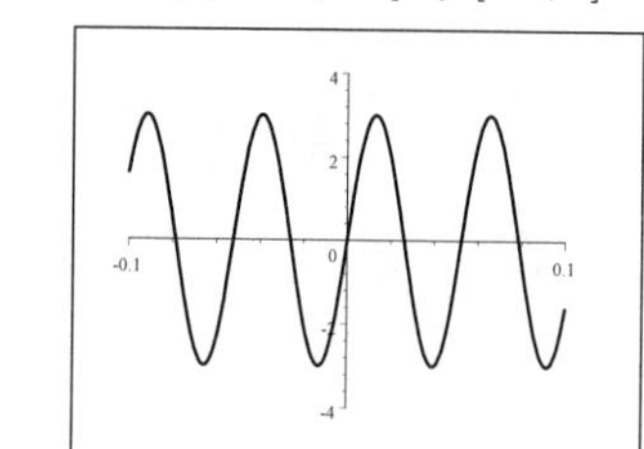


**51.** $f(x) = \sin\dfrac{x}{40}$, $[-250, 250]$ by $[-1.5, 1.5]$

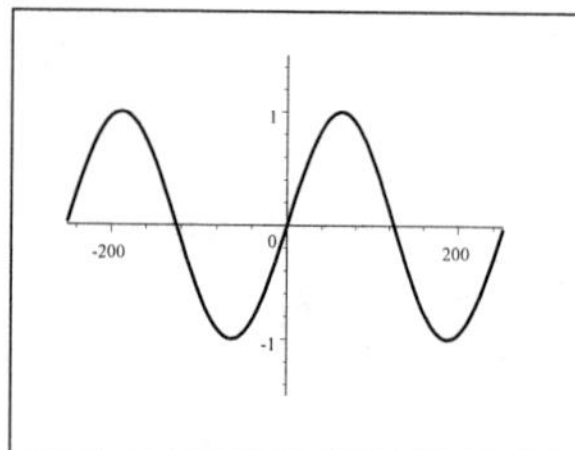


**52.** $f(x) = \cos\dfrac{x}{80}$, $[0, 500]$ by $[-1, 1]$

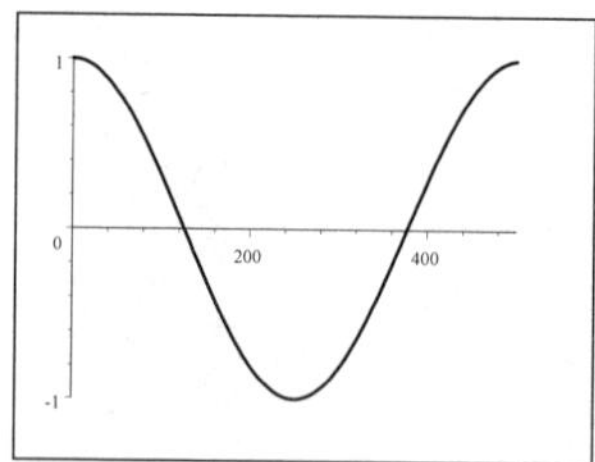


**53.** $y = \tan 25x$, $[-0.2, 0.2]$ by $[-3, 3]$

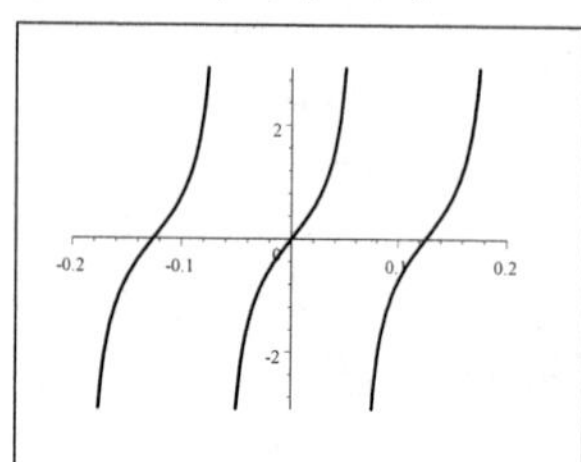


**54.** $y = \csc 40x$, $[-0.1, 0.1]$ by $[-10, 10]$

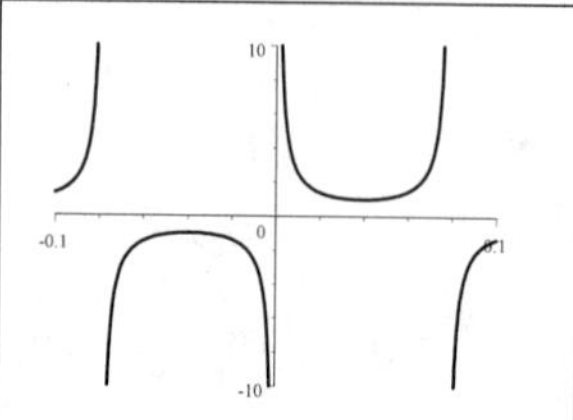

**55.** $y = \sin^2 20x$, $[-0.5, 0.5]$ by $[-0.2, 1.2]$

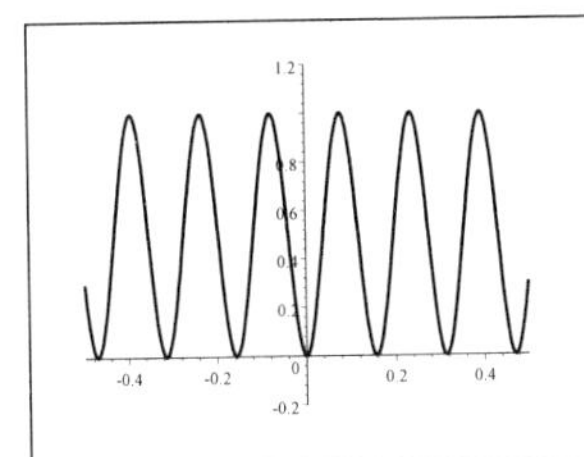

**56.** $y = \sqrt{\tan 10\pi x}$, $[-0.2, 0.2]$ by $[-1, 4]$

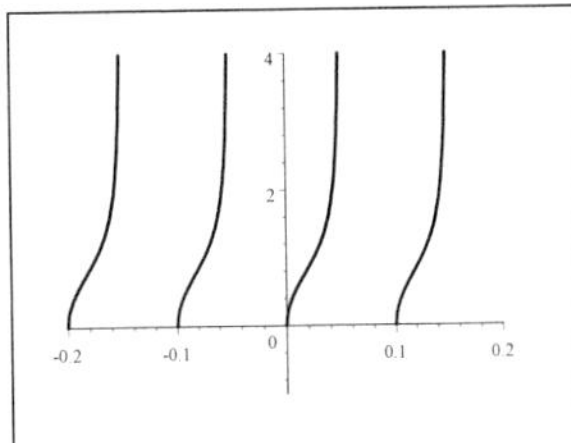

**57.** $f(x) = x$, $g(x) = \sin x$

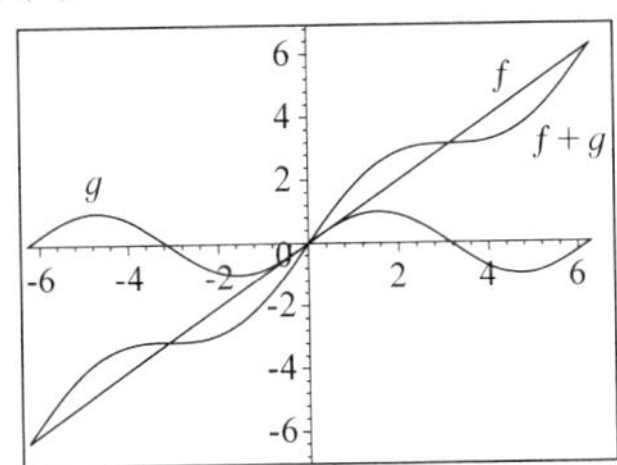


**58.** $f(x) = \sin x$, $g(x) = \sin 2x$

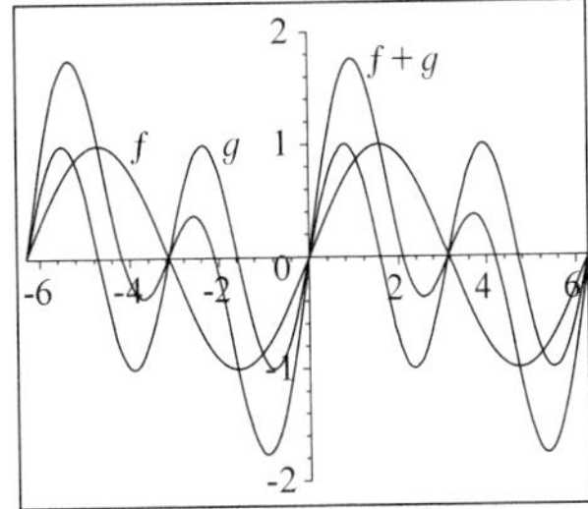


**59.** $y = x^2 \sin x$ is a sine curve that lies between the graphs of $y = x^2$ and $y = -x^2$.

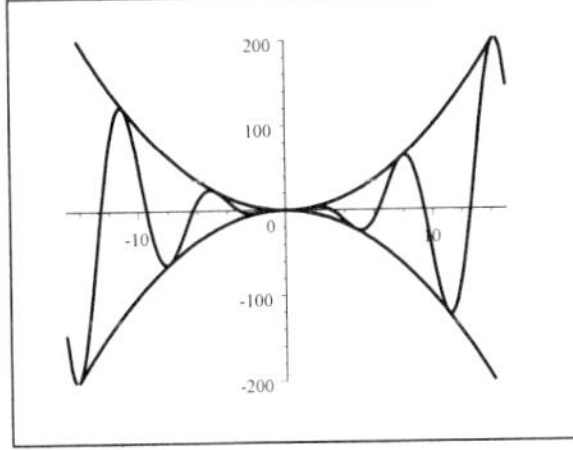

**60.** $y = x \cos x$ is a cosine curve that lies between the graphs of $y = x$ and $y = -x$.

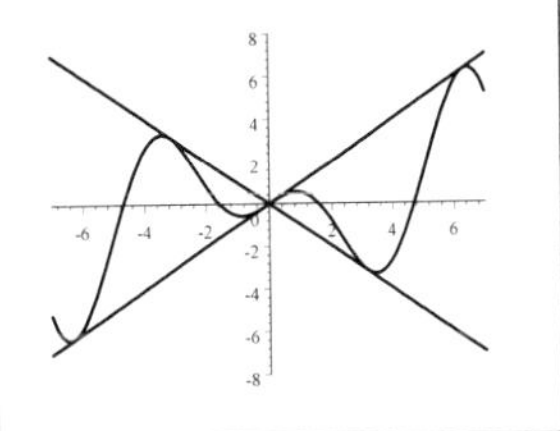

**61.** $y = \sqrt{x} \sin 5\pi x$ is a sine curve that lies between the graphs of $y = \sqrt{x}$ and $y = -\sqrt{x}$.

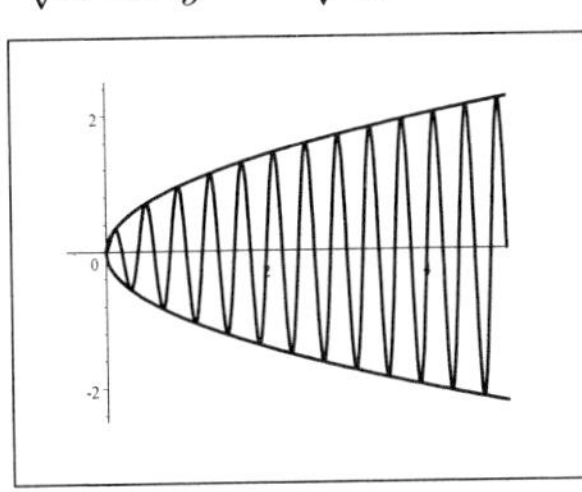

**62.** $y = \dfrac{\cos 2\pi x}{1 + x^2}$ is a cosine curve that lies between the graphs of $y = \dfrac{1}{1 + x^2}$ and $y = -\dfrac{1}{1 + x^2}$.

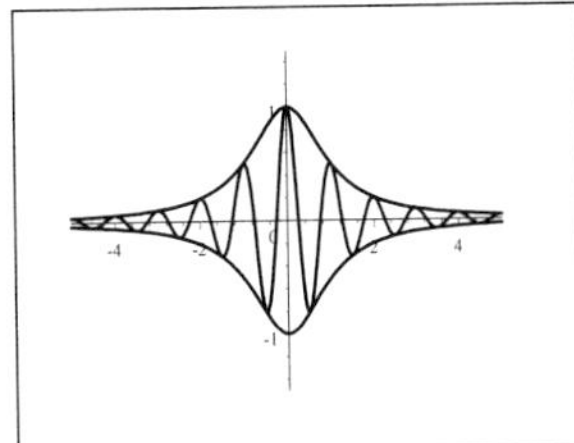

**63.** $y = \cos 3\pi x \cos 21\pi x$ is a cosine curve that lies between the graphs of $y = \cos 3\pi x$ and $y = -\cos 3\pi x$.

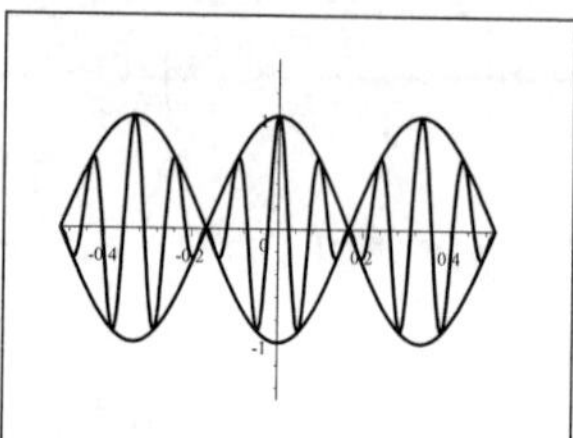

**64.** $y = \sin 2\pi x \sin 10\pi x$ is a sine curve that lies between the graphs of $y = \sin 2\pi x$ and $y = -\sin 2\pi x$.

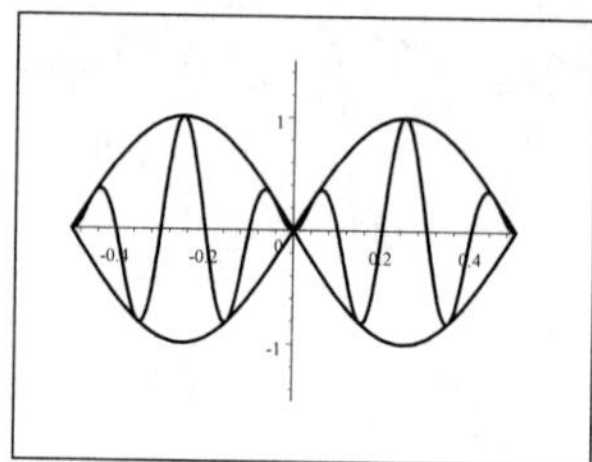

**65.** $y = \sin x + \sin 2x$. The period is $2\pi$, so we graph the function over one period, $(-\pi, \pi)$. Maximum value 1.76 when $x \approx 0.94 + 2n\pi$, minimum value $-1.76$ when $x \approx -0.94 + 2n\pi$, $n$ any integer.

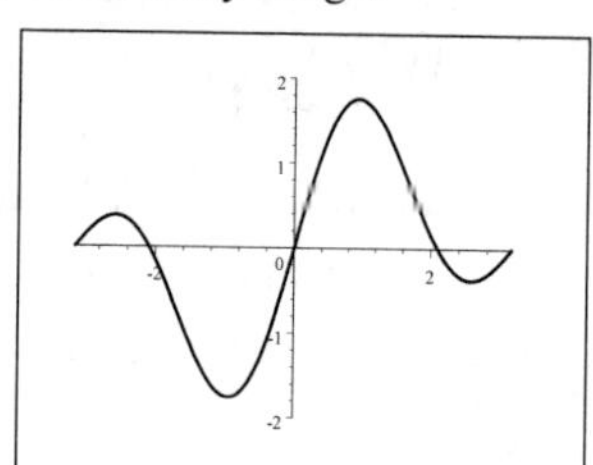

**66.** $y = x - 2\sin x$, $0 \le x \le 2\pi$. Maximum value 6.97 when $x \approx 5.24$, minimum value $-0.68$ when $x \approx 1.05$.

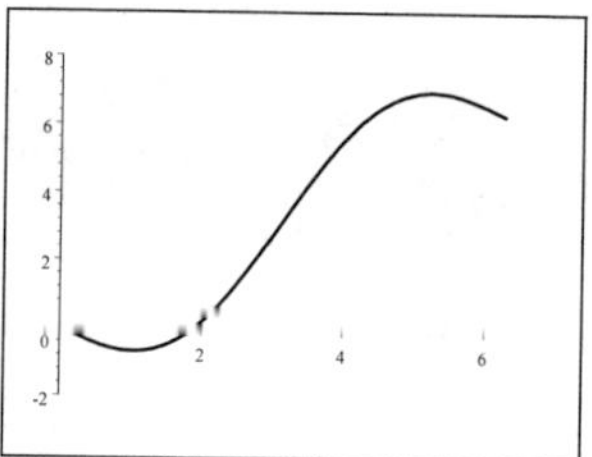

**67.** $y = 2\sin x + \sin^2 x$. The period is $2\pi$, so we graph the function over one period, $(-\pi, \pi)$. Maximum value 3.00 when $x \approx 1.57 + 2n\pi$, minimum value $-1.00$ when $x \approx -1.57 + 2n\pi$, $n$ any integer.

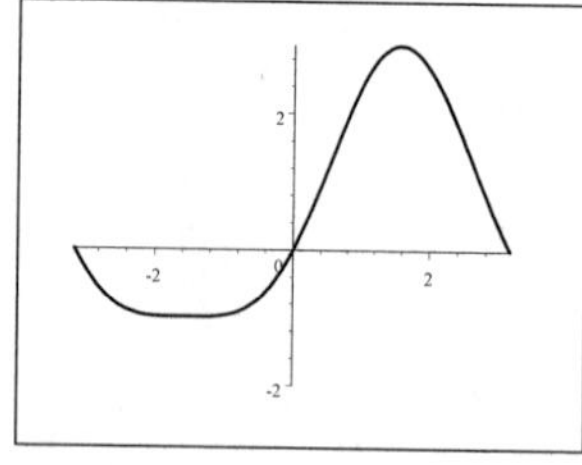

**68.** $y = \dfrac{\cos x}{2 + \sin x}$. The period is $2\pi$, so we graph the function over one period. Maximum value 0.58 when $x \approx 5.76 + 2n\pi$ (exact value $x = \frac{11\pi}{6} + 2n\pi$); Minimum value $-0.58$ when $x \approx 3.67 + 2n\pi$ (exact value $x = \frac{7\pi}{6} + 2n\pi$) for any integer $n$.

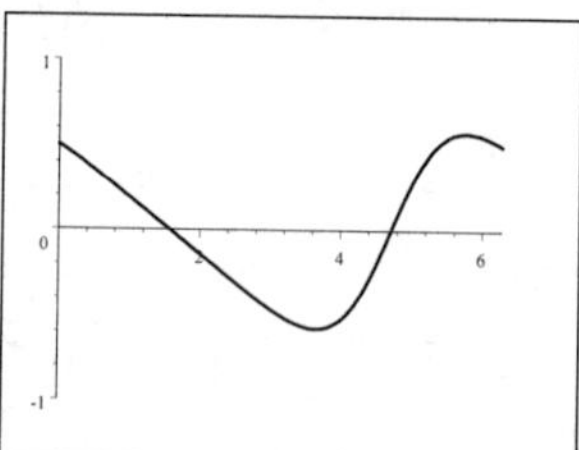

**69.** $\cos x = 0.4$, $x \in [0, \pi]$. The solution is $x \approx 1.16$.

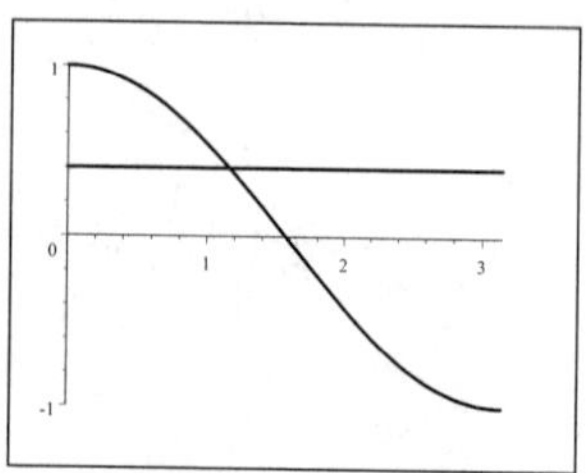

**70.** $\tan x = 2$, $x \in [0, \pi]$. The solution is $x \approx 1.11$.

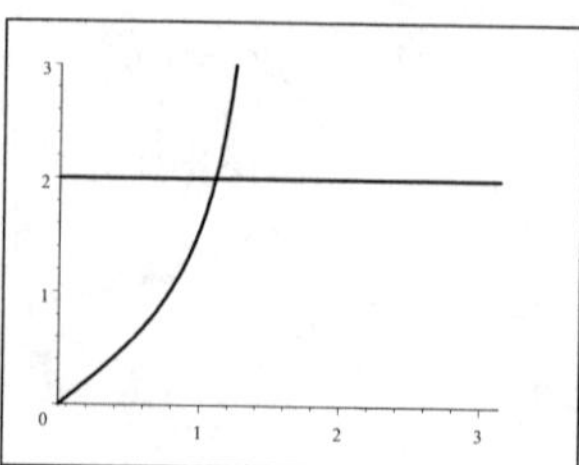

**71.** $\csc x = 3$, $x \in [0, \pi]$. The solutions are $x \approx 0.34, 2.80$.

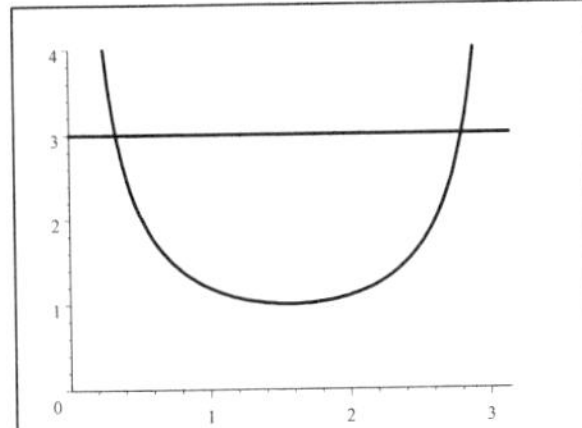

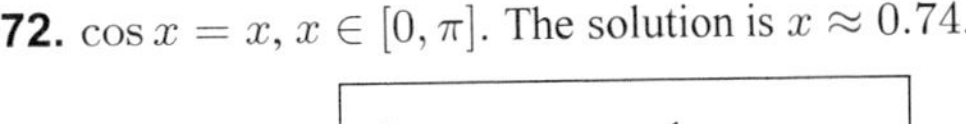

**72.** $\cos x = x$, $x \in [0, \pi]$. The solution is $x \approx 0.74$.

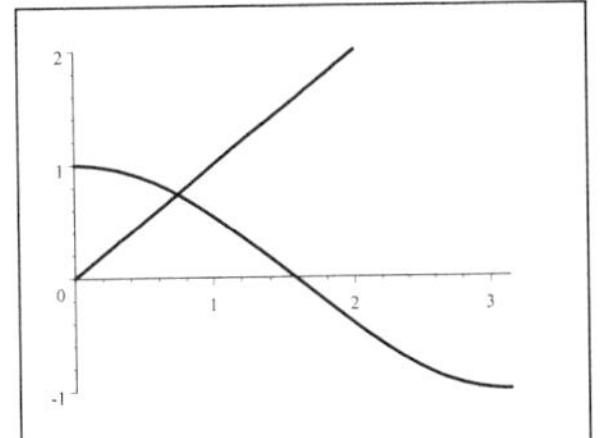

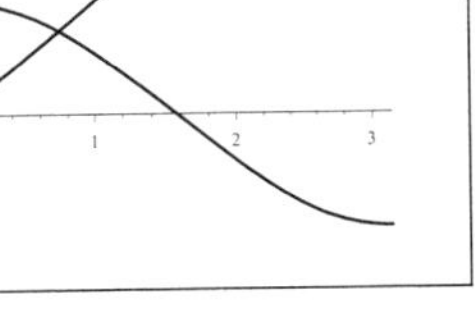

**73.** $f(x) = \dfrac{1 - \cos x}{x}$

**(a)** Since $f(-x) = \dfrac{1 - \cos(-x)}{-x} = \dfrac{1 - \cos x}{-x} = -f(x)$, the function is odd.

**(b)** The $x$-intercepts occur when $1 - \cos x = 0 \quad \Leftrightarrow \quad \cos x = 1 \quad \Leftrightarrow$ $x = 0, \pm 2\pi, \pm 4\pi, \pm 6\pi, \ldots$

**(c)**

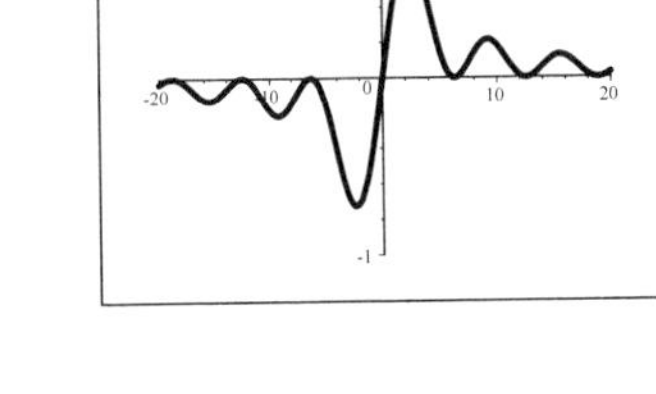

**(d)** As $x \to \pm\infty$, $f(x) \to 0$.

**(e)** As $x \to 0$, $f(x) \to 0$.

**74.** $f(x) = \dfrac{\sin 4x}{2x}$

**(a)** $f(-x) = \dfrac{\sin(-4x)}{2(-x)} = \dfrac{-\sin 4x}{-2x} = f(x)$, so the function is even.

**(b)** The $x$-intercepts occur when $\sin 4x = 0$, $x \neq 0 \Leftrightarrow 4x = n\pi \quad \Leftrightarrow$ $x = \frac{1}{4}n\pi$, $n$ any nonzero integer.

**(c)**

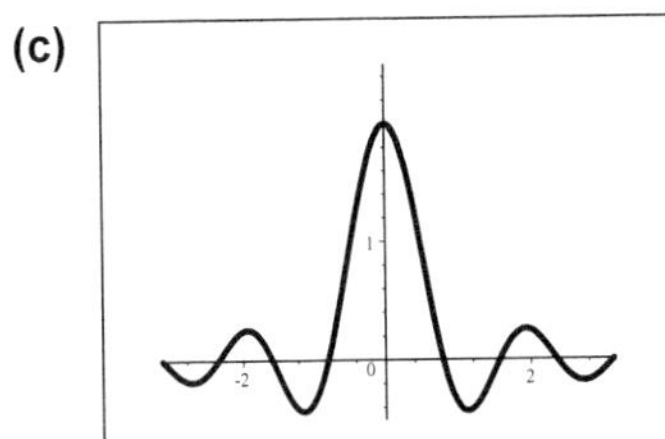

**(d)** As $x \to \pm\infty$, $f(x) \to 0$.

**(e)** As $x \to 0$, $f(x) \to 2$.

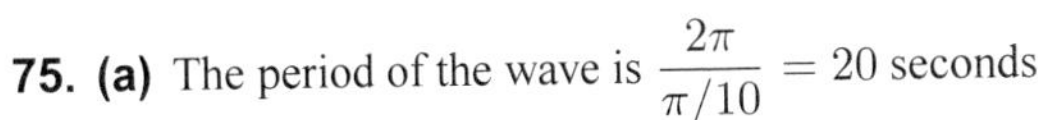

**75. (a)** The period of the wave is $\dfrac{2\pi}{\pi/10} = 20$ seconds.

**(b)** Since $h(0) = 3$ and $h(10) = -3$, the wave height is $3 - (-3) = 6$ feet.

**76. (a)** The period of the vibration is $\dfrac{2\pi}{880\pi} = \dfrac{1}{440}$ second.

**(b)** Since each vibration takes $\dfrac{1}{440}$ of a second, there are 440 vibrations per second.

**(c)**

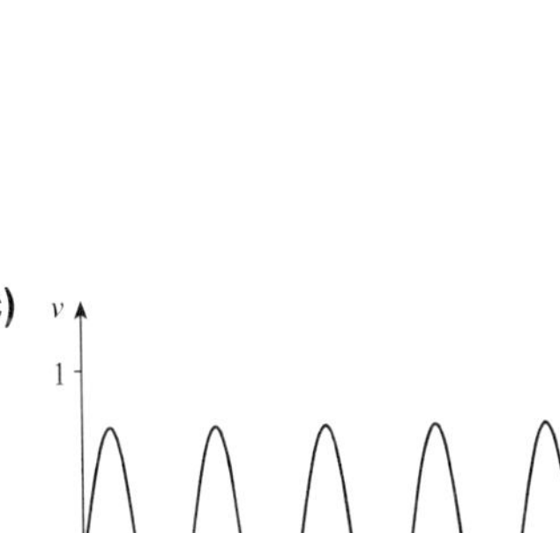


**77. (a)** The period of $p$ is $\dfrac{2\pi}{160\pi} = \dfrac{1}{80}$ minute.

**(b)** Since each period represents a heart beat, there are 80 heart beats per minute.

**(c)**

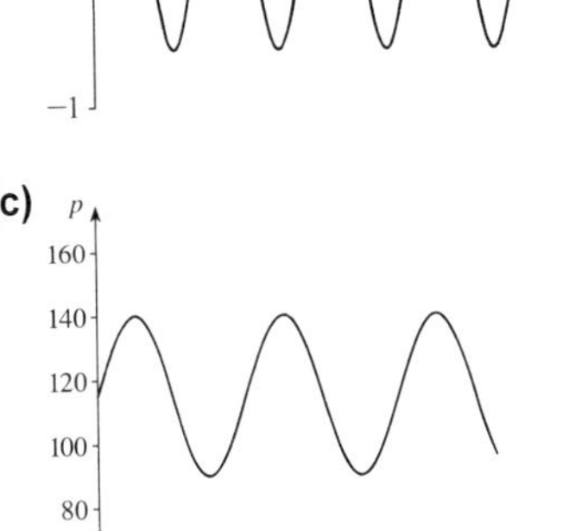


**(d)** The maximum (or systolic) is $115 + 25 = 140$ and the minimum (or diastolic) is $115 - 25 = 90$. The read would be $140/90$ which higher than normal.

**78. (a)** The period of R Leonis is $\dfrac{2\pi}{\pi/156} = 312$ days.

**(b)** The maximum brightness is $7.9 + 2.1 = 10$; the minimum brightness is $7.9 - 2.1 = 5.8$.

**(c)**

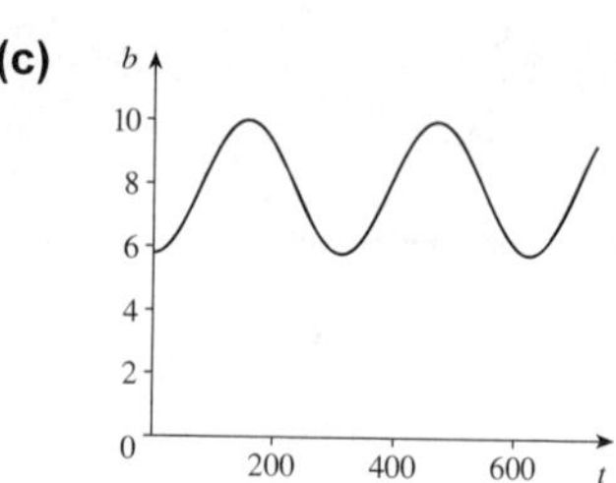

**79. (a)** $y = \sin(\sqrt{x})$. This graph looks like a sine function which has been stretched horizontally (stretched more for larger values of $x$). It is defined only for $x \geq 0$, so it is neither even nor odd.

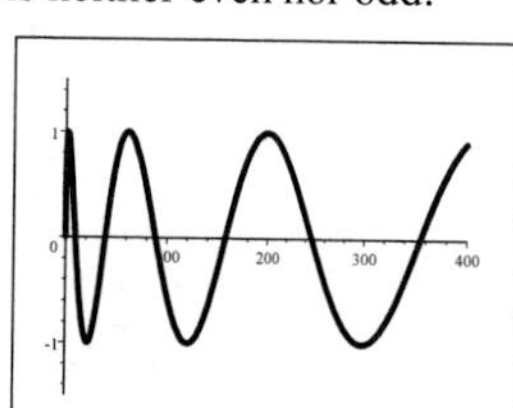

**(b)** $y = \sin(x^2)$. This graph looks like a graph of $\sin|x|$ which has been shrunk for $|x| > 1$ (shrunk more for larger values of $x$) and stretched for $|x| < 1$. It is an even function, whereas $\sin x$ is odd.

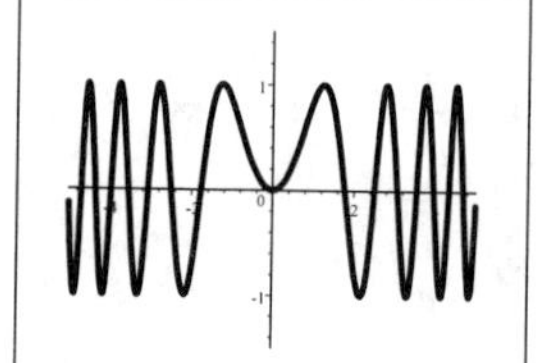

**80. (a)** This function is periodic with period 2.

**(b)** This function is not periodic. $f(2.5) \neq f(4.5)$.

**(c)** This function is not periodic; its graph on the interval $[-2, 0]$ is different from its graph on the interval $[0, 2]$.

**(d)** This function is periodic with period 3.

**81. (a)** The graph of $y = |\sin x|$ is shown in the viewing rectangle $[-6.28, 6.28]$ by $[-0.5, 1.5]$. This function is periodic with period$\pi$.

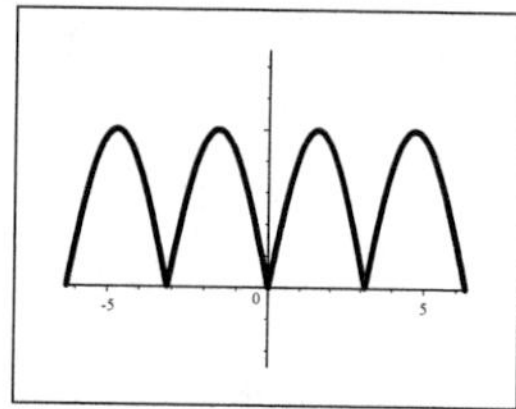

**(b)** The graph of $y = \sin|x|$ is shown in the viewing rectangle $[-10, 10]$ by $[-1.5, 1.5]$. The function is not periodic. Note that while $\sin|x + 2\pi| = \sin|x|$ for many values of $x$, it is false for $x \in (-2\pi, 0)$. For example $\sin\left|-\frac{\pi}{2}\right| = \sin\frac{\pi}{2} = 1$ while $\sin\left|-\frac{\pi}{2} + 2\pi\right| = \sin\frac{3\pi}{2} = -1$.

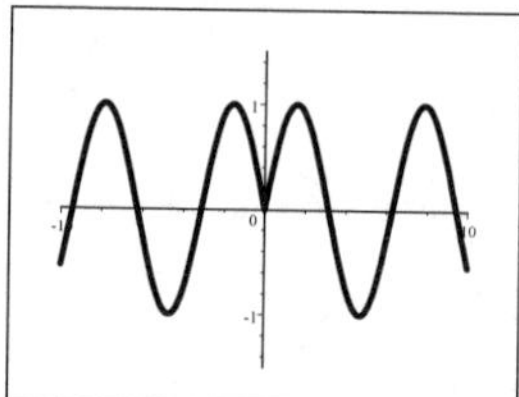

**(c)** The graph of $y = 2^{\cos x}$ is shown in the viewing rectangle $[-10, 10]$ by $[-1, 3]$. This function is periodic with period $= 2\pi$.

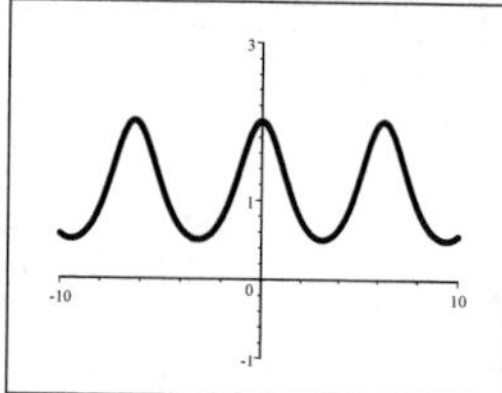

**(d)** The graph of $y = x - [\![x]\!]$ is shown in the viewing rectangle $[-7.5, 7.5]$ by $[-0.5, 1.5]$. This function is periodic with period1. Be sure to turn off "connected" mode when graphing functions with gaps in their graph.

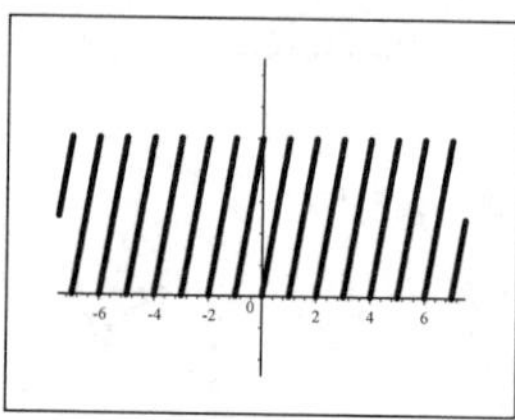

**82.** From the graph we see that the amplitude is 5 and the period is $2\pi$. In Exercises 25 and 26 in Section 8.2, we will prove the reduction formulas $\sin(x-\pi) = -\sin x$ and $\cos(x-\pi) = -\cos x$. So starting with $y = -5\sin x$ we may also represent the curve by $y = -5\sin(x-b)$ where $b = 2n\pi$, where $n$ is an integer; $y = 5\sin(x-b)$ where $b = (2n+1)\pi$, where $n$ is an integer. Then starting with $y = -5\cos\left(x - \frac{\pi}{2}\right)$ we may also represent the curve by $y = -5\cos(x-b)$ where $b = \frac{\pi}{2} + 2n\pi$, where $n$ is an integer; $y = 5\cos(x-b)$ where $b = \frac{\pi}{2} + (2n+1)\pi$, where $n$ is an integer.

## 7.4 More Trigonometric Graphs

**1.** $f(x) = \tan\left(x + \frac{\pi}{4}\right)$ corresponds to Graph II. $f$ is undefined at $x = \frac{\pi}{4}$ and $x = \frac{3\pi}{4}$, and Graph II has the shape of a graph of a tangent function.

**2.** $f(x) = \sec 2x$ corresponds to Graph III. $f$ is undefined at $x = \frac{\pi}{4}$ and $x = \frac{3\pi}{4}$, and Graph III has the shape of a graph of a secant function.

**3.** $f(x) = \cot 2x$ corresponds to Graph VI.

**4.** $f(x) = -\tan x$ corresponds to Graph I.

**5.** $f(x) = 2\sec x$ corresponds to Graph IV.

**6.** $f(x) = 1 + \csc x$ corresponds to Graph V.

**7.** $y = 4\tan x$ has period $\pi$.

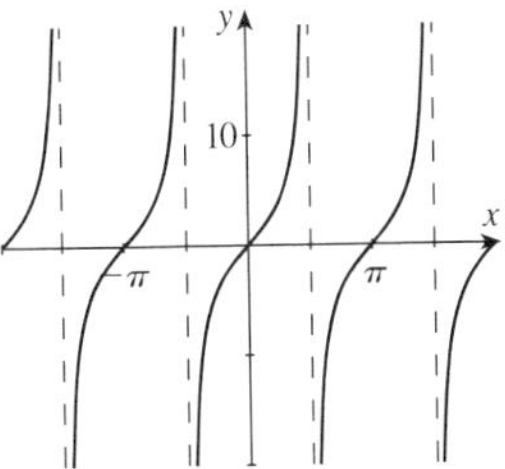


**8.** $y = -4\tan x$ has period $\pi$.

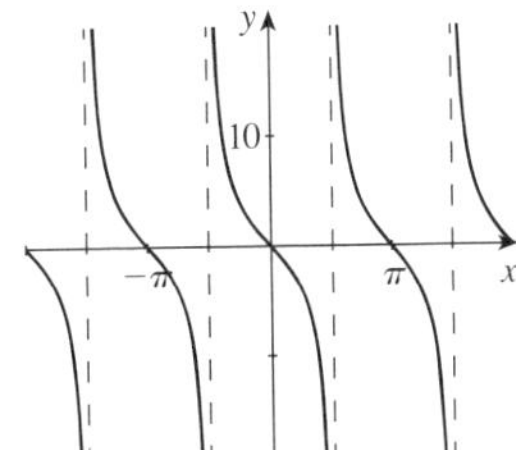


**9.** $y = -\frac{1}{2}\tan x$ has period $\pi$.

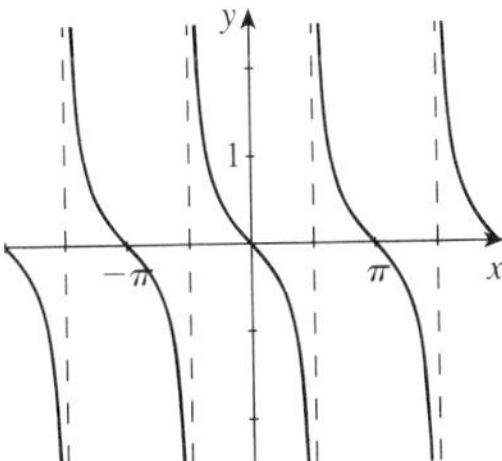


**10.** $y = \frac{1}{2}\tan x$ has period $\pi$.

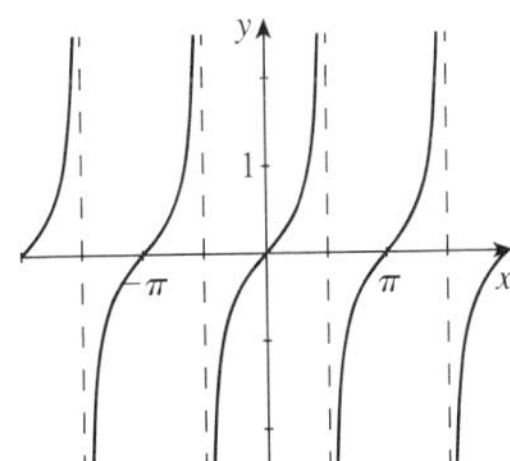


**11.** $y = -\cot x$ has period $\pi$.

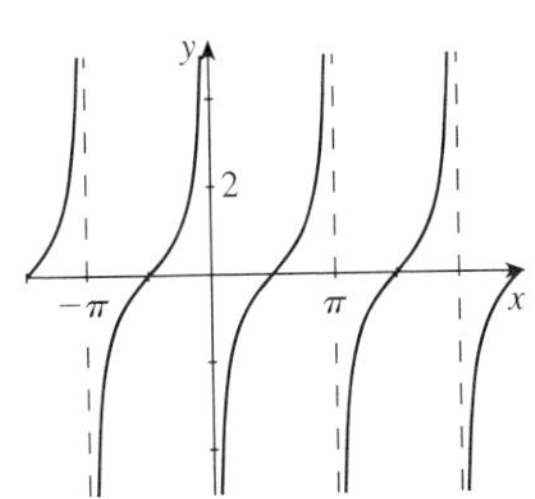


**12.** $y = 2\cot x$ has period $\pi$.

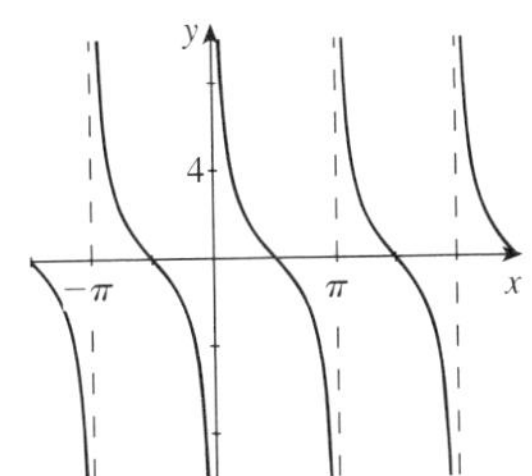

**13.** $y = 2\csc x$ has period $2\pi$.

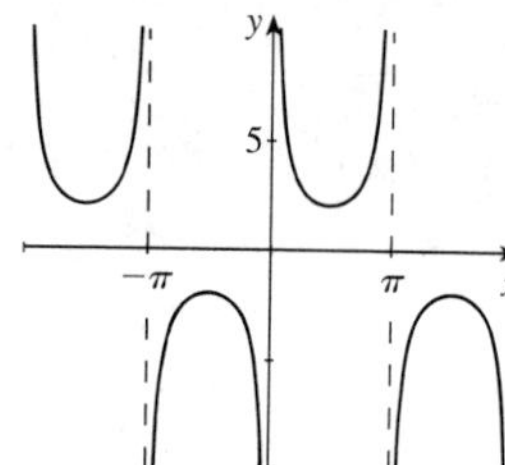


**14.** $y = \frac{1}{2}\csc x$ has period $2\pi$.

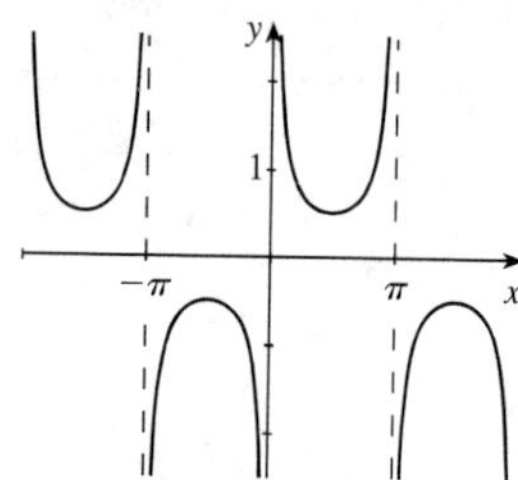


**15.** $y = 3\sec x$ has period $2\pi$.

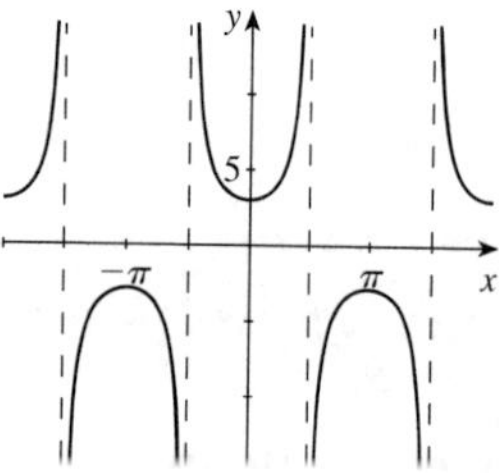


**16.** $y = -3\sec x$ has period $2\pi$.

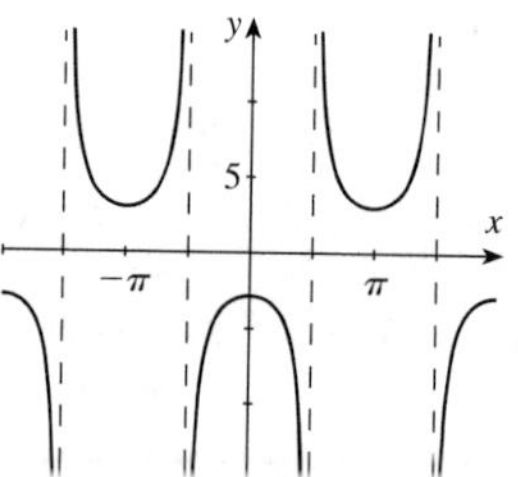


**17.** $y = \tan\left(x + \frac{\pi}{2}\right)$ has period $\pi$.

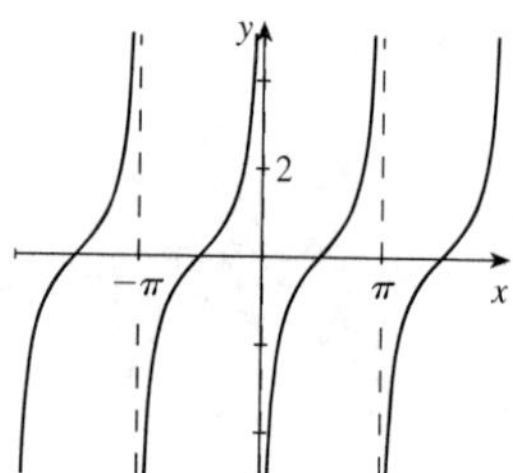


**18.** $y = \tan\left(x - \frac{\pi}{4}\right)$ has period $\pi$.

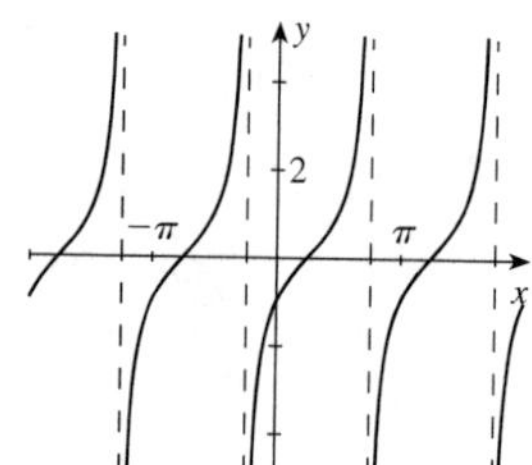


**19.** $y = \csc\left(x - \frac{\pi}{2}\right)$ has period $2\pi$.

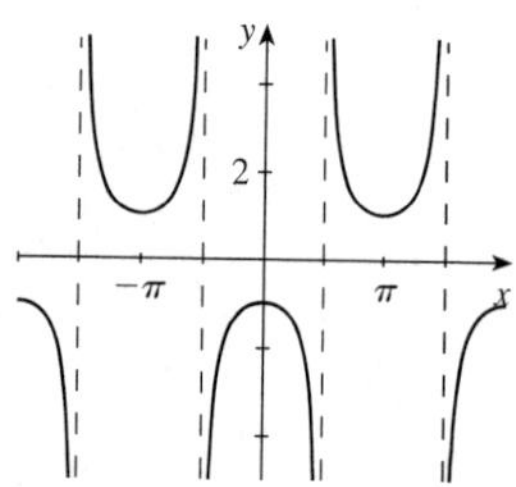


**20.** $y = \sec\left(x + \frac{\pi}{4}\right)$ has period $2\pi$.

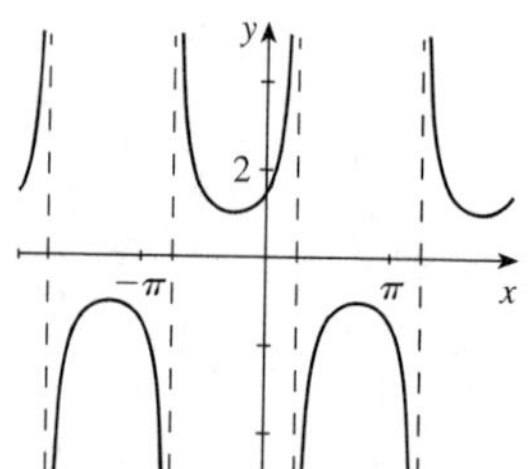


**21.** $y = \cot\left(x + \frac{\pi}{4}\right)$ has period $\pi$.

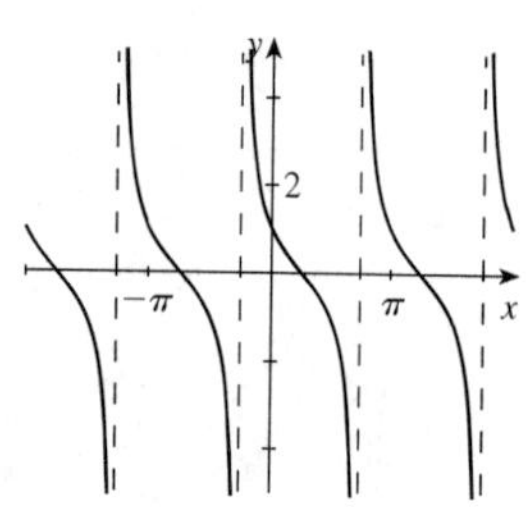


**22.** $y = 2\csc\left(x - \frac{\pi}{3}\right)$ has period $2\pi$.

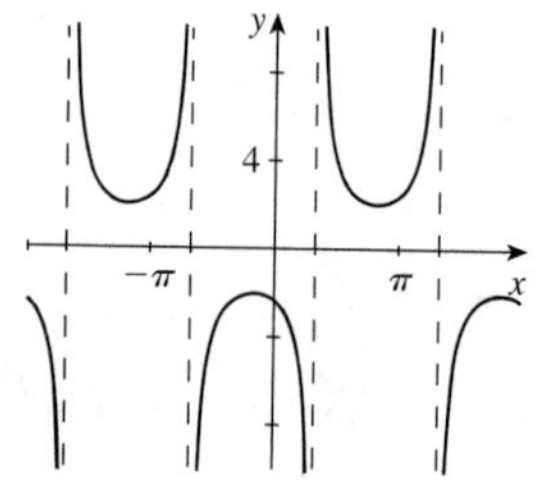

**23.** $y = \frac{1}{2}\sec\left(x - \frac{\pi}{6}\right)$ has period $2\pi$.

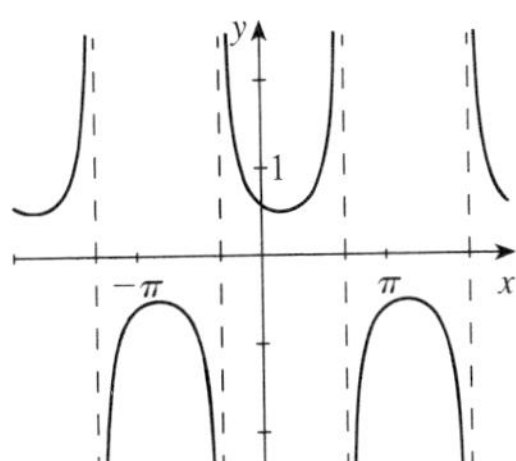


**24.** $y = 3\csc\left(x + \frac{\pi}{2}\right)$ has period $2\pi$.

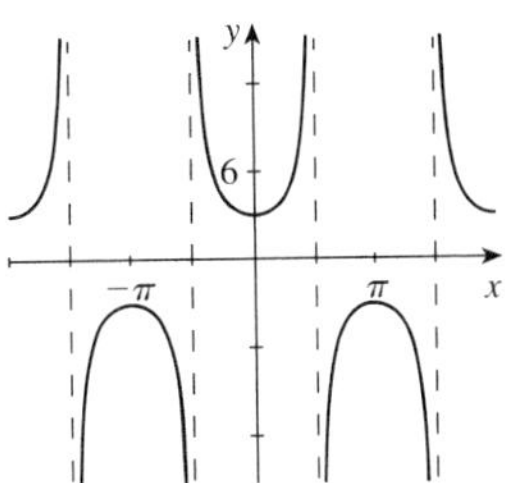


**25.** $y = \tan 2x$ has period $\frac{\pi}{2}$.

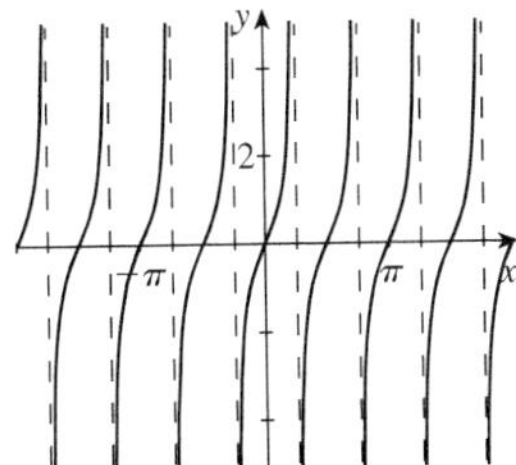


**26.** $y = \tan\left(\frac{1}{2}x\right)$ has period $\dfrac{\pi}{\left(\frac{1}{2}\right)} = 2\pi$.

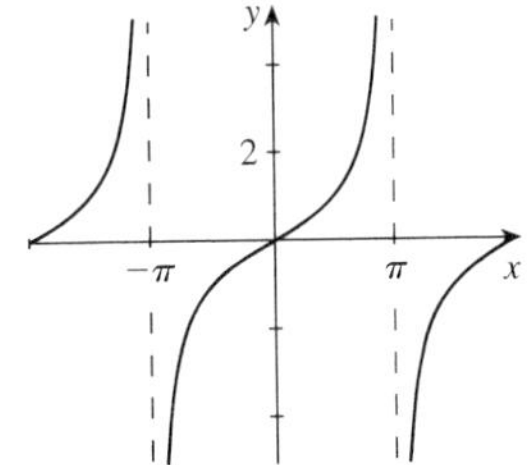


**27.** $y = \tan\left(\frac{\pi}{4}x\right)$ has period $\dfrac{\pi}{\frac{\pi}{4}} = 4$.

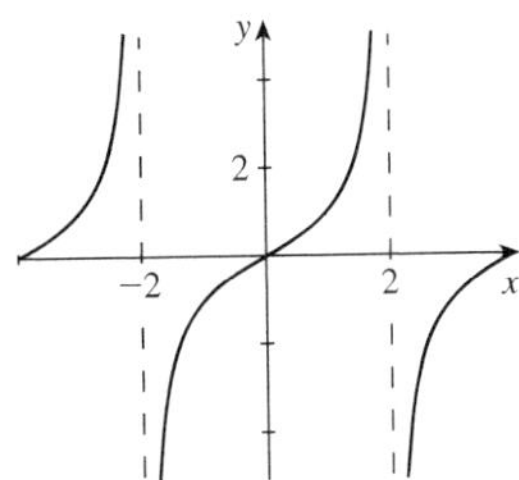


**28.** $y = \cot\left(\frac{\pi}{2}x\right)$ has period $\dfrac{\pi}{\left(\frac{\pi}{2}\right)} = 2$.

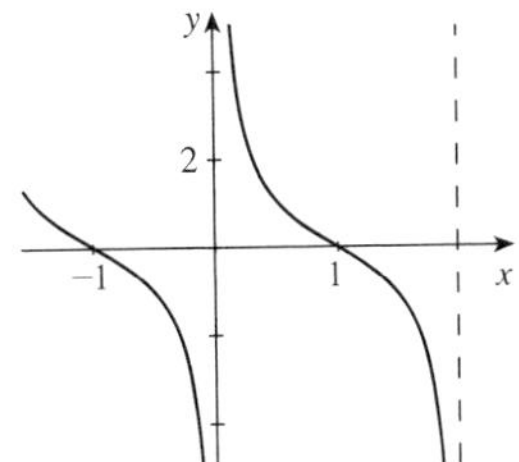


**29.** $y = \sec 2x$ has period $\dfrac{2\pi}{2} = \pi$.

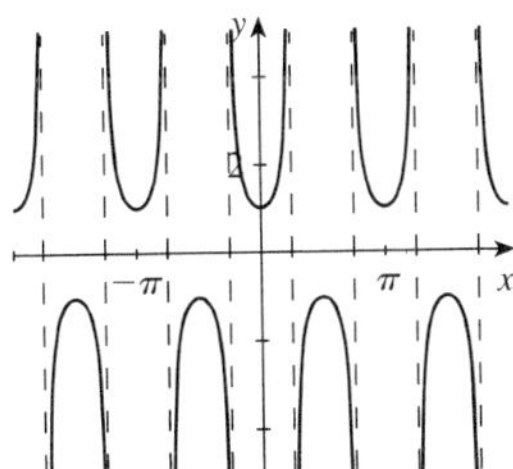


**30.** $y = 5\csc(3x)$ has period $\frac{2\pi}{3}$.

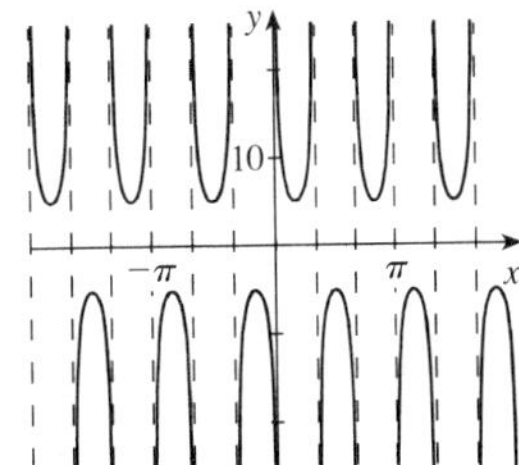


**31.** $y = \csc 2x$ has period $\dfrac{2\pi}{2} = \pi$.

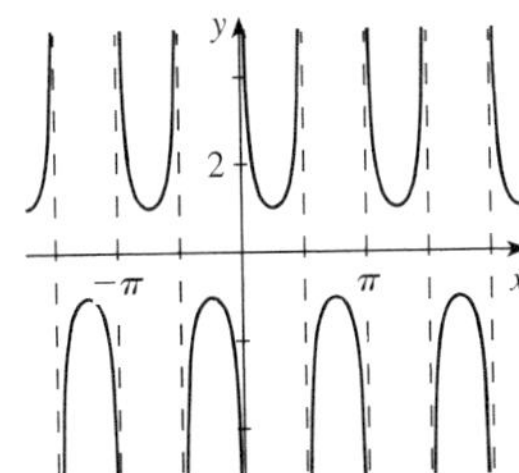


**32.** $y = \csc\left(\frac{1}{2}x\right)$ has period $\dfrac{2\pi}{\left(\frac{1}{2}\right)} = 4\pi$.

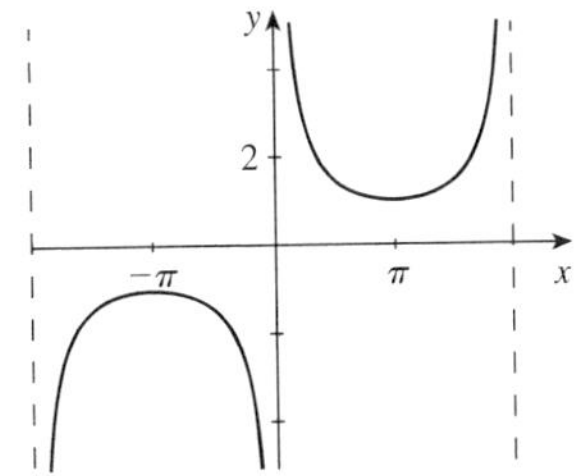

**33.** $y = 2\tan 3\pi x$ has period $\frac{1}{3}$.

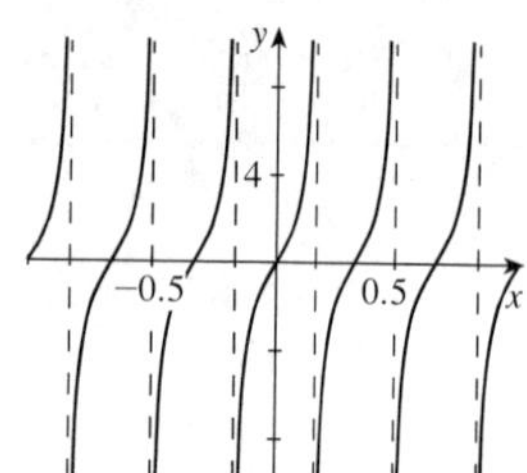


**34.** $y = 2\tan\left(\frac{\pi}{2}x\right)$ has period $\frac{\pi}{\left(\frac{\pi}{2}\right)} = 2$.

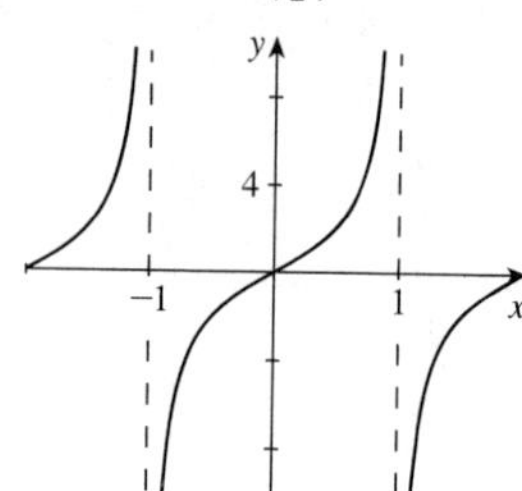


**35.** $y = 5\csc\frac{3\pi}{2}x$ has period $\dfrac{2\pi}{\frac{3\pi}{2}} = \dfrac{4}{3}$.

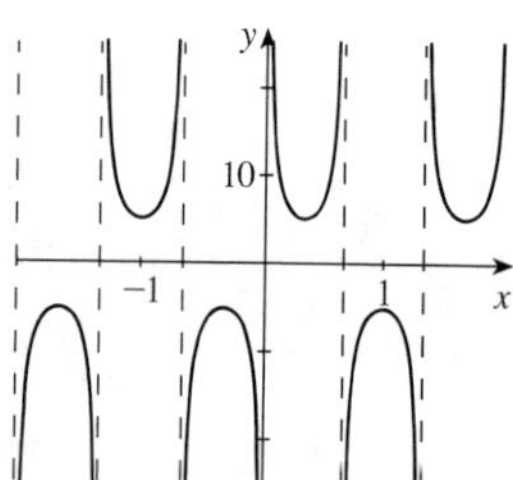


**36.** $y = 5\sec(2\pi x)$ has period $\frac{2\pi}{2\pi} = 1$.

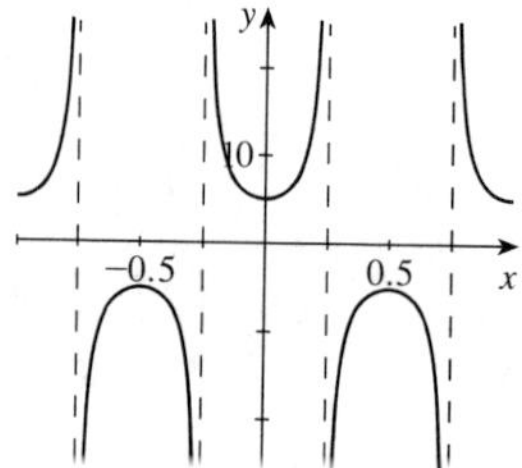


**37.** $y = \tan 2\left(x + \frac{\pi}{2}\right)$ has period $\frac{\pi}{2}$.

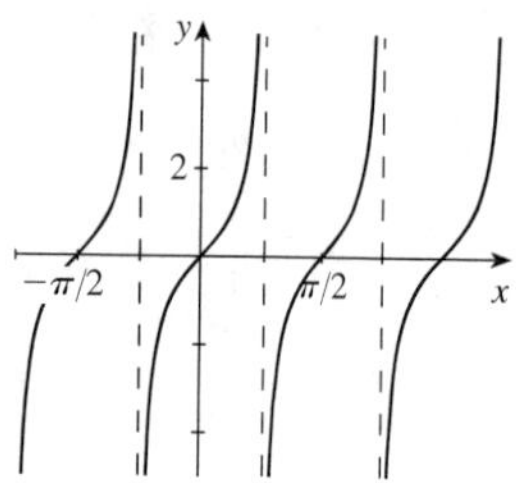


**38.** $y = \csc 2\left(x + \frac{\pi}{2}\right)$ has period $\frac{2\pi}{2} = \pi$.

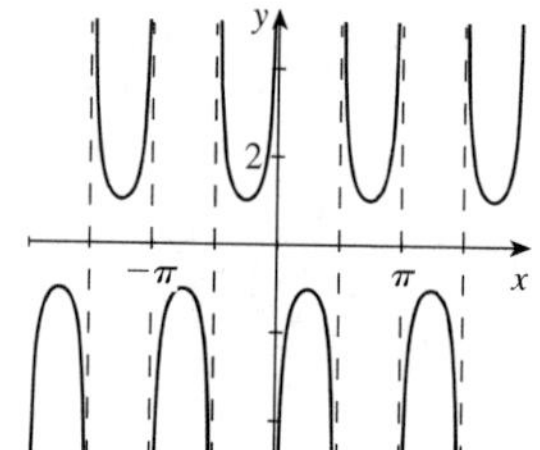


**39.** $y = \tan 2(x - \pi) = \tan 2x$ has period $\frac{\pi}{2}$.

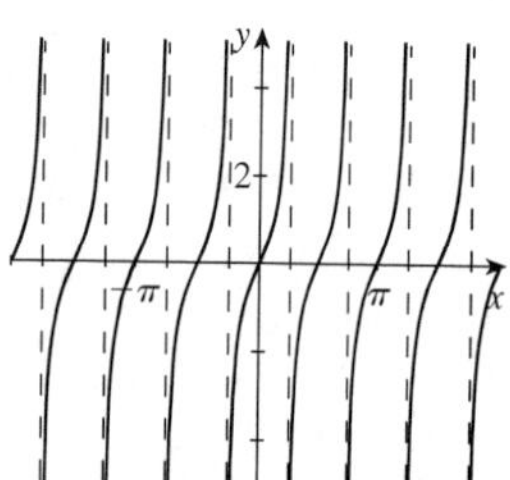


**40.** $y = \sec 2\left(x - \frac{\pi}{2}\right)$ has period $\frac{2\pi}{2} = \pi$.

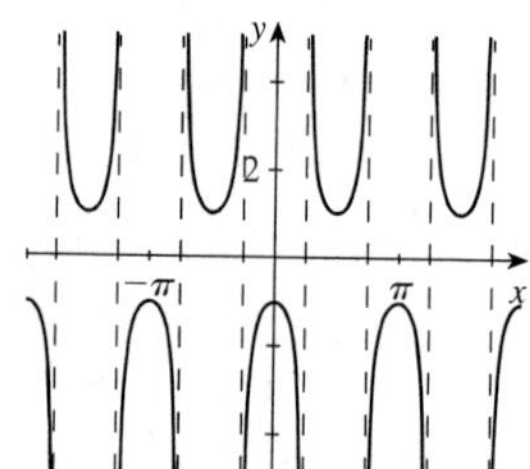


**41.** $y = \cot\left(2x - \frac{\pi}{2}\right) = \cot 2\left(x - \frac{\pi}{4}\right)$ has period $\frac{\pi}{2}$.

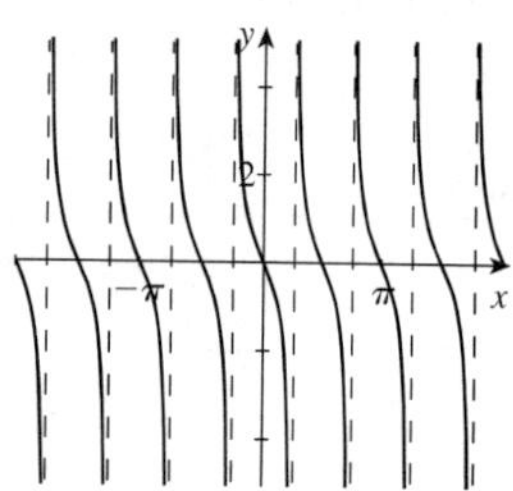


**42.** $y = \frac{1}{2}\tan(\pi x - \pi) = \frac{1}{2}\tan\pi(x - 1)$ has period $\frac{\pi}{\pi} = 1$.

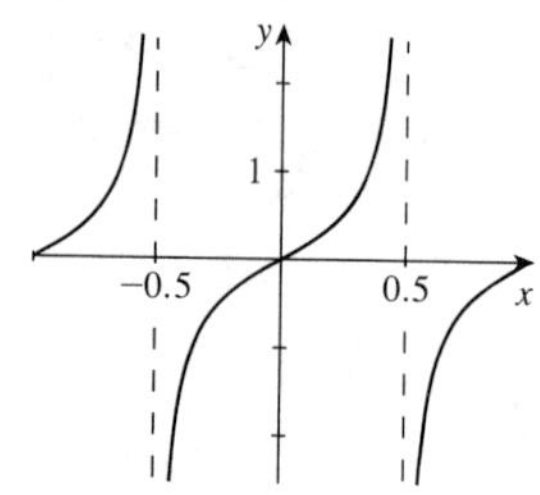

**43.** $y = 2\csc\left(\pi x - \frac{\pi}{3}\right) = 2\csc\pi\left(x - \frac{1}{3}\right)$ has period $\frac{2\pi}{\pi} = 2$.

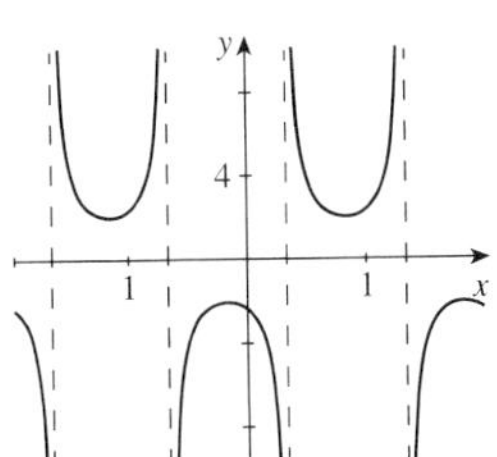


**44.** $y = 2\sec\left(\frac{1}{2}x - \frac{\pi}{3}\right) = 2\sec\frac{1}{2}\left(x - \frac{2\pi}{3}\right)$ has period $2\pi/\left(\frac{1}{2}\right) = 4\pi$.

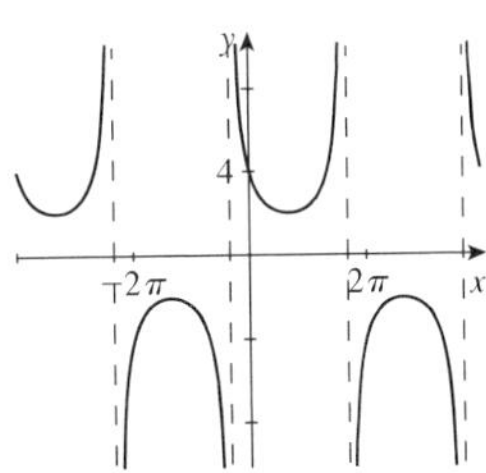


**45.** $y = 5\sec\left(3x - \frac{\pi}{2}\right) = 5\sec 3\left(x - \frac{\pi}{6}\right)$ has period $\frac{2\pi}{3}$.

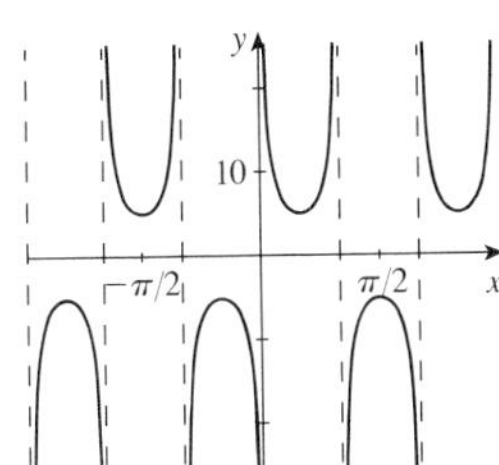


**46.** $y = \frac{1}{2}\sec(2\pi x - \pi) = \frac{1}{2}\sec 2\pi\left(x - \frac{1}{2}\right)$ has period $\frac{2\pi}{2\pi} = 1$.

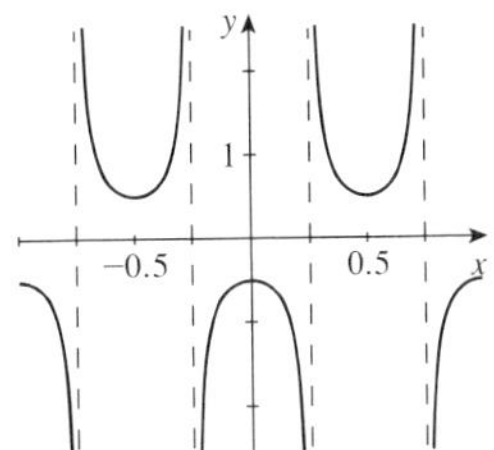


**47.** $y = \tan\left(\frac{2}{3}x - \frac{\pi}{6}\right) = \tan\frac{2}{3}\left(x - \frac{\pi}{4}\right)$ has period $\pi/\left(\frac{2}{3}\right) = \frac{3\pi}{2}$.

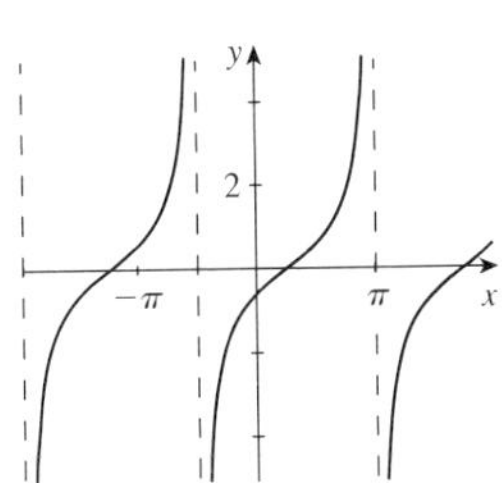


**48.** $y = \tan\frac{1}{2}\left(x + \frac{\pi}{4}\right)$ has period $\dfrac{\pi}{\left(\frac{1}{2}\right)} = 2\pi$.

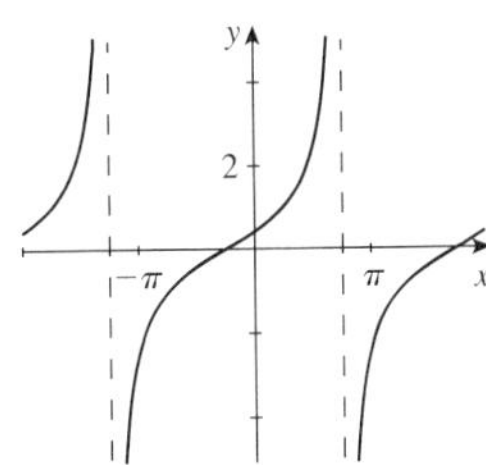


**49.** $y = 3\sec\pi\left(x + \frac{1}{2}\right)$ has period $\frac{2\pi}{\pi} = 2$.

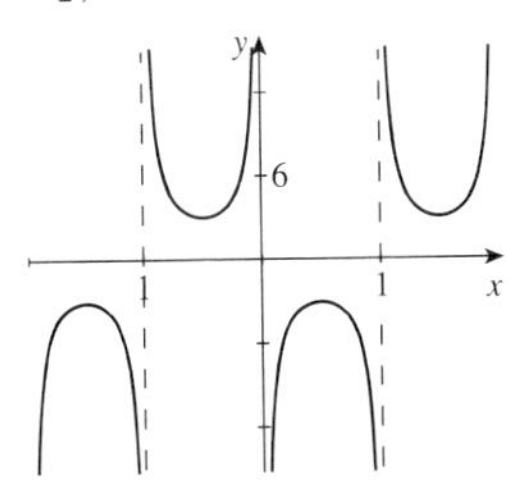


**50.** $y = \sec\left(3x + \frac{\pi}{2}\right) = \sec 3\left(x + \frac{\pi}{6}\right)$ has period $\frac{2\pi}{3}$.

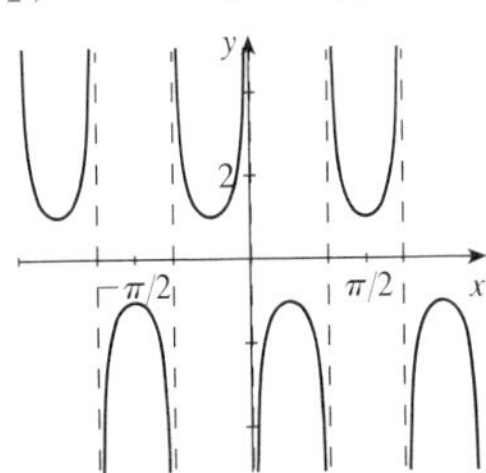


**51.** $y = -2\tan\left(2x - \frac{\pi}{3}\right) = -2\tan 2\left(x - \frac{\pi}{6}\right)$ has period $\frac{\pi}{2}$.

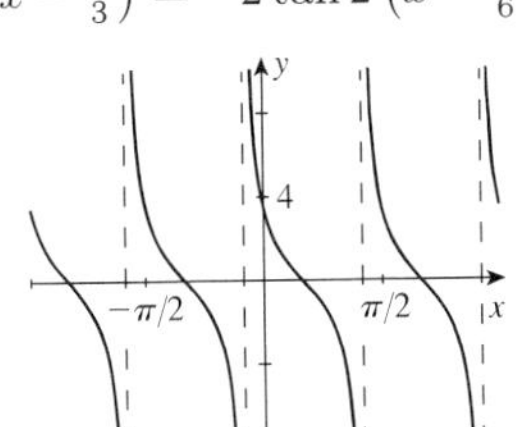


**52.** $y = 2\csc(3x + 3) = 2\csc 3(x + 1)$ has period $\frac{2\pi}{3}$.

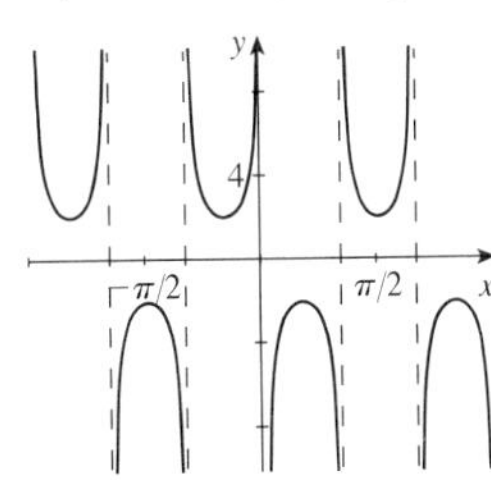

**53. (a)** If $f$ is periodic with period $p$, then by the definition of a period, $f(x+p) = f(x)$ for all $x$ in the domain of $f$. Therefore, $\dfrac{1}{f(x+p)} = \dfrac{1}{f(x)}$ for all $f(x) \neq 0$. Thus, $\dfrac{1}{f}$ is also periodic with period $p$.

**(b)** Since $\sin x$ has period $2\pi$, it follows from part (a) that $\csc x = \dfrac{1}{\sin x}$ also has period $2\pi$. Similarly, since $\cos x$ has period $2\pi$, we conclude $\sec x = \dfrac{1}{\cos x}$ also has period $2\pi$.

**54.** If $f$ and $g$ are periodic with period $p$, then $f(x+p) = f(x)$ for all $x$ and $g(x+p) = g(x)$ for all $x$. Thus $\dfrac{f(x+p)}{g(x+p)} = \dfrac{f(x)}{g(x)}$ for all $x$ [unless $g(x)$ is undefined, in which case both are undefined.] But consider $f(x) = \sin x$ (period $2\pi$) and $g(x) = \cos x$ (period $2\pi$). Their quotient is $\dfrac{f(x)}{g(x)} = \dfrac{\sin x}{\cos x} = \tan x$, whose period is $\pi$.

**55. (a)** $d(t) = 3\tan \pi t$, so $d(0.15) \approx 1.53$, $d(0.25) \approx 3.00$, and $d(0.45) \approx 18.94$.

**(b)**

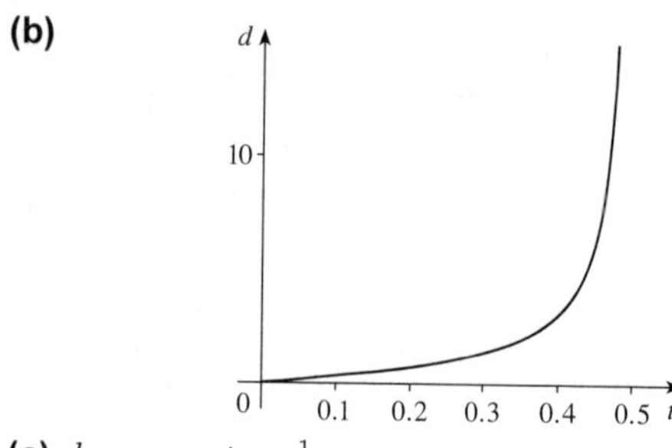


**(c)** $d \to \infty$ as $t \to \frac{1}{2}$.

**56. (a)** $S(t) = 6\left|\cot \frac{\pi}{12}t\right|$, so $S(2) \approx 10.39$, $S(6) = 0$, $S(8) \approx 3.46$, $S(11.75) \approx 91.54$, and $S(12)$ is undefined.

**(b)**



**(c)** From the graph, it appears that $S(t) = 6$ at approximately $t = 3$ and $t = 9$, corresponding to 9 A.M. and 3 P.M.

**(d)** As $t \to 12^-$, the sun approaches the horizon and the man's shadow grows longer and longer.

**57.** The graph of $y = -\cot x$ is the same as the graph of $y = \tan x$ shifted $\frac{\pi}{2}$ units to the right, and the graph of $y = \csc x$ is the same as the graph of $y = \sec x$ shifted $\frac{\pi}{2}$ units to the right.

## 7.5 Modeling Harmonic Motion

**1.** $y = 2\sin 3t$

**(a)** Amplitude 2, period $\frac{2\pi}{3}$, frequency $\dfrac{1}{\text{period}} = \dfrac{3}{2\pi}$.

**(b)**

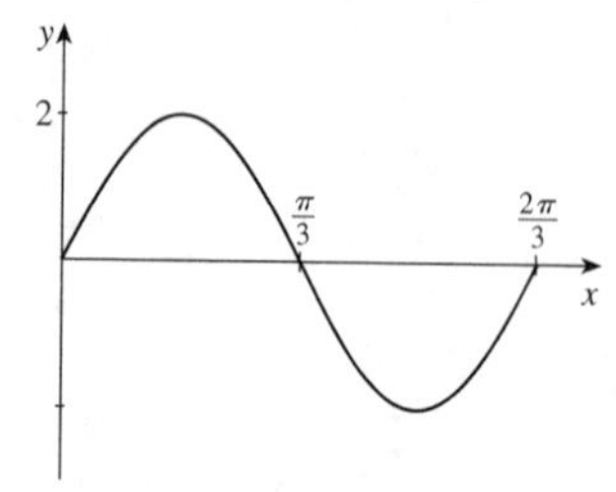


**2.** $y = 3\cos \frac{1}{2}t$

**(a)** Amplitude 3, period $\dfrac{2\pi}{\left(\frac{1}{2}\right)} = 4\pi$, frequency $\dfrac{1}{\text{period}} = \dfrac{1}{4\pi}$.

**(b)**

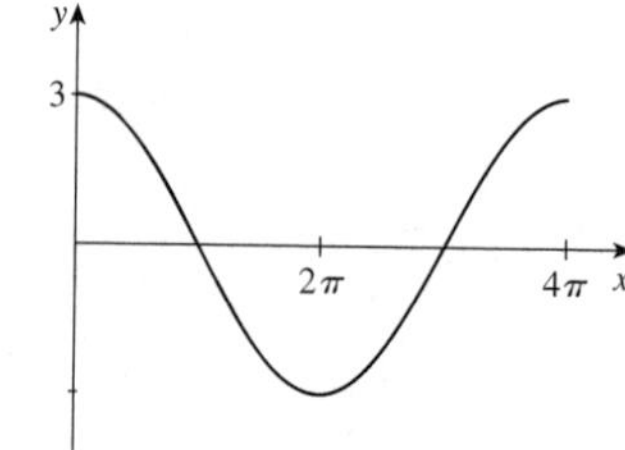

**3.** $y = -\cos 0.3t$

**(a)** Amplitude 1, period $\frac{2\pi}{0.3} = \frac{20\pi}{3}$, frequency $\frac{3}{20\pi}$.

**(b)**

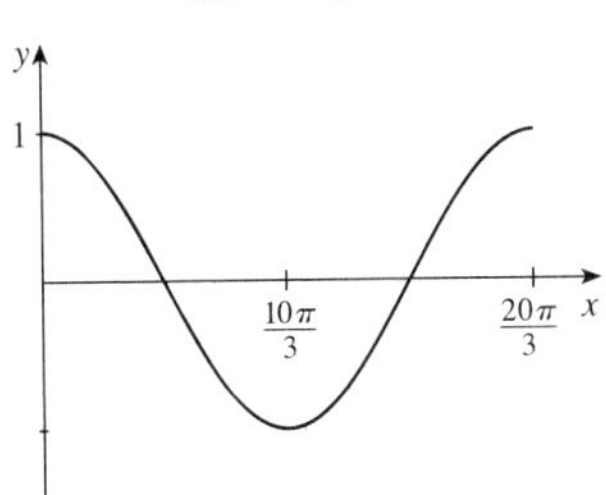

**4.** $y = 2.4 \sin 3.6t$

**(a)** Amplitude 2.4, period $\frac{2\pi}{3.6} = \frac{5\pi}{9}$, frequency $\frac{9}{5\pi}$.

**(b)**

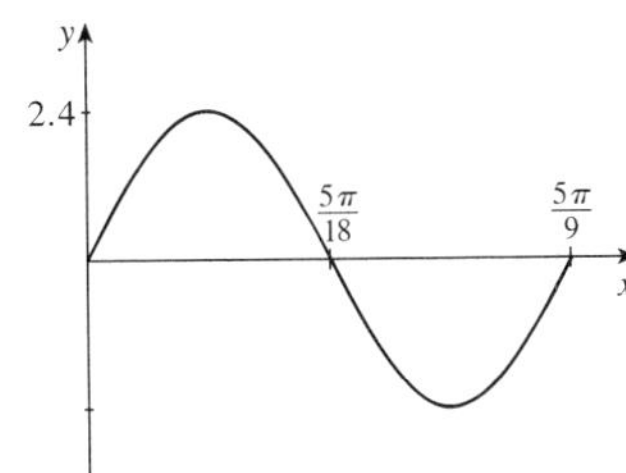

**5.** $y = -0.25 \cos\left(1.5t - \frac{\pi}{3}\right) = -0.25 \cos\left(\frac{3}{2}t - \frac{\pi}{3}\right)$
$= -0.25 \cos \frac{3}{2}\left(t - \frac{2\pi}{9}\right)$

**(a)** Amplitude 0.25, period $\dfrac{2\pi}{\frac{3}{2}} = \frac{4\pi}{3}$, frequency $\frac{3}{4\pi}$.

**(b)**

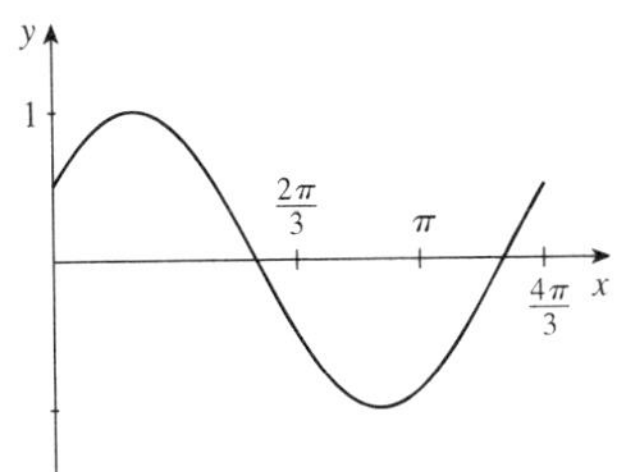

**6.** $y = -\frac{3}{2} \sin(0.2t + 1.4) = -\frac{3}{2} \sin 0.2(t + 7)$

**(a)** Amplitude $\frac{3}{2}$, period $\dfrac{2\pi}{0.2} = 10\pi$, frequency $\frac{1}{10\pi}$.

**(b)**

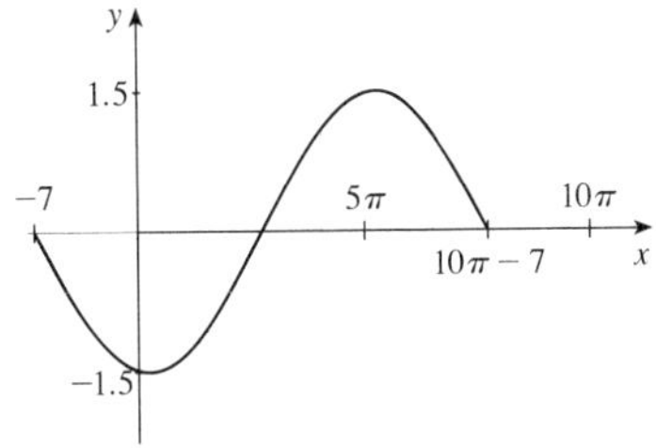

**7.** $y = 5 \cos\left(\frac{2}{3}t + \frac{3}{4}\right) = 5 \cos \frac{2}{3}\left(t + \frac{9}{8}\right)$

**(a)** Amplitude 5, period $\dfrac{2\pi}{\frac{2}{3}} = 3\pi$, frequency $\frac{1}{3\pi}$.

**(b)**

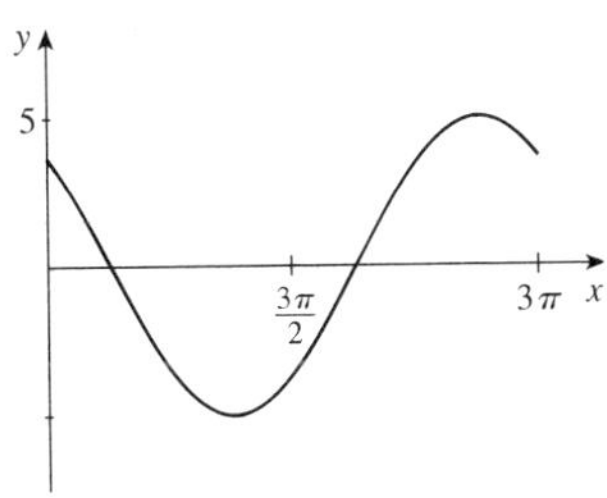

**8.** $y = 1.6 \sin(t - 1.8)$

**(a)** Amplitude 1.6, period $2\pi$, frequency $\frac{1}{2\pi}$.

**(b)**

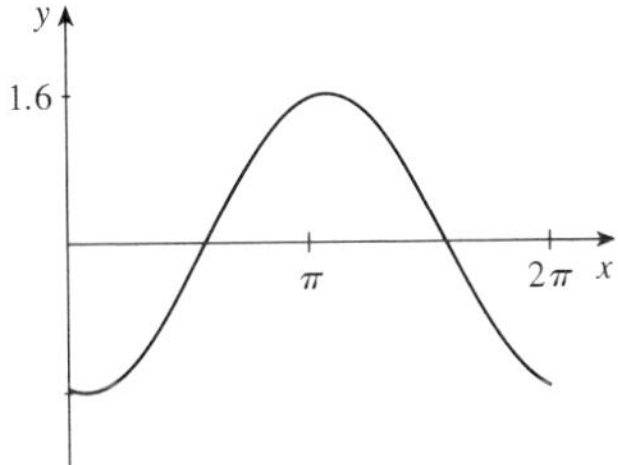

**9.** The amplitude is $a = 10$ cm, the period is $\dfrac{2\pi}{k} = 3$ s, and $f(0) = 0$, so $f(x) = 10 \sin \frac{2\pi}{3}t$.

**10.** The amplitude is 24 ft, the period is $\dfrac{2\pi}{k} = 2$ min, and $f(0) = 0$, so $f(x) = 24 \sin \pi t$.

**11.** The amplitude is 6 in., the frequency is $\dfrac{k}{2\pi} = \dfrac{5}{\pi}$ Hz, and $f(0) = 0$, so $f(x) = 6 \sin 10t$.

**12.** The amplitude is 1.2 m, the frequency is $\dfrac{k}{2\pi} = 0.5$ Hz, and $f(0) = 0$, so $f(x) = 1.2 \sin \pi t$.

**13.** The amplitude is 60 ft, the period is $\dfrac{2\pi}{k} = 0.5$ min, and $f(0) = 60$, so $f(x) = 60 \cos 4\pi t$.

**14.** The amplitude is 35 cm, the period is $\dfrac{2\pi}{k} = 8$ s, and $f(0) = 35$, so $f(x) = 35 \cos \frac{\pi}{4}t$.

**15.** The amplitude is 2.4 m, the frequency is $\dfrac{k}{2\pi} = 750$ Hz, and $f(0) = 2.4$, so $f(x) = 2.4 \cos 1500\pi t$.

**16.** The amplitude is 6.25 in., the frequency is $\dfrac{k}{2\pi} = 60$ Hz, and $f(0) = 6.25$, so $f(x) = 6.25 \cos 120\pi t$.

**17. (a)** $k = 2$, $c = 1.5$, and $f = 3 \Rightarrow \omega = 6\pi$, so we have $y = 2e^{-1.5t} \cos 6\pi t$.

**(b)**

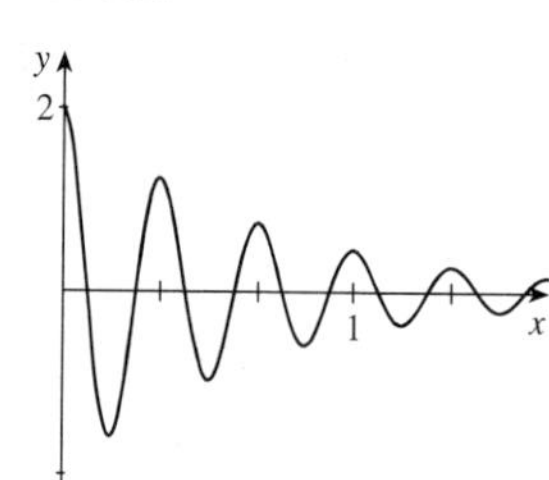


**18. (a)** $k = 15$, $c = 0.25$, and $f = 0.6 \Rightarrow \omega = 1.2\pi$, so we have $y = 15e^{-0.25t} \cos 1.2\pi t$.

**(b)**

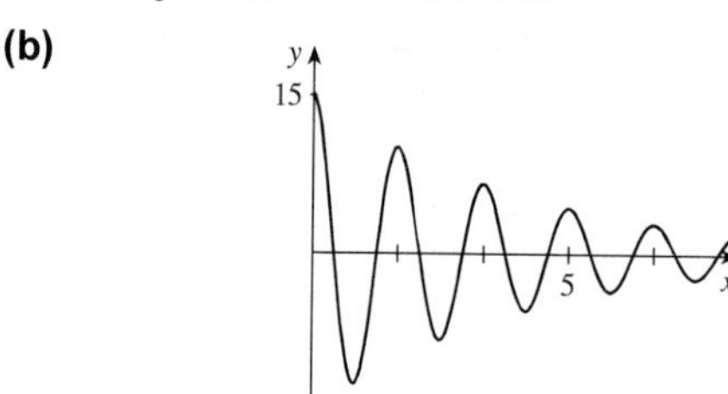


**19. (a)** $k = 100$, $c = 0.05$, and $p = 4 \Rightarrow \omega = \frac{\pi}{2}$, so we have $y = 100e^{-0.05t} \cos \frac{\pi}{2}t$.

**(b)**

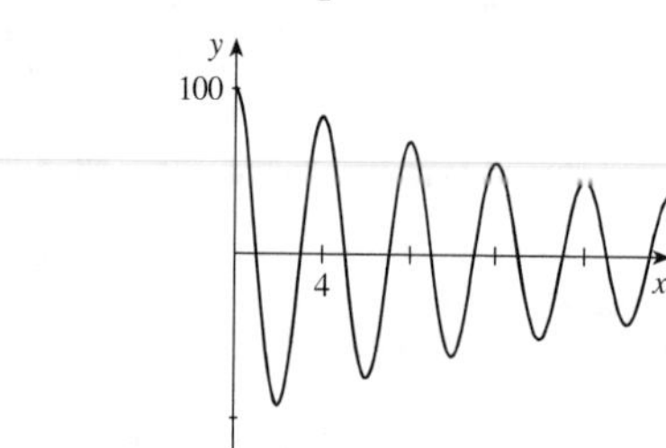


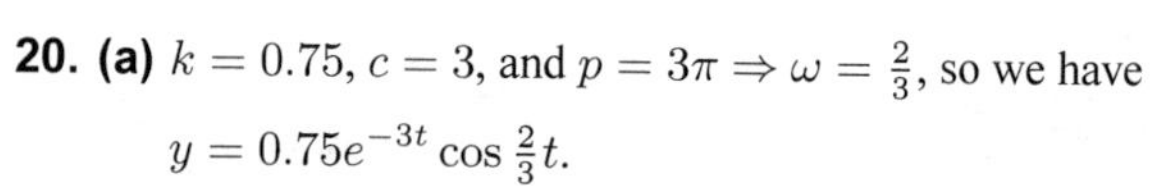

**20. (a)** $k = 0.75$, $c = 3$, and $p = 3\pi \Rightarrow \omega = \frac{2}{3}$, so we have $y = 0.75e^{-3t} \cos \frac{2}{3}t$.

**(b)**

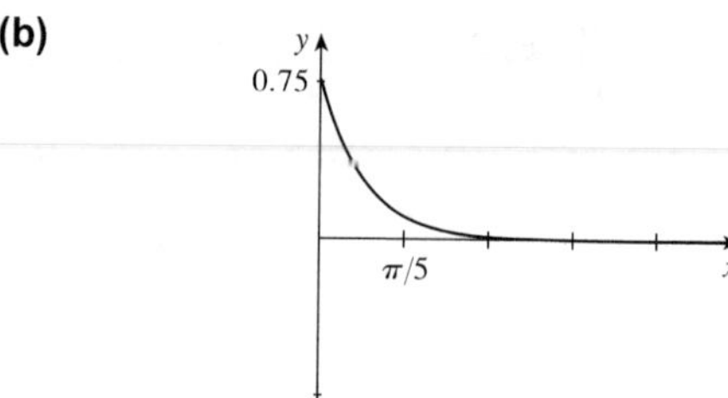


**21. (a)** $k = 7$, $c = 10$, and $p = \frac{\pi}{6} \Rightarrow \omega = 12$, so we have $y = 7e^{-10t} \sin 12t$.

**(b)**

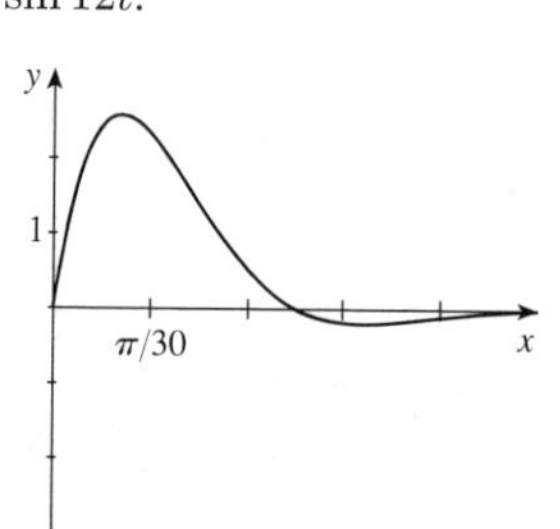


**22. (a)** $k = 1$, $c = 1$, and $p = 1 \Rightarrow \omega = 2\pi$, so we have $y = e^{-t} \sin 2\pi t$.

**(b)**

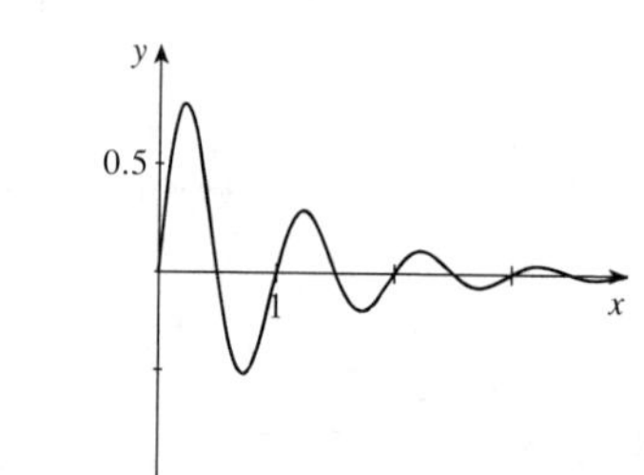


**23. (a)** $k = 0.3$, $c = 0.2$, and $f = 20 \Rightarrow \omega = 40\pi$, so we have $y = 0.3e^{-0.2t} \sin 40\pi t$.

**(b)**

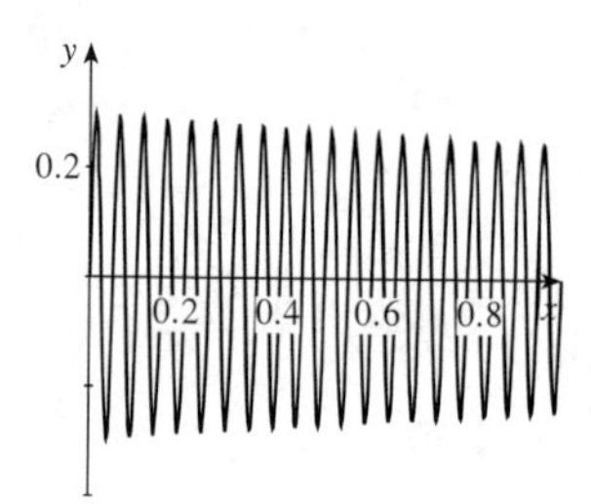


**24. (a)** $k = 12$, $c = 0.01$, and $f = 8 \Rightarrow \omega = 16\pi$, so we have $y = 12e^{-0.01t} \sin 16\pi t$.

**(b)**

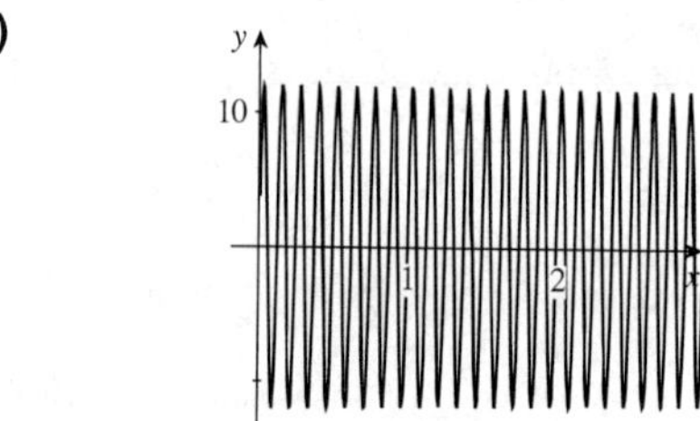

**25.** $y = 0.2\cos 20\pi t + 8$

**(a)** The frequency is $\frac{20\pi}{2\pi} = 10$ cycles/min.

**(b)**

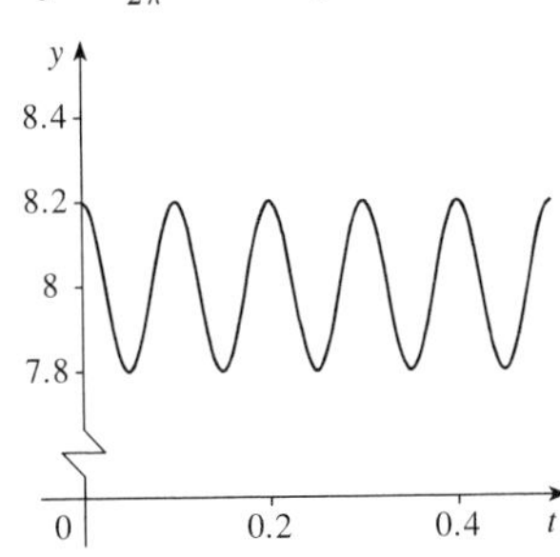


**(c)** Since $y = 0.2\cos 20\pi t + 8 \leq 0.2(1) + 8 = 8.2$ and when $t = 0$, $y = 8.2$, the maximum displacement is 8.2 m.

**26.** $y = a\sin\left(2\pi\left(9.15 \times 10^7\right)t\right)$. The period is $\dfrac{2\pi}{2\pi\left(9.15 \times 10^7\right)} = \dfrac{1}{9.15} \times 10^{-7} = 1.09 \times 10^{-8}$ and the frequency is $\dfrac{2\pi\left(9.15 \times 10^7\right)}{2\pi} = 9.15 \times 10^7$.

**27. (a)** When $t = 0, \pi, 2\pi, \ldots$ we have $\cos 2t = 1$ so $y$ is maximized at $y = 8900$.

**(b)** The length of time between successive periods of maximum population is the length of a period which is $\frac{2\pi}{2} = \pi \approx 3.14$ years.

**28.** $p(t) = 115 + 25\sin(160\pi t)$

**(a)** Amplitude 25, period $\frac{2\pi}{160\pi} = \frac{1}{80} = 0.0125$, frequency $\dfrac{1}{\text{period}} = 80$.

**(b)**

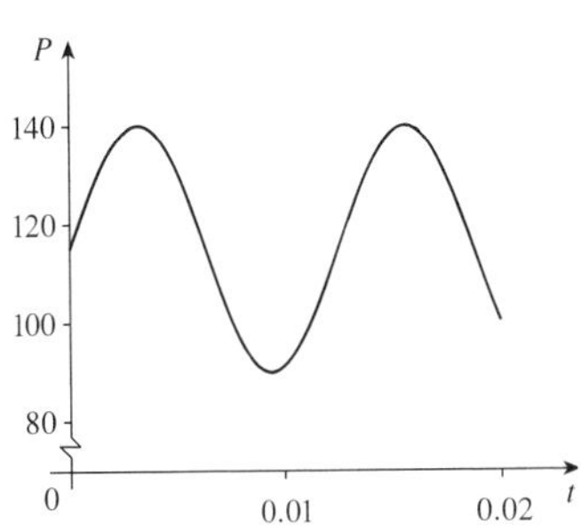


**(c)** The period decreases and the frequency increases.

**29.** The graph resembles a sine wave with an amplitude of 5, a period of $\frac{2}{5}$, and no phase shift. Therefore, $a = 5$, $\dfrac{2\pi}{\omega} = \frac{2}{5} \quad\Leftrightarrow\quad \omega = 5\pi$, and a formula is $d(t) = 5\sin 5\pi t$.

**30.** From the graph we see that the amplitude is 6 feet and the period is 12 hours. Also, the sine curve is shifted 6 hours (exactly half its period) to the right, which is equivalent to reflecting the curve about the $t$-axis. Thus, the equation is $y = -(6)\sin\left(\frac{2\pi}{12}\right)t = -6\sin\frac{\pi}{6}t$.

**31.** $a = 21$, $f = \frac{1}{12}$ cycle/hour $\quad\Rightarrow\quad \dfrac{\omega}{2\pi} = \frac{1}{12} \quad\Leftrightarrow\quad \omega = \frac{\pi}{6}$. So, $y = 21\sin\left(\frac{\pi}{6}t\right)$ (assuming the tide is at mean level and rising when $t = 0$).

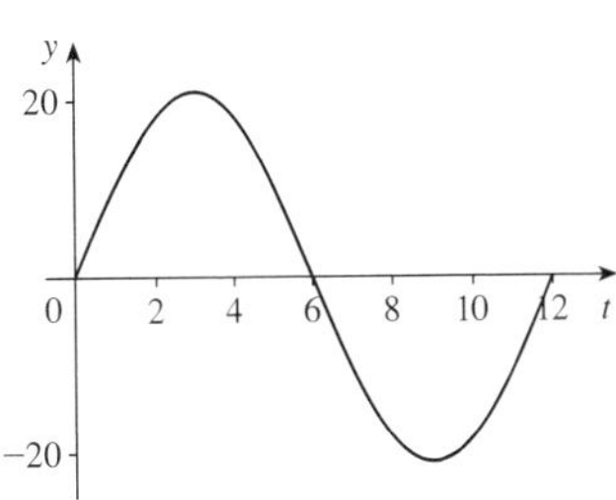


**32.** $a = 2$, $\dfrac{2\pi}{\omega} = 1 \quad\Leftrightarrow\quad \omega = 2\pi$. So $y = -2\cos 2\pi t$.

**33.** Since the mass travels from its highest point (compressed spring) to its lowest point in $\frac{1}{2}$ s, it completes half a period in $\frac{1}{2}$ s. So, $\frac{1}{2}$ (one period) $= \frac{1}{2}$ s $\quad\Rightarrow\quad \frac{1}{2} \cdot \dfrac{2\pi}{\omega} = \frac{1}{2} \quad\Leftrightarrow\quad \omega = 2\pi$. Also, $a = 5$. So $y = 5\cos 2\pi t$.

**34. (a)** $m = 10$ g, $k = 3$, $a = 5$ cm. Then $f(t) = 5\cos\left(\sqrt{\frac{3}{10}}t\right)$.

**(b)** The function $f(t) = a\cos\sqrt{k/m}t$ describes an object oscillating in simple harmonic motion, so by comparing it with the general equation $y = a\cos\omega t$ we see that $\omega = \sqrt{k/m}$. This means the frequency is $f = \frac{\omega}{2\pi} = \frac{\sqrt{k/m}}{2\pi} = \frac{1}{2\pi}\cdot\sqrt{\frac{k}{m}}$.

**(c)** If the mass is increased, then the denominator of $\sqrt{\frac{k}{m}}$ increases, so overall the frequency decreases. If the frequency has decreased, then by definition the oscillations are slower.

**(d)** If a stiffer spring is used, $k$ is larger and so the numerator of $\sqrt{\frac{k}{m}}$ increases. Thus, overall the frequency increases. If the frequency has increased, then by definition the oscillations are faster.

**35.** Since the Ferris wheel has a radius of 10 m and the bottom of the wheel is 1 m above the ground, the minimum height is 1 m and the maximum height is 21 m. Then $a = 10$ and $\frac{2\pi}{\omega} = 20$ s $\Leftrightarrow$ $\omega = \frac{\pi}{10}$, and so $y = 11 + 10\sin\left(\frac{\pi}{10}t\right)$, where $t$ is in seconds.

**36.** Let $f(t)$ represent the measure of the angle $\theta$ at time $t$. The amplitude of this motion is 10. Since the period is 2s, we have $\frac{2\pi}{\omega} = 2$ $\Leftrightarrow$ $\omega = \pi$. Thus $f(t) = 10\sin\pi t$.

**37.** $a = 0.2$, $\frac{2\pi}{\omega} = 10$ $\Leftrightarrow$ $\omega = \frac{\pi}{5}$. Then $y = 3.8 + 0.2\sin\left(\frac{\pi}{5}t\right)$.

**38.** $a = 1.5$, $\frac{\omega}{2\pi} = \frac{1}{5.4} \Leftrightarrow \omega = \frac{2\pi}{5.4}$. So $R(t) = 20 + 1.5\sin\left(\frac{2\pi}{5.4}t\right)$, where $R$ is in millions of miles and $t$ is in days.

**39.** $E_0 = 310$, frequency is 100 $\Rightarrow$ $\frac{\omega}{2\pi} = 100$ $\Leftrightarrow$ $\omega = 200\pi$. Then, $E(t) = 310\cos 200\pi t$. The maximum voltage produced occurs when $\cos 2\pi t = 1$, and hence is $E_{\max} = 310$ V. The rms voltage is $\frac{310}{\sqrt{2}} \approx 219$ V.

**40.** The amplitude is $\frac{1}{2}(100 - 80) = 10$ mmHG, the period is 24 hours, and the phase shift is 8 hours, so $f(t) = 10\sin\left(\frac{\pi}{12}(t-8)\right) + 90$.

**41. (a)** The maximum voltage is the amplitude, that is, $V_{\max} = a = 45$ V.

**(b)** From the graph we see that 4 cycles are completed every 0.1 seconds, or equivalently, 40 cycles are completed every second, so $f = 40$.

**(c)** The number of revolutions per second of the armature is the frequency, that is, $\frac{\omega}{2\pi} = f = 40$.

**(d)** $a = 45$, $f = \frac{\omega}{2\pi} = 40$ $\Leftrightarrow$ $\omega = 80\pi$. Then $V(t) = 45\cos 80\pi t$.

**42. (a)** As the car approaches, the perceived frequency is $f = 500\left(\frac{1130}{1130-110}\right) \approx 553.9$ Hz. As it moves away, the perceived frequency is $f = 500\left(\frac{1130}{1130+110}\right) \approx 455.6$ Hz.

**(b)** The frequency is $\frac{\omega}{2\pi} = 500\left(\frac{1130}{1130\pm110}\right)$, so $\omega = 1000\pi\left(\frac{1130}{1130\pm110}\right) \approx 1107.8\pi$ (approaching) or $911.3\pi$ (receding). Thus, models are $y = A\sin 1107.8\pi t$ and $A\sin 911.3\pi t$.

**43.** $k = 1$, $c = 0.9$, and $\frac{\omega}{2\pi} = \frac{1}{2} \Leftrightarrow \omega = \pi$. Since $f(0) = 0$, $f(t) = e^{-0.9t}\sin\pi t$.

**44.** $k = 6$, $c = 2.8$, and $\frac{\omega}{2\pi} = 2$ $\Leftrightarrow$ $\omega = 4\pi$.

**(a)** Since $f(0) = 6$, $f(t) = 6e^{-2.8t}\cos 4\pi t$.

**(b)** The amplitude of the vibration is 0.5 when $6e^{-2.8t} = 0.5$ $\Rightarrow$ $e^{-2.8t} = \frac{1}{12}$ $\Rightarrow$ $-2.8t = \ln\frac{1}{12}$ $\Rightarrow$ $t = -\frac{1}{2}\ln\frac{1}{12} = \frac{\ln 12}{2.8} \approx 0.88$ s.

**45.** $\dfrac{ke^{-ct}}{ke^{-c(t+3)}} = 4 \Leftrightarrow e^{-ct+c(t+3)} = 4 \quad \Leftrightarrow \quad e^{3c} = 4 \quad \Leftrightarrow \quad 3c = \ln 4 \quad \Leftrightarrow \quad c = \frac{1}{3}\ln 4 \approx 0.46.$

**46. (a)** $\dfrac{ke^{-c\cdot 0}}{ke^{-c\cdot 2}} = \dfrac{3}{0.6} = 5 \quad \Leftrightarrow \quad e^{2c} = 5 \quad \Leftrightarrow \quad 2c = \ln 5 \quad \Leftrightarrow \quad c = \frac{1}{2}\ln 5 \approx 0.80.$

**(b)** $f(t) = 3e^{-0.8t}\cos\omega t$. If the frequency is 165 cycles per second, then $\dfrac{\omega}{2\pi} = 165 \quad \Leftrightarrow \quad \omega = 330\pi$. Thus, $f(t) = 3e^{-0.8t}\cos 330\pi t$.

# Chapter 7 Review

**1. (a)** Since $\left(-\frac{\sqrt{3}}{2}\right)^2 + \left(\frac{1}{2}\right)^2 = \frac{3}{4} + \frac{1}{4} = 1$, the point $P\left(-\frac{\sqrt{3}}{2}, \frac{1}{2}\right)$ lies on the unit circle.

**(b)** $\sin t = \frac{1}{2}$, $\cos t = -\frac{\sqrt{3}}{2}$, $\tan t = \dfrac{\frac{1}{2}}{-\frac{\sqrt{3}}{2}} = -\frac{\sqrt{3}}{3}$.

**2. (a)** Since $\left(\frac{3}{5}\right)^2 + \left(-\frac{4}{5}\right)^2 = \frac{9}{25} + \frac{1}{2} = 1$, the point $P\left(\frac{3}{5}, -\frac{4}{5}\right)$ lies on the unit circle.

**(b)** $\sin t = -\frac{4}{5}$, $\cos t = \frac{3}{5}$, $\tan t = \dfrac{-\frac{4}{5}}{\frac{3}{5}} = -\frac{4}{3}$.

**3.** $t = \frac{2\pi}{3}$

**(a)** $\bar{t} = \pi - \frac{2\pi}{3} = \frac{\pi}{3}$
**(b)** $P\left(-\frac{1}{2}, \frac{\sqrt{3}}{2}\right)$
**(c)** $\sin t = \frac{\sqrt{3}}{2}$, $\cos t = -\frac{1}{2}$, $\tan t = -\sqrt{3}$, $\csc t = \frac{2\sqrt{3}}{3}$, $\sec t = -2$, and $\cot t = -\frac{\sqrt{3}}{3}$.

**4.** $t = \frac{5\pi}{3}$

**(a)** $\bar{t} = 2\pi - \frac{5\pi}{3} = \frac{\pi}{3}$
**(b)** $P\left(\frac{1}{2}, -\frac{\sqrt{3}}{2}\right)$
**(c)** $\sin t = -\frac{\sqrt{3}}{2}$, $\cos t = \frac{1}{2}$, $\tan t = -\sqrt{3}$, $\csc t = -\frac{2\sqrt{3}}{3}$, $\sec t = 2$, and $\cot t = -\frac{\sqrt{3}}{3}$.

**5.** $t = -\frac{11\pi}{4}$

**(a)** $\bar{t} = 3\pi + \left(-\frac{11\pi}{4}\right) = \frac{\pi}{4}$
**(b)** $P\left(-\frac{\sqrt{2}}{2}, -\frac{\sqrt{2}}{2}\right)$
**(c)** $\sin t = -\frac{\sqrt{2}}{2}$, $\cos t = -\frac{\sqrt{2}}{2}$, $\tan t = 1$, $\csc t = -\sqrt{2}$, $\sec t = -\sqrt{2}$, and $\cot t = 1$.

**6.** $t = -\frac{7\pi}{6}$

**(a)** $\bar{t} = \frac{7\pi}{6} - \pi = \frac{\pi}{6}$
**(b)** $P\left(-\frac{\sqrt{3}}{2}, \frac{1}{2}\right)$
**(c)** $\sin t = \frac{1}{2}$, $\cos t = -\frac{\sqrt{3}}{2}$, $\tan t = -\frac{\sqrt{3}}{3}$, $\csc t = 2$, $\sec t = -\frac{2\sqrt{3}}{3}$, and $\cot t = -\sqrt{3}$.

**7. (a)** $\sin\frac{3\pi}{4} = \sin\frac{\pi}{4} = \frac{\sqrt{2}}{2}$
**(b)** $\cos\frac{3\pi}{4} = -\cos\frac{\pi}{4} = -\frac{\sqrt{2}}{2}$

**8. (a)** $\tan\frac{\pi}{3} = \sqrt{3}$
**(b)** $\tan\left(-\frac{\pi}{3}\right) = -\sqrt{3}$

**9. (a)** $\sin 1.1 \approx 0.89121$
**(b)** $\cos 1.1 \approx 0.45360$

**10. (a)** $\cos\frac{\pi}{5} \approx 0.80902$
**(b)** $\cos\left(-\frac{\pi}{5}\right) \approx 0.80902$

**11. (a)** $\cos\frac{9\pi}{2} = \cos\frac{\pi}{2} = 0$
**(b)** $\sec\frac{9\pi}{2}$ is undefined

**12. (a)** $\sin\frac{\pi}{7} \approx 0.43388$
**(b)** $\csc\frac{\pi}{7} \approx 2.30476$

**13. (a)** $\tan\frac{5\pi}{2}$ is undefined
**(b)** $\cot\frac{5\pi}{2} = \cot\frac{\pi}{2} = 0$

**14. (a)** $\sin 2\pi = 0$
**(b)** $\csc 2\pi$ is undefined

**15. (a)** $\tan\frac{5\pi}{6} = -\frac{\sqrt{3}}{3}$
**(b)** $\cot\frac{5\pi}{6} = -\sqrt{3}$

**16. (a)** $\cos\frac{\pi}{3} = \frac{1}{2}$
**(b)** $\sin\frac{\pi}{6} = \frac{1}{2}$

**17.** $\dfrac{\tan t}{\cos t} = \dfrac{\frac{\sin t}{\cos t}}{\cos t} = \dfrac{\sin t}{\cos^2 t} = \dfrac{\sin t}{1 - \sin^2 t}$

**18.** $\tan^2 t \cdot \sec t = \tan^2 t \cdot \dfrac{1}{\cos t} = \dfrac{\sin^2 t}{\cos^2 t} \cdot \dfrac{1}{\cos t} = \dfrac{1 - \cos^2 t}{\cos^3 t}$

**19.** $\tan t = \dfrac{\sin t}{\cos t} = \dfrac{\sin t}{\pm\sqrt{1 - \sin^2 t}} = \dfrac{\sin t}{\sqrt{1 - \sin^2 t}}$ (since $t$ is in quadrant IV, $\cos t$ is positive)

**20.** $\sec t = \dfrac{1}{\cos t} = \dfrac{1}{\pm\sqrt{1 - \sin^2 t}} = \dfrac{1}{-\sqrt{1 - \sin^2 t}}$ (since $t$ is in quadrant II $\Rightarrow$ $\cos t$ is negative)

**21.** $\sin t = \frac{5}{13}$, $\cos t = -\frac{12}{13}$. Then $\tan t = \dfrac{\frac{5}{13}}{-\frac{12}{13}} = -\frac{5}{12}$, $\csc t = \frac{13}{5}$, $\sec t = -\frac{13}{12}$, and $\cot t = -\frac{12}{5}$.

**22.** $\sin t = -\frac{1}{2}$, $\cos t > 0$. Since $\sin t$ is negative and $\cos t$ is positive, $t$ is in quadrant IV. Thus, $t$ determines the terminal point $P\left(\frac{\sqrt{3}}{2}, -\frac{1}{2}\right)$, and $\cos t = \frac{\sqrt{3}}{2}$, $\tan t = -\dfrac{\sqrt{3}}{3}$, $\csc t = -2$, $\sec t = \dfrac{2\sqrt{3}}{3}$, $\cot t = -\sqrt{3}$.

**23.** $\cot t = -\frac{1}{2}$, $\csc t = \frac{\sqrt{5}}{2}$. Since $\csc t = \dfrac{1}{\sin t}$, we know $\sin t = \frac{2}{\sqrt{5}} = \frac{2\sqrt{5}}{5}$. Now $\cot t = \dfrac{\cos t}{\sin t}$, so $\cos t = \sin t \cdot \cot t = \frac{2\sqrt{5}}{5} \cdot \left(-\frac{1}{2}\right) = -\frac{\sqrt{5}}{5}$, and $\tan t = \dfrac{1}{\left(-\frac{1}{2}\right)} = -2$ while $\sec t = \dfrac{1}{\cos t} = \dfrac{1}{\left(-\frac{\sqrt{5}}{5}\right)} = -\frac{5}{\sqrt{5}} = -\sqrt{5}$.

**24.** $\cos t = -\frac{3}{5}$, $\tan t < 0$. Since $\cos t$ and $\tan t$ are both negative, $t$ is in quadrant II. Thus, $t$ determines the terminal point $P\left(-\frac{3}{5}, \frac{4}{5}\right)$, and $\sin t = \frac{4}{5}$, $\tan t = -\frac{4}{3}$, $\csc t = \frac{5}{4}$, $\sec t = -\dfrac{5}{3}$, $\cot t = -\frac{3}{4}$.

**25.** $\tan t = \frac{1}{4}$, $t$ is in quadrant III $\Rightarrow$

$$\sec t + \cot t = -\sqrt{\tan^2 t + 1} + \frac{1}{\tan t} = -\sqrt{\left(\tfrac{1}{4}\right)^2 + 1} + 4 = -\sqrt{\tfrac{17}{16}} + 4 = 4 - \tfrac{\sqrt{17}}{4} = \frac{16 - \sqrt{17}}{4}$$

**26.** $\sin t = -\frac{8}{17}$, $t$ is in quadrant IV $\Rightarrow$

$$\begin{aligned}\csc t + \sec t &= \frac{1}{\sin t} + \frac{1}{\cos t} = \frac{\cos t + \sin t}{\sin t \cos t} = \frac{\sqrt{1 - \sin^2 t} + \sin t}{\sin t \sqrt{1 - \sin^2 t}} = \frac{\frac{15}{17} + \left(-\frac{8}{17}\right)}{\left(-\frac{8}{17}\right) \cdot \frac{15}{17}} \\ &= \tfrac{7}{17} \cdot \tfrac{17^2}{-8 \cdot 15} = -\tfrac{7 \cdot 17}{8 \cdot 15} = -\tfrac{119}{120} \approx -0.99167 \text{ (since } \cos t \text{ is positive in quadrant IV)}\end{aligned}$$

**27.** $\cos t = \frac{3}{5}$, $t$ is in quadrant I $\Rightarrow$

$$\tan t + \sec t = \frac{\sin t}{\cos t} + \frac{1}{\cos t} = \frac{\sqrt{1 - \cos^2 t}}{\cos t} + \frac{1}{\cos t} = \frac{\sqrt{1 - \left(\frac{3}{5}\right)^2}}{\frac{3}{5}} + \frac{5}{3} = \frac{\sqrt{\frac{1}{2}}}{\frac{3}{5}} + \frac{5}{3} = \frac{4}{5} \cdot \frac{5}{3} + \frac{5}{3} = \frac{9}{3} = 3$$

**28.** The value of $\sec t$ is irrelevant, since $\sin^2 t + \cos^2 t = 1$ for all $t$.

**29.** $y = 10 \cos \frac{1}{2}x$

**(a)** This function has amplitude10, period $\dfrac{2\pi}{\frac{1}{2}} = 4\pi$, and phase shift 0.

**(b)**

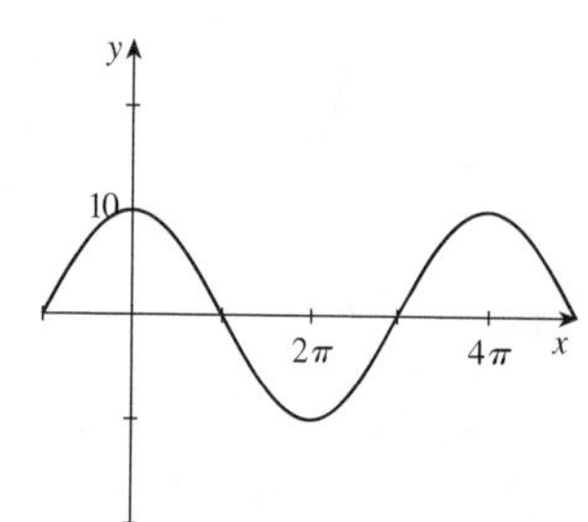


**30.** $y = 4 \sin 2\pi x$

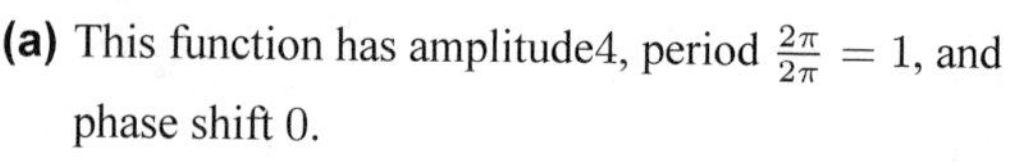

**(a)** This function has amplitude4, period $\frac{2\pi}{2\pi} = 1$, and phase shift 0.

**(b)**

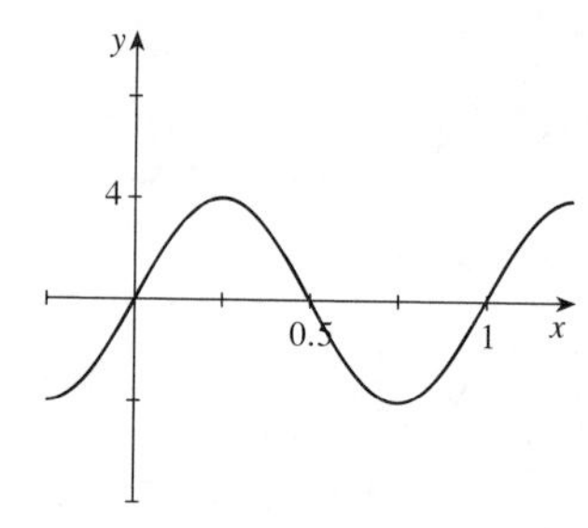

**31.** $y = -\sin\frac{1}{2}x$

**(a)** This function has amplitude1, period $\dfrac{2\pi}{\left(\frac{1}{2}\right)} = 4\pi$, and phase shift 0.

**(b)**

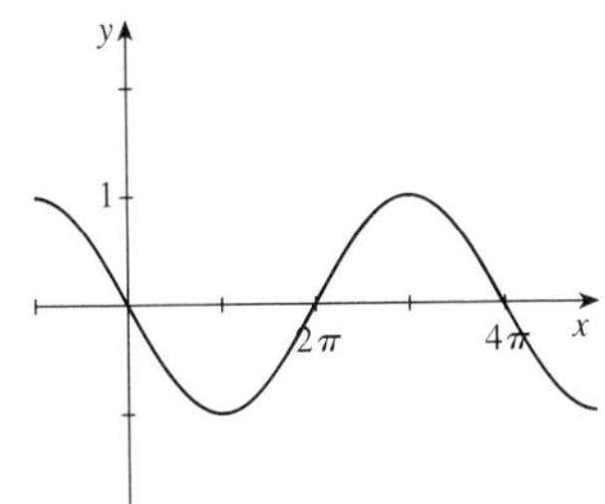


**32.** $y = 2\sin\left(x - \frac{\pi}{4}\right)$

**(a)** This function has amplitude2, period $2\pi$, and phase shift $\frac{\pi}{4}$.

**(b)**

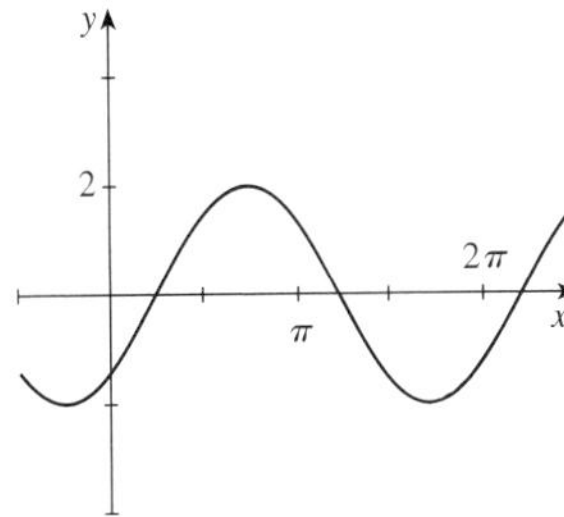


**33.** $y = 3\sin(2x - 2) = 3\sin 2(x - 1)$

**(a)** This function has amplitude3, period $\frac{2\pi}{2} = \pi$, and phase shift 1.

**(b)**

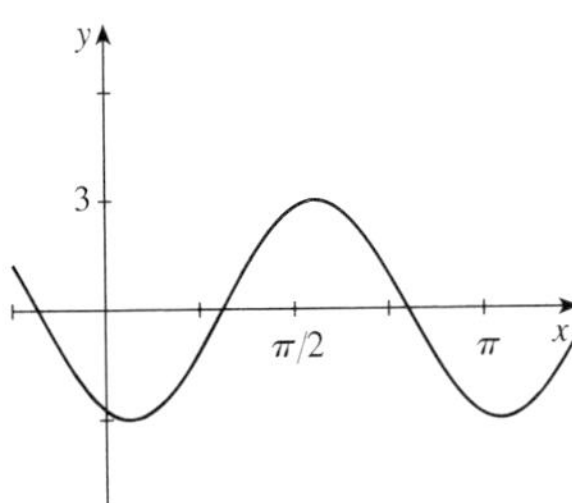


**34.** $y = \cos 2\left(x - \frac{\pi}{2}\right)$

**(a)** This function has amplitude1, period $\frac{2\pi}{2} = \pi$, and phase shift $\frac{\pi}{2}$.

**(b)**

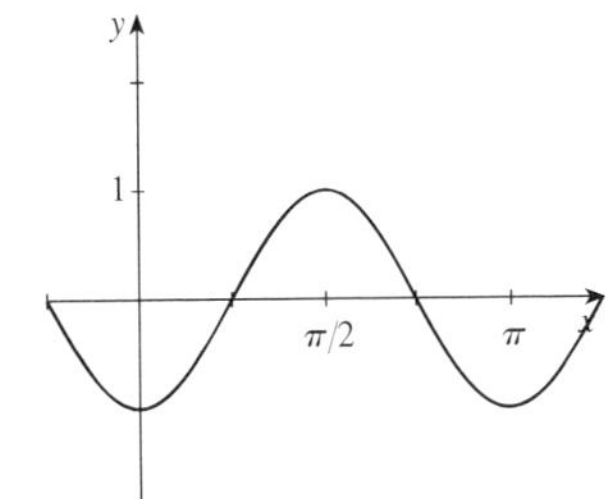


**35.** $y = -\cos\left(\frac{\pi}{2}x + \frac{\pi}{6}\right) = -\cos\frac{\pi}{2}\left(x + \frac{1}{3}\right)$

**(a)** This function has amplitude1, period $\dfrac{2\pi}{\left(\frac{\pi}{2}\right)} = 4$, and phase shift $-\frac{1}{3}$.

**(b)**

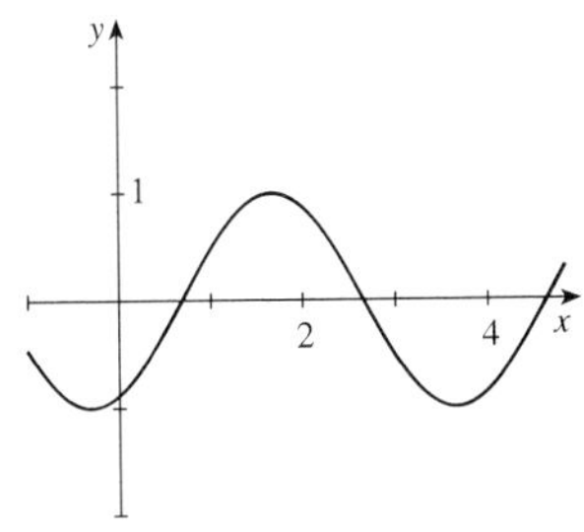


**36.** $y = 10\sin\left(2x - \frac{\pi}{2}\right) = 10\sin 2\left(x - \frac{\pi}{4}\right)$

**(a)** This function has amplitude10, period $\frac{2\pi}{2} = \pi$, and phase shift $\frac{\pi}{4}$.

**(b)**

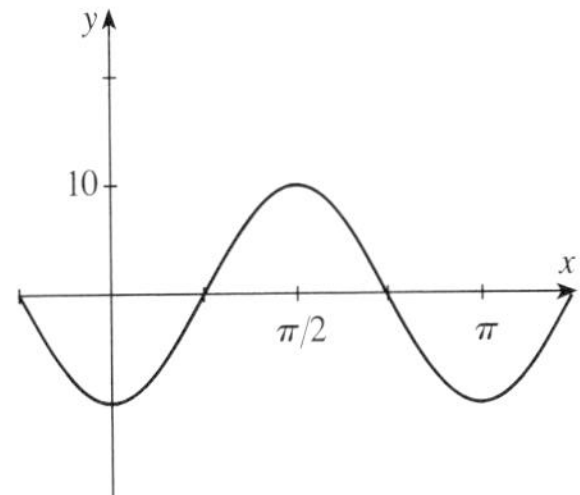


**37.** From the graph we see that the amplitude is 5, the period is $\frac{\pi}{2}$, and there is no phase shift. Therefore, the function is $y = 5\sin 4x$.

**38.** From the graph we see that the amplitude is 2, the period is 4 since $\frac{1}{4}$of the period has been completed at $(1, 2)$, and there is no phase shift. Thus, $\dfrac{2\pi}{k} = 4 \Leftrightarrow k = \frac{\pi}{2}$. Therefore, the function is $y = 2\sin\frac{\pi}{2}x$.

**39.** From the graph we see that the amplitude is $\frac{1}{2}$, the period is 1, and there is a phase shift of $-\frac{1}{3}$. Therefore, the function is $y = \frac{1}{2}\sin 2\pi\left(x + \frac{1}{3}\right)$.

**40.** From the graph we see that the amplitude is 4, the period is $\frac{4\pi}{3}$. Thus, $\dfrac{2\pi}{k} = \frac{4\pi}{3} \Leftrightarrow k = \frac{3}{2}$. The phase shift is $-\frac{\pi}{3}$. Therefore, the function is $y = 4\sin\frac{3}{2}\left(x + \frac{\pi}{3}\right)$.

**41.** $y = 3\tan x$ has period $\pi$.

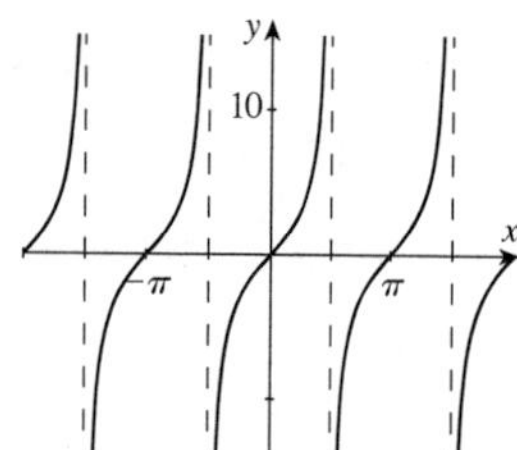


**42.** $y = \tan \pi x$ has period $\frac{\pi}{\pi} = 1$.

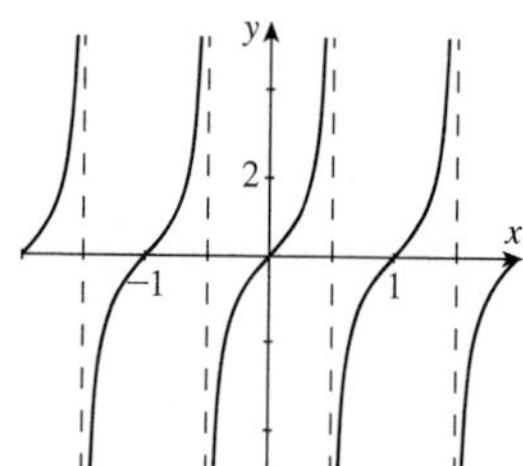


**43.** $y = 2\cot\left(x - \frac{\pi}{2}\right)$ has period $\pi$.

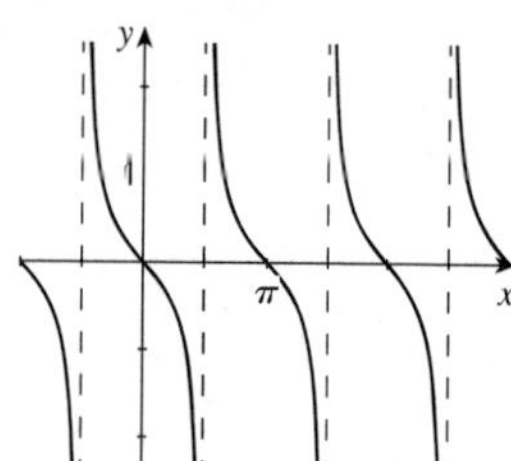


**44.** $y = \sec\left(\frac{1}{2}x - \frac{\pi}{2}\right) = \sec\frac{1}{2}(x - \pi)$ has period $\dfrac{2\pi}{\left(\frac{1}{2}\right)} = 4\pi$.

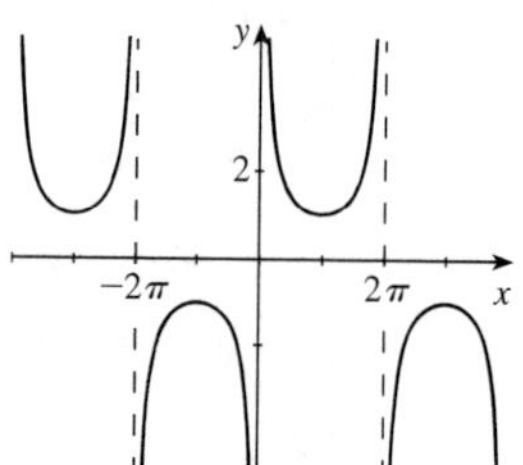


**45.** $y = 4\csc(2x + \pi) = 4\csc 2\left(x + \frac{\pi}{2}\right)$ has period $\frac{2\pi}{2} = \pi$.

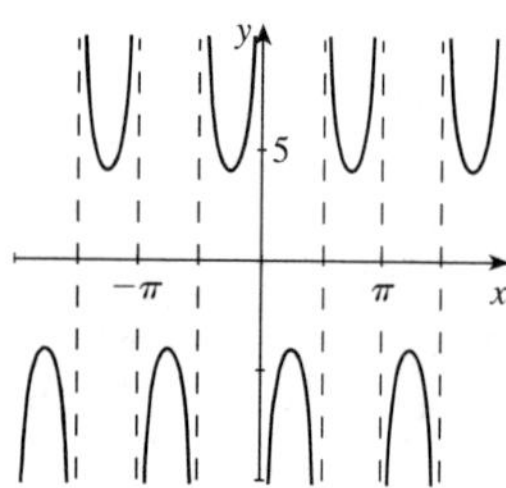


**46.** $y = \tan\left(x + \frac{\pi}{6}\right)$ has period $\pi$.

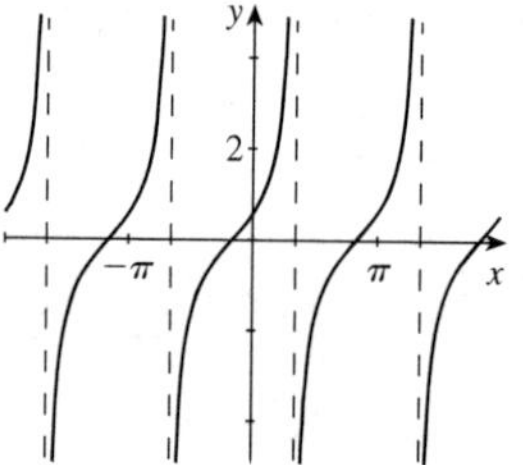


**47.** $y = \tan\left(\frac{1}{2}x - \frac{\pi}{8}\right) = \tan\frac{1}{2}\left(x - \frac{\pi}{4}\right)$ has period $\dfrac{\pi}{\frac{1}{2}} = 2\pi$.

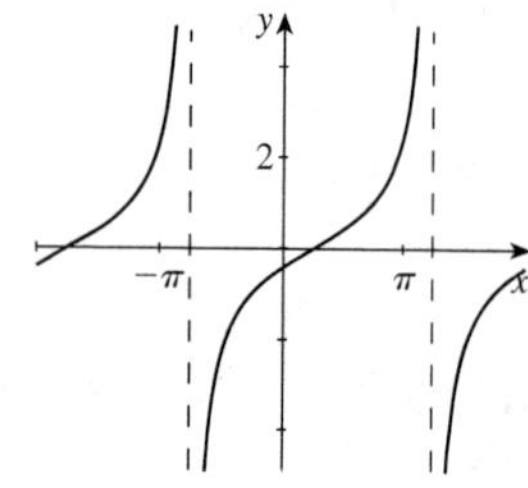


**48.** $y = -4\sec(4\pi x)$ has period $\frac{2\pi}{4\pi} = \frac{1}{2}$.

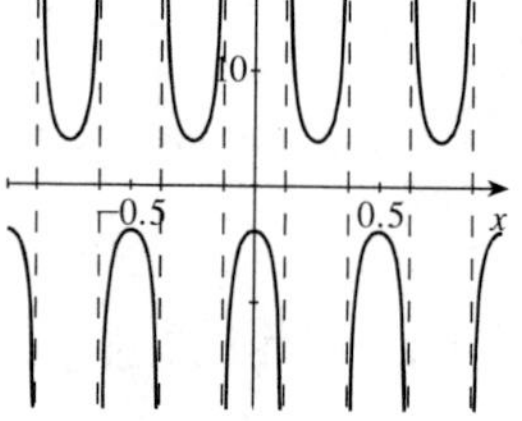

**49. (a)** $y = |\cos x|$

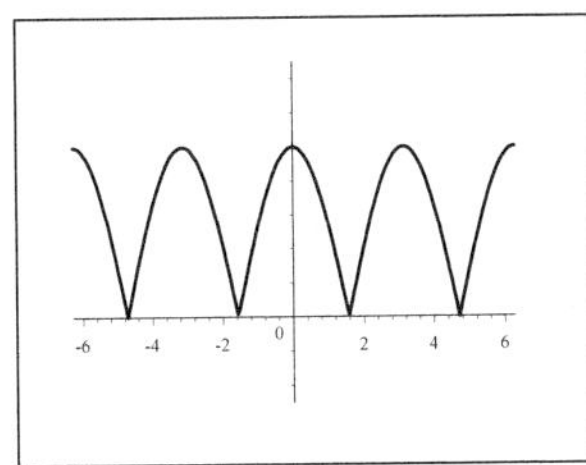

**(b)** This function has period $\pi$.

**(c)** This function is even.

**50. (a)** $y = \sin(\cos x)$

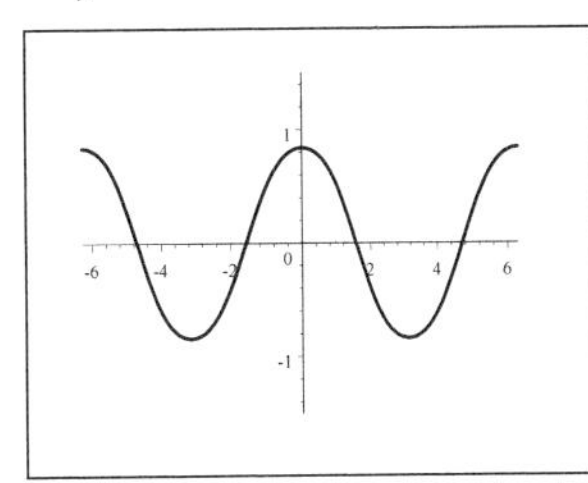

**(b)** This function has period $2\pi$.

**(c)** This function is even.

**51. (a)** $y = \cos\left(2^{0.1x}\right)$

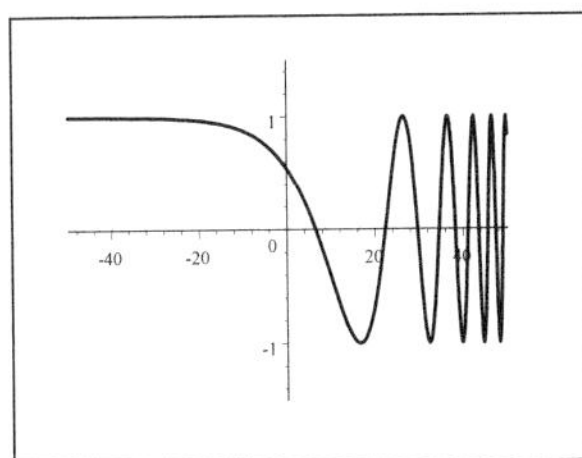

**(b)** This function is not periodic.

**(c)** This function is neither even nor odd.

**52. (a)** $y = 1 + 2^{\cos x}$

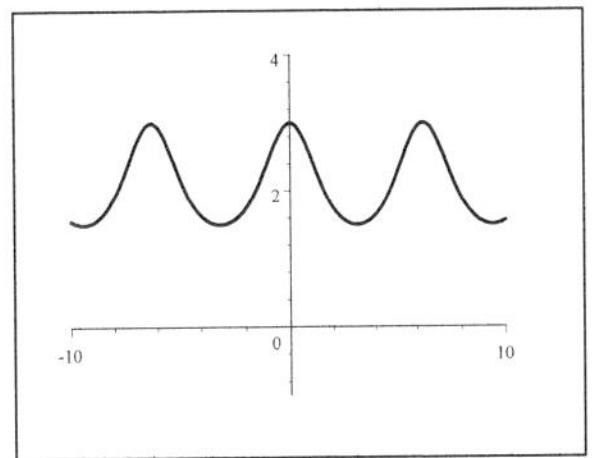

**(b)** This function has period $2\pi$.

**(c)** This function is even.

**53. (a)** $y = |x| \cos 3x$

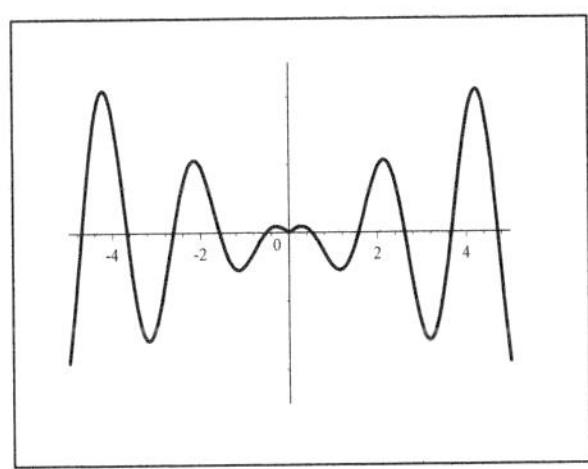

**(b)** This function is not periodic.

**(c)** This function is even.

**54. (a)** $y = \sqrt{x} \sin 3x$ $(x > 0)$

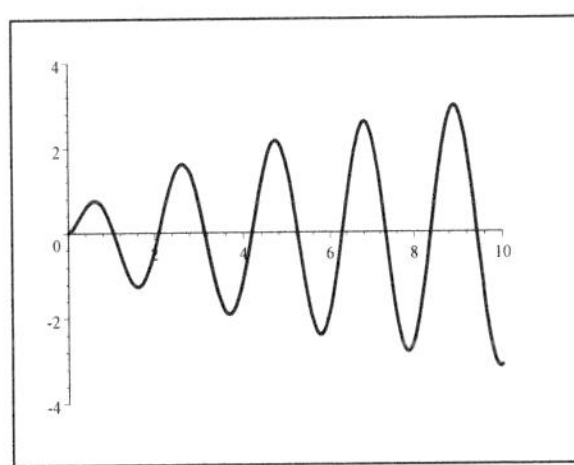

**(b)** This function is not periodic.

**(c)** This function is neither even nor odd.

**55.** $y = x \sin x$ is a sine function whose graph lies between those of $y = x$ and $y = -x$.

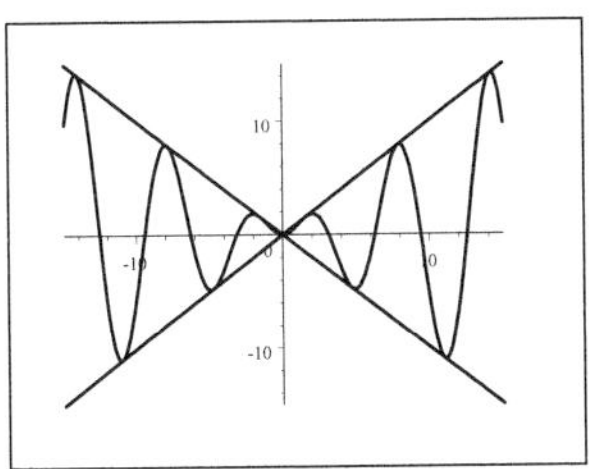

**56.** $y = 2^{-x} \cos 4\pi x$ is a cosine function whose graph lies between the graphs of $y = 2^{-x}$ and $y = -2^{-x}$.

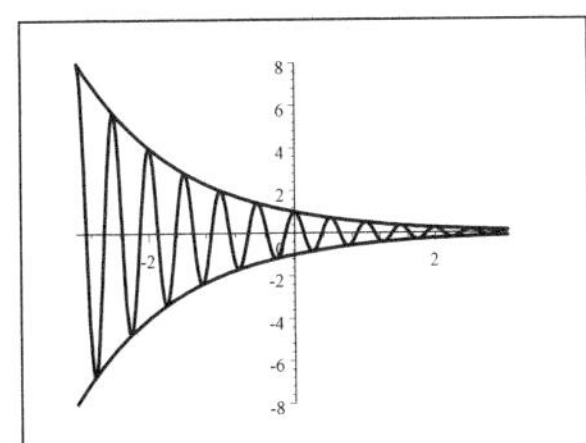

**57.** $y = x + \sin 4x$ is the sum of the two functions $y = x$ and $y = \sin 4x$.

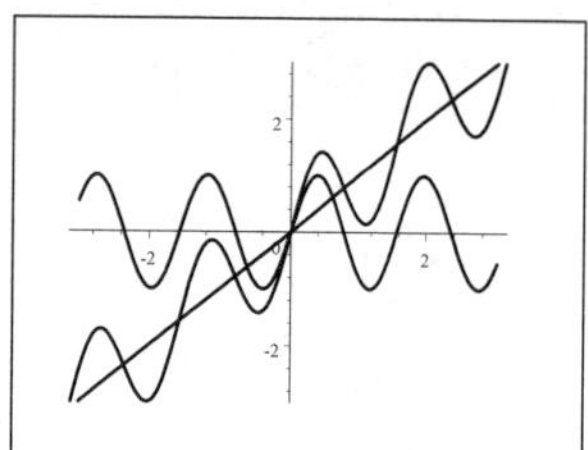

**58.** $y = \sin^2 x + \cos^2 x$ is the sum of the two functions $y = \sin^2 x$ and $y = \cos^2 x$. Note that $\sin^2 x + \cos^2 x = 1$ for all $x$.

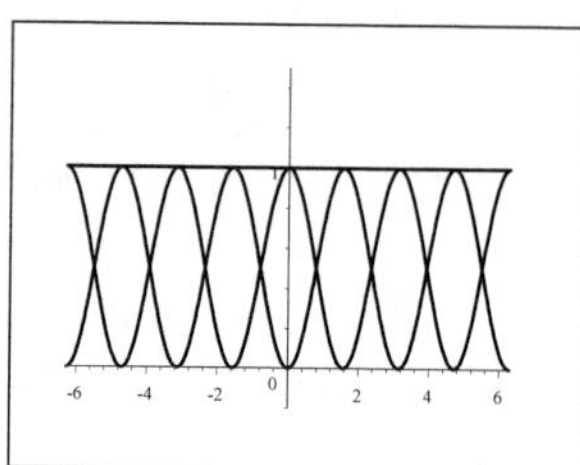

**59.** $y = \cos x + \sin 2x$. Since the period is $2\pi$, we graph over the interval $[-\pi, \pi]$. The maximum value is $1.76$ when $x \approx 0.63 \pm 2n\pi$, the minimum value is $-1.76$ when $x \approx 2.51 \pm 2n\pi$, $n$ an integer.

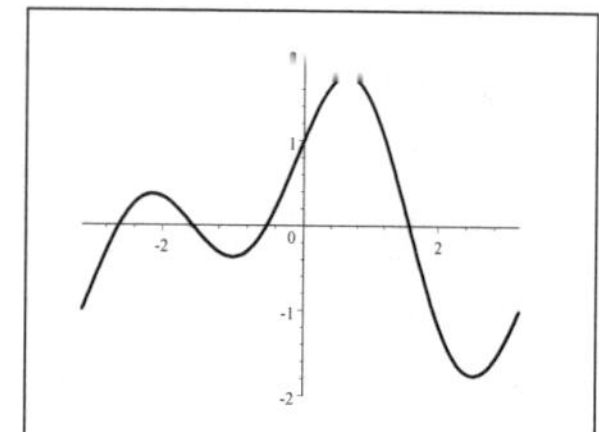

**60.** $y = \cos x + \sin^2 x$. Since the period of this function is $2\pi$, we graph this function over the interval $-\pi$ to $\pi$. The maximum value is $1.25$ when $x \approx 1.05 \pm 2n\pi$, the minimum value is $-1$ when $x = \frac{\pi}{2} \pm 2n\pi \approx 1.57 \pm 2n\pi$, $n$ an integer.

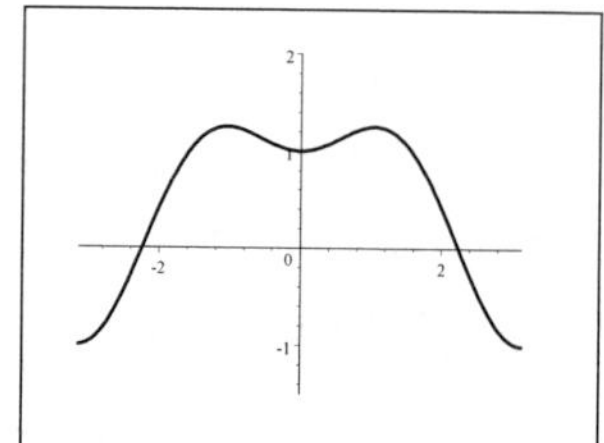

**61.** We want to find solutions to $\sin x = 0.3$ in the interval $[0, 2\pi]$, so we plot the functions $y = \sin x$ and $y = 0.3$ and look for their intersection. We see that $x \approx 0.305$ or $x \approx 2.837$.

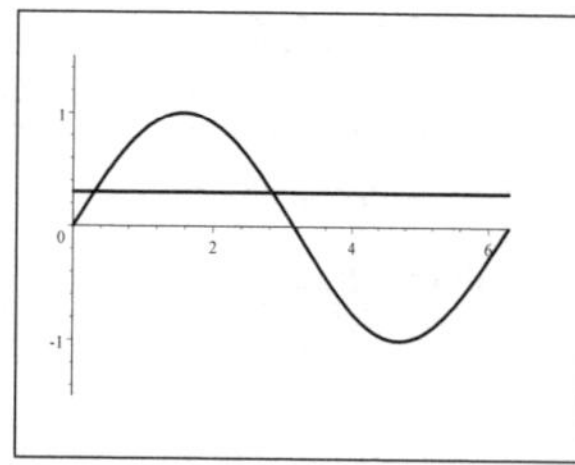

**62.** We want to find solutions to $\cos 3x = x$ in the interval $[0, \pi]$, so we plot the functions $y = \cos 3x$ and $y = x$ and look for their intersection. We see that $x \approx 0.390$.

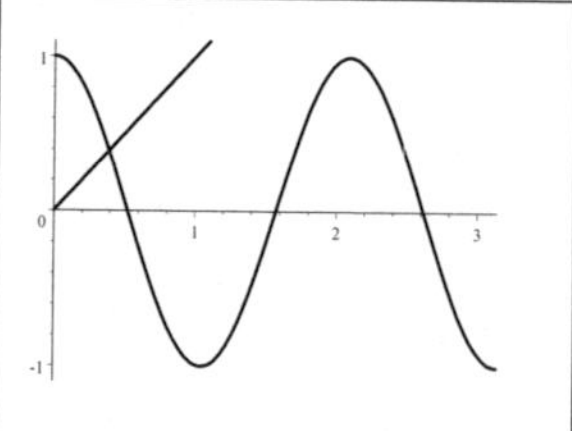

**63.** $f(x) = \dfrac{\sin^2 x}{x}$

**(a)** The function is odd.

**(b)** The graph intersects the $x$-axis at $x = 0, \pm\pi, \pm 2\pi, \pm 3\pi, \ldots$

**(c)**

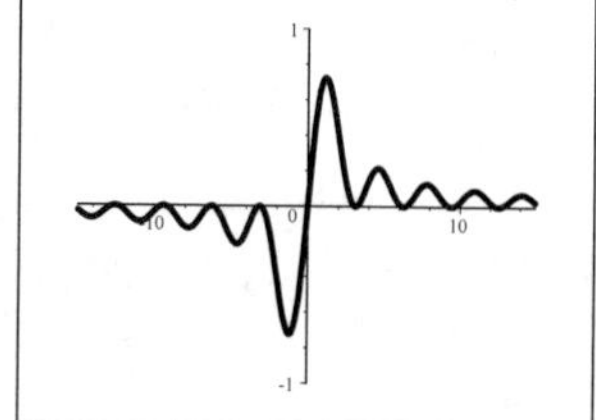

**(d)** As $x \to \pm\infty$, $f(x) \to 0$.

**(e)** As $x \to 0$, $f(x) \to 0$.

**64.** $y_1 = \cos(\sin x)$, $y_2 = \sin(\cos x)$

**(a)**

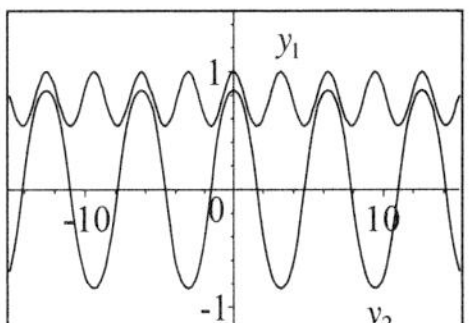


**(b)** $y_1$ has period $\pi$, while $y_2$ has period $2\pi$.

**(c)** $\sin(\cos x) < \cos(\sin x)$ for all $x$.

**65.** The amplitude is $a = 50$ cm. The frequency is 8 Hz, so $\omega = 8(2\pi) = 16\pi$. Since the mass is at its maximum displacement when $t = 0$, the motion follows a cosine curve. So a function describing the motion of $P$ is $f(t) = 50\cos 16\pi t$.

**66.** The amplitude is $\frac{1}{2}(100) = 50$ cm, the frequency is 4 Hz, so $\omega = 4(2\pi) = 8\pi$. Since the mass is at its lowest point when $t = 0$, a function describing the distance of the mass from its rest position is $f(t) = -50\cos 8\pi t$.

**67.** From the graph, we see that the amplitude is 4 ft, the period is 12 hours, and there is no phase shift. Thus, the variation in water level is described by $y = 4\cos\frac{\pi}{6}t$.

**68. (a)** The initial amplitude is 16 cm and the frequency is 1.4 Hz, so a function describing the motion is $y = 16e^{-0.72t}\cos 2.8\pi t$.

**(b)**

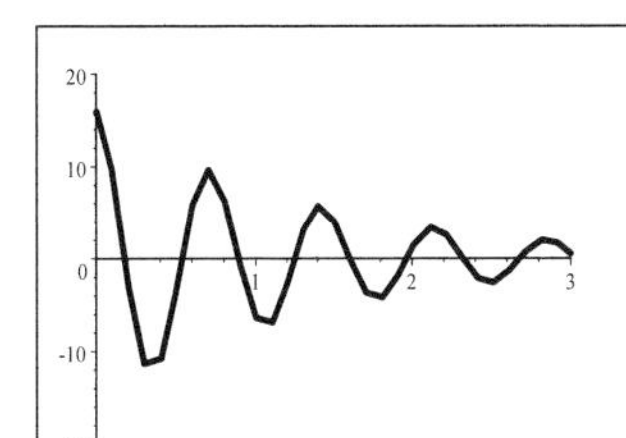


**(c)** When $t = 10$ s, $y \approx 16e^{-7.2}\cos 28\pi \approx 0.0119$ cm.

# Chapter 7 Test

**1.** Since $P(x, y)$ lies on the unit circle, $x^2 + y^2 = 1 \quad \Rightarrow \quad y = \pm\sqrt{1 - \left(\frac{\sqrt{11}}{6}\right)^2} = \pm\sqrt{\frac{25}{36}} = \pm\frac{5}{6}$. But $P(x, y)$ lies in the fourth quadrant. Therefore $y$ is negative $\quad \Rightarrow \quad y = -\frac{5}{6}$.

**2.** Since $P$ is on the unit circle, $x^2 + y^2 = 1 \Leftrightarrow x^2 = 1 - y^2$. Thus, $x^2 = 1 - \left(\frac{4}{5}\right)^2 = \frac{9}{25}$, and so $x = \pm\frac{3}{5}$. From the diagram, $x$ is clearly negative, so $x = -\frac{3}{5}$. Therefore, $P$ is the point $\left(-\frac{3}{5}, \frac{4}{5}\right)$.

**(a)** $\sin t = \frac{4}{5}$

**(b)** $\cos t = -\frac{3}{5}$

**(c)** $\tan t = \dfrac{\frac{4}{5}}{-\frac{3}{5}} = -\frac{4}{3}$

**(d)** $\sec(t) = -\frac{5}{3}$

**3. (a)** $\sin\frac{7\pi}{6} = -0.5$

**(b)** $\cos\dfrac{13\pi}{4} = -\frac{\sqrt{2}}{2}$

**(c)** $\tan\left(-\frac{5\pi}{3}\right) = \sqrt{3}$

**(d)** $\csc\left(\frac{3\pi}{2}\right) = -1$

**4.** $\tan t = \dfrac{\sin t}{\cos t} = \dfrac{\sin t}{\pm\sqrt{1 - \sin^2 t}}$. But $t$ is in quadrant II $\quad \Rightarrow \quad \cos t$ is negative, so we choose the negative square root.

Thus, $\tan t = \dfrac{\sin t}{-\sqrt{1 - \sin^2 t}}$.

**5.** $\cos t = -\frac{8}{17}$, $t$ in quadrant III $\quad \Rightarrow \quad \tan t \cdot \cot t + \csc t = 1 + \dfrac{1}{-\sqrt{1 - \cos^2 t}}$ (since $t$ is in quadrant III)

$= 1 - \dfrac{1}{-\sqrt{1 - \frac{64}{289}}} = 1 - \dfrac{1}{\frac{15}{17}} = -\frac{2}{15}$.

**6.** $y = -5\cos 4x$

**(a)** This function has amplitude5, period $\frac{2\pi}{4} = \frac{\pi}{2}$, and phase shift 0.

**(b)**

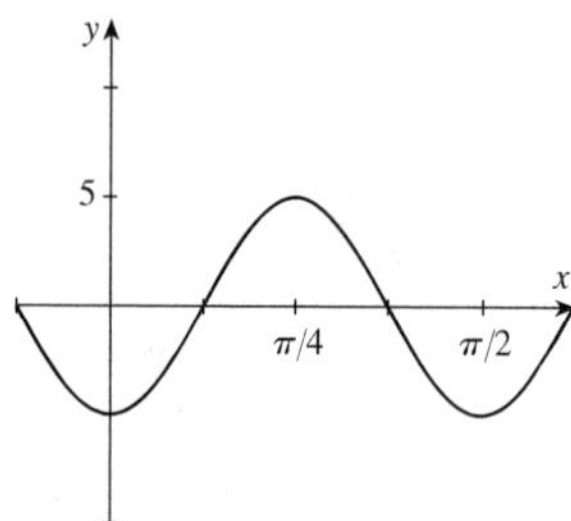


**7.** $y = 2\sin\left(\frac{1}{2}x - \frac{\pi}{6}\right) = \sin\frac{1}{2}\left(x - \frac{\pi}{3}\right)$

**(a)** This function has amplitude2, period $\dfrac{2\pi}{\frac{1}{2}} = 4\pi$, and phase shift $\frac{\pi}{3}$.

**(b)**

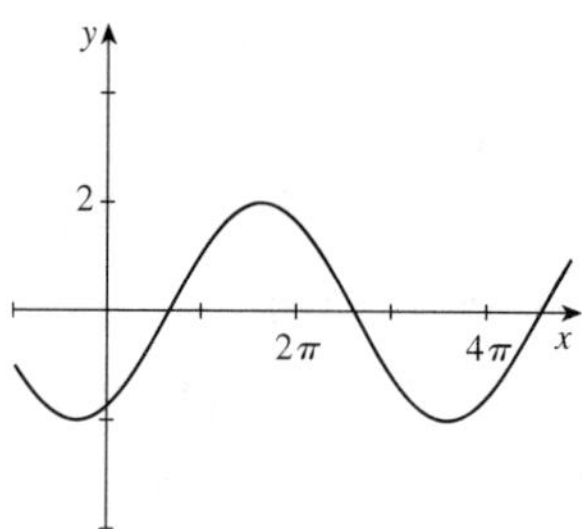


**8.** $y = -\csc 2x$ has period $\frac{2\pi}{2} = \pi$.

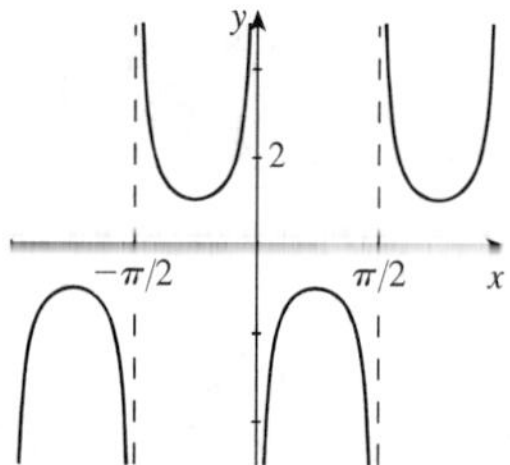


**9.** $y = \tan 2\left(x - \frac{\pi}{4}\right)$ has period $\frac{\pi}{2}$.

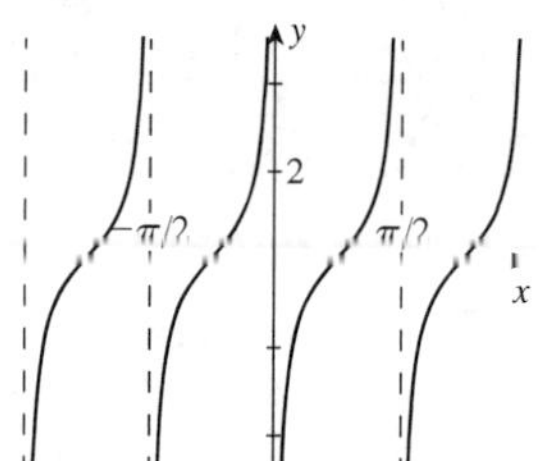


**10.** From the graph, we see that the amplitude is 2 and the phase shift is $-\frac{\pi}{3}$. Also, the period is $\pi$, so $\frac{2\pi}{k} = \pi \Rightarrow k = \frac{2\pi}{\pi} = 2$. Thus, the function is $y = 2\sin 2\left(x + \frac{\pi}{3}\right)$.

**11.** $y = \dfrac{\cos x}{1 + x^2}$

**(a)**

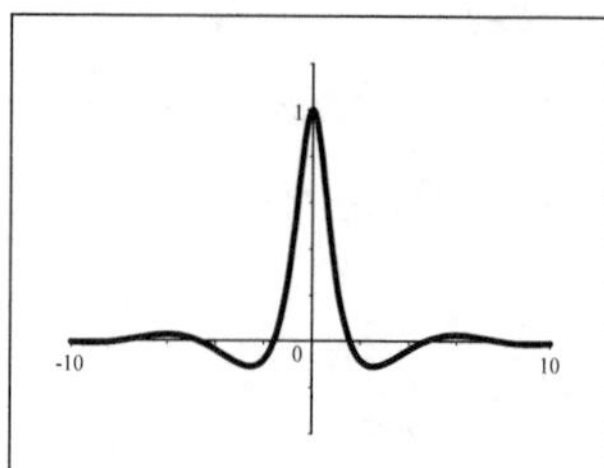


**(b)** The function is even.

**(c)** The function has a minimum value of approximately $-0.11$ when $x \approx \pm 2.54$ and a maximum value of 1 when $x = 0$.

**12.** The amplitude is $\frac{1}{2}(10) = 5$ cm and the frequency is 2 Hz. Assuming that the mass is at its rest position and moving upward when $t = 0$, a function describing the distance of the mass from its rest position is $f(t) = 5\sin 4\pi t$.

**13. (a)** The initial amplitude is 16 in and the frequency is 12 Hz, so a function describing the motion is $y = 16e^{-0.1t}\cos 24\pi t$.

**(b)**

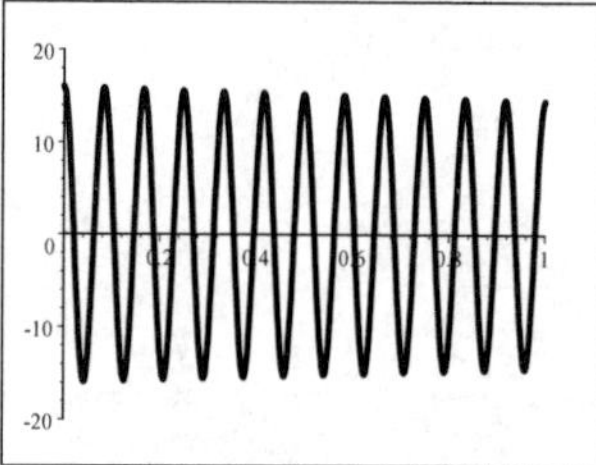

# Focus on Modeling: Fitting Sinusoidal Curves to Data

**1. (a)** See the graph in part (c).

**(b)** Using the method of Example 1, we find the vertical shift $b = \frac{1}{2}(\text{maximum value} + \text{minimum value}) = \frac{1}{2}(2.1 - 2.1) = 0$, the amplitude $a = \frac{1}{2}(\text{maximum value} - \text{minimum value}) = \frac{1}{2}(2.1 - (-2.1)) = 2.1$, the period $\frac{2\pi}{\omega} = 2(6 - 0) = 12$ (so $\omega \approx 0.5236$), and the phase shift $c = 0$. Thus, our model is $y = 2.1 \cos \frac{\pi}{6} t$.

**(c)** The curve fits the data quite well.

**(d)** Using the SinReg command on the TI-83, we find $y = 2.048714222 \sin(0.5030795477t + 1.551856108) - 0.0089616507$.

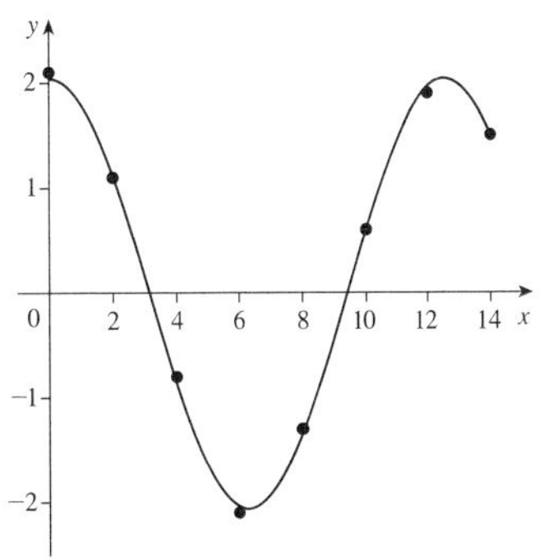

**(e)** Our model from part (b) is equivalent to $y = 2.1 \sin(0.5236t + 1.5708)$, which is close to the model in part (d).

**2. (a)** See the graph in part (c).

**(b)** Using the method of Example 1, we find the vertical shift $b = \frac{1}{2}(\text{maximum value} + \text{minimum value}) = \frac{1}{2}(200 + 95) = 147.5$, the amplitude $a = \frac{1}{2}(\text{maximum value} - \text{minimum value}) = \frac{1}{2}(200 - 95) = 52.5$, the period $\frac{2\pi}{\omega} = 2(275 - 125) = 300$ (so $\omega \approx 0.02094$), and the phase shift $c = -25$. Thus, our model is $y = 52.5 \cos(0.02094(t + 25)) + 147.5$.

**(c)**

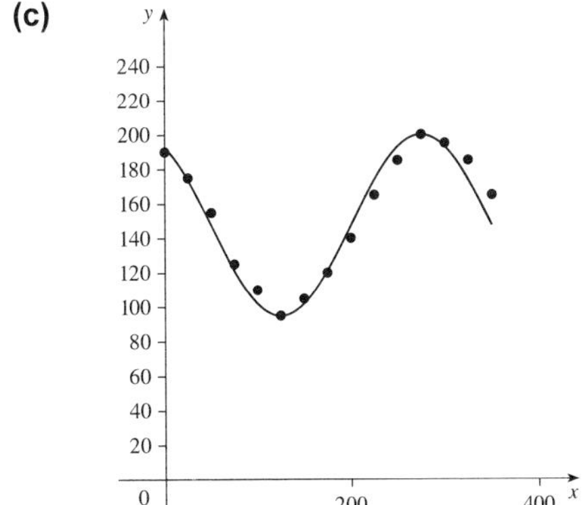

The curve fits the data quite well.

**(d)** Using the SinReg command on the TI-83, we find $y = 49.70329025 \sin(0.0200800412x + 2.093611012) + 149.1294508$.

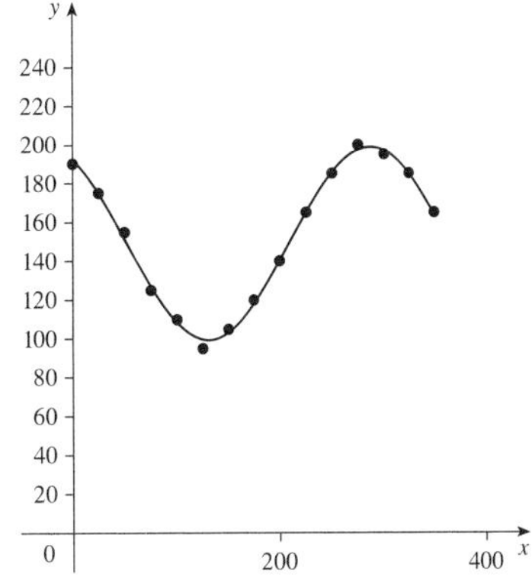

**(e)** Our model from part (b) is equivalent to $y = 52.5 \sin(0.02094t + 2.094) + 147.5$, which is close to the model in part (d).

**3. (a)** See the graph in part (c).

**(b)** Using the method of Example 1, we find the vertical shift $b = \frac{1}{2}(\text{maximum value} + \text{minimum value}) = \frac{1}{2}(25.1 + 1.0) = 13.05$, the amplitude $a = \frac{1}{2}(\text{maximum value} - \text{minimum value}) = \frac{1}{2}(25.1 - 1.0) = 12.05$, the period $\frac{2\pi}{\omega} = 2(1.5 - 0.9) = 1.2$ (so $\omega \approx 5.236$), and the phase shift $c = 0.3$. Thus, our model is $y = 12.05\cos(5.236(t - 0.3)) + 13.05$.

**(c)**

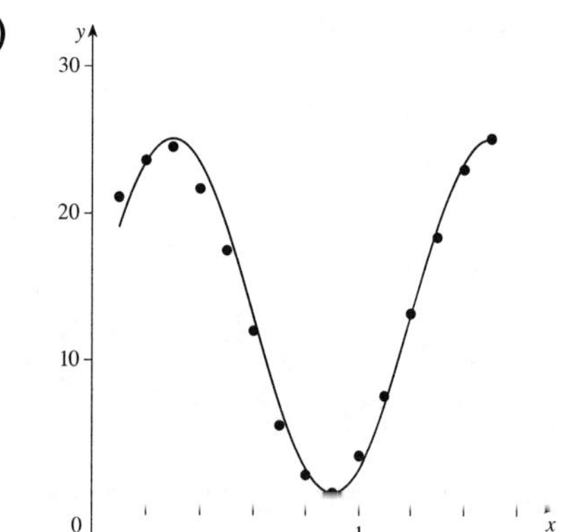

The curve fits the data fairly well.

**(d)** Using the `SinReg` command on the TI-83, we find

$$y = 11.71905062\sin(5.048853286t + 0.2388957877) + 12.96070536.$$

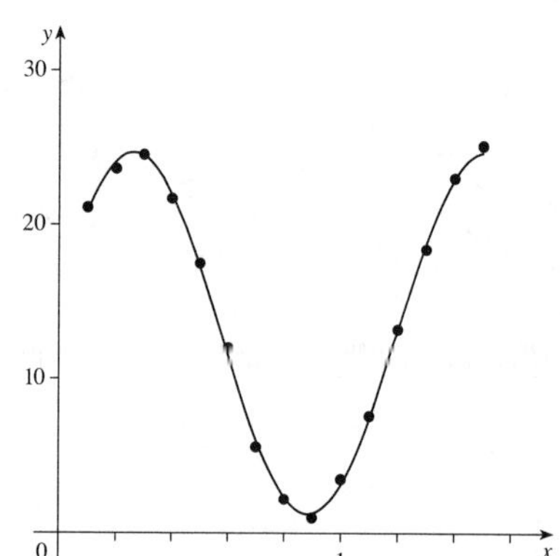

**(e)** Our model from part (b) is equivalent to $y = 12.05\sin 5.236t + 13.05$, which is close to the model in part (d).

**4. (a)** See the graph in part (c).

**(b)** Using the method of Example 1, we find the vertical shift $b = \frac{1}{2}(\text{maximum value} + \text{minimum value}) = \frac{1}{2}(0.63 - 0.10) = 0.265$, the amplitude $a = \frac{1}{2}(\text{maximum value} - \text{minimum value}) = \frac{1}{2}(0.63 - (-0.10)) = 0.365$, the period $\frac{2\pi}{\omega} = 2(5.5 - 2.5) = 6$ (so $\omega \approx 1.047$), and the phase shift $c = -0.5$. Thus, our model is $y = 0.365\cos(1.047(t + 0.5)) + 0.265$.

**(c)**

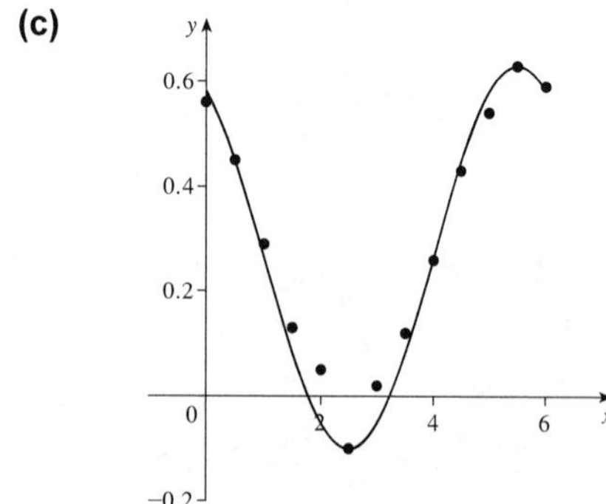

The curve fits the data reasonably well.

**(d)** Using the `SinReg` command on the TI-83, we find

$$y = 0.327038879\sin(1.021164911t + 2.118186963) + 0.2896017397.$$

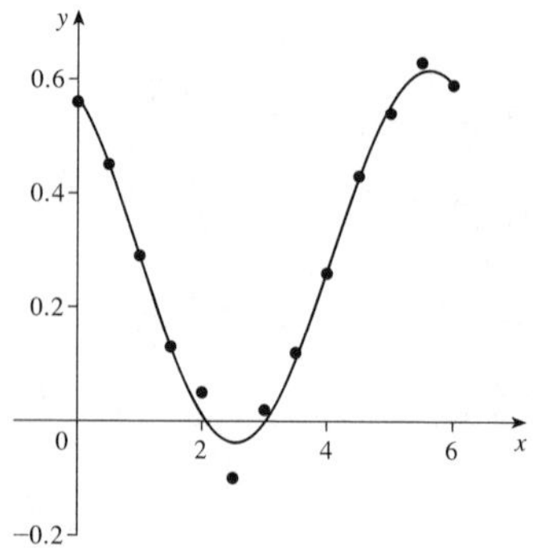

**(e)** Our model from part (b) is equivalent to $y = 0.365\cos(1.047t + 2.094) + 0.265$, which is close to the model in part (d). Our model in part (b) gives undue weight to the minimum value of $-0.10$.

**5. (a)** See the graph in part (c).

**(b)** Let $t$ be the time (in months) from January. We find a function of the form $y = a\cos\omega(t - c) + b$, where $y$ is temperature in °F.
$a = \frac{1}{2}(85.8 - 40) = 22.9$. The period is $2(\text{Jul} - \text{Jan}) = 2(6 - 0) = 12$ and so $\omega = \frac{2\pi}{12} \approx 0.52$.
$b = \frac{1}{2}(85.8 + 40) = 62.9$. Since the maximum value occurs in July, $c = 6$. Thus the function is $y = 22.9\cos 0.52(t - 6) + 62.9$.

**(c)**

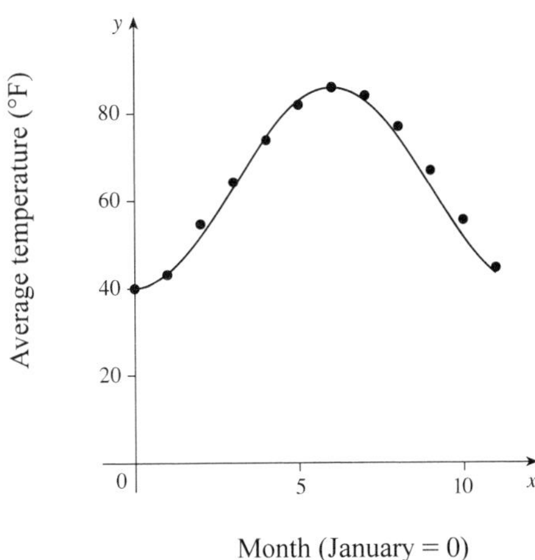


**(d)** Using the `SinReg` command on the TI-83 we find that for the function $y = a\sin(bt + c) + d$, where $a = 23.4$, $b = 0.48$, $c = -1.36$, and $d = 62.2$. Thus we get the model $y = 23.4\sin(0.48t - 1.36) + 62.2$.

**6. (a)** See the graph in part (c).

**(b)** Let $t$ be the time since midnight. We find a function of the form $y = a\sin\omega(t - c) + b$. $a = \frac{1}{2}(37.4 - 36.6) = 0.4$. The period is 24 and so $\omega = \frac{2\pi}{24} \approx 0.26$. $b = \frac{1}{2}(37.4 + 36.6) = 37$. Since the maximum value occurs at $t = 16$, we get $c = 16$. Thus the function is $y = 0.4\cos 0.26(t - 16) + 37$.

**(c)**

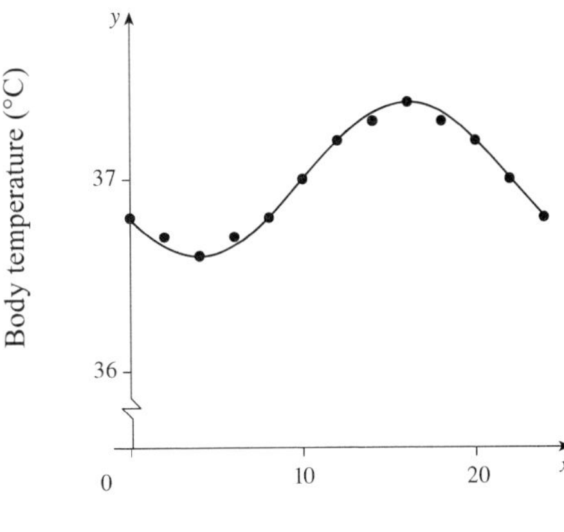


**(d)** Using the `SinReg` command on the TI-83 we obtain the function $y = a\sin(bt + c) + d$, where $a = 0.4$, $b = 0.26$, $c = -2.64$, and $d = 37.0$. Thus we get the model $y = 0.4\sin(0.26t - 2.64) + 37.0$.

**7. (a)** See the graph in part (c).

**(b)** Let $t$ be the time years. We find a function of the form $y = a \sin \omega (t - c) + b$, where $y$ is the owl population. $a = \frac{1}{2}(80 - 20) = 30$. The period is $2(9 - 3) = 12$ and so $\omega = \frac{2\pi}{12} \approx 0.52$. $b = \frac{1}{2}(80 + 20) = 50$. Since the values start at the middle we have $c = 0$. Thus the function is $y = 30 \sin 0.52t + 50$.

**(c)**

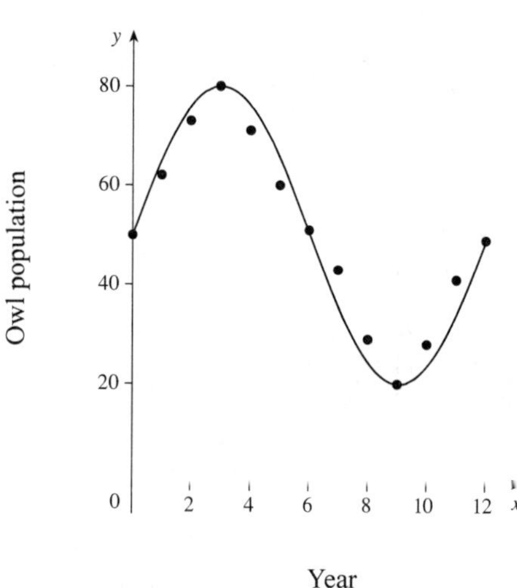


**(d)** Using the `SinReg` command on the TI–83 we find that for the function $y = a \sin(bt + c) + d$, where $a = 25.8$, $b = 0.52$, $c = -0.02$, and $d = 50.6$. Thus we get the model $y = 25.8 \sin(0.52t - 0.02) + 50.6$.

**8. (a)** See the graph in part (c).

**(b)** Let $t$ be the time since 1985. We find a function of the form $y = a \sin \omega (t - c) + b$. $a = \frac{1}{2}(63 - 22) = 20.5$. The periodis $2(15 - 9) = 12$ and so $\omega = \frac{2\pi}{20} \approx 0.52$. $b = \frac{1}{2}(63 + 22) = 42.5$. The average value occurs in the ninth year, so $c = 6$. Thus, our model is $y = 20.5 \sin 0.52(t - 6) + 42.5$.

**(c)**

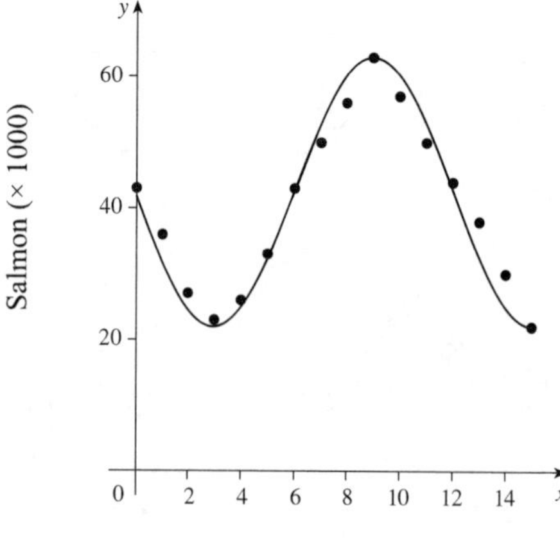


**(d)** Using the `SinReg` command on the TI-83, we find that for the function $y = a \sin(bt + c) + d$, where $a = 17.8$, $b = 0.52$, $c = 3.11$, and $d = 42.4$. Thus we get the model $y = 17.8 \sin(0.52t + 3.11) + 42.4$.

**9. (a)** See the graph in part (c).

**(b)** Let $t$ be the time since 1975. We find a function of the form $y = a\cos\omega(t - c) + b$. $a = \frac{1}{2}(158 - 9) = 74.5$. The first period seems to stretch from 1980 to 1989 (maximum to maximum) and the second period from 1989 to 2000, so the period is about 10 years and so $\omega = \frac{2\pi}{10} \approx 0.628$. $b = \frac{1}{2}(158 + 9) = 83.5$. The first maximum value occurs in the fifth year, so $c = 5$. Thus, our model is $y = 74.5\cos 0.628(t - 5) + 83.5$.

**(c)**

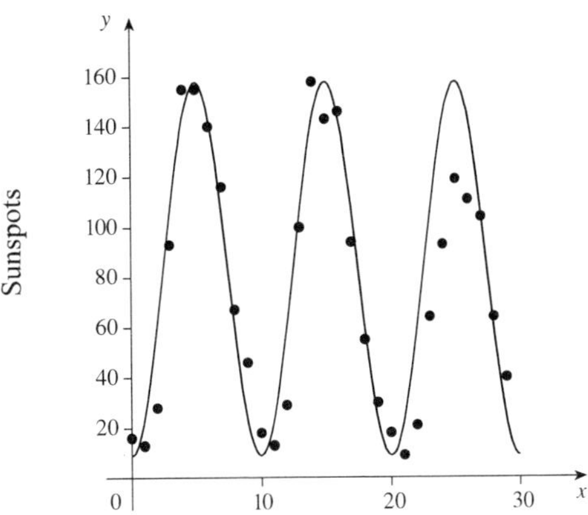


Year since 1975

**(d)** Using the `SinReg` command on the TI-83, we find that for the function $y = a\sin(bt + c) + d$, where $a = 67.65094323$, $b = 0.6205550572$, $c = -1.654463632$, and $d = 74.50460325$. Thus we get the model $y = 67.65\sin(0.62t - 1.65) + 74.5$. This model is more accurate for more recent years than the one we found in part (b).

# 8 Analytic Trigonometry

## 8.1 Trigonometric Identities

**1.** $\cos t \tan t = \cos t \cdot \dfrac{\sin t}{\cos t} = \sin t$

**2.** $\cos t \csc t = \cos t \cdot \dfrac{1}{\sin t} = \cot t$

**3.** $\sin\theta \sec\theta = \sin\theta \cdot \dfrac{1}{\cos\theta} = \tan\theta$

**4.** $\tan\theta \csc\theta = \dfrac{\sin\theta}{\cos\theta} \cdot \dfrac{1}{\sin\theta} = \dfrac{1}{\cos\theta} = \sec\theta$

**5.** $\tan^2 x - \sec^2 x = \dfrac{\sin^2 x}{\cos^2 x} - \dfrac{1}{\cos^2 x} = \dfrac{\sin^2 x - 1}{\cos^2 x} = \dfrac{-\cos^2 x}{\cos^2 x} = -1$

**6.** $\dfrac{\sec x}{\csc x} = \dfrac{\frac{1}{\cos x}}{\frac{1}{\sin x}} = \dfrac{\sin x}{\cos x} = \tan x$

**7.** $\sin u + \cot u \cos u = \sin u + \dfrac{\cos u}{\sin u} \cdot \cos u = \dfrac{\sin^2 u + \cos^2 u}{\sin u} = \dfrac{1}{\sin u} = \csc u$

**8.** $\cos^2\theta\left(1 + \tan^2\theta\right) = \cos^2\theta\left(1 + \dfrac{\sin^2\theta}{\cos^2\theta}\right) = \cos^2\theta + \sin^2\theta = 1$

**9.** $\dfrac{\sec\theta - \cos\theta}{\sin\theta} = \dfrac{\frac{1}{\cos\theta} - \cos\theta}{\sin\theta} = \dfrac{1 - \cos^2\theta}{\sin\theta\cos\theta} = \dfrac{\sin^2\theta}{\sin\theta\cos\theta} = \dfrac{\sin\theta}{\cos\theta} = \tan\theta$

**10.** $\dfrac{\cot\theta}{\csc\theta - \sin\theta} = \dfrac{\frac{\cos\theta}{\sin\theta}}{\frac{1}{\sin\theta} - \sin\theta} = \dfrac{\cos\theta}{1 - \sin^2\theta} = \dfrac{\cos\theta}{\cos^2\theta} = \sec\theta$

**11.** $\dfrac{\sin x \sec x}{\tan x} = \dfrac{\sin x \cdot \frac{1}{\cos x}}{\frac{\cos x}{\sin x}} = 1$

**12.** $\cos^3 x + \sin^2 x \cos x = \cos x\left(\cos^2 x + \sin^2 x\right) = \cos x$

**13.** $\dfrac{1 + \cos y}{1 + \sec y} = \dfrac{1 + \cos y}{1 + \frac{1}{\cos y}} = \dfrac{1 + \cos y}{\frac{\cos y + 1}{\cos y}} = \dfrac{1 + \cos y}{1} \cdot \dfrac{\cos y}{\cos y + 1} = \cos y$

**14.** $\dfrac{\tan x}{\sec(-x)} = \dfrac{\tan x}{\sec x} = \dfrac{\sin x}{\cos x} \cdot \cos x = \sin x$

**15.** $\dfrac{\sec^2 x - 1}{\sec^2 x} = \dfrac{\tan^2 x}{\sec^2 x} = \dfrac{\sin^2 x}{\cos^2 x} \cdot \cos^2 x = \sin^2 x$. *Another method:* $\dfrac{\sec^2 x - 1}{\sec^2 x} = 1 - \dfrac{1}{\sec^2 x} = 1 - \cos^2 x = \sin^2 x$

**16.** $\dfrac{\sec x - \cos x}{\tan x} = \dfrac{\frac{1}{\cos x} - \frac{\cos^2 x}{\cos x}}{\frac{\sin x}{\cos x}} = \dfrac{1 - \cos^2 x}{\sin x} = \dfrac{\sin^2 x}{\sin x} = \sin x$

**17.** $\dfrac{1 + \csc x}{\cos x + \cot x} = \dfrac{1 + \frac{1}{\sin x}}{\cos x + \frac{\cos x}{\sin x}} = \dfrac{1 + \frac{1}{\sin x}}{\cos x + \frac{\cos x}{\sin x}} \cdot \dfrac{\sin x}{\sin x} = \dfrac{\sin x + 1}{\cos x\left(\sin x + 1\right)} = \dfrac{1}{\cos x} = \sec x$

**18.** $\dfrac{\sin x}{\csc x} + \dfrac{\cos x}{\sec x} = \sin x \sin x + \cos x \cos x = \sin^2 x + \cos^2 x = 1$

**19.** $\dfrac{1 + \sin u}{\cos u} + \dfrac{\cos u}{1 + \sin u} = \dfrac{(1 + \sin u)^2 + \cos^2 u}{\cos u\,(1 + \sin u)} = \dfrac{1 + 2\sin u + \sin^2 u + \cos^2 u}{\cos u\,(1 + \sin u)} = \dfrac{1 + 2\sin u + 1}{\cos u\,(1 + \sin u)}$

$= \dfrac{2 + 2\sin u}{\cos u\,(1 + \sin u)} = \dfrac{2\,(1 + \sin u)}{\cos u\,(1 + \sin u)} = \dfrac{2}{\cos u} = 2\sec u$

**20.** $\tan x \cos x \csc x = \dfrac{\sin x}{\cos x} \cdot \cos x \cdot \dfrac{1}{\sin x} = 1$

**21.** $\dfrac{2 + \tan^2 x}{\sec^2 x} - 1 = \dfrac{1 + 1 + \tan^2 x}{\sec^2 x} - 1 = \dfrac{1}{\sec^2 x} + \dfrac{1 + \tan^2 x}{\sec^2 x} - 1 = \dfrac{1}{\sec^2 x} + \dfrac{\sec^2 x}{\sec^2 x} - 1$

$= \dfrac{1}{\sec^2 x} + 1 - 1 = \dfrac{1}{\sec^2 x} = \cos^2 x$

**22.** $\dfrac{1 + \cot A}{\csc A} = (1 + \cot A)\sin A = \sin A + \sin A \cot A = \sin A + \sin A \dfrac{\cos A}{\sin A} = \sin A + \cos A$

**23.** $\tan\theta + \cos(-\theta) + \tan(-\theta) = \tan\theta + \cos\theta - \tan\theta = \cos\theta$

**24.** $\dfrac{\cos x}{\sec x + \tan x} = \dfrac{\cos x}{\dfrac{1}{\cos x} + \dfrac{\sin x}{\cos x}} = \dfrac{\cos^2 x}{1 + \sin x} = \dfrac{1 - \sin^2 x}{1 + \sin x} = \dfrac{(1 + \sin x)(1 - \sin x)}{1 + \sin x} = 1 - \sin x$

**25.** $\dfrac{\sin\theta}{\tan\theta} = \dfrac{\sin\theta}{\dfrac{\sin\theta}{\cos\theta}} = \sin\theta \cdot \dfrac{\cos\theta}{\sin\theta} = \cos\theta$

**26.** $\dfrac{\tan x}{\sec x} = \dfrac{\sin x}{\cos x} \cdot \cos x = \sin x$

**27.** $\dfrac{\cos u \sec u}{\tan u} = \cos u \dfrac{1}{\cos u} \cot u = \cot u$

**28.** $\dfrac{\cot x \sec x}{\csc x} = \dfrac{\cos x}{\sin x} \cdot \dfrac{1}{\cos x} \cdot \sin x = 1$

**29.** $\dfrac{\tan y}{\csc y} = \dfrac{\sin y}{\cos y} \sin y = \dfrac{\sin^2 y}{\cos y} = \dfrac{1 - \cos^2 y}{\cos y} = \sec y - \cos y$

**30.** $\dfrac{\cos v}{\sec v \sin v} = \cos v \cdot \dfrac{\cos v}{\sin v} = \dfrac{1 - \sin^2 v}{\sin v} = \csc v - \sin v$

**31.** $\sin B + \cos B \cot B = \sin B + \cos B \dfrac{\cos B}{\sin B} = \dfrac{\sin^2 B + \cos^2 B}{\sin B} = \dfrac{1}{\sin B} = \csc B$

**32.** $\cos(-x) - \sin(-x) = \cos x - (-\sin x) = \cos x + \sin x$

**33.** $\cot(-\alpha)\cos(-\alpha) + \sin(-\alpha) = -\dfrac{\cos\alpha}{\sin\alpha}\cos\alpha - \sin\alpha = \dfrac{-\cos^2\alpha - \sin^2\alpha}{\sin\alpha} = \dfrac{-1}{\sin\alpha} = -\csc\alpha$

**34.** $\csc x\,[\csc x + \sin(-x)] = \csc^2 x - \sin x \csc x = \csc^2 x - 1 = \cot^2 x$

**35.** $\tan\theta + \cot\theta = \dfrac{\sin\theta}{\cos\theta} + \dfrac{\cos\theta}{\sin\theta} = \dfrac{\sin^2\theta + \cos^2\theta}{\cos\theta\sin\theta} = \dfrac{1}{\cos\theta\sin\theta} = \sec\theta\csc\theta$

**36.** $(\sin x + \cos x)^2 = \sin^2 x + 2\sin x\cos x + \cos^2 x = 1 + 2\sin x\cos x$

**37.** $(1 - \cos\beta)(1 + \cos\beta) = 1 - \cos^2\beta = \sin^2\beta = \dfrac{1}{\csc^2\beta}$

**38.** $\dfrac{\cos x}{\sec x} + \dfrac{\sin x}{\csc x} = \cos^2 x + \sin^2 x = 1$

**39.** $\dfrac{(\sin x + \cos x)^2}{\sin^2 x - \cos^2 x} = \dfrac{(\sin x + \cos x)^2}{(\sin x + \cos x)(\sin x - \cos x)} = \dfrac{\sin x + \cos x}{\sin x - \cos x} = \dfrac{(\sin x + \cos x)(\sin x - \cos x)}{(\sin x - \cos x)(\sin x - \cos x)}$

$= \dfrac{\sin^2 x - \cos^2 x}{(\sin x - \cos x)^2}$

**40.** $(\sin x + \cos x)^4 = \left[(\sin x + \cos x)^2\right]^2 = \left(\sin^2 x + 2\sin x\cos x + \cos^2 x\right)^2 = (1 + 2\sin x\cos x)^2$

**41.** $\dfrac{\sec t-\cos t}{\sec t}=\dfrac{\dfrac{1}{\cos t}-\cos t}{\dfrac{1}{\cos t}}=\dfrac{\dfrac{1}{\cos t}-\cos t}{\dfrac{1}{\cos t}}\cdot\dfrac{\cos t}{\cos t}=\dfrac{1-\cos^2 t}{1}=\sin^2 t$

**42.** $\dfrac{1-\sin x}{1+\sin x}=\dfrac{1-\sin x}{1+\sin x}\cdot\dfrac{1-\sin x}{1-\sin x}=\dfrac{1-2\sin x+\sin^2 x}{1-\sin^2 x}=\dfrac{1-2\sin x+\sin^2 x}{\cos^2 x}=\dfrac{1}{\cos^2 x}-\dfrac{2\sin x}{\cos^2 x}+\dfrac{\sin^2 x}{\cos^2 x}$

$=\sec^2 x-2\sec x\tan x+\tan^2 x=(\sec x-\tan x)^2$

**43.** $\dfrac{1}{1-\sin^2 y}=\dfrac{1}{\cos^2 y}=\sec^2 y=1+\tan^2 y$

**44.** $\csc x-\sin x=\dfrac{1}{\sin x}-\sin x=\dfrac{1-\sin^2 x}{\sin x}=\dfrac{\cos^2 x}{\sin x}=\cos x\cot x$

**45.** $(\cot x-\csc x)(\cos x+1)=\cot x\cos x+\cot x-\csc x\cos x-\csc x=\dfrac{\cos^2 x}{\sin x}+\dfrac{\cos x}{\sin x}-\dfrac{\cos x}{\sin x}-\dfrac{1}{\sin x}$

$=\dfrac{\cos^2 x-1}{\sin x}=\dfrac{-\sin^2 x}{\sin x}=-\sin x$

**46.** $\sin^4\theta-\cos^4\theta=\left(\sin^2\theta\right)^2-\left(\cos^2\theta\right)^2=\left(\sin^2\theta-\cos^2\theta\right)\left(\sin^2\theta+\cos^2\theta\right)=\sin^2\theta-\cos^2\theta$

**47.** $\left(1-\cos^2 x\right)\left(1+\cot^2 x\right)=\sin^2 x\left(1+\dfrac{\cos^2 x}{\sin^2 x}\right)=\sin^2 x+\cos^2 x=1$

**48.** $\cos^2 x-\sin^2 x=\cos^2 x-\left(1-\cos^2 x\right)=2\cos^2 x-1$

**49.** $2\cos^2 x-1=2\left(1-\sin^2 x\right)-1=2-2\sin^2 x-1=1-2\sin^2 x$

**50.** $\tan y+\cot y=\dfrac{\sin y}{\cos y}+\dfrac{\cos y}{\sin y}=\dfrac{\sin^2 y+\cos^2 y}{\sin y\cos y}=\dfrac{1}{\sin y\cos y}=\sec y\csc y$

**51.** $\dfrac{1-\cos\alpha}{\sin\alpha}=\dfrac{1-\cos\alpha}{\sin\alpha}\cdot\dfrac{1+\cos\alpha}{1+\cos\alpha}=\dfrac{1-\cos^2\alpha}{\sin\alpha\,(1+\cos\alpha)}=\dfrac{\sin^2\alpha}{\sin\alpha\,(1+\cos\alpha)}=\dfrac{\sin\alpha}{1+\cos\alpha}$

**52.** $\sin^2\alpha+\cos^2\alpha+\tan^2\alpha=1+\tan^2\alpha=\sec^2\alpha$

**53.** $\tan^2\theta\sin^2\theta=\tan^2\theta\left(1-\cos^2\theta\right)=\tan^2\theta-\dfrac{\sin^2\theta}{\cos^2\theta}\cos^2\theta=\tan^2\theta-\sin^2\theta$

**54.** $\cot^2\theta\cos^2\theta=\cot^2\theta\left(1-\sin^2\theta\right)=\cot^2\theta-\dfrac{\cos^2\theta}{\sin^2\theta}\sin^2\theta=\cot^2\theta-\cos^2\theta$

**55.** $\dfrac{\sin x-1}{\sin x+1}=\dfrac{\sin x-1}{\sin x+1}\cdot\dfrac{\sin x+1}{\sin x+1}=\dfrac{\sin^2 x-1}{(\sin x+1)^2}=\dfrac{-\cos^2 x}{(\sin x+1)^2}$

**56.** $\dfrac{\sin w}{\sin w+\cos w}=\dfrac{\sin w}{\sin w+\cos w}\cdot\dfrac{\dfrac{1}{\cos w}}{\dfrac{1}{\cos w}}=\dfrac{\tan w}{1+\tan w}$

**57.** $\dfrac{(\sin t+\cos t)^2}{\sin t\cos t}=\dfrac{\sin^2 t+2\sin t\cos t+\cos^2 t}{\sin t\cos t}=\dfrac{\sin^2 t+\cos^2 t}{\sin t\cos t}+\dfrac{2\sin t\cos t}{\sin t\cos t}=\dfrac{1}{\sin t\cos t}+2=2+\sec t\csc t$

**58.** $\sec t\csc t\,(\tan t+\cot t)=\dfrac{1}{\cos t}\dfrac{1}{\sin t}\left(\dfrac{\sin t}{\cos t}+\dfrac{\cos t}{\sin t}\right)=\dfrac{1}{\cos^2 t}+\dfrac{1}{\sin^2 t}=\sec^2 t+\csc^2 t$

**59.** $\dfrac{1+\tan^2 u}{1-\tan^2 u}=\dfrac{1+\dfrac{\sin^2 u}{\cos^2 u}}{1-\dfrac{\sin^2 u}{\cos^2 u}}=\dfrac{1+\dfrac{\sin^2 u}{\cos^2 u}}{1-\dfrac{\sin^2 u}{\cos^2 u}}\cdot\dfrac{\cos^2 u}{\cos^2 u}=\dfrac{\cos^2 u+\sin^2 u}{\cos^2 u-\sin^2 u}=\dfrac{1}{\cos^2 u-\sin^2 u}$

**60.** $\dfrac{1+\sec^2 x}{1+\tan^2 x}=\dfrac{1+\sec^2 x}{\sec^2 x}=\dfrac{1}{\sec^2 x}+1=\cos^2 x+1$

**61.** $\dfrac{\sec x}{\sec x-\tan x}=\dfrac{\sec x}{\sec x-\tan x}\cdot\dfrac{\sec x+\tan x}{\sec x+\tan x}=\dfrac{\sec x\,(\sec x+\tan x)}{\sec^2 x-\tan^2 x}=\dfrac{\sec x\,(\sec x+\tan x)}{1}$

$=\sec x\,(\sec x+\tan x)$

**62.** $\dfrac{\sec x+\csc x}{\tan x+\cot x}=\dfrac{\dfrac{1}{\cos x}+\dfrac{1}{\sin x}}{\dfrac{\sin x}{\cos x}+\dfrac{\cos x}{\sin x}}=\dfrac{\dfrac{1}{\cos x}+\dfrac{1}{\sin x}}{\dfrac{\sin x}{\cos x}+\dfrac{\cos x}{\sin x}}\cdot\dfrac{\sin x\cos x}{\sin x\cos x}=\dfrac{\sin x+\cos x}{\sin^2 x+\cos^2 x}=\sin x+\cos x$

**63.** $\sec v-\tan v=(\sec v-\tan v)\cdot\dfrac{\sec v+\tan v}{\sec v+\tan v}=\dfrac{\sec^2 v-\tan^2 v}{\sec v+\tan v}=\dfrac{1}{\sec v+\tan v}$

**64.** $\dfrac{\sin A}{1-\cos A}-\cot A=\dfrac{\sin A}{1-\cos A}\cdot\dfrac{1+\cos A}{1+\cos A}-\cot A=\dfrac{\sin A\,(1+\cos A)}{1-\cos^2 A}-\cot A=\dfrac{\sin A\,(1+\cos A)}{\sin^2 A}-\dfrac{\cos A}{\sin A}$

$=\dfrac{1}{\sin A}+\dfrac{\cos A}{\sin A}-\dfrac{\cos A}{\sin A}=\dfrac{1}{\sin A}=\csc A$

**65.** $\dfrac{\sin x+\cos x}{\sec x+\csc x}=\dfrac{\sin x+\cos x}{\dfrac{1}{\cos x}+\dfrac{1}{\sin x}}=\dfrac{\sin x+\cos x}{\dfrac{\sin x+\cos x}{\cos x\sin x}}=(\sin x+\cos x)\dfrac{\cos x\sin x}{\sin x+\cos x}=\cos x\sin x$

**66.** $\dfrac{1-\cos x}{\sin x}+\dfrac{\sin x}{1-\cos x}=\dfrac{1-\cos x}{\sin x}\cdot\dfrac{1-\cos x}{1-\cos x}+\dfrac{\sin x}{1-\cos x}\cdot\dfrac{\sin x}{\sin x}=\dfrac{1-2\cos x+\cos^2 x+\sin^2 x}{\sin x\,(1-\cos x)}$

$=\dfrac{2-2\cos x}{\sin x\,(1-\cos x)}=\dfrac{2\,(1-\cos x)}{\sin x\,(1-\cos x)}=2\csc x$

**67.** $\dfrac{\csc x-\cot x}{\sec x-1}=\dfrac{\dfrac{1}{\sin x}-\dfrac{\cos x}{\sin x}}{\dfrac{1}{\cos x}-1}=\dfrac{\dfrac{1}{\sin x}-\dfrac{\cos x}{\sin x}}{\dfrac{1}{\cos x}-1}\cdot\dfrac{\sin x\cos x}{\sin x\cos x}=\dfrac{\cos x\,(1-\cos x)}{\sin x\,(1-\cos x)}=\dfrac{\cos x}{\sin x}=\cot x$

**68.** $\dfrac{\csc^2 x-\cot^2 x}{\sec^2 x}=\dfrac{1}{\sec^2 x}=\cos^2 x$

**69.** $\tan^2 u-\sin^2 u=\dfrac{\sin^2 u}{\cos^2 u}-\dfrac{\sin^2 u\cos^2 u}{\cos^2 u}=\dfrac{\sin^2 u}{\cos^2 u}\left(1-\cos^2 u\right)=\tan^2 u\sin^2 u$

**70.** $\dfrac{\tan v\sin v}{\tan v+\sin v}=\dfrac{\tan v\sin v}{\tan v+\sin v}\cdot\dfrac{\tan v-\sin v}{\tan v-\sin v}=\dfrac{\tan v\sin v\,(\tan v-\sin v)}{\tan^2 v-\sin^2 v}=\dfrac{\tan v\sin v\,(\tan v-\sin v)}{\sin^2 v\,(\sec^2 v-1)}$

$=\dfrac{\tan v\sin v\,(\tan v-\sin v)}{\tan^2 v\sin^2 v}=\dfrac{\tan v-\sin v}{\tan v\sin v}$

**71.** $\sec^4 x-\tan^4 x=\left(\sec^2 x-\tan^2 x\right)\left(\sec^2 x+\tan^2 x\right)=1\left(\sec^2 x+\tan^2 x\right)=\sec^2 x+\tan^2 x$

**72.** $\dfrac{\cos\theta}{1-\sin\theta}=\dfrac{\cos\theta}{1-\sin\theta}\cdot\dfrac{1+\sin\theta}{1+\sin\theta}=\dfrac{\cos\theta\,(1+\sin\theta)}{1-\sin^2\theta}=\dfrac{\cos\theta\,(1+\sin\theta)}{\cos^2\theta}=\dfrac{1}{\cos\theta}+\dfrac{\sin\theta}{\cos\theta}=\sec\theta+\tan\theta$

**73.** $\dfrac{\sin\theta-\csc\theta}{\cos\theta-\cot\theta}=\dfrac{\sin\theta-\dfrac{1}{\sin\theta}}{\cos\theta-\dfrac{\cos\theta}{\sin\theta}}=\dfrac{\dfrac{\sin^2\theta-1}{\sin\theta}}{\dfrac{\cos\theta\sin\theta-\cos\theta}{\sin\theta}}=\dfrac{\cos^2\theta}{\cos\theta\,(\sin\theta-1)}=\dfrac{\cos\theta}{\sin\theta-1}$

**74.** $\dfrac{1+\tan x}{1-\tan x}=\dfrac{1+\dfrac{\sin x}{\cos x}}{1-\dfrac{\sin x}{\cos x}}\cdot\dfrac{\cos x}{\cos x}=\dfrac{\cos x+\sin x}{\cos x-\sin x}$

**75.** $\dfrac{\cos^2 t+\tan^2 t-1}{\sin^2 t}=\dfrac{-\sin^2 t+\tan^2 t}{\sin^2 t}=-1+\dfrac{\sin^2 t}{\cos^2 t}\cdot\dfrac{1}{\sin^2 t}=-1+\sec^2 t=\tan^2 t$

**76.** $\dfrac{1}{1-\sin x}-\dfrac{1}{1+\sin x}=\dfrac{(1+\sin x)-(1-\sin x)}{(1-\sin x)(1+\sin x)}=\dfrac{2\sin x}{1-\sin^2 x}=\dfrac{2\sin x}{\cos^2 x}=2\dfrac{\sin x}{\cos x}\cdot\dfrac{1}{\cos x}=2\tan x\sec x$

**77.** $\dfrac{1}{\sec x+\tan x}+\dfrac{1}{\sec x-\tan x}=\dfrac{\sec x-\tan x+\sec x+\tan x}{(\sec x+\tan x)(\sec x-\tan x)}=\dfrac{2\sec x}{\sec^2 x-\tan^2 x}=\dfrac{2\sec x}{1}=2\sec x$

**78.** $\dfrac{1+\sin x}{1-\sin x}-\dfrac{1-\sin x}{1+\sin x}=\dfrac{(1+\sin x)^2-(1-\sin x)^2}{(1-\sin x)(1+\sin x)}=\dfrac{1+2\sin x+\sin^2 x-1+2\sin x-\sin^2 x}{1-\sin^2 x}$

$=\dfrac{4\sin x}{\cos^2 x}=4\dfrac{\sin x}{\cos x}\cdot\dfrac{1}{\cos x}=4\tan x\sec x$

**79.** $(\tan x+\cot x)^2=\tan^2 x+2\tan x\cot x+\cot^2 x=\tan^2 x+2+\cot^2 x=\left(\tan^2 x+1\right)+\left(\cot^2 x+1\right)=\sec^2 x+\csc^2 x$

**80.** $\tan^2 x - \cot^2 x = \left(\sec^2 x - 1\right) - \left(\csc^2 x - 1\right) = \sec^2 x - \csc^2 x$

**81.** $\dfrac{\sec u - 1}{\sec u + 1} = \dfrac{\dfrac{1}{\cos u} - 1}{\dfrac{1}{\cos u} + 1} \cdot \dfrac{\cos u}{\cos u} = \dfrac{1 - \cos u}{1 + \cos u}$

**82.** $\dfrac{\cot x + 1}{\cot x - 1} = \dfrac{\cot x + 1}{\cot x - 1} \cdot \dfrac{\tan x}{\tan x} = \dfrac{\tan x \cot x + \tan x}{\tan x \cot x - \tan x} = \dfrac{1 + \tan x}{1 - \tan x}$

**83.** $\dfrac{\sin^3 x + \cos^3 x}{\sin x + \cos x} = \dfrac{(\sin x + \cos x)\left(\sin^2 x - \sin x \cos x + \cos^2 x\right)}{\sin x + \cos x} = \sin^2 - \sin x \cos x + \cos^2 x = 1 - \sin x \cos x$

**84.** $\dfrac{\tan v - \cot v}{\tan^2 v - \cot^2 v} = \dfrac{\tan v - \cot v}{(\tan v - \cot v)(\tan v + \cot v)} = \dfrac{1}{\tan v + \cot v} = \dfrac{1}{\dfrac{\sin v}{\cos v} + \dfrac{\cos v}{\sin v}} \cdot \dfrac{\sin v \cos v}{\sin v \cos v}$

$= \dfrac{\sin v \cos v}{\sin^2 v + \cos^2 v} = \sin v \cos v$

**85.** $\dfrac{1 + \sin x}{1 - \sin x} = \dfrac{1 + \sin x}{1 - \sin x} \cdot \dfrac{1 + \sin x}{1 + \sin x} = \dfrac{(1 + \sin x)^2}{1 - \sin^2 x} = \dfrac{(1 + \sin x)^2}{\cos^2 x} = \left(\dfrac{1 + \sin x}{\cos x}\right)^2 = (\tan x + \sec x)^2$

**86.** $\dfrac{\tan x + \tan y}{\cot x + \cot y} = \dfrac{\dfrac{\sin x}{\cos x} + \dfrac{\sin y}{\cos y}}{\dfrac{\cos x}{\sin x} + \dfrac{\cos y}{\sin y}} = \dfrac{\dfrac{\sin x \cos y + \cos x \sin y}{\cos x \cos y}}{\dfrac{\cos x \sin y + \sin x \cos y}{\sin x \sin y}}$

$= \left(\dfrac{\sin x \cos y + \cos x \sin y}{\cos x \cos y}\right)\left(\dfrac{\sin x \sin y}{\cos x \sin y + \sin x \cos y}\right) = \dfrac{\sin x \sin y}{\cos x \cos y} = \tan x \tan y$

**87.** $(\tan x + \cot x)^4 = \left(\dfrac{\sin x}{\cos x} + \dfrac{\cos x}{\sin x}\right)^4 = \left(\dfrac{\sin^2 x + \cos^2 x}{\sin x \cos x}\right)^4 = \left(\dfrac{1}{\sin x \cos x}\right)^4 = \sec^4 x \csc^4 x$

**88.** $(\sin\alpha - \tan\alpha)(\cos\alpha - \cot\alpha) = \left(\sin\alpha - \dfrac{\sin\alpha}{\cos\alpha}\right)\left(\cos\alpha - \dfrac{\cos\alpha}{\sin\alpha}\right) = \sin\alpha\left(1 - \dfrac{1}{\cos\alpha}\right) \cdot \cos\alpha\left(1 - \dfrac{1}{\sin\alpha}\right)$

$= \cos\alpha\left(1 - \dfrac{1}{\cos\alpha}\right)\sin\alpha\left(1 - \dfrac{1}{\sin\alpha}\right) = (\cos\alpha - 1)(\sin\alpha - 1)$

**89.** $x = \sin\theta$; then $\dfrac{x}{\sqrt{1 - x^2}} = \dfrac{\sin\theta}{\sqrt{1 - \sin^2\theta}} = \dfrac{\sin\theta}{\sqrt{\cos^2\theta}} = \dfrac{\sin\theta}{\cos\theta} = \tan\theta$ (since $\cos\theta \ge 0$ for $0 \le \theta \le \frac{\pi}{2}$).

**90.** $x = \tan\theta$; then $\sqrt{1 + x^2} = \sqrt{1 + \tan^2\theta} = \sqrt{1 + (\sec^2\theta - 1)} = \sqrt{\sec^2\theta} = \sec\theta$(since $\sec\theta \ge 0$ for $0 \le \theta \le \frac{\pi}{2}$).

**91.** $x = \sec\theta$; then $\sqrt{x^2 - 1} = \sqrt{\sec^2\theta - 1} = \sqrt{(\tan^2\theta + 1) - 1} = \sqrt{\tan^2\theta} = \tan\theta$ (since $\tan\theta \ge 0$ for $0 \le \theta < \frac{\pi}{2}$)

**92.** $x = 2\tan\theta$; then

$\dfrac{1}{x^2\sqrt{4 + x^2}} = \dfrac{1}{(2\tan\theta)^2\sqrt{4 + (2\tan\theta)^2}} = \dfrac{1}{4\tan^2\theta \cdot \sqrt{4\left(1 + \tan^2\theta\right)}} = \dfrac{1}{4\tan^2\theta \cdot 2\sqrt{\sec^2\theta}}$

$= \dfrac{1}{8\,\tan^2\theta \cdot \sec\theta} = \frac{1}{8} \cdot \dfrac{\cos^3\theta}{\sin^2\theta}$ (since $\sec\theta > 0$ for $0 \le \theta < \frac{\pi}{2}$).

**93.** $x = 3\sin\theta$; then $\sqrt{9 - x^2} = \sqrt{9 - (3\sin\theta)^2} = \sqrt{9 - 9\sin^2\theta} = \sqrt{9\left(1 - \sin^2\theta\right)} = 3\sqrt{\cos^2\theta} = 3\cos\theta$ (since $\cos\theta \ge 0$ for $0 \le \theta < \frac{\pi}{2}$).

**94.** $x = 5\sec\theta$; then $\dfrac{\sqrt{x^2 - 25}}{x} = \dfrac{\sqrt{(5\sec\theta)^2 - 25}}{5\sec\theta} = \dfrac{\sqrt{25\left(\sec^2\theta - 1\right)}}{5\sec\theta} = \dfrac{5\sqrt{\tan^2\theta}}{5\sec\theta} = \dfrac{\tan\theta}{\sec\theta} = \dfrac{\dfrac{\sin\theta}{\cos\theta}}{\dfrac{1}{\cos\theta}} = \sin\theta$ (since $\tan\theta > 0$ for $0 \le \theta < \frac{\pi}{2}$).

**95.** $f(x) = \cos^2 x - \sin^2 x$, $g(x) = 1 - 2\sin^2 x$. From the graph, $f(x) = g(x)$ this appears to be an identity. *Proof:*
$f(x) = \cos^2 x - \sin^2 x = \cos^2 x + \sin^2 x - 2\sin^2 x = 1 - 2\sin^2 x = g(x)$.
Since $f(x) = g(x)$ for all $x$, this is an identity.

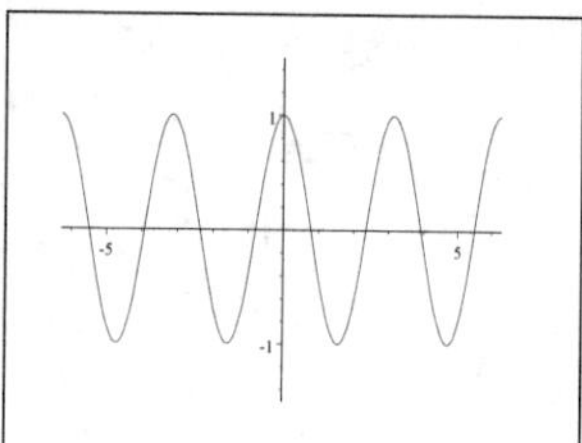

**96.** $f(x) = \tan x\,(1 + \sin x)$, $g(x) = \dfrac{\sin x \cos x}{1 + \sin x}$. From the graph, $f(x) = g(x)$ does not appear to be an identity. In order to show this, let $x = \frac{\pi}{4}$. Then,
$f\left(\frac{\pi}{4}\right) = (1) \cdot \left[1 + \left(\frac{1}{\sqrt{2}}\right)\right] = \frac{\sqrt{2}+1}{\sqrt{2}}$. However,
$g\left(\frac{\pi}{4}\right) = \dfrac{\frac{1}{\sqrt{2}} \cdot \frac{1}{\sqrt{2}}}{\left[1 + \left(\frac{1}{\sqrt{2}}\right)\right]} = \dfrac{1}{2} \cdot \dfrac{1}{\frac{\sqrt{2}+1}{\sqrt{2}}} = \frac{\sqrt{2}}{2}$. Since $f\left(\frac{\pi}{4}\right) \neq g\left(\frac{\pi}{4}\right)$, this is not an identity.

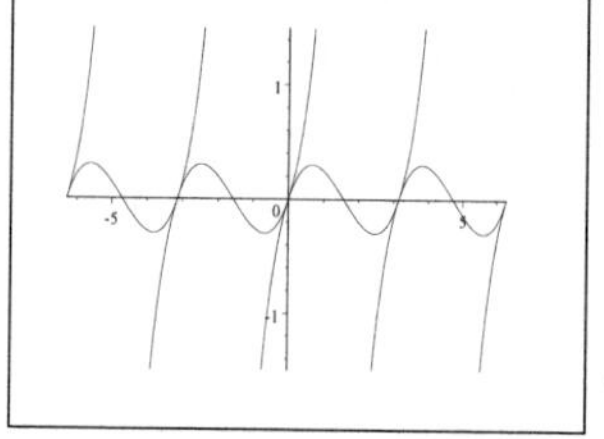

**97.** $f(x) = (\sin x + \cos x)^2$, $g(x) = 1$. From the graph, $f(x) = g(x)$ does not appear to be an identity. In order to show this, we can set $x = \frac{\pi}{4}$. Then we have
$f\left(\frac{\pi}{4}\right) = \left(\frac{1}{\sqrt{2}} + \frac{1}{\sqrt{2}}\right)^2 = \left(\frac{2}{\sqrt{2}}\right)^2 = \left(\sqrt{2}\right)^2 = 2 \neq 1 = g\left(\frac{\pi}{4}\right)$. Since
$f\left(\frac{\pi}{4}\right) \neq g\left(\frac{\pi}{4}\right)$, this is not an identity.

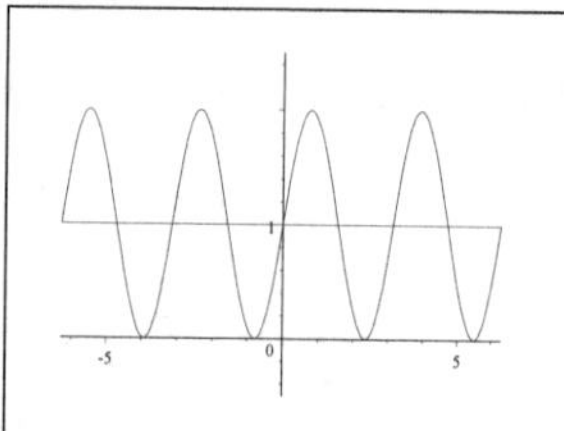

**98.** $f(x) = \cos^4 x - \sin^4 x$, $g(x) = 2\cos^2 x - 1$. From the graph, $f(x) = g(x)$ appears to be an identity. In order to prove this, simplify the expression $f(x)$:

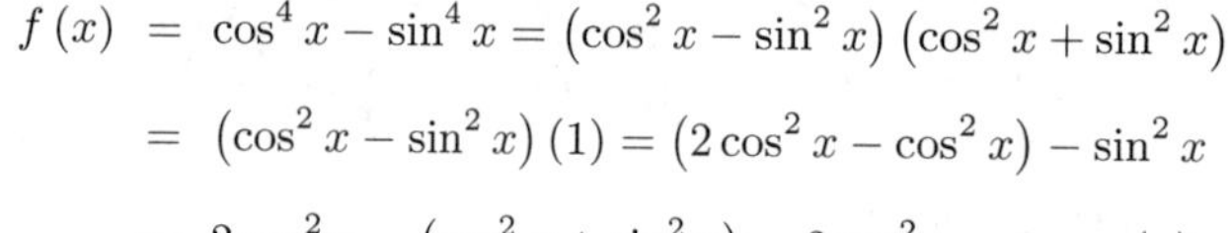

$$\begin{aligned} f(x) &= \cos^4 x - \sin^4 x = \left(\cos^2 x - \sin^2 x\right)\left(\cos^2 x + \sin^2 x\right) \\ &= \left(\cos^2 x - \sin^2 x\right)(1) = \left(2\cos^2 x - \cos^2 x\right) - \sin^2 x \\ &= 2\cos^2 x - \left(\cos^2 x + \sin^2 x\right) = 2\cos^2 x - 1 = g(x) \end{aligned}$$

Since $f(x) = g(x)$ for all $x$, this is an identity.

**99. (a)** Choose $x = \frac{\pi}{2}$. Then $\sin 2x = \sin \pi = 0$ whereas $2\sin x = 2\sin\frac{\pi}{2} = 2$.

**(b)** Choose $x = \frac{\pi}{4}$ and $y = \frac{\pi}{4}$. Then $\sin(x + y) = \sin\frac{\pi}{2} = 1$ whereas $\sin x + \sin y = \sin\frac{\pi}{4} + \sin\frac{\pi}{4} = \frac{1}{\sqrt{2}} + \frac{1}{\sqrt{2}} = \frac{2}{\sqrt{2}}$. Since these are not equal, the equation is not an identity.

**(c)** Choose $\theta = \frac{\pi}{4}$. Then $\sec^2\theta + \csc^2\theta = \left(\sqrt{2}\right)^2 + \left(\sqrt{2}\right)^2 = 4 \neq 1$.

**(d)** Choose $x = \frac{\pi}{4}$. Then $\dfrac{1}{\sin x + \cos x} = \dfrac{1}{\sin\frac{\pi}{4} + \cos\frac{\pi}{4}} = \dfrac{1}{\frac{1}{\sqrt{2}} + \frac{1}{\sqrt{2}}} = \frac{1}{\sqrt{2}}$ whereas $\csc x + \sec x = \csc\frac{\pi}{4} + \sec\frac{\pi}{4} = \sqrt{2} + \sqrt{2}$. Since these are not equal, the equation is not an identity.

**100.** Label $a$ the side opposite $v$, $b$ the side opposite $u$, and $c$ the hypotenuse. Since $u + v + \frac{\pi}{2} = \pi$, we must have $u + v = \frac{\pi}{2} \Leftrightarrow v = \frac{\pi}{2} - u$. Next we express all six trigonometric function for each angle: $\cos u = \frac{a}{c} = \sin v \Rightarrow \cos u = \sin\left(\frac{\pi}{2} - u\right)$, $\sin u = \frac{b}{c} = \cos v \Rightarrow \sin u = \cos\left(\frac{\pi}{2} - u\right)$, $\tan u = \frac{b}{a} = \cot v \Rightarrow \tan u = \cot\left(\frac{\pi}{2} - u\right)$, $\cot u = \frac{a}{b} = \tan v \Rightarrow \cot u = \tan\left(\frac{\pi}{2} - u\right)$, $\sec u = \frac{c}{a} = \csc v \Rightarrow \sec u = \csc\left(\frac{\pi}{2} - u\right)$, and $\csc u = \frac{c}{b} = \sec v \Rightarrow \csc u = \sec\left(\frac{\pi}{2} - u\right)$.

**101.** No. All this proves is that $f(x) = g(x)$ for $x$ in the range of the viewing rectangle. It does not prove that these functions are equal for all values of $x$. For example, let $f(x) = 1 - \frac{x^2}{2} + \frac{x^4}{24} - \frac{x^6}{720}$ and $g(x) = \cos x$. In the first viewing rectangle the graphs of these two functions appear identical. However, when the domain is expanded in the second viewing rectangle, you can see that these two functions are not identical.

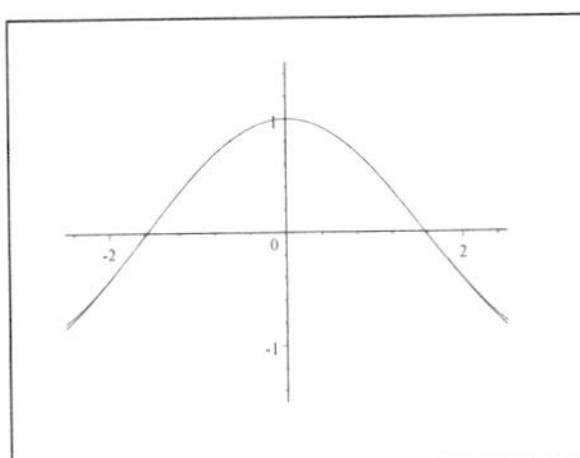

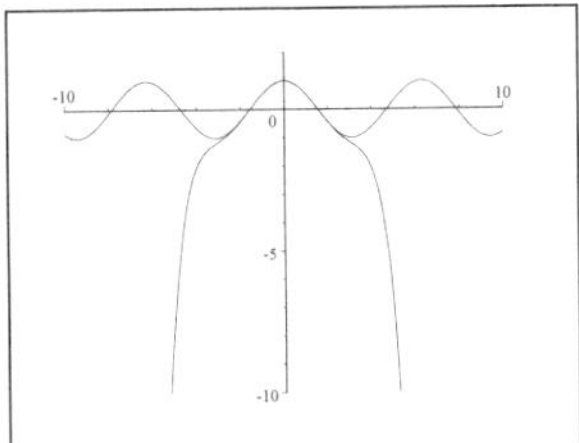

**102.** Answers will vary.

## 8.2 Addition and Subtraction Formulas

**1.** $\sin 75^\circ = \sin(45^\circ + 30^\circ) = \sin 45^\circ \cos 30^\circ + \cos 45^\circ \sin 30^\circ = \frac{\sqrt{2}}{2} \cdot \frac{\sqrt{3}}{2} + \frac{\sqrt{2}}{2} \cdot \frac{1}{2} = \frac{\sqrt{6}+\sqrt{2}}{4}$

**2.** $\sin 15^\circ = \sin(45^\circ - 30^\circ) = \sin 45^\circ \cos 30^\circ - \cos 45^\circ \sin 30^\circ = \frac{\sqrt{2}}{2} \cdot \frac{\sqrt{3}}{2} - \frac{\sqrt{2}}{2} \cdot \frac{1}{2} = \frac{\sqrt{6}-\sqrt{2}}{4}$

**3.** $\cos 105^\circ = \cos(60^\circ + 45^\circ) = \cos 60^\circ \cos 45^\circ - \sin 60^\circ \sin 45^\circ = \frac{1}{2} \cdot \frac{\sqrt{2}}{2} - \frac{\sqrt{3}}{2} \cdot \frac{\sqrt{2}}{2} = \frac{\sqrt{2}-\sqrt{6}}{4}$

**4.** $\cos 195^\circ = -\cos 15^\circ = -\cos(45^\circ - 30^\circ) = -(\cos 45^\circ \cos 30^\circ + \sin 45^\circ \sin 30^\circ) = -\left(\frac{\sqrt{2}}{2} \cdot \frac{\sqrt{3}}{2} + \frac{\sqrt{2}}{2} \cdot \frac{1}{2}\right)$
$= -\frac{\sqrt{6}+\sqrt{2}}{4}$

**5.** $\tan 15^\circ = \tan(45^\circ - 30^\circ) = \dfrac{\tan 45^\circ - \tan 30^\circ}{1 + \tan 45^\circ \tan 30^\circ} = \dfrac{1 - \frac{\sqrt{3}}{3}}{1 + 1 \cdot \frac{\sqrt{3}}{3}} = \dfrac{3 - \sqrt{3}}{3 + \sqrt{3}} = 2 - \sqrt{3}$

**6.** $\tan 165^\circ = -\tan 15^\circ = -\tan(45^\circ - 30^\circ) = -\dfrac{\tan 45^\circ - \tan 30^\circ}{1 + \tan 45^\circ \tan 30^\circ} = \dfrac{\frac{\sqrt{3}}{3} - 1}{1 + 1 \cdot \frac{\sqrt{3}}{3}} = \dfrac{\sqrt{3} - 3}{3 + \sqrt{3}} = \sqrt{3} - 2$

**7.** $\sin \frac{19\pi}{12} = -\sin \frac{7\pi}{12} = -\sin\left(\frac{\pi}{4} + \frac{\pi}{3}\right) = -\sin \frac{\pi}{4} \cos \frac{\pi}{3} - \cos \frac{\pi}{4} \sin \frac{\pi}{3} = -\frac{\sqrt{2}}{2} \cdot \frac{1}{2} - \frac{\sqrt{2}}{2} \cdot \frac{\sqrt{3}}{2} = -\frac{\sqrt{6}+\sqrt{2}}{4}$

**8.** $\cos \frac{17\pi}{12} = -\cos \frac{5\pi}{12} = -\cos\left(\frac{\pi}{4} + \frac{\pi}{6}\right) = -\cos \frac{\pi}{4} \cos \frac{\pi}{6} + \sin \frac{\pi}{4} \sin \frac{\pi}{6} = -\frac{\sqrt{2}}{2} \cdot \frac{\sqrt{3}}{2} + \frac{\sqrt{2}}{2} \cdot \frac{1}{2} = -\frac{\sqrt{6}-\sqrt{2}}{4}$

**9.** $\tan\left(-\frac{\pi}{12}\right) = -\tan \frac{\pi}{12} = -\tan\left(\frac{\pi}{3} - \frac{\pi}{4}\right) = -\dfrac{\tan \frac{\pi}{3} - \tan \frac{\pi}{4}}{1 + \tan \frac{\pi}{3} \tan \frac{\pi}{4}} = \dfrac{1 - \sqrt{3}}{1 + \sqrt{3}} = \sqrt{3} - 2$

**10.** $\sin\left(-\frac{5\pi}{12}\right) = -\sin \frac{5\pi}{12} = -\sin\left(\frac{\pi}{4} + \frac{\pi}{6}\right) = -\sin \frac{\pi}{4} \cos \frac{\pi}{6} - \cos \frac{\pi}{4} \sin \frac{\pi}{6} = -\frac{\sqrt{2}}{2} \cdot \frac{\sqrt{3}}{2} - \frac{\sqrt{2}}{2} \cdot \frac{1}{2} = -\frac{\sqrt{6}+\sqrt{2}}{4}$

**11.** $\cos \frac{11\pi}{12} = -\cos \frac{\pi}{12} = -\cos\left(\frac{\pi}{3} - \frac{\pi}{4}\right) = -\cos \frac{\pi}{3} \cos \frac{\pi}{4} - \sin \frac{\pi}{3} \sin \frac{\pi}{4} = -\frac{\sqrt{3}}{2} \cdot \frac{\sqrt{2}}{2} - \frac{1}{2} \cdot \frac{\sqrt{2}}{2} = -\frac{\sqrt{6}+\sqrt{2}}{4}$

**12.** $\tan \frac{7\pi}{12} = -\tan \frac{5\pi}{12} = -\tan\left(\frac{\pi}{4} + \frac{\pi}{6}\right) = -\dfrac{\tan \frac{\pi}{4} + \tan \frac{\pi}{6}}{1 - \tan \frac{\pi}{4} \tan \frac{\pi}{6}} = -\dfrac{1 + \frac{\sqrt{3}}{3}}{1 - 1 \cdot \frac{\sqrt{3}}{3}} = -\dfrac{3 + \sqrt{3}}{3 - \sqrt{3}} = -2 - \sqrt{3}$

**13.** $\sin 18^\circ \cos 27^\circ + \cos 18^\circ \sin 27^\circ = \sin(18^\circ + 27^\circ) = \sin 45^\circ = \frac{1}{\sqrt{2}} = \frac{\sqrt{2}}{2}$

**14.** $\cos 10^\circ \cos 80^\circ - \sin 10^\circ \sin 80^\circ = \cos(10^\circ + 80^\circ) = \cos 90^\circ = 0$

**15.** $\cos \frac{3\pi}{7} \cos \frac{2\pi}{21} + \sin \frac{3\pi}{7} \sin \frac{2\pi}{21} = \cos\left(\frac{3\pi}{7} - \frac{2\pi}{21}\right) = \cos \frac{7\pi}{21} = \cos \frac{\pi}{3} = \frac{1}{2}$

**16.** $\dfrac{\tan \frac{\pi}{18} + \tan \frac{\pi}{9}}{1 - \tan \frac{\pi}{18} \tan \frac{\pi}{9}} = \tan\left(\frac{\pi}{18} + \frac{\pi}{9}\right) = \tan \frac{3\pi}{18} = \tan \frac{\pi}{6} = \frac{\sqrt{3}}{3}$

**17.** $\dfrac{\tan 73^\circ - \tan 13^\circ}{1 + \tan 73^\circ \tan 13^\circ} = \tan(73^\circ - 13^\circ) = \tan 60^\circ = \sqrt{3}$

**18.** $\cos \frac{13\pi}{15} \cos\left(-\frac{\pi}{5}\right) - \sin \frac{13\pi}{15} \sin\left(-\frac{\pi}{5}\right) = \cos\left(\frac{13\pi}{15} + \left(-\frac{\pi}{5}\right)\right) = \cos \frac{10\pi}{15} = \cos \frac{2\pi}{3} = -\frac{1}{2}$

**19.** $\tan\left(\frac{\pi}{2}-u\right) = \dfrac{\sin\left(\frac{\pi}{2}-u\right)}{\cos\left(\frac{\pi}{2}-u\right)} = \dfrac{\sin\frac{\pi}{2}\cos u - \cos\frac{\pi}{2}\sin u}{\cos\frac{\pi}{2}\cos u + \sin\frac{\pi}{2}\sin u} = \dfrac{1\cdot\cos u - 0\cdot\sin u}{0\cdot\cos u + 1\cdot\sin u} = \dfrac{\cos u}{\sin u} = \cot u$

**20.** $\cot\left(\frac{\pi}{2}-u\right) = \dfrac{\cos\left(\frac{\pi}{2}-u\right)}{\sin\left(\frac{\pi}{2}-u\right)} = \dfrac{\cos\frac{\pi}{2}\cos u + \sin\frac{\pi}{2}\sin u}{\sin\frac{\pi}{2}\cos u - \cos\frac{\pi}{2}\sin u} = \dfrac{0\cdot\cos u + 1\cdot\sin u}{1\cdot\cos u - 0\cdot\sin u} = \dfrac{\sin u}{\cos u} = \tan u$

**21.** $\sec\left(\frac{\pi}{2}-u\right) = \dfrac{1}{\cos\left(\frac{\pi}{2}-u\right)} = \dfrac{1}{\cos\frac{\pi}{2}\cos u + \sin\frac{\pi}{2}\sin u} = \dfrac{1}{0\cdot\cos u + 1\cdot\sin u} = \dfrac{1}{\sin u} = \csc u$

**22.** $\csc\left(\frac{\pi}{2}-u\right) = \dfrac{1}{\sin\left(\frac{\pi}{2}-u\right)} = \dfrac{1}{\sin\frac{\pi}{2}\cos u - \cos\frac{\pi}{2}\sin u} = \dfrac{1}{1\cdot\cos u - 0\cdot\sin u} = \dfrac{1}{\cos u} = \sec u$

**23.** $\sin\left(x-\frac{\pi}{2}\right) = \sin x\cos\frac{\pi}{2} - \cos x\sin\frac{\pi}{2} = 0\cdot\sin x - 1\cdot\cos x = -\cos x$

**24.** $\cos\left(x-\frac{\pi}{2}\right) = \cos x\cos\frac{\pi}{2} + \sin x\sin\frac{\pi}{2} = 0\cdot\cos x + 1\cdot\sin x = \sin x$

**25.** $\sin(x-\pi) = \sin x\cos\pi - \cos x\sin\pi = -1\cdot\sin x - 0\cdot\cos x = -\sin x$

**26.** $\cos(x-\pi) = \cos x\cos\pi + \sin x\sin\pi = -1\cdot\cos x + 0\cdot\sin x = -\cos x$

**27.** $\tan(x-\pi) = \dfrac{\tan x - \tan\pi}{1+\tan x\tan\pi} = \dfrac{\tan x - 0}{1+\tan x\cdot 0} = \tan x$

**28.** LHS $= \sin\left(\frac{\pi}{2}-x\right) = \sin\frac{\pi}{2}\cos x - \cos\frac{\pi}{2}\sin x = 1\cdot\cos x - 0\cdot\sin x = \cos x$ and RHS $= \sin\left(\frac{\pi}{2}+x\right) = \sin\frac{\pi}{2}\cos x + \cos\frac{\pi}{2}\sin x = 1\cdot\cos x + 0\cdot\sin x = \cos x$. Therefore, LHS = RHS.

**29.** $\cos\left(x+\frac{\pi}{6}\right) + \sin\left(x-\frac{\pi}{3}\right) = \cos x\cos\frac{\pi}{6} - \sin x\sin\frac{\pi}{6} + \sin x\cos\frac{\pi}{3} - \cos x\sin\frac{\pi}{3}$

**30.** $\tan\left(x-\frac{\pi}{4}\right) = \dfrac{\tan x - \tan\frac{\pi}{4}}{1+\tan x\tan\frac{\pi}{4}} = \dfrac{\tan x - 1}{1+\tan x\cdot 1} = \dfrac{\tan x - 1}{\tan x + 1}$

**31.** $\sin(x+y) - \sin(x-y) = \sin x\cos y + \cos x\sin y - (\sin x\cos y - \cos x\sin y) = 2\cos x\sin y$

**32.** $\cos(x+y) + \cos(x-y) = \cos x\cos y - \sin x\sin y + \cos x\cos y + \sin x\sin y = 2\cos x\cos y$

**33.** $\cot(x-y) = \dfrac{1}{\tan(x-y)} = \dfrac{1+\tan x\tan y}{\tan x - \tan y} = \dfrac{1+\dfrac{1}{\cot x}\dfrac{1}{\cot y}}{\dfrac{1}{\cot x} - \dfrac{1}{\cot y}} \cdot \dfrac{\cot x\cot y}{\cot x\cot y} = \dfrac{\cot x\cot y + 1}{\cot y - \cot x}$

**34.** $\cot(x+y) = \dfrac{1}{\tan(x+y)} = \dfrac{1-\tan x\tan y}{\tan x + \tan y} = \dfrac{1-\dfrac{1}{\cot x}\dfrac{1}{\cot y}}{\dfrac{1}{\cot x} + \dfrac{1}{\cot y}} \cdot \dfrac{\cot x\cot y}{\cot x\cot y} = \dfrac{\cot x\cot y - 1}{\cot x + \cot y}$

**35.** $\tan x - \tan y = \dfrac{\sin x}{\cos x} - \dfrac{\sin y}{\cos y} = \dfrac{\sin x\cos y - \cos x\sin y}{\cos x\cos y} = \dfrac{\sin(x-y)}{\cos x\cos y}$

**36.** $1 - \tan x\tan y = 1 - \dfrac{\sin x\sin y}{\cos x\cos y} = \dfrac{\cos x\cos y - \sin x\sin y}{\cos x\cos y} = \dfrac{\cos(x+y)}{\cos x\cos y}$

**37.** $\dfrac{\sin(x+y) - \sin(x-y)}{\cos(x+y) + \cos(x-y)} = \dfrac{\sin x\cos y + \cos x\sin y - (\sin x\cos y - \cos x\sin y)}{\cos x\cos y - \sin x\sin y + \cos x\cos y + \sin x\sin y} = \dfrac{2\cos x\sin y}{2\cos x\cos y} = \tan y$

**38.** $\cos(x+y)\cos(x-y) = (\cos x\cos y - \sin x\sin y)(\cos x\cos y + \sin x\sin y) = \cos^2 x\cos^2 y - \sin^2 x\sin^2 y$
$= \cos^2 x\left(1-\sin^2 y\right) - \left(1-\cos^2 x\right)\sin^2 y = \cos^2 x - \sin^2 y\cos^2 x + \sin^2 y\cos^2 x - \sin^2 y$
$= \cos^2 x - \sin^2 y$

**39.** $\sin(x+y+z) = \sin((x+y)+z) = \sin(x+y)\cos z + \cos(x+y)\sin z$
$= \cos z(\sin x\cos y + \cos x\sin y) + \sin z(\cos x\cos y - \sin x\sin y)$
$= \sin x\cos y\cos z + \cos x\sin y\cos z + \cos x\cos y\sin z - \sin x\sin y\sin z$

**40.** The addition formula for the tangent function can be written as $\tan A + \tan B = \tan(A+B)(1-\tan A\tan B)$. Also note that $\tan(-A) = -\tan A$. Using these facts, we get

$$\begin{aligned}\tan(x-y)+\tan(y-z)+\tan(z-x) &= \tan(x-y+y-z)\left[1-\tan(x-y)\tan(y-z)\right]+\tan(z-x)\\ &= \tan(x-z)\left[1-\tan(x-y)\tan(y-z)\right]+\tan(z-x)\\ &= \tan(x-z)+\tan(z-x)-\tan(x-y)\tan(y-z)\tan(x-z)\\ &= \tan(x-z)-\tan(x-z)-\tan(x-y)\tan(y-z)\tan(x-z)\\ &= 0-\tan(x-y)\tan(y-z)\tan(x-z)\\ &= \tan(x-y)\tan(y-z)\tan(z-x)\end{aligned}$$

**41.** $k=\sqrt{A^2+B^2}=\sqrt{\left(-\sqrt{3}\right)^2+1^2}=\sqrt{4}=2$. $\sin\phi=\frac{1}{2}$ and $\cos\phi=\dfrac{-\sqrt{3}}{2}$ $\Rightarrow$ $\phi=\frac{5\pi}{6}$. Therefore, $-\sqrt{3}\sin x+\cos x=k\sin(x+\phi)=2\sin\left(x+\frac{5\pi}{6}\right)$.

**42.** $k=\sqrt{A^2+B^2}=\sqrt{1^2+1^2}=\sqrt{2}$ and $\phi$ satisfies $\sin\phi=\frac{1}{\sqrt{2}}$, $\cos\phi=\frac{1}{\sqrt{2}}$ $\Leftrightarrow$ $\phi=\frac{\pi}{4}$. Thus, $\sin x+\cos x=k\sin(x+\phi)=\sqrt{2}\sin\left(x+\frac{\pi}{4}\right)=\frac{\sqrt{3}}{2}\cos x-\frac{1}{2}\sin x+\frac{1}{2}\sin x-\frac{\sqrt{3}}{2}\cos x=0$

**43.** $k=\sqrt{A^2+B^2}=\sqrt{5^2+(-5)^2}=\sqrt{50}=5\sqrt{2}$. $\sin\phi=-\frac{5}{5\sqrt{2}}=-\frac{1}{\sqrt{2}}$ and $\cos\phi=\frac{5}{5\sqrt{2}}=\frac{1}{\sqrt{2}}$ $\Rightarrow$ $\phi=\frac{7\pi}{4}$. Therefore, $5(\sin 2x-\cos 2x)=k\sin(2x+\phi)=5\sqrt{2}\sin\left(2x+\frac{7\pi}{4}\right)$.

**44.** $k=\sqrt{A^2+B^2}=\sqrt{3^2+\left(3\sqrt{3}\right)^2}=\sqrt{36}=6$ and $\phi$ satisfies $\sin\phi=\frac{3\sqrt{3}}{6}=\frac{\sqrt{3}}{2}$, $\cos\phi=\frac{3}{6}=\frac{1}{2}$ $\Leftrightarrow$ $\phi=\frac{\pi}{3}$. Thus, $3\sin\pi x+3\sqrt{3}\cos\pi x=k\sin(\pi x+\phi)=6\sin\left(\pi x+\frac{\pi}{3}\right)=6\sin\pi\left(x+\frac{1}{3}\right)$.

**45. (a)** $f(x)=\sin x+\cos x\Rightarrow k=\sqrt{1^2+1^2}=\sqrt{2}$, and $\phi$ satisfies $\sin\phi=\cos\phi=\frac{1}{\sqrt{2}}\Rightarrow\phi=\frac{\pi}{4}$. Thus, we can write $f(x)=k\sin(x+\phi)=\sqrt{2}\sin\left(x+\frac{\pi}{4}\right)$.

**(b)** This is a sine curve with amplitude $\sqrt{2}$, period $2\pi$, and phase shift $-\frac{\pi}{4}$.

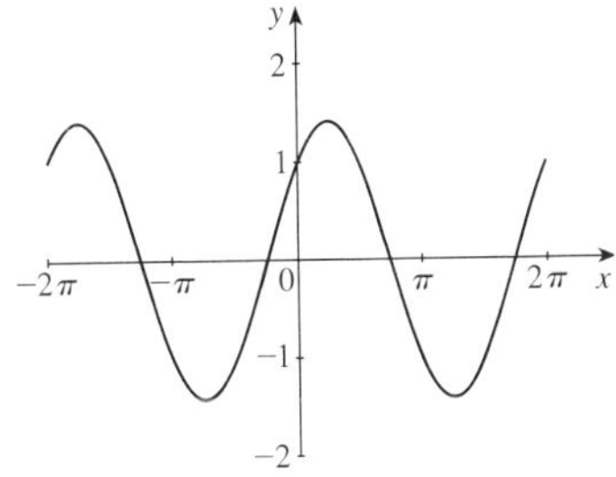


**46. (a)** $g(x)=\cos 2x+\sqrt{3}\sin 2x\Rightarrow$ $k=\sqrt{1^2+\left(\sqrt{3}\right)^2}=\sqrt{4}=2$, and $\phi$ satisfies $\sin\phi=\frac{1}{2}$, $\cos\phi=\frac{\sqrt{3}}{2}\Rightarrow\phi=\frac{\pi}{6}$. Thus, we can write $g(x)=k\sin(2x+\phi)=2\sin\left(2x+\frac{\pi}{6}\right)=2\sin 2\left(x+\frac{\pi}{12}\right)$.

**(b)** This is a sine curve with amplitude 2, period $\pi$, and phase shift $-\frac{\pi}{12}$.

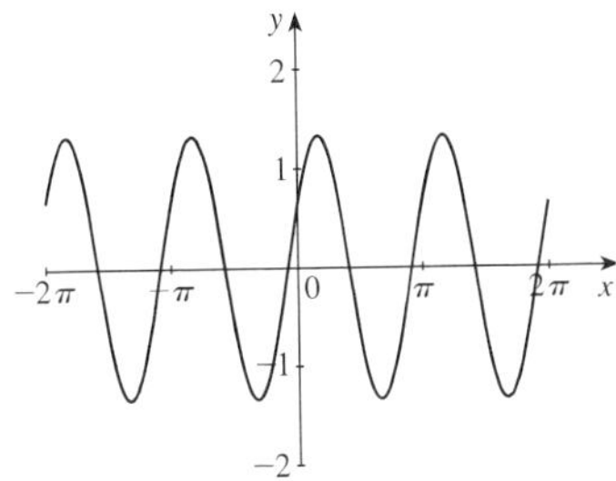


**47.** If $\beta-\alpha=\frac{\pi}{2}$, then $\beta=\alpha+\frac{\pi}{2}$. Now if we let $y=x+\alpha$, then $\sin(x+\alpha)+\cos\left(x+\alpha+\frac{\pi}{2}\right)=\sin y+\cos\left(y+\frac{\pi}{2}\right)=\sin y+(-\sin y)=0$. Therefore, $\sin(x+\alpha)+\cos(x+\beta)=0$.

**48.** $g(x) = \cos x$. Now

$$\frac{g(x+h)-g(x)}{h} = \frac{\cos(x+h)-\cos x}{h} = \frac{\cos x\cos h - \sin x \sin h - \cos x}{h}$$
$$= \frac{-\cos x\,(1-\cos h) - \sin h\,(\sin x)}{h} = -\cos x\left(\frac{1-\cos h}{h}\right) - \left(\frac{\sin h}{h}\right)\sin x$$

**49.** Let $\angle A$ and $\angle B$ be the two angles shown in the diagram. Then

$180^\circ = \gamma + A + B$, $90^\circ = \alpha + A$, and $90^\circ = \beta + B$.

Subtracting the second and third equation from the first, we get

$180^\circ - 90^\circ - 90^\circ = \gamma + A + B - (\alpha + A) - (\beta + B) \quad \Leftrightarrow$

$\alpha + \beta = \gamma$. Then

$$\tan\gamma = \tan(\alpha+\beta) = \frac{\tan\alpha + \tan\beta}{1-\tan\alpha\tan\beta} = \frac{\frac{4}{6}+\frac{3}{4}}{1-\frac{4}{6}\cdot\frac{3}{4}} = \frac{\frac{8}{12}+\frac{9}{12}}{1-\frac{1}{2}} = 2\cdot\tfrac{17}{12} = \tfrac{17}{6}.$$

**50. (a)** By definition, $m = \dfrac{\Delta y}{\Delta x}$ and $\tan\theta = \dfrac{\Delta y}{\Delta x}$. Thus, $m = \tan\theta$.

**(b)** $\tan\psi = \tan(\theta_2 - \theta_1) = \dfrac{\tan\theta_2 - \tan\theta_1}{1+\tan\theta_2\tan\theta_1}$. From part (a), we have $m_1 = \tan\theta_1$ and $m_2 = \tan\theta_2$. Then by substitution, $\tan\psi = \dfrac{m_2 - m_1}{1+m_1m_2}$.

**(c)** Let $\psi$ be the unknown angle as in part (b). Since $m_1 = \frac{1}{3}$ and $m_2 = -\frac{1}{2}$, $\tan\psi = \dfrac{m_2-m_1}{1+m_1m_2} = \dfrac{-\frac{1}{2}-\frac{1}{3}}{1+\frac{1}{3}\left(-\frac{1}{2}\right)} = \dfrac{-\frac{5}{6}}{\frac{5}{6}} = -1$ $\Leftrightarrow \quad \psi = \frac{3\pi}{4}$.

**(d)** From part (b), we have $\cot\psi = \dfrac{1+m_1m_2}{m_2-m_1}$. If the two lines are perpendicular then $\psi = 90^\circ$ and so $\cot\psi = 0$. Thus we have $0 = \dfrac{1+m_1m_2}{m_2-m_1} \quad\Leftrightarrow\quad 0 = 1 + m_1m_2 \quad\Leftrightarrow\quad m_1m_2 = -1 \quad\Leftrightarrow\quad m_2 = -1/m_1$. Thus $m_2$ is the negative reciprocal of $m_1$.

**51. (a)** $y = \sin^2\left(x+\frac{\pi}{4}\right) + \sin^2\left(x-\frac{\pi}{4}\right)$. From the graph we see that the value of $y$ seems to always be equal to 1.

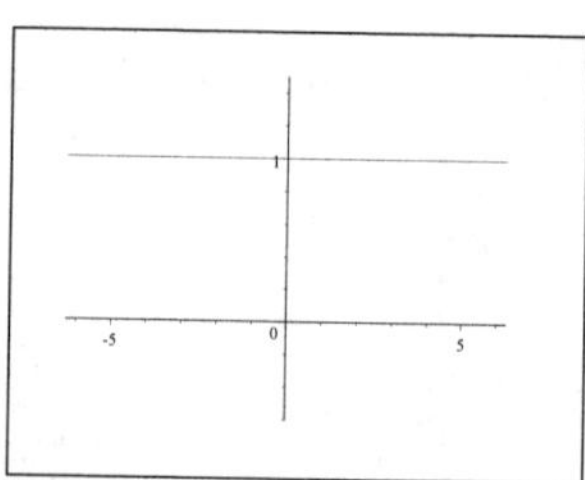


**(b)** $y = \sin^2\left(x+\frac{\pi}{4}\right) + \sin^2\left(x-\frac{\pi}{4}\right) = \left(\sin x\cos\frac{\pi}{4} + \cos x\sin\frac{\pi}{4}\right)^2 + \left(\sin x\cos\frac{\pi}{4} - \cos x\sin\frac{\pi}{4}\right)^2$

$= \left[\frac{1}{\sqrt{2}}(\sin x + \cos x)\right]^2 + \left[\frac{1}{\sqrt{2}}(\sin x - \cos x)\right]^2 = \frac{1}{2}\left[(\sin x+\cos x)^2 + (\sin x - \cos x)^2\right]$

$= \frac{1}{2}\left[\left(\sin^2 x + 2\sin x\cos x + \cos^2 x\right) + \left(\sin^2 x - 2\sin x\cos x + \cos^2 x\right)\right]$

$= \frac{1}{2}\left[(1+2\sin x\cos x) + (1-2\sin x\cos x)\right] = \frac{1}{2}\cdot 2 = 1$

**52. (a)** $y = -\frac{1}{2}\left[\cos(x+\pi) + \cos(x-\pi)\right]$. The graph of $y$ appears to be the same as that of $\cos x$.

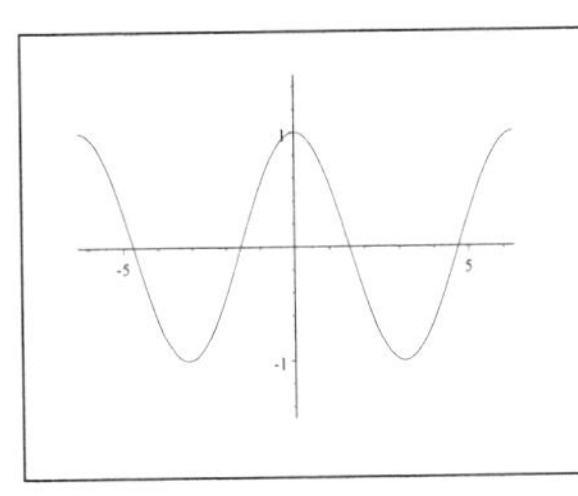

**(b)** $y = -\frac{1}{2}\left[\cos(x+\pi) + \cos(x-\pi)\right]$
$= -\frac{1}{2}\left[(\cos x \cos\pi - \sin x \sin\pi) + (\cos x \cos\pi + \sin x\ \sin\pi)\right]$
$= -\frac{1}{2}\left[(-\cos x - 0) + (-\cos x + 0)\right] = -\frac{1}{2}(-2\cos x) = \cos x$

**53.** Clearly $C = \frac{\pi}{4}$. Now $\tan(A+B) = \dfrac{\tan A + \tan B}{1 - \tan A \tan B} = \dfrac{\frac{1}{3}+\frac{1}{2}}{1 - \frac{1}{3}\cdot\frac{1}{2}} = 1$. Thus $A+B = \frac{\pi}{4}$, so $A+B+C = \frac{\pi}{4}+\frac{\pi}{4} = \frac{\pi}{2}$.

**54. (a)** $y = f_1(t) + f_2(t) = 5\sin t + 5\cos t = 5(\sin t + \cos t)$

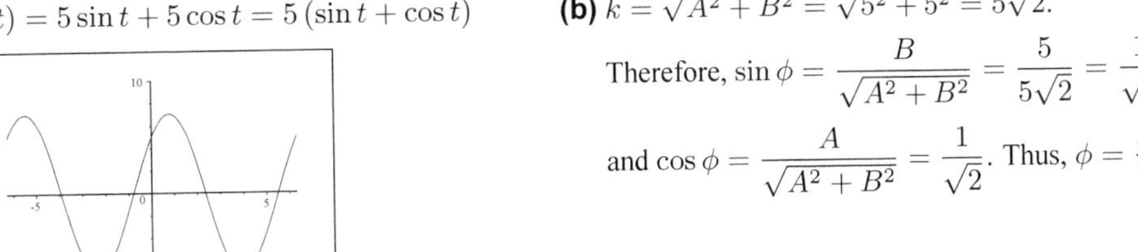

**(b)** $k = \sqrt{A^2+B^2} = \sqrt{5^2+5^2} = 5\sqrt{2}$.

Therefore, $\sin\phi = \dfrac{B}{\sqrt{A^2+B^2}} = \dfrac{5}{5\sqrt{2}} = \dfrac{1}{\sqrt{2}}$

and $\cos\phi = \dfrac{A}{\sqrt{A^2+B^2}} = \dfrac{1}{\sqrt{2}}$. Thus, $\phi = \frac{\pi}{4}$.

**55. (a)** $f(t) = C\sin\omega t + C\sin(\omega t + \alpha) = C\sin\omega t + C(\sin\omega t\cos\alpha + \cos\omega t\sin\alpha)$
$= C(1+\cos\alpha)\sin\omega t + C\sin\alpha\cos\omega t = A\sin\omega t + B\cos\omega t$

where $A = C(1+\cos\alpha)$ and $B = C\sin\alpha$.

**(b)** In this case, $f(t) = 10\left(1+\cos\frac{\pi}{3}\right)\sin\omega t + 10\sin\frac{\pi}{3}\cos\omega t = 15\sin\omega t + 5\sqrt{3}\cos\omega t$. Thus $k = \sqrt{15^2 + \left(5\sqrt{3}\right)^2} = 10\sqrt{3}$ and $\phi$ has $\cos\phi = \frac{15}{10\sqrt{3}} = \frac{\sqrt{3}}{2}$ and $\sin\phi = \frac{5\sqrt{3}}{10\sqrt{3}} = \frac{1}{2}$, so $\phi = \frac{\pi}{6}$. Therefore, $f(t) = 10\sqrt{3}\sin\left(\omega t + \frac{\pi}{6}\right)$.

**56.** $\sin(s+t) = \cos\left[\frac{\pi}{2} - (s+t)\right] = \cos\left[\left(\frac{\pi}{2} - s\right) - t\right] = \cos\left(\frac{\pi}{2} - s\right)\cos t + \sin\left(\frac{\pi}{2} - s\right)\sin t = \sin s\cos t + \cos s\sin t$. The last equality comes from again applying the cofunction identities.

**57.** $\tan(s+t) = \dfrac{\sin(s+t)}{\cos(s+t)} = \dfrac{\sin s\cos t + \cos s\sin t}{\cos s\cos t + \sin s\sin t}$

$$= \frac{\sin s\cos t + \cos s\sin t}{\cos s\cos t - \sin s\sin t}\cdot\frac{\dfrac{1}{\cos s\cos t}}{\dfrac{1}{\cos s\cos t}} = \frac{\dfrac{\sin s}{\cos s} + \dfrac{\sin t}{\cos t}}{1 - \dfrac{\sin s}{\cos s}\cdot\dfrac{\sin t}{\cos t}} = \frac{\tan s + \tan t}{1 - \tan s\tan t}$$

# 8.3 Double-Angle, Half-Angle, and Product-Sum Formulas

**1.** $\sin x = \frac{5}{13}$, $x$ in quadrant I $\Rightarrow$ $\cos x = \frac{12}{13}$ and $\tan x = \frac{5}{12}$. Thus, $\sin 2x = 2\sin x\cos x = 2\left(\frac{5}{13}\right)\left(\frac{12}{13}\right) = \frac{120}{169}$, $\cos 2x = \cos^2 x - \sin^2 x = \left(\frac{12}{13}\right)^2 - \left(\frac{5}{13}\right)^2 = \frac{144-25}{169} = \frac{119}{169}$, and $\tan 2x = \dfrac{\sin 2x}{\cos 2x} = \dfrac{\frac{120}{169}}{\frac{119}{169}} = \frac{120}{169} \cdot \frac{169}{119} = \frac{120}{119}$.

**2.** $\tan x = -\frac{4}{3}$. Then $\sin x = \frac{4}{5}$ and $\cos x = -\frac{3}{5}$ ($x$ is in quadrant II). Thus, $\sin 2x = 2\sin x\cos x = 2 \cdot \frac{4}{5}\left(-\frac{3}{5}\right) = -\frac{24}{25}$, $\cos 2x = \cos^2 x - \sin^2 x = \left(-\frac{3}{5}\right)^2 - \left(\frac{4}{5}\right)^2 = \frac{9-16}{25} = -\frac{7}{25}$, and $\tan 2x = \dfrac{\sin 2x}{\cos 2x} = \dfrac{-\frac{24}{25}}{-\frac{7}{25}} = \frac{24}{25} \cdot \frac{25}{7} = \frac{24}{7}$.

**3.** $\cos x = \frac{4}{5}$. Then $\sin x = -\frac{3}{5}$ ($\csc x < 0$) and $\tan x = -\frac{3}{4}$. Thus, $\sin 2x = 2\sin x\cos x = 2\left(-\frac{3}{5}\right) \cdot \frac{4}{5} = -\frac{24}{25}$, $\cos 2x = \cos^2 x - \sin^2 x = \left(\frac{4}{5}\right)^2 - \left(-\frac{3}{5}\right)^2 = \frac{16-9}{25} = \frac{7}{25}$, and $\tan 2x = \dfrac{\sin 2x}{\cos 2x} = \dfrac{-\frac{24}{25}}{\frac{7}{25}} = -\frac{24}{25} \cdot \frac{25}{7} = -\frac{24}{7}$.

**4.** $\csc x = 4$. Then $\sin x = \frac{1}{4}$, $\cos x = -\frac{\sqrt{15}}{4}$, and $\tan x = -\frac{1}{\sqrt{15}}$ ($\tan x < 0$). Thus, $\sin 2x = 2\sin x\cos x = 2 \cdot \frac{1}{4}\left(-\frac{\sqrt{15}}{4}\right) = -\frac{\sqrt{15}}{8}$, $\cos 2x = \cos^2 x - \sin^2 x = \left(-\frac{\sqrt{15}}{4}\right)^2 - \left(\frac{1}{4}\right)^2 = \frac{15-1}{16} = \frac{7}{8}$, and $\tan 2x = \dfrac{\sin 2x}{\cos 2x} = \dfrac{-\frac{\sqrt{15}}{8}}{\frac{7}{8}} = -\frac{\sqrt{15}}{8} \cdot \frac{8}{7} = -\frac{\sqrt{15}}{7}$.

**5.** $\sin x = -\frac{3}{5}$. Then, $\cos x = -\frac{4}{5}$ and $\tan x = \frac{3}{4}$ ($x$ is in quadrant III). Thus, $\sin 2x = 2\sin x\cos x = 2\left(-\frac{3}{5}\right)\left(-\frac{4}{5}\right) = \frac{24}{25}$, $\cos 2x = \cos^2 x - \sin^2 x = \left(-\frac{4}{5}\right)^2 - \left(-\frac{3}{5}\right)^2 = \frac{16-9}{25} = \frac{7}{25}$, and $\tan 2x = \dfrac{\sin 2x}{\cos 2x} = \dfrac{\frac{24}{25}}{\frac{7}{25}} = \frac{24}{25} \cdot \frac{25}{7} = \frac{24}{7}$.

**6.** $\sec x = 2$. Then $\cos x = \frac{1}{2}$, $\sin x = -\frac{\sqrt{3}}{2}$, and $\tan x = -\sqrt{3}$ ($x$ is in quadrant IV). Thus, $\sin 2x = 2\sin x\cos x = 2\left(-\frac{\sqrt{3}}{2}\right)\left(\frac{1}{2}\right) = -\frac{\sqrt{3}}{2}$, $\cos 2x = \cos^2 x - \sin^2 x = \left(\frac{1}{2}\right)^2 - \left(-\frac{\sqrt{3}}{2}\right)^2 = -\frac{1}{2}$, and $\tan 2x = \dfrac{\sin 2x}{\cos 2x} = \dfrac{-\frac{\sqrt{3}}{2}}{-\frac{1}{2}} = \sqrt{3}$.

**7.** $\tan x = -\frac{1}{3}$ and $\cos x > 0$, so $\sin x < 0$. Thus, $\sin x = -\frac{1}{\sqrt{10}}$ and $\cos x = \frac{3}{\sqrt{10}}$. Thus, $\sin 2x = 2\sin x\cos x = 2\left(-\frac{1}{\sqrt{10}}\right)\left(\frac{3}{\sqrt{10}}\right) = -\frac{6}{10} = -\frac{3}{5}$, $\cos 2x = \cos^2 x - \sin^2 x = \left(\frac{3}{\sqrt{10}}\right)^2 - \left(-\frac{1}{\sqrt{10}}\right)^2 = \frac{8}{10} = \frac{4}{5}$, and $\tan 2x = \dfrac{\sin 2x}{\cos 2x} = \dfrac{-\frac{3}{5}}{\frac{4}{5}} = -\frac{3}{5} \cdot \frac{5}{4} = -\frac{3}{4}$.

**8.** $\cot x = \frac{2}{3}$. Then $\tan x = \frac{3}{2}$, $\sin x = \frac{3}{\sqrt{13}}$ ($\sin x > 0$), and $\cos x = \frac{2}{\sqrt{13}}$. Thus, $\sin 2x = 2\sin x\cos x = 2\left(\frac{3}{\sqrt{13}}\right)\left(\frac{2}{\sqrt{13}}\right) = \frac{12}{13}$, $\cos 2x = \cos^2 x - \sin^2 x = \left(\frac{2}{\sqrt{13}}\right)^2 - \left(\frac{3}{\sqrt{13}}\right)^2 = \frac{4-9}{13} = -\frac{5}{13}$, and $\tan 2x = \dfrac{\sin 2x}{\cos 2x} = \dfrac{\frac{12}{13}}{-\frac{5}{13}} = \frac{12}{13} \cdot \left(-\frac{13}{5}\right) = -\frac{12}{5}$.

**9.** $\sin^4 x = \left(\sin^2 x\right)^2 = \left(\dfrac{1-\cos 2x}{2}\right)^2 = \frac{1}{4} - \frac{1}{2}\cos 2x + \frac{1}{4}\cos^2 2x$
$= \frac{1}{4} - \frac{1}{2}\cos 2x + \frac{1}{4} \cdot \dfrac{1+\cos 4x}{2} = \frac{1}{4} - \frac{1}{2}\cos 2x + \frac{1}{8} + \frac{1}{8}\cos 4x = \frac{3}{8} - \frac{1}{2}\cos 2x + \frac{1}{8}\cos 4x$
$= \frac{1}{2}\left(\frac{3}{4} - \cos 2x + \frac{1}{4}\cos 4x\right)$

**10.** $\cos^4 x = \left(\cos^2 x\right)^2 = \left(\dfrac{1+\cos 2x}{2}\right)^2 = \frac{1}{4} + \frac{1}{2}\cos 2x + \frac{1}{4}\cos^2 2x$
$= \frac{1}{4} + \frac{1}{2}\cos 2x + \frac{1}{4} \cdot \dfrac{1+\cos 4x}{2} = \frac{1}{4} + \frac{1}{2}\cos 2x + \frac{1}{8} + \frac{1}{8}\cos 4x = \frac{3}{8} + \frac{1}{2}\cos 2x + \frac{1}{8}\cos 4x$
$= \frac{1}{2}\left(\frac{3}{4} + \cos 2x + \frac{1}{4}\cos 4x\right)$

**11.** We use the result of Example 4 to get

$$\begin{aligned}\cos^2 x \sin^4 x &= \left(\sin^2 x \cos^2 x\right)\sin^2 x = \left(\tfrac{1}{8} - \tfrac{1}{8}\cos 4x\right)\cdot\left(\tfrac{1}{2} - \tfrac{1}{2}\cos 2x\right) \\ &= \tfrac{1}{16}\left(1 - \cos 2x - \cos 4x + \cos 2x \cos 4x\right)\end{aligned}$$

**12.** Using Example 4, we have

$$\begin{aligned}\cos^4 x \sin^2 x &= \cos^2 x\left(\cos^2 x \sin^2 x\right) = \tfrac{1}{2}\left(1 + \cos 2x\right)\cdot\tfrac{1}{8}\left(1 - \cos 4x\right) \\ &= \tfrac{1}{16}\left(1 - \cos 4x + \cos 2x - \cos 2x \cos 4x\right)\end{aligned}$$

**13.** Since $\sin^4 x \cos^4 x = \left(\sin^2 x \cos^2 x\right)^2$ we can use the result of Example 4 to get

$$\begin{aligned}\sin^4 x \cos^4 x &= \left(\tfrac{1}{8} - \tfrac{1}{8}\cos 4x\right)^2 = \tfrac{1}{64} - \tfrac{1}{32}\cos 4x + \tfrac{1}{64}\cos^2 4x \\ &= \tfrac{1}{64} - \tfrac{1}{32}\cos 4x + \tfrac{1}{64}\cdot\tfrac{1}{2}\left(1 + \cos 8x\right) = \tfrac{1}{64} - \tfrac{1}{32}\cos 4x + \tfrac{1}{128} + \tfrac{1}{128}\cos 8x \\ &= \tfrac{3}{128} - \tfrac{1}{32}\cos 4x + \tfrac{1}{128}\cos 8x = \tfrac{1}{32}\left(\tfrac{3}{4} - \cos 4x + \tfrac{1}{4}\cos 8x\right)\end{aligned}$$

**14.** Using the result of Exercise 10, we have

$$\begin{aligned}\cos^6 x &= \left(\cos^2 x\right)^2\cdot\cos^2 x = \left(\tfrac{3}{8} + \tfrac{1}{2}\cos 2x + \tfrac{1}{8}\cos 4x\right)\left(\tfrac{1}{2} + \tfrac{1}{2}\cos 2x\right) \\ &= \tfrac{3}{16} + \tfrac{1}{4}\cos 2x + \tfrac{1}{16}\cos 4x + \tfrac{3}{16}\cos 2x + \tfrac{1}{16}\cos 2x \cos 4x \\ &= \tfrac{3}{16} + \tfrac{7}{16}\cos 2x + \tfrac{1}{16}\cos 4x + \tfrac{1}{16}\cos 2x \cos 4x = \tfrac{1}{16}\left(3 + 7\cos 2x + \cos 4x + \cos 2x \cos 4x\right)\end{aligned}$$

**15.** $\sin 15^\circ = \sqrt{\tfrac{1}{2}\left(1 - \cos 30^\circ\right)} = \sqrt{\tfrac{1}{2}\left(1 - \tfrac{\sqrt{3}}{2}\right)} = \sqrt{\tfrac{1}{4}\left(2 - \sqrt{3}\right)} = \tfrac{1}{2}\sqrt{2 - \sqrt{3}}$

**16.** $\tan 15^\circ = \dfrac{1 - \cos 30^\circ}{\sin 30^\circ} = \dfrac{1 - \frac{\sqrt{3}}{2}}{\frac{1}{2}} = 2 - \sqrt{3}$

**17.** $\tan 22.5^\circ = \dfrac{1 - \cos 45^\circ}{\sin 45^\circ} = \dfrac{1 - \frac{\sqrt{2}}{2}}{\frac{\sqrt{2}}{2}} = \sqrt{2} - 1$

**18.** $\sin 75^\circ = \sqrt{\tfrac{1}{2}\left(1 - \cos 150^\circ\right)} = \sqrt{\tfrac{1}{2}\left(1 - \left[-\tfrac{\sqrt{3}}{2}\right]\right)} = \sqrt{\tfrac{1}{4}\left(2 + \sqrt{3}\right)} = \tfrac{1}{2}\sqrt{2 + \sqrt{3}}$

**19.** $\cos 165^\circ = -\sqrt{\tfrac{1}{2}\left(1 + \cos 330^\circ\right)} = -\sqrt{\tfrac{1}{2}\left(1 + \cos 30^\circ\right)} = -\sqrt{\tfrac{1}{2}\left(1 + \tfrac{\sqrt{3}}{2}\right)} = -\tfrac{1}{2}\sqrt{2 + \sqrt{3}}$

**20.** $\cos 112.5^\circ = -\sqrt{\tfrac{1}{2}\left(1 + \cos 225^\circ\right)} = -\sqrt{\tfrac{1}{2}\left(1 - \cos 45^\circ\right)} = -\sqrt{\tfrac{1}{2}\left(1 - \tfrac{\sqrt{2}}{2}\right)} = -\tfrac{1}{2}\sqrt{2 - \sqrt{2}}$

**21.** $\tan\frac{\pi}{8} = \dfrac{1 - \cos\frac{\pi}{4}}{\sin\frac{\pi}{4}} = \dfrac{1 - \frac{\sqrt{2}}{2}}{\frac{\sqrt{2}}{2}} = \sqrt{2} - 1$

**22.** $\cos\frac{3\pi}{8} = \cos\left(\frac{1}{2}\cdot\frac{3\pi}{4}\right) = \sqrt{\dfrac{1 + \cos\frac{3\pi}{4}}{2}} = \sqrt{\dfrac{1 - \frac{\sqrt{2}}{2}}{2}} = \sqrt{\dfrac{2 - \sqrt{2}}{4}} = \frac{1}{2}\sqrt{2 - \sqrt{2}}$. Note that we have chosen the positive root because $\frac{3\pi}{8}$ is in quadrant I, so $\cos\frac{3\pi}{8} > 0$.

**23.** $\cos\frac{\pi}{12} = \sqrt{\tfrac{1}{2}\left(1 + \cos\frac{\pi}{6}\right)} = \sqrt{\tfrac{1}{2}\left(1 + \tfrac{\sqrt{3}}{2}\right)} = \tfrac{1}{2}\sqrt{2 + \sqrt{3}}$

**24.** $\tan\frac{5\pi}{12} = \dfrac{1 - \cos\frac{5\pi}{6}}{\sin\frac{5\pi}{6}} = \dfrac{1 + \frac{\sqrt{3}}{2}}{\frac{1}{2}} = 2 + \sqrt{3}$

**25.** $\sin\frac{9\pi}{8} = -\sqrt{\tfrac{1}{2}\left(1 - \cos\frac{9\pi}{4}\right)} = -\sqrt{\tfrac{1}{2}\left(1 - \tfrac{\sqrt{2}}{2}\right)} = -\tfrac{1}{2}\sqrt{2 - \sqrt{2}}$. We have chosen the negative root because $\frac{9\pi}{8}$ is in quadrant III, so $\sin\frac{9\pi}{8} < 0$.

**26.** $\sin\frac{11\pi}{12} = \sqrt{\tfrac{1}{2}\left(1 - \cos\frac{11\pi}{6}\right)} = \sqrt{\tfrac{1}{2}\left(1 - \tfrac{\sqrt{3}}{2}\right)} = \tfrac{1}{2}\sqrt{2 - \sqrt{3}}$. We have chosen the positive root because $\frac{11\pi}{12}$ is in quadrant II, so $\sin\frac{11\pi}{12} > 0$.

**27. (a)** $2\sin 18^\circ \cos 18^\circ = \sin 36^\circ$

**(b)** $2\sin 3\theta \cos 3\theta = \sin 6\theta$

**28. (a)** $\dfrac{2\tan 7^\circ}{1-\tan^2 7^\circ} = \tan 14^\circ$

**(b)** $\dfrac{2\tan 7\theta}{1-\tan^2 7\theta} = \tan 14\theta$

**29. (a)** $\cos^2 34^\circ - \sin^2 34^\circ = \cos 68^\circ$

**(b)** $\cos^2 5\theta - \sin^2 5\theta = \cos 10\theta$

**30. (a)** $\cos^2 \dfrac{\theta}{2} - \sin^2 \dfrac{\theta}{2} = \cos\theta$

**(b)** $2\sin\dfrac{\theta}{2}\cos\dfrac{\theta}{2} = \sin\theta$

**31. (a)** $\dfrac{\sin 8^\circ}{1+\cos 8^\circ} = \tan\dfrac{8^\circ}{2} = \tan 4^\circ$

**(b)** $\dfrac{1-\cos 4\theta}{\sin 4\theta} = \tan\dfrac{4\theta}{2} = \tan 2\theta$

**32. (a)** $\sqrt{\dfrac{1-\cos 30^\circ}{2}} = \sin 15^\circ$

**(b)** $\sqrt{\dfrac{1-\cos 8\theta}{2}} = \sin 4\theta$

**33.** $\sin(x+x) = \sin x\cos x + \cos x\sin x = 2\sin x\cos x$

**34.** $\tan(x+x) = \dfrac{\tan x + \tan x}{1-\tan x\tan x} = \dfrac{2\tan x}{1-\tan^2 x}$

**35.** $\sin x = \frac{3}{5}$. Since $x$ is in quadrant I, $\cos x = \frac{4}{5}$ and $\frac{x}{2}$ is also in quadrant I. Thus, $\sin\frac{x}{2} = \sqrt{\frac{1}{2}(1-\cos x)} = \sqrt{\frac{1}{2}\left(1-\frac{4}{5}\right)} = \frac{1}{\sqrt{10}} = \frac{\sqrt{10}}{10}$, $\cos\frac{x}{2} = \sqrt{\frac{1}{2}(1+\cos x)} = \sqrt{\frac{1}{2}\left(1+\frac{4}{5}\right)} = \frac{3}{\sqrt{10}} = \frac{3\sqrt{10}}{10}$, and $\tan\frac{x}{2} = \dfrac{\sin\frac{x}{2}}{\cos\frac{x}{2}} = \frac{1}{\sqrt{10}}\cdot\frac{\sqrt{10}}{3} = \frac{1}{3}$.

**36.** $\cos x = -\frac{4}{5}$. Since $x$ is in quadrant III, $\sin x = -\frac{3}{5}$ and $\tan x = \frac{3}{4}$. Also, since $180^\circ \le x \le 270^\circ$, $90^\circ \le \frac{x}{2} \le 135^\circ$ and so $\frac{x}{2}$ is in quadrant II. Thus, $\sin\frac{x}{2} = \sqrt{\frac{1}{2}(1-\cos x)} = \sqrt{\frac{1}{2}\left(1+\frac{4}{5}\right)} = \frac{3}{\sqrt{10}} = \frac{3\sqrt{10}}{10}$, $\cos\frac{x}{2} = -\sqrt{\frac{1}{2}(1+\cos x)} = -\sqrt{\frac{1}{2}\left(1-\frac{4}{5}\right)} = -\frac{1}{\sqrt{10}} = -\frac{\sqrt{10}}{10}$, and $\tan\frac{x}{2} = \dfrac{\sin\frac{x}{2}}{\cos\frac{x}{2}} = \frac{3}{\sqrt{10}}\cdot\frac{\sqrt{10}}{-1} = -3$.

**37.** $\csc x = 3$. Then, $\sin x = \frac{1}{3}$ and since $x$ is in quadrant II, $\cos x = -\dfrac{2\sqrt{2}}{3}$. Since $90^\circ \le x \le 180^\circ$, we have $45^\circ \le \frac{x}{2} \le 90^\circ$ and so $\frac{x}{2}$ is in quadrant I. Thus, $\sin\frac{x}{2} = \sqrt{\frac{1}{2}(1-\cos x)} = \sqrt{\frac{1}{2}\left(1+\frac{2\sqrt{2}}{3}\right)} = \sqrt{\frac{1}{6}(3+2\sqrt{2})}$, $\cos\frac{x}{2} = \sqrt{\frac{1}{2}(1+\cos x)} = \sqrt{\frac{1}{2}\left(1-\frac{2\sqrt{2}}{3}\right)} = \sqrt{\frac{1}{6}(3-2\sqrt{2})}$, and $\tan\frac{x}{2} = \dfrac{\sin\frac{x}{2}}{\cos\frac{x}{2}} = \sqrt{\frac{3+2\sqrt{2}}{3-2\sqrt{2}}} = 3+2\sqrt{2}$.

**38.** $\tan x = 1$. Then $\sin x = \frac{\sqrt{2}}{2}$ and $\cos x = \frac{\sqrt{2}}{2}$, since $x$ is in quadrant I. Also, since $0^\circ \le x \le 90^\circ$, $0^\circ \le \frac{x}{2} \le 45^\circ$ and so $\frac{x}{2}$ is also in quadrant I. Thus, $\sin\frac{x}{2} = \sqrt{\frac{1}{2}(1-\cos x)} = \sqrt{\frac{1}{2}\left(1-\frac{\sqrt{2}}{2}\right)} = \frac{1}{2}\sqrt{2-\sqrt{2}}$, $\cos\frac{x}{2} = \sqrt{\frac{1}{2}(1+\cos x)} = \sqrt{\frac{1}{2}\left(1+\frac{\sqrt{2}}{2}\right)} = \frac{1}{2}\sqrt{2+\sqrt{2}}$, and $\tan\frac{x}{2} = \dfrac{1-\cos x}{\sin x} = \dfrac{1-\frac{\sqrt{2}}{2}}{\frac{\sqrt{2}}{2}} = \sqrt{2}-1$.

**39.** $\sec x = \frac{3}{2}$. Then $\cos x = \frac{2}{3}$ and since $x$ is in quadrant IV, $\sin x = -\frac{\sqrt{5}}{3}$. Since $270^\circ \le x \le 360^\circ$, we have $135^\circ \le \frac{x}{2} \le 180^\circ$ and so $\frac{x}{2}$ is in quadrant II. Thus, $\sin\frac{x}{2} = \sqrt{\frac{1}{2}(1-\cos x)} = \sqrt{\frac{1}{2}\left(1-\frac{2}{3}\right)} = \frac{1}{\sqrt{6}} = \frac{\sqrt{6}}{6}$, $\cos\frac{x}{2} = -\sqrt{\frac{1}{2}(1+\cos x)} = -\sqrt{\frac{1}{2}\left(1+\frac{2}{3}\right)} = -\frac{\sqrt{5}}{\sqrt{6}} = -\dfrac{\sqrt{30}}{6}$, and $\tan\frac{x}{2} = \dfrac{\sin\frac{x}{2}}{\cos\frac{x}{2}} = \frac{1}{\sqrt{6}}\cdot\frac{\sqrt{6}}{-\sqrt{5}} = -\frac{1}{\sqrt{5}} = -\frac{\sqrt{5}}{5}$.

**40.** $\cot x = 5$. Then, $\cos x = -\frac{5}{\sqrt{26}}$ and $\sin x = -\frac{1}{\sqrt{2}}$ ($\csc x < 0$). Since $\cot x > 0$ and $\csc x < 0$, it follows that $x$ is in quadrant III. Thus $180^\circ \le x \le 270^\circ$ and so $90^\circ \le \frac{x}{2} \le 135^\circ$. Thus $\frac{x}{2}$ is in quadrant II. $\sin\frac{x}{2} = \sqrt{\frac{1}{2}(1-\cos x)} = \sqrt{\frac{1}{2}\left(1+\frac{5}{\sqrt{26}}\right)} = \frac{1}{2}\sqrt{\frac{26+5\sqrt{26}}{13}}$, $\cos\frac{x}{2} = -\sqrt{\frac{1}{2}(1+\cos x)} = -\sqrt{\frac{1}{2}\left(1-\frac{5}{\sqrt{26}}\right)} = -\frac{1}{2}\sqrt{\frac{26-5\sqrt{26}}{13}}$, and $\tan\frac{x}{2} = \dfrac{1-\cos x}{\sin x} = \dfrac{1+\frac{5}{\sqrt{26}}}{-\frac{1}{\sqrt{2}}} = -5-\sqrt{26}$.

**41.** $\sin 2x \cos 3x = \frac{1}{2}[\sin(2x+3x)+\sin(2x-3x)] = \frac{1}{2}(\sin 5x - \sin x)$

**42.** $\sin x \sin 5x = \frac{1}{2}[\cos(x-5x)-\cos(x+5x)] = \frac{1}{2}(\cos 4x - \cos 6x)$

**43.** $\cos x \sin 4x = \frac{1}{2}[\sin(4x+x)+\sin(4x-x)] = \frac{1}{2}(\sin 5x + \sin 3x)$

**44.** $\cos 5x \cos 3x = \frac{1}{2}[\cos(5x+3x)+\cos(5x-3x)] = \frac{1}{2}(\cos 8x + \cos 2x)$

**45.** $3\cos 4x \cos 7x = 3 \cdot \frac{1}{2}[\cos(4x+7x)+\cos(4x-7x)] = \frac{3}{2}(\cos 11x + \cos 3x)$

**46.** $11 \sin\frac{x}{2}\cos\frac{x}{4} = 11 \cdot \frac{1}{2}\left[\sin\left(\frac{x}{2}+\frac{x}{4}\right)+\sin\left(\frac{x}{2}-\frac{x}{4}\right)\right] = \frac{11}{2}\left(\sin\frac{3x}{4}+\sin\frac{x}{4}\right)$

**47.** $\sin 5x + \sin 3x = 2\sin\left(\frac{5x+3x}{2}\right)\cos\left(\frac{5x-3x}{2}\right) = 2\sin 4x \cos x$

**48.** $\sin x - \sin 4x = 2\cos\left(\frac{x+4x}{2}\right)\sin\left(\frac{x-4x}{2}\right) = 2\cos\frac{5x}{2}\sin\left(-\frac{3x}{2}\right) = -2\cos\frac{5x}{2}\sin\frac{3x}{2}$

**49.** $\cos 4x - \cos 6x = -2\sin\left(\frac{4x+6x}{2}\right)\sin\left(\frac{4x-6x}{2}\right) = -2\sin 5x \sin(-x) = 2\sin 5x \sin x$

**50.** $\cos 9x + \cos 2x = 2\cos\left(\frac{9x+2x}{2}\right)\cos\left(\frac{9x-2x}{2}\right) = 2\cos\frac{11x}{2}\cos\frac{7x}{2}$

**51.** $\sin 2x - \sin 7x = 2\cos\left(\frac{2x+7x}{2}\right)\sin\left(\frac{2x-7x}{2}\right) = 2\cos\frac{9x}{2}\sin\left(-\frac{5x}{2}\right) = -2\cos\frac{9x}{2}\sin\frac{5x}{2}$

**52.** $\sin 3x + \sin 4x = 2\sin\left(\frac{3x+4x}{2}\right)\cos\left(\frac{3x-4x}{2}\right) = 2\sin\frac{7x}{2}\cos\left(-\frac{x}{2}\right) = 2\sin\frac{7x}{2}\cos\frac{x}{2}$

**53.** $2\sin 52.5^\circ \sin 97.5^\circ = 2 \cdot \frac{1}{2}[\cos(52.5^\circ - 97.5^\circ) - \cos(52.5^\circ + 97.5^\circ)] = \cos(-45^\circ) - \cos 150^\circ$
$= \cos 45^\circ - \cos 150^\circ = \frac{\sqrt{2}}{2} + \frac{\sqrt{3}}{2} = \frac{1}{2}\left(\sqrt{2}+\sqrt{3}\right)$

**54.** $3\cos 37.5^\circ \cos 7.5^\circ = \frac{3}{2}(\cos 45^\circ + \cos 30^\circ) = \frac{3}{2}\left(\frac{\sqrt{2}}{2}+\frac{\sqrt{3}}{2}\right) = \frac{3}{4}\left(\sqrt{2}+\sqrt{3}\right)$

**55.** $\cos 37.5^\circ \sin 7.5^\circ = \frac{1}{2}(\sin 45^\circ - \sin 30^\circ) = \frac{1}{2}\left(\frac{\sqrt{2}}{2}-\frac{1}{2}\right) = \frac{1}{4}\left(\sqrt{2}-1\right)$

**56.** $\sin 75^\circ + \sin 15^\circ = 2\sin\left(\frac{75^\circ+15^\circ}{2}\right)\cos\left(\frac{75^\circ-15^\circ}{2}\right) = 2\sin 45^\circ \cos 30^\circ = 2 \cdot \frac{\sqrt{2}}{2} \cdot \frac{\sqrt{3}}{2} = \frac{\sqrt{6}}{2}$

**57.** $\cos 255^\circ - \cos 195^\circ = -2\sin\left(\frac{255^\circ+195^\circ}{2}\right)\sin\left(\frac{255^\circ-195^\circ}{2}\right) = -2\sin 225^\circ \sin 30^\circ = -2\left(-\frac{\sqrt{2}}{2}\right)\frac{1}{2} = \frac{\sqrt{2}}{2}$

**58.** $\cos\frac{\pi}{12} + \cos\frac{5\pi}{12} = 2\cos\left[\frac{1}{2}\left(\frac{\pi}{12}+\frac{5\pi}{12}\right)\right]\cos\left[\frac{1}{2}\left(\frac{\pi}{12}-\frac{5\pi}{12}\right)\right] = 2\cos\frac{\pi}{4}\cos\frac{\pi}{6} = 2 \cdot \frac{\sqrt{2}}{2} \cdot \frac{\sqrt{3}}{2} = \frac{\sqrt{6}}{2}$

**59.** $\cos^2 5x - \sin^2 5x = \cos(2 \cdot 5x) = \cos 10x$

**60.** $\sin 8x = \sin(2 \cdot 4x) = 2\sin 4x \cos 4x$

**61.** $(\sin x + \cos x)^2 = \sin^2 x + 2\sin x \cos x + \cos^2 x = 1 + 2\sin x \cos x = 1 + \sin 2x$

**62.** $\dfrac{2\tan x}{1+\tan^2 x} = \dfrac{2\tan x}{\sec^2 x} = 2 \cdot \dfrac{\sin x}{\cos x}\cos^2 x = 2\sin x \cos x = \sin 2x$

**63.** $\dfrac{\sin 4x}{\sin x} = \dfrac{2\sin 2x \cos 2x}{\sin x} = \dfrac{2(2\sin x \cos x)(\cos 2x)}{\sin x} = 4\cos x \cos 2x$

**64.** $\dfrac{1+\sin 2x}{\sin 2x} = \dfrac{1+2\sin x \cos x}{2\sin x \cos x} = 1 + \dfrac{1}{2\sin x \cos x} = 1 + \frac{1}{2}\csc x \sec x$

**65.** $\dfrac{2(\tan x - \cot x)}{\tan^2 x - \cot^2 x} = \dfrac{2(\tan x - \cot x)}{(\tan x + \cot x)(\tan x - \cot x)} = \dfrac{2}{\tan x + \cot x} = \dfrac{2}{\dfrac{\sin x}{\cos x}+\dfrac{\cos x}{\sin x}}$

$= \dfrac{2}{\dfrac{\sin x}{\cos x}+\dfrac{\cos x}{\sin x}} \cdot \dfrac{\sin x \cos x}{\sin x \cos x} = \dfrac{2\sin x \cos x}{\sin^2 x + \cos^2 x} = 2\sin x \cos x = \sin 2x$

**66.** $\cot 2x = \dfrac{1}{\tan 2x} = \dfrac{1}{\dfrac{2\tan x}{1-\tan^2 x}} = \dfrac{1-\tan^2 x}{2\tan x}$

**67.** $\tan 3x = \tan(2x + x) = \dfrac{\tan 2x + \tan x}{1 - \tan 2x \tan x} = \dfrac{\dfrac{2\tan x}{1 - \tan^2 x} + \tan x}{1 - \dfrac{2\tan x}{1 - \tan^2 x}\tan x} = \dfrac{2\tan x + \tan x\left(1 - \tan^2 x\right)}{1 - \tan^2 x - 2\tan x \tan x}$

$= \dfrac{3\tan x - \tan^3 x}{1 - 3\tan^2 x}$

**68.** $4\left(\sin^6 x + \cos^6 x\right) = 4\left[\left(\sin^2 x + \cos^2 x\right)^3 - 3\left(\sin^4 x \cos^2 x + \sin^2 x \cos^4 x\right)\right]$

$= 4\left[1 - 3\sin^2 x \cos^2 x\left(\sin^2 x + \cos^2 x\right)\right] = 4 - 12\sin^2 x \cos^2 x$

$= 4 - 3\left(2\sin x \cos x\right)^2 = 4 - 3\sin^2 2x$

**69.** $\cos^4 x - \sin^4 x = \left(\cos^2 x + \sin^2 x\right)\left(\cos^2 x - \sin^2 x\right) = \cos^2 x - \sin^2 x = \cos 2x$

**70.** Let $y = \dfrac{x}{2} + \dfrac{\pi}{4} \Leftrightarrow 2y = x + \dfrac{\pi}{2}$. Then

$\tan^2\left(\dfrac{x}{2} + \dfrac{\pi}{4}\right) = \tan^2 y = \dfrac{1 - \cos 2y}{1 + \cos 2y} = \dfrac{1 - \cos\left(x + \frac{\pi}{2}\right)}{1 + \cos\left(x + \frac{\pi}{2}\right)} = \dfrac{1 - (-\sin x)}{1 + (-\sin x)} = \dfrac{1 + \sin x}{1 - \sin x}$

**71.** $\dfrac{\sin x + \sin 5x}{\cos x + \cos 5x} = \dfrac{2\sin 3x \cos 2x}{2\cos 3x \cos 2x} = \dfrac{\sin 3x}{\cos 3x} = \tan 3x$

**72.** $\dfrac{\sin 3x + \sin 7x}{\cos 3x - \cos 7x} = \dfrac{2\sin 5x \cos 2x}{-2\sin 5x \sin(-2x)} = \dfrac{\cos 2x}{\sin 2x} = \cot 2x$

**73.** $\dfrac{\sin 10x}{\sin 9x + \sin x} = \dfrac{2\sin 5x \cos 5x}{2\sin 5x \cos 4x} = \dfrac{\cos 5x}{\cos 4x}$

**74.** $\dfrac{\sin x + \sin 3x + \sin 5x}{\cos x + \cos 3x + \cos 5x} = \dfrac{\sin x + \sin 5x + \sin 3x}{\cos x + \cos 5x + \cos 3x} = \dfrac{2\sin 3x \cos 2x + \sin 3x}{2\cos 3x \cos 2x + \cos 3x} = \dfrac{\sin 3x\,(2\cos 2x + 1)}{\cos 3x\,(2\cos 2x + 1)} = \tan 3x$

**75.** $\dfrac{\sin x + \sin y}{\cos x + \cos y} = \dfrac{2\sin\left(\dfrac{x + y}{2}\right)\cos\left(\dfrac{x - y}{2}\right)}{2\cos\left(\dfrac{x + y}{2}\right)\cos\left(\dfrac{x - y}{2}\right)} = \dfrac{\sin\left(\dfrac{x + y}{2}\right)}{\cos\left(\dfrac{x + y}{2}\right)} = \tan\left(\dfrac{x + y}{2}\right)$

**76.** $\dfrac{\sin(x + y) - \sin(x - y)}{\cos(x + y) + \cos(x - y)} = \dfrac{2\sin\left(\dfrac{x + y + x - y}{2}\right)\cos\left(\dfrac{x + y - x + y}{2}\right)}{2\cos\left(\dfrac{x + y + x - y}{2}\right)\cos\left(\dfrac{x + y - x + y}{2}\right)} = \dfrac{\sin x}{\cos x} = \tan x$

**77.** $\sin 130^\circ - \sin 110^\circ = 2\cos\dfrac{130^\circ + 110^\circ}{2}\sin\dfrac{130^\circ - 110^\circ}{2} = 2\cos 120^\circ \sin 10^\circ = 2\left(-\frac{1}{2}\right)\sin 10^\circ = -\sin 10^\circ$

**78.** $\cos 100^\circ - \cos 200^\circ = -2\sin\dfrac{100^\circ + 200^\circ}{2}\sin\dfrac{100^\circ - 200^\circ}{2} = -2\sin 150^\circ \sin(-50^\circ)$

$= -2\left(\frac{1}{2}\right)(-\sin 50^\circ) = \sin 50^\circ$

**79.** $\sin 45^\circ + \sin 15^\circ = 2\sin\left(\dfrac{45^\circ + 15^\circ}{2}\right)\cos\left(\dfrac{45^\circ - 15^\circ}{2}\right) = 2\sin 30^\circ \cos 15^\circ = 2 \cdot \frac{1}{2} \cdot \cos 15^\circ$

$= \cos 15^\circ = \sin(90^\circ - 15^\circ) = \sin 75^\circ$ (applying the cofunction identity)

**80.** $\cos 87^\circ + \cos 33^\circ = 2\cos\dfrac{87^\circ + 33^\circ}{2}\cos\dfrac{87^\circ - 33^\circ}{2} = 2\cos 60^\circ \cos 27^\circ$

$= 2 \cdot \frac{1}{2}\cos 27^\circ = \cos 27^\circ = \sin(90^\circ - 27^\circ) = \sin 63^\circ$

**81.** $\dfrac{\sin x + \sin 2x + \sin 3x + \sin 4x + \sin 5x}{\cos x + \cos 2x + \cos 3x + \cos 4x + \cos 5x} = \dfrac{(\sin x + \sin 5x) + (\sin 2x + \sin 4x) + \sin 3x}{(\cos x + \cos 5x) + (\cos 2x + \cos 4x) + \cos 3x}$

$= \dfrac{2\sin 3x \cos 2x + 2\sin 3x \cos x + \sin 3x}{2\cos 3x \cos 2x + 2\cos 3x \cos x + \cos 3x} = \dfrac{\sin 3x\,(2\cos 2x + 2\cos x + 1)}{\cos 3x\,(2\cos 2x + 2\cos x + 1)} = \tan 3x$

**82.** $n = 1$: $\sin\left(2^1 x\right) = 2\sin x\cos x = 2^1 \sin x\cos\left(2^0 x\right)$

$n = 2$: $\sin\left(2^2 x\right) = \sin 4x = 2\sin 2x\cos 2x = 2\left(2\sin x\cos x\right)\cos 2x = 4\sin x\cos x\cos 2x = 2^2\sin x\cos x\cos\left(2^1 x\right)$

$n = 3$: $\sin\left(2^3 x\right) = \sin 8x = 2\sin 4x\cos 4x = 2\left(4\sin x\cos x\cos 2x\right)\cos 4x = 8\sin x\cos x\cos 2x\cos 4x$
$= 2^3\sin x\cos x\cos 2x\cos\left(2^2 x\right)$

In general, for $n > 0$ we have

$$\begin{aligned}\sin\left(2^n x\right) &= \sin 2\left(2^{n-1}x\right) = 2\sin\left(2^{n-1}x\right)\cos\left(2^{n-1}x\right)\\ &= 2\left[2^{n-1}\sin x\cos x\cos 2x\cos 4x\cos 8x\cdots\cos\left(2^{n-2}x\right)\right]\cos\left(2^{n-1}x\right)\\ &= 2^n\sin x\cos x\cos 2x\cos 4x\cos 8x\cdots\cos\left(2^{n-2}x\right)\cos\left(2^{n-1}x\right)\end{aligned}$$

**83. (a)** $f(x) = \dfrac{\sin 3x}{\sin x} - \dfrac{\cos 3x}{\cos x}$

The function appears to have a constant value of 2 wherever it is defined.

**(b)** $$\begin{aligned}f(x) &= \frac{\sin 3x}{\sin x} - \frac{\cos 3x}{\cos x}\\ &= \frac{\sin 3x\cos x - \cos 3x\sin x}{\sin x\cos x} = \frac{\sin\left(3x - x\right)}{\sin x\cos x}\\ &= \frac{\sin 2x}{\sin x\cos x} = \frac{2\sin x\cos x}{\sin x\cos x} = 2\end{aligned}$$

for all $x$ for which the function is defined.

**84. (a)** $f(x) = \cos 2x + 2\sin^2 x$

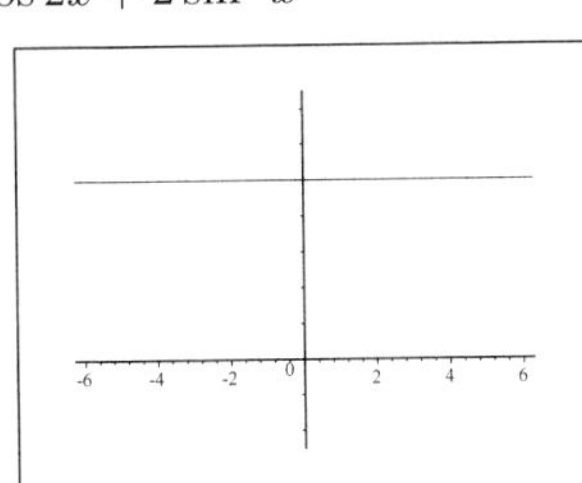

The function appears to have a constant value of 1.

**(b)** $$\begin{aligned}f(x) &= \cos 2x + 2\sin^2 x\\ &= \left(\cos^2 x - \sin^2 x\right) + 2\sin^2 x\\ &= \cos^2 x + \sin^2 x = 1\end{aligned}$$

**85. (a)** $y = \sin 6x + \sin 7x$

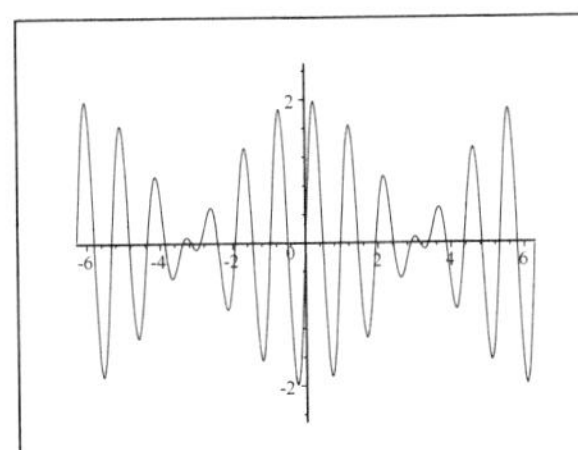

**(b)** By a sum-to-product formula,

$$\begin{aligned}y &= \sin 6x + \sin 7x\\ &= 2\sin\left(\frac{6x + 7x}{2}\right)\cos\left(\frac{6x - 7x}{2}\right)\\ &= 2\sin\left(\tfrac{13}{2}x\right)\cos\left(-\tfrac{1}{2}x\right)\\ &= 2\sin\tfrac{13}{2}x\cos\tfrac{1}{2}x\end{aligned}$$

**(c)** We graph $y = \sin 6x + \sin 7x$, $y = 2\cos\left(\frac{1}{2}x\right)$, and $y = -2\cos\left(\frac{1}{2}x\right)$.

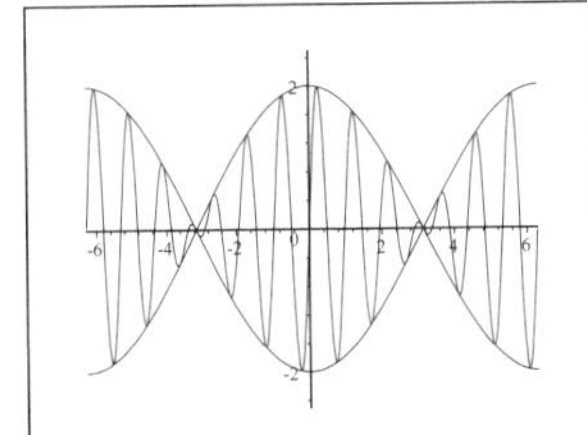

The graph of $y = f(x)$ lies between the other two graphs.

**86.** From Example 2, we have $\cos 3x = 4\cos^3 x - 3\cos x$. If $3x = \frac{\pi}{3}$, then $\cos\frac{\pi}{3} = \frac{1}{2} = 4\cos^3 x - 3\cos x \Leftrightarrow 1 = 8\cos^3 x - 6\cos x \Leftrightarrow 8\cos^3 x - 6\cos x - 1 = 0$. Substituting $y = \cos x$ gives $8y^3 - 6y - 1 = 0$.

**87. (a)** $\cos 4x = \cos(2x+2x) = 2\cos^2 2x - 1 = 2\left(2\cos^2 x - 1\right)^2 - 1 = 8\cos^4 x - 8\cos^2 x + 1$. Thus the desired polynomial is $P(t) = 8t^4 - 8t^2 + 1$.

**(b)** $\cos 5x = \cos(4x+x) = \cos 4x\cos x - \sin 4x\sin x = \cos x\left(8\cos^4 x - 8\cos^2 x + 1\right) - 2\sin 2x\cos 2x\sin x$

$$= 8\cos^5 x - 8\cos^3 x + \cos x - 4\sin x\cos x\left(2\cos^2 x - 1\right)\sin x \quad \text{[from part (a)]}$$
$$= 8\cos^5 x - 8\cos^3 x + \cos x - 4\cos x\left(2\cos^2 x - 1\right)\sin^2 x$$
$$= 8\cos^5 x - 8\cos^3 x + \cos x - 4\cos x\left(2\cos^2 x - 1\right)\left(1-\cos^2 x\right)$$
$$= 8\cos^5 x - 8\cos^3 x + \cos x + 8\cos^5 x - 12\cos^3 x + 4\cos x = 16\cos^5 x - 20\cos^3 x + 5\cos x$$

Thus, the desired polynomial is $P(t) = 16t^5 - 20t^3 + 5t$.

**88.** Let $c_1$ and $c_2$ be the lengths of the segments shown in the figure. By the Law of Sines applied to $\triangle ABC$, we have $\dfrac{c}{\sin 2x} = \dfrac{b}{\sin B}$ or $c = \dfrac{b\sin 2x}{\sin B}$. Also by the Law of Sines applied to $\triangle BCD$ and $\triangle ACD$ we have $\dfrac{s}{\sin B} = \dfrac{c_1}{\sin x}$ and $\dfrac{s}{\sin A} = \dfrac{c_2}{\sin x}$. So $c_1 = \dfrac{s\sin x}{\sin B}$ and $c_2 = \dfrac{s\sin x}{\sin A}$. Since $c = c_1 + c_2$, we have

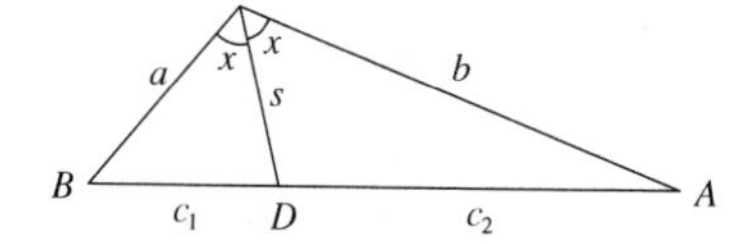


$\dfrac{b\sin 2x}{\sin B} = \dfrac{s\sin x}{\sin B} + \dfrac{s\sin x}{\sin A} = s\sin x\left(\dfrac{1}{\sin B} + \dfrac{1}{\sin A}\right)$. By applying the Law of Sines to $\triangle ABC$, $\dfrac{b}{\sin B} = \dfrac{a}{\sin A}$ $\Leftrightarrow$ $\dfrac{1}{\sin A} = \dfrac{b}{a\sin B}$. Substituting we have $\dfrac{b\sin 2x}{\sin B} = s\sin x\left(\dfrac{1}{\sin B} + \dfrac{b}{a\sin B}\right) = s\dfrac{\sin x}{\sin B}\left(1+\dfrac{b}{a}\right)$ $\Leftrightarrow$ $b\sin 2x = s\sin x\left(1+\dfrac{b}{a}\right)$ $\Leftrightarrow$ $2b\sin x\cos x = \sin x\left(1+\dfrac{b}{a}\right)$ $\Leftrightarrow$ $2b\cos x = s\left(1+\dfrac{b}{a}\right) = s\left(\dfrac{a+b}{a}\right)$ $\Leftrightarrow$ $s = \dfrac{2ab\cos x}{a+b}$.

**89.** Using a product-to-sum formula,
RHS $= 4\sin A\sin B\sin C = 4\sin A\left\{\frac{1}{2}\left[\cos(B-C) - \cos(B+C)\right]\right\} = 2\sin A\cos(B-C) - 2\sin A\cos(B+C)$.
Using another product-to-sum formula, this is equal to
$$2\left\{\tfrac{1}{2}\left[\sin(A+B-C) + \sin(A-B+C)\right]\right\} - 2\left\{\tfrac{1}{2}\left[\sin(A+B+C) + \sin(A-B-C)\right]\right\}$$
$$= \sin(A+B-C) + \sin(A-B+C) - \sin(A+B+C) - \sin(A-B-C)$$
Now $A+B+C = \pi$, so $A+B-C = \pi - 2C$, $A-B+C = \pi - 2B$, and $A-B-C = 2A - \pi$. Thus our expression simplifies to
$$\sin(A+B-C) + \sin(A-B+C) - \sin(A+B+C) - \sin(A-B-C)$$
$$= \sin(\pi-2C) + \sin(\pi-2B) + 0 - \sin(2A-\pi) = \sin 2C + \sin 2B + \sin 2A = \text{LHS}$$

**90. (a)** The length of the base of the inscribed rectangle twice the length of the adjacent side which is $2(5\cos\theta)$ and the length of the opposite side is $5\sin\theta$. Thus the area of the rectangle is modeled by $A(\theta) = 2(5\cos\theta)(5\sin\theta) = 25(2\sin\theta\cos\theta) = 25\sin 2\theta$.

**(b)** The function $y = \sin u$ is maximized when $u = \frac{\pi}{2}$. So $2\theta = \frac{\pi}{2}$ $\Leftrightarrow$ $\theta = \frac{\pi}{4}$. Thus the maximum cross-sectional area is $A\left(\frac{\pi}{4}\right) = 25\sin 2\left(\frac{\pi}{4}\right) = 25\text{ cm}^2$.

**(c)** The length of the base is $2\left(5\cos\frac{\pi}{4}\right) = 5\sqrt{2} \approx 7.07$ cm and the width of the rectangle is $5\sin\frac{\pi}{4} = \frac{5\sqrt{2}}{2} = 3.54$ cm.

**91. (a)** In both logs the length of the adjacent side is $20\cos\theta$ and the length of the opposite side is $20\sin\theta$. Thus the cross-sectional area of the beam is modeled by
$A(\theta) = (20\cos\theta)(20\sin\theta) = 400\sin\theta\cos\theta = 200(2\sin\theta\cos\theta) = 200\sin 2\theta$.

**(b)** The function $y = \sin u$ is maximized when $u = \frac{\pi}{2}$. So $2\theta = \frac{\pi}{2}$ $\Leftrightarrow$ $\theta = \frac{\pi}{4}$. Thus the maximum cross-sectional area is $A\left(\frac{\pi}{4}\right) = 200\sin 2\left(\frac{\pi}{4}\right) = 200$.

**92.** We first label the figure as shown. Because the sheet of paper is folded over, $\angle EAC = \angle CAB = \theta$. Thus $\angle BCA = \angle ACE = 90^\circ - \theta$. It follows that $\angle ECD = 180^\circ - \angle BCA - \angle ACE = 180^\circ - (90^\circ - \theta) - (90^\circ - \theta) = 2\theta$. Also, from the figure we see that $BC = L\sin\theta$ and $CE = L\sin\theta$, so $DC = EC\cos 2\theta = L\sin\theta\cos 2\theta$. Thus $6 = DB = DC + CB = L\sin\theta\cos 2\theta + L\sin\theta = L\sin\theta\,(1 + \cos 2\theta) = L\sin\theta \cdot 2\cos^2\theta$. So $L = \dfrac{6}{2\sin\theta\cos^2\theta} = \dfrac{3}{\sin\theta\cos^2\theta}$.

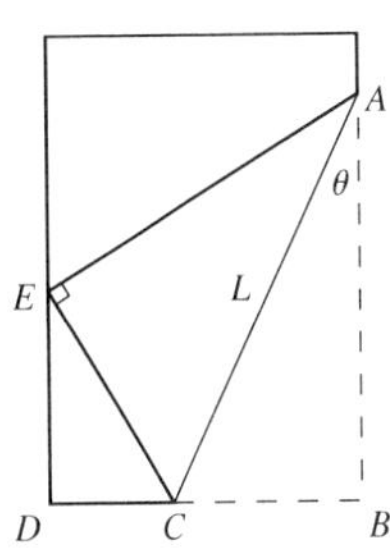


**93. (a)** $y = f_1(t) + f_2(t) = \cos 11t + \cos 13t$

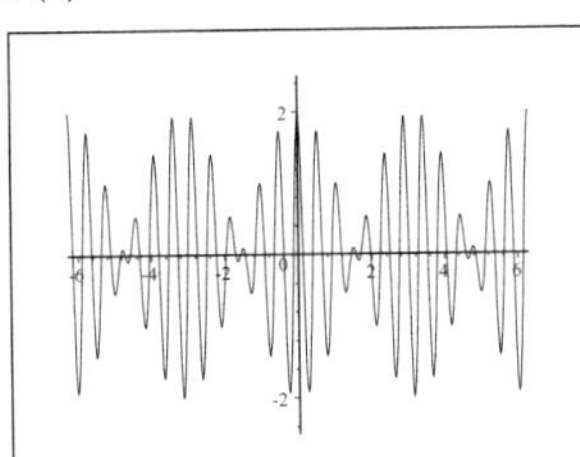

**(b)** Using the identity

$$\cos\alpha + \cos y = 2\cdot\cos\left(\frac{\alpha + y}{2}\right)\cos\left(\frac{\alpha - y}{2}\right), \text{ we have}$$

$$f(t) = \cos 11t + \cos 13t = 2\cdot\cos\left(\frac{11t + 13t}{2}\right)\cos\left(\frac{11t - 13t}{2}\right)$$
$$= 2\cdot\cos 12t\cdot\cos(-t) = 2\cos 12t\cos t$$

**(c)** We graph $y = \cos 11t + \cos 13t$, $y = 2\cos t$, and $y = -2\cos t$.

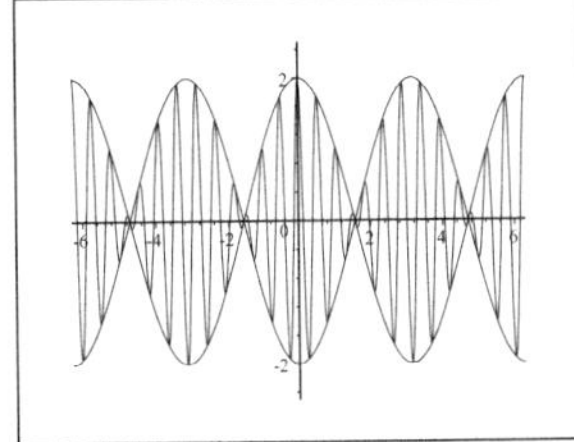

The graph of $f$ lies between the graphs of $y = 2\cos t$ and $y = -2\cos t$. Thus, the loudness of the sound varies between $y = \pm 2\cos t$.

**94. (a)** $f_1 = 770$ Hz and $f_2 = 1209$ Hz, so

$$y = \sin(2\pi\cdot 770t) + \sin(2\pi\cdot 1209t)$$
$$= \sin 1540\pi t + \sin 2418\pi t$$

**(b)** Using a sum-to-product formula, we have

$$y = 2\sin\frac{1540\pi t + 2418\pi t}{2}\cos\frac{1540\pi t - 2418\pi t}{2}$$
$$= 2\sin 1979\pi t\cos 439\pi t$$

**(c)**

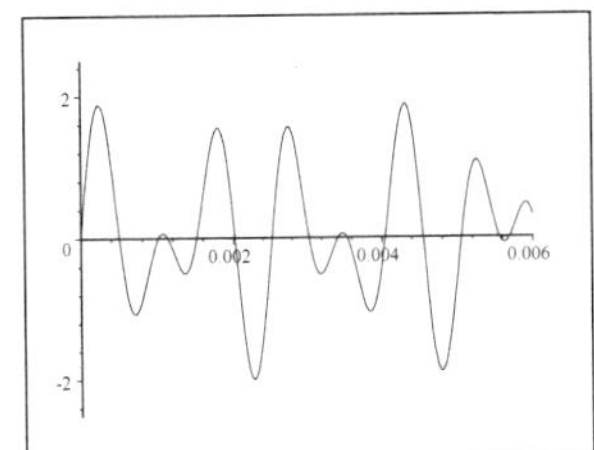

**95.** We find the area of $\triangle ABC$ in two different ways. First, let $AB$ be the base and $CD$ be the height. Since $\angle BOC = 2\theta$ we see that $CD = \sin 2\theta$. So the area is $\frac{1}{2}$ (base) (height) $= \frac{1}{2}\cdot 2\cdot\sin 2\theta = \sin 2\theta$. On the other hand, in $\triangle ABC$ we see that $\angle C$ is a right angle. So $BC = 2\sin\theta$ and $AC = 2\cos\theta$, and the area is $\frac{1}{2}$ (base) (height) $= \frac{1}{2}\cdot(2\sin\theta)(2\cos\theta) = 2\sin\theta\cos\theta$. Equating the two expressions for the area of $\triangle ABC$, we get $\sin 2\theta = 2\sin\theta\cos\theta$.

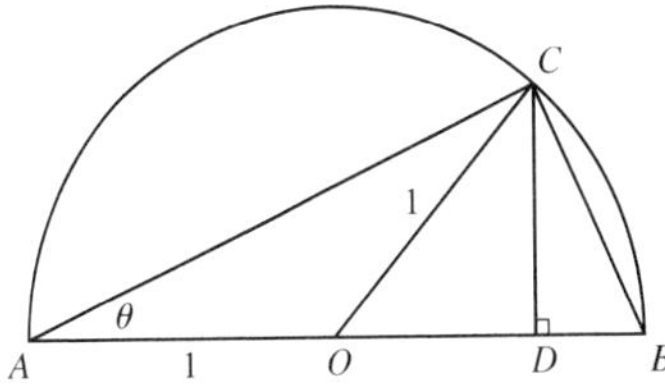

# 8.4 Inverse Trigonometric Functions

**1. (a)** $\sin^{-1}\frac{1}{2}=\frac{\pi}{6}$
**(b)** $\cos^{-1}\frac{1}{2}=\frac{\pi}{3}$
**(c)** $\cos^{-1}2$ is not defined.

**2. (a)** $\sin^{-1}\frac{\sqrt{3}}{2}=\frac{\pi}{3}$
**(b)** $\cos^{-1}\frac{\sqrt{3}}{2}=\frac{\pi}{6}$
**(c)** $\cos^{-1}\left(-\frac{\sqrt{3}}{2}\right)=\frac{5\pi}{6}$

**3. (a)** $\sin^{-1}\frac{\sqrt{2}}{2}=\frac{\pi}{4}$
**(b)** $\cos^{-1}\frac{\sqrt{2}}{2}=\frac{\pi}{4}$
**(c)** $\sin^{-1}\left(-\frac{\sqrt{2}}{2}\right)=-\frac{\pi}{4}$

**4. (a)** $\tan^{-1}\sqrt{3}=\frac{\pi}{3}$
**(b)** $\tan^{-1}\left(-\sqrt{3}\right)=-\frac{\pi}{3}$
**(c)** $\sin^{-1}\sqrt{3}$ is not defined.

**5. (a)** $\sin^{-1}1=\frac{\pi}{2}$
**(b)** $\cos^{-1}1=0$
**(c)** $\cos^{-1}(-1)=\pi$

**6. (a)** $\tan^{-1}1=\frac{\pi}{4}$
**(b)** $\tan^{-1}(-1)=-\frac{\pi}{4}$
**(c)** $\tan^{-1}0=0$

**7. (a)** $\tan^{-1}\frac{\sqrt{3}}{3}=\frac{\pi}{6}$
**(b)** $\tan^{-1}\left(-\frac{\sqrt{3}}{3}\right)=-\frac{\pi}{6}$
**(c)** $\sin^{-1}(-2)$ is not defined.

**8. (a)** $\sin^{-1}0=0$
**(b)** $\cos^{-1}0=\frac{\pi}{2}$
**(c)** $\cos^{-1}\left(-\frac{1}{2}\right)=\frac{2\pi}{3}$

**9. (a)** $\sin^{-1}(0.13844)\approx 0.13889$
**(b)** $\cos^{-1}(-0.92761)\approx 2.75876$

**10. (a)** $\cos^{-1}(0.31187)\approx 1.25364$
**(b)** $\tan^{-1}(26.23110)\approx 1.53269$

**11. (a)** $\tan^{-1}(1.23456)\approx 0.88998$
**(b)** $\sin^{-1}(1.23456)$ is not defined.

**12. (a)** $\cos^{-1}(-0.25713)\approx 1.83085$
**(b)** $\tan^{-1}(-0.25713)\approx -0.25168$

**13.** $\sin\left(\sin^{-1}\frac{1}{4}\right)=\frac{1}{4}$

**14.** $\cos\left(\cos^{-1}\frac{3}{2}\right)=\frac{3}{2}$

**15.** $\tan\left(\tan^{-1}5\right)=5$

**16.** $\sin\left(\sin^{-1}5\right)$ does not exist because 5 is not in the domain of the function $\sin^{-1}$.

**17.** $\cos^{-1}\left(\cos\frac{\pi}{3}\right)=\frac{\pi}{3}$

**18.** $\tan^{-1}\left(\tan\frac{\pi}{6}\right)=\frac{\pi}{6}$

**19.** $\sin^{-1}\left(\sin\left(-\frac{\pi}{6}\right)\right)=-\frac{\pi}{6}$

**20.** Since $\sin\frac{5\pi}{6}=\sin\frac{\pi}{6}$, $\sin^{-1}\left(\sin\frac{5\pi}{6}\right)=\frac{\pi}{6}$

**21.** $\tan^{-1}\left(\tan\frac{2\pi}{3}\right)=\tan^{-1}\left(\tan\left(\frac{2\pi}{3}-\pi\right)\right)$
$=\tan^{-1}\left(\tan\left(-\frac{\pi}{3}\right)\right)=-\frac{\pi}{3}$ since $\frac{2\pi}{3}>\frac{\pi}{2}$.

**22.** Since $\cos\left(-\frac{\pi}{4}\right)=\cos\frac{\pi}{4}$, $\cos^{-1}\left(\cos\left(-\frac{\pi}{4}\right)\right)=\frac{\pi}{4}$.

**23.** $\tan\left(\sin^{-1}\frac{1}{2}\right)=\tan\frac{\pi}{6}=\frac{\sqrt{3}}{3}$

**24.** $\sin\left(\sin^{-1}0\right)=0$

**25.** $\cos\left(\sin^{-1}\frac{\sqrt{3}}{2}\right)=\cos\frac{\pi}{3}=\frac{1}{2}$

**26.** $\tan\left(\sin^{-1}\frac{\sqrt{2}}{2}\right)=\tan\frac{\pi}{4}=1$

**27.** $\tan^{-1}\left(2\sin\frac{\pi}{3}\right)=\tan^{-1}\left(2\cdot\frac{\sqrt{3}}{2}\right)=\tan^{-1}\sqrt{3}=\frac{\pi}{3}$

**28.** $\cos^{-1}\left(\sqrt{3}\sin\frac{\pi}{6}\right)=\cos^{-1}\left(\sqrt{3}\cdot\frac{1}{2}\right)=\frac{\pi}{6}$

**29.** Let $u=\cos^{-1}\frac{3}{5}$, so $\cos u=\frac{3}{5}$. Then from the triangle, $\sin\left(\cos^{-1}\frac{3}{5}\right)=\sin u=\frac{4}{5}$.

**30.** Let $u=\sin^{-1}\frac{4}{5}$, so $\sin u=\frac{4}{5}$. Then from the triangle, $\tan\left(\sin^{-1}\frac{4}{5}\right)=\tan u=\frac{4}{3}$.

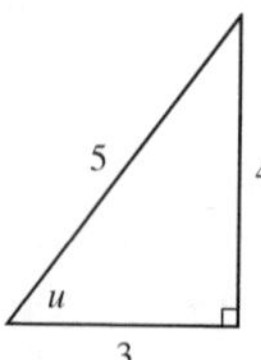

**31.** Let $u = \tan^{-1}\frac{12}{5}$, so $\tan u = \frac{12}{5}$. Then from the triangle, $\sin\left(\tan^{-1}\frac{12}{5}\right) = \sin u = \frac{12}{13}$.

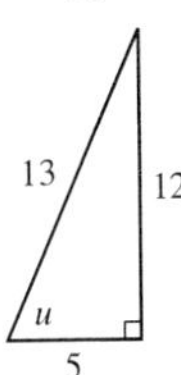


**32.** Let $u = \tan^{-1} 5$, so $\tan u = 5$. Then from the triangle, $\cos\left(\tan^{-1} 5\right) = \cos u = \frac{\sqrt{26}}{26}$.

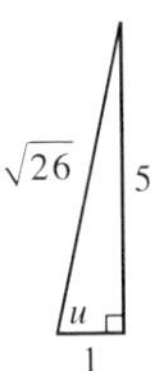


**33.** Let $\theta = \sin^{-1}\frac{12}{13}$, so $\sin\theta = \frac{12}{13}$. Then from the triangle, $\sec\left(\sin^{-1}\frac{12}{13}\right) = \sec\theta = \frac{13}{5}$.

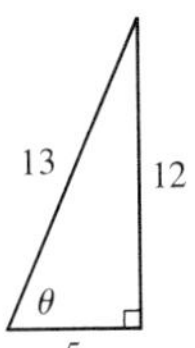


**34.** Let $\alpha = \cos^{-1}\frac{7}{25}$, so $\cos\alpha = \frac{7}{25}$. Then from the triangle, $\csc\left(\cos^{-1}\frac{7}{25}\right) = \csc\alpha = \frac{25}{24}$.

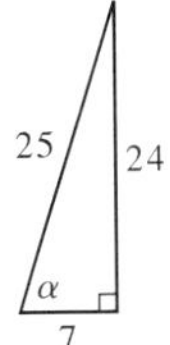


**35.** Let $u = \tan^{-1} 2$, so $\tan u = 2$. Then from the triangle, $\cos\left(\tan^{-1} 2\right) = \cos u = \frac{1}{\sqrt{5}} = \frac{\sqrt{5}}{5}$.

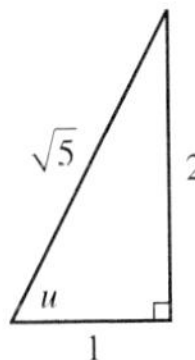


**36.** Let $u = \sin^{-1}\frac{2}{3}$, so $\sin u = \frac{2}{3}$. Then from the triangle, $\cot\left(\sin^{-1}\frac{2}{3}\right) = \cot u = \frac{\sqrt{5}}{2}$.

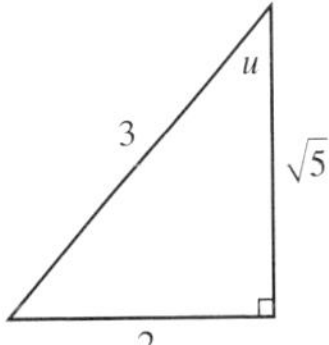


**37.** Let $u = \cos^{-1}\frac{3}{5}$, so $\cos u = \frac{3}{5}$. From the triangle, $\sin u = \frac{4}{5}$, so

$$\sin\left(2\cos^{-1}\frac{3}{5}\right) = \sin(2u) = 2\sin u\cos u = 2\cdot\frac{4}{5}\cdot\frac{3}{5} = \frac{24}{25}$$

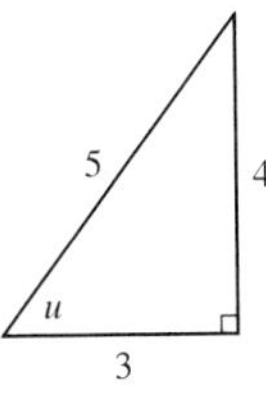


**38.** Let $u = \tan^{-1}\frac{5}{13}$, so $\tan u = \frac{5}{13}$. Then from the triangle,

$$\tan\left(2\tan^{-1}\frac{5}{13}\right) = \tan 2u = \frac{2\tan u}{1-\tan^2 u} = \frac{2\cdot\frac{5}{13}}{1-\frac{25}{169}} = \frac{10}{13}\cdot\frac{169}{144} = \frac{65}{72}$$

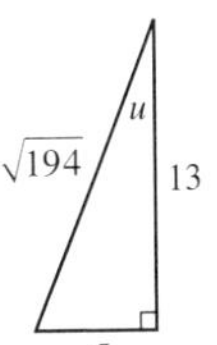

**39.** Let $u = \cos^{-1}\frac{1}{2}$ and $v = \sin^{-1}\frac{1}{2}$, so $\cos u = \frac{1}{2}$ and $\sin v = \frac{1}{2}$. From the triangle,

$$\sin\left(\sin^{-1}\tfrac{1}{2} + \cos^{-1}\tfrac{1}{2}\right) = \sin(v+u)$$
$$= \sin v \cos u + \cos v \sin u = \tfrac{1}{2}\cdot\tfrac{1}{2} + \tfrac{\sqrt{3}}{2}\cdot\tfrac{\sqrt{3}}{2} = 1$$

*Another method:*

$$\sin\left(\sin^{-1}\tfrac{1}{2} + \cos^{-1}\tfrac{1}{2}\right) = \sin\left(\tfrac{\pi}{6} + \tfrac{\pi}{3}\right) = \sin\tfrac{\pi}{2} = 1$$

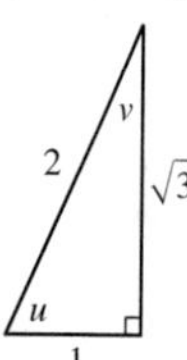


**40.** Let $\theta = \sin^{-1}\frac{3}{5}$ and $\gamma = \cos^{-1}\frac{3}{5}$, so $\sin\theta = \frac{3}{5}$ and $\cos\gamma = \frac{3}{5}$. From the triangle,

$$\cos\left(\sin^{-1}\tfrac{3}{5} - \cos^{-1}\tfrac{3}{5}\right)$$
$$= \cos\left(\sin^{-1}\tfrac{3}{5}\right)\cos\left(\cos^{-1}\tfrac{3}{5}\right) + \sin\left(\sin^{-1}\tfrac{3}{5}\right)\sin\left(\cos^{-1}\tfrac{3}{5}\right)$$
$$= \cos\theta\cos\gamma + \sin\theta\sin\gamma = \tfrac{4}{5}\cdot\tfrac{3}{5} + \tfrac{3}{5}\cdot\tfrac{4}{5} = \tfrac{24}{25}$$

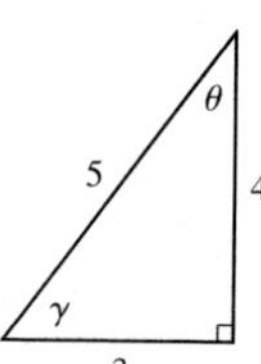


**41.** Let $u = \sin^{-1} x$, so $\sin u = x$. From the triangle, $\cos\left(\sin^{-1} x\right) = \cos u = \sqrt{1-x^2}$.

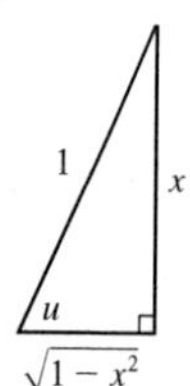


**42.** Let $u = \tan^{-1} x$, so $\tan u = x$. Then from the triangle, $\sin\left(\tan^{-1} x\right) = \sin u = \dfrac{x}{\sqrt{1+x^2}}$

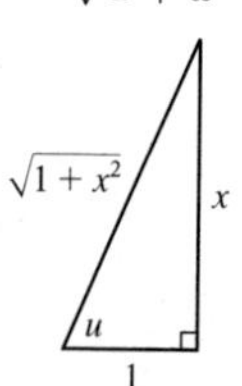


**43.** Let $u = \sin^{-1} x$, so $\sin u = x$. From the triangle, $\tan\left(\sin^{-1} x\right) = \tan u = \dfrac{x}{\sqrt{1-x^2}}$.

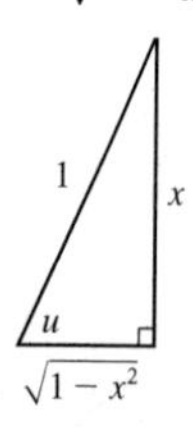


**44.** Let $u = \tan^{-1} x$, so $\tan u = x$. Then from the triangle, $\cos\left(\tan^{-1} x\right) = \cos u = \dfrac{1}{\sqrt{1+x^2}}$.

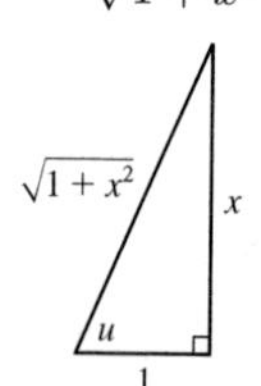


**45.** Let $u = \tan^{-1} x$, so $\tan u = x$. From the triangle,

$$\cos\left(2\tan^{-1} x\right) = \cos 2u = 2\cos^2 u - 1$$
$$= \frac{2}{1+x^2} - 1 = \frac{1-x^2}{1+x^2}$$

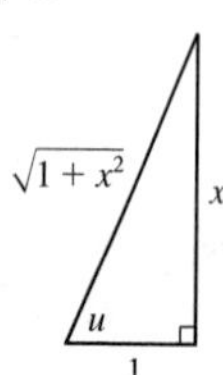


**46.** Let $u = \sin^{-1} x$, so $\sin u = x$. From the triangle, $\sin\left(2\sin^{-1} x\right) = \sin 2u = 2\sin u\cos u = 2x\sqrt{1-x^2}$.

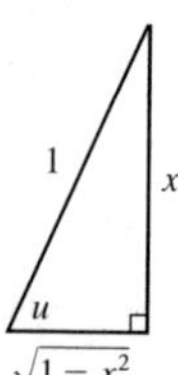

**47.** Let $u = \cos^{-1} x$ and $v = \sin^{-1} x$, so $\cos u = x$ and $\sin v = x$. From the triangle,

$$\cos\left(\cos^{-1} x + \sin^{-1} x\right) = \cos\left(\cos^{-1} x\right)\cos\left(\sin^{-1} x\right) - \sin\left(\cos^{-1} x\right)\sin\left(\sin^{-1} x\right)$$
$$= \cos u \cos v - \sin u \sin v = x\sqrt{1-x^2} - x\sqrt{1-x^2} = 0$$

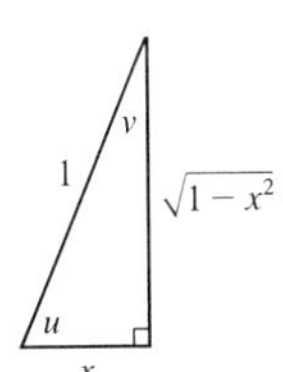


**48.** Let $u = \tan^{-1} x$ and $v = \sin^{-1} x$, so $\tan u = x$ and $\sin v = x$. From the triangles we have

$$\sin\left(\tan^{-1} x - \sin^{-1} x\right) = \sin(u - v) = \sin u \cos v - \cos u \sin v$$
$$= \frac{x}{\sqrt{1+x^2}} \cdot \sqrt{1-x^2} - \frac{1}{\sqrt{1+x^2}} \cdot x$$
$$= x\frac{\sqrt{1-x^2} - 1}{\sqrt{1+x^2}}$$

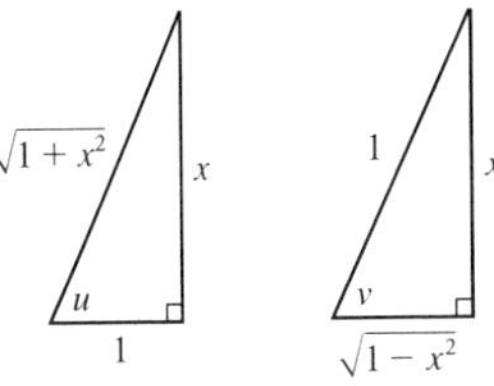


**49. (a)** $y = \sin^{-1} x + \cos^{-1} x$. Note the domain of both $\sin^{-1} x$ and $\cos^{-1} x$ is $[-1, 1]$.

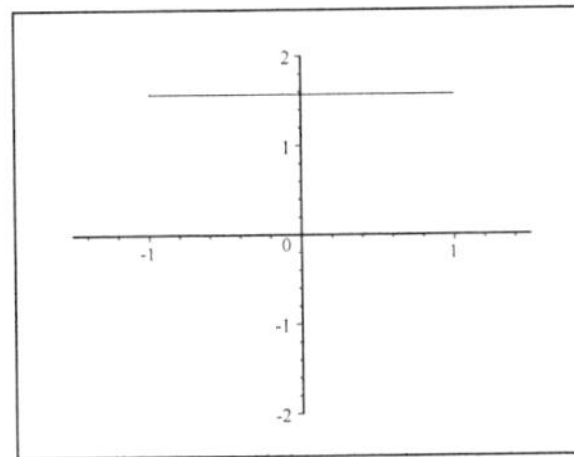


*Conjecture:* $y = \frac{\pi}{2}$ for $-1 \le x \le 1$.

**(b)** To prove this conjecture, let $\sin^{-1} x = u \quad\Leftrightarrow\quad \sin u = x$, $-\frac{\pi}{2} \le u \le \frac{\pi}{2}$. Using the cofunction identity $\sin u = \cos\left(\frac{\pi}{2} - u\right) = x$, $\cos^{-1} x = \frac{\pi}{2} - u$. Therefore, $\sin^{-1} x + \cos^{-1} x = u + \left(\frac{\pi}{2} - u\right) = \frac{\pi}{2}$.

**50. (a)** $y = \tan^{-1} x + \tan^{-1}\left(\dfrac{1}{x}\right)$

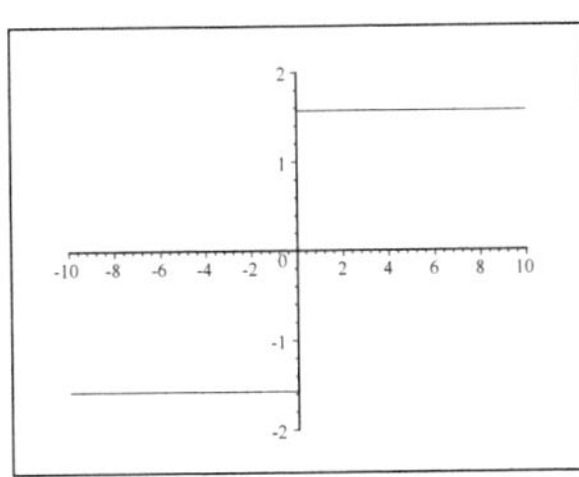


*Conjecture:* $y = \begin{cases} \frac{\pi}{2} & \text{if } x > 0 \\ -\frac{\pi}{2} & \text{if } x < 0 \end{cases}$

**(b)** To prove this conjecture, let $\tan^{-1} x = u \quad\Leftrightarrow\quad \tan u = x$, for $-\frac{\pi}{2} < u < \frac{\pi}{2}$. Then, $\dfrac{1}{x} = \dfrac{1}{\tan u} = \cot u = \tan\left(\frac{\pi}{2} - u\right)$ by the cofunction identity. So

$$\tan^{-1} x + \tan^{-1}\left(\frac{1}{x}\right) = u + \tan^{-1}\left(\tan\left(\tfrac{\pi}{2} - u\right)\right).$$

If $x > 0$, then $\frac{\pi}{2} > u > 0$, so $0 < \frac{\pi}{2} - u < \frac{\pi}{2} \Rightarrow \frac{\pi}{2} - u$ is in quadrant I. Therefore, $\tan^{-1}\left(\tan\left(\frac{\pi}{2} - u\right)\right) = \frac{\pi}{2} - u$, so

$$\tan^{-1} x + \tan^{-1}\left(\frac{1}{x}\right) = u + \left(\tfrac{\pi}{2} - u\right) = \tfrac{\pi}{2}.$$

If $x < 0$, then $-\frac{\pi}{2} < u < 0$, so $\pi > \frac{\pi}{2} - u > \frac{\pi}{2} \Rightarrow \frac{\pi}{2} - u$ is in quadrant II. Thus, $\tan^{-1}\left(\tan\left(\frac{\pi}{2} - u\right)\right) = \left(\frac{\pi}{2} - u\right) - \pi = -\frac{\pi}{2} - u$, and hence $\tan^{-1} x + \tan^{-1}\left(\dfrac{1}{x}\right) = u + \left(-\frac{\pi}{2} - u\right) = -\frac{\pi}{2}$.

**51. (a)** $\tan^{-1} x + \tan^{-1} 2x = \frac{\pi}{4}$

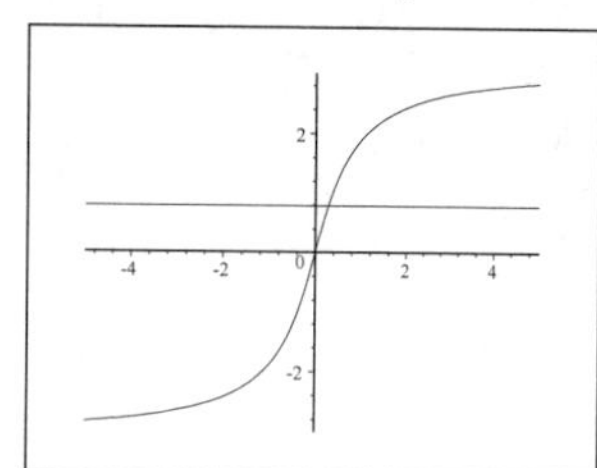


From the graph, the solution is $x \approx 0.28$.

**(b)** To solve the equation exactly, we set $\tan^{-1} x + \tan^{-1} 2x = \frac{\pi}{4} \quad \Leftrightarrow$ $\tan^{-1} x = \frac{\pi}{4} - \tan^{-1} 2x \Rightarrow \tan\left(\tan^{-1} x\right) = \tan\left(\frac{\pi}{4} - \tan^{-1} 2x\right)$

$$\Rightarrow x = \frac{\tan\frac{\pi}{4} - \tan\left(\tan^{-1} 2x\right)}{1 + \tan\frac{\pi}{4} \cdot \tan\left(\tan^{-1} 2x\right)} = \frac{1 - 2x}{1 + (1)\,2x} \Rightarrow$$

$x\,(1 + 2x) = 1 - 2x \quad \Leftrightarrow \quad x + 2x^2 = 1 - 2x$. Thus, $2x^2 + 3x - 1 = 0$. We use the quadratic formula to solve for $x$:

$x = \frac{-3 \pm \sqrt{3^2 - 4(2)(-1)}}{2(2)} = \frac{-3 \pm \sqrt{17}}{4}$. Substituting into the original equation, we see that $x = \frac{-3 - \sqrt{17}}{4}$ is not a solution, and so $x = \frac{-3 + \sqrt{17}}{4} \approx 0.28$ is the only root.

**52. (a)** $y = \sin^{-1} x - \cos^{-1} x$

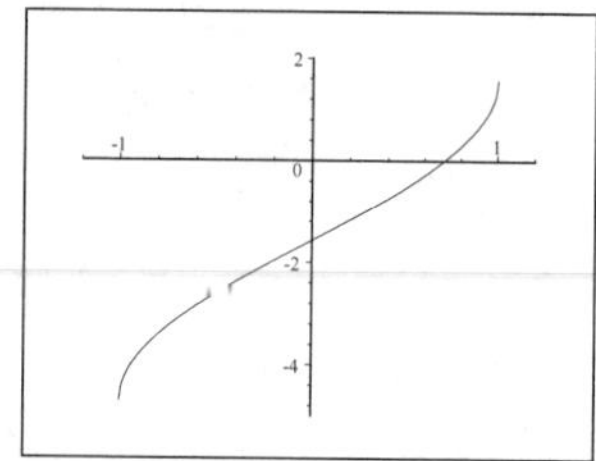


From the graph, we see that $y = 0$ when $x \approx 0.71$.

**(b)** To solve the equation exactly, we set $\tan^{-1} x + \tan^{-1} 2x = \frac{\pi}{4} \quad \Leftrightarrow$ $\tan^{-1} x = \frac{\pi}{4} - \tan^{-1} 2x \Rightarrow \tan\left(\tan^{-1} x\right) = \tan\left(\frac{\pi}{4} - \tan^{-1} 2x\right)$

$$\Rightarrow x = \frac{\tan\frac{\pi}{4} - \tan\left(\tan^{-1} 2x\right)}{1 + \tan\frac{\pi}{4} \cdot \tan\left(\tan^{-1} 2x\right)} = \frac{1 - 2x}{1 + (1)\,2x} \Rightarrow$$

$x\,(1 + 2x) = 1 - 2x \quad \Leftrightarrow \quad x + 2x^2 = 1 - 2x$. Thus, $2x^2 + 3x - 1 = 0$. We use the quadratic formula to solve for $x$:

$x = \frac{-3 \pm \sqrt{3^2 - 4(2)(-1)}}{2(2)} = \frac{-3 \pm \sqrt{17}}{4}$. Substituting into the original equation, we see that $x = \frac{-3 - \sqrt{17}}{4}$ is not a solution, and so $x = \frac{-3 + \sqrt{17}}{4} \approx 0.28$ is the only root.To find the exact solution, we set $\sin^{-1} x - \cos^{-1} x = 0 \Leftrightarrow \sin^{-1} x = \cos^{-1} x \Rightarrow$ $\sin\left(\sin^{-1} x\right) = \sin\left(\cos^{-1} x\right) \Rightarrow x = \sqrt{1 - x^2} \Rightarrow x^2 = 1 - x^2$ $\Leftrightarrow \quad 2x^2 = 1 \Rightarrow x = \pm\frac{1}{\sqrt{2}}$. But $x = \sqrt{1 - x^2} \geq 0$, so $x = \frac{1}{\sqrt{2}}$.

**53. (a)** Solving $\tan\theta = h/2$ for $h$, we have $h = 2\tan\theta$.

**(b)** Solving $\tan\theta = h/2$ for $\theta$ we have $\theta = \tan^{-1}(h/2)$.

**54. (a)** Solving $\tan\theta = 50/s$. Solving for $\theta$, we have $\theta = \tan^{-1}(50/s)$.

**(b)** Set $s = 20$ ft to get $\tan\theta = \frac{50}{20} = \frac{5}{2}$. Solving for $\theta$, we have $\theta = \tan^{-1}\frac{5}{2} \approx 68.2^\circ$.

**55. (a)** Solving $\sin\theta = h/680$ for $\theta$ we have $\theta = \sin^{-1}(h/680)$.

**(b)** Set $h = 500$ to get $\theta = \sin^{-1}\left(\frac{500}{680}\right) \approx 0.826$ rad.

**56. (a)** Since the radius of the earth is 3960 miles, we have the relationship $\cos\theta = \frac{3960}{h + 3960}$. Solving for $\theta$, we get $\theta = \cos^{-1}\left(\frac{3960}{h + 3960}\right)$.

**(b)** The arc length is $s = (\text{radius})\,(\text{included angle}) = (3960)\,(2\theta) = 7920\theta$.

**(c)** $s = 7920\cos^{-1}\left(\frac{3960}{h + 3960}\right)$.

**(d)** When $h = 100$ we have $s = 7920\cos^{-1}\left(\frac{3960}{100 + 3960}\right) = 7920\cos^{-1}\left(\frac{3960}{4060}\right) \approx 1761.5$ miles.

**(e)** When $s = 2450$ we have $2450 = 7920\cos^{-1}\left(\frac{3960}{h + 3960}\right) \quad \Leftrightarrow \quad \frac{2450}{7920} = \cos^{-1}\left(\frac{3960}{h + 3960}\right) \quad \Leftrightarrow$ $\frac{3960}{h + 3960} = \cos\left(\frac{2450}{7920}\right) \quad \Leftrightarrow \quad h + 3960 = 3960\sec\left(\frac{2450}{7920}\right) \quad \Leftrightarrow \quad h = 3960\sec\left(\frac{2450}{7920}\right) - 3960 \approx 197.3$ mi.

**57. (a)** $\theta = \sin^{-1}\left(\dfrac{1}{(2\cdot 3+1)\tan 10^\circ}\right) = \sin^{-1}\left(\dfrac{1}{7\tan 10^\circ}\right) \approx \sin^{-1} 0.8102 \approx 54.1^\circ$

**(b)** For $n = 2$, $\theta = \sin^{-1}\left(\dfrac{1}{5\tan 15^\circ}\right) \approx 48.3^\circ$. For $n = 3$, $\theta = \sin^{-1}\left(\dfrac{1}{7\tan 15^\circ}\right) \approx 32.2^\circ$. For $n = 4$, $\theta = \sin^{-1}\left(\dfrac{1}{9\tan 15^\circ}\right) \approx 24.5^\circ$. $n = 0$ and $n = 1$ are outside of the domain for $\beta = 15^\circ$, because $\dfrac{1}{\tan 15^\circ} \approx 3.732$ and $\dfrac{1}{3\tan 15^\circ} \approx 1.244$, neither of which is in the domain of $\sin^{-1}$.

**58.** $f(x) = \sin\left(\sin^{-1} x\right) = x$ has domain $[-1,1]$ and range $[-1,1]$.

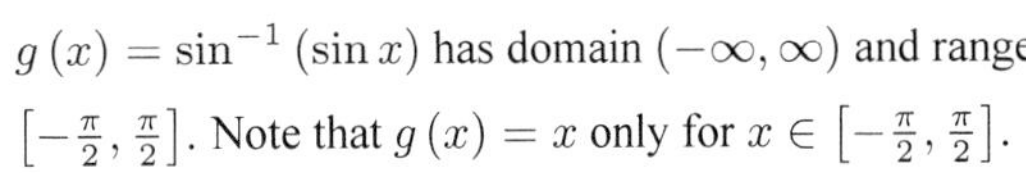
$g(x) = \sin^{-1}(\sin x)$ has domain $(-\infty,\infty)$ and range $\left[-\frac{\pi}{2},\frac{\pi}{2}\right]$. Note that $g(x) = x$ only for $x \in \left[-\frac{\pi}{2},\frac{\pi}{2}\right]$.

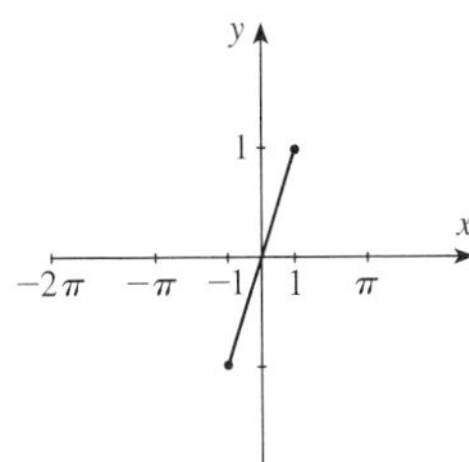


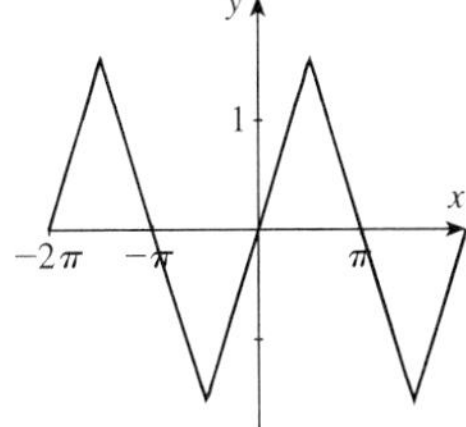


**59. (a)** Let $\theta = \sec^{-1} x$. Then $\sec\theta = x$, as shown in the figure. Then $\cos\theta = \dfrac{1}{x}$, so $\theta = \cos^{-1}\left(\dfrac{1}{x}\right)$. Thus $\sec^{-1} x = \cos^{-1}\left(\dfrac{1}{x}\right)$, $x \geq 1$.

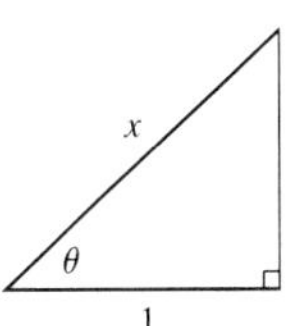


**(b)** Let $\gamma = \csc^{-1} x$. Then $\csc\gamma = x$, as shown in the figure. Then $\sin\gamma = \dfrac{1}{x}$, so $\gamma = \sin^{-1}\left(\dfrac{1}{x}\right)$. Thus $\csc^{-1} x = \sin^{-1}\left(\dfrac{1}{x}\right)$, $x \geq 1$.

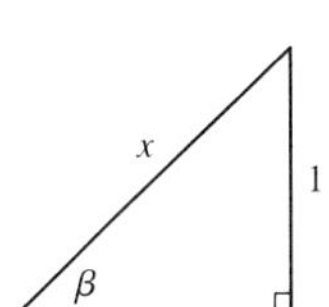


**(c)** Let $\beta = \cot^{-1} x$. Then $\cot\beta = x$, as shown in the figure. Then $\tan\beta = \dfrac{1}{x}$, so $\beta = \tan^{-1}\left(\dfrac{1}{x}\right)$. Thus $\cot^{-1} x = \tan^{-1}\left(\dfrac{1}{x}\right)$, $x \geq 1$.

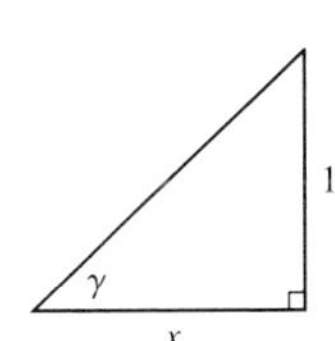


# 8.5 Trigonometric Equations

**1.** $\cos x + 1 = 0 \quad\Leftrightarrow\quad \cos x = -1$. In the interval $[0, 2\pi)$ the only solution is $x = \pi$.Thus the solutions are $x = (2k+1)\pi$ for any integer $k$.

**2.** $\sin x + 1 = 0 \quad\Leftrightarrow\quad \sin x = -1$. In the interval $[0, 2\pi)$ the only solution is $x = \frac{3\pi}{2}$. Therefore, the solutions are $x = \frac{3\pi}{2} + 2k\pi$ for any integer $k$.

**3.** $2\sin x - 1 = 0 \quad\Leftrightarrow\quad 2\sin x = 1 \quad\Leftrightarrow\quad \sin x = \frac{1}{2}$. In the interval $[0, 2\pi)$ the solutions are $x = \frac{\pi}{6}, \frac{5\pi}{6}$ . Therefore, the solutions are $x = \frac{\pi}{6} + 2k\pi$, $\frac{5\pi}{6} + 2k\pi$, $k = 0, \pm 1, \pm 2, \ldots$.

**4.** $\sqrt{2}\cos x - 1 = 0 \quad\Leftrightarrow\quad \sqrt{2}\cos x = 1 \quad\Leftrightarrow\quad \cos x = \frac{1}{\sqrt{2}}$. The solutions in the interval $[0, 2\pi)$ are $x = \frac{\pi}{4}, \frac{7\pi}{4}$. Thus the solutions are $x = \frac{\pi}{4} + 2k\pi$, $\frac{7\pi}{4} + 2k\pi$ for any integer $k$.

**5.** $\sqrt{3}\tan x + 1 = 0 \iff \sqrt{3}\tan x = -1 \iff \tan x = -\frac{\sqrt{3}}{3}$. In the interval $[0, \pi)$ the only solution is $x = \frac{5\pi}{6}$. Therefore, the solutions are $x = \frac{5\pi}{6} + k\pi$, $k = 0, \pm1, \pm2, \ldots$.

**6.** $\cot x + 1 = 0 \iff \cot x = -1$. The solution in the interval $(0, \pi)$ is $x = \frac{3\pi}{4}$. Thus, the solutions are $x = \frac{3\pi}{4} + k\pi$ for any integer $k$.

**7.** $4\cos^2 x - 1 = 0 \iff 4\cos^2 x = 1 \iff 4\cos^2 x = 1 \iff \cos^2 x = \frac{1}{4} \iff \cos x = \pm\frac{1}{2}$. In the interval $[0, 2\pi)$ the solutions are $x = \frac{\pi}{3}, \frac{2\pi}{3}, \frac{4\pi}{3}, \frac{5\pi}{3}$. So the solutions are $x = \frac{\pi}{3} + 2k\pi, \frac{2\pi}{3} + 2k\pi, \frac{4\pi}{3} + 2k\pi, \frac{5\pi}{3} + 2k\pi$, which can be expressed more simply as $x = \frac{\pi}{3} + k\pi, \frac{2\pi}{3} + k\pi$ for any integer $k$.

**8.** $2\cos^2 x - 1 = 0 \iff \cos^2 x = \frac{1}{2} \iff \cos x = \pm\frac{1}{\sqrt{2}} \iff x = \frac{\pi}{4}, \frac{3\pi}{4}, \frac{5\pi}{4}, \frac{7\pi}{4}$ in $[0, 2\pi)$. Thus, the solutions are $x = \frac{\pi}{4} + k\pi, \frac{3\pi}{4} + k\pi$ for any integer $k$.

**9.** $\sec^2 x - 2 = 0 \iff \sec^2 x = 2 \iff \sec x = \pm\sqrt{2}$. In the interval $[0, 2\pi)$ the solutions are $x = \frac{\pi}{4}, \frac{3\pi}{4}, \frac{5\pi}{4}, \frac{7\pi}{4}$. Thus, the solutions are $x = (2k+1)\frac{\pi}{4}$ for any integer $k$.

**10.** $\csc^2 x - 4 = 0 \iff \csc^2 x = 4 \iff \csc x = \pm2$. In the interval $[0, 2\pi)$ the solutions are $x = \frac{\pi}{6}, \frac{5\pi}{6}, \frac{7\pi}{6}, \frac{11\pi}{6}$. So the solutions are $x = \frac{\pi}{6} + k\pi, \frac{5\pi}{6} + k\pi$ for any integer $k$.

**11.** $3\csc^2 x - 4 = 0 \iff \csc^2 x = \frac{4}{3} \iff \csc x = \pm\frac{2}{\sqrt{3}} \iff \sin x = \pm\frac{\sqrt{3}}{2} \iff x = \frac{\pi}{3}, \frac{2\pi}{3}, \frac{4\pi}{3}, \frac{5\pi}{3}$ for $x$ in $[0, 2\pi)$. Thus, the solutions are $x = \frac{\pi}{3} + k\pi, \frac{2\pi}{3} + k\pi$ for any integer $k$.

**12.** $1 - \tan^2 x = 0 \iff \tan^2 x = 1 \iff \tan x = \pm1 \iff x = \frac{\pi}{4}, \frac{3\pi}{4}$ in $(0, \pi)$. Thus, the solutions are $\frac{\pi}{4} + k\pi$, $\frac{3\pi}{4} + k\pi$ for any integer $k$.

**13.** $\cos x\,(2\sin x + 1) = 0 \iff \cos x = 0$ or $2\sin x + 1 = 0 \iff \sin x = -\frac{1}{2}$. On $[0, 2\pi)$, $\cos x = 0 \iff x = \frac{\pi}{2}, \frac{3\pi}{2}$ and $\sin x = -\frac{1}{2} \iff x = \frac{7\pi}{6}, \frac{11\pi}{6}$. Thus the solutions are $x = \frac{\pi}{2} + k\pi$, $x = \frac{7\pi}{6} + 2k\pi$, $x = \frac{11\pi}{6} + 2k\pi$, $k = 0, \pm1, \pm2, \ldots$.

**14.** $\sec x\left(2\cos x - \sqrt{2}\right) = 0 \iff \sec x = 0$ or $2\cos x - \sqrt{2} = 0$. Since $|\sec x| \geq 1$, $\sec x = 0$ has no solution. Thus $2\cos x - \sqrt{2} = 0 \iff 2\cos x = \sqrt{2} \iff \cos x = \frac{\sqrt{2}}{2} \iff x = \frac{\pi}{4}, \frac{7\pi}{4}$ in $[0, 2\pi)$. Thus $x = \frac{\pi}{4} + 2k\pi, \frac{7\pi}{4} + 2k\pi$ for any integer $k$.

**15.** $\left(\tan x + \sqrt{3}\right)(\cos x + 2) = 0 \iff \tan x + \sqrt{3} = 0$ or $\cos x + 2 = 0$. Since $|\cos x| \leq 1$ for all $x$, there is no solution for $\cos x + 2 = 0$. Hence, $\tan x + \sqrt{3} = 0 \iff \tan x = -\sqrt{3} \iff x = -\frac{\pi}{3}$ on $\left(-\frac{\pi}{2}, \frac{\pi}{2}\right)$. Thus the solutions are $x = -\frac{\pi}{3} + k\pi$, $k = 0, \pm1, \pm2, \ldots$.

**16.** $\left(2\cos x + \sqrt{3}\right)(2\sin x - 1) = 0 \iff 2\cos x + \sqrt{3} = 0$ or $2\sin x - 1 = 0 \iff \cos x = -\frac{\sqrt{3}}{2}$ or $\sin x = \frac{1}{2} \iff x = \frac{\pi}{6}, \frac{5\pi}{6}, \frac{7\pi}{6}$ in $[0, 2\pi)$. Thus, $x = \frac{\pi}{6} + k\pi, \frac{5\pi}{6} + 2k\pi$ for any integer $k$.

**17.** $\cos x\sin x - 2\cos x = 0 \iff \cos x\,(\sin x - 2) = 0 \iff \cos x = 0$ or $\sin x - 2 = 0$. Since $|\sin x| \leq 1$ for all $x$, there is no solution for $\sin x - 2 = 0$. Hence, $\cos x = 0 \iff x = \frac{\pi}{2} + 2k\pi, \frac{3\pi}{2} + 2k\pi \iff x = \frac{\pi}{2} + k\pi$, $k = 0, \pm1, \pm2, \ldots$.

**18.** $\tan x\sin x + \sin x = 0 \iff \sin x\,(\tan x + 1) = 0 \iff \sin x = 0$ or $\tan x + 1 = 0$. Now $\sin x = 0$ when $x = k\pi$ and $\tan x + 1 = 0 \iff \tan x = -1 \Rightarrow x = \frac{3\pi}{4} + k\pi$. Thus, the solutions are $x = k\pi, \frac{3\pi}{4} + k\pi$ for any integer $k$.

**19.** $4\cos^2 x - 4\cos x + 1 = 0 \iff (2\cos x - 1)^2 = 0 \iff 2\cos x - 1 = 0 \iff \cos x = \frac{1}{2} \iff x = \frac{\pi}{3} + 2k\pi$, $\frac{5\pi}{3} + 2k\pi$, $k = 0, \pm1, \pm2, \ldots$.

**20.** $2\sin^2 x - \sin x - 1 = 0 \iff (2\sin x + 1)(\sin x - 1) = 0 \iff 2\sin x + 1 = 0$ or $\sin x - 1 = 0$. Since $2\sin x + 1 = 0 \iff 2\sin x = -\frac{1}{2} \iff x = \frac{7\pi}{6}, \frac{11\pi}{6}$ in $[0, 2\pi)$ and $\sin x = 1 \iff x = \frac{\pi}{2}$ in $[0, 2\pi)$. Thus the solutions are $x = \frac{7\pi}{6} + 2k\pi, \frac{11\pi}{6} + 2k\pi, \frac{\pi}{2} + 2k\pi$ for any integer $k$.

**21.** $\sin^2 x = 2\sin x + 3 \iff \sin^2 x - 2\sin x - 3 = 0 \iff (\sin x - 3)(\sin x + 1) = 0 \iff \sin x - 3 = 0$ or $\sin x + 1 = 0$. Since $|\sin x| \leq 1$ for all $x$, there is no solution for $\sin x - 3 = 0$. Hence $\sin x + 1 = 0 \iff \sin x = -1 \iff x = \frac{3\pi}{2} + 2k\pi$, $k = 0, \pm1, \pm2, \ldots$.

**22.** $3\tan^3 x = \tan x \quad\Leftrightarrow\quad 3\tan^3 x - \tan x = 0 \quad\Leftrightarrow\quad \tan x\left(3\tan^2 x - 1\right) = 0 \quad\Leftrightarrow\quad \tan x = 0$ or $3\tan^2 x - 1 = 0$. Now $\tan x = 0 \quad\Rightarrow\quad x = k\pi$ and $3\tan^2 x - 1 = 0 \quad\Leftrightarrow\quad \tan^2 x = \frac{1}{3} \quad\Rightarrow\quad \tan x = \pm\frac{1}{\sqrt{3}} \quad\Rightarrow\quad x = \frac{\pi}{6} + k\pi$, $\frac{5\pi}{6} + k\pi$. Thus the solutions are $x = k\pi$, $\frac{\pi}{6} + k\pi$, $\frac{5\pi}{6} + k\pi$ for any integer $k$.

**23.** $\sin^2 x = 4 - 2\cos^2 x \quad\Leftrightarrow\quad \sin^2 x + \cos^2 x + \cos^2 x = 4 \quad\Leftrightarrow\quad 1 + \cos^2 x = 4 \quad\Leftrightarrow\quad \cos^2 x = 3$ Since $|\cos x| \le 1$ for all $x$, it follows that $\cos^2 x \le 1$ and so there is no solution for $\cos^2 x = 3$.

**24.** $2\cos^2 x + \sin x = 1 \quad\Leftrightarrow\quad 2\left(1 - \sin^2 x\right) + \sin x - 1 = 0 \quad\Leftrightarrow\quad -2\sin^2 x + \sin x + 1 = 0 \quad\Leftrightarrow$ $2\sin^2 x - \sin x - 1 = 0$ which we solved in Exercise 20. Thus $x = \frac{7\pi}{6} + 2k\pi$, $\frac{11\pi}{6} + 2k\pi$, $\frac{\pi}{2} + 2k\pi$ for any integer $k$.

**25.** $2\sin 3x + 1 = 0 \quad\Leftrightarrow\quad 2\sin 3x = -1 \quad\Leftrightarrow\quad \sin 3x = -\frac{1}{2}$. In the interval $[0, 6\pi)$ the solution are $3x = \frac{7\pi}{6}, \frac{11\pi}{6}, \frac{19\pi}{6}, \frac{23\pi}{6}, \frac{31\pi}{6}, \frac{35\pi}{6} \quad\Leftrightarrow\quad x = \frac{7\pi}{18}, \frac{11\pi}{18}, \frac{19\pi}{18}, \frac{23\pi}{18}, \frac{25\pi}{18}, \frac{29\pi}{18}$. So $x = \frac{7\pi}{18} + 2k\frac{\pi}{3}$, $\frac{11\pi}{18} + 2k\frac{\pi}{3}$ for any integer $k$.

**26.** $2\cos 2x + 1 = 0 \quad\Leftrightarrow\quad \cos 2x = -\frac{1}{2} \quad\Leftrightarrow\quad 2x = \frac{2\pi}{3} + 2k\pi, \frac{4\pi}{3} + 2k\pi \quad\Leftrightarrow\quad x = \frac{\pi}{3} + k\pi$, $\frac{2\pi}{3} + k\pi$ for any integer $k$.

**27.** $\sec 4x - 2 = 0 \quad\Leftrightarrow\quad \sec 4x = 2 \quad\Leftrightarrow\quad 4x = \frac{\pi}{3} + 2k\pi, \frac{5\pi}{3} + 2k\pi \quad\Leftrightarrow\quad x = \frac{\pi}{12} + \frac{1}{2}k\pi$, $-\frac{\pi}{12} + \frac{1}{2}k\pi$ for any integer $k$.

**28.** $\sqrt{3}\tan 3x + 1 = 0 \quad\Leftrightarrow\quad \tan 3x = -\frac{1}{\sqrt{3}} \quad\Leftrightarrow\quad 3x = \frac{5\pi}{6} + k\pi \quad\Leftrightarrow\quad x = \frac{5\pi}{18} + \frac{1}{3}k\pi$ for any integer $k$.

**29.** $\sqrt{3}\sin 2x = \cos 2x \quad\Leftrightarrow\quad \tan 2x = \frac{1}{\sqrt{3}}$ (if $\cos 2x \neq 0$) $\quad\Leftrightarrow\quad 2x = \frac{\pi}{6} + k\pi \quad\Leftrightarrow\quad x = \frac{\pi}{12} + \frac{1}{2}k\pi$ for any integer $k$.

**30.** $\csc 3x = \sin 3x \quad\Leftrightarrow\quad \dfrac{1}{\sin 3x} = \sin 3x \quad\Leftrightarrow\quad \sin^2 3x = 1 \quad\Leftrightarrow\quad \sin 3x = \pm 1 \quad\Leftrightarrow\quad 3x = \frac{\pi}{2} + k\pi \quad\Leftrightarrow$ $x = \frac{\pi}{6} + k\frac{\pi}{3}$ for any integer $k$. Notice that multiplying by $\sin 3x$ in the first step did not introduce extraneous roots.

**31.** $\cos\frac{x}{2} - 1 = 0 \quad\Leftrightarrow\quad \cos\frac{x}{2} = 1 \quad\Leftrightarrow\quad \frac{x}{2} = 2k\pi \quad\Leftrightarrow\quad x = 4k\pi$ for any integer $k$.

**32.** $2\sin\frac{x}{3} + \sqrt{3} = 0 \quad\Leftrightarrow\quad 2\sin\frac{x}{3} = -\sqrt{3} \quad\Leftrightarrow\quad \sin\frac{x}{3} = -\frac{\sqrt{3}}{2} \quad\Leftrightarrow\quad \frac{x}{3} = \frac{4\pi}{3} + 2k\pi, \frac{5\pi}{3} + 2k\pi \quad\Leftrightarrow\quad x = 4\pi + 6k\pi$, $5\pi + 6k\pi$ for any integer $k$.

**33.** $\tan\frac{x}{4} + \sqrt{3} = 0 \quad\Leftrightarrow\quad \tan\frac{x}{4} = -\sqrt{3} \quad\Leftrightarrow\quad \frac{x}{4} = \frac{2\pi}{3} + k\pi \quad\Leftrightarrow\quad x = \frac{8\pi}{3} + 4k\pi$ for any integer $k$.

**34.** $\sec\frac{x}{2} = \cos\frac{x}{2} \quad\Leftrightarrow\quad \cos^2\frac{x}{2} = 1 \quad\Rightarrow\quad \cos\frac{x}{2} = \pm 1 \quad\Leftrightarrow\quad \frac{x}{2} = k\pi \quad\Leftrightarrow\quad x = 2k\pi$ for any integer $k$.

**35.** $\tan^5 x - 9\tan x = 0 \quad\Leftrightarrow\quad \tan x\left(\tan^4 x - 9\right) = 0 \quad\Leftrightarrow\quad \tan x = 0$ or $\tan^4 x = 9 \quad\Leftrightarrow\quad \tan x = 0$ or $\tan x = \pm\sqrt{3} \quad\Leftrightarrow\quad x = 0, \pi, \frac{\pi}{3}, \frac{2\pi}{3}, \frac{4\pi}{3}, \frac{5\pi}{3}$ in $[0, 2\pi)$. Thus, $x = \frac{k\pi}{3}$ for any integer $k$.

**36.** $3\tan^3 x - 3\tan^2 x - \tan x + 1 = 0 \quad\Leftrightarrow\quad (\tan x - 1)\left(3\tan^2 x - 1\right) = 0 \quad\Leftrightarrow\quad \tan x = 1$ or $3\tan^2 x = 1 \quad\Leftrightarrow$ $\tan x = 1$ or $\tan x = \pm\frac{1}{\sqrt{3}} \quad\Leftrightarrow\quad x = \frac{\pi}{6}, \frac{\pi}{4}, \frac{5\pi}{6}$, for $x \in [0, \pi)$. Thus $x = \frac{\pi}{6} + k\pi$, $\frac{\pi}{4} + k\pi$, $\frac{5\pi}{6} + k\pi$ for any integer $k$.

**37.** $4\sin x\cos x + 2\sin x - 2\cos x - 1 = 0 \quad\Leftrightarrow\quad (2\sin x - 1)(2\cos x + 1) = 0 \quad\Leftrightarrow\quad 2\sin x - 1 = 0$ or $2\cos x + 1 = 0$ $\quad\Leftrightarrow\quad \sin x = \frac{1}{2}$ or $\cos x = -\frac{1}{2} \quad\Leftrightarrow\quad x = \frac{\pi}{6}, \frac{5\pi}{6}, \frac{2\pi}{3}, \frac{4\pi}{3}$ in $[0, 2\pi)$. Thus $x = \frac{\pi}{6} + 2k\pi$, $\frac{5\pi}{6} + 2k\pi$, $\frac{2\pi}{3} + 2k\pi$, $\frac{4\pi}{3} + 2k\pi$ for any integer $k$.

**38.** $\sin 2x = 2\tan 2x \quad\Leftrightarrow\quad \sin 2x = \dfrac{2\sin 2x}{\cos 2x} \quad\Leftrightarrow\quad \sin 2x\cos 2x - 2\sin 2x = 0 \quad\Leftrightarrow\quad \sin 2x(\cos 2x - 2) = 0 \quad\Leftrightarrow$ $\cos 2x = 2$ (which is impossible since $|\cos u| \le 1$) or $\sin 2x = 0 \quad\Leftrightarrow\quad 2x = k\pi \quad\Leftrightarrow\quad x = \frac{1}{2}k\pi$ for any integer $k$.

**39.** $\cos^2 2x - \sin^2 2x = 0 \quad\Leftrightarrow\quad (\cos 2x - \sin 2x)(\cos 2x + \sin 2x) = 0 \quad\Leftrightarrow\quad \cos 2x = \pm\sin 2x \quad\Leftrightarrow\quad \tan 2x = \pm 1$ $\quad\Leftrightarrow\quad 2x = \frac{\pi}{4}, \frac{3\pi}{4}, \frac{5\pi}{4}, \frac{7\pi}{4}, \frac{9\pi}{4}, \frac{11\pi}{4}, \frac{13\pi}{4}, \frac{15\pi}{4}$ in $[0, 4\pi) \quad\Leftrightarrow\quad x = \frac{\pi}{8}, \frac{3\pi}{8}, \frac{5\pi}{8}, \frac{7\pi}{8}, \frac{9\pi}{8}, \frac{11\pi}{8}, \frac{13\pi}{8}, \frac{15\pi}{8}$ in $[0, 2\pi)$. So the solution can be expressed as the odd multiples of $\frac{\pi}{8}$ which are $x = \frac{\pi}{8} + \frac{k\pi}{4}$ for any integer $k$.

**40.** $\sec x - \tan x = \cos x \quad\Leftrightarrow\quad \cos x(\sec x - \tan x) = \cos x(\cos x) \quad\Leftrightarrow\quad 1 - \sin x = \cos^2 x \quad\Leftrightarrow$ $1 - \sin x = 1 - \sin^2 x \quad\Leftrightarrow\quad \sin x = \sin^2 x \quad\Leftrightarrow\quad \sin^2 x - \sin x = 0 \quad\Leftrightarrow\quad \sin x(\sin x - 1) = 0 \quad\Leftrightarrow\quad \sin x = 0$ or $\sin x = 1 \quad\Leftrightarrow\quad x = 0, \pi$ or $x = \frac{\pi}{2}$ in $[0, 2\pi)$. However, since the equation is undefined when $x = \frac{\pi}{2}$, the solutions are $x = k\pi$ for any integer $k$.

**41.** $2\cos 3x = 1 \quad\Leftrightarrow\quad \cos 3x = \frac{1}{2} \quad\Rightarrow\quad 3x = \frac{\pi}{3}, \frac{5\pi}{3}, \frac{7\pi}{3}, \frac{11\pi}{3}, \frac{13\pi}{3}, \frac{17\pi}{3}$ on $[0, 6\pi) \quad\Leftrightarrow\quad x = \frac{\pi}{9}, \frac{5\pi}{9}, \frac{7\pi}{9}, \frac{11\pi}{9}, \frac{13\pi}{9}, \frac{17\pi}{9}$ on $[0, 2\pi)$.

**42.** $3\csc^2 x = 4 \quad\Leftrightarrow\quad \csc^2 x = \frac{4}{3} \quad\Leftrightarrow\quad \csc x = \pm\frac{2}{\sqrt{3}} \quad\Leftrightarrow\quad \sin x = \pm\frac{\sqrt{3}}{2} \quad\Leftrightarrow\quad x = \frac{\pi}{3}, \frac{2\pi}{3}, \frac{4\pi}{3}, \frac{5\pi}{3}$ in $[0, 2\pi)$.

**43.** $2\sin x\tan x - \tan x = 1 - 2\sin x \Leftrightarrow 2\sin x\tan x - \tan x + 2\sin x - 1 = 0 \Leftrightarrow (2\sin x - 1)(\tan x + 1) = 0 \Leftrightarrow 2\sin x - 1 = 0$ or $\tan x + 1 = 0 \Leftrightarrow \sin x = \frac{1}{2}$ or $\tan x = -1 \Leftrightarrow x = \frac{\pi}{6}, \frac{5\pi}{6}$ or $x = \frac{3\pi}{4}, \frac{7\pi}{4}$. Thus, the solutions in $[0, 2\pi)$ are $\frac{\pi}{6}, \frac{3\pi}{4}, \frac{5\pi}{6}, \frac{7\pi}{4}$.

**44.** $\sec x\tan x - \cos x\cot x = \sin x \Leftrightarrow \frac{1}{\cos x}\frac{\sin x}{\cos x} - \cos x \cdot \frac{\cos x}{\sin x} = \sin x \Leftrightarrow \frac{\sin x}{\cos^2 x} - \frac{\cos^2 x}{\sin x} = \sin x$. Multiplying both sides by the common denominator $\cos^2 x\sin x$ gives $\sin^2 x - \cos^4 x = \sin^2 x\cos^2 x \Leftrightarrow \sin^2 x - \cos^4 x = (1 - \cos^2 x)\cos^2 x \Leftrightarrow \sin^2 x - \cos^4 x = \cos^2 x - \cos^4 x \Leftrightarrow \sin^2 x = \cos^2 x \Leftrightarrow \sin^2 x = 1 - \sin^2 x \Leftrightarrow 2\sin^2 x = 1 \Leftrightarrow \sin x = \pm\frac{1}{\sqrt{2}} \Leftrightarrow x = \frac{\pi}{4}, \frac{3\pi}{4}, \frac{5\pi}{4}, \frac{7\pi}{4}$. Since we multiplied the above equation by $\cos^2 x\sin x$ (which could be zero) we must check to see if we have introduced extraneous solutions. However, each of the values of $x$ satisfies the original equation and so the solutions on $[0, 2\pi)$ are $x = \frac{\pi}{4}, \frac{3\pi}{4}, \frac{5\pi}{4}, \frac{7\pi}{4}$.

**45.** $\tan x - 3\cot x = 0 \Leftrightarrow \frac{\sin x}{\cos x} - \frac{3\cos x}{\sin x} = 0 \Leftrightarrow \frac{\sin^2 x - 3\cos^2 x}{\cos x\sin x} = 0 \Leftrightarrow \frac{\sin^2 x + \cos^2 x - 4\cos^2 x}{\cos x\sin x} = 0 \Leftrightarrow \frac{1 - 4\cos^2 x}{\cos x\sin x} = 0 \Leftrightarrow 1 - 4\cos^2 x = 0 \Leftrightarrow 4\cos^2 x = 1 \Leftrightarrow \cos x = \pm\frac{1}{2} \Leftrightarrow x = \frac{\pi}{3}, \frac{2\pi}{3}, \frac{4\pi}{3}, \frac{5\pi}{3}$ in $[0, 2\pi)$.

**46.** $2\sin^2 x - \cos x = 1 \Leftrightarrow 2(1 - \cos^2 x) - \cos x - 1 = 0 \Leftrightarrow -2\cos^2 x - \cos x + 1 = 0 \Leftrightarrow (2\cos x - 1)(\cos x + 1) = 0 \Leftrightarrow 2\cos x - 1 = 0$ or $\cos x + 1 = 0 \Leftrightarrow \cos x = \frac{1}{2}$ or $\cos x = -1 \Leftrightarrow x = \frac{\pi}{3}, \frac{5\pi}{3}, \pi$ in $[0, 2\pi)$.

**47.** $\tan 3x + 1 = \sec 3x \Rightarrow (\tan 3x + 1)^2 = \sec^2 3x \Leftrightarrow \tan^2 3x + 2\tan 3x + 1 = \sec^2 3x \Leftrightarrow \sec^2 3x + 2\tan 3x = \sec^2 3x \Leftrightarrow 2\tan 3x = 0 \Leftrightarrow 3x = 0, \pi, 2\pi, 3\pi, 4\pi, 5\pi$ in $[0, 6\pi) \Leftrightarrow x = 0, \frac{\pi}{3}, \frac{2\pi}{3}, \pi, \frac{4\pi}{3}, \frac{5\pi}{3}$ in $[0, 2\pi)$ .Since squaring both sides is an operation that can introduce extraneous solutions, we must check each of the possible solution in the original equation. We see that $x = \frac{\pi}{3}, \pi, \frac{5\pi}{3}$ are not solutions. Hence the only solutions are $x = 0, \frac{2\pi}{3}, \frac{4\pi}{3}$.

**48.** $3\sec^2 x + 8\cos^2 x = 7 \Leftrightarrow \frac{3}{\cos^2 x} + 4\cos^2 x - 7 = 0$ (for $\cos x \neq 0$) $\Leftrightarrow 3 + 4\cos^4 x - 7\cos^2 x = 0 \Leftrightarrow 4\cos^4 x - 7\cos^2 x + 3 = 0 \Leftrightarrow (4\cos^2 x - 3)(\cos^2 x - 1) = 0 \Leftrightarrow 4\cos^2 x - 3 = 0$ or $\cos^2 x - 1 = 0 \Leftrightarrow \cos x = \pm\frac{\sqrt{3}}{2}$ or $\cos x = \pm 1 \Leftrightarrow x = \frac{\pi}{6}, \frac{5\pi}{6}, \frac{7\pi}{6}, \frac{11\pi}{6}$ or $x = 0, \pi$ in $[0, 2\pi)$. So the solutions are $x = 0, \frac{\pi}{6}, \frac{5\pi}{6}, \pi, \frac{7\pi}{6}, \frac{11\pi}{6}$.

**49. (a)** Since the period of cosine is $2\pi$ the solutions are $x \approx 1.15928 + 2k\pi$ and $x \approx 5.12391 + 2k\pi$ for any integer $k$.

**(b)** $\cos x = 0.4 \Rightarrow x = \cos^{-1} 0.4 \approx 1.15928$. The other solution is $2\pi - \cos^{-1} 0.4 \approx 5.12391$.

**50. (a)** Since the period is $\pi$, the solutions are of the form $x \approx 1.41815 + k\pi$ for any integer $k$.

**(b)** $2\tan x = 13 \Leftrightarrow \tan x = \frac{13}{2} \Rightarrow x \approx 1.41815$. Since the period for tangent is $\pi$, the other solution in $[0, 2\pi]$ is $x \approx 1.41815 + \pi \approx 4.55974$.

**51. (a)** Since the period of secant is $2\pi$ the solutions are $x \approx 1.36943 \pm 2k\pi$ and $\approx 4.91375 \pm 2k\pi$ for any integer $k$.

**(b)** $\sec x = 5 \Leftrightarrow \cos x = \frac{1}{5} \Rightarrow x = \cos^{-1}\frac{1}{5} \approx 1.36944$. The other solution is $2\pi - \cos^{-1}\frac{1}{5} \approx 4.91375$.

**52. (a)** Since the period is $\pi$, the solutions are of the form $x \approx 1.16590 + k\pi$ for any integer $k$.

**(b)** Since $\cos x = 0$ is not a solution, we divide both sides by $\cos x$. $3\sin x = 7\cos x \Leftrightarrow \tan x = \frac{7}{3} \Leftrightarrow x \approx 1.16590$. Since the period for tangent is $\pi$, the other solution in $[0, 2\pi]$ is $x \approx 1.16590 + \pi \approx 4.30750$.

**53. (a)** Since $\sin(\pi + x) = -\sin x$, we can express the general solutions as $x \approx 0.46365 + k\pi$ and $x \approx 2.67795 + k\pi$ for any integer $k$.

**(b)** $5\sin^2 x - 1 = 0 \Rightarrow \sin^2 x = \frac{1}{5} \Rightarrow \sin x = \pm\frac{1}{\sqrt{5}}$. Now $\sin x = \frac{1}{\sqrt{5}} \Rightarrow x = \sin^{-1}\left(\frac{1}{\sqrt{5}}\right) \approx 0.46365$,and $x \approx \pi - 0.46365 \approx 2.67795$. Also, $\sin x = -\frac{1}{\sqrt{5}} \Rightarrow x = \sin^{-1}\left(-\frac{1}{\sqrt{5}}\right) \approx -0.46365$, so in the interval $[0, 2\pi)$, $x \approx \pi - (-0.46365) \approx 3.60524$ and $x \approx 2\pi - 0.46365 \approx 5.81954$.

**54. (a)** Since the period is $2\pi$, the solutions are of the form $x \approx 1.57080 + k\pi$,$x \approx 0.25268 + 2k\pi$, $x \approx 2.88891 + 2k\pi$ for any integer $k$.

**(b)** $2\sin 2x - \cos x = 0 \Leftrightarrow 2(2\sin x\cos x) - \cos x = 0 \Leftrightarrow 4\sin x\cos x - \cos x = 0 \Leftrightarrow \cos x(4\sin x - 1) = 0 \Leftrightarrow \cos x = 0$ or $4\sin x - 1 = 0$. Now $\cos x = 0 \Leftrightarrow x = \frac{\pi}{2}, \frac{3\pi}{2}$ and $4\sin x - 1 = 0 \Leftrightarrow \sin x = \frac{1}{4} \Rightarrow x \approx 0.25268$ or $x \approx \pi - 0.25268 \approx 2.88891$. Thus the solutions in $[0, 2\pi)$ are $x \approx 0.25268, 1.57080, 2.88891, 4.71239$.

**55. (a)** Since the period for sine is $2\pi$, the general solution is of the form $x \approx 0.33984 + 2k\pi$ and $x \approx 2.80176 + 2k\pi$ for any integer $k$.

**(b)** $3\sin^2 x - 7\sin x + 2 = 0 \Rightarrow (3\sin x - 1)(\sin x - 2) = 0 \Rightarrow 3\sin x - 1 = 0$ or $\sin x - 2 = 0$. Since $|\sin x| \le 1$, $\sin x - 2 = 0$ has no solution. Thus $3\sin x - 1 = 0 \Rightarrow \sin x = \frac{1}{3} \Rightarrow x \approx 0.33984$ and $x \approx \pi - 0.33984 \approx 2.80176$.

**56. (a)** Since the period is $\pi$, the solutions are of the form $x \approx 1.10715 + k\pi$, $x \approx 2.03444 + k\pi$, $x \approx 1.24905 + k\pi$, $x \approx 1.89255 + k\pi$ for any integer $k$.

**(b)** $\tan^4 x - 13\tan^2 x + 36 = 0 \Leftrightarrow (\tan^2 x - 4)(\tan^2 x - 9) = 0 \Leftrightarrow \tan x = \pm 2$ or $\tan x = \pm 3 \Rightarrow x \approx 1.10715, -1.10715, 1.24905, -1.24905$ in $\left(-\frac{\pi}{2}, \frac{\pi}{2}\right)$. Since the period for tangent is $\pi$, the solutions in $[0, 2\pi)$ are $x \approx 1.10715, 1.24905, 1.89255, 2.03444, 4.24874, 4.39064, 5.03414, 5.17604$.

**57.** $f(x) = 3\cos x + 1$; $g(x) = \cos x - 1$. $f(x) = g(x)$ when $3\cos x + 1 = \cos x - 1 \Leftrightarrow 2\cos x = -2 \Leftrightarrow \cos x = -1 \Leftrightarrow x = \pi + 2k\pi = (2k+1)\pi$. The points of intersection are $((2k+1)\pi, -2)$ for any integer $k$.

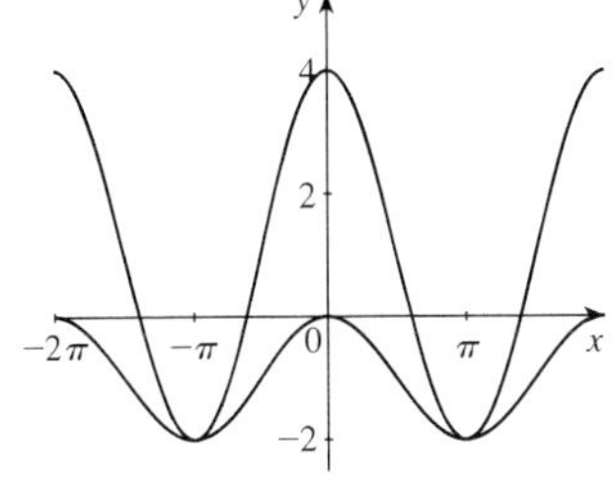


**58.** $f(x) = \sin 2x$; $g(x) = 2\sin 2x + 1$. $f(x) = g(x)$ when $\sin 2x = 2\sin 2x + 1 \Leftrightarrow \sin 2x = -1 \Leftrightarrow 2x = \frac{3\pi}{2} + 2k\pi \Leftrightarrow x = \frac{3\pi}{4} + k\pi$ for any integer $k$. So the intersection points are $\left(\frac{3\pi}{4} + k\pi, -1\right)$, for any integer $k$.

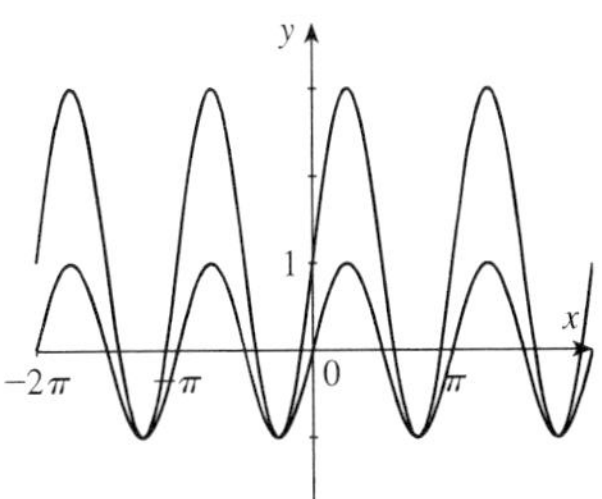


**59.** $f(x) = \tan x$; $g(x) = \sqrt{3}$. $f(x) = g(x)$ when $\tan x = \sqrt{3} \Leftrightarrow x = \frac{\pi}{3} + k\pi$.The intersection points are $\left(\frac{\pi}{3} + k\pi, \sqrt{3}\right)$ for any integer $k$.

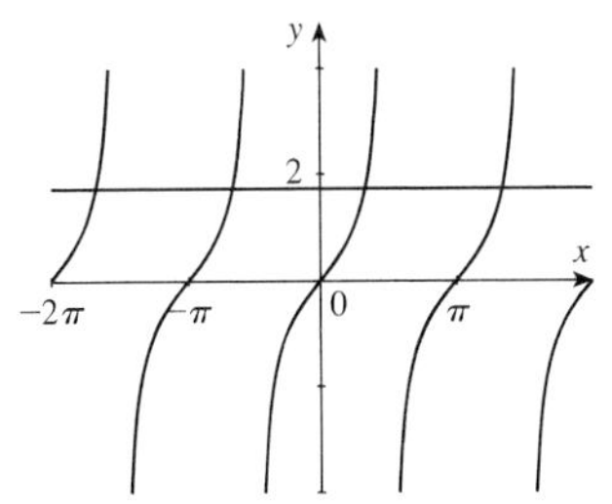

**60.** $f(x) = \sin x - 1$; $g(x) = \cos x$. $f(x) = g(x)$ when $\sin x - 1 = \cos x \Rightarrow$ $(\sin x - 1)^2 = \cos^2 x \Leftrightarrow \sin^2 x - 2\sin x + 1 = \cos^2 x \Leftrightarrow$ $\sin^2 x - 2\sin x + 1 - \cos^2 x = 0 \Leftrightarrow 2\sin^2 x - 2\sin x = 0 \Leftrightarrow$ $2\sin x(\sin x - 1) = 0 \Leftrightarrow \sin x = 0$ or $\sin x = 1 \Leftrightarrow x = k\pi, \frac{\pi}{2} + 2k\pi$. However, $x = k\pi$ is not a solution when $k$ is even. (The extraneous solutions were introduced by squaring both sides.) So the solutions are $x = (2k+1)\pi, \frac{\pi}{2} + 2k\pi$, and the intersection points are $(\pi + 2k\pi, -1), \left(\frac{\pi}{2} + 2k\pi, 0\right)$ for any integer $k$.

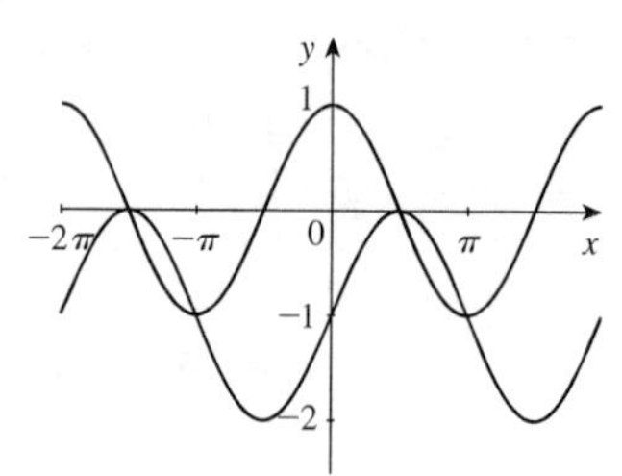


*Another method:* $\sin x - 1 = \cos x \Leftrightarrow \sin x - \cos x = 1 \Leftrightarrow \sqrt{2}\left(\frac{1}{\sqrt{2}}\sin x - \frac{1}{\sqrt{2}}\cos x\right) = 1 \Leftrightarrow \frac{\sqrt{2}}{2}\sin x - \frac{\sqrt{2}}{2}\cos x = \frac{\sqrt{2}}{2}$. Since $\sin\frac{\pi}{4} = \cos\frac{\pi}{4} = \frac{\sqrt{2}}{2}$, we may write $\frac{\sqrt{2}}{2}\sin x - \frac{\sqrt{2}}{2}\cos x = \cos\frac{\pi}{4}\sin x - \sin\frac{\pi}{4}\cos x = \frac{\sqrt{2}}{2} \Leftrightarrow$ $\sin\left(x - \frac{\pi}{4}\right) = \frac{\sqrt{2}}{2} \Leftrightarrow x - \frac{\pi}{4} = \frac{\pi}{4}, \frac{3\pi}{4} \Leftrightarrow x = \frac{\pi}{2}, \pi$ in $[0, 2\pi)$. So the solutions are $x = (2k+1)\pi, \frac{\pi}{2} + 2k\pi$, and the intersection points are $(\pi + 2k\pi, -1), \left(\frac{\pi}{2} + 2k\pi, 0\right)$ for any integer $k$.

**61.** $\cos x\cos 3x - \sin x\sin 3x = 0 \Leftrightarrow \cos(x + 3x) = 0 \Leftrightarrow \cos 4x = 0 \Leftrightarrow 4x = \frac{\pi}{2}, \frac{3\pi}{2}, \frac{5\pi}{2}, \frac{7\pi}{2}, \frac{9\pi}{2}, \frac{11\pi}{2}, \frac{13\pi}{2}, \frac{15\pi}{2}$ in $[0, 8\pi) \Leftrightarrow x = \frac{\pi}{8}, \frac{3\pi}{8}, \frac{5\pi}{8}, \frac{7\pi}{8}, \frac{9\pi}{8}, \frac{11\pi}{8}, \frac{13\pi}{8}, \frac{15\pi}{8}$ in $[0, 2\pi)$.

**62.** $\cos x\cos 2x + \sin x\sin 2x = \frac{1}{2} \Leftrightarrow \cos(x - 2x) = \frac{1}{2} \Leftrightarrow \cos(-x) = \frac{1}{2} \Leftrightarrow \cos x = \frac{1}{2} \Leftrightarrow x = \frac{\pi}{3}, \frac{5\pi}{3}$ in $[0, 2\pi)$.

**63.** $\sin 2x\cos x + \cos 2x\sin x = \frac{\sqrt{3}}{2} \Leftrightarrow \sin(2x + x) = \frac{\sqrt{3}}{2} \Leftrightarrow \sin 3x = \frac{\sqrt{3}}{2} \Leftrightarrow 3x = \frac{\pi}{3}, \frac{2\pi}{3}, \frac{7\pi}{3}, \frac{8\pi}{3}, \frac{13\pi}{3}, \frac{14\pi}{3}$ in $[0, 6\pi) \Leftrightarrow x = \frac{\pi}{9}, \frac{2\pi}{9}, \frac{7\pi}{9}, \frac{8\pi}{9}, \frac{13\pi}{9}, \frac{14\pi}{9}$ in $[0, 2\pi)$.

**64.** $\sin 3x\cos x - \cos 3x\sin x = 0 \Leftrightarrow \sin(3x - x) = 0 \Leftrightarrow \sin 2x = 0 \Leftrightarrow 2x = 0, \pi, 2\pi, 3\pi, 4\pi$ in $[0, 4\pi)$ $\Leftrightarrow x = 0, \frac{\pi}{2}, \pi, \frac{3\pi}{2}$ in $[0, 2\pi)$.

**65.** $\sin 2x + \cos x = 0 \Leftrightarrow 2\sin x\cos x + \cos x = 0 \Leftrightarrow \cos x(2\sin x + 1) = 0 \Leftrightarrow \cos x = 0$ or $\sin x = -\frac{1}{2}$ $\Leftrightarrow x = \frac{\pi}{2}, \frac{7\pi}{6}, \frac{3\pi}{2}, \frac{11\pi}{6}$.

**66.** $\tan\dfrac{x}{2} - \sin x = 0 \Leftrightarrow \dfrac{\sin x}{1 + \cos x} - \sin x = 0 \Leftrightarrow \sin x - \sin x(1 + \cos x) = 0$ (and $\cos x \neq -1 \Leftrightarrow x \neq \pi$) $\Leftrightarrow \sin x(-\cos x) = 0 \Leftrightarrow \sin x = 0$ or $\cos x = 0 \Leftrightarrow x = 0, \frac{\pi}{2}, \frac{3\pi}{2}$ in $[0, 2\pi)$ ($x = \pi$ is inadmissible).

**67.** $\cos 2x + \cos x = 2 \Leftrightarrow 2\cos^2 x - 1 + \cos x - 2 = 0 \Leftrightarrow 2\cos^2 x + \cos x - 3 = 0 \Leftrightarrow$ $(2\cos x + 3)(\cos x - 1) = 0 \Leftrightarrow 2\cos x + 3 = 0$ or $\cos x - 1 = 0 \Leftrightarrow \cos x = -\frac{3}{2}$ (which is impossible) or $\cos x = 1 \Leftrightarrow x = 0$ in $[0, 2\pi)$.

**68.** $\tan x + \cot x = 4\sin 2x \Leftrightarrow \dfrac{\sin x}{\cos x} + \dfrac{\cos x}{\sin x} = 8\sin x\cos x \Leftrightarrow \left(\dfrac{\sin x}{\cos x} + \dfrac{\cos x}{\sin x}\right)\cdot\sin x\cos x = (8\sin x\cos x)\cdot$ $\sin x\cos x \Leftrightarrow \sin^2 x + \cos^2 x = 8\sin^2 x\cos^2 x \Leftrightarrow 1 = 2(2\sin x\cos x)^2 \Leftrightarrow (\sin 2x)^2 = \frac{1}{2} \Leftrightarrow$ $\sin 2x = \pm\frac{1}{\sqrt{2}}$. Therefore, $2x = \frac{\pi}{4} + k\pi$ or $2x = \frac{3\pi}{4} + k\pi \Leftrightarrow x = \frac{\pi}{8} + \frac{k\pi}{2}$ or $x = \frac{3\pi}{8} + \frac{k\pi}{2}$. Thus on the interval $[0, 2\pi)$ the solutions are $x = \frac{\pi}{8}, \frac{5\pi}{8}, \frac{9\pi}{8}, \frac{13\pi}{8}$ and $\frac{3\pi}{8}, \frac{7\pi}{8}, \frac{11\pi}{8}, \frac{15\pi}{8}$. Together we write the solutions as $x = \frac{\pi}{8}, \frac{3\pi}{8}, \frac{5\pi}{8}, \frac{7\pi}{8}, \frac{9\pi}{8}, \frac{11\pi}{8}, \frac{13\pi}{8}, \frac{15\pi}{8}$, which are odd multiples of $\frac{\pi}{8}$. So we can express the general solution as $x = \frac{\pi}{8} + k\frac{\pi}{4}$ where $k$ is any integer.

**69.** $\sin x + \sin 3x = 0 \Leftrightarrow 2\sin 2x\cos(-x) = 0 \Leftrightarrow 2\sin 2x\cos x = 0 \Leftrightarrow \sin 2x = 0$ or $\cos x = 0 \Leftrightarrow$ $2x = k\pi$ or $x = k\frac{\pi}{2} \Leftrightarrow x = k\frac{\pi}{2}$ for any integer $k$.

**70.** $\cos 5x - \cos 7x = 0 \Leftrightarrow -2\sin 6x\sin(-x) = 0 \Leftrightarrow \sin 6x\sin x = 0 \Leftrightarrow \sin 6x = 0$ or $\sin x = 0 \Leftrightarrow$ $6x = k\pi$ or $x = k\pi \Leftrightarrow x = k\frac{\pi}{6}$ for any integer $k$.

**71.** $\cos 4x + \cos 2x = \cos x \Leftrightarrow 2\cos 3x\cos x = \cos x \Leftrightarrow \cos x(2\cos 3x - 1) = 0 \Leftrightarrow \cos x = 0$ or $\cos 3x = \frac{1}{2}$ $\Leftrightarrow x = \frac{\pi}{2}$ or $3x = \frac{\pi}{3} + 2k\pi, \frac{5\pi}{3} + 2k\pi, \frac{7\pi}{3} + 2k\pi, \frac{11\pi}{3} + 2k\pi, \frac{13\pi}{3} + 2k\pi, \frac{17\pi}{3} + 2k\pi \Leftrightarrow x = \frac{\pi}{2} + k\pi, \frac{\pi}{9} + \frac{2}{3}k\pi,$ $\frac{5\pi}{9} + \frac{2}{3}k\pi$.

**72.** $\sin 5x - \sin 3x = \cos 4x \Leftrightarrow 2\cos 4x \sin x = \cos 4x \Leftrightarrow \cos 4x\,(2\sin x - 1) = 0 \Leftrightarrow \cos 4x = 0$ or $\sin x = \frac{1}{2} \Leftrightarrow 4x = \frac{\pi}{2} + k\pi$ or $x = \frac{\pi}{6} + 2k\pi, \frac{5\pi}{6} + 2k\pi \Leftrightarrow x = \dfrac{\pi}{8} + k\frac{\pi}{4}, \frac{\pi}{6} + 2k\pi, \frac{5\pi}{6} + 2k\pi$.

**73.** $\sin 2x = x$

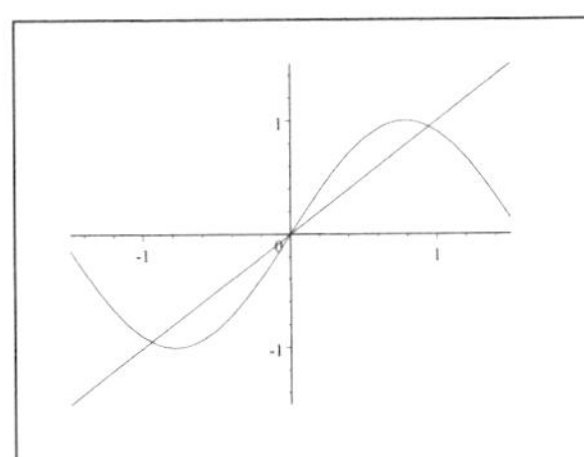

The three solutions are $x = 0$ and $x \approx \pm 0.95$.

**74.** $\cos x = \dfrac{x}{3}$

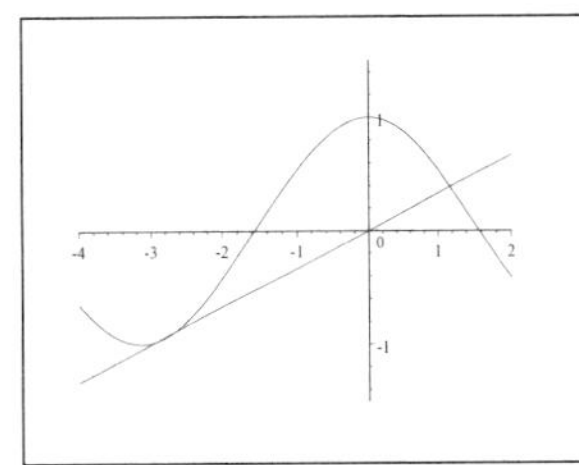

The three solutions are $x \approx 1.17, -2.66$, and $-2.94$.

**75.** $2^{\sin x} = x$

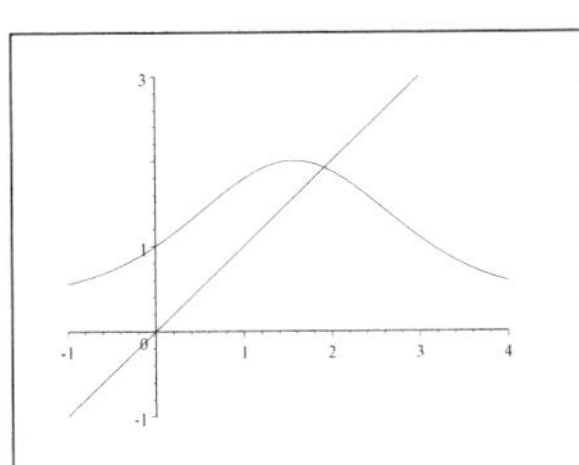

The only solution is $x \approx 1.92$.

**76.** $\sin x = x^3$

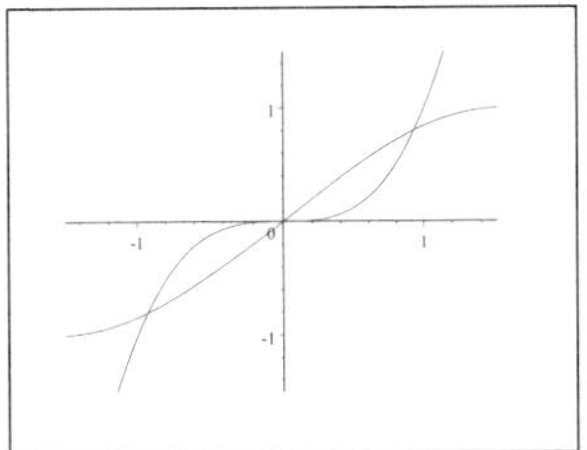

The three solutions are $x = 0$ and $x \approx \pm 0.93$.

**77.** $\dfrac{\cos x}{1 + x^2} = x^2$

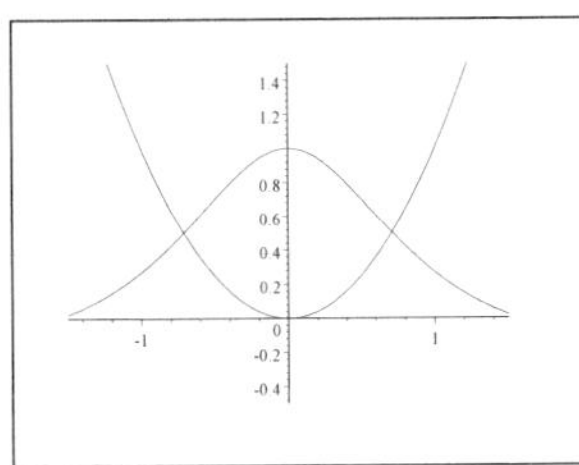

The two solutions are $x \approx \pm 0.71$.

**78.** $\sin x = \frac{1}{2}\left(e^x + e^{-x}\right)$

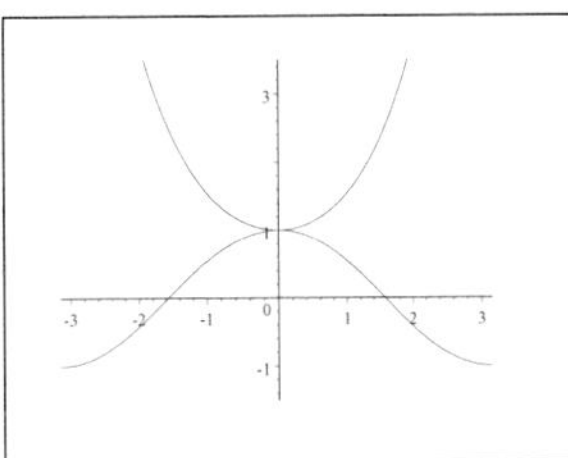

The only solution is $x = 0$.

**79.** We substitute $v_0 = 2200$ and $R(\theta) = 5000$ and solve for $\theta$. So $5000 = \dfrac{(2200)^2 \sin 2\theta}{32} \Leftrightarrow 5000 = 151250 \sin 2\theta \Leftrightarrow \sin 2\theta = 0.03308 \Rightarrow 2\theta = 1.89442^\circ$ or $2\theta = 180^\circ - 1.89442^\circ = 178.10558^\circ$. If $2\theta = 1.89442^\circ$, then $\theta = 0.94721^\circ$, and if $2\theta = 178.10558^\circ$, then $\theta = 89.05279^\circ$.

**80.** Since $4e^{-3t} > 0$, we have $0 = 4e^{-3t} \sin 2\pi t \Leftrightarrow 0 = \sin 2\pi t \Leftrightarrow 2\pi t = 0, \pi, 2\pi, \ldots \Leftrightarrow t = 0, \frac{1}{2}, 1, \frac{3}{2}, \ldots$.

**81.** We substitute $\theta_1 = 70^\circ$ and $\dfrac{v_1}{v_2} = 1.33$ into Snell's Law to get $\dfrac{\sin 70^\circ}{\sin \theta_2} = 1.33 \Leftrightarrow \sin \theta_2 = \dfrac{\sin 70^\circ}{1.33} = 0.7065 \Rightarrow \theta_2 \approx 44.95^\circ$.

**82.** The index of refraction from glass to air is $\dfrac{1}{1.52} \approx 0.658$, so we substitute $\theta_2 = 90^\circ$ into Snell's Law to get $\dfrac{\sin \theta_1}{\sin 90^\circ} = 0.658 \Rightarrow \sin \theta_1 \approx 0.658 \Rightarrow \theta_1 \approx 41.1^\circ$.

**83. (a)** $10 = 12 + 2.83\sin\left(\frac{2\pi}{3}(t-80)\right) \quad \Leftrightarrow \quad 2.83\sin\left(\frac{2\pi}{3}(t-80)\right) = -2 \quad \Leftrightarrow \quad \sin\left(\frac{2\pi}{3}(t-80)\right) = -0.70671$. Now $\sin\theta = -0.70671$ and $\theta = -0.78484$. If $\frac{2\pi}{3}(t-80) = -0.78484 \quad \Leftrightarrow \quad t - 80 = 45.6 \quad \Leftrightarrow \quad t = 34.4$. Now in the interval $[0, 2\pi)$, we have $\theta = \pi + 0.78484 \approx 3.92644$ and $\theta = 2\pi - 0.78484 \approx 5.49834$. If $\frac{2\pi}{3}(t-80) = 3.92644 \quad \Leftrightarrow \quad t - 80 = 228.1 \quad \Leftrightarrow \quad t = 308.1$. And if $\frac{2\pi}{3}(t-80) = 5.49834 \quad \Leftrightarrow$ $t - 80 = 319.4 \quad \Leftrightarrow \quad t = 399.4 \quad (399.4 - 365 = 34.4)$. So according to this model, there should be 10 hours of sunshine on the 34th day (February 3) and on the 308th day (November 4).

**(b)** Since $L(t) = 12 + 2.83\sin\left(\frac{2\pi}{3}(t-80)\right) \geq 10$ for $t \in [34, 308]$, the number of days with more than 10 hours of daylight is $308 - 34 + 1 = 275$ days.

**84. (a)** $F = \frac{1}{2}(1 - \cos\theta) = 0 \quad \Rightarrow \quad \cos\theta = 1 \quad \Rightarrow \quad \theta = 0$

**(b)** $F = \frac{1}{2}(1 - \cos\theta) = 0.25 \quad \Rightarrow \quad 1 - \cos\theta = 0.5 \quad \Rightarrow \quad \cos\theta = 0.5 \quad \Rightarrow \quad \theta = 60^\circ$ or $120^\circ$

**(c)** $F = \frac{1}{2}(1 - \cos\theta) = 0.5 \quad \Rightarrow \quad 1 - \cos\theta = 1 \quad \Rightarrow \quad \cos\theta = 0 \quad \Rightarrow \quad \theta = 90^\circ$ or $270^\circ$

**(d)** $F = \frac{1}{2}(1 - \cos\theta) = 1 \quad \Rightarrow \quad 1 - \cos\theta = 2 \quad \Rightarrow \quad \cos\theta = -1 \quad \Rightarrow \quad \theta = 180^\circ$

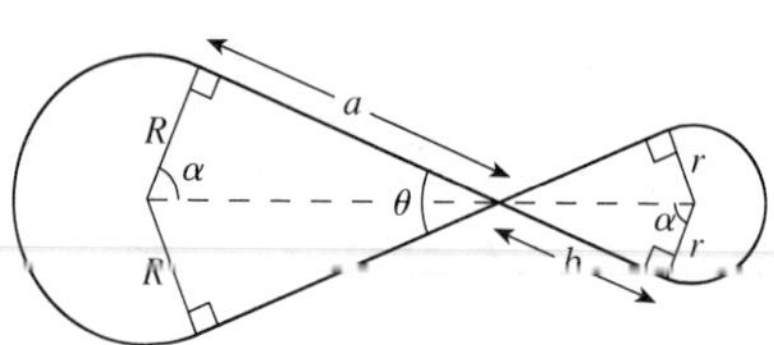


**85. (a)** First note that $\alpha = \frac{\pi}{2} - \frac{\theta}{2}$. The part of the belt touching the larger pulley has length $2(\pi - \alpha)R = (\theta + \pi)R$ and similarly the part touching the smaller belt has length $(\theta + \pi)r$. To calculate $a$ and $b$, we write $\cot\frac{\theta}{2} = \frac{a}{R} = \frac{b}{r} \Rightarrow a = R\cot\frac{\theta}{2}$ and $b = r\cot\frac{\theta}{2}$, so the length of the straight parts of the belt is $2a + 2b = 2(R + r)\cot\frac{\theta}{2}$. Thus, the total length of the belt is $L = (\theta + \pi)R + (\theta + \pi)r + 2(R + r)\cot\frac{\theta}{2} = (R + r)\left(\theta + \pi + 2\cot\frac{\theta}{2}\right)$ and so $\theta + \pi + 2\cot\frac{\theta}{2} = \frac{L}{R + r} \quad \Leftrightarrow$ $\theta + 2\cot\frac{\theta}{2} = \frac{L}{R + r} - \pi$.

**(b)** We plot $\theta + 2\cot\frac{\theta}{2}$ and $\frac{L}{R + r} - \pi = \frac{27.78}{2.42 + 1.21} - \pi \approx 4.5113$ in the same viewing rectangle. The solution is $\theta \approx 1.047$ rad $\approx 60^\circ$.

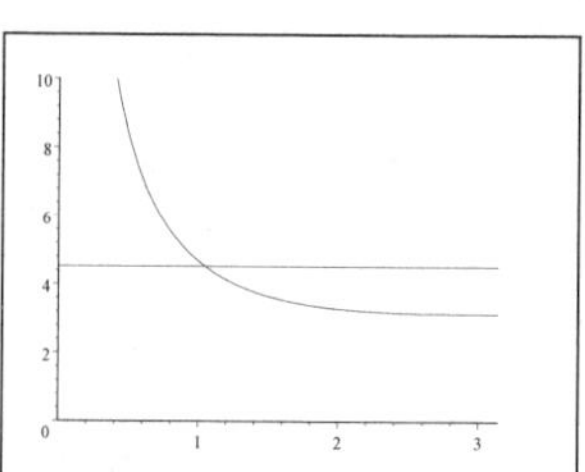


**86.** Statement A is true: every identity is an equation. However, Statement B is false: not every equation is an identity. The difference between an identity and an equation is that an identity is true for all values in the domain, whereas an equation may only be true for certain values in the domain and false for others. For example, $x = 0$ is an equation but not an identity, because it is true for only one value of $x$.

**87.** $\sin(\cos x)$ is a function of a function, that is, a composition of trigonometric functions (see Section 2.7). Most of the other equations involve sums, products, differences, or quotients of trigonometric functions.

$\sin(\cos x) = 0 \quad \Leftrightarrow \quad \cos x = 0$ or $\cos x = \pi$. However, since $|\cos x| \leq 1$, the only solution is $\cos x = 0 \Rightarrow x = \frac{\pi}{2} + k\pi$. The graph of $f(x) = \sin(\cos x)$ is shown.

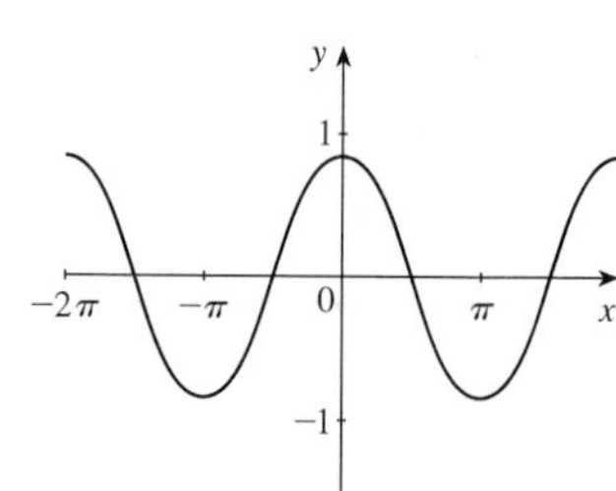

# Chapter 8 Review

**1.** $\sin\theta\,(\cot\theta+\tan\theta)=\sin\theta\left(\dfrac{\cos\theta}{\sin\theta}+\dfrac{\sin\theta}{\cos\theta}\right)=\cos\theta+\dfrac{\sin^2\theta}{\cos\theta}=\dfrac{\cos^2\theta+\sin^2\theta}{\cos\theta}=\dfrac{1}{\cos\theta}=\sec\theta$

**2.** $(\sec\theta-1)(\sec\theta+1)=\sec^2\theta-1=\tan^2\theta$

**3.** $\cos^2 x\csc x-\csc x=\left(1-\sin^2 x\right)\csc x-\csc x=\csc x-\sin^2 x\csc x-\csc x=-\sin^2 x\cdot\dfrac{1}{\sin x}=-\sin x$

**4.** $\dfrac{1}{1-\sin^2 x}=\dfrac{1}{\cos^2 x}=\sec^2 x=1+\tan^2 x$

**5.** $\dfrac{\cos^2 x-\tan^2 x}{\sin^2 x}=\dfrac{\cos^2 x}{\sin^2 x}-\dfrac{\tan^2 x}{\sin^2 x}=\cot^2 x-\dfrac{1}{\cos^2 x}=\cot^2 x-\sec^2 x$

**6.** $\dfrac{1+\sec x}{\sec x}=\dfrac{1}{\sec x}+1=1+\cos x=(1+\cos x)\cdot\dfrac{1-\cos x}{1-\cos x}=\dfrac{1-\cos^2 x}{1-\cos x}=\dfrac{\sin^2 x}{1-\cos x}$

**7.** $\dfrac{\cos^2 x}{1-\sin x}=\dfrac{\cos x}{\dfrac{1}{\cos x}(1-\sin x)}=\dfrac{\cos x}{\dfrac{1}{\cos x}-\dfrac{\sin x}{\cos x}}=\dfrac{\cos x}{\sec x-\tan x}$

**8.** $(1-\tan x)(1-\cot x)=1-\cot x-\tan x+\tan x\cot x=2-(\cot x+\tan x)=2-\left(\dfrac{\cos x}{\sin x}+\dfrac{\sin x}{\cos x}\right)$

$=2-\dfrac{\cos^2 x+\sin^2 x}{\cos x\sin x}=2-\dfrac{1}{\cos x\sin x}=2-\sec x\csc x$

**9.** $\sin^2 x\cot^2 x+\cos^2 x\tan^2 x=\sin^2 x\cdot\dfrac{\cos^2 x}{\sin^2 x}+\cos^2 x\cdot\dfrac{\sin^2 x}{\cos^2 x}=\cos^2 x+\sin^2 x=1$

**10.** $(\tan x+\cot x)^2=\left(\dfrac{\sin x}{\cos x}+\dfrac{\cos x}{\sin x}\right)^2=\left(\dfrac{\sin^2 x+\cos^2 x}{\cos x\sin x}\right)^2=\left(\dfrac{1}{\cos x\sin x}\right)^2=(\sec x\csc x)^2=\csc^2 x\sec^2 x$

**11.** $\dfrac{\sin 2x}{1+\cos 2x}=\dfrac{2\sin x\cos x}{1+2\cos^2 x-1}=\dfrac{2\sin x\cos x}{2\cos^2 x}=\dfrac{2\sin x}{2\cos x}=\tan x$

**12.** $\dfrac{\cos(x+y)}{\cos x\sin y}=\dfrac{\cos x\cos y-\sin x\sin y}{\cos x\sin y}=\dfrac{\cos x\cos y}{\cos x\sin y}-\dfrac{\sin x\sin y}{\cos x\sin y}=\dfrac{\cos y}{\sin y}-\dfrac{\sin x}{\cos x}=\cot y-\tan x$

**13.** $\tan\dfrac{x}{2}=\dfrac{1-\cos x}{\sin x}=\dfrac{1}{\sin x}-\dfrac{\cos x}{\sin x}=\csc x-\cot x$

**14.** $\dfrac{\sin(x+y)+\sin(x-y)}{\cos(x+y)+\cos(x-y)}=\dfrac{2\sin\left(\dfrac{(x+y)+(x-y)}{2}\right)\cos\left(\dfrac{(x+y)-(x-y)}{2}\right)}{2\cos\left(\dfrac{(x+y)+(x-y)}{2}\right)\cos\left(\dfrac{(x+y)-(x-y)}{2}\right)}=\dfrac{2\sin x\cos y}{2\cos x\cos y}=\dfrac{\sin x}{\cos x}=\tan x$

**15.** $\sin(x+y)\sin(x-y)=\frac{1}{2}\left[\cos((x+y)-(x-y))-\cos((x+y)+(x-y))\right]=\frac{1}{2}(\cos 2y-\cos 2x)$

$=\frac{1}{2}\left[1-2\sin^2 y-\left(1-2\sin^2 x\right)\right]=\frac{1}{2}\left(2\sin^2 x-2\sin^2 y\right)=\sin^2 x-\sin^2 y$

**16.** $\csc x-\tan\dfrac{x}{2}=\csc x-\dfrac{1-\cos x}{\sin x}=\csc x-(\csc x-\cot x)=\cot x$

**17.** $1+\tan x\tan\dfrac{x}{2}=1+\dfrac{\sin x}{\cos x}\cdot\dfrac{1-\cos x}{\sin x}=1+\dfrac{1-\cos x}{\cos x}=1+\dfrac{1}{\cos x}-1=\dfrac{1}{\cos x}=\sec x$

**18.** $\dfrac{\sin 3x+\cos 3x}{\cos x-\sin x}=\dfrac{\sin(x+2x)+\cos(x+2x)}{\cos x-\sin x}=\dfrac{\sin x\cos 2x+\cos x\sin 2x+\cos x\cos 2x-\sin x\sin 2x}{\cos x-\sin x}$

$=\dfrac{\cos 2x(\sin x+\cos x)+\sin 2x(\cos x-\sin x)}{\cos x-\sin x}=\dfrac{\cos 2x(\sin x+\cos x)}{\cos x-\sin x}+\sin 2x$

$=\dfrac{\left(\cos^2 x-\sin^2 x\right)(\sin x+\cos x)}{\cos x-\sin x}+\sin 2x=(\cos x+\sin x)(\sin x+\cos x)+\sin 2x$

$=\sin^2 x+\cos^2 x+2\sin x\cos x+\sin 2x=1+\sin 2x+\sin 2x=1+2\sin 2x$

**19.** $\left(\cos\dfrac{x}{2}-\sin\dfrac{x}{2}\right)^2=\cos^2\dfrac{x}{2}-2\sin\dfrac{x}{2}\cos\dfrac{x}{2}+\sin^2\dfrac{x}{2}=\sin^2\dfrac{x}{2}+\cos^2\dfrac{x}{2}-2\sin\dfrac{x}{2}\cos\dfrac{x}{2}=1-\sin\left(2\cdot\dfrac{x}{2}\right)=1-\sin x$

**20.** $\dfrac{\cos 3x - \cos 7x}{\sin 3x + \sin 7x} = \dfrac{-2\sin\left(\dfrac{3x+7x}{2}\right)\sin\left(\dfrac{3x-7x}{2}\right)}{2\sin\left(\dfrac{3x+7x}{2}\right)\cos\left(\dfrac{3x-7x}{2}\right)} = \dfrac{-2\sin 5x\sin(-2x)}{2\sin 5x\cos(-2x)} = \dfrac{2\sin 5x\sin 2x}{2\sin 5x\cos 2x} = \dfrac{\sin 2x}{\cos 2x} = \tan 2x$

**21.** $\dfrac{\sin 2x}{\sin x} - \dfrac{\cos 2x}{\cos x} = \dfrac{2\sin x\cos x}{\sin x} - \dfrac{2\cos^2 x - 1}{\cos x} = 2\cos x - 2\cos x + \dfrac{1}{\cos x} = \sec x$

**22.** $(\cos x + \cos y)^2 + (\sin x - \sin y)^2 = \cos^2 x + 2\cos x\cos y + \cos^2 y + \sin^2 x - 2\sin x\sin y + \sin^2 y = \left(\cos^2 x + \sin^2 x\right) + \left(\sin^2 y + \cos^2 y\right) + 2(\cos x\cos y - \sin x\sin y) = 2 + 2\cos(x+y)$

**23.** $\tan\left(x + \frac{\pi}{4}\right) = \dfrac{\tan x + \tan\frac{\pi}{4}}{1 - \tan x\tan\frac{\pi}{4}} = \dfrac{1+\tan x}{1-\tan x}$

**24.** $\dfrac{\sec x - 1}{\sin x\sec x} = \dfrac{\dfrac{1}{\cos x} - 1}{\sin x\dfrac{1}{\cos x}} = \dfrac{\dfrac{1}{\cos x} - 1}{\sin x\dfrac{1}{\cos x}} \cdot \dfrac{\cos x}{\cos x} = \dfrac{1-\cos x}{\sin x} = \tan\dfrac{x}{2}$

**25. (a)** $f(x) = 1 - \left(\cos\frac{x}{2} - \sin\frac{x}{2}\right)^2, g(x) = \sin x$

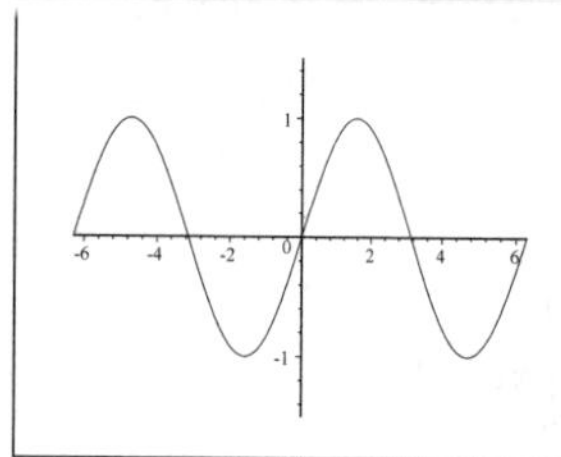

**(b)** The graphs suggest that $f(x) = g(x)$ is an identity. To prove this, expand $f(x)$ and simplify, using the double-angle formula for sine:

$$\begin{aligned} f(x) &= 1 - \left(\cos\tfrac{x}{2} - \sin\tfrac{x}{2}\right)^2 \\ &= 1 - \left(\cos^2\tfrac{x}{2} - 2\cos\tfrac{x}{2}\sin\tfrac{x}{2} + \sin^2\tfrac{x}{2}\right) \\ &= 1 + 2\cos\tfrac{x}{2}\sin\tfrac{x}{2} - \left(\cos^2\tfrac{x}{2} + \sin^2\tfrac{x}{2}\right) \\ &= 1 + \sin x - (1) = \sin x = g(x) \end{aligned}$$

**26. (a)** $f(x) = \sin x + \cos x$,

$g(x) = \sqrt{\sin^2 x + \cos^2 x}$

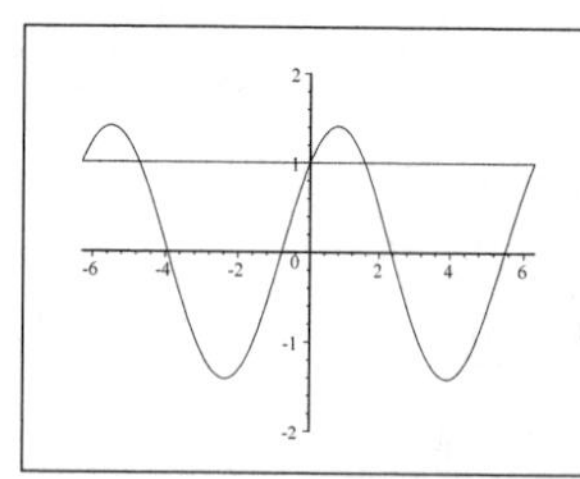

**(b)** The graphs suggest that $f(x) \neq g(x)$ in general. For example, choose $x = \frac{\pi}{6}$ and evaluate the functions: $f\left(\frac{\pi}{6}\right) = \frac{1}{2} + \frac{\sqrt{3}}{2} = \frac{1+\sqrt{3}}{2}$, whereas $g\left(\frac{\pi}{6}\right) = \sqrt{\frac{1}{4} + \frac{3}{4}} = \sqrt{1} = 1$, so $f(x) \neq g(x)$.

**27. (a)** $f(x) = \tan x\tan\frac{x}{2}, g(x) = \dfrac{1}{\cos x}$

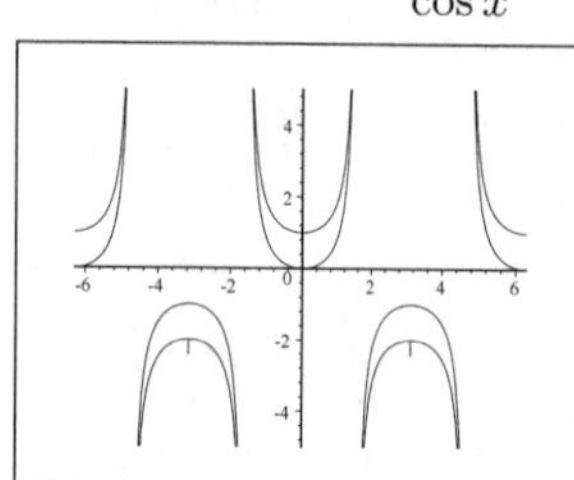

**(b)** The graphs suggest that $f(x) \neq g(x)$ in general. For example, choose $x = \frac{\pi}{3}$ and evaluate: $f\left(\frac{\pi}{3}\right) = \tan\frac{\pi}{3}\tan\frac{\pi}{6} = \sqrt{3}\cdot\frac{1}{\sqrt{3}} = 1$, whereas $g\left(\frac{\pi}{3}\right) = \dfrac{1}{\frac{1}{2}} = 2$, so $f(x) \neq g(x)$.

**28. (a)** $f(x) = 1 - 8\sin^2 x + 8\sin^4 x$, $g(x) = \cos 4x$

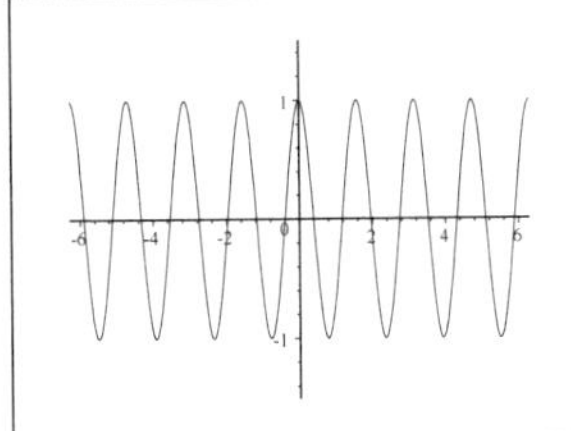

**(b)** The graphs suggest that $f(x) = g(x)$ is an identity. To show this, expand $g(x)$ by using double-angle identities:

$$\begin{aligned} g(x) &= \cos 4x = \cos(2(2x)) = 1 - 2\sin^2 2x \\ &= 1 - 2(2\sin x\cos x)^2 = 1 - 2\left(4\sin^2 x\cos^2 x\right) \\ &= 1 - 8\sin^2 x \cdot \left(1 - \sin^2 x\right) \\ &= 1 - 8\sin^2 x + 8\sin^4 x = f(x) \end{aligned}$$

**29. (a)** $f(x) = 2\sin^2 3x + \cos 6x$

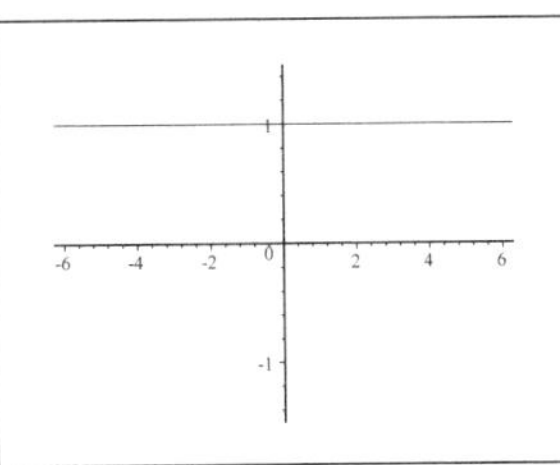

**(b)** The graph suggests that $f(x) = 1$ for all $x$. To prove this, we use the double angle formula to note that $\cos 6x = \cos(2(3x)) = 1 - 2\sin^2 3x$, so $f(x) = 2\sin^2 3x + \left(1 - 2\sin^2 3x\right) = 1$.

**30. (a)** $f(x) = \sin x \cot \dfrac{x}{2}$, $g(x) = \cos x$

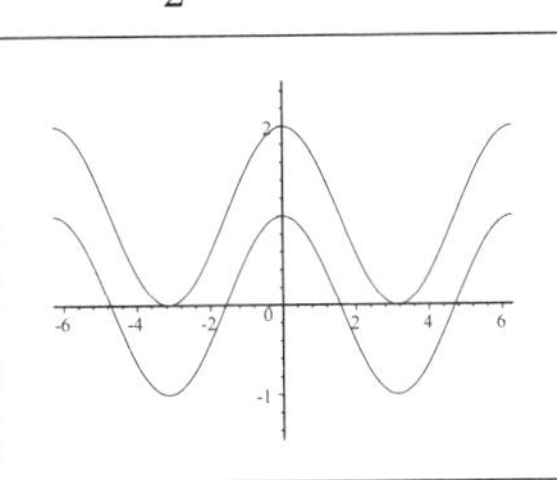

*Conjecture:* $f(x) = g(x) + 1$

**(b)** *Proof:*

$$\begin{aligned} f(x) &= \sin x \cot \tfrac{x}{2} = \sin x \cdot \left(\frac{\cos \frac{x}{2}}{\sin \frac{x}{2}}\right) \\ &= \left(2\sin \tfrac{x}{2}\cos \tfrac{x}{2}\right) \cdot \left(\frac{\cos \frac{x}{2}}{\sin \frac{x}{2}}\right) = 2\cos^2 \tfrac{x}{2} \end{aligned}$$

Now subtract and add 1, so $f(x) = 2\cos^2 \dfrac{x}{2} - 1 + 1 = \cos x + 1 = g(x) + 1$.

**31.** $\cos x \sin x - \sin x = 0 \quad\Leftrightarrow\quad \sin x(\cos x - 1) = 0 \quad\Leftrightarrow\quad \sin x = 0$ or $\cos x = 1 \quad\Leftrightarrow\quad x = 0, \pi$ or $x = 0$. Therefore, the solutions are $x = 0$ and $\pi$.

**32.** $\sin x - 2\sin^2 x = 0 \quad\Leftrightarrow\quad \sin x(1 - 2\sin x) = 0 \quad\Leftrightarrow\quad \sin x = 0$ or $\sin x = \frac{1}{2} \quad\Leftrightarrow\quad x = 0, \pi$ or $x = \frac{\pi}{6}, \frac{5\pi}{6}$. Therefore, the solutions in $[0, 2\pi)$ are $x = 0, \frac{\pi}{6}, \frac{5\pi}{6}, \pi$.

**33.** $2\sin^2 x - 5\sin x + 2 = 0 \quad\Leftrightarrow\quad (2\sin x - 1)(\sin x - 2) = 0 \quad\Leftrightarrow\quad \sin x = \frac{1}{2}$ or $\sin x = 2$ (which is inadmissible) $\Leftrightarrow\quad x = \frac{\pi}{6}, \frac{5\pi}{6}$. Thus, the solutions in $[0, 2\pi)$ are $x = \frac{\pi}{6}$ and $\frac{5\pi}{6}$.

**34.** $\sin x - \cos x - \tan x = -1 \quad\Leftrightarrow\quad \sin x\cos x - \cos^2 x - \sin x = -\cos x \quad\Leftrightarrow\quad \sin x\cos x - \sin x - \cos^2 x + \cos x = 0$ $\Leftrightarrow\quad \sin x(\cos x - 1) - \cos x(\cos x - 1) = 0 \quad\Leftrightarrow\quad (\sin x - \cos x)(\cos x - 1) = 0 \quad\Leftrightarrow\quad \sin x = \cos x$ or $\cos x = 1 \quad\Leftrightarrow\quad \tan x = 1$ or $\cos x = 1 \quad\Leftrightarrow\quad x = \frac{\pi}{4}, \frac{5\pi}{4}$ or $x = 0$. Therefore, the solutions in $[0, 2\pi)$ are $x = 0, \frac{\pi}{4}, \frac{5\pi}{4}$.

**35.** $2\cos^2 x - 7\cos x + 3 = 0 \quad\Leftrightarrow\quad (2\cos x - 1)(\cos x - 3) = 0 \quad\Leftrightarrow\quad \cos x = \frac{1}{2}$ or $\cos x = 3$ (which is inadmissible) $\Leftrightarrow\quad x = \frac{\pi}{3}, \frac{5\pi}{3}$. Therefore, the solutions in $[0, 2\pi)$ are $x = \frac{\pi}{3}, \frac{5\pi}{3}$.

**36.** $4\sin^2 x + 2\cos^2 x = 3 \quad\Leftrightarrow\quad 2\sin^2 x + 2\left(\sin^2 x + \cos^2 x\right) - 3 = 0 \quad\Leftrightarrow\quad 2\sin^2 x + 2 - 3 = 0 \quad\Leftrightarrow\quad 2\sin^2 x = 1$ $\Leftrightarrow\quad \sin x = \pm\frac{1}{\sqrt{2}}$. So the solutions in $[0, 2\pi)$ are $x = \frac{\pi}{4}, \frac{3\pi}{4}, \frac{5\pi}{4}, \frac{7\pi}{4}$.

**37.** Note that $x = \pi$ is not a solution because the denominator is zero. $\dfrac{1 - \cos x}{1 + \cos x} = 3 \quad\Leftrightarrow\quad 1 - \cos x = 3 + 3\cos x \quad\Leftrightarrow$ $-4\cos x = 2 \quad\Leftrightarrow\quad \cos x = -\frac{1}{2} \quad\Leftrightarrow\quad x = \frac{2\pi}{3}, \frac{4\pi}{3}$ in $[0, 2\pi)$.

**38.** $\sin x = \cos 2x \quad \Leftrightarrow \quad \sin x = 1 - 2\sin^2 x \quad \Leftrightarrow \quad 2\sin^2 x + \sin x - 1 = 0 \quad \Leftrightarrow \quad (2\sin x - 1)(\sin x + 1) = 0 \quad \Leftrightarrow$ $\sin x = \frac{1}{2}$ or $\sin x = -1 \quad \Leftrightarrow \quad x = \frac{\pi}{6}, \frac{5\pi}{6}$ or $x = \frac{3\pi}{2}$. Thus, the solutions in $[0, 2\pi)$ are $x = \frac{\pi}{6}, \frac{3\pi}{2}, \frac{5\pi}{6}$.

**39.** Factor by grouping: $\tan^3 x + \tan^2 x - 3\tan x - 3 = 0 \quad \Leftrightarrow \quad (\tan x + 1)\left(\tan^2 x - 3\right) = 0 \quad \Leftrightarrow \quad \tan x = -1$ or $\tan x = \pm\sqrt{3} \quad \Leftrightarrow \quad x = \frac{3\pi}{4}, \frac{7\pi}{4}$ or $x = \frac{\pi}{3}, \frac{2\pi}{3}, \frac{4\pi}{3}, \frac{5\pi}{3}$. Therefore, the solutions in $[0, 2\pi)$ are $x = \frac{\pi}{3}, \frac{2\pi}{3}, \frac{3\pi}{4}, \frac{4\pi}{3}, \frac{5\pi}{3}, \frac{7\pi}{4}$.

**40.** $\cos 2x \csc^2 x = 2\cos 2x \quad \Leftrightarrow \quad \cos 2x \csc^2 x - 2\cos 2x = 0 \quad \Leftrightarrow \quad \cos 2x\left(\csc^2 x - 2\right) = 0 \quad \Leftrightarrow \quad \cos 2x = 0$ or $\csc^2 x = 2 \quad \Leftrightarrow \quad \cos 2x = 0$or $\sin^2 x = \frac{1}{2} \quad \Leftrightarrow \quad \cos 2x = 0$ or $\sin x = \pm\frac{1}{\sqrt{2}}$.

For $\cos 2x = 0$, the solutions in $[0, 4\pi)$ are $2x = \frac{\pi}{2}, \frac{3\pi}{2}, \frac{5\pi}{2}, \frac{7\pi}{2} \quad \Leftrightarrow$ the solutions in $[0, 2\pi)$ are $x = \frac{\pi}{4}, \frac{3\pi}{4}, \frac{5\pi}{4}, \frac{7\pi}{4}$.
For $\sin x = \pm\frac{1}{\sqrt{2}}$, the solutions in $[0, 2\pi)$ are $x = \frac{\pi}{4}, \frac{3\pi}{4}, \frac{5\pi}{4}, \frac{7\pi}{4}$.

Thus, the solutions of the equation in $[0, 2\pi)$ are $x = \frac{\pi}{4}, \frac{3\pi}{4}, \frac{5\pi}{4}, \frac{7\pi}{4}$.

**41.** $\tan\frac{1}{2}x + 2\sin 2x = \csc x \quad \Leftrightarrow \quad \dfrac{1 - \cos x}{\sin x} + 4\sin x \cos x = \dfrac{1}{\sin x} \quad \Leftrightarrow \quad 1 - \cos x + 4\sin^2 x \cos x = 1 \quad \Leftrightarrow$ $4\sin^2 x \cos x - \cos x = 0 \quad \Leftrightarrow \quad \cos x\left(4\sin^2 x - 1\right) = 0 \quad \Leftrightarrow \quad \cos x = 0$ or $\sin x = \pm\frac{1}{2} \quad \Leftrightarrow \quad x = \frac{\pi}{2}, \frac{3\pi}{2}$ or $x = \frac{\pi}{6}, \frac{5\pi}{6}, \frac{7\pi}{6}, \frac{11\pi}{6}$. Thus, the solutions in $[0, 2\pi)$ are $x = \frac{\pi}{6}, \frac{\pi}{2}, \frac{5\pi}{6}, \frac{7\pi}{6}, \frac{3\pi}{2}, \frac{11\pi}{6}$.

**42.** $\cos 3x + \cos 2x + \cos x = 0 \quad \Leftrightarrow \quad \cos 2x \cos x - \sin 2x \sin x + \cos 2x + \cos x = 0 \quad \Leftrightarrow$
$\cos 2x \cos x + \cos 2x - \sin 2x \sin x + \cos x = 0 \quad \Leftrightarrow \quad \cos 2x(\cos x + 1) - 2\sin^2 x \cos x + \cos x = 0$
$\Leftrightarrow \quad \cos 2x(\cos x + 1) + \cos x\left(1 - 2\sin^2 x\right) = 0 \quad \Leftrightarrow \quad \cos 2x(\cos x + 1) + \cos x \cos 2x = 0 \quad \Leftrightarrow$ $\cos 2x(\cos x + 1 + \cos x) = 0 \quad \Leftrightarrow \quad \cos 2x(2\cos x + 1) = 0 \quad \Leftrightarrow \quad \cos 2x = 0$ or $\cos x = -\frac{1}{2} \quad \Leftrightarrow \quad 2x = \frac{\pi}{2}, \frac{3\pi}{2}, \frac{5\pi}{2}, \frac{7\pi}{2}$ (in $[0, 4\pi)$) or $x = \frac{2\pi}{3}, \frac{4\pi}{3}$(in $[0, 2\pi)$). Thus, the solutions in $[0, 2\pi)$ are $x = \frac{\pi}{4}, \frac{2\pi}{3}, \frac{3\pi}{4}, \frac{5\pi}{4}, \frac{4\pi}{3}, \frac{7\pi}{4}$.

**43.** $\tan x + \sec x = \sqrt{3} \quad \Leftrightarrow \quad \dfrac{\sin x}{\cos x} + \dfrac{1}{\cos x} = \sqrt{3} \quad \Leftrightarrow \quad \sin x + 1 = \sqrt{3}\cos x \quad \Leftrightarrow \quad \sqrt{3}\cos x - \sin x = 1$
$\Leftrightarrow \quad \frac{\sqrt{3}}{2}\cos x - \frac{1}{2}\sin x = \frac{1}{2} \quad \Leftrightarrow \quad \cos\frac{\pi}{6}\cos x - \sin\frac{\pi}{6}\sin x = \frac{1}{2} \quad \Leftrightarrow \quad \cos\left(x + \frac{\pi}{6}\right) = \frac{1}{2} \quad \Leftrightarrow \quad x + \frac{\pi}{6} = \frac{\pi}{3}, \frac{5\pi}{3} \quad \Leftrightarrow$ $x = \frac{\pi}{6}, \frac{3\pi}{2}$. However, $x = \frac{3\pi}{2}$ is inadmissible because $\sec\frac{3\pi}{2}$ is undefined. Thus, the only solution in $[0, 2\pi)$ is $x = \frac{\pi}{6}$.

**44.** $2\cos x - 3\tan x = 0 \quad \Leftrightarrow \quad 2\cos x - 3\dfrac{\sin x}{\cos x} = 0 \quad \Leftrightarrow \quad 2\cos^2 x - 3\sin x = 0 \; (\cos x \neq 0) \quad \Leftrightarrow$ $2\left(1 - \sin^2 x\right) - 3\sin x = 0 \quad \Leftrightarrow \quad 2\sin^2 x + 3\sin x - 2 = 0 \quad \Leftrightarrow \quad (2\sin x - 1)(\sin x + 2) = 0 \quad \Leftrightarrow \quad \sin x = \frac{1}{2}$ or $\sin x = -2$ (which has no solution) $\Leftrightarrow \quad x = \frac{\pi}{6}, \frac{5\pi}{6}$.

**45.** We graph $f(x) = \cos x$ and $g(x) = x^2 - 1$ in the viewing rectangle $[0, 6.5]$ by $[-2, 2]$. The two functions intersect at only one point, $x \approx 1.18$.

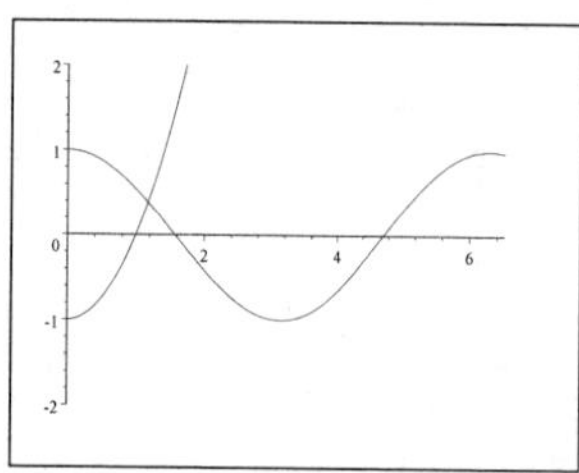

**46.** We graph $f(x) = e^{\sin x}$ and $g(x) = x$ in the viewing rectangle $[0, 6.5]$ by $[-1, 3]$. The two functions intersect at only one point, $x \approx 2.22$.

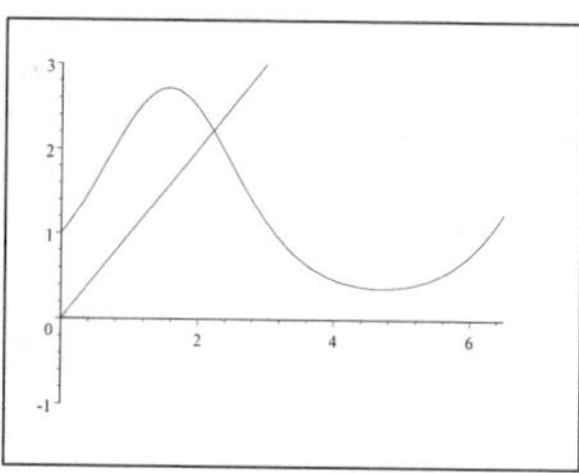

**47. (a)** $2000 = \dfrac{(400)^2 \sin^2\theta}{64} \quad \Leftrightarrow \quad \sin^2\theta = 0.8 \quad \Leftrightarrow \quad \sin\theta \approx 0.8944 \quad \Leftrightarrow \quad \theta \approx 63.4^\circ$

**(b)** $\dfrac{(400)^2 \sin^2\theta}{64} = 2500\sin^2\theta \leq 2500$. Therefore it is impossible for the projectile to reach a height of 3000 ft.

**(c)** The function $M(\theta) = 2500\sin^2\theta$ is maximized when $\sin^2\theta = 1$, so $\theta = 90^\circ$. The projectile will travel the highest when it is shot straight up.

**48.** Since $e^{-0.2t} > 0$ we have $f(t) = e^{-0.2t}\sin 4\pi t = 0 \quad\Leftrightarrow\quad \sin 4\pi t = 0 \quad\Leftrightarrow\quad 4\pi t = k\pi \quad\Leftrightarrow\quad t = \dfrac{k}{4}$ where $k$ is any integer. Thus the shock absorber is at equilibrium position every quarter second.

**49.** Since $15^\circ$ is in quadrant I, $\cos 15^\circ = \sqrt{\dfrac{1+\cos 30^\circ}{2}} = \sqrt{\dfrac{2+\sqrt{3}}{4}} = \frac{1}{2}\sqrt{2+\sqrt{3}}$.

**50.** $\sin\frac{5\pi}{12} = \sin\left(\frac{\pi}{4}+\frac{\pi}{6}\right) = \sin\frac{\pi}{4}\cos\frac{\pi}{6} + \cos\frac{\pi}{4}\sin\frac{\pi}{6} = \frac{1}{\sqrt{2}}\cdot\frac{\sqrt{3}}{2} + \frac{1}{\sqrt{2}}\cdot\frac{1}{2} = \frac{1+\sqrt{3}}{2\sqrt{2}} = \frac{\sqrt{2}+\sqrt{6}}{4}$

**51.** $\tan\frac{\pi}{8} = \dfrac{1-\cos\frac{\pi}{4}}{\sin\frac{\pi}{4}} = \dfrac{1-\frac{1}{\sqrt{2}}}{\frac{1}{\sqrt{2}}} = \left(1-\frac{1}{\sqrt{2}}\right)\sqrt{2} = \sqrt{2}-1$

**52.** $2\sin\frac{\pi}{12}\cos\frac{\pi}{12} = \sin\left(2\cdot\frac{\pi}{12}\right) = \sin\frac{\pi}{6} = \frac{1}{2}$

**53.** $\sin 5^\circ\cos 40^\circ + \cos 5^\circ\sin 40^\circ = \sin(5^\circ+40^\circ) = \sin 45^\circ = \frac{1}{\sqrt{2}} = \frac{\sqrt{2}}{2}$

**54.** $\dfrac{\tan 66^\circ - \tan 6^\circ}{1+\tan 66^\circ\tan 6^\circ} = \tan(66^\circ - 6^\circ) = \tan 60^\circ = \sqrt{3}$

**55.** $\cos^2\frac{\pi}{8} - \sin^2\frac{\pi}{8} = \cos\left(2\left(\frac{\pi}{8}\right)\right) = \cos\frac{\pi}{4} = \frac{1}{\sqrt{2}} = \frac{\sqrt{2}}{2}$

**56.** $\frac{1}{2}\cos\frac{\pi}{12} + \frac{\sqrt{3}}{2}\sin\frac{\pi}{12} = \sin\frac{\pi}{6}\cos\frac{\pi}{12} + \cos\frac{\pi}{6}\sin\frac{\pi}{12} = \sin\left(\frac{\pi}{6}+\frac{\pi}{12}\right) = \sin\frac{\pi}{4} = \frac{1}{\sqrt{2}} = \frac{\sqrt{2}}{2}$

**57.** We use a product-to-sum formula: $\cos 37.5^\circ\cos 7.5^\circ = \frac{1}{2}(\cos 45^\circ + \cos 30^\circ) = \frac{1}{2}\left(\frac{\sqrt{2}}{2}+\frac{\sqrt{3}}{2}\right) = \frac{1}{4}(\sqrt{2}+\sqrt{3})$.

**58.** $\cos 67.5^\circ + \cos 22.5^\circ = 2\cos\left(\dfrac{67.5^\circ+22.5^\circ}{2}\right)\cos\left(\dfrac{67.5^\circ-22.5^\circ}{2}\right) = 2\cos 45^\circ\cos 22.5^\circ = 2\cdot\dfrac{1}{\sqrt{2}}\cdot\sqrt{\dfrac{1+\cos 45^\circ}{2}}$

$= \sqrt{1+\cos 45^\circ} = \sqrt{1+\frac{1}{\sqrt{2}}} = \sqrt{\frac{2+\sqrt{2}}{2}}$

**In Solutions 59–64, $x$ and $y$ are in quadrant I, so we know that $\sec x = \frac{3}{2} \Rightarrow \cos x = \frac{2}{3}$, so $\sin x = \frac{\sqrt{5}}{3}$ and $\tan x = \frac{\sqrt{5}}{2}$. Also, $\csc y = 3 \Rightarrow \sin y = \frac{1}{3}$, and so $\cos y = \frac{2\sqrt{2}}{3}$, and $\tan y = \frac{1}{2\sqrt{2}} = \frac{\sqrt{2}}{4}$.**

**59.** $\sin(x+y) = \sin x\cos y + \cos x\sin y = \frac{\sqrt{5}}{3}\cdot\frac{2\sqrt{2}}{3} + \frac{2}{3}\cdot\frac{1}{3} = \frac{2}{9}(1+\sqrt{10})$.

**60.** $\cos(x-y) = \cos x\cos y + \sin x\sin y = \frac{2}{3}\cdot\frac{2\sqrt{2}}{3} + \frac{\sqrt{5}}{3}\cdot\frac{1}{3} = \frac{1}{9}(4\sqrt{2}+\sqrt{5})$.

**61.** $\tan(x+y) = \dfrac{\tan x + \tan y}{1-\tan x\tan y} = \dfrac{\frac{\sqrt{5}}{2}+\frac{\sqrt{2}}{4}}{1-\left(\frac{\sqrt{5}}{2}\right)\left(\frac{\sqrt{2}}{4}\right)} = \dfrac{\frac{\sqrt{5}}{2}+\frac{\sqrt{2}}{4}}{1-\left(\frac{\sqrt{5}}{2}\right)\left(\frac{\sqrt{2}}{4}\right)}\cdot\dfrac{8}{8} = \dfrac{2(2\sqrt{5}+\sqrt{2})}{8-\sqrt{10}}\cdot\dfrac{8+\sqrt{10}}{8+\sqrt{10}}$

$= \frac{2}{3}(\sqrt{2}+\sqrt{5})$

**62.** $\sin 2x = 2\sin x\cos x = 2\cdot\frac{\sqrt{5}}{3}\cdot\frac{2}{3} = \frac{4\sqrt{5}}{9}$.

**63.** $\cos\dfrac{y}{2} = \sqrt{\dfrac{1+\cos y}{2}} = \sqrt{\dfrac{1+\left(\frac{2\sqrt{2}}{3}\right)}{2}} = \sqrt{\dfrac{3+2\sqrt{2}}{6}}$ (since cosine is positive in quadrant I)

**64.** $\tan\dfrac{y}{2} = \dfrac{1-\cos y}{\sin y} = \dfrac{1-\frac{2\sqrt{2}}{3}}{\frac{1}{3}} = \dfrac{3-2\sqrt{2}}{1} = 3-2\sqrt{2}$

**65.** $\sin^{-1}\frac{\sqrt{3}}{2} = \frac{\pi}{3}$

**66.** $\tan^{-1}\frac{\sqrt{3}}{3} = \frac{\pi}{6}$

**67.** $\cos\left(\tan^{-1}\sqrt{3}\right) = \cos\frac{\pi}{3} = \frac{1}{2}$

**68.** $\sin\left(\cos^{-1}\frac{\sqrt{3}}{2}\right) = \sin\frac{\pi}{6} = \frac{1}{2}$

**69.** Let $u = \sin^{-1}\frac{2}{5}$ and so $\sin u = \frac{2}{5}$. Then from the triangle, $\tan\left(\sin^{-1}\frac{2}{5}\right) = \tan u = \frac{2}{\sqrt{21}}$.

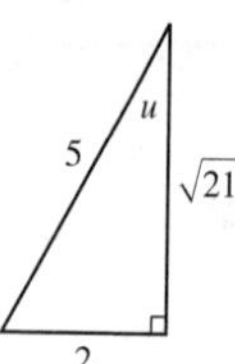


**70.** Let $u = \cos^{-1}\frac{3}{8}$ then $\cos u = \frac{3}{8}$. From the triangle, we have $\sin\left(\cos^{-1}\frac{3}{8}\right) = \sin u = \frac{\sqrt{55}}{8}$.

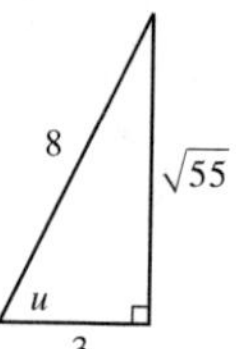


**71.** $\cos\left(2\sin^{-1}\frac{1}{3}\right) = 1 - 2\sin^2\left(\sin^{-1}\frac{1}{3}\right) = 1 - 2\left(\frac{1}{3}\right)^2 = \frac{7}{9}$

**72.** Let $u = \sin^{-1}\frac{5}{13}$ and $v = \cos^{-1}\frac{4}{5}$. Then $\sin u = \frac{5}{13}$ and $\cos v = \frac{4}{5}$. So from the triangles, we have

$$\begin{aligned}\cos\left(\sin^{-1}\tfrac{5}{13} - \cos^{-1}\tfrac{4}{5}\right) &= \cos(u - v)\\ &= \cos u\cos v + \sin u\sin v\\ &= \tfrac{12}{13}\cdot\tfrac{4}{5} + \tfrac{5}{13}\cdot\tfrac{3}{5} = \tfrac{63}{65}\end{aligned}$$

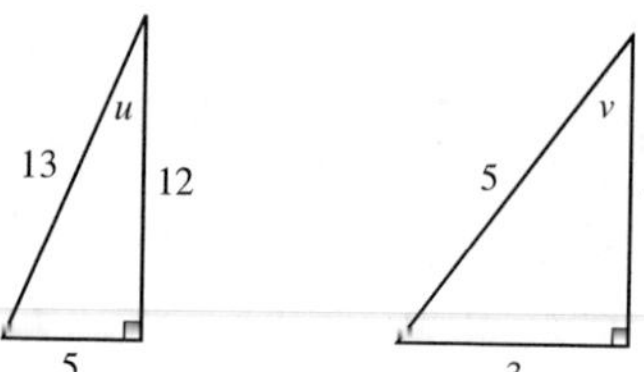


**73.** Let $\theta = \tan^{-1}x \quad\Leftrightarrow\quad \tan\theta = x$. Then from the triangle, we have $\sin\left(\tan^{-1}x\right) = \sin\theta = \dfrac{x}{\sqrt{1+x^2}}$.

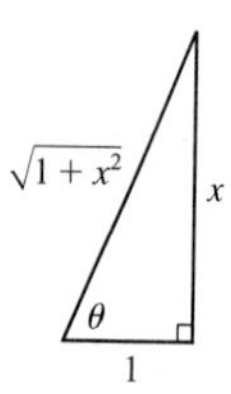


**74.** Let $\theta = \sin^{-1}x$. Then $\sin\theta = x$. From the triangle, we have $\sec\left(\sin^{-1}x\right) = \sec\theta = \dfrac{1}{\sqrt{1-x^2}}$.

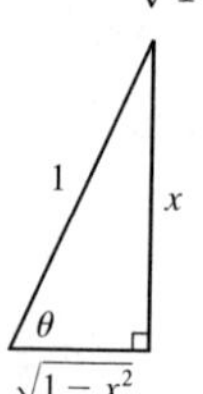


**75.** $\cos\theta = \dfrac{x}{3} \quad\Rightarrow\quad \theta = \cos^{-1}\left(\dfrac{x}{3}\right)$

**76.** $\tan\theta = \dfrac{x}{2} \quad\Rightarrow\quad \theta = \tan^{-1}\left(\dfrac{x}{2}\right)$

**77. (a)** $\tan\theta = \dfrac{10}{x} \quad\Leftrightarrow\quad \theta = \tan^{-1}\left(\dfrac{10}{x}\right)$

**(b)** $\theta = \tan^{-1}\left(\dfrac{10}{x}\right)$, for $x > 0$. Since the road sign can first be seen when $\theta = 2°$, we have $2° = \tan^{-1}\left(\dfrac{10}{x}\right) \quad\Leftrightarrow\quad x = \dfrac{10}{\tan 2°} \approx 286.4$ ft. Thus, the sign can first be seen at a height of 286.4 ft.

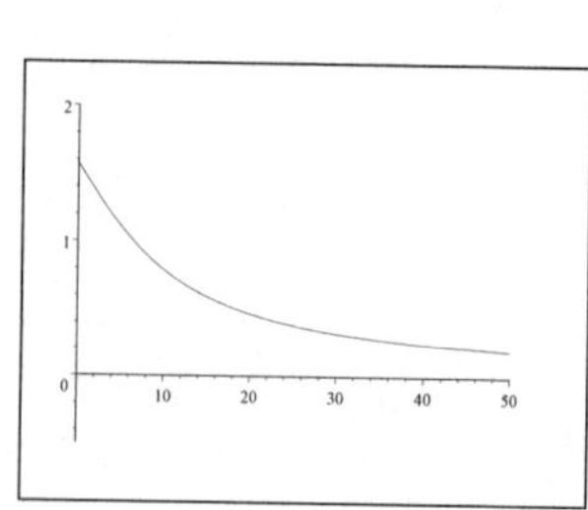

**78. (a)** Let $\gamma$ be the angle formed by the top of the tower, the car, and the base of the building, and let $\beta$ be the angle formed by base of the tower, the car, and the base of the building, as shown in the diagram. Then $\tan\gamma = \dfrac{420}{x} \Leftrightarrow$ $\gamma = \tan^{-1}\left(\dfrac{420}{x}\right)$ and $\tan\beta = \dfrac{380}{x} \Leftrightarrow \beta = \tan^{-1}\left(\dfrac{380}{x}\right)$. Thus, $\theta = \gamma - \beta = \tan^{-1}\left(\dfrac{420}{x}\right) - \tan^{-1}\left(\dfrac{380}{x}\right)$.

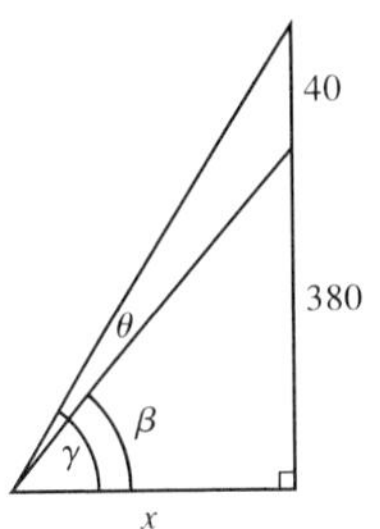


**(b)** We graph $\theta(x)$ and find that $\theta$ is maximized when $x = 400$.

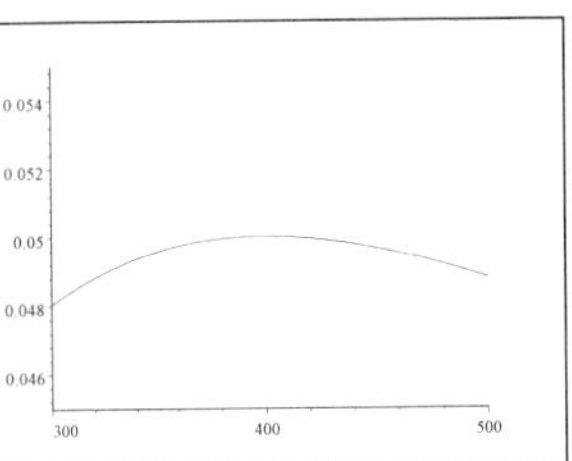


# Chapter 8 Test

**1. (a)** $\tan\theta\sin\theta + \cos\theta = \dfrac{\sin\theta}{\cos\theta}\sin\theta + \cos\theta = \dfrac{\sin^2\theta}{\cos\theta} + \dfrac{\cos^2\theta}{\cos\theta} = \dfrac{1}{\cos\theta} = \sec\theta$

**(b)** $\dfrac{\tan x}{1-\cos x} = \dfrac{\tan x}{1-\cos x}\cdot\dfrac{1+\cos x}{1+\cos x} = \dfrac{\tan x\,(1+\cos x)}{1-\cos^2 x} = \dfrac{\dfrac{\sin x}{\cos x}(1+\cos x)}{\sin^2 x} = \dfrac{1}{\sin x}\cdot$ $\dfrac{1+\cos x}{\cos x} = \csc x\,(1+\sec x)$

**(c)** $\dfrac{2\tan x}{1+\tan^2 x} = \dfrac{2\tan x}{\sec^2 x} = \dfrac{2\sin x}{\cos x}\cdot\cos^2 x = 2\sin x\cos x = \sin 2x$

**2.** $\dfrac{x}{\sqrt{4-x^2}} = \dfrac{2\sin\theta}{\sqrt{4-(2\sin\theta)^2}} = \dfrac{2\sin\theta}{\sqrt{4-4\sin^2\theta}} = \dfrac{2\sin\theta}{2\sqrt{1-\sin^2\theta}} = \dfrac{\sin\theta}{\sqrt{\cos^2\theta}} = \dfrac{\sin\theta}{|\cos\theta|} = \dfrac{\sin\theta}{\cos\theta} = \tan\theta$ (because $\cos\theta > 0$ for $-\frac{\pi}{2} < \theta < \frac{\pi}{2}$)

**3. (a)** $\sin 8^\circ\cos 22^\circ + \cos 8^\circ\sin 22^\circ = \sin(8^\circ + 22^\circ) = \sin 30^\circ = \frac{1}{2}$

**(b)** $\sin 75^\circ = \sin(45^\circ + 30^\circ) = \sin 45^\circ\cos 30^\circ + \cos 45^\circ\sin 30^\circ = \frac{\sqrt{2}}{2}\cdot\frac{\sqrt{3}}{2} + \frac{\sqrt{2}}{2}\cdot\frac{1}{2} = \frac{1}{4}\left(\sqrt{6}+\sqrt{2}\right)$

**(c)** $\sin\frac{\pi}{12} = \sin\left(\frac{\pi}{2}\right) = \sqrt{\dfrac{1-\cos\frac{\pi}{6}}{2}} = \sqrt{\dfrac{1-\frac{\sqrt{3}}{2}}{2}} = \sqrt{\dfrac{2-\sqrt{3}}{4}} = \frac{1}{2}\sqrt{2-\sqrt{3}}$

**4.** From the figures, we have $\cos(\alpha+\beta) = \cos\alpha\cos\beta - \sin\alpha\sin\beta = \frac{2}{\sqrt{5}}\cdot\frac{\sqrt{5}}{3} - \frac{1}{\sqrt{5}}\cdot\frac{2}{3} = \frac{2\sqrt{5}-2}{3\sqrt{5}} = \frac{10-2\sqrt{5}}{15}$.

**5. (a)** $\sin 3x\cos 5x = \frac{1}{2}\left[\sin(3x+5x) + \sin(3x-5x)\right] = \frac{1}{2}(\sin 8x - \sin 2x)$

**(b)** $\sin 2x - \sin 5x = 2\cos\left(\dfrac{2x+5x}{2}\right)\sin\left(\dfrac{2x-5x}{2}\right) = -2\cos\dfrac{7x}{2}\sin\dfrac{3x}{2}$

**6.** $\sin\theta = -\frac{4}{5}$. Since $\theta$ is in quadrant III, $\cos\theta = -\frac{3}{5}$. Then $\tan\dfrac{\theta}{2} = \dfrac{1-\cos\theta}{\sin\theta} = \dfrac{1-\left(-\frac{3}{5}\right)}{-\frac{4}{5}} = -\dfrac{1+\frac{3}{5}}{\frac{4}{5}} = -\dfrac{5+3}{4} = -2$.

**7.** $y = \sin x$ has domain $(-\infty, \infty)$ and $y = \sin^{-1} x$ has domain $[-1, 1]$.

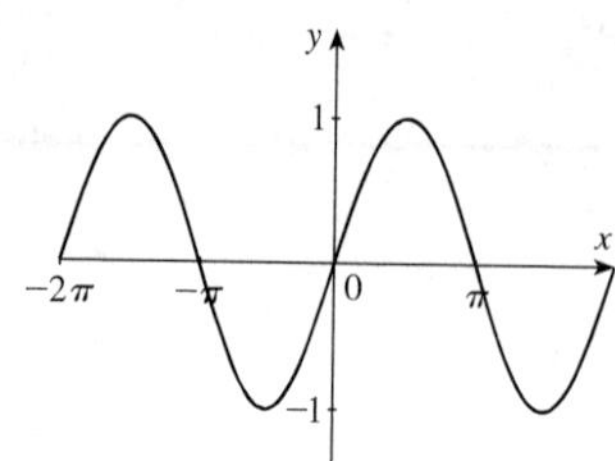


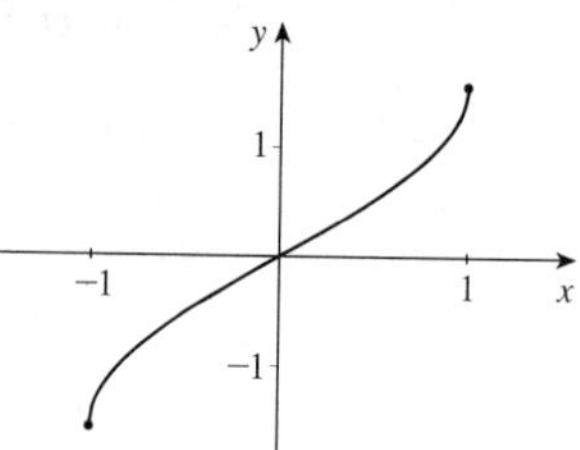


**8. (a)** $\tan\theta = \dfrac{x}{4} \quad\Rightarrow\quad \theta = \tan^{-1}\left(\dfrac{x}{4}\right)$

**(b)** $\cos\theta = \dfrac{3}{x} \quad\Rightarrow\quad \theta = \cos^{-1}\left(\dfrac{3}{x}\right)$

**9. (a)** $2\cos^2 x + 5\cos x + 2 = 0 \quad\Leftrightarrow\quad (2\cos x + 1)(\cos x + 2) = 0 \quad\Leftrightarrow\quad \cos x = -\frac{1}{2}$ or $\cos x = -2$ (which is impossible). So in the interval $[0, 2\pi)$, the solutions are $x = \frac{2\pi}{3}, \frac{4\pi}{3}$.

**(b)** $\sin 2x - \cos x = 0 \quad\Leftrightarrow\quad 2\sin x\cos x - \cos x = 0 \quad\Leftrightarrow\quad \cos x\,(2\sin x - 1) = 0 \quad\Leftrightarrow\quad \cos x = 0$ or $\sin x = \frac{1}{2}$ $\Leftrightarrow\quad x = \frac{\pi}{2}, \frac{3\pi}{2}$ or $x = \frac{\pi}{6}, \frac{5\pi}{6}$. Therefore, the solutions in $[0, 2\pi)$ are $x = \frac{\pi}{6}, \frac{\pi}{2}, \frac{5\pi}{6}, \frac{3\pi}{2}$.

**10.** $5\cos 2x = 2 \quad\Leftrightarrow\quad \cos 2x = \frac{2}{5} \quad\Leftrightarrow\quad 2x = \cos^{-1} 0.4 \approx 1.159279$. The solutions in $[0, 4\pi)$ are $2x \approx 1.159279$, $2\pi - 1.159279$, $2\pi + 1.159279$, $4\pi - 1.159279 \quad\Leftrightarrow\quad 2x \approx 1.159279, 5.123906, 7.442465, 11.407091 \quad\Leftrightarrow$ $x \approx 0.57964, 2.56195, 3.72123, 5.70355$ in $[0, 2\pi)$.

**11.** Let $u = \tan^{-1}\frac{9}{40}$ so $\tan u = \frac{9}{40}$. From the triangle, $r = \sqrt{9^2 + 40^2} = 41$. So $\cos\left(\tan^{-1}\frac{9}{40}\right) = \cos u = \frac{40}{41}$.

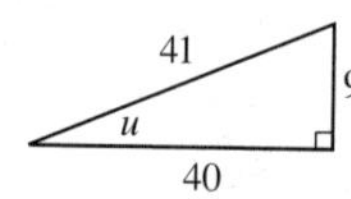

## Focus on Modeling: Traveling and Standing Waves

**1. (a)** Substituting $x = 0$, we get $y(0,t) = 5\sin\left(2 \cdot 0 - \frac{\pi}{2}t\right) = 5\sin\left(-\frac{\pi}{2}t\right) = -5\sin\frac{\pi}{2}t$.

**(b)**

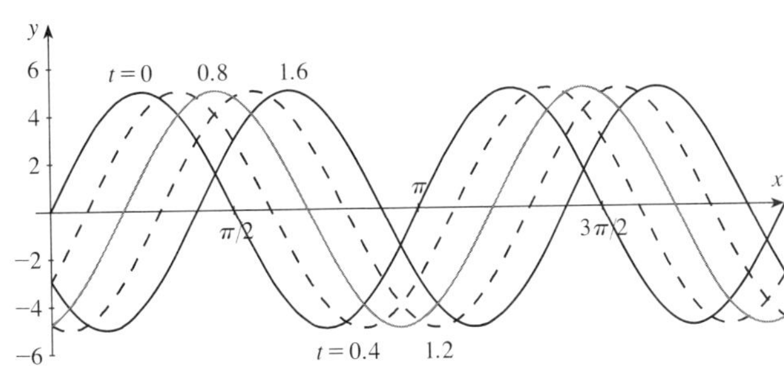


**(c)** We express the function in the standard form $y(x,t) = A\sin k(x - vt)$: $y(x,t) = 5\sin\left(2x - \frac{\pi}{2}t\right) = 5\sin 2\left(x - \frac{\pi}{4}t\right)$. Comparing this to the standard form, we see that the velocity of the wave is $v = \frac{\pi}{4}$.

**2. (a)** $y = 0.2\sin(1.047x - 0.524t) + 0.2\sin(1.047x + 0.524t) = 2(0.2)\sin 1.047x\cos 0.524t = 0.4\sin 1.047x\cos 0.524t$. The nodes occur when $1.047x = m\pi \quad\Leftrightarrow\quad x = \frac{\pi}{1.047}m \approx 3m$. So the nodes are at 3, 6, 9, 12, ....

**(b)**

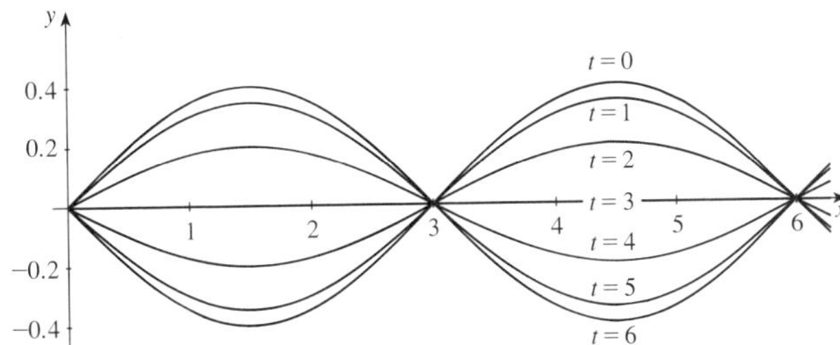


Note that when $t = 3$, $\cos(0.524 \cdot 3) \approx -0.001$, so $|y(x,3)| \leq 0.0004$. Thus, the graph of $y(x,3)$ cannot be distinguished from the $x$-axis in the diagram. Yes, this is a standing wave.

**3.** From the graph, we see that the amplitude is $A = 2.7$ and the period is 9.2, so $k = \frac{2\pi}{9.2} \approx 0.68$. Since $v = 6$, we have $kv = \frac{2\pi}{9.2} \cdot 6 \approx 4.10$, so the equation we seek is $y(x,t) = 2.7\sin(0.68x - 4.10t)$.

**4. (a)** We are given $A = 5$, period $\frac{2\pi}{3}$, and $v = 0.5$. Since the period is $\frac{2\pi}{3}$, we have $k = \frac{2\pi}{2\pi/3} = 3$. Thus, expressing the function in standard form, we have $y(x,t) = 5\sin 3(x - 0.5t) = 5\sin(3x - 1.5t)$.

**(b)**

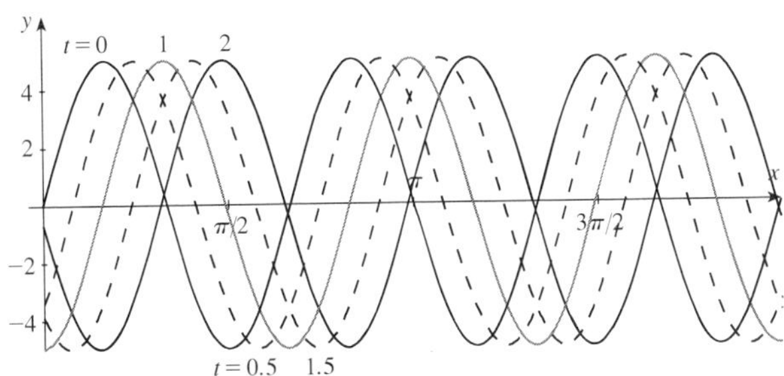


**5.** From the graphs, we see that the amplitude is $A = 0.6$. The nodes occur at $x = 0, 1, 2, 3$. Since $\sin\alpha x = 0$ when $\alpha x = k\pi$ ($k$ any integer), we have $\alpha = \pi$. Then since the frequency is $\beta/2\pi$, we get $20 = \beta/2\pi \quad\Leftrightarrow\quad \beta = 40\pi$. Thus, an equation for this model is $f(x,t) = 0.6\sin\pi x\cos 40\pi t$.

**6.** From the graph, we see that the amplitude is $A = 7$. Now $\sin\alpha x = 0$ when $\alpha x = k\pi$ ($k$ an integer). So for $k = 1$, we must have $\alpha\left(\frac{\pi}{2}\right) = \pi \quad\Leftrightarrow\quad \alpha = 2$. Then since the period is $4\pi$, we have $2\pi/\beta = 4\pi \quad\Leftrightarrow\quad \beta = \frac{1}{2}$. Thus, an equation for this model is $f(x,t) = 7\sin 2x\cos\frac{1}{2}t$.

**7.** **(a)** The first standing wave has $\alpha = 1$, the second has $\alpha = 2$, the third has $\alpha = 3$, and the fourth has $\alpha = 4$.

**(b)** $\alpha$ is equal to the number of nodes minus 1. The first string has two nodes and $\alpha = 1$; the second string has three nodes and $\alpha = 2$, and so forth.

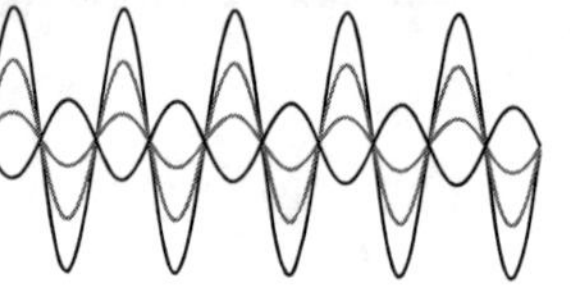

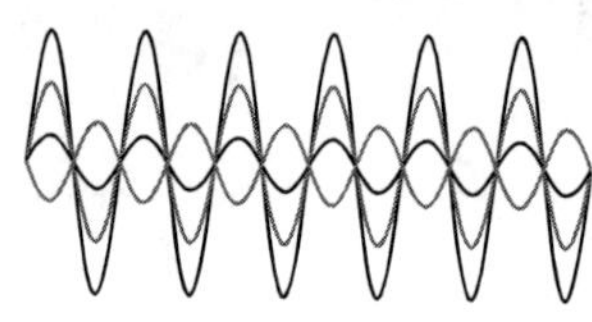

**(c)** Since the frequency is $\beta/2\pi$, we have $440 = \beta/2\pi \quad \Leftrightarrow \quad \beta = 880\pi$.

**(d)** The first standing wave has equation $y = \sin x \cos 880\pi t$, the second has equation $y = \sin 2x \cos 880\pi t$, the third has equation $y = \sin 3x \cos 880\pi t$, and the fourth has equation $y = \sin 4x \cos 880\pi t$.

**8.** **(a)** The nodes of the tube occur when $\cos \frac{1}{2}x = 0$ and $0 \le x \le 37.7$. So $\frac{1}{2}x = (2k+1)\frac{\pi}{2} \Leftrightarrow x = (2k+1)\pi$. Thus, the nodes are at $x = \pi, 3\pi, 5\pi, 7\pi, 9\pi$, and $11\pi$. We stop there since $13\pi > 37.7$. Note that the endpoints of the tube ($x = 0$ and $x = 37.7$) are not nodes.

**(b)** In the function $y = A \cos \alpha x \cos \beta t$, the frequency is $\beta/2\pi$. In this case, the frequency is $\frac{50\pi}{2\pi} = 25$ Hz.

# 9 Polar Coordinates and Vectors

## 9.1 Polar Coordinates

1. 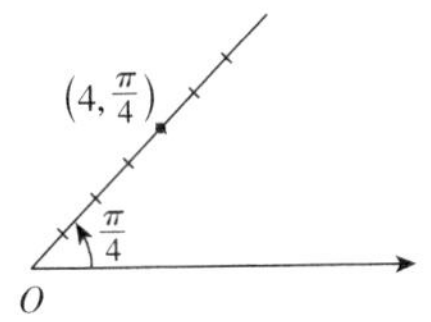


2. 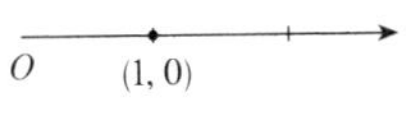


3. 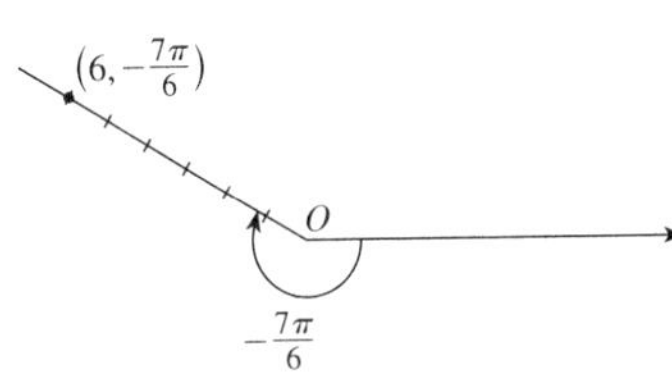


4. 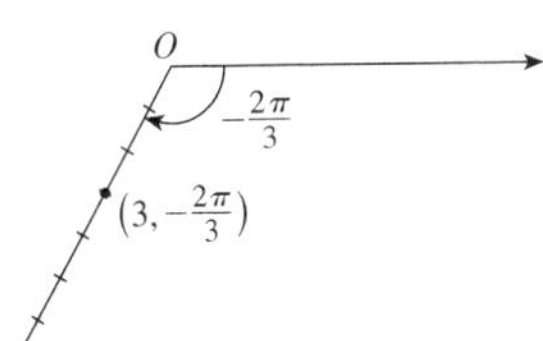


5. 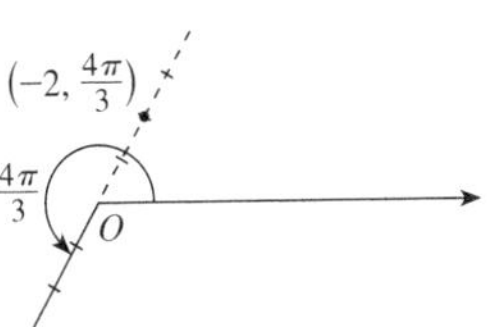


6. 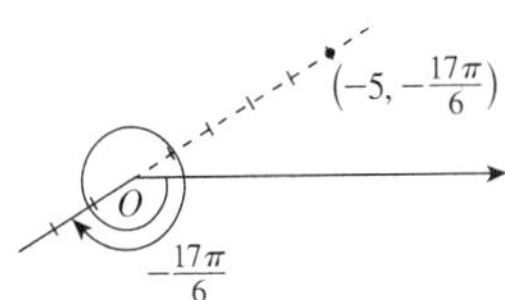


**Answers to Exercises 7–12 will vary.**

7. $\left(3, \frac{\pi}{2}\right)$ also has polar coordinates $\left(3, \frac{5\pi}{2}\right)$ or $\left(-3, \frac{3\pi}{2}\right)$

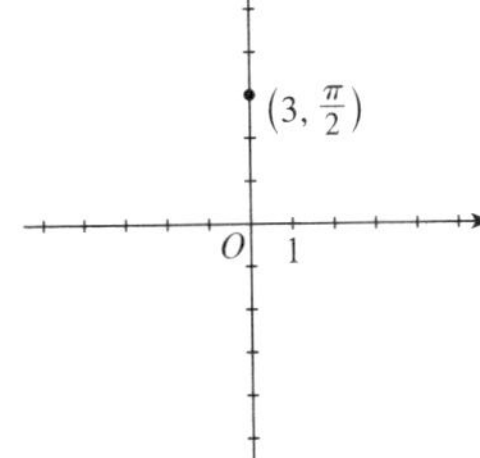


8. $\left(2, \frac{3\pi}{4}\right)$ also has polar coordinates $\left(2, \frac{11\pi}{4}\right)$ or $\left(-2, \frac{7\pi}{4}\right)$.

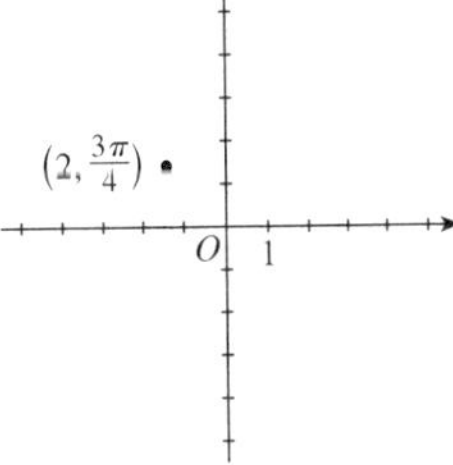


9. $\left(-1, \frac{7\pi}{6}\right)$ also has polar coordinates $\left(1, \frac{\pi}{6}\right)$ or $\left(-1, -\frac{5\pi}{6}\right)$.

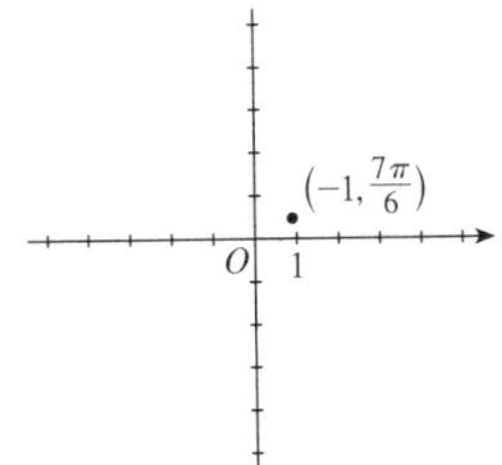


10. $\left(-2, -\frac{\pi}{3}\right)$ also has polar coordinates $\left(2, \frac{2\pi}{3}\right)$ or $\left(-2, \frac{5\pi}{3}\right)$.

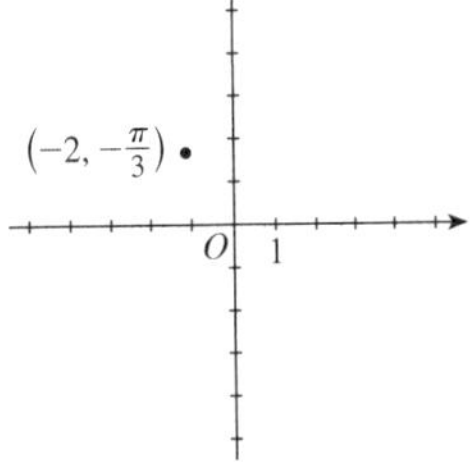

**11.** $(-5, 0)$ also has polar coordinates $(5, \pi)$ or $(-5, 2\pi)$.

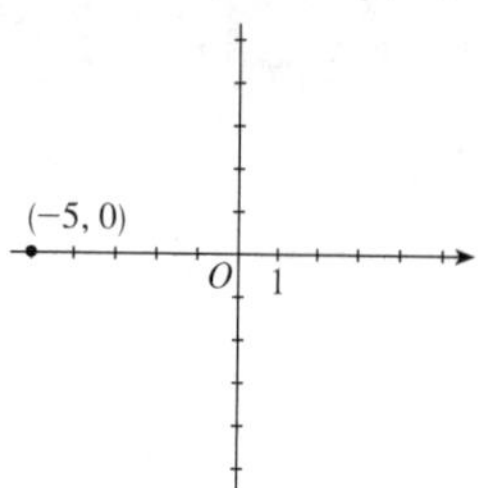


**12.** $(3, 1)$ also has polar coordinates $(3, 1 + 2\pi)$ or $(-3, 1 + \pi)$.

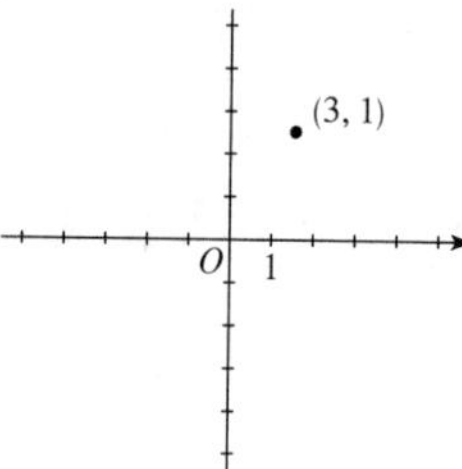


**13.** $Q$ has coordinates $\left(4, \frac{3\pi}{4}\right)$.

**14.** $R$ has coordinates $\left(4, -\frac{3\pi}{4}\right) = \left(4, \frac{5\pi}{4}\right)$.

**15.** $Q$ has coordinates $\left(-4, -\frac{\pi}{4}\right) = \left(4, \frac{3\pi}{4}\right)$.

**16.** $P$ has coordinates $\left(-4, \frac{13\pi}{4}\right) = \left(-4, \frac{5\pi}{4}\right) = \left(4, \frac{\pi}{4}\right)$.

**17.** $P$ has coordinates $\left(4, -\frac{23\pi}{4}\right) = \left(4, \frac{\pi}{4}\right)$.

**18.** $Q$ has coordinates $\left(-4, \frac{23\pi}{4}\right) = (-4, -\pi/4) = \left(4, \frac{3\pi}{4}\right)$.

**19.** $P$ has coordinates $\left(-4, \frac{101\pi}{4}\right) = \left(-4, \frac{5\pi}{4}\right) = \left(4, \frac{\pi}{4}\right)$.

**20.** $S$ has coordinates $\left(4, \frac{103\pi}{4}\right) = \left(4, \frac{7\pi}{4}\right)$.

**21.** $P = (-3, 3)$ in rectangular coordinates, so $r^2 = x^2 + y^2 = (-3)^2 + 3^2 = 18$ and we can take $r = 3\sqrt{2}$. $\tan\theta = \dfrac{y}{x} = \dfrac{3}{-3} = -1$, so since $P$ is in quadrant 2 we take $\theta = \frac{3\pi}{4}$. Thus, polar coordinates for $P$ are $\left(3\sqrt{2}, \frac{3\pi}{4}\right)$.

**22.** $Q = (0, -3)$ in rectangular coordinates, so $r = 3$ and $\theta = \frac{3\pi}{2}$. Polar coordinates for $Q$ are $\left(3, \frac{3\pi}{2}\right)$.

**23.** Here $r = 5$ and $\theta = -\frac{2\pi}{3}$, so $x = r\cos\theta = 5\cos\left(-\frac{2\pi}{3}\right) = -\frac{5}{2}$ and $y = r\sin\theta = 5\sin\left(-\frac{2\pi}{3}\right) = -\frac{5\sqrt{3}}{2}$. $R$ has rectangular coordinates $\left(-\frac{5}{2}, -\frac{5\sqrt{3}}{2}\right)$.

**24.** $r = 2$ and $\theta = \frac{5\pi}{6}$, so $S$ has rectangular coordinates $(r\cos\theta, r\sin\theta) = \left(2\cos\frac{5\pi}{6}, 2\sin\frac{5\pi}{6}\right) = (-\sqrt{3}, 1)$.

**25.** $(r, \theta) = \left(4, \frac{\pi}{6}\right)$. So $x = r\cos\theta = 4\cos\frac{\pi}{6} = 4 \cdot \frac{\sqrt{3}}{2} = 2\sqrt{3}$ and $y = r\sin\theta = 4\sin\frac{\pi}{6} = 4 \cdot \frac{1}{2} = 2$. Thus, the rectangular coordinates are $(2\sqrt{3}, 2)$.

**26.** $(r, \theta) = \left(6, \frac{2\pi}{3}\right)$. So $x = r\cos\theta = 6\cos\frac{2\pi}{3} = -6 \cdot \frac{1}{2} = -3$ and $y = r\sin\theta = 6\sin\frac{2\pi}{3} = 6 \cdot \frac{\sqrt{3}}{2} = 3\sqrt{3}$. Thus, the rectangular coordinates are $(-3, 3\sqrt{3})$.

**27.** $(r, \theta) = \left(\sqrt{2}, -\frac{\pi}{4}\right)$. So $x = r\cos\theta = \sqrt{2}\cos\left(-\frac{\pi}{4}\right) = \sqrt{2} \cdot \frac{1}{\sqrt{2}} = 1$, and $y = r\sin\theta = \sqrt{2}\sin\left(-\frac{\pi}{4}\right) = \sqrt{2}\left(-\frac{1}{\sqrt{2}}\right) = -1$. Thus, the rectangular coordinates are $(1, -1)$.

**28.** $(r, \theta) = \left(-1, \frac{5\pi}{2}\right)$. So $x = r\cos\theta = -1\cos\frac{5\pi}{2} = -1 \cdot 0 = 0$ and $y = r\sin\theta = -1\sin\frac{5\pi}{2} = -1 \cdot 1 = -1$. Thus, the rectangular coordinates are $(0, -1)$.

**29.** $(r, \theta) = (5, 5\pi)$. So $x = r\cos\theta = 5\cos 5\pi = -5$, and $y = r\sin\theta = 5\sin 5\pi = 0$. Thus, the rectangular coordinates are $(-5, 0)$.

**30.** $(r, \theta) = (0, 13\pi)$. So $(x, y) = (0, 0)$ because $r = 0$.

**31.** $(r, \theta) = \left(6\sqrt{2}, \frac{11\pi}{6}\right)$. So $x = r\cos\theta = 6\sqrt{2}\cos\frac{11\pi}{6} = 3\sqrt{6}$ and $y = r\sin\theta = 6\sqrt{2}\sin\frac{11\pi}{6} = -3\sqrt{2}$. Thus, the rectangular coordinates are $(3\sqrt{6}, -3\sqrt{2})$.

**32.** $(r, \theta) = \left(\sqrt{3}, -\frac{5\pi}{3}\right)$. So $x = r\cos\theta = \sqrt{3}\cos\left(-\frac{5\pi}{3}\right) = \frac{\sqrt{3}}{2}$ and $y = r\sin\theta = \sqrt{3}\sin\left(-\frac{5\pi}{3}\right) = \frac{3}{2}$. Thus, the rectangular coordinates are $\left(\frac{\sqrt{3}}{2}, \frac{3}{2}\right)$.

**33.** $(x, y) = (-1, 1)$. Since $r^2 = x^2 + y^2$, we have $r^2 = (-1)^2 + 1^2 = 2$, so $r = \sqrt{2}$. Now $\tan\theta = \dfrac{y}{x} = \dfrac{1}{-1} = -1$, so, since the point is in the second quadrant, $\theta = \frac{3\pi}{4}$. Thus, polar coordinates are $\left(\sqrt{2}, \frac{3\pi}{4}\right)$.

**34.** $(x,y)=(3\sqrt{3},-3)$. Since $r^2=x^2+y^2$, we have $r^2=(3\sqrt{3})^2+(-3)^2=36$, so $r=6$. Now $\tan\theta=\dfrac{y}{x}=-\frac{3}{3\sqrt{3}}=-\dfrac{1}{\sqrt{3}}$, so, since the point is in the fourth quadrant, $\theta=\frac{11\pi}{6}$. Thus, polar coordinates are $\left(6,\frac{11\pi}{6}\right)$.

**35.** $(x,y)=(\sqrt{8},\sqrt{8})$. Since $r^2=x^2+y^2$, we have $r^2=(\sqrt{8})^2+(\sqrt{8})^2=16$, so $r=4$. Now $\tan\theta=\dfrac{y}{x}=\frac{\sqrt{8}}{\sqrt{8}}=1$, so, since the point is in the first quadrant, $\theta=\frac{\pi}{4}$. Thus, polar coordinates are $\left(4,\frac{\pi}{4}\right)$.

**36.** $(x,y)=(-\sqrt{6},-\sqrt{2})$. Since $r^2=x^2+y^2$, we have $r^2=(-\sqrt{6})^2+(-\sqrt{2})^2=8$, so $r=2\sqrt{2}$. Now $\tan\theta=\dfrac{y}{x}=\dfrac{-\sqrt{2}}{-\sqrt{6}}=\frac{1}{\sqrt{3}}$, so, since the point is in the third quadrant, $\theta=\frac{7\pi}{6}$. Thus, polar coordinates are $\left(2\sqrt{2},\frac{7\pi}{6}\right)$.

**37.** $(x,y)=(3,4)$. Since $r^2=x^2+y^2$, we have $r^2=3^2+4^2=25$, so $r=5$. Now $\tan\theta=\dfrac{y}{x}=\frac{4}{3}$, so, since the point is in the first quadrant, $\theta=\tan^{-1}\frac{4}{3}$. Thus, polar coordinates are $\left(5,\tan^{-1}\frac{4}{3}\right)$.

**38.** $(x,y)=(1,-2)$. Since $r^2=x^2+y^2$, we have $r^2=1^2+(-2)^2=2$, so $r=\sqrt{5}$. Now, $\tan\theta=\dfrac{y}{x}=\dfrac{-2}{1}=-2$, and since the point is in the fourth quadrant, $\theta=2\pi+\tan^{-1}(-2)$ (since we need $0\le\theta<2\pi$). Thus, polar coordinates are $\left(\sqrt{5},2\pi+\tan^{-1}(-2)\right)$.

**39.** $(x,y)=(-6,0)$. $r^2=(-6^2)=36$, so $r=6$. Now $\tan\theta=\dfrac{y}{x}=0$, so since the point is on the negative $x$-axis, $\theta=\pi$. Thus, polar coordinates are $(6,\pi)$.

**40.** $(x,y)=(0,-\sqrt{3})$. $r=\sqrt{3}$ and since the point is on the negative $y$-axis, $\theta=\frac{3\pi}{2}$. Thus, polar coordinates are $\left(\sqrt{3},\frac{3\pi}{2}\right)$.

**41.** $x=y \quad\Leftrightarrow\quad r\cos\theta=r\sin\theta \quad\Leftrightarrow\quad \tan\theta=1$, and so $\theta=\frac{\pi}{4}$.

**42.** $x^2+y^2=9$. By substitution, $(r\cos\theta)^2+(r\sin\theta)^2=9 \quad\Leftrightarrow\quad r^2\left(\cos^2\theta+\sin^2\theta\right)=9 \quad\Leftrightarrow\quad r^2=9 \quad\Leftrightarrow\quad r=3$.

**43.** $y=x^2$. We substitute and then solve for $r$: $r\sin\theta=(r\cos\theta)^2=r^2\cos^2\theta \quad\Leftrightarrow\quad \sin\theta=r\cos^2\theta \quad\Leftrightarrow\quad r=\dfrac{\sin\theta}{\cos^2\theta}=\tan\theta\sec\theta$.

**44.** $y=5$. By substitution, $r\sin\theta=5 \quad\Leftrightarrow\quad r=\dfrac{5}{\sin\theta}=5\csc\theta$.

**45.** $x=4$. We substitute and then solve for $r$: $r\cos\theta=4 \quad\Leftrightarrow\quad r=\dfrac{4}{\cos\theta}=4\sec\theta$.

**46.** $x^2-y^2=1$. By substitution, $(r\cos\theta)^2-(r\sin\theta)^2=1 \quad\Leftrightarrow\quad r^2\left(\cos^2\theta-\sin^2\theta\right)=1 \quad\Leftrightarrow\quad r^2\cos2\theta=1 \quad\Leftrightarrow\quad r^2=\dfrac{1}{\cos2\theta}=\sec2\theta$.

**47.** $r=7$. But $r^2=x^2+y^2$, so $x^2+y^2=r^2=49$. Hence, the equivalent equation in rectangular coordinates is $x^2+y^2=49$.

**48.** $\theta=\pi \quad\Rightarrow\quad \tan\theta=0 \quad\Rightarrow\quad \dfrac{y}{x}=0 \quad\Rightarrow\quad y=0$.

**49.** $r\cos\theta=6$. But $x=r\cos\theta$, and so $x=6$ is the equation.

**50.** $r=6\cos\theta \quad\Leftrightarrow\quad r^2=6r\cos\theta$. By substitution, $x^2+y^2=6x \quad\Leftrightarrow\quad \left(x^2-6x+9\right)+y^2=9 \quad\Leftrightarrow\quad (x-3)^2+y^2=9$.

**51.** $r^2=\tan\theta$. Substituting $r^2=x^2+y^2$ and $\tan\theta=\dfrac{y}{x}$, we get $x^2+y^2=\dfrac{y}{x}$.

**52.** $r^2=\sin2\theta=2\sin\theta\cos\theta \quad\Leftrightarrow\quad r^4=2r^2\sin\theta\cos\theta=2(r\cos\theta)(r\sin\theta)$. By substitution, $\left(x^2+y^2\right)^2=2xy \quad\Leftrightarrow\quad x^4+2x^2y^2+y^4-2xy=0$.

**53.** $r=\dfrac{1}{\sin\theta-\cos\theta} \quad\Rightarrow\quad r(\sin\theta-\cos\theta)=1 \quad\Leftrightarrow\quad r\sin\theta-r\cos\theta=1$, and since $r\cos\theta=x$ and $r\sin\theta=y$, we get $y-x=1$.

**54.** $r = \dfrac{1}{1+\sin\theta} \quad\Leftrightarrow\quad r(1+\sin\theta) = 1 \quad\Leftrightarrow\quad r + r\sin\theta = 1$. Thus $r = 1 - r\sin\theta$, and squaring both sides gives $r^2 = (1 - r\sin\theta)^2 \quad\Leftrightarrow\quad x^2 + y^2 = (1-y)^2 = 1 - 2y + y^2 \quad\Leftrightarrow\quad x^2 + 2y - 1 = 0$.

**55.** $r = 1 + \cos\theta$. If we multiply both sides of this equation by $r$ we get $r^2 = r + r\cos\theta$. Thus $r^2 - r\cos\theta = r$, and squaring both sides gives $\left(r^2 - r\cos\theta\right)^2 = r^2 \quad\Leftrightarrow\quad \left(x^2 + y^2 - x\right)^2 = x^2 + y^2$

**56.** $r = \dfrac{4}{1+2\sin\theta} \quad\Leftrightarrow\quad r(1+2\sin\theta) = 4 \quad\Leftrightarrow\quad r + 2r\sin\theta = 4$. Thus $r = 4 - 2r\sin\theta$. Squaring both sides, we get $r^2 = (4 - 2r\sin\theta)^2$. Substituting, $x^2 + y^2 = (4-2y)^2 \quad\Leftrightarrow\quad x^2 + y^2 = 16 - 16y + 4y^2 \quad\Leftrightarrow\quad x^2 - 3y^2 + 16y - 16 = 0$.

**57.** $r = 2\sec\theta \quad\Leftrightarrow\quad r = 2\cdot\dfrac{1}{\cos\theta} \quad\Leftrightarrow\quad r\cos\theta = 2 \quad\Leftrightarrow\quad x = 2$.

**58.** $r = 2 - \cos\theta \quad\Leftrightarrow\quad r^2 = 2r - r\cos\theta \quad\Leftrightarrow\quad r^2 + r\cos\theta = 2r \quad\Leftrightarrow\quad \left(r^2 + r\cos\theta\right)^2 = (2r)^2 \quad\Leftrightarrow\quad \left(r^2 + r\cos\theta\right)^2 = 4r^2 \quad\Rightarrow\quad \left(x^2 + y^2 - x\right)^2 = 4\left(x^2 + y^2\right)$

**59.** $\sec\theta = 2 \quad\Leftrightarrow\quad \cos\theta = \frac{1}{2} \quad\Leftrightarrow\quad \theta = \pm\frac{\pi}{3} \quad\Leftrightarrow\quad \tan\theta = \pm\sqrt{3} \quad\Leftrightarrow\quad \dfrac{y}{x} = \pm\sqrt{3} \quad\Leftrightarrow\quad y = \pm\sqrt{3}x$.

**60.** $\cos 2\theta = 1$ means that $2\theta = 0 \quad\Rightarrow\quad \theta = 0 \quad\Rightarrow\quad \tan\theta = 0 \quad\Rightarrow\quad \dfrac{y}{x} = 0 \quad\Rightarrow\quad y = 0$

**61. (a)** In rectangular coordinates, the points $(r_1, \theta_1)$ and $(r_2, \theta_2)$ are $(x_1, y_1) = (r_1\cos\theta_1, r_1\sin\theta_1)$ and $(x_2, y_2) = (r_2\cos\theta_2, r_2\sin\theta_2)$. Then, the distance between the points is

$$\begin{aligned} D &= \sqrt{(x_1 - x_2)^2 + (y_1 - y_2)^2} = \sqrt{(r_1\cos\theta_1 - r_2\cos\theta_2)^2 + (r_1\sin\theta_1 - r_2\sin\theta_2)^2} \\ &= \sqrt{r_1^2\left(\cos^2\theta_1 + \sin^2\theta_1\right) + r_2^2\left(\cos^2\theta_2 + \sin^2\theta_2\right) - 2r_1r_2\left(\cos\theta_1\cos\theta_2 + \sin\theta_1\sin\theta_2\right)} \\ &= \sqrt{r_1^2 + r_2^2 - 2r_1r_2\cos\left(\theta_2 - \theta_1\right)} \end{aligned}$$

**(b)** The distance between the points $\left(3, \frac{3\pi}{4}\right)$ and $\left(-1, \frac{7\pi}{6}\right)$ is

$$D = \sqrt{3^2 + (-1)^2 - 2(3)(-1)\cos\left(\tfrac{7\pi}{6} - \tfrac{3\pi}{4}\right)} = \sqrt{9 + 1 + 6\cos\tfrac{5\pi}{12}} \approx 3.40$$

## 9.2 Graphs of Polar Equations

**1.** VI **2.** III **3.** II **4.** IV **5.** I **6.** V

**7.** Polar axis: $2 - \sin(-\theta) = 2 + \sin\theta \neq r$, so the graph is not symmetric about the polar axis.
Pole: $2 - \sin(\theta + \pi) = 2 - (\sin\pi\cos\theta + \cos\pi\sin\theta) = 2 - (-\sin\theta) = 2 + \sin\theta \neq r$, so the graph is not symmetric about the pole.
Line $\theta = \frac{\pi}{2}$: $2 - \sin(\pi - \theta) = 2 - (\sin\pi\cos\theta - \cos\pi\sin\theta) = 2 - \sin\theta = r$, so the graph is symmetric about $\theta = \frac{\pi}{2}$.

**8.** Polar axis: $4 + 8\cos(-\theta) = 4 + 8\cos\theta = r$, so the graph is symmetric about the polar axis.
Pole: $4 + 8\cos(\theta + \pi) = 4 + 8(\cos\pi\cos\theta - \sin\pi\sin\theta) = 4 + 8(-\cos\theta) \neq r$, so the graph is not symmetric about the pole.
Line $\theta = \frac{\pi}{2}$: $4 + 8\cos(\pi - \theta) = 4 + 8(\cos\pi\cos\theta + \sin\pi\sin\theta) = 4 + 8(-\cos\theta) \neq r$, so the graph is not symmetric about $\theta = \frac{\pi}{2}$.

**9.** Polar axis: $3\sec(-\theta) = 3\sec\theta = r$, so the graph is symmetric about the polar axis.

Pole: $3\sec(\theta+\pi) = \dfrac{3}{\cos(\theta+\pi)} = \dfrac{1}{\cos\pi\cos\theta - \sin\pi\sin\theta} = \dfrac{3}{-\cos\theta} = -3\sec\theta \neq r$, so the graph is not symmetric about the pole.

Line $\theta = \frac{\pi}{2}$: $3\sec(\pi-\theta) = \dfrac{3}{\cos(\pi-\theta)} = \dfrac{1}{\cos\pi\cos\theta + \sin\pi\sin\theta} = \dfrac{3}{-\cos\theta} = -3\sec\theta \neq r$, so the graph is not symmetric about $\theta = \frac{\pi}{2}$.

**10.** Polar axis: $5\cos(-\theta)\csc(-\theta) = 5\cos\theta(-\csc\theta) \neq r$, so the graph is not symmetric about the polar axis.

Pole: $5\cos(\theta+\pi)\csc(\theta+\pi) = 5\cos(\theta+\pi)\dfrac{1}{\sin(\theta+\pi)}$

$$= 5(\cos\pi\cos\theta - \sin\pi\sin\theta)\frac{1}{\sin\pi\cos\theta + \cos\pi\sin\theta} = 5(-\cos\theta)\frac{1}{-\sin\theta} = 5\cos\theta\csc\theta = r,$$

so the graph is symmetric about the pole.

Line $\theta = \frac{\pi}{2}$: $5\cos(\pi-\theta)\csc(\pi-\theta) = 5\cos(\pi-\theta)\dfrac{1}{\sin(\pi-\theta)}$

$$= 5(\cos\pi\cos\theta + \sin\pi\sin\theta)\frac{1}{\sin\pi\cos\theta - \cos\pi\sin\theta} = 5(-\cos\theta)\frac{1}{\sin\theta} = -5\cos\theta\csc\theta \neq r,$$

so the graph is not symmetric about $\theta = \frac{\pi}{2}$.

**11.** Polar axis: $\dfrac{4}{3-2\sin(-\theta)} = \dfrac{4}{3+2\sin\theta} \neq r$, so the graph is not symmetric about the polar axis.

Pole: $\dfrac{4}{3-2\sin(\theta+\pi)} = \dfrac{4}{3-2(\sin\pi\cos\theta+\cos\pi\sin\theta)} = \dfrac{4}{3-2(-\sin\theta)} = \dfrac{4}{3+2\sin\theta} \neq r$, so the graph is not symmetric about the pole.

Line $\theta = \frac{\pi}{2}$: $\dfrac{4}{3-2\sin(\pi-\theta)} = \dfrac{4}{3-2(\sin\pi\cos\theta - \cos\pi\sin\theta)} = \dfrac{4}{3-2\sin\theta} = r$, so the graph is symmetric about $\theta = \frac{\pi}{2}$.

**12.** Polar axis: $\dfrac{5}{1+3\cos(-\theta)} = \dfrac{5}{1+3\cos\theta} = r$, so the graph is symmetric about the polar axis.

Pole: $\dfrac{5}{1+3\cos(\theta+\pi)} = \dfrac{5}{1+3(\cos\pi\cos\theta - \sin\pi\sin\theta)} = \dfrac{5}{1+3(-\cos\theta)} = \dfrac{5}{1-3\cos\theta} \neq r$, so the graph is not symmetric about the pole.

Line $\theta = \frac{\pi}{2}$: $\dfrac{5}{1+3\cos(\pi-\theta)} = \dfrac{5}{1+3(\cos\pi\cos\theta + \sin\pi\sin\theta)} = \dfrac{5}{1-3\cos\theta} \neq r$, so the graph is not symmetric about $\theta = \frac{\pi}{2}$.

**13.** Polar axis: $4\cos 2(-\theta) = 4\cos 2\theta = r^2$, so the graph is symmetric about the polar axis.

Pole: $(-r)^2 = r^2$, so the graph is symmetric about the pole.

Line $\theta = \frac{\pi}{2}$: $4\cos 2(\pi-\theta) = 4\cos(2\pi - 2\theta) = 4\cos(-2\theta) = 4\cos 2\theta = r^2$, so the graph is symmetric about $\theta = \frac{\pi}{2}$.

**14.** Polar axis: $9\sin(-\theta) = -9\sin\theta \neq r^2$, so the graph is not symmetric about the polar axis.

Pole: $(-r)^2 = r^2$, so the graph is symmetric about the pole.

Line $\theta = \frac{\pi}{2}$: $9\sin(\pi-\theta) = 9(\sin\pi\cos\theta - \cos\pi\sin\theta) = 9\sin\theta = r^2$, so the graph is symmetric about $\theta = \frac{\pi}{2}$.

**15.** $r = 2$. Circle.

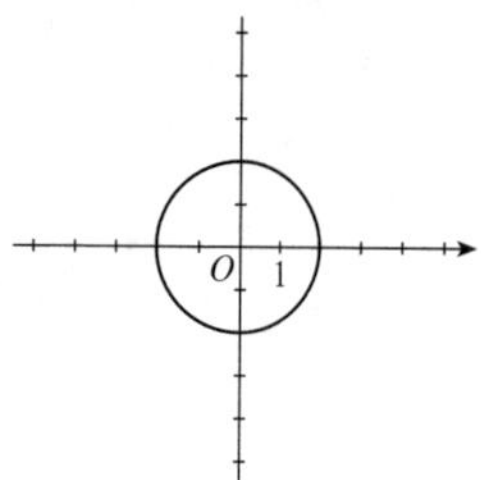


**16.** $r = -1$. Circle.

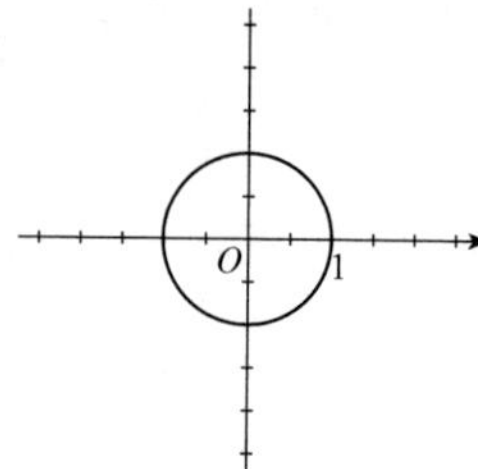


**17.** $\theta = -\frac{\pi}{2}$. Line.

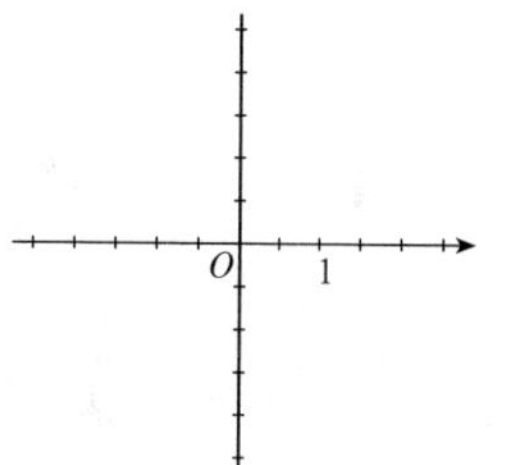


**18.** $\theta = \frac{5\pi}{6}$. Line.

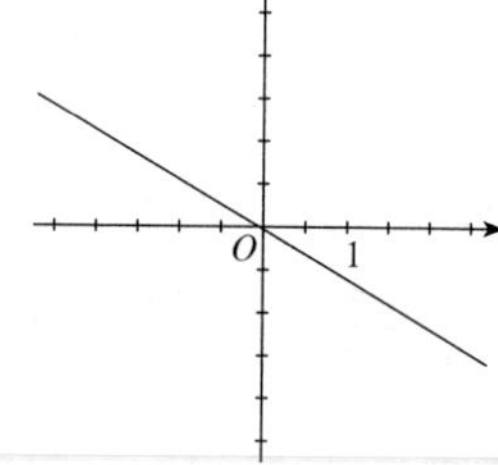


**19.** $r = 6\sin\theta$. Circle.

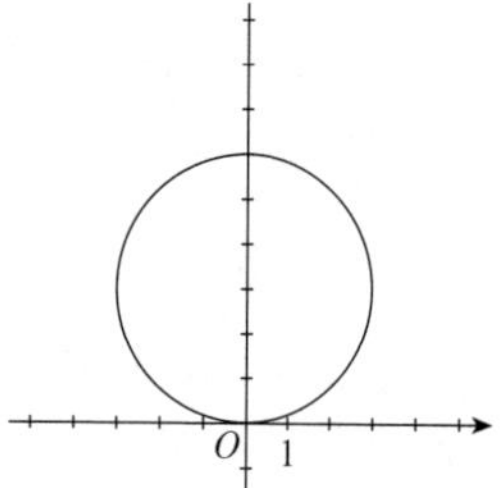


**20.** $r = \cos\theta$. Circle.

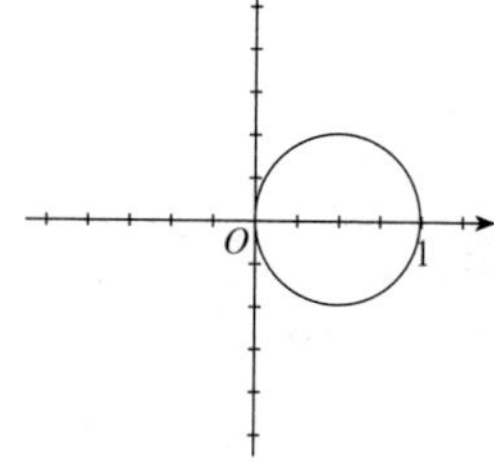


**21.** $r = -2\cos\theta$. Circle.

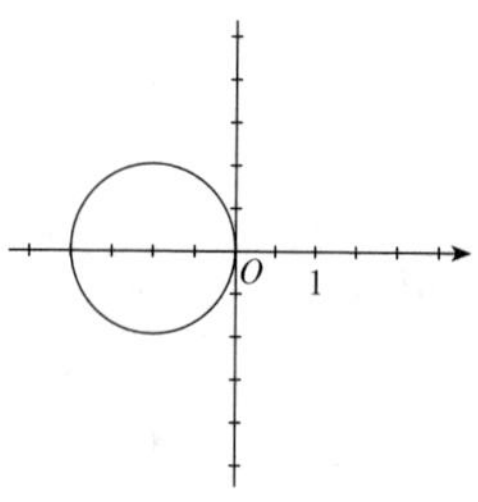


**22.** $r = 2\sin\theta + 2\cos\theta \quad \Leftrightarrow \quad r^2 = 2r\sin\theta + 2r\cos\theta$
$\Leftrightarrow \quad x^2 + y^2 = 2x + 2y \quad \Leftrightarrow$
$(x-1)^2 + (y-1)^2 = 2$. Circle.

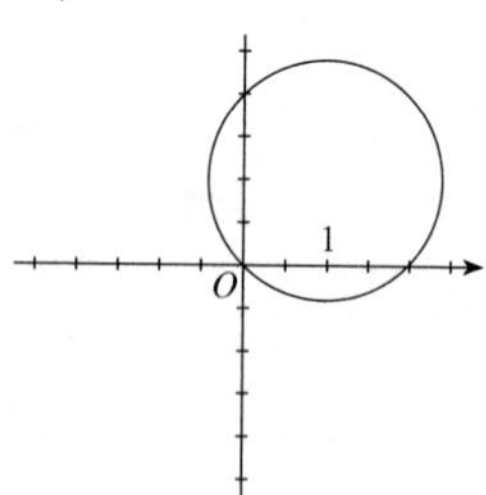


**23.** $r = 2 - 2\cos\theta$. Cardioid.

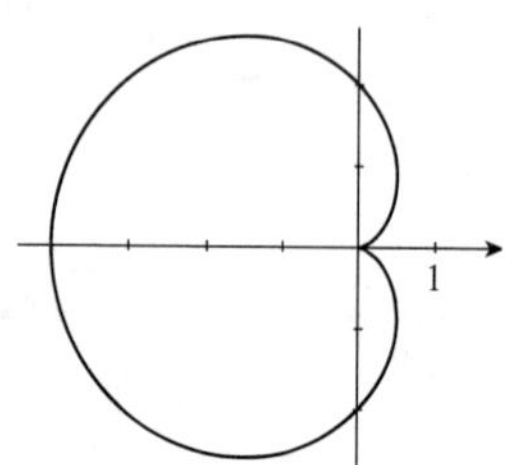


**24.** $r = 1 + \sin\theta$. Cardioid.

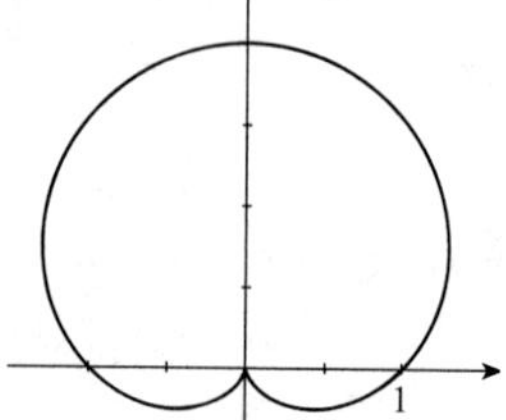

**25.** $r = -3(1 + \sin\theta)$. Cardioid.

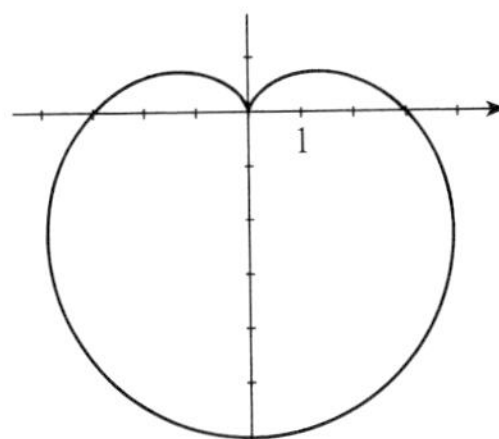


**26.** $r = \cos\theta - 1$. Cardioid.

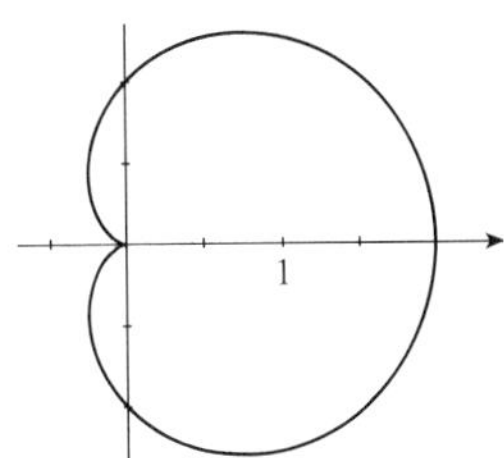


**27.** $r = \theta, \theta \geq 0$

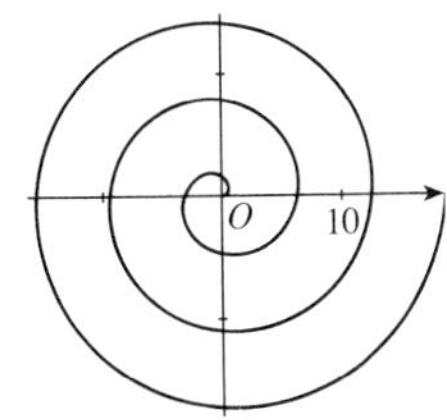


**28.** $r\theta = 1, \theta > 0$

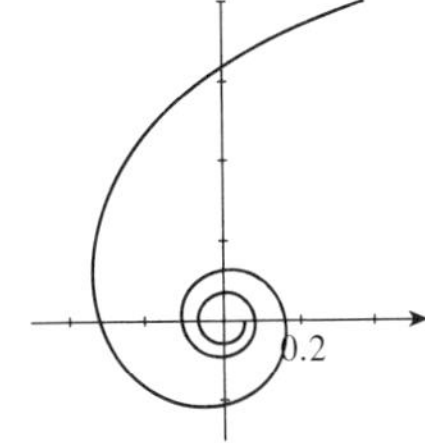


**29.** $r = \sin 2\theta$

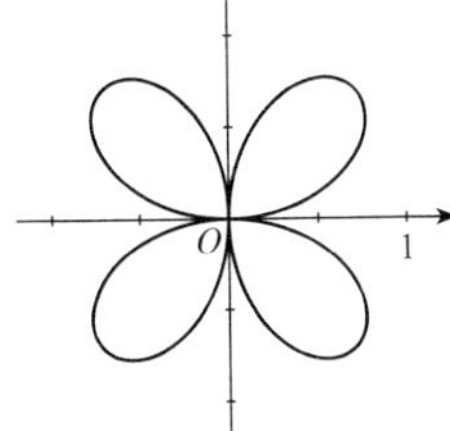


**30.** $r = 2\cos 3\theta$

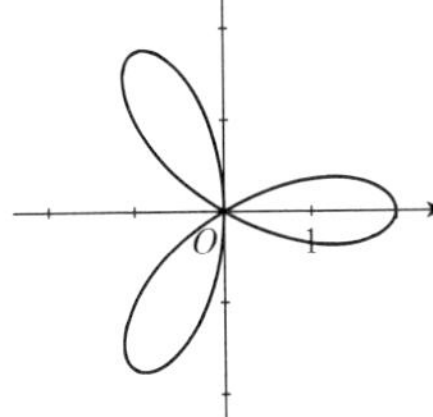


**31.** $r^2 = \cos 2\theta$

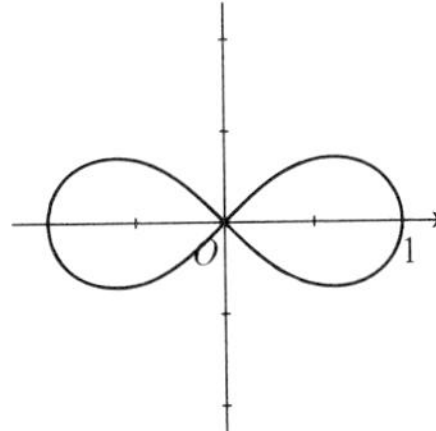


**32.** $r^2 = 4\sin 2\theta$

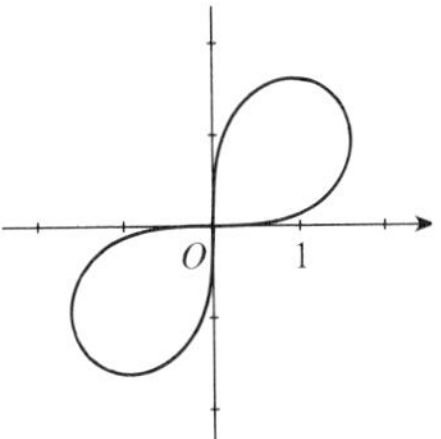


**33.** $r = 2 + \sin\theta$

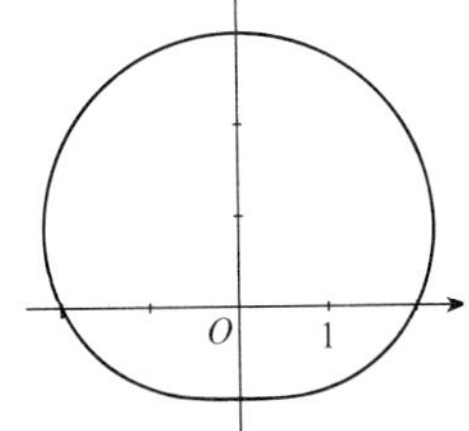


**34.** $r = 1 - 2\cos\theta$

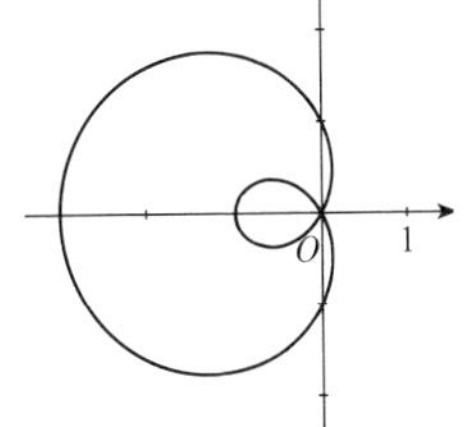

**35.** $r = 2 + \sec\theta$

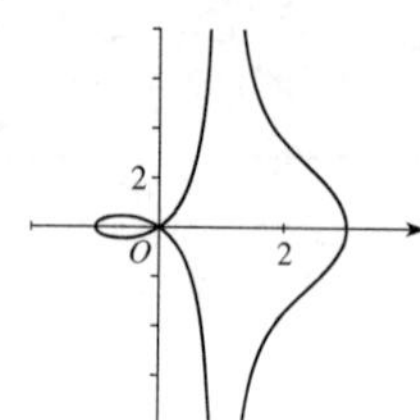

**36.** $r = \sin\theta\tan\theta$

**37.** $r = \cos\left(\dfrac{\theta}{2}\right), \theta \in [0, 4\pi]$

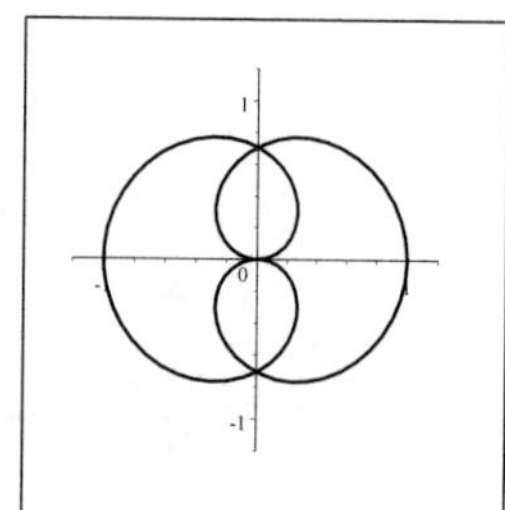

**38.** $r = \sin\left(\dfrac{8\theta}{5}\right), \theta \in [0, 10\pi]$

**39.** $r = 1 + 2\sin\left(\dfrac{\theta}{2}\right), \theta \in [0, 4\pi]$

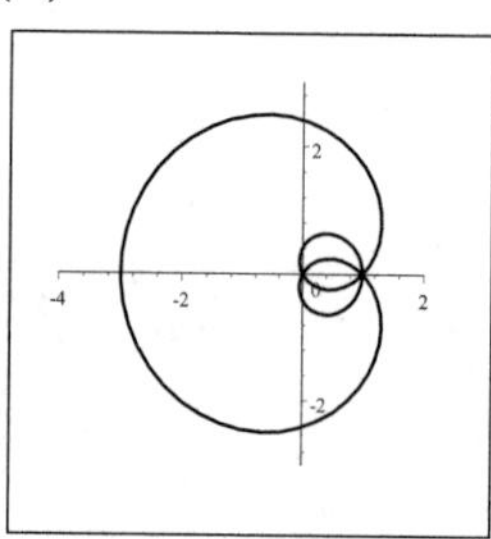

**40.** $r = \sqrt{1 - 0.8\sin^2\theta}, \theta \in [0, 2\pi]$

**41.** $r = 1 + \sin n\theta$. The number of loops is $n$.

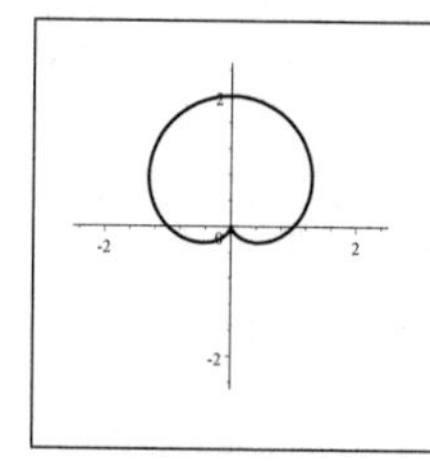

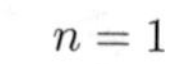

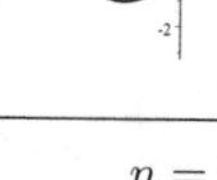

$n = 2$

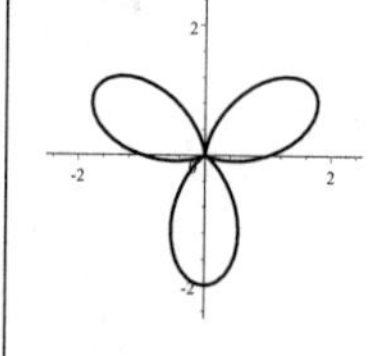

$n = 3$

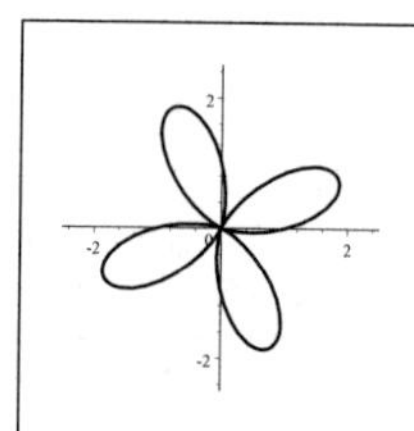

$n = 4$

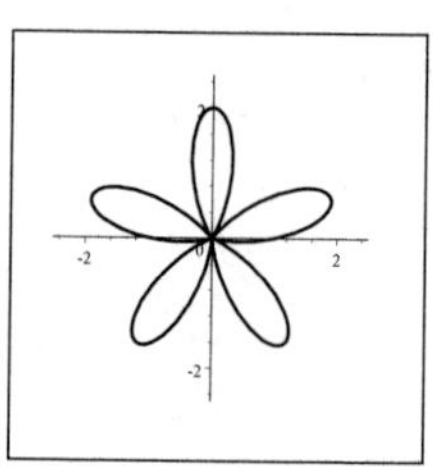

$n = 5$

**42.** $r = 1 + c\sin 2\theta$

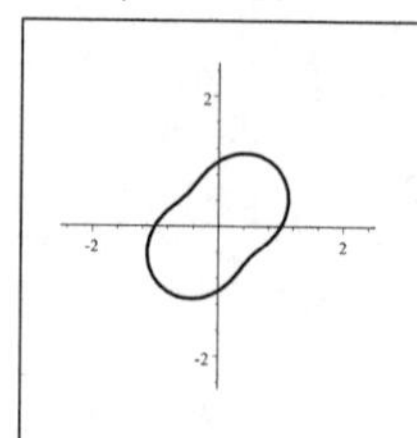

$c = 0.3$

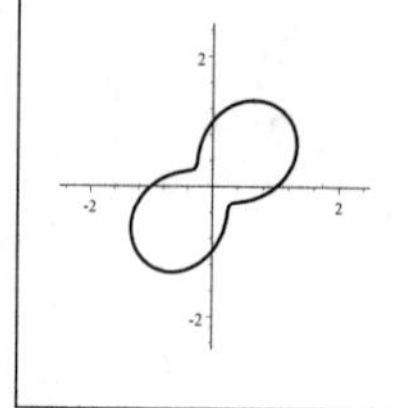

$c = 0.6$

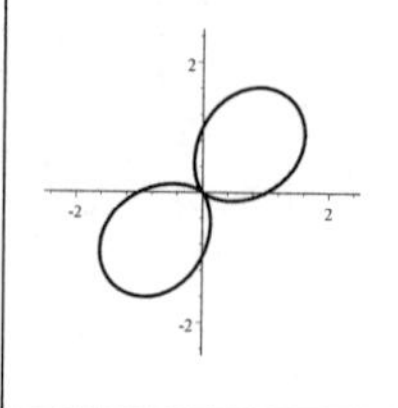

$c = 1$

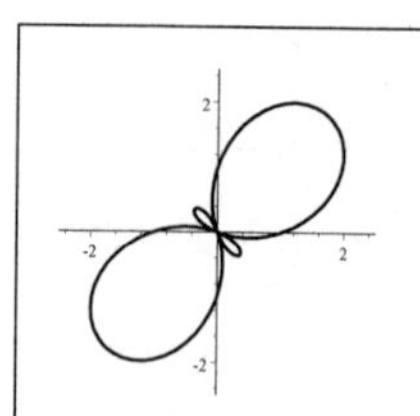

$c = 1.5$

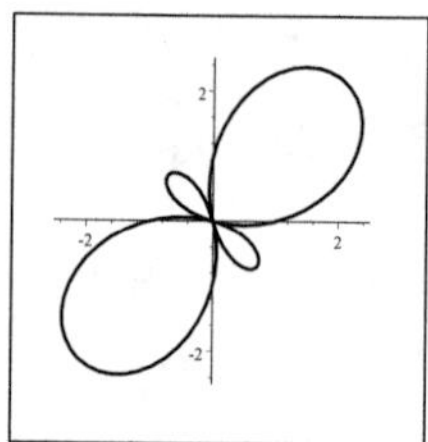

$c = 2$

As $c$ increases, the graph becomes "pinched" along the line $\theta = \frac{3\pi}{4}$, eventually growing more "leaves" with that line as their axis.

**43.** The graph of $r = \sin\left(\dfrac{\theta}{2}\right)$ is IV, since the graph must contain the points $(0,0)$, $\left(\dfrac{1}{\sqrt{2}}, \frac{\pi}{2}\right)$, $(1,\pi)$, and so on.

**44.** The graph of $r = \dfrac{1}{\sqrt{\theta}}$ is I, since as $\theta$ increase, $r$ decreases ($\theta > 0$). So this is a spiral into the origin.

**45.** The graph of $r = \theta \sin\theta$ is III, since for $\theta = \frac{\pi}{2}, \frac{5\pi}{2}, \frac{7\pi}{2}, \ldots$ the values of $r$ are also $\frac{\pi}{2}, \frac{5\pi}{2}, \frac{7\pi}{2}, \ldots$. Thus the graph must cross the vertical axis at an infinite number of points.

**46.** The graph of $r = 1 + 3\cos(3\theta)$ is II, since when $\theta = 0, r = 1 + 3 = 4$ and when $\theta = \pi, r = 1 - 3 = -2$, so there should be two intercepts on the positive-axis.

**47.** $(x^2+y^2)^3 = 4x^2y^2 \quad \Leftrightarrow \quad (r^2)^3 = 4(r\cos\theta)^2(r\sin\theta)^2 \quad \Leftrightarrow$
$r^6 = 4r^4\cos^2\theta\sin^2\theta \quad \Leftrightarrow \quad r^2 = 4\cos^2\theta\sin^2\theta \quad \Leftrightarrow$
$r = 2\cos\theta\sin\theta = \sin 2\theta$. The equation is $r = \sin 2\theta$, a rose.

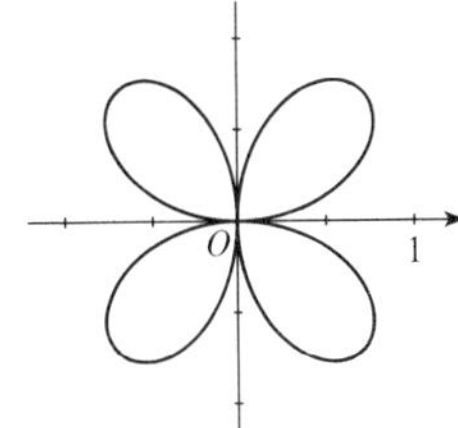


**48.** $(x^2+y^2)^3 = (x^2-y^2)^2 \quad \Leftrightarrow \quad (r^2)^3 = \left[(r\cos\theta)^2 - (r\sin\theta)^2\right]^2 \quad \Leftrightarrow$
$r^6 = (r^2\cos^2\theta - r^2\sin^2\theta)^2 \quad \Leftrightarrow \quad r^6 = \left[r^2(\cos^2\theta - \sin^2\theta)\right]^2 \quad \Leftrightarrow$
$r^6 = r^4(\cos^2\theta - \sin^2\theta)^2 \quad \Leftrightarrow \quad r^2 = (\cos 2\theta)^2 \quad \Leftrightarrow \quad r = \pm\cos 2\theta$. The equations $r = \cos 2\theta$ and $r = -\cos 2\theta$ have the same graph, a rose.

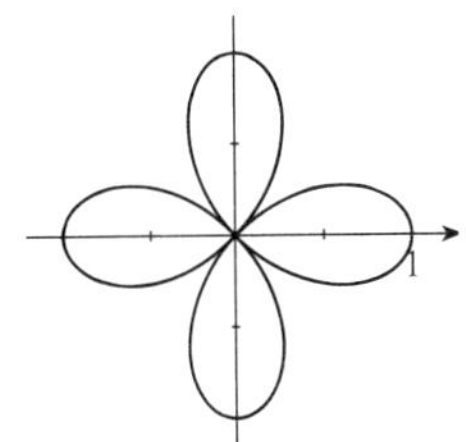


**49.** $(x^2+y^2)^2 = x^2 - y^2 \quad \Leftrightarrow \quad (r^2)^2 = (r\cos\theta)^2 - (r\sin\theta)^2 \quad \Leftrightarrow$
$r^4 = r^2\cos^2\theta - r^2\sin^2\theta \quad \Leftrightarrow \quad r^4 = r^2(\cos^2\theta - \sin^2\theta) \quad \Leftrightarrow$
$r^2 = \cos^2\theta - \sin^2\theta = \cos 2\theta$. The graph is $r^2 = \cos 2\theta$, a leminiscate.

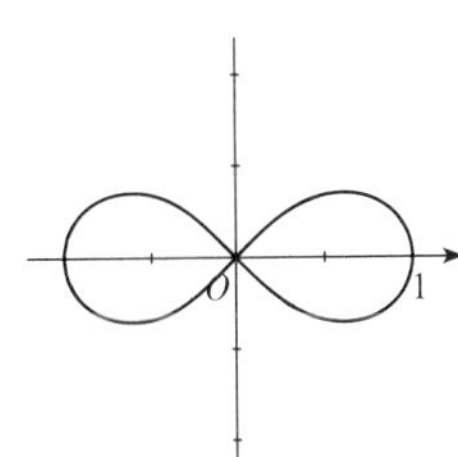


**50.** $x^2+y^2 = (x^2+y^2-x)^2 \quad \Leftrightarrow \quad r^2 = (r^2 - r\cos\theta)^2 \quad \Leftrightarrow$
$r^2 = [r(r-\cos\theta)]^2 \quad \Leftrightarrow \quad r^2 = r^2(r-\cos\theta)^2 \quad \Leftrightarrow \quad 1 = (r-\cos\theta)^2$
$\Leftrightarrow \quad \pm 1 = r - \cos\theta \quad \Leftrightarrow \quad r = \cos\theta \pm 1$. Both $r = \cos\theta + 1$ and $r = \cos\theta - 1$ give the same graph. To see this, replace $\theta$ with $\theta + \pi$:
$\cos(\theta+\pi) - 1 = -\cos\theta - 1 = -(\cos\theta + 1)$. This is a cardioid.

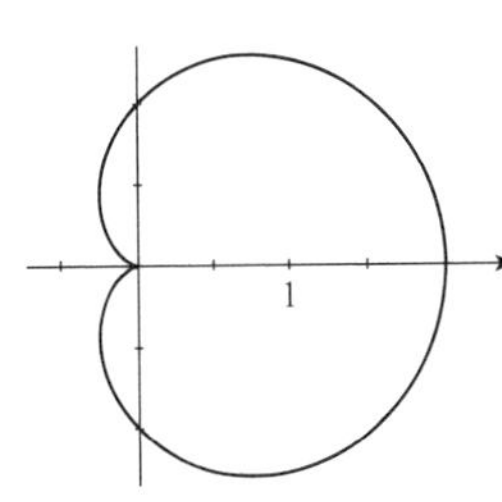


**51.** $r = a\cos\theta + b\sin\theta \quad \Leftrightarrow \quad r^2 = ar\cos\theta + br\sin\theta \quad \Leftrightarrow \quad x^2+y^2 = ax+by \quad \Leftrightarrow \quad x^2 - ax + y^2 - by = 0 \quad \Leftrightarrow$
$x^2 - ax + \frac{1}{4}a^2 + y^2 - by + \frac{1}{4}b^2 = \frac{1}{4}a^2 + \frac{1}{4}b^2 \quad \Leftrightarrow$
$\left(x - \frac{1}{2}a\right)^2 + \left(y - \frac{1}{2}b\right)^2 = \frac{1}{4}(a^2+b^2)$. Thus, in rectangular coordinates the center is $\left(\frac{1}{2}a, \frac{1}{2}b\right)$ and the radius is $\frac{1}{2}\sqrt{a^2+b^2}$.

**52. (a)**

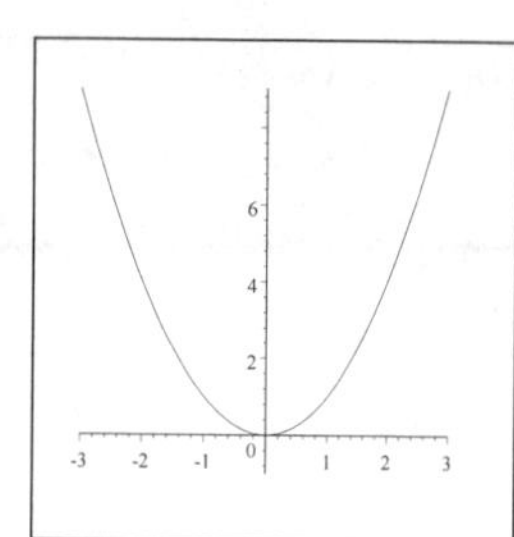

**(b)** $r = \tan\theta \sec\theta = \dfrac{\sin\theta}{\cos\theta}\left(\dfrac{1}{\cos\theta}\right) = \dfrac{y/r}{(x/r)^2} = \dfrac{ry}{x^2} \quad\Leftrightarrow\quad \dfrac{y}{x^2} = 1$

$\Leftrightarrow \quad y = x^2.$

**53. (a)**

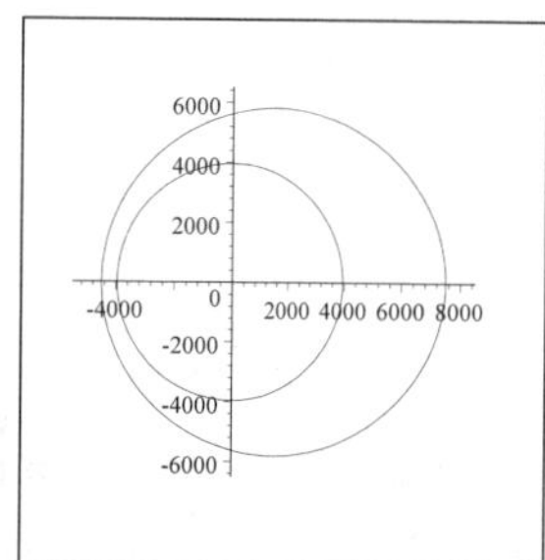

At $\theta = 0$, the satellite is at the "rightmost" point in its orbit, $(5625, 0)$. As $\theta$ increases, it travels counterclockwise. Note that it is moving fastest when $\theta = \pi$.

**(b)** The satellite is closest to earth when $\theta = \pi$. Its height above the earth's surface at this point is $22500/(4 - \cos\pi) - 3960 = 4500 - 3960 = 540$ mi.

**54. (a)**

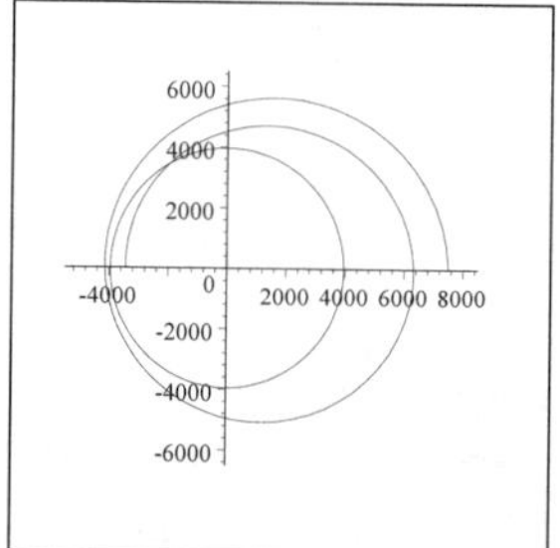

The satellite starts out at the same point, $(5625, 0)$, but its orbit decays. It narrowly misses the earth at its first perigee (closest approach to the earth), and crashes during its second orbit.

**(b)** The satellite crashes when $\theta \approx 8.37$ rad $\approx 480°$.

**55.** The graphs of $r = 1 + \sin\left(\theta - \frac{\pi}{6}\right)$ and $r = 1 + \sin\left(\theta - \frac{\pi}{3}\right)$ have the same shape as $r = 1 + \sin\theta$, rotated through angles of $\frac{\pi}{6}$ and $\frac{\pi}{3}$, respectively. Similarly, the graph of $r = f(\theta - \alpha)$ is the graph of $r = f(\theta)$ rotated by the angle $\alpha$.

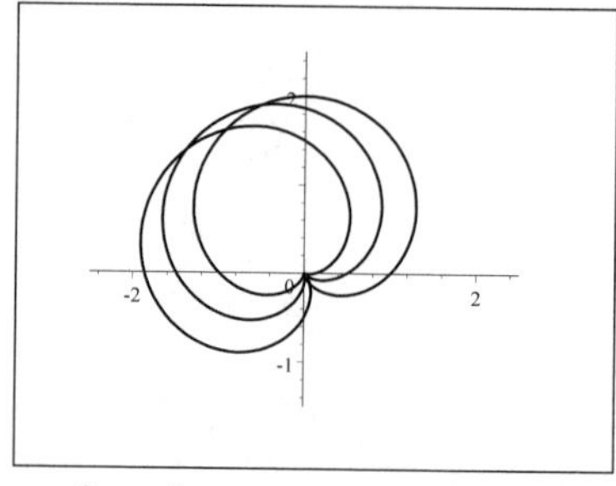

**56.** The circle $r = 2$ (in polar coordinates) has rectangular coordinate equation $x^2 + y^2 = 4$. The polar coordinate form is simpler. The graph of the equation $r = \sin 2\theta$ is a four-leafed rose. Multiplying both sides by $r^2$, we get $r^3 = r^2 2\cos\theta\sin\theta = 2(r\cos\theta)(r\sin\theta)$ which is $\left(x^2 + y^2\right)^{3/2} = 2xy$ in rectangular form. The polar form is definitely simpler.

**57.** $y = 2 \quad\Leftrightarrow\quad r\sin\theta = 2 \quad\Leftrightarrow\quad r = 2\csc\theta$. The rectangular coordinate system gives the simpler equation here. It is easier to study lines in rectangular coordinates.

# 9.3 Polar Form of Complex Numbers; DeMoivre's Theorem

**1.** $|4i| = \sqrt{0^2 + 4^2} = 4$

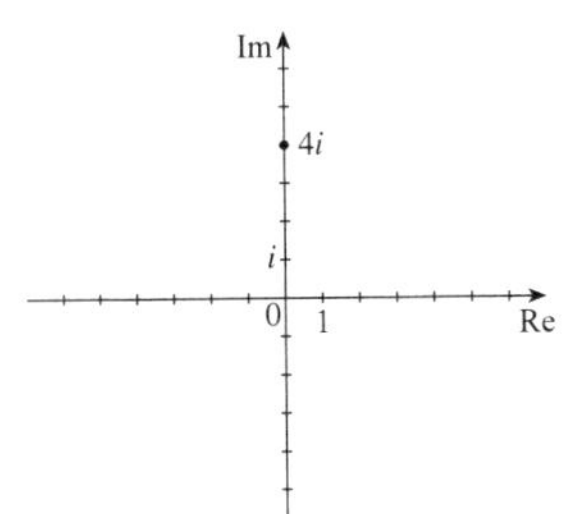


**2.** $|-3i| = \sqrt{0 + 9} = 3$

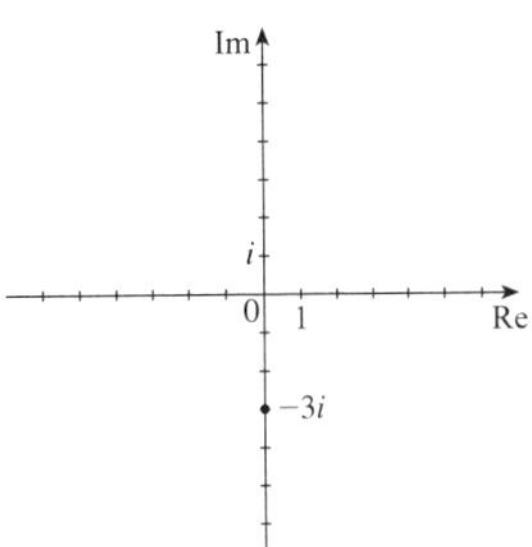


**3.** $|-2| = \sqrt{4 + 0} = 2$

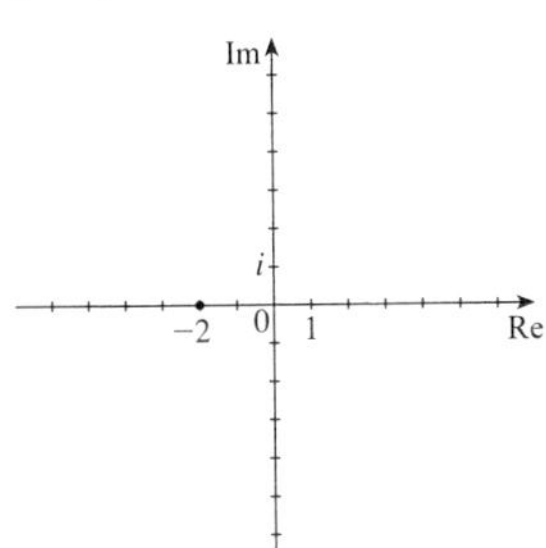


**4.** $|6| = 6$

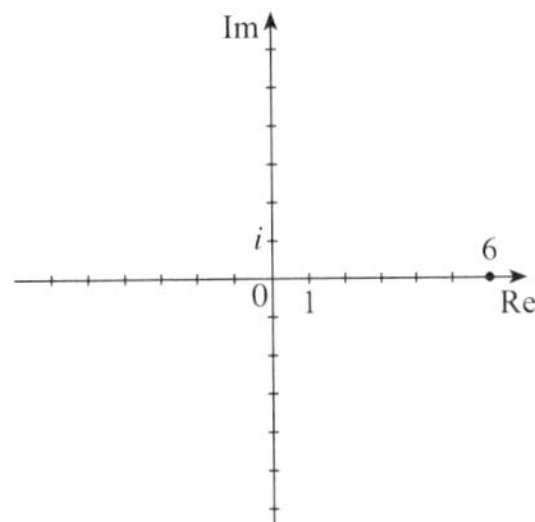


**5.** $|5 + 2i| = \sqrt{5^2 + 2^2} = \sqrt{29}$

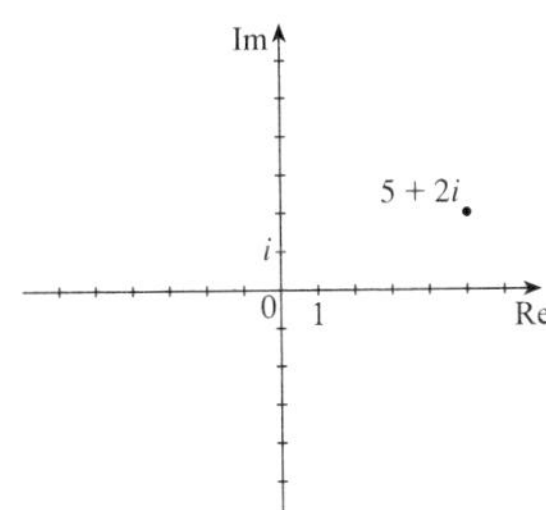


**6.** $|7 - 3i| = \sqrt{49 + 9} = \sqrt{58}$

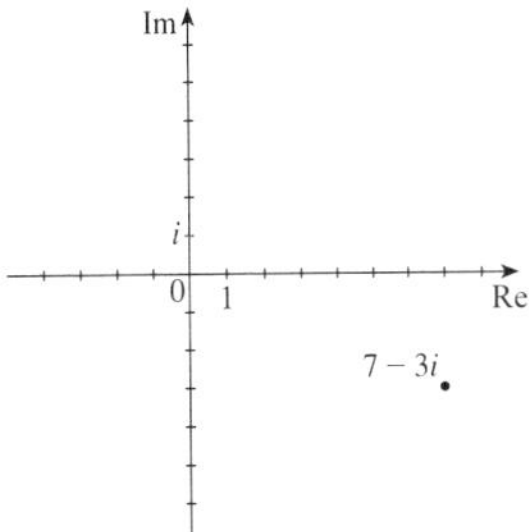


**7.** $\left|\sqrt{3} + i\right| = \sqrt{3 + 1} = 2$

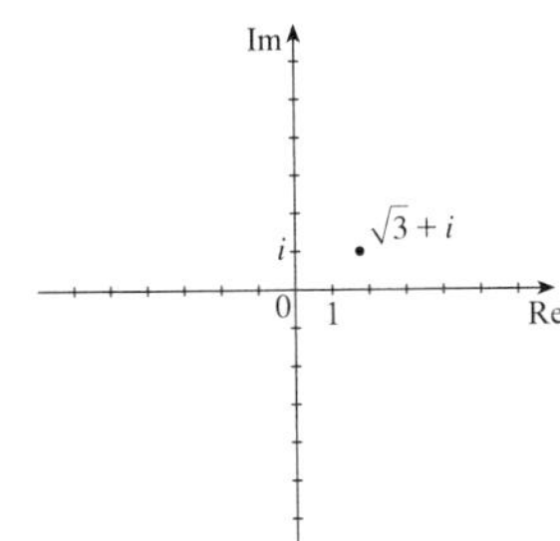


**8.** $\left|-1 - \frac{\sqrt{3}}{3}i\right| = \sqrt{1 + \frac{1}{3}} = \sqrt{\frac{4}{3}} = \frac{2\sqrt{3}}{3}$

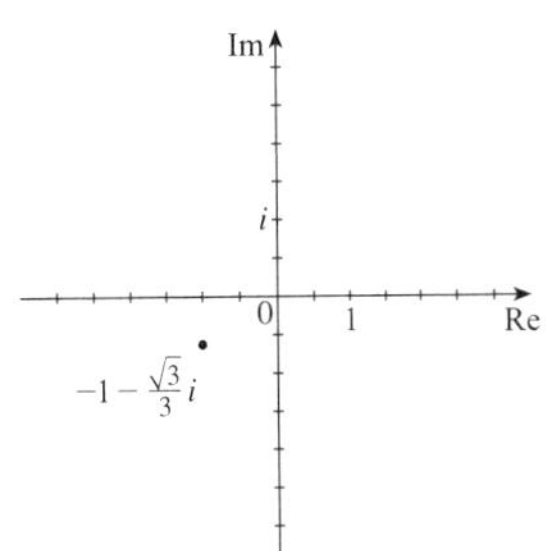

**9.** $\left|\dfrac{3+4i}{5}\right| = \sqrt{\frac{9}{25}+\frac{16}{25}} = 1$

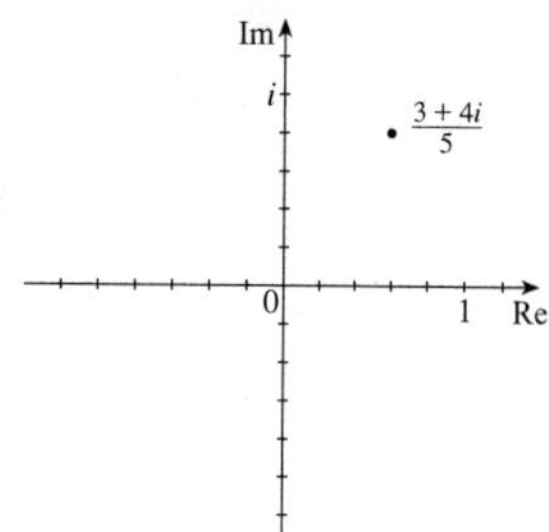


**10.** $\left|\dfrac{-\sqrt{2}+i\sqrt{2}}{2}\right| = \sqrt{\frac{1}{2}+\frac{1}{2}} = 1$

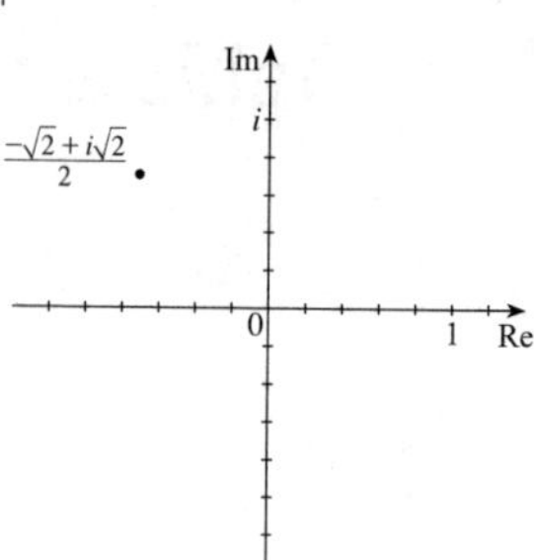


**11.** $z = 1+i$, $2z = 2+2i$, $-z = -1-i$, $\frac{1}{2}z = \frac{1}{2}+\frac{1}{2}i$

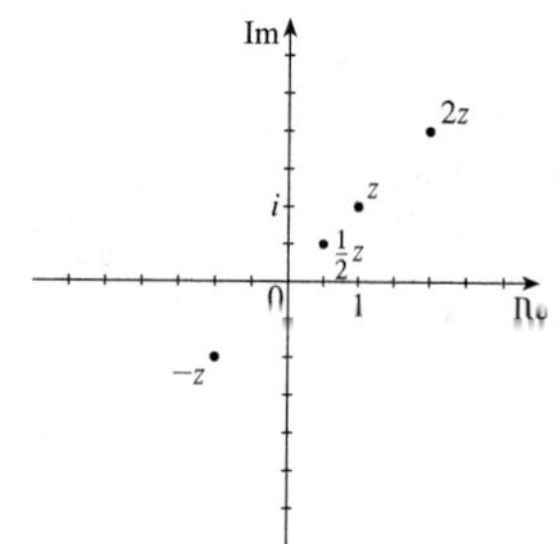


**12.** $z = -1+\sqrt{3}i$, $2z = -2+2\sqrt{3}i$, $-z = 1-\sqrt{3}i$, $\frac{1}{2}z = -\frac{1}{2}+\frac{\sqrt{3}}{2}i$

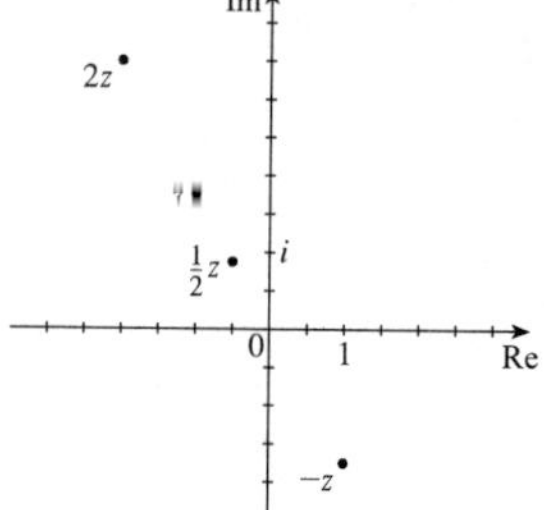


**13.** $z = 8+2i$, $\overline{z} = 8-2i$

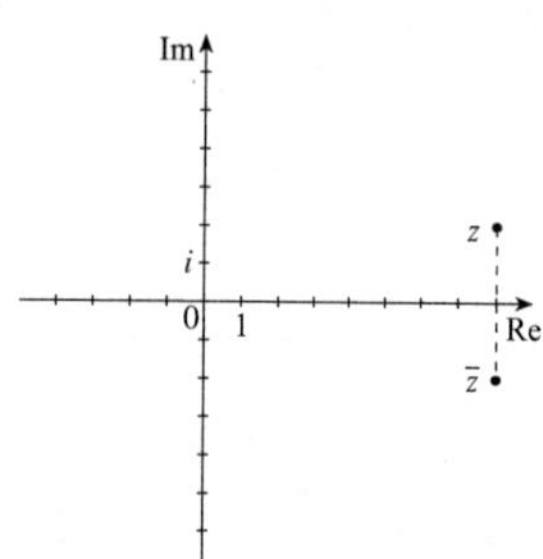


**14.** $z = -5+6i$, $\overline{z} = -5-6i$

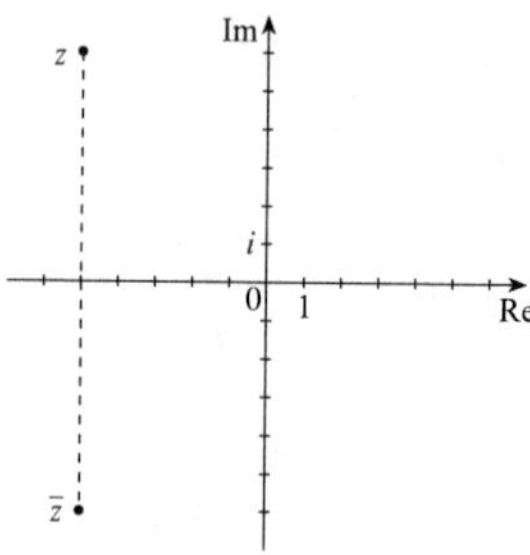


**15.** $z_1 = 2-i$, $z_2 = 2+i$, $z_1+z_2 = 2-i+2+i = 4$,
$z_1z_2 = (2-i)(2+i) = 4-i^2 = 5$

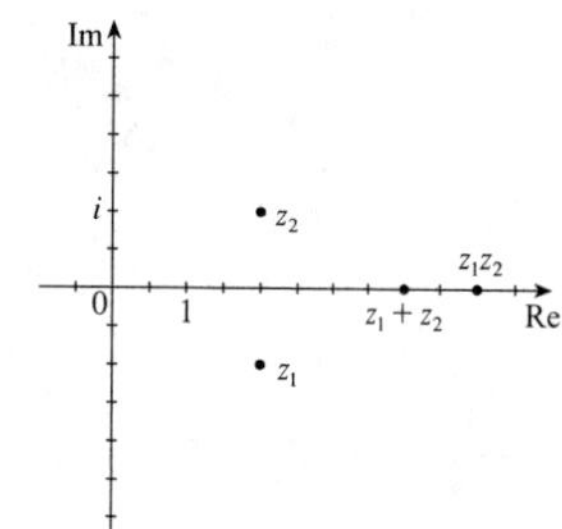


**16.** $z_1 = -1+i$, $z_2 = 2-3i$,
$z_1+z_2 = -1+i+2-3i = 1-2i$,
$z_1z_2 = (-1+i)(2-3i) = -2+3i+2i-3i^2 = 1+5i$

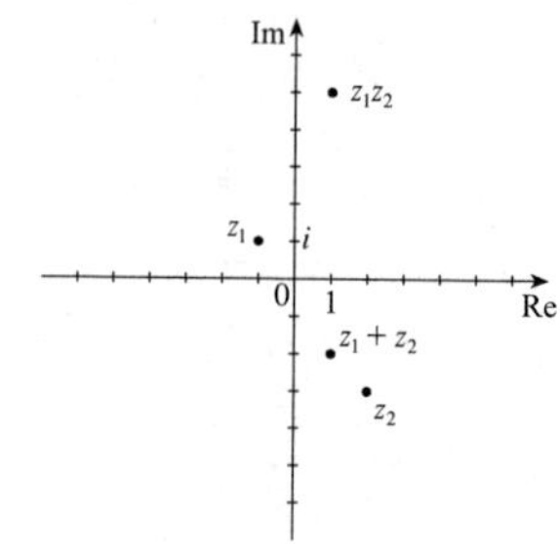

**17.** $\{z = a + bi \mid a \leq 0, b \geq 0\}$

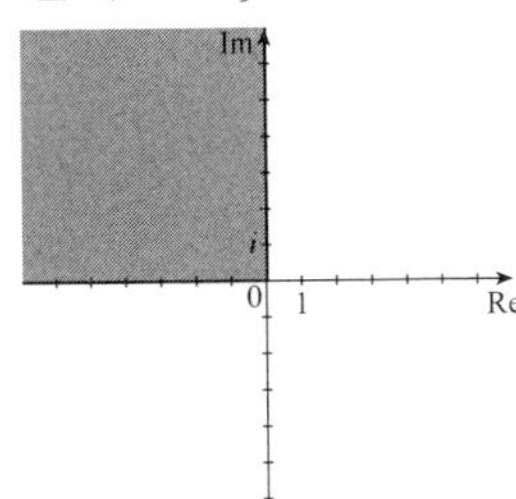


**18.** $\{z = a + bi \mid a > 1, b > 1\}$

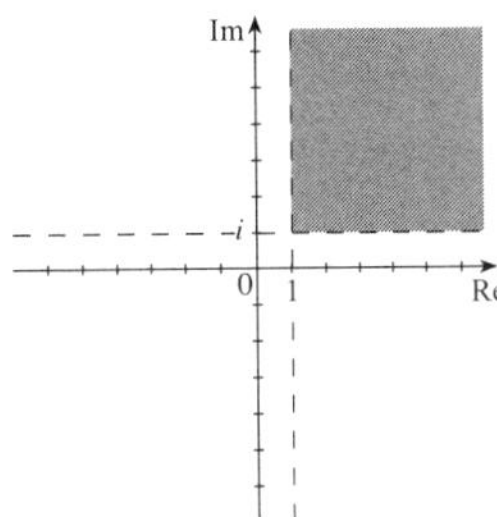


**19.** $\{z \mid |z| = 3\}$

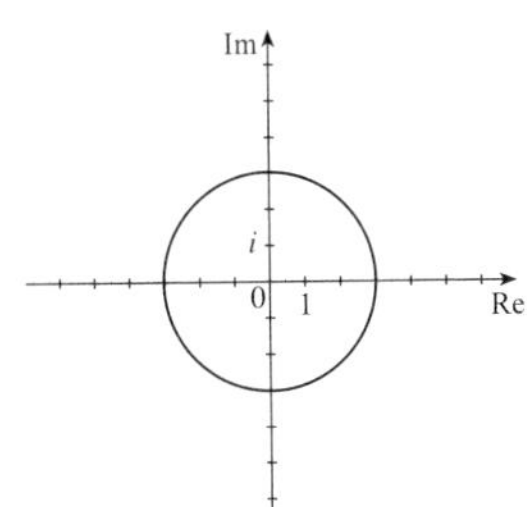


**20.** $\{z \mid |z| \geq 1\}$

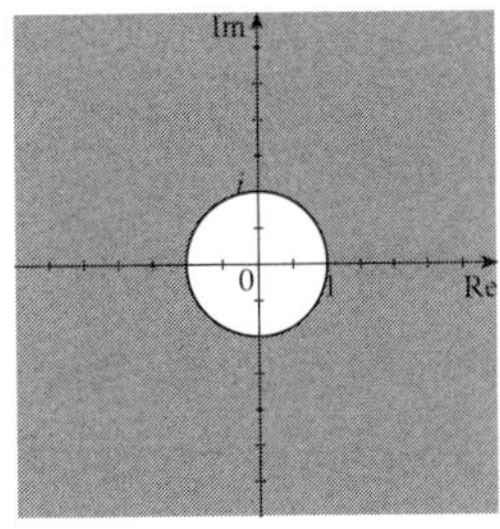


**21.** $\{z \mid |z| < 2\}$

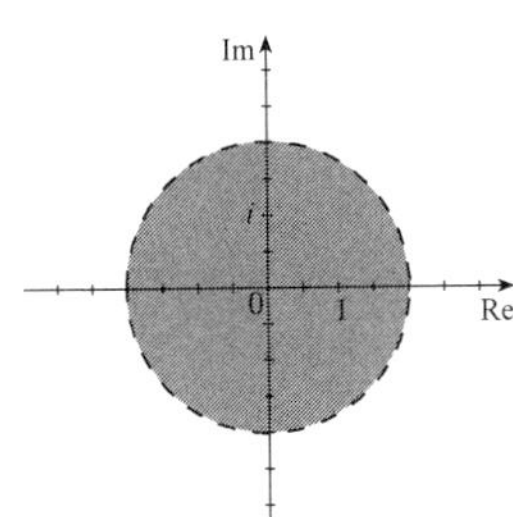


**22.** $\{z \mid 2 \leq |z| \leq 5\}$

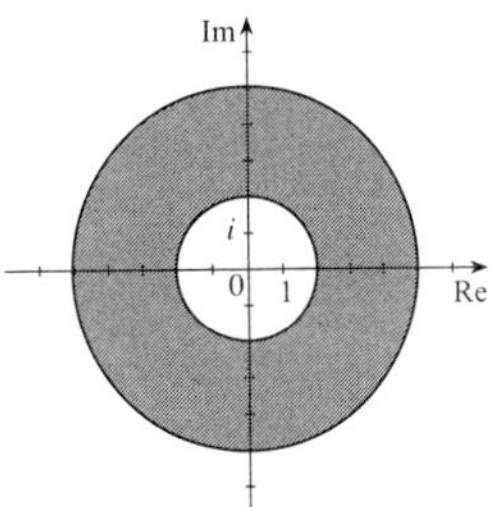


**23.** $\{z = a + bi \mid a + b < 2\}$

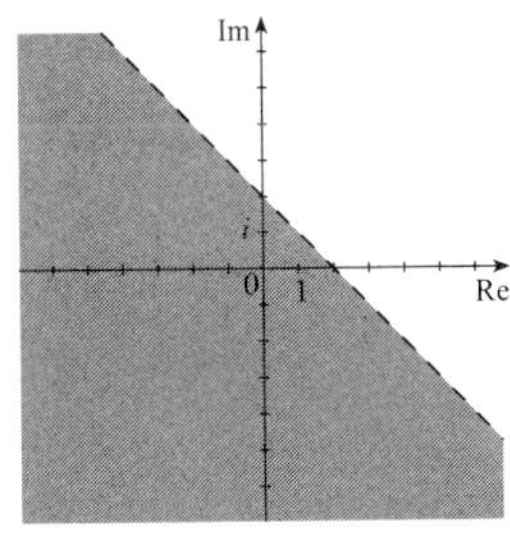


**24.** $\{z = a + bi \mid a \geq b\}$

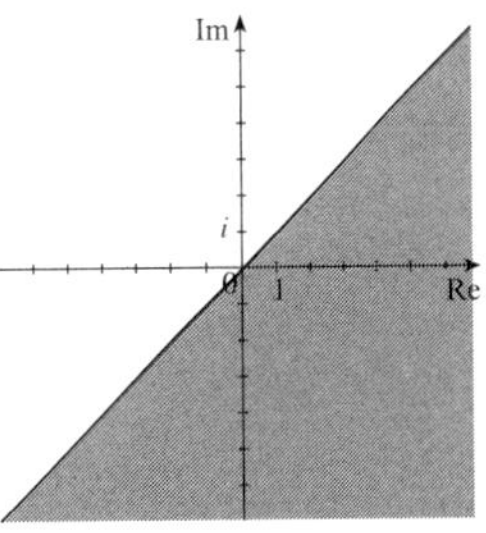


**25.** $1 + i$. Then $\tan\theta = \frac{1}{1} = 1$ with $\theta$ in quadrant I $\Rightarrow$ $\theta = \frac{\pi}{4}$, and $r = \sqrt{1^2 + 1^2} = \sqrt{2}$. Hence, $1 + i = \sqrt{2}\left(\cos\frac{\pi}{4} + i\sin\frac{\pi}{4}\right)$.

**26.** $1 + \sqrt{3}i$. Then $\tan\theta = \sqrt{3}$ with $\theta$ in quadrant I $\Rightarrow$ $\theta = \frac{\pi}{3}$, and $r = \sqrt{1 + 3} = 2$. Hence, $1 - \sqrt{3}i = 2\left(\cos\frac{\pi}{3} + i\sin\frac{\pi}{3}\right)$.

**27.** $\sqrt{2} - \sqrt{2}i$. Then $\tan\theta = \frac{\sqrt{2}}{\sqrt{2}} = -1$ with $\theta$ in quadrant IV $\Rightarrow$ $\theta = \frac{7\pi}{4}$, and $r = \sqrt{2 + 2} = 2$. Hence, $\sqrt{2} - \sqrt{2}i = 2\left(\cos\frac{7\pi}{4} + i\sin\frac{7\pi}{4}\right)$.

**28.** $1 - i$. Then $\tan\theta = -1$ with $\theta$ in quadrant IV $\Rightarrow$ $\theta = \frac{7\pi}{4}$, and $r = \sqrt{1 + 1} = \sqrt{2}$. Hence, $1 - i = \sqrt{2}\left(\cos\frac{7\pi}{4} + i\sin\frac{7\pi}{4}\right)$.

**29.** $2\sqrt{3}-2i$. Then $\tan\theta = \frac{-2}{2\sqrt{3}} = -\frac{1}{\sqrt{3}}$ with $\theta$ in quadrant IV $\Rightarrow$ $\theta = \frac{11\pi}{6}$, and $r = \sqrt{12+4} = 4$. Hence, $2\sqrt{3}-2i = 4\left(\cos\frac{11\pi}{6} + i\sin\frac{11\pi}{6}\right)$.

**30.** $-1+i$. Then $\tan\theta = -1$ with $\theta$ in quadrant II $\Rightarrow$ $\theta = \frac{3\pi}{4}$, and $r = \sqrt{1+1} = \sqrt{2}$. Hence, $-1+i = \sqrt{2}\left(\cos\frac{3\pi}{4} + i\sin\frac{3\pi}{4}\right)$.

**31.** $-3i$. Then $\theta = \frac{3\pi}{2}$, and $r = \sqrt{0+9} = 3$. Hence, $3i = 3\left(\cos\frac{3\pi}{2} + i\sin\frac{3\pi}{2}\right)$.

**32.** $-3-3\sqrt{3}i$. Then $\tan\theta = \sqrt{3}$ with $\theta$ in quadrant III $\Rightarrow$ $\theta = \frac{4\pi}{3}$, and $r = \sqrt{9+27} = 6$. Hence, $-3-3\sqrt{3}i = 6\left(\cos\frac{4\pi}{3} + i\sin\frac{4\pi}{3}\right)$.

**33.** $5+5i$. Then $\tan\theta = \frac{5}{5} = 1$ with $\theta$ in quadrant I $\Rightarrow$ $\theta = \frac{\pi}{4}$, and $r = \sqrt{25+25} = 5\sqrt{2}$. Hence, $5+5i = 5\sqrt{2}\left(\cos\frac{\pi}{4} + i\sin\frac{\pi}{4}\right)$.

**34.** $4$. Then $\theta = 0$, and $r = \sqrt{16+0} = 4$. Hence, $4 = 4\left(\cos 0 + i\sin 0\right)$.

**35.** $4\sqrt{3}-4i$. Then $\tan\theta = \frac{-4}{4\sqrt{3}} = -\frac{1}{\sqrt{3}}$ with $\theta$ in quadrant IV $\Rightarrow$ $\theta = \frac{11\pi}{6}$, and $r = \sqrt{48+16} = 8$. Hence, $4\sqrt{3}-4i = 8\left(\cos\frac{11\pi}{6} + i\sin\frac{11\pi}{6}\right)$.

**36.** $8i$. Then $\theta = \frac{\pi}{2}$, and $r = 8$. Hence, $8i = 8\left(\cos\frac{\pi}{2} + i\sin\frac{\pi}{2}\right)$.

**37.** $-20$. Then $\theta = \pi$, and $r = 20$. Hence, $-20 = 20\left(\cos\pi + i\sin\pi\right)$.

**38.** $\sqrt{3}+i$. Then $\tan\theta = \frac{1}{\sqrt{3}}$ with $\theta$ in quadrant I $\Rightarrow$ $\theta = \frac{\pi}{6}$, and $r = \sqrt{3+1} = 2$. Hence, $\sqrt{3}+i = 2\left(\cos\frac{\pi}{6} + i\sin\frac{\pi}{6}\right)$.

**39.** $3+4i$. Then $\tan\theta = \frac{4}{3}$ with $\theta$ in quadrant I $\Rightarrow$ $\theta = \tan^{-1}\frac{4}{3}$, and $r = \sqrt{9+16} = 5$. Hence, $3+4i = 5\left[\cos\left(\tan^{-1}\frac{4}{3}\right) + i\sin\left(\tan^{-1}\frac{4}{3}\right)\right]$.

**40.** $i(2-2i) = 2+2i$. Then $\tan\theta = 1$ with $\theta$ in quadrant I $\Rightarrow$ $\theta = \frac{\pi}{4}$, and $r = \sqrt{4+4} = 2\sqrt{2}$. Hence, $i(2-2i) = 2\sqrt{2}\left(\cos\frac{\pi}{4} + i\sin\frac{\pi}{4}\right)$.

**41.** $3i(1+i) = -3+3i$. Then $\tan\theta = \frac{3}{-3} = -1$ with $\theta$ in quadrant II $\Rightarrow$ $\theta = \frac{3\pi}{4}$, and $r = \sqrt{9+9} = 3\sqrt{2}$. Hence, $3i(1+i) = 3\sqrt{2}\left(\cos\frac{3\pi}{4} + i\sin\frac{3\pi}{4}\right)$.

**42.** $2(1-i) = 2-2i$. Then $\tan\theta = -1$ with $\theta$ in quadrant IV $\Rightarrow$ $\theta = \frac{7\pi}{4}$, and $r = \sqrt{4+4} = 2\sqrt{2}$. Hence, $2(1-i) = 2\sqrt{2}\left(\cos\frac{7\pi}{4} + i\sin\frac{7\pi}{4}\right)$.

**43.** $4\left(\sqrt{3}+i\right) = 4\sqrt{3}+4i$. Then $\tan\theta = \frac{4}{4\sqrt{3}} = \frac{1}{\sqrt{3}}$ with $\theta$ in quadrant I $\Rightarrow$ $\theta = \frac{\pi}{6}$, and $r = \sqrt{48+16} = 8$. Hence, $4\left(\sqrt{3}+i\right) = 8\left(\cos\frac{\pi}{6} + i\sin\frac{\pi}{6}\right)$.

**44.** $-3-3i$. Then $\tan\theta = 1$ with $\theta$ in quadrant III $\Rightarrow$ $\theta = \frac{5\pi}{4}$, and $r = \sqrt{9+9} = 3\sqrt{2}$. Hence, $-3-3i = 3\sqrt{2}\left(\cos\frac{5\pi}{4} + i\sin\frac{5\pi}{4}\right)$.

**45.** $2+i$. Then $\tan\theta = \frac{1}{2}$ with $\theta$ in quadrant I $\Rightarrow$ $\theta = \tan^{-1}\frac{1}{2}$, and $r = \sqrt{4+1} = \sqrt{5}$. Hence, $2+i = \sqrt{5}\left[\cos\left(\tan^{-1}\frac{1}{2}\right) + i\sin\left(\tan^{-1}\frac{1}{2}\right)\right]$.

**46.** $3+\sqrt{3}i$. Then $\tan\theta = \frac{1}{\sqrt{3}}$ with $\theta$ in quadrant I $\Rightarrow$ $\theta = \frac{\pi}{6}$, and $r = \sqrt{9+3} = 2\sqrt{3}$. Hence, $3+\sqrt{3}i = 2\sqrt{3}\left(\cos\frac{\pi}{6} + i\sin\frac{\pi}{6}\right)$.

**47.** $\sqrt{2}+\sqrt{2}i$. Then $\tan\theta = \frac{\sqrt{2}}{\sqrt{2}} = 1$ with $\theta$ in quadrant I $\Rightarrow$ $\theta = \frac{\pi}{4}$, and $r = \sqrt{2+2} = 2$. Hence, $2+\sqrt{2}i = 2\left(\cos\frac{\pi}{4} + i\sin\frac{\pi}{4}\right)$.

**48.** $-\pi i$. Then $\theta = \frac{3\pi}{2}$, and $r = \pi$. Hence $-\pi i = \pi\left(\cos\frac{3\pi}{2} + i\sin\frac{3\pi}{2}\right)$.

**49.** $z_1 = \cos\pi + i\sin\pi$, $z_2 = \cos\frac{\pi}{3} + i\sin\frac{\pi}{3}$, $z_1z_2 = \cos\left(\pi + \frac{\pi}{3}\right) + i\sin\left(\pi + \frac{\pi}{3}\right) = \cos\frac{4\pi}{3} + i\sin\frac{4\pi}{3}$,
$z_1/z_2 = \cos\left(\pi - \frac{\pi}{3}\right) + i\sin\left(\pi - \frac{\pi}{3}\right) = \cos\frac{2\pi}{3} + i\sin\frac{2\pi}{3}$

**50.** $z_1 = \cos\frac{\pi}{4} + i\sin\frac{\pi}{4}$, $z_2 = \cos\frac{3\pi}{4} + i\sin\frac{3\pi}{4}$, $z_1z_2 = \cos\left(\frac{\pi}{4} + \frac{3\pi}{4}\right) + i\sin\left(\frac{\pi}{4} + \frac{3\pi}{4}\right) = \cos\pi + i\sin\pi$,
$z_1/z_2 = \cos\left(\frac{\pi}{4} - \frac{3\pi}{4}\right) + i\sin\left(\frac{\pi}{4} - \frac{3\pi}{4}\right) = \cos\left(-\frac{\pi}{2}\right) + i\sin\left(-\frac{\pi}{2}\right) = \cos\frac{\pi}{2} - i\sin\frac{\pi}{2}$

**51.** $z_1 = 3\left(\cos\frac{\pi}{6} + i\sin\frac{\pi}{6}\right)$, $z_2 = 5\left(\cos\frac{4\pi}{3} + i\sin\frac{4\pi}{3}\right)$,
$z_1z_2 = 3\cdot 5\left[\cos\left(\frac{\pi}{6} + \frac{4\pi}{3}\right) + i\sin\left(\frac{\pi}{6} + \frac{4\pi}{3}\right)\right] = 15\left(\cos\frac{9\pi}{6} + i\sin\frac{9\pi}{6}\right) = 15\left(\cos\frac{3\pi}{2} + i\sin\frac{3\pi}{2}\right)$,
$z_1/z_2 = \frac{3}{5}\left[\cos\left(\frac{\pi}{6} - \frac{4\pi}{3}\right) + i\sin\left(\frac{\pi}{6} - \frac{4\pi}{3}\right)\right] = \frac{3}{5}\left[\cos\left(-\frac{7\pi}{6}\right) + i\sin\left(-\frac{7\pi}{6}\right)\right] = \frac{3}{5}\left[\cos\left(\frac{7\pi}{6}\right) - i\sin\left(\frac{7\pi}{6}\right)\right]$

**52.** $z_1 = 7\left(\cos\frac{9\pi}{8} + i\sin\frac{9\pi}{8}\right)$, $z_2 = 2\left(\cos\frac{\pi}{8} + i\sin\frac{\pi}{8}\right)$,
$z_1z_2 = 7\cdot 2\left[\cos\left(\frac{9\pi}{8} + \frac{\pi}{8}\right) + i\sin\left(\frac{9\pi}{8} + \frac{\pi}{8}\right)\right] = 14\left(\cos\frac{5\pi}{4} + i\sin\frac{5\pi}{4}\right)$,
$z_1/z_2 = \frac{7}{2}\left[\cos\left(\frac{9\pi}{8} - \frac{\pi}{8}\right) + i\sin\left(\frac{9\pi}{8} - \frac{\pi}{8}\right)\right] = \frac{7}{2}\left(\cos\pi + i\sin\pi\right)$

**53.** $z_1 = 4\left(\cos 120^\circ + i\sin 120^\circ\right)$, $z_2 = 2\left(\cos 30^\circ + i\sin 30^\circ\right)$,
$z_1z_2 = 4\cdot 2\left[\cos\left(120^\circ + 30^\circ\right) + i\sin\left(120^\circ + 30^\circ\right)\right] = 8\left(\cos 150^\circ + i\sin 150^\circ\right)$,
$z_1/z_2 = \frac{4}{2}\left[\cos\left(120^\circ - 30^\circ\right) + i\sin\left(120^\circ - 30^\circ\right)\right] = 2\left(\cos 90^\circ + i\sin 90^\circ\right)$

**54.** $z_1 = \sqrt{2}\left(\cos 75^\circ + i\sin 75^\circ\right)$, $z_2 = 3\sqrt{2}\left(\cos 60^\circ + i\sin 60^\circ\right)$,
$z_1z_2 = \sqrt{2}\cdot 3\sqrt{2}\cdot\left[\cos\left(75^\circ + 60^\circ\right) + i\sin\left(75^\circ + 60^\circ\right)\right] = 6\left(\cos 135^\circ + i\sin 135^\circ\right)$,
$z_1/z_2 = \frac{\sqrt{2}}{3\sqrt{2}}\left[\cos\left(75^\circ - 60^\circ\right) + i\sin\left(75^\circ - 60^\circ\right)\right] = \frac{1}{3}\left(\cos 15^\circ + i\sin 15^\circ\right)$

**55.** $z_1 = 4\left(\cos 200^\circ + i\sin 200^\circ\right)$, $z_2 = 25\left(\cos 150^\circ + i\sin 150^\circ\right)$,
$z_1z_2 = 4\cdot 25\left[\cos\left(200^\circ + 150^\circ\right) + i\sin\left(200^\circ + 150^\circ\right)\right] = 100\left(\cos 350^\circ + i\sin 350^\circ\right)$,
$z_1/z_2 = \frac{4}{25}\left[\cos\left(200^\circ - 150^\circ\right) + i\sin\left(200^\circ - 150^\circ\right)\right] = \frac{4}{25}\left(\cos 50^\circ + i\sin 50^\circ\right)$

**56.** $z_1 = \frac{4}{5}\left(\cos 25^\circ + i\sin 25^\circ\right)$, $z_2 = \frac{1}{5}\left(\cos 155^\circ + i\sin 155^\circ\right)$,
$z_1z_2 = \frac{4}{5}\cdot\frac{1}{5}\cdot\left[\cos\left(25^\circ + 155^\circ\right) + i\sin\left(25^\circ + 155^\circ\right)\right] = \frac{4}{25}\left(\cos 180^\circ + i\sin 180^\circ\right)$,
$z_1/z_2 = \frac{4/5}{1/5}\left[\cos\left(25^\circ - 155^\circ\right) + i\sin\left(25^\circ - 155^\circ\right)\right] = 4\left[\cos\left(-130^\circ\right) + i\sin\left(-130^\circ\right)\right] = 4\left(\cos 130^\circ - i\sin 130^\circ\right)$

**57.** $z_1 = \sqrt{3} + i$, so $\tan\theta_1 = \frac{1}{\sqrt{3}}$ with $\theta_1$ in quadrant I $\Rightarrow$ $\theta_1 = \frac{\pi}{6}$, and $r_1 = \sqrt{3+1} = 2$.
$z_2 = 1 + \sqrt{3}i$, so $\tan\theta_2 = \sqrt{3}$ with $\theta_2$ in quadrant I $\Rightarrow$ $\theta_2 = \frac{\pi}{3}$, and $r_1 = \sqrt{1+3} = 2$.
Hence, $z_1 = 2\left(\cos\frac{\pi}{6} + i\sin\frac{\pi}{6}\right)$ and $z_2 = 2\left(\cos\frac{\pi}{3} + i\sin\frac{\pi}{3}\right)$.
Thus, $z_1z_2 = 2\cdot 2\left[\cos\left(\frac{\pi}{6} + \frac{\pi}{3}\right) + i\sin\left(\frac{\pi}{6} + \frac{\pi}{3}\right)\right] = 4\left(\cos\frac{\pi}{2} + i\sin\frac{\pi}{2}\right)$,
$z_1/z_2 = \frac{2}{2}\left[\cos\left(\frac{\pi}{6} - \frac{\pi}{3}\right) + i\sin\left(\frac{\pi}{6} - \frac{\pi}{3}\right)\right] = \cos\left(-\frac{\pi}{6}\right) + i\sin\left(-\frac{\pi}{6}\right) = \cos\frac{\pi}{6} - i\sin\frac{\pi}{6}$, and
$1/z_1 = \frac{1}{2}\left[\cos\left(-\frac{\pi}{6}\right) + i\sin\left(-\frac{\pi}{6}\right)\right] = \frac{1}{2}\left(\cos\frac{\pi}{6} - i\sin\frac{\pi}{6}\right)$.

**58.** $z_1 = \sqrt{2} - \sqrt{2}i$, so $\tan\theta_1 = -1$ with $\theta_1$ in quadrant IV $\Rightarrow$ $\theta_1 = \frac{7\pi}{4}$, and $r_1 = \sqrt{2+2} = 2$.
$z_2 = 1 - i$, so $\tan\theta_2 = -1$ with $\theta_2$ in quadrant IV $\Rightarrow$ $\theta_2 = \frac{7\pi}{4}$, and $r_2 = \sqrt{1+1} = \sqrt{2}$.
Hence, $z_1 = 2\left(\cos\frac{7\pi}{4} + i\sin\frac{7\pi}{4}\right)$ and $z_2 = \sqrt{2}\left(\cos\frac{7\pi}{4} + i\sin\frac{7\pi}{4}\right)$.
Thus, $z_1z_2 = 2\cdot\sqrt{2}\left[\cos\left(\frac{7\pi}{4} + \frac{7\pi}{4}\right) + i\sin\left(\frac{7\pi}{4} + \frac{7\pi}{4}\right)\right] = 2\sqrt{2}\left(\cos\frac{7\pi}{2} + i\sin\frac{7\pi}{2}\right) = 2\sqrt{2}\left(\cos\frac{\pi}{2} - i\sin\frac{\pi}{2}\right)$,
$z_1/z_2 = \frac{2}{\sqrt{2}}\left[\cos\left(\frac{7\pi}{4} - \frac{7\pi}{4}\right) + i\sin\left(\frac{7\pi}{4} - \frac{7\pi}{4}\right)\right] = \sqrt{2}\left(\cos 0 + i\sin 0\right)$, and
$1/z_1 = \frac{1}{2}\left[\cos\left(-\frac{7\pi}{4}\right) + i\sin\left(-\frac{7\pi}{4}\right)\right] = \frac{1}{2}\left(\cos\frac{7\pi}{4} - i\sin\frac{7\pi}{4}\right)$.

**59.** $z_1 = 2\sqrt{3} - 2i$, so $\tan\theta_1 = \frac{-2}{2\sqrt{3}} = -\frac{1}{\sqrt{3}}$ with $\theta_1$ in quadrant IV $\Rightarrow$ $\theta_1 = \frac{11\pi}{6}$, and $r_1 = \sqrt{12+4} = 4$.
$z_2 = -1 + i$, so $\tan\theta_2 = -1$ with $\theta_2$ in quadrant II $\Rightarrow$ $\theta_2 = \frac{3\pi}{4}$, and $r_2 = \sqrt{1+1} = \sqrt{2}$.
Hence, $z_1 = 4\left(\cos\frac{11\pi}{6} + i\sin\frac{11\pi}{6}\right)$ and $z_2 = \sqrt{2}\left(\cos\frac{3\pi}{4} + i\sin\frac{3\pi}{4}\right)$.
Thus, $z_1 z_2 = 4 \cdot \sqrt{2}\left[\cos\left(\frac{11\pi}{6} + \frac{3\pi}{4}\right) + i\sin\left(\frac{11\pi}{6} + \frac{3\pi}{4}\right)\right] = 4\sqrt{2}\left(\cos\frac{7\pi}{12} + i\sin\frac{7\pi}{12}\right)$,
$z_1/z_2 = \dfrac{4}{\sqrt{2}}\left[\cos\left(\frac{11\pi}{6} - \frac{3\pi}{4}\right) + i\sin\left(\frac{11\pi}{6} - \frac{3\pi}{4}\right)\right] = 2\sqrt{2}\left(\cos\frac{13\pi}{12} + i\sin\frac{13\pi}{12}\right)$, and
$1/z_1 = \frac{1}{4}\left(\cos\left(-\frac{11\pi}{6}\right) + i\sin\left(-\frac{11\pi}{6}\right)\right) = \frac{1}{4}\left(\cos\frac{11\pi}{6} - i\sin\frac{11\pi}{6}\right)$.

**60.** $z_1 = -\sqrt{2}i$, so $\theta_1 = \frac{3\pi}{2}$, and $r_1 = \sqrt{2}$.
$z_2 = -3 - 3\sqrt{3}i$, so $\tan\theta_2 = \sqrt{3}$ with $\theta_2$ in quadrant IV $\Rightarrow$ $\theta_2 = \frac{4\pi}{3}$, and $r_2 = \sqrt{9+27} = 6$.
Hence, $z_1 = \sqrt{2}\left(\cos\frac{3\pi}{2} + i\sin\frac{3\pi}{2}\right)$ and $z_2 = 6\left(\cos\frac{4\pi}{3} + i\sin\frac{4\pi}{3}\right)$.
Thus, $z_1 z_2 = \sqrt{2} \cdot 6\left[\cos\left(\frac{3\pi}{2} + \frac{4\pi}{3}\right) + i\sin\left(\frac{3\pi}{2} + \frac{4\pi}{3}\right)\right] = 6\sqrt{2}\left(\cos\frac{5\pi}{6} + i\sin\frac{5\pi}{6}\right)$,
$z_1/z_2 = \frac{\sqrt{2}}{6}\left[\cos\left(\frac{3\pi}{2} - \frac{4\pi}{3}\right) + i\sin\left(\frac{3\pi}{2} - \frac{4\pi}{3}\right)\right] = \frac{\sqrt{2}}{6}\left(\cos\frac{\pi}{6} + i\sin\frac{\pi}{6}\right)$, and
$1/z_1 = \frac{1}{\sqrt{2}}\left[\cos\left(-\frac{3\pi}{2}\right) + i\sin\left(-\frac{3\pi}{2}\right)\right] = \frac{1}{\sqrt{2}}\left(\cos\frac{3\pi}{2} - i\sin\frac{3\pi}{2}\right)$.

**61.** $z_1 = 5 + 5i$, so $\tan\theta_1 = \frac{5}{5} = 1$ with $\theta_1$ in quadrant I $\Rightarrow$ $\theta_1 = \frac{\pi}{4}$, and $r_1 = \sqrt{25+25} = 5\sqrt{2}$.
$z_2 = 4$, so $\theta_2 = 0$, and $r_2 = 4$.
Hence, $z_1 = 5\sqrt{2}\left(\cos\frac{\pi}{4} + i\sin\frac{\pi}{4}\right)$ and $z_2 = 4\left(\cos 0 + i\sin 0\right)$.
Thus, $z_1 z_2 = 5\sqrt{2} \cdot 4\left[\cos\left(\frac{\pi}{4} + 0\right) + i\sin\left(\frac{\pi}{4} + 0\right)\right] = 20\sqrt{2}\left(\cos\frac{\pi}{4} + i\sin\frac{\pi}{4}\right)$, $z_1/z_2 = \frac{5\sqrt{2}}{4}\left(\cos\frac{\pi}{4} + i\sin\frac{\pi}{4}\right)$, and
$1/z_1 = \frac{1}{5\sqrt{2}}\left(\cos\left(-\frac{\pi}{4}\right) + i\sin\left(-\frac{\pi}{4}\right)\right) = \frac{\sqrt{2}}{10}\left(\cos\frac{\pi}{4} - i\sin\frac{\pi}{4}\right)$.

**62.** $z_1 = 4\sqrt{3} - 4i$, so $\tan\theta_1 = \frac{-4}{4\sqrt{3}} = -\frac{\sqrt{3}}{3}$ with $\theta_1$ in quadrant III $\Rightarrow$ $\theta_1 = \frac{11\pi}{6}$, and $r_1 = \sqrt{48+16} = 8$.
$z_2 = 8i$, so $\theta_2 = \frac{\pi}{2}$, and $r_2 = 8$.
Hence, $z_1 = 8\left(\cos\frac{11\pi}{6} + i\sin\frac{11\pi}{6}\right)$ and $z_2 = 8\left(\cos\frac{\pi}{2} + i\sin\frac{\pi}{2}\right)$.
Thus, $z_1 z_2 = 8 \cdot 8\left[\cos\left(\frac{11\pi}{6} + \frac{\pi}{2}\right) + i\sin\left(\frac{11\pi}{6} + \frac{\pi}{2}\right)\right] = 64\left(\cos\frac{\pi}{3} + i\sin\frac{\pi}{3}\right)$,
$z_1/z_2 = \frac{8}{8}\left[\cos\left(\frac{11\pi}{6} - \frac{\pi}{2}\right) + i\sin\left(\frac{11\pi}{6} - \frac{\pi}{2}\right)\right] = \cos\frac{4\pi}{3} + i\sin\frac{4\pi}{3}$, and $1/z_1 = \frac{1}{8}\left(\cos\frac{11\pi}{6} - i\sin\frac{11\pi}{6}\right)$.

**63.** $z_1 = -20$, so $\theta_1 = \pi$, and $r_1 = 20$.
$z_2 = \sqrt{3} + i$, so $\tan\theta_2 = \frac{1}{\sqrt{3}}$ with $\theta_2$ in quadrant I $\Rightarrow$ $\theta_2 = \frac{\pi}{6}$, and $r_2 = \sqrt{3+1} = 2$.
Hence, $z_1 = 20\left(\cos\pi + i\sin\pi\right)$ and $z_2 = 2\left(\cos\frac{\pi}{6} + i\sin\frac{\pi}{6}\right)$.
Thus, $z_1 z_2 = 20 \cdot 2\left[\cos\left(\pi + \frac{\pi}{6}\right) + i\sin\left(\pi + \frac{\pi}{6}\right)\right] = 40\left(\cos\frac{7\pi}{6} + i\sin\frac{7\pi}{6}\right)$,
$z_1/z_2 = \frac{20}{2}\left[\cos\left(\pi - \frac{\pi}{6}\right) + i\sin\left(\pi - \frac{\pi}{6}\right)\right] = 10\left(\cos\frac{5\pi}{6} + i\sin\frac{5\pi}{6}\right)$, and
$1/z_1 = \frac{1}{20}\left[\cos(-\pi) + i\sin(-\pi)\right] = \frac{1}{20}\left(\cos\pi - i\sin\pi\right)$.

**64.** $z_1 = 3 + 4i$, so $\tan\theta_1 = \frac{4}{3}$ with $\theta_1$ in quadrant I $\Rightarrow$ $\theta_1 = \tan^{-1}\frac{4}{3}$, and $r_1 = \sqrt{9+16} = 5$.
$z_2 = 2 - 2i$, so $\tan\theta_2 = -1$ with $\theta_2$ in quadrant IV $\Rightarrow$ $\theta_2 = \frac{7\pi}{4}$, and $r_2 = \sqrt{4+4} = 2\sqrt{2}$.
Hence, $z_1 = 5\left[\cos\left(\tan^{-1}\frac{4}{3}\right) + i\sin\left(\tan^{-1}\frac{4}{3}\right)\right] \approx 5\left(\cos 0.927 + i\sin 0.927\right)$ and $z_2 = 2\sqrt{2}\left(\cos\frac{7\pi}{4} + i\sin\frac{7\pi}{4}\right)$.
Thus, $z_1 z_2 = 5 \cdot 2\sqrt{2}\left[\cos\left(\tan^{-1}\frac{4}{3} + \frac{7\pi}{4}\right) + i\sin\left(\tan^{-1}\frac{4}{3} + \frac{7\pi}{4}\right)\right] = 10\sqrt{2}\left(\cos 0.142 + i\sin 0.142\right)$,
$z_1/z_2 = \frac{5}{2\sqrt{2}}\left[\cos\left(\tan^{-1}\frac{4}{3} - \frac{7\pi}{4}\right) + i\sin\left(\tan^{-1}\frac{4}{3} - \frac{7\pi}{4}\right)\right] = \frac{5\sqrt{2}}{4}\left[\cos\left(\tan^{-1}\frac{4}{3} - \frac{7\pi}{4}\right) + i\sin\left(\tan^{-1}\frac{4}{3} - \frac{7\pi}{4}\right)\right]$
$= \frac{5\sqrt{2}}{4}\left(\cos 4.57 + i\sin 4.57\right)$, and
$1/z_1 = \frac{1}{5}\left[\cos\left(\tan^{-1}\frac{4}{3}\right) - i\sin\left(\tan^{-1}\frac{4}{3}\right)\right] = \frac{1}{5}\left(\cos 0.927 + i\sin 0.927\right)$.

**65.** From Exercise 25, $1+i=\sqrt{2}\left(\cos\frac{\pi}{4}+i\sin\frac{\pi}{4}\right)$. Thus, $(1+i)^{20}=\left(\sqrt{2}\right)^{20}\left[\cos 20\left(\frac{\pi}{4}\right)+i\sin 20\left(\frac{\pi}{4}\right)\right]=\left(2^{1/2}\right)^{20}\left(\cos 5\pi+i\sin 5\pi\right)=2^{10}\left(-1+0i\right)=-1024$.

**66.** $1-\sqrt{3}i=2\left(\cos\frac{5\pi}{3}+i\sin\frac{5\pi}{3}\right)$, so $\left(1-\sqrt{3}i\right)^5=2^5\left(\cos\frac{25\pi}{3}+i\sin\frac{25\pi}{3}\right)=32\left(\cos\frac{\pi}{3}+i\sin\frac{\pi}{3}\right)=32\left(\frac{1}{2}+i\frac{\sqrt{3}}{2}\right)=16+16\sqrt{3}i$.

**67.** $r=\sqrt{12+4}=4$ and $\tan\theta=\frac{2}{2\sqrt{3}}=\frac{1}{\sqrt{3}}\Rightarrow\theta=\frac{\pi}{6}$. Thus, $2\sqrt{3}+2i=4\left(\cos\frac{\pi}{6}+i\sin\frac{\pi}{6}\right)$. So $\left(2\sqrt{3}+2i\right)^5=4^5\left(\cos\frac{5\pi}{6}+i\sin\frac{5\pi}{6}\right)=1024\left(-\frac{\sqrt{3}}{2}+\frac{1}{2}i\right)=512\left(-\sqrt{3}+i\right)$.

**68.** From Exercise 28, $1-i=\sqrt{2}\left(\cos\frac{7\pi}{4}+i\sin\frac{7\pi}{4}\right)$. Thus, $(1-i)^8=\left(\sqrt{2}\right)^8\left(\cos 14\pi+i\sin 14\pi\right)=16\left(1+0i\right)=16$.

**69.** $r=\sqrt{\frac{1}{2}+\frac{1}{2}}=1$ and $\tan\theta=1\Rightarrow\theta=\frac{\pi}{4}$. Thus $\frac{\sqrt{2}}{2}+\frac{\sqrt{2}}{2}i=\cos\frac{\pi}{4}+i\sin\frac{\pi}{4}$. Therefore, $\left(\frac{\sqrt{2}}{2}+\frac{\sqrt{2}}{2}i\right)^{12}=\cos 12\left(\frac{\pi}{4}\right)+i\sin 12\left(\frac{\pi}{4}\right)=\cos 3\pi+i\sin 3\pi=-1$.

**70.** $r=\sqrt{3+1}=2$ and $\tan\theta=\frac{-1}{\sqrt{3}}$ with $\theta$ in quadrant IV $\Rightarrow\theta=\frac{11\pi}{6}$. Thus $\sqrt{3}-i=2\left(\cos\frac{11\pi}{6}+i\sin\frac{11\pi}{6}\right)$, so $\left(\sqrt{3}-i\right)^{-10}=\left(\frac{1}{2}\right)^{10}\left(\cos\frac{-110\pi}{6}+i\sin\frac{-110\pi}{6}\right)=\frac{1}{1024}\left(\cos\frac{2\pi}{3}-i\sin\frac{2\pi}{3}\right)=\frac{1}{1024}\left(-\frac{1}{2}+\frac{\sqrt{3}}{2}i\right)=\frac{1}{2048}\left(-1+\sqrt{3}i\right)$.

**71.** $r=\sqrt{4+4}=4\sqrt{2}$ and $\tan\theta=-1$ with $\theta$ in quadrant IV $\Rightarrow\theta=\frac{7\pi}{4}$. Thus $2-2i=2\sqrt{2}\left(\cos\frac{7\pi}{4}+i\sin\frac{7\pi}{4}\right)$, so $(2-2i)^8=\left(2\sqrt{2}\right)^8\left(\cos 14\pi+i\sin 14\pi\right)=4096\left(1-0i\right)=4096$.

**72.** $r=\sqrt{\frac{1}{4}+\frac{3}{4}}=1$ and $\tan\theta=\sqrt{3}$ with $\theta$ in quadrant III $\Rightarrow\theta=-\frac{4\pi}{3}$. Thus $-\frac{1}{2}-\frac{\sqrt{3}}{2}i=\cos\frac{4\pi}{3}+i\sin\frac{4\pi}{3}$, so $\left(-\frac{1}{2}-\frac{\sqrt{3}}{2}i\right)^{15}=\cos\frac{60\pi}{3}+i\sin\frac{60\pi}{3}=\cos 20\pi+i\sin 20\pi=1$.

**73.** $r=\sqrt{1+1}=\sqrt{2}$ and $\tan\theta=1$ with $\theta$ in quadrant III $\Rightarrow\theta=\frac{5\pi}{4}$. Thus $-1-i=\sqrt{2}\left(\cos\frac{5\pi}{4}+i\sin\frac{5\pi}{4}\right)$, so $(-1-i)^7=\left(\sqrt{2}\right)^7\left(\cos\frac{35\pi}{4}+i\sin\frac{35\pi}{4}\right)=8\sqrt{2}\left(\cos\frac{3\pi}{4}+i\sin\frac{3\pi}{4}\right)=8\sqrt{2}\left(\frac{1}{\sqrt{2}}-i\frac{1}{\sqrt{2}}\right)=8\left(-1+i\right)$.

**74.** $r=\sqrt{9+3}=2\sqrt{3}$ and $\tan\theta=\frac{\sqrt{3}}{3}$ with $\theta$ in quadrant I $\Rightarrow\theta=\frac{\pi}{6}$. Thus $3+\sqrt{3}i=2\sqrt{3}\left(\cos\frac{\pi}{6}+i\sin\frac{\pi}{6}\right)$, so $\left(3+\sqrt{3}i\right)^4=144\left(\cos\frac{2\pi}{3}+i\sin\frac{2\pi}{3}\right)=144\left(-\frac{1}{2}+\frac{\sqrt{3}}{2}i\right)=72\left(-1+\sqrt{3}i\right)$.

**75.** $r=\sqrt{12+4}=4$ and $\tan\theta=\frac{2}{2\sqrt{3}}=\frac{1}{\sqrt{3}}\Rightarrow\theta=\frac{\pi}{6}$. Thus $2\sqrt{3}+2i=4\left(\cos\frac{\pi}{6}+i\sin\frac{\pi}{6}\right)$, so $\left(2\sqrt{3}+2i\right)^{-5}=\left(\frac{1}{4}\right)^5\left(\cos\frac{-5\pi}{6}+i\sin\frac{-5\pi}{6}\right)=\frac{1}{1024}\left(-\frac{\sqrt{3}}{2}-\frac{1}{2}i\right)=\frac{1}{2048}\left(-\sqrt{3}-i\right)$

**76.** $r=\sqrt{1+1}=\sqrt{2}$ and $\tan\theta=-1$ with $\theta$ in quadrant IV $\Rightarrow\theta=\frac{7\pi}{4}$. Thus $1-i=\sqrt{2}\left(\cos\frac{7\pi}{4}+i\sin\frac{7\pi}{4}\right)$, so $(1-i)^{-8}=\frac{1}{16}\left(\cos\frac{-56\pi}{4}+i\sin\frac{-56\pi}{4}\right)=\frac{1}{16}\left(\cos 14\pi-i\sin 14\pi\right)=\frac{1}{16}$.

**77.** $r=\sqrt{48+16}=8$ and $\tan\theta=\frac{4}{4\sqrt{3}}=\frac{1}{\sqrt{3}}\Rightarrow\theta=\frac{\pi}{6}$. Thus $4\sqrt{3}+4i=8\left(\cos\frac{\pi}{6}+i\sin\frac{\pi}{6}\right)$. So,

$$\left(4\sqrt{3}+4i\right)^{1/2}=\sqrt{8}\left[\cos\left(\frac{\pi/6+2k\pi}{2}\right)+i\sin\left(\frac{\pi/6+2k\pi}{2}\right)\right]\text{ for }k=0,1.$$

Thus the two roots are $w_0=2\sqrt{2}\left(\cos\frac{\pi}{12}+i\sin\frac{\pi}{12}\right)$ and $w_1=2\sqrt{2}\left(\cos\frac{13\pi}{12}+i\sin\frac{13\pi}{12}\right)$.

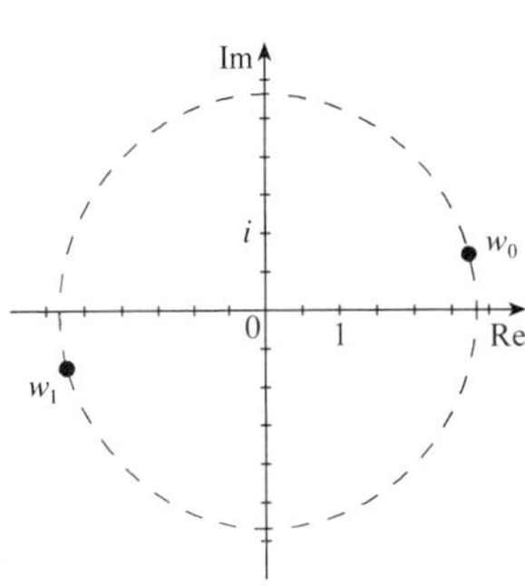

**78.** $r = \sqrt{48+16} = 8$ and $\tan\theta = \frac{4}{4\sqrt{3}} = \frac{1}{\sqrt{3}} \Rightarrow \theta = \frac{\pi}{6}$. Thus $4\sqrt{3}+4i = 8\left(\cos\frac{\pi}{6} + i\sin\frac{\pi}{6}\right)$. So

$$\left(4\sqrt{3}+4i\right)^{1/3} = 2\left[\cos\left(\frac{\pi/6+2k\pi}{3}\right) + i\sin\left(\frac{\pi/6+2k\pi}{3}\right)\right] \text{ for } k = 0, 1, 2.$$

Thus the three roots are $w_0 = 2\left(\cos\frac{\pi}{18} + i\sin\frac{\pi}{18}\right)$, $w_1 = 2\left(\cos\frac{13\pi}{18} + i\sin\frac{13\pi}{8}\right)$, and $w_2 = 2\left(\cos\frac{25\pi}{18} + i\sin\frac{25\pi}{18}\right)$.

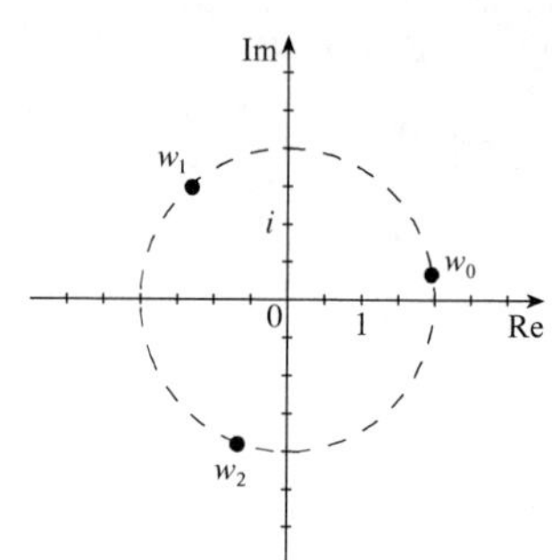


**79.** $-81i = 81\left(\cos\frac{3\pi}{2} + i\sin\frac{3\pi}{2}\right)$. Thus,

$$(-81i)^{1/4} = 81^{1/4}\left[\cos\left(\frac{3\pi/2+2k\pi}{4}\right) + i\sin\left(\frac{3\pi/2+2k\pi}{4}\right)\right] \text{ for } k = 0, 1,$$

2, 3. The four roots are $w_0 = 3\left(\cos\frac{3\pi}{8} + i\sin\frac{3\pi}{8}\right)$, $w_1 = 3\left(\cos\frac{7\pi}{8} + i\sin\frac{7\pi}{8}\right)$, $w_2 = 3\left(\cos\frac{11\pi}{8} + i\sin\frac{11\pi}{8}\right)$, and $w_3 = 3\left(\cos\frac{15\pi}{8} + i\sin\frac{15\pi}{8}\right)$.

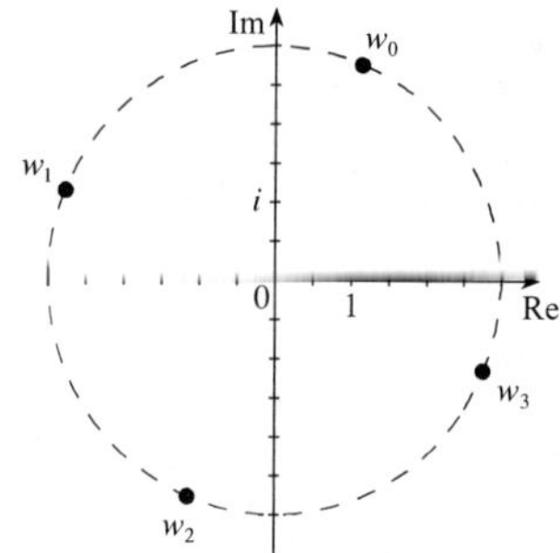


**80.** $32 = 32(\cos 0 + i\sin 0)$. Thus, $32^{1/5}\left(\cos\dfrac{2k\pi}{5} + i\sin\dfrac{2k\pi}{5}\right)$ for $k = 0, 1, 2, 3,$ 4. Thus the five roots are $w_0 = 2(\cos 0 + i\sin 0)$, $w_1 = 2\left(\cos\frac{2\pi}{5} + i\sin\frac{2\pi}{5}\right)$, $w_2 = 2\left(\cos\frac{4\pi}{5} + i\sin\frac{4\pi}{5}\right)$, $w_3 = 2\left(\cos\frac{6\pi}{5} + i\sin\frac{6\pi}{5}\right)$, and $w_4 = 2\left(\cos\frac{8\pi}{5} + i\sin\frac{8\pi}{5}\right)$.

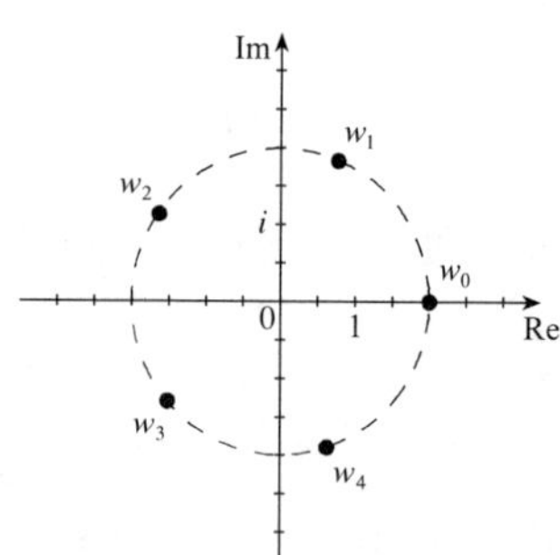


**81.** $1 = \cos 0 + i\sin 0$. Thus, $1^{1/8} = \cos\dfrac{2k\pi}{8} + i\sin\dfrac{2k\pi}{8}$, for $k = 0, 1, 2, 3, 4, 5, 6,$ 7. So the eight roots are $w_0 = \cos 0 + i\sin 0 = 1$,
$w_1 = \cos\frac{\pi}{4} + i\sin\frac{\pi}{4} = \frac{\sqrt{2}}{2} + i\frac{\sqrt{2}}{2}$, $w_2 = \cos\frac{\pi}{2} + i\sin\frac{\pi}{2} = i$,
$w_3 = \cos\frac{3\pi}{4} + i\sin\frac{3\pi}{4} = -\frac{\sqrt{2}}{2} + i\frac{\sqrt{2}}{2}$, $w_4 = \cos\pi + i\sin\pi = -1$,
$w_5 = \cos\frac{5\pi}{4} + i\sin\frac{5\pi}{4} = -\frac{\sqrt{2}}{2} - i\frac{\sqrt{2}}{2}$, $w_6 = \cos\frac{3\pi}{2} + i\sin\frac{3\pi}{2} = -i$, and
$w_7 = \cos\frac{7\pi}{4} + i\sin\frac{7\pi}{4} = \frac{\sqrt{2}}{2} - i\frac{\sqrt{2}}{2}$.

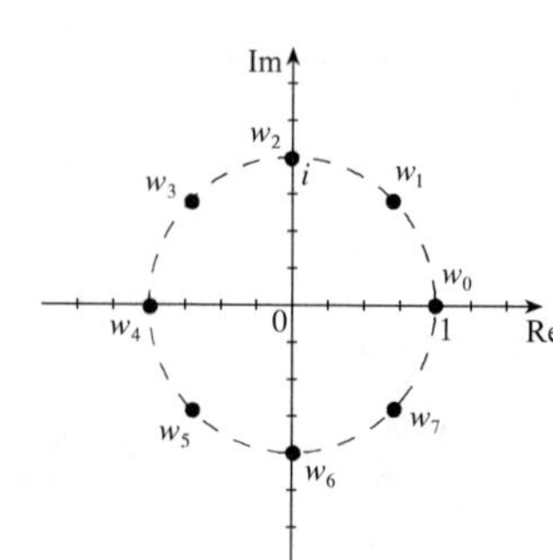

**82.** From Exercise 25, $1+i=\sqrt{2}\left(\cos\frac{\pi}{4}+i\sin\frac{\pi}{4}\right)$. So

$(1+i)^{1/3}=\left(\sqrt{2}\right)^{1/3}\left[\cos\left(\frac{\pi/4+2k\pi}{3}\right)+i\sin\left(\frac{\pi/4+2k\pi}{3}\right)\right]$ for $k=0$, 1, 2. Thus the three roots are $w_0=2^{1/6}\left(\cos\frac{\pi}{12}+i\sin\frac{\pi}{12}\right)$, $w_1=2^{1/6}\left(\cos\frac{9\pi}{12}+i\sin\frac{9\pi}{12}\right)$, and $w_2=2^{1/6}\left(\cos\frac{17\pi}{12}+i\sin\frac{17\pi}{12}\right)$.

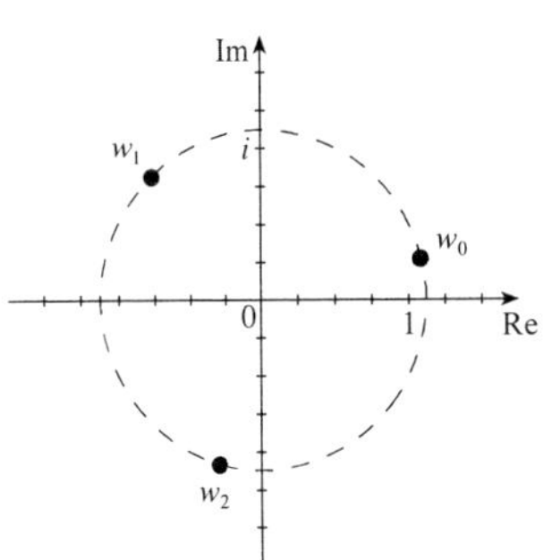


**83.** $i=\cos\frac{\pi}{2}+i\sin\frac{\pi}{2}$, so $i^{1/3}=\cos\left(\frac{\pi/2+2k\pi}{3}\right)+i\sin\left(\frac{\pi/2+2k\pi}{3}\right)$ for $k=0$, 1, 2. Thus the three roots are $w_0=\cos\frac{\pi}{6}+i\sin\frac{\pi}{6}=\frac{\sqrt{3}}{2}+\frac{1}{2}i$, $w_1=\cos\frac{5\pi}{6}+i\sin\frac{5\pi}{6}=-\frac{\sqrt{3}}{2}+\frac{1}{2}i$, and $w_2=\cos\frac{3\pi}{2}+i\sin\frac{3\pi}{2}=-i$.

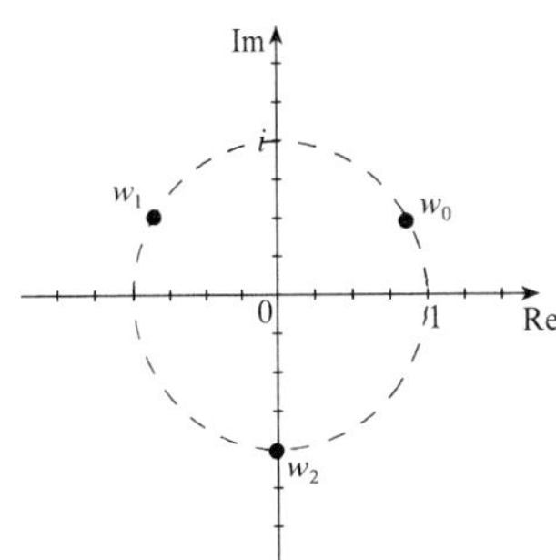


**84.** $i=\cos\frac{\pi}{2}+i\sin\frac{\pi}{2}$, so $i^{1/5}=\cos\left(\frac{\pi/2+2k\pi}{5}\right)+i\sin\left(\frac{\pi/2+2k\pi}{5}\right)$ for $k=0$, 1, 2, 3, 4. Thus the five roots are $w_0=\cos\frac{\pi}{10}+i\sin\frac{\pi}{10}$, $w_1=\cos\frac{\pi}{2}+i\sin\frac{\pi}{2}$, $w_2=\cos\frac{9\pi}{10}+i\sin\frac{9\pi}{10}$, $w_3=\cos\frac{13\pi}{10}+i\sin\frac{13\pi}{10}$, and $w_4=\cos\frac{17\pi}{10}+i\sin\frac{17\pi}{10}$.

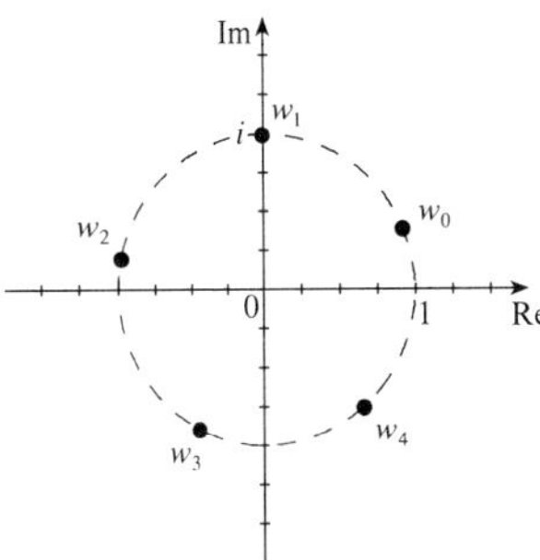


**85.** $-1=\cos\pi+i\sin\pi$. Then $(-1)^{1/4}=\cos\left(\frac{\pi+2k\pi}{4}\right)+i\sin\left(\frac{\pi+2k\pi}{4}\right)$ for $k=0$, 1, 2, 3. So the four roots are $w_0=\cos\frac{\pi}{4}+i\sin\frac{\pi}{4}=\frac{\sqrt{2}}{2}+i\frac{\sqrt{2}}{2}$, $w_1=\cos\frac{3\pi}{4}+i\sin\frac{3\pi}{4}=-\frac{\sqrt{2}}{2}+i\frac{\sqrt{2}}{2}$, $w_2=\cos\frac{5\pi}{4}+i\sin\frac{5\pi}{4}=-\frac{\sqrt{2}}{2}-i\frac{\sqrt{2}}{2}$, and $w_3=\cos\frac{7\pi}{4}+i\sin\frac{7\pi}{4}=\frac{\sqrt{2}}{2}-i\frac{\sqrt{2}}{2}$.

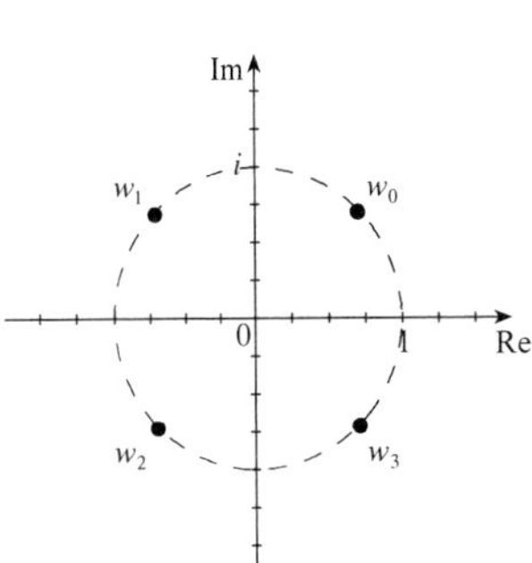

**86.** $r = \sqrt{16^2(1+3)} = 32$ and $\tan\theta = \frac{16\sqrt{3}}{16} = \sqrt{3}$ with $\theta$ in quadrant III $\Rightarrow$ $\theta = \frac{4\pi}{3}$. Thus $-16 - 16\sqrt{3}i = 32\left(\cos\frac{4\pi}{3} + i\sin\frac{4\pi}{3}\right)$. So $\left(-16 - 16\sqrt{3}i\right)^{1/5} = 32^{1/5}\left[\cos\left(\dfrac{4\pi/3 + 2k\pi}{5}\right) + i\sin\left(\dfrac{4\pi/3 + 2k\pi}{5}\right)\right]$ for $k = 0, 1, 2, 3, 4$. The five roots are $w_0 = 2\left(\cos\frac{4\pi}{15} + i\sin\frac{4\pi}{15}\right)$, $w_1 = 2\left(\cos\frac{2\pi}{3} + i\sin\frac{2\pi}{3}\right)$, $w_2 = 2\left(\cos\frac{16\pi}{15} + i\sin\frac{16\pi}{15}\right)$, $w_3 = 2\left(\cos\frac{22\pi}{15} + i\sin\frac{22\pi}{15}\right)$, and $w_4 = 2\left(\cos\frac{28\pi}{15} + i\sin\frac{28\pi}{15}\right)$.

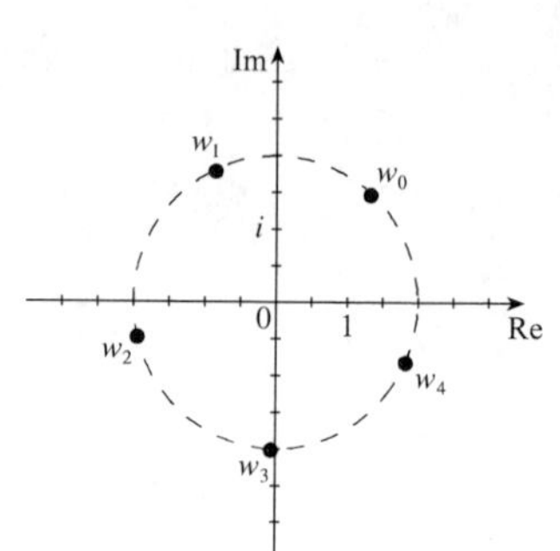


**87.** $z^4 + 1 = 0 \quad\Leftrightarrow\quad z = (-1)^{1/4} = \frac{\sqrt{2}}{2}(\pm 1 \pm i)$ (from Exercise 85)

**88.** $z^8 - i = 0 \quad\Rightarrow\quad x = i^{1/8}$. Since $i = \cos\frac{\pi}{2} + i\sin\frac{\pi}{2}$, then, $z = i^{1/8} = \cos\left(\dfrac{\pi/2 + 2k\pi}{8}\right) + i\sin\left(\dfrac{\pi/2 + 2k\pi}{8}\right)$ for $k = 0, 1, 2, 3, 4, 5, 6, 7$. Thus there are eight solutions: $z = \cos\frac{\pi}{16} + i\sin\frac{\pi}{16}$, $\cos\frac{5\pi}{16} + i\sin\frac{5\pi}{16}$, $\cos\frac{9\pi}{16} + i\sin\frac{9\pi}{16}$, $\cos\frac{13\pi}{16} + i\sin\frac{13\pi}{16}$, $\cos\frac{17\pi}{16} + i\sin\frac{17\pi}{16}$, $\cos\frac{21\pi}{16} + i\sin\frac{21\pi}{16}$, $\cos\frac{25\pi}{16} + i\sin\frac{25\pi}{16}$, and $\cos\frac{29\pi}{16} + i\sin\frac{29\pi}{16}$.

**89.** $z^3 - 4\sqrt{3} - 4i = 0 \quad\Leftrightarrow\quad z = \left(4\sqrt{3} + 4i\right)^{1/3}$. Since $4\sqrt{3} + 4i = 8\left(\cos\frac{\pi}{6} + i\sin\frac{\pi}{6}\right)$, $\left(4\sqrt{3} + 4i\right)^{1/3} = 8^{1/3}\left[\cos\left(\dfrac{\pi/6 + 2k\pi}{3}\right) + i\sin\left(\dfrac{\pi/6 + 2k\pi}{3}\right)\right]$, for $k = 0, 1, 2$. Thus the three roots are $z = 2\left(\cos\frac{\pi}{18} + i\sin\frac{\pi}{18}\right)$, $z = 2\left(\cos\frac{13\pi}{18} + i\sin\frac{13\pi}{8}\right)$, and $z = 2\left(\cos\frac{25\pi}{18} + i\sin\frac{25\pi}{18}\right)$.

**90.** $z^6 - 1 = 0 \quad\Leftrightarrow\quad z = 1^{1/6}$. Since $1 = \cos 0 + i\sin 0$, $z = 1^{1/6} = \cos\dfrac{2k\pi}{6} + i\sin\dfrac{2k\pi}{6}$ for $k = 0, 1, 2, 3, 4, 5$. Thus there are six solutions: $z = \pm 1, \pm\frac{1}{2} \pm \frac{\sqrt{3}}{2}i$.

**91.** $z^3 + 1 = -i \quad\Rightarrow\quad z = (-1 - i)^{1/3}$. Since $-1 - i = \sqrt{2}\left(\cos\frac{5\pi}{4} + i\sin\frac{5\pi}{4}\right)$, $z = (-1 - i)^{1/3} = 2^{1/6}\left[\cos\left(\dfrac{5\pi/4 + 2k\pi}{3}\right) + i\sin\left(\dfrac{5\pi/4 + 2k\pi}{3}\right)\right]$ for $k = 0, 1, 2$. Thus the three solutions to this equation are $z = 2^{1/6}\left(\cos\frac{5\pi}{12} + i\sin\frac{5\pi}{12}\right)$, $2^{1/6}\left(\cos\frac{13\pi}{12} + i\sin\frac{13\pi}{12}\right)$, and $2^{1/6}\left(\cos\frac{21\pi}{12} + i\sin\frac{21\pi}{12}\right)$.

**92.** $z^3 - 1 = 0 \quad\Rightarrow\quad z = 1^{1/3}$. Since $1 = \cos 0 + i\sin 0$, $z = 1^{1/3} = \cos\dfrac{2k\pi}{3} + i\sin\dfrac{2k\pi}{3}$ for $k = 0, 1, 2$. Thus the three solutions to this equation are $z = \cos 0 + i\sin 0$, $\cos\frac{2\pi}{3} + i\sin\frac{2\pi}{3}$, and $\cos\frac{4\pi}{3} + i\sin\frac{4\pi}{3}$ or $z = 1, -\frac{1}{2} + \frac{\sqrt{3}}{2}i, -\frac{1}{2} - \frac{\sqrt{3}}{2}i$.

**93. (a)** $w = \cos\dfrac{2\pi}{n} + i\sin\dfrac{2\pi}{n}$ for a positive integer $n$. Then, $w^k = \cos\dfrac{2k\pi}{n} + i\sin\dfrac{2k\pi}{n}$. Now $w^0 = \cos 0 + i\sin 0 = 1$ and for $k \neq 0$, $\left(w^k\right)^n = \cos 2k\pi + i\sin 2k\pi = 1$.So the $n$th roots of 1 are $\cos\dfrac{2k\pi}{n} + i\sin\dfrac{2k\pi}{n} = w^k$ for $k = 0, 1, 2, \ldots, n-1$. In other words, the $n$th roots of 1 are $w^0, w^1, w^2, w^3, \ldots, w^{n-1}$ or $1, w, w^2, w^3, \ldots, w^{n-1}$.

**(b)** For $k = 0, 1, \ldots, n-1$, we have $\left(sw^k\right)^n = s^n\left(w^k\right)^n = z \cdot 1 = z$, so $sw^k$ are $n$th roots of $z$ for $k = 0, 1, \ldots, n-1$.

**94.** The cube roots of 1 are $w^0 = 1$, $w^1 = \cos\frac{2\pi}{3} + i\sin\frac{2\pi}{3} = -\frac{1}{2} + \frac{\sqrt{3}}{2}i$, and $w^2 = \cos\frac{4\pi}{3} + i\sin\frac{4\pi}{3} = \frac{1}{2} - \frac{\sqrt{3}}{2}i$, so their sum is $w^0 + w^1 + w^2 = 1 + \left(-\frac{1}{2} + \frac{\sqrt{3}}{2}i\right) + \left(\frac{1}{2} - \frac{\sqrt{3}}{2}i\right) = 0$.

The fourth roots of 1 are $w^0 = 1$, $w^1 = i$, $w^2 = -1$, and $w^3 = -i$, so their sum is $w^0 + w^1 + w^2 + w^3 = 1 + i - 1 - i = 0$.

The fifth roots of 1 are $w^0 = 1$, $w^1 = \cos\frac{2\pi}{5} + i\sin\frac{2\pi}{5}$, $w^2 = \cos\frac{4\pi}{5} + i\sin\frac{4\pi}{5}$, $w^3 = \cos\frac{6\pi}{5} + i\sin\frac{6\pi}{5}$, and $w^4 = \cos\frac{8\pi}{5} + i\sin\frac{8\pi}{5}$, so their sum is

$$1 + \left(\cos\tfrac{2\pi}{5} + i\sin\tfrac{2\pi}{5}\right) + \left(\cos\tfrac{4\pi}{5} + i\sin\tfrac{4\pi}{5}\right) + \left(\cos\tfrac{6\pi}{5} + i\sin\tfrac{6\pi}{5}\right) + \left(\cos\tfrac{8\pi}{5} + i\sin\tfrac{8\pi}{5}\right)$$

$$= 1 + 2\cos\tfrac{2\pi}{5} + 2\cos\tfrac{6\pi}{5} \quad \text{(most terms cancel)} \quad = 1 + 2\left[\left(\tfrac{\sqrt{5}}{4} - \tfrac{1}{4}\right) + \left(-\tfrac{\sqrt{5}}{4} - \tfrac{1}{4}\right)\right] = 0$$

The sixth roots of 1 are $w^0 = 1$, $w^1 = \cos\frac{\pi}{3} + i\sin\frac{\pi}{3} = \frac{1}{2} + \frac{\sqrt{3}}{2}i$, $w^2 = \cos\frac{2\pi}{3} + i\sin\frac{2\pi}{3} = -\frac{1}{2} + \frac{\sqrt{3}}{2}i$, $w^3 = -1$, $w^4 = \cos\frac{4\pi}{3} + i\sin\frac{4\pi}{3} = -\frac{1}{2} - \frac{\sqrt{3}}{2}i$, and $w^5 = \cos\frac{5\pi}{3} + i\sin\frac{5\pi}{3} = \frac{1}{2} - \frac{\sqrt{3}}{2}i$, so their sum is $1 + \left(\frac{1}{2} + \frac{\sqrt{3}}{2}i\right) + \left(-\dfrac{1}{2} + \frac{\sqrt{3}}{2}i\right) - 1 + \left(-\frac{1}{2} - \frac{\sqrt{3}}{2}i\right) + \left(\frac{1}{2} - \frac{\sqrt{3}}{2}i\right) = 0$.

The eight roots of 1 are $w^0 = 1$, $w^1 = \cos\frac{\pi}{4} + i\sin\frac{\pi}{4} = \frac{\sqrt{2}}{2} + \frac{\sqrt{2}}{2}i$, $w^2 = i$, $w^3 = \cos\frac{3\pi}{4} + i\sin\frac{3\pi}{4} = -\frac{\sqrt{2}}{2} + \frac{\sqrt{2}}{2}i$, $w^4 = -1$, $w^5 = \cos\frac{5\pi}{4} + i\sin\frac{5\pi}{4} = -\frac{\sqrt{2}}{2} - \frac{\sqrt{2}}{2}i$, $w^6 = -i$, and $w^7 = \cos\frac{7\pi}{4} + i\sin\frac{7\pi}{4} = \frac{\sqrt{2}}{2} - \frac{\sqrt{2}}{2}i$, so their sum is $1 + \left(\frac{\sqrt{2}}{2} + \frac{\sqrt{2}}{2}i\right) + i + \left(-\frac{\sqrt{2}}{2} + \frac{\sqrt{2}}{2}\right) - 1 + \left(-\frac{\sqrt{2}}{2} - \frac{\sqrt{2}}{2}i\right) - i + \left(\frac{\sqrt{2}}{2} - \frac{\sqrt{2}}{2}i\right) = 0$.

It seems that the sum of any set of $n$th roots is 0.

To prove this, factor $w^n - 1 = \left(w^1 - 1\right)\left(1 + w^1 + w^2 + w^3 + \cdots + w^{n-1}\right)$. Since this is 0 and $w^1 \neq 0$, we must have $1 + w^1 + w^2 + w^3 + \cdots + w^{n-1} = 0$.

**95.** The cube roots of 1 are $w^0 = 1$, $w^1 = \cos\frac{2\pi}{3} + i\sin\frac{2\pi}{3}$, and $w^2 = \cos\frac{4\pi}{3} + i\sin\frac{4\pi}{3}$, so their product is $w^0 \cdot w^1 \cdot w^2 = (1)\left(\cos\frac{2\pi}{3} + i\sin\frac{2\pi}{3}\right)\left(\cos\frac{4\pi}{3} + i\sin\frac{4\pi}{3}\right) = \cos 2\pi + i\sin 2\pi = 1$.

The fourth roots of 1 are $w^0 = 1$, $w^1 = i$, $w^2 = -1$, and $w^3 = -i$, so their product is $w^0 \cdot w^1 \cdot w^2 \cdot w^3 = (1) \cdot (i) \cdot (-1) \cdot (-i) = i^2 = -1$.

The fifth roots of 1 are $w^0 = 1$, $w^1 = \cos\frac{2\pi}{5} + i\sin\frac{2\pi}{5}$, $w^2 = \cos\frac{4\pi}{5} + i\sin\frac{4\pi}{5}$, $w^3 = \cos\frac{6\pi}{5} + i\sin\frac{6\pi}{5}$, and $w^4 = \cos\frac{8\pi}{5} + i\sin\frac{8\pi}{5}$, so their product is $1\left(\cos\frac{2\pi}{5} + i\sin\frac{2\pi}{5}\right)\left(\cos\frac{4\pi}{5} + i\sin\frac{4\pi}{5}\right)\left(\cos\frac{6\pi}{5} + i\sin\frac{6\pi}{5}\right)\left(\cos\frac{8\pi}{5} + i\sin\frac{8\pi}{5}\right) = \cos 4\pi + i\sin 4\pi = 1$.

The sixth roots of 1 are $w^0 = 1$, $w^1 = \cos\frac{\pi}{3} + i\sin\frac{\pi}{3}$, $w^2 = \cos\frac{2\pi}{3} + i\sin\frac{2\pi}{3} = -\frac{1}{2} + \frac{\sqrt{3}}{2}i$, $w^3 = -1$, $w^4 = \cos\frac{4\pi}{3} + i\sin\frac{4\pi}{3} = -\frac{1}{2} - \frac{\sqrt{3}}{2}i$, and $w^5 = \cos\frac{5\pi}{3} + i\sin\frac{5\pi}{3} = \frac{1}{2} - \frac{\sqrt{3}}{2}i$, so their product is $1\left(\cos\frac{\pi}{3} + i\sin\frac{\pi}{3}\right)\left(\cos\frac{2\pi}{3} + i\sin\frac{2\pi}{3}\right)(-1)\left(\cos\frac{4\pi}{3} + i\sin\frac{4\pi}{3}\right)\left(\cos\frac{5\pi}{3} + i\sin\frac{5\pi}{3}\right) = \cos 5\pi + i\sin 5\pi = -1$.

The eight roots of 1 are $w^0 = 1$, $w^1 = \cos\frac{\pi}{4} + i\sin\frac{\pi}{4}$, $w^2 = i$, $w^3 = \cos\frac{3\pi}{4} + i\sin\frac{3\pi}{4}$, $w^4 = -1$, $w^5 = \cos\frac{5\pi}{4} + i\sin\frac{5\pi}{4}$, $w^6 = -i$, $w^7 = \cos\frac{7\pi}{4} + i\sin\frac{7\pi}{4}$, so their product is $1\left(\cos\frac{\pi}{4} + i\sin\frac{\pi}{4}\right) i \left(\cos\frac{3\pi}{4} + i\sin\frac{3\pi}{4}\right)(-1)\left(\cos\frac{5\pi}{4} + i\sin\frac{5\pi}{4}\right)(-i)\left(\cos\frac{7\pi}{4} + i\sin\frac{7\pi}{4}\right) = i^2 \cdot (\cos 2\pi + i\sin 2\pi) = -1$.

The product of the $n$th roots of 1 is $-1$ if $n$ is even and 1 if $n$ is odd.

The proof requires the fact that the sum of the first $m$ integers is $\dfrac{m(m+1)}{2}$.

Let $w = \cos\dfrac{2\pi}{n} + i\sin\dfrac{2\pi}{n}$. Then $w^k = \cos\dfrac{2k\pi}{n} + i\sin\dfrac{2k\pi}{n}$ for $k = 0, 1, 2, \ldots, n-1$. The argument of the product of the $n$ roots of unity can be found by adding the arguments of each $w^k$. So the argument of the product is

$$\theta = 0 + \frac{2(1)\pi}{n} + \frac{2(2)\pi}{n} + \frac{2(3)\pi}{n} + \cdots + \frac{2(n-2)\pi}{n} + \frac{2(n-1)\pi}{n} = \frac{2\pi}{n}\left[0 + 1 + 2 + 3 + \cdots + (n-2) + (n-1)\right].$$

Since this is the sum of the first $n-1$ integers, this sum is $\dfrac{2\pi}{n} \cdot \dfrac{(n-1)n}{2} = (n-1)\pi$. Thus the product of the $n$ roots of unity is $\cos((n-1)\pi) + i\sin((n-1)\pi) = -1$ if $n$ is even and 1 if $n$ is odd.

**96. (a)** $z = \frac{-(1+i)\pm\sqrt{(1+i)^2-4(1)(i)}}{2(1)} = \frac{-1-i\pm\sqrt{2i-4i}}{2} = \frac{-1-i\pm\sqrt{-2i}}{2}$. Since $-2i = 2\left(\cos\frac{3\pi}{2} + i\sin\frac{3\pi}{2}\right)$, we apply DeMoivre's Theorem to find $\sqrt{-2i}$, we have $w_1 = \sqrt{2}\left(\cos\frac{3\pi}{4} + i\sin\frac{3\pi}{4}\right) = \sqrt{2}\left(-\frac{\sqrt{2}}{2} + \frac{\sqrt{2}}{2}i\right) = -1+i$ and $w_2 = \sqrt{2}\left(\cos\frac{7\pi}{4} + i\sin\frac{7\pi}{4}\right) = \sqrt{2}\left(\frac{\sqrt{2}}{2} - \frac{\sqrt{2}}{2}i\right) = 1-i$. (Notice that $w_2 = -w_1$.) So

$z = \frac{-1-i+(-1+i)}{2} = \frac{-1-i-1+i}{2} = \frac{-2}{2} = -1$ or $z = \frac{-1-i+(1-i)}{2} = -\frac{-1-i+1-i}{2} - \frac{-2i}{2} = -i$.

Checking: $(-1)^2 + (1+i)(-1) + i = 1 - 1 - i + i = 0$ and $(-i)^2 + (1+i)(-i) + i = -1 - i - i^2 + i = 0$.

Alternatively we could have factored

$$z^2 + (1+i)z + i = z^2 + z + zi + i = z^2 + zi + z + i = \left(z^2 + zi\right) + (z+i) = z(z+i) + (z+i)$$
$$= (z+i)(z+1) = 0$$

so $z + i = 0 \quad\Leftrightarrow\quad z = -i$ or $z + 1 = 0 \quad\Leftrightarrow\quad z = -1$.

**(b)** $z = \frac{-(-i)\pm\sqrt{(-i)^2-4(1)(1)}}{2(1)} = \frac{i\pm\sqrt{i^2-4}}{2} = \frac{i\pm\sqrt{-5}}{2} = \frac{i\pm\sqrt{5}i}{2}$.

So $z = \dfrac{i+\sqrt{5}i}{2} = \dfrac{1+\sqrt{5}}{2}i$ or $z = \dfrac{i-\sqrt{5}i}{2} = \dfrac{1-\sqrt{5}}{2}i$. Checking:

$\left(\frac{1+\sqrt{5}}{2}i\right)^2 - i\left(\frac{1+\sqrt{5}}{2}i\right) + 1 = -\frac{1+2\sqrt{5}+5}{4} + \frac{1+\sqrt{5}}{2} + 1 = -\frac{6+2\sqrt{5}}{4} + \frac{1+\sqrt{5}}{2} + \frac{2}{2} = -\frac{3+\sqrt{5}}{2} + \frac{3+\sqrt{5}}{2} = 0$

$\left(\frac{1-\sqrt{5}}{2}i\right)^2 - i\left(\frac{1-\sqrt{5}}{2}i\right) + 1 = -\frac{1-2\sqrt{5}+5}{4} + \frac{1-\sqrt{5}}{2} + 1 = -\frac{6-2\sqrt{5}}{4} + \frac{1-\sqrt{5}}{2} + \frac{2}{2} = -\frac{3-\sqrt{5}}{2} + \frac{3-\sqrt{5}}{2} = 0$

This polynomial does not factor as nicely as the polynomial in part (a).

**(c)** $z = \frac{(2-i)\pm\sqrt{(2-i)^2-4(1)\left(-\frac{1}{4}i\right)}}{2(1)} = \frac{2-i\pm\sqrt{4-4i+i^2+i}}{2} = \frac{2-i\pm\sqrt{3-3i}}{2}$.

To apply DeMoivre's Theorem, we have $3 - 3i = 3\sqrt{2}\left(\cos\frac{7\pi}{4} + i\sin\frac{7\pi}{4}\right)$, so one root of $\sqrt{3-3i}$ is $w_1 = \sqrt{3\sqrt{2}}\left(\cos\frac{7\pi}{8} + i\sin\frac{7\pi}{8}\right)$. Now using the half-angle formulas and the fact that $\cos\frac{7\pi}{8} < 0$ and $\sin\frac{7\pi}{8} > 0$, we have $\cos\frac{7\pi}{8} = \cos\dfrac{\frac{7\pi}{4}}{2} = -\sqrt{\dfrac{1+\cos\frac{7\pi}{4}}{2}} = -\sqrt{\dfrac{1+\frac{\sqrt{2}}{2}}{2}} = -\sqrt{\frac{2+\sqrt{2}}{4}}$ and

$\sin\frac{7\pi}{8} = \sin\dfrac{\frac{7\pi}{4}}{2} = \sqrt{\dfrac{1-\cos\frac{7\pi}{4}}{2}} = \sqrt{\dfrac{1-\frac{\sqrt{2}}{2}}{2}} = \sqrt{\dfrac{2-\sqrt{2}}{4}}$. So

$$w_1 = \sqrt{3\sqrt{2}}\left(\cos\tfrac{7\pi}{8} + i\sin\tfrac{7\pi}{8}\right) = \sqrt{3\sqrt{2}}\left(-\sqrt{\tfrac{2+\sqrt{2}}{4}}\right) + \sqrt{3\sqrt{2}}\sqrt{\tfrac{2-\sqrt{2}}{4}}\,i$$
$$= -\sqrt{\tfrac{3\sqrt{2}+3}{2}} + \sqrt{\tfrac{3\sqrt{2}-3}{2}}\,i$$

$w_2 = -w_1 = \sqrt{\frac{3\sqrt{2}+3}{2}} - \sqrt{\frac{3\sqrt{2}-3}{2}}\,i$. Then

$$z = \frac{2 - i + \left(-\sqrt{\frac{3\sqrt{2}+3}{2}} + \sqrt{\frac{3\sqrt{2}-3}{2}}\,i\right)}{2} = 1 - \tfrac{1}{2}i - \sqrt{\tfrac{3\sqrt{2}+3}{8}} + \sqrt{\tfrac{3\sqrt{2}-3}{8}}\,i$$
$$= \left(1 - \sqrt{\tfrac{3\sqrt{2}+3}{8}}\right) + \left(-\tfrac{1}{2} + \sqrt{\tfrac{3\sqrt{2}-3}{8}}\right)i$$

or

$$z = \frac{2 - i + \left(\sqrt{\frac{3\sqrt{2}+3}{2}} - \sqrt{\frac{3\sqrt{2}-3}{2}}\,i\right)}{2} = 1 - \tfrac{1}{2}i + \sqrt{\tfrac{3\sqrt{2}+3}{8}} + \sqrt{\tfrac{3\sqrt{2}-3}{8}}\,i$$
$$= \left(1 + \sqrt{\tfrac{3\sqrt{2}+3}{8}}\right) + \left(-\frac{1}{2} - \sqrt{\tfrac{3\sqrt{2}-3}{8}}\right)i$$

I have discovered a most wondrous proof of this fact, but unfortunately the margin is too small to contain it.

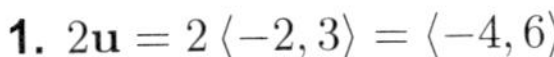

# 9.4 Vectors

**1.** $2\mathbf{u} = 2\langle -2, 3\rangle = \langle -4, 6\rangle$

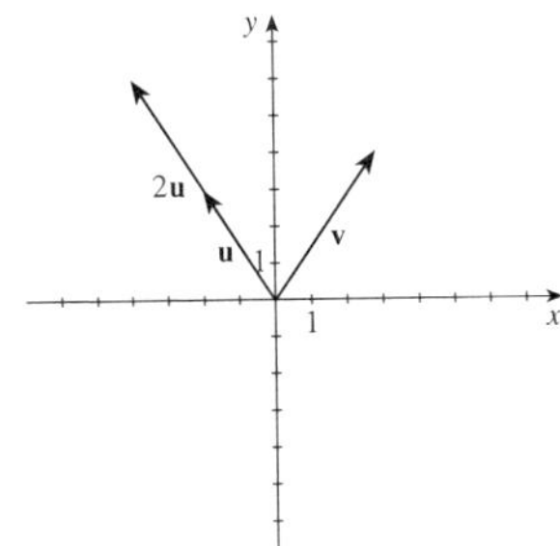


**2.** $-\mathbf{v} = -\langle 3, 4\rangle = \langle -3, -4\rangle$

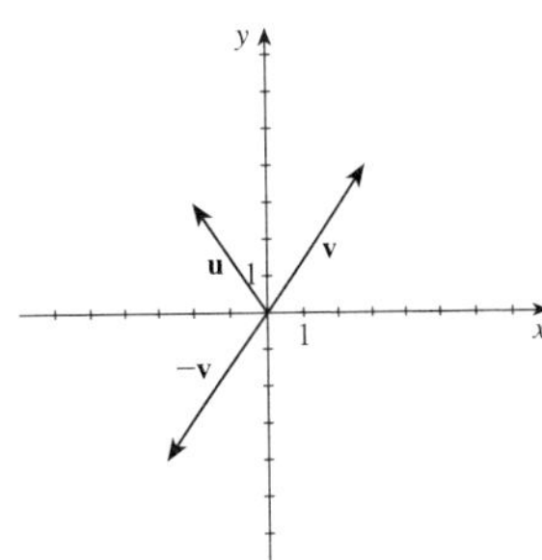


**3.** $\mathbf{u} + \mathbf{v} = \langle -2, 3\rangle + \langle 3, 4\rangle = \langle -2 + 3, 3 + 4\rangle = \langle 1, 7\rangle$

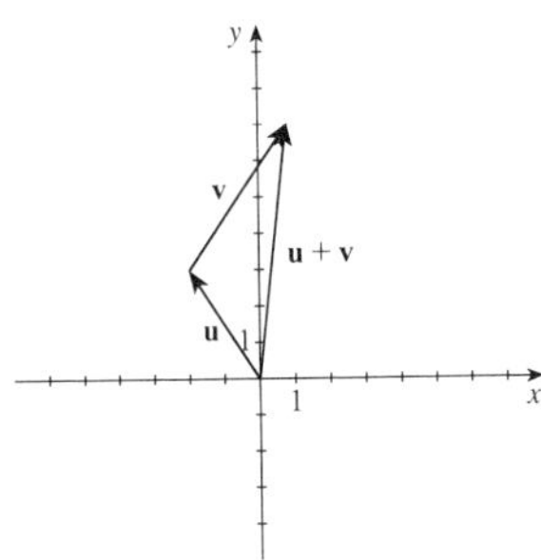


**4.** $\mathbf{u} - \mathbf{v} = \langle -2, 3\rangle - \langle 3, 4\rangle = \langle -2 - 3, 3 - 4\rangle = \langle -5, -1\rangle$

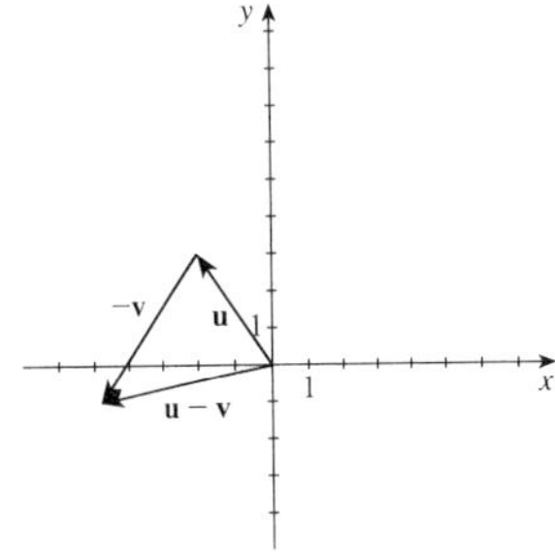


**5.** $\mathbf{v} - 2\mathbf{u} = \langle 3, 4\rangle - 2\langle -2, 3\rangle = \langle 3 - 2(-2), 4 - 2(3)\rangle = \langle 7, -2\rangle$

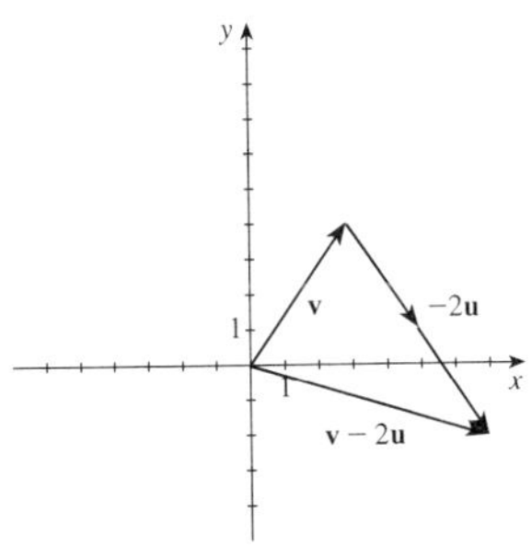


**6.** $2\mathbf{u} + \mathbf{v} = 2\langle -2, 3\rangle + \langle 3, 4\rangle = \langle 2(-2) + 3, 2(3) + 4\rangle = \langle -1, 10\rangle$

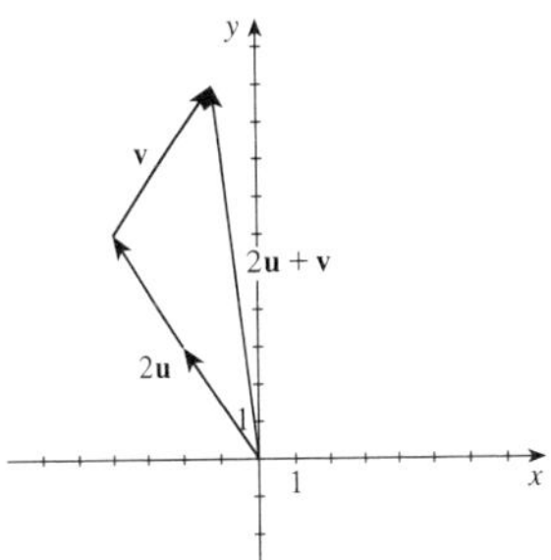


**In Solutions 7–16, v represents the vector with initial point $P$ and terminal point $Q$.**

**7.** $P(2, 1)$, $Q(5, 4)$. $\mathbf{v} = \langle 5 - 2, 4 - 1\rangle = \langle 3, 3\rangle$

**8.** $P(2, 1)$, $Q(-3, 4)$. $\mathbf{v} = \langle -3 - 2, 4 - 1\rangle = \langle -5, 3\rangle$

**9.** $P(1, 2)$, $Q(4, 1)$. $\mathbf{v} = \langle 4 - 1, 1 - 2\rangle = \langle 3, -1\rangle$

**10.** $P(3, 1)$, $Q(1, -2)$. $\mathbf{v} = \langle 1 - 3, -2 - 1\rangle = \langle -2, -3\rangle$

**11.** $P(3, 2)$, $Q(8, 9)$. $\mathbf{v} = \langle 8 - 3, 9 - 2\rangle = \langle 5, 7\rangle$

**12.** $P(1, 1)$, $Q(9, 9)$. $\mathbf{v} = \langle 9 - 1, 9 - 1\rangle = \langle 8, 8\rangle$

**13.** $P(5, 3)$, $Q(1, 0)$. $\mathbf{v} = \langle 1 - 5, 0 - 3\rangle = \langle -4, -3\rangle$

**14.** $P(-1, 3)$, $Q(-6, -1)$. $\mathbf{v} = \langle -6 - (-1), -1 - 3\rangle = \langle -5, -4\rangle$

**15.** $P(-1, -1)$, $Q(-1, 1)$. $\mathbf{v} = \langle -1 - (-1), 1 - (-1)\rangle = \langle 0, 2\rangle$

**16.** $P(-8, -6)$, $Q(-1, -1)$. $\mathbf{v} = \langle -1 - (-8), -1 - (-6)\rangle = \langle 7, 5\rangle$

**17.** $\mathbf{u} = \langle 2, 7 \rangle$, $\mathbf{v} = \langle 3, 1 \rangle$. $2\mathbf{u} = 2 \cdot \langle 2, 7 \rangle = \langle 4, 14 \rangle$; $-3\mathbf{v} = -3 \cdot \langle 3, 1 \rangle = \langle -9, -3 \rangle$; $\mathbf{u} + \mathbf{v} = \langle 2, 7 \rangle + \langle 3, 1 \rangle = \langle 5, 8 \rangle$; $3\mathbf{u} - 4\mathbf{v} = \langle 6, 21 \rangle - \langle 12, 4 \rangle = \langle -6, 17 \rangle$

**18.** $\mathbf{u} = \langle -2, 5 \rangle$, $\mathbf{v} = \langle 2, -8 \rangle$. $2\mathbf{u} = 2 \cdot \langle -2, 5 \rangle = \langle -4, 10 \rangle$; $-3\mathbf{v} = -3 \cdot \langle 2, -8 \rangle = \langle -6, 24 \rangle$; $\mathbf{u} + \mathbf{v} = \langle -2, 5 \rangle + \langle 2, -8 \rangle = \langle 0, -3 \rangle$; $3\mathbf{u} - 4\mathbf{v} = \langle -6, 15 \rangle - \langle 8, -32 \rangle = \langle -14, 47 \rangle$

**19.** $\mathbf{u} = \langle 0, -1 \rangle$, $\mathbf{v} = \langle -2, 0 \rangle$. $2\mathbf{u} = 2 \cdot \langle 0, -1 \rangle = \langle 0, -2 \rangle$; $-3\mathbf{v} = -3 \cdot \langle -2, 0 \rangle = \langle 6, 0 \rangle$; $\mathbf{u} + \mathbf{v} = \langle 0, -1 \rangle + \langle -2, 0 \rangle = \langle -2, -1 \rangle$; $3\mathbf{u} - 4\mathbf{v} = \langle 0, -3 \rangle - \langle -8, 0 \rangle = \langle 8, -3 \rangle$

**20.** $\mathbf{u} = \mathbf{i}$, $\mathbf{v} = -2\mathbf{j}$. $2\mathbf{u} = 2\mathbf{i}$; $-3\mathbf{v} = -3(-2\mathbf{j}) = 6\mathbf{j}$; $\mathbf{u} + \mathbf{v} = \mathbf{i} - 2\mathbf{j}$; $3\mathbf{u} - 4\mathbf{v} = 3\mathbf{i} + 8\mathbf{j}$

**21.** $\mathbf{u} = 2\mathbf{i}$, $\mathbf{v} = 3\mathbf{i} - 2\mathbf{j}$. $2\mathbf{u} = 2 \cdot 2\mathbf{i} = 4\mathbf{i}$; $-3\mathbf{v} = -3(3\mathbf{i} - 2\mathbf{j}) = -9\mathbf{i} + 6\mathbf{j}$; $\mathbf{u} + \mathbf{v} = 2\mathbf{i} + 3\mathbf{i} - 2\mathbf{j} = 5\mathbf{i} - 2\mathbf{j}$; $3\mathbf{u} - 4\mathbf{v} = 3 \cdot 2\mathbf{i} - 4(3\mathbf{i} - 2\mathbf{j}) = -6\mathbf{i} + 8\mathbf{j}$

**22.** $\mathbf{u} = \mathbf{i} + \mathbf{j}$, $\mathbf{v} = \mathbf{i} - \mathbf{j}$. $2\mathbf{u} = 2\mathbf{i} + 2\mathbf{j}$, $-3\mathbf{v} = -3\mathbf{i} + 3\mathbf{j}$, $\mathbf{u} + \mathbf{v} = \mathbf{i} + \mathbf{j} + \mathbf{i} - \mathbf{j} = 2\mathbf{i}$; $3\mathbf{u} - 4\mathbf{v} = 3(\mathbf{i} + \mathbf{j}) - 4(\mathbf{i} - \mathbf{j}) = -\mathbf{i} + 7\mathbf{j}$

**23.** $\mathbf{u} = 2\mathbf{i} + \mathbf{j}$, $\mathbf{v} = 3\mathbf{i} - 2\mathbf{j}$. Then $|\mathbf{u}| = \sqrt{2^2 + 1^2} = \sqrt{5}$; $|\mathbf{v}| = \sqrt{3^2 + 2^2} = \sqrt{13}$; $2\mathbf{u} = 4\mathbf{i} + 2\mathbf{j}$; $|2\mathbf{u}| = \sqrt{4^2 + 2^2} = 2\sqrt{5}$; $\frac{1}{2}\mathbf{v} = \frac{3}{2}\mathbf{i} - \mathbf{j}$; $\left|\frac{1}{2}\mathbf{v}\right| = \sqrt{\left(\frac{3}{2}\right)^2 + 1^2} = \frac{1}{2}\sqrt{13}$; $\mathbf{u} + \mathbf{v} = 5\mathbf{i} - \mathbf{j}$; $|\mathbf{u} + \mathbf{v}| = \sqrt{5^2 + 1^2} = \sqrt{26}$; $\mathbf{u} - \mathbf{v} = 2\mathbf{i} + \mathbf{j} - 3\mathbf{i} + 2\mathbf{j} = -\mathbf{i} + 3\mathbf{j}$; $|\mathbf{u} - \mathbf{v}| = \sqrt{1^2 + 3^2} = \sqrt{10}$; $|\mathbf{u}| - |\mathbf{v}| = \sqrt{5} - \sqrt{13}$

**24.** $\mathbf{u} = -2\mathbf{i} + 3\mathbf{j}$, $\mathbf{v} = \mathbf{i} - 2\mathbf{j}$. Then $|\mathbf{u}| = \sqrt{4 + 9} = \sqrt{13}$; $|\mathbf{v}| = \sqrt{1 + 4} = \sqrt{5}$; $2\mathbf{u} = -4\mathbf{i} + 6\mathbf{j}$; $|2\mathbf{u}| = \sqrt{16 + 36} = 2\sqrt{13}$; $\frac{1}{2}\mathbf{v} = \frac{1}{2}\mathbf{i} - \mathbf{j}$; $\left|\frac{1}{2}\mathbf{v}\right| = \sqrt{\frac{1}{4} + 1} = \frac{1}{2}\sqrt{5}$; $\mathbf{u} + \mathbf{v} = -\mathbf{i} + \mathbf{j}$; $|\mathbf{u} + \mathbf{v}| = \sqrt{1 + 1} = \sqrt{2}$; $\mathbf{u} - \mathbf{v} = -3\mathbf{i} + 5\mathbf{j}$; $|\mathbf{u} - \mathbf{v}| = \sqrt{9 + 25} = \sqrt{34}$; $|\mathbf{u}| - |\mathbf{v}| = \sqrt{13} - \sqrt{5}$

**25.** $\mathbf{u} = \langle 10, -1 \rangle$, $\mathbf{v} = \langle -2, -2 \rangle$. Then $|\mathbf{u}| = \sqrt{10^2 + 1^2} = \sqrt{101}$; $|\mathbf{v}| = \sqrt{(-2)^2 + (-2)^2} = 2\sqrt{2}$; $2\mathbf{u} = \langle 20, -2 \rangle$; $|2\mathbf{u}| = \sqrt{20^2 + 2^2} = \sqrt{404} = 2\sqrt{101}$; $\frac{1}{2}\mathbf{v} = \langle -1, -1 \rangle$; $\left|\frac{1}{2}\mathbf{v}\right| = \sqrt{(-1)^2 + (-1)^2} = \sqrt{2}$; $\mathbf{u} + \mathbf{v} = \langle 8, -3 \rangle$; $|\mathbf{u} + \mathbf{v}| = \sqrt{8^2 + 3^2} = \sqrt{73}$; $\mathbf{u} - \mathbf{v} = \langle 12, 1 \rangle$; $|\mathbf{u} - \mathbf{v}| = \sqrt{12^2 + 1^2} = \sqrt{145}$; $|\mathbf{u}| - |\mathbf{v}| = \sqrt{101} - 2\sqrt{2}$

**26.** $\mathbf{u} = \langle -6, 6 \rangle$, $\mathbf{v} = \langle -2, -1 \rangle$. Then $|\mathbf{u}| = \sqrt{36 + 36} = 6\sqrt{2}$; $|\mathbf{v}| = \sqrt{4 + 1} = \sqrt{5}$; $2\mathbf{u} = \langle -12, 12 \rangle$; $|2\mathbf{u}| = \sqrt{144 + 144} = 12\sqrt{2}$; $\frac{1}{2}\mathbf{v} = \langle -1, -\frac{1}{2} \rangle$; $\left|\frac{1}{2}\mathbf{v}\right| = \sqrt{1 + \frac{1}{4}} = \frac{1}{2}\sqrt{5}$; $\mathbf{u} + \mathbf{v} = \langle -8, 5 \rangle$; $|\mathbf{u} + \mathbf{v}| = \sqrt{64 + 25} = \sqrt{89}$; $\mathbf{u} - \mathbf{v} = \langle -4, 7 \rangle$; $|\mathbf{u} - \mathbf{v}| = \sqrt{16 + 49} = \sqrt{65}$; $|\mathbf{u}| - |\mathbf{v}| = 6\sqrt{2} - \sqrt{5}$

**In Solutions 27–32, $x$ represents the horizontal component and $y$ the vertical component.**

**27.** $|\mathbf{v}| = 40$, direction $\theta = 30°$. $x = 40 \cos 30° = 20\sqrt{3}$ and $y = 40 \sin 30° = 20$. Thus, $\mathbf{v} = x\mathbf{i} + y\mathbf{j} = 20\sqrt{3}\mathbf{i} + 20\mathbf{j}$.

**28.** $|\mathbf{v}| = 50$, direction $\theta = 120°$. $x = 50 \cos 120° = -25$ and $y = 50 \sin 120° = 25\sqrt{3}$. Thus, $\mathbf{v} = x\mathbf{i} + y\mathbf{j} = -25\mathbf{i} + 25\sqrt{3}\mathbf{j}$.

**29.** $|\mathbf{v}| = 1$, direction $\theta = 225°$. $x = \cos 225° = -\frac{1}{\sqrt{2}}$ and $y = \sin 225° = -\frac{1}{\sqrt{2}}$. Thus, $\mathbf{v} = x\mathbf{i} + y\mathbf{j} = -\frac{1}{\sqrt{2}}\mathbf{i} - \frac{1}{\sqrt{2}}\mathbf{j} = -\frac{\sqrt{2}}{2}\mathbf{i} - \frac{\sqrt{2}}{2}\mathbf{j}$.

**30.** $|\mathbf{v}| = 800$, direction $\theta = 125°$. $x = 800 \cos 125° \approx -458.86$ and $y = 800 \sin 125° \approx 655.32$. Thus, $\mathbf{v} = x\mathbf{i} + y\mathbf{j} = (800 \cos 125°)\,\mathbf{i} + (800 \sin 125°)\,\mathbf{j} \approx -458.86\mathbf{i} + 655.32\mathbf{j}$.

**31.** $|\mathbf{v}| = 4$, direction $\theta = 10°$. $x = 4 \cos 10° \approx 3.94$ and $y = 4 \sin 10° \approx 0.69$. Thus, $\mathbf{v} = x\mathbf{i} + y\mathbf{j} = (4 \cos 10°)\,\mathbf{i} + (4 \sin 10°)\,\mathbf{j} \approx 3.94\mathbf{i} + 0.69\mathbf{j}$.

**32.** $|\mathbf{v}| = \sqrt{3}$, direction $\theta = 300°$. $x = \sqrt{3} \cos 300° = \frac{\sqrt{3}}{2}$ and $y = \sqrt{3} \sin 300° = -\frac{3}{2}$. Thus, $\mathbf{v} = x\mathbf{i} + y\mathbf{j} = \frac{\sqrt{3}}{2}\mathbf{i} - \frac{3}{2}\mathbf{j}$.

**33.** $\mathbf{v} = \langle 3, 4 \rangle$. The magnitude is $|\mathbf{v}| = \sqrt{3^2 + 4^2} = 5$. The direction is $\theta$ where $\tan \theta = \frac{4}{3}$ $\Leftrightarrow$ $\theta = \tan^{-1}\left(\frac{4}{3}\right) \approx 53.13°$.

**34.** $\mathbf{v} = \left\langle -\frac{\sqrt{2}}{2}, -\frac{\sqrt{2}}{2} \right\rangle$. The magnitude is $|\mathbf{v}| = \sqrt{\frac{1}{2} + \frac{1}{2}} = 1$. The direction is $\theta$ where $\tan \theta = 1$ with $\theta$ in quadrant III $\Leftrightarrow$ $\theta = \tan^{-1} 1 = 225°$.

**35.** $\mathbf{v} = \langle -12, 5 \rangle$. The magnitude is $|\mathbf{v}| = \sqrt{(-12)^2 + 5^2} = \sqrt{169} = 13$. The direction is $\theta$ where $\tan \theta = -\frac{5}{12}$ with $\theta$ in quadrant II $\Leftrightarrow$ $\theta = \tan^{-1}\left(-\frac{5}{12}\right) \approx 157.38°$.

**36.** $\mathbf{v} = \langle 40, 9 \rangle$. The magnitude is $|\mathbf{v}| = \sqrt{1600 + 81} = 41$. The direction is $\theta$ where $\tan \theta = \frac{9}{40}$ with $\theta$ in quadrant I $\Leftrightarrow$ $\theta = \tan^{-1} \frac{9}{40} \approx 12.68°$.

**37.** $\mathbf{v} = \mathbf{i} + \sqrt{3}\mathbf{j}$. The magnitude is $|\mathbf{v}| = \sqrt{1^2 + (\sqrt{3})^2} = 2$. The direction is $\theta$ where$\tan\theta = \sqrt{3}$ with $\theta$ in quadrant I $\Leftrightarrow$ $\theta = \tan^{-1}\sqrt{3} = 60°$.

**38.** $\mathbf{v} = \mathbf{i} + \mathbf{j}$. The magnitude is $|\mathbf{v}| = \sqrt{1+1} = \sqrt{2}$. The direction is $\theta$ where $\tan\theta = 1$ with $\theta$ in quadrant I $\Leftrightarrow$ $\theta = \tan^{-1} 1 = 45°$.

**39.** $|\mathbf{v}| = 30$, direction $\theta = 30°$. $x = 30\cos 30° = 30 \cdot \frac{\sqrt{3}}{2} \approx 25.98$, $y = 30 \sin 30° = 15$. So the horizontal component of force is $15\sqrt{3}$ lb and the vertical component is $-15$ lb.

**40.** $|\mathbf{v}| = 500$, direction $\theta = 70°$. $x = 500\cos 70° \approx 171.01$, $y = 500 \sin 70° \approx 469.85$. So the east component of the velocity is 171.01 mi/h and the north component is 469.85 mi/h.

**41.** The flow of the river can be represented by the vector $\mathbf{v} = -3\mathbf{j}$ and the swimmer can be represented by the vector $\mathbf{u} = 2\mathbf{i}$. Therefore the true velocity is $\mathbf{u} + \mathbf{v} = 2\mathbf{i} - 3\mathbf{j}$.

**42.** The ocean currents can be represented by the vector $\mathbf{v} = 3\mathbf{i}$ and the salmon can be represented by the vector $\mathbf{u} = \frac{5\sqrt{2}}{2}\mathbf{i} + \frac{5\sqrt{2}}{2}\mathbf{j}$. Therefore $\mathbf{u} + \mathbf{v} = \left(\frac{5\sqrt{2}}{2} + 3\right)\mathbf{i} + \frac{5\sqrt{2}}{2}\mathbf{j}$ represents the true velocity of the fish.

**43.** **(a)** The velocity of the wind is $40\mathbf{j}$.

**(b)** The velocity of the jet relative to the air is $425\mathbf{i}$.

**(c)** The true velocity of the jet is $\mathbf{v} = 425\mathbf{i} + 40\mathbf{j} = \langle 425, 40\rangle$.

**(d)** The true speed of the jet is $|\mathbf{v}| = \sqrt{425^2 + 40^2} \approx 427$ mi/h, and the true direction is $\theta = \tan^{-1}\left(\frac{40}{425}\right) \approx 5.4°$ $\Rightarrow$ $\theta$ is N 84.6° E.

**44.** **(a)** The velocity of the wind is $\mathbf{v} = \langle 55\cos 60°, 55 \sin 60°\rangle = \left\langle \frac{55}{2}, \frac{55\sqrt{3}}{2}\right\rangle$.

**(b)** The velocity of the jet is $\mathbf{w} = \langle 765\cos 45°, 765\sin 45°\rangle = \left\langle \frac{765\sqrt{2}}{2}, \frac{765\sqrt{2}}{2}\right\rangle$.

**(c)** The true velocity of the jet is $\mathbf{v} + \mathbf{w} = \left\langle \frac{55}{2} + \frac{765\sqrt{2}}{2}, \frac{55\sqrt{3}}{2} + \frac{765\sqrt{2}}{2}\right\rangle \approx \langle 568.44, 588.57\rangle$.

**(d)** The true speed of the jet is $|\mathbf{v} + \mathbf{w}| \approx \sqrt{568.44^2 + 588.57^2} \approx 818$ mi/h. The direction of the vector $\mathbf{w} + \mathbf{v}$ is $\theta \approx \tan^{-1}\left(\frac{588.57}{568.44}\right) \approx 46°$. Thus the true direction of the jet is approximately N 44° E.

**45.** If the direction of the plane is N 30° W, the airplane's velocity is $\mathbf{u} = \langle \mathbf{u}_x, \mathbf{u}_y\rangle$ where $\mathbf{u}_x = -765\cos 60° = -382.5$, and $\mathbf{u}_y = 765 \sin 60° \approx 662.51$. If the direction of the wind is N 30° E, the wind velocity is $\mathbf{w} = \langle w_x, w_y\rangle$ where $w_x = 55\cos 60° = 27.5$, and $w_y = 55\sin 60° \approx 47.63$. Thus, the actual flight path is $\mathbf{v} = \mathbf{u} + \mathbf{w} = \langle -382.5 + 27.5, 662.51 + 47.63\rangle = \langle -355, 710.14\rangle$, and so the true speed is $|\mathbf{v}| = \sqrt{355^2 + 710.14^2} \approx 794$ mi/h, and the true direction is $\theta = \tan^{-1}\left(-\frac{710.14}{355}\right) \approx 116.6°$ so $\theta$ is N 26.6° W.

**46.** Let $\mathbf{v}$ be the velocity vector of the jet and let $\theta$ be the direction of this vector. Thus $\mathbf{v} = (765\cos\theta)\,\mathbf{i} + (765 \sin\theta)\,\mathbf{j}$. If w is the velocity vector of the wind then the true course of the jet is $\mathbf{v} + \mathbf{w} = \left(765\cos\theta + \frac{55}{2}\right)\mathbf{i} + \left(765\sin\theta + \frac{55\sqrt{3}}{2}\right)\mathbf{j}$. To achieve a course of true north, the east-west component (the $\mathbf{i}$ component) of the jet's velocity vector must be 0. That is $765\cos\theta + \frac{55}{2} = 0$ $\Leftrightarrow$ $\cos\theta = -\frac{27.5}{765}$ $\Leftrightarrow$ $\theta = \cos^{-1}\left(-\frac{27.5}{765}\right) \approx 92.1°$. Thus the pilot should head his plane in the direction N 2.1° W.

**47.** **(a)** The velocity of the river is represented by the vector $\mathbf{r} = \langle 10, 0\rangle$.

**(b)** Since the boater direction is 60° from the shore at 20 mi/h, the velocity of the boat is represented by the vector $\mathbf{b} = \langle 20\cos 60°, 20\sin 60°\rangle \approx \langle 10, 17.32\rangle$.

**(c)** $\mathbf{w} = \mathbf{r} + \mathbf{b} = \langle 10 + 10, 0 + 17.32\rangle = \langle 20, 17.32\rangle$

**(d)** The true speed of the boat is $|\mathbf{w}| = \sqrt{20^2 + 17.32^2} \approx 26.5$ mi/h, and the true direction is $\theta = \tan^{-1}\left(\frac{17.32}{20}\right) \approx 40.9° \approx$ N 49.1° E.

**48.** Let $\mathbf{w}$ be the velocity vector of the water, let $\mathbf{v}$ be the velocity vector of the boat, and let $\theta$ be the direction of $\mathbf{v}$. Then $\mathbf{v} = (20\cos\theta)\,\mathbf{i} + (20\sin\theta)\,\mathbf{j}$ and $\mathbf{w} = 10\mathbf{i}$. The true course of the boat is $\mathbf{v} + \mathbf{w} = (20\cos\theta + 10)\,\mathbf{i} + (20\sin\theta)\,\mathbf{j}$. To achieve a course of true north, the east-west component of the boat's velocity vector must be 0. Thus, $20\cos\theta + 10 = 0$ $\Leftrightarrow$ $\cos\theta = \dfrac{-10}{20} = -\frac{1}{2}$ $\Leftrightarrow$ $\theta = \cos^{-1}\left(-\frac{1}{2}\right) = 120^\circ$. Thus the boater should head her boat in the direction N $30^\circ$ W.

**49. (a)** Let $\mathbf{b} = \langle b_x, b_y \rangle$ represent the velocity of the boat relative to the water. Then $\mathbf{b} = \langle 24\cos 18^\circ, 24\sin 18^\circ \rangle$.

**(b)** Let $\mathbf{w} = \langle w_x, w_y \rangle$ represent the velocity of the water. Then $\mathbf{w} = \langle 0, w \rangle$ where $w$ is the speed of the water. So the true velocity of the boat is $\mathbf{b} + \mathbf{w} = \langle 24\cos 18^\circ, 24\sin 18^\circ - w \rangle$. For the direction to be due east, we must have $24\sin 18^\circ - w = 0$ $\Leftrightarrow$ $w = 7.42$ mi/h. Therefore, the true speed of the water is 7.4 mi/h. Since $\mathbf{b} + \mathbf{w} = \langle 24\cos 18^\circ, 0 \rangle$, the true speed of the boat is $|\mathbf{b} + \mathbf{w}| = 24\cos 18^\circ \approx 22.8$ mi/h.

**50.** Let w represent the velocity of the woman and l the velocity of the ocean liner. Then $\mathbf{w} = \langle -2, 0 \rangle$, and $\mathbf{l} = \langle 0, 25 \rangle$, and so $\mathbf{r} = \langle -2 + 0, 0 + 25 \rangle$. Hence, relative to the water, the woman's speed is $|\mathbf{r}| = \sqrt{4 + 625} \approx 25.08$ mi/h, and her direction is $\theta = \tan^{-1}\frac{25}{-2} \approx 94.57^\circ$ or approximately N $4.57^\circ$ W.

**51.** $\mathbf{F}_1 = \langle 2, 5 \rangle$ and $\mathbf{F}_2 = \langle 3, -8 \rangle$.

**(a)** $\mathbf{F}_1 + \mathbf{F}_2 = \langle 2 + 3, 5 - 8 \rangle = \langle 5, -3 \rangle$

**(b)** The additional force required is $\mathbf{F}_3 = \langle 0, 0 \rangle - \langle 5, -3 \rangle = \langle -5, 3 \rangle$.

**52.** $\mathbf{F}_1 = \langle 3, -7 \rangle$, $\mathbf{F}_2 = \langle 4, -2 \rangle$, and $\mathbf{F}_3 = \langle -7, 9 \rangle$.

**(a)** $\mathbf{F}_1 + \mathbf{F}_2 + \mathbf{F}_3 = \langle 3 + 4 - 7, -7 - 2 + 9 \rangle = \langle 0, 0 \rangle$

**(b)** No additional force is required.

**53.** $\mathbf{F}_1 = 4\mathbf{i} - \mathbf{j}$, $\mathbf{F}_2 = 3\mathbf{i} - 7\mathbf{j}$, $\mathbf{F}_3 = -8\mathbf{i} + 3\mathbf{j}$, and $\mathbf{F}_4 = \mathbf{i} + \mathbf{j}$.

**(a)** $\mathbf{F}_1 + \mathbf{F}_2 + \mathbf{F}_3 + \mathbf{F}_4 = (4 + 3 - 8 + 1)\,\mathbf{i} + (-1 - 7 + 3 + 1)\,\mathbf{j} = 0\mathbf{i} - 4\mathbf{j}$

**(b)** The additional force required is $\mathbf{F}_5 = 0\mathbf{i} + 0\mathbf{j} - (0\mathbf{i} - 4\mathbf{j}) = 4\mathbf{j}$.

**54.** $\mathbf{F}_1 = \mathbf{i} - \mathbf{j}$, $\mathbf{F}_2 = \mathbf{i} + \mathbf{j}$, and $\mathbf{F}_3 = -2\mathbf{i} + \mathbf{j}$.

**(a)** $\mathbf{F}_1 + \mathbf{F}_2 + \mathbf{F}_3 = (1 + 1 - 2)\,\mathbf{i} + (-1 + 1 + 1)\,\mathbf{j} = 0\mathbf{i} + \mathbf{j}$

**(b)** The additional force required is $\mathbf{F}_4 = 0\mathbf{i} + 0\mathbf{j} - (0\mathbf{i} + \mathbf{j}) = -\mathbf{j}$.

**55.** $\mathbf{F}_1 = \langle 10\cos 60^\circ, 10\sin 60^\circ \rangle = \langle 5, 5\sqrt{3} \rangle$, $\mathbf{F}_2 = \langle -8\cos 30^\circ, 8\sin 30^\circ \rangle = \langle -4\sqrt{3}, 4 \rangle$, and $\mathbf{F}_3 = \langle -6\cos 20^\circ, -6\sin 20^\circ \rangle \approx \langle -5.638, -2.052 \rangle$.

**(a)** $\mathbf{F}_1 + \mathbf{F}_2 + \mathbf{F}_3 = \langle 5 - 4\sqrt{3} - 5.638, 5\sqrt{3} + 4 - 2.052 \rangle \approx \langle -7.57, 10.61 \rangle$.

**(b)** The additional force required is $\mathbf{F}_4 = \langle 0, 0 \rangle - \langle -7.57, 10.61 \rangle = \langle 7.57, -10.61 \rangle$.

**56.** $\mathbf{F}_1 = \langle 3, -1 \rangle$, $\mathbf{F}_2 = \langle 1, 2 \rangle$, $\mathbf{F}_3 = \langle -2, -1 \rangle$, and $\mathbf{F}_4 = \langle 0, -4 \rangle$.

**(a)** $\mathbf{F}_1 + \mathbf{F}_2 + \mathbf{F}_3 + \mathbf{F}_4 = \langle 3 + 1 - 2 + 0, -1 + 2 - 1 - 4 \rangle = \langle 2, -4 \rangle$

**(b)** The additional force required is $\mathbf{F}_5 = \langle 0, 0 \rangle - \langle 2, -4 \rangle = \langle -2, 4 \rangle$.

**57.** From the figure we see that $\mathbf{T}_1 = -|\mathbf{T}_1|\cos 50^\circ\mathbf{i} + |\mathbf{T}_1|\sin 50^\circ\mathbf{j}$ and $\mathbf{T}_2 = |\mathbf{T}_2|\cos 30^\circ\mathbf{i} + |\mathbf{T}_2|\sin 30^\circ\mathbf{j}$. Since $\mathbf{T}_1 + \mathbf{T}_2 = 100\mathbf{j}$ we get $-|\mathbf{T}_1|\cos 50^\circ + |\mathbf{T}_2|\cos 30^\circ = 0$ and $|\mathbf{T}_1|\sin 50^\circ + |\mathbf{T}_2|\sin 30^\circ = 100$. From the first equation, $|\mathbf{T}_2| = |\mathbf{T}_1|\dfrac{\cos 50^\circ}{\cos 30^\circ}$, and substituting into the second equation gives $|\mathbf{T}_1|\sin 50^\circ + |\mathbf{T}_1|\dfrac{\cos 50^\circ \sin 30^\circ}{\cos 30^\circ} = 100$ $\Leftrightarrow$ $|\mathbf{T}_1|(\sin 50^\circ\cos 30^\circ + \cos 50^\circ \sin 30^\circ) = 100\cos 30^\circ$ $\Leftrightarrow$ $|\mathbf{T}_1|\sin(50^\circ + 30^\circ) = 100\cos 30^\circ$ $\Leftrightarrow$ $|\mathbf{T}_1| = 100\dfrac{\cos 30^\circ}{\sin 80^\circ} \approx 87.9385$.

Similarly, solving for $|\mathbf{T}_1|$ in the first equation gives $|\mathbf{T}_1| = |\mathbf{T}_2|\dfrac{\cos 30^\circ}{\cos 50^\circ}$ and substituting gives

$|\mathbf{T}_2|\dfrac{\cos 30^\circ \sin 50^\circ}{\cos 50^\circ} + |\mathbf{T}_2|\sin 30^\circ = 100$ $\Leftrightarrow$ $|\mathbf{T}_2|(\cos 30^\circ\sin 50^\circ + \cos 50^\circ\sin 30^\circ) = 100\cos 50^\circ$ $\Leftrightarrow$

$|\mathbf{T}_2| = \dfrac{100\cos 50^\circ}{\sin 80^\circ} \approx 65.2704$. Thus, $\mathbf{T}_1 \approx (-87.9385\cos 50^\circ)\,\mathbf{i} + (87.9385\sin 50^\circ)\,\mathbf{j} \approx -56.5\mathbf{i} + 67.4\mathbf{j}$ and $\mathbf{T}_2 \approx (65.2704\cos 30^\circ)\,\mathbf{i} + (65.2704\sin 30^\circ)\,\mathbf{j} \approx 56.5\mathbf{i} + 32.6\mathbf{j}$.

**58.** From the figure we see that $\mathbf{T}_1 = -|\mathbf{T}_1|\cos 22.3^\circ\mathbf{i} + |\mathbf{T}_1|\sin 22.3^\circ\mathbf{j}$ and $\mathbf{T}_2 = |\mathbf{T}_2|\cos 41.5^\circ\mathbf{i} + |\mathbf{T}_2|\sin 41.5^\circ\mathbf{j}$. Since $\mathbf{T}_1 + \mathbf{T}_2 = 18{,}278\mathbf{j}$, we get $-|\mathbf{T}_1|\cos 22.3^\circ + |\mathbf{T}_2|\cos 41.5^\circ = 0$ and $|\mathbf{T}_1|\sin 22.3^\circ + |\mathbf{T}_2|\sin 41.5^\circ = 18{,}278$. From the first equation, $|\mathbf{T}_2| = |\mathbf{T}_1|\dfrac{\cos 22.3^\circ}{\cos 41.5^\circ}$, and substituting into the second equation gives

$|\mathbf{T}_1|\sin 22.3^\circ + |\mathbf{T}_1|\dfrac{\cos 22.3^\circ\sin 41.5^\circ}{\cos 41.5^\circ} = 18{,}278$ $\Leftrightarrow$

$|\mathbf{T}_1|(\sin 22.3^\circ\cos 41.5^\circ + \cos 22.3^\circ\sin 41.5^\circ) = 18{,}278\cos 41.5^\circ$ $\Leftrightarrow$

$|\mathbf{T}_1|\sin(22.3^\circ + 41.5^\circ) = 18{,}278\cos 41.5^\circ$ $\Leftrightarrow$ $|\mathbf{T}_1| = 18{,}278\dfrac{\cos 41.5^\circ}{\sin 63.8^\circ} \approx 15{,}257$.

Similarly, solving for $|\mathbf{T}_1|$ in the first equation gives $|\mathbf{T}_1| = |\mathbf{T}_2|\dfrac{\cos 41.5^\circ}{\cos 22.3^\circ}$ and substituting gives $|\mathbf{T}_2|\dfrac{\cos 41.5^\circ\sin 22.3^\circ}{\cos 22.3^\circ} + |\mathbf{T}_2|\sin 41.5^\circ = 18{,}278$ $\Leftrightarrow$

$|\mathbf{T}_2|(\sin 22.3^\circ\cos 41.5^\circ + \sin 41.5^\circ\cos 22.3^\circ) = 18{,}278\cos 22.3^\circ$ $\Leftrightarrow$ $|\mathbf{T}_2| = \dfrac{18{,}278\cos 22.3^\circ}{\sin 63.8^\circ} \approx 18{,}847$.

Thus, $\mathbf{T}_1 \approx (-15{,}257\cos 22.3^\circ)\,\mathbf{i} + (15{,}257\sin 22.3^\circ)\,\mathbf{j} \approx -14{,}116\mathbf{i} + 5{,}789\mathbf{j}$ and $\mathbf{T}_2 \approx (18{,}847\cos 41.5^\circ)\,\mathbf{i} + (18{,}847\sin 41.5^\circ)\,\mathbf{j} \approx 14{,}116\mathbf{i} + 12{,}488\mathbf{j}$.

**59.** When we add two (or more vectors), the resultant vector can be found by first placing the initial point of the second vector at the terminal point of the first vector. The resultant vector can then found by using the new terminal point of the second vector and the initial point of the first vector. When the $n$ vectors are placed head to tail in the plane so that they form a polygon, the initial point and the terminal point are the same. Thus the sum of these $n$ vectors is the zero vector.

## 9.5 The Dot Product

**1. (a)** $\mathbf{u}\cdot\mathbf{v} = \langle 2,0\rangle\cdot\langle 1,1\rangle = 2 + 0 = 2$

**(b)** $\cos\theta = \dfrac{\mathbf{u}\cdot\mathbf{v}}{|\mathbf{u}|\,|\mathbf{v}|} = \frac{2}{2\cdot\sqrt{2}} = \frac{1}{\sqrt{2}} \Rightarrow \theta = 45^\circ$

**2. (a)** $\mathbf{u}\cdot\mathbf{v} = \langle 1,\sqrt{3}\rangle\cdot\langle -\sqrt{3},1\rangle = -\sqrt{3} + \sqrt{3} = 0$

**(b)** $\cos\theta = \dfrac{\mathbf{u}\cdot\mathbf{v}}{|\mathbf{u}|\,|\mathbf{v}|} = 0 \Rightarrow \theta = 90^\circ$

**3. (a)** $\mathbf{u}\cdot\mathbf{v} = \langle 2,7\rangle\cdot\langle 3,1\rangle = 6 + 7 = 13$

**(b)** $\cos\theta = \dfrac{\mathbf{u}\cdot\mathbf{v}}{|\mathbf{u}|\,|\mathbf{v}|} = \frac{13}{\sqrt{53}\cdot\sqrt{10}} \Rightarrow \theta \approx 56^\circ$

**4. (a)** $\mathbf{u}\cdot\mathbf{v} = \langle -6,6\rangle\cdot\langle 1,-1\rangle = -6 + (-6) = -12$

**(b)** $\cos\theta = \dfrac{\mathbf{u}\cdot\mathbf{v}}{|\mathbf{u}|\,|\mathbf{v}|} = \frac{-12}{6\sqrt{2}\cdot\sqrt{2}} = -1 \Rightarrow \theta = 180^\circ$

**5. (a)** $\mathbf{u}\cdot\mathbf{v} = \langle 3,-2\rangle\cdot\langle 1,2\rangle = 3+(-4) = -1$

**(b)** $\cos\theta = \dfrac{\mathbf{u}\cdot\mathbf{v}}{|\mathbf{u}|\,|\mathbf{v}|} = \frac{-1}{\sqrt{13}\cdot\sqrt{5}} \Rightarrow \theta \approx 97^\circ$

**6. (a)** $\mathbf{u}\cdot\mathbf{v} = \langle 2,1\rangle\cdot\langle 3,-2\rangle = 6+(-2) = 4$

**(b)** $\cos\theta = \dfrac{\mathbf{u}\cdot\mathbf{v}}{|\mathbf{u}|\,|\mathbf{v}|} = \frac{4}{\sqrt{5}\cdot\sqrt{13}} \Rightarrow \theta \approx 60.3^\circ$

**7. (a)** $\mathbf{u}\cdot\mathbf{v} = \langle 0,-5\rangle\cdot\langle -1,-\sqrt{3}\rangle = 0+5\sqrt{3} = 5\sqrt{3}$

**(b)** $\cos\theta = \dfrac{\mathbf{u}\cdot\mathbf{v}}{|\mathbf{u}|\,|\mathbf{v}|} = \frac{5\sqrt{3}}{5\cdot 2} = \frac{\sqrt{3}}{2} \Rightarrow \theta = 30^\circ$

**8. (a)** $\mathbf{u}\cdot\mathbf{v} = \langle 1,1\rangle\cdot\langle 1,-1\rangle = 1+(-1) = 0$

**(b)** $\cos\theta = \dfrac{\mathbf{u}\cdot\mathbf{v}}{|\mathbf{u}|\,|\mathbf{v}|} = \frac{0}{\sqrt{2}\cdot\sqrt{2}} = 0 \Rightarrow \theta = 90^\circ$

**9.** $\mathbf{u}\cdot\mathbf{v} = -12+12 = 0 \Rightarrow$ vectors are orthogonal

**10.** $\mathbf{u}\cdot\mathbf{v} = 0+0 = 0 \Rightarrow$ vectors are orthogonal

**11.** $\mathbf{u}\cdot\mathbf{v} = -8+12 = 4 \neq 0 \Rightarrow$ vectors are not orthogonal

**12.** $\mathbf{u}\cdot\mathbf{v} = 0+0 = 0 \Rightarrow$ vectors are orthogonal

**13.** $\mathbf{u}\cdot\mathbf{v} = -24+24 = 0 \Rightarrow$ vectors are orthogonal

**14.** $\mathbf{u}\cdot\mathbf{v} = -4+0 = -4 \Rightarrow$ vectors are not orthogonal

**15.** $\mathbf{u}\cdot\mathbf{v}+\mathbf{u}\cdot\mathbf{w} = \langle 2,1\rangle\cdot\langle 1,-3\rangle + \langle 2,1\rangle\cdot\langle 3,4\rangle$
$= 2-3+6+4 = 9$

**16.** $\mathbf{u}\cdot(\mathbf{v}+\mathbf{w}) = \langle 2,1\rangle\cdot[\langle 1,-3\rangle+\langle 3,4\rangle]$
$= \langle 2,1\rangle\cdot\langle 4,1\rangle = 8+1 = 9$

**17.** $(\mathbf{u}+\mathbf{v})\cdot(\mathbf{u}-\mathbf{v}) = [\langle 2,1\rangle+\langle 1,-3\rangle]\cdot[\langle 2,1\rangle-\langle 1,-3\rangle]$
$= \langle 3,-2\rangle\cdot\langle 1,4\rangle = 3-8 = -5$

**18.** $(\mathbf{u}\cdot\mathbf{v})(\mathbf{u}\cdot\mathbf{w}) = (\langle 2,1\rangle\cdot\langle 1,-3\rangle)(\langle 2,1\rangle\cdot\langle 3,4\rangle)$
$= (-1)(10) = -10$

**19.** $x = \dfrac{\mathbf{u}\cdot\mathbf{v}}{|\mathbf{v}|} = \dfrac{12-24}{5} = -\dfrac{12}{5}$

**20.** $x = \dfrac{\mathbf{u}\cdot\mathbf{v}}{|\mathbf{v}|} = \dfrac{-\frac{3}{\sqrt{2}}+\frac{5}{\sqrt{2}}}{1} = \dfrac{2}{\sqrt{2}} = \sqrt{2}$

**21.** $x = \dfrac{\mathbf{u}\cdot\mathbf{v}}{|\mathbf{v}|} = \dfrac{0-24}{1} = -24$

**22.** $x = \dfrac{\mathbf{u}\cdot\mathbf{v}}{|\mathbf{v}|} = \dfrac{56+0}{10} = \frac{28}{5}$

**23. (a)** $\mathbf{u}_1 = \operatorname{proj}_{\mathbf{v}}\mathbf{u} = \left(\dfrac{\mathbf{u}\cdot\mathbf{v}}{|\mathbf{v}|^2}\right)\mathbf{v} = \left(\dfrac{\langle -2,4\rangle\cdot\langle 1,1\rangle}{1^2+1^2}\right)\langle 1,1\rangle = \langle 1,1\rangle.$

**(b)** $\mathbf{u}_2 = \mathbf{u}-\mathbf{u}_1 = \langle -2,4\rangle - \langle 1,1\rangle = \langle -3,3\rangle$

**24. (a)** $\mathbf{u}_1 = \operatorname{proj}_{\mathbf{v}}\mathbf{u} = \left(\dfrac{\mathbf{u}\cdot\mathbf{v}}{|\mathbf{v}|^2}\right)\mathbf{v} = \left(\dfrac{\langle 7,-4\rangle\cdot\langle 2,1\rangle}{2^2+1^2}\right)\langle 2,1\rangle = 2\langle 2,1\rangle = \langle 4,2\rangle$

**(b)** $\mathbf{u}_2 = \mathbf{u}-\mathbf{u}_1 = \langle 7,-4\rangle - \langle 4,2\rangle = \langle 3,-6\rangle$

**25. (a)** $\mathbf{u}_1 = \operatorname{proj}_{\mathbf{v}}\mathbf{u} = \left(\dfrac{\mathbf{u}\cdot\mathbf{v}}{|\mathbf{v}|^2}\right)\mathbf{v} = \left(\dfrac{\langle 1,2\rangle\cdot\langle 1,-3\rangle}{1^2+(-3)^2}\right)\langle 1,-3\rangle = -\frac{1}{2}\langle 1,-3\rangle = \left\langle -\frac{1}{2},\frac{3}{2}\right\rangle$

**(b)** $\mathbf{u}_2 = \mathbf{u}-\mathbf{u}_1 = \langle 1,2\rangle - \left\langle -\frac{1}{2},\frac{3}{2}\right\rangle = \left\langle \frac{3}{2},\frac{1}{2}\right\rangle$

**26. (a)** $\mathbf{u}_1 = \operatorname{proj}_{\mathbf{v}}\mathbf{u} = \left(\dfrac{\mathbf{u}\cdot\mathbf{v}}{|\mathbf{v}|^2}\right)\mathbf{v} = \left(\dfrac{\langle 11,3\rangle\cdot\langle -3,-2\rangle}{(-3)^2+(-2)^2}\right)\langle -3,-2\rangle = -3\langle -3,-2\rangle = \langle 9,6\rangle$

**(b)** $\mathbf{u}_2 = \mathbf{u}-\mathbf{u}_1 = \langle 11,3\rangle - \langle 9,6\rangle = \langle 2,-3\rangle$

**27. (a)** $\mathbf{u}_1 = \operatorname{proj}_{\mathbf{v}}\mathbf{u} = \left(\dfrac{\mathbf{u}\cdot\mathbf{v}}{|\mathbf{v}|^2}\right)\mathbf{v} = \left(\dfrac{\langle 2,9\rangle\cdot\langle -3,4\rangle}{(-3)^3+4^2}\right)\langle -3,4\rangle = \frac{6}{5}\langle -3,4\rangle = \left\langle -\frac{18}{5},\frac{24}{5}\right\rangle$

**(b)** $\mathbf{u}_2 = \mathbf{u}-\mathbf{u}_1 = \langle 2,9\rangle - \left\langle -\frac{18}{5},\frac{24}{5}\right\rangle = \left\langle \frac{28}{5},\frac{21}{5}\right\rangle$

**28. (a)** $\mathbf{u}_1 = \operatorname{proj}_{\mathbf{v}}\mathbf{u} = \left(\dfrac{\mathbf{u}\cdot\mathbf{v}}{|\mathbf{v}|^2}\right)\mathbf{v} = \left(\dfrac{\langle 1,1\rangle\cdot\langle 2,-1\rangle}{2^2+(-1)^2}\right)\langle 2,-1\rangle = \frac{1}{5}\langle 2,-1\rangle = \left\langle \frac{2}{5},-\frac{1}{5}\right\rangle$

**(b)** $\mathbf{u}_2 = \mathbf{u}-\mathbf{u}_1 = \langle 1,1\rangle - \left\langle \frac{2}{5},-\frac{1}{5}\right\rangle = \left\langle \frac{3}{5},\frac{6}{5}\right\rangle$

**29.** $W = \mathbf{F}\cdot\mathbf{d} = \langle 4,-5\rangle\cdot\langle 3,8\rangle = -28$

**30.** $W = \mathbf{F}\cdot\mathbf{d} = \langle 400,50\rangle\cdot\langle 201,0\rangle = 80,400$

**31.** $W = \mathbf{F}\cdot\mathbf{d} = \langle 10,3\rangle\cdot\langle 4,-5\rangle = 25$

**32.** $W = \mathbf{F}\cdot\mathbf{d} = \langle -4,20\rangle\cdot\langle 5,15\rangle = 280$

**33.** Let $\mathbf{u} = \langle \mathbf{u}_1,\mathbf{u}_2\rangle$ and $\mathbf{v} = \langle \mathbf{v}_1,\mathbf{v}_2\rangle$. Then
$\mathbf{u}\cdot\mathbf{v} = \langle u_1,u_2\rangle\cdot\langle v_1,v_2\rangle = u_1v_1+u_2v_2 = v_1u_1+v_2u_2 = \langle v_1,v_2\rangle\cdot\langle u_1,u_2\rangle = \mathbf{v}\cdot\mathbf{u}$

**34.** Let $\mathbf{u} = \langle \mathbf{u}_1,\mathbf{u}_2\rangle$ and $\mathbf{v} = \langle \mathbf{v}_1,\mathbf{v}_2\rangle$. Then

$(a\mathbf{u})\cdot\mathbf{v} = (a\langle u_1,u_2\rangle)\cdot\langle v_1,v_2\rangle = \langle au_1,au_2\rangle\cdot\langle v_1,v_2\rangle = au_1v_1+au_2v_2 = a(u_1v_1+u_2v_2) = a(\mathbf{u}\cdot\mathbf{v})$
$= u_1av_1+u_2av_2 = \langle u_1,u_2\rangle\cdot\langle av_1,av_2\rangle = \mathbf{u}\cdot(a\mathbf{v})$

**35.** Let $\mathbf{u} = \langle u_1, u_2 \rangle$, $\mathbf{v} = \langle v_1, v_2 \rangle$, and $\mathbf{w} = \langle w_1, w_2 \rangle$. Then

$$\begin{aligned}(\mathbf{u}+\mathbf{v})\cdot\mathbf{w} &= (\langle u_1, u_2\rangle + \langle v_1, v_2\rangle)\cdot\langle w_1, w_2\rangle = \langle u_1+v_1, u_2+v_2\rangle\cdot\langle w_1, w_2\rangle \\ &= u_1w_1 + v_1w_1 + u_2w_2 + v_2w_2 = u_1w_1 + u_2w_2 + v_1w_1 + v_2w_2 \\ &= \langle u_1, u_2\rangle\cdot\langle w_1, w_2\rangle + \langle v_1, v_2\rangle\cdot\langle w_1, w_2\rangle = \mathbf{u}\cdot\mathbf{w} + \mathbf{v}\cdot\mathbf{w}\end{aligned}$$

**36.** Let $\mathbf{u} = \langle u_1, u_2 \rangle$ and $\mathbf{v} = \langle v_1, v_2 \rangle$. Then

$$\begin{aligned}(\mathbf{u}-\mathbf{v})\cdot(\mathbf{u}+\mathbf{v}) &= (\langle u_1, u_2\rangle - \langle v_1, v_2\rangle)\cdot(\langle u_1, u_2\rangle + \langle v_1, v_2\rangle) = \langle u_1 - v_1, u_2 - v_2\rangle\cdot\langle u_1+v_1, u_2+, v_2\rangle \\ &= u_1^2 - v_1^2 + u_2^2 - v_2^2 = (u_1^2 + u_2^2) - (v_1^2 + v_2^2) = \left(\sqrt{u_1^2+u_2^2}\right)^2 - \left(\sqrt{v_1^2+v_2^2}\right)^2 = |\mathbf{u}|^2 - |\mathbf{v}|^2\end{aligned}$$

**37.** We use the definition that $\operatorname{proj}_{\mathbf{v}}\mathbf{u} = \left(\dfrac{\mathbf{u}\cdot\mathbf{v}}{|\mathbf{v}|^2}\right)\mathbf{v}$. Then

$$\begin{aligned}\operatorname{proj}_{\mathbf{v}}\mathbf{u}\cdot(\mathbf{u} - \operatorname{proj}_{\mathbf{v}}\mathbf{u}) &= \left(\frac{\mathbf{u}\cdot\mathbf{v}}{|\mathbf{v}|^2}\right)\mathbf{v}\cdot\left[\mathbf{u} - \left(\frac{\mathbf{u}\cdot\mathbf{v}}{|\mathbf{v}|^2}\right)\mathbf{v}\right] = \left(\frac{\mathbf{u}\cdot\mathbf{v}}{|\mathbf{v}|^2}\right)(\mathbf{v}\cdot\mathbf{u}) - \left(\frac{\mathbf{u}\cdot\mathbf{v}}{|\mathbf{v}|^2}\right)\mathbf{v}\cdot\left(\frac{\mathbf{u}\cdot\mathbf{v}}{|\mathbf{v}|^2}\right)\mathbf{v} \\ &= \frac{(\mathbf{u}\cdot\mathbf{v})^2}{|\mathbf{v}|^2} - \frac{(\mathbf{u}\cdot\mathbf{v})^2}{|\mathbf{v}|^4}|\mathbf{v}|^2 = \frac{(\mathbf{u}\cdot\mathbf{v})^2}{|\mathbf{v}|^2} - \frac{(\mathbf{u}\cdot\mathbf{v})^2}{|\mathbf{v}|^2} = 0\end{aligned}$$

Thus $\mathbf{u}$ and $\mathbf{u} - \operatorname{proj}_{\mathbf{v}}\mathbf{u}$ are orthogonal.

**38.** $\mathbf{v}\cdot\operatorname{proj}_{\mathbf{v}}\mathbf{u} = \mathbf{v}\cdot\left(\dfrac{\mathbf{u}\cdot\mathbf{v}}{|\mathbf{v}|^2}\right)\mathbf{v} = \dfrac{(\mathbf{u}\cdot\mathbf{v})(\mathbf{v}\cdot\mathbf{v})}{|\mathbf{v}|^2} = \mathbf{u}\cdot\mathbf{v}$.

**39.** $W = \mathbf{F}\cdot\mathbf{d} = \langle 4, -7\rangle\cdot\langle 4, 0\rangle = 16$ ft-lb

**40.** The displacement of the object is $\mathbf{D} = \langle 11, 13\rangle - \langle 2, 5\rangle = \langle 9, 8\rangle$. Hence, the work done is $W = \mathbf{F}\cdot\mathbf{D} = \langle 2, 8\rangle\cdot\langle 9, 8\rangle = 18 + 64 = 82$ ft-lb.

**41.** The distance vector is $\mathbf{D} = \langle 200, 0\rangle$ and the force vector is $\mathbf{F} = \langle 50\cos 30^\circ, 50\sin 30^\circ\rangle$. Hence, the work done is $W = \mathbf{F}\cdot\mathbf{D} = \langle 200, 0\rangle\cdot\langle 50\cos 30^\circ, 50\sin 30^\circ\rangle = 200\cdot 50\cos 30^\circ \approx 8660$ ft-lb.

**42.** $W = \mathbf{F}\cdot\mathbf{d}$, and in this problem $\mathbf{F} = \langle 0, -2500\rangle$, $\mathbf{d} = \langle 500\cos 12^\circ, 500\sin 12^\circ\rangle$ Thus, $W = \langle 0, -2500\rangle\cdot\langle 500\cos 12^\circ, 500\sin 12^\circ\rangle \approx 0 - (2500)(104.0) = -260{,}000$ ft-lb. This is the work done by gravity; the car does (positive) work in overcoming the force of gravity. So the work done by the car is 260,000 ft-lb.

**43.** Since the weight of the car is 2755 lb, the force exerted perpendicular to the earth is 2755 lb. Resolving this into a force $\mathbf{u}$ perpendicular to the driveway gives $|\mathbf{u}| = 2766\cos 65^\circ \approx 1164$ lb. Thus, a force of about 1164 lb is required.

**44.** **(a)** Since the force parallel to the driveway is $490 = |\mathbf{w}|\sin 10^\circ \quad\Leftrightarrow\quad |\mathbf{w}| = \dfrac{490}{\sin 10^\circ} \approx 2821.8$, and thus the weight of the car is about 2822 lb.

**(b)** The force exerted against the driveway is $2821.8\cos 10^\circ \approx 2779$ lb.

**45.** Since the force required parallel to the plane is 80 lb and the weight of the package is 200 lb, it follows that $80 = 200\sin\theta$, where $\theta$ is the angle of inclination of the plane. Then $\theta = \sin^{-1}\left(\frac{80}{200}\right) \approx 23.58^\circ$, and so the angle of inclination is approximately $23.6^\circ$.

**46.** Let $\mathbf{R}$ represent the force exerted by the rope and $\mathbf{d}$ the force causing the cart to roll down the ramp. Gravity acting on the cart exerts a force $\mathbf{w}$ of 40 lb directly downward. So the magnitude of the part of that force causing the cart to roll down the ramp is $|\mathbf{d}| = 40\sin 15^\circ$. The angle between $\mathbf{R}$ and $\mathbf{d}$ is $45^\circ$, so the magnitude of the force holding the cart up is $|\mathbf{R}|\cos 45^\circ$. Equating these two, we have $|\mathbf{R}|\cos 45^\circ = 40\sin 15^\circ$, so $|\mathbf{R}| = \dfrac{40\sin 15^\circ}{\cos 45^\circ} \approx 14.64$ lb.

**47. (a)** $2(0) + 4(2) = 8$, so $Q(0,2)$ lies on $L$. $2(2) + 4(1) = 4 + 4 = 8$, so $R(2,1)$ lies on $L$.

**(b)** $\mathbf{u} = \overrightarrow{QP} = \langle 0,2\rangle - \langle 3,4\rangle = \langle -3,-2\rangle$.

$\mathbf{v} = \overrightarrow{QR} = \langle 0,2\rangle - \langle 2,1\rangle = \langle -2,1\rangle$.

$\mathbf{w} = \text{proj}_{\mathbf{v}}\,\mathbf{u} = \left(\frac{\mathbf{u}\cdot\mathbf{v}}{|\mathbf{v}|^2}\right)\mathbf{v} = \frac{\langle -3,-2\rangle\cdot\langle 2,1\rangle}{(-2)^2+1^2}\langle -2,1\rangle$
$= -\frac{8}{5}\langle -2,1\rangle = \left\langle \frac{16}{5}, -\frac{8}{5}\right\rangle$

**(c)** From the graph, we can see that $\mathbf{u} - \mathbf{w}$ is orthogonal to $\mathbf{v}$ (and thus to $L$). Thus, the distance from $P$ to $L$ is $|\mathbf{u} - \mathbf{w}|$.

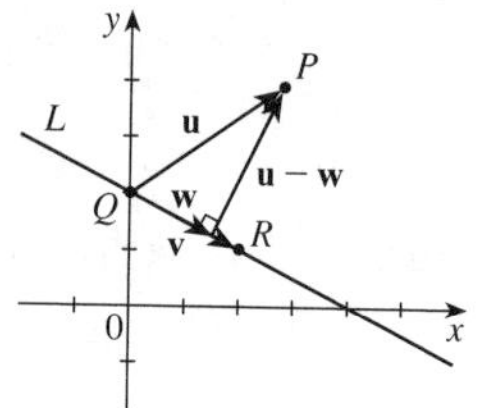


# Chapter 9 Review

**1. (a)**

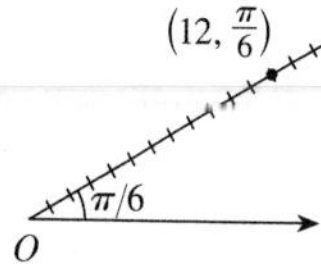


**(b)** $x = 12\cos\frac{\pi}{6} = 12\cdot\frac{\sqrt{3}}{2} = 6\sqrt{3}$,

$y = 12\sin\frac{\pi}{6} = 12\cdot\frac{1}{2} = 6$. Thus, the rectangular coordinates of $P$ are $(6\sqrt{3}, 6)$.

**2. (a)**

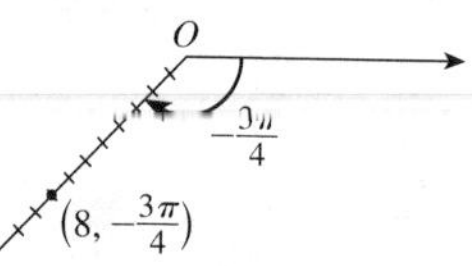


**(b)** $x = 8\cos\left(-\frac{3\pi}{4}\right) = 8\left(-\frac{\sqrt{2}}{2}\right) = -4\sqrt{2}$,

$y = 8\sin\left(-\frac{3\pi}{4}\right) = 8\left(-\frac{\sqrt{2}}{2}\right) = -4\sqrt{2}$. Thus, the rectangular coordinates of $P$ are $(-4\sqrt{2}, -4\sqrt{2})$.

**3. (a)**

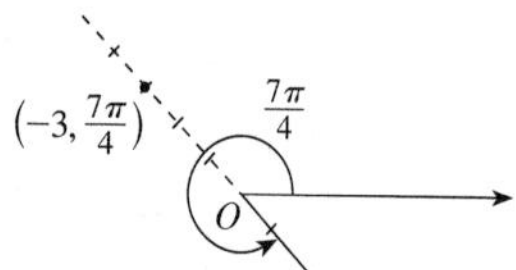


**(b)** $x = -3\cos\frac{7\pi}{4} = -3\left(\frac{\sqrt{2}}{2}\right) = -\frac{3\sqrt{2}}{2}$,

$y = -3\sin\frac{7\pi}{4} = -3\left(-\frac{\sqrt{2}}{2}\right) = \frac{3\sqrt{2}}{2}$. Thus, the rectangular coordinates of $P$ are $\left(-\frac{3\sqrt{2}}{2}, \frac{3\sqrt{2}}{2}\right)$.

**4. (a)**

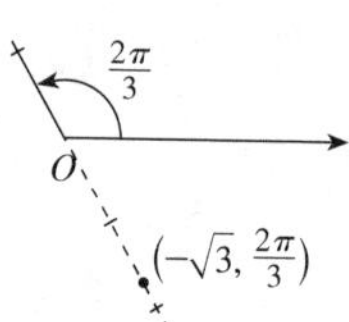


**(b)** $x = -\sqrt{3}\cos\frac{2\pi}{3} = -\sqrt{3}\left(-\frac{1}{2}\right) = \frac{\sqrt{3}}{2}$,

$y = -\sqrt{3}\sin\frac{2\pi}{3} = -\sqrt{3}\left(\frac{\sqrt{3}}{2}\right) = -\frac{3}{2}$. Thus, the rectangular coordinates of $P$ are $\left(\frac{\sqrt{3}}{2}, -\frac{3}{2}\right)$.

**5. (a)**

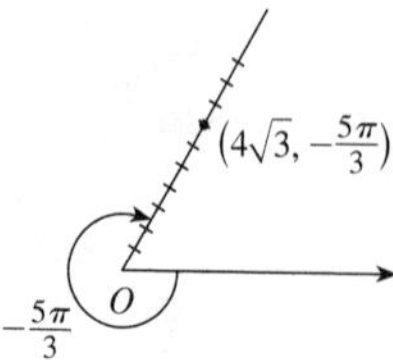


**(b)** $x = 4\sqrt{3}\cos\left(-\frac{5\pi}{3}\right) = 4\sqrt{3}\left(\frac{1}{2}\right) = 2\sqrt{3}$,

$y = 4\sqrt{3}\sin\left(-\frac{5\pi}{3}\right) = 4\sqrt{3}\left(\frac{\sqrt{3}}{2}\right) = 6$. Thus, the rectangular coordinates of $P$ are $(2\sqrt{3}, 6)$.

**6. (a)**

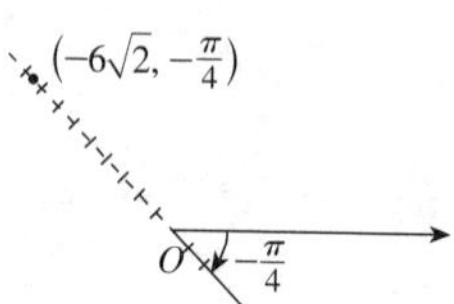


**(b)** $x = -6\sqrt{2}\cos\left(-\frac{\pi}{4}\right) = -6\sqrt{2}\left(\frac{\sqrt{2}}{2}\right) = -6$,

$y = -6\sqrt{2}\sin\left(-\frac{\pi}{4}\right) = -6\sqrt{2}\left(-\frac{\sqrt{2}}{2}\right) = 6$. Thus, the rectangular coordinates of $P$ are $(-6, 6)$.

**7. (a)**

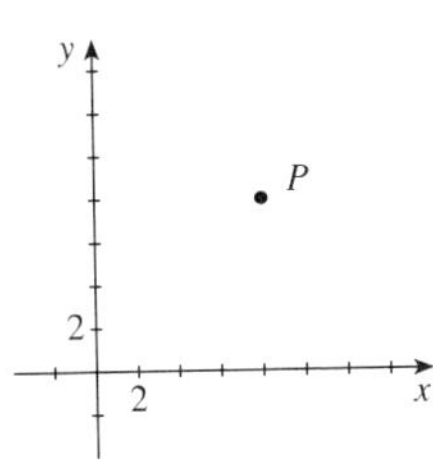


**(b)** $r = \sqrt{8^2 + 8^2} = \sqrt{128} = 8\sqrt{2}$ and $\overline{\theta} = \tan^{-1} \frac{8}{8}$. Since $P$ is in quadrant I, $\theta = \frac{\pi}{4}$. Polar coordinates for $P$ are $\left(8\sqrt{2}, \frac{\pi}{4}\right)$.

**(c)** $\left(-8\sqrt{2}, \frac{5\pi}{4}\right)$

**8. (a)**

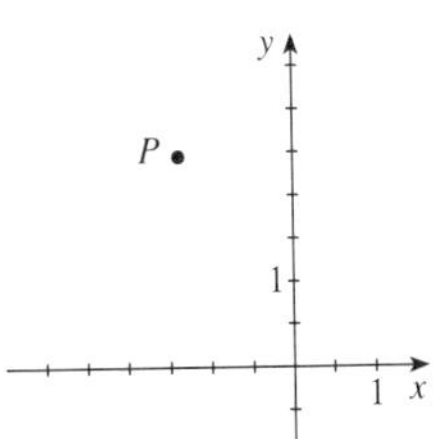


**(b)** $r = \sqrt{\left(-\sqrt{2}\right)^2 + \left(\sqrt{6}\right)^2} = \sqrt{8} = 2\sqrt{2}$ and $\overline{\theta} = \tan^{-1} \frac{\sqrt{6}}{-\sqrt{2}} = \tan^{-1}\left(-\sqrt{3}\right) = -\frac{\pi}{3}$. Since $x$ is negative and $y$ is positive, $\theta = \frac{2\pi}{3}$. Polar coordinates for $P$ are $\left(2\sqrt{2}, \frac{2\pi}{3}\right)$.

**(c)** $\left(-2\sqrt{2}, \frac{5\pi}{3}\right)$

**9. (a)**

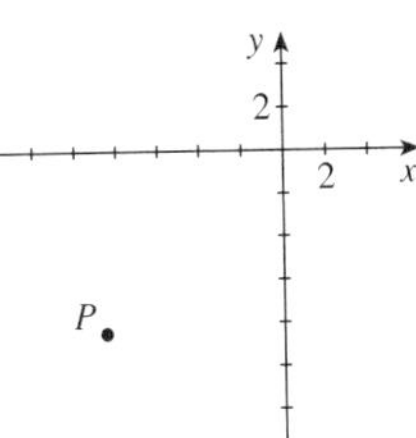


**(b)** $r = \sqrt{\left(-6\sqrt{2}\right)^2 + \left(-6\sqrt{2}\right)^2} = \sqrt{144} = 12$ and $\overline{\theta} = \tan^{-1} \frac{-6\sqrt{2}}{-6\sqrt{2}} = \frac{\pi}{4}$. Since $P$ is in quadrant III, $\theta = \frac{5\pi}{4}$. Polar coordinates for $P$ are $\left(12, \frac{5\pi}{4}\right)$.

**(c)** $\left(-12, \frac{\pi}{4}\right)$

**10. (a)**

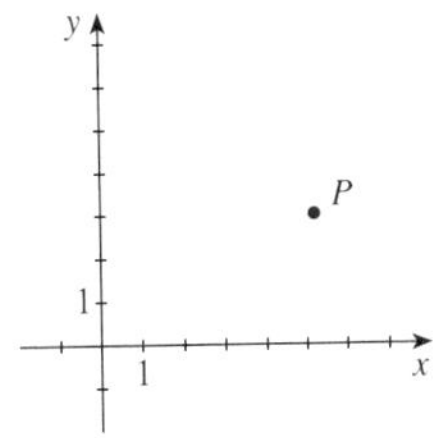


**(b)** $r = \sqrt{\left(3\sqrt{3}\right)^2 + 3^2} = \sqrt{36} = 6$ and $\overline{\theta} = \tan^{-1} \frac{3}{3\sqrt{3}} = \tan^{-1} \frac{\sqrt{3}}{3} = \frac{\pi}{6}$. Since $P$ is in quadrant I, $\theta = \frac{\pi}{6}$. Polar coordinates for $P$ are $\left(6, \frac{\pi}{6}\right)$.

**(c)** $\left(-6, \frac{7\pi}{6}\right)$

**11. (a)**

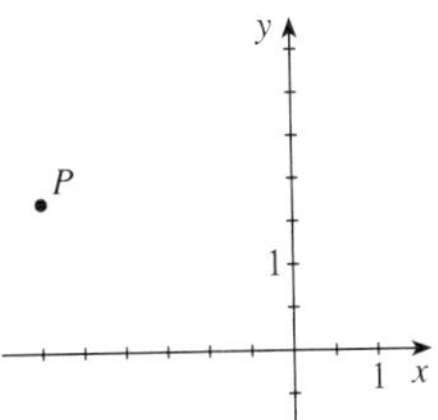


**(b)** $r = \sqrt{(-3)^2 + \left(\sqrt{3}\right)^2} = \sqrt{12} = 2\sqrt{3}$ and $\overline{\theta} = \tan^{-1} \frac{\sqrt{3}}{-3}$. Since $P$ is in quadrant II, $\theta = \frac{5\pi}{6}$. Polar coordinates for $P$ are $\left(2\sqrt{3}, \frac{5\pi}{6}\right)$.

**(c)** $\left(-2\sqrt{3}, -\frac{\pi}{6}\right)$

**12. (a)**

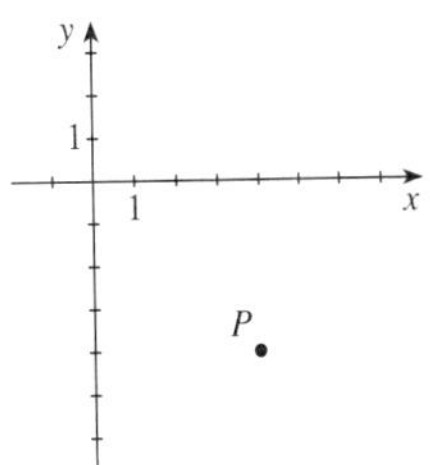


**(b)** $r = \sqrt{4^2 + (-4)^2} = \sqrt{32} = 4\sqrt{2}$ and $\overline{\theta} = \tan^{-1} \frac{-4}{4}$. Since $P$ is in quadrant IV, $\theta = \frac{7\pi}{4}$. Polar coordinates for $P$ are $\left(4\sqrt{2}, \frac{7\pi}{4}\right)$.

**(c)** $\left(-4\sqrt{2}, \frac{3\pi}{4}\right)$

**13. (a)** $x+y=4 \quad \Leftrightarrow \quad r\cos\theta + r\sin\theta = 4 \quad \Leftrightarrow$
$r(\cos\theta+\sin\theta)=4 \quad \Leftrightarrow \quad r = \dfrac{4}{\cos\theta+\sin\theta}$

**(b)** The rectangular equation is easier to graph.

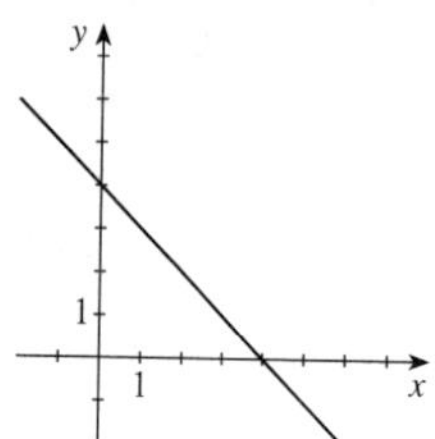


**14. (a)** $xy=1 \quad \Leftrightarrow \quad (r\cos\theta)(r\sin\theta)=1 \quad \Leftrightarrow$
$r^2(\cos\theta\sin\theta)=1 \quad \Leftrightarrow$
$r^2 = \dfrac{1}{\cos\theta\sin\theta} = \dfrac{2}{\sin 2\theta}$ or $r^2 = 2\csc 2\theta$.

**(b)** The rectangular equation is easier to graph.

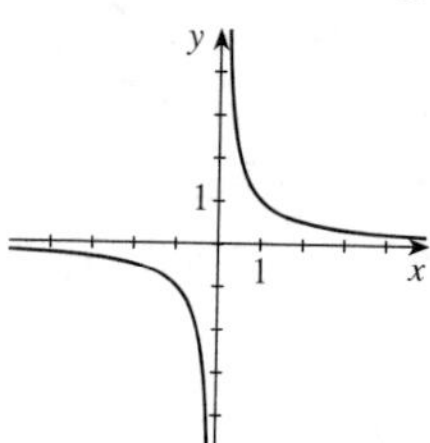


**15. (a)** $x^2+y^2=4x+4y \quad \Leftrightarrow \quad r^2 = 4r\cos\theta + 4r\sin\theta$
$\Leftrightarrow \quad r^2 = r(4\cos\theta+4\sin\theta) \quad \Leftrightarrow$
$r = 4\cos\theta + 4\sin\theta$

**(b)** The polar equation is easier to graph.

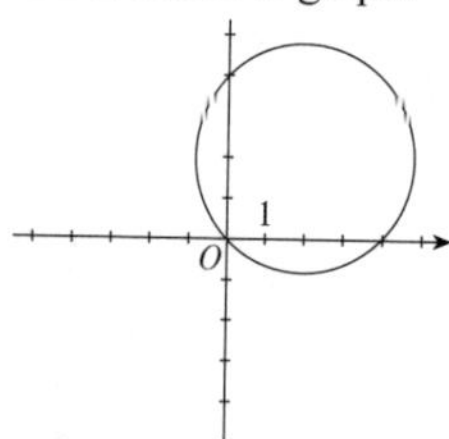
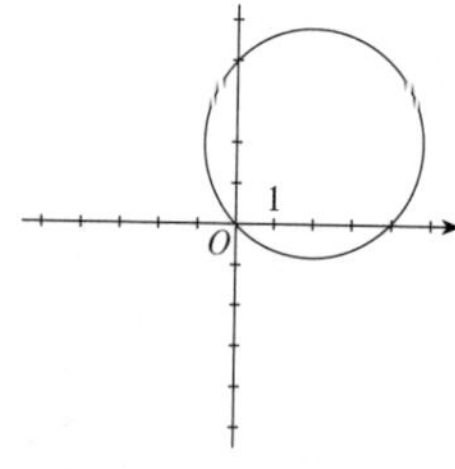


**16. (a)** $(x^2+y^2)^2 = 2xy \Leftrightarrow (r^2)^2 = 2(r\cos\theta)(r\sin\theta)$
$\Leftrightarrow r^4 = 2r^2\cos\theta\sin\theta \Leftrightarrow r^2 = 2\cos\theta\sin\theta \Leftrightarrow$
$r^2 = \sin 2\theta$

**(b)** The polar equation is easier to graph.

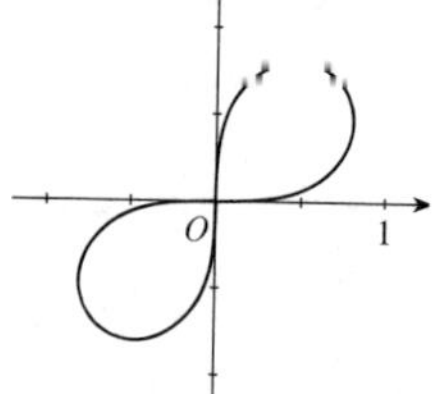


**17. (a)**

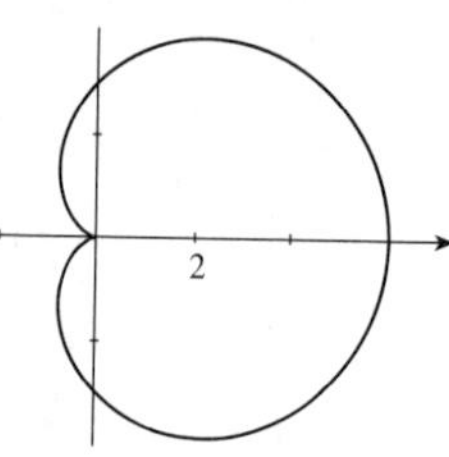


**(b)** $r = 3+3\cos\theta \quad \Leftrightarrow \quad r^2 = 3r + 3r\cos\theta$, which gives $x^2+y^2 = 3\sqrt{x^2+y^2}+3x \quad \Leftrightarrow$
$x^2-3x+y^2 = 3\sqrt{x^2+y^2}$. Squaring both sides gives $(x^2-3x+y^2)^2 = 9(x^2+y^2)$.

**18. (a)**

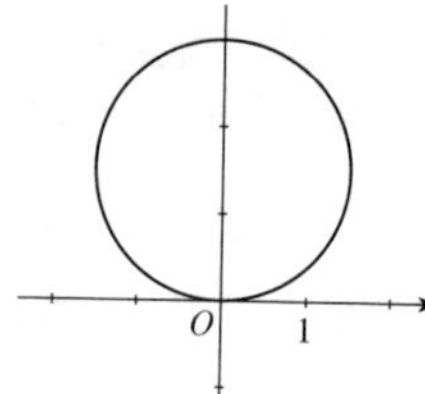


**(b)** $r = 3\sin\theta \quad \Leftrightarrow \quad r^2 = 3r\sin\theta$, so $x^2+y^2=3y$
$\Leftrightarrow \quad x^2 + \left(y^2 - 3y + \frac{9}{4}\right) = \frac{9}{4} \quad \Leftrightarrow$
$x^2 + \left(y-\frac{3}{2}\right)^2 = \frac{9}{4}$.

**19. (a)**

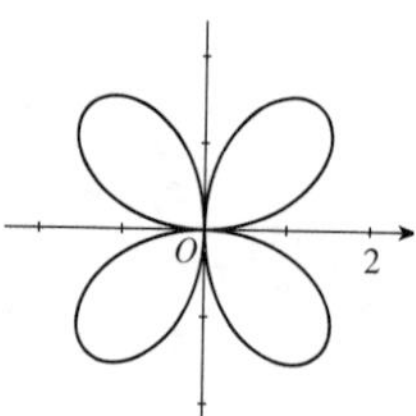


**(b)** $r = 2\sin 2\theta \quad \Leftrightarrow \quad r = 2\cdot 2\sin\theta\cos\theta \quad \Leftrightarrow$
$r^3 = 4r^2\sin\theta\cos\theta \quad \Leftrightarrow$
$(r^2)^{3/2} = 4(r\sin\theta)(r\cos\theta)$ and so, since $x = r\cos\theta$ and $y = r\sin\theta$, we get $(x^2+y^2)^3 = 16x^2y^2$.

**20. (a)**

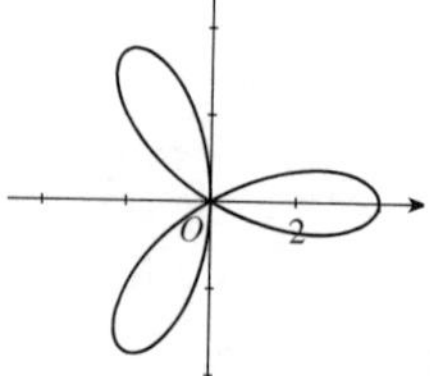


**(b)** $r = 4\cos 3\theta = 4(\cos 2\theta\cos\theta - \sin 2\theta\sin\theta)$
$= 4\left[\cos\theta\left(\cos^2\theta - \sin^2\theta\right) - 2\sin^2\theta\cos\theta\right] \quad \Leftrightarrow$
$r = 4\left(\cos^3\theta - 3\sin^2\theta\cos\theta\right) \quad \Leftrightarrow$
$r^4 = 4r^3\left(\cos^3\theta - 3\sin^2\theta\cos\theta\right)$, which gives $(x^2+y^2)^2 = 4x^3 - 12xy^2$.

**21. (a)**

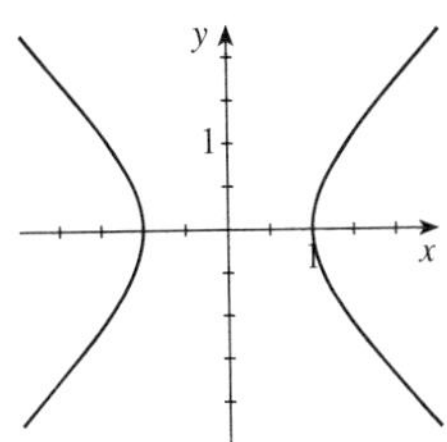

**(b)** $r^2 = \sec 2\theta = \dfrac{1}{\cos 2\theta} = \dfrac{1}{\cos^2\theta - \sin^2\theta} \quad \Leftrightarrow$

$r^2\left(\cos^2\theta - \sin^2\theta\right) = 1 \quad \Leftrightarrow$

$r^2\cos^2\theta - r^2\sin^2\theta = 1 \quad \Leftrightarrow$

$(r\cos\theta)^2 - (r\sin\theta)^2 = 1 \quad \Leftrightarrow \quad x^2 - y^2 = 1.$

**22. (a)**

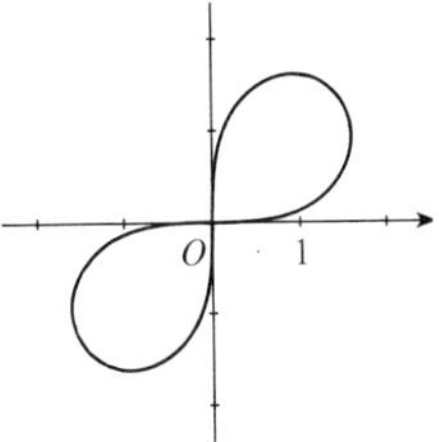

**(b)** $r^2 = 4\sin 2\theta = 8\sin\theta\cos\theta \quad \Leftrightarrow$

$r^4 = 8r^2\sin\theta\cos\theta$, so $\left(x^2+y^2\right)^2 = 8xy.$

**23. (a)**

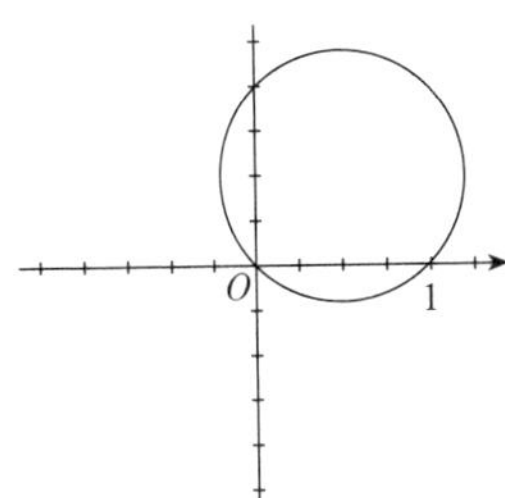

**(b)** $r = \sin\theta + \cos\theta \quad \Leftrightarrow \quad r^2 = r\sin\theta + r\cos\theta$, so $x^2 + y^2 = y + x$

$\Leftrightarrow \quad \left(x^2 - x + \frac{1}{4}\right) + \left(y^2 - y + \frac{1}{4}\right) = \frac{1}{2} \quad \Leftrightarrow$

$\left(x - \frac{1}{2}\right)^2 + \left(y - \frac{1}{2}\right)^2 = \frac{1}{2}.$

**24. (a)**

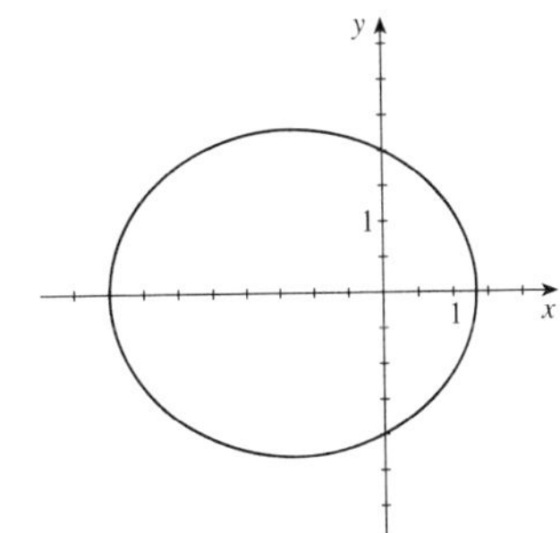

**(b)** $r = \dfrac{4}{2 + \cos\theta} \quad \Leftrightarrow \quad 2r + r\cos\theta = 4$, so $2\sqrt{x^2+y^2} + x = 4 \quad \Leftrightarrow$

$2\sqrt{x^2+y^2} = 4 - x$. Squaring both sides gives

$4\left(x^2+y^2\right) = (4-x)^2 = 16 - 8x + x^2 \quad \Leftrightarrow \quad 3x^2 + 8x + 4y^2 = 16$

$\Leftrightarrow \quad 3\left(x + \frac{4}{3}\right)^2 + 4y^2 = 16 + \frac{16}{3} \quad \Leftrightarrow \quad \dfrac{\left(x+\frac{4}{3}\right)^2}{\frac{64}{9}} + \dfrac{y^2}{\frac{16}{3}} = 1.$

**25.** $r = \cos(\theta/3)$, $\theta \in [0, 3\pi]$.

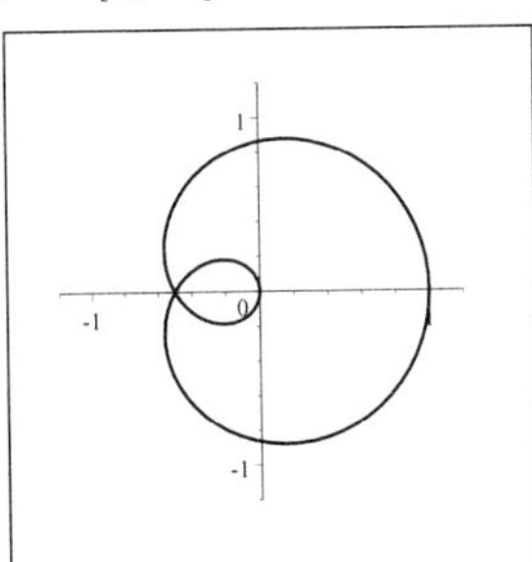

**26.** $r = \sin(9\theta/4)$, $\theta \in [0, 8\pi]$

**27.** $r = 1 + 4\cos(\theta/3)$, $\theta \in [0, 6\pi]$.

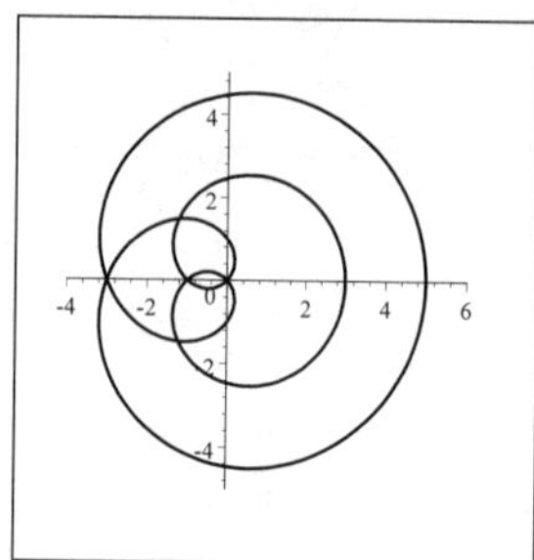

**28.** $r = \theta \sin\theta$, $\theta \in [-3\pi, 3\pi]$. Note that this graph is not bounded.

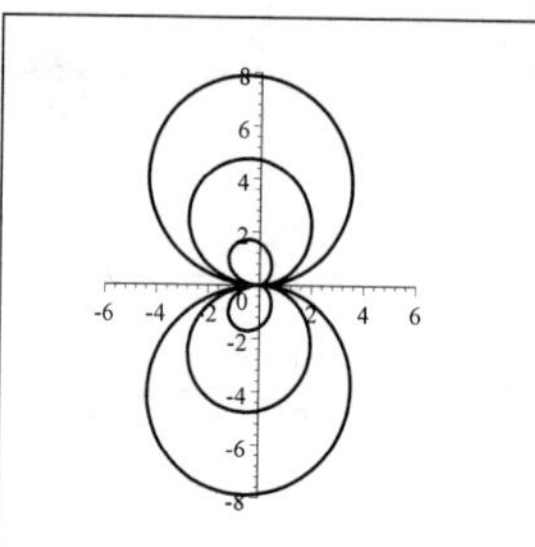

**29. (a)**

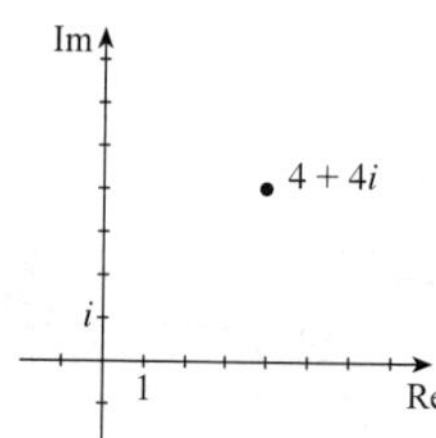


**(b)** $4 + 4i$ has $r = \sqrt{16 + 16} = 4\sqrt{2}$, and $\theta = \tan^{-1} \frac{4}{4} = \frac{\pi}{4}$ (in quadrant I).

**(c)** $4 + 4i = 4\sqrt{2}\left(\cos\frac{\pi}{4} + i\sin\frac{\pi}{4}\right)$

**30. (a)**

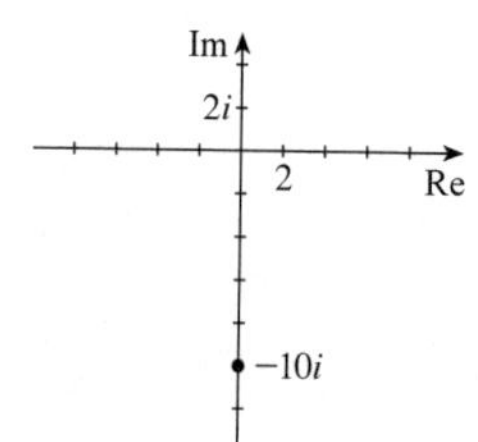


**(b)** $-10i$ has $r = 10$ and $\theta = \frac{3\pi}{2}$.

**(c)** $-10i = 10\left(\cos\frac{3\pi}{2} + i\sin\frac{3\pi}{2}\right)$

**31. (a)**

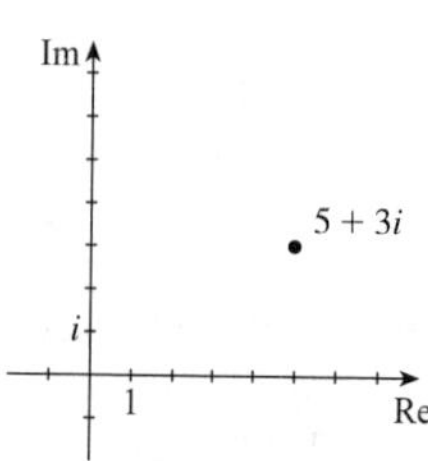


**(b)** $5 + 3i$. Then $r = \sqrt{25 + 9} = \sqrt{34}$, and $\theta = \tan^{-1} \frac{3}{5}$.

**(c)** $5 + 3i = \sqrt{34}\left[\cos\left(\tan^{-1}\frac{3}{5}\right) + i\sin\left(\tan^{-1}\frac{3}{5}\right)\right]$

**32. (a)**

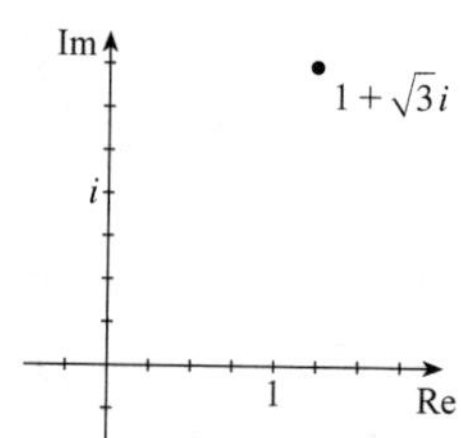


**(b)** $1 + \sqrt{3}i$ has $r = \sqrt{1 + 3} = 2$, and $\theta = \tan^{-1}\sqrt{3} = \frac{\pi}{3}$ (since $\theta$ is in quadrant I).

**(c)** $1 + \sqrt{3}i = 2\left(\cos\frac{\pi}{3} + i\sin\frac{\pi}{3}\right)$

**33. (a)**

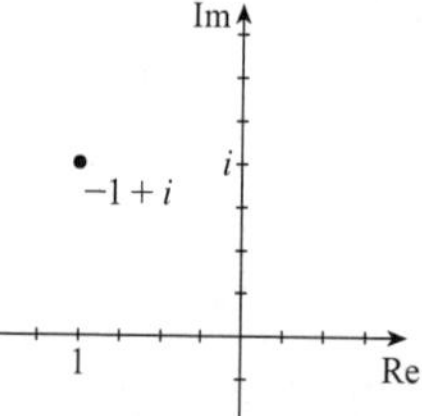


**(b)** $-1 + i$ has $r = \sqrt{1 + 1} = \sqrt{2}$ and $\tan\theta = \dfrac{1}{-1}$ with $\theta$ in quadrant II $\Leftrightarrow$ $\theta = \frac{3\pi}{4}$.

**(c)** $-1 + i = \sqrt{2}\left(\cos\frac{3\pi}{4} + i\sin\frac{3\pi}{4}\right)$

**34. (a)**

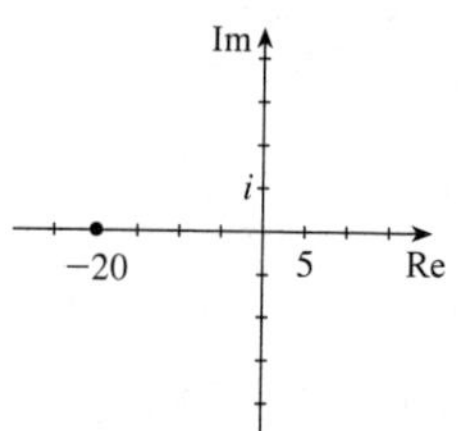


**(b)** $-20$ has $r = 20$ and $\theta = \pi$.

**(c)** $-20 = 20(\cos\pi + i\sin\pi)$

**35.** $1 - \sqrt{3}i$ has $r = \sqrt{1+3} = 2$ and $\tan\theta = \frac{-\sqrt{3}}{1} = -\sqrt{3}$ with $\theta$ in quadrant III $\Leftrightarrow$ $\theta = \frac{5\pi}{3}$. Therefore, $1 - \sqrt{3}i = 2\left(\cos\frac{5\pi}{3} + i\sin\frac{5\pi}{3}\right)$, and so $\left(1 - \sqrt{3}i\right)^4 = 2^4\left(\cos\frac{20\pi}{3} + i\sin\frac{20\pi}{3}\right) = 16\left(\cos\frac{2\pi}{3} + i\sin\frac{2\pi}{3}\right) = 16\left(-\frac{1}{2} + i\frac{\sqrt{3}}{2}\right) = 8\left(-1 + i\sqrt{3}\right)$.

**36.** $1+i$ has $r=\sqrt{1+1}=\sqrt{2}$ and $\theta=\frac{\pi}{4}$. Thus, $1+i=\sqrt{2}\left(\cos\frac{\pi}{4}+i\sin\frac{\pi}{4}\right)$, and so $(1+i)^8=\left(\sqrt{2}\right)^8(\cos 2\pi+i\sin 2\pi)=16(1+i\cdot 0)=16$.

**37.** $\sqrt{3}+i$ has $r=\sqrt{3+1}=2$ and $\tan\theta=\frac{1}{\sqrt{3}}$ with $\theta$ in quadrant I $\Leftrightarrow$ $\theta=\frac{\pi}{6}$. Therefore, $\sqrt{3}+i=2\left(\cos\frac{\pi}{6}+i\sin\frac{\pi}{6}\right)$, and so

$$\left(\sqrt{3}+i\right)^{-4}=2^{-4}\left(\cos\tfrac{-4\pi}{6}+i\sin\tfrac{-4\pi}{6}\right)=\tfrac{1}{16}\left(\cos\tfrac{2\pi}{3}-i\sin\tfrac{2\pi}{3}\right)=\tfrac{1}{16}\left(-\tfrac{1}{2}-i\tfrac{\sqrt{3}}{2}\right)$$
$$=\tfrac{1}{32}\left(-1-i\sqrt{3}\right)=-\tfrac{1}{32}\left(1+i\sqrt{3}\right)$$

**38.** $\frac{1}{2}+\frac{\sqrt{3}}{2}i$ has $r=\sqrt{\frac{1}{4}+\frac{3}{4}}=1$ and $\theta=\tan^{-1}\frac{\sqrt{3}/2}{1/2}=\frac{\pi}{3}$ (since $\theta$ is in quadrant I). So $\frac{1}{2}+\frac{\sqrt{3}}{2}i=\cos\frac{\pi}{3}+i\sin\frac{\pi}{3}$. Thus

$$\left(\tfrac{1}{2}+\tfrac{\sqrt{3}}{2}i\right)^{20}=1^{20}\left(\cos\tfrac{20\pi}{3}+i\sin\tfrac{20\pi}{3}\right)=\cos\tfrac{2\pi}{3}+i\sin\tfrac{2\pi}{3}=-\tfrac{1}{2}+\tfrac{\sqrt{3}}{2}i=\tfrac{1}{2}\left(-1+i\sqrt{3}\right)$$

**39.** $-16i$ has $r=16$ and $\theta=\frac{3\pi}{2}$. Thus, $-16i=16\left(\cos\frac{3\pi}{2}+i\sin\frac{3\pi}{2}\right)$ and so $(-16i)^{1/2}=16^{1/2}\left[\cos\left(\dfrac{3\pi+4k\pi}{2}\right)+i\sin\left(\dfrac{3\pi+4k\pi}{2}\right)\right]$ for $k=0,1$. Thus the roots are $w_0=4\left(\cos\frac{3\pi}{4}+i\sin\frac{3\pi}{4}\right)=4\left(-\frac{1}{\sqrt{2}}+i\frac{1}{\sqrt{2}}\right)=2\sqrt{2}(-1+i)$ and $w_1=4\left(\cos\frac{7\pi}{4}+i\sin\frac{7\pi}{4}\right)=4\left(\frac{1}{\sqrt{2}}-i\frac{1}{\sqrt{2}}\right)=2\sqrt{2}(1-i)$.

**40.** $4+4\sqrt{3}i$ has $r=\sqrt{16+48}=8$ and $\theta=\tan^{-1}\frac{4\sqrt{3}}{4}=\frac{\pi}{3}$. Therefore, $4+4\sqrt{3}i=8\left(\cos\frac{\pi}{3}+i\sin\frac{\pi}{3}\right)$. Thus $\left(4+4\sqrt{3}i\right)^{1/3}=\sqrt[3]{8}\left[\cos\left(\dfrac{\pi+6k\pi}{9}\right)+i\sin\left(\dfrac{\pi+6k\pi}{9}\right)\right]$ for $k=0,1,2$. Thus the three roots are $w_0=2\left(\cos\frac{\pi}{9}+i\sin\frac{\pi}{9}\right)$, $w_1=2\left(\cos\frac{7\pi}{9}+i\sin\frac{7\pi}{9}\right)$, and $w_2=2\left(\cos\frac{13\pi}{9}+i\sin\frac{13\pi}{9}\right)$.

**41.** $1=\cos 0+i\sin 0$. Then $1^{1/6}=1\left(\cos\dfrac{2k\pi}{6}+i\sin\dfrac{2k\pi}{6}\right)$ for $k=0,1,2,3,4,5$. Thus the six roots are $w_0=1(\cos 0+i\sin 0)=1$, $w_1=1\left(\cos\frac{\pi}{3}+i\sin\frac{\pi}{3}\right)=\frac{1}{2}+i\frac{\sqrt{3}}{2}$, $w_2=1\left(\cos\frac{2\pi}{3}+i\sin\frac{2\pi}{3}\right)=-\frac{1}{2}+i\frac{\sqrt{3}}{2}$, $w_3=1(\cos\pi+i\sin\pi)=-1$, $w_4=1\left(\cos\frac{4\pi}{3}+i\sin\frac{4\pi}{3}\right)=-\frac{1}{2}-i\frac{\sqrt{3}}{2}$, and $w_5=1\left(\cos\frac{5\pi}{3}+i\sin\frac{5\pi}{3}\right)=\frac{1}{2}-i\frac{\sqrt{3}}{2}$.

**42.** $i=\cos\frac{\pi}{2}+i\sin\frac{\pi}{2}$. Then $i^{1/8}=\cos\left(\dfrac{\pi/2+2k\pi}{8}\right)+i\sin\left(\dfrac{\pi/2+2k\pi}{8}\right)=\cos\left(\dfrac{\pi+4k\pi}{16}\right)+i\sin\left(\dfrac{\pi+4k\pi}{16}\right)$ for $k=0,1,2,3,4,5,6,7$. Thus the eight roots are $w_0=\cos\frac{\pi}{16}+i\sin\frac{\pi}{16}$, $w_1=\cos\frac{5\pi}{16}+i\sin\frac{5\pi}{16}$, $w_2=\cos\frac{9\pi}{16}+i\sin\frac{9\pi}{16}$, $w_3=\cos\frac{13\pi}{16}+i\sin\frac{13\pi}{16}$, $w_4=\cos\frac{17\pi}{16}+i\sin\frac{17\pi}{16}$, $w_5=\cos\frac{21\pi}{16}+i\sin\frac{21\pi}{6}$, $w_6=\cos\frac{25\pi}{16}+i\sin\frac{25\pi}{16}$, and $w_7=\cos\frac{29\pi}{16}+i\sin\frac{29\pi}{16}$.

**43.** $\mathbf{u}=\langle -2,3\rangle$ and $\mathbf{v}=\langle 8,1\rangle$. Then $|\mathbf{u}|=\sqrt{(-2)^2+3^2}=\sqrt{13}$, $\mathbf{u}+\mathbf{v}=\langle -2+8,3+1\rangle=\langle 6,4\rangle$, $\mathbf{u}-\mathbf{v}=\langle -2-8,3-1\rangle=\langle -10,2\rangle$, $2\mathbf{u}=\langle -4,6\rangle$, and $3\mathbf{u}-2\mathbf{v}=\langle -6,9\rangle-\langle 16,2\rangle=\langle -6-16,9-2\rangle=\langle -22,7\rangle$.

**44.** $\mathbf{u}=2\mathbf{i}+\mathbf{j}$ and $\mathbf{v}=\mathbf{i}-2\mathbf{j}$. Then $|\mathbf{u}|=\sqrt{2^2+1^2}=\sqrt{5}$, $\mathbf{u}+\mathbf{v}=(2+1)\mathbf{i}+(1-2)\mathbf{j}=3\mathbf{i}-\mathbf{j}$, $\mathbf{u}-\mathbf{v}=(2-1)\mathbf{i}+(1+2)\mathbf{j}=\mathbf{i}+3\mathbf{j}$, $2\mathbf{u}=4\mathbf{i}+2\mathbf{j}$, and $3\mathbf{u}-2\mathbf{v}=(6-2)\mathbf{i}+(3+4)\mathbf{j}=4\mathbf{i}+7\mathbf{j}$.

**45.** $P(0,3)$ and $Q(3,-1)$. $\mathbf{u}=\langle 3-0,-1-3\rangle=\langle 3,-4\rangle=3\mathbf{i}-4\mathbf{j}$

**46.** $\mathbf{u}=(|\mathbf{u}|\cos\theta)\,\mathbf{i}+(|\mathbf{u}|\sin\theta)\,\mathbf{j}=20\cos 60^\circ\mathbf{i}+20\sin 60^\circ\mathbf{j}=10\mathbf{i}+10\sqrt{3}\mathbf{j}$

**47.** Let $Q(x,y)$ be the terminal point. Then $\langle x-5,y-6\rangle=\langle 5,-8\rangle$ $\Leftrightarrow$ $x-5=5$ and $y-6=-8$ $\Leftrightarrow$ $x=10$ and $y=-2$. Therefore, the terminal point is $Q(10,-2)$.

**48.** $\tan\theta=\dfrac{-5}{2}$ with $\theta$ in quadrant IV $\Leftrightarrow$ $\theta=\tan^{-1}(-2.5)\approx -68.2^\circ$.

**49. (a)** The resultant force $\mathbf{r}$ is the sum of the forces; to find this we first resolve the two vectors:
$\mathbf{v}_1 = 2.0 \times 10^4$ lb N 50° E $= \left\langle 2.0 \times 10^4 \cdot \cos 40°, 2.0 \times 10^4 \cdot \sin 40° \right\rangle \approx \langle 15321, 12856 \rangle$ and
$\mathbf{v}_2 = 3.4 \times 10^4$ lb S 75° E $= \left\langle 3.4 \times 10^4 \cdot \cos(-15°), 3.4 \times 10^4 \cdot \sin(-15°) \right\rangle \approx \langle 32841, -8800 \rangle$. Thus
$\mathbf{r} = \mathbf{v}_1 + \mathbf{v}_2 \approx \langle 15321, 12856 \rangle + \langle 32841, -8800 \rangle = \langle 48163, 4056 \rangle \approx \langle 4.82, 0.41 \rangle \times 10^4 = (4.82\mathbf{i} + 0.41\mathbf{j}) \times 10^4$.

**(b)** The magnitude is $|\mathbf{r}| = \sqrt{48163^2 + 4056^2} \approx 4.8 \times 10^4$ lb. We have $\theta \approx \tan^{-1}\left(\frac{4056}{48163}\right) \approx 4.8°$. Thus the direction of $\mathbf{r}$ is approximately N 85.2° E.

**50. (a)** The initial velocity of the airplane is $\mathbf{v} = 600 \cos 30° \, \mathbf{i} + 600 \sin 30° \, \mathbf{j} = \left\langle 300\sqrt{3}, 300 \right\rangle$. The wind has a velocity $\mathbf{w} = 50 \cos 120° \, \mathbf{i} + 50 \sin 120° \, \mathbf{j} = \left\langle -25, 25\sqrt{3} \right\rangle$. Thus, the resultant velocity is $\mathbf{r} = \left\langle 300\sqrt{3} - 25, 300 + 25\sqrt{3} \right\rangle$.

**(b)** The true speed of the plane is $|\mathbf{r}| = \sqrt{\left(300\sqrt{3} - 25\right)^2 + \left(300 + 25\sqrt{3}\right)^2} \approx 602.1$ mi/h. Since

$\tan\theta = \dfrac{300 + 25\sqrt{3}}{300\sqrt{3} - 25} \approx 0.69408$, $\theta \approx \tan^{-1}(0.69408) \approx 34.8°$, and the true direction is N 55.2° E.

**51.** $\mathbf{u} = \langle 4, -3 \rangle$ and $\mathbf{v} = \langle 9, -8 \rangle$. Then $|\mathbf{u}| = \sqrt{(4)^2 + (-3)^2} = \sqrt{25} = 5$,
$\mathbf{u} \cdot \mathbf{u} = \langle 4, -3 \rangle \cdot \langle 4, -3 \rangle = (4)(4) + (-3)(-3) = 16 + 9 = 25$, and
$\mathbf{u} \cdot \mathbf{v} = \langle 4, -3 \rangle \cdot \langle 9, -8 \rangle = (4)(9) + (-3)(-8) = 36 + 24 = 60$.

**52.** $\mathbf{u} = \langle 5, 12 \rangle$ and $\mathbf{v} = \langle 10, -4 \rangle$. Then $|\mathbf{u}| = \sqrt{(5)^2 + (12)^2} = \sqrt{169} = 13$,
$\mathbf{u} \cdot \mathbf{u} = \langle 5, 12 \rangle \cdot \langle 5, 12 \rangle = (5)(5) + (12)(12) = 25 + 144 = 169$, and
$\mathbf{u} \cdot \mathbf{v} = \langle 5, 12 \rangle \cdot \langle 10, -4 \rangle = (5)(10) + (12)(-4) = 50 - 48 = 2$.

**53.** $\mathbf{u} = -2\mathbf{i} + 2\mathbf{j}$ and $\mathbf{v} = \mathbf{i} + \mathbf{j}$. Then $|\mathbf{u}| = \sqrt{(-2)^2 + (2)^2} = \sqrt{8} = 2\sqrt{2}$,
$\mathbf{u} \cdot \mathbf{u} = -2\mathbf{i} + 2\mathbf{j} \cdot -2\mathbf{i} + 2\mathbf{j} = (-2)(-2) + (2)(2) = 4 + 4 = 8$, and
$\mathbf{u} \cdot \mathbf{v} = -2\mathbf{i} + 2\mathbf{j} \cdot \mathbf{i} + \mathbf{j} = (-2)(1) + (2)(1) = -2 + 2 = 0$.

**54.** $\mathbf{u} = 10\mathbf{j}$ and $\mathbf{v} = 5\mathbf{i} - 3\mathbf{j}$. Then $|\mathbf{u}| = \sqrt{(0)^2 + (10)^2} = \sqrt{100} = 10$,
$\mathbf{u} \cdot \mathbf{u} = 10\mathbf{j} \cdot 10\mathbf{j} = (0)(0) + (10)(10) = 100$, and $\mathbf{u} \cdot \mathbf{v} = 10\mathbf{j} \cdot 5\mathbf{i} - 3\mathbf{j} = (0)(5) + (10)(-3) = -30$.

**55.** $\mathbf{u} = \langle -4, 2 \rangle$ and $\mathbf{v} = \langle 3, 6 \rangle$. Since $\mathbf{u} \cdot \mathbf{v} = -12 + 12 = 0$, the vectors are orthogonal.

**56.** $\mathbf{u} = \langle 5, 3 \rangle$ and $\mathbf{v} = \langle -2, 6 \rangle$. Then $\cos\theta = \dfrac{\mathbf{u} \cdot \mathbf{v}}{|\mathbf{u}|\,|\mathbf{v}|} = \dfrac{-10 + 18}{\sqrt{34} \cdot \sqrt{40}} = \dfrac{5}{2\sqrt{340}} = \dfrac{5}{4\sqrt{85}}$. Thus, $\theta = \cos^{-1} \frac{5}{4\sqrt{85}} \approx 82.2°$, so the vectors are not orthogonal.

**57.** $\mathbf{u} = 2\mathbf{i} + \mathbf{j}$ and $\mathbf{v} = \mathbf{i} + 3\mathbf{j}$. Then $\cos\theta = \dfrac{\mathbf{u} \cdot \mathbf{v}}{|\mathbf{u}|\,|\mathbf{v}|} = \dfrac{2 + 3}{\sqrt{5}\sqrt{10}} = \dfrac{\sqrt{2}}{2}$. Thus, $\theta = \cos^{-1} \frac{\sqrt{2}}{2} = 45°$, so the vectors are not orthogonal.

**58.** $\mathbf{u} = \mathbf{i} - \mathbf{j}$ and $\mathbf{v} = \mathbf{i} + \mathbf{j}$. Then $\cos\theta = \dfrac{\mathbf{u} \cdot \mathbf{v}}{|\mathbf{u}|\,|\mathbf{v}|} = \dfrac{1 - 1}{\sqrt{2} \cdot \sqrt{2}} = 0$, so the vectors are orthogonal.

**59. (a)** $\mathbf{u} = \langle 3, 1 \rangle$ and $\mathbf{v} = \langle 6, -1 \rangle$. Then the component of $\mathbf{u}$ along $\mathbf{v}$ is $\dfrac{\mathbf{u} \cdot \mathbf{v}}{|\mathbf{v}|} = \dfrac{\langle 3, 1 \rangle \cdot \langle 6, -1 \rangle}{\sqrt{\langle 6, -1 \rangle \cdot \langle 6, -1 \rangle}} = \dfrac{18 - 1}{\sqrt{36 + 1}} = \dfrac{17\sqrt{37}}{37}$.

**(b)** $\mathbf{u}_1 = \text{proj}_{\mathbf{v}}\,\mathbf{u} = \left(\dfrac{\mathbf{u} \cdot \mathbf{v}}{|\mathbf{v}|^2}\right)\mathbf{v} = \left(\dfrac{\langle 3, 1 \rangle \cdot \langle 6, -1 \rangle}{6^2 + (-1)^2}\right)\langle 6, -1 \rangle = \frac{17}{37}\langle 6, -1 \rangle = \left\langle \frac{102}{37}, -\frac{17}{37} \right\rangle$.

**(c)** $\mathbf{u}_2 = \mathbf{u} - \mathbf{u}_1 = \langle 3, 1 \rangle - \left\langle \frac{102}{37}, -\frac{17}{37} \right\rangle = \left\langle \frac{9}{37}, \frac{54}{37} \right\rangle$.

**60. (a)** $\mathbf{u} = \langle -8, 6 \rangle$ and $\mathbf{v} = \langle 20, 20 \rangle$. Then the component of $\mathbf{u}$ along $\mathbf{v}$ is

$$\frac{\mathbf{u} \cdot \mathbf{v}}{|\mathbf{v}|} = \frac{\langle -8, 6 \rangle \cdot \langle 20, 20 \rangle}{\sqrt{\langle 20, 20 \rangle \cdot \langle 20, 20 \rangle}} = \frac{-160 + 120}{\sqrt{400 + 400}} = \frac{-40}{\sqrt{800}} = \frac{-40}{20\sqrt{2}} = -\sqrt{2}.$$

**(b)** $\mathbf{u}_1 = \text{proj}_{\mathbf{v}}\,\mathbf{u} = \left(\dfrac{\mathbf{u} \cdot \mathbf{v}}{|\mathbf{v}|^2}\right)\mathbf{v} = \left(\dfrac{\langle -8, 6 \rangle \cdot \langle 20, 20 \rangle}{20^2 + 20^2}\right)\langle 20, 20 \rangle = -\frac{1}{20}\langle 20, 20 \rangle = \langle -1, -1 \rangle$.

**(c)** $\mathbf{u}_2 = \mathbf{u} - \mathbf{u}_1 = \langle -8, 6 \rangle - \langle -1, -1 \rangle = \langle -7, 7 \rangle$.

**61.** The displacement is $\mathbf{D} = \langle 7-1, -1-1 \rangle = \langle 6, -2 \rangle$. Thus, $W = \mathbf{F} \cdot \mathbf{D} = \langle 2, 9 \rangle \cdot \langle 6, -2 \rangle = -6$.

**62.** $W = \mathbf{F} \cdot \mathbf{D} = 3800$ ft-lb. But $\mathbf{F} \cdot \mathbf{D} = |\mathbf{F}|\,|\mathbf{D}| \cos\theta \quad \Leftrightarrow \quad \cos\theta = \dfrac{\mathbf{F} \cdot \mathbf{D}}{|\mathbf{F}|\,|\mathbf{D}|} = \dfrac{3800}{(250)\,(20)} = 0.76$. Therefore, $\theta = \cos^{-1}(0.76) \approx 40.5°$.

# Chapter 9 Test

**1. (a)** $x = 8\cos\frac{5\pi}{4} = 8\left(-\frac{\sqrt{2}}{2}\right) = -4\sqrt{2}$, $y = 8\sin\frac{5\pi}{4} = 8\left(-\frac{\sqrt{2}}{2}\right) = -4\sqrt{2}$. So the point has rectangular coordinates $(-4\sqrt{2}, -4\sqrt{2})$.

**(b)** $P = (-6, 2\sqrt{3})$ in rectangular coordinates. So $\tan\theta = \frac{2\sqrt{3}}{-6}$ and the reference angle is $\overline{\theta} = \frac{\pi}{6}$. Since $P$ is in quadrant II, we have $\theta = \frac{5\pi}{6}$. Next, $r^2 = (-6)^2 + \left(2\sqrt{3}\right)^2 = 36 + 12 = 48$, so $r = 4\sqrt{3}$. Thus, polar coordinates for the point are $\left(4\sqrt{3}, \frac{5\pi}{6}\right)$ or $\left(-4\sqrt{3}, \frac{11\pi}{6}\right)$.

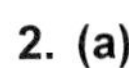

**2. (a)**

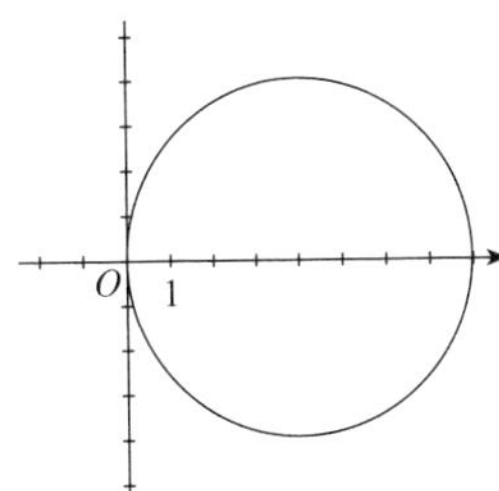


**(b)** $r = 8\cos\theta \quad \Leftrightarrow \quad r^2 = 8r\cos\theta \quad \Leftrightarrow \quad x^2 + y^2 = 8x \quad \Leftrightarrow$
$x^2 - 8x + y^2 = 0 \quad \Leftrightarrow \quad x^2 - 8x + 16 + y^2 = 16 \quad \Leftrightarrow$
$(x-4)^2 + y^2 = 16$

**3. (a)**

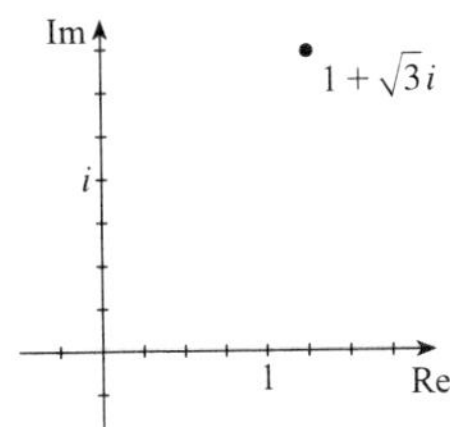


**(b)** $1 + \sqrt{3}i$ has $r = \sqrt{1+3} = 2$ and $\theta = \tan^{-1}\left(\sqrt{3}\right) = \frac{\pi}{3}$. So, in trigonometric form, $1 + \sqrt{3}i = 2\left(\cos\frac{\pi}{3} + i\sin\frac{\pi}{3}\right)$.

(c) $z = 1 + \sqrt{3}i = 2\left(\cos\frac{\pi}{3} + i\sin\frac{\pi}{3}\right) \quad \Rightarrow$

$$z^9 = 2^9\left(\cos\tfrac{9\pi}{3} + i\sin\tfrac{9\pi}{3}\right) = 512\left(\cos 3\pi + i\sin 3\pi\right) = 512\left(-1 + i\,(0)\right) = -512$$

**4.** $z_1 = 4\left(\cos\frac{7\pi}{12} + i\sin\frac{7\pi}{12}\right)$ and $z_2 = 2\left(\cos\frac{5\pi}{12} + i\sin\frac{5\pi}{12}\right)$.

Then $z_1 z_2 = 4 \cdot 2\left[\cos\left(\dfrac{7\pi + 5\pi}{12}\right) + i\sin\left(\dfrac{7\pi + 5\pi}{12}\right)\right] = 8\left(\cos\pi + i\sin\pi\right) = -8$ and

$$z_1/z_2 = \tfrac{4}{2}\left[\cos\left(\frac{7\pi - 5\pi}{12}\right) + i\sin\left(\frac{7\pi - 5\pi}{12}\right)\right] = 2\left(\cos\tfrac{\pi}{6} + i\sin\tfrac{\pi}{6}\right) = 2\left(\tfrac{\sqrt{3}}{2} + \tfrac{1}{2}i\right) = \sqrt{3} + i.$$

**5.** $27i$ has $r = 27$ and $\theta = \frac{\pi}{2}$, so $27i = 27\left(\cos\frac{\pi}{2} + i\sin\frac{\pi}{2}\right)$. Thus,

$$(27i)^{1/3} = \sqrt[3]{27}\left[\cos\left(\frac{\frac{\pi}{2} + 2k\pi}{3}\right) + i\sin\left(\frac{\frac{\pi}{2} + 2k\pi}{3}\right)\right] = 3\left[\cos\left(\frac{\pi + 4k\pi}{6}\right) + i\sin\left(\frac{\pi + 4k\pi}{6}\right)\right]$$

for $k = 0, 1, 2$. Thus, the three roots are

$w_0 = 3\left(\cos\frac{\pi}{6} + i\sin\frac{\pi}{6}\right) = 3\left(\frac{\sqrt{3}}{2} + \frac{1}{2}i\right) = \frac{3}{2}\left(\sqrt{3} + i\right)$,

$w_1 = 3\left(\cos\frac{5\pi}{6} + i\sin\frac{5\pi}{6}\right) = 3\left(-\frac{\sqrt{3}}{2} + \frac{1}{2}i\right) = \frac{3}{2}\left(-\sqrt{3} + i\right)$, and

$w_2 = 3\left(\cos\frac{9\pi}{6} + i\sin\frac{9\pi}{6}\right) = -3i$.

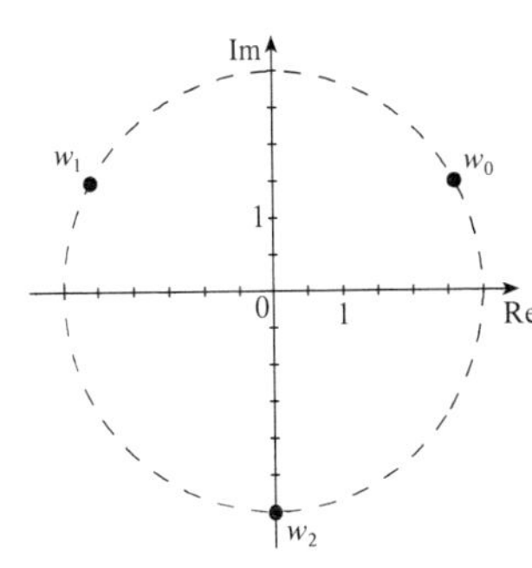

**6. (a)** $\mathbf{u}$ has initial point $P(3,-1)$ and terminal point $Q(-3,9)$. Therefore,
$\mathbf{u} = \langle -3-3, 9-(-1)\rangle = \langle -6, 10\rangle = -6\mathbf{i} + 10\mathbf{j}$.

**(b)** $\mathbf{u} = \langle -6, 10\rangle \quad\Rightarrow\quad |\mathbf{u}| = \sqrt{(-6)^2 + 10^2} = \sqrt{4\cdot(9+25)} = 2\sqrt{34}$

**7.** $\mathbf{u} = \langle 1, 3\rangle$ and $\mathbf{v} = \langle -6, 2\rangle$.

**(a)** $\mathbf{u} - 3\mathbf{v} = \langle 1,3\rangle - \langle -18, 6\rangle = \langle 19, -3\rangle$

**(b)** $|\mathbf{u}+\mathbf{v}| = |\langle 1-6, 3+2\rangle| = |\langle -5,5\rangle| = \sqrt{(-5)^2 + 5^2} = 5\sqrt{2}$

**(c)** $\mathbf{u}\cdot\mathbf{v} = (1)(-6) + (3)(2) = -6+6 = 0$

**(d)** Since $\mathbf{u}\cdot\mathbf{v} = 0$, it is true that $\mathbf{u}$ and $\mathbf{v}$ are perpendicular.

**8. (a)**

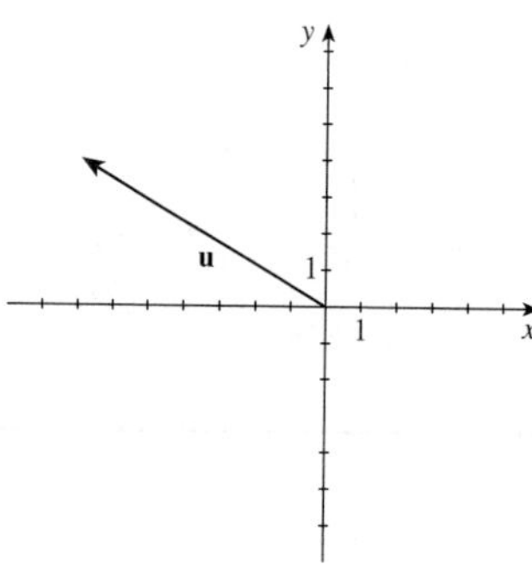


**(b)** $|\mathbf{u}| = \sqrt{\left(-4\sqrt{3}\right)^2 + 4^2} = \sqrt{48+16} = 8$ and $\tan\theta = \frac{4}{-4\sqrt{3}} = -\frac{\sqrt{3}}{3}$, so $\theta = \frac{5\pi}{6}$ (since the terminal point of $\mathbf{u}$ is in quadrant III).

**9.** Let $\mathbf{r}$ represent the velocity of the river and $\mathbf{b}$ the velocity of the boat. Then $\mathbf{r} = \langle 8, 0\rangle$ and $\mathbf{b} = \langle 12\sin 30°, 12\cos 30°\rangle = \left\langle 6, 6\sqrt{3}\right\rangle$.

**(a)** The resultant (or true) velocity is $\mathbf{v} = \mathbf{r} + \mathbf{b} = \left\langle 14, 6\sqrt{3}\right\rangle = 14\mathbf{i} + 6\sqrt{3}\mathbf{j}$.

**(b)** $|\mathbf{v}| = \sqrt{14^2 + \left(6\sqrt{3}\right)^2} = \sqrt{196+108} = \sqrt{304} \approx 17.44$ and $\theta = \tan^{-1}\left(\frac{6\sqrt{3}}{14}\right) = \tan^{-1}\left(\frac{3\sqrt{3}}{7}\right) \approx 36.6°$.

Therefore, the true speed of the boat is approximately 17.4 mi/h and the true direction is approximately N 53.4° E.

**10. (a)** $\cos\theta = \dfrac{\langle 3,2\rangle\cdot\langle 5,-1\rangle}{|\langle 3,2\rangle|\,|\langle 5,-1\rangle|} = \dfrac{15-2}{\sqrt{13}\sqrt{26}} = \dfrac{13}{13\sqrt{2}} = \dfrac{1}{\sqrt{2}} \quad\Rightarrow\quad \theta = \frac{\pi}{4}\text{ rad} = 45°$.

**(b)** The component of $\mathbf{u}$ along $\mathbf{v}$ is $\dfrac{\mathbf{u}\cdot\mathbf{v}}{|\mathbf{v}|} = \dfrac{\langle 3,2\rangle\cdot\langle 5,-1\rangle}{\sqrt{26}} = \dfrac{15-2}{\sqrt{26}} = \dfrac{13}{\sqrt{26}} = \dfrac{\sqrt{26}}{2}$.

**(c)** $\text{proj}_{\mathbf{v}}\,\mathbf{u} = \left(\dfrac{\mathbf{u}\cdot\mathbf{v}}{|\mathbf{v}|^2}\right)\mathbf{v} = \left(\dfrac{\langle 3,2\rangle\cdot\langle 5,-1\rangle}{5^2+(-1)^2}\right)\langle 5,-1\rangle = \frac{13}{26}\langle 5,-1\rangle = \frac{1}{2}\langle 5,-1\rangle = \left\langle \frac{5}{2}, -\frac{1}{2}\right\rangle$.

**11.** The displacement vector is $\mathbf{D} = \langle 7,-13\rangle - \langle 2,2\rangle = \langle 5,-15\rangle$. Therefore, the work done is $W = \mathbf{F}\cdot\mathbf{D} = \langle 3,-5\rangle\cdot\langle 5,-15\rangle = 15+75 = 90$.

## Focus on Modeling: Mapping the World

**1. (a)** Since we want $x = 36$ when $\alpha = 360°$, we substitute these values and solve for $R$: $36 = \frac{\pi}{180} \cdot 360R \quad \Leftrightarrow$ $R = \frac{36}{2\pi} \approx 5.73$ inches.

**(b)** The circumference of the earth at the equator is $2\pi\,(3960)$. Since the length of the equator on the map is 36 inches, each inch represents $\frac{1}{36}$ of that length, or $\frac{1}{36} \cdot 7920\pi \approx 691.2$ miles.

**2.** The distance from the equator to $70°$ N is $y = 5.73 \tan 70° \approx 15.74$. So the map should be $2 \cdot 15.74 = 31.48$ inches tall.

**3. (a)** $x = \frac{\pi}{180}\,(-122.3)\,(5.73) \approx -12.33$; $y = 5.73 \tan 47.6° \approx 6.28$.

**(b)** $x = \frac{\pi}{180}\,(37.6)\,(5.73) \approx 3.76$; $y = 5.73 \tan 55.8° \approx 8.43$

**(c)** $x = \frac{\pi}{180}\,(151.2)\,(5.73) \approx 15.12$; $y = 5.73 \tan(-33.9°) \approx -3.85$

**(d)** $x = \frac{\pi}{180}\,(-43.1)\,(5.73) \approx -4.31$; $y = 5.73 \tan(-22.9°) \approx -2.42$

**4. (a)** Since the outside of the map corresponds to the equator, we use $\beta = 0$ in the formula, with $r = 20$. Thus,

$$20 = 2R \tan 45° \quad \Leftrightarrow \quad R = \frac{10}{\tan 45°} = 10.$$

**(b)** $\theta = 151.2°$ and $r = 2\,(10) \tan\left(\frac{-33.9°}{2} + 45°\right) = 20 \tan 28.05° \approx 10.66$. So Sydney, Australia corresponds to the point $(10.66, 151.2°)$.

**5.** The formula for the length of a circular arc is $s = r\theta$, where $\theta$ is measured in radians. Using this formula, we find that the distance along a meridian between the latitudes $\beta°$ and $(\beta + 1)°$ is $\frac{\pi}{180} R \approx 0.01745R$.

**(a)** The projected distance on the cylinder is $R \tan 21° - R \tan 10° \approx 0.01989R$. Thus, the ratio is $\dfrac{0.01989R}{0.01745R} \approx 1.14$.

**(b)** The projected distance on the cylinder is $R \tan 41° - R \tan 40° \approx 0.03019R$. Thus, the ratio is $\dfrac{0.03019R}{0.01745R} \approx 1.73$.

**(c)** The projected distance on the cylinder is $R \tan 81° - R \tan 80° \approx 0.64247R$. Thus, the ratio is $\dfrac{0.64247R}{0.01745R} \approx 36.82$.

**6.** We first need to determine the radius $r$ of the circle at latitude $\beta$.

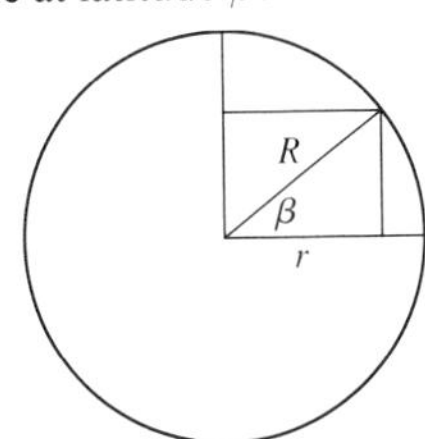


From the figure, we see that $\cos\beta = r/R$, so $r = R\cos\beta$. Using the formula for the length of a circular arc, we find that the distance between two points on the sphere at latitude $\beta$ that are $1°$ longitude apart is $\frac{\pi}{180} r = \frac{\pi}{180} R\cos\beta$. On the cylinder, the distance between two points that are $1°$ longitude apart is $\frac{\pi}{180} R$. Thus, the ratio of the projected distance on the cylinder to the distance on the sphere is $\dfrac{\frac{\pi}{180} R}{\frac{\pi}{180} R\cos\beta} = \sec\beta$.

**(a)** At $20°$ N, the ratio is $\sec 20° \approx 1.06$.

**(b)** At $40°$ N, the ratio is $\sec 40° \approx 1.31$.

**(c)** At $80°$ N, the ratio is $\sec 80° \approx 5.76$.

**7.** The projected distance on the plane between latitude $\beta_1^\circ$ and $\beta_2^\circ$ is

$$2R\tan\left(\tfrac{1}{2}\beta_1 + 45^\circ\right) - 2R\tan\left(\tfrac{1}{2}\beta_2 + 45^\circ\right) = 2R\left[\tan\left(\tfrac{1}{2}\beta_1 + 45^\circ\right) - \tan\left(\tfrac{1}{2}\beta_2 + 45^\circ\right)\right]$$

In Problem 5 we found that on the sphere, the distance along a meridian between the latitudes $\beta^\circ$ and $(\beta + 1)^\circ$ is $\frac{\pi}{180}R \approx 0.01745R$.

**(a)** The projected distance on the plane is

$$\begin{aligned} 2R\left[\tan\left(\tfrac{1}{2}(-20^\circ) + 45^\circ\right) - \tan\left(\tfrac{1}{2}(-21^\circ) + 45^\circ\right)\right] &= 2R(\tan 35^\circ - R\tan 34.5^\circ) \\ &\approx 0.02583R \end{aligned}$$

Thus, the ratio is $\dfrac{0.02583R}{0.01745R} \approx 1.48$.

**(b)** The projected distance on the plane is

$$\begin{aligned} 2R\left[\tan\left(\tfrac{1}{2}(-40) + 45^\circ\right) - \tan\left(\tfrac{1}{2}(-41) + 45^\circ\right)\right] &= 2R(\tan 25^\circ - R\tan 24.5^\circ) \\ &\approx 0.02116R \end{aligned}$$

Thus, the ratio is $\dfrac{0.02116R}{0.01745R} \approx 1.21$.

**(c)** The projected distance on the plane is

$$\begin{aligned} 2R\left[\tan\left(\tfrac{1}{2}(-80) + 45^\circ\right) - \tan\left(\tfrac{1}{2}(-81) + 45^\circ\right)\right] &= 2R(\tan 5^\circ - R\tan 4.5^\circ) \\ &\approx 0.01757R \end{aligned}$$

Thus, the ratio is $\dfrac{0.01757R}{0.01745R} \approx 1.01$.

**8.** Using the formula for the length of a circular arc, we find that the projected distance on the plane between two points that are $1^\circ$ longitude apart (at latitude $\beta^\circ$) is $\frac{\pi}{180}r$, where $r = 2R\tan\left(\frac{1}{2}\beta + 45^\circ\right)$. Thus, the ratio of the projected distance on the plane to the distance on the sphere is $\dfrac{\frac{\pi}{180}2R\tan\left(\frac{1}{2}\beta + 45^\circ\right)}{\frac{\pi}{180}R\cos\beta} = \dfrac{2\tan\left(\frac{1}{2}\beta + 45^\circ\right)}{\cos\beta}$.

**(a)** At $20^\circ$ S, the ratio is $\dfrac{2\tan\left(\frac{1}{2}(-20^\circ) + 45^\circ\right)}{\cos 20^\circ} = \dfrac{2\tan 35^\circ}{\cos 20^\circ} \approx 1.49$.

**(b)** At $40^\circ$ S, the ratio is $\dfrac{2\tan\left(\frac{1}{2}(-40^\circ) + 45^\circ\right)}{\cos 40^\circ} = \dfrac{2\tan 25^\circ}{\cos 40^\circ} \approx 1.22$.

**(c)** At $80^\circ$ S, the ratio is $\dfrac{2\tan\left(\frac{1}{2}(-80^\circ) + 45^\circ\right)}{\cos 40^\circ} = \dfrac{2\tan 5^\circ}{\cos 80^\circ} \approx 1.01$.

# 10 Systems of Equations and Inequalities

## 10.1 Systems of Equations

**1.** $\begin{cases} x - \ \ y = 2 \\ 2x + 3y = 9 \end{cases}$ Solving the first equation for $x$, we get $x = y + 2$, and substituting this into the second equation gives $2(y+2) + 3y = 9 \Leftrightarrow 5y + 4 = 9 \Leftrightarrow 5y = 5 \Leftrightarrow y = 1$. Substituting for $y$ we get $x = y + 2 = (1) + 2 = 3$. Thus, the solution is $(3, 1)$.

**2.** $\begin{cases} 2x + \ \ y = 7 \\ x + 2y = 2 \end{cases}$ Solving the first equation for $y$, we get $y = 7 - 2x$, and substituting this into the second equation gives $x + 2(7 - 2x) = 2 \Leftrightarrow x + 14 - 4x = 2 \Leftrightarrow -3x = -12 \Leftrightarrow x = 4$. Substituting for $x$ we get $y = 7 - 2x = 7 - 2(4) = -1$. Thus, the solution is $(4, -1)$.

**3.** $\begin{cases} y = x^2 \\ y = x + 12 \end{cases}$ Substituting $y = x^2$ into the second equation gives $x^2 = x + 12 \Leftrightarrow 0 = x^2 - x - 12 = (x - 4)(x + 3) \Rightarrow x = 4$ or $x = -3$. So since $y = x^2$, the solutions are $(-3, 9)$ and $(4, 16)$.

**4.** $\begin{cases} x^2 + y^2 = 25 \\ y = 2x \end{cases}$ Substituting for $y$ in the first equation gives $x^2 + (2x)^2 = 25 \Leftrightarrow 5x^2 = 25 \Leftrightarrow x^2 = 5 \Rightarrow x = \pm\sqrt{5}$ When $x = \sqrt{5}$ then $y = 2\sqrt{5}$, and when $x = -\sqrt{5}$ then $y = 2(-\sqrt{5}) = -2\sqrt{5}$. Thus the solutions are $(\sqrt{5}, 2\sqrt{5})$ and $(-\sqrt{5}, -2\sqrt{5})$.

**5.** $\begin{cases} x^2 + y^2 = 8 \\ x + \ \ y = 0 \end{cases}$ Solving the second equation for $y$ gives $y = -x$, and substituting this into the first equation gives $x^2 + (-x)^2 = 8 \Leftrightarrow 2x^2 = 8 \Leftrightarrow x = \pm 2$. So since $y = -x$, the solutions are $(2, -2)$ and $(-2, 2)$.

**6.** $\begin{cases} x^2 + y = 9 \\ x - y + 3 = 0 \end{cases}$ Solving the first equation for $y$, we get $y = 9 - x^2$. Substituting this into the second equation gives $x - (9 - x^2) + 3 = 0 \Leftrightarrow x^2 + x - 6 = 0 \Leftrightarrow (x + 3)(x - 2) = 0 \Leftrightarrow x = -3$ or $x = 2$. If $x = -3$, then $y = 9 - (-3)^2 = 0$, and if $x = 2$, then $y = 9 - (2)^2 = 5$. Thus the solutions are $(-3, 0)$ and $(2, 5)$.

**7.** $\begin{cases} x + \ \ y^2 = \ \ 0 \\ 2x + 5y^2 = 75 \end{cases}$ Solving the first equation for $x$ gives $x = -y^2$, and substituting this into the second equation gives $2(-y^2) + 5y^2 = 75 \Leftrightarrow 3y^2 = 75 \Leftrightarrow y^2 = 25 \Leftrightarrow y = \pm 5$. So since $x = -y^2$, the solutions are $(-25, -5)$ and $(-25, 5)$.

**8.** $\begin{cases} x^2 - \ \ y - \ \ 1 \\ 2x^2 + 3y = 17 \end{cases}$ Solving the first equation for $y$, we get $y = x^2 - 1$. Substituting this into the second equation gives $2x^2 + 3(x^2 - 1) = 17 \Leftrightarrow 2x^2 + 3x^2 - 3 = 17 \Leftrightarrow 5x^2 = 20 \Leftrightarrow x^2 = 4 \Leftrightarrow x = \pm 2$. If $x = -2$, then $y = (-2)^2 - 1 = 3$, and if $x = 2$, then $y = (2)^2 - 1 = 3$. Thus the solutions are $(-2, 3)$ and $(2, 3)$.

**9.** $\begin{cases} x + 2y = 5 \\ 2x + 3y = 8 \end{cases}$ Multiplying the first equation by 2 and the second by $-1$ gives the system

$\begin{cases} 2x + 4y = 10 \\ -2x - 3y = -8 \end{cases}$ Adding, we get $y = 2$, and substituting into the first equation in the original system gives $x + 2(2) = 5 \Leftrightarrow x + 4 = 5 \quad \Leftrightarrow \quad x = 1$. The solution is $(1, 2)$.

**10.** $\begin{cases} 4x - 3y = 11 \\ 8x + 4y = 12 \end{cases}$ Multiplying the first equation by 2 gives the system $\begin{cases} 8x - 6y = 22 \\ 8x + 4y = 12 \end{cases}$ Subtracting the equations gives $-10y = 10 \Leftrightarrow y = -1$. Substituting this value into the second equation gives $8x + 4(-1) = 12 \quad \Leftrightarrow$ $8x = 16 \Leftrightarrow x = 2$. Thus, the solution is $(2, -1)$.

**11.** $\begin{cases} x^2 - 2y = 1 \\ x^2 + 5y = 29 \end{cases}$ Subtracting the first equation from the second equation gives $7y = 28 \Rightarrow y = 4$. Substituting $y = 4$ into the first equation of the original system gives $x^2 - 2(4) = 1 \quad \Leftrightarrow \quad x^2 = 9 \quad \Leftrightarrow \quad x = \pm 3$. The solutions are $(3, 4)$ and $(-3, 4)$.

**12.** $\begin{cases} 3x^2 + 4y = 17 \\ 2x^2 + 5y = 2 \end{cases}$ Multiplying the first equation by 2 and the second by 3 gives the system

$\begin{cases} 6x^2 + 8y = 34 \\ -6x^2 - 15y = -6 \end{cases}$ Adding we get $-7y = 28 \quad \Leftrightarrow \quad y = -4$. Substituting this value into the second equation gives $2x^2 + 5(-4) = 2 \Rightarrow 2x^2 = 22 \quad \Leftrightarrow \quad x^2 = 11 \Leftrightarrow x = \pm\sqrt{11}$. Thus the solutions are $\left(\sqrt{11}, -4\right)$ and $\left(-\sqrt{11}, -4\right)$.

**13.** $\begin{cases} 3x^2 - y^2 = 11 \\ x^2 + 4y^2 = 8 \end{cases}$ Multiplying the first equation by 4 gives the system $\begin{cases} 12x^2 - 4y^2 = 44 \\ x^2 + 4y^2 = 8 \end{cases}$ Adding the equations gives $13x^2 = 52 \Leftrightarrow x = \pm 2$. Substituting into the first equation we get $3(4) - y^2 = 11 \quad \Leftrightarrow \quad y = \pm 1$. Thus, the solutions are $(2, 1)$, $(2, -1)$, $(-2, 1)$, and $(-2, -1)$.

**14.** $\begin{cases} 2x^2 + 4y = 13 \\ x^2 - y^2 = \frac{7}{2} \end{cases}$ Multiplying the second equation by 2 gives the system $\begin{cases} 2x^2 + 4y = 13 \\ 2x^2 - 2y^2 = 7 \end{cases}$ Subtracting the equations gives $4y + 2y^2 = 6 \Leftrightarrow y^2 + 2y - 3 = 0 \quad \Leftrightarrow \quad (y + 3)(y - 1) = 0 \quad \Leftrightarrow \quad y = -3, y = 1$. If $y = -3$, then $2x^2 + 4(-3) = 13 \quad \Leftrightarrow \quad x^2 = \frac{25}{2} \Leftrightarrow x = \pm\dfrac{5\sqrt{2}}{2}$. If $y = 1$, then $2x^2 + 4(1) = 13 \quad \Leftrightarrow \quad x^2 = \frac{9}{2} \Leftrightarrow x = \pm\frac{3\sqrt{2}}{2}$. Hence, the solutions are $\left(\pm\frac{5\sqrt{2}}{2}, -3\right)$ and $\left(\pm\frac{3\sqrt{2}}{2}, 1\right)$.

**15.** $\begin{cases} x - y^2 + 3 = 0 \\ 2x^2 + y^2 - 4 = 0 \end{cases}$ Adding the two equations gives $2x^2 + x - 1 = 0$. Using the quadratic formula we have

$x = \dfrac{-1 \pm \sqrt{1 - 4(2)(-1)}}{2(2)} = \dfrac{-1 \pm \sqrt{9}}{4} = \dfrac{-1 \pm 3}{4}$. So $x = \dfrac{-1 - 3}{4} = -1$ or $x = \dfrac{-1 + 3}{4} = \frac{1}{2}$. Substituting $x = -1$ into the first equation gives $-1 - y^2 + 3 = 0 \Leftrightarrow y^2 = 2 \quad \Leftrightarrow \quad y = \pm\sqrt{2}$. Substituting $x = \frac{1}{2}$ into the first equation gives $\frac{1}{2} - y^2 + 3 = 0 \quad \Leftrightarrow \quad y^2 = \frac{7}{2} \quad \Leftrightarrow \quad y = \pm\sqrt{\frac{7}{2}}$. Thus the solutions are $(-1, \pm\sqrt{2})$ and $\left(\frac{1}{2}, \pm\sqrt{\frac{7}{2}}\right)$.

**16.** $\begin{cases} x^2 - y^2 = 1 \\ 2x^2 - y^2 = x + 3 \end{cases}$ Subtracting the first equation from the second equation gives $x^2 = x + 2 \Leftrightarrow$ $x^2 - x - 2 = 0 \Leftrightarrow (x-2)(x+1) = 0 \Leftrightarrow x = 2, x = -1$. Solving the first equation for $y^2$ we have $y^2 = x^2 - 1$. When $x = -1$, $y^2 = (-1)^2 - 1 = 0$ so $y = 0$ and when $x = 2$, $y^2 = (2)^2 - 1 = 3$ so $y = \pm\sqrt{3}$. Thus, the solutions are $(-1, 0)$, $(2, -\sqrt{3})$, and $(2, \sqrt{3})$.

**17.** $\begin{cases} 2x + y = -1 \\ x - 2y = -8 \end{cases}$ By inspection of the graph, it appears that $(-2, 3)$ is the solution to the system. We check this in both equations to verify that it is a solution. $2(-2) + 3 = -4 + 3 = -1$ and $-2 - 2(3) = -2 - 6 = -8$. Since both equations are satisfied, the solution is $(-2, 3)$.

**18.** $\begin{cases} x + y = 2 \\ 2x + y = 5 \end{cases}$ By inspection of the graph, it appears that $(3, -1)$ is the solution to the system. We check this in both equations to verify that it is a solution. $3 + (-1) = 2$ and $2(3) + (-1) = 6 - 1 = 5$. Since both equations are satisfied, the solution is $(3, -1)$ .

**19.** $\begin{cases} x^2 + y = 8 \\ x - 2y = -6 \end{cases}$ By inspection of the graph, it appears that $(2, 4)$ is a solution, but is difficult to get accurate values for the other point. Multiplying the first equation by 2 gives the system $\begin{cases} 2x^2 + 2y = 16 \\ x - 2y = -6 \end{cases}$ Adding the equations gives $2x^2 + x = 10 \Leftrightarrow 2x^2 + x - 10 = 0 \Leftrightarrow (2x+5)(x-2) = 0$. So $x = -\frac{5}{2}$ or $x = 2$. If $x = -\frac{5}{2}$, then $-\frac{5}{2} - 2y = -6 \Leftrightarrow -2y = -\frac{7}{2} \Leftrightarrow y = \frac{7}{4}$, and if $x = 2$, then $2 - 2y = -6 \Leftrightarrow -2y = -8 \Leftrightarrow y = 4$. Hence, the solutions are $\left(-\frac{5}{2}, \frac{7}{4}\right)$ and $(2, 4)$.

**20.** $\begin{cases} x - y^2 = -4 \\ x - y = 2 \end{cases}$ By inspection of the graph, it appears that $(0, -2)$ and $(5, 3)$ are solutions to the system. We check each point in both equations to verify that it is a solution. For $(0, -2)$: $(0) - (-2)^2 = -4$ and $(0) - (-2) = 2$. For $(5, 3)$: $5 - 3^2 = 5 - 9 = -4$ and $5 - 3 = 2$. Thus, the solutions are $(0, -2)$ and $(5, 3)$.

**21.** $\begin{cases} x^2 + y = 0 \\ x^3 - 2x - y = 0 \end{cases}$ By inspection of the graph, it appears that $(-2, -4)$, $(0, 0)$, and $(1, -1)$ are solutions to the system. We check each point in both equations to verify that it is a solution.
For $(-2, -4)$: $(-2)^2 + (-4) = 4 - 4 = 0$ and $(-2)^3 - 2(-2) - (-4) = -8 + 4 + 4 = 0$.
For $(0, 0)$: $(0)^2 + (0) = 0$ and $(0)^3 - 2(0) - (0) = 0$.
For $(1, -1)$: $(1)^2 + (-1) = 1 - 1 = 0$ and $(1)^3 - 2(1) - (-1) = 1 - 2 + 1 = 0$.
Thus, the solutions are $(-2, -4)$, $(0, 0)$, and $(1, -1)$.

**22.** $\begin{cases} x^2 + y^2 = 4x \\ x = y^2 \end{cases}$ By inspection of the graph, it appears that $(0, 0)$ is a solution, but is difficult to get accurate values for the other points. Substituting for $y^2$ we have $x^2 + x = 4x \Leftrightarrow x^2 - 3x = 0 \Leftrightarrow x(x - 3) = 0$. So $x = 0$ or $x = 3$. If $x = 0$, then $y^2 = 0$ so $y = 0$. And is $x = 3$ then $y^2 = 3$ so $y = \pm\sqrt{3}$. Hence, the solutions are $(0, 0)$, $(3, -\sqrt{3})$, and $(3, \sqrt{3})$.

**23.** $\begin{cases} y + x^2 = 4x \\ y + 4x = 16 \end{cases}$ Subtracting the second equation from the first equation gives $x^2 - 4x = 4x - 16 \Leftrightarrow x^2 - 8x + 16 = 0 \Leftrightarrow (x-4)^2 = 0 \Leftrightarrow x = 4$. Substituting this value for $x$ into either of the original equations gives $y = 0$. Therefore, the solution is $(4, 0)$.

**24.** $\begin{cases} x - y^2 = 0 \\ y - x^2 = 0 \end{cases}$ Solving the first equation for $x$ and the second equation for $y$ gives $\begin{cases} x = y^2 \\ y = x^2 \end{cases}$ Substituting for $y$ in the first equation gives $x = x^4 \Leftrightarrow x\left(x^3 - 1\right) = 0 \Leftrightarrow x = 0, x = 1$. Thus, the solutions are $(0, 0)$ and $(1, 1)$.

**25.** $\begin{cases} x - 2y = 2 \\ y^2 - x^2 = 2x + 4 \end{cases}$ Now $x - 2y = 2 \Leftrightarrow x = 2y + 2$. Substituting for $x$ gives $y^2 - x^2 = 2x + 4 \Leftrightarrow y^2 - (2y+2)^2 = 2(2y+2) + 4 \Leftrightarrow y^2 - 4y^2 - 8y - 4 = 4y + 4 + 4 \Leftrightarrow y^2 + 4y + 4 = 0 \Leftrightarrow (y+2)^2 = 0 \Leftrightarrow y = -2$. Since $x = 2y + 2$, we have $x = 2(-2) + 2 = -2$. Thus, the solution is $(-2, -2)$.

**26.** $\begin{cases} y = 4 - x^2 \\ y = x^2 - 4 \end{cases}$ Setting the two equations equal, we get $4 - x^2 = x^2 - 4 \Leftrightarrow 2x^2 = 8 \Leftrightarrow x = \pm 2$. Therefore, the solutions are $(2, 0)$ and $(-2, 0)$.

**27.** $\begin{cases} x - y = 4 \\ xy = 12 \end{cases}$ Now $x - y = 4 \Leftrightarrow x = 4 + y$. Substituting for $x$ gives $xy = 12 \Leftrightarrow (4+y)y = 12 \Leftrightarrow y^2 + 4y - 12 = 0 \Leftrightarrow (y+6)(y-2) = 0 \Leftrightarrow y = -6, y = 2$. Since $x = 4 + y$, the solutions are $(-2, -6)$ and $(6, 2)$.

**28.** $\begin{cases} xy = 24 \\ 2x^2 - y^2 + 4 = 0 \end{cases}$ Since $x = 0$ is not a solution, from the first equation we get $y = \dfrac{24}{x}$. Substituting into the second equation, we get $2x^2 + 4 = \left(\dfrac{24}{x}\right)^2 \Rightarrow 2x^4 + 4x^2 = 576 \Leftrightarrow x^4 + 2x^2 - 288 = 0 \Leftrightarrow \left(x^2 + 18\right)\left(x^2 - 16\right) = 0$. Since $x^2 + 18$ cannot be 0 if $x$ is real, we have $x^2 - 16 = 0 \Leftrightarrow x = \pm 4$. When $x = 4$, we have $y = \frac{24}{4} = 6$ and when $x = -4$, we have $y = \dfrac{24}{-4} = -6$. Thus the solutions are $(4, 6)$ and $(-4, -6)$.

**29.** $\begin{cases} x^2 y = 16 \\ x^2 + 4y + 16 = 0 \end{cases}$ Now $x^2 y = 16 \Leftrightarrow x^2 = \dfrac{16}{y}$. Substituting for $x^2$ gives $\dfrac{16}{y} + 4y + 16 = 0 \Rightarrow 4y^2 + 16y + 16 = 0 \Leftrightarrow y^2 + 4y + 4 = 0 \Leftrightarrow (y+2)^2 = 0 \Leftrightarrow y = -2$. Therefore, $x^2 = \dfrac{16}{-2} = -8$, which has no real solution, and so the system has no solution.

**30.** $\begin{cases} x + \sqrt{y} = 0 \\ y^2 - 4x^2 = 12 \end{cases}$ Solving the first equation for $x$, we get $x = -\sqrt{y}$. Substituting for $x$ gives $y^2 - 4\left(-\sqrt{y}\right)^2 = 12 \Leftrightarrow y^2 - 4y - 12 = 0 \Leftrightarrow (y-6)(y+2) = 0 \Rightarrow y = 6, y = -2$. Since $x = -\sqrt{-2}$ is not a real solution, the only solution is $\left(-\sqrt{6}, 6\right)$.

**31.** $\begin{cases} x^2 + y^2 = 9 \\ x^2 - y^2 = 1 \end{cases}$ Adding the equations gives $2x^2 = 10 \Leftrightarrow x^2 = 5 \Leftrightarrow x = \pm\sqrt{5}$. Now $x = \pm\sqrt{5} \Rightarrow y^2 = 9 - 5 = 4 \Leftrightarrow y = \pm 2$, and so the solutions are $\left(\sqrt{5}, 2\right)$, $\left(\sqrt{5}, -2\right)$, $\left(-\sqrt{5}, 2\right)$, and $\left(-\sqrt{5}, -2\right)$.

**32.** $\begin{cases} x^2 + 2y^2 = 2 \\ 2x^2 - 3y = 15 \end{cases}$ Multiplying the first equation by 2 gives the system $\begin{cases} 2x^2 + 4y^2 = 4 \\ 2x^2 - 3y = 15 \end{cases}$ Subtracting the two equations gives $4y^2 + 3y = -11 \Leftrightarrow 4y^2 + 3y + 11 = 0 \Rightarrow y = \dfrac{-3 \pm \sqrt{9 - 4(4)(11)}}{2(4)}$ which is not a real number. Therefore, there are no real solutions.

**33.** $\begin{cases} 2x^2 - 8y^3 = 19 \\ 4x^2 + 16y^3 = 34 \end{cases}$ Multiplying the first equation by 2 gives the system $\begin{cases} 4x^2 - 16y^3 = 38 \\ 4x^2 + 16y^3 = 34 \end{cases}$ Adding the two equations gives $8x^2 = 72 \Leftrightarrow x = \pm 3$, and then substituting into the first equation we have $2(9) - 8y^3 = 19 \Leftrightarrow y^3 = -\frac{1}{8} \Leftrightarrow y = -\frac{1}{2}$. Therefore, the solutions are $\left(3, -\frac{1}{2}\right)$ and $\left(-3, -\frac{1}{2}\right)$.

**34.** $\begin{cases} x^4 - y^3 = 15 \\ 3x^4 + 5y^3 = 53 \end{cases}$ Multiplying the first equation by 3 gives the system $\begin{cases} 3x^4 - 3y^3 = 45 \\ 3x^4 + 5y^3 = 53 \end{cases}$ Subtracting the equations gives $8y^3 = 8 \Leftrightarrow y^3 = 1 \Rightarrow y = 1$, and then $x^4 - 1 = 15 \Leftrightarrow x = \pm 2$. Therefore, the solutions are $(2, 1)$ and $(-2, 1)$.

**35.** $\begin{cases} \dfrac{2}{x} - \dfrac{3}{y} = 1 \\ -\dfrac{4}{x} + \dfrac{7}{y} = 1 \end{cases}$ If we let $u = \dfrac{1}{x}$ and $v = \dfrac{1}{y}$,the system is equivalent to $\begin{cases} 2u - 3v = 1 \\ -4u + 7v = 1 \end{cases}$ Multiplying the first equation by 4 gives the system $\begin{cases} 4u - 6v = 2 \\ -4u + 7v = 1 \end{cases}$ Adding the equations gives $v = 3$, and then substituting into the first equation gives $2u - 9 = 1 \Leftrightarrow u = 5$. Thus, the solution is $\left(\frac{1}{5}, \frac{1}{3}\right)$.

**36.** $\begin{cases} \dfrac{4}{x^2} + \dfrac{6}{y^4} = \dfrac{7}{2} \\ \dfrac{1}{x^2} - \dfrac{2}{y^4} = 0 \end{cases}$ If we let $u = \dfrac{1}{x^2}$ and $v = \dfrac{1}{y^4}$, the system is equivalent to $\begin{cases} 4u + 6v = \frac{7}{2} \\ u - 2v = 0 \end{cases}$, and multiplying the second equation by 3, gives $\begin{cases} 4u + 6v = \frac{7}{2} \\ 3u - 6v = 0 \end{cases}$ Adding the equations gives $7u = \frac{7}{2} \Leftrightarrow u = \frac{1}{2}$, and $v = \frac{1}{4}$. Therefore, $x^2 = \dfrac{1}{u} = 2 \Leftrightarrow x = \pm\sqrt{2}$, and $y^4 = \dfrac{1}{v} = 4 \Leftrightarrow y = \pm\sqrt{2}$. Thus, the solutions are $\left(\sqrt{2}, \sqrt{2}\right)$, $\left(\sqrt{2}, -\sqrt{2}\right)$, $\left(-\sqrt{2}, \sqrt{2}\right)$, and $\left(-\sqrt{2}, -\sqrt{2}\right)$.

**37.** $\begin{cases} y = 2x + 6 \\ y = -x + 5 \end{cases}$

The solution is approximately $(-0.33, 5.33)$.

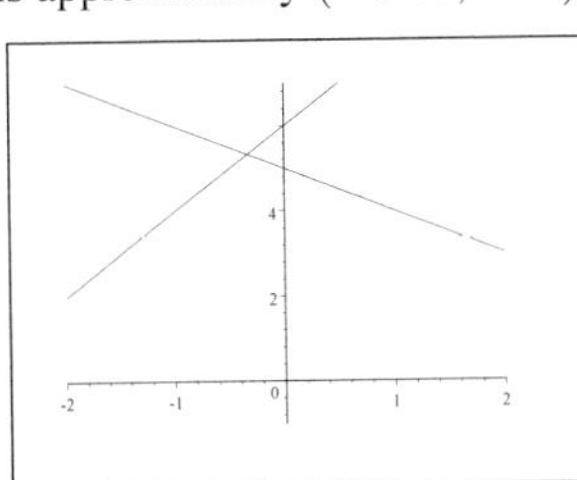


**38.** $\begin{cases} y = -2x + 12 \\ y = x + 3 \end{cases}$

The solution is $(3, 6)$.

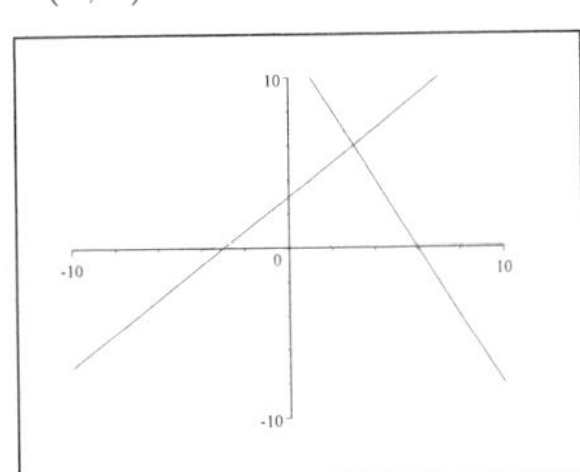

**39.** $\begin{cases} y = x^2 + 8x \\ y = 2x + 16 \end{cases}$

The solutions are $(-8, 0)$ and $(2, 20)$.

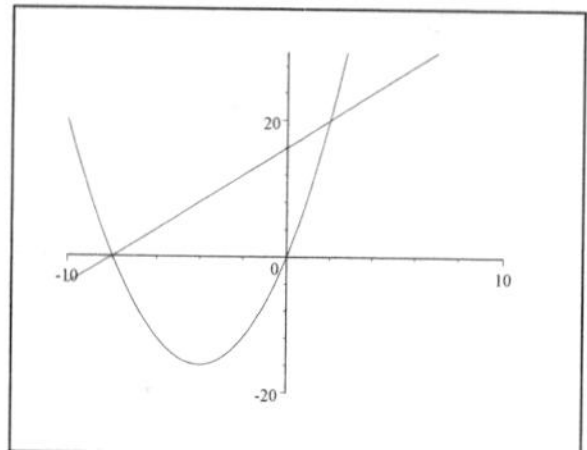

**40.** $\begin{cases} y = x^2 - 4x \\ 2x - y = 2 \end{cases} \Leftrightarrow \begin{cases} y = x^2 - 4x \\ y = 2x - 2 \end{cases}$

The solutions are approximately $(0.35, -1.30)$ and $(5.65, 9.30)$.

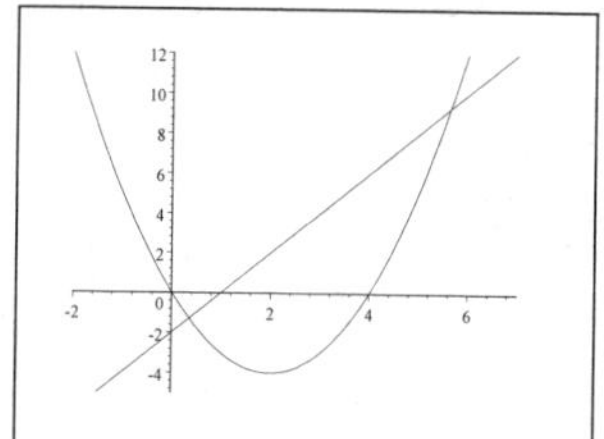

**41.** $\begin{cases} x^2 + y^2 = 25 \\ x + 3y = 2 \end{cases} \Leftrightarrow \begin{cases} y = \pm\sqrt{25 - x^2} \\ y = -\frac{1}{3}x + \frac{2}{3} \end{cases}$

The solutions are $(-4.51, 2.17)$ and $(4.91, -0.97)$.

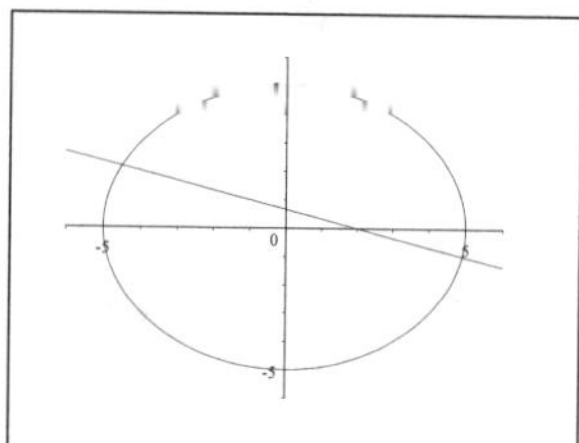

**42.** $\begin{cases} x^2 + y^2 = 17 \\ x^2 - 2x + y^2 = 13 \end{cases} \Leftrightarrow \begin{cases} y = \pm\sqrt{17 - x^2} \\ y = \pm\sqrt{13 + 2x - x^2} \end{cases}$

The solutions are approximately $(2, \pm 3.61)$.

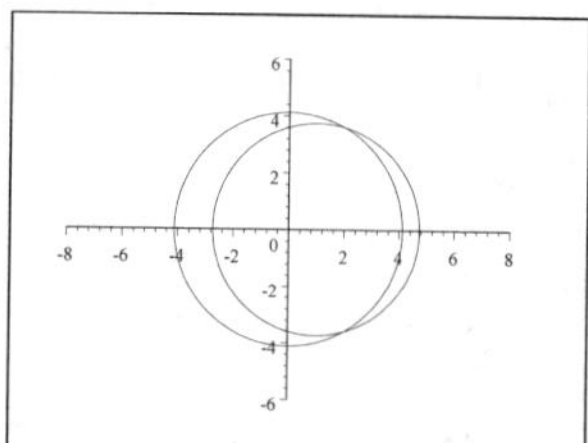

**43.** $\begin{cases} \dfrac{x^2}{9} + \dfrac{y^2}{18} = 1 \\ y = -x^2 + 6x - 2 \end{cases} \Leftrightarrow \begin{cases} y = \pm\sqrt{18 - 2x^2} \\ y = -x^2 + 6x - 2 \end{cases}$

The solutions are $(1.23, 3.87)$ and $(-0.35, -4.21)$.

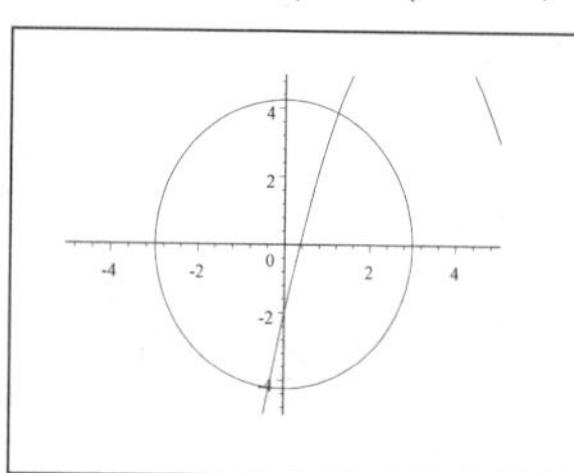

**44.** $\begin{cases} x^2 - y^2 = 3 \\ y = x^2 - 2x - 8 \end{cases} \Leftrightarrow \begin{cases} y = \pm\sqrt{x^2 - 3} \\ y = x^2 - 2x - 8 \end{cases}$

The solutions are approximately $(-2.22, 1.40)$, $(-1.88, -0.72)$, $(3.45, -2.99)$, and $(4.65, 4.31)$.

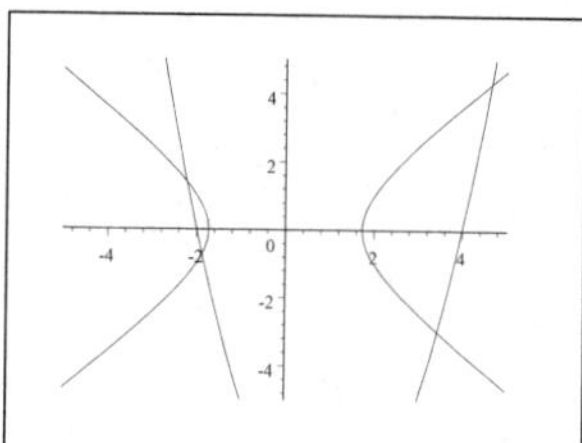

**45.** $\begin{cases} x^4 + 16y^4 = 32 \\ x^2 + 2x + y = 0 \end{cases} \Leftrightarrow \begin{cases} y = \pm\dfrac{\sqrt[4]{32 - x^4}}{2} \\ y = -x^2 - 2x \end{cases}$

The solutions are $(-2.30, -0.70)$ and $(0.48, -1.19)$.

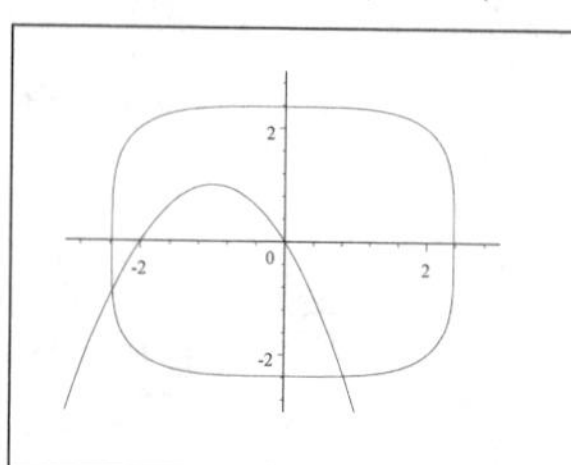

**46.** $\begin{cases} y = e^x + e^{-x} \\ y = 5 - x^2 \end{cases}$ The solution are approximately $(1.19, 3.59)$ and $(-1.19, 3.59)$.

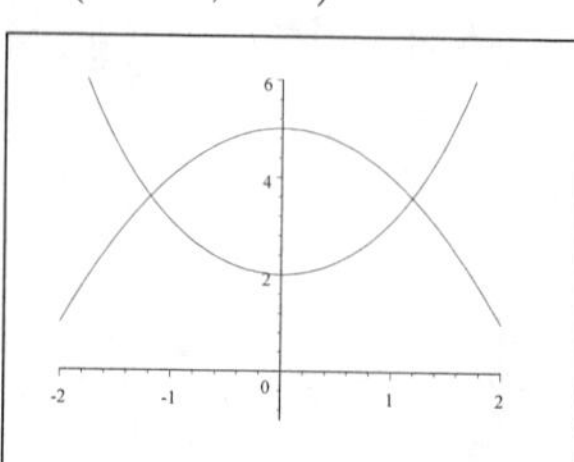

**47.** Let $w$ and $l$ be the lengths of the sides, in cm. Then we have the system $\begin{cases} lw = 180 \\ 2l + 2w = 54 \end{cases}$ We solve the second equation for $w$ giving, $w = 27 - l$, and substitute into the first equation to get $l\,(27 - l) = 180 \Leftrightarrow l^2 - 27l + 180 = 0 \quad \Leftrightarrow$ $(l - 15)\,(l - 12) = 0 \Rightarrow l = 15$ or $l = 12$. If $l = 15$, then $w = 27 - 15 = 12$, and if $l = 12$, then $w = 27 - 12 = 15$. Therefore, the dimensions of the rectangle are 12 cm by 15 cm.

**48.** Let $b$ be the length of the base of the triangle, in feet, and $h$ be the height of the triangle, in feet. Then $\begin{cases} \frac{1}{2}bh = 84 \\ b^2 + h^2 = 25^2 = 625 \end{cases}$ The first equation gives $b = \dfrac{168}{h}$. By substitution, $\left(\dfrac{168}{h}\right)^2 + h^2 = 625 \Leftrightarrow h^4 - 625h^2 + 168^2 = 0 \quad \Leftrightarrow \quad \left(h^2 - 49\right)\left(h^2 - 576\right) = 0 \Rightarrow h = 7$ or $h = 24$. Thus, the lengths of the other two sides are 7 ft and 24 ft.

**49.** Let $l$ and $w$ be the length and width, respectively, of the rectangle. Then, the system of equations is $\begin{cases} 2l + 2w = 70 \\ \sqrt{l^2 + w^2} = 25 \end{cases}$ Solving the first equation for $l$, we have $l = 35 - w$, and substituting into the second gives $\sqrt{l^2 + w^2} = 25 \Leftrightarrow l^2 + w^2 = 625 \quad \Leftrightarrow \quad (35 - w)^2 + w^2 = 625 \quad \Leftrightarrow$ $1225 - 70w + w^2 + w^2 = 625 \Leftrightarrow 2w^2 - 70w + 600 = 0 \quad \Leftrightarrow \quad (w - 15)\,(w - 20) = 0 \Rightarrow w = 15$ or $w = 20$. So the dimensions of the rectangle are 15 and 20.

**50.** Let $w$ be the width and $l$ be the length of the rectangle, in inches. From the figure, the diagonals of the rectangle are simply diameters of the circle. Then, $\begin{cases} wl = 160 \\ w^2 + l^2 = 20^2 = 400 \end{cases} \quad \Leftrightarrow \quad w = \dfrac{160}{l}$. By substitution, $\dfrac{160^2}{l^2} + l^2 = 400 \quad \Leftrightarrow$ $l^4 - 400l^2 + 160^2 = 0 \Leftrightarrow \left(l^2 - 80\right)\left(l^2 - 320\right) = 0 \Rightarrow l = \sqrt{80} = 4\sqrt{5}$ or $l = \sqrt{320} = 8\sqrt{5}$. Therefore, the dimensions of the rectangle are $4\sqrt{5}$ in. and $8\sqrt{5}$ in..

**51.** At the points where the rocket path and the hillside meet, we have $\begin{cases} y = \frac{1}{2}x \\ y = -x^2 + 401x \end{cases}$ Substituting for $y$ in the second equation gives $\frac{1}{2}x = -x^2 + 401x \quad \Leftrightarrow \quad x^2 - \frac{801}{2}x = 0 \Leftrightarrow x\left(x - \frac{801}{2}\right) = 0 \Rightarrow x = 0,\ x = \frac{801}{2}$. When $x = 0$, the rocket has not left the pad. When $x = \frac{801}{2}$, then $y = \frac{1}{2}\left(\frac{801}{2}\right) = \frac{801}{4}$. So the rocket lands at the point $\left(\frac{801}{2}, \frac{801}{4}\right)$. The distance from the base of the hill is $\sqrt{\left(\frac{801}{2}\right)^2 + \left(\frac{801}{4}\right)^2} \approx 447.77$ meters.

**52.** Let $x$ be the circumference and $y$ be length of the stove pipe. Using the circumference we can determine the radius, $2\pi r = x \Leftrightarrow r = \dfrac{x}{2\pi}$. Thus the volume is $\pi\left(\dfrac{x}{2\pi}\right)^2 y = \dfrac{1}{4\pi}x^2y$. So the system is given by $\begin{cases} xy = 1200 \\ \dfrac{1}{4\pi}x^2y = 600 \end{cases}$ Substituting for $xy$ in the second equation gives $\dfrac{1}{4\pi}x^2y = \dfrac{1}{4\pi}x\,(xy) = \dfrac{1}{4\pi}x\,(1200) = 600 \quad \Leftrightarrow$ $x = 2\pi$. So $y = \dfrac{1200}{x} = \dfrac{1200}{2\pi} = \dfrac{600}{\pi}$. Thus the dimensions of the sheet metal are $2\pi \approx 6.3$ in and $\dfrac{600}{\pi} \approx 191.0$ in.

**53.** The point $P$ is at an intersection of the circle of radius 26 centered at $A\,(22,32)$ and the circle of radius 20 centered at $B\,(28,20)$. We have the system

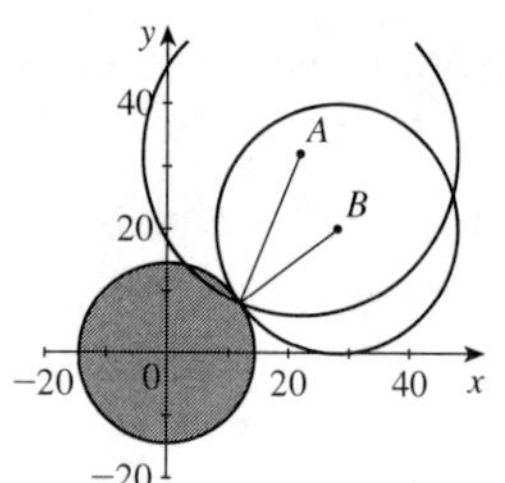


$$\begin{cases} (x-22)^2+(y-32)^2=26^2 \\ (x-28)^2+(y-20)^2=20^2 \end{cases} \Leftrightarrow$$

$$\begin{cases} x^2-44x+484+y^2-64y+1024=676 \\ x^2-56x+784+y^2-40y+400=400 \end{cases} \Leftrightarrow$$

$\begin{cases} x^2-44x+y^2-64y=-832 \\ x^2-56x+y^2-40y=-784 \end{cases}$ Subtracting the two equations, we get $12x-24y=-48 \Leftrightarrow x-2y=-4$, which is the equation of a line. Solving for $x$, we have $x=2y-4$. Substituting into the first equation gives $(2y-4)^2-44(2y-4)+y^2-64y=-832 \quad\Leftrightarrow\quad 4y^2-16y+16-88y+176+y^2-64y=-832$ $\Leftrightarrow \quad 5y^2-168y+192=-832 \quad\Leftrightarrow\quad 5y^2-168y+1024=0$. Using the quadratic formula, we have $y=\frac{168\pm\sqrt{168^2-4(5)(1024)}}{2(5)}=\frac{168\pm\sqrt{7744}}{10}=\frac{168\pm 88}{10} \quad\Leftrightarrow\quad y=8$ or $y=25.60$. Since the $y$-coordinate of the point $P$ must be less than that of point $A$, we have $y=8$. Then $x=2(8)-4=12$. So the coordinates of $P$ are $(12,8)$. To solve graphically, we must solve each equation for $y$. This gives $(x-22)^2+(y-32)^2=26^2$ $\Leftrightarrow (y-32)^2=26^2-(x-22)^2 \Rightarrow y-32=\pm\sqrt{676-(x-22)^2} \Leftrightarrow y=32\pm\sqrt{676-(x-22)^2}$. We use the function $y=32-\sqrt{676-(x-22)^2}$ because the intersection we at interested in is below the point $A$. Likewise, solving the second equation for $y$, we would get the function $y=20-\sqrt{400-(x-28)^2}$. In a three-dimensional situation, you would need a minimum of three satellites, since a point on the earth can be uniquely specified as the intersection of three spheres centered at the satellites.

**54.** The graphs of $y=x^2$ and $y=x+k$ for various values of $k$ are shown. If we solve the system $\begin{cases} y=x^2 \\ y=x+k \end{cases}$ we get $x^2-x-k=0$. Using the quadratic formula, we have $x=\dfrac{-1\pm\sqrt{1+4k}}{2}$. So there is no solution if $\sqrt{1+4k}$ is undefined, that is, if $1+4k<0 \quad\Leftrightarrow\quad k<-\frac{1}{4}$. There is be exactly one solution if $1+4k=0$ $\Leftrightarrow \quad k<-\frac{1}{4}$, and there are two solutions if $1+4k>0 \quad\Leftrightarrow\quad k>-\frac{1}{4}$.

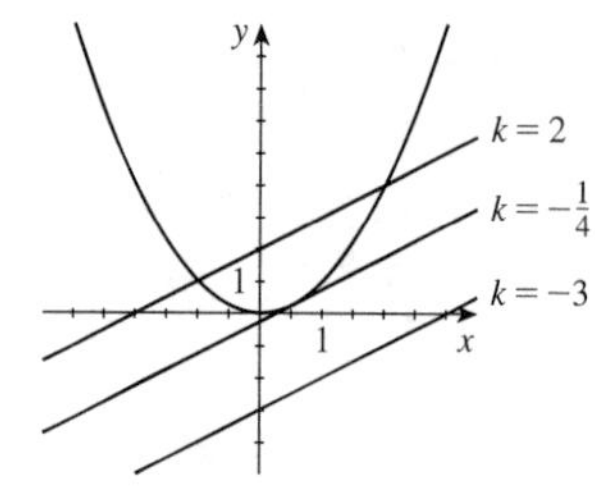


**55. (a)** $\begin{cases} \log x+\log y=\frac{3}{2} \\ 2\log x-\log y=0 \end{cases}$ Adding the two equations gives $3\log x=\frac{3}{2} \Leftrightarrow \log x=\frac{1}{2} \quad\Leftrightarrow\quad x=\sqrt{10}$. Substituting into the second equation we get $2\log 10^{1/2}-\log y=0 \Leftrightarrow \log 10-\log y=0 \quad\Leftrightarrow\quad \log y=1 \Leftrightarrow y=10$. Thus, the solution is $(\sqrt{10},10)$.

**(b)** $\begin{cases} 2^x+2^y=10 \\ 4^x+4^y=68 \end{cases} \Leftrightarrow \begin{cases} 2^x+2^y=10 \\ 2^{2x}+2^{2y}=68 \end{cases}$ If we let $u=2^x$ and $v=2^y$, the system becomes $\begin{cases} u+v=10 \\ u^2+v^2=68 \end{cases}$ Solving the first equation for $u$, and substituting this into the second equation gives $u+v=10 \quad\Leftrightarrow\quad u=10-v$, so $(10-v)^2+v^2=68 \quad\Leftrightarrow\quad 100-20v+v^2+v^2=68 \quad\Leftrightarrow\quad v^2-10v+16=0 \Leftrightarrow (v-8)(v-2)=0 \Rightarrow v=2$ or $v=8$. If $v=2$, then $u=8$, and so $y=1$ and $x=3$. If $v=8$, then $u=2$, and so $y=3$ and $x=1$. Thus, the solutions are $(1,3)$ and $(3,1)$.

**(c)** $\begin{cases} x - y = 3 \\ x^3 - y^3 = 387 \end{cases}$ Solving the first equation for $x$ gives $x = 3 + y$ and using the hint, $x^3 - y^3 = 387 \Leftrightarrow (x - y)(x^2 + xy + y^2) = 387$. Next, substituting for $x$, we get $3\left[(3+y)^2 + y(3+y) + y^2\right] = 387 \Leftrightarrow 9 + 6y + y^2 + 3y + y^2 + y^2 = 129 \Leftrightarrow 3y^2 + 9y + 9 = 129 \Leftrightarrow$ $(y+8)(y-5) = 0 \Rightarrow y = -8$ or $y = 5$. If $y = -8$, then $x = 3 + (-8) = -5$, and if $y = 5$, then $x = 3 + 5 = 8$. Thus the solutions are $(-5, -8)$ and $(8, 5)$.

**(d)** $\begin{cases} x^2 + xy = 1 \\ xy + y^2 = 3 \end{cases}$ Adding the equations gives $x^2 + xy + xy + y^2 = 4 \Leftrightarrow x^2 + 2xy + y^2 = 4 \Leftrightarrow$ $(x+y)^2 = 4 \Rightarrow x + y = \pm 2$. If $x + y = 2$, then from the first equation we get $x(x+y) = 1 \Rightarrow x \cdot 2 = 1 \Rightarrow x = \frac{1}{2}$, and so $y = 2 - \frac{1}{2} = \frac{3}{2}$. If $x + y = -2$, then from the first equation we get $x(x+y) = 1 \Rightarrow$ $x \cdot (-2) = 1 \Rightarrow x = -\frac{1}{2}$, and so $y = -2 - \left(-\frac{1}{2}\right) = -\frac{3}{2}$. Thus the solutions are $\left(\frac{1}{2}, \frac{3}{2}\right)$ and $\left(-\frac{1}{2}, -\frac{3}{2}\right)$.

# 10.2 Systems of Linear Equations in Two Variables

**1.** $\begin{cases} x + y = 4 \\ 2x - y = 2 \end{cases}$

The solution is $x = 2$, $y = 2$.

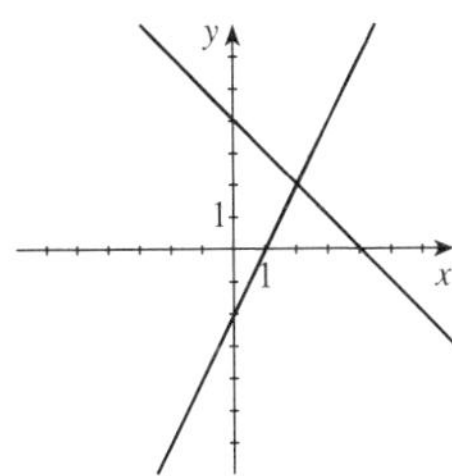


**2.** $\begin{cases} 2x + y = 11 \\ x - 2y = 4 \end{cases}$

The solution is $x = 5.2$, $y = 0.6$.

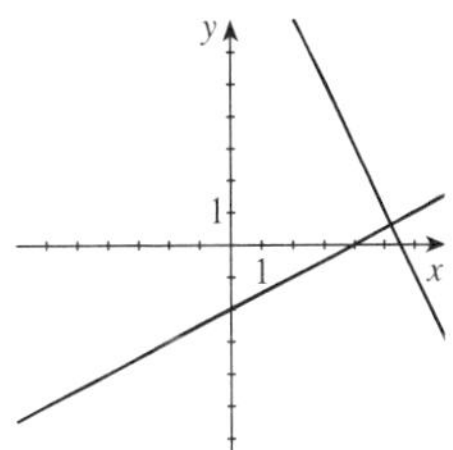


**3.** $\begin{cases} 2x - 3y = 12 \\ -x + \frac{3}{2}y = 4 \end{cases}$

The lines are parallel, so there is no intersection and hence no solution.

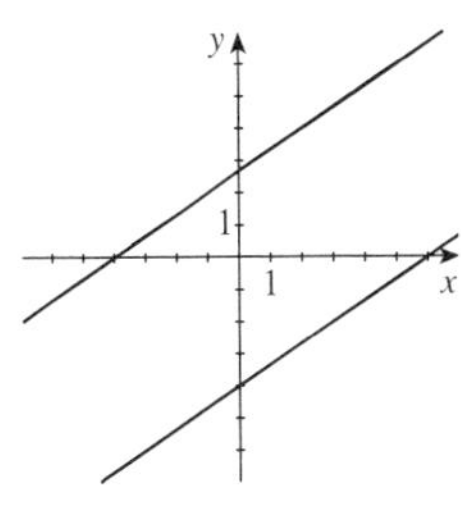


**4.** $\begin{cases} 2x + 6y = 0 \\ -3x - 9y = 18 \end{cases}$

The lines are parallel, so there is no intersection and hence no solution.

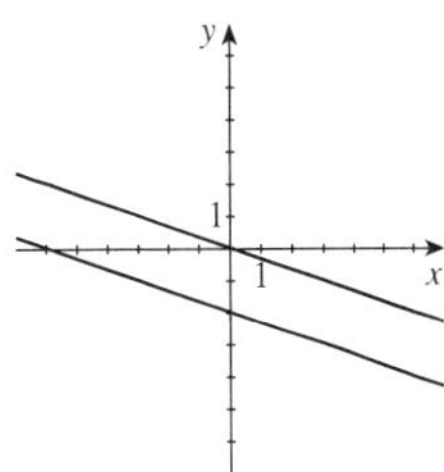

**5.** $\left\{\begin{array}{rr|r} -x + \frac{1}{2}y & = & -5 \\ 2x - \quad y & = & 10 \end{array}\right.$

There are infinitely many solutions.

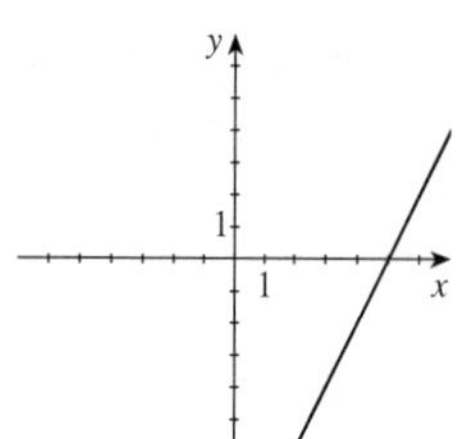


**6.** $\left\{\begin{array}{rl} 12x + 15y = & -18 \\ 2x + \frac{5}{2}y = & -3 \end{array}\right.$

There are infinitely many solutions. The lines are the same.

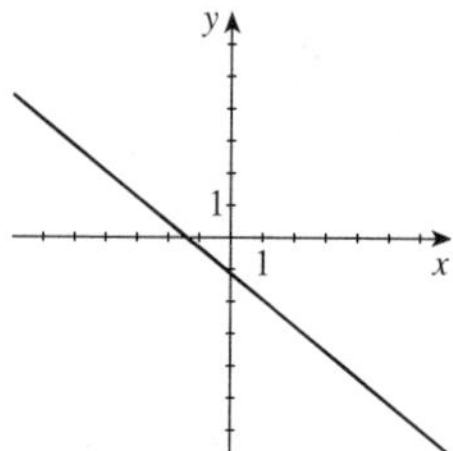


**7.** $\left\{\begin{array}{r} x + y = 4 \\ -x + y = 0 \end{array}\right.$ Adding the two equations gives $2y = 4 \Leftrightarrow y = 2$. Substituting for $y$ in the first equation gives $x + 2 = 4 \Leftrightarrow x = 2$. Hence, the solution is $(2, 2)$.

**8.** $\left\{\begin{array}{r} x - \quad y = 3 \\ x + 3y = 7 \end{array}\right.$ Subtracting the first equation from the second equation gives $4y = 4 \Leftrightarrow y = 1$. Substituting, we have $x - 1 = 3 \Leftrightarrow x = 4$. Hence, the solution is $(4, 1)$.

**9.** $\left\{\begin{array}{r} 2x - 3y = 9 \\ 4x + 3y = 9 \end{array}\right.$ Adding the two equations gives $6x = 18 \Leftrightarrow x = 3$. Substituting for $x$ in the second equation gives $4(3) + 3y = 9 \Leftrightarrow 12 + 3y = 9 \Leftrightarrow 3y = -3 \Leftrightarrow x = -1$. Hence, the solution is $(3, -1)$.

**10.** $\left\{\begin{array}{r} 3x + 2y = 0 \\ -x - 2y = 8 \end{array}\right.$ Adding the two equations gives $2x = 8 \Leftrightarrow x = 4$. Substituting for $x$ in the second equation gives $3(4) + 2y = 0 \Leftrightarrow 12 + 2y = 0 \Leftrightarrow y = -6$. Hence, the solution is $(4, -6)$.

**11.** $\left\{\begin{array}{r} x + 3y = 5 \\ 2x - \quad y = 3 \end{array}\right.$ Solving the first equation for $x$ gives $x = -3y + 5$. Substituting for $x$ in the second equation gives $2(-3y + 5) - y = 3 \Leftrightarrow -6y + 10 - y = 3 \Leftrightarrow -7y = -7 \Leftrightarrow y = 1$. Then $x = -3(1) + 5 = 2$. Hence, the solution is $(2, 1)$.

**12.** $\left\{\begin{array}{r} x + \quad y = \quad 7 \\ 2x - 3y = -1 \end{array}\right.$ Adding 3 times the first equation to the second equation gives $5x = 20 \Leftrightarrow x = 4$. So $4 + y = 7 \Leftrightarrow y = 3$, and the solution is $(4, 3)$.

**13.** $-x + y = 2 \Leftrightarrow y = x + 2$. Substituting for $y$ into $4x - 3y = -3$ gives $4x - 3(x + 2) = -3 \Leftrightarrow 4x - 3x - 6 = -3 \Leftrightarrow x = 3$, and so $y = (3) + 2 = 5$. Hence, the solution is $(3, 5)$.

**14.** $9x - y = -6 \Leftrightarrow y = 9x + 6$. Substituting for $y$ into $4x - 3y = 28$ gives $4x - 3(9x + 6) = 28 \Leftrightarrow -23x = 46 \Leftrightarrow x = -2$, and so $y = 9(-2) + 6 = -12$. Thus, the solution is $(-2, -12)$.

**15.** $x + 2y = 7 \Leftrightarrow x = 7 - 2y$. Substituting for $x$ into $5x - y = 2$ gives $5(7 - 2y) - y = 2 \Leftrightarrow 35 - 10y - y = 2 \Leftrightarrow -11y = -33 \Leftrightarrow y = 3$, and so $x = 7 - 2(3) = 1$. Hence, the solution is $(1, 3)$.

**16.** $-4x + 12y = 0 \Leftrightarrow x = 3y$. Substituting for $x$ into $12x + 4y = 160$ gives $12(3y) + 4y = 160 \Leftrightarrow 40y = 160 \Leftrightarrow y = 4$, and so $x = 3(4) = 12$. Therefore, the solution is $(12, 4)$.

**17.** $\frac{1}{2}x + \frac{1}{3}y = 2 \quad \Leftrightarrow \quad x + \frac{2}{3}y = 4 \quad \Leftrightarrow \quad x = 4 - \frac{2}{3}y$. Substituting for $x$ into $\frac{1}{5}x - \frac{2}{3}y = 8$ gives $\frac{1}{5}\left(4 - \frac{2}{3}y\right) - \frac{2}{3}y = 8$ $\Leftrightarrow \quad \frac{4}{5} - \frac{2}{15}y - \frac{10}{15}y = 8 \Leftrightarrow 12 - 2y - 10y = 120 \quad \Leftrightarrow \quad y = -9$, and so $x = 4 - \frac{2}{3}(-9) = 10$. Hence, the solution is $(10, -9)$.

**18.** $0.2x - 0.2y = -1.8 \quad \Leftrightarrow \quad x = y - 9$. Substituting for $x$ into $-0.3x + 0.5y = 3.3$ gives $-0.3(y - 9) + 0.5y = 3.3 \Leftrightarrow 0.2y = 0.6 \quad \Leftrightarrow \quad y = 3$, and so $x = (3) - 9 = -6$. Hence, the solution is $(-6, 3)$.

**19.** Adding twice the first equation to the second gives $0 = 32$, which is false. Thus, the system has no solution.

**20.** $\begin{cases} 4x + 2y = 16 \\ x - 5y = 70 \end{cases}$ Adding the first equation to $-4$ times the second equation gives $22y = -264 \quad \Leftrightarrow \quad y = -12$, so $4x + 2(-12) = 16 \quad \Leftrightarrow \quad x = 10$, and the solution is $(10, -12)$.

**21.** $\begin{cases} x + 4y = 8 \\ 3x + 12y = 2 \end{cases}$ Adding $-3$ times the first equation to the second equation gives $0 = -22$, which is never true. Thus, the system has no solution.

**22.** $\begin{cases} -3x + 5y = 2 \\ 9x - 15y = 6 \end{cases}$ Adding 3 times the first equation to the second equation gives $0 = 12$, which is false. Therefore, there is no solution to this system.

**23.** $\begin{cases} 2x - 6y = 10 \\ -3x + 9y = -15 \end{cases}$ Adding 3 times the first equation to 2 times the second equation gives $0 = 0$. Writing the equation in slope-intercept form, we have $2x - 6y = 10 \quad \Leftrightarrow \quad -6y = -2x + 10 \quad \Leftrightarrow \quad y = \frac{1}{3}x - \frac{5}{3}$, so the solutions are all pairs of the form $\left(x, \frac{1}{3}x - \frac{5}{3}\right)$ where $x$ is a real number.

**24.** $\begin{cases} 2x - 3y = -8 \\ 14x - 21y = 3 \end{cases}$ Adding 7 times the first equation to $-1$ times the second equation gives $0 = -59$, which is false. Therefore, there is no solution to this system.

**25.** $\begin{cases} 6x + 4y = 12 \\ 9x + 6y = 18 \end{cases}$ Adding 3 times the first equation to $-2$ times the second equation gives $0 = 0$. Writing the equation in slope-intercept form, we have $6x + 4y = 12 \quad \Leftrightarrow \quad 4y = -6x + 12 \quad \Leftrightarrow \quad y = -\frac{3}{2}x + 3$, so the solutions are all pairs of the form $\left(x, -\frac{3}{2}x + 3\right)$ where $x$ is a real number.

**26.** $\begin{cases} 25x - 75y = 100 \\ -10x + 30y = -40 \end{cases}$ Adding $\frac{1}{25}$ times the first equation to $\frac{1}{10}$ times the second equation gives $0 = 0$, which is always true. We now put the equation in slope-intercept form. We have $x - 3y = 4 \Leftrightarrow -3y = -x + 4 \quad \Leftrightarrow \quad y = \frac{1}{3}x - \frac{4}{3}$, so the solutions are all pairs of the form $\left(x, \frac{1}{3}x - \frac{4}{3}\right)$ where $x$ is a real number.

**27.** $\begin{cases} 8s - 3t = -3 \\ 5s - 2t = -1 \end{cases}$ Adding 2 times the first equation to 3 times the second equation gives $s = -3$, so $8(-3) - 3t = -3 \Leftrightarrow -24 - 3t = -3 \quad \Leftrightarrow \quad t = -7$. Thus, the solution is $(-3, -7)$.

**28.** $\begin{cases} u - 30v = -5 \\ -3u + 80v = 5 \end{cases}$ Adding 3 times the first equation to the second equation gives $-10v = -10 \quad \Leftrightarrow \quad v = 1$, so $u - 30(1) = -5 \quad \Leftrightarrow \quad u = 25$. Thus, the solution is $(u, v) = (25, 1)$.

**29.** $\begin{cases} \frac{1}{2}x + \frac{3}{5}y = 3 \\ \frac{5}{3}x + 2y = 10 \end{cases}$ Adding 10 times the first equation to $-3$ times the second equation gives $0 = 0$. Writing the equation in slope-intercept form, we have $\frac{1}{2}x + \frac{3}{5}y = 3 \quad \Leftrightarrow \quad \frac{3}{5}y = -\frac{1}{2}x + 3 \quad \Leftrightarrow \quad y = -\frac{5}{6}x + 5$, so the solutions are all pairs of the form $\left(x, -\frac{5}{6}x + 5\right)$ where $x$ is a real number.

**30.** $\begin{cases} \frac{3}{2}x - \frac{1}{3}y = \frac{1}{2} \\ 2x - \frac{1}{2}y = -\frac{1}{2} \end{cases}$ Adding $-6$ times the first equation to 4 times the second equation gives $-x = -5 \quad \Leftrightarrow \quad x = 5$. So $9(5) - 2y = 3 \quad \Leftrightarrow \quad y = 21$. Thus, the solution is $(5, 21)$.

**31.** $\begin{cases} 0.4x + 1.2y = 14 \\ 12x - 5y = 10 \end{cases}$ Adding 30 times the first equation to $-1$ times the second equation gives $41y = 410 \quad \Leftrightarrow$ $y = 10$, so $12x - 5(10) = 10 \quad \Leftrightarrow \quad 12x = 60 \quad \Leftrightarrow \quad x = 5$. Thus, the solution is $(5, 10)$.

**32.** $\begin{cases} 26x - 10y = -4 \\ -0.6x + 1.2y = 3 \end{cases}$ Adding 3 times the first equation to 25 times the second equation gives $63x = 63 \quad \Leftrightarrow$ $x = 1$, so $26(1) - 10y = -4 \quad \Leftrightarrow \quad -10y = -30 \quad \Leftrightarrow \quad y = 3$. Thus, the solution is $(1, 3)$.

**33.** $\begin{cases} \frac{1}{3}x - \frac{1}{4}y = 2 \\ -8x + 6y = 10 \end{cases}$ Adding 24 times the first equation to the second equation gives $0 = 58$, which is never true. Thus, the system has no solution.

**34.** $\begin{cases} -\frac{1}{10}x + \frac{1}{2}y = 4 \\ 2x - 10y = -80 \end{cases}$ Adding 20 times the first equation to the second equation gives $0 = 0$, which is always true. We now put the equation in slope-intercept form. We have $2x - 10y = -80 \quad \Leftrightarrow \quad -10y = -2x - 80 \Leftrightarrow y = \frac{1}{5}x + 8$, so the solutions are all pairs of the form $\left(x, \frac{1}{5}x + 8\right)$ where $x$ is a real number.

**35.** $\begin{cases} 0.21x + 3.17y = 9.51 \\ 2.35x - 1.17y = 5.89 \end{cases}$

The solution is approximately $(3.87, 2.74)$.

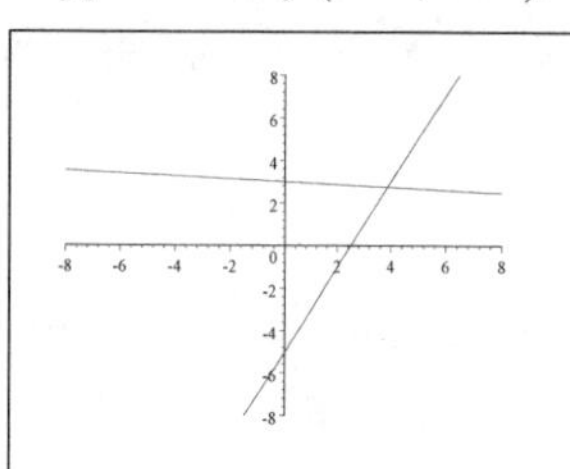

**36.** $\begin{cases} 18.72x - 14.91y = 12.33 \\ 6.21x - 12.92y = 17.82 \end{cases}$

The solution is approximately $(-0.71, -1.72)$.

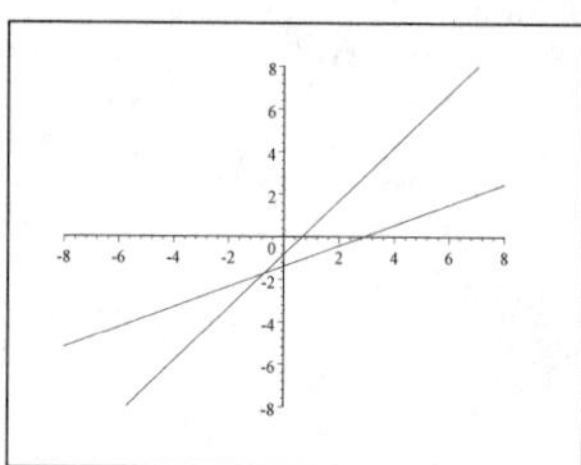

**37.** $\begin{cases} 2371x - 6552y = 13{,}591 \\ 9815x + 992y = 618{,}555 \end{cases}$

The solution is approximately $(61.00, 20.00)$.

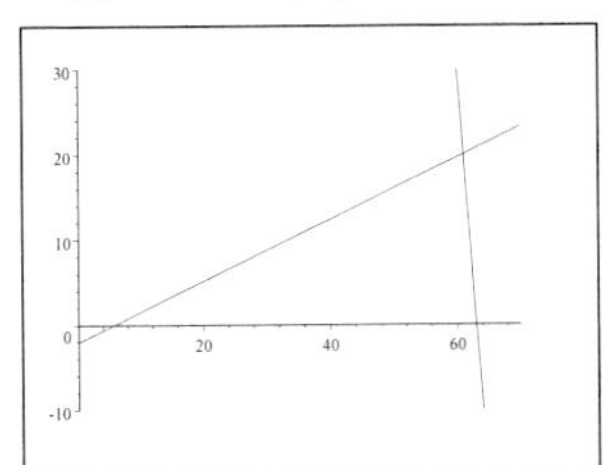

**38.** $\begin{cases} -435x + 912y = 0 \\ 132x + 455y = 994 \end{cases}$

The solution is approximately $(2.85, 1.36)$.

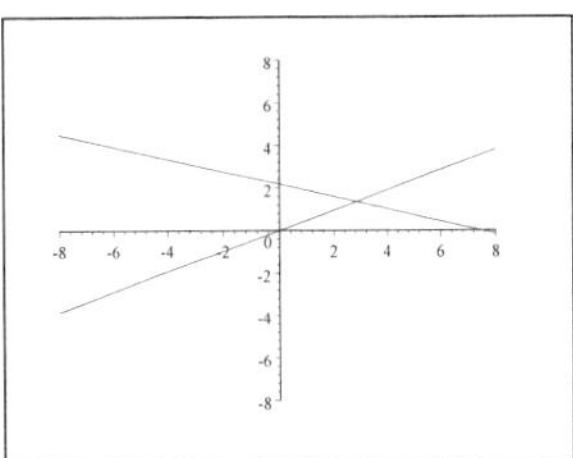

**39.** Subtracting the first equation from the second, we get $ay - y = 1 \Leftrightarrow y(a-1) = 1 \quad \Leftrightarrow \quad y = \dfrac{1}{a-1}$, $a \neq 1$. So $x + \left(\dfrac{1}{a-1}\right) = 0 \quad \Leftrightarrow \quad x = \dfrac{1}{1-a} = -\dfrac{1}{a-1}$. Thus, the solution is $\left(-\dfrac{1}{a-1}, \dfrac{1}{a-1}\right)$.

**40.** Subtracting the first equation from $a$ times the second, we get $(a-b)\,y = a \quad \Leftrightarrow \quad y = \dfrac{a}{a-b}$, $a \neq b$. So $x + \left(\dfrac{a}{a-b}\right) = 1 \quad \Leftrightarrow \quad x = \dfrac{b}{b-a}$. Hence, the solution is $\left(\dfrac{b}{b-a}, \dfrac{a}{a-b}\right)$.

**41.** Subtracting $b$ times the first equation from $a$ times the second, we get $\left(a^2 - b^2\right) y = a - b \quad \Leftrightarrow \quad y = \dfrac{a-b}{a^2-b^2} = \dfrac{1}{a+b}$, $a^2 - b^2 \neq 0$. So $ax + \dfrac{b}{a+b} = 1 \quad \Leftrightarrow \quad ax = \dfrac{a}{a+b} \quad \Leftrightarrow \quad x = \dfrac{1}{a+b}$. Thus, the solution is $\left(\dfrac{1}{a+b}, \dfrac{1}{a+b}\right)$.

**42.** Subtracting $a$ times the first equation from the second, we get $b\,(b-a)\,y = 1 \quad \Leftrightarrow \quad y = \dfrac{1}{b\,(b-a)}$. So $ax + \dfrac{b}{b\,(b-a)} = 0 \quad \Leftrightarrow \quad x = -\dfrac{1}{a\,(b-a)}$. Hence, the solution is $\left(-\dfrac{1}{a\,(b-a)}, \dfrac{1}{b\,(b-a)}\right) = \left(\dfrac{1}{a^2-ab}, \dfrac{1}{b^2-ab}\right)$.

**43.** Let the two numbers be $x$ and $y$. Then $\begin{cases} x + y = 34 \\ x - y = 10 \end{cases}$ Adding these two equations gives $2x = 44 \Leftrightarrow x = 22$. So $22 + y = 34 \quad \Leftrightarrow \quad y = 12$. Therefore, the two numbers are 22 and 12.

**44.** Let $x$ be the larger number and $y$ be the other number. This gives $\begin{cases} x + y = 2\,(x-y) \\ x = 6 + 2y \end{cases} \Rightarrow$ $\begin{cases} -x + 3y = 0 \\ x - 2y = 6 \end{cases}$ Adding these two equations gives $y = 6$, so $x = 6 + 2\,(6) = 18$. Therefore, the two numbers are 18 and 6.

**45.** Let $d$ be the number of dimes and $q$ be the number of quarters. This gives $\begin{cases} d + q = 14 \\ 0.10d + 0.25q = 2.75 \end{cases}$ Subtracting the first equation from 10 times the second gives $1.5q = 13.5 \quad \Leftrightarrow \quad q = 9$. So $d + 9 = 14 \Leftrightarrow d = 5$. Thus, the number of dimes is 5 and the number of quarters is 9.

**46.** Let $c$ be the number of children and $a$ be the number of adults. This gives $\begin{cases} c + a = 2200 \\ 1.50c + 4.00a = 5050 \end{cases}$ Subtracting 3 times the first equation from 2 times the second gives $\Rightarrow 5a = 3500 \quad \Leftrightarrow \quad a = 700$, so $c + 700 = 2200 \quad \Leftrightarrow \quad c = 1500$. Therefore, the number of children admitted was 1500 and the number of adults was 700.

**47.** Let $x$ be the speed of the plane in still air and $y$ be the speed of the wind. This gives $\begin{cases} 2x - \phantom{1.}2y = 180 \\ 1.2x + 1.2y = 180 \end{cases}$ Subtracting 6 times the first equation from 10 times the second gives $24x = 2880 \quad \Leftrightarrow \quad x = 120$, so $2(120) - 2y = 180 \quad \Leftrightarrow$ $-2y = -60 \quad \Leftrightarrow \quad y = 30$. Therefore, the speed of the plane is 120 mi/h and the wind speed is 30 mi/h.

**48.** Let $x$ be speed of the boat in still water and $y$ be speed of the river flow. $\begin{cases} \text{Downriver:} \quad x + y = 20 \\ \text{Upriver: } 2.5x - 2.5y = 20 \end{cases}$ Adding 5 times the first equation to 2 times the second gives $10x = 140 \quad \Leftrightarrow \quad x = 14$, so $14 + y = 20 \Leftrightarrow y = 6$. Therefore, the boat's speed is 14 mph and the current in the river flows at 6 mph.

**49.** Let $x$ be the cycling speed and $y$ be the running speed. (Remember to divide by 60 to convert minutes to decimal hours.) We have $\begin{cases} 0.5x + 0.5y = 12.5 \\ 0.75x + 0.2y = \phantom{1}16 \end{cases}$ Subtracting 2 times the first equation from 5 times the second, we get $2.75x = 55$ $\Leftrightarrow \quad x = 20$, so $20 + y = 25 \quad \Leftrightarrow \quad y = 5$. Thus, the cycling speed is 20 mi/h and the running speed is 5 mi/h.

**50.** Let $x$ and $y$ be the number of milliliters of the two brine solutions. $\begin{cases} \text{Quantity:} \quad x + y = 1000 \\ \text{Concentrations: } 0.05x + 0.20y = 0.14 \end{cases}$ Subtracting the first equation from 20 times the second gives $3y = 1800$ $\Leftrightarrow \quad y = 600$, so $x + 600 = 1000 \Leftrightarrow x = 400$. Therefore, 400 milliliters of the 5% solution and 600 milliliters of the 20% solution should be mixed.

**51.** Let $a$ and $b$ be the number of grams of food A and food B. Then $\begin{cases} 0.12a + 0.20b = \phantom{22,0}32 \\ 100a + 50b = 22{,}000 \end{cases}$ Subtracting 250 times the first equation from the second, we get $70a = 14{,}000 \quad \Leftrightarrow \quad a = 200$, so $0.12(200) + 0.20b = 32 \quad \Leftrightarrow \quad 0.20b = 8$ $\Leftrightarrow \quad b = 40$. Thus, she should use 200 grams of food A and 40 grams of food B.

**52.** Let $x$ be the number of pounds of Kenyan coffee and $y$ be the number of pounds of Sri Lankan coffee. This gives $\begin{cases} 3.50x + 5.60y = 11.55 \\ x + y = 3 \end{cases}$ Adding $-10$ times the first equation and $-35$ times the second gives $21y = 10.5 \quad \Leftrightarrow$ $y = 0.5$, so $x + (2.5) = 3 \quad \Leftrightarrow \quad x = 2.5$. Thus, 2.5 pounds of Kenyan coffee and 0.5 pounds of Sri Lankan coffee should be mixed.

**53.** Let $x$ and $y$ be the sulfuric acid concentrations in the first and second containers. $\begin{cases} 300x + 600y = 900(0.15) \\ 100x + 500y = 600(0.125) \end{cases}$ Subtracting the first equation from 3 times the second gives $900y = 90 \quad \Leftrightarrow$ $y = 0.10$, so $100x + 500(0.10) = 75 \quad \Leftrightarrow \quad x = 0.25$. Thus, the concentrations of sulfuric acid are 25% in the first container and 10% in the second.

**54.** Let $x$ be the amount invested at 5% and $y$ the amount invested at 8%. $\begin{cases} \text{Total invested:} \quad x + y = 20{,}000 \\ \text{Interest earned: } 0.05x + 0.08y = 1180 \end{cases}$ Subtracting 5 times the first equation from 100 times the second gives $3y = 18{,}000 \quad \Leftrightarrow \quad y = 6{,}000$, so $x + 6{,}000 = 20{,}000 \quad \Leftrightarrow \quad x = 14{,}000$. She invests \$14,000 at 5% and \$6,000 at 8%.

**55.** Let $x$ be the amount invested at 6% and $y$ the amount invested at 10%. The ratio of the amounts invested gives $x = 2y$. Then the interest earned is $0.06x + 0.10y = 3520 \quad \Leftrightarrow \quad 6x + 10y = 352{,}000$. Substituting gives $6(2y) + 10y = 352{,}000 \Leftrightarrow 22y = 352{,}000 \quad \Leftrightarrow \quad y = 16{,}000$. Then $x = 2(16{,}000) = 32{,}000$. Thus, he invests \$32,000 at 6% and \$16,000 at 10%.

**56.** Let $x$ be the length of time John drives and $y$ be the length of time Mary drives. Then $y = x + 0.25$, so $-x + y = 0.25$, and multiplying by 40, we get $-40x + 40y = 10$. Comparing the distances, we get $60x = 40y + 35$, or $60x - 40y = 35$.

This gives the system $\begin{cases} -40x + 40y = 10 \\ 60x - 40y = 35 \end{cases}$ Adding, we get $20x = 45 \Leftrightarrow x = 2.25$, so $y = 2.25 + 0.25 = 2.5$.

Thus, John drives for $2\frac{1}{4}$ hours and Mary drives for $2\frac{1}{2}$ hours.

**57.** Let $x$ be the tens digit and $y$ be the ones digit of the number. $\begin{cases} x + y = 7 \\ 10y + x = 27 + 10x + y \end{cases}$ Adding 9 times the first equation to the second gives $18x = 36 \Leftrightarrow x = 2$, so $2 + y = 7 \Leftrightarrow y = 5$. Thus, the number is 25.

**58.** First let us find the intersection point of the two lines. The $y$-coordinate of the intersection point is the height of the triangle.

We have $\begin{cases} y = 2x - 4 \\ y = -4x + 20 \end{cases}$ Adding 2 times the first equation to the second gives $3y = 12$, so the triangle has height 4.

Furthermore, $y = 2x - 4$ intersects the $x$-axis at $x = 2$, and $y = -4x + 20$ intersects the $x$-axis at $x = 5$. Thus the base has length $5 - 2 = 3$. Therefore, the area of the triangle is $A = \frac{1}{2}bh = \frac{1}{2} \cdot 3 \cdot 4 = 6$.

**59.** $n = 5$, so $\sum_{k=1}^{n} x_k = 1 + 2 + 3 + 5 + 7 = 18$,

$\sum_{k=1}^{n} y_k = 3 + 5 + 6 + 6 + 9 = 29$,

$\sum_{k=1}^{n} x_k y_k = 1(3) + 2(5) + 3(6) + 5(6) + 7(9) = 124$, and

$\sum_{k=1}^{n} x_k^2 = 1^2 + 2^2 + 3^2 + 5^2 + 7^2 = 88$. Thus we get the system

$\begin{cases} 18a + 5b = 29 \\ 88a + 18b = 124 \end{cases}$ Subtracting 18 times the first equation from 5 times the second, we get $116a = 98 \Leftrightarrow a \approx 0.845$. Then

$b = \frac{1}{5}[-18(0.845) + 29] \approx 2.758$. So the regression line is $y = 0.845x + 2.758$.

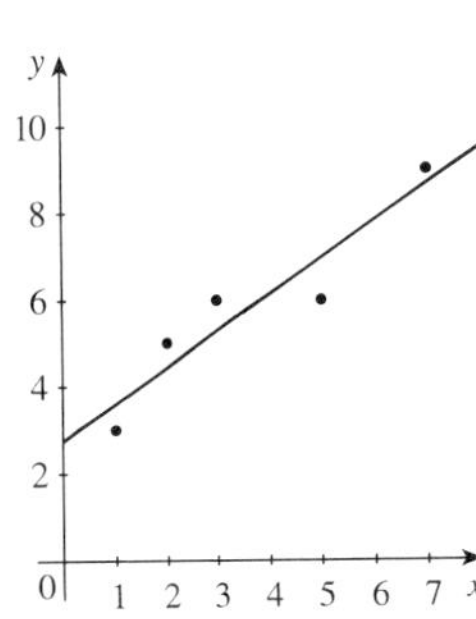


## 10.3 Systems of Linear Equations in Several Variables

**1.** The equation $6x - \sqrt{3}y + \frac{1}{2}z = 0$ is linear.

**2.** The equation $x^2 + y^2 + z^2 = 4$ is not linear, since it contains squares of variables.

**3.** The system $\begin{cases} xy - 3y + z = 5 \\ x - y^2 + 5z = 0 \\ 2x + yz = 3 \end{cases}$ is not a linear system, since the first equation contains a product of variables. In fact both the second and the third equation are not linear.

**4.** The system $\begin{cases} x - 2y + 3z = 10 \\ 2x + 5y = 2 \\ y + 2z = 4 \end{cases}$ is linear.

**5.** $\begin{cases} x - 2y + 4z = 3 \\ y + 2z = 7 \\ z = 2 \end{cases}$ Substituting $z = 2$ into the second equation gives $y + 2(2) = 7 \Leftrightarrow y = 3$. Substituting $z = 2$ and $y = 3$ into the first equation gives $x - 2(3) + 4(2) = 3 \Leftrightarrow x = 1$. Thus, the solution is $(1, 3, 2)$.

**6.** $\begin{cases} x + y - 3z = 8 \\ y - 3z = 5 \\ z = -1 \end{cases}$ Substituting $z = -1$ into the second equation gives $y - 3(-1) = 5 \Leftrightarrow y = 2$. Substituting $z = -1$ and $y = 2$ into the first equation gives $x + 2 - 3(-1) = 8 \Leftrightarrow x = 3$. Thus, the solution is $(3, 2, -1)$.

**7.** $\begin{cases} x + 2y + z = 7 \\ -y + 3z = 9 \\ 2z = 6 \end{cases}$ Solving we get $2z = 6 \Leftrightarrow z = 3$. Substituting $z = 3$ into the second equation gives $-y + 3(3) = 9 \Leftrightarrow y = 0$. Substituting $z = 3$ and $y = 0$ into the first equation gives $x + 2(0) + 3 = 7 \Leftrightarrow x = 4$. Thus, the solution is $(4, 0, 3)$.

**8.** $\begin{cases} x - 2y + 3z = 10 \\ 2y - z = 2 \\ 3z = 12 \end{cases}$ Solving we get $3z = 12 \Leftrightarrow z = 4$. Substituting $z = 4$ into the second equation gives $2y - 4 = 2 \Leftrightarrow y = 3$. Substituting $z = 4$ and $y = 3$ into the first equation gives $x - 2(3) + 3(4) = 10 \Leftrightarrow x = 4$. Thus, the solution is $(4, 3, 4)$.

**9.** $\begin{cases} 2x - y + 6z = 5 \\ y + 4z = 0 \\ -2z = 1 \end{cases}$ Solving we get $-2z = 1 \Leftrightarrow z = -\frac{1}{2}$. Substituting $z = -\frac{1}{2}$ into the second equation gives $y + 4\left(-\frac{1}{2}\right) = 0 \Leftrightarrow y = 2$. Substituting $z = -\frac{1}{2}$ and $y = 2$ into the first equation gives $2x - (2) + 6\left(-\frac{1}{2}\right) = 5 \Leftrightarrow x = 5$. Thus, the solution is $\left(5, 2, -\frac{1}{2}\right)$.

**10.** $\begin{cases} 4x + 3z = 10 \\ 2y - z = -6 \\ \frac{1}{2}z = 2 \end{cases}$ Solving we get $\frac{1}{2}z = 4 \Leftrightarrow z = 8$ . Substituting $z = 8$ into the second equation gives $2y - 8 = -6 \Leftrightarrow 2y = 2 \Leftrightarrow y = 1$. Substituting $z = 8$ into the first equation gives $4x + 3(8) = 10 \Leftrightarrow 4x = -14 \Leftrightarrow x = -\frac{7}{2}$. Thus, the solution is $\left(-\frac{7}{2}, 1, 8\right)$.

**11.** $\begin{cases} x - 2y - z = 4 \\ x - y + 3z = 0 \\ 2x + y + z = 0 \end{cases}$ Subtract the first equation from the second equation: $\begin{cases} x - 2y - z = 4 \\ y + 4z = -4 \\ 2x + y + z = 0 \end{cases}$

Or, subtract $\frac{1}{2}$ times the third equation from the second equation: $\begin{cases} x - 2y - z = 4 \\ -\frac{3}{2}y + \frac{5}{2}z = 0 \\ 2x + y + z = 0 \end{cases}$

**12.** $\begin{cases} x + y - 3z = 3 \\ -2x + 3y + z = 2 \\ x - y + 2z = 0 \end{cases}$ Add 2 times the first equation to the second equation: $\begin{cases} x + y - 3z = 3 \\ 5y - 5z = 8 \\ x - y + 2z = 0 \end{cases}$

Or, add 2 times the third equation to the second equation: $\begin{cases} x + y - 3z = 3 \\ y + 5z = 2 \\ x - y + 2z = 0 \end{cases}$

**13.** $\begin{cases} 2x - y + 3z = 2 \\ x + 2y - z = 4 \\ -4x + 5y + z = 10 \end{cases}$ Add 2 times the first equation to the third equation: $\begin{cases} 2x - y + 3z = 2 \\ x + 2y - z = 4 \\ 3y + 7z = 14 \end{cases}$

Or, add 4 times the second equation to the third equation: $\begin{cases} 2x - y + 3z = 2 \\ x + 2y - z = 4 \\ 13y - 3z = 26 \end{cases}$

**14.** $\begin{cases} x - 4y + z = 3 \\ y - 3z = 10 \\ 3y - 8z = 24 \end{cases}$ Add $\frac{3}{4}$ times the first equation to the third equation: $\begin{cases} x - 4y + z = 3 \\ y - 3z = 10 \\ \frac{3}{4}x - \frac{29}{4}z = \frac{105}{4} \end{cases}$

Or, subtract 3 times the second equation from the third equation: $\begin{cases} x - 4y + z = 3 \\ y - 3z = 10 \\ z = -6 \end{cases}$

**15.** $\begin{cases} x + y + z = 4 \\ x + 3y + 3z = 10 \\ 2x + y - z = 3 \end{cases} \Leftrightarrow \begin{cases} x + y + z = 4 \\ 2y + 2z = 6 & (-1) \times \text{Eq. } 1 + \text{Eq. } 2 \\ y + 3z = 5 & 2 \times \text{Eq. } 1 + (-1) \times \text{Eq. } 3 \end{cases} \Leftrightarrow \begin{cases} x + y + z = 4 \\ y + 3z = 5 & \text{Eq. } 3 \\ 2y + 2z = 6 & \text{Eq. } 2 \end{cases}$

$\Leftrightarrow \begin{cases} x + y + z = 4 \\ y + 3z = 5 \\ -4z = -4 & (-2) \times \text{Eq. } 2 + \text{Eq. } 3 \end{cases} \Leftrightarrow$ $z = 1$ and $y + 3(1) = 8 \Leftrightarrow y = 2$. Then $x + 2 + 1 = 4 \Leftrightarrow$ $x = 1$. So the solution is $(1, 2, 1)$.

**16.** $\begin{cases} x + y + z = 0 \\ -x + 2y + 5z = 3 \\ 3x - y = 6 \end{cases} \Leftrightarrow \begin{cases} x + y + z = 0 \\ y + 2z = 1 & \frac{1}{3} \times \text{Eq. } 1 + \frac{1}{3} \times \text{Eq. } 2 \\ -4y - 3z = 6 & (-3) \times \text{Eq. } 1 + \text{Eq. } 3 \end{cases} \Leftrightarrow \begin{cases} x + y + z = 0 \\ y + 2z = 1 \\ 5z = 10 & 4 \times \text{Eq. } 2 + \text{Eq. } 3 \end{cases}$

So $z = 2$ and $y + 2(2) = 1 \Leftrightarrow y = -3$. Then $x + (-3) + 2 = 0 \Leftrightarrow x = 1$. So the solution is $(1, -3, 2)$.

**17.** $\begin{cases} x - 4z = 1 \\ 2x - y - 6z = 4 \\ 2x + 3y - 2z = 8 \end{cases} \Leftrightarrow \begin{cases} x - 4z = 1 \\ -y + 2z = 2 & (-2) \times \text{Eq. } 1 + \text{Eq. } 2 \\ 3y + 6z = 6 & (-2) \times \text{Eq. } 1 + \text{Eq. } 3 \end{cases} \Leftrightarrow \begin{cases} x - 4z = 1 \\ -y + 2z = 2 \\ 12z = 12 & 3 \times \text{Eq. } 2 + \text{Eq. } 3 \end{cases}$

So $z = 1$ and $-y + 2(1) = 2 \Leftrightarrow y = 0$. Then $x - 4(1) = 1 \Leftrightarrow x = 5$. So the solution is $(5, 0, 1)$.

**18.** $\begin{cases} x - y + 2z = 2 \\ 3x + y + 5z = 8 \\ 2x - y - 2z = -7 \end{cases} \Leftrightarrow \begin{cases} x - y + 2z = 2 \\ y - 6z = -11 & (-3) \times \text{Eq. } 1 + \text{Eq. } 3 \\ 4y - z = 2 & (-2) \times \text{Eq. } 1 + \text{Eq. } 2 \end{cases} \Leftrightarrow \begin{cases} x - y + 2z = 2 \\ y - 6z = -11 \\ 23z = 46 & (-4) \times \text{Eq. } 2 + \text{Eq. } 2 \end{cases}$

So $z = 2$ and $y - 6(2) = -11 \Leftrightarrow y = 1$. Then $x + 1 + 2 = 2 \Leftrightarrow x = -1$. So the solution is $(-1, 1, 2)$.

**19.** $\begin{cases} 2x + 4y - z = 2 \\ x + 2y - 3z = -4 \\ 3x - y + z = 1 \end{cases} \Leftrightarrow \begin{cases} x + 2y - 3z = -4 \\ 2x + 4y - z = 2 \\ 3x - y + z = 1 \end{cases} \Leftrightarrow \begin{cases} x + 2y - 3z = -4 \\ -7y + 10z = 13 \\ 5z = 10 \end{cases}$ So $z = 2$ and $-7y + 10(2) = 13 \Leftrightarrow y = 1$. Then $x + 2(1) - 3(2) = -4 \Leftrightarrow x = 0$. So the solution is $(0, 1, 2)$.

**20.** $\begin{cases} 2x + y - z = -8 \\ -x + y + z = 3 \\ -2x + 4z = 18 \end{cases} \Leftrightarrow \begin{cases} -x + y + z = 3 \\ 2x + y - z = -8 \\ -2x + 4z = 18 \end{cases} \Leftrightarrow \begin{cases} -x + y + z = 3 \\ y - z = -6 \\ 3y + z = -2 \end{cases} \Leftrightarrow \begin{cases} -x + y + z = 3 \\ y - z = -6 \\ 4z = 16 \end{cases}$

So $z = 4$ and $y - 4 = -6 \Leftrightarrow y = -2$. Then $-x + (-2) + 4 = 3 \Leftrightarrow x = -1$. So the solution is $(-1, -2, 4)$.

**21.** $\begin{cases} y - 2z = 0 \\ 2x + 3y = 2 \\ -x - 2y + z = -1 \end{cases} \Leftrightarrow \begin{cases} -x - 2y + z = -1 \\ y - 2z = 0 \\ 2x + 3y = 2 \end{cases} \Leftrightarrow \begin{cases} -x - 2y + z = -1 \\ y - 2z = 0 \\ -y + 2z = 0 \end{cases} \Leftrightarrow \begin{cases} -x - 2y + z = -1 \\ y - 2z = 0 \\ 0 = 0 \end{cases}$

The system is dependent, so when $z = t$ we solve for $y$ to get $y - 2t = 0 \Leftrightarrow y = 2t$. Then $-x - 2(2t) + t = -1 \Leftrightarrow -x - 3t = -1 \Leftrightarrow x = -3t + 1$. So the solutions are $(-3t + 1, 2t, t)$, where $t$ is any real number.

**22.** $\begin{cases} 2y + z = 3 \\ 5x + 4y + 3z = -1 \\ x - 3y = -2 \end{cases} \Leftrightarrow \begin{cases} x - 3y = -2 \\ 2y + z = 3 \\ 5x + 4y + 3z = -1 \end{cases} \Leftrightarrow \begin{cases} x - 3y = -2 \\ 2y + z = 3 \\ 19y + 3z = 9 \end{cases} \Leftrightarrow \begin{cases} x - 3y = -2 \\ 2y + z = 3 \\ -13z = -39 \end{cases}$

So $z = 3$ and $2y + 3 = 3 \Leftrightarrow y = 0$. Then $x - 3(0) = -2 \Leftrightarrow x = -2$. So the solution is $(-2, 0, 3)$.

**23.** $\begin{cases} x + 2y - z = 1 \\ 2x + 3y - 4z = -3 \\ 3x + 6y - 3z = 4 \end{cases} \Leftrightarrow \begin{cases} x + 2y - z = 1 \\ -y - 2z = -5 \\ 0 = 1 \end{cases}$ Since $0 = 1$ is false, this system is inconsistent.

**24.** $\begin{cases} -x + 2y + 5z = 4 \\ x - 2z = 0 \\ 4x - 2y - 11z = 2 \end{cases} \Leftrightarrow \begin{cases} x - 2z = 0 \\ -x + 2y + 5z = 4 \\ 4x - 2y - 11z = 2 \end{cases} \Leftrightarrow \begin{cases} x - 2z = 0 \\ 2y + 3z = 4 \\ -2y - 3z = 2 \end{cases} \Leftrightarrow \begin{cases} x - 2z = 0 \\ 2y + 3z = 4 \\ 0 = 6 \end{cases}$

Since $0 = 6$ is false, this system is inconsistent.

**25.** $\begin{cases} 2x + 3y - z = 1 \\ x + 2y = 3 \\ x + 3y + z = 4 \end{cases} \Leftrightarrow \begin{cases} x + 2y = 3 \\ 2x + 3y - z = 1 \\ x + 3y + z = 4 \end{cases} \Leftrightarrow \begin{cases} x + 2y = 3 \\ -y - z = -5 \\ y + z = 1 \end{cases} \Leftrightarrow \begin{cases} x + 2y = 3 \\ -y - z = -5 \\ 0 = -4 \end{cases}$

Since $0 = -4$ is false, this system is inconsistent.

**26.** $\begin{cases} x - 2y - 3z = 5 \\ 2x + y - z = 5 \\ 4x - 3y - 7z = 5 \end{cases} \Leftrightarrow \begin{cases} x - 2y - 3z = 5 \\ 5y + 5z = -5 \\ 5y + 5z = -15 \end{cases} \Leftrightarrow \begin{cases} x - 2y - 3z = 5 \\ 5y + 5z = -5 \\ 0 = -10 \end{cases}$

Since $0 = -10$ is false, this system is inconsistent.

**27.** $\begin{cases} x + y - z = 0 \\ x + 2y - 3z = -3 \\ 2x + 3y - 4z = -3 \end{cases} \Leftrightarrow \begin{cases} x + y - z = 0 \\ y - 2z = -3 \\ y - 2z = -3 \end{cases} \Leftrightarrow \begin{cases} x + y - z = 0 \\ y - 2z = -3 \\ 0 = 0 \end{cases}$

So $z = t$ and $y - 2t = -3 \Leftrightarrow y = 2t - 3$. Then $x + (2t - 3) - t = 0 \Leftrightarrow x = -t + 3$. So the solutions are $(-t + 3, 2t - 3, t)$, where $t$ is any real number.

**28.** $\begin{cases} x - 2y + z = 3 \\ 2x - 5y + 6z = 7 \\ 2x - 3y - 2z = 5 \end{cases} \Leftrightarrow \begin{cases} x - 2y + z = 3 \\ -y + 4z = 1 \\ y - 4z = -1 \end{cases} \Leftrightarrow \begin{cases} x - 2y + z = 3 \\ -y + 4z = 1 \\ 0 = 0 \end{cases}$

Then we have $-y + 4t = 1 \Leftrightarrow y = 4t - 1$. Substituting into the first equation, we have $x - 2(4t - 1) + t = 3 \Leftrightarrow x = 7t + 1$. So the solutions are $(7t + 1, 4t - 1, t)$, where $t$ is any real number.

**29.** $\begin{cases} x + 3y - 2z = 0 \\ 2x \qquad + 4z = 4 \\ 4x + 6y \qquad = 4 \end{cases} \Leftrightarrow \begin{cases} x + 3y - 2z = 0 \\ -6y + 8z = 4 \\ -6y + 8z = 4 \end{cases} \Leftrightarrow \begin{cases} x + 3y - 2z = 0 \\ -6y + 8z = 4 \\ 0 = 0 \end{cases}$

So $z = t$ and $-6y + 8t = 4 \Leftrightarrow -6y = -8t + 4 \Leftrightarrow y = \frac{4}{3}t - \frac{2}{3}$. Then $x + 3\left(\frac{4}{3}t - \frac{2}{3}\right) - 2t = 0 \Leftrightarrow x = -2t + 2$. So the solutions are $\left(-2t + 2, \frac{4}{3}t - \frac{2}{3}, t\right)$, where $t$ is any real number.

**30.** $\begin{cases} 2x + 4y - z = 3 \\ x + 2y + 4z = 6 \\ x + 2y - 2z = 0 \end{cases} \Leftrightarrow \begin{cases} x + 2y - 2z = 0 \\ 2x + 4y - z = 3 \\ x + 2y + 4z = 6 \end{cases} \Leftrightarrow \begin{cases} x + 2y - 2z = 0 \\ 3z = 3 \\ 6z = 6 \end{cases} \Leftrightarrow \begin{cases} x + 2y - 2z = 0 \\ 3z = 3 \\ 0 = 0 \end{cases}$

So $z = 1$, and substituting into the first equation we have $x + 2t - 2(1) = 0 \Leftrightarrow x = -2t + 2$. Thus, the solutions are $(-2t + 2, t, 1)$, where $t$ is any real number.

**31.** $\begin{cases} x \qquad + z + 2w = 6 \\ y - 2z \qquad = -3 \\ x + 2y - z \qquad = -2 \\ 2x + y + 3z - 2w = 0 \end{cases} \Leftrightarrow \begin{cases} x \qquad + z + 2w = 6 \\ y - 2z \qquad = -3 \\ 2y - 2z - 2w = -8 \\ y + z - 6w = -12 \end{cases} \Leftrightarrow \begin{cases} x \qquad + z + 2w = 6 \\ y - 2z \qquad = -3 \\ z - w = -1 \\ 3z - 6w = -9 \end{cases} \Leftrightarrow$

$\begin{cases} x \qquad + z + 2w = 6 \\ y - 2z \qquad = -3 \\ z - w = -1 \\ -3w = -6 \end{cases}$ So $w = 2$ and $z - 2 = -1 \Leftrightarrow z = 1$. Then $y - 2(1) = -3 \Leftrightarrow y = -1$ and $x + 1 + 2(2) = 6 \Leftrightarrow x = 1$. Thus, the solution is $(1, -1, 1, 2)$.

**32.** $\begin{cases} x + y + z + w = 0 \\ x + y + 2z + 2w = 0 \\ 2x + 2y + 3z + 4w = 1 \\ 2x + 3y + 4z + 5w = 2 \end{cases} \Leftrightarrow \begin{cases} x + y + z + w = 0 \\ z + w = 0 \\ z + 2w = 1 \\ y + 2z + 3w = 2 \end{cases} \Leftrightarrow \begin{cases} x + y + z + w = 0 \\ y + 2z + 3w = 2 \\ z + w = 0 \\ z + 2w = 1 \end{cases} \Leftrightarrow$

$\begin{cases} x + y + z + w = 0 \\ y + 2z + 3w = 2 \\ z + w = 0 \\ w = 1 \end{cases}$ So $w = 1$ and $z + 1 = 0 \Leftrightarrow z = -1$. Then $y + 2(-1) + 3(1) = 2 \Leftrightarrow y = 1$ and $x + 1 + (-1) + 1 = 0 \Leftrightarrow x = -1$. So the solution is $(-1, 1, -1, 1)$.

**33.** Let $x$ be the amount invested at 4%, $y$ the amount invested at 5%, and $z$ the amount invested at 6%. We set up a model and get the following equations: $\begin{cases} \text{Total money:} & x + y + z = 100{,}000 \\ \text{Annual income:} & 0.04x + 0.05y + 0.06z = 0.051\,(100{,}000) \\ \text{Equal amounts:} & x = y \end{cases} \Leftrightarrow$

$\begin{cases} x + y + z = 100{,}000 \\ 4x + 5y + 6z = 510{,}000 \\ x - y \qquad = 0 \end{cases} \Leftrightarrow \begin{cases} x + y + z = 100{,}000 \\ y + 2z = 110{,}000 \\ -2y - z = -100{,}000 \end{cases} \Leftrightarrow \begin{cases} x + y + z = 100{,}000 \\ y + 2z = 110{,}000 \\ 3z = 120{,}000 \end{cases}$ So $z = 40{,}000$ and $y + 2(40{,}000) = 110{,}000 \Leftrightarrow y = 30{,}000$. Since $x = y$, $x = 30{,}000$. She must invest \$30,000 in short-term bonds, \$30,000 in intermediate-term bonds, and \$40,000 in long-term bonds.

**34.** Let $x$ be the amount invested at 4%, $y$ the amount invested at 6%, and $z$ the amount invested at 8%. We set up a model and get the following equations: $\begin{cases} \text{Total money:} & x+y+z=100{,}000 \\ \text{Annual income:} & 0.04x+0.06y+0.08z=6{,}700 \\ \text{Equal amounts:} & y=z \end{cases}$ $\Leftrightarrow$ $\begin{cases} x+y+z=100{,}000 \\ 4x+6y+8z=670{,}000 \\ y-z=0 \end{cases}$

$\Leftrightarrow$ $\begin{cases} x+y+z=100{,}000 \\ 2y+4z=270{,}000 \\ y-z=0 \end{cases}$ $\Leftrightarrow$ $\begin{cases} x+y+z=100{,}000 \\ 2y+4z=270{,}000 \\ -6z=-270{,}000 \end{cases}$ So $z=45{,}000$ and $y=z=45{,}000$. Since $x+45{,}000+45{,}000=100{,}000$ $\Leftrightarrow$ $x=10{,}000$. She must invest \$10,000 in short-term bonds, \$45,000 in intermediate-term bonds, and \$45,000 in long-term bonds.

**35.** Let $a$, $b$, and $c$ be the number of ounces of Type A, Type B, and Type C pellets used. The requirements for the different vitamins gives the following system: $\begin{cases} 2a+3b+c=9 \\ 3a+b+3c=14 \\ 8a+5b+7c=32 \end{cases}$ $\Leftrightarrow$ $\begin{cases} 2a+3b+c=9 \\ -7b+3c=1 \\ -7b+3c=-4 \end{cases}$

Equations 2 and 3 are inconsistent, so there is no solution.

**36.** $\begin{cases} I_1+I_2-I_3=0 \\ 16I_1-8I_2=4 \\ 8I_2+4I_3=5 \end{cases}$ $\Leftrightarrow$ $\begin{cases} I_1+I_2-I_3=0 \\ -24I_2+16I_3=4 \\ 8I_2+4I_3=5 \end{cases}$ $\Leftrightarrow$ $\begin{cases} I_1+I_2-I_3=0 \\ 8I_2+4I_3=5 \\ 28I_3=19 \end{cases}$

So $I_3=\frac{19}{28}\approx 0.68$ and $8I_2+4\left(\frac{19}{28}\right)=5$ $\Leftrightarrow$ $I_2=\frac{2}{7}\approx 0.29$. Then $I_1+\frac{2}{7}-\frac{19}{28}=0$ $\Leftrightarrow$ $I_1=\frac{11}{28}\approx 0.39$.

**37.** Let $x$, $y$, and $z$ be the number of acres of land planted with corn, wheat, and soybeans. We set up a model and get the following equations: $\begin{cases} \text{Total acres:} & x+y+z=1200 \\ \text{Market demand:} & 2x=y \\ \text{Total cost:} & 45x+60y+50z=63{,}750 \end{cases}$ Substituting $2x$ for $y$, we get

$\begin{cases} x+2x+z=1200 \\ 2x=y \\ 45x+60(2x)+50z=63{,}750 \end{cases}$ $\Leftrightarrow$ $\begin{cases} 3x+z=1200 \\ 2x-y=0 \\ 165x+50z=63{,}750 \end{cases}$ $\Leftrightarrow$ $\begin{cases} 3x+z=1200 \\ 2x-y=0 \\ 15x=3750 \end{cases}$ So $15x=3{,}750$ $\Leftrightarrow$ $x=250$ and $y=2(250)=500$. Substituting into the original equation, we have $250+500+z=1200$ $\Leftrightarrow$ $z=450$. Thus the farmer should plant 250 acres of corn, 500 acres of wheat, and 450 acres of soybeans.

**38.** Let $a$, $b$, and $c$ be the number of shares of Stock A, Stock B, and Stock C in the investor's portfolio. Since the total value remains unchanged, we get the following system:

$\begin{cases} 10a+25b+29c=74{,}000 \\ 12a+20b+32c=74{,}000 \\ 16a+15b+32c=74{,}000 \end{cases}$ $\Leftrightarrow$ $\begin{cases} 10a+25b+29c=74{,}000 \\ 50b+14c=74{,}000 \\ 125b+72c=222{,}000 \end{cases}$ $\Leftrightarrow$ $\begin{cases} 10a+25b+29c=74{,}000 \\ 50b+14c=74{,}000 \\ -74c=-74{,}000 \end{cases}$

So $c=1{,}000$. Back-substituting we have $50b+14(1000)=74{,}000 \Leftrightarrow 50b=60{,}000$ $\Leftrightarrow$ $b=1{,}200$. And finally $10a+25(1200)+29(1000)=74{,}000$ $10a+30{,}000+29{,}000=74{,}000$ $\Leftrightarrow$ $10a=15{,}000 \Leftrightarrow a=1{,}500$. Thus the portfolio consists of 1,500 shares of Stock A, 1,200 shares of Stock B, and 1,000 shares of Stock C.

**39. (a)** We begin by substituting $\frac{x_0+x_1}{2}$, $\frac{y_0+y_1}{2}$, and $\frac{z_0+z_1}{2}$ into the left-hand side of the first equation:

$$a_1\left(\frac{x_0+x_1}{2}\right)+b_1\left(\frac{y_0+y_1}{2}\right)+c_1\left(\frac{z_0+z_1}{2}\right)=\tfrac{1}{2}\left[(a_1x_0+b_1y_0+c_1z_0)+(a_1x_1+b_1y_1+c_1z_1)\right]$$
$$=\tfrac{1}{2}\left[d_1+d_1\right]=d_1$$

Thus the given ordered triple satisfies the first equation. We can show that it satisfies the second and the third in exactly the same way. Thus it is a solution of the system.

**(b)** We have shown in part (a) that if the system has two different solutions, we can find a third one by averaging the two solutions. But then we can find a fourth and a fifth solution by averaging the new one with each of the previous two. Then we can find four more by repeating this process with these new solutions, and so on. Clearly this process can continue indefinitely, so there are infinitely many solutions.

# 10.4 Systems of Linear Equations: Matrices

**1.** $3\times 2$ **2.** $2\times 4$ **3.** $2\times 1$ **4.** $3\times 1$ **5.** $1\times 3$ **6.** $2\times 2$

**7. (a)** Yes, this matrix is in row-echelon form.

**(b)** Yes, this matrix is in reduced row-echelon form.

**(c)** $\begin{cases} x=-3 \\ y=\ \ 5 \end{cases}$

**8. (a)** Yes, this matrix is in row-echelon form.

**(b)** No, this matrix not in reduced row-echelon form. The entry above the leading 1 in the second row is not 0.

**(c)** $\begin{cases} x+3y=-3 \\ \qquad y=\ \ 5 \end{cases}$

**9. (a)** Yes, this matrix is in row-echelon form.

**(b)** No, this matrix is not in reduced row-echelon form, since the leading 1 in the second row does not have a zero above it.

**(c)** $\begin{cases} x+2y+8z=0 \\ \qquad y+3z=2 \\ \qquad\qquad 0=0 \end{cases}$

**10. (a)** Yes, the matrix is in row-echelon form.

**(b)** Yes, the matrix is in reduced row-echelon form.

**(c)** $\begin{cases} x \qquad -7z=0 \\ \qquad y+3z=0 \\ \qquad\qquad 0=1 \end{cases}$

**11. (a)** No, this matrix is not in row-echelon form, since the row of zeros is not at the bottom.

**(b)** No, this matrix is not in reduced row-echelon form.

**(c)** $\begin{cases} x \qquad\qquad =0 \\ \qquad\qquad 0=0 \\ \quad y+5z=1 \end{cases}$

**12. (a)** Yes, the matrix is in row-echelon form.

**(b)** Yes, the matrix is in reduced row-echelon form.

**(c)** $\begin{cases} x \qquad =1 \\ \quad y \quad =2 \\ \qquad z=3 \end{cases}$

**13. (a)** Yes, this matrix is in row-echelon form.

**(b)** Yes, this matrix is in reduced row-echelon form.

**(c)** $\begin{cases} x+3y \qquad -w=0 \\ \qquad\qquad z+2w=2 \\ \qquad\qquad\qquad 0=1 \\ \qquad\qquad\qquad 0=0 \end{cases}$

Notice that this system has no solution.

**14. (a)** No, this matrix is not in row-echelon form, since the fourth column has the leading 1 of two rows.

**(b)** No, this matrix is not in reduced row-echelon form.

**(c)** $\begin{cases} x+3y \ +\ w \qquad =0 \\ \qquad y\ +4w \qquad =0 \\ \qquad\qquad w+u=2 \\ \qquad\qquad w \qquad =0 \end{cases}$

**15.** $\begin{bmatrix} 1 & -2 & 1 & 1 \\ 0 & 1 & 2 & 5 \\ 1 & 1 & 3 & 8 \end{bmatrix} \xrightarrow{R_3 - R_1 \to R_3} \begin{bmatrix} 1 & -2 & 1 & 1 \\ 0 & 1 & 2 & 5 \\ 0 & 3 & 2 & 7 \end{bmatrix} \xrightarrow{R_3 - 3R_2 \to R_3} \begin{bmatrix} 1 & -2 & 1 & 1 \\ 0 & 1 & 2 & 5 \\ 0 & 0 & -4 & -8 \end{bmatrix}$. Thus, $-4z = -8 \Leftrightarrow z = 2$; $y + 2(2) = 5 \Leftrightarrow y = 1$; and $x - 2(1) + (2) = 1 \Leftrightarrow x = 1$. Therefore, the solution is $(1, 1, 2)$.

**16.** $\begin{bmatrix} 1 & 1 & 6 & 3 \\ 1 & 1 & 3 & 3 \\ 1 & 2 & 4 & 7 \end{bmatrix} \xrightarrow{R_2 - R_1 \to R_2} \begin{bmatrix} 1 & 1 & 6 & 3 \\ 0 & 0 & -3 & 0 \\ 1 & 2 & 4 & 7 \end{bmatrix} \xrightarrow[R_3 - R_1 \to R_3]{-\frac{1}{3}R_2} \begin{bmatrix} 1 & 1 & 6 & 3 \\ 0 & 0 & 1 & 0 \\ 0 & 1 & -2 & 4 \end{bmatrix} \xrightarrow{R_3 \leftrightarrow R_2} \begin{bmatrix} 1 & 1 & 6 & 3 \\ 0 & 1 & -2 & 4 \\ 0 & 0 & 1 & 0 \end{bmatrix}$.
Thus, $z = 0$; $y - 2z = y - 0 = y = 4$; and $x + y + 6z = 3 \Leftrightarrow x + 4 - 0 = 3 \Leftrightarrow x = -1$. Therefore, the solution is $(-1, 4, 0)$.

**17.** $\begin{bmatrix} 1 & 1 & 1 & 2 \\ 2 & -3 & 2 & 4 \\ 4 & 1 & -3 & 1 \end{bmatrix} \xrightarrow[R_3 - 4R_1 \to R_3]{R_2 - 2R_1 \to R_2} \begin{bmatrix} 1 & 1 & 1 & 2 \\ 0 & -5 & 0 & 0 \\ 0 & -3 & -7 & -7 \end{bmatrix} \xrightarrow{R_3 - \frac{3}{5}R_2 \to R_3} \begin{bmatrix} 1 & 1 & 1 & 2 \\ 0 & -5 & 0 & 0 \\ 0 & 0 & -7 & -7 \end{bmatrix}$. Thus, $-7z = -7 \Leftrightarrow z = 1$; $-5y = 0 \Leftrightarrow y = 0$; and $x + 0 + 1 = 2 \Leftrightarrow x = 1$. Therefore, the solution is $(1, 0, 1)$.

**18.** $\begin{bmatrix} 1 & 1 & 1 & 4 \\ -1 & 2 & 3 & 17 \\ 2 & -1 & 0 & -7 \end{bmatrix} \xrightarrow[R_3 - 2R_1 \to R_3]{R_2 + R_1 \to R_2} \begin{bmatrix} 1 & 1 & 1 & 4 \\ 0 & 3 & 4 & 21 \\ 0 & -3 & -2 & -15 \end{bmatrix} \xrightarrow[R_3 - 2R_1 \to R_3]{R_3 + R_2 \to R_3} \begin{bmatrix} 1 & 1 & 1 & 4 \\ 0 & 3 & 4 & 21 \\ 0 & 0 & 2 & 6 \end{bmatrix}$. Thus, $2z = 6 \Leftrightarrow z = 3$; $3y + 4(3) = 21 \Leftrightarrow y = 3$; and $x + (3) + (3) = 4 \Leftrightarrow x = -2$. The solution is $(-2, 3, 3)$.

**19.** $\begin{bmatrix} 1 & 2 & -1 & -2 \\ 1 & 0 & 1 & 0 \\ 2 & -1 & -1 & -3 \end{bmatrix} \xrightarrow[R_3 - 2R_1 \to R_3]{R_2 - R_1 \to R_2} \begin{bmatrix} 1 & 2 & -1 & -2 \\ 0 & -2 & 2 & 2 \\ 0 & -5 & 1 & 1 \end{bmatrix} \xrightarrow{-\frac{1}{2}R_2} \begin{bmatrix} 1 & 2 & -1 & -2 \\ 0 & 1 & 1 & 1 \\ 0 & -5 & 1 & 1 \end{bmatrix} \xrightarrow{R_3 + 5R_2 \to R_3}$
$\begin{bmatrix} 1 & 2 & -1 & -2 \\ 0 & 1 & 1 & 1 \\ 0 & 0 & 6 & 6 \end{bmatrix}$. Thus, $6z = 6 \Leftrightarrow z = 1$; $y + (1) = 1 \Leftrightarrow y = 0$; and $x + 2(0) - (1) = -2 \Leftrightarrow x = -1$.
Therefore, the solution is $(-1, 0, 1)$.

**20.** $\begin{bmatrix} 0 & 2 & 1 & 4 \\ 1 & 1 & 0 & 4 \\ 3 & 3 & -1 & 10 \end{bmatrix} \xrightarrow{R_1 \leftrightarrow R_2} \begin{bmatrix} 1 & 1 & 0 & 4 \\ 0 & 2 & 1 & 4 \\ 3 & 3 & -1 & 10 \end{bmatrix} \xrightarrow{R_3 - 3R_1 \to R_3} \begin{bmatrix} 1 & 1 & 0 & 4 \\ 0 & 2 & 1 & 4 \\ 0 & 0 & -1 & -2 \end{bmatrix}$. Thus, $-z = -2 \Leftrightarrow z = 2$; $2y + (2) = 4 \Leftrightarrow y = 1$; and $x + (1) = 4 \Leftrightarrow x = 3$. The solution is $(3, 1, 2)$

**21.** $\begin{bmatrix} 1 & 2 & -1 & 9 \\ 2 & 0 & -1 & -2 \\ 3 & 5 & 2 & 22 \end{bmatrix} \xrightarrow[R_3 - 3R_1 \to R_3]{R_2 - 2R_1 \to R_2} \begin{bmatrix} 1 & 2 & -1 & 9 \\ 0 & -4 & 1 & -20 \\ 0 & -1 & 5 & -5 \end{bmatrix} \xrightarrow{4R_3 - R_2 \to R_3} \begin{bmatrix} 1 & 2 & -1 & 9 \\ 0 & -4 & 1 & -20 \\ 0 & 0 & 19 & 0 \end{bmatrix}$ Thus, $19x_3 = 0 \Leftrightarrow x_3 = 0$; $-4x_2 = -20 \Leftrightarrow x_2 = 5$; and $x_1 + 2(5) = 9 \Leftrightarrow x_1 = -1$. Therefore, the solution is $(-1, 5, 0)$.

**22.** $\begin{bmatrix} 2 & 1 & 0 & 7 \\ 2 & -1 & 1 & 6 \\ 3 & -2 & 4 & 11 \end{bmatrix} \xrightarrow[2R_3 - 3R_1 \to R_3]{R_2 - R_1 \to R_2} \begin{bmatrix} 2 & 1 & 0 & 7 \\ 0 & -2 & 1 & -1 \\ 0 & -7 & 8 & 1 \end{bmatrix} \xrightarrow{2R_3 - 7R_2 \to R_3} \begin{bmatrix} 2 & 1 & 0 & 7 \\ 0 & -2 & 1 & -1 \\ 0 & 0 & 9 & 9 \end{bmatrix}$ Then, $9x_3 = 9 \Leftrightarrow x_3 = 1$; $-2x_2 + 1 = -1 \Leftrightarrow x_2 = 1$; and $2x_1 + 1 = 7 \Leftrightarrow x_1 = 3$. Therefore, the solution is $(x_1, x_2, x_3) = (3, 1, 1)$.

**23.** $\begin{bmatrix} 2 & -3 & -1 & 13 \\ -1 & 2 & -5 & 6 \\ 5 & -1 & -1 & 49 \end{bmatrix} \xrightarrow[2R_3 - 5R_1 \to R_3]{2R_2 + R_1 \to R_2} \begin{bmatrix} 2 & -3 & -1 & 13 \\ 0 & 1 & -11 & 25 \\ 0 & 13 & 3 & 33 \end{bmatrix} \xrightarrow{R_3 - 13R_2 \to R_3} \begin{bmatrix} 2 & -3 & -1 & 13 \\ 0 & 1 & -11 & 25 \\ 0 & 0 & 146 & -292 \end{bmatrix}$ Thus, $146z = -292 \quad \Leftrightarrow \quad z = -2$; $y - 11(-2) = 25 \quad \Leftrightarrow \quad y = 3$; and $2x - 3 \cdot 3 + 2 = 13 \Leftrightarrow x = 10$. Therefore, the solution is $(10, 3, -2)$.

**24.** $\begin{bmatrix} 10 & 10 & -20 & 60 \\ 15 & 20 & 30 & -25 \\ -5 & 30 & -10 & 45 \end{bmatrix} \xrightarrow[2R_3 + R_1 \to R_3]{2R_2 - 3R_1 \to R_2} \begin{bmatrix} 10 & 10 & -20 & 60 \\ 0 & 10 & 120 & -230 \\ 0 & 70 & -40 & 150 \end{bmatrix} \xrightarrow{R_3 - 7R_2 \to R_3} \begin{bmatrix} 10 & 10 & -20 & 60 \\ 0 & 10 & 120 & -230 \\ 0 & 0 & -880 & 1760 \end{bmatrix}$ Thus, $-880z = 1760 \quad \Leftrightarrow \quad z = -2$; $10y + 120(-2) = -230 \quad \Leftrightarrow \quad y = 1$; and $10x + 10 + 40 = 60 \Leftrightarrow x = 1$. Therefore, the solution is $(1, 1, -2)$.

**25.** $\begin{bmatrix} 1 & 1 & 1 & 2 \\ 0 & 1 & -3 & 1 \\ 2 & 1 & 5 & 0 \end{bmatrix} \xrightarrow{R_3 - 2R_1 \to R_3} \begin{bmatrix} 1 & 1 & 1 & 2 \\ 0 & 1 & -3 & 1 \\ 0 & -1 & 3 & -4 \end{bmatrix} \xrightarrow{R_3 + R_2 \to R_3} \begin{bmatrix} 1 & 1 & 1 & 3 \\ 0 & 1 & -3 & 1 \\ 0 & 0 & 0 & -3 \end{bmatrix}$. The third row of the matrix states $0 = -3$, which is impossible. Hence, the system is inconsistent, and there is no solution.

**26.** $\begin{bmatrix} 1 & 0 & 3 & 3 \\ 2 & 1 & -2 & 5 \\ 0 & -1 & 8 & 8 \end{bmatrix} \xrightarrow{R_2 - 2R_1 \to R_2} \begin{bmatrix} 1 & 0 & 3 & 3 \\ 0 & 1 & -8 & -1 \\ 0 & -1 & 8 & 8 \end{bmatrix} \xrightarrow{R_3 + R_2 \to R_3} \begin{bmatrix} 1 & 0 & 3 & 3 \\ 0 & 1 & -8 & -1 \\ 0 & 0 & 0 & 7 \end{bmatrix}$. The system is inconsistent, and there is no solution.

**27.** $\begin{bmatrix} 2 & -3 & -9 & -5 \\ 1 & 0 & 3 & 2 \\ -3 & 1 & -4 & -3 \end{bmatrix} \xrightarrow{R_1 \leftrightarrow R_2} \begin{bmatrix} 1 & 0 & 3 & 2 \\ 2 & -3 & -9 & -5 \\ -3 & 1 & -4 & -3 \end{bmatrix} \xrightarrow[R_3 + 3R_1 \to R_3]{R_2 - 2R_1 \to R_2} \begin{bmatrix} 1 & 0 & 3 & 2 \\ 0 & -3 & -15 & -9 \\ 0 & 1 & 5 & 3 \end{bmatrix} \xrightarrow{-\frac{1}{3}R_2}$

$\begin{bmatrix} 1 & 0 & 3 & 2 \\ 0 & 1 & 5 & 3 \\ 0 & 1 & 5 & 3 \end{bmatrix} \xrightarrow{R_3 - R_2 \to R_3} \begin{bmatrix} 1 & 0 & 3 & 2 \\ 0 & 1 & 5 & 3 \\ 0 & 0 & 0 & 0 \end{bmatrix}$. Therefore, this system has infinitely many solutions, given by $x + 3t = 2$ $\Leftrightarrow \quad x = 2 - 3t$, and $y + 5t = 3 \quad \Leftrightarrow \quad y = 3 - 5t$. Hence, the solutions are $(2 - 3t, 3 - 5t, t)$, where $t$ is any real number.

**28.** $\begin{bmatrix} 1 & -2 & 5 & 3 \\ -2 & 6 & -11 & 1 \\ 3 & -16 & 20 & -26 \end{bmatrix} \xrightarrow[R_3 - 3R_1 \to R_3]{R_2 + 2R_1 \to R_2} \begin{bmatrix} 1 & -2 & 5 & 3 \\ 0 & 2 & -1 & 7 \\ 0 & -10 & 5 & -35 \end{bmatrix} \xrightarrow{R_3 + 5R_2 \to R_3} \begin{bmatrix} 1 & -2 & 5 & 3 \\ 0 & 2 & -1 & 7 \\ 0 & 0 & 0 & 0 \end{bmatrix}$ The system is dependent; there are infinitely many solutions, given by $2y - t = 7 \quad \Leftrightarrow \quad 2y = t + 7 \quad \Leftrightarrow \quad y = \frac{1}{2}t + \frac{7}{2}$; and $x - 2\left(\frac{1}{2}t + \frac{7}{2}\right) + 5t = 3 \Leftrightarrow x - t - 7 + 5t = 3 \quad \Leftrightarrow \quad x = -4t + 10$. The solutions are $\left(-4t + 10, \frac{1}{2}t + \frac{7}{2}, t\right)$, where $t$ is any real number.

**29.** $\begin{bmatrix} 1 & -1 & 3 & 3 \\ 4 & -8 & 32 & 24 \\ 2 & -3 & 11 & 4 \end{bmatrix} \xrightarrow[R_3 - 2R_1 \to R_3]{R_2 - 4R_1 \to R_2} \begin{bmatrix} 1 & -1 & 3 & 3 \\ 0 & -4 & 20 & 12 \\ 0 & -1 & 5 & -2 \end{bmatrix} \xrightarrow[R_3 + R_2 \to R_3]{-\frac{1}{4}R_2} \begin{bmatrix} 1 & -1 & 3 & 3 \\ 0 & 1 & -5 & -3 \\ 0 & 0 & 0 & -5 \end{bmatrix}$ The third row of the matrix states $0 = -5$, which is impossible. Hence, the system is inconsistent, and there is no solution.

**30.** $$\begin{bmatrix} -2 & 6 & -2 & -12 \\ 1 & -3 & 2 & 10 \\ -1 & 3 & 2 & 6 \end{bmatrix} \xrightarrow{-\frac{1}{2}R_1} \begin{bmatrix} 1 & -3 & 1 & 6 \\ 1 & -3 & 2 & 10 \\ -1 & 3 & 2 & 6 \end{bmatrix} \xrightarrow[R_3+R_1 \to R_3]{R_2-R_1 \to R_2} \begin{bmatrix} 1 & -3 & 1 & 6 \\ 0 & 0 & 1 & 4 \\ 0 & 0 & 3 & 12 \end{bmatrix} \xrightarrow{R_3-3R_2 \to R_3} \begin{bmatrix} 1 & -3 & 1 & 6 \\ 0 & 0 & 1 & 4 \\ 0 & 0 & 0 & 0 \end{bmatrix}.$$

The system is dependent; there are infinitely many solutions, given by $z = 4$, and $x - 3y + z = 6$, so $x - 3t + 4 = 6 \Leftrightarrow x = 3t + 2$. The solutions are $(3t + 2, t, 4)$, where $t$ is any real number.

**31.** $$\begin{bmatrix} 1 & 4 & -2 & -3 \\ 2 & -1 & 5 & 12 \\ 8 & 5 & 11 & 30 \end{bmatrix} \xrightarrow[R_3-8R_1 \to R_3]{R_2-2R_1 \to R_2} \begin{bmatrix} 1 & 4 & -2 & -3 \\ 0 & -9 & 9 & 18 \\ 0 & -27 & 27 & 54 \end{bmatrix} \xrightarrow{R_3-3R_2 \to R_3} \begin{bmatrix} 1 & 4 & -2 & -3 \\ 0 & -9 & 9 & 18 \\ 0 & 0 & 0 & 0 \end{bmatrix}.$$

Therefore, this system has infinitely many solutions, given by $-9y + 9t = 18 \Leftrightarrow y = -2 + t$, and $x + 4(-2 + t) - 2t = -3 \Leftrightarrow x = 5 - 2t$. Hence, the solutions are $(5 - 2t, -2 + t, t)$, where $t$ is any real number.

**32.** $$\begin{bmatrix} 3 & 2 & -3 & 10 \\ 1 & -1 & -1 & -5 \\ 1 & 4 & -1 & 20 \end{bmatrix} \xrightarrow{R_1 \leftrightarrow R_2} \begin{bmatrix} 1 & -1 & -1 & -5 \\ 3 & 2 & -3 & 10 \\ 1 & 4 & -1 & 20 \end{bmatrix} \xrightarrow[R_3-R_1 \to R_3]{R_2-3R_1 \to R_2} \begin{bmatrix} 1 & -1 & -1 & -5 \\ 0 & 5 & 0 & 25 \\ 0 & 5 & 0 & 25 \end{bmatrix} \xrightarrow{R_3-R_2 \to R_3}$$

$$\begin{bmatrix} 1 & -1 & -1 & -5 \\ 0 & 5 & 0 & 25 \\ 0 & 0 & 0 & 0 \end{bmatrix}.$$ The system is dependent, there are infinitely many solutions, given by $5s = 25 \Leftrightarrow s = 5$, and $r - 5 - t = -5 \Leftrightarrow r = t$. Hence, the solutions are $(t, 5, t)$, where $t$ is any real number.

**33.** $$\begin{bmatrix} 2 & 1 & -2 & 12 \\ -1 & -\frac{1}{2} & 1 & -6 \\ 3 & \frac{3}{2} & -3 & 18 \end{bmatrix} \xrightarrow[-R_1]{R_1 \leftrightarrow R_2} \begin{bmatrix} 1 & \frac{1}{2} & -1 & 6 \\ 2 & 1 & -2 & 12 \\ 3 & \frac{3}{2} & -3 & 18 \end{bmatrix} \xrightarrow[R_3-3R_1 \to R_3]{R_2-2R_1 \to R_2} \begin{bmatrix} 1 & \frac{1}{2} & -1 & 6 \\ 0 & 0 & 0 & 0 \\ 0 & 0 & 0 & 0 \end{bmatrix}$$ Therefore, this system has infinitely many solutions, given by $x + \frac{1}{2}s - t = 6 \Leftrightarrow x = 6 - \frac{1}{2}s + t$. Hence, the solutions are $(6 - \frac{1}{2}s + t, s, t)$, where $s$ and $t$ are any real numbers.

**34.** $$\begin{bmatrix} 0 & 1 & -5 & 7 \\ 3 & 2 & 0 & 12 \\ 3 & 0 & 10 & 80 \end{bmatrix} \xrightarrow{R_1 \leftrightarrow R_2} \begin{bmatrix} 3 & 2 & 0 & 12 \\ 0 & 1 & -5 & 7 \\ 3 & 0 & 10 & 80 \end{bmatrix} \xrightarrow{R_3-R_1 \to R_3} \begin{bmatrix} 3 & 2 & 0 & 12 \\ 0 & 1 & -5 & 7 \\ 0 & -2 & 10 & 68 \end{bmatrix} \xrightarrow{R_3+2R_2 \to R_3} \begin{bmatrix} 3 & 2 & 0 & 12 \\ 0 & 1 & -5 & 7 \\ 0 & 0 & 0 & 82 \end{bmatrix}.$$

Therefore, the system is inconsistent and has no solution.

**35.** $$\begin{bmatrix} 4 & -3 & 1 & -8 \\ -2 & 1 & -3 & -4 \\ 1 & -1 & 2 & 3 \end{bmatrix} \xrightarrow{R_1 \leftrightarrow R_3} \begin{bmatrix} 1 & -1 & 2 & 3 \\ -2 & 1 & -3 & -4 \\ 4 & -3 & 1 & -8 \end{bmatrix} \xrightarrow[R_3-4R_1 \to R_3]{R_2+2R_1 \to R_2} \begin{bmatrix} 1 & -1 & 2 & 3 \\ 0 & -1 & 1 & 2 \\ 0 & 1 & -7 & -20 \end{bmatrix} \xrightarrow{-R_2}$$

$$\begin{bmatrix} 1 & -1 & 2 & 3 \\ 0 & 1 & -1 & -2 \\ 0 & 1 & -7 & -20 \end{bmatrix} \xrightarrow{R_3-R_2 \to R_3} \begin{bmatrix} 1 & -1 & 2 & 3 \\ 0 & 1 & -1 & -2 \\ 0 & 0 & -6 & -18 \end{bmatrix}.$$ Therefore, $-6z = -18 \Leftrightarrow z = 3$; $y - (3) = -2 \Leftrightarrow y = 1$; and $x - (1) + 2(3) = 3 \Leftrightarrow x = -2$. Hence, the solution is $(-2, 1, 3)$.

**36.** $$\begin{bmatrix} 2 & -3 & 5 & 14 \\ 4 & -1 & -2 & -17 \\ -1 & -1 & 1 & 3 \end{bmatrix} \xrightarrow[-R_1]{R_1 \leftrightarrow R_3} \begin{bmatrix} 1 & 1 & -1 & -3 \\ 4 & -1 & -2 & -17 \\ 2 & -3 & 5 & 14 \end{bmatrix} \xrightarrow[R_3-2R_1 \to R_3]{R_2-4R_1 \to R_2} \begin{bmatrix} 1 & 1 & -1 & -3 \\ 0 & -5 & 2 & -5 \\ 0 & -5 & 7 & 20 \end{bmatrix} \xrightarrow{R_3-R_2 \to R_3} \begin{bmatrix} 1 & 1 & -1 & -3 \\ 0 & -5 & 2 & -5 \\ 0 & 0 & 5 & 25 \end{bmatrix}.$$

Thus $5z = 25 \Leftrightarrow z = 5$; $-5y + 2(5) = -5 \Leftrightarrow -5y = -15 \Leftrightarrow y = 3$; and $x + (3) - (5) = -3 \Leftrightarrow x = -1$. Hence, the solution is $(-1, 3, 5)$.

**37.** $\begin{bmatrix} 1 & 2 & -3 & -5 \\ -2 & -4 & -6 & 10 \\ 3 & 7 & -2 & -13 \end{bmatrix} \xrightarrow[\mathrm{R}_3 - 3\mathrm{R}_1 \to \mathrm{R}_3]{\mathrm{R}_2 + 2\mathrm{R}_1 \to \mathrm{R}_2} \begin{bmatrix} 1 & 2 & -3 & -5 \\ 0 & 0 & -12 & 0 \\ 0 & 1 & 7 & 2 \end{bmatrix} \xrightarrow{\mathrm{R}_2 \leftrightarrow \mathrm{R}_3} \begin{bmatrix} 1 & 2 & -3 & -5 \\ 0 & 1 & 7 & 2 \\ 0 & 0 & -12 & 0 \end{bmatrix}$. Therefore, $-12z = 0$ $\Leftrightarrow$ $z = 0; y + 7(0) = 2$ $\Leftrightarrow$ $y = 2$; and $x + 2(2) - 3(0) = -5 \Leftrightarrow x = -9$. Hence, the solution is $(-9, 2, 0)$.

**38.** $\begin{bmatrix} 3 & -1 & 2 & -1 \\ 4 & -2 & 1 & -7 \\ -1 & 3 & -2 & -1 \end{bmatrix} \xrightarrow[-\mathrm{R}_1]{\mathrm{R}_1 \leftrightarrow \mathrm{R}_3} \begin{bmatrix} 1 & -3 & 2 & 1 \\ 4 & -2 & 1 & -7 \\ 3 & -1 & 2 & -1 \end{bmatrix} \xrightarrow[\mathrm{R}_3 - 3\mathrm{R}_1 \to \mathrm{R}_3]{\mathrm{R}_2 - 4\mathrm{R}_1 \to \mathrm{R}_2} \begin{bmatrix} 1 & -3 & 2 & 1 \\ 0 & 10 & -7 & -11 \\ 0 & 8 & -4 & -4 \end{bmatrix} \xrightarrow{\mathrm{R}_2 \leftrightarrow \frac{1}{4}\mathrm{R}_3}$

$\begin{bmatrix} 1 & -3 & 2 & 1 \\ 0 & 2 & -1 & -1 \\ 0 & 10 & -7 & -11 \end{bmatrix} \xrightarrow{\mathrm{R}_3 - 5\mathrm{R}_2 \to \mathrm{R}_3} \begin{bmatrix} 1 & -3 & 2 & 1 \\ 0 & 2 & -1 & -1 \\ 0 & 0 & -2 & -6 \end{bmatrix}$. Thus $-2z = -6$ $\Leftrightarrow$ $z = 3$; $2y - (3) = -1 \Leftrightarrow 2y = 2$ $\Leftrightarrow$ $y = 1$; and $x - 3(1) + 2(3) = 1$ $\Leftrightarrow$ $x = -2$. Hence, the solution is $(-2, 1, 3)$.

**39.** $\begin{bmatrix} -1 & 2 & 1 & -3 & 3 \\ 3 & -4 & 1 & 1 & 9 \\ -1 & -1 & 1 & 1 & 0 \\ 2 & 1 & 4 & -2 & 3 \end{bmatrix} \xrightarrow{-\mathrm{R}_1} \begin{bmatrix} 1 & -2 & -1 & 3 & -3 \\ 3 & -4 & 1 & 1 & 9 \\ -1 & -1 & 1 & 1 & 0 \\ 2 & 1 & 4 & -2 & 3 \end{bmatrix} \xrightarrow[\mathrm{R}_4 - 2\mathrm{R}_1 \to \mathrm{R}_4]{\substack{\mathrm{R}_2 - 3\mathrm{R}_1 \to \mathrm{R}_2 \\ \mathrm{R}_3 + \mathrm{R}_1 \to \mathrm{R}_3}} \begin{bmatrix} 1 & -2 & -1 & 3 & -3 \\ 0 & 2 & 4 & -8 & 18 \\ 0 & -3 & 0 & 4 & -3 \\ 0 & 5 & 6 & -8 & 9 \end{bmatrix} \xrightarrow{\frac{1}{2}\mathrm{R}_2}$

$\begin{bmatrix} 1 & -2 & -1 & 3 & -3 \\ 0 & 1 & 2 & -4 & 9 \\ 0 & -3 & 0 & 4 & -3 \\ 0 & 5 & 6 & -8 & 9 \end{bmatrix} \xrightarrow[\mathrm{R}_4 - 5\mathrm{R}_2 \to \mathrm{R}_4]{\mathrm{R}_3 + 3\mathrm{R}_2 \to \mathrm{R}_3} \begin{bmatrix} 1 & -2 & -1 & 3 & -3 \\ 0 & 1 & 2 & -4 & 9 \\ 0 & 0 & 6 & -8 & 24 \\ 0 & 0 & -4 & 12 & -36 \end{bmatrix} \xrightarrow{3\mathrm{R}_4 + 2\mathrm{R}_3 \to \mathrm{R}_4} \begin{bmatrix} 1 & -2 & -1 & 3 & -3 \\ 0 & 1 & 2 & -4 & 9 \\ 0 & 0 & 6 & -8 & 24 \\ 0 & 0 & 0 & 20 & -60 \end{bmatrix}$.

Therefore, $20w = -60$ $\Leftrightarrow$ $w = -3; 6z + 24 = 24 \Leftrightarrow z = 0$. Then $y + 12 = 9$ $\Leftrightarrow$ $y = -3$ and $x + 6 - 9 = -3$ $\Leftrightarrow$ $x = 0$. Hence, the solution is $(0, -3, 0, -3)$.

**40.** $\begin{bmatrix} 1 & 1 & -1 & -1 & 6 \\ 2 & 0 & 1 & -3 & 8 \\ 1 & -1 & 0 & 4 & -10 \\ 3 & 5 & -1 & -1 & 20 \end{bmatrix} \xrightarrow[\mathrm{R}_4 - 3\mathrm{R}_1 \to \mathrm{R}_4]{\substack{\mathrm{R}_2 - 2\mathrm{R}_1 \to \mathrm{R}_2 \\ \mathrm{R}_3 - \mathrm{R}_1 \to \mathrm{R}_3}} \begin{bmatrix} 1 & 1 & -1 & -1 & 6 \\ 0 & -2 & 3 & -1 & -4 \\ 0 & -2 & 1 & 5 & -16 \\ 0 & 2 & 2 & 2 & 2 \end{bmatrix} \xrightarrow[\frac{1}{2}\mathrm{R}_4]{\substack{\mathrm{R}_2 + \mathrm{R}_4 \to \mathrm{R}_2 \\ \mathrm{R}_3 + \mathrm{R}_4 \to \mathrm{R}_3}} \begin{bmatrix} 1 & 1 & -1 & -1 & 6 \\ 0 & 0 & 5 & 1 & -2 \\ 0 & 0 & 3 & 7 & -14 \\ 0 & 1 & 1 & 1 & 1 \end{bmatrix}$

$\xrightarrow[\mathrm{R}_3 \leftrightarrow \mathrm{R}_4]{\mathrm{R}_4 \leftrightarrow \mathrm{R}_2} \begin{bmatrix} 1 & 1 & -1 & -1 & 6 \\ 0 & 1 & 1 & 1 & 1 \\ 0 & 0 & 5 & 1 & -2 \\ 0 & 0 & 3 & 7 & -14 \end{bmatrix} \xrightarrow{5\mathrm{R}_4 - 3\mathrm{R}_3 \to \mathrm{R}_4} \begin{bmatrix} 1 & 1 & -1 & -1 & 6 \\ 0 & 1 & 1 & 1 & 1 \\ 0 & 0 & 5 & 1 & -2 \\ 0 & 0 & 0 & 32 & -64 \end{bmatrix}$. Thus $32w = -64$ $\Leftrightarrow$ $w = -2$;

$5z + (-2) = -2$ $\Leftrightarrow$ $5z = 0$ $\Leftrightarrow$ $z = 0; y + 0 + (-2) = 1$ $\Leftrightarrow$ $y = 3$; and $x + 3 - 0 - (-2) = 6 \Leftrightarrow x = 1$.

Hence the solution is $(1, 3, 0, -2)$.

**41.** $\begin{bmatrix} 1 & 1 & 2 & -1 & -2 \\ 0 & 3 & 1 & 2 & 2 \\ 1 & 1 & 0 & 3 & 2 \\ -3 & 0 & 1 & 2 & 5 \end{bmatrix} \xrightarrow[\text{R}_4 + 3\text{R}_1 \to \text{R}_4]{\text{R}_3 - \text{R}_1 \to \text{R}_3} \begin{bmatrix} 1 & 1 & 2 & -1 & -2 \\ 0 & 3 & 1 & 2 & 2 \\ 0 & 0 & -2 & 4 & 4 \\ 0 & 3 & 7 & -1 & -1 \end{bmatrix} \xrightarrow{\text{R}_4 - \text{R}_2 \to \text{R}_4} \begin{bmatrix} 1 & 1 & 2 & -1 & -2 \\ 0 & 3 & 1 & 2 & 2 \\ 0 & 0 & -2 & 4 & 4 \\ 0 & 0 & 6 & -3 & -3 \end{bmatrix}$

$\xrightarrow{\text{R}_4 + 3\text{R}_3 \to \text{R}_4} \begin{bmatrix} 1 & 1 & 2 & -1 & -2 \\ 0 & 3 & 1 & 2 & 2 \\ 0 & 0 & -2 & 4 & 4 \\ 0 & 0 & 0 & 9 & 9 \end{bmatrix}$. Therefore, $9w = 9 \quad \Leftrightarrow \quad w = 1; -2z + 4(1) = 4 \quad \Leftrightarrow \quad z = 0$. Then $3y + (0) + 2(1) = 2 \quad \Leftrightarrow \quad y = 0$and $x + (0) + 2(0) - (1) = -2 \quad \Leftrightarrow \quad x = -1$. Hence, the solution is $(-1, 0, 0, 1)$.

**42.** $\begin{bmatrix} 1 & -3 & 2 & 1 & -2 \\ 1 & -2 & 0 & -2 & -10 \\ 0 & 0 & 1 & 5 & 15 \\ 3 & 0 & 2 & 1 & -3 \end{bmatrix} \xrightarrow[\text{R}_4 - 3\text{R}_1 \to \text{R}_4]{\text{R}_2 - \text{R}_1 \to \text{R}_2} \begin{bmatrix} 1 & -3 & 2 & 1 & -2 \\ 0 & 1 & -2 & -3 & -8 \\ 0 & 0 & 1 & 5 & 15 \\ 0 & 9 & -4 & -2 & 3 \end{bmatrix} \xrightarrow{\text{R}_4 - 9\text{R}_2 \to \text{R}_4} \begin{bmatrix} 1 & -3 & 2 & 1 & -2 \\ 0 & 1 & -2 & -3 & -8 \\ 0 & 0 & 1 & 5 & 15 \\ 0 & 0 & 14 & 25 & 75 \end{bmatrix}$

$\xrightarrow{\text{R}_4 - 14\text{R}_3 \to \text{R}_4} \begin{bmatrix} 1 & -3 & 2 & 1 & -2 \\ 0 & 1 & -2 & -3 & -8 \\ 0 & 0 & 1 & 5 & 15 \\ 0 & 0 & 0 & -45 & -135 \end{bmatrix}$. Thus $-45w = -135 \quad \Leftrightarrow \quad w = 3; z + 5(3) = 15 \quad \Leftrightarrow \quad z = 0;$ $y - 2(0) - 3(3) = -8 \Leftrightarrow y = 1$; and $x - 3(1) + 2(0) + (3) = -2 \quad \Leftrightarrow \quad x = -2$. Hence, the solution is $(-2, 1, 0, 3)$.

**43.** $\begin{bmatrix} 1 & 0 & 1 & 1 & 4 \\ 0 & 1 & -1 & 0 & -4 \\ 1 & -2 & 3 & 1 & 12 \\ 2 & 0 & -2 & 5 & -1 \end{bmatrix} \xrightarrow[\text{R}_4 - 2\text{R}_1 \to \text{R}_4]{\text{R}_3 - \text{R}_1 \to \text{R}_3} \begin{bmatrix} 1 & 0 & 1 & 1 & 4 \\ 0 & 1 & -1 & 0 & -4 \\ 0 & -2 & 2 & 0 & 8 \\ 0 & 0 & -4 & 3 & -9 \end{bmatrix} \xrightarrow{\text{R}_3 + 2\text{R}_2 \to \text{R}_3} \begin{bmatrix} 1 & 0 & 1 & 1 & 4 \\ 0 & 1 & -1 & 0 & -4 \\ 0 & 0 & 0 & 0 & 0 \\ 0 & 0 & -4 & 3 & -9 \end{bmatrix} \xrightarrow{\text{R}_3 \leftrightarrow -\text{R}_4}$

$\begin{bmatrix} 1 & 0 & 1 & 1 & 4 \\ 0 & 1 & -1 & 0 & -4 \\ 0 & 0 & 4 & -3 & 9 \\ 0 & 0 & 0 & 0 & 0 \end{bmatrix}$. Therefore, $4z - 3t = 9 \quad \Leftrightarrow \quad 4z = 9 + 3t \Leftrightarrow z = \frac{9}{4} + \frac{3}{4}t$. Then we have $y - \left(\frac{9}{4} + \frac{3}{4}t\right) = -4$ $\Leftrightarrow \quad y = \frac{-7}{4} + \frac{3}{4}t$ and $x + \left(\frac{9}{4} + \frac{3}{4}t\right) + t = 4 \quad \Leftrightarrow \quad x = \frac{7}{4} - \frac{7}{4}t$. Hence, the solutions are $\left(\frac{7}{4} - \frac{7}{4}t, -\frac{7}{4} + \frac{3}{4}t, \frac{9}{4} + \frac{3}{4}t, t\right)$, where $t$ is any real number.

**44.** $\begin{bmatrix} 0 & 1 & -1 & 2 & 0 \\ 3 & 2 & 0 & 1 & 0 \\ 2 & 0 & 0 & 4 & 12 \\ -2 & 0 & -2 & 5 & 6 \end{bmatrix} \xrightarrow[\frac{1}{2}\text{R}_3]{\substack{\text{R}_2 - \text{R}_3 \to \text{R}_2 \\ \text{R}_4 + \text{R}_3 \to \text{R}_4}} \begin{bmatrix} 0 & 1 & -1 & 2 & 0 \\ 1 & 2 & 0 & -3 & -12 \\ 1 & 0 & 0 & 2 & 6 \\ 0 & 0 & -2 & 9 & 18 \end{bmatrix} \xrightarrow[\text{R}_2 - \text{R}_1 \to \text{R}_2]{\text{R}_1 \leftrightarrow \text{R}_3} \begin{bmatrix} 1 & 0 & 0 & 2 & 6 \\ 0 & 2 & 0 & -5 & -18 \\ 0 & 1 & -1 & 2 & 0 \\ 0 & 0 & -2 & 9 & 18 \end{bmatrix}$

$\xrightarrow[\text{R}_2 \leftrightarrow \text{R}_3]{\text{R}_2 - 2\text{R}_3 \to \text{R}_2} \begin{bmatrix} 1 & 0 & 0 & 2 & 6 \\ 0 & 1 & -1 & 2 & 0 \\ 0 & 0 & 2 & -9 & -18 \\ 0 & 0 & -2 & 9 & 18 \end{bmatrix} \xrightarrow{\text{R}_4 + \text{R}_3 \to \text{R}_4} \begin{bmatrix} 1 & 0 & 0 & 2 & 6 \\ 0 & 1 & -1 & 2 & 0 \\ 0 & 0 & 2 & -9 & -18 \\ 0 & 0 & 0 & 0 & 0 \end{bmatrix}$. Thus, the system has infinitely many solutions given by $-2z + 9t = 18 \quad \Leftrightarrow \quad z = \frac{9}{2}t - 9$; $y - \left(\frac{9}{2}t - 9\right) + 2t = 0 \quad \Leftrightarrow \quad y = \frac{5}{2}t - 9$; and $x + 2t = 6 \Leftrightarrow x = -2t + 6$. So the solutions are $\left(-2t + 6, \frac{5}{2}t - 9, \frac{9}{2}t - 9, t\right)$, where $t$ is any real number.

**45.** $\begin{bmatrix} 1 & -1 & 0 & 1 & 0 \\ 3 & 0 & -1 & 2 & 0 \\ 1 & -4 & 1 & 2 & 0 \end{bmatrix} \xrightarrow[\text{R}_3 - \text{R}_1 \to \text{R}_3]{\text{R}_2 - 3\text{R}_1 \to \text{R}_2} \begin{bmatrix} 1 & -1 & 0 & 1 & 0 \\ 0 & 3 & -1 & -1 & 0 \\ 0 & -3 & 1 & 1 & 0 \end{bmatrix} \xrightarrow{\text{R}_3 + \text{R}_2 \to \text{R}_3} \begin{bmatrix} 1 & -1 & 0 & 1 & 0 \\ 0 & 3 & -1 & -1 & 0 \\ 0 & 0 & 0 & 0 & 0 \end{bmatrix}$.

Therefore, the system has infinitely many solutions, given by $3y - s - t = 0 \Leftrightarrow y = \frac{1}{3}(s + t)$ and $x - \frac{1}{3}(s+t) + t = 0$ $\Leftrightarrow x = \frac{1}{3}(s - 2t)$. So the solutions are $\left(\frac{1}{3}(s-2t), \frac{1}{3}(s+t), s, t\right)$, where $s$ and $t$ are any real numbers.

**46.** $\begin{bmatrix} 2 & -1 & 2 & 1 & 5 \\ -1 & 1 & 4 & -1 & 3 \\ 3 & -2 & -1 & 0 & 0 \end{bmatrix} \xrightarrow[-\text{R}_1]{\text{R}_1 \leftrightarrow \text{R}_2} \begin{bmatrix} 1 & -1 & -4 & 1 & -3 \\ 2 & -1 & 2 & 1 & 5 \\ 3 & -2 & -1 & 0 & 0 \end{bmatrix} \xrightarrow[\text{R}_3 - 3\text{R}_1 \to \text{R}_3]{\text{R}_2 - 2\text{R}_1 \to \text{R}_2} \begin{bmatrix} 1 & -1 & -4 & 1 & -3 \\ 0 & 1 & 10 & -1 & 11 \\ 0 & 1 & 11 & -3 & 9 \end{bmatrix}$

$\xrightarrow{\text{R}_3 - \text{R}_2 \to \text{R}_3} \begin{bmatrix} 1 & -1 & -4 & 1 & -3 \\ 0 & 1 & 10 & -1 & 11 \\ 0 & 0 & 1 & -2 & -2 \end{bmatrix}$. Thus, the system has infinitely many solutions, given by $z - 2t = -2 \Leftrightarrow$ $z = -2 + 2t$; $y + 10(-2 + 2t) - t = 11 \Leftrightarrow y = 31 - 19t$; and $x - (31 - 19t) - 4(-2 + 2t) + t = -3 \Leftrightarrow$ $x = 20 - 12t$. Hence, the solutions are $(20 - 12t, 31 - 19t, -2 + 2t, t)$, where $t$ is any real number.

**47.** Let $x$, $y$, $z$ represent the number of VitaMax, Vitron, and VitaPlus pills taken daily. The matrix representation for the system of equations is

$\begin{bmatrix} 5 & 10 & 15 & 50 \\ 15 & 20 & 0 & 50 \\ 10 & 10 & 10 & 50 \end{bmatrix} \xrightarrow[\frac{1}{5}\text{R}_3]{\frac{1}{5}\text{R}_1,\ \frac{1}{5}\text{R}_2} \begin{bmatrix} 1 & 2 & 3 & 10 \\ 3 & 4 & 0 & 10 \\ 2 & 2 & 2 & 10 \end{bmatrix} \xrightarrow[\text{R}_3 - 2\text{R}_1 \to \text{R}_3]{\text{R}_2 - 3\text{R}_1 \to \text{R}_2} \begin{bmatrix} 1 & 2 & 3 & 10 \\ 0 & -2 & -9 & -20 \\ 0 & -2 & -4 & -10 \end{bmatrix} \xrightarrow{\text{R}_3 - \text{R}_2 \to \text{R}_3} \begin{bmatrix} 1 & 2 & 3 & 10 \\ 0 & -2 & -9 & -20 \\ 0 & 0 & 5 & 10 \end{bmatrix}$.

Thus, $5z = 10 \Leftrightarrow z = 2$; $-2y - 18 = -20 \Leftrightarrow y = 1$; and $x + 2 + 6 = 10 \Leftrightarrow x = 2$. Hence, he should take 2 VitaMax, 1 Vitron, and 2 VitaPlus pills daily.

**48.** Let $x$ be the quantity, in mL, of 10% acid, $y$ the quantity of 20% acid, and $z$ the quantity of 40% acid. Then $\begin{cases} 0.1x + 0.2y + 0.4z = 18 \\ x + y + z = 100 \\ x - 4z = 0 \end{cases}$ Writing this equation in matrix form, we get $\begin{bmatrix} 1 & 1 & 1 & 100 \\ 0.1 & 0.2 & 0.4 & 18 \\ 1 & 0 & -4 & 0 \end{bmatrix} \xrightarrow[\text{R}_3 - \text{R}_1 \to \text{R}_3]{10\text{R}_2,\ \text{R}_2 - \text{R}_1 \to \text{R}_2}$

$\begin{bmatrix} 1 & 1 & 1 & 100 \\ 0 & 1 & 3 & 80 \\ 0 & -1 & -5 & -100 \end{bmatrix} \xrightarrow[-\text{R}_3]{\text{R}_3 + \text{R}_2 \to \text{R}_3} \begin{bmatrix} 1 & 1 & 1 & 100 \\ 0 & 1 & 3 & 80 \\ 0 & 0 & 2 & 20 \end{bmatrix}$. Thus $2z = 20 \Leftrightarrow z = 10$; $y + 30 = 80 \Leftrightarrow y = 50$; and $x + 50 + 10 = 100 \Leftrightarrow x = 40$. So he should mix together 40 mL of 10% acid, 50 mL of 20% acid, and 10 mL of 40% acid.

**49.** Let $x$, $y$, and $z$ represent the distance, in miles, of the run, swim, and cycle parts of the race respectively. Then, since $\text{time} = \dfrac{\text{distance}}{\text{speed}}$, we get the following equations from the three contestants' race times:

$\begin{cases} \left(\frac{x}{10}\right) + \left(\frac{y}{4}\right) + \left(\frac{z}{20}\right) = 2.5 \\ \left(\frac{x}{7.5}\right) + \left(\frac{y}{6}\right) + \left(\frac{z}{15}\right) = 3 \\ \left(\frac{x}{15}\right) + \left(\frac{y}{3}\right) + \left(\frac{z}{40}\right) = 1.75 \end{cases} \Leftrightarrow \begin{cases} 2x + 5y + z = 50 \\ 4x + 5y + 2z = 90 \\ 8x + 40y + 3z = 210 \end{cases}$ which has the following matrix representation:

$\begin{bmatrix} 2 & 5 & 1 & 50 \\ 4 & 5 & 2 & 90 \\ 8 & 40 & 3 & 210 \end{bmatrix} \xrightarrow[\text{R}_3 - 4\text{R}_1 \to \text{R}_3]{\text{R}_2 - 2\text{R}_1 \to \text{R}_2} \begin{bmatrix} 2 & 5 & 1 & 50 \\ 0 & -5 & 0 & -10 \\ 0 & 20 & -1 & 10 \end{bmatrix} \xrightarrow{\text{R}_3 + 4\text{R}_2 \to \text{R}_3} \begin{bmatrix} 2 & 5 & 1 & 50 \\ 0 & -5 & 0 & -10 \\ 0 & 0 & -1 & -30 \end{bmatrix}$.

Thus, $-z = -30 \Leftrightarrow z = 30$; $-5y = -10 \Leftrightarrow y = 2$; and $2x + 10 + 30 = 50 \Leftrightarrow x = 5$. So the race has a 5 mile run, 2 mile swim, and 30 mile cycle.

**50.** Let $a$, $b$, and $c$ be the number of students in classrooms A, B, and C, respectively, where $a, b, c \geq 0$. Then, the system of equations is $\begin{cases} a+b+c=100 \\ \frac{1}{2}a=\frac{1}{5}b \\ \frac{1}{5}b=\frac{1}{3}c \end{cases} \Leftrightarrow \begin{cases} a+b+c=100 \\ b=\frac{5}{2}a \\ c=\frac{3}{5}b=\frac{3}{2}a \end{cases}$ By substitution, $a+\frac{5}{2}a+\frac{3}{2}a=100 \Leftrightarrow a=20$; $b=\frac{5}{2}(20)=50$; and $c=\frac{3}{2}(20)=30$. So there are 20 students in classroom A, 50 students in classroom B, and 30 students in classroom C.

**51.** Let $t$ be the number of tables produced, $c$ the number of chairs, and $a$ the number of armoires. Then, the system of equations is $\begin{cases} \frac{1}{2}t+c+a=300 \\ \frac{1}{2}t+\frac{3}{2}c+a=400 \\ t+\frac{3}{2}c+2a=590 \end{cases} \Leftrightarrow \begin{cases} t+2c+2a=600 \\ t+3c+2a=800 \\ 2t+3c+4a=1180 \end{cases}$ and a matrix representation is

$$\begin{bmatrix} 1 & 2 & 2 & 600 \\ 1 & 3 & 2 & 800 \\ 2 & 3 & 4 & 1180 \end{bmatrix} \xrightarrow[\text{R}_3-2\text{R}_1\to\text{R}_3]{\text{R}_2-\text{R}_1\to\text{R}_2} \begin{bmatrix} 1 & 2 & 2 & 600 \\ 0 & 1 & 0 & 200 \\ 0 & -1 & 0 & -20 \end{bmatrix} \xrightarrow{\text{R}_3+\text{R}_2\to\text{R}_3} \begin{bmatrix} 1 & 2 & 2 & 600 \\ 0 & 1 & 0 & 200 \\ 0 & 0 & 0 & 180 \end{bmatrix}.$$

The third row states $0=180$, which is impossible, and so the system is inconsistent. Therefore, it is impossible to use all of the available labor-hours.

**52.** The number of cars entering each intersection must equal the number of cars leaving that intersection. This leads to the following equations: $\begin{cases} 200+180=x+z \\ x+70=20+w \\ w+200=y+30 \\ y+z=400+200 \end{cases}$ Simplifying and writing this in matrix form, we get

$$\begin{bmatrix} 1 & 0 & 1 & 0 & 380 \\ 1 & 0 & 0 & -1 & -50 \\ 0 & 1 & 0 & -1 & 170 \\ 0 & 1 & 1 & 0 & 600 \end{bmatrix} \xrightarrow[\text{R}_4-\text{R}_3\to\text{R}_4]{\text{R}_2-\text{R}_1\to\text{R}_2} \begin{bmatrix} 1 & 0 & 1 & 0 & 380 \\ 0 & 0 & -1 & -1 & -430 \\ 0 & 1 & 0 & -1 & 170 \\ 0 & 0 & 1 & 1 & 430 \end{bmatrix} \xrightarrow{\text{R}_2\leftrightarrow\text{R}_3} \begin{bmatrix} 1 & 0 & 1 & 0 & 380 \\ 0 & 1 & 0 & -1 & 170 \\ 0 & 0 & -1 & -1 & -430 \\ 0 & 0 & 1 & 1 & 430 \end{bmatrix}.$$

Therefore, $z+t=430 \Leftrightarrow z=430-t$; $y-t=170 \Leftrightarrow y=170+t$; and $x+430-t=380 \Leftrightarrow x=t-50$. Since $x$, $y$, $z$, $w \geq 0$, it follows that $50 \leq t \leq 430$, and so the solutions are $(t-50, 170+t, 430-t, t)$, where $50 \leq t \leq 430$.

**53.** *Line containing the points* $(0, 0)$ *and* $(1, 12)$*:* Using the general form of a line, $y=ax+b$, we substitute for $x$ and $y$ and solve for $a$ and $b$. The point $(0,0)$ gives $0=a(0)+b \Rightarrow b=0$; the point $(1,12)$ gives $12=a(1)+b \Rightarrow a=12$. Since $a=12$ and $b=0$, the equation of the line is $y=12x$.

*Quadratic containing the points* $(0,0)$, $(1,12)$, *and* $(3,6)$*:* Using the general form of a quadratic, $y=ax^2+bx+c$, we substitute for $x$ and $y$ and solve for $a$, $b$, and $c$. The point $(0,0)$ gives $0=a(0)^2+b(0)+c \Rightarrow c=0$; the point $(1,12)$ gives $12=a(1)^2+b(1)+c \Rightarrow a+b=12$; the point $(3,6)$ gives $6=a(3)^2+b(3)+c \Rightarrow 9a+3b=6$. Subtracting the third equation from $-3$ times the third gives $6a=-30 \Leftrightarrow a=-5$. So $a+b=12 \Leftrightarrow b=12-a \Rightarrow b=17$. Since $a=-5$, $b=17$, and $c=0$, the equation of the quadratic is $y=-5x^2+17x$.

*Cubic containing the points* $(0,\ 0)$, $(1,\ 12)$, $(2, 40)$, *and* $(3, 6)$: Using the general form of a cubic, $y = ax^3 + bx^2 + cx + d$, we substitute for $x$ and $y$ and solve for $a$, $b$, $c$, and $d$. The point $(0, 0)$ gives $0 = a(0)^3 + b(0)^2 + c(0) + d \quad \Rightarrow \quad d = 0$; the point the point $(1, 12)$ gives $12 = a(1)^3 + b(1)^2 + c(1) + d \quad \Rightarrow \quad a + b + c + d = 12$; the point $(2, 40)$ gives $40 = a(2)^3 + b(2)^2 + c(2) + d \quad \Rightarrow \quad 8a + 4b + 2c + d = 40$; the point $(3, 6)$ gives $6 = a(3)^3 + b(3)^2 + c(3) + d$ $\Rightarrow \quad 27a + 9b + 3c + d = 6$. Since $d = 0$, the system reduces to $\begin{cases} a + b + c = 12 \\ 8a + 4b + 2c = 40 \\ 27a + 9b + 3c = 6 \end{cases}$ which has representation

$$\begin{bmatrix} 1 & 1 & 1 & 12 \\ 8 & 4 & 2 & 40 \\ 27 & 9 & 3 & 6 \end{bmatrix} \xrightarrow[\mathrm{R}_3 - 27\mathrm{R}_1 \to \mathrm{R}_3]{\mathrm{R}_2 - 8\mathrm{R}_1 \to \mathrm{R}_2} \begin{bmatrix} 1 & 1 & 1 & 12 \\ 0 & -4 & -6 & -56 \\ 0 & -18 & -24 & -318 \end{bmatrix} \xrightarrow[-\frac{1}{6}\mathrm{R}_3]{-\frac{1}{2}\mathrm{R}_2} \begin{bmatrix} 1 & 1 & 1 & 12 \\ 0 & 2 & 3 & 28 \\ 0 & 3 & 4 & 53 \end{bmatrix} \xrightarrow{2\mathrm{R}_3 - 3\mathrm{R}_2 \to \mathrm{R}_3} \begin{bmatrix} 1 & 1 & 1 & 12 \\ 0 & 2 & 3 & 28 \\ 0 & 0 & -1 & 22 \end{bmatrix}.$$

So $c = -22$ and back-substituting we have $2b + 3(-22) = 28 \quad \Leftrightarrow \quad b = 47$ and $a + 47 + (-22) = 0 \Leftrightarrow a = -13$. So the cubic is $y = -13x^3 + 47x^2 - 22x$.

*Fourth-degree polynomial containing the points* $(0, 0)$, $(1, 12)$, $(2, 40)$, $(3, 6)$, *and* $(-1, -14)$: Using the general form of a fourth-degree polynomial, $y = ax^4 + bx^3 + cx^2 + dx + e$, we substitute for $x$ and $y$ and solve for $a$, $b$, $c$, $d$, and $e$. The point $(0, 0)$ gives $0 = a(0)^4 + b(0)^3 + c(0)^2 + d(0) + e \quad \Rightarrow \quad e = 0$; the point $(1, 12)$ gives $12 = a(1)^4 + b(1)^3 + c(1)^2 + d(1) + e$; the point $(2, 40)$ gives $40 = a(2)^4 + b(2)^3 + c(2)^2 + d(2) + e$; the point $(3, 6)$ gives $6 = a(3)^4 + b(3)^3 + c(3)^2 + d(3) + e$; the point $(-1, -14)$ gives $-14 = a(-1)^4 + b(-1)^3 + c(-1)^2 + d(-1) + e$. Since the first equation is $e = 0$, we eliminate $e$ from the other equations to get the system $\begin{cases} a + b + c + d = 12 \\ 16a + 8b + 4c + 2d = 40 \\ 81a + 27b + 9c + 3d = 6 \\ a - b + c - d = -14 \end{cases}$

which has the matrix representation

$$\begin{bmatrix} 1 & 1 & 1 & 1 & 12 \\ 16 & 8 & 4 & 2 & 40 \\ 81 & 27 & 9 & 3 & 6 \\ 1 & -1 & 1 & -1 & -14 \end{bmatrix} \xrightarrow[\mathrm{R}_4 - \mathrm{R}_1 \to \mathrm{R}_4]{\substack{\mathrm{R}_2 - 16\mathrm{R}_1 \to \mathrm{R}_2 \\ \mathrm{R}_3 - 81\mathrm{R}_1 \to \mathrm{R}_3}} \begin{bmatrix} 1 & 1 & 1 & 1 & 12 \\ 0 & -8 & -12 & -14 & -152 \\ 0 & -54 & -72 & -78 & -966 \\ 0 & -2 & 0 & -2 & -26 \end{bmatrix} \xrightarrow[\mathrm{R}_3 \to \mathrm{R}_4]{\substack{-\frac{1}{2}\mathrm{R}_4 \to \mathrm{R}_2 \\ \mathrm{R}_2 \to \mathrm{R}_3}}$$

$$\begin{bmatrix} 1 & 1 & 1 & 1 & 12 \\ 0 & 1 & 0 & 1 & 13 \\ 0 & -8 & -12 & -14 & -152 \\ 0 & -54 & -72 & -78 & -966 \end{bmatrix} \xrightarrow[\mathrm{R}_4 + 54\mathrm{R}_2 \to \mathrm{R}_4]{\mathrm{R}_3 + 8\mathrm{R}_2 \to \mathrm{R}_3} \begin{bmatrix} 1 & 1 & 1 & 1 & 12 \\ 0 & 1 & 0 & 1 & 13 \\ 0 & 0 & -12 & -6 & -48 \\ 0 & 0 & -72 & -24 & -264 \end{bmatrix} \xrightarrow{\mathrm{R}_4 - 6\mathrm{R}_3 \to \mathrm{R}_4} \begin{bmatrix} 1 & 1 & 1 & 1 & 12 \\ 0 & 1 & 0 & 1 & 13 \\ 0 & 0 & -12 & -6 & -48 \\ 0 & 0 & 0 & 12 & 24 \end{bmatrix}.$$

So $d = 2$. Then $-12c - 6(2) = -48 \quad \Leftrightarrow \quad c = 3$and $b + 2 = 13 \quad \Leftrightarrow$ $b = 11$. Finally, $a + 11 + 3 + 2 = 12 \quad \Leftrightarrow \quad a = -4$. So the fourth-degree polynomial containing these points is $y = -4x^4 + 11x^3 + 3x^2 + 2x$.

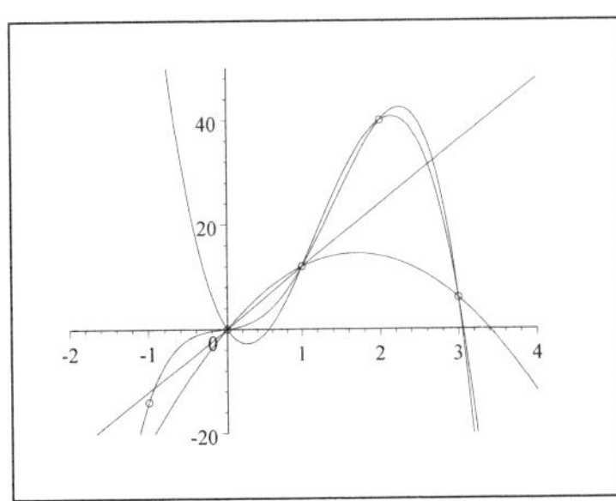

# 10.5 The Algebra of Matrices

**1.** The matrices have different dimensions, so they cannot be equal.

**2.** Since $\frac{1}{4} = 0.25$, $\ln 1 = 0$, $2 = \sqrt{4}$, and $3 = \frac{6}{2}$, the corresponding entries are equal, so the matrices are equal.

**3.** $\begin{bmatrix} 2 & 6 \\ -5 & 3 \end{bmatrix} + \begin{bmatrix} -1 & -3 \\ 6 & 2 \end{bmatrix} = \begin{bmatrix} 1 & 3 \\ 1 & 5 \end{bmatrix}$

**4.** $\begin{bmatrix} 0 & 1 & 1 \\ 1 & 1 & 0 \end{bmatrix} - \begin{bmatrix} 2 & 1 & -1 \\ 1 & 3 & -2 \end{bmatrix} = \begin{bmatrix} -2 & 0 & 2 \\ 0 & -2 & 2 \end{bmatrix}$

**5.** $3\begin{bmatrix} 1 & 2 \\ 4 & -1 \\ 1 & 0 \end{bmatrix} = \begin{bmatrix} 3 & 6 \\ 12 & -3 \\ 3 & 0 \end{bmatrix}$

**6.** $2\begin{bmatrix} 1 & 1 & 0 \\ 1 & 0 & 1 \\ 0 & 1 & 1 \end{bmatrix} + \begin{bmatrix} 1 & 1 \\ 2 & 1 \\ 3 & 1 \end{bmatrix}$ is undefined because these matrices have incompatible dimensions.

**7.** $\begin{bmatrix} 2 & 6 \\ 1 & 3 \\ 2 & 4 \end{bmatrix}\begin{bmatrix} 1 & -2 \\ 3 & 6 \\ -2 & 0 \end{bmatrix}$ is undefined because these matrices have incompatible dimensions.

**8.** $\begin{bmatrix} 2 & 1 & 2 \\ 6 & 3 & 4 \end{bmatrix}\begin{bmatrix} 1 & -2 \\ 3 & 6 \\ -2 & 0 \end{bmatrix} = \begin{bmatrix} 1 & 2 \\ 7 & 64 \end{bmatrix}$

**9.** $\begin{bmatrix} 1 & 2 \\ -1 & 4 \end{bmatrix}\begin{bmatrix} 1 & -2 & 3 \\ 2 & 2 & -1 \end{bmatrix} = \begin{bmatrix} 5 & 2 & 1 \\ 7 & 10 & -7 \end{bmatrix}$

**10.** $\begin{bmatrix} 2 & -3 \\ 0 & 1 \\ 1 & 2 \end{bmatrix}\begin{bmatrix} 5 \\ 1 \end{bmatrix} = \begin{bmatrix} 7 \\ 1 \\ 7 \end{bmatrix}$

**11.** $2X + A = B \quad \Leftrightarrow \quad X = \frac{1}{2}(B - A) = \dfrac{1}{2}\left(\begin{bmatrix} 2 & 5 \\ 3 & 7 \end{bmatrix} - \begin{bmatrix} 4 & 6 \\ 1 & 3 \end{bmatrix}\right) = \dfrac{1}{2}\begin{bmatrix} -2 & -1 \\ 2 & 4 \end{bmatrix} = \begin{bmatrix} -1 & -\frac{1}{2} \\ 1 & 2 \end{bmatrix}$.

**12.** $3X - B = C \quad \Leftrightarrow \quad 3X = C + B$. but $C$ is a $3 \times 2$ matrix and $B$ is a $2 \times 2$ matrix, so $C + B$ is impossible. Thus, there is no solution.

**13.** $2(B - X) = D$. Since $B$ is a $2 \times 2$ matrix, $B - X$ is defined only when $X$ is a $2 \times 2$ matrix, so $2(B - X)$ is a $2 \times 2$ matrix. But $D$ is a $3 \times 2$ matrix. Thus, there is no solution.

**14.** $5(X - C) = D \quad \Leftrightarrow \quad X - C = \frac{1}{5}D \quad \Leftrightarrow$

$$X = \tfrac{1}{5}D + C = \frac{1}{5}\begin{bmatrix} 10 & 20 \\ 30 & 20 \\ 10 & 0 \end{bmatrix} + \begin{bmatrix} 2 & 3 \\ 1 & 0 \\ 0 & 2 \end{bmatrix} = \begin{bmatrix} 2 & 4 \\ 6 & 4 \\ 2 & 0 \end{bmatrix} + \begin{bmatrix} 2 & 3 \\ 1 & 0 \\ 0 & 2 \end{bmatrix} = \begin{bmatrix} 4 & 7 \\ 7 & 4 \\ 2 & 2 \end{bmatrix}.$$

**15.** $\frac{1}{5}(X + D) = C \quad \Leftrightarrow \quad X + D = 5C \quad \Leftrightarrow$

$$X = 5C - D = 5\begin{bmatrix} 2 & 3 \\ 1 & 0 \\ 0 & 2 \end{bmatrix} - \begin{bmatrix} 10 & 20 \\ 30 & 20 \\ 10 & 0 \end{bmatrix} = \begin{bmatrix} 10 & 15 \\ 5 & 0 \\ 0 & 10 \end{bmatrix} - \begin{bmatrix} 10 & 20 \\ 30 & 20 \\ 10 & 0 \end{bmatrix} = \begin{bmatrix} 0 & -5 \\ -25 & -20 \\ -10 & 10 \end{bmatrix}.$$

**16.** $2A = B - 3X \quad \Leftrightarrow \quad 2A - B = -3X \quad \Leftrightarrow$

$X = -\frac{1}{3}(2A - B)$

$$= -\frac{1}{3}\left(2\begin{bmatrix} 4 & 6 \\ 1 & 3 \end{bmatrix} - \begin{bmatrix} 2 & 5 \\ 3 & 7 \end{bmatrix}\right) = -\frac{1}{3}\left(\begin{bmatrix} 8 & 12 \\ 2 & 6 \end{bmatrix} - \begin{bmatrix} 2 & 5 \\ 3 & 7 \end{bmatrix}\right) = -\frac{1}{3}\begin{bmatrix} 6 & 7 \\ -1 & -1 \end{bmatrix} = \begin{bmatrix} -2 & -\frac{7}{3} \\ \frac{1}{3} & \frac{1}{3} \end{bmatrix}$$

**In Solutions 17–38, the matrices $A$, $B$, $C$, $D$, $E$, $F$, and $G$ are defined as follows:**

$$A = \begin{bmatrix} 2 & -5 \\ 0 & 7 \end{bmatrix} \qquad B = \begin{bmatrix} 3 & \frac{1}{2} & 5 \\ 1 & -1 & 3 \end{bmatrix} \qquad C = \begin{bmatrix} 2 & -\frac{5}{2} & 0 \\ 0 & 2 & -3 \end{bmatrix}$$

$$D = \begin{bmatrix} 7 & 3 \end{bmatrix} \qquad E = \begin{bmatrix} 1 \\ 2 \\ 0 \end{bmatrix} \qquad F = \begin{bmatrix} 1 & 0 & 0 \\ 0 & 1 & 0 \\ 0 & 0 & 1 \end{bmatrix} \qquad G = \begin{bmatrix} 5 & -3 & 10 \\ 6 & 1 & 0 \\ -5 & 2 & 2 \end{bmatrix}$$

**17.** $B + C = \begin{bmatrix} 3 & \frac{1}{2} & 5 \\ 1 & -1 & 3 \end{bmatrix} + \begin{bmatrix} 2 & -\frac{5}{2} & 0 \\ 0 & 2 & -3 \end{bmatrix} = \begin{bmatrix} 5 & -2 & 5 \\ 1 & 1 & 0 \end{bmatrix}$

**18.** $B + F$ is undefined because $B$ $(2 \times 3)$ and $F$ $(3 \times 3)$ have incompatible dimensions.

**19.** $C - B = \begin{bmatrix} 2 & -\frac{5}{2} & 0 \\ 0 & 2 & -3 \end{bmatrix} - \begin{bmatrix} 3 & \frac{1}{2} & 5 \\ 1 & -1 & 3 \end{bmatrix} = \begin{bmatrix} -1 & -3 & -5 \\ -1 & 3 & -6 \end{bmatrix}$

**20.** $5A = 5\begin{bmatrix} 2 & -5 \\ 0 & 7 \end{bmatrix} = \begin{bmatrix} 10 & -25 \\ 0 & 35 \end{bmatrix}$

**21.** $3B + 2C = 3\begin{bmatrix} 3 & \frac{1}{2} & 5 \\ 1 & -1 & 3 \end{bmatrix} + 2\begin{bmatrix} 2 & -\frac{5}{2} & 0 \\ 0 & 2 & -3 \end{bmatrix} = \begin{bmatrix} 13 & -\frac{7}{2} & 15 \\ 3 & 1 & 3 \end{bmatrix}$

**22.** $C - 5A$ is undefined because $C$ $(2 \times 3)$ and $A$ $(2 \times 2)$ have incompatible dimensions.

**23.** $2C - 6B = 2\begin{bmatrix} 2 & -\frac{5}{2} & 0 \\ 0 & 2 & -3 \end{bmatrix} - 6\begin{bmatrix} 3 & \frac{1}{2} & 5 \\ 1 & -1 & 3 \end{bmatrix} = \begin{bmatrix} -14 & -8 & -30 \\ -6 & 10 & -24 \end{bmatrix}$

**24.** $DA = \begin{bmatrix} 7 & 3 \end{bmatrix}\begin{bmatrix} 2 & -5 \\ 0 & 7 \end{bmatrix} = \begin{bmatrix} 14 & -14 \end{bmatrix}$

**25.** $AD$ is undefined because $A$ $(2 \times 2)$ and $D$ $(1 \times 2)$ have incompatible dimensions.

**26.** $BC$ is undefined because $B$ $(2 \times 3)$ and $C$ $(2 \times 3)$ have incompatible dimensions.

**27.** $BF = \begin{bmatrix} 3 & \frac{1}{2} & 5 \\ 1 & -1 & 3 \end{bmatrix}\begin{bmatrix} 1 & 0 & 0 \\ 0 & 1 & 0 \\ 0 & 0 & 1 \end{bmatrix} = \begin{bmatrix} 3 & \frac{1}{2} & 5 \\ 1 & -1 & 3 \end{bmatrix}$

**28.** $GF = \begin{bmatrix} 5 & -3 & 10 \\ 6 & 1 & 0 \\ -5 & 2 & 2 \end{bmatrix}\begin{bmatrix} 1 & 0 & 0 \\ 0 & 1 & 0 \\ 0 & 0 & 1 \end{bmatrix} = \begin{bmatrix} 5 & -3 & 10 \\ 6 & 1 & 0 \\ -5 & 2 & 2 \end{bmatrix}$

**29.** $(DA)\,B = \begin{bmatrix} 7 & 3 \end{bmatrix}\begin{bmatrix} 2 & -5 \\ 0 & 7 \end{bmatrix}\begin{bmatrix} 3 & \frac{1}{2} & 5 \\ 1 & -1 & 3 \end{bmatrix} = \begin{bmatrix} 14 & -14 \end{bmatrix}\begin{bmatrix} 3 & \frac{1}{2} & 5 \\ 1 & -1 & 3 \end{bmatrix} = \begin{bmatrix} 28 & 21 & 28 \end{bmatrix}$

**30.** $D(AB) = \begin{bmatrix} 7 & 3 \end{bmatrix} \begin{bmatrix} 2 & -5 \\ 0 & 7 \end{bmatrix} \begin{bmatrix} 3 & \frac{1}{2} & 5 \\ 1 & -1 & 3 \end{bmatrix} = \begin{bmatrix} 7 & 3 \end{bmatrix} \begin{bmatrix} 1 & 6 & -5 \\ 7 & -7 & 21 \end{bmatrix} = \begin{bmatrix} 28 & 21 & 28 \end{bmatrix}$

**31.** $GE = \begin{bmatrix} 5 & -3 & 10 \\ 6 & 1 & 0 \\ -5 & 2 & 2 \end{bmatrix} \begin{bmatrix} 1 \\ 2 \\ 0 \end{bmatrix} = \begin{bmatrix} -1 \\ 8 \\ -1 \end{bmatrix}$

**32.** $A^2 = \begin{bmatrix} 2 & -5 \\ 0 & 7 \end{bmatrix} \begin{bmatrix} 2 & -5 \\ 0 & 7 \end{bmatrix} = \begin{bmatrix} 4 & -45 \\ 0 & 49 \end{bmatrix}$

**33.** $A^3 = \begin{bmatrix} 2 & -5 \\ 0 & 7 \end{bmatrix} \begin{bmatrix} 2 & -5 \\ 0 & 7 \end{bmatrix} \begin{bmatrix} 2 & -5 \\ 0 & 7 \end{bmatrix} = \begin{bmatrix} 4 & -45 \\ 0 & 49 \end{bmatrix} \begin{bmatrix} 2 & -5 \\ 0 & 7 \end{bmatrix} = \begin{bmatrix} 8 & -335 \\ 0 & 343 \end{bmatrix}$

**34.** $DB+DC = D(B+C) = \begin{bmatrix} 7 & 3 \end{bmatrix} \left( \begin{bmatrix} 3 & \frac{1}{2} & 5 \\ 1 & -1 & 3 \end{bmatrix} + \begin{bmatrix} 2 & -\frac{5}{2} & 0 \\ 0 & 2 & -3 \end{bmatrix} \right) = \begin{bmatrix} 7 & 3 \end{bmatrix} \begin{bmatrix} 5 & -2 & 5 \\ 1 & 1 & 0 \end{bmatrix} = \begin{bmatrix} 38 & -11 & 35 \end{bmatrix}$

**35.** $B^2$ is undefined because the dimensions $2 \times 3$ and $2 \times 3$ are incompatible.

**36.** $F^2 = \begin{bmatrix} 1 & 0 & 0 \\ 0 & 1 & 0 \\ 0 & 0 & 1 \end{bmatrix} \begin{bmatrix} 1 & 0 & 0 \\ 0 & 1 & 0 \\ 0 & 0 & 1 \end{bmatrix} = \begin{bmatrix} 1 & 0 & 0 \\ 0 & 1 & 0 \\ 0 & 0 & 1 \end{bmatrix}$

**37.** $BF + FE$ is undefined because the dimensions $(2 \times 3) \cdot (3 \times 3) = (2 \times 3)$ and $(3 \times 3) \cdot (3 \times 1) = (3 \times 1)$ are incompatible.

**38.** $ABE = \begin{bmatrix} 2 & -5 \\ 0 & 7 \end{bmatrix} \begin{bmatrix} 3 & \frac{1}{2} & 5 \\ 1 & -1 & 3 \end{bmatrix} \begin{bmatrix} 1 \\ 2 \\ 0 \end{bmatrix} = \begin{bmatrix} 1 & 6 & -5 \\ 7 & -7 & 21 \end{bmatrix} \begin{bmatrix} 1 \\ 2 \\ 0 \end{bmatrix} = \begin{bmatrix} 13 \\ -7 \end{bmatrix}$

**39.** $\begin{bmatrix} x & 2y \\ 4 & 6 \end{bmatrix} = \begin{bmatrix} 2 & -2 \\ 2x & -6y \end{bmatrix}$. Thus we must solve the system $\begin{cases} x = 2 \\ 2y = -2 \\ 4 = 2x \\ 6 = -6y \end{cases}$ So $x = 2$ and $2y = -2 \quad \Leftrightarrow \quad y = -1$.

Since these values for $x$ and $y$ also satisfy the last two equations, the solution is $x = 2$, $y = -1$.

**40.** $3 \cdot \begin{bmatrix} x & y \\ y & x \end{bmatrix} = \begin{bmatrix} 6 & -9 \\ -9 & 6 \end{bmatrix}$. Since $3 \cdot \begin{bmatrix} x & y \\ y & x \end{bmatrix} = \begin{bmatrix} 3x & 3y \\ 3y & 3x \end{bmatrix}$, we must solve the system $\begin{cases} 3x = 6 \\ 3y = -9 \\ 3x = 6 \\ 3y = -9 \end{cases}$ So $3x = 6 \quad \Leftrightarrow$

$x = 2$ and $3y = -9 \quad \Leftrightarrow \quad y = -3$. Thus, the solution is $x = 2$, $y = -3$.

**41.** $2\begin{bmatrix} x & y \\ x+y & x-y \end{bmatrix} = \begin{bmatrix} 2 & -4 \\ -2 & 6 \end{bmatrix}$. Since $2\begin{bmatrix} x & y \\ x+y & x-y \end{bmatrix} = \begin{bmatrix} 2x & 2y \\ 2(x+y) & 2(x-y) \end{bmatrix}$, Thus we must solve the system $\begin{cases} 2x = 2 \\ 2y = -4 \\ 2(x+y) = -2 \\ 2(x-y) = 6 \end{cases}$ So $x = 1$ and $y = -2$. Since these values for $x$ and $y$ also satisfy the last two equations, the solution is $x = 1$, $y = -2$.

**42.** $\begin{bmatrix} x & y \\ -y & x \end{bmatrix} - \begin{bmatrix} y & x \\ x & -y \end{bmatrix} = \begin{bmatrix} 4 & -4 \\ -6 & 6 \end{bmatrix} \Leftrightarrow \begin{bmatrix} x-y & y-x \\ -x-y & x+y \end{bmatrix} = \begin{bmatrix} 4 & -4 \\ -6 & 6 \end{bmatrix}$. Thus we must solve the system $\begin{cases} x-y = 4 \\ y-x = -4 \\ -x-y = -6 \\ x+y = 6 \end{cases}$ Adding the first equation to the last equation gives $2x = 10 \Leftrightarrow x = 5$ and $5 + y = 6$ $\Leftrightarrow y = 1$. So the solution is $x = 5$, $y = 1$.

**43.** $\begin{cases} 2x - 5y = 7 \\ 3x + 2y = 4 \end{cases}$ written as a matrix equation is $\begin{bmatrix} 2 & -5 \\ 3 & 2 \end{bmatrix}\begin{bmatrix} x \\ y \end{bmatrix} = \begin{bmatrix} 7 \\ 4 \end{bmatrix}$.

**44.** $\begin{cases} 6x - y + z = 12 \\ 2x + z = 7 \\ y - 2z = 4 \end{cases}$ written as a matrix equation is $\begin{bmatrix} 6 & -1 & 1 \\ 2 & 0 & 1 \\ 0 & 1 & -2 \end{bmatrix}\begin{bmatrix} x \\ y \\ z \end{bmatrix} = \begin{bmatrix} 12 \\ 7 \\ 4 \end{bmatrix}$.

**45.** $\begin{cases} 3x_1 + 2x_2 - x_3 + x_4 = 0 \\ x_1 - x_3 = 5 \\ 3x_2 + x_3 - x_4 = 4 \end{cases}$ written as a matrix equation is $\begin{bmatrix} 3 & 2 & -1 & 1 \\ 1 & 0 & -1 & 0 \\ 0 & 3 & 1 & -1 \end{bmatrix}\begin{bmatrix} x_1 \\ x_2 \\ x_3 \\ x_4 \end{bmatrix} = \begin{bmatrix} 0 \\ 5 \\ 4 \end{bmatrix}$.

**46.** $\begin{cases} x - y + z = 2 \\ 4x - 2y - z = 2 \\ x + y + 5z = 2 \\ -x - y - z = 2 \end{cases}$ written as a matrix equation is $\begin{bmatrix} 1 & -1 & 1 \\ 4 & -2 & -1 \\ 1 & 1 & 5 \\ -1 & -1 & -1 \end{bmatrix}\begin{bmatrix} x \\ y \\ z \end{bmatrix} = \begin{bmatrix} 2 \\ 2 \\ 2 \\ 2 \end{bmatrix}$.

**47.** $A = \begin{bmatrix} 1 & 0 & 6 & -1 \\ 2 & \frac{1}{2} & 4 & 0 \end{bmatrix}$, $B = \begin{bmatrix} 1 & 7 & -9 & 2 \end{bmatrix}$, and $C = \begin{bmatrix} 1 \\ 0 \\ -1 \\ -2 \end{bmatrix}$. $ABC$ is undefined because the dimensions of $A$ $(2 \times 4)$ and $B$ $(1 \times 4)$ are not compatible. $ACB = \begin{bmatrix} -3 \\ -2 \end{bmatrix}\begin{bmatrix} 1 & 7 & -9 & 2 \end{bmatrix} = \begin{bmatrix} -3 & -21 & 27 & -6 \\ -2 & -14 & 18 & -4 \end{bmatrix}$. $BAC$ is undefined because the dimensions of $B$ $(1 \times 4)$ and $A$ $(2 \times 4)$ are not compatible. $BCA$ is undefined because the dimensions of $C$ $(4 \times 1)$ and $A$ $(2 \times 4)$ are not compatible. $CAB$ is undefined because the dimensions of $C$ $(4 \times 1)$ and $A$ $(2 \times 4)$ are not compatible. $CBA$ is undefined because the dimensions of $B$ $(1 \times 4)$ and $A$ $(2 \times 4)$ are not compatible.

**48. (a)** Let $A = \begin{bmatrix} a & b \\ c & d \end{bmatrix}$ and $B = \begin{bmatrix} e & f \\ g & h \end{bmatrix}$. Then $A + B = \begin{bmatrix} a+e & b+f \\ c+g & d+h \end{bmatrix}$, and

$$(A+B)^2 = \begin{bmatrix} a+e & b+f \\ c+g & d+h \end{bmatrix} \begin{bmatrix} a+e & b+f \\ c+g & d+h \end{bmatrix}$$
$$= \begin{bmatrix} (a+e)^2 + (b+f)(c+g) & (a+e)(b+f) + (b+f)(d+h) \\ (c+g)(a+e) + (d+h)(c+g) & (c+g)(b+f) + (d+h)^2 \end{bmatrix};$$

$$A^2 = \begin{bmatrix} a & b \\ c & d \end{bmatrix} \begin{bmatrix} a & b \\ c & d \end{bmatrix} = \begin{bmatrix} a^2+bc & ab+bd \\ ac+cd & bc+d^2 \end{bmatrix}; B^2 = \begin{bmatrix} e & f \\ g & h \end{bmatrix} \begin{bmatrix} e & f \\ g & h \end{bmatrix} = \begin{bmatrix} e^2+fg & ef+fh \\ eg+gh & fg+h^2 \end{bmatrix};$$

$$AB = \begin{bmatrix} a & b \\ c & d \end{bmatrix} \begin{bmatrix} e & f \\ g & h \end{bmatrix} = \begin{bmatrix} ae+bg & af+bh \\ ce+dg & cf+dh \end{bmatrix}; BA = \begin{bmatrix} e & f \\ g & h \end{bmatrix} \begin{bmatrix} a & b \\ c & d \end{bmatrix} = \begin{bmatrix} ae+cf & be+df \\ ag+ch & bg+dh \end{bmatrix}. \text{ Then}$$

$$A^2 + AB + BA + B^2 = \begin{bmatrix} a^2+bc+ae+bg+ae+cf+e^2+fg & ab+bd+ef+fh+af+bh+be+df \\ ac+cd+eg+gh+ce+dg+ag+ch & bc+d^2+fg+h^2+cf+dh+bg+dh \end{bmatrix}$$
$$= \begin{bmatrix} a^2+2ae+e^2+b(c+g)+f(c+g) & a(b+f)+e(b+f)+b(d+h)+f(d+h) \\ c(a+e)+g(a+e)+d(c+g)+h(c+g) & c(b+f)+g(b+f)+d^2+2dh+h^2 \end{bmatrix}$$
$$= \begin{bmatrix} (a+e)^2+(b+f)(c+g) & (a+e)(b+f)+(b+f)(d+h) \\ (c+g)(a+e)+(d+h)(c+g) & (c+g)(b+f)+(d+h)^2 \end{bmatrix} = (A+B)^2$$

**(b)** No. From Exercise 46, $(A+B)^2 = (A+B)(A+B) = A^2 + AB + BA + B^2 \neq A^2 + 2AB + B^2$ unless $AB = BA$ which is, in general, not true, as we saw in Example 3.

**49. (a)** $BA = \begin{bmatrix} \$0.90 & \$0.80 & \$1.10 \end{bmatrix} \begin{bmatrix} 4000 & 1000 & 3500 \\ 400 & 300 & 200 \\ 700 & 500 & 9000 \end{bmatrix} = \begin{bmatrix} \$4690 & \$1690 & \$13{,}210 \end{bmatrix}$

**(b)** The entries in the product matrix represent the total food sales in Santa Monica, Long Beach, and Anaheim, respectively.

**50. (a)** $AB = \begin{bmatrix} 12 & 10 & 0 \\ 4 & 4 & 20 \\ 8 & 9 & 12 \end{bmatrix} \begin{bmatrix} \$1000 & \$500 \\ \$2000 & \$1200 \\ \$1500 & \$1000 \end{bmatrix} = \begin{bmatrix} \$32{,}000 & \$18{,}000 \\ \$42{,}000 & \$26{,}800 \\ \$44{,}000 & \$26{,}800 \end{bmatrix}$

**(b)** The daily profit in January from the Biloxi plant is the $(2, 1)$ matrix entry, namely \$42,000.

**(c)** The total daily profit from all three plants in February was \$18,000 + \$26,800 + \$26,800 = \$71,600.

**51. (a)** $AB = \begin{bmatrix} 6 & 10 & 14 & 28 \end{bmatrix} \begin{bmatrix} 2000 & 2500 \\ 3000 & 1500 \\ 2500 & 1000 \\ 1000 & 500 \end{bmatrix} = \begin{bmatrix} 105{,}000 & 58{,}000 \end{bmatrix}$

**(b)** That day they canned 105,000 ounces of tomato sauce and 58,000 ounces of tomato paste.

**52. (a)** $AC = \begin{bmatrix} 120 & 50 & 60 \\ 40 & 25 & 30 \\ 60 & 30 & 20 \end{bmatrix} \begin{bmatrix} 0.10 \\ 0.50 \\ 1.00 \end{bmatrix} = \begin{bmatrix} 97.00 \\ 46.50 \\ 41.00 \end{bmatrix}$ Amy's stand sold \$97 worth of produce on Saturday, Beth's stand sold \$46.50 worth, and Chad's stand sold \$41 worth.

**(b)** $BC = \begin{bmatrix} 100 & 60 & 30 \\ 35 & 20 & 20 \\ 60 & 25 & 30 \end{bmatrix} \begin{bmatrix} 0.10 \\ 0.50 \\ 1.00 \end{bmatrix} = \begin{bmatrix} 70.00 \\ 33.50 \\ 48.50 \end{bmatrix}$ Amy's stand sold \$70 worth of produce on Sunday, Beth's stand sold \$33.50 worth, and Chad's stand sold \$48.50 worth.

**(c)** $A + B = \begin{bmatrix} 120 & 50 & 60 \\ 40 & 25 & 30 \\ 60 & 30 & 20 \end{bmatrix} + \begin{bmatrix} 100 & 60 & 30 \\ 35 & 20 & 20 \\ 60 & 25 & 30 \end{bmatrix} = \begin{bmatrix} 220 & 110 & 90 \\ 75 & 45 & 50 \\ 120 & 55 & 50 \end{bmatrix}$ This represents the melons, squash, and tomatoes they sold during the weekend.

**(d)** $(A + B) C = \left( \begin{bmatrix} 120 & 50 & 60 \\ 40 & 25 & 30 \\ 60 & 30 & 20 \end{bmatrix} + \begin{bmatrix} 100 & 60 & 30 \\ 35 & 20 & 20 \\ 60 & 25 & 30 \end{bmatrix} \right) \begin{bmatrix} 0.10 \\ 0.50 \\ 1.00 \end{bmatrix} = \begin{bmatrix} 220 & 110 & 90 \\ 75 & 45 & 50 \\ 120 & 55 & 50 \end{bmatrix} \begin{bmatrix} 0.10 \\ 0.50 \\ 1.00 \end{bmatrix} = \begin{bmatrix} 167.00 \\ 80.00 \\ 89.50 \end{bmatrix}$

During the weekend, Amy's stand sold \$167 worth, Beth's stand sold \$80 worth, and Chad's stand sold \$89.50 worth of produce. Notice that $(A + B) C = AC + BC = \begin{bmatrix} 97.00 \\ 46.50 \\ 41.00 \end{bmatrix} + \begin{bmatrix} 70.00 \\ 33.50 \\ 48.50 \end{bmatrix} = \begin{bmatrix} 167.00 \\ 80.00 \\ 89.50 \end{bmatrix}$.

**53. (a)** $\begin{bmatrix} 1 & 0 & 1 & 0 & 1 & 1 \\ 0 & 3 & 0 & 1 & 2 & 1 \\ 1 & 2 & 0 & 0 & 3 & 0 \\ 1 & 3 & 2 & 3 & 2 & 0 \\ 0 & 3 & 0 & 0 & 2 & 1 \\ 1 & 2 & 0 & 1 & 3 & 1 \end{bmatrix}$ **(b)** $\begin{bmatrix} 2 & 1 & 2 & 1 & 2 & 2 \\ 1 & 3 & 1 & 2 & 3 & 2 \\ 2 & 3 & 1 & 1 & 3 & 1 \\ 2 & 3 & 3 & 3 & 3 & 1 \\ 1 & 3 & 1 & 1 & 3 & 2 \\ 2 & 3 & 1 & 2 & 3 & 2 \end{bmatrix}$ **(c)** $\begin{bmatrix} 2 & 3 & 2 & 3 & 2 & 2 \\ 3 & 0 & 3 & 2 & 1 & 2 \\ 2 & 1 & 3 & 3 & 0 & 3 \\ 2 & 0 & 1 & 0 & 1 & 3 \\ 3 & 0 & 3 & 3 & 1 & 2 \\ 2 & 1 & 3 & 2 & 0 & 2 \end{bmatrix}$

**(d)** 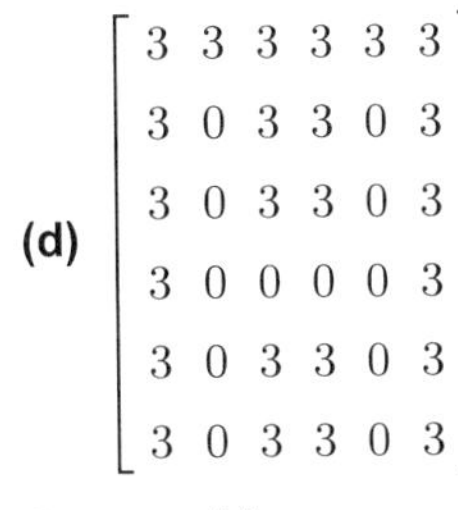

$\begin{bmatrix} 3 & 3 & 3 & 3 & 3 & 3 \\ 3 & 0 & 3 & 3 & 0 & 3 \\ 3 & 0 & 3 & 3 & 0 & 3 \\ 3 & 0 & 0 & 0 & 0 & 3 \\ 3 & 0 & 3 & 3 & 0 & 3 \\ 3 & 0 & 3 & 3 & 0 & 3 \end{bmatrix}$

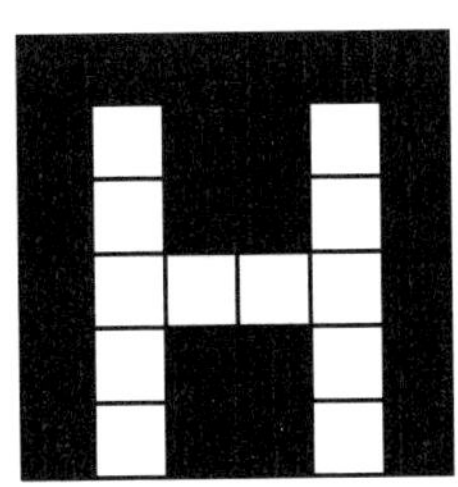

**(e)** 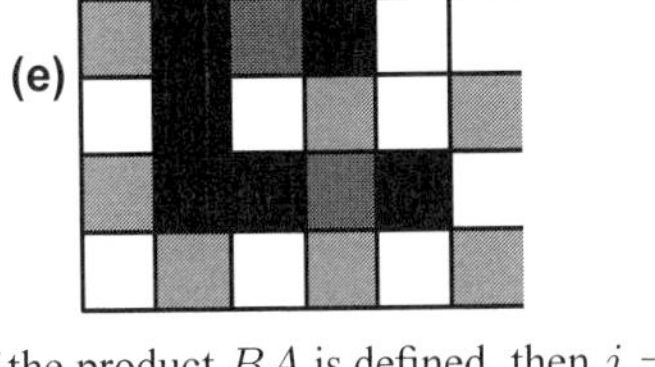

**54.** Suppose $A$ is $n \times m$ and $B$ is $i \times j$. If the product $AB$ is defined, then $m = i$. If the product $BA$ is defined, then $j = n$. Thus if both products are defined and if $A$ is $n \times m$, then $B$ must be $m \times n$.

**55.** $A = \begin{bmatrix} 1 & 1 \\ 0 & 1 \end{bmatrix}$; $A^2 = \begin{bmatrix} 1 & 1 \\ 0 & 1 \end{bmatrix} \begin{bmatrix} 1 & 1 \\ 0 & 1 \end{bmatrix} = \begin{bmatrix} 1 & 2 \\ 0 & 1 \end{bmatrix}$; $A^3 = A \cdot A^2 = \begin{bmatrix} 1 & 1 \\ 0 & 1 \end{bmatrix} \begin{bmatrix} 1 & 2 \\ 0 & 1 \end{bmatrix} = \begin{bmatrix} 1 & 3 \\ 0 & 1 \end{bmatrix}$;

$A^4 = A \cdot A^3 = \begin{bmatrix} 1 & 1 \\ 0 & 1 \end{bmatrix} \begin{bmatrix} 1 & 3 \\ 0 & 1 \end{bmatrix} = \begin{bmatrix} 1 & 4 \\ 0 & 1 \end{bmatrix}$. Therefore, it seems that $A^n = \begin{bmatrix} 1 & n \\ 0 & 1 \end{bmatrix}$.

**56.** $A = \begin{bmatrix} 1 & 1 \\ 1 & 1 \end{bmatrix}$; $A^2 = \begin{bmatrix} 1 & 1 \\ 1 & 1 \end{bmatrix}\begin{bmatrix} 1 & 1 \\ 1 & 1 \end{bmatrix} = \begin{bmatrix} 2 & 2 \\ 2 & 2 \end{bmatrix}$; $A^3 = A \cdot A^2 = \begin{bmatrix} 1 & 1 \\ 1 & 1 \end{bmatrix}\begin{bmatrix} 2 & 2 \\ 2 & 2 \end{bmatrix} = \begin{bmatrix} 4 & 4 \\ 4 & 4 \end{bmatrix}$;

$A^4 = A \cdot A^3 = \begin{bmatrix} 1 & 1 \\ 1 & 1 \end{bmatrix}\begin{bmatrix} 4 & 4 \\ 4 & 4 \end{bmatrix} = \begin{bmatrix} 8 & 8 \\ 8 & 8 \end{bmatrix}$. From this pattern, we see that $A^n = \begin{bmatrix} 2^{n-1} & 2^{n-1} \\ 2^{n-1} & 2^{n-1} \end{bmatrix}$.

**57.** Let $A = \begin{bmatrix} a & b \\ c & d \end{bmatrix}$. For the first matrix, we have $A^2 = \begin{bmatrix} a & b \\ c & d \end{bmatrix}$.

$$\begin{bmatrix} a & b \\ c & d \end{bmatrix} = \begin{bmatrix} a^2+bc & ab+bd \\ ac+cd & bc+d^2 \end{bmatrix} = \begin{bmatrix} a^2+bc & b(a+d) \\ c(a+d) & bc+d^2 \end{bmatrix}. \text{ So } A^2 = \begin{bmatrix} 4 & 0 \\ 0 & 9 \end{bmatrix} \Leftrightarrow \begin{cases} a^2+bc=4 \\ b(a+d)=0 \\ c(a+d)=0 \\ bc+d^2=9 \end{cases}$$

If $a+d=0$, then $a=-d$, so $4=a^2+bc=(-d)^2+bc=d^2+bc=9$, which is a contradiction. Thus $a+d \neq 0$. Since $b(a+d)=0$ and $c(a+d)=0$, we must have $b=0$ and $c=0$. So the first equation becomes $a^2=4 \Rightarrow a=\pm 2$, and the fourth equation becomes $d^2=9 \Rightarrow d=\pm 3$.

Thus the square roots of $\begin{bmatrix} 4 & 0 \\ 0 & 9 \end{bmatrix}$ are $A_1 = \begin{bmatrix} 2 & 0 \\ 0 & 3 \end{bmatrix}$, $A_2 = \begin{bmatrix} 2 & 0 \\ 0 & -3 \end{bmatrix}$, $A_3 = \begin{bmatrix} -2 & 0 \\ 0 & 3 \end{bmatrix}$, and $A_4 = \begin{bmatrix} -2 & 0 \\ 0 & -3 \end{bmatrix}$.

For the second matrix, we have $A^2 = \begin{bmatrix} 1 & 5 \\ 0 & 9 \end{bmatrix} \Leftrightarrow \begin{cases} a^2+bc=1 \\ b(a+d)=5 \\ c(a+d)=0 \\ bc+d^2=9 \end{cases}$ Since $a+d \neq 0$ and $c(a+d)=0$, we must have $c=0$. The equations then simplify into the system $\begin{cases} a^2=1 \\ b(a+d)=5 \\ d^2=9 \end{cases} \Rightarrow \begin{cases} a=\pm 1 \\ b(a+d)=5 \\ d=\pm 3 \end{cases}$

We consider the four possible values of $a$ and $d$. If $a=1$ and $d=3$, then $b(a+d)=5 \Rightarrow b(4)=5 \Rightarrow b=\frac{5}{4}$. If $a=1$ and $d=-3$, then $b(a+d)=5 \Rightarrow b(-2)=5 \Rightarrow b=-\frac{5}{2}$. If $a=-1$ and $d=3$, then $b(a+d)=5 \Rightarrow b(2)=5 \Rightarrow b=\frac{5}{2}$. If $a=-1$ and $d=-3$, then $b(a+d)=5 \Rightarrow b(-4)=5 \Rightarrow b=-\frac{5}{4}$. Thus, the square roots of $\begin{bmatrix} 1 & 5 \\ 0 & 9 \end{bmatrix}$ are $A_1 = \begin{bmatrix} 1 & \frac{5}{4} \\ 0 & 3 \end{bmatrix}$, $A_2 = \begin{bmatrix} 1 & -\frac{5}{2} \\ 0 & -3 \end{bmatrix}$, $A_3 = \begin{bmatrix} -1 & \frac{5}{2} \\ 0 & 3 \end{bmatrix}$, and $A_4 = \begin{bmatrix} -1 & -\frac{5}{4} \\ 0 & -3 \end{bmatrix}$.

## 10.6 Inverses of Matrices and Matrix Equations

**1.** $A = \begin{bmatrix} 4 & 1 \\ 7 & 2 \end{bmatrix}$; $B = \begin{bmatrix} 2 & -1 \\ -7 & 4 \end{bmatrix}$.

$AB = \begin{bmatrix} 4 & 1 \\ 7 & 2 \end{bmatrix}\begin{bmatrix} 2 & -1 \\ -7 & 4 \end{bmatrix} = \begin{bmatrix} 1 & 0 \\ 0 & 1 \end{bmatrix}$ and $BA = \begin{bmatrix} 2 & -1 \\ -7 & 4 \end{bmatrix}\begin{bmatrix} 4 & 1 \\ 7 & 2 \end{bmatrix} = \begin{bmatrix} 1 & 0 \\ 0 & 1 \end{bmatrix}$.

**2.** $A = \begin{bmatrix} 2 & -3 \\ 4 & -7 \end{bmatrix}$; $B = \begin{bmatrix} \frac{7}{2} & -\frac{3}{2} \\ 2 & -1 \end{bmatrix}$. $AB = \begin{bmatrix} 2 & -3 \\ 4 & -7 \end{bmatrix} \begin{bmatrix} \frac{7}{2} & -\frac{3}{2} \\ 2 & -1 \end{bmatrix} = \begin{bmatrix} 1 & 0 \\ 0 & 1 \end{bmatrix}$ and

$BA = \begin{bmatrix} \frac{7}{2} & -\frac{3}{2} \\ 2 & -1 \end{bmatrix} \begin{bmatrix} 2 & -3 \\ 4 & -7 \end{bmatrix} = \begin{bmatrix} 1 & 0 \\ 0 & 1 \end{bmatrix}$.

**3.** $A = \begin{bmatrix} 1 & 3 & -1 \\ 1 & 4 & 0 \\ -1 & -3 & 2 \end{bmatrix}$; $B = \begin{bmatrix} 8 & -3 & 4 \\ -2 & 1 & -1 \\ 1 & 0 & 1 \end{bmatrix}$. $AB = \begin{bmatrix} 1 & 3 & -1 \\ 1 & 4 & 0 \\ -1 & -3 & 2 \end{bmatrix} \begin{bmatrix} 8 & -3 & 4 \\ -2 & 1 & -1 \\ 1 & 0 & 1 \end{bmatrix} = \begin{bmatrix} 1 & 0 & 0 \\ 0 & 1 & 0 \\ 0 & 0 & 1 \end{bmatrix}$ and

$BA = \begin{bmatrix} 8 & -3 & 4 \\ -2 & 1 & -1 \\ 1 & 0 & 1 \end{bmatrix} \begin{bmatrix} 1 & 3 & -1 \\ 1 & 4 & 0 \\ -1 & -3 & 2 \end{bmatrix} = \begin{bmatrix} 1 & 0 & 0 \\ 0 & 1 & 0 \\ 0 & 0 & 1 \end{bmatrix}$

**4.** $A = \begin{bmatrix} 3 & 2 & 4 \\ 1 & 1 & -6 \\ 2 & 1 & 12 \end{bmatrix}$; $B = \begin{bmatrix} 9 & -10 & -8 \\ -12 & 14 & 11 \\ -\frac{1}{2} & \frac{1}{2} & \frac{1}{2} \end{bmatrix}$. $AB = \begin{bmatrix} 3 & 2 & 4 \\ 1 & 1 & -6 \\ 2 & 1 & 12 \end{bmatrix} \begin{bmatrix} 9 & -10 & -8 \\ -12 & 14 & 11 \\ -\frac{1}{2} & \frac{1}{2} & \frac{1}{2} \end{bmatrix} = \begin{bmatrix} 1 & 0 & 0 \\ 0 & 1 & 0 \\ 0 & 0 & 1 \end{bmatrix}$ and

$BA = \begin{bmatrix} 9 & -10 & -8 \\ -12 & 14 & 11 \\ -\frac{1}{2} & \frac{1}{2} & \frac{1}{2} \end{bmatrix} \begin{bmatrix} 3 & 2 & 4 \\ 1 & 1 & -6 \\ 2 & 1 & 12 \end{bmatrix} = \begin{bmatrix} 1 & 0 & 0 \\ 0 & 1 & 0 \\ 0 & 0 & 1 \end{bmatrix}$.

**5.** $A = \begin{bmatrix} 7 & 4 \\ 3 & 2 \end{bmatrix} \Leftrightarrow A^{-1} = \dfrac{1}{14-12} \begin{bmatrix} 2 & -4 \\ -3 & 7 \end{bmatrix} = \begin{bmatrix} 1 & -2 \\ -\frac{3}{2} & \frac{7}{2} \end{bmatrix}$. Then,

$AA^{-1} = \begin{bmatrix} 7 & 4 \\ 3 & 2 \end{bmatrix} \begin{bmatrix} 1 & -2 \\ -\frac{3}{2} & \frac{7}{2} \end{bmatrix} = \begin{bmatrix} 1 & 0 \\ 0 & 1 \end{bmatrix}$ and $A^{-1}A = \begin{bmatrix} 1 & -2 \\ -\frac{3}{2} & \frac{7}{2} \end{bmatrix} \begin{bmatrix} 7 & 4 \\ 3 & 2 \end{bmatrix} = \begin{bmatrix} 1 & 0 \\ 0 & 1 \end{bmatrix}$.

**6.** $B = \begin{bmatrix} 1 & 3 & 2 \\ 0 & 2 & 2 \\ -2 & -1 & 0 \end{bmatrix}$. We begin with a $3 \times 6$ matrix whose left half is $B$ and whose right half is $I_3$.

$$\begin{bmatrix} 1 & 3 & 2 & 1 & 0 & 0 \\ 0 & 2 & 2 & 0 & 1 & 0 \\ -2 & -1 & 0 & 0 & 0 & 1 \end{bmatrix} \xrightarrow{R_3 + 2R_1 \to R_3} \begin{bmatrix} 1 & 3 & 2 & 1 & 0 & 0 \\ 0 & 2 & 2 & 0 & 1 & 0 \\ 0 & 5 & 4 & 2 & 0 & 1 \end{bmatrix} \xrightarrow{2R_3 - 5R_2 \to R_3} \begin{bmatrix} 1 & 3 & 2 & 1 & 0 & 0 \\ 0 & 2 & 2 & 0 & 1 & 0 \\ 0 & 0 & -2 & 4 & -5 & 2 \end{bmatrix} \xrightarrow[-\frac{1}{2}R_3]{\frac{1}{2}R_2}$$

$$\begin{bmatrix} 1 & 3 & 2 & 1 & 0 & 0 \\ 0 & 1 & 1 & 0 & \frac{1}{2} & 0 \\ 0 & 0 & 1 & -2 & \frac{5}{2} & -1 \end{bmatrix} \xrightarrow[R_2 - R_3 \to R_2]{R_1 - 3R_2 \to R_1} \begin{bmatrix} 1 & 0 & -1 & 1 & -\frac{3}{2} & 0 \\ 0 & 1 & 0 & 2 & -2 & 1 \\ 0 & 0 & 1 & -2 & \frac{5}{2} & -1 \end{bmatrix} \xrightarrow{R_1 + R_3 \to R_1} \begin{bmatrix} 1 & 0 & 0 & -1 & 1 & -1 \\ 0 & 1 & 0 & 2 & -2 & 1 \\ 0 & 0 & 1 & -2 & \frac{5}{2} & -1 \end{bmatrix}.$$

Then $B^{-1} = \begin{bmatrix} -1 & 1 & -1 \\ 2 & -2 & 1 \\ -2 & \frac{5}{2} & -1 \end{bmatrix}$; $B^{-1}B = \begin{bmatrix} -1 & 1 & -1 \\ 2 & -2 & 1 \\ -2 & \frac{5}{2} & -1 \end{bmatrix} \begin{bmatrix} 1 & 3 & 2 \\ 0 & 2 & 2 \\ -2 & -1 & 0 \end{bmatrix} = \begin{bmatrix} 1 & 0 & 0 \\ 0 & 1 & 0 \\ 0 & 0 & 1 \end{bmatrix}$; and

$BB^{-1} = \begin{bmatrix} 1 & 3 & 2 \\ 0 & 2 & 2 \\ -2 & -1 & 0 \end{bmatrix} \begin{bmatrix} -1 & 1 & -1 \\ 2 & -2 & 1 \\ -2 & \frac{5}{2} & -1 \end{bmatrix} = \begin{bmatrix} 1 & 0 & 0 \\ 0 & 1 & 0 \\ 0 & 0 & 1 \end{bmatrix}$.

**7.** $\begin{bmatrix} 5 & 3 \\ 3 & 2 \end{bmatrix}^{-1} = \dfrac{1}{10-9}\begin{bmatrix} 2 & -3 \\ -3 & 5 \end{bmatrix} = \begin{bmatrix} 2 & -3 \\ -3 & 5 \end{bmatrix}$

**8.** $\begin{bmatrix} 3 & 4 \\ 7 & 9 \end{bmatrix}^{-1} = \dfrac{1}{27-28}\begin{bmatrix} 9 & -4 \\ -7 & 3 \end{bmatrix} = -\begin{bmatrix} 9 & -4 \\ -7 & 3 \end{bmatrix} = \begin{bmatrix} -9 & 4 \\ 7 & -3 \end{bmatrix}$

**9.** $\begin{bmatrix} 2 & 5 \\ -5 & -13 \end{bmatrix}^{-1} = \dfrac{1}{-26+25}\begin{bmatrix} -13 & -5 \\ 5 & 2 \end{bmatrix} = \begin{bmatrix} 13 & 5 \\ -5 & -2 \end{bmatrix}$

**10.** $\begin{bmatrix} -7 & 4 \\ 8 & -5 \end{bmatrix}^{-1} = \dfrac{1}{35-32}\begin{bmatrix} -5 & -4 \\ -8 & -7 \end{bmatrix} = \begin{bmatrix} -\frac{5}{3} & -\frac{4}{3} \\ -\frac{8}{3} & -\frac{7}{3} \end{bmatrix}$

**11.** $\begin{bmatrix} 6 & -3 \\ -8 & 4 \end{bmatrix}^{-1} = \dfrac{1}{24-24}\begin{bmatrix} 4 & 3 \\ 8 & 6 \end{bmatrix}$, which is not defined, and so there is no inverse.

**12.** $\begin{bmatrix} \frac{1}{2} & \frac{1}{3} \\ 5 & 4 \end{bmatrix}^{-1} = \dfrac{1}{2-\frac{5}{3}}\begin{bmatrix} 4 & -\frac{1}{3} \\ -5 & \frac{1}{2} \end{bmatrix} = \begin{bmatrix} 12 & -1 \\ -15 & \frac{3}{2} \end{bmatrix}$

**13.** $\begin{bmatrix} 0.4 & -1.2 \\ 0.3 & 0.6 \end{bmatrix}^{-1} = \dfrac{1}{0.24+0.36}\begin{bmatrix} 0.6 & 1.2 \\ -0.3 & 0.4 \end{bmatrix} = \begin{bmatrix} 1 & 2 \\ -\frac{1}{2} & \frac{2}{3} \end{bmatrix}$

**14.** $\left[\begin{array}{ccc|ccc} 4 & 2 & 3 & 1 & 0 & 0 \\ 3 & 3 & 2 & 0 & 1 & 0 \\ 1 & 0 & 1 & 0 & 0 & 1 \end{array}\right] \xrightarrow[4R_3 - R_1 \to R_3]{4R_2 - 3R_1 \to R_2} \left[\begin{array}{ccc|ccc} 4 & 2 & 3 & 1 & 0 & 0 \\ 0 & 6 & -1 & -3 & 4 & 0 \\ 0 & -2 & 1 & -1 & 0 & 4 \end{array}\right] \xrightarrow[3R_3 + R_2 \to R_3]{R_1 + R_3 \to R_1} \left[\begin{array}{ccc|ccc} 4 & 0 & 4 & 0 & 0 & 4 \\ 0 & 6 & -1 & -3 & 4 & 0 \\ 0 & 0 & 2 & -6 & 4 & 12 \end{array}\right] \xrightarrow[\frac{1}{2}R_3]{\frac{1}{4}R_1}$

$\left[\begin{array}{ccc|ccc} 1 & 0 & 1 & 0 & 0 & 1 \\ 0 & 6 & -1 & -3 & 4 & 0 \\ 0 & 0 & 1 & -3 & 2 & 6 \end{array}\right] \xrightarrow[R_2 + R_3 \to R_2]{R_1 - R_3 \to R_1} \left[\begin{array}{ccc|ccc} 1 & 0 & 0 & 3 & -2 & -5 \\ 0 & 6 & 0 & -6 & 6 & 6 \\ 0 & 0 & 1 & -3 & 2 & 6 \end{array}\right] \xrightarrow{\frac{1}{6}R_2} \left[\begin{array}{ccc|ccc} 1 & 0 & 0 & 3 & -2 & -5 \\ 0 & 1 & 0 & -1 & 1 & 1 \\ 0 & 0 & 1 & -3 & 2 & 6 \end{array}\right].$

Therefore, the inverse matrix is $\begin{bmatrix} 3 & -2 & -5 \\ -1 & 1 & 1 \\ -3 & 2 & 6 \end{bmatrix}$.

**15.** $\left[\begin{array}{ccc|ccc} 2 & 4 & 1 & 1 & 0 & 0 \\ -1 & 1 & -1 & 0 & 1 & 0 \\ 1 & 4 & 0 & 0 & 0 & 1 \end{array}\right] \xrightarrow[2R_3 - R_1 \to R_3]{2R_2 + R_1 \to R_2} \left[\begin{array}{ccc|ccc} 2 & 4 & 1 & 1 & 0 & 0 \\ 0 & 6 & -1 & 1 & 2 & 0 \\ 0 & 4 & -1 & -1 & 0 & 2 \end{array}\right] \xrightarrow[3R_1 - 2R_2 \to R_1]{3R_3 - 2R_2 \to R_3} \left[\begin{array}{ccc|ccc} 6 & 0 & 5 & 1 & -4 & 0 \\ 0 & 6 & -1 & 1 & 2 & 0 \\ 0 & 0 & -1 & -5 & -4 & 6 \end{array}\right]$

$\xrightarrow[R_2 - R_3 \to R_2]{R_1 + 5R_3 \to R_1} \left[\begin{array}{ccc|ccc} 6 & 0 & 0 & -24 & -24 & 30 \\ 0 & 6 & 0 & 6 & 6 & -6 \\ 0 & 0 & -1 & -5 & -4 & 6 \end{array}\right] \xrightarrow[\frac{1}{6}R_2,\ -R_3]{\frac{1}{6}R_1} \left[\begin{array}{ccc|ccc} 1 & 0 & 0 & -4 & -4 & 5 \\ 0 & 1 & 0 & 1 & 1 & -1 \\ 0 & 0 & 1 & 5 & 4 & -6 \end{array}\right].$

Therefore, the inverse matrix is $\begin{bmatrix} -4 & -4 & 5 \\ 1 & 1 & -1 \\ 5 & 4 & -6 \end{bmatrix}$.

**16.** $\left[\begin{array}{rrrrrr} 5 & 7 & 4 & 1 & 0 & 0 \\ 3 & -1 & 3 & 0 & 1 & 0 \\ 6 & 7 & 5 & 0 & 0 & 1 \end{array}\right] \xrightarrow[R_1 \leftrightarrow R_3]{R_3 - R_1 \to R_3} \left[\begin{array}{rrrrrr} 1 & 0 & 1 & -1 & 0 & 1 \\ 3 & -1 & 3 & 0 & 1 & 0 \\ 5 & 7 & 4 & 1 & 0 & 0 \end{array}\right] \xrightarrow[R_3 - 5R_1 \to R_3]{R_2 - 3R_1 \to R_2} \left[\begin{array}{rrrrrr} 1 & 0 & 1 & -1 & 0 & 1 \\ 0 & -1 & 0 & 3 & 1 & -3 \\ 0 & 7 & -1 & 6 & 0 & -5 \end{array}\right]$

$\xrightarrow{R_3 + 7R_2 \to R_3} \left[\begin{array}{rrrrrr} 1 & 0 & 1 & -1 & 0 & 1 \\ 0 & -1 & 0 & 3 & 1 & -3 \\ 0 & 0 & -1 & 27 & 7 & -26 \end{array}\right] \xrightarrow[-R_2,\, -R_3]{R_1 + R_3 \to R_1} \left[\begin{array}{rrrrrr} 1 & 0 & 0 & 26 & 7 & -25 \\ 0 & 1 & 0 & -3 & -1 & 3 \\ 0 & 0 & 1 & -27 & -7 & 26 \end{array}\right].$

Therefore, the inverse matrix is $\left[\begin{array}{rrr} 26 & 7 & -25 \\ -3 & -1 & 3 \\ -27 & -7 & 26 \end{array}\right]$.

**17.** $\left[\begin{array}{rrrrrr} 1 & 2 & 3 & 1 & 0 & 0 \\ 4 & 5 & -1 & 0 & 1 & 0 \\ 1 & -1 & -10 & 0 & 0 & 1 \end{array}\right] \xrightarrow[R_3 - R_1 \to R_3]{R_2 - 4R_1 \to R_2} \left[\begin{array}{rrrrrr} 1 & 2 & 3 & 1 & 0 & 0 \\ 0 & -3 & -13 & -4 & 1 & 0 \\ 0 & -3 & -13 & -1 & 0 & 1 \end{array}\right] \xrightarrow{R_3 - R_2 \to R_3} \left[\begin{array}{rrrrrr} 1 & 2 & 3 & 1 & 0 & 0 \\ 0 & -3 & -13 & -4 & 1 & 0 \\ 0 & 0 & 0 & 3 & -1 & 1 \end{array}\right].$

Since the left half of the last row consists entirely of zeros, there is no inverse matrix.

**18.** $\left[\begin{array}{rrrrrr} 2 & 1 & 0 & 1 & 0 & 0 \\ 1 & 1 & 4 & 0 & 1 & 0 \\ 2 & 1 & 2 & 0 & 0 & 1 \end{array}\right] \xrightarrow{R_1 \leftrightarrow R_2} \left[\begin{array}{rrrrrr} 1 & 1 & 4 & 0 & 1 & 0 \\ 2 & 1 & 0 & 1 & 0 & 0 \\ 2 & 1 & 2 & 0 & 0 & 1 \end{array}\right] \xrightarrow[R_3 - 2R_1 \to R_3]{R_2 - 2R_1 \to R_2} \left[\begin{array}{rrrrrr} 1 & 1 & 4 & 0 & 1 & 0 \\ 0 & -1 & -8 & 1 & -2 & 0 \\ 0 & -1 & -6 & 0 & -2 & 1 \end{array}\right] \xrightarrow[R_3 - R_2 \to R_3]{R_1 + R_2 \to R_1}$

$\left[\begin{array}{rrrrrr} 1 & 0 & -4 & 1 & -1 & 0 \\ 0 & -1 & -8 & 1 & -2 & 0 \\ 0 & 0 & 2 & -1 & 0 & 1 \end{array}\right] \xrightarrow[R_2 + 4R_3 \to R_2]{R_1 + 2R_3 \to R_1} \left[\begin{array}{rrrrrr} 1 & 0 & 0 & -1 & -1 & 2 \\ 0 & -1 & 0 & -3 & -2 & 4 \\ 0 & 0 & 2 & -1 & 0 & 1 \end{array}\right] \xrightarrow[\frac{1}{2}R_3]{-R_2} \left[\begin{array}{rrrrrr} 1 & 0 & 0 & -1 & -1 & 2 \\ 0 & 1 & 0 & 3 & 2 & -4 \\ 0 & 0 & 1 & -\frac{1}{2} & 0 & \frac{1}{2} \end{array}\right].$

Therefore, the inverse matrix is $\left[\begin{array}{rrr} -1 & -1 & 2 \\ 3 & 2 & -4 \\ -\frac{1}{2} & 0 & \frac{1}{2} \end{array}\right]$.

**19.** $\left[\begin{array}{rrrrrr} 0 & -2 & 2 & 1 & 0 & 0 \\ 3 & 1 & 3 & 0 & 1 & 0 \\ 1 & -2 & 3 & 0 & 0 & 1 \end{array}\right] \xrightarrow{R_1 \leftrightarrow R_3} \left[\begin{array}{rrrrrr} 1 & -2 & 3 & 0 & 0 & 1 \\ 3 & 1 & 3 & 0 & 1 & 0 \\ 0 & -2 & 2 & 1 & 0 & 0 \end{array}\right] \xrightarrow{R_2 - 3R_1 \to R_2} \left[\begin{array}{rrrrrr} 1 & -2 & 3 & 0 & 0 & 1 \\ 0 & 7 & -6 & 0 & 1 & -3 \\ 0 & -2 & 2 & 1 & 0 & 0 \end{array}\right]$

$\xrightarrow[R_2 + 3R_3 \to R_2]{R_1 - R_3 \to R_1} \left[\begin{array}{rrrrrr} 1 & 0 & 1 & -1 & 0 & 1 \\ 0 & 1 & 0 & 3 & 1 & -3 \\ 0 & -2 & 2 & 1 & 0 & 0 \end{array}\right] \xrightarrow{R_3 + 2R_2 \to R_3} \left[\begin{array}{rrrrrr} 1 & 0 & 1 & -1 & 0 & 1 \\ 0 & 1 & 0 & 3 & 1 & -3 \\ 0 & 0 & 2 & 7 & 2 & -6 \end{array}\right] \xrightarrow{\frac{1}{2}R_3}$

$\left[\begin{array}{rrrrrr} 1 & 0 & 1 & -1 & 0 & 1 \\ 0 & 1 & 0 & 3 & 1 & -3 \\ 0 & 0 & 1 & \frac{7}{2} & 1 & -3 \end{array}\right] \xrightarrow{R_1 - R_3 \to R_1} \left[\begin{array}{rrrrrr} 1 & 0 & 0 & -\frac{9}{2} & -1 & 4 \\ 0 & 1 & 0 & 3 & 1 & -3 \\ 0 & 0 & 1 & \frac{7}{2} & 1 & -3 \end{array}\right]$. Therefore, the inverse matrix is $\left[\begin{array}{rrr} -\frac{9}{2} & -1 & 4 \\ 3 & 1 & -3 \\ \frac{7}{2} & 1 & -3 \end{array}\right]$.

**20.** $\left[\begin{array}{rrrrrr} 3 & -2 & 0 & 1 & 0 & 0 \\ 5 & 1 & 1 & 0 & 1 & 0 \\ 2 & -2 & 0 & 0 & 0 & 1 \end{array}\right] \xrightarrow{R_1 - R_3 \to R_1} \left[\begin{array}{rrrrrr} 1 & 0 & 0 & 1 & 0 & -1 \\ 5 & 1 & 1 & 0 & 1 & 0 \\ 2 & -2 & 0 & 0 & 0 & 1 \end{array}\right] \xrightarrow[R_3 - 2R_1 \to R_3]{R_2 - 5R_1 \to R_2} \left[\begin{array}{rrrrrr} 1 & 0 & 0 & 1 & 0 & -1 \\ 0 & 1 & 1 & -5 & 1 & 5 \\ 0 & -2 & 0 & -2 & 0 & 3 \end{array}\right] \xrightarrow{R_3 + 2R_2 \to R_3}$

$\left[\begin{array}{rrrrrr} 1 & 0 & 0 & 1 & 0 & -1 \\ 0 & 1 & 1 & -5 & 1 & 5 \\ 0 & 0 & 2 & -12 & 2 & 13 \end{array}\right] \xrightarrow[R_2 - R_3 \to R_2]{\frac{1}{2}R_3} \left[\begin{array}{rrrrrr} 1 & 0 & 0 & 1 & 0 & -1 \\ 0 & 1 & 0 & 1 & 0 & -\frac{3}{2} \\ 0 & 0 & 1 & -6 & 1 & \frac{13}{2} \end{array}\right]$. Therefore, the inverse matrix is $\left[\begin{array}{rrr} 1 & 0 & -1 \\ 1 & 0 & -\frac{3}{2} \\ -6 & 1 & \frac{13}{2} \end{array}\right]$.

**21.** $\begin{bmatrix} 1 & 2 & 0 & 3 & 1 & 0 & 0 & 0 \\ 0 & 1 & 1 & 1 & 0 & 1 & 0 & 0 \\ 0 & 1 & 0 & 1 & 0 & 0 & 1 & 0 \\ 1 & 2 & 0 & 2 & 0 & 0 & 0 & 1 \end{bmatrix} \xrightarrow[\text{R}_4 - \text{R}_1 \to \text{R}_4]{\text{R}_3 - \text{R}_2 \to \text{R}_3} \begin{bmatrix} 1 & 2 & 0 & 3 & 1 & 0 & 0 & 0 \\ 0 & 1 & 1 & 1 & 0 & 1 & 0 & 0 \\ 0 & 0 & -1 & 0 & 0 & -1 & 1 & 0 \\ 0 & 0 & 0 & -1 & -1 & 0 & 0 & 1 \end{bmatrix} \xrightarrow[-\text{R}_4]{-\text{R}_3} \begin{bmatrix} 1 & 2 & 0 & 3 & 1 & 0 & 0 & 0 \\ 0 & 1 & 1 & 1 & 0 & 1 & 0 & 0 \\ 0 & 0 & 1 & 0 & 0 & 1 & -1 & 0 \\ 0 & 0 & 0 & 1 & 1 & 0 & 0 & -1 \end{bmatrix}$

$\xrightarrow[\text{R}_2 - \text{R}_3 \to \text{R}_2]{\text{R}_1 - 2\text{R}_2 \to \text{R}_1} \begin{bmatrix} 1 & 0 & -2 & 1 & 1 & -2 & 0 & 0 \\ 0 & 1 & 0 & 1 & 0 & 0 & 1 & 0 \\ 0 & 0 & 1 & 0 & 0 & 1 & -1 & 0 \\ 0 & 0 & 0 & 1 & 1 & 0 & 0 & -1 \end{bmatrix} \xrightarrow[\text{R}_2 - \text{R}_4 \to \text{R}_2]{\text{R}_1 + 2\text{R}_3 \to \text{R}_1} \begin{bmatrix} 1 & 0 & 0 & 1 & 1 & 0 & -2 & 0 \\ 0 & 1 & 0 & 0 & -1 & 0 & 1 & 1 \\ 0 & 0 & 1 & 0 & 0 & 1 & -1 & 0 \\ 0 & 0 & 0 & 1 & 1 & 0 & 0 & -1 \end{bmatrix} \xrightarrow{\text{R}_1 \to \text{R}_1 - \text{R}_4}$

$\begin{bmatrix} 1 & 0 & 0 & 0 & 0 & 0 & -2 & 1 \\ 0 & 1 & 0 & 0 & -1 & 0 & 1 & 1 \\ 0 & 0 & 1 & 0 & 0 & 1 & -1 & 0 \\ 0 & 0 & 0 & 1 & 1 & 0 & 0 & -1 \end{bmatrix}$. Therefore, the inverse matrix is $\begin{bmatrix} 0 & 0 & -2 & 1 \\ -1 & 0 & 1 & 1 \\ 0 & 1 & -1 & 0 \\ 1 & 0 & 0 & -1 \end{bmatrix}$.

**22.** $\begin{bmatrix} 1 & 0 & 1 & 0 & 1 & 0 & 0 & 0 \\ 0 & 1 & 0 & 1 & 0 & 1 & 0 & 0 \\ 1 & 1 & 1 & 0 & 0 & 0 & 1 & 0 \\ 1 & 1 & 1 & 1 & 0 & 0 & 0 & 1 \end{bmatrix} \xrightarrow[\text{R}_3 - \text{R}_1 \to \text{R}_3]{\text{R}_4 - \text{R}_1 \to \text{R}_4} \begin{bmatrix} 1 & 0 & 1 & 0 & 1 & 0 & 0 & 0 \\ 0 & 1 & 0 & 1 & 0 & 1 & 0 & 0 \\ 0 & 1 & 0 & 0 & -1 & 0 & 1 & 0 \\ 0 & 1 & 0 & 1 & -1 & 0 & 0 & 1 \end{bmatrix} \xrightarrow[\text{R}_3 - \text{R}_2 \to \text{R}_3]{\text{R}_1 - \text{R}_2 \to \text{R}_1} \begin{bmatrix} 1 & 0 & 1 & 0 & 1 & 0 & 0 & 0 \\ 0 & 1 & 0 & 1 & 0 & 1 & 0 & 0 \\ 0 & 0 & 0 & -1 & -1 & -1 & 1 & 0 \\ 0 & 0 & 0 & 0 & -1 & -1 & 0 & 1 \end{bmatrix}$.

Therefore, there is no inverse matrix.

**23.** $\begin{cases} 5x + 3y = 4 \\ 3x + 2y = 0 \end{cases}$ is equivalent to the matrix equation $\begin{bmatrix} 5 & 3 \\ 3 & 2 \end{bmatrix} \begin{bmatrix} x \\ y \end{bmatrix} = \begin{bmatrix} 4 \\ 0 \end{bmatrix}$. Using the inverse from Exercise 3, $\begin{bmatrix} x \\ y \end{bmatrix} = \begin{bmatrix} 2 & -3 \\ -3 & 5 \end{bmatrix} \begin{bmatrix} 4 \\ 0 \end{bmatrix} = \begin{bmatrix} 8 \\ -12 \end{bmatrix}$. Therefore, $x = 8$ and $y = -12$.

**24.** $\begin{cases} 3x + 4y = 10 \\ 7x + 9y = 20 \end{cases}$ is equivalent to the matrix equation $\begin{bmatrix} 3 & 4 \\ 7 & 9 \end{bmatrix} \begin{bmatrix} x \\ y \end{bmatrix} = \begin{bmatrix} 10 \\ 20 \end{bmatrix}$ $\Leftrightarrow$ $\begin{bmatrix} x \\ y \end{bmatrix} = \begin{bmatrix} -9 & 4 \\ 7 & -3 \end{bmatrix} \begin{bmatrix} 10 \\ 20 \end{bmatrix} = \begin{bmatrix} -10 \\ 10 \end{bmatrix}$. Therefore, $x = -10$ and $y = 10$.

**25.** $\begin{cases} 2x + 5y = 2 \\ -5x - 13y = 20 \end{cases}$ is equivalent to the matrix equation $\begin{bmatrix} 2 & 5 \\ -5 & -13 \end{bmatrix} \begin{bmatrix} x \\ y \end{bmatrix} = \begin{bmatrix} 2 \\ 20 \end{bmatrix}$. Using the inverse from Exercise 5, $\begin{bmatrix} x \\ y \end{bmatrix} = \begin{bmatrix} 13 & 5 \\ -5 & -2 \end{bmatrix} \begin{bmatrix} 2 \\ 20 \end{bmatrix} = \begin{bmatrix} 126 \\ -50 \end{bmatrix}$. Therefore, $x = 126$ and $y = -50$.

**26.** $\begin{cases} -7x + 4y = 0 \\ 8x - 5y = 100 \end{cases}$ is equivalent to the matrix equation $\begin{bmatrix} -7 & 4 \\ 8 & -5 \end{bmatrix} \begin{bmatrix} x \\ y \end{bmatrix} = \begin{bmatrix} 0 \\ 100 \end{bmatrix}$ $\Leftrightarrow$ $\begin{bmatrix} x \\ y \end{bmatrix} = \begin{bmatrix} -\frac{5}{3} & -\frac{4}{3} \\ -\frac{8}{3} & -\frac{7}{3} \end{bmatrix} \begin{bmatrix} 0 \\ 100 \end{bmatrix} = \begin{bmatrix} -\frac{400}{3} \\ -\frac{700}{3} \end{bmatrix}$. Therefore, $x = -\frac{400}{3}$ and $y = -\frac{700}{3}$.

**27.** $\begin{cases} 2x+4y+z=7 \\ -x+y-z=0 \\ x+4y=-2 \end{cases}$ is equivalent to the matrix equation $\begin{bmatrix} 2 & 4 & 1 \\ -1 & 1 & -1 \\ 1 & 4 & 0 \end{bmatrix}\begin{bmatrix} x \\ y \\ z \end{bmatrix} = \begin{bmatrix} 7 \\ 0 \\ -2 \end{bmatrix}$. Using the inverse from Exercise 11, $\begin{bmatrix} x \\ y \\ z \end{bmatrix} = \begin{bmatrix} -4 & -4 & 5 \\ 1 & 1 & -1 \\ 5 & 4 & -6 \end{bmatrix}\begin{bmatrix} 7 \\ 0 \\ -2 \end{bmatrix} = \begin{bmatrix} -38 \\ 9 \\ 47 \end{bmatrix}$. Therefore, $x=-38$, $y=9$, and $z=47$.

**28.** $\begin{cases} 5x+7y+4z=1 \\ 3x-y+3z=1 \\ 6x+7y+5z=1 \end{cases}$ is equivalent to the matrix equation $\begin{bmatrix} 5 & 7 & 4 \\ 3 & -1 & 3 \\ 6 & 7 & 5 \end{bmatrix}\begin{bmatrix} x \\ y \\ z \end{bmatrix} = \begin{bmatrix} 1 \\ 1 \\ 1 \end{bmatrix} \Leftrightarrow$ $\begin{bmatrix} x \\ y \\ z \end{bmatrix} = \begin{bmatrix} 26 & 7 & -25 \\ -3 & -1 & 3 \\ -27 & -7 & 26 \end{bmatrix}\begin{bmatrix} 1 \\ 1 \\ 1 \end{bmatrix} = \begin{bmatrix} 8 \\ -1 \\ -8 \end{bmatrix}$. Therefore, $x=8$, $y=-1$, and $z=-8$.

**29.** $\begin{cases} -2y+2z=12 \\ 3x+y+3z=-2 \\ x-2y+3z=8 \end{cases}$ is equivalent to the matrix equation $\begin{bmatrix} 0 & -2 & 2 \\ 3 & 1 & 3 \\ 1 & -2 & 3 \end{bmatrix}\begin{bmatrix} x \\ y \\ z \end{bmatrix} = \begin{bmatrix} 12 \\ -2 \\ 8 \end{bmatrix}$. Using the inverse from Exercise 15, $\begin{bmatrix} x \\ y \\ z \end{bmatrix} = \begin{bmatrix} -\frac{9}{2} & -1 & 4 \\ 3 & 1 & -3 \\ \frac{7}{2} & 1 & -3 \end{bmatrix}\begin{bmatrix} 12 \\ -2 \\ 8 \end{bmatrix} = \begin{bmatrix} -20 \\ 10 \\ 16 \end{bmatrix}$. Therefore, $x=-20$, $y=10$, and $z=16$.

**30.** $\begin{cases} x+2y+3w=0 \\ y+z+w=1 \\ y+w=2 \\ x+2y+2w=3 \end{cases}$ is equivalent to the matrix equation $\begin{bmatrix} 1 & 2 & 0 & 3 \\ 0 & 1 & 1 & 1 \\ 0 & 1 & 0 & 1 \\ 1 & 2 & 0 & 2 \end{bmatrix}\begin{bmatrix} x \\ y \\ z \\ w \end{bmatrix} = \begin{bmatrix} 0 \\ 1 \\ 2 \\ 3 \end{bmatrix} \Leftrightarrow$ $\begin{bmatrix} x \\ y \\ z \\ w \end{bmatrix} = \begin{bmatrix} 0 & 0 & -2 & 1 \\ -1 & 0 & 1 & 1 \\ 0 & 1 & -1 & 0 \\ 1 & 0 & 0 & -1 \end{bmatrix}\begin{bmatrix} 0 \\ 1 \\ 2 \\ 3 \end{bmatrix} = \begin{bmatrix} -1 \\ 5 \\ -1 \\ -3 \end{bmatrix}$. Therefore, $x=-1$, $y=5$, $z=-1$, and $w=-3$.

**31.** Using a calculator, we get the result $(3,2,1)$.

**32.** Using a calculator, we get the result $(-1,2,3)$.

**33.** Using a calculator, we get the result $(3,-2,2)$.

**34.** Using a calculator, we get the result $(6,12,24)$.

**35.** Using a calculator, we get the result $(8,1,0,3)$.

**36.** Using a calculator, we get the result $(8,4,2,1)$.

**37.** This has the form $MX=C$, so $M^{-1}(MX)=M^{-1}C$ and $M^{-1}(MX)=(M^{-1}M)X=X$. Now $M^{-1}=\begin{bmatrix} 3 & -2 \\ -4 & 3 \end{bmatrix}^{-1} = \frac{1}{9-8}\begin{bmatrix} 3 & 2 \\ 4 & 3 \end{bmatrix} = \begin{bmatrix} 3 & 2 \\ 4 & 3 \end{bmatrix}$. Since $X=M^{-1}C$, we get $\begin{bmatrix} x & y & z \\ u & v & w \end{bmatrix} = \begin{bmatrix} 3 & 2 \\ 4 & 3 \end{bmatrix}\begin{bmatrix} 1 & 0 & -1 \\ 2 & 1 & 3 \end{bmatrix} = \begin{bmatrix} 7 & 2 & 3 \\ 10 & 3 & 5 \end{bmatrix}$.

**38.** Using the inverse matrix from Exercise 19, we see that $\begin{bmatrix} -\frac{9}{2} & -1 & 4 \\ 3 & 1 & -3 \\ \frac{7}{2} & 1 & -3 \end{bmatrix}\begin{bmatrix} 3 & 6 \\ 6 & 12 \\ 0 & 0 \end{bmatrix} = \begin{bmatrix} -\frac{39}{2} & -39 \\ 15 & 30 \\ \frac{33}{2} & 33 \end{bmatrix}$.

Hence, $\begin{bmatrix} x & u \\ y & v \\ z & w \end{bmatrix} = \begin{bmatrix} -\frac{39}{2} & -39 \\ 15 & 30 \\ \frac{33}{2} & 33 \end{bmatrix}$.

**39.** $\begin{bmatrix} a & -a \\ a & a \end{bmatrix}^{-1} = \dfrac{1}{a^2-(-a^2)}\begin{bmatrix} a & a \\ -a & a \end{bmatrix} = \dfrac{1}{2a^2}\begin{bmatrix} a & a \\ -a & a \end{bmatrix} = \dfrac{1}{2a}\begin{bmatrix} 1 & 1 \\ -1 & 1 \end{bmatrix}$

**40.** $\begin{bmatrix} a & 0 & 0 & 0 & 1 & 0 & 0 & 0 \\ 0 & b & 0 & 0 & 0 & 1 & 0 & 0 \\ 0 & 0 & c & 0 & 0 & 0 & 1 & 0 \\ 0 & 0 & 0 & d & 0 & 0 & 0 & 1 \end{bmatrix} \xrightarrow[\frac{1}{c}R_3,\ \frac{1}{d}R_4]{\frac{1}{a}R_1,\ \frac{1}{b}R_2} \begin{bmatrix} 1 & 0 & 0 & 0 & 1/a & 0 & 0 & 0 \\ 0 & 1 & 0 & 0 & 0 & 1/b & 0 & 0 \\ 0 & 0 & 1 & 0 & 0 & 0 & 1/c & 0 \\ 0 & 0 & 0 & 1 & 0 & 0 & 0 & 1/d \end{bmatrix}$.

Thus the matrix $\begin{bmatrix} a & 0 & 0 & 0 \\ 0 & b & 0 & 0 \\ 0 & 0 & c & 0 \\ 0 & 0 & 0 & d \end{bmatrix}$ has inverse $\begin{bmatrix} 1/a & 0 & 0 & 0 \\ 0 & 1/b & 0 & 0 \\ 0 & 0 & 1/c & 0 \\ 0 & 0 & 0 & 1/d \end{bmatrix}$.

**41.** $\begin{bmatrix} 2 & x \\ x & x^2 \end{bmatrix}^{-1} = \dfrac{1}{2x^2-x^2}\begin{bmatrix} x^2 & -x \\ -x & 2 \end{bmatrix} = \dfrac{1}{x^2}\begin{bmatrix} x^2 & -x \\ -x & 2 \end{bmatrix} = \begin{bmatrix} 1 & -1/x \\ -1/x & 2/x^2 \end{bmatrix}$. The inverse does not exist when $x = 0$.

**42.** $\begin{bmatrix} e^x & -e^{2x} \\ e^{2x} & e^{3x} \end{bmatrix}^{-1} = \dfrac{1}{e^{4x}+e^{4x}}\begin{bmatrix} e^{3x} & e^{2x} \\ -e^{2x} & e^x \end{bmatrix} = \frac{1}{2}\begin{bmatrix} e^{-x} & e^{-2x} \\ -e^{-2x} & e^{-3x} \end{bmatrix}$. The inverse exists for all $x$.

**43.** $\begin{bmatrix} 1 & e^x & 0 & 1 & 0 & 0 \\ e^x & -e^{2x} & 0 & 0 & 1 & 0 \\ 0 & 0 & 2 & 0 & 0 & 1 \end{bmatrix} \xrightarrow{R_2 - e^x R_1 \to R_2} \begin{bmatrix} 1 & e^x & 0 & 1 & 0 & 0 \\ 0 & -2e^{2x} & 0 & -e^x & 1 & 0 \\ 0 & 0 & 2 & 0 & 0 & 1 \end{bmatrix} \xrightarrow[\frac{1}{2}R_3]{-\frac{1}{2}e^{-2x}R_2}$

$\begin{bmatrix} 1 & e^x & 0 & 1 & 0 & 0 \\ 0 & 1 & 0 & \frac{1}{2}e^{-x} & -\frac{1}{2}e^{-2x} & 0 \\ 0 & 0 & 1 & 0 & 0 & \frac{1}{2} \end{bmatrix} \xrightarrow{R_1 - e^x R_2 \to R_1} \begin{bmatrix} 1 & 0 & 0 & \frac{1}{2} & \frac{1}{2}e^{-x} & 0 \\ 0 & 1 & 0 & \frac{1}{2}e^{-x} & -\frac{1}{2}e^{-2x} & 0 \\ 0 & 0 & 1 & 0 & 0 & \frac{1}{2} \end{bmatrix}$.

Therefore, the inverse matrix is $\begin{bmatrix} \frac{1}{2} & \frac{1}{2}e^{-x} & 0 \\ \frac{1}{2}e^{-x} & -\frac{1}{2}e^{-2x} & 0 \\ 0 & 0 & \frac{1}{2} \end{bmatrix}$. The inverse exists for all $x$.

**44.** $\begin{bmatrix} x & 1 \\ -x & \dfrac{1}{x-1} \end{bmatrix}^{-1} = \dfrac{1}{\dfrac{x}{x-1}+x}\begin{bmatrix} \dfrac{1}{x-1} & -1 \\ x & x \end{bmatrix} = \dfrac{x-1}{x^2}\begin{bmatrix} \dfrac{1}{x-1} & -1 \\ x & x \end{bmatrix} = \begin{bmatrix} \dfrac{1}{x^2} & -\dfrac{x-1}{x^2} \\ \dfrac{x-1}{x} & \dfrac{x-1}{x} \end{bmatrix}$.

The inverse exists for $x \neq 0, 1$.

**45.** $\begin{bmatrix} \cos x & \sin x \\ -\sin x & \cos x \end{bmatrix}^{-1} = \dfrac{1}{\cos^2 x + \sin^2 x}\begin{bmatrix} \cos x & -\sin x \\ \sin x & \cos x \end{bmatrix} = \begin{bmatrix} \cos x & -\sin x \\ \sin x & \cos x \end{bmatrix}$. The inverse exists for all $x$.

**46.** $\begin{bmatrix} \sec x & \tan x \\ \tan x & \sec x \end{bmatrix}^{-1} = \dfrac{1}{\sec^2 x - \tan^2 x} \begin{bmatrix} \sec x & -\tan x \\ -\tan x & \sec x \end{bmatrix} = \begin{bmatrix} \sec x & -\tan x \\ -\tan x & \sec x \end{bmatrix}$. The inverse does not exist for $x = \frac{\pi}{2} + k\pi$, $k$ an integer.

**47. (a)** $\left[\begin{array}{ccc|ccc} 3 & 1 & 3 & 1 & 0 & 0 \\ 4 & 2 & 4 & 0 & 1 & 0 \\ 3 & 2 & 4 & 0 & 0 & 1 \end{array}\right] \xrightarrow[R_1 \leftrightarrow R_2]{R_3 - R_1 \to R_3} \left[\begin{array}{ccc|ccc} 4 & 2 & 4 & 0 & 1 & 0 \\ 3 & 1 & 3 & 1 & 0 & 0 \\ 0 & 1 & 1 & -1 & 0 & 1 \end{array}\right] \xrightarrow{R_1 - R_2 \to R_1} \left[\begin{array}{ccc|ccc} 1 & 1 & 1 & -1 & 1 & 0 \\ 3 & 1 & 3 & 1 & 0 & 0 \\ 0 & 1 & 1 & -1 & 0 & 1 \end{array}\right]$

$\xrightarrow{R_2 - 3R_1 \to R_2} \left[\begin{array}{ccc|ccc} 1 & 1 & 1 & -1 & 1 & 0 \\ 0 & -2 & 0 & 4 & -3 & 0 \\ 0 & 1 & 1 & -1 & 0 & 1 \end{array}\right] \xrightarrow[R_3 + \frac{1}{2}R_2 \to R_3]{R_1 + \frac{1}{2}R_2 \to R_1,\ -\frac{1}{2}R_2} \left[\begin{array}{ccc|ccc} 1 & 0 & 1 & 1 & -\frac{1}{2} & 0 \\ 0 & 1 & 0 & -2 & \frac{3}{2} & 0 \\ 0 & 0 & 1 & 1 & -\frac{3}{2} & 1 \end{array}\right] \xrightarrow{R_1 - R_3 \to R_1}$

$\left[\begin{array}{ccc|ccc} 1 & 0 & 0 & 0 & 1 & -1 \\ 0 & 1 & 0 & -2 & \frac{3}{2} & 0 \\ 0 & 0 & 1 & 1 & -\frac{3}{2} & 1 \end{array}\right]$. Therefore, the inverse of the matrix is $\begin{bmatrix} 0 & 1 & -1 \\ -2 & \frac{3}{2} & 0 \\ 1 & -\frac{3}{2} & 1 \end{bmatrix}$.

**(b)** $\begin{bmatrix} A \\ B \\ C \end{bmatrix} = \begin{bmatrix} 0 & 1 & -1 \\ -2 & \frac{3}{2} & 0 \\ 1 & -\frac{3}{2} & 1 \end{bmatrix} \begin{bmatrix} 10 \\ 14 \\ 13 \end{bmatrix} = \begin{bmatrix} 1 \\ 1 \\ 2 \end{bmatrix}$.

Therefore, he should feed the rats 1 oz of food A, 1 oz of food B, and 2 oz of food C.

**(c)** $\begin{bmatrix} A \\ B \\ C \end{bmatrix} = \begin{bmatrix} 0 & 1 & -1 \\ -2 & \frac{3}{2} & 0 \\ 1 & -\frac{3}{2} & 1 \end{bmatrix} \begin{bmatrix} 9 \\ 12 \\ 10 \end{bmatrix} = \begin{bmatrix} 2 \\ 0 \\ 1 \end{bmatrix}$.

Therefore, he should feed the rats 2 oz of food A, no food B, and 1 oz of food C.

**(d)** $\begin{bmatrix} A \\ B \\ C \end{bmatrix} = \begin{bmatrix} 0 & 1 & -1 \\ -2 & \frac{3}{2} & 0 \\ 1 & -\frac{3}{2} & 1 \end{bmatrix} \begin{bmatrix} 2 \\ 4 \\ 11 \end{bmatrix} = \begin{bmatrix} -7 \\ 2 \\ 7 \end{bmatrix}$.

Since $A < 0$, there is no combination of foods giving the required supply.

**48.** $\left[\begin{array}{ccc|ccc} 3 & 1 & 4 & 1 & 0 & 0 \\ 4 & 2 & 6 & 0 & 1 & 0 \\ 3 & 2 & 5 & 0 & 0 & 1 \end{array}\right] \xrightarrow[R_3 - R_1 \to R_3]{R_2 - R_1 \to R_2} \left[\begin{array}{ccc|ccc} 3 & 1 & 4 & 1 & 0 & 0 \\ 1 & 1 & 2 & -1 & 1 & 0 \\ 0 & 1 & 1 & -1 & 0 & 1 \end{array}\right] \xrightarrow{R_1 \leftrightarrow R_2} \left[\begin{array}{ccc|ccc} 1 & 1 & 2 & -1 & 1 & 0 \\ 3 & 1 & 4 & 1 & 0 & 0 \\ 0 & 1 & 1 & -1 & 0 & 1 \end{array}\right] \xrightarrow{R_2 - 3R_1 \to R_2}$

$\left[\begin{array}{ccc|ccc} 1 & 1 & 2 & -1 & 1 & 0 \\ 0 & -2 & -2 & 4 & -3 & 0 \\ 0 & 1 & 1 & -1 & 0 & 1 \end{array}\right] \xrightarrow{R_3 + \frac{1}{2}R_2 \to R_3} \left[\begin{array}{ccc|ccc} 1 & 1 & 2 & -1 & 1 & 0 \\ 0 & -2 & -2 & 4 & -3 & 0 \\ 0 & 0 & 0 & 1 & -\frac{3}{2} & 1 \end{array}\right]$.

Since the inverse matrix does not exist, it would not be possible to use matrix inversion in the solutions of parts (b), (c), and (d).

**49. (a)** $\begin{cases} x + y + 2z = 675 \\ 2x + y + z = 600 \\ x + 2y + z = 625 \end{cases}$ **(b)** $\begin{bmatrix} 1 & 1 & 2 \\ 2 & 1 & 1 \\ 1 & 2 & 1 \end{bmatrix} \begin{bmatrix} x \\ y \\ z \end{bmatrix} = \begin{bmatrix} 675 \\ 600 \\ 625 \end{bmatrix}$

**(c)** $\begin{bmatrix} 1 & 1 & 2 & 1 & 0 & 0 \\ 2 & 1 & 1 & 0 & 1 & 0 \\ 1 & 2 & 1 & 0 & 0 & 1 \end{bmatrix} \xrightarrow[R_3 - R_1 \to R_3]{R_2 - 2R_1 \to R_2} \begin{bmatrix} 1 & 1 & 2 & 1 & 0 & 0 \\ 0 & -1 & -3 & -2 & 1 & 0 \\ 0 & 1 & -1 & -1 & 0 & 1 \end{bmatrix} \xrightarrow[R_3 + R_2 \to R_3]{R_1 + R_2 \to R_1} \begin{bmatrix} 1 & 0 & -1 & -1 & 1 & 0 \\ 0 & -1 & -3 & -2 & 1 & 0 \\ 0 & 0 & -4 & -3 & 1 & 1 \end{bmatrix}$

$\xrightarrow[-\frac{1}{4}R_3]{-R_2} \begin{bmatrix} 1 & 0 & -1 & -1 & 1 & 0 \\ 0 & 1 & 3 & 2 & -1 & 0 \\ 0 & 0 & 1 & \frac{3}{4} & -\frac{1}{4} & -\frac{1}{4} \end{bmatrix} \xrightarrow[R_2 - 3R_3 \to R_2]{R_1 + R_3 \to R_1} \begin{bmatrix} 1 & 0 & 0 & -\frac{1}{4} & \frac{3}{4} & -\frac{1}{4} \\ 0 & 1 & 0 & -\frac{1}{4} & -\frac{1}{4} & \frac{3}{4} \\ 0 & 0 & 1 & \frac{3}{4} & -\frac{1}{4} & -\frac{1}{4} \end{bmatrix}$. Therefore, the inverse of the matrix is $\begin{bmatrix} -\frac{1}{4} & \frac{3}{4} & -\frac{1}{4} \\ -\frac{1}{4} & -\frac{1}{4} & \frac{3}{4} \\ \frac{3}{4} & -\frac{1}{4} & -\frac{1}{4} \end{bmatrix}$ and $\begin{bmatrix} x \\ y \\ z \end{bmatrix} = \begin{bmatrix} -\frac{1}{4} & \frac{3}{4} & -\frac{1}{4} \\ -\frac{1}{4} & -\frac{1}{4} & \frac{3}{4} \\ \frac{3}{4} & -\frac{1}{4} & -\frac{1}{4} \end{bmatrix} \begin{bmatrix} 675 \\ 600 \\ 625 \end{bmatrix} = \begin{bmatrix} 125 \\ 150 \\ 200 \end{bmatrix}$. Thus, he earns \$125 on a standard set, \$150 on a deluxe set, and \$200 on a leather-bound set.

**50.** No. Consider the following counterexample: $A = \begin{bmatrix} 0 & 1 \\ 0 & 0 \end{bmatrix}$ and $B = \begin{bmatrix} 0 & 2 \\ 0 & 0 \end{bmatrix}$. Then, $AB = O$, but neither $A = O$ nor $B = O$. There are infinitely many matrices for which $A^2 = O$. One example is $A = \begin{bmatrix} 0 & 1 \\ 0 & 0 \end{bmatrix}$. Then, $A^2 = O$, but $A \neq O$.

## 10.7 Determinants and Cramer's Rule

**1.** The matrix $\begin{bmatrix} 2 & 0 \\ 0 & 3 \end{bmatrix}$ has determinant $|D| = (2)(3) - (0)(0) = 6$.

**2.** The matrix $\begin{bmatrix} 0 & -1 \\ 2 & 0 \end{bmatrix}$ has determinant $|D| = (0)(0) - (-1)(2) = 2$.

**3.** The matrix $\begin{bmatrix} 4 & 5 \\ 0 & -1 \end{bmatrix}$ has determinant $|D| = (4)(-1) - (5)(0) = -4$.

**4.** The matrix $\begin{bmatrix} -2 & 1 \\ 3 & -2 \end{bmatrix}$ has determinant $|D| = (-2)(-2) - (1)(3) = 1$.

**5.** The matrix $\begin{bmatrix} 2 & 5 \end{bmatrix}$ does not have a determinant because the matrix is not square.

**6.** The matrix $\begin{bmatrix} 3 \\ 0 \end{bmatrix}$ does not have a determinant because the matrix is not square.

**7.** The matrix $\begin{bmatrix} \frac{1}{2} & \frac{1}{8} \\ 1 & \frac{1}{2} \end{bmatrix}$ has determinant $|D| = \frac{1}{2} \cdot \frac{1}{2} - 1 \cdot \frac{1}{8} = \frac{1}{4} - \frac{1}{8} = \frac{1}{8}$.

**8.** The matrix $\begin{bmatrix} 2.2 & -1.4 \\ 0.5 & 1.0 \end{bmatrix}$ has determinant $|D| = (2.2)(1.0) - (0.5)(-1.4) = 2.2 + 0.7 = 2.9$.

**In Solutions 9–14,** $A = \begin{bmatrix} 1 & 0 & \frac{1}{2} \\ -3 & 5 & 2 \\ 0 & 0 & 4 \end{bmatrix}$.

**9.** $M_{11} = 5 \cdot 4 - 0 \cdot 2 = 20$, $A_{11} = (-1)^2 M_{11} = 20$

**10.** $M_{33} = 1 \cdot 5 + 3 \cdot 0 = 5$, $A_{33} = (-1)^6 M_{33} = 5$

**11.** $M_{12} = -3 \cdot 4 - 0 \cdot 2 = -12$, $A_{12} = (-1)^3 M_{12} = 12$

**12.** $M_{13} = -3 \cdot 0 - 0 \cdot 5 = 0$, $A_{13} = (-1)^4 M_{13} = 0$

**13.** $M_{23} = 1 \cdot 0 - 0 \cdot 0 = 0$, $A_{23} = (-1)^5 M_{23} = 0$

**14.** $M_{32} = 1 \cdot 2 + 3 \cdot \frac{1}{2} = \frac{7}{2}$, $A_{32} = (-1)^5 M_{32} = -\frac{7}{2}$

**15.** $M = \begin{bmatrix} 2 & 1 & 0 \\ 0 & -2 & 4 \\ 0 & 1 & -3 \end{bmatrix}$. Therefore, expanding by the first column, $|M| = 2\begin{vmatrix} -2 & 4 \\ 1 & -3 \end{vmatrix} = 2(6-4) = 4$. Since $|M| \neq 0$, the matrix has an inverse.

**16.** $M = \begin{bmatrix} 0 & -1 & 0 \\ 2 & 6 & 4 \\ 1 & 0 & 3 \end{bmatrix}$. Therefore, expanding by the first row, $|M| = -(-1)\begin{vmatrix} 2 & 4 \\ 1 & 3 \end{vmatrix} = 6 - 4 = 2$. Since $|M| \neq 0$, the matrix has an inverse.

**17.** $M = \begin{bmatrix} 1 & 3 & 7 \\ 2 & 0 & -1 \\ 0 & 2 & 6 \end{bmatrix}$. Therefore, expanding by the third row,

$|M| = -2\begin{vmatrix} 1 & 7 \\ 2 & -1 \end{vmatrix} + 6\begin{vmatrix} 1 & 3 \\ 2 & 0 \end{vmatrix} = -2(-1-14) + 6(0-6) = 30 - 36 = -6$. Since $|M| \neq 0$, the matrix has an inverse.

**18.** $M = \begin{bmatrix} -2 & -\frac{3}{2} & \frac{1}{2} \\ 2 & 4 & 0 \\ \frac{1}{2} & 2 & 1 \end{bmatrix}$. Therefore, $|M| = \frac{1}{2}\begin{vmatrix} 2 & 4 \\ \frac{1}{2} & 2 \end{vmatrix} + 1\begin{vmatrix} -2 & -\frac{3}{2} \\ 2 & 4 \end{vmatrix} = \frac{1}{2}(4-2) + (-8+3) = 1 - 5 = -4$, and the matrix has an inverse.

**19.** $M = \begin{bmatrix} 30 & 0 & 20 \\ 0 & -10 & -20 \\ 40 & 0 & 10 \end{bmatrix}$. Therefore, expanding by the first row,

$|M| = 30\begin{vmatrix} -10 & -20 \\ 0 & 10 \end{vmatrix} + 20\begin{vmatrix} 0 & -10 \\ 40 & 0 \end{vmatrix} = 30(-100+0) + 20(0+400) = -3000 + 8000 = 5000$, and so $M^{-1}$ exists.

**20.** $M = \begin{bmatrix} 1 & 2 & 5 \\ -2 & -3 & 2 \\ 3 & 5 & 3 \end{bmatrix}$. Therefore,

$|M| = 1\begin{vmatrix} -3 & 2 \\ 5 & 3 \end{vmatrix} - 2\begin{vmatrix} -2 & 2 \\ 3 & 3 \end{vmatrix} + 5\begin{vmatrix} -2 & -3 \\ 3 & 5 \end{vmatrix} = (-9-10) - 2(-6-6) + 5(-10+9) = -19 + 24 - 5 = 0$, and so the matrix does not have an inverse.

**21.** $M=\begin{bmatrix}1&3&3&0\\0&2&0&1\\-1&0&0&2\\1&6&4&1\end{bmatrix}$. Therefore, expanding by the third row,

$$|M|=-1\begin{vmatrix}3&3&0\\2&0&1\\6&4&1\end{vmatrix}-2\begin{vmatrix}1&3&3\\0&2&0\\1&6&4\end{vmatrix}=1\begin{vmatrix}3&3\\6&4\end{vmatrix}-1\begin{vmatrix}3&3\\2&0\end{vmatrix}-4\begin{vmatrix}1&3\\1&4\end{vmatrix}=-6+6-4=-4,$$ and so $M^{-1}$exists.

**22.** $M=\begin{bmatrix}1&2&0&2\\3&-4&0&4\\0&1&6&0\\1&0&2&0\end{bmatrix}$. Therefore, $|M|=-1\begin{vmatrix}2&0&2\\-4&0&4\\1&6&0\end{vmatrix}-2\begin{vmatrix}1&2&2\\3&-4&4\\0&1&0\end{vmatrix}=6\begin{vmatrix}2&2\\-4&4\end{vmatrix}+2\begin{vmatrix}1&2\\3&4\end{vmatrix}=6\cdot16-2\cdot2=92,$

and so $M^{-1}$exists.

**23.** $|M|=\begin{vmatrix}0&0&4&6\\2&1&1&3\\2&1&2&3\\3&0&1&7\end{vmatrix}=\begin{vmatrix}0&0&4&6\\2&1&1&3\\0&0&1&0\\3&0&1&7\end{vmatrix}$, by replacing $R_3$ with $R_3-R_2$. Then, expanding by the third row,

$$|M|=1\begin{vmatrix}0&0&6\\2&1&3\\3&0&7\end{vmatrix}=6\begin{vmatrix}2&1\\3&0\end{vmatrix}=6(2\cdot0-3\cdot1)=-18.$$

**24.** $M=\begin{bmatrix}-2&3&-1&7\\4&6&-2&3\\7&7&0&5\\3&-12&4&0\end{bmatrix}$. Then $|M|=\begin{vmatrix}-2&3&-1&7\\4&6&-2&3\\7&7&0&5\\3&-12&4&0\end{vmatrix}=\begin{vmatrix}-2&0&-1&7\\4&0&-2&3\\7&7&0&5\\3&0&4&0\end{vmatrix}$,

by replacing $C_2$ with $C_2+3C_3$. So expanding about the second column,

$$|M|=-7\begin{vmatrix}-2&-1&7\\4&-2&3\\3&4&0\end{vmatrix}=-7\left(7\begin{vmatrix}4&-2\\3&4\end{vmatrix}-3\begin{vmatrix}-2&-1\\3&4\end{vmatrix}\right)=-7(7\cdot22+3\cdot5)=-1183.$$

**25.** $M=\begin{bmatrix}1&2&3&4&5\\0&2&4&6&8\\0&0&3&6&9\\0&0&0&4&8\\0&0&0&0&5\end{bmatrix}$, so $|M|=5\begin{vmatrix}1&2&3&4\\0&2&4&6\\0&0&3&6\\0&0&0&4\end{vmatrix}=5\cdot4\begin{vmatrix}1&2&3\\0&2&4\\0&0&3\end{vmatrix}=20\cdot3\begin{vmatrix}1&2\\0&2\end{vmatrix}=60\cdot2=120.$

**26.** $M=\begin{bmatrix}2&-1&6&4\\7&2&-2&5\\4&-2&10&8\\6&1&1&4\end{bmatrix}$. Then, $|M|=\begin{vmatrix}2&-1&6&4\\7&2&-2&5\\4&-2&10&8\\6&1&1&4\end{vmatrix}=\begin{vmatrix}0&-1&6&4\\11&2&-2&5\\0&-2&10&8\\8&1&1&4\end{vmatrix}$, by replacing $C_1$with $C_1+2C_2$. So

$$|M|=-11\begin{vmatrix}-1&6&4\\-2&10&8\\1&1&4\end{vmatrix}-8\begin{vmatrix}-1&6&4\\2&-2&5\\-2&10&8\end{vmatrix}=-11\begin{vmatrix}-1&6&0\\-2&10&0\\1&1&8\end{vmatrix}-8\begin{vmatrix}-1&6&0\\2&-2&13\\-2&10&0\end{vmatrix}=-88\begin{vmatrix}-1&6\\-2&10\end{vmatrix}+104\begin{vmatrix}-1&6\\-2&10\end{vmatrix}$$

$$=-88\cdot2+104\cdot2=32$$

**27.** $B = \begin{bmatrix} 4 & 1 & 0 \\ -2 & -1 & 1 \\ 4 & 0 & 3 \end{bmatrix}$

**(a)** $|B| = 2\begin{vmatrix} 1 & 0 \\ 0 & 3 \end{vmatrix} - 1\begin{vmatrix} 4 & 0 \\ 4 & 3 \end{vmatrix} - 1\begin{vmatrix} 4 & 1 \\ 4 & 0 \end{vmatrix} = 6 - 12 + 4 = -2$

**(b)** $|B| = -1\begin{vmatrix} 4 & 1 \\ 4 & 0 \end{vmatrix} + 3\begin{vmatrix} 4 & 1 \\ -2 & -1 \end{vmatrix} = 4 - 6 = -2$

**(c)** Yes, as expected, the results agree.

**28.** $\begin{cases} x + 2y + 6z = 5 \\ -3x - 6y + 5z = 8 \\ 2x + 6y + 9z = 7 \end{cases}$

**(a)** If $x = -1$, $y = 0$, and $z = 1$, then $x+2y+6z = (-1)+2(0)+6(1) = 5$, $-3x-6y+5z = -3(-1)-6(0)+5(1) = 8$, and $2x + 6y + 9z = 2(-1) + 6(0) + 9(1) = 7$. Therefore, $x = -1$, $y = 0$, $z = 1$ is a solution of the system.

**(b)** $M = \begin{bmatrix} 1 & 2 & 6 \\ -3 & -6 & 5 \\ 2 & 6 & 9 \end{bmatrix}$. Then, $|M| = \begin{vmatrix} 1 & 2 & 6 \\ -3 & -6 & 5 \\ 2 & 6 & 9 \end{vmatrix} = \begin{vmatrix} 1 & 0 & 6 \\ -3 & 0 & 5 \\ 2 & 2 & 9 \end{vmatrix}$ (replacing $C_2$ with $C_2 - 2C_1$), so

$|M| = -2\begin{vmatrix} 1 & 6 \\ -3 & 5 \end{vmatrix} = -2(5 + 18) = -46.$

**(c)** We can write the system as a matrix equation: $\begin{bmatrix} 1 & 2 & 6 \\ -3 & -6 & 5 \\ 2 & 6 & 9 \end{bmatrix}\begin{bmatrix} x \\ y \\ z \end{bmatrix} = \begin{bmatrix} 5 \\ 8 \\ 7 \end{bmatrix}$ or $MX = B$. Since $|M| \neq 0$, $M$ has an inverse. If we multiply both sides of the matrix equation by $M^{-1}$, then we get a unique solution for $X$, given by $X = M^{-1}B$. Thus, the equation has no other solution.

**(d)** Yes, since $|M| \neq 0$.

**29.** $\begin{cases} 2x - y = -9 \\ x + 2y = 8 \end{cases}$ Then $|D| = \begin{vmatrix} 2 & -1 \\ 1 & 2 \end{vmatrix} = 5$, $|D_x| = \begin{vmatrix} -9 & -1 \\ 8 & 2 \end{vmatrix} = -10$, and $|D_y| = \begin{vmatrix} 2 & -9 \\ 1 & 8 \end{vmatrix} = 25$.

Hence, $x = \dfrac{|D_x|}{|D|} = \dfrac{-10}{5} = -2$, $y = \dfrac{|D_y|}{|D|} = \dfrac{25}{5} = 5$, and so the solution is $(-2, 5)$.

**30.** $\begin{cases} 6x + 12y = 33 \\ 4x + 7y = 20 \end{cases}$ Then $|D| = \begin{vmatrix} 6 & 12 \\ 4 & 7 \end{vmatrix} = -6$, $|D_x| = \begin{vmatrix} 33 & 12 \\ 20 & 7 \end{vmatrix} = -9$, and $|D_y| = \begin{vmatrix} 6 & 33 \\ 4 & 20 \end{vmatrix} = -12$.

Hence, $x = \dfrac{|D_x|}{|D|} = \dfrac{-9}{-6} = \dfrac{3}{2}$ and $y = \dfrac{|D_y|}{|D|} = \dfrac{-12}{-6} = 2$, and so the solution is $\left(\frac{3}{2}, 2\right)$.

**31.** $\begin{cases} x - 6y = 3 \\ 3x + 2y = 1 \end{cases}$ Then, $|D| = \begin{vmatrix} 1 & -6 \\ 3 & 2 \end{vmatrix} = 20$, $|D_x| = \begin{vmatrix} 3 & -6 \\ 1 & 2 \end{vmatrix} = 12$, and $|D_y| = \begin{vmatrix} 1 & 3 \\ 3 & 1 \end{vmatrix} = -8$.

Hence, $x = \dfrac{|D_x|}{|D|} = \frac{12}{20} = 0.6$, $y = \dfrac{|D_y|}{|D|} = \dfrac{-8}{20} = -0.4$,and so the solution is $(0.6, -0.4)$.

**32.** $\begin{cases} \frac{1}{2}x + \frac{1}{3}y = 1 \\ \frac{1}{4}x - \frac{1}{6}y = -\frac{3}{2} \end{cases}$ Then, $|D| = \begin{vmatrix} \frac{1}{2} & \frac{1}{3} \\ \frac{1}{4} & -\frac{1}{6} \end{vmatrix} = -\frac{1}{6}$, $|D_x| = \begin{vmatrix} 1 & \frac{1}{3} \\ -\frac{3}{2} & -\frac{1}{6} \end{vmatrix} = \frac{1}{3}$, and $|D_y| = \begin{vmatrix} \frac{1}{2} & 1 \\ \frac{1}{4} & -\frac{3}{2} \end{vmatrix} = -1$.

Hence, $x = \dfrac{|D_x|}{|D|} = \dfrac{\frac{1}{3}}{-\frac{1}{6}} = -2$, $y = \dfrac{|D_y|}{|D|} = \dfrac{-1}{-\frac{1}{6}} = 6$, and so the solution is $(-2, 6)$.

**33.** $\begin{cases} 0.4x + 1.2y = 0.4 \\ 1.2x + 1.6y = 3.2 \end{cases}$ Then, $|D| = \begin{vmatrix} 0.4 & 1.2 \\ 1.2 & 1.6 \end{vmatrix} = -0.8$, $|D_x| = \begin{vmatrix} 0.4 & 1.2 \\ 3.2 & 1.6 \end{vmatrix} = -3.2$, and $|D_y| = \begin{vmatrix} 0.4 & 0.4 \\ 1.2 & 3.2 \end{vmatrix} = 0.8$.

Hence, $x = \dfrac{|D_x|}{|D|} = \dfrac{-3.2}{-0.8} = 4$, $y = \dfrac{|D_y|}{|D|} = \dfrac{0.8}{-0.8} = -1$,and so the solution is $(4, -1)$.

**34.** $\begin{cases} 10x - 17y = 21 \\ 20x - 31y = 39 \end{cases}$. Then, $|D| = \begin{vmatrix} 10 & -17 \\ 20 & -31 \end{vmatrix} = 30$, $|D_x| = \begin{vmatrix} 21 & -17 \\ 39 & -31 \end{vmatrix} = 12$, and $|D_y| = \begin{vmatrix} 10 & 21 \\ 20 & 39 \end{vmatrix} = -30$.

Hence, $x = \dfrac{|D_x|}{|D|} = \frac{12}{30} = \frac{2}{5}$, $y = \dfrac{|D_y|}{|D|} = \dfrac{-30}{30} = -1$, and so the solution is $\left(\frac{2}{5}, -1\right)$.

**35.** $\begin{cases} x - y + 2z = 0 \\ 3x + z = 11 \\ -x + 2y = 0 \end{cases}$ Then expanding by the second row,

$$|D| = \begin{vmatrix} 1 & -1 & 2 \\ 3 & 0 & 1 \\ -1 & 2 & 0 \end{vmatrix} = -3 \begin{vmatrix} -1 & 2 \\ 2 & 0 \end{vmatrix} - 1 \begin{vmatrix} 1 & -1 \\ -1 & 2 \end{vmatrix} = 12 - 1 = 11, \ |D_x| = \begin{vmatrix} 0 & -1 & 2 \\ 11 & 0 & 1 \\ 0 & 2 & 0 \end{vmatrix} = -11 \begin{vmatrix} -1 & 2 \\ 2 & 0 \end{vmatrix} = 44,$$

$$|D_y| = \begin{vmatrix} 1 & 0 & 2 \\ 3 & 11 & 1 \\ -1 & 0 & 0 \end{vmatrix} = 11 \begin{vmatrix} 1 & 2 \\ -1 & 0 \end{vmatrix} = 22, \text{ and } |D_z| = \begin{vmatrix} 1 & -1 & 0 \\ 3 & 0 & 11 \\ -1 & 2 & 0 \end{vmatrix} = -11 \begin{vmatrix} 1 & -1 \\ -1 & 2 \end{vmatrix} = -11.$$

Therefore, $x = \frac{44}{11} = 4$, $y = \frac{22}{11} = 2$, $z = \frac{-11}{11} = -1$, and so the solution is $(4, 2, -1)$.

**36.** $\begin{cases} 5x - 3y + z = 6 \\ 4y - 6z = 22 \\ 7x + 10y = -13 \end{cases}$ Then $|D| = \begin{vmatrix} 5 & -3 & 1 \\ 0 & 4 & -6 \\ 7 & 10 & 0 \end{vmatrix} = 1 \begin{vmatrix} 0 & 4 \\ 7 & 10 \end{vmatrix} + 6 \begin{vmatrix} 5 & -3 \\ 7 & 10 \end{vmatrix} = -28 + 426 = 398,$

$$|D_x| = \begin{vmatrix} 6 & -3 & 1 \\ 22 & 4 & -6 \\ -13 & 10 & 0 \end{vmatrix} = 1 \begin{vmatrix} 22 & 4 \\ -13 & 10 \end{vmatrix} + 6 \begin{vmatrix} 6 & -3 \\ -13 & 10 \end{vmatrix} = 272 + 126 = 398,$$

$$|D_y| = \begin{vmatrix} 5 & 6 & 1 \\ 0 & 22 & -6 \\ 7 & -13 & 0 \end{vmatrix} = 1 \begin{vmatrix} 0 & 22 \\ 7 & -13 \end{vmatrix} + 6 \begin{vmatrix} 5 & 6 \\ 7 & -13 \end{vmatrix} = -154 - 642 = -796, \text{ and}$$

$$|D_z| = \begin{vmatrix} 5 & -3 & 6 \\ 0 & 4 & 22 \\ 7 & 10 & -13 \end{vmatrix} = 4 \begin{vmatrix} 5 & 6 \\ 7 & -13 \end{vmatrix} - 22 \begin{vmatrix} 5 & -3 \\ 7 & 10 \end{vmatrix} = -428 - 1562 = -1990.$$

Therefore, $x = \frac{398}{398} = 1$, $y = \frac{-796}{398} = -2$, and $z = \frac{-1990}{398} = -5$, and so the solution is $(1, -2, -5)$.

**37.** $\begin{cases} 2x_1 + 3x_2 - 5x_3 = 1 \\ x_1 + x_2 - x_3 = 2 \\ 2x_2 + x_3 = 8 \end{cases}$

Then, expanding by the third row,

$$|D| = \begin{vmatrix} 2 & 3 & -5 \\ 1 & 1 & -1 \\ 0 & 2 & 1 \end{vmatrix} = -2 \begin{vmatrix} 2 & -5 \\ 1 & -1 \end{vmatrix} + \begin{vmatrix} 2 & 3 \\ 1 & 1 \end{vmatrix} = -6 - 1 = -7,$$

$$|D_{x_1}| = \begin{vmatrix} 1 & 3 & -5 \\ 2 & 1 & -1 \\ 8 & 2 & 1 \end{vmatrix} = \begin{vmatrix} 1 & -1 \\ 2 & 1 \end{vmatrix} - 3 \begin{vmatrix} 2 & -1 \\ 8 & 1 \end{vmatrix} - 5 \begin{vmatrix} 2 & 1 \\ 8 & 2 \end{vmatrix} = 3 - 30 + 20 = -7,$$

$$|D_{x_2}| = \begin{vmatrix} 2 & 1 & -5 \\ 1 & 2 & -1 \\ 0 & 8 & 1 \end{vmatrix} = -8 \begin{vmatrix} 2 & -5 \\ 1 & -1 \end{vmatrix} + \begin{vmatrix} 2 & 1 \\ 1 & 2 \end{vmatrix} = -24 + 3 = -21, \text{ and}$$

$$|D_{x_3}| = \begin{vmatrix} 2 & 3 & 1 \\ 1 & 1 & 2 \\ 0 & 2 & 8 \end{vmatrix} = -2 \begin{vmatrix} 2 & 1 \\ 1 & 2 \end{vmatrix} + 8 \begin{vmatrix} 2 & 3 \\ 1 & 1 \end{vmatrix} = -6 - 8 = -14.$$

Thus, $x_1 = \frac{-7}{-7} = 1$, $x_2 = \frac{-21}{-7} = 3$, $x_3 = \frac{-14}{-7} = 2$, and so the solution is $(1, 3, 2)$.

**38.** $\begin{cases} -2a + c = 2 \\ a + 2b - c = 9 \\ 3a + 5b + 2c = 22 \end{cases}$

Then $|D| = \begin{vmatrix} -2 & 0 & 1 \\ 1 & 2 & -1 \\ 3 & 5 & 2 \end{vmatrix} = -2 \begin{vmatrix} 2 & -1 \\ 5 & 2 \end{vmatrix} + 1 \begin{vmatrix} 1 & 2 \\ 3 & 5 \end{vmatrix} = -18 - 1 = -19,$

$$|D_a| = \begin{vmatrix} 2 & 0 & 1 \\ 9 & 2 & -1 \\ 22 & 5 & 2 \end{vmatrix} = 2 \begin{vmatrix} 2 & -1 \\ 5 & 2 \end{vmatrix} + 1 \begin{vmatrix} 9 & 2 \\ 22 & 5 \end{vmatrix} = 18 + 1 = 19,$$

$$|D_b| = \begin{vmatrix} -2 & 2 & 1 \\ 1 & 9 & -1 \\ 3 & 22 & 2 \end{vmatrix} = -2 \begin{vmatrix} 9 & -1 \\ 22 & 2 \end{vmatrix} - 2 \begin{vmatrix} 1 & -1 \\ 3 & 2 \end{vmatrix} + 1 \begin{vmatrix} 1 & 9 \\ 3 & 22 \end{vmatrix} = -80 - 10 - 5 = -95, \text{ and}$$

$$|D_c| = \begin{vmatrix} -2 & 0 & 2 \\ 1 & 2 & 9 \\ 3 & 5 & 22 \end{vmatrix} = -2 \begin{vmatrix} 2 & 9 \\ 5 & 22 \end{vmatrix} + 2 \begin{vmatrix} 1 & 2 \\ 3 & 5 \end{vmatrix} = 2 - 2 = 0.$$

Hence, $a = -1$, $b = 5$, and $c = 0$, and so the solution is $(-1, 5, 0)$.

**39.** $\begin{cases} \frac{1}{3}x - \frac{1}{5}y + \frac{1}{2}z = \frac{7}{10} \\ -\frac{2}{3}x + \frac{2}{5}y + \frac{3}{2}z = \frac{11}{10} \\ x - \frac{4}{5}y + z = \frac{9}{5} \end{cases} \Leftrightarrow \begin{cases} 10x - 6y + 15z = 21 \\ -20x + 12y + 45z = 33 \\ 5x - 4y + 5z = 9 \end{cases}$ Then

$$|D| = \begin{vmatrix} 10 & -6 & 15 \\ -20 & 12 & 45 \\ 5 & -4 & 5 \end{vmatrix} = 10\begin{vmatrix} 12 & 45 \\ -4 & 5 \end{vmatrix} + 6\begin{vmatrix} -20 & 45 \\ 5 & 5 \end{vmatrix} + 15\begin{vmatrix} -20 & 12 \\ 5 & -4 \end{vmatrix} = 2400 - 1950 + 300 = 750,$$

$$|D_x| = \begin{vmatrix} 21 & -6 & 15 \\ 33 & 12 & 45 \\ 9 & -4 & 5 \end{vmatrix} = 21\begin{vmatrix} 12 & 45 \\ -4 & 5 \end{vmatrix} + 6\begin{vmatrix} 33 & 45 \\ 9 & 5 \end{vmatrix} + 15\begin{vmatrix} 33 & 12 \\ 9 & -4 \end{vmatrix} = 5040 - 1440 - 3600 = 0,$$

$$|D_y| = \begin{vmatrix} 10 & 21 & 15 \\ -20 & 33 & 45 \\ 5 & 9 & 5 \end{vmatrix} = 10\begin{vmatrix} 33 & 45 \\ 9 & 5 \end{vmatrix} - 21\begin{vmatrix} -20 & 45 \\ 5 & 5 \end{vmatrix} + 15\begin{vmatrix} -20 & 33 \\ 5 & 9 \end{vmatrix} = -2400 + 6825 - 5175 = -750, \text{ and}$$

$$|D_z| = \begin{vmatrix} 10 & -6 & 21 \\ -20 & 12 & 33 \\ 5 & -4 & 9 \end{vmatrix} = 10\begin{vmatrix} 12 & 33 \\ -4 & 9 \end{vmatrix} + 6\begin{vmatrix} -20 & 33 \\ 5 & 9 \end{vmatrix} + 21\begin{vmatrix} -20 & 12 \\ 5 & -4 \end{vmatrix} = 2400 - 2070 + 420 = 750.$$

Therefore, $x = 0$, $y = -1$, $z = 1$, and so the solution is $(0, -1, 1)$.

**40.** $\begin{cases} 2x - y = 5 \\ 5x + 3z = 19 \\ 4y + 7z = 17 \end{cases}$ Then $|D| = \begin{vmatrix} 2 & -1 & 0 \\ 5 & 0 & 3 \\ 0 & 4 & 7 \end{vmatrix} = 2\begin{vmatrix} 0 & 3 \\ 4 & 7 \end{vmatrix} + 1\begin{vmatrix} 5 & 3 \\ 0 & 7 \end{vmatrix} = -24 + 35 = 11,$

$$|D_x| = \begin{vmatrix} 5 & -1 & 0 \\ 19 & 0 & 3 \\ 17 & 4 & 7 \end{vmatrix} = 5\begin{vmatrix} 0 & 3 \\ 4 & 7 \end{vmatrix} + 1\begin{vmatrix} 19 & 3 \\ 17 & 7 \end{vmatrix} = -60 + 82 = 22,$$

$$|D_y| = \begin{vmatrix} 2 & 5 & 0 \\ 5 & 19 & 3 \\ 0 & 17 & 7 \end{vmatrix} = 2\begin{vmatrix} 19 & 3 \\ 17 & 7 \end{vmatrix} - 5\begin{vmatrix} 5 & 3 \\ 0 & 7 \end{vmatrix} = 164 - 175 = -11, \text{ and}$$

$|D_z| = \begin{vmatrix} 2 & -1 & 5 \\ 5 & 0 & 19 \\ 0 & 4 & 17 \end{vmatrix} = -4\begin{vmatrix} 2 & 5 \\ 5 & 19 \end{vmatrix} + 17\begin{vmatrix} 2 & -1 \\ 5 & 0 \end{vmatrix} = -52 + 85 = 33$. Thus, $x = 2$, $y = -1$, and $z = 3$, and so the solution is $(2, -1, 3)$.

**41.** $\begin{cases} 3y + 5z = 4 \\ 2x - z = 10 \\ 4x + 7y = 0 \end{cases}$ Then $|D| = \begin{vmatrix} 0 & 3 & 5 \\ 2 & 0 & -1 \\ 4 & 7 & 0 \end{vmatrix} = -3\begin{vmatrix} 2 & -1 \\ 4 & 0 \end{vmatrix} + 5\begin{vmatrix} 2 & 0 \\ 4 & 7 \end{vmatrix} = -12 + 70 = 58,$

$$|D_x| = \begin{vmatrix} 4 & 3 & 5 \\ 10 & 0 & -1 \\ 0 & 7 & 0 \end{vmatrix} = -7\begin{vmatrix} 4 & 5 \\ 10 & -1 \end{vmatrix} = 378, \quad |D_y| = \begin{vmatrix} 0 & 4 & 5 \\ 2 & 10 & -1 \\ 4 & 0 & 0 \end{vmatrix} = 4\begin{vmatrix} 4 & 5 \\ 10 & -1 \end{vmatrix} = -216, \text{ and}$$

$$|D_z| = \begin{vmatrix} 0 & 3 & 4 \\ 2 & 0 & 10 \\ 4 & 7 & 0 \end{vmatrix} = 4\begin{vmatrix} 3 & 4 \\ 0 & 10 \end{vmatrix} - 7\begin{vmatrix} 0 & 4 \\ 2 & 10 \end{vmatrix} = 120 + 56 = 176.$$

Thus, $x = \frac{189}{29}$, $y = -\frac{108}{29}$, and $z = \frac{88}{29}$, and so the solution is $\left(\frac{189}{29}, -\frac{108}{29}, \frac{88}{29}\right)$.

**42.** $\begin{cases} 2x - 5y = 4 \\ x + y - z = 8 \\ 3x + 5z = 0 \end{cases}$ Then $|D| = \begin{vmatrix} 2 & -5 & 0 \\ 1 & 1 & -1 \\ 3 & 0 & 5 \end{vmatrix} = 2\begin{vmatrix} 1 & -1 \\ 0 & 5 \end{vmatrix} + 5\begin{vmatrix} 1 & -1 \\ 3 & 5 \end{vmatrix} = 10 + 40 = 50,$

$|D_x| = \begin{vmatrix} 4 & -5 & 0 \\ 8 & 1 & -1 \\ 0 & 0 & 5 \end{vmatrix} = 5\begin{vmatrix} 4 & -5 \\ 8 & 1 \end{vmatrix} = 220$, $|D_y| = \begin{vmatrix} 2 & 4 & 0 \\ 1 & 8 & -1 \\ 3 & 0 & 5 \end{vmatrix} = 2\begin{vmatrix} 8 & -1 \\ 0 & 5 \end{vmatrix} - 4\begin{vmatrix} 1 & -1 \\ 3 & 5 \end{vmatrix} = 80 - 32 = 48$, and

$|D_z| = \begin{vmatrix} 2 & -5 & 4 \\ 1 & 1 & 8 \\ 3 & 0 & 0 \end{vmatrix} = 3\begin{vmatrix} -5 & 4 \\ 1 & 8 \end{vmatrix} = -132$. Thus, $x = \frac{22}{5}$, $y = \frac{24}{25}$, and $z = -\frac{66}{25}$, and so the solution is $\left(\frac{22}{5}, \frac{24}{25}, -\frac{66}{25}\right)$.

**43.** $\begin{cases} x + y + z + w = 0 \\ 2z + w = 0 \\ y - z = 0 \\ x + 2z = 1 \end{cases}$ Then

$$|D| = \begin{vmatrix} 1 & 1 & 1 & 1 \\ 2 & 0 & 0 & 1 \\ 0 & 1 & -1 & 0 \\ 1 & 0 & 2 & 0 \end{vmatrix} = -1\begin{vmatrix} 2 & 0 & 1 \\ 0 & -1 & 0 \\ 1 & 2 & 0 \end{vmatrix} - 1\begin{vmatrix} 1 & 1 & 1 \\ 2 & 0 & 1 \\ 1 & 2 & 0 \end{vmatrix} = -\left(2\begin{vmatrix} -1 & 0 \\ 2 & 0 \end{vmatrix} + 1\begin{vmatrix} 0 & 1 \\ -1 & 0 \end{vmatrix}\right) - \left(-1\begin{vmatrix} 2 & 1 \\ 1 & 0 \end{vmatrix} - 2\begin{vmatrix} 1 & 1 \\ 2 & 1 \end{vmatrix}\right)$$

$$= -2(0) - 1(1) + 1(-1) + 2(-1) = -4,$$

$$|D_x| = \begin{vmatrix} 0 & 1 & 1 & 1 \\ 0 & 0 & 0 & 1 \\ 0 & 1 & -1 & 0 \\ 1 & 0 & 2 & 0 \end{vmatrix} = -1\begin{vmatrix} 1 & 1 & 1 \\ 0 & 0 & 1 \\ 1 & -1 & 0 \end{vmatrix} = -1(-1)\begin{vmatrix} 1 & 1 \\ 1 & -1 \end{vmatrix} = -2,$$

$$|D_y| = \begin{vmatrix} 1 & 0 & 1 & 1 \\ 2 & 0 & 0 & 1 \\ 0 & 0 & -1 & 0 \\ 1 & 1 & 2 & 0 \end{vmatrix} = 1\begin{vmatrix} 1 & 1 & 1 \\ 2 & 0 & 1 \\ 0 & -1 & 0 \end{vmatrix} = 1\begin{vmatrix} 0 & 1 \\ -1 & 0 \end{vmatrix} - 2\begin{vmatrix} 1 & 1 \\ -1 & 0 \end{vmatrix} = 1 - 2(1) = -1,$$

$$|D_z| = \begin{vmatrix} 1 & 1 & 0 & 1 \\ 2 & 0 & 0 & 1 \\ 0 & 1 & 0 & 0 \\ 1 & 0 & 1 & 0 \end{vmatrix} = -1\begin{vmatrix} 1 & 1 & 1 \\ 2 & 0 & 1 \\ 0 & 1 & 0 \end{vmatrix} = -1\begin{vmatrix} 0 & 1 \\ 1 & 0 \end{vmatrix} + 2\begin{vmatrix} 1 & 1 \\ 1 & 0 \end{vmatrix} = -1(-1) + 2(-1) = -1, \text{ and}$$

$$|D_w| = \begin{vmatrix} 1 & 1 & 1 & 0 \\ 2 & 0 & 0 & 0 \\ 0 & 1 & -1 & 0 \\ 1 & 0 & 2 & 1 \end{vmatrix} = 1\begin{vmatrix} 1 & 1 & 1 \\ 2 & 0 & 0 \\ 0 & 1 & -1 \end{vmatrix} = -2\begin{vmatrix} 1 & 1 \\ 1 & -1 \end{vmatrix} = -2(-2) = 4.$$ Hence, we have $x = \frac{|D_x|}{|D|} = \frac{-2}{-4} = \frac{1}{2}$,

$y = \frac{|D_y|}{|D|} = \frac{-1}{-4} = \frac{1}{4}$, $z = \frac{|D_z|}{|D|} = \frac{-1}{-4} = \frac{1}{4}$, and $w = \frac{|D_w|}{|D|} = \frac{4}{-4} = -1$, and the solution is $\left(\frac{1}{2}, \frac{1}{4}, \frac{1}{4}, -1\right)$.

**44.** $\begin{cases} x+y=1 \\ y+z=2 \\ z+w=3 \\ w-x=4 \end{cases}$ Then $|D| = \begin{vmatrix} 1 & 1 & 0 & 0 \\ 0 & 1 & 1 & 0 \\ 0 & 0 & 1 & 1 \\ -1 & 0 & 0 & 1 \end{vmatrix} = 1\begin{vmatrix} 1 & 1 & 0 \\ 0 & 1 & 1 \\ 0 & 0 & 1 \end{vmatrix} - 1\begin{vmatrix} 0 & 1 & 0 \\ 0 & 1 & 1 \\ -1 & 0 & 1 \end{vmatrix} = 1\begin{vmatrix} 1 & 1 \\ 0 & 1 \end{vmatrix} - (-1)\begin{vmatrix} 1 & 0 \\ 1 & 1 \end{vmatrix} = 1+1 = 2,$

$$|D_x| = \begin{vmatrix} 1 & 1 & 0 & 0 \\ 2 & 1 & 1 & 0 \\ 3 & 0 & 1 & 1 \\ 4 & 0 & 0 & 1 \end{vmatrix} = 1\begin{vmatrix} 1 & 1 & 0 \\ 0 & 1 & 1 \\ 0 & 0 & 1 \end{vmatrix} - 1\begin{vmatrix} 2 & 1 & 0 \\ 3 & 1 & 1 \\ 4 & 0 & 1 \end{vmatrix} = \begin{vmatrix} 1 & 1 \\ 0 & 1 \end{vmatrix} - \left(2\begin{vmatrix} 1 & 1 \\ 0 & 1 \end{vmatrix} - 1\begin{vmatrix} 3 & 1 \\ 4 & 1 \end{vmatrix}\right) = 1 - 3 = -2,$$

$$|D_y| = \begin{vmatrix} 1 & 1 & 0 & 0 \\ 0 & 2 & 1 & 0 \\ 0 & 3 & 1 & 1 \\ -1 & 4 & 0 & 1 \end{vmatrix} = 1\begin{vmatrix} 2 & 1 & 0 \\ 3 & 1 & 1 \\ 4 & 0 & 1 \end{vmatrix} - 1\begin{vmatrix} 0 & 1 & 0 \\ 0 & 1 & 1 \\ -1 & 0 & 1 \end{vmatrix} = \left(2\begin{vmatrix} 1 & 1 \\ 0 & 1 \end{vmatrix} - 1\begin{vmatrix} 3 & 1 \\ 4 & 1 \end{vmatrix}\right) - (-1)\begin{vmatrix} 1 & 0 \\ 1 & 1 \end{vmatrix} = 3+1 = 4,$$

$$|D_z| = \begin{vmatrix} 1 & 1 & 1 & 0 \\ 0 & 1 & 2 & 0 \\ 0 & 0 & 3 & 1 \\ -1 & 0 & 4 & 1 \end{vmatrix} = 1\begin{vmatrix} 1 & 2 & 0 \\ 0 & 3 & 1 \\ 0 & 4 & 1 \end{vmatrix} + 1\begin{vmatrix} 1 & 1 & 0 \\ 1 & 2 & 0 \\ 0 & 3 & 1 \end{vmatrix} = 1\begin{vmatrix} 3 & 1 \\ 4 & 1 \end{vmatrix} + 1\begin{vmatrix} 1 & 1 \\ 1 & 2 \end{vmatrix} = -1 + 1 = 0,$$

$$|D_w| = \begin{vmatrix} 1 & 1 & 0 & 1 \\ 0 & 1 & 1 & 2 \\ 0 & 0 & 1 & 3 \\ -1 & 0 & 0 & 4 \end{vmatrix} = 1\begin{vmatrix} 1 & 1 & 2 \\ 0 & 1 & 3 \\ 0 & 0 & 4 \end{vmatrix} + 1\begin{vmatrix} 1 & 0 & 1 \\ 1 & 1 & 2 \\ 0 & 1 & 3 \end{vmatrix} = \begin{vmatrix} 1 & 3 \\ 0 & 4 \end{vmatrix} + \left(1\begin{vmatrix} 1 & 2 \\ 1 & 3 \end{vmatrix} - 1\begin{vmatrix} 0 & 1 \\ 1 & 3 \end{vmatrix}\right) = 4+2 = 6.$$

Hence, the solution is $x = \dfrac{|D_x|}{|D|} = \dfrac{-2}{2} = -1$, $y = \dfrac{|D_y|}{|D|} = \dfrac{4}{2} = 2$, $z = \dfrac{|D_z|}{|D|} = \dfrac{0}{2} = 0$, and $w = \dfrac{|D_w|}{|D|} = \dfrac{6}{2} = 3$.

**45.** $\begin{vmatrix} a & 0 & 0 & 0 & 0 \\ 0 & b & 0 & 0 & 0 \\ 0 & 0 & c & 0 & 0 \\ 0 & 0 & 0 & d & 0 \\ 0 & 0 & 0 & 0 & e \end{vmatrix} = a\begin{vmatrix} b & 0 & 0 & 0 \\ 0 & c & 0 & 0 \\ 0 & 0 & d & 0 \\ 0 & 0 & 0 & e \end{vmatrix} = ab\begin{vmatrix} c & 0 & 0 \\ 0 & d & 0 \\ 0 & 0 & e \end{vmatrix} = abc\begin{vmatrix} d & 0 \\ 0 & e \end{vmatrix} = abcde$

**46.** $\begin{vmatrix} a & a & a & a & a \\ 0 & a & a & a & a \\ 0 & 0 & a & a & a \\ 0 & 0 & 0 & a & a \\ 0 & 0 & 0 & 0 & a \end{vmatrix} = a\begin{vmatrix} a & a & a & a \\ 0 & a & a & a \\ 0 & 0 & a & a \\ 0 & 0 & 0 & a \end{vmatrix} = a^2\begin{vmatrix} a & a & a \\ 0 & a & a \\ 0 & 0 & a \end{vmatrix} = a^3\begin{vmatrix} a & a \\ 0 & a \end{vmatrix} = a^5$

**47.** $\begin{vmatrix} x & 12 & 13 \\ 0 & x-1 & 23 \\ 0 & 0 & x-2 \end{vmatrix} = 0 \quad\Leftrightarrow\quad (x-2)\begin{vmatrix} x & 12 \\ 0 & x-1 \end{vmatrix} = 0 \quad\Leftrightarrow\quad (x-2)\cdot x\,(x-1) = 0 \quad\Leftrightarrow\quad x = 0, 1, \text{ or } 2$

**48.** $\begin{vmatrix} x & 1 & 1 \\ 1 & 1 & x \\ x & 1 & x \end{vmatrix} = x\begin{vmatrix} 1 & x \\ 1 & x \end{vmatrix} - \begin{vmatrix} 1 & x \\ x & x \end{vmatrix} + \begin{vmatrix} 1 & 1 \\ x & 1 \end{vmatrix} = x\,(0) - (x - x^2) + 1 - x = x^2 - 2x + 1 = 0 \Leftrightarrow (x-1)^2 = 0 \Leftrightarrow x = 1$

**49.** $\begin{vmatrix} 1 & 0 & x \\ x^2 & 1 & 0 \\ x & 0 & 1 \end{vmatrix} = 0 \quad\Leftrightarrow\quad 1\begin{vmatrix} 1 & 0 \\ 0 & 1 \end{vmatrix} + x\begin{vmatrix} x^2 & 1 \\ x & 0 \end{vmatrix} = 0 \quad\Leftrightarrow\quad 1 - x^2 = 0 \quad\Leftrightarrow\quad x^2 = 1 \quad\Leftrightarrow\quad x = \pm 1$

**50.** $\begin{vmatrix} a & b & x-a \\ x & x+b & x \\ 0 & 1 & 1 \end{vmatrix} = -1\begin{vmatrix} a & x-a \\ x & x \end{vmatrix} + \begin{vmatrix} a & b \\ x & x+b \end{vmatrix} = -1\left[ax - x(x-a)\right] + a(x+b) - bx$

$= -ax + x^2 - ax + ax + ab - bx = x^2 - ax - bx + ab = (x-a)(x-b) = 0 \Leftrightarrow x = a$ or $x = b$

**51.** Area $= \pm\frac{1}{2}\begin{vmatrix} 0 & 0 & 1 \\ 6 & 2 & 1 \\ 3 & 8 & 1 \end{vmatrix} = \pm\frac{1}{2}\begin{vmatrix} 6 & 2 \\ 3 & 8 \end{vmatrix} = \pm\frac{1}{2}(48-6) = \frac{1}{2}(42) = 21$

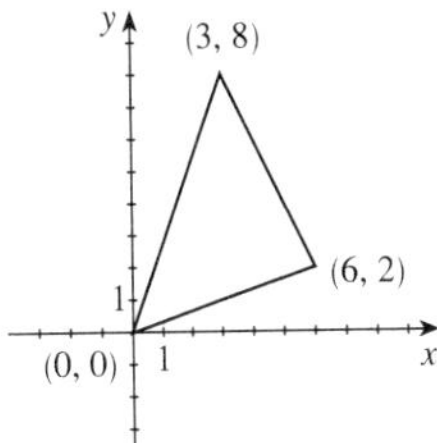


**52.** Area $= \pm\frac{1}{2}\begin{vmatrix} 1 & 0 & 1 \\ 3 & 5 & 1 \\ -2 & 2 & 1 \end{vmatrix} = \pm\frac{1}{2}\left(1\begin{vmatrix} 5 & 1 \\ 2 & 1 \end{vmatrix} + 1\begin{vmatrix} 3 & 5 \\ -2 & 2 \end{vmatrix}\right)$

$= \pm\frac{1}{2}\left[(5-2) + (6+10)\right] = \pm\frac{1}{2}\left[3+16\right] = \frac{1}{2}(19) = \frac{19}{2}$

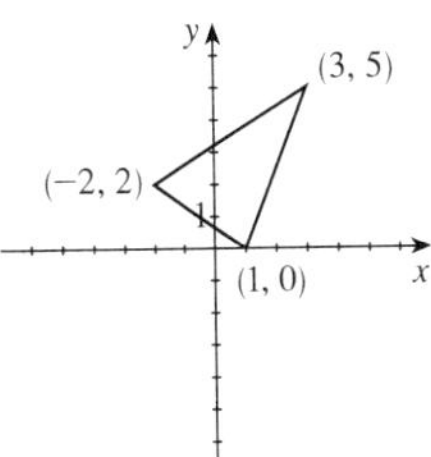


**53.** Area $= \pm\frac{1}{2}\begin{vmatrix} -1 & 3 & 1 \\ 2 & 9 & 1 \\ 5 & -6 & 1 \end{vmatrix} = \pm\frac{1}{2}\left[-1\begin{vmatrix} 9 & 1 \\ -6 & 1 \end{vmatrix} - 3\begin{vmatrix} 2 & 1 \\ 5 & 1 \end{vmatrix} + 1\begin{vmatrix} 2 & 9 \\ 5 & -6 \end{vmatrix}\right]$

$= \pm\frac{1}{2}\left[-1(9+6) - 3(2-5) + 1(-12-45)\right]$

$= \pm\frac{1}{2}\left[-15 - 3(-3) + (-57)\right] = \pm\frac{1}{2}(-63) = \frac{63}{2}$

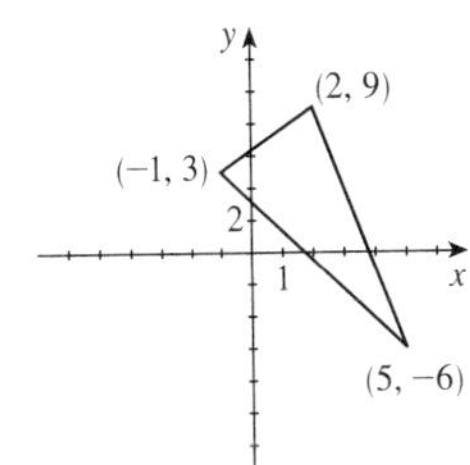


**54.** Area $= \pm\frac{1}{2}\begin{vmatrix} -2 & 5 & 1 \\ 7 & 2 & 1 \\ 3 & -4 & 1 \end{vmatrix} = \pm\frac{1}{2}\left(-2\begin{vmatrix} 2 & 1 \\ -4 & 1 \end{vmatrix} - 5\begin{vmatrix} 7 & 1 \\ 3 & 1 \end{vmatrix} + 1\begin{vmatrix} 7 & 2 \\ 3 & -4 \end{vmatrix}\right)$

$= \pm\frac{1}{2}\left[-2(2+4) - 5(7-3) + (-28-6)\right] = \pm\frac{1}{2}\left[-2(6) - 5(4) + (-34)\right]$

$= \pm\frac{1}{2}(-12-20-34) = \frac{1}{2}(66) = 33$

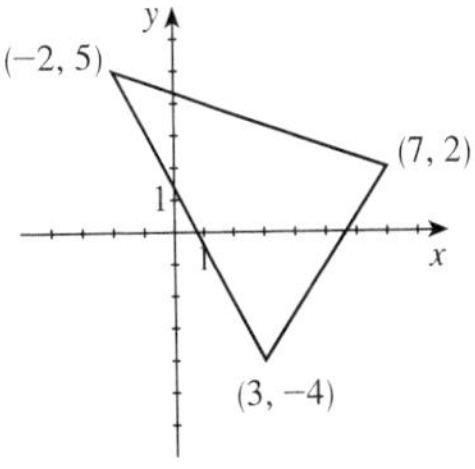


**55.** $\begin{vmatrix} 1 & x & x^2 \\ 1 & y & y^2 \\ 1 & z & z^2 \end{vmatrix} = 1\begin{vmatrix} y & y^2 \\ z & z^2 \end{vmatrix} - 1\begin{vmatrix} x & x^2 \\ z & z^2 \end{vmatrix} + 1\begin{vmatrix} x & x^2 \\ y & y^2 \end{vmatrix} = yz^2 - y^2z - \left(xz^2 - x^2z\right) + \left(xy^2 - xy^2\right)$

$= yz^2 - y^2z - xz^2 - x^2z + xy^2 - xy^2 + xyz - xyz = xyz - xz^2 - y^2z + yz^2 - x^2y + x^2z + zy^2 - xyz$

$= z\left(xy - xz - y^2 + yz\right) - x\left(xy - xz - y^2 + yz\right) = (z-x)\left(xy - xz - y^2 + yz\right)$

$= (z-x)\left[x(y-z) - y(y-z)\right] = (z-x)(x-y)(y-z)$

**56. (a)** Let $x$ be the amount of apples, $y$ the amount of peaches, and $z$ the amount of pears (in pounds).

We get the model $\begin{cases} x + \quad y + \quad z = 18 \\ 0.75x + 0.90y + 0.60z = 13.80 \\ -0.75x + 0.90y + 0.60z = 1.80 \end{cases}$

**(b)** $|D| = \begin{vmatrix} 1 & 1 & 1 \\ 0.75 & 0.90 & 0.60 \\ -0.75 & 0.90 & 0.60 \end{vmatrix} = 1 \cdot \begin{vmatrix} 0.90 & 0.60 \\ 0.90 & 0.60 \end{vmatrix} - 1 \cdot \begin{vmatrix} 0.75 & 0.60 \\ -0.75 & 0.60 \end{vmatrix} + 1 \cdot \begin{vmatrix} 0.75 & 0.90 \\ -0.75 & 0.90 \end{vmatrix} = 0 - 0.90 + 1.35 = 0.45,$

$|D_x| = \begin{vmatrix} 18 & 1 & 1 \\ 13.80 & 0.90 & 0.60 \\ 1.80 & 0.90 & 0.60 \end{vmatrix} = 18 \cdot \begin{vmatrix} 0.90 & 0.60 \\ 0.90 & 0.60 \end{vmatrix} - 1 \cdot \begin{vmatrix} 13.80 & 0.60 \\ 1.80 & 0.60 \end{vmatrix} + 1 \cdot \begin{vmatrix} 13.80 & 0.90 \\ 1.80 & 0.90 \end{vmatrix} = 0 - 7.2 + 10.8 = 3.6,$

$|D_y| = \begin{vmatrix} 1 & 18 & 1 \\ 0.75 & 13.80 & 0.60 \\ -0.75 & 1.80 & 0.60 \end{vmatrix} = 1 \cdot \begin{vmatrix} 13.80 & 0.60 \\ 1.80 & 0.60 \end{vmatrix} - 18 \cdot \begin{vmatrix} 0.75 & 0.60 \\ -0.75 & 0.60 \end{vmatrix} + 1 \cdot \begin{vmatrix} 0.75 & 13.80 \\ -0.75 & 1.80 \end{vmatrix}$

$= 7.2 - 16.2 + 11.7 = 2.7$, and

$|D_z| = \begin{vmatrix} 1 & 1 & 18 \\ 0.75 & 0.90 & 13.80 \\ -0.75 & 0.90 & 1.80 \end{vmatrix} = 1 \cdot \begin{vmatrix} 0.90 & 13.80 \\ 0.90 & 1.80 \end{vmatrix} - 1 \cdot \begin{vmatrix} 0.75 & 13.80 \\ -0.75 & 1.80 \end{vmatrix} + 18 \cdot \begin{vmatrix} 0.75 & 0.90 \\ -0.75 & 0.90 \end{vmatrix}$

$= -10.8 - 11.7 + 24.3 = 1.8.$

So $x = \dfrac{|D_x|}{|D|} = \dfrac{3.6}{0.45} = 8$; $y = \dfrac{|D_y|}{|D|} = \dfrac{27}{0.45} = 6$; and $z = \dfrac{|D_z|}{|D|} = \dfrac{1.8}{0.45} = 4$.

Thus, Muriel buys 8 pounds of apples, 6 pounds of peaches, and 4 pounds of pears.

**57. (a)** Using the points $(10, 25)$, $(15, 33.75)$, and $(40, 40)$,we substitute for $x$ and $y$ and get the system

$\begin{cases} 100a + 10b + c = 25 \\ 225a + 15b + c = 33.75 \\ 1600a + 40b + c = 40 \end{cases}$

**(b)** $|D| = \begin{vmatrix} 100 & 10 & 1 \\ 225 & 15 & 1 \\ 1600 & 40 & 1 \end{vmatrix} = 1 \cdot \begin{vmatrix} 225 & 15 \\ 1600 & 40 \end{vmatrix} - 1 \cdot \begin{vmatrix} 100 & 10 \\ 1600 & 40 \end{vmatrix} + 1 \cdot \begin{vmatrix} 100 & 10 \\ 225 & 15 \end{vmatrix}$

$= (9000 - 24{,}000) - (4000 - 16{,}000) + (1500 - 2250) = -15{,}000 + 12{,}000 - 750 = -3750,$

$|D_a| = \begin{vmatrix} 25 & 10 & 1 \\ 33.75 & 15 & 1 \\ 40 & 40 & 1 \end{vmatrix} = 1 \cdot \begin{vmatrix} 33.75 & 15 \\ 40 & 40 \end{vmatrix} - 1 \cdot \begin{vmatrix} 25 & 10 \\ 40 & 40 \end{vmatrix} + 1 \cdot \begin{vmatrix} 25 & 10 \\ 33.75 & 15 \end{vmatrix}$

$= (1350 - 600) - (1000 - 400) + (375 - 337.5) = 750 - 600 + 37.5 = 187.5,$

$|D_b| = \begin{vmatrix} 100 & 25 & 1 \\ 225 & 33.75 & 1 \\ 1600 & 40 & 1 \end{vmatrix} = 1 \cdot \begin{vmatrix} 225 & 33.75 \\ 1600 & 40 \end{vmatrix} - 1 \cdot \begin{vmatrix} 100 & 25 \\ 1600 & 40 \end{vmatrix} + 1 \cdot \begin{vmatrix} 100 & 25 \\ 225 & 33.75 \end{vmatrix}$

$= (9000 - 54{,}000) - (4000 - 40{,}000) + (3375 - 5625) = -45{,}000 + 36{,}000 - 2250 = -11{,}250$, and

$$|D_c| = \begin{vmatrix} 100 & 10 & 25 \\ 225 & 15 & 33.75 \\ 1600 & 40 & 40 \end{vmatrix} = 25 \cdot \begin{vmatrix} 225 & 15 \\ 1600 & 40 \end{vmatrix} - 33.75 \cdot \begin{vmatrix} 100 & 10 \\ 1600 & 40 \end{vmatrix} + 40 \cdot \begin{vmatrix} 100 & 10 \\ 225 & 15 \end{vmatrix}$$

$$= 25 \cdot (9{,}000 - 24{,}000) - 33.75 \cdot (4{,}000 - 16{,}000) + 40 \cdot (1{,}500 - 2{,}250)$$

$$= 25 \cdot (-15{,}000) + 33.75 \cdot 12{,}000 + 40 \cdot (-750) = -375{,}000 + 405{,}000 - 30{,}000 = 0.$$

Thus, $a = \frac{|D_a|}{|D|} = \frac{187.5}{-3750} = 0.05$, $b = \frac{|D_b|}{|D|} = \frac{-11{,}250}{-3{,}750} = 3$, and $c = \frac{|D_c|}{|D|} = \frac{0}{-3{,}750} = 0$. Thus, the model is $y = 0.05x^2 + 3x$.

**58.** Using the determinant formula for the area of a triangle (Exercise 59), we have

$$\text{Area} = \pm\frac{1}{2} \begin{vmatrix} 1000 & 2000 & 1 \\ 5000 & 4000 & 1 \\ 2000 & 6000 & 1 \end{vmatrix} = \pm\frac{1}{2}\left(1 \cdot \begin{vmatrix} 5000 & 4000 \\ 2000 & 6000 \end{vmatrix} - 1 \cdot \begin{vmatrix} 1000 & 2000 \\ 2000 & 6000 \end{vmatrix} + 1 \cdot \begin{vmatrix} 1000 & 2000 \\ 5000 & 4000 \end{vmatrix}\right)$$

$$= \pm\tfrac{1}{2} \cdot (22{,}000{,}000 - 2{,}000{,}000 - 6{,}000{,}000) = \pm\tfrac{1}{2} \cdot 14{,}000{,}000 = \pm 7{,}000{,}000$$

Thus, the area is 7,000,000 ft$^2$.

**59. (a)** The coordinates of the vertices of the surrounding rectangle are $(a_1, b_1)$, $(a_2, b_1)$, $(a_2, b_3)$, and $(a_1, b_3)$. The area of the surrounding rectangle is given by $(a_2 - a_1) \cdot (b_3 - b_1) = a_2b_3 + a_1b_1 - a_2b_1 - a_1b_3 = a_1b_1 + a_2b_3 - a_1b_3 - a_2b_1$.

**(b)** The area of the three blue triangles are as follows:

Area of $\triangle\left((a_1, b_1), (a_2, b_1), (a_2, b_2)\right)$: $\frac{1}{2}(a_2 - a_1) \cdot (b_2 - b_1) = \frac{1}{2}(a_2b_2 + a_1b_1 - a_2b_1 - a_1b_2)$

Area of $\triangle\left((a_2, b_2), (a_2, b_3), (a_3, b_3)\right)$: $\frac{1}{2}(a_2 - a_3) \cdot (b_3 - b_2) = \frac{1}{2}(a_2b_3 + a_3b_2 - a_2b_2 - a_3b_3)$

Area of $\triangle\left((a_1, b), (a_1, b_3), (a_3, b_3)\right)$: $\frac{1}{2}(a_3 - a_1) \cdot (b_3 - b_1) = \frac{1}{2}(a_3b_3 + a_1b_1 - a_3b_1 - a_1b_3)$.

Thus the sum of the areas of the blue triangles, $B$, is

$$B = \tfrac{1}{2}(a_2b_2 + a_1b_1 - a_2b_1 - a_1b_2) + \tfrac{1}{2}(a_2b_3 + a_3b_2 - a_2b_2 - a_3b_3) + \tfrac{1}{2}(a_3b_3 + a_1b_1 - a_3b_1 - a_1b_3)$$

$$= \tfrac{1}{2}(a_1b_1 + a_1b_1 + a_2b_2 + a_2b_3 + a_3b_2 + a_3b_3) - \tfrac{1}{2}(a_1b_2 + a_1b_3 + a_2b_1 + a_2b_2 + a_3b_1 + a_3b_3)$$

$$= a_1b_1 + \tfrac{1}{2}(a_2b_3 + a_3b_2) - \tfrac{1}{2}(a_1b_2 + a_1b_3 + a_2b_1 + a_3b_1)$$

So the area of the red triangle $A$ is the area of the rectangle minus the sum of the areas of the blue triangles, that is,

$$A = (a_1b_1 + a_2b_3 - a_1b_3 - a_2b_1) - \left[a_1b_1 + \tfrac{1}{2}(a_2b_3 + a_3b_2) - \tfrac{1}{2}(a_1b_2 + a_1b_3 + a_2b_1 + a_3b_1)\right]$$

$$= a_1b_1 + a_2b_3 - a_1b_3 - a_2b_1 - a_1b_1 - \tfrac{1}{2}(a_2b_3 + a_3b_2) + \tfrac{1}{2}(a_1b_2 + a_1b_3 + a_2b_1 + a_3b_1)$$

$$= \tfrac{1}{2}(a_1b_2 + a_2b_3 + a_3b_1) - \tfrac{1}{2}(a_1b_3 + a_2b_1 + a_3b_2)$$

**(c)** We first find $Q = \begin{vmatrix} a_1 & b_1 & 1 \\ a_2 & b_2 & 1 \\ a_3 & b_3 & 1 \end{vmatrix}$ by expanding about the third column.

$$Q = 1 \begin{vmatrix} a_2 & b_2 \\ a_3 & b_3 \end{vmatrix} - 1 \begin{vmatrix} a_1 & b_1 \\ a_3 & b_3 \end{vmatrix} + 1 \begin{vmatrix} a_1 & b_1 \\ a_2 & b_2 \end{vmatrix} = a_2b_3 - a_3b_2 - (a_1b_3 - a_3b_1) + a_1b_2 - a_2b_1$$

$$= a_1b_2 + a_2b_3 + a_3b_1 - a_1b_3 - a_2b_1 - a_3b_2$$

So $\frac{1}{2}Q = \frac{1}{2}(a_1b_2 + a_2b_3 + a_3b_1) - \frac{1}{2}(a_1b_3 - a_2b_1 - a_3b_2)$, the area of the red triangle. Since $\frac{1}{2}Q$ is not always positive, the area is $\pm\frac{1}{2}Q$.

**60. (a)** If three points lie on a line then the area of the "triangle" they determine is 0, that is, $\pm\frac{1}{2}Q = \pm\frac{1}{2}\begin{vmatrix} a_1 & b_1 & 1 \\ a_2 & b_2 & 1 \\ a_3 & b_3 & 1 \end{vmatrix} = 0$ $\Leftrightarrow$ $Q = 0$. If the points are not collinear, then the point form a triangle, and the area of the triangle determined by these points is nonzero. If $Q = 0$, then $\pm\frac{1}{2}Q = \pm\frac{1}{2}(0) = 0$, so the "triangle" has no area, and the points are collinear.

**(b) (i)** 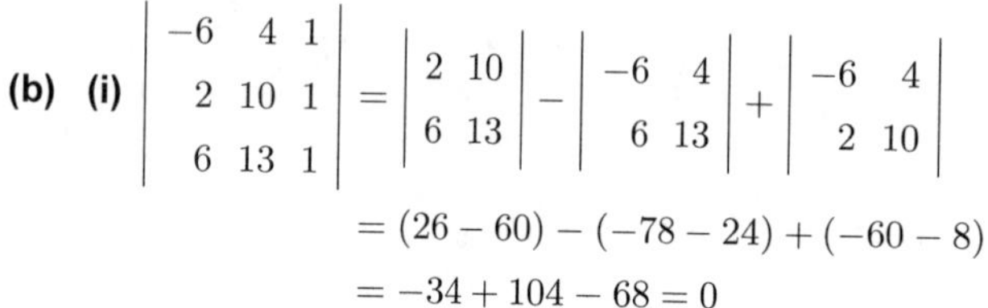

$$\begin{vmatrix} -6 & 4 & 1 \\ 2 & 10 & 1 \\ 6 & 13 & 1 \end{vmatrix} = \begin{vmatrix} 2 & 10 \\ 6 & 13 \end{vmatrix} - \begin{vmatrix} -6 & 4 \\ 6 & 13 \end{vmatrix} + \begin{vmatrix} -6 & 4 \\ 2 & 10 \end{vmatrix}$$

$$= (26 - 60) - (-78 - 24) + (-60 - 8)$$
$$= -34 + 104 - 68 = 0$$

Thus, these points are collinear.

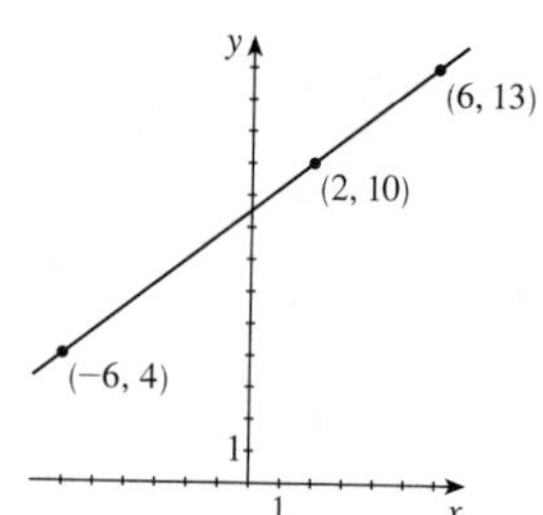


**(ii)**

$$\begin{vmatrix} -5 & 10 & 1 \\ 2 & 6 & 1 \\ 15 & -2 & 1 \end{vmatrix} = \begin{vmatrix} 2 & 6 \\ 15 & -2 \end{vmatrix} - \begin{vmatrix} -5 & 10 \\ 15 & -2 \end{vmatrix} + \begin{vmatrix} -5 & 10 \\ 2 & 6 \end{vmatrix}$$

$$= (-4 - 90) - (10 - 150) + (-30 - 20)$$
$$= -94 + 140 - 50 = -4$$

These points are not collinear. Note that this is difficult to determine from the diagram.

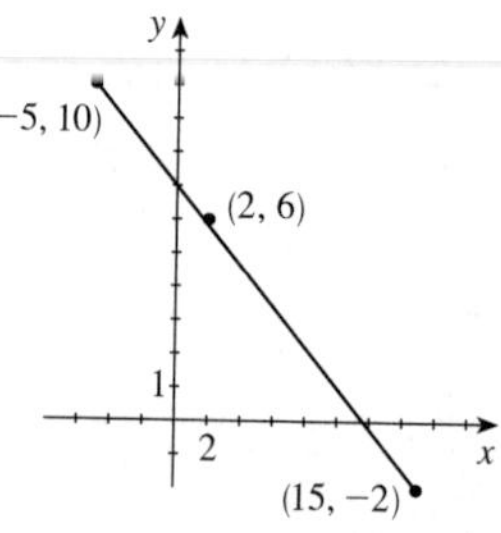


**61. (a)** Let $|M| = \begin{vmatrix} x & y & 1 \\ x_1 & y_1 & 1 \\ x_2 & y_2 & 1 \end{vmatrix}$. Then, expanding by the third column,

$$|M| = \begin{vmatrix} x_1 & y_1 \\ x_2 & y_2 \end{vmatrix} - \begin{vmatrix} x & y \\ x_2 & y_2 \end{vmatrix} + \begin{vmatrix} x & y \\ x_1 & y_1 \end{vmatrix} = (x_1y_2 - x_2y_1) - (xy_2 - x_2y) + (xy_1 - x_1y)$$

$$= x_1y_2 - x_2y_1 - xy_2 + x_2y + xy_1 - x_1y = x_2y - x_1y - xy_2 + xy_1 + x_1y_2 - x_2y_1$$

$$= (x_2 - x_1)\,y - (y_2 - y_1)\,x + x_1y_2 - x_2y_1$$

So $|M| = 0$ $\Leftrightarrow$ $(x_2 - x_1)\,y - (y_2 - y_1)\,x + x_1y_2 - x_2y_1 = 0 \Leftrightarrow (x_2 - x_1)\,y = (y_2 - y_1)\,x - x_1y_2 + x_2y_1$ $\Leftrightarrow$ $(x_2 - x_1)\,y = (y_2 - y_1)\,x - x_1y_2 + x_1y_1 - x_1y_1 + x_2y_1$ $\Leftrightarrow$ $y = \dfrac{y_2 - y_1}{x_2 - x_1}x - \dfrac{x_1(y_2 - y_1)}{x_2 - x_1} + \dfrac{y_1(x_2 - x_1)}{x_2 - x_1}$ $\Leftrightarrow$ $y = \dfrac{y_2 - y_1}{x_2 - x_1}(x - x_1) + y_1$ $\Leftrightarrow$ $y - y_1 = \dfrac{y_2 - y_1}{x_2 - x_1}(x - x_1)$, which is the "two-point" form of the equation for the line passing through the points $(x_1, y_1)$ and $(x_2, y_2)$.

**(b)** Using the result of part (a), the line has equation

$$\begin{vmatrix} x & y & 1 \\ 20 & 50 & 1 \\ -10 & 25 & 1 \end{vmatrix} = 0 \Leftrightarrow \begin{vmatrix} 20 & 50 \\ -10 & 25 \end{vmatrix} - \begin{vmatrix} x & y \\ -10 & 25 \end{vmatrix} + \begin{vmatrix} x & y \\ 20 & 50 \end{vmatrix} = 0 \Leftrightarrow$$

$(500 + 500) - (25x + 10y) + (50x - 20y) = 0$ $\Leftrightarrow$ $25x - 30y + 1000 = 0$ $\Leftrightarrow$ $5x - 6y + 200 = 0$.

**62. (a)** If $A$ is a matrix with a row or column consisting entirely of zeros, then if we expand the determinant by this row or column, we get $|A| = 0 \cdot |A_{1j}| - 0 \cdot |A_{2j}| + \cdots - 0 \cdot |A_{nj}| = 0$.

**(b)** Use the principle that if matrix $B$ is a square matrix obtained from $A$ by adding a multiple of one row to another, or a multiple of one column to another, then $|A| = |B|$. If we let $B$ be the matrix obtained by subtracting the two rows (or columns) that are the same, then matrix $B$ will have a row or column that consists entirely of zeros. So $|B| = 0 \Rightarrow |A| = 0$.

**(c)** Again use the principle that if matrix $B$ is a square matrix obtained from $A$ by adding a multiple of one row to another, or a multiple of one column to another, then $|A| = |B|$. If we let $B$ be the matrix obtained by subtracting the proper multiple of the row (or column) from the other similar row (or column), then matrix $B$ will have a row or column that consists entirely of zeros. So $|B| = 0 \Rightarrow |A| = 0$.

**63.** Gaussian elimination is superior, since it takes much longer to evaluate six $5 \times 5$ determinants than it does to perform one five-equation Gaussian elimination.

## 10.8 Partial Fractions

**1.** $\dfrac{1}{(x-1)(x+2)} = \dfrac{A}{x-1} + \dfrac{B}{x+2}$

**2.** $\dfrac{x}{x^2+3x-4} = \dfrac{x}{(x-1)(x+4)} = \dfrac{A}{(x-1)} + \dfrac{B}{(x+4)}$

**3.** $\dfrac{x^2-3x+5}{(x-2)^2(x+4)} = \dfrac{A}{x-2} + \dfrac{B}{(x-2)^2} + \dfrac{C}{x+4}$

**4.** $\dfrac{1}{x^4-x^3} = \dfrac{1}{x^3(x-1)} = \dfrac{A}{x} + \dfrac{B}{x^2} + \dfrac{C}{x^3} + \dfrac{D}{x-1}$

**5.** $\dfrac{x^2}{(x-3)(x^2+4)} = \dfrac{A}{x-3} + \dfrac{Bx+C}{x^2+4}$

**6.** $\dfrac{1}{x^4-1} = \dfrac{1}{(x^2-1)(x^2+1)} = \dfrac{1}{(x-1)(x+1)(x^2+1)} = \dfrac{A}{x-1} + \dfrac{B}{x+1} + \dfrac{Cx+D}{x^2+1}$

**7.** $\dfrac{x^3-4x^2+2}{(x^2+1)(x^2+2)} = \dfrac{Ax+B}{x^2+1} + \dfrac{Cx+D}{x^2+2}$

**8.** $\dfrac{x^4+x^2+1}{x^2(x^2+4)^2} = \dfrac{A}{x} + \dfrac{B}{x^2} + \dfrac{Cx+D}{x^2+4} + \dfrac{Ex+F}{(x^2+4)^2}$

**9.** $\dfrac{x^3+x+1}{x(2x-5)^3(x^2+2x+5)^2} = \dfrac{A}{x} + \dfrac{B}{2x-5} + \dfrac{C}{(2x-5)^2} + \dfrac{D}{(2x-5)^3} + \dfrac{Ex+F}{x^2+2x+5} + \dfrac{Gx+H}{(x^2+2x+5)^2}$

**10.** Since $(x^3-1)(x^2-1) = (x-1)(x^2+x+1)(x-1)(x+1) = (x-1)^2(x+1)(x^2+x+1)$, we have $\dfrac{1}{(x^3-1)(x^2-1)} = \dfrac{1}{(x-1)^2(x+1)(x^2+x+1)} = \dfrac{A}{x-1} + \dfrac{B}{(x-1)^2} + \dfrac{C}{x+1} + \dfrac{Dx+E}{x^2+x+1}$.

**11.** $\dfrac{2}{(x-1)(x+1)} = \dfrac{A}{x-1} + \dfrac{B}{x+1}$. Multiplying by $(x-1)(x+1)$, we get $2 = A(x+1) + B(x-1) \Leftrightarrow$ $2 = Ax + A + Bx - B$. Thus $\begin{cases} A+B=0 \\ A-B=2 \end{cases}$ Adding we get $2A = 2 \Leftrightarrow A = 1$. Now $A + B = 0 \Leftrightarrow$ $B = -A$, so $B = -1$. Thus, the required partial fraction decomposition is $\dfrac{2}{(x-1)(x+1)} = \dfrac{1}{x-1} - \dfrac{1}{x+1}$.

**12.** $\dfrac{2x}{(x-1)(x+1)} = \dfrac{A}{x-1} + \dfrac{B}{x+1}$. Multiplying by $(x-1)(x+1)$, we get $2x = A(x+1) + B(x-1) \Leftrightarrow$ $2x = Ax + A + Bx - B$. Thus $\begin{cases} A+B=2 \\ A-B=0 \end{cases}$ Adding we $2A = 2 \Leftrightarrow A = 1$. Since $A - B = 0 \Leftrightarrow B = A$, $B = 1$. The required partial fraction decomposition is $\dfrac{2x}{(x-1)(x+1)} = \dfrac{1}{x-1} + \dfrac{1}{x+1}$.

**13.** $\dfrac{5}{(x-1)(x+4)} = \dfrac{A}{x-1} + \dfrac{B}{x+4}$. Multiplying by $(x-1)(x+4)$, we get $5 = A(x+4) + B(x-1) \Leftrightarrow$ $5 = Ax + 4A + Bx - B$. Thus $\begin{cases} A + B = 0 \\ 4A - B = 5 \end{cases}$ Now $A + B = 0 \Leftrightarrow B = -A$, so substituting,we get $4A - (-A) = 5 \Leftrightarrow 5A = 5 \Leftrightarrow A = 1$and $B = -1$. The required partial fraction decomposition is $\dfrac{5}{(x-1)(x+4)} = \dfrac{1}{x-1} - \dfrac{1}{x+4}$.

**14.** $\dfrac{x+6}{x(x+3)} = \dfrac{A}{x} + \dfrac{B}{x+3}$. Multiplying by $x(x+3)$we get $x + 6 = A(x+3) + xB \Leftrightarrow$ $x + 6 = Ax + 3A + Bx = (A+B)x + 3A$. Thus $\begin{cases} A + B = 1 \\ 3A \qquad = 6 \end{cases}$ Now $3A = 6 \Leftrightarrow A = 2$, and $2 + B = 1 \Leftrightarrow B = -1$. The required partial fraction decomposition is $\dfrac{x+6}{x(x+3)} = \dfrac{2}{x} - \dfrac{1}{x+3}$.

**15.** $\dfrac{12}{x^2-9} = \dfrac{12}{(x-3)(x+3)} = \dfrac{A}{x-3} + \dfrac{B}{x+3}$. Multiplying by $(x-3)(x+3)$, we get $12 = A(x+3) + B(x-3)$ $\Leftrightarrow 12 = Ax + 3A + Bx - 3B$. Thus $\begin{cases} A + B = 0 \\ 3A - 3B = 12 \end{cases} \Leftrightarrow \begin{cases} A + B = 0 \\ A - B = 4 \end{cases}$ Adding, we get $2A = 4 \Leftrightarrow$ $A = 2$. So $2 + B = 0 \Leftrightarrow B = -2$. The required partial fraction decomposition is $\dfrac{12}{x^2-9} = \dfrac{2}{x-3} - \dfrac{2}{x+3}$.

**16.** $\dfrac{x-12}{x^2-4x} = \dfrac{x-12}{x(x-4)} = \dfrac{A}{x} + \dfrac{B}{x-4}$. Multiplying by $x(x-4)$, we get $x - 12 = A(x-4) + Bx = Ax - 4A + Bx = (A+B)x - 4A$. Thus we must solve the system $\begin{cases} A + B = 1 \\ -4A \qquad = -12 \end{cases}$ This gives $-4A = -12 \Leftrightarrow A = 3$, and $3 + B = 1 \Leftrightarrow B = -2$. Thus $\dfrac{x-12}{x^2-4x} = \dfrac{3}{x} - \dfrac{2}{x-4}$.

**17.** $\dfrac{4}{x^2-4} = \dfrac{4}{(x-2)(x+2)} = \dfrac{A}{x-2} + \dfrac{B}{x+2}$. Multiplying by $x^2 - 4$, we get $4 = A(x+2) + B(x-2) = (A+B)x + (2A-2B)$, and so $\begin{cases} A + B = 0 \\ 2A - 2B = 4 \end{cases} \Leftrightarrow \begin{cases} A + B = 0 \\ A - B = 2 \end{cases}$ Adding we get $2A = 2 \Leftrightarrow A = 1$, and $B = -1$. Therefore, $\dfrac{4}{x^2-4} = \dfrac{1}{x-2} - \dfrac{1}{x+2}$.

**18.** $\dfrac{2x+1}{x^2+x-2} = \dfrac{2x+1}{(x+2)(x-1)} = \dfrac{A}{x+2} + \dfrac{B}{x-1}$. Thus, $2x+1 = A(x-1) + B(x+2) = (A+B)x + (-A+2B)$, and so $\begin{cases} A + B = 2 \\ -A + 2B = 1 \end{cases}$ Adding the two equations, we get $3B = 3 \Leftrightarrow B = 1$. Thus $A + 1 = 2 \Leftrightarrow A = 1$. Therefore, $\dfrac{2x+1}{x^2+x-2} = \dfrac{1}{x+2} + \dfrac{1}{x-1}$.

**19.** $\dfrac{x+14}{x^2-2x-8} = \dfrac{x+14}{(x-4)(x+2)} = \dfrac{A}{x-4} + \dfrac{B}{x+2}$. Hence, $x+14 = A(x+2) + B(x-4) = (A+B)x + (2A-4B)$, and so $\begin{cases} A + B = 1 \\ 2A - 4B = 14 \end{cases} \Leftrightarrow \begin{cases} 2A + 2B = 2 \\ A - 2B = 7 \end{cases}$ Adding, we get $3A = 9 \Leftrightarrow A = 3$. So $(3) + B = 1 \Leftrightarrow$ $B = -2$. Therefore, $\dfrac{x+14}{x^2-2x-8} = \dfrac{3}{x-4} - \dfrac{2}{x+2}$.

**20.** $\dfrac{8x-3}{2x^2-x} = \dfrac{8x-3}{x(2x-1)} = \dfrac{A}{x} + \dfrac{B}{2x-1}$. Hence, $8x-3 = A(2x-1)+Bx = (2A+B)x+(-A)$,

giving $\begin{cases} 2A+B = 8 \\ -A = -3 \end{cases}$ So $-A = -3 \Leftrightarrow A = 3$, and $2(3)+B = 8 \Leftrightarrow B = 2$. Therefore,

$\dfrac{8x-3}{2x^2-x} = \dfrac{3}{x} + \dfrac{2}{2x-1}$.

**21.** $\dfrac{x}{8x^2-10x+3} = \dfrac{x}{(4x-3)(2x-1)} = \dfrac{A}{4x-3} + \dfrac{B}{2x-1}$. Hence,

$x = A(2x-1)+B(4x-3) = (2A+4B)x+(-A-3B)$, and so $\begin{cases} 2A+4B = 1 \\ -A-3B = 0 \end{cases} \Leftrightarrow \begin{cases} 2A+4B = 1 \\ -2A-6B = 0 \end{cases}$

Adding, we get $-2B = 1 \Leftrightarrow B = -\frac{1}{2}$, and $A = \frac{3}{2}$. Therefore, $\dfrac{x}{8x^2-10x+3} = \dfrac{\frac{3}{2}}{4x-3} - \dfrac{\frac{1}{2}}{2x-1}$.

**22.** $\dfrac{7x-3}{x^3+2x^2-3x} = \dfrac{7x-3}{x(x+3)(x-1)} = \dfrac{A}{x} + \dfrac{B}{x+3} + \dfrac{C}{x-1}$. Hence,

$$\begin{aligned} 7x-3 &= A(x+3)(x-1)+Bx(x-1)+Cx(x+3) = A\left(x^2+2x-3\right)+B\left(x^2-x\right)+C\left(x^2+3x\right) \\ &= (A+B+C)x^2+(2A-B+3C)x-3A \end{aligned}$$

Thus $\begin{cases} A+B+C = 0 & \text{Coefficients of } x^2 \\ 2A-B+3C = 7 & \text{Coefficients of } x \\ -3A = -3 & \text{Constant terms} \end{cases}$ So $-3A = -3 \Leftrightarrow A = 1$. Substituting, the system reduces

to $\begin{cases} 1+B+C = 0 \\ 2-B+3C = 7 \end{cases}$ Adding these two equations, we get $3+4C = 7 \Leftrightarrow C = 1$. Thus $1+B+1 = 0 \Leftrightarrow$

$B = -2$. Therefore, $\dfrac{7x-3}{x^3+2x^2-3x} = \dfrac{1}{x} - \dfrac{2}{x+3} + \dfrac{1}{x-1}$.

**23.** $\dfrac{9x^2-9x+6}{2x^3-x^2-8x+4} = \dfrac{9x^2-9x+6}{(x-2)(x+2)(2x-1)} = \dfrac{A}{x-2} + \dfrac{B}{x+2} + \dfrac{C}{2x-1}$. Thus,

$$\begin{aligned} 9x^2-9x+6 &= A(x+2)(2x-1)+B(x-2)(2x-1)+C(x-2)(x+2) \\ &= A\left(2x^2+3x-2\right)+B\left(2x^2-5x+2\right)+C\left(x^2-4\right) \\ &= (2A+2B+C)x^2+(3A-5B)x+(-2A+2B-4C) \end{aligned}$$

This leads to the system $\begin{cases} 2A+2B+C = 9 & \text{Coefficients of } x^2 \\ 3A-5B = -9 & \text{Coefficients of } x \\ -2A+2B-4C = 6 & \text{Constant terms} \end{cases} \Leftrightarrow \begin{cases} 2A+2B+C = 9 \\ 16B+3C = 45 \\ 4B-3C = 15 \end{cases} \Leftrightarrow$

$\begin{cases} 2A+2B+C = 9 \\ 16B+3C = 45 \\ 15C = -15 \end{cases}$ Hence, $-15C = 15 \Leftrightarrow C = -1$; $16B-3 = 45 \Leftrightarrow B = 3$; and $2A+6-1 = 9$

$\Leftrightarrow A = 2$. Therefore, $\dfrac{9x^2-9x+6}{2x^3-x^2-8x+4} = \dfrac{2}{x-2} + \dfrac{3}{x+2} - \dfrac{1}{2x-1}$.

**24.** $\dfrac{-3x^2-3x+27}{(x+2)(2x^2+3x-9)}=\dfrac{-3x^2-3x+27}{(x+2)(2x-3)(x+3)}=\dfrac{A}{x+2}+\dfrac{B}{2x-3}+\dfrac{C}{x+3}$. Thus,

$$\begin{aligned}-3x^2-3x+27&=A(2x-3)(x+3)+B(x+2)(x+3)+C(x+2)(2x-3)\\&=A\left(2x^2+3x-9\right)+B\left(x^2+5x+6\right)+C\left(2x^2+x-6\right)\\&=(2A+B+2C)x^2+(3A+5B+C)x+(-9A+6B-6C)\end{aligned}$$

So $\begin{cases}2A+\phantom{5}B+2C=-3 & \text{Coefficients of } x^2\\ 3A+5B+\phantom{2}C=-3 & \text{Coefficients of } x\\ -9A+6B-6C=\phantom{-}27 & \text{Constant terms}\end{cases}\Leftrightarrow\begin{cases}2A+B+2C=-3\\ 7B-4C=\phantom{-}3\\ 21B+6C=\phantom{-}27\end{cases}\Leftrightarrow\begin{cases}2A+B+2C=-3\\ 7B-6C=\phantom{-}3\\ 24C=\phantom{-}24\end{cases}$

Hence, $24C=24\quad\Leftrightarrow\quad C=1$; then $7B-4=3\quad\Leftrightarrow\quad B=1$; and $2A+1+2=-3\quad\Leftrightarrow\quad A=-3$.

Therefore, $\dfrac{-3x^2-3x+27}{(x+2)(2x^2+3x-9)}=\dfrac{-3}{x+2}+\dfrac{1}{2x-3}+\dfrac{1}{x+3}$.

**25.** $\dfrac{x^2+1}{x^3+x^2}=\dfrac{x^2+1}{x^2(x+1)}=\dfrac{A}{x}+\dfrac{B}{x^2}+\dfrac{C}{x+1}$. Hence,

$x^2+1=Ax(x+1)+B(x+1)+Cx^2=(A+C)x^2+(A+B)x+B$, and so $B=1$; $A+1=0\quad\Leftrightarrow\quad A=-1$;

and $-1+C=1\quad\Leftrightarrow\quad C=2$. Therefore, $\dfrac{x^2+1}{x^3+x^2}=\dfrac{-1}{x}+\dfrac{1}{x^2}+\dfrac{2}{x+1}$.

**26.** $\dfrac{3x^2+5x-13}{(3x+2)(x^2-4x+4)}=\dfrac{3x^2+5x-13}{(3x+2)(x-2)^2}=\dfrac{A}{3x+2}+\dfrac{B}{x-2}+\dfrac{C}{(x-2)^2}$. Thus,

$$\begin{aligned}3x^2+5x-13&=A(x-2)^2+B(3x+2)(x-2)+C(3x+2)\\&=A\left(x^2-4x+4\right)+B\left(3x^2-4x-4\right)+C(3x+2)\\&=(A+3B)x^2+(-4A-4B+3C)x+(4A-4B+2C)\end{aligned}$$

This leads to the following system: $\begin{cases}A+3B=3 & \text{Coefficients of } x^2\\ -4A-4B+3C=5 & \text{Coefficients of } x\\ 4A-4B+2C=-13 & \text{Constant terms}\end{cases}\Leftrightarrow$

$\begin{cases}A+3B=3\\ 8B+3C=17\\ -8B+5C=-8\end{cases}\Leftrightarrow\begin{cases}A+3B=3\\ 8B+3C=17\\ 8C=9\end{cases}$ Hence, $8C=9\Leftrightarrow C=\frac{9}{8}$;

$8B+3\left(\frac{9}{8}\right)=8B+\frac{27}{8}=17\quad\Leftrightarrow\quad B=\frac{109}{64}$; and $A+3\left(\frac{109}{64}\right)=A+\frac{327}{64}=3\quad\Leftrightarrow\quad A=-\frac{135}{64}$. Therefore,

$\dfrac{3x^2+5x-13}{(3x+2)(x-2)^2}=\dfrac{\frac{-135}{64}}{3x+2}+\dfrac{\frac{109}{64}}{x-2}+\dfrac{\frac{9}{8}}{(x-2)^2}$.

**27.** $\dfrac{2x}{4x^2+12x+9}=\dfrac{2x}{(2x+3)^2}=\dfrac{A}{2x+3}+\dfrac{B}{(2x+3)^2}$. Hence, $2x=A(2x+3)+B=2Ax+(3A+B)$. So $2A=2$

$\Leftrightarrow\quad A=1$; and $3(1)+B=0\quad\Leftrightarrow\quad B=-3$. Therefore, $\dfrac{2x}{4x^2+12x+9}=\dfrac{1}{2x+3}-\dfrac{3}{(2x+3)^2}$.

**28.** $\dfrac{x-4}{(2x-5)^2}=\dfrac{A}{2x-5}+\dfrac{B}{(2x-5)^2}$. Hence, $x-4=A(2x-5)+B=2Ax+(-5A+B)$, and so $\begin{cases}2A=1\\ -5A+B=-4\end{cases}$

$\Rightarrow\quad A=\frac{1}{2}$ and $-5\left(\frac{1}{2}\right)+B=-4\Leftrightarrow B=-\frac{3}{2}$. Therefore, $\dfrac{x-4}{(2x-5)^2}=\dfrac{\frac{1}{2}}{2x-5}-\dfrac{\frac{3}{2}}{(2x-5)^2}$.

**29.** $\dfrac{4x^2-x-2}{x^4+2x^3}=\dfrac{4x^2-x-2}{x^3(x+2)}=\dfrac{A}{x}+\dfrac{B}{x^2}+\dfrac{C}{x^3}+\dfrac{D}{x+2}$. Hence,

$$\begin{aligned}4x^2-x-2&=Ax^2(x+2)+Bx(x+2)+C(x+2)+Dx^3\\&=(A+D)x^3+(2A+B)x^2+(2B+C)x+2C\end{aligned}$$

So $2C=-2$ $\Leftrightarrow$ $C=-1$; $2B-1=-1$ $\Leftrightarrow$ $B=0$; $2A+0=4$ $\Leftrightarrow$ $A=2$; and $2+D=0$ $\Leftrightarrow$ $D=-2$. Therefore, $\dfrac{4x^2-x-2}{x^4+2x^3}=\dfrac{2}{x}-\dfrac{1}{x^3}-\dfrac{2}{x+2}$.

**30.** $\dfrac{x^3-2x^2-4x+3}{x^4}=\dfrac{A}{x}+\dfrac{B}{x^2}+\dfrac{C}{x^3}+\dfrac{D}{x^4}$. Hence, $x^3-2x^2-4x+3=Ax^3+Bx^2+Cx+D$. Thus $A=1$; $B=-2$; $C=-4$; and $D=3$. Therefore, $\dfrac{x^3-2x^2-4x+3}{x^4}=\dfrac{1}{x}-\dfrac{2}{x^2}-\dfrac{4}{x^3}+\dfrac{3}{x^4}$.

**31.** $\dfrac{-10x^2+27x-14}{(x-1)^3(x+2)}=\dfrac{A}{x+2}+\dfrac{B}{x-1}+\dfrac{C}{(x-1)^2}+\dfrac{D}{(x-1)^3}$. Thus,

$$\begin{aligned}-10x^2+27x-14&=A(x-1)^3+B(x+2)(x-1)^2+C(x+2)(x-1)+D(x+2)\\&=A\left(x^3-3x^2+3x-1\right)+B(x+2)\left(x^2-2x+1\right)+C\left(x^2+x-2\right)+D(x+2)\\&=A\left(x^3-3x^2+3x-1\right)+B\left(x^3-3x+2\right)+C\left(x^2+x-2\right)+D(x+2)\\&=(A+B)x^3+(-3A+C)x^2+(3A-3B+C+D)x+(-A+2B-2C+2D)\end{aligned}$$

which leads to the system

$$\left\{\begin{array}{rl}A+B=0 & \text{Coefficients of } x^3\\ -3A+C=-10 & \text{Coefficients of } x^2\\ 3A-3B+C+D=27 & \text{Coefficients of } x\\ -A+2B-2C+2D=-14 & \text{Constant terms}\end{array}\right. \Leftrightarrow \left\{\begin{array}{r}A+B=0\\ 3B+C=-10\\ -3B+2C+D=17\\ 3B-5C+7D=-15\end{array}\right. \Leftrightarrow$$

$$\left\{\begin{array}{r}A+B=0\\ 3B+C=-10\\ 3C+D=7\\ -3C+8D=2\end{array}\right. \Leftrightarrow \left\{\begin{array}{r}A+B=0\\ 3B+C=-10\\ 3C+D=7\\ 9D=9\end{array}\right.$$

Hence, $9D=9$ $\Leftrightarrow$ $D=1$, $3C+1=7$ $\Leftrightarrow$ $C=2$, $3B+2=-10$ $\Leftrightarrow$ $B=-4$, and $A-4=0$ $\Leftrightarrow$ $A=4$. Therefore, $\dfrac{-10x^2+27x-14}{(x-1)^3(x+2)}=\dfrac{4}{x+2}-\dfrac{4}{x-1}+\dfrac{2}{(x-1)^2}+\dfrac{1}{(x-1)^3}$.

**32.** $\dfrac{-2x^2+5x-1}{x^4-2x^3+2x-1} = \dfrac{-2x^2+5x-1}{(x-1)(x^3-x^2-x+1)} = \dfrac{-2x^2+5x-1}{(x-1)^3(x+1)} = \dfrac{A}{x+1} + \dfrac{B}{x-1} + \dfrac{C}{(x-1)^2} + \dfrac{D}{(x-1)^3}$.

Thus,

$$\begin{aligned} -2x^2+5x-1 &= A(x-1)^3 + B(x+1)(x-1)^2 + C(x+1)(x-1) + D(x+1) \\ &= A(x^3-3x^2+3x-1) + B(x+1)(x^2-2x+1) + C(x^2-1) + D(x+1) \\ &= A(x^3-3x^2+3x-1) + B(x^3-x^2-x+1) + C(x^2-1) + D(x+1) \\ &= (A+B)x^3 + (-3A-B+C)x^2 + (3A-B+D)x + (-A+B-C+D) \end{aligned}$$

which leads to the system $\begin{cases} A+B = 0 & \text{Coefficients of } x^3 \\ -3A-B+C = -2 & \text{Coefficients of } x^2 \\ 3A-B+D = 5 & \text{Coefficients of } x \\ -A+B-C+D = -1 & \text{Constant terms} \end{cases} \Leftrightarrow \begin{cases} A+B = 0 \\ 2B+C = -2 \\ -2B+C+D = 3 \\ 2B-3C+4D = 2 \end{cases}$

$\Leftrightarrow \begin{cases} A+B = 0 \\ 2B+C = -2 \\ 2C+D = 1 \\ -2C+5D = 5 \end{cases} \Leftrightarrow \begin{cases} A+B = 0 \\ 2B+C = -2 \\ 2C+D = 1 \\ 6D = 6 \end{cases}$ Hence, $6D = 6 \Leftrightarrow D = 1$,

$2C+1 = 1 \Leftrightarrow C = 0$, $2B+0 = -2 \Leftrightarrow B = -1$, and $A-1 = 0 \Leftrightarrow A = 1$. Therefore, $\dfrac{-2x^2+5x-1}{x^4-2x^3+2x-1} = \dfrac{1}{x+1} - \dfrac{1}{x-1} + \dfrac{1}{(x-1)^3}$.

**33.** $\dfrac{3x^3+22x^2+53x+41}{(x+2)^2(x+3)^2} = \dfrac{A}{x+2} + \dfrac{B}{(x+2)^2} + \dfrac{C}{x+3} + \dfrac{D}{(x+3)^2}$. Thus,

$$\begin{aligned} 3x^3+22x^2+53x+41 &= A(x+2)(x+3)^2 + B(x+3)^2 + C(x+2)^2(x+3) + D(x+2)^2 \\ &= A(x^3+8x^2+21x+18) + B(x^2+6x+9) \\ &\qquad + C(x^3+7x^2+16x+12) + D(x^2+4x+4) \\ &= (A+C)x^3 + (8A+B+7C+D)x^2 \\ &\qquad + (21A+6B+16C+4D)x + (18A+9B+12C+4D) \end{aligned}$$

so we must solve the system $\begin{cases} A + C = 3 & \text{Coefficients of } x^3 \\ 8A+B+7C+D = 22 & \text{Coefficients of } x^2 \\ 21A+6B+16C+4D = 53 & \text{Coefficients of } x \\ 18A+9B+12C+4D = 41 & \text{Constant terms} \end{cases} \Leftrightarrow$

$\begin{cases} A + C = 3 \\ B - C + D = -2 \\ 6B - 5C + 4D = -10 \\ 9B - 6C + 4D = -13 \end{cases} \Leftrightarrow \begin{cases} A + C = 3 \\ B - C + D = -2 \\ C - 2D = 2 \\ 3C - 5D = 5 \end{cases} \Leftrightarrow \begin{cases} A + C = 3 \\ B - C + D = -2 \\ C - 2D = 2 \\ D = -1 \end{cases}$ Hence,

$D = -1$, $C+2 = 2 \Leftrightarrow C = 0$, $B-0-1 = -2 \Leftrightarrow B = -1$, and $A+0 = 3 \Leftrightarrow A = 3$. Therefore, $\dfrac{3x^3+22x^2+53x+41}{(x+2)^2(x+3)^2} = \dfrac{3}{x+2} - \dfrac{1}{(x+2)^2} - \dfrac{1}{(x+3)^2}$.

**34.** $\dfrac{3x^2+12x-20}{x^4-8x^2+16} = \dfrac{3x^2+12x-20}{(x^2-4)^2} = \dfrac{3x^2+12x-20}{(x+2)^2(x-2)^2} = \dfrac{A}{x+2} + \dfrac{B}{(x+2)^2} + \dfrac{C}{x-2} + \dfrac{D}{(x-2)^2}$. Thus,

$$\begin{aligned} 3x^2+12x-20 &= A(x+2)(x-2)^2 + B(x-2)^2 + C(x+2)^2(x-2) + D(x+2)^2 \\ &= A(x^3-2x^2-4x+8) + B(x^2-4x+4) + C(x^3+2x^2-4x-8) + D(x^2+4x+4) \\ &= (A+C)x^3 + (-2A+B+2C+D)x^2 + (-4A-4B-4C+4D)x + (8A+4B-8C+4D) \end{aligned}$$

which leads to the system $\begin{cases} A + C = 0 & \text{Coefficients of } x^3 \\ -2A + B + 2C + D = 3 & \text{Coefficients of } x^2 \\ -4A - 4B - 4C + 4D = 12 & \text{Coefficients of } x \\ 8A + 4B - 8C + 4D = -20 & \text{Constant terms} \end{cases} \Leftrightarrow$

$$\begin{cases} A + C = 0 \\ B + 4C + D = 3 \\ 6B + 8C - 2D = -6 \\ -4B - 16C + 12D = 4 \end{cases} \Leftrightarrow \begin{cases} A + C = 0 \\ B + 4C + D = 3 \\ 16C + 8D = 24 \\ -32C + 32D = 0 \end{cases} \Leftrightarrow \begin{cases} A + C = 0 \\ B + 4C + D = 3 \\ 16C + 8D = 24 \\ 48D = 48 \end{cases}$$

Hence, $48D = 48 \Leftrightarrow D = 1$, $16C + 8 = 24 \Leftrightarrow C = 1$; $B + 4 + 1 = 3 \Leftrightarrow B = -2$, and $A + 1 = 0 \Leftrightarrow$ $A = -1$. Therefore, $\dfrac{3x^2+12x-20}{x^4-8x^2+16} = -\dfrac{1}{x+2} - \dfrac{2}{(x+2)^2} + \dfrac{1}{x-2} + \dfrac{1}{(x-2)^2}$.

**35.** $\dfrac{x-3}{x^3+3x} = \dfrac{x-3}{x(x^2+3)} = \dfrac{A}{x} + \dfrac{Bx+C}{x^2+3}$. Hence, $x - 3 = A(x^2+3) + Bx^2 + Cx = (A+B)x^2 + Cx + 3A$. So $3A = -3 \Leftrightarrow A = -1$; $C = 1$; and $-1 + B = 0 \Leftrightarrow B = 1$. Therefore, $\dfrac{x-3}{x^3+3x} = -\dfrac{1}{x} + \dfrac{x+1}{x^2+3}$.

**36.** $\dfrac{3x^2-2x+8}{x^3-x^2+2x-2} = \dfrac{3x^2-2x+8}{(x^2+2)(x-1)} = \dfrac{Ax+B}{x^2+2} + \dfrac{C}{x-1}$. Thus, $3x^2 - 2x + 8 = (Ax+B)(x-1) + C(x^2+2) = (A+C)x^2 + (-A+B)x + (-B+2C)$, which leads to the system $\begin{cases} A + C = 3 & \text{Coefficients of } x^2 \\ -A + B = -2 & \text{Coefficients of } x \\ -B + 2C = 8 & \text{Constant terms} \end{cases} \Leftrightarrow \begin{cases} A + C = 3 & \text{Coefficients of } x^2 \\ B + C = 1 & \text{Coefficients of } x \\ 2C = 6 & \text{Constant terms} \end{cases}$

Hence, $2C = 6 \Leftrightarrow C = 3$, $B + 3 = 1 \Leftrightarrow B = -2$; and $A + 3 = 3 \Leftrightarrow A = 0$. Therefore, $\dfrac{3x^2-2x+8}{x^3-x^2+2x-2} = -\dfrac{2}{x^2+2} + \dfrac{3}{x-1}$.

**37.** $\dfrac{2x^3+7x+5}{(x^2+x+2)(x^2+1)} = \dfrac{Ax+B}{x^2+x+2} + \dfrac{Cx+D}{x^2+1}$. Thus,

$$\begin{aligned} 2x^3+7x+5 &= (Ax+B)(x^2+1) + (Cx+D)(x^2+x+2) \\ &= Ax^3 + Ax + Bx^2 + B + Cx^3 + Cx^2 + 2Cx + Dx^2 + Dx + 2D \\ &= (A+C)x^3 + (B+C+D)x^2 + (A+2C+D)x + (B+2D) \end{aligned}$$

We must solve the system

$$\begin{cases} A + C = 2 & \text{Coefficients of } x^3 \\ B + C + D = 0 & \text{Coefficients of } x^2 \\ A + 2C + D = 7 & \text{Coefficients of } x \\ B + 2D = 5 & \text{Constant terms} \end{cases} \Leftrightarrow \begin{cases} A + C = 2 \\ B + C + D = 0 \\ C + D = 5 \\ C - D = -5 \end{cases} \Leftrightarrow \begin{cases} A + C = 2 \\ B + C + D = 0 \\ C + D = 5 \\ 2D = 10 \end{cases}$$

Hence, $2D = 10 \Leftrightarrow D = 5$, $C + 5 = 5 \Leftrightarrow C = 0$, $B + 0 + 5 = 0 \Leftrightarrow B = -5$, and $A + 0 = 2 \Leftrightarrow$ $A = 2$. Therefore, $\dfrac{2x^3+7x+5}{(x^2+x+2)(x^2+1)} = \dfrac{2x-5}{x^2+x+2} + \dfrac{5}{x^2+1}$.

**38.** $\dfrac{x^2+x+1}{2x^4+3x^2+1}=\dfrac{x^2+x+1}{(2x^2+1)(x^2+1)}=\dfrac{Ax+B}{2x^2+1}+\dfrac{Cx+D}{x^2+1}$. Thus,

$$\begin{aligned} x^2+x+1 &= (Ax+B)(x^2+1)+(Cx+D)(2x^2+1)=Ax^3+Ax+Bx^2+B+2Cx^3+2Dx^2+Cx+D \\ &= (A+2C)x^3+(B+2D)x^2+(A+C)x+(B+D) \end{aligned}$$

which leads to the system

$$\begin{cases} A + 2C = 0 & \text{Coefficients of } x^3 \\ B + 2D = 1 & \text{Coefficients of } x^2 \\ A + C = 1 & \text{Coefficients of } x \\ B + D = 1 & \text{Constant terms} \end{cases} \Leftrightarrow \begin{cases} A + 2C = 0 \\ B + 2D = 1 \\ C = -1 \\ D = 0 \end{cases}$$

Hence, $D=0$, $C=-1$, $B+0=1 \Leftrightarrow B=1$, and $A-2=0 \Leftrightarrow A=2$. Therefore, $\dfrac{x^2+x+1}{2x^4+3x^2+1}=\dfrac{2x+1}{2x^2+1}-\dfrac{x}{x^2+1}$.

**39.** $\dfrac{x^4+x^3+x^2-x+1}{x(x^2+1)^2}=\dfrac{A}{x}+\dfrac{Bx+C}{x^2+1}+\dfrac{Dx+E}{(x^2+1)^2}$. Hence,

$$\begin{aligned} x^4+x^3+x^2-x+1 &= A(x^2+1)^2+(Bx+C)x(x^2+1)+x(Dx+E) \\ &= A(x^4+2x^2+1)+(Bx^2+Cx)(x^2+1)+Dx^2+Ex \\ &= A(x^4+2x^2+1)+Bx^4+Bx^2+Cx^3+Cx+Dx^2+Ex \\ &= (A+B)x^4+Cx^3+(2A+B+D)x^2+(C+E)x+A \end{aligned}$$

So $A=1$, $1+B=1 \Leftrightarrow B=0$; $C=1$; $2+0+D=1 \Leftrightarrow D=-1$; and $1+E=-1 \Leftrightarrow E=-2$. Therefore, $\dfrac{x^4+x^3+x^2-x+1}{x(x^2+1)^2}=\dfrac{1}{x}+\dfrac{1}{x^2+1}-\dfrac{x+2}{(x^2+1)^2}$.

**40.** $\dfrac{2x^2-x+8}{(x^2+4)^2}=\dfrac{Ax+B}{x^2+4}+\dfrac{Cx+D}{(x^2+4)^2}$. Thus,

$$\begin{aligned} 2x^2-x+8 &= (Ax+B)(x^2+4)+Cx+D=Ax^3+4Ax+Bx^2+4B+Cx+D \\ &= Ax^3+Bx^2+(4A+C)x+(4B+D) \end{aligned}$$

and so $A=0$, $B=2$, $0+C=-1 \Leftrightarrow C=-1$, and $8+D=8 \Leftrightarrow D=0$. Therefore, $\dfrac{2x^2-x+8}{(x^2+4)^2}=\dfrac{2}{x^2+4}-\dfrac{1}{(x^2+4)^2}$.

**41.** We must first get a proper rational function. Using long division, we find that $\dfrac{x^5-2x^4+x^3+x+5}{x^3-2x^2+x-2}=x^2+\dfrac{2x^2+x+5}{x^3-2x^2+x-2}=x^2+\dfrac{2x^2+x+5}{(x-2)(x^2+1)}=x^2+\dfrac{A}{x-2}+\dfrac{Bx+C}{x^2+1}$. Hence,

$$\begin{aligned} 2x^2+x+5 &= A(x^2+1)+(Bx+C)(x-2)=Ax^2+A+Bx^2+Cx-2Bx-2C \\ &= (A+B)x^2+(C-2B)x+(A-2C) \end{aligned}$$

Equating coefficients, we get the system

$$\begin{cases} A + B = 2 & \text{Coefficients of } x^2 \\ -2B + C = 1 & \text{Coefficients of } x \\ A - 2C = 5 & \text{Constant terms} \end{cases} \Leftrightarrow \begin{cases} A + B = 2 \\ -2B + C = 1 \\ B + 2C = -3 \end{cases} \Leftrightarrow \begin{cases} A + B = 2 \\ -2B + C = 1 \\ 5C = -5 \end{cases}$$

Therefore, $5C=-5 \Leftrightarrow C=-1$, $-2B-1=1 \Leftrightarrow B=-1$, and $A-1=2 \Leftrightarrow A=3$. Therefore, $\dfrac{x^5-2x^4+x^3+x+5}{x^3-2x^2+x-2}=x^2+\dfrac{3}{x-2}-\dfrac{x+1}{x^2+1}$.

**42.** $\dfrac{x^5-3x^4+3x^3-4x^2+4x+12}{(x-2)^2(x^2+2)}=\dfrac{x^5-3x^4+3x^3-4x^2+4x+12}{x^4-4x^3+6x^2-8x+8}$. We use long division to get a proper rational function:

$$\begin{array}{r|l} & x+1 \\ \hline x^4-4x^3+6x^2-8x+8 & x^5-3x^4+3x^3-4x^2+4x+12 \\ & x^5-4x^4+6x^3-8x^2+8x \\ \hline & x^4-3x^3+4x^2-4x+12 \\ & x^4-4x^3+6x^2-8x+8 \\ \hline & x^3-2x^2+4x+4 \end{array}$$

Thus, $\dfrac{x^5-3x^4+3x^3-4x^2+4x+12}{(x-2)^2(x^2+2)}=x+1+\dfrac{x^3-2x^2+4x+4}{(x-2)^2(x^2+2)}=x+1+\dfrac{A}{x-2}+\dfrac{B}{(x-2)^2}+\dfrac{Cx+D}{x^2+2}$, so

$$\begin{aligned} x^3-2x^2+4x+4 &= A(x-2)(x^2+2)+B(x^2+2)+(Cx+D)(x-2)^2 \\ &= A(x^3-2x^2+2x-4)+B(x^2+2)+(Cx+D)(x^2-4x+4) \\ &= Ax^3-2Ax^2+2Ax-4A+Bx^2+2B+Cx^3-4Cx^2+4Cx+Dx^2-4Dx+4D \\ &= (A+C)x^3+(-2A+B-4C+D)x^2+(2A+4C-4D)x+(-4A+2B+4D) \end{aligned}$$

which leads to the system $\left\{\begin{array}{rlr} A+C &= 1 & \text{Coefficients of } x^3 \\ -2A+B-4C+D &= -2 & \text{Coefficients of } x^2 \\ 2A+4C-4D &= 4 & \text{Coefficients of } x \\ -4A+2B+4D &= 4 & \text{Constant terms} \end{array}\right. \Leftrightarrow$

$\left\{\begin{array}{l} A+C=1 \\ B-2C+D=0 \\ B-3D=2 \\ 2B+8C-4D=12 \end{array}\right. \Leftrightarrow \left\{\begin{array}{l} A+C=1 \\ B-2C+D=0 \\ 2C-4D=2 \\ 8C+2D=8 \end{array}\right. \Leftrightarrow \left\{\begin{array}{l} A+C=1 \\ B-2C+D=0 \\ 2C-4D=2 \\ 18D=0 \end{array}\right.$ Then $D=0$; $2C=2 \Leftrightarrow C=1$; $B-2=0 \Leftrightarrow B=2$; and $A+1=1 \Leftrightarrow A=0$. Therefore, $\dfrac{x^5-3x^4+3x^3-4x^2+4x+12}{(x-2)^2(x^2+2)}=x+1+\dfrac{2}{(x-2)^2}+\dfrac{x}{x^2+2}$.

**43.** $\dfrac{ax+b}{x^2-1}=\dfrac{A}{x-1}+\dfrac{B}{x+1}$. Hence, $ax+b=A(x+1)+B(x-1)=(A+B)x+(A-B)$.

So $\left\{\begin{array}{l} A+B=a \\ A-B=b \end{array}\right.$ Adding, we get $2A=a+b \Leftrightarrow A=\dfrac{a+b}{2}$.

Substituting, we get $B=a-A=\dfrac{2a}{2}-\dfrac{a+b}{2}=\dfrac{a-b}{2}$. Therefore, $A=\dfrac{a+b}{2}$ and $B=\dfrac{a-b}{2}$.

**44.** $\dfrac{ax^3+bx^2}{(x^2+1)^2}=\dfrac{Ax+B}{x^2+1}+\dfrac{Cx+D}{(x^2+1)^2}$. Hence,

$$\begin{aligned} ax^3+bx^2 &= (Ax+B)(x^2+1)+Cx+D=Ax^3+Ax+Bx^2+B+Cx+D \\ &= Ax^3+Bx^2+(A+C)x+(B+D) \end{aligned}$$

and so $A=a$, $B=b$, $a+C=0 \Leftrightarrow C=-a$, and $b+D=0 \Leftrightarrow D=-b$. Therefore, $A=a$, $B=b$, $C=-a$, and $D=-b$.

**45. (a)** The expression $\dfrac{x}{x^2+1}+\dfrac{1}{x+1}$ is already a partial fraction decomposition. The denominator in the first term is a quadratic which cannot be factored and the degree of the numerator is less than 2. The denominator of the second term is linear and the numerator is a constant.

**(b)** The term $\dfrac{x}{(x+1)^2}$ can be decomposed further, since the numerator and denominator both have linear factors. $\dfrac{x}{(x+1)^2}=\dfrac{A}{x+1}+\dfrac{B}{(x+1)^2}$. Hence, $x=A(x+1)+B=Ax+(A+B)$. So $A=1$, $B=-1$, and $\dfrac{x}{(x+1)^2}=\dfrac{1}{x+1}+\dfrac{-1}{(x+1)^2}$.

**(c)** The expression $\dfrac{1}{x+1}+\dfrac{2}{(x+1)^2}$ is already a partial fraction decomposition, since each numerator is constant.

**(d)** The expression $\dfrac{x+2}{(x^2+1)^2}$ is already a partial fraction decomposition, since the denominator is the square of a quadratic which cannot be factored,and the degree of the numerator is less than 2.

**46.** Combining the terms, we have

$$\begin{aligned}\frac{2}{x-1}+\frac{1}{(x-1)^2}+\frac{1}{x+1}&=\frac{2(x^2-1)}{(x-1)^2(x+1)}+\frac{1(x+1)}{(x-1)^2(x+1)}+\frac{1(x^2-2x+1)}{(x-1)^2(x+1)}\\&=\frac{2x^2-2+x+1+x^2-2x+1}{(x-1)^2(x+1)}=\frac{3x^2-x}{(x-1)^2(x+1)}\end{aligned}$$

Now to find the partial fraction decomposition, we have $\dfrac{3x^2-x}{(x-1)^2(x+1)}=\dfrac{A}{x-1}+\dfrac{B}{(x-1)^2}+\dfrac{C}{x+1}$, and so

$$\begin{aligned}3x^2-x&=A(x-1)(x+1)+B(x+1)+C(x-1)^2=A(x^2-1)+B(x+1)+C(x^2-2x+1)\\&=Ax^2-A+Bx+B+Cx^2-2Cx+C=(A+C)x^2+(B-2C)x+(-A+B+C)\end{aligned}$$

which result in the system $\begin{cases}A+C=3 & \text{Coefficients of } x^2\\ B-2C=-1 & \text{Coefficients of } x\\ -A+B+C=0 & \text{Constant terms}\end{cases}$ $\Leftrightarrow$ $\begin{cases}A+C=3\\ B-2C=-1\\ B+2C=3\end{cases}$ $\Leftrightarrow$

$\begin{cases}A+C=3\\ B-2C=-1\\ 4C=4\end{cases}$ so $4C=4$ $\Leftrightarrow$ $C=1$, $B-2(1)=-1$ $\Leftrightarrow$ $B=1$, and $A+1=3$ $\Leftrightarrow$ $A=2$.

Therefore, we get back the same expression: $\dfrac{3x^2-x}{(x-1)^2(x+1)}=\dfrac{2}{x-1}+\dfrac{1}{(x-1)^2}+\dfrac{1}{x+1}$.

# 10.9 Systems of Inequalities

**1.** $x < 3$

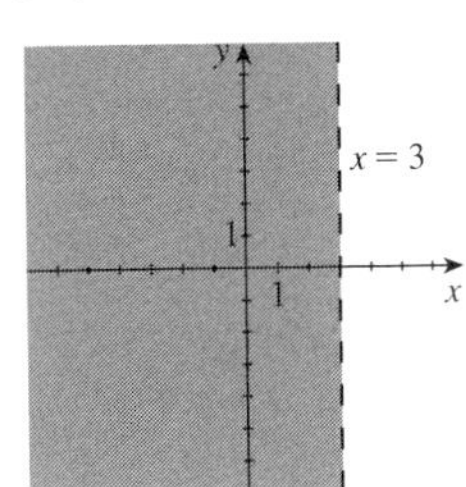


**2.** $y \geq -2$

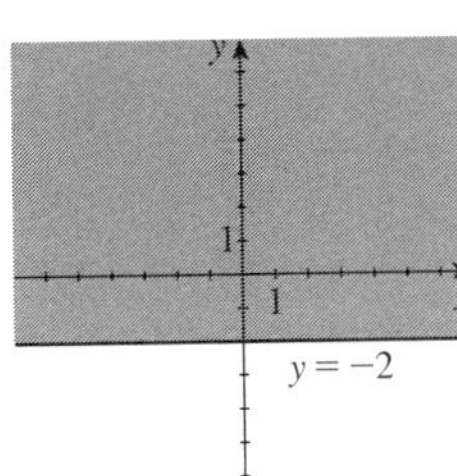


**3.** $y > x$

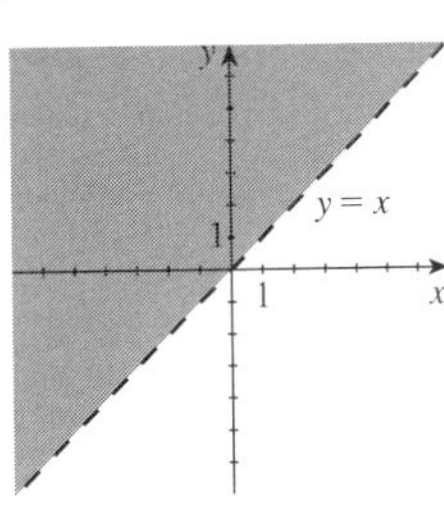


**4.** $y < x + 2$

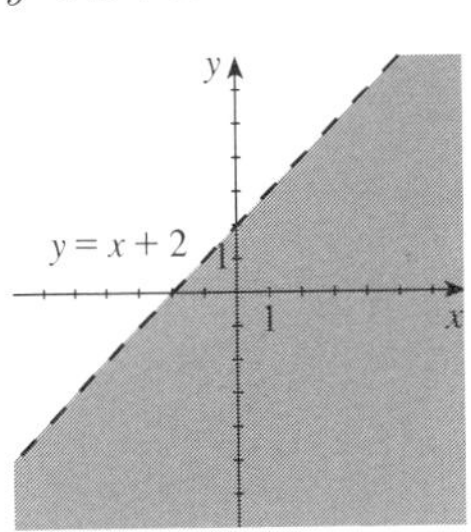


**5.** $y \leq 2x + 2$

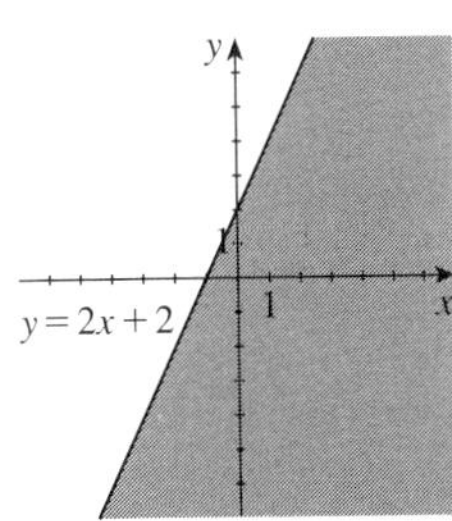


**6.** $y < -x + 5$

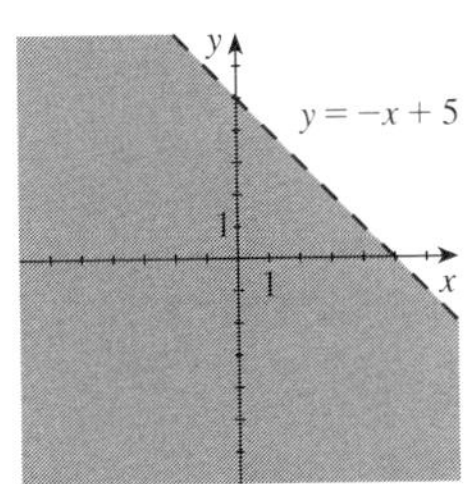


**7.** $2x - y \leq 8$

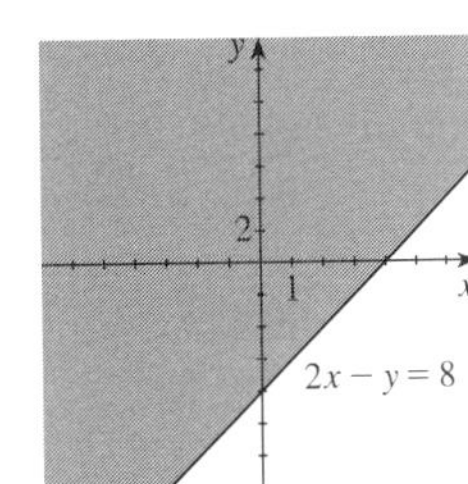


**8.** $3x + 4y + 12 > 0$

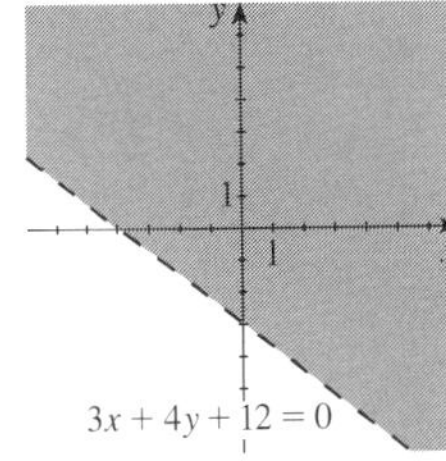


**9.** $4x + 5y < 20$

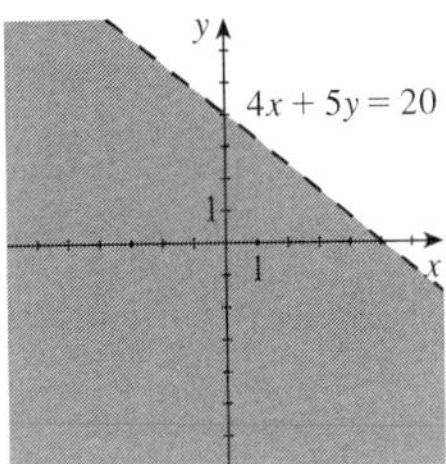


**10.** $-x^2 + y \geq 10$

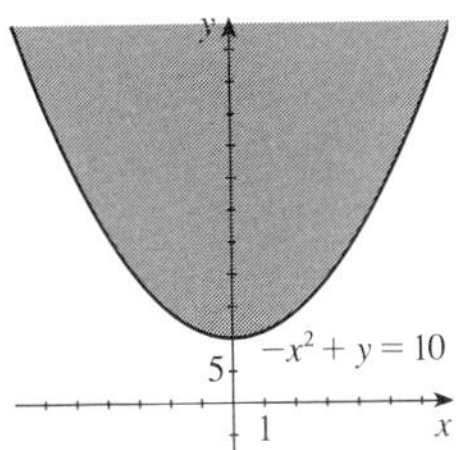


**11.** $y > x^2 + 1$

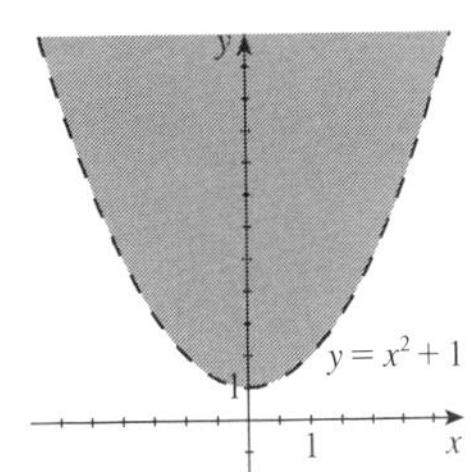


**12.** $x^2 + y^2 \geq 9$

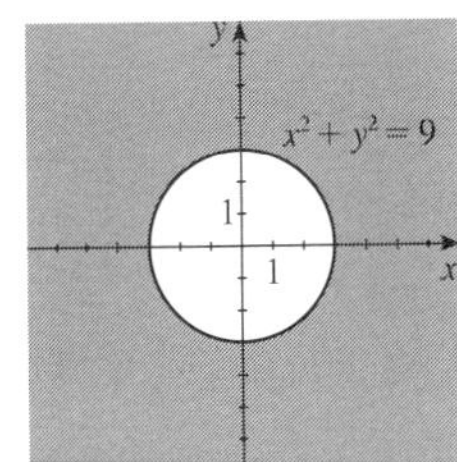


**13.** $x^2 + y^2 \leq 25$

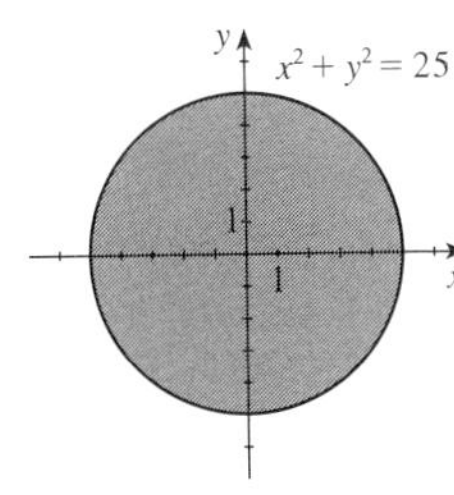


**14.** $x^2 + (y - 1)^2 \leq 1$

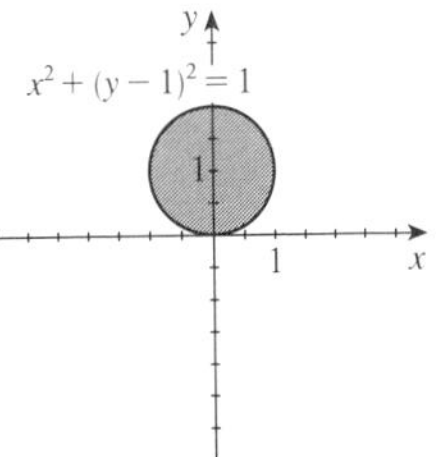


**15.** The boundary is a solid curve, so we have the inequality $y \leq \frac{1}{2}x - 1$. We take the test point $(0, -2)$ and verify that it satisfies the inequality: $-2 \leq \frac{1}{2}(0) - 1$.

**16.** The boundary is a solid curve, so we have the inequality $y \leq x^2 + 2$. We take the test point $(0, 0)$ and verify that it satisfies the inequality: $0 \leq 0^2 + 2$.

**17.** The boundary is a broken curve, so we have the inequality $x^2 + y^2 > 4$. We take the test point $(0, 4)$ and verify that it satisfies the inequality: $0^2 + 4^2 > 4$.

**18.** The boundary is a solid curve, so we have the inequality $y \geq x^3 - 4x$. We take the test point $(1, 1)$ and verify that it satisfies the inequality: $1 \geq 1^3 - 4(1)$.

**19.** $\begin{cases} x+y \le 4 \\ y \ge x \end{cases}$ The vertices occur where $\begin{cases} x+y=4 \\ y=x \end{cases}$ Substituting, we have $2x=4 \quad \Leftrightarrow \quad x=2$. Since $y=x$, the vertex is $(2,2)$, and the solution set is not bounded.

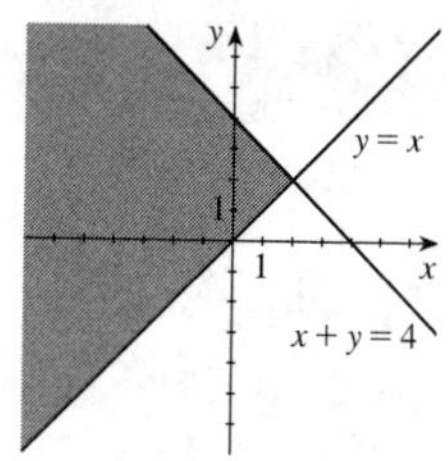


**20.** $\begin{cases} 2x+3y > 12 \\ 3x - \phantom{3}y < 21 \end{cases}$ The vertices occur where $\begin{cases} 2x+3y=12 \\ 3x - \phantom{3}y = 21 \end{cases} \Leftrightarrow$ $\begin{cases} 2x+3y=12 \\ 9x-3y=63 \end{cases}$ and adding the two equations gives $11x=75 \quad \Leftrightarrow$ $x=\frac{75}{11}$. Then $2\left(\frac{75}{11}\right)+3y=12 \quad \Leftrightarrow \quad 3y=\frac{132-150}{11}=-\frac{18}{11} \quad \Leftrightarrow$ $y=-\frac{6}{11}$. Therefore, the vertex is $\left(\frac{75}{11},-\frac{6}{11}\right)$, and the solution set is not bounded.

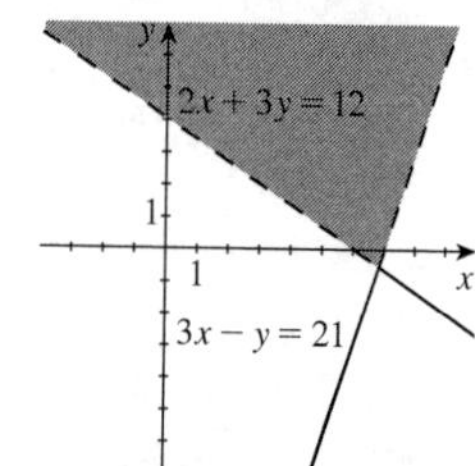


**21.** $\begin{cases} y < \frac{1}{4}x+2 \\ y \ge 2x-5 \end{cases}$ The vertex occurs where $\begin{cases} y=\frac{1}{4}x+2 \\ y=2x-5 \end{cases}$ Substituting for $y$ gives $\frac{1}{4}x+2=2x-5 \quad \Leftrightarrow \quad \frac{7}{4}x=7 \quad \Leftrightarrow \quad x=4$, so $y=3$. Hence, the vertex is $(4,3)$, and the solution is not bounded.

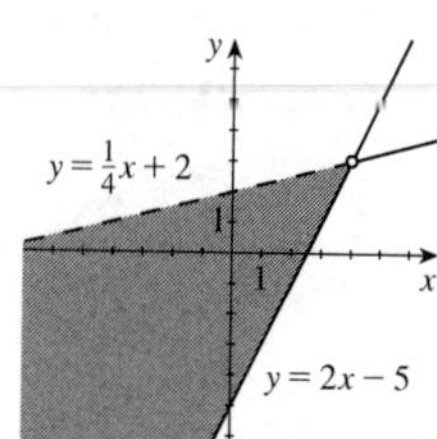


**22.** $\begin{cases} x-y>0 \\ 4+y \le 2x \end{cases}$ The vertices occur where $\begin{cases} y=x \\ 4+y=2x \end{cases}$ Substituting for $y$ gives $4+x=2x \quad \Leftrightarrow \quad x=4$, so $y=4$. Hence, the vertex is $(4,4)$, and the solution set is not bounded.

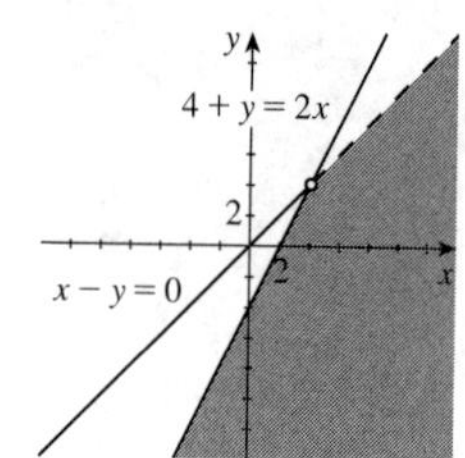


**23.** $\begin{cases} x \ge 0 \\ y \ge 0 \\ 3x+5y \le 15 \\ 3x+2y \le 9 \end{cases}$ From the graph, the points $(3,0)$, $(0,3)$ and $(0,0)$ are vertices, and the fourth vertex occurs where the lines $3x+5y=15$ and $3x+2y=9$ intersect. Subtracting these two equations gives $3y=6 \quad \Leftrightarrow$ $y=2$, and so $x=\frac{5}{3}$. Thus, the fourth vertex is $\left(\frac{5}{3},2\right)$, and the solution set is bounded.

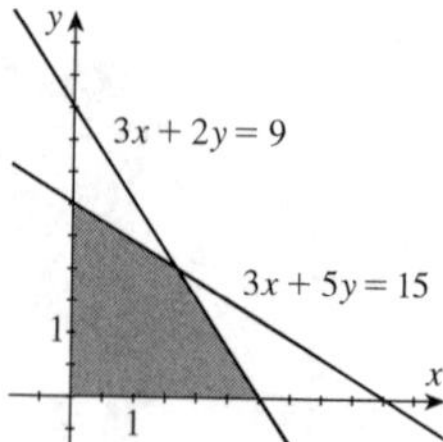


**24.** $\begin{cases} x > 2 \\ y < 12 \\ 2x-4y > 8 \end{cases}$ From the graph, the vertices occur at $(2,-1)$ and $(28,12)$. The solution set is not bounded.

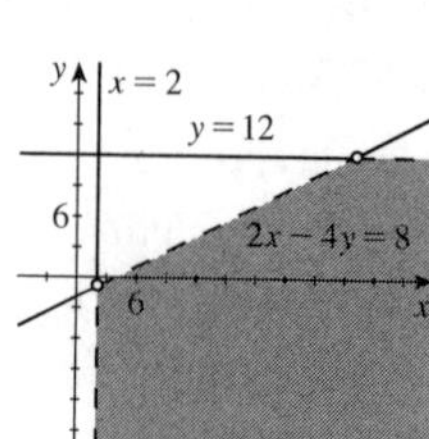

**25.** $\begin{cases} y < 9 - x^2 \\ y \geq x + 3 \end{cases}$ The vertices occur where $\begin{cases} y = 9 - x^2 \\ y = x + 3 \end{cases}$ Substituting for $y$ gives $9 - x^2 = x + 3 \quad \Leftrightarrow \quad x^2 + x - 6 = 0 \quad \Leftrightarrow$ $(x - 2)(x + 3) = 0 \Rightarrow x = -3, x = 2$. Therefore, the vertices are $(-3, 0)$ and $(2, 5)$, and the solution set is bounded.

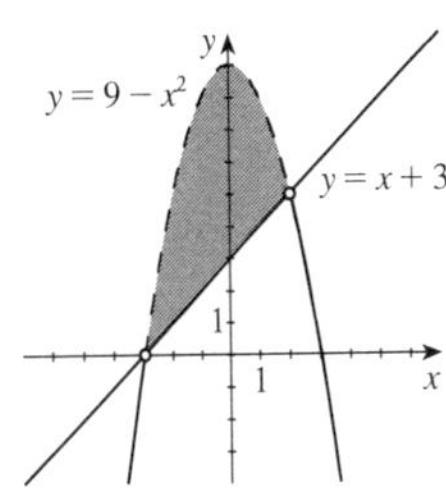


**26.** $\begin{cases} y \geq x^2 \\ x + y \geq 6 \end{cases}$ The vertices occur where $\begin{cases} y = x^2 \\ x + y = 6 \end{cases}$ Substituting for $y$ gives $x^2 + x = 6 \quad \Leftrightarrow \quad x^2 + x - 6 = 0 \Leftrightarrow (x + 3)(x - 2) = 0 \Rightarrow x = -3$, $x = 2$. Since $y = x^2$, the vertices are $(-3, 9)$ and $(2, 4)$, and the solution set is not bounded.

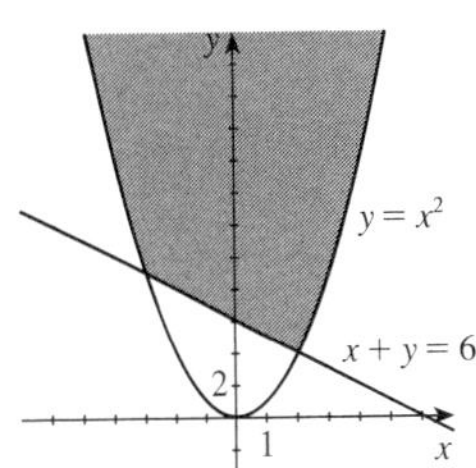


**27.** $\begin{cases} x^2 + y^2 \leq 4 \\ x - y > 0 \end{cases}$ The vertices occur where $\begin{cases} x^2 + y^2 = 4 \\ x - y = 0 \end{cases}$ Since $x - y = 0$ $\Leftrightarrow \quad x = y$, substituting for $x$ gives $y^2 + y^2 = 4 \quad \Leftrightarrow \quad y^2 = 2 \Rightarrow y = \pm\sqrt{2}$, and $x = \pm\sqrt{2}$. Therefore, the vertices are $\left(-\sqrt{2}, -\sqrt{2}\right)$ and $\left(\sqrt{2}, \sqrt{2}\right)$, and the solution set is bounded.

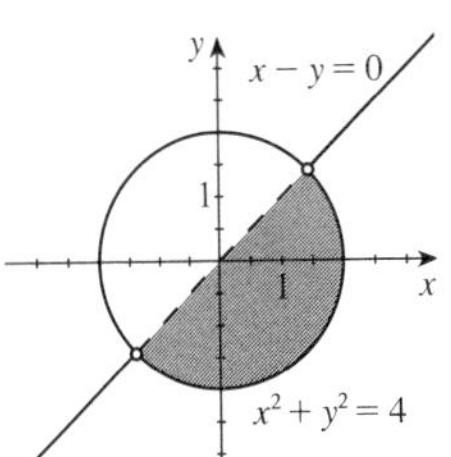


**28.** $\begin{cases} x > 0 \\ y > 0 \\ x + y < 10 \\ x^2 + y^2 > 9 \end{cases}$ From the graph, the vertices are $(0, 3)$, $(0, 10)$, $(3, 0)$, and $(10, 0)$. The solution set is bounded.

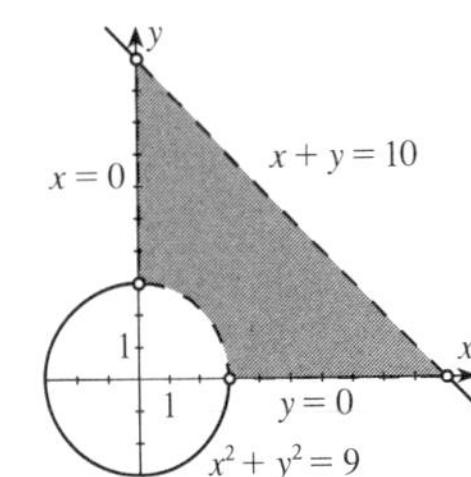


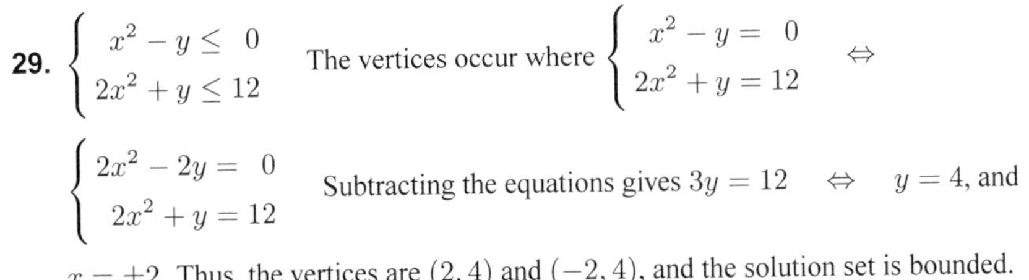

**29.** $\begin{cases} x^2 - y \leq 0 \\ 2x^2 + y \leq 12 \end{cases}$ The vertices occur where $\begin{cases} x^2 - y = 0 \\ 2x^2 + y = 12 \end{cases} \Leftrightarrow$ $\begin{cases} 2x^2 - 2y = 0 \\ 2x^2 + y = 12 \end{cases}$ Subtracting the equations gives $3y = 12 \quad \Leftrightarrow \quad y = 4$, and $x = \pm 2$. Thus, the vertices are $(2, 4)$ and $(-2, 4)$, and the solution set is bounded.

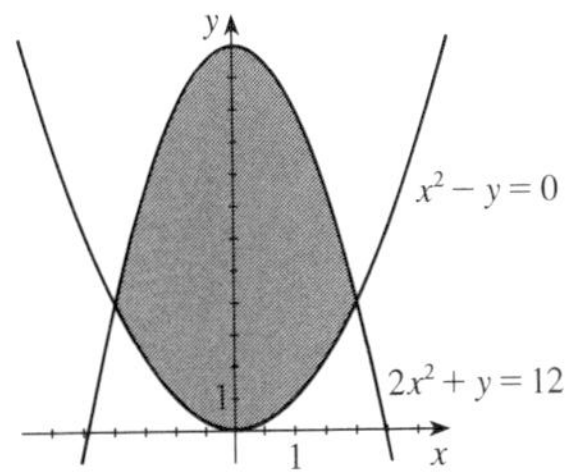


**30.** $\begin{cases} x^2 + y^2 < 9 \\ 2x + y^2 \geq 1 \end{cases}$ The vertices occur where $\begin{cases} x^2 + y^2 = 9 \\ 2x + y^2 = 1 \end{cases}$ Subtracting the equations gives $x^2 - 2x = 8 \quad \Leftrightarrow \quad x^2 - 2x - 8 = 0 \quad \Leftrightarrow$ $(x - 4)(x + 2) = 0 \Leftrightarrow x = -2, x = 4$. Therefore, the vertices are $\left(-2, -\sqrt{5}\right)$ and $\left(-2, \sqrt{5}\right)$, since $x = 4$ does not give a real solution for $y$. The solution set is bounded.

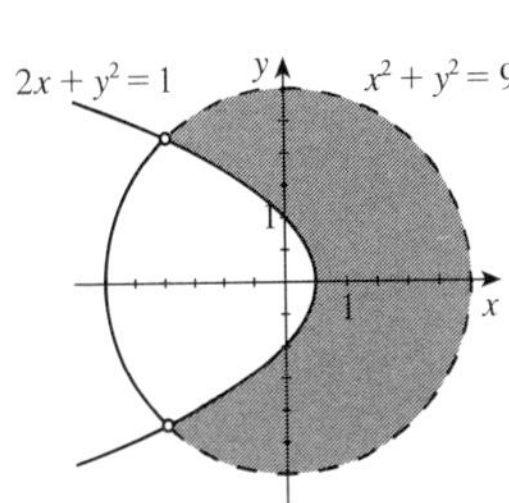

**31.** $\begin{cases} x + 2y \le 14 \\ 3x - \phantom{2}y \ge 0 \\ x - \phantom{2}y \ge 2 \end{cases}$ We find the vertices of the region by solving pairs of the corresponding equations: $\begin{cases} x + 2y = 14 \\ x - \phantom{2}y = 2 \end{cases} \Leftrightarrow \begin{cases} x + 2y = 14 \\ 3y = 12 \end{cases} \Leftrightarrow y = 4$ and $x = 6$. $\begin{cases} 3x - y = 0 \\ x - y = 2 \end{cases} \Leftrightarrow \begin{cases} 3x - y = 0 \\ 2x - y = -2 \end{cases} \Leftrightarrow x = -1$ and $y = -3$. Therefore, the vertices are $(6, 4)$ and $(-1, -3)$, and the solution set is not bounded.

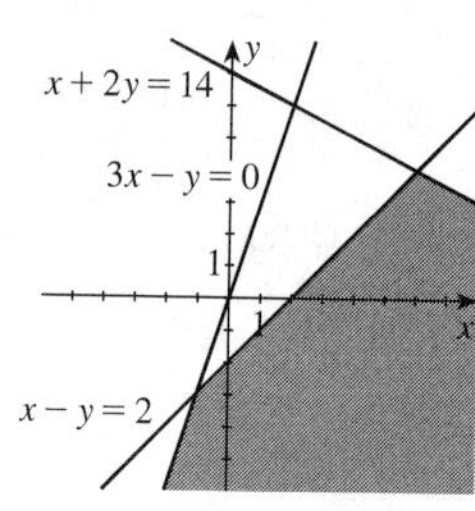


**32.** $\begin{cases} y < x + 6 \\ 3x + 2y \ge 12 \\ x - 2y \le 2 \end{cases}$ To find where the line $y = x + 6$ intersects the lines $3x + 2y = 12$ and $x - 2y = 2$, we substitute for $y$: $3x + 2(x + 6) = 12 \Leftrightarrow x = 0$ and $y = 6$; $x - 2(x + 6) = 2 \Leftrightarrow x = -14$ and $y = -8$. Next, adding the equations $3x + 2y = 12$ and $x - 2y = 2$ yields $4x = 14 \Leftrightarrow x = \frac{7}{2}$. So these lines intersect at the point $\left(\frac{7}{2}, \frac{3}{4}\right)$. Since the vertex $(-14, -8)$ is not part of the solution set, the vertices are $(0, 6)$ and $\left(\frac{7}{2}, \frac{3}{4}\right)$, and the solution set is not bounded.

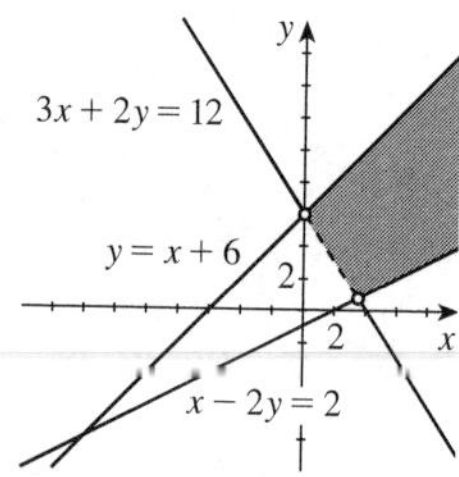


**33.** $\begin{cases} x \ge 0, y \ge 0 \\ x \le 5, x + y \le 7 \end{cases}$ The points of intersection are $(0, 7)$, $(0, 0)$, $(7, 0)$, $(5, 2)$, and $(5, 0)$. However, the point $(7, 0)$ is not in the solution set. Therefore, the vertices are $(0, 7)$, $(0, 0)$, $(5, 0)$, and $(5, 2)$, and the solution set is bounded.

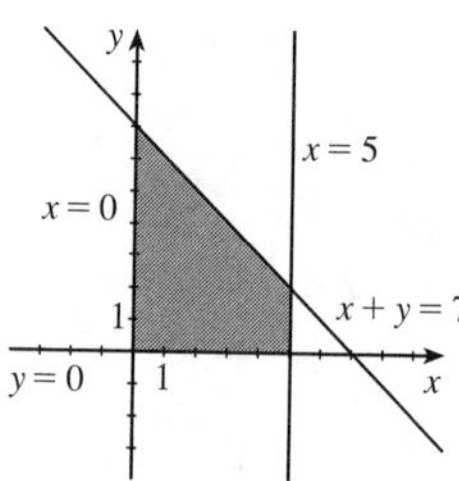


**34.** $\begin{cases} x \ge 0, y \ge 0 \\ y \le 4, 2x + y \le 8 \end{cases}$ The points of intersection are $(0, 8)$, $(0, 4)$, $(4, 0)$, $(2, 4)$, and $(3, 2)$. However, the point $(0, 8)$ is not in the solution set. Therefore, the vertices are $(0, 4)$, $(2, 4)$, $(4, 0)$, and $(0, 0)$. The solution set is bounded.

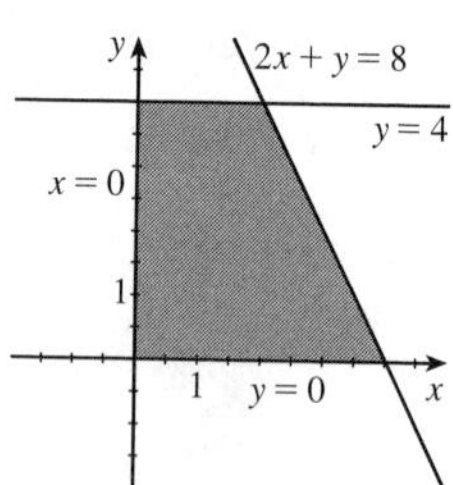


**35.** $\begin{cases} y > x + 1 \\ x + 2y \le 12 \\ x + 1 > 0 \end{cases}$ We find the vertices of the region by solving pairs of the corresponding equations. Using $x = -1$ and substituting for $x$ in the line $y = x + 1$ gives the point $(-1, 0)$. Substituting for $x$ in the line $x + 2y = 12$ gives the point $\left(-1, \frac{13}{2}\right)$. $\begin{cases} y = x + 1 \\ x + 2y = 12 \end{cases} \Leftrightarrow x = y - 1$ and $y - 1 + 2y = 12$ $\Leftrightarrow 3y = 13 \Leftrightarrow y = \frac{13}{3}$ and $x = \frac{10}{3}$. So the vertices are $(-1, 0)$, $\left(-1, \frac{13}{2}\right)$, and $\left(\frac{10}{3}, \frac{13}{3}\right)$, and none of these vertices is in the solution set. The solution set is bounded.

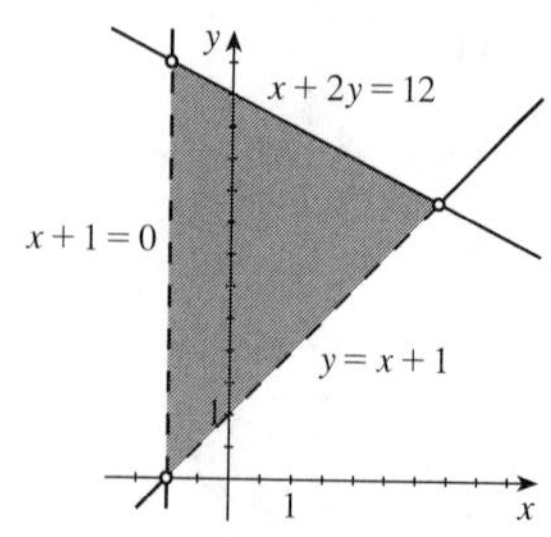

**36.** $\begin{cases} x + y > 12 \\ y < \frac{1}{2}x - 6 \\ 3x + y < 6 \end{cases}$ Graphing these inequalities, we see that there are no points that satisfy all three, and hence the solution set is empty.

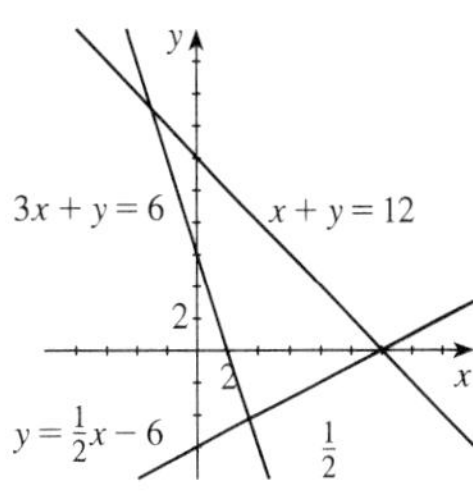


**37.** $\begin{cases} x^2 + y^2 \le 8 \\ x \ge 2,\ y \ge 0 \end{cases}$ The intersection points are $(2, \pm 2)$, $(2, 0)$, and $\left(2\sqrt{2}, 0\right)$. However, since $(2, -2)$ is not part of the solution set, the vertices are $(2, 2)$, $(2, 0)$, and $\left(2\sqrt{2}, 0\right)$. The solution set is bounded.

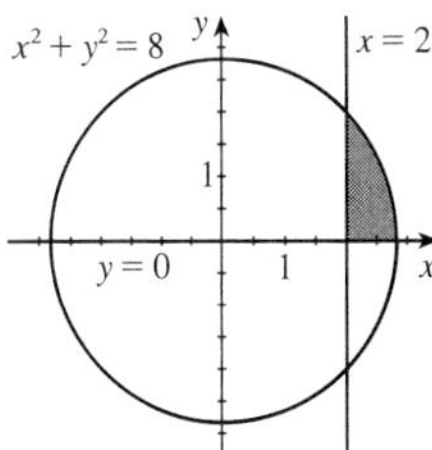


**38.** $\begin{cases} x^2 - y \ge 0 \\ x + y < 6 \\ x - y < 6 \end{cases}$ Adding the equations $x + y = 6$ and $x - y = 6$ yields $2x = 12$ $\Leftrightarrow$ $x = 6$. So these curves intersect at the point $(6, 0)$. To find where $x^2 - y = 0$ and $x + y = 6$ intersect, we solve the first for $y$, giving $y = x^2$, and then substitute into the second equation to get $x + x^2 = 6$ $\Leftrightarrow$ $x^2 + x - 6 = 0$ $\Leftrightarrow$ $(x + 3)(x - 2) = 0$ $\Leftrightarrow$ $x = -3$ or $x = 2$. When $x = -3$, we have $y = 9$, and when $x = 2$, we have $y = 4$, so the points of intersection are $(-3, 9)$ and $(2, 4)$. Substituting $y = x^2$ into the equation $x - y = 6$ gives $x - x^2 = 6$, which has no solution. Thus the vertices (which are not in the solution set) are $(6, 0)$, $(-3, 9)$, and $(2, 4)$. The solution set is not bounded.

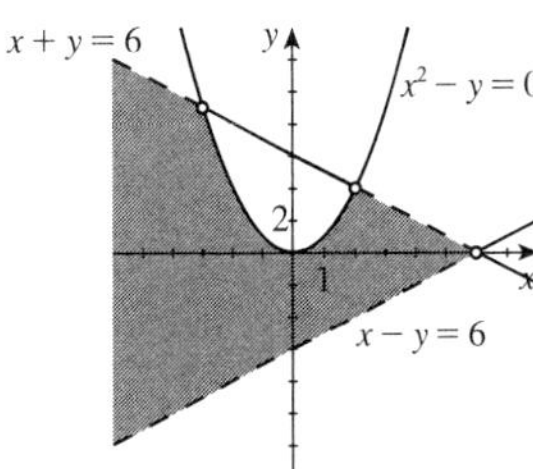


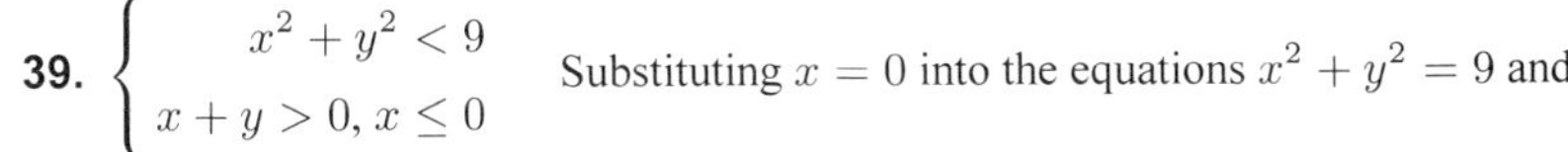

**39.** $\begin{cases} x^2 + y^2 < 9 \\ x + y > 0,\ x \le 0 \end{cases}$ Substituting $x = 0$ into the equations $x^2 + y^2 = 9$ and $x + y = 0$ gives the vertices $(0, \pm 3)$ and $(0, 0)$. To find the points of intersection for the equations $x^2 + y^2 = 9$ and $x + y = 0$, we solve for $x = -y$ and substitute into the first equation. This gives $(-y)^2 + y^2 = 9 \Rightarrow y = \pm\frac{3\sqrt{2}}{2}$. The points $(0, -3)$and $\left(\frac{3\sqrt{2}}{2}, -\frac{3\sqrt{2}}{2}\right)$ lie away from the solution set, so the vertices are $(0, 0)$, $(0, 3)$, and $\left(-\frac{3\sqrt{2}}{2}, \frac{3\sqrt{2}}{2}\right)$. Note that the vertices are not solutions in this case. The solution set is bounded.

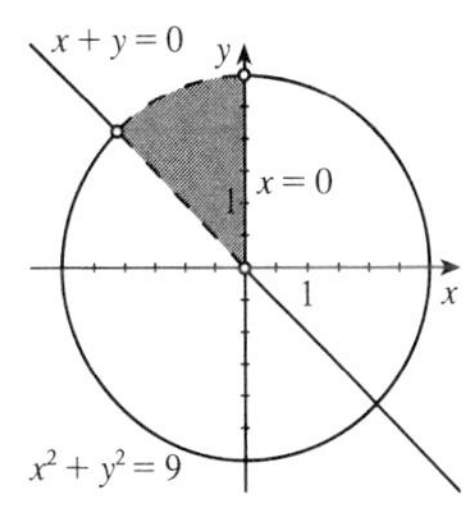


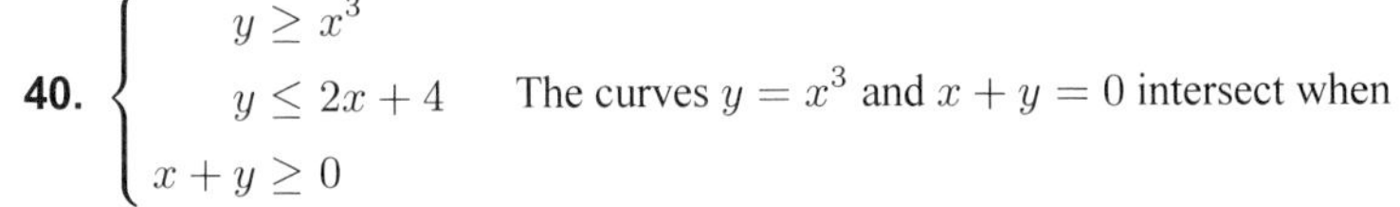

**40.** $\begin{cases} y \ge x^3 \\ y \le 2x + 4 \\ x + y \ge 0 \end{cases}$ The curves $y = x^3$ and $x + y = 0$ intersect when $x^3 + x = 0$ $\Leftrightarrow$ $x = 0$ $\Leftrightarrow$ $y = 0$. The lines $x + y = 0$ and $y = 2x + 4$ intersect when $-x = 2x + 4$ $\Leftrightarrow$ $3x = -4$ $\Leftrightarrow$ $x = -\frac{4}{3}$ $\Leftrightarrow$ $y = \frac{4}{3}$. To find where $y = x^3$ and $y = 2x + 4$ intersect, we substitute for $y$ and get $x^3 = 2x + 4$ $\Leftrightarrow$ $x^3 - 2x - 4 = 0 \Leftrightarrow (x - 2)\left(x^2 + 2x + 2\right) = 0$ $\Rightarrow$ $x = 2$ (the other factor has no real solution). When $x = 2$ we have $y = 8$.Thus, the vertices of the region are $(0, 0)$, $\left(-\frac{4}{3}, \frac{4}{3}\right)$, and $(2, 8)$. The solution set is bounded.

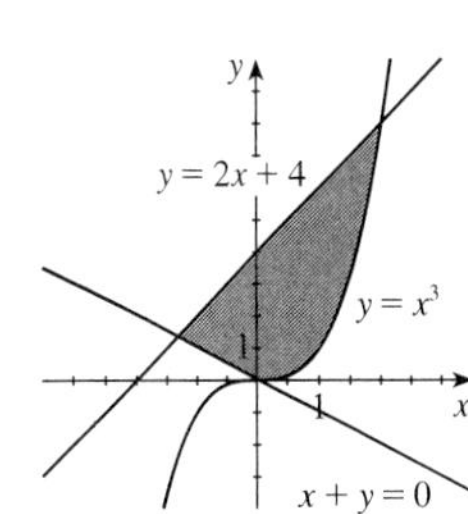

**41.** $\begin{cases} y \geq x - 3 \\ y \geq -2x + 6 \\ y \leq 8 \end{cases}$ Using a graphing calculator, we find the region shown. The vertices are $(3, 0)$, $(-1, 8)$, and $(11, 8)$.

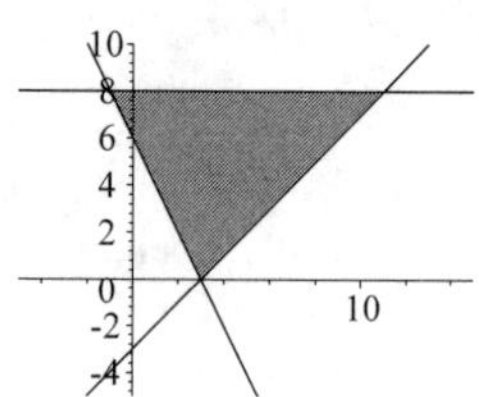

**42.** $\begin{cases} x + y \geq 12 \\ 2x + y \leq 24 \\ x - y \geq -6 \end{cases}$ Using the graphing calculator we find the region shown. The vertices are $(3, 9)$, $(6, 12)$, and $(12, 0)$.

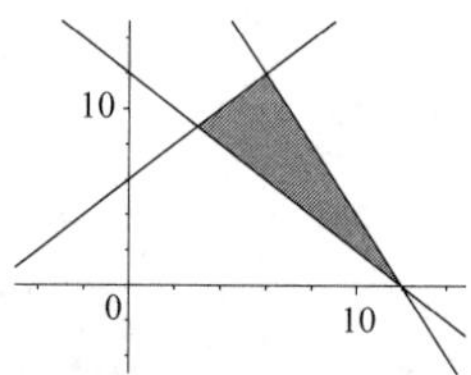

**43.** $\begin{cases} y \leq 6x - x^2 \\ x + y \geq 4 \end{cases}$ Using a graphing calculator, we find the region shown. The vertices are $(0.6, 3.4)$ and $(6.4, -2.4)$.

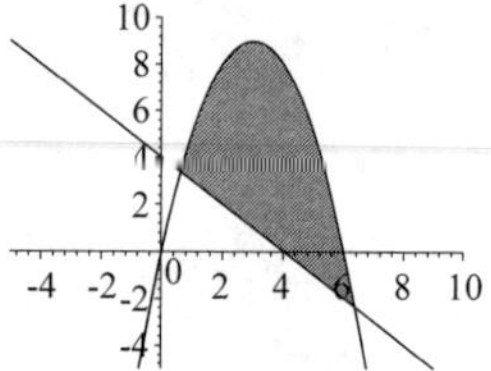

**44.** $\begin{cases} y \geq x^3 \\ 2x + y \geq 0 \\ y \leq 2x + 6 \end{cases}$ Using a graphing calculator, we find the region shown. The vertices are $(0, 0)$, $(2.2, 10.3)$ and $(-1.5, 3)$.

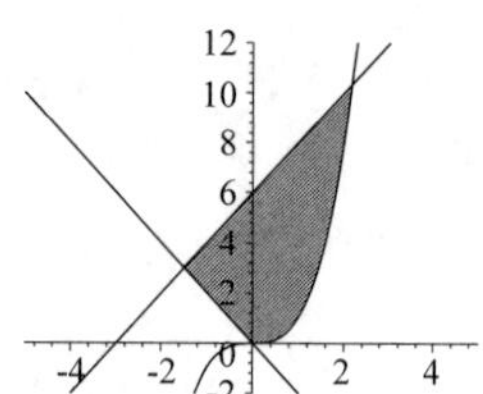

**45.** Let $x$ be the number of fiction books published in a year and $y$ the number of nonfiction books. Then the following system of inequalities holds:

$\begin{cases} x \geq 0, y \geq 0 \\ x + y \leq 100 \\ y \geq 20, x \geq y \end{cases}$ From the graph, we see that the vertices are $(50, 50)$, $(80, 20)$ and $(20, 20)$.

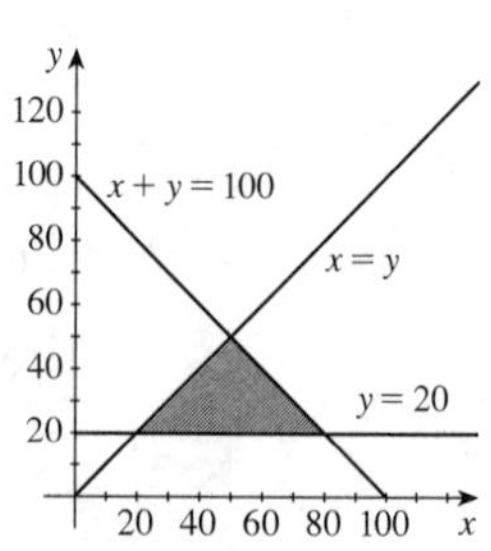


**46.** Let $x$ be the number of chairs made and $y$ the number of tables. Then the following system of inequalities holds: $\begin{cases} 2x + 3y \leq 12 \\ 2x + y \leq 8 \\ x \geq 0, y \geq 0 \end{cases}$ The intersection points are $(0, 4)$, $(0, 8)$, $(6, 0)$, $(4, 0)$, $(0, 0)$, and $(3, 2)$. Since the points $(0, 8)$ and $(6, 0)$ are not in the solution set, the vertices are $(0, 4)$, $(0, 0)$, $(4, 0)$, and $(3, 2)$.

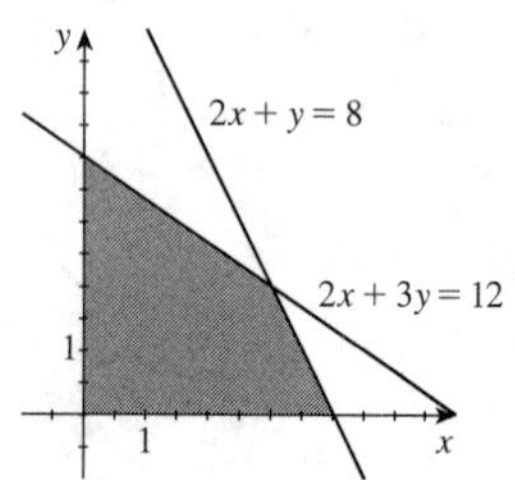

**47.** Let $x$ be the number of Standard Blend packages and $y$ be the number of Deluxe Blend packages. Since there are 16 ounces per pound, we get the following system of inequalities holds:

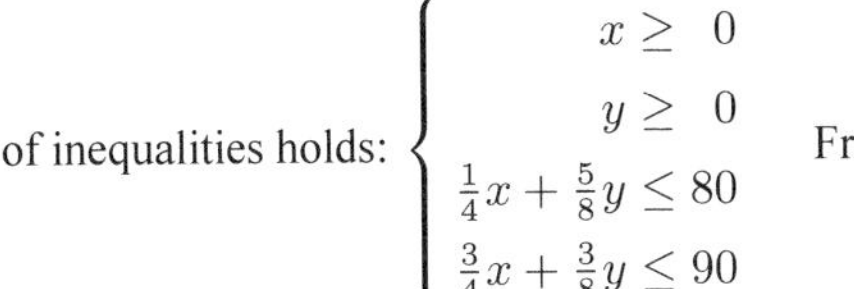

$$\begin{cases} x \geq 0 \\ y \geq 0 \\ \frac{1}{4}x + \frac{5}{8}y \leq 80 \\ \frac{3}{4}x + \frac{3}{8}y \leq 90 \end{cases}$$

From the graph, we see that the vertices are $(0, 0)$, $(120, 0)$, $(70, 100)$ and $(0, 128)$.

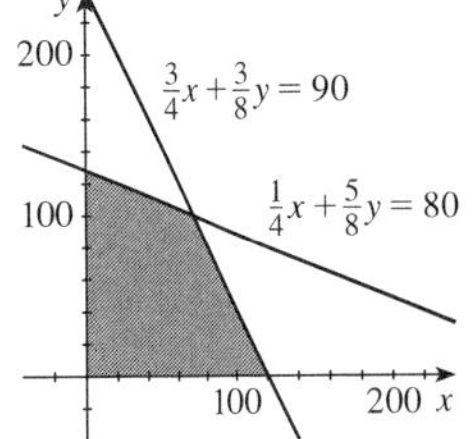


**48.** Let $x$ be the amount of fish and $y$ the amount of beef (in ounces) in each can. Then the following system of inequalities holds: $\begin{cases} 12x + 6y \geq 60 \\ 3x + 9y \geq 45 \\ x \geq 0, \ y \geq 0 \end{cases}$ From the graph, we see that the vertices are $(15, 0)$, $(3, 4)$, and $(0, 10)$.

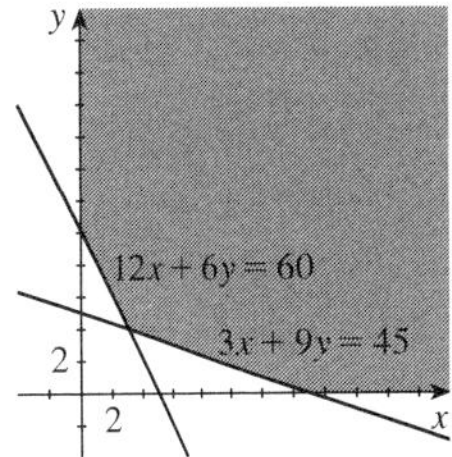


**49.** $x + 2y > 4$, $-x + y < 1$, $x + 3y \leq 9$, $x < 3$.

*Method 1:* We shade the solution to each inequality with lines perpendicular to the boundary. As you can see, as the number of inequalities in the system increases, it gets harder to locate the region where *all* of the shaded parts overlap.

*Method 2:* Here, if a region is shaded then it fails to satisfy at least one inequality. As a result, the region that is left unshaded satisfies each inequality, and is the solution to the system of inequalities. In this case, this method makes it easier to identify the solution set.

To finish, we find the vertices of the solution set. The line $x = 3$ intersects the line $x + 2y = 4$ at $\left(3, \frac{1}{2}\right)$ and the line $x + 3y = 9$ at $(3, 2)$. To find where the lines $-x + y = 1$ and $x + 2y = 4$ intersect, we add the two equations, which gives $3y = 5 \quad \Leftrightarrow \quad y = \frac{5}{3}$, and $x = \frac{2}{3}$. To find where the lines $-x + y = 1$ and $x + 3y = 9$ intersect, we add the two equations, which gives $4y = 10 \quad \Leftrightarrow \quad y = \frac{10}{4} = \frac{5}{2}$, and $x = \frac{3}{2}$. The vertices are $\left(3, \frac{1}{2}\right)$, $(3, 2)$, $\left(\frac{2}{3}, \frac{5}{3}\right)$, and $\left(\frac{3}{2}, \frac{5}{2}\right)$, and the solution set is bounded.

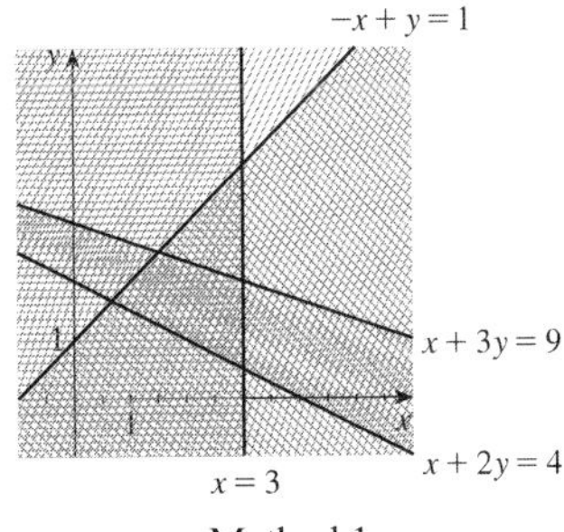

Method 1

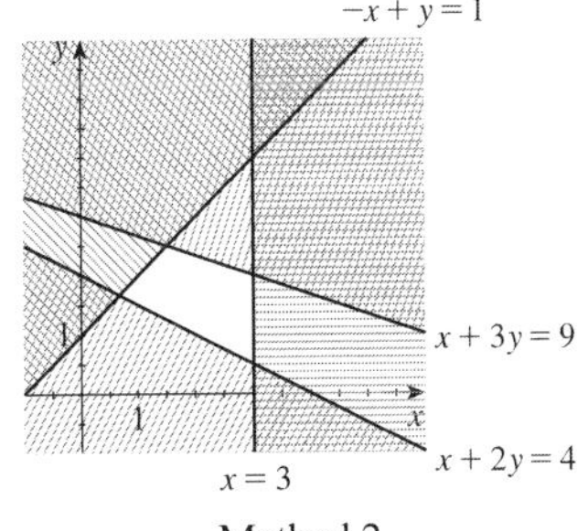

Method 2

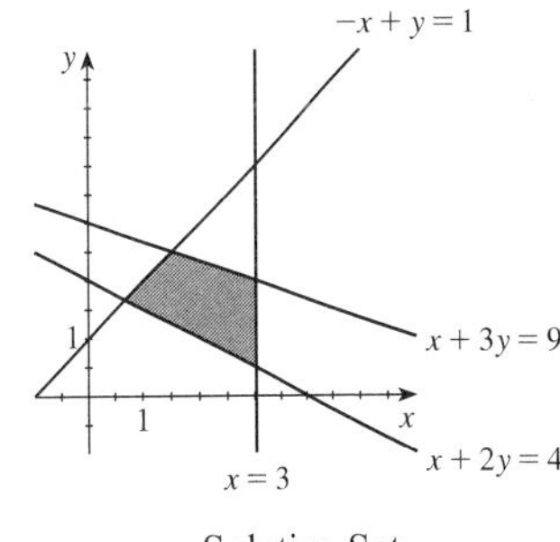

Solution Set

# Chapter 10 Review

**1.** $\begin{cases} 2x + 3y = 7 \\ x - 2y = 0 \end{cases}$ By inspection of the graph, it appears that $(2, 1)$ is the solution to the system. We check this in both equations to verify that it is the solution. $2(2) + 3(1) = 4 + 3 = 7$ and $2 - 2(1) = 2 - 2 = 0$. Since both equations are satisfied, the solution is indeed $(2, 1)$.

**2.** $\begin{cases} 3x + y = 8 \\ y = x^2 - 5x \end{cases}$ By inspection of the graph, it appears that $(-2, 14)$ and $(4, -4)$ are solutions to the system. We check each possible solution in both equations to verify that it is a solution: $3(-2) + (14) = -6 + 14 = 8$ and $(-2)^2 - 5(-2) = 4 + 10 = 14$; also $3(4) + (-4) = 12 - 4 = 8$ and $(4)^2 - 5(4) = 16 - 20 = -4$. Since both points satisfy both equations, the solutions are $(-2, 14)$ and $(4, -4)$.

**3.** $\begin{cases} x^2 + y = 2 \\ x^2 - 3x - y = 0 \end{cases}$ By inspection of the graph, it appears that $(2, -2)$ is a solution to the system, but is difficult to get accurate values for the other point. Adding the equations, we get $2x^2 - 3x = 2 \Leftrightarrow 2x^2 - 3x - 2 = 0 \Leftrightarrow (2x + 1)(x - 2) = 0$. So $2x + 1 = 0 \Leftrightarrow x = -\frac{1}{2}$ or $x = 2$. If $x = -\frac{1}{2}$, then $\left(-\frac{1}{2}\right)^2 + y = 2 \Leftrightarrow y = \frac{7}{4}$. If $x = 2$, then $2^2 + y = 2 \Leftrightarrow y = -2$. Thus, the solutions are $\left(-\frac{1}{2}, \frac{7}{4}\right)$ and $(2, -2)$.

**4.** $\begin{cases} x - y = -2 \\ x^2 + y^2 - 4y = 4 \end{cases}$ By inspection of the graph, it appears that $(-2, 0)$ and $(2, 4)$ are solutions to the system. We check each possible solution in both equations to verify that it is a solution: $(-2) - (0) = -2$ and $(-2)^2 + (0)^2 - 4(0) = 4$; also $(2) - (4) = -2$ and $(2)^2 + (4)^2 - 4(4) = 4 + 16 - 16 = 4$. Since both points satisfy both equations, the solutions are $(-2, 0)$ and $(2, 4)$.

**5.** $\begin{cases} 3x - y = 5 \\ 2x + y = 5 \end{cases}$ Adding, we get $5x = 10 \Leftrightarrow x = 2$. So $2(2) + y = 5 \Leftrightarrow y = 1$. Thus, the solution is $(2, 1)$.

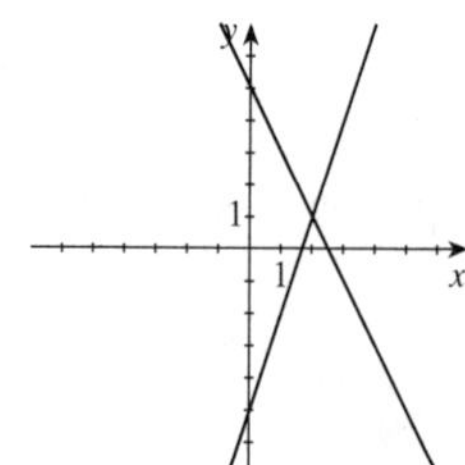


**6.** $\begin{cases} y = 2x + 6 \\ y = -x + 3 \end{cases}$ Subtracting the second equation from the first, we get $0 = 3x + 3 \Leftrightarrow x = -1$. So $y = -(-1) + 3 = 4$. Thus, the solution is $(-1, 4)$.

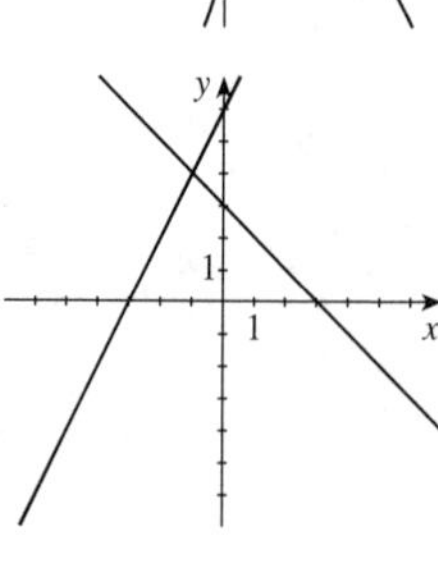


**7.** $\begin{cases} 2x - 7y = 28 \\ y = \frac{2}{7}x - 4 \end{cases} \Leftrightarrow \begin{cases} 2x - 7y = 28 \\ 2x - 7y = 28 \end{cases}$ Since these equations represent the same line, any point on this line will satisfy the system. Thus the solution are $\left(x, \frac{2}{7}x - 4\right)$, where $x$ is any real number.

**8.** $\begin{cases} 6x - 8y = 15 \\ -\frac{3}{2}x + 2y = -4 \end{cases} \Leftrightarrow \begin{cases} 6x - 8y = 15 \\ -6x + 8y = -16 \end{cases}$ Adding gives $0 = -1$ which is false. Hence, there is no solution. The lines are parallel.

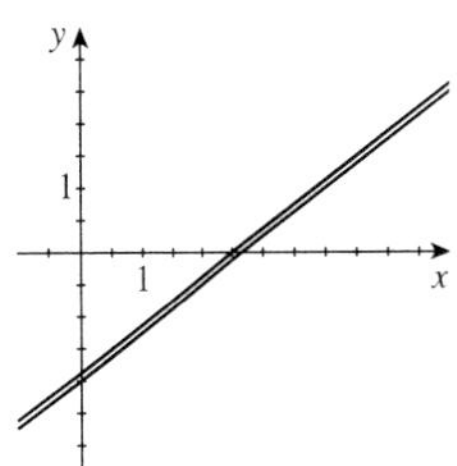


**9.** $\begin{cases} 2x - \phantom{3}y = 1 \\ x + 3y = 10 \\ 3x + 4y = 15 \end{cases}$ Solving the first equation for $y$, we get $y = -2x + 1$.

Substituting into the second equation gives $x + 3(-2x + 1) = 10 \quad \Leftrightarrow \quad -5x = 7 \quad \Leftrightarrow \quad x = -\frac{7}{5}$. So $y = -\left(-\frac{7}{5}\right) + 1 = \frac{12}{5}$.Checking the point $\left(-\frac{7}{5}, \frac{12}{5}\right)$ in the third equation we have $3\left(-\frac{7}{5}\right) + 4\left(\frac{12}{5}\right) \stackrel{?}{=} 15$ but $-\frac{21}{5} + \frac{48}{5} \neq 15$. Thus, there is no solution, and the lines do not intersect at one point.

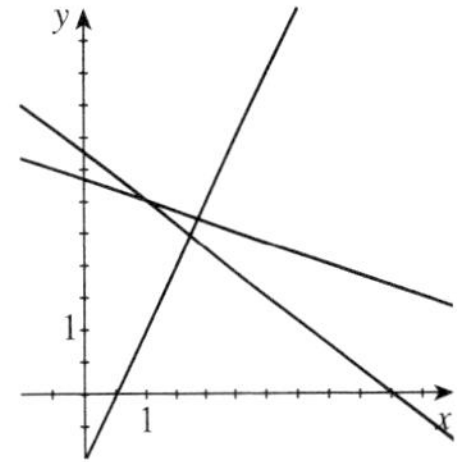


**10.** $\begin{cases} 2x + 5y = 9 \\ -x + 3y = 1 \\ 7x - 2y = 14 \end{cases}$ Adding the first equation to twice the second equation gives

$11y = 11 \quad \Leftrightarrow \quad y = 1$. Substituting back into the second equation, we get $-x + 3(1) = 1 \quad \Leftrightarrow \quad x = 4$. Checking point $(4, 1)$ in the third equation gives $7(4) - 2(1) = 26 \neq 14$. Thus there is no solution, and the lines do not intersect at one point.

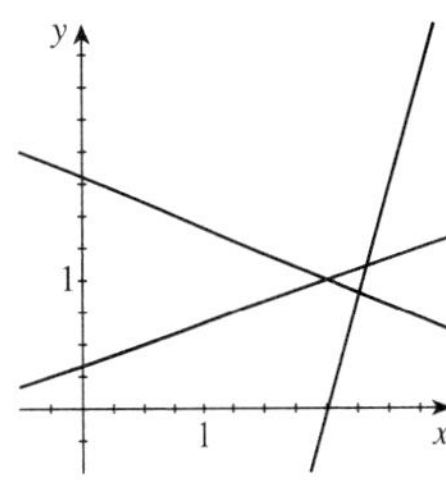


**11.** $\begin{cases} y = x^2 + 2x \\ y = 6 + x \end{cases}$ Substituting for $y$ gives $6 + x = x^2 + 2x \quad \Leftrightarrow \quad x^2 + x - 6 = 0$. Factoring, we have $(x - 2)(x + 3) = 0$. Thus $x = 2$ or $-3$. If $x = 2$, then $y = 8$, and if $x = -3$, then $y = 3$. Thus the solutions are $(-3, 3)$ and $(2, 8)$.

**12.** $\begin{cases} x^2 + y^2 = 8 \\ y = x + 2 \end{cases}$ Substituting for $y$ in the first equation gives $x^2 + (x + 2)^2 = 8 \quad \Leftrightarrow \quad 2x^2 + 4x - 4 = 0 \Leftrightarrow 2\left(x^2 + 2x - 2\right) = 0$. Using the quadratic formula, we have $x = \dfrac{-2 \pm \sqrt{4+8}}{2} = \dfrac{-2 \pm 2\sqrt{3}}{2} = -1 \pm \sqrt{3}$. If $x = -1 - \sqrt{3}$. then $y = \left(-1 - \sqrt{3}\right) + 2 = 1 - \sqrt{3}$, and if $x = -1 + \sqrt{3}$, then $y = \left(-1 + \sqrt{3}\right) + 2 = 1 + \sqrt{3}$. Thus, the solutions are $\left(-1 - \sqrt{3}, 1 - \sqrt{3}\right)$ and $\left(-1 + \sqrt{3}, 1 + \sqrt{3}\right)$.

**13.** $\begin{cases} 3x + \dfrac{4}{y} = 6 \\ x - \dfrac{8}{y} = 4 \end{cases}$ Adding twice the first equation to the second gives $7x = 16 \quad \Leftrightarrow \quad x = \frac{16}{7}$. So $\dfrac{16}{7} - \dfrac{8}{y} = 4 \quad \Leftrightarrow \quad 16y - 56 = 28y \quad \Leftrightarrow \quad -12y = 56 \quad \Leftrightarrow \quad y = -\frac{14}{3}$. Thus, the solution is $\left(\frac{16}{7}, -\frac{14}{3}\right)$.

**14.** $\begin{cases} x^2 + y^2 = 10 \\ x^2 + 2y^2 - 7y = 0 \end{cases}$ Subtracting the first equation from the second gives $y^2 - 7y = -10 \Leftrightarrow$ $y^2 - 7y + 10 = 0 \Leftrightarrow (y-2)(y-5) = 0 \Rightarrow y = 2, y = 5$. If $y = 2$, then $x^2 + 4 = 10 \Leftrightarrow x^2 = 6 \Rightarrow x = \pm\sqrt{6}$, and if $y = 5$, then $x^2 + 25 = 10 \Leftrightarrow x^2 = -15$, which leads to no real solution. Thus the solutions are $(\sqrt{6}, 2)$ and $(-\sqrt{6}, 2)$.

**15.** $\begin{cases} 0.32x + 0.43y = 0 \\ 7x - 12y = 341 \end{cases} \Leftrightarrow \begin{cases} y = -\dfrac{32x}{43} \\ y = \dfrac{7x - 341}{12} \end{cases}$

The solution is approximately $(21.41, -15.93)$.

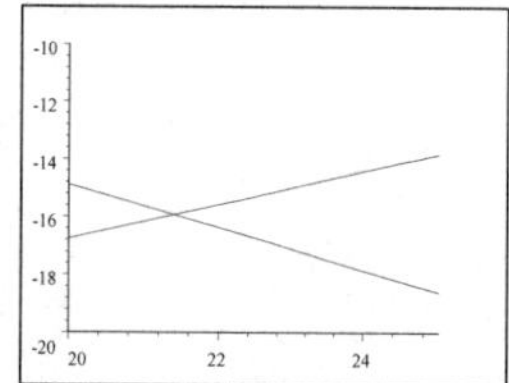

**16.** 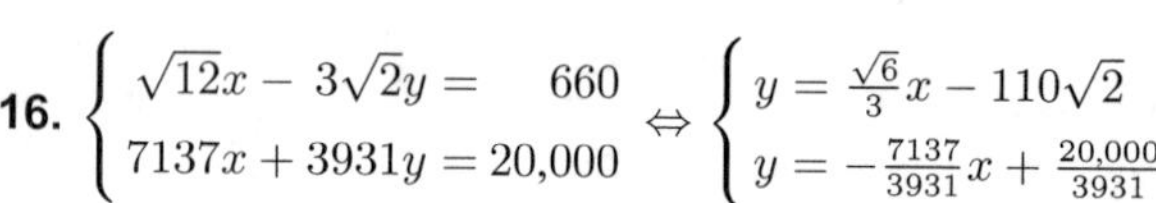

$\begin{cases} \sqrt{12}x - 3\sqrt{2}y = 660 \\ 7137x + 3931y = 20{,}000 \end{cases} \Leftrightarrow \begin{cases} y = \frac{\sqrt{6}}{3}x - 110\sqrt{2} \\ y = -\frac{7137}{3931}x + \frac{20{,}000}{3931} \end{cases}$

The solution is approximately $(61.04, -105.73)$.

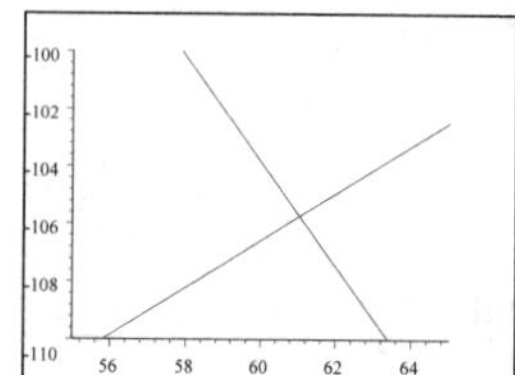

**17.** $\begin{cases} x - y^2 = 10 \\ x = \frac{1}{22}y + 12 \end{cases} \Leftrightarrow \begin{cases} y = \pm\sqrt{x - 10} \\ y = 22(x - 12) \end{cases}$

The solutions are $(11.94, -1.39)$ and $(12.07, 1.44)$.

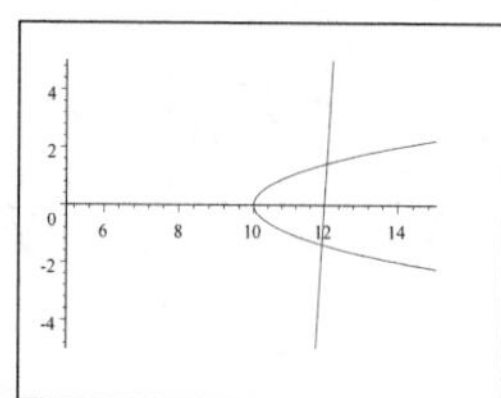

**18.** $\begin{cases} y = 5^x + x \\ y = x^5 + 5 \end{cases}$ The solutions are approximately $(-1.45, -1.35)$, $(1, 6)$, and $(1.51, 12.93)$.

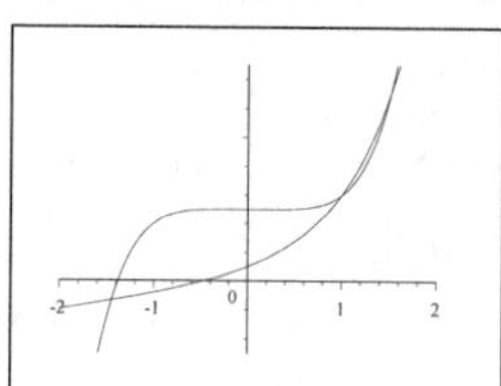

**19. (a)** $2 \times 3$

**(b)** Yes, this matrix is in row-echelon form.

**(c)** No, this matrix is not in reduced row-echelon form, since the leading 1 in the second row does not have a 0 above it.

**(d)** $\begin{cases} x + 2y = -5 \\ y = 3 \end{cases}$

**20. (a)** $2 \times 3$

**(b)** Yes, this matrix is in row-echelon form.

**(c)** Yes, this matrix is in reduced row-echelon form.

**(d)** $\begin{cases} x = 6 \\ y = 0 \end{cases}$

**21. (a)** $3 \times 4$

**(b)** Yes, this matrix is in row-echelon form.

**(c)** Yes, this matrix is in reduced row-echelon form.

**(d)** $\begin{cases} x + 8z = 0 \\ y + 5z = -1 \\ 0 = 0 \end{cases}$

**22. (a)** $3 \times 4$

**(b)** No, this matrix is not in row-echelon form, since the leading 1 in the second row is not to the left of the one above it.

**(c)** Since this matrix is not in row-echelon form, it is not in reduced row-echelon form.

**(d)** $\begin{cases} x + 3y + 6z = 2 \\ 2x + y = 5 \\ z = 0 \end{cases}$

**23.** **(a)** $3 \times 4$

**(b)** No, this matrix is not in row-echelon form. The leading 1 in the second row is not to the left of the one above it.

**(c)** No, this matrix is not in reduced row-echelon form.

**(d)** $\begin{cases} \phantom{x+{}} y - 3z = 4 \\ x + y \phantom{{}+z} = 7 \\ x + 2y + z = 2 \end{cases}$

**24.** **(a)** $4 \times 4$

**(b)** No, this matrix is not in row-echelon form. The leading 1 in the fourth row is not to the left of the one above it.

**(c)** No, this matrix is not in reduced row-echelon form.

**(d)** $\begin{cases} x + 8y + 6z = -4 \\ y - 3z = 5 \\ 2z = -7 \\ x + y + z = 0 \end{cases}$

**25.** $\begin{cases} x + y + 2z = 6 \\ 2x + 5z = 12 \\ x + 2y + 3z = 9 \end{cases} \Leftrightarrow \begin{cases} x + y + 2z = 6 \\ 2y - z = 0 \\ 4y + z = 6 \end{cases} \Leftrightarrow \begin{cases} x + y + 2z = 6 \\ 2y - z = 0 \\ 3z = 6 \end{cases}$ Therefore, $3z = 6 \Leftrightarrow$ $z = 2$, $2y - 2 = 0 \Leftrightarrow y = 1$, and $x + 1 + 2(2) = 6 \Leftrightarrow x = 1$. Hence, the solution is $(1, 1, 2)$.

**26.** $\begin{cases} x - 2y + 3z = 1 \\ x - 3y - z = 0 \\ 2x - 6z = 6 \end{cases} \Leftrightarrow \begin{cases} x - 2y + 3z = 1 \\ y + 4z = 1 \\ 6y - 4z = 6 \end{cases} \Leftrightarrow \begin{cases} x - 2y + 3z = 1 \\ y + 4z = 1 \\ 28z = 0 \end{cases}$ Thus, $z = 0$, $y = 1$, and $x - 2(1) + 3(0) = 1 \Leftrightarrow x = 3$, and so the solution is $(3, 1, 0)$.

**27.** $\begin{cases} x - 2y + 3z = 1 \\ 2x - y + z = 3 \\ 2x - 7y + 11z = 2 \end{cases} \Leftrightarrow \begin{cases} x - 2y + 3z = 1 \\ 3y - 5z = 1 \\ 6y - 10z = 1 \end{cases} \Leftrightarrow \begin{cases} x - 2y + 3z = 1 \\ 3y - 5z = 1 \\ 0 = -1 \end{cases}$ which is impossible. Therefore, the system has no solution.

**28.** $\begin{cases} x + y + z + w = 2 \\ 2x - 3z = 5 \\ x - 2y + 4w = 9 \\ x + y + 2z + 3w = 5 \end{cases} \Leftrightarrow \begin{cases} x + y + z + w = 2 \\ 2y + 5z + 2w = -1 \\ 3y + z - 3w = -7 \\ z + 2w = 3 \end{cases} \Leftrightarrow \begin{cases} x + y + z + w = 2 \\ 2y + 5z + 2w = -1 \\ 13z + 12w = 11 \\ z + 2w = 3 \end{cases} \Leftrightarrow$

$\begin{cases} x + y + z + w = 2 \\ 2y + 5z + 2w = -1 \\ 13z + 12w = 11 \\ 14w = 28 \end{cases}$ Therefore, $14w = 28 \Leftrightarrow w = 2$, $13z + 12(2) = 11 \Leftrightarrow z = -1$; $2y + 5(-1) + 2(2) = -1 \Leftrightarrow 2y - 1 = -1 \Leftrightarrow y = 0$; and $x - 1 + 2 = 2 \Leftrightarrow x = 1$. So the solution is $(1, 0, -1, 2)$.

**29.** $\begin{bmatrix} 1 & 2 & 2 & 6 \\ 1 & -1 & 0 & -1 \\ 2 & 1 & 3 & 7 \end{bmatrix} \xrightarrow{R_2 \leftrightarrow R_1} \begin{bmatrix} 1 & -1 & 0 & -1 \\ 1 & 2 & 2 & 6 \\ 2 & 1 & 3 & 7 \end{bmatrix} \xrightarrow[R_3 - 2R_1 \to R_3]{R_2 - R_1 \to R_2} \begin{bmatrix} 1 & -1 & 0 & -1 \\ 0 & 3 & 2 & 7 \\ 0 & 3 & 3 & 9 \end{bmatrix} \xrightarrow{R_3 - R_2 \to R_3} \begin{bmatrix} 1 & -1 & 0 & -1 \\ 0 & 3 & 2 & 7 \\ 0 & 0 & 1 & 2 \end{bmatrix}$.

Thus, $z = 2$, $3y + 2(2) = 7 \Leftrightarrow 3y = 3 \Leftrightarrow y = 1$, and $x - (1) = -1 \Leftrightarrow x = 0$, and so the solution is $(0, 1, 2)$.

**30.** $\begin{bmatrix} 1 & -1 & 1 & 2 \\ 1 & 1 & 3 & 6 \\ 0 & 2 & 3 & 5 \end{bmatrix} \xrightarrow{R_2 - R_1 \to R_2} \begin{bmatrix} 1 & -1 & 1 & 2 \\ 0 & 2 & 2 & 4 \\ 0 & 2 & 3 & 5 \end{bmatrix} \xrightarrow{R_3 - R_2 \to R_3} \begin{bmatrix} 1 & -1 & 1 & 2 \\ 0 & 2 & 2 & 4 \\ 0 & 0 & 1 & 1 \end{bmatrix}$. Thus, $z = 1$; $2y + 2(1) = 4$ $\Leftrightarrow y = 1$; and $x - (1) + (1) = 2 \Leftrightarrow x = 2$, and so the solution is $(2, 1, 1)$.

**31.** $\begin{bmatrix} 1 & -2 & 3 & -2 \\ 2 & -1 & 1 & 2 \\ 2 & -7 & 11 & -9 \end{bmatrix} \xrightarrow[R_3 - 2R_1 \to R_3]{R_2 - 2R_1 \to R_2} \begin{bmatrix} 1 & -2 & 3 & -2 \\ 0 & 3 & -5 & 6 \\ 0 & -3 & 5 & -5 \end{bmatrix} \xrightarrow{R_3 + R_2 \to R_3} \begin{bmatrix} 1 & -2 & 3 & -2 \\ 0 & 3 & -5 & 6 \\ 0 & 0 & 0 & 1 \end{bmatrix}$.

The last row corresponds to the equation $0 = 1$, which is always false. Thus, there is no solution.

**32.** $\begin{bmatrix} 1 & -1 & 1 & 2 \\ 1 & 1 & 3 & 6 \\ 3 & -1 & 5 & 10 \end{bmatrix} \xrightarrow[R_3 - 3R_1 \to R_3]{R_2 - R_1 \to R_2} \begin{bmatrix} 1 & -1 & 1 & 2 \\ 0 & 2 & 2 & 4 \\ 0 & 2 & 2 & 4 \end{bmatrix} \xrightarrow{R_3 - R_2 \to R_3} \begin{bmatrix} 1 & -1 & 1 & 2 \\ 0 & 2 & 2 & 4 \\ 0 & 0 & 0 & 0 \end{bmatrix} \xrightarrow{\frac{1}{2}R_2} \begin{bmatrix} 1 & -1 & 1 & 2 \\ 0 & 1 & 1 & 2 \\ 0 & 0 & 0 & 0 \end{bmatrix}$

$\xrightarrow{R_1 + R_2 \to R_1} \begin{bmatrix} 1 & 0 & 2 & 4 \\ 0 & 1 & 1 & 2 \\ 0 & 0 & 0 & 0 \end{bmatrix}$. Let $z = t$. Then $y + t = 2 \Leftrightarrow y = 2 - t$ and $x + 2t = 4 \Leftrightarrow x = 4 - 2t$, and so the solutions are $(4 - 2t, 2 - t, t)$, where $t$ is any real number.

**33.** $\begin{bmatrix} 1 & 1 & 1 & 1 & 0 \\ 1 & -1 & -4 & -1 & -1 \\ 1 & 2 & 0 & 4 & -7 \\ 2 & 2 & 3 & 4 & -3 \end{bmatrix} \xrightarrow[R_4 - 2R_1 \to R_4]{\substack{R_2 - R_1 \to R_2 \\ R_3 - R_1 \to R_3}} \begin{bmatrix} 1 & 1 & 1 & 1 & 0 \\ 0 & -2 & -5 & -2 & -1 \\ 0 & -3 & -1 & 3 & -7 \\ 0 & 0 & 1 & 2 & -3 \end{bmatrix} \xrightarrow{-R_3 + R_2 \to R_3} \begin{bmatrix} 1 & 1 & 1 & 1 & 0 \\ 0 & 1 & -4 & -5 & 6 \\ 0 & -3 & -1 & 3 & -7 \\ 0 & 0 & 1 & 2 & -3 \end{bmatrix}$

$\xrightarrow{R_3 + 3R_2 \to R_3} \begin{bmatrix} 1 & 1 & 1 & 1 & 0 \\ 0 & 1 & -4 & -5 & 6 \\ 0 & 0 & -13 & -12 & 11 \\ 0 & 0 & 1 & 2 & -3 \end{bmatrix} \xrightarrow{R_3 \leftrightarrow R_4} \begin{bmatrix} 1 & 1 & 1 & 1 & 0 \\ 0 & 1 & -4 & -5 & 6 \\ 0 & 0 & 1 & 2 & -3 \\ 0 & 0 & -13 & -12 & 11 \end{bmatrix} \xrightarrow{R_4 + 13R_3 \to R_4} \begin{bmatrix} 1 & 1 & 1 & 1 & 0 \\ 0 & 1 & -4 & -5 & 6 \\ 0 & 0 & 1 & 2 & -3 \\ 0 & 0 & 0 & 14 & -28 \end{bmatrix}$.

Therefore, $14w = -28 \Leftrightarrow w = -2$, $z + 2(-2) = -3 \Leftrightarrow z = 1$, $y - 4(1) - 5(-2) = 6 \Leftrightarrow y = 0$, and $x + 0 + 1 + (-2) = 0 \Leftrightarrow x = 1$. So the solution is $(1, 0, 1, -2)$.

**34.** $\begin{bmatrix} 1 & 0 & 3 & 0 & -1 \\ 0 & 1 & 0 & -4 & 5 \\ 0 & 2 & 1 & 1 & 0 \\ 2 & 1 & 5 & -4 & 4 \end{bmatrix} \xrightarrow{R_4 - 2R_1 \to R_4} \begin{bmatrix} 1 & 0 & 3 & 0 & -1 \\ 0 & 1 & 0 & -4 & 5 \\ 0 & 2 & 1 & 1 & 0 \\ 0 & 1 & -1 & -4 & 6 \end{bmatrix} \xrightarrow[R_4 - R_2 \to R_4]{R_3 - 2R_2 \to R_3} \begin{bmatrix} 1 & 0 & 3 & 0 & -1 \\ 0 & 1 & 0 & -4 & 5 \\ 0 & 0 & 1 & 9 & -10 \\ 0 & 0 & -1 & 0 & 1 \end{bmatrix} \xrightarrow{R_3 \leftrightarrow R_4}$

$\begin{bmatrix} 1 & 0 & 3 & 0 & -1 \\ 0 & 1 & 0 & -4 & 5 \\ 0 & 0 & -1 & 0 & 1 \\ 0 & 0 & 1 & 9 & -10 \end{bmatrix} \xrightarrow{R_4 + 3R_3 \to R_4} \begin{bmatrix} 1 & 0 & 3 & 0 & -1 \\ 0 & 1 & 0 & -4 & 5 \\ 0 & 0 & -1 & 0 & 1 \\ 0 & 0 & 0 & 9 & -9 \end{bmatrix} \xrightarrow[\frac{1}{9}R_4]{-R_3} \begin{bmatrix} 1 & 0 & 3 & 0 & -1 \\ 0 & 1 & 0 & -4 & 5 \\ 0 & 0 & 1 & 0 & -1 \\ 0 & 0 & 0 & 1 & -1 \end{bmatrix} \xrightarrow[R_2 + 4R_4 \to R_2]{R_1 - 3R_3 \to R_1}$

$\begin{bmatrix} 1 & 0 & 0 & 0 & 2 \\ 0 & 1 & 0 & 0 & 1 \\ 0 & 0 & 1 & 0 & -1 \\ 0 & 0 & 0 & 1 & -1 \end{bmatrix}$. Therefore, the solution is $(2, 1, -1, -1)$.

**35.** $\begin{cases} x - 3y + z = 4 \\ 4x - y + 15z = 5 \end{cases} \Leftrightarrow \begin{cases} x - 3y + z = 4 \\ y + z = -1 \end{cases}$ Thus, the system has infinitely many solutions given by $z = t$, $y + t = -1 \Leftrightarrow y = -1 - t$, and $x + 3(1 + t) + t = 4 \Leftrightarrow x = 1 - 4t$. Therefore, the solutions are $(1 - 4t, -1 - t, t)$, where $t$ is any real number.

**36.** $\begin{cases} 2x - 3y + 4z = 3 \\ 4x - 5y + 9z = 13 \\ 2x + 7z = 0 \end{cases} \Leftrightarrow \begin{cases} 2x - 3y + 4z = 3 \\ y + z = 7 \\ 3y + 3z = -3 \end{cases} \Leftrightarrow \begin{cases} 2x - 3y + 4z = 3 \\ y + z = 7 \\ 0 = -24 \end{cases}$

Since this last equation is impossible, the system is inconsistent and has no solution.

**37.** $\begin{cases} -x + 4y + z = 8 \\ 2x - 6y + z = -9 \\ x - 6y - 4z = -15 \end{cases} \Leftrightarrow \begin{cases} -x + 4y + z = 8 \\ 2y + 3z = 7 \\ 6y + 9z = 21 \end{cases} \Leftrightarrow \begin{cases} -x + 4y + z = 8 \\ 2y + 3z = 7 \\ 0 = 0 \end{cases}$

Thus, the system has infinitely many solutions. Letting $z = t$, we find $2y + 3t = 7 \Leftrightarrow y = \frac{7}{2} - \frac{3}{2}t$, and $-x + 4\left(\frac{7}{2} - \frac{3}{2}t\right) + t = 8 \Leftrightarrow x = 6 - 5t$. Therefore, the solutions are $\left(6 - 5t, \frac{7}{2} - \frac{3}{2}t, t\right)$, where $t$ is any real number.

**38.** $\begin{cases} x - z + w = 2 \\ 2x + y - 2w = 12 \\ 3y + z + w = 4 \\ x + y - z = 10 \end{cases} \Leftrightarrow \begin{cases} x - z + w = 2 \\ y + 2z - 4w = 8 \\ 3y + z + w = 4 \\ y - w = 8 \end{cases} \Leftrightarrow \begin{cases} x - z + w = 2 \\ y + 2z - 4w = 8 \\ 5z - 13w = 20 \\ z + 4w = -20 \end{cases} \Leftrightarrow$

$\begin{cases} x - z + w = 2 \\ y + 2z - 4w = 8 \\ 5z - 13w = 20 \\ 33w = -120 \end{cases}$ Therefore, $33w = -120 \Leftrightarrow w = -\frac{40}{11}$, $5z - 13\left(-\frac{40}{11}\right) = 20 \Leftrightarrow z = -\frac{60}{11}$,

$y + 2\left(-\frac{60}{11}\right) - 4\left(-\frac{40}{11}\right) = 8 \Leftrightarrow y = \frac{48}{11}$; and $x - \left(-\frac{60}{11}\right) + \left(-\frac{40}{11}\right) = 2 \Leftrightarrow x = \frac{2}{11}$. Hence, the solution is $\left(\frac{2}{11}, \frac{48}{11}, -\frac{60}{11}, -\frac{40}{11}\right)$.

**39.** $\begin{bmatrix} 1 & -1 & 3 & 2 \\ 2 & 1 & 1 & 2 \\ 3 & 0 & 4 & 4 \end{bmatrix} \xrightarrow[R_3 - 3R_1 \to R_3]{R_2 - 2R_1 \to R_2} \begin{bmatrix} 1 & -1 & 3 & 2 \\ 0 & 3 & -5 & -2 \\ 0 & 3 & -5 & -2 \end{bmatrix} \xrightarrow{R_3 - R_2 \to R_3} \begin{bmatrix} 1 & -1 & 3 & 2 \\ 0 & 3 & -5 & -2 \\ 0 & 0 & 0 & 0 \end{bmatrix} \xrightarrow{\frac{1}{3}R_2}$

$\begin{bmatrix} 1 & -1 & 3 & 2 \\ 0 & 1 & -\frac{5}{3} & -\frac{2}{3} \\ 0 & 0 & 0 & 0 \end{bmatrix} \xrightarrow{R_1 + R_2 \to R_1} \begin{bmatrix} 1 & 0 & \frac{4}{3} & \frac{4}{3} \\ 0 & 1 & -\frac{5}{3} & -\frac{2}{3} \\ 0 & 0 & 0 & 0 \end{bmatrix}$. The system is dependent, so let $z = t$: $y - \frac{5}{3}t = -\frac{2}{3} \Leftrightarrow$

$y = \frac{5}{3}t - \frac{2}{3}$ and $x + \frac{4}{3}t = \frac{4}{3} \Leftrightarrow x = -\frac{4}{3}t + \frac{4}{3}$. So the solution is $\left(-\frac{4}{3}t + \frac{4}{3}, \frac{5}{3}t - \frac{2}{3}, t\right)$, where $t$ is any real number.

**40.** $\begin{bmatrix} 1 & -1 & 0 & 1 \\ 1 & 1 & 2 & 3 \\ 1 & -3 & -2 & -1 \end{bmatrix} \xrightarrow[R_3 - R_1 \to R_3]{R_2 - R_1 \to R_2} \begin{bmatrix} 1 & -1 & 0 & 1 \\ 0 & 2 & 2 & 2 \\ 0 & -2 & -2 & -2 \end{bmatrix} \xrightarrow{R_3 + R_2 \to R_3} \begin{bmatrix} 1 & -1 & 0 & 1 \\ 0 & 2 & 2 & 2 \\ 0 & 0 & 0 & 0 \end{bmatrix} \xrightarrow{\frac{1}{2}R_4}$

$\begin{bmatrix} 1 & -1 & 0 & 1 \\ 0 & 1 & 1 & 1 \\ 0 & 0 & 0 & 0 \end{bmatrix} \xrightarrow{R_1 + R_2 \to R_1} \begin{bmatrix} 1 & 0 & 1 & 2 \\ 0 & 1 & 1 & 1 \\ 0 & 0 & 0 & 0 \end{bmatrix}$. Since the system is dependent, let $z = t$. Then $y + t = 1 \Leftrightarrow$

$y = -t + 1$ and $x + t = 2 \Leftrightarrow x = -t + 2$. So, the solution is $(-t + 2, -t + 1, t)$ where $t$ is any real number.

**41.** $\begin{bmatrix} 1 & -1 & 1 & -1 & 0 \\ 3 & -1 & -1 & -1 & 2 \end{bmatrix} \xrightarrow{R_2 - 3R_1 \to R_2} \begin{bmatrix} 1 & -1 & 1 & -1 & 0 \\ 0 & 2 & -4 & 2 & 2 \end{bmatrix} \xrightarrow{\frac{1}{2}R_4} \begin{bmatrix} 1 & -1 & 1 & -1 & 0 \\ 0 & 1 & -2 & 1 & 1 \end{bmatrix} \xrightarrow{R_1 + R_2 \to R_1}$

$\begin{bmatrix} 1 & 0 & -1 & 0 & 1 \\ 0 & 1 & -2 & 1 & 1 \end{bmatrix}$. Since the system is dependent, Let $z = s$ and $w = t$. Then $y - 2s + t = 1 \Leftrightarrow y = 2s - t + 1$

and $x - s = 1 \Leftrightarrow x = s + 1$. So the solution is $(s + 1, 2s - t + 1, s, t)$, where $s$ and $t$ are any real numbers.

**42.** $\begin{bmatrix} 1 & -1 & 3 \\ 2 & 1 & 6 \\ 1 & -2 & 9 \end{bmatrix} \xrightarrow[\text{R}_3 - \text{R}_1 \to \text{R}_3]{\text{R}_2 - 2\text{R}_1 \to \text{R}_2} \begin{bmatrix} 1 & -1 & 3 \\ 0 & 3 & 0 \\ 0 & -1 & 6 \end{bmatrix} \xrightarrow{\text{R}_2 + 3\text{R}_3 \to \text{R}_2} \begin{bmatrix} 1 & -1 & 3 \\ 0 & 0 & 18 \\ 0 & -1 & 6 \end{bmatrix}$. Since the second row corresponds to the equation $0 = 18$, which is always false, this system has no solution.

**43.** $\begin{bmatrix} 1 & -1 & 1 & 0 \\ 3 & 2 & -1 & 6 \\ 1 & 4 & -3 & 3 \end{bmatrix} \xrightarrow[\text{R}_3 - \text{R}_1 \to \text{R}_3]{\text{R}_2 - 3\text{R}_1 \to \text{R}_2} \begin{bmatrix} 1 & -1 & 1 & 0 \\ 0 & 5 & -4 & 6 \\ 0 & 5 & -4 & 3 \end{bmatrix} \xrightarrow{\text{R}_3 - \text{R}_2 \to \text{R}_3} \begin{bmatrix} 1 & -1 & 1 & 0 \\ 0 & 5 & -4 & 6 \\ 0 & 0 & 0 & 3 \end{bmatrix}$. The last row of this matrix corresponds to the equation $0 = 3$, which is always false. Hence there is no solution.

**44.** $\begin{bmatrix} 1 & 2 & 3 & 2 \\ 2 & -1 & -5 & 1 \\ 4 & 3 & 1 & 6 \end{bmatrix} \xrightarrow[\text{R}_3 - 4\text{R}_1 \to \text{R}_3]{\text{R}_2 - 2\text{R}_1 \to \text{R}_2} \begin{bmatrix} 1 & 2 & 3 & 2 \\ 0 & -5 & -11 & -3 \\ 0 & -5 & -11 & -2 \end{bmatrix} \xrightarrow{\text{R}_3 - \text{R}_2 \to \text{R}_3} \begin{bmatrix} 1 & 2 & 3 & 2 \\ 0 & -5 & -11 & -3 \\ 0 & 0 & 0 & 1 \end{bmatrix}$. Since the third row corresponds to the equation $0 = 1$, which is always false, this system has no solution.

**45.** $\begin{bmatrix} 1 & 1 & -1 & -1 & 2 \\ 1 & -1 & 1 & -1 & 0 \\ 2 & 0 & 0 & 2 & 2 \\ 2 & 4 & -4 & -2 & 6 \end{bmatrix} \xrightarrow{\text{R}_1 \leftrightarrow \frac{1}{2}\text{R}_3} \begin{bmatrix} 1 & 0 & 0 & 1 & 1 \\ 1 & -1 & 1 & -1 & 0 \\ 1 & 1 & -1 & -1 & 2 \\ 2 & 4 & -4 & -2 & 6 \end{bmatrix} \xrightarrow[\text{R}_4 - 2\text{R}_1 \to \text{R}_4]{\substack{\text{R}_2 - \text{R}_1 \to \text{R}_2 \\ \text{R}_3 - \text{R}_1 \to \text{R}_3}} \begin{bmatrix} 1 & 0 & 0 & 1 & 1 \\ 0 & -1 & 1 & -2 & -1 \\ 0 & 1 & -1 & -2 & 1 \\ 0 & 4 & -4 & -4 & 4 \end{bmatrix}$

$\xrightarrow[\text{R}_4 + 4\text{R}_2 \to \text{R}_4]{\text{R}_3 + \text{R}_2 \to \text{R}_3} \begin{bmatrix} 1 & 0 & 0 & 1 & 1 \\ 0 & -1 & 1 & -2 & -1 \\ 0 & 0 & 0 & -4 & 0 \\ 0 & 0 & 0 & -12 & 0 \end{bmatrix} \xrightarrow[-\frac{1}{12}\text{R}_4]{-\frac{1}{4}\text{R}_3} \begin{bmatrix} 1 & 0 & 0 & 1 & 1 \\ 0 & -1 & 1 & -2 & -1 \\ 0 & 0 & 0 & 0 & 0 \\ 0 & 0 & 0 & 1 & 0 \end{bmatrix} \xrightarrow[\text{R}_4 - \text{R}_3 \to \text{R}_4]{\substack{\text{R}_1 - \text{R}_3 \to \text{R}_1 \\ \text{R}_2 + 2\text{R}_3 \to \text{R}_2}} \begin{bmatrix} 1 & 0 & 0 & 0 & 1 \\ 0 & -1 & 1 & 0 & -1 \\ 0 & 0 & 0 & 1 & 0 \\ 0 & 0 & 0 & 0 & 0 \end{bmatrix}$.

This system is dependent. Let $z = t$, so $-y + t = -1 \Leftrightarrow y = t + 1$; $x = 1 \Leftrightarrow x = 1$. So the solution is $(1, t + 1, t, 0)$, where $t$ is any real number.

**46.** $\begin{bmatrix} 1 & -1 & -2 & 3 & 0 \\ 0 & 1 & -1 & 1 & 1 \\ 3 & -2 & -7 & 10 & 2 \end{bmatrix} \xrightarrow{\text{R}_3 - 3\text{R}_1 \to \text{R}_3} \begin{bmatrix} 1 & -1 & -2 & 3 & 0 \\ 0 & 1 & -1 & 1 & 1 \\ 0 & 1 & -1 & 1 & 2 \end{bmatrix} \xrightarrow{\text{R}_3 - \text{R}_2 \to \text{R}_3} \begin{bmatrix} 1 & -1 & -2 & 3 & 0 \\ 0 & 1 & -1 & 1 & 1 \\ 0 & 0 & 0 & 0 & 1 \end{bmatrix}$. Since the third row corresponds to the equation $0 = 1$, which is always false, this system has no solution.

**47.** Let $x$ be the amount in the 6% account and $y$ the amount in the 7% account. The system is $\begin{cases} y = 2x \\ 0.06x + 0.07y = 600 \end{cases}$

Substituting gives $0.06x + 0.07(2x) = 600 \Leftrightarrow 0.2x = 600 \Leftrightarrow x = 3000$, so $y = 2(3000) = 6000$. Hence, the man has \$3,000 invested at 6% and \$6,000 invested at 7%.

**48.** Let $n$ be the number of nickels, $d$ the number of dimes, and $q$ the number of quarter in the piggy bank. We get the following system: $\begin{cases} n + d + q = 50 \\ 5n + 10d + 25z = 560 \\ 10d = 5(5n) \end{cases}$ Since $10d = 25n$, we have $d = \frac{5}{2}n$, so substituting into the first equation we get $n + \frac{5}{2}n + q = 50 \Leftrightarrow \frac{7}{2}n + q = 50 \Leftrightarrow q = 50 - \frac{7}{2}n$. Now substituting this into the second equation we have $5n + 10\left(\frac{5}{2}n\right) + 25\left(50 - \frac{7}{2}n\right) = 560 \Leftrightarrow 5n + 25n + 1250 - \frac{175}{2}n = 560 \Leftrightarrow 1250 - \frac{115}{2}n = 560$ $\Leftrightarrow \frac{115}{2}n = 650 \Leftrightarrow n = 12$. Then $d = \frac{5}{2}(12) = 30$ and $q = 50 - n - d = 50 - 12 - 30 = 8$. Thus the piggy bank contains 12 nickels, 30 dimes, and 8 quarters.

**49.** Let $x$ be the amount invested in Bank A, $y$ the amount invested in Bank B, and $z$ the amount invested in Bank C.

We get the following system: $\begin{cases} x + y + z = 60{,}000 \\ 0.02x + 0.025y + 0.03z = 1575 \\ 2x + 2z = y \end{cases} \Leftrightarrow \begin{cases} x + y + z = 60{,}000 \\ 2x + 2.5y + 3z = 157{,}500 \\ 2x - y + 2z = 0 \end{cases}$

which has matrix representation $\begin{bmatrix} 1 & 1 & 1 & 60{,}000 \\ 2 & 2.5 & 3 & 157{,}500 \\ 2 & -1 & 2 & 0 \end{bmatrix} \xrightarrow[\text{R}_3 - 2\text{R}_1 \to \text{R}_3]{\text{R}_2 - 2\text{R}_1 \to \text{R}_2} \begin{bmatrix} 1 & 1 & 1 & 60{,}000 \\ 0 & 0.5 & 1 & 37{,}500 \\ 0 & -3 & 0 & -120{,}000 \end{bmatrix} \xrightarrow{\text{R}_2 \leftrightarrow -\frac{1}{3}\text{R}_3}$

$\begin{bmatrix} 1 & 1 & 1 & 60{,}000 \\ 0 & 1 & 0 & 40{,}000 \\ 0 & 0.5 & 1 & 37{,}500 \end{bmatrix} \xrightarrow[\text{R}_3 - 0.5\text{R}_2 \to \text{R}_3]{\text{R}_1 - \text{R}_2 \to \text{R}_1} \begin{bmatrix} 1 & 0 & 1 & 20{,}000 \\ 0 & 1 & 0 & 40{,}000 \\ 0 & 0 & 1 & 17{,}500 \end{bmatrix} \xrightarrow{\text{R}_1 - \text{R}_3 \to \text{R}_1} \begin{bmatrix} 1 & 0 & 1 & 2{,}500 \\ 0 & 1 & 0 & 40{,}000 \\ 0 & 0 & 1 & 17{,}500 \end{bmatrix}$. Thus, she invests $2,500 in Bank A, $40,000 in Bank B, and $17,500 in Bank C.

**50.** Let $x$ be the amount of haddock, $y$ the amount of sea bass, and $z$ the amount of red snapper, in pounds. Our system

has the matrix representation $\begin{bmatrix} 1 & 1 & 1 & 560 \\ 1.25 & 0.75 & 2.00 & 575 \\ 1.25 & 0 & 2.00 & 320 \end{bmatrix} \xrightarrow{\text{R}_2 - \text{R}_3 \to \text{R}_2} \begin{bmatrix} 1 & 1 & 1 & 560 \\ 0 & 0.75 & 0 & 255 \\ 1.25 & 0 & 2.00 & 320 \end{bmatrix} \xrightarrow{\text{R}_3 - 1.25\text{R}_1 \to \text{R}_3}$

$\begin{bmatrix} 1 & 1 & 1 & 560 \\ 0 & 0.75 & 0 & 255 \\ 0 & -1.25 & 0.75 & -380 \end{bmatrix} \xrightarrow{\frac{4}{3}\text{R}_2} \begin{bmatrix} 1 & 1 & 1 & 560 \\ 0 & 1 & 0 & 340 \\ 0 & -1.25 & 0.75 & -380 \end{bmatrix} \xrightarrow{\text{R}_3 + 1.25\text{R}_2 \to \text{R}_3} \begin{bmatrix} 1 & 1 & 1 & 560 \\ 0 & 1 & 0 & 340 \\ 0 & 0 & 0.75 & 45 \end{bmatrix} \xrightarrow{\frac{4}{3}\text{R}_3}$

$\begin{bmatrix} 1 & 1 & 1 & 560 \\ 0 & 1 & 0 & 340 \\ 0 & 0 & 1 & 60 \end{bmatrix} \xrightarrow{\text{R}_1 - \text{R}_3 \to \text{R}_1} \begin{bmatrix} 1 & 1 & 0 & 500 \\ 0 & 1 & 0 & 340 \\ 0 & 0 & 1 & 60 \end{bmatrix} \xrightarrow{\text{R}_1 - \text{R}_2 \to \text{R}_1} \begin{bmatrix} 1 & 0 & 0 & 160 \\ 0 & 1 & 0 & 340 \\ 0 & 0 & 1 & 60 \end{bmatrix}$. Thus, he caught 160 lb of haddock, 340 lb of sea bass, and 60 lb of red snapper.

**In Solutions 51–62, the matrices $A$, $B$, $C$, $D$, $E$, $F$, and $G$ are defined as follows:**

$$A = \begin{bmatrix} 2 & 0 & -1 \end{bmatrix} \qquad B = \begin{bmatrix} 1 & 2 & 4 \\ -2 & 1 & 0 \end{bmatrix} \qquad C = \begin{bmatrix} \frac{1}{2} & 3 \\ 2 & \frac{3}{2} \\ -2 & 1 \end{bmatrix}$$

$$D = \begin{bmatrix} 1 & 4 \\ 0 & -1 \\ 2 & 0 \end{bmatrix} \qquad E = \begin{bmatrix} 2 & -1 \\ -\frac{1}{2} & 1 \end{bmatrix} \qquad F = \begin{bmatrix} 4 & 0 & 2 \\ -1 & 1 & 0 \\ 7 & 5 & 0 \end{bmatrix} \qquad G = \begin{bmatrix} 5 \end{bmatrix}$$

**51.** $A + B$ is not defined because the matrix dimensions $1 \times 3$ and $2 \times 3$ are not compatible.

**52.** $C - D = \begin{bmatrix} \frac{1}{2} & 3 \\ 2 & \frac{3}{2} \\ -2 & 1 \end{bmatrix} - \begin{bmatrix} 1 & 4 \\ 0 & -1 \\ 2 & 0 \end{bmatrix} = \begin{bmatrix} -\frac{1}{2} & -1 \\ 2 & \frac{5}{2} \\ -4 & 1 \end{bmatrix}$

**53.** $2C + 3D = 2\begin{bmatrix} \frac{1}{2} & 3 \\ 2 & \frac{3}{2} \\ -2 & 1 \end{bmatrix} + 3\begin{bmatrix} 1 & 4 \\ 0 & -1 \\ 2 & 0 \end{bmatrix} = \begin{bmatrix} 1 & 6 \\ 4 & 3 \\ -4 & 2 \end{bmatrix} + \begin{bmatrix} 3 & 12 \\ 0 & -3 \\ 6 & 0 \end{bmatrix} = \begin{bmatrix} 4 & 18 \\ 4 & 0 \\ 2 & 2 \end{bmatrix}$

**54.** $5B - 2C$ is not defined because the matrix dimensions $2 \times 3$ and $3 \times 2$ are not compatible.

**55.** $GA = \begin{bmatrix} 5 \end{bmatrix}\begin{bmatrix} 2 & 0 & -1 \end{bmatrix} = \begin{bmatrix} 10 & 0 & -5 \end{bmatrix}$

**56.** $AG$ is undefined because the matrix dimensions $1 \times 3$ and $1 \times 1$ are not compatible.

**57.** $BC = \begin{bmatrix} 1 & 2 & 4 \\ -2 & 1 & 0 \end{bmatrix} \begin{bmatrix} \frac{1}{2} & 3 \\ 2 & \frac{3}{2} \\ -2 & 1 \end{bmatrix} = \begin{bmatrix} -\frac{7}{2} & 10 \\ 1 & -\frac{9}{2} \end{bmatrix}$

**58.** $CB = \begin{bmatrix} \frac{1}{2} & 3 \\ 2 & \frac{3}{2} \\ -2 & 1 \end{bmatrix} \begin{bmatrix} 1 & 2 & 4 \\ -2 & 1 & 0 \end{bmatrix} = \begin{bmatrix} -\frac{11}{2} & 4 & 2 \\ -1 & \frac{11}{2} & 8 \\ -4 & -3 & -8 \end{bmatrix}$

**59.** $BF = \begin{bmatrix} 1 & 2 & 4 \\ -2 & 1 & 0 \end{bmatrix} \begin{bmatrix} 4 & 0 & 2 \\ -1 & 1 & 0 \\ 7 & 5 & 0 \end{bmatrix} = \begin{bmatrix} 30 & 22 & 2 \\ -9 & 1 & -4 \end{bmatrix}$

**60.** $FC = \begin{bmatrix} 4 & 0 & 2 \\ -1 & 1 & 0 \\ 7 & 5 & 0 \end{bmatrix} \begin{bmatrix} \frac{1}{2} & 3 \\ 2 & \frac{3}{2} \\ -2 & 1 \end{bmatrix} = \begin{bmatrix} -2 & 14 \\ \frac{3}{2} & -\frac{3}{2} \\ \frac{27}{2} & \frac{57}{2} \end{bmatrix}$

**61.** $(C + D) E = \left( \begin{bmatrix} \frac{1}{2} & 3 \\ 2 & \frac{3}{2} \\ -2 & 1 \end{bmatrix} + \begin{bmatrix} 1 & 4 \\ 0 & -1 \\ 2 & 0 \end{bmatrix} \right) \begin{bmatrix} 2 & -1 \\ -\frac{1}{2} & 1 \end{bmatrix} = \begin{bmatrix} \frac{3}{2} & 7 \\ 2 & \frac{1}{2} \\ 0 & 1 \end{bmatrix} \begin{bmatrix} 2 & -1 \\ -\frac{1}{2} & 1 \end{bmatrix} = \begin{bmatrix} -\frac{1}{2} & \frac{11}{2} \\ \frac{15}{4} & -\frac{3}{2} \\ -\frac{1}{2} & 1 \end{bmatrix}$

**62.** $F (2C - D) = \begin{bmatrix} 4 & 0 & 2 \\ -1 & 1 & 0 \\ 7 & 5 & 0 \end{bmatrix} \left( \begin{bmatrix} 1 & 6 \\ 4 & 3 \\ -4 & 2 \end{bmatrix} - \begin{bmatrix} 1 & 4 \\ 0 & -1 \\ 2 & 0 \end{bmatrix} \right) = \begin{bmatrix} 4 & 0 & 2 \\ -1 & 1 & 0 \\ 7 & 5 & 0 \end{bmatrix} \begin{bmatrix} 0 & 2 \\ 4 & 4 \\ -6 & 2 \end{bmatrix} = \begin{bmatrix} -12 & 12 \\ 4 & 2 \\ 20 & 34 \end{bmatrix}$

**63.** $AB = \begin{bmatrix} 2 & -5 \\ -2 & 6 \end{bmatrix} \begin{bmatrix} 3 & \frac{5}{2} \\ 1 & 1 \end{bmatrix} = \begin{bmatrix} 1 & 0 \\ 0 & 1 \end{bmatrix}$ and $BA = \begin{bmatrix} 3 & \frac{5}{2} \\ 1 & 1 \end{bmatrix} \begin{bmatrix} 2 & -5 \\ -2 & 6 \end{bmatrix} = \begin{bmatrix} 1 & 0 \\ 0 & 1 \end{bmatrix}$

**64.** $AB = \begin{bmatrix} 2 & -1 & 3 \\ 2 & -2 & 1 \\ 0 & 1 & 1 \end{bmatrix} \begin{bmatrix} -\frac{3}{2} & 2 & \frac{5}{2} \\ -1 & 1 & 2 \\ 1 & -1 & -1 \end{bmatrix} = \begin{bmatrix} 1 & 0 & 0 \\ 0 & 1 & 0 \\ 0 & 0 & 1 \end{bmatrix}$ and $BA = \begin{bmatrix} -\frac{3}{2} & 2 & \frac{5}{2} \\ -1 & 1 & 2 \\ 1 & -1 & -1 \end{bmatrix} \begin{bmatrix} 2 & -1 & 3 \\ 2 & -2 & 1 \\ 0 & 1 & 1 \end{bmatrix} = \begin{bmatrix} 1 & 0 & 0 \\ 0 & 1 & 0 \\ 0 & 0 & 1 \end{bmatrix}$.

**In Solutions 65–70,** $A = \begin{bmatrix} 2 & 1 \\ 3 & 2 \end{bmatrix}$, $B = \begin{bmatrix} 1 & -2 \\ -2 & 4 \end{bmatrix}$, **and** $C = \begin{bmatrix} 0 & 1 & 3 \\ -2 & 4 & 0 \end{bmatrix}$.

**65.** $A + 3X = B \Leftrightarrow 3X = B - A \Leftrightarrow X = \frac{1}{3}(B - A)$. Thus,

$$X = \frac{1}{3} \left( \begin{bmatrix} 1 & -2 \\ -2 & 4 \end{bmatrix} - \begin{bmatrix} 2 & 1 \\ 3 & 2 \end{bmatrix} \right) = \frac{1}{3} \begin{bmatrix} -1 & -3 \\ -5 & 2 \end{bmatrix}.$$

**66.** $\frac{1}{2}(X - 2B) = A \Leftrightarrow X - 2B = 2A \Leftrightarrow X = 2A + 2B = 2(A + B)$. Thus,

$$X = 2 \left( \begin{bmatrix} 2 & 1 \\ 3 & 2 \end{bmatrix} + \begin{bmatrix} 1 & -2 \\ -2 & 4 \end{bmatrix} \right) = 2 \begin{bmatrix} 3 & -1 \\ 1 & 6 \end{bmatrix} = \begin{bmatrix} 6 & -2 \\ 2 & 12 \end{bmatrix}.$$

**67.** $2(X - A) = 3B \Leftrightarrow X - A = \frac{3}{2}B \Leftrightarrow X = A + \frac{3}{2}B$. Thus,

$$X = \begin{bmatrix} 2 & 1 \\ 3 & 2 \end{bmatrix} + \frac{3}{2} \begin{bmatrix} 1 & -2 \\ -2 & 4 \end{bmatrix} = \begin{bmatrix} 2 & 1 \\ 3 & 2 \end{bmatrix} + \begin{bmatrix} \frac{3}{2} & -3 \\ -3 & 6 \end{bmatrix} = \begin{bmatrix} \frac{7}{2} & -2 \\ 0 & 8 \end{bmatrix}.$$

**68.** $2X + C = 5A \Leftrightarrow 2X = 5A - C$, but the difference $5A - C$ is not defined because the dimensions of $5A$ and $C$ are not the same.

**69.** $AX = C \Leftrightarrow A^{-1}AX = X = A^{-1}C$. Now

$$A^{-1} = \frac{1}{4 - 3} \begin{bmatrix} 2 & -1 \\ -3 & 2 \end{bmatrix} = \begin{bmatrix} 2 & -1 \\ -3 & 2 \end{bmatrix}. \text{ Thus, } X = A^{-1}C = \begin{bmatrix} 2 & -1 \\ -3 & 2 \end{bmatrix} \begin{bmatrix} 0 & 1 & 3 \\ -2 & 4 & 0 \end{bmatrix} = \begin{bmatrix} 2 & -2 & 6 \\ -4 & 5 & -9 \end{bmatrix}.$$

**70.** $AX = B \Leftrightarrow A^{-1}AX = X = A^{-1}B$. From Exercise 69,

$$A^{-1} = \begin{bmatrix} 2 & -1 \\ -3 & 2 \end{bmatrix}. \text{ Thus, } X = A^{-1}B = \begin{bmatrix} 2 & -1 \\ -3 & 2 \end{bmatrix} \begin{bmatrix} 1 & -2 \\ -2 & 4 \end{bmatrix} = \begin{bmatrix} 4 & -8 \\ -7 & 14 \end{bmatrix}.$$

**71.** $D = \begin{bmatrix} 1 & 4 \\ 2 & 9 \end{bmatrix}$. Then $|D| = 1(9) - 2(4) = 1$, and so $D^{-1} = \begin{bmatrix} 9 & -4 \\ -2 & 1 \end{bmatrix}$.

**72.** $D = \begin{bmatrix} 2 & 2 \\ 1 & -3 \end{bmatrix}$. Then $|D| = 2(-3) - 1(2) = -8$, and so $D^{-1} = -\frac{1}{8}\begin{bmatrix} -3 & -2 \\ -1 & 2 \end{bmatrix} = \begin{bmatrix} \frac{3}{8} & \frac{1}{4} \\ \frac{1}{8} & -\frac{1}{4} \end{bmatrix}$.

**73.** $D = \begin{bmatrix} 4 & -12 \\ -2 & 6 \end{bmatrix}$. Then $|D| = 4(6) - 2(12) = 0$, and so $D$ has no inverse.

**74.** $D = \begin{bmatrix} 2 & 4 & 0 \\ -1 & 1 & 2 \\ 0 & 3 & 2 \end{bmatrix}$. Then $|D| = 2\begin{vmatrix} 1 & 2 \\ 3 & 2 \end{vmatrix} - 4\begin{vmatrix} -1 & 2 \\ 0 & 2 \end{vmatrix} = 2(2-6) - 4(-2) = 0$, and so $D$ has no inverse.

**75.** $D = \begin{bmatrix} 3 & 0 & 1 \\ 2 & -3 & 0 \\ 4 & -2 & 1 \end{bmatrix}$. Then, $|D| = 1\begin{vmatrix} 2 & -3 \\ 4 & -2 \end{vmatrix} + 1\begin{vmatrix} 3 & 0 \\ 2 & -3 \end{vmatrix} = -4 + 12 - 9 = -1$. So $D^{-1}$ exists.

$$\begin{bmatrix} 3 & 0 & 1 & 1 & 0 & 0 \\ 2 & -3 & 0 & 0 & 1 & 0 \\ 4 & -2 & 1 & 0 & 0 & 1 \end{bmatrix} \xrightarrow{R_1 - R_2 \to R_1} \begin{bmatrix} 1 & 3 & 1 & 1 & -1 & 0 \\ 2 & -3 & 0 & 0 & 1 & 0 \\ 4 & -2 & 1 & 0 & 0 & 1 \end{bmatrix} \xrightarrow[R_3 - 4R_1 \to R_3]{R_2 - 2R_1 \to R_2} \begin{bmatrix} 1 & 3 & 1 & 1 & -1 & 0 \\ 0 & -9 & -2 & -2 & 3 & 0 \\ 0 & -14 & -3 & -4 & 4 & 1 \end{bmatrix} \xrightarrow[-2R_3]{-3R_2}$$

$$\begin{bmatrix} 1 & 3 & 1 & 1 & -1 & 0 \\ 0 & 27 & 6 & 6 & -9 & 0 \\ 0 & 28 & 6 & 8 & -8 & -2 \end{bmatrix} \xrightarrow{R_3 - R_2 \to R_3} \begin{bmatrix} 1 & 3 & 1 & 1 & -1 & 0 \\ 0 & 27 & 6 & 6 & -9 & 0 \\ 0 & 1 & 0 & 2 & 1 & -2 \end{bmatrix} \xrightarrow[\frac{1}{3}R_3]{R_3 \leftrightarrow R_2} \begin{bmatrix} 1 & 3 & 1 & 1 & -1 & 0 \\ 0 & 1 & 0 & 2 & 1 & -2 \\ 0 & 9 & 2 & 2 & -3 & 0 \end{bmatrix} \xrightarrow[R_1 - 3R_2 \to R_1]{R_3 - 9R_2 \to R_3}$$

$$\begin{bmatrix} 1 & 0 & 1 & -5 & -4 & 6 \\ 0 & 1 & 0 & 2 & 1 & -2 \\ 0 & 0 & 2 & -16 & -12 & 18 \end{bmatrix} \xrightarrow[R_1 - R_3 \to R_1]{\frac{1}{2}R_3} \begin{bmatrix} 1 & 0 & 0 & 3 & 2 & -3 \\ 0 & 1 & 0 & 2 & 1 & -2 \\ 0 & 0 & 1 & -8 & -6 & 9 \end{bmatrix}. \text{ Thus, } D^{-1} = \begin{bmatrix} 3 & 2 & -3 \\ 2 & 1 & -2 \\ -8 & -6 & 9 \end{bmatrix}.$$

**76.** $D = \begin{bmatrix} 1 & 2 & 3 \\ 2 & 4 & 5 \\ 2 & 5 & 6 \end{bmatrix}$. Then, $|D| = 1\begin{vmatrix} 4 & 5 \\ 5 & 6 \end{vmatrix} - 2\begin{vmatrix} 2 & 5 \\ 2 & 6 \end{vmatrix} + 3\begin{vmatrix} 2 & 4 \\ 2 & 5 \end{vmatrix} = -1 - 4 + 6 = 1$. So $D^{-1}$exists.

$$\begin{bmatrix} 1 & 2 & 3 & 1 & 0 & 0 \\ 2 & 4 & 5 & 0 & 1 & 0 \\ 2 & 5 & 6 & 0 & 0 & 1 \end{bmatrix} \xrightarrow[R_3 - 2R_1 \to R_3]{R_2 - 2R_1 \to R_2} \begin{bmatrix} 1 & 2 & 3 & 1 & 0 & 0 \\ 0 & 0 & -1 & -2 & 1 & 0 \\ 0 & 1 & 0 & -2 & 0 & 1 \end{bmatrix} \xrightarrow{-R_2 \leftrightarrow R_3} \begin{bmatrix} 1 & 2 & 3 & 1 & 0 & 0 \\ 0 & 1 & 0 & -2 & 0 & 1 \\ 0 & 0 & 1 & 2 & -1 & 0 \end{bmatrix}$$

$$\xrightarrow{R_1 - 3R_3 \to R_1} \begin{bmatrix} 1 & 2 & 0 & -5 & 3 & 0 \\ 0 & 1 & 0 & -2 & 0 & 1 \\ 0 & 0 & 1 & 2 & -1 & 0 \end{bmatrix} \xrightarrow{R_1 - 2R_2 \to R_1} \begin{bmatrix} 1 & 0 & 0 & -1 & 3 & -2 \\ 0 & 1 & 0 & -2 & 0 & 1 \\ 0 & 0 & 1 & 2 & -1 & 0 \end{bmatrix}. \text{ Thus, } D^{-1} = \begin{bmatrix} -1 & 3 & -2 \\ -2 & 0 & 1 \\ 2 & -1 & 0 \end{bmatrix}.$$

**77.** $D = \begin{bmatrix} 1 & 0 & 0 & 1 \\ 0 & 2 & 0 & 2 \\ 0 & 0 & 3 & 3 \\ 0 & 0 & 0 & 4 \end{bmatrix}$. Thus, $|D| = \begin{vmatrix} 2 & 0 & 2 \\ 0 & 3 & 3 \\ 0 & 0 & 4 \end{vmatrix} = 2\begin{vmatrix} 3 & 3 \\ 0 & 4 \end{vmatrix} = 24$ and $D^{-1}$exists. $\begin{bmatrix} 1 & 0 & 0 & 1 & 1 & 0 & 0 & 0 \\ 0 & 2 & 0 & 2 & 0 & 1 & 0 & 0 \\ 0 & 0 & 3 & 3 & 0 & 0 & 1 & 0 \\ 0 & 0 & 0 & 4 & 0 & 0 & 0 & 1 \end{bmatrix} \xrightarrow[\frac{1}{4}R_4]{\frac{1}{2}R_2,\ \frac{1}{3}R_3}$

$$\begin{bmatrix} 1 & 0 & 0 & 1 & 1 & 0 & 0 & 0 \\ 0 & 1 & 0 & 1 & 0 & \frac{1}{2} & 0 & 0 \\ 0 & 0 & 1 & 1 & 0 & 0 & \frac{1}{3} & 0 \\ 0 & 0 & 0 & 1 & 0 & 0 & 0 & \frac{1}{4} \end{bmatrix} \xrightarrow[R_3 - R_4 \to R_3]{R_1 - R_4 \to R_1,\ R_2 - R_4 \to R_2} \begin{bmatrix} 1 & 0 & 0 & 0 & 1 & 0 & 0 & -\frac{1}{4} \\ 0 & 1 & 0 & 0 & 0 & \frac{1}{2} & 0 & -\frac{1}{4} \\ 0 & 0 & 1 & 0 & 0 & 0 & \frac{1}{3} & -\frac{1}{4} \\ 0 & 0 & 0 & 1 & 0 & 0 & 0 & \frac{1}{4} \end{bmatrix}$$

. Therefore, $D^{-1} = \begin{bmatrix} 1 & 0 & 0 & -\frac{1}{4} \\ 0 & \frac{1}{2} & 0 & -\frac{1}{4} \\ 0 & 0 & \frac{1}{3} & -\frac{1}{4} \\ 0 & 0 & 0 & \frac{1}{4} \end{bmatrix}$.

**78.** $D = \begin{bmatrix} 1 & 0 & 1 & 0 \\ 0 & 1 & 0 & 1 \\ 1 & 1 & 1 & 2 \\ 1 & 2 & 1 & 2 \end{bmatrix}$. Thus,

$$|D| = \begin{vmatrix} 1 & 0 & 1 \\ 1 & 1 & 2 \\ 2 & 1 & 2 \end{vmatrix} + \begin{vmatrix} 0 & 1 & 1 \\ 1 & 1 & 2 \\ 1 & 2 & 2 \end{vmatrix} = \left(\begin{vmatrix} 1 & 2 \\ 1 & 2 \end{vmatrix} + \begin{vmatrix} 1 & 1 \\ 2 & 1 \end{vmatrix}\right) + \left(-\begin{vmatrix} 1 & 2 \\ 1 & 2 \end{vmatrix} + \begin{vmatrix} 1 & 1 \\ 1 & 2 \end{vmatrix}\right) = (0 - 1) + (0 + 1) = 0.$$

Hence, $D^{-1}$ does not exist.

**79.** $\begin{bmatrix} 12 & -5 \\ 5 & -2 \end{bmatrix}\begin{bmatrix} x \\ y \end{bmatrix} = \begin{bmatrix} 10 \\ 17 \end{bmatrix}$. If we let $A = \begin{bmatrix} 12 & -5 \\ 5 & -2 \end{bmatrix}$, then $A^{-1} = \dfrac{1}{-24+25}\begin{bmatrix} -2 & 5 \\ -5 & 12 \end{bmatrix} = \begin{bmatrix} -2 & 5 \\ -5 & 12 \end{bmatrix}$, and so $\begin{bmatrix} x \\ y \end{bmatrix} = \begin{bmatrix} -2 & 5 \\ -5 & 12 \end{bmatrix}\begin{bmatrix} 10 \\ 17 \end{bmatrix} = \begin{bmatrix} 65 \\ 154 \end{bmatrix}$. Therefore, the solution is $(65, 154)$.

**80.** $\begin{bmatrix} 6 & -5 \\ 8 & -7 \end{bmatrix}\begin{bmatrix} x \\ y \end{bmatrix} = \begin{bmatrix} 1 \\ -1 \end{bmatrix}$. If we let $A = \begin{bmatrix} 6 & -5 \\ 8 & -7 \end{bmatrix}$, then

$A^{-1} = \dfrac{1}{-42+40}\begin{bmatrix} -7 & 5 \\ -8 & 6 \end{bmatrix} = -\frac{1}{2}\begin{bmatrix} -7 & 5 \\ -8 & 6 \end{bmatrix} = \begin{bmatrix} \frac{7}{2} & -\frac{5}{2} \\ 4 & -3 \end{bmatrix}$, and so $\begin{bmatrix} x \\ y \end{bmatrix} = \begin{bmatrix} \frac{7}{2} & -\frac{5}{2} \\ 4 & -3 \end{bmatrix}\begin{bmatrix} 1 \\ -1 \end{bmatrix} = \begin{bmatrix} 6 \\ 7 \end{bmatrix}$.

Therefore, the solution is $(6, 7)$.

**81.** $\begin{bmatrix} 2 & 1 & 5 \\ 1 & 2 & 2 \\ 1 & 0 & 3 \end{bmatrix}\begin{bmatrix} x \\ y \\ z \end{bmatrix} = \begin{bmatrix} \frac{1}{3} \\ \frac{1}{4} \\ \frac{1}{6} \end{bmatrix}$. Let $A = \begin{bmatrix} 2 & 1 & 5 \\ 1 & 2 & 2 \\ 1 & 0 & 3 \end{bmatrix}$. Then $\begin{bmatrix} 2 & 1 & 5 & 1 & 0 & 0 \\ 1 & 2 & 2 & 0 & 1 & 0 \\ 1 & 0 & 3 & 0 & 0 & 1 \end{bmatrix} \xrightarrow{R_1 \leftrightarrow R_2} \begin{bmatrix} 1 & 2 & 2 & 0 & 1 & 0 \\ 2 & 1 & 5 & 1 & 0 & 0 \\ 1 & 0 & 3 & 0 & 0 & 1 \end{bmatrix}$

$$\xrightarrow[R_3 - R_1 \to R_3]{R_2 - 2R_1 \to R_2} \begin{bmatrix} 1 & 2 & 2 & 0 & 1 & 0 \\ 0 & -3 & 1 & 1 & -2 & 0 \\ 0 & -2 & 1 & 0 & -1 & 1 \end{bmatrix} \xrightarrow{R_2 - 2R_3 \to R_2} \begin{bmatrix} 1 & 2 & 2 & 0 & 1 & 0 \\ 0 & 1 & -1 & 1 & 0 & -2 \\ 0 & -2 & 1 & 0 & -1 & 1 \end{bmatrix} \xrightarrow[R_3 \to R_3 + 2R_2]{R_1 - 2R_2 \to R_1}$$

$$\begin{bmatrix} 1 & 0 & 4 & -2 & 1 & 4 \\ 0 & 1 & -1 & 1 & 0 & -2 \\ 0 & 0 & -1 & 2 & -1 & -3 \end{bmatrix} \xrightarrow{-R_3} \begin{bmatrix} 1 & 0 & 4 & -2 & 1 & 4 \\ 0 & 1 & -1 & 1 & 0 & -2 \\ 0 & 0 & 1 & -2 & 1 & 3 \end{bmatrix} \xrightarrow[R_2 + R_3 \to R_2]{R_1 - 4R_3 \to R_1} \begin{bmatrix} 1 & 0 & 0 & 6 & -3 & -8 \\ 0 & 1 & 0 & -1 & 1 & 1 \\ 0 & 0 & 1 & -2 & 1 & 3 \end{bmatrix}$$

. Hence,

$A^{-1} = \begin{bmatrix} 6 & -3 & -8 \\ -1 & 1 & 1 \\ -2 & 1 & 3 \end{bmatrix}$ and $\begin{bmatrix} x \\ y \\ z \end{bmatrix} = \begin{bmatrix} 6 & -3 & -8 \\ -1 & 1 & 1 \\ -2 & 1 & 3 \end{bmatrix}\begin{bmatrix} \frac{1}{3} \\ \frac{1}{4} \\ \frac{1}{6} \end{bmatrix} = \begin{bmatrix} -\frac{1}{12} \\ \frac{1}{12} \\ \frac{1}{12} \end{bmatrix}$, and so the solution is $\left(-\frac{1}{12}, \frac{1}{12}, \frac{1}{12}\right)$.

**82.** $\begin{bmatrix} 2 & 0 & 3 \\ 1 & 1 & 6 \\ 3 & -1 & 1 \end{bmatrix}\begin{bmatrix} x \\ y \\ z \end{bmatrix} = \begin{bmatrix} 5 \\ 0 \\ 5 \end{bmatrix}$. Let $A = \begin{bmatrix} 2 & 0 & 3 \\ 1 & 1 & 6 \\ 3 & -1 & 1 \end{bmatrix}$. Then $\left[\begin{array}{ccc|ccc} 2 & 0 & 3 & 1 & 0 & 0 \\ 1 & 1 & 6 & 0 & 1 & 0 \\ 3 & -1 & 1 & 0 & 0 & 1 \end{array}\right] \xrightarrow{R_1 \leftrightarrow R_2} \left[\begin{array}{ccc|ccc} 1 & 1 & 6 & 0 & 1 & 0 \\ 2 & 0 & 3 & 1 & 0 & 0 \\ 3 & -1 & 1 & 0 & 0 & 1 \end{array}\right]$

$\xrightarrow[R_3 - 3R_1 \to R_3]{R_2 - 2R_1 \to R_2} \left[\begin{array}{ccc|ccc} 1 & 1 & 6 & 0 & 1 & 0 \\ 0 & -2 & -9 & 1 & -2 & 0 \\ 0 & -4 & -17 & 0 & -3 & 1 \end{array}\right] \xrightarrow{R_2 - 2R_3 \to R_2} \left[\begin{array}{ccc|ccc} 1 & 1 & 6 & 0 & 1 & 0 \\ 0 & -2 & -9 & 1 & -2 & 0 \\ 0 & 0 & 1 & -2 & 1 & 1 \end{array}\right] \xrightarrow[R_2 - 6R_3 \to 2R_2]{R_1 + 9R_3 \to R_1}$

$\left[\begin{array}{ccc|ccc} 1 & 1 & 0 & 12 & -5 & -6 \\ 0 & -2 & 0 & -17 & 7 & 9 \\ 0 & 0 & 1 & -2 & 1 & 1 \end{array}\right] \xrightarrow{-\frac{1}{2}R_2} \left[\begin{array}{ccc|ccc} 1 & 1 & 0 & 12 & -5 & -6 \\ 0 & 1 & 0 & \frac{17}{2} & -\frac{7}{2} & -\frac{9}{2} \\ 0 & 0 & 1 & -2 & 1 & 1 \end{array}\right] \xrightarrow{R_1 - R_2 \to R_1} \left[\begin{array}{ccc|ccc} 1 & 0 & 0 & \frac{7}{2} & -\frac{3}{2} & -\frac{3}{2} \\ 0 & 1 & 0 & \frac{17}{2} & -\frac{7}{2} & -\frac{9}{2} \\ 0 & 0 & 1 & -2 & 1 & 1 \end{array}\right]$.

Hence, $A^{-1} = \begin{bmatrix} \frac{7}{2} & -\frac{3}{2} & -\frac{3}{2} \\ \frac{17}{2} & -\frac{7}{2} & -\frac{9}{2} \\ -2 & 1 & 1 \end{bmatrix}$ and $\begin{bmatrix} x \\ y \\ z \end{bmatrix} = \begin{bmatrix} \frac{7}{2} & -\frac{3}{2} & -\frac{3}{2} \\ \frac{17}{2} & -\frac{7}{2} & -\frac{9}{2} \\ -2 & 1 & 1 \end{bmatrix}\begin{bmatrix} 5 \\ 0 \\ 5 \end{bmatrix} = \begin{bmatrix} 10 \\ 20 \\ -5 \end{bmatrix}$, and so the solution is $(10, 20, -5)$.

**83.** $|D| = \begin{vmatrix} 2 & 7 \\ 6 & 16 \end{vmatrix} = 32 - 42 = -10$, $|D_x| = \begin{vmatrix} 13 & 7 \\ 30 & 16 \end{vmatrix} = 208 - 210 = -2$, and $|D_y| = \begin{vmatrix} 2 & 13 \\ 6 & 30 \end{vmatrix} = 60 - 78 = -18$.

Therefore, $x = \frac{-2}{-10} = \frac{1}{5}$ and $y = \frac{-18}{-10} = \frac{9}{5}$, and so the solution is $\left(\frac{1}{5}, \frac{9}{5}\right)$.

**84.** $|D| = \begin{vmatrix} 12 & -11 \\ 7 & 9 \end{vmatrix} = 108 + 77 = 185$, $|D_x| = \begin{vmatrix} 140 & -11 \\ 20 & 9 \end{vmatrix} = 1260 + 220 = 1480$, and

$|D_y| = \begin{vmatrix} 12 & 140 \\ 7 & 20 \end{vmatrix} = 240 - 980 = -740$. Therefore, $x = \frac{1480}{185} = 8$ and $y = \frac{-740}{185} = -4$, and so the solution is $(8, -4)$.

**85.** $|D| = \begin{vmatrix} 2 & -1 & 5 \\ -1 & 7 & 0 \\ 5 & 4 & 3 \end{vmatrix} = 5\begin{vmatrix} -1 & 7 \\ 5 & 4 \end{vmatrix} + 3\begin{vmatrix} 2 & -1 \\ -1 & 7 \end{vmatrix} = -195 + 39 = -156$,

$|D_x| = \begin{vmatrix} 0 & -1 & 5 \\ 9 & 7 & 0 \\ -9 & 4 & 3 \end{vmatrix} = 5\begin{vmatrix} 9 & 7 \\ -9 & 4 \end{vmatrix} + 3\begin{vmatrix} 0 & -1 \\ 9 & 7 \end{vmatrix} = 495 + 27 = 522$,

$|D_y| = \begin{vmatrix} 2 & 0 & 5 \\ -1 & 9 & 0 \\ 5 & -9 & 3 \end{vmatrix} = 5\begin{vmatrix} -1 & 9 \\ 5 & -9 \end{vmatrix} + 3\begin{vmatrix} 2 & 0 \\ -1 & 9 \end{vmatrix} = -180 + 54 = -126$, and

$|D_z| = \begin{vmatrix} 2 & -1 & 0 \\ -1 & 7 & 9 \\ 5 & 4 & -9 \end{vmatrix} = -9\begin{vmatrix} 2 & -1 \\ 5 & 4 \end{vmatrix} - 9\begin{vmatrix} 2 & -1 \\ -1 & 7 \end{vmatrix} = -117 - 117 = -234$.

Therefore, $x = \frac{522}{-156} = -\frac{87}{26}$, $y = \frac{-126}{-156} = \frac{21}{26}$, and $z = \frac{-234}{-156} = \frac{3}{2}$, and so the solution is $\left(-\frac{87}{26}, \frac{21}{26}, \frac{3}{2}\right)$.

**86.** $|D| = \begin{vmatrix} 3 & 4 & -1 \\ 1 & 0 & -4 \\ 2 & 1 & 5 \end{vmatrix} = -4\begin{vmatrix} 1 & -4 \\ 2 & 5 \end{vmatrix} - 1\begin{vmatrix} 3 & -1 \\ 1 & -4 \end{vmatrix} = -52 + 11 = -41,$

$|D_x| = \begin{vmatrix} 10 & 4 & -1 \\ 20 & 0 & -4 \\ 30 & 1 & 5 \end{vmatrix} = -4\begin{vmatrix} 20 & -4 \\ 30 & 5 \end{vmatrix} - 1\begin{vmatrix} 10 & -1 \\ 20 & -4 \end{vmatrix} = -880 + 20 = -860,$

$|D_y| = \begin{vmatrix} 3 & 10 & -1 \\ 1 & 20 & -4 \\ 2 & 30 & 5 \end{vmatrix} = 3\begin{vmatrix} 20 & -4 \\ 30 & 5 \end{vmatrix} - 1\begin{vmatrix} 10 & -1 \\ 30 & 5 \end{vmatrix} + 2\begin{vmatrix} 10 & -1 \\ 20 & -4 \end{vmatrix} = 660 - 80 - 40 = 540$, and

$|D_z| = \begin{vmatrix} 3 & 4 & 10 \\ 1 & 0 & 20 \\ 2 & 1 & 30 \end{vmatrix} = -4\begin{vmatrix} 1 & 20 \\ 2 & 30 \end{vmatrix} - 1\begin{vmatrix} 3 & 10 \\ 1 & 20 \end{vmatrix} = 40 - 50 = -10.$

Therefore, $x = \frac{-860}{-41} = \frac{860}{41}$, $y = -\frac{540}{41}$, and $z = \frac{10}{41}$, and so the solution is $\left(\frac{860}{41}, -\frac{540}{41}, \frac{10}{41}\right)$.

**87.** The area is $\pm\frac{1}{2}\begin{vmatrix} -1 & 3 & 1 \\ 3 & 1 & 1 \\ -2 & -2 & 1 \end{vmatrix} = \pm\frac{1}{2}\left(\begin{vmatrix} 3 & 1 \\ -2 & -2 \end{vmatrix} - \begin{vmatrix} -1 & 3 \\ -2 & -2 \end{vmatrix} + \begin{vmatrix} -1 & 3 \\ 3 & 1 \end{vmatrix}\right) = \pm\frac{1}{2}(-4 - 8 - 10) = 11.$

**88.** The area is $\pm\frac{1}{2}\begin{vmatrix} 5 & -2 & 1 \\ 1 & 5 & 1 \\ -4 & 1 & 1 \end{vmatrix} = \pm\frac{1}{2}\left(\begin{vmatrix} 1 & 5 \\ -4 & 1 \end{vmatrix} - \begin{vmatrix} 5 & -2 \\ -4 & 1 \end{vmatrix} + \begin{vmatrix} 5 & -2 \\ 1 & 5 \end{vmatrix}\right) = \pm\frac{1}{2}(21 + 3 + 27) = \frac{51}{2}.$

**89.** $\frac{3x+1}{x^2-2x-15} = \frac{3x+1}{(x-5)(x+3)} = \frac{A}{x-5} + \frac{B}{x+3}$. Thus, $3x+1 = A(x+3) + B(x-5) = x(A+B) + (3A-5B)$,

and so $\begin{cases} A + B = 3 \\ 3A - 5B = 1 \end{cases} \Leftrightarrow \begin{cases} -3A - 3B = -9 \\ 3A - 5B = 1 \end{cases}$ Adding, we have $-8B = -8 \Leftrightarrow B = 1$, and $A = 2$.

Hence, $\frac{3x+1}{x^2-2x-15} = \frac{2}{x-5} + \frac{1}{x+3}$.

**90.** $\frac{8}{x^3-4x} = \frac{8}{x(x^2-4)} = \frac{8}{x(x+2)(x-2)} = \frac{A}{x} + \frac{B}{x-2} + \frac{C}{x+2}$. Then $8 = A(x^2-4) + Bx(x+2) + Cx(x-2) = x^2(A+B+C) + x(2B-2C) - 4A$. Thus, $-4A = 8 \Leftrightarrow A = -2$, $2B - 2C = 0 \Leftrightarrow B = C$, and $-2 + 2B = 0 \Leftrightarrow B = 1$, so $C = 1$. Therefore, $\frac{8}{x^3-4x} = -\frac{2}{x} + \frac{1}{x-2} + \frac{1}{x+2}$.

**91.** $\frac{2x-4}{x(x-1)^2} = \frac{A}{x} + \frac{B}{x-1} + \frac{C}{(x-1)^2}$. Then $2x - 4 = A(x-1)^2 + Bx(x-1) + Cx = Ax^2 - 2Ax + A + Bx^2 - Bx + Cx = x^2(A+B) + x(-2A-B+C) + A$. So $A = -4$, $-4 + B = 0 \Leftrightarrow B = 4$, and $8 - 4 + C = 2 \Leftrightarrow C = -2$. Therefore, $\frac{2x-4}{x(x-1)^2} = -\frac{4}{x} + \frac{4}{x-1} - \frac{2}{(x-1)^2}$.

**92.** $\dfrac{x+6}{x^3-2x^2+4x-8}=\dfrac{x+6}{(x^2+4)(x-2)}=\dfrac{Ax+B}{x^2+4}+\dfrac{C}{x-2}$. Thus,

$$\begin{aligned} x+6 &= (Ax+B)(x-2)+C(x^2+4)=Ax^2-2Ax+Bx-2B+Cx^2+4C \\ &= x^2(A+C)+x(-2A+B)+(-2B+4C) \end{aligned}$$

and we get the system $\begin{cases} A \quad + C = 0 \\ -2A + B \quad = 1 \\ \quad -2B+4C=6 \end{cases} \Leftrightarrow \begin{cases} A \quad + C = 0 \\ B+2C=1 \\ -2B+4C=6 \end{cases} \Leftrightarrow \begin{cases} A \quad + C = 0 \\ B+2C=1 \\ 8C=8 \end{cases}$

Thus, $8C=8 \Leftrightarrow C=1,\ B+2=1 \Leftrightarrow B=-1$, and $A+1=0 \Leftrightarrow A=-1$. So $\dfrac{x+6}{x^3-2x^2+4x-8}=-\dfrac{x+1}{x^2+4}+\dfrac{1}{x-2}$.

**93.** $\dfrac{2x-1}{x^3+x}=\dfrac{2x-1}{x(x^2+1)}=\dfrac{A}{x}+\dfrac{Bx+C}{x^2+1}$. Then $2x-1=A(x^2+1)+(Bx+C)x=Ax^2+A+Bx^2+Cx=(A+B)x^2+Cx+A$. So $A=-1$, $C=2$, and $A+B=0$ gives us $B=1$. Thus $\dfrac{2x-1}{x^3+x}=-\dfrac{1}{x}+\dfrac{x+2}{x^2+1}$.

**94.** Since $x^4+x^2-2=(x^2-1)(x^2+2)=(x-1)(x+1)(x^2+2)$, we have $\dfrac{5x^2-3x+10}{x^4+x^2-2}=\dfrac{A}{x-1}+\dfrac{B}{x+1}+\dfrac{Cx+D}{x^2+2}$. Thus

$$\begin{aligned} 5x^2-3x+10 &= (x+1)(x^2+2)A+(x-1)(x^2+2)B+(x^2-1)(Cx+D) \\ &= Ax^3+Ax^2+2Ax+2A+Bx^3-Bx^2+2Bx-2B+Cx^3+Dx^2-Cx-D \\ &= (A+B+C)x^3=(A-B+D)x^2+(2A+2B-C)x+(2A-2B-D) \end{aligned}$$

This leads to the system $\begin{cases} A+B+C \quad = 0 & \text{Coefficients of } x^3 \\ A-B \quad +D= 5 & \text{Coefficients of } x^2 \\ 2A+2B-C \quad =-3 & \text{Coefficients of } x \\ 2A-2B \quad -D= 10 & \text{Constant terms} \end{cases} \Leftrightarrow \begin{cases} A+B+C \quad = 0 \\ 2B+C-D=-5 \\ 3C \quad = 3 \\ 4B-C+D=-13 \end{cases}$

$\Leftrightarrow \begin{cases} A+B+C \quad = 0 \\ 2B+C-D=-5 \\ C \quad = 1 \\ 3C-3D= 3 \end{cases}$ Thus $C=1,\ 3(1)-3D=3 \Leftrightarrow D=0,\ 2B+1-0=-5 \Leftrightarrow$

$B=-3$, and $A-3+1=0 \Leftrightarrow A=2$. Thus, $\dfrac{5x^2-3x+10}{x^4+x^2-2}=\dfrac{2}{x-1}-\dfrac{3}{x+1}+\dfrac{x}{x^2+2}$.

**95.** The boundary is a solid curve, so we have the inequality $x+y^2\le 4$. We take the test point $(0,0)$ and verify that it satisfies the inequality: $0+0^2\le 4$.

**96.** The boundary is a solid curve, so we have the inequality $x^2+y^2\ge 8$. We take the test point $(0,3)$ and verify that it satisfies the inequality: $0^2+3^2\ge 8$.

**97.** $3x + y \le 6$

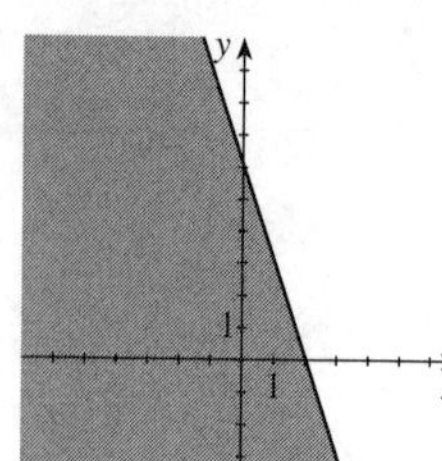

**98.** $y \ge x^2 - 3$

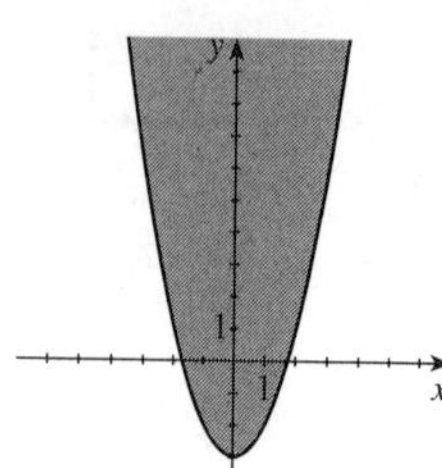

**99.** $x^2 + y^2 > 9$

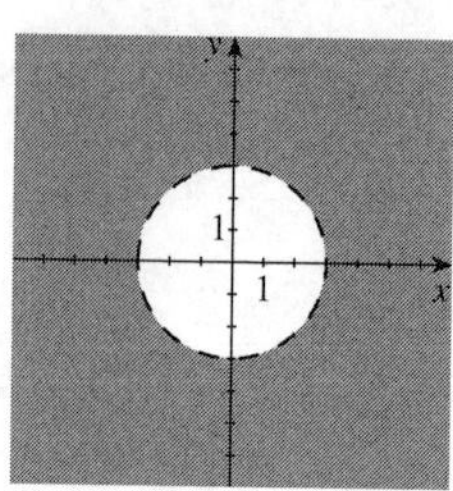

**100.** $x - y^4 < 4$

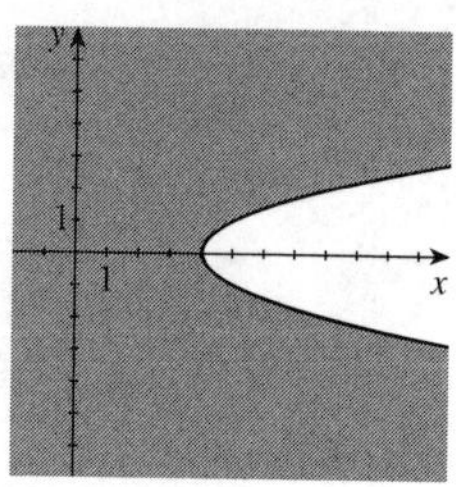

**101.** $\begin{cases} y \ge x^2 - 3x \\ y \le \frac{1}{3}x - 1 \end{cases}$

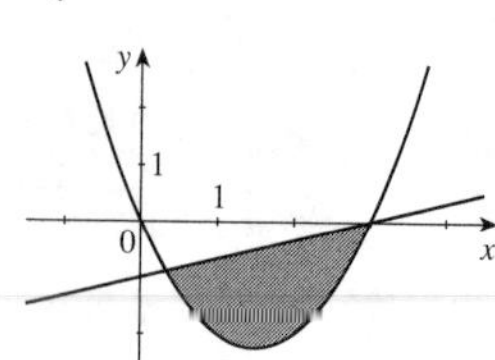

**102.** $\begin{cases} y \ge x - 1 \\ x^2 + y^2 \le 1 \end{cases}$

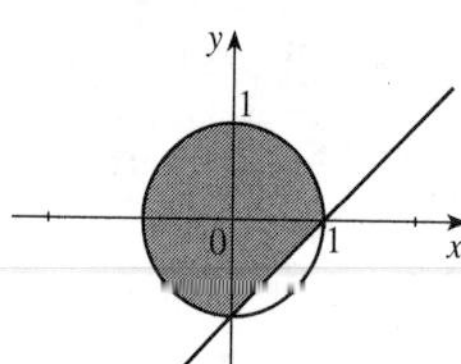

**103.** $\begin{cases} x + y \ge 2 \\ y - x \le 2 \\ x \le 3 \end{cases}$

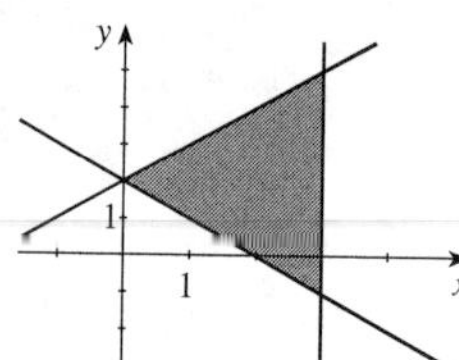

**104.** $\begin{cases} y \ge -2x \\ y \le 2x \\ y \le -\frac{1}{2}x + 2 \end{cases}$

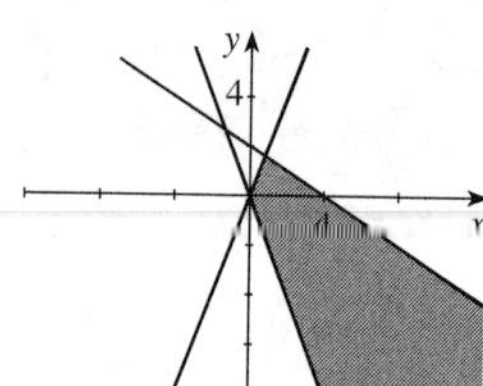

**105.** $\begin{cases} x^2 + y^2 < 9 \\ x + y < 0 \end{cases}$ The vertices occur where $y = -x$. By substitution, $x^2 + x^2 = 9 \quad \Leftrightarrow \quad x = \pm\frac{3}{\sqrt{2}}$, and so $y = \mp\frac{3}{\sqrt{2}}$. Therefore, the vertices are $\left(\frac{3}{\sqrt{2}}, -\frac{3}{\sqrt{2}}\right)$ and $\left(-\frac{3}{\sqrt{2}}, \frac{3}{\sqrt{2}}\right)$ and the solution set is bounded.

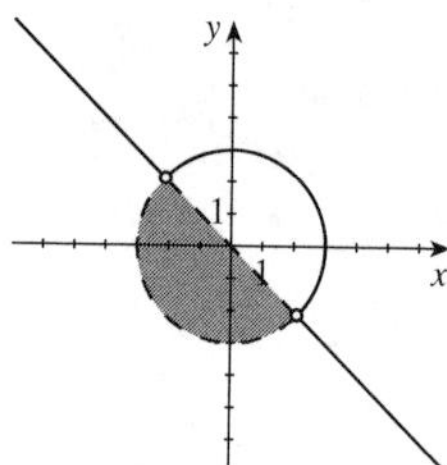

**106.** $\begin{cases} y - x^2 \ge 4 \\ y < 20 \end{cases}$ The vertices occur where $y = x^2 + 4$ and $y = 2$. By substitution, $x^2 + 4 = 20 \quad \Leftrightarrow \quad x^2 = 16 \quad \Leftrightarrow \quad x = \pm 4$ and $y = 20$. Thus, the vertices are $(\pm 4, 20)$ and the solution set is bounded.

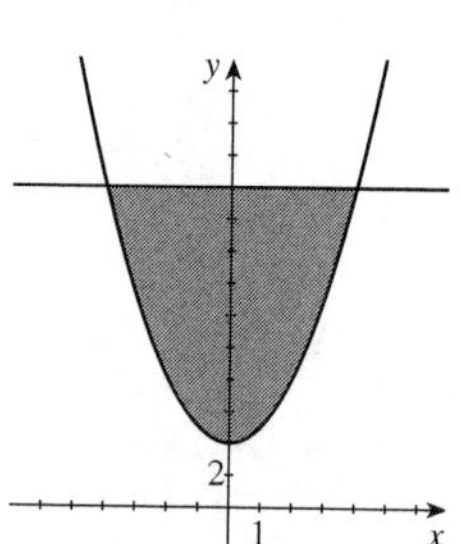

**107.** $\begin{cases} x \ge 0, y \ge 0 \\ x + 2y \le 12 \\ y \le x + 4 \end{cases}$ The intersection points are $(-4, 0)$, $(0, 4)$, $\left(\frac{4}{3}, \frac{16}{3}\right)$, $(0, 6)$, $(0, 0)$, and $(12, 0)$. Since the points $(-4, 0)$ and $(0, 6)$ are not in the solution set, the vertices are $(0, 4)$, $\left(\frac{4}{3}, \frac{16}{3}\right)$, $(12, 0)$, and $(0, 0)$. The solution set is bounded.

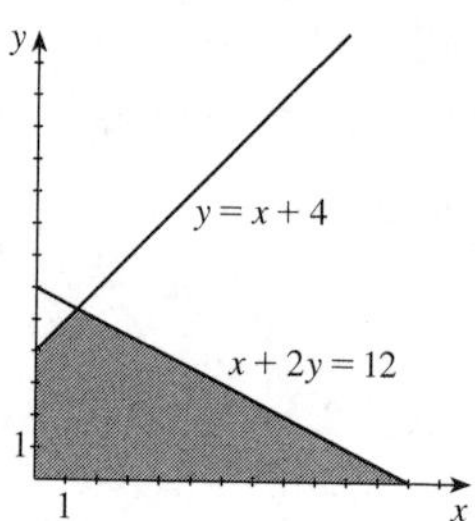

**108.** $\begin{cases} x \geq 4 \\ x + y \geq 24 \\ x \leq 2y + 12 \end{cases}$ The lines $x + y = 24$ and $x = 2y + 12$ intersect at the point $(20, 4)$. The lines $x = 4$ and $x = 2y + 12$ intersect at the point $(-4, 4)$, however, this vertex does not satisfy the other inequality. The lines $x + y = 24$ and $x = 4$ intersect at the point $(4, 20)$. Hence, the vertices are $(4, 20)$ and $(20, 4)$. The solution set is not bounded.

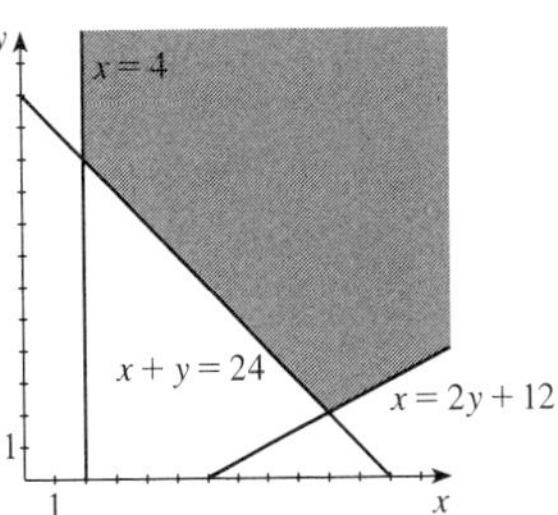


**109.** $\begin{cases} -x + y + z = a \\ x - y + z = b \\ x + y - z = c \end{cases} \Leftrightarrow \begin{cases} -x + y + z = a \\ 2z = a + b \\ 2y = a + c \end{cases}$ Thus, $y = \dfrac{a+c}{2}$, $z = \dfrac{a+b}{2}$, and $-x + \dfrac{a+c}{2} + \dfrac{a+b}{2} = a$ $\Leftrightarrow$ $x = \dfrac{b+c}{2}$. The solution is $\left(\dfrac{b+c}{2}, \dfrac{a+c}{2}, \dfrac{a+b}{2}\right)$.

**110.** $\begin{cases} ax + by + cz = a - b + c \\ bx + by + cz = c \\ cx + cy + cz = c \end{cases}$ $a \neq b \neq c \neq 0$ $\Leftrightarrow$ $\begin{cases} ax + by + cz = a - b + c \\ bx + by + cz = c \\ x + y + z = 1 \end{cases}$ Subtracting the second equation from the first gives $(a - b)x = a - b \Leftrightarrow x = 1$. Subtracting the third equation from the second, $(b - c)x + (b - c)y = 0 \Leftrightarrow y = -x = -1$. So $1 - 1 + z = 1 \Leftrightarrow z = 1$, and the solution is $(1, -1, 1)$.

**111.** Solving the second equation for $y$, we have $y = kx$. Substituting for $y$in the first equation gives us $x + kx = 12 \Leftrightarrow (1 + k)x = 12 \Leftrightarrow x = \dfrac{12}{k+1}$. Substituting for $y$ in the third equation gives us $kx - x = 2k \Leftrightarrow (k - 1)x = 2k \Leftrightarrow x = \dfrac{2k}{k-1}$. These points of intersection are the same when the $x$-values are equal. Thus, $\dfrac{12}{k+1} = \dfrac{2k}{k-1} \Leftrightarrow 12(k - 1) = 2k(k + 1) \Leftrightarrow 12k - 12 = 2k^2 + 2k \Leftrightarrow$ $0 = 2k^2 - 10k + 12 = 2(k^2 - 5k + 6) = 2(k - 3)(k + 2)$. Hence, $k = 2$ or $k = 3$.

**112.** The system will have infinitely many solutions when the system has solutions other than $(0, 0, 0)$. So we solve the system with $x \neq 0$, $y \neq 0$, and $z \neq 0$:

$$\begin{cases} kx + y + z = 0 \\ x + 2y + kz = 0 \\ -x + 3z = 0 \end{cases} \Leftrightarrow \begin{cases} kx + y + z = 0 \\ 2y + (k+3)z = 0 \\ y + (3k+1)z = 0 \end{cases} \Leftrightarrow \begin{cases} kx + y + z = 0 \\ + 2y + (k+3)z = 0 \\ [(k+3) - 2(3k+1)]z = 0 \end{cases}$$

Since $z \neq 0$, we must have $(k + 3) - 2(3k + 1) = -5k + 1 = 0 \Leftrightarrow k = \frac{1}{5}$.

# Chapter 10 Test

**1. (a)** The system is linear.

**(b)** $\begin{cases} x + 3y = 7 \\ 5x + 2y = -4 \end{cases}$ Multiplying the first equation by $-5$ and then adding gives $-13y = -39 \quad \Leftrightarrow \quad y = 3$. So $x + 3(3) = 7 \quad \Leftrightarrow \quad x = -2$. Thus, the solution is $(-2, 3)$.

**2. (a)** The system is nonlinear.

**(b)** $\begin{cases} 6x + y^2 = 10 \\ 3x - y = 5 \end{cases} \quad \Leftrightarrow \quad \begin{cases} 6x + y^2 = 10 \\ y^2 + 2y = 0 \end{cases}$ Thus $y^2 + 2y = y(y+2) = 0$, so either $y = 0$ or $y = -2$. If $y = 0$, then $3x = 5 \quad \Leftrightarrow \quad x = \frac{5}{3}$ and if $y = -2$, then $3x - (-2) = 5 \quad \Leftrightarrow \quad 3x = 3 \quad \Leftrightarrow \quad x = 1$. Thus the solutions are $\left(\frac{5}{3}, 0\right)$ and $(1, -2)$.

**3.** $\begin{cases} x - 2y = 1 \\ y = x^3 - 2x^2 \end{cases}$

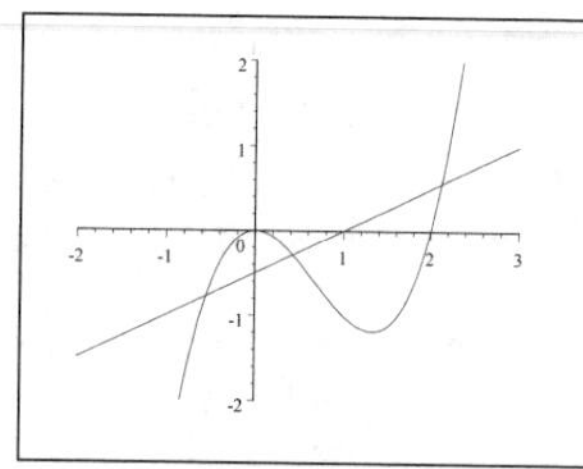


The solutions are approximately $(-0.55, -0.78)$, $(0.43, -0.29)$, and $(2.12, 0.56)$.

**4.** Let $w$ be the speed of the wind and $a$ the speed of the airplane in still air, in kilometers per hour. Then the speed of the of the plane flying against the wind is $a - w$ and the speed of the plane flying with the wind is $a + w$. Using distance = rate × time, we get the system $\begin{cases} 600 = 2.5(a - w) \\ 300 = \frac{50}{60}(a + w) \end{cases} \quad \Leftrightarrow \quad \begin{cases} 240 = a - w \\ 360 = a + w \end{cases}$ Adding the two equations, we get $600 = 2a \quad \Leftrightarrow \quad a = 300$. So $360 = 300 + w \quad \Leftrightarrow \quad w = 60$.Thus the speed of the airplane in still air is 300 km/h and the speed of the wind is 60 km/h.

**5. (a)** This matrix is in row-echelon form, but not in reduced row-echelon form, since the leading 1 in the second row does not have a 0 above it.

**(b)** This matrix is in reduced row-echelon form.

**(c)** This matrix is in neither row-echelon form nor reduced row-echelon form.

**6. (a)** $\begin{cases} x - y + 2z = 0 \\ 2x - 4y + 5z = -5 \\ 2y - 3z = 5 \end{cases}$ has the matrix representation $\begin{bmatrix} 1 & -1 & 2 & 0 \\ 2 & -4 & 5 & -5 \\ 0 & 2 & -3 & 5 \end{bmatrix} \xrightarrow{R_2 - R_1 \to R_2} \begin{bmatrix} 1 & -1 & 2 & 0 \\ 0 & -2 & 1 & -5 \\ 0 & 2 & -3 & 5 \end{bmatrix}$

$\xrightarrow{R_3 + R_2 \to R_3} \begin{bmatrix} 1 & -1 & 2 & 0 \\ 0 & -2 & 1 & -5 \\ 0 & 0 & -2 & 0 \end{bmatrix} \xrightarrow{-\frac{1}{2}R_3} \begin{bmatrix} 1 & -1 & 2 & 0 \\ 0 & -2 & 1 & -5 \\ 0 & 0 & 1 & 0 \end{bmatrix}$. Thus $z = 0$, $-2y + 0 = -5 \quad \Leftrightarrow \quad y = \frac{5}{2}$, and $x - \frac{5}{2} + 2(0) = 0 \quad \Leftrightarrow \quad x = \frac{5}{2}$. Thus, the solution is $\left(\frac{5}{2}, \frac{5}{2}, 0\right)$.

**(b)** $\begin{cases} 2x - 3y + z = 3 \\ x + 2y + 2z = -1 \\ 4x + y + 5z = 4 \end{cases}$ has the matrix representation $\begin{bmatrix} 2 & -3 & 1 & 3 \\ 1 & 2 & 2 & -1 \\ 4 & 1 & 5 & 4 \end{bmatrix} \xrightarrow{R_1 \leftrightarrow R_2} \begin{bmatrix} 1 & 2 & 2 & -1 \\ 2 & -3 & 1 & 3 \\ 4 & 1 & 5 & 4 \end{bmatrix}$

$$\xrightarrow[R_3 - 4R_1 \to R_3]{R_2 - 2R_1 \to R_2} \begin{bmatrix} 1 & 2 & 2 & -1 \\ 0 & -7 & -3 & 5 \\ 0 & -7 & -3 & 8 \end{bmatrix} \xrightarrow{R_3 - R_2 \to R_3} \begin{bmatrix} 1 & 2 & 2 & -1 \\ 0 & -7 & -3 & 5 \\ 0 & 0 & 0 & 3 \end{bmatrix}.$$

Since the last row corresponds to the equation $0 = 3$, this system has no solution.

**7.** $\begin{cases} x + 3y - z = 0 \\ 3x + 4y - 2z = -1 \\ -x + 2y = 1 \end{cases}$ has the matrix representation $\begin{bmatrix} 1 & 3 & -1 & 0 \\ 3 & 4 & -2 & -1 \\ -1 & 2 & 0 & 1 \end{bmatrix} \xrightarrow[R_3 + R_1 \to R_3]{R_2 - 3R_1 \to R_2} \begin{bmatrix} 1 & 3 & -1 & 0 \\ 0 & -5 & 1 & -1 \\ 0 & 5 & -1 & 1 \end{bmatrix}$

$$\xrightarrow{R_3 - R_2 \to R_3} \begin{bmatrix} 1 & 3 & -1 & 0 \\ 0 & -5 & 1 & -1 \\ 0 & 0 & 0 & 0 \end{bmatrix} \xrightarrow{-\frac{1}{5}R_2} \begin{bmatrix} 1 & 3 & -1 & 0 \\ 0 & 1 & -\frac{1}{5} & \frac{1}{5} \\ 0 & 0 & 0 & 0 \end{bmatrix} \xrightarrow{R_1 - 3R_2 \to R_1} \begin{bmatrix} 1 & 0 & -\frac{2}{5} & -\frac{3}{5} \\ 0 & 1 & -\frac{1}{5} & \frac{1}{5} \\ 0 & 0 & 0 & 0 \end{bmatrix}.$$

Since this system is dependent, let $z = t$. Then $y - \frac{1}{5}t = \frac{1}{5} \Leftrightarrow y = \frac{1}{5}t + \frac{1}{5}$ and $x - \frac{2}{5}t = -\frac{3}{5} \Leftrightarrow x = \frac{2}{5}t - \frac{3}{5}$. Thus, the solution is $\left(\frac{2}{5}t - \frac{3}{5}, \frac{1}{5}t + \frac{1}{5}, t\right)$.

**8.** Let $x$, $y$, and $z$ represent the price in dollars for coffee, juice, and donuts respectively. Then the system of equations is

$$\begin{cases} 2x + y + 2z = 6.25 & \text{Anne} \\ x + 3z = 3.75 & \text{Barry} \\ 3x + y + 4z = 9.25 & \text{Cathy} \end{cases} \Leftrightarrow \begin{cases} 2x + y + 2z = 6.25 \\ y - 4z = -1.25 \\ y - 5z = -2.00 \end{cases} \Leftrightarrow \begin{cases} 2x + y + 2z = 6.25 \\ y - 4z = -1.25 \\ z = 0.75 \end{cases}$$

Thus, $z = 0.75$, $y - 4(0.75) = -1.25 \Leftrightarrow y = 1.75$, and $2x + 1.75 + 2(0.75) = 6.25 \Leftrightarrow x = 1.5$. Thus coffee costs \$1.50, juice costs \$1.75, and donuts cost \$0.75.

**9.** $A = \begin{bmatrix} 2 & 3 \\ 2 & 4 \end{bmatrix}$, $B = \begin{bmatrix} 2 & 4 \\ -1 & 1 \\ 3 & 0 \end{bmatrix}$, and $C = \begin{bmatrix} 1 & 0 & 4 \\ -1 & 1 & 2 \\ 0 & 1 & 3 \end{bmatrix}$.

**(a)** $A + B$ is undefined because $A$ is $2 \times 2$ and $B$ is $3 \times 2$, so they have incompatible dimensions.

**(b)** $AB$ is undefined because $A$ is $2 \times 2$ and $B$ is $3 \times 2$, so they have incompatible dimensions.

**(c)** $BA - 3B = \begin{bmatrix} 2 & 4 \\ -1 & 1 \\ 3 & 0 \end{bmatrix} \begin{bmatrix} 2 & 3 \\ 2 & 4 \end{bmatrix} - 3 \begin{bmatrix} 2 & 4 \\ -1 & 1 \\ 3 & 0 \end{bmatrix} = \begin{bmatrix} 12 & 22 \\ 0 & 1 \\ 6 & 9 \end{bmatrix} - \begin{bmatrix} 6 & 12 \\ -3 & 3 \\ 9 & 0 \end{bmatrix} = \begin{bmatrix} 6 & 10 \\ 3 & -2 \\ -3 & 9 \end{bmatrix}$

**(d)** $CBA = \begin{bmatrix} 1 & 0 & 4 \\ -1 & 1 & 2 \\ 0 & 1 & 3 \end{bmatrix} \begin{bmatrix} 2 & 4 \\ -1 & 1 \\ 3 & 0 \end{bmatrix} \begin{bmatrix} 2 & 3 \\ 2 & 4 \end{bmatrix} = \begin{bmatrix} 14 & 4 \\ 3 & -3 \\ 8 & 1 \end{bmatrix} \begin{bmatrix} 2 & 3 \\ 2 & 4 \end{bmatrix} = \begin{bmatrix} 36 & 58 \\ 0 & -3 \\ 18 & 28 \end{bmatrix}$

**(e)** $A = \begin{bmatrix} 2 & 3 \\ 2 & 4 \end{bmatrix} \Leftrightarrow A^{-1} = \dfrac{1}{8 - 6} \begin{bmatrix} 4 & -3 \\ -2 & 2 \end{bmatrix} = \begin{bmatrix} 2 & -\frac{3}{2} \\ -1 & 1 \end{bmatrix}$

**(f)** $B^{-1}$ does not exist because $B$ is not a square matrix. **(g)** $\det(B)$ is not defined because $B$ is not a square matrix.

**(h)** $\det(C) = \begin{vmatrix} 1 & 0 & 4 \\ -1 & 1 & 2 \\ 0 & 1 & 3 \end{vmatrix} = 1 \begin{vmatrix} 1 & 2 \\ 1 & 3 \end{vmatrix} + 4 \begin{vmatrix} -1 & 1 \\ 0 & 1 \end{vmatrix} = 1 - 4 = -3$

**10. (a)** The system $\begin{cases} 4x - 3y = 10 \\ 3x - 2y = 30 \end{cases}$ is equivalent to the matrix equation $\begin{bmatrix} 4 & -3 \\ 3 & -2 \end{bmatrix} \begin{bmatrix} x \\ y \end{bmatrix} = \begin{bmatrix} 10 \\ 30 \end{bmatrix}$.

**(b)** We have $|D| = \begin{vmatrix} 4 & -3 \\ 3 & -2 \end{vmatrix} = 4(-2) - 3(-3) = 1$. So $D^{-1} = \begin{bmatrix} -2 & 3 \\ -3 & 4 \end{bmatrix}$ and $\begin{bmatrix} x \\ y \end{bmatrix} = \begin{bmatrix} -2 & 3 \\ -3 & 4 \end{bmatrix} \begin{bmatrix} 10 \\ 30 \end{bmatrix} = \begin{bmatrix} 70 \\ 90 \end{bmatrix}$.

Therefore, $x = 70$ and $y = 90$.

**11.** $|A| = \begin{vmatrix} 1 & 4 & 1 \\ 0 & 2 & 0 \\ 1 & 0 & 1 \end{vmatrix} = 2 \begin{vmatrix} 1 & 1 \\ 1 & 1 \end{vmatrix} = 0$, $|B| = \begin{vmatrix} 1 & 4 & 0 \\ 0 & 2 & 0 \\ -3 & 0 & 1 \end{vmatrix} = 2 \begin{vmatrix} 1 & 0 \\ -3 & 1 \end{vmatrix} = 2$. Since $|A| = 0$, $A$ does not have an inverse, and since $|B| \neq 0$, $B$ does have an inverse. $\left[\begin{array}{ccc|ccc} 1 & 4 & 0 & 1 & 0 & 0 \\ 0 & 2 & 0 & 0 & 1 & 0 \\ -3 & 0 & 1 & 0 & 0 & 1 \end{array}\right] \xrightarrow{R_3 + 3R_1 \to R_3} \left[\begin{array}{ccc|ccc} 1 & 4 & 0 & 1 & 0 & 0 \\ 0 & 2 & 0 & 0 & 1 & 0 \\ 0 & 12 & 1 & 3 & 0 & 1 \end{array}\right] \xrightarrow[R_3 - 6R_2 \to R_3]{R_1 - 2R_2 \to R_1}$

$\left[\begin{array}{ccc|ccc} 1 & 0 & 0 & 1 & -2 & 0 \\ 0 & 2 & 0 & 0 & 1 & 0 \\ 0 & 0 & 1 & 3 & -6 & 1 \end{array}\right] \xrightarrow{\frac{1}{2}R_2} \left[\begin{array}{ccc|ccc} 1 & 0 & 0 & 1 & -2 & 0 \\ 0 & 1 & 0 & 0 & \frac{1}{2} & 0 \\ 0 & 0 & 1 & 3 & -6 & 1 \end{array}\right]$. Therefore, $B^{-1} = \begin{bmatrix} 1 & -2 & 0 \\ 0 & \frac{1}{2} & 0 \\ 3 & -6 & 1 \end{bmatrix}$.

**12.** $\begin{cases} 2x \phantom{{}-y} - z = 14 \\ 3x - y + 5z = 0 \\ 4x + 2y + 3z = -2 \end{cases}$ Then $|D| = \begin{vmatrix} 2 & 0 & -1 \\ 3 & -1 & 5 \\ 4 & 2 & 3 \end{vmatrix} = 2 \begin{vmatrix} -1 & 5 \\ 2 & 3 \end{vmatrix} - 1 \begin{vmatrix} 3 & -1 \\ 4 & 2 \end{vmatrix} = -26 - 10 = -36$,

$|D_x| = \begin{vmatrix} 14 & 0 & -1 \\ 0 & -1 & 5 \\ -2 & 2 & 3 \end{vmatrix} = 14 \begin{vmatrix} -1 & 5 \\ 2 & 3 \end{vmatrix} - 1 \begin{vmatrix} 0 & -1 \\ 4 & 2 \end{vmatrix} = -182 + 2 = -180$,

$|D_y| = \begin{vmatrix} 2 & 14 & -1 \\ 3 & 0 & 5 \\ 4 & -2 & 3 \end{vmatrix} = -3 \begin{vmatrix} 14 & -1 \\ -2 & 3 \end{vmatrix} - 5 \begin{vmatrix} 2 & 14 \\ 4 & -2 \end{vmatrix} = -120 + 300 = 180$, and

$|D_z| = \begin{vmatrix} 2 & 0 & 14 \\ 3 & -1 & 0 \\ 4 & 2 & -2 \end{vmatrix} = 2 \begin{vmatrix} -1 & 0 \\ 2 & -2 \end{vmatrix} + 14 \begin{vmatrix} 3 & -1 \\ 2 & 2 \end{vmatrix} = 4 + 140 = 144$.

Therefore, $x = \frac{-180}{-36} = 5$, $y = \frac{180}{-36} = -5$, $z = \frac{144}{-36} = -4$, and so the solution is $(5, -5, -4)$.

**13. (a)** $\dfrac{4x-1}{(x-1)^2(x+2)} = \dfrac{A}{x-1} + \dfrac{B}{(x-1)^2} + \dfrac{C}{x+2}$. Thus,

$$\begin{aligned} 4x-1 &= A(x-1)(x+2) + B(x+2) + C(x-1)^2 = A(x^2+x-2) + B(x+2) + C(x^2-2x+1) \\ &= (A+C)x^2 + (A+B-2C)x + (-2A+2B+C) \end{aligned}$$

which leads to the system of equations

$$\begin{cases} A \quad\quad + C = 0 \\ A + B - 2C = 4 \\ -2A + 2B + C = -1 \end{cases} \Leftrightarrow \begin{cases} A \quad + C = 0 \\ B - 3C = 4 \\ 2B + 3C = -1 \end{cases} \Leftrightarrow \begin{cases} A \quad + C = 0 \\ B - 3C = 4 \\ 9C = -9 \end{cases}$$

Therefore, $9C = -9 \quad \Leftrightarrow \quad C = -1$, $B - 3(-1) = 4 \quad \Leftrightarrow \quad B = 1$, and $A + (-1) = 0 \quad \Leftrightarrow \quad A = 1$.

Therefore, $\dfrac{4x-1}{(x-1)^2(x+2)} = \dfrac{1}{x-1} + \dfrac{1}{(x-1)^2} - \dfrac{1}{x+2}$.

**(b)** $\dfrac{2x-3}{x^3+3x} = \dfrac{2x-3}{x(x^2+3)} = \dfrac{A}{x} + \dfrac{Bx+C}{x^2+3}$. Then

$2x - 3 = A(x^2+3) + (Bx+C)x = Ax^2 + 3A + Bx^2 + Cx = (A+B)x^2 + Cx + 3A$.

So $3A = -3 \quad \Leftrightarrow \quad A = -1$, $C = 2$ and $A + B = 0$ gives us $B = 1$. Thus $\dfrac{2x-3}{x^3+x} = -\dfrac{1}{x} + \dfrac{x+2}{x^2+3}$.

**14. (a)** $\begin{cases} 2x + y \le 8 \\ x - y \ge -2 \\ x + 2y \ge 4 \end{cases}$ From the graph, the points $(4, 0)$ and $(0, 2)$ are vertices. The third vertex occurs where the lines $2x + y = 8$ and $x - y = -2$ intersect. Adding these two equations gives $3x = 6 \quad \Leftrightarrow \quad x = 2$, and so $y = 8 - 2(2) = 4$. Thus, the third vertex is $(2, 4)$.

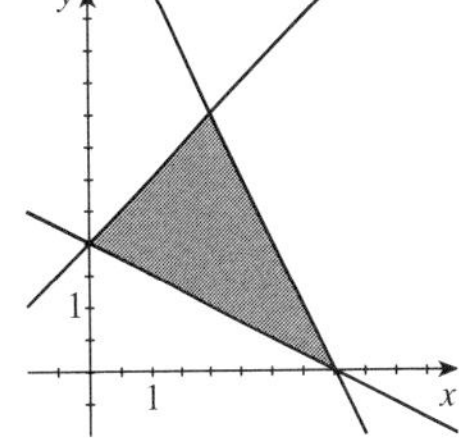


**(b)** $\begin{cases} x^2 - 5 + y \le 0 \\ y \le 5 + 2x \end{cases}$ Substituting $y = 5 + 2x$ into the first equation gives $x^2 - 5 + (5+2x) = 0 \quad \Leftrightarrow \quad x^2 + 2x = 0 \Leftrightarrow x(x+2) = 0 \quad \Leftrightarrow \quad x = 0$ or $x = -2$. If $x = 0$, then $y = 5 + 2(0) = 5$, and if $x = -2$, then $y = 5 + 2(-2) = 1$. Thus, the vertices are $(0, 5)$ and $(-2, 1)$.

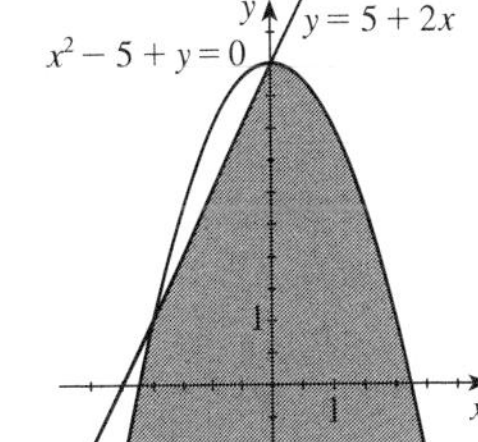

# Focus on Modeling: Linear Programming

**1.**

| Vertex | $M = 200 - x - y$ |
|---|---|
| $(0,2)$ | $200 - (0) - (2) = 198$ |
| $(0,5)$ | $200 - (0) - (5) = 195$ |
| $(4,0)$ | $200 - (4) - (0) = 196$ |

Thus, the maximum value is 198 and the minimum value is 195.

**2.**

| Vertex | $N = \frac{1}{2}x + \frac{1}{4}y + 40$ |
|---|---|
| $(1,0)$ | $\frac{1}{2}(1) + \frac{1}{4}(0) + 40 = 40.5$ |
| $\left(\frac{1}{2},\frac{1}{2}\right)$ | $\frac{1}{2}\left(\frac{1}{2}\right) + \frac{1}{4}\left(\frac{1}{2}\right) + 40 = 40.375$ |
| $(2,2)$ | $\frac{1}{2}(2) + \frac{1}{4}(2) + 40 = 41.5$ |
| $(4,0)$ | $\frac{1}{2}(4) + \frac{1}{4}(0) + 40 = 42$ |

Thus, the maximum value is 42 and the minimum value is 40.375.

**3.** $\begin{cases} x \geq 0, y \geq 0 \\ 2x + y \leq 10 \\ 2x + 4y \leq 28 \end{cases}$ The objective function is $P = 140 - x + 3y$. From the graph, the vertices are $(0,0)$, $(5,0)$, $(2,6)$, and $(0,7)$.

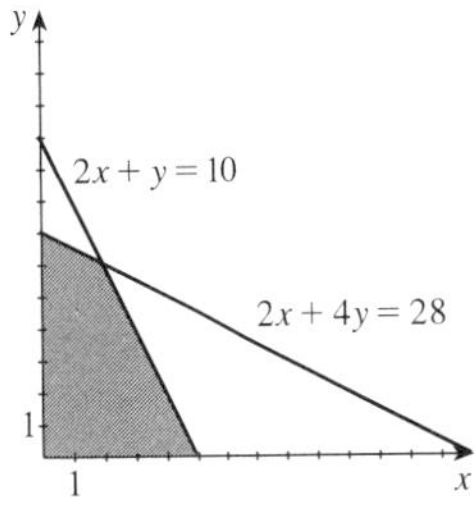


| Vertex | $P = 140 - x + 3y$ |
|---|---|
| $(0,0)$ | $140 - (0) + 3(0) = 140$ |
| $(5,0)$ | $140 - (5) + 3(0) = 135$ |
| $(2,6)$ | $140 - (2) + 3(6) = 156$ |
| $(0,7)$ | $140 - (0) + 3(7) = 161$ |

Thus the maximum value is 161, and the minimum value is 135.

**4.** $\begin{cases} x \geq 0, y \geq 0 \\ x \leq 10, y \leq 20 \\ x + y \geq 5 \\ x + 2y \leq 18 \end{cases}$ The objective function is $Q = 70x + 82y$. From the graph, the vertices are at $(0,9)$, $(0,5)$, $(5,0)$, $(10,0)$, and $(10,4)$. Note that the restriction $y \leq 20$ is irrelevant, superseded by $x + 2y \leq 18$ and $x \geq 0$.

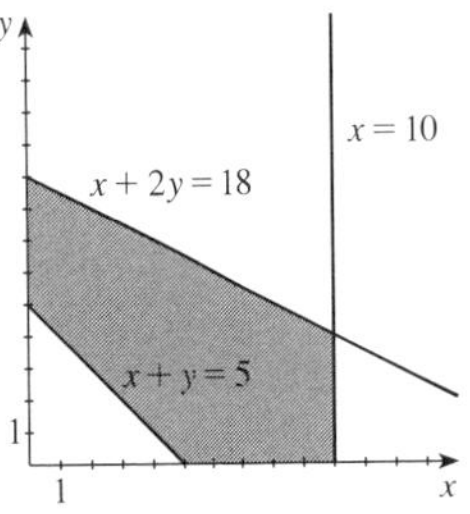


| Vertex | $Q = 70x + 82y$ |
|---|---|
| $(0,9)$ | $70(0) + 82(9) = 738$ |
| $(0,5)$ | $70(0) + 82(5) = 410$ |
| $(5,0)$ | $70(5) + 82(0) = 350$ |
| $(10,0)$ | $70(10) + 82(0) = 700$ |
| $(10,4)$ | $70(10) + 82(4) = 1028$ |

Thus, the maximum value of $Q$ is 1028 and the minimum value is 350.

**5.** Let $t$ be the number of tables made daily and $c$ be the number of chairs made daily. Then the data given can be summarized by the following table:

| | Tables $t$ | Chairs $c$ | Available time |
|---|---|---|---|
| Carpentry | 2 h | 3 h | 108 h |
| Finishing | 1 h | $\frac{1}{2}$ h | 20 h |
| Profit | \$35 | \$20 | |

Thus we wish to maximize the total profit $P = 35t + 20c$ subject to the constraints $\begin{cases} 2t + 3c \leq 108 \\ t + \frac{1}{2}c \leq 20 \\ t \geq 0, c \geq 0 \end{cases}$

From the graph, the vertices occur at $(0,0)$, $(20,0)$, $(0,36)$, and $(3,34)$.

| Vertex | $P = 35t + 20c$ |
|---|---|
| $(0,0)$ | $35(0) + 20(0) = 0$ |
| $(20,0)$ | $35(20) + 20(0) = 700$ |
| $(0,36)$ | $35(0) + 20(36) = 720$ |
| $(3,34)$ | $35(3) + 20(34) = 785$ |

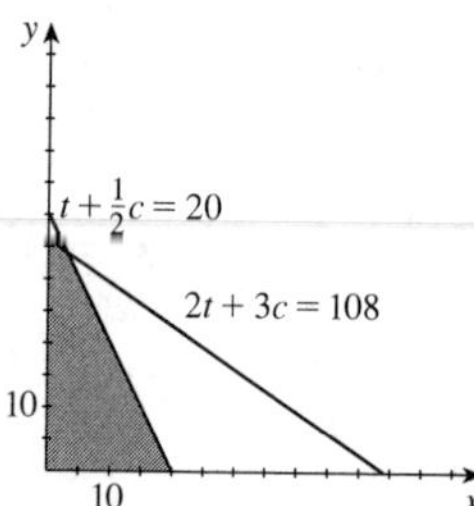


Hence, 3 tables and 34 chairs should be produced daily for a maximum profit of \$785.

**6.** Let $c$ be the number of colonial homes built and $r$ the number of ranch homes built. Since there are 100 lots available, $c + r \leq 100$. From the capital restriction, we get $30{,}000c + 40{,}000r \leq 3{,}600{,}000$, or $3c + 4r \leq 360$. Thus, we wish to maximize the profit $P = 4000c + 8000r$ subject to the constraints $\begin{cases} c \geq 0, r \geq 0 \\ c + r \leq 100 \\ 3c + 4r \leq 360 \end{cases}$

From the graph, the vertices occur at $(0,0)$, $(100,0)$, $(40,60)$, and $(0,90)$.

| Vertex | $P = 4000c + 8000r$ |
|---|---|
| $(0,0)$ | $4000(0) + 8000(0) = 0$ |
| $(100,0)$ | $4000(100) + 8000(0) = 400{,}000$ |
| $(40,60)$ | $4000(40) + 8000(60) = 640{,}000$ |
| $(0,90)$ | $4000(0) + 8000(90) = 720{,}000$ |

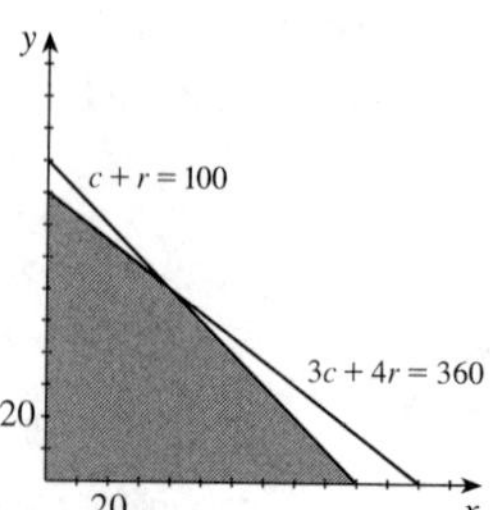


Therefore, he should build no colonial style and 90 ranch style houses for a maximum profit of \$720,000. Note that ten of the lots will be left vacant.

**7.** Let $x$ be the number of crates of oranges and $y$ the number of crates of grapefruit. Then the data given can be summarized by the following table:

| | Oranges | Grapefruit | Available |
|---|---|---|---|
| Volume | 4 ft$^3$ | 6 ft$^3$ | 300 ft$^3$ |
| Weight | 80 lb | 100 lb | 5600 lb |
| Profit | \$2.50 | \$4.00 | |

In addition, $x \geq y$. Thus we wish to maximize the total profit $P = 2.5x + 4y$ subject to the constraints

$$\begin{cases} x \geq 0, y \geq 0, x \geq y \\ 4x + 6y \leq 300 \\ 80x + 100y \leq 5600 \end{cases}$$

From the graph, the vertices occur at $(0, 0)$, $(30, 30)$, $(45, 20)$, and $(70, 0)$.

| Vertex | $P = 2.5x + 4y$ |
|---|---|
| $(0, 0)$ | $2.5(0) + 4(0) = 0$ |
| $(30, 30)$ | $2.5(30) + 4(30) = 195$ |
| $(45, 20)$ | $2.5(45) + 4(20) = 192.5$ |
| $(70, 0)$ | $2.5(70) + 4(0) = 175$ |

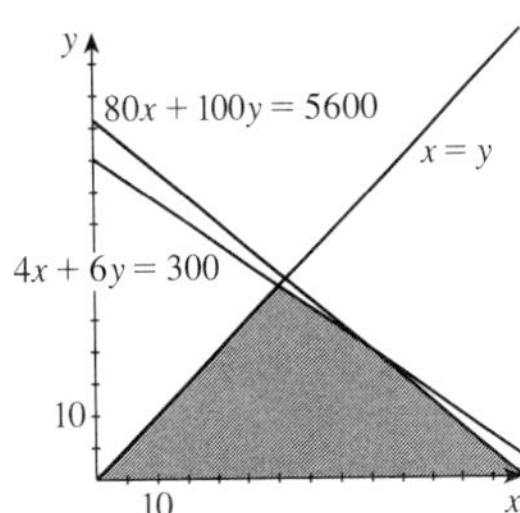


Thus, she should carry 30 crates of oranges and 30 crates of grapefruit for a maximum profit of \$195.

**8.** Let $x$ be the daily production of standard calculators and $y$ the daily production of scientific calculators. Then the inequalities $\begin{cases} x \geq 100, y \geq 80 \\ x \leq 200, y \leq 170 \\ x + y \geq 200 \end{cases}$ describe the constraints. From the graph, the vertices occur at $(100, 100)$, $(100, 170)$, $(200, 170)$, $(200, 80)$, and $(120, 80)$.

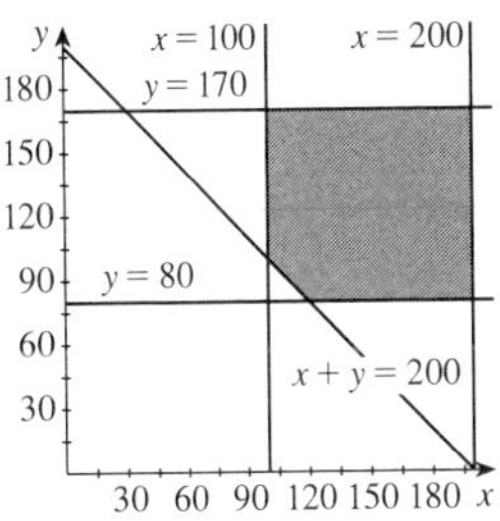


**(a)** Minimize the objective function $C = 5x + 7y$:

| Vertex | $C = 5x + 7y$ |
|---|---|
| $(100, 100)$ | $5(100) + 7(100) = 1200$ |
| $(100, 170)$ | $5(100) + 7(170) = 1690$ |
| $(200, 170)$ | $5(200) + 7(170) = 2190$ |
| $(200, 80)$ | $5(200) + 7(80) = 1560$ |
| $(120, 80)$ | $5(120) + 7(80) = 1160$ |

So to minimize cost, they should produce 120 standard and 80 scientific calculators.

**(b)** Maximize the objective function $P = -2x + 5y$:

| Vertex | $P = -2x + 5y$ |
|---|---|
| $(100, 100)$ | $-2(100) + 5(100) = 300$ |
| $(100, 170)$ | $-2(100) + 5(170) = 650$ |
| $(200, 170)$ | $-2(200) + 5(170) = 450$ |
| $(200, 80)$ | $-2(200) + 5(80) = 0$ |
| $(120, 80)$ | $-2(120) + 5(80) = 160$ |

So to maximize profit, they should produce 100 standard and 170 scientific calculators.

**9.** Let $x$ be the number of stereo sets shipped from Long Beach to Santa Monica and $y$ the number of stereo sets shipped from Long Beach to El Toro. Thus, $15 - x$ sets must be shipped to Santa Monica from Pasadena and $19 - y$ sets to El Toro from Pasadena. Thus, $x \geq 0$, $y \geq 0$, $15 - x \geq 0$, $19 - y \geq 0$, $x + y \leq 24$, and $(15 - x) + (19 - y) \leq 18$. Simplifying, we get

the constraints $\begin{cases} x \geq 0, y \geq 0 \\ x \leq 15, y \leq 19 \\ x + y \leq 24 \\ x + y \geq 16 \end{cases}$

The objective function is the cost $C = 5x + 6y + 4(15 - x) + 5.5(19 - y) = x + 0.5y + 164.5$, which we wish to minimize. From the graph, the vertices occur at $(0, 16)$, $(0, 19)$, $(5, 19)$, $(15, 9)$, and $(15, 1)$.

| Vertex | $C = x + 0.5y + 164.5$ |
|---|---|
| $(0, 16)$ | $(0) + 0.5(16) + 164.5 = 172.5$ |
| $(0, 19)$ | $(0) + 0.5(19) + 164.5 = 174$ |
| $(5, 19)$ | $(5) + 0.5(19) + 164.5 = 179$ |
| $(15, 9)$ | $(15) + 0.5(9) + 164.5 = 184$ |
| $(15, 1)$ | $(15) + 0.5(1) + 164.5 = 180$ |

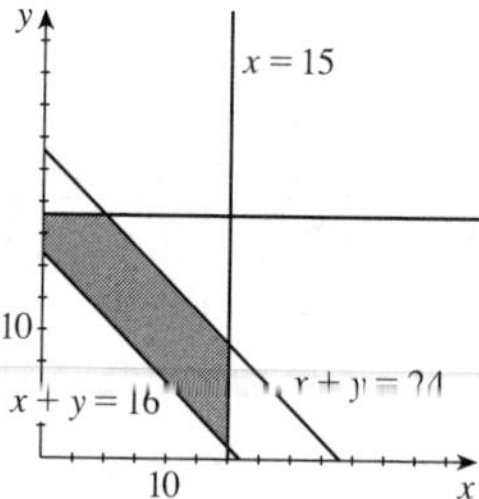


The minimum cost is \$172.50 and occurs when $x = 0$ and $y = 16$. Hence, no stereo should be shipped from Long Beach to Santa Monica, 16 from Long Beach to El Toro, 15 from Pasadena to Santa Monica, and 3 from Pasadena to El Toro.

**10.** Let $x$ be the number of sheets shipped from the east side store to customer A and $y$ be the number of sheets shipped from the east side store to customer B. Then $50 - x$ sheets must be shipped to customer A from the west side store and $70 - y$ sheets must be shipped to customer B from the west side store. Thus, we obtain the constraints

$$\begin{cases} x \geq 0, y \geq 0 \\ x \leq 50, y \leq 70 \\ x + y \leq 80 \\ 50 - x + 70 - y \leq 45 \end{cases} \Leftrightarrow \begin{cases} x \geq 0, y \geq 0 \\ x \leq 50, y \leq 70 \\ x + y \leq 80 \\ x + y \geq 75 \end{cases}$$

The objective function is the cost $C = 0.5x + 0.6y + 0.4(50 - x) + 0.55(70 - y) = 0.1x + 0.05y + 58.5$, which we wish to minimize. From the graph, the vertices occur at $(5, 70)$, $(10, 70)$, $(50, 30)$, and $(50, 25)$.

| Vertex | $C = 0.1x + 0.05y + 58.5$ |
|---|---|
| $(5, 70)$ | $0.1(5) + 0.05(70) + 58.5 = 62.5$ |
| $(10, 70)$ | $0.1(10) + 0.05(70) + 58.5 = 63$ |
| $(50, 30)$ | $0.1(50) + 0.05(30) + 58.5 = 65$ |
| $(50, 25)$ | $0.1(50) + 0.05(25) + 58.5 = 64.75$ |

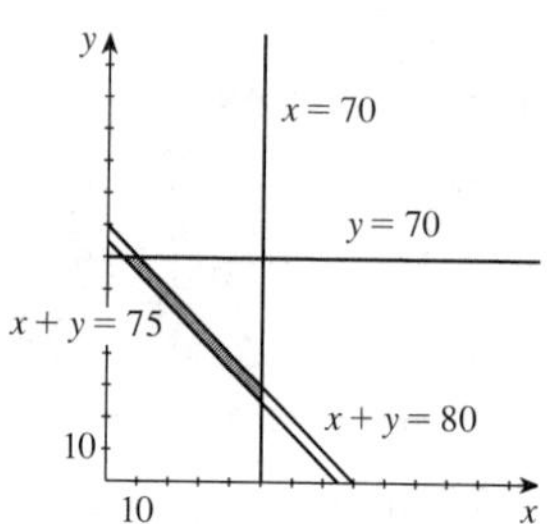


Therefore, the minimum cost is \$62.50 and occurs when $x = 5$ and $y = 70$. So 5 sheets should be shipped from the east side store to customer A, 70 sheets from the east side store to customer B, 45 sheets from the west side store to customer A, and no sheet from the west side store to customer B.

**11.** Let $x$ be the number of bags of standard mixtures and $y$ be the number of bags of deluxe mixtures. Then the data can be summarized by the following table:

| | Standard | Deluxe | Available |
|---|---|---|---|
| Cashews | 100 g | 150 g | 15 kg |
| Peanuts | 200 g | 50 g | 20 kg |
| Selling price | \$1.95 | \$2.20 | |

Thus the total revenue, which we want to maximize, is given by $R = 1.95x + 2.25y$.We have the constraints

$$\begin{cases} x \geq 0, y \geq 0, x \geq y \\ 0.1x + 0.15y \leq 15 \\ 0.2x + 0.05y \leq 20 \end{cases} \Leftrightarrow \begin{cases} x \geq 0, y \geq 0, x \geq y \\ 10x + 15y \leq 1500 \\ 20x + 5y \leq 2000 \end{cases}$$

From the graph, the vertices occur at $(0,0)$, $(60,60)$, $(90,40)$, and $(100,0)$.

| Vertex | $R = 1.95x + 2.25y$ |
|---|---|
| $(0,0)$ | $1.96\,(0) + 2.25\,(0) = 0$ |
| $(60,60)$ | $1.95\,(60) + 2.25\,(60) = 252$ |
| $(90,40)$ | $1.95\,(90) + 2.25\,(40) = 265.5$ |
| $(100,0)$ | $1.95\,(100) + 2.25\,(0) = 195$ |

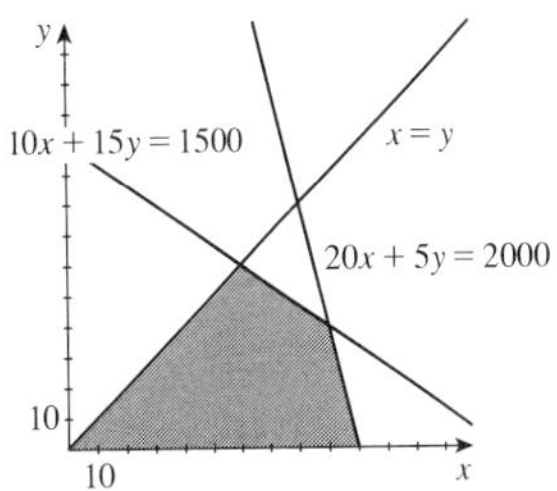


Hence, he should pack 90 bags of standard and 40 bags of deluxe mixture for a maximum revenue of \$265.50.

**12.** Let $x$ be the quantity of type I food and $y$ the quantity of type II food, in ounces. Then the data can be summarized by the following table:

| | Type I | Type II | Required |
|---|---|---|---|
| Fat | 8 g | 12 g | 24 g |
| Carbohydrate | 12 g | 12 g | 36 g |
| Protein | 2 g | 1 g | 4 g |
| Cost | \$0.20 | \$0.30 | |

Also, the total amount of food must be no more than 5 oz. Thus, the constraints are $\begin{cases} x \geq 0, y \geq 0, \\ x + y \leq 5, 8x + 12y \geq 24 \\ 12x + 12y \geq 36, 2x + y \geq 4 \end{cases}$

The objective function is the cost $C = 0.2x + 0.3y$, which we wish to minimize. From the graph, the vertices occur at $(1,2)$, $(0,4)$, $(0,5)$, $(5,0)$, and $(3,0)$.

| Vertex | $C = 0.2x + 0.3y$ |
|---|---|
| $(1,2)$ | $0.2\,(1) + 0.3\,(2) = 0.8$ |
| $(0,4)$ | $0.2\,(0) + 0.3\,(4) = 1.2$ |
| $(0,5)$ | $0.2\,(0) + 0.3\,(5) = 1.5$ |
| $(5,0)$ | $0.2\,(5) + 0.3\,(0) = 1.0$ |
| $(3,0)$ | $0.2\,(3) + 0.3\,(0) = 0.6$ |

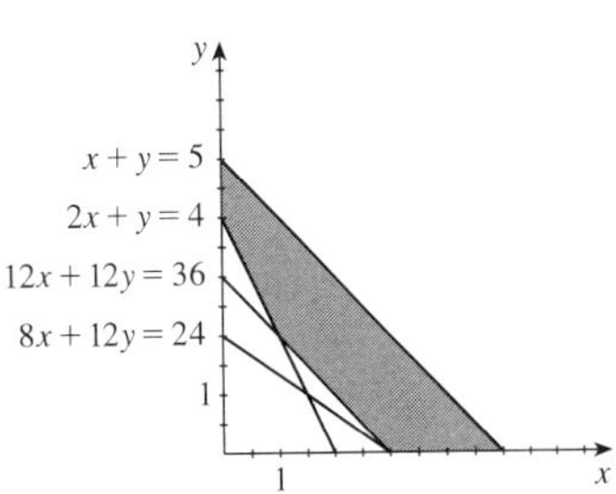


Hence, the rabbits should be fed 3 oz of type I food and no type II food, for a minimum cost of \$0.60.

**13.** Let $x$ be the amount in municipal bonds and $y$ the amount in bank certificates, both in dollars. Then $12000 - x - y$ is the amount in high-risk bonds. So our constraints can be stated as

$$\begin{cases} x \geq 0,\ y \geq 0,\ x \geq 3y \\ 12{,}000 - x - y \geq 0 \\ 12{,}000 - x - y \leq 2000 \end{cases} \Leftrightarrow \begin{cases} x \geq 0,\ y \geq 0,\ x \geq 3y \\ x + y \leq 12{,}000 \\ x + y \geq 10{,}000 \end{cases}$$

From the graph, the vertices occur at $(7500, 2500)$, $(10000, 0)$, $(12000, 0)$, and $(9000, 3000)$. The objective function is $P = 0.07x + 0.08y + 0.12(12000 - x - y) = 1440 - 0.05x - 0.04y$, which we wish to maximize.

| Vertex | $P = 1440 - 0.05x - 0.04y$ |
|---|---|
| $(7500, 2500)$ | $1440 - 0.05(7500) - 0.04(2500) = 965$ |
| $(10000, 0)$ | $1440 - 0.05(10{,}000) - 0.04(0) = 940$ |
| $(12000, 0)$ | $1440 - 0.05(12{,}000) - 0.04(0) = 840$ |
| $(9000, 3000)$ | $1440 - 0.05(9000) - 0.04(3000) = 870$ |

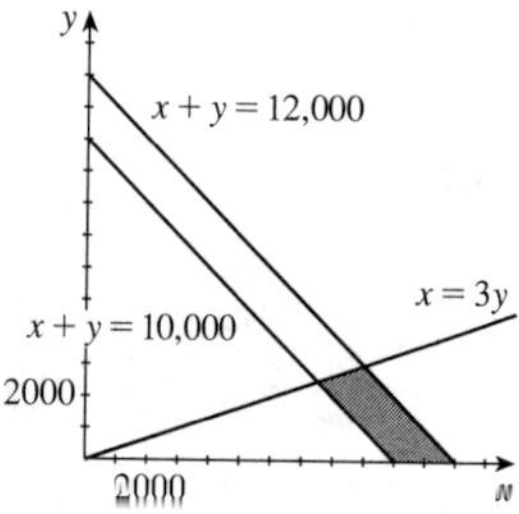


Hence, she should invest \$7500 in municipal bonds, \$2500 in bank certificates, and the remaining \$2000 in high-risk bonds for a maximum yield of \$965.

**14.** The only change that needs to be made to the constraints in Exercise 13 is that the 2000 in the last inequality becomes 3000. Then we have

$$\begin{cases} x \geq 0,\ y \geq 0,\ x \geq 3y \\ 12{,}000 - x - y \geq 0 \\ 12{,}000 - x - y \leq 3000 \end{cases} \Leftrightarrow \begin{cases} x \geq 0,\ y \geq 0,\ x \geq 3y \\ x + y \leq 12{,}000 \\ x + y \geq 9000 \end{cases}$$

From the graph, the vertices occur at $(6750, 2250)$, $(9000, 0)$, $(12000, 0)$, and $(9000, 3000)$. The objective function is $Y = 0.07x + 0.08y + 0.12(12000 - x - y) = 1440 - 0.05x - 0.04y$, which we wish to maximize.

| Vertex | $Y = 1440 - 0.05x - 0.04y$ |
|---|---|
| $(6750, 2250)$ | $1440 - 0.05(6750) - 0.04(2250) = 1012.5$ |
| $(9000, 0)$ | $1440 - 0.05(9000) - 0.04(0) = 990$ |
| $(12000, 0)$ | $1440 - 0.05(12{,}000) - 0.04(0) = 840$ |
| $(9000, 3000)$ | $1440 - 0.05(9000) - 0.04(3000) = 870$ |

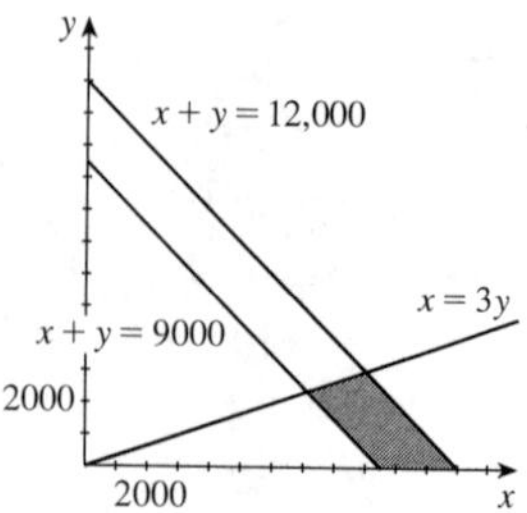


Hence, she should invest \$6750 in municipal bonds, \$2250 in bank certificates, and \$3000 in high-risk bonds for a maximum yield of \$1012.50, which is an increase of \$47.50, over her yield in Exercise 13.

**15.** Let $g$ be the number of games published and $e$ be the number of educational programs published. Then the number of utility programs published is $36 - g - e$. Hence we wish to maximize profit, $P = 5000g + 8000e + 6000(36 - g - e) = 216{,}000 - 1000g + 2000e$, subject to the constraints

$$\begin{cases} g \geq 4,\ e \geq 0 \\ 36 - g - e \geq 0 \\ 36 - g - e \leq 2e \end{cases} \Leftrightarrow \begin{cases} g \geq 4,\ e \geq 0 \\ g + e \leq 36 \\ g + 3e \geq 36. \end{cases}$$

From the graph, the vertices are at $\left(4, \frac{32}{3}\right)$, $(4, 32)$, and $(36, 0)$. The objective function is $P = 216{,}000 - 1000g + 2000e$.

| Vertex | $P = 216{,}000 - 1000g + 2000e$ |
|---|---|
| $\left(4, \frac{32}{3}\right)$ | $216{,}000 - 1000(4) + 2000\left(\frac{32}{3}\right) = 233{,}333.33$ |
| $(4, 32)$ | $216{,}000 - 1000(4) + 2000(32) = 276{,}000$ |
| $(36, 0)$ | $216{,}000 - 1000(36) + 2000(0) = 180{,}000$ |

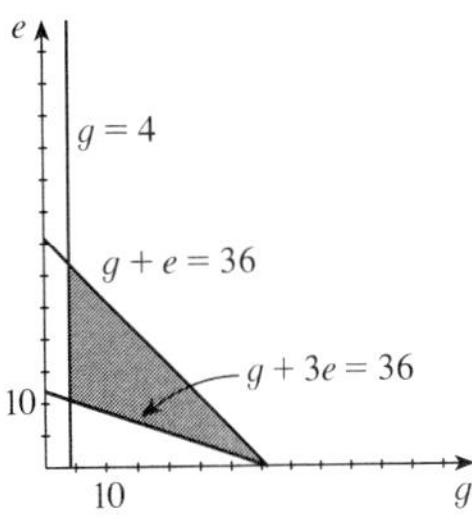


So, they should publish 4 games, 32 educational programs, and no utility program for a maximum profit of \$276,000 annually.

**16.** $\begin{cases} x \geq 0,\ x \geq y \\ x + 2y \leq 12 \\ x + y \leq 10 \end{cases}$

**(a)**

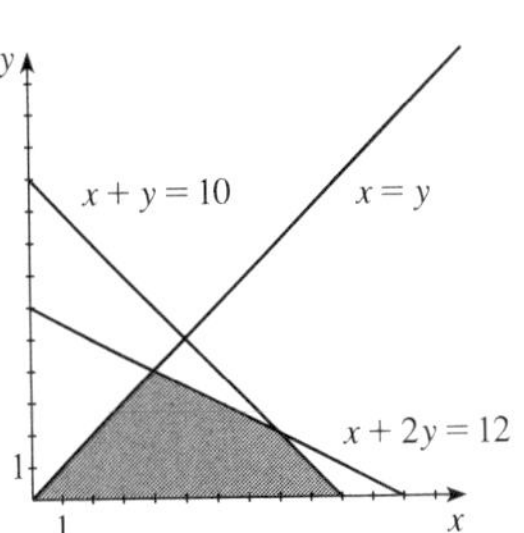


**(b)**

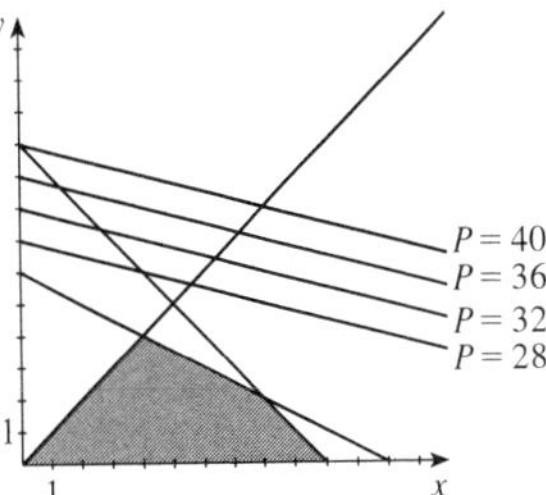


**(c)** These lines will first touch the feasible region at the top vertex, $(4, 4)$.

**(d)**

| Vertex | $P = x + 4y$ |
|---|---|
| $(0, 0)$ | $(0) + 4(0) = 0$ |
| $(10, 0)$ | $(10) + 4(0) = 10$ |
| $(4, 4)$ | $(4) + 4(4) = 20$ |
| $(8, 2)$ | $(8) + 4(2) = 16$ |

The maximum value of the objective function occurs at the vertex $(4, 4)$.

# 11 Analytic Geometry

## 11.1 Parabolas

**1.** $y^2 = 2x$ is Graph III, which opens to the right and is not as wide as the graph for Exercise 5.

**2.** $y^2 = -\frac{1}{4}x$ is Graph V, the only graph that opens downward.

**3.** $x^2 = -6y$ is Graph II, which opens downward and is narrower than the graph for Exercise 6.

**4.** $2x^2 = y$ is Graph I, the only graph that opens to the right.

**5.** $y^2 - 8x = 0$ is Graph VI, which opens to the right and is wider than the graph for Exercise 1.

**6.** $12y + x^2 = 0$ is Graph IV, which opens downward and is wider than the graph for Exercise 3.

**7.** $y^2 = 4x$. Then $4p = 4 \Leftrightarrow p = 1$. The focus is $(1, 0)$, the directrix is $x = -1$, and the focal diameter is 4.

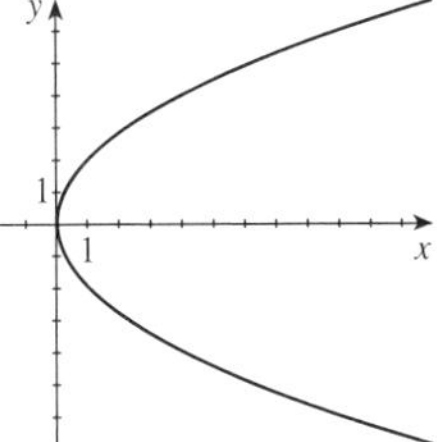


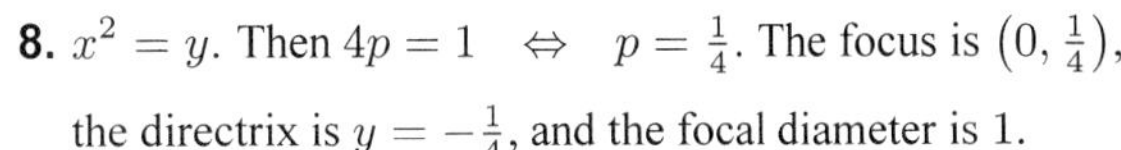

**8.** $x^2 = y$. Then $4p = 1 \Leftrightarrow p = \frac{1}{4}$. The focus is $\left(0, \frac{1}{4}\right)$, the directrix is $y = -\frac{1}{4}$, and the focal diameter is 1.

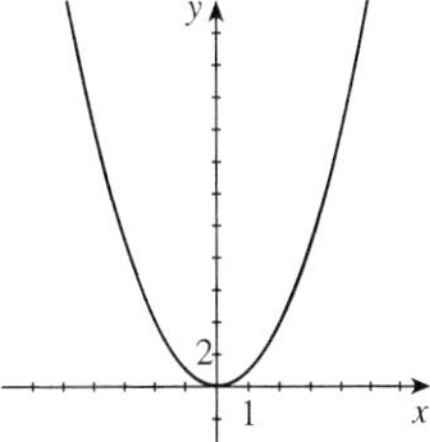


**9.** $x^2 = 9y$. Then $4p = 9 \Leftrightarrow p = \frac{9}{4}$. The focus is $\left(0, \frac{9}{4}\right)$, the directrix is $y = -\frac{9}{4}$, and the focal diameter is 9.

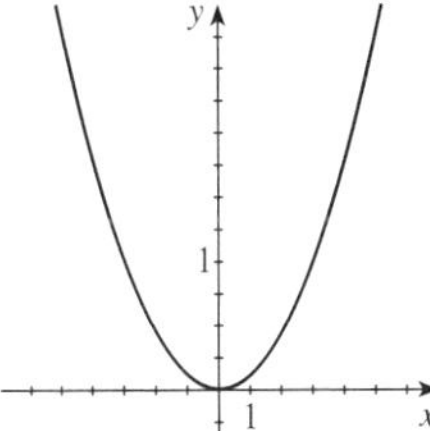


**10.** $y^2 = 3x$. Then $4p = 3 \Leftrightarrow p = \frac{3}{4}$. The focus is $\left(\frac{3}{4}, 0\right)$, the directrix is $x = -\frac{3}{4}$, and the focal diameter is 3.

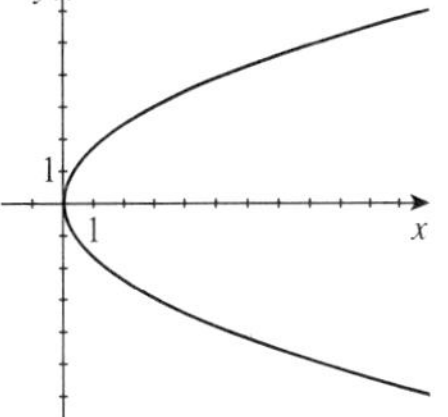


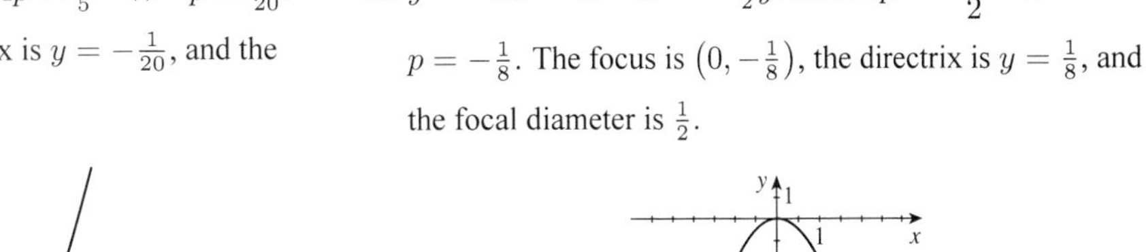

**11.** $y = 5x^2 \Leftrightarrow x^2 = \frac{1}{5}y$. Then $4p = \frac{1}{5} \Leftrightarrow p = \frac{1}{20}$. The focus is $\left(0, \frac{1}{20}\right)$, the directrix is $y = -\frac{1}{20}$, and the focal diameter is $\frac{1}{5}$.

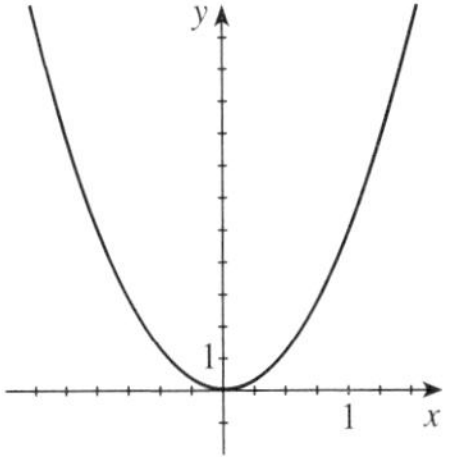


**12.** $y = -2x^2 \Leftrightarrow x^2 = -\frac{1}{2}y$. Then $4p = -\dfrac{1}{2} \Leftrightarrow p = -\frac{1}{8}$. The focus is $\left(0, -\frac{1}{8}\right)$, the directrix is $y = \frac{1}{8}$, and the focal diameter is $\frac{1}{2}$.

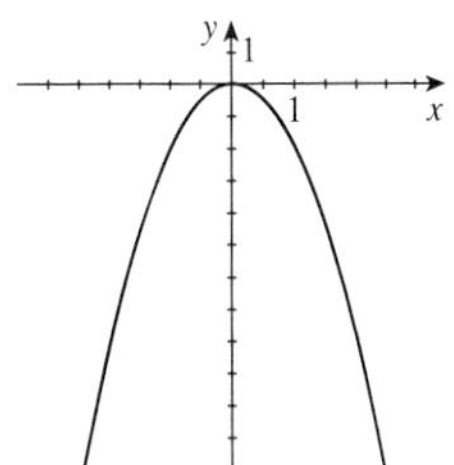

**13.** $x = -8y^2 \Leftrightarrow y^2 = -\frac{1}{8}x$. Then $4p = -\dfrac{1}{8} \Leftrightarrow$ $p = -\frac{1}{32}$. The focus is $\left(-\frac{1}{32}, 0\right)$, the directrix is $x = \frac{1}{32}$, and the focal diameter is $\frac{1}{8}$.

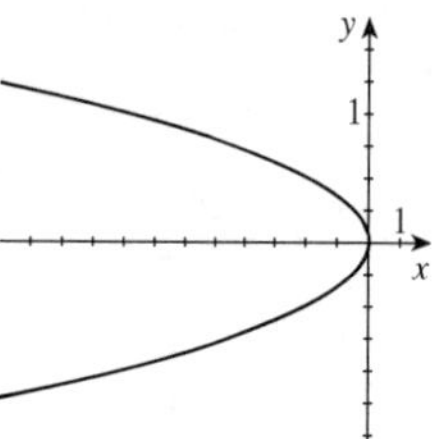

**14.** $x = \frac{1}{2}y^2 \Leftrightarrow y^2 = 2x$. Then $4p = 2 \Leftrightarrow p = \frac{1}{2}$. The focus is $\left(\frac{1}{2}, 0\right)$, the directrix is $x = -\frac{1}{2}$, and the focal diameter is 2.

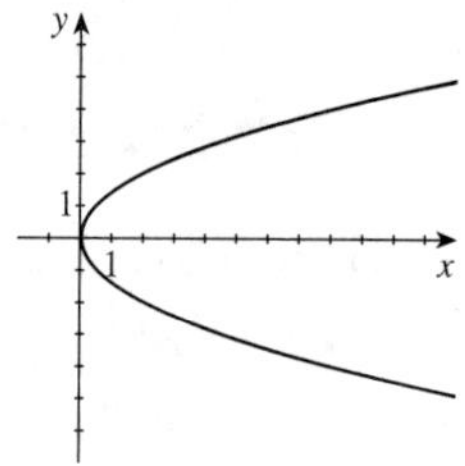

**15.** $x^2 + 6y = 0 \Leftrightarrow x^2 = -6y$. Then $4p = -6 \Leftrightarrow$ $p = -\frac{3}{2}$. The focus is $\left(0, -\frac{3}{2}\right)$, the directrix is $y = \frac{3}{2}$, and the focal diameter is 6.

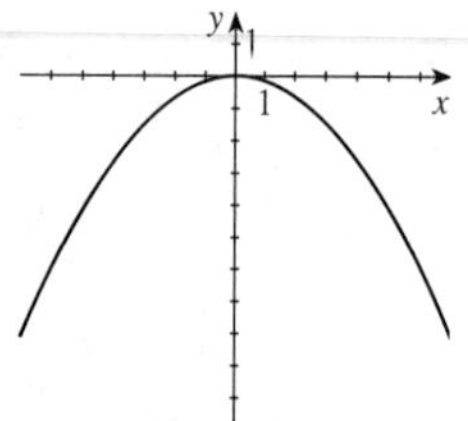

**16.** $x - 7y^2 = 0 \Leftrightarrow y^2 = \frac{1}{7}x$. Then $4p = \dfrac{1}{7} \Leftrightarrow$ $p = \frac{1}{28}$. The focus is $\left(\frac{1}{28}, 0\right)$, the directrix is $x = -\frac{1}{28}$, and the focal diameter is $\frac{1}{7}$.

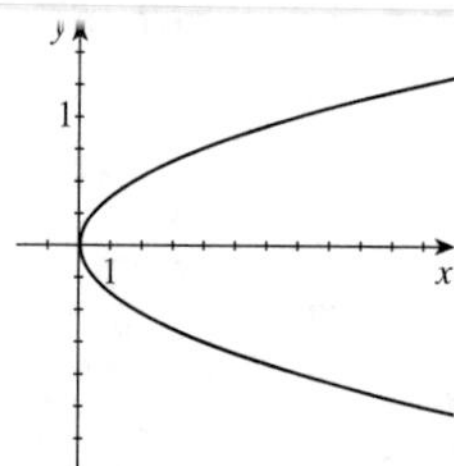

**17.** $5x + 3y^2 = 0 \Leftrightarrow y^2 = -\frac{5}{3}x$. Then $4p = -\frac{5}{3} \Leftrightarrow$ $p = -\frac{5}{12}$. The focus is $\left(-\frac{5}{12}, 0\right)$, the directrix is $x = \frac{5}{12}$, and the focal diameter is $\frac{5}{3}$.

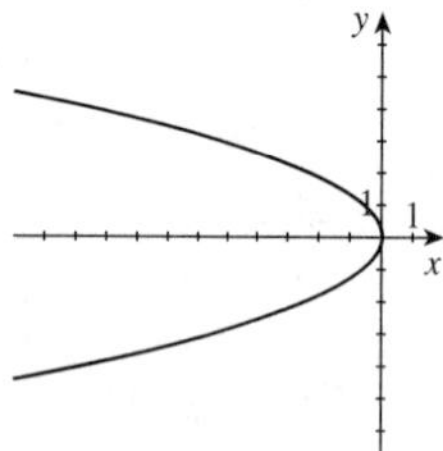

**18.** $8x^2 + 12y = 0 \Leftrightarrow x^2 = -\frac{3}{2}y$. Then $4p = -\frac{3}{2} \Leftrightarrow$ $p = -\frac{3}{8}$. The focus is $\left(0, -\frac{3}{8}\right)$, the directrix is $y = \frac{3}{8}$, and the focal diameter is $\frac{3}{2}$.

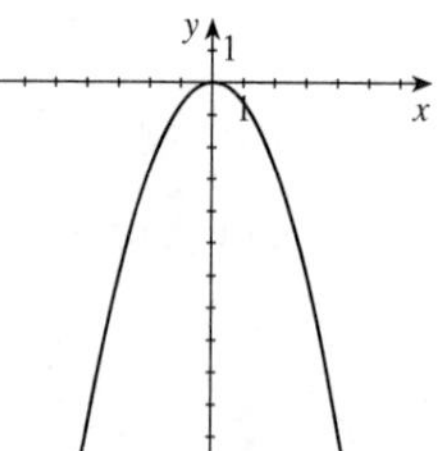

**19.** $x^2 = 16y$

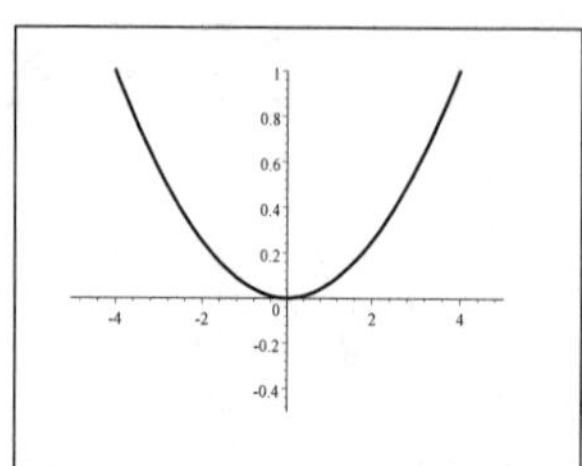

**20.** $x^2 = -8y$

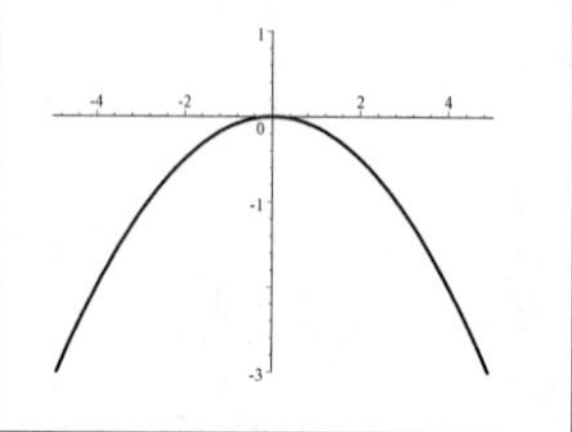

**21.** $y^2 = -\frac{1}{3}x$

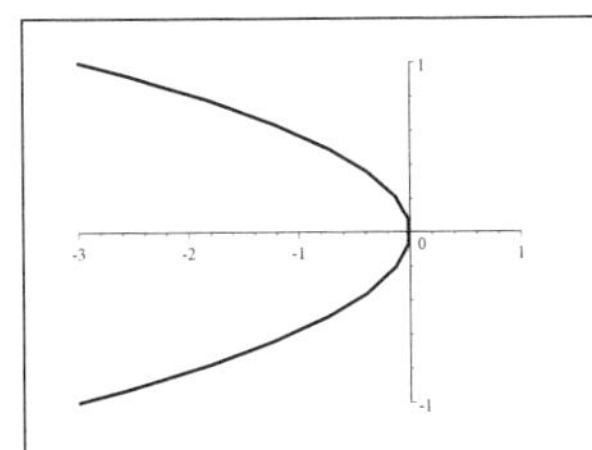

**22.** $8y^2 = x$

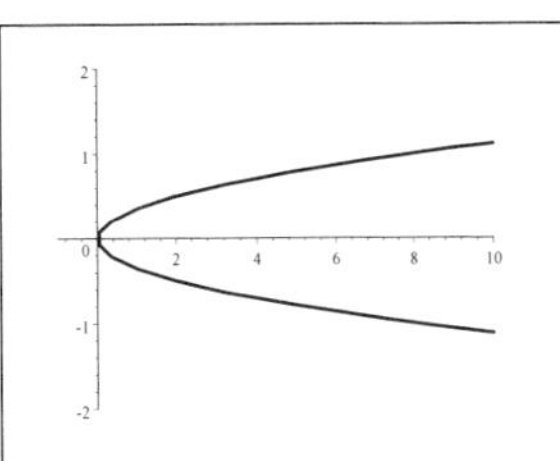

**23.** $4x + y^2 = 0$

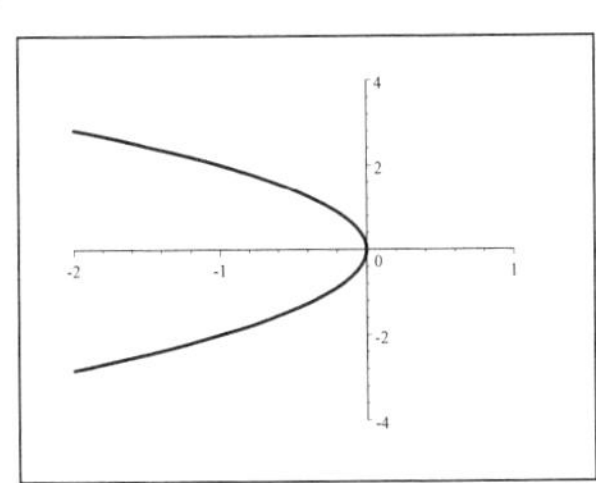

**24.** $x - 2y^2 = 0$

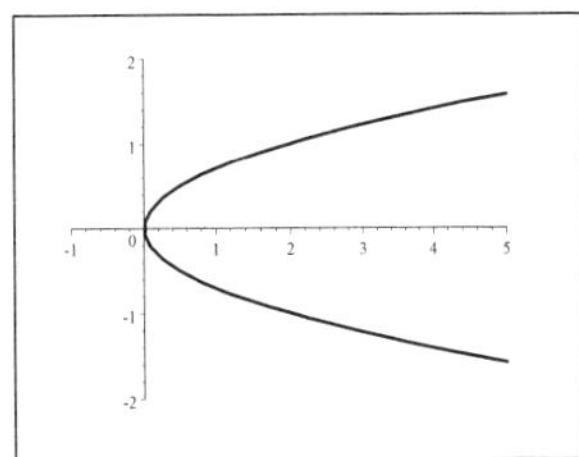

**25.** Since the focus is $(0, 2)$, $p = 2 \Leftrightarrow 4p = 8$. Hence, an equation of the parabola is $x^2 = 8y$.

**26.** Since the focus is $\left(0, -\frac{1}{2}\right)$, $p = -\frac{1}{2} \Leftrightarrow 4p = -2$. So an equation of the parabola is $x^2 = -2y$.

**27.** Since the focus is $(-8, 0)$, $p = -8 \Leftrightarrow 4p = -32$. Hence, an equation of the parabola is $y^2 = -32x$.

**28.** Since the focus is $(5, 0)$, $p = 5 \Leftrightarrow 4p = 20$. Hence, an equation of the parabola is $y^2 = 20x$.

**29.** Since the directrix is $x = 2$, $p = -2 \Leftrightarrow 4p = -8$. Hence, an equation of the parabola is $y^2 = -8x$.

**30.** Since the directrix is $y = 6$, $p = -6 \Leftrightarrow 4p = -24$. Hence, an equation of the parabola is $x^2 = -24y$.

**31.** Since the directrix is $y = -10$, $p = 10 \Leftrightarrow 4p = 40$. Hence, an equation of the parabola is $x^2 = 40y$.

**32.** Since the directrix is $x = -\frac{1}{8}$, $p = \frac{1}{8} \Leftrightarrow 4p = \frac{1}{2}$. Hence, an equation of the parabola is $y^2 = \frac{1}{2}x$.

**33.** The focus is on the positive $x$-axis, so the parabola opens horizontally with $2p = 2 \Leftrightarrow 4p = 4$. So an equation of the parabola is $y^2 = 4x$.

**34.** The directrix has $y$-intercept 6, and so $p = -6 \Leftrightarrow 4p = -24$. Therefore, an equation of the parabola is $x^2 = -24y$.

**35.** Since the parabola opens upward with focus 5 units from the vertex, the focus is $(5, 0)$. So $p = 5 \Leftrightarrow 4p = 20$. Thus an equation of the parabola is $x^2 = 20y$.

**36.** Since the focal diameter is 8 and the focus is on the negative $y$-axis, $4p = -8$. So the equation is $x^2 = -8y$.

**37.** $p = 2 \Leftrightarrow 4p = 8$. Since the parabola opens upward, its equation is $x^2 = 8y$.

**38.** The directrix is $x = -2$, and so $p = 2 \Leftrightarrow 4p = 8$. Since the parabola opens to the left, its equation is $y^2 = 8x$.

**39.** $p = 4 \Leftrightarrow 4p = 16$. Since the parabola opens to the left, its equation is $y^2 = -16x$.

**40.** $p = 3 \Leftrightarrow 4p = 12$. Since the parabola opens downward, its equation is $x^2 = -12y$.

**41.** The focal diameter is $4p = \frac{3}{2} + \frac{3}{2} = 3$. Since the parabola opens to the left, its equation is $y^2 = -3x$.

**42.** The focal diameter is $4p = 2(5) = 10$. Since the parabola opens upward, its equation is $x^2 = 10y$.

**43.** The equation of the parabola has the form $y^2 = 4px$. Since the parabola passes through the point $(4, -2)$, $(-2)^2 = 4p(4)$ $\Leftrightarrow 4p = 1$, and so an equation is $y^2 = x$.

**44.** Since the directrix is $x = -p$, we have $p^2 = 16$, so $p = 4$, and an equation is $y^2 = 4px$ or $y^2 = 16x$.

**45.** The area of the shaded region is width $\times$ height $= 4p \cdot p = 8$, and so $p^2 = 2 \Leftrightarrow p = -\sqrt{2}$ (because the parabola opens downward). Therefore, an equation is $x^2 = 4py = -4\sqrt{2}y \Leftrightarrow x^2 = -4\sqrt{2}y$.

**46.** The focus is $(0, p)$. Since the line has slope $\frac{1}{2}$, an equation of the line is $y = \frac{1}{2}x + p$. Therefore, the point where the line intersects the parabola has $y$-coordinate $\frac{1}{2}(2) + p = p + 1$. The parabola's equation is of the form $x^2 = 4py$, so $(2)^2 = 4p(p+1) \Leftrightarrow p^2 + p - 1 = 0 \Leftrightarrow p = \dfrac{-1+\sqrt{5}}{2}$ (since $p > 0$). Hence, an equation of the parabola is $x^2 = 2(\sqrt{5}-1)y$.

**47. (a)** A parabola with directrix $y = -p$ has equation $x^2 = 4py$. If the directrix is $y = \frac{1}{2}$, then $p = -\frac{1}{2}$, so an equation is $x^2 = 4\left(-\frac{1}{2}\right)y \Leftrightarrow x^2 = -2y$. If the directrix is $y = 1$, then $p = -1$, so an equation is $x^2 = 4(-1)y \Leftrightarrow x^2 = -4y$. If the directrix is $y = 4$, then $p = -4$, so an equation is $x^2 = 4(-4)y \Leftrightarrow x^2 = -16y$. If the directrix is $y = 8$, then $p = -8$, so an equation is $x^2 = 4(-8)y \Leftrightarrow x^2 = -32y$.

**(b)**

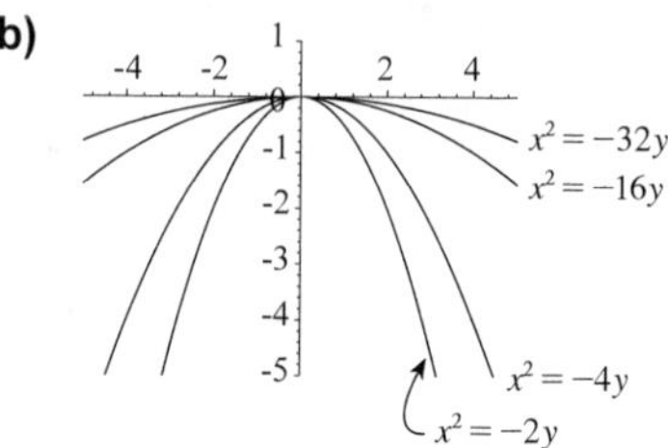


As the directrix moves further from the vertex, the parabolas get flatter.

**48. (a)** If the focal diameter of a parabola is $4p$, it has equation $x^2 = 4py$. If the focal diameter is $4p = 1$, an equation is $x^2 = y$. If the focal diameter is $4p = 2$, an equation is $x^2 = 2y$. If the focal diameter is $4p = 4$, an equation is $x^2 = 4y$. If the focal diameter is $4p = 8$, an equation is $x^2 = 8y$.

**(b)**

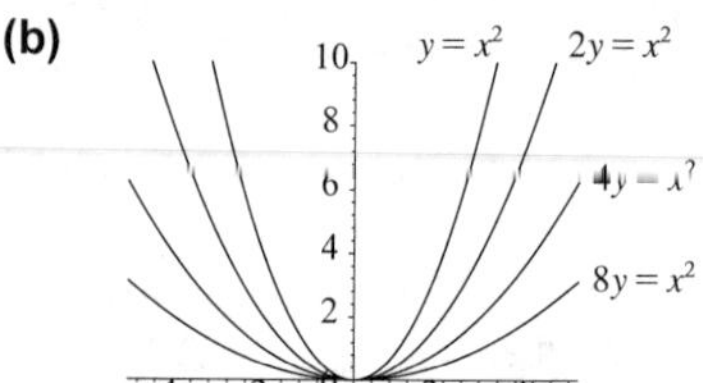


As the focal diameter increases, the parabolas get flatter.

**49. (a)** Since the focal diameter is 12 cm, $4p = 12$. Hence, the parabola has equation $y^2 = 12x$.

**(b)** At a point 20 cm horizontally from the vertex, the parabola passes through the point $(20, y)$, and hence from part (a), $y^2 = 12(20) \Leftrightarrow y^2 = 240 \Leftrightarrow y = \pm 4\sqrt{15}$. Thus, $|CD| = 8\sqrt{15} \approx 31$ cm.

**50.** The equation of the parabola has the form $x^2 = 4py$. From the diagram, the parabola passes through the point $(10, 1)$, and so $(10)^2 = 4p(1) \Leftrightarrow 4p = 100$ and $p = 25$. Therefore, the receiver is 25 ft from the vertex.

**51.** With the vertex at the origin, the top of one tower will be at the point $(300, 150)$. Inserting this point into the equation $x^2 = 4py$ gives $(300)^2 = 4p(150) \Leftrightarrow 90000 = 600p \Leftrightarrow p = 150$. So an equation of the parabolic part of the cables is $x^2 = 4(150)y \Leftrightarrow x^2 = 600y$.

**52.** The equation of the parabola has the form $x^2 = 4py$. From the diagram, the parabola passes through the point $(100, 3.79)$, and so $(100)^2 = 4p(3.79) \Leftrightarrow 15.16p = 10000$ and $p \approx 659.63$. Therefore, the receiver is about 659.63 inches $\approx$ 55 feet from the vertex.

**53.** Many answers are possible: satellite dish TV antennas, sound surveillance equipment, solar collectors for hot water heating or electricity generation, bridge pillars, etc.

**54.** Yes. If a cone intersects a plane that is parallel to a line on the cone, the resulting curve is a parabola, as shown in the text.

# 11.2 Ellipses

**1.** $\dfrac{x^2}{16}+\dfrac{y^2}{4}=1$ is Graph II. The major axis is horizontal and the vertices are $(\pm4,0)$.

**2.** $x^2+\dfrac{y^2}{9}=1$ is Graph IV. The major axis is vertical and the vertices are $(0,\pm3)$.

**3.** $4x^2+y^2=4$ is Graph I. The major axis is vertical and the vertices are $(0,\pm2)$.

**4.** $16x^2+25y^2=400$ is Graph III. The major axis is horizontal and the vertices are $(\pm5,0)$.

**5.** $\dfrac{x^2}{25}+\dfrac{y^2}{9}=1$. This ellipse has $a=5$, $b=3$, and so $c^2=a^2-b^2=16$ $\Leftrightarrow$ $c=4$. The vertices are $(\pm5,0)$, the foci are $(\pm4,0)$, the eccentricity is $e=\dfrac{c}{a}=\frac{4}{5}=0.8$, the length of the major axis is $2a=10$, and the length of the minor axis is $2b=6$.

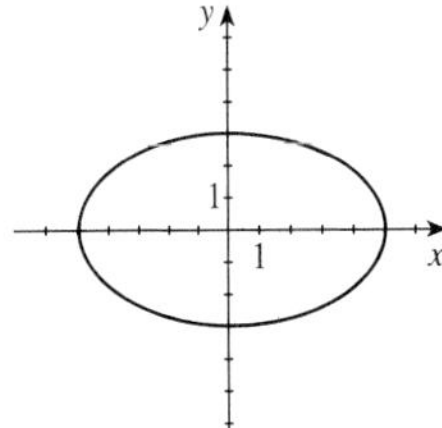

**6.** $\dfrac{x^2}{16}+\dfrac{y^2}{25}=1$. This ellipse has $a=5$, $b=4$, and so $c^2=25-16=9$ $\Leftrightarrow$ $c=3$. The vertices are $(0,\pm5)$, the foci are $(0,\pm3)$, the eccentricity is $e=\dfrac{c}{a}=\frac{3}{5}=0.6$, the length of the major axis is $2a=10$, and the length of the minor axis is $2b=8$.

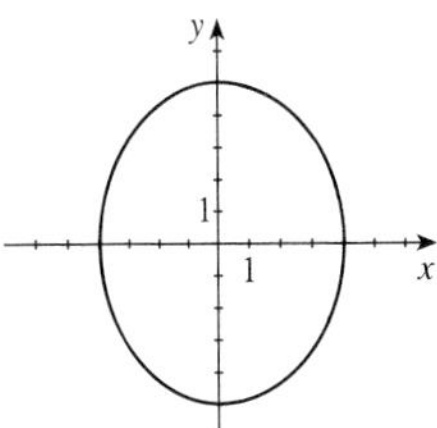

**7.** $9x^2+4y^2=36$ $\Leftrightarrow$ $\dfrac{x^2}{4}+\dfrac{y^2}{9}=1$. This ellipse has $a=3$, $b=2$, and so $c^2=9-4=5$ $\Leftrightarrow$ $c=\sqrt{5}$. The vertices are $(0,\pm3)$, the foci are $(0,\pm\sqrt{5})$, the eccentricity is $e=\dfrac{c}{a}=\dfrac{\sqrt{5}}{3}$, the length of the major axis is $2a=6$, and the length of the minor axis is $2b=4$.

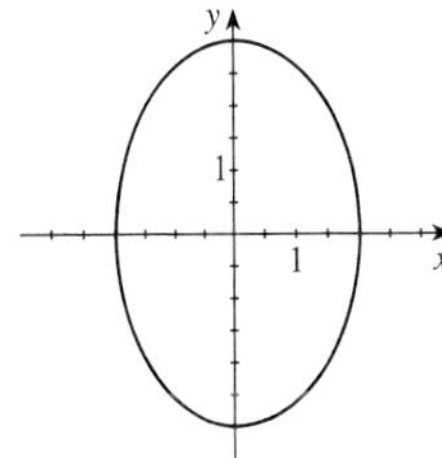

**8.** $4x^2+25y^2=100$ $\Leftrightarrow$ $\dfrac{x^2}{25}+\dfrac{y^2}{4}=1$. This ellipse has $a=5$, $b=2$, and so $c^2=25-4=21$ $\Leftrightarrow$ $c=\sqrt{21}$. The vertices are $(\pm5,0)$, the foci are $(\pm\sqrt{21},0)$, the eccentricity is $e=\dfrac{c}{a}=\dfrac{\sqrt{21}}{5}$, the length of the major axis is $2a=10$, and the length of the minor axis is $2b=4$.

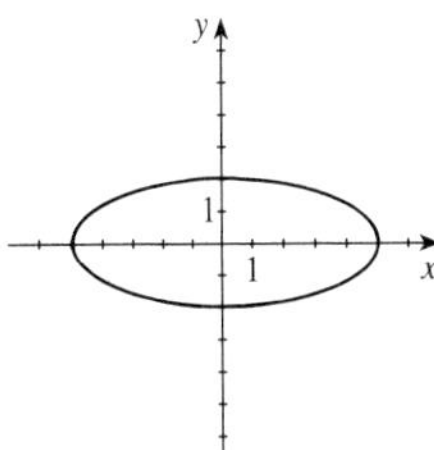

**9.** $x^2+4y^2=16$ $\Leftrightarrow$ $\dfrac{x^2}{16}+\dfrac{y^2}{4}=1$. This ellipse has $a=4$, $b=2$, and so $c^2=16-4=12$ $\Leftrightarrow$ $c=2\sqrt{3}$.The vertices are $(\pm4,0)$, the foci are $(\pm2\sqrt{3},0)$, the eccentricity is $e=\dfrac{c}{a}=\dfrac{2\sqrt{3}}{4}=\frac{\sqrt{3}}{2}$, the length of the major axis is $2a=8$, and the length of the minor axis is $2b=4$.

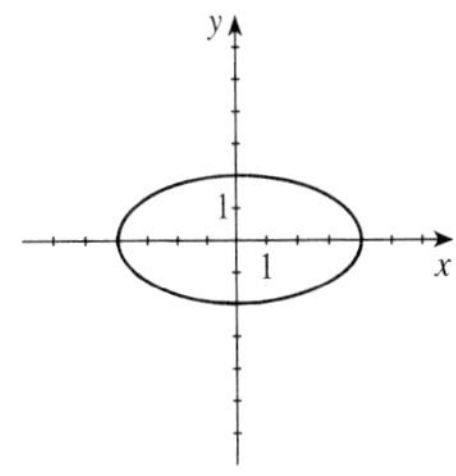

**10.** $4x^2 + y^2 = 16 \Leftrightarrow \dfrac{x^2}{4} + \dfrac{y^2}{16} = 1$. This ellipse has $a = 4$, $b = 2$, and so $c^2 = 16 - 4 = 12 \Leftrightarrow c = 2\sqrt{3}$. The vertices are $(0, \pm 4)$, the foci are $(0, \pm 2\sqrt{3})$, the eccentricity is $e = \dfrac{c}{a} = \dfrac{2\sqrt{3}}{4} = \frac{\sqrt{3}}{2}$, the length of the major axis is $2a = 8$, and the length of the minor axis is $2b = 4$.

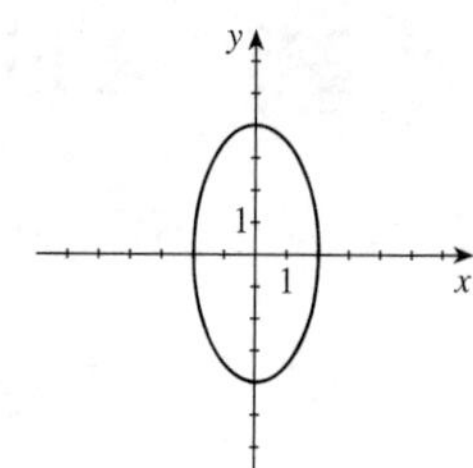

**11.** $2x^2 + y^2 = 3 \Leftrightarrow \dfrac{x^2}{\frac{3}{2}} + \dfrac{y^2}{3} = 1$. This ellipse has $a = \sqrt{3}$, $b = \sqrt{\frac{3}{2}}$, and so $c^2 = 3 - \frac{3}{2} = \frac{3}{2} \Leftrightarrow c = \sqrt{\frac{3}{2}} = \frac{\sqrt{6}}{2}$. The vertices are $(0, \pm\sqrt{3})$, the foci are $\left(0, \pm\frac{\sqrt{6}}{2}\right)$, the eccentricity is $e = \dfrac{c}{a} = \dfrac{\frac{\sqrt{6}}{2}}{\sqrt{3}} = \frac{\sqrt{2}}{2}$, the length of the major axis is $2a = 2\sqrt{3}$, and the length of the minor axis is $2b = 2 \cdot \frac{\sqrt{6}}{2} = \sqrt{6}$.

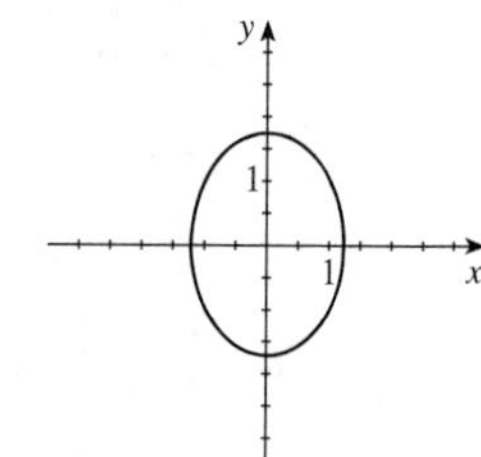

**12.** $5x^2 + 6y^2 = 30 \Leftrightarrow \dfrac{x^2}{6} + \dfrac{y^2}{5} = 1$. This ellipse has $a = \sqrt{6}$, $b = \sqrt{5}$, and so $c^2 = 6 - 5 = 1 \Leftrightarrow c = 1$. The vertices are $(\pm\sqrt{6}, 0)$, the foci are $(\pm 1, 0)$, the eccentricity is $e = \dfrac{c}{a} = \frac{1}{\sqrt{6}} = \frac{\sqrt{6}}{6}$: length of the major axis: $2a = 2\sqrt{6}$, and the length of the minor axis is $2b = 2\sqrt{5}$.

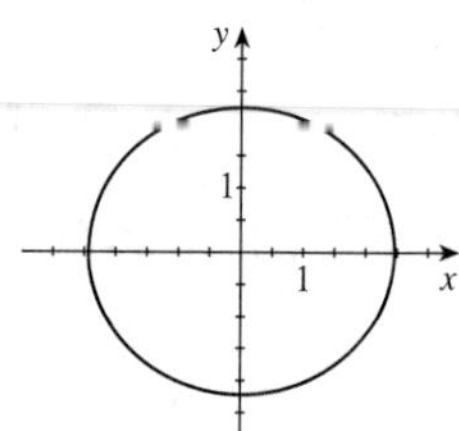

**13.** $x^2 + 4y^2 = 1 \Leftrightarrow \dfrac{x^2}{1} + \dfrac{y^2}{\frac{1}{4}} = 1$. This ellipse has $a = 1$, $b = \frac{1}{2}$, and so $c^2 = 1 - \frac{1}{4} = \frac{3}{4} \Leftrightarrow c = \frac{\sqrt{3}}{2}$. The vertices are $(\pm 1, 0)$, the foci are $\left(\pm\frac{\sqrt{3}}{2}, 0\right)$, the eccentricity is $e = \dfrac{c}{a} = \frac{\sqrt{3}/2}{1} = \frac{\sqrt{3}}{2}$, the length of the major axis is $2a = 2$, and the length of the minor axis is $2b = 1$.

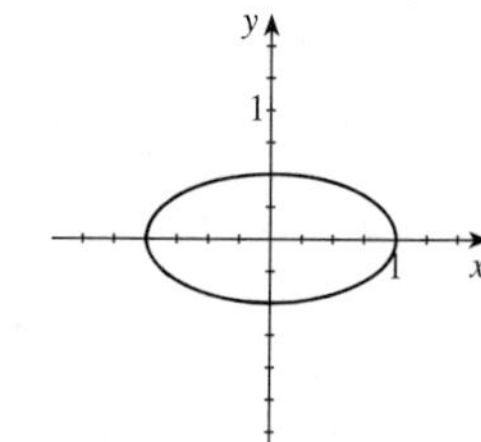

**14.** $9x^2 + 4y^2 = 1 \Leftrightarrow \dfrac{x^2}{1/9} + \dfrac{y^2}{1/4} = 1$. This ellipse has $a = \frac{1}{2}$, and so $c^2 = \frac{1}{4} - \frac{1}{9} = \frac{5}{36} \Leftrightarrow c = \frac{\sqrt{5}}{6}$. The vertices are $\left(0, \pm\frac{1}{2}\right)$, the foci are $\left(0, \pm\frac{\sqrt{5}}{6}\right)$, the eccentricity is $e = \dfrac{c}{a} = \frac{\sqrt{5}/6}{1/2} = \frac{\sqrt{5}}{3}$, the length of the major axis is $2a = 1$, and the length of the minor axis is $2b = \frac{2}{3}$.

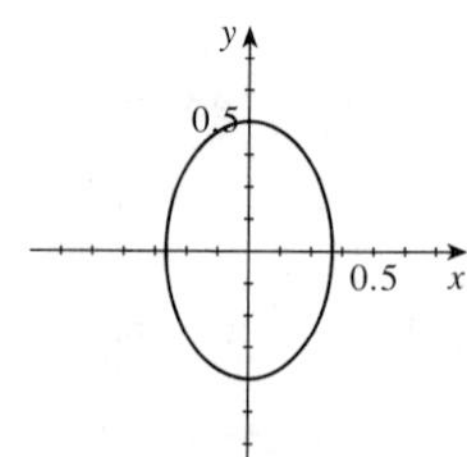

**15.** $\frac{1}{2}x^2 + \frac{1}{8}y^2 = \frac{1}{4} \Leftrightarrow 2x^2 + \frac{1}{2}y^2 = 1 \Leftrightarrow \dfrac{x^2}{\frac{1}{2}} + \dfrac{y^2}{2} = 1$. This ellipse has $a = \sqrt{2}$, $b = \frac{1}{\sqrt{2}}$, and so $c^2 = 2 - \frac{1}{2} = \frac{3}{2} \Leftrightarrow c = \sqrt{\frac{3}{2}} = \frac{\sqrt{6}}{2}$. The vertices are $(0, \pm\sqrt{2})$, the foci are $\left(0, \pm\frac{\sqrt{6}}{2}\right)$, the eccentricity is $e = \dfrac{c}{a} = \dfrac{\frac{\sqrt{6}}{2}}{\sqrt{2}} = \frac{\sqrt{3}}{2}$, the length of the major axis is $2a = 2\sqrt{2}$, and length of the minor axis is $2b = \sqrt{2}$.

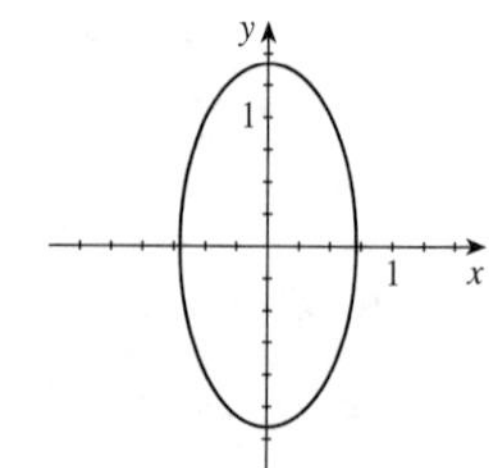

**16.** $x^2 = 4 - 2y^2 \quad \Leftrightarrow \quad x^2 + 2y^2 = 4 \quad \Leftrightarrow \quad \dfrac{x^2}{4} + \dfrac{y^2}{2} = 1$. This ellipse has $a = 2$, $b = \sqrt{2}$, and so $c^2 = 4 - 2 = 2 \quad \Leftrightarrow \quad c = \sqrt{2}$. The vertices are $(\pm 2, 0)$, the foci are $(\pm\sqrt{2}, 0)$, the eccentricity is $e = \dfrac{c}{a} = \dfrac{\sqrt{2}}{2}$, the length of the major axis is $2a = 4$, and the length of the minor axis is $2b = 2\sqrt{2}$.

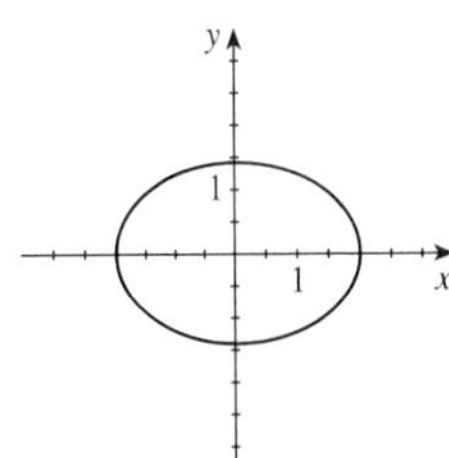


**17.** $y^2 = 1 - 2x^2 \quad \Leftrightarrow \quad 2x^2 + y^2 = 1 \quad \Leftrightarrow \quad \dfrac{x^2}{\frac{1}{2}} + \dfrac{y^2}{1} = 1$. This ellipse has $a = 1$, $b = \frac{\sqrt{2}}{2}$, and so $c^2 = 1 - \frac{1}{2} = \frac{1}{2} \quad \Leftrightarrow \quad c = \frac{\sqrt{2}}{2}$. The vertices are $(0, \pm 1)$, the foci are $\left(0, \pm\frac{\sqrt{2}}{2}\right)$, the eccentricity is $e = \dfrac{c}{a} = \frac{1/\sqrt{2}}{1} = \frac{\sqrt{2}}{2}$, the length of the major axis is $2a = 2$, and the length of the minor axis is $2b = \sqrt{2}$.

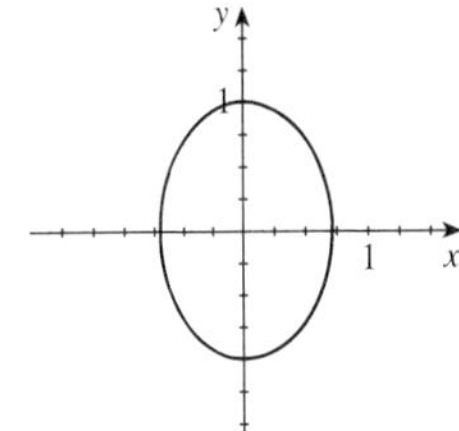


**18.** $20x^2 + 4y^2 = 5 \quad \Leftrightarrow \quad \dfrac{x^2}{1/4} + \dfrac{y^2}{5/4} = 1$. This ellipse has $a = \frac{\sqrt{5}}{2}$, $b = \frac{1}{2}$, and so $c^2 = \frac{5}{4} - \frac{1}{4} = 1 \quad \Leftrightarrow \quad c = 1$. The vertices are $\left(0, \pm\frac{\sqrt{5}}{2}\right)$, the foci are $(0, \pm 1)$, the eccentricity is $e = \dfrac{c}{a} = \frac{1}{\sqrt{5}/2} = \frac{2\sqrt{5}}{5}$, the length of the major axis is $2a = \sqrt{5}$, and the length of the minor axis is $2b = 1$.

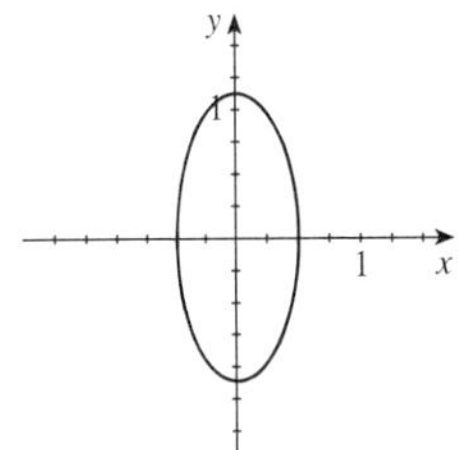


**19.** This ellipse has a horizontal major axis with $a = 5$ and $b = 4$, so an equation is $\dfrac{x^2}{(5)^2} + \dfrac{y^2}{(4)^2} = 1 \quad \Leftrightarrow \quad \dfrac{x^2}{25} + \dfrac{y^2}{16} = 1$.

**20.** This ellipse has a vertical major axis with $a = 5$ and $b = 2$. Thus an equation is $\dfrac{x^2}{2^2} + \dfrac{y^2}{5^2} = 1 \quad \Leftrightarrow \quad \dfrac{x^2}{4} + \dfrac{y^2}{25} = 1$.

**21.** This ellipse has a vertical major axis with $c = 2$ and $b = 2$. So $a^2 = c^2 + b^2 = 2^2 + 2^2 = 8 \quad \Leftrightarrow \quad a = 2\sqrt{2}$. So an equation is $\dfrac{x^2}{(2)^2} + \dfrac{y^2}{\left(2\sqrt{2}\right)^2} = 1 \quad \Leftrightarrow \quad \dfrac{x^2}{4} + \dfrac{y^2}{8} = 1$.

**22.** This ellipse has a vertical major axis with $a = 4$ and $c = 3$. So $c^2 = a^2 - b^2 \quad \Leftrightarrow \quad 9 = 16 - b^2 \quad \Leftrightarrow \quad b^2 = 7$. Thus an equation is $\dfrac{x^2}{7} + \dfrac{y^2}{4^2} = 1 \quad \Leftrightarrow \quad \dfrac{x^2}{7} + \dfrac{y^2}{16} = 1$.

**23.** This ellipse has a horizontal major axis with $a = 16$, so an equation of the ellipse is of the form $\dfrac{x^2}{16^2} + \dfrac{y^2}{b^2} = 1$. Substituting the point $(8, 6)$ into the equation, we get $\frac{64}{256} + \dfrac{36}{b^2} = 1 \quad \Leftrightarrow \quad \dfrac{36}{b^2} = 1 - \frac{1}{4} \quad \Leftrightarrow \quad \dfrac{36}{b^2} = \dfrac{3}{4} \quad \Leftrightarrow \quad b^2 = \dfrac{4\,(36)}{3} = 48$. Thus, an equation of the ellipse is $\dfrac{x^2}{256} + \dfrac{y^2}{48} = 1$.

**24.** This ellipse has a vertical major axis with $b = 2$, so an equation of the ellipse is of the form $\dfrac{x^2}{2^2} + \dfrac{y^2}{a^2} = 1$. Substituting the point $(-1, 2)$ into the equation, we get $\frac{1}{4} + \dfrac{4}{a^2} = 1 \quad \Leftrightarrow \quad \dfrac{4}{a^2} = 1 - \dfrac{1}{4} \quad \Leftrightarrow \quad \dfrac{4}{a^2} = \dfrac{3}{4} \quad \Leftrightarrow \quad a^2 = \dfrac{4\,(4)}{3} = \dfrac{16}{3}$. Thus, an equation of the ellipse is $\dfrac{x^2}{4} + \dfrac{y^2}{16/3} = 1 \quad \Leftrightarrow \quad \dfrac{x^2}{4} + \dfrac{3y^2}{16} = 1$.

**25.** $\dfrac{x^2}{25} + \dfrac{y^2}{20} = 1 \quad \Leftrightarrow \quad \dfrac{y^2}{20} = 1 - \dfrac{x^2}{25} \quad \Leftrightarrow$

$y^2 = 20 - \dfrac{4x^2}{5} \quad \Rightarrow \quad y = \pm\sqrt{20 - \dfrac{4x^2}{5}}.$

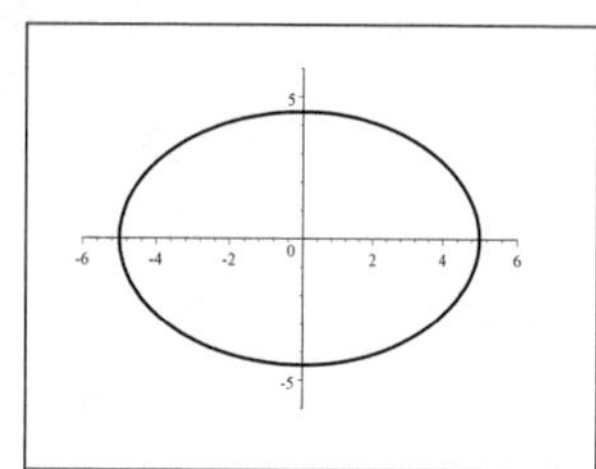

**26.** $x^2 + \dfrac{y^2}{12} = 1 \quad \Leftrightarrow \quad \dfrac{y^2}{12} = 1 - x^2 \quad \Leftrightarrow \quad y^2 = 12 - 12x^2$

$\Rightarrow \quad y = \pm\sqrt{12 - 12x^2}.$

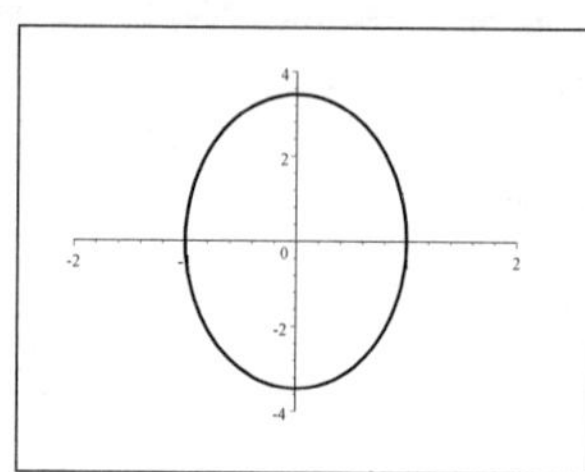

**27.** $6x^2 + y^2 = 36 \quad \Leftrightarrow \quad y^2 = 36 - 6x^2 \quad \Rightarrow$

$y = \pm\sqrt{36 - 6x^2}.$

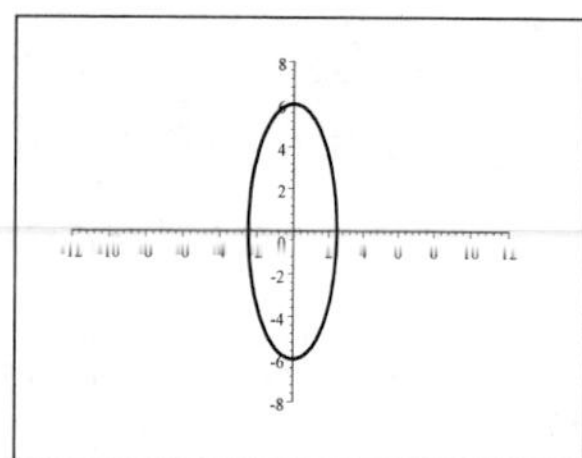

**28.** $x^2 + 2y^2 = 8 \quad \Leftrightarrow \quad 2y^2 = 8 - x^2 \quad \Leftrightarrow \quad y^2 = 4 - \dfrac{x^2}{2}$

$\Rightarrow \quad y = \pm\sqrt{4 - \dfrac{x^2}{2}}.$

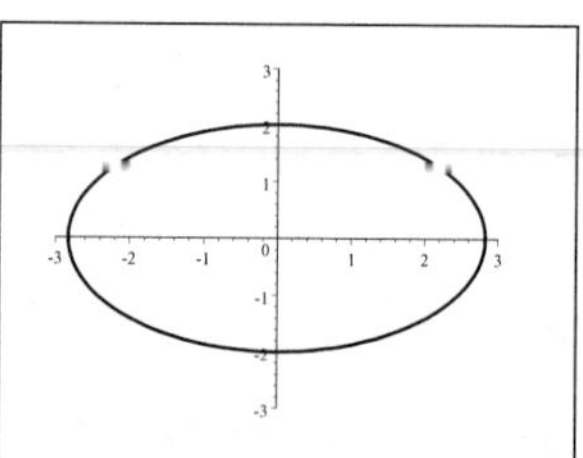

**29.** The foci are $(\pm 4, 0)$, and the vertices are $(\pm 5, 0)$. Thus, $c = 4$ and $a = 5$, and so $b^2 = 25 - 16 = 9$. Therefore, an equation of the ellipse is $\dfrac{x^2}{25} + \dfrac{y^2}{9} = 1$.

**30.** The foci are $(0, \pm 3)$ and the vertices are $(0, \pm 5)$. Thus, $c = 3$ and $a = 5$, and so $c^2 = a^2 - b^2 \quad \Leftrightarrow \quad 9 = 25 - b^2 \quad \Leftrightarrow$ $b^2 = 25 - 9 = 16$. Therefore, an equation of the ellipse is $\dfrac{x^2}{16} + \dfrac{y^2}{25} = 1$.

**31.** The length of the major axis is $2a = 4 \quad \Leftrightarrow \quad a = 2$, the length of the minor axis is $2b = 2 \quad \Leftrightarrow \quad b = 1$, and the foci are on the $y$-axis. Therefore, an equation of the ellipse is $x^2 + \dfrac{y^2}{4} = 1$.

**32.** The length of the major axis is $2a = 6 \quad \Leftrightarrow \quad a = 3$, the length of the minor axis is $2b = 4 \quad \Leftrightarrow \quad b = 2$, and the foci are on the $x$-axis. Therefore, an equation of the ellipse is $\dfrac{x^2}{9} + \dfrac{y^2}{4} = 1$.

**33.** The foci are $(0, \pm 2)$, and the length of the minor axis is $2b = 6 \quad \Leftrightarrow \quad b = 3$. Thus, $a^2 = 4 + 9 = 13$. Since the foci are on the $y$-axis, an equation is $\dfrac{x^2}{9} + \dfrac{y^2}{13} = 1$.

**34.** The foci are $(\pm 5, 0)$, and the length of the major axis is $2a = 12 \quad \Leftrightarrow \quad a = 6$. Thus, $c^2 = a^2 - b^2 \quad \Leftrightarrow \quad 25 = 36 - b^2$ $\Leftrightarrow \quad b^2 = 36 - 25 = 11$. Since the foci are on the $x$-axis, an equation is $\dfrac{x^2}{36} + \dfrac{y^2}{11} = 1$.

**35.** The endpoints of the major axis are $(\pm 10, 0) \quad \Leftrightarrow \quad a = 10$, and the distance between the foci is $2c = 6 \quad \Leftrightarrow \quad c = 3$. Therefore, $b^2 = 100 - 9 = 91$, and so an equation of the ellipse is $\dfrac{x^2}{100} + \dfrac{y^2}{91} = 1$.

**36.** Since the endpoints of the minor axis are $(0, \pm 3)$, we have $b = 3$. The distance between the foci is $2c = 8$, so $c = 4$. Thus, $a^2 = b^2 + c^2 = 9 + 16 = 25$, and an equation of the ellipse is $\dfrac{x^2}{25} + \dfrac{y^2}{9} = 1$.

**37.** The length of the major axis is 10, so $2a = 10 \Leftrightarrow a = 5$, and the foci are on the $x$-axis, so the form of the equation is $\dfrac{x^2}{25} + \dfrac{y^2}{b^2} = 1$. Since the ellipse passes through $(\sqrt{5}, 2)$, we know that $\dfrac{(\sqrt{5})^2}{25} + \dfrac{(2)^2}{b^2} = 1 \Leftrightarrow \dfrac{5}{25} + \dfrac{4}{b^2} = 1 \Leftrightarrow \dfrac{4}{b^2} = \frac{4}{5} \Leftrightarrow b^2 = 5$, and so an equation is $\dfrac{x^2}{25} + \dfrac{y^2}{5} = 1$.

**38.** Since the foci are $(0, \pm 2)$, we have $c = 2$. The eccentricity is $\frac{1}{9} = \dfrac{c}{a} \Leftrightarrow a = 9c = 9(2) = 18$, and so $b^2 = a^2 - c^2 = 18^2 - 2^2 = 320$. Since the foci lie on the $y$-axis, an equation of the ellipse is $\dfrac{x^2}{320} + \dfrac{y^2}{324} = 1$.

**39.** Since the foci are $(\pm 1.5, 0)$, we have $c = \dfrac{3}{2}$. Since the eccentricity is $0.8 = \dfrac{c}{a}$, we have $a = \dfrac{\frac{3}{2}}{\frac{4}{5}} = \dfrac{15}{8}$, and so $b^2 = \dfrac{225}{64} - \dfrac{9}{4} = \dfrac{225 - 16 \cdot 9}{64} = \dfrac{81}{64}$. Therefore, an equation of the ellipse is $\dfrac{x^2}{(15/8)^2} + \dfrac{y^2}{81/64} = 1 \Leftrightarrow \dfrac{64x^2}{225} + \dfrac{64y^2}{81} = 1$.

**40.** Since the length of the major axis is $2a = 4$, we have $a = 2$. The eccentricity is $\frac{\sqrt{3}}{2} = \dfrac{c}{a} = \dfrac{c}{2}$, so $c = \sqrt{3}$. Then $b^2 = a^2 - c^2 = 4 - 3 = 1$, and since the foci are on the $y$-axis, an equation of the ellipse is $x^2 + \dfrac{y^2}{4} = 1$.

**41.** $\begin{cases} 4x^2 + y^2 = 4 \\ 4x^2 + 9y^2 = 36 \end{cases}$ Subtracting the first equation from the second gives $8y^2 = 32 \Leftrightarrow y^2 = 4 \Leftrightarrow y = \pm 2$. Substituting $y = \pm 2$ in the first equation gives $4x^2 + (\pm 2)^2 = 4 \Leftrightarrow x = 0$, and so the points of intersection are $(0, \pm 2)$.

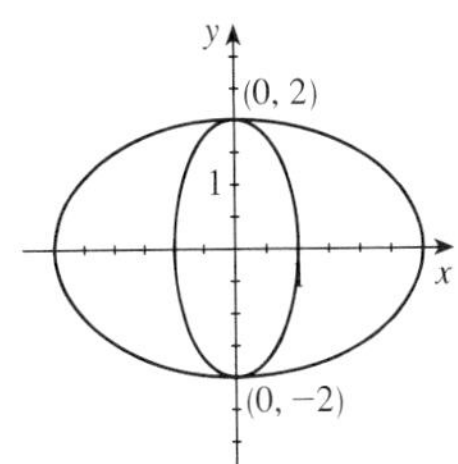


**42.** $\begin{cases} \dfrac{x^2}{16} + \dfrac{y^2}{9} = 1 \\ \dfrac{x^2}{9} + \dfrac{y^2}{16} = 1 \end{cases} \Leftrightarrow \begin{cases} 9x^2 + 16y^2 = 144 \\ 16x^2 + 9y^2 = 144 \end{cases} \Leftrightarrow \begin{cases} 144x^2 + 256y^2 = 2304 \\ -144x^2 - 81y^2 = -1296 \end{cases}$

Adding gives $175y^2 = 1008 \Rightarrow y = \pm\frac{12}{5}$. Substituting for $y$ gives $9x^2 + 16\left(\frac{12}{5}\right)^2 = 144 \Leftrightarrow 9x^2 = 144 - \frac{2304}{25} = \dfrac{1296}{25} \Leftrightarrow x = \pm\frac{12}{5}$, and so the four points of intersection are $\left(\pm\frac{12}{5}, \pm\frac{12}{5}\right)$.

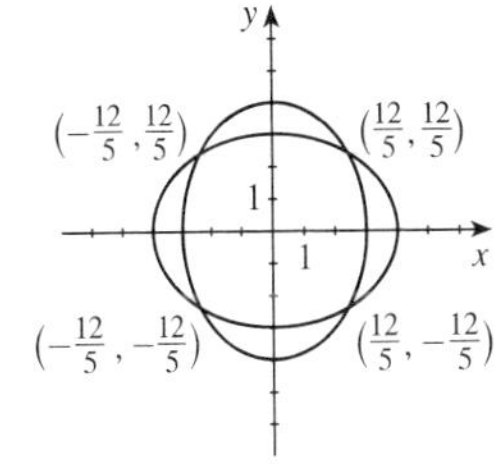


**43.** $\begin{cases} 100x^2 + 25y^2 = 100 \\ x^2 + \dfrac{y^2}{9} = 1 \end{cases}$ Dividing the first equation by 100 gives $x^2 + \dfrac{y^2}{4} = 1$.

Subtracting this equation from the second equation gives $\dfrac{y^2}{9} - \dfrac{y^2}{4} = 0 \Leftrightarrow \left(\frac{1}{9} - \frac{1}{4}\right) y^2 = 0 \Leftrightarrow y = 0$. Substituting $y = 0$ in the second equation gives $x^2 + (0)^2 = 1 \Leftrightarrow x = \pm 1$, and so the points of intersection are $(\pm 1, 0)$.

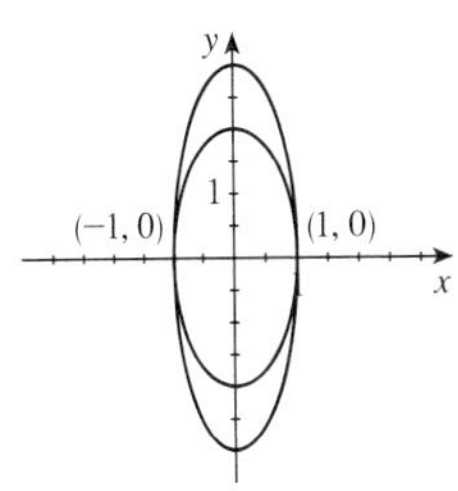


**44. (a)** The ellipse $x^2 + 4y^2 = 16 \Leftrightarrow \dfrac{x^2}{16} + \dfrac{y^2}{4} = 1$ has $a = 4$ and $b = 2$. Thus, an equation of the ancillary circle is $x^2 + y^2 = 4$.

**(b)** If $(s, t)$ is a point on the ancillary circle, then $s^2 + t^2 = 4 \Leftrightarrow 4s^2 + 4t^2 = 16 \Leftrightarrow (2s)^2 + 4(t)^2 = 16$, which implies that $(2s, t)$ is a point on the ellipse.

**45. (a)** $x^2 + ky^2 = 100 \Leftrightarrow ky^2 = 100 - x^2 \Leftrightarrow y = \pm\frac{1}{k}\sqrt{100 - x^2}$.

For the top half, we graph $y = \frac{1}{k}\sqrt{100 - x^2}$ for $k = 4$, 10, 25, and 50.

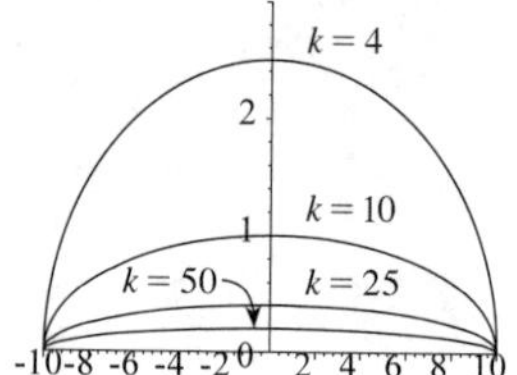


**(b)** This family of ellipses have common major axes and vertices, and the eccentricity increases as $k$ increases.

**46.** $\frac{x^2}{k} + \frac{y^2}{4+k} = 1$ is an ellipse for $k > 0$. Then $a^2 = 4 + k$, $b^2 = k$, and so $c^2 = 4 + k - k = 4 \Leftrightarrow c = \pm 2$. Therefore, all of the ellipses' foci are $(0, \pm 2)$ regardless of the value of $k$.

**47.** Using the perihelion, $a - c = 147{,}000{,}000$, while using the aphelion, $a + c = 153{,}000{,}000$. Adding, we have $2a = 300{,}000{,}000 \Leftrightarrow a = 150{,}000{,}000$. So $b^2 = a^2 - c^2 = (150 \times 10^6)^2 - (3 \times 10^6)^2 = 22{,}491 \times 10^{12} = 2.2491 \times 10^{16}$. Thus, an equation of the orbit is $\frac{x^2}{2.2500 \times 10^{16}} + \frac{y^2}{2.2491 \times 10^{16}} = 1$.

**48.** Using the eccentricity, $e = 0.25 = \frac{c}{a} \Leftrightarrow c = 0.25a$. Using the length of the minor axis, $2b = 10{,}000{,}000{,}000 \Leftrightarrow b = 5 \times 10^9$. Since $a^2 = c^2 + b^2$, $a^2 = (0.25a)^2 + 25 \times 10^{18} \Leftrightarrow \frac{15}{16}a^2 = 25 \times 10^{18} \Leftrightarrow a^2 = \frac{80}{3} \times 10^{18} \Leftrightarrow a = \sqrt{\frac{80}{3}} \times 10^9 = 4\sqrt{\frac{5}{3}} \times 10^9$. Then $c = 0.25\left(4\sqrt{\frac{5}{3}} \times 10^9\right) = \sqrt{\frac{5}{3}} \times 10^9$. Since the Sun is at one focus of the ellipse, the distance from Pluto to the Sun at perihelion is $a - c = 4\sqrt{\frac{5}{3}} \times 10^9 - \sqrt{\frac{5}{3}} \times 10^9 = 3\sqrt{\frac{5}{3}} \times 10^9 \approx 3.87 \times 10^9$ km; the distance from Pluto to the Sun at aphelion is $a + c = 4\sqrt{\frac{5}{3}} \times 10^9 + \sqrt{\frac{5}{3}} \times 10^9 = 5\sqrt{\frac{5}{3}} \times 10^9 \approx 6.45 \times 10^9$ km.

**49.** Using the perilune, $a - c = 1075 + 68 = 1143$, and using the apolune, $a + c = 1075 + 195 = 1270$. Adding, we get $2a = 2413 \Leftrightarrow a = 1206.5$. So $c = 1270 - 1206.5 \Leftrightarrow c = 63.5$. Therefore, $b^2 = (1206.5)^2 - (63.5)^2 = 1{,}451{,}610$. Since $a^2 \approx 1{,}455{,}642$, an equation of Apollo 11's orbit is $\frac{x^2}{1{,}455{,}642} + \frac{y^2}{1{,}451{,}610} = 1$.

**50.** Placing the origin at the center of the sheet of plywood and letting the $x$-axis be the long central axis, we have $2a = 8$, so that $a = 4$, and $2b = 4$, so that $b = 2$. So $c^2 = a^2 - b^2 = 4^2 - 2^2 = 12 \Rightarrow c = 2\sqrt{3} \approx 3.46$. So the tacks should be located $2(3.46) = 6.92$ feet apart and the string should be $2a = 8$ feet long.

**51.** From the diagram, $a = 40$ and $b = 20$, and so an equation of the ellipse whose top half is the window is $\frac{x^2}{1600} + \frac{y^2}{400} = 1$. Since the ellipse passes through the point $(25, h)$, by substituting, we have $\frac{25^2}{1600} + \frac{h^2}{400} = 1 \Leftrightarrow 625 + 4y^2 = 1600 \Leftrightarrow y = \frac{\sqrt{975}}{2} = \frac{5\sqrt{39}}{2} \approx 15.61$ in. Therefore, the window is approximately 15.6 inches high at the specified point.

**52.** Have each friend hold one end of the string on the blackboard. These fixed points will be the foci. Then, keeping the string taut with the chalk, draw the ellipse.

**53.** We start with the flashlight perpendicular to the wall; this shape is a circle. As the angle of elevation increases, the shape of the light changes to an ellipse. When the flashlight is angled so that the outer edge of the light cone is parallel to the wall, the shape of the light is a parabola. Finally, as the angle of elevation increases further, the shape of the light is hyperbolic.

**54.** The foci are $(\pm c, 0)$, where $c^2 = a^2 - b^2$. The endpoints of one *latus rectum* are the points $(c, \pm k)$, and the length is $2k$. Substituting this point into the equation, we get $\frac{c^2}{a^2} + \frac{k^2}{b^2} = 1 \Leftrightarrow \frac{k^2}{b^2} = 1 - \frac{c^2}{a^2} = \frac{a^2 - c^2}{a^2} \Leftrightarrow k^2 = \frac{b^2(a^2 - c^2)}{a^2}$. Since $b^2 = a^2 - c^2$, the last equation becomes $k^2 = \frac{b^4}{a^2} \Rightarrow k = \frac{b^2}{a}$. Thus, the length of the *latus rectum* is $2k = 2\left(\frac{b^2}{a}\right) = \frac{2b^2}{a}$.

**55.** The shape drawn on the paper is almost, but not quite, an ellipse. For example, when the bottle has radius 1 unit and the compass legs are set 1 unit apart, then it can be shown that an equation of the resulting curve is $1 + y^2 = 2\cos x$. The graph of this curve differs very slightly from the ellipse with the same major and minor axis. This example shows that in mathematics, things are not always as they appear to be.

# 11.3 Hyperbolas

**1.** $\frac{x^2}{4} - y^2 = 1$ is Graph III, which opens horizontally and has vertices at $(\pm 2, 0)$.

**2.** $y^2 - \frac{x^2}{9} = 1$ is Graph IV, which opens vertically and has vertices at $(0, \pm 1)$.

**3.** $16y^2 - x^2 = 144$ is Graph II, which pens vertically and has vertices at $(0, \pm 3)$.

**4.** $9x^2 - 25y^2 = 225$ is Graph I, which opens horizontally and has vertices at $(\pm 5, 0)$.

**5.** The hyperbola $\frac{x^2}{4} - \frac{y^2}{16} = 1$ has $a = 2$, $b = 4$, and $c^2 = 16 + 4 \quad \Rightarrow \quad c = 2\sqrt{5}$. The vertices are $(\pm 2, 0)$, the foci are $\left(\pm 2\sqrt{5}, 0\right)$, and the asymptotes are $y = \pm \frac{4}{2} x \quad \Leftrightarrow \quad y = \pm 2x$.

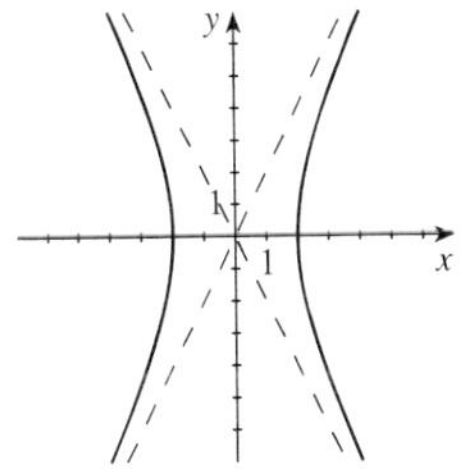


**6.** The hyperbola $\frac{y^2}{9} - \frac{x^2}{16} = 1$ has $a = 3$, $b = 4$, and $c^2 = 9 + 16 = 25 \quad \Rightarrow \quad c = 5$. The vertices are $(0, \pm 3)$, the foci are $(0, \pm 5)$, and the asymptotes are $y = \pm \frac{3}{4} x$.

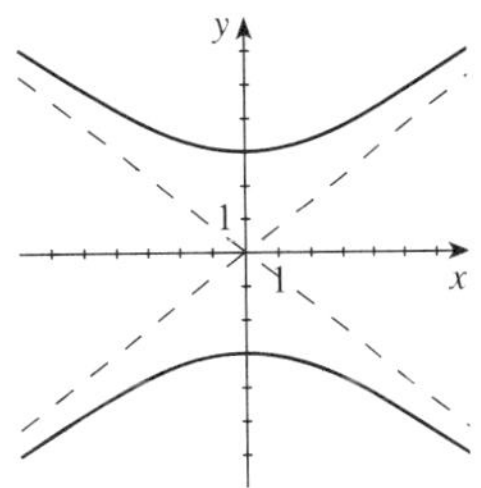


**7.** The hyperbola $\frac{y^2}{1} - \frac{x^2}{25} = 1$ has $a = 1$, $b = 5$, and $c^2 = 1 + 25 = 26 \quad \Rightarrow \quad c = \sqrt{26}$. The vertices are $(0, \pm 1)$, the foci are $\left(0, \pm \sqrt{26}\right)$, and the asymptotes are $y = \pm \frac{1}{5} x$.

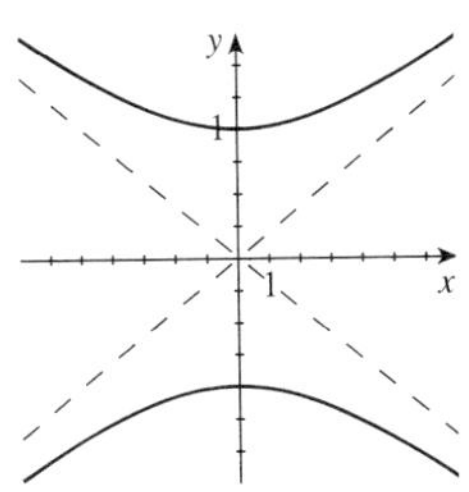


**8.** The hyperbola $\frac{x^2}{2} - \frac{y^2}{1} = 1$ has $a = \sqrt{2}$, $b = 1$, and $c^2 = 2 + 1 = 3 \quad \Rightarrow \quad c = \sqrt{3}$. The vertices are $\left(\pm \sqrt{2}, 0\right)$, the foci are $\left(\pm \sqrt{3}, 0\right)$, and the asymptotes are $y = \pm \frac{\sqrt{2}}{2} x$.

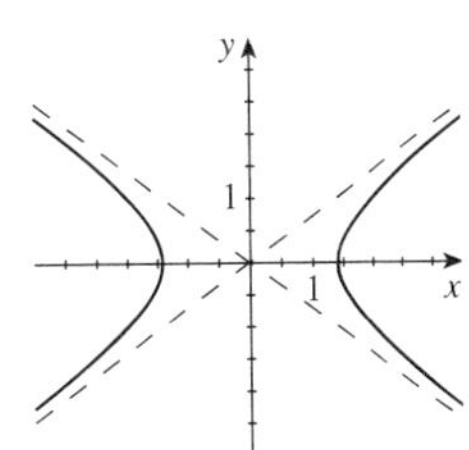

**9.** The hyperbola $x^2 - y^2 = 1$ has $a = 1$, $b = 1$, and $c^2 = 1 + 1 = 2 \Rightarrow c = \sqrt{2}$. The vertices are $(\pm 1, 0)$, the foci are $(\pm\sqrt{2}, 0)$, and the asymptotes are $y = \pm x$.

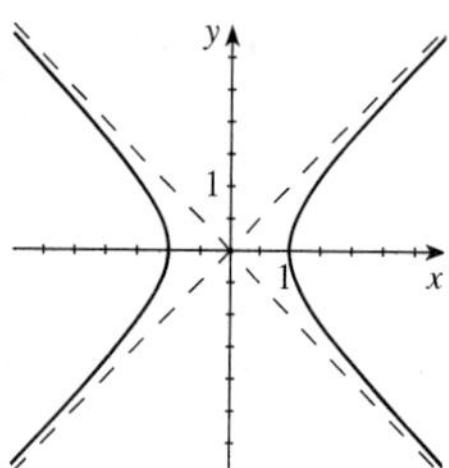


**10.** The hyperbola $9x^2 - 4y^2 = 36 \Leftrightarrow \dfrac{x^2}{4} - \dfrac{y^2}{9} = 1$ has $a = 2$, $b = 3$, and $c^2 = 4 + 9 = 13 \Rightarrow c = \sqrt{13}$. The vertices are $(\pm 2, 0)$, the foci are $(\pm\sqrt{13}, 0)$, and the asymptotes are $y = \pm\frac{3}{2}x$.

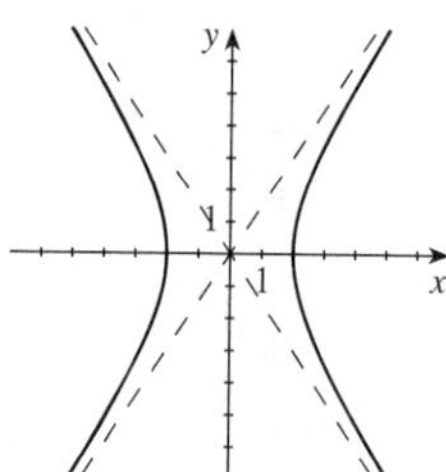


**11.** The hyperbola $25y^2 - 9x^2 = 225 \Leftrightarrow \dfrac{y^2}{9} - \dfrac{x^2}{25} = 1$ has $a = 3$, $b = 5$, and $c^2 = 25 + 9 = 34 \Rightarrow c = \sqrt{34}$. The vertices are $(0, \pm 3)$, the foci are $(0, \pm\sqrt{34})$, and the asymptotes are $y = \pm\frac{3}{5}x$.

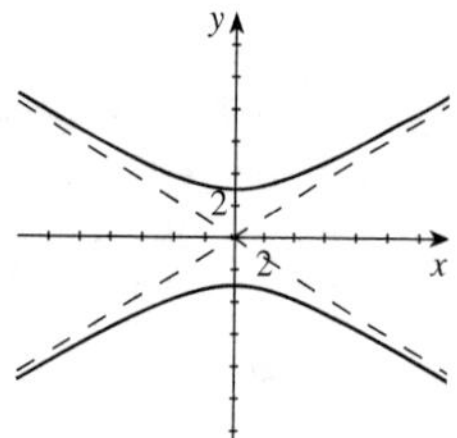


**12.** The hyperbola $x^2 - y^2 + 4 = 0 \Leftrightarrow y^2 - x^2 = 4 \Leftrightarrow \dfrac{y^2}{4} - \dfrac{x^2}{4} = 1$ has $a = 2$, $b = 2$, and $c^2 = 4 + 4 = 8 = 2\sqrt{2}$. The vertices are $(0, \pm 2)$, the foci are $(0, \pm 2\sqrt{2})$, and the asymptotes are $y = \pm x$.

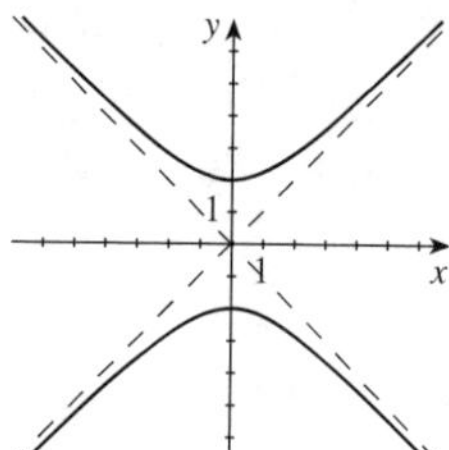


**13.** The hyperbola $x^2 - 4y^2 - 8 = 0 \Leftrightarrow \dfrac{x^2}{8} - \dfrac{y^2}{2} = 1$ has $a = \sqrt{8}$, $b = \sqrt{2}$, and $c^2 = 8 + 2 = 10 \Rightarrow c = \sqrt{10}$. The vertices are $(\pm 2\sqrt{2}, 0)$, the foci are $(\pm\sqrt{10}, 0)$, and the asymptotes are $y = \pm\frac{\sqrt{2}}{\sqrt{8}}x = \pm\frac{1}{2}x$.

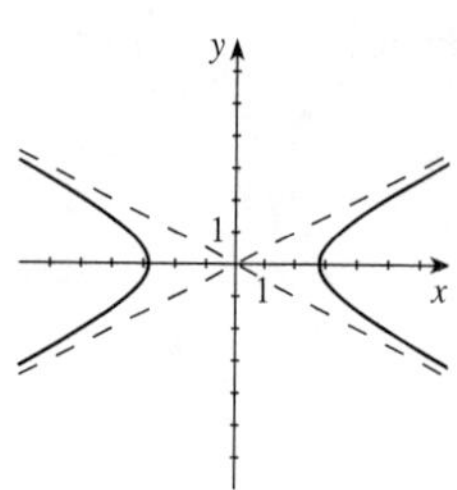


**14.** The hyperbola $x^2 - 2y^2 = 3 \Leftrightarrow \dfrac{x^2}{3} - \dfrac{y^2}{3/2} = 1$ has $a = \sqrt{3}$, $b = \sqrt{3/2} = \frac{\sqrt{6}}{2}$, and $c^2 = 3 + \frac{3}{2} = \frac{9}{2} \Rightarrow c = \frac{3\sqrt{2}}{2}$. The vertices are $(\pm\sqrt{3}, 0)$, the foci are $\left(\pm\frac{3\sqrt{2}}{2}, 0\right)$, and the asymptotes are

$$y = \pm\frac{\frac{\sqrt{6}}{2}}{\sqrt{3}}x = \pm\frac{\sqrt{2}}{2}x.$$

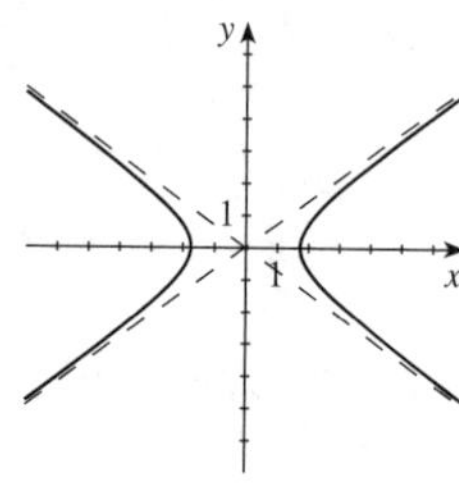

**15.** The hyperbola $4y^2 - x^2 = 1 \quad \Leftrightarrow \quad \dfrac{y^2}{\frac{1}{4}} - x^2 = 1$ has $a = \frac{1}{2}$, $b = 1$, and $c^2 = \frac{1}{4} + 1 = \frac{5}{4} \quad \Rightarrow \quad c = \frac{\sqrt{5}}{2}$. The vertices are $\left(0, \pm\frac{1}{2}\right)$, the foci are $\left(0, \pm\frac{\sqrt{5}}{2}\right)$, and the asymptotes are $y = \pm\frac{1/2}{1}x = \pm\frac{1}{2}x$.

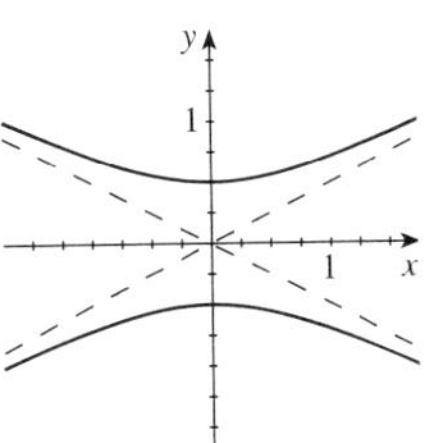


**16.** The hyperbola $9x^2 - 16y^2 = 1 \quad \Leftrightarrow \quad \dfrac{x^2}{1/9} - \dfrac{y^2}{1/16} = 1$ has $a = \frac{1}{3}$, $b = \frac{1}{4}$, and $c^2 = \frac{1}{9} + \frac{1}{16} = \frac{25}{144} \quad \Rightarrow$ $c = \frac{5}{12}$. The vertices are $\left(\pm\frac{1}{3}, 0\right)$, the foci are $\left(\pm\frac{5}{12}, 0\right)$, and the asymptotes are $y = \pm\frac{1/4}{1/3}x = \pm\frac{3}{4}x$.

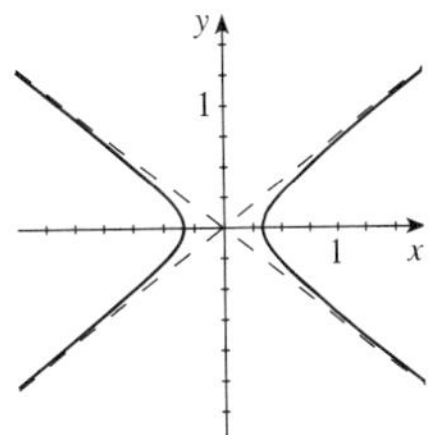


**17.** From the graph, the foci are $(\pm 4, 0)$, and the vertices are $(\pm 2, 0)$, so $c = 4$ and $a = 2$. Thus, $b^2 = 16 - 4 = 12$, and since the vertices are on the $x$-axis, an equation of the hyperbola is $\dfrac{x^2}{4} - \dfrac{y^2}{12} = 1$.

**18.** From the graph, the foci are $(0, \pm 13)$ and the vertices are $(0, \pm 12)$, so $c = 13$ and $a = 12$. Then $b^2 = c^2 - a^2 = 169 - 144 = 25$, and since the vertices are on the $y$-axis, an equation of the hyperbola is $\dfrac{y^2}{144} - \dfrac{x^2}{25} = 1$.

**19.** From the graph, the vertices are $(0, \pm 4)$, the foci are on the $y$-axis, and the hyperbola passes through the point $(3, -5)$. So the equation is of the form $\dfrac{y^2}{16} - \dfrac{x^2}{b^2} = 1$. Substituting the point $(3, -5)$, we have $\dfrac{(-5)^2}{16} - \dfrac{(3)^2}{b^2} = 1 \quad \Leftrightarrow \quad \dfrac{25}{16} - 1 = \dfrac{9}{b^2}$ $\Leftrightarrow \quad \dfrac{9}{16} = \dfrac{9}{b^2} \quad \Leftrightarrow \quad b^2 = 16$. Thus, an equation of the hyperbola is $\dfrac{y^2}{16} - \dfrac{x^2}{16} = 1$.

**20.** The vertices are $\left(\pm 2\sqrt{3}, 0\right)$, so $a = 2\sqrt{3}$, so an equation of the hyperbola is of the form $\dfrac{x^2}{\left(2\sqrt{3}\right)^2} - \dfrac{y^2}{b^2} = 1$. Substituting the point $(4, 4)$ into the equation, we get $\dfrac{16}{12} - \dfrac{16}{b^2} = 1 \quad \Leftrightarrow \quad \dfrac{16}{b^2} = \dfrac{16}{12} - 1 \quad \Leftrightarrow \quad \dfrac{16}{b^2} = \dfrac{4}{12} \quad \Leftrightarrow \quad b^2 = 48$. Thus, an equation of the hyperbola is $\dfrac{x^2}{12} - \dfrac{y^2}{48} = 1$.

**21.** The vertices are $(\pm 3, 0)$, so $a = 3$. Since the asymptotes are $y = \pm\frac{1}{2}x = \pm\dfrac{b}{a}x$, we have $\dfrac{b}{3} = \dfrac{1}{2} \quad \Leftrightarrow \quad b = \dfrac{3}{2}$. Since the vertices are on the $x$-axis, an equation is $\dfrac{x^2}{3^2} - \dfrac{y^2}{(3/2)^2} = 1 \quad \Leftrightarrow \quad \dfrac{x^2}{9} - \dfrac{4y^2}{9} = 1$.

**22.** From the graph, the vertices are $(0, \pm 3)$, so $a = 3$. Since the asymptotes are $y = \pm 3x = \pm\dfrac{a}{b}x$, we have $\dfrac{3}{b} = 3 \quad \Leftrightarrow$ $b = 1$. Since the vertices are on the $x$-axis, an equation is $\dfrac{y^2}{3^2} - \dfrac{x^2}{1^2} = 1 \quad \Leftrightarrow \quad \dfrac{y^2}{9} - x^2 = 1$.

**23.** $x^2 - 2y^2 = 8 \quad \Leftrightarrow \quad 2y^2 = x^2 - 8 \Leftrightarrow y^2 = \frac{1}{2}x^2 - 4$ $\Rightarrow y = \pm\sqrt{\frac{1}{2}x^2 - 4}$

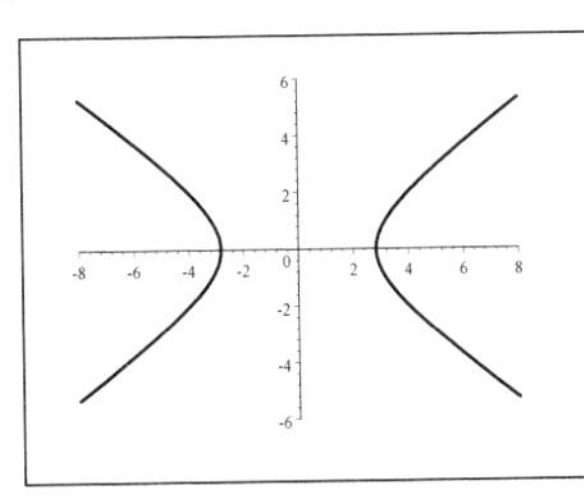


**24.** $3y^2 - 4x^2 = 24 \quad \Leftrightarrow \quad 3y^2 = 4x^2 + 24 \Leftrightarrow$ $y^2 = \frac{4}{3}x^2 + 8 \quad \Rightarrow y = \pm\sqrt{\frac{4}{3}x^2 + 8}$

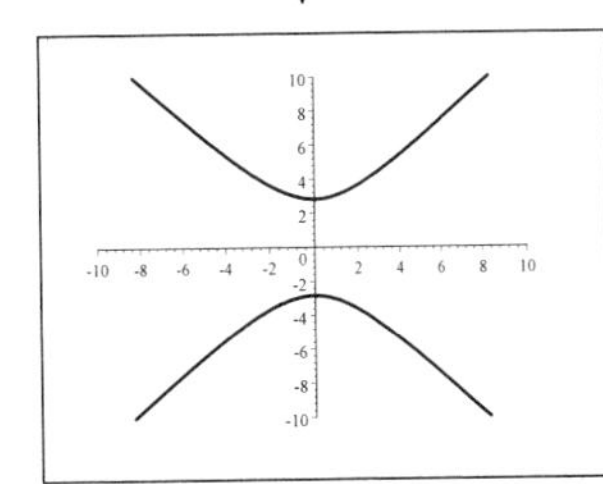

**25.** $\dfrac{y^2}{2} - \dfrac{x^2}{6} = 1 \Leftrightarrow \dfrac{y^2}{2} = \dfrac{x^2}{6} + 1 \Leftrightarrow y^2 = \dfrac{x^2}{3} + 2 \Rightarrow$

$y = \pm\sqrt{\dfrac{x^2}{3} + 2}$

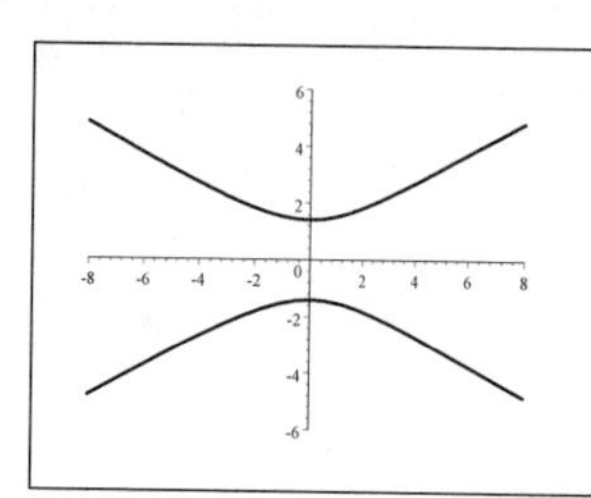

**26.** $\dfrac{x^2}{100} - \dfrac{y^2}{64} = 1 \Leftrightarrow 16x^2 - 25y^2 = 1600 \Leftrightarrow$

$25y^2 = 16x^2 - 1600 \Leftrightarrow y^2 = \frac{16}{25}x^2 - 64 \Rightarrow$

$y = \pm\sqrt{\frac{16}{25}x^2 - 64}$

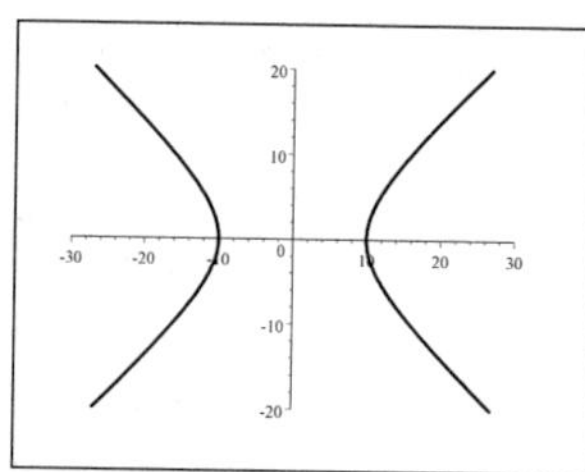

**27.** The foci are $(\pm5, 0)$ and the vertices are $(\pm3, 0)$, so $c = 5$ and $a = 3$. Then $b^2 = 25 - 9 = 16$, and since the vertices are on the $x$-axis, an equation of the hyperbola is $\dfrac{x^2}{9} - \dfrac{y^2}{16} = 1$.

**28.** The foci are $(0, \pm10)$ and the vertices are $(0, \pm8)$, so $c = 10$ and $a = 8$. Then $b^2 = c^2 - a^2 = 100 - 64 = 36$, and since the vertices are on the $y$-axis, an equation of the hyperbola is $\dfrac{y^2}{64} - \dfrac{x^2}{36} = 1$.

**29.** The foci are $(0, \pm2)$ and the vertices are $(0, \pm1)$, so $c = 2$ and $a = 1$. Then $b^2 = 4 - 1 = 3$, and since the vertices are on the $y$-axis, an equation is $y^2 - \dfrac{x^2}{3} = 1$.

**30.** The foci are $(\pm6, 0)$ and the vertices are $(\pm2, 0)$, so $c = 6$ and $a = 2$. Then $b^2 = c^2 - a^2 = 36 - 4 = 32$, and since the vertices are on the $x$-axis, an equation is $\dfrac{x^2}{4} - \dfrac{y^2}{32} = 1$.

**31.** The vertices are $(\pm1, 0)$ and the asymptotes are $y = \pm5x$, so $a = 1$. The asymptotes are $y = \pm\dfrac{b}{a}x$, so $\dfrac{b}{1} = 5 \Leftrightarrow b = 5$. Therefore, an equation of the hyperbola is $x^2 - \dfrac{y^2}{25} = 1$.

**32.** The vertices are $(0, \pm6)$, so $a = 6$. The asymptotes are $y = \pm\frac{1}{3}x = \pm\dfrac{a}{b}x \Leftrightarrow \dfrac{6}{b} = \frac{1}{3} \Leftrightarrow b = 18$. Since the vertices are on the $y$-axis, an equation of the hyperbola is $\dfrac{y^2}{36} - \dfrac{x^2}{324} = 1$.

**33.** The foci are $(0, \pm8)$, and the asymptotes are $y = \pm\frac{1}{2}x$, so $c = 8$. The asymptotes are $y = \pm\dfrac{a}{b}x$, so $\dfrac{a}{b} = \dfrac{1}{2}$ and $b = 2a$. Since $a^2 + b^2 = c^2 = 64$, we have $a^2 + 4a^2 = 64 \Leftrightarrow a^2 = \frac{64}{5}$ and $b^2 = 4a^2 = \frac{256}{5}$. Thus, an equation of the hyperbola is $\dfrac{y^2}{64/5} - \dfrac{x^2}{256/5} = 1 \Leftrightarrow \dfrac{5y^2}{64} - \dfrac{5x^2}{256} = 1$.

**34.** The vertices are $(0, \pm6)$, so $a = 6$. Since the vertices are on the $y$-axis, the hyperbola has an equation of the form $\dfrac{y^2}{36} - \dfrac{x^2}{b^2} = 1$. Since the hyperbola passes through the point $(-5, 9)$, we have $\dfrac{81}{36} - \dfrac{25}{b^2} = 1 \Leftrightarrow \dfrac{25}{b^2} = \dfrac{45}{36} \Leftrightarrow$ $b^2 = 20$. Thus, an equation is $\dfrac{y^2}{36} - \dfrac{x^2}{20} = 1$.

**35.** The asymptotes of the hyperbola are $y = \pm x$, so $b = a$. Since the hyperbola passes through the point $(5, 3)$, its foci are on the $x$-axis, and an equation is of the form, $\frac{x^2}{a^2} - \frac{y^2}{a^2} = 1$, so it follows that $\frac{25}{a^2} - \frac{9}{a^2} = 1 \Leftrightarrow a^2 = 16 = b^2$. Therefore, an equation of the hyperbola is $\frac{x^2}{16} - \frac{y^2}{16} = 1$.

**36.** The foci are $(\pm 3, 0)$, so $c = 3$. Since the vertices are on the $x$-axis, an equation of the hyperbola is of the form $\frac{x^2}{a^2} - \frac{y^2}{b^2} = 1$. Since the hyperbola passes through the point $(4, 1)$, we have $\frac{16}{a^2} - \frac{1}{b^2} = 1$. Using the foci, $c^2 = a^2 + b^2 = 9 \Leftrightarrow a^2 = 9 - b^2$, and substituting gives $\frac{16}{9 - b^2} - \frac{1}{b^2} = 1 \Leftrightarrow 16b^2 - (9 - b^2) = b^2(9 - b^2) \Leftrightarrow 17b^2 - 9 = 9b^2 - b^4 \Leftrightarrow b^4 + 8b^2 - 9 = 0 \Leftrightarrow (b^2 + 9)(b^2 - 1) = 0 \Leftrightarrow b^2 = 1$ or $b^2 = -9$ (never). Then we have $a^2 = 9 - b^2 = 8$. Thus, an equation of the hyperbola is $\frac{x^2}{8} - y^2 = 1$.

**37.** The foci are $(\pm 5, 0)$, and the length of the transverse axis is 6, so $c = 5$ and $2a = 6 \Leftrightarrow a = 3$. Thus, $b^2 = 25 - 9 = 16$, and an equation is $\frac{x^2}{9} - \frac{y^2}{16} = 1$.

**38.** The foci are $(0, \pm 1)$, and the length of the transverse axis is 1, so $c = 1$ and $2a = 1 \Leftrightarrow a = \frac{1}{2}$. Then $b^2 = c^2 - a^2 = 1 - \frac{1}{4} = \frac{3}{4}$, and since the foci are on the $y$-axis, an equation is $\frac{y^2}{1/4} - \frac{x^2}{3/4} = 1 \Leftrightarrow 4y^2 - \frac{4x^2}{3} = 1$.

**39. (a)** The hyperbola $x^2 - y^2 = 5 \Leftrightarrow \frac{x^2}{5} - \frac{y^2}{5} = 1$ has $a = \sqrt{5}$ and $b = \sqrt{5}$. Thus, the asymptotes are $y = \pm x$, and their slopes are $m_1 = 1$ and $m_2 = -1$. Since $m_1 \cdot m_2 = -1$, the asymptotes are perpendicular.

**(b)** Since the asymptotes are perpendicular, they must have slopes $\pm 1$, so $a = b$. Therefore, $c^2 = 2a^2 \Leftrightarrow a^2 = \frac{c^2}{2}$, and since the vertices are on the $x$-axis, an equation is $\frac{x^2}{\frac{1}{2}c^2} - \frac{y^2}{\frac{1}{2}c^2} = 1 \Leftrightarrow x^2 - y^2 = \frac{c^2}{2}$.

**40.** The hyperbolas $\frac{x^2}{a^2} - \frac{y^2}{b^2} = 1$ and $\frac{x^2}{a^2} - \frac{y^2}{b^2} = -1$ are conjugate to each other.

**(a)** $x^2 - 4y^2 + 16 = 0 \Leftrightarrow \frac{x^2}{16} - \frac{y^2}{4} = -1$ and $4y^2 - x^2 + 16 = 0 \Leftrightarrow \frac{x^2}{16} - \frac{y^2}{4} = 1$. So the hyperbolas are conjugate to each other.

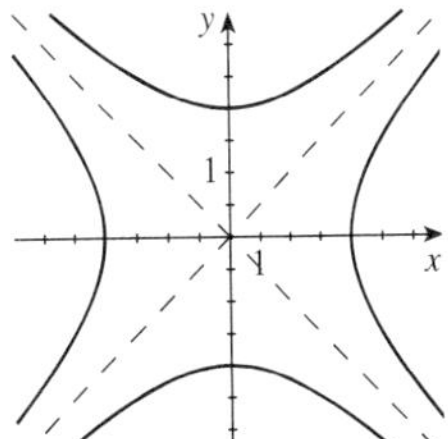


**(b)** They both have the same asymptotes, $y = \pm\frac{1}{2}x$.

**(c)** The two general conjugate hyperbolas both have asymptotes $y = \pm\frac{b}{a}x$.

**41.** $\sqrt{(x + c)^2 + y^2} - \sqrt{(x - c)^2 + y^2} = \pm 2a$. Let us consider the positive case only. Then $\sqrt{(x + c)^2 + y^2} = 2a + \sqrt{(x - c)^2 + y^2}$, and squaring both sides gives $x^2 + 2cx + c^2 + y^2 = 4a^2 + 4a\sqrt{(x - c)^2 + y^2} + x^2 - 2cx + c^2 + y^2 \Leftrightarrow 4a\sqrt{(x - c)^2 + y^2} = 4cx - 4a^2$. Dividing by 4 and squaring both sides gives $a^2(x^2 - 2cx + c^2 + y^2) = c^2x^2 - 2a^2cx + a^4 \Leftrightarrow a^2x^2 - 2a^2cx + a^2c^2 + a^2y^2 = c^2x^2 - 2a^2cx + a^4 \Leftrightarrow a^2x^2 + a^2c^2 + a^2y^2 = c^2x^2 + a^4$. Rearranging the order, we have $c^2x^2 - a^2x^2 - a^2y^2 = a^2c^2 - a^4 \Leftrightarrow (c^2 - a^2)x^2 - a^2y^2 = a^2(c^2 - a^2)$. The negative case gives the same result.

**42. (a)** The hyperbola $\frac{x^2}{9} - \frac{y^2}{16} = 1$ has $a = 3$, $b = 4$, so $c^2 = 9 + 16 = 25$, and $c = 5$. Therefore, the foci are $F_1(5, 0)$ and $F_2(-5, 0)$.

**(b)** Substituting $P\left(5, \frac{16}{3}\right)$ into an equation of the hyperbola, we get $\frac{25}{9} - \frac{256/9}{16} = \frac{25}{9} - \frac{16}{9} = 1$, so $P$ lies on the hyperbola.

**(c)** $d(P, F_1) = \sqrt{(5-5)^2 + (16/3 - 0)^2} = \frac{16}{3}$, and $d(P, F_2) = \sqrt{(-5-5)^2 + (16/3 - 0)^2} = \frac{1}{3}\sqrt{900 + 256} = \frac{34}{3}$.

**(d)** $d(P, F_2) - d(P, F_1) = \frac{34}{3} - \frac{16}{3} = 6 = 2(3) = 2a$.

**43. (a)** From the equation, we have $a^2 = k$ and $b^2 = 16 - k$. Thus, $c^2 = a^2 + b^2 = k + 16 - k = 16 \Rightarrow c = \pm 4$. Thus the foci of the family of hyperbolas are $(0, \pm 4)$.

**(b)** $\frac{y^2}{k} - \frac{x^2}{16 - k} = 1 \Leftrightarrow y^2 = k\left(1 + \frac{x^2}{16 - k}\right) \Rightarrow$

$y = \pm\sqrt{k + \frac{kx^2}{16 - k}}$. For the top branch, we graph $y = \sqrt{k + \frac{kx^2}{16 - k}}$, $k = 1, 4, 8, 12$. As $k$ increases, the asymptotes get steeper and the vertices move further apart.

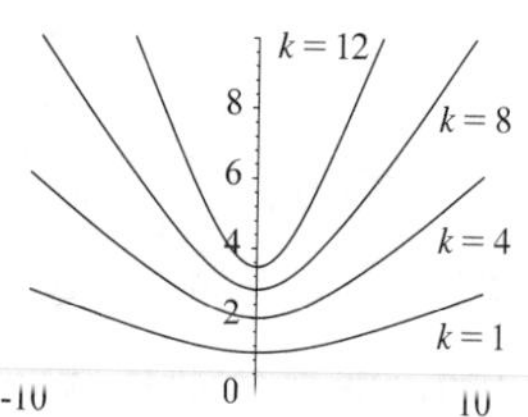


**44.** $d(AB) = 500 = 2c \Leftrightarrow c = 250$.

**(a)** Since $\Delta t = 2640$ and $v = 980$, we have $\Delta d = d(PA) - d(PB) = v\Delta t = (980 \text{ ft}/\mu\text{s}) \cdot (2640\ \mu\text{s}) = 2{,}587{,}200 \text{ ft} = 490\text{mi}$.

**(b)** $c = 250$, $2a = 490 \Leftrightarrow a = 245$ and the foci are on the $y$-axis. Then $b^2 = 250^2 - 245^2 = 2475$. Hence, an equation is $\frac{y^2}{60{,}025} - \frac{x^2}{2475} = 1$.

**(c)** Since $P$ is due east of $A$, $c = 250$ is the $y$-coordinate of $P$. Therefore, $P$ is at $(x, 250)$, and so $\frac{250^2}{245^2} - \frac{x^2}{2475} = 1 \Leftrightarrow$ $x^2 = 2475\left(\frac{250^2}{245^2} - 1\right) \approx 102.05$. Then $x \approx 10.1$, and so $P$ is approximately 10.1 miles from $A$.

**45.** Since the asymptotes are perpendicular, $a = b$. Also, since the sun is a focus and the closest distance is $2 \times 10^9$, it follows that $c - a = 2 \times 10^9$. Now $c^2 = a^2 + b^2 = 2a^2$, and so $c = \sqrt{2}a$. Thus, $\sqrt{2}a - a = 2 \times 10^9 \Rightarrow a = \frac{2 \times 10^9}{\sqrt{2} - 1}$ and $a^2 = b^2 = \frac{4 \times 10^{18}}{3 - 2\sqrt{2}} \approx 2.3 \times 10^{19}$. Therefore, an equation of the hyperbola is $\frac{x^2}{2.3 \times 10^{19}} - \frac{y^2}{2.3 \times 10^{19}} = 1 \Leftrightarrow$ $x^2 - y^2 = 2.3 \times 10^{19}$.

**46. (a)** These equally spaced concentric circles can be used as a kind of measure where we count the number of rings. In the case of the red dots, the sum of the number of wave crests from each center is a constant, in this case 17. As you move out one wave crest from the left stone you move in one wave crest from the right stone. Therefore this satisfies the geometric definition of an ellipse.

**(b)** Similarly, in the case of the blue dots, the difference of the number of wave crests from each center is a constant. As you move out one wave crest from the left center you also move out one wave crest from the right stone. Therefore this satisfies the geometric definition of a hyperbola.

**47.** Some possible answers are: as cross-sections of nuclear power plant cooling towers, or as reflectors for camouflaging the location of secret installations.

**48.** The wall is parallel to the axis of the cone of light coming from the top of the shade, so the intersection of the wall and the cone of light is a hyperbola. In the case of the flashlight, hold it parallel to the ground to form a hyperbola.

# 11.4 Shifted Conics

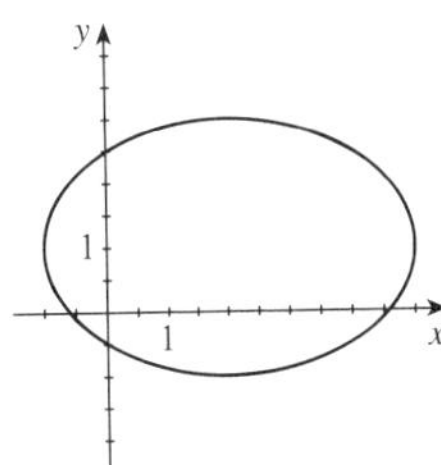

**1.** The ellipse $\dfrac{(x-2)^2}{9}+\dfrac{(y-1)^2}{4}=1$ is obtained from the ellipse $\dfrac{x^2}{9}+\dfrac{y^2}{4}=1$ by shifting it 2 units to the right and 1 unit upward. So $a=3$, $b=2$, and $c=\sqrt{9-4}=\sqrt{5}$. The center is $(2,1)$, the foci are $\left(2\pm\sqrt{5},1\right)$, the vertices are $(2\pm3,1)=(-1,1)$ and $(5,1)$, the length of the major axis is $2a=6$, and the length of the minor axis is $2b=4$.

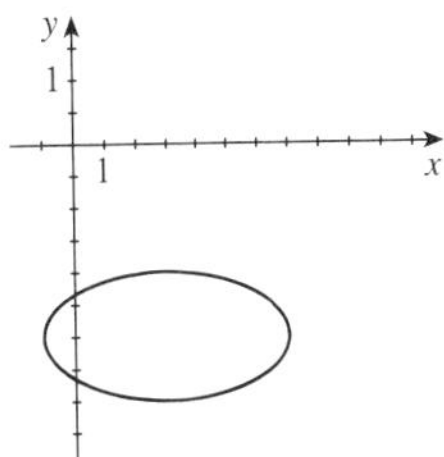

**2.** The ellipse $\dfrac{(x-3)^2}{16}+(y+3)^2=1$ is obtained from the ellipse $\dfrac{x^2}{16}+y^2=1$ by shifting to the right 3 units and downward 3 units. So $a=4$, $b=1$, and $c=\sqrt{16-1}=\sqrt{15}$. The center is $(3,-3)$, the foci are $\left(3\pm\sqrt{15},-3\right)$, the vertices are $(3\pm4,-3)=(-1,-3)$ and $(7,-3)$, the length of the major axis is $2a=8$, and the length of the minor axis is $2b=2$.

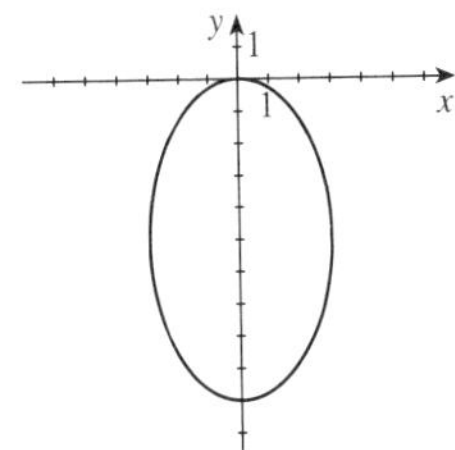

**3.** The ellipse $\dfrac{x^2}{9}+\dfrac{(y+5)^2}{25}=1$ is obtained from the ellipse $\dfrac{x^2}{9}+\dfrac{y^2}{25}=1$ by shifting it 5 units downward. So $a=5$, $b=3$, and $c=\sqrt{25-9}=4$. The center is $(0,-5)$, the foci are $(0,-5\pm4)=(0,-9)$ and $(0,-1)$, the vertices are $(0,-5\pm5)=(0,-10)$ and $(0,0)$, the length of the major axis is $2a=10$, and the length of the minor axis is $2b=6$.

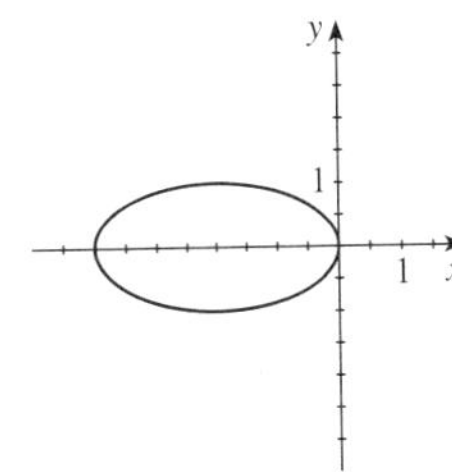

**4.** The ellipse $\dfrac{(x+2)^2}{4}+y^2=1$ is obtained from the ellipse $\dfrac{x^2}{4}+\dfrac{y^2}{1}=1$ by shifting to the left 2 units. So $a=2$, $b=1$, and $c=\sqrt{4-1}=\sqrt{3}$. The center is $(-2,0)$, the foci are $\left(-2\pm\sqrt{3},0\right)$, the vertices are $(-2\pm2,0)=(-4,0)$ and $(0,0)$, the length of the major axis is $2a=4$, and the length of the minor axis is $2b=2$.

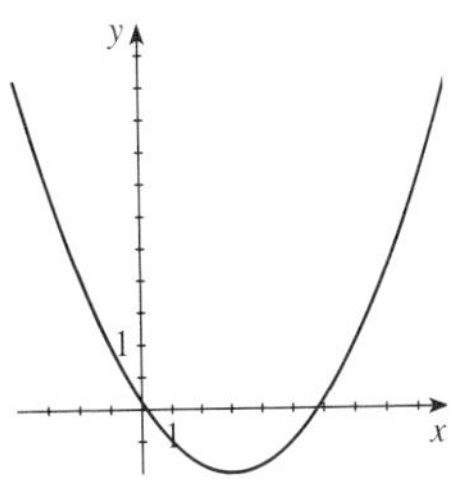

**5.** The parabola $(x-3)^2=8(y+1)$ is obtained from the parabola $x^2=8y$ by shifting it 3 units to the right and 1 unit down. So $4p=8 \Leftrightarrow p=2$. The vertex is $(3,-1)$, the focus is $(3,-1+2)=(3,1)$, and the directrix is $y=-1-2=-3$.

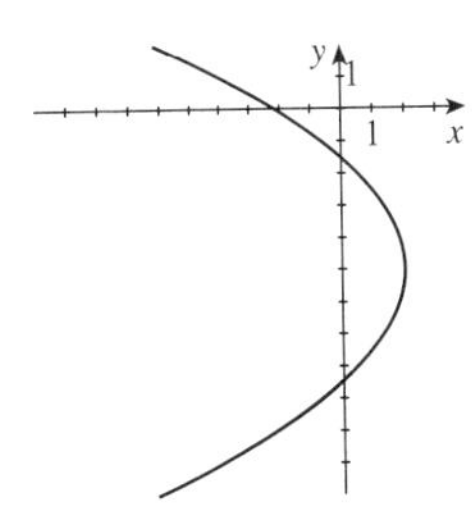

**6.** The parabola $(y+5)^2=-6x+12=-6(x-2)$ is obtained from the parabola $y^2=-6x$ by shifting to the right 2 units and down 5 units. So $4p=-6 \Leftrightarrow p=-\frac{3}{2}$. The vertex is $(2,-5)$, the focus is $\left(2-\frac{3}{2},-5\right)=\left(\frac{1}{2},-5\right)$, and the directrix is $x=2+\frac{3}{2}=\frac{7}{2}$.

**7.** The parabola $-4\left(x+\frac{1}{2}\right)^2 = y \quad \Leftrightarrow \quad \left(x+\frac{1}{2}\right)^2 = -\frac{1}{4}y$ is obtained from the parabola $x^2 = -\frac{1}{4}y$ by shifting it $\frac{1}{2}$ unit to the left. So $4p = -\frac{1}{4} \quad \Leftrightarrow \quad p = -\frac{1}{16}$. The vertex is $\left(-\frac{1}{2}, 0\right)$, the focus is $\left(-\frac{1}{2}, 0 - \frac{1}{16}\right) = \left(-\frac{1}{2}, -\frac{1}{16}\right)$, and the directrix is $y = 0 + \frac{1}{16} = \frac{1}{16}$.

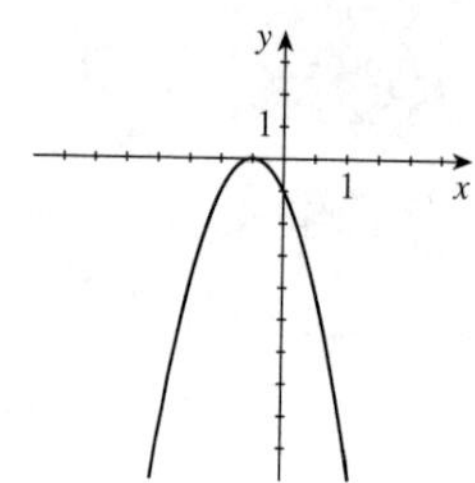


**8.** The parabola $y^2 = 16x - 8 = 16(x - \frac{1}{2})$ is obtained from the parabola $y^2 = 16x$ by shifting to the right $\frac{1}{2}$ unit. So $4p = 16 \quad \Leftrightarrow \quad p = 4$. The vertex is $\left(\frac{1}{2}, 0\right)$, the focus is $\left(\frac{1}{2} + 4, 0\right) = \left(\frac{9}{2}, 0\right)$, and the directrix is $x = \frac{1}{2} - 4 = -\frac{7}{2}$.

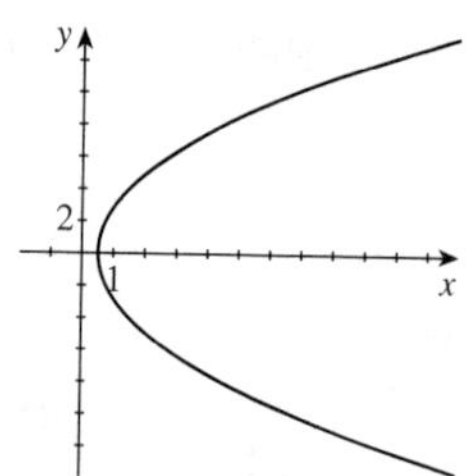


**9.** The hyperbola $\dfrac{(x+1)^2}{9} - \dfrac{(y-3)^2}{16} = 1$ is obtained from the hyperbola $\dfrac{x^2}{9} - \dfrac{y^2}{16} = 1$ by shifting it 1 unit to the left and 3 units up. So $a = 3$, $b = 4$, and $c = \sqrt{9+16} = 5$. The center is $(-1, 3)$, the foci are $(-1 \pm 5, 3) = (-6, 3)$ and $(4, 3)$, the vertices are $(-1 \pm 3, 3) = (-4, 3)$ and $(2, 3)$, and the asymptotes are $(y - 3) = \pm\frac{4}{3}(x+1) \quad \Leftrightarrow \quad y = \pm\frac{4}{3}(x+1) + 3 \quad \Leftrightarrow \quad 3y = 4x + 13$ and $3y = -4x + 5$.

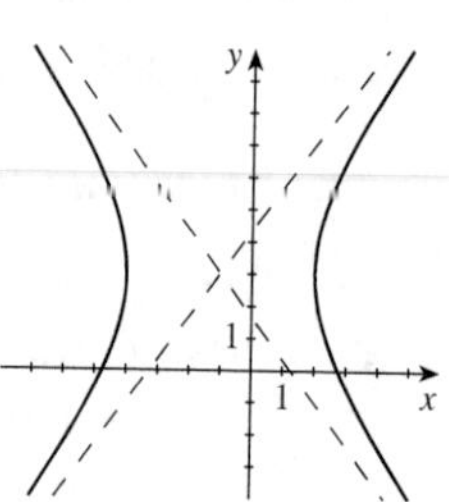


**10.** The hyperbola $(x-8)^2 - (y+6)^2 = 1$ is obtained from the hyperbola $x^2 - y^2 = 1$ by shifting to the right 8 units and downward 6 units. So $a = 1$, $b = 1$, and $c = \sqrt{1+1} = \sqrt{2}$. The center is $(8, -6)$, the foci are $\left(8 \pm \sqrt{2}, -6\right)$, the vertices are $(8 \pm 1, -6) = (7, -6)$ and $(9, -6)$, and the asymptotes are $y + 6 = \pm(x - 8) \quad \Leftrightarrow \quad y = x - 14$ and $y = -x + 2$.

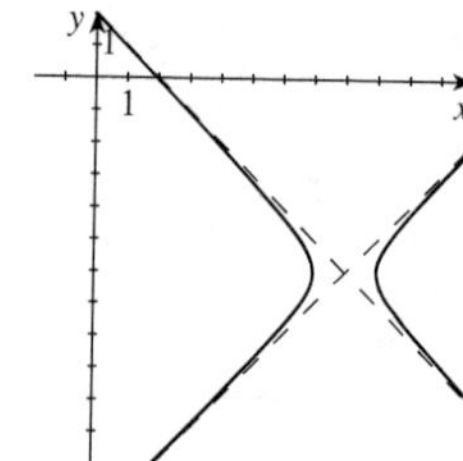


**11.** The hyperbola $y^2 - \dfrac{(x+1)^2}{4} = 1$ is obtained from the hyperbola $y^2 - \dfrac{x^2}{4} = 1$ by shifting it 1 unit to the left. So $a = 1$, $b = 2$, and $c = \sqrt{1+4} = \sqrt{5}$. The center is $(-1, 0)$, the foci are $\left(-1, \pm\sqrt{5}\right) = \left(-1, -\sqrt{5}\right)$ and $\left(-1, \sqrt{5}\right)$, the vertices are $(-1, \pm 1) = (-1, -1)$ and $(-1, 1)$, and the asymptotes are $y = \pm\frac{1}{2}(x+1) \quad \Leftrightarrow$ $y = \frac{1}{2}(x+1)$ and $y = -\frac{1}{2}(x+1)$.

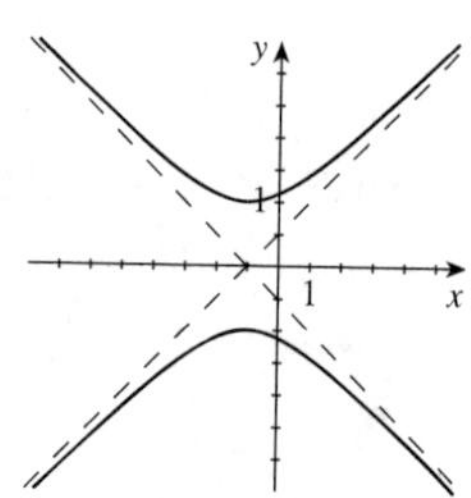


**12.** The hyperbola $\dfrac{(y-1)^2}{25} - (x+3)^2 = 1$ is obtained from the hyperbola $\dfrac{y^2}{25} - x^2 = 1$ by shifting to the left 3 units and upward 1 unit. So $a = 5$, $b = 1$, and $c = \sqrt{25+1} = \sqrt{26}$. The center is $(-3, 1)$, the foci are $\left(-3, 1 \pm \sqrt{26}\right)$, the vertices are $(-3, 1 \pm 5) = (-3, -4)$ and $(-3, 6)$, and the asymptotes are $(y - 1) = \pm 5(x + 3) \quad \Leftrightarrow \quad y = 5x + 16$ and $y = -5x - 14$.

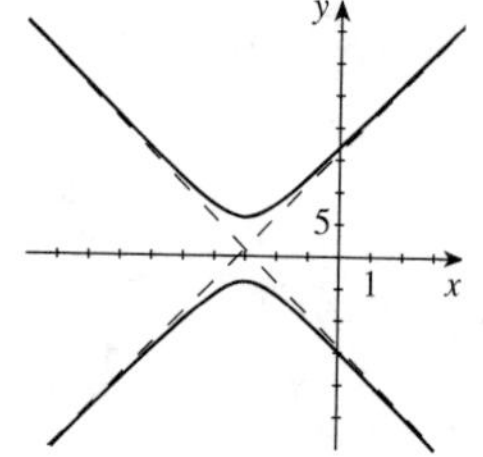

**13.** This is a parabola that opens down with its vertex at $(0,4)$, so its equation is of the form $x^2 = a(y-4)$. Since $(1,0)$ is a point on this parabola, we have $(1)^2 = a(0-4) \Leftrightarrow 1 = -4a \Leftrightarrow a = -\frac{1}{4}$. Thus, an equation is $x^2 = -\frac{1}{4}(y-4)$.

**14.** This is a parabola that opens to the right with its vertex at $(-6,0)$, so its equation is of the form $y^2 = 4p(x+6)$, with $p > 0$. Since the distance from the vertex to the directrix is $p = -6-(-12) = 6$, an equation is $y^2 = 4(6)(x+6) \Leftrightarrow y^2 = 24(x+6)$

**15.** This is an ellipse with the major axis parallel to the $x$-axis, with one vertex at $(0,0)$, the other vertex at $(10,0)$, and one focus at $(8,0)$. The center is at $\left(\frac{0+10}{2},0\right) = (5,0)$, $a = 5$, and $c = 3$ (the distance from one focus to the center). So $b^2 = a^2 - c^2 = 25 - 9 = 16$. Thus, an equation is $\dfrac{(x-5)^2}{25} + \dfrac{y^2}{16} = 1$.

**16.** This is an ellipse with the major axis parallel to the $y$-axis. From the graph the center is $(2,-3)$, with $a = 3$ and $b = 2$. Thus, an equation is $\dfrac{(x-2)^2}{4} + \dfrac{(y+3)^2}{9} = 1$.

**17.** This is a hyperbola with center $(0,1)$ and vertices $(0,0)$ and $(0,2)$. Since $a$ is the distance form the center to a vertex, we have $a = 1$. The slope of the given asymptote is 1, so $\dfrac{a}{b} = 1 \Leftrightarrow b = 1$. Thus, an equation of the hyperbola is $(y-1)^2 - x^2 = 1$.

**18.** From the graph, the vertices are $(2,0)$ and $(6,0)$. The center is the midpoint between the vertices, so the center is $\left(\dfrac{2+6}{2}, \dfrac{0+0}{2}\right) = (4,0)$. Since $a$ is the distance from the center to a vertex, $a = 2$. Since the vertices are on the $x$-axis, the equation is of the form $\dfrac{(x-4)^2}{4} - \dfrac{y^2}{b^2} = 1$. Since the point $(0,4)$ lies on the hyperbola, we have $\dfrac{(0-4)^2}{4} - \dfrac{(4)^2}{b^2} = 1$ $\Leftrightarrow \frac{16}{4} - \dfrac{16}{b^2} = 1 \Leftrightarrow 3 = \dfrac{16}{b^2} \Leftrightarrow b^2 = \frac{16}{3}$. Thus, an equation of the hyperbola is $\dfrac{(x-4)^2}{4} - \dfrac{y^2}{16/3} = 1 \Leftrightarrow$ $\dfrac{(x-4)^2}{4} - \dfrac{3y^2}{16} = 1$.

**19.** $9x^2 - 36x + 4y^2 = 0 \Leftrightarrow 9(x^2 - 4x + 4) - 36 + 4y^2 = 0 \Leftrightarrow$ $9(x-2)^2 + 4y^2 = 36 \Leftrightarrow \dfrac{(x-2)^2}{4} + \dfrac{y^2}{9} = 1$. This is an ellipse with $a = 3$, $b = 2$, and $c = \sqrt{9-4} = \sqrt{5}$. The center is $(2,0)$, the foci are $\left(2, \pm\sqrt{5}\right)$, the vertices are $(2,\pm 3)$, the length of the major axis is $2a = 6$, and the length of the minor axis is $2b = 4$.

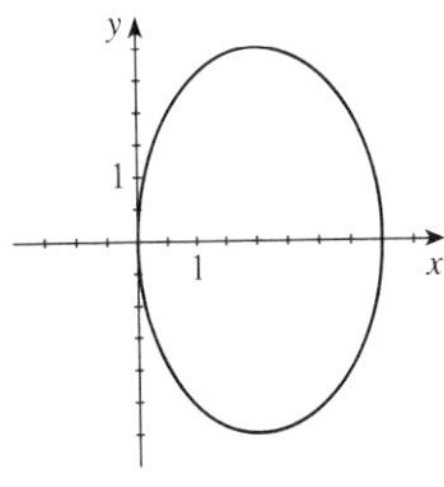


**20.** $y^2 = 4(x+2y) \Leftrightarrow y^2 - 8y = 4x \Leftrightarrow y^2 - 8y + 16 = 4x + 16 \Leftrightarrow$ $(y-4)^2 = 4(x+4)$. This is a parabola with $4p = 4 \Leftrightarrow p = 1$. The vertex is $(-4,4)$, the focus is $(-4+1,4) = (-3,4)$, and the directrix is $x = -4 - 1 = -5$.

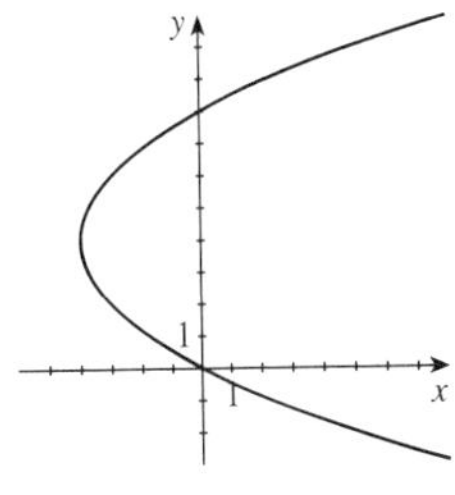

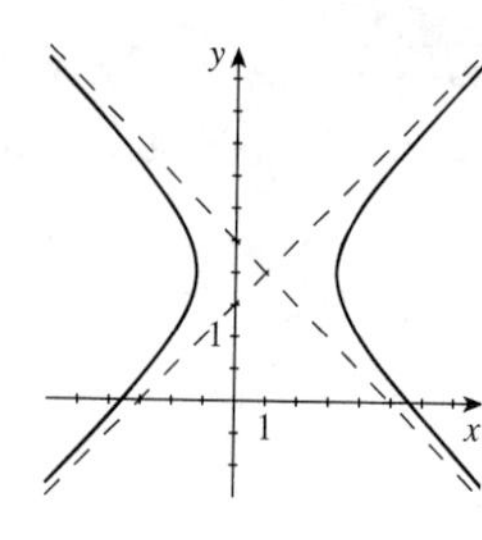


**21.** $x^2 - 4y^2 - 2x + 16y = 20 \quad \Leftrightarrow$

$(x^2 - 2x + 1) - 4(y^2 - 4y + 4) = 20 + 1 - 16 \quad \Leftrightarrow \quad (x-1)^2 - 4(y-2)^2 = 5$

$\Leftrightarrow \quad \dfrac{(x-1)^2}{5} - \dfrac{(y-2)^2}{\frac{5}{4}} = 1$. This is a hyperbola with $a = \sqrt{5}$, $b = \frac{1}{2}\sqrt{5}$, and $c = \sqrt{5 + \frac{5}{4}} = \frac{5}{2}$. The center is $(1,2)$, the foci are $\left(1 \pm \frac{5}{2}, 2\right) = \left(-\frac{3}{2}, 2\right)$ and $\left(\frac{7}{2}, 2\right)$, the vertices are $\left(1 \pm \sqrt{5}, 2\right)$, and the asymptotes are $y - 2 = \pm\frac{1}{2}(x-1)$ $\Leftrightarrow \quad y = \pm\frac{1}{2}(x-1) + 2 \quad \Leftrightarrow \quad y = \frac{1}{2}x + \frac{3}{2}$ and $y = -\frac{1}{2}x + \frac{5}{2}$.

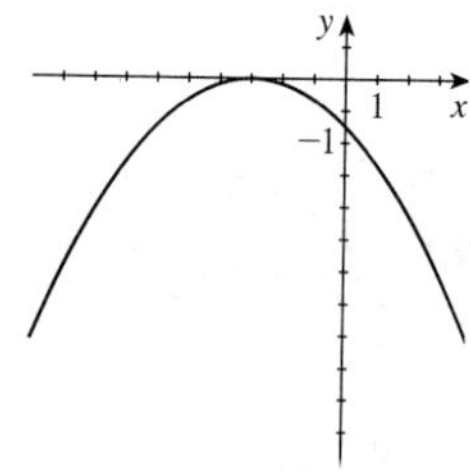


**22.** $x^2 + 6x + 12y + 9 = 0 \quad \Leftrightarrow \quad x^2 + 6x + 9 = -12y \quad \Leftrightarrow \quad (x+3)^2 = -12y$. This is a parabola with $4p = -12 \quad \Leftrightarrow \quad p = -3$. The vertex is $(-3, 0)$, the focus is $(-3, -3)$, and the directrix is $y = 3$.

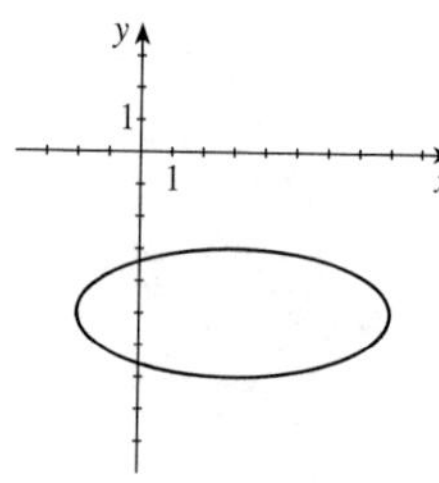


**23.** $4x^2 + 25y^2 - 24x + 250y + 561 = 0 \quad \Leftrightarrow \quad 4\left(x^2 - 6x + 9\right) +$ $25\left(y^2 + 10y + 25\right) = -561 + 36 + 625 \Leftrightarrow \quad 4(x-3)^2 + 25(y+5)^2 = 100$

$\Leftrightarrow \quad \dfrac{(x-3)^2}{25} + \dfrac{(y+5)^2}{4} = 1$. This is an ellipse with $a = 5$, $b = 2$, and $c = \sqrt{25 - 4} = \sqrt{21}$. The center is $(3, -5)$, the foci are $\left(3 \pm \sqrt{21}, -5\right)$, the vertices are $(3 \pm 5, -5) = (-2, -5)$ and $(8, -5)$, the length of the major axis is $2a = 10$, and the length of the minor axis is $2b = 4$.

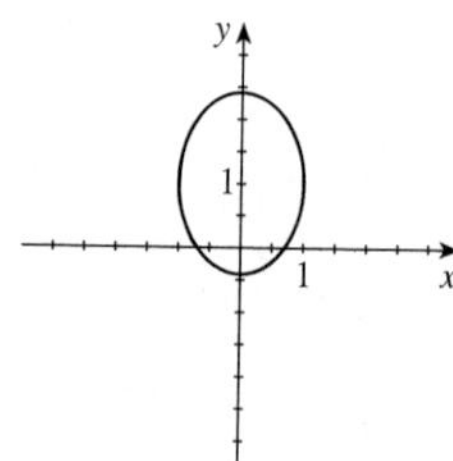


**24.** $2x^2 + y^2 = 2y + 1 \quad \Leftrightarrow \quad 2x^2 + \left(y^2 - 2y + 1\right) = 1 + 1 \quad \Leftrightarrow$

$2x^2 + (y-1)^2 = 2 \quad \Leftrightarrow \quad x^2 + \dfrac{(y-1)^2}{2} = 1$. This is an ellipse with $a = \sqrt{2}$, $b = 1$, and $c = \sqrt{2 - 1} = 1$. The center is $(0, 1)$, the foci are $(0, 1 \pm 1) = (0, 0)$ and $(0, 2)$, the vertices are $\left(0, 1 \pm \sqrt{2}\right)$, the length of the major axis is $2a = 2\sqrt{2}$, and the length of the minor axis is $2b = 2$.

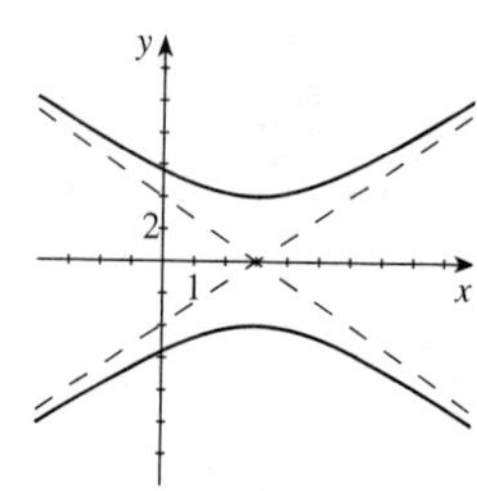


**25.** $16x^2 - 9y^2 - 96x + 288 = 0 \quad \Leftrightarrow \quad 16\left(x^2 - 6x\right) - 9y^2 + 288 = 0 \quad \Leftrightarrow$

$16\left(x^2 - 6x + 9\right) - 9y^2 = 144 - 288 \quad \Leftrightarrow \quad 16(x-3)^2 - 9y^2 = -144 \quad \Leftrightarrow$

$\dfrac{y^2}{16} - \dfrac{(x-3)^2}{9} = 1$. This is a hyperbola with $a = 4$, $b = 3$, and $c = \sqrt{16 + 9} = 5$. The center is $(3, 0)$, the foci are $(3, \pm 5)$, the vertices are $(3, \pm 4)$, and the asymptotes are $y = \pm\frac{4}{3}(x-3) \quad \Leftrightarrow \quad y = \frac{4}{3}x - 4$ and $y = 4 - \frac{4}{3}x$.

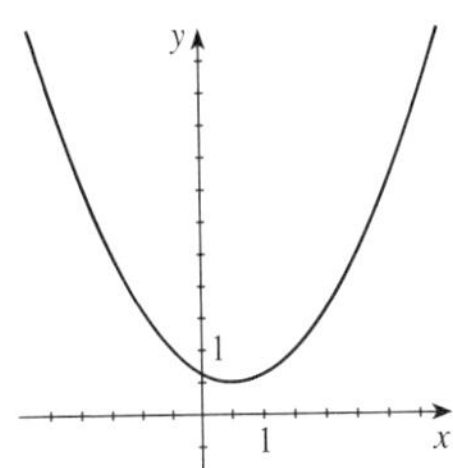


**26.** $4x^2 - 4x - 8y + 9 = 0 \quad \Leftrightarrow \quad 4\left(x^2 - x\right) - 8y + 9 = 0 \quad \Leftrightarrow$
$4\left(x^2 - x + \frac{1}{4}\right) = 8y - 9 + 1 \quad \Leftrightarrow \quad 4\left(x - \frac{1}{2}\right)^2 = 8\left(y - 1\right) \quad \Leftrightarrow$
$\left(x - \frac{1}{2}\right)^2 = 2\left(y - 1\right)$. This is a parabola with $4p = 2 \quad \Leftrightarrow \quad p = \frac{1}{2}$. The vertex is $\left(\frac{1}{2}, 1\right)$, the focus is $\left(\frac{1}{2}, \frac{3}{2}\right)$, and the directrix is $y = -\frac{1}{2} + 1 = \frac{1}{2}$.

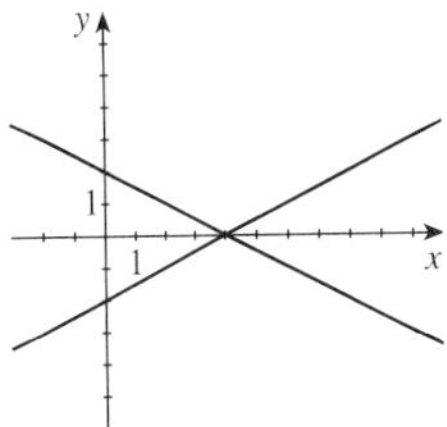


**27.** $x^2 + 16 = 4\left(y^2 + 2x\right) \quad \Leftrightarrow \quad x^2 - 8x - 4y^2 + 16 = 0 \quad \Leftrightarrow$
$\left(x^2 - 8x + 16\right) - 4y^2 = -16 + 16 \quad \Leftrightarrow \quad 4y^2 = \left(x - 4\right)^2 \quad \Leftrightarrow$
$y = \pm\frac{1}{2}\left(x - 4\right)$. Thus, the conic is degenerate, and its graph is the pair of lines $y = \frac{1}{2}\left(x - 4\right)$ and $y = -\frac{1}{2}\left(x - 4\right)$.

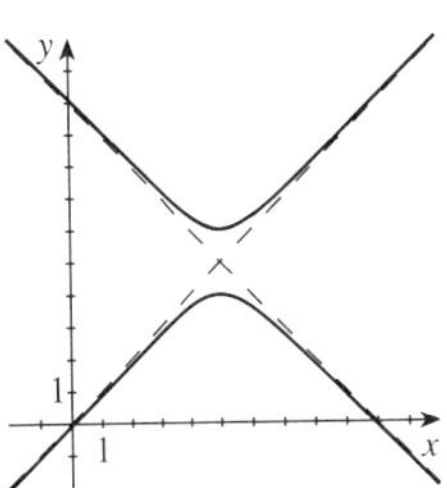


**28.** $x^2 - y^2 = 10\left(x - y\right) + 1 \quad \Leftrightarrow \quad x^2 - 10x - y^2 + 10y = 1 \quad \Leftrightarrow$
$\left(x^2 - 10x + 25\right) - \left(y^2 - 10y + 25\right) = 1 + 25 - 25 \quad \Leftrightarrow$
$\left(x - 5\right)^2 - \left(y - 5\right)^2 = 1$. This is a hyperbola with $a = 1$, $b = 1$, and $c = \sqrt{1 + 1} = \sqrt{2}$. The center is $(5, 5)$, the foci are $\left(5 \pm \sqrt{2}, 5\right)$, the vertices are $(5 \pm 1, 5) = (4, 5)$ and $(6, 5)$, and the asymptotes are $y - 5 = \pm\left(x - 5\right) \quad \Leftrightarrow$ $y = x$ and $y = -x + 10$.

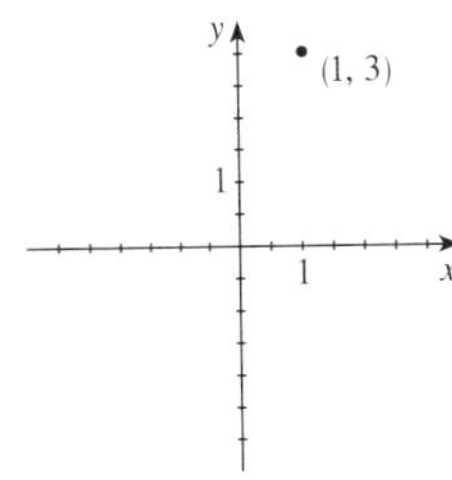


**29.** $3x^2 + 4y^2 - 6x - 24y + 39 = 0 \quad \Leftrightarrow \quad 3\left(x^2 - 2x\right) + 4\left(y^2 - 6y\right) = -39 \quad \Leftrightarrow$
$3\left(x^2 - 2x + 1\right) + 4\left(y^2 - 6y + 9\right) = -39 + 3 + 36 \quad \Leftrightarrow$
$3\left(x - 1\right)^2 + 4\left(y - 3\right)^2 = 0 \quad \Leftrightarrow \quad x = 1$ and $y = 3$. This is a degenerate conic whose graph is the point $(1, 3)$.

**30.** $x^2 + 4y^2 + 20x - 40y + 300 = 0 \quad \Leftrightarrow \quad \left(x^2 + 20x\right) + 4\left(y^2 - 10y\right) = -300 \quad \Leftrightarrow \quad \left(x^2 + 20x + 100\right) + 4\left(y^2 - 10y + 25\right) = -300 + 100 + 100 \quad \Leftrightarrow \quad \left(x + 10\right)^2 + 4\left(y - 5\right)^2 = -100$. Since $u^2 + v^2 \geq 0$ for all $u, v \in \mathbb{R}$, there is no $(x, y)$ such that $\left(x + 10\right)^2 + 4\left(y - 5\right)^2 = -100$. So there is no solution, and the graph is empty.

**31.** $2x^2 - 4x + y + 5 = 0 \quad \Leftrightarrow \quad y = -2x^2 + 4x - 5$.

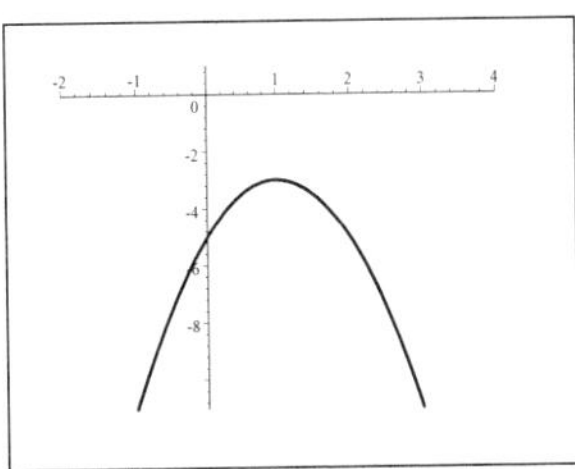


**32.** $4x^2 + 9y^2 - 36y = 0 \quad \Leftrightarrow \quad 4x^2 + 9\left(y^2 - 4y\right) = 0 \quad \Leftrightarrow$
$4x^2 + 9\left(y^2 - 4y + 4\right) = 36 \quad \Leftrightarrow \quad 9\left(y^2 - 4y + 4\right) = 36 - 4x^2 \quad \Leftrightarrow$
$y^2 - 4y + 4 = 4 - \frac{4}{9}x^2 \quad \Leftrightarrow \quad y - 2 = \pm\sqrt{4 - \frac{4}{9}x^2} \quad \Leftrightarrow \quad y = 2 \pm \sqrt{4 - \frac{4}{9}x^2}$

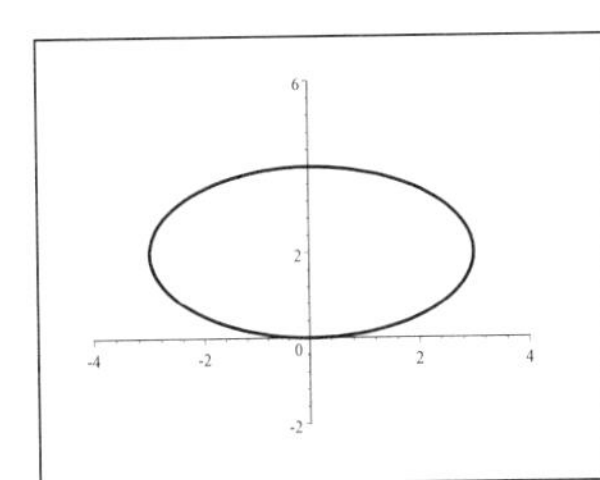

**33.** $9x^2 + 36 = y^2 + 36x + 6y \Leftrightarrow x^2 - 36x + 36 = y^2 + 6y \Leftrightarrow$
$9x^2 - 36x + 45 = y^2 + 6y + 9 \Leftrightarrow 9\left(x^2 - 4x + 5\right) = (y+3)^2 \Leftrightarrow$
$y + 3 = \pm\sqrt{9\left(x^2 - 4x + 5\right)} \Leftrightarrow y = -3 \pm 3\sqrt{x^2 - 4x + 5}$

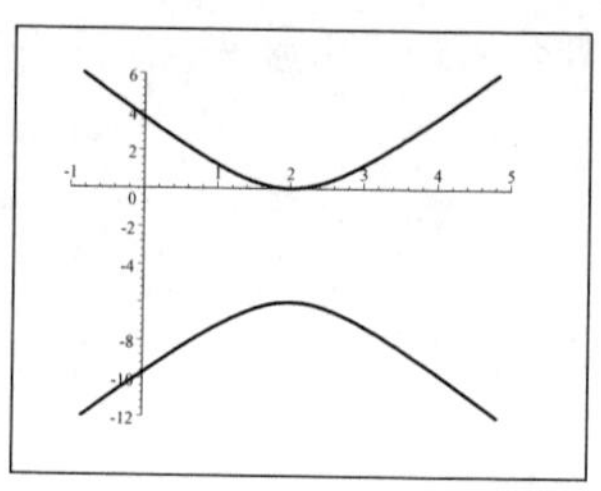

**34.** $x^2 - 4y^2 + 4x + 8y = 0 \Leftrightarrow x^2 + 4x - 4\left(y^2 - 2y\right) = 0 \Leftrightarrow$
$\left(x^2 + 4x + 4\right) - 4\left(y^2 - 2y + 1\right) = 0 \Leftrightarrow 4(y-1)^2 = (x+2)^2 \Leftrightarrow$
$2(y-1) = \pm(x+2) \Leftrightarrow y - 1 = \pm\frac{1}{2}(x+2) \Leftrightarrow y = 1 \pm \frac{1}{2}(x+2)$.So
$y = 1 + \frac{1}{2}(x+2) = \frac{1}{2}x + 2$ and $y = 1 - \frac{1}{2}(x+2) = -\frac{1}{2}x$. This is a degenerate conic.

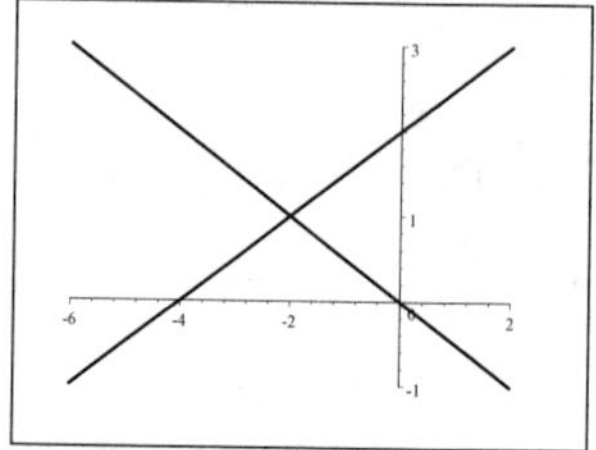

**35.** $4x^2 + y^2 + 4(x - 2y) + F = 0 \Leftrightarrow 4\left(x^2 + x\right) + \left(y^2 - 8y\right) = -F \Leftrightarrow$
$4\left(x^2 + x + \frac{1}{4}\right) + \left(y^2 - 8y + 16\right) = 16 + 1 - F \Leftrightarrow 4\left(x + \frac{1}{2}\right)^2 + (y-1)^2 = 17 - F$

**(a)** For an ellipse, $17 - F > 0 \Leftrightarrow F < 17$.

**(b)** For a single point, $17 - F = 0 \Leftrightarrow F = 17$.

**(c)** For the empty set, $17 - F < 0 \Leftrightarrow F > 17$.

**36.** The parabola $x^2 + y = 100 \Leftrightarrow x^2 = -(y - 100)$ has $4p = -1 \Leftrightarrow p = -\frac{1}{4}$. The vertex is $(0, 100)$ and so the focus is $\left(0, 100 - \frac{1}{4}\right) = \left(0, \frac{399}{4}\right)$. Thus, one vertex of the ellipse is $(0, 100)$, and one focus is $\left(0, \frac{399}{4}\right)$. Since the second focus of the ellipse is $(0, 0)$, the second vertex is $\left(0, -\frac{1}{4}\right)$. So $2a = 100 + \frac{1}{4} = \frac{401}{4} \Leftrightarrow a = \frac{401}{8}$. Since $2c$ is the distance between the foci of the ellipse, $2c = \frac{399}{4}$, $c = \frac{399}{8}$, and then $b^2 = a^2 - c^2 = \frac{401^2}{64} - \frac{399^2}{64} = 25$. The center of the ellipse is $\left(0, \frac{399}{8}\right)$ and so its equation is $\dfrac{x^2}{25} + \dfrac{\left(y - \frac{399}{8}\right)^2}{\left(\frac{401}{8}\right)^2} = 1$, which simplifies to $\dfrac{x^2}{25} + \dfrac{(8y - 399)^2}{160{,}801} = 1$.

**37. (a)** $x^2 = 4p(y + p)$, for
$p = -2, -\frac{3}{2}, -1, -\frac{1}{2}, \frac{1}{2}, 1, \frac{3}{2}, 2$.

**(b)** The graph of $x^2 = 4p(y + p)$ is obtained by shifting the graph of $x^2 = 4py$ vertically $-p$ units so that the vertex is at $(0, -p)$. The focus of $x^2 = 4py$ is at $(0, p)$, so this point is also shifted $-p$ units vertically to the point $(0, p - p) = (0, 0)$. Thus, the focus is located at the origin.

**(c)** The parabolas become narrower as the vertex moves toward the origin.

**38.** Since $(0, 0)$ and $(1600, 0)$ are both points on the parabola, the $x$-coordinate of the vertex is 800. And since the highest point it reaches is 3200, the $y$-coordinate of the vertex is 3200. Thus the vertex is $(800, 3200)$, and the equation is of the form $(x - 800)^2 = 4p(y - 3200)$. Substituting the point $(0, 0)$, we get $(0 - 800)^2 = 4p(0 - 3200)$ $\Leftrightarrow 640{,}000 = -12{,}800p \Leftrightarrow p = -50$. So an equation is $(x - 800)^2 = 4(-50)(y - 3200) \Leftrightarrow$ $(x - 800)^2 = -200(y - 3200)$.

**39.** Since the height of the satellite above the earth varies between 140 and 440, the length of the major axis is $2a = 140 + 2(3960) + 440 = 8500 \Leftrightarrow a = 4250$. Since the center of the earth is at one focus, we have $a - c = (\text{earth radius}) + 140 = 3960 + 140 = 4100 \Leftrightarrow$ $c = a - 4100 = 4250 - 4100 = 150$. Thus, the center of the ellipse is $(-150, 0)$. So $b^2 = a^2 - c^2 = 4250^2 - 150^2 = 18{,}062{,}500 - 22500 = 18{,}040{,}000$. Hence, an equation is $\dfrac{(x+150)^2}{18{,}062{,}500} + \dfrac{y^2}{18{,}040{,}000} = 1$.

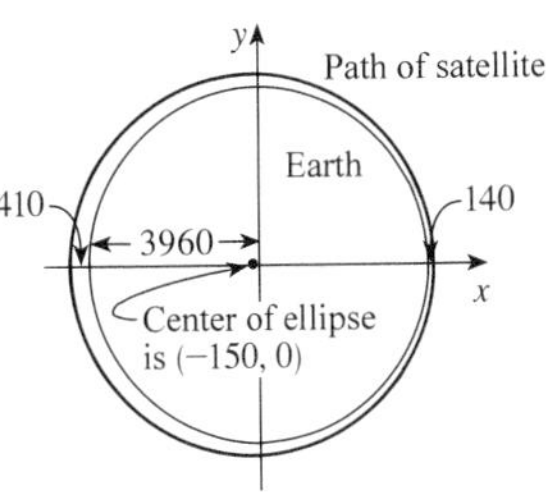


**40.** **(a)** We assume that $(0, 1)$ is the focus closer to the vertex $(0, 0)$, as shown in the figure in the text. Then the center of the ellipse is $(0, a)$ and $1 = a - c$. So $c = a - 1$ and $(a-1)^2 = a^2 - b^2 \Leftrightarrow a^2 - 2a + 1 = a^2 - b^2 \Leftrightarrow$ $b^2 = 2a - 1$. Thus, the equation is $\dfrac{x^2}{2a-1} + \dfrac{(y-a)^2}{a^2} = 1$. If we choose $a = 2$, then we get $\dfrac{x^2}{3} + \dfrac{(y-2)^2}{4} = 1$. If we choose $a = 5$, then we get $\dfrac{x^2}{9} + \dfrac{(y-5)^2}{25} = 1$. (Answers will vary, depending on your choices of $a > 1$.)

**(b)** Since a vertex is at $(0, 0)$ and a focus is at $(0, 1)$, we must have $c - a = 1$ ($a > 0$), and the center of the hyperbola is $(0, -a)$. So $c = a + 1$ and $(a+1)^2 = a^2 + b^2 \Leftrightarrow a^2 + 2a + 1 = a^2 + b^2 \Leftrightarrow b^2 = 2a + 1$. Thus the equation is $\dfrac{(y+a)^2}{a^2} - \dfrac{x^2}{2a+1} = 1$. If we let $a = 2$, then we get $\dfrac{(y+2)^2}{4} - \dfrac{x^2}{5} = 1$. If we let $a = 5$, then we get $\dfrac{(y+5)^2}{25} - \dfrac{x^2}{11} = 1$. (Answers will vary, depending on your choices of $a$.)

**(c)** Since the vertex is at $(0, 0)$ and the focus is at $(0, 1)$, we must have $p = 1$. So $(x-0)^2 = 4(1)(y-0) \Leftrightarrow x^2 = 4y$, and there is no other possible parabola.

**(d)** Graphs will vary, depending on the choices of $a$ in parts (a) and (b).

**(e)** The ellipses are inside the parabola and the hyperbolas are outside the parabola. All touch at the origin.

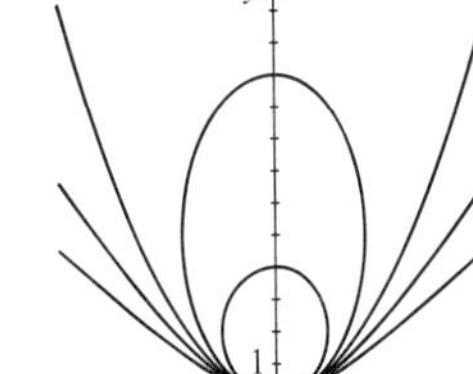


# 11.5 Rotation of Axes

**1.** $(x, y) = (1, 1)$, $\phi = 45°$. Then $X = x\cos\phi + y\sin\phi = 1 \cdot \frac{1}{\sqrt{2}} + 1 \cdot \frac{1}{\sqrt{2}} = \sqrt{2}$ and $Y = -x\sin\phi + y\cos\phi = -1 \cdot \frac{1}{\sqrt{2}} + 1 \cdot \frac{1}{\sqrt{2}} = 0$. Therefore, the $XY$-coordinates of the given point are $(X, Y) = (\sqrt{2}, 0)$.

**2.** $(x, y) = (-2, 1)$, $\phi = 30°$. Then $X = x\cos\phi + y\sin\phi = -2 \cdot \frac{\sqrt{3}}{2} + 1 \cdot \frac{1}{2} = \frac{1}{2}(1 - 2\sqrt{3})$ and $Y = -x\sin\phi + y\cos\phi = 2 \cdot \frac{1}{2} + 1 \cdot \frac{\sqrt{3}}{2} = \frac{1}{2}(2 + \sqrt{3})$. Therefore, the $XY$-coordinates of the given point are $(X, Y) = \left(\frac{1}{2}(1 - 2\sqrt{3}), \frac{1}{2}(2 + \sqrt{3})\right)$.

**3.** $(x, y) = (3, -\sqrt{3})$, $\phi = 60°$. Then $X = x\cos\phi + y\sin\phi = 3 \cdot \frac{1}{2} - \sqrt{3} \cdot \frac{\sqrt{3}}{2} = 0$ and $Y = -x\sin\phi + y\cos\phi = -3 \cdot \frac{\sqrt{3}}{2} - \sqrt{3} \cdot \frac{1}{2} = -2\sqrt{3}$. Therefore, the $XY$-coordinates of the given point are $(X, Y) = (0, -2\sqrt{3})$.

**4.** $(x, y) = (2, 0)$, $\phi = 15°$. Then $X = x\cos\phi + y\sin\phi = 2\cos 15° + 0\sin 15° \approx 1.9319$ and $Y = -x\sin\phi + y\cos\phi = -2\sin 15° + 0\cos 15° \approx -0.5176$. Therefore, the $XY$-coordinates of the given point are approximately $(X, Y) = (1.9319, -0.5176)$.

**5.** $(x, y) = (0, 2)$, $\phi = 55°$. Then $X = x\cos\phi + y\sin\phi = 0\cos 55° + 2\sin 55° \approx 1.6383$ and $Y = -x\sin\phi + y\cos\phi = -0\sin 55° + 2\cos 55° \approx 1.1472$. Therefore, the $XY$-coordinates of the given point are approximately $(X, Y) = (1.6383, 1.1472)$.

**6.** $(x, y) = (\sqrt{2}, 4\sqrt{2})$, $\phi = 45°$. Then $X = x\cos\phi + y\sin\phi = \sqrt{2}\cdot\frac{1}{\sqrt{2}} + 4\sqrt{2}\cdot\frac{1}{\sqrt{2}} = 5$ and $Y = -x\sin\phi + y\cos\phi = -\sqrt{2}\cdot\frac{1}{\sqrt{2}} + 4\sqrt{2}\cdot\frac{1}{\sqrt{2}} = 3$. Therefore, the $XY$-coordinates of the given point are $(X, Y) = (5, 3)$.

**7.** $x^2 - 3y^2 = 4$, $\phi = 60°$. Then $x = X\cos 60° - Y\sin 60° = \frac{1}{2}X - \frac{\sqrt{3}}{2}Y$ and $y = X\sin 60° + Y\cos 60° = \frac{\sqrt{3}}{2}X + \frac{1}{2}Y$. Substituting these values into the equation, we get $\left(\frac{1}{2}X - \frac{\sqrt{3}}{2}Y\right)^2 - 3\left(\frac{\sqrt{3}}{2}X + \frac{1}{2}Y\right)^2 = 4 \Leftrightarrow$ $\frac{X^2}{4} - \frac{\sqrt{3}XY}{2} + \frac{3Y^2}{4} - 3\left(\frac{3X^2}{4} + \frac{\sqrt{3}XY}{2} + \frac{Y^2}{4}\right) = 4 \Leftrightarrow \frac{X^2}{4} - \frac{9}{4}X^2 + \frac{3Y^2}{4} - \frac{3Y^2}{4} - \frac{\sqrt{3}XY}{2} - \frac{3\sqrt{3}XY}{2} = 4$ $\Leftrightarrow -2X^2 - 2\sqrt{3}XY = 4 \Leftrightarrow X^2 + \sqrt{3}XY = -2$.

**8.** $y = (x-1)^2$, $\phi = 45°$. Then $x = X\cos 45° - Y\sin 45° = \frac{\sqrt{2}}{2}X - \frac{\sqrt{2}}{2}Y$ and $y = X\sin 45° + Y\cos 45° = \frac{\sqrt{2}}{2}X + \frac{\sqrt{2}}{2}Y$. Substituting these values into the equation, we get

$$\left(\frac{\sqrt{2}}{2}X + \frac{\sqrt{2}}{2}Y\right) = \left(\frac{\sqrt{2}}{2}X - \frac{\sqrt{2}}{2}Y - 1\right)^2 = \frac{X^2}{2} - 2\frac{\sqrt{2}}{2}X\left(\frac{\sqrt{2}}{2}Y + 1\right) + \left(\frac{\sqrt{2}}{2}Y + 1\right)^2$$

$$= \frac{X^2}{2} - 2\frac{\sqrt{2}}{2}X\left(\frac{\sqrt{2}}{2}Y + 1\right) + \frac{Y^2}{2} + \sqrt{2}Y + 1 \Leftrightarrow$$

$$\frac{X^2}{2} - \frac{\sqrt{2}}{2}X - \sqrt{2}X\left(\frac{\sqrt{2}}{2}Y + 1\right) + \frac{Y^2}{2} + \sqrt{2}Y + 1 - \frac{\sqrt{2}}{2}Y = 0 \Leftrightarrow X^2 + Y^2 - 2XY - 3\sqrt{2}X + \sqrt{2}Y + 2 = 0.$$

**9.** $x^2 - y^2 = 2y$, $\phi = \cos^{-1}\left(\frac{3}{5}\right)$. So $\cos\phi = \frac{3}{5}$ and $\sin\phi = \frac{4}{5}$. Then $(X\cos\phi - Y\sin\phi)^2 - (X\sin\phi + Y\cos\phi)^2 = 2(X\sin\phi + Y\cos\phi) \Leftrightarrow \left(\frac{3}{5}X - \frac{4}{5}Y\right)^2 -$ $\left(\frac{4}{5}X + \frac{3}{5}Y\right)^2 = 2\left(\frac{4}{5}X + \frac{3}{5}Y\right) \Leftrightarrow \frac{9X^2}{25} - \frac{24XY}{25} + \frac{16Y^2}{25} - \frac{16X^2}{25} - \frac{24XY}{25} - \frac{9Y^2}{25} = \frac{8X}{5} + \frac{6Y}{5} \Leftrightarrow$ $-\frac{7X^2}{25} - \frac{48XY}{25} + \frac{7Y^2}{25} - \frac{8X}{5} - \frac{6Y}{5} = 0 \Leftrightarrow 7Y^2 - 48XY - 7X^2 - 40X - 30Y = 0$.

**10.** $x^2 + 2y^2 = 16$, $\phi = \sin^{-1}\left(\frac{3}{5}\right)$. So $\sin\phi = \frac{3}{5}$ and $\cos\phi = \frac{4}{5}$. Then $x = \frac{4}{5}X - \frac{3}{5}Y$ and $y = \frac{3}{5}X + \frac{4}{5}Y$. Substituting these values into the equation, we get $\left(\frac{4}{5}X - \frac{3}{5}Y\right)^2 + 2\left(\frac{3}{5}X + \frac{4}{5}Y\right)^2 = 16 \Leftrightarrow \frac{16}{25}X^2 - \frac{24}{25}XY + \frac{9}{25}Y^2 + 2\left(\frac{9}{25}X^2 + \frac{24}{25}XY + \frac{16}{25}Y^2\right) = 16$ $\Leftrightarrow \frac{16}{25}X^2 + \frac{18}{25}X^2 + \frac{9}{25}Y^2 + \frac{32}{25}Y^2 - \frac{24}{25}XY + \frac{48}{25}XY = 16 \Leftrightarrow \frac{34}{25}X^2 + \frac{41}{25}Y^2 + \frac{24}{25}XY = 16 \Leftrightarrow$ $34X^2 + 41Y^2 + 24XY = 400$.

**11.** $x^2 + 2\sqrt{3}xy - y^2 = 4$, $\phi = 30°$. Then $x = X\cos 30° - Y\sin 30° = \frac{\sqrt{3}}{2}X - \frac{1}{2}Y = \frac{1}{2}\left(\sqrt{3}X - Y\right)$ and $y = X\sin 30° + Y\cos 30° = \frac{1}{2}X + \frac{\sqrt{3}}{2}Y = \frac{1}{2}\left(X + \sqrt{3}Y\right)$. Substituting these values into the equation, we get $\left[\frac{1}{2}\left(\sqrt{3}X - Y\right)\right]^2 + 2\sqrt{3}\left[\frac{1}{2}\left(\sqrt{3}X - Y\right)\right]\left[\frac{1}{2}\left(X + \sqrt{3}Y\right)\right] - \left[\frac{1}{2}\left(X + \sqrt{3}Y\right)\right]^2 = 4 \Leftrightarrow$ $\left(\sqrt{3}X - Y\right)^2 + 2\sqrt{3}\left(\sqrt{3}X - Y\right)\left(X + \sqrt{3}Y\right) - \left(X + \sqrt{3}Y\right)^2 = 16 \Leftrightarrow$ $\left(3X^2 - 2\sqrt{3}XY + Y^2\right) + \left(6X^2 + 4\sqrt{3}XY - 6Y^2\right) - \left(X^2 + 2\sqrt{3}XY + 3Y^2\right) = 16 \Leftrightarrow 8X^2 - 8Y^2 = 16 \Leftrightarrow$ $\frac{X^2}{2} - \frac{Y^2}{2} = 1$. This is a hyperbola.

**12.** $xy = x + y$, $\phi = \frac{\pi}{4}$. Therefore, $x = X\cos\frac{\pi}{4} - Y\sin\frac{\pi}{4} = \frac{1}{\sqrt{2}}X - \frac{1}{\sqrt{2}}Y = \frac{1}{\sqrt{2}}(X - Y)$ and $y = X\sin\frac{\pi}{4} + Y\cos\frac{\pi}{4} = \frac{1}{\sqrt{2}}X + \frac{1}{\sqrt{2}}Y = \frac{1}{\sqrt{2}}(X + Y)$. Substituting into the equation gives $\left[\frac{1}{\sqrt{2}}(X - Y)\right]\left[\frac{1}{\sqrt{2}}(X + Y)\right] = \frac{1}{\sqrt{2}}(X - Y) + \frac{1}{\sqrt{2}}(X + Y) \Leftrightarrow \frac{1}{2}(X^2 - Y^2) = \frac{2}{\sqrt{2}}X \Leftrightarrow$
$X^2 - Y^2 = 2\sqrt{2}X \Leftrightarrow \dfrac{(X - \sqrt{2})^2}{2} - \dfrac{Y^2}{2} = 1$. This is a hyperbola.

**13. (a)** $xy = 8 \Leftrightarrow 0x^2 + xy + 0y^2 = 8$. So $A = 0$, $B = 1$, and $C = 0$, and so the discriminant is $B^2 - 4AC = 1^2 - 4(0)(0) = 1$. Since the discriminant is positive, the equation represents a hyperbola.

**(b)** $\cot 2\phi = \dfrac{A - C}{B} = 0 \Rightarrow 2\phi = 90° \Leftrightarrow \phi = 45°$. Therefore, $x = \frac{\sqrt{2}}{2}X - \frac{\sqrt{2}}{2}Y$ and $y = \frac{\sqrt{2}}{2}X + \frac{\sqrt{2}}{2}Y$. After substitution, the original equation becomes $\left(\frac{\sqrt{2}}{2}X - \frac{\sqrt{2}}{2}Y\right)\left(\frac{\sqrt{2}}{2}X + \frac{\sqrt{2}}{2}Y\right) = 8 \Leftrightarrow$
$\dfrac{(X - Y)(X + Y)}{2} = 8 \Leftrightarrow \dfrac{X^2}{16} - \dfrac{Y^2}{16} = 1$. This is a hyperbola with $a = 4$, $b = 4$, and $c = 4\sqrt{2}$. Hence, the vertices are $V(\pm 4, 0)$ and the foci are $F(\pm 4\sqrt{2}, 0)$.

**(c)**

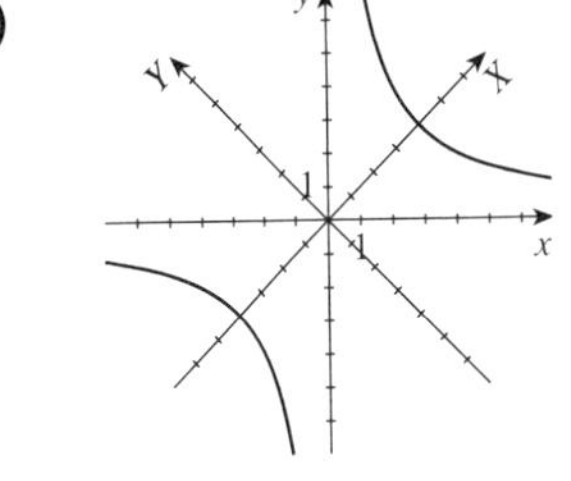


**14. (a)** $xy + 4 = 0 \Leftrightarrow 0x^2 + xy + 0y^2 + 4 = 0$. So $A = 0$, $B = 1$, and $C = 0$, and so the discriminant is $B^2 - 4AC = 1^2 - 4(0)(0) = 1$. Since the discriminant is positive, the equation represents a hyperbola.

**(b)** $\cot 2\phi = \dfrac{A - C}{B} = 0 \Rightarrow 2\phi = 90° \Leftrightarrow \phi = 45°$. Therefore, $x = \frac{\sqrt{2}}{2}X - \frac{\sqrt{2}}{2}Y$ and $y = \frac{\sqrt{2}}{2}X + \frac{\sqrt{2}}{2}Y$. After substitution, the original equation becomes $\left(\frac{\sqrt{2}}{2}X - \frac{\sqrt{2}}{2}Y\right)\left(\frac{\sqrt{2}}{2}X + \frac{\sqrt{2}}{2}Y\right) + 4 = 0 \Leftrightarrow$
$\dfrac{(X - Y)(X + Y)}{2} = -4 \Leftrightarrow \dfrac{Y^2}{8} - \dfrac{X^2}{8} = 1$. This is a hyperbola with $a = 2\sqrt{2}$, $b = 2\sqrt{2}$, and $c = 4$. Hence, the vertices are $V(0, \pm 2\sqrt{2})$ and the foci are $F(0, \pm 4)$.

**(c)**

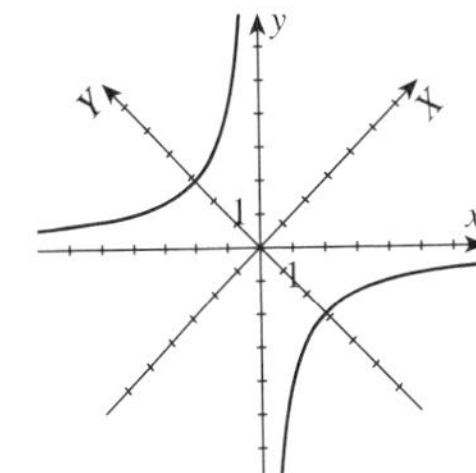


**15. (a)** $x^2 + 2xy + y^2 + x - y = 0$. So $A = 1$, $B = 2$, and $C = 1$, and so the discriminant is $B^2 - 4AC = 2^2 - 4(1)(1) = 0$. Since the discriminant is zero, the equation represents a parabola.

**(b)** $\cot 2\phi = \dfrac{A - C}{B} = 0 \Rightarrow 2\phi = 90° \Leftrightarrow \phi = 45°$. Therefore, $x = \frac{\sqrt{2}}{2}X - \frac{\sqrt{2}}{2}Y$ and $y = \frac{\sqrt{2}}{2}X + \frac{\sqrt{2}}{2}Y$. After substitution, the original equation becomes

$$\left(\tfrac{\sqrt{2}}{2}X - \tfrac{\sqrt{2}}{2}Y\right)^2 + 2\left(\tfrac{\sqrt{2}}{2}X - \tfrac{\sqrt{2}}{2}Y\right)\left(\tfrac{\sqrt{2}}{2}X + \tfrac{\sqrt{2}}{2}Y\right) + \left(\tfrac{\sqrt{2}}{2}X + \tfrac{\sqrt{2}}{2}Y\right)^2 + \left(\tfrac{\sqrt{2}}{2}X - \tfrac{\sqrt{2}}{2}Y\right) - \left(\tfrac{\sqrt{2}}{2}X + \tfrac{\sqrt{2}}{2}Y\right) = 0 \Leftrightarrow$$

$\frac{1}{2}X^2 - XY + \frac{1}{2}Y^2 + X^2 - Y^2 + \frac{1}{2}X^2 + XY + Y^2 - \sqrt{2}Y = 0 \Leftrightarrow 2X^2 - \sqrt{2}Y = 0 \Leftrightarrow X^2 = \frac{\sqrt{2}}{2}Y$.

This is a parabola with $4p = \frac{1}{\sqrt{2}}$ and hence the focus is $F\left(0, \frac{1}{4\sqrt{2}}\right)$.

**(c)**

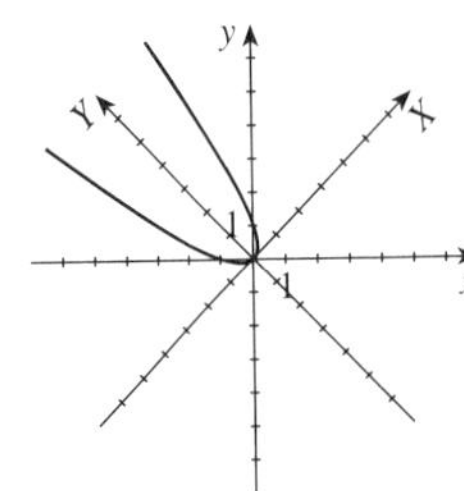

**16. (a)** $13x^2 + 6\sqrt{3}xy + 7y^2 = 16$. Then $A = 13$, $B = 6\sqrt{3}$, and $C = 7$, and so the discriminant is $B^2 - 4AC = \left(6\sqrt{3}\right)^2 - 4\,(13)\,(7) = -256$. Since the discriminant is negative, the equation represents an ellipse.

**(b)** $\cot 2\phi = \dfrac{A-C}{B} = \dfrac{13-7}{6\sqrt{3}} = \dfrac{1}{\sqrt{2}} \quad\Rightarrow\quad 2\phi = 60^\circ \quad\Leftrightarrow\quad \phi = 30^\circ$.

Therefore, $x = \frac{\sqrt{3}}{2}X - \frac{1}{2}Y$ and $y = \frac{1}{2}X + \frac{\sqrt{3}}{2}Y$. After substitution, the original equation becomes

$$13\left(\tfrac{\sqrt{3}}{2}X - \tfrac{1}{2}Y\right)^2 + 6\sqrt{3}\left(\tfrac{\sqrt{3}}{2}X - \tfrac{1}{2}Y\right)\left(\tfrac{1}{2}X + \tfrac{\sqrt{3}}{2}Y\right) + 7\left(\tfrac{1}{2}X + \tfrac{\sqrt{3}}{2}Y\right)^2 = 16 \quad\Leftrightarrow$$

$$\tfrac{13}{4}\left(3X^2 - 2\sqrt{3}XY + Y^2\right) + \tfrac{3\sqrt{3}}{2}\left(\sqrt{3}X^2 + 2XY - \sqrt{3}Y^2\right) + \tfrac{7}{4}\left(X^2 + 2\sqrt{3}XY + 3Y^2\right) = 16 \quad\Leftrightarrow$$

$$X^2\left(\tfrac{39}{4} + \tfrac{9}{2} + \tfrac{7}{4}\right) + XY\left(-\tfrac{13\sqrt{3}}{2} + \tfrac{6\sqrt{3}}{2} + \tfrac{7\sqrt{3}}{2}\right) + Y^2\left(\tfrac{13}{4} - \tfrac{9}{2} + \tfrac{21}{4}\right) = 16 \quad\Leftrightarrow\quad 16X^2 + 4Y^2 = 16 \quad\Leftrightarrow$$

$X^2 + \dfrac{Y^2}{4} = 1$. This is an ellipse with $a = 2$, $b = 1$, and $c = \sqrt{4-1} = \sqrt{3}$. Thus, the vertices are $V\left(0, \pm 2\right)$ and the foci are $F\left(0, \pm\sqrt{3}\right)$.

**(c)**

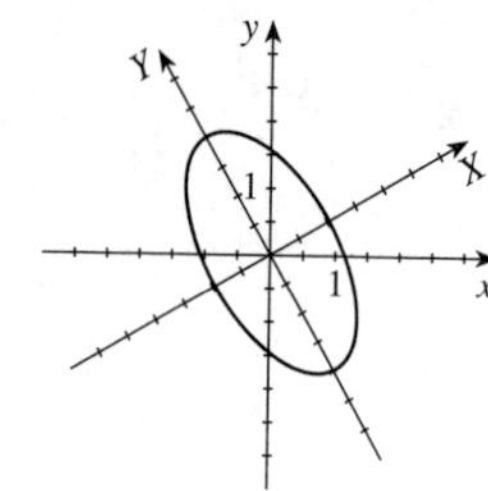


**17. (a)** $x^2 + 2\sqrt{3}xy - y^2 + 2 = 0$. So $A = 1$, $B = 2\sqrt{3}$, and $C = -1$, and so the discriminant is $B^2 - 4AC = \left(2\sqrt{3}\right)^2 - 4\,(1)\,(-1) > 0$. Since the discriminant is positive, the equation represents a hyperbola.

**(b)** $\cot 2\phi = \dfrac{A-C}{B} = \dfrac{1+1}{2\sqrt{3}} = \dfrac{1}{\sqrt{3}} \quad\Rightarrow\quad 2\phi = 60^\circ \quad\Leftrightarrow\quad \phi = 30^\circ$.

Therefore, $x = \frac{\sqrt{3}}{2}X - \frac{1}{2}Y$ and $y = \frac{1}{2}X + \frac{\sqrt{3}}{2}Y$. After substitution, the original equation becomes

$$\left(\tfrac{\sqrt{3}}{2}X - \tfrac{1}{2}Y\right)^2 + 2\sqrt{3}\left(\tfrac{\sqrt{3}}{2}X - \tfrac{1}{2}Y\right)\left(\tfrac{1}{2}X + \tfrac{\sqrt{3}}{2}Y\right) - \left(\tfrac{1}{2}X + \tfrac{\sqrt{3}}{2}Y\right)^2 + 2 = 0 \quad\Leftrightarrow$$

$$\tfrac{3}{4}X^2 - \tfrac{\sqrt{3}}{2}XY + \tfrac{1}{4}Y^2 + \tfrac{\sqrt{3}}{2}\left(\sqrt{3}X^2 + 2XY - \sqrt{3}Y^2\right) - \tfrac{1}{4}X^2 - \tfrac{\sqrt{3}}{2}XY - \tfrac{3}{4}Y^2 + 2 = 0 \quad\Leftrightarrow$$

$$X^2\left(\tfrac{3}{4} + \tfrac{3}{2} - \tfrac{1}{4}\right) + XY\left(-\tfrac{\sqrt{3}}{2} + \sqrt{3} - \tfrac{\sqrt{3}}{2}\right) + Y^2\left(\tfrac{1}{4} - \tfrac{3}{2} - \tfrac{3}{4}\right) = -2 \quad\Leftrightarrow\quad 2X^2 - 2Y^2 = -2 \quad\Leftrightarrow$$

$Y^2 - X^2 = 1$.

**(c)**

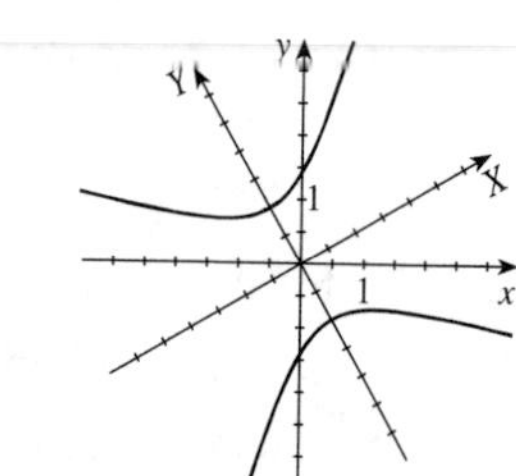


**18. (a)** $21x^2 + 10\sqrt{3}xy + 31y^2 = 144$. Then $A = 21$, $B = 10\sqrt{3}$, and $C = 31$, and so the discriminant is $B^2 - 4AC = \left(10\sqrt{3}\right)^2 - 4\,(21)\,(31) = -2304$. Since the discriminant is negative, the equation represents an ellipse.

**(b)** $\cot 2\phi = \dfrac{A-C}{B} = \dfrac{21-31}{10\sqrt{3}} = -\dfrac{1}{\sqrt{3}} \Rightarrow 2\phi = 120^\circ \quad\Leftrightarrow\quad \phi = 60^\circ$.

Therefore, $x = \frac{1}{2}X - \frac{\sqrt{3}}{2}Y$ and $y = \frac{\sqrt{3}}{2}X + \frac{1}{2}Y$. After substitution, the original equation becomes

$$21\left(\tfrac{1}{2}X - \tfrac{\sqrt{3}}{2}Y\right)^2 + 10\sqrt{3}\left(\tfrac{1}{2}X - \tfrac{\sqrt{3}}{2}Y\right)\left(\tfrac{\sqrt{3}}{2}X + \tfrac{1}{2}Y\right) + 31\left(\tfrac{\sqrt{3}}{2}X + \tfrac{1}{2}Y\right)^2 = 144 \quad\Leftrightarrow$$

$$\tfrac{21}{4}\left(X^2 - 2\sqrt{3}XY + 3Y^2\right) + \tfrac{5\sqrt{3}}{2}\left(\sqrt{3}X^2 - 2XY - \sqrt{3}Y^2\right) + \tfrac{31}{4}\left(3X^2 + 2\sqrt{3}XY + Y^2\right) = 144 \quad\Leftrightarrow$$

$$X^2\left(\tfrac{21}{4} + \tfrac{15}{2} + \tfrac{93}{4}\right) + XY\left(-\tfrac{21\sqrt{3}}{2} - \tfrac{10\sqrt{3}}{2} + \tfrac{31\sqrt{3}}{2}\right) + Y^2\left(\tfrac{63}{4} - \tfrac{15}{2} + \tfrac{31}{4}\right) = 144 \quad\Leftrightarrow$$

$36X^2 + 16Y^2 = 144 \quad\Leftrightarrow\quad \frac{1}{4}X^2 + \frac{1}{9}Y^2 = 1$. This is an ellipse with $a = 3$, $b = 2$, and $c = \sqrt{9-4} = \sqrt{5}$.

**(c)**

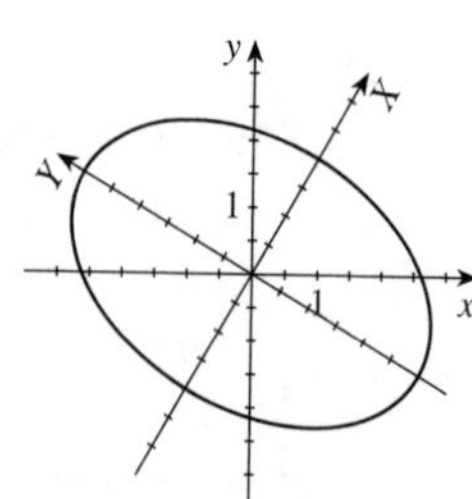

**19. (a)** $11x^2 - 24xy + 4y^2 + 20 = 0$. So $A = 11$, $B = -24$, and $C = 4$, and so the discriminant is $B^2 - 4AC = (-24)^2 - 4(11)(4) > 0$. Since the discriminant is positive, the equation represents a hyperbola.

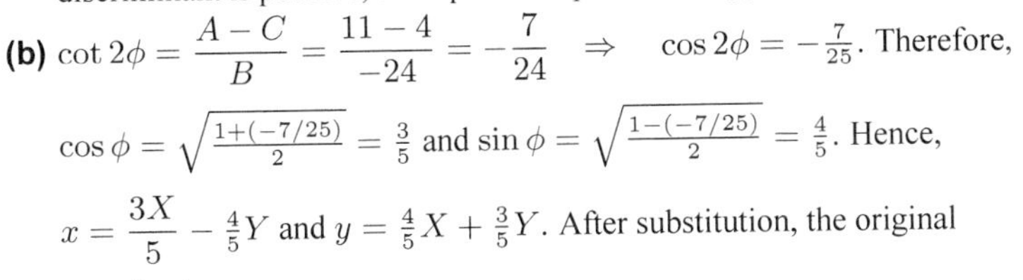

**(b)** $\cot 2\phi = \dfrac{A-C}{B} = \dfrac{11-4}{-24} = -\dfrac{7}{24} \quad \Rightarrow \quad \cos 2\phi = -\frac{7}{25}$. Therefore,

$\cos\phi = \sqrt{\frac{1+(-7/25)}{2}} = \frac{3}{5}$ and $\sin\phi = \sqrt{\frac{1-(-7/25)}{2}} = \frac{4}{5}$. Hence,

$x = \dfrac{3X}{5} - \frac{4}{5}Y$ and $y = \frac{4}{5}X + \frac{3}{5}Y$. After substitution, the original equation becomes

$$11\left(\tfrac{3}{5}X - \tfrac{4}{5}Y\right)^2 - 24\left(\tfrac{3}{5}X - \tfrac{4}{5}Y\right)\left(\tfrac{4}{5}X + \tfrac{3}{5}Y\right) + 4\left(\tfrac{4}{5}X + \tfrac{3}{5}Y\right)^2 + 20 = 0 \quad \Leftrightarrow$$
$$\tfrac{11}{25}\left(9X^2 - 24XY + 16Y^2\right) - \tfrac{24}{25}\left(12X^2 - 7XY - 12Y^2\right) + \tfrac{4}{25}\left(16X^2 + 24XY + 9Y^2\right) + 20 = 0 \quad \Leftrightarrow$$
$$X^2(99 - 288 + 64) + XY(-264 + 168 + 96) + Y^2(176 + 288 + 36) = -500 \quad \Leftrightarrow$$
$$-125X^2 + 500Y^2 = -500 \quad \Leftrightarrow \quad \tfrac{1}{4}X^2 - Y^2 = 1.$$

**(c)**

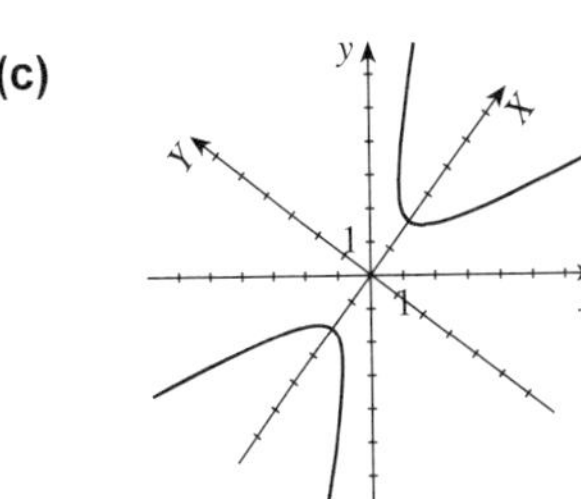


Since $\cos 2\phi = -\frac{7}{25}$, we have $2\phi \approx 106.26°$, so $\phi \approx 53°$.

**20. (a)** $25x^2 - 120xy + 144y^2 - 156x - 65y = 0$. Then $A = 25$, $B = -120$, and $C = 144$, and so the discriminant is $B^2 - 4AC = (-120)^2 - 4(25)(144) = 0$. Since the discriminant is zero, the equation represents a parabola.

**(b)** $\cot 2\phi = \dfrac{A-C}{B} = \dfrac{25-144}{-120} = \dfrac{-119}{-120} \quad \Rightarrow \quad \cos 2\phi = \frac{119}{169}$.

Therefore, $\cos\phi = \sqrt{\frac{1+119/169}{2}} = \frac{12}{13}$ and $\sin\phi = \sqrt{\frac{1-119/169}{2}} = \frac{5}{13}$.

Hence, $x = \dfrac{12X}{13} - \dfrac{5Y}{13}$ and $y = \dfrac{5X}{13} + \dfrac{12Y}{13}$. Substituting gives

$$25\left(\frac{12X}{13} - \frac{5Y}{13}\right)^2 - 120\left(\frac{12X}{13} - \frac{5Y}{13}\right)\left(\frac{5X}{13} + \frac{12Y}{13}\right) + 144\left(\frac{5X}{13} + \frac{12Y}{13}\right)^2 - 156\left(\frac{12X}{13} - \frac{5Y}{13}\right) -$$
$$65\left(\frac{5X}{13} + \frac{12Y}{13}\right) = 0 \quad \Leftrightarrow \quad \frac{X^2}{169}\left(25 \cdot 12^2 - 120 \cdot 12 \cdot 5 + 144 \cdot 5^2\right) +$$
$$\frac{Y^2}{169}\left(25 \cdot 5^2 - 120(-5)(12) + 144 \cdot 12^2\right) - \frac{X}{13}(156 \cdot 12 + 65 \cdot 5) + \frac{Y}{13}(156 \cdot 5 - 65 \cdot 12) = 0 \quad \Leftrightarrow$$
$$169Y^2 - 169X = 0 \quad \Leftrightarrow \quad X = Y^2.$$ This is a parabola with $4p = 1$.

**(c)**

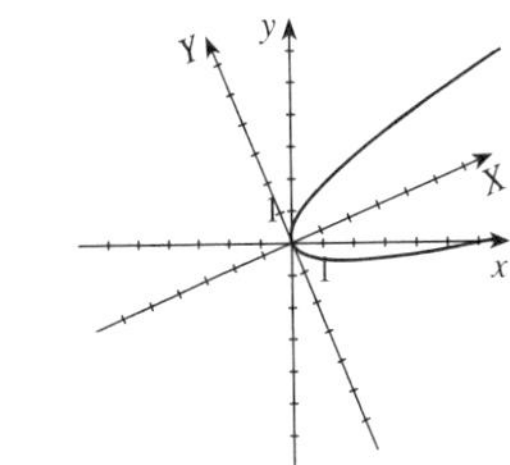


Since $\cos 2\phi = \frac{119}{169}$, we have $2\phi \approx 45.2°$, so $\phi \approx 23°$.

**21. (a)** $\sqrt{3}x^2 + 3xy = 3$. So $A = \sqrt{3}$, $B = 3$, and $C = 0$, and so the discriminant is $B^2 - 4AC = (3)^2 - 4\left(\sqrt{3}\right)(0) = 9$. Since the discriminant is positive, the equation represents a hyperbola.

**(b)** $\cot 2\phi = \dfrac{A-C}{B} = \dfrac{1}{\sqrt{3}} \quad \Rightarrow \quad 2\phi = 60° \quad \Leftrightarrow \quad \phi = 30°$. Therefore,

$x = \frac{\sqrt{3}}{2}X - \frac{1}{2}Y$ and $y = \frac{1}{2}X + \frac{\sqrt{3}}{2}Y$. After substitution, the equation becomes $\sqrt{3}\left(\frac{\sqrt{3}}{2}X - \frac{1}{2}Y\right)^2 + 3\left(\frac{\sqrt{3}}{2}X - \frac{1}{2}Y\right)\left(\frac{1}{2}X + \frac{\sqrt{3}}{2}Y\right) = 3$

$$\Leftrightarrow \quad \tfrac{\sqrt{3}}{4}\left(3X^2 - 2\sqrt{3}XY + Y^2\right) + \tfrac{3}{4}\left(\sqrt{3}X^2 + 2XY - \sqrt{3}Y^2\right) = 3 \quad \Leftrightarrow$$
$$X^2\left(\tfrac{3\sqrt{3}}{4} + \tfrac{3\sqrt{3}}{4}\right) + XY\left(\tfrac{-6}{4} + \tfrac{6}{4}\right) + Y^2\left(\tfrac{\sqrt{3}}{4} - \tfrac{3\sqrt{3}}{4}\right) = 3 \quad \Leftrightarrow \quad \tfrac{3\sqrt{3}}{2}X^2 - \tfrac{\sqrt{3}}{2}Y^2 = 3 \quad \Leftrightarrow$$
$$\tfrac{\sqrt{3}}{2}X^2 - \tfrac{1}{2\sqrt{3}}Y^2 = 1.$$ This is a hyperbola with $a = \sqrt{\frac{2}{\sqrt{3}}}$ and $b = \sqrt{2\sqrt{3}}$.

**(c)**

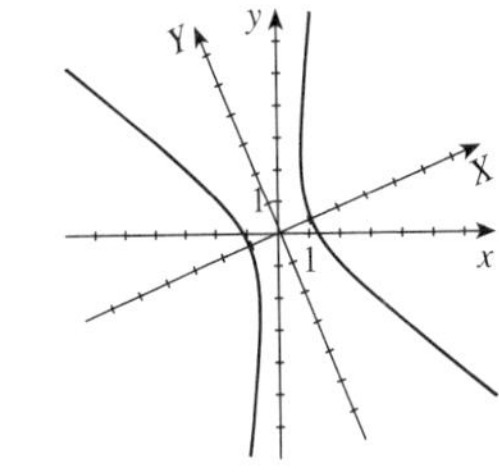

**22. (a)** $153x^2 + 192xy + 97y^2 = 225$. Then $A = 153$, $B = 192$, and $C = 97$, and so the discriminant is $B^2 - 4AC = (192)^2 - 4(153)(97) = -22500$. Since the discriminant is negative, the equation represents an ellipse.

**(b)** $\cot 2\phi = \dfrac{A-C}{B} = \dfrac{153-97}{192} = \dfrac{56}{192} \Rightarrow \cos 2\phi = \dfrac{56}{200}$. Therefore, $\cos\phi = \sqrt{\frac{1+56/200}{2}} = \frac{16}{20} = \frac{4}{5}$ and $\sin\phi = \sqrt{\frac{1-56/200}{2}} = \frac{12}{20} = \frac{3}{5} \Leftrightarrow$ $\phi = \cos^{-1}\frac{4}{5} \approx 36.9°$. Substituting gives

$153\left(\frac{4}{5}X - \frac{3}{5}Y\right)^2 + 192\left(\frac{4}{5}X - \frac{3}{5}Y\right)\left(\frac{3}{5}X + \frac{4}{5}Y\right) + 97\left(\frac{3}{5}X + \frac{4}{5}Y\right)^2 = 225 \quad \Leftrightarrow$

$\frac{153}{25}\left(16X^2 - 24XY + 9Y^2\right) + \frac{192}{25}\left(12X^2 + 7XY - 12Y^2\right) + \frac{97}{25}\left(9X^2 + 24XY + 16Y^2\right) = 225 \quad \Leftrightarrow$

$X^2(2448 + 2304 + 873) + XY(-3672 + 1344 + 2328) + Y^2(1377 - 2304 + 1552) = 5625 \quad \Leftrightarrow$

$5625X^2 + 625Y^2 = 5625 \quad \Leftrightarrow \quad X^2 + \frac{1}{9}Y^2 = 1$. This is an ellipse with $a = 3$, $b = 1$, and $c = \sqrt{9-1} = 2\sqrt{2}$.

**(c)**

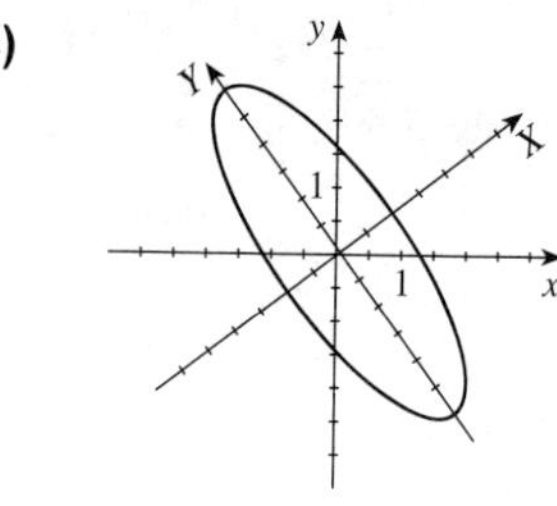

**23. (a)** $2\sqrt{3}x^2 - 6xy + \sqrt{3}x + 3y = 0$. So $A = 2\sqrt{3}$, $B = -6$, and $C = 0$, and so the discriminant is $B^2 - 4AC = (-6)^2 - 4\left(2\sqrt{3}\right)(0) = 36$. Since the discriminant is positive, the equation represents a hyperbola.

**(b)** $\cot 2\phi = \dfrac{A-C}{B} = \dfrac{2\sqrt{3}}{-6} = -\dfrac{1}{\sqrt{3}} \quad \Rightarrow \quad 2\phi = 120° \quad \Leftrightarrow \quad \phi = 60°$.

Therefore, $x = \frac{1}{2}X - \frac{\sqrt{3}}{2}Y$ and $y = \frac{\sqrt{3}}{2}X + \frac{1}{2}Y$, and substituting gives

$2\sqrt{3}\left(\frac{1}{2}X - \frac{\sqrt{3}}{2}Y\right)^2 - 6\left(\frac{1}{2}X - \frac{\sqrt{3}}{2}Y\right)\left(\frac{\sqrt{3}}{2}X + \frac{1}{2}Y\right) + \sqrt{3}\left(\frac{1}{2}X - \frac{\sqrt{3}}{2}Y\right) + 3\left(\frac{\sqrt{3}}{2}X + \frac{1}{2}Y\right) = 0 \quad \Leftrightarrow$

$\frac{\sqrt{3}}{2}\left(X^2 - 2\sqrt{3}XY + 3Y^2\right) - \frac{3}{2}\left(\sqrt{3}X^2 - 2XY - \sqrt{3}Y^2\right) + \frac{\sqrt{3}}{2}\left(X - \sqrt{3}Y\right) + \frac{3}{2}\left(\sqrt{3}X + Y\right) = 0 \quad \Leftrightarrow$

$X^2\left(\frac{\sqrt{3}}{2} - \frac{3\sqrt{3}}{2}\right) + X\left(\frac{\sqrt{3}}{2} + \frac{3\sqrt{3}}{2}\right) + XY(-3+3) + Y^2\left(\frac{3\sqrt{3}}{2} + \frac{3\sqrt{3}}{2}\right) + Y\left(-\frac{3}{2} + \frac{3}{2}\right) = 0 \quad \Leftrightarrow$

$-\sqrt{3}X^2 + 2\sqrt{3}X + 3\sqrt{3}Y^2 = 0 \quad \Leftrightarrow \quad -X^2 + 2X + 3Y^2 = 0 \quad \Leftrightarrow \quad 3Y^2 - \left(X^2 - 2X + 1\right) = -1 \quad \Leftrightarrow$

$(X-1)^2 - 3Y^2 = 1$. This is a hyperbola with $a = 1$, $b = \frac{\sqrt{3}}{3}$, $c = \sqrt{1 + \frac{1}{3}} = \frac{2}{\sqrt{3}}$, and $C(1, 0)$.

**(c)**

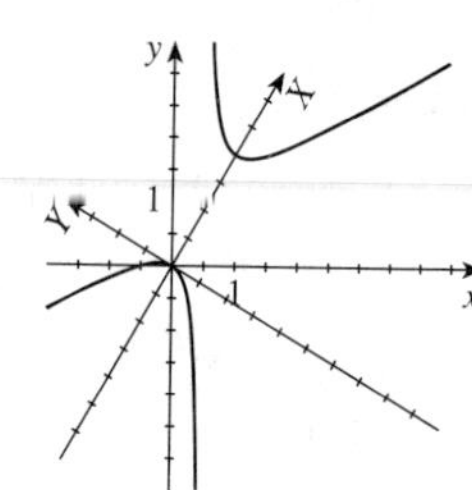

**24. (a)** $9x^2 - 24xy + 16y^2 = 100(x - y - 1)$. Then $A = 9$, $B = -24$, and $C = 16$, and so the discriminant is $B^2 - 4AC = (-24)^2 - 4(9)(16) = 0$. Since the discriminant is zero, the equation represents a parabola.

**(b)** $\cot 2\phi = \dfrac{A-C}{B} = \dfrac{9-16}{-24} = \dfrac{7}{24} \quad \Rightarrow \quad \cos 2\phi = \frac{7}{25} \quad \Leftrightarrow \quad \phi \approx 36.9°$.

Now $\cos\phi = \sqrt{\frac{1+7/25}{2}} = \frac{4}{5}$ and $\sin\phi = \sqrt{\frac{1-7/25}{2}} = \frac{3}{5}$, and so $x = \frac{4}{5}X - \frac{3}{5}Y$, $y = \frac{3}{5}X + \frac{4}{5}Y$. By substitution,

$9\left(\frac{4}{5}X - \frac{3}{5}Y\right)^2 - 24\left(\frac{4}{5}X - \frac{3}{5}Y\right)\left(\frac{3}{5}X + \frac{4}{5}Y\right) + 16\left(\frac{3}{5}X + \frac{4}{5}Y\right)^2 = 100\left(\frac{4}{5}X - \frac{3}{5}Y - \frac{3}{5}X - \frac{4}{5}Y - 1\right) \quad \Leftrightarrow$

$\frac{9}{25}\left(16X^2 - 24XY + 9Y^2\right) - \frac{24}{25}\left(12X^2 + 7XY - 12Y^2\right) + \frac{16}{25}\left(9X^2 + 24XY + 16Y^2\right) = 100\left(\frac{1}{5}X - \frac{7}{5}Y - 1\right)$

$\Leftrightarrow \quad 625Y^2 = 500X - 3500Y - 2500 \quad \Leftrightarrow \quad 5Y^2 + 28Y = 4X - 20 \quad \Leftrightarrow$

$5\left(Y^2 + \frac{28}{5}Y + \frac{196}{25}\right) = 4X - 20 + \frac{196}{5} = 4X + \frac{96}{5} = 4\left(X + \frac{24}{5}\right) \quad \Leftrightarrow \quad \left(Y + \frac{14}{5}\right)^2 = \frac{4}{5}\left(X + \frac{24}{5}\right)$. This is a parabola with $4p = \frac{4}{5}$ and $V\left(-\frac{24}{5}, -\frac{14}{5}\right)$.

**(c)**

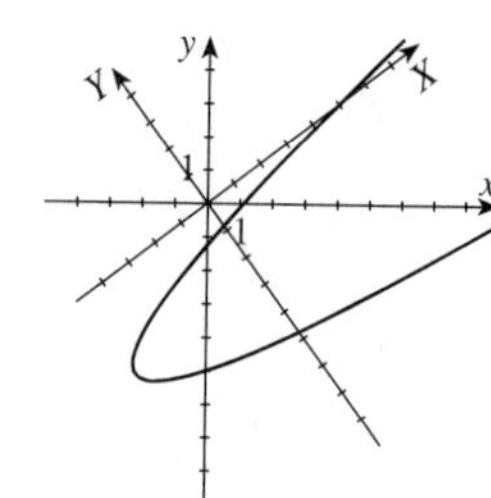

**25. (a)** $52x^2 + 72xy + 73y^2 = 40x - 30y + 75$. So $A = 52$, $B = 72$, and $C = 73$, and so the discriminant is $B^2 - 4AC = (72)^2 - 4(52)(73) = -10{,}000$. Since the discriminant is decidedly negative, the equation represents an ellipse.

**(b)** $\cot 2\phi = \dfrac{A - C}{B} = \dfrac{52 - 73}{72} = -\dfrac{7}{24}$. Therefore, as in Exercise 19(b), we get $\cos\phi = \frac{3}{5}$, $\sin\phi = \frac{4}{5}$, and $x = \frac{3}{5}X - \frac{4}{5}Y$, $y = \frac{4}{5}X + \frac{3}{5}Y$. By substitution,

$$52\left(\tfrac{3}{5}X - \tfrac{4}{5}Y\right)^2 + 72\left(\tfrac{3}{5}X - \tfrac{4}{5}Y\right)\left(\tfrac{4}{5}X + \tfrac{3}{5}Y\right) + 73\left(\tfrac{4}{5}X + \tfrac{3}{5}Y\right)^2 = 40\left(\tfrac{3}{5}X - \tfrac{4}{5}Y\right) - 30\left(\tfrac{4}{5}X + \tfrac{3}{5}Y\right) + 75 \quad\Leftrightarrow$$

$$\tfrac{52}{25}\left(9X^2 - 24XY + 16Y^2\right) + \tfrac{72}{25}\left(12X^2 - 7XY - 12Y^2\right) + \tfrac{73}{25}\left(16X^2 + 24XY + 9Y^2\right) = 24X - 32Y - 24X - 18Y + 75 \quad\Leftrightarrow$$

$468X^2 + 832Y^2 + 864X^2 - 864Y^2 + 1168X^2 + 657Y^2 = -1250Y + 1875 \quad\Leftrightarrow\quad 2500X^2 + 625Y^2 + 1250Y = 1875$ $\Leftrightarrow\quad 100X^2 + 25Y^2 + 50Y = 75 \quad\Leftrightarrow\quad X^2 + \frac{1}{4}(Y+1)^2 = 1$. This is an ellipse with $a = 2$, $b = 1$, $c = \sqrt{4 - 1} = \sqrt{3}$, and center $C(0, -1)$.

**(c)**

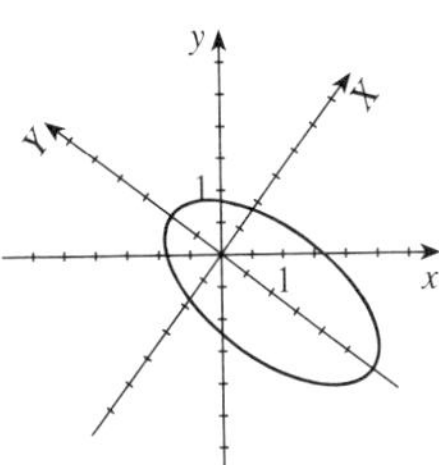


Since $\cos 2\phi = -\frac{7}{25}$, we have $2\phi = \cos^{-1}\left(-\frac{7}{25}\right) \approx 106.26°$ and so $\phi \approx 53°$.

**26. (a)** $(7x + 24y)^2 = 49x^2 + 336xy + 576y^2 = 600x - 175y + 25$. So $A = 49$, $B = 336$, and $C = 576$, and so the discriminant is $B^2 - 4AC = (336)^2 - 4(49)(576) = 0$. Since the discriminant is zero, the equation represents a parabola.

**(b)** $\cot 2\phi = \dfrac{A - C}{B} = \dfrac{49 - 576}{336} = -\dfrac{527}{336} \quad\Rightarrow\quad \cos 2\phi = -\frac{527}{625}$. Therefore, $\cos\phi = \sqrt{\frac{1 - 527/625}{2}} = \frac{7}{25}$ and $\sin\phi = \sqrt{\frac{1 + 527/625}{2}} = \frac{24}{25}$. Substituting $x = \frac{7}{25}X - \frac{24}{25}Y$ and $y = \frac{24}{25}X + \frac{7}{25}Y$ gives

$$49\left(\tfrac{7}{25}X - \tfrac{24}{25}Y\right)^2 + 336\left(\tfrac{7}{25}X - \tfrac{24}{25}Y\right)\left(\tfrac{24}{25}X + \tfrac{7}{25}Y\right) + 576\left(\tfrac{24}{25}X + \tfrac{7}{25}Y\right)^2 = 600\left(\tfrac{7}{25}X - \tfrac{24}{25}Y\right) - 175\left(\tfrac{24}{25}X + \tfrac{7}{25}Y\right) + 25 \quad\Leftrightarrow$$

$$\tfrac{49}{625}\left(49X^2 - 336XY + 576Y^2\right) + \tfrac{336}{625}\left(168X^2 - 527XY - 168Y^2\right) + \tfrac{576}{625}\left(576X^2 + 336XY + 49Y^2\right) = 168X - 576Y - 168X - 49Y + 25 \quad\Leftrightarrow$$

$$X^2(2401 + 56{,}448 + 331{,}776) + XY(-16{,}464 - 177{,}072 + 193{,}536) + Y^2(28{,}224 - 56{,}448 + 28{,}224) = -625^2Y + 15{,}625 \quad\Leftrightarrow$$

$390{,}625X^2 + 390{,}625Y = 15{,}625 \quad\Leftrightarrow\quad 25X^2 + 25Y = 1 \quad\Leftrightarrow\quad X^2 = -Y + \frac{1}{25} = -\left(Y - \frac{1}{25}\right)$. This is a parabola with $4p = -1$ and vertex $\left(0, \frac{1}{25}\right)$.

**(c)**

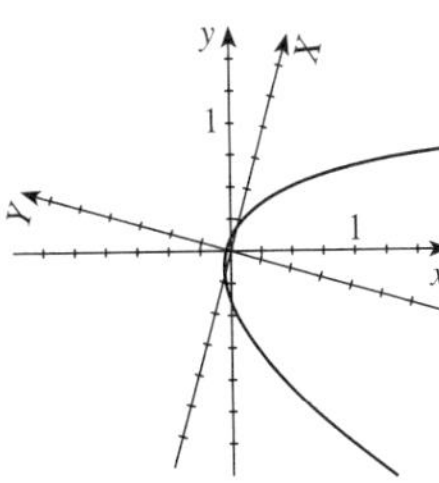


Since $\cos 2\phi = -\frac{527}{625}$, we have $2\phi = \cos^{-1}\left(-\frac{527}{625}\right) \approx 147.48°$ and so $\phi \approx 74°$.

**27. (a)** The discriminant is $B^2 - 4AC = (-4)^2 + 4(2)(2) = 0$. Since the discriminant is 0, the equation represents a parabola.

**(b)** $2x^2 - 4xy + 2y^2 - 5x - 5 = 0 \quad \Leftrightarrow \quad 2y^2 - 4xy = -2x^2 + 5x + 5$
$\Leftrightarrow \quad 2\left(y^2 - 2xy\right) = -2x^2 + 5x + 5 \quad \Leftrightarrow$
$2\left(y^2 - 2xy + x^2\right) = -2x^2 + 5x + 5 + 2x^2 \quad \Leftrightarrow$
$2(y-x)^2 = 5x + 5 \quad \Leftrightarrow$
$(y-x)^2 = \frac{5}{2}x + \frac{5}{2} \quad \Rightarrow y - x = \pm\sqrt{\frac{5}{2}x + \frac{5}{2}} \quad \Leftrightarrow$
$y = x \pm \sqrt{\frac{5}{2}x + \frac{5}{2}}$

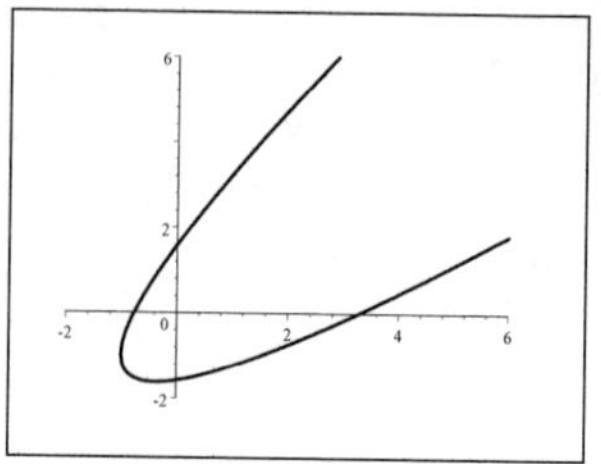

**28. (a)** The discriminant is $B^2 - 4AC = (-2)^2 + 4(1)(3) = -8 < 0$. Since the discriminant is negative, the equation represents an ellipse.

**(b)** $x^2 - 2xy + 3y^2 = 8 \quad \Leftrightarrow \quad 3y^2 - 2xy = 8 - x^2 \quad \Leftrightarrow$
$3\left(y^2 - \frac{2}{3}xy\right) = 8 - x^2 \quad \Leftrightarrow \quad 3\left(y^2 - \frac{2}{3}xy + \frac{1}{9}x^2\right) = 8 - x^2 + \frac{1}{3}x^2$
$\Leftrightarrow \quad 3\left(y - \frac{1}{3}x\right)^2 = 8 - \frac{2}{3}x^2 \quad \Leftrightarrow \quad \left(y - \frac{1}{3}x\right)^2 = \frac{8}{3} - \frac{2}{9}x^2 \quad \Rightarrow$
$y - \frac{1}{3}x = \pm\sqrt{\frac{8}{3} - \frac{2}{9}x^2} \Leftrightarrow y = \frac{1}{3}x \pm \sqrt{\frac{8}{3} - \frac{2}{9}x^2}$

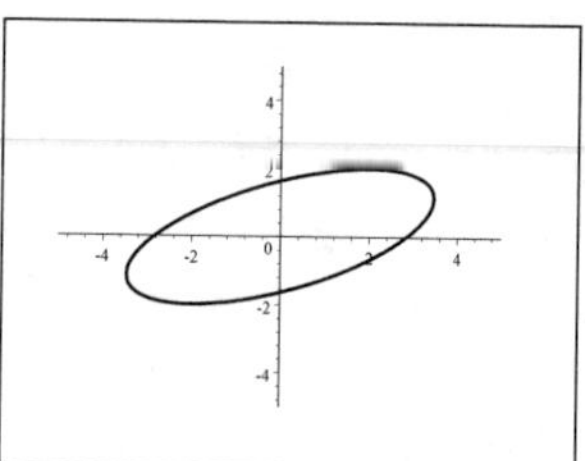

**29. (a)** The discriminant is $B^2 - 4AC = 10^2 + 4(6)(3) = 28 > 0$. Since the discriminant is positive, the equation represents a hyperbola.

**(b)** $6x^2 + 10xy + 3y^2 - 6y = 36 \quad \Leftrightarrow \quad 3y^2 + 10xy - 6y = 36 - 6x^2 \quad \Leftrightarrow$
$3y^2 + 2(5x - 3)y = 36 - 6x^2 \quad \Leftrightarrow \quad y^2 + 2\left(\frac{5}{3}x - 1\right)y = 12 - 2x^2$
$\Leftrightarrow \quad y^2 + 2\left(\frac{5}{3}x - 1\right)y + \left(\frac{5}{3}x - 1\right)^2 = \left(\frac{5}{3}x - 1\right)^2 + 12 - 2x^2 \quad \Leftrightarrow$
$\left[y + \left(\frac{5}{3}x - 1\right)\right]^2 = \frac{25}{9}x^2 - \frac{10}{3}x + 1 + 12 - 2x^2 \quad \Leftrightarrow$
$\left[y + \left(\frac{5}{3}x - 1\right)\right]^2 = \frac{7}{9}x^2 - \frac{10}{3}x + 13 \quad \Leftrightarrow$
$y + \left(\frac{5}{3}x - 1\right) = \pm\sqrt{\frac{7}{9}x^2 - \frac{10}{3}x + 13} \quad \Leftrightarrow$
$y = -\frac{5}{3}x + 1 \pm \sqrt{\frac{7}{9}x^2 - \frac{10}{3}x + 13}$

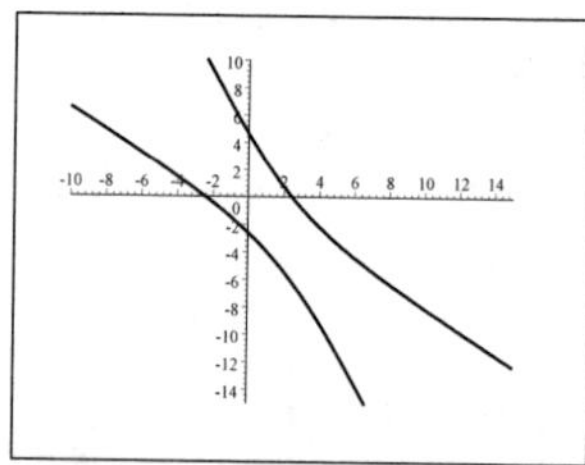

**30. (a)** The discriminant is $B^2 - 4AC = (-6)^2 + 4(9)(1) = 0$. Since the discriminant is 0, the equation represents a parabola.

**(b)** $9x^2 - 6xy + y^2 + 6x - 2y = 0 \quad \Leftrightarrow \quad y^2 - 6xy - 2y = -9x^2 - 6x$
$\Leftrightarrow \quad y^2 - 2(3x + 1)y = -9x^2 - 6x \quad \Leftrightarrow$
$y^2 - 2(3x + 1)y + (3x + 1)^2 = (3x + 1)^2 - 9x^2 - 6x \quad \Leftrightarrow$
$[y - (3x + 1)]^2 = 1 \quad \Leftrightarrow \quad y - (3x + 1) = \pm 1 \quad \Leftrightarrow$
$y = 3x + 1 \pm 1$. So $y = 3x$ or $y = 3x + 2$. This is a degenerate conic.

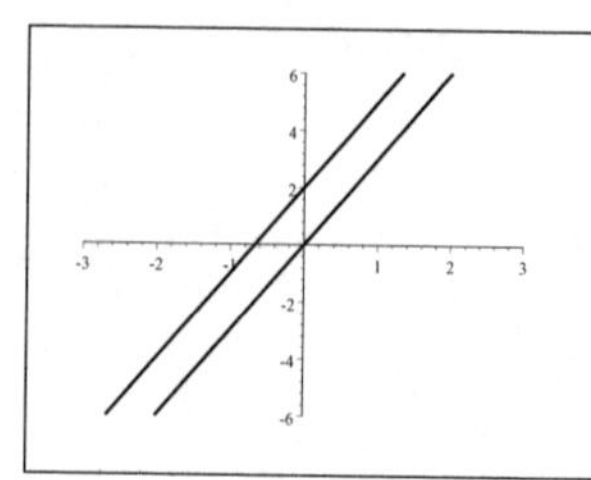

**31. (a)** $7x^2 + 48xy - 7y^2 - 200x - 150y + 600 = 0$. Then $A = 7$, $B = 48$, and $C = -7$, and so the discriminant is $B^2 - 4AC = (48)^2 - 4(7)(7) > 0$. Since the discriminant is positive, the equation represents a hyperbola. We now find the equation in terms of $XY$-coordinates. We have $\cot 2\phi = \dfrac{A-C}{B} = \dfrac{7}{24} \quad\Rightarrow\quad \cos\phi = \frac{4}{5}$ and $\sin\phi = \frac{3}{5}$. Therefore, $x = \frac{4}{5}X - \frac{3}{5}Y$ and $y = \frac{3}{5}X + \frac{4}{5}Y$, and substitution gives

$$7\left(\tfrac{4}{5}X - \tfrac{3}{5}Y\right)^2 + 48\left(\tfrac{4}{5}X - \tfrac{3}{5}Y\right)\left(\tfrac{3}{5}X + \tfrac{4}{5}Y\right) - 7\left(\tfrac{3}{5}X + \tfrac{4}{5}Y\right)^2 - 200\left(\tfrac{4}{5}X - \tfrac{3}{5}Y\right) - 150\left(\tfrac{3}{5}X + \tfrac{4}{5}Y\right) + 600 = 0$$

$$\Leftrightarrow \quad \tfrac{7}{25}\left(16X^2 - 24XY + 9Y^2\right) + \tfrac{48}{25}\left(12X^2 + 7XY - 12Y^2\right) - \tfrac{7}{25}\left(9X^2 + 24XY + 16Y^2\right) - 160X + 120Y - 90X - 120Y + 600 = 0$$

$$\Leftrightarrow \quad 112X^2 - 168XY + 63Y^2 + 576X^2 + 336XY - 576Y^2 - 63X^2 - 168XY - 112Y^2 - 6250X + 15{,}000 = 0$$

$\Leftrightarrow \quad 25X^2 - 25Y^2 - 250X + 600 = 0 \quad\Leftrightarrow\quad 25\left(X^2 - 10X + 25\right) - 25Y^2 = -600 + 625 \quad\Leftrightarrow$ $(X-5)^2 - Y^2 = 1$. This is a hyperbola with $a = 1$, $b = 1$, $c = \sqrt{1+1} = \sqrt{2}$, and center $C(5,0)$.

**(b)** In the $XY$-plane, the center is $C(5,0)$, the vertices are $V(5 \pm 1, 0) = V_1(4,0)$ and $V_2(6,0)$, and the foci are $F\left(5 \pm \sqrt{2}, 0\right)$. In the $xy$-plane, the center is $C\left(\frac{4}{5}\cdot 5 - \frac{3}{5}\cdot 0, \frac{3}{5}\cdot 5 + \frac{4}{5}\cdot 0\right) = C(4,3)$, the vertices are $V_1\left(\frac{4}{5}\cdot 4 - \frac{3}{5}\cdot 0, \frac{3}{5}\cdot 4 + \frac{4}{5}\cdot 0\right) = V_1\left(\frac{16}{5}, \frac{12}{5}\right)$ and $V_2\left(\frac{4}{5}\cdot 6 - \frac{3}{5}\cdot 0, \frac{3}{5}\cdot 6 + \frac{4}{5}\cdot 0\right) = V_2\left(\frac{24}{5}, \frac{18}{5}\right)$, and the foci are $F_1\left(4 + \frac{4}{5}\sqrt{2}, 3 + \frac{3}{5}\sqrt{2}\right)$ and $F_2\left(4 - \frac{4}{5}\sqrt{2}, 3 - \frac{3}{5}\sqrt{2}\right)$.

**(c)** In the $XY$-plane, the equations of the asymptotes are $Y = X - 5$ and $Y = -X + 5$. In the $xy$-plane, these equations become $-x\cdot\frac{3}{5} + y\cdot\frac{4}{5} = x\cdot\frac{4}{5} + y\cdot\frac{3}{5} - 5 \quad\Leftrightarrow\quad 7x - y - 25 = 0$. Similarly, $-x\cdot\frac{3}{5} + y\cdot\frac{4}{5} = -x\cdot\frac{4}{5} - y\cdot\frac{3}{5} + 5$ $\Leftrightarrow \quad x + 7y - 25 = 0$.

**32. (a)** $2\sqrt{2}(x+y)^2 = 7x + 9y \quad\Leftrightarrow\quad 2\sqrt{2}x^2 + 4\sqrt{2}xy + 2\sqrt{2}y^2 = 7x + 9y$. Therefore, $A = 2\sqrt{2}$, $B = 4\sqrt{2}$, and $C = 2\sqrt{2}$, and so $\cot 2\phi = \dfrac{A-C}{B} = 0 \quad\Rightarrow\quad 2\phi = 90^\circ \quad\Rightarrow\quad \phi = 45^\circ$. Thus $x = \dfrac{X-Y}{\sqrt{2}}$, $y = \dfrac{X+Y}{\sqrt{2}}$ $\Rightarrow \quad x + y = \sqrt{2}X$. Thus the equation becomes $2\sqrt{2}\left(\sqrt{2}X\right)^2 = \dfrac{16X + 2Y}{\sqrt{2}} \quad\Leftrightarrow\quad 8X^2 = 16X + 2Y \quad\Leftrightarrow$ $4\left(X^2 + 2X + 1\right) = Y + 4 \quad\Leftrightarrow\quad 4(X+1)^2 = Y + 4$. This is a parabola.

**(b)** $4p = \frac{1}{4} \quad\Rightarrow\quad p = \frac{1}{16}$. Thus, in $XY$-coordinates the vertex is $V(-1,-4)$ and the focus is $F\left(-1, -\frac{63}{16}\right)$. In $xy$-coordinates, $V\left(\frac{3}{\sqrt{2}}, \frac{-5}{\sqrt{2}}\right)$ and $F\left(\frac{47\sqrt{2}}{32}, \frac{-79\sqrt{2}}{32}\right)$.

**(c)** The directrix is $Y = -\frac{65}{16}$. Thus $\dfrac{-x+y}{\sqrt{2}} = -\dfrac{65}{16}$ or $y = x - \frac{65\sqrt{2}}{16}$.

**33.** We use the hint and eliminate $Y$ by adding: $x = X\cos\phi - Y\sin\phi \quad\Leftrightarrow\quad x\cos\phi = X\cos^2\phi - Y\sin\phi\cos\phi$ and $y = X\sin\phi + Y\cos\phi \quad\Leftrightarrow\quad y\sin\phi = X\sin^2\phi + Y\sin\phi\cos\phi$, and adding these two equations gives $x\cos\phi + y\sin\phi = X\left(\cos^2\phi + \sin^2\phi\right) \quad\Leftrightarrow\quad x\cos\phi + y\sin\phi = X$. In a similar manner, we eliminate $X$ by subtracting: $x = X\cos\phi - Y\sin\phi \quad\Leftrightarrow\quad -x\sin\phi = -X\cos\phi\sin\phi + Y\sin^2\phi$ and $y = X\sin\phi + Y\cos\phi \quad\Leftrightarrow$ $y\cos\phi = X\sin\phi\cos\phi + Y\cos^2\phi$, so $-x\sin\phi + y\cos\phi = Y\left(\cos^2\phi + \sin^2\phi\right) \quad\Leftrightarrow\quad -x\sin\phi + y\cos\phi = Y$. Thus, $X = x\cos\phi + y\sin\phi$ and $Y = -x\sin\phi + y\cos\phi$.

**34.** $\sqrt{x}+\sqrt{y}=1$. Squaring both sides gives $x+2\sqrt{xy}+y=1 \Leftrightarrow 2\sqrt{xy}=1-x-y$, and squaring both sides again gives $4xy=(1-x-y)^2=1-x-y-x+x^2+xy-y+xy+y^2 \Leftrightarrow 4xy=x^2+y^2+2xy-2x-2y+1$ $\Leftrightarrow x^2+y^2-2xy-2x-2y+1=0$. Then $A=1$, $B=-2$, and $C=1$, and so $\cot 2\phi=\dfrac{A-C}{B}=0 \Rightarrow 2\phi=90^\circ$ $\Leftrightarrow \phi=45^\circ$. Therefore, $x=\frac{\sqrt{2}}{2}X-\frac{\sqrt{2}}{2}Y$ and $y=\frac{\sqrt{2}}{2}X+\frac{\sqrt{2}}{2}Y$, and substituting gives

$$\left(\tfrac{\sqrt{2}}{2}X-\tfrac{\sqrt{2}}{2}Y\right)^2-2\left(\tfrac{\sqrt{2}}{2}X-\tfrac{\sqrt{2}}{2}Y\right)\left(\tfrac{\sqrt{2}}{2}X+\tfrac{\sqrt{2}}{2}Y\right)$$
$$+\left(\tfrac{\sqrt{2}}{2}X+\tfrac{\sqrt{2}}{2}Y\right)^2-2\left(\tfrac{\sqrt{2}}{2}X-\tfrac{\sqrt{2}}{2}Y\right)-2\left(\tfrac{\sqrt{2}}{2}X+\tfrac{\sqrt{2}}{2}Y\right)+1=0 \quad \Leftrightarrow$$

$\frac{1}{2}\left(X^2-2XY+Y^2\right)-\left(X^2-Y^2\right)+\frac{1}{2}\left(X^2+2XY+Y^2\right)-2\sqrt{2}X+1=0 \quad \Leftrightarrow \quad X^2\left(\frac{1}{2}-1+\frac{1}{2}\right)+XY(-1+1)+Y^2\left(\frac{1}{2}+1+\frac{1}{2}\right)-2\sqrt{2}X+1=0 \Leftrightarrow 2Y^2=2\sqrt{2}X-1 \quad \Leftrightarrow \quad Y^2=\sqrt{2}X-\frac{1}{2}=\sqrt{2}\left(X-\frac{1}{2\sqrt{2}}\right)$.

This is a parabola with $4p=\sqrt{2}$ and vertex $V\left(\dfrac{1}{2\sqrt{2}},0\right)$. However, in the original equation we must have $x\geq 0$ and $y\geq 0$, so we get only the part of the parabola that lies in the first quadrant.

**35.** $Z=\begin{bmatrix} x \\ y \end{bmatrix}$, $Z'=\begin{bmatrix} X \\ Y \end{bmatrix}$, and $R=\begin{bmatrix} \cos\phi & -\sin\phi \\ \sin\phi & \cos\phi \end{bmatrix}$.

Thus $Z=RZ' \quad \Leftrightarrow \quad \begin{bmatrix} x \\ y \end{bmatrix}=\begin{bmatrix} \cos\phi & -\sin\phi \\ \sin\phi & \cos\phi \end{bmatrix}\begin{bmatrix} X \\ Y \end{bmatrix}=\begin{bmatrix} X\cos\phi-Y\sin\phi Y \\ X\sin\phi+Y\cos\phi \end{bmatrix}$. Equating the entries in this matrix equation gives the first pair of rotation of axes formulas. Now

$R^{-1}=\dfrac{1}{\cos^2\phi+\sin^2\phi}\begin{bmatrix} \cos\phi & \sin\phi \\ -\sin\phi & \cos\phi \end{bmatrix}=\begin{bmatrix} \cos\phi & \sin\phi \\ -\sin\phi & \cos\phi \end{bmatrix}$ and so $Z'=R^{-1}Z \quad \Leftrightarrow$

$\begin{bmatrix} X \\ Y \end{bmatrix}=\begin{bmatrix} \cos\phi & \sin\phi \\ -\sin\phi & \cos\phi \end{bmatrix}\begin{bmatrix} x \\ y \end{bmatrix}=\begin{bmatrix} x\cos\phi+y\sin\phi \\ -x\sin\phi+y\cos\phi \end{bmatrix}$. Equating the entries in this matrix equation gives the second pair of rotation of axes formulas.

**36. (a)** Using $A'=A\cos^2\phi+B\sin\phi\cos\phi+C\sin^2\phi$, $B'=2(C-a)\sin\phi\cos\phi+B\left(\cos^2\phi-\sin^2\phi\right)$, and $C'=A\sin^2\phi-B\sin\phi\cos\phi+C\cos^2\phi$, we first expand the terms.

$$\begin{aligned}
\left(B'\right)^2 &= \left[2(C-a)\sin\phi\cos\phi+B\left(\cos^2\phi-\sin^2\phi\right)\right]^2 \\
&= 4(C-A)^2\sin^2\phi\cos^2\phi+4B(C-A)\sin\phi\cos\phi\left(\cos^2\phi-\sin^2\phi\right)+B^2\left(\cos^2\phi-\sin^2\phi\right)^2 \\
&= 4(C-A)^2\sin^2\phi\cos^2\phi+4B(C-A)\sin\phi\cos^3\phi-4B(C-A)\sin^3\phi\cos\phi \\
&\qquad +B^2\cos^4\phi-2B^2\sin^2\phi\cos^2\phi+B^2\sin^4\phi
\end{aligned}$$

$$\begin{aligned}
4A'C' &= 4\left(A\cos^2\phi+B\sin\phi\cos\phi+C\sin^2\phi\right)\left(A\sin^2\phi-B\sin\phi\cos\phi+C\cos^2\phi\right) \\
&= 4A^2\sin^2\phi\cos^2\phi-4AB\sin\phi\cos^3\phi+4AC\cos^4\phi+4AB\sin^3\phi\cos\phi-4B^2\sin^2\phi\cos^2\phi \\
&\qquad +4BC\sin\phi\cos^3\phi+4AC\sin^4\phi-4BC\sin^3\phi\cos\phi+4C^2\sin^2\phi\cos^2\phi
\end{aligned}$$

Since there are many terms in this expansion, we find the coefficients of like trignometric terms.

| Term | $(B')^2$ | $-4A'C'$ | Sum |
|---|---|---|---|
| $\cos^4\phi$ | $B^2$ | $-4AC$ | $B^2-4AC$ |
| $\sin\phi\cos^3\phi$ | $4B(C-A)$ | $4AB-4BC$ | $4BC-4AB+4AB-4BC=0$ |
| $\sin^2\phi\cos^2\phi$ | $4(C-A)^2-2B^2$ | $-4A^2+4B^2-4C^2$ | $-8AC+2B^2=2(B^2-4AC)$ |
| $\sin^3\phi\cos\phi$ | $-4B(C-A)$ | $-4AB+4BC$ | $-4BC+AB-4AB+4BC=0$ |
| $\sin^4\phi$ | $B^2$ | $-4AC$ | $B^2-4AC$ |

So

$$\begin{aligned} B'^2-4A'C' &= (B^2-4AC)\cos^4\phi+2(B^2-4AC)\sin^2\phi\cos^2\phi+(B^2-4AC)\sin^4\phi \\ &= (B^2-4AC)(\cos^4\phi+2\sin^2\phi\cos^2\phi+\sin^4\phi)=(B^2-4AC)(\cos^2\phi+\sin^2\phi)^2 \\ &= (B^2-4AC)(1^2)=B^2-4AC \end{aligned}$$

So $B^2-4AC=B'^2-4A'C'$.

**(b)** Using $A'=A\cos^2\phi+B\sin\phi\cos\phi+C\sin^2\phi$ and $C'=A\sin^2\phi-B\sin\phi\cos\phi+C\cos^2\phi$, we have

$$\begin{aligned} A'+C' &= A\cos^2\phi+B\sin\phi\cos\phi+C\sin^2\phi+A\sin^2\phi-B\sin\phi\cos\phi+C\cos^2\phi \\ &= A(\sin^2\phi+\cos^2\phi)+C(\sin^2\phi+\cos^2\phi) \\ &= A+C \end{aligned}$$

**(c)** Since $F'=F$, $F$ is also invariant under rotation.

**37.** Let $P$ be the point $(x_1,y_1)$ and $Q$ be the point $(x_2,y_2)$ and let $P'(X_1,Y_1)$ and $Q'(X_2,Y_2)$ be the images of $P$ and $Q$ under the rotation of $\phi$. So $X_1=x_1\cos\phi+y_1\sin\phi$, $Y_1=-x_1\sin\phi+y_1\cos\phi$, $X_2=x_2\cos\phi+y_2\sin\phi$,and $Y_2=-x_2\sin\phi+y_2\cos\phi$. Thus $d(P',Q')=\sqrt{(X_2-X_1)^2+(Y_2-Y_1)^2}$, where

$$\begin{aligned} (X_2-X_1)^2 &= [(x_2\cos\phi+y_2\sin\phi)-(x_1\cos\phi+y_1\sin\phi)]^2=[(x_2-x_1)\cos\phi+(y_2-y_1)\sin\phi]^2 \\ &= (x_2-x_1)^2\cos^2\phi+(x_2-x_1)(y_2-y_1)\sin\phi\cos\phi+(y_2-y_1)^2\sin^2\phi \end{aligned}$$

and

$$\begin{aligned} (Y_2-Y_1)^2 &= [(-x_2\sin\phi+y_2\cos\phi)-(-x_1\sin\phi+y_1\cos\phi)]^2=[-(x_2-x_1)\sin\phi+(y_2-y_1)\cos\phi]^2 \\ &= (x_2-x_1)^2\sin^2\phi-(x_2-x_1)(y_2-y_1)\sin\phi\cos\phi+(y_2-y_1)^2\cos^2\phi \end{aligned}$$

So

$$\begin{aligned} (X_2-X_1)^2+(Y_2-Y_1)^2 &= (x_2-x_1)^2\cos^2\phi+(x_2-x_1)(y_2-y_1)\sin\phi\cos\phi+(y_2-y_1)^2\sin^2\phi \\ &\quad +(x_2-x_1)^2\sin^2\phi-(x_2-x_1)(y_2-y_1)\sin\phi\cos\phi+(y_2-y_1)^2\cos^2\phi \\ &= (x_2-x_1)^2\cos^2\phi+(y_2-y_1)^2\sin^2\phi+(x_2-x_1)^2\sin^2\phi+(y_2-y_1)^2\cos^2\phi \\ &= (x_2-x_1)^2(\cos^2\phi+\sin^2\phi)+(y_2-y_1)^2(\sin^2\phi+\cos^2\phi)=(x_2-x_1)^2+(y_2-y_1)^2 \end{aligned}$$

Putting these equations together gives
$d(P',Q')=\sqrt{(X_2-X_1)^2+(Y_2-Y_1)^2}=\sqrt{(x_2-x_1)^2+(y_2-y_1)^2}=d(P,Q)$.

# 11.6 Polar Equations of Conics

**1.** Substituting $e = \frac{2}{3}$ and $d = 3$ into the general equation of a conic with vertical directrix, we get $r = \dfrac{\frac{2}{3} \cdot 3}{1 + \frac{2}{3}\cos\theta} \Leftrightarrow r = \dfrac{6}{3 + 2\cos\theta}$.

**2.** Substituting $e = \frac{4}{3}$ and $d = 3$ into the general equation of a conic with vertical directrix, we get $r = \dfrac{\frac{4}{3} \cdot 3}{1 - \frac{4}{3}\cos\theta} \Leftrightarrow r = \dfrac{12}{3 - 4\cos\theta}$.

**3.** Substituting $e = 1$ and $d = 2$ into the general equation of a conic with horizontal directrix, we get $r = \dfrac{1 \cdot 2}{1 + \sin\theta} \Leftrightarrow r = \dfrac{2}{1 + \sin\theta}$.

**4.** Substituting $e = \frac{1}{2}$ and $d = 4$ into the general equation of a conic with horizontal directrix, we get $r = \dfrac{\frac{1}{2} \cdot 4}{1 - \frac{1}{2}\sin\theta} \Leftrightarrow r = \dfrac{4}{2 - \sin\theta}$.

**5.** $r = 5\sec\theta \Leftrightarrow r\cos\theta = 5 \Leftrightarrow x = 5$. So $d = 5$ and $e = 4$ gives $r = \dfrac{4 \cdot 5}{1 + 4\cos\theta} \Leftrightarrow r = \dfrac{20}{1 + 4\cos\theta}$.

**6.** $r = 2\csc\theta \Leftrightarrow r\sin\theta = 2 \Leftrightarrow y = 2$. So $d = 2$ and $e = 0.6$ gives $r = \dfrac{0.6 \cdot 2}{1 + 0.6\sin\theta} \Leftrightarrow r = \dfrac{1.2}{1 + 0.6\sin\theta}$.

**7.** Since this is a parabola whose focus is at the origin and vertex at $(5, \pi/2)$, the directrix must be $y = 10$. So $d = 10$ and $e = 1$ gives $r = \dfrac{1 \cdot 10}{1 + \sin\theta} = \dfrac{10}{1 + \sin\theta}$.

**8.** Since the vertex is at $(2, 0)$ we have $d(P, F) = 2$. Now, since $\dfrac{d(P, F)}{d(P, \ell)} = e$ we get $\dfrac{2}{d(P, \ell)} = 0.4 \Leftrightarrow d(P, \ell) = \dfrac{2}{0.4} = 5$. The directrix is $x = 7$, so substituting $e = 0.4$ and $d = 7$ we get $r = \dfrac{0.4 \cdot 7}{1 + 0.4\cos\theta} \Leftrightarrow r = \dfrac{2.8}{1 + 0.4\cos\theta}$.

**9.** $r = \dfrac{6}{1 + \cos\theta}$ is Graph II. The eccentricity is 1, so this is a parabola. When $\theta = 0$, we have $r = 3$ and when $\theta = \frac{\pi}{2}$, we have $r = 6$.

**10.** $r = \dfrac{2}{2 - \cos\theta}$ is Graph III. $r = \dfrac{1}{1 - \frac{1}{2}\cos\theta}$, so $e = \frac{1}{2}$ and this is an ellipse. When $\theta = 0$, $r = 2$, and when $\theta = \pi$, $r = \frac{2}{3}$.

**11.** $r = \dfrac{3}{1 - 2\sin\theta}$ is Graph VI. $e = 2$, so this is a hyperbola. When $\theta = 0$, $r = 3$, and when $\theta = \pi$, $r = 3$.

**12.** $r = \dfrac{5}{3 - 3\sin\theta}$ is Graph I. $r = \dfrac{\frac{5}{3}}{1 - \sin\theta}$, so $e = 1$ and this is a parabola. When $\theta = 0$, $r = \frac{5}{3}$ and when $\theta = \pi$, $r = \frac{5}{3}$.

**13.** $r = \dfrac{12}{3 + 2\sin\theta}$ is Graph IV. $r = \dfrac{4}{1 + \frac{2}{3}\sin\theta}$, so $e = \frac{2}{3}$ and this is an ellipse. When $\theta = 0$, $r = 4$, and when $\theta = \pi$, $r = 4$.

**14.** $r = \dfrac{12}{2 + 3\cos\theta}$ is Graph V. $r = \dfrac{6}{1 + \frac{3}{2}\cos\theta}$, so $e = \frac{3}{2}$ and this is a hyperbola. When $\theta = 0$, $r = \frac{12}{5}$, and when $\theta = \pi$, we have $r = 12$.

**15. (a)** $r = \dfrac{4}{1 + 3\cos\theta} \quad \Rightarrow \quad e = 3$, so the conic is a hyperbola.

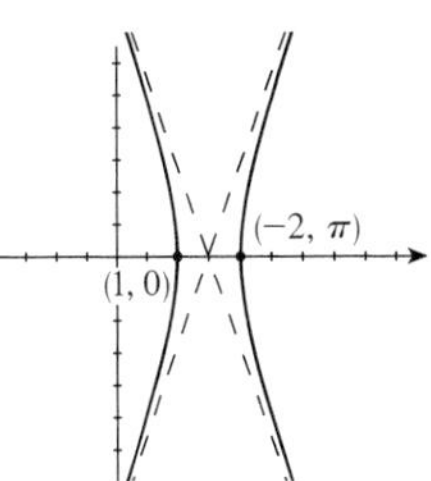


**(b)** The vertices occur where $\theta = 0$ and $\theta = \pi$. Now $\theta = 0 \quad \Rightarrow$
$r = \dfrac{4}{1 + 3\cos 0} = 1$, and $\theta = \pi \quad \Rightarrow \quad r = \dfrac{4}{1 + 3\cos\pi} = \dfrac{4}{-2} = -2$. Thus the vertices are $(1, 0)$ and $(-2, \pi)$.

**16. (a)** $r = \dfrac{8}{3 + 3\cos\theta} \quad \Leftrightarrow \quad r = \dfrac{\frac{8}{3}}{1 + \cos\theta} \quad \Rightarrow \quad e = 1$, so the conic is a parabola.

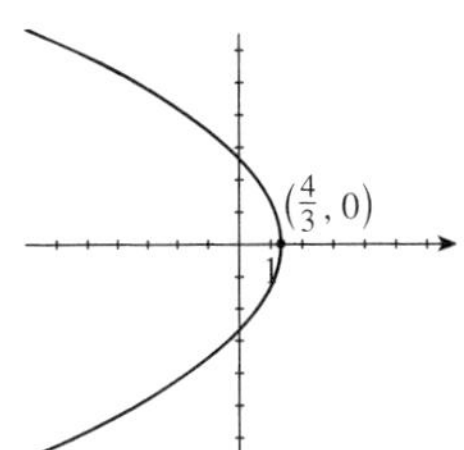


**(b)** Substituting $\theta = 0$, we have $r = \dfrac{8}{3 + 3\cos 0} = \dfrac{8}{6} = \dfrac{4}{3}$. Thus the vertex is $\left(\frac{4}{3}, 0\right)$.

**17. (a)** $r = \dfrac{2}{1 - \cos\theta} \quad \Rightarrow \quad e = 1$, so the conic is a parabola.

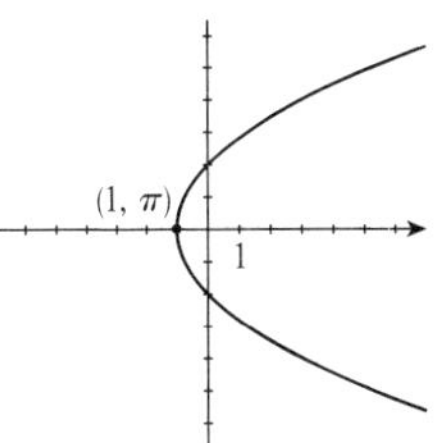


**(b)** Substituting $\theta = \pi$, we have $r = \dfrac{2}{1 - \cos\pi} = \frac{2}{2} = 1$. Thus the vertex is $(1, \pi)$.

**18. (a)** $r = \dfrac{10}{3 - 2\sin\theta} \quad \Leftrightarrow \quad r = \dfrac{\frac{10}{3}}{1 - \frac{2}{3}\sin\theta} \quad \Rightarrow \quad e = \frac{2}{3}$, so the conic is an ellipse.

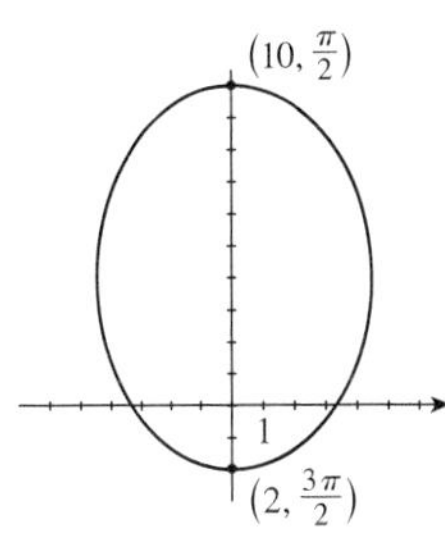


**(b)** The vertices occur where $\theta = \frac{\pi}{2}$ and $\theta = \frac{3\pi}{2}$. Now $\theta = \frac{\pi}{2} \quad \Rightarrow$
$r = \dfrac{10}{3 - 2\sin\frac{\pi}{2}} = \dfrac{10}{1} = 10$ and $\theta = \frac{3\pi}{2} \quad \Rightarrow \quad r = \dfrac{10}{3 - 2\sin\frac{3\pi}{2}} = \dfrac{10}{5} = 2$.
Thus, the vertices are $\left(10, \frac{\pi}{2}\right)$ and $\left(2, \frac{3\pi}{2}\right)$.

**19. (a)** $r = \dfrac{6}{2 + \sin\theta} \quad \Leftrightarrow \quad r = \dfrac{\frac{1}{2} \cdot 6}{1 + \frac{1}{2}\sin\theta} \quad \Rightarrow \quad e = \frac{1}{2}$, so the conic is an ellipse.

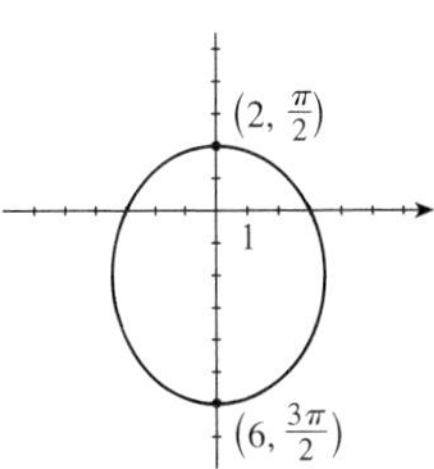


**(b)** The vertices occur where $\theta = \frac{\pi}{2}$ and $\theta = \frac{3\pi}{2}$. Now $\theta = \frac{\pi}{2} \quad \Rightarrow$
$r = \dfrac{6}{2 + \sin\frac{\pi}{2}} = \dfrac{6}{3} = 2$ and $\theta = \frac{3\pi}{2} \quad \Rightarrow \quad r = \dfrac{6}{2 + \sin\frac{3\pi}{2}} = \dfrac{6}{1} = 6$. Thus, the vertices are $\left(2, \frac{\pi}{2}\right)$ and $\left(6, \frac{3\pi}{2}\right)$.

**20. (a)** $r = \dfrac{5}{2 - 3\sin\theta} \quad \Leftrightarrow \quad r = \dfrac{\frac{5}{2}}{1 - \frac{3}{2}\sin\theta} \quad \Rightarrow \quad e = \frac{3}{2}$, so the conic is a hyperbola.

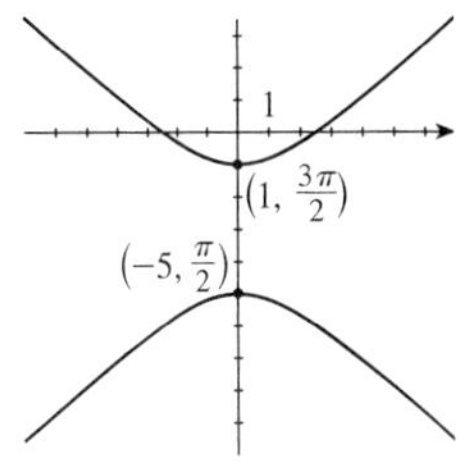


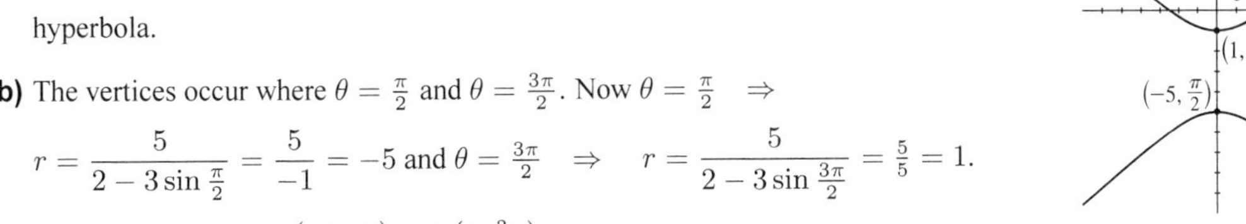

**(b)** The vertices occur where $\theta = \frac{\pi}{2}$ and $\theta = \frac{3\pi}{2}$. Now $\theta = \frac{\pi}{2} \quad \Rightarrow$
$r = \dfrac{5}{2 - 3\sin\frac{\pi}{2}} = \dfrac{5}{-1} = -5$ and $\theta = \frac{3\pi}{2} \quad \Rightarrow \quad r = \dfrac{5}{2 - 3\sin\frac{3\pi}{2}} = \frac{5}{5} = 1$.
Thus, the vertices are $\left(-5, \frac{\pi}{2}\right)$ and $\left(1, \frac{3\pi}{2}\right)$.

**21. (a)** $r = \dfrac{7}{2 - 5\sin\theta} \quad\Leftrightarrow\quad r = \dfrac{\frac{7}{2}}{1 - \frac{5}{2}\sin\theta} \quad\Rightarrow\quad e = \frac{5}{2}$, so the conic is a hyperbola.

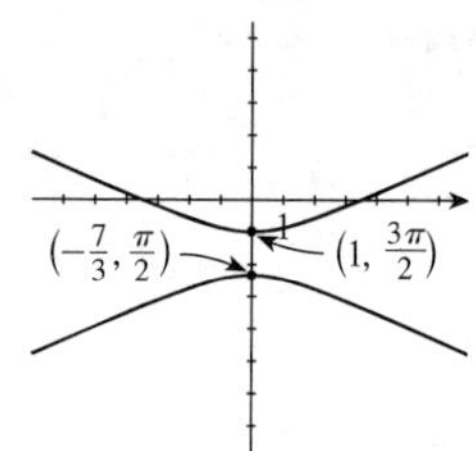


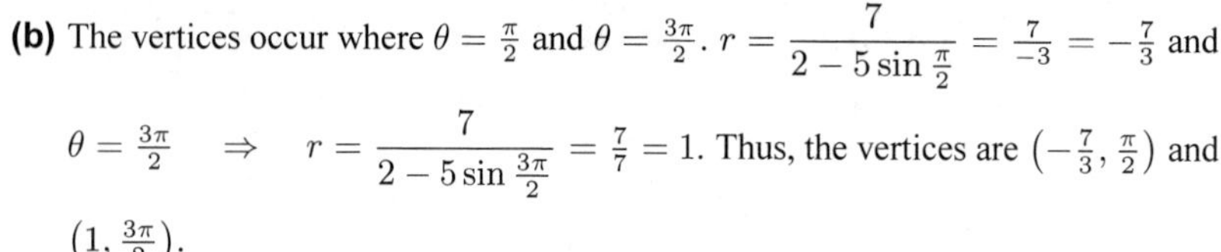

**(b)** The vertices occur where $\theta = \frac{\pi}{2}$ and $\theta = \frac{3\pi}{2}$. $r = \dfrac{7}{2 - 5\sin\frac{\pi}{2}} = \frac{7}{-3} = -\frac{7}{3}$ and $\theta = \frac{3\pi}{2} \quad\Rightarrow\quad r = \dfrac{7}{2 - 5\sin\frac{3\pi}{2}} = \frac{7}{7} = 1$. Thus, the vertices are $\left(-\frac{7}{3}, \frac{\pi}{2}\right)$ and $\left(1, \frac{3\pi}{2}\right)$.

**22. (a)** $r = \dfrac{8}{3 + \cos\theta} \quad\Leftrightarrow\quad r = \dfrac{\frac{8}{3}}{1 + \frac{1}{3}\cos\theta} \quad\Rightarrow\quad e = \frac{1}{3}$, so the conic is an ellipse.

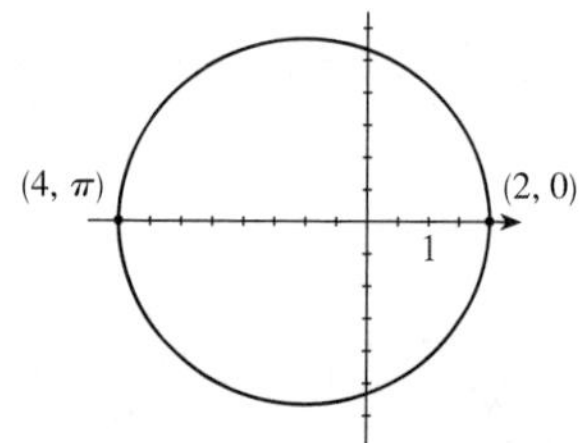


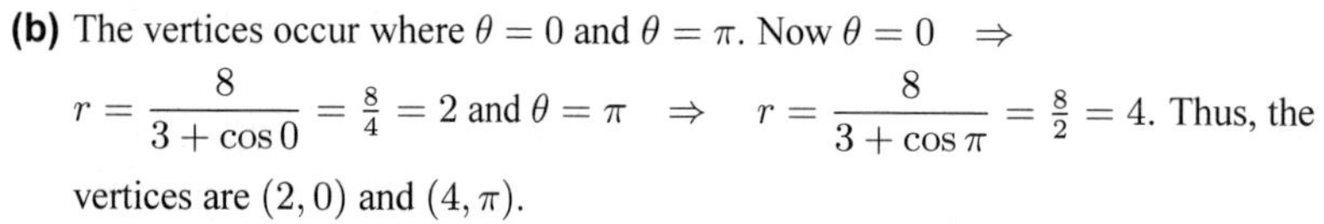

**(b)** The vertices occur where $\theta = 0$ and $\theta = \pi$. Now $\theta = 0 \quad\Rightarrow$ $r = \dfrac{8}{3 + \cos 0} = \frac{8}{4} = 2$ and $\theta = \pi \quad\Rightarrow\quad r = \dfrac{8}{3 + \cos\pi} = \frac{8}{2} = 4$. Thus, the vertices are $(2, 0)$ and $(4, \pi)$.

**23. (a)** $r = \dfrac{1}{4 - 3\cos\theta} \quad\Leftrightarrow\quad r = \dfrac{\frac{1}{4}}{1 - \frac{3}{4}\cos\theta} \quad\Rightarrow\quad e = \frac{3}{4}$. Since $ed = \frac{1}{4}$ we have $d = \dfrac{\frac{1}{4}}{e} = \frac{4}{3} \cdot \frac{1}{4} = \frac{1}{3}$. Therefore the eccentricity is $\frac{3}{4}$ and the directrix is $x = -\frac{1}{3} \quad\Leftrightarrow\quad r = -\frac{1}{3}\sec\theta$.

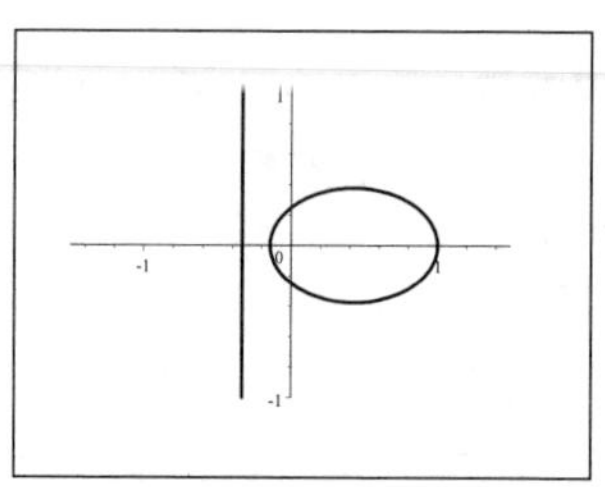

**(b)** We replace $\theta$ by $\theta - \frac{\pi}{3}$ to get $r = \dfrac{1}{4 - 3\cos\left(\theta - \frac{\pi}{3}\right)}$.

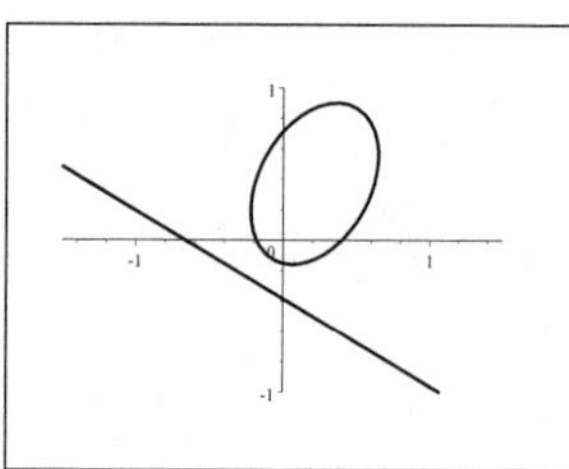

**24.** $r = \dfrac{5}{2 + 2\sin\theta} \quad\Leftrightarrow\quad r = \dfrac{\frac{5}{2}}{1 + \sin\theta}$. Thus $d = \frac{5}{2}$ and the directrix is $y = \frac{5}{2}$ $\quad\Leftrightarrow\quad r = \frac{5}{2}\csc\theta$. The equation of the conic obtained by rotating this parabola about its focus through an angle of $\frac{\pi}{6}$ is $r = \dfrac{5}{2 + 2\sin\left(\theta - \frac{\pi}{6}\right)}$.

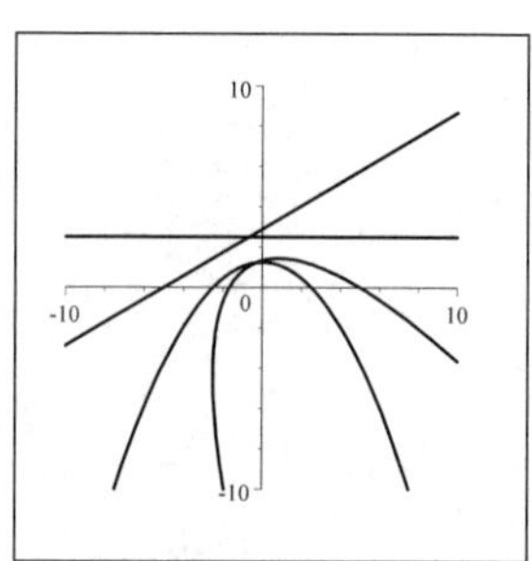


**25.** The ellipse is nearly circular when $e$ is close to 0 and becomes more elongated as $e \to 1^-$. At $e = 1$, the curve becomes a parabola.

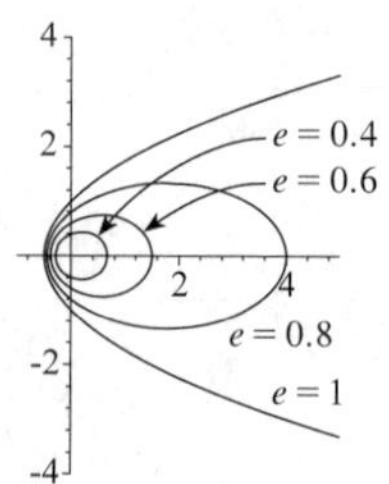

**26. (a)** Shown are the graphs of the conics $r = \dfrac{d}{1+\sin\theta}$ where $d = \frac{1}{2}$, $d = 2$, and $d = 10$. As $d$ increases, the parabolas get flatter while the vertex moves further from the focus at the origin.

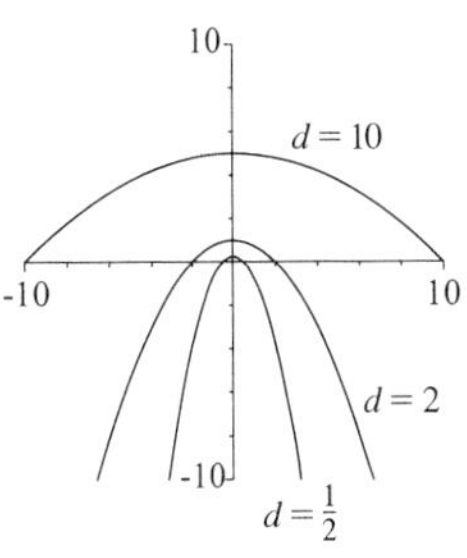


**(b)** Shown are the graphs of the conics $r = \dfrac{e}{1+e\sin\theta}$ where $e = 0.5$, $e = 1$, and $e = 10$. As $e$ increases, the conic changes from an ellipse to a parabola, and finally to a hyperbola. The vertex gets closer to the directrix (shown as a dashed line in the figure).

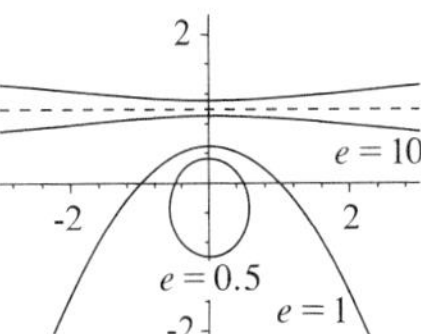


**27. (a)** Since the polar form of an ellipse with directrix $x = -d$ is $r = \dfrac{ed}{1-e\cos\theta}$ we need to show that $ed = a\left(1-e^2\right)$. From the proof of the Equivalent Description of Conics we have $a^2 = \dfrac{e^2d^2}{(1-e^2)^2}$. Since the conic is an ellipse, $e < 1$ and so the quantities $a$, $d$, and $\left(1-e^2\right)$ are all positive. Thus we can take the square roots of both sides and maintain equality. Thus $a^2 = \dfrac{e^2d^2}{(1-e^2)^2} \quad\Leftrightarrow\quad a = \dfrac{ed}{1-e^2} \quad\Leftrightarrow\quad ed = a\left(1-e^2\right)$. As a result, $r = \dfrac{ed}{1-e\cos\theta} \quad\Leftrightarrow$ $r = \dfrac{a\left(1-e^2\right)}{1-e\cos\theta}$.

**(b)** Since $2a = 2.99\times10^8$ we have $a = 1.495\times10^8$, so a polar equation for the earth's orbit (using $e \approx 0.017$) is $r = \dfrac{1.495\times10^8\left[1-(0.017)^2\right]}{1-0.017\cos\theta} \approx \dfrac{1.49\times10^8}{1-0.017\cos\theta}$.

**28. (a)** Using the form of the equation from Exercise 27, the perihelion distance occurs when $\theta = \pi$ and the aphelion distance occurs when $\theta = 0$. Since the focus is at the origin, the perihelion distance is $\dfrac{a\left(1-e^2\right)}{1+e} = \dfrac{a(1-e)(1+e)}{1+e} = a(1-e)$ and the aphelion distance is $\dfrac{a\left(1-e^2\right)}{1-e} = \dfrac{a(1-e)(1+e)}{1-e} = a(1+e)$.

**(b)** Given $e = 0.017$ and $2a = 2.99\times10^8$ we have $a = 1.495\times10^8$. Thus the perihelion distance is $1.495\times10^8\,[1-(0.017)] \approx 1.468\times10^8$ km and the aphelion distance is $1.495\times10^8\,(1+0.017) \approx 1.520\times10^8$ km.

**29.** From Exercise 28, we know that at perihelion $r = 4.43\times10^9 = a(1-e)$ and at aphelion $r = 7.37\times10^9 = a(1+e)$. Dividing these equations gives $\dfrac{7.37\times10^9}{4.43\times10^9} = \dfrac{a(1+e)}{a(1-e)} \quad\Leftrightarrow\quad 1.664 = \dfrac{1+e}{1-e} \quad\Leftrightarrow\quad 1.664(1-e) = 1+e \quad\Leftrightarrow$ $1.664 - 1 = e + 1.664 \quad\Leftrightarrow\quad 0.664 = 2.664e \quad\Leftrightarrow\quad e = \dfrac{0.664}{2.664} \approx 0.25$.

**30.** Since the focus is at the origin, the distance from the focus to any point on the conic is the absolute value of the $r$-coordinate of that point, which we can obtain from the polar equation.

**31.** The $r$-coordinate of the satellite will be its distance from the focus (the center of the earth). From the $r$-coordinate we can easily calculate the height of the satellite.

# 11.7 Plane Curves and Parametric Equations

**1. (a)** $x = 2t$, $y = t + 6$

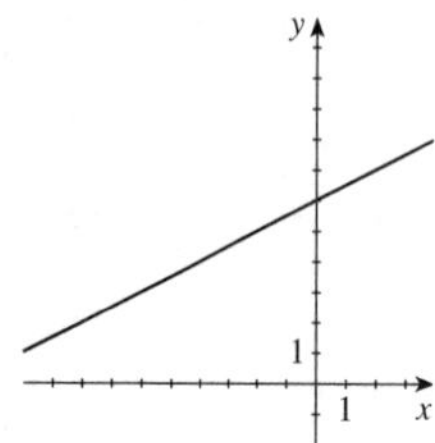

**(b)** Since $x = 2t$, $t = \dfrac{x}{2}$ and so $y = \dfrac{x}{2} + 6 \Leftrightarrow$ $x - 2y + 12 = 0$.

**2. (a)** $x = 6t - 4$, $y = 3t$, $t \geq 0$

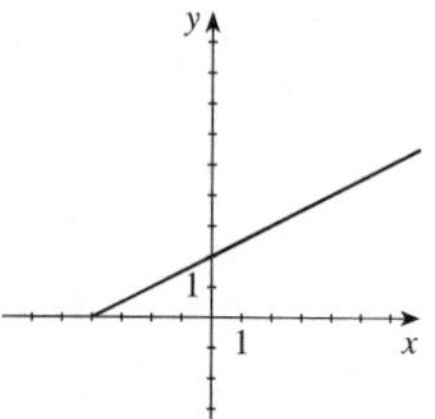

**(b)** Since $y = 3t$, $t = \dfrac{y}{3}$ and so $x = 6\left(\dfrac{y}{3}\right) - 4 = 2y - 4$ $\Leftrightarrow$ $2y = x + 4$, $y \geq 0$.

**3. (a)** $x = t^2$, $y = t - 2$, $2 \leq t \leq 4$

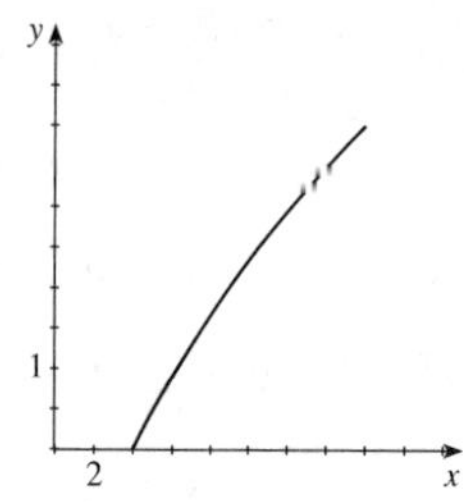

**(b)** Since $y = t - 2 \Leftrightarrow t = y + 2$, we have $x = t^2$ $\Leftrightarrow$ $x = (y + 2)^2$, and since $2 \leq t \leq 4$, we have $4 \leq x \leq 16$.

**4. (a)** $x = 2t + 1$, $y = \left(t + \frac{1}{2}\right)^2$

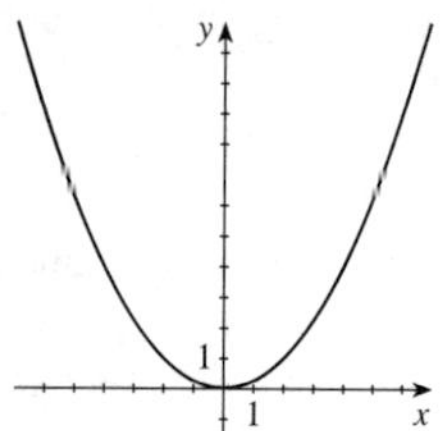

**(b)** Since $y = \left(t + \frac{1}{2}\right)^2$, $4y = 4\left(t + \frac{1}{2}\right)^2 = (2t + 1)^2$, and since $x = 2t + 1$, we have $x^2 = (2t + 1)^2 = 4y$ $\Leftrightarrow$ $y = \frac{1}{4}x^2$.

**5. (a)** $x = \sqrt{t}$, $y = 1 - t \Rightarrow t \geq 0$

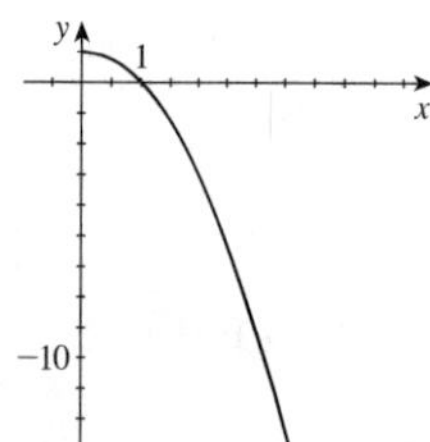

**(b)** Since $x = \sqrt{t}$, we have $x^2 = t$, and so $y = 1 - x^2$ with $x \geq 0$.

**6. (a)** $x = t^2$, $y = t^4 + 1$

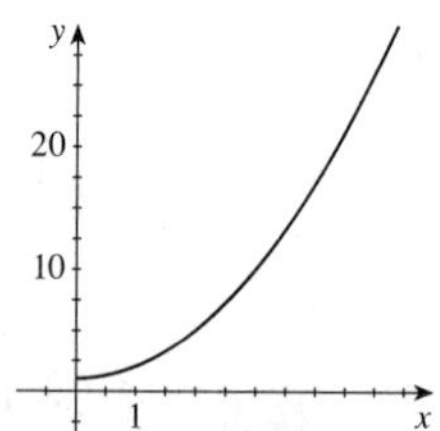

**(b)** Since $x = t^2$, we have $x^2 = t^4$ and so $y = x^2 + 1$, $x \geq 0$.

**7. (a)** $x = \dfrac{1}{t}$, $y = t + 1$

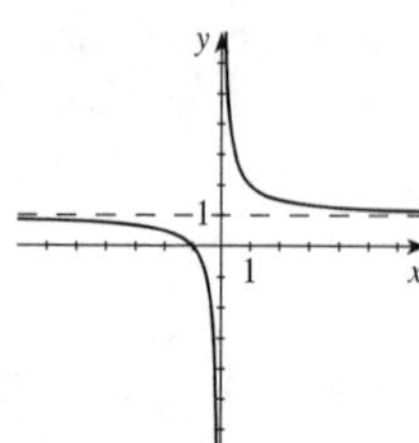

**(b)** Since $x = \dfrac{1}{t}$ we have $t = \dfrac{1}{x}$ and so $y = \dfrac{1}{x} + 1$.

**8. (a)** $x = t + 1$, $y = \dfrac{t}{t + 1}$

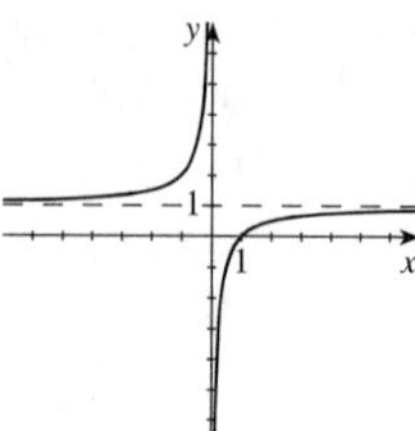

**(b)** Since $x = t + 1$, we have $t = x - 1$, so $y = \dfrac{x - 1}{x}$.

**9. (a)** $x = 4t^2$, $y = 8t^3$

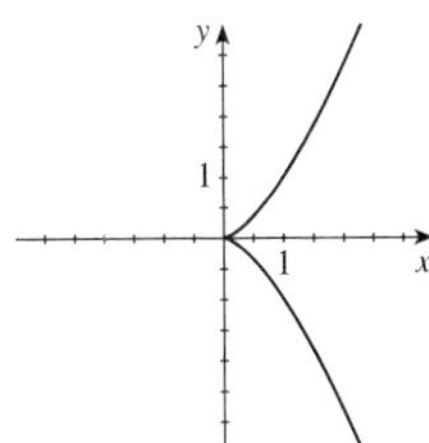


**(b)** Since $y = 8t^3 \quad \Leftrightarrow \quad y^2 = 64t^6 = \left(4t^2\right)^3 = x^3$, we have $y^2 = x^3$.

**10. (a)** $x = |t|$, $y = |1 - |t||$

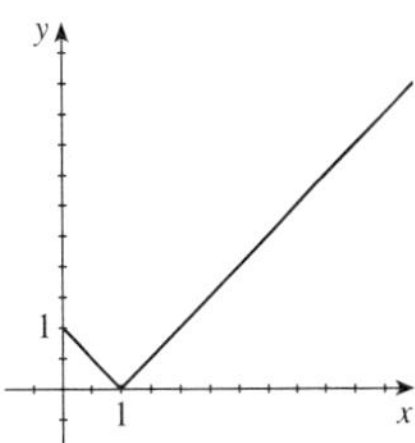


**(b)** Since $x = |t|$, we have $y = |1 - x|$, where $x \geq 0$.

**11. (a)** $x = 2\sin t$, $y = 2\cos t$, $0 \leq t \leq \pi$

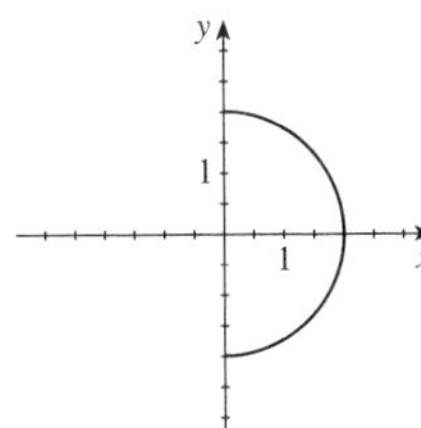


**(b)** $x^2 = (2\sin t)^2 = 4\sin^2 t$ and $y^2 = 4\cos^2 t$. Hence, $x^2 + y^2 = 4\sin^2 t + 4\cos^2 t = 4 \quad \Leftrightarrow$ $x^2 + y^2 = 4$, where $x \geq 0$.

**12. (a)** $x = 2\cos t$, $y = 3\sin t$, $0 \leq t \leq 2\pi$

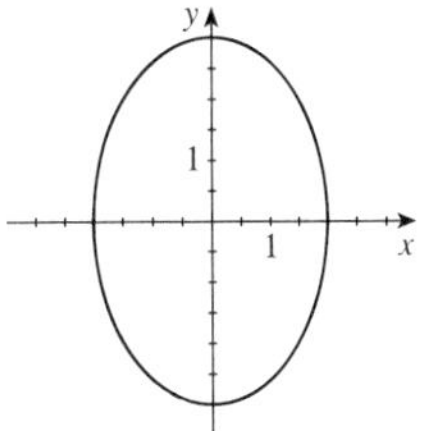


**(b)** We have $\cos t = x/2$ and $\sin t = y/3$, so $(x/2)^2 + (y/3)^2 = \cos^2 t + \sin^2 t = 1 \quad \Leftrightarrow$ $\frac{1}{4}x^2 + \frac{1}{9}y^2 = 1$.

**13. (a)** $x = \sin^2 t$, $y = \sin^4 t$

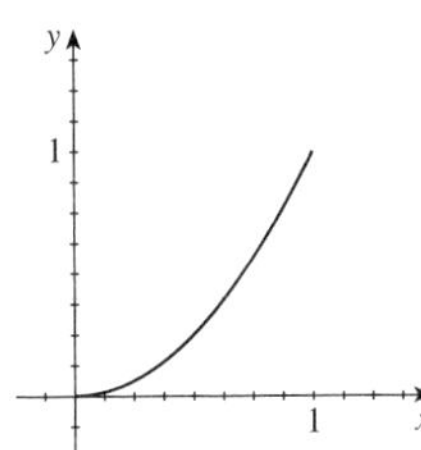


**(b)** Since $x = \sin^2 t$ we have $x^2 = \sin^4 t$ and so $y = x^2$. But since $0 \leq \sin^2 t \leq 1$ we only get the part of this parabola for which $0 \leq x \leq 1$.

**14. (a)** $x = \sin^2 t$, $y = \cos t$

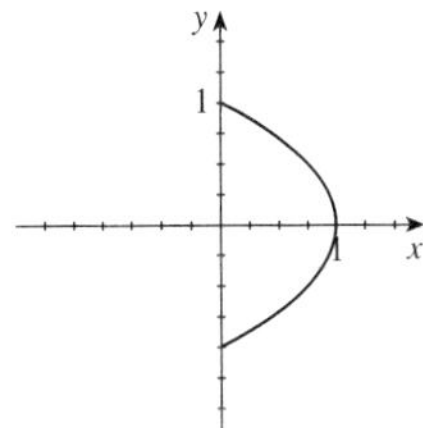


**(b)** Since $y = \cos t$, we have $y^2 = \cos^2 t$ and so $x + y^2 = \sin^2 t + \cos^2 t = 1$ or $x = 1 - y^2$ (actually $y = \pm\sqrt{1 - x}$), where $0 \leq x \leq 1$.

**15. (a)** $x = \cos t$, $y = \cos 2t$

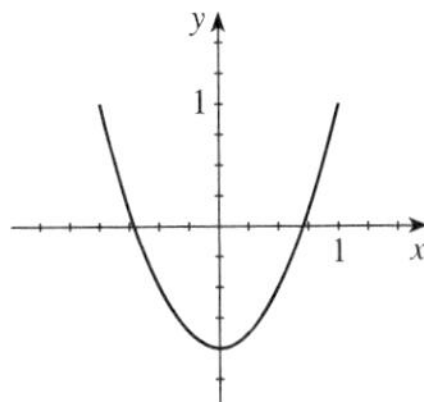


**(b)** Since $x = \cos t$ we have $x^2 = \cos^2 t$, so $2x^2 - 1 = 2\cos^2 t - 1 = \cos 2t = y$. Hence, the rectangular equation is $y = 2x^2 - 1$, $-1 \leq x \leq 1$.

**16. (a)** $x = \cos 2t$, $y = \sin 2t$

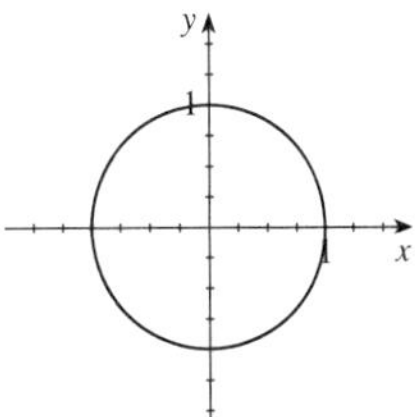


**(b)** $x^2 = \cos^2 2t$ and $y^2 = \sin^2 2t$. Then $x^2 + y^2 = \cos^2 2t + \sin^2 2t = 1 \quad \Leftrightarrow$ $x^2 + y^2 = 1$.

**17.** **(a)** $x = \sec t$, $y = \tan t$, $0 \le t < \frac{\pi}{2}$ $\Rightarrow$ $x \ge 1$ and $y \ge 0$.

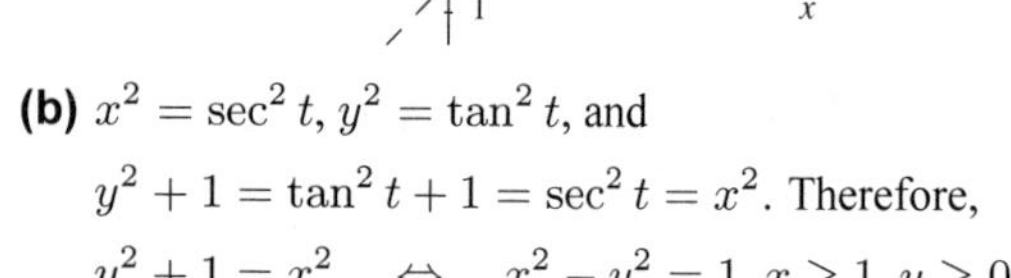

**(b)** $x^2 = \sec^2 t$, $y^2 = \tan^2 t$, and $y^2 + 1 = \tan^2 t + 1 = \sec^2 t = x^2$. Therefore, $y^2 + 1 = x^2$ $\Leftrightarrow$ $x^2 - y^2 = 1$, $x \ge 1$, $y \ge 0$.

**18.** **(a)** $x = \cot t$, $y = \csc t$, $0 < t < \pi$ so $y \ge 1$.

**(b)** $x^2 = \cot^2 t$, $y^2 = \csc^2 t$, and so $x^2 + 1 = \cot^2 t + 1 = \csc^2 t = y^2$. Therefore, $y^2 = x^2 + 1$, with $y \ge 1$. This is the top half of the hyperbola $y^2 - x^2 = 1$.

**19.** **(a)** $x = e^t$, $y = e^{-t}$ $\Rightarrow$ $x > 0, y > 0$.

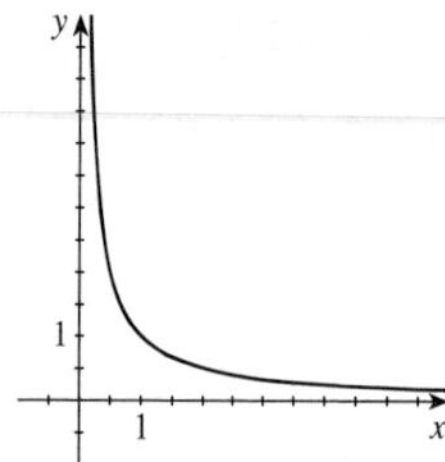

**(b)** $xy = e^t \cdot e^{-t} = e^0 = 1$. Hence, the equation is $xy = 1$, with $x > 0$, $y > 0$.

**20.** **(a)** $x = \sec t$, $y = \tan^2 t$, $0 \le t \le \frac{\pi}{2}$.

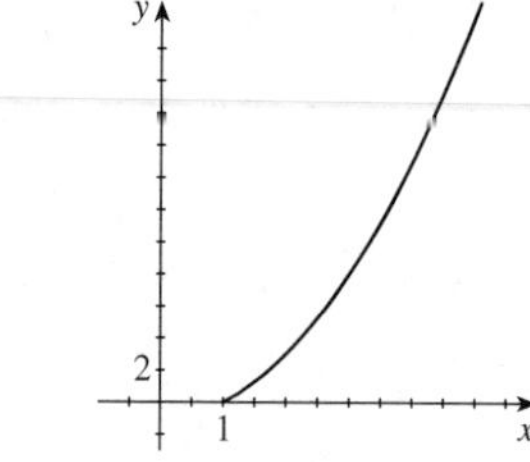

**(b)** $x^2 = \sec^2 t = 1 + \tan^2 t$, so $x^2 = 1 + y$, $x \ge 1$, $y \ge 0$.

**21.** **(a)** $x = \cos^2 t$, $y = \sin^2 t$

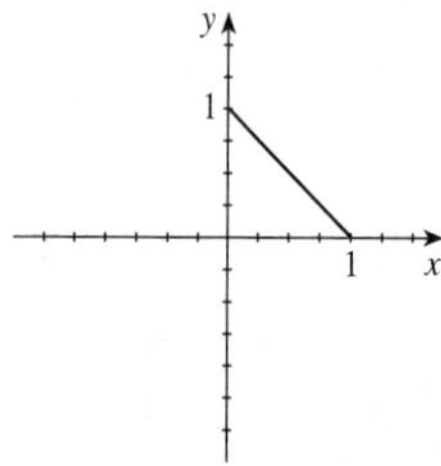

**(b)** $x + y = \cos^2 t + \sin^2 t = 1$. Hence, the equation is $x + y = 1$ with $0 \le x, y \le 1$.

**22.** **(a)** $x = \cos^3 t$, $y = \sin^3 t$, $0 \le t \le 2\pi$

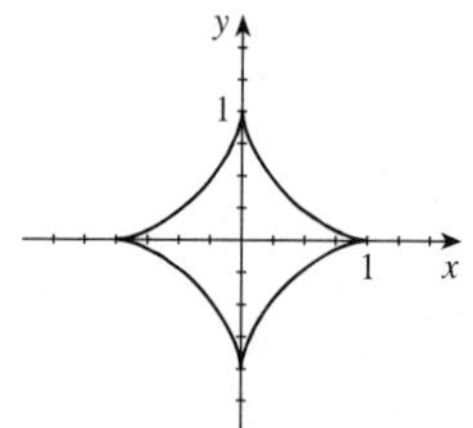

**(b)** $x^{2/3} = \cos^2 t$ and $y^{2/3} = \sin^2 t$, and so $x^{2/3} + y^{2/3} = 1$.

**23.** Since the line passes through the point $(4, -1)$ and has slope $\frac{1}{2}$, parametric equations for the line are $x = 4 + t$, $y = -1 + \frac{1}{2}t$.

**24.** Since the line passes through the point $(-10, -20)$ and has slope $-2$, parametric equations for the line are $x = -10 + t$, $y = -20 - 2t$.

**25.** Since the line passes through the points $(6, 7)$ and $(7, 8)$, its slope is $\dfrac{8-7}{7-6} = 1$. Thus, parametric equations for the line are $x = 6 + t$, $y = 7 + t$.

**26.** Since the line passes through the points $(0, 0)$ and $(12, 7)$, its slope is $\dfrac{7-0}{12-0} = \dfrac{7}{12}$. Thus, parametric equations for the line are $x = t$, $y = \frac{7}{12}t$.

**27.** Since $\cos^2 t + \sin^2 t = 1$, we have $a^2\cos^2 t + a^2\sin^2 t = a^2$. If we let $x = a\cos t$ and $y = a\sin t$, then $x^2 + y^2 = a^2$. Hence, parametric equations for the circle are $x = a\cos t$, $y = a\sin t$.

**28.** Since $\cos^2 t + \sin^2 t = 1$, we have $\dfrac{a^2\cos^2 t}{a^2} + \dfrac{b^2\sin^2 t}{b^2} = 1$. If we let $x = a\cos t$ and $y = b\sin t$, then $\dfrac{x^2}{a^2} + \dfrac{y^2}{b^2} = 1$. Hence, parametric equations for the ellipse are $x = a\cos t$, $y = b\sin t$.

**29.** $x = a\tan\theta \quad\Leftrightarrow\quad \tan\theta = \dfrac{x}{a} \quad\Rightarrow\quad \tan^2\theta = \dfrac{x^2}{a^2}$. Also, $y = b\sec\theta \quad\Leftrightarrow\quad \sec\theta = \dfrac{y}{b} \quad\Rightarrow\quad \sec^2\theta = \dfrac{y^2}{b^2}$. Since $\tan^2\theta = \sec^2\theta - 1$, we have $\dfrac{x^2}{a^2} = \dfrac{y^2}{b^2} - 1 \quad\Leftrightarrow\quad \dfrac{y^2}{b^2} - \dfrac{x^2}{a^2} = 1$, which is the equation of a hyperbola.

**30.** Substituting the given values for $x$ and $y$ into the equation we derived in Exercise 29, we get $\dfrac{\left(b\sqrt{t+1}\right)^2}{b^2} - \dfrac{\left(a\sqrt{t}\right)^2}{a^2} = (t+1) - t = 1$. Thus the points on this curve satisfy the equation, which is that of a hyperbola. However, this hyperbola is only the part of $\dfrac{y^2}{b^2} - \dfrac{x^2}{a^2} = 1$ for which $x \geq 0$ and $y \geq 0$.

**31.** $x = t\cos t$, $y = t\sin t$, $t \geq 0$

| $t$ | $x$ | $y$ |
|---|---|---|
| $0$ | $0$ | $0$ |
| $\frac{\pi}{4}$ | $\frac{\pi\sqrt{2}}{8}$ | $\frac{\pi\sqrt{2}}{8}$ |
| $\frac{\pi}{2}$ | $0$ | $\frac{\pi}{2}$ |
| $\frac{3\pi}{4}$ | $-\frac{3\pi\sqrt{2}}{8}$ | $\frac{3\pi\sqrt{2}}{8}$ |
| $\pi$ | $-\pi$ | $0$ |

| $t$ | $x$ | $y$ |
|---|---|---|
| $\frac{5\pi}{4}$ | $-\frac{5\pi\sqrt{2}}{8}$ | $-\frac{5\pi\sqrt{2}}{8}$ |
| $\frac{3\pi}{2}$ | $0$ | $-\frac{3\pi}{2}$ |
| $\frac{7\pi}{4}$ | $\frac{7\pi\sqrt{2}}{8}$ | $-\frac{7\pi\sqrt{2}}{8}$ |
| $2\pi$ | $2\pi$ | $0$ |

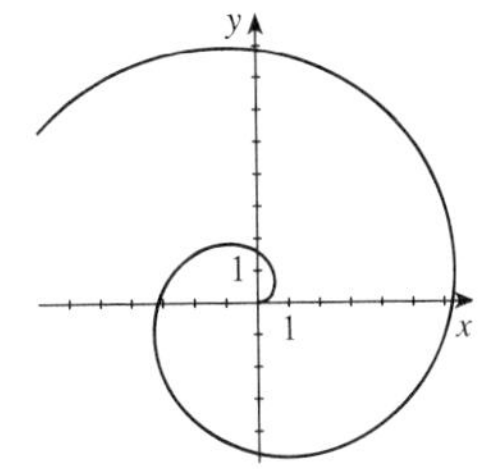


**32.** $x = \sin t$, $y = \sin 2t$

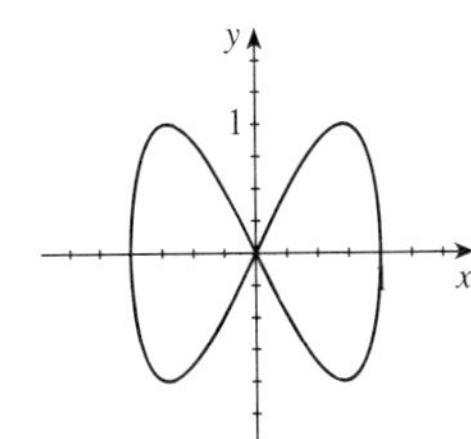


**33.** $x = \dfrac{3t}{1+t^3}$, $y = \dfrac{3t^2}{1+t^3}$, $t \neq -1$

| $t$ | $x$ | $y$ |
|---|---|---|
| $-0.9$ | $-9.96$ | $8.97$ |
| $-0.75$ | $-3.89$ | $2.92$ |
| $-0.5$ | $-1.71$ | $0.86$ |
| $0$ | $0$ | $0$ |
| $0.5$ | $1.33$ | $0.67$ |
| $1$ | $1.5$ | $1.5$ |
| $1.5$ | $1.03$ | $1.54$ |

| $t$ | $x$ | $y$ |
|---|---|---|
| $2$ | $0.67$ | $1.33$ |
| $2.5$ | $0.45$ | $1.13$ |
| $3$ | $0.32$ | $0.96$ |
| $4$ | $0.18$ | $0.74$ |
| $5$ | $0.12$ | $0.60$ |
| $6$ | $0.08$ | $0.50$ |

| $t$ | $x$ | $y$ |
|---|---|---|
| $-1.1$ | $9.97$ | $-10.97$ |
| $-1.25$ | $3.93$ | $-4.92$ |
| $-1.5$ | $1.89$ | $-2.84$ |
| $-2$ | $0.86$ | $-1.71$ |
| $-2.5$ | $0.51$ | $-1.28$ |
| $-3$ | $0.35$ | $-1.04$ |
| $-3.5$ | $0.25$ | $-0.88$ |

| $t$ | $x$ | $y$ |
|---|---|---|
| $-4$ | $0.19$ | $-0.76$ |
| $-4.5$ | $0.15$ | $-0.67$ |
| $-5$ | $0.12$ | $-0.60$ |
| $-6$ | $0.08$ | $-0.50$ |
| $-7$ | $0.06$ | $-0.43$ |
| $-8$ | $0.05$ | $-0.38$ |

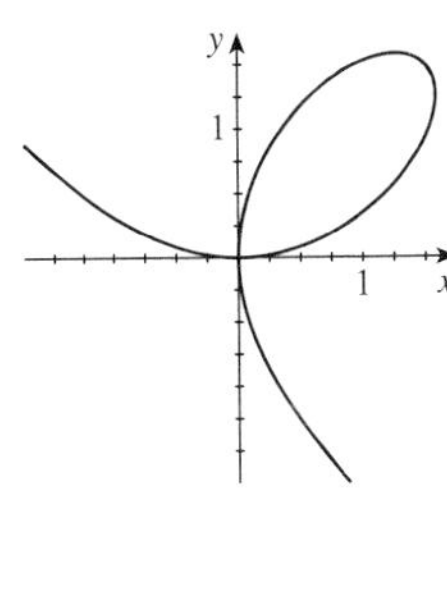


As $t \to -1^-$ we have $x \to \infty$ and $y \to -\infty$. As $t \to -1^+$ we have $x \to -\infty$ and $y \to \infty$. As $t \to \infty$ we have $x \to 0^+$ and $y \to 0^+$. As $t \to -\infty$ we have $x \to 0^+$ and $y \to 0^-$.

**34.** $x = \cot t$, $y = 2\sin^2 t$, $0 < t < \pi$

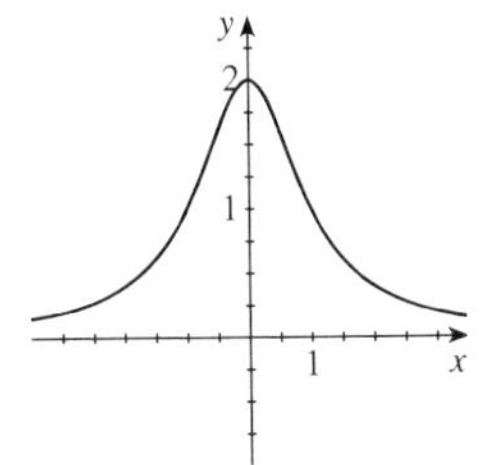


**35.** $x = (v_0\cos\alpha)\,t$, $y = (v_0\sin\alpha)\,t - 16t^2$. From the equation for $x$, $t = \dfrac{x}{v_0\cos\alpha}$. Substituting into the equation for $y$ gives

$$y = (v_0\sin\alpha)\,\frac{x}{v_0\cos\alpha} - 16\left(\frac{x}{v_0\cos\alpha}\right)^2 = x\tan\alpha - \frac{16x^2}{v_0^2\cos^2\alpha}.$$

Thus the equation is of the form $y = c_1 x - c_2 x^2$, where $c_1$ and $c_2$ are constants, so its graph is a parabola.

**36.** $v_0 = 2048$ ft/s and $\alpha = 30°$.

**(a)** The bullet will hit the ground when $y = 0$. Since $y = (v_0 \sin\alpha)\,t - 16t^2$, we have $0 = (2048 \cdot \sin 30°)\,t - 16t^2$ $\Leftrightarrow$ $0 = 1024t - 16t^2$ $\Leftrightarrow$ $16t\,(t - 64) = 0$ $\Leftrightarrow$ $t = 0$ or $t = 64$. Hence, the bullet will hit the ground after 64 seconds.

**(b)** After 64 seconds, the bullet will hit the ground at $x = (2048 \cdot \cos 30°) \cdot 64 = 65{,}536\sqrt{3} \approx 113{,}511.7$ ft $\approx 21.5$ miles.

**(c)** The maximum height attained by the bullet is the maximum value of $y$. Since $y = 1024t - 16t^2 = -16\left(t^2 - 64t\right) = -16\left(t^2 - 64t + 1024\right) + 16{,}384 = -16\,(t - 32)^2 + 16{,}384$, $y$ is maximized when $t = 32$, and therefore the maximum height is 16,384 ft $\approx 3.1$ miles.

**37.** $x = \sin t$, $y = 2\cos 3t$

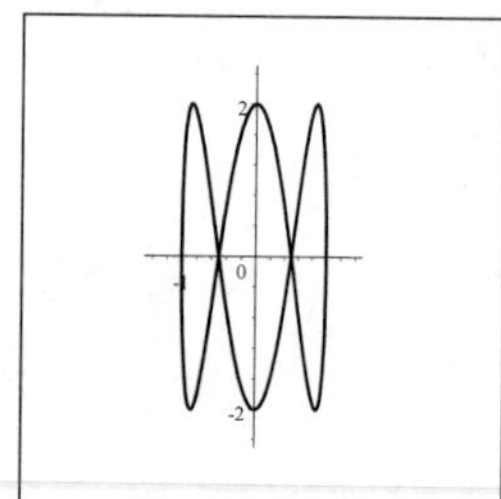

**38.** $x = 2\sin t$, $y = \cos 4t$

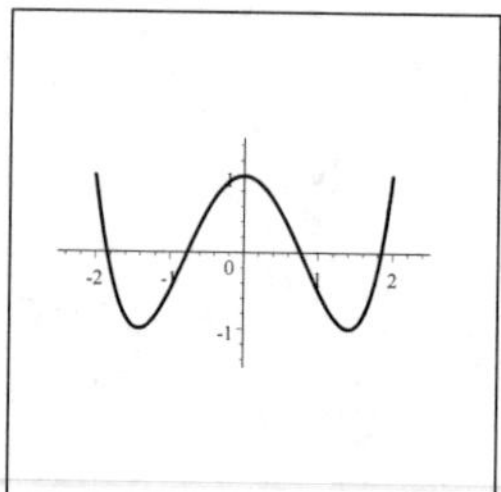

**39.** $x = 3\sin 5t$, $y = 5\cos 3t$

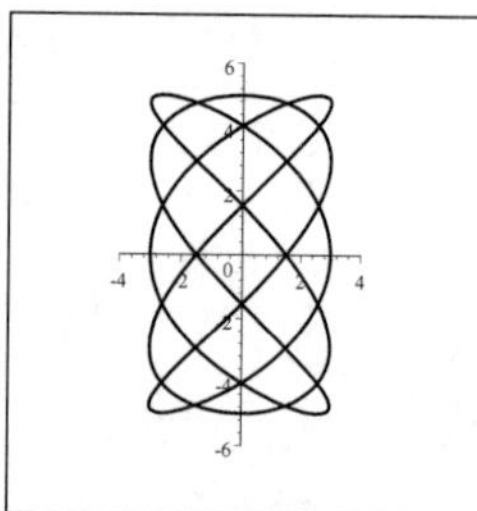

**40.** $x = \sin 4t$, $y = \cos 3t$

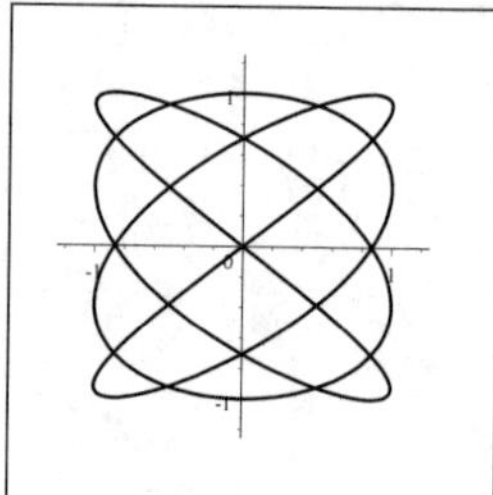

**41.** $x = \sin(\cos t)$, $y = \cos t^{3/2}$, $0 \le t \le 2\pi$

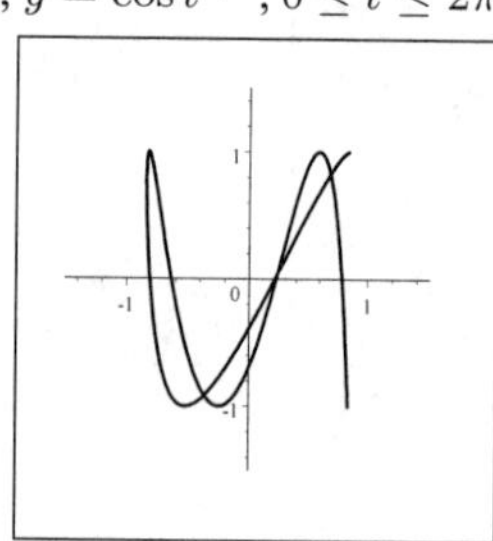

**42.** $x = 2\cos t + \cos 2t$, $y = 2\sin t - \sin 2t$

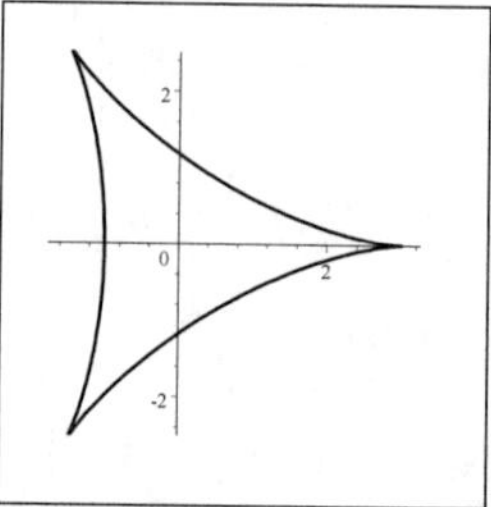

**43.** **(a)** $r = e^{\theta/12}$, $0 \le \theta \le 4\pi$ $\Rightarrow$ $x = e^{t/12}\cos t$, $y = e^{t/12}\sin t$

**(b)**

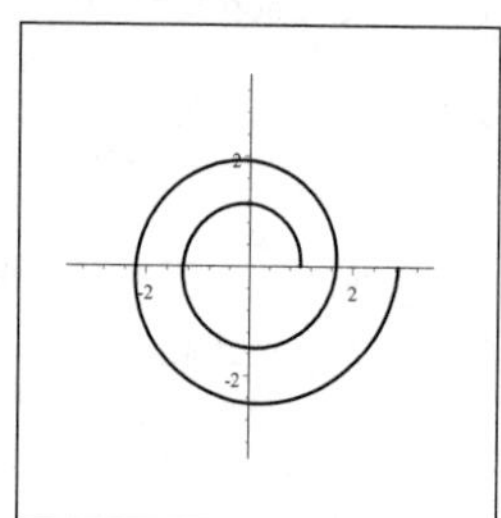

**44.** **(a)** $r = \sin\theta + 2\cos\theta \Leftrightarrow x = (\sin\theta + 2\cos\theta)\cos\theta$, $y = (\sin\theta + 2\cos\theta)\sin\theta$

**(b)**

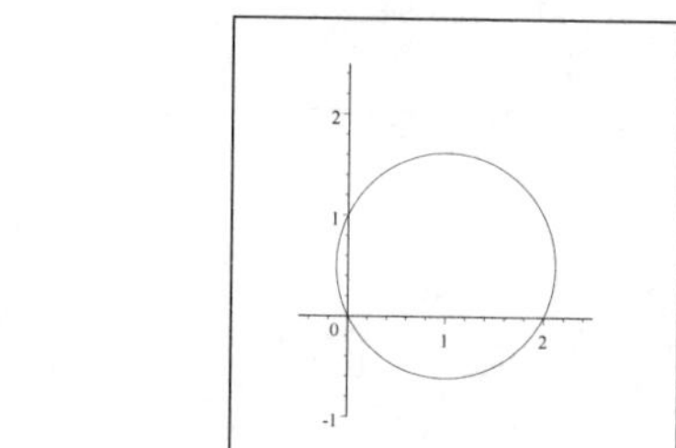

**45. (a)** $r = \dfrac{4}{2 - \cos\theta} \quad \Leftrightarrow \quad x = \dfrac{4\cos t}{2 - \cos t},\ y = \dfrac{4\sin t}{2 - \cos t}$

**(b)**

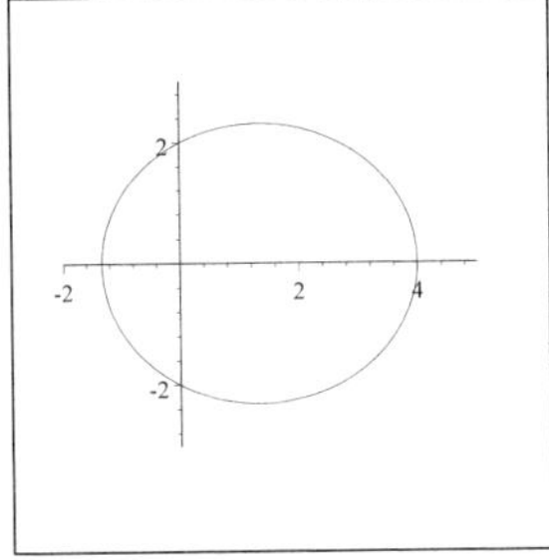

**46. (a)** $r = 2^{\sin\theta} \quad \Rightarrow \quad x = 2^{\sin\theta}\cos\theta,\ y = 2^{\sin\theta}\sin\theta$

**(b)**

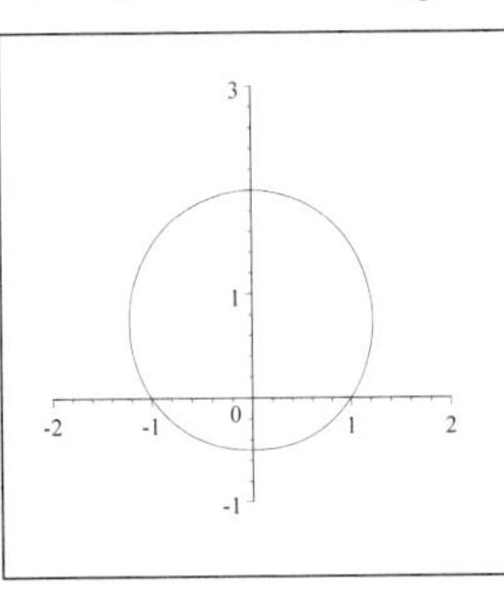

**47.** $x = t^3 - 2t,\ y = t^2 - t$ is Graph III, since $y = t^2 - t = \left(t^2 - t + \frac{1}{4}\right) - \frac{1}{4} = \left(t - \frac{1}{2}\right)^2 - \frac{1}{4}$, and so $y \geq -\frac{1}{4}$ on this curve, while $x$ is unbounded.

**48.** $x = \sin 3t,\ y = \sin 4t$ is Graph IV, since when $t = 0$ and $t = \pi$ the curve passes through $(0, 0)$. Thus this curve must pass through $(0, 0)$ twice.

**49.** $x = t + \sin 2t,\ y = t + \sin 3t$ is Graph II, since the values of $x$ and $y$ oscillate about their values on the line $x = t,\ y = t$ $\Leftrightarrow$ $y = x$.

**50.** $x = \sin(t + \sin t),\ y = \cos(t + \cos t)$ is Graph I, since this curve does not pass through the point $(0, 0)$.

**51. (a)** If we modify Figure 8 so that $|PC| = b$, then by the same reasoning as in Example 6, we see that $x = |OT| - |PQ| = a\theta - b\sin\theta$ and $y = |TC| - |CQ| = a - b\cos\theta$. We graph the case where $a = 3$ and $b = 2$.

**(b)**

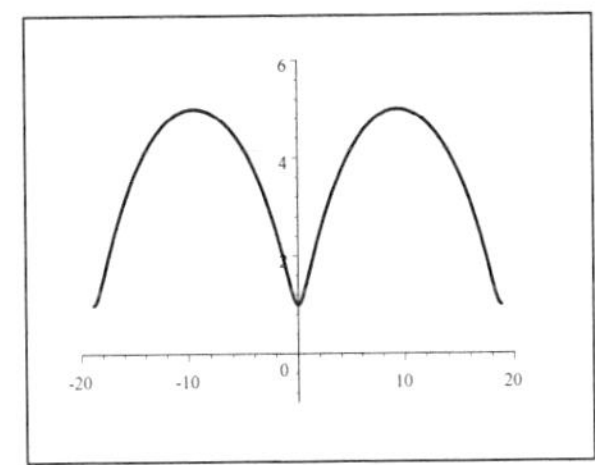

**52.** From the figure, we see that $x = |OT| - |PQ| = a\theta - b\sin\theta$ and $y = |TC| - |CQ| = a - b\cos\theta$. When $a = 1$ and $b = 2$, the parametric equations become $x = \theta - 2\sin\theta,\ y = 1 - 2\cos\theta$. The curve is graphed for $0 \leq \theta \leq 4\pi$.

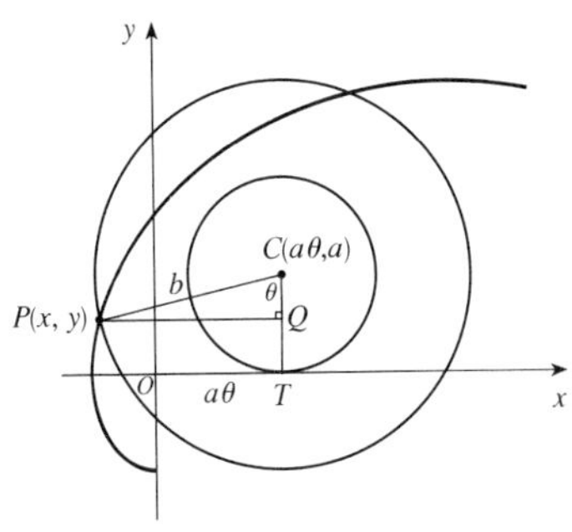


**(b)**

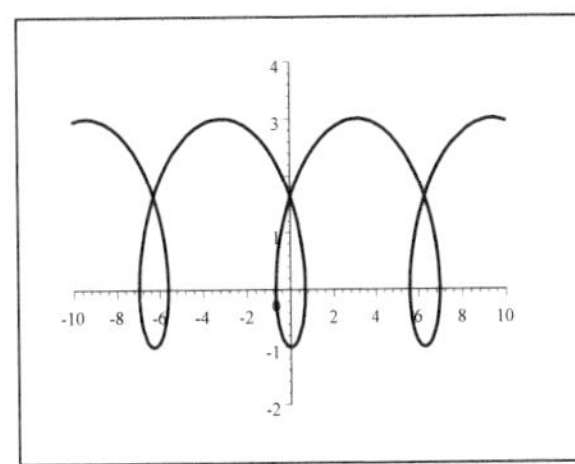

**53. (a)** We first note that the center of circle $C$ (the small circle) has coordinates $([a-b]\cos\theta, [a-b]\sin\theta)$. Now the arc $PQ$ has the same length as the arc $P'Q$, so $b\phi = a\theta \Leftrightarrow \phi = \frac{a}{b}\theta$, and so $\phi - \theta = \frac{a}{b}\theta - \theta = \frac{a-b}{b}\theta$. Thus the $x$-coordinate of $P$ is the $x$-coordinate of the center of circle $C$ plus $b\cos(\phi - \theta) = b\cos\left(\frac{a-b}{b}\theta\right)$, and the $y$-coordinate of $P$ is the $y$-coordinate of the center of circle $C$ minus $b \cdot \sin(\phi - \theta) = b\sin\left(\frac{a-b}{b}\theta\right)$. So $x = (a-b)\cos\theta + b\cos\left(\frac{a-b}{b}\theta\right)$ and $y = (a-b)\sin\theta - b\sin\left(\frac{a-b}{b}\theta\right)$.

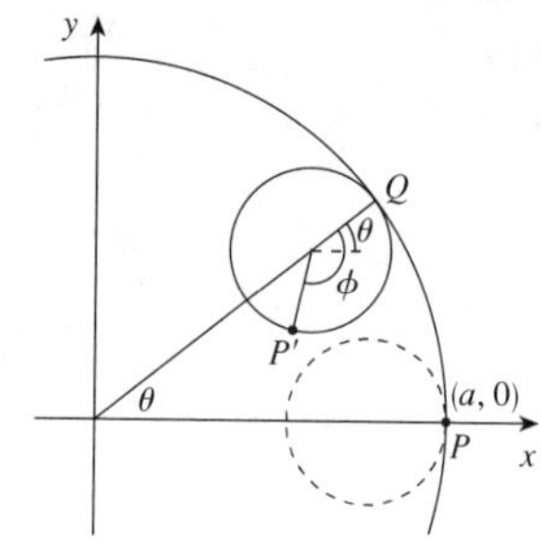


**(b)** If $a = 4b$, $b = \frac{a}{4}$, and $x = \frac{3}{4}a\cos\theta + \frac{1}{4}a\cos 3\theta$, $y = \frac{3}{4}a\sin\theta - \frac{1}{4}a\sin 3\theta$. From Example 2 in Section 7.3, $\cos 3\theta = 4\cos^3\theta - 3\cos\theta$. Similarly, one can prove that $\sin 3\theta = 3\sin\theta - 4\sin^3\theta$. Substituting, we get

$x = \frac{3}{4}a\cos\theta + \frac{1}{4}a\left(4\cos^3\theta - 3\cos\theta\right) = a\cos^3\theta$

$y = \frac{3}{4}a\sin\theta - \frac{1}{4}a\left(3\sin\theta - 4\sin^3\theta\right) = a\sin^3\theta$. Thus,

$x^{2/3} + y^{2/3} = a^{2/3}\cos^2\theta + a^{2/3}\sin^2\theta = a^{2/3}$, so $x^{2/3} + y^{2/3} = a^{2/3}$.

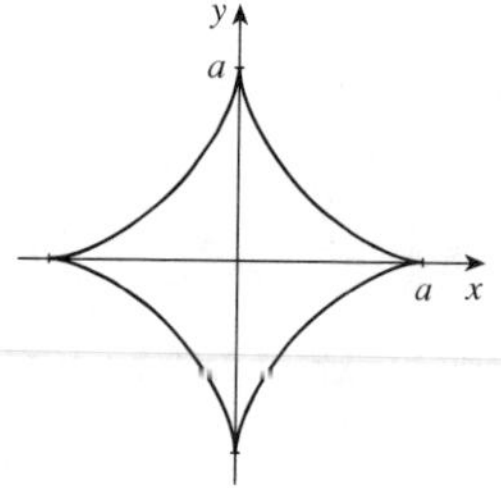


**54.** The coordinates of $C$ are $([a+b]\cos\theta, [a+b]\sin\theta)$. Let $\phi$ be $\angle OCP$ as shown in the figure. Then the arcs traced out by rolling the circle along the outside are equal, that is, $b\phi = a\theta \Leftrightarrow \phi = \frac{a}{b}\theta$. $P$ is displaced from $C$ by amounts equal to the legs of the right triangle $CPT$, where $CP$ is the hypotenuse. Since $\triangle OCQ$ is a right triangle, it follows that $\angle OCQ = \frac{\pi}{2} - \theta$. Thus, $\alpha = \phi - \left(\frac{\pi}{2} - \theta\right) = \frac{a}{b}\theta - \frac{\pi}{2} + \theta = \left(\frac{a+b}{b}\right)\theta - \frac{\pi}{2}$. So

$$\begin{aligned} x &= (a+b)\cos\theta + b\sin\left[\left(\frac{a+b}{b}\right)\theta - \frac{\pi}{2}\right] = (a+b)\cos\theta - b\sin\left[\frac{\pi}{2} - \left(\frac{a+b}{b}\right)\theta\right] \\ &= (a+b)\cos\theta - b\cos\left(\frac{a+b}{b}\right)\theta \end{aligned}$$

and

$$\begin{aligned} y &= (a+b)\sin\theta - b\cos\left[\left(\frac{a+b}{b}\right)\theta - \frac{\pi}{2}\right] = (a+b)\sin\theta - b\cos\left[\frac{\pi}{2} - \left(\frac{a+b}{b}\right)\theta\right] \\ &= (a+b)\sin\theta - b\sin\left(\frac{a+b}{b}\right)\theta \end{aligned}$$

Therefore, parametric equations for the epicycloid are

$x = (a+b)\cos\theta - b\cos\left(\frac{a+b}{b}\right)\theta$ and $y = (a+b)\sin\theta - b\sin\left(\frac{a+b}{b}\right)\theta$.

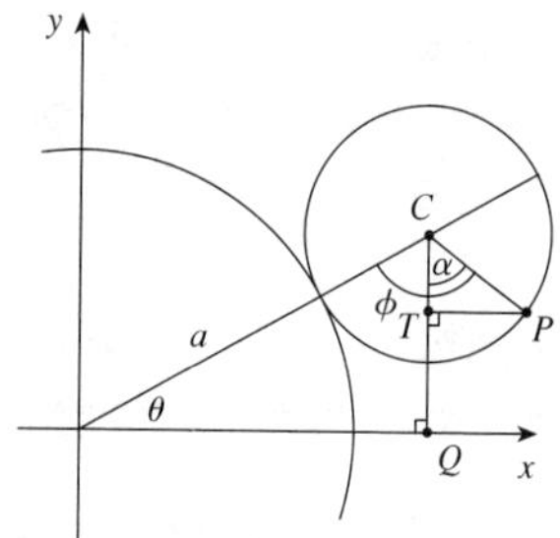


**55.** A polar equation for the circle is $r = 2a\sin\theta$. Thus the coordinates of $Q$ are $x = r\cos\theta = 2a\sin\theta\cos\theta$ and $y = r\sin\theta = 2a\sin^2\theta$. The coordinates of $R$ are $x = 2a\cot\theta$ and $y = 2a$. Since $P$ is the midpoint of $QR$, we use the midpoint formula to get $x = a(\sin\theta\cos\theta + \cot\theta)$ and $y = a\left(1 + \sin^2\theta\right)$.

**56. (a)** It is apparent that $x = |OQ|$ and $y = |QP| = |ST|$. From the diagram, $x = |OQ| = a\cos\theta$ and $y = |ST| = b\sin\theta$. Thus, parametric equations are $x = a\cos\theta$ and $y = b\sin\theta$.

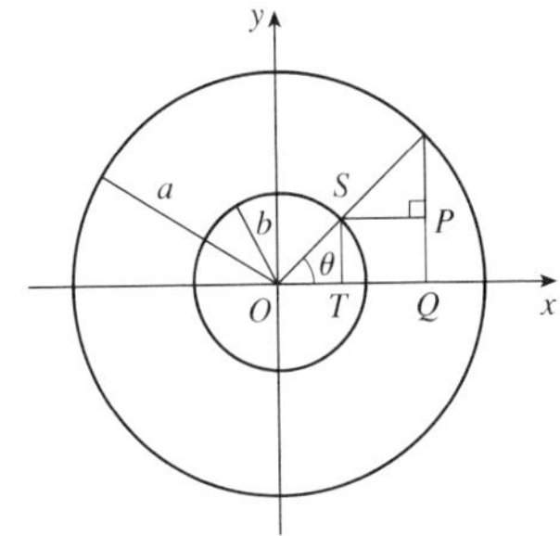


**(b)** The curve is graphed with $a = 3$ and $b = 2$.

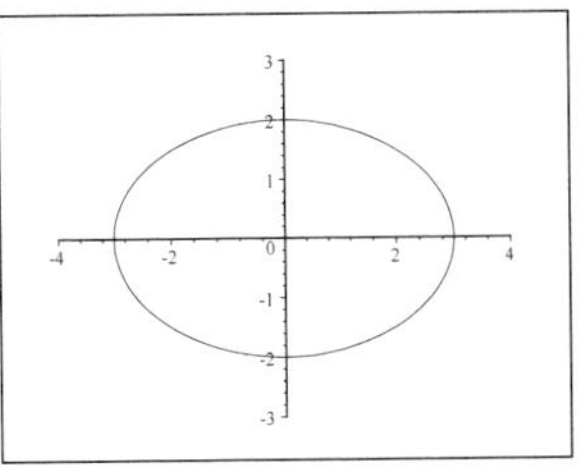

**(c)** To eliminate $\theta$ we rearrange: $\sin\theta = y/b \Rightarrow \sin^2\theta = (y/b)^2$ and $\cos\theta = x/a \Rightarrow \cos^2\theta = (x/a)^2$. Adding the two equations: $\sin^2\theta + \cos^2\theta = 1 = x^2/a^2 + y^2/b^2$. As indicated in part (b), the curve is an ellipse.

**57. (a)** $A$ has coordinates $(a\cos\theta, a\sin\theta)$. Since $OA$ is perpendicular to $AB$, $\Delta OAB$ is a right triangle and $B$ has coordinates $(a\sec\theta, 0)$. It follows that $P$ has coordinates $(a\sec\theta, b\sin\theta)$. Thus, the parametric equations are $x = a\sec\theta$, $y = b\sin\theta$.

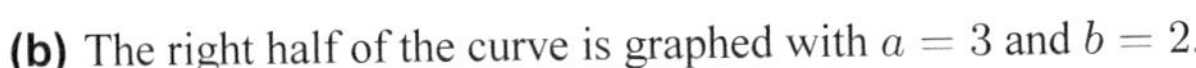

**(b)** The right half of the curve is graphed with $a = 3$ and $b = 2$.

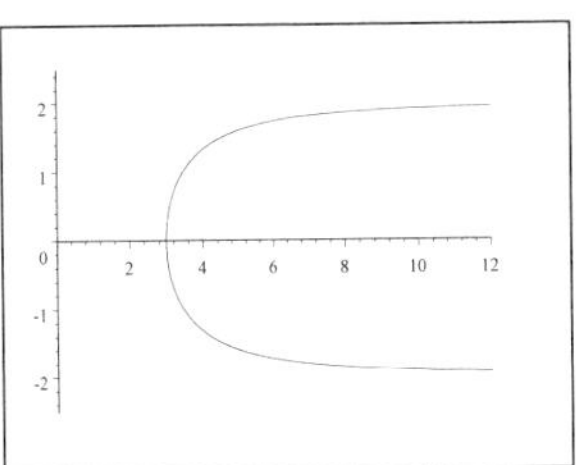

**58. (a)** $C = (2a\cot\theta, 2a)$, so the $x$-coordinate of $P$ is $x = 2a\cot\theta$. Let $B = (0, 2a)$. Then $\angle OAB$ is a right angle and $\angle OBA = \theta$, so $|OA| = 2a\sin\theta$ and $A = \left(2a\sin\theta\cos\theta, 2a\sin^2\theta\right)$. Thus, the $y$-coordinate of $P$ is $y = 2a\sin^2\theta$.

**(b)** The curve is graphed with $a = 1$.

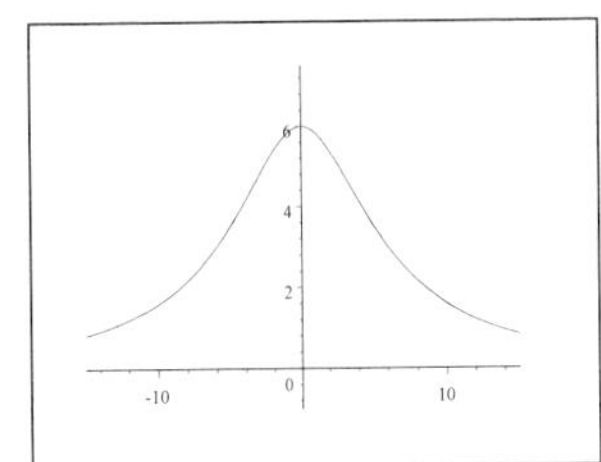

**59.** We use the equation for $y$ from Example 6 and solve for $\theta$. Thus for $0 \le \theta \le \pi$, $y = a(1 - \cos\theta) \Leftrightarrow \dfrac{a-y}{a} = \cos\theta$ $\Leftrightarrow \quad \theta = \cos^{-1}\left(\dfrac{a-y}{a}\right)$. Substituting into the equation for $x$, we get $x = a\left[\cos^{-1}\left(\dfrac{a-y}{a}\right) - \sin\left(\cos^{-1}\left(\dfrac{a-y}{a}\right)\right)\right]$.

However, $\sin\left(\cos^{-1}\left(\dfrac{a-y}{a}\right)\right) = \sqrt{1 - \left(\dfrac{a-y}{a}\right)^2} = \dfrac{\sqrt{2ay - y^2}}{a}$. Thus, $x = a\left[\cos^{-1}\left(\dfrac{a-y}{a}\right) - \dfrac{\sqrt{2ay - y^2}}{a}\right]$,

and we have $\dfrac{\sqrt{2ay - y^2} + x}{a} = \cos^{-1}\left(\dfrac{a-y}{a}\right) \quad\Rightarrow\quad 1 - \dfrac{y}{a} = \cos\left(\dfrac{\sqrt{2ay - y^2} + x}{a}\right) \quad\Rightarrow$

$$y = a\left[1 - \cos\left(\frac{\sqrt{2ay - y^2} + x}{a}\right)\right].$$

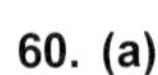

**60. (a)**

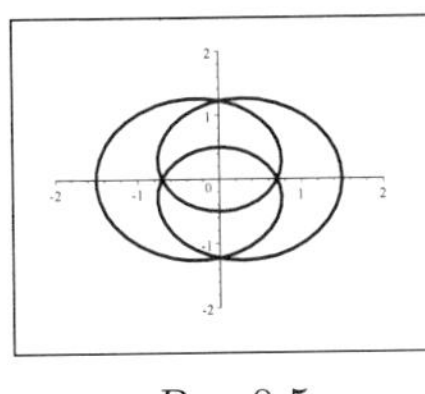

$R = 0.5$ $\qquad R = 1$ $\qquad R = 3$ $\qquad R = 5$

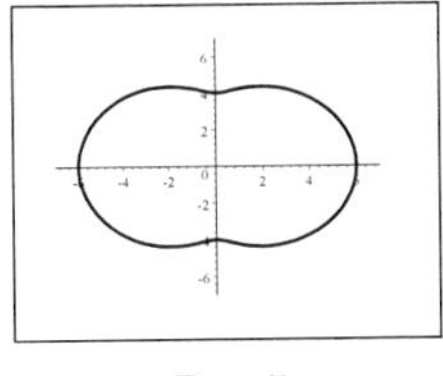

**(b)** The graph with $R = 5$ seems to most closely resemble the profile of the engine housing.

**61. (a)** In the figure, since $OQ$ and $QT$ are perpendicular and $OT$ and $TD$ are perpendicular, the angles formed by their intersections are equal, that is, $\theta = \angle DTQ$. Now the coordinates of $T$ are $(\cos\theta, \sin\theta)$. Since $|TD|$ is the length of the string that has been unwound from the circle, it must also have arc length $\theta$, so $|TD| = \theta$. Thus the $x$-displacement from $T$ to $D$ is $\theta \cdot \sin\theta$ while the $y$-displacement from $T$ to $D$ is $\theta \cdot \cos\theta$. So the coordinates of $D$ are $x = \cos\theta + \theta\sin\theta$ and $y = \sin\theta - \theta\cos\theta$.

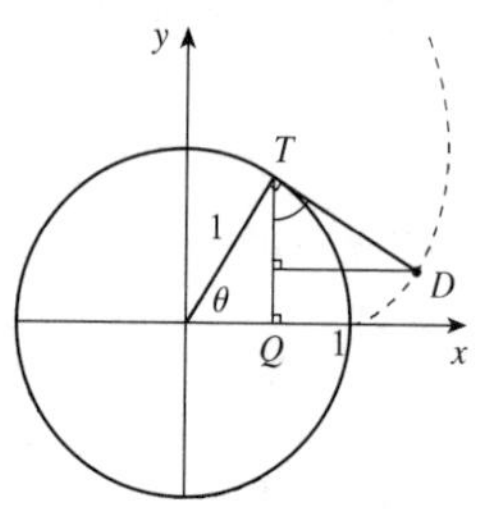


**(b)**

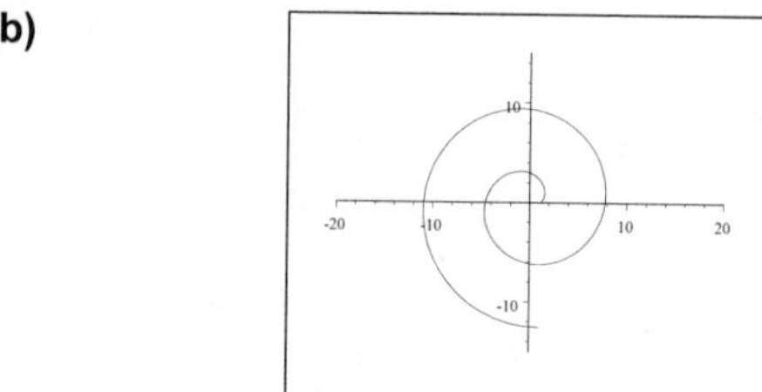


**62. (a)** It takes $2\pi$ units of time. The parametric equations of the particle that moves twice as fast around the circle are $x = \sin 2t$, $y = \cos 2t$.

**(b)** From the first table, we see that the particle travels counterclockwise. If we want the particle to travel in a clockwise direction, then we want the second table to apply. Possible parametric equations for clockwise traversal of the unit circle are $x = -\sin t$, $y = \cos t$.

| $t$ | $x = \sin t$ | $y = \cos t$ |
|---|---|---|
| 0 | 0 | 1 |
| $\frac{\pi}{2}$ | 1 | 0 |
| $\pi$ | 0 | $-1$ |
| $\frac{3\pi}{2}$ | $-1$ | 0 |
| $2\pi$ | 0 | 1 |

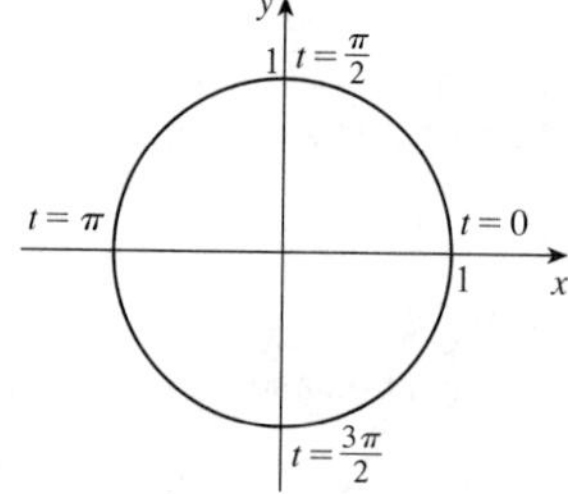


| $t$ | $x$ | $y$ |
|---|---|---|
| 0 | 0 | 1 |
| $\frac{\pi}{2}$ | $-1$ | 0 |
| $\pi$ | 0 | $-1$ |
| $\frac{3\pi}{2}$ | 1 | 0 |
| $2\pi$ | 0 | 1 |

**63.** $C$: $x = t, y = t^2$; $D$: $x = \sqrt{t}, y = t, t \geq 0$ $E$: $x = \sin t, y = 1 - \cos^2 t$ $F$: $x = e^t, y = e^{2t}$

**(a)** For $C$, $x = t, y = t^2 \Rightarrow y = x^2$.

For $D$, $x = \sqrt{t}, y = t \Rightarrow y = x^2$.

For $E$, $x = \sin t \Rightarrow x^2 = \sin^2 t = 1 - \cos^2 t = y$ and so $y = x^2$.

For $F$, $x = e^t \Rightarrow x^2 = e^{2t} = y$ and so $y = x^2$. Therefore, the points on all four curves satisfy the same rectangular equation.

**(b)** Curve $C$ is the entire parabola $y = x^2$. Curve $D$ is the right half of the parabola because $t \geq 0$ and so $x \geq 0$. Curve $E$ is the portion of the parabola for $-1 \leq x \leq 1$. Curve $F$ is the portion of the parabola where $x > 0$, since $e^t > 0$ for all $t$.

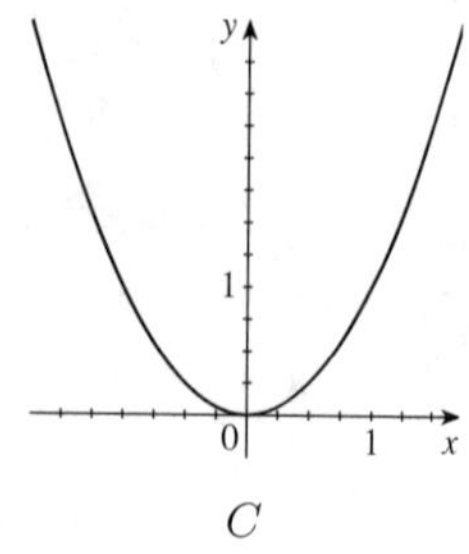


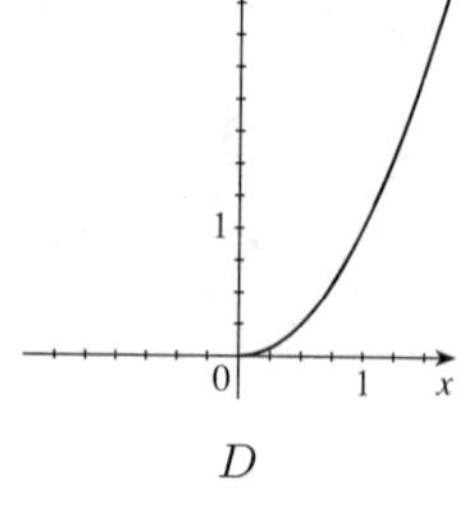


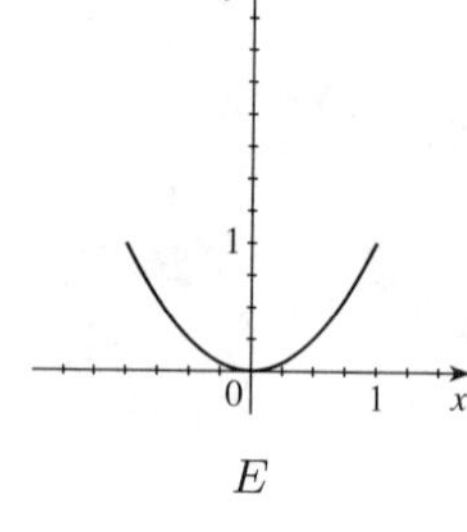


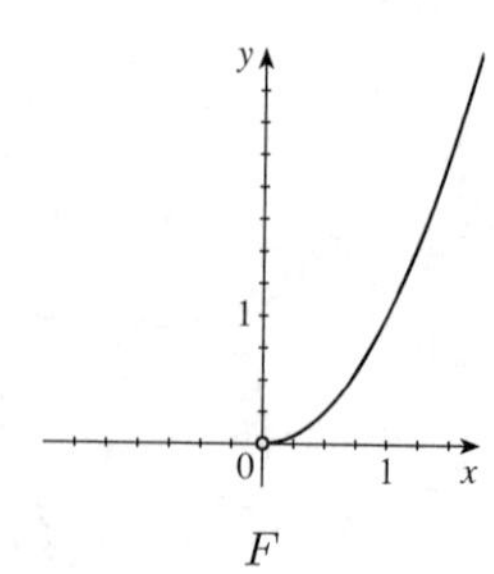

# Chapter 11 Review

**1.** $y^2 = 4x$. This is a parabola with $4p = 4 \quad \Leftrightarrow \quad p = 1$. The vertex is $(0,0)$, the focus is $(1,0)$, and the directrix is $x = -1$.

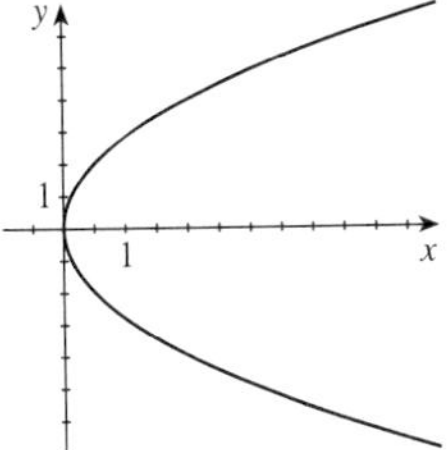

**2.** $x = \frac{1}{12}y^2 \quad \Leftrightarrow \quad y^2 = 12x$. This is a parabola with $4p = 12 \quad \Leftrightarrow \quad p = 3$. The vertex is $(0,0)$, the focus is $(3,0)$, and the directrix is $x = -3$.

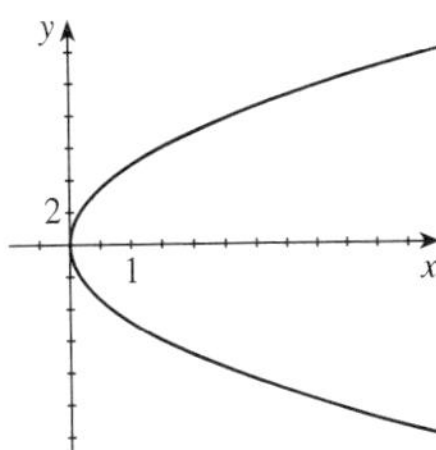

**3.** $x^2 + 8y = 0 \quad \Leftrightarrow \quad x^2 = -8y$. This is a parabola with $4p = -8 \quad \Leftrightarrow \quad p = -2$. The vertex is $(0,0)$, the focus is $(0,-2)$, and the directrix is $y = 2$.

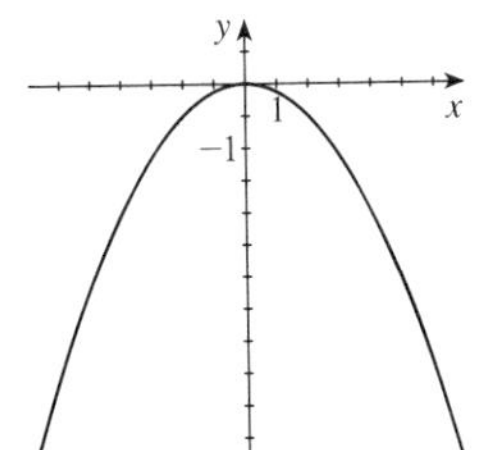

**4.** $2x - y^2 = 0 \quad \Leftrightarrow \quad y^2 = 2x$. This is a parabola with $4p = 2 \quad \Leftrightarrow \quad p = \frac{1}{2}$. The vertex is $(0,0)$, the focus is $\left(\frac{1}{2}, 0\right)$, and the directrix is $x = -\frac{1}{2}$.

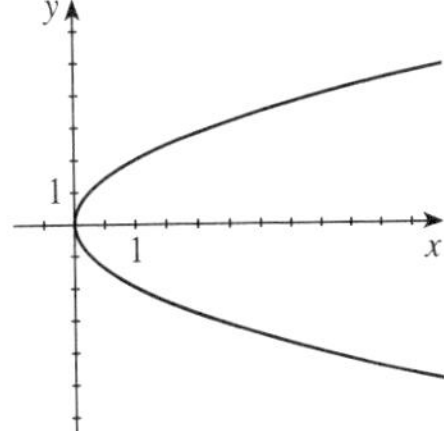

**5.** $x - y^2 + 4y - 2 = 0 \quad \Leftrightarrow \quad x - \left(y^2 - 4y + 4\right) - 2 = -4 \quad \Leftrightarrow$ $x - (y-2)^2 = -2 \quad \Leftrightarrow \quad (y-2)^2 = x + 2$. This is a parabola with $4p = 1 \quad \Leftrightarrow$ $p = \frac{1}{4}$. The vertex is $(-2,2)$, the focus is $\left(-2 + \frac{1}{4}, 2\right) = \left(-\frac{7}{4}, 2\right)$, and the directrix is $x = -2 - \frac{1}{4} = -\frac{9}{4}$.

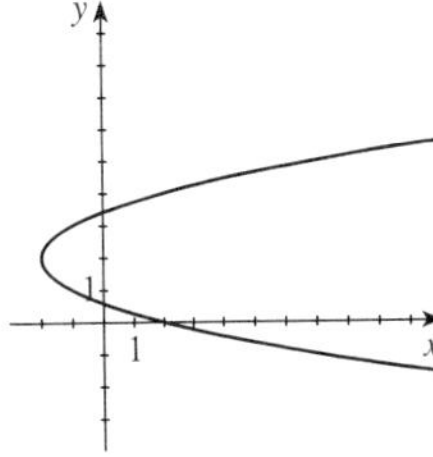

**6.** $2x^2 + 6x + 5y + 10 = 0 \quad \Leftrightarrow \quad 2\left(x^2 + 3x + \frac{9}{4}\right) + 5y + 10 = \frac{9}{2} \quad \Leftrightarrow$ $2\left(x + \frac{3}{2}\right)^2 = -5y - \frac{11}{2} \quad \Leftrightarrow \quad 2\left(x + \frac{3}{2}\right)^2 = -5\left(y + \frac{11}{10}\right) \quad \Leftrightarrow$ $\left(x + \frac{3}{2}\right)^2 = -\frac{5}{2}\left(y + \frac{11}{10}\right)$. This is a parabola with $4p = -\frac{5}{2} \quad \Leftrightarrow \quad p = -\frac{5}{8}$. The vertex is $\left(-\frac{3}{2}, -\frac{11}{10}\right)$, the focus is $\left(-\frac{3}{2}, -\frac{11}{10} - \frac{5}{8}\right) = \left(-\frac{3}{2}, -\frac{44+25}{40}\right) = \left(-\frac{3}{2}, -\frac{69}{20}\right)$, and the directrix is $y = -\frac{11}{10} + \frac{5}{8} = \frac{-44+25}{40} = -\frac{19}{40}$.

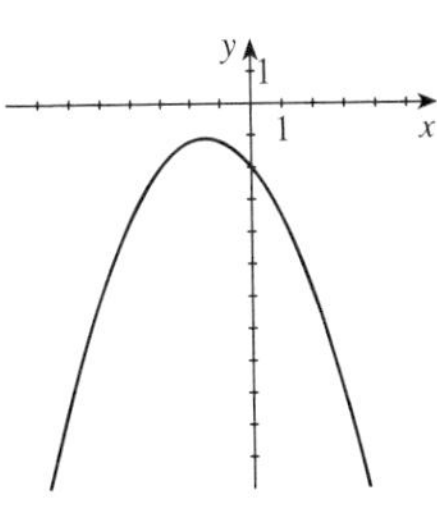

**7.** $\frac{1}{2}x^2 + 2x = 2y + 4 \Leftrightarrow x^2 + 4x = 4y + 8 \Leftrightarrow x^2 + 4x + 4 = 4y + 8$ $\Leftrightarrow (x+2)^2 = 4(y+3)$. This is a parabola with $4p = 4 \Leftrightarrow p = 1$. The vertex is $(-2, -3)$, the focus is $(-2, -3+1) = (-2, -2)$, and the directrix is $y = -3 - 1 = -4$.

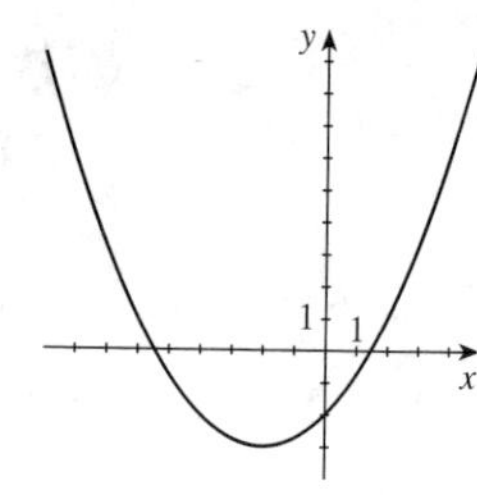


**8.** $x^2 = 3(x+y) \Leftrightarrow x^2 - 3x = 3y \Leftrightarrow x^2 - 3x + \frac{9}{4} = 3y + \frac{9}{4} \Leftrightarrow$ $\left(x - \frac{3}{2}\right)^2 = 3\left(y + \frac{3}{4}\right)$. This is a parabola with $4p = 3 \Leftrightarrow p = \frac{3}{4}$. The vertex is $\left(\frac{3}{2}, -\frac{3}{4}\right)$, the focus is $\left(\frac{3}{2}, -\frac{3}{4} + \frac{3}{4}\right) = \left(\frac{3}{2}, 0\right)$, and the directrix is $y = -\frac{3}{4} - \frac{3}{4} = -\frac{3}{2}$.

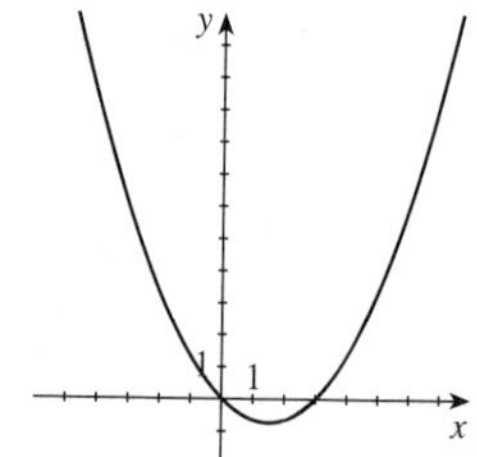


**9.** $\dfrac{x^2}{9} + \dfrac{y^2}{25} = 1$. This is an ellipse with $a = 5$, $b = 3$, and $c = \sqrt{25 - 9} = 4$. The center is $(0, 0)$, the vertices are $(0, \pm 5)$, the foci are $(0, \pm 4)$, the length of the major axis is $2a = 10$, and the length of the minor axis is $2b = 6$.

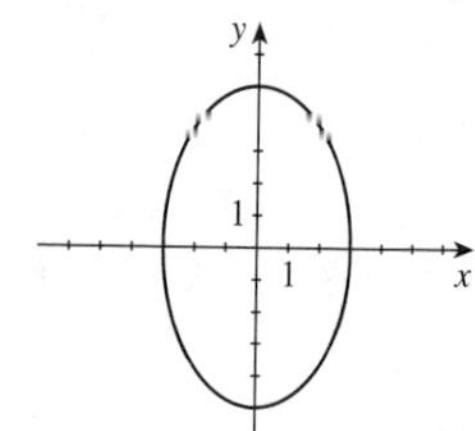


**10.** $\dfrac{x^2}{49} + \dfrac{y^2}{9} = 1$. This is an ellipse with $a = 7$, $b = 3$, and $c = \sqrt{49 - 9} = 2\sqrt{10}$. The center is $(0, 0)$, the vertices are $(\pm 7, 0)$, the foci are $\left(\pm 2\sqrt{10}, 0\right)$, the length of the major axis is $2a = 14$, and the length of the minor axis is $2b = 6$.

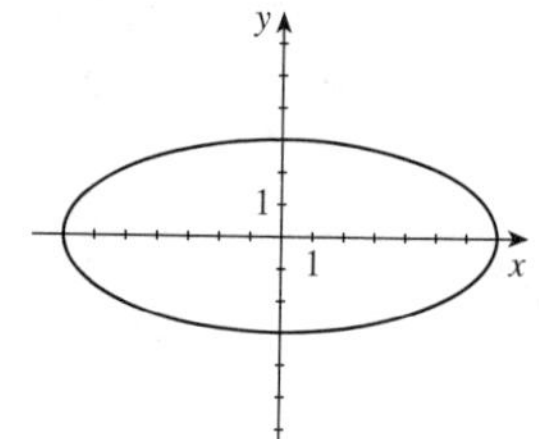


**11.** $x^2 + 4y^2 = 16 \Leftrightarrow \dfrac{x^2}{16} + \dfrac{y^2}{4} = 1$. This is an ellipse with $a = 4$, $b = 2$, and $c = \sqrt{16 - 4} = 2\sqrt{3}$. The center is $(0, 0)$, the vertices are $(\pm 4, 0)$, the foci are $\left(\pm 2\sqrt{3}, 0\right)$, the length of the major axis is $2a = 8$, and the length of the minor axis is $2b = 4$.

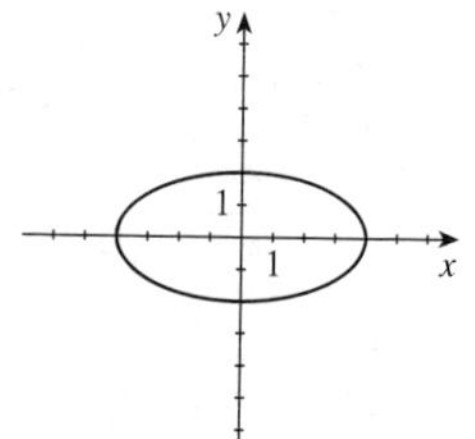


**12.** $9x^2 + 4y^2 = 1 \Leftrightarrow \dfrac{x^2}{1/9} + \dfrac{y^2}{1/4} = 1$. This is an ellipse with $a = \frac{1}{2}$, $b = \frac{1}{3}$, and $c = \sqrt{\frac{1}{4} - \frac{1}{9}} = \frac{1}{6}\sqrt{5}$. The center is $(0, 0)$, the vertices are $\left(0, \pm\frac{1}{2}\right)$, the foci are $\left(0, \pm\frac{1}{6}\sqrt{5}\right)$, the length of the major axis is $2a = 1$, and the length of the minor axis is $2b = \frac{2}{3}$.

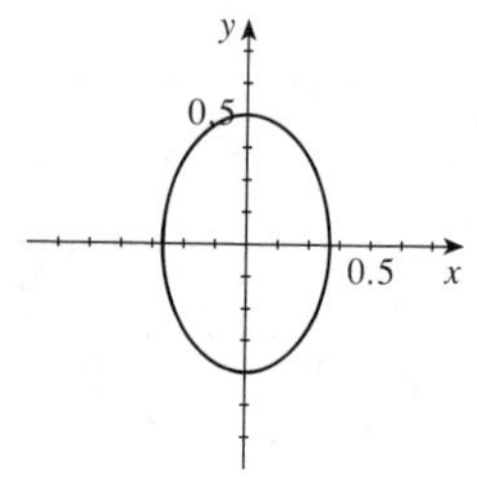

**13.** $\dfrac{(x-3)^2}{9}+\dfrac{y^2}{16}=1$.This is an ellipse with $a=4$, $b=3$, and $c=\sqrt{16-9}=\sqrt{7}$. The center is $(3,0)$, the vertices are $(3,\pm 4)$, the foci are $\left(3,\pm\sqrt{7}\right)$, the length of the major axis is $2a=8$, and the length of the minor axis is $2b=6$.

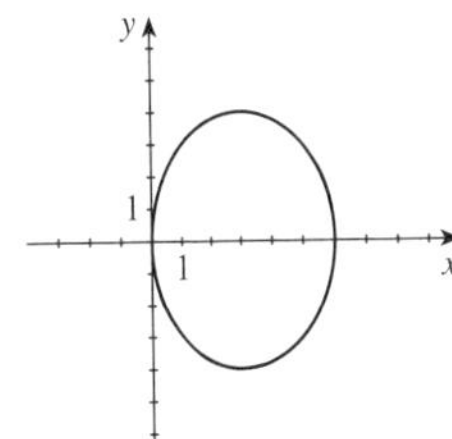


**14.** $\dfrac{(x-2)^2}{25}+\dfrac{(y+3)^2}{16}=1$. This is an ellipse with $a=5$, $b=4$, and $c=\sqrt{25-16}=3$. The center is $(2,-3)$, the vertices are $(2\pm 5,-3)=(-3,-3)$ and $(7,-3)$, the foci are $(2\pm 3,-3)=(-1,-3)$ and $(5,-3)$, the length of the major axis is $2a=10$, and the length of the minor axis is $2b=8$.

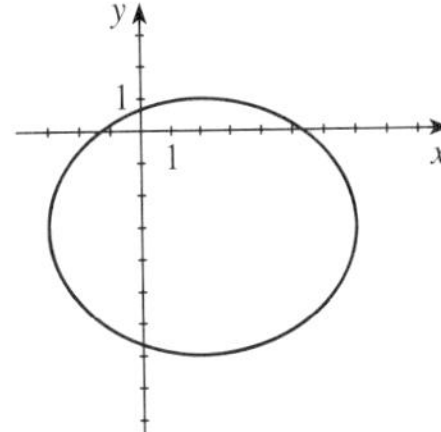


**15.** $4x^2+9y^2=36y \quad\Leftrightarrow\quad 4x^2+9\left(y^2-4y+4\right)=36 \quad\Leftrightarrow$ $4x^2+9(y-2)^2=36 \quad\Leftrightarrow\quad \dfrac{x^2}{9}+\dfrac{(y-2)^2}{4}=1$. This is an ellipse with $a=3$, $b=2$, and $c=\sqrt{9-4}=\sqrt{5}$. The center is $(0,2)$, the vertices are $(\pm 3,2)$, the foci are $\left(\pm\sqrt{5},2\right)$, the length of the major axis is $2a=6$, and the length of the minor axis is $2b=4$.

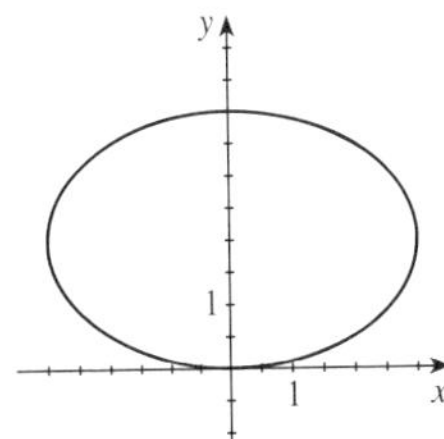


**16.** $2x^2+y^2=2+4(x-y) \quad\Leftrightarrow\quad 2x^2-4x+y^2+4y=2 \quad\Leftrightarrow$

$2\left(x^2-2x+1\right)+\left(y^2+4y+4\right)=2+2+4 \quad\Leftrightarrow$

$2(x-1)^2+(y+2)^2=8 \quad\Leftrightarrow\quad \dfrac{(x-1)^2}{4}+\dfrac{(y+2)^2}{8}=1$. This is an ellipse with $a=2\sqrt{2}$, $b=2$, and $c=\sqrt{8-4}=2$. The center is $(1,-2)$, the vertices are $\left(1,-2\pm 2\sqrt{2}\right)$, the foci are $(1,-2\pm 2)$, so the foci are $(1,0)$ and $(1,-4)$, the length of the major axis is $2a=4\sqrt{2}$, and the length of the minor axis is $2b=4$.

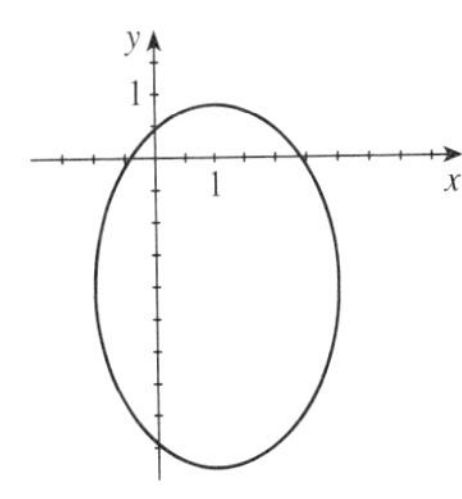


**17.** $-\dfrac{x^2}{9}+\dfrac{y^2}{16}=1 \quad\Leftrightarrow\quad \dfrac{y^2}{16}-\dfrac{x^2}{9}=0$. This is a hyperbola with $a=4$, $b=3$, and $c=\sqrt{16+9}=\sqrt{25}=5$. The center is $(0,0)$, the vertices are $(0,\pm 4)$, the foci are $(0,\pm 5)$, and the asymptotes are $y=\pm\frac{4}{3}x$.

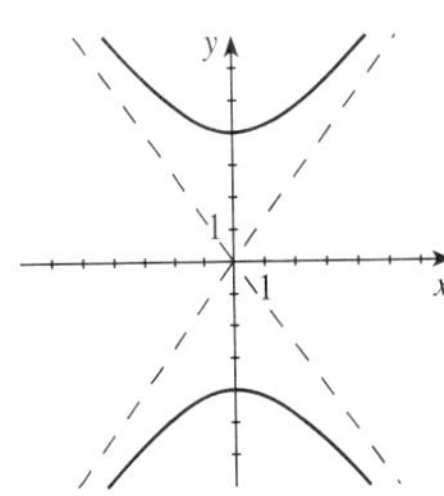


**18.** $\dfrac{x^2}{49}-\dfrac{y^2}{32}=1$. This is a hyperbola with $a=7$, $b=4\sqrt{2}$, and $c=\sqrt{49+32}=9$. The center is $(0,0)$, the vertices are $(\pm 7,0)$, the foci are $(\pm 9,0)$, and the asymptotes are $y=\pm\frac{4\sqrt{2}}{7}x$.

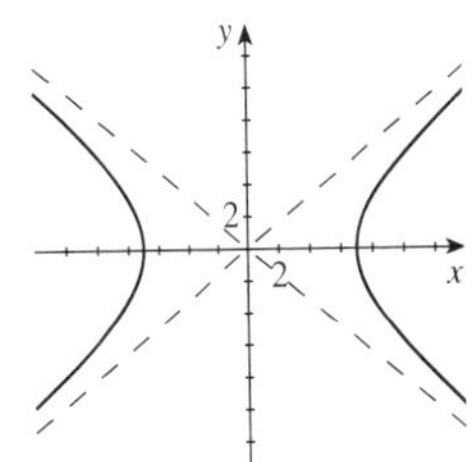

**19.** $x^2 - 2y^2 = 16 \quad \Leftrightarrow \quad \dfrac{x^2}{16} - \dfrac{y^2}{8} = 1$. This is a hyperbola with $a = 4$, $b = 2\sqrt{2}$, and $c = \sqrt{16+8} = \sqrt{24} = 2\sqrt{6}$. The center is $(0,0)$, the vertices are $(\pm 4, 0)$, the foci are $(\pm 2\sqrt{6}, 0)$, and the asymptotes are $y = \pm \frac{2\sqrt{2}}{4} x \quad \Leftrightarrow \quad y = \pm \frac{1}{\sqrt{2}} x$.

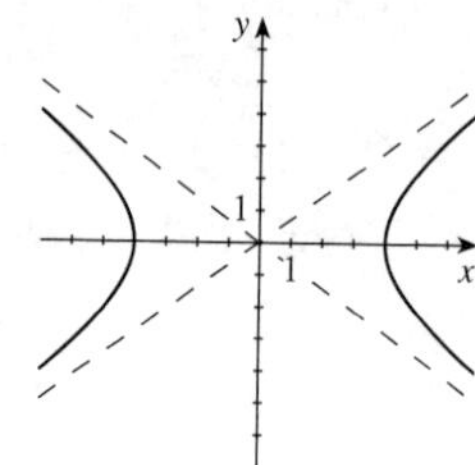


**20.** $x^2 - 4y^2 + 16 = 0 \quad \Leftrightarrow \quad 4y^2 - x^2 = 16 \quad \Leftrightarrow \quad \dfrac{y^2}{4} - \dfrac{x^2}{16} = 1$. This is a hyperbola with $a = 2$, $b = 4$, and $c = \sqrt{4+16} = 2\sqrt{5}$. The center is $(0,0)$, the vertices are $(0, \pm 2)$, the foci are $(0, \pm 2\sqrt{5})$, and the asymptotes are $y = \pm \frac{2}{4} x$ $\Leftrightarrow \quad y = \pm \frac{1}{2} x$.

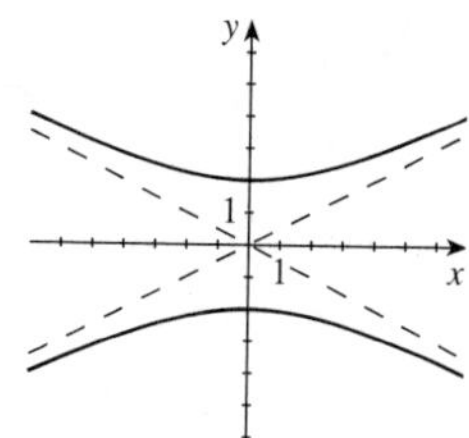


**21.** $\dfrac{(x+4)^2}{16} - \dfrac{y^2}{16} = 1$. This is a hyperbola with $a = 4$, $b = 4$ and $c = \sqrt{16+16} = 4\sqrt{2}$. The center is $(-4, 0)$, the vertices are $(-4 \pm 4, 0)$ which are $(-8, 0)$ and $(0, 0)$, the foci are $(-4 \pm 4\sqrt{2}, 0)$, and the asymptotes are $y = \pm (x + 4)$.

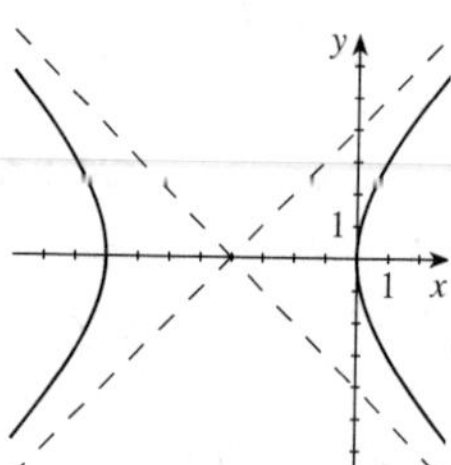


**22.** $\dfrac{(x-2)^2}{8} - \dfrac{(y+2)^2}{8} = 1$. This is a hyperbola with $a = 2\sqrt{2}$, $b = 2\sqrt{2}$ and $c = \sqrt{8+8} = 4$. The center is $(2, -2)$, the vertices are $(2 \pm 2\sqrt{2}, -2)$, the foci are $(2 \pm 4, -2) = (-2, -2)$ and $(6, -2)$, and the asymptotes are $y + 2 = \pm (x - 2) \quad \Leftrightarrow \quad y = -x$ and $y = x - 4$.

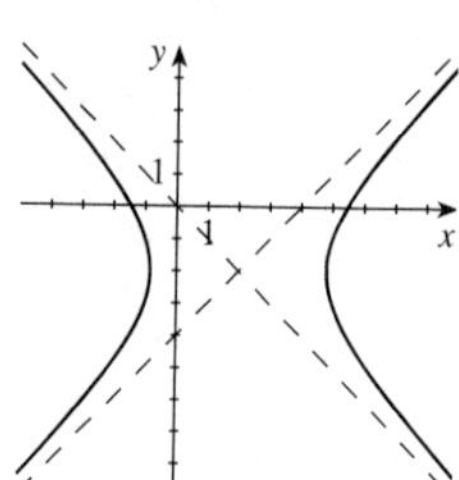


**23.** $9y^2 + 18y = x^2 + 6x + 18 \quad \Leftrightarrow$
$9(y^2 + 2y + 1) = (x^2 + 6x + 9) + 9 - 9 + 18 \quad \Leftrightarrow$
$9(y+1)^2 - (x+3)^2 = 18 \quad \Leftrightarrow \quad \dfrac{(y+1)^2}{2} - \dfrac{(x+3)^2}{18} = 1$. This is a hyperbola with $a = \sqrt{2}$, $b = 3\sqrt{2}$, and $c = \sqrt{2+18} = 2\sqrt{5}$.The center is $(-3, -1)$, the vertices are $(-3, -1 \pm \sqrt{2})$, the foci are $(-3, -1 \pm 2\sqrt{5})$, and the asymptotes are $y + 1 = \pm \frac{1}{3}(x + 3) \quad \Leftrightarrow \quad y = \frac{1}{3}x$ and $y = -\frac{1}{3}x - 2$.

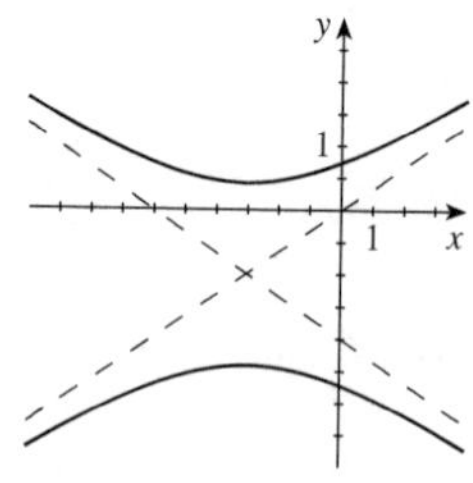


**24.** $y^2 = x^2 + 6y \quad \Leftrightarrow \quad (y^2 - 6y + 9) - x^2 = 9 \quad \Leftrightarrow \quad (y-3)^2 - x^2 = 9 \quad \Leftrightarrow$
$\dfrac{(y-3)^2}{9} - \dfrac{x^2}{9} = 1$. This is a hyperbola with $a = 3$, $b = 3$, and $c = \sqrt{9+9} = 3\sqrt{2}$. The center is $(0, 3)$, the vertices are $(0, 3 \pm 3) = (0, 6)$ and $(0, 0)$, the foci are $(0, 3 \pm 3\sqrt{2})$, and the asymptotes are $y - 3 = \pm x \quad \Leftrightarrow$
$y = x + 3$ and $y = -x + 3$.

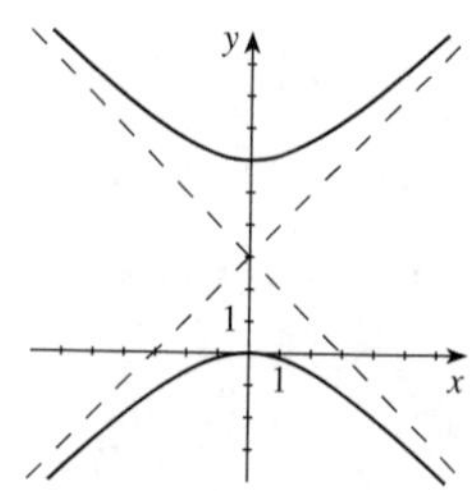

**25.** This is a parabola that opens to the right with its vertex at $(0,0)$ and the focus at $(2,0)$. So $p = 2$, and the equation is $y^2 = 4(2)x \quad \Leftrightarrow \quad y^2 = 8x$.

**26.** This is an ellipse with the center at $(0,0)$, $a = 12$, and $b = 5$. The equation is then $\dfrac{x^2}{12^2} + \dfrac{y^2}{5^2} = 1 \quad \Leftrightarrow \quad \dfrac{x^2}{144} + \dfrac{y^2}{25} = 1$.

**27.** From the graph, the center is $(0,0)$, and the vertices are $(0,-4)$ and $(0,4)$. Since $a$ is the distance from the center to a vertex, we have $a = 4$. Because one focus is $(0,5)$, we have $c = 5$, and since $c^2 = a^2 + b^2$, we have $25 = 16 + b^2 \quad \Leftrightarrow$ $b^2 = 9$. Thus an equation of the hyperbola is $\dfrac{y^2}{16} - \dfrac{x^2}{9} = 1$.

**28.** This is a parabola that opens to the left with its vertex at $(4,4)$, so its equation is of the form $(y-4)^2 = 4p(x-4)$ with $p < 0$. Since $(0,0)$ is a point on this hyperbola, we must have $(0-4)^2 = 4p(0-4) \quad \Leftrightarrow \quad 16 = -16p \quad \Leftrightarrow \quad p = -1$. Thus the equation is $(y-4)^2 = -4(x-4)$.

**29.** From the graph, the center of the ellipse is $(4,2)$, and so $a = 4$ and $b = 2$. The equation is $\dfrac{(x-4)^2}{4^2} + \dfrac{(y-2)^2}{2^2} = 1 \Leftrightarrow \quad \dfrac{(x-4)^2}{16} + \dfrac{(y-2)^2}{4} = 1$.

**30.** From the graph, the center is at $(1,0)$, and the vertices are $(0,0)$ and $(2,0)$. Since $a$ is the distance form the center to a vertex, $a = 1$. From the graph, the slope of one of the asymptotes is $\dfrac{b}{a} = 1 \quad \Leftrightarrow \quad b = 1$. Thus an equation of the hyperbola is $(x-1)^2 - y^2 = 1$.

**31.** $\dfrac{x^2}{12} + y = 1 \quad \Leftrightarrow \quad \dfrac{x^2}{12} = -(y-1) \quad \Leftrightarrow \quad x^2 = -12(y-1)$. This is a parabola with $4p = -12 \quad \Leftrightarrow \quad p = -3$. The vertex is $(0,1)$ and the focus is $(0, 1-3) = (0,-2)$.

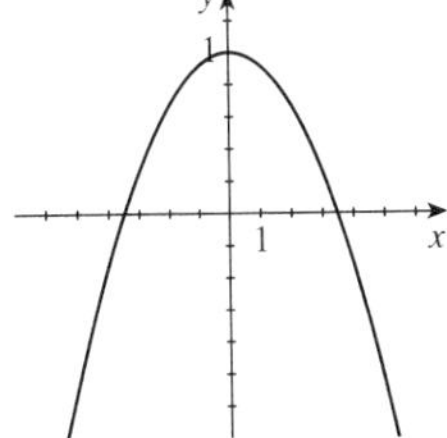


**32.** $\dfrac{x^2}{12} + \dfrac{y^2}{144} = \dfrac{y}{12} \quad \Leftrightarrow \quad 12x^2 + y^2 - 12y = 0 \quad \Leftrightarrow$

$12x^2 + \left(y^2 - 12y + 36\right) = 36 \quad \Leftrightarrow \quad \dfrac{x^2}{3} + \dfrac{(y-6)^2}{36} = 1$. This is an ellipse with $a = 6$, $b = \sqrt{3}$, and $c = \sqrt{36-3} = \sqrt{33}$. The vertices are $(0, 6 \pm 6) = (0,0)$ and $(0,12)$, and the foci are $\left(0, 6 \pm \sqrt{33}\right)$.

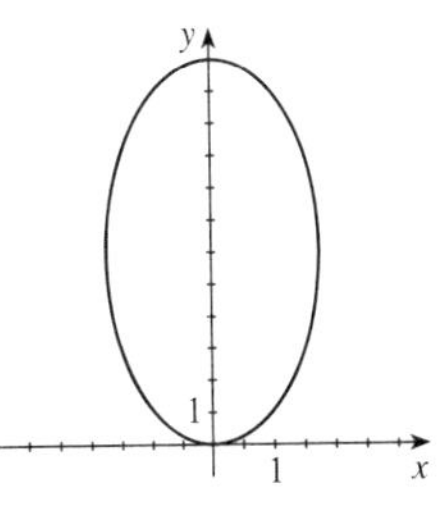


**33.** $x^2 - y^2 + 144 = 0 \quad \Leftrightarrow \quad \dfrac{y^2}{144} - \dfrac{x^2}{144} = 1$. This is a hyperbola with $a = 12$, $b = 12$, and $c = \sqrt{144 + 144} = 12\sqrt{2}$. The vertices are $(0, \pm 12)$ and the foci are $\left(0, \pm 12\sqrt{2}\right)$.

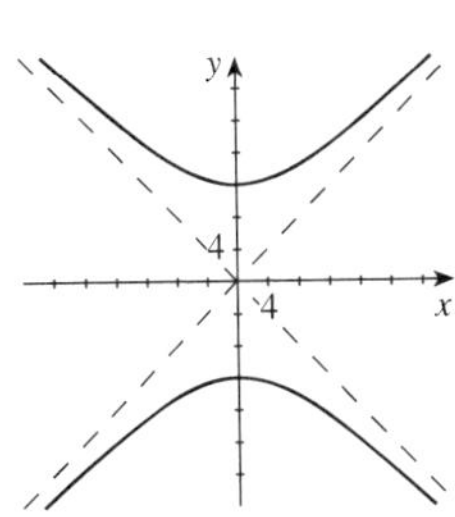

**34.** $x^2 + 6x = 9y^2 \quad \Leftrightarrow \quad (x^2 + 6x + 9) - 9y^2 = 9 \quad \Leftrightarrow \quad \dfrac{(x+3)^2}{9} - y^2 = 1$. This is a hyperbola with $a = 3$, $b = 1$, and $c = \sqrt{9+1} = \sqrt{10}$. The vertices are $(-3 \pm 3, 0) = (-6, 0)$ and $(0, 0)$, and the foci are $\left(-3 \pm \sqrt{10}, 0\right)$.

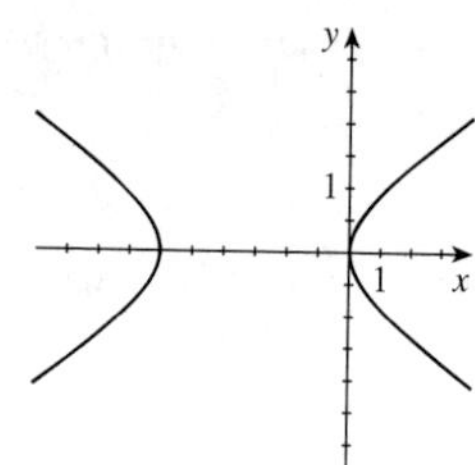


**35.** $4x^2 + y^2 = 8(x + y) \quad \Leftrightarrow \quad 4(x^2 - 2x) + (y^2 - 8y) = 0 \quad \Leftrightarrow$
$4(x^2 - 2x + 1) + (y^2 - 8y + 16) = 4 + 16 \quad \Leftrightarrow \quad 4(x-1)^2 + (y-4)^2 = 20$
$\Leftrightarrow \quad \dfrac{(x-1)^2}{5} + \dfrac{(y-4)^2}{20} = 1$. This is an ellipse with $a = 2\sqrt{5}$, $b = \sqrt{5}$, and $c = \sqrt{20-5} = \sqrt{15}$. The vertices are $\left(1, 4 \pm 2\sqrt{5}\right)$ and the foci are $\left(1, 4 \pm \sqrt{15}\right)$.

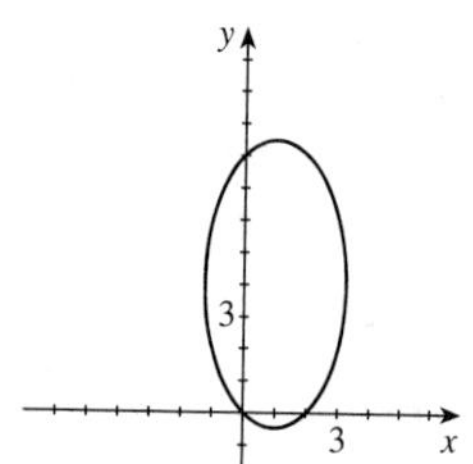


**36.** $3x^2 - 6(x + y) = 10 \quad \Leftrightarrow \quad 3x^2 - 6x = 6y + 10 \quad \Leftrightarrow$
$3(x^2 - 2x + 1) = 6y + 10 + 3 \quad \Leftrightarrow \quad 3(x-1)^2 = 6y + 13 \quad \Leftrightarrow$
$3(x-1)^2 = 6\left(y + \frac{13}{6}\right) \quad \Leftrightarrow \quad (x-1)^2 = 2\left(y + \frac{13}{6}\right)$. This is a parabola with $4p = 2 \quad \Leftrightarrow \quad p = \frac{1}{2}$. The vertex is $\left(1, -\frac{13}{6}\right)$ and the focus is $\left(1, -\frac{13}{6} + \frac{1}{2}\right) = \left(1, -\frac{10}{6}\right) = \left(1, -\frac{5}{3}\right)$.

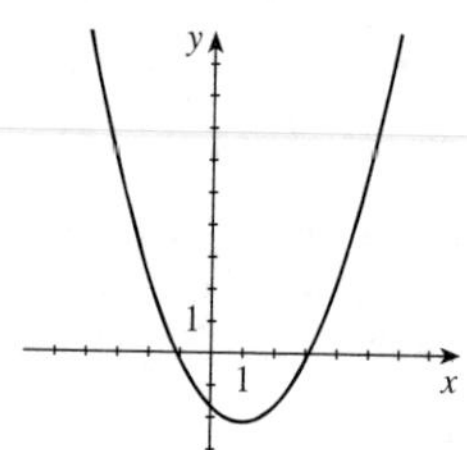


**37.** $x = y^2 - 16y \quad \Leftrightarrow \quad x + 64 = y^2 - 16y + 64 \quad \Leftrightarrow \quad (y-8)^2 = x + 64$. This is a parabola with $4p = 1 \quad \Leftrightarrow \quad p = \frac{1}{4}$. The vertex is $(-64, 8)$ and the focus is $\left(-64 + \frac{1}{4}, 8\right) = \left(-\frac{255}{4}, 8\right)$.

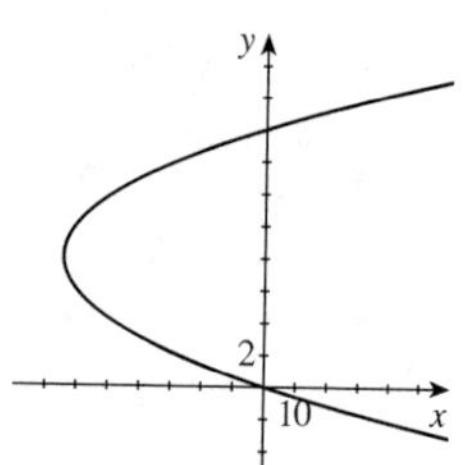


**38.** $2x^2 + 4 = 4x + y^2 \quad \Leftrightarrow \quad y^2 - 2x^2 + 4x = 4 \quad \Leftrightarrow$
$y^2 - 2(x^2 - 2x + 1) = 4 - 2 \quad \Leftrightarrow \quad y^2 - 2(x-1)^2 = 2 \quad \Leftrightarrow$
$\dfrac{y^2}{2} - (x-1)^2 = 1$. This is a hyperbola with $a = \sqrt{2}$, $b = 1$, and $c = \sqrt{2+1} = \sqrt{3}$. The vertices are $\left(1, \pm\sqrt{2}\right)$ and the foci are $\left(1, \pm\sqrt{3}\right)$.

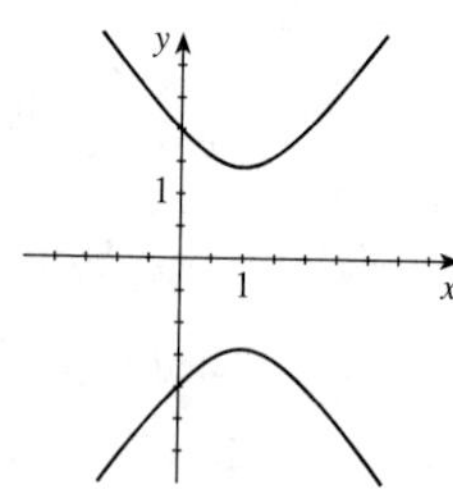


**39.** $2x^2 - 12x + y^2 + 6y + 26 = 0 \quad \Leftrightarrow \quad 2(x^2 - 6x) + (y^2 + 6y) = -26 \quad \Leftrightarrow$
$2(x^2 - 6x + 9) + (y^2 + 6y + 9) = -26 + 18 + 9 \quad \Leftrightarrow$
$2(x-3)^2 + (y+3)^2 = 1 \quad \Leftrightarrow \quad \dfrac{(x-3)^2}{\frac{1}{2}} + (y+3)^2 = 1$. This is an ellipse with $a = 1$, $b = \frac{\sqrt{2}}{2}$, and $c = \sqrt{1 - \frac{1}{2}} = \frac{\sqrt{2}}{2}$. The vertices are $(3, -3 \pm 1) = (3, -4)$ and $(3, -2)$, and the foci are $\left(3, -3 \pm \frac{\sqrt{2}}{2}\right)$.

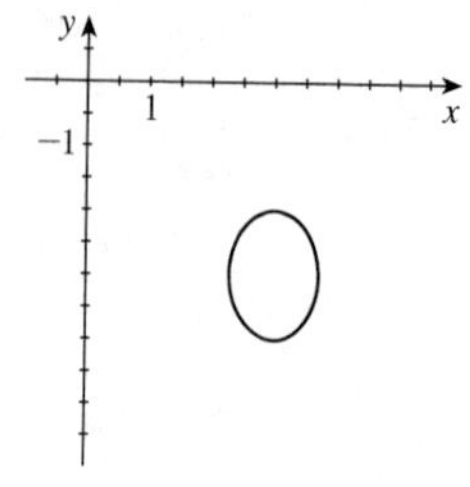

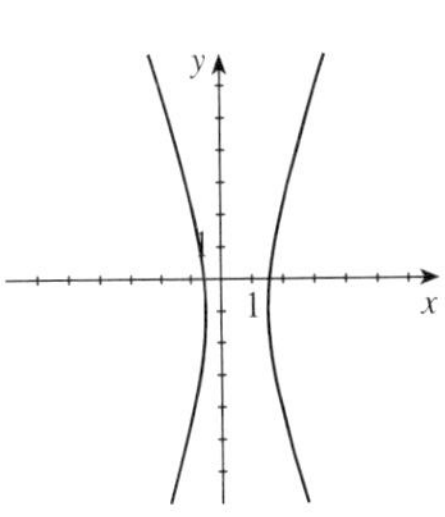


**40.** $36x^2 - 4y^2 - 36x - 8y = 31 \quad \Leftrightarrow \quad 36\left(x^2 - x\right) - 4\left(y^2 + 2y\right) = 31 \quad \Leftrightarrow$
$36\left(x^2 - x + \frac{1}{4}\right) - 4\left(y^2 + 2y + 1\right) = 31 + 9 - 4 \quad \Leftrightarrow$
$36\left(x - \frac{1}{2}\right)^2 - 4\left(y + 1\right)^2 = 36 \quad \Leftrightarrow \quad \left(x - \frac{1}{2}\right)^2 - \dfrac{\left(y+1\right)^2}{9} = 1$. This is a hyperbola with $a = 1$, $b = 3$, and $c = \sqrt{1+9} = \sqrt{10}$. The vertices are $\left(\frac{1}{2} \pm 1, -1\right) = \left(-\frac{1}{2}, -1\right)$ and $\left(\frac{3}{2}, -1\right)$, and the foci are $\left(\frac{1}{2} \pm \sqrt{10}, -1\right)$.

**41.** $9x^2 + 8y^2 - 15x + 8y + 27 = 0 \quad \Leftrightarrow \quad 9\left(x^2 - \frac{5}{3}x + \frac{25}{36}\right) + 8\left(y^2 + y + \frac{1}{4}\right) = -27 + \frac{25}{4} + 2 \quad \Leftrightarrow$
$9\left(x - \frac{5}{6}\right)^2 + 8\left(y + \frac{1}{2}\right)^2 = -\frac{75}{4}$. However, since the left-hand side of the equation is greater than or equal to 0, there is no point that satisfies this equation. The graph is empty.

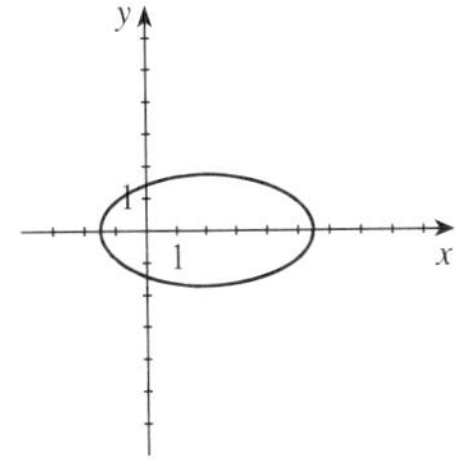


**42.** $x^2 + 4y^2 = 4x + 8 \quad \Leftrightarrow \quad \left(x^2 - 4x + 4\right) + 4y^2 = 8 + 4 \quad \Leftrightarrow$
$(x-2)^2 + 4y^2 = 12 \quad \Leftrightarrow \quad \dfrac{(x-2)^2}{12} + \dfrac{y^2}{3} = 1$. This is an ellipse with $a = 2\sqrt{3}$, $b = \sqrt{3}$, and $c = \sqrt{12 - 3} = 3$. The vertices are $\left(2 \pm 2\sqrt{3}, 0\right)$ and the foci are $(2 \pm 3, 0) = (-1, 0)$ and $(5, 0)$.

**43.** The parabola has focus $(0, 1)$ and directrix $y = -1$. Therefore, $p = 1$ and so $4p = 4$. Since the focus is on the $y$-axis and the vertex is $(0, 0)$, an equation of the parabola is $x^2 = 4y$.

**44.** The ellipse has center $C\,(0, 4)$, foci $F_1\,(0, 0)$ and $F_2\,(0, 8)$, and major axis of length 10. Then $2c = 8 - 0 \quad \Leftrightarrow \quad c = 4$. Also, since the length of the major axis is 10, $2a = 10 \quad \Leftrightarrow \quad a = 5$. Therefore, $b^2 = a^2 - c^2 = 25 - 16 = 9$. Since the foci are on the $y$-axis, the vertices are on the $y$-axis, and an equation of the ellipse is $\dfrac{x^2}{9} + \dfrac{(y-4)^2}{25} = 1$.

**45.** The hyperbola has vertices $(0, \pm 2)$ and asymptotes $y = \pm\frac{1}{2}x$. Therefore, $a = 2$, and the foci are on the $y$-axis. Since the slopes of the asymptotes are $\pm\frac{1}{2} = \pm\dfrac{a}{b} \quad \Leftrightarrow \quad b = 2a = 4$, an equation of the hyperbola is $\dfrac{y^2}{4} - \dfrac{x^2}{16} = 1$.

**46.** The hyperbola has center $C\,(2, 4)$, foci $F_1\,(2, 7)$ and $F_2\,(2, 1)$, and vertices $V_1\,(2, 6)$ and $V_2\,(2, 2)$. Thus, $2a = 6 - 2 = 4$ $\Leftrightarrow \quad a = 2$. Also, $2c = 7 - 1 = 6 \quad \Leftrightarrow \quad c = 3$. So $b^2 = 9 - 4 = 5$. Since the hyperbola has center $C\,(2, 4)$, its equation is $\dfrac{(y-4)^2}{4} - \dfrac{(x-2)^2}{5} = 1$.

**47.** The ellipse has foci $F_1\,(1, 1)$ and $F_2\,(1, 3)$, and one vertex is on the $x$-axis. Thus, $2c = 3 - 1 = 2 \quad \Leftrightarrow \quad c = 1$, and so the center of the ellipse is $C\,(1, 2)$. Also, since one vertex is on the $x$-axis, $a = 2 - 0 = 2$, and thus $b^2 = 4 - 1 = 3$. So an equation of the ellipse is $\dfrac{(x-1)^2}{3} + \dfrac{(y-2)^2}{4} = 1$.

**48.** The parabola has vertex $V\,(5, 5)$ and directrix the $y$-axis. Therefore, $-p = 0 - 5 \quad \Leftrightarrow \quad p = 5 \quad \Leftrightarrow \quad 4p = 20$. Since the parabola opens to the right, its equation is $(y-5)^2 = 20\,(x - 5)$.

**49.** The ellipse has vertices $V_1\,(7, 12)$ and $V_2\,(7, -8)$ and passes through the point $P\,(1, 8)$. Thus, $2a = 12 - (-8) = 20 \quad \Leftrightarrow$ $a = 10$, and the center is $\left(7, \dfrac{-8+12}{2}\right) = (7, 2)$. Thus an equation of the ellipse has the form $\dfrac{(x-7)^2}{b^2} + \dfrac{(y-2)^2}{100} = 1$. Since the point $P\,(1, 8)$ is on the ellipse, $\dfrac{(1-7)^2}{b^2} + \dfrac{(8-2)^2}{100} = 1 \quad \Leftrightarrow \quad 3600 + 36b^2 = 100b^2 \quad \Leftrightarrow \quad 64b^2 = 3600 \quad \Leftrightarrow$ $b^2 = \frac{225}{4}$. Therefore, an equation of the ellipse is $\dfrac{(x-7)^2}{225/4} + \dfrac{(y-2)^2}{100} = 1 \quad \Leftrightarrow \quad \dfrac{4\,(x-7)^2}{225} + \dfrac{(y-2)^2}{100} = 1$.

**50.** The parabola has vertex $V(-1,0)$, horizontal axis of symmetry, and crosses the $y$-axis where $y=2$. Since the parabola has a horizontal axis of symmetry and $V(-1,0)$, its equation is of the form $y^2=4p(x+1)$. Also, since the parabola crosses the $y$-axis where $y=2$, it passes through the point $(0,2)$. Substituting this point gives $(2)^2=4p(0+1) \Leftrightarrow 4p=4$. Therefore, an equation of the parabola is $y^2=4(x+1)$.

**51.** The length of the major axis is $2a=186{,}000{,}000 \Leftrightarrow a=93{,}000{,}000$. The eccentricity is $e=c/a=0.017$, and so $c=0.017(93{,}000{,}000)=1{,}581{,}000$.

**(a)** The earth is closest to the sun when the distance is $a-c=93{,}000{,}000-1{,}581{,}000=91{,}419{,}000$.

**(b)** The earth is furthest from the sun when the distance is $a+c=93{,}000{,}000+1{,}581{,}000=94{,}581{,}000$.

**52.** We sketch the LORAN station on the $y$-axis and place the $x$-axis halfway between them as suggested in the exercise. This gives us the general form $\dfrac{y^2}{a^2}-\dfrac{x^2}{b^2}=1$. Since the ship is 80 miles closer to $A$ than to $B$ we have $2a=80 \Leftrightarrow a=40$.Since the foci are $(0,\pm150)$, we have $c=150$. Thus $b^2=c^2-a^2=150^2-40^2=20900$. So this places the ship on the hyperbola given by the equation $\dfrac{y^2}{1600}-\dfrac{x^2}{20{,}900}=1$. When $x=40$, we get $\dfrac{y^2}{1600}-\dfrac{1600}{20{,}900}=1 \Leftrightarrow$ $\dfrac{y^2}{1600}=\dfrac{225}{209} \Leftrightarrow y\approx 41.5$. (Note that $y>0$, since $A$ is on the positive $y$-axis.) Thus, the ship's position is approximately $(40,41.5)$.

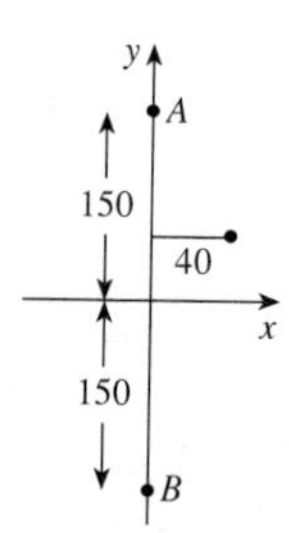


**53. (a)** The graphs of $\dfrac{x^2}{16+k^2}+\dfrac{y^2}{k^2}=1$ for $k=1$, 2, 4, and 8 are shown in the figure.

**(b)** $c^2=(16+k^2)-k^2=16 \Rightarrow c=\pm4$. Since the center is $(0,0)$, the foci of each of the ellipses are $(\pm4,0)$.

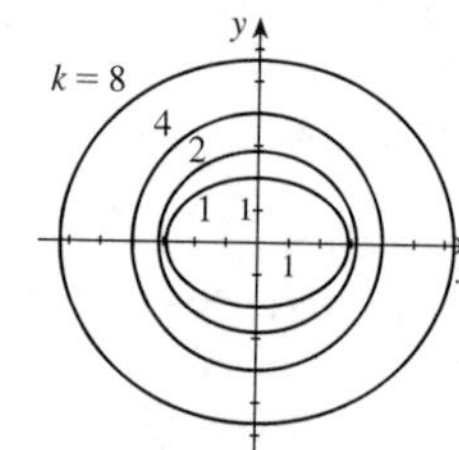


**54. (a)** The graphs of $y=kx^2$ for $k=\frac{1}{2}$, 1, 2, and 4 are shown in the figure.

**(b)** $y=kx^2 \Leftrightarrow x^2=\dfrac{1}{k}y=4\left(\dfrac{1}{4k}\right)y$. Thus the foci are $\left(0,\dfrac{1}{4k}\right)$.

**(c)** As $k$ increases, the focus gets closer to the vertex.

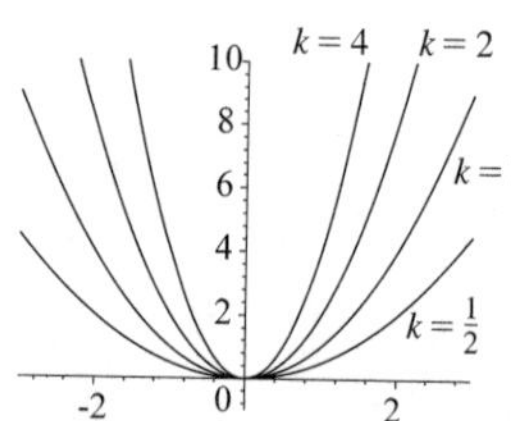


**55. (a)** $x^2+4xy+y^2=1$. Then $A=1$, $B=4$, and $C=1$, so the discriminant is $4^2-4(1)(1)=12$. Since the discriminant is positive, the equation represents a hyperbola.

**(b)** $\cot 2\phi=\dfrac{A-C}{B}=\dfrac{1-1}{4}=0 \Rightarrow 2\phi=90^\circ \Leftrightarrow \phi=45^\circ$. Therefore, $x=\frac{\sqrt{2}}{2}X-\frac{\sqrt{2}}{2}Y$ and $y=\frac{\sqrt{2}}{2}X+\frac{\sqrt{2}}{2}Y$. Substituting into the original equation gives

$\left(\frac{\sqrt{2}}{2}X-\frac{\sqrt{2}}{2}Y\right)^2+4\left(\frac{\sqrt{2}}{2}X-\frac{\sqrt{2}}{2}Y\right)\left(\frac{\sqrt{2}}{2}X+\frac{\sqrt{2}}{2}Y\right)+\left(\frac{\sqrt{2}}{2}X+\frac{\sqrt{2}}{2}Y\right)^2=1 \Leftrightarrow$

$\frac{1}{2}(X^2-2XY+Y^2)+2(X^2+XY-XY-Y^2)+\frac{1}{2}(X^2+2XY+Y^2)=1$

$\Leftrightarrow 3X^2-Y^2=1 \Leftrightarrow 3X^2-Y^2=1$. This is a hyperbola with $a=\frac{1}{\sqrt{3}}$, $b=1$, and $c=\sqrt{\frac{1}{3}+1}=\frac{2}{\sqrt{3}}$. Therefore, the hyperbola has vertices $V\left(\pm\frac{1}{\sqrt{3}},0\right)$ and foci $F\left(\pm\frac{2}{\sqrt{3}},0\right)$, in $XY$-coordinates.

**(c)**

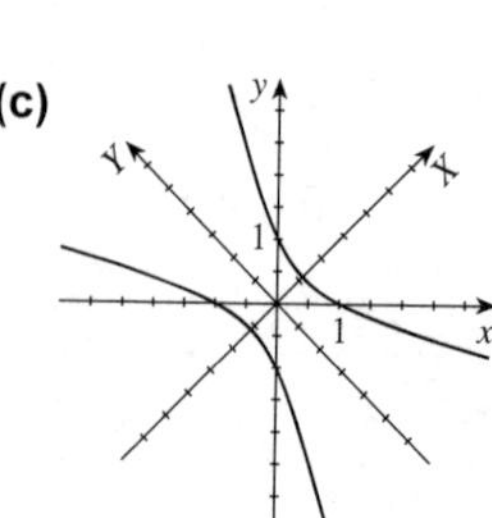

**56. (a)** $5x^2 - 6xy + 5y^2 - 8x + 8y - 8 = 0$. Then $A = 5$, $B = -6$, and $C = 5$, so the discriminant is $(-6)^2 - 4(5)(5) = 64$. Since the discriminant is negative, the equation represents an ellipse.

**(b)** $\cot 2\phi = \dfrac{A - C}{B} = 0 \quad \Rightarrow \quad 2\phi = 90^\circ \quad \Leftrightarrow \quad \phi = 45^\circ$. Therefore, $x = \frac{\sqrt{2}}{2}X - \frac{\sqrt{2}}{2}Y$ and $y = \frac{\sqrt{2}}{2}X + \frac{\sqrt{2}}{2}Y$.

Substituting into the original equation gives

$$5\left(\tfrac{\sqrt{2}}{2}X - \tfrac{\sqrt{2}}{2}Y\right)^2 - 6\left(\tfrac{\sqrt{2}}{2}X - \tfrac{\sqrt{2}}{2}Y\right)\left(\tfrac{\sqrt{2}}{2}X + \tfrac{\sqrt{2}}{2}Y\right)$$
$$+ 5\left(\tfrac{\sqrt{2}}{2}X + \tfrac{\sqrt{2}}{2}Y\right)^2 - 8\left(\tfrac{\sqrt{2}}{2}X - \tfrac{\sqrt{2}}{2}Y\right) + 8\left(\tfrac{\sqrt{2}}{2}X + \tfrac{\sqrt{2}}{2}Y\right) - 8 = 0 \quad \Leftrightarrow$$

$$\tfrac{5}{2}\left(X^2 - 2XY + Y^2\right) - 3\left(X^2 - Y^2\right) + \tfrac{5}{2}\left(X^2 + 2XY + Y^2\right) - 4\sqrt{2}X + 4\sqrt{2}Y + 4\sqrt{2}X + 4\sqrt{2}Y - 8 = 0 \quad \Leftrightarrow$$

$5X^2 - 3X^2 + 3Y^2 + 5Y^2 + 8\sqrt{2}Y - 8 = 0 \quad \Leftrightarrow$

$2X^2 + 8\left(Y^2 + \sqrt{2}Y + \frac{1}{2}\right) = 8 + 4 \quad \Leftrightarrow \quad \dfrac{X^2}{6} + \dfrac{(Y+1)^2}{3/2} = 1$. This ellipse has $a = \sqrt{6}$, $b = \sqrt{3/2}$, and $c = \sqrt{6 - \frac{3}{2}} = \frac{3\sqrt{2}}{2}$. Therefore, the vertices are $V\left(\pm\sqrt{6}, -1\right)$ and the foci are $F\left(\pm\frac{3\sqrt{2}}{2}, -1\right)$.

**(c)**

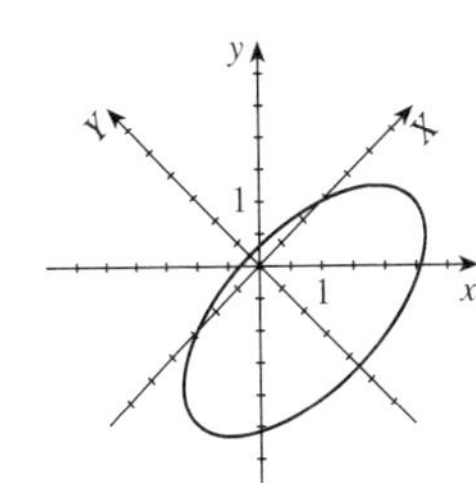


**57. (a)** $7x^2 - 6\sqrt{3}xy + 13y^2 - 4\sqrt{3}x - 4y = 0$. Then $A = 7$, $B = -6\sqrt{3}$, and $C = 13$, so the discriminant is $\left(-6\sqrt{3}\right)^2 - 4(7)(13) = -256$. Since the discriminant is negative, the equation represents an ellipse.

**(b)** $\cot 2\phi = \dfrac{A - C}{B} = \dfrac{7 - 13}{-6\sqrt{3}} = \dfrac{1}{\sqrt{3}} \quad \Rightarrow \quad 2\phi = 60^\circ \quad \Leftrightarrow \quad \phi = 30^\circ$. Therefore, $x = \frac{\sqrt{3}}{2}X - \frac{1}{2}Y$ and $y = \frac{1}{2}X + \frac{\sqrt{3}}{2}Y$. Substituting into the original equation gives

$$7\left(\tfrac{\sqrt{3}}{2}X - \tfrac{1}{2}Y\right)^2 - 6\sqrt{3}\left(\tfrac{\sqrt{3}}{2}X - \tfrac{1}{2}Y\right)\left(\tfrac{1}{2}X + \tfrac{\sqrt{3}}{2}Y\right)$$
$$+ 13\left(\tfrac{1}{2}X + \tfrac{\sqrt{3}}{2}Y\right)^2 - 4\sqrt{3}\left(\tfrac{\sqrt{3}}{2}X - \tfrac{1}{2}Y\right) - 4\left(\tfrac{1}{2}X + \tfrac{\sqrt{3}}{2}Y\right) = 0 \quad \Leftrightarrow$$

$$\tfrac{7}{4}\left(3X^2 - 2\sqrt{3}XY + Y^2\right) - \tfrac{3\sqrt{3}}{2}\left(\sqrt{3}X^2 + 3XY - XY - \sqrt{3}Y^2\right)$$
$$+ \tfrac{13}{4}\left(X^2 + 2\sqrt{3}XY + 3Y^2\right) - 6X + 2\sqrt{3}Y - 2X - 2\sqrt{3}Y = 0 \quad \Leftrightarrow$$

$X^2\left(\frac{21}{4} - \frac{9}{2} + \frac{13}{4}\right) - 8X + Y^2\left(\frac{7}{4} + \frac{9}{2} + \frac{39}{4}\right) = 0 \quad \Leftrightarrow$

$4X^2 - 8X + 16Y^2 = 0 \quad \Leftrightarrow \quad 4\left(X^2 - 2X + 1\right) + 16Y^2 = 4 \quad \Leftrightarrow$

$(X-1)^2 + 4Y^2 = 1$. This ellipse has $a = 1$, $b = \frac{1}{2}$, and $c = \sqrt{1 - \frac{1}{4}} = \frac{1}{2}\sqrt{3}$.

Therefore, the vertices are $V(1 \pm 1, 0) = V_1(0,0)$ and $V_2(2,0)$ and the foci are $F\left(1 \pm \frac{1}{2}\sqrt{3}, 0\right)$.

**(c)**

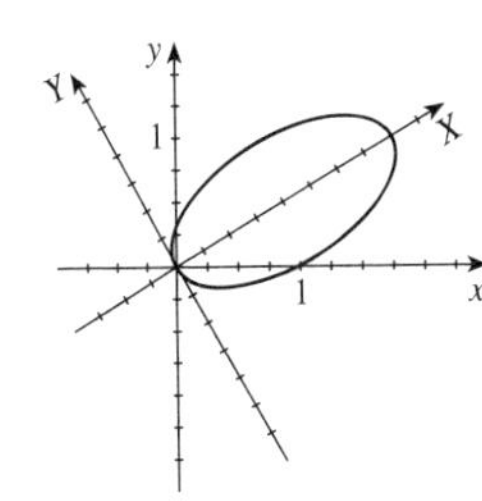

**58. (a)** $9x^2 + 24xy + 16y^2 = 25$. Then $A = 9$, $B = 24$, and $C = 16$, so the discriminant is $24^2 - 4\,(9)\,(16) = 0$. Since the discriminant is zero, the equation represents a parabola.

**(b)** $\cot 2\phi = \dfrac{A-C}{B} = \dfrac{9-16}{24} = -\dfrac{7}{24} \quad\Rightarrow\quad \cos 2\phi = -\dfrac{7}{25}$, so $\cos\phi = \sqrt{\frac{1-(7/25)}{2}} = \frac{3}{5}$, $\sin\phi = \sqrt{\frac{1+(7/25)}{2}} = \frac{4}{5}$

$\Leftrightarrow\quad \phi \approx 53^\circ$, and thus $x = \frac{3}{5}X - \frac{4}{5}Y$ and $y = \frac{4}{5}X + \frac{3}{5}Y$. Substituting,

$9x^2 + 24xy + 16y^2 = 25 \quad\Leftrightarrow$

$9\left(\frac{3}{5}X - \frac{4}{5}Y\right)^2 + 24\left(\frac{3}{5}X - \frac{4}{5}Y\right)\left(\frac{4}{5}X + \frac{3}{5}Y\right) + 16\left(\frac{4}{5}X + \frac{3}{5}Y\right)^2 = 25 \quad\Leftrightarrow$

$25X^2 = 25 \quad\Leftrightarrow\quad X^2 = 1 \quad\Leftrightarrow\quad X = \pm 1$. Thus the graph is a degenerate conic that consists of two lines. Converting back to $xy$-coordinates, we see that $X = \frac{3}{5}x + \frac{4}{5}y$, so $\frac{3}{5}x + \frac{4}{5}y = \pm 1 \quad\Leftrightarrow\quad 3x + 4y = \pm 5$.

**(c)**

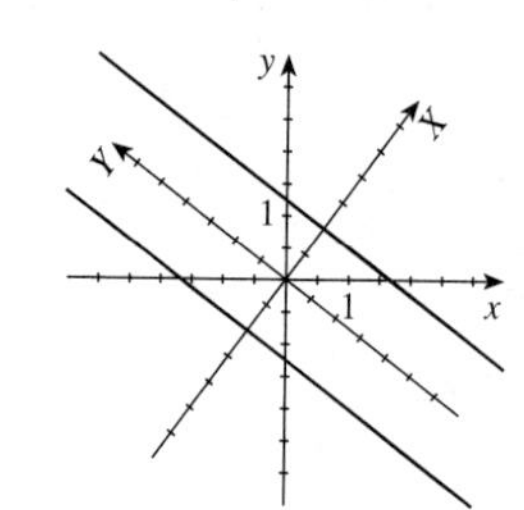


**59.** $5x^2 + 3y^2 = 60 \quad\Leftrightarrow\quad 3y^2 = 60 - 5x^2 \quad\Leftrightarrow\quad y^2 = 20 - \frac{5}{3}x^2$. This conic is an ellipse.

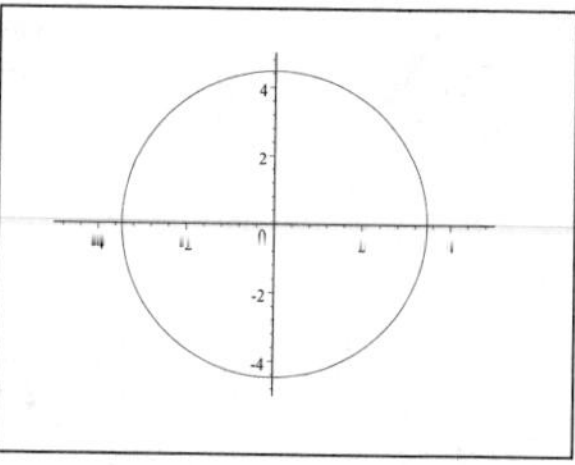

**60.** $9x^2 - 12y^2 + 36 = 0 \quad\Leftrightarrow\quad 12y^2 = 9x^2 + 36 \quad\Leftrightarrow\quad y^2 = \frac{3}{4}x^2 + 3 \quad\Leftrightarrow$ $y = \pm\sqrt{\frac{3}{4}x^2 + 3}$. This conic is a hyperbola.

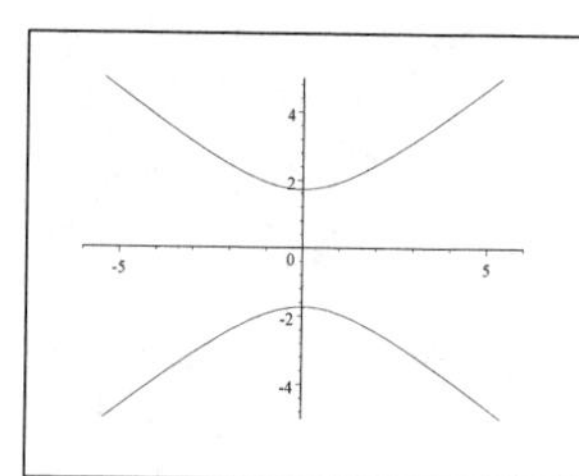

**61.** $6x + y^2 - 12y = 30 \quad\Leftrightarrow\quad y^2 - 12y = 30 - 6x \quad\Leftrightarrow$
$y^2 - 12y + 36 = 66 - 6x \quad\Leftrightarrow\quad (y-6)^2 = 66 - 6x \quad\Leftrightarrow$
$y - 6 = \pm\sqrt{66 - 6x} \quad\Leftrightarrow\quad y = 6 \pm \sqrt{66 - 6x}$. This conic is a parabola.

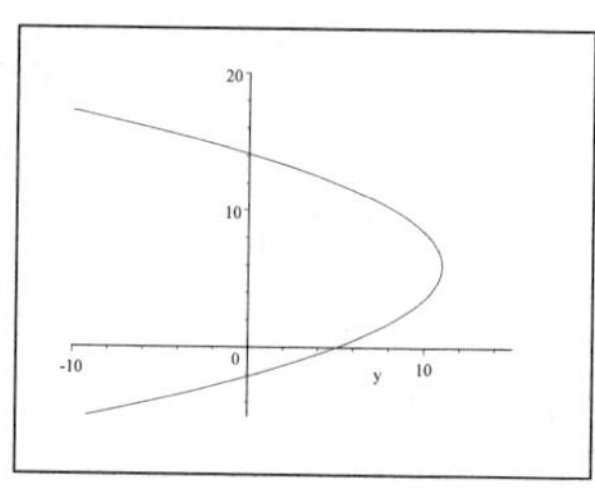

**62.** $52x^2 - 72xy + 73y^2 = 100 \Leftrightarrow 73y^2 - 72xy + 52x^2 - 100 = 0$. Using the quadratic formula,

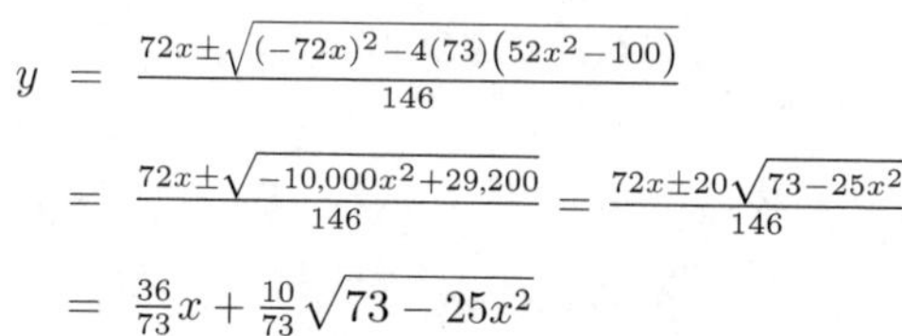

$$
\begin{aligned}
y &= \frac{72x \pm \sqrt{(-72x)^2 - 4(73)(52x^2 - 100)}}{146} \\
&= \frac{72x \pm \sqrt{-10{,}000x^2 + 29{,}200}}{146} = \frac{72x \pm 20\sqrt{73 - 25x^2}}{146} \\
&= \frac{36}{73}x + \frac{10}{73}\sqrt{73 - 25x^2}
\end{aligned}
$$

This conic is an ellipse.

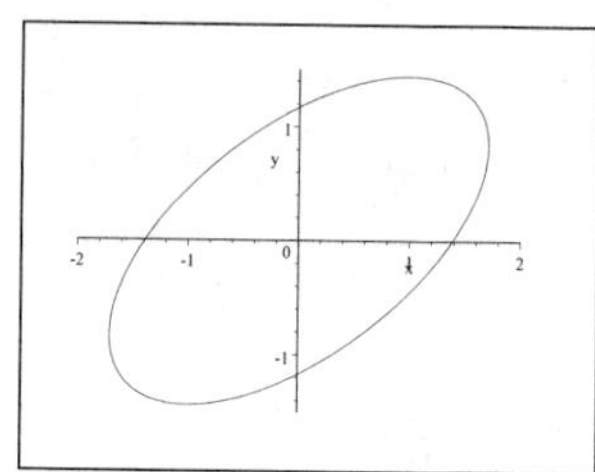

**63. (a)** $r = \dfrac{1}{1 - \cos\theta} \quad \Rightarrow \quad e = 1$. Therefore, this is a parabola.

**(b)**

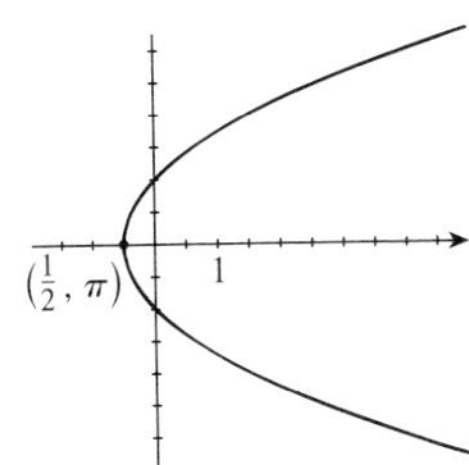


**64. (a)** $r = \dfrac{2}{3 + 2\sin\theta} \quad \Rightarrow \quad e = \frac{2}{3}$. Therefore, this is an ellipse.

**(b)**

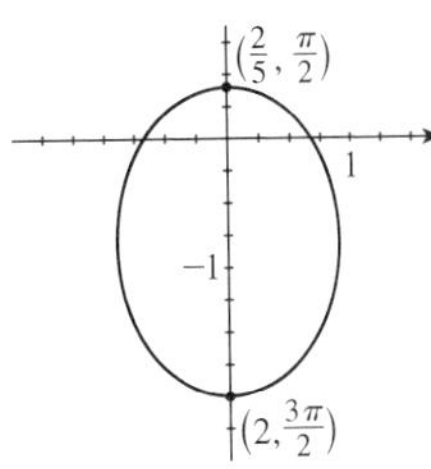


**65. (a)** $r = \dfrac{4}{1 + 2\sin\theta} \quad \Rightarrow \quad e = 2$. Therefore, this is a hyperbola.

**(b)**

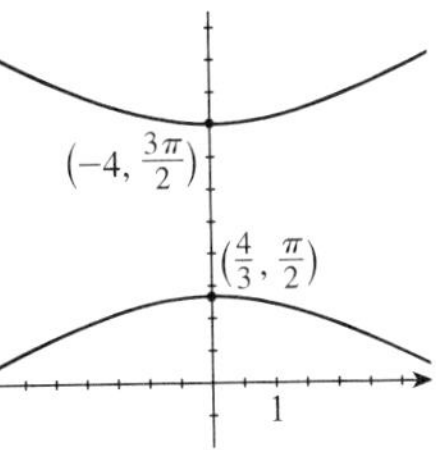


**66. (a)** $r = \dfrac{12}{1 - 4\cos\theta} \quad \Rightarrow \quad e = 4$. Therefore, this is a hyperbola.

**(b)**

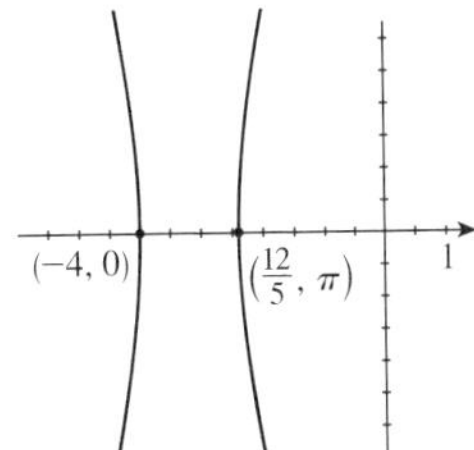


**67. (a)**

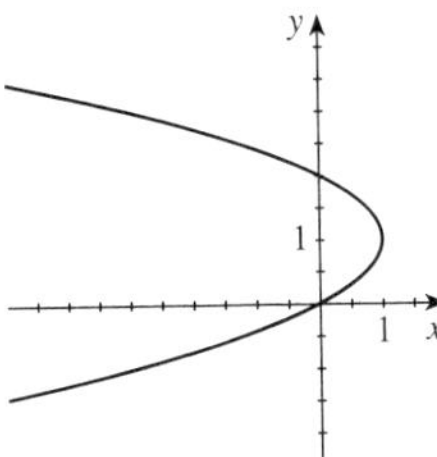


**(b)** $x = 1 - t^2, y = 1 + t \quad \Leftrightarrow \quad t = y - 1$. Substituting for $t$ gives $x = 1 - (y - 1)^2 \quad \Leftrightarrow$ $(y - 1)^2 = 1 - x$ which is the rectangular coordinate equation.

**68. (a)**

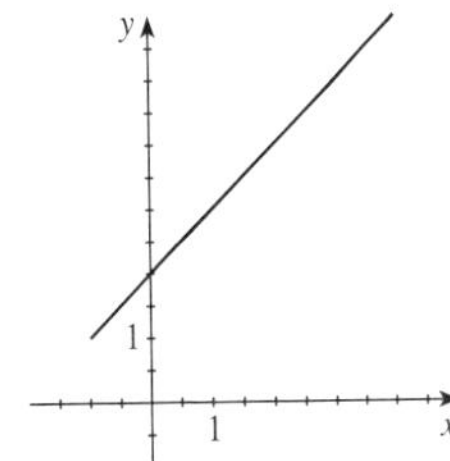


**(b)** $x = t^2 - 1, y = t^2 + 1 \quad \Leftrightarrow \quad t^2 = y - 1$. Substituting for $t^2$ gives $x = (y - 1) - 1 \quad \Leftrightarrow$ $y = x + 2$ where $x \geq -1$ and $y \geq 1$.

**69. (a)**

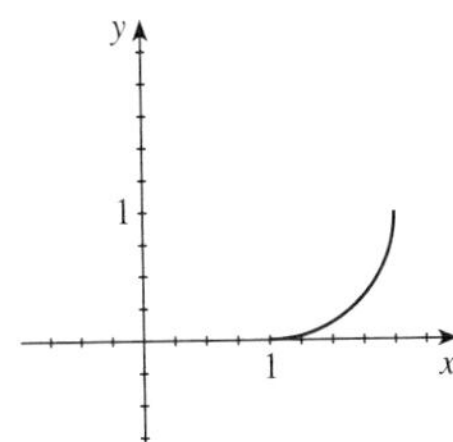


**(b)** $x = 1 + \cos t \quad \Leftrightarrow \quad \cos t = x - 1$, and $y = 1 - \sin t \quad \Leftrightarrow \quad \sin t = 1 - y$. Since $\cos^2 t + \sin^2 t = 1$, it follows that $(x - 1)^2 + (1 - y)^2 = 1 \quad \Leftrightarrow$ $(x - 1)^2 + (y - 1)^2 = 1$. Since $t$ is restricted by $0 \leq t \leq \frac{\pi}{2}$, $1 + \cos 0 \leq x \leq 1 + \cos\frac{\pi}{2} \quad \Leftrightarrow$ $1 \leq x \leq 2$, and similarly, $0 \leq y \leq 1$. (This is the lower right quarter of the circle.)

**70. (a)**

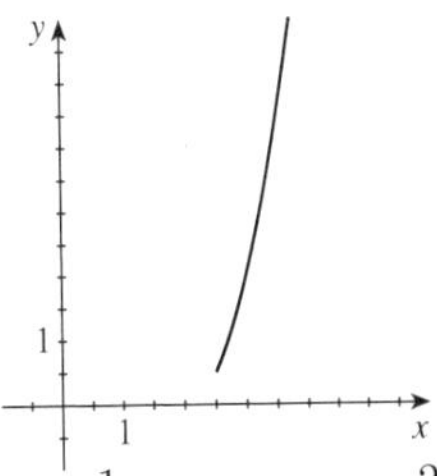


**(b)** $x = \dfrac{1}{t} + 2 \quad \Leftrightarrow \quad \dfrac{1}{t} = x - 2$, $y = \dfrac{2}{t^2}$. Substituting for $\dfrac{1}{t}$ gives $y = 2(x - 2)^2$. Since $t$ is restricted by $0 < t \leq 2$, we have $\dfrac{1}{t} \geq \dfrac{1}{2}$, so $x \geq \frac{5}{2}$ and $y \geq \frac{1}{2}$. The rectangular coordinate equation is $y = 2(x - 2)^2$, $x \geq \frac{5}{2}$.

**71.** $x = \cos 2t$, $y = \sin 3t$

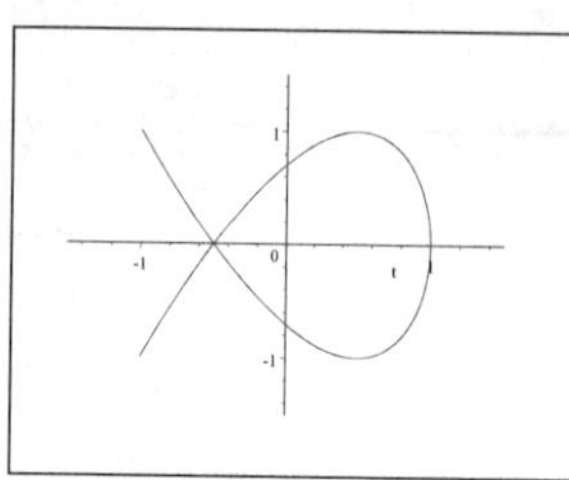

**72.** $x = \sin(t + \cos 2t)$, $y = \cos(t + \sin 3t)$

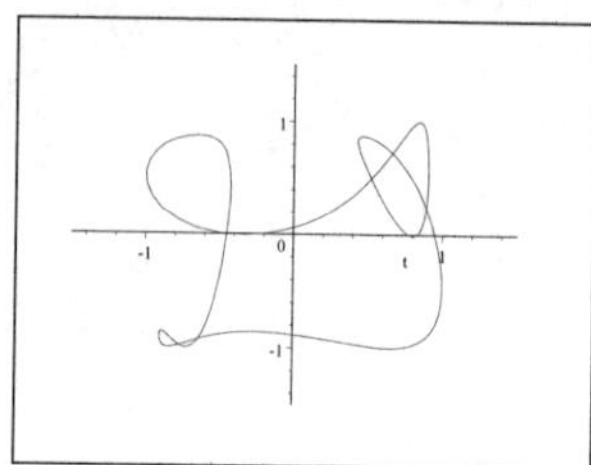

**73.** The coordinates of $Q$ are $x = \cos\theta$ and $y = \sin\theta$. The coordinates of $R$ are $x = 1$ and $y = \tan\theta$. Hence, the midpoint $P$ is $\left(\dfrac{1+\cos\theta}{2}, \dfrac{\sin\theta+\tan\theta}{2}\right)$, so parametric equations for the curve are $x = \dfrac{1+\cos\theta}{2}$ and $y = \dfrac{\sin\theta+\tan\theta}{2}$.

# Chapter 11 Test

**1.** $x^2 = -12y$. This is a parabola with $4p = -12 \Leftrightarrow p = -3$. The focus is $(0, -3)$ and the directrix is $y = 3$.

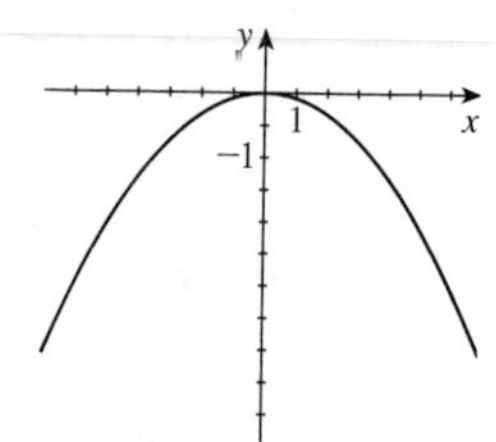


**2.** $\dfrac{x^2}{16} + \dfrac{y^2}{4} = 1$. This is an ellipse with $a = 4$, $b = 2$, and $c = \sqrt{16-4} = 2\sqrt{3}$. The vertices are $(\pm 4, 0)$, the foci are $(\pm 2\sqrt{3}, 0)$, the length of the major axis is $2a = 8$, and the length of the minor axis is $2b = 4$.

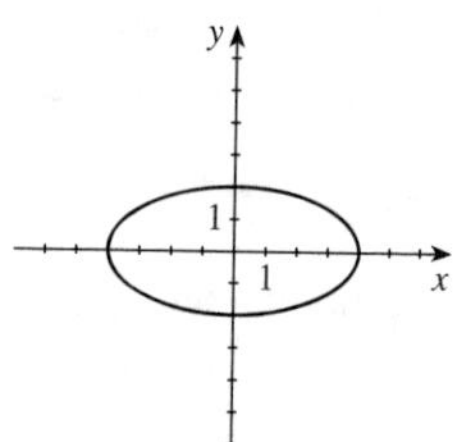


**3.** $\dfrac{y^2}{9} - \dfrac{x^2}{16} = 1$. This is a hyperbola with $a = 3$, $b = 4$, and $c = \sqrt{9+16} = 5$. The vertices are $(0, \pm 3)$, the foci are $(0, \pm 5)$, and the asymptotes are $y = \pm\frac{3}{4}x$.

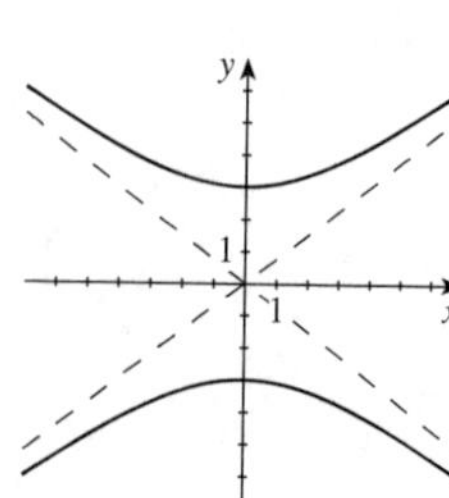


**4.** This is a parabola that opens to the left with its vertex at $(0, 0)$. So its equation is of the form $y^2 = 4px$ with $p < 0$. Substituting the point $(-4, 2)$, we have $2^2 = 4p(-4) \Leftrightarrow 4 = -16p \Leftrightarrow p = -\frac{1}{4}$. So an equation is $y^2 = 4\left(-\frac{1}{4}\right)x \Leftrightarrow y^2 = -x$.

**5.** This is an ellipse tangent to the $x$-axis at $(0, 0)$ and with one vertex at the point $(4, 3)$. The center is $(0, 3)$, and $a = 4$ and $b = 3$. Thus the equation is $\dfrac{x^2}{16} + \dfrac{(y-3)^2}{9} = 1$.

**6.** This a hyperbola with a horizontal transverse axis, vertices at $(1,0)$ and $(3,0)$, and foci at $(0,0)$ and $(4,0)$. Thus the center is $(2,0)$, and $a = 3 - 2 = 1$ and $c = 4 - 2 = 2$. Thus $b^2 = 2^2 - 1^2 = 3$. So an equation is $\dfrac{(x-2)^2}{1^2} - \dfrac{y^2}{3} = 1 \quad \Leftrightarrow$ $(x-2)^2 - \dfrac{y^2}{3} = 1$.

**7.** $16x^2 + 36y^2 - 96x + 36y + 9 = 0 \quad \Leftrightarrow \quad 16\left(x^2 - 6x\right) + 36\left(y^2 + y\right) = -9$ $\Leftrightarrow \quad 16\left(x^2 - 6x + 9\right) + 36\left(y^2 + y + \frac{1}{4}\right) = -9 + 144 + 9 \quad \Leftrightarrow$ $16(x-3)^2 + 36\left(y + \frac{1}{2}\right)^2 = 144 \quad \Leftrightarrow \quad \dfrac{(x-3)^2}{9} + \dfrac{\left(y+\frac{1}{2}\right)^2}{4} = 1$. This is an ellipse with $a = 3$, $b = 2$, and $c = \sqrt{9-4} = \sqrt{5}$. The center is $\left(3, -\frac{1}{2}\right)$, the vertices are $\left(3 \pm 3, -\frac{1}{2}\right) = \left(0, -\frac{1}{2}\right)$ and $\left(6, -\frac{1}{2}\right)$, and the foci are $(h \pm c, k) = \left(3 \pm \sqrt{5}, -\frac{1}{2}\right) = \left(3 + \sqrt{5}, -\frac{1}{2}\right)$ and $\left(3 - \sqrt{5}, -\frac{1}{2}\right)$.

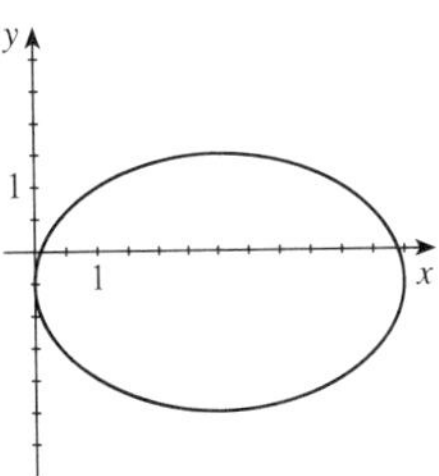


**8.** $9x^2 - 8y^2 + 36x + 64y = 164 \quad \Leftrightarrow \quad 9\left(x^2 + 4x\right) - 8\left(y^2 - 8y\right) = 164 \quad \Leftrightarrow$ $9\left(x^2 + 4x + 4\right) - 8\left(y^2 - 8y + 16\right) = 164 + 36 - 128 \quad \Leftrightarrow$ $9(x+2)^2 - 8(y-4)^2 = 72 \quad \Leftrightarrow \quad \dfrac{(x+2)^2}{8} - \dfrac{(y-4)^2}{9} = 1$. This conic is a hyperbola.

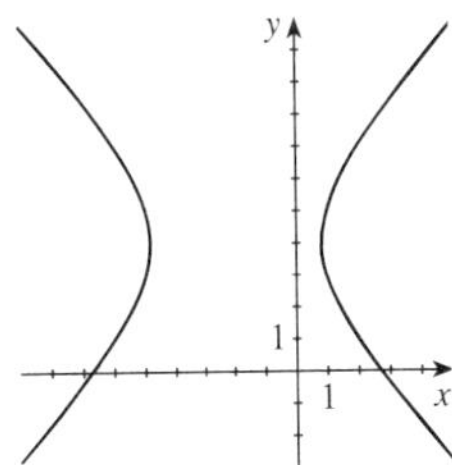


**9.** $2x + y^2 + 8y + 8 = 0 \quad \Leftrightarrow \quad y^2 + 8y + 16 = -2x - 8 + 16 \quad \Leftrightarrow$ $(y+4)^2 = -2(x-4)$. This is a parabola with $4p = -2 \quad \Leftrightarrow \quad p = -\frac{1}{2}$. The vertex is $(4,-4)$ and the focus is $\left(4 - \frac{1}{2}, -4\right) = \left(\frac{7}{2}, -4\right)$.

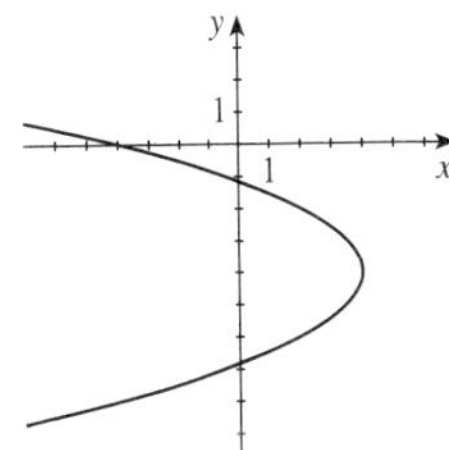


**10.** The hyperbola has foci $(0, \pm 5)$ and asymptotes $y = \pm\frac{3}{4}x$. Since the foci are $(0, \pm 5)$, $c = 5$, the foci are on the $y$-axis, and the center is $(0,0)$. Also, since $y = \pm\frac{3}{4}x = \pm\dfrac{a}{b}x$, it follows that $\dfrac{a}{b} = \dfrac{3}{4} \quad \Leftrightarrow \quad a = \frac{3}{4}b$. Then $c^2 = 5^2 = 25 = a^2 + b^2 = \left(\frac{3}{4}b\right)^2 + b^2 = \frac{25}{16}b^2 \quad \Leftrightarrow \quad b^2 = 16$, and by substitution, $a = \frac{3}{4}(4) = 3$. Therefore, an equation of the hyperbola is $\dfrac{y^2}{9} - \dfrac{x^2}{16} = 1$.

**11.** The parabola has focus $(2,4)$ and directrix the $x$-axis $(y = 0)$. Therefore, $2p = 4 - 0 = 4 \quad \Leftrightarrow \quad p = 2 \quad \Leftrightarrow$ $4p = 8$, and the vertex is $(2, 4 - p) = (2,2)$. Hence, an equation of the parabola is $(x-2)^2 = 8(y-2) \quad \Leftrightarrow$ $x^2 - 4x + 4 = 8y - 16 \quad \Leftrightarrow \quad x^2 - 4x - 8y + 20 = 0$.

**12.** We place the vertex of the parabola at the origin, so the parabola contains the points $(3, \pm 3)$, and the equation is of the form $y^2 = 4px$. Substituting the point $(3,3)$, we get $3^2 = 4p(3) \quad \Leftrightarrow \quad 9 = 12p \quad \Leftrightarrow \quad p = \frac{3}{4}$. So the focus is $\left(\frac{3}{4}, 0\right)$, and we should place the light bulb $\frac{3}{4}$ inch from the vertex.

**13. (a)** $5x^2 + 4xy + 2y^2 = 18$. Then $A = 5$, $B = 4$, and $C = 2$, so the discriminant is $(4)^2 - 4\,(5)\,(2) = -24$. Since the discriminant is negative, the equation represents an ellipse.

**(b)** $\cot 2\phi = \dfrac{A-C}{B} = \dfrac{5-2}{4} = \dfrac{3}{4}$. Thus, $\cos 2\phi = \frac{3}{5}$ and so $\cos\phi = \sqrt{\frac{1+(3/5)}{2}} = \frac{2\sqrt{5}}{5}$, $\sin\phi = \sqrt{\frac{1-(3/5)}{2}} = \frac{\sqrt{5}}{5}$. It follows that $x = \frac{2\sqrt{5}}{5}X - \frac{\sqrt{5}}{5}Y$ and $y = \frac{\sqrt{5}}{5}X + \frac{2\sqrt{5}}{5}Y$. By substitution, $5\left(\frac{2\sqrt{5}}{5}X - \frac{\sqrt{5}}{5}Y\right)^2 + 4\left(\frac{2\sqrt{5}}{5}X - \frac{\sqrt{5}}{5}Y\right)\left(\frac{\sqrt{5}}{5}X + \frac{2\sqrt{5}}{5}Y\right) + 2\left(\frac{\sqrt{5}}{5}X + \frac{2\sqrt{5}}{5}Y\right)^2 = 18$ $\Leftrightarrow$ $4X^2 - 4XY + Y^2 + \frac{4}{5}\left(2X^2 + 4XY - XY - 2Y^2\right) + \frac{2}{5}\left(X^2 + 4XY + 4Y^2\right) = 18$ $\Leftrightarrow$ $X^2\left(4 + \frac{8}{5} + \frac{2}{5}\right) + XY\left(-4 + \frac{12}{5} + \frac{8}{5}\right) + Y^2\left(1 - \frac{8}{5} + \frac{4}{5}\right) = 18$ $\Leftrightarrow$ $6X^2 + Y^2 = 18$ $\Leftrightarrow$ $\dfrac{X^2}{3} + \dfrac{Y^2}{18} = 1$. This is an ellipse with $a = 3\sqrt{2}$ and $b = \sqrt{3}$.

**(c)**

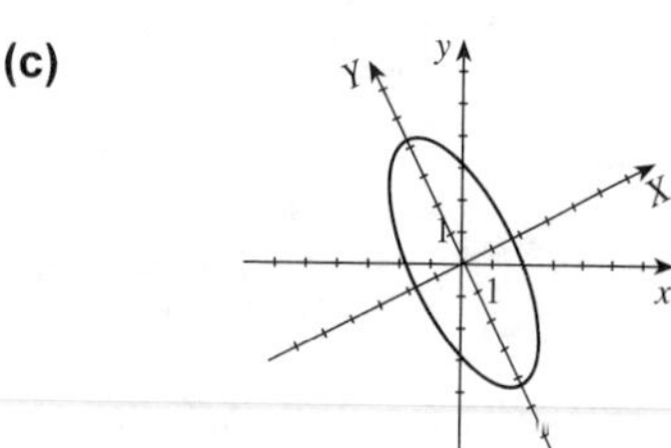


Since $\cos 2\phi = \frac{3}{5}$ we have $2\phi = \cos^{-1}\frac{3}{5} \approx 53.13°$, so $\phi \approx 27°$.

**(d)** In $XY$-coordinates, the vertices are $V\left(0, \pm 3\sqrt{2}\right)$. Therefore, in $xy$-coordinates, the vertices are $x = -\frac{3\sqrt{2}}{\sqrt{5}}$ and $y = \frac{6\sqrt{2}}{\sqrt{5}}$ $\Rightarrow$ $V_1\left(-\frac{3\sqrt{2}}{\sqrt{5}}, \frac{6\sqrt{2}}{\sqrt{5}}\right)$, and $x = \frac{3\sqrt{2}}{\sqrt{5}}$ and $y = -\frac{6\sqrt{2}}{\sqrt{5}}$ $\Rightarrow$ $V_2\left(\frac{3\sqrt{2}}{\sqrt{5}}, \frac{-6\sqrt{2}}{\sqrt{5}}\right)$.

**14. (a)** Since the focus of this conic is the origin and the directrix is $x = 2$, the equation has the form $r = \dfrac{ed}{1 + e\cos\theta}$. Subsituting $e = \frac{1}{2}$ and $d = 2$ we get $r = \dfrac{1}{1 + \frac{1}{2}\cos\theta}$ $\Leftrightarrow$ $r = \dfrac{2}{2 + \cos\theta}$.

**(b)** $r = \dfrac{3}{2 - \sin\theta}$ $\Leftrightarrow$ $r = \dfrac{\frac{3}{2}}{1 - \frac{1}{2}\sin\theta}$. So $e = \frac{1}{2}$ and the conic is an ellipse.

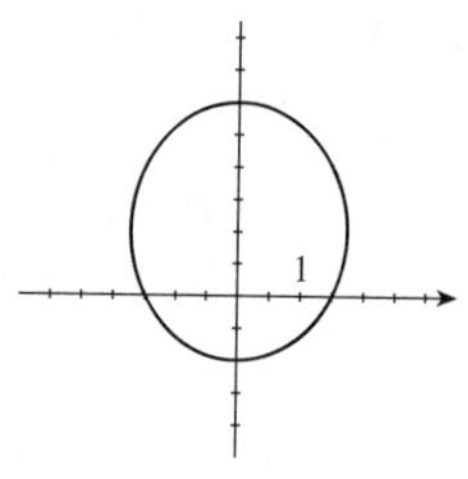


**15. (a)** $x = 3\sin\theta + 3$, $y = 2\cos\theta$, $0 \le \theta \le \pi$. From the work of part (b), we see that this is the half-ellipse shown.

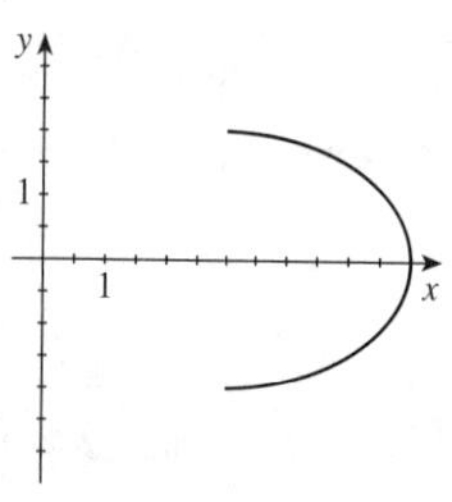


**(b)** $x = 3\sin\theta + 3 \Leftrightarrow x - 3 = 3\sin\theta \Leftrightarrow \dfrac{x-3}{3} = \sin\theta$. Squaring both sides gives $\dfrac{(x-3)^2}{9} = \sin^2\theta$. Similarly, $y = 2\cos\theta \Leftrightarrow \dfrac{y}{2} = \cos\theta$, and squaring both sides gives $\dfrac{y^2}{4} = \cos^2\theta$. Since $\sin^2\theta + \cos^2\theta = 1$, it follows that $\dfrac{(x-3)^2}{9} + \dfrac{y^2}{4} = 1$. Since $0 \le \theta \le \pi$, $\sin\theta \ge 0$, so $3\sin\theta \ge 0 \Rightarrow 3\sin\theta + 3 \ge 3$, and so $x \ge 3$. Thus the curve consists of only the right half of the ellipse.

# Focus on Modeling: The Path of a Projectile

**1.** From $x = (v_0 \cos\theta)\, t$, we get $t = \dfrac{x}{v_0 \cos\theta}$. Substituting this value for $t$ into the equation for $y$, we get

$$y = (v_0 \sin\theta)\, t - \tfrac{1}{2} g t^2 \quad \Leftrightarrow \quad y = (v_0 \sin\theta) \left( \frac{x}{v_0 \cos\theta} \right) - \tfrac{1}{2} g \left( \frac{x}{v_0 \cos\theta} \right)^2 \quad \Leftrightarrow \quad y = (\tan\theta)\, x - \frac{g}{2v_0^2 \cos^2\theta} x^2.$$

This shows that $y$ is a quadratic function of $x$, so its graph is a parabola as long as $\theta \neq 90^\circ$. When $\theta = 90^\circ$, the path of the projectile is a straight line up (then down).

**2. (a)** Applying the given values, we get $x = (v_0 \cos\theta)\, t = 15t$ and $y = 4 + (v_0 \sin\theta)\, t - \frac{1}{2} g t^2 = 4 + 25.98t - 16t^2$ as the parametric equations for the path of the baseball.

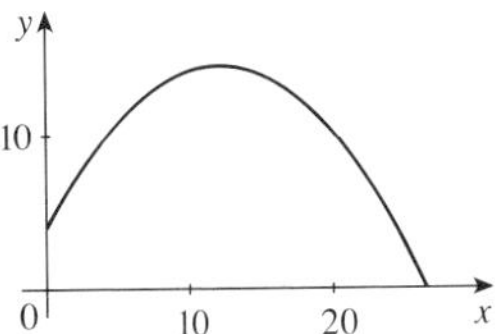


**(b)** The baseball will hit the ground when $y = 0 = 4 + 25.98t - 16t^2$. Using the quadratic formula, $t = \frac{-25.98 \pm \sqrt{(25.98)^2 + 4(-16)(4)}}{-32} \approx 1.77$ seconds (since $t$ must be positive). So the baseball travels $x = 15 \cdot 1.77 = 26.5$ ft before hitting the ground after 1.77 s.

**3. (a)** We use the equation $t = \frac{2v_0 \sin\theta}{g}$. Substituting $g \approx 32$ ft/s$^2$, $\theta = 5^\circ$, and $v_0 = 1000$, we get $t = \frac{2 \cdot 1000 \cdot \sin 5^\circ}{32} \approx 5.447$ seconds.

**(b)** Substituting the given values into $y = (v_0 \sin\theta)\, t - \frac{1}{2} g t^2$, we get $y = 87.2t - 16t^2$. The maximum value of $y$ is attained at the vertex of the parabola; thus $y = 87.2t - 16t^2 = -16\left(t^2 - 5.45t\right) \quad \Leftrightarrow$ $y = -16\left[t^2 - 2\,(2.725)\, t + 7.425625\right] + 118.7$. Thus the greatest height is 118.7 ft.

**(c)** The rocket hits the ground after 5.447 s, so substituting this into the expression for the horizontal distance gives $x = (1000 \cos 5^\circ)\, 5.447 = 5426$ ft.

**(d)**

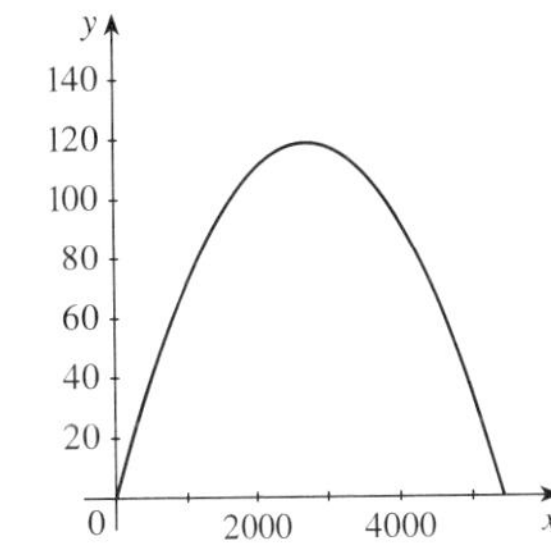


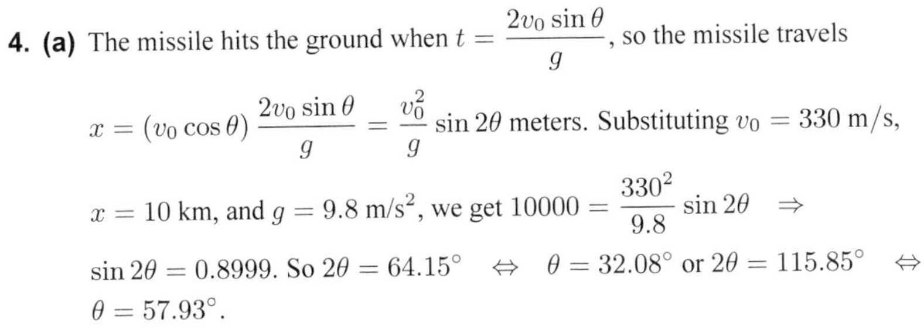

**4. (a)** The missile hits the ground when $t = \dfrac{2v_0 \sin\theta}{g}$, so the missile travels

$$x = (v_0 \cos\theta) \frac{2v_0 \sin\theta}{g} = \frac{v_0^2}{g} \sin 2\theta \text{ meters. Substituting } v_0 = 330 \text{ m/s},$$

$x = 10$ km, and $g = 9.8$ m/s$^2$, we get $10000 = \dfrac{330^2}{9.8} \sin 2\theta \quad \Rightarrow$

$\sin 2\theta = 0.8999$. So $2\theta = 64.15^\circ \quad \Leftrightarrow \quad \theta = 32.08^\circ$ or $2\theta = 115.85^\circ \quad \Leftrightarrow$ $\theta = 57.93^\circ$.

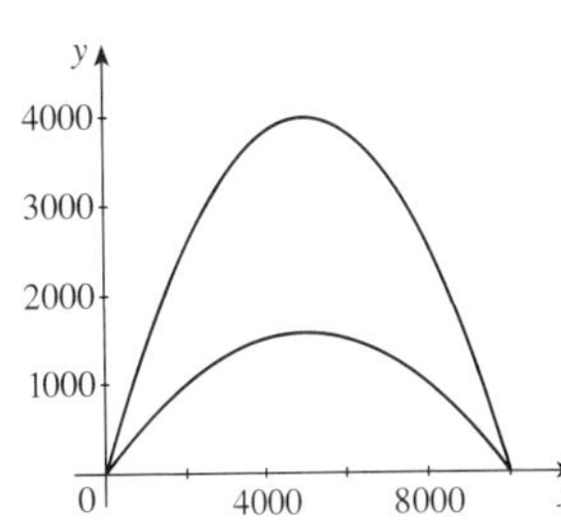


**(b)** The missile fired at $32.08^\circ$ will hit the target in 35.7 seconds, while the missile fired at $57.93^\circ$ reaches the target in 57.1 seconds.

**5.** We use the equation of the parabola from Exercise 1 and find its vertex:

$$y = (\tan\theta)\,x - \frac{g}{2v_0^2\cos^2\theta}x^2 \Leftrightarrow y = -\frac{g}{2v_0^2\cos^2\theta}\left[x^2 - \frac{2v_0^2\sin\theta\cos\theta x}{g}\right] \quad \Leftrightarrow$$

$$y = -\frac{g}{2v_0^2\cos^2\theta}\left[x^2 - \frac{2v_0^2\sin\theta\cos\theta x}{g} + \left(\frac{v_0^2\sin\theta\cos\theta}{g}\right)^2\right] + \frac{g}{2v_0^2\cos^2\theta}\cdot\left(\frac{v_0^2\sin\theta\cos\theta}{g}\right)^2 \quad \Leftrightarrow$$

$y = -\dfrac{g}{2v_0^2\cos^2\theta}\left[x - \dfrac{v_0^2\sin\theta\cos\theta}{g}\right]^2 + \dfrac{v_0^2\sin^2\theta}{2g}$. Thus the vertex is at $\left(\dfrac{v_0^2\sin\theta\cos\theta}{g}, \dfrac{v_0^2\sin^2\theta}{2g}\right)$, so the maximum height is $\dfrac{v_0^2\sin^2\theta}{2g}$.

**6.** Since the horizontal component of the projectile's velocity has been reduced by $w$, the parametric equations become $x = (v_0\cos\theta - w)\,t$, $y = (v_0\sin\theta)\,t - \frac{1}{2}gt^2$.

**7.** In Exercise 6 we derived the equations $x = (v_0\cos\theta - w)\,t$, $y = (v_0\sin\theta)\,t - \frac{1}{2}gt^2$. We plot the graphs for the given values of $v_0$, $w$, and $\theta$ in the figure to the right. The projectile will be blown backwards if the horizontal component of its velocity is less than the speed of the wind, that is, $32\cos\theta < 24 \quad \Leftrightarrow \quad \cos\theta < \frac{3}{4} \quad \Rightarrow \quad \theta > 41.4°$.

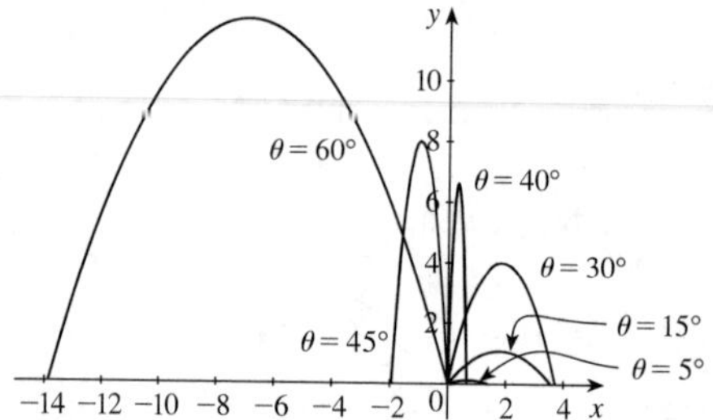


The optimal firing angle appears to be between 15° and 30°. We graph the trajectory for $\theta = 20°$, $\theta = 23°$, and $\theta = 25°$. The solution appears to be close to 23°.

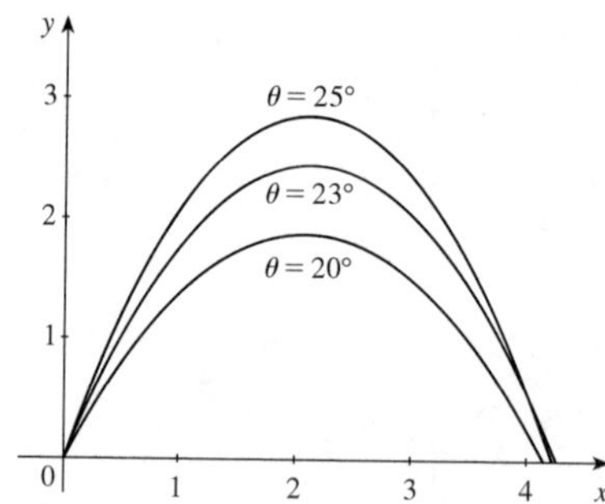


**8. (a)** Answers will vary.

**(b)** Both projectiles land at the same spot. The projectile fired with $\theta = 30°$ flies lower and lands first.

**(c)** When $\theta = 90°$, we have $y = v_0 t - 4.9t^2 = -4.9\left(t^2 - \dfrac{v_0}{4.9}t\right) = -4.9\left[t^2 - \dfrac{v_0}{4.9}t + \left(\dfrac{v_0}{9.8}\right)^2\right] + \dfrac{v_0^2}{19.6}$. Thus, when $v_0$ is doubled the maximum height of the ball increases by a factor of 4.

# 12 Sequences and Series

## 12.1 Sequences and Summation Notation

**1.** $a_n = n + 1$. Then $a_1 = 1 + 1 = 2$, $a_2 = 2 + 1 = 3$, $a_3 = 3 + 1 = 4$, $a_4 = 4 + 1 = 5$, and $a_{100} = 100 + 1 = 101$.

**2.** $a_n = 2n + 3$. Then $a_1 = 2(1) + 3 = 5$, $a_2 = 2(2) + 3 = 7$, $a_3 = 2(3) + 3 = 9$, $a_4 = 2(4) + 3 = 11$, and $a_{100} = 2(100) + 3 = 203$.

**3.** $a_n = \dfrac{1}{n+1}$. Then $a_1 = \dfrac{1}{1+1} = \dfrac{1}{2}$, $a_2 = \dfrac{1}{2+1} = \dfrac{1}{3}$, $a_3 = \dfrac{1}{3+1} = \dfrac{1}{4}$, $a_4 = \dfrac{1}{4+1} = \dfrac{1}{5}$, and $a_{100} = \dfrac{1}{100+1} = \dfrac{1}{101}$.

**4.** $a_n = n^2 + 1$. Then $a_1 = (1)^2 + 1 = 2$, $a_2 = (2)^2 + 1 = 5$, $a_3 = (3)^2 + 1 = 10$, $a_4 = (4)^2 + 1 = 17$, and $a_{100} = (100)^2 + 1 = 10{,}001$.

**5.** $a_n = \dfrac{(-1)^n}{n^2}$. Then $a_1 = \dfrac{(-1)^1}{1^2} = -1$, $a_2 = \dfrac{(-1)^2}{2^2} = \dfrac{1}{4}$, $a_3 = \dfrac{(-1)^3}{3^2} = -\dfrac{1}{9}$, $a_4 = \dfrac{(-1)^4}{4^2} = \dfrac{1}{16}$, and $a_{100} = \dfrac{(-1)^{100}}{100^2} = \dfrac{1}{10{,}000}$.

**6.** $a_n = \dfrac{1}{n^2}$. Then $a_1 = \dfrac{1}{(1)^2} = 1$, $a_2 = \dfrac{1}{(2)^2} = \dfrac{1}{4}$, $a_3 = \dfrac{1}{(3)^2} = \dfrac{1}{9}$, $a_4 = \dfrac{1}{(4)^2} = \dfrac{1}{16}$, and $a_{100} = \dfrac{1}{(100)^2} = \dfrac{1}{10{,}000}$.

**7.** $a_n = 1 + (-1)^n$. Then $a_1 = 1 + (-1)^1 = 0$, $a_2 = 1 + (-1)^2 = 2$, $a_3 = 1 + (-1)^3 = 0$, $a_4 = 1 + (-1)^4 = 2$, and $a_{100} = 1 + (-1)^{100} = 2$.

**8.** $a_n = \dfrac{(-1)^{n+1} n}{n+1}$. Then $a_1 = \dfrac{(-1)^2 \cdot 1}{1+1} = \dfrac{1}{2}$, $a_2 = \dfrac{(-1)^3 \cdot 2}{2+1} = -\dfrac{2}{3}$, $a_3 = \dfrac{(-1)^4 \cdot 3}{3+1} = \dfrac{3}{4}$, $a_4 = \dfrac{(-1)^5 \cdot 4}{4+1} = -\dfrac{4}{5}$, and $a_{100} = \dfrac{(-1)^{101} \cdot 100}{101} = -\dfrac{100}{101}$.

**9.** $a_n = n^n$. Then $a_1 = 1^1 = 1$, $a_2 = 2^2 = 4$, $a_3 = 3^3 = 27$, $a_4 = 4^4 = 256$, and $a_{100} = 100^{100} = 10^{200}$.

**10.** $a_n = 3$. Then $a_1 = 3$, $a_2 = 3$, $a_3 = 3$, $a_4 = 3$, and $a_{100} = 3$.

**11.** $a_n = 2(a_{n-1} - 2)$ and $a_1 = 3$. Then $a_2 = 2[(3) - 2] = 2$, $a_3 = 2[(2) - 2] = 0$, $a_4 = 2[(0) - 2] = -4$, and $a_5 = 2[(-4) - 2] = -12$.

**12.** $a_n = \dfrac{a_{n-1}}{2}$ and $a_1 = -8$. Then $a_2 = \dfrac{-8}{2} = -4$, $a_3 = \dfrac{-4}{2} = -2$, $a_4 = \dfrac{-2}{2} = -1$, and $a_5 = \dfrac{-1}{2} = -\dfrac{1}{2}$.

**13.** $a_n = 2a_{n-1} + 1$ and $a_1 = 1$. Then $a_2 = 2(1) + 1 = 3$, $a_3 = 2(3) + 1 = 7$, $a_4 = 2(7) + 1 = 15$, and $a_5 = 2(15) + 1 = 31$.

**14.** $a_n = \dfrac{1}{1 + a_{n-1}}$ and $a_1 = 1$. Then $a_2 = \dfrac{1}{1+1} = \dfrac{1}{2}$, $a_3 = \dfrac{1}{1+\frac{1}{2}} = \dfrac{2}{3}$, $a_4 = \dfrac{1}{1+\frac{2}{3}} = \dfrac{3}{5}$, and $a_5 = \dfrac{1}{1+\frac{3}{5}} = \dfrac{5}{8}$.

**15.** $a_n = a_{n-1} + a_{n-2}$, $a_1 = 1$, and $a_2 = 2$. Then $a_3 = 2 + 1 = 3$, $a_4 = 3 + 2 = 5$, and $a_5 = 5 + 3 = 8$.

**16.** $a_n = a_{n-1} + a_{n-2} + a_{n-3}$ and $a_1 = 1$, $a_2 = 1$, and $a_3 = 1$. Then $a_4 = 1 + 1 + 1 = 3$ and $a_5 = 3 + 1 + 1 = 5$.

**17. (a)** $a_1 = 7, a_2 = 11, a_3 = 15, a_4 = 19, a_5 = 23,$
$a_6 = 27, a_7 = 31, a_8 = 35, a_9 = 39, a_{10} = 43$

**(b)**

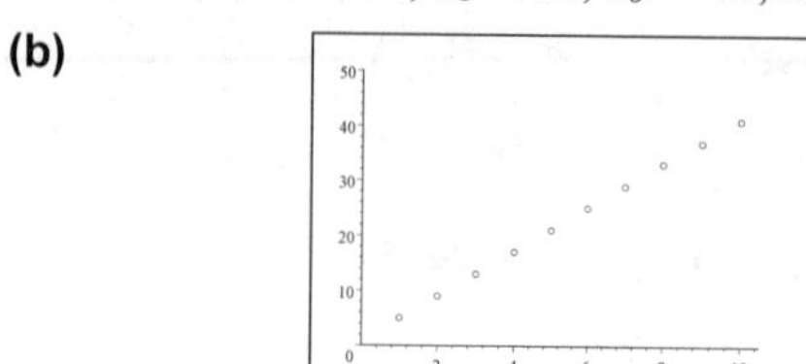

**18. (a)** $a_1 = 2, a_2 = 6, a_3 = 12, a_4 = 20, a_5 = 30,$
$a_6 = 42, a_7 = 56, a_8 = 72, a_9 = 90, a_{10} = 110$

**(b)**

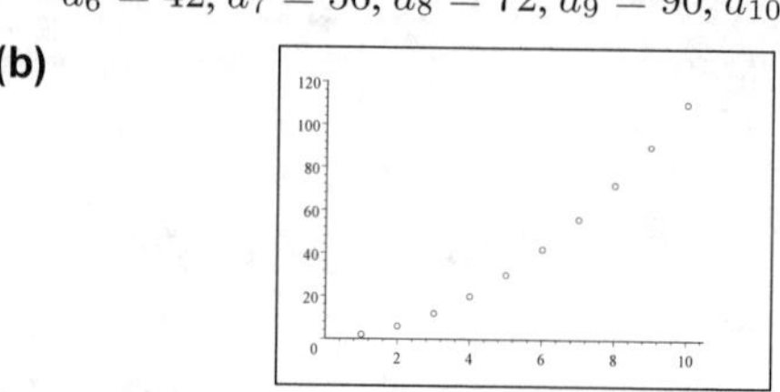

**19. (a)** $a_1 = \frac{12}{1} = 12, a_2 = \frac{12}{2} = 6, a_3 = \frac{12}{3} = 4,$
$a_4 = \frac{12}{4} = 3, a_5 = \frac{12}{5}, a_6 = \frac{12}{6} = 2, a_7 = \frac{12}{7},$
$a_8 = \frac{12}{8} = \frac{3}{2}, a_9 = \frac{12}{9} = \frac{4}{3}, a_{10} = \frac{12}{10} = \frac{6}{5}$

**(b)**

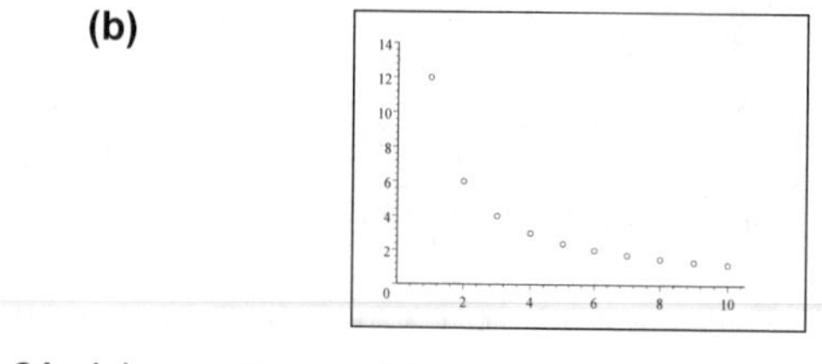

**20. (a)** $a_1 = 6, a_2 = 2, a_3 = 6, a_4 = 2, a_5 = 6, a_6 = 2,$
$a_7 = 6, a_8 = 2, a_9 = 6, a_{10} = 2$

**(b)**

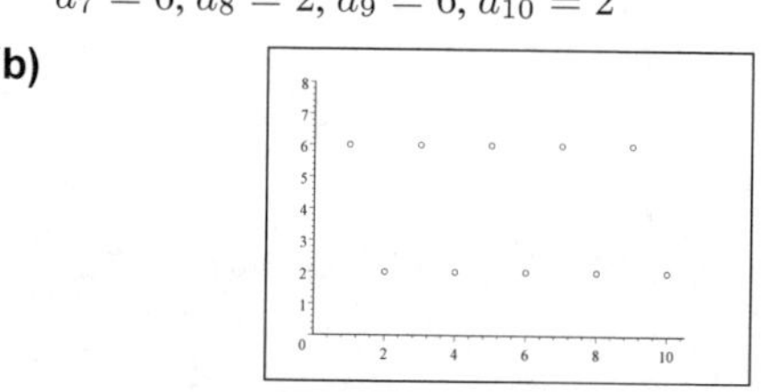

**21. (a)** $a_1 = 2, a_2 = 0.5, a_3 = 2, a_4 = 0.5, a_5 = 2,$
$a_6 = 0.5, a_7 = 2, a_8 = 0.5, a_9 = 2, a_{10} = 0.5$

**(b)**

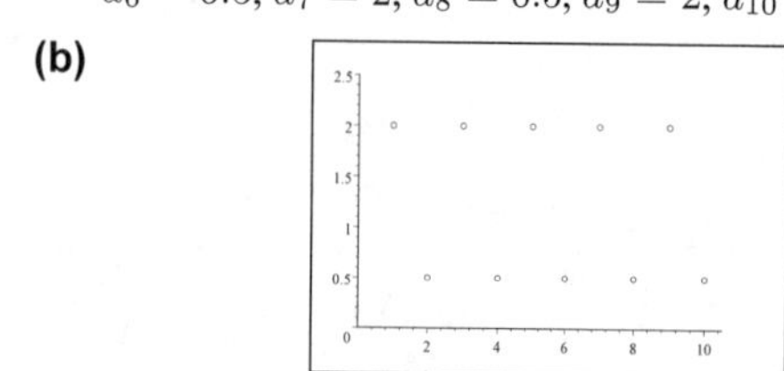

**22. (a)** $a_1 = 1, a_2 = 3, a_3 = 2, a_4 = -1, a_5 = -3,$
$a_6 = -2, a_7 = 1, a_8 = 3, a_9 = 2, a_{10} = -1$

**(b)**

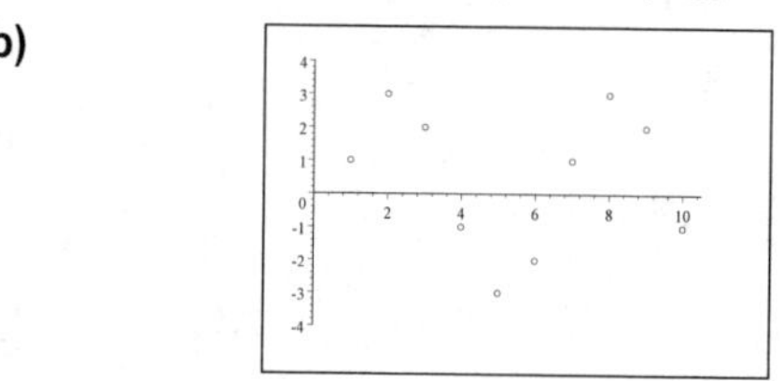

**23.** 2, 4, 8, 16, .... All are powers of 2, so $a_1 = 2, a_2 = 2^2, a_3 = 2^3, a_4 = 2^4, \ldots$ Thus $a_n = 2^n$.

**24.** $-\frac{1}{3}, \frac{1}{9}, -\frac{1}{27}, \frac{1}{81}, \ldots$. The denominators are all powers of 3, and the terms alternate in sign. Thus $a_1 = \dfrac{(-1)^1}{3^1}$, $a_2 = \dfrac{(-1)^2}{3^2}, a_3 = \dfrac{(-1)^3}{3^3}, a_4 = \dfrac{(-1)^4}{3^4}, \ldots$. So $a_n = \dfrac{(-1)^n}{3^n}$.

**25.** 1, 4, 7, 10, .... The difference between any two consecutive terms is 3, so $a_1 = 3(1) - 2, a_2 = 3(2) - 2, a_3 = 3(3) - 2,$ $a_4 = 3(4) - 2, \ldots$ Thus $a_n = 3n - 2$.

**26.** 5, −25, 125, −625, .... These terms are powers of 5, and the terms alternate in sign. So $a_1 = (-1)^2 \cdot 5^1, a_2 = (-1)^3 \cdot 5^2,$ $a_3 = (-1)^4 \cdot 5^3, a_4 = (-1)^5 \cdot 5^4, \ldots$. Thus $a_n = (-1)^{n+1} \cdot 5^n$.

**27.** $1, \frac{3}{4}, \frac{5}{9}, \frac{7}{16}, \frac{9}{25}, \ldots$. We consider the numerator separately from the denominator. The numerators of the terms differ by 2, and the denominators are perfect squares. So $a_1 = \dfrac{2(1) - 1}{1^2}, a_2 = \dfrac{2(2) - 1}{2^2}, a_3 = \dfrac{2(3) - 1}{3^2}, a_4 = \dfrac{2(4) - 1}{4^2},$ $a_5 = \dfrac{2(5) - 1}{5^2}, \ldots$. Thus $a_n = \dfrac{2n - 1}{n^2}$.

**28.** $\frac{3}{4}, \frac{4}{5}, \frac{5}{6}, \frac{6}{7}, \ldots$. Both the numerator and the denominator increase by 1, so $a_1 = \dfrac{1+2}{1+3}, a_2 = \dfrac{2+2}{2+3}, a_3 = \dfrac{3+2}{3+3},$ $a_4 = \dfrac{4+2}{4+3}, \ldots$. Thus $a_n = \dfrac{n+2}{n+3}$.

**29.** 0, 2, 0, 2, 0, 2, .... These terms alternate between 0 and 2. So $a_1 = 1 - 1, a_2 = 1 + 1, a_3 = 1 - 1, a_4 = 1 + 1,$ $a_5 = 1 - 1, a_6 = 1 + 1, \ldots$ Thus $a_n = 1 + (-1)^n$.

**30.** $1, \frac{1}{2}, 3, \frac{1}{4}, 5, \frac{1}{6}, \ldots$. So $a_1 = 1, a_2 = 2^{-1}, a_3 = 3^1, a_4 = 4^{-1}, \ldots$. Thus $a_n = n^{(-1)^{n+1}}$.

**31.** $a_1 = 1$, $a_2 = 3$, $a_3 = 5$, $a_4 = 7$, .... Therefore, $a_n = 2n - 1$. So $S_1 = 1$, $S_2 = 1 + 3 = 4$, $S_3 = 1 + 3 + 5 + = 9$, $S_4 = 1 + 3 + 5 + 7 = 16$, $S_5 = 1 + 3 + 5 + 7 + 9 = 25$, and $S_6 = 1 + 3 + 5 + 7 + 9 + 11 = 36$.

**32.** $a_1 = 1^2$, $a_2 = 2^2$, $a_3 = 3^2$, $a_4 = 4^2$, .... Therefore, $a_n = n^2$. So $S_1 = 1^2 = 1$, $S_2 = 1 + 2^2 = 5$, $S_3 = 5 + 3^2 = 5 + 9 = 14$, $S_4 = 14 + 4^2 = 14 + 16 = 30$, $S_5 = 30 + 5^2 = 30 + 25 = 55$, and $S_6 = 55 + 6^2 = 55 + 36 = 91$.

**33.** $a_1 = \frac{1}{3}$, $a_2 = \dfrac{1}{3^2}$, $a_3 = \dfrac{1}{3^3}$, $a_4 = \dfrac{1}{3^4}$, .... Therefore, $a_n = \dfrac{1}{3^n}$. So $S_1 = \frac{1}{3}$, $S_2 = \frac{1}{3} + \dfrac{1}{3^2} = \dfrac{4}{9}$, $S_3 = \frac{1}{3} + \dfrac{1}{3^2} + \dfrac{1}{3^3} = \frac{13}{27}$, $S_4 = \frac{1}{3} + \dfrac{1}{3^2} + \dfrac{1}{3^3} + \dfrac{1}{3^4} = \frac{40}{81}$, and $S_5 = \frac{1}{3} + \dfrac{1}{3^2} + \dfrac{1}{3^3} + \dfrac{1}{3^4} + \dfrac{1}{3^5} = \frac{121}{243}$, $S_6 = \frac{1}{3} + \dfrac{1}{3^2} + \dfrac{1}{3^3} + \dfrac{1}{3^4} + \dfrac{1}{3^5} + \dfrac{1}{3^6} = \frac{364}{729}$.

**34.** $a_1 = -1$, $a_2 = 1$, $a_3 = -1$, $a_4 = 1$, .... Therefore, $a_n = (-1)^n$. So $S_1 = -1$, $S_2 = -1 + 1 = 0$, $S_3 = 0 - 1 = -1$, $S_4 = -1 + 1 = 0$, $S_5 = 0 - 1 = -1$, and $S_6 = -1 + 1 = 0$.

**35.** $a_n = \dfrac{2}{3^n}$. So $S_1 = \frac{2}{3}$, $S_2 = \dfrac{2}{3} + \dfrac{2}{3^2} = \frac{8}{9}$, $S_3 = \frac{2}{3} + \dfrac{2}{3^2} + \dfrac{2}{3^3} = \frac{26}{27}$, and $S_4 = \frac{2}{3} + \dfrac{2}{3^2} + \dfrac{2}{3^3} + \dfrac{2}{3^4} = \frac{80}{81}$. Therefore, $S_n = \dfrac{3^n - 1}{3^n}$.

**36.** $a_n = \dfrac{1}{n+1} - \dfrac{1}{n+2}$. So $S_1 = \frac{1}{2} - \frac{1}{3}$, $S_2 = \left(\frac{1}{2} - \frac{1}{3}\right) + \left(\frac{1}{3} - \frac{1}{4}\right) = \frac{1}{2} + \left(-\frac{1}{3} + \frac{1}{3}\right) - \frac{1}{4} = \frac{1}{2} - \frac{1}{4}$,
$S_3 = \left(\frac{1}{2} - \frac{1}{3}\right) + \left(\frac{1}{3} - \frac{1}{4}\right) + \left(\frac{1}{4} - \frac{1}{5}\right) = \frac{1}{2} + \left(-\frac{1}{3} + \frac{1}{3}\right) + \left(-\frac{1}{4} + \frac{1}{4}\right) - \frac{1}{5} = \frac{1}{2} - \frac{1}{5}$, and
$S_4 = \left(\frac{1}{2} - \frac{1}{3}\right) + \left(\frac{1}{3} - \frac{1}{4}\right) + \left(\frac{1}{4} - \frac{1}{5}\right) + \left(\frac{1}{5} - \frac{1}{6}\right) = \frac{1}{2} + \left(-\frac{1}{3} + \frac{1}{3}\right) + \left(-\frac{1}{4} + \frac{1}{4}\right) + \left(-\frac{1}{5} + \frac{1}{5}\right) - \frac{1}{6} = \frac{1}{2} - \frac{1}{6}$. Therefore,
$S_n = \left(\frac{1}{2} - \frac{1}{3}\right) + \left(\frac{1}{3} - \frac{1}{4}\right) + \cdots + \left(\frac{1}{n+1} - \frac{1}{n+2}\right) = \frac{1}{2} + \left(-\frac{1}{3} + \frac{1}{3}\right) + \cdots + \left(-\frac{1}{n+1} + \frac{1}{n+1}\right) - \frac{1}{n+2} = \frac{1}{2} - \frac{1}{n+2}$.

**37.** $a_n = \sqrt{n} - \sqrt{n+1}$. So $S_1 = \sqrt{1} - \sqrt{2} = 1 - \sqrt{2}$, $S_2 = \left(\sqrt{1} - \sqrt{2}\right) + \left(\sqrt{2} - \sqrt{3}\right) = 1 + \left(-\sqrt{2} + \sqrt{2}\right) - \sqrt{3} = 1 - \sqrt{3}$,
$S_3 = \left(\sqrt{1} - \sqrt{2}\right) + \left(\sqrt{2} - \sqrt{3}\right) + \left(\sqrt{3} - \sqrt{4}\right) = 1 + \left(-\sqrt{2} + \sqrt{2}\right) + \left(-\sqrt{3} + \sqrt{3}\right) - \sqrt{4} = 1 - \sqrt{4}$,

$$S_4 = \left(\sqrt{1} - \sqrt{2}\right) + \left(\sqrt{2} - \sqrt{3}\right) + \left(\sqrt{3} - \sqrt{4}\right) + \left(\sqrt{4} - \sqrt{5}\right)$$
$$= 1 + \left(-\sqrt{2} + \sqrt{2}\right) + \left(-\sqrt{3} + \sqrt{3}\right) + \left(-\sqrt{4} + \sqrt{4}\right) - \sqrt{5} = 1 - \sqrt{5}$$

Therefore,

$$S_n = \left(\sqrt{1} - \sqrt{2}\right) + \left(\sqrt{2} - \sqrt{3}\right) + \cdots + \left(\sqrt{n} - \sqrt{n+1}\right)$$
$$= 1 + \left(-\sqrt{2} + \sqrt{2}\right) + \left(-\sqrt{3} + \sqrt{3}\right) + \cdots + \left(-\sqrt{n} + \sqrt{n}\right) - \sqrt{n+1} = 1 - \sqrt{n+1}$$

**38.** $a_n = \log\left(\dfrac{k}{k+1}\right) = \log k - \log(k+1)$. So $S_1 = \log 1 - \log 2 = -\log 2$, $S_2 = (-\log 2) + (\log 2 - \log 3) = -\log 3$,
$S_3 = -\log 2 + (\log 2 - \log 3) + (\log 3 - \log 4) = (-\log 2 + \log 2) + (-\log 3 + \log 3) - \log 4 = -\log 4$, and

$$S_4 = -\log 2 + (\log 2 - \log 3) + (\log 3 - \log 4) + (\log 4 - \log 5)$$
$$= (-\log 2 + \log 2) + (-\log 3 + \log 3) + (-\log 4 + \log 4) - \log 5 = -\log 5$$

Therefore, $S_n = -\log(n+1)$.

**39.** $\sum_{k=1}^{4} k = 1 + 2 + 3 + 4 = 10$

**40.** $\sum_{k=1}^{4} k^2 = 1 + 2^2 + 3^2 + 4^2 = 1 + 4 + 9 + 16 = 30$

**41.** $\sum_{k=1}^{3} \dfrac{1}{k} = 1 + \frac{1}{2} + \frac{1}{3} = \frac{6}{6} + \frac{3}{6} + \frac{2}{6} = \frac{11}{6}$

**42.** $\sum_{j=1}^{100} (-1)^j = (-1)^1 + (-1)^2 + (-1)^3 + (-1)^4 + \cdots + (-1)^{99} + (-1)^{100} = -1 + 1 - 1 + 1 - 1 + \cdots - 1 + 1 = 0$

**43.** $\sum_{i=1}^{8} \left[1 + (-1)^i\right] = 0 + 2 + 0 + 2 + 0 + 2 + 0 + 2 = 8$

**44.** $\sum_{i=4}^{12} 10 = 10 + 10 + 10 + 10 + 10 + 10 + 10 + 10 + 10 = 90$

**45.** $\sum_{k=1}^{5} 2^{k-1} = 2^0 + 2^1 + 2^2 + 2^3 + 2^4 = 1 + 2 + 4 + 8 + 16 = 31$

**46.** $\sum_{i=1}^{3} i2^i = 1 \cdot 2^1 + 2 \cdot 2^2 + 3 \cdot 2^3 = 2 + 8 + 24 = 34$

**47.** 385 **48.** 15,550 **49.** 46,438 **50.** 0.153146 **51.** 22 **52.** −0.688172

**53.** $\sum_{k=1}^{5} \sqrt{k} = \sqrt{1} + \sqrt{2} + \sqrt{3} + \sqrt{4} + \sqrt{5}$

**54.** $\sum_{i=0}^{4} \frac{2i-1}{2i+1} = \frac{2\cdot 0-1}{2\cdot 0+1} + \frac{2\cdot 1-1}{2\cdot 1+1} + \frac{2\cdot 2-1}{2\cdot 2+1} + \frac{2\cdot 3-1}{2\cdot 3+1} + \frac{2\cdot 4-1}{2\cdot 4+1} = -1 + \frac{1}{3} + \frac{3}{5} + \frac{5}{7} + \frac{7}{9}$

**55.** $\sum_{k=0}^{6} \sqrt{k+4} = \sqrt{4} + \sqrt{5} + \sqrt{6} + \sqrt{7} + \sqrt{8} + \sqrt{9} + \sqrt{10}$

**56.** $\sum_{k=6}^{9} k(k+3) = 6\cdot 9 + 7\cdot 10 + 8\cdot 11 + 9\cdot 12 = 54 + 70 + 88 + 108$

**57.** $\sum_{k=3}^{100} x^k = x^3 + x^4 + x^5 + \cdots + x^{100}$

**58.** $\sum_{j=1}^{n} (-1)^{j+1} x^j = (-1)^2 x + (-1)^3 x^2 + (-1)^4 x^3 + \cdots + (-1)^{n+1} x^n = x - x^2 + x^3 - \cdots + (-1)^{n+1} x^n$

**59.** $1 + 2 + 3 + 4 + \cdots + 100 = \sum_{k=1}^{100} k$

**60.** $2 + 4 + 6 + \cdots + 20 = \sum_{k=1}^{10} 2k$

**61.** $1^2 + 2^2 + 3^2 + \cdots + 10^2 = \sum_{k=1}^{10} k^2$

**62.** $\frac{1}{2\ln 2} - \frac{1}{3\ln 3} + \frac{1}{4\ln 4} - \frac{1}{5\ln 5} + \cdots + \frac{1}{100\ln 100} = \sum_{k=2}^{100} \frac{(-1)^k}{k\ln k}$

**63.** $\frac{1}{1\cdot 2} + \frac{1}{2\cdot 3} + \frac{1}{3\cdot 4} + \cdots + \frac{1}{999\cdot 1000} = \sum_{k=1}^{999} \frac{1}{k(k+1)}$

**64.** $\frac{\sqrt{1}}{1^2} + \frac{\sqrt{2}}{2^2} + \frac{\sqrt{3}}{3^2} + \cdots + \frac{\sqrt{n}}{n^2} = \sum_{k=1}^{n} \frac{\sqrt{k}}{k^2}$

**65.** $1 + x + x^2 + x^3 + \cdots + x^{100} = \sum_{k=0}^{100} x^k$

**66.** $1 - 2x + 3x^2 - 4x^3 + 5x^4 + \cdots - 100x^{99} = \sum_{k=1}^{100} (-1)^{k+1} \cdot k \cdot x^{k-1}$

**67.** $\sqrt{2}, \sqrt{2\sqrt{2}}, \sqrt{2\sqrt{2\sqrt{2}}}, \sqrt{2\sqrt{2\sqrt{2\sqrt{2}}}}, \ldots$. We simplify each term in an attempt to determine a formula for $a_n$. So $a_1 = 2^{1/2}$, $a_2 = \sqrt{2\cdot 2^{1/2}} = \sqrt{2^{3/2}} = 2^{3/4}$, $a_3 = \sqrt{2\cdot 2^{3/4}} = \sqrt{2^{7/4}} = 2^{7/8}$, $a_4 = \sqrt{2\cdot 2^{7/8}} = \sqrt{2^{15/8}} = 2^{15/16}, \ldots$. Thus $a_n = 2^{(2^n-1)/2^n}$.

**68.** $G_1 = 1$, $G_2 = 1$, $G_3 = 2$, $G_4 = 3$, $G_5 = 5$, $G_6 = 8$, $G_7 = 13$, $G_8 = 21$, $G_9 = 34$, $G_{10} = 55$

**69.** **(a)** $A_1 = \$2004$, $A_2 = \$2008.01$, $A_3 = \$2012.02$, $A_4 = \$2016.05$, $A_5 = \$2020.08$, $A_6 = \$2024.12$

**(b)** Since 3 years is 36 months, we get $A_{36} = \$2149.16$.

**70.** **(a)** $I_1 = 0$, $I_2 = \$0.50$, $I_3 = \$1.50$, $I_4 = \$3.01$, $I_5 = \$5.03$, $I_6 = \$7.55$

**(b)** Since 5 years is 60 months, we have $I_{60} = \$977.00$.

**71.** **(a)** $P_1 = 35{,}700$, $P_2 = 36{,}414$, $P_3 = 37{,}142$, $P_4 = 37{,}885$, $P_5 = 38{,}643$

**(b)** Since 2014 is 10 years after 2004, $P_{10} = 42{,}665$.

**72.** **(a)** The amount she owes at the end of the month, $A_n$, is the amount she owes at the beginning of the month, $A_{n-1}$, plus the interest, $0.005A_{n-1}$, minus the \$200 she repay her uncle. Thus $A_n = A_{n-1} + 0.005A_{n-1} - 200 \Leftrightarrow A_n = 1.005A_{n-1} - 200$.

**(b)** $A_1 = 9850$, $A_2 = 9699.25$, $A_3 = 9547.74$, $A_4 = 9395.48$, $A_5 = 9242.46$, $A_6 = 9088.67$. Thus she owes her uncle \$9088.67 after six months.

**73.** **(a)** The number of catfish at the end of the month, $P_n$, is the population at the start of the month, $P_{n-1}$, plus the increase in population, $0.08P_{n-1}$, minus the 300 catfish harvested. Thus $P_n = P_{n-1} + 0.08P_{n-1} - 300 \Leftrightarrow P_n = 1.08P_{n-1} - 300$.

**(b)** $P_1 = 5100$, $P_2 = 5208$, $P_3 = 5325$, $P_4 = 5451$, $P_5 = 5587$, $P_6 = 5734$, $P_7 = 5892$, $P_8 = 6064$, $P_9 = 6249$, $P_{10} = 6449$, $P_{11} = 6665$, $P_{12} = 6898$. Thus there should be 6898 catfish in the pond at the end of 12 months.

**74.** **(a)** Let $n$ be the years since 2002. So $P_0 = \$240{,}000$. Each month the median price of a house in Orange County increases to 1.06 the previous months. Thus $P_n = 1.06^n P_0$.

**(b)** Since $2010 - 2002 = 8$. Thus $P_8 = 382{,}523.53$. Thus in 2010, the median price of a house in Orange County should be \$382,524.

**75. (a)** Let $S_n$ be his salary in the $n$th year. Then $S_1 = \$30{,}000$. Since his salary increase by 2000 each year, $S_n = S_{n-1} + 2000$. Thus $S_1 = \$30{,}000$ and $S_n = S_{n-1} + 2000$.

**(b)** $S_5 = S_4 + 2000 = (S_3 + 2000) + 2000 = (S_2 + 2000) + 4000 = (S_1 + 2000) + 6000 = \$38{,}000$.

**76. (a)** Let $n$ be the days after she starts, so $C_0 = 4$. And each day the concentration increases by 10%. Thus after $n$ daysthe brine solution is $C_n = 1.10C_{n-1}$ with $C_0 = 4$.

**(b)** $C_8 = 1.10C_7 = 1.10\,(1.10C_6) = \cdots = 1.10^8 \cdot C_0 = 1.10^8 \cdot 4 \approx 8.6$. Thus the brine solution is 8.6 g/L of salt.

**77.** Let $F_n$ be the number of pairs of rabbits in the $n$th month. Clearly $F_1 = F_2 = 1$. In the $n$th month each pair that is two or more months old (that is, $F_{n-2}$ pairs) will add a pair of offspring to the $F_{n-1}$ pairs already present. Thus $F_n = F_{n-1} + F_{n-2}$. So $F_n$ is the Fibonacci sequence.

**78. (a)** $a_n = n^2$. Then $a_1 = 1^2 = 1$, $a_2 = 2^2 = 4$, $a_3 = 3^2 = 9$, $a_4 = 4^2 = 16$.

**(b)** $a_n = n^2 + (n-1)(n-2)(n-3)(n-4)$, $a_1 = 1^2 + (1-1)(1-2)(1-3)(1-4) = 1 + 0(-1)(-2)(-3) = 1$,
$a_2 = 2^2 + (2-1)(2-2)(2-3)(2-4) = 4 + 1 \cdot 0(-1)(-2) = 4$,
$a_3 = 3^2 + (3-1)(3-2)(3-3)(3-4) = 9 + 2 \cdot 1 \cdot 0(-1) = 9$,
$a_4 = 4^2 + (4-1)(4-2)(4-3)(4-4) = 16 + 3 \cdot 2 \cdot 1 \cdot 0 = 16$.
Hence, the sequences agree in the first four terms. However, for the second sequence, $a_5 = 5^2 + (5-1)(5-2)(5-3)(5-4) = 25 + 4 \cdot 3 \cdot 2 \cdot 1 = 49$, and for the first sequence, $a_5 = 5^2 = 25$, and thus the sequences disagree from the fifth term on.

**(c)** $a_n = n^2 + (n-1)(n-2)(n-3)(n-4)(n-5)(n-6)$ agrees with $a_n = n^2$ in the first six terms only.

**(d)** $a_n = 2^n$ and $b_n = 2^n + (n-1)(n-2)(n-3)(n-4)$.

**79.** $a_{n+1} = \begin{cases} \dfrac{a_n}{2} & \text{if } a_n \text{ is even} \\ 3a_n + 1 & \text{if } a_n \text{ is odd} \end{cases}$ With $a_1 = 11$, we have $a_2 = 34$, $a_3 = 17$, $a_4 = 52$, $a_5 = 26$, $a_6 = 13$, $a_7 = 40$, $a_8 = 20$, $a_9 = 10$, $a_{10} = 5$, $a_{11} = 16$, $a_{12} = 8$, $a_{13} = 4$, $a_{14} = 2$, $a_{15} = 1$, $a_{16} = 4$, $a_{17} = 2$, $a_{18} = 1, \ldots$ (with 4, 2, 1 repeating). So $a_{3n+1} = 4$, $a_{3n+2} = 2$, and $a_{3n} = 1$, for $n \geq 5$. With $a_1 = 25$, we have $a_2 = 76$, $a_3 = 38$, $a_4 = 19$, $a_5 = 58$, $a_6 = 29$, $a_7 = 88$, $a_8 = 44$, $a_9 = 22$, $a_{10} = 11$, $a_{11} = 34$, $a_{12} = 17$, $a_{13} = 52$, $a_{14} = 26$, $a_{15} = 13$, $a_{16} = 40$, $a_{17} = 20$, $a_{18} = 10$, $a_{19} = 5$, $a_{20} = 16$, $a_{21} = 8$, $a_{22} = 4$, $a_{23} = 2$, $a_{24} = 1$, $a_{25} = 4$, $a_{26} = 2$, $a_{27} = 1, \ldots$ (with 4, 2, 1 repeating). So $a_{3n+1} = 4$, $a_{3n+2} = 2$, and $a_{3n+3} = 1$ for $n \geq 7$.
We conjecture that the sequence will always return to the numbers 4, 2, 1 repeating.

**80.** $a_n = a_{n-a_{n-1}} + a_{n-a_{n-2}}$, $a_1 = 1$, and $a_2 = 1$. So $a_3 = a_{3-1} + a_{3-1} = a_2 + a_2 = 1 + 1 = 2$, $a_4 = a_{4-2} + a_{4-1} = a_2 + a_3 = 1 + 2 = 3$, $a_5 = a_{5-3} + a_{5-2} = a_2 + a_3 = 1 + 2 = 3$, $a_6 = a_{6-3} + a_{6-3} = a_3 + a_3 = 2 + 2 = 4$, $a_7 = a_{7-4} + a_{7-3} = a_3 + a_4 = 2 + 3 = 5$, $a_8 = a_{8-5} + a_{8-4} = a_3 + a_4 = 2 + 3 = 5$, $a_9 = a_{9-5} + a_{9-5} = a_4 + a_4 = 3 + 3 = 6$, and $a_{10} = a_{10-6} + a_{10-5} = a_4 + a_5 = 3 + 3 = 6$. The definition of $a_n$ depends on the values of certain preceding terms. So $a_n$ is the sum of two preceding terms whose choice depends on the *values* of $a_{n-1}$ and $a_{n-2}$ (not on $n-1$ and $n-2$).

# 12.2 Arithmetic Sequences

**1. (a)** $a_1 = 5 + 2(1 - 1) = 5$,
$a_2 = 5 + 2(2 - 1) = 5 + 2 = 7$,
$a_3 = 5 + 2(3 - 1) = 5 + 4 = 9$,
$a_4 = 5 + 2(4 - 1) = 5 + 6 = 11$,
$a_5 = 5 + 2(5 - 1) = 5 + 8 = 13$

**(b)** The common difference is 2.

**(c)**

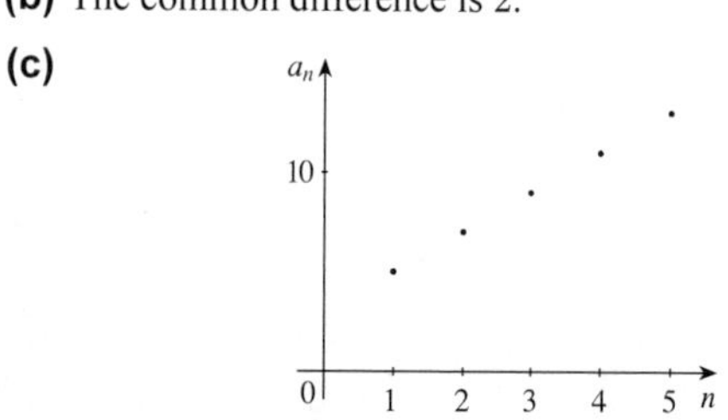

**2. (a)** $a_1 = 3 - 4(1 - 1) = 3$, $a_2 = 3 - 4(2 - 1) = -1$,
$a_3 = 3 - 4(3 - 1) = -5$, $a_4 = 3 - 4(4 - 1) = -9$,
$a_5 = 3 - 4(5 - 1) = -13$

**(b)** The common difference is $-4$.

**(c)**

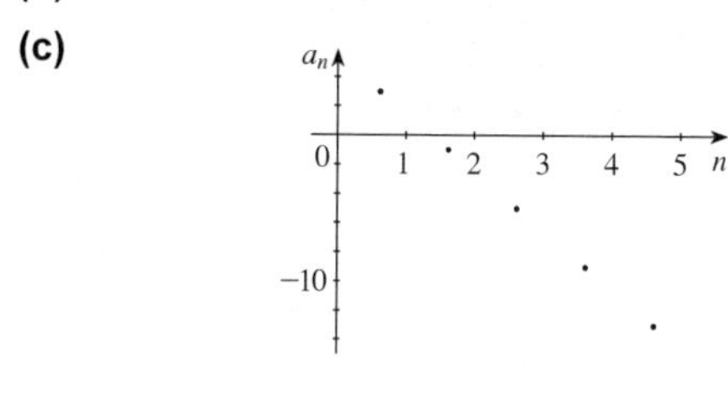

**3. (a)** $a_1 = \frac{5}{2} - (1 - 1) = \frac{5}{2}$, $a_2 = \frac{5}{2} - (2 - 1) = \frac{3}{2}$,
$a_3 = \frac{5}{2} - (3 - 1) = \frac{1}{2}$, $a_4 = \frac{5}{2} - (4 - 1) = -\frac{1}{2}$,
$a_5 = \frac{5}{2} - (5 - 1) = -\frac{3}{2}$

**(b)** The common difference is $-1$.

**(c)**

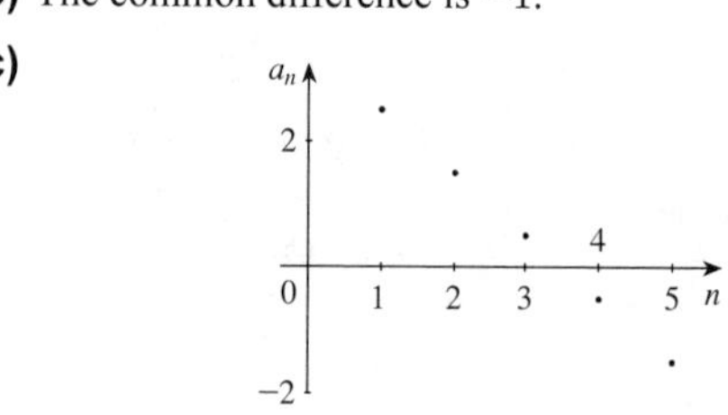

**4. (a)** $a_1 = \frac{1}{2}(1 - 1) = 0$, $a_2 = \frac{1}{2}(2 - 1) = \frac{1}{2}$,
$a_3 = \frac{1}{2}(3 - 1) = 1$, $a_4 = \frac{1}{2}(4 - 1) = \frac{3}{2}$,
$a_5 = \frac{1}{2}(5 - 1) = 2$

**(b)** The common difference is $\frac{1}{2}$.

**(c)**

**5.** $a = 3$, $d = 5$, $a_n = a + d(n - 1) = 3 + 5(n - 1)$. So $a_{10} = 3 + 5(10 - 1) = 48$.

**6.** $a = -6$, $d = 3$, $a_n = a + d(n - 1) = -6 + 3(n - 1)$. So $a_{10} = -6 + 3(10 - 1) = 21$.

**7.** $a = \frac{5}{2}$, $d = -\frac{1}{2}$, $a_n = a + d(n - 1) = \frac{5}{2} - \frac{1}{2}(n - 1)$. So $a_{10} = \frac{5}{2} - \frac{1}{2}(10 - 1) = -2$.

**8.** $a = \sqrt{3}$, $d = \sqrt{3}$, $a_n = a + d(n - 1) = \sqrt{3} + \sqrt{3}(n - 1)$. So $a_{10} = \sqrt{3} + \sqrt{3}(10 - 1) = 10\sqrt{3}$.

**9.** $a_4 - a_3 = 14 - 11 = 3$, $a_3 - a_2 = 11 - 8 = 3$, $a_2 - a_1 = 8 - 5 = 3$. This sequence is arithmetic with common difference 3.

**10.** Since $a_2 - a_1 = 6 - 3 = 3$ and $a_4 - a_3 = 13 - 9 = 4$, the terms of the sequence do not have a common difference. This sequence is not arithmetic.

**11.** Since $a_2 - a_1 = 4 - 2 = 2$ and $a_4 - a_3 = 16 - 8 = 8$, the terms of the sequence do not have a common difference. This sequence is not arithmetic.

**12.** $a_4 - a_3 = a_3 - a_2 = a_2 - a_1 = 2$. This sequence is arithmetic with common difference 2.

**13.** $a_4 - a_3 = -\frac{3}{2} - 0 = -\frac{3}{2}$, $a_3 - a_2 = 0 - \dfrac{3}{2} = -\frac{3}{2}$, $a_2 - a_1 = \frac{3}{2} - 3 = -\frac{3}{2}$. This sequence is arithmetic with common difference $-\frac{3}{2}$.

**14.** $a_4 - a_3 = \ln 16 - \ln 8 = \ln \frac{16}{8} = \ln 2$, $a_3 - a_2 = \ln 8 - \ln 4 = \ln \frac{8}{4} = \ln 2$, $a_2 - a_1 = \ln 4 - \ln 2 = \ln \frac{4}{2} = \ln 2$. This sequence is arithmetic with common difference $\ln 2$.

**15.** $a_4 - a_3 = 7.7 - 6.0 = 1.7$, $a_3 - a_2 = 6.0 - 4.3 = 1.7$, $4. - a_1 = 4.3 - 2.6 = 1.7$. This sequence is arithmetic with common difference 1.7.

**16.** Since $a_4 - a_3 = \frac{1}{5} - \frac{1}{4} = -\frac{1}{20}$ and $a_3 - a_2 = \frac{1}{4} - \frac{1}{3} = -\frac{1}{12}$, the terms of the sequence do not have a common difference. This sequence is not arithmetic.

**17.** $a_1 = 4 + 7(1) = 11$, $a_2 = 4 + 7(2) = 18$, $a_3 = 4 + 7(3) = 25$, $a_4 = 4 + 7(4) = 32$, $a_5 = 4 + 7(5) = 39$. This sequence is arithmetic, the common difference is $d = 7$ and $a_n = 4 + 7n = 4 + 7n - 7 + 7 = 11 + 7(n-1)$.

**18.** $a_1 = 4 + 2^1 = 6$, $a_2 = 4 + 2^2 = 8$, $a_3 = 4 + 2^3 = 12$, $a_4 = 4 + 2^4 = 20$, $a_5 = 4 + 2^5 = 36$. Since $a_4 - a_3 = 8$ and $a_3 - a_2 = 4$, the terms of the sequence do not have a common difference. This sequence is not arithmetic.

**19.** $a_1 = \dfrac{1}{1+2(1)} = \dfrac{1}{3}$, $a_2 = \dfrac{1}{1+2(2)} = \dfrac{1}{5}$, $a_3 = \dfrac{1}{1+2(3)} = \dfrac{1}{7}$, $a_4 = \dfrac{1}{1+2(4)} = \dfrac{1}{9}$, $a_5 = \dfrac{1}{1+2(5)} = \dfrac{1}{11}$. Since $a_4 - a_3 = \frac{1}{9} - \frac{1}{7} = -\frac{2}{63}$ and $a_3 - a_2 = \frac{1}{7} - \frac{1}{3} = -\frac{2}{21}$, the terms of the sequence do not have a common difference. This sequence is not arithmetic.

**20.** $a_1 = 1 + \frac{1}{2} = \frac{3}{2}$, $a_2 = 1 + \frac{2}{2} = 2$, $a_3 = 1 + \frac{3}{2} = \frac{5}{2}$, $a_4 = 1 + \frac{4}{2} = 3$, $a_5 = 1 + \frac{5}{2} = \frac{7}{2}$. This sequence is arithmetic, the common difference is $d = \frac{1}{2}$ and $a_n = 1 + \frac{n}{2} = 1 + \frac{1}{2}n - \frac{1}{2} + \frac{1}{2} = \frac{3}{2} + \frac{1}{2}(n-1)$.

**21.** $a_1 = 6(1) - 10 = -4$, $a_2 = 6(2) - 10 = 2$, $a_3 = 6(3) - 10 = 8$, $a_4 = 6(4) - 10 = 14$, $a_5 = 6(5) - 10 = 20$. This sequence is arithmetic, the common difference is $d = 6$ and $a_n = 6n - 10 = 6n - 6 + 6 - 10 = -4 + 6(n-1)$.

**22.** $a_1 = 3 + (-1)^1(1) = 2$, $a_2 = 3 + (-1)^2(2) = 5$, $a_3 = 3 + (-1)^3(3) = 0$, $a_4 = 3 + (-1)^4(4) = 7$, $a_5 = 3 + (-1)^5(5) = -2$. Since $a_4 - a_3 = 7$ and $a_3 - a_2 = -5$, the terms of the sequence do not have a common difference. This sequence is not arithmetic.

**23.** $2, 5, 8, 11, \ldots$. Then $d = a_2 - a_1 = 5 - 2 = 3$, $a_5 = a_4 + 3 = 11 + 3 = 14$, $a_n = 2 + 3(n-1)$, and $a_{100} = 2 + 3(99) = 299$.

**24.** $1, 5, 9, 13, \ldots$. Then $d = a_2 - a_1 = 5 - 1 = 4$, $a_5 = a_4 + 4 = 13 + 4 = 17$, $a_n = 1 + 4(n-1)$, $a_{100} = 1 + 4(99) = 397$.

**25.** $4, 9, 14, 19, \ldots$. Then $d = a_2 - a_1 = 9 - 4 = 5$, $a_5 = a_4 + 5 = 19 + 5 = 24$, $a_n = 4 + 5(n-1)$, and $a_{100} = 4 + 5(99) = 499$.

**26.** $11, 8, 5, 2, \ldots$. Then $d = a_2 - a_1 = 8 - 11 = -3$, $a_5 = a_4 - 3 = 2 - 3 = -1$, $a_n = 11 - 3(n-1)$, and $a_{100} = 11 - 3(99) = -286$.

**27.** $-12, -8, -4, 0, \ldots$. Then $d = a_2 - a_1 = -8 - (-12) = 4$, $a_5 = a_4 + 4 = 0 + 4 = 4$, $a_n = -12 + 4(n-1)$, and $a_{100} = -12 + 4(99) = 384$.

**28.** $\frac{7}{6}, \frac{5}{3}, \frac{13}{6}, \frac{8}{3}, \ldots$. Then $d = a_2 - a_1 = \frac{10}{6} - \frac{7}{6} = \frac{1}{2}$, $a_5 = a_4 + \frac{1}{2} = \frac{8}{3} + \frac{1}{2} = \frac{19}{6}$, $a_n = \frac{7}{6} + \frac{1}{2}(n-1)$, $a_{100} = \frac{7}{6} + \frac{1}{2}(99) = \frac{7+297}{6} = \frac{152}{3}$.

**29.** $25, 26.5, 28, 29.5, \ldots$. Then $d = a_2 - a_1 = 26.5 - 25 = 1.5$, $a_5 = a_4 + 1.5 = 29.5 + 1.5 = 31$, $a_n = 25 + 1.5(n-1)$, $a_{100} = 25 + 1.5(99) = 173.5$.

**30.** $15, 12.3, 9.6, 6.9, \ldots$. Then $d = a_2 - a_1 = 12.3 - 15 = -2.7$, $a_5 = a_4 - 2.7 = 6.9 - 2.7 = 4.2$, $a_n = 15 - 2.7(n-1)$, and $a_{100} = 15 - 2.7(99) = -252.3$.

**31.** $2, 2+s, 2+2s, 2+3s, \ldots$. Then $d = a_2 - a_1 = 2 + s - 2 = s$, $a_5 = a_4 + s = 2 + 3s + s = 2 + 4s$, $a_n = 2 + (n-1)s$, and $a_{100} = 2 + 99s$.

**32.** $-t, -t+3, -t+6, -t+9, \ldots$. Then $d = a_2 - a_1 = (-t+3) - (-t) = 3$, $a_5 = a_4 + 3 = -t + 9 + 3 = -t + 12$, $a_n = -t + 3(n-1)$, and $a_{100} = -t + 3(99) = -t + 297$.

**33.** $a_{10} = \frac{55}{2}$, $a_2 = \frac{7}{2}$, and $a_n = a + d(n-1)$. Then $a_2 = a + d = \frac{7}{2} \quad\Leftrightarrow\quad d = \frac{7}{2} - a$. Substituting into $a_{10} = a + 9d = \frac{55}{2}$ gives $a + 9\left(\frac{7}{2} - a\right) = \frac{55}{2} \quad\Leftrightarrow\quad a = \frac{1}{2}$. Thus, the first term is $a_1 = \frac{1}{2}$.

**34.** $a_{12} = 32$, $a_5 = 18$, and $a_n = a + d(n-1)$. Then $a_5 = a + 4d = 18 \quad\Leftrightarrow\quad a = 18 - 4d$. Substituting for $a$, we have $a_{12} = a + 11d = 32 \quad\Leftrightarrow\quad (18 - 4d) + 11d = 32 \quad\Leftrightarrow\quad 7d = 14 \quad\Leftrightarrow\quad d = 2$. Then we have $a = 18 - 4 \cdot 2 = 10$, and hence, the 20th term is $a_{20} = 10 + 19(2) = 48$.

**35.** $a_{100} = 98$ and $d = 2$. Note that $a_{100} = a + 99d = a + 99(2) = a + 198$. Since $a_{100} = 98$, we have $a + 198 = a_{100} = 98 \quad\Leftrightarrow\quad a = -100$. Hence, $a_1 = -100$, $a_2 = -100 + 2 = -98$, and $a_3 = -100 + 4 = -96$.

**36.** $a_{20} = 101$, and $d = 3$. Note that $a_{20} = a + 19d = a + 19(3) = a + 57$. Since $a_{20} = 101$ we have $a + 57 = a_{20} = 101 \quad\Leftrightarrow\quad a = 44$. Hence the $n$th term is $a_n = 44 + 3(n-1) = 41 + 3n$.

**37.** The arithmetic sequence is $1, 4, 7, \ldots$. So $d = 4 - 1 = 3$ and $a_n = 1 + 3(n-1)$. Then $a_n = 88 \Leftrightarrow 1 + 3(n-1) = 88 \Leftrightarrow 3(n-1) = 87 \Leftrightarrow n - 1 = 29 \Leftrightarrow n = 30$. So 88 is the 30th term.

**38.** If 11,937 is a term of an arithmetic sequence with $a_1 = 1$ and common difference 4, then $11937 = 1 + 4(n-1)$ for some integer $n$. Solving for $n$, we have $11937 = 1 + 4(n-1) = -3 + 4n \Leftrightarrow 11940 = 4n \Leftrightarrow n = 2985$. Thus 11,937 is the 2985th term of this sequence.

**39.** $a = 1, d = 2, n = 10$. Then $S_{10} = \frac{10}{2}[2a + (10-1)d] = \frac{10}{2}[2 \cdot 1 + 9 \cdot 2] = 100$.

**40.** $a = 3, d = 2, n = 12$. Then $S_{12} = \frac{12}{2}[2a + (12-1)d] = \frac{12}{2}[2 \cdot 3 + 11 \cdot 2] = 168$.

**41.** $a = 4, d = 2, n = 20$. Then $S_{20} = \frac{20}{2}[2a + (20-1)d] = \frac{20}{2}[2 \cdot 4 + 19 \cdot 2] = 460$.

**42.** $a = 100, d = -5, n = 8$. Then $S_8 = \frac{8}{2}[2a + (8-1)d] = \frac{8}{2}[2 \cdot 100 - 7 \cdot 5] = 660$.

**43.** $a_1 = 55, d = 12, n = 10$. Then $S_{10} = \frac{10}{2}[2a + (10-1)d] = \frac{10}{2}[2 \cdot 55 + 9 \cdot 12] = 1090$.

**44.** $a_2 = 8, a_5 = 9.5, n = 15$. Thus $a_2 = a + d = 8$ and $a_5 = a + 4d = 9.5$. Subtracting the first equation from the second gives $3d = 1.5 \Leftrightarrow d = 0.5$. Substituting for $d$ in the first equation gives $a + 0.5 = 8 \Leftrightarrow a = 7.5$. Thus $S_{15} = \frac{15}{2}[2 \cdot 7.5 + 14 \cdot 0.5] = 165$.

**45.** $1 + 5 + 9 + \cdots + 401$ is a partial sum of an arithmetic series, where $a = 1$ and $d = 5 - 1 = 4$. The last term is $401 = a_n = 1 + 4(n-1)$, so $n - 1 = 100 \Leftrightarrow n = 101$. So the partial sum is $S_{101} = \frac{101}{2}(1 + 401) = 101 \cdot 201 = 20{,}301$.

**46.** $-3 + \left(-\frac{3}{2}\right) + 0 + \frac{3}{2} + 3 + \cdots + 30$ is a partial sum of an arithmetic sequence, where $a = -3$ and $d = \left(-\frac{3}{2}\right) - (-3) = \frac{3}{2}$. The last term is $30 = a_n = -3 + \frac{3}{2}(n-1)$, so $22 = n - 1 \Leftrightarrow n = 23$. So the partial sum is $S_{23} = \frac{23}{2}(-3 + 30) = \frac{621}{2} = 310.5$.

**47.** $0.7 + 2.7 + 4.7 + \cdots + 56.7$ is a partial sum of an arithmetic series, where $a = 0.7$ and $d = 2.7 - 0.7 = 2$. The last term is $56.7 = a_n = 0.7 + 2(n-1) \Leftrightarrow 28 = n - 1 \Leftrightarrow n = 29$. So the partial sum is $S_{29} = \frac{29}{2}(0.7 + 56.7) = 832.3$.

**48.** $-10 + (-9.9) + (-9.8) + \cdots + (-0.1)$ is a partial sum of an arithmetic sequence, where $a = -10$ and $d = 0.1$. The last term is $-0.1 = a_n = -10 + 0.1(n-1)$, so $99 = n - 1 \Leftrightarrow n = 100$. So the partial sum is $S_{100} = \frac{100}{2}(-10 - 0.1) = -505$.

**49.** $\sum_{k=0}^{10}(3 + 0.25k)$ is a partial sum of an arithmetic series where $a = 3 + 0.25 \cdot 0 = 3$ and $d = 0.25$. The last term is $a_{11} = 3 + 0.25 \cdot 10 = 5.5$. So the partial sum is $S_{11} = \frac{11}{2}(3 + 5.5) = 46.75$.

**50.** $\sum_{n=0}^{20}(1 - 2n)$ is a partial sum of an arithmetic sequence where $a = 1 - 2 \cdot 0 = 1$, $d = -2$, and the last term is $a_{21} = 1 - 2 \cdot 20 = -39$. So the partial sum is $S_{21} = \frac{21}{2}(1 - 39) = -399$.

**51.** Let $x$ denote the length of the side between the length of the other two sides. Then the lengths of the three sides of the triangle are $x - a$, $x$, and $x + a$, for some $a > 0$. Since $x + a$ is the longest side, it is the hypotenuse, and by the Pythagorean Theorem, we know that $(x-a)^2 + x^2 = (x+a)^2 \Leftrightarrow x^2 - 2ax + a^2 + x^2 = x^2 + 2ax + a^2 \Leftrightarrow x^2 - 4ax = 0 \Leftrightarrow x(x - 4a) = 0 \Rightarrow x = 4a$ ($x = 0$ is not a possible solution). Thus, the lengths of the three sides are $x - a = 4a - a = 3a$, $x = 4a$, and $x + a = 4a + a = 5a$. The lengths $3a$, $4a$, $5a$ are proportional to 3, 4, 5, and so the triangle is similar to a 3-4-5 triangle.

**52.** $P = 10^{1/10} \cdot 10^{2/10} \cdot 10^{3/10} \cdot \cdots \cdot 10^{19/10} = 10^{(1+2+3+\cdots+19)/10}$. Now, $1 + 2 + 3 + \cdots + 19$ is an arithmetic series with $a = 1$, $d = 1$, and $n = 19$. Thus, $1 + 2 + 3 + \cdots + 19 = S_{19} = 19\dfrac{(1+19)}{2} = 190$, and so $P = 10^{190/10} = 10^{19}$.

**53.** The sequence $1, \frac{3}{5}, \frac{3}{7}, \frac{1}{3}, \ldots$ is harmonic if $1, \frac{5}{3}, \frac{7}{3}, 3, \ldots$ forms an arithmetic sequence. Since $\frac{5}{3} - 1 = \frac{7}{3} - \frac{5}{3} = 3 - \frac{7}{3} = \frac{2}{3}$, the sequence of reciprocals is arithmetic and thus the original sequence is harmonic.

**54.** The two original numbers are 3 and 5. Thus, the reciprocals are $\frac{1}{3}$ and $\frac{1}{5}$, and their average is $\frac{1}{2}\left(\frac{1}{3} + \frac{1}{5}\right) = \frac{1}{2}\left(\frac{5}{15} + \frac{3}{15}\right) = \frac{4}{15}$. Therefore, the harmonic mean is $\frac{15}{4}$.

**55.** We have an arithmetic sequence with $a = 5$ and $d = 2$. We seek $n$ such that $2700 = S_n = \frac{n}{2}[2a + (n-1)d]$. Solving for $n$, we have $2700 = \frac{n}{2}[10 + 2(n-1)] \Leftrightarrow 5400 = 10n + 2n^2 - 2n \Leftrightarrow n^2 + 4n - 2700 = 0 \Leftrightarrow (n-50)(n+54) = 0 \Leftrightarrow n = 50$ or $n = -54$. Since $n$ is a positive integer, 50 terms of the sequence must be added to get 2700.

**56.** Since the sequence is arithmetic, $a_4 - a_1 = 3d = 16 - 1 = 15 \Leftrightarrow 3d = 15 \Leftrightarrow d = 5$. Now since $S_n = \frac{n}{2}[2a + (n-1)d]$ we have $2356 = \frac{n}{2}[2(1) + (n-1)5] \Leftrightarrow 4712 = n(2 + 5n - 5) = 5n^2 - 3n \Leftrightarrow 5n^2 - 3n - 4712 = 0$. Using the quadratic equation we have $n = \dfrac{3 \pm \sqrt{3^2 + 4(5)(4712)}}{2(5)} = \dfrac{3 \pm \sqrt{94{,}249}}{10} = \dfrac{3 \pm 307}{10}$. Since $n$ is positive, we have $n = \frac{310}{10} = 31$. Thus 31 terms must be added.

**57.** The diminishing values of the computer form an arithmetic sequence with $a_1 = 12{,}500$ and common difference $d = -1875$. Thus the value of the computer after 6 years is $a_7 = 12{,}500 + (7-1)(-1875) = \$1250$.

**58.** The number of poles in a layer can be viewed as an arithmetic sequence, where $a_1 = 25$ and the common difference is $-1$. The number of poles in the first 12 layers is $S_{12} = \frac{12}{2}[2(25) + 11(-1)] = 6 \cdot 39 = 234$.

**59.** The increasing values of the man's salary form an arithmetic sequence with $a_1 = 30{,}000$ and common difference $d = 2300$. Then his total earnings for a ten-year period are $S_{10} = \frac{10}{2}[2(30{,}000) + 9(2300)] = 403{,}500$. Thus his total earnings for the 10 year period are \$403,500.

**60.** The number of cars that can park in a row can be viewed as an arithmetic sequence, where $a_1 = 20$ and the common difference is 2. Thus the number of cars that can park in the 21 rows is $S_{21} = \frac{21}{2}[2(20) + 20(2)] = 10.5 \cdot 80 = 840$.

**61.** The number of seats in the $n$th row is given by the $n$th term of an arithmetic sequence with $a_1 = 15$ and common difference $d = 3$. We need to find $n$ such that $S_n = 870$. So we solve $870 = S_n = \frac{n}{2}[2(15) + (n-1)3]$ for $n$. We have $870 = \frac{n}{2}(27 + 3n) \Leftrightarrow 1740 = 3n^2 + 27n \Leftrightarrow 3n^2 + 27n - 1740 = 0 \Leftrightarrow n^2 + 9n - 580 = 0 \Leftrightarrow (x-20)(x+29) = 0 \Rightarrow n = 20$ or $n = -29$. Since the number of rows is positive, the theater must have 20 rows.

**62.** The sequence is 16, 48, 80, .... This is an arithmetic sequence with $a = 16$ and $d = 48 - 16 = 32$.

**(a)** The total distance after 6 seconds is $S_6 = \frac{6}{2}(32 + 5 \cdot 32) = 3 \cdot 192 = 576$ ft.

**(b)** The total distance after $n$ seconds is $S_n = \frac{n}{2}[32 + 32(n-1)] = 16n^2$ ft.

**63.** The number of gifts on the 12th day is $1 + 2 + 3 + 4 + \cdots + 12$. Since $a_2 - a_1 = a_3 - a_2 = a_4 - a_3 = \cdots = 1$, the number of gifts on the 12th day is the partial sum of an arithmetic sequence with $a = 1$ and $d = 1$. So the sum is $S_{12} = 12\left(\dfrac{1+12}{2}\right) = 6 \cdot 13 = 78$.

**64. (a)** We want an arithmetic sequence with4 terms, so let $a_1 = 10$ and $a_4 = 18$. Since the sequence is arithmetic, $a_4 - a_1 = 3d = 18 - 10 = 8 \Leftrightarrow 3d = 8 \Leftrightarrow d = \frac{8}{3}$. Therefore, $a_2 = 10 + \frac{8}{3} = \frac{38}{3}$ and $a_3 = 10 + 2\left(\frac{8}{3}\right) = \frac{46}{3}$ are the two arithmetic means between 10 and 18.

**(b)** We want an arithmetic sequence with5 terms, so let $a_1 = 10$ and $a_5 = 18$. Since the sequence is arithmetic, $a_5 - a_1 = 4d = 18 - 10 = 8 \Leftrightarrow 4d = 8 \Leftrightarrow d = 2$. Therefore, $a_2 = 10 + 2 = 12$, $a_3 = 10 + 2(2) = 14$, and $a_4 = 10 + 3(2) = 16$ are the three arithmetic means between 10 and 18.

**(c)** We want an arithmetic sequence with6 terms, with the starting dosage $a_1 = 100$ and the final dosage $a_6 = 300$. Since the sequence is arithmetic, $a_6 - a_1 = 5d = 300 - 100 = 200 \Leftrightarrow 5d = 200 \Leftrightarrow d = 40$. Therefore, $a_2 = 140$, $a_3 = 180$, $a_4 = 220$, $a_5 = 260$, and $a_6 = 300$. The patient should take 140 mg, then 180 mg, then 220 mg, then 260 mg, and finally arrive at 300 mg.

# 12.3 Geometric Sequences

**1. (a)** $a_1 = 5(2)^0 = 5$, $a_2 = 5(2)^1 = 10$,

$a_3 = 5(2)^2 = 20$, $a_4 = 5(2)^3 = 40$,

$a_5 = 5(2)^4 = 80$

**(b)** The common ratio is 2.

**(c)**

**2. (a)** $a_1 = 3(-4)^0 = 3$, $a_2 = 3(-4)^1 = -12$,

$a_3 = 3(-4)^2 = 48$, $a_4 = 3(-4)^3 = -192$,

$a_5 = 3(-4)^4 = 768$

**(b)** The common ratio is $-4$.

**(c)**

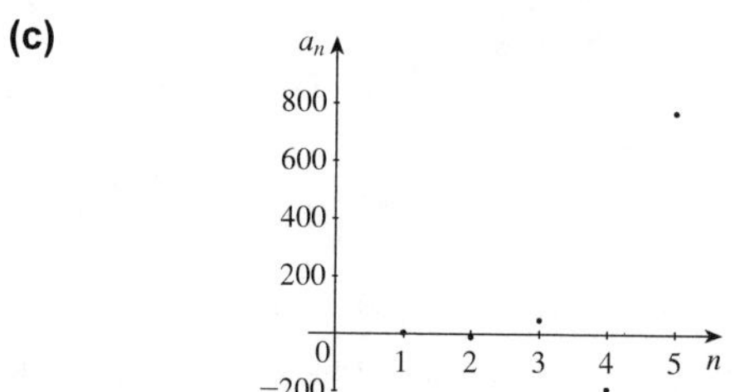

**3. (a)** $a_1 = \frac{5}{2}\left(-\frac{1}{2}\right)^0 = \frac{5}{2}$, $a_2 = \frac{5}{2}\left(-\frac{1}{2}\right)^1 = -\frac{5}{4}$,

$a_3 = \frac{5}{2}\left(-\frac{1}{2}\right)^2 = \frac{5}{8}$, $a_4 = \frac{5}{2}\left(-\frac{1}{2}\right)^3 = -\frac{5}{16}$,

$a_5 = \frac{5}{2}\left(-\frac{1}{2}\right)^4 = \frac{5}{32}$

**(b)** The common ratio is $-\frac{1}{2}$.

**(c)**

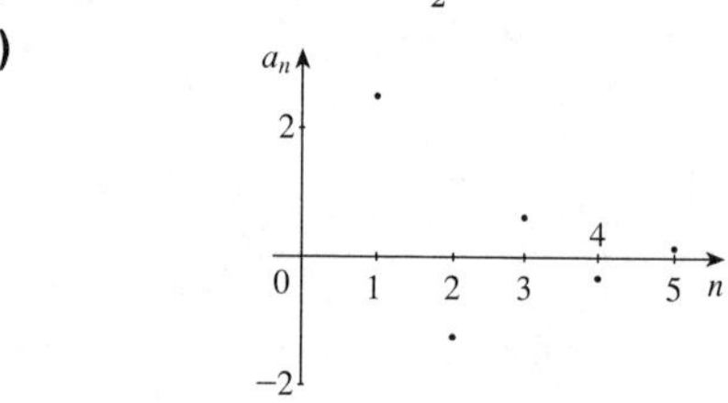

**4. (a)** $a_1 = 3^0 = 1$, $a_2 = 3^1 = 3$, $a_3 = 3^2 = 9$,

$a_4 = 3^3 = 27$, $a_5 = 3^4 = 81$

**(b)** The common ratio is 3.

**(c)**

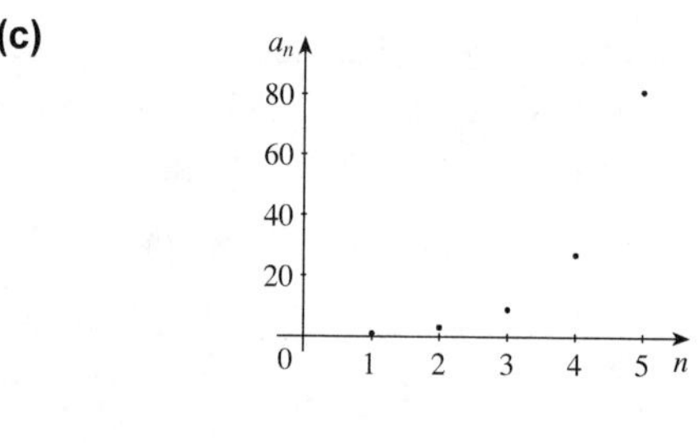

**5.** $a = 3$, $r = 5$. So $a_n = ar^{n-1} = 3(5)^{n-1}$ and $a_4 = 3 \cdot 5^3 = 375$.

**6.** $a = -6$, $r = 3$. So $a_n = ar^{n-1} = -6(3)^{n-1}$ and $a_4 = -6 \cdot 3^3 = -162$.

**7.** $a = \frac{5}{2}$, $r = -\frac{1}{2}$. So $a_n = ar^{n-1} = \frac{5}{2}\left(-\frac{1}{2}\right)^{n-1}$ and $a_4 = \frac{5}{2} \cdot \left(-\frac{1}{2}\right)^3 = -\frac{5}{16}$.

**8.** $a = \sqrt{3}$, $r = \sqrt{3}$. So $a_n = ar^{n-1} = \sqrt{3}\left(\sqrt{3}\right)^{n-1} = \left(\sqrt{3}\right)^n$ and $a_4 = \sqrt{3} \cdot \left(\sqrt{3}\right)^3 = 9$.

**9.** $\dfrac{a_2}{a_1} = \dfrac{4}{2} = 2$, $\dfrac{a_3}{a_2} = \dfrac{8}{4} = 2$, $\dfrac{a_4}{a_3} = \dfrac{16}{8} = 2$. Since these ratios are the same, the sequence is geometric with the common ratio 2.

**10.** $\dfrac{a_2}{a_1} = \dfrac{6}{2} = 3$, $\dfrac{a_4}{a_3} = \dfrac{36}{18} = 2$. Since these ratios are not the same, this is not a geometric sequence.

**11.** $\dfrac{a_2}{a_1} = \dfrac{3/2}{3} = \dfrac{1}{2}$, $\dfrac{a_3}{a_2} = \dfrac{3/4}{3/2} = \dfrac{1}{2}$, $\dfrac{a_4}{a_3} = \dfrac{3/8}{3/4} = \dfrac{1}{2}$. Since these ratios are the same, the sequence is geometric with the common ratio $\frac{1}{2}$.

**12.** $\dfrac{a_2}{a_1} = \dfrac{-9}{27} = -\dfrac{1}{3}$, $\dfrac{a_3}{a_2} = \dfrac{3}{-9} = -\dfrac{1}{3}$, $\dfrac{a_4}{a_3} = -\dfrac{1}{3}$. Since these ratios are the same, the sequence is geometric with the common ratio $-\frac{1}{3}$.

**13.** $\dfrac{a_2}{a_1} = \dfrac{1/3}{1/2} = \dfrac{2}{3}$, $\dfrac{a_4}{a_3} = \dfrac{1/5}{1/4} = \dfrac{4}{5}$. Since these ratios are not the same, this is not a geometric sequence.

**14.** $\dfrac{a_2}{a_1} = \dfrac{e^4}{e^2} = e^2$, $\dfrac{a_3}{a_2} = \dfrac{e^6}{e^4} = e^2$, $\dfrac{a_4}{a_3} = \dfrac{e^8}{e^6} = e^2$. Since these ratios are the same, the sequence is geometric with the common ratio $e^2$.

**15.** $\dfrac{a_2}{a_1} = \dfrac{1.1}{1.0} = 1.1$, $\dfrac{a_3}{a_2} = \dfrac{1.21}{1.1} = 1.1$, $\dfrac{a_4}{a_3} = \dfrac{1.331}{1.21} = 1.1$. Since these ratios are the same, the sequence is geometric with the common ratio 1.1.

**16.** $\dfrac{a_2}{a_1} = \dfrac{\frac{1}{4}}{\frac{1}{2}} = \dfrac{1}{2}$, $\dfrac{a_4}{a_3} = \dfrac{\frac{1}{8}}{\frac{1}{6}} = \dfrac{3}{4}$. Since these ratios are not the same, this is not a geometric sequence.

**17.** $a_1 = 2(3)^1 = 6$, $a_2 = 2(3)^2 = 18$, $a_3 = 2(3)^3 = 54$, $a_4 = 2(3)^4 = 162$, $a_5 = 2(3)^5 = 486$. This sequence is geometric, the common ratio is $r = 3$ and $a_n = a_1 r^{n-1} = 6(3)^{n-1}$.

**18.** $a_1 = 4 + (3)^1 = 7$, $a_2 = 4 + (3)^2 = 13$, $a_3 = 4 + (3)^3 = 31$, $a_4 = 4 + (3)^4 = 85$, $a_5 = 4 + (3)^5 = 247$. Since $\dfrac{a_2}{a_1} = \frac{13}{7}$ and $\dfrac{a_3}{a_2} = \frac{31}{13}$ are different ratios, this is not a geometric sequence.

**19.** $a_1 = \dfrac{1}{4}$, $a_2 = \dfrac{1}{4^2} = \dfrac{1}{16}$, $a_3 = \dfrac{1}{4^3} = \dfrac{1}{64}$, $a_4 = \dfrac{1}{4^4} = \dfrac{1}{256}$, $a_5 = \dfrac{1}{4^5} = \dfrac{1}{1024}$. This sequence is geometric, the common ratio is $r = \frac{1}{4}$ and $a_n = a_1 r^{n-1} = \frac{1}{4}\left(\frac{1}{4}\right)^{n-1}$.

**20.** $a_1 = (-1)^1(2)^1 = -2$, $a_2 = (-1)^2(2)^2 = 4$, $a_3 = (-1)^3(2)^3 = -8$, $a_4 = (-1)^4(2)^4 = 16$, $a_5 = (-1)^5(2)^5 = -32$. This sequence is geometric, the common ratio is $r = -2$ and $a_n = a_1 r^{n-1} = -2(-2)^{n-1}$.

**21.** Since $\ln a^b = b \ln a$, we have $a_1 = \ln(5^0) = \ln 1 = 0$, $a_2 = \ln(5^1) = \ln 5$, $a_3 = \ln(5^2) = 2\ln 5$, $a_4 = \ln(5^3) = 3\ln 5$, $a_5 = \ln(5^4) = 4\ln 5$. Since $a_1 = 0$ and $a_2 \neq 0$, this sequence is not geometric.

**22.** $a_1 = 1^1 = 1$, $a_2 = 2^2 = 4$, $a_3 = 3^3 = 27$, $a_4 = 4^4 = 256$, $a_5 = 5^5 = 3125$. Since $\dfrac{a_2}{a_1} = \dfrac{4}{1} = 4$ and $\dfrac{a_3}{a_2} = \dfrac{27}{4}$ are different, this is not a geometric sequence.

**23.** $2, 6, 18, 54, \ldots$. Then $r = \dfrac{a_2}{a_1} = \frac{6}{2} = 3$, $a_5 = a_4 \cdot 3 = 54(3) = 162$, and $a_n = 2 \cdot 3^{n-1}$.

**24.** $7, \frac{14}{3}, \frac{28}{9}, \frac{56}{27}, \ldots$. Then $r = \dfrac{a_2}{a_1} = \dfrac{\frac{14}{3}}{7} = \frac{2}{3}$, $a_5 = a_4 \cdot \frac{2}{3} = \frac{56}{27} \cdot \frac{2}{3} = \frac{112}{81}$, and $a_n = 7\left(\frac{2}{3}\right)^{n-1}$.

**25.** $0.3, -0.09, 0.027, -0.0081, \ldots$. Then $r = \dfrac{a_2}{a_1} = \dfrac{-0.09}{0.3} = -0.3$, $a_5 = a_4 \cdot (-0.3) = -0.0081(-0.3) = 0.00243$, and $a_n = 0.3(-0.3)^{n-1}$.

**26.** $1, \sqrt{2}, 2, 2\sqrt{2}, \ldots$. Then $r = \dfrac{a_2}{a_1} = \dfrac{\sqrt{2}}{1} = \sqrt{2}$, $a_5 = a_4 \cdot \sqrt{2} = 2\sqrt{2} \cdot \sqrt{2} = 4$, and $a_n = \left(\sqrt{2}\right)^{n-1}$.

**27.** $144, -12, 1, -\frac{1}{12}, \ldots$. Then $r = \dfrac{a_2}{a_1} = \dfrac{-12}{144} = -\frac{1}{12}$, $a_5 = a_4 \cdot \left(-\frac{1}{12}\right) = -\frac{1}{12}\left(-\frac{1}{12}\right) = \frac{1}{144}$, $a_n = 144\left(-\frac{1}{12}\right)^{n-1}$.

**28.** $-8, -2, -\frac{1}{2}, -\frac{1}{8}, \ldots$. Then $r = \dfrac{a_2}{a_1} = \dfrac{-2}{-8} = \frac{1}{4}$, $a_5 = a_4 \cdot \dfrac{1}{4} = -\frac{1}{8} \cdot \frac{1}{4} = -\frac{1}{32}$, and $a_n = -8\left(\frac{1}{4}\right)^{n-1}$.

**29.** $3, 3^{5/3}, 3^{7/3}, 27, \ldots$. Then $r = \dfrac{a_2}{a_1} = \dfrac{3^{5/3}}{3} = 3^{2/3}$, $a_5 = a_4 \cdot \left(3^{2/3}\right) = 27 \cdot 3^{2/3} = 3^{11/3}$, and $a_n = 3\left(3^{2/3}\right)^{n-1} = 3 \cdot 3^{(2n-2)/3} = 3^{(2n+1)/3}$.

**30.** $t, \dfrac{t^2}{2}, \dfrac{t^3}{4}, \dfrac{t^4}{8}, \ldots$. Then $r = \dfrac{a_2}{a_1} = \dfrac{\frac{t^2}{2}}{t} = \dfrac{t}{2}$, $a_5 = a_4 \cdot \dfrac{t}{2} = \dfrac{t^4}{8} \cdot \dfrac{t}{2} = \dfrac{t^5}{16}$, and $a_n = t\left(\dfrac{t}{2}\right)^{n-1}$.

**31.** $1, s^{2/7}, s^{4/7}, s^{6/7}, \ldots$. Then $r = \dfrac{a_2}{a_1} = \dfrac{s^{2/7}}{1} = s^{2/7}$, $a_5 = a_4 \cdot s^{2/7} = s^{6/7} \cdot s^{2/7} = s^{8/7}$, and $a_n = \left(s^{2/7}\right)^{n-1} = s^{(2n-2)/7}$.

**32.** $5, 5^{c+1}, 5^{2c+1}, 5^{3c+1}, \ldots$. Then $r = \dfrac{a_2}{a_1} = \dfrac{5^{c+1}}{5} = 5^c$, $a_5 = a_4 \cdot 5^c = 5^{3c+1} \cdot 5^c = 5^{4c+1}$, and $a_n = 5(5^c)^{n-1} = 5 \cdot 5^{cn-c} = 5^{cn-c+1}$.

**33.** $a_1 = 8$, $a_2 = 4$. Thus $r = \dfrac{a_2}{a_1} = \dfrac{4}{8} = \dfrac{1}{2}$ and $a_5 = a_1 r^{5-1} = 8\left(\frac{1}{2}\right)^4 = \dfrac{8}{16} = \dfrac{1}{2}$.

**34.** $a_1 = 3$, $a_3 = \frac{4}{3}$. Thus $r^2 = \dfrac{\frac{4}{3}}{3} = \dfrac{4}{9} \quad \Leftrightarrow \quad r = \pm\dfrac{2}{3}$. If $r = \frac{2}{3}$, then $a_5 = 3\left(\frac{2}{3}\right)^4 = \frac{16}{27}$, and if $r = -\frac{2}{3}$, then $a_5 = 3\left(-\frac{2}{3}\right)^4 = \frac{16}{27}$. Therefore, $a_5 = \frac{16}{27}$.

**35.** $r = \frac{2}{5}$, $a_4 = \frac{5}{2}$. Since $r = \dfrac{a_4}{a_3}$, we have $a_3 = \dfrac{a_4}{r} = \dfrac{5/2}{2/5} = \dfrac{25}{4}$.

**36.** $r = \frac{3}{2}$, $a_5 = 1$. Then $1 = a_5 = a \cdot r^4 = a\left(\frac{3}{2}\right)^4 \quad \Leftrightarrow \quad 1 = a\left(\frac{3}{2}\right)^4 \quad \Leftrightarrow \quad a = \left(\frac{2}{3}\right)^4 = \frac{16}{81}$. Therefore, $a_1 = \frac{16}{81}$, $a_2 = \frac{16}{81}\left(\frac{3}{2}\right) = \frac{8}{27}$, and $a_3 = \frac{8}{27}\left(\frac{3}{2}\right) = \frac{4}{9}$.

**37.** The geometric sequence is 2, 6, 18, .... Thus $r = \dfrac{a_2}{a_1} = \frac{6}{2} = 3$. We need to find $n$ so that $a_n = 2 \cdot 3^{n-1} = 118{,}098 \quad \Leftrightarrow$ $3^{n-1} = 59{,}049 \quad \Leftrightarrow \quad n - 1 = \log_3 59{,}049 = 10 \quad \Leftrightarrow \quad n = 11$. Therefore, 118,098 is the 11th term of the geometric sequence.

**38.** $a_2 = 10$, $a_5 = 1250$. Thus, $\frac{1250}{10} = \dfrac{a_5}{a_2} = \dfrac{ar^4}{ar} = r^3 \quad \Leftrightarrow \quad r = 5$. Then $a_1 = \dfrac{a_2}{r} = \frac{10}{5} = 2$, so $a_n = 2 \cdot 5^{n-1}$. We seek $n$ such that $a_n = 31{,}250 \quad \Leftrightarrow \quad 2 \cdot 5^{n-1} = 31{,}250 \quad \Leftrightarrow \quad 5^{n-1} = 15{,}625 \quad \Leftrightarrow \quad n - 1 = \log_5 15{,}625 = 6 \quad \Leftrightarrow \quad n = 7$. Therefore, 31,250 is the 7th term of the sequence.

**39.** $a = 5$, $r = 2$, $n = 6$. Then $S_6 = 5\dfrac{1 - 2^6}{1 - 2} = (-5)(-63) = 315$.

**40.** $a = \frac{2}{3}$, $r = \frac{1}{3}$, $n = 4$. Then $S_4 = \left(\dfrac{2}{3}\right)\dfrac{1 - \left(\frac{1}{3}\right)^4}{1 - \frac{1}{3}} = \left(\frac{2}{3}\right)\dfrac{\frac{80}{81}}{\frac{2}{3}} = \frac{80}{81}$.

**41.** $a_3 = 28$, $a_6 = 224$, $n = 6$. So $\dfrac{a_6}{a_3} = \dfrac{ar^5}{ar^2} = r^3$. So we have $r^3 = \dfrac{a_6}{a_3} = \frac{224}{28} = 8$, and hence $r = 2$. Since $a_3 = a \cdot r^2$, we get $a = \dfrac{a_3}{r^2} = \dfrac{28}{2^2} = 7$. So $S_6 = 7\dfrac{1 - 2^6}{1 - 2} = (-7)(-63) = 441$.

**42.** $a_2 = 0.12$, $a_5 = 0.00096$, $n = 4$. So $r^3 = \dfrac{a_5}{a_2} = \dfrac{0.00096}{0.12} = 0.008 \quad \Leftrightarrow \quad r = 0.2$, and thus $a_1 = \dfrac{a_2}{r} = \dfrac{0.12}{0.2} = 0.6$.

Therefore, $S_4 = (0.6)\,\dfrac{1 - (0.2)^4}{1 - 0.2} = 0.7488$.

**43.** $1 + 3 + 9 + \cdots + 2187$ is a partial sum of a geometric sequence, where $a = 1$ and $r = \dfrac{a_2}{a_1} = \frac{3}{1} = 3$. Then the last term is $2187 = a_n = 1 \cdot 3^{n-1} \quad \Leftrightarrow \quad n - 1 = \log_3 2187 = 7 \quad \Leftrightarrow \quad n = 8$. So the partial sum is $S_8 = (1)\,\dfrac{1 - 3^8}{1 - 3} = 3280$.

**44.** $1 - \frac{1}{2} + \frac{1}{4} - \frac{1}{8} + \cdots - \frac{1}{512}$ is a partial sum of a geometric sequence, where $a = 1$ and $r = \dfrac{a_2}{a_1} = \dfrac{-\frac{1}{2}}{1} = -\frac{1}{2}$. The last term is $a_n$, where $\dfrac{-1}{512} = a_n = 1\left(-\frac{1}{2}\right)^{n-1}$, so $n = 10$. So the partial sum is $S_{10} = (1)\,\dfrac{1 - \left(-\frac{1}{2}\right)^{10}}{1 - \left(-\frac{1}{2}\right)} = \frac{341}{512}$.

**45.** $\sum_{k=0}^{10} 3\left(\frac{1}{2}\right)^k$ is a partial sum of a geometric sequence, where $a = 3$, $r = \frac{1}{2}$, and $n = 11$. So the partial sum is $S_{11} = (3)\,\dfrac{1 - \left(\frac{1}{2}\right)^{11}}{1 - \left(\frac{1}{2}\right)} = 6\left[1 - \left(\frac{1}{2}\right)^{11}\right] = \frac{6141}{1024} \approx 5.997070313$.

**46.** $\sum_{j=0}^{5} 7\left(\frac{3}{2}\right)^j$ is a partial sum of a geometric sequence, where $a = 7$, $r = \frac{3}{2}$, and $n = 6$. So the partial sum is $S_6 = (7)\,\dfrac{1 - \left(\frac{3}{2}\right)^6}{1 - \frac{3}{2}} = (7)\,\dfrac{1 - \left(\frac{729}{64}\right)}{-\frac{1}{2}} = -14\left(1 - \frac{729}{64}\right) = \dfrac{4655}{32}$.

**47.** $1 + \frac{1}{3} + \frac{1}{9} + \frac{1}{27} + \cdots$ is an infinite geometric series with $a = 1$ and $r = \frac{1}{3}$. Therefore, the sum of the series is $S = \dfrac{a}{1 - r} = \dfrac{1}{1 - \left(\frac{1}{3}\right)} = \dfrac{3}{2}$.

**48.** $1-\frac{1}{2}+\frac{1}{4}-\frac{1}{8}+\cdots$ is an infinite geometric series with $a=1$ and $r=-\frac{1}{2}$. Therefore, the sum of the series is $S=\dfrac{1}{1-\left(-\frac{1}{2}\right)}=\dfrac{1}{\frac{3}{2}}=\dfrac{2}{3}$.

**49.** $1-\frac{1}{3}+\frac{1}{9}-\frac{1}{27}+\cdots$ is an infinite geometric series with $a=1$ and $r=-\frac{1}{3}$. Therefore, the sum of the series is $S=\dfrac{a}{1-r}=\dfrac{1}{1-\left(-\frac{1}{3}\right)}=\dfrac{3}{4}$.

**50.** $\frac{2}{5}+\frac{4}{25}+\frac{8}{125}+\cdots$ is an infinite geometric series with $a=\frac{2}{5}$ and $r=\frac{2}{5}$. Therefore, the sum of the series is $S=\dfrac{\frac{2}{5}}{1-\frac{2}{5}}=\dfrac{\frac{2}{5}}{\frac{3}{5}}=\dfrac{2}{3}$.

**51.** $\dfrac{1}{3^6}+\dfrac{1}{3^8}+\dfrac{1}{3^{10}}+\dfrac{1}{3^{12}}+\cdots$ is an infinite geometric series with $a=\dfrac{1}{3^6}$ and $r=\dfrac{1}{3^2}=\frac{1}{9}$. Therefore, the sum of the series is $S=\dfrac{a}{1-r}=\dfrac{\frac{1}{3^6}}{1-\left(\frac{1}{9}\right)}=\dfrac{1}{3^6}\cdot\dfrac{9}{8}=\dfrac{1}{648}$.

**52.** $3-\frac{3}{2}+\frac{3}{4}-\frac{3}{8}+\cdots$ is an infinite geometric series with $a=3$ and $r=-\frac{1}{2}$. Therefore, the sum of the series is $S=\dfrac{3}{1-\left(-\frac{1}{2}\right)}=2$.

**53.** $-\frac{100}{9}+\frac{10}{3}-1+\frac{3}{10}-\cdots$ is an infinite geometric series with $a=-\dfrac{100}{9}$ and $r=\dfrac{\frac{10}{3}}{-\frac{100}{9}}=\dfrac{3}{10}$. Therefore, the sum of the series is $S=\dfrac{a}{1-r}=\dfrac{-\frac{100}{9}}{1-\left(-\frac{3}{10}\right)}=\dfrac{-\frac{100}{9}}{\frac{13}{10}}=-\frac{100}{9}\cdot\frac{10}{13}=-\frac{1000}{117}$.

**54.** $\frac{1}{\sqrt{2}}+\frac{1}{2}+\frac{1}{2\sqrt{2}}+\frac{1}{4}+\cdots$ is an infinite geometric series with $a=\frac{1}{\sqrt{2}}$ and $r=\frac{1}{\sqrt{2}}$. Therefore, the sum of the series is $S=\dfrac{\frac{1}{\sqrt{2}}}{1-\frac{1}{\sqrt{2}}}=\dfrac{1}{\sqrt{2}-1}=\sqrt{2}+1$.

**55.** $0.777\ldots=\frac{7}{10}+\frac{7}{100}+\frac{7}{1000}+\cdots$ is an infinite geometric series with $a=\frac{7}{10}$ and $r=\frac{1}{10}$. Thus $0.777\ldots=\dfrac{a}{1-r}=\dfrac{\frac{7}{10}}{1-\frac{1}{10}}=\dfrac{7}{9}$.

**56.** $0.2535353\ldots=0.2+\frac{53}{1000}+\frac{53}{100{,}000}+\frac{53}{10{,}000{,}000}+\cdots$ is an infinite geometric series (after the first term) with $a=\frac{53}{1000}$ and $r=\frac{1}{100}$. Thus $0.05353\ldots=\dfrac{\frac{53}{1000}}{1-\frac{1}{100}}=\frac{53}{1000}\cdot\frac{100}{99}=\frac{53}{990}$, and so $0.2535353\ldots=\dfrac{2}{10}+\dfrac{53}{990}=\dfrac{2\cdot 99+53}{990}=\dfrac{251}{990}$.

**57.** $0.030303\ldots=\frac{3}{100}+\frac{3}{10{,}000}+\frac{3}{1{,}000{,}000}+\cdots$ is an infinite geometric series with $a=\frac{3}{100}$ and $r=\frac{1}{100}$. Thus $0.030303\ldots=\dfrac{a}{1-r}=\dfrac{\frac{3}{100}}{1-\frac{1}{100}}=\dfrac{3}{99}=\dfrac{1}{33}$.

**58.** $2.11252525\ldots=2.11+\frac{25}{10{,}000}+\frac{25}{1{,}000{,}000}+\frac{25}{100{,}000{,}000}+\cdots$ is an infinite geometric series (after the first term) with $a=\frac{25}{10{,}000}$ and $r=\frac{1}{100}$. Thus $0.00252525\ldots=\dfrac{\frac{25}{10{,}000}}{1-\frac{1}{100}}=\dfrac{25}{9900}$, and so $2.11252525\ldots=\dfrac{211}{100}+\dfrac{25}{9900}=\dfrac{211\cdot 99+25}{9900}=\dfrac{20{,}914}{9900}=\dfrac{10{,}457}{4950}$.

**59.** $0.\overline{112}=0.112112112\ldots=\frac{112}{1000}+\frac{112}{1{,}000{,}000}+\frac{112}{1{,}000{,}000{,}000}+\cdots$ is an infinite geometric series with $a=\frac{112}{1000}$ and $r=\frac{1}{1000}$. Thus $0.112112112\ldots=\dfrac{a}{1-r}=\dfrac{\frac{112}{1000}}{1-\frac{1}{1000}}=\frac{112}{999}$.

**60.** $0.123123123\ldots=\frac{123}{1000}+\frac{123}{1{,}000{,}000}+\frac{123}{1{,}000{,}000{,}000}+\cdots$ is an infinite geometric series with $a=\frac{123}{1000}$ and $r=\frac{1}{1000}$. Thus $0.123123123\ldots=\dfrac{\frac{123}{1000}}{1-\frac{1}{1000}}=\dfrac{123}{999}$.

**61.** Since we have 5 terms, let us denote $a_1 = 5$ and $a_5 = 80$. Also, $\dfrac{a_5}{a_1} = r^4$ because the sequence is geometric, and so $r^4 = \frac{80}{5} = 16 \Leftrightarrow r = \pm 2$. If $r = 2$, the three geometric means are $a_2 = 10$, $a_3 = 20$, and $a_4 = 40$. (If $r = -2$, the three geometric means are $a_2 = -10$, $a_3 = 20$, and $a_4 = -40$, but these are not between 5 and 80.)

**62.** The sum is given by

$$\begin{aligned}(a+b)+(a^2+2b)+(a^3+3b)+\cdots+(a^{10}+10b) &= (a+a^2+a^3+\cdots+a^{10})+(b+2b+3b+\cdots+10b)\\ &= a\cdot\frac{1-a^{10}}{1-a}+\frac{10}{2}\cdot(b+10b) = a\cdot\frac{1-a^{10}}{1-a}+55b\end{aligned}$$

**63.** **(a)** The value at the end of the year is equal to the the value at beginning less the depreciation, so $V_n = V_{n-1} - 0.2V_{n-1} = 0.8V_{n-1}$ with $V_1 = 160{,}000$. Thus $V_n = 160{,}000 \cdot 0.8^{n-1}$.

**(b)** $V_n < 100{,}000 \Leftrightarrow 0.8^{n-1} \cdot 160{,}000 < 100{,}000 \Leftrightarrow 0.8^{n-1} < 0.625 \Leftrightarrow (n-1)\log 0.8 < \log 0.625 \Leftrightarrow n - 1 > \dfrac{\log 0.625}{\log 0.8} = 2.11$. Thus it will depreciate to below \$100,000 during the fourth year.

**64.** Let $a_n$ denote the number of ancestors a person has $n$ generations back. Then $a_1 = 2$, $a_2 = 4$, $a_3 = 8$, .... Since $\frac{4}{2} = \frac{8}{4} = \cdots = 2$, this is a geometric sequence with $r = 2$. Therefore, $a_{15} = 2 \cdot 2^{14} = 2^{15} = 32{,}768$.

**65.** Since the ball is dropped from a height of 80 feet, $a = 80$. Also since the ball rebounds three-fourths of the distance fallen, $r = \frac{3}{4}$. So on the $n$th bounce, the ball attains a height of $a_n = 80\left(\frac{3}{4}\right)^n$. Hence, on the fifth bounce, the ball goes $a_5 = 80\left(\frac{3}{4}\right)^5 = \frac{80\cdot 243}{1024} \approx 19$ ft high.

**66.** $a = 5000$, $r = 1.08$. After 1 hour, there are $5000 \cdot 1.08 = 5400$, after 2 hours, $5400 \cdot 1.08 = 5832$, after 3 hours, $5832 \cdot 1.08 = 6298.56$, after 4 hours, $6298.56 \cdot 1.08 = 6802.4448$, and after 5 hours, $6802.4448 \cdot 1.08 \approx 7347$ bacteria. After $n$ hours, the number of bacteria is $a_n = 5000 \cdot (1.08)^n$.

**67.** Let $a_n$ be the amount of water remaining at the $n$th stage. We start with 5 gallons, so $a = 5$. When 1 gallon (that is, $\frac{1}{5}$ of the mixture) is removed, $\frac{4}{5}$ of the mixture (and hence $\frac{4}{5}$ of the water in the mixture) remains. Thus, $a_1 = 5 \cdot \frac{4}{5}$, $a_2 = 5 \cdot \frac{4}{5} \cdot \frac{4}{5}$, ..., and in general, $a_n = 5\left(\frac{4}{5}\right)^n$. The amount of water remaining after 3 repetitions is $a_3 = 5\left(\frac{4}{5}\right)^3 = \frac{64}{25}$, and after 5 repetitions it is $a_5 = 5\left(\frac{4}{5}\right)^5 = \frac{1024}{625}$.

**68.** Let $a_C$ denote the term of the geometric series that is the frequency of middle C. Then $a_C = 256$ and $a_{C+1} = 512$. Since this is a geometric sequence, $r = \frac{512}{256} = 2$, and so $a_{C-2} = \dfrac{a_C}{r^2} = \dfrac{256}{2^2} = 64$.

**69.** Let $a_n$ be the height the ball reaches on the $n$th bounce. From the given information, $a_n$ is the geometric sequence $a_n = 9 \cdot \left(\frac{1}{3}\right)^n$. (Notice that the ball hits the ground for the fifth time after the fourth bounce.)

**(a)** $a_0 = 9$, $a_1 = 9 \cdot \frac{1}{3} = 3$, $a_2 = 9 \cdot \left(\frac{1}{3}\right)^2 = 1$, $a_3 = 9 \cdot \left(\frac{1}{3}\right)^3 = \frac{1}{3}$, and $a_4 = 9 \cdot \left(\frac{1}{3}\right)^4 = \frac{1}{9}$. The total distance traveled is $a_0 + 2a_1 + 2a_2 + 2a_3 + 2a_4 = 9 + 2 \cdot 3 + 2 \cdot 1 + 2 \cdot \frac{1}{3} + 2 \cdot \frac{1}{9} = \frac{161}{9} = 17\frac{8}{9}$ ft.

**(b)** The total distance traveled at the instant the ball hits the ground for the $n$th time is

$$\begin{aligned}D_n &= 9+2\cdot9\cdot\tfrac{1}{3}+2\cdot9\cdot\left(\tfrac{1}{3}\right)^2+2\cdot9\cdot\left(\tfrac{1}{3}\right)^3+2\cdot9\cdot\left(\tfrac{1}{3}\right)^4+\cdots+2\cdot9\cdot\left(\tfrac{1}{3}\right)^{n-1}\\ &= 2\left[9+9\cdot\tfrac{1}{3}+9\cdot\left(\tfrac{1}{3}\right)^2+9\cdot\left(\tfrac{1}{3}\right)^3+9\cdot\left(\tfrac{1}{3}\right)^4+\cdots+9\cdot\left(\tfrac{1}{3}\right)^{n-1}\right]-9\\ &= 2\left[9\cdot\frac{1-\left(\frac{1}{3}\right)^n}{1-\frac{1}{3}}\right]-9 = 27\left[1-\left(\tfrac{1}{3}\right)^n\right]-9 = 18-\left(\tfrac{1}{3}\right)^{n-3}\end{aligned}$$

**70.** We have a geometric sequence with $a = 1$ and $r = 2$. Then $S_n = (1)\dfrac{1-2^n}{1-2} = 2^n - 1$. At the end of 30 days, she will have $S_{30} = 2^{30} - 1 = 1{,}073{,}741{,}823$ cents $= \$10{,}737{,}418.23$. To become a billionaire, we want $2^n - 1 = 10^{11}$ or approximately $2^n = 10^{11}$. So $\log 2^n = \log 10^{11} \Leftrightarrow n = \dfrac{11}{\log 2} \approx 36.5$. Thus it will take 37 days.

**71.** Let $a_1 = 1$ be the man with 7 wives. Also, let $a_2 = 7$ (the wives), $a_3 = 7a_2 = 7^2$ (the sacks), $a_4 = 7a_3 = 7^3$ (the cats), and $a_5 = 7a_4 = 7^4$ (the kits). The total is $a_1 + a_2 + a_3 + a_4 + a_5 = 1 + 7 + 7^2 + 7^3 + 7^4$, which is a partial sum of a geometric sequence with $a = 1$ and $r = 7$. Thus, the number in the party is $S_5 = 1 \cdot \dfrac{1 - 7^5}{1 - 7} = 2801$.

**72. (a)** $\sum_{k=1}^{10} 50\left(\frac{1}{2}\right)^{k-1} = 50 \cdot \dfrac{1 - \left(\frac{1}{2}\right)^{10}}{1 - \frac{1}{2}} \approx 100\,(1 - 0.0009765) = 99.9023$ mg

**(b)** $\sum_{k=1}^{\infty} 50\left(\frac{1}{2}\right)^{k-1} = \dfrac{50}{1 - \frac{1}{2}} = 100$ mg

**73.** Let $a_n$ be the height the ball reaches on the $n$th bounce. We have $a_0 = 1$ and $a_n = \frac{1}{2}a_{n-1}$. Since the total distance $d$ traveled includes the bounce up as well and the distance down, we have

$$\begin{aligned} d &= a_0 + 2 \cdot a_1 + 2 \cdot a_2 + \cdots = 1 + 2\left(\tfrac{1}{2}\right) + 2\left(\tfrac{1}{2}\right)^2 + 2\left(\tfrac{1}{2}\right)^3 + 2\left(\tfrac{1}{2}\right)^4 + \cdots \\ &= 1 + 1 + \tfrac{1}{2} + \left(\tfrac{1}{2}\right)^2 + \left(\tfrac{1}{2}\right)^3 + \cdots = 1 + \sum_{i=0}^{\infty}\left(\tfrac{1}{2}\right)^i = 1 + \frac{1}{1 - \frac{1}{2}} = 3 \end{aligned}$$

Thus the total distance traveled is about 3 m.

**74.** The time required for the ball to stop bouncing is $t = 1 + \frac{1}{\sqrt{2}} + \left(\frac{1}{\sqrt{2}}\right)^2 + \cdots$ which is an infinite geometric series with $a = 1$ and $r = \frac{1}{\sqrt{2}}$. The sum of this series is $t = \dfrac{1}{1 - \frac{1}{\sqrt{2}}} = \dfrac{\sqrt{2}}{\sqrt{2} - 1} = \dfrac{\sqrt{2}}{\sqrt{2} - 1} \cdot \dfrac{\sqrt{2} + 1}{\sqrt{2} + 1} = \dfrac{2 + \sqrt{2}}{2 - 1} = 2 + \sqrt{2}$. Thus the time required for the ball to stop is $2 + \sqrt{2} \approx 3.41$ s.

**75. (a)** If a square has side $x$, then by the Pythagorean Theorem the length of the side of the square formed by joining the midpoints is, $\sqrt{\left(\frac{x}{2}\right)^2 + \left(\frac{x}{2}\right)^2} = \sqrt{\frac{x^2}{4} + \frac{x^2}{4}} = \frac{x}{\sqrt{2}}$. In our case, $x = 1$ and the side of the first inscribed square is $\frac{1}{\sqrt{2}}$, the side of the second inscribed square is $\frac{1}{\sqrt{2}} \cdot \frac{1}{\sqrt{2}} = \left(\frac{1}{\sqrt{2}}\right)^2$, the side of the third inscribed square is $\left(\frac{1}{\sqrt{2}}\right)^3$, and so on. Since this pattern continues, the total area of all the squares is $A = 1^2 + \left(\frac{1}{\sqrt{2}}\right)^2 + \left(\frac{1}{\sqrt{2}}\right)^4 + \left(\frac{1}{\sqrt{2}}\right)^6 + \cdots = 1 + \frac{1}{2} + \left(\frac{1}{2}\right)^2 + \left(\frac{1}{2}\right)^3 + \cdots = \dfrac{1}{1 - \frac{1}{2}} = 2$.

**(b)** As in part (a), the sides of the squares are $1, \frac{1}{\sqrt{2}}, \left(\frac{1}{\sqrt{2}}\right)^2, \left(\frac{1}{\sqrt{2}}\right)^3, \ldots$. Thus the sum of the perimeters is $S = 4 \cdot 1 + 4 \cdot \frac{1}{\sqrt{2}} + 4 \cdot \left(\frac{1}{\sqrt{2}}\right)^2 + 4 \cdot \left(\frac{1}{\sqrt{2}}\right)^3 + \cdots$, which is an infinite geometric series with $a = 4$ and $r = \frac{1}{\sqrt{2}}$. Thus the sum of the perimeters is $S = \dfrac{4}{1 - \frac{1}{\sqrt{2}}} = \dfrac{4\sqrt{2}}{\sqrt{2} - 1} = \dfrac{4\sqrt{2}}{\sqrt{2} - 1} \cdot \dfrac{\sqrt{2} + 1}{\sqrt{2} + 1} = \dfrac{4 \cdot 2 + 4\sqrt{2}}{2 - 1} = 8 + 4\sqrt{2}$.

**76.** Let $A_n$ be the area of the disks of paper placed at the $n$th stage. Then $A_1 = \pi R^2$, $A_2 = 2 \cdot \pi\left(\frac{1}{2}R\right)^2 = \frac{\pi}{2}R^2$, $A_3 = 4 \cdot \pi\left(\frac{1}{4}R\right)^2 = \frac{\pi}{4}R^2, \ldots$. We see from this pattern that the total area is $A = \pi R^2 + \frac{1}{2}\pi R^2 + \frac{1}{4}\pi R^2 + \cdots$. Thus, the total area, $A$, is an infinite geometric series with $a_1 = \pi R^2$ and $r = \frac{1}{2}$. So $A = \dfrac{\pi R^2}{1 - \frac{1}{2}} = 2\pi R^2$.

**77.** Let $a_n$ denote the area colored blue at $n$th stage. Since only the middle squares are colored blue, $a_n = \frac{1}{9} \times$ (area remaining yellow at the $(n-1)$th stage). Also, the area remaining yellow at the $n$th stage is $\frac{8}{9}$ of the area remaining yellow at the preceding stage. So $a_1 = \frac{1}{9}$, $a_2 = \frac{1}{9}\left(\frac{8}{9}\right)$, $a_3 = \frac{1}{9}\left(\frac{8}{9}\right)^2$, $a_4 = \frac{1}{9}\left(\frac{8}{9}\right)^3, \ldots$. Thus the total area colored blue $A = \frac{1}{9} + \frac{1}{9}\left(\frac{8}{9}\right) + \frac{1}{9}\left(\frac{8}{9}\right)^2 + \frac{1}{9}\left(\frac{8}{9}\right)^3 + \cdots$ is an infinite geometric series with $a = \frac{1}{9}$ and $r = \frac{8}{9}$. So the total area is $A = \dfrac{\frac{1}{9}}{1 - \frac{8}{9}} = 1$.

**78. (a)** $5, -3, 5, -3, \ldots$. Now $a_2 - a_1 = -3 - 5 = -8$, but $a_3 - a_2 = 5 - (-3) = 8$, and $\dfrac{a_2}{a_1} = -\frac{3}{5}$, but $\dfrac{a_3}{a_2} = -\frac{5}{3}$. Thus, the sequence is neither arithmetic nor geometric.

**(b)** $\frac{1}{3}, 1, \frac{5}{3}, \frac{7}{3}, \ldots$. Now $a_2 - a_1 = 1 - \frac{1}{3} = \frac{2}{3}$, $a_3 - a_2 = \frac{5}{3} - 1 = \frac{2}{3}$, and $a_4 - a_3 = \frac{7}{3} - \frac{5}{3} = \frac{2}{3}$. Therefore, the sequence is arithmetic with $d = \frac{2}{3}$ and $a_5 = \frac{7}{3} + \frac{2}{3} = 3$.

**(c)** $\sqrt{3}, 3, 3\sqrt{3}, 9, \ldots$. Now $\dfrac{a_2}{a_1} = \dfrac{3}{\sqrt{3}} = \sqrt{3}$, $\dfrac{a_3}{a_2} = \dfrac{3\sqrt{3}}{3} = \sqrt{3}$, and $\dfrac{a_4}{a_3} = \dfrac{9}{3\sqrt{3}} = \sqrt{3}$. Therefore, the sequence is geometric with $r = \sqrt{3}$ and $a_5 = 9\sqrt{3}$.

**(d)** $1, -1, 1, -1, \ldots$. Now $\dfrac{a_2}{a_1} = \dfrac{-1}{1} = -1$, $\dfrac{a_3}{a_2} = \dfrac{1}{-1} = -1$, and $\dfrac{a_4}{a_3} = \dfrac{-1}{1} = -1$. Therefore, the sequence is geometric with $r = -1$ and $a_5 = (-1)(-1) = 1$.

**(e)** $2, -1, \frac{1}{2}, 2, \ldots$. Now $a_2 - a_1 = -1 - 2 = -3$, but $a_3 - a_2 = \frac{1}{2} + 1 = \frac{3}{2}$, and so the sequence is not arithmetic. Also, $\dfrac{a_2}{a_1} = -\dfrac{1}{2}$, but $\dfrac{a_4}{a_3} = \dfrac{2}{\frac{1}{2}} = 4$, and so the sequence is not geometric. Thus, the sequence is neither arithmetic nor geometric.

**(f)** $x - 1, x, x + 1, x + 2, \ldots$. Now $a_2 - a_1 = x - (x - 1) = 1$, $a_3 - a_2 = (x + 1) - x = 1$, and $a_4 - a_3 = (x + 2) - (x + 1) = 1$. Therefore, the sequence is arithmetic with $d = 1$ and $a_5 = (x + 2) + 1 = x + 3$.

**(g)** $-3, -\frac{3}{2}, 0, \frac{3}{2}, \ldots$. Now $a_2 - a_1 = -\frac{3}{2} - (-3) = \frac{3}{2}$, $a_3 - a_2 = 0 - \left(-\frac{3}{2}\right) = \dfrac{3}{2}$, and $a_4 - a_3 = \frac{3}{2} - 0 = \frac{3}{2}$. Therefore, the sequence is arithmetic with $d = \frac{3}{2}$, and $a_5 = \frac{3}{2} + \frac{3}{2} = 3$.

**(h)** $\sqrt{5}, \sqrt[3]{5}, \sqrt[6]{5}, 1, \ldots$. Now $a_2 - a_1 = \sqrt[3]{5} - \sqrt{5}$, but $a_3 - a_2 = \sqrt[6]{5} - \sqrt[3]{5}$. Thus the sequence is not arithmetic. However, $\dfrac{a_2}{a_1} = \dfrac{\sqrt[3]{5}}{\sqrt{5}} = \dfrac{5^{1/3}}{5^{1/2}} = 5^{-1/6}$, $\dfrac{a_3}{a_2} = \dfrac{\sqrt[6]{5}}{\sqrt[3]{5}} = \dfrac{5^{1/6}}{5^{1/3}} = 5^{-1/6}$, and $\dfrac{a_4}{a_3} = \dfrac{1}{\sqrt[6]{5}} = \dfrac{1}{5^{1/6}} = 5^{-1/6}$. Therefore, the sequence is geometric with $r = 5^{-1/6}$ and $a_5 = 1 \cdot 5^{-1/6} = \dfrac{1}{\sqrt[6]{5}}$.

**79.** Let $a_1, a_2, a_3, \ldots$ be a geometric sequence with common ratio $r$. Thus $a_2 = a_1 r$, $a_3 = a_1 \cdot r^2, \ldots, a_n = a_1 \cdot r^{n-1}$. Hence, $\dfrac{1}{a_2} = \dfrac{1}{a_1 \cdot r} = \dfrac{1}{a_1} \cdot \dfrac{1}{r}$, $\dfrac{1}{a_3} = \dfrac{1}{a_1 \cdot r^2} = \dfrac{1}{a_1} \cdot \dfrac{1}{r^2} = \dfrac{1}{a_1} \cdot \left(\dfrac{1}{r}\right)^2, \ldots \dfrac{1}{a_n} = \dfrac{1}{a_1 \cdot r^{n-1}} = \dfrac{1}{a_1} \cdot \dfrac{1}{r^{n-1}} = \dfrac{1}{a_1}\left(\dfrac{1}{r}\right)^{n-1}$, and so $\dfrac{1}{a_1}, \dfrac{1}{a_2}, \dfrac{1}{a_3}, \ldots$ is a geometric sequence with common ratio $\dfrac{1}{r}$.

**80.** $a_1, a_2, a_3, \ldots$ is a geometric sequence with common ratio $r$. Thus $a_2 = a_1 r$, $a_3 = a_1 \cdot r^2, \ldots, a_n = a_1 \cdot r^{n-1}$. Hence $\log a_2 = \log(a_1 r) = \log a_1 + \log r$, $\log a_3 = \log(a_1 \cdot r^2) = \log a_1 + \log(r^2) = \log a_1 + 2\log r, \ldots$, $\log a_n = \log(a_1 \cdot r^{n-1}) = \log a_1 + \log(r^{n-1}) = \log a_1 + (n-1)\log r$, and so $\log a_1, \log a_2, \log a_3, \ldots$ is an arithmetic sequence with common difference $\log r$.

**81.** Since $a_1, a_2, a_3, \ldots$ is an arithmetic sequence with common difference $d$, the terms can be expressed as $a_2 = a_1 + d$, $a_3 = a_1 + 2d, \ldots, a_n = a_1 + (n-1)d$. So $10^{a_2} = 10^{a_1 + d} = 10^{a_1} \cdot 10^d$, $10^{a_3} = 10^{a_1 + 2d} = 10^{a_1} \cdot (10^d)^2, \ldots$, $10^{a_n} = 10^{a_1 + (n-1)d} = 10^{a_1} \cdot (10^d)^{n-1}$, and so $10^{a_1}, 10^{a_2}, 10^{a_3}, \ldots$ is a geometric sequence with common ratio $r = 10^d$.

# 12.4 Mathematics of Finance

**1.** $n = 10$, $R = \$1000$, $i = 0.06$. So $A_f = R\dfrac{(1+i)^n - 1}{i} = 1000\dfrac{(1+0.06)^{10} - 1}{0.06} = \$13{,}180.79$.

**2.** $R = 500$, $n = 24$, $i = \dfrac{0.08}{12} \approx 0.0066667$. So $A_f = R\dfrac{(1+i)^n - 1}{i} = 500\dfrac{(1.0066667)^{24} - 1}{0.0066667} = \$12{,}966.59$.

**3.** $n = 20$, $R = \$5000$, $i = 0.12$. So $A_f = R\dfrac{(1+i)^n - 1}{i} = 5000\dfrac{(1+0.12)^{20} - 1}{0.12} = \$360{,}262.21$.

**4.** $R = 500$, $n = 20$, $i = \dfrac{0.06}{2} = 0.03$. So $A_f = R\dfrac{(1+i)^n - 1}{i} = 500\dfrac{(1.03)^{20} - 1}{0.03} = \$13{,}435.19$.

**5.** $n = 16$, $R = \$300$, $i = \dfrac{0.08}{4} = 0.02$. So $A_f = R\dfrac{(1+i)^n - 1}{i} = 300\dfrac{(1+0.02)^{16} - 1}{0.02} = \$5{,}591.79$.

**6.** $A_f = 5000$, $n = 4 \cdot 2 = 8$, $i = \dfrac{0.10}{4} = 0.025$. So $R = \dfrac{iA_f}{(1+i)^n - 1} = \dfrac{(0.025)(5000)}{(1.025)^8 - 1} = \$572.34$.

**7.** $A_f = \$2000$, $i = \dfrac{0.06}{12} = 0.005$, $n = 8$. Then $R = \dfrac{iA_f}{(1+i)^n - 1} = \dfrac{(0.005)(2000)}{(1+0.005)^8 - 1} = \$245.66$.

**8.** $R = 1000$, $n = 20$, $i = \dfrac{0.09}{2} = 0.045$. So $A_p = R\dfrac{1-(1+i)^{-n}}{i} = 1000\dfrac{1-(1.045)^{-20}}{0.045} = \$13{,}007.94$.

**9.** $R = \$200$, $n = 20$, $i = \dfrac{0.09}{2} = 0.045$. So $A_p = R\dfrac{1-(1+i)^{-n}}{i} = (200)\dfrac{1-(1+0.045)^{-20}}{0.045} = \$2601.59$.

**10.** $A_p = 50{,}000$, $n = 10(2) = 20$, $i = \dfrac{0.08}{2} = 0.04$. So $R = \dfrac{iA_p}{1-(1+i)^{-n}} = \dfrac{(0.04)(50{,}000)}{1-(1.04)^{-20}} = \$3679.09$.

**11.** $A_p = \$12,000$, $i = \dfrac{0.105}{12} = 0.00875$, $n = 48$. Then $R = \dfrac{iA_p}{1-(1+i)^{-n}} = \dfrac{(0.00875)(12000)}{1-(1+0.00875)^{-48}} = \$307.24$.

**12.** $A_p = 80{,}000$, $i = \dfrac{0.09}{12} = 0.0075$. Over a 30 year period, $n = 30(12) = 360$, and the monthly payment is $R = \dfrac{iA_p}{1-(1+i)^{-n}} = \dfrac{(0.0075)(80{,}000)}{1-(1.0075)^{-360}} = \$643.70$. Over a 15 year period, $n = 15 \cdot 12 = 180$, and so the monthly payment is $R = \dfrac{80{,}000 \cdot 0.0075}{1-(1.0075)^{-180}} = \$811.41$.

**13.** $A_p = \$100{,}000$, $i = \dfrac{0.08}{12} \approx 0.006667$, $n = 360$. Then $R = \dfrac{iA_p}{1-(1+i)^{-n}} = \dfrac{(0.006667)(100{,}000)}{1-(1+0.006667)^{-360}} = \$733.76$.
Therefore, the total amount paid on this loan over the 30 year period is $(360)(733.76) = \$264{,}153.60$.

**14.** $R = 650$, $n = 12(30) = 360$, $i = \dfrac{0.09}{12} = 0.0075$. So $A_p = R\dfrac{1-(1+i)^{-n}}{i} = 650\dfrac{1-(1.0075)^{-360}}{0.0075} = \$80{,}783.21$.
Therefore, the couple can afford a loan of \$80,783.21.

**15.** $A_p = 100{,}000$, $n = 360$, $i = \dfrac{0.0975}{12} = 0.008125$.

**(a)** $R = \dfrac{iA_p}{1-(1+i)^{-n}} = \dfrac{(0.008125)(100{,}000)}{1-(1+0.008125)^{-360}} = \$859.15$.

**(b)** The total amount that will be paid over the 30 year period is $(360)(859.15) = \$309{,}294.00$.

**(c)** $R = \$859.15$, $i = \dfrac{0.0975}{12} = 0.008125$, $n = 360$. So $A_f = 859.15\dfrac{(1+0.008125)^{360} - 1}{0.008125} = \$1{,}841{,}519.29$.

**16.** $R = 220$, $n = 12(3) = 36$, $i = \dfrac{0.08}{12} \approx 0.00667$. The amount borrowed is $A_p = R\dfrac{1-(1+i)^{-n}}{i} = 220\dfrac{1-(1.00667)^{-36}}{0.00667} = \$7{,}020.60$. So she purchased the car for $\$7{,}020.60 + \$2000 = \$9020.60$.

**17.** $R = \$30$, $i = \dfrac{0.10}{12} \approx 0.008333$, $n = 12$. Then $A_p = R\dfrac{1-(1+i)^{-n}}{i} = 30\dfrac{1-(1+0.008333)^{-12}}{0.008333} = \$341.24$.

**18.** $A_p = \$12{,}500$, $R = \$420$, $n = 36$. We want to solve for the interest rate using the equation $R = \dfrac{iA_P}{1-(1+i)^{-n}}$. Let $x$ be the interest rate, then $i = \dfrac{x}{12}$. So we can express $R$ as a function of $x$ as follows: $R(x) = \dfrac{\frac{x}{12}\cdot 12{,}500}{1-\left(1+\frac{x}{12}\right)^{-36}}$. We graph $R(x)$ and $y = 420$ in the rectangle $[0.12, 0.13]$ by $[415, 425]$. The $x$-coordinate of the intersection is about $0.1280$, which corresponds to an interest rate of $12.80\%$.

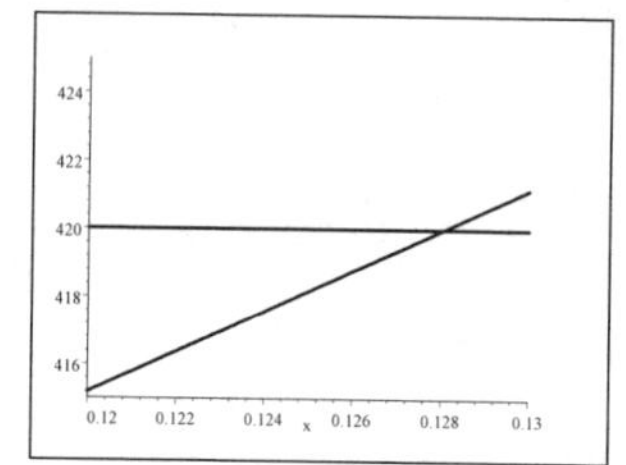


**19.** $A_p = \$640$, $R = \$32$, $n = 24$. We want to solve the equation $R = \dfrac{iA_p}{1-(1+i)^{n}}$ for the interest rate $i$. Let $x$ be the interest rate, then $i = \dfrac{x}{12}$. So we can express $R$ as a function of $x$ by $R(x) = \dfrac{\frac{x}{12}\cdot 640}{1-\left(1+\frac{x}{12}\right)^{-24}}$. We graph $R(x)$ and $y = 32$ in the rectangle $[0.12, 0.22] \times [30, 34]$. The $x$-coordinate of the intersection is about $0.1816$, which corresponds to an interest rate of $18.16\%$.

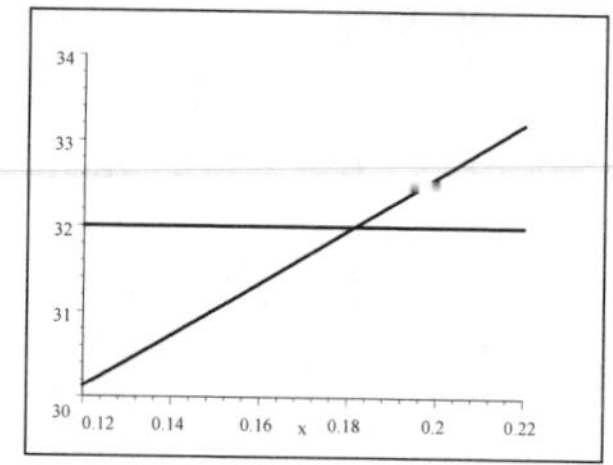


**20.** $A_p = \$2000 - \$200 = \$1800$, $R = \$88$, $n = 24$. We want to solve for the interest rate using the equation $R = \dfrac{iA_P}{1-(1+i)^{-n}}$. Let $x$ be the interest rate, then $i = \dfrac{x}{12}$. So we can express $R$ as a function of $x$ as follows:

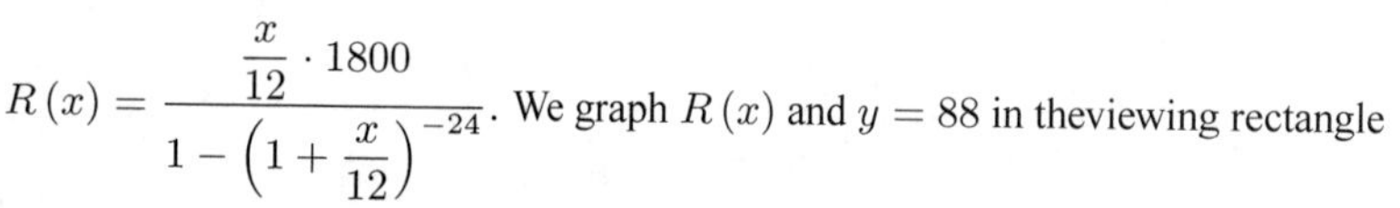

$R(x) = \dfrac{\frac{x}{12}\cdot 1800}{1-\left(1+\frac{x}{12}\right)^{-24}}$. We graph $R(x)$ and $y = 88$ in theviewing rectangle $[0.14, 0.16]$ by $[87, 89]$. The $x$-coordinate of the intersection is about $0.1584$, which corresponds to an interest rate of $15.84\%$.

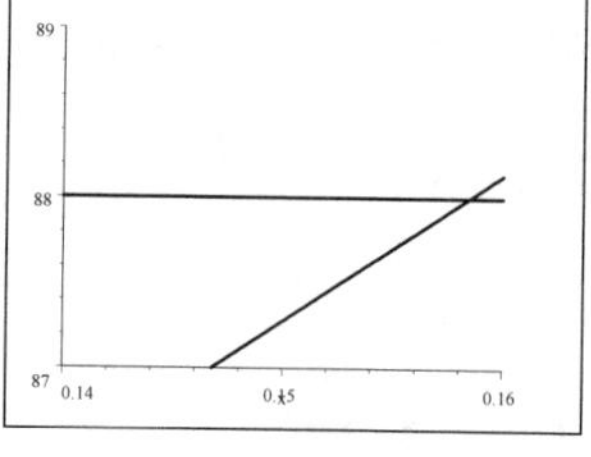


**21.** $A_p = \$189.99$, $R = \$10.50$, $n = 20$. We want to solve the equation $R = \dfrac{iA_p}{1-(1+i)^{n}}$ for the interest rate $i$. Let $x$ be the interest rate, then $i = \dfrac{x}{12}$. So we can express $R$ as a function of $x$ by $R(x) = \dfrac{\frac{x}{12}\cdot 189.99}{1-\left(1+\frac{x}{12}\right)^{-20}}$. We graph $R(x)$ and $y = 10.50$ in the rectangle $[0.10, 0.18] \times [10, 11]$. The $x$-coordinate of the intersection is about $0.1168$, which corresponds to an interest rate of $11.68\%$.

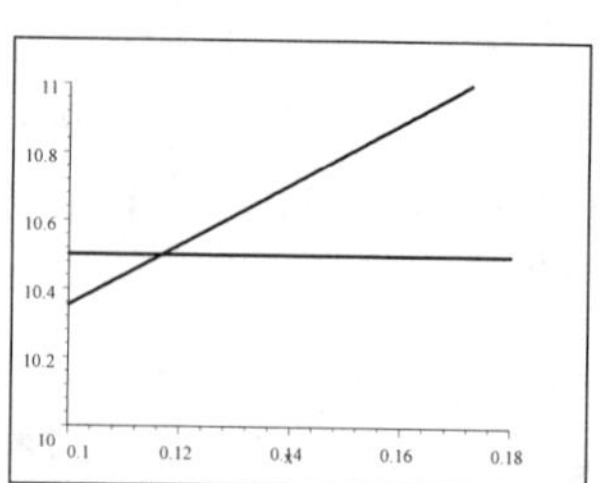

**22. (a)** The present value of the $k$th payment is $PV = R(1+i)^{-k} = \dfrac{R}{(1+i)^k}$. The present value of an annuity is the sum of the present values of each of the payments of $R$ dollars, as shown in the time line.

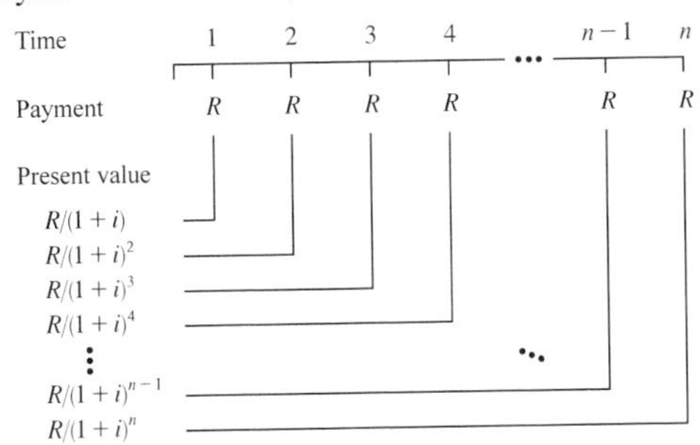


**(b)** $A_p = \dfrac{R}{1+i} + \dfrac{R}{(1+i)^2} + \dfrac{R}{(1+i)^3} + \cdots + \dfrac{R}{(1+i)^n}$

$= \dfrac{R}{1+i} + \left(\dfrac{R}{1+i}\right)\left(\dfrac{1}{1+i}\right) + \left(\dfrac{R}{1+i}\right)\left(\dfrac{1}{1+i}\right)^2 + \cdots + \left(\dfrac{R}{1+i}\right)\left(\dfrac{1}{1+i}\right)^{n-1}$

This is a geometric series with $a = \dfrac{R}{1+i}$ and $r = \dfrac{1}{1+i}$. Since $S_n = a\dfrac{1-r^n}{1-r}$, we have

$$A_p = \left(\frac{R}{1+i}\right)\frac{1-\left[\dfrac{1}{(1+i)}\right]^n}{1-\left(\dfrac{1}{1+i}\right)} = R\frac{1-(1+i)^{-n}}{(1+i)\left[1-\left(\dfrac{1}{1+i}\right)\right]} = R\frac{1-(1+i)^{-n}}{(1+i)-1} = R\frac{1-(1+i)^{-n}}{i}.$$

**23. (a)** The present value of the $k$th payment is $PV = \dfrac{R}{(1+i)^k}$. The amount of money to be invested now ($A_p$) to ensure an annuity in perpetuity is the (infinite) sum of the present values of each of the payments, as shown in the time line.

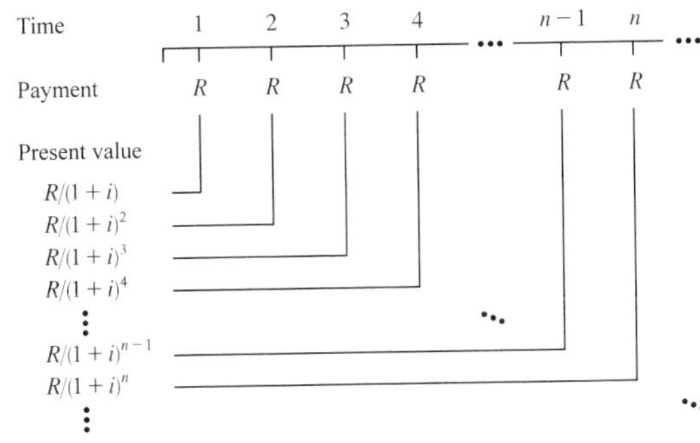


**(b)** $A_p = \dfrac{R}{1+i} + \dfrac{R}{(1+i)^2} + \dfrac{R}{(1+i)^3} + \cdots + \dfrac{R}{(1+i)^n} + \cdots$. This is an infinite geometric series with $a = \dfrac{R}{1+i}$ and $r = \dfrac{1}{1+i}$. Therefore, $A_p = \dfrac{\dfrac{R}{1+i}}{1-\dfrac{1}{1+i}} = \dfrac{R}{1+i}\cdot\dfrac{1+i}{i} = \dfrac{R}{i}$.

**(c)** Using the result from part (b), we have $R = 5000$ and $i = 0.10$. Then $A_p = \dfrac{R}{i} = \dfrac{5000}{0.10} = \$50{,}000$.

**(d)** We are given two different time periods: the interest is compounded quarterly, while the annuity is paid yearly. In order to use the formula in part (b), we need to find the effective annual interest rate produced by 8% interest compounded quarterly, which is $(1+i/n)^n - 1$. Now $i = 8\%$ and it is compounded quarterly, $n = 4$, so the effective annual yield is $(1+0.02)^4 - 1 \approx 0.08243216$. Thus by the formula in part (b), the amount that must be invested is $A_p = \frac{3000}{0.08243216} = \$36{,}393.56$.

**24. (a)** Using the hint, we calculate the present value of the remaining 240 payments with $R = 724.17$, $i = 0.0075$, and $n = 240$. Since $A_p = R\frac{1-(1+i)^{-n}}{i} = (724.17)\frac{1-(1.0075)^{-240}}{0.0075} = 80{,}487.84$, they still owe \$80,487.84 on their mortgage.

**(b)** On their next payment, $0.0075\,(80{,}487.84) = \$603.66$ is interest and $\$724.17 - 603.66 = \$120.51$ goes toward the principal.

## 12.5 Mathematical Induction

**1.** Let $P(n)$ denote the statement $2+4+6+\cdots+2n = n(n+1)$.
*Step 1:* $P(1)$ is the statement that $2 = 1(1+1)$, which is true.
*Step 2:* Assume that $P(k)$ is true; that is, $2+4+6+\cdots+2k = k(k+1)$. We want to use this to show that $P(k+1)$ is true. Now

$$\begin{aligned} 2+4+6+\cdots+2k+2(k+1) &= k(k+1)+2(k+1) && \text{induction hypothesis} \\ &= (k+1)(k+2) = (k+1)[(k+1)+1] \end{aligned}$$

Thus, $P(k+1)$ follows from $P(k)$. So by the Principle of Mathematical Induction, $P(n)$ is true for all $n$.

**2.** Let $P(n)$ denote the statement $1+4+7+10+\cdots+(3n-2) = \frac{n(3n-1)}{2}$.

*Step 1:* $P(1)$ is the statement that $1 = \frac{1[3(1)-1]}{2} = \frac{1\cdot 2}{2}$, which is true.

*Step 2:* Assume that $P(k)$ is true; that is. $1+4+7+\cdots+(3k-2) = \frac{k(3k-1)}{2}$. We want to use this to show that $P(k+1)$ is true. Now

$$\begin{aligned} 1+4+7+10+\cdots+(3k-2)+[3(k+1)-2] &= \frac{k(3k-1)}{2}+3k+1 && \text{induction hypothesis} \\ &= \frac{k(3k-1)}{2}+\frac{6k+2}{2} = \frac{3k^2-k+6k+2}{2} = \frac{3k^2+5k+2}{2} \\ &= \frac{(k+1)(3k+2)}{2} = \frac{(k+1)[3(k+1)-1]}{2} \end{aligned}$$

Thus, $P(k+1)$ follows from $P(k)$. So by the Principle of Mathematical Induction, $P(n)$ is true for all $n$.

**3.** Let $P(n)$ denote the statement $5+8+11+\cdots+(3n+2) = \frac{n(3n+7)}{2}$.

*Step 1:* We need to show that $P(1)$ is true. But $P(1)$ says that $5 = \frac{1\cdot(3\cdot 1+7)}{2}$, which is true.

*Step 2:* Assume that $P(k)$ is true; that is, $5+8+11+\cdots+(3k+2) = \frac{k(3k+7)}{2}$. We want to use this to show that $P(k+1)$ is true. Now

$$\begin{aligned} 5+8+11+\cdots+(3k+2)+[3(k+1)+2] &= \frac{k(3k+7)}{2}+(3k+5) && \text{induction hypothesis} \\ &= \frac{3k^2+7k}{2}+\frac{6k+10}{2} = \frac{3k^2+13k+10}{2} \\ &= \frac{(3k+10)(k+1)}{2} = \frac{(k+1)[3(k+1)+7]}{2} \end{aligned}$$

Thus, $P(k+1)$ follows from $P(k)$. So by the Principle of Mathematical Induction, $P(n)$ is true for all $n$.

**4.** Let $P(n)$ denote the statement $1^2 + 2^2 + 3^2 + \cdots + n^2 = \frac{n(n+1)(2n+1)}{6}$.

*Step 1:* $P(1)$ is the statement that $1^2 = \frac{1 \cdot 2 \cdot 3}{6}$, which is true.

*Step 2:* Assume that $P(k)$ is true; that is, $1^2 + 2^2 + 3^2 + \cdots + k^2 = \frac{k(k+1)(2k+1)}{6}$. We want to use this to show that $P(k+1)$ is true. Now,

$$\begin{aligned} 1^2 + 2^2 + 3^2 + \cdots + k^2 + (k+1)^2 &= \frac{k(k+1)(2k+1)}{6} + (k+1)^2 \qquad \text{induction hypothesis} \\ &= (k+1)\left[\frac{k(2k+1) + 6(k+1)}{6}\right] = (k+1)\left[\frac{2k^2 + k + 6k + 6}{6}\right] \\ &= (k+1)\left[\frac{2k^2 + 7k + 6}{6}\right] = \frac{(k+1)(k+2)(2k+3)}{6} \\ &= \frac{(k+1)[(k+1)+1][2(k+1)+1]}{6}. \end{aligned}$$

Thus $P(k+1)$ follows from $P(k)$. So by the Principle of Mathematical Induction, $P(n)$ is true for all $n$.

**5.** Let $P(n)$ denote the statement $1 \cdot 2 + 2 \cdot 3 + 3 \cdot 4 + \cdots + n(n+1) = \frac{n(n+1)(n+2)}{3}$.

*Step 1:* $P(1)$ is the statement that $1 \cdot 2 = \frac{1 \cdot (1+1) \cdot (1+2)}{3}$, which is true.

*Step 2:* Assume that $P(k)$ is true; that is, $1 \cdot 2 + 2 \cdot 3 + 3 \cdot 4 + \cdots + k(k+1) = \frac{k(k+1)(k+2)}{3}$. We want to use this to show that $P(k+1)$ is true. Now

$$\begin{aligned} &1 \cdot 2 + 2 \cdot 3 + 3 \cdot 4 + \cdots + k(k+1) + (k+1)[(k+1)+1] \\ &\quad = \frac{k(k+1)(k+2)}{3} + (k+1)(k+2) \qquad \text{induction hypothesis} \\ &\quad = \frac{k(k+1)(k+2)}{3} + \frac{3(k+1)(k+2)}{3} = \frac{(k+1)(k+2)(k+3)}{3} \end{aligned}$$

Thus, $P(k+1)$ follows from $P(k)$. So by the Principle of Mathematical Induction, $P(n)$ is true for all $n$.

**6.** Let $P(n)$ denote the statement $1 \cdot 3 + 2 \cdot 4 + 3 \cdot 5 + \cdots + n(n+2) = \frac{n(n+1)(2n+7)}{6}$.

*Step 1:* $P(1)$ is the statement that $1 \cdot 3 = \frac{1 \cdot 2 \cdot 9}{6}$, which is true.

*Step 2:* Assume that $P(k)$ is true; that is, $1 \cdot 3 + 2 \cdot 4 + 3 \cdot 5 + \cdots + k(k+2) = \frac{k(k+1)(2k+7)}{6}$. We want to use this to show that $P(k+1)$ is true. Now

$$\begin{aligned} &1 \cdot 3 + 2 \cdot 4 + 3 \cdot 5 + \cdots + k(k+2) + (k+1)[(k+1)+2] \\ &\quad = \frac{k(k+1)(2k+7)}{6} + (k+1)(k+3) \qquad \text{induction hypothesis} \\ &\quad = (k+1)\left[\frac{k(2k+7)}{6} + \frac{6(k+3)}{6}\right] = (k+1)\left(\frac{2k^2 + 7k + 6k + 18}{6}\right) \\ &\quad = \frac{(k+1)[(k+1)+1][2(k+1)+7]}{6} \end{aligned}$$

Thus $P(k+1)$ follows from $P(k)$. So by the Principle of Mathematical Induction, $P(n)$ is true for all $n$.

**7.** Let $P(n)$ denote the statement $1^3 + 2^3 + 3^3 + \cdots + n^3 = \dfrac{n^2(n+1)^2}{4}$.

*Step 1:* $P(1)$ is the statement that $1^3 = \dfrac{1^2 \cdot (1+1)^2}{4}$, which is clearly true.

*Step 2:* Assume that $P(k)$ is true; that is, $1^3 + 2^3 + 3^3 + \cdots + k^3 = \dfrac{k^2(k+1)^2}{4}$. We want to use this to show that $P(k+1)$ is true. Now

$$\begin{aligned} 1^3 + 2^3 + 3^3 + \cdots + k^3 + (k+1)^3 &= \frac{k^2(k+1)^2}{4} + (k+1)^3 \qquad \text{induction hypothesis} \\ &= \frac{(k+1)^2\left[k^2 + 4(k+1)\right]}{4} = \frac{(k+1)^2\left[k^2+4k+4\right]}{4} \\ &= \frac{(k+1)^2(k+2)^2}{4} = \frac{(k+1)^2\left[(k+1)+1\right]^2}{4} \end{aligned}$$

Thus, $P(k+1)$ follows from $P(k)$. So by the Principle of Mathematical Induction, $P(n)$ is true for all $n$.

**8.** Let $P(n)$ denote the statement $1^3 + 3^3 + 5^3 + \cdots + (2n-1)^3 = n^2\left(2n^2 - 1\right)$.
*Step 1:* $P(1)$ is the statement that $1^3 = 1^2\left(2 \cdot 1^2 - 1\right)$, which is clearly true.
*Step 2:* Assume that $P(k)$ is true; that is, $1^3 + 3^3 + 5^3 + \cdots + (2k-1)^3 = k^2\left(2k^2-1\right)$. We want to use this to show that $P(k+1)$ is true. Now

$$\begin{aligned} 1^3 + 3^3 + 5^3 + \cdots + (2k-1)^3 + (2k+1)^3 &= k^2\left(2k^2-1\right) + (2k+1)^3 \qquad \text{induction hypothesis} \\ &= 2k^4 - k^2 + 8k^3 + 12k^2 + 6k + 1 = 2k^4 + 8k^3 + 11k^2 + 6k + 1 \\ &= \left(k^2+2k+1\right)\left(2k^2+4k+1\right) = (k+1)^2\left[2(k+1)^2 - 1\right] \end{aligned}$$

Thus $P(k+1)$ follows from $P(k)$. So by the Principle of Mathematical Induction, $P(n)$ is true for all $n$.

**9.** Let $P(n)$ denote the statement $2^3 + 4^3 + 6^3 + \cdots + (2n)^3 = 2n^2(n+1)^2$.
*Step 1:* $P(1)$ is true since $2^3 = 2(1)^2(1+1)^2 = 2 \cdot 4 = 8$.
*Step 2:* Assume that $P(k)$ is true; that is, $2^3 + 4^3 + 6^3 + \cdots + (2k)^3 = 2k^2(k+1)^2$. We want to use this to show that $P(k+1)$ is true. Now

$$\begin{aligned} 2^3 + 4^3 + 6^3 + \cdots + (2k)^3 + [2(k+1)]^3 &= 2k^2(k+1)^2 + [2(k+1)]^3 \qquad \text{induction hypothesis} \\ &= 2k^2(k+1)^2 + 8(k+1)(k+1)^2 = (k+1)^2\left(2k^2+8k+8\right) \\ &= 2(k+1)^2(k+2)^2 = 2(k+1)^2\left[(k+1)+1\right]^2 \end{aligned}$$

Thus, $P(k+1)$ follows from $P(k)$. So by the Principle of Mathematical Induction, $P(n)$ is true for all $n$.

**10.** Let $P(n)$ denote the statement $\dfrac{1}{1 \cdot 2} + \dfrac{1}{2 \cdot 3} + \dfrac{1}{3 \cdot 4} + \cdots + \dfrac{1}{n(n+1)} = \dfrac{n}{n+1}$.

*Step 1:* $P(1)$ is the statement that $\dfrac{1}{1 \cdot 2} = \frac{1}{2}$, which is clearly true.

*Step 2:* Assume that $P(k)$ is true; that is $\dfrac{1}{1 \cdot 2} + \dfrac{1}{2 \cdot 3} + \dfrac{1}{3 \cdot 4} + \cdots + \dfrac{1}{k(k+1)} = \dfrac{k}{k+1}$. We want to use this to show that $P(k+1)$ is true. Now

$$\begin{aligned} \frac{1}{1 \cdot 2} + \frac{1}{2 \cdot 3} + \cdots + \frac{1}{k(k+1)} + \frac{1}{(k+1)(k+2)} &= \frac{k}{k+1} + \frac{1}{(k+1)(k+2)} \qquad \text{induction hypothesis} \\ &= \frac{k(k+2)+1}{(k+1)(k+2)} = \frac{k^2+2k+1}{(k+1)(k+2)} = \frac{(k+1)^2}{(k+1)(k+2)} \\ &= \frac{k+1}{k+2} = \frac{k+1}{(k+1)+1} \end{aligned}$$

Thus $P(k+1)$ follows from $P(k)$. So by the Principle of Mathematical Induction, $P(n)$ is true for all $n$.

**11.** Let $P(n)$ denote the statement $1 \cdot 2 + 2 \cdot 2^2 + 3 \cdot 2^3 + 4 \cdot 2^4 + \cdots + n \cdot 2^n = 2[1 + (n-1)2^n]$.

*Step 1:* $P(1)$ is the statement that $1 \cdot 2 = 2[1 + 0]$, which is clearly true.

*Step 2:* Assume that $P(k)$ is true; that is, $1 \cdot 2 + 2 \cdot 2^2 + 3 \cdot 2^3 + 4 \cdot 2^4 + \cdots + k \cdot 2^k = 2\left[1 + (k-1)2^k\right]$. We want to use this to show that $P(k+1)$ is true. Now

$$\begin{aligned} 1 \cdot 2 + 2 \cdot 2^2 + 3 \cdot 2^3 + 4 \cdot 2^4 + \cdots + k \cdot 2^k + (k+1) \cdot 2^{(k+1)} & \\ = 2\left[1 + (k-1)2^k\right] + (k+1) \cdot 2^{k+1} & \quad \text{induction hypothesis} \\ = 2\left[1 + (k-1) \cdot 2^k + (k+1) \cdot 2^k\right] = 2\left[1 + 2k \cdot 2^k\right] & \\ = 2\left[1 + k \cdot 2^{k+1}\right] = 2\left\{1 + [(k+1) - 1]\, 2^{k+1}\right\} & \end{aligned}$$

Thus $P(k+1)$ follows from $P(k)$. So by the Principle of Mathematical Induction, $P(n)$ is true for all $n$.

**12.** Let $P(n)$ denote the statement $1 + 2 + 2^2 + \cdots + 2^{n-1} = 2^n - 1$.

*Step 1:* $P(1)$ is the statement that $1 = 2^1 - 1$, which is clearly true.

*Step 2:* Assume that $P(k)$ is true; that is, $1 + 2 + 2^2 + \cdots + 2^{k-1} = 2^k - 1$. We want to use this to show that $P(k+1)$ is true. Now

$$\begin{aligned} 1 + 2 + 2^2 + \cdots + 2^{k-1} + 2^k &= 2^k - 1 + 2^k && \text{induction hypothesis} \\ &= 2 \cdot 2^k - 1 = 2^{k+1} - 1 \end{aligned}$$

Thus $P(k+1)$ follows from $P(k)$. So by the Principle of Mathematical Induction, $P(n)$ is true for all $n$.

**13.** Let $P(n)$ denote the statement $n^2 + n$ is divisible by 2.

*Step 1:* $P(1)$ is the statement that $1^2 + 1 = 2$ is divisible by 2, which is clearly true.

*Step 2:* Assume that $P(k)$ is true; that is, $k^2 + k$ is divisible by 2. Now $(k+1)^2 + (k+1) = k^2 + 2k + 1 + k + 1 = \left(k^2 + k\right) + 2k + 2 = \left(k^2 + k\right) + 2(k+1)$. By the induction hypothesis, $k^2 + k$ is divisible by 2, and clearly $2(k+1)$ is divisible by 2. Thus, the sum is divisible by 2, so $P(k+1)$ is true. Therefore, $P(k+1)$ follows from $P(k)$. So by the Principle of Mathematical Induction, $P(n)$ is true for all $n$.

**14.** Let $P(n)$ denote the statement that $5^n - 1$ is divisible by 4.

*Step 1:* $P(1)$ is the statement that $5^1 - 1 = 4$ is divisible by 4, which is clearly true.

*Step 2:* Assume that $P(k)$ is true; that is, $5^k - 1$ is divisible by 4. We want to use this to show that $P(k+1)$ is true. Now, $5^{(k+1)} - 1 = 5 \cdot 5^k - 1 = 5 \cdot 5^k - 5 + 4 = 5\left(5^k - 1\right) + 4$ which is divisible by 4 since $5\left(5^k - 1\right)$ is divisible by 4 by the induction hypothesis. Thus $P(k+1)$ follows from $P(k)$. So by the Principle of Mathematical Induction, $P(n)$ is true for all $n$.

**15.** Let $P(n)$ denote the statement that $n^2 - n + 41$ is odd.

*Step 1:* $P(1)$ is the statement that $1^2 - 1 + 41 = 41$ is odd, which is clearly true.

*Step 2:* Assume that $P(k)$ is true; that is, $k^2 - k + 41$ is odd. We want to use this to show that $P(k+1)$ is true. Now, $(k+1)^2 - (k+1) + 41 = k^2 + 2k + 1 - k - 1 + 41 = \left(k^2 - k + 41\right) + 2k$, which is also odd because $k^2 - k + 41$ is odd by the induction hypothesis, $2k$ is always even, and an odd number plus an even number is always odd. Therefore, $P(k+1)$ follows from $P(k)$. So by the Principle of Mathematical Induction, $P(n)$ is true for all $n$.

**16.** Let $P(n)$ denote the statement that $n^3 - n + 3$ is divisible by 3.

*Step 1:* $P(1)$ is the statement that $1^3 - 1 + 3 = 3$ is divisible by 3, which is true.

*Step 2:* Assume that $P(k)$ is true; that is, $k^3 - k + 3$ is divisible by 3. We want to use this to show that $P(k+1)$ is true. Now, $(k+1)^3 - (k+1) + 3 = k^3 + 3k^2 + 3k + 1 - k - 1 + 3 = k^3 - k + 3 + 3k^2 + 3k = (k^3 - k + 3) + 3(k^2 + k)$, which is divisible by 3, since $k^3 - k + 3$ is divisible by 3 by the induction hypothesis, and $3(k^2 + k)$ is divisible by 3. Thus $P(k+1)$ follows from $P(k)$. So by the Principle of Mathematical Induction, $P(n)$ is true for all $n$.

**17.** Let $P(n)$ denote the statement that $8^n - 3^n$ is divisible by 5.

*Step 1:* $P(1)$ is the statement that $8^1 - 3^1 = 5$ is divisible by 5, which is clearly true.

*Step 2:* Assume that $P(k)$ is true; that is, $8^k - 3^k$ is divisible by 5. We want to use this to show that $P(k+1)$ is true. Now, $8^{k+1} - 3^{k+1} = 8 \cdot 8^k - 3 \cdot 3^k = 8 \cdot 8^k - (8-5) \cdot 3^k = 8 \cdot (8^k - 3^k) + 5 \cdot 3^k$, which is divisible by 5 because $8^k - 3^k$ is divisible by 5 by our induction hypothesis, and $5 \cdot 3^k$ is divisible by 5. Thus $P(k+1)$ follows from $P(k)$. So by the Principle of Mathematical Induction, $P(n)$ is true for all $n$.

**18.** Let $P(n)$ denote the statement that $3^{2n} - 1$ is divisible by 8.

*Step 1:* $P(1)$ is the statement that $3^2 - 1 = 8$ is divisible by 8, which is clearly true.

*Step 2:* Assume that $P(k)$ is true; that is, $3^{2k} - 1$ is divisible by 8. We want to use this to show that $P(k+1)$ is true. Now, $3^{2(k+1)} - 1 = 9 \cdot 3^{2k} - 1 = 9 \cdot 3^{2k} - 9 + 8 = 9(3^{2k} - 1) + 8$, which is divisible by 8, since $3^{2k} - 1$ is divisible by 8 by the induction hypothesis. Thus $P(k+1)$ follows from $P(k)$. So by the Principle of Mathematical Induction, $P(n)$ is true for all $n$.

**19.** Let $P(n)$ denote the statement $n < 2^n$.

*Step 1:* $P(1)$ is the statement that $1 < 2^1 = 2$, which is clearly true.

*Step 2:* Assume that $P(k)$ is true; that is, $k < 2^k$. We want to use this to show that $P(k+1)$ is true. Adding 1 to both sides of $P(k)$ we have $k + 1 < 2^k + 1$. Since $1 < 2^k$ for $k \geq 1$, we have $2^k + 1 < 2^k + 2^k = 2 \cdot 2^k = 2^{k+1}$. Thus $k + 1 < 2^{k+1}$, which is exactly $P(k+1)$. Therefore, $P(k+1)$ follows from $P(k)$. So by the Principle of Mathematical Induction, $P(n)$ is true for all $n$.

**20.** Let $P(n)$ denote the statement $(n+1)^2 < 2n^2$, for all $n \geq 3$.

*Step 1:* $P(3)$ is the statement that $(3+1)^2 < 2 \cdot 3^2$ or $16 < 18$, which is true.

*Step 2:* Assume that $P(k)$ is true; that is, $(k+1)^2 < 2k^2$, $k \geq 3$. We want to use this to show that $P(k+1)$ is true. Now

$$\begin{aligned}
(k+2)^2 &= k^2 + 4k + 4 = (k^2 + 2k + 1) + (2k + 3) = (k+1)^2 + (2k+1) \\
&< 2k^2 + (2k+3) && \text{induction hypothesis} \\
&< 2k^2 + (2k+3) + (2k-1) && \text{because } 2k - 1 > 0 \text{ for } k \geq 3 \\
&= 2k^2 + 4k + 2 = 2(k+1)^2
\end{aligned}$$

Thus $P(k+1)$ follows from $P(k)$. So by the Principle of Mathematical Induction, $P(n)$ is true for all $n \geq 3$.

**21.** Let $P(n)$ denote the statement $(1+x)^n \geq 1+nx$, if $x > -1$.

*Step 1:* $P(1)$ is the statement that $(1+x)^1 \geq 1+1x$, which is clearly true.

*Step 2:* Assume that $P(k)$ is true; that is, $(1+x)^k \geq 1+kx$. Now, $(1+x)^{k+1} = (1+x)(1+x)^k \geq (1+x)(1+kx)$, by the induction hypothesis. Since $(1+x)(1+kx) = 1+(k+1)x+kx^2 \geq 1+(k+1)x$ (since $kx^2 \geq 0$), we have $(1+x)^{k+1} \geq 1+(k+1)x$, which is $P(k+1)$. Thus $P(k+1)$ follows from $P(k)$. So the Principle of Mathematical Induction, $P(n)$ is true for all $n$.

**22.** Let $P(n)$ denote the statement $100n \leq n^2$, for all $n \geq 100$.

*Step 1:* $P(100)$ is the statement that $100(100) \leq (100)^2$, which is true.

*Step 2:* Assume that $P(k)$ is true; that is, $100k \leq k^2$. We want to use this to show that $P(k+1)$ is true. Now

$$\begin{aligned} 100(k+1) &= 100k + 100 \leq k^2 + 100 && \text{induction hypothesis} \\ &\leq k^2 + 2k + 1 = (k+1)^2 && \text{because } 2k+1 \geq 100 \text{ for } k \geq 100 \end{aligned}$$

Thus $P(k+1)$ follows from $P(k)$. So by the Principle of Mathematical Induction, $P(n)$ is true for all $n \geq 100$.

**23.** Let $P(n)$ be the statement that $a_n = 5 \cdot 3^{n-1}$.

*Step 1:* $P(1)$ is the statement that $a_1 = 5 \cdot 3^0 = 5$, which is true.

*Step 2:* Assume that $P(k)$ is true; that is, $a_k = 5 \cdot 3^{k-1}$. We want to use this to show that $P(k+1)$ is true. Now, $a_{k+1} = 3a_k = 3 \cdot (5 \cdot 3^{k-1})$, by the induction hypothesis. Therefore, $a_{k+1} = 3 \cdot (5 \cdot 3^{k-1}) = 5 \cdot 3^k$, which is exactly $P(k+1)$. Thus, $P(k+1)$ follows from $P(k)$. So by the Principle of Mathematical Induction, $P(n)$ is true for all $n$.

**24.** $a_{n+1} = 3a_n - 8$ and $a_1 = 4$. Then $a_2 = 3 \cdot 4 - 8 = 4$, $a_3 = 3 \cdot 4 - 8 = 4$, $a_4 = 3 \cdot 4 - 8 = 4, \ldots$, and the conjecture is that $a_n = 4$. Let $P(n)$ denote the statement that $a_n = 4$.

*Step 1:* $P(1)$ is the statement that $a_1 = 4$, which is true.

*Step 2:* Assume that $P(k)$ is true; that is, $a_k = 4$. We want to use this to show that $P(k+1)$ is true. Now, $a_{k+1} = 3 \cdot a_k - 8 = 3 \cdot 4 - 8 = 4$, by the induction hypothesis. This is exactly $P(k+1)$, so by the Principle of Mathematical Induction, $P(n)$ is true for all $n$.

**25.** Let $P(n)$ be the statement that $x-y$ is a factor of $x^n - y^n$ for all natural numbers $n$.

*Step 1:* $P(1)$ is the statement that $x-y$ is a factor of $x^1 - y^1$, which is clearly true.

*Step 2:* Assume that $P(k)$ is true; that is, $x-y$ is a factor of $x^k - y^k$. We want to use this to show that $P(k+1)$ is true. Now, $x^{k+1} - y^{k+1} = x^{k+1} - x^k y + x^k y - y^{k+1} = x^k(x-y) + (x^k - y^k)y$, for which $x-y$ is a factor because $x-y$ is a factor of $x^k(x-y)$, and $x-y$ is a factor of $(x^k - y^k)y$, by the induction hypothesis. Thus $P(k+1)$ follows from $P(k)$. So by the Principle of Mathematical Induction, $P(n)$ is true for all $n$.

**26.** Let $P(n)$ be the statement that $x+y$ is a factor of $x^{2n-1} + y^{2n-1}$.

*Step 1:* $P(1)$ is the statement that $x+y$ is a factor of $x^1 + y^1$, which is clearly true.

*Step 2:* Assume that $P(k)$ is true; that is, $x+y$ is a factor of $x^{2k-1} + y^{2k-1}$. We want to use this to show that $P(k+1)$ is true. Now

$$\begin{aligned} x^{2(k+1)-1} + y^{2(k+1)-1} &= x^{2k+1} + y^{2k+1} = x^{2k+1} - x^{2k-1}y^2 + x^{2k-1}y^2 + y^{2k+1} \\ &= x^{2k-1}(x^2 - y^2) + (x^{2k-1} + y^{2k-1})y^2 \end{aligned}$$

for which $x+y$ is a factor. This is because $x+y$ is a factor of $x^2 - y^2 = (x+y)(x-y)$ and $(x+y)$ is a factor of $x^{2k-1} + y^{2k-1}$ by our induction hypothesis. Thus $P(k+1)$ follows from $P(k)$. So by the Principle of Mathematical Induction, $P(n)$ is true for all $n$.

**27.** Let $P(n)$ denote the statement that $F_{3n}$ is even for all natural numbers $n$.

*Step 1:* $P(1)$ is the statement that $F_3$ is even. Since $F_3 = F_2 + F_1 = 1 + 1 = 2$, this statement is true.

*Step 2:* Assume that $P(k)$ is true; that is, $F_{3k}$ is even. We want to use this to show that $P(k+1)$ is true. Now, $F_{3(k+1)} = F_{3k+3} = F_{3k+2} + F_{3k+1} = F_{3k+1} + F_{3k} + F_{3k+1} = F_{3k} + 2 \cdot F_{3k+1}$, which is even because $F_{3k}$ is even by the induction hypothesis, and $2 \cdot F_{3k+1}$ is even. Thus $P(k+1)$ follows from $P(k)$. So by the Principle of Mathematical Induction, $P(n)$ is true for all $n$.

**28.** Let $P(n)$ denote the statement that $F_1 + F_2 + F_3 + \cdots + F_n = F_{n+2} - 1$.

*Step 1:* $P(1)$ is the statement that $F_1 = F_3 - 1$. But $F_1 = 1 = 2 - 1 = F_3 - 1$, which is true.

*Step 2:* Assume that $P(k)$ is true; that is, $F_1 + F_2 + F_3 + \cdots + F_k = F_{k+2} - 1$. We want to use this to show that $P(k+1)$ is true. Now $F_{(k+1)+2} - 1 = F_{k+3} - 1 = F_{k+2} + F_{k+1} - 1 = (F_{k+2} - 1) + F_{k+1} = F_1 + F_2 + F_3 + \cdots + F_k + F_{k+1}$ by the induction hypothesis. Thus $P(k+1)$ follows from $P(k)$. So by the Principle of Mathematical Induction, $P(n)$ is true for all $n$.

**29.** Let $P(n)$ denote the statement that $F_1^2 + F_2^2 + F_3^2 + \cdots + F_n^2 = F_n \cdot F_{n+1}$.

*Step 1:* $P(1)$ is the statement that $F_1^2 = F_1 \cdot F_2$ or $1^2 = 1 \cdot 1$, which is true.

*Step 2:* Assume that $P(k)$ is true, that is, $F_1^2 + F_2^2 + F_3^2 + \cdots + F_k^2 = F_k \cdot F_{k+1}$. We want to use this to show that $P(k+1)$ is true. Now

$$\begin{aligned} F_1^2 + F_2^2 + F_3^2 + \cdots + F_k^2 + F_{k+1}^2 &= F_k \cdot F_{k+1} + F_{k+1}^2 && \text{induction hypothesis} \\ &= F_{k+1}(F_k + F_{k+1}) \\ &= F_{k+1} \cdot F_{k+2} && \text{by definition of the Fibonacci sequence} \end{aligned}$$

Thus $P(k+1)$ follows from $P(k)$. So by the Principle of Mathematical Induction, $P(n)$ is true for all $n$.

**30.** Let $P(n)$ denote the statement that $F_1 + F_3 + \cdots + F_{2n-1} = F_{2n}$.

*Step 1:* $P(1)$ is the statement that $F_1 = F_2$, which is true since $F_1 = 1$ and $F_2 = 1$.

*Step 2:* Assume that $P(k)$ is true; that is, $F_1 + F_3 + \cdots + F_{2k-1} = F_{2k}$, for some $k \geq 1$. We want to use this to show that $P(k+1)$ is true; that is, $F_1 + F_3 + \cdots + F_{2(k+1)-1} = F_{2(k+1)}$. Now

$$\begin{aligned} F_1 + F_3 + \cdots + F_{2k-1} + F_{2k+1} &= F_{2k} + F_{2k+1} && \text{induction hypothesis} \\ &= F_{2k+2} && \text{definition of } F_{2k+2} \\ &= F_{2(k+1)} \end{aligned}$$

Therefore, $P(k+1)$ is true. So by the Principle of Mathematical Induction, $P(n)$ is true for all $n$.

**31.** Let $P(n)$ denote the statement $\begin{bmatrix} 1 & 1 \\ 1 & 0 \end{bmatrix}^n = \begin{bmatrix} F_{n+1} & F_n \\ F_n & F_{n-1} \end{bmatrix}$.

*Step 1:* Since $\begin{bmatrix} 1 & 1 \\ 1 & 0 \end{bmatrix}^2 = \begin{bmatrix} 1 & 1 \\ 1 & 0 \end{bmatrix}\begin{bmatrix} 1 & 1 \\ 1 & 0 \end{bmatrix} = \begin{bmatrix} 2 & 1 \\ 1 & 1 \end{bmatrix} = \begin{bmatrix} F_3 & F_2 \\ F_2 & F_1 \end{bmatrix}$, it follows that $P(2)$ is true.

*Step 2:* Assume that $P(k)$ is true; that is, $\begin{bmatrix} 1 & 1 \\ 1 & 0 \end{bmatrix}^k = \begin{bmatrix} F_{k+1} & F_k \\ F_k & F_{k-1} \end{bmatrix}$. We show that $P(k+1)$ follows from this. Now,

$$\begin{aligned}\begin{bmatrix} 1 & 1 \\ 1 & 0 \end{bmatrix}^{k+1} &= \begin{bmatrix} 1 & 1 \\ 1 & 0 \end{bmatrix}^k \begin{bmatrix} 1 & 1 \\ 1 & 0 \end{bmatrix} = \begin{bmatrix} F_{k+1} & F_k \\ F_k & F_{k-1} \end{bmatrix}\begin{bmatrix} 1 & 1 \\ 1 & 0 \end{bmatrix} && \text{induction hypothesis} \\ &= \begin{bmatrix} F_{k+1}+F_k & F_{k+1} \\ F_k+F_{k-1} & F_k \end{bmatrix} = \begin{bmatrix} F_{k+2} & F_{k+1} \\ F_{k+1} & F_k \end{bmatrix} && \text{by definition of the Fibonacci sequence}\end{aligned}$$

Thus $P(k+1)$ follows from $P(k)$. So by the Principle of Mathematical Induction, $P(n)$ is true for all $n \geq 2$.

**32.** Let $a_1 = 1$ and $a_{n+1} = \dfrac{1}{1+a_n}$, for $n \geq 1$. Let $P(n)$ be the statement that $a_n = \dfrac{F_n}{F_{n+1}}$, for all $n \geq 1$.

*Step 1:* $P(1)$ is the statement that $a_1 = \dfrac{F_1}{F_2}$, which is true since $a_1 = 1$ and $\dfrac{F_1}{F_2} = \frac{1}{1} = 1$.

*Step 2:* Assume that $P(k)$ is true; that is, $a_k = \dfrac{F_k}{F_{k+1}}$. We want to use this to show that $P(k+1)$ is true. Now

$$a_{k+1} = \frac{1}{1+a_k} \text{ (definition of } a_{k+1}) = \frac{1}{1+\dfrac{F_k}{F_{k+1}}} \text{ (induction hypothesis)} = \frac{F_{k+1}}{F_k+F_{k+1}} = \frac{F_{k+1}}{F_{k+2}} \text{ (definition of } F_{k+2}).$$

Thus $P(k+1)$ follows from $P(k)$. So by the Principle of Mathematical Induction, $P(n)$ is true for all $n \geq 100$.

**33.** Since $F_1 = 1$, $F_2 = 1$, $F_3 = 2$, $F_4 = 3$, $F_5 = 5$, $F_6 = 8$, $F_7 = 13$, ... our conjecture is that $F_n \geq n$, for all $n \geq 5$. Let $P(n)$ denote the statement that $F_n \geq n$.

*Step 1:* $P(5)$ is the statement that $F_5 = 5 \geq 5$, which is clearly true.

*Step 2:* Assume that $P(k)$ is true; that is, $F_k \geq k$, for some $k \geq 5$. We want to use this to show that $P(k+1)$ is true. Now, $F_{k+1} = F_k + F_{k-1} \geq k + F_{k-1}$ (by the induction hypothesis)$\geq k+1$ (because $F_{k-1} \geq 1$). Thus $P(k+1)$ follows from $P(k)$. So by the Principle of Mathematical Induction, $P(n)$ is true for all $n \geq 5$.

**34.** Since $100 \cdot 10 = 10^3$, $100 \cdot 11 < 11^3$, $100 \cdot 12 < 12^3$, ..., our conjecture is that $100n \leq n^3$, for all natural numbers $n \geq 10$. Let $P(n)$ denote the statement that $100n \leq n^3$, for $n \geq 10$.

*Step 1:* $P(10)$ is the statement that $100 \cdot 10 = 1{,}000 \leq 10^3 = 1{,}000$, which is true.

*Step 2:* Assume that $P(k)$ is true; that is, $100k \leq k^3$. We want to use this to show that $P(k+1)$ is true. Now

$$\begin{aligned}100(k+1) &= 100k + 100 \leq k^2 + 100 && \text{induction hypothesis} \\ &\leq k^3 + k^2 && \text{because } k \geq 10 \\ &\leq k^3 + 3k^2 + 3k + 1 = (k+1)^3\end{aligned}$$

Thus $P(k+1)$ follows from $P(k)$. So by the Principle of Mathematical Induction, $P(n)$ is true for all $n \geq 10$.

**35.** **(a)** $P(n) = n^2 - n + 11$ is prime for all $n$. This is false as the case for $n = 11$ demonstrates: $P(11) = 11^2 - 11 + 11 = 121$, which is not prime since $11^2 = 121$.

**(b)** $n^2 > n$, for all $n \geq 2$. This is true. Let $P(n)$ denote the statement that $n^2 > n$.

*Step 1:* $P(2)$ is the statement that $2^2 = 4 > 2$, which is clearly true.

*Step 2:* Assume that $P(k)$ is true; that is, $k^2 > k$. We want to use this to show that $P(k+1)$ is true. Now $(k+1)^2 = k^2 + 2k + 1$. Using the induction hypothesis (to replace $k^2$), we have $k^2 + 2k + 1 > k + 2k + 1 = 3k + 1 > k + 1$, since $k \geq 2$. Therefore, $(k+1)^2 > k + 1$, which is exactly $P(k+1)$. Thus $P(k+1)$ follows from $P(k)$. So by the Principle of Mathematical Induction, $P(n)$ is true for all $n$.

**(c)** $2^{2n+1} + 1$ is divisible by 3, for all $n \geq 1$. This is true. Let $P(n)$ denote the statement that $2^{2n+1} + 1$ is divisible by 3.

*Step 1:* $P(1)$ is the statement that $2^3 + 1 = 9$ is divisible by 3, which is clearly true.

*Step 2:* Assume that $P(k)$ is true; that is, $2^{2k+1} + 1$ is divisible by 3. We want to use this to show that $P(k+1)$ is true. Now, $2^{2(k+1)+1} + 1 = 2^{2k+3} + 1 = 4 \cdot 2^{2k+1} + 1 = (3+1)\,2^{2k+1} + 1 = 3 \cdot 2^{2k+1} + \left(2^{2k+1} + 1\right)$, which is divisible by 3 since $2^{2k+1} + 1$ is divisible by 3 by the induction hypothesis, and $3 \cdot 2^{2k+1}$ is clearly divisible by 3. Thus $P(k+1)$ follows from $P(k)$. So by the Principle of Mathematical Induction, $P(n)$ is true for all $n$.

**(d)** The statement $n^3 \geq (n+1)^2$ for all $n \geq 2$ is false. The statement fails when $n = 2$: $2^3 = 8 < (2+1)^2 = 9$.

**(e)** $n^3 - n$ is divisible by 3, for all $n \geq 2$. This is true. Let $P(n)$ denote the statement that $n^3 - n$ is divisible by 3.

*Step 1:* $P(2)$ is the statement that $2^3 - 2 = 6$ is divisible by 3, which is clearly true.

*Step 2:* Assume that $P(k)$ is true; that is, $k^3 - k$ is divisible by 3. We want to use this to show that $P(k+1)$ is true. Now $(k+1)^3 - (k+1) = k^3 + 3k^2 + 3k + 1 - (k+1) = k^3 + 3k^2 + 2k = k^3 - k + 3k^2 + 2k + k = \left(k^3 - k\right) + 3\left(k^2 + k\right)$. The term $k^3 - k$ is divisible by 3 by our induction hypothesis, and the term $3\left(k^2 + k\right)$ is clearly divisible by 3. Thus $(k+1)^3 - (k+1)$ is divisible by 3, which is exactly $P(k+1)$. So by the Principle of Mathematical Induction, $P(n)$ is true for all $n$.

**(f)** $n^3 - 6n^2 + 11n$ is divisible by 6, for all $n \geq 1$. This is true. Let $P(n)$ denote the statement that $n^3 - 6n^2 + 11n$ is divisible by 6.

*Step 1:* $P(1)$ is the statement that $(1)^3 - 6(1)^2 + 11(1) = 6$ is divisible by 6, which is clearly true.

*Step 2:* Assume that $P(k)$ is true; that is, $k^3 - 6k^2 + 11k$ is divisible by 6. We show that $P(k+1)$ is then also true. Now

$$\begin{aligned}(k+1)^3 - 6(k+1)^2 + 11(k+1) &= k^3 + 3k^2 + 3k + 1 - 6k^2 - 12k - 6 + 11k + 11 \\ &= k^3 - 3k^2 + 2k + 6 = k^3 - 6k^2 + 11k + \left(3k^2 - 9k + 6\right) \\ &= \left(k^3 - 6k^2 + 11k\right) + 3\left(k^2 - 3k + 2\right) = \left(k^3 - 6k^2 + 11k\right) + 3(k-1)(k-2)\end{aligned}$$

In this last expression, the first term is divisible by 6 by our induction hypothesis. The second term is also divisible by 6. To see this, notice that $k - 1$ and $k - 2$ are consecutive natural numbers, and so one of them must be even (divisible by 2). Since 3 also appears in this second term, it follows that this term is divisible by 2 and 3 and so is divisible by 6. Thus $P(k+1)$ follows from $P(k)$. So by the Principle of Mathematical Induction, $P(n)$ is true for all $n$.

**36.** The induction step fails when $k = 2$, that is, $P(2)$ does not follow from $P(1)$. If there are only two cats, Midnight and Sparky, and we remove Sparky, then only Midnight remains. So at this point, we still know only that Midnight is black. Now removing Midnight and putting Sparky back leaves Midnight alone. So the induction hypothesis does not allow us to conclude that Sparky is black.

# 12.6 The Binomial Theorem

**1.** $(x+y)^6 = x^6 + 6x^5y + 15x^4y^2 + 20x^3y^3 + 15x^2y^4 + 6xy^5 + y^6$

**2.** $(2x+1)^4 = (2x)^4 + 4(2x)^3 + 6(2x)^2 + 4 \cdot 2x + 1 = 16x^4 + 32x^3 + 24x^2 + 8x + 1$

**3.** $\left(x + \frac{1}{x}\right)^4 = x^4 + 4x^3 \cdot \frac{1}{x} + 6x^2\left(\frac{1}{x}\right)^2 + 4x\left(\frac{1}{x}\right)^3 + \left(\frac{1}{x}\right)^4 = x^4 + 4x^2 + 6 + \frac{4}{x^2} + \frac{1}{x^4}$

**4.** $(x-y)^5 = x^5 - 5x^4y + 10x^3y^2 - 10x^2y^3 + 5xy^4 - y^5$

**5.** $(x-1)^5 = x^5 - 5x^4 + 10x^3 - 10x^2 + 5x - 1$

**6.** $$\left(\sqrt{a} + \sqrt{b}\right)^6 = a^3 + 6a^2\sqrt{a}\sqrt{b} + 15a^2b + 20a\sqrt{a}b\sqrt{b} + 15ab^2 + 6\sqrt{a}b^2\sqrt{b} + b^3$$
$$= a^3 + 6a^2\sqrt{ab} + 15a^2b + 20ab\sqrt{ab} + 15ab^2 + 6b^2\sqrt{ab} + b^3$$

or $a^3 + 6a^{5/2}b^{1/2} + 15a^2b + 20a^{3/2}b^{3/2} + 15ab^2 + 6a^{1/2}b^{5/2} + b^3$.

**7.** $(x^2y-1)^5 = (x^2y)^5 - 5(x^2y)^4 + 10(x^2y)^3 - 10(x^2y)^2 + 5x^2y - 1 = x^{10}y^5 - 5x^8y^4 + 10x^6y^3 - 10x^4y^2 + 5x^2y - 1$

**8.** $$\left(1+\sqrt{2}\right)^6 = 1^6 + 6 \cdot 1^5 \cdot \sqrt{2} + 15 \cdot 1^4 \cdot 2 + 20 \cdot 1^3 \cdot 2\sqrt{2} + 15 \cdot 1^2 \cdot 4 + 6 \cdot 1 \cdot 4\sqrt{2} + 2^3$$
$$= 1 + 6\sqrt{2} + 30 + 40\sqrt{2} + 60 + 24\sqrt{2} + 8 = 99 + 70\sqrt{2}$$

**9.** $(2x-3y)^3 = (2x)^3 - 3(2x)^2 3y + 3 \cdot 2x(3y)^2 - (3y)^3 = 8x^3 - 36x^2y + 54xy^2 - 27y^3$

**10.** $(1+x^3)^3 = 1^3 + 3 \cdot 1^2 \cdot x^3 + 3 \cdot 1(x^3)^2 + (x^3)^3 = 1 + 3x^3 + 3x^6 + x^9$

**11.** $$\left(\frac{1}{x} - \sqrt{x}\right)^5 = \left(\frac{1}{x}\right)^5 - 5\left(\frac{1}{x}\right)^4\sqrt{x} + 10\left(\frac{1}{x}\right)^3 x - 10\left(\frac{1}{x}\right)^2 x\sqrt{x} + 5\left(\frac{1}{x}\right)x^2 - x^2\sqrt{x}$$
$$= \frac{1}{x^5} - \frac{5}{x^{7/2}} + \frac{10}{x^2} - \frac{10}{x^{1/2}} + 5x - x^{5/2}$$

**12.** $\left(2 + \frac{x}{2}\right)^5 = (2)^5 + 5(2)^4\frac{x}{2} + 10(2)^3\left(\frac{x}{2}\right)^2 + 10(2)^2\left(\frac{x}{2}\right)^3 + 5(2)\left(\frac{x}{2}\right)^4 + \left(\frac{x}{2}\right)^5 = 32 + 40x + 20x^2 + 5x^3 + \frac{5}{8}x^4 + \frac{1}{32}x^5$

**13.** $\binom{6}{4} = \frac{6!}{4!\,2!} = \frac{6 \cdot 5 \cdot 4!}{2 \cdot 1 \cdot 4!} = 15$

**14.** $\binom{8}{3} = \frac{8!}{3!\,5!} = \frac{8 \cdot 7 \cdot 6 \cdot 5!}{3 \cdot 2 \cdot 1 \cdot 5!} = 8 \cdot 7 = 56$

**15.** $\binom{100}{98} = \frac{100!}{98!\,2!} = \frac{100 \cdot 99 \cdot 98!}{98! \cdot 2 \cdot 1} = 4950$

**16.** $\binom{10}{5} = \frac{10!}{5!\,5!} = \frac{10 \cdot 9 \cdot 8 \cdot 7 \cdot 6 \cdot 5!}{5 \cdot 4 \cdot 3 \cdot 2 \cdot 1 \cdot 5!} = 3 \cdot 2 \cdot 7 \cdot 6 = 252$

**17.** $\binom{3}{1}\binom{4}{2} = \frac{3!}{1!\,2!}\frac{4!}{2!\,2!} = \frac{3 \cdot 2! \cdot 4 \cdot 3 \cdot 2!}{1 \cdot 2! \cdot 2 \cdot 1 \cdot 2!} = 18$

**18.** $\binom{5}{2}\binom{5}{3} = \frac{5!}{2!\,3!} \cdot \frac{5!}{3!\,2!} = \frac{5 \cdot 4 \cdot 3!}{2 \cdot 1 \cdot 3!} \cdot \frac{5 \cdot 4 \cdot 3!}{3! \cdot 2 \cdot 1} = 10 \cdot 10 = 100$

**19.** $\binom{5}{0} + \binom{5}{1} + \binom{5}{2} + \binom{5}{3} + \binom{5}{4} + \binom{5}{5} = (1+1)^5 = 2^5 = 32$

**20.** $\binom{5}{0} - \binom{5}{1} + \binom{5}{2} - \binom{5}{3} + \binom{5}{4} - \binom{5}{5} = 1 - \frac{5!}{1!\,4!} + \frac{5!}{2!\,3!} - \frac{5!}{3!\,2!} + \frac{5!}{4!\,1!} - 1 = 0$. Notice that the first and sixth terms cancel, as do the second and fifth terms and the third and fourth terms.

**21.** $(x+2y)^4 = \binom{4}{0}x^4 + \binom{4}{1}x^3 \cdot 2y + \binom{4}{2}x^2 \cdot 4y^2 + \binom{4}{3}x \cdot 8y^3 + \binom{4}{4}16y^4 = x^4 + 8x^3y + 24x^2y^2 + 32xy^3 + 16y^4$

**22.** $(1-x)^5 = \binom{5}{0}(1)^5 - \binom{5}{1}(1)^4x + \binom{5}{2}(1)^3x^2 - \binom{5}{3}(1)^2x^3 + \binom{5}{4}(1)x^4 - \binom{5}{5}x^5 = 1 - 5x + 10x^2 - 10x^3 + 5x^4 - x^5$

**23.** $\left(1+\frac{1}{x}\right)^6 = \binom{6}{0}1^6 + \binom{6}{1}1^5\left(\frac{1}{x}\right) + \binom{6}{2}1^4\left(\frac{1}{x}\right)^2 + \binom{6}{3}1^3\left(\frac{1}{x}\right)^3 + \binom{6}{4}1^2\left(\frac{1}{x}\right)^4 + \binom{6}{5}1\left(\frac{1}{x}\right)^5 + \binom{6}{6}\left(\frac{1}{x}\right)^6$
$= 1 + \frac{6}{x} + \frac{15}{x^2} + \frac{20}{x^3} + \frac{15}{x^4} + \frac{6}{x^5} + \frac{1}{x^6}$

**24.** $(2A+B^2)^4 = \binom{4}{0}(2A)^4 + \binom{4}{1}(2A)^3(B^2) + \binom{4}{2}(2A)^2(B^2)^2 + \binom{4}{3}(2A)(B^2)^3 + \binom{4}{4}(B^2)^4$
$= 16A^4 + 32A^3B^2 + 24A^2B^4 + 8AB^6 + B^8$

**25.** The first three terms in the expansion of $(x+2y)^{20}$ are $\binom{20}{0}x^{20} = x^{20}$, $\binom{20}{1}x^{19}\cdot 2y = 40x^{19}y$, and $\binom{20}{2}x^{18}\cdot(2y)^2 = 760x^{18}y^2$.

**26.** The first four terms in the expansion of $\left(x^{1/2}+1\right)^{30}$ are $\binom{30}{0}\left(x^{1/2}\right)^{30} = x^{15}$, $\binom{30}{1}\left(x^{1/2}\right)^{29}(1) = 30x^{29/2}$, $\binom{30}{2}\left(x^{1/2}\right)^{28}(1)^2 = 435x^{14}$, and $\binom{30}{3}\left(x^{1/2}\right)^{27}(1)^3 = 4060x^{27/2}$.

**27.** The last two terms in the expansion of $\left(a^{2/3}+a^{1/3}\right)^{25}$ are $\binom{25}{24}a^{2/3}\cdot\left(a^{1/3}\right)^{24} = 25a^{26/3}$, and $\binom{25}{25}a^{25/3} = a^{25/3}$.

**28.** The first three terms in the expansion of $\left(x+\frac{1}{x}\right)^{40}$ are $\binom{40}{0}x^{40} = x^{40}$, $\binom{40}{1}x^{39}\left(\frac{1}{x}\right) = 40x^{38}$, and $\binom{40}{2}x^{38}\left(\frac{1}{x}\right)^2 = 780x^{36}$

**29.** The middle term in the expansion of $(x^2+1)^{18}$ occurs when both terms are raised to the 9th power. So this term is $\binom{18}{9}(x^2)^9 1^9 = 48{,}620x^{18}$.

**30.** The fifth term in the expansion of $(ab-1)^{20}$ is $\binom{20}{4}(ab)^{16}(-1)^4 = 4845a^{16}b^{16}$.

**31.** The 24th term in the expansion of $(a+b)^{25}$ is $\binom{25}{23}a^2b^{23} = 300a^2b^{23}$.

**32.** The 28th term in the expansion of $(A-B)^{30}$ is $\binom{30}{27}A^3(-B)^{27} = -4060A^3B^{27}$.

**33.** The 100th term in the expansion of $(1+y)^{100}$ is $\binom{100}{99}1^1\cdot y^{99} = 100y^{99}$.

**34.** The second term in the expansion of $\left(x^2-\frac{1}{x}\right)^{25}$ is $\binom{25}{1}(x^2)^{24}\left(-\frac{1}{x}\right) = -25x^{47}$.

**35.** The term that contains $x^4$ in the expansion of $(x+2y)^{10}$ has exponent $r = 4$. So this term is $\binom{10}{4}x^4\cdot(2y)^{10-4} = 13{,}440x^4y^6$.

**36.** The $r$th term in the expansion of $(\sqrt{2}+y)^{12}$ is $\binom{12}{r}(\sqrt{2})^r y^{12-r}$. The term that contains $y^3$ occurs when $12-r=3 \Leftrightarrow r=9$. Therefore, the term is $\binom{12}{9}(\sqrt{2})^9 y^3 = 3520\sqrt{2}y^3$.

**37.** The $r$th term is $\binom{12}{r}a^r(b^2)^{12-r} = \binom{12}{r}a^r b^{24-2r}$. Thus the term that contains $b^8$ occurs where $24-2r=8 \Leftrightarrow r=8$. So the term is $\binom{12}{8}a^8b^8 = 495a^8b^8$.

**38.** The $r$th term is $\binom{8}{r}(8x)^r\left(\frac{1}{2x}\right)^{8-r} = \binom{8}{r}\frac{8^r}{2^{8-r}}\cdot\frac{x^r}{x^{8-r}} = \binom{8}{r}\frac{8^r}{2^{8-r}}\cdot x^{2r-8}$. So the term that does not contain $x$ occurs when $2r-8=0 \Leftrightarrow r=4$. Thus, the term is $\binom{8}{4}(8x)^4\left(\frac{1}{2x}\right)^4 = 17{,}920$.

**39.** $x^4+4x^3y+6x^2y^2+4xy^3+y^4 = (x+y)^4$

**40.** $(x-1)^5+5(x-1)^4+10(x-1)^3+10(x-1)^2+5(x-1)+1 = [(x-1)+1]^5 = x^5$

**41.** $8a^3+12a^2b+6ab^2+b^3 = \binom{3}{0}(2a)^3+\binom{3}{1}(2a)^2b+\binom{3}{2}2ab^2+\binom{3}{3}b^3 = (2a+b)^3$

**42.** $x^8+4x^6y+6x^4y^2+4x^2y^3+y^4 = \binom{4}{0}(x^2)^4+\binom{4}{1}(x^2)^3y+\binom{4}{2}(x^2)^2y^2+\binom{4}{3}x^2y^3+\binom{4}{4}y^4 = (x^2+y)^4$

**43.** $\frac{(x+h)^3-x^3}{h} = \frac{x^3+3x^2h+3xh^2+h^3-x^3}{h} = \frac{3x^2h+3xh^2+h^3}{h} = \frac{h(3x^2+3xh+h^2)}{h} = 3x^2+3xh+h^2$

**44.** $$\frac{(x+h)^4-x^4}{h} = \frac{\binom{4}{0}x^4+\binom{4}{1}x^3h+\binom{4}{2}x^2h^2+\binom{4}{3}xh^3+\binom{4}{4}h^4-x^4}{h} = \frac{x^4+4x^3h+6x^2h^2+4xh^3+h^4-x^4}{h}$$
$$= \frac{4x^3h+6x^2h^2+4xh^3+h^4}{h} = \frac{h\left(4x^3+6x^2h+4xh^2+h^3\right)}{h} = 4x^3+6x^2h+4xh^2+h^3$$

**45.** $(1.01)^{100} = (1+0.01)^{100}$. Now the first term in the expansion is $\binom{100}{0}1^{100} = 1$, the second term is $\binom{100}{1}1^{99}(0.01) = 1$, and the third term is $\binom{100}{2}1^{98}(0.01)^2 = 0.495$. Now each term is nonnegative, so $(1.01)^{100} = (1+0.01)^{100} > 1+1+.0.495 > 2$. Thus $(1.01)^{100} > 2$.

**46.** $\binom{n}{0} = \frac{n!}{0!n!} = \frac{n!}{1\cdot n!} = 1$. $\binom{n}{n} = \frac{n!}{n!0!} = \frac{n!}{n!\cdot 1} = 1$. Therefore, $\binom{n}{0} = \binom{n}{n}$.

**47.** $\binom{n}{1} = \frac{n!}{1!\,(n-1)!} = \frac{n\,(n-1)!}{1\,(n-1)!} = \frac{n}{1} = n$. $\binom{n}{n-1} = \frac{n!}{(n-1)!\,1!} = \frac{n\,(n-1)!}{(n-1)!\,1} = n$. Therefore, $\binom{n}{1} = \binom{n}{n-1} = n$.

**48.** $\binom{n}{r} = \frac{n!}{r!\,(n-r)!} = \frac{n!}{(n-r)!\,r!} = \binom{n}{n-r}$ for $0 \le r \le n$.

**49. (a)** $\binom{n}{r-1} + \binom{n}{r} = \frac{n!}{(r-1)!\,[n-(r-1)]!} + \frac{n!}{r!\,(n-r)!}$.

**(b)** $$\frac{n!}{(r-1)!\,[n-(r-1)]!} + \frac{n!}{r!\,(n-r)!} = \frac{r\cdot n!}{r\cdot(r-1)!\,(n-r+1)!} + \frac{(n-r+1)\cdot n!}{r!\,(n-r+1)\,(n-r)!}$$
$$= \frac{r\cdot n!}{r!\,(n-r+1)!} + \frac{(n-r+1)\cdot n!}{r!\,(n-r+1)!}$$

Thus a common denominator is $r!\,(n-r+1)!$.

**(c)** Therefore, using the results of parts (a) and (b),

$$\binom{n}{r-1} + \binom{n}{r} = \frac{n!}{(r-1)!\,[n-(r-1)]!} + \frac{n!}{r!\,(n-r)!} = \frac{r\cdot n!}{r!\,(n-r+1)!} + \frac{(n-r+1)\cdot n!}{r!\,(n-r+1)!}$$
$$= \frac{r\cdot n! + (n-r+1)\cdot n!}{r!\,(n-r+1)!} = \frac{n!\,(r+n-r+1)}{r!\,(n-r+1)!} = \frac{n!\,(n+1)}{r!\,(n+1-r)!} = \frac{(n+1)!}{r!\,(n+1-r)!} = \binom{n+1}{r}$$

**50.** Let $P(n)$ be the proposition that $\binom{n}{r}$ is an integer for the number $n$, $0 \le r \le n$.

*Step 1:* Suppose $n = 0$. If $0 \le r \le n$, then $r = 0$, and so $\binom{n}{r} = \binom{0}{0} = 1$, which is obviously an integer. Therefore, $P(0)$ is true.

*Step 2:* Suppose that $P(k)$ is true. We want to use this to show that $P(k+1)$ must also be true; that is, $\binom{k+1}{r}$ is an integer for $0 \le r \le k+1$. But we know that $\binom{k+1}{r} = \binom{k}{r-1} + \binom{k}{r}$ by the key property of binomial coefficients (see Exercise 49). Furthermore, $\binom{k}{r-1}$ and $\binom{k}{r}$ are both integers by the induction hypothesis. Since the sum of two integers is always an integer, $\binom{k+1}{r}$ must be an integer. Thus, $P(k+1)$ is true if $P(k)$ is true. So by the Principal of Mathematical induction, $\binom{n}{r}$ is an integer for all $n \ge 0$, $0 \le r \le n$.

**51.** Notice that $(100!)^{101} = (100!)^{100}\cdot 100!$ and $(101!)^{100} = (101\cdot 100!)^{100} = 101^{100}\cdot(100!)^{100}$. Now $100! = 1\cdot 2\cdot 3\cdot 4\cdot\cdots\cdot 99\cdot 100$ and $101^{100} = 101\cdot 101\cdot 101\cdot\cdots\cdot 101$. Thus each of these last two expressions consists of 100 factors multiplied together, and since each factor in the product for $101^{100}$ is larger than each factor in the product for $100!$, it follows that $100! < 101^{100}$. Thus $(100!)^{100}\cdot 100! < (100!)^{100}\cdot 101^{100}$. So $(100!)^{101} < (101!)^{100}$.

**52.**

$$1+1=2$$
$$1+2+1=4$$
$$1+3+3+1=8$$
$$1+4+6+4+1=16$$
$$1+5+10+10+5+1=32$$

*Conjecture:* The sum is $2^n$.

*Proof:*

$$\begin{aligned} 2^n &= (1+1)^n = \binom{n}{0} 1^0 \cdot 1^n + \binom{n}{1} 1^1 \cdot 1^{n-1} + \binom{n}{2} 1^2 \cdot 1^{n-2} + \cdots + \binom{n}{n} 1^n \cdot 1^0 \\ &= \binom{n}{0} + \binom{n}{1} + \binom{n}{2} + \cdots + \binom{n}{n} \end{aligned}$$

**53.** $0 = 0^n = (-1+1)^n = \binom{n}{0}(-1)^0 (1)^n + \binom{n}{1}(-1)^1 (1)^{n-1} + \binom{n}{2}(-1)^2 (1)^{n-2} + \cdots + \binom{n}{n}(-1)^n (1)^0$
$= \binom{n}{0} - \binom{n}{1} + \binom{n}{2} - \cdots + (-1)^k \binom{n}{k} + \cdots + (-1)^n \binom{n}{n}$

# Chapter 12 Review

**1.** $a_n = \dfrac{n^2}{n+1}$. Then $a_1 = \dfrac{1^2}{1+1} = \dfrac{1}{2}$, $a_2 = \dfrac{2^2}{2+1} = \dfrac{4}{3}$, $a_3 = \dfrac{3^2}{3+1} = \dfrac{9}{4}$, $a_4 = \dfrac{4^2}{4+1} = \dfrac{16}{5}$, and $a_{10} = \dfrac{10^2}{10+1} = \dfrac{100}{11}$.

**2.** $a_n = (-1)^n \dfrac{2^n}{n}$. Then $a_1 = (-1)^1 \dfrac{2^1}{1} = -2$, $a_2 = (-1)^2 \dfrac{2^2}{2} = 2$, $a_3 = (-1)^3 \dfrac{2^3}{3} = -\dfrac{8}{3}$, $a_4 = (-1)^4 \dfrac{2^4}{4} = 4$, and $a_{10} = (-1)^{10} \dfrac{2^{10}}{10} = \dfrac{1024}{10} = \dfrac{512}{5}$.

**3.** $a_n = \dfrac{(-1)^n + 1}{n^3}$. Then $a_1 = \dfrac{(-1)^1 + 1}{1^3} = 0$, $a_2 = \dfrac{(-1)^2 + 1}{2^3} = \dfrac{2}{8} = \dfrac{1}{4}$, $a_3 = \dfrac{(-1)^3 + 1}{3^3} = 0$, $a_4 = \dfrac{(-1)^4 + 1}{4^3} = \dfrac{2}{64} = \dfrac{1}{32}$, and $a_{10} = \dfrac{(-1)^{10} + 1}{10^3} = \dfrac{1}{500}$.

**4.** $a_n = \dfrac{n(n+1)}{2}$. Then $a_1 = \dfrac{1(1+1)}{2} = 1$, $a_2 = \dfrac{2(2+1)}{2} = 3$, $a_3 = \dfrac{3(3+1)}{2} = 6$, $a_4 = \dfrac{4(4+1)}{2} = 10$, and $a_{10} = \dfrac{10(10+1)}{2} = 55$.

**5.** $a_n = \dfrac{(2n)!}{2^n n!}$. Then $a_1 = \dfrac{(2\cdot 1)!}{2^1 \cdot 1!} = 1$, $a_2 = \dfrac{(2\cdot 2)!}{2^2 \cdot 2!} = 3$, $a_3 = \dfrac{(2\cdot 3)!}{2^3 \cdot 3!} = \dfrac{6\cdot 5\cdot 4}{8} = 15$, $a_4 = \dfrac{(2\cdot 4)!}{2^4 \cdot 4!} = \dfrac{8\cdot 7\cdot 6\cdot 5}{16} = 105$, and $a_{10} = \dfrac{(2\cdot 10)!}{2^{10}\cdot 10!} = 654{,}729{,}075$.

**6.** $a_n = \dbinom{n+1}{2}$. Then $a_1 = \dbinom{1+1}{2} = 1$, $a_2 = \dbinom{2+1}{2} = \dfrac{3!}{2!\,1!} = 3$, $a_3 = \dbinom{3+1}{2} = \dfrac{4!}{2!\,2!} = 6$, $a_4 = \dbinom{4+1}{2} = \dfrac{5!}{2!\,3!} = 10$, and $a_{10} = \dbinom{10+1}{2} = \dfrac{11!}{2!\,9!} = 55$.

**7.** $a_n = a_{n-1} + 2n - 1$ and $a_1 = 1$. Then $a_2 = a_1 + 4 - 1 = 4$, $a_3 = a_2 + 6 - 1 = 9$, $a_4 = a_3 + 8 - 1 = 16$, $a_5 = a_4 + 10 - 1 = 25$, $a_6 = a_5 + 12 - 1 = 36$, and $a_7 = a_6 + 14 - 1 = 49$.

**8.** $a_n = \dfrac{a_{n-1}}{n}$ and $a_1 = 1$. Then $a_2 = \dfrac{a_1}{2} = \dfrac{1}{2}$, $a_3 = \dfrac{a_2}{3} = \dfrac{1}{6}$, $a_4 = \dfrac{a_3}{4} = \dfrac{1}{24}$, $a_5 = \dfrac{a_4}{5} = \dfrac{1}{120}$, $a_6 = \dfrac{a_5}{6} = \dfrac{1}{720}$, and $a_7 = \dfrac{a_6}{7} = \dfrac{1}{5040}$.

**9.** $a_n = a_{n-1} + 2a_{n-2}$, $a_1 = 1$ and $a_2 = 3$. Then $a_3 = a_2 + 2a_1 = 5$, $a_4 = a_3 + 2a_2 = 11$, $a_5 = a_4 + 2a_3 = 21$, $a_6 = a_5 + 2a_4 = 43$, and $a_7 = a_6 + 2a_5 = 85$.

**10.** $a_n = \sqrt{3a_{n-1}}$ and $a_1 = \sqrt{3} = 3^{1/2}$. Then $a_2 = \sqrt{3a_1} = \sqrt{3\sqrt{3}} = \left(3 \cdot 3^{1/2}\right)^{1/2} = 3^{3/4}$, $a_3 = \sqrt{3a_2} = \sqrt{3 \cdot 3^{3/4}} = \sqrt{3^{7/4}} = 3^{7/8}$, $a_4 = \sqrt{3a_3} = \sqrt{3 \cdot 3^{7/8}} = \sqrt{3^{15/8}} = 3^{15/16}$, $a_5 = \sqrt{3a_4} = \sqrt{3 \cdot 3^{15/16}} = \sqrt{3^{31/16}} = 3^{31/32}$, $a_6 = \sqrt{3a_5} = \sqrt{3 \cdot 3^{31/32}} = \sqrt{3^{63/32}} = 3^{63/64}$, $a_7 = \sqrt{3a_6} = \sqrt{3 \cdot 3^{63/64}} = \sqrt{3^{127/64}} = 3^{127/128}$.

**11. (a)** $a_1 = 2(1) + 5 = 7$, $a_2 = 2(2) + 5 = 9$, $a_3 = 2(3) + 5 = 11$, $a_4 = 2(4) + 5 = 13$, $a_5 = 2(5) + 5 = 15$

**(b)**

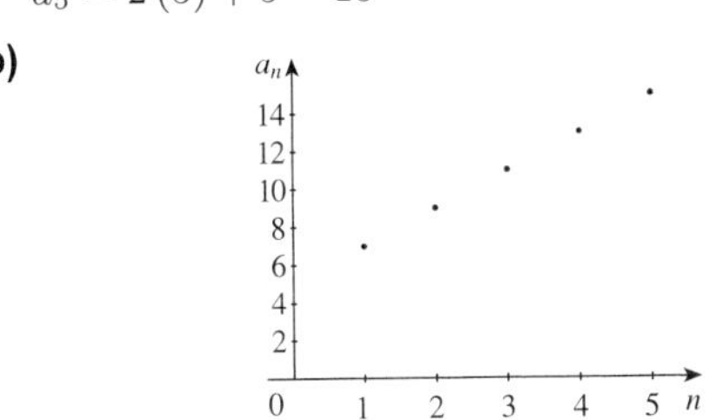


**(c)** This sequence is arithmetic with common difference 2.

**12. (a)** $a_1 = \dfrac{5}{2^1} = \dfrac{5}{2}$, $a_2 = \dfrac{5}{2^2} = \dfrac{5}{4}$, $a_3 = \dfrac{5}{2^3} = \dfrac{5}{8}$, $a_4 = \dfrac{5}{2^4} = \dfrac{5}{16}$, $a_5 = \dfrac{5}{2^5} = \dfrac{5}{32}$

**(b)**

**(c)** This sequence is geometric with common ratio $\frac{1}{2}$.

**13. (a)** $a_1 = \dfrac{3^1}{2^2} = \dfrac{3}{4}$, $a_2 = \dfrac{3^2}{2^3} = \dfrac{9}{8}$, $a_3 = \dfrac{3^3}{2^4} = \dfrac{27}{16}$, $a_4 = \dfrac{3^4}{2^5} = \dfrac{81}{32}$, $a_5 = \dfrac{3^5}{2^6} = \dfrac{243}{64}$

**(b)**

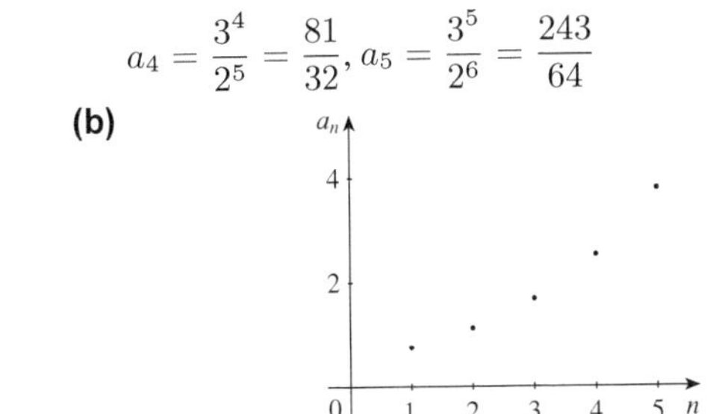


**(c)** This sequence is geometric with common ratio $\frac{3}{2}$.

**14. (a)** $a_1 = 4 - \dfrac{1}{2} = \dfrac{7}{2}$, $a_2 = 4 - \dfrac{2}{2} = 3$, $a_3 = 4 - \dfrac{3}{2} = \dfrac{5}{2}$, $a_4 = 4 - \dfrac{4}{2} = 2$, $a_5 = 4 - \dfrac{5}{2} = \dfrac{3}{2}$

**(b)**

**(c)** This sequence is arithmetic with common difference $-\frac{1}{2}$.

**15.** $5, 5.5, 6, 6.5, \ldots$. Since $5.5 - 5 = 6 - 5.5 = 6.5 - 6 = 0.5$, this is an arithmetic sequence with $a_1 = 5$ and $d = 0.5$. Then $a_5 = a_4 + 0.5 = 7$.

**16.** $1, -\frac{3}{2}, 2, -\frac{5}{2}, \ldots$. Since $-\frac{3}{2} - 1 \neq 2 + \frac{3}{2}$, and $\dfrac{-\frac{3}{2}}{1} \neq \dfrac{2}{-\frac{3}{2}}$, the series is neither arithmetic nor geometric.

**17.** $\sqrt{2}, 2\sqrt{2}, 3\sqrt{2}, 4\sqrt{2}, \ldots$. Since $2\sqrt{2} - \sqrt{2} = 3\sqrt{2} - 2\sqrt{2} = 4\sqrt{2} - 3\sqrt{2} = \sqrt{2}$, this is an arithmetic sequence with $a_1 = \sqrt{2}$ and $d = \sqrt{2}$. Then $a_5 = a_4 + \sqrt{2} = 4\sqrt{2} + \sqrt{2} = 5\sqrt{2}$.

**18.** $\sqrt{2}, 2, 2\sqrt{2}, 4, \ldots$. Since $\frac{2}{\sqrt{2}} = \frac{2\sqrt{2}}{2} = \frac{4}{2\sqrt{2}} = \sqrt{2}$, this is a geometric sequence with $a_1 = \sqrt{2}$ and $r = \sqrt{2}$. Then $a_5 = a_4 \cdot r = 4 \cdot \sqrt{2} = 4\sqrt{2}$.

**19.** $t - 3, t - 2, t - 1, t, \ldots$. Since $(t - 2) - (t - 3) = (t - 1) - (t - 2) = t - (t - 1) = 1$, this is an arithmetic sequence with $a_1 = t - 3$ and $d = 1$. Then $a_5 = a_4 + 1 = t + 1$.

**20.** $t^3, t^2, t, 1, \ldots$. Since $\dfrac{t^2}{t^3} = \dfrac{t}{t^2} = \dfrac{1}{t}$, this is a geometric sequence with $a_1 = t^3$ and $r = \dfrac{1}{t}$. Then $a_5 = a_4 \cdot r = 1 \cdot \dfrac{1}{t} = \dfrac{1}{t}$.

**21.** $\frac{3}{4}, \frac{1}{2}, \frac{1}{3}, \frac{2}{9}, \ldots$. Since $\dfrac{\frac{1}{2}}{\frac{3}{4}} = \dfrac{\frac{1}{3}}{\frac{1}{2}} = \dfrac{\frac{2}{9}}{\frac{1}{3}} = \frac{2}{3}$, this is a geometric sequence with $a_1 = \frac{3}{4}$ and $r = \frac{2}{3}$. Then $a_5 = a_4 \cdot r = \frac{2}{9} \cdot \frac{2}{3} = \frac{4}{27}$.

**22.** $a, 1, \frac{1}{a}, \frac{1}{a^2}, \ldots$ Since $\frac{1}{a} = \frac{\frac{1}{a}}{1}$ and $\frac{\frac{1}{a^2}}{\frac{1}{a}} = \frac{1}{a}$, this is a geometric sequence with $a_1 = a$ and $r = \frac{1}{a}$. Then $a_5 = a_4 \cdot r = \frac{1}{a^2} \cdot \frac{1}{a} = \frac{1}{a^3}$.

**23.** $3, 6i, -12, -24i, \ldots$ Since $\frac{6i}{3} = 2i$, $\frac{-12}{6i} = \frac{-2}{i} = \frac{-2i}{i^2} = 2i$, $\frac{-24i}{-12} = 2i$, this is a geometric sequence with common ratio $r = 2i$.

**24.** The sequence $2,\ 2+2i,\ 4i,\ -4+4i,\ -8, \ldots$ is geometric (where $i^2 = -1$), since $\frac{a_2}{a_1} = \frac{2+2i}{2} = 1+i$, $\frac{a_3}{a_2} = \frac{4i}{2+2i} = \frac{4i}{2+2i} \cdot \frac{2-2i}{2-2i} = \frac{8+8i}{8} = 1+i$, $\frac{a_4}{a_3} = \frac{-4+4i}{4i} = -\frac{1}{i} + 1 = -\frac{1}{i} \cdot \frac{i}{i} + 1 = \frac{-i}{i^2} + 1 = 1+i$, $\frac{a_5}{a_4} = \frac{-8}{-4+4i} = \frac{-8}{-4+4i} \cdot \frac{-4-4i}{-4-4i} = \frac{32+32i}{32} = 1+i$. Thus the common ratio is $i+1$ and the first term is 2. So the $n$th term is $a_n = a_1 r^{n-1} = 2(1+i)^{n-1}$.

**25.** $a_6 = 17 = a + 5d$ and $a_4 = 11 = a + 3d$. Then, $a_6 - a_4 = 17 - 11 \Leftrightarrow (a+5d) - (a+3d) = 6 \Leftrightarrow 6 = 2d \Leftrightarrow d = 3$. Substituting into $11 = a + 3d$ gives $11 = a + 3 \cdot 3$, and so $a = 2$. Thus $a_2 = a + (2-1)d = 2 + 3 = 5$.

**26.** $a_{20} = 96$ and $d = 5$. Then $96 = a_{20} = a + 19 \cdot 5 = a + 95 \Leftrightarrow a = 96 - 95 = 1$. Therefore, $a_n = 1 + 5(n-1)$.

**27.** $a_3 = 9$ and $r = \frac{3}{2}$. Then $a_5 = a_3 \cdot r^2 = 9 \cdot \left(\frac{3}{2}\right)^2 = \frac{81}{4}$.

**28.** $a_2 = 10$ and $a_5 = \frac{1250}{27}$. Then $r^3 = \frac{a_5}{a_2} = \frac{\frac{1250}{27}}{10} = \frac{125}{27} \Leftrightarrow r = \frac{5}{3}$ and $a = a_1 = \frac{a_2}{r} = \frac{10}{\frac{5}{3}} = 6$. Therefore, $a_n = ar^{n-1} = 6\left(\frac{5}{3}\right)^{n-1}$.

**29. (a)** $A_n = 32{,}000 \cdot 1.05^{n-1}$

**(b)** $A_1 = \$32{,}000$, $A_2 = 32{,}000 \cdot 1.05^1 = \$33{,}600$, $A_3 = 32{,}000 \cdot 1.05^2 = \$35{,}280$, $A_4 = 32{,}000 \cdot 1.05^3 = \$37{,}044$, $A_5 = 32{,}000 \cdot 1.05^4 = \$38{,}896.20$, $A_6 = 32{,}000 \cdot 1.05^5 = \$40{,}841.01$, $A_7 = 32{,}000 \cdot 1.05^6 = \$42{,}883.06$, $A_8 = 32{,}000 \cdot 1.05^7 = \$45{,}027.21$

**30. (a)** $A_n = 35{,}000 + 1{,}200(n-1)$

**(b)** $A_8 = 35{,}000 + 1{,}200(7) = \$43{,}400$. The salary for this teacher is higher for the first 6 years is higher, but from the 7th year on, the salary of the teacher in Exercise 29 is higher.

**31.** Let $a_n$ be the number of bacteria in the dish at the end of $5n$ seconds. So $a_0 = 3$, $a_1 = 3 \cdot 2$, $a_2 = 3 \cdot 2^2$, $a_3 = 3 \cdot 2^3, \ldots$. Then, clearly, $a_n$ is a geometric sequence with $r = 2$ and $a = 3$. Thus at the end of $60 = 5(12)$ seconds, the number of bacteria is $a_{12} = 3 \cdot 2^{12} = 12{,}288$.

**32.** Let $d$ be the common difference in the arithmetic sequence $a_1, a_2, a_3, \ldots$, so that $a_n = a_1 + (n-1)d$, $n = 1, 2, 3, \ldots$, and let $e$ be the common difference for $b_1, b_2, b_3, \ldots$, so that $b_n = b_1 + (n-1)e$. Then $a_n + b_n = [a_1 + (n-1)a] + [b_1 + (n-1)e] = (a_1 + b_1) + (n-1)(d+e)$, $n = 1, 2, 3, \ldots$. Thus $a_1 + b_1, a_2 + b_2, \ldots$ is an arithmetic sequence with first term $a_1 + b_1$ and common difference $d + e$.

**33.** Suppose that the common ratio in the sequence $a_1, a_2, a_3, \ldots$ is $r$. Also, suppose that the common ratio in the sequence $b_1, b_2, b_3, \ldots$ is $s$. Then $a_n = a_1 r^{n-1}$ and $b_n = b_1 s^{n-1}$, $n = 1, 2, 3, \ldots$. Thus $a_n b_n = a_1 r^{n-1} \cdot b_1 s^{n-1} = (a_1 b_1)(rs)^{n-1}$. So the sequence $a_1 b_1, a_2 b_2, a_3 b_3, \ldots$ is geometric with first term $a_1 b_1$ and common ratio $rs$.

**34. (a)** Yes. If the common difference is $d$, then $a_n = a_1 + (n-1)d$. So $a_n + 2 = a_1 + 2 + (n-1)d$, and thus the sequence $a_1 + 2, a_2 + 2, a_3 + 2, \ldots$ is an arithmetic sequence with the common difference $d$, but with the first term $a_1 + 2$.

**(b)** Yes. If the common ratio is $r$, then $a_n = a_1 \cdot r^{n-1}$. So $5a_n = (5a_1) \cdot r^{n-1}$, and the sequence $5a_1, 5a_2, 5a_3, \ldots$ is also geometric, with common ratio $r$, but with the first term $5a_1$.

**35. (a)** $6, x, 12, \ldots$ is arithmetic if $x - 6 = 12 - x \Leftrightarrow 2x = 18 \Leftrightarrow x = 9$.

**(b)** $6, x, 12, \ldots$ is geometric if $\frac{x}{6} = \frac{12}{x} \Leftrightarrow x^2 = 72 \Leftrightarrow x = \pm 6\sqrt{2}$.

**36.** **(a)** $2, x, y, 17, \ldots$ is arithmetic. Therefore, $15 = 17 - 2 = a_4 - a_1 = a + 3d - a = 3d$. So $d = 5$, and hence, $x = a + d = 2 + 5 = 7$ and $y = a + 2d = 2 + 2 \cdot 5 = 12$.

**(b)** $2, x, y, 17, \ldots$ is geometric. Therefore, $\frac{17}{2} = \dfrac{a_4}{a_1} = \dfrac{ar^3}{a} = r^3 \Leftrightarrow r = \left(\frac{17}{2}\right)^{1/3}$. So $x = a_2 = ar = 2\left(\frac{17}{2}\right)^{1/3} = 2^{2/3}17^{1/3}$ and $y = a_3 = ar^2 = 2\left[\left(\frac{17}{2}\right)^{1/3}\right]^2 = 2\left(\frac{17}{2}\right)^{2/3} = 2^{1/3}17^{2/3}$.

**37.** $\sum_{k=3}^{6}(k+1)^2 = (3+1)^2 + (4+1)^2 + (5+1)^2 + (6+1)^2 = 16 + 25 + 36 + 49 = 126$

**38.** $\sum_{i=1}^{4}\dfrac{2i}{2i-1} = \dfrac{2\cdot 1}{2\cdot 1-1} + \dfrac{2\cdot 2}{2\cdot 2-1} + \dfrac{2\cdot 3}{2\cdot 3-1} + \dfrac{2\cdot 4}{2\cdot 4-1} = 2 + \dfrac{4}{3} + \dfrac{6}{5} + \dfrac{8}{7} = \dfrac{210+140+126+120}{105} = \dfrac{596}{105}$

**39.** $\sum_{k=1}^{6}(k+1)2^{k-1} = 2\cdot 2^0 + 3\cdot 2^1 + 4\cdot 2^2 + 5\cdot 2^3 + 6\cdot 2^4 + 7\cdot 2^5 = 2 + 6 + 16 + 40 + 96 + 224 = 384$

**40.** $\sum_{m=1}^{5}3^{m-2} = 3^{-1} + 3^0 + 3^1 + 3^2 + 3^3 = \frac{1}{3} + 1 + 3 + 9 + 27 = \frac{121}{3}$

**41.** $\sum_{k=1}^{10}(k-1)^2 = 0^2 + 1^2 + 2^2 + 3^2 + 4^2 + 5^2 + 6^2 + 7^2 + 8^2 + 9^2$

**42.** $\sum_{j=2}^{100}\dfrac{1}{j-1} = \frac{1}{1} + \frac{1}{2} + \frac{1}{3} + \frac{1}{4} + \frac{1}{5} + \cdots + \frac{1}{98} + \frac{1}{99}$

**43.** $\sum_{k=1}^{50}\dfrac{3^k}{2^{k+1}} = \dfrac{3}{2^2} + \dfrac{3^2}{2^3} + \dfrac{3^3}{2^4} + \dfrac{3^4}{2^5} + \cdots + \dfrac{3^{49}}{2^{50}} + \dfrac{3^{50}}{2^{51}}$

**44.** $\sum_{n=1}^{10}n^2 2^n = 1^2\cdot 2^1 + 2^2\cdot 2^2 + 3^2\cdot 2^3 + \cdots + 9^2\cdot 2^9 + 10^2\cdot 2^{10}$

**45.** $3 + 6 + 9 + 12 + \cdots + 99 = 3(1) + 3(2) + 3(3) + \cdots + 3(33) = \sum_{k=1}^{33}3k$

**46.** $1^2 + 2^2 + 3^3 + \cdots + 100^2 = \sum_{k=1}^{100}k^2$

**47.** $1\cdot 2^3 + 2\cdot 2^4 + 3\cdot 2^5 + 4\cdot 2^6 + \cdots + 100\cdot 2^{102}$

$$= (1)\,2^{(1)+2} + (2)\,2^{(2)+2} + (3)\,2^{(3)+2} + (4)\,2^{(4)+2} + \cdots + (100)\,2^{(100)+2} = \sum_{k=1}^{100}k\cdot 2^{k+2}$$

**48.** $\dfrac{1}{1\cdot 2} + \dfrac{1}{2\cdot 3} + \dfrac{1}{3\cdot 4} + \cdots + \dfrac{1}{999\cdot 1000} = \sum_{k=1}^{999}\dfrac{1}{k(k+1)}$

**49.** $1 + 0.9 + (0.9)^2 + \cdots + (0.9)^5$ is a geometric series with $a = 1$ and $r = \dfrac{0.9}{1} = 0.9$. Thus, the sum of the series is $S_6 = \dfrac{1-(0.9)^6}{1-0.9} = \dfrac{1-0.531441}{0.1} = 4.68559$.

**50.** $3 + 3.7 + 4.4 + \cdots + 10$ is an arithmetic series with $a = 3$ and $d = 0.7$. Then $10 = a_n = 3 + 0.7(n-1) \Leftrightarrow 0.7(n-1) = 7 \Leftrightarrow n = 11$. So the sum of the series is $S_{11} = \frac{11}{2}(3+10) = \frac{143}{2} = 71.5$.

**51.** $\sqrt{5} + 2\sqrt{5} + 3\sqrt{5} + \cdots + 100\sqrt{5}$ is an arithmetic series with $a = \sqrt{5}$ and $d = \sqrt{5}$. Then $100\sqrt{5} = a_n = \sqrt{5} + \sqrt{5}(n-1) \Leftrightarrow n = 100$. So the sum is $S_{100} = \frac{100}{2}\left(\sqrt{5} + 100\sqrt{5}\right) = 50\left(101\sqrt{5}\right) = 5050\sqrt{5}$.

**52.** $\frac{1}{3} + \frac{2}{3} + 1 + \frac{4}{3} + \cdots + 33$ is an arithmetic series with $a = \frac{1}{3}$ and $d = \frac{1}{3}$. Then $a_n = 33 = \frac{1}{3} + \frac{1}{3}(n-1) \Leftrightarrow n = 99$. So the sum is $S_{99} = \frac{99}{2}\left(\frac{2}{3} + \dfrac{99-1}{3}\right) = \frac{99}{2}\cdot\frac{100}{3} = 1650$.

**53.** $\sum_{n=0}^{6}3\cdot(-4)^n$ is a geometric series with $a = 3$, $r = -4$, and $n = 7$. Therefore, the sum of the series is $S_7 = 3\cdot\dfrac{1-(-4)^7}{1-(-4)} = \frac{3}{5}\left(1+4^7\right) = 9831$.

**54.** $\sum_{k=0}^{8}7\cdot 5^{k/2}$ is a geometric series with $a = 7$, $r = 5^{1/2}$, and $n = 9$. Thus, the sum of the series is

$$S_9 = 7\cdot\dfrac{1-5^{9/2}}{1-\sqrt{5}} = 7\cdot\dfrac{1-5^{9/2}}{1-\sqrt{5}}\,\dfrac{1+\sqrt{5}}{1+\sqrt{5}} = 7\cdot\dfrac{1-5^{9/2}+\sqrt{5}-5^5}{1-5}$$
$$= -\tfrac{7}{4}\left(1 - 625\sqrt{5} + \sqrt{5} - 3125\right) = 5467 + 1092\sqrt{5} \approx 7908.8$$

**55.** We have an arithmetic sequence with $a = 7$ and $d = 3$. Then $S_n = 325 = \dfrac{n}{2}[2a + (n-1)d] = \dfrac{n}{2}[14 + 3(n-1)] = \dfrac{n}{2}(11+3n) \Leftrightarrow 650 = 3n^2 + 11n \Leftrightarrow (3n+50)(n-13) = 0 \Leftrightarrow n = 13$ (because $n = -\frac{50}{3}$ is inadmissible). Thus, 13 terms must be added.

**56.** We have a geometric series with $S_3 = 52$ and $r = 3$. Then $52 = S_3 = a + 3a + 9a = 13a \Leftrightarrow a = 4$, and so the first term is 4.

**57.** This is a geometric sequence with $a = 2$ and $r = 2$. Then $S_{15} = 2 \cdot \dfrac{1 - 2^{15}}{1 - 2} = 2\left(2^{15} - 1\right) = 65{,}534$, and so the total number of ancestors is 65,534.

**58.** $R = 1000$, $i = 0.08$, and $n = 16$. Thus, $A = 1000\dfrac{1.08^{16} - 1}{1.08 - 1} = 12{,}500\left[1.08^{16} - 1\right] = \$30{,}324.28$.

**59.** $A = 10{,}000$, $i = 0.03$, and $n = 4$. Thus, $10{,}000 = R\dfrac{(1.03)^4 - 1}{0.03} \Leftrightarrow R = \dfrac{10{,}000 \cdot 0.03}{(1.03)^4 - 1} = \$2390.27$.

**60.** $A = 60{,}000$ and $i = \dfrac{0.09}{12} = 0.0075$.

**(a)** If the period is 30 years, $n = 360$ and $R = \dfrac{60{,}000 \cdot 0.0075}{1 - 1.0075^{-360}} = \$482.77$.

**(b)** If the period is 15 years, $n = 180$ and $R = \dfrac{60{,}000 \cdot 0.0075}{1 - 1.0075^{-180}} = \$608.56$.

**61.** $1 - \frac{2}{5} + \frac{4}{25} - \frac{8}{125} + \cdots$ is a geometric series with $a = 1$ and $r = -\frac{2}{5}$. Therefore, the sum is $S = \dfrac{a}{1 - r} = \dfrac{1}{1 - \left(-\frac{2}{5}\right)} = \dfrac{5}{7}$.

**62.** $0.1 + 0.01 + 0.001 + 0.0001 + \cdots$ is an infinite geometric series with $a = 0.1$ and $r = 0.1$. Therefore, the sum is $S = \dfrac{0.1}{1 - 0.1} = \dfrac{1}{9}$.

**63.** $1 + \dfrac{1}{3^{1/2}} + \frac{1}{3} + \dfrac{1}{3^{3/2}} + \cdots$ is an infinite geometric series with $a = 1$ and $r = \dfrac{1}{\sqrt{3}}$. Thus, the sum is $S = \dfrac{1}{1 - \frac{1}{\sqrt{3}}} = \dfrac{\sqrt{3}}{\sqrt{3} - 1} = \frac{1}{2}\left(3 + \sqrt{3}\right)$.

**64.** $a + ab^2 + ab^4 + ab^6 + \cdots$ is an infinite geometric series with first term $a$ and common ratio $b^2$. Thus, the sum is $S = \dfrac{a}{1 - b^2}$.

**65.** Let $P(n)$ denote the statement that $1 + 4 + 7 + \cdots + (3n - 2) = \dfrac{n(3n - 1)}{2}$.

*Step 1:* $P(1)$ is the statement that $1 = \dfrac{1[3(1) - 1]}{2} = \dfrac{1 \cdot 2}{2}$, which is true.

*Step 2:* Assume that $P(k)$ is true; that is, $1 + 4 + 7 + \cdots + (3k - 2) = \dfrac{k(3k - 1)}{2}$. We want to use this to show that $P(k + 1)$ is true. Now

$$
\begin{aligned}
1 + 4 + 7 + 10 + \cdots + (3k - 2) + [3(k + 1) - 2] &= \frac{k(3k - 1)}{2} + 3k + 1 \quad \text{induction hypothesis} \\
&= \frac{k(3k - 1)}{2} + \frac{6k + 2}{2} = \frac{3k^2 - k + 6k + 2}{2} \\
&= \frac{3k^2 + 5k + 2}{2} = \frac{(k + 1)(3k + 2)}{2} \\
&= \frac{(k + 1)[3(k + 1) - 1]}{2}
\end{aligned}
$$

Thus, $P(k + 1)$ follows from $P(k)$. So by the Principle of Mathematical Induction, $P(n)$ is true for all $n$.

**66.** Let $P(n)$ denote the statement that $\frac{1}{1\cdot 3}+\frac{1}{3\cdot 5}+\frac{1}{5\cdot 7}+\cdots+\frac{1}{(2n-1)(2n+1)}=\frac{n}{2n+1}$.

*Step 1:* $P(1)$ is the statement that $\frac{1}{1\cdot 3}=\frac{1}{2\cdot 1+1}$, which is true.

*Step 2:* Assume that $P(k)$ is true; that is, $\frac{1}{1\cdot 3}+\frac{1}{3\cdot 5}+\frac{1}{5\cdot 7}+\cdots+\frac{1}{(2k-1)(2k+1)}=\frac{k}{2k+1}$. We want to use this to show that $P(k+1)$ is true. Now

$$\begin{aligned}
&\frac{1}{1\cdot 3}+\frac{1}{3\cdot 5}+\frac{1}{5\cdot 7}+\cdots+\frac{1}{(2k-1)(2k+1)}+\frac{1}{(2k+1)(2k+3)}\\
&\quad=\frac{k}{2k+1}+\frac{1}{(2k+1)(2k+3)} && \text{induction hypothesis}\\
&\quad=\frac{k(2k+3)+1}{(2k+1)(2k+3)}=\frac{2k^2+3k+1}{(2k+1)(2k+3)}\\
&\quad=\frac{(k+1)(2k+1)}{(2k+1)(2k+3)}=\frac{k+1}{2k+3}=\frac{k+1}{2(k+1)+1}
\end{aligned}$$

Thus $P(k+1)$ follows from $P(k)$. So by the Principle of Mathematical Induction, $P(n)$ is true for all $n$.

**67.** Let $P(n)$ denote the statement that $\left(1+\frac{1}{1}\right)\left(1+\frac{1}{2}\right)\left(1+\frac{1}{3}\right)\cdot\cdots\cdot\left(1+\frac{1}{n}\right)=n+1$.

*Step 1:* $P(1)$ is the statement that $1+\frac{1}{1}=1+1$, which is clearly true.

*Step 2:* Assume that $P(k)$ is true; that is, $\left(1+\frac{1}{1}\right)\left(1+\frac{1}{2}\right)\left(1+\frac{1}{3}\right)\cdot\cdots\cdot\left(1+\frac{1}{k}\right)=k+1$. We want to use this to show that $P(k+1)$ is true. Now

$$\begin{aligned}
&\left(1+\frac{1}{1}\right)\left(1+\frac{1}{2}\right)\left(1+\frac{1}{3}\right)\cdot\cdots\cdot\left(1+\frac{1}{k}\right)\left(1+\frac{1}{k+1}\right)\\
&\quad=\left[\left(1+\frac{1}{1}\right)\left(1+\frac{1}{2}\right)\left(1+\frac{1}{3}\right)\cdot\cdots\cdot\left(1+\frac{1}{k}\right)\right]\left(1+\frac{1}{k+1}\right)\\
&\quad=(k+1)\left(1+\frac{1}{k+1}\right) && \text{induction hypothesis}\\
&\quad=(k+1)+1
\end{aligned}$$

Thus, $P(k+1)$ follows from $P(k)$. So by the Principle of Mathematical Induction, $P(n)$ is true for all $n$.

**68.** Let $P(n)$ denote the statement that $7^n-1$ is divisible by 6.

*Step 1:* $P(1)$ is the statement that $7^1-1=6$ is divisible by 6, which is clearly true.

*Step 2:* Assume that $P(k)$ is true; that is, $7^k-1$ is divisible by 6. We want to use this to show that $P(k+1)$ is true. Now $7^{k+1}-1=7\cdot 7^k-1=7\cdot 7^k-7+6=7\left(7^k-1\right)+6$, which is divisible by 6. This is because $7^k-1$ is divisible by 6 by the induction hypothesis, and clearly 6 is divisible by 6. Thus $P(k+1)$ follows from $P(k)$. So by the Principle of Mathematical Induction, $P(n)$ is true for all $n$.

**69.** $a_{n+1}=3a_n+4$ and $a_1=4$. Let $P(n)$ denote the statement that $a_n=2\cdot 3^n-2$.

*Step 1:* $P(1)$ is the statement that $a_1=2\cdot 3^1-2=4$, which is clearly true.

*Step 2:* Assume that $P(k)$ is true; that is, $a_k=2\cdot 3^k-2$. We want to use this to show that $P(k+1)$ is true. Now

$$\begin{aligned}
a_{k+1}&=3a_k+4 && \text{definition of } a_{k+1}\\
&=3\left(2\cdot 3^k-2\right)+4 && \text{induction hypothesis}\\
&=2\cdot 3^{k+1}-6+4=2\cdot 3^{k+1}-2
\end{aligned}$$

Thus $P(k+1)$ follows from $P(k)$. So by the Principle of Mathematical Induction, $P(n)$ is true for all $n$.

**70.** Let $P(n)$ denote the statement that $F_{4n}$ is divisible by 3.
*Step 1:* Show that $P(1)$ is true, but $P(1)$ is true since $F_4 = 3$ is divisible by 3.
*Step 2:* Assume that $P(k)$ is true; that is, $F_{4k}$ is divisible by 3. We want to use this to show that $P(k+1)$ is true. Now, $F_{4(k+1)} = F_{4k+4} = F_{4k+2} + F_{4k+3} = (F_{4k} + F_{4k+1}) + (F_{4k+1} + F_{4k+2}) = F_{4k} + F_{4k+1} + F_{4k+1} + (F_{4k} + F_{4k+1}) = 2 \cdot F_{4k} + 3 \cdot F_{4k+1}$, which is divisible by 3 because $F_{4k}$ is divisible by 3 by our induction hypothesis, and $3 \cdot F_{4k+1}$ is clearly divisible by 3. Thus, $P(k+1)$ follows from $P(k)$. So by the Principle of Mathematical Induction, $P(n)$ is true for all $n$.

**71.** Let $P(n)$ denote the statement that $n! > 2^n$, for all natural numbers $n \geq 4$.
*Step 1:* $P(4)$ is the statement that $4! = 24 > 2^4 = 16$, which is true.
*Step 2:* Assume that $P(k)$ is true; that is, $k! > 2^k$. We want to use this to show that $P(k+1)$ is true. Now

$$\begin{aligned} (k+1)! &= (k+1)\,k! > (k+1) \cdot 2^k && \text{induction hypothesis} \\ &> 2 \cdot 2^k && \text{because } k+1 > 2 \text{ for } k \geq 4 \\ &= 2^{k+1} \end{aligned}$$

Thus $P(k+1)$ follows from $P(k)$. So by the Principle of Mathematical Induction, $P(n)$ is true for all $n \geq 4$.

**72.** $\binom{5}{2}\binom{5}{3} = \dfrac{5!}{2!\,3!} \cdot \dfrac{5!}{3!\,2!} = \dfrac{5 \cdot 4}{2} \cdot \dfrac{5 \cdot 4}{2} = 10 \cdot 10 = 100$

**73.** $\binom{10}{2} + \binom{10}{6} = \dfrac{10!}{2!\,8!} + \dfrac{10!}{6!\,4!} = \dfrac{10 \cdot 9}{2} + \dfrac{10 \cdot 9 \cdot 8 \cdot 7}{4 \cdot 3 \cdot 2} = 45 + 210 = 255.$

**74.** $\sum_{k=0}^{5} \binom{5}{k} = \binom{5}{0} + \binom{5}{1} + \binom{5}{2} + \binom{5}{3} + \binom{5}{4} + \binom{5}{5} = 2\left(\frac{5!}{0!\,5!} + \frac{5!}{1!\,4!} + \frac{5!}{2!\,3!}\right) = 2(1 + 5 + 10) = 32$

**75.** $\sum_{k=0}^{8} \binom{8}{k}\binom{8}{8-k} = 2\binom{8}{0}\binom{8}{8} + 2\binom{8}{1}\binom{8}{7} + 2\binom{8}{2}\binom{8}{6} + 2\binom{8}{3}\binom{8}{5} + \binom{8}{4}\binom{8}{4} = 2 + 2 \cdot 8^2 + 2 \cdot (28)^2 + 2 \cdot (56)^2 + (70)^2 = 12{,}870.$

**76.** $(1 - x^2)^6 = \binom{6}{0}1^6 - \binom{6}{1}1^5x^2 + \binom{6}{2}1^4x^4 - \binom{6}{3}1^3x^6 + \binom{6}{4}1^2x^8 - \binom{6}{5}x^{10} + \binom{6}{6}x^{12}$
$= 1 - 6x^2 + 15x^4 - 20x^6 + 15x^8 - 6x^{10} + x^{12}$

**77.** $(2x + y)^4 = \binom{4}{0}(2x)^4 + \binom{4}{1}(2x)^3 y + \binom{4}{2}(2x)^2 y^2 + \binom{4}{3} \cdot 2xy^3 + \binom{4}{4}y^4 = 16x^4 + 32x^3y + 24x^2y^2 + 8xy^3 + y^4$

**78.** The 20th term is $\binom{22}{19}a^3b^{19} = 1540a^3b^{19}$.

**79.** The first three terms in the expansion of $\left(b^{-2/3} + b^{1/3}\right)^{20}$ are $\binom{20}{0}\left(b^{-2/3}\right)^{20} = b^{-40/3}$, $\binom{20}{1}\left(b^{-2/3}\right)^{19}\left(b^{1/3}\right) = 20b^{-37/3}$, and $\binom{20}{2}\left(b^{-2/3}\right)^{18}\left(b^{1/3}\right)^2 = 190b^{-34/3}$.

**80.** The $r$th term in the expansion of $(A + 3B)^{10}$ is $\binom{10}{r}A^r(3B)^{10-r}$. The term that contains $A^6$ occurs when $r = 6$. Thus, the term is $\binom{10}{6}A^6(3B)^4 = 210A^6 81B^4 = 17{,}010A^6B^4$.

# Chapter 12 Test

**1.** $a_1 = 1^2 - 1 = 0$, $a_2 = 2^2 - 1 = 3$, $a_3 = 3^2 - 1 = 8$, $a_4 = 4^2 - 1 = 15$, and $a_{10} = 10^2 - 1 = 99$.

**2.** $a_{n+2} = (a_n)^2 - a_{n+1}$, $a_1 = 1$ and $a_2 = 1$. Then $a_3 = a_1^2 - a_2 = 1^2 - 1 = 0$, $a_4 = a_2^2 - a_3 = 1^2 - 0 = 1$, and $a_5 = a_3^2 - a_4 = 0^2 - 1 = -1$.

**3. (a)** The common difference is $d = 5 - 2 = 3$.

**(b)** $a_n = 2 + (n-1)\,3$

**(c)** $a_{35} = 2 + 3(35 - 1) = 104$

**4. (a)** The common ratio is $r = \frac{3}{12} = \frac{1}{4}$.

**(b)** $a_n = a_1 r^{n-1} = 12\left(\frac{1}{4}\right)^{n-1}$

**(c)** $a_{10} = 12\left(\frac{1}{4}\right)^{10-1} = \frac{3}{4^8} = \frac{3}{65{,}536}$

**5. (a)** $a_1 = 25$, $a_4 = \frac{1}{5}$. Then $r^3 = \dfrac{\frac{1}{5}}{25} = \dfrac{1}{125} \quad \Leftrightarrow \quad r = \frac{1}{5}$, so $a_5 = ra_4 = \frac{1}{25}$.

**(b)** $S_8 = 25\dfrac{1-\left(\frac{1}{5}\right)^8}{1-\frac{1}{5}} = \dfrac{5^8-1}{12{,}500} = \dfrac{97{,}656}{3125}$

**6. (a)** $a_1 = 10$ and $a_{10} = 2$, so $9d = -8 \quad \Leftrightarrow \quad d = -\frac{8}{9}$ and $a_{100} = a_1 + 99d = 10 - 88 = -78$.

**(b)** $S_{10} = \frac{10}{2}\left[2a + (10-1)d\right] = \frac{10}{2}\left(2 \cdot 10 + 9\left(-\frac{8}{9}\right)\right) = 60$

**7.** Let the common ratio for the geometric series $a_1, a_2, a_3, \ldots$ be $r$, so that $a_n = a_1 r^{n-1}$, $n = 1, 2, 3, \ldots$. Then $a_n^2 = \left(a_1 r^{n-1}\right)^2 = \left(a_1^2\right)\left(r^2\right)^{n-1}$. Therefore, the sequence $a_1^2, a_2^2, a_3^2, \ldots$ is geometric with common ratio $r^2$.

**8. (a)** $\sum_{n=1}^{5}\left(1-n^2\right) = \left(1-1^2\right) + \left(1-2^2\right) + \left(1-3^2\right) + \left(1-4^2\right) + \left(1-5^2\right) = 0 - 3 - 8 - 15 - 24 = -50$

**(b)** $\sum_{n=3}^{6}(-1)^n 2^{n-2} = (-1)^3 2^{3-2} + (-1)^4 2^{4-2} + (-1)^5 2^{5-2} + (-1)^6 2^{6-2} = -2 + 4 - 8 + 16 = 10$

**9. (a)** The geometric sum $\frac{1}{3} + \frac{2}{3^2} + \frac{2^2}{3^3} + \frac{2^3}{3^4} + \cdots + \frac{2^9}{3^{10}}$ has $a = \frac{1}{3}$, $r = \frac{2}{3}$, and $n = 10$. So $S_{10} = \frac{1}{3} \cdot \frac{1-(2/3)^{10}}{1-(2/3)} = \frac{1}{3} \cdot 3\left(1 - \frac{1024}{59{,}049}\right) = \frac{58{,}025}{59{,}049}$.

**(b)** The infinite geometric series $1 + \frac{1}{2^{1/2}} + \frac{1}{2} + \frac{1}{2^{3/2}} + \cdots$ has $a = 1$ and $r = 2^{-1/2} = \frac{1}{\sqrt{2}}$. Thus, $S = \frac{1}{1-1/\sqrt{2}} = \frac{\sqrt{2}}{\sqrt{2}-1} = \frac{\sqrt{2}}{\sqrt{2}-1} \cdot \frac{\sqrt{2}+1}{\sqrt{2}+1} = 2 + \sqrt{2}$.

**10.** Let $P(n)$ denote the statement that $1^2 + 2^2 + 3^2 + \cdots + n^2 = \dfrac{n(n+1)(2n+1)}{6}$.

*Step 1:* Show that $P(1)$ is true. But $P(1)$ says that $1^2 = \dfrac{1 \cdot 2 \cdot 3}{6}$, which is true.

*Step 2:* Assume that $P(k)$ is true; that is, $1^2 + 2^2 + 3^2 + \cdots + k^2 = \dfrac{k(k+1)(2k+1)}{6}$. We want to use this to show that $P(k+1)$ is true. Now

$$\begin{aligned} 1^2 + 2^2 + 3^2 + \cdots + k^2 + (k+1)^2 &= \frac{k(k+1)(2k+1)}{6} + (k+1)^2 \qquad \text{induction hypothesis} \\ &= \frac{k(k+1)(2k+1) + 6(k+1)^2}{6} = \frac{(k+1)\left(2k^2+k\right) + (6k+6)(k+1)}{6} \\ &= \frac{(k+1)\left(2k^2+k+6k+6\right)}{6} = \frac{(k+1)\left(2k^2+7k+6\right)}{6} \\ &= \frac{(k+1)(k+2)(2k+3)}{6} = \frac{(k+1)\left[(k+1)+1\right]\left[2(k+1)+1\right]}{6} \end{aligned}$$

Thus $P(k+1)$ follows from $P(k)$. So by the Principle of Mathematical Induction, $P(n)$ is true for all $n$.

**11.** $$\begin{aligned} \left(2x+y^2\right)^5 &= \tbinom{5}{0}(2x)^5 + \tbinom{5}{1}(2x)^4 y^2 + \tbinom{5}{2}(2x)^3\left(y^2\right)^2 + \tbinom{5}{3}(2x)^2\left(y^2\right)^3 + \tbinom{5}{4}(2x)\left(y^2\right)^4 + \tbinom{5}{5}\left(y^2\right)^5 \\ &= 32x^5 + 80x^4y^2 + 80x^3y^4 + 40x^2y^6 + 10xy^8 + y^{10} \end{aligned}$$

**12.** $\binom{10}{3}(3x)^3(-2)^7 = 120 \cdot 27x^3(-128) = -414{,}720x^3$

**13. (a)** Each week he gains 24% in weight, that is, $0.24a_n$. Thus, $a_{n+1} = a_n + 0.24a_n = 1.24a_n$ for $n \geq 1$. $a_0$ is given to be 0.85 lb.

**(b)** $a_6 = 1.24a_5 = 1.24(1.24a_4) = \cdots = 1.24^6 a_0 = 1.24^6(0.85) \approx 3.1$ lb

**(c)** The sequence $a_1, a_2, a_3, \ldots$ is geometric with common ratio 1.24.

## Focus on Modeling: Modeling with Recursive Sequences

**1. (a)** Since there are 365 days in a year, the interest earned per day is $\frac{0.0365}{365} = 0.0001$. Thus the amount in the account at the end of the $n$th day is $A_n = 1.0001A_{n-1}$ with $A_0 = \$275{,}000$.

**(b)** $A_0 = \$275{,}000$, $A_1 = 1.0001A_0 = 1.0001 \cdot 275{,}000 = \$275{,}027.50$,
$A_2 = 1.0001A_1 = 1.0001\,(1.0001A_0) = 1.0001^2 A_0 = \$275{,}055.00$,
$A_3 = 1.0001A_2 = 1.0001\left(1.0001^2 A_0\right) = 1.0001^3 A_0 = \$275{,}082.51$,
$A_4 = 1.0001A_3 = 1.0001\left(1.0001^3 A_0\right) = 1.0001^4 A_0 = \$275{,}110.02$,
$A_5 = 1.0001A_4 = 1.0001\left(1.0001^4 A_0\right) = 1.0001^5 A_0 = \$275{,}137.53$,
$A_6 = 1.0001A_5 = 1.0001\left(1.0001^5 A_0\right) = 1.0001^6 A_0 = \$275{,}165.04$,
$A_7 = 1.0001A_6 = 1.0001\left(1.0001^6 A_0\right) = 1.0001^7 A_0 = \$275{,}192.56$

**(c)** $A_n = 1.0001^n \cdot 275{,}000$

**2. (a)** $T_n = T_{n-1} + 1.5$ with $T_1 = 5$.

**(b)** $T_1 = 5$, $T_2 = T_1 + 1.5 = 5 + 1.5 = 6.5$, $T_3 = T_2 + 1.5 = (5 + 1.5) + 1.5 = 5 + 2 \cdot 1.5 = 8.0$,
$T_4 = T_3 + 1.5 = (5 + 2 \cdot 1.5) + 1.5 = 5 + 3 \cdot 1.5 = 9.5$, $T_5 = T_4 + 1.5 = (5 + 3 \cdot 1.5) + 1.5 = 5 + 4 \cdot 1.5 = 11.0$,
$T_6 = T_5 + 1.5 = (5 + 4 \cdot 1.5) + 1.5 = 5 + 5 \cdot 1.5 = 12.5$

**(c)** This is an arithmetic sequence with $T_n = 5 + 1.5\,(n-1)$.

**(d)** $T_n = 65 = 5 + 1.5n - 1.5 \;\Leftrightarrow\; 61.5 = 1.5n \;\Leftrightarrow\; n = 41$.So Sheila swims 65 minutes on the 41st day.

**(e)** Using the partial sum of an arithmetic sequence, she swims for $\frac{30}{2}\,[2\,(5) + (30-1)\,1.5] = 15 \cdot 53.5 = 802.5$ minutes $= 13$ hours $22.5$ minutes.

**3. (a)** Since there are 12 months in a year, the interest earned per day is $\frac{0.03}{12} = 0.0025$. Thus the amount in the account at the end of the $n$th month is $A_n = 1.0025A_{n-1} + 100$ with $A_0 = \$100$.

**(b)** $A_0 = \$100$, $A_1 = 1.0025A_0 + 100 = 1.0025 \cdot 100 + 100 = \$200.25$, $A_2 = 1.0025A_1 + 100 = 1.0025\,(1.0025 \cdot 100 + 100) + 100 = 1.0025^2 \cdot 100 + 1.0025 \cdot 100 + 100 = \$300.75$, $A_3 = 1.0025A_2 + 100 = 1.0025\left(1.0025^2 \cdot 100 + 1.0025 \cdot 100 + 100\right) + 100 = 1.0025^3 \cdot 100 + 1.0025^2 \cdot 100 + 1.0025 \cdot 100 + 100 = \$401.50$, $A_4 = 1.0025A_3 + 100 = 1.0025\left(1.0025^3 \cdot 100 + 1.0025^2 \cdot 100 + 1.0025 \cdot 100 + 100\right) + 100 = 1.0025^4 \cdot 100 + 1.0025^3 \cdot 100 + 1.0025^2 \cdot 100 + 1.0025 \cdot 100 + 100 = \$502.51$

**(c)** $A_n = 1.0025^n \cdot 100 + \cdots + 1.0025^2 \cdot 100 + 1.0025 \cdot 100 + 100$, the partial sum of a geometric series, so
$$A_n = 100 \cdot \frac{1 - 1.0025^{n+1}}{1 - 1.0025} = 100 \cdot \frac{1.0025^{n+1} - 1}{0.0025}.$$

**(d)** Since 5 years is 60 months, we have $A_{60} = 100 \cdot \dfrac{1.0025^{61} - 1}{0.0025} \approx \$6{,}580.83$.

**4.** In each case, $P_0 = 4000$.

**(a)** $P_n = 1.2P_{n-1}.P_5 = 1.2^5\,(4000) \approx 9953$.

**(b)** $P_n = 1.2P_{n-1} - 600$. $P_1 = 1.2\,(4000) - 600 = 4200$, $P_2 = 1.2\,(4200) - 600 = 4440$,
$P_3 = 1.2\,(4440) - 600 = 4728$, $P_4 = 1.2\,(4728) - 600 = 5073$, $P_5 = 1.2\,(5073) - 600 = 5488$.

**(c)** $P_n = 1.2P_{n-1} + 250$. $P_1 = 1.2\,(4000) + 250 = 5050$, $P_2 = 1.2\,(5050) + 250 = 6310$,
$P_3 = 1.2\,(6310) + 250 = 7822$, $P_4 = 1.2\,(7822) + 250 = 9636$, $P_5 = 1.2\,(9636) + 250 = 11{,}814$.

**(d)** Since 10% is have $P_n = 1.2P_{n-1} - 0.1P_{n-1} + 300 = 1.1P_{n-1} + 300$. $P_1 = 1.1\,(4000) + 300 = 4700$,
$P_2 = 1.1\,(4700) + 300 = 5470$, $P_3 = 1.1\,(5470) + 300 = 6317$, $P_4 = 1.1\,(6317) + 300 = 7249$,
$P_5 = 1.1\,(7249) + 300 = 8{,}274$.

**5. (a)** The amount $A_n$ of pollutants in the lake in the $n$th year is 30% of the amount from the preceding year $(0.30A_{n-1})$ plus the amount discharged that year (2400 tons). Thus $A_n = 0.30A_{n-1} + 2400$.

**(b)** $A_0 = 2400$, $A_1 = 0.30\,(2400) + 2400 = 3120$,

$A_2 = 0.30\,[0.30\,(2400) + 2400] + 2400 = 0.30^2\,(2400) + 2400\,(2400) + 2400 = 3336$,

$$A_3 = 0.30\left[0.30^2\,(2400) + 2400\,(2400) + 2400\right] + 2400$$
$$= 0.03^3\,(2400) + 0.30^2\,(2400) + 2400\,(2400) + 2400 = 3400.8,$$

$$A_4 = 0.30\left[0.03^3\,(2400) + 0.30^2\,(2400) + 2400\,(2400) + 2400\right] + 2400$$
$$= 0.03^4\,(2400) + 0.03^3\,(2400) + 0.30^2\,(2400) + 2400\,(2400) + 2400 = 3420.2$$

**(c)** $A_n$ is the partial sum of a geometric series, so

$$A_n = 2400 \cdot \frac{1 - 0.30^{n+1}}{1 - 0.30} = 2400 \cdot \frac{1 - 0.30^{n+1}}{0.70}$$
$$\approx 3428.6\left(1 - 0.30^{n+1}\right)$$

**(d)** $A_6 = 2400 \cdot \dfrac{1 - 0.30^7}{0.70} = 3427.8$ tons. The sum of a geometric series, is $A = 2400 \cdot \dfrac{1}{0.70} = 3428.6$ tons.

**(e)**

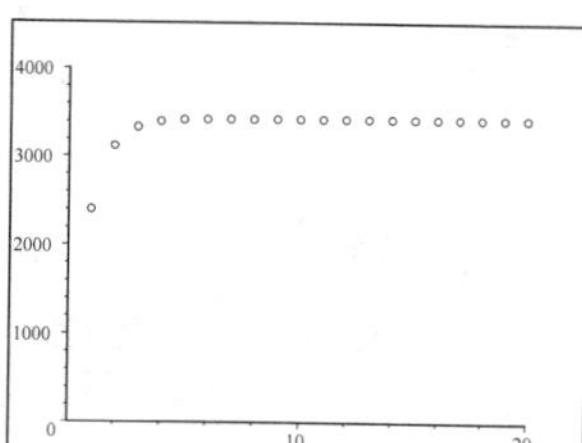

**6. (a)** $U_n = 1.05U_{n-1} + 0.10\,(1.05U_{n-1}) = 1.10\,(1.05)\,U_{n-1} = 1.155U_{n-1}$ with $U_0 = 5000$.

**(b)** $U_0 = \$5000$, $U_1 = 1.155U_0 = \$5775$, $U_2 = 1.155U_1 = 1.155\,(1.155 \cdot 5000) = 1.155^2 \cdot 5000 = \$6670.13$, $U_3 = 1.155U_2 = 1.155\left(1.155^2 \cdot 5000\right) = 1.155^3 \cdot 5000 = \$7703.99$, $U_4 = 1.155U_3 = 1.155\left(1.155^3 \cdot 5000\right) = 1.155^4 \cdot 5000 = \$8898.11$.

**(c)** Using the pattern we found in part (b), we have $U_n = 1.155^n \cdot 5000$.

**(d)** $U_{10} = 1.155^{10} \cdot 5000 = \$21{,}124.67$.

**7. (a)** In the $n$th year since Victoria's initial deposit the amount $V_n$ in her CD is the amount from the preceding year $(V_{n-1})$, plus the 5% interest earned on that amount $(0.05V_{n-1})$, plus \$500 times the number of years since her initial deposit $(500n)$. Thus $V_n = 1.05V_{n-1} + 500n$.

**(b)** Ursula's savings surpass Victoria's savings in the 35th year.

**8. (a)** $T_n = T_{n-1} - 0.03\,(T_{n-1} - 70) = 0.97T_{n-1} + 2.1$, with $T_0 = 170$.

**(b)** $T_0 = 170\ ^\circ\text{F}$; $T_{10} = 143.7\ ^\circ\text{F}$; $T_{20} = 124.4\ ^\circ\text{F}$; $T_{30} = 110.1\ ^\circ\text{F}$; $T_{40} = 99.6\ ^\circ\text{F}$; $T_{50} = 91.8\ ^\circ\text{F}$; $T_{60} = 86.1\ ^\circ\text{F}$.

**(c)**

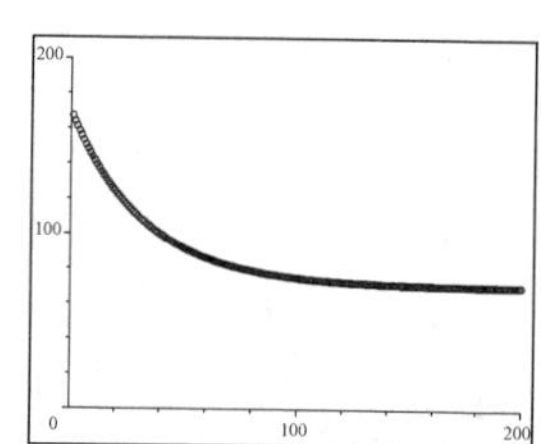

The temperature of the soup approaches 70 °F.

**9. (a)** $R_1 = 104$, $R_2 = 108.0$, $R_3 = 112.0$, $R_4 = 115.9$, $R_5 = 119.8$, $R_6 = 123.7$, $R_7 = 127.4$

**(b)**

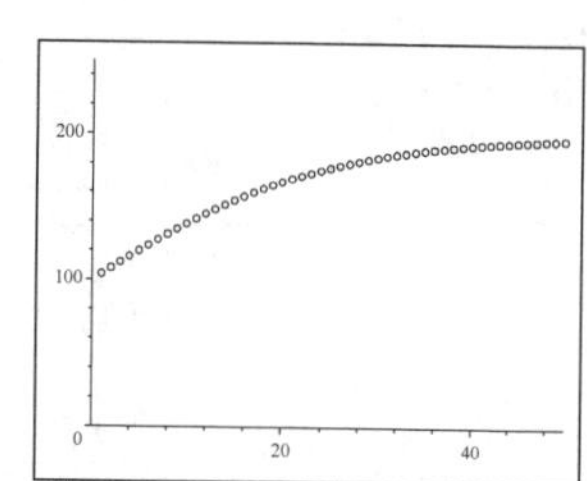

As $n$ becomes large the raccoon population approaches 200.

# 13 Counting and Probability

## 13.1 Counting Principles

**1.** By the Fundamental Counting Principle, the number of possible single-scoop ice cream cones is

$$\begin{pmatrix}\text{number of ways to}\\ \text{choose the flavor}\end{pmatrix}\cdot\begin{pmatrix}\text{number of ways to}\\ \text{choose the type of cone}\end{pmatrix}=4\cdot 3=12.$$

**2.** By the Fundamental Counting Principle, the possible number of 3-letter words is

$$\begin{pmatrix}\text{number of ways to}\\ \text{choose the 1st letter}\end{pmatrix}\cdot\begin{pmatrix}\text{number of ways to}\\ \text{choose the 2nd letter}\end{pmatrix}\cdot\begin{pmatrix}\text{number of ways to}\\ \text{choose the 3rd letter}\end{pmatrix}.$$

**(a)** Since repetitions are allowed, we have 26 choices for each letter. Thus, there are $26\cdot 26\cdot 26=17{,}576$ words.

**(b)** Since repetitions are not allowed, we have 26 choices for the 1st letter, 25 choices for the 2nd letter, and 24 choices for the 3rd letter. Thus there are $26\cdot 25\cdot 24=15{,}600$ words.

**3.** By the Fundamental Counting Principle, the possible number of 3-letter words is

$$\begin{pmatrix}\text{number of ways to}\\ \text{choose the 1st letter}\end{pmatrix}\cdot\begin{pmatrix}\text{number of ways to}\\ \text{choose the 2nd letter}\end{pmatrix}\cdot\begin{pmatrix}\text{number of ways to}\\ \text{choose the 3rd letter}\end{pmatrix}.$$

**(a)** Since repetitions are allowed, we have 4 choices for each letter. Thus there are $4\cdot 4\cdot 4=64$ words.

**(b)** Since repetitions are not allowed, we have 4 choices for the 1st letter, 3 choices for the 2nd letter, and 2 choices for the 3rd letter. Thus there are $4\cdot 3\cdot 2=24$ words.

**4. (a)** By the Fundamental Counting Principle, the possible number of ways 8 horses can complete a race, assuming no ties in any position, is

$$\begin{pmatrix}\text{number of ways to}\\ \text{choose the 1st finisher}\end{pmatrix}\cdot\begin{pmatrix}\text{number of ways to}\\ \text{choose the 2nd finisher}\end{pmatrix}\cdot\cdots\cdot\begin{pmatrix}\text{number of ways to}\\ \text{choose the 8th finisher}\end{pmatrix}=8\cdot 7\cdot 6\cdot 5\cdot 4\cdot 3\cdot 2\cdot 1$$
$$=8!=40{,}320$$

**(b)** By the Fundamental Counting Principle, the possible number of ways the first, second, and third place can be decided, assuming no ties, is $\begin{pmatrix}\text{number of ways to}\\ \text{choose the 1st finisher}\end{pmatrix}\cdot\begin{pmatrix}\text{number of ways to}\\ \text{choose the 2nd finisher}\end{pmatrix}\cdot\begin{pmatrix}\text{number of ways to}\\ \text{choose the 3th finisher}\end{pmatrix}=8\cdot 7\cdot 6=336.$

**5.** Since there are four choices for each of the five questions, by the Fundamental Counting Principle there are $4\cdot 4\cdot 4\cdot 4\cdot 4=1024$ different ways the test can be completed.

**6.** The number of possible seven-digit phone numbers is

$\begin{pmatrix}\text{number of ways to}\\ \text{choose the 1st digit}\end{pmatrix}\cdot\begin{pmatrix}\text{number of ways to}\\ \text{choose the 2nd digit}\end{pmatrix}\cdot\cdots\cdot\begin{pmatrix}\text{number of ways to}\\ \text{choose the 7th digit}\end{pmatrix}$. Since the first digit cannot be a 0 or a 1, there are only 8 digits to choose from, while there are 10 digits to choose from for the other six digits in the phone number. Thus the number of possible seven-digit phone numbers is $8\cdot 10\cdot 10\cdot 10\cdot 10\cdot 10\cdot 10=8{,}000{,}000$.

**7.** Since a runner can only finish once, there are no repetitions. And since we are assuming that there is no tie, the number of different finishes is $\begin{pmatrix}\text{number of ways to}\\ \text{choose the 1st runner}\end{pmatrix}\cdot\begin{pmatrix}\text{number of ways to}\\ \text{choose the 2nd runner}\end{pmatrix}\cdot\cdots\cdot\begin{pmatrix}\text{number of ways to}\\ \text{choose the 5th runner}\end{pmatrix}=5\cdot 4\cdot 3\cdot 2\cdot 1=120.$

**8.** By the Fundamental Counting Principle, the number of ways of seating 5 people in a row of 5 seats is $5\cdot 4\cdot 3\cdot 2\cdot 1=5!=120$.

**9.** Since there are 4 main courses, there are 6 ways to choose a main course. Likewise, there are 5 drinks and 3 desserts so there are 5 ways to choose a drink and 3 ways to choose a dessert. So the number of different meals consisting of a main course, a drink, and a dessert is $\begin{pmatrix}\text{number of ways to}\\\text{choose the main course}\end{pmatrix}\cdot\begin{pmatrix}\text{number of ways to}\\\text{choose a drink}\end{pmatrix}\cdot\begin{pmatrix}\text{number of ways to}\\\text{choose a dessert}\end{pmatrix}=(4)(5)(3)=60.$

**10.** By the Fundamental Counting Principle, the number of ways of arranging 5 mathematics books is $5\cdot4\cdot3\cdot2\cdot1=5!=120$.

**11.** By the Fundamental Counting Principle, the number of different routes from town A to town D via towns B and C is $\begin{pmatrix}\text{number of routes}\\\text{from A to B}\end{pmatrix}\cdot\begin{pmatrix}\text{number of routes}\\\text{from B to C}\end{pmatrix}\cdot\begin{pmatrix}\text{number of routes}\\\text{from C to D}\end{pmatrix}=(4)(5)(6)=120.$

**12.** The number of different boy-girl combinations of 4 children is

$$\begin{pmatrix}\text{number of ways to}\\\text{choose a boy or a}\\\text{girl for the 1st child}\end{pmatrix}\cdot\begin{pmatrix}\text{number of ways to}\\\text{choose a boy or a}\\\text{girl for the 2nd child}\end{pmatrix}\cdot\begin{pmatrix}\text{number of ways to}\\\text{choose a boy or a}\\\text{girl for the 3rd child}\end{pmatrix}\cdot\begin{pmatrix}\text{number of ways to}\\\text{choose a boy or a}\\\text{girl for the 4th child}\end{pmatrix}=2^4=16.$$

**13.** The number of possible sequences of heads and tails when a coin is flipped 5 times is

$$\begin{pmatrix}\text{number of possible}\\\text{outcomes on the 1st flip}\end{pmatrix}\cdot\begin{pmatrix}\text{number of possible}\\\text{outcomes on the 2nd flip}\end{pmatrix}\cdot\dots\cdot\begin{pmatrix}\text{number of possible}\\\text{outcomes on the 5th flip}\end{pmatrix}=(2)(2)(2)(2)(2)$$
$$=2^5=32$$

Here there are only two choices, heads or tails, for each flip.

**14.** Since each die has six different faces, the number of different outcomes when rolling a red and a white die is $6\cdot6=36$.

**15.** Since there are six different faces on each die, the number of possible outcomes when a red die and a blue die and a white die are rolled is

$$\begin{pmatrix}\text{number of possible}\\\text{outcomes on the red die}\end{pmatrix}\cdot\begin{pmatrix}\text{number of possible}\\\text{outcomes on the blue die}\end{pmatrix}\cdot\begin{pmatrix}\text{number of possible}\\\text{outcomes on the white die}\end{pmatrix}=(6)(6)(6)=6^3=216.$$

**16. (a)** Since there are 13 spades and 13 hearts, the number of ways for which the first card is a spade and the second is a heart is $(13)(13)=169$.

**(b)** Since the card is not replaced, the number of ways for which the first card is a spade and the second is a spade is $(13)(12)=156$.

**17.** The number of possible skirt-blouse-shoe outfits is

$$\begin{pmatrix}\text{number of ways}\\\text{to choose a skirt}\end{pmatrix}\cdot\begin{pmatrix}\text{number of ways}\\\text{to choose a blouse}\end{pmatrix}\cdot\begin{pmatrix}\text{number of ways}\\\text{to choose shoes}\end{pmatrix}=(5)(8)(12)=480.$$

**18.** The number of possible ID numbers consisting of one letter followed by 3 digits is $(26)(10)(10)(10)=26{,}000$.

**19.** The number of possible ID numbers of one letter followed by two digits is

$$\begin{pmatrix}\text{number of ways}\\\text{to choose a letter}\end{pmatrix}\cdot\begin{pmatrix}\text{number of ways}\\\text{to choose a digit}\end{pmatrix}\cdot\begin{pmatrix}\text{number of ways}\\\text{to choose a digit}\end{pmatrix}=(26)(10)(10)=2600.$$

Since $2600<2844$, it is not possible to give each of the company's employees a different ID number using this scheme.

**20.** The number of possible pairs of a pitcher and a catcher is $\begin{pmatrix}\text{number of ways to}\\\text{choose a pitcher}\end{pmatrix}\cdot\begin{pmatrix}\text{number of ways to}\\\text{choose a catcher}\end{pmatrix}=(7)(3)=21.$

**21.** The number of different California license plates possible is

$$\begin{pmatrix}\text{number of ways to}\\\text{choose a nonzero digit}\end{pmatrix}\cdot\begin{pmatrix}\text{number of ways}\\\text{to choose 3 letters}\end{pmatrix}\cdot\begin{pmatrix}\text{number of ways}\\\text{to choose 3 digits}\end{pmatrix}=(9)\left(26^3\right)\left(10^3\right)=158{,}184{,}000.$$

**22.** Since successive numbers cannot be the same, the number of possible choices for the second number in the combination is only 59. The third number in the combination cannot be the same as the second in the combination, but it can be the same as the first number, so the number of possible choices for the third number in the combination is also 59. So the number of possible combinations consisting of a number in the clockwise direction, a number in the counterclockwise direction, and then a number in the clockwise direction is $(60)(59)(59) = 208{,}860$.

**23.** There are two possible ways to answer each question on a true-false test. Thus the possible number of different ways to complete a true-false test containing 10 questions is

$$\begin{pmatrix}\text{number of ways}\\\text{to answer question 1}\end{pmatrix}\cdot\begin{pmatrix}\text{number of ways}\\\text{to answer question 2}\end{pmatrix}\cdot\cdots\cdot\begin{pmatrix}\text{number of ways}\\\text{to answer question 10}\end{pmatrix} = 2^{10} = 1024.$$

**24.** The possible number of different cars is

$$\begin{pmatrix}\text{number of}\\\text{models}\end{pmatrix}\cdot\begin{pmatrix}\text{number}\\\text{of colors}\end{pmatrix}\cdot\begin{pmatrix}\text{number of}\\\text{types of stereos}\end{pmatrix}\cdot\begin{pmatrix}\text{air cond.}\\\text{or none}\end{pmatrix}\cdot\begin{pmatrix}\text{sunroof}\\\text{or none}\end{pmatrix} = (5)(4)(3)(2)(2) = 240.$$

**25.** The number of different classifications possible is

$$\begin{pmatrix}\text{number of ways}\\\text{to choose a major}\end{pmatrix}\cdot\begin{pmatrix}\text{number of ways to}\\\text{choose a minor}\end{pmatrix}\cdot\begin{pmatrix}\text{number of ways}\\\text{to choose a year}\end{pmatrix}\cdot\begin{pmatrix}\text{number of ways}\\\text{to choose a sex}\end{pmatrix} = (32)(32)(4)(2) = 8192.$$

To see that there are 32 ways to choose a minor, we start with 32 fields, subtract the major (you cannot major and minor in the same subject), and then add one for the possibility of no minor.

**26.** The possible number of three initial monograms is

$$\begin{pmatrix}\text{number of ways to}\\\text{chose the 1st initial}\end{pmatrix}\cdot\begin{pmatrix}\text{number of ways to}\\\text{chose the 2nd initial}\end{pmatrix}\cdot\begin{pmatrix}\text{number of ways to}\\\text{chose the 3rd initial}\end{pmatrix} = (26)(26)(26) = 17{,}576.$$

**27.** The number of possible license plates of two letters followed by three digits is

$\begin{pmatrix}\text{number of ways to}\\\text{choose 2 letters}\end{pmatrix}\cdot\begin{pmatrix}\text{number of ways}\\\text{to choose 3 digits}\end{pmatrix} = (26^2)(10^3) = 676{,}000$. Since $676{,}000 < 8{,}000{,}000$, there will not be enough different license plates for the state's 8 million registered cars.

**28. (a)** The number of different plates consisting of 1 letter followed by 5 digits is $(26)(10^5) = 2{,}600{,}000 = 2.6$ million. Since this is less than the 17 million registered cars, there are not enough different plates to go around.

**(b)** The system with two letters followed by 4 digits gives $(26^2)(10^4) = 6{,}760{,}000 < 17$ million which is not enough. The system consisting of 3 letters followed by 3 digits gives $(26^3)(10^3) = 17{,}576{,}000 > 17$ million. Thus the smallest number of letters that will provide enough different plates is 3.

**29.** Since a student can hold only one office, the number of ways that a president, a vice-president and a secretary can be chosen from a class of 30 students is

$$\begin{pmatrix}\text{number of ways}\\\text{to choose a president}\end{pmatrix}\cdot\begin{pmatrix}\text{number of ways to}\\\text{choose a vice-president}\end{pmatrix}\cdot\begin{pmatrix}\text{number of ways}\\\text{to choose a secretary}\end{pmatrix} = (30)(29)(28) = 24{,}360.$$

**30.** The number of ways that a president, a vice-president and a secretary can be chosen is

$$\begin{pmatrix}\text{number of ways to}\\\text{choose a female}\\\text{from the 20 females}\end{pmatrix}\cdot\begin{pmatrix}\text{number of ways to}\\\text{choose a male}\\\text{from the 30 males}\end{pmatrix}\cdot\begin{pmatrix}\text{number of ways to choose}\\\text{the secretary from the}\\\text{remaining 48 students}\end{pmatrix} = (20)(30)(48) = 28{,}800.$$

**31.** The number of ways a chairman, vice-chairman, and a secretary can be chosen if the chairman must be a Democrat and the vice-chairman must be a Republican is

$$\begin{pmatrix}\text{number of ways to choose}\\\text{a Democratic chairman}\\\text{from the 10 Democrats}\end{pmatrix}\begin{pmatrix}\text{number of ways to choose}\\\text{a Republican vice-chairman}\\\text{from the 10 Republicans}\end{pmatrix}\cdot\begin{pmatrix}\text{number of ways to choose}\\\text{a secretary from}\\\text{the remaining 15 members}\end{pmatrix} = (10)(7)(15) = 1050.$$

**32.** We have seven choices for the first digit and 10 choices for each of the other 8 digits. Thus, the number of Social Security numbers is $7 \cdot 10^8 = 700{,}000{,}000$.

**33.** The possible number of 5-letter words formed using the letters A, B, C, D, E, F, and G is calculated as follows:

**(a)** If repetition of letters is allowed, then there are 7 ways to pick each letter in the word. Thus the number of 5-letter words is $\begin{pmatrix}\text{number of ways to}\\ \text{choose the 1st letter}\end{pmatrix} \cdot \begin{pmatrix}\text{number of ways to}\\ \text{choose the 2nd letter}\end{pmatrix} \cdot \cdots \cdot \begin{pmatrix}\text{number of ways to}\\ \text{choose the 5th letter}\end{pmatrix} = 7^5 = 16{,}807$.

**(b)** If no letter can be repeated in a word, then the number of 5-letter words is

$$\begin{pmatrix}\text{number of ways to}\\ \text{choose the 1st letter}\end{pmatrix} \cdot \begin{pmatrix}\text{number of ways to}\\ \text{choose the 2nd letter}\end{pmatrix} \cdot \cdots \cdot \begin{pmatrix}\text{number of ways to}\\ \text{choose the 5th letter}\end{pmatrix} = (7)(6)(5)(4)(3) = 2520$$

**(c)** If each word must begin with the letter A, then the number of 5-letter words (letters can be repeated) is

$$\begin{pmatrix}\text{number of ways to}\\ \text{choose the letter A}\end{pmatrix} \cdot \begin{pmatrix}\text{number of ways to}\\ \text{choose the 2nd letter}\end{pmatrix} \cdot \cdots \cdot \begin{pmatrix}\text{number of ways to}\\ \text{choose the 5th letter}\end{pmatrix} = (1)(7)(7)(7)(7) = 7^4 = 2401.$$

**(d)** If the letter C must be in the middle, then the number of 5-letter words (letters can be repeated) is

$$\begin{pmatrix}\text{number of ways to}\\ \text{choose the 1st letter}\end{pmatrix} \cdot \begin{pmatrix}\text{number of ways to}\\ \text{choose the 2nd letter}\end{pmatrix} \cdot \begin{pmatrix}\text{number of ways to}\\ \text{choose the letter C}\end{pmatrix} \cdot \begin{pmatrix}\text{number of ways to}\\ \text{choose the 4th letter}\end{pmatrix} \cdot \begin{pmatrix}\text{number of ways to}\\ \text{choose the 5th letter}\end{pmatrix}$$

$$= (7)(7)(1)(7)(7) = 7^4 = 2401$$

**(e)** If the middle letter must be a vowel, then the number of 5-letter words is $(7)(7)(2)(7)(7) = 2 \cdot 7^4 = 4802$.

**34.** Since the fourth letter must be the same as the second letter and the fifth letter must be the same as the first letter, the possible number of 5-letter palindromes is $(26)(26)(26)(1)(1) = 17{,}576$.

**35.** The number of possible variable names is $\begin{pmatrix}\text{number of ways to}\\ \text{choose a letter}\end{pmatrix} \cdot \begin{pmatrix}\text{number of ways to choose}\\ \text{a letter or a digit}\end{pmatrix} = (26)(36) = 936$.

**36. (a)** The number of possible different 3-character code words with a letter as the first entry is $26 \cdot 36 \cdot 36 = 33{,}696$.

**(b)** The number of possible different 3-character code words in which zero is not the first entry is $35 \cdot 36 \cdot 36 = 45{,}360$.

**37.** The number of ways 4 men and 4 women may be seated in a row of 8 seats is calculated as follows:

**(a)** If the women are to be seated together and the men together, then the number of ways they can be seated is

$$\begin{pmatrix}\text{number of ways to seat the}\\ \text{women first and the men second}\end{pmatrix} + \begin{pmatrix}\text{number of ways to seat the}\\ \text{men first and the women second}\end{pmatrix}$$

$$= (4)(3)(2)(1)(4)(3)(2)(1) + (4)(3)(2)(1)(4)(3)(2)(1) = (4!)(4!) + (4!)(4!) = 576 + 576 = 1152$$

**(b)** If they are to be seated alternately by gender, then the number of ways they can be seated is

$$\begin{pmatrix}\text{number of ways to seat them}\\ \text{alternately if a woman is first}\end{pmatrix} + \begin{pmatrix}\text{number of ways to seat them}\\ \text{alternately if a man is first}\end{pmatrix}$$

$$= (4)(4)(3)(3)(2)(2)(1)(1) + (4)(4)(3)(3)(2)(2)(1)(1) = 1152$$

**38.** Since the two algebra books must be next to each other, we first consider them as one object. So we now have four objects to arrange and there are 4! ways to arrange these four objects. Now there are 2 ways to arrange the two algebra books. Thus the number of ways that 5 mathematics books may be placed on a shelf if the two algebra books are to be next to each other is $2 \cdot 4! = 48$.

**39.** The number of ways that 8 mathematics books and 3 chemistry books may be placed on a shelf if the math books are to be next to each other and the chemistry books next to each other is

$$\begin{pmatrix}\text{number of ways to place them}\\ \text{if the math books are first}\end{pmatrix} + \begin{pmatrix}\text{number of ways to place them}\\ \text{if the chemistry books are first}\end{pmatrix} = (8!)(3!) + (3!)(8!) = 483{,}840.$$

**40. (a)** Since the numbers to be formed are less than 700, the first digit must be either 2, 4, or 5. Thus there are 3 possible ways to choose the first digit. The remaining 2 digits can be 2, 4, 5, or 7. Therefore the number of 3-digit numbers that can be formed using the digits 2, 4, 5, and 7 if the numbers are less than 700 is $(3)(4)(4) = 48$.

**(b)** Since the numbers to be formed must be even, the last digit must be either 2 or 4. The first 2 digits can be 2, 4, 5, or 7. Therefore the number of even 3-digit numbers that can be formed using the digits 2, 4, 5, and 7 is $(4)(4)(2) = 32$.

**(c)** Since the numbers to be formed must be divisible by 5, the last digit must be 5. The first 2 digits can be 2, 4, 5, or 7. Therefore the number of 3-digit numbers divisible by 5 that can be formed using the digits 2, 4, 5, and 7 is $(4)(4)(1) = 16$.

**41.** In order for a number to be odd, the last digit must be odd. In this case, it must be a 1. Thus, the other two digits must be chosen from the three digits 2, 4, and 6. So, the number of 3-digit odd numbers that can be formed using the digits 1, 2, 4, and 6, if no digit may be used more than once is

$$\begin{pmatrix}\text{number of ways to}\\\text{choose a digit}\\\text{from the 3 digits}\end{pmatrix}\cdot\begin{pmatrix}\text{number of ways to}\\\text{choose a digit from}\\\text{the 2 remaining digits}\end{pmatrix}\begin{pmatrix}\text{number of ways}\\\text{to choose}\\\text{an odd digits}\end{pmatrix}=(3)(2)(1)=6.$$

**42.** Since there are 26 letters, the possible number of combinations of the first and the last initials is $(26)(26) = 676$. Since $677 > 676$, there must be at least 2 people that have the same first and last initials in any group of 677 people.

**43. (a)** From Exercise 6, the number of seven-digit phone numbers is 8,000,000. Under the old rules, the possible number of telephone area codes consisting of 3 digits with the given conditions is $(8)(2)(10) = 160$. So the possible number of area code + telephone number combinations under the old rules is $160 \cdot 8{,}000{,}000 = 1{,}280{,}000{,}000$.

**(b)** Under the new rules, where the second digit can be any number, the number of area codes possible is $(8)(10)(10) = 800$. Thus the possible number of area code + telephone number combinations under the new rules is $800 \cdot 8{,}000{,}000 = 6{,}400{,}000{,}000$.

**(c)** Although over $1\frac{1}{4}$ billion telephone numbers may seem to be enough for a population of 300,000,000, so many individuals and businesses now have multiple lines for fax, Internet access, and cellular phones that this number has proven insufficient.

**(d)** There are now more "new" area codes than "old" area codes. Numerous answers are possible.

## 13.2 Permutations and Combinations

**1.** $P(8,3) = \dfrac{8!}{(8-3)!} = \dfrac{8!}{5!} = 8 \cdot 7 \cdot 6 = 336$

**2.** $P(9,2) = \dfrac{9!}{(9-2)!} = \dfrac{9!}{7!} = 9 \cdot 8 = 72$

**3.** $P(11,4) = \dfrac{11!}{(11-4)!} = \dfrac{11!}{7!} = 11 \cdot 10 \cdot 9 \cdot 8 = 7920$

**4.** $P(10,5) = \dfrac{10!}{(10-5)!} = \dfrac{10!}{5!} = 30{,}240$

**5.** $P(100,1) = \dfrac{100!}{(100-1)!} = \dfrac{100!}{99!} = 100$

**6.** $P(99,3) == \dfrac{99!}{(99-3)!} = \dfrac{99!}{96!} = 941{,}094$

**7.** Here we have 6 letters, of which 3 are A, 2 are B, and 1 is C. Thus the number of distinguishable permutations is $\dfrac{6!}{3!\,2!\,1!} = \dfrac{(6)(5)(4)(3!)}{(2)(3!)} = 60$.

**8.** Here we have 9 letters, of which 3 are A, 3 are B, and 3 are C. Thus the number of distinguishable permutations is $\dfrac{9!}{3!\,3!\,3!} = \dfrac{9 \cdot 8 \cdot 7 \cdot 6 \cdot 5 \cdot 4 \cdot 3!}{3 \cdot 2 \cdot 1 \cdot 3 \cdot 2 \cdot 1 \cdot 3!} = 1680$.

**9.** Here we have 5 letters, of which 2 are A, 1 is B, 1 is C, and 1 is D. Thus the number of distinguishable permutations is $\dfrac{5!}{2!\,1!\,1!\,1!} = \dfrac{(5)(4)(3)(2!)}{2!} = 60$.

**10.** Here we have 8 letters, of which 1 is A, 1 is B, 1 is a C, 3 are D, and 2 are E. Thus the number of distinguishable permutations is $\dfrac{8!}{1!\,1!\,1!\,3!\,2!} = 3360$.

**11.** Here we have 9 letters, of which 2 are X, 3 are Y, and 4 are Z. Thus the number of distinguishable permutations is $\dfrac{9!}{2!\,3!\,4!} = \dfrac{9 \cdot 8 \cdot 7 \cdot 6 \cdot 5 \cdot 4!}{2 \cdot 3 \cdot 2 \cdot 4!} = 1260.$

**12.** Here we have 7 letters, of which 2 are X, 2 are Y, and 3 are Z. Thus the number of distinguishable permutations is $\dfrac{7!}{2!\,2!\,3!} = \dfrac{7 \cdot 6 \cdot 5 \cdot 4 \cdot 3!}{2 \cdot 2 \cdot 3!} = 210.$

**13.** $C(8,3) = \dfrac{8!}{3!\,(8-3)!} = \dfrac{8!}{3!\,5!} = \dfrac{8 \cdot 7 \cdot 6}{3 \cdot 2 \cdot 1} = 56$

**14.** $C(9,2) = \dfrac{9!}{2!\,(9-2)!} = \dfrac{9!}{2!\,7!} = \dfrac{9 \cdot 8}{2 \cdot 1} = 36$

**15.** $C(11,4) = \dfrac{11!}{4!\,7!} = \dfrac{11 \cdot 10 \cdot 9 \cdot 8}{4 \cdot 3 \cdot 2 \cdot 1} = 330$

**16.** $C(10,5) = \dfrac{10!}{5!\,(10-5)!} = \dfrac{10!}{5!\,5!} = 252$

**17.** $C(100,1) = \dfrac{100!}{1!\,99!} = \frac{100}{1} = 100$

**18.** $C(99,3) = \dfrac{99!}{3!\,96!} = \dfrac{99 \cdot 98 \cdot 97}{3 \cdot 2 \cdot 1} = 156{,}849$

**19.** We need the number of ways of selecting three students in order for the positions of president, vice president, and secretary from the 15 students in the class. Since a student cannot hold more than one position, this number is $P(15,3) = \dfrac{15!}{(15-3)!} = \dfrac{15!}{12!} = 15 \cdot 14 \cdot 13 = 2730.$

**20.** Since the order of finish is important, we want the number of permutations of 8 objects (the contestants) taken three at a time, which is $P(8,3) = \dfrac{8!}{(8-3)!} = \dfrac{8!}{5!} = 8 \cdot 7 \cdot 6 = 336.$

**21.** Since a person cannot occupy more than one chair at a time, the number of ways of seating 6 people from 10 people in a row of 6 chairs is $P(10,6) = 10 \cdot 9 \cdot 8 \cdot 7 \cdot 6 \cdot 5 = 151{,}200.$

**22.** The number of ways of ordering 6 distinct objects (the 6 people) is $P(6,6) = 6! = 720.$

**23.** The number of ways of selecting 3 objects in order (a 3-letter word) from 6 distinct objects (the 6 letters) assuming that the letters cannot be repeated is $P(6,3) = 6 \cdot 5 \cdot 4 = 120.$

**24.** Since the word permutations is used in this exercise, we are looking for the number of ways to arrange these 4 letters without repeats. Thus, the number of ways is $P(4,4) = 4! = 24.$

**25.** The number of ways of selecting 3 objects in order (a 3-digit number) from 4 distinct objects (the 4 digits) with no repetition of the digits is $P(4,3) = 4 \cdot 3 \cdot 2 = 24.$

**26.** The number of ways of ordering 8 pieces in order (without repeats) is $P(8,8) = 8! = 40{,}320.$

**27.** The number of ways of ordering 9 distinct objects (the contestants) is $P(9,9) = 9! = 362{,}880$. Here a runner cannot finish more than once, so no repetitions are allowed, and order is important.

**28.** The number of ways of ordering three of the five distinct flags is $P(5,3) = 60.$

**29.** The number of ways of ordering 1000 distinct objects (the contestants) taking 3 at a time is $P(1000,3) = 1000 \cdot 999 \cdot 998 = 997{,}002{,}000$. We are assuming that a person cannot win more than once, that is, there are no repetitions.

**30.** In selecting these officers, order is important and repeats are not allowed, so the number of ways of choosing 4 officers from 30 students$= P(30,4) = 657{,}720.$

**31.** We first place Jack in the first seat, and then seat the remaining 4 students. Thus the number of these arrangements is $\left(\begin{array}{c}\text{number of ways to}\\ \text{seat Jack in the 1st seat}\end{array}\right) \cdot \left(\begin{array}{c}\text{number of ways to seat}\\ \text{the remaining 4 students}\end{array}\right) = P(1,1) \cdot P(4,4) = 1!\,4! = 24.$

**32.** We start by first placing Jack in the middle seat, and then we place the remaining 4 students in the remaining 4 seats. Thus the number of a these arrangements is $\left(\begin{array}{c}\text{number of ways}\\ \text{to seat Jack}\end{array}\right) \cdot \left(\begin{array}{c}\text{number of ways to seat}\\ \text{the remaining 4 students}\end{array}\right) = 1 \cdot P(4,4) = 1!\,4! = 24.$

**33.** Here we have 6 objects, of which 2 are blue marbles and 4 are red marbles. Thus the number of distinguishable permutations is $\dfrac{6!}{2!\,4!} = \dfrac{6 \cdot 5 \cdot 4!}{2 \cdot 4!} = 15.$

**34.** Here we have 14 objects (the 14 balls) of which 5 are red balls, 2 are white balls, and 7 are blue balls. So the number of distinguishable permutations is $\dfrac{14!}{5!\,2!\,7!} = 72{,}072$.

**35.** The number of distinguishable permutations of 12 objects (the 12 coins), from like groups of size 4 (the pennies), of size 3 (the nickels), of size 2 (the dimes) and of size 3 (the quarters) is $\dfrac{12!}{4!\,3!\,2!\,3!} = 277{,}200$.

**36.** The word ELEEMOSYNARY has 12 letters of which 3 are E, 2 are Y, and the remaining letters are distinct. So we wish to find the number of distinguishable permutations of 12 objects (the 12 letters) from like groups of size 3, 2, and seven of size 1. We get $\dfrac{12!}{3!\,2!\,1!\,1!\,1!\,1!\,1!\,1!\,1!} = 39{,}916{,}800$.

**37.** The number of distinguishable permutations of 12 objects (the 12 ice cream cones) from like groups of size 3 (the vanilla cones), of size 2 (the chocolate cones), of size 4 (the strawberry cones), and of size 5 (the butterscotch cones) is $\dfrac{14!}{3!\,2!\,4!\,5!} = 2{,}522{,}520$.

**38.** This is the number of distinguishable permutations of 7 objects (the students) from like groups of size 3 (the ones who stay in the 3-person room), size 2 (the ones who stay in the 2-person room), size 1 (the one who stays in the 1-person room), and size 1 (the one who sleeps in the car). This number is $\dfrac{7!}{3!\,2!\,1!\,1!} = 420$.

**39.** The number of distinguishable permutations of 8 objects (the 8 cleaning tasks) from like groups of size 5, 2, and 1 workers, respectively is $\dfrac{8!}{5!\,2!\,1!} = 168$.

**40.** The number of distinguishable permutations of 13 objects (the total number of blocks he must travel) which can be partitioned into like groups of size 8 (the east blocks) and of size 5 (the north blocks) is $\dfrac{13!}{8!\;5!} = 1{,}287$.

**41.** We want the number of ways of choosing a group of three from a group of six. This number is $C(6,3) = \dfrac{6!}{3!\,3!} = 20$.

**42.** In this exercise, we assume that the pizza toppings cannot be repeated, so we are interested in the number of ways to select a subset of 3 toppings from a set of 12 toppings. The number of ways this can occur is $C(12,3) = \dfrac{12!}{3!\,9!} = \dfrac{12 \cdot 11 \cdot 10}{3 \cdot 2 \cdot 1} = 220$.

**43.** We want the number of ways of choosing a group of six people from a group of ten people. The number of combinations of 10 objects (10 people) taken 6 at a time is $C(10,6) = \dfrac{10!}{6!\,4!} = 210$.

**44.** Here we are interested in the number of ways of choosing 3 objects (the 3 members of the committee) from a set of 25 objects (the 25 members). The number of combinations of 25 objects taken 3 at a time is $C(25,3) = \dfrac{25!}{3!\,22!} = 2300$.

**45.** We want the number of ways of choosing a group (the 5-card hand) where order of selection is not important. The number of combinations of 52 objects (the 52 cards) taken 5 at a time is $C(52,5) = \dfrac{52!}{5!\,47!} = 2{,}598{,}960$.

**46.** Since order is not important in a 7-card hand, the number of combinations of 52 objects (the 52 cards) taken 7 at a time is $C(52,7) = \dfrac{52!}{7!\,45!} = 133{,}784{,}560$.

**47.** The order of selection is not important, hence we must calculate the number of combinations of 10 objects (the 10 questions) taken 7 at a time, this gives $C(10,7) = \dfrac{10!}{7!\,3!} = 120$.

**48.** In this exercise, we assume that the pizza toppings cannot be repeated, so we are interested in the number of ways to select a subset of 3 toppings from a set of 16 toppings. The number of ways this can occur is $C(16,3) = \dfrac{16!}{3!\,13!} = 560$.

**49.** We assume that the order in which he plays the pieces in the recital is not important, so the number of combinations of 12 objects (the 12 pieces) taken 8 at a time is $C(12,8) = \dfrac{12!}{8!\,4!} = 495$.

**50.** The order the skirts are selected is not important and no skirt is repeated. So the number of combinations of 8 skirts taken 5 at a time is $C(8,5) = \dfrac{8!}{5!\,3!} = 56$.

**51.** We are just interested in the group of seven students taken from the class of 30 students, not the order in which they are picked. Thus the number is $C(30,7) = \dfrac{30!}{7!\,23!} = 2{,}035{,}800$.

**52.** Since Jack must go on the field trip, we first pick Jack to go on the field trip, and then select the six other students from the remaining 29 students. Since $C(29,6) = \dfrac{29!}{6!\,23!} = 475{,}020$, there are 475,020ways to select the students to go on the field trip with Jack.

**53.** We first take Jack out of the class of 30 students and select the 7 students from the remaining 29 students. Thus there are $C(29,7) = \dfrac{29!}{7!\,22!} = 1{,}560{,}780$ ways to pick the 7 students for the field trip.

**54.** Since the order in which the numbers are selected is not important, the number of combinations of 49 numbers taken 6 at a time is $C(49,6) = \dfrac{49!}{6!\,43!} = 13{,}983{,}816$.

**55.** We must count the number of different ways to choose the group of 6 numbers from the 53 numbers. There are $C(53,6) = \dfrac{53!}{6!\,47!} = 22{,}957{,}480$ possible tickets, so it would cost \$22,957,480.

**56.** **(a)** The number of ways of choosing 5 students from the 20 students is $C(20,5) = \dfrac{20!}{5!\,15!} = 15{,}504$.

**(b)** The number of ways of choosing 5 students for the committee from the 12 females is $C(12,5) = \dfrac{12!}{5!\,7!} = 792$.

**(c)** We use the Fundamental Counting Principle to count the number of possible committees with 3 females and 2 males. Thus, we get

$$\begin{pmatrix}\text{number of ways to choose the}\\ \text{3 females from the 12 females}\end{pmatrix} \cdot \begin{pmatrix}\text{number of ways to choose the}\\ \text{2 males from the 8 males}\end{pmatrix} = C(12,3)\cdot C(8,2) = (220)(28) = 6160.$$

**57.** **(a)** The number of ways to select 5 of the 8 objects is $C(8,5) = \dfrac{8!}{5!\,3!} = 56$.

**(b)** A set with 8 elements has $2^8 = 256$ subsets.

**58.** We may choose any subset of the 8 available brochures. There are $2^8 = 256$ ways to do this.

**59.** Each subset of toppings constitutes a different way a hamburger can be ordered. Since a set with 10 elements has $2^{10} = 1024$ subsets, there are 1024 different ways to order a hamburger.

**60.** We consider a set of 20 objects (the shoppers in the mall) and a subset that corresponds to those shoppers that enter the store. Since a set of 20 objects has $2^{20} = 1{,}048{,}576$ subsets, there are 1,048,576 outcomes to their decisions.

**61.** We pick the two men from the group of ten men, and we pick the two women from the group of ten women. So the number of ways 2 men and 2 women can be chosen is

$$\begin{pmatrix}\text{number of ways to pick}\\ \text{2 of the 10 men}\end{pmatrix} \cdot \begin{pmatrix}\text{number of ways to pick}\\ \text{2 of the 10 women}\end{pmatrix} = C(10,2)\cdot C(10,2) = (45)(45) = 2025.$$

**62.** The number of ways the committee can be chosen is

$$\begin{pmatrix}\text{number of ways to}\\ \text{choose 2 of 6 freshmen}\end{pmatrix} \cdot \begin{pmatrix}\text{number of ways to}\\ \text{choose 3 of 8 sophomores}\end{pmatrix} \cdot \begin{pmatrix}\text{number of ways to}\\ \text{choose 4 of 12 juniors}\end{pmatrix} \cdot \begin{pmatrix}\text{number of ways to}\\ \text{choose 5 of 10 seniors}\end{pmatrix}$$

$$= C(6,2)\cdot C(8,3)\cdot C(12,4)\cdot C(10,5) = 15\cdot 56\cdot 495\cdot 252 = 104{,}781{,}600$$

**63.** The leading and supporting roles are different (order counts), while the extra roles are not (order doesn't count). Also, the male roles must be filled by the male actors and the female roles filled by the female actresses. Thus, number of ways the actors and actresses can be chosen is

$$\left(\begin{array}{c}\text{number of ways the}\\ \text{leading and the supporting}\\ \text{actors can be chosen}\end{array}\right)\cdot\left(\begin{array}{c}\text{number of ways the}\\ \text{leading and the supporting}\\ \text{actresses can be chosen}\end{array}\right)\cdot\left(\begin{array}{c}\text{number of ways to}\\ \text{choose 5 of 8 male extras}\end{array}\right)\cdot\left(\begin{array}{c}\text{number of ways to}\\ \text{choose 3 of 10 female extras}\end{array}\right)$$

$$= P(10,2)\cdot P(12,2)\cdot C(8,5)\cdot C(10,3) = (90)(132)(56)(120) = 79{,}833{,}600.$$

**64.** We choose 3 forwards from the forwards, 2 defense players from the defense men, and the goalie from the two goalies. Thus the number of ways to pick the 6 starting players is

$$\left(\begin{array}{c}\text{number of ways to}\\ \text{pick 3 of 12 forwards}\end{array}\right)\cdot\left(\begin{array}{c}\text{number of ways to}\\ \text{pick 2 of 6 defenders}\end{array}\right)\cdot\left(\begin{array}{c}\text{number of ways to}\\ \text{pick 1 of 2 goalies}\end{array}\right) = C(12,3)\cdot C(6,2)\cdot C(2,1)$$

$$= (220)(15)(2) = 6600$$

**65.** To order a pizza, we must make several choices. First the size (4 choices), the type of crust (2 choices), and then the toppings. Since there are 14 toppings, the number of possible choices is the number of subsets of the 14 toppings, that is $2^{14}$ choices. So by the Fundamental Counting Principle, the number of possible pizzas is $(4)(2)\cdot 2^{14} = 131{,}072$.

**66. (a)** We choose 2 girls from the girls on the camping trip and the 4 boys from the boys on the camping trip. Thus the number of ways to pick the 6 to gather firewood is $\left(\begin{array}{c}\text{number of ways to}\\ \text{choose 2 of 9 girls}\end{array}\right)\cdot\left(\begin{array}{c}\text{number of ways to}\\ \text{choose 4 of 16 boys}\end{array}\right) = C(9,2)\cdot$ $C(16,4) = (36)(1820) = 65{,}520$.

**(b)** *Method 1:* We consider the number of ways of selecting the group of 6 from the 25 campers and subtract off the groups that contain no girls and those that contain one girls. Thus the number groups that contain at least two girls is $C(25,6) - C(9,0)\cdot C(16,6) - C(9,1)\cdot C(16,5) = 177{,}100 - (1)(8008) - (9)(4368) = 129{,}780$.

*Method 2:* In the method we construct all the groups that are possible. Thus the number of groups is

$$\left(\begin{array}{c}\text{groups with 2 girls}\\ \text{and 4 boys}\end{array}\right)+\left(\begin{array}{c}\text{groups with 3 girls}\\ \text{and 3 boys}\end{array}\right)+\left(\begin{array}{c}\text{groups with 4 girls}\\ \text{and 2 boys}\end{array}\right)+\left(\begin{array}{c}\text{groups with 5 girls}\\ \text{and 1 boy}\end{array}\right)+\left(\text{groups with 6 girls}\right)$$

$$= C(9,2)\cdot C(16,4) + C(9,3)\cdot C(16,3) + C(9,4)\cdot C(16,2) + C(9,5)\cdot C(16,1) + C(9,6)\cdot C(16,0)$$

$$= (36)(1820) + (84)(560) + (126)(120) + (126)(16) + (84)(1)$$

$$= 65{,}520 + 47{,}040 + 15{,}120 + 2{,}016 + 84 = 129{,}780.$$

**67.** We treat John and Jane as one object and Mike and Molly as one object, so we need the number of ways of permuting 8 objects. We then multiply this by the number of ways of arranging John and Jane within their group and arranging Mike and Molly within their group. Thus the number of possible arrangements is $P(8,8)\cdot P(2,2)\cdot P(2,2) = 8!\cdot 2!\cdot 2! = 161{,}280$.

**68.** We count the number of arrangements where Mike and Molly are standing together and subtract the arrangements in which John and Jane are also standing together. If we consider Mike and Molly as one object, then we need the number of ways of arranging 9 objects. We then need to arrange Mike and Molly. Thus the number of arrangements where Mike and Molly are standing together is $P(9,9)\cdot P(2,2) = 9!\cdot 2! = 725{,}760$. To count the number of arrangements where Mike and Molly stand together and John and Jane also stand together, we treat John and Jane as one object and Mike and Molly as one object and arrange the 8 objects. So there are $P(8,8)\cdot P(2,2)\cdot P(2,2) = 8!\cdot 2!\cdot 2! = 161{,}280$ ways to do this. (Remember, we still needed to arrange John and Jane within their group and Mike and Molly within their group.) Hence the number of ways to arrange the students is $725{,}760 - 161{,}280 = 564{,}480$.

**69. (a)** To find the number of ways the men and women can be seated we first select and place a man in the first seat and then arrange the other 7 people. Thus we get $\binom{\text{select 1 of}}{\text{the 4 men}} \cdot \binom{\text{arrange the}}{\text{remaining 7 people}} = C(4,1) \cdot P(7,7) = 4 \cdot 7! = 20{,}160$.

**(b)** To find the number of ways the men and women can be seated we first select and place a woman in the first and last seats and then arrange the other 6 people. Thus we get

$$\binom{\text{arrange 2 of}}{\text{the 4 women}} \cdot \binom{\text{arrange the}}{\text{remaining 6 people}} = P(4,2) \cdot P(6,6) = 6 \cdot 6! = 8{,}640.$$

**70. (a)** We treat the women as one object and find the number of ways of permuting 5 objects. We then permute the 4 women. Thus the number of arrangements is $P(5,5) \cdot P(4,4) = 5! \cdot 4! = 2880$.

**(b)** There are two ways to solve this exercise.

*Method 1:* The number of ways the men and women can be seated is

$$\binom{\text{number of ways the}}{\text{men can be arranged}} \cdot \binom{\text{number of ways the}}{\text{women can be arranged}} \cdot \binom{\text{number of ways to select}}{\text{the gender of the first seat}} = P(4,4) \cdot P(4,4) \cdot C(2,1)$$
$$= 4! \cdot 4! \cdot 2 = 1152$$

*Method 2:* There are 8 choices for the first seat, 4 for the second, 3 for the third, 3 for the fourth, 2 for the fifth, 2 for the sixth, 1 for the seventh, and 1 for the eighth. Thus the number of ways of seating the men and women in the required fashion is $8 \cdot 4 \cdot 3 \cdot 3 \cdot 2 \cdot 2 \cdot 1 \cdot 1 = 1152$

**71.** The number of ways the top finalist can be chosen is

$$\begin{pmatrix}\text{number of ways}\\ \text{to choose the 6}\\ \text{semifinalists from the 30}\end{pmatrix} \cdot \begin{pmatrix}\text{number of ways}\\ \text{to choose the 2}\\ \text{finalists from the 6}\end{pmatrix} \cdot \begin{pmatrix}\text{number of ways}\\ \text{to choose the winner}\\ \text{from the 2 finalists}\end{pmatrix} = C(30,6) \cdot C(6,2) \cdot C(2,1)$$
$$= (593{,}775)(15)(2) = 17{,}813{,}250$$

**72.** There are many different possibilities here, so we consider the complement where *no* professor is chosen for the delegates and subtract this number from the way to select 3 people from the group of 8 people, which is $C(8,3)$. If the professor cannot to be selected, then we must select 3 people from a group of 5 and this can be done in $C(5,3)$ ways. Thus the number of delegations that contain a professor is $C(8,3) - C(5,3) = 56 - 10 = 46$.

**73.** We find the number of committees that can be formed and subtract those committees that contain the two picky people. There are $C(10,4)$ ways to select the committees without restrictions. The number of committees that contain the two picky people is $C(8,2)$ because once we place these two people on the committee, then we need only choose two more members from the remaining eight candidates to complete the committee. Thus the number of committees that do not contain these two people is $C(10,4) - C(8,2) = 210 - 28 = 182$.

**74.** Since two points determine a line and no three points are collinear, the exercise is how many ways can you select 2 of the 12 dots. Thus the number of possible lines is $C(12,2) = 66$.

**75.** We show two methods to solve this exercise.

*Method 1:* We consider the number of 5 member committees that can be formed from the 26 interested people and subtract off those that contain no teacher and those that contain no students. Thus the number of committees is $C(26,5) - C(12,5) - C(14,5) = 65{,}780 - 792 - 2002 = 62{,}986$.

*Method 2:* In the method we construct all the committees that are possible. Thus the number of committees is

$$\binom{\text{committees with}}{\text{1 student and 4 teachers}} + \binom{\text{committees with}}{\text{2 students and 3 teachers}} + \binom{\text{committees with}}{\text{3 students and 2 teachers}} + \binom{\text{committees with}}{\text{4 students and 1 teacher}}$$
$$= C(14,1) \cdot C(12,4) + C(14,2) \cdot C(12,3) + C(14,3) \cdot C(12,2) + C(14,4) \cdot C(12,1)$$
$$= (14)(495) + (91)(220) + (364)(66) + (1001)(12)$$
$$= 6{,}930 + 20{,}020 + 24{,}024 + 12{,}012 = 62{,}986$$

**76.** When two objects are chosen from ten objects, it determines a unique set of eight objects, those not chosen. So choosing two objects from ten objects is the same a choosing eight objects from ten objects. In general, every subset of $r$ objects chosen from a set of $n$ objects determines a corresponding set of $(n-r)$ objects, namely, those not chosen. Therefore, the total number of combinations for each type are equal.

**77.** We are only interested in selecting a set of three marbles to give to Luke and a set of two marbles to give to Mark, not the order in which we hand out the marbles. Since both $C(10,3) \cdot C(7,2)$ and $C(10,2) \cdot C(8,3)$ count the number of ways this can be done, these numbers must be equal. (Calculating these values shows that they are indeed equal.) In general, if we wish to find two distinct sets of $k$ and $r$ objects selected from $n$ objects ($k + r \le n$), then we can either first select the $k$ objects from the $n$ objects and then select the $r$ objects from the $n-k$ remaining objects, or we can first select the $r$ objects from the $n$ objects and then the $k$ objects from the $n-r$ remaining objects. Thus $\binom{n}{r} \cdot \binom{n-r}{k} = \binom{n}{k} \cdot \binom{n-k}{r}$.

**78.** **(a)**

$$\begin{aligned}(x+y)^5 &= (x+y)(x+y)(x+y)(x+y)(x+y)\\ &= (x+y)(x+y)(x+y)(xx+xy+yx+yy)\\ &= (x+y)(x+y)(xxx+xxy+xyx+xyy+yxx+yxy+yyx+yyy)\\ &= (x+y)(xxxx+xxxy+xxyx+xxyy+xyxx+xyxy+xyyx+xyyy\\ &\qquad +yxxx+yxxy+yxyx+yxyy+yyxx+yyxy+yyyx+yyyy)\\ &= xxxxx+xxxxy+xxxyx+xxxyy+xxyxx+xxyxy+xxyyx+xxyyy\\ &\quad +xyxxx+xyxxy+xyxyx+xyxyy+xyyxx+xyyxy+xyyyx+xyyyy\\ &\quad +yxxxx+yxxxy+yxxyx+yxxyy+yxyxx+yxyxy+yxyyx+yxyyy\\ &\quad +yyxxx+yyxxy+yyxyx+yyxyy+yyyxx+yyyxy+yyyyx+yyyyy\end{aligned}$$

**(b)** There are ten terms that contain two $x$'s and three $y$'s. Their sum is

$$xxyyy+xyxyy+xyyxy+xyyyx+yxxyy+yxyxy+yxyyx+yyxxy+yyxyx+yyyxx$$

**(c)** To count the number of terms with two $x$'s, we must count the number of ways to pick two of the five positions to contain an $x$. This number is $C(5,2)$.

**(d)** In the Binomial Theorem, the coefficient $\binom{n}{r}$ is the number of ways of picking $r$ positions in a term with $n$ factors to contain an $x$. By definition, this is $C(n,r)$.

## 13.3 Probability

**1.** Let $H$ stand for head and $T$ for tails.

**(a)** The sample space is $S = \{HH, HT, TH, TT\}$.

**(b)** Let $E$ be the event of getting exactly two heads, so $E = \{HH\}$. Then $P(E) = \frac{n(E)}{n(S)} = \frac{1}{4}$.

**(c)** Let $F$ be the event of getting at least one head. Then $F = \{HH, HT, TH\}$, and $P(F) = \frac{n(F)}{n(S)} = \frac{3}{4}$.

**(d)** Let $G$ be the event of getting exactly one head, that is, $G = \{HT, TH\}$. Then $P(G) = \frac{n(G)}{n(S)} = \frac{2}{4} = \frac{1}{2}$.

**2.** Let $H$ stand for heads and $T$ for tails; the numbers 1, 2, . . ., 6 are the faces of the die.

**(a)** $S = \{H1, H2, H3, H4, H5, H6, T1, T2, T3, T4, T5, T6\}$

**(b)** Let $E$ be the event of getting heads and rolling an even number. So $E = \{H2, H4, H6\}$, and $P(E) = \frac{n(E)}{n(S)} = \frac{3}{12} = \frac{1}{4}$.

**(c)** Let $F$ be the event of getting heads and rolling a number greater than 4. So $F = \{H5, H6\}$, and $P(F) = \frac{n(F)}{n(S)} = \frac{2}{12} = \frac{1}{6}$.

**(d)** Let $G$ be the event of getting tails and rolling an odd number. So $G = \{T1, T3, T5\}$, and $P(G) = \frac{n(G)}{n(S)} = \frac{3}{12} = \frac{1}{4}$.

**3. (a)** Let $E$ be the event of rolling a six. Then $P(E) = \frac{n(E)}{n(S)} = \frac{1}{6}$.

**(b)** Let $F$ be the event of rolling an even number. Then $F = \{2, 4, 6\}$. So $P(F) = \frac{n(F)}{n(S)} = \frac{3}{6} = \frac{1}{2}$.

**(c)** Let $G$ be the event of rolling a number greater than 5. Since 6 is the only face greater than 5, $P(G) = \frac{n(G)}{n(S)} = \frac{1}{6}$.

**4. (a)** Let $E$ be the event of rolling a two or a three. Then $P(E) = \frac{n(E)}{n(S)} = \frac{2}{6} = \frac{1}{3}$.

**(b)** Let $F$ be the event of rolling an odd number. So $F = \{1, 3, 5\}$, and $P(F) = \frac{n(F)}{n(S)} = \frac{3}{6} = \frac{1}{2}$.

**(c)** Let $G$ be the event of rolling a number divisible by 3. So $G = \{3, 6\}$, and $P(G) = \frac{n(G)}{n(S)} = \frac{1}{3}$.

**5. (a)** Let $E$ be the event of choosing a king. Since a deck has 4 kings, $P(E) = \frac{n(E)}{n(S)} = \frac{4}{52} = \frac{1}{13}$.

**(b)** Let $F$ be the event of choosing a face card. Since there are 3 face cards per suit and 4 suits, $P(F) = \frac{n(F)}{n(S)} = \frac{12}{52} = \frac{3}{13}$.

**(c)** Let $F$ be the event of choosing a face card. Then $P(F') = 1 - P(F) = 1 - \frac{3}{13} = \frac{10}{13}$.

**6. (a)** Let $E$ be the event of choosing a heart. Since there are 13 hearts, $P(E) = \frac{n(E)}{n(S)} = \frac{13}{52} = \frac{1}{4}$.

**(b)** Let $F$ be the event of choosing a heart or a spade. Since there are 13 hearts and 13 spades, $P(E) = \frac{n(E)}{n(S)} = \frac{26}{52} = \frac{1}{2}$.

**(c)** Let $G$ be the event of choosing a heart, a diamond or a spade. Since there are 13 cards in each suit, $P(G) = \frac{n(G)}{n(S)} = \frac{39}{52} = \frac{3}{4}$.

**7. (a)** Let $E$ be the event of selecting a red ball. Since the jar contains 5 red balls, $P(E) = \frac{n(E)}{n(S)} = \frac{5}{8}$.

**(b)** Let $F$ be the event of selecting a yellow ball. Since there is only one yellow ball, $P(F') = 1 - P(F) = 1 - \frac{n(F)}{n(S)} = 1 - \frac{1}{8} = \frac{7}{8}$.

**(c)** Let $G$ be the event of selecting a black ball. Since there are no black balls in the jar, $P(G) = \frac{n(G)}{n(S)} = \frac{0}{8} = 0$.

**8. (a)** Let $E$ be the event of selecting a white or a yellow ball. Since there are 2 white balls and 1 yellow ball, $P(E') = 1 - P(E) = 1 - \frac{n(E)}{n(S)} = 1 - \frac{3}{8} = \frac{5}{8}$.

**(b)** Let $F$ be the event of selecting a red, a white or a yellow ball. Since all the types of balls are in the jar, $P(E) = 1$.

**(c)** Let $G$ be the event of selecting a white ball. Since there are 2 white balls, $P(E') = 1 - P(E) = 1 - \frac{n(E)}{n(S)} = 1 - \frac{2}{8} = \frac{6}{8} = \frac{3}{4}$.

**9. (a)** Let $E$ be the event of drawing a red sock. Since 3 pairs are red, the drawer contains 6 red socks, and so $P(E) = \frac{n(E)}{n(S)} = \frac{6}{18} = \frac{1}{3}$.

**(b)** Let $F$ be the event of drawing another red sock. Since there are 17 socks left of which 5 are red, $P(F) = \frac{n(F)}{n(S)} = \frac{5}{17}$.

**10.** The spinner has 9 equal sized regions numbered 1 through 9.

**(a)** Let $E$ be the event that the spinner stops at an even number. Since there are 4 even numbers, $P(E) = \frac{4}{9}$.

**(b)** Let $F$ be the event that the spinner stops at an odd number or a number greater than 3. Since the only number not greater than 3 and not odd is 2, $P(E) = 1 - P(E') = 1 - \frac{n(E')}{n(S)} = 1 - \frac{1}{9} = \frac{8}{9}$.

**11. (a)** Let $E$ be the event of choosing a "T". Since 3 of the 16 letters are T's, $P(E) = \frac{3}{16}$.

**(b)** Let $F$ be the event of choosing a vowel. Since there are 6 vowels, $P(F) = \frac{6}{16} = \frac{3}{8}$.

**(c)** Let $F$ be the event of choosing a vowel. Then $P(F') = 1 - \frac{3}{8} = \frac{5}{8}$.

**12.** Let $E$ be the event of dealing 5 hearts. Since there are 13 hearts, $P(E) = \dfrac{C(13,5)}{C(52,5)} = \dfrac{1287}{2{,}598{,}960} \approx 0.000495$.

**13.** Let $E$ be the event of choosing 5 cards of the same suit. Since there are 4 suits and 13 cards in each suit, $n(E) = 4 \cdot C(13,5)$. Also, $n(S) = C(52,5)$. Therefore, $P(E) = \dfrac{4 \cdot C(13,5)}{C(52,5)} = \dfrac{5{,}148}{2{,}598{,}960} \approx 0.00198$.

**14.** Let $E$ be the event of dealing 5 face cards. Since there are 3 face cards for each suit and 4 suits, $P(E) = \dfrac{C(12,5)}{C(52,5)} = \dfrac{792}{2{,}598{,}960} \approx 0.000305$.

**15.** Let $E$ be the event of dealing a royal flush (ace, king, queen, jack, and 10 of the same suit). Since there is only one such sequence for each suit, there are only 4 royal flushes, so $P(E) = \dfrac{4}{C(52,5)} = \dfrac{4}{2{,}598{,}960} \approx 1.53908 \times 10^{-6}$.

**16. (a)** $S = \{(1,1), (1,2), (1,3), (1,4), (1,5), (1,6), (2,1), (2,2), (2,3), (2,4), (2,5), (2,6),$
$(3,1), (3,2), (3,3), (3,4), (3,5), (3,6), (4,1), (4,2), (4,3), (4,4), (4,5), (4,6),$
$(5,1), (5,2), (5,3), (5,4), (5,5), (5,6), (6,1), (6,2), (6,3), (6,4), (6,5), (6,6)\}$

**(b)** Let $E$ be the event of getting a sum of 7. Then $E = \{(1,6), (2,5), (3,4), (4,3), (5,2), (6,1)\}$, and $P(E) = \frac{6}{36} = \frac{1}{6}$.

**(c)** Let $F$ be the event of getting a sum of 9. Then $F = \{(3,6), (4,5), (5,4), (6,3)\}$, and $P(F) = \frac{4}{36} = \frac{1}{9}$.

**(d)** Let $E$ be the event that the two dice show the same number. Then $P(E) = \frac{6}{36} = \frac{1}{6}$.

**(e)** Let $E$ be the event that the two dice show different numbers. Then $E'$ is the event that the two dice show the same number. Thus, $P(E) = 1 - P(E') = 1 - \frac{1}{6} = \frac{5}{6}$.

**(f)** Let $E$ be the event of getting a sum of 9 or higher. Then $P(E) = \frac{10}{36} = \frac{5}{18}$.

**17. (a)** Let B stand for "boy"and G stand for "girl". Then
$S = \{BBBB, GBBB, BGBB, BBGB, BBBG, GGBB, GBGB, GBBG,$
$BGGB, BGBG, BBGG, BGGG, GBGG, GGBG, GGGB, GGGG\}$

**(b)** Let $E$ be the event that the couple has only boys. Then $E = \{BBBB\}$ and $P(E) = \frac{1}{16}$.

**(c)** Let $F$ be the event that the couple has 2 boys and 2 girls. Then $F = \{GGBB, GBGB, GBBG, BGGB, BGBG, BBGG\}$, so $P(F) = \frac{6}{16} = \frac{3}{8}$.

**(d)** Let $G$ be the event that the couple has 4 children of the same sex. Then $G = \{BBBB, GGGG\}$, and $P(G) = \frac{2}{16} = \frac{1}{8}$.

**(e)** Let $H$ be the event that the couple has at least 2 girls. Then $H'$ is the event that the couple has fewer than two girls. Thus, $H' = \{BBBB, GBBB, BGBB, BBGB, BBBG\}$, so $n(H') = 5$, and $P(H) = 1 - P(H') = 1 - \frac{5}{16} = \frac{11}{16}$.

**18.** Let $E$ be the event that a 13-card bridge hand consists of all cards from the same suit. Since there are exactly 4 such hands (one for each suit), $P(E) = \dfrac{4}{C(52,13)} \approx 6.3 \times 10^{-12}$.

**19.** Let $E$ be the event that the ball lands in an odd numbered slot. Since there are 18 odd numbers between 1 and 36, $P(E) = \frac{18}{38} = \frac{9}{19}$.

**20. (a)** Let $E$ be the event that the toddler arranges the word "FRENCH". Since the letters are distinct, there are $P(6,6)$ ways of arranging the blocks of which only one spells the word "FRENCH". Thus $P(E) = \dfrac{1}{P(6,6)} = \dfrac{1}{720} \approx 0.0014$.

**(b)** Let $E$ be the event that the toddler arranges the letters in alphabetical order. Since there are $P(6,6)$ ways of arranging the blocks of which only one is in alphabetical order, $P(E) = \dfrac{1}{P(6,6)} = \dfrac{1}{720} \approx 0.0014$.

**21.** Let $E$ be the event of picking the 6 winning numbers. Since there is only one way to pick these, $P(E) = \dfrac{1}{C(49,6)} = \dfrac{1}{13{,}983{,}816} \approx 7.15 \times 10^{-8}$.

**22.** Let $E$ be the event that no women are chosen. The number of ways that no women are chosen is the same as the number of ways that only men are chosen, which is $C(11,6)$. Thus $P(E) = \dfrac{C(11,6)}{C(30,6)} \approx 0.00078$.

**23.** The sample space consist of all possible True-False combinations, so $n(S) = 2^{10}$.

**(a)** Let $E$ be the event that the student answers all 10 questions correctly. Since there is only one way to answer all 10 questions correctly, $P(E) = \dfrac{1}{2^{10}} = \dfrac{1}{1024}$.

**(b)** Let $F$ be the event that the student answers exactly 7 questions correctly. The number of ways to answer exactly 7 of the 10 questions correctly is the number of ways to choose 7 of the 10 questions, so $n(F) = C(10,7)$. Therefore, $P(E) = \dfrac{C(10,7)}{2^{10}} = \dfrac{120}{1024} = \dfrac{15}{128}$.

**24.** Let $E$ be the event that the batch will be discarded. Thus, $E$ is the event that at least one defective bulb is found. It is easier to find $E'$, the event that no defective bulbs are found. Since there are 10 bulbs in the batch of which 8 are non-defective, $P(E') = \dfrac{C(8,3)}{C(10,3)}$. Thus $P(E) = 1 - P(E') = 1 - \dfrac{C(8,3)}{C(10,3)} \approx 1 - 0.4667 = 0.5333$.

**25. (a)** Let $E$ be the event that the monkey types "Hamlet" as his first word. Since "Hamlet" contains 6 letters and there are 48 typewriter keys, $P(E) = \dfrac{1}{48^6} \approx 8.18 \times 10^{-11}$.

**(b)** Let $F$ be the event that the monkey types "to be or not to be" as his first words. Since this phrase has 18 characters (including the blanks), $P(F) = \dfrac{1}{48^{18}} \approx 5.47 \times 10^{-31}$.

**26.** Let $E$ be the event that the monkey will arrange the 6 blocks to spell HAMLET. Then $P(E) = \dfrac{1}{6!} = \frac{1}{720} \approx 0.0014$.

**27.** Let $E$ be the event that the monkey will arrange the 11 blocks to spell PROBABILITY as his first word. The number of ways of arranging these blocks is the number of distinguishable permutations of 11 blocks. Since there are two blocks labeled B and two blocks labeled I, the number of distinguishable permutations is $\dfrac{11!}{2!\,2!}$. Only one of these arrangements spells the word PROBABILITY. Thus $P(E) = \dfrac{1}{\frac{11!}{2!\,2!}} = \dfrac{2!\,2!}{11!} \approx 1.00 \times 10^{-7}$.

**28.** Let $E$ be the event that you predict the correct order for the horses to finish the race. Since there are eight horses, there are $P(8,8) = 8!$ ways that the horses could finish, with only one being the correct order. Thus, $P(E) = \dfrac{1}{P(8,8)} = \dfrac{1}{40{,}320} \approx 2.48 \times 10^{-5}$.

**29. (a)** Let $E$ be the event that the pea is tall. Since tall is dominant, $E = \{TT, Tt, tT\}$. So $P(E) = \frac{3}{4}$.

**(b)** $E'$ is the event that the pea is short. So $P(E') = 1 - P(E) = 1 - \frac{3}{4} = \frac{1}{4}$.

**30.**

| | | Parent 2 | |
|---|---|---|---|
| | | t | t |
| Parent 1 | T | Tt | Tt |
| | t | tt | tt |

**(a)** Let $E$ be the event that the offspring will be tall. Since only offspring with genotype Tt will be tall, $P(E) = \frac{2}{4} = \frac{1}{2}$.

**(b)** $E'$ is the event that the offspring will not be tall (thus, the offspring is short). So $P(E') = 1 - P(E) = 1 - \frac{1}{2} = \frac{1}{2}$.

**31. (a)** Yes, the events are mutually exclusive since a person cannot be both male and female.

**(b)** No, the events are not mutually exclusive since a person can be both tall and blond.

**32. (a)** No, the events are not mutually exclusive since a student can be female and wear glasses.

**(b)** No, the events are not mutually exclusive since a student can be male and have long hair.

**33. (a)** Yes, the events are mutually exclusive since the number cannot be both even and odd. So $P(E \cup F) = P(E) + P(F) = \frac{3}{6} + \frac{3}{6} = 1$.

**(b)** No, the events are not mutually exclusive since 6 is both even and greater than 4. So $P(E \cup F) = P(E) + P(F) - P(E \cap F) = \frac{3}{6} + \frac{2}{6} - \frac{1}{6} = \frac{2}{3}$.

**34. (a)** No, the events are not mutually exclusive since 4 is greater than 3 and also less than 5. So $P(E \cup F) = P(E) + P(F) - P(E \cap F) = \frac{3}{6} + \frac{4}{6} - \frac{1}{6} = 1$.

**(b)** Yes, the events are mutually exclusive since there are only 2 numbers less than 3, namely 1 and 2, but they are not divisible by 3. So $P(E \cup F) = P(E) + P(F) = \frac{2}{6} + \frac{2}{6} = \frac{2}{3}$.

**35. (a)** No, the events $E$ and $F$ are not mutually exclusive since the Jack, Queen, and King of spades are both face cards and spades. So $P(E \cup F) = P(E) + P(F) - P(E \cap F) = \frac{13}{52} + \frac{12}{52} - \frac{3}{52} = \frac{11}{26}$.

**(b)** Yes, the events $E$ and $F$ are mutually exclusive since the card cannot be both a heart and a spade. So $P(E \cup F) = P(E) + P(F) = \frac{13}{52} + \frac{13}{52} = \frac{1}{2}$.

**36. (a)** No, events $E$ and $F$ are not mutually exclusive since a king can be a club. So $P(E \cup F) = P(E) + P(F) - P(E \cap F) = \frac{13}{52} + \frac{4}{52} - \frac{1}{52} = \frac{4}{13}$.

**(b)** No, events $E$ and $F$ are not mutually exclusive since an ace can be a spade. So $P(E \cup F) = P(E) + P(F) - P(E \cap F) = \frac{4}{52} + \frac{13}{52} - \frac{1}{52} = \frac{4}{13}$.

**37. (a)** Let $E$ be the event that the spinner stops on red. Since 12 of the regions are red, $P(E) = \frac{12}{16} = \frac{3}{4}$.

**(b)** Let $F$ be the event that the spinner stops on an even number. Since 8 of the regions are even-numbered, $P(F) = \frac{8}{16} = \frac{1}{2}$.

**(c)** Since 4 of the even-numbered regions are red, $P(E \cup F) = P(E) + P(F) - P(E \cap F) = \frac{3}{4} + \frac{1}{2} - \frac{4}{16} = 1$.

**38. (a)** Let $E$ be the event that the spinner stops on blue. Since only 4 of the regions are blue, $P(E) = \frac{4}{16} = \frac{1}{4}$.

**(b)** Let $F$ be the event that the spinner stops on an odd number. Since 8 of the regions are odd-numbered, $P(F) = \frac{8}{16} = \frac{1}{2}$.

**(c)** Since none of the odd-numbered regions are blue, $P(E \cup F) = P(E) + P(F) = \frac{1}{4} + \frac{1}{2} = \frac{3}{4}$.

**39.** Let $E$ be the event that the ball lands in an odd numbered slot and $F$ be the event that it lands in a slot with a number higher than 31. Since there are two odd-numbered slots with numbers greater than 31, $P(E \cup F) = P(E) + P(F) - P(E \cap F) = \frac{18}{38} + \frac{5}{38} - \frac{2}{38} = \frac{21}{38}$.

**40.** Let $E$ be the event of arranging the letters to spell TRIANGLE and $F$ be the event of arranging them to spell INTEGRAL. There are $P(8,8)$ possible ways of arranging the blocks and only one way to spell each of these words. Also, these events are mutually exclusive since the blocks cannot spell both words at the same time, so

$$P(E \cup F) = P(E) + P(F) = \frac{1}{P(8,8)} + \frac{1}{P(8,8)} = \frac{2}{8!} = \frac{1}{20{,}160} \approx 4.96 \times 10^{-5}.$$

**41.** Let $E$ be the event that the committee is all male and $F$ the event it is all female. The sample space is the set of all ways that 5 people can be chosen from the group of 14. These events are mutually exclusive, so

$$P(E \cup F) = P(E) + P(F) = \frac{C(6,5)}{C(14,5)} + \frac{C(8,5)}{C(14,5)} = \frac{6+56}{2002} = \frac{31}{1001}.$$

**42.** Let $E$ be the event that a player has exactly 5 winning numbers and $F$ be the event that a player has all 6 winning numbers. These events are mutually exclusive. For a players to have exactly 5 winning numbers means that the player has 5 of the 6 winning numbers and 1 number that was not selected in the lottery. So $n(E) = C(6,5) \cdot C(43,1)$. Thus,

$$P(\text{at least 5 winning numbers}) = P(E \cup F) = P(E) + P(F) = \frac{C(6,5) \cdot C(43,1)}{C(49,6)} + \frac{1}{C(49,6)} \approx 0.0000185.$$

**43.** Let $E$ be the event that the marble is red and $F$ be the event that the number is odd-numbered. Then $E'$ is the event that the marble is blue, and $F'$ is the event that the marble is even-numbered.

**(a)** $P(E) = \frac{6}{16} = \frac{3}{8}$

**(b)** $P(F) = \frac{8}{16} = \frac{1}{2}$

**(c)** $P(E \cup F) = P(E) + P(F) - P(E \cap F) = \frac{6}{16} + \frac{8}{16} - \frac{3}{16} = \frac{11}{16}$

**(d)** $P(E' \cup F') = P(E') + P(F') - P(E' \cap F') = \frac{10}{16} + \frac{8}{16} - \frac{5}{16} = \frac{13}{16}$.

**44. (a)** Yes, because the first flip does not influence the outcome of the second flip.

**(b)** The probability of showing heads on both tosses is $P(E \cap F) = P(E) \cdot P(F) = \left(\frac{1}{2}\right)\left(\frac{1}{2}\right) = \frac{1}{4}$.

**45. (a)** Yes, the first roll does not influence the outcome of the second roll.

**(b)** The probability of getting a six on both rolls is $P(E \cap F) = P(E) \cdot P(F) = \left(\frac{1}{6}\right)\left(\frac{1}{6}\right) = \frac{1}{36}$.

**46. (a)** Yes. What happens on spinner A does not influence what happens on spinner B.

**(b)** The probability that A stops on red and B stops on yellow is $P(E \cap F) = P(E) \cdot P(F) = \left(\frac{2}{4}\right)\left(\frac{2}{8}\right) = \frac{1}{8}$.

**47. (a)** Let $E_{\text{A}}$ and $E_{\text{B}}$ be the event that the respective spinners stop on a purple region. Since these events are independent, $P(E_{\text{A}} \cap E_{\text{B}}) = P(E_{\text{A}}) \cdot P(E_{\text{B}}) = \frac{1}{4} \cdot \frac{2}{8} = \frac{1}{16}$.

**(b)** Let $F_{\text{A}}$ and $F_{\text{B}}$ be the event that the respective spinners stop on a blue region. Since these events are independent, $P(F_{\text{A}} \cap F_{\text{B}}) = P(F_{\text{A}}) \cdot P(F_{\text{B}}) = \frac{1}{4} \cdot \frac{1}{8} = \frac{1}{32}$.

**48.** Let $E$ be the event of getting a 1 on the first toss, and let $F$ be the event of getting a 1 on the second toss. Since the events are independent, $P(E \cap F) = P(E) \cdot P(F) = \frac{1}{6} \cdot \frac{1}{6} = \frac{1}{36}$.

**49.** Let $E$ be the event of getting a 1 on the first roll, and let $F$ be the event of getting an even number on the second roll. Since these events are independent, $P(E \cap F) = P(E) \cdot P(F) = \frac{1}{6} \cdot \frac{3}{6} = \frac{1}{12}$.

**50.** Since the card is replaced, the selection of the first card does not influence the selection of the second card, so the events are independent.

**(a)** Let $E$ be the event of getting an ace on the first draw and $F$ be the event of getting an ace on the second draw. Then $P(E \cap F) = P(E) \cdot P(F) = \left(\frac{4}{52}\right)\left(\frac{4}{52}\right) = \frac{1}{169}$.

**(b)** Let $E$ be the event of getting an ace on the first draw and $F$ be the event of getting a spade on the second draw. Then $P(E \cap F) = P(E) \cdot P(F) = \left(\frac{4}{52}\right)\left(\frac{13}{52}\right) = \frac{1}{52}$.

**51.** Let $E$ be the event that the player wins on spin 1, and let $F$ be the event that the player wins on spin 2. What happens on the first spin does not influence what happens on the second spin, so the events are independent. Thus, $P(E \cap F) = P(E) \cdot P(F) = \frac{1}{38} \cdot \frac{1}{38} = \frac{1}{1444}$.

**52.** The number of ways to arrange the letters E, O, K, M, N, and Y is $P(6,6) = 720$. So the probability that the monkey can arrange the letters correctly (if he is merely arranging the blocks randomly) is $\frac{1}{720}$. Also, since the monkey is arranging the blocks randomly, each arrangement is independent of every other arrangement. Hence, the probability that the monkey will be able to spell the word correctly three consecutive times is $\left(\frac{1}{720}\right)^3 = \frac{1}{373{,}248{,}000} \approx 2.68 \times 10^{-9}$.

**53.** Let $E$, $F$ and $G$ denote the events of rolling two ones on the first, second, and third rolls, respectively, of a pair of dice. The events are independent, so $P(E \cap F \cap G) = P(E) \cdot P(F) \cdot P(G) = \frac{1}{36} \cdot \frac{1}{36} \cdot \frac{1}{36} = \frac{1}{36^3} \approx 2.14 \times 10^{-5}$.

**54.** The number of ways a set of six numbers can be selected from a group of 49 numbers is $C(49,6)$. Since the games are independent, the probability of winning the lottery two times in a row is $\left[\frac{1}{C(49,6)}\right]^2 \approx 5.11 \times 10^{-15}$.

**55.** The probability of getting 2 red balls by picking from jar B is $\left(\frac{5}{7}\right)\left(\frac{4}{6}\right) = \frac{10}{21}$. The probability of getting 2 red balls by picking one ball from each jar is $\left(\frac{3}{7}\right)\left(\frac{5}{7}\right) = \frac{15}{49}$. The probability of getting 2 red balls after putting all balls in one jar is $\left(\frac{8}{14}\right)\left(\frac{7}{13}\right) = \frac{4}{13}$. Hence, picking both balls from jar B gives the greatest probability.

**56. (a)** The probability that the first wheel has a bar is $\frac{1}{11}$, and is the same for the second and third wheels. The events are independent, and so the probability of getting 3 bars is $\frac{1}{11} \cdot \frac{1}{11} \cdot \frac{1}{11} = \frac{1}{1331}$.

**(b)** The probability of getting a number on the first wheel is $\frac{10}{11}$, the probability of getting the same number on the second wheel is $\frac{1}{11}$, and the probability of getting the same number on the third wheel is $\frac{1}{11}$. Thus, the probability of getting the same number on each wheel is $\frac{10}{11} \cdot \frac{1}{11} \cdot \frac{1}{11} = \frac{10}{1331}$.

**(c)** We use the complement, no bar, to determine the probability of at least one bar. The probability that the first wheel does not have a bar is $\frac{10}{11}$, and is the same for the second and third wheels. Since the events are independent, the probability of getting no bar is $\frac{10}{11} \cdot \frac{10}{11} \cdot \frac{10}{11} = \frac{10^3}{11^3} = \frac{1000}{1331}$, and so $P(\text{at least one bar}) = 1 - \frac{1000}{1331} = \frac{331}{1331}$.

**57.** Let $E$ be the event that two of the students have the same birthday. It is easier to consider the complementary event $E'$: no two students have the same birthday. Then

$$\begin{aligned} P(E') &= \frac{\text{number of ways to assign 8 different birthdays}}{\text{number of ways to assign 8 birthdays}} = \frac{P(365, 8)}{365^8} \\ &= \frac{365 \cdot 364 \cdot 363 \cdot 362 \cdot 361 \cdot 360 \cdot 359 \cdot 358}{365 \cdot 365 \cdot 365 \cdot 365 \cdot 365 \cdot 365 \cdot 365 \cdot 365} \approx 0.92566 \end{aligned}$$

So $P(E) = 1 - P(E') \approx 0.07434$.

**58.** Let $E$ be the event that at least two have a birthday in the same month, so that $E'$ is the event that no two have a birthday in the same month. So

$$\begin{aligned} P(E') &= \frac{\text{number of ways to assign 6 distinct birth months}}{\text{number of ways to assign 6 birth months}} = \frac{P(12, 6)}{12^6} \\ &= \frac{12 \cdot 11 \cdot 10 \cdot 9 \cdot 8 \cdot 7}{12^6} = \frac{385}{1728} \end{aligned}$$

Thus $P(E) = 1 - P(E') = 1 - \frac{385}{1728} = \frac{1343}{1728}$.

**59.** Let $E$ be the event that she opens the lock within an hour. The number of combinations she can try in one hour is $10 \cdot 60 = 600$. The number of possible combinations is $P(40, 3)$ assuming that no number can be repeated. Thus $P(E) = \dfrac{600}{P(40, 3)} = \dfrac{600}{59{,}280} = \dfrac{5}{494} \approx 0.010$ .

**60. (a)** Let $E$ be the event that the curriculum committee consists of 2 women and 4 men. So $P(E) = \dfrac{\text{number of committees with 2 women and 4 men}}{\text{number of ways to select 6 member committee}} = \dfrac{C(8, 2) \cdot C(10, 4)}{C(18, 6)} = \dfrac{28 \cdot 210}{18564} = \dfrac{490}{1547}$.

**(b)** Let $F$ be the event that curriculum committee consists of 2 or fewer women. Then $F$ is the event that the committee has no woman or 1 woman or 2 women. Then

$$\begin{aligned} P(F) &= \frac{C(8, 0) \cdot C(10, 6)}{C(18, 6)} + \frac{C(8, 1) \cdot C(10, 5)}{C(18, 6)} + \frac{C(8, 2) \cdot C(10, 4)}{C(18, 6)} = \frac{1 \cdot 210 + 8 \cdot 252 + 28 \cdot 210}{18{,}564} \\ &= \frac{8106}{18{,}564} = \frac{193}{442} \end{aligned}$$

**(c)** So $F'$ is the event the at curriculum committee has more that 2 women. The $P(F') = 1 - P(F) = 1 - \frac{193}{442} = \frac{249}{442}$.

**61.** Let $E$ be the event that Paul stands next to Phyllis. To find $n(E)$ we treat Paul and Phyllis as one object and find the number of ways to arrange the 19 objects and then multiply the result by the number of ways to arrange Paul and Phyllis. So $n(E) = 19! \cdot 2!$. The sample space is all the ways that 20 people can be arranged. Thus $P(E) = \dfrac{19! \cdot 2!}{20!} = \dfrac{2}{20} = 0.10$.

**62.** Let $E$ be the event that all the boys are standing together and all the girls are standing together. If we consider all 8 boys as one object and all 12 girls as one object, then there are 2! ways to arrange these 2 objects (8 boys and 12 girls). There are 8! possible ways to arrange 8 boys and 12! possible ways to arrange 12 girls, thus there are $2! \cdot 8! \cdot 12!$ possible ways to arrange 8 boys and 12 girls in which all the boys are standing together and all the girls are standing together. Therefore $P(E) = \dfrac{2! \cdot 8! \cdot 12!}{20!} \approx 1.59 \times 10^{-5}$.

**63.** The sample space for the genders of Mrs. Smith's children is $S_S = \{(b,g),(b,b)\}$, where we list the firstborn first. The sample space for the genders of Mrs. Jones' children is $S_J = \{(b,g),(g,b),(b,b)\}$. Thus, the probability that Mrs. Smith's other child is a boy is $\frac{1}{2}$, while the probability that Mrs. Jones' other child is a boy is $\frac{1}{3}$. The reason there is a difference is that we know that Mrs. Smith's firstborn is a boy.

**64.** Since most families in the US have three or less children, we construct a table in which we calculate the probability that a randomly selected child is the oldest son or daughter.

| Number of children in the family | Sample space | Probability that the child is the oldest son or daughter |
|---|---|---|
| 1 | $(b)$<br>$(g)$ | $\left.\begin{array}{r} 0.5 \cdot 1 = 0.5 \\ 0.5 \cdot 1 = 0.5 \end{array}\right\} = 1$ |
| 2 | $(b,b)$<br>$(b,g)$<br>$(g,b)$<br>$(g,g)$ | $\left.\begin{array}{r} 0.25 \cdot 0.5 = 0.125 \\ 0.25 \cdot 1 = 0.25 \\ 0.25 \cdot 1 = 0.25 \\ 0.25 \cdot 0.5 = 0.125 \end{array}\right\} = 0.75$ |
| 3 | $(b,b,b)$<br>$(b,b,g)$<br>$(b,g,b)$<br>$(g,b,b)$<br>$(b,g,g)$<br>$(g,b,g)$<br>$(g,g,b)$<br>$(g,g,g)$ | $\left.\begin{array}{r} 0.125 \cdot 0.333 = 0.041625 \\ 0.125 \cdot 0.667 = 0.08325 \\ 0.125 \cdot 0.667 = 0.08325 \\ 0.125 \cdot 0.667 = 0.08325 \\ 0.125 \cdot 0.667 = 0.08325 \\ 0.125 \cdot 0.667 = 0.08325 \\ 0.125 \cdot 0.667 = 0.08325 \\ 0.125 \cdot 0.333 = 0.041625 \end{array}\right\} = 0.58275$ |

As the table shows, for families of this size, the relevant probability is always greater than 0.5.

## 13.4 Binomial Probability

**1.** $P(2 \text{ successes in } 5) = C(5,2) \cdot (0.7^2)(0.3^3) = 0.13230$

**2.** $P(3 \text{ successes in } 5) = C(5,3) \cdot (0.7^3)(0.3^2) = 0.30870$

**3.** $P(0 \text{ success in } 5) = C(5,0) \cdot (0.7^0)(0.3^5) = 0.00243$

**4.** $P(5 \text{ successes in } 5) = C(5,5) \cdot (0.7^5)(0.3^0) = 0.16807$

**5.** $P(1 \text{ success in } 5) = C(5,1) \cdot (0.7^1)(0.3^4) = 0.02835$

**6.** $P(1 \text{ failure in } 5) = C(5,1) \cdot (0.3^1)(0.7^4) = 0.36015$. Note that exactly one failure is the same as exactly 4 successes, and $P(4 \text{ successes in } 5) = C(5,4) \cdot (0.7^4)(0.3^1) = 0.36015$.

**7.** $P\,(\text{at least 4 successes}) = P\,(\text{4 successes}) + P\,(\text{5 successes}) = 0.36015 + 0.16807 = 0.52822$

**8.** $P\,(\text{at least 3 successes}) = P\,(\text{3 successes}) + P\,(\text{4 successes}) + P\,(\text{5 successes})$
$= 0.30870 + 0.36015 + 0.16807 = 0.83692$

**9.** $P\,(\text{at most 1 failure}) = P\,(\text{0 failure}) + P\,(\text{1 failure}) = P\,(\text{5 successes}) + P\,(\text{4 successes}) = 0.52822$

**10.** $P\,(\text{at most 2 failures}) = P\,(\text{0 failure}) + P\,(\text{1 failure}) + P\,(\text{2 failure})$
$= P\,(\text{5 successes}) + P\,(\text{4 successes}) + P\,(\text{3 successes})$
$= 0.16807 + 0.36015 + 0.30870 = 0.83692$

**11.** $P\,(\text{at least 2 successes}) = P\,(\text{2 successes}) + P\,(\text{3 successes}) + P\,(\text{4 successes}) + P\,(\text{5 successes})$
$= 0.13230 + 0.30870 + 0.36015 + 0.16807 = 0.96922$

**12.** $P\,(\text{at most 3 failures}) = P\,(\text{0 failure}) + P\,(\text{1 failure}) + P\,(\text{2 failures}) + P\,(\text{3 failures})$
$= P\,(\text{5 successes}) + P\,(\text{4 successes}) + P\,(\text{3 successes}) + P\,(\text{2 successes})$
$= 0.16807 + 0.36015 + 0.30870 + 0.13230 = 0.96922$

**13.** Here "success" is "face is 4" and $P\,(\text{face is 4}) = \frac{1}{6}$. Then $P\,(\text{2 successes in 6}) = C\,(6,2) \cdot \left(\frac{1}{6}\right)^2 \left(\frac{5}{6}\right)^4 = 0.20094$.

**14.** Here $P\,(\text{success}) = 0.8$ and $P\,(\text{failure}) = 0.2$.

**(a)** $P\,(\text{0 success in 7}) = C\,(7,0) \cdot (0.8^0)\,(0.2^7) = 0.0000128$

**(b)** $P\,(\text{7 successes in 7}) = C\,(7,7) \cdot (0.8^7)\,(0.2^0) \approx 0.209715$

**(c)** $P\,(\text{he hits target more than once}) = 1 - [P\,(\text{0 success in 7}) + P\,(\text{1 success in 7})]$
$= 1 - \left[C\,(7,0) \cdot (0.8^0)\,(0.2^7) + C\,(7,1) \cdot (0.8^1)\,(0.2^6)\right] \approx 0.99963$

**(d)** $P\,(\text{at least 5 successes}) = P\,(\text{5 successes}) + P\,(\text{6 successes}) + P\,(\text{7 successes})$
$= C\,(7,5) \cdot (0.8^5)\,(0.2^2) + C\,(7,6) \cdot (0.8^6)\,(0.2^1) + C\,(7,7) \cdot (0.8^7)\,(0.2^0) \approx 0.85197$

**15.** $P\,(\text{4 successes in 10}) = C\,(10,4) \cdot (0.4^4)\,(0.6^6) \approx 0.25082$

**16.** The complement is that none of the raccoons had rabies.
$P\,(\text{at least 1 had rabies}) = 1 - P\,(\text{none had rabies}) = 1 - C\,(4,0) \cdot (0.1^0)\,(0.9^4) = 1 - 0.6561 = 0.3439$

**17. (a)** $P\,(\text{5 in 10}) = C\,(10,5) \cdot (0.45^5)\,(0.55^5) \approx 0.23403$

**(b)** $P\,(\text{at least 3}) = 1 - P\,(\text{at most 2}) = P\,(\text{0 in 10}) + P\,(\text{1 in 10}) + P\,(\text{2 in 10})$
$= 1 - \left[C\,(10,0) \cdot (0.45^0)\,(0.55^{10}) + C\,(10,1) \cdot (0.45^1)\,(0.55^9) + C\,(10,2) \cdot (0.45^2)\,(0.55^8)\right]$
$\approx 0.90044$

**18. (a)** $P\,(\text{12 in 15}) = C\,(15,12) \cdot (0.1^{12})\,(0.9^3) \approx 3.31695 \times 10^{-10}$

**(b)** $P\,(\text{at least 12}) = P\,(\text{12 in 15}) + P\,(\text{13 in 15}) + P\,(\text{14 in 15}) + P\,(\text{15 in 15})$
$= C\,(15,12) \cdot (0.1^{12})\,(0.9^3) + C\,(15,13) \cdot (0.1^{13})\,(0.9^2)$
$+ C\,(15,14) \cdot (0.1^{14})\,(0.9^1) + C\,(15,15) \cdot (0.1^{15})\,(0.9^0) \approx 3.40336 \times 10^{-10}$

**19. (a)** The complement of at least 1 germinating is no seed germinating, so
$P\,(\text{at least 1 germinates}) = 1 - P\,(\text{0 germinates}) = 1 - C\,(4,0) \cdot (0.75^0)\,(0.25^4) \approx 0.99609$.

**(b)** $P\,(\text{at least 2 germinate}) = P\,(\text{2 germinates}) + P\,(\text{3 germinates}) + P\,(\text{4 germinates})$
$= C\,(4,2) \cdot (0.75^2)\,(0.25^2) + C\,(4,3) \cdot (0.75^3)\,(0.25^1) + C\,(4,4) \cdot (0.75^4)\,(0.25^0)$
$\approx 0.94922$

**(c)** $P\,(\text{4 germinates}) = C\,(4,4) \cdot (0.75^4)\,(0.25^0) \approx 0.31641$

**20. (a)** $P\,(\text{at least 3 boys}) = P\,(3\text{ boys}) + P\,(4\text{ boys}) + P\,(5\text{ boys})$
$= C\,(5,3)\cdot\left(0.5^3\right)\left(0.5^4\right) + C\,(5,4)\cdot\left(0.5^4\right)\left(0.5^1\right) + C\,(5,5)\cdot\left(0.5^5\right)\left(0.5^0\right) = 0.5$

**(b)** $P\,(\text{at least 4 girls}) = P\,(4\text{ girls}) + P\,(5\text{ girls}) + P\,(6\text{ girls}) + P\,(7\text{ girls})$
$= C\,(7,4)\cdot\left(0.5^4\right)\left(0.5^3\right) + C\,(7,5)\cdot\left(0.5^5\right)\left(0.5^2\right) + C\,(7,6)\cdot\left(0.5^6\right)\left(0.5^1\right) + C\,(7,7)\cdot\left(0.5^7\right)\left(0.5^0\right) = 0.5$

**21. (a)** $P\,(\text{all 10 are boys}) = C\,(10,10)\cdot\left(0.52^{10}\right)\left(0.48^0\right) \approx 0.0014456.$

**(b)** $P\,(\text{all 10 are girls}) = C\,(10,0)\cdot\left(0.52^0\right)\left(0.48^{10}\right) \approx 0.00064925.$

**(c)** $P\,(5\text{ in 10 are boys}) = C\,(10,5)\cdot\left(0.52^5\right)\left(0.48^5\right) \approx 0.24413.$

**22. (a)** $P\,(2\text{ in }12) = C\,(12,2)\cdot\left(0.2^2\right)\left(0.8^3\right) \approx 0.28347.$

**(b)** The complement of "at least 3" is "at most 2", so

$$P\,(\text{at least 3 in 12}) = 1 - P\,(\text{at most 2 in 12}) = 1 - \left[P\,(0\text{ in }12) + P\,(1\text{ in }12) + P\,(2\text{ in }12)\right]$$
$$= 1 - \left[C\,(12,0)\cdot\left(0.2^0\right)\left(0.8^{12}\right) + C\,(12,1)\cdot\left(0.2^1\right)\left(0.8^{11}\right) + C\,(12,2)\cdot\left(0.2^2\right)\left(0.8^{10}\right)\right]$$
$$\approx 0.44165$$

**23. (a)** $P\,(3\text{ in }3) = C\,(3,3)\cdot\left(0.005^3\right)\left(0.995^0\right) \approx 0.000000125.$

**(b)** The complement of "one or more bulbs is defective" is "none of the bulbs is defective." So
$P\,(\text{at least 1 defective}) = 1 - P\,(\text{none is defective}) = 1 - C\,(3,0)\cdot\left(0.005^0\right)\left(0.995^0\right) \approx 0.014925.$

**24.** $P\,(\text{at least 1 in 10}) = 1 - P\,(0\text{ in }10) = 1 - C\,(10,0)\cdot\left(0.05^0\right)\left(0.95^{10}\right) \approx 0.40126$

**25.** The complement of "2 or more workers call in sick" is "0 or 1 worker calls in sick." So

$$P\,(2\text{ or more}) = 1 - \left[P\,(0\text{ in }8) + P\,(1\text{ in }8)\right]$$
$$= 1 - \left[C\,(8,0)\cdot\left(0.04^0\right)\left(0.96^8\right) + C\,(8,1)\cdot\left(0.04^1\right)\left(0.96^7\right)\right] \approx 0.038147$$

**26.** $P\,(3\text{ in 5 favor}) = C\,(5,3)\cdot\left(0.6^3\right)\left(0.4^2\right) = 0.3456$

**27. (a)** $P\,(6\text{ in }6) = C\,(6,6)\cdot\left(0.75^6\right)\left(0.25^0\right) \approx 0.17798$

**(b)** $P\,(0\text{ in }6) = C\,(6,0)\cdot\left(0.75^0\right)\left(0.25^6\right) \approx 0.00024414$

**(c)** $P\,(3\text{ in }6) = C\,(6,3)\cdot\left(0.75^3\right)\left(0.25^3\right) \approx 0.13184$

**(d)** $P\,(\text{at least 2 seasick}) = 1 - P\,(\text{at most 1 seasick}) = 1 - \left[P\,(6\text{ in 6 OK}) + P\,(5\text{ in 6 OK})\right]$
$= 1 - \left[C\,(6,6)\cdot\left(0.75^6\right)\left(0.25^0\right) + C\,(6,5)\cdot\left(0.75^5\right)\left(0.25^1\right)\right] \approx 0.46606$

**28. (a)** $P\,(\text{machine breaks}) = P\,(\text{at least 1 component fails}) = 1 - P\,(0\text{ component fails})$
$= 1 - C\,(4,0)\cdot\left(0.01^0\right)\left(0.99^4\right) \approx 0.039404$

**(b)** $P\,(0\text{ component fails}) = C\,(4,0)\cdot\left(0.01^0\right)\left(0.99^4\right) \approx 0.960596$

**(c)** $P\,(3\text{ components fail}) = C\,(4,3)\cdot\left(0.01^3\right)\left(0.99^1\right) \approx 0.00000396$

**29. (a)** The complement of "at least one gets the disease" is "none gets the disease." Then
$P\,(\text{at least 1 gets the disease}) = 1 - P\,(0\text{ gets the disease}) = 1 - C\,(4,0)\cdot\left(0.25^0\right)\left(0.75^4\right) \approx 0.68359.$

**(b)** $P\,(\text{at least 3 get the disease}) = P\,(3\text{ get the disease}) + P\,(4\text{ get the disease})$
$= C\,(4,3)\cdot\left(0.25^3\right)\left(0.75^1\right) + C\,(4,4)\cdot\left(0.25^4\right)\left(0.75^0\right) \approx 0.05078$

**30.** There are 52 cards in the deck of which 13 belong to any one suit, so
$P\,(\text{heart}) = P\,(\text{spade}) = P\,(\text{diamond}) = P\,(\text{club}) = 0.25.$

**(a)** $P\,(3\text{ in 3 are hearts}) = C\,(3,3)\cdot\left(0.25^3\right)\left(0.75^0\right) = 0.015625$

**(b)** $P\,(2\text{ in 3 are spades}) = C\,(3,2)\cdot\left(0.25^2\right)\left(0.75^1\right) = 0.140625$

**(c)** $P\,(0\text{ in 3 are diamonds}) = C\,(3,0)\cdot\left(0.25^0\right)\left(0.75^3\right) = 0.421875$

**(d)** $P\,(\text{at least 1 is a club}) = 1 - P\,(\text{none is a club}) = 1 - C\,(3,0)\cdot\left(0.25^0\right)\left(0.75^3\right) \approx 0.578125$

**31.** Fred (a nonsmoker) is already in the room, so this exercise concerns the remaining 4 participants assigned to the room.

**(a)** $P(1 \text{ in } 4 \text{ is a smoker}) = C(4,1) \cdot (0.3^1)(0.7^3) = 0.4116$

**(b)** $P(\text{at least 1 smoker}) = 1 - P(0 \text{ in } 4 \text{ is a smoker}) = 1 - C(4,0) \cdot (0.3^0)(0.7^4) = 0.7599$

**32. (a)** $P(2 \text{ or more}) = 1 - [P(0 \text{ in } 100) + P(1 \text{ in } 100)]$
$= 1 - \left[C(100,0) \cdot (0.02^0)(0.98^{100}) + C(100,1) \cdot (0.02^1)(0.98^{99})\right] \approx 0.59673$

**(b)** Since $P(\text{at least 1 interested}) = 1 - P(0 \text{ interested})$ and $0.98^{35} \approx 0.507$, the telephone consultant needs to make at least 35 calls to ensure at least a 0.5 probability of reaching one or more interested parties.

**33. (a)**

| Number of heads | Probability |
|---|---|
| 0 | 0.001953 |
| 1 | 0.017578 |
| 2 | 0.070313 |
| 3 | 0.164063 |
| 4 | 0.246094 |
| 5 | 0.246094 |
| 6 | 0.164063 |
| 7 | 0.070313 |
| 8 | 0.017578 |
| 9 | 0.001953 |

**(b)** Between 4 and 5 heads.

**(c)** If the coin is flipped 101 times, then 50 and 51 heads has the greatest probability of occurring. If the coin is flipped 100 times, then 50 heads has the greatest probability of occurring.

**34. (a)**

| Number of Heads | Probability |
|---|---|
| 0 | 0.000020 |
| 1 | 0.000413 |
| 2 | 0.003858 |
| 3 | 0.021004 |
| 4 | 0.073514 |
| 5 | 0.171532 |
| 6 | 0.266828 |
| 7 | 0.266828 |
| 8 | 0.155650 |
| 9 | 0.040354 |

**(b)** Between 5 and 6 heads.

## 13.5 Expected Value

**1.** Mike gets \$2 with probability $\frac{1}{2}$ and \$1 with probability $\frac{1}{2}$. Thus, $E = (2)\left(\frac{1}{2}\right) + (1)\left(\frac{1}{2}\right) = 1.5$, and so his expected winnings are \$1.50 per game.

**2.** The probability that Jane gets \$10 is $\frac{1}{6}$, and the probability that she loses \$1 is $\frac{5}{6}$. Thus $E = (10)\left(\frac{1}{6}\right) + (-1)\left(\frac{5}{6}\right) \approx 0.833$, and so her expectation is \$0.833.

**3.** Since the probability of drawing the ace of spades is $\frac{1}{52}$, the expected value of this game is $E = (100)\left(\frac{1}{52}\right) + (-1)\left(\frac{51}{52}\right) = \frac{49}{52} \approx 0.94$. So your expected winnings are \$0.94 per game.

**4.** The expected value of this game is $E = (3)\left(\frac{1}{2}\right) + (2)\left(\frac{1}{2}\right) = \frac{5}{2} = 2.5$. So Tim's expected winnings are \$2.50 per game.

**5.** Since the probability that Carol rolls a six is $\frac{1}{6}$, the expected value of this game is $E = (3)\left(\frac{1}{6}\right) + (0.50)\left(\frac{5}{6}\right) = \frac{5.5}{6} \approx 0.9167$. So Carol expects to win \$0.92 per game.

**6.** The probability that Albert gets two tails is $\left(\frac{1}{2}\right)^2 = \frac{1}{4}$, the probability that Albert gets one tail and one head is $C(2,1)\left(\frac{1}{2}\right)^2 = \frac{1}{2}$, and the probability that Albert gets two heads is $\left(\frac{1}{2}\right)^2 = \frac{1}{4}$. If Albert gets two heads, he will receive \$4, if he get one head and one tail, he will get \$2− \$1 = \$1, and if he get two tails, he will lose \$2. Thus the expected value of this game is $E = (4)\left(\frac{1}{4}\right) + (1)\left(\frac{1}{2}\right) + (-2)\left(\frac{1}{4}\right) = 1$. So Albert's expected winnings are \$1 per game.

**7.** Since the probability that the die shows an even number equals the probability that that die shows an odd number, the expected value of this game is $E = (2)\left(\frac{1}{2}\right) + (-2)\left(\frac{1}{2}\right) = 0$. So Tom should expect to break even after playing this game many times.

**8.** Since there are 4 aces, 12 face cards and only one 8 of clubs, the expected value of this game is $E = (104)\left(\frac{4}{52}\right) + (26)\left(\frac{12}{52}\right) + (13)\left(\frac{1}{52}\right) = \$14.25$.

**9.** Since it costs \$0.50 to play, if you get a silver dollar, you win only $1 - 0.50 = \$0.50$. Thus the expected value of this game is $E = (0.50)\left(\frac{2}{10}\right) + (-0.50)\left(\frac{8}{10}\right) = -0.30$. So your expected winnings are −\$0.30 per game. In other words, you should expect to lose \$0.30 per game.

**10.** The probability of choosing 2 white balls (that is, no black ball) is $\frac{8}{10} \cdot \frac{7}{9} = \frac{56}{90}$, and the probability of not choosing 2 white balls (at least one black) is $\left(1 - \frac{56}{90}\right) = \frac{34}{90}$. Therefore, the expected value of this game is $E = (5)\left(\frac{56}{90}\right) + (0)\left(\frac{34}{90}\right) \approx 3.111$. Thus, John's expected winnings are \$3.11 per game.

**11.** You can either win \$35 or lose \$1, so the expected value of this game is $E = (35)\left(\frac{1}{38}\right) + (-1)\left(\frac{37}{38}\right) = -\frac{2}{38} = -0.0526$. Thus the expected value is −\$0.0526 per game.

**12. (a)** We have $P(\text{winning the first prize}) = \dfrac{1}{2 \times 10^6}$. After the first prize winner is selected, then $P(\text{winning the second prize}) = \dfrac{1}{2 \times 10^6 - 1}$. Similarly, $P(\text{winning the third prize}) = \dfrac{1}{2 \times 10^6 - 2}$. So the expected value of this game is $E = (10^6)\left(\dfrac{1}{2 \times 10^6}\right) + (10^5)\left(\dfrac{1}{2 \times 10^6 - 1}\right) + (10^4)\left(\dfrac{1}{2 \times 10^6 - 2}\right) = \$0.555$.

**(b)** Since we expect to win \$0.555 on the average per game, if we pay \$1.00, then our net outcome is a loss of \$0.445 per game. Hence, it is not worth playing, because on average you will lose \$0.445 per game.

**13.** By the rules of the game, a player can win \$10 or \$5, break even, or lose \$100. Thus the expected value of this game is $E = (10)\left(\frac{10}{100}\right) + (5)\left(\frac{10}{100}\right) + (-100)\left(\frac{2}{100}\right) + (0)\left(\frac{78}{100}\right) = -0.50$. So the expected winnings per game are −\$0.50.

**14.** Since the safe has a six digit combination, there are $10^6$ possible combinations to the safe, of which only one is correct. The expected value of this game is $E = (10^6 - 1)\left(\dfrac{1}{10^6}\right) + (-1)\left(\dfrac{10^6 - 1}{10^6}\right) = 0$.

**15.** If the stock goes up to \$20, she expects to make \$20 − \$5 = \$15. And if the stock falls to \$1, then she has lost \$5 − \$1 = \$4. So the expected value of her profit is $E = (15)(0.1) - (4)(0.9) = -2.1$. Thus, her expected profit per share is −\$2.10, that is, she should expect to lose \$2.10 per share. She did not make a wise investment.

**16.** Since the wheels of the slot machine are independent, the probability that you get three watermelons is $\left(\frac{1}{11}\right)^3$. So the expected value of this game is $E = (4.75)\left(\frac{1}{11^3}\right) + (-0.25)\left(1 - \frac{1}{11^3}\right) \approx -\$0.246$.

**17.** There are $C(49,6)$ ways to select a group of six numbers from the group of 49 numbers, of which only one is a winning set. Thus the expected value of this game is $E = (10^6 - 1)\left(\dfrac{1}{C(49,6)}\right) + (-1)\left(1 - \dfrac{1}{C(49,6)}\right) \approx -\$0.93$.

**18.** Let $x$ be the fair price to pay to play this game. Then if you win, you gain $1 - x$, and if you lose, your loss will be $-x$. So $E = 0 \Leftrightarrow (1-x)\left(\frac{2}{8}\right) + (-x)\left(\frac{6}{8}\right) = 0 \Leftrightarrow 1 - 4x = 0 \Leftrightarrow x = \frac{1}{4} = 0.25$. Thus, a fair price to play this game is \$0.25.

**19.** Let $x$ be the fair price to pay to play this game. Then the game is fair whenever $E = 0 \Leftrightarrow (13 - x)\left(\frac{4}{52}\right) + (-x)\left(\frac{48}{52}\right) = 0 \Leftrightarrow 52 - 52x = 0 \Leftrightarrow x = 1$. Thus, a fair price to pay to play this game is \$1.

**20.** If you win, you win \$1 million minus the price of the stamp. If you lose, you lose only the price of the stamp (currently 34 cents). So the expected value of this game is $(999{,}999.67) \cdot \frac{1}{20 \times 10^6} + (-0.34) \cdot \frac{20 \times 10^6 - 1}{20 \times 10^6} \approx -0.29$. Thus you expect to lose 29 cents on each entry, and so it's not worth it.

# Chapter 13 Review

**1.** The number of possible outcomes is

$$\binom{\text{number of outcomes}}{\text{when a coin is tossed}} \cdot \binom{\text{number of outcomes}}{\text{a die is rolled}} \cdot \binom{\text{number of ways}}{\text{to draw a card}} = (2)(6)(52) = 624.$$

**2. (a)** The number of 3-digit numbers that can be formed using the digits 1–6 if a digit can be used any number of times is $(6)(6)(6) = 216$.

**(b)** The number of 3-digit numbers that can be formed using the digits 1–6 if a digit can be used only once is $(6)(5)(4) = 120$.

**3. (a)** Order is not important, and there are no repetitions, so the number of different two-element subsets is $C(5,2) = \frac{5!}{2!\,3!} = \frac{5 \cdot 4}{2} = 10$.

**(b)** Order is important, and there are no repetitions, so the number of different two-letter words is $P(5,2) = \frac{5!}{3!} = 20$.

**4.** Since the order in which the people are chosen is not important and a person cannot be bumped more than once (no repetitions), the number of ways that 7 passengers can be bumped is $C(120,7) \approx 5.9488 \times 10^{10}$.

**5.** You earn a score of 70% by answering exactly 7 of the 10 questions correctly. The number of different ways to answer the questions correctly is $C(10,7) = \frac{10!}{7!\,3!} = 120$.

**6.** There 2 ways to answer each of the 10 true-false questions and 4 ways to answer each of the 5 multiple choice questions. So the number of ways that this test can be completed is $(2^{10})(4^5) = 1{,}048{,}576$.

**7.** You must choose two of the ten questions to omit, and the number of ways of choosing these two questions is $C(10,2) = \frac{10!}{2!\,8!} = 45$.

**8.** Since the order of the scoops of ice cream is not important and the scoops cannot be repeated, the number of ways to have a banana split is $C(15,4) = 1365$.

**9.** The maximum number of employees using this security system is

$$\binom{\text{number of choices}}{\text{for the first letter}} \cdot \binom{\text{number of choices}}{\text{for the second letter}} \cdot \binom{\text{number of choices}}{\text{for the third letter}} = (26)(26)(26) = 17{,}576.$$

**10.** Since there are $n!$ ways to arrange a group of size $n$ and $5! = 120$, there are 5 students in this class.

**11.** We could count the number of ways of choosing 7 of the flips to be heads; equivalently we could count the number of ways of choosing 3 of the flips to be tails. Thus, the number of different ways this can occur is $C(10,7) = C(10,3) = \frac{10!}{3!\,7!} = 120$.

**12.** The number ways to form a license plate consisting of 2 letters followed by 3 numbers is $(26)(26)(10)(10)(10) = 676{,}000$. Since there are fewer possible license plates than 700,000, there must be fewer than 700,000 licensed cars in the Yukon.

**13.** Let $x$ be the number of people in the group. Then $C(x,2)=10 \Leftrightarrow \dfrac{x!}{2!\,(x-2)!}=10 \Leftrightarrow \dfrac{x!}{(x-2)!}=20 \Leftrightarrow x(x-1)=20$ $\Leftrightarrow x^2-x-20=0 \Leftrightarrow (x-5)(x+4)=0 \Leftrightarrow x=5$ or $x=-4$. So there are 5 people in this group.

**14.** Each topping corresponds to a subset of a set with $n$ elements. Since a set with $n$ elements has $2^n$ subsets and $2^{11}=2048$, there are 11 toppings that the pizza parlor offers.

**15.** A letter can be represented by a sequence of length 1, a sequence of length 2, or a sequence of length 3. Since each symbol is either a dot or a dash, the possible number of letters is $\begin{pmatrix}\text{number of letters}\\ \text{using 3 symbols}\end{pmatrix}+\begin{pmatrix}\text{number of letters}\\ \text{using 2 symbols}\end{pmatrix}+\begin{pmatrix}\text{number of letters}\\ \text{using 1 symbol}\end{pmatrix}=2^3+2^2+2=14.$

**16.** Since the nucleotides can be repeated, the number of possible words of length $n$ is $4^n$. Since $4^2=16<20$ and $4^3=64$, the minimum length of word needed is 3.

**17. (a)** Since we cannot choose a major and a minor in the same subject, the number of ways a student can select a major and a minor is $P(16,2)=16\cdot 15=240$.

**(b)** Again, since we cannot have repetitions and the order of selection is important, the number of ways to select a major, a first minor, and a second minor is $P(16,3)=16\cdot 15\cdot 14=3360$.

**(c)** When we select a major and 2 minors, the order in which we choose the minors is not important. Thus the number of ways to select a major and 2 minors is $\begin{pmatrix}\text{number of ways}\\ \text{to select a major}\end{pmatrix}\cdot\begin{pmatrix}\text{number of ways to}\\ \text{select two minors}\end{pmatrix}=16\cdot C(15,2)=16\cdot 105=1680.$

**18. (a)** *Solution 1:* Since the leftmost digit of a three-digit number cannot be zero, there are 9 choices for this first digit and 10 choices for each of the other two digits. Thus, the number of three-digit numbers is $(9)(10)(10)=900$.

*Solution 2:* Since there are 999 numbers between 1 and 999, of which the numbers between 1 and 99 do not have three digits, there are $999-99=900$ three-digit numbers.

**(b)** There are 1001 numbers from 0–1000. From part (a), there are 900 three-digit numbers. Therefore the probability that the number chosen is a three-digit number is $P(E)=\frac{900}{1001}\approx 0.899$.

**19.** Since the letters are distinct, the number of anagrams of the word TRIANGLE is $8!=40{,}320$.

**20.** Since MISSISSIPPI has 4 I's, 4 S's, 2 P's, and 1 M, the number of distinguishable anagrams of the word MISSISSIPPI is $\dfrac{11!}{4!\,4!\,2!\,1!}=34{,}650$.

**21. (a)** The possible number of committees is $C(18,7)=31{,}824$.

**(b)** Since we must select the 4 men from the group of 10 men and the 3 women from the group of 8 women, the possible number of committees is $\begin{pmatrix}\text{number of ways to}\\ \text{choose 4 of 10 men}\end{pmatrix}\cdot\begin{pmatrix}\text{number of ways to}\\ \text{choose 3 of 8 women}\end{pmatrix}=C(10,4)\cdot C(8,3)=210\cdot 56=11{,}760.$

**(c)** We remove Susie from the group of 18,so the possible number of committees is $C(17,7)=19{,}448$.

**(d)** The possible number of committees is

$$\begin{pmatrix}\text{possible number of}\\ \text{committees with 5 women}\end{pmatrix}+\begin{pmatrix}\text{possible number of}\\ \text{committees with 6 women}\end{pmatrix}+\begin{pmatrix}\text{possible number of}\\ \text{committees with 7 women}\end{pmatrix}$$
$$=C(8,5)\cdot C(10,2)+C(8,6)\cdot C(10,1)+C(8,7)\cdot C(10,0)=56\cdot 45+28\cdot 10+8\cdot 1=2808$$

**(e)** Since the committee is to have 7 members, "at most two men" is the same as "at least five women," which we found in part (d). So the number is also 2808.

**(f)** We select the specific offices first, then complete the committee from the remaining members of the group. So the number of possible committees is

$$\begin{pmatrix}\text{number of ways to choose}\\ \text{a chairman, a vice-chairman,}\\ \text{and a secretary}\end{pmatrix}\cdot\begin{pmatrix}\text{number of ways to choose}\\ \text{4 other members}\end{pmatrix}=P(18,3)\cdot C(15,4)=4896\cdot 1365=6{,}683{,}040.$$

**22.** **(a)** The probability that the ball is red is $\frac{10}{15} = \frac{2}{3}$.

**(b)** The probability that the ball is even numbered is $\frac{8}{15}$.

**(c)** The probability that the ball is white and an odd number is $\frac{2}{15}$.

**(d)** The probability that the ball is red or odd numbered is $P(\text{red}) + P(\text{odd}) - P(\text{red} \cap \text{odd}) = \frac{10}{15} + \frac{7}{15} - \frac{5}{15} = \frac{12}{15} = \frac{4}{5}$.

**23.** Let $R_n$ denote the event that the $n$th ball is red and let $W_n$ denote the event that the $n$th ball is white.

**(a)** $P(\text{both balls are red}) = P(R_1 \cap R_2) = P(R_1) \cdot P(R_2 \mid R_1) = \frac{10}{15} \cdot \frac{9}{14} = \frac{3}{7}$.

**(b)** *Solution 1:* The probability that one is white and that the other is red is

$$\frac{\text{number of ways to select one white and one red}}{\text{number of ways to select two balls}} = \frac{C(10,1) \cdot C(5,1)}{C(15,2)} = \frac{10}{21}.$$

*Solution 2:*

$$\begin{aligned} P(\text{one white and one red}) &= P(W_1 \cap R_2) + P(R_1 \cap W_2) = P(W_1) \cdot P(R_2 \mid W_1) + P(R_1) \cdot P(W_2 \mid R_1) \\ &= \tfrac{5}{15} \cdot \tfrac{10}{14} + \tfrac{10}{15} \cdot \tfrac{5}{14} = \tfrac{10}{21} \end{aligned}$$

**(c)** *Solution 1:* Let $E$ be the event "at least one is red". Then $E'$ is the event "both are white". $P(E') = P(W_1 \cap W_2) = P(W_1) \cdot P(W_2 \mid W_1) = \frac{5}{15} \cdot \frac{4}{14} = \frac{2}{21}$. Thus $P(E) = 1 - \frac{2}{21} = \frac{19}{21}$.

*Solution 2:* $P(\text{at least one is red}) = P(\text{one red and one white}) + P(\text{both red}) = \frac{10}{21} + \frac{9}{21} = \frac{19}{21}$ (from (a) and (b)).

**(d)** Since 5 of the 15 balls are both red and even-numbered, the probability that both balls are red and even-numbered is $\frac{5}{15} \cdot \frac{4}{14} = \frac{2}{21}$.

**(e)** Since 2 of the 15 balls are both white and odd-numbered, the probability that both are white is $\frac{2}{15} \cdot \frac{1}{14} = \frac{1}{105}$.

**24.** **(a)** $S = \{HHH, HHT, HTH, HTT, THH, THT, TTH, TTT\}$.

**(b)** $P(HHH) = \frac{1}{8}$

**(c)** $P(\text{2 or more heads}) = P(\text{exactly 2 heads}) + P(\text{3 heads}) = \frac{3}{8} + \frac{1}{8} = \frac{4}{8} = \frac{1}{2}$

**(d)** $P(\text{tails on the first toss}) = \frac{4}{8} = \frac{1}{2}$

**25.** The probability that you select a mathematics book is $\dfrac{\text{number of ways to select a mathematics book}}{\text{number of ways to select a book}} = \dfrac{4}{10} = \dfrac{2}{5} = 0.4$.

**26.** Since rolling a die and selecting a card is independent,
$P(\text{both show a six}) = P(\text{die shows a six}) \cdot P(\text{card is a six}) = \frac{1}{6} \cdot \frac{4}{52} = \frac{1}{78}$.

**27.** **(a)** $P(\text{ace}) = \frac{4}{52} = \frac{1}{13}$

**(b)** Let $E$ be the event the card chosen is an ace, and let $F$ be the event the card chosen is a jack. Then $P(E \cup F) = P(E) + P(F) = \frac{4}{52} + \frac{4}{52} = \frac{2}{13}$.

**(c)** Let $E$ be the event the card chosen is an ace, and let $F$ be the event the card chosen is a spade. Then $P(E \cup F) = P(E) + P(F) - P(E \cap F) = \frac{4}{52} + \frac{13}{52} - \frac{1}{52} = \frac{4}{13}$.

**(d)** Let $E$ be the event the card chosen is an ace, and let $F$ be the event the card chosen is a red card. Then

$$P(E \cap F) = \frac{n(E \cap F)}{n(S)} = \tfrac{2}{52} = \tfrac{1}{26}.$$

**28.** **(a)** Since these events are independent, the probability of getting the ace of spades, a six, and a head is $\frac{1}{52} \cdot \frac{1}{6} \cdot \frac{1}{2} = \frac{1}{624}$.

**(b)** The probability of getting a spade, a six, and a head is $\frac{13}{52} \cdot \frac{1}{6} \cdot \frac{1}{2} = \frac{1}{48}$.

**(c)** The probability of getting a face card, a number greater than 3, and a head is $\frac{12}{52} \cdot \frac{3}{6} \cdot \frac{1}{2} = \frac{3}{52}$.

**29. (a)** The probability the first die shows some number is 1, and the probability the second die shows the same number is $\frac{1}{6}$. So the probability each die shows the same number is $1 \cdot \frac{1}{6} = \frac{1}{6}$.

**(b)** By part (a), the event of showing the same number has probability of $\frac{1}{6}$, and the complement of this event is that the dice show different numbers. Thus the probability that the dice show different numbers is $1 - \frac{1}{6} = \frac{5}{6}$.

**30. (a)** Since there are four kings in a standard deck, $P(4\text{ kings}) = \dfrac{C(4,4)}{C(52,4)} = \dfrac{1}{\frac{52\cdot51\cdot50\cdot49}{4\cdot3\cdot2\cdot1}} = \dfrac{1}{270{,}725} \approx 3.69 \times 10^{-6}$.

**(b)** Since there are 13 spades in a standard deck, $P(4\text{ spades}) = \dfrac{C(13,4)}{C(52,4)} = \dfrac{\frac{13\cdot12\cdot11\cdot10}{4\cdot3\cdot2\cdot1}}{\frac{52\cdot51\cdot50\cdot49}{4\cdot3\cdot2\cdot1}} = \frac{11}{4165} \approx 0.00264$.

**(c)** Since there are 26 red cards and 26 black cards, $P(\text{all same color}) = \dfrac{2\cdot C(26,4)}{C(52,4)} = \dfrac{2\cdot\frac{26\cdot25\cdot24\cdot23}{4\cdot3\cdot2\cdot1}}{\frac{52\cdot51\cdot50\cdot49}{4\cdot3\cdot2\cdot1}} = \frac{92}{833} \approx 0.11044$.

**31.** In the numbers game lottery, there are 1000 possible "winning" numbers.

**(a)** The probability that John wins \$500 is $\frac{1}{1000}$.

**(b)** There are $P(3,3) = 6$ ways to arrange the digits 1, 5, 9. However, if John wins only \$50, it means that his number 159 was not the winning number. Thus the probability is $\frac{5}{1000} = \frac{1}{200}$.

**32.** She knows the first digit and must arrange the other four digits. Since only one of the $P(4,4) = 24$ arrangements is correct, the probability that she guesses correctly is $\frac{1}{24}$.

**33.** There are 36 possible outcomes in rolling two dice and 6 ways in which both dice show the same numbers, namely, $(1,1)$, $(2,2)$, $(3,3)$, $(4,4)$, $(5,5)$, and $(6,6)$. So the expected value of this game is $E = (5)\left(\frac{6}{36}\right) + (-1)\left(\frac{30}{36}\right) = 0$.

**34.** Using the same logic as in Exercise 29(a), the probability that all three dice show the same number is $1 \cdot \frac{1}{6} \cdot \frac{1}{6} = \frac{1}{36}$, while the probability they are not all the same is $1 - \frac{1}{36} = \frac{35}{36}$. Thus, the expected value of this game is $E = (5)\left(\frac{1}{36}\right) + (-1)\left(\frac{35}{36}\right) = -\frac{30}{36} = -0.83$. So John's expected winnings per game are $-\$0.83$, that is, he expects to lose \$0.83 per game.

**35.** Since Mary makes a guess as to the order of ratification of the 13 original states, the number of such guesses is $P(13,13) = 13!$, while the probability that she guesses the correct order is $\dfrac{1}{13!}$. Thus the expected value is $E = (1{,}000{,}000)\left(\dfrac{1}{13!}\right) + (0)\left(\dfrac{13!-1}{13!}\right) = 0.00016$. So Mary's expected winnings are \$0.00016.

**36.** The number of different pizzas is the number of subsets of the set of 12 toppings, that is, $2^{12} = 4096$. The number of pizzas with anchovies is the number of ways of choosing anchovies and then choosing a subset of the 11 remaining toppings, that is, $(1)\cdot 2^{11} = 2048$. Thus, $P(\text{getting anchovies}) = \frac{2048}{4096} = \frac{1}{2}$.

Note that this makes intuitive sense: for each pizza combination without anchovies there is a corresponding one with anchovies, so half will have anchovies and half will not.

**37. (a)** Since there are only two colors of socks, any 3 socks must contain a matching pair.

**(b)** *Method 1:* If the two socks drawn form a matching pair then they are either both red or both blue. So

$$P\left(\begin{array}{c}\text{choosing a}\\\text{matching pair}\end{array}\right) = P\left(\begin{array}{c}\text{both red or}\\\text{both blue}\end{array}\right) = P(\text{both red}) + P(\text{both blue}) = \frac{C(20,2)}{C(50,2)} + \frac{C(30,2)}{C(50,2)} \approx 0.51.$$

*Method 2:* The complement of choosing a matching pair is choosing one sock of each color. So

$$P(\text{choosing a matching pair}) = 1 - P(\text{different colors}) = 1 - \frac{C(20,1)\cdot C(30,1)}{C(50,2)} \approx 1 - 0.49 = 0.51.$$

**38.** *Method 1:* First choose the 2 forward players, then choose the 3 defense players from the remaining 7 players. Thus the number of ways of choosing a starting lineup is $C(9,2)\cdot C(7,3) = 1260$.

*Method 2:* First choose the 3 defense players, then choose the 2 forwards from the remaining 6 players. Thus the number of ways of choosing a starting lineup is $C(9,3)\cdot C(6,2) = 1260$.

**39. (a)** (number of codes) $=$ (choices for 1st digit) $\cdot$ (choices for 2nd digit) $\cdot \cdots \cdot$ (choices for 5th digit)
$= 10 \cdot 10 \cdot 10 \cdot 10 \cdot 10 = 10^5 = 100{,}000$

**(b)** Since there are five numbers (0, 1, 6, 8 and 9) that can be read upside down, we have (number of codes) $=$ (choices for 1st digit) $\cdot$ (choices for 2nd digit) $\cdot \cdots \cdot$ (choices for 5th digit) $= 5^5 = 3125$.

**(c)** Let $E$ be the event that a zip code can be read upside down. Then by parts (a) and (b), $P(E) = \dfrac{n(E)}{n(S)} = \dfrac{5^5}{10^5} = \dfrac{1}{32}$.

**(d)** Suppose a zip code is turned upside down. Then the middle digit remains the middle digit, so it must be a digit that reads the same when turned upside down, that is, a 0, 1 or 8. Also, the last digit becomes the first digit, and the next to last digit becomes the second digit. Thus, once the first two digits are chosen, the last two are determined. Therefore, the number of zip codes that read the same upside down as right side up is
(number of codes) $=$ (choices for 1st digit) $\cdot$ (choices for 2nd digit) $\cdot \cdots \cdot$ (choices for 5th digit) $= 5 \cdot 5 \cdot 3 \cdot 1 \cdot 1 = 75$.

**40. (a)** Order is important, and repeats are possible. Thus there are 10 choices for each digit. So the number of different Zip+4 codes is $10 \cdot 10 \cdot \cdots \cdot 10 = 10^9$.

**(b)** If a Zip+4 code is to be a palindrome, the first 5 digits can be chosen arbitrarily. But once chosen, the last 4 digits are determined. Since there are 10 ways to choose each of the first 5 digits, there are $10^5$ palindromes.

**(c)** By parts (a) and (b), the probability that a randomly chosen Zip+4 code is a palindrome is $\frac{10^5}{10^9} = 10^{-4}$.

**41. (a)** Using the rule for the number of distinguishable combinations, the number of divisors of $N$ is $(7+1)(2+1)(5+1) = 144$.

**(b)** An even divisor of $N$ must contain 2 as a factor. Thus we place a 2 as one of the factors and count the number of distinguishable combinations of $M = 2^6 3^2 5^5$. So using the rule for the number of distinguishable combinations, the number of even divisors of $N$ is $(6+1)(2+1)(5+1) = 126$.

**(c)** A divisor is a multiple of 6 if 2 is a factor and 3 is a factor. Thus we place a 2 as one of the factors and a 3 as one of the factors. Then we count the number of distinguishable combinations of $K = 2^6 3^1 5^5$. So using the rule for the number of distinguishable combinations, the number of even divisors of $N$ is $(6+1)(1+1)(5+1) = 84$.

**(d)** Let $E$ be the event that the divisor is even. Then using parts (a) and (b), $P(E) = \frac{n(E)}{n(S)} = \frac{126}{144} = \frac{7}{8}$.

**42.** *Method 1:* We choose the 5 states first and then one of the two senators from each state. Thus the number of committees is $C(50,5) \cdot 2^5 = 67{,}800{,}320$.
*Method 2:* We choose one of 100 senators, then choose one of the remaining 98 senators (deleting the chosen senator and the other senator form that state), then choose one of the remaining 96 senators, continuing this way until the 5 senators are chosen. Finally, we need to divide by the number of ways to arrange the 5 senators. Thus the number of committees is $\frac{100 \cdot 98 \cdot 96 \cdot 94 \cdot 92}{5!} = 67{,}800{,}320$.

**43. (a)** $P(4 \text{ sixes in 8 rolls}) = C(8,4) \cdot \left(\frac{1}{6}\right)^4 \left(\frac{5}{6}\right)^4 \approx 0.026048$.

**(b)** There are three even numbers on a die and three odds numbers, so $P(\text{even}) = P(\text{odd}) = 0.5$. Thus

$$\begin{aligned} P(2 \text{ or more evens in 8 rolls}) &= 1 - P(\text{fewer than 2 evens}) = 1 - [P(0 \text{ evens}) + P(1 \text{ even})] \\ &= 1 - \left[C(8,0) \cdot (0.5^0)(0.5^8) + C(8,1) \cdot (0.5^1)(0.5^7)\right] \approx 0.96484 \end{aligned}$$

**44. (a)** $P(5 \text{ in 5 are white flesh}) = C(5,5) \cdot (0.3^5)(0.7^0) = 0.00243$

**(b)** $P(0 \text{ in 5 are white flesh}) = C(5,0) \cdot (0.3^0)(0.7^5) \approx 0.16807$

**(c)** $P(2 \text{ in 5 are white flesh}) = C(5,2) \cdot (0.3^2)(0.7^3) = 0.3087$

**(d)**
$$\begin{aligned} P(3 \text{ or more are red flesh}) &= P(3 \text{ in 5 are red flesh}) + P(4 \text{ in 5 are red flesh}) + P(5 \text{ in 5 are red flesh}) \\ &= C(5,3) \cdot (0.3^2)(0.7^3) + C(5,4) \cdot (0.3^1)(0.7^4) + C(5,5) \cdot (0.3^0)(0.7^5) \\ &= 0.83692 \end{aligned}$$

# Chapter 13 Test

**1. (a)** If repetition is allowed, then each letter of the word can be chosen in 10 ways since there are 10 letters. Thus the number of possible five-letter words is $10 \cdot 10 \cdot 10 \cdot 10 \cdot 10 = 10^5 = 100{,}000$.

**(b)** If repetition is not allowed, then since order is important, we need the number of permutations of 10 objects (the 10 letters) taken 5 at a time. Therefore, the number of possible five-letter words is $P(10,5) = 10 \cdot 9 \cdot 8 \cdot 7 \cdot 6 = 30{,}240$.

**2.** There are three choices to be made: one choice each of a main course, a dessert, and a drink. Since a main course can be chosen in one of five ways, a dessert in one of three ways, and a drink in one of four ways, there are $5 \cdot 3 \cdot 4 = 60$ possible ways that a customer could order a meal.

**3. (a)** Order is important in the arrangement, therefore the number of ways to arrange $P(30,4) = 657{,}720$.

**(b)** Here we are interested in the group of books to be taken on vacation so order is not important, therefore the number of ways to choose these books is $C(30,4) = 27{,}405$.

**4.** There are two choices to be made: choose a road to travel from Ajax to Barrie, and then choose a different road from Barrie to Ajax. Since there are 4 roads joining the two cities, we need the number of permutations of 4 objects (the roads) taken 2 at a time (the road there and the road back). This number is $P(4,2) = 4 \cdot 3 = 12$.

**5.** A customer must choose a size of pizza and must make a choice of toppings. There are 4 sizes of pizza, and each choice of toppings from the 14 available corresponds to a subset of the 14 objects. Since a set with 14 objects has $2^{14}$ subsets, the number of different pizzas this parlor offers is $4 \cdot 2^{14} = 65{,}536$.

**6. (a)** We want the number of ways of arranging 4 distinct objects (the letters L, O, V, E). This is the number of permutations of 4 objects taken 4 at a time. Therefore, the number of anagrams of the word LOVE is $P(4,4) = 4! = 24$.

**(b)** We want the number of distinguishable permutations of 6 objects (the letters K, I, S, S, E, S) consisting of three like groups of size 1 and a like group of size 3 (the S's). Therefore, the number of different anagrams of the word KISSES is $\dfrac{6!}{1!\,1!\,1!\,3!} = \dfrac{6!}{3!} = 120$.

**7.** We choose the officers first. Here order is important, because the officers are different. Thus there are $P(30,3)$ ways to do this. Next we choose the other 5 members from the remaining 27 members. Here order is not important, so there are $C(27,5)$ ways to do this. Therefore the number of ways that the board of directors can be chosen is $P(30,3) \cdot C(27,5) = 30 \cdot 29 \cdot 28 \cdot \dfrac{27!}{5!\,22!} = 1{,}966{,}582{,}800$.

**8.** One card is drawn from a deck.

**(a)** Since there are 26 red cards, the probability that the card is red is $\frac{26}{52} = \frac{1}{2}$.

**(b)** Since there are 4 kings, the probability that the card is a king is $\frac{4}{52} = \frac{1}{13}$.

**(c)** Since there are 2 red kings, the probability that the card is a red king is $\frac{2}{52} = \frac{1}{26}$.

**9.** Let $R$ be the event that the ball chosen is red. Let $E$ be the event that the ball chosen is even-numbered.

**(a)** Since 5 of the 13 balls are red, $P(R) = \frac{5}{13} \approx 0.3846$.

**(b)** Since 6 of the 13 balls are even-numbered, $P(E) = \frac{6}{13} \approx .4615$.

**(c)** $P(R \text{ or } E) = P(R) + P(E) - P(R \cap E) = \frac{5}{13} + \frac{6}{13} - \frac{2}{13} = \frac{9}{13} \approx 0.6923$.

**10.** Let $E$ be the event of choosing 3 men. Then $P(E) = \dfrac{n(E)}{n(S)} = \dfrac{\text{number of ways to choose 3 men}}{\text{number of ways to choose 3 people}} = \dfrac{C(5,3)}{C(15,3)} \approx 0.022$.

**11.** Two dice are rolled. Let $E$ be the event of getting doubles. Since a double may occur in 6 ways, $P(E) = \dfrac{n(E)}{n(S)} = \dfrac{6}{36} = \dfrac{1}{6}$.

**12.** There are 4 students and 12 astrological signs. Let $E$ be the event that at least 2 have the same astrological sign. Then $E'$ is the event that no 2 have the same astrological sign. It is easier to find $E'$. So

$$P(E') = \frac{\text{number of ways to assign 4 different astrological signs}}{\text{number of ways to assign 4 astrological signs}} = \frac{P(12,4)}{12^4} = \frac{12 \cdot 11 \cdot 10 \cdot 9}{12 \cdot 12 \cdot 12 \cdot 12} = \frac{55}{96}.$$

Therefore, $P(E) = 1 - P(E') = 1 - \frac{55}{96} = \frac{41}{96} \approx 0.427$.

**13.** **(a)** $P(\text{6 heads in 10 tosses}) = C(10,6) \cdot (0.55^6)(0.45^4) = 0.23837$.

**(b)** "Fewer than 3 heads" is the same as "0, 1, or 2 heads." So

$$\begin{aligned} P(\text{fewer than 3 heads}) &= P(\text{0 head in 10}) + P(\text{1 head in 10}) + P(\text{2 heads in 10}) \\ &= C(10,0) \cdot (0.55^0)(0.45^{10}) + C(10,1) \cdot (0.55^1)(0.45^9) + C(10,2) \cdot (0.55^2)(0.45^8) \\ &= 0.02739 \end{aligned}$$

**14.** A deck of cards contains 4 aces, 12 face cards, and 36 other cards. So the probability of an ace is $\frac{4}{52} = \frac{1}{13}$, the probability of a face card is $\frac{12}{52} = \frac{3}{13}$, and the probability of a non-ace, non-face card is $\frac{36}{52} = \frac{9}{13}$. Thus the expected value of this game is $E = (10)\left(\frac{1}{13}\right) + (1)\left(\frac{3}{13}\right) + (-.5)\left(\frac{9}{13}\right) = \frac{8.5}{13} \approx 0.654$, that is, about \$0.65.

## Focus on Modeling: The Monte Carlo Method

**1. (a)** You should find that with the switching strategy, you win about 90% of the time. The more games you play, the closer to 90% your winning ratio will be.

**(b)** The probability that the contestant has selected the winning door to begin with is $\frac{1}{10}$, since there are ten doors and only one is a winner. So the probability that he has selected a losing door is $\frac{9}{10}$. If the contestant switches, he exchanges a losing door for a winning door (and vice versa), so the probability that he loses is now $\frac{1}{10}$, and the probability that he wins is now $\frac{9}{10}$.

**2. (a)** You should find that you get a combination consisting of one head and one tail about 50% of the time.

**(b)** The possible gender combinations are $\{BB, BG, GB, GG\}$. Thus, the probability of having one child of each sex is $\frac{2}{4} = \frac{1}{2}$.

**3. (a)** You should find that player A wins about $\frac{7}{8}$of the time. That is, if you play this game 80 times, player A should win approximately 70 times.

**(b)** The game will end when either player A gets one more head or player B gets three more tails. Each toss is independent, and both heads and tails have probability $\frac{1}{2}$, so we obtain the following probabilities.

| Outcome | Probability |
|---|---|
| H | $\frac{1}{2}$ |
| TH | $\frac{1}{2} \cdot \frac{1}{2} = \frac{1}{4}$ |
| TTH | $\frac{1}{2} \cdot \frac{1}{2} \cdot \frac{1}{2} = \frac{1}{8}$ |
| TTT | $\frac{1}{2} \cdot \frac{1}{2} \cdot \frac{1}{2} = \frac{1}{8}$ |

Since Player A wins for any outcome that ends in heads, the probability that he wins is $\frac{1}{2} + \frac{1}{4} + \frac{1}{8} = \frac{7}{8}$.

**4. (a)** If you simulate 80 World Series with coin tosses, you should expect the series to end in 4 games about 10 times, in 5 games about 20 times, in 6 games about 25 times, and in 7 games about 25 times.

**(b)** We first calculate the number of ways that the series can end with team A winning. (Note that a team must win the final game plus three of the preceding games to win the series.) To win in 4 games, team A must win 4 games right off the bat, and there is only 1 way this can happen. To win in 5 games, team A must win the final game plus 3 of the first 4 games, so this can happen in $C(4, 3) = 4$ ways. To win in 6 games, team A must win the final game plus 3 of the first 5 games, so this can happen in $C(5, 3) = \dfrac{5 \cdot 4}{2 \cdot 1} = 10$ ways. To win in 7 games, team A must win the final game plus 3 of the first 6 games, so this can happen in $C(6, 3) = \dfrac{6 \cdot 5 \cdot 4}{3 \cdot 2 \cdot 1} = 20$ ways. By symmetry, it is also true for team B that they can win in 4 games 1 way, in 5 games 4 ways, in 6 games 10 ways, and in 7 games 20 ways. The probability that any particular team wins a given game is $\frac{1}{2}$; this fact, together with the assumption that the games are independent allows us to calculate the probabilities in the following table.

| Series | Number of ways | Probability |
|---|---|---|
| 4 games | 2 | $2 \cdot \left(\frac{1}{2} \cdot \frac{1}{2} \cdot \frac{1}{2} \cdot \frac{1}{2}\right) = \frac{1}{8}$ |
| 5 games | 8 | $8 \cdot \left(\frac{1}{2} \cdot \frac{1}{2} \cdot \frac{1}{2} \cdot \frac{1}{2} \cdot \frac{1}{2}\right) = \frac{1}{4}$ |
| 6 games | 20 | $20 \cdot \left(\frac{1}{2} \cdot \frac{1}{2} \cdot \frac{1}{2} \cdot \frac{1}{2} \cdot \frac{1}{2} \cdot \frac{1}{2}\right) = \frac{5}{16}$ |
| 7 games | 40 | $40 \cdot \left(\frac{1}{2} \cdot \frac{1}{2} \cdot \frac{1}{2} \cdot \frac{1}{2} \cdot \frac{1}{2} \cdot \frac{1}{2} \cdot \frac{1}{2}\right) = \frac{5}{16}$ |

**(c)** The expected value is $\frac{1}{8} \cdot 4 + \frac{1}{4} \cdot 5 + \frac{5}{16} \cdot 6 + \frac{5}{16} \cdot 7 = 5\frac{13}{16} \approx 5.8$. Thus, on average, we expect a World Series to end in about 5.8 games.

**5.** With 1000 trials, you are likely to obtain an estimate for $\pi$ that is between 3.1 and 3.2.

**6.** Modify the TI-83 program in Problem 5 to the following:

```
PROGRAM:PROB6
:0→P
:For(N,1,1000)
:rand→X:rand→Y
:P+(X²>Y)→P
:End
:Disp "PROBABILITY IS APPROX",P/1000
```

You should find that the probability is very close to $\frac{1}{3}$.

**7. (a)** We can use the following TI-83 program to model this experiment. It is a minor modification of the one given in Problem 5.

```
PROGRAM:PROB7
:0→P
:For(N,1,1000)
:rand→X:rand→Y
:P+((X+Y)<1)→P
:End
:Disp "PROBABILITY IS APPROX",P/1000
```

You should find that the probability is very close to $\frac{1}{2}$.

**(b)** Following the hint, the points in the square for which $x + y < 1$ are the ones that lie below the line $x + y = 1$. This triangle has area $\frac{1}{2}$ (it takes up half the square), so the probability that $x + y < 1$ is $\frac{1}{2}$.